Physical State at Room Temperature

**Black** = Solid
**Blue** = Liquid
**Red** = Gas

| | IB | IIB | IIIA | IVA | VA | VIA | VIIA | Rare gases VIIIA |
|---|---|---|---|---|---|---|---|---|
| | | | | | | | | 2 He 4.00260 |
| | | | 5 B 10.811 | 6 C 12.011 | 7 N 14.00674 | 8 O 15.9994 | 9 F 18.99840 | 10 Ne 20.1792 |
| | | | 13 Al 26.98154 | 14 Si 28.0855 | 15 P 30.97376 | 16 S 32.066 | 17 Cl 35.4527 | 18 Ar 39.948 |
| 28 Ni 58.6934 | 29 Cu 63.546 | 30 Zn 65.39 | 31 Ga 69.723 | 32 Ge 72.61 | 33 As 74.92159 | 34 Se 78.96 | 35 Br 79.904 | 36 Kr 83.80 |
| 46 Pd 106.42 | 47 Ag 107.8682 | 48 Cd 112.411 | 49 In 114.82 | 50 Sn 118.710 | 51 Sb 121.757 | 52 Te 127.60 | 53 I 126.09447 | 54 Xe 131.29 |
| 78 Pt 195.08 | 79 Au 196.96654 | 80 Hg 200.59 | 81 Tl 204.3833 | 82 Pb 207.2 | 83 Bi 208.98037 | 84 Po (209)[b] | 85 At (210)[b] | 86 Rn (222)[b] |

| | | | | | | | | |
|---|---|---|---|---|---|---|---|---|
| 63 Eu 151.965 | 64 Gd 157.25 | 65 Tb 158.92534 | 66 Dy 162.50 | 67 Ho 164.93032 | 68 Er 167.26 | 69 Tm 168.93421 | 70 Yb 173.04 | 71 Lu 174.967 |
| 95 Am 243.0614 | 96 Cm (247)[b] | 97 Bk (248)[b] | 98 Cf (250)[b] | 99 Es 252.083 | 100 Fm 257.0951 | 101 Md (257)[b] | 102 No 259.1009 | 103 Lr 262.11 |

# CHEMISTRY

## AN EXPERIMENTAL SCIENCE 2/e

**GEORGE M. BODNER**
Purdue University

**HARRY L. PARDUE**
Purdue University

**JOHN WILEY & SONS, INC.**
New York • Chichester • Brisbane • Toronto • Singapore

ACQUISITIONS EDITOR Nedah Rose
DEVELOPMENTAL EDITOR Kathleen Dolan
MARKETING MANAGER Catherine Faduska
SENIOR PRODUCTION EDITOR Marcia Craig
PRODUCTION SERVICE York Production Services
TEXT DESIGNER Ann Marie Renzi
MANUFACTURING MANAGER Susan Stetzer
PHOTO EDITORS Charles A. Hamilton
Lisa Passmore
PHOTO RESEARCHER Ramón Rivera Moret
SENIOR ILLUSTRATION COORDINATOR Sigmund Malinowski
ELECTRONIC ILLUSTRATIONS Precision Graphics
COVER: DESIGN AND ILLUSTRATION Kenny Beck
ART DIRECTOR Ann Marie Renzi
PHOTOGRAPHY Andy Washnik
PHOTOGRAPHY ART DIRECTION Charles A. Hamilton

This book was set in Times Roman by York Graphic Services, Inc. and printed and bound by Von Hoffman Press. The cover was printed by Phoenix Color Corp.

Recognizing the importance of preserving what has been written, it is a policy of John Wiley & Sons, Inc. to have books of enduring value published in the United States printed on acid-free paper, and we exert our best efforts to that end.

The paper on this book was manufactured by a mill whose forest management programs include sustained yield harvesting of its timberlands. Sustained yield harvesting principles ensure that the number of trees cut each year does not exceed the amount of new growth.



**Library of Congress Cataloging-in-Publication Data**

Bodner, George M.
Chemistry, an experimental science / George M. Bodner, Harry L. Pardue. — 2nd ed.
p. cm.
Includes index.
ISBN 0-471-59386-9
1. Chemistry. I. Pardue, Harry L. II. Title.
QD33.B684 1995
540—dc20 94-22161
CIP

## HOW TO LOOK AT THE ILLUSTRATIONS

- Color is used in this text to identify, associate, and relate items dealt with in various contexts. Drawings of orbitals, for example, are color coded so that each category of orbital (*s, p, d, sp,* etc.) has its own identifying color. These colors not only appear in three-dimensional pictorial depictions of the orbitals but also in the diagrams and charts that deal with these orbitals.
- The relative sizes and shapes of the various orbitals are drawn to scale in accordance with the known mathematical data defining these orbitals to ensure an authentic and consistent appearance.
- Various rays such as infrared rays, alpha, beta, and gamma rays, ultraviolet rays, and x-rays have been given characteristic colors in both illustrations of the laboratory equipment used to generate these rays and the charts and diagrams in which data are reported.
- Elements in ball-and-stick or space-filling molecular models are identified by conventional colors throughout the book. Thus hydrogen is white, oxygen red, chlorine green, sulfur yellow, and so on.
- Space-filling molecular models are drawn to a scale that reflects the relative sizes of their atoms. In figures of the water molecule, for instance, the radii of the hydrogen and oxygen atoms are drawn so as to reflect the experimental ratio of 0.37 to 0.66.
- Bond angles in figures of ball-and-stick and space-filling models are drawn to reflect experimental data. The angle between the hydrogen atoms in a water molecule, for example, is shown at a true $104° 28'$. The result is an authentic depiction and not a casual impression.
- Three-dimensional formula structures are drawn to emphasize their spatial geometry with clarity.
- Laboratory equipment (flasks, beakers, bunsen burners, etc.) has been drawn with reference to lab equipment catalogues to ensure accuracy in appearance.
- Classic experiments are depicted with careful attention to key aspects concerning the physical equipment used and its set-up.
- Crystal structures are drawn to ensure that the positions of the atoms in space relative to each other and thus the geometries of the structures are unequivocally clear.

# PREFACE

## TO THE INSTRUCTOR

The title of this text—*Chemistry: An Experimental Science*—reflects our beliefs about the direction in which general chemistry courses should evolve. We agree with the Committee on Professional Training of the American Chemical Society, which recommends putting the ''chemistry'' back into the introductory chemistry course. To us, however, this means more than just adding more inorganic chemistry. It means returning to an experimental perspective, in which observations are made before they are explained.

## GOALS

The second edition of this text was written with the following goals in mind.

- **Teach chemistry as a process, not a product.** Many texts presume that we have to develop all the theory needed to *explain* why the intensity of the color of a sealed tube of $NO_2$ decreases when it is cooled before we can *show* students that this happens. The result is a course that talks about science, but doesn't show it being done. As much as possible, we have structured this text around an experimental approach. We start by noting what happens when chemical systems are observed. We then try to develop explanations for this behavior. As a result, chemistry is portrayed as a process students can use to probe the structure of the world in which they live, rather than a recitation of the product of work by previous generations of scientists.
- **Recognize that it is the process, not the product, of science that is exciting.** Students often leave introductory courses without any understanding of how the information they have been asked to learn was discovered. This is unfortunate because it is the process of science, not its product, that is exciting. We have tried to combat this by raising epistemological questions such as ''Where do equilibrium constants come from?'' and then refusing to accept answers such as: ''From the back of the textbook!'' Students who read this book should understand how the value of $K_a$ for a weak acid can be extracted from a titration curve, how an equilibrium constant can be obtained from electrochemical data (even if the reaction doesn't seem to involve the transfer of electrons), and how equilibrium constants can be extracted from thermochemical data.
- **Provide reasons why someone might decide to become a chemist.** Students often leave chemistry courses with the impression that there is no reason for anyone to become a chemist because there is nothing left to do. That the topics

discussed in general chemistry are "dead," or at least dying. And, perhaps worst of all, there is no way that they can compare with the famous scientists whose work is described in the textbook. We have tried to combat these beliefs with descriptions of research being done in the 90s that relates to the material the students have been asked to learn. The goal is to portray chemistry as an active field, in which students can participate—even those who are not going to be chemists themselves.

- **Introduce chemistry in terms of major themes, not 25 different topics.** The major themes in this text are introduced by separate sections called *Intersections,* which are designed to help students build connections between the next series of topics under consideration and previous chapters. This text is organized around the following major themes:

  The Nature of Science

  Nature on the Macroscopic Scale

  Nature on an Atomic Scale

  Descriptive Chemistry of the Elements

  The Structure of Solids and Liquids

  Reactions Do Not Go to Completion

  Oxidation-Reduction Reactions

  The Forces That Control a Chemical Reaction

  Application of Chemical Knowledge

- **Recognize that knowledge is constructed in the mind of the learner.** Developmental biologists have argued that "ontogeny recapitulates phylogeny." (The process by which an individual organism develops reflects the process by which the species evolved.) There is abundant evidence that the same can be said about the learning process. Students have to pass through a stage in which they believe the caloric theory of heat on their way to an understanding of the kinetic theory, for example. We therefore start many chapters with a section labeled *The Process of Discovery,* which explicitly recognizes the basis upon which knowledge will be built in that chapter.

- **Help students develop problem-solving skills.** This text assumes that the skills students will retain, long after the facts of chemistry have been forgotten, are the problem-solving skills they develop in the introductory chemistry course. Explicit attention is therefore paid to developing problem-solving strategies. Some of what we have learned by doing research on problem solving in chemistry is summarized below, in the section entitled *To the Student*. Other aspects of this research have been built into problem-solving strategies that are discussed within the body of the textbook.

Several years ago, we noted that there is a fundamental difference between what expert problem solvers do when they are working a novel problem and what they do when they work routine exercises.[1] This difference is important because many of the tasks our students face are novel problems to them but routine exercises to us. To illustrate the difference, consider how you would attack the following question.

[1] G. M. Bodner, "A View From Chemistry," in *Toward A Unified Theory of Problem Solving: Views From the Content Domains,* M. U. Smith, Ed., Lawrence Erlbaum Associates, Hillsdale, NJ, **1991**, pp. 21–34.

> Two trains are stopped on adjacent tracks. The engine of one train is 1000 yards ahead of the engine of the other. The end of the caboose of the first train is 400 yards ahead of the end of the caboose of the other. The first train is three times as long as the second. How long are the trains?

Virtually every chemist we've talked to acknowledges that they would start with a drawing that summarized some of the relevant information in this question. We get a very different response, however, when we ask them to consider how they would attack the following question.

> What is the molarity of an acetic acid solution if 34.57 ml of this solution is needed to neutralize 25.19 ml of 0.1025 M sodium hydroxide?

$$CH_3CO_2H(aq) + NaOH(aq) \longrightarrow Na^+(aq) + CH_3CO_2^-(aq) + H_2O(l)$$

Very few testify that they would start by drawing a beaker containing one of the solutions and a buret filled with the other, and labeling the contents of the two pieces of glassware. To recognize the role that simple drawings can play in the task of solving novel problems, we have incorporated these diagrams into many of the exercises in this textbook.

## ORGANIZATION

The nine major themes around which this text was constructed dictate the order in which the chapters appear.

*The Nature of Science* (Chapter 1): This section acknowledges the difference between mathematics—where numbers are exact—and science—where the process of taking a measurement introduces a finite element of error.

*Nature on the Macroscopic Scale* (Chapters 2–5): It is easier to deal with objects and events we can see than those we can't. We therefore begin our discussion of chemistry by examining what we can deduce from measurements of the mass of a pure substance or the volume of a solution. As our first step away from the world that can be observed toward the world whose structures must be inferred, we examine the properties of gases. This section concludes with a chapter on thermochemistry, which some instructors will skip for the moment.

*Nature on an Atomic Scale* (Chapters 7 and 8): This section begins with a discussion of the structure of the atom. It then uses patterns in the physical properties of these atoms to introduce the periodic table. Once these trends are established, the chemistry of ionic and covalent compounds is introduced.

*Descriptive Chemistry of the Elements* (Chapters 9–12): Because we believe that chemistry is an experimental science, in which questions precede answers, the descriptive chemistry of the elements is introduced as soon as possible. The chemistry of the main-group metals is used to introduce the concepts of oxidation and reduction. The chapter on nonmetals uses the chemistry of oxygen, sulfur, nitrogen, and phosphorus to compare and contrast the behavior of elements with similar electron configurations. The chapter on acids and bases provides the basis upon which later discussions of acid-base equilibria will be built. This section concludes with a chapter on transition-metal chemistry, which some instructors will delay until the end of the second semester.

*The Structure of Liquids and Solids* (Chapters 13–15): General chemistry courses have historically spent far more time on gases than on solids. The balance between these topics has begun to change because so much work in science and engineering requires an understanding of the solid phase. To increase the flexibility of this section, the discussion of liquids and solutions has been divided into separate chapters in this edition.

*Reactions That Do Not Go to Completion* (Chapters 16–18): This section begins with gas-phase reactions because they are easier to understand. We then introduce interactions with the solvent, examine heterogeneous equilibria, and conclude with a brief discussion of systems in which many equilibria exist simultaneously.

*Oxidation-Reduction Reactions* (Chapters 19 and 20): Our belief that chemistry is an experimental science determined the order in which electrochemistry and thermodynamics are introduced. We start with something that can be experienced—electrochemical cells—to introduce the concept of spontaneous reactions.

*The Forces That Control a Chemical Reaction* (Chapters 21 and 22): We now use thermodynamics to explain what makes a chemical reaction spontaneous. We then introduce the notion of kinetic versus thermodynamic control to remind students that many reactions that thermodynamics predicts should be spontaneous occur at an infinitesimal rate, in the absence of a spark, a source of heat, or a catalyst.

*Application of Chemical Knowledge* (Chapters 23–25): Although most of this material could be covered at any point in the course, we use these topics as illustrations of how some of the knowledge of chemistry constructed in this text can be applied.

## PEDAGOGICAL FEATURES

A number of new pedagogical features have been added in this edition to make the material more accessible to students.

**Intersection Essays:** The goal of this book is to provide students with an easy-to-read roadmap for their study of chemistry. Each major step in this journey is introduced with an *Intersection* that helps students see the connections between a series of individual chapters, and thereby begin the process of building bridges between these chapters.

**Bridge Icons:** Connections between chapters are reinforced by bridge icons that refer students to previous discussions of a concept.

**Conceptual Questions:** Each chapter begins with a set of conceptual questions designed to stimulate the reader's curiosity and provide clues to the chapter's content. These questions are similar in character to those described in ''I Have Found You An Argument: The Conceptual Knowledge of Beginning Chemistry Graduate Students,'' *J. Chem. Ed., 68,* 385–388 (1991).

**Research in the 90s:** Most chapters contain a discussion of recent research related to one of the concepts introduced in the chapter. These sections were designed to combat the impression that the material we ask the students to learn is of nothing more than historical interest—everything important has either been done or can be predicted from thermodynamic data. These sections should help students understand why someone might want to become a chemist, even if they might not. And, more importantly, they should combat the belief that there is no room in science for people, like ourselves, who are not another Rutherford or another Pauling.

**Special Topics Sections:** To broaden the population who can successfully use this textbook, certain advanced topics have been moved to *Special Topics* sections at the end of the chapter. Instructors who want to include discussions of the van der Waals equation, or molecular orbital theory, or combined equilibria, or who want to show how the integrated rate equations can be derived can assign these sections.

**Lecture Demonstrations:** Many instructors have neither the time nor the facilities to show students what happens when you cool a sealed tube of $NO_2$ gas in liquid nitrogen or to collect the data that demonstrate Avogadro's hypothesis. We have therefore included photographs of some lecture demonstrations and tables of data that would be collected in other demonstrations.

**Checkpoints and Exercises:** Review questions that allow students to test their understanding of the material they have read are included as *Checkpoints. Exercises* are then used to help students learn how to apply what they have read.

**Problem-Solving Strategies:** Because the development of problem-solving strategies is an important component of this text, a problem-solving strategy icon is used to draw student's attention to portions of an exercise in which these strategies are discussed.

**Key Terms and Key Equations:** Key terms and equations are indicated in boldface type and then listed at the end of each chapter. Explicit attention is paid to chemical etymology, to convince students that the terms they encounter in chemistry are rational, not nonsense syllables to be memorized. (Students are reminded, for example, that the term *aldehyde* was introduced to describe the product of an *al*cohol *dehyd*rogenation reaction.)

## SUPPLEMENTS

The following supplements to this textbook are available:

**Student Solutions Manual** by Denise Magnuson, Baylor University, and Roy Garvey, North Dakota State University. Contains worked-out solutions to approximately half of the text's end of chapter problems.

**Study Guide** by Thomas J. Greenbowe, Iowa State University, and Jeff Pribyl, Mankato State University. This thorough student guide contains further discussions of difficult concepts, chapter objectives, additional worked-out examples with problem-solving strategies, as well as extensive review exercises.

**Instructor's Manual** by George M. Bodner. In addition to the author's discussion of *Eternal Verities* (or Things That Will Be True 'Til the End of Time), this rich instructor's tool contains chapter goals, lecture focus sections, section notes, as well as worked-out solutions to those problems not solved in the Student Solutions Manual.

**Test Bank** by George M. Bodner. Contains over 1,500 multiple-choice, short-answer, and discussion questions.

**Computerized Test Bank.** IBM and Macintosh versions of the entire Test Bank are available with full editing features to help you customize tests.

**Four-color Overhead Transparencies.** Approximately 200 four-color illustrations are provided in a form suitable for projection in the classroom.

**Lecture Demonstration Manual** by George M. Bodner, Paul E. Smith, Kurt L. Keyes, Purdue University, and Tom Greenbowe, Iowa State University. Contains over 150 lecture demonstrations that provide concrete examples of abstract concepts in general chemistry. Safety and safe handling and disposal of chemicals are stressed throughout the manual.

**Chemistry: An Experimental Science Videotape.** This 72-minute videotape contains 38 chemical demonstrations (each 1 to 4 minutes) that illustrate the principles of general chemistry. The narration explains the demonstration, its techniques, and its underlying principles. Proper safety precautions are observed and emphasized throughout the videotape.

## TO THE STUDENT

Students enroll in general chemistry courses for many reasons. For some, it is a direct result of their choice of major because the language and critical-thinking skills chemists use has been judged to be a valuable tool for success in that field. Others take chemistry to fulfill a science elective or because it is a required course for processional schools in medicine, dentistry, or veterinary medicine. Some of you might even be considering a career in one of the chemical sciences. No matter why you are taking general chemistry, this course is designed to meet the following objectives.

- To introduce you to the language that chemists use to describe the world around us—a language that has been adopted by professionals in such diverse fields as political science and astronomy.
- To introduce you to concepts and skills that are needed in later courses in your major.
- To help you improve problem-solving skills that can be transferred to your profession or to life in general.

Developing problem-solving skills is such an important component of chemistry courses that it is useful to summarize the results of research on what makes someone a good problem solver.

***Good Problem Solvers:***

- Believe they can solve almost any problem if they work long enough.
- Are persistent; they don't give up easily.
- Read carefully, and reread a problem, until they understand what information is given and what they are asked to solve for.
- Break problems into small steps, while they solve one at a time.
- Organize their work so that they don't lose sight of what they've accomplished, and can follow the steps they've taken so far.
- Check their work, not only at the end of the problem, but at various points along the way.
- Build representations of the problem, which can take the form of a list of relevant information, a picture of the system under consideration, or a concrete example.
- Try to solve a simpler, related problem when faced with a problem they can't solve.
- Try out several approaches to a problem until they are successful.

This text contains worked examples designed to help you: (1) determine what the problem asks for, (2) select relevant information, (3) keep track of this information,

(4) check the results of calculations, (5) work problems that contain too much information, (6) work problems that don't seem to contain enough information, (7) work backwards, and (8) make assumptions or approximations that turn complex problems into simpler ones.

As you learn how to work the examples in this text, it might be useful to distinguish between two closely related concepts: **problems** and **exercises.** Hayes[2] defined a problem as follows.

> Whenever there is a gap between where you are now and where you want to be, and you don't know how to find a way to cross that gap, you have a problem.

If you know what to do when you read a question, it's an exercise not a problem.

Status as a problem is not an innate characteristic of a question, it is a subtle interaction between the question and the individual trying to answer the question. It reflects experience with that type of question more than intellectual ability. When you go to class, you may find that your instructor has developed an impressive repertoire of techniques that can be used to turn problems into exercises. We call these techniques *algorithms,* which are defined as ''rules for calculating something, especially by machine.'' Algorithms are useful for solving routine exercises. In fact, the existence of an algorithm constructed from prior experience may be what turns a question from a problem into an exercise.

Students who have not built algorithms for at least some of the steps in a problem will have difficulty solving the problem. There is more to working problems, however, than applying algorithms in the correct order. Problem solving has been defined as ''What you do, when you don't know what to do.''[3] By definition, there is no clear cut answer to the question: ''What should I do when faced with a novel problem?'' We believe, however, that successful problem solvers follow a general pattern when faced with a novel problem. This pattern consists of a series of steps that cycle repeatedly until the answer is obtained.

1. Read the problem.
2. Now read the problem again.
3. Write down what you hope is the relevant information.
4. Read the problem again.
5. Draw a picture or make a list to help you build a model, or representation, of the problem.
6. Try something.
7. Try something else.
8. See where this gets you.
9. Read the problem again.
10. Try something else.
11. See where that gets you.
12. Inevitably, this process generates an intermediate result. Test this intermediate result to see if you are making any progress toward the answer.
13. Read the problem again.
14. Repeat this process until you get ''an'' answer, which is not necessarily ''the'' answer.

[2] J. Hayes, ''The Complete Problem Solver,'' Franklin Institute Press, Philadelphia, PA, 1980.

[3] G. H. Wheatley, Problem solving in school mathematics. MEPS Technical Report 84.01, School Mathematics and Science Center, Purdue University, West Lafayette, IN.

15. Test the answer to see if it makes sense.
16. Start over if you have to, celebrate if you don't.

We have two suggestions for improving your problem-solving skills. First, and foremost, recognize the importance of practice. The more problems you work, the better you will become at solving problems. Second, recognize the importance of working with other students. Take turns working problems out loud, explaining each step in the problem to the others in your group. While you do this, the other members of the group should listen carefully to what you say, to make sure they understand each step you take, to check each step to make sure that you aren't making any errors, to identify errors when they perceive them (without giving any hints about what they believe is the correct answer), and to ensure that you vocalize each of the major steps in the problem. Research has shown that this approach can significantly improve the problem-solving skills of each member of the group.

## ACKNOWLEDGMENTS

The authors would like to express their appreciation to the people who made this book possible. To Beverly Peavler and Lindsey Ardwin, who struggled to teach us the difference between "that" and "which." To John Balbalis, who took simplistic, hand-drawn figures and turned them into beautiful artwork, and to Sigmund Malinowski who brought the artwork into the computer age.

We are indebted to Paul Smith, who set up the demonstrations photographed in this text, to Charles Hamilton, who understood what had to be done to capture these demonstrations in a static image, and to Andy Washnik, who shot the photographs for this edition. It is also a pleasure to acknowledge the contributions of Ramón Rivera Moret, Lisa Passmore, and Stella Kupferberg, who incorporated these photographs and so many more into the text.

We owe a debt of gratitude to Kathleen Dolan, who saw the second edition through its development, and to Marcia Craig, who supervised its production. We would like to thank Ann Marie Renzi, who designed this text, Sandy Schnetzka, who turned a well-thumbed manuscript into a finished product, and Jackie Price and Nicole Carlson, who helped with the arduous task of proofreading galleys. We are grateful to Cliff Mills, who offered us the initial contract for this book, to Glenn Turner, who orchestrated the revision of the first edition, and to Nedah Rose, whose patience was sorely tested during the evolution of this edition. We would like to thank Joan Kalkut who coordinated the development of the supplements that accompany this text.

We wish to thank the following list of individuals who were involved in the reviewing process for this text. Their knowledge of chemistry and their insight into the process of teaching chemistry helped prevent us from making many errors of both omission and commission. Errors that remain are solely our fault, and we apologize for them.

Rathindra Bose
*Kent State University*

A. Wallace Cordes
*University of Arkansas*

Roberta Day
*University of Massachusetts*

Michael Davis
*University of Texas at El Paso*

Mauri Ditzler
*Holy Cross College*

Thomas Eckman
*University of North Alabama*

C. Dan Foote
*Eastern Illinois University*

Anne Harmon
*Lamar University*

David Harris
*University of California-Santa Barbara*
Alton Hassell
*Baylor University*
Richard Kiefer
*College of William and Mary*
Joanna Kirvaitis
*Moraine Valley Community College*
Larry Krannich
*University of Alabama*
Michael Mackey
*Memorial University of Newfoundland*
John McCracken
*Michigan State University*
John Melton
*Messiah College*
Christopher Ott
*Assumption College*
George Page
*Central Connecticut State University*
Gary Pfeiffer
*Ohio University*
Robert Pinnell
*The Claremont Colleges*
Robin Rogers
*Northern Illinois University*
Bradley Saville
*University of Toronto*
Barbara Sawrey
*University of California-San Diego*
Amy Schacter
*Santa Clara University*
Mahesh B. Sharma
*Columbus College*
William Shirley
*Pittsburgh State University*
Donald Titus
*Temple University*

Finally, we would like to thank our families and friends, who know too well the cost of writing a textbook, and our coworkers who carried much of the burden we neglected during the development of this book.

George M. Bodner
Harry L. Pardue

# FEATURES OF THIS BOOK

### Intersection Essays:

The goal of this book is to provide students with an easy-to-read road-map for their study of chemistry. Each major step in this journey is introduced with an essay that helps students see the connections between chapters, thereby introducing chemistry in terms of major themes, rather than 25 separate topics.

### Conceptual Questions:

Each chapter begins with a set of "Points of Interest" conceptual questions designed to stimulate the reader's curiosity about the leg of the journey to be encountered in the chapter to follow.

---

**Intersection**

## THE STRUCTURE OF SOLIDS AND LIQUIDS

The kinetic molecular theory explains the characteristic properties of gases by assuming that gas particles are in a state of constant, random motion and that the diameter of these particles is very much smaller than the distance between the particles. Because most of the volume of a gas is empty space, the simplest analogy compares the particles of a gas to fruit flies in a jar.

Many of the properties of solids have been captured in the way the term *solid* is used in English. It describes something that holds its shape, such as a solidly constructed house. It implies continuity, as in a movie that runs for three solid hours. It implies the absence of empty space, as in a solid chocolate Easter bunny. Finally, it describes things that occupy three dimensions, as in solid geometry. A solid may be compared to a brick wall in which the individual bricks form a regular structure and the amount of empty space is kept to a minimum.

Liquids have properties between the extremes of gases and solids. Like gases, they flow to conform to the shape of their containers. Like solids, they cannot expand to fill their containers, and they are very difficult to compress. The structure of a liquid may be compared to a collection of marbles in a bag being shaken back and forth.

Because water is the only substance that we encounter routinely as a solid, a liquid, and a gas, it is useful to consider what happens to water as we change the temperature. At low temperatures, it is a solid in which the individual molecules are locked into a rigid structure. As we raise the temperature, the average kinetic energy of the molecules increases, which increases the rate at which these molecules move.

There are three ways in which a water molecule can move: (1) vibration, (2) rotation, and (3) translation, as shown in art below. Water molecules *vibrate* when H—O bonds are stretched or bent. *Rotation* involves the motion of a molecule around its center of gravity. *Translation* literally means to change from one place to another. It therefore describes the motion of molecules through space.

To understand the effect of this motion, we need to differentiate between intramolecular and intermolecular bonds. The covalent bonds between the hydrogen and oxygen atoms in a water molecule are called **intramolecular bonds.** (The prefix *intra-* comes from the Latin stem meaning "within or inside." Thus, intramural sports match teams from within the same institution.) The bonds between the neighboring water molecules in ice are called **intermolecular bonds,** from the Latin stem *inter* meaning "between." (This far more common prefix is used in words such as *interact, intermediate,* and *international.*)

The *intramolecular* bonds that hold the atoms in $H_2O$ molecules together are almost 25 times as strong as the *intermolecular* bonds between water molecules. (It takes

---

CHAPTER 13

## THE STRUCTURE OF SOLIDS

Gases are so disordered that their physical properties usually don't depend on the identity of the gas. Solids represent the other extreme. They are so organized that we have to specify the structure of each substance before we can understand its physical properties. Our discussion of solids will provide the basis for answering questions such as the following:

**POINTS OF INTEREST**

- Why does Saran Wrap cling, whereas other plastic wraps do not?
- Why does the metal of a car left in the summer sun feel so much hotter than the grass next to which it was parked?
- Why is steel so much stronger than the iron from which it is made?
- Why do salts with similar formulas—such as NaCl, ZnS, and CsCl—crystallize in different structures?
- How do we know that the radius of a nickel atom is 0.1246 nm, or that the radius of a $Cl^-$ ion is 0.181 nm, if these particles are far too small to be seen?
- Why do metals become better conductors of electricity when cooled? Why do semiconductors become better conductors when heated?
- What is superconductivity? Why do we have to cool substances to very low temperatures before they become superconductors? What progress has been made in recent years toward a "high-temperature" superconductor?

475

17.13 BASES **621**

The terms on the right side of this equation should look familiar. The first is the inverse of the $K_b$ expression, the second is the expression for $K_w$:

$$K_a = \frac{1}{K_b} \times K_w$$

Rearranging this equation gives the following result:

$$K_a \times K_b = K_w$$

According to this equation, the value of $K_b$ for the reaction between the benzoate ion and water can be calculated from $K_a$ for benzoic acid:

$$K_b = \frac{K_w}{K_a} = \frac{1.0 \times 10^{-14}}{6.3 \times 10^{-5}} = 1.6 \times 10^{-10}$$

**CHECKPOINT**

Use the relationship between the values of $K_a$ for an acid and $K_b$ for its conjugate base to explain the following rules from Chapter 11.

Sec. 11.8

**Strong acids have weak conjugate bases.**
**Strong bases have weak conjugate acids.**

Now that we know $K_b$ for the benzoate ion, we can calculate the pH of an 0.030 *M* NaOBz solution with the techniques used to handle weak-acid equilibria. We start, once again, by building a representation for the problem:

**Problem-Solving Strategy**

| | $OBz^-(aq) + H_2O(l) \rightleftharpoons$ | $HOBz(aq)$ + | $OH^-(aq)$ | $K_b = 1.6 \times 10^{-10}$ |
|---|---|---|---|---|
| Initial: | 0.030 *M* | 0 | ≈0 | |
| Equilibrium: | 0.030 − Δ | Δ | Δ | |

We then substitute this information into the $K_b$ expression:

$$K_b = \frac{[HOBz][OH^-]}{[OBz^-]} = \frac{[\Delta][\Delta]}{[0.030 - \Delta]} = 1.6 \times 10^{-10}$$

Because $K_b$ is relatively small, we assume that Δ is small compared with 0.030:

$$\frac{[\Delta][\Delta]}{[0.030]} \approx 1.6 \times 10^{-10}$$

9.6 REACTIONS WITH $H_2O$ AND $NH_3$ **323**

that electrons are neither created nor destroyed, we can obtain an overall equation for the reaction:

$$2\,[Na \longrightarrow Na^+ + e^-]$$
$$2\,H_2O + 2\,e^- \longrightarrow H_2 + 2\,OH^-$$
$$2\,Na + 2\,H_2O \longrightarrow 2\,Na^+ + 2\,OH^- + H_2$$

The balanced equation for this reaction can be written as follows:

$$2\,Na(s) + 2\,H_2O(l) \longrightarrow 2\,Na^+(aq) + 2\,OH^-(aq) + H_2(g)$$

When heated, magnesium metal bursts into flame as it reacts with the $O_2$ and $N_2$ in the atmosphere to form MgO and a trace of $Mg_3N_2$. The reaction gives off both heat and a brilliant white light.

**CHECKPOINT**

Magnesium metal burns rapidly in air to form magnesium oxide. This reaction gives off an enormous amount of energy in the form of light and is used in both flares and fireworks. Use the fact that magnesium is an active metal to predict what happens when someone tries to extinguish a magnesium flare by pouring water on the reaction.

**ACTIVE METALS AND AMMONIA**

The same line of reasoning can be used to predict what happens when an alkali metal, such as potassium, reacts with liquid ammonia ($NH_3$). We can start by noting that the potassium atoms will lose one electron each:

$$K \longrightarrow K^+ + e^-$$

Figure 9.4 shows what happens to these electrons.

Hydrogen atoms in a +1 oxidation state gain electrons to form neutral hydrogen atoms, which combine to form $H_2$ molecules:

$$2\,H^+ + 2\,e^- \longrightarrow H_2$$

Removing an $H^+$ ion from an $NH_3$ molecule leaves a negatively charged $NH_2^-$ ion. The following equation therefore describes what happens when ammonia gains electrons:

$$2\,NH_3 + 2\,e^- \longrightarrow H_2 + 2\,NH_2^-$$

Combining the two halves of this reaction so that electrons are conserved gives the overall equation for the reaction:

$$2\,[K \longrightarrow K^+ + e^-]$$
$$2\,NH_3 + 2\,e^- \longrightarrow H_2 + 2\,NH_2^-$$
$$2\,K + 2\,NH_3 \longrightarrow 2\,K^+ + 2\,NH_2^- + H_2$$

The balanced equation for this reaction is written:

$$2\,K(s) + 2\,NH_3(l) \longrightarrow 2\,K^+(sol) + 2\,NH_2^-(sol) + H_2(g)$$

(The symbol *sol* in this equation indicates that the $K^+$ and $NH_2^-$ ions in this solution are solvated by neighboring $NH_3$ molecules.)

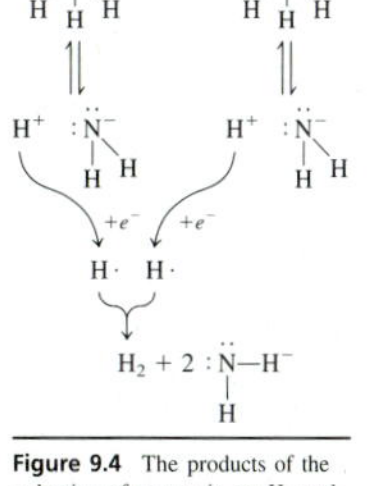


**Figure 9.4** The products of the reduction of ammonia are $H_2$ molecules and $NH_2^-$ ions.

**CHECKPOINT**

Write a balanced equation for the reaction between calcium metal and liquid ammonia.

## Bridge Icons:

Connections between chapters are reinforced by bridge icons that refer students to previous discussions of a concept.

## Checkpoints:

Review questions that allow students to test their understanding of the material they have just read are integrated throughout each chapter. Answers are located at the end of the book.

## Lecture Demonstrations:

Many instructors lack the time or facilities to show students what happens, for example, when magnesium metal is heated in air. Therefore numerous photographs of lecture demonstrations are included, as well as tables of data that would be collected in other demonstrations.

# FEATURES OF THIS BOOK

### Exercises:

Thorough, step-by-step explanations in worked examples help students learn how to apply what they have read.

### Problem-solving Strategy:

Included in many exercises, problem-solving strategy signposts in the margin draw students' attention to key strategies that will help develop their problem-solving skills.

98 Chapter 3 STOICHIOMETRY: COUNTING ATOMS AND MOLECULES

Sec. 1.13

The amount of solute or solvent in a solution is an extensive property. So is the amount of solution formed when the solute and solvent are mixed. The ratio of the amount of solute to the amount either of the solvent or the solution, however, is an intensive property. This ratio, which is known as the **concentration** of the solution, does not depend on the size of the sample.

$$\text{Concentration} = \frac{\text{amount of solute}}{\text{amount of solvent or solution}}$$

The concept of concentration is a common one. We talk about *concentrated* orange juice, which must be *diluted* with water. We even describe certain laundry products as *concentrated,* which means that we don't have to use as much.

**3.18 MOLARITY AS A WAY OF COUNTING ATOMS IN SOLUTIONS**

All concentration units have one thing in common: they describe the ratio of the amount of solute to the amount either of solvent or solution. Chemists use one concentration unit more than any other: **molarity** ($M$). The molarity of a solution is defined as the number of moles of solute per liter of solution. Molarity is calculated by dividing the number of moles of solute in the solution by the volume of the solution in liters:

$$\textbf{Molarity } (M) = \frac{\textbf{moles of solute}}{\textbf{liters of solution}}$$

**EXERCISE 3.15**

Copper sulfate is available as blue crystals that contain water molecules coordinated to the $Cu^{2+}$ ions in this crystal. Because these crystals contain five water molecules per $Cu^{2+}$ ion, the compound is called a pentahydrate, and the formula is written $CuSO_4 \cdot 5\,H_2O$. Calculate the molarity of a solution prepared by dissolving 1.25 grams of this compound in enough water to give 50.0 mL of solution.

Problem-Solving Strategy

**SOLUTION** A useful strategy for solving problems involves looking at the goal of the problem and asking: What information do we need to reach this goal? The molarity of a solution is calculated by dividing the number of moles of solute by the volume of the solution. We therefore need two pieces of information to reach the goal of this exercise—the number of moles of solute and the volume of the solution in liters.

The volume of the solution is easy to calculate:

$$50.0\,\cancel{\text{mL}} \times \frac{1\text{ L}}{1000\,\cancel{\text{mL}}} = 0.0500\text{ L}$$

The number of moles of solute can be calculated from the mass of solute used to prepare the solution and the mass of a mole of this compound:

$$1.25\,\cancel{\text{g CuSO}_4 \cdot 5\,\text{H}_2\text{O}} \times \frac{1\text{ mol}}{249.6\,\cancel{\text{g CuSO}_4 \cdot 5\,\text{H}_2\text{O}}} = 0.00500\text{ mol CuSO}_4 \cdot 5\,\text{H}_2\text{O}$$

The solution therefore has a concentration of 0.100 $M$:

$$\frac{0.00500\text{ mol CuSO}_4 \cdot 5\,\text{H}_2\text{O}}{0.0500\text{ L}} = \mathbf{0.100}\ \boldsymbol{M}\ \mathbf{CuSO_4}$$

Crystals of $CuSO_4 \cdot 5\,H_2O$.

**RESEARCH IN THE 90s**

### THE SEARCH TO UNDERSTAND SALTINESS

In the 1950s Lloyd Beidler of Florida State University showed that the $Na^+$ ions in table salt play a more important role than the $Cl^-$ ions in producing a "salty" taste when they encounter the taste cells on the surface of the tongue. The $Na^+$ ions apparently enter these cells through special channels in the cell membrane. This reduces the negative charge within the taste cell, which changes the voltage across the cell membrane, thereby exciting the cell. The excited cells release a neurotransmitter, which stimulates an adjacent nerve cell that begins the process of carrying the signal to the brain.

Beidler found that the chloride ion has an effect on the magnitude of the response of the taste cells to $Na^+$ ions. In 1987, Harry Harper, a physiologist at Stauffer Chemical Company, proposed a possible explanation for the effect of the $Cl^-$ ion on the perception of saltiness. He suggested that the $Cl^-$ ions diffuse through junctions *between* taste cells in the epithelium on the surface of the tongue, as shown in Figure 9.2. By altering the concentration of negative ions that surround the taste cells, these ions modify the cells' response to the $Na^+$ ion.

John DeSimone, Gerald Heck, and Qing Ye of Virginia Commonwealth University recently reported experimental evidence to support Harper's hypothesis [*Science,* **254,** 654 (1991)]. They developed an apparatus that could simultaneously deliver solutions to a rat's tongue and measure the voltage across the surface layer of cells that line the tongue. They found that sodium acetate (NaOAc) increased the voltage across the surface layer of cells but that sodium chloride (NaCl) produced only a small change in this voltage.

These results suggest that both $Na^+$ and $Cl^-$ ions can diffuse across the layer of cells on the tongue into the epithelium. The $OAc^-$ ion, on the other hand, is too large to diffuse into the epithelium and surround the taste cell the way the $Cl^-$ ion can.

When $Na^+$ ions diffuse into the taste cell and $Cl^-$ ions diffuse through the junction between these cells, we get a voltage *across the cell membrane* that excites the cell. When $Na^+$ ions diffuse into the taste cell and $OAc^-$ ions remain behind, we get a voltage *across the epithelium* that interferes with the electrical signal that excites the cell. It therefore takes significantly larger amounts of NaOAc to produce the response to the $Na^+$ ion obtained with a sample of NaCl.

**Figure 9.2** Chloride ions pass easily through the junction between taste cells on the surface of the tongue. Larger ions, such as the acetate ion, cannot pass through this junction.

## Research in the 90s Boxes:

These boxes, included in most chapters, discuss research being done today and show the relevance of the material students have been asked to learn.



**SPECIAL TOPIC**

### LIQUID CRYSTALS

Molecules that are large, rigid, and linear can form an intermediate phase during the transition between the liquid and solid states. Because it has some of the structure of solids and some of the freedom of motion associated with liquids, this phase is best described as a **liquid crystal.**

Liquid crystals were discovered in 1888, but they were primarily a laboratory curiosity until about 30 years ago. They are now used routinely in the displays of electrical devices such as digital watches, calculators, and computers. These LCD devices take advantage of the fact that the weak bonds that hold molecules together in a liquid crystal are easily affected by changes in pressure, temperature, or electromagnetic fields.

Liquid crystals are divided into three categories: smectic, nematic, and cholesteric. *Smectic* liquid crystals have a structure that resembles a handful of cigars, as shown in Figure 13.26*a*. Not only do the molecules all point in the same direction, they are so well ordered that they form planes perpendicular to the axes of the molecules. *Nematic* liquid crystals are slightly less well ordered. The molecules still point in the same direction, but they start and stop at different positions within the liquid, as shown in Figure 13.26*b*.

Advances in liquid crystal displays have allowed portable computers to compete with desk-top models.

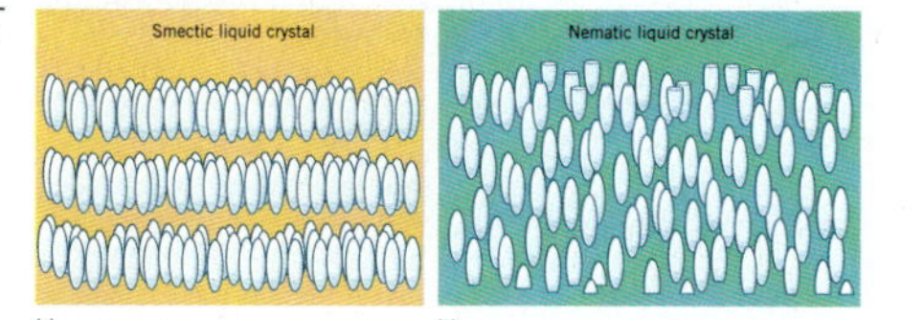


**Figure 13.26** The structures of (*a*) smectic and (*b*) nematic liquid crystals.

*Cholesteric* liquid crystals have a structure similar to nematic liquid crystals, but each plane of molecules is twisted slightly in relation to the plane above or below. These liquid crystals received their name from the fact that many derivatives of cholesterol form this structure. The slight twist in the planes these structures form tends to make these liquid crystals colored, and the fact that changes in the amount of twisting lead to changes in color make these crystals sensitive indicators of changes in temperature or pressure. The most sensitive cholesteric liquid crystals show a detectable color change with temperature changes as small as 0.001°C.

## Special Topics:

To allow for greater flexibility of use, some chapters include Special Topic sections at the end of the chapter that discusses certain advanced topics, such as the van der Waals equation or molecular orbital theory.

# FEATURES OF THIS BOOK

**Summary:**

A clear review of the key points of the chapter is presented in numbered paragraphs at the end of each chapter.

**Key Terms and Key Equations:**

Key terms and equations are indicated in boldface type within the chapter and then listed at the end of each chapter with page references.

**Problems:**

Comprehensive problem sets at the ends of chapters provide students with practice in solving both conceptual and numerical problems. In addition to problems classified by topic, most chapters also provide Integrated Problems that are unclassified and often require knowledge of two or more concepts.

**504** **Chapter 13** THE STRUCTURE OF SOLIDS

## SUMMARY

**1.** Some solids are **crystalline;** they have structures that repeat in a regular pattern. Some are **polycrystalline;** they contain small regions where the structure fits a regular pattern. Others are **amorphous;** their structures have little, if any, regularity.

**2.** **Ionic compounds** have structures that maximize the force of attraction between the ions of opposite charge. **Covalent solids,** such as diamond, are held together by extended arrays of relatively strong covalent bonds. **Metals** form solids that maximize the number of metallic bonds that can form. **Molecular solids** contain individual molecules, held together by relatively weak intermolecular bonds.

**3.** There are two ways of describing the regular patterns in a crystal. We can look at the crystal as an extended array of planes of atoms, ions, or molecules, or we can focus attention on the simplest repeating unit in the crystal. From the first perspective, solids are described in terms of **simple cubic, body-centered cubic, hexagonal closest packed,** and **cubic closest packed** structures. The other perspective classifies solids in terms of unit cells, such as **simple cubic, body-centered cubic,** and **face-centered cubic** unit cells.

**4.** Ionic compounds can be thought of as extended arrays of relatively large negative ions, with smaller positive ions packed in the holes between the planes of negative ions. When the negative ions form a closest packed structure, the positive ions can occupy either tetrahedral or octahedral holes, depending on their size. When the positive ions are too big to fit into these holes, the negative ions often pack in a simple cubic structure in which the positive ions occupy cubic holes.

## KEY TERMS

**Amorphous solid** (p. 476)
**Body-centered cubic packing** (p. 479)
**Body-centered cubic unit cell** (p. 490)
**Bragg equation** (p. 494)
**Closest packed structures** (p. 480)
**Conduction band** (p. 499)
**Coordination number** (p. 479)
**Covalent solid** (p. 477)
**Crystalline solid** (p. 476)
**Cubic closest packed** (p. 481)
**Cubic hole** (p. 487)
**Face-centered cubic unit cell** (p. 490)
**Hexagonal closest packed** (p. 480)
**Ionic solid** (p. 477)
**Lattice points** (p. 488)
**Liquid crystal** (p. 498)
**Metallic solid** (p. 477)
**Molecular solid** (p. 476)
**Octahedral hole** (p. 486)
**Polycrystalline solids** (p. 476)
**Radius ratio** (p. 487)
**Semiconductor** (p. 499)
**Simple cubic packing** (p. 479)
**Simple cubic unit cell** (p. 490)
**Superconductor** (p. 502)
**Tetrahedral hole** (p. 486)
**Unit cell** (p. 488)

## KEY EQUATIONS

$$\text{Radius ratio} = \frac{r_+}{r_-}$$

$$n\lambda = 2d \sin \Theta$$

## PROBLEMS

**Gases, Liquids, and Solids**

**13-1** Describe the differences in the properties of gases, liquids, and solids on the atomic scale. Explain how these differences give rise to the observed differences in the macroscopic properties of these three states of matter.

**13-2** Describe the difference between *intermolecular* and *intramolecular* bonds, giving examples of each. Which is stronger?

**van der Waals Forces**

**13-3** Which term in the van der Waals equation corrects for the force of attraction between molecules?

**13-4** Describe the difference between the three forms of van der Waals forces.

**13-5** Propose an explanation for the fact that induced dipole–induced dipole forces increase as the number of electrons on an atom increases.

**Crystals, Amorphous Solids, and Polycrystalline Substances**

**13-6** Describe the difference between *crystalline* and *amorphous* solids. Give three common examples of both kinds of solids.

**13-7** Disposable plastic glasses made from polystyrene are

▶ **Intersection** THE NATURE OF SCIENCE **1**

**CHAPTER 1**
THE FUNDAMENTALS OF MEASUREMENT 4

▶ **Intersection** NATURE ON THE MACROSCOPIC SCALE **40**

**CHAPTER 2**
ELEMENTS AND COMPOUNDS 42

**CHAPTER 3**
STOICHIOMETRY: COUNTING ATOMS AND MOLECULES 70

**CHAPTER 4**
GASES 110

**CHAPTER 5**
THERMOCHEMISTRY 152

▶ **Intersection** NATURE ON AN ATOMIC SCALE **193**

**CHAPTER 6**
THE STRUCTURE OF THE ATOM 194

**CHAPTER 7**
THE PERIODIC TABLE: AN INTRODUCTION TO IONIC COMPOUNDS 238

**CHAPTER 8**
THE COVALENT BOND 270

▶ **Intersection** DESCRIPTIVE CHEMISTRY OF THE ELEMENTS **312**

**CHAPTER 9**
THE MAIN-GROUP METALS AND THEIR SALTS 314

**CHAPTER 10**
THE CHEMISTRY OF THE NONMETALS 348

**CHAPTER 11**
ACIDS AND BASES 400

**CHAPTER 12**
TRANSITION-METAL CHEMISTRY 438

▶ **Intersection** THE STRUCTURE OF SOLIDS AND LIQUIDS 471

**CHAPTER 13**
THE STRUCTURE OF SOLIDS 474

**CHAPTER 14**
LIQUIDS 508

**CHAPTER 15**
SOLUTIONS 530

▶ **Intersection** REACTIONS DO NOT GO TO COMPLETION 559

**CHAPTER 16**
GAS-PHASE REACTIONS: AN INTRODUCTION TO KINETICS AND EQUILIBRIA 562

**CHAPTER 17**
ACID-BASE EQUILIBRIA 598

**CHAPTER 18**
SOLUBILITY AND COMPLEX-ION EQUILIBRIA 646

▶ **Intersection** OXIDATION-REDUCTION REACTIONS 692

**CHAPTER 19**
OXIDATION-REDUCTION REACTIONS 694

**CHAPTER 20**
ELECTROCHEMISTRY 724

▶ **Intersection** THE FORCES THAT CONTROL A CHEMICAL REACTION 762

**CHAPTER 21**
CHEMICAL THERMODYNAMICS 764

**CHAPTER 22**
KINETICS 800

▶ **Intersection** APPLICATION OF CHEMICAL KNOWLEDGE 839

**CHAPTER 23**
NUCLEAR CHEMISTRY 840

**CHAPTER 24**
THE ORGANIC CHEMISTRY OF CARBON 882

**CHAPTER 25**
POLYMERS: SYNTHETIC AND NATURAL 935

# CONTENTS


▶ **Intersection:** THE NATURE OF SCIENCE 1

**CHAPTER 1**
THE FUNDAMENTALS OF MEASUREMENT 4

**1.1 THE PROCESS OF DISCOVERY: AN INTRODUCTION TO MEASUREMENT 6**
**1.2 ENGLISH UNITS OF MEASUREMENT 8**
**1.3 SIMPLE UNIT CONVERSIONS 8**
**1.4 THE METRIC SYSTEM 10**
**1.5 MASS VERSUS WEIGHT 11**
**1.6 SI UNITS OF MEASUREMENT 12**
**1.7 UNCERTAINTY IN MEASUREMENT 13**
**1.8 SYSTEMATIC AND RANDOM ERRORS—ACCURACY AND PRECISION 14**

▶ **Research in the 90s:** MEASURING RISK 16

**1.9 SIGNIFICANT FIGURES 19**
**1.10 ROUNDING OFF 22**
**1.11 THE DIMENSIONAL ANALYSIS APPROACH TO UNIT CONVERSIONS 22**
**1.12 SCIENTIFIC NOTATION 25**
**1.13 EXTENSIVE AND INTENSIVE QUANTITIES 27**
**1.14 DENSITY AS AN EXAMPLE OF AN INTENSIVE QUANTITY 27**
**1.15 TEMPERATURE AS AN INTENSIVE QUANTITY 29**

▶ **Special Topic:** THE GRAPHICAL TREATMENT OF DATA 32

▶ **Intersection:** NATURE OF THE MACROSCOPIC SCALE 40

**CHAPTER 2**
ELEMENTS AND COMPOUNDS 42

**2.1 THE PROCESS OF DISCOVERY: THE GREEK ELEMENTS 44**
**2.2 ELEMENTS, MIXTURES, AND COMPOUNDS 45**
**2.3 EVIDENCE FOR THE EXISTENCE OF ATOMS 46**
**2.4 ATOMS AND MOLECULES 48**

▶ **Research in the 90s:** SCANNING TUNNELING MICROSCOPY 49

**2.5 THE MACROSCOPIC, ATOMIC, AND SYMBOLIC WORLDS OF CHEMISTRY 50**

2.6 THE CHEMISTRY OF THE ELEMENTS 51
2.7 METALS, NONMETALS, AND SEMIMETALS 52
2.8 IONIC AND COVALENT COMPOUNDS 54
2.9 THE STRUCTURE OF ATOMS 56
2.10 THE DIFFERENCE BETWEEN ATOMS AND IONS 56
2.11 FORMULAS OF COMMON IONIC COMPOUNDS, OR SALTS 57
2.12 POLYATOMIC IONS 58
2.13 OXIDATION NUMBERS 59

▶ **Special Topic:** NOMENCLATURE 62

**CHAPTER 3**
STOICHIOMETRY: COUNTING ATOMS AND MOLECULES 70

3.1 THE PROCESS OF DISCOVERY: THE ORIGINS OF STOICHIOMETRY 72
3.2 THE RELATIVE MASSES OF ATOMS 73
3.3 THE DIFFERENCE BETWEEN ATOMIC MASS AND ATOMIC WEIGHT 74
3.4 THE MOLE AS A COLLECTION OF ATOMS 75
3.5 THE MOLE AS A COLLECTION OF MOLECULES 76
3.6 AVOGADRO'S CONSTANT: THE NUMBER OF PARTICLES IN A MOLE 77
3.7 CONVERTING GRAMS INTO MOLES 78
3.8 DETERMINING EMPIRICAL FORMULAS 80
3.9 EMPIRICAL VERSUS MOLECULAR FORMULAS 81
3.10 CHEMICAL REACTIONS AND THE LAW OF CONSERVATION OF MATTER 83
3.11 CHEMICAL EQUATIONS AS A REPRESENTATION OF CHEMICAL REACTIONS 83
3.12 TWO VIEWS OF CHEMICAL EQUATIONS: MOLECULES VERSUS MOLES 84
3.13 BALANCING CHEMICAL EQUATIONS 85
3.14 MOLE RATIOS AND CHEMICAL EQUATIONS 87
3.15 PREDICTING THE MASS OF REACTANTS CONSUMED OR PRODUCTS GIVEN OFF IN A CHEMICAL REACTION 88
3.16 THE NUTS AND BOLTS OF LIMITING REAGENTS 90
3.17 SOLUTIONS AND CONCENTRATION 94

▶ **Research in the 90s:** THE STOICHIOMETRY OF THE BREATHALYZER 95

3.18 MOLARITY AS A WAY OF COUNTING ATOMS IN SOLUTIONS 98
3.19 DILUTION AND TITRATION CALCULATIONS 101

▶ **Special Topic:** ELEMENTAL ANALYSIS 103

**CHAPTER 4**
GASES 110

4.1 THE STATES OF MATTER 112
4.2 ELEMENTS OR COMPOUNDS THAT ARE GASES AT ROOM TEMPERATURE 112
4.3 THE PROPERTIES OF GASES 112
4.4 PRESSURE VERSUS FORCE 115
4.5 ATMOSPHERIC PRESSURE 116
4.6 BOYLE'S LAW 119
4.7 AMONTONS' LAW 120
4.8 CHARLES' LAW 122
4.9 GAY-LUSSAC'S LAW 125
4.10 AVOGADRO'S HYPOTHESIS 126
4.11 THE IDEAL GAS EQUATION 128

4.12 IDEAL GAS CALCULATIONS: PART I 129
4.13 IDEAL GAS CALCULATIONS: PART II 132
4.14 DALTON'S LAW OF PARTIAL PRESSURES 133

▶ **Research in the 90s:** GAS-PHASE ION CHEMISTRY 136

4.15 THE KINETIC MOLECULAR THEORY 138
4.16 HOW THE KINETIC MOLECULAR THEORY EXPLAINS THE GAS LAWS 139
4.17 GRAHAM'S LAWS OF DIFFUSION AND EFFUSION 140

▶ **Special Topic:** DEVIATIONS FROM IDEAL GAS LAW BEHAVIOR: VAN DER WAALS' EQUATION 143

▶ **Special Topic:** ANALYSIS OF THE VAN DER WAALS CONSTANTS 146

**CHAPTER 5**
THERMOCHEMISTRY 152

5.1 TEMPERATURE 154
5.2 TEMPERATURE AS AN INTENSIVE PROPERTY OF MATTER 154
5.3 HEAT AND HEAT CAPACITY 155
5.4 LATENT HEAT 157
5.5 THE CALORIC THEORY 159
5.6 HEAT AND THE KINETIC MOLECULAR THEORY 159
5.7 WORK 161
5.8 HEAT FROM WORK AND VICE VERSA 163
5.9 THE FIRST LAW OF THERMODYNAMICS: CONSERVATION OF ENERGY 164
5.10 THE FIRST LAW OF THERMODYNAMICS: INTERCONVERSION OF HEAT AND WORK 164
5.11 STATE FUNCTIONS 166
5.12 MEASURING HEAT WITH A CALORIMETER 168
5.13 ENTHALPY VERSUS INTERNAL ENERGY 171
5.14 ENTHALPIES OF REACTION 174
5.15 STANDARD-STATE ENTHALPIES OF REACTION 176
5.16 HESS'S LAW 177
5.17 ENTHALPIES OF FORMATION 178
5.18 BOND-DISSOCIATION ENTHALPIES 182

▶ **Research in the 90s:** BIOLOGICAL MICROCALORIMETRY 186

▶ **Intersection:** NATURE ON AN ATOMIC SCALE 193

**CHAPTER 6**
THE STRUCTURE OF THE ATOM 194

6.1 THE PROCESS OF DISCOVERY: ELECTRICITY 196
6.2 THE PROCESS OF DISCOVERY: THE ELECTRON 196
6.3 THE DISCOVERY OF X-RAYS 198
6.4 THE DISCOVERY OF RADIOACTIVITY 198
6.5 MILLIKAN'S OIL DROP EXPERIMENT 199
6.6 THOMSON'S RAISIN PUDDING MODEL OF THE ATOM 201
6.7 THE RUTHERFORD MODEL OF THE ATOM 202
6.8 THE DISCOVERY OF THE PROTON 203
6.9 THE DISCOVERY OF THE NEUTRON 205
6.10 PARTICLES AND WAVES 206
6.11 LIGHT AND OTHER FORMS OF ELECTROMAGNETIC RADIATION 208
6.12 ATOMIC SPECTRA 209

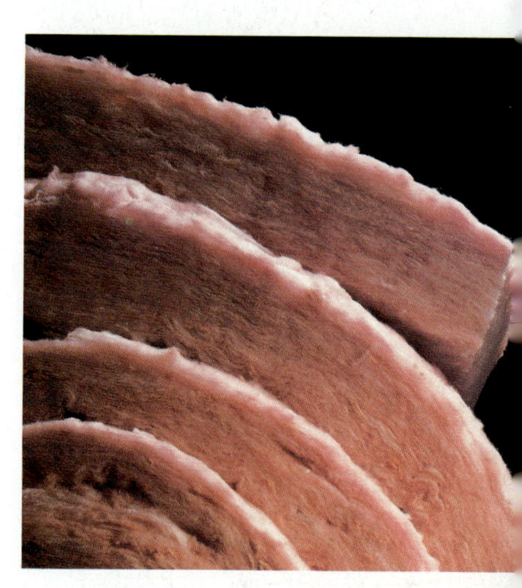

**6.13 QUANTIZATION OF ENERGY 211**
**6.14 THE BOHR MODEL OF THE ATOM 213**
**6.15 THE SUCCESS AND THE FAILURE OF THE BOHR MODEL 216**
**6.16 WAVE–PARTICLE DUALITY 216**
**6.17 CONSEQUENCES OF THE WAVE PROPERTIES OF ELECTRONS 217**
**6.18 QUANTUM NUMBERS 217**
**6.19 SHELLS AND SUBSHELLS OF ORBITALS 220**

▶ **Research in the 90s:** CONSEQUENCES OF THE SPIN OF SUBATOMIC PARTICLES 223

**6.20 THE RELATIVE ENERGIES OF ATOMIC ORBITALS 226**
**6.21 ELECTRON CONFIGURATIONS 227**
**6.22 EXCEPTIONS TO PREDICTED ELECTRON CONFIGURATIONS 230**
**6.23 ELECTRON CONFIGURATIONS AND THE PERIODIC TABLE 230**

**CHAPTER 7**
THE PERIODIC TABLE: AN INTRODUCTION TO IONIC COMPOUNDS 238

**7.1 THE PROCESS OF DISCOVERY: THE ELEMENTS 240**
**7.2 THE DEVELOPMENT OF THE PERIODIC TABLE 241**
**7.3 MODERN VERSIONS OF THE PERIODIC TABLE 243**
**7.4 THE SIZE OF ATOMS: METALLIC RADII 245**
**7.5 THE SIZE OF ATOMS: COVALENT RADII 246**
**7.6 THE SIZE OF ATOMS: IONIC RADII 247**
**7.7 THE RELATIVE SIZE OF ATOMS AND THEIR IONS 247**
**7.8 PATTERNS IN IONIC RADII 249**
**7.9 THE FIRST IONIZATION ENERGY 250**
**7.10 PATTERNS IN THE FIRST IONIZATION ENERGIES 251**
**7.11 EXCEPTIONS TO THE GENERAL PATTERN OF FIRST IONIZATION ENERGIES 252**
**7.12 SECOND, THIRD, FOURTH, AND HIGHER IONIZATION ENERGIES 253**
**7.13 ELECTRON AFFINITY 255**

▶ **Research in the 90s:** MEASURING FUNDAMENTAL PHYSICAL CONSTANTS 258

**7.14 CONSEQUENCES OF THE RELATIVE SIZE OF IONIZATION ENERGIES AND ELECTRON AFFINITIES 259**
**7.15 LATTICE ENERGIES AND THE STRENGTH OF THE IONIC BOND 259**
**7.16 LATTICE ENERGIES AND SOLUBILITY 260**
**7.17 WHY DOES SODIUM FORM NaCl? 261**
**7.18 WHY DOES MAGNESIUM FORM $MgCL_2$? 263**
**7.19 WHY DO SEMIMETALS EXIST? 263**

**CHAPTER 8**
THE COVALENT BOND 270

**8.1 THE PROCESS OF DISCOVERY VALENCE ELECTRONS 272**
**8.2 THE PROCESS OF DISCOVERY THE COVALENT BOND 273**
**8.3 HOW DOES SHARING OF ELECTRONS BOND ATOMS? 274**
**8.4 SIMILARITIES AND DIFFERENCES BETWEEN IONIC AND COVALENT COMPOUNDS 275**
**8.5 ELECTRONEGATIVITY AND POLARITY 277**
**8.6 LIMITATIONS OF THE ELECTRONEGATIVITY CONCEPT 281**
**8.7 THE DIFFERENCE BETWEEN POLAR BONDS AND POLAR MOLECULES 282**
**8.8 WRITING LEWIS STRUCTURES BY TRIAL AND ERROR 282**
**8.9 DRAWING SKELETON STRUCTURES 283**
**8.10 A STEP-BY-STEP APPROACH TO WRITING LEWIS STRUCTURES 283**

**8.11 MOLECULES THAT DON'T SEEM TO SATISFY THE OCTET RULE 284**
**8.12 RESONANCE HYBRIDS 286**
**8.13 FORMAL CHARGE 288**
**8.14 THE SHAPES OF MOLECULES 288**

▶ **Research in the 90s:** THE SHAPES OF MOLECULES 289

**8.15 PREDICTING THE SHAPES OF MOLECULES 291**
**8.16 INCORPORATING DOUBLE AND TRIPLE BONDS INTO THE VSEPR THEORY 292**
**8.17 THE ROLE OF NONBONDING ELECTRONS IN THE VSEPR THEORY 293**

▶ **Special Topic:** HYBRID ATOMIC ORBITALS 297

▶ **Special Topic:** MOLECULAR ORBITAL THEORY 301

▶ **Intersection:** DESCRIPTIVE CHEMISTRY OF THE ELEMENTS 312

**CHAPTER 9**
THE MAIN-GROUP METALS 314

**9.1 THE ACTIVE METALS 316**
**9.2 GROUP IA: THE ALKALI METALS 317**
**9.3 HALIDES, HYDRIDES, SULFIDES, NITRIDES, AND PHOSPHIDES 318**

▶ **Research in the 90s:** THE SEARCH TO UNDERSTAND SALTINESS 319

**9.4 PREDICTING THE PRODUCT OF MAIN-GROUP METAL REACTIONS 319**
**9.5 OXIDES, PEROXIDES, AND SUPEROXIDES 320**
**9.6 REACTIONS WITH $H_2O$ AND $NH_3$ 321**
**9.7 GROUP IIA: THE ALKALINE-EARTH METALS 324**
**9.8 GROUP IIIA: THE CHEMISTRY OF ALUMINUM 325**
**9.9 MAGNESIUM AND ALUMINUM AS STRUCTURAL METALS 326**
**9.10 GROUP IVA: TIN AND LEAD 326**
**9.11 OXIDATION–REDUCTION REACTIONS 327**
**9.12 THE ROLE OF OXIDATION NUMBERS IN OXIDATION–REDUCTION REACTIONS 328**
**9.13 OXIDATION NUMBERS VERSUS THE TRUE CHARGE ON IONS 329**
**9.14 OXIDIZING AGENTS AND REDUCING AGENTS 330**
**9.15 CONJUGATE OXIDIZING AGENT–REDUCING AGENT PAIRS 331**
**9.16 THE RELATIVE STRENGTHS OF METALS AS REDUCING AGENTS 333**
**9.17 THE PREPARATION OF METALS: CHEMICAL MEANS 336**
**9.18 THE PREPARATION OF ACTIVE METALS: ELECTROLYSIS 339**

▶ **Special Topic:** SALTS OF THE MAIN-GROUP METALS 341

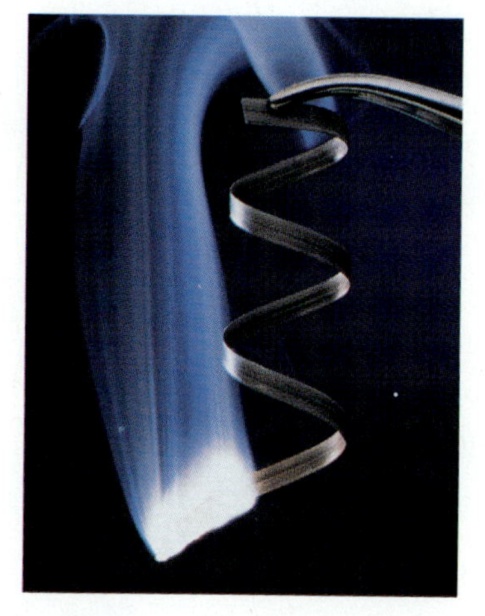

**CHAPTER 10**
THE CHEMISTRY OF THE NONMETALS 348

**10.1 THE NONMETALS 350**
**10.2 THE CHEMISTRY OF HYDROGEN 352**
**10.3 THE CHEMISTRY OF OXYGEN 356**

▶ **Research in the 90s:** THE CHEMISTRY OF THE ATMOSPHERE 362

**10.4 THE CHEMISTRY OF SULFUR 364**
**10.5 THE CHEMISTRY OF NITROGEN 370**
**10.6 THE CHEMISTRY OF PHOSPHORUS 377**
**10.7 THE CHEMISTRY OF THE HALOGENS 382**
**10.8 THE CHEMISTRY OF THE RARE GASES 387**

▶ **Special Topic:** THE INORGANIC CHEMISTRY OF CARBON 390

▶ **Research in the 90s:** FULLERENES 395

**CHAPTER 11**
ACIDS AND BASES 400

**11.1 THE PROCESS OF DISCOVERY: ACIDS AND BASES 402**
**11.2 THE ARRHENIUS DEFINITION OF ACIDS AND BASES 403**
**11.3 AN OPERATIONAL DEFINITION OF ACIDS AND BASES 404**
**11.4 TYPICAL ACIDS AND BASES 406**
**11.5 WHY ARE METAL HYDROXIDES BASES AND NONMETAL HYDROXIDES ACIDS? 409**
**11.6 THE BRØNSTED DEFINITION OF ACIDS AND BASES 410**
**11.7 THE ROLE OF WATER IN THE BRØNSTED THEORY 413**
**11.8 CONJUGATE ACID-BASE PAIRS 414**
**11.9 THE RELATIVE STRENGTHS OF ACIDS AND BASES 415**
**11.10 THE RELATIVE STRENGTHS OF CONJUGATE ACID-BASE PAIRS 417**
**11.11 THE RELATIVE STRENGTHS OF PAIRS OF ACIDS AND BASES 417**
**11.12 THE ADVANTAGES OF THE BRØNSTED DEFINITION 420**

▶ **Research in the 90s:** SOLID-STATE ACIDS 421

**11.13 pH AS A MEASURE OF THE CONCENTRATION OF THE $H_3O^+$ ION 422**
**11.14 FACTORS THAT CONTROL THE RELATIVE STRENGTHS OF ACIDS AND BASES 426**

▶ **Special Topic:** TRANSITION-METAL IONS AS BRØNSTED ACIDS 429

▶ **Special Topic:** THE LEWIS DEFINITIONS OF ACIDS AND BASES 430

**CHAPTER 12**
TRANSITION-METAL CHEMISTRY 438

**12.1 THE TRANSITION METALS 440**
**12.2 THE PROCESS OF DISCOVERY: WERNER'S MODEL OF COORDINATION COMPLEXES 441**
**12.3 TYPICAL COORDINATION NUMBERS 444**
**12.4 THE ELECTRON CONFIGURATION OF TRANSITION-METAL IONS 445**
**12.5 OXIDATION STATES OF THE TRANSITION METALS 446**
**12.6 LEWIS ACID-LEWIS BASE APPROACH TO BONDING IN COMPLEXES 446**
**12.7 TYPICAL LIGANDS 448**

▶ **Research in the 90s:** NITROGEN FIXATION 450

**12.8 COORDINATION COMPLEXES IN NATURE 451**
**12.9 NOMENCLATURE OF COMPLEXES 453**
**12.10 ISOMERS 454**
**12.11 THE VALENCE-BOND APPROACH TO BONDING IN COMPLEXES 456**
**12.12 CRYSTAL-FIELD THEORY 457**
**12.13 THE SPECTROCHEMICAL SERIES 461**
**12.14 HIGH-SPIN VERSUS LOW-SPIN OCTAHEDRAL COMPLEXES 461**
**12.15 THE COLORS OF TRANSITION-METAL COMPLEXES 463**

▶ **Special Topic:** LIGAND-FIELD THEORY 466

▶ **Intersection:** THE STRUCTURE OF SOLIDS AND LIQUIDS 471

**CHAPTER 13**
THE STRUCTURE OF SOLIDS 474

**13.1 SOLIDS 476**
**13.2 THE STRUCTURE OF METALS AND OTHER MONATOMIC SOLIDS 478**

**18.3 SOLUBILITY RULES 651**
**18.4 THE SOLUBILITY PRODUCT EXPRESSION 652**
**18.5 THE RELATIONSHIP BETWEEN $K_{SP}$ AND THE SOLUBILITY OF A SALT 654**
**18.6 COMMON MISCONCEPTIONS ABOUT SOLUBILITY PRODUCT CALCULATIONS 656**
**18.7 USING $K_{SP}$ AS A MEASURE OF THE SOLUBILITY OF A SALT 657**
**18.8 THE ROLE OF THE ION PRODUCT ($Q_{SP}$) IN SOLUBILITY CALCULATIONS 658**
**18.9 THE COMMON-ION EFFECT 660**
**18.10 HOW TO KEEP A SALT FROM PRECIPITATING 662**
**18.11 HOW TO SEPARATE IONS BY SELECTIVE PRECIPITATION 664**
**18.12 HOW TO ADJUST THE CONCENTRATION OF AN ION 666**

▶ **Research in the 90s:** SOLUBILITY EQUILIBRIA 668

**18.13 COMPLEX IONS 668**
**18.14 THE STEPWISE FORMATION OF COMPLEX IONS 670**
**18.15 COMPLEX-DISSOCIATION EQUILIBRIUM CONSTANTS 673**
**18.16 APPROXIMATE COMPLEX ION CALCULATIONS 674**
**18.17 USING COMPLEX-ION EQUILIBRIA TO DISSOLVE AN INSOLUBLE SALT 676**

▶ **Research in the 90s:** BIOCHEMICAL COMPLEXES 678

▶ **Special Topic:** A QUALITATIVE VIEW OF COMBINED EQUILIBRIA 680

▶ **Special Topic:** A QUANTITATIVE VIEW OF COMBINED EQUILIBRIA 683

▶ **Intersection:** OXIDATION–REDUCTION REACTIONS 692

**CHAPTER 19**
OXIDATION–REDUCTION REACTIONS 694

**19.1 THE PROCESS OF DISCOVERY: OXIDATION AND REDUCTION 696**
**19.2 OXIDATION–REDUCTION REACTIONS 698**
**19.3 ASSIGNING OXIDATION NUMBERS 699**
**19.4 RECOGNIZING OXIDATION–REDUCTION REACTIONS 701**
**19.5 BALANCING OXIDATION–REDUCTION EQUATIONS 702**
**19.6 THE HALF-REACTION METHOD OF BALANCING REDOX EQUATIONS 703**
**19.7 REDOX REACTIONS IN ACIDIC SOLUTIONS 704**
**19.8 REDOX REACTIONS IN BASIC SOLUTIONS 710**
**19.9 MOLECULAR REDOX REACTIONS 713**
**19.10 COMMON OXIDIZING AGENTS AND REDUCING AGENTS 715**
**19.11 THE RELATIVE STRENGTHS OF OXIDIZING AND REDUCING AGENTS 717**

**CHAPTER 20**
ELECTROCHEMISTRY 724

**20.1 ELECTROCHEMICAL REACTIONS 726**
**20.2 ELECTRICAL WORK FROM SPONTANEOUS OXIDATION–REDUCTION REACTIONS 726**
**20.3 VOLTAIC CELLS 728**
**20.4 STANDARD-STATE CELL POTENTIALS FOR VOLTAIC CELLS 729**
**20.5 PREDICTING SPONTANEOUS REDOX REACTIONS FROM THE SIGN OF $E°$ 730**
**20.6 STANDARD-STATE REDUCTION HALF-CELL POTENTIALS 731**
**20.7 PREDICTING STANDARD-STATE CELL POTENTIALS 733**

20.8 LINE NOTATION FOR VOLTAIC CELLS 735
20.9 THE NERNST EQUATION 736
20.10 USING THE NERNST EQUATION TO MEASURE EQUILIBRIUM CONSTANTS 740
20.11 ELECTROLYTIC CELLS 743
20.12 THE ELECTROLYSIS OF MOLTEN NaCl 743
20.13 THE ELECTROLYSIS OF AQUEOUS NaCl 745
20.14 THE ELECTROLYSIS OF WATER 747
20.15 FARADAY'S LAWS 748

▶ **Special Topic:** BATTERIES 750

▶ **Special Topic:** GALVANIC CORROSION AND CATHODIC PROTECTION 754

▶ **Intersection:** THE FORCES THAT CONTROL A CHEMICAL REACTION 762

**CHAPTER 21**
CHEMICAL THERMODYNAMICS 764

21.1 CHEMICAL THERMODYNAMICS 766
21.2 THE FIRST LAW OF THERMODYNAMICS 766
21.3 ENTHALPY VERSUS INTERNAL ENERGY 768
21.4 SPONTANEOUS CHEMICAL REACTIONS 771
21.5 ENTROPY AS A MEASURE OF DISORDER 772
21.6 ENTROPY AND THE SECOND LAW OF THERMODYNAMICS 774
21.7 THE THIRD LAW OF THERMODYNAMICS 776
21.8 STANDARD-STATE ENTROPIES OF REACTION 776
21.9 THE DIFFERENCE BETWEEN ENTHALPY OF REACTION AND ENTROPY OF REACTION CALCULATIONS 778
21.10 GIBBS FREE ENERGY 779
21.11 THE EFFECT OF TEMPERATURE ON THE FREE ENERGY OF A REACTION 782
21.12 BEWARE OF OVERSIMPLIFICATION 783
21.13 STANDARD-STATE FREE ENERGIES OF REACTION 784
21.14 INTERPRETING STANDARD-STATE FREE ENERGY OF REACTION DATA 784
21.15 THE RELATIONSHIP BETWEEN FREE ENERGY AND EQUILIBRIUM CONSTANTS 785

▶ **Research in the 90s:** THE THERMODYNAMICS OF BIOLOGICAL SYSTEMS 789

21.16 THE TEMPERATURE DEPENDENCE OF EQUILIBRIUM CONSTANTS 790
21.17 THE RELATIONSHIP BETWEEN FREE ENERGY AND CELL POTENTIALS 792

**CHAPTER 22**
KINETICS 800

22.1 CHEMICAL KINETICS 802
22.2 INSTANTANEOUS RATES OF REACTION AND THE RATE LAW FOR A REACTION 804
22.3 RATE LAWS AND RATE CONSTANTS 805
22.4 DIFFERENT WAYS OF EXPRESSING THE RATE OF REACTION 806
22.5 THE RATE LAW VERSUS THE STOICHIOMETRY OF A REACTION 807
22.6 ORDER AND MOLECULARITY 808
22.7 A COLLISION THEORY MODEL OF CHEMICAL REACTIONS 809
22.8 THE MECHANISMS OF CHEMICAL REACTIONS 811

22.9 THE RELATIONSHIP BETWEEN THE RATE CONSTANTS AND THE EQUILIBRIUM CONSTANT FOR A REACTION 812
22.10 DETERMINING THE ORDER OF A REACTION FROM RATE OF REACTION DATA 813
22.11 THE INTEGRATED FORM OF FIRST-ORDER AND SECOND-ORDER RATE LAWS 815
22.12 DETERMINING THE ORDER OF A REACTION WITH THE INTEGRATED FORM OF RATE LAWS 818
22.13 REACTIONS THAT ARE FIRST-ORDER IN TWO REACTANTS 820

▶ **Research in the 90s:** DETERMINING THE MECHANISM OF CHEMICAL REACTIONS 821

22.14 THE ACTIVATION ENERGY OF CHEMICAL REACTIONS 823
22.15 CATALYSTS AND THE RATES OF CHEMICAL REACTIONS 825
22.16 DETERMINING THE ACTIVATION ENERGY OF A REACTION 826

▶ **Special Topic:** DERIVING THE INTEGRATED RATE LAWS 829

▶ **Special Topic:** THE KINETICS OF ENZYME-CATALYZED REACTIONS 830

▶ **Intersection:** APPLICATION OF CHEMICAL KNOWLEDGE 839

**CHAPTER 23**
NUCLEAR CHEMISTRY 840

23.1 THE PROCESS OF DISCOVERY: RADIOACTIVITY 842
23.2 THE STRUCTURE OF THE ATOM 844
23.3 MODES OF RADIOACTIVE DECAY 845
23.4 NEUTRON-RICH VERSUS NEUTRON-POOR NUCLIDES 848
23.5 BINDING ENERGY CALCULATIONS 850
23.6 THE KINETICS OF RADIOACTIVE DECAY 854
23.7 DATING BY RADIOACTIVE DECAY 856
23.8 IONIZING VERSUS NONIONIZING RADIATION 857
23.9 BIOLOGICAL EFFECTS OF IONIZING RADIATION 859
23.10 NATURAL VERSUS INDUCED RADIOACTIVITY 862
23.11 NUCLEAR FISSION 868
23.12 NUCLEAR FUSION 872
23.13 NUCLEAR SYNTHESIS 874

▶ **Research in the 90s:** NUCLEAR MEDICINE 876

**CHAPTER 24**
THE ORGANIC CHEMISTRY OF CARBON 882

24.1 ORGANIC COMPOUNDS 884
24.2 THE SATURATED HYDROCARBONS: ALKANES AND CYCLOALKANES 885
24.3 THE UNSATURATED HYDROCARBONS: ALKENES AND ALKYNES 890
24.4 THE REACTIONS OF ALKANES, ALKENES AND ALKYNES 892
24.5 NATURALLY OCCURRING HYDROCARBONS AND THEIR DERIVATIVES 893
24.6 THE AROMATIC HYDROCARBONS AND THEIR DERIVATIVES 894
24.7 THE CHEMISTRY OF PETROLEUM PRODUCTS 897
24.8 THE CHEMISTRY OF COAL 900
24.9 FUNCTIONAL GROUPS 903
24.10 ALKYL HALIDES 904
24.11 ALCOHOLS AND ETHERS 905
24.12 SUBSTITUTION REACTIONS 908
24.13 OXIDATION–REDUCTION REACTIONS 910
24.14 ALDEHYDES AND KETONES 912

**24.15 CARBOXYLIC ACIDS AND CARBOXYLATE IONS 914**
**24.16 ESTERS 916**
**24.17 AMINES, ALKALOIDS, AND AMIDES 917**
**24.18 GRIGNARD REAGENTS 920**
**24.19 OPTICAL ACTIVITY 924**

▶ **Research in the 90s:** THE CHEMISTRY OF GARLIC 925

**CHAPTER 25**
POLYMERS: SYNTHETIC AND NATURAL 935

**25.1 THE PROCESS OF DISCOVERY: POLYMERS 936**
**25.2 DEFINITIONS OF TERMS 936**
**25.3 ELASTOMERS 940**
**25.4 FREE-RADICAL POLYMERIZATION REACTIONS 941**
**25.5 IONIC AND COORDINATION POLYMERIZATION REACTIONS 942**
**25.6 ADDITION POLYMERS 944**
**25.7 CONDENSATION POLYMERS 948**
**25.8 NATURAL POLYMERS 951**
**25.9 THE AMINO ACIDS 952**
**25.10 PEPTIDES AND PROTEINS 955**
**25.11 THE STRUCTURE OF PROTEINS 958**
**25.12 CARBOHYDRATES: THE MONOSACCHARIDES 962**
**25.13 CARBOHYDRATES: THE DISACCHARIDES AND POLYSACCHARIDES 965**
**25.14 LIPIDS 968**

▶ **Research in the 90s:** MEDICINAL CHEMISTRY 970

**25.15 NUCLEIC ACIDS 973**

▶ **Special Topic:** PROTEIN BIOSYNTHESIS 976

**APPENDIX A:** TABLES A-1
**Table A.1** The SI System A-2
**Table A.2** Values of Selected Fundamental Constants A-2
**Table A.3** Selected Conversion Factors A-3
**Table A.4** The Vapor Pressure of Water A-3
**Table A.5** Radii of Atoms and Ions A-4
**Table A.6** Ionization Energies A-5
**Table A.7** Electron Affinities A-7
**Table A.8** Electronegativities A-8
**Table A.9** Acid-Dissociation Equilibrium Constants A-9
**Table A.10** Base-Ionization Equilibrium Constants A-10
**Table A.11** Solubility Product Equilibrium Constants A-11
**Table A.12** Complex Formation Equilibrium Constants A-12
**Table A.13** Standard-State Reduction Potentials A-12
**Table A.14** Bond-Dissociation Enthalpies A-15
**Table A.15** Standard-State Enthalpy of Formation, Free Energy of Formation, and Absolute Entropy Data A-15

**APPENDIX B:** ANSWERS TO CHECKPOINT QUESTIONS B-1

**APPENDIX C:** ANSWERS TO SELECTED PROBLEMS C-1

**GLOSSARY G-1**

**PHOTO CREDITS P-1**

**INDEX I-1**

# Intersection

## THE NATURE OF SCIENCE

It seems logical to start a book of this nature with the question: What is chemistry? Most dictionaries define chemistry as the science that deals with the composition and properties of substances and the reactions by which one substance is converted into another. Knowing the definition of chemistry, however, is not the same as understanding what it means.

Perhaps the best way to understand the nature of chemistry is to examine examples of what it is not. In 1921, a group from the American Museum of Natural History began excavations at an archaeological site on Dragon-Bone Hill, near the town of Chou-k'outien, 34 miles southwest of Beijing, China. Fossils found at this site were assigned to a new species, *Homo erectus pekinensis,* commonly known as Peking man. Material found at these excavations suggested that for at least 500,000 years, people have known enough about the properites of stone to make tools, and they have been able to take advantage of the chemical reactions involved in combustion in order to cook food. But even the most liberal interpretation would not allow us to call this chemistry because of the absence of any proof that anything more than trial and error was involved in the control over these reactions or processes.

The ability to control the transformation of one substance into another can be traced back to the origin of two different technologies: brewing and metallurgy. People have been brewing beer for at least 12,000 years, since the time when the first cereal grains were cultivated, and the process of extracting metals from their ores has been practiced for at least 6000 years, since copper was first produced by heating the ore malachite.

But brewing beer by burying barley until it germinates and then allowing the barley to ferment in the open air still was not chemistry. Neither was extracting copper metal from one of its ores because this process was carried out without any understanding of what was happening, or why. Even the discovery around 3500 B.C. that copper mixed with 10% to 12% tin gave a new metal that was harder than copper, and yet easier to melt and cast, was not chemistry. The preparation of bronze was a major technical breakthrough in metallurgy, but it did not provide an understanding of how the proccss could be used to make other metals.

Between the sixth century B.C. and the third century B.C., the Greek philosophers tried to build a theoretical model for the behavior of the natural world. They argued that the world was made up of four primary, or *elementary,* substances: fire, air, earth, and water. These substances differed in two properties: hot versus cold, and dry versus wet. Fire was hot and dry; air was hot and wet; earth was cold and dry; water was cold and wet.

This model was the first step toward the goal of understanding the properties and compositions of different substances and the reactions that convert one substance to another. But some elements of modern chemistry were still missing. This model could explain certain observations of how the natural world behaved, but it couldn't predict new observations or behaviors. It was also based on pure speculation. In fact, its proponents were not interested in using the results of experiments to test the model.

Modern chemistry is based on certain general principles:

**1. One of the goals of chemistry is to recognize patterns in the behaviors of different substances.**
An example of this might be Lavoisier's discovery that every substance that burns in air gains weight.

**2. Once a pattern is recognized, it should be possible to develop a model that explains these observations.**
Lavoisier concluded that any substance that burns in air combines with the oxygen in the air to form a product that weighs more than the starting material.

**3. These models should allow us to predict the behavior of other substances.**
In 1869, Dmitri Mendeléeff used his model for the behavior of the known elements to predict the properties of elements that had not yet been discovered.

**4. When possible, the models should be quantitative.**
They should predict not only what happens, but by how much.

**5. The models should be able to make predictions that can be tested experimentally.**
Mendeléeff's periodic table was accepted by other chemists because of the agreement between his predictions and the results of experiments based on these predictions.

Sociologists and philosophers of science have identified certain normal practices that scientists follow. In his book, *False Prophets: Fraud and Error in Science and Medicine,* Alexander Kohn summarized these practices as follows.

1. Scientific facts and theories should be judged in terms of intellectual criteria valid in that branch of science. They should not be accepted or rejected because of the personal attributes of their author.
2. Scientists should direct their activities and efforts toward an extension of scientific knowledge, not the personal interests of an individual or group of scientists.
3. Science is a collaborative effort, which requires the open exchange of information.
4. Science should be approached rationally.
5. The validity of a scientific fact or theory should be judged from an emotionally neutral perspective.
6. Scientists must be honest, objective, tolerant, and unselfish.

In a discussion of the ethics of science, Hans Mohr has summarized these rules as follows. "Be honest; never manipulate data; be precise; be fair with regard to priority of ideas; be without bias with regard to data and ideas of your rival; do not make compromises in trying to solve a problem."

Young children are inherently curious. They want to know answers to questions ranging from why the sky is blue to why they have to go to bed. This book is dedicated to those who have retained some of this innate curiosity. It is written for people who at one time or another have wondered why a hot-air balloon rises, or why water evaporates at room temperature but doesn't boil until it reaches 212°F. It is written for those who may not have noticed that the sand on a damp beach dries out for an instant when they step on it, but who want to know why this happens when it is brought to their attention.

An etching of Robert Boyle (1627–1691), who has been called the father of modern chemistry.

Explanations for these observations are all based on models scientists construct to explain the results of quantitative experiments. The first step in this process is often the discovery of an **experimental law,** such as Boyle's law. The next step involves the development of a theoretical model that explains this experimental law. These models exist at three different levels of sophistication. At the first level, a **hypothesis** is proposed. Once a hypothesis exists, experiments can be designed to test it. A common cliché states that "the exception *proves* the rule." Nothing could be further from the truth. The following is a more accurate translation of the source of this cliché: "The exception *probes* the rule."

Once the hypothesis has been tested, and modified to fit new experimental results, it may reach the level of a **theory.** Most of the time, development of the model stops at this stage. Every once in a while, a model reaches the level of sophistication at which it becomes a **scientific law.** Examples of this rare phenomenon are the laws of thermodynamics and the law of conservation of mass in a chemical reaction.

**Definitions:**

**Experimental law:** A perceived regularity in the results of experiments that can be represented by an equation or formula.

**Hypothesis:** A suggested explanation that has not been subjected to extensive testing.

**Theory:** A systematic statement of a principle that has been verified by repeated experimentation.

**Scientific law:** A statement to which there are no known exceptions.

The height of Mt. Everest was estimated in 1856 by a team of surveyors working from observation stations as much as 100 miles from the mountain.

# THE FUNDAMENTALS OF MEASUREMENT

This chapter examines the role of measurement in science and provides the basis for answering questions such as the following:

## POINTS OF INTEREST

- Which travels faster: lightning at 87,000 miles per hour or an $O_2$ molecule traveling at 440 meters per second?
- Why do salad dressings made from oil and vinegar separate into two layers? Which liquid floats on top of the other, and why?
- What happens to the volume of the system when the ice in a glass of tea melts? Does it increase, decrease, or remain the same?
- Why were both the Titanic and the iceberg it hit able to float until they collided?

## 1.1 THE PROCESS OF DISCOVERY: AN INTRODUCTION TO MEASUREMENT

All measurements contain three elements: a **number** that indicates the magnitude of the quantity being measured, **units** that provide a basis for comparing this quantity with a standard reference, and some **uncertainty** or **error.**

The importance and complexity of accurate measurements are illustrated by the first estimate of the height of Mt. Everest. In 1818, a 28-year-old engineer named George Everest joined the Great Trigonometrical Survey of India—a project on which he eventually became surveyor general. The starting point of this survey was a carefully measured 7.5-mile baseline near Madras, on the southeast coast of India. A theodolite, which is a high-resolution telescope that measures horizontal and vertical angles, was used to measure the angle from both ends of the baseline to a tower, 30 miles from the baseline.

These three measurements—one length and two angles—provided the information necessary to calculate the length of the other two sides of the triangle. The surveyors then moved the theodolite to the apex of the triangle and took a new set of measurements. Figure 1.1 is a reproduction of the chart, first published in 1876, that shows how this laborious process was extended across the Indian subcontinent.

By 1849 Everest's successor, Sir Andrew Waugh, had brought the survey to the plains at the northeastern border of India. Access to Nepal was forbidden, however, so the surveying team resorted to shooting measurements of the peaks of the Himalaya mountains from observation stations about 100 miles from the mountain range.

In 1852 a team of "human computers" finally got around to calculating the height of the peaks in this mountain range. By their reckoning, one peak, which had been given the name "XV," was 29,002 feet above mean sea level. In 1856 after the surveying team was convinced that no peak in the Himalayas was taller than XV, Waugh announced the discovery of the tallest mountain on the planet. Although this mountain was known to the Tibetans as Chomolungma ("Goddess Mother of the Land"), the British named it Mt. Everest. In 1954 the height of Mt. Everest was revised by another team of British surveyors, who reported a value of 29,028 feet.

Measurement of the height of Mt. Everest raises several important issues discussed in this chapter. First, we must recognize that there is an element of error in any measurement. Second, we must understand the various sources of error in a measurement, so that the amount of error can be kept to a minimum. Finally, we must find a way to report the result of a measurement that reflects the amount of error involved in its determination.

The following sources of error can affect measurements of the location and height of the Himalaya mountains.

1. The starting point for the Great Trigonometrical Survey was more than 1000 miles from Mt. Everest.
2. The quality of all other data from this survey rests on the quality of the measurement of the baseline near Madras. (This was well understood by the surveying team, who used 6-foot bars with lengths known to within $\pm 0.001$ inch to establish a 7.5-mile baseline in which the total error was less than $\pm 0.5$ inches, or 1 part in 1,000,000.)
3. Near the Himalayas, measurements of vertical angles were affected by plumb-line deflection because the enormous mass of the mountains had a tendency to tug the leveling bubbles in the surveyors' instruments toward the mountains.
4. Atmospheric refraction tends to bend light rays as they pass through the atmosphere, causing the mountain to appear taller than it is. The amount of error this produces depends on the temperature and density of every layer of the atmosphere through which the light passes.

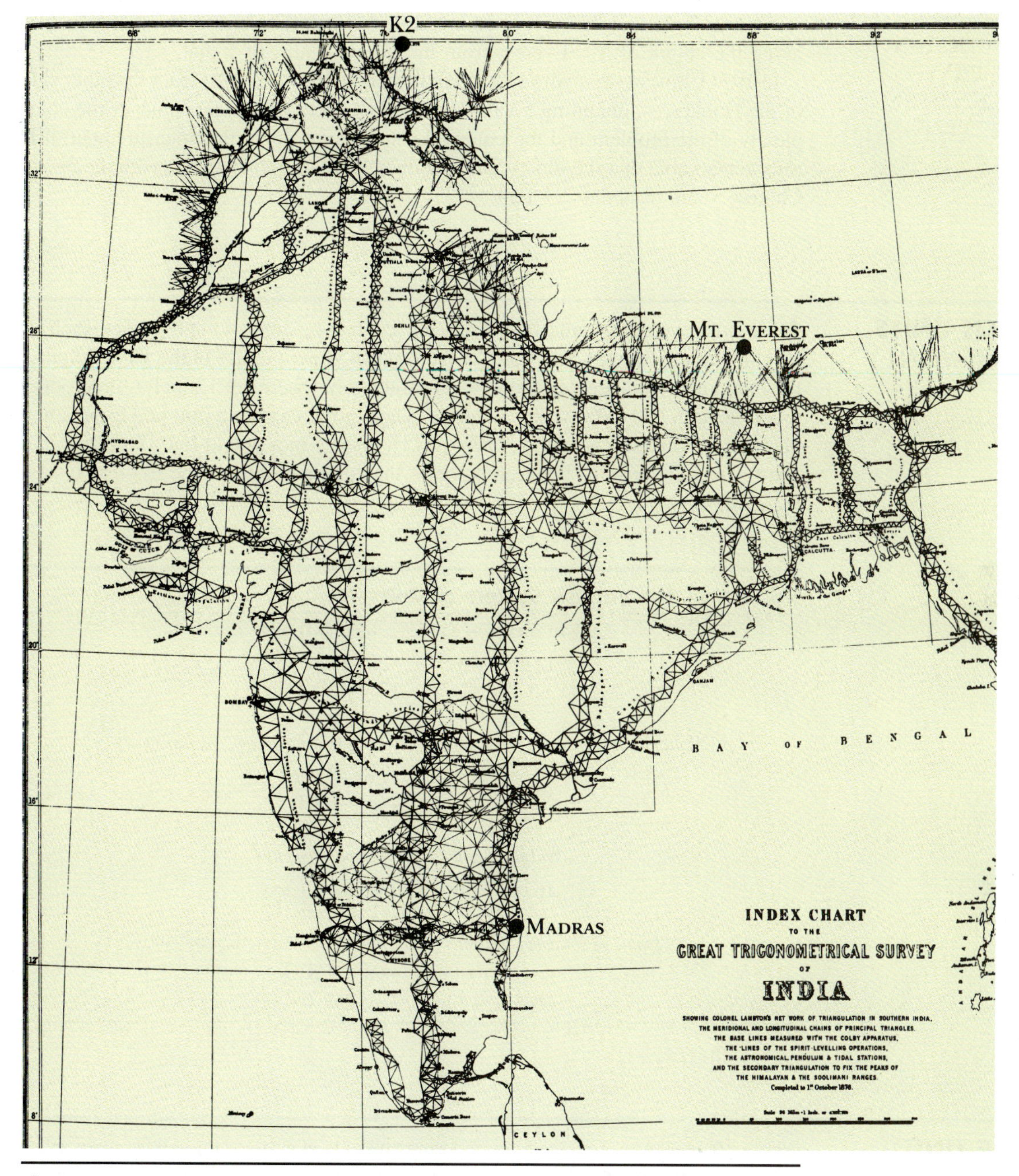


**Figure 1.1** This 1876 map shows the results of the elaborate triangulations involved in the survey of India which started in Madras and extended more than a thousand miles to Mt. Everest.

Even from only 1 mile, the effect of atmospheric refraction is large enough to make the mountain appear to shrink by several hundred feet when measurements taken at dawn are compared with measurements later in the day, after the atmosphere has warmed. The effect of atmospheric refraction is magnified, of course, by the distance between the theodolite and the mountain. Surveyors have therefore used corrections as large as 1,375 feet to compensate for atmosphere refraction when calculating the height of Mt. Everest from the distant plains of India.

In 1974 Chinese surveyors estimated the height of Mt. Everest from the other side of the Himalayas, obtaining a value of 29,029.4 feet. When you consider the complexity of the problem and the potential sources of error in this measurement, it is truly remarkable that the older value obtained by the English agrees with the newer Chinese value to about ±1 foot.

## 1.2 ENGLISH UNITS OF MEASUREMENT

There are several systems of units, each containing units for properties such as length, volume, weight, and time. In the English system in use in the United States, the individual units are defined in an arbitrary way. There are 12 inches in a foot, 3 feet in a yard, and 1760 yards in a mile. There are 2 cups in a pint and 2 pints in a quart but 4 quarts in a gallon. There are 16 ounces in a pound but 32 ounces in a quart. The relationships between some of the common units in the English system are given in Table 1.1.

**TABLE 1.1 The English System of Units**

| | |
|---|---|
| *Length: inch (in) foot (ft), yard (yd), mile (mi)* | |
| 12 in = 1 ft | 5280 ft = 1 mi |
| 3 ft = 1 yd | 1760 yd = 1 mi |
| *Volume: fluid ounce (oz), cup (c), pint (pt), quart (qt), gallon (gal)* | |
| 2 c = 1 pt | 32 oz = 1 qt |
| 2 pt = 1 qt | 4 qt = 1 gal |
| *Weight: ounce (oz), pound (lb), ton* | |
| 16 oz = 1 lb | 2000 lb = 1 ton |
| *Time: second (s), minute (min), hour (h), day (d), year (y)* | |
| 60 s = 1 min | 24 h = 1 d |
| 60 min = 1 h | $365\frac{1}{4}$ d = 1 y |

## 1.3 SIMPLE UNIT CONVERSIONS

About 10 years ago, first-year engineering students at a major university were asked to translate the following sentence into an equation:

**"There are six times as many students as professors at this university."**

The most common wrong answer was:

$$6\,S = P$$

You can prove that this equation is wrong by using a concrete example. If you substitute $S = 2$ into the equation, you will come to the conclusion that there are 12 professors for every 2 students, which is absurd.

The source of this error is confusion between *equations* and *equalities*, which are often written in the same way but mean different things. The following is an example of an **equation:**

$$S = 6\,P \qquad \text{(an equation)}$$

It states that the number of students can be calculated by multiplying the number of professors by 6.

The relationship between students and professors also can be described in terms of the following **equality,** which states that one professor is equivalent to six students.

$$1\,P = 6\,S \qquad \text{(an equality)}$$

It is often useful to write the relationships between different sets of units in the English system in terms of equalities, such as the following (see Figure 1.2):

$$1 \text{ ft} = 12 \text{ in}$$

It is important to recognize, however, that this is not an equation for converting from feet into inches, or vice versa. It is an equality that can be used to construct a **unit factor** that allows us to do this conversion. The importance of appropriate unit factors is illustrated in Exercise 1.1.

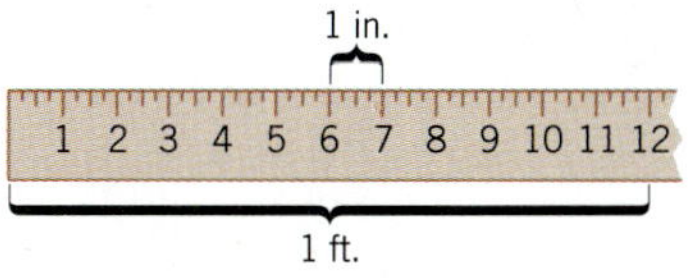

**Figure 1.2** The equality: 1 ft = 12 in is merely a statement of a fact that is obvious to anyone who has used a common ruler.

## EXERCISE 1.1

Convert 6.5 feet into inches.

**SOLUTION** We can start by writing an equality for the relationship between feet and inches:

$$12 \text{ in} = 1 \text{ ft}$$

We then divide both sides of the equality by 1 foot:

$$\frac{12 \text{ in}}{1 \text{ ft}} = \frac{1 \cancel{\text{ft}}}{1 \cancel{\text{ft}}}$$

The result is a ratio, or factor, that is equal to 1. This ratio is therefore a *unit* factor.

$$\frac{12 \text{ in}}{1 \text{ ft}} = 1$$

This is not the only unit factor that can be constructed. We can divide both sides of the equality by 12 inches to create another unit factor:

$$\frac{1 \text{ ft}}{12 \text{ in}} = 1$$

**Problem-Solving Strategy**

Both unit factors are valid, but only one is useful. If we choose the right unit factor, the units of feet will cancel, or *factor* out, when we multiply the original measurement by the unit factor:

$$6.5 \cancel{\text{ft}} \times \frac{12 \text{ in}}{1 \cancel{\text{ft}}} = 78 \text{ in}$$

## 1.4 THE METRIC SYSTEM

More than 300 years ago, the Royal Society of London discussed replacing the irregular English system of units with one based on decimals. It was not until the French Revolution, however, that a decimal-based system of units was adopted. This **metric system** is based on the fundamental units of measurement for length, volume, and mass shown in Table 1.2 and Figure 1.3.

**TABLE 1.2 The Fundamental Units of the Metric System**

*Length: meter (m)*

1 m = 1.094 yd
1 yd = 0.9144 m

*Volume: liter (L)*

1 L = 1.057 qt
1 qt = 0.9464 L

*Mass: gram (g)*

1 g = 0.002205 lb
1 lb = 453.6 g

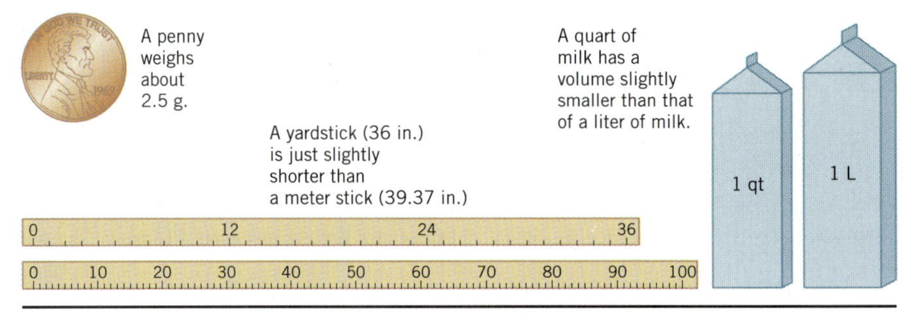


**Figure 1.3** The fundamental units in the metric system are the gram for mass, the meter for length, and the liter for volume. A penny has a mass of about 2.5 grams, a meter is about 3 inches longer than a yard, and a liter is about 5% larger than a quart.

A standard sample that has a mass of one kilogram.

The principal advantage of the metric system is the ease with which the base units can be converted into a unit that is more appropriate for the quantity being measured. This is done by adding a prefix to the name of the base unit. The prefix *kilo-* (abbreviated k), for example, stands for multiplication by a factor of 1000. Thus, a kilometer is equal to 1000 meters:

$$1 \text{ km} = 1000 \text{ m}$$

The prefix *milli-* (m), on the other hand, means division by a factor of 1000. A milliliter (mL) is equal to 0.001 liters:

$$1 \text{ mL} = 0.001 \text{ L}$$

The common metric prefixes are given in Table 1.3. The prefixes you will encounter most often in chemistry are highlighted in color.

**TABLE 1.3 Metric System Prefixes**

| Prefix | Symbol | Meaning |
|---|---|---|
| Femto- | f | $\times 1/1{,}000{,}000{,}000{,}000{,}000$ ($10^{-15}$) |
| Pico- | p | $\times 1/1{,}000{,}000{,}000{,}000$ ($10^{-12}$) |
| **Nano-** | **n** | **$\times 1/1{,}000{,}000{,}000$ ($10^{-9}$)** |
| **Micro-** | **$\mu$** | **$\times 1/1{,}000{,}000$ ($10^{-6}$)** |
| **Milli-** | **m** | **$\times 1/1{,}000$ ($10^{-3}$)** |
| **Centi-** | **c** | **$\times 1/100$ ($10^{-2}$)** |
| Deci- | d | $\times 1/10$ ($10^{-1}$) |
| **Kilo-** | **k** | **$\times 1{,}000$ ($10^{3}$)** |
| Mega- | M | $\times 1{,}000{,}000$ ($10^{6}$) |
| Giga- | G | $\times 1{,}000{,}000{,}000$ ($10^{9}$) |
| Tera- | T | $\times 1{,}000{,}000{,}000{,}000$ ($10^{12}$) |

**EXERCISE 1.2**

Convert 0.135 kilometers into meters.

**SOLUTION** We start with the definition of a kilometer:

$$1 \text{ km} = 1000 \text{ m}$$

In theory, we can turn this equality into two unit factors:

$$\frac{1 \text{ km}}{1000 \text{ m}} = 1 \qquad \frac{1000 \text{ m}}{1 \text{ km}} = 1$$

By choosing the correct unit factor, we can get the units of kilometers in the original measurement to cancel:

$$0.135 \cancel{\text{km}} \times \frac{1000 \text{ m}}{1 \cancel{\text{km}}} = 135 \text{ m}$$

A second advantage of the metric system is the link between the base units of length and volume (see Figure 1.4). By definition, a liter is equal to the volume of

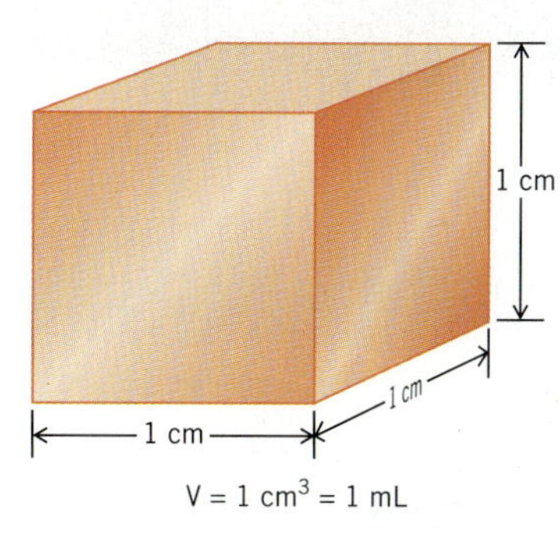


**Figure 1.4** The liter was defined so that a cube exactly 1 cm × 1 cm × 1 cm would have a volume of 1 milliliter: 1 $cm^3$ = 1 mL.

a cube exactly 10 cm tall, 10 cm long, and 10 cm wide. Because the volume of this cube is 1000 cubic centimeters and a liter contains 1000 milliliters, 1 milliliter is equivalent to 1 cubic centimeter:

$$1 \text{ mL} = 1 \text{ cm}^3$$

The third advantage of the metric system is the link between the base units of volume and weight. The gram was originally defined as the mass of 1 mL of water at 4°C. (It was important to specify the temperature in this definition because water expands or contracts as the temperature changes.)

## 1.5 MASS VERSUS WEIGHT

There is a subtle but important difference between *weight* and *mass*. **Mass** is a measure of the amount of matter in an object, so the mass of an object is constant. **Weight** is a measure of the force of attraction of the earth acting on an object. The weight of an object is not constant. There is a slight difference between the weight of an object resting on the earth's surface at sea level and the weight of the same object at the top of the Himalaya mountains. Because the mass of an object is constant, it is a more fundamental quantity than weight.

Most balances are designed to measure mass by comparing the object with samples of known mass. Unfortunately there is no English equivalent to the verb *weigh* that can be used to describe what happens when the mass of an object is measured. You are therefore likely to encounter the terms *weigh* and *weight* for operations and quantities that are more accurately associated with the term *mass*.

## 1.6 SI UNITS OF MEASUREMENT

A series of international conferences on weights and measures has been held periodically since 1875 to refine the metric system. At the 11th conference, in 1960, a new system of units known as the **International System of Units** (abbreviated **SI** in all languages) was proposed as a replacement for the metric system. The seven base units for the SI system are given in Table 1.4.

**TABLE 1.4 SI Base Units**

| Physical Quantity | Name of Unit | Symbol |
|---|---|---|
| Length | meter | m |
| Mass | kilogram | kg |
| Time | second | s |
| Temperature | kelvin | K |
| Electric current | ampere | A |
| Amount of substance | mole | mol |
| Luminous intensity | candela | cd |

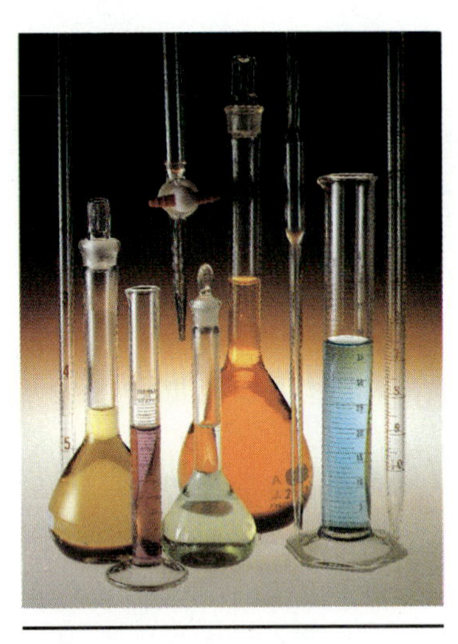

Although the SI system would recommend reporting the volume of these volumetric flasks as $0.00050\ m^3$, it is more useful to refer to them as 500-mL flasks.

### DERIVED SI UNITS

The units of every measurement in the SI system, no matter how simple or complex, must be derived from one or more of the seven base units. The preferred unit for volume is the cubic meter, for example, because volume has units of length cubed and the SI unit for length is the meter. The preferred unit for speed is meters per second because speed is the distance traveled divided by the time it takes to cover this distance.

SI unit of volume: $m^3$
SI unit of speed: m/s

Some of the common derived SI units are given in Table 1.5.

**TABLE 1.5 Common Derived SI Units in Chemistry**

| Physical Quantity | Name of Unit | Symbol |
|---|---|---|
| Density | | $kg/m^3$ |
| Electric charge | coulomb | C (A · s) |
| Electric potential | volt | V (J/C) |
| Energy | joule | J ($kg\text{-}m^2/s^2$) |
| Force | newton | N ($kg\text{-}m/s^2$) |
| Frequency | hertz | Hz ($s^{-1}$) |
| Pressure | pascal | Pa ($N/m^2$) |
| Velocity (speed) | meters per second | m/s |
| Volume | cubic meter | $m^3$ |

## NON-SI UNITS

Strict adherence to SI units would require changing directions such as "add 250 mL of water to a 1-L beaker" to "add 0.00025 cubic meters of water to an 0.001-$m^3$ container." Because of this, a number of units that are not srictly acceptable under the SI convention are still in use. Some of these non-SI units are given in Table 1.6.

**TABLE 1.6 Non-SI Units in Common Use**

| Physical Quantity | Name of Unit | Symbol |
|---|---|---|
| Volume | liter | L ($10^{-3}$ $m^3$) |
| Length | angstrom | Å (0.1 nm) |
| Pressure | atmosphere | atm (101.325 kPa) |
| | torr | mmHg (133.32 Pa) |
| Energy | electron volt | eV ($1.601 \times 10^{-19}$ J) |
| Temperature | degree Celsius | °C (K − 273.15) |
| Concentration | molarity | *M* (mol/L) |

## 1.7 UNCERTAINTY IN MEASUREMENT

There is a fundamental difference between stating that there are 12 inches in a foot and stating that the circumference of the earth at the equator is 24,903.01 miles. The first relationship is based on a *definition*. By convention, there are exactly 12 inches in 1 foot. The second relationship is based on a *measurement*. It reports the circumference of the earth to within the limits of experimental error in an actual measurement.

Many unit factors are based on definitions. There are exactly 5280 feet in a mile and 2.54 centimeters in an inch, for example. Unit factors based on definitions are known with complete certainty. (There is no error or uncertainty associated with these numbers.) Measurements, however, are always accompanied by a finite amount of error or uncertainty, which reflects limitations in the techniques used to make them.

The first measurement of the circumference of the Earth, in the third century B.C., for example, gave a value of 250,000 stadia—29,000 miles (see Figure 1.5). As the

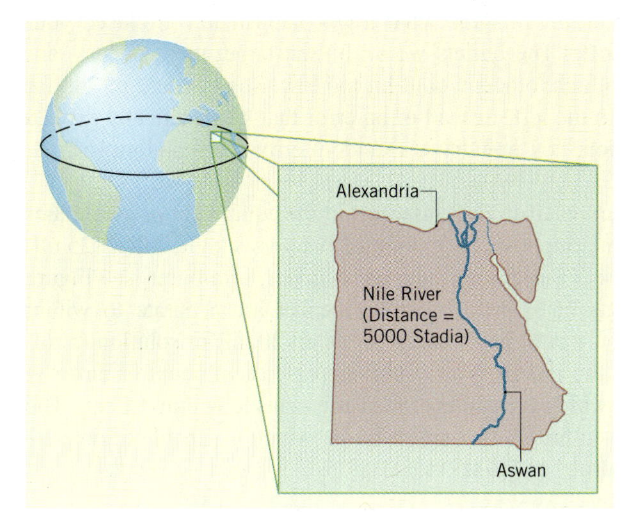


**Figure 1.5** The first estimate of the circumference of the Earth was based on the assumption that the distance between Alexandria and Aswan, where the measurements were made, was 5000 stadia.

quality of the instruments used to make this measurement has improved, the amount of error has gradually decreased. But it has never disappeared. Regardless of how carefully measurements are made, they always contain an element of uncertainty.

## 1.8 SYSTEMATIC AND RANDOM ERRORS—ACCURACY AND PRECISION

There are two sources of error in a measurement: (1) limitations in the sensitivity of the instruments used, and (2) imperfections in the techniques used to make the measurement. These errors can be divided into two classes: systematic and random.

The idea of **systematic error** can be understood in terms of the bull's-eye analogy shown in Figure 1.6. Imagine what would happen if you aimed at a target with a rifle whose sights were not properly adjusted. Instead of hitting the bull's-eye, you would systematically hit the target at another point. Your results would be influenced by a systematic error caused by an imperfection in the equipment being used.

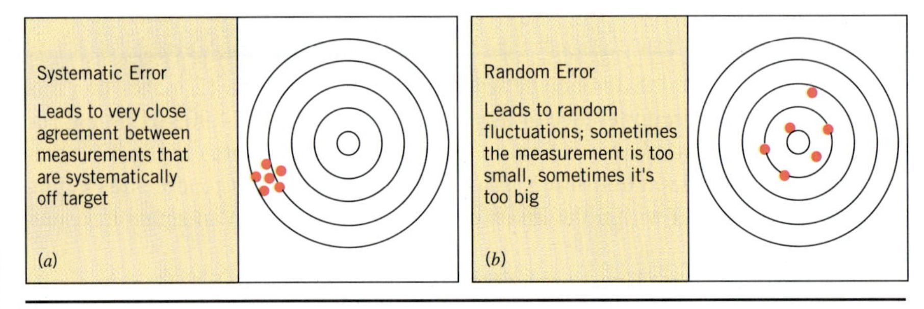


**Figure 1.6** (*a*) Systematic errors give results that are systematically too small or too large. (*b*) Random errors give results that fluctuate between being too small and too large.

Because the weight of the water is known to only ±0.1 grams, the total weight of this solution can be known to only ±0.1 grams, regardless of how carefully the weight of the salt is measured.

Systematic error can also result from mistakes the individual makes while taking the measurement. In the bull's-eye analogy, a systematic error of this kind might occur if you flinched and pulled the rifle toward you each time it was fired.

To understand **random error,** imagine what might happen if you closed your eyes before you fired the rifle. The bullets would hit the target more or less randomly. Some would hit too high, others would hit too low. Some would hit too far to the right, others too far to the left. Instead of an error that systematically gives a result too far in one direction, you now have a random error with random fluctuations.

Random errors most often result from limitations in the equipment or techniques used to make a measurement. Suppose, for example, that you want to collect 25 mL of a solution. You could use a beaker, a graduated cylinder, or a buret (see Figure 1.7). Volume measurements made with a 50-mL beaker are accurate to within ±5 mL. In other words, you would be as likely to obtain 20 mL of solution (5 mL too little) as 30 mL (5 mL too much). You could decrease the amount of error by using a graduated cylinder, which is capable of measurements to within ±1 mL. The error could be decreased even further by using a buret, which is capable of delivering a volume to within 1 drop, or ±0.05 mL.

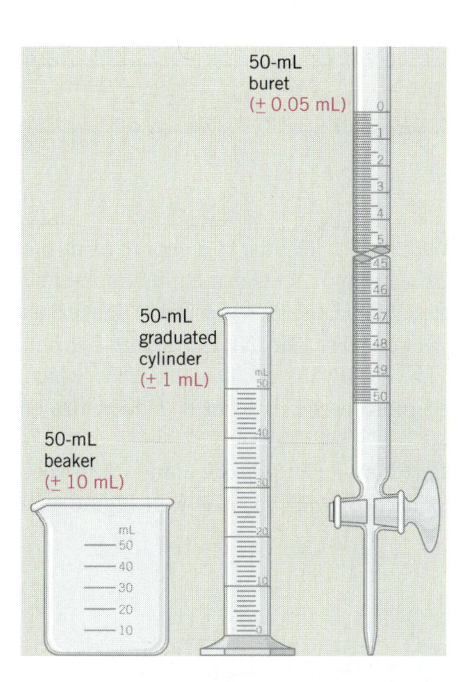


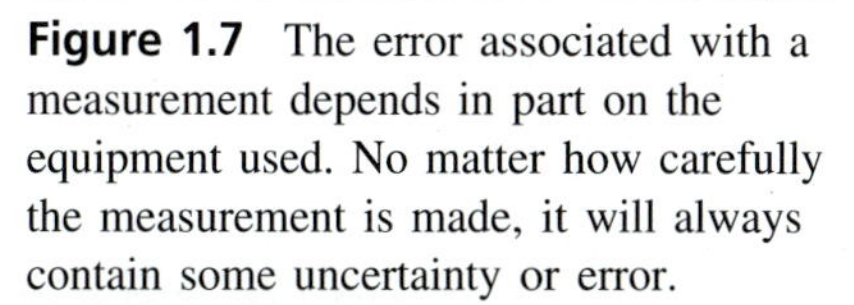

**Figure 1.7** The error associated with a measurement depends in part on the equipment used. No matter how carefully the measurement is made, it will always contain some uncertainty or error.

**EXERCISE 1.3**

Which of the following procedures would lead to systematic errors, and which would produce random errors?

(a) Using a 1-quart milk carton to measure 1-liter samples of milk.

(b) Using a balance that is sensitive to ±0.1 gram to obtain 250 milligrams of vitamin C.

(c) Using a 100-mL graduated cylinder to measure 2.5 mL of solution.

**SOLUTION** Procedure (a) would result in a systematic error. The volume would always be too small because a quart is slightly smaller than a liter. Procedures (b) and (c) would lead to random errors because the equipment used to make the measurements is not sensitive enough.

## ACCURACY AND PRECISION

To most people, *accuracy* and *precision* are synonyms, words that have the same or nearly the same meaning. In the physical sciences, there is an important difference between these terms. Measurements are **accurate** when they agree with the true value of the quantity being measured. They are **precise** when individual measurements of the same quantity agree.

The difference between accuracy and precision is similar to the difference between two terms from statistics: *validity* and *reliability*. Precision means the same thing as reliability. A measurement is precise, or reliable, if we get essentially the same result each time we make the measurement. Measurements are therefore precise when they are reproducible. Accuracy is the same as validity. A measurement is accurate, or valid, only if we get the correct answer.

## RESEARCH IN THE 90s

### MEASURING RISK

On February 26, 1989, ''60 Minutes'' broadcast the results of a report from the National Resource Defense Council (NRDC) warning parents that children might be at a high risk for cancer and neurological damage because of exposure to dangerous levels of pesticide residues on fruits and vegetables. The NRDC report focused particular attention on a growth regulator known by the trade name Alar, which kept apples on the tree longer and produced more perfectly shaped, redder, firmer fruit.

Researchers at the NRDC noted that Alar (daminozide) can decompose to succinic acid and unsymmetric dimethylhydrazine (UDMH), as shown in Figure 1.8.

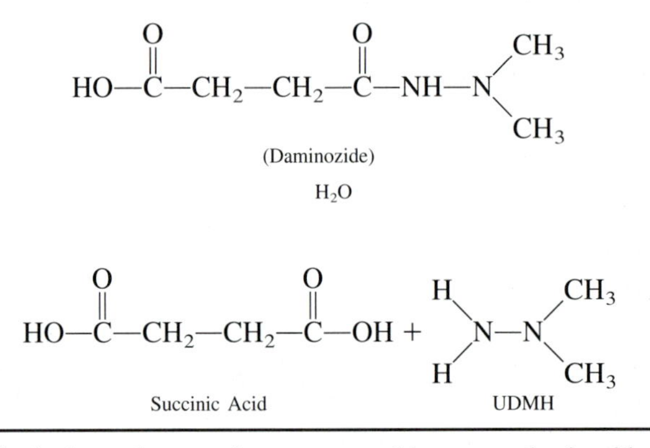


**Figure 1.8** Alar is the trade name for a compound known as daminozide that decomposes to form succinic acid and unsymmetric dimethylhydrazine (UDMH).

They then based their assessment of the risk of exposure to daminozide on research done by Bela Toth on the effect on rodents of high-dose exposure to UDMH.

The reaction to the ''60 Minutes'' broadcast and the subsequent media attention has been described as follows.

> Fears over the potential presence in apples of residues of a possibly carcinogenic growth regulator led schools across the U.S. to ban lunchroom use of apples and apple products . . . Mothers, nurses, and school officials, some ''scared to the point of hysteria,'' deluged the Environmental Protection Agency with calls. Parents poured apple juice down the drain. (*Chemical and Engineering News,* March 20, 1989, p. 7)

This reaction led the EPA, the Food and Drug Administration (FDA), and the Department of Agriculture to issue an extraordinary joint statement declaring that ''The federal government believes that it is safe for Americans to eat apples. There is not an imminent hazard posed to children, despite claims to the contrary.'' The damage, however, was done. Apple growers and processors experienced a 25% decrease in sales of apples and a 30% decrease in sales of apple juice.

Pesticides being sprayed onto a field of crops.

For our purposes, the question of who was right—the NRDC or the EPA/FDA/ Department of Agriculture—is not relevant. Two important questions arise: Why did the EPA and other agencies discount Toth's work on UDMH in their assessment of the risk of exposure to Alar? Second, why is it so difficult to measure the risk stemming from exposure to pesticides and pesticide residues?

The key to the first question is the difference between advances that have been made in measurements of exposure to potentially hazardous substances (detection) and measurements of their biological effect (risk). Advances in technology have enabled us to detect substances at levels as low as a few picograms ($10^{-12}$ g). Unfortunately the bioassays used to determine whether these substances are hazardous are insensitive and subject to large experimental errors. As a result, studies of the effect of exposure to *reasonable* amounts of these substances can't be done. High doses are therefore used, and the data are then extrapolated to low-dosage situations.

As a rule, toxicologists try to limit the dose in bioassays to the Maximum Tolerated Dose (MTD) that causes no more than a 10% weight loss and few early deaths within the test animal population. In Toth's work, the top dose of UDMH was 29 milligrams per kilogram of body weight per day. Because the dose used in this study was so high that many animals died early, the EPA concluded that it was not valid to estimate the risk of exposure to UDMH in humans. They noted that the average citizen's exposure to UDMH is only 0.000047 milligrams per kilogram, and other studies showed no significant increase in tumors at exposure levels up to 3 milligrams per kilogram.

A major theme for research in the 90s will be the attempt to improve our ability to assess the risks of exposure to potentially hazardous substances. In their book *In Search of Safety: Chemicals and Cancer Risk* (Harvard University Press, 1988), John Graham, Laura Green, and Marc Roberts wrote "as scientists discover more of the causal mechanism underlying the health effects of chemicals . . . they will be able to make quantitative risk assessments with the reliability that the legal/political system pretends is possible today."

The present state of risk assessment for exposure to potential carcinogens was summarized by Bruce Ames and Lois Swirsky Gold (*Chemical and Engineering News,* January 7, 1991, p. 28):

> The attempt to prevent cancer by regulating low levels of synthetic chemicals by "risk assessment," using worst-case, one-in-a-million risk scenarios is not scientifically justified. Testing chemicals for carcinogenicity at near-toxic doses in rodents does not provide enough information to predict the excess numbers of human cancers that might occur at low-dose exposures. In addition, this cancer prevention strategy is . . . counterproductive because it diverts resources from much more important risks, and, in the case of synthetic pesticides, makes fruits and vegetables more expensive, thus serving to decrease consumption of foods that help prevent cancer.

Ames and Gold note that there is no fundamental difference between the response of test animals to high-dose exposures to synthetic and natural chemicals. About one-half of the chemicals tested seem to be carcinogens, regardless of whether they are natural or synthetic.

Ames and Gold argue that natural pesticides account for 99.99% of the average human exposure to pesticides in food. They base this conclusion on a comparison of the results of FDA analyses of chemical residues in food and estimates of the average consumption of natural pesticides. The FDA tested for 200 chemicals in food, including the most important synthetic pesticides and industrial chemicals. They found residues for 105 of these substances and concluded that the average American consumes a total of 0.09 mg of these chemicals per day. Ames and Gold note, however, that the average American eats about 1.5 grams of natural pesticides each day, "which is 10,000 times more than they eat of synthetic pesticides residues." They also note that the cooking of food produces about 2 grams per person per day of burnt material that contains known carcinogens. Furthermore, they note that a typical cup of coffee contains at least 10 mg of known carcinogens. A single cup of coffee therefore contains the equivalent of a year's worth of synthetic pesticide residues.

Some argue that the risk of exposure to hazardous chemicals involves more than the threat of cancer. Many chemicals are also *teratogens,* which lead to birth defects. Much attention has been paid in recent years to the risk of exposure to an industrial by-product known as dioxin, or, more accurately, 2,3,7,8-tetrachlorodibenzo-*p*-dioxin (TCDD). TCDD has been shown to be both carcinogenic and teratogenic in rodents at low doses. Ames and Gold argue, however, that TCDD should be of minor concern as a teratogen, when compared with alcohol. They note that the potential for birth defects due to the estimated average exposure of TCDD in the United States is equivalent to the threat associated with the consumption of 0.000036 ounces of beer each day.

To understand how systematic and random errors affect accuracy and precision, let's return to the bull's-eye analogy (see Figure 1.9). Systematic errors influence the accuracy, but not necessarily the precision, of a measurement. It is possible to get measurements that are consistently the same but systematically wrong. Random errors influence both the accuracy and the precision of the measurement.

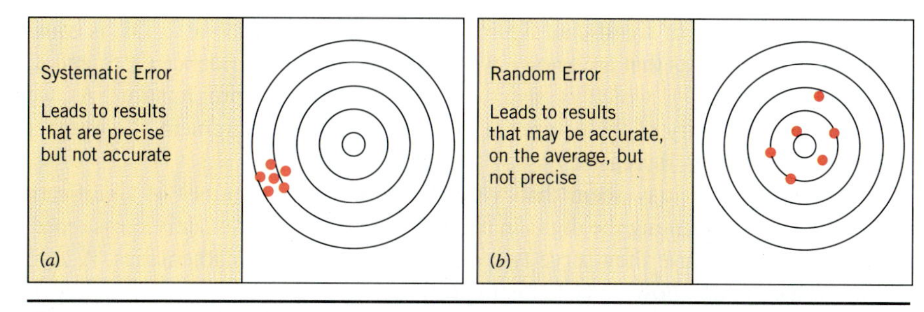


**Figure 1.9** (*a*) Systematic errors affect the accuracy of a measurement. The measurement may still be precise, but the results are systematically different from the correct answer. (*b*) Random errors can affect both the accuracy and precision of a measurement.

Systematic errors can be reduced by increasing the care and patience of the individual making the measurement or by improving the sensitivity of the equipment used. Random errors can be reduced by averaging the results of many measurements of the same quantity because the precision of a series of measurements increases with the square root of the number of measurements.

### CHECKPOINT

Describe a source of random error and a source of systematic error other than those identified in Exercise 1.3. Explain why these errors are random or systematic and describe the effect of these errors on accuracy and/or precision of the measurement.

## 1.9 SIGNIFICANT FIGURES

It is important to be honest when reporting a measurement, so that it does not appear to be more accurate than the equipment used to make the measurement allows. We can achieve this goal by controlling the number of digits, or **significant figures,** used to report the measurement.

Imagine what would happen if you determined the mass of an old copper penny on a postage scale, a two-pan laboratory balance, and an analytical balance (see Figure 1.10). The postage scale might give a mass of about 3 grams, which means that the penny is closer to 3 grams than either 2 grams or 4 grams. The two-pan balance is inherently more sensitive; it can give results to within the nearest hundredth of a gram (±0.01 g). When this balance is used, the penny might have a mass of 2.53 g. The analytical balance is even more sensitive. It can measure the mass of an object to the nearest ±0.001 g, in which case you might find that the penny has a mass of 2.531 g.

| | |
|---|---|
| Postage scale: | 3 ± 1 g |
| Two-pan balance: | 2.53 ± 0.01 g |
| Analytical balance: | 2.531 ± 0.001 g |

**Figure 1.10** If the price of copper is known to only two significant figures, the value of the copper in this penny would be reported to two significant figures, regardless of how accurately the weight of the penny was measured.

The postage scale gives a measurement that has only one reliable digit. This measurement is therefore said to be good to only one significant figure. The two-pan balance gives three significant figures (2.53) and the analytical balance gives four

(2.531). The number of significant figures in a measurement, such as 2.531, is equal to the number of digits that are known with some degree of confidence (2, 5, and 3) plus the last digit (1), which is generally an estimate or approximation. As we improve the sensitivity of the equipment used to make a measurement, the number of significant figures increases.

At first glance, it may seem that we can determine the number of significant figures by simply counting the digits in the measurement. Unfortunately zeros represent a problem because they serve two different functions in mathematics. A zero can act as a counter, allowing us to distinguish between 304 and 324, for example. It also can be used to set the decimal point, so that we can distinguish between 0.045, 0.45, 4.5, 45, 450, and so on. In the first case, the zero is a significant figure; in the second, it isn't.

To understand why zeros used to set the decimal point are not significant, think about the relative accuracy of measurements of 0.045 inches and 45 inches. Both measurements are known to an accuracy of $\pm 1$ part in 45. Both measurements must therefore contain the same number of significant figures. The keys to counting significant figures are summarized below.

- Zeros *within* a number are always significant. Both 4308 and 40.05 contain four significant figures.
- Zeros that do nothing but set the decimal point are not significant. Thus, 470,000 has two significant figures.
- Trailing zeros that aren't needed to hold the decimal point are significant. For example, 4.00 has three significant figures.

**CHECKPOINT**

Explain why the zero in 407 is a significant figure but not the zero in 470.

If you are not sure whether a digit is significant, assume that it isn't. For example, if the directions for an experiment read: "Add the sample to 400 mL of water," assume the volume of water is known to one significant figure.

**EXERCISE 1.4**

Determine the number of significant figures in the following measurements.

(a) 23.07 in (b) 0.003210 m (c) 2000 lb (d) $400.00

**SOLUTION**

(a) Four significant figures, because the zero is a counter.
(b) Four significant figures (3210) because the last zero is a counter.
(c) One significant figure because the zeros are needed to set the decimal point.
(d) Five significant figures because this number has more zeros than needed to set the decimal point.

### ADDITION AND SUBTRACTION WITH SIGNIFICANT FIGURES

What is the mass of a solution prepared by adding 0.507 gram of salt to 150.0 grams of water? If we attacked this problem without considering significant figures, we would simply add the two measurements.

$$\begin{array}{rl} & 150.0 \text{ g } H_2O \\ + & 0.507 \text{ g salt} \\ \hline & 150.507 \text{ g solution} \end{array} \qquad \text{(without using significant figures)}$$

But this answer doesn't make sense. We only know the mass of the water to the nearest tenth of a gram, so we can only know the total mass of the solution to within $\pm 0.1$ g. Taking significant figures into account, we find that adding 0.507 grams of salt to 150.0 grams of water gives a solution with a mass of 150.5 grams.

$$\begin{array}{rl} & 150.0 \text{ g } H_2O \\ + & 0.507 \text{ g salt} \\ \hline & 150.5 \text{ g solution} \end{array} \qquad \text{(using significant figures)}$$

Many of the calculations in this book are done by combining measurements with different degrees of accuracy and precision. The guiding principle in carrying out these calculations is easily stated:

**The accuracy of the final answer can be no greater than the least accurate measurement.**

This principle can be translated into a simple rule for addition and subtraction as follows:

**When measurements are added or subtracted, the answer can contain no more *decimal places* than the least accurate measurement.**

### MULTIPLICATION AND DIVISION WITH SIGNIFICANT FIGURES

The same principle governs the use of significant figures in multiplication and division: the final result can be no more accurate than the least accurate measurement. In this case, however, we count the significant figures in each measurement, not the number of decimal places:

**When measurements are multiplied or divided, the answer can contain no more *significant figures* than the least accurate measurement.**

To illustrate this rule, let's calculate the cost of the copper in an old penny that is pure copper. Let's assume that the penny has a mass of 2.531 grams, that it is essentially pure copper, and that the price of copper is 67 cents per pound. We can start by converting from grams to pounds:

$$2.531 \not{g} \times \frac{1 \text{ lb}}{453.6 \not{g}} = 0.005580 \text{ lb}$$

We then use the price of a pound of copper to calculate the cost of the copper metal:

$$0.005580 \not{\text{lb}} \times \frac{67¢}{1 \not{\text{lb}}} = 0.37¢$$

There are four significant figures in both the mass of the penny (2.531) and the number of grams in a pound (453.6). But there are only two significant figures in the price of copper, so the final answer can only have two significant figures. Exercise 1.5 illustrates the effect of defined conversion factors on the number of significant figures.

**EXERCISE 1.5**

Calculate the length in inches of a piece of wood 1.245 feet long.

**SOLUTION** This problem is easy to set up:

$$1.245 \not{ft} \times \frac{12 \text{ in}}{1 \not{ft}} = 14.94 \text{ in}$$

It is not as easy to decide how many significant figures the final answer should have. The original measurement (1.254 feet) has four significant figures, but there seem to be only two signficant figures in the number of inches in a foot. Thus, it might seem that the answer should contain only two significant figures.

We can clear up this confusion by remembering that only *measurements* involve error or uncertainty. Many unit factors are based on *definitions*. For example, 1 foot is defined as exactly 12 inches. *Unit factors based on definitions have an infinite number of significant figures*. The answer to this problem therefore contains four significant figures.

## 1.10 ROUNDING OFF

When the answer to a calculation contains too many significant figures, it must be rounded off. Assume that the answer to a calculation is 1.247 and that the least accurate measurement has only three significant figures. The simplest way to round off this number would be to ignore the final digit and report the first three: 1.24. This approach has the disadvantage of introducing a systematic error into our calculations. Each time we round off, we would underestimate the value of the final answer.

There are 10 digits that can occur in the last decimal place in a calculation. One way of rounding off involves underestimating the answer for five of these digits (0, 1, 2, 3, and 4) and overestimating the answer for the other five (5, 6, 7, 8, and 9). This approach to rounding off is summarized as follows.

- If the digit is smaller than 5, drop this digit and leave the remaining number unchanged. Thus, 1.684 becomes 1.68.
- If the digit is 5 or larger, drop this digit and add 1 to the preceding digit. Thus, 1.247 becomes 1.25.

## 1.11 THE DIMENSIONAL ANALYSIS APPROACH TO UNIT CONVERSIONS

Section 1.3 used unit factors as a tool for converting from one set of units to another. This section introduces a technique known as **dimensional analysis** that can be used to extend unit factors to similar tasks. Dimensional analysis involves watching what happens to the units (or dimensions) in a calculation to make sure it is done correctly. If the units cancel as expected, the calculation has been set up properly. The following exercises illustrate possible pitfalls.

Secretariat, in the Belmont Stakes, as he became the first triple crown winner in 25 years.

## EXERCISE 1.6

The record for the Kentucky Derby is held by Secretariat, who ran the 10 furlongs in 1 minute, 59.4 seconds. Calculate his average speed in miles per hour.

**Problem-Solving Strategy**

**SOLUTION** The key to this calculation is keeping track of the units used to express the dimensions of the problem. The problem gives the distance for the Kentucky Derby in furlongs and the length of the race in minutes and seconds. It then asks us to convert these data into units of miles and hours. We can start by looking up the definition of a furlong: 1/8 mile.

Two unit factors can be constructed from the definition of a furlong. One describes the number of furlongs per mile. The other tells us the number of miles in a furlong.

$$\frac{8 \text{ furlongs}}{1 \text{ mi}} = 1 \qquad \frac{1 \text{ mi}}{8 \text{ furlongs}} = 1$$

If we multiply 10 furlongs by the appropriate unit factor, the units of furlongs cancel and the distance comes out in the desired units, miles.

$$10 \cancel{\text{ furlongs}} \times \frac{1 \text{ mi}}{8 \cancel{\text{ furlongs}}} = 1.25 \text{ miles}$$

(There are an endless number of significant figures in this result because none of the quantities in this calculation are measurements. The distance of the Kentucky Derby is defined as 10 furlongs, and a furlong is defined as 1/8 of a mile.)

To complete this problem we must convert 1 minute, 59.4 seconds into units of hours. We might start by calculating the total number of seconds:

$$\left[1 \cancel{\text{ min}} \times \frac{60 \text{ s}}{1 \cancel{\text{ min}}}\right] + 59.4 \text{ s} = 119.4 \text{ s}$$

We can then use the appropriate unit factors to convert seconds into minutes and minutes into hours:

$$119.4\ \text{s} \times \frac{1\ \text{min}}{60\ \text{s}} \times \frac{1\ \text{h}}{60\ \text{min}} = 0.03317\ \text{h}$$

To calculate the average speed in miles per hour we divide the distance traveled in miles by the time it took in hours:

$$\frac{1.25\ \text{mi}}{0.03317\ \text{h}} = 37.68\ \text{mi/h}$$

## EXERCISE 1.7

Calculate the volume in liters of a cubic container 0.500 meter tall.

**SOLUTION** We can start by calculating the volume of the container in units of cubic meters, which is the product of its height times its length times its width:

$$V = (0.500\ \text{m} \times 0.500\ \text{m} \times 0.500\ \text{m}) = 0.125\ \text{m}^3$$

**Problem-Solving Strategy**

It isn't obvious what should be done next, so let's look at the goal of the calculation and work backwards.

Our goal is the volume of the container in liters. We might therefore ask: What do we know about this unit? We know that a liter contains 1000 mL. But a milliliter is equivalent to a cubic centimeter. Thus, 1 liter contains 1000 cubic centimeters:

$$1\ \text{L} = 1000\ \text{cm}^3$$

If we can convert the volume of the container from cubic meters to cubic centimeters, we can then convert cubic centimeters to liters.

The conversion between cubic meters and cubic centimeters is based on the equality:

$$1\ \text{m} = 100\ \text{cm}$$

The unit factor used to do this conversion must be cubed, however, because we have to convert from cubic meters to cubic centimeters:

$$0.125\ \text{m}^3 \times \left[\frac{100\ \text{cm}}{1\ \text{m}}\right]^3 = 1.25 \times 10^5\ \text{cm}^3$$

Now that we know the volume in cubic centimeters, we can calculate the volume in liters:

$$1.25 \times 10^5\ \text{cm}^3 \times \left[\frac{1\ \text{L}}{1000\ \text{cm}^3}\right] = 125\ \text{L}$$

## EXERCISE 1.8

What is the value of a gold ingot 20.0 cm long by 8.5 cm wide by 6.0 cm tall, if the mass of a cubic centimeter of gold is 19.3 grams and the price of gold is $356 per ounce?

**SOLUTION** Because the problem gives the length, height, and width of the gold ingot, we might start by calculating the volume of the ingot.

$$V = (20.0 \text{ cm} \times 8.5 \text{ cm} \times 6.0 \text{ cm}) = 1020 \text{ cm}^3$$

This calculation is useful because we know something about the mass of a cubic centimeter of gold:

$$1 \text{ cm}^3 = 19.3 \text{ g}$$

We can therefore calculate the mass of the ingot in grams:

$$1020 \cancel{\text{cm}^3} \times \frac{19.3 \text{ g}}{1 \cancel{\text{cm}^3}} = 19{,}700 \text{ g}$$

We can now convert from grams into pounds:

$$19{,}700 \cancel{\text{g}} \times \frac{1 \text{ lb}}{453.6 \cancel{\text{g}}} = 43.4 \text{ lb}$$

We can then calculate the number of ounces of gold in the ingot:

$$43.4 \cancel{\text{lb}} \times \frac{16 \text{ oz}}{1 \cancel{\text{lb}}} = 694 \text{ oz}$$

We are now ready to use the price of gold to calculate the value of the ingot in dollars:

$$694 \cancel{\text{oz}} \times \frac{\$356}{1 \cancel{\text{oz}}} = \$250{,}000$$

The final answers in Exercises 1.7 and 1.8 have been rounded off to the number of significant figures that reflect the accuracy of the least accurate measurements. An extra significant figure was carried in intermediate steps in these calculations, however, to avoid the error that can creep in if each step is rounded off to the number of significant figures allowed in the final answer.

## 1.12 SCIENTIFIC NOTATION

Chemists routinely work with numbers that are either extremely large or extremely small. There are 10,300,000,000,000,000,000,000 carbon atoms in a 1-carat diamond, for example, each of which has a mass of 0.000,000,000,000,000,000,000,020 gram. In theory, the product of these numbers is the mass of a 1-carat diamond. It is impossible to multiply these numbers with most calculators, however, because they can't accept either number as it is written here. To do a calculation like this, it is necessary to express these numbers in **scientific notation:** as a number between 1 and 10 multiplied by 10 raised to some exponent.

Before we discuss how to translate numbers into scientific notation, it might be useful to review some of the basics of exponential mathematics.

- Any number raised to the zero power is equal to 1:

$$1^0 = 1 \qquad 10^0 = 1$$

- Any number raised to the first power is equal to itself:

$$1^1 = 1 \qquad 10^1 = 10$$

- Any number raised to the $n$th power is equal to the product of that number times itself $n - 1$ times:

$$2^2 = 2 \times 2 = 4$$
$$10^5 = 10 \times 10 \times 10 \times 10 \times 10 = 100{,}000$$

- Dividing by a number raised to some exponent is the same as multiplying by that number raised to an exponent of the opposite sign:

$$\frac{1}{10^2} = 10^{-2} \qquad \frac{1}{10^{-3}} = 10^3$$

The following rule can be used to convert numbers into scientific notation:

**The exponent in scientific notation is equal to the number of times the decimal point must be moved to produce a number between 1 and 10.**

According to the 1990 census, for example, the population of Chicago was 6,070,000 ± 1000. To convert this number to scientific notation, we move the decimal point to the left six times:

$$6\,070000 = 6.070 \times 10^6$$

To translate 10,300,000,000,000,000,000,000 carbon atoms into scientific notation, we move the decimal point to the left 22 times:

$$1\,0300000000000000000000 = 1.03 \times 10^{22}$$

To convert numbers smaller than 1 into scientific notation, we have to move the decimal point to the right. The decimal point in 0.000985, for example, must be moved to the right four times:

$$0.0009\,85 = 9.85 \times 10^{-4}$$

Converting 0.000,000,000,000,000,000,000,020 grams per carbon atom into scientific notation involves moving the decimal point to the right 23 times:

$$0.00000000000000000000002\,0 = 2.0 \times 10^{-23}$$

The primary reason for converting numbers into scientific notation is to make calculations with unusually large or small numbers less cumbersome. But there is another important advantage to scientific notation. Because zeros are no longer used to set the decimal point, all of the digits in a number in scientific notation are significant, as shown by the following examples:

| | |
|---|---|
| $1.03 \times 10^{22}$ | 3 significant figures |
| $9.85 \times 10^{-4}$ | 3 significant figures |
| $2.0 \times 10^{-23}$ | 4 significant figures |

**EXERCISE 1.9**

Convert the following numbers into scientific notation.

(a) 0.004694 (b) 1.98 (c) 4,679,000

**SOLUTION**

(a) $4.694 \times 10^{-3}$ (b) $1.98 \times 10^0$ (c) $4.679 \times 10^6$

## 1.13 EXTENSIVE AND INTENSIVE QUANTITIES

Many of the quantities whose measurements are discussed in this chapter are **extensive properties,** which depend on the size of the sample. Mass and volume, for example, are extensive properties because the larger the sample, the bigger the mass or volume. Other extensive properties include distance and time.

At first glance, it might seem that anything you measure would be an extensive property. But that isn't true. There are also **intensive properties,** which do not depend on the size of the sample. Temperature and pressure are both examples of intensive properties. On the average, the temperature of the air in your immediate vicinity is the same as the temperature of the air throughout the room. The same can be said about the pressure of the gas in one portion of the room compared with the pressure of the gas throughout the room.

Many intensive properties are the ratios of a pair of extensive properties that are proportional to each other. If you divide the number of times your heart beats by the length of time during which you counted, the result is an intensive quantity. It doesn't matter whether you count 35 beats in 30 seconds or 175 beats in 2.5 minutes. In either case, the heart rate would be 70 beats per minute. Speed is another example of an intensive property that is the ratio of a pair of extensive properties that are proportional to each other. Regardless of whether you drive 3.90 miles in 216 seconds or 45.5 miles in 42 minutes, your speed is 65 miles per hour.

## 1.14 DENSITY AS AN EXAMPLE OF AN INTENSIVE QUANTITY

If you ask children, "Why does ice float on water?" the most common answer is, "Ice is lighter than water." If you ask them to compare pieces of copper and lead, they often say that lead is heavier than copper. These responses result from confusion between extensive and intensive properties.

The key to creating an intensive property is to find a pair of extensive properties that are proportional to each other. The mass and volume of a substance offer a perfect example of this phenomenon. Figure 1.11 shows how the mass of a sample of sugar depends on its volume. Both mass and volume are extensive properties. (The bigger the sample, the more it weighs and the more space it occupies.) But the

Ice floats because it is not as dense as water.

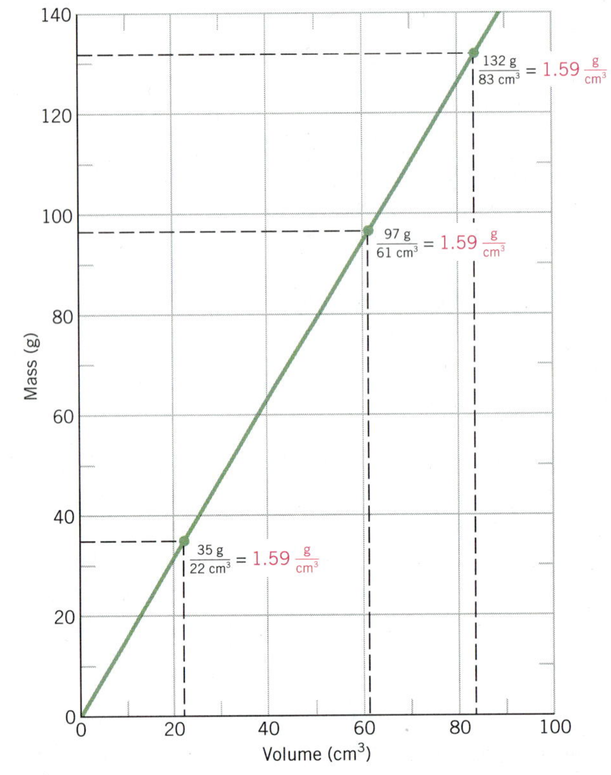


**Figure 1.11** Mass and volume are both extensive properties. The larger the sample, the larger the mass or volume. Density is an intensive property. No matter how large or small the sample, the ratio of mass to volume is the same. This graph shows that we get the same density for sugar regardless of the size of the sample we analyze.

ratio of the mass of the sample to its volume is constant. This ratio is an intensive property known as **density,** which is a characteristic property of the substance:

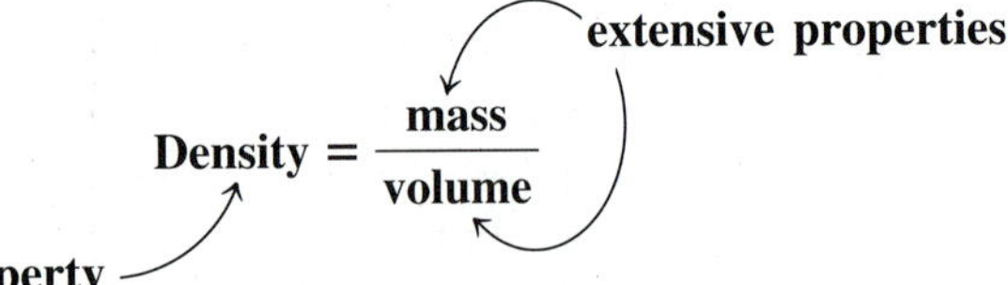


The density of water, for example, is about 1 gram per milliliter, regardless of whether we measure water by the drop or by the gallon. Intensive properties, such as density, are therefore characteristic properties of a substance that can be used to identify the substance.

Densities of solids and liquids are given in units of grams per milliliter (g/mL) or grams per cubic centimeter (g/$cm^3$). Most substances have densities that fall in the range between 0.10 and 10 g/$cm^3$. The densities of certain common substances are given in Table 1.7, and the densities of some metals are given in Table 1.8.

Ice doesn't float on water because it is lighter than water; it floats because it is less dense. A given volume of ice weighs less than an equivalent volume of water, so the water settles to the bottom, and the ice floats on top. A piece of lead may feel heavy when we pick it up. But it isn't so much the weight that we notice as the difference between what we feel and what we expect for an object that size. Lead isn't heavier than copper, it is more dense.

Because lead is more dense than copper, a piece of lead seems ''heavy'' when compared with a piece of copper.

## ▶ CHECKPOINT

The difference between the density of vinegar (1 g/$cm^3$) and oil (0.9 g/$cm^3$) explains why salad dressings that contain both ingredients often separate to form two liquids. Which liquid floats on top? Is it the oil, or the vinegar?

**TABLE 1.7 Densities of Some Common Substances**

| Substance | Density ($g/cm^3$) | Substance | Density ($g/cm^3$) |
|---|---|---|---|
| Air at 25°C | 0.0408 | Limestone | 2.7 |
| Bone | 1.7 | Olive oil | 0.918 |
| Cement | 2.7–3 | Salt | 2.18 |
| Charcoal | 0.28–0.57 | Sugar | 1.59 |
| Coal | 1.2–1.8 | Water at 4°C | 1.00 |
| Cork | 0.22–0.26 | Wood | |
| The Earth | 5.519 | Balsa | 0.11–0.14 |
| Gasoline | 0.66–0.69 | Oak | 0.60–0.90 |
| Ice | 0.917 | | |

**TABLE 1.8 Densities of Some Common Metals**

| Substance | Density ($g/cm^3$) | Substance | Density ($g/cm^3$) |
|---|---|---|---|
| Aluminum | 2.70 | Nickel | 8.90 |
| Brass | 8.40 | Platinum | 21.45 |
| Chromium | 7.20 | Potassium | 0.856 |
| Copper | 8.92 | Silver | 10.5 |
| Gold | 19.3 | Sodium | 0.97 |
| Iron | 7.86 | Steel | 7.8 |
| Lead | 11.34 | Titanium | 4.5 |
| Magnesium | 1.74 | Tungsten | 19.35 |
| Mercury | 13.60 | Zinc | 7.14 |

**CHECKPOINT**

A ship can be described as a ''steel balloon.'' Use this to explain how the Titanic was able to float until it hit the iceberg, even though steel is almost eight times as dense as water.

**EXERCISE 1.10**

Calculate the density of concentrated sulfuric acid if a 175-mL sample of this acid has a mass of 322 g.

**SOLUTION** Density is calculated by dividing the mass of a sample by its volume. The density of concentrated sulfuric acid is therefore 1.84 g/mL, or 1.84 $g/cm^3$:

$$\text{Density} = \frac{322 \text{ g}}{175 \text{ mL}} = 1.84 \text{ g/mL}$$

The key to this calculation, of course, is recognizing that the measurements of mass and volume must be made on the same sample.

**CHECKPOINT**

Archaeologists use a technique they call *flotation* to isolate small fragments of charcoal or bone. They start by dissolving large quantities of a soluble salt in water to produce a solution with a density of about 1.8 $g/cm^3$. The mixture of sand, soil, small rocks, and fragments of charcoal and bone they wish to separate is then added to this solution. Explain why charcoal and bone float, but sand (2.5 $g/cm^3$), soil (3 $g/cm^3$), and rocks ($\approx$5 $g/cm^3$) sink to the bottom.

## 1.15 TEMPERATURE AS AN INTENSIVE QUANTITY

The origin of the thermometer can be traced to the work of Galileo, who constructed the first ''thermoscope'' in 1592 by trapping air in a large glass bulb with a long narrow neck inverted over a container of water or wine. Changes in the **temperature** of the room cause the air in the bulb to expand or contract, forcing the liquid to travel up or down the tube. It was not until almost 20 years later, however, that a colleague of Galileo suggested adding a scale to the thermoscope to make the first thermometer.

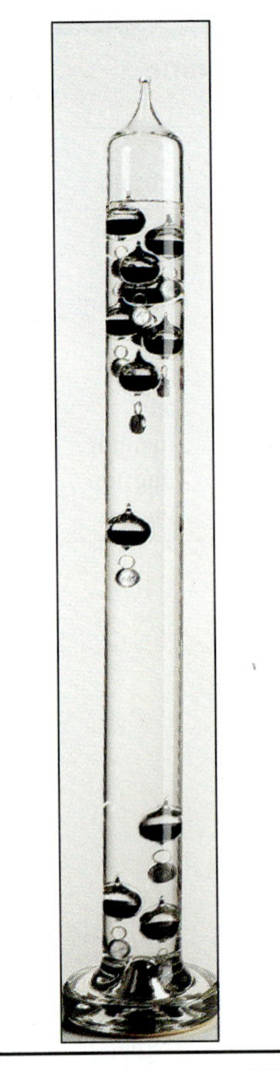

This sculpture recreates a 17th-century experiment by Galileo, in which hand-blown glass spheres of different weights float upward or downward depending on temperature.

## RELATIVE TEMPERATURE SCALES

By the early 1700s, at least 35 different temperature scales had been proposed. At that time, a Dutch instrument maker by the name of Daniel Gabriel Fahrenheit became famous for his mercury thermometers. The Fahrenheit scale he developed is still the most widely used temperature scale in the United States.

The problem that Fahrenheit faced is still a common one. It is easy to make *relative measurements,* such as determining that one object is hotter than another. But it is much harder to make *absolute measurements,* such as determining the temperature of an object on an absolute scale.

The approach Fahrenheit took to this problem is still used for most instruments today. Fahrenheit calibrated his thermometer against a pair of references: body temperature and the lowest temperature he could achieve by adding salt to ice water. On this scale, the temperature of ice by itself is 32°F and the temperature of boiling water is 212°F.

As early as the second century A.D., the Greek physician Galen suggested using ice and boiling water as the basis of a temperature scale. This was not done until 1742, however, when the Swedish astronomer Anders Celsius introduced the temperature scale now used in almost every country in the world. Celsius defined the temperature of ice and boiling water as 0°C and 100°C, respectively. His scale was known as the centigrade scale because a degree on this scale is 1/100 of the difference between the temperatures of ice and boiling water. In 1948, the name was officially changed to the Celsius scale.

The Fahrenheit and Celsius scales are both relative temperature scales. They define two reference points, divide the range of temperatures between these points into degrees, and then compare all temperatures with these arbitrary references.

## ABSOLUTE TEMPERATURE SCALES

At the beginning of the 1800s, a relationship was discovered between the volume and the temperature of a gas. This relationship suggests that the volume of a gas should become zero at a temperature of −273.15°C. In 1848 the British physicist William Thompson, who later became Lord Kelvin, suggested that this observation could be used as the basis for an absolute temperature scale. On the Kelvin scale, absolute zero (0 K) is the temperature at which the volume of a gas becomes zero. It is therefore the lowest possible temperature, or the absolute zero on any temperature scale. Zero on the Kelvin scale is therefore −273.15°C.

$$0 \text{ K} = -273.15°\text{C}$$

Each unit on this scale, or each *kelvin,* is equal to 1 degree on the Celsius scale. There is a subtle difference between the units on these scales, however. Because the Celsius scale is based on two arbitrary reference points, the difference between the temperatures of these two points is divided into *degrees.* The Kelvin scale, however, is an absolute scale. Zero is not arbitrarily defined; it is the lowest possible temperature that can be achieved. Thus, temperatures on the Kelvin scale are not divided into degrees. Temperatures on this scale are reported in units of "kelvin," not in "degrees kelvin."

## TEMPERATURE CONVERSIONS

Figure 1.12 shows the relationship between the Fahrenheit, Celsius, and Kelvin scales. The simplest temperature scales to interconvert are the Celsius and Kelvin

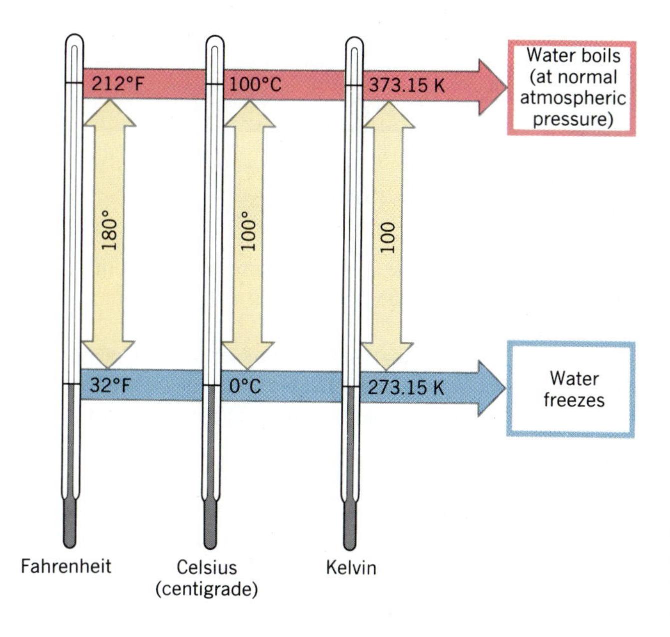


**Figure 1.12** The three common temperature scales. Note that water boils at 212°F, 100°C, or 373.15 K and freezes at 32°F, 0°C, or 273.15 K.

scales. All we have to remember is that zero on the Kelvin scale is equivalent to −273.15°C, and that 1 K is equal to 1°C. Thus, the temperature on the Kelvin scale is equal to the temperature in degrees Celsius plus 273.15:

$$T_{K} = T_{°C} + 273.15$$

Converting from degrees Celsius to degrees Fahrenheit, or vice versa, is a bit more difficult. We can start by noting that the difference between the temperatures of ice and boiling water is either 180°F or 100°C. This means that a degree on the Fahrenheit scale is smaller than a degree on the Celsius scale. Their relationship can be expressed as a factor of 180°/100°, or 9/5.

There is another problem, however. The temperature at which the two scales reach zero is not the same; zero on the Celsius scale is equivalent to 32°F. To convert from Celsius into Fahrenheit we therefore have to multiply by 9/5 and then add 32°.

$$T_{°F} = \frac{9}{5} T_{°C} + 32$$

This equation can be rearranged to convert from Fahrenheit to Celsius.

$$T_{°C} = \frac{5}{9} [T_{°F} - 32]$$

These equations are difficult to remember. There are two ways of overcoming this problem. You can remember how they are derived, in which case you no longer have to remember the equations. Likewise, you can check to make sure the equation you think you remember gives the right results. The equations are correct if they convert 212°F into 100°C and 32°F into 0°C.

**CHECKPOINT**

Show that the °F and °C scales converge at −40°.

−40°F = −40°C

**SPECIAL TOPIC**

## THE GRAPHICAL TREATMENT OF DATA

If the basis of science is a natural curiosity about the world that surrounds us, an important step in doing science is trying to find patterns in the observations and measurements that result from this curiosity.

Anyone who has played with a prism, or seen a rainbow, has watched what happens when white light is split into a spectrum of different colors. The difference between the blue and red light in this spectrum is the result of differences in the frequencies and wavelengths of the light, as shown in Table 1.9.

**TABLE 1.9 Characteristic Wavelengths and Frequencies of Light of Different Colors**

| Color | Wavelength ($\lambda$) | Frequency ($v$) |
|---|---|---|
| Violet | $4.100 \times 10^{-7}$ m | $7.312 \times 10^{14}$ s$^{-1}$ |
| Blue | $4.700 \times 10^{-7}$ m | $6.379 \times 10^{14}$ s$^{-1}$ |
| Green | $5.200 \times 10^{-7}$ m | $5.765 \times 10^{14}$ s$^{-1}$ |
| Yellow | $5.800 \times 10^{-7}$ m | $5.169 \times 10^{14}$ s$^{-1}$ |
| Orange | $6.000 \times 10^{-7}$ m | $4.997 \times 10^{14}$ s$^{-1}$ |
| Red | $6.500 \times 10^{-7}$ m | $4.612 \times 10^{14}$ s$^{-1}$ |

There is an obvious pattern in these data: As the wavelength of the light increases, the frequency decreases. But recognizing this pattern is not enough. It would be even more useful to construct a mathematical equation that fits these data. This would allow us to calculate the frequency of light of any wavelength, such as blue-green light with a wavelengh of $5.00 \times 10^{-7}$ m, or to calculate the wavelength of light of a known frequency, such as blue-violet light with a frequency of $7.00 \times 10^{14}$ cycles per second.

The first step toward constructing an equation that fits these data involves plotting the data in different ways until we get a straight line. We might decide, for example, to plot the wavelengths on the vertical axis and the frequencies on the horizontal axis, as shown in Figure 1.13. When we construct a graph, we should keep the following points in mind.

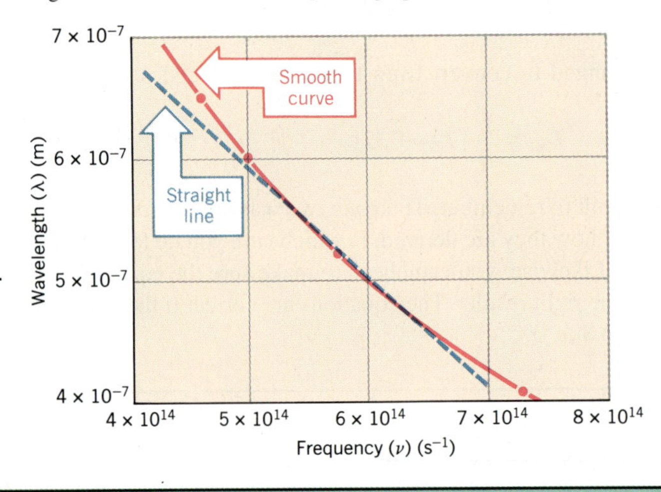


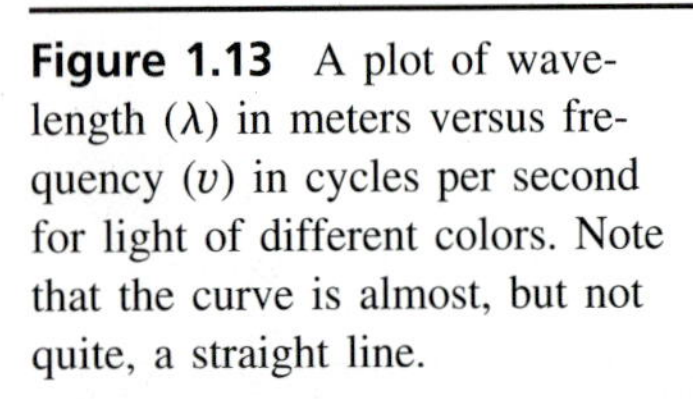
**Figure 1.13** A plot of wavelength ($\lambda$) in meters versus frequency ($v$) in cycles per second for light of different colors. Note that the curve is almost, but not quite, a straight line.

- The scales of the graph should be chosen so that the data fill as much of the available space as possible.
- It isn't necessary to include the origin (0,0) on the graph. In fact, it may be more efficient to leave off the origin, so that the data fill the available space.
- Once the scales have been chosen and labeled, the data are plotted one point at a time.
- A straight line or a smooth curve is then drawn through as many points as possible. Because of experimental error, the line or curve may not pass through every data point.

If this process gives a straight line, we can conclude that the quantity plotted on the vertical axis ($v$) is directly proportional to the quantity on the horizontal axis ($\lambda$). We can then fit the graph to the equation for a straight line, $y = mx + b$:

$$\lambda = mv + b$$

The graph in Figure 1.13 is almost, but not quite, a straight line. Because the data are known to four significant figures, the deviation from a straight line is not the result of experimental error. We must therefore conclude that the wavelength and frequency of light are not *directly proportional*. We might therefore look for an *inverse proportionality* between the two sets of measurements. In other words, we might try to fit the data to the following equation:

$$\lambda = m\left[\frac{1}{v}\right] + b$$

We can test this relationship by plotting the wavelengths in Table 1.9 versus the inverse of the frequencies, as shown in Figure 1.14.

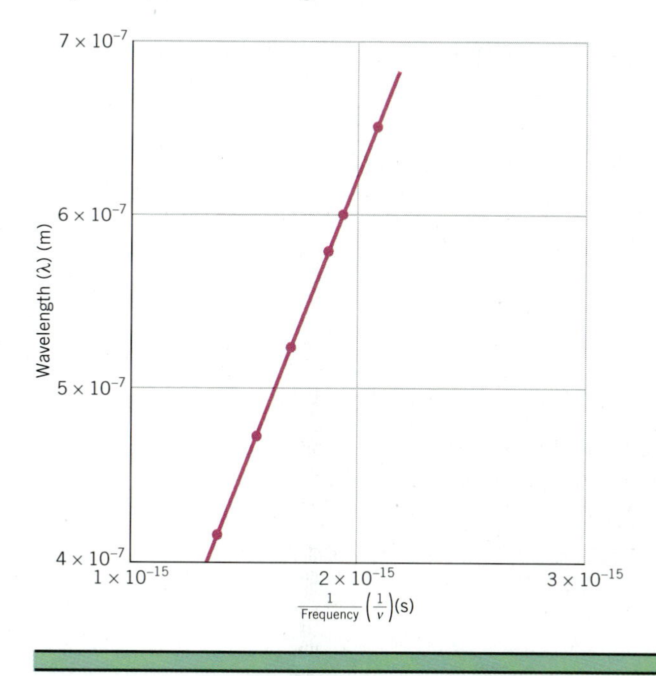


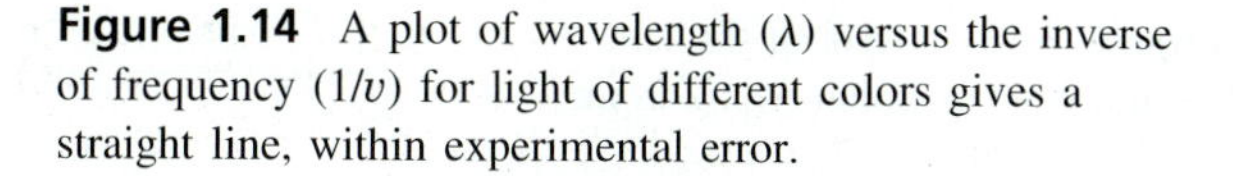
**Figure 1.14** A plot of wavelength ($\lambda$) versus the inverse of frequency ($1/v$) for light of different colors gives a straight line, within experimental error.

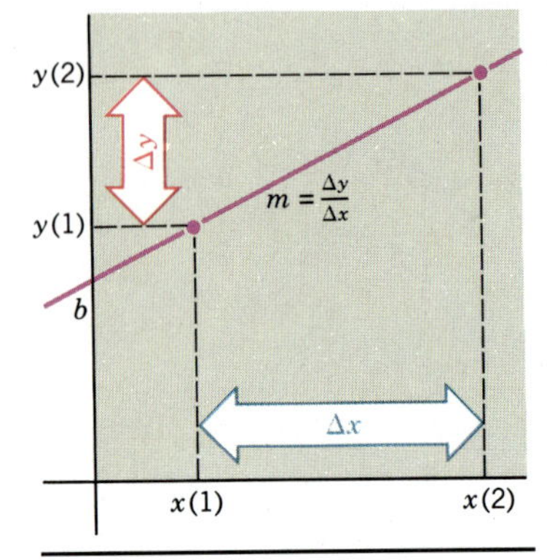


**Figure 1.15** To calculate the slope ($m$) of a straight line, divide the distance between two points on the vertical axis by the corresponding distance between two points along the horizontal axis.

This graph gives a beautiful straight-line relationship. We can now calculate the slope of the line ($m$) and the intercept ($b$), as shown in Figure 1.15. To minimize the error in our calculations, it is important to choose points for this calculation that are *not* original data points. When this is done, the slope of the line in Figure 1.14 is found to be equal to the speed of light: $2.998 \times 10^8$ meters per second.

When the graph includes the origin, the intercept can be determined by reading the value on the vertical axis that corresponds to a value of zero on the horizontal axis. This isn't possible for the data in Figure 1.14, however, because the origin was left off this graph. Another approach to determining the intercept starts by selecting a point on the straight line. We then combine estimates of the values of $y$ and $x$ for this point with the slope of the line to calculate the value of the intercept. When this approach is applied to the data in Figure 1.14, we find that the intercept is equal to zero. The data in Figure 1.14 therefore fit the following equation:

$$\lambda = (2.998 \times 10^8)\left[\frac{1}{v}\right]$$

Rearranging this equation, we find that the product of the frequency times the wavelength of light is equal to the speed of light.

$$v\lambda = 2.998 \times 10^8 \text{ m/s}$$

## EXERCISE 1.11

The following data were obtained from a study of the relationship between the volume ($V$) and temperature ($T$) of a gas at constant pressure:

| | | | | | | |
|---|---|---|---|---|---|---|
| Volume (mL): | 273.0 | 277.4 | 282.7 | 287.8 | 293.1 | 298.1 |
| Temperature (°C): | 0.0 | 5.0 | 10.0 | 15.0 | 20.0 | 25.0 |

Determine the temperature at which the volume of the gas should become equal to zero.

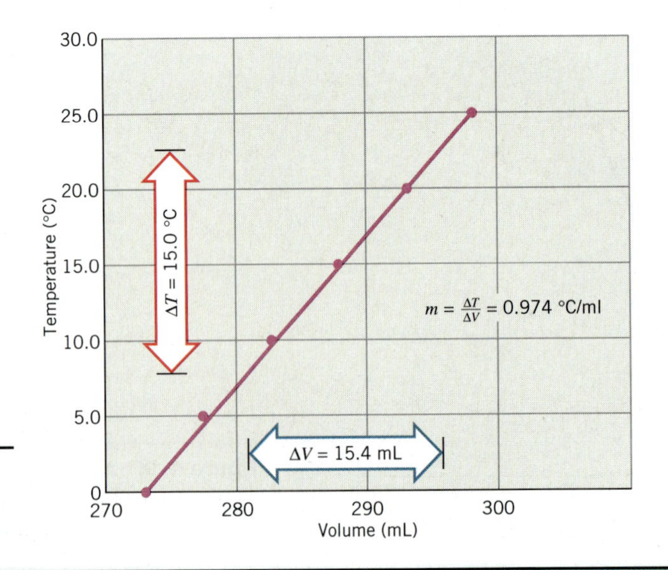


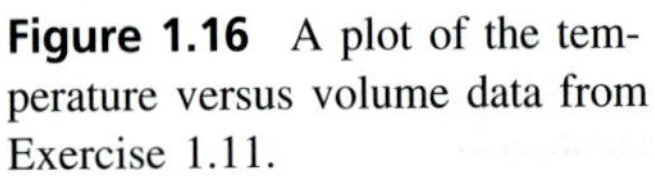
**Figure 1.16** A plot of the temperature versus volume data from Exercise 1.11.

**SOLUTION** A plot of these data gives a straight line, as shown in Figure 1.16. The equation for this straight line can be written as follows:

$$T = mV + b$$

By choosing any two points on this curve, such as $T = 7.5°C$ and 22.5°C, and estimating the volumes at these temperatures from the straight line that passes through the points, we can calculate the slope ($m$) of the line:

$$m = \frac{\Delta T}{\Delta V} = \frac{(22.5°C - 7.5°C)}{(295.5 \text{ mL} - 280.1 \text{ mL})} = 0.974°C/\text{mL}$$

The value of $b$, the $y$-axis intercept, for this equation can be calculated from the data for any point on the straight line, such as the point at which the temperature is 22.5°C and the volume is 295.5 mL:

$$b = T - mV = (22.5°C) - (0.974°C/\cancel{\text{mL}})(295.5 \cancel{\text{mL}}) = -265°C$$

According to these data, the volume of the gas becomes zero when the gas is cooled until the temperature reaches about −265°C.

## SUMMARY

**1.** All measurements contain at least three elements: a **number** that indicates the size of the quantity being measured, **units** that provide a basis for comparing this quantity with a standard reference, and some **uncertainty** or **error.**

**2.** The error or uncertainty associated with a measurement requires that we: (1) pay attention to the accuracy and/or precision of the measurement, (2) try to minimize random and systematic errors involved in the measurement, and (3) estimate the magnitude of the error involved in a measurement through the use of significant figures.

**3.** The quantities measured in this chapter can be divided into two categories: **extensive** and **intensive** properties. Extensive properties, such as mass and volume, depend on the size of the sample. Intensive properties, such as density and temperature, do not depend on the size of the sample. Intensive properties are therefore useful for characterizing substances.

**4.** An intensive property is often the ratio of a pair of extensive properties that are proportional to each other. The density of a substance, for example, is equal to the ratio of the mass of a sample of the substance divided by the volume of the sample.

## KEY TERMS

**Accuracy** (p. 15)
**Density** (p. 28)
**Dimensional analysis** (p. 22)
**Error** (p. 6)
**Equality** (p. 9)
**Equation** (p. 9)
**Extensive property** (p. 27)
**Intensive property** (p. 27)
**Mass** (p. 11)
**Metric system** (p. 10)
**Number** (p. 6)
**Precision** (p. 15)
**Random error** (p. 14)
**Scientific notation** (p. 25)
**SI units** (p. 12)
**Significant figures** (p. 19)
**Systematic error** (p. 14)
**Temperature** (p. 29)
**Unit factor** (p. 9)
**Units** (p. 6)
**Uncertainty** (p. 6)
**Weight** (p. 11)

## KEY EQUATIONS

$$\text{Density} = \frac{\text{mass}}{\text{volume}}$$

$$T_K = T_{°C} + 273.15$$

$$T_{°F} = \frac{9}{5} T_{°C} + 32$$

$$T_{°C} = \frac{5}{9} [T_{°F} - 32]$$

## PROBLEMS

### English Units of Measurement

**1-1** Describe the three elements that are present in all measurements.

**1-2** Find a source that explains the origin of one or more units in the English system of units, and describe why the system seemed rational to its inventors.

**1-3** A gallon in the United States contains 128 ounces. Canada once used the imperial gallon, which contains 160 ounces. If there are four imperial quarts in an imperial gallon, how many ounces are there in an imperial quart?

### Unit Conversions in the English System

**1-4** Calculate the number of seconds in a year and the number of ounces in a ton.

**1-5** A typical aspirin tablet weighs 5.0 grains. Calculate the number of grains in a pound if there are 16 ounces in a pound, 16 drams in an ounce, and 437.5 grains in a dram.

**1-6** Calculate the number of ounces by weight in a fluid ounce if a pint of water weighs 1.04 pounds.

**1-7** A British gallon contains 0.027777 barrel, 0.125 bushel, 0.111111 firkin, 0.5 peck, 0.019063 hogshead, or 0.05555 kilderkin. Calculate the number of: (a) hogsheads in a peck, (b) kilderkins in a firkin, (c) barrels in a bushel.

**1-8** A bushel is a dry measure equal to 4 pecks. There are 8 dry quarts in a peck and 2 dry pints in a quart. If a dry pint contains 33.6 cubic inches, what is the volume of a bushel in cubic inches?

**1-9** A light-year is the distance light travels in a year. If a light-year is $5.87851 \times 10^{12}$ miles, what is the length of a distance of 1 foot in units of light years?

### The Metric System

**1-10** Describe three advantages of the metric system.

**1-11** Define the following prefixes from the metric system:

(a) nano- (b) micro- (c) milli- (d) centi- (e) kilo-

### Unit Conversions in the Metric System

**1-12** Make the following conversions:

(a) 0.043 kg into grams (b) 2.45 L into mL
(c) 0.00814 L into mL (d) 346.8 mm into cm

**1-13** The diameter of a helium atom is 1.9 angstroms (Å), where an angstrom is $10^{-8}$ centimeters. Calculate the diameter of a helium atom in units of centimeters, meters, nanometers, and picometers.

**1-14** Light is the small portion of the electromagnetic spectrum visible to the naked eye. It has wavelengths between about $4 \times 10^{-5}$ and $7 \times 10^{-5}$ centimeters. Calculate the range of wavelengths of light in units of ''microns''—micrometers—and in units of millimicrons, nanometers, and angstroms.

### SI Units of Measurement

**1-15** Determine the derived SI base unit for the following quantities:

(a) speed (b) area (c) volume (d) density

**1-16** Explain why the liter, angstrom, atmosphere, and calorie are exceptions to the rules for derived SI units.

**1-17** The same prefixes—kilo-, milli-, nano-, and so on—are used in both the SI and the metric system. There is only one difference. In SI, the prefixes are applied to the unit of grams—to give milligrams or micrograms—instead of the base unit of mass in this system. Explain why.

### Unit Conversions Between the English and Metric Systems

**1-18** Calculate the number of pounds in a kilogram, assuming a kilogram weighs 2.205 pounds.

**1-19** Calculate the number of liters in a gallon, assuming a liter is 0.2542 gallons.

**1-20** Make the following conversions:

(a) 32.0 liters into gallons (b) 8.76 pounds into grams
(c) 135 meters into miles (d) 6500 grams of water into liters

**1-21** Calculate the mass in milligrams of the aspirin in a tablet that contains 5.0 grains of aspirin. Assume 1 pound contains 7000 grains.

**1-22** Calculate the volume in milliliters of a bottle of typewriting correction fluid that contains 0.6 fluid ounces.

**1-23** Which takes longer to run at a constant speed, a 100-meter dash or 100-yard dash?

**1-24** A barrel of liquor contains 31 gallons, but a barrel of oil or gasoline contains 42 gallons. Calculate the number of liters per barrel for both liquor and gasoline.

**1-25** Liquor, which used to be sold in "fifths," is now sold in 750 mL bottles. If a fifth is one-fifth of a gallon, which is the better buy: a fifth of Scotch selling for $12.50 or a 750 mL bottle selling for the same price?

**1-26** Determine the number of gallons in 1.00 $m^3$ of water and the mass in pounds of 1.00 $ft^3$ of water.

**1-27** Calculate the number of cubic feet of peat moss in a bag marked 113 cubic decimeters.

**1-28** The largest Harley-Davidson motorcycle has an engine with a displacement of 80 cubic inches. The largest motorcycle Honda makes has a displacement of 1500 cubic centimeters. Which motorcycle has the larger engine?

**1-29** Calculate the number of cubic inches in a liter and the number of cubic meters in a cubic foot.

### Unit Conversions Involving Ratios of Two Quantities

**1-30** Estimate the speed of sound in air in miles per hour, assuming the sound of thunder takes 5 seconds to travel 1 mile.

**1-31** Calculate the speed limit on the nation's interstate highways in units of meters per second, kilometers per hour, and feet per second.

**1-32** The speed of light is $2.9979 \times 10^8$ meters per second. Calculate the speed of light in centimeters per second, miles per hour, and miles per year.

**1-33** What is the speed in miles per hour of an electron traveling across a cathode ray tube if the electron travels 1 yard in 12 nanoseconds?

**1-34** If your heart beats an average of 70 times per minute, how many times will it beat in a 75-year lifetime? The Jarvik-7 artificial heart beats 40,000,000 times per year. How does this compare with a normal human heart?

**1-35** Air flow is measured in units of cubic feet per minute (CFM). Convert 100 CFM into units of cubic meters per second.

**1-36** Convert atmospheric pressure from 14.7 pounds per square inch into units of tons per square foot.

**1-37** Lightning travels at a speed of 87,000 miles per hour. How does this compare with the average velocity of an oxygen molecule at room temperature, which is 440 meters per second, or the speed of light, which is $3 \times 10^8$ meters per second?

**1-38** Isopropyl alcohol, or rubbing alcohol, has a density of 0.786 g/$cm^3$. Calculate the mass of the rubbing alcohol in a quart bottle.

**1-39** Calculate the density of corn oil in grams per cubic centimeter if 1 liter of this oil has a mass of 0.91 kg.

**1-40** Calculate the density of water in units of kilograms per cubic meter.

### Uncertainty in Measurement

**1-41** Describe the difference between systematic and random error, and give examples of each.

**1-42** Describe the difference between accuracy and precision. Give examples of measurements that are accurate but not precise, precise but not accurate, and both accurate and precise.

**1-43** Four nickels selected at random were found to have masses of 5.0601 grams, 4.8881 grams, 4.9238 grams, and 5.0603 grams, respectively. Calculate the average mass of the four nickels, and then calculate the difference between the average mass and the mass of each nickel. Use the largest of these differences to estimate the error involved in counting nickels with a balance. How many nickels can we have at a time before the error involved in the measurement of their mass is equal to the mass of the average nickel?

### Significant Figures

**1-44** Determine the number of significant figures in the following numbers:

(a) 0.00641 (b) 0.07850 (c) 500 (d) 50,003

**1-45** Determine the number of significant figures in the following numbers:

(a) $3.4 \times 10^{-2}$ (b) $5.98521 \times 10^3$ (c) $8.709 \times 10^{-6}$
(d) $7.00 \times 10^{-5}$

**1-46** Round off the following numbers to three significant figures:

(a) 474.53 (b) 0.067981 (c) $9.463 \times 10^{10}$
(d) 30.0974

### Dimensional Analysis

**1-47** A 20-pound bag of 20:10:5 fertilizer is 20% nitrogen by weight. Calculate the number of grams of nitrogen per square foot if this bag of fertilizer covers 5000 square feet of lawn.

**1-48** Calculate the fraction of the body weight that is water in an average human body, which has a mass of 70 kilograms and contains 12.5 gallons of water.

**1-49** The Pacific Ocean has a volume of $6.96189 \times 10^{23}$ L and a surface area of $1.66241 \times 10^{11}$ $km^2$. Calculate the average depth of the Pacific Ocean.

**1-50** Calculate the mass of $CO_2$ in the atmosphere if the atmosphere is 338 parts per million $CO_2$ by mass and the mass of the atmosphere is $5.2 \times 10^{15}$ tons.

**1-51** The $LD_{50}$ for a drug is the dose that would be lethal for 50% of the population. The $LD_{50}$ for aspirin in rats is 1.75 grams per kilogram of body weight. Calculate the number of tablets containing 325 mg of aspirin a 70-kg human would have to consume to achieve this dose.

**1-52** The $LD_{50}$ for sodium cyclamate in mice is 17.0 grams per kilogram of body weight. Calculate the number of cans of diet soda a 70-kg human would have to drink to achieve this dose if each can contains 0.096 grams of sodium cyclamate.

**1-53** Ipecac syrup is used to induce vomiting in people who have swallowed a poison. The syrup contains 7 grams of ipecac per 100 mL, and the recommended dosage is 1 tablespoon, or 15 mL. Calculate the amount of ipecac in the average dose.

**1-54** The typical bar buys a 750-mL bottle of Jack Daniels bourbon for about $8.00 and pours drinks that use a jigger, or 1.5 ounces, of the whiskey. Estimate the profit per drink if the bar sells a jigger of bourbon for $2.50.

**1-55** In 1773 Benjamin Franklin observed that one teaspoon of oil spilled on a pond near London spread out to form a film that covered an area of about 22,000 square feet. If a teaspoon of oil has a volume of about 5 $cm^3$ and the oil spread out to form a film roughly one molecule tall, what is the average height of an oil molecule?

**1-56** A gardener wants to improve the soil in her vegetable garden. The Purdue extension agent recommends incorporating 6 inches of loamy topsoil into the existing soil. How many cubic yards of topsoil will she need for a garden 30 feet long and 20 feet wide? What is the density of topsoil if 1 cubic yard has a mass between 1.1 and 1.5 tons, depending on how wet?

**1-57** The height-and-weight charts that used to hang in doctors' offices are being replaced by formulas for estimating when someone has "medically significant obesity." For men, the mass in kilograms is divided by the square of the height in meters. For women, the mass in kilograms is divided by the height in meters raised to the 1.5 power. If this ratio is larger than 30, the person is obese. Assume that one of the authors of this text is 6 feet, 1 inch tall. What is the maximum weight in pounds he could carry before he would be considered obese? How tall would he have to be if he had a mass of 250 pounds to avoid being obese?

**1-58** One thousand cubic feet of natural gas delivers $1 \times 10^6$ BTU of energy at an average cost of $6.63. If a gallon of # 2 fuel oil delivers $0.139 \times 10^6$ BTU of energy at an average cost of $1.01, which is the cheaper fuel for heating a home: gas or oil?

**1-59** It takes 1000 $ft^3$ of natural gas or 293 kilowatt hours of electricity to generate $1.0 \times 10^6$ BTU of energy. What is the ratio of the cost of electricity to the cost of natural gas if electricity sells for $0.058 per kilowatt hour and natural gas sells for $0.663 per 100 cubic feet?

**1-60** Calculate the energy released per gram of solar fuel burned if the sun releases $3.6 \times 10^{26}$ joules of energy per second and loses $1.25 \times 10^{14}$ metric tons per year. If the earth absorbs $5.4 \times 10^{24}$ joules per year of energy from the sun, what fraction of the sun's energy is absorbed by the earth?

**1-61** Gold is the most malleable element. An ounce of gold can be beaten into a thin sheet that covers an area of 300 square feet. Calculate the thickness of this sheet in centimeters, assuming the density of gold is 19.3 g/$cm^3$. Calculate the thickness of the sheet in terms of the number of gold atoms if the radius of an individual gold atom is $1.4 \times 10^{-8}$ cm.

### Scientific Notation

**1-62** Convert the following numbers to scientific notation:
(a) 11.98 (b) 0.0046940 (c) 4,679,000

**1-63** Convert the following numbers to scientific notation:
(a) 212.6 (b) 0.189 (c) 16,221

**1-64** Convert the following numbers from scientific notation:
(a) $5.60 \times 10^{-3}$ (b) $7.025 \times 10^5$ (c) $8.216 \times 10^{-2}$

**1-65** Do the following calculations. (Keep track of significant figures.)
(a) $132.76 \times 21.16071$ (b) $32 + 0.9767$ (c) $3.02 \times 10^4 + 1.69 \times 10^3$ (d) $4.18 \times 10^{-2} + 1.29 \times 10^{-3}$

**1-66** Do the following calculations. (Keep track of significant figures.)
(a) $28 \times 4.80$ (b) $32.1/0.75$ (c) $(8.16 \times 10^{-4}) \times (4.78 \times 10^{15})$ (d) $(5.00 \times 10^4)/(1.60 \times 10^{-2})$ (e) $(1.39 \times 10^7)/(1.10 \times 10^{-3})$

### Extensive and Intensive Quantities

**1-67** Describe why intensive properties are more useful than extensive properties for comparing two systems.

**1-68** Describe why intensive properties are characteristic properties of a system, whereas extensive properties are not.

**1-69** Use the concept of intensive quantities to explain what is wrong with signs that say "open 24 hours." What should these signs say?

**1-70** List five quantities whose measurements were discussed in this chapter that are extensive properties. List as many intensive properties as you can.

### Density

**1-71** Use the density of ice and water to predict whether the total volume of the system increases, decreases, or remains the same when the ice in a glass of tea melts.

**1-72** A 55-gallon drum holds 309.8 pounds of gasoline or 459.0 pounds of water. Which is more dense: gasoline or water?

**1-73** Calculate the mass of a cubic yard of peat moss, assuming the density of peat is 0.84 g/$cm^3$.

**1-74** Calculate the mass of a gallon of milk if the density of milk is 1.032 g/$cm^3$.

**1-75** Liquid mercury is sold in units of flasks, which weigh 76 pounds each. What is the volume of a flask of mercury?

**1-76** Calculate the mass of mercury that would fill a barometer tube 760 mm tall and 1.00 cm in diameter.

**1-77** What is the radius of the moon if it has a mass of $7.4 \times 10^{22}$ kg and its density is 3.3 g/$cm^3$? (Hint: Assume the moon is spherical and that the volume of a sphere is given by $V = \frac{4}{3}\pi r^3$.)

**1-78** The density of Earth is 5.5 g/$cm^3$ and its radius averages $6.4 \times 10^6$ meters. Calculate the mass of the planet.

**1-79** Neutron stars have been estimated to have an average radius of 6 miles and a density of 1 billion tons per cubic inch. How does this compare with the density of water? How does it compare with the density of the sun ($1.41 \text{ g/cm}^3$)?

**1-80** A 2-$\text{ft}^3$ bag of mulch sold at lawn-care centers weighs 30 lb. Will the mulch float on water?

**1-81** Which of the following liquids would float on top of water and which would sink?

(a) benzene ($C_6H_6$; density = $0.877 \text{ g/cm}^3$)
(b) carbon disulfide ($CS_2$; density = $1.263 \text{ g/cm}^3$)
(c) chloroform ($CHCl_3$; density = $1.483 \text{ g/cm}^3$)
(d) ether [$(C_2H_5)_2O$; density = $0.714 \text{ g/cm}^3$]
(e) gasoline (density = $0.675 \text{ g/cm}^3$)

**1-82** If a beaker that contains 250 mL of chloroform (density = $1.483 \text{ g/cm}^3$) has a mass of 524 grams, what is the mass of the beaker?

**1-83** A student was given two metal cubes that looked similar. One was 1.05 cm on an edge and had a mass of 8.33 grams. The other was 2.84 cm on an edge and had a mass of 164.9 grams. Were the cubes made of the same metal?

**1-84** An irregularly shaped piece of metal has a mass of 54.6 grams and displaces $2.83 \text{ cm}^3$ of water when dropped into a graduated cylinder filled with water. What is the metal?

**1-85** The amount of air trapped in a can when it is filled with beer varies from one can to another. Use this to explain why some cans of beer will float in a cooler in which the ice has melted.

### Temperature

**1-86** Explain why we need two reference points (such as 32°F and 212°F or 0°C and 100°C) to define a temperature scale.

**1-87** Normal body temperature is 98.6°F. What is the equivalent temperature in degrees Celsius? In kelvins?

**1-88** At what temperature are the readings on the Celsius and Fahrenheit scales the same? At what temperature are the magnitudes the same but the signs different?

**1-89** Fahrenheit defined zero on his scale as the lowest temperature that could be achieved by adding salt to ice and found that ice by itself freezes at 32° on this scale. Unfortunately, Fahrenheit's measurements were wrong. The lowest temperature that can be achieved by adding salt to ice is −21.2°C. Calculate the equivalent temperature on the Fahrenheit scale.

### The Graphical Treatment of Data

**1-90** The following data show how the pressure ($P$) of a gas depends on its volume ($V$) when the amount of gas and the temperature of the gas are held constant. Plot these data in terms of $P$ versus $V$, $P$ versus $1/V$, and $V$ versus $1/P$. Write an equation of the form $y = mx + b$ for each straight-line relationship you find.

| Pressure (atm): | 0.0586 | 0.0856 | 0.142 | 0.289 |
|---|---|---|---|---|
| Volume (L): | 1.28 | 0.876 | 0.528 | 0.259 |

**1-91** $^{14}C$ nuclei are radioactive, decaying to $^{14}N$ with a half-life of 5730 years. The following data show the change in the activity ($A$) of a sample containing $^{14}C$ in units of counts per minute (cpm) versus the age of the sample in years. Plot these data in terms of $A$ versus $t$, $1/A$ versus $t$, and log $A$ versus $t$. Write an equation of the form $y = mx + b$ for each straight-line relationship you find.

| Activity (cpm): | 1000 | 800 | 600 | 400 | 200 | 100 |
|---|---|---|---|---|---|---|
| Age (year): | 0 | 1845 | 4224 | 7576 | 13,307 | 19,039 |

**1-92** Some of the light that shines on a colored solution is absorbed as it passes through the solution. The intensity of the light that passes through the sample ($I$) is therefore lower than the intensity of the light that shines on the sample ($I_0$). The following data show how the ratio of $I_0/I$ depends on the concentration of the light-absorbing compoound. Plot the $I_0/I$ ratio versus concentration, the $I_0/I$ ratio versus the log of the concentration, and log $I_0/I$ versus concentration for the following data. Write an equation of the form $y = mx + b$ for each straight-line relationship you find.

| $I_0/I$: | 2 | 3 | 4 | 5 | 6 |
|---|---|---|---|---|---|
| Concentration: | 0.250 | 0.396 | 0.500 | 0.581 | 0.646 |

# Intersection

## NATURE ON THE MACROSCOPIC SCALE

In the 1930s, John Dickson Carr wrote a mystery story in which a magic trick was described. Although the trick originally referred to water and wine, chemists would prefer the following version, which uses water and methanol.

> Start with a glass of water, a glass of methanol, and a teaspoon. Remove one teaspoon of water from the glass of water and add it to the glass of methanol. Stir the resulting mixture of methanol and water until the two liquids are thoroughly mixed. Then transfer one teaspoon of this mixture back to the water. Which of the following statements is true? (1) The volume of water that ends up in the methanol is *larger* than the volume of methanol that ends up in the water. (2) The volume of water that ends up in the methanol is *smaller* than the volume of methanol that ends up in the water. (3) The volume of water added to the methanol is *the same as* the volume of methanol added to the water.

Many people conclude that the first statement is true. They argue that pure water was transferred into the methanol, and a mixture of water and methanol was transferred back. It therefore seems logical to assume that more water ends up in the methanol than methanol in the water.

Unfortunately, the most popular answer is wrong. Exactly the same volume of water and methanol is transferred in this process. When told they are wrong, many people who have chosen the first answer vehemently defend that answer. Often, the only way to convince them they are wrong is to ask them to consider a concrete example, such as the following:

> Assume that you start with two beakers that contain different colored marbles. Assume, for the sake of argument, that there are 100 white marbles in one beaker and 90 black marbles in the other. Furthermore, assume that we use a spoon that transfers 10 marbles at a time. In the first step, we will transfer 10 white marbles into the other beaker. (This leaves us with 90 white marbles in one beaker and a mixture of 90 black marbles and *10 white marbles* in the other.) We then use the spoon to transfer a representative sample of the mixture back into the first beaker. The spoon will pick up 10 marbles, *nine of which will be black.*

The first step toward understanding the structure of the infinitesimally small molecules that collide with our bodies when we experience hurricane-force winds involves describing the properties of nature on the macroscopic scale, the scale of objects such as this hurricane that are visible to the naked eye.

The first beaker now contains 91 white marbles (100 − 10 + 1) and nine black marbles. The second beaker contains 81 black marbles (90 − 9) and nine white marbles. Thus, nine white marbles are transferred in one direction,

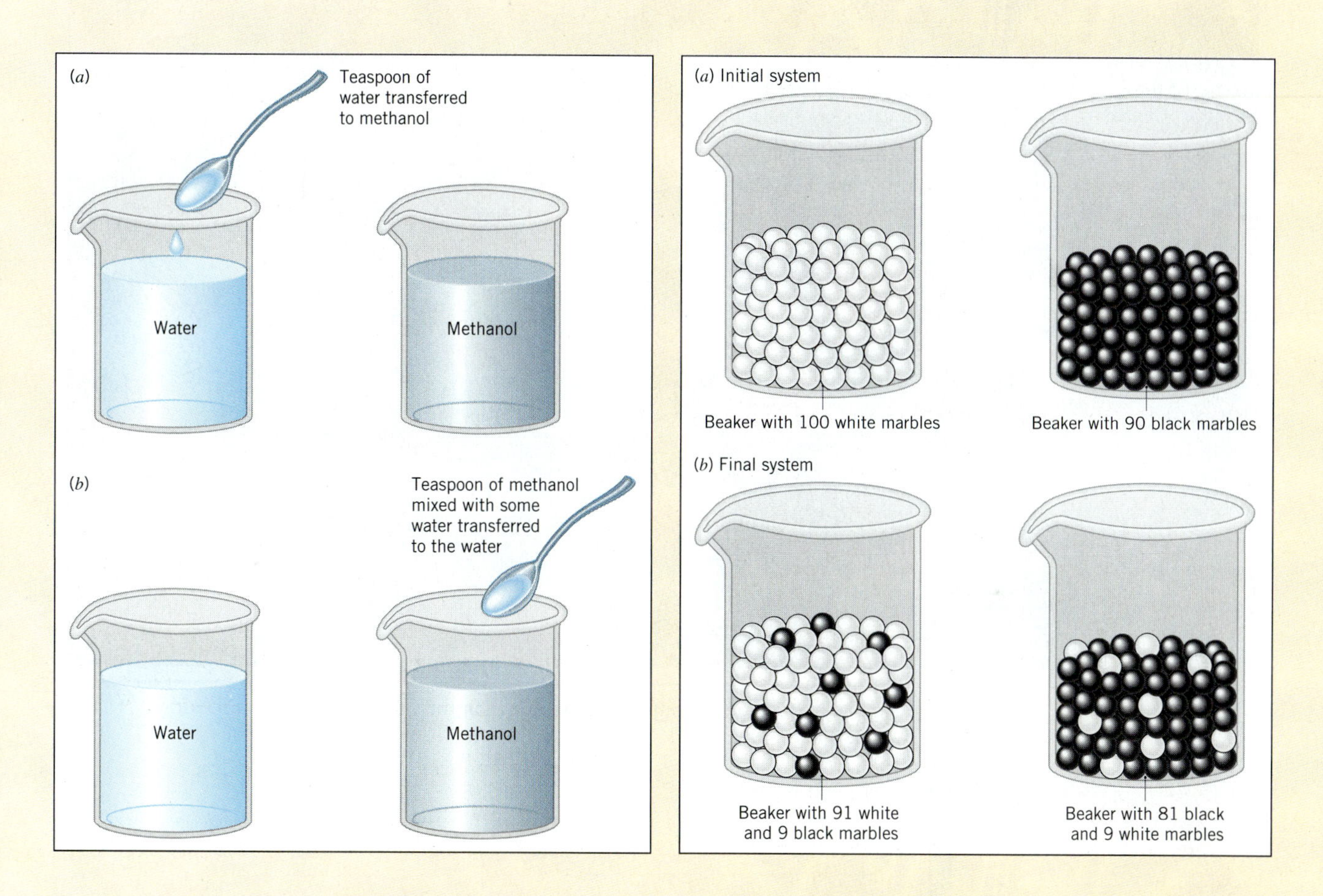


while nine black marbles are transferred in the other. In other words, exactly the same number of marbles is transferred in each direction. After considering one or more concrete examples, such as this, many people conclude that their original answer was wrong—equal volumes of water and methanol are in fact transferred in the original experiment.

This trick has important implications for anyone trying to learn chemistry. The first step in building an understanding of any abstract field must rely on concrete examples—things we can sense, that we can touch, that exist on a scale large enough to be observed. As a result, the next few chapters will focus on the properties of objects on the scale of objects visible to the naked eye. In the course of this discussion, we will inevitably talk about atoms and molecules. But most of our attention will focus on properties of nature that exist on the *macroscopic scale.* This section examines the consequences of measurements of the weight of a solid, the volume of a liquid, or the pressure and volume of a gas. It concludes with a discussion of the concepts of temperature, heat, and work, which are primarily macroscopic phenomena.

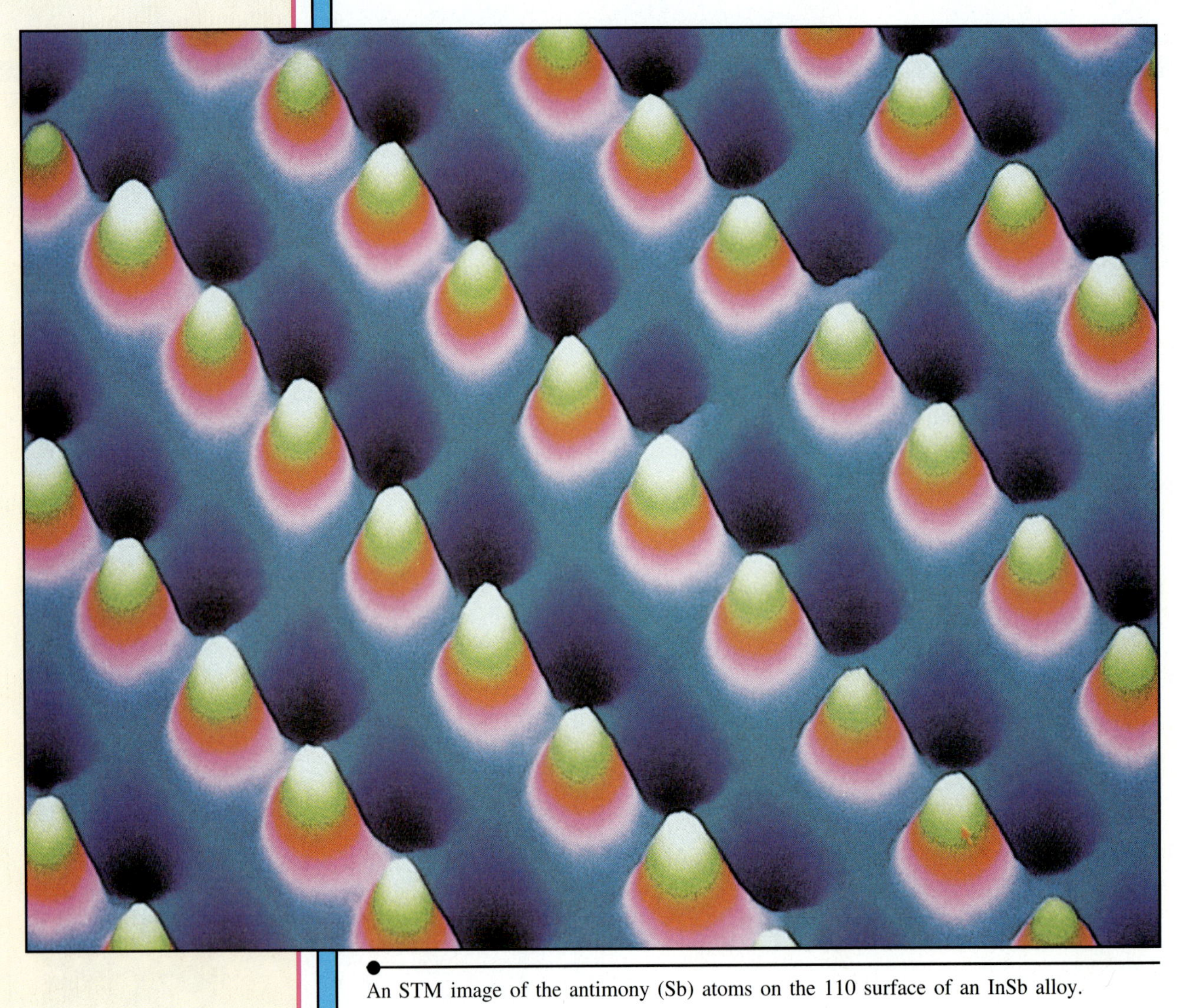

An STM image of the antimony (Sb) atoms on the 110 surface of an InSb alloy.

CHAPTER 2

# ELEMENTS AND COMPOUNDS

By examining what chemists mean when they use the words *elements, compounds, mixtures, atoms,* and *molecules,* this chapter provides the basis for answering questions such as the following:

## POINTS OF INTEREST

- Why do we believe that matter is made up of atoms, despite clear evidence from our senses that the world in which we live is continuous?
- What do we mean when we say that the oxygen in the atmosphere or the aluminum in aluminum foil is an *element?*
- What is the difference between the mixture of oxygen and nitrogen in the atmosphere and compounds formed when nitrogen and oxygen combine?
- What is the formula of the "stannous fluoride" found in some fluoride toothpastes?

## 2.1 THE PROCESS OF DISCOVERY: THE GREEK ELEMENTS

The histories of many cultures contain myths that explain the creation of the Earth, the origin of men and women, or the acquisition of tools, food, and fire. The main characters in these myths are gods or cultural heroes who overcome other gods, monsters, or mythical beings to bring one of the essentials of life to humankind. As such they explain much about our relationship to the natural world, but much less about that world itself.

The first step toward the development of modern science in the Western world occurred in the sixth century B.C. For the first time, the search for knowledge was not inspired by religion or day-to-day practical applications; it was based entirely on the desire to understand the world in which we live. This process of speculating about nature began among a group of philosophers in Miletus, an Ionian city on the coast of what is now Turkey.

The Ionian philosophers agreed on one thing: There was some boundless and eternal material, which was indestructible and unchangeable, out of which the world was created. Thales (ca. 600 B.C.) argued that water played a central role in matter. He envisioned the Earth as a disk that floated on water and received water in the form of rain from above.

Anaximander (610–545 B.C.) believed in a single, boundless, eternal substance that contained the opposite properties of matter: hot versus cold and wet versus dry. He argued that conflict within this primordial substance brought about separation of these properties and resulted in the materials of the known world, which appeared to contain fire, air, water, and earth.

Anaximenes (570–500 B.C.) believed that air was the primordial substance. When it is compressed, it becomes colder, heavier, and darker—first wind, then cloud, then water, and eventually earth. When air is dilated, on the other hand, it becomes hotter, lighter, and brighter, until it eventually takes on the properties of fire.

Heraclitus focused on fire as the primary substance because he saw change and motion in everything he encountered. Parmenides disagreed. He argued that change is impossible, and that all that can occur is the appearance of change. The debate over whether matter is eternal—unchangeable—or characterized by constant changes was resolved by a group of philosophers known as the pluralists. This school rejected the philosophies of Thales, Anaximander, and Anaximenes, which assumed that there was a single primary substance, and introduced a model based on the existence of more than one primary substance.

Empedocles, for example, argued that matter was made up of four primary substances—fire, air, earth, and water—which were indestructible, as Parmenides argued. But change was constant, as Heraclitus believed, because the elements were constantly being mixed or separated by the forces that acted on them: love (the force of attraction) and hate (the force of repulsion).

Plato adopted Empedocles' theory and coined the term *element* to describe these four substances. His successor, Aristotle, also adopted the concept of four elements and introduced the idea that the elements can be differentiated on the basis of properties such as hot versus cold and wet versus dry, as shown in Figure 2.1.

*Fire:* hot and dry
*Air:* hot and wet
*Water:* cold and wet
*Earth:* cold and dry

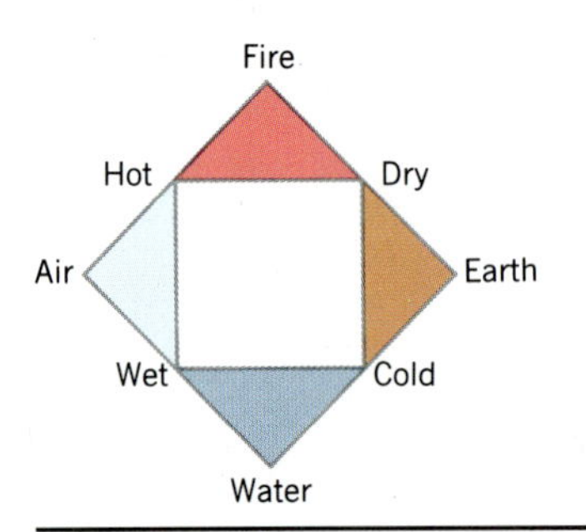


**Figure 2.1** The four Greek elements differed in two properties: hot versus cold and dry versus wet.

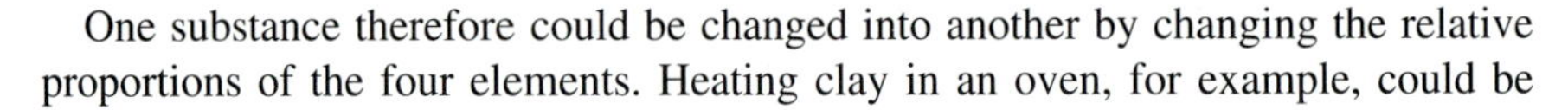
One substance therefore could be changed into another by changing the relative proportions of the four elements. Heating clay in an oven, for example, could be

thought of as driving off water and adding fire, thereby transforming the clay into a pot. Alternatively, substances could be changed into one another by changing one of the properties that made the elements different from each other. Water (cold and wet) falls from the sky as rain, for example, when the air (hot and wet) cools down. A piece of wood, on the other hand, which is obviously rich in earth (cold and dry), bursts into flame (hot and dry) when heated.

## 2.2 ELEMENTS, MIXTURES, AND COMPOUNDS

The Greek concept of elements was popular for almost 2200 years and was the guiding force behind the alchemists' search for ways to turn cheap metals such as lead into gold. In 1661, however, the English scientist Robert Boyle raised an important objection to this model. In his book, *The Sceptical Chymist,* Boyle noted that it was impossible to combine the four Greek elements to form any substance and it was equally impossible to extract these elements from a substance. He therefore proposed a new definition of an element that became the basis for the modern definition of this concept.

Boyle's definition of an element was based on the observation that many substances can be decomposed into simpler substances. Water, for example, decomposes into a mixture of hydrogen and oxygen when an electric current is passed through the liquid. Hydrogen and oxygen, on the other hand, cannot be decomposed into simpler substances. They are therefore examples of the elementary, or simplest, chemical substances. Thus, as Boyle pointed out, an **element** is any substance that cannot be decomposed into a simpler substance.

It isn't obvious from their appearance that the liquid on the left is a compound (water) and the liquid on the right (mercury metal) is an element.

This raises an interesting question: What is the difference between a mixture of elements and a compound formed by a reaction between these elements? The best way to answer this question is to compare the properties of chemical compounds and mixtures of elements.

Between 1798 and 1808, Joseph Louis Proust analyzed different sources of several compounds. He found that they always contained the same ratio by weight of their elements. Table salt, for example, always contained 1.5 times as much chlorine as sodium. He also found that water always contained eight times as much oxygen as hydrogen. Finally, he found that the mineral known as cinnabar always contained 6.25 times as much mercury as sulfur. These results led Proust to propose the **law of constant composition,** which states that the ratio by mass of the elements in a chemical compound is always the same, regardless of the source of the compound.

The law of constant composition can be used to distinguish between compounds and mixtures of elements:

**Compounds have a constant composition; mixtures do not.**

Brass is an example of a mixture of two elements: copper and zinc. It can contain as little as 10%, or as much as 45%, zinc. Bronze is another mixture of two elements, this time copper and tin. Once again, the composition of this mixture can vary over a wide range.

Another difference between compounds and mixtures of elements is the ease with which the elements can be separated. Mixtures, such as the atmosphere, contain two or more substances that are relatively easy to separate. At 20°C, the atmosphere is 76.2% nitrogen, 20.5% oxygen, 2.4% water, and 0.9% argon, by mass. The individual components of this mixture can be physically separated from each other. The

Cinnabar (shown on the right) always contains 6.25 times as much mercury as sulfur by weight. Brass (shown on the left) can be made with widely differing ratios of copper and zinc. Cinnabar is therefore a compound, whereas brass is a mixture.

water can be removed, for example, by passing the gas over a dehydrating agent. Oxygen can be separated from nitrogen and argon by cooling the gas until it becomes a liquid and then boiling off the more volatile nitrogen and argon.

Chemical compounds are very different from mixtures:

**The elements in a chemical compound can be separated only by destroying the compound.**

Some of the differences between chemical compounds and mixtures of elements are illustrated by the following exercise.

**EXERCISE 2.1**

Raisin bran can be thought of as a mixture of "elements"—raisins and bran flakes. Crispix, on the other hand, can be thought of as a "compound" in which each particle consists of a rice flake fused with a corn flake.

At breakfast one morning, one of the authors was faced with a choice between two cereals. One was raisin bran and the other was "Crispix," which contains flakes of rice fused with flakes of corn. Describe the characteristic properties of these cereals that makes one an analogy for a mixture of elements and the other an analogy for a chemical compound.

**SOLUTION** Raisin bran has the following characteristic properties of a *mixture.*

1. The cereal does not have a constant composition; the ratio of raisins to bran flakes changes from sample to sample.
2. It is easy to physically separate the two "elements," to pick out the raisins, for example, and eat them separately.

Crispix has some of the characteristic properties of a *compound.*

1. The ratio of rice flakes to corn flakes is constant; it is 1 : 1 in every sample.
2. There is no way to separate the "elements" without breaking the bonds that hold them together.

## 2.3 EVIDENCE FOR THE EXISTENCE OF ATOMS

Our senses suggest that matter is continuous. The air that surrounds us, for example, feels like a continuous fluid. (We do not feel bombarded by individual particles in the air.) The water we drink seems to be a continuous fluid. We can take a glass of water, divide it in halves, and repeat this process again and again, without appearing to reach the point at which it is impossible to divide it one more time.

Because our senses suggest that matter is continuous, it isn't surprising that the debate about the existence of atoms goes back as far as we can trace and has continued well into this century. The first proponents of an atomic theory were the Greek philosophers Leucippus and Democritus, who proposed the following model in the fifth century B.C.

1. Matter is composed of atoms separated by empty space through which the atoms move.
2. Atoms are solid, homogeneous, indivisible, and unchangeable.
3. All apparent changes in matter result from changes in the groupings of atoms.
4. There are different kinds of atoms that differ in size and shape.
5. The properties of matter reflect the properties of the atoms the matter contains.

This model attracted few supporters among later generations of Greek philosophers. Aristotle, in particular, refused to accept the idea that the natural world could be reduced to a random assortment of atoms moving through a vacuum.

The first indication that matter might be composed of indivisible particles came from Boyle's work on chemical elements. Boyle argued that the existence of elements that cannot be decomposed into simpler substances is consistent with the idea that there are elementary particles (or corpuscles, as he called them) that can combine to form compounds but cannot be subdivided.

Further evidence for the existence of atoms was the **law of definite proportions** proposed by Jeremias Benjamin Richter in 1792. Richter found that the ratio by weight of the compounds consumed in a chemical reaction was always the same. It took 615 parts by weight of magnesia (MgO), for example, to neutralize 1000 parts by weight of sulfuric acid. A few years later, when Proust reported his work on the constant composition of chemical compounds, the time was ripe for the reinvention of an atomic theory. The laws of definite proportions and constant composition do not prove that atoms exist, but they are difficult to explain without assuming that chemical compounds are formed when atoms combine in constant proportions.

Experiments with gases that first became possible at the turn of the nineteenth century led John Dalton in 1803 to propose a modern theory of the atom based on the following assumptions.

1. Matter is made up of atoms that are indivisible and indestructible.
2. All atoms of an element are identical.
3. Atoms of different elements have different weights and different chemical properties.
4. Atoms of different elements combine in simple whole numbers to form compounds.
5. Atoms cannot be created or destroyed. When a compound decomposes, the atoms are recovered unchanged.

Although arguments about the existence of atoms continued well into the twentieth century, active debate has now ended. The existence of atoms is now universally accepted.

### CHECKPOINT

Which assumptions of Dalton's model of the atom are still accepted today? Which of his assumptions must be modified? Describe how they could be modified to bring this model up to date.

The concept of elementary particles or atoms provides a better way of distinguishing between elements and compounds. **Elements** are substances that contain only one kind of atom. **Compounds** are substances that contain atoms of more than one element.

## 2.4 ATOMS AND MOLECULES

Imagine cutting a piece of gold metal in half and then repeating this process again and again and again. In theory, we should eventually end up with a single **atom,** which is smallest particle that has any of the properties of the element.

What would happen if we did the same thing to a compound, such as water? Eventually we would end up with a single molecule of water, which contains two hydrogen atoms and an oxygen atom. A **molecule** is therefore the smallest particle that has any of the properties of a compound, as shown in Figure 2.2. It is important to note that elements can also occur as molecules, but these molecules are composed of identical atoms, as shown in Figure 2.3.

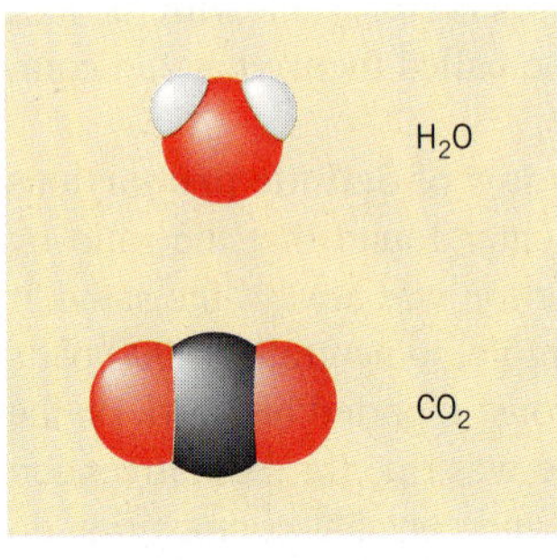


**Figure 2.2** A molecule is the smallest particle that has the properties of a chemical compound. Water consists of molecules that contain two hydrogen atoms bound to an oxygen atom. Carbon dioxide molecules contain two oxygen atoms bound to a carbon atom.

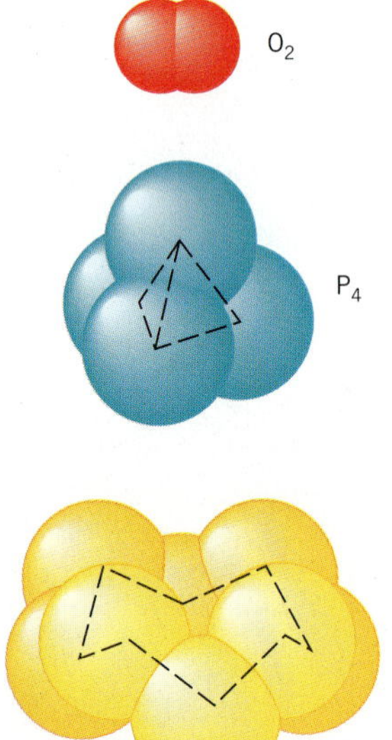


**Figure 2.3** Some nonmetallic elements also form molecules. At room temperature, oxygen exists as linear $O_2$ molecules, phosphorus forms tetrahedral $P_4$ molecules, and sulfur forms cyclic $S_8$ molecules.

Chemists use a shorthand notation to save both time and space when describing molecules. Each element is represented by a unique symbol. Most of these symbols make sense because they are derived from the name of the element. Symbols that don't seem to make sense usually come from the Latin or Greek names for the elements. Fortunately there are only a handful of elements in this category.

| | |
|---|---|
| Ag = silver (argentum) | Na = sodium (natrium) |
| Au = gold (aurum) | Pb = lead (plumbum) |
| Cu = copper (cuprum) | Sb = antimony (stibium) |
| Fe = iron (ferrum) | Sn = tin (stannum) |
| Hg = mercury (hydrargyrum) | W = tungsten (wolfram) |
| K = potassium (kalium) | |

The shorthand notation for a compound also describes the number of atoms of each element, which is indicated by a subscript written after the symbol for the element. By convention, no subscript is written when a molecule contains only one atom of an element. Thus, water is $H_2O$ and carbon dioxide is $CO_2$.

The advantage of this shorthand notation becomes obvious when we look at complex molecules, such as vitamin $B_{12}$, which has the formula $C_{63}H_{88}N_{14}O_{14}CoP$. Although a single $B_{12}$ molecule seems large, because it contains 181 atoms, it is useful to keep the size of these molecules in perspective. A teaspoon of vitamin $B_{12}$ contains about $2.5 \times 10^{21}$—or 2,500,000,000,000,000,000,000—molecules.

## RESEARCH IN THE 90s

### SCANNING TUNNELING MICROSCOPY

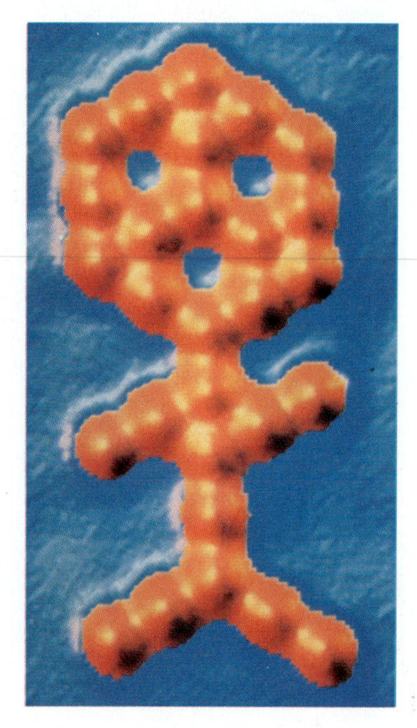

In 1982 a paper describing a new technique known as *scanning tunneling microscopy* (STM) was published by scientists from the IBM Research Laboratory in Zürich [G. Binning, H. Rohrer, Ch. Geber, and E. Weibel, *Physical Review Letters,* **49,** 57 (1982)]. The research reported in this paper was built on prior work in the same laboratory that showed that the current that flowed between the tip of a sharp piece of tungsten metal and the surface of platinum metal over which it was moved was very sensitive to the distance between the tip and the metal surface, so sensitive that variations on the order of atomic sizes could be distinguished. In their 1982 paper, the IBM researchers showed how this information could be used to study the surface of the metal. They described the STM technique as follows.

> The principle of the STM is straightforward. It consists essentially in scanning a metal tip over the surface at *constant* tunnel current. . . . The displacements of the metal tip . . . yield a topographic picture of the surface.

In other words, as the tungsten tip moves over the surface being studied, it is raised or lowered as needed to give a constant current. Measurements of the motion of the tip are then recorded and analyzed.

The photographs at the beginning of this chapter are STM images of individual iodine atoms absorbed onto a platinum metal surface [B. C. Schardt, S.-L. Yau, and F. Rinaldi, *Science,* **243,** 1050 (1989)]. They give the clear impression that scientists have finally developed a technique that can ''see'' (or at least ''feel'') individual atoms. It therefore isn't surprising that Binning and Rohrer received the Nobel Prize in physics in 1986 for the development of STM. Nor is it surprising that the number of papers describing the use of STM is increasing exponentially.

Scanning tunneling microscopy is ideally suited to probing the structure of metal surfaces. It has been used to study the mechanism by which metals nucleate and grow, the way metal films develop on atomically flat surfaces, and the process by which metals can be deposited on a semiconductor substrate. STM has also provided useful information about the correlation between electron density on the surface of semiconductors and their structures, and it has been used to study the effect of dislocations on the growth of superconductors.

Research in the 1990s will find STM being used as a routine probe of the structure of molecules absorbed onto the surface of a solid. A recent paper, for example, reported the use of STM to study the structure of a small segment of synthetic DNA that contained 12 base pairs [Y. Kim, E. C. Long, J. K. Barton, and C. M. Lieber, *Langmuir,* **8,** 496 (1992)]. The STM images not only showed individual molecules of the double-stranded DNA, they were able to resolve the two strands. When an $Ru^{3+}$ complex was added to the oligonucleotide, the STM images provided an estimate of the distance between the site at which this complex binds and the end of the nucleotide fragment that is consistent with the position expected from studies of the site at which this complex cuts DNA.

## 2.5 THE MACROSCOPIC, ATOMIC, AND SYMBOLIC WORLDS OF CHEMISTRY

Chemists simultaneously work in three very different worlds (see Figure 2.4). Most measurements are done in the **macroscopic world,** with objects that are visible to the naked eye. When you walk into a chemical laboratory, you find a variety of bottles, tubes, flasks, and beakers designed to study samples of liquids and solids large enough to be seen. You may also find sophisticated instruments that can be used to detect very small quantities of materials, but the samples injected into these instruments are still large enough to be seen.

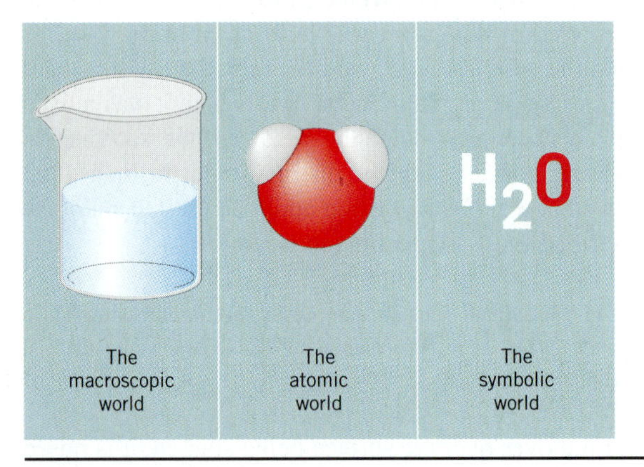


**Figure 2.4** Chemists simultaneously work in three different worlds: (1) the macroscopic world of objects visible to the naked eye, which may be represented by a beaker of water, (2) the atomic world, in which water is thought of as molecules that contain two hydrogen atoms bound to an oxygen atom, and (3) the symbolic world, in which water is represented as $H_2O$.

Although their experiments are done on the macroscopic scale, chemists think about the behavior of matter in terms of a world of atoms and molecules. In this **atomic world,** water is no longer a liquid that freezes at 0°C and boils at 100°C but individual molecules that contain two hydrogen atoms and an oxygen atom. One of the challenges students face when they encounter chemistry for the first time is understanding the process by which chemists perform experiments on the macroscopic scale that can be interpreted in terms of the structure of matter on the atomic scale.

The task of bridging the gap between the atomic and macroscopic worlds is made more difficult by the fact that chemists also work in a **symbolic world,** in which they represent water as $H_2O$ and write equations such as the following to represent what happens when hydrogen and oxygen react to form water:

$$2\,H_2 + O_2 \longrightarrow 2\,H_2O$$

The problem with the symbolic world is that chemists use the same symbols to describe what happens on both the macroscopic and the atomic scales. The symbol "$H_2O$," for example, is used to represent both a single water molecule and a beaker full of water.

The link between the symbols chemists use to represent reactions and the particles involved in these reactions must be kept in mind. Figure 2.5 provides an example of how you might envision the reaction described in the chemical equation written above. The reaction starts with a mixture of $H_2$ and $O_2$ molecules, each of which contains a pair of atoms. It produces water molecules, which contain two hydrogen atoms and an oxygen atom.

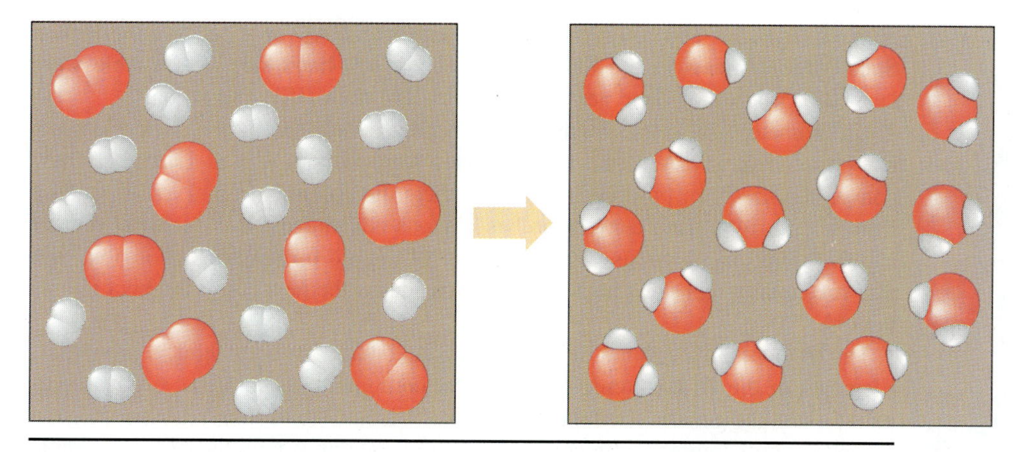

**Figure 2.5** Mechanical models such as this may help you remember that chemical reactions involve particles in a state of constant motion that collide and then react.

## 2.6 THE CHEMISTRY OF THE ELEMENTS

In 1661 when Boyle introduced the modern definition of an element, only 13 elements were known. As the number of elements increased, chemists began to look for ways to group these elements on the basis of their chemical and physical properties. The process that led to the periodic table in Figure 2.6 will be described in Chapter 7. For now, we can concentrate on the fact that this table is divided into 18 vertical columns and 7 horizontal rows.

The vertical columns are known as **groups,** or **families.** Traditionally these groups have been distinguished by a shorthand notation consisting of a Roman

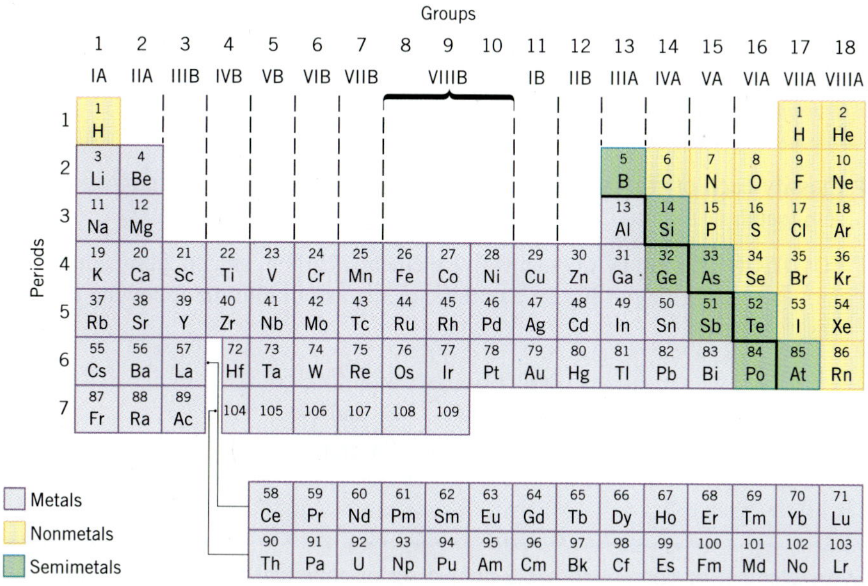

Groups

| Periods | 1 IA | 2 IIA | 3 IIIB | 4 IVB | 5 VB | 6 VIB | 7 VIIB | 8 VIIIB | 9 VIIIB | 10 VIIIB | 11 IB | 12 IIB | 13 IIIA | 14 IVA | 15 VA | 16 VIA | 17 VIIA | 18 VIIIA |
|---|---|---|---|---|---|---|---|---|---|---|---|---|---|---|---|---|---|---|
| 1 | 1 H | | | | | | | | | | | | | | | | 1 H | 2 He |
| 2 | 3 Li | 4 Be | | | | | | | | | | | 5 B | 6 C | 7 N | 8 O | 9 F | 10 Ne |
| 3 | 11 Na | 12 Mg | | | | | | | | | | | 13 Al | 14 Si | 15 P | 16 S | 17 Cl | 18 Ar |
| 4 | 19 K | 20 Ca | 21 Sc | 22 Ti | 23 V | 24 Cr | 25 Mn | 26 Fe | 27 Co | 28 Ni | 29 Cu | 30 Zn | 31 Ga | 32 Ge | 33 As | 34 Se | 35 Br | 36 Kr |
| 5 | 37 Rb | 38 Sr | 39 Y | 40 Zr | 41 Nb | 42 Mo | 43 Tc | 44 Ru | 45 Rh | 46 Pd | 47 Ag | 48 Cd | 49 In | 50 Sn | 51 Sb | 52 Te | 53 I | 54 Xe |
| 6 | 55 Cs | 56 Ba | 57 La | 72 Hf | 73 Ta | 74 W | 75 Re | 76 Os | 77 Ir | 78 Pt | 79 Au | 80 Hg | 81 Tl | 82 Pb | 83 Bi | 84 Po | 85 At | 86 Rn |
| 7 | 87 Fr | 88 Ra | 89 Ac | 104 | 105 | 106 | 107 | 108 | 109 | | | | | | | | | |

| 58 Ce | 59 Pr | 60 Nd | 61 Pm | 62 Sm | 63 Eu | 64 Gd | 65 Tb | 66 Dy | 67 Ho | 68 Er | 69 Tm | 70 Yb | 71 Lu |
|---|---|---|---|---|---|---|---|---|---|---|---|---|---|
| 90 Th | 91 Pa | 92 U | 93 Np | 94 Pu | 95 Am | 96 Cm | 97 Bk | 98 Cf | 99 Es | 100 Fm | 101 Md | 102 No | 103 Lr |

Metals
Nonmetals
Semimetals

**Figure 2.6** The elements can be organized into a periodic table in which elements with similar chemical properties are placed in vertical columns or groups. More than 75% of the known elements are metals. Another 15%, clustered primarily in the upper right corner of the table, are nonmetals. Along the dividing line between these categories is a handful of semimetals, which have properties that lie between the extremes of metals and nonmetals.

This photograph shows samples of various elements.

numeral (I, II, III, and so on) followed by either an ''A'' or a ''B.'' In the United States, the elements in the first column on the left-hand side of the table were labeled ''Group IA.'' The next column was IIA, then IIIB, and so on.

Unfortunately the same notation was not used in all countries. The elements known as Group VIA in the United States were Group VIB in Europe. A new convention has been proposed, which numbers the columns from 1 to 18, reading from left to right. This convention has obvious advantages: it is perfectly regular and therefore unambiguous. The advantages of the old format are less obvious, but they are equally real. This book therefore introduces the new convention but retains the old.

The elements in a column of the periodic table have similar chemical properties. Elements in the first column, for example, combine with chlorine to form compounds with the generic formula **MCl** (HCl, LiCl, NaCl, and so on). Elements in the second column combine with chlorine to form compounds with the generic formula $MCl_2$ ($BeCl_2$, $MgCl_2$, $CaCl_2$, and so on).

The horizontal rows in this table are called **periods.** The first period contains only two elements: H and He. The second period contains eight elements (Li, Be, B, C, N, O, F, and Ne). Although there are nine horizontal rows in the periodic table in Figure 2.6, there are only seven periods—the two rows at the bottom of the table belong in the sixth and seventh periods.

The elements in the periodic table can be divided into three categories: **metals, nonmetals,** and **semimetals** or **metalloids.** As you can see from Figure 2.6, more than 75% of the elements are metals, which are found toward the bottom and the left side of the periodic table.

Only 17 elements are nonmetals. With only one exception—hydrogen, which appears on both sides of the table—these elements are clustered in the upper right corner of the table. The dividing line between the metals and the nonmetals in Figure 2.6 is marked with a heavy diagonal line. A cluster of elements that are neither metals nor nonmetals can be found on either side of this line. These elements are the semimetals or metalloids.

### EXERCISE 2.2

Classify each of the elements in Group IVA as either a metal, a nonmetal, or a semimetal.

**SOLUTION** Group IVA contains five elements: carbon, silicon, germanium, tin, and lead. According to Figure 2.6, these elements fall into the following categories.

| | |
|---|---|
| Nonmetal: | C |
| Semimetal: | Si and Ge |
| Metal: | Sn and Pb |

## 2.7 METALS, NONMETALS, AND SEMIMETALS

Every element or compound has a unique set of chemical and physical properties that can be used to identify that substance. **Physical properties,** such as the melting point or boiling point, are characteristics of the substance itself. **Chemical properties** describe the way the substance interacts with other elements and compounds.

The tendency to divide elements into metals, nonmetals, and semimetals is based on differences between the chemical and physical properties of these three categories of elements.

### PHYSICAL PROPERTIES

**Metals** have the following *physical properties.*

1. They have a metallic shine or luster. (They look like metals!)
2. They are usually solids at room temperature.
3. They are *malleable.* They can be hammered, pounded, or pressed into different shapes without breaking.
4. They are *ductile.* They can be drawn into thin sheets or wires without breaking.
5. They conduct heat and electricity.

**Nonmetals** have the opposite physical properties.

1. They seldom have a metallic luster. They tend to be colorless, like the $O_2$ and $N_2$ in the atmosphere, or brilliantly colored, like bromine.
2. They are often gases at room temperature.
3. The nonmetallic elements (such as carbon, phosphorus, sulfur, and iodine) that are solids at room temperature are neither malleable nor ductile. They cannot be shaped with a hammer or drawn into sheets or wires.
4. They are poor conductors either of heat or electricity. Nonmetals tend to be insulators, not conductors.

Samples of bromine (a nonmetal), copper (a metal), and silicon (a semimetal).

The **semimetals** have properties that lie between these extremes. They often look like metals but are brittle like nonmetals. They are neither conductors nor insulators but make excellent semiconductors.

### CHEMICAL PROPERTIES

The *chemical properties* of metals and nonmetals are also very different. Chemical reactions that involve metals or nonmetals can be divided into three general classes, each of which leads to a different kind of product.

Metals combine with other metals to form **alloys.** Brass, for example, is an alloy of copper and zinc, whereas bronze is an alloy of copper and tin. Alloys retain the physical properties of a metal. They are usually solids at room temperature, with a metallic shine or luster. They are usually malleable and ductile, and they conduct both heat and electricity. Because they can be made with a wide range of compositions, alloys are mixtures, not chemical compounds.

Mercury is the only metal that is a liquid at room temperature, but gallium metal melts at about 30°C.

Alloys, such as brass and bronze, are mixtures of metals that retain the characteristic properties of a metal.

Aluminum metal reacts vigorously with liquid bromine to form a compound that has the properties of neither a metal nor a nonmetal.

Metals often react with nonmetals to form **ionic compounds** or **salts.** Sodium, for example, reacts with chlorine to form sodium chloride, or table salt. NaCl is a white crystalline solid that readily dissolves in water. It therefore has none of the properties of either the metal or the nonmetal from which it is made.

As a rule, nonmetals combine with each other to form **covalent compounds** such as water ($H_2O$) and carbon dioxide ($CO_2$), which have many of the properties of the nonmetallic elements. Like the nonmetallic elements, they exist as molecules. They are often gases at room temperature, or low-boiling liquids. They are neither malleable nor ductile, and they are poor conductors of either heat or electricity.

**EXERCISE 2.3**

For each of the following compounds, predict whether you would expect it to be ionic or covalent.

(a) chromium(III) oxide, $Cr_2O_3$ (b) carbon tetrachloride, $CCl_4$
(c) methanol, $CH_3OH$ (d) strontium fluoride, $SrF_2$

**SOLUTION** $Cr_2O_3$ and $SrF_2$ are likely to be ionic because they contain a metal and a nonmetal. $CCl_4$ and $CH_3OH$ are more likely to be covalent compounds because they contain only nonmetallic elements.

## 2.8 IONIC AND COVALENT COMPOUNDS

There are important differences between ionic and covalent compounds. As evidence of this, let's compare the best known examples of these classes of chemical compounds: table salt and sugar.

Table salt is an ionic compound known as sodium chloride, with the formula NaCl. It is a white, crystalline solid with a very high melting point (801°C) and an even higher boiling point (1465°C). Sugar is a covalent compound, known as sucrose, that has the formula $C_{12}H_{22}O_{11}$. It also is a white, crystalline solid. But as anyone who has made candy knows, it melts at a fairly low temperature (185°C).

**EXERCISE 2.4**

Use the following data to propose a way of distinguishing between ionic and covalent compounds.

| *Compound* | *Melting Point (°C)* | *Boiling Point (°C)* |
|---|---|---|
| $Cr_2O_3$ | 2266 | 4000 |
| $SrF_2$ | 1470 | 2489 |
| $CCl_4$ | −22.9 | 76.6 |
| $CH_3OH$ | −97.8 | 64.7 |

**SOLUTION** In general, ionic compounds (such as $Cr_2O_3$ and $SrF_2$) have much higher melting points and boiling points than covalent compounds (such as $CCl_4$ and $CH_3OH$).

One of the key differences between ionic and covalent compounds becomes apparent when these compounds are dissolved in water. Table salt is fairly soluble in water; up to 36 grams of NaCl will dissolve in 100 grams of water. Sugar is even more soluble; 180 grams of this compound will dissolve in 100 grams of water. These solutions differ significantly, however, in their ability to conduct an electric current.

A simple conductivity apparatus, which consists of a light bulb connected to a pair of thin electrical wires that do not touch, is shown in Figure 2.7. Nothing happens when the light bulb is plugged into a wall socket because there is a gap in the electrical circuit. When something that conducts a current is used to bridge this gap, the light bulb begins to glow.

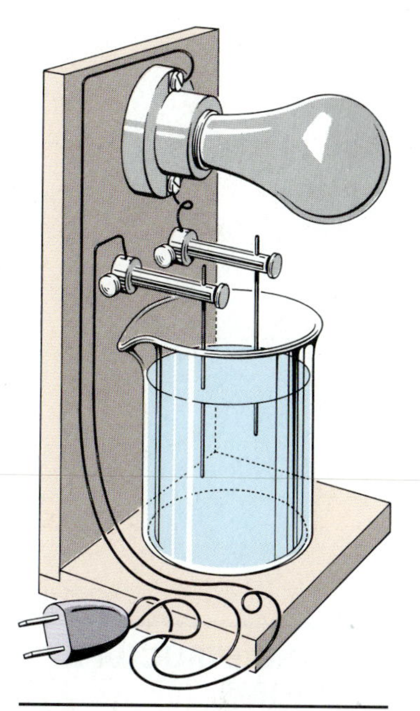

**Figure 2.7** A conductivity apparatus, such as the one shown here, can be used to determine whether a compound can conduct an electric current when it is dissolved in water.

The light bulb does not glow when the wires are immersed in a solution of sugar dissolved in water. But it glows brightly when the wires are immersed in a solution of table salt dissolved in water. Repeating this experiment with solutions of other compounds dissolved in water gives the following general rule of thumb:

**Ionic compounds, or salts, that dissolve in water give solutions that conduct an electric current. Solutions of covalent compounds in water usually do not conduct an electric current.**

An explanation for this behavior was proposed by Svante Arrhenius while he was a graduate student at the University of Uppsala in Sweden. In 1884 Arrhenius suggested that NaCl **dissociates** into positively charged $Na^+$ ions and negatively charged $Cl^-$ ions when it dissolves in water. Although his professors questioned this theory and only reluctantly granted him his degree, we now know that his theory is correct.

Sodium chloride consists of positively charged $Na^+$ ions and negatively charged $Cl^-$ ions that are released into the solution when NaCl dissolves in water. It is the presence of these charged ions that allows the solution to conduct an electric current. Solutions of sugar dissolved in water don't conduct electricity because this compound consists of neutral $C_{12}H_{22}O_{11}$ molecules that do not dissociate into positive and negative ions when they dissolve in water.

The light bulb in this conductivity apparatus glows brightly because the apparatus was immersed in an aqueous solution of NaCl. NaCl dissociates into $Na^+$ and $Cl^-$ ions when it dissolves in water, which can carry the electric current through the solution.

**EXERCISE 2.5**

Which of the following compounds should conduct an electric current when dissolved in water?

(a) methanol, $CH_3OH$ (b) strontium fluoride, $SrF_2$

**SOLUTION**

(a) $CH_3OH$ is a covalent compound that consists of neutral $CH_3OH$ molecules that do not dissociate into ions when the compound dissolves in water. As a result, a solution of this compound in water should not conduct an electric current.

(b) $SrF_2$ is an ionic compound, which should dissociate into positive and negative ions when it dissolves in water to form a solution that conducts an electric current.

## 2.9 THE STRUCTURE OF ATOMS

To understand what it means to say that NaCl consists of positively charged $Na^+$ ions and negatively charged $Cl^-$ ions, it is necessary to take a closer look at the assumption that atoms are indivisible. Today, we recognize that atoms are composed of the three fundamental subatomic particles listed in Table 2.1: *electrons, protons,* and *neutrons.*

**TABLE 2.1 Fundamental Subatomic Particles**

| Particle | Symbol | Charge |
|---|---|---|
| Electron | $e^-$ | −1 |
| Proton | $p^+$ | +1 |
| Neutron | $n^0$ | 0 |

The charge on a proton has the same magnitude as the charge on an electron, but it has the opposite sign. One proton therefore exactly balances the charge on an electron, and vice versa. Thus, atoms are electrically neutral when they contain the same number of electrons and protons. Each atom in the periodic table has been assigned an **atomic number** between 1 and 109, which is equal to the number of protons and electrons in a neutral atom of that element.

## 2.10 THE DIFFERENCE BETWEEN ATOMS AND IONS

We are now ready to propose a model that explains the difference between a neutral sodium atom and a positively charged $Na^+$ ion or a neutral chlorine atom and a negatively charged $Cl^-$ ion. By definition, an **ion** is an electrically charged particle produced by either removing electrons from a neutral atom to give a positive ion or adding electrons to a neutral atom to give a negative ion.

Neutral atoms can be turned into positively charged ions by removing one or more electrons, as shown in Figure 2.8. By removing an electron from a sodium

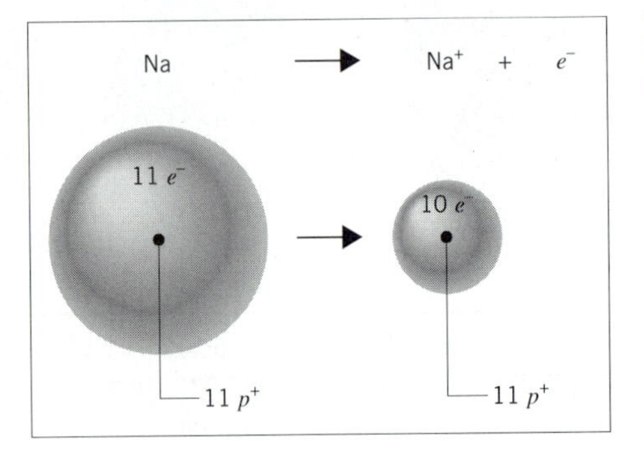


**Figure 2.8** Removing an electron from a neutral sodium atom produces a $Na^+$ ion that has a net charge of +1.

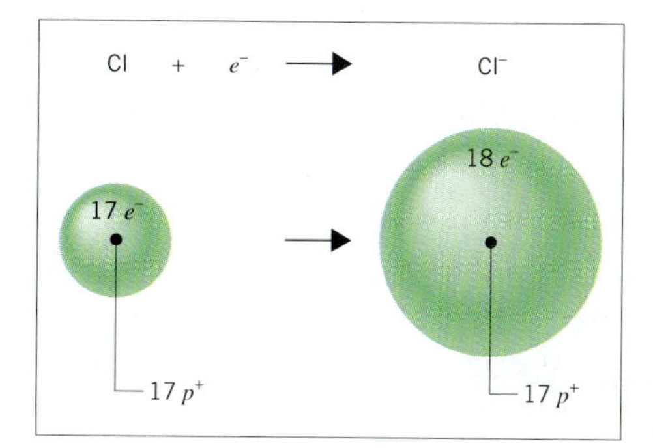


**Figure 2.9** Adding an extra electron to a neutral chlorine atom produces a $Cl^-$ ion that has a net charge of $-1$.

There is an enormous difference between Na and $Na^+$. The first represents a metal that bursts into flame in the presence of water, the other is an ion found in table salt that dissolves readily in water.

atom with 11 protons and 11 electrons, we can produce an $Na^+$ ion with 11 protons and 10 electrons that has a net charge of $+1$. Atoms that gain extra electrons become negatively charged, as shown in Figure 2.9. By adding one more electron to a neutral chlorine atom, for example, we can produce a $Cl^-$ ion that has a net charge of $-1$.

The gain or loss of electrons by an atom to form negative or positive ions has an enormous impact on the chemical and physical properties of the atom. Sodium metal, for example, which consists of neutral sodium atoms, bursts into flame when it comes in contact with water. Positively charged $Na^+$ ions are so unreactive they are essentially inert. Neutral chlorine atoms instantly combine to form $Cl_2$ molecules, which are so reactive that entire communities are evacuated when trains carrying chlorine gas derail. Negatively charged $Cl^-$ ions are essentially inert to chemical reactions.

The enormous difference between the chemistry of neutral atoms and their ions means that you will have to pay particular attention to the formulas you read, to make sure that you don't confuse one with the other.

## 2.11 FORMULAS OF COMMON IONIC COMPOUNDS, OR SALTS

The existence of positive and negative ions raises two important questions. How do we decide whether an element forms a positive ion or a negative ion? And, how do we know how many electrons are gained or lost when an atom forms an ion? It will take a substantial portion of Chapters 6, 7, 8, 9, and 10 to develop a satisfactory answer to these questions. For now, we can note that every column in the periodic table is identified with a **group number** (IA, IIA, and so on). We can use the following general rules to predict the charge on common ions that contain a single atom.

- Metals form positive ions, such as the $Na^+$ and $Cr^{3+}$ ions.
- Nonmetals tend to form negative ions, such as the $Cl^-$ or $O^{2-}$ ions.
- Metals often form positive ions by losing the number of electrons equal to the group number. Thus, sodium in Group IA forms $Na^+$ ions; magnesium in Group IIA forms $Mg^{2+}$ ions; and aluminum in Group IIIA forms $Al^{3+}$ ions.
- Nonmetals often form negative ions by gaining electrons until the charge on the ion is equal to the group number minus 8. Thus, chlorine in Group VIIA forms $Cl^-$ ions ($7 - 8 = -1$); and sulfur in Group VIA forms $S^{2-}$ ions ($6 - 8 = -2$).

The charges on positive and negative ions dictate the formulas of the ionic compounds they form. Ionic compounds have no net electric charge, which means they must contain just as many positive as negative charges. If a salt containing the $Na^+$ and $Co(NO_2)_6^{3-}$ ions is isolated from solution, for example, the formula of the compound would have to be $Na_3Co(NO_2)_6$.

**EXERCISE 2.6**

Predict the formula of the compound that forms when magnesium metal reacts with nitrogen to form magnesium nitride.

**SOLUTION** Magnesium is in Group IIA and nitrogen is in Group VA. The magnesium atoms therefore lose two electrons to form $Mg^{2+}$ ions and the nitrogen atoms gain three electrons to form $N^{3-}$ ions $(5 - 8 = -3)$. To balance the charge, we need three $Mg^{2+}$ ions for every two $N^{3-}$ ions. The product of this reaction is therefore $Mg_3N_2$.

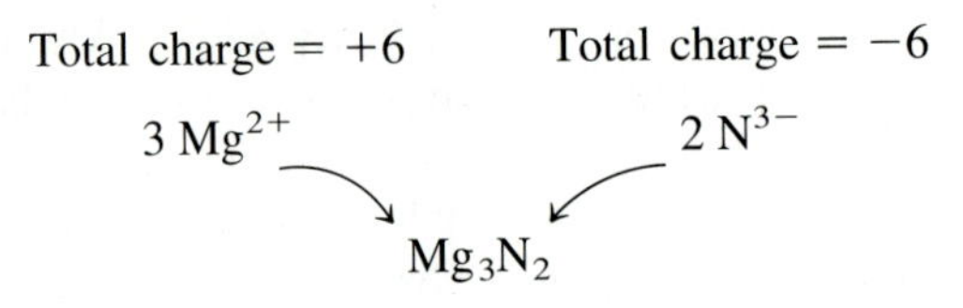


Exercise 2.6 explains why both conventions for labeling the groups in the periodic table are used in this text. The new convention, which numbers the columns from 1 to 18, is both logical and unambiguous. But the old convention allows us to predict the number of electrons gained or lost when an atom forms ions.

**EXERCISE 2.7**

Aluminum chlorhydrate is added to antiperspirants to stop people from sweating. This compound contains $Al^{3+}$, $OH^-$, and $Cl^-$ ions. What is the value of *x*, if the formula for this compound is $Al_x(OH)_5Cl$?

**SOLUTION** The formula for this compound contains five $OH^-$ ions and one $Cl^-$ ion, for a net charge of $-6$. The total charge on the $Al^{3+}$ ions must therefore be $+6$, which means that there must be two $Al^{3+}$ ions in the formula of the compound: $Al_2(OH)_5Cl$.

## 2.12 POLYATOMIC IONS

Simple ions, such as the $Mg^{2+}$ and $N^{3-}$ ions, are formed by adding or subtracting electrons from neutral atoms. It is also possible to add or subtract electrons from neutral molecules to produce **polyatomic ions.** Polyatomic negative ions, such as the $CO_3^{2-}$ or $SO_4^{2-}$ ions, are far more common than polyatomic positive ions, such as the $NH_4^+$ and $H_3O^+$ ion. Some of the important polyatomic negative ions are given in Table 2.2.

**EXERCISE 2.8**

The bone and tooth enamel in your body contains ionic compounds such as calcium phosphate and hydroxyapatite. Predict the formula of calcium phosphate, which contains $Ca^{2+}$ and $PO_4^{3-}$ ions. Calculate the value of *x*, if the formula of hydroxyapatite is $Ca_x(PO_4)_3(OH)$.

**SOLUTION** The formula for calcium phosphate is $Ca_3(PO_4)_2$ because it takes three $Ca^{2+}$ ions to balance the charge on two $PO_4^{3-}$ ions. The formula for hydroxyapatite is $Ca_5(PO_4)_3(OH)$ because it takes five $Ca^{2+}$ ions to balance the charge on three $PO_4^{3-}$ ions and one $OH^-$ ion.

**TABLE 2.2 Common Polyatomic Negative Ions**

| | | | |
|---|---|---|---|
| | −1 ions | | |
| $HCO_3^-$ | bicarbonate | $HSO_4^-$ | hydrogen sulfate (bisulfate) |
| $CH_3CO_2^-$ | acetate | $ClO_4^-$ | perchlorate |
| $NO_3^-$ | nitrate | $ClO_3^-$ | chlorate |
| $NO_2^-$ | nitrite | $ClO_2^-$ | chlorite |
| $MnO_4^-$ | permanganate | $ClO^-$ | hypochlorite |
| $CN^-$ | cyanide | $OH^-$ | hydroxide |
| | −2 ions | | |
| $CO_3^{2-}$ | carbonate | $O_2^{2-}$ | peroxide |
| $SO_4^{2-}$ | sulfate | $CrO_4^{2-}$ | chromate |
| $SO_3^{2-}$ | sulfite | $Cr_2O_7^{2-}$ | dichromate |
| $S_2O_3^{2-}$ | thiosulfate | $HPO_4^{2-}$ | hydrogen phosphate |
| | −3 ions | | |
| $PO_4^{3-}$ | phosphate | $AsO_4^{3-}$ | arsenate |
| $BO_3^{3-}$ | borate | | |

## 2.13 OXIDATION NUMBERS

The **oxidation number** or **oxidation state** of an atom is defined as the charge that atom would carry if the compound were composed of ions. The strontium atom in $SrF_2$ and the carbon atom in CO, for example, are both assigned oxidation numbers of +2 or described as being in the +2 oxidation state. Oxidation numbers are assigned as follows.

- The oxidation number of an atom is zero in a neutral substance that contains atoms of only one element. Thus, the atoms in $O_2$, $O_3$, $P_4$, $S_8$, and aluminum metal all have an oxidation number of 0.
- The oxidation number of simple ions is equal to the charge on the ion. The oxidation number of sodium in the $Na^+$ ion is +1, for example, and the oxidation number of chlorine in the $Cl^-$ ion is −1.
- The oxidation number of hydrogen is +1 when it is combined with a *nonmetal.* Hydrogen is therefore in the +1 oxidation state in $CH_4$, $NH_3$, $H_2O$, and HCl.
- The oxidation number of hydrogen is −1 when it is combined with a *metal.* Hydrogen is therefore in the −1 oxidation state in LiH, NaH, $CaH_2$, and $LiAlH_4$.
- The metals in Group IA form compounds (such as $Li_3N$ and $Na_2S$) in which the metal atom is in the +1 oxidation state.

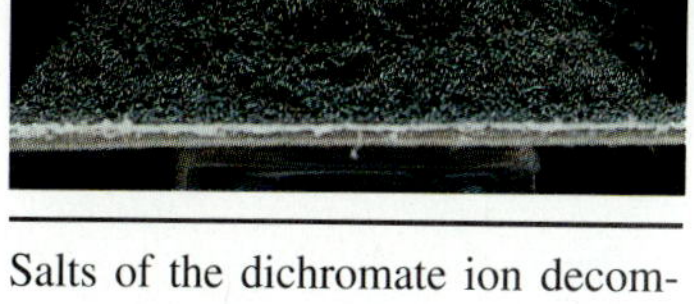

Salts of the dichromate ion decompose violently when heated to form a dark green solid ($Cr_2O_3$).

- The elements in Group IIA form compounds (such as $Mg_3N_2$ and $CaCO_3$) in which the metal atom is in the +2 oxidation state.
- Oxygen usually has an oxidation number of −2. Exceptions include molecules and polyatomic ions that contain O–O bonds, such as $O_2$, $O_3$, $H_2O_2$, and the $O_2^{2-}$ ion.
- The elements in Group VIIA often form compounds (such as $AlF_3$, HCl, and $ZnBr_2$) in which the nonmetal is in the −1 oxidation state.
- The sum of the oxidation numbers in a neutral compound is zero:

$$H_2O: \quad 2(+1) + (-2) = 0$$

The sum of the oxidation numbers in a polyatomic ion is equal to the charge on the ion. The oxidation number of the sulfur atom in the $SO_4^{2-}$ ion must be +6, for example, because the sum of the oxidation numbers of the atoms in this ion must equal −2:

$$SO_4^{2-}: \quad (+6) + 4(-2) = -2$$

- Elements toward the bottom left corner of the periodic table are more likely to form positive oxidation states than those toward the upper right corner of the table. Sulfur carries a positive oxidation state in $SO_2$, for example, because it is below oxygen in the periodic table:

$$SO_2: \quad (+4) + 2(-2) = 0$$

## EXERCISE 2.9

Assign the oxidation numbers of the atoms in the following compounds:

(a) $Al_2O_3$ (b) $XeF_4$ (c) $K_2Cr_2O_7$

**SOLUTION**

(a) The sum of the oxidation numbers in $Al_2O_3$ must be zero because the compound is neutral. If we assume that oxygen is present in the −2 oxidation state, the oxidation state of the aluminum must be +3:

$$Al_2O_3: \quad 2(+3) + 3(-2) = 0$$

(b) Because the oxidation number of the fluorine is −1, the xenon atom must be present in the +4 oxidation state:

$$XeF_4: \quad (+4) + 4(-1) = 0$$

(c) Assigning oxidation numbers in this compound is simplified if you recognize that $K_2Cr_2O_7$ is an ionic compound that contains $K^+$ and $Cr_2O_7^{2-}$ ions. The oxidation state of the potassium is +1 in the $K^+$ ion. Because the oxidation number of oxygen is usually −2, the oxidation state of the chromium in the $Cr_2O_7^{2-}$ ion is +6:

$$Cr_2O_7^{2-}: \quad 2(+6) + 7(-2) = -2$$

## EXERCISE 2.10

Arrange the following compounds in order of increasing oxidation state for the carbon atom.

(a) $CO$ (b) $CO_2$ (c) $H_2CO$ (d) $CH_3OH$ (e) $CH_4$

**SOLUTION**

$$CH_4 < CH_3OH < H_2CO < CO < CO_2$$

$$-4 \qquad -2 \qquad\qquad 0 \qquad +2 \qquad +4$$

Figure 2.10 shows the common oxidation numbers for many of the elements in the periodic table. There are several clear patterns in these data.

1. Elements in the same group often have the same oxidation numbers.
2. The largest, or most positive, oxidation state of an atom is often equal to the group number of the element. The largest oxidation number for phosphorus, in Group VA, for example, is +5.
3. The smallest, or most negative, oxidation number of a nonmetal can be found by subtracting 8 from the group number. The most negative oxidation state of phosphorus, for example, is 5 − 8, or −3.

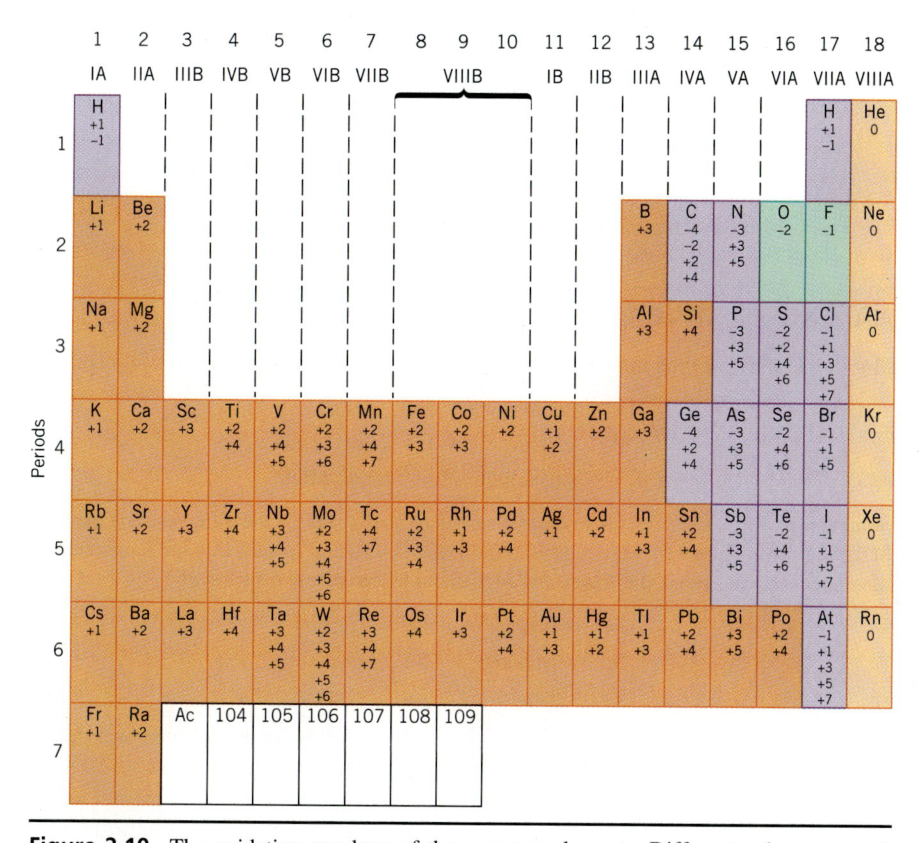


**Figure 2.10** The oxidation numbers of the common elements. Different colors are used to distinguish between elements that form only positive or negative oxidation states and those that exhibit both positive and negative oxidation states.

## SPECIAL TOPIC

# NOMENCLATURE

Long before chemists knew the formulas for chemical compounds, they developed a system of **nomenclature** (from the Latin *nomen,* name, and *calare,* to call) that gave each compound a unique name. Today we often use chemical formulas, such as NaCl, $C_{12}H_{22}O_{11}$, and $Co(NH_3)_6(ClO_4)_3$, to describe chemical compounds. But we still need unique names that unambiguously identify each compound.

### COMMON NAMES

Some compounds have been known for so long that a systematic nomenclature cannot compete with well-established common names. Water ($H_2O$), ammonia ($NH_3$), and methane ($CH_4$) are examples of compounds for which common names are used.

### NAMING IONIC COMPOUNDS, OR SALTS

The names of ionic compounds are written by listing the name of the positive ion followed by the name of the negative ion:

| | |
|---|---|
| NaCl | sodium chloride |
| $(NH_4)_2SO_4$ | ammonium sulfate |
| $NaHCO_3$ | sodium bicarbonate |

We therefore need a series of rules that allow us to unambiguously name positive and negative ions before we can name the salts these ions form.

### NAMING POSITIVE IONS

Monatomic positive ions have the name of the element from which they are formed:

| | | | |
|---|---|---|---|
| $Na^+$ | sodium | $Zn^{2+}$ | zinc |
| $Ca^{2+}$ | calcium | $H^+$ | hydrogen |
| $K^+$ | potassium | $Sr^{2+}$ | strontium |

Some metals form positive ions in more than one oxidation state. One of the earliest methods of distinguishing between these ions used the suffixes *-ous* and *-ic* added to the Latin name of the element to represent the lower and higher oxidation states, respectively:

| | | | |
|---|---|---|---|
| $Fe^{2+}$ | ferrous | $Fe^{3+}$ | ferric |
| $Sn^{2+}$ | stannous | $Sn^{4+}$ | stannic |
| $Cu^+$ | cuprous | $Cu^{2+}$ | cupric |

Chemists now use a simpler method, in which the charge on the ion is indicated by a Roman numeral in parentheses immediately after the name of the element:

| | | | |
|---|---|---|---|
| $Fe^{2+}$ | iron(II) | $Fe^{3+}$ | iron(III) |
| $Sn^{2+}$ | tin(II) | $Sn^{4+}$ | tin(IV) |
| $Cu^+$ | copper(I) | $Cu^{2+}$ | copper(II) |

Polyatomic positive ions often have common names ending with the suffix *-onium:*

| | |
|---|---|
| $H_3O^+$ | hydronium |
| $NH_4^+$ | ammonium |

## NAMING NEGATIVE IONS

Negative ions that consist of a single atom are named by adding the suffix *-ide* to the stem of the name of the element:

| | | | |
|---|---|---|---|
| $F^-$ | fluoride | $O^{2-}$ | oxide |
| $Cl^-$ | chloride | $S^{2-}$ | sulfide |
| $Br^-$ | bromide | $N^{3-}$ | nitride |
| $I^-$ | iodide | $P^{3-}$ | phosphide |
| $H^-$ | hydride | $C^{4-}$ | carbide |

At first glance, the nomenclature of the polyatomic negative ions given in Table 2.2 (Section 2.12) seems hopelessly confusing. There are several general rules, however, that can bring some order out of this apparent chaos:

- The name of the ion usually ends in either *-ite* or *-ate.*
- The *-ite* ending indicates a low oxidation state. Thus, the $NO_2^-$ ion is the nitrite ion.
- The *-ate* ending indicates a high oxidation state. The $NO_3^-$ ion, for example, is the nitrate ion.
- The prefix *hypo-* is used to indicate the very lowest oxidation state. The $ClO^-$ ion, for example, is the hypochlorite ion.
- The prefix *per-* (as in hyper-) is used to indicate the very highest oxidation state. The $ClO_4^-$ ion is therefore the perchlorate ion.

There are only a handful of exceptions to these generalizations. The names of the hydroxide ($OH^-$), cyanide ($CN^-$), and peroxide ($O_2^{2-}$) ions, for example, have the *-ide* ending because they were once thought to be monatomic ions.

### EXERCISE 2.11

Name the following ionic compounds:

(a) $Fe(NO_3)_3$ (b) $SrCO_3$ (c) $Na_2SO_3$ (d) $Ca(ClO)_2$

**SOLUTION**

(a) According to Figure 2.10, iron exists in two oxidation states: +2 or +3. It is therefore important to specify that this compound contains the $Fe^{3+}$ and $NO_3^-$ ions. The compound is therefore known as iron(III) nitrate.

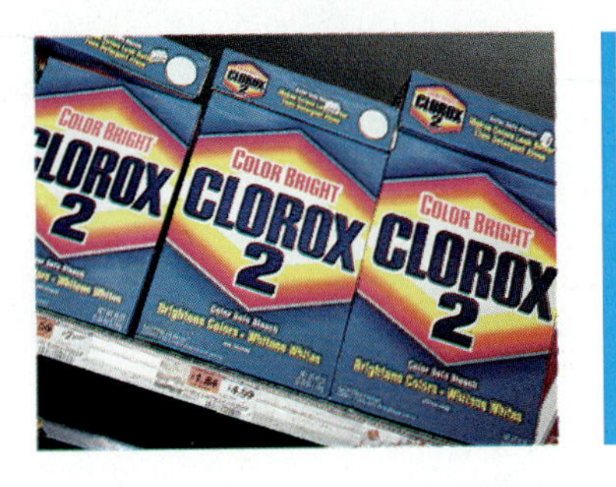


(b) There is no need to specify the oxidation state of strontium because strontium is in the +2 oxidation state in all of its compounds. $SrCO_3$ is therefore known as strontium carbonate.

(c) There is no need to specify the oxidation state of sodium in this compound because sodium is always in the +1 oxidation state in its compounds. The negative ion in this compound is the $SO_3^{2-}$, or sulfite ion, so the compound is sodium sulfite.

(d) This compound consists of $Ca^{2+}$ and $ClO^-$ ions. Once again, the metal is only found in one oxidation state, so this is simply a calcium salt. This compound, which is the solid bleach used in Clorox 2, is known as calcium hypochlorite.

## NAMING SIMPLE COVALENT COMPOUNDS

Oxidation states also play an important role in naming simple covalent compounds. The name of the atom in the positive oxidation state is listed first. The suffix *-ide* is then added to the stem of the name of the atom in the negative oxidation state:

| | |
|---|---|
| HCl | hydrogen chloride |
| NO | nitrogen oxide |
| BrCl | bromine chloride |

As a rule, chemists write formulas in which the element in the positive oxidation state is written first, followed by the element(s) with negative oxidation numbers.

The number of atoms of an element in simple covalent compounds is indicated by adding one of the following Greek prefixes to the name of the element:

| | | | |
|---|---|---|---|
| 1 | mono- | 6 | hexa- |
| 2 | di- | 7 | hepta- |
| 3 | tri- | 8 | octa- |
| 4 | tetra- | 9 | nona- |
| 5 | penta- | 10 | deca- |

The prefix *mono-* is seldom used because it is redundant. The principal exception to this rule is carbon monoxide (CO).

**EXERCISE 2.12**

Name the following covalent compounds:

(a) $N_2O$ (b) NO (c) $NO_2$ (d) $N_2O_3$ (e) $N_2O_4$ (f) $N_2O_5$

**SOLUTION** Nitrogen is present in a positive oxidation state in all these compounds. We therefore add the *-ide* ending to the stem of the name of the element oxygen and specify the number of nitrogen and oxygen atoms in each of these compounds as follows:

(a) $N_2O$, dinitrogen oxide (b) NO, nitrogen oxide (c) $NO_2$, nitrogen dioxide
(d) $N_2O_3$, dinitrogen trioxide (e) $N_2O_4$, dinitrogen tetroxide
(f) $N_2O_5$, dinitrogen pentoxide

## NAMING ACIDS

Simple covalent compounds that contain hydrogen, such as HCl, HBr, and HCN, often dissolve in water to produce acids. These solutions are named by adding the prefix *hydro-* to the name of the compound and then replacing the suffix *-ide* with *-ic*. For example, hydrogen chloride (HCl) is a gas that dissolves in water to form hydrochloric acid. Hydrogen bromide (HBr) forms hydrobromic acid; and hydrogen cyanide (HCN) forms hydrocyanic acid.

Many of the oxygen-rich polyatomic negative ions in Table 2.2 form acids that are named by replacing the suffix *-ate* with *-ic* and the suffix *-ite* with *-ous:*

| | | | |
|---|---|---|---|
| $CH_3CO_2^-$ | acetate | $CH_3CO_2H$ | acetic acid |
| $CO_3^{2-}$ | carbonate | $H_2CO_3$ | carbonic acid |
| $BO_3^{3-}$ | borate | $H_3BO_3$ | boric acid |
| $NO_3^-$ | nitrate | $HNO_3$ | nitric acid |
| $NO_2^-$ | nitrite | $HNO_2$ | nitrous acid |
| $SO_4^{2-}$ | sulfate | $H_2SO_4$ | sulfuric acid |
| $SO_3^{2-}$ | sulfite | $H_2SO_3$ | sulfurous acid |
| $ClO_4^-$ | perchlorate | $HClO_4$ | perchloric acid |
| $ClO_3^-$ | chlorate | $HClO_3$ | chloric acid |
| $ClO_2^-$ | chlorite | $HClO_2$ | chlorous acid |
| $ClO^-$ | hypochlorite | $HClO$ | hypochlorous acid |
| $PO_4^{3-}$ | phosphate | $H_3PO_4$ | phosphoric acid |
| $MnO_4^-$ | permanganate | $HMnO_4$ | permanganic acid |
| $CrO_4^{2-}$ | chromate | $H_2CrO_4$ | chromic acid |

Complex acids can be named by indicating the presence of an acidic hydrogen as follows:

| | |
|---|---|
| $NaHCO_3$ | sodium hydrogen carbonate (also known as sodium bicarbonate) |
| $NaHSO_3$ | sodium hydrogen sulfite (also known as sodium bisulfite) |
| $KH_2PO_4$ | potassium dihydrogen phosphate |

### EXERCISE 2.13

Name the following compounds:

(a) $NaClO_3$ (b) $Al_2(SO_4)_3$
(c) $P_4S_3$ (d) $SCl_4$

**SOLUTION** The key to naming these compounds is recognizing that the first two are ionic compounds and the last two are covalent compounds:

(a) sodium chlorate (b) aluminum sulfate (c) tetraphosphorus trisulfide
(d) sulfur tetrachloride

## SUMMARY

**1.** An element is any substance that contains only one kind of atom. Because atoms cannot be created or destroyed in a chemical reaction, elements cannot be broken down into simpler substances by these reactions.

**2.** The elements can be divided into three categories: **metals, nonmetals,** and **semimetals** that have characteristic properties. Most elements are metals, which are found on the left and toward the bottom of the periodic table. A handful of nonmetals is clustered in the upper right corner of the periodic table. The semimetals can be found along the dividing line between the metals and the nonmetals.

**3.** Elements combine to form chemical compounds that are often divided into two categories. **Ionic compounds,** or salts, are usually formed when metals react with nonmetals. These compounds are composed of positive and negative ions formed by adding or subtracting electrons from neutral atoms and molecules. **Covalent compounds,** which exist as neutral molecules, are formed when two or more nonmetals combine.

**4.** It is often useful to follow chemical reactions by looking at changes in the oxidation numbers of the atoms in each compound during the reaction. By definition, the **oxidation number** of an atom is the charge that atom would have if the compound was composed of ions. Oxidation numbers also play an important role in the systematic nomenclature of chemical compounds.

## KEY TERMS

**Alloy** (p. 53)
**Atom** (p. 48)
**Atomic number** (p. 56)
**Chemical properties** (p. 52)
**Compound** (p. 48)
**Covalent compound** (p. 54)
**Dissociate** (p. 55)
**Element** (p. 45)
**Groups** (p. 51)
**Group number** (p. 57)
**Ion** (p. 56)
**Ionic compound** (p. 54)
**Law of constant composition** (p. 45)
**Law of definite proportions** (p. 47)
**Metal** (p. 52)
**Metalloid** (p. 52)
**Molecule** (p. 48)
**Nomenclature** (p. 62)
**Nonmetal** (p. 52)
**Oxidation number** (p. 59)
**Period** (p. 52)
**Physical properties** (p. 52)
**Polyatomic ion** (p. 58)
**Salt** (p. 54)
**Semimetal** (p. 52)

## PROBLEMS

### Elements and Compounds

**2-1** Define the following terms: *element, compound,* and *mixture.* Give an example of each.

**2-2** Describe the difference between elements and compounds on the macroscopic scale and on the atomic scale.

**2-3** Classify the following substances into the categories of elements, compounds, mixtures, metals, nonmetals, and semimetals. Use as many labels as necessary to classify each substance. Use whatever reference you need to identify each substance.

(a) diamond (b) brass (c) soil (d) glass (e) cotton (f) milk of magnesia (g) salt (h) iron (i) steel

**2-4** Granite consists of four minerals: feldspar, magnetite, mica, and quartz. If one of these minerals can be physically separated from the others, is granite an element, a compound, or a mixture?

**2-5** List the symbols for the following elements:

(a) antimony (b) gold (c) iron (d) mercury (e) potassium (f) silver (g) tin (h) tungsten

**2-6** Name the elements with the following symbols:

(a) Na (b) Mg (c) Al (d) Si (e) P (f) Cl (g) Ar

**2-7** Name the elements with the following symbols:

(a) Ti (b) V (c) Cr (d) Mn (e) Fe (f) Co (g) Ni (h) Cu (i) Zn

**2-8** Name the elements with the following symbols:

(a) Mo (b) W (c) Rh (d) Ir (e) Pd (f) Pt (g) Ag (h) Au (i) Hg

### The Law of Constant Composition

**2-9** Describe the law of constant composition and show how it can be used to distinguish between compounds and mixtures.

**2-10** Explain why the law of constant composition is consistent with Dalton's theory of the atom.

### Evidence for the Existence of Atoms

**2-11** Describe some of the evidence for the existence of atoms and some of the evidence from our senses that seems to deny the existence of atoms.

**2-12** Compare Dalton's theory of the atom with the theory proposed by Leucippus and Democritus 2300 years earlier. In what ways are these theories the same? In what ways are they different?

**2-13** What changes would have to be made in the five postulates of Dalton's atomic theory to bring this theory up to date?

**2-14** Describe some of the evidence for the assumption that atoms are small.

### Atoms and Molecules

**2-15** Describe what the formula $P_4S_3$ tells us about this compound.

**2-16** Describe the difference between the following pairs of symbols:

(a) Co and CO (b) Cs and $CS_2$ (c) Ho and $H_2O$ (d) 4 P and $P_4$

**2-17** Describe the difference between the following pairs of symbols:

(a) H and $H^+$ (b) H and $H^-$ (c) 2 H and $H_2$ (d) $H^+$ and $H^-$

### The Macroscopic, Atomic, and Symbolic Worlds of Chemistry

**2-18** Explain the difference between $H^+$ ions, H atoms, and $H_2$ molecules on both the atomic and the macroscopic scales.

**2-19** Describe some of the properties of water on both the macroscopic and the atomic scale.

**2-20** Which of the following properties describe elements and compounds on the macroscopic scale? Which describe elements and compounds on the atomic scale? Which can be used to describe them on both scales?

(a) temperature (b) pressure (c) volume (d) kinetic energy (e) the shape of molecules (f) elemental symbols such as Au or Pt (g) weight (h) color (i) melting point (j) boiling point

### The Chemistry of the Elements

**2-21** Classify each of the following elements as a metal, a nonmetal, or a semimetal:

(a) Na (b) Mg (c) Al (d) Si (e) P (f) S (g) Cl (h) Ar

**2-22** Classify each of the following elements as a metal, nonmetal, or semimetal:

(a) $S_8$ (b) Sb (c) Sc (d) Se (e) Si (f) Sm (g) Sn (h) Sr

**2-23** Describe the difference between a group and a period.

**2-24** Classify the elements in Group VA as either metals, nonmetals, or semimetals. Describe what happens to the properties of these elements as we go down this column of elements.

**2-25** Classify the elements in the third period as either metals, nonmetals, or semimetals. Describe what happens to the properties of these elements as we go from left to right across this period.

### Metals, Nonmetals, and Semimetals

**2-26** Describe what happens to the physical properties of a nonmetal when it combines with another nonmetal to form a covalent compound.

**2-27** Describe what happens to the physical properties of a metal and a nonmetal when they combine to form an ionic compound, or salt.

**2-28** Describe what happens to the physical properties of a metal when it combines with another metal to form an alloy.

**2-29** Describe the differences between the chemical and physical properties of covalent compounds such as $CO_2$ and ionic compounds such as NaCl.

**2-30** Which of the following compounds should be ionic?

(a) ZnS (b) $AlCl_3$ (c) $SnF_2$ (d) $BH_3$ (e) $H_2S$

**2-31** Which of the following compounds should be covalent?

(a) $CH_4$ (b) $CO_2$ (c) $SrCl_2$ (d) NaH (e) $SF_4$

**2-32** Which of the following pairs of elements should combine to give ionic compounds?

(a) $Mg + O_2$ (b) $S_8 + F_2$ (c) $P_4 + Na$ (d) Na + Hg (e) $K + I_2$

**2-33** Which of the following pairs of elements should combine to give covalent compounds?

(a) $N_2 + O_2$ (b) $Cl_2 + F_2$ (c) $Cl_2 + Cr$ (d) $S_8 + Na$ (e) Cu + Sn

### Ionic and Covalent Compounds

**2-34** Which of the following compounds should conduct an electric current when dissolved in water?

(a) $MgCl_2$ (b) $CO_2$ (c) $CH_3OH$ (d) $KNO_3$ (e) $Ca_3P_2$

**2-35** Many covalent compounds have very distinctive odors, such as the $H_2S$ given off by rotten eggs, or the $CH_3CO_2H$ in vinegar. Ionic compounds such as NaCl or $Al_2O_3$ have no odor. What is the difference between the physical properties of ionic and covalent compounds that is responsible for this characteristic?

**2-36** One of the simplest ways of distinguishing between two covalent compounds is to measure their melting points or boiling points. Naphthalene melts at 80.5°C and camphor melts at 178.8°C, for example. Explain why this procedure is not as useful when distinguishing between ionic compounds such as NaCl and $Al_2O_3$.

**2-37** Hydrogen chloride is a covalent compound that is a gas at room temperature. When cooled, it condenses to form a liquid. Neither HCl gas nor liquid HCl conducts electricity. But when HCl is dissolved in water, the result is hydrochloric acid, which conducts electricity very well. Explain why.

**2-38** Which of the following substances would you expect to conduct an electric current?

(a) solid Na metal (b) liquid Na metal (c) solid NaCl (d) liquid NaCl (e) NaCl dissolved in water

### Formulas of Common Ionic Compounds, or Salts

**2-39** Just as there is a difference between an orange and an orange peel, there is a difference between Fe metal and $Fe^{3+}$ ions. If you are told that you have a shortage of iron, calcium, or zinc in your diet, which would you add to your diet, pieces of the metal or salts of their ions?

**2-40** Calculate the charge on the positive ions formed by the following elements:

(a) Mg (b) Al (c) Si (d) Cs (e) Ba

**2-41** Calculate the charge on the negative ions formed by the following elements:

(a) C (b) P (c) S (d) I

**2-42** Fluoride toothpastes convert the mineral apatite in tooth enamel into fluorapatite, $Ca_5(PO_4)_3F$. If fluorapatite contains $Ca^{2+}$ and $PO_4^{3-}$ ions, what is the charge on the fluoride ion in this compound?

**2-43** Verdigris is a green pigment used in paint. The simplest formula for this compound is $Cu_3(OH)_2(CH_3CO_2)_4$. What is the charge on the copper ions in this compound, if the other ions both carry a charge of $-1$?

**2-44** Predict the formulas for neutral compounds containing the following pairs of ions:

(a) $Mg^{2+}$ and $NO_3^-$ (b) $Fe^{3+}$ and $SO_4^{2-}$ (c) $Na^+$ and $CO_3^{2-}$

**2-45** Predict the formulas for neutral compounds containing the following pairs of ions:

(a) $Na^+$ and $O_2^{2-}$ (b) $Zn^{2+}$ and $PO_4^{3-}$ (c) $K^+$ and $PtCl_6^{2-}$

**2-46** Predict the formulas for potassium nitride and aluminum nitride, if the formula for magnesium nitride is $Mg_3N_2$.

**2-47** Compounds that contain the $O^{2-}$ ion are called oxides. Those that contain the $O_2^{2-}$ ion are called peroxides. If the formula for potassium oxide is $K_2O$, what is the formula for potassium peroxide?

**2-48** What is the value of $x$ in the $[Co(NO_2)_x]^{3-}$ ion if this complex ion contains $Co^{3+}$ and $NO_2^-$ ions?

**2-49** Magnetic iron oxide has the formula $Fe_3O_4$. Explain this formula by assuming that $Fe_3O_4$ contains both $Fe^{2+}$ and $Fe^{3+}$ ions.

### Oxidation Numbers

**2-50** An area of active research interest in recent years has involved compounds such as $Re_2Cl_8^{2-}$, $Cr_2Cl_9^{3-}$, and $Mo_2Cl_8^{4-}$ that contain bonds between metal atoms. Calculate the oxidation number of the metal atom in each of these compounds.

**2-51** Determine the oxidation state of phosphorus in the following compounds:

(a) $K_3P$ (b) $AlPO_4$ (c) $PO_3^{3-}$ (d) $P_2Cl_4$

**2-52** Calculate the oxidation number of the aluminum atom in the following compounds:

(a) $LiAlH_4$ (b) $Al(H_2O)_6^{3+}$ (c) $Al(OH)_4^-$

**2-53** The active ingredient in Rolaids has the formula $NaAl(OH)_2CO_3$. Calculate the oxidation state of the aluminum atom in this compound.

**2-54** Which of the following compounds contain hydrogen in a negative oxidation state?

(a) $H_2S$ (b) $H_2O$ (c) $NH_3$ (d) $PH_4^-$ (e) $LiAlH_4$
(f) HF (g) $CaH_2$ (h) $CH_4$

**2-55** Calculate the oxidation number of the chlorine atom in the following compounds:

(a) $Cl_2$ (b) $Cl^-$ (c) $ClO^-$ (d) $ClO_2^-$ (e) $ClO_3^-$
(f) $ClO_4^-$

**2-56** Calculate the oxidation number of the iodine atom in the following compounds. Group together any compounds in which iodine has the same oxidation number.

(a) HI (b) KI (c) $I_2$ (d) HOI (e) $KIO_3$ (f) $I_2O_5$
(g) $KIO_4$ (h) $H_5IO_6$

**2-57** Calculate the oxidation number of the xenon atom in the following compounds and describe any trends in these oxidation states:

(a) $XeF_2$ (b) $XeF_4$ (c) $XeOF_2$ (d) $XeF_6$
(e) $XeOF_4$ (f) $XeO_3$ (g) $XeO_4$ (h) $XeO_6^{4-}$

**2-58** Calculate the oxidation number of barium in $BaO_2$. Does your answer make sense? If it doesn't, determine the consequence of assuming that barium is present in its usual oxidation state in this compound.

**2-59** Carbon can have any oxidation number between $-4$ and $+4$. Calculate the oxidation number of carbon in the following compounds. Group compounds with the same oxidation number and describe any trends.

(a) $CCl_4$ (b) $COCl_2$ (c) CO (d) $CO_2$ (e) $CS_2$
(f) $CH_3Li$ (g) $CH_4$ (h) $H_2CO$ (i) $Na_2CO_3$
(j) $HCO_2H$

**2-60** Sulfur can have any oxidation number between $+6$ and $-2$. Calculate the oxidation number of sulfur in the following compounds. Group compounds with the same oxidation number and describe any trends.

(a) $S_8$ (b) $H_2S$ (c) ZnS (d) $SF_4$ (e) $SF_6$ (f) $SO_2$
(g) $SO_3$ (h) $SO_3^{2-}$ (i) $SO_4^{2-}$ (j) $H_2SO_3$
(k) $H_2SO_4$

**2-61** Calculate the oxidation number of manganese in the following compounds. Group compounds with the same oxidation number and describe any trends.

(a) MnO (b) $Mn_2O_3$ (c) $MnO_2$ (d) $MnO_3$
(e) $Mn_2O_7$ (f) $Mn(OH)_2$ (g) $Mn(OH)_3$ (h) $H_2MnO_4$
(i) $HMnO_4$ (j) $CaMnO_3$ (k) $MnSO_4$

**2-62** Calculate the oxidation number of titanium in the following compounds. Group compounds with the same oxidation number and describe any trends.

(a) TiO (b) $TiO_2$ (c) $Ti_2O_3$ (d) $Ti_3O_7$ (e) $TiCl_3$
(f) $TiCl_4$ (g) $K_2TiO_3$ (h) $H_2TiCl_6$ (i) $Ti(SO_4)_2$

**2-63** Prussian blue is a pigment with the formula $Fe_4[Fe(CN)_6]_3$. If this compound contains the $Fe(CN)_6^{4-}$ ion, what is the oxidation state of the other four iron atoms? Turnbull's blue is a pigment with the formula $Fe_3[Fe(CN)_6]_2$. If this compound contains the $Fe(CN)_6^{3-}$ ion, what is the oxidation state of the other three iron atoms?

### Nomenclature

**2-64** Describe what is wrong with the common names for the following compounds and write a better name for each compound.

(a) phosphorus pentoxide ($P_2O_5$) (b) iron oxide ($Fe_2O_3$) (c) chlorine monoxide ($Cl_2O$) (d) copper bromide ($CuBr_2$)

**2-65** Explain why calcium bromide is a satisfactory name for $CaBr_2$, but $FeBr_2$ is called iron(II) bromide.

**2-66** Write the formulas for the following compounds:

(a) tetraphosphorus trisulfide (b) silicon dioxide (c) carbon disulfide (d) carbon tetrachloride (e) phosphorus pentafluoride

**2-67** Write the formulas for the following compounds:

(a) silicon tetrafluoride (b) sulfur hexafluoride (c) oxygen difluoride (d) dichlorine heptoxide (e) chlorine trifluoride

**2-68** Write the formulas for the following compounds:

(a) tin(II) chloride (b) mercury(II) nitrate (c) tin(IV) sulfide (d) chromium(III) oxide (e) iron(II) phosphide

**2-69** Write the formulas for the following compounds:

(a) beryllium fluoride (b) magnesium nitride (c) calcium carbide (d) barium peroxide (e) potassium carbonate

**2-70** Write the formulas for the following compounds:

(a) cobalt(III) nitrate (b) iron(III) sulfate (c) gold(III) chloride (d) manganese(IV) oxide (e) tungsten(VI) chloride

**2-71** Name the following compounds:

(a) $KNO_3$ (b) $Li_2CO_3$ (c) $BaSO_4$ (d) $PbI_2$

**2-72** Name the following compounds:

(a) $AlCl_3$ (b) $Na_3N$ (c) $Ca_3P_2$ (d) $Li_2S$ (e) $MgO$

**2-73** Name the following compounds:

(a) $NH_4OH$ (b) $H_2O_2$ (c) $Mg(OH)_2$ (d) $Ca(OCl)_2$ (e) $NaCN$

**2-74** Name the following compounds:

(a) $Sb_2S_3$ (b) $SnCl_2$ (c) $SF_4$ (d) $SrBr_2$ (e) $SiCl_4$

**2-75** Write the formulas of the following common acids:

(a) acetic acid (b) hydrochloric acid (c) sulfuric acid (d) phosphoric acid (e) nitric acid

**2-76** Write the formulas of the following less common acids:

(a) carbonic acid (b) hydrocyanic acid (c) boric acid (d) phosphorus acid (e) nitrous acid

**2-77** If sodium carbonate is $Na_2CO_3$ and sodium bicarbonate is $NaHCO_3$, what are the formulas for sodium bisulfide and sodium bisulfite?

**2-78** The prefix *thio-* describes compounds in which sulfur replaces oxygen—for example, cyanate ($OCN^-$), and thiocyanate ($SCN^-$). If $SO_4^{2-}$ is the sulfate ion, what is the formula for the thiosulfate ion?

**2-79** Name the compound in each of the following minerals:

(a) fluorite ($CaF_2$) (b) galena ($PbS$) (c) rutile ($TiO_2$) (d) hematite ($Fe_2O_3$)

**2-80** Name the compound in each of the following minerals:

(a) magnetite ($Fe_3O_4$) (b) calcite ($CaCO_3$) (c) barite ($BaSO_4$) (d) quartz ($SiO_2$)

This chapter describes how the number of molecules in a drop of water can be calculated from measurements of either its mass or its volume.

# STOICHIOMETRY: COUNTING ATOMS AND MOLECULES

This chapter provides the basis for answering the following questions:

## POINTS OF INTEREST

- What are the consequences of the fact that matter exists as infinitesimally small atoms that cannot be created or destroyed?
- What happens to the weight of an iron bar when it rusts? Does it increase, decrease, or remain the same?
- On a per gram basis, what is the most efficient source of $Ca^{2+}$ ions to prevent osteoporosis: calcium carbonate ($CaCO_3$), calcium sulfate ($CaSO_4$), or calcium phosphate [$Ca_3(PO_4)_2$]?
- How much oxygen do we have to breathe to burn 10.0 grams of sugar?
- What happens when *concentrated* orange juice is *diluted* with water?

## 3.1 THE PROCESS OF DISCOVERY: THE ORIGINS OF STOICHIOMETRY

Sec. 2.3

Between 1792 and 1794 Jeremias Benjamin Richter, who was fascinated with the role of mathematics in chemistry, published a three-volume summary of his work on the law of definite proportions. In this book Richter introduced the term *stoichiometry,* which he defined as the ''art of chemical measurements, which has to deal with the laws according to which substances unite to form chemical compounds.''

Unfortunately Richter's writing style is best described as ''obscure and clumsy.'' His work therefore had little impact until 1802, when it was summarized by Ernst Gottfried Fischer in terms of tables, such as Table 3.1. According to this table, it takes 615 parts by weight of magnesia to neutralize either 1000 parts by weight of sulfuric acid or 1405 parts by weight of nitric acid.

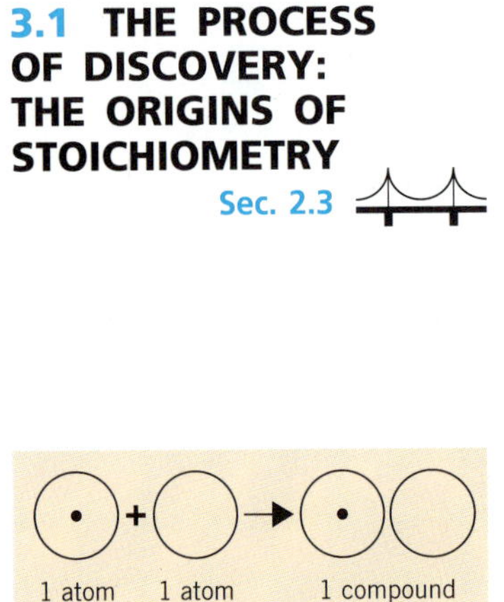


**Figure 3.1** Dalton assumed that water contains one atom of hydrogen and one atom of oxygen and concluded that the relative weight of the oxygen atom must be 5.6 times as large as the hydrogen atom.

**TABLE 3.1 Weights of Acids and Bases That Are Chemically Equivalent**

| Bases | | Acids | |
|---|---|---|---|
| Alumina | 525 | Carbonic | 577 |
| Magnesia | 615 | Muriatic | 712 |
| Lime | 793 | Phosphoric | 979 |
| Soda | 859 | Sulfuric | 1000 |
| Potash | 1605 | Nitric | 1405 |
| | | Acetic | 1480 |

John Dalton was not familiar with Richter's work when he developed his atomic theory in 1803 (see Section 2.3). By 1807, however, references to this work appeared in Dalton's notebooks, and Dalton's contemporaries viewed his atomic theory as a way of explaining why compounds combine in definite proportions.

Consider water, for example. In his famous textbook, *Traité Élémentaire de Chimie,* which was published in 1789, Lavoisier reported that water was roughly 85% oxygen and 15% hydrogen by weight. Water therefore seemed to contain 5.6 times more oxygen by weight than hydrogen. Dalton assumed that water contains one atom of hydrogen and one atom of oxygen, as shown in Figure 3.1, and concluded that an oxygen atom must weight 5.6 times more than a hydrogen atom. On the basis of such reasoning, Dalton constructed a table of the relative atomic weights of a handful of elements.

Jöns Jacob Berzelius (1779–1848) was famous for his contributions to both the theoretical and experimental aspects of the field that would become chemistry.

Jöns Jacob Berzelius was so impressed with the work of Richter and Dalton that he analyzed 2000 compounds to provide the experimental basis for the atomic theory. Working in a laboratory with facilities no more elaborate than a kitchen, Berzelius prepared and purified the necessary reagents, developed the techniques to perform the analyses, and collected data on the relative weights of atoms of 43 elements.

Berzelius also introduced the symbolism with which chemical formulas are still written, although he wrote the numbers that specify the ratio of the elements as superscripts ($H^2O$) rather than as subscripts ($H_2O$). Table 3.2 illustrates the power of Berzelius's work by comparing some of his results with the most recently published data. In most cases, Berzelius's data agree with those obtained with modern instrumentation. Where significant differences exist, they are the result of assumptions that Berzelius had to make about the formulas of the compounds he analyzed.

**TABLE 3.2 Comparison of Berzelius's Atomic Weights with Modern Values**

| Element | Berzelius's Atomic Weights (1826) | Modern Atomic Weights (1983) |
|---|---|---|
| Hydrogen | 0.998 | 1.008 |
| Carbon | 12.25 | 12.01 |
| Nitrogen | 14.16 | 14.01 |
| Oxygen | 16.00 | 16.00 |
| Sulfur | 32.19 | 32.06 |
| Chlorine | 35.41 | 35.45 |
| Calcium | 40.96 | 40.08 |
| Sodium | 46.54 | 22.99 |
| Iron | 54.27 | 55.85 |
| Chromium | 56.29 | 52.01 |
| Copper | 63.31 | 63.54 |
| Potassium | 78.39 | 39.10 |
| Strontium | 87.56 | 87.62 |
| Iodine | 123.00 | 126.90 |
| Barium | 137.10 | 137.34 |
| Gold | 198.88 | 196.97 |
| Mercury | 202.53 | 200.59 |
| Lead | 207.12 | 207.19 |
| Silver | 216.26 | 107.87 |

## 3.2 THE RELATIVE MASSES OF ATOMS

Atoms are so small that a sliver of copper metal just big enough to be detected on a good analytical balance contains about 100,000,000,000,000,000 ($1 \times 10^{17}$) atoms. As a result, it is impossible to measure the absolute mass of a single atom. We can, however, measure the relative masses of different atoms.

Figure 3.2 shows a block diagram of a *mass spectrometer,* which can be used to determine the relative mass of an atom or molecule. The sample enters the instru-

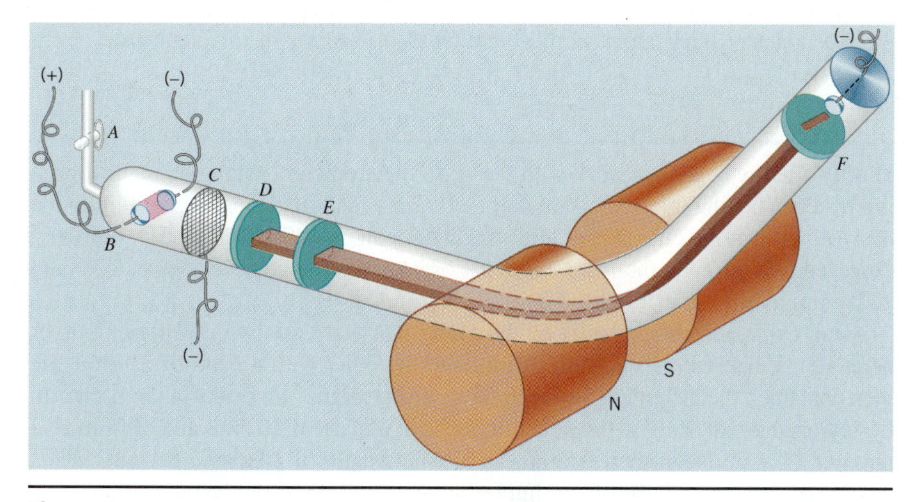


**Figure 3.2** A block diagram of a mass spectrometer.

A mass spectrometer.

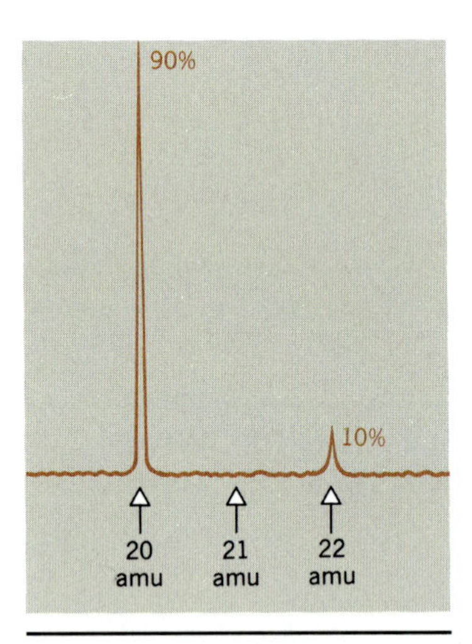


**Figure 3.3** The mass spectrum of neon shows two lines corresponding to the two isotopes of neon.

ment through a valve (*A*) into an evacuated chamber. The particles in the sample flow past a filament (*B*), where they collide with high-energy electrons. As a result of these collisions, the neutral atoms or molecules in the sample lose an electron to form positively charged ions. These ions are then accelerated by a high-voltage accelerator (*C*), focused by a series of slits (*D* and *E*), and passed between the poles of a magnet. The interaction between the magnetic field and the charge on the ion bends the path along which the ions travel. The larger the mass of the ion, the smaller the angle through which its path is bent before it enters the detector (*F*).

The mass spectrometer was invented by F. W. Aston shortly after World War I. By 1927 Aston had built an instrument that was accurate to more than 1 part in 10,000, and mass spectrometry became the method of choice for measuring the relative mass of an atom or molecule. It can tell us, for example, that the mass of a single fluorine atom is 2.010808 times the mass of a beryllium atom.

The logical units for reporting the relative mass of an atom are **atomic mass units,** or **amu.** In theory, the lightest atom could be assigned a mass of 1 amu, and the mass of any other atom would be expressed in terms of this standard.

## 3.3 THE DIFFERENCE BETWEEN ATOMIC MASS AND ATOMIC WEIGHT

Shortly after he began his work with the mass spectrometer, Aston obtained surprising results when he injected a sample of neon into his instrument. According to the best estimates at the time, the atomic weight of neon was 20.2 amu. Aston, however, observed two peaks in the *mass spectrum* of neon with relative masses of 20.0 and 22.0 amu, as shown in Figure 3.3.

Aston explained these results by assuming that there are two different kinds of neon atoms, one with a mass of 20 amu and the other with a mass of 22 amu. He then explained the difference between the intensity of the two peaks in the spectrum by suggesting that 90% of the neon atoms have a mass of 20 amu and 10% have a mass of 22 amu. Neon therefore has an atomic weight of 20.2 amu because it is a weighted average of 90% of the lighter atom and 10% of the heavier atom.

$$\left(20 \text{ amu} \times \frac{90}{100}\right) + \left(22 \text{ amu} \times \frac{10}{100}\right) = 20.2 \text{ amu}$$

Aston called these different neon atoms **isotopes** (literally, ''same place''), using a term introduced by Frederick Soddy to describe different atoms that occupy the same place in the periodic table.

**TABLE 3.3 Common Isotopes of the Lighter Elements**

| Isotope | Mass | Natural Abundance (%) |
|---|---|---|
| $^{1}H$ | 1.007825 | 99.985 |
| $^{2}H$ | 2.0140 | 0.015 |
| $^{6}Li$ | 6.01512 | 7.42 |
| $^{7}Li$ | 7.01600 | 92.58 |
| $^{10}B$ | 10.0129 | 19.7 |
| $^{11}B$ | 11.00931 | 80.3 |
| $^{12}C$ | 12.00000 | 98.89 |
| $^{13}C$ | 13.00335 | 1.11 |
| $^{16}O$ | 15.99491 | 99.76 |
| $^{17}O$ | 16.99913 | 0.04 |
| $^{18}O$ | 17.99916 | 0.20 |
| $^{20}Ne$ | 19.99244 | 90.51 |
| $^{21}Ne$ | 20.99395 | 0.27 |
| $^{22}Ne$ | 21.99138 | 9.22 |

**EXERCISE 3.1**

Calculate the atomic weight of chlorine if 75.77% of the atoms have a mass of 34.97 amu and 24.23% have a mass of 36.97 amu.

**SOLUTION** *Percent* literally means ''per hundred.'' Chlorine is therefore a mixture of atoms for which 75.77 parts per hundred have a mass of 34.97 amu and 24.23 parts per hundred have a mass of 36.97 amu. The atomic weight of chlorine is therefore 35.45 amu (to four significant figures).

$$\left[34.97 \text{ amu} \times \frac{75.77}{100}\right] + \left[36.97 \text{ amu} \times \frac{24.23}{100}\right] = 35.45 \text{ amu}$$

The isotopes of an atom are identified by writing the mass number of the isotope in the upper left corner of the symbol for the element. Thus, we can talk about the $^{20}Ne$ and $^{22}Ne$ isotopes of neon or the $^{35}Cl$ and $^{37}Cl$ isotopes of chlorine. Common isotopes of the lighter elements are given in Table 3.3.

The discovery of isotopes led to a small, but embarrassing, discrepancy between the tables of atomic weights used by chemists and those used by physicists. Physicists used atomic weights based on the assumption that the most common isotope of oxygen ($^{16}0$) had a mass of exactly 16 amu. The tables used by chemists were based on a weighted average of $^{16}O$, $^{17}O$, and $^{18}O$ atoms. Differences between the two scales were small, corresponding to discrepancies in the fifth or sixth significant figures, but it was obvious that a new standard for atomic masses was needed.

This discrepancy was resolved when the SI units were created. By definition, atomic weights are now based on the assumption that the mass of the $^{12}C$ isotope of carbon is exactly 12 amu. Note, however, that naturally occurring carbon is a mixture of three isotopes, $^{12}C$ (98.89%), $^{13}C$ (1.11%), and infinitesimal amounts of $^{14}C$. As a result, the average mass of a carbon atom is 12.011 amu. No individual carbon atom ever has a mass of 12.011 amu, but the weighted average of the different naturally occurring isotopes of carbon gives an **atomic weight** for this element of 12.011 amu.

## 3.4 THE MOLE AS A COLLECTION OF ATOMS

Atoms are so small that it takes an enormous number of atoms to give a sample large enough to be seen with the naked eye. It is therefore useful to create a unit that represents a collection of atoms or molecules that can serve as bridge between chemistry on the atomic and macroscopic scale. This unit is called a **mole** (from the Latin meaning ''a huge mass''). By definition,

**A mole is the amount of any substance that contains as many elementary particles as there are atoms in exactly 12 grams of the $^{12}C$ isotope of carbon.**

Note that a single $^{12}C$ atom has a mass of exactly 12 amu and a mole of these atoms has a mass of exactly 12 grams.

$$1\ ^{12}\text{C atom} = 12 \text{ amu}$$
$$1 \text{ mol of } ^{12}\text{C atoms} = 12 \text{ g}$$

Samples that contain one mole of the following elements: mercury, zinc, silicon, aluminum, sulfur and bromine.

**A mole of atoms of any element therefore has a mass in grams equal to the atomic weight of the element.**

**EXERCISE 3.2**

Predict the mass of a mole of magnesium atoms.

**SOLUTION** The atomic weight of magnesium is 24.305 amu, which means that a mole of magnesium atoms would have a mass of 24.305 grams.

## 3.5 THE MOLE AS A COLLECTION OF MOLECULES

Samples that contain one mole of the following compounds—HgS, $PbI_2$, $CuSO_4 \cdot 5\ H_2O$ and $(NH_4)_2Cr_2O_7$.

The sugar we add to food consists of $C_{12}H_{22}O_{11}$ molecules.

The unit *mole* (abbreviated mol) can be applied to any particle. We can talk about a mole of Mg atoms, a mole of $Na^+$ ions, a mole of electrons, or a mole of $CO_2$ molecules. Each time we use the term, we refer to a number of particles equal to the number of atoms in exactly 12 grams of the $^{12}C$ isotope of carbon.

Before we can apply the concept of the mole to covalent compounds such as carbon dioxide ($CO_2$) or sugar ($C_{12}H_{22}O_{11}$), we need to be able to calculate the **molecular weight** of a compound, which is the sum of the atomic weights of the atoms in the molecules that form these compounds.

**EXERCISE 3.3**

Calculate the molecular weights of carbon dioxide and sugar and the mass of a mole of each compound.

**SOLUTION** The molecular weight of carbon dioxide is the sum of the atomic weights of the three atoms in a $CO_2$ molecule:

$$\begin{array}{lrl} CO_2: & 1\text{ C atom} = 1(12.011)\text{ amu} & = 12.011\text{ amu} \\ & 2\text{ O atoms} = 2(15.9994)\text{ amu} & = \underline{31.9988\text{ amu}} \\ & & \mathbf{44.010\text{ amu}} \end{array}$$

$CO_2$ therefore has a molecular weight of 44.010 amu, and a mole of carbon dioxide would have a mass of 44.010 grams.

The molecular weight of sugar is the sum of the atomic weights of the 45 atoms in a $C_{12}H_{22}O_{11}$ molecule. Thus, sugar has a molecular weight of 343.30 amu, and a mole of $C_{12}H_{22}O_{11}$ molecules would have a mass of 342.30 grams.

$$\begin{array}{lrl} C_{12}H_{22}O_{11}: & 12\text{ C atoms} = 12(12.011)\text{ amu} & = 144.13\text{ amu} \\ & 22\text{ H atoms} = 22(1.00794)\text{ amu} & = 22.1747\text{ amu} \\ & 11\text{ O atoms} = 11(15.9994)\text{ amu} & = \underline{175.993\text{ amu}} \\ & & \mathbf{342.30\text{ amu}} \end{array}$$

**CHECKPOINT**

Explain why none of the $C_{12}H_{22}O_{11}$ molecules in a box of sugar has a mass of 342.30 amu, even though the molecular weight of this compound is 342.30 amu.

**EXERCISE 3.4**

Describe the difference between the mass of a mole of oxygen atoms and the mass of a mole of $O_2$ molecules.

**SOLUTION** Because the atomic weight of oxygen is 16.0 amu, a mole of oxygen atoms has a mass of 16.0 grams. Each $O_2$ molecule has two atoms, however, so the molecular weight of these molecules is twice as large as the atomic weight of the element. The molecular weight of an $O_2$ molecule is 32.0 amu, and a mole of these molecules would have a mass of 32.0 grams.

$$1 \text{ mol O} = 16.0 \text{ g} \qquad 1 \text{ mol } O_2 = 32.0 \text{ g}$$

The mass of a mole of any substance is often called the **molar mass.** Chemists who don't feel comfortable using the term ''weight'' to describe measurements of mass often use the term *molar mass* in place of the terms *atomic weight* or *molecular weight.*

## 3.6 AVOGADRO'S CONSTANT: THE NUMBER OF PARTICLES IN A MOLE

The only way to determine the number of particles in a mole is to measure the same quantity on both the atomic and macroscopic scales. In 1910 Robert Millikan measured the charge on an electron for the first time. Because the charge on a mole of electrons was already known, it was possible to estimate the number of particles in a mole. Using more recent (and more accurate) data, we get the following results:

$$\frac{96{,}484.56 \not{C}}{1 \text{ mol}} \times \frac{1 \text{ electron}}{1.6021892 \times 10^{-19} \not{C}} = 6.022045 \times 10^{23} \frac{\text{electrons}}{\text{mol}}$$

The number of particles in a mole is called **Avogadro's number** or, more accurately, **Avogadro's constant.** For most calculations, three ($6.02 \times 10^{23}$) or at most four ($6.022 \times 10^{23}$) significant figures for Avogadro's number are enough.

**EXERCISE 3.5**

Calculate the mass in grams of a single $^{12}C$ atom.

**SOLUTION** We can start with the fact that a mole of $^{12}C$ has a mass of exactly 12 grams. We then construct a unit factor based on the fact that a mole of any substance contains $6.022 \times 10^{23}$ atoms:

$$\frac{12.00 \text{ g } ^{12}C}{\cancel{1 \text{ mol C}}} \times \frac{\cancel{1 \text{ mol C}}}{6.022 \times 10^{23} \text{ atoms}} = 1.993 \times 10^{-23} \text{ g/atom}$$

Thus, to four significant figures, the mass of a single $^{12}C$ atom is $1.993 \times 10^{-23}$ g.

We now know the mass of a single $^{12}C$ atom in units of both grams and atomic mass units. We can therefore calculate the number of grams per amu:

$$\frac{1.993 \times 10^{-23} \text{ g}}{12.00 \text{ amu}} = 1.661 \times 10^{-24} \text{ g/amu}$$

By inverting this ratio, we can also calculate the number of amu per gram:

$$\frac{12.00 \text{ amu}}{1.993 \times 10^{-23} \text{ g}} = 6.022 \times 10^{23} \text{ amu/g}$$

Avogadro's number is therefore equal to the number of atomic mass units in a gram, as might be expected from the definition of a mole.

Avogadro's number provides us with another way to define the concept of a mole:

**A mole of any substance contains Avogadro's number of elementary particles.**

It doesn't matter whether we talk about a mole of atoms, a mole of molecules, a mole of electrons, or a mole of ions. By definition, a mole always contains $6.022 \times 10^{23}$ elementary particles.

There are $6.022 \times 10^{23}$ oxygen atoms in a mole of oxygen.

There are $6.022 \times 10^{23}$ $H_2O$ molecules in a mole of water.

There are $6.022 \times 10^{23}$ $Na^+$ ions in a mole of sodium ions.

There are $6.022 \times 10^{23}$ $Cl^-$ ions in a mole of chlorine ions.

There are $6.022 \times 10^{23}$ electrons in a mole of electrons.

Avogadro's number is so large it is difficult to comprehend. It would take 6 trillion galaxies the size of the Milky Way to yield $6 \times 10^{23}$ stars. At the speed of light, it would take 102 billion years to travel $6 \times 10^{23}$ miles. When you consider the size of Avogadro's number, it isn't surprising that it took so long to prove that atoms exist.

## 3.7 CONVERTING GRAMS INTO MOLES

The mole is a powerful tool, which enables chemists armed with nothing more than a table of atomic weights and a balance to determine the number of atoms, ions, or molecules in a sample. To show how this is done, let's calculate the number of $C_{12}H_{22}O_{11}$ molecules in a pound of sugar.

We start by noting that a pound of sugar has a mass of 453.6 g. We then use the molecular weight of sugar calculated in Exercise 3.3 to construct a pair of unit factors:

$$\frac{1 \text{ mol } C_{12}H_{22}O_{11}}{342.3 \text{ g } C_{12}H_{22}O_{11}} = 1 \qquad \frac{342.3 \text{ g } C_{12}H_{22}O_{11}}{1 \text{ mol } C_{12}H_{22}O_{11}} = 1$$

By paying attention to the dimensions or units during the calculation, it isn't difficult to choose the correct unit factor to convert grams of sugar into moles of sugar:

$$453.6 \cancel{\text{ g } C_{12}H_{22}O_{11}} \times \frac{1 \text{ mol } C_{12}H_{22}O_{11}}{342.3 \cancel{\text{ g } C_{12}H_{22}O_{11}}} = \mathbf{1.325 \text{ mol } C_{12}H_{22}O_{11}}$$

We then use Avogadro's number to determine the number of $C_{12}H_{22}O_{11}$ molecules in our sample:

$$1.325 \cancel{\text{ mol } C_{12}H_{22}O_{11}} \times \frac{6.022 \times 10^{23} \text{ molecules}}{1 \cancel{\text{ mol } C_{12}H_{22}O_{11}}} = \mathbf{6.218 \times 10^{23} \text{ molecules}}$$

**EXERCISE 3.6**

What is the formula of magnesium chloride if 2.55 grams of magnesium combine with 7.45 grams of chlorine to form 10.0 grams of this compound?

**SOLUTION** The first step toward answering of an unfamiliar problem often involves drawing a diagram that helps us organize the information in the problem.

We could start, for example, with the simple diagram in Figure 3.4, which summarizes the relationship between the mass of magnesium metal and chlorine gas consumed in this reaction.

The next step in any problem of this kind is to try to convert grams into moles. To do this, we need to know the relationship between the number of grams and the number of moles of the substance. It doesn't matter which element we start with because we eventually have to work with both, so let's arbitrarily start with magnesium.

The atomic weight of magnesium is 24.31 amu, which means that a mole of magnesium has a mass of 24.31 grams. We can use this information to construct two unit factors:

$$\frac{1 \text{ mol Mg}}{24.31 \text{ g Mg}} = 1 \qquad \frac{24.31 \text{ g Mg}}{1 \text{ mol Mg}} = 1$$

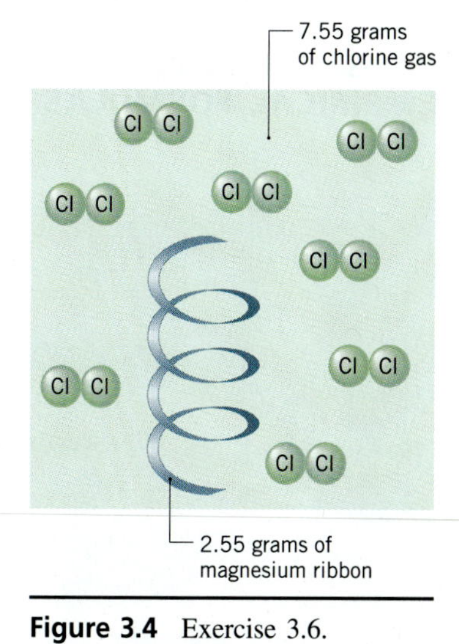


**Figure 3.4** Exercise 3.6.

Converting grams of magnesium into moles requires a unit factor that has units of moles in the numerator and grams in the denominator. Dimensional analysis therefore suggests that the problem should be set up as follows:

$$2.55 \cancel{\text{g Mg}} \times \frac{1 \text{ mol Mg}}{24.31 \cancel{\text{g Mg}}} = \mathbf{0.1049 \text{ mol Mg}}$$

The same format can be used to convert grams of chlorine into moles of chlorine atoms:

$$7.45 \cancel{\text{g Cl}} \times \frac{1 \text{ mol Cl}}{35.45 \cancel{\text{g Cl}}} = \mathbf{0.2102 \text{ mol Cl}}$$

Everything we have done so far is strictly mechanical, using a well-defined set of rules to convert from one set of units (grams) into another (moles). Once this is done, we write down the information we have obtained so far.

The product of this reaction contains:
- 0.1049 mol Mg atoms
- 0.2102 mol Cl atoms

We then reread the problem and ask: ''Have we made any progress toward the answer?''

**Problem-Solving Strategy**

In this case, we are trying to find the formula for magnesium chloride, which gives us the ratio of magnesium atoms to chlorine atoms. (If the formula is MgCl, there are just as many magnesium atoms as chlorine atoms in the compound. If the formula is $MgCl_2$, there are twice as many chlorine atoms in the compound.) The next step in the problem might therefore involve determining the relationship between the number of moles of magnesium and the number of moles of chlorine in our sample.

$$\frac{0.2102 \text{ mol Cl}}{0.1049 \text{ mol Mg}} = \mathbf{2.003}$$

Within experimental error, there are twice as many moles of chlorine atoms as moles of magnesium atoms in this sample. Because a mole of atoms always contains the same number of atoms, the only possible conclusion is that there are twice as many chlorine atoms as magnesium atoms in the compound. In other words, the formula for magnesium chloride must be $MgCl_2$.

## 3.8 DETERMINING EMPIRICAL FORMULAS

Methane was once known as marsh gas, because it was discovered bubbling out of the marshes in England.

Exercise 3.6 showed one way to determine the formula of a compound. By carefully measuring the amount of magnesium and chlorine that combine to form magnesium chloride, it was possible to show that the formula for this compound is $MgCl_2$. Let's look at another way to approach this problem. This time we will examine a compound once know as "marsh gas" because it was first collected above certain swamps, or marshes, in Britain.

Marsh gas, or methane as it is now known, is 74.9% carbon and 25.1% hydrogen by mass. A 100-gram sample of this gas would therefore contain 74.9 grams of carbon and 25.1 grams of hydrogen:

$$100 \text{ g methane} \times \left[\frac{74.9 \text{ g C}}{100 \text{ g sample}}\right] = 74.9 \text{ g C}$$

$$100 \text{ g methane} \times \left[\frac{25.1 \text{ g H}}{100 \text{ g sample}}\right] = 25.1 \text{ g H}$$

We can now use the atomic weights of these elements to convert grams of carbon and hydrogen in this sample into moles of these elements:

$$74.9 \text{ g C} \times \frac{1 \text{ mol C}}{12.01 \text{ g C}} = \mathbf{6.236 \text{ mol C}}$$

$$25.1 \text{ g H} \times \frac{1 \text{ mol H}}{1.008 \text{ g H}} = \mathbf{24.90 \text{ mol H}}$$

Let's stop at this point to review the information obtained so far: 100 grams of methane contains 6.24 moles of carbon and 24.9 moles of hydrogen. Because we know the number of moles of carbon and hydrogen in the same sample, it might be useful to look at the ratio of the moles of these elements in the sample.

$$\frac{24.9 \text{ mol H}}{6.24 \text{ mol C}} = \mathbf{3.99}$$

Within experimental error, our 100-gram sample of methane contains four times as many moles of hydrogen atoms as moles of carbon atoms. This means that there are four times as many hydrogen atoms as carbon atoms in this sample. We can explain these results by assuming that the **empirical formula** for this gas is $CH_4$. (The term *empirical* comes from the Greek stem that means "to experience." The empirical formula is therefore literally the formula that comes from the data we experience.) By definition,

**The empirical formula of a compound is the simplest whole-number ratio of the atoms of each element in the compound.**

This experiment does not tell us the formula of a methane molecule. These results are consistent with molecules that contain one carbon atom and four hydrogen atoms: $CH_4$. But they are also consistent with formulas of $C_2H_8$, $C_3H_{12}$, $C_4H_{16}$, and so on. All we know at this point is that the formula for the molecule is some multiple of an empirical formula that can be written as $CH_4$.

We can use the results of the calculations for $CH_4$ to generate a sequence of steps that can be followed more or less automatically to determine the empirical formula for a compound from percent-by-mass data for the various elements.

1. Start with a 100-g sample of the compound.
2. Use the percent-by-mass data to calculate the number of grams of each element in the 100-g sample.
3. Convert these data into the number of moles of atoms of each element.
4. Divide through by the element with the smallest number of moles of atoms.
5. Multiply the ratio of the number of atoms by small whole numbers until all of the coefficients in the formula are integers.

**EXERCISE 3.7**

Calculate the empirical formula of aspirin, which is 60.0% C, 4.48% H, and 35.5% O by mass.

**SOLUTION** We start by calculating the number of grams of each element in a 100-gram sample of aspirin:

$$\begin{aligned} 100\text{ g} \times 60.0\%\text{ C} &= 60.0\text{ g C} \\ 100\text{ g} \times 4.48\%\text{ H} &= 4.48\text{ g H} \\ 100\text{ g} \times 35.5\%\text{ O} &= 35.5\text{ g O} \end{aligned}$$

We then convert the number of grams of each element into the number of moles of atoms of that element:

$$60.0\ \cancel{\text{g C}} \times \frac{1\text{ mol C}}{12.01\ \cancel{\text{g C}}} = 5.00\text{ mol C}$$

$$4.48\ \cancel{\text{g H}} \times \frac{1\text{ mol H}}{1.008\ \cancel{\text{g H}}} = 4.44\text{ mol H}$$

$$35.5\ \cancel{\text{g O}} \times \frac{1\text{ mol O}}{16.00\ \cancel{\text{g O}}} = 2.22\text{ mol O}$$

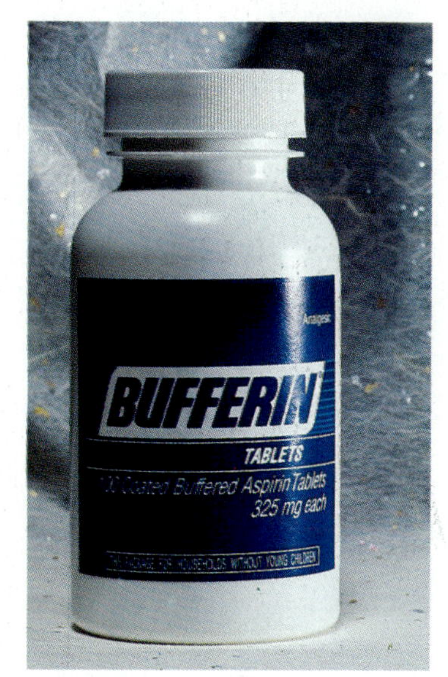


Because we are interested in the simplest whole-number ratio of these elements, we then divide through by the element with the smallest number of moles of atoms:

$$\frac{5.00\text{ mol C}}{2.22\text{ mol O}} = 2.25 \qquad \frac{4.44\text{ mol H}}{2.22\text{ mol O}} = 2.00$$

It doesn't make sense to write the ratio of atoms as $C_{2.25}H_2O$, because there is no such thing as one-fourth of a carbon atom. We therefore multiply this ratio by small whole numbers until we get a formula in which all of the coefficients are integers:

$$\begin{aligned} 2(C_{2.25}H_2O) &= C_{4.5}H_4O_2 \\ 3(C_{2.25}H_2O) &= C_{6.75}H_6O_3 \\ 4(C_{2.25}H_2O) &= \mathbf{C_9H_8O_4} \end{aligned}$$

Thus, the empirical formula for aspirin is $C_9H_8O_4$.

## 3.9 EMPIRICAL VERSUS MOLECULAR FORMULAS

The preceding section showed how percent-by-mass data can be used to determine the empirical formula of a compound. Let's look at the results of an analysis of vitamin C.

## EXERCISE 3.8

Calculate the empirical formula for vitamin C if this compound is 40.9% C, 54.5% O, and 4.58% H by mass.

**SOLUTION** A 100-gram sample of vitamin C would contain 40.9 grams of carbon, 54.5 grams of oxygen, and 4.58 grams of hydrogen. This corresponds to 3.41 moles of carbon, 3.41 moles of oxygen, and 4.54 moles of hydrogen:

$$40.9 \not{g}\,\not{C} \times \frac{1 \text{ mol C}}{12.01 \not{g}\,\not{C}} = 3.41 \text{ mol C}$$

$$54.5 \not{g}\,\not{O} \times \frac{1 \text{ mol O}}{16.00 \not{g}\,\not{O}} = 3.41 \text{ mol O}$$

$$4.58 \not{g}\,\not{H} \times \frac{1 \text{ mol H}}{1.008 \not{g}\,\not{H}} = 4.54 \text{ mol H}$$

Dividing through by 3.41 moles of C atoms gives the following results:

$$\frac{3.41 \text{ mol O}}{3.41 \text{ mol C}} = 1 \qquad \frac{4.54 \text{ mol H}}{3.41 \text{ mol C}} = 1.33$$

The ratio of C to H to O atoms in vitamin C is therefore 1:1.33:1. Multiplying this ratio by three gives an empirical formula of $C_3H_4O_3$ for vitamin C.

The **empirical weight** of a compound is the sum of the atomic weights of the atoms in its empirical formula. If the empirical formula for vitamin C is $C_3H_4O_3$, the empirical weight is 88.06 grams per mole.

$$\begin{array}{llcl} C_3H_4O_3: & 3 \text{ C atoms} = 3(12.01) \text{ amu} & = & 36.03 \text{ amu} \\ & 4 \text{ H atoms} = 4(1.008) \text{ amu} & = & 4.032 \text{ amu} \\ & 3 \text{ O atoms} = 3(16.00) \text{ amu} & = & \underline{48.00 \text{ amu}} \\ & & & 88.06 \text{ amu} \end{array}$$

This is as far as we can go with percent-by-mass data. It is possible to determine the **molecular weight** of a compound in a separate experiment, however. When this is done, the molecular weight of vitamin C is found to be about 176 grams per mole.

How do we explain the difference between the molecular weight and the empirical weight of vitamin C? We can start by comparing the two measurements.

$$\frac{\text{Molecular weight}}{\text{Empirical weight}} = \frac{176 \not{g}/\not{mol}}{88.1 \not{g}/\not{mol}} = 2$$

Within experimental error, the molecular weight of this compound is twice as large as the empirical weight. The only possible conclusion is that a molecule of vitamin C is twice as large as the empirical formula. In other words, the **molecular formula** of vitamin C is $C_6H_8O_6$.

---

### ▶ CHECKPOINT

Describe the difference between the following terms: *empirical formula, molecular formula, empirical weight,* and *molecular weight.*

---

## 3.10 CHEMICAL REACTIONS AND THE LAW OF CONSERVATION OF MATTER

We have focused so far on individual compounds such as methane ($CH_4$), aspirin ($C_9H_8O_4$), and vitamin C ($C_6H_8O_6$). Much of the fascination of chemistry, however, revolves around chemical reactions. The first breakthrough in the study of chemical reactions resulted from the work of French chemist Antoine Lavoisier between 1772 and 1794. Lavoisier found that mass is conserved in a chemical reaction. The total mass of the products of a chemical reaction is always the same as the total mass of the starting materials consumed in the reaction. His results led to one of the fundamental laws of chemical behavior: the **law of conservation of matter,** which states that matter is conserved in a chemical reaction.

We now understand why matter is conserved—atoms are neither created nor destroyed in a chemical reaction. Hydrogen atoms in an $H_2$ molecule can combine with oxygen atoms in an $O_2$ molecule to form $H_2O$, for example, as shown in Figure 3.5. But the number of hydrogen and oxygen atoms before and after the reaction is the same. The total mass of the products of a reaction therefore must be the same as the total mass of the reactants.

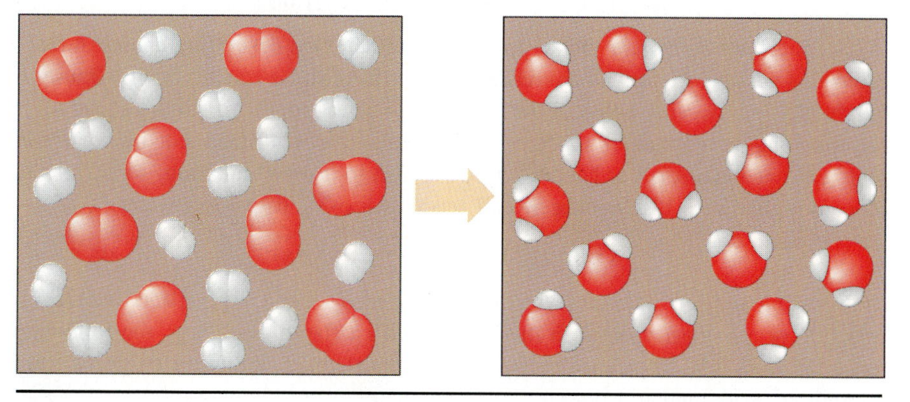

**Figure 3.5** Chemical reactions occur on the atomic scale. This figure represents what happens in a small corner of a container in which $H_2$ and $O_2$ molecules react to form $H_2O$ molecules.

## 3.11 CHEMICAL EQUATIONS AS A REPRESENTATION OF CHEMICAL REACTIONS

It is possible to describe a chemical reaction in words, but it is much easier to describe it with a **chemical equation.** Chemical equations are always written from left to right, even in languages that do not read from the left to right. The formulas of the starting materials, or **reactants,** are written on the left side of the equation and the formulas of the **products** are written on the right. Instead of an equal sign, the reactants and products are separated by an arrow that shows the direction of the reaction. The reaction between hydrogen and oxygen to form water is represented by the following equation.

$$2\,H_2 + O_2 \longrightarrow 2\,H_2O$$

The reaction between aluminum and iron oxide could be written as follows.

$$2\,Al + Fe_2O_3 \longrightarrow Al_2O_3 + 2\,Fe$$

It is often useful to indicate whether the reactants or products are solids, liquids, or gases by writing an *s*, *l*, or *g* in parentheses after the symbol for the reactants or products, as shown in the following equations.

The thermite reaction used to weld iron rails while building railroad tracks involves the reaction between powdered aluminum and iron oxide to form aluminum oxide and molten iron metal.

$$2\,H_2(g) + O_2(g) \longrightarrow 2\,H_2O(g)$$
$$2\,Al(s) + Fe_2O_3(s) \longrightarrow Al_2O_3(s) + 2\,Fe(l)$$

Reactions between gases are often so fast they are dangerous. Reactions between solids are usually so slow that they seem to take forever. Reactions between liquids are often fast enough to give products in a reasonable amount of time, without being so fast they produce explosions. Reactions between compounds that are not liquids at room temperature are therefore often carried out by first dissolving the compound in a liquid.

Most of the reactions you will encounter in this course will occur when solutions of two substances dissolved in water are mixed. These **aqueous** solutions (from the Latin *aqua,* ''water'') are so important that we use the special symbol *aq* to describe them. This way we can distinguish between sugar as a solid, $C_{12}H_{22}O_{11}(s)$, and solutions of sugar dissolved in water, $C_{12}H_{22}O_{11}(aq)$. Or between salt as a solid, $NaCl(s)$, and solutions of salt dissolved in water, $NaCl(aq)$. The process in which a sample dissolves in water will be indicated by equations such as the following:

$$C_{12}H_{22}O_{11}(s) \xrightarrow{H_2O} C_{12}H_{22}O_{11}(aq)$$

Chemical equations are such a powerful shorthand for describing chemical reactions that we tend to think about reactions in terms of these equations. This causes a slight problem because some of the grammar of a reaction is lost in the equation. It is important to remember that a chemical equation is a statement of what *can happen,* not necessarily what *will happen.* The following equation, for example, does not guarantee that hydrogen will react with oxygen to form water:

$$2\,H_2(g) + O_2(g) \longrightarrow 2\,H_2O(g)$$

It is possible to fill a balloon with a mixture of hydrogen and oxygen and find that no reaction occurs until you touch the balloon with a flame. Similarly, the following equation does not say that aluminum metal must react with iron oxide to form aluminum oxide and iron metal:

$$2\,Al(s) + Fe_2O_3(s) \longrightarrow Al_2O_3(s) + 2\,Fe(l)$$

All the equation tells us is what would happen if, or when, the reaction occurs.

## 3.12 TWO VIEWS OF CHEMICAL EQUATIONS: MOLECULES VERSUS MOLES

Chemical equations such as the following can be used to represent what happens on either the atomic or macroscopic scale:

$$2\,H_2(g) + O_2(g) \longrightarrow 2\,H_2O(g)$$

This equation can be read in either of the following ways.

- If, or when, hydrogen reacts with oxygen, two molecules of hydrogen and one molecule of oxygen are consumed for every two molecules of water produced.
- If, or when, hydrogen reacts with oxygen, 2 moles of hydrogen and 1 mole of oxygen are consumed for every 2 moles of water produced.

Regardless of whether we think of the reaction in terms of molecules or moles, chemical equations must be balanced—they must have the same number of atoms of each element on both sides of the equation. As a result, the mass of the reactants must be equal to the mass of the products of the reaction. On the atomic scale, the following equation is balanced because the total mass of the reactants is equal to the mass of the products:

$$2\,H_2(g) + O_2(g) \longrightarrow 2\,H_2O(g)$$

$$\underbrace{2 \times 2 \text{ amu} + 32 \text{ amu}}_{36 \text{ amu}} \qquad \underbrace{2 \times 18 \text{ amu}}_{36 \text{ amu}}$$

On the macroscopic scale, it is balanced because the mass of 2 moles of hydrogen and 1 mole of oxygen is equal to the mass of 2 moles of water:

$$2\,H_2(g) + O_2(g) \longrightarrow 2\,H_2O(g)$$

$$\underbrace{2 \times 2 \text{ g} + 32 \text{ g}}_{36 \text{ g}} \qquad \underbrace{2 \times 18 \text{ g}}_{36 \text{ g}}$$

## 3.13 BALANCING CHEMICAL EQUATIONS

There is no sequence of rules that can be followed blindly to get a balanced chemical equation. All we can do is manipulate the coefficients written in front of the formulas of the reactants and products until the number of atoms of each element on both sides of the equation is the same.

The first thing to look for when balancing equations are relationships between the two sides of an equation. In the following equation, for example, we can see that at least three lithium atoms will be needed to make lithium nitride, $Li_3N$. It therefore seems reasonable to start with three lithium atoms on the left-hand side of the equation and one $Li_3N$ on the right:

$$\mathbf{3\,Li} + \underline{\qquad}\, N_2 \longrightarrow 1\,\mathbf{Li_3}N$$

Half the problem is now solved; the lithium atoms are balanced.

We can now try to balance the nitrogen atoms. There are two nitrogen atoms on the left and only one on the right, so we have to assume that the reaction produces two $Li_3N$ for each $N_2$ consumed. But doubling the amount of $Li_3N$ produced in this reaction means that we must double the amount of Li metal consumed:

$$\mathbf{6\,Li} + 1\,N_2 \longrightarrow \mathbf{2\,Li_3}N$$

Both sides of the equation now contain six lithium atoms and two nitrogen atoms. The balanced equation for this reaction is therefore written as follows:

$$6\,Li(s) + N_2(g) \longrightarrow 2\,Li_3N(s)$$

The flame of a propane torch.

The key to balancing chemical equations is persistence. Keep exploring the equation until you have the same number of atoms of each element on both sides. While doing this, keep in mind that it is usually a good idea to tackle the easiest part of a problem first. Consider, for example, the equation for the reaction that fuels the propane burners that campers use. Propane ($C_3H_8$) burns in air to form $CO_2$ and $H_2O$:

$$\underline{\qquad}\, C_3H_8 + \underline{\qquad}\, O_2 \longrightarrow \underline{\qquad}\, CO_2 + \underline{\qquad}\, H_2O$$

If you look at this equation carefully, you may conclude that it is going to be easier to balance the carbon and hydrogen atoms than the oxygen atoms in this reaction. All of the carbon atoms in propane end up in $CO_2$ and all of the hydrogen atoms end up in $H_2O$, but some of the oxygen atoms end up in each compound. This means that there is no way to predict the number of $O_2$ molecules that are consumed in this reaction until you know how many $CO_2$ and $H_2O$ molecules are produced.

We can start the process of balancing this equation by noting that there are three carbon atoms in each $C_3H_8$ molecule. Thus, three $CO_2$ molecules are formed for every $C_3H_8$ molecule consumed:

$$1\ C_3H_8 + \_\_\_\_\ O_2 \longrightarrow 3\ CO_2 + \_\_\_\_\ H_2O$$

If there are eight hydrogen atoms in each $C_3H_8$ molecule, there must be eight hydrogen atoms, or four $H_2O$ molecules, on the right-hand side of the equation:

$$1\ C_3H_8 + \_\_\_\_\ O_2 \longrightarrow 3\ CO_2 + 4\ H_2O$$

Now that the carbon and hydrogen atoms are balanced, we can try to balance the oxygen atoms. There are six oxygen atoms in three $CO_2$ molecules and four oxygen atoms in four $H_2O$ molecules. To balance the 10 oxygen atoms in the products of this reaction, we need five $O_2$ molecules among the reactants:

$$1\ C_3H_8 + 5\ O_2 \longrightarrow 3\ CO_2 + 4\ H_2O$$

There are now three carbon atoms, eight hydrogen atoms, and 10 oxygen atoms on each side of the equation. The balanced equation for this reaction is therefore written:

$$C_3H_8(g) + 5\ O_2(g) \longrightarrow 3\ CO_2(g) + 4\ H_2O(g)$$

## EXERCISE 3.9

Write a balanced equation for the reaction that occurs when ammonia burns in air to form nitrogen oxide and water:

$$\_\_\_\_\ NH_3 + \_\_\_\_\ O_2 \longrightarrow \_\_\_\_\ NO + \_\_\_\_\ H_2O$$

**SOLUTION** We might start by balancing the nitrogen atoms. If we start with one molecule of ammonia and form one molecule of NO, the nitrogen atoms are balanced.

$$1\ NH_3 + \_\_\_\_\ O_2 \longrightarrow 1\ NO + \_\_\_\_\ H_2O$$

We can then turn to the hydrogen atoms. We have three hydrogen atoms on the left and two hydrogen atoms on the right in this equation. One way of balancing the hydrogen atoms is to look for the lowest common denominator: $2 \times 3 = 6$. We therefore set up this equation so that there are six hydrogen atoms on both sides. Doing this doubles the amount of $NH_3$ consumed in the reaction, so we have to double the amount of NO produced.

$$2\ NH_3 + \_\_\_\_\ O_2 \longrightarrow 2\ NO + 3\ H_2O$$

Because the nitrogen and hydrogen atoms are both balanced, the only task left is to balance the oxygen atoms. There are five oxygen atoms on the right side of this equation, so we need five oxygen atoms on the left. But that's impossible because these atoms come from $O_2$ molecules that each contain two oxygen atoms. If we insist that chemical equations work on both the atomic and the macroscopic scales, we cannot leave the equation as follows:

$$2\ NH_3 + 2\tfrac{1}{2}\ O_2 \longrightarrow 2\ NO + 3\ H_2O$$

Because there is no such thing as one-half of an $O_2$ molecule, the only alternative is to multiply the equation by 2:

$$4\ NH_3 + 5\ O_2 \longrightarrow 4\ NO + 6\ H_2O$$

The balanced equation for this reaction is therefore written:

$$4\ NH_3(g) + 5\ O_2(g) \longrightarrow 4\ NO(g) + 6\ H_2O(g)$$

## 3.14 MOLE RATIOS AND CHEMICAL EQUATIONS

There are two fundamental goals of science: (1) explaining observations about the world around us, and (2) predicting what will happen under a particular set of conditions. Any chemical equation explains something about the world, but a balanced chemical equation has the added advantage of allowing us to predict what happens when a reaction takes place.

Let's see, for example, how a balanced chemical equation can predict the amount of $O_2$ consumed when one of the fireworks that brightens the sky each Fourth of July is ignited. These fireworks are based on the reaction between magnesium and oxygen to form magnesium oxide:

$$2\ Mg(s) + O_2(g) \longrightarrow 2\ MgO(s)$$

If one of these flares contains 0.40 moles of magnesium metal, how much $O_2$ is consumed when the magnesium burns?

The white light emitted during fireworks displays is produced by burning magnesium metal.

The balanced equation for this reaction can be used to construct two unit factors that describe the relationship between the amounts of magnesium and oxygen consumed in this reaction:

$$\frac{1\ \text{mol}\ O_2}{2\ \text{mol Mg}} = 1 \qquad \frac{2\ \text{mol Mg}}{1\ \text{mol}\ O_2} = 1$$

By focusing on the units of this problem, we can select the correct **mole ratio** to convert moles of magnesium into an equivalent number of moles of oxygen:

$$0.40\ \cancel{\text{mol Mg}} \times \frac{1\ \text{mol}\ O_2}{2\ \cancel{\text{mol Mg}}} = \mathbf{0.20\ mol\ O_2}$$

A can of butane lighter fluid.

**EXERCISE 3.10**

A can of butane lighter fluid contains 1.20 moles of butane ($C_4H_{10}$). Calculate the number of moles of carbon dioxide given off when this butane is burned.

**SOLUTION** This calculation involves two steps: writing a balanced equation for the reaction, and then using the balanced equation to predict the number of moles of carbon dioxide produced when 1.20 moles of $C_4H_{10}$ burn. Balancing the equation is fairly easy. Each butane molecule contains four carbon atoms and 10 hydrogen atoms. For each butane molecule consumed, four $CO_2$ molecules and five $H_2O$ molecules are produced.

$$1\ C_4H_{10} + \underline{\qquad}\ O_4 \longrightarrow 4\ CO_2 + 5\ H_2O$$

There are 13 oxygen atoms among the products of this reaction, so we need 13 oxygen atoms among the reactants. Because 13 oxygen atoms corresponds to $6\frac{1}{2}\ O_2$ molecules, we multiply the equation by 2 and try again. Now there are 8 carbon atoms, 20 hydrogen atoms, and 26 oxygen atoms on each side of the equation:

$$2\ C_4H_{10}(g) + 13\ O_2(g) \longrightarrow 8\ CO_2(g) + 10\ H_2O(g)$$

According to this equation, we get 8 moles of $CO_2$ for every 2 moles of butane that burn. We can therefore convert 1.20 moles of butane into an equivalent number of moles of carbon dioxide as follows:

$$1.20\ \cancel{\text{mol } C_4H_{10}} \times \frac{8\ \text{mol } CO_2}{2\ \cancel{\text{mol } C_4H_{10}}} = \mathbf{4.80\ mol\ CO_2}$$

Thus, 4.80 moles of carbon dioxide will be given off when the contents of the typical can of butane lighter fluid burn.

## 3.15 PREDICTING THE MASS OF REACTANTS CONSUMED OR PRODUCTS GIVEN OFF IN A CHEMICAL REACTION

By now, you have encountered all the steps necessary to do calculations of the sort that are grouped under the heading **stoichiometry.** The word *stoichiometry* comes from the Greek stems meaning *quantities* and *to measure*. The goal of these calculations is to use a balanced chemical equation to predict the relationships between the amounts of the reactants and products of a chemical reaction.

To illustrate how these calculations are done, let's predict the amount of oxygen that must be inhaled to digest 10.0 grams of sugar. We can assume that the sugar in our diet comes to us as $C_{12}H_{22}O_{11}$ molecules and that our bodies burn this sugar according to the following equation:

$$C_{12}H_{22}O_{11}(s) + 12\ O_2(g) \longrightarrow 12\ CO_2(g) + 11\ H_2O(l)$$

There are many ways of doing this calculation. The following list introduces an approach based on a few relatively easy ground rules.

- Identify the goal of the problem.
- Write down the key elements of the problem, or draw a simple picture that summarizes the key information in the problem.
- Try to do what can be done.

- Don't try to do the impossible.
- Never lose sight of what you have accomplished.
- Never lose sight of your goal.
- Don't try to work the problem in your head. Write down all the intermediate steps.
- If you get lost, go back and read the question again.
- Don't give up; explore the problem until you get an answer.

**Problem-Solving Strategy**

Let's apply these rules to the calculation described above. We can start by asking "What are we trying to find?" We then summarize what we think are the important pieces of information in the problem.

*Goal*: Find out how many grams of $O_2$ are consumed when 10.0 grams of sugar are burned.
*Fact*: We start with 10.0 grams of sugar.
*Fact*: Sugar has the formula $C_{12}H_{22}O_{11}$.
*Fact*: The balanced equation for this reaction can be written as follows:

$$C_{12}H_{22}O_{11} + 12\ O_2 \longrightarrow 12\ CO_2 + 11\ H_2O$$

We then turn to the task of "doing what can be done." Because we know the molecular weight of sugar, we can convert the known mass of the sugar into an equivalent number of moles:

$$10.0\ \cancel{\text{g}\ C_{12}H_{22}O_{11}} \times \frac{1\ \text{mol}\ C_{12}H_{22}O_{11}}{342.3\ \cancel{\text{g}\ C_{12}H_{22}O_{11}}} = \mathbf{0.02921\ mol\ C_{12}H_{22}O_{11}}$$

Before we continue, let's take a look at the fourth of our general rules: Don't try to do the impossible. There is no way to get from moles of sugar directly to grams of oxygen in one step, so instead, we look for something that we can do. In other words, we continue exploring the problem.

What else can we do? We have a balanced chemical equation, and we know the number of *moles of sugar* in the sample. As a step toward the goal of the problem we can calculate the number of *moles of oxygen* consumed in the reaction. The equation for this reaction suggests that 12 moles of $O_2$ are consumed for every mole of sugar in this reaction. We can therefore calculate the number of moles of oxygen needed to burn 0.02921 mole of sugar as follows:

$$0.02921\ \cancel{\text{mol}\ C_{12}H_{22}O_{11}} \times \frac{12\ \text{mol}\ O_2}{1\ \cancel{\text{mol}\ C_{12}H_{22}O_{11}}} = \mathbf{0.3505\ mol\ O_2}$$

We now have the necessary information to get to the goal of our calculation. We know the amount of $O_2$ consumed in this reaction in units of moles, and we can calculate the mass of 0.3505 mole of $O_2$ from the mass of a mole of oxygen:

$$0.3505\ \cancel{\text{mol}\ O_2} \times \frac{32.00\ \text{g}\ O_2}{1\ \cancel{\text{mol}\ O_2}} = \mathbf{11.22\ g\ O_2}$$

According to this calculation, it takes 11.2 grams of $O_2$ to burn 10.0 grams of sugar.

**EXERCISE 3.11**

Calculate the amount of ammonia and oxygen needed to prepare 3.00 grams of nitrogen oxide by the following reaction:

$$4\ NH_3(g) + 5\ O_2(g) \longrightarrow 4\ NO(g) + 6\ H_2O(g)$$

**SOLUTION** By now, the first step in this calculation should be almost automatic. We start by converting 3.00 grams of NO into an equivalent number of moles of this compound. To do this, we need to calculate the mass of a mole of NO, which is 30.0 grams per mole. The number of moles of NO formed in this reaction therefore can be calculated as follows:

$$3.00\ \cancel{g\ NO} \times \frac{1\ mol\ NO}{30.0\ \cancel{g\ NO}} = 0.100\ mol\ NO$$

We then use the balanced equation for this reaction to determine the mole ratios that allow us to calculate the number of moles of $NH_3$ and $O_2$ needed to produce 0.100 mole of NO:

$$0.100\ \cancel{mol\ NO} \times \frac{4\ mol\ NH_3}{4\ \cancel{mol\ NO}} = 0.100\ mol\ NH_3$$

$$0.100\ \cancel{mol\ NO} \times \frac{5\ mol\ O_2}{4\ \cancel{mol\ NO}} = 0.125\ mol\ O_2$$

We then use the mass of a mole of $NH_3$ and $O_2$ to calculate the mass of each of the reactants consumed in this reaction:

$$0.100\ \cancel{mol\ NH_3} \times \frac{17.0\ g\ NH_3}{1\ \cancel{mol\ NH_3}} = \mathbf{1.70\ g\ NH_3}$$

$$0.125\ \cancel{mol\ O_2} \times \frac{32.0\ g\ O_2}{1\ \cancel{mol\ O_2}} = \mathbf{4.00\ g\ O_2}$$

We can check this calculation by predicting the amount of $H_2O$ formed in the reaction:

$$0.100\ \cancel{mol\ NH_3} \times \frac{6\ mol\ H_2O}{4\ \cancel{mol\ NH_3}} = 0.150\ mol\ H_2O$$

$$0.150\ \cancel{mol\ H_2O} \times \frac{18.0\ g\ H_2O}{1\ \cancel{mol\ H_2O}} = \mathbf{2.70\ g\ H_2O}$$

Our calculation should be correct because the sum of the masses of the reactants (1.70 g $NH_3$ + 4.00 g $O_2$) is equal to the sum of the masses of the products (3.00 g NO + 2.70 g $H_2O$).

## 3.16 THE NUTS AND BOLTS OF LIMITING REAGENTS

According to Exercise 3.11, it takes 1.70 grams of ammonia and 4.00 grams of oxygen to make 3.00 grams of nitrogen oxide by the following reaction:

$$4\ NH_3(g) + 5\ O_2(g) \longrightarrow 4\ NO(g) + 6\ H_2O(g)$$

What would happen to the amount of NO produced in this reaction if we kept the amount of $O_2$ the same (4.00 g) but changed the amount of $NH_3$ that was present initially? Figure 3.6 shows how the amount of NO produced in this reaction would depend on the amount of $NH_3$ with which we started.

At first, the amount of NO produced would be directly proportional to the amount of $NH_3$ present when the reaction began. At some point, however, the yield of the reaction would become constant. No matter how much $NH_3$ we add to the system, no more NO is produced. The problem is simple. We eventually reach a point at which the reaction runs out of $O_2$ before all of the $NH_3$ is consumed. When this happens, the reaction must stop. No matter how much $NH_3$ is added to the system, we can't get more than 3.00 grams of NO from 4.00 grams of oxygen.

When there isn't enough $O_2$ to consume all the $NH_3$ in the reaction, the amount of $O_2$ limits the amount of NO that can be produced. Oxygen is therefore the **limiting reagent** in this reaction. Because there is more $NH_3$ than we need, it is the **excess reagent.**

The concept of limiting reagent is important because chemists frequently run reactions in which only a limited amount of one of the reactants is present. An analogy may help clarify what goes on in limiting reagent problems.

Let's start with exactly 10 nuts and 10 bolts, as shown in Figure 3.7. How many *NB* "molecules" can be made by screwing one nut (*N*) onto each bolt (*B*)? The answer is obvious—we can make exactly 10. After that, we run out of both nuts and bolts. Because we run out of both nuts and bolts at the same time, neither is a limiting reagent.

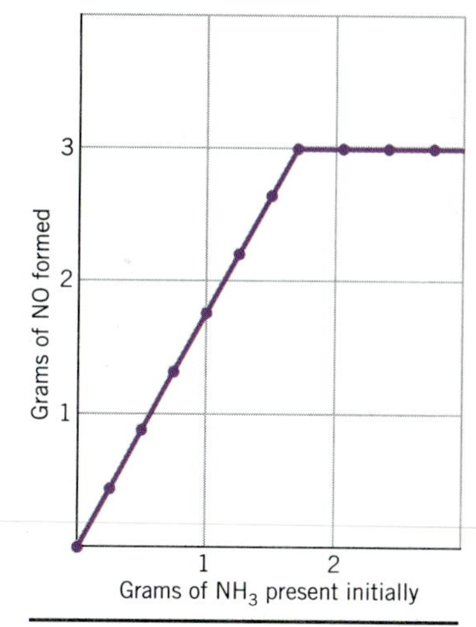


**Figure 3.6** A graph of the amount of NO that can be produced by adding different amounts of $NH_3$ to 4.00 grams of $O_2$.

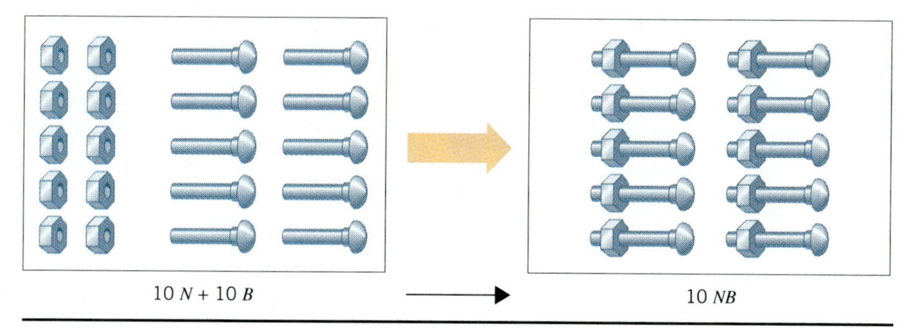


**Figure 3.7** Starting with 10 nuts (*N*) and 10 bolts (*B*) we can make 10 *NB* molecules, with no nuts or bolts left over.

Now let's assemble $N_2B$ molecules by screwing two nuts onto each bolt. Starting with 10 nuts and 10 bolts, we can only make five $N_2B$ molecules, as shown in Figure 3.8. Because we run out of nuts, they must be the limiting reagent. Because five bolts are left over, they are the excess reagent.

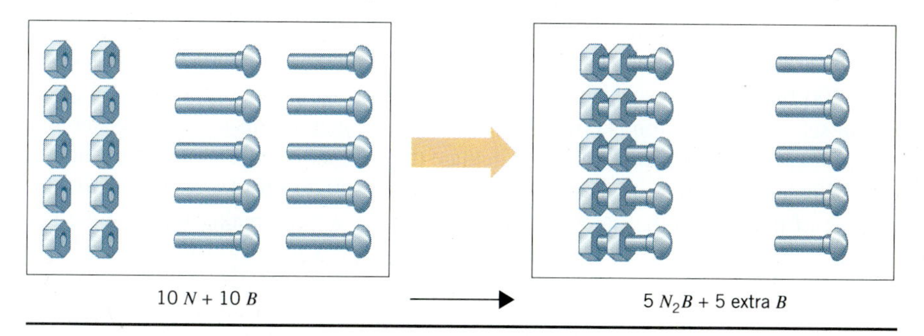


**Figure 3.8** Starting with 10 nuts and 10 bolts, we can make only five $N_2B$ molecules and we will have five bolts left over. Because the number of $N_2B$ molecules is limited by the number of nuts, the nuts are the limiting reagent in this analogy.

It is also possible to put two bolts into a single nut. Starting with 10 nuts and 10 bolts, we can only make five $NB_2$ molecles, as shown in Figure 3.9. This time, the bolts are the limiting reagent and the nuts are present in excess.

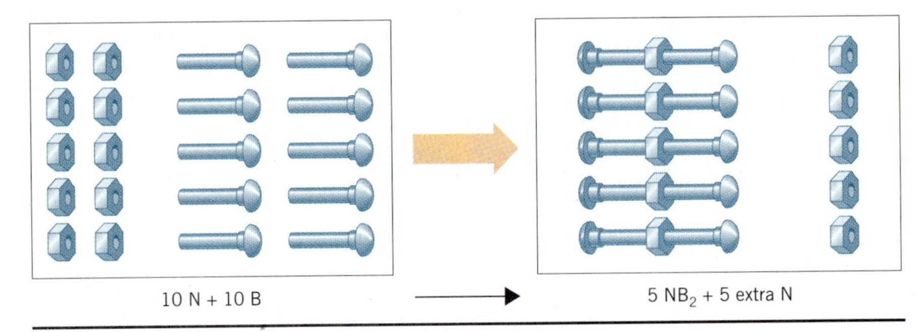


**Figure 3.9** Starting with 10 nuts and 10 bolts, we can make only five $NB_2$ molecules, and we will have five nuts left over. In this case, the bolts are the limiting reagent.

This analogy can be extended to a slightly more difficult problem in which we assemble as many $N_2B$ molecules as possible from a collection of 300 nuts and 200 bolts. There are three possibilities: (1) we have too many nuts and not enough bolts, (2) we have too many bolts and not enough nuts, or (3) we have just the right number of both nuts and bolts.

**Problem-Solving Strategy**

One way to approach this problem is to pick one of these alternatives and test it. Because there are more nuts (300) than bolts (200), let's assume that we have too many nuts and not enough bolts. In other words, let's assume that bolts are the limiting reagent in this problem. Now let's test this assumption. According to the formula, $N_2B$, we need two nuts for every bolt. Thus, we need 400 nuts to consume 200 bolts.

$$200\ \cancel{\text{bolts}} \times \frac{2\ \text{nuts}}{1\ \cancel{\text{bolt}}} = 400\ \text{nuts}$$

According to this calculation, we need more nuts (400) than we have (300). Thus, our original assumption is wrong. We don't run out of bolts, we run out of nuts.

Because our original assumption is wrong, let's turn it around and try again. Now let's assume that nuts are the limiting reagent and calculate the number of bolts we need.

$$300\ \cancel{\text{nuts}} \times \frac{1\ \text{bolt}}{2\ \cancel{\text{nuts}}} = 150\ \text{bolts}$$

Do we have enough bolts to use up all the nuts? Yes, we only need 150 bolts and we have 200 bolts to choose from.

Our second assumption is correct. The limiting reagent in this case is nuts and the excess reagent is bolts. We can now calculate the number of $N_2B$ molecules that can be assembled from 300 nuts and 200 bolts. Because the limiting reagent is nuts, the number of nuts limits the number of $N_2B$ molecules we can make. Because we get one $N_2B$ molecule for every two nuts, we can make a total of 150 of these $N_2B$ molecules:

$$300\ \cancel{\text{nuts}} \times \frac{1\ N_2B\ \text{molecule}}{2\ \cancel{\text{nuts}}} = 150\ N_2B\ \text{molecules}$$

The key to limiting reagent problems is the following sequence of steps:

- Recognize that you have a limiting reagent problem, or at least consider the possibility that there may be a limiting amount of one of the reactants.
- Assume that one of the reactants is the limiting reagent.
- See if you have enough of the other reactant to consume the material you have assumed to be the limiting reactant.
- If you do, your original assumption was correct.
- If you don't, assume that another reagent is the limiting reagent and test this assumption.
- Once you have identified the limiting reagent, calculate the amount of product formed.

**EXERCISE 3.12**

How much magnesium oxide (MgO) is formed when 10.0 grams of magnesium reacts with 10.0 grams of $O_2$?

**SOLUTION** It might be useful to start with a simple diagram, such as Figure 3.10, that summarizes the relevant information in the problem. The next step toward solving this problem is to write a balanced equation for the reaction:

**Problem-Solving Strategy**

$$2\ \text{Mg}(s) + \text{O}_2(g) \rightarrow 2\ \text{MgO}(s)$$

We then pick one of the reactants and assume it is the limiting reagent. For the sake of argument, we'll assume that magnesium is the limiting reagent and $O_2$ is present in excess. Our immediate goal is to test the validity of this assumption. If it is correct, we have more $O_2$ than we need to burn 10.0 grams of magnesium. If it is wrong, $O_2$ is the limiting reagent.

We start by converting grams of magnesium into moles of magnesium:

$$10.0\ \cancel{\text{g Mg}} \times \frac{1\ \text{mol Mg}}{24.31\ \cancel{\text{g Mg}}} = 0.4114\ \text{mol Mg}$$

We then use the balanced equation to predict the number of moles of $O_2$ needed to burn this much magnesium. According to the equation for this reaction, it takes 1 mole of $O_2$ to burn 2 moles of magnesium. We therefore need 0.2057 mole of $O_2$ to consume all of the magnesium:

$$0.4114\ \cancel{\text{mol Mg}} \times \frac{1\ \text{mol O}_2}{2\ \cancel{\text{mol Mg}}} = 0.2057\ \text{mol O}_2$$

We now calculate the mass of the $O_2$:

$$0.2057\ \cancel{\text{mol O}_2} \times \frac{32.00\ \text{g O}_2}{1\ \cancel{\text{mol O}_2}} = \mathbf{6.582\ g\ O_2}$$

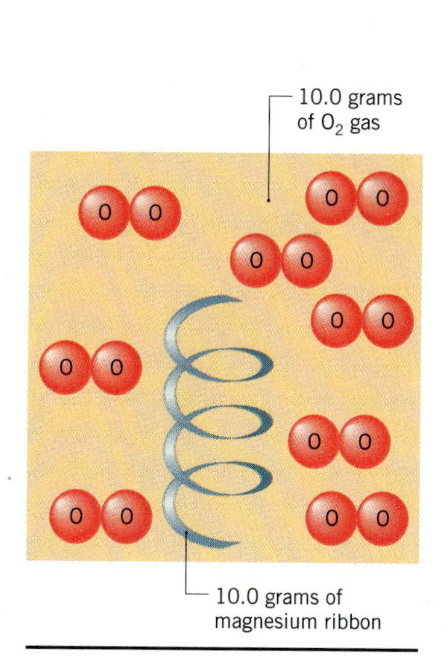


**Figure 3.10** Exercise 3.12.

According to this calculation, we need 6.58 grams of $O_2$ to burn all the magnesium. Because we have 10.0 grams of $O_2$, our original assumption was correct. We have more than enough $O_2$ and only a limited amount of magnesium.

We can now calculate the amount of magnesium oxide formed when all of the limiting reagent is consumed. The balanced equation suggests that 2 moles of

MgO are produced for every 2 moles of magnesium consumed. Thus, 0.4114 mole of MgO can be formed in this reaction:

$$0.4114 \text{ mol Mg} \times \frac{2 \text{ mol MgO}}{2 \text{ mol Mg}} = 0.4114 \text{ mol MgO}$$

We can now use the molar mass of MgO to calculate the number of grams of MgO that can be formed:

$$0.4114 \text{ mol MgO} \times \frac{40.31 \text{ g MgO}}{1 \text{ mol MgO}} = \mathbf{16.6 \text{ g MgO}}$$

We can check the result of this calculation by noting that 10.0 grams of magnesium combine with 6.6 grams of $O_2$ to form 16.6 grams of MgO. (3.4 g of $O_2$ remain unused.) Mass is therefore conserved and we can feel confident that our calculation is correct.

## 3.17 SOLUTIONS AND CONCENTRATION

The calculations done so far all have been based on the mass of a sample, which is a convenient measurement for solids. When working with liquids, it is often easier to measure the volume. We therefore need a way to determine the number of atoms, ions, or molecules in a sample from measurements of its volume. For a pure substance, we can do this by measuring the volume and density of the sample.

### EXERCISE 3.13

Use the density of mercury (13.60 g/cm$^3$) to calculate the number of atoms in a liter of this liquid.

**SOLUTION** The volume of mercury is given in liters and the density is given in grams per cubic centimeter. We might therefore start by calculating the volume of the mercury in cubic centimeters:

$$1.000 \text{ L} \times \frac{1000 \text{ mL}}{1 \text{ L}} \times \frac{1 \text{ cm}^3}{1 \text{ mL}} = 1000 \text{ cm}^3$$

We then combine the volume of the sample with the density of mercury to calculate the mass of the sample:

$$1000 \text{ cm}^3 \times \frac{13.60 \text{ g Hg}}{1 \text{ cm}^3} = 1.360 \times 10^4 \text{ g Hg}$$

We then convert grams of mercury into moles of mercury:

$$1.360 \times 10^4 \text{ g Hg} \times \frac{1 \text{ mol Hg}}{200.6 \text{ g Hg}} = 67.80 \text{ mol Hg}$$

We can then use the number of atoms per mole to calculate the number of mercury atoms in a liter of liquid mercury:

$$67.80 \text{ mol Hg} \times \frac{6.022 \times 10^{23} \text{ atoms}}{1 \text{ mol Hg}} = \mathbf{4.083 \times 10^{25} \text{ Hg atoms}}$$

**RESEARCH IN THE 90s**

## THE STOICHIOMETRY OF THE BREATHALYZER

A patent was issued to R. F. Borkenstein in 1958 for the Breathalyzer, which still remains the method of choice for determining whether an individual is DUI—driving under the influence—or DWI—driving while intoxicated. The chemistry behind the Breathalyzer is described by the following equation:

$$3\,CH_3CH_2OH(g) + 2\,Cr_2O_7^{2-}(aq) + 16\,H^+(aq) \longrightarrow 3\,CH_3CO_2H(aq) + 4\,Cr^{3+}(aq) + 11\,H_2O(l)$$

The instrument contains two ampules that hold 0.75 mg of potassium dichromate ($K_2Cr_2O_7$) dissolved in 3 mL of 9 *M* $H_2SO_4$. One of these ampules is used as a reference. The other is opened, and the breath sample to be analyzed is added to this ampule. If alcohol is present in the breath, it reduces the yellow-orange $Cr_2O_7^{2-}$ ion to form a green $Cr^{3+}$ ion. The extent to which the color balance between the two ampules is disturbed is a direct measure of the amount of alcohol in the breath sample.

Measurements of the alcohol on the breath are then converted into estimates of the concentration of the alcohol in the blood. The link between these quantities is the assumption that 2100 mL of the air exhaled from the lungs contains the same amount of alcohol as 1 mL of blood.

Measurements taken with the Breathalyzer are reported in units of percent blood-alcohol concentration (BAC) from 0 to 0.40%. In most states, a BAC of 0.10% is sufficient for a DWI conviction. (This corresponds to a BAC of 0.10 grams of alcohol per 100 mL of blood.)

The breathalyzer analyzes for alcohol exhaled by the lungs, which is directly proportional to the concentration of the alcohol in the blood.

Between January 1989 and December 1990, almost 50 papers were published that described research related to the measurement of blood-alcohol concentration. Several studies have probed the implications of the fact that the ratio of alcohol in the breath to the blood-alcohol concentration varies from one individual to another. This research has shown that breathalyzers are far more likely to *underestimate* BAC than to overestimate it [G. Simpson, *Journal of Analytical Toxicology,* **13**(2), 120 (1989)]. Research has also shown that there is no significant difference in the rate at which an individual metabolizes alcohol with age [P. M. Hein and R. Volk, *Blutalkohol,* **26**(4), 276 (1990).] This research has also led to the development of more accurate methods for determining blood-alcohol concentration, particularly during autopsies that are done on those who drink and drive [J. V. Maracini, T. Carroll, S. Grant, S. Halleran, and J. A. Benz, *Journal of Forensic Science,* **41,** 181 (1989).]

Work with general chemistry students recently revealed an interesting misconception. Some students believe they can ''cheat'' on a Breathalyzer test by placing a copper penny in their mouths. (Modern folklore apparently suggests that this will decrease the amount of alcohol on the breath.)

Copper metal will, in fact, catalyze the following reaction, in which ethyl alcohol is oxidized to acetaldehyde:

$$CH_3CH_2OH \xrightarrow{Cu} CH_3CHO + H_2$$

There is only one minor problem—*the copper penny has to be heated until it glows red hot before it will do this.*

## SOLUTE, SOLVENT, AND SOLUTION

Before we apply the same techniques to mixtures, it is useful to define three closely related terms: *solute, solvent,* and *solution.* Suppose we dissolve a solid, such as copper(II) sulfate pentahydrate ($CuSO_4 \cdot 5\ H_2O$) in water as shown in Figure 3.11. The substance that dissolves ($CuSO_4 \cdot 5\ H_2O$) is called the **solute.** The substance in which the solute dissolves (water) is called the **solvent.** The mixture of solute and solvent is called the **solution.**

The following rules can be used to decide which component of a solution is the solute and which is the solvent.

- There are three states of matter: solids, liquids, and gases. Any reagent that undergoes a change in state when it forms a solution is the *solute.*
- If neither reagent changes state, the reagent present in the smallest quantity is the *solute.*

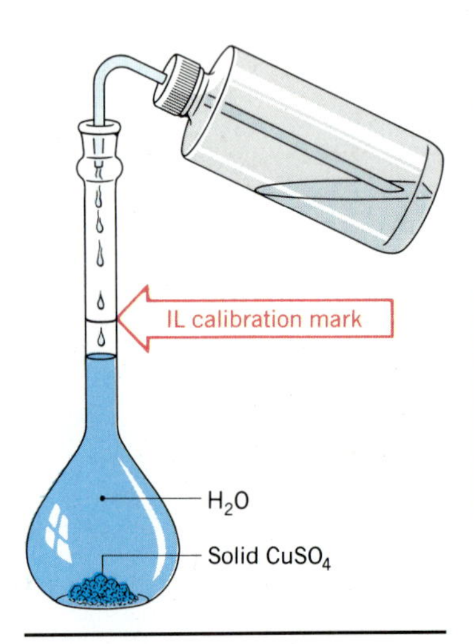


**Figure 3.11** A *solution* of $CuSO_4$ in water is made by dissolving the *solute* ($CuSO_4$) in a *solvent* ($H_2O$).

Table 3.4 gives examples of different kinds of solutions. $CuSO_4 \cdot 5\ H_2O$ is a dark blue solid; because it changes state when it dissolves in water, it is the solute in Figure 3.11. When $H_2$ gas dissolves in platinum metal to form a solid solution, $H_2$ is the solute. When solid sodium metal reacts with liquid mercury to form a solid solution, mercury is the solute. Wine that is 12% alcohol by volume is a solution of a small quantity of alcohol (the solute) in a larger volume of water (the solvent).

When these rules fail, such as a 50:50 mixture of alcohol and water, either component of the mixture can be thought of as the solute.

It is possible to determine the number of atoms, ions, or molecules in a mixture using measurements of density and volume if the percent by mass of each of the components in the mixture is known.

**TABLE 3.4 Examples of Solutions**

| Solute | Solvent | Solution |
|---|---|---|
| $CuSO_4(s)$ | $H_2O(l)$ | $CuSO_4(aq)$ |
| $H_2(g)$ | $Pt(s)$ | $H_2/Pt(s)$ |
| $Hg(l)$ | $Na(s)$ | $Na/Hg(s)$ |
| Alcohol | Water | Wine |

## EXERCISE 3.14

Concentrated sulfuric acid is 96.0% $H_2SO_4$ by mass. (The remaining 4.0% is water.) Calculate the number of moles of $H_2SO_4$ in a liter of concentrated sulfuric acid if the density of this solution is 1.84 g/cm$^3$.

**SOLUTION** Problems like this are best solved by trying to relate what we know to the goal of the problem.

*Fact*: The solution has a density of 1.84 g/cm$^3$.
*Fact*: We have a liter of this solution.
*Fact*: 96.0% of the mass of this solution is sulfuric acid.
*Fact*: The rest of the mass is water.
*Goal*: Calculate the number of moles of $H_2SO_4$ in a liter of this solution.

This information is summarized in the drawing in Figure 3.12.

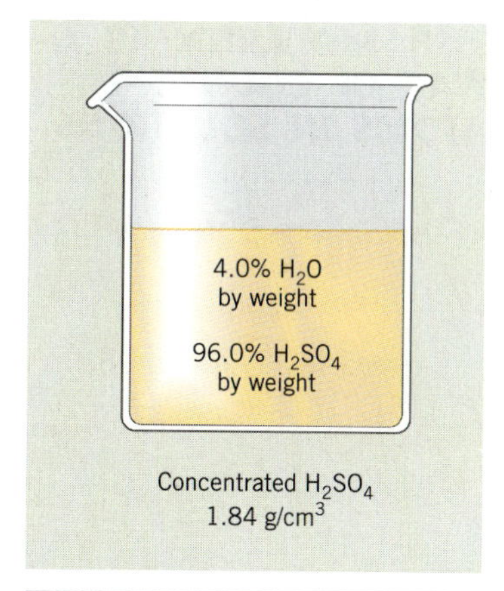


**Figure 3.12** Exercise 3.14.

**Problem-Solving Strategy**

Because we know the density in units of grams per cubic centimeter, we might start by calculating the volume of the solution in cubic centimeters:

$$1.00 \cancel{L} \times \frac{1000 \cancel{mL}}{1 \cancel{L}} \times \frac{1 \text{ cm}^3}{1 \cancel{mL}} = 1000 \text{ cm}^3$$

We can then use the density to calculate the mass of this solution:

$$1000 \cancel{cm^3} \times \frac{1.84 \text{ g}}{1 \cancel{cm^3}} = 1840 \text{ g}$$

At this point it may be useful to ask: Are we making any progress toward the answer to this exercise? Our goal is to determine the number of moles of $H_2SO_4$ in this solution and we know the total mass of the solution. We might therefore use the fact that this solution is 96.0% $H_2SO_4$ by mass to calculate the number of grams of $H_2SO_4$ in a liter of the solution:

$$1840 \cancel{\text{g soln}} \times \frac{96.0 \text{ g } H_2SO_4}{100 \cancel{\text{g soln}}} = 1770 \text{ g } H_2SO_4$$

We can then use the molecular weight of $H_2SO_4$ to calculate the number of moles of this compound in a liter of the solution:

$$1770 \cancel{\text{g } H_2SO_4} \times \frac{1 \text{ mol } H_2SO_4}{98.1 \cancel{\text{g } H_2SO_4}} = \mathbf{18.0 \text{ mol } H_2SO_4}$$

According to this calculation, concentrated sulfuric acid contains 18.0 moles of $H_2SO_4$ per liter.

### CHECKPOINT

Explain why chemists may find it more useful to describe concentrated $H_2SO_4$ as a solution that contains 18.0 moles of solute per liter instead of as a solution that is 96.0% $H_2SO_4$ by mass.

Sec. 1.13

The amount of solute or solvent in a solution is an extensive property. So is the amount of solution formed when the solute and solvent are mixed. The ratio of the amount of solute to the amount either of the solvent or the solution, however, is an intensive property. This ratio, which is known as the **concentration** of the solution, does not depend on the size of the sample.

$$\text{Concentration} = \frac{\text{amount of solute}}{\text{amount of solvent or solution}}$$

The concept of concentration is a common one. We talk about *concentrated* orange juice, which must be *diluted* with water. We even describe certain laundry products as *concentrated,* which means that we don't have to use as much.

## 3.18 MOLARITY AS A WAY OF COUNTING ATOMS IN SOLUTIONS

All concentration units have one thing in common: they describe the ratio of the amount of solute to the amount either of solvent or solution. Chemists use one concentration unit more than any other: **molarity** (*M*). The molarity of a solution is defined as the number of moles of solute per liter of solution. Molarity is calculated by dividing the number of moles of solute in the solution by the volume of the solution in liters:

$$\textbf{Molarity } (M) = \frac{\textbf{moles of solute}}{\textbf{liters of solution}}$$

**EXERCISE 3.15**

Copper sulfate is available as blue crystals that contain water molecules coordinated to the $Cu^{2+}$ ions in this crystal. Because these crystals contain five water molecules per $Cu^{2+}$ ion, the compound is called a pentahydrate, and the formula is written $CuSO_4 \cdot 5\,H_2O$. Calculate the molarity of a solution prepared by dissolving 1.25 grams of this compound in enough water to give 50.0 mL of solution.

**Problem-Solving Strategy**

**SOLUTION** A useful strategy for solving problems involves looking at the goal of the problem and asking: What information do we need to reach this goal? The molarity of a solution is calculated by dividing the number of moles of solute by the volume of the solution. We therefore need two pieces of information to reach the goal of this exercise—the number of moles of solute and the volume of the solution in liters.

The volume of the solution is easy to calculate:

$$50.0\ \cancel{\text{mL}} \times \frac{1\ \text{L}}{1000\ \cancel{\text{mL}}} = 0.0500\ \text{L}$$

The number of moles of solute can be calculated from the mass of solute used to prepare the solution and the mass of a mole of this compound:

$$1.25\ \cancel{\text{g CuSO}_4 \cdot 5\,\text{H}_2\text{O}} \times \frac{1\ \text{mol}}{249.6\ \cancel{\text{g CuSO}_4 \cdot 5\,\text{H}_2\text{O}}} = 0.00500\ \text{mol CuSO}_4 \cdot 5\,\text{H}_2\text{O}$$

The solution therefore has a concentration of 0.100 *M*:

$$\frac{0.00500\ \text{mol CuSO}_4 \cdot 5\,\text{H}_2\text{O}}{0.0500\ \text{L}} = 0.100\ M\ \text{CuSO}_4$$

Crystals of $CuSO_4 \cdot 5\,H_2O$.

## EXERCISE 3.16

Calculate the volume of 1.50 *M* HCl that would react with 25.0 grams of $CaCO_3$ according to the following balanced equation:

$$CaCO_3(s) + 2\,HCl(aq) \longrightarrow Ca^{2+}(aq) + 2\,Cl^-(aq) + CO_2(g) + H_2O(l)$$

**SOLUTION** We have only one piece of information about the HCl solution (its concentration is 1.50 *M*) and only one piece of information about $CaCO_3$ (it has a mass of 25.0 grams), as shown in Figure 3.13. Which of these numbers do we start with? The answer is based on two points that were made in Section 3.15: Do what can be done, and don't try to do the impossible.

**Problem-Solving Strategy**

There is nothing we can do with the concentration of the HCl solution unless we know either the number of moles of HCl consumed in the reaction or the volume of the solution. We can work with the mass of $CaCO_3$ consumed in the reaction, however. We can start the calculation by converting grams of $CaCO_3$ into moles of $CaCO_3$:

$$25.0\ \cancel{g\ CaCO_3} \times \frac{1\ \text{mol}\ CaCO_3}{100.1\ \cancel{g\ CaCO_3}} = 0.250\ \text{mol}\ CaCO_3$$

**Figure 3.13** Exercise 3.16.

We then ask: Does this information get us any closer to our goal of calculating the volume of HCl consumed in this reaction? We know the number of moles of $CaCO_3$ consumed in the reaction, and we have a balanced equation that states that two moles of HCl are consumed for every mole of $CaCO_3$. We can therefore calculate the number of moles of HCl consumed in the reaction:

$$0.250\ \cancel{\text{mol}\ CaCO_3} \times \frac{2\ \text{mol HCl}}{1\ \cancel{\text{mol}\ CaCO_3}} = 0.500\ \text{mol HCl}$$

We now have the number of moles of HCl (0.500 mol) and the molarity of the solution (1.50 *M*). How do we put these two facts together? We can start by remembering that molarity has units of moles per liter. The product of the molarity of a solution times its volume in liters is therefore equal to the number of moles of solute dissolved in the solution.

$$\frac{\text{mol}}{\cancel{L}} \times \cancel{L} = \text{mol}$$

We can write this as a generic equation as follows:

$$M \times V = n$$

In this equation, *M* stands for the concentration of the solution in units of moles per liter, *V* is the volume of the solution in liters, and *n* is the number of moles of solute. Substituting the known values of the concentration of the HCl solution and the number of moles of HCl into this equation gives the following result:

$$\frac{1.50\ \text{mol}}{1\ \text{L}} \times V = 0.500\ \text{mol HCl}$$

We then solve this equation for the volume of the solution that would contain this amount of HCl:

$$V = 0.333\ \text{L}$$

According to this calculation, we need 333 mL of 1.50 *M* HCl to consume 25.0 grams of $CaCO_3$.

With a little imagination, the concept of concentration can be used to do far more interesting calculations.

**EXERCISE 3.17**

Assume that a metal (M) reacts with hydrochloric acid according to the following balanced equation.

$$M(s) + 2\ HCl(aq) \longrightarrow M^{2+}(aq) + 2\ Cl^{-}(aq) + H_2(g)$$

Calculate the atomic weight of the metal if 125 milliliters of 0.200 *M* HCl reacts with 0.304 gram of the metal.

**SOLUTION** At first glance, it seems that we don't have enough information to solve this problem (see Figure 3.14). The only way to proceed with a question like this is to follow the rough guidelines for problem solving outlined in Section 3.15. Start by identifying what you know, do what can be done, and see where this leads you. In other words, start by exploring the problem. What do we know?

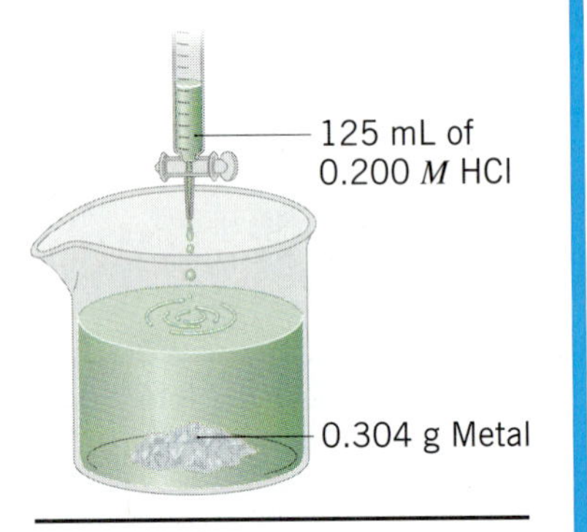


**Figure 3.14** Exercise 3.17.

*Fact*: The metal reacts with hydrochloric acid according to the following balanced equation: $M + 2\ HCl \longrightarrow MCl_2 + H_2$

*Fact*: We start with 0.304 g of the metal.

*Fact*: It takes 125 mL of 0.200 *M* HCl to consume the metal.

*Goal*: Find the atomic weight of a metal.

What can we do? We know the volume (125 mL) and the concentration (0.200 *M*) of a solution. We might therefore start by calculating the number of moles of solute in this solution:

$$\frac{0.200\ \text{mol HCl}}{1\ \cancel{L}} \times 0.125\ \cancel{L} = 0.0250\ \text{mol HCl}$$

Now what? We know the number of moles of HCl and we have a balanced equation. Furthermore, we are interested in one of the properties of the metal. It seems reasonable to convert moles of HCl consumed in this reaction into moles of metal consumed:

$$0.0250\ \cancel{\text{mol HCl}} \times \frac{1\ \text{mol M}}{2\ \cancel{\text{mol HCl}}} = 0.0125\ \text{mol M}$$

**Problem-Solving Strategy**

It is important to never lose sight of the goal of the problem. In this case, the problem asks for the atomic weight of the metal. It may be useful to go to the end of the problem and work backward. Atomic weight has units of grams per mole. If we knew both the number of grams and the number of moles of metal in a sample, we could calculate the atomic weight of the metal.

But we already have that information. We know the number of moles of metal (0.0125 mol) in a sample of known mass (0.304 g). The ratio of these numbers is the atomic weight of the metal:

$$\frac{0.304\ \text{g M}}{0.0125\ \text{mol M}} = \mathbf{24.3\ g/mol}$$

By looking at a table of atomic weights, we can deduce that the metal is magnesium.

## 3.19 DILUTION AND TITRATION CALCULATIONS

Adding more solvent to a solution to decrease the concentration is known as **dilution.** Starting with a known volume of a solution of known molarity, we should be able to prepare a more dilute solution of any desired concentration.

### EXERCISE 3.18

Describe how you would prepare 2.50 liters of an 0.360 *M* solution of sulfuric acid starting with concentrated sulfuric acid that is 18.0 *M*.

**SOLUTION** We have one piece of information about the concentrated $H_2SO_4$ solution (the concentration is 18.0 moles per liter) and two pieces of information about the dilute solution (the volume is 2.50 liters and the concentration is 0.360 mole per liter), as shown in Figure 3.15. It seems reasonable to start with the solution about which we know the most. If we know the concentration (0.360 *M*) and the volume (2.50 liters) of the dilute $H_2SO_4$ solution we are trying to prepare, we can calculate the number of moles of $H_2SO_4$ it contains:

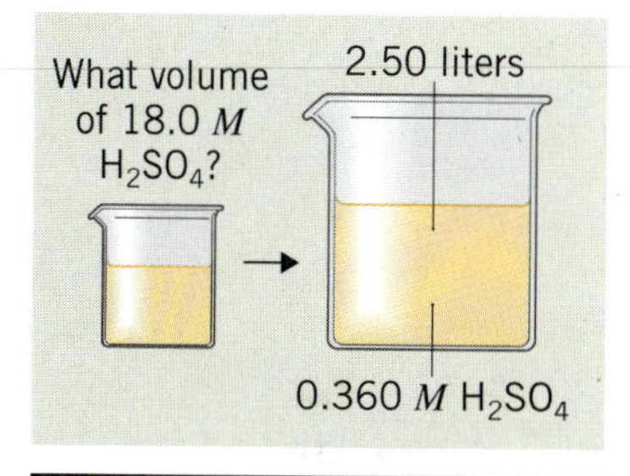


**Figure 3.15** Exercise 3.18.

$$\frac{0.360 \text{ mol } H_2SO_4}{1 \cancel{L}} \times 2.50 \cancel{L} = 0.900 \text{ mol } H_2SO_4$$

What volume of concentrated $H_2SO_4$ contains the same number of moles of $H_2SO_4$? We can start with the equation that describes the relationship between the molarity ($M$) of a solution, the volume of the solution ($V$), and the number of moles of solute in the solution ($n$):

$$M \times V = n$$

We then substitute into this equation the molarity of the concentrated sulfuric acid solution and the number of moles of $H_2SO_4$ molecules needed to prepare the dilute solution:

$$\frac{18.0 \text{ mol } H_2SO_4}{1 \text{ L}} \times V = 0.900 \text{ mol } H_2SO_4$$

We then solve this equation for the volume of the solution.

$$V = 0.0500 \text{ L}$$

According to this calculation, we can prepare 2.50 liters of an 0.360 *M* $H_2SO_4$ solution by adding 50.0 mL of concentrated sulfuric acid to enough water to give a total volume of 2.50 liters.

We can measure the concentration of a solution by a technique known as a **titration.** The solution being studied is slowly added to a known quantity of a reagent with which it reacts until we observe something that tells us that exactly equivalent numbers of moles of the reagents are present. Titrations are therefore dependent on the existence of a class of compounds known as **indicators.**

One of the indicators commonly used in acid-base titrations is phenolphthalein, which turns from colorless to pink in the presence of excess base.

Phenolphthalein is an example of an indicator for acid–base reactions. When added to solutions that contain an acid, phenolphthalein is colorless. In the presence of a base, it has a light pink color. Phenolphthalein can therefore indicate when enough base has been added to consume the acid that was initially present in the solution, because it turns from colorless to pink *just after* the point when equivalent numbers of moles of acid and base have been added to the solution, which is called the **equivalence point.**

## ▶ CHECKPOINT

The **endpoint** of an acid–base titration is the point at which the indicator turns color. The equivalence point is the point at which exactly enough base has been added to neutralize the acid. Explain why the endpoint of a titration can differ slightly from the equivalence point. Explain why chemists stop acid–base titrations that use phenolphthalein as the indicator one drop before the indicator turns a permanent color.

## EXERCISE 3.19

The reaction between oxalic acid ($H_2C_2O_4$) and sodium hydroxide (NaOH) can be described by the following equation:

$$H_2C_2O_4(aq) + 2\,NaOH(aq) \longrightarrow 2\,Na^+(aq) + C_2O_4^{2-}(aq) + 2\,H_2O(l)$$

Calculate the concentration of an oxalic acid solution if it takes 34.0 mL of an 0.200 *M* NaOH solution to consume the acid in a 25.0-mL sample of this oxalic acid solution.

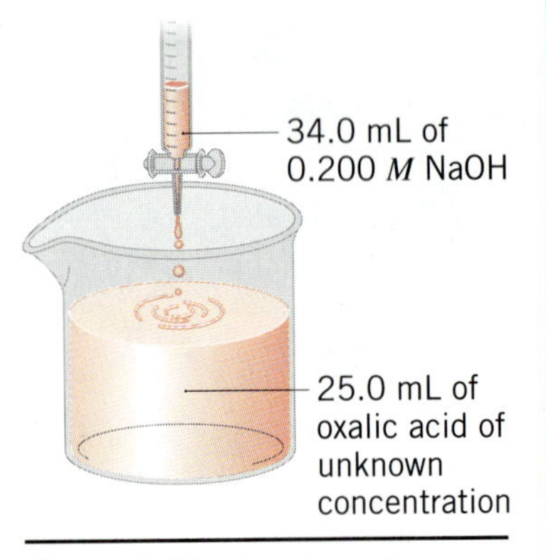


**Figure 3.16** Exercise 3.19.

**SOLUTION** We only know the volume of the oxalic acid solution, but we know both the volume and the concentration of the NaOH solution, as shown in Figure 3.16. It therefore seems reasonable to start by calculating the number of moles of NaOH in this solution:

$$\frac{0.200 \text{ mol NaOH}}{1 \cancel{\text{L}}} \times 0.0340 \cancel{\text{L}} = 6.80 \times 10^{-3} \text{ mol NaOH}$$

We now know the number of moles of NaOH consumed in the reaction, and we have a balanced chemical equation for the reaction that occurs when the two solutions are mixed. We can therefore calculate the number of moles of $H_2C_2O_4$ needed to consume this much NaOH:

$$6.80 \times 10^{-3} \cancel{\text{mol NaOH}} \times \frac{1 \text{ mol } H_2C_2O_4}{2 \cancel{\text{mol NaOH}}} = 3.40 \times 10^{-3} \text{ mol } H_2C_2O_4$$

We now know the number of moles of $H_2C_2O_4$ in the original oxalic acid solution and the volume of this solution. We can therefore calculate the number of moles of oxalic acid per liter, or the molarity of the solution.

$$\frac{3.40 \times 10^{-3} \text{ mol } H_2C_2O_4}{0.0250 \text{ L}} = \mathbf{0.136}\ \boldsymbol{M}\ \mathbf{H_2C_2O_4}$$

The oxalic acid solution therefore has a concentration of 0.136 mole per liter.

SPECIAL TOPIC

## ELEMENTAL ANALYSIS

Percent-by-mass data for a compound are obtained by a process known as **elemental analysis.** If the compound contains carbon and hydrogen, a small sample is burned in the microanalysis apparatus shown in Figure 3.17. A few milligrams of the compound are added to a tiny platinum boat, which is placed in a furnace heated to about 850°C, and a stream of oxygen ($O_2$) gas is passed over the sample.

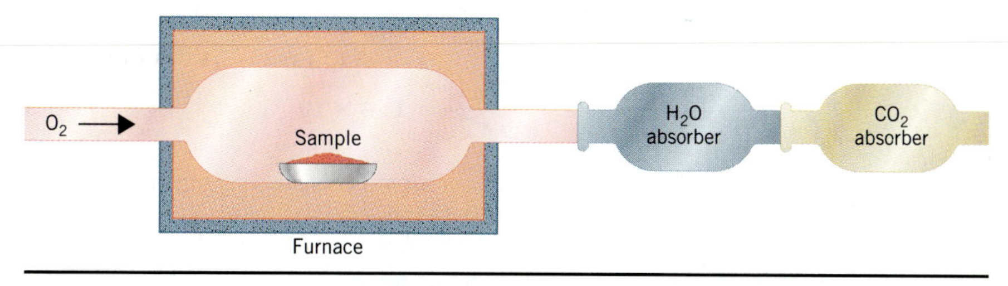


**Figure 3.17** A block diagram of the microanalysis apparatus used to determine the percent by mass of carbon and hydrogen in a compound.

Compounds that contain only carbon and hydrogen burn to form a mixture of $CO_2$ and $H_2O$. If elements besides carbon and hydrogen are present, other gases may be formed as well. The $CO_2$ and $H_2O$ produced in this combustion reaction are swept out of the furnace by the stream of oxygen gas and trapped in a pair of absorbers. The water vapor is absorbed onto a sample of magnesium perchlorate [$Mg(ClO_4)_2$] of known mass. The carbon dioxide is absorbed onto a known mass of the mineral ascharite ($Mg_2B_2O_4 \cdot 2\,H_2O$).

Let's assume that we analyzed a 3.00-mg sample of aspirin, which is known to contain three elements: carbon, hydrogen, and oxygen. Furthermore, let's assume that 6.60 mg of $CO_2$ and 1.20 mg of $H_2O$ were formed. Our goal is to convert this information into data that describe the percent by mass of carbon and hydrogen in aspirin. We start by converting the number of grams of $CO_2$ and $H_2O$ given off in this reaction into moles of these compounds:

$$0.00660\ \cancel{\text{g CO}_2} \times \frac{1\ \text{mol CO}_2}{44.01\ \cancel{\text{g CO}_2}} = 1.50 \times 10^{-4}\ \text{mol CO}_2$$

$$0.00120\ \cancel{\text{g H}_2\text{O}} \times \frac{1\ \text{mol H}_2\text{O}}{18.02\ \cancel{\text{g H}_2\text{O}}} = 6.66 \times 10^{-5}\ \text{mol H}_2\text{O}$$

We then note that there is one carbon atom in each $CO_2$ molecule, which means there is a mole of carbon atoms in a mole of $CO_2$ molecules:

$$1.50 \times 10^{-4}\ \cancel{\text{mol CO}_2} \times \frac{1\ \text{mol C}}{1\ \cancel{\text{mol CO}_2}} = \mathbf{1.50 \times 10^{-4}\ mol\ C}$$

Because all of the carbon came from the original sample, the aspirin sample must have contained $1.50 \times 10^{-4}$ moles of carbon atoms.

There are two hydrogen atoms in each $H_2O$ molecule and therefore 2 moles of hydrogen atoms in a mole of $H_2O$ molecules:

$$6.66 \times 10^{-5}\ \cancel{\text{mol H}_2\text{O}} \times \frac{2\ \text{mol H}}{1\ \cancel{\text{mol H}_2\text{O}}} = \mathbf{1.33 \times 10^{-4}\ mol\ H}$$

The aspirin therefore contained $1.33 \times 10^{-4}$ moles of hydrogen atoms.

Let's summarize our results so far:

A 3.00-mg sample of aspirin contains $1.50 \times 10^{-4}$ moles of carbon and $1.33 \times 10^{-4}$ moles of hydrogen.

Because we know the number of moles of carbon and hydrogen in this sample, we can calculate the number of grams of these elements in the sample:

$$1.50 \times 10^{-4} \text{ mol C} \times \frac{12.01 \text{ g C}}{1 \text{ mol C}} = \mathbf{1.80 \times 10^{-3} \text{ g C}}$$

$$1.33 \times 10^{-4} \text{ mol H} \times \frac{1.008 \text{ g H}}{1 \text{ mol H}} = \mathbf{1.34 \times 10^{-4} \text{ g H}}$$

According to this calculation, a 3.00-mg sample of aspirin contains 1.80 mg of carbon and 0.134 mg of hydrogen. Aspirin is therefore 60.0% C and 4.48% H by mass.

$$\frac{1.80 \text{ mg C}}{3.00 \text{ mg aspirin}} \times 100\% = \mathbf{60.0\% \text{ C}}$$

$$\frac{0.134 \text{ mg H}}{3.00 \text{ mg aspirin}} \times 100\% = \mathbf{4.48\% \text{ H}}$$

Carbon and hydrogen add up to only 64.5% of the total mass of the aspirin. The remaining mass (35.5%) therefore must be due to the third element: oxygen. Microanalysis therefore suggests that aspirin is 60.0% C, 4.48% H, and 35.5% O by mass. As we saw in Exercise 3.7, these data can be used to determine that the empirical formula for aspirin is $C_9H_8O_4$.

## SUMMARY

**1.** Atoms are so small that we can only determine the relative mass of one atom compared with another. The mass of an atom is therefore reported in atomic mass units, or amu. Because isotopes exist, the average mass of the atoms of an element—the atomic weight—is different from the mass of a single atom of most elements.

**2.** Atoms are so small they can't be counted individually. We therefore need a unit that describes a collection of atoms large enough to be visible to the naked eye. By convention, this goal is achieved by defining a mole of atoms as the sample that contains enough of these atoms to have a mass in grams equal to the atomic weight. Because the atomic weight of carbon is 12.011 amu, a mole of carbon atoms has a mass of 12.011 grams.

**3.** The concept of a mole can be extended to other elementary particles, such as molecules. Because the molecular weight of $CO_2$ is 44.01 amu, a mole of $CO_2$ has a mass of 44.01 grams.

**4.** The concept of a mole allows us to predict the relationship between the masses of the reactants and products of a chemical reaction. We start by converting grams of one of the components of the reaction into moles of that component. We then use a balanced equation to convert from moles of that component to moles of another. We then use the mole concept to convert from moles of the second component into grams of this material.

**5.** The mole concept allows us to determine the number of atoms, ions, or molecules from the mass of the sample, which is a convenient property as long as the sample is a solid. But volume is a more convenient measurement when working with solutions. We therefore introduce the concept of concentration, and one particular unit of concentration known as the molarity of a solution, to enable us to determine the number of atoms, ions, or molecules in a solution from measurements of the volume of the solution.

## KEY TERMS

**Aqueous** (p. 84)
**Atomic mass units (amu)** (p. 74)
**Atomic weight** (p. 75)
**Avogadro's number** (p. 77)
**Chemical equation** (p. 83)
**Concentration** (p. 98)
**Dilution** (p. 101)
**Elemental analysis** (p. 103)
**Empirical formula** (p. 80)
**Empirical weight** (p. 82)
**Endpoint** (p. 102)
**Equivalence point** (p. 102)
**Excess reagent** (p. 91)
**Indicators** (p. 101)
**Isotopes** (p. 75)
**Law of conservation of matter** (p. 83)
**Limiting reagent** (p. 91)
**Molar mass** (p. 77)
**Molarity** (p. 98)
**Mole** (p. 75)
**Mole ratio** (p. 87)
**Molecular formula** (p. 82)
**Molecular weight** (p. 82)
**Products** (p. 83)
**Reactants** (p. 83)
**Stoichiometry** (p. 88)
**Solute** (p. 96)
**Solution** (p. 96)
**Solvent** (p. 96)
**Titration** (p. 101)

## KEY EQUATIONS

$$\text{Molarity } (M) = \frac{\text{moles of solute}}{\text{liters of solution}}$$

$$M \times V = n$$

## PROBLEMS

### The Relative Masses of Atoms

**3-1** Identify the element that has an atomic weight 4.33 times as large as carbon.

**3-2** What would be the atomic weight of neon on a scale on which the mass of a $^{12}C$ is defined as exactly 1 amu?

**3-3** Identify the element that contains atoms that have an average mass of $4.48 \times 10^{-23}$ grams.

**3-4** Calculate the atomic weight of bromine if naturally occurring bromine is 50.69% $^{79}Br$ atoms that have a mass of 78.9183 amu and 49.31% $^{81}Br$ atoms that have a mass of 80.9163 amu.

**3-5** Naturally occurring zinc is 48.6% $^{64}Zn$ atoms (63.9291 amu), 27.9% $^{66}Zn$ atoms (65.9260 amu), 4.1% $^{67}Zn$ atoms (66.9721 amu), 18.8% $^{68}Zn$ atoms (67.9249 amu), and 0.6% $^{70}Zn$ atoms (69.9253 amu). Calculate the atomic weight of zinc.

### The Mole as a Collection of Atoms

**3-6** If the average mass of a chromium atom is 51.9961 amu, what is the mass of a mole of chromium atoms?

**3-7** If the average sulfur atom is approximately twice as heavy as the average oxygen atom, what is the ratio of the mass of a mole of sulfur atoms to the mass of a mole of oxygen atoms?

**3-8** Calculate the mass in grams of a mole of atoms of the following elements:

(a) C (b) Ni (c) Hg

**3-9** If eggs sell for 90¢ a dozen, what does it cost to buy 2.5 dozen eggs? If the molar mass of carbon is 12.011 grams, what is the mass of 2.5 moles of carbon atoms?

### Avogadro's Number

**3-10** What would be the value of Avogardro's number if a mole were defined as the number of $^{12}C$ atoms in 12 pounds of $^{12}C$?

**3-11** Calculate the number of atoms in 16 grams of $O_2$, 31 grams of $P_4$, and 32 grams of $S_8$.

**3-12** Calculate the number of chlorine atoms in 0.756 gram of $K_2PtCl_6$.

**3-13** Calculate the number of oxygen atoms in each of the following:

(a) 0.100 mole of potassium permanganate, $KMnO_4$
(b) 0.25 mole of dinitrogen pentoxide, $N_2O_5$
(c) 0.45 mole of penicillin, $C_{16}H_{17}N_2O_5SK$

**3-14** Benzaldehyde has the pleasant, distinctive odor of almonds. What is the molecular weight of benzaldehyde if a single molecule has a mass of $1.762 \times 10^{-22}$ gram?

### The Mole as a Collection of Molecules

**3-15** Which pair of samples contains the same number of hydrogen atoms?

(a) 1 mole of $NH_3$ and 1 mole of $N_2H_4$ (b) 2 moles of $NH_3$ and 1 mole of $N_2H_4$ (c) 2 moles of $NH_3$ and 3 moles of $N_2H_4$ (d) 4 moles of $NH_3$ and 3 moles of $N_2H_4$

**3-16** Which of the following contains the largest number of carbon atoms?

(a) 0.10 mole of acetic acid, $CH_3CO_2H$ (b) 0.25 mole of carbon dioxide, $CO_2$ (c) 0.050 mole of glucose, $C_6H_{12}O_6$ (d) 0.0010 mole of sucrose, $C_{12}H_{22}O_{11}$

**3-17** Calculate the molecular weights of formic acid, $HCO_2H$, and formaldehyde, $H_2CO$.

**3-18** Calculate the molecular weights of the following compounds:

(a) methane, $CH_4$ (b) glucose, $C_6H_{12}O_6$ (c) diethyl ether, $(CH_3CH_2)_2O$ (d) thioacetamide, $CH_3CSNH_2$

**3-19** Calculate the molar mass of the following compounds:

(a) tetraphosphorus decasulfide, $P_4S_{10}$ (b) nitrogen dioxide, $NO_2$ (c) zinc sulfide, ZnS (d) potassium permanganate, $KMnO_4$

**3-20** Calculate the molar mass of the following compounds:

(a) chromium hexacarbonyl, $Cr(CO)_6$ (b) iron(III) nitrate, $Fe(NO_3)_3$ (c) potassium dichromate, $K_2Cr_2O_7$ (d) calcium phosphate, $Ca_3(PO_4)_2$

**3-21** Root beer hasn't tasted the same since the FDA outlawed the use of sassafras oil as a food additive because sassafras oil is 80% safrole, which has been shown to cause cancer in rats and mice. Calculate the molecular weight of safrole, $C_{10}H_{10}O_2$.

**3-22** MSG ($C_5H_8NNaO_4$) is a spice commonly used in Chinese cooking that causes some people to feel light headed. Calculate the molecular weight of MSG.

**3-23** Calculate the molecular weights of the active ingredients in the following prescription drugs:

(a) Darvon, $C_{22}H_{30}ClNO_2$ (b) Valium, $C_{16}H_{13}ClN_2O$ (c) tetracycline, $C_{22}H_{24}N_2O_8$

### Using Measurements of Mass to Count Atoms and Molecules

**3-24** Calculate the number of moles of carbon in 0.244 gram of calcium carbide, $CaC_2$.

**3-25** Calculate the number of moles of phosphorus in 15.95 grams of tetraphosphorus decaoxide, $P_4O_{10}$.

**3-26** Calculate the atomic weight of platinum, if 0.8170 mole of this metal has a mass of 159.4 grams.

**3-27** Calculate the mass of 0.0582 mole of carbon tetrachloride, $CCl_4$.

**3-28** Calculate the density of carbon tetrachloride, if 1 mole of $CCl_4$ occupies a volume of 96.94 cubic centimeters.

**3-29** Calculate the volume of a mole of aluminum, if the density of this metal is 2.70 $g/cm^3$.

### Percent-by-Mass Calculations

**3-30** Calculate the percent by mass of chromium in each of the following oxides:

(a) CrO (b) $Cr_2O_3$ (c) $CrO_3$

**3-31** Calculate the percent by mass of nitrogen in the following fertilizers:

(a) $(NH_4)_2SO_4$ (b) $KNO_3$ (c) $NaNO_3$ (d) $(H_2N)_2CO$

**3-32** Calculate the percent by mass of carbon, hydrogen, and chlorine in DDT, $C_{14}H_9Cl_5$.

**3-33** Emeralds are gem-quality forms of the mineral beryl, $Be_3Al_2(SiO_3)_6$. Calculate the percent by mass of silicon in beryl.

**3-34** Osteoporosis is a disease common in older women who have not had enough calcium in their diets. Calcium can be added to the diet by tablets that contain either calcium carbonate ($CaCO_3$), calcium sulfate ($CaSO_4$), or calcium phosphate [$Ca_3(PO_4)_2$]. On a per gram basis, which is the most efficient way of getting $Ca^{2+}$ ions into the body?

### Empirical Formulas

**3-35** Stannous fluoride, or "Fluoristan," is added to toothpaste to help prevent tooth decay. What is the empirical formula for stannous fluoride if this compound is 24.25% F and 75.75% Sn by mass?

**3-36** Iron reacts with oxygen to form three compounds: FeO, $Fe_2O_3$, and $Fe_3O_4$. One of these compounds, known as magnetite, is 72.36% Fe and 27.64% O by mass. What is the formula of magnetite?

**3-37** The chief ore of manganese is an oxide known as pyrolusite, which is 36.8% O and 63.2% Mn by mass. Which of the following oxides of manganese is pyrolusite?

(a) MnO (b) $MnO_2$ (c) $Mn_2O_3$ (d) $MnO_3$ (e) $Mn_2O_7$

**3-38** Nitrogen combines with oxygen to form a variety of compounds, including $N_2O$, NO, $NO_2$, $N_2O_3$, $N_2O_4$, and $N_2O_5$. One of these compounds is called nitrous oxide, or "laughing gas." What is the formula of nitrous oxide if this compound is 63.65% N and 36.35% O by mass?

**3-39** Chalcopyrite is a bronze-colored mineral that is 34.67% Cu, 30.43% Fe, and 34.94% S by mass. Calculate the empirical formula for this mineral.

**3-40** A compound of xenon and fluorine is found to be 53.5% xenon by mass. What is the oxidation number of the xenon atom in this compound?

**3-41** In 1914 E. Merck and Company synthesized and patented a compound known as MDMA as an appetite suppressant. Although it was never marketed, it has reappeared in recent years as a street drug known as "ecstasy." What is the empirical formula of this compound if it contains 68.4% C, 7.8% H, 7.2% N, and 16.6% O by mass?

**3-42** What is the empirical formula of the compound that contains 0.483 grams of nitrogen and 1.104 grams of oxygen?

(a) $N_2O$ (b) NO (c) $NO_2$ (d) $N_2O_3$ (e) $N_2O_4$

**3-43** What is the empirical formula of the compound formed when 9.33 grams of copper metal react with excess chlorine to give 14.54 grams of this compound?

### Empirical and Molecular Formulas

**3-44** $\beta$-Carotene is the protovitamin from which nature builds vitamin A. It is widely distributed in the plant and animal kingdoms, always occurring in plants together with chlorophyll. Calculate the molecular formula for $\beta$-carotene if this compound is 89.49% C and 10.51% H by mass and its molecular weight is 536.89 grams per mole.

**3-45** The phenolphthalein used as an indicator in acid–base titrations is also the active ingredient in laxatives such as ExLax. Calculate the molecular formula for phenolphthalein if this compound is 75.46% C, 4.43% H, and 20.10% O by mass and it has a molecular weight of 318.31 grams per mole.

**3-46** Caffeine is a central nervous system stimulant found in coffee, tea, and cola nuts. Calculate the molecular formula of caffeine if this compound is 49.48% C, 5.19% H, 28.85% N, and 16.48% O by mass and it has a molecular weight of 194.2 grams per mole.

**3-47** Aspartame, also known as NutraSweet, is 160 times sweeter than sugar when dissolved in water. The true name for this artificial sweetener is *N*-L-$\alpha$-aspartyl-L-phenylalanine methyl ester. Calculate the molecular formula of aspartame if this compound is 57.14% C, 6.16% H, 9.52% N, and 27.18% O by mass and it has a molecular weight of 294.30 grams per mole.

### Balancing Chemical Equations

**3-48** Balance the following chemical equations:

(a) $Cr(s) + O_2(g) \longrightarrow Cr_2O_3(s)$

(b) $SiH_4(g) \longrightarrow Si(s) + H_2(g)$

(c) $SO_3(g) \longrightarrow SO_2(g) + O_2(g)$

**3-49** Balance the following chemical equations:

(a) $Pb(NO_3)_2(s) \longrightarrow PbO(s) + NO_2(g) + O_2(g)$

(b) $NH_4NO_2(s) \longrightarrow N_2(g) + H_2O(g)$

(c) $(NH_4)_2Cr_2O_7(s) \longrightarrow N_2(g) + Cr_2O_3(s) + H_2O(g)$

**3-50** Balance the following chemical equations:

(a) $CH_4(g) + O_2(g) \longrightarrow CO_2(g) + H_2O(g)$

(b) $H_2S(g) + O_2(g) \longrightarrow H_2O(g) + SO_2(g)$

(c) $B_5H_9(g) + O_2(g) \longrightarrow B_2O_3(s) + H_2O(g)$

**3-51** Balance the following chemical equations:

(a) $PF_3(g) + H_2O(l) \longrightarrow H_3PO_3(aq) + HF(aq)$

(b) $P_4O_{10}(s) + H_2O(l) \longrightarrow H_3PO_4(aq)$

**3-52** Balance the following chemical equations:

(a) $C_3H_8(g) + O_2(g) \longrightarrow CO_2(g) + H_2O(g)$

(b) $C_2H_5OH(l) + O_2(g) \longrightarrow CO_2(g) + H_2O(g)$

(c) $C_6H_{12}O_6(s) + O_2(g) \longrightarrow CO_2(g) + H_2O(aq)$

### Mole Ratios and Chemical Equations

**3-53** Carbon disulfide burns in oxygen to form carbon dioxide and sulfur dioxide:

$$CS_2(l) + 3\,O_2(g) \longrightarrow CO_2(g) + 2\,SO_2(g)$$

Calculate the number of $O_2$ molecules it would take to consume 500 molecules of $CS_2$. Calculate the number of moles of $O_2$ molecules it would take to consume 5.00 moles of $CS_2$.

**3-54** Calculate the number of moles of oxygen produced when 6.75 moles of manganese dioxide decomposes to $Mn_3O_4$ and $O_2$:

$$3\,MnO_2(s) \longrightarrow Mn_3O_4(s) + O_2(g)$$

**3-55** Calculate the number of moles of carbon monoxide needed to reduce 3.00 moles of iron(III) oxide to iron metal:

$$Fe_2O_3(s) + 3\,CO(g) \longrightarrow 2\,Fe(s) + 3\,CO_2(g)$$

### Stoichiometry

**3-56** Calculate the mass of oxygen released when enough mercury(II) oxide decomposes to give 25 grams of liquid mercury:

$$2\,HgO(s) \longrightarrow 2\,Hg(l) + O_2(g)$$

**3-57** Calculate the mass of $CO_2$ produced and the mass of oxygen consumed when 10.0 grams of methane ($CH_4$) are burned.

**3-58** How many pounds of sulfur will react with 10.0 pounds of zinc to form zinc sulfide, ZnS?

**3-59** Calculate the mass of oxygen that can be prepared by decomposing 25.0 grams of potassium chlorate.

$$2\,KClO_3(s) \longrightarrow 2\,KCl(s) + 3\,O_2(g)$$

**3-60** Predict the formula of the compound produced when 1.00 gram of chromium metal reacts with 0.923 gram of oxygen.

**3-61** Ethanol, or ethyl alcohol, is produced by the fermentation of sugars such as glucose:

$$C_6H_{12}O_6(aq) \longrightarrow 2\,C_2H_5OH(aq) + 2\,CO_2(g)$$

Calculate the number of kilograms of alcohol that can be produced from a kilogram of glucose.

**3-62** Calculate the number of pounds of aluminum metal that can be obtained from a ton of bauxite, $Al_2O_3 \cdot 2\,H_2O$.

**3-63** Calculate the amount of phosphine, $PH_3$, that can be prepared when 10.0 grams of calcium phosphide, $Ca_3P_2$, react with water.

$$Ca_3P_2(s) + 6\,H_2O(l) \longrightarrow 3\,Ca(OH)_2(aq) + 2\,PH_3(g)$$

**3-64** Hydrogen chloride can be made by reacting phosphorus trichloride with water and then boiling the HCl gas out of the solution.

$$PCl_3(g) + 3\,H_2O(l) \longrightarrow 3\,HCl(aq) + H_3PO_3(aq)$$

Calculate the mass of HCl gas that can be prepared from 15.0 grams of $PCl_3$.

**3-65** Nitrogen reacts with hydrogen to form ammonia:

$$N_2(g) + 3\,H_2(g) \longrightarrow 2\,NH_3(g)$$

which burns in the presence of oxygen to form nitrogen oxide:

$$4\,NH_3(g) + 5\,O_2(g) \longrightarrow 4\,NO(g) + 6\,H_2O(g)$$

which reacts with excess oxygen to form nitrogen dioxide:

$$2\,NO(g) + O_2(g) \longrightarrow 2\,NO_2(g)$$

which dissolves in water to give nitric acid:

$$3\,NO_2(g) + H_2O(l) \longrightarrow 2\,HNO_3(aq) + NO(g)$$

Calculate the mass of nitrogen needed to make 150 grams of nitric acid.

### Limiting Reagents

**3-66** Calculate the number of water molecules that can be prepared from 500 $H_2$ molecules and 500 $O_2$ molecules:

$$2\,H_2(g) + O_2(g) \longrightarrow 2\,H_2O(l)$$

What would happen to the potential yield of water molecules if the amount of $O_2$ was doubled? What if the amount of $H_2$ was doubled?

**3-67** Calculate the number of moles of $P_4S_{10}$ that can be produced from 0.500 mole of $P_4$ and 0.500 mole of $S_8$:

$$4\ P_4(s) + 5\ S_8(s) \longrightarrow 4\ P_4S_{10}(s)$$

What would happen to the potential yield of $P_4S_{10}$ if the amount of $P_4$ were doubled? What if the amount of $S_8$ were doubled?

**3-68** Calculate the number of moles of nitrogen dioxide that could be prepared from 0.35 mole of nitrogen oxide and 0.25 mole of oxygen:

$$2\ NO(g) + O_2(g) \longrightarrow 2\ NO_2(g)$$

Identify the limiting reagent and the excess reagent in this reaction. What would happen to the potential yield of $NO_2$ if the amount of NO were increased? What if the amount of $O_2$ were increased?

**3-69** Calculate the mass of hydrogen chloride that can be produced from 10.0 grams of hydrogen and 10.0 grams of chlorine.

$$H_2(g) + Cl_2(g) \longrightarrow 2\ HCl(g)$$

What would have to be done to increase the amount of hydrogen chloride produced in this reaction?

**3-70** Calculate the mass of calcium nitride, $Ca_3N_2$, that can be prepared from 54.9 grams of calcium and 43.2 grams of nitrogen.

$$3\ Ca(s) + N_2(g) \longrightarrow Ca_3N_2(s)$$

**3-71** $PF_3$ reacts with $XeF_4$ to give $PF_5$.

$$2\ PF_3(g) + XeF_4(s) \longrightarrow 2\ PF_5(g) + Xe(g)$$

How many moles of $PF_5$ can be produced from 100.0 g of $PF_3$ and 50.0 g of $XeF_4$?

**3-72** Trimethyl aluminum, $Al(CH_3)_3$, must be handled in an apparatus from which oxygen has been rigorously excluded because it bursts into flame in the presence of oxygen. Calculate the mass of trimethyl aluminum that can be prepared from 5.00 grams of aluminum metal and 25.0 grams of dimethyl mercury.

$$2\ Al(s) + 3\ Hg(CH_3)_2(l) \longrightarrow 2\ Al(CH_3)_3(l) + 3\ Hg(l)$$

**3-73** The thermite reaction, used to weld rails together in the building of railroads, is described by the following equation:

$$Fe_2O_3(s) + 2\ Al(s) \longrightarrow Al_2O_3(s) + 2\ Fe(l)$$

Calculate the mass of iron metal that can be prepared from 150 grams of aluminum and 250 grams of iron(III) oxide.

### Solutions and Concentration

**3-74** Explain why the molarity of a solution cannot be calculated from percent by mass data unless the density of the solution is known.

**3-75** Describe in detail the steps you would take to prepare 125 mL of 0.745 *M* oxalic acid, starting with solid oxalic acid dihydrate ($H_2C_2O_4 \cdot 2\ H_2O$). Describe the glassware you would need, the chemicals, the amounts of each chemical, and the sequence of steps you would take.

**3-76** "Muriatic acid" is sold in many hardware stores for cleaning bricks and tile. What is the molarity of this solution if 125 mL of this solution contain 27.3 grams of HCl?

**3-77** Silver chloride is only marginally soluble in water; only 0.00019 gram of AgCl dissolves in 100 mL of water. Calculate the molarity of this solution.

**3-78** Ammonia ($NH_3$) is relatively soluble in water. Calculate the molarity of a solution that contains 252 grams of $NH_3$ per liter.

**3-79** Ethyl alcohol, or ethanol, $C_2H_5OH$, is infinitely soluble in water. The concentration of a solution of this alcohol in water is often expressed in units of "proof." Pure alcohol is 200 proof, and a 50:50 mixture is 100 proof. Calculate the molarity of a 90-proof solution of ethyl alcohol in water. Assume that the density of this solution is 0.894 $g/cm^3$.

**3-80** During a physical exam, one of the authors was found to have a cholesterol level of 160 milligrams per deciliter. If the molecular weight of cholesterol is 386.67 grams per mole, what is the cholesterol level in his blood in units of moles per liter?

**3-81** Concentrated hydrochloric acid is 38.0% HCl by mass. Calculate the molarity of this solution if it has a density of 1.1977 $g/cm^3$.

**3-82** At 25°C, 5.77 grams of chlorine gas dissolve in a liter of water. Calculate the molarity of $Cl_2$ in this solution.

**3-83** Calculate the mass of $Na_2SO_4$ needed to prepare 0.500 liter of a 0.150 *M* solution.

**3-84** Calculate the concentration of an aqueous KCl solution if 25.00 mL of this solution give 0.430 gram of AgCl when treated with excess $AgNO_3$.

$$KCl(aq) + AgNO_3(aq) \longrightarrow AgCl(s) + KNO_3(aq)$$

**3-85** Calculate the volume of 0.25 *M* NaI that would be needed to precipitate all of the $Hg^{2+}$ ion from 45 mL of a 0.10 *M* $Hg(NO_3)_2$ solution.

$$2\ NaI(aq) + Hg(NO_3)_2(aq) \longrightarrow HgI_2(s) + 2\ NaNO_3(aq)$$

### Dilution Calculations

**3-86** Calculate the volume of 17.4 *M* acetic acid needed to prepare 1.00 liter of 3.00 *M* acetic acid.

**3-87** Calculate the concentration of the solution formed when 15.0 mL of 6.00 *M* HCl are diluted with 25.0 mL of water.

**3-88** Describe how you would prepare 0.200 liter of 1.25 *M* nitric acid from a solution that is 5.94 *M* $HNO_3$.

### Titration Calculations

**3-89** Calculate the molarity of an acetic acid solution if 34.57 mL of this solution are needed to neutralize 25.19 mL of 0.1025 *M* sodium hydroxide.

$$CH_3CO_2H(aq) + NaOH(aq) \longrightarrow Na^+(aq) + CH_3CO_2^-(aq) + H_2O(l)$$

**3-90** Calculate the molarity of a sodium hydroxide solution if 10.42 mL of this solution are needed to neutralize 25.00 mL of 0.2043 *M* oxalic acid.

$$H_2C_2O_4(aq) + 2\ NaOH(aq) \longrightarrow Na_2C_2O_4(aq) + 2\ H_2O(l)$$

**3-91** Calculate the volume of 0.0985 *M* sulfuric acid that would be needed to neutralize 10.89 mL of a 0.01043 *M* aqueous ammonia solution.

$$H_2SO_4(aq) + 2\ NH_3(aq) \longrightarrow (NH_4)_2SO_4(aq)$$

**3-92** α-D-Glucopyranose reacts with the periodate ion as follows.

$$C_6H_{12}O_6(aq) + 5\ IO_4^-(aq) \longrightarrow 5\ IO_3^-(aq) + 5\ HCO_2H(aq) + H_2CO(aq)$$

Calculate the molarity of the glucopyranose solution if 25.0 mL of 0.750 *M* $IO_4^-$ are required to consume 10.0 mL of this solution.

**3-93** Oxalic acid reacts with the chromate ion in acidic solution as follows.

$$3\ H_2C_2O_4(aq) + 2\ CrO_4^{2-}(aq) + 10\ H^+(aq) \longrightarrow 6\ CO_2(g) + 2\ Cr^{3+}(aq) + 8\ H_2O(l)$$

Calculate the molarity of the oxalic acid solution if 10.0 mL of this solution consume 40.0 mL of 0.0250 *M* $CrO_4^{2-}$.

### Elemental Analysis: Experimental Results

**3-94** How accurately would you have to measure the percent by mass of carbon and hydrogen to tell the difference between diazepam (Valium), with the formula $C_{16}H_{13}ClN_2O$, and chlordiazepoxide (Librium), with the formula $C_{16}H_{14}ClN_3O$?

**3-95** The methane in natural gas, the propane used in camping stoves, and the butane used in butane lighters are all members of a family of compounds known as the alkanes, which have the generic formula $C_nH_{2n+2}$. Calculate the value of *n* for butane if 3.15 mg of butane burn in air to form 9.54 mg of $CO_2$ and 4.88 mg of $H_2O$.

**3-96** In small quantities, the nicotine in tobacco is addictive. In large quantities, it is a deadly poison. Calculate the molecular formula of nicotine, $C_xH_yN_z$, if the molecular weight of nicotine is 162.2 grams per mole and 4.38 mg of this compound burn to form 11.9 mg of $CO_2$ and 3.41 mg of water.

### Integrated Problems

**3-97** Calculate the atomic weight of the metal (M) that forms a compound with the formula $MCl_2$ that is 74.5% Cl by mass.

**3-98** Halothane is an anaesthetic that is 12.17% C, 0.51% H, 40.48% Br, 17.96% Cl, and 28.87% F by mass. What is the molecular formula of this compound if each molecule contains one hydrogen atom?

**3-99** A compound that is 31.9% K and 28.9% Cl by mass decomposes when heated to give $O_2$ and a compound that is 52.4% K and 47.6% Cl by mass. Write a balanced chemical equation for this reaction.

**3-100** Calculate the number of water molecules in the formula for the hydrate of copper sulfate if 2.47 grams of $CuSO_4$ pick up water to form 3.86 grams of a compound with the formula $CuSO_4 \cdot x\ H_2O$.

**3-101** Predict the formula of the compound produced when 1.00 gram of chromium metal reacts with 0.923 gram of oxygen.

**3-102** A 3.500-gram sample of an oxide of manganese contains 1.288 grams of oxygen. What is the empirical formula of this compound?

**3-103** Cocaine is a naturally occurring substance that can be extracted from the leaves of the *coca* plant, which grows in South America. (Not to be confused with chocolate, or *cocoa,* which is extracted from the seeds of another plant that grows in South America.) If the chemical formula for cocaine is $C_{17}H_{21}O_4N$, what is the percent by mass of carbon, hydrogen, oxygen, and nitrogen in this compound? Comment on the ease with which elemental analysis of the carbon and hydrogen in a compound can be used to distinguish between the white, crystalline powder known as aspirin ($C_9H_8O_4$), which is used to cure headaches, and the white, crystalline powder known as cocaine, which is more likely to cause headaches.

**3-104** The oxygen-carrier protein known as hemoglobin is 0.335% Fe by mass. Calculate the molecular weight of this protein if there are four iron atoms in each molecule of hemoglobin.

**3-105** Metal carbonates decompose when they are heated to form metal oxides and carbon dioxide:

$$MCO_3(s) \longrightarrow MO(s) + CO_2(g)$$

Which of the following metal carbonates would lose 35.1% of its mass when it decomposes?

(a) $Li_2CO_3$ (b) $MgCO_3$ (c) $CaCO_3$ (d) $ZnCO_3$
(e) $BaCO_3$

**3-106** A crucible and sample of $CaCO_3$ weighing 42.670 grams were heated until they decomposed to form CaO and $CO_2$:

$$CaCO_3(s) \longrightarrow CaO(s) + CO_2(g)$$

The crucible had a mass of 35.351 g. What is the theoretical mass of the crucible and residue after the decomposition is complete?

**3-107** Nitrogen reacts with red-hot magnesium to form magnesium nitride

$$3\ Mg(s) + N_2(g) \longrightarrow Mg_3N_2(s)$$

which reacts with water to form magnesium hydroxide and ammonia.

$$Mg_3N_2(s) + 6\ H_2O(l) \longrightarrow 3\ Mg(OH)_2(aq) + 2\ NH_3(aq)$$

Calculate the number of grams of magnesium that would be needed to prepare 15.0 grams of ammonia.

**3-108** A 2.50-gram sample of bronze was dissolved in sulfuric acid. The copper in this alloy reacted with sulfuric acid as follows:

$$Cu(s) + 2\ H_2SO_4(aq) \longrightarrow CuSO_4(aq) + SO_2(g) + 2\ H_2O(l)$$

The $CuSO_4$ formed in this reaction was mixed with KI to form CuI:

$$2\ CuSO_4(aq) + 5\ I^-(aq) \longrightarrow 2\ CuI(s) + I_3^-(aq) + 2\ SO_4^{2-}(aq)$$

The $I_3^-$ formed in this reaction was then titrated with $S_2O_3^{2-}$:

$$I_3^-(aq) + 2\ S_2O_3^{2-}(aq) \longrightarrow 3\ I^-(aq) + S_4O_6^{2-}(aq)$$

Calculate the percent by mass of copper in the original sample if 31.5 mL of 1.00 *M* $S_2O_3^{2-}$ were consumed in this titration.

For many years, chemists struggled to find a way to determine whether all gases are the same. Today we recognize that the air in the atmosphere is a mixture of different gases that is about 80% $N_2$, 20% $O_2$, and 1% Ar, with varying amounts of the water vapor that forms these clouds.

# GASES

The term *gas* comes from the Greek word for chaos because gases consist of a chaotic collection of particles in a state of constant, random motion. In the course of our discussion of gases we will provide the basis for answering the following questions.

## POINTS OF INTEREST

- Why does popcorn "pop" when we heat it?
- Why does a hot-air balloon rise when the air in the balloon is heated?
- Why does a balloon filled with helium rise? Why does a balloon filled with $CO_2$ sink?
- At 25°C and 1 atmosphere pressure, which weighs more: dry air or air saturated with water vapor?
- Is the volume of the $O_2$ in your room the same as the volume of the $N_2$? Is the pressure of the $O_2$ the same as the pressure of the $N_2$?

## 4.1 THE STATES OF MATTER

The term *state* can be defined as a set of conditions that describes a person or thing at a given time. It is in this sense of the word that scientists divide matter into the three **states** shown in Figure 4.1.

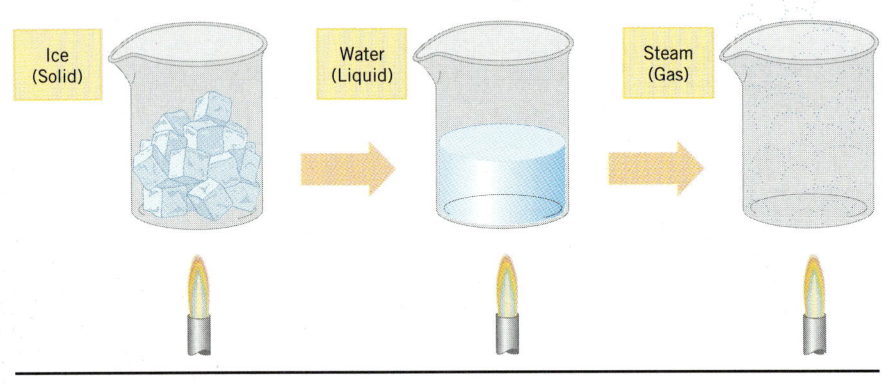


**Figure 4.1** The three states of matter have characteristic properties. Solids have a distinct shape. When they melt, the resulting liquid conforms to the shape of its container. Gases expand to fill their containers.

Because it exists as relatively light covalent molecules, $NO_2$ is a gas at room temperature.

There are two reasons for studying gases before liquids or solids. First, the behavior of gases is easier to describe because most of the properties of gases do not depend on the identity of the gas. We can therefore develop a model for a gas without worrying about whether the gas is $O_2$, $N_2$, $H_2$, or a mixture of these gases. Second, a relatively simple, yet powerful, model known as the *kinetic molecular theory* is available, which explains most of the behavior of gases.

## 4.2 ELEMENTS OR COMPOUNDS THAT ARE GASES AT ROOM TEMPERATURE

Before we examine the chemical and physical properties of gases, it may be useful to ask: What kinds of elements or compounds are gases at room temperature? To help answer this question, a list of some common compounds that are gases at room temperature is given in Table 4.1.

There are several patterns in Table 4.1.

1. Common gases at room temperature include both elements (such as $H_2$ and $O_2$) and compounds (such as $CO_2$ and $NH_3$).
2. Elements that are gases at room temperature are all *nonmetals* (such as He, Ar, $N_2$, and $O_2$).
3. Compounds that are gases at room temperature are all *covalent compounds* (such as $CO_2$, $SO_2$, and $NH_3$) that contain two or more nonmetals.
4. With only rare exception, these gases have relatively small molar masses.

As a general rule, compounds that consist of relatively light, covalent molecules are most likely to be gases at room temperature.

## 4.3 THE PROPERTIES OF GASES

Gases have three characteristic properties: (1) they are easy to compress, (2) they expand to fill their containers, and (3) they occupy far more space than the liquids or solids from which they form.

**TABLE 4.1 Common Gases at Room Temperature**

| Element or Compound | Molar Mass |
| --- | --- |
| $H_2$ (hydrogen) | 2.02 |
| He (helium) | 4.00 |
| $CH_4$ (methane) | 16.04 |
| $NH_3$ (ammonia) | 17.03 |
| Ne (neon) | 20.18 |
| HCN (hydrogen cyanide) | 27.03 |
| CO (carbon monoxide) | 28.01 |
| $N_2$ (nitrogen) | 28.01 |
| NO (nitrogen oxide) | 30.01 |
| $C_2H_6$ (ethane) | 30.07 |
| $O_2$ (oxygen) | 32.00 |
| $PH_3$ (phosphine) | 34.00 |
| $H_2S$ (hydrogen sulfide) | 34.08 |
| HCl (hydrogen chloride) | 36.46 |
| $F_2$ (fluorine) | 38.00 |
| Ar (argon) | 39.95 |
| $CO_2$ (carbon dioxide) | 44.01 |
| $N_2O$ (dinitrogen oxide) | 44.01 |
| $C_3H_8$ (propane) | 44.10 |
| $NO_2$ (nitrogen dioxide) | 46.01 |
| $O_3$ (ozone) | 48.00 |
| $C_4H_{10}$ (butane) | 58.12 |
| $SO_2$ (sulfur dioxide) | 64.06 |
| $BF_3$ (boron trifluoride) | 67.80 |
| $Cl_2$ (chlorine) | 70.91 |
| Kr (krypton) | 83.80 |
| $CF_2Cl_2$ (dichlorodifluoromethane) | 120.91 |
| Xe (xenon) | 131.30 |
| $SF_6$ (sulfur hexafluoride) | 146.05 |

## COMPRESSIBILITY

The internal combustion engine provides a good example of the ease with which gases can be compressed. In a typical four-stroke engine, the piston is first pulled out of the cylinder to create a partial vacuum, which draws a mixture of gasoline vapor and air into the cylinder (see Figure 4.2). The piston is then pushed into the cylinder, compressing the gasoline–air mixture to a fraction of its original volume.

The ratio of the volume of the gas in the cylinder after the first stroke to its volume after the second stroke is the *compression ratio* of the engine. Modern cars run at compression ratios of about 9:1, which means the gasoline–air mixture in the cylinder is compressed by a factor of nine in the second stroke. After the gasoline–air mixture is compressed, the spark plug at the top of the cylinder fires and the resulting explosion pushes the piston out of the cylinder in the third stroke. Finally, the piston is pushed back into the cylinder in the fourth stroke, clearing out the exhaust gases.

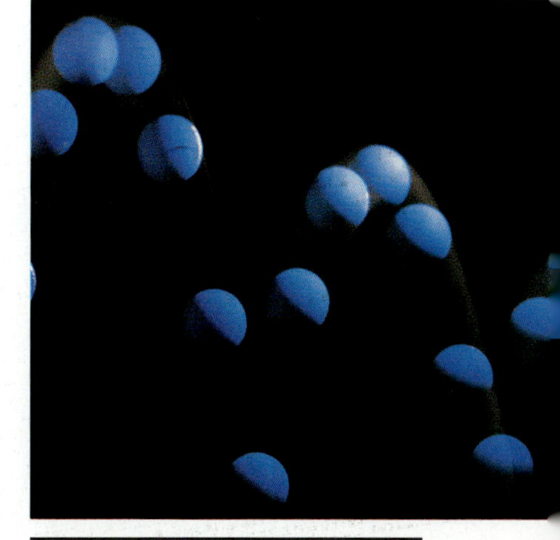

Most balls that bounce are hollow. It is the ease with which the gas inside the ball can be compressed and then return to its initial volume that explains why they bounce.

Liquids are much harder to compress than gases. They are so hard to compress that the hydraulic brake systems used in most cars operate on the principle that there is essentially no change in the volume of the brake fluid when pressure is applied to this liquid. Most solids are even harder to compress. The only exceptions belong to

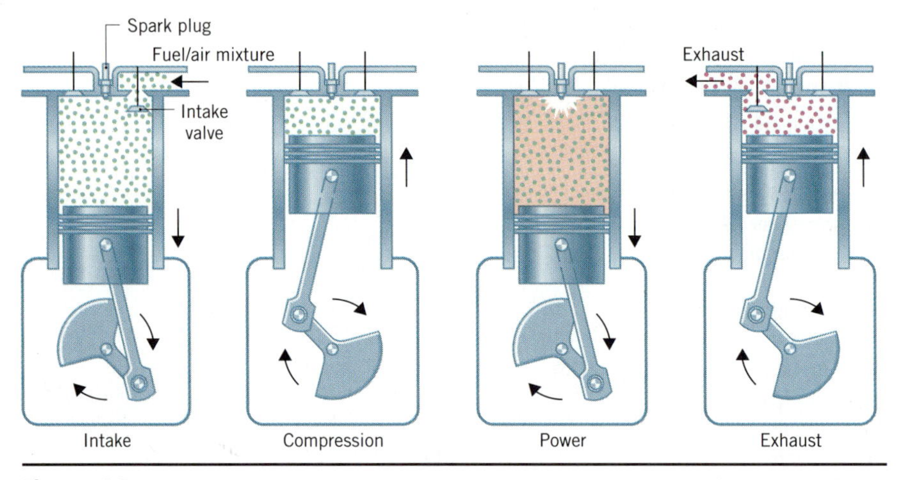


**Figure 4.2** The operation of a four-stroke engine can be divided into four cycles: intake, compression, power, and exhaust stages.

a small class of compounds that includes natural and synthetic rubber. Most rubber balls that seem easy to compress, such as a racquetball, are filled with air, which is compressed when the ball is squeezed.

## EXPANDABILITY

Anyone who has walked into a kitchen where bread is baking has experienced the fact that gases expand to fill their containers, as the air in the kitchen becomes filled with wonderful odors. Unfortunately the same thing happens when someone breaks open a rotten egg and the characteristic odor of hydrogen sulfide ($H_2S$) rapidly diffuses through the room. Because gases expand to fill their containers, it is safe to assume that the volume of a gas is equal to the volume of its container.

## VOLUMES OF GASES VERSUS VOLUMES OF LIQUIDS OR SOLIDS

The difference between the volume of a gas and the volume of the liquid or solid from which it forms can be illustrated with the following examples. One gram of liquid oxygen at its boiling point (−183°C) has a volume of 0.894 mL. The same amount of $O_2$ gas at 0°C and atmospheric pressure has a volume of 700 mL, which is almost 800 times larger. Similar results are obtained when the volumes of solids and gases are compared. One gram of solid $CO_2$ has a volume of 0.641 mL. At 0°C and atmospheric pressure, the same amount of $CO_2$ gas has a volume of 556 mL, which is more than 850 times as large. As a general rule, the volume of a liquid or solid increases by a factor of about 800 when it forms a gas.

### ▶ CHECKPOINT

Ammonia ($NH_3$) is a gas at room temperature that condenses to become a liquid at temperatures below −33°C. The density of the gas is 0.719 g/L at 20°C and atmospheric pressure. The density of the liquid at its boiling point is 0.648 g/mL. Show that $NH_3$ as a gas has a volume more than 900 times as large as an equivalent amount of liquid $NH_3$.

This enormous change in volume is frequently used to do work. The steam engine, which brought about the industrial revolution, is based on the fact that water boils to form a gas (steam) that has a much larger volume. The gas therefore escapes from the container in which it has been generated, and the escaping steam can be made to do work. The same principle is at work when dynamite is used to blast rocks. In 1867, the Swedish chemist Alfred Nobel discovered that the highly dangerous liquid explosive nitroglycerin could be absorbed onto clay or sawdust to produce a solid that was much more stable and therefore safer to use. When dynamite is detonated, the nitroglycerin decomposes to produce a mixture of $CO_2$, $H_2O$, $N_2$, and $O_2$ gases:

$$4\ C_3H_5N_3O_9(l) \longrightarrow 12\ CO_2(g) + 10\ H_2O(g) + 6\ N_2(g) + O_2(g)$$

Because 29 moles of gas are produced for every 4 moles of liquid that decompose, and because each mole of gas occupies a volume roughly 800 times larger than a mole of liquid, this reaction produces a shock wave that destroys anything in its vicinity.

The same phenomenon occurs on a much smaller scale when we pop popcorn. When kernels of popcorn are heated in oil, the liquids inside the kernel turn into gases. The pressure that builds up inside the kernel is enormous and the kernel eventually explodes.

Popcorn ''pops'' because of the enormous difference between the volume of the oils inside the kernel and the volume of the gases these oils produce when they boil.

## 4.4 PRESSURE VERSUS FORCE

The volume of a gas is one of its characteristic properties. Another characteristic property is the **pressure** the gas exerts on its surroundings. Many of us got our first exposure to the concept of the pressure of a gas when we rode to the neighborhood gas station to check the pressure of our bicycle tires. Depending on the kind of bicycle we had, we added air to the tires until the pressure gauge read between 30 and 70 pounds per square inch ($lb/in^2$ or psi). Two important properties of pressure can be obtained from this example:

1. The pressure of a gas increases as more gas is added to the container.
2. Pressure is measured in units (such as $lb/in^2$) that describe the **force** exerted by the gas divided by the **area** over which this force is distributed.[1]

The first conclusion can be summarized in the following relationship, where $P$ is the pressure of the gas and $n$ is the amount of gas in the container.

$$P \propto n$$

Because the pressure increases as gas is added to the container, $P$ is directly proportional to $n$.

The second conclusion describes the relationship between pressure and force. Pressure is defined as the force exerted on an object divided by the area over which the force is distributed.

$$\textbf{Pressure} = \frac{\textbf{force}}{\textbf{area}}$$

The difference between pressure and force can be illustrated with an analogy based on a 10-penny nail, a hammer, and a piece of wood, as shown in Figure 4.3. By resting the nail on its point, and hitting the head with the hammer, we can drive the nail into the wood. But what happens if we turn the nail over and rest the head of the nail against the wood? If we hit the nail with the same force, we can't get the nail to stick into the wood.

Most of us are familiar with measurements of pressures in units of pounds per square inch ($lb/in^2$).

[1] The SI unit for pressure is the pascal, which is defined as a force of 1 newton averaged over an area of 1 $m^2$.

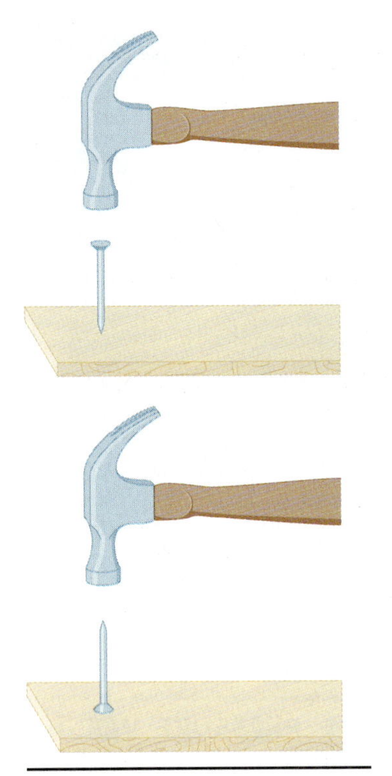

**Figure 4.3** The *force* exerted by a hammer hitting a nail is the same regardless of whether the hammer hits the nail on the head or on the point. But the *pressure* exerted on the wood is very different.

When we hit the nail on the head, the force of this blow is applied to the very small area of the wood in contact with the sharp point of the nail, and the nail slips easily into the wood. But when we turn the nail over, and hit it on the point, the force is distributed over a much larger area. The force is now distributed over the surface of the wood that touches any part of the nail head. As a result, the pressure applied to the wood is much smaller and the nail just bounces off the wood.

**EXERCISE 4.1**

(a) Calculate the pressure exerted by a 200-lb man wearing size 10 shoes, if the area of each shoe in contact with the floor is 20 square inches.

(b) Calculate the pressure exerted by the heels of a 100-lb woman in high heels, if the area beneath the heel of each shoe is 0.25 square inch.

**SOLUTION**

(a) The pressure can be calculated by dividing the force by the area over which it is distributed. Because one-half of the weight of the man is applied to each shoe, the pressure in this case is 5 lb/in$^2$.

$$\text{Pressure} = \frac{\text{force}}{\text{area}} = \frac{100 \text{ lb}}{20 \text{ in}^2} = \mathbf{5 \text{ lb/in}^2}$$

(b) We can assume that about one-fourth of the weight of the woman is applied to each heel, if her weight is evenly divided between the heel and sole of each shoe:

$$\text{Pressure} = \frac{\text{force}}{\text{area}} = \frac{25 \text{ lb}}{0.25 \text{ in}^2} = \mathbf{100 \text{ lb/in}^2}$$

The pressure exerted by the heels of this 100-lb woman is 20 times larger than the pressure exerted by the man, even though he weighs twice as much.

## 4.5 ATMOSPHERIC PRESSURE

What would happen if we bent a long piece of glass tubing into the shape of the letter U and then carefully filled one arm of this U-tube with water and the other arm with ethyl alcohol? Most people expect the height of the columns of liquid in the two arms of the tube to be the same. Experimentally, we find the results shown in Figure 4.4. A 100-cm column of water balances a 127-cm column of ethyl alcohol, regardless of the diameter of the glass tubing.

We can explain this observation by comparing the densities of water (1.00 g/cm$^3$) and ethyl alcohol (0.789 g/cm$^3$). A column of water 100 cm tall exerts a pressure of 100 grams per square centimeter:

$$100 \cancel{\text{cm}} \times \frac{1.00 \text{ g}}{1 \cancel{\text{cm}^3}} = 100 \text{ g/cm}^2$$

A column of ethyl alcohol 127 cm tall exerts the same pressure:

$$127 \cancel{\text{cm}} \times \frac{0.789 \text{ g}}{1 \cancel{\text{cm}^3}} = 100 \text{ g/cm}^2$$

Because the pressure of the water pushing down on one arm of the U-tube is equal to the pressure of the alcohol pushing down on the other arm of the tube, the system is in balance. This demonstration provides the basis for understanding how a mercury barometer can be used to measure the pressure of the atmosphere.

## THE DISCOVERY OF THE BAROMETER

In the early 1600s, Galileo argued that suction pumps were able to draw water from a well because of the "force of vacuum" inside the pump. After Galileo's death, Italian mathematician and physicist Evangelista Torricelli (1608–1647) proposed another explanation. He suggested that the air in our atmosphere has weight and that the force of the atmosphere pushing down on the surface of the water drives the water into the suction pump when it is evacuated.

In 1646 Torricelli described an experiment in which a glass tube about a meter long was sealed at one end, filled with mercury, and then inverted into a dish filled with mercury, as shown in Figure 4.5. Some, but not all, of the mercury drained out of the glass tube into the dish. Torricelli explained this by assuming that mercury drains from the glass tube until the force of the column of mercury pushing down on the *inside* of the tube exactly balances the force of the atmosphere pushing down on the surface of the liquid *outside* the tube.

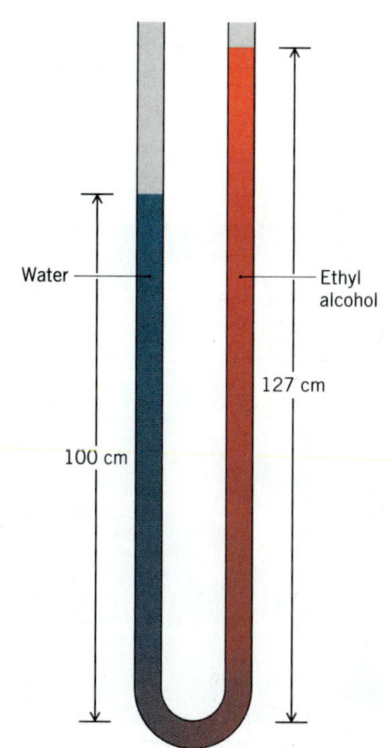


**Figure 4.4** Because of the difference between the densities of water and ethyl alcohol, the weight of a column of water 100-cm long balances the weight of a column of ethyl alcohol 127-cm long in the other arm of a U-tube.

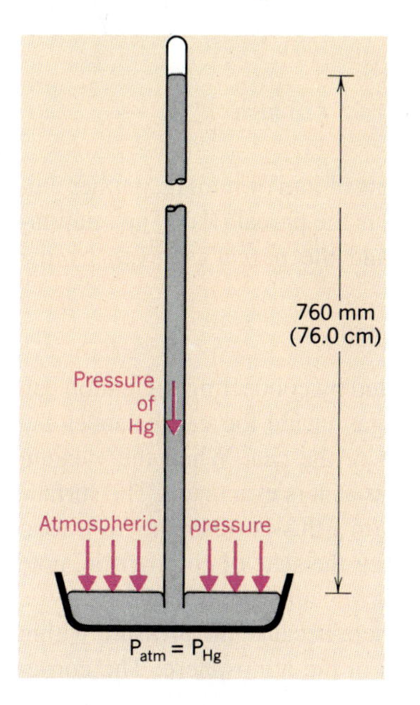


**Figure 4.5** On a sunny day, at sea level, the weight of a 760-mm column of mercury inside a glass tube balances the weight of the atmosphere pushing down on the pool of mercury that surrounds the tube. The pressure of the atmosphere is therefore said to be equivalent to 760 mmHg.

### CHECKPOINT

Use the results of the demonstration in Figure 4.4 to explain how a Torricelli barometer reflects atmospheric pressure.

Torricelli predicted that the height of the mercury column would change from day to day as the pressure of the atmosphere changed. Today, his apparatus is known as a *barometer,* from the Greek *baros,* meaning "weight," because it literally measures the weight of the atmosphere. Repeated experiments showed that the average pressure of the atmosphere at sea level is equal to the pressure of a column of mercury 760 mm tall. Thus, a standard unit of pressure known as the *atmosphere* was defined as follows.

$$1 \text{ atm} = 760 \text{ mmHg}$$

To recognize Torricelli's contributions, some scientists describe pressure in units of "torr," which are defined as follows:

$$1 \text{ torr} = 1 \text{ mmHg}$$

**CHECKPOINT**

What would happen to the height of the mercury in the barometer in Figure 4.5 if the tube was tilted 30° away from the vertical?

The weight of the atmosphere can be calculated from measurements of atmospheric pressure and the surface area of the planet.

**EXERCISE 4.2**

On the day this exercise was written, atmospheric pressure was 745.8 mmHg. Calculate the pressure in units of atmospheres.

**SOLUTION** The conversion between mmHg and atmospheres is based on the following definition:

$$1 \text{ atm} = 760 \text{ mmHg}$$

Using this equality to generate an appropriate unit factor gives the following result:

$$745.8 \cancel{\text{mmHg}} \times \frac{1 \text{ atm}}{760 \cancel{\text{mmHg}}} = 0.9813 \text{ atm}$$

Although chemists still work with pressures in units of atm or mmHg, neither unit is accepted in the SI system. The SI unit of pressure is the pascal (Pa). The relationship between one standard atmosphere pressure and the pascal is given by the following equalities:

$$1 \text{ atm} = 101{,}325 \text{ Pa} = 101.325 \text{ kPa}$$

The pressure of the atmosphere can be demonstrated by connecting a 1-gallon can to a vacuum pump. Normally the pressure of the gas inside the can balances the pressure of the atmosphere pushing on the outside of the can. When the vacuum pump is turned on, however, the can rapidly collapses as it is evacuated. The surface area of a 1-gallon can is about 250 $\text{in}^2$. At 14.7 $\text{lb/in}^2$, this corresponds to a total force over the surface of the can of about 3700 lb. For the sake of comparison, note that each of the 18 wheels of a 70,000-lb truck carries only about 3900 lb.

We usually don't feel the pressure of the atmosphere because the pressure within our bodies counterbalances the pressure of the gas in the atmosphere. The conse-

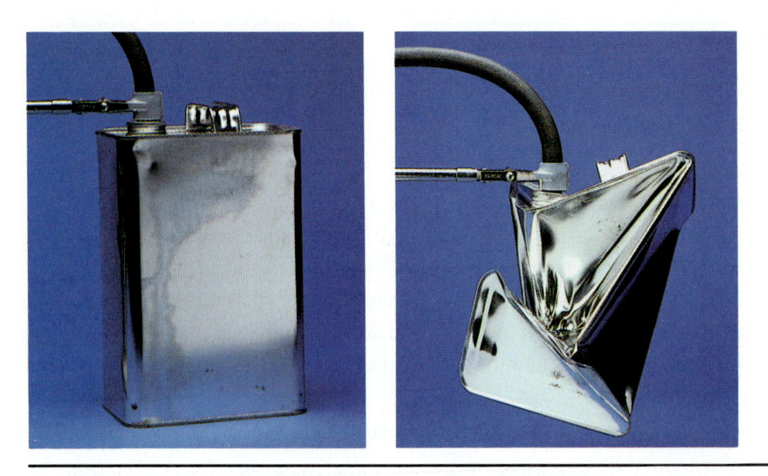

The pressure of the atmosphere can be demonstrated by connecting an empty paint-thinner can to a vacuum pump. Within seconds of turning on the vacuum pump, the can collapses.

quences of this inner pressure have been shown quite graphically in several movies. The puncture of a space suit in the vacuum of outer space immediately leads to the rupture of the body because there is nothing outside to balance the body's inner pressure.

### THE DIFFERENCE BETWEEN PRESSURE OF A GAS AND PRESSURE DUE TO WEIGHT

There is an important difference between the pressure of a gas and the other examples of pressure discussed in this section. The pressure exerted by a woman in high heels or a 70,000-pound truck is directional. A truck, for example, exerts all of its pressure on the surface beneath its wheels. In contrast, gas pressure is the same in all directions. To demonstrate this, we can fill a glass cylinder with water and rest a glass plate on top of the cylinder. When we turn the cylinder over, the plate doesn't fall to the floor because the pressure of the air outside the cylinder pushing up on the bottom of the plate is larger than the pressure exerted by the water in the cylinder pushing down on the plate. It would take a column of water 33.9 feet tall to produce a pressure equivalent to the pressure of the gas in the atmosphere.

**CHECKPOINT**

Use the densities of water (1.00 g/cm$^3$) and mercury (13.6 g/cm$^3$) to explain why atmospheric pressure can balance a column of mercury 760 mm tall or a column of water 33.9 feet tall.

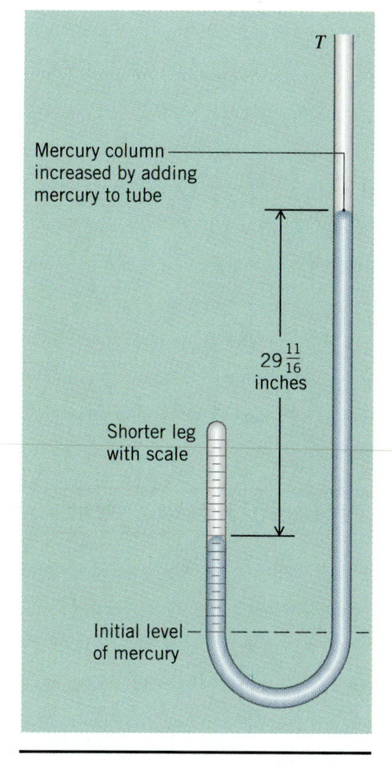


**Figure 4.6** Boyle's law is based on data obtained with a J-tube apparatus such as this.

## 4.6 BOYLE'S LAW

Torricelli's experiment did more than just show that air has weight; it also provided a way of creating a vacuum because the space above the column of mercury at the top of Figure 4.5 is almost completely empty. (It is free of air or other gases except a negligible amount of mercury vapor.) Torricelli's work with a vacuum soon caught the attention of British scientist Robert Boyle.

Boyle's most famous experiments with gases dealt with what he called the "spring of air." These experiments were based on the observation that gases are *elastic.* (They return to their original size and shape after being stretched or squeezed.) Boyle studied the elasticity of gases in a J-tube similar to the apparatus shown in Figure 4.6. By adding mercury to the open end of the tube, he trapped a small volume of air in the sealed end.

Boyle studied what happened to the volume of the gas in the sealed end of the tube as he added mercury to the open end. Table 4.2 contains some of the experimental data he reported in his book, *New Experiments Physico-Mechanicall, Touching the Spring of Air, and its Effects . . . ,* published in 1662. The first column is the volume of the gas in the sealed end of the J-tube, in arbitrary units. The second column is the difference between the height of the mercury in the sealed and open arms of the J-tube, to the nearest $\pm\frac{1}{16}$ inch. The third column is the product of the volume of the gas times the pressure.

Boyle noticed that the product of the pressure times the volume for any measurement in this table was equal to the product of the pressure times the volume for any other measurement, within experimental error:

$$P_1V_1 = P_2V_2$$

**TABLE 4.2 Boyle's Data**

| Volume | Pressure | P × V |
|---|---|---|
| 48 | $29\frac{2}{16}$ | 1398 |
| 46 | $30\frac{9}{16}$ | 1406 |
| 44 | $31\frac{15}{16}$ | 1405 |
| 42 | $33\frac{8}{16}$ | 1407 |
| 40 | $35\frac{5}{16}$ | 1413 |
| 38 | 37 | 1406 |
| 36 | $39\frac{4}{16}$ | 1413 |
| 34 | $41\frac{10}{16}$ | 1415 |
| 32 | $44\frac{3}{16}$ | 1414 |
| 30 | $47\frac{1}{16}$ | 1412 |
| 28 | $50\frac{5}{16}$ | 1409 |
| 26 | $54\frac{5}{16}$ | 1412 |
| 24 | $58\frac{13}{16}$ | 1412 |
| 22 | $64\frac{1}{16}$ | 1409 |
| 20 | $70\frac{11}{16}$ | 1414 |
| 18 | $77\frac{14}{16}$ | 1402 |
| 16 | $87\frac{14}{16}$ | 1406 |
| 14 | $100\frac{7}{16}$ | 1406 |
| 12 | $117\frac{9}{16}$ | 1411 |

This expression, or its equivalent,

$$P \propto \frac{1}{V}$$

is now known as **Boyle's Law.**

**EXERCISE 4.3**

Calculate the pressure in atmospheres in a motorcycle engine at the end of the compression stroke. Assume that at the start of this stroke, the pressure of the mixture of gasoline and air in the cylinder is 745.8 mmHg and the volume of each cylinder is 246.8 mL. Assume that the volume of the cylinder is 24.2 mL at the end of the compression stroke.

**Problem-Solving Strategy**

**SOLUTION** The most important step in this problem is recognizing what we know and what we don't know. We know the pressure and the volume at the start of the compression stroke and the volume at the end of the stroke:

| *Before compression* | *After compression* |
|---|---|
| $P_1 = 745.8$ mmHg | $P_2 = ?$ |
| $V_1 = 246.8$ mL | $V_2 = 24.2$ mL |

According to Boyle's law, the product of the pressure times the volume at the start of the compression stroke is equal to the product of the pressure times the volume at the end of this stroke:

$$P_1V_1 = P_2V_2$$

Before we substitute the known information into this equation, we have to make sure that the initial and final pressures are expressed in a consistent set of units:

$$745.8 \text{ mmHg} \times \frac{1 \text{ atm}}{760 \text{ mmHg}} = 0.9813 \text{ atm}$$

We can now substitute the known information into Boyle's law:

$$(0.9813 \text{ atm})(246.8 \text{ mL}) = (P_2)(24.2 \text{ mL})$$

We then rearrange this equation and solve for the unknown pressure:

$$P_2 = \frac{(0.9813 \text{ atm})(246.8 \text{ mL})}{(24.2 \text{ mL})} = \mathbf{10.0 \text{ atm}}$$

## 4.7 AMONTONS' LAW

Toward the end of the 1600s, French physicist Guillaume Amontons built a thermometer based on the fact that the pressure of a gas is directly proportional to its temperature. The relationship between the pressure and the temperature of a gas is therefore known as **Amontons' law:**

$$P \propto T$$

Amontons' law explains why car manufacturers recommend adjusting the pressure of your car's tires before you start on a trip. The flexing of the tire as you drive inevitably raises the temperature of the air in the tire. When this happens, the pressure of the gas inside the tires increases.

Amontons' law can be demonstrated with the apparatus shown in Figure 4.7. Data obtained with this apparatus at various temperatures are given in Table 4.3.

**Figure 4.7** The apparatus for demonstrating Amontons' law consists of a pressure gauge connected to a metal sphere of constant volume, which is immersed in solutions that have different temperatures.

In 1779 Joseph Lambert proposed a definition for absolute zero on the temperature scale that was based on the straight-line relationship between the temperature and pressure of a gas shown in Figure 4.8. He defined **absolute zero** as the temperature

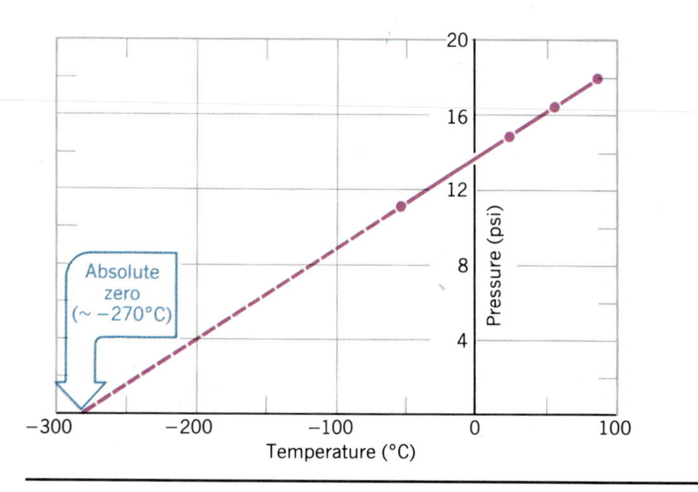


**Figure 4.8** When data obtained with the Amontons' law apparatus in Figure 4.7 are extrapolated, the pressure of the gas approaches zero when the temperature of the gas is approximately −270°C.

at which the pressure of a gas becomes zero when a plot of pressure versus temperature for a gas is extrapolated. According to the data in Table 4.3, the pressure of a gas approaches zero when the temperature is about −270°C. When more accurate measurements are made, the pressure of a gas extrapolates to zero when the temperature is −273.15°C. Absolute zero on the Celsius scale is therefore −273.15°C.

Sec. 1.15

The relationship between the temperature and pressure data in Table 4.3 can be greatly simplified by converting the temperatures from the Celsius to the Kelvin scale:

$$T_K = T_{°C} + 273.15$$

When this is done, a plot of the temperature versus the pressure of a gas gives a straight line that passes through the origin. Any two points along the line therefore fit the following equation:

$$\frac{P_1}{P_2} = \frac{T_1}{T_2}$$

It is important to remember that this equation is only valid if the temperatures are converted from the Celsius to the Kelvin scale before calculations are done.

**TABLE 4.3 The Dependence of the Pressure of a Gas on Its Temperature**

| Temperature (°C) | Pressure (lb/in$^2$) |
|---|---|
| 100 | 18.1 |
| 74 | 16.7 |
| 24 | 14.5 |
| 0 | 13.2 |
| −47 | 10.8 |

### EXERCISE 4.4

Assume that the pressure in the tires of your car is 32 lb/in$^2$ at 20°C. What is the pressure when the gas in these tires heats up to a temperature of 40°C?

**SOLUTION** We start by listing what we know and what we don't about this problem:

| *Initial conditions* | *Final conditions* |
|---|---|
| $P_1 = 32\ \text{lb/in}^2$ | $P_2 = ?$ |
| $T_1 = 20°\text{C}$ | $T_2 = 40°\text{C}$ |

This information suggests that the problem involves Amontons' law. We therefore start with one of the equations for this law:

$$\frac{P_1}{P_2} = \frac{T_1}{T_2}$$

Before we can use this equation, we have to convert the temperatures from °C to K:

| *Initial conditions* | *Final conditions* |
|---|---|
| $P_1 = 32\ \text{lb/in}^2$ | $P_2 = ?$ |
| $\mathbf{T_1 = 293\ K}$ | $\mathbf{T_2 = 313\ K}$ |

We then substitute the known information into the equation:

$$\frac{(32\ \text{lb/in}^2)}{P_2} = \frac{(293\ \text{K})}{(313\ \text{K})}$$

and solve the equation for the unknown pressure:

$$P_2 = \frac{(32\ \text{lb/in}^2)(313\ \cancel{\text{K}})}{(293\ \cancel{\text{K}})} = \mathbf{34\ lb/in^2}$$

This answer is consistent with Amontons' law, which suggests that the pressure should increase as the tires become warmer.

**CHECKPOINT**

Describe how to use Amontons' law to build a thermometer. Include a description of how the thermometer could be calibrated.

## 4.8 CHARLES' LAW

On June 5, 1783, Joseph and Étienne Montgolfier used a fire to inflate a spherical balloon about 30 feet in diameter that traveled about a mile and one-half before it came back to earth. News of this remarkable achievement spread throughout France, and Jacques-Alexandre-César Charles immediately tried to duplicate this performance. As a result of his work with balloons, Charles noticed that the volume of a gas is directly proportional to its temperature:

$$V \propto T$$

This relationship between the temperature and volume of a gas, which became known as **Charles' law,** provides an explanation of how hot-air balloons work. Ever

An etching of the first Montgolfier ascent at Versailles on September 19, 1783.

since the third century B.C., it has been known that an object floats when it weighs less than the fluid it displaces. If a gas expands when heated, then a given weight of hot air occupies a larger volume than the same weight of cold air. The weight of the gas in the balloon therefore decreases when it is heated. Once the air in a balloon gets hot enough, the net weight of the balloon plus this hot air is less than the weight of an equivalent volume of cold air, and the balloon starts to rise.

Charles' law can be demonstrated with the apparatus shown in Figure 4.9. A 30-mL syringe and a thermometer are inserted through a rubber stopper into a flask that has been cooled to 0°C. The ice bath is then removed and the flask is immersed in a warm-water bath. The gas in the flask expands as it warms, slowly pushing the piston out of the syringe. The total volume of the gas in the system is equal to the volume of the flask plus the volume of the syringe. Table 4.4 contains typical data obtained with this apparatus.

Figure 4.10 shows a plot of the data in Table 4.4. This graph provides us with another way of defining absolute zero on the temperature scale. **Absolute zero** is the

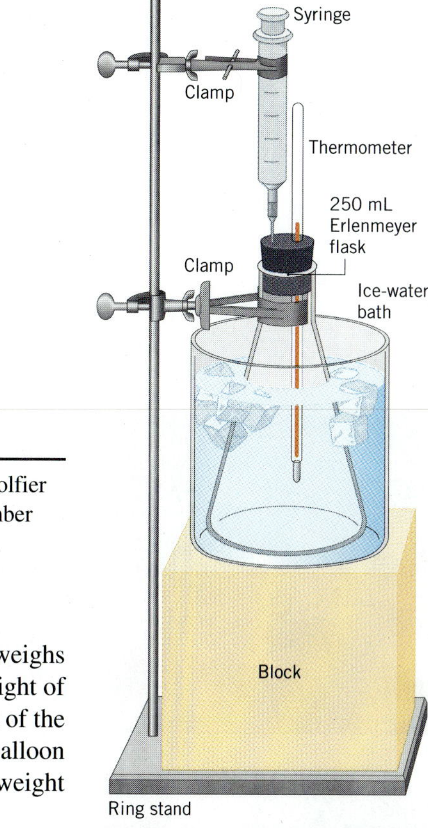


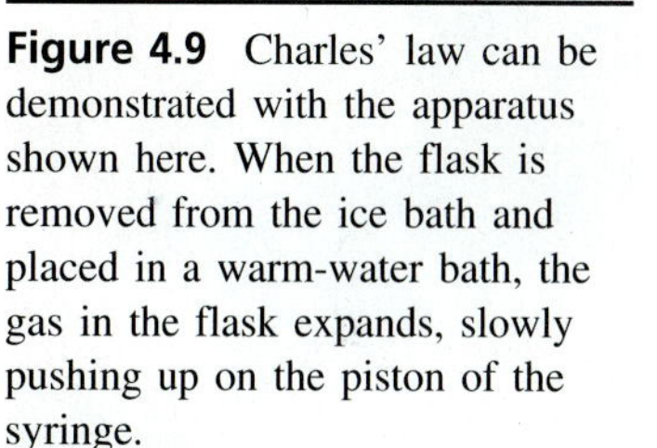

**Figure 4.9** Charles' law can be demonstrated with the apparatus shown here. When the flask is removed from the ice bath and placed in a warm-water bath, the gas in the flask expands, slowly pushing up on the piston of the syringe.

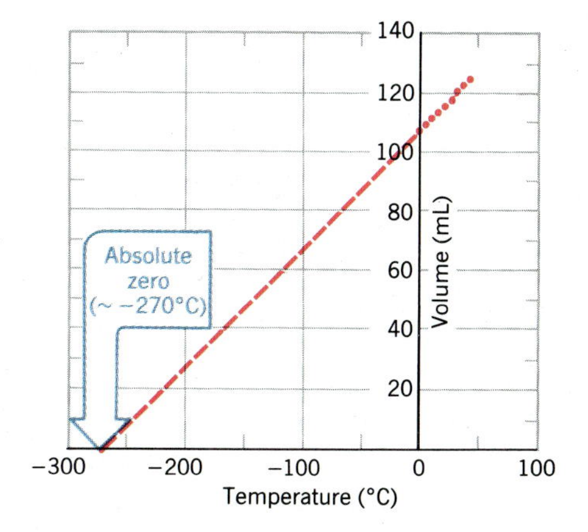


**Figure 4.10** When data obtained with the Charles' law apparatus in Figure 4.9 are extrapolated, the volume of the gas approaches zero when the temperature of the gas is approximately −270°C.

**TABLE 4.4 The Dependence of the Volume of a Gas on Its Temperature**

| Temperature (°C) | Volume (mL) |
|---|---|
| 0 | 107.9 |
| 5 | 109.7 |
| 10 | 111.7 |
| 15 | 113.6 |
| 20 | 115.5 |
| 25 | 117.5 |
| 30 | 119.4 |
| 35 | 121.3 |
| 40 | 123.2 |

temperature at which the volume of a gas becomes zero when the plot of the volume versus temperature for a gas are extrapolated. As expected, the value of absolute zero obtained by extrapolating the data in Table 4.4 is essentially the same as the value obtained from the graph of pressure versus temperature in the preceding section. Absolute zero therefore can be more accurately defined as the temperature at which the pressure and the volume of a gas extrapolate to zero.

When the temperatures in Table 4.4 are converted from the Celsius to the Kelvin scale, a plot of the volume versus the temperature of a gas becomes a straight line that passes through the origin. Any two points along this line therefore can be used to construct the following equation, which is known as **Charles' law:**

$$\frac{V_1}{V_2} = \frac{T_1}{T_2}$$

Because most of the volume of a gas is empty space, a balloon filled with $H_2$ collapses when immersed in liquid nitrogen at a temperature of −195.8°C.

Before you use this equation, it is important to remember to convert temperatures from °C to K.

## EXERCISE 4.5

Assume that the volume of a balloon filled with $H_2$ is 1.00 L at 25°C. Calculate the volume of the balloon when it is cooled to −78°C in a low-temperature bath made by adding dry ice to acetone.

**SOLUTION** We start, once again, by listing what we know and what we don't know:

| *Initial conditions* | *Final conditions* |
|---|---|
| $V_1 = 1.00$ L | $V_2 = ?$ |
| $T_1 = 25$°C | $T_2 = -78$°C |

This leads us to conclude that the problem involves Charles' law:

$$\frac{V_1}{V_2} = \frac{T_1}{T_2}$$

Before we use this equation, we have to convert the temperatures from °C to K:

| *Initial conditions* | *Final conditions* |
|---|---|
| $V_1 = 1.00$ L | $V_2 = ?$ |
| $T_1 = 298$ K | $T_2 = 195$ K |

We can now substitute this information into the Charles' law equation:

$$\frac{(1.00 \text{ L})}{V_2} = \frac{(298 \text{ K})}{(195 \text{ K})}$$

and then solve for the unknown volume:

$$V_2 = \frac{(1.00 \text{ L})(195 \cancel{\text{K}})}{(298 \cancel{\text{K}})} = \mathbf{0.654 \text{ L}}$$

According to this calculation, the volume of the balloon will shrink by about 35%. This is consistent with what we would expect from Charles' law; the volume of the gas should decrease as the gas is cooled.

## 4.9 GAY-LUSSAC'S LAW

Joseph Louis Gay-Lussac (1778–1850) began his career in 1801 by very carefully showing the validity of Charles' law for a number of different gases. Gay-Lussac's most important contributions to the study of gases, however, were experiments he performed on the ratio of the volumes of gases involved in a chemical reaction.

Gay-Lussac was interested in the reaction between hydrogen and oxygen to form water. He argued that measurements of the *weights* of hydrogen and oxygen consumed in this reaction could be influenced by the moisture present in the reaction flask, but this moisture would not affect the *volumes* of hydrogen and oxygen gases consumed in the reaction.

Much to his surprise, Gay-Lussac found that 199.89 parts by volume of hydrogen were consumed for every 100 parts by volume of oxygen. Thus, hydrogen and oxygen seemed to combine in a simple 2:1 ratio by volume:

$$\underset{\text{2 volumes}}{\text{hydrogen}} + \underset{\text{1 volume}}{\text{oxygen}} \longrightarrow \text{water}$$

Gay-Lussac found similar whole-number ratios for the reactions between other pairs of gases. The compound we now know as hydrogen chloride (HCl) combined with ammonia ($NH_3$) in a simple 1:1 ratio by volume:

$$\underset{\text{1 volume}}{\text{hydrogen chloride}} + \underset{\text{1 volume}}{\text{ammonia}} \longrightarrow \text{ammonium chloride}$$

Carbon monoxide combined with oxygen in a 2:1 ratio by volume:

$$\underset{\text{2 volumes}}{\text{carbon monoxide}} + \underset{\text{1 volume}}{\text{oxygen}} \longrightarrow \text{carbon dioxide}$$

Gay-Lussac obtained similar results when he analyzed the volumes of gases given off when compounds decomposed. Ammonia, for example, decomposes to give three times as much hydrogen by volume as nitrogen:

$$\text{ammonia} \longrightarrow \underset{\text{1 volume}}{\text{nitrogen}} + \underset{\text{3 volumes}}{\text{hydrogen}}$$

On December 31, 1808, Gay-Lussac announced his **law of combining volumes** to a meeting of the Societé Philomatique in Paris. Today, **Gay-Lussac's law** is stated as follows:

**The ratio of the volumes of gases consumed or produced in a chemical reaction is equal to the ratio of simple whole numbers.**

**EXERCISE 4.6**

Use the following balanced chemical equations to explain the results of Gay-Lussac's experiments:

$$2\ H_2(g) + O_2(g) \longrightarrow 2\ H_2O(g)$$
$$HCl(g) + NH_3(g) \longrightarrow NH_4Cl(s)$$
$$2\ CO(g) + O_2(g) \longrightarrow 2\ CO_2(g)$$
$$2\ NH_3(g) \longrightarrow N_2(g) + 3\ H_2(g)$$

**SOLUTION** In each case, the ratio of the number of moles of gases consumed or produced in this reaction is the same as the ratio of the volumes of gases involved in the reaction. For example, Gay-Lussac found a 2:1 ratio by volume of the hydrogen and oxygen that react to form water, which is consistent with the 2:1 ratio of moles of $H_2$ and $O_2$ in the balanced equation for this reaction.

**EXERCISE 4.7**

The invigorating odor that accompanies summer thunderstorms is due to the formation of trace quantities of ozone ($O_3$) when lightning passes through the atmosphere. Calculate the volume of oxygen formed when 1.00 L of ozone decomposes back to oxygen according to the following equation:

$$2\,O_3(g) \longrightarrow 3\,O_2(g)$$

$O_2$ reacts to form traces of $O_3$ in the presence of a stroke of lightning.

**Problem-Solving Strategy**

**SOLUTION** The key to this problem is remembering that equal volumes of different gases at the same temperature and pressure contain the same number of particles. This means that we can translate the balanced equation into a relationship between the volumes of gases involved in the reaction. The balanced equation suggests that we get 3 moles of $O_2$ for every 2 moles of $O_3$ that decompose. In other words, 1.5 times as many moles of $O_2$ are produced in this reaction as moles of $O_3$ consumed. This means that 1.50 liters of $O_2$ will be produced in this reaction for every 1.00 liter of $O_3$ consumed.

## 4.10 AVOGADRO'S HYPOTHESIS

Gay-Lussac's law of combining volumes was announced ony a few years after John Dalton proposed his atomic theory. The link between these two ideas was first recognized by Italian physicist Amadeo Avogadro three years later, in 1811. Avogadro argued that Gay-Lussac's law of combining volumes could be explained by assuming that equal volumes of different gases collected under similar conditions contain the same number of particles.

HCl and $NH_3$ therefore combine in a 1:1 ratio by volume because one molecule of HCl is consumed for every molecule of $NH_3$ in this reaction and equal volumes of these gases contain the same number of molecules:

$$NH_3(g) + HCl(g) \longrightarrow NH_4Cl(s)$$

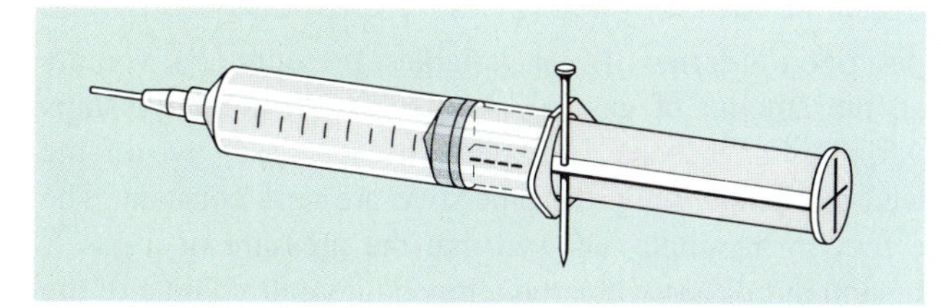

**Figure 4.11** The apparatus used to demonstrate Avogadro's hypothesis.

Anyone who has blown up a balloon will have noticed that the volume of a gas is proportional to the number of particles in the gas:

$$V \propto n$$

The more air you add to a balloon, the bigger it gets. However, this example does not test **Avogadro's hypothesis** that equal volumes of *different gases* contain the same number of particles. The best way to probe the validity of this hypothesis is to measure the number of molecules in a given volume of different gases, which can be done with the apparatus shown in Figure 4.11.

A small hole is drilled through the plunger of a 50-mL plastic syringe. The plunger is then pushed into the syringe and the syringe is sealed with a syringe cap. The plunger is then pulled out of the syringe until the volume reads 50 mL, and a nail is inserted through the hole in the plunger so that the plunger is not sucked back into the barrel of the syringe. The "empty" syringe is then weighed, the syringe is filled with 50 mL of a gas, and the syringe is reweighed. The difference between these measurements is the mass of 50 mL of the gas.

The results of experiments with six gases are given in Table 4.5. The number of molecules in a 50-mL sample of any one of these gases can be calculated from the mass of the sample, the molar mass of the gas, and the number of molecules in a mole. Consider the following calculation of the number of $H_2$ molecules in 50 mL of hydrogen gas, for example:

$$0.005 \cancel{\text{g H}_2} \times \frac{1 \cancel{\text{mol}}\ \text{H}_2}{2.02 \cancel{\text{g H}_2}} \times \frac{6.02 \times 10^{23}\ \text{molecules}}{1 \cancel{\text{mol}}} = \mathbf{1 \times 10^{21}\ H_2\ molecules}$$

The last column in Table 4.5 summarizes the results obtained when this calculation is repeated for each gas. The number of significant figures in the answer changes from one calculation to the next, but the number of molecules in each sample is the same, within experimental error. We therefore conclude that equal volumes of different gases collected under the same conditions of temperature and pressure do in fact contain the same number of particles.

**TABLE 4.5 Experimental Data for the Mass of 50-mL Samples of Different Gases**

| Compound | Mass of 50 mL of Gas (g) | Molar Mass of Gas (g/mol) | Number of Molecules |
|---|---|---|---|
| $H_2$ | 0.005 | 2.02 | $1 \times 10^{21}$ |
| $N_2$ | 0.055 | 28.01 | $1.2 \times 10^{21}$ |
| $O_2$ | 0.061 | 32.00 | $1.1 \times 10^{21}$ |
| $CO_2$ | 0.088 | 44.01 | $1.2 \times 10^{21}$ |
| $C_4H_{10}$ | 0.111 | 58.12 | $1.15 \times 10^{21}$ |
| $CCl_2F_2$ | 0.228 | 120.91 | $1.14 \times 10^{21}$ |

## 4.11 THE IDEAL GAS EQUATION

So far, gases have been described in terms of four variables: pressure ($P$), volume ($V$), temperature ($T$), and the amount of gas ($n$). And so far, five relationships between pairs of these variables have been discussed. In each case, two of the variables have been allowed to change while the other two are held constant. The discussion of the bicycle tire, for example, showed that the pressure of a gas is directly proportional to the amount of gas when the temperature and volume of the gas are held constant:

$$P \propto n \qquad (T \text{ and } V \text{ constant})$$

Other relationships between pairs of these variables include the following:

| | | |
|---|---|---|
| Boyle's law: | $P \propto 1/V$ | ($T$ and $n$ constant) |
| Amontons' law: | $P \propto T$ | ($V$ and $n$ constant) |
| Charles' law: | $V \propto T$ | ($P$ and $n$ constant) |
| Avogadro's hypothesis: | $V \propto n$ | ($P$ and $T$ constant) |

Each of these relationships is a special case of a more general relationship known as the **ideal gas equation:**

$$PV = nRT$$

In this equation, $R$ is a proportionality constant known as the *ideal gas constant* and $T$ is the absolute temperature. The value of $R$ depends on the units used to express the four variables $P$, $V$, $n$, and $T$. By convention, most chemists use the following set of units:

| | |
|---|---|
| $P$: atmospheres | $T$: kelvin |
| $V$: liters | $n$: moles |

### EXERCISE 4.8

Calculate the value of the ideal gas constant, $R$, if exactly 1 mole of an ideal gas occupies a volume of 22.414 liters at 0°C and 1 atmosphere pressure.

**SOLUTION** According to the ideal gas law, the product of the pressure times the volume of an ideal gas divided by the product of the amount of gas times the absolute temperature is a constant:

$$\frac{PV}{nT} = R$$

We can calculate the value of $R$ for any set of units of $P$, $V$, $n$, and $T$ by simply substituting the known values of these quantities into this equation:

$$\frac{(1.0000 \text{ atm})(22.414 \text{ L})}{(1.0000 \text{ mol})(273.15 \text{ K})} = \mathbf{0.082057 \text{ L-atm/mol-K}}$$

For most ideal gas calculations, four significant figures are sufficient. The standard value of the ideal gas constant is therefore 0.08206 L-atm/mol-K.

### ▶ CHECKPOINT

Calculate the value of the ideal gas constant in units of mL-torr/mol-K.

## 4.12 IDEAL GAS CALCULATIONS: PART I

The ideal gas equation can be used to predict the value of any one of the variables that describe a gas from known values of the other three.

Compressed gas cylinders are used to store gases at high pressures.

### EXERCISE 4.9

Many gases are available for use in the laboratory in compressed gas cylinders, in which they are stored at high pressures. Calculate the mass of $O_2$ that can be stored at 21°C and 170 atm in a cylinder with a volume of 60.0 L.

**SOLUTION** The key to this problem is recognizing that we know three of the four variables in the ideal gas equation:

$$P = 170 \text{ atm} \qquad V = 60.0 \text{ L} \qquad T = 21°\text{C}$$

This suggests that the ideal gas equation will play an important role in solving this problem:

$$PV = nRT$$

Before we can use this equation, we have to convert the temperature from °C to K:

$$T_K = T_{°C} + 273 = 294 \text{ K}$$

We can then substitute the known information into the ideal gas equation:

$$(170 \text{ atm})(60.0 \text{ L}) = (n)(0.08206 \text{ L-atm/mol-K})(294 \text{ K})$$

and solve this equation for the number of moles of gas in the container:

$$n = \frac{(170 \cancel{\text{atm}})(60.0 \cancel{\text{L}})}{(0.08206 \cancel{\text{L}}\text{-}\cancel{\text{atm}}\text{/mol-}\cancel{\text{K}})(294 \cancel{\text{K}})} = 422.8 \text{ mol}$$

We then use the mass of a mole of $O_2$ to calculate the number of grams of oxygen that can be stored in this cylinder:

$$422.8 \cancel{\text{mol O}_2} \times \frac{32.00 \text{ g O}_2}{1 \cancel{\text{mol O}_2}} = 1.35 \times 10^4 \text{ g O}_2$$

According to this calculation, 13.5 kilograms of $O_2$ can be stored in a 60-L compressed gas cylinder at 21°C and 170 atm.

The key to solving ideal gas problems often involves recognizing what is known and deciding how to use this information.

### EXERCISE 4.10

Calculate the mass of the air in a hot-air balloon that has a volume of $4.00 \times 10^5$ liters when the temperature of the gas is 30°C and the pressure is 748 mmHg. Assume that the average molar mass of air is 29.0 grams per mole.

**SOLUTION** In this problem, we know the pressure, volume, and temperature of a gas:

$$P = 748 \text{ mmHg} \qquad V = 4.00 \times 10^5 \text{ L} \qquad T = 30°\text{C}$$

We might therefore consider using the ideal gas equation to calculate the number of moles of gas in this balloon:

$$PV = nRT$$

Before we do this, however, we have to convert the pressure and temperature into appropriate units:

$$748 \text{ mmHg} \times \frac{1 \text{ atm}}{760 \text{ mmHg}} = 0.9842 \text{ atm}$$

$$T_K = T_{°C} + 273 = 303 \text{ K}$$

We can then substitute this information into the ideal gas equation and solve for the variable we don't know:

$$n = \frac{PV}{RT} = \frac{(0.9842 \text{ atm})(3.993 \times 10^5 \text{ L})}{(0.08206 \text{ L-atm/mol-K})(303 \text{ K})} = \mathbf{1.58 \times 10^4 \text{ mol}}$$

**Problem-Solving Strategy**

We then go back and reread the problem to see whether we are any closer to its solution.

The problem asks for the mass of the air in this balloon, which can be calculated from the number of moles of gas and the mass of a mole of air:

$$1.58 \times 10^4 \text{ mol} \times \frac{29.0 \text{ g}}{1 \text{ mol}} = \mathbf{4.58 \times 10^5 \text{ g}}$$

This balloon contains just over 1000 pounds of air.

The ideal gas equation can be applied even to problems that don't seem to ask for one of the variables in this equation.

## EXERCISE 4.11

Calculate the molecular weight of butane if 0.5813 gram of this gas fills a 250.0-mL flask at a temperature of 24.4°C and a pressure of 742.6 mmHg.

**SOLUTION** We know something about the pressure, volume, and temperature of the gas:

$$P = 742.6 \text{ mmHg} \qquad V = 250.0 \text{ mL} \qquad T = 24.4°\text{C}$$

We might therefore start by calculating the number of moles of gas in the sample. To do this, we need to convert the pressure, volume, and temperature into appropriate units:

$$742.6 \text{ mmHg} \times \frac{1 \text{ atm}}{760 \text{ mmHg}} = 0.9771 \text{ atm}$$

$$250.0 \text{ mL} \times \frac{1 \text{ L}}{1000 \text{ mL}} = 0.2500 \text{ L}$$

$$T_K = T_{°C} + 273.15 = 297.6 \text{ K}$$

Solving the ideal gas equation for the number of moles of gas and substituting the known values of the pressure, volume, and temperature into this equation gives the following result:

$$n = \frac{PV}{RT} = \frac{(0.9771 \text{ atm})(0.2500 \text{ L})}{(0.08206 \text{ L-atm/mol-K})(297.6 \text{ K})} = \mathbf{0.01000 \text{ mol}}$$

**Problem-Solving Strategy**

The key to solving this problem involves recognizing what we have achieved so far. The problem asks for the molecular weight of the gas. At this point, we know the mass of the gas in a sample and the number of moles of gas in the sample. We can therefore calculate the molecular weight of butane by dividing the mass of the sample by the number of moles of gas in the sample.

$$\frac{0.5813 \text{ g}}{0.01000 \text{ mol}} = \mathbf{58.13 \text{ g/mol}}$$

The molecular weight of butane is therefore 58.13 grams per mole.

The ideal gas equation can be used even to solve problems that don't seem to contain enough information.

## EXERCISE 4.12

Calculate the density in grams per liter of $O_2$ gas at 0°C and 1.00 atm.

**SOLUTION** This time we have information about only two of the variables in the ideal gas equation:

$$P = 1.00 \text{ atm} \qquad T = 0°\text{C}$$

What can we calculate from this information? One way to answer this question is to rearrange the ideal gas equation, putting the knowns on one side and the unknowns on the other:

**Problem-Solving Strategy**

$$\frac{n}{V} = \frac{P}{RT}$$

We now have enough information to calculate the number of moles of gas per liter, so let's do that:

$$\frac{n}{V} = \frac{P}{RT} = \frac{(1.00 \text{ atm})}{(0.08206 \text{ L-atm/mol-K})(273 \text{ K})} = 0.0446 \text{ mol } O_2/\text{L}$$

The problem asks for the density of the gas in grams per liter. Because we know the number of moles of $O_2$ per liter, we can use the mass of a mole of $O_2$ to calculate the number of grams per liter:

$$\frac{0.0446 \text{ mol } O_2}{1 \text{ L}} \times \frac{32.00 \text{ g}}{1 \text{ mol } O_2} = \mathbf{1.43 \text{ g/L}}$$

The density of $O_2$ gas at 0°C and 1.00 atm is therefore 1.43 grams per liter.

## EXERCISE 4.13

Explain why balloons filled with helium at room temperature (21°C) and 1 atm pressure float, whereas balloons filled with $CO_2$ sink.

**SOLUTION** The number of moles of gas per liter at 21°C and 1 atm does not depend on the identity of the gas:

$$\frac{n}{V} = \frac{P}{RT} = \frac{(1.00 \text{ atm})}{(0.08206 \text{ L-atm/mol-K})(294 \text{ K})} = 0.0415 \text{ mol/L}$$

Air saturated with water vapor.

Helium therefore must be less dense than air, whereas $CO_2$ is more dense than air:

$$d_{He} = \frac{0.0415 \text{ mol}}{1 \text{ L}} \times \frac{4.00 \text{ g He}}{1 \text{ mol}} = 0.166 \text{ g He/L}$$

$$d_{air} = \frac{0.0415 \text{ mol}}{1 \text{ L}} \times \frac{29.0 \text{ g air}}{1 \text{ mol}} = 1.20 \text{ g air/L}$$

$$d_{CO_2} = \frac{0.0415 \text{ mol}}{1 \text{ L}} \times \frac{44.0 \text{ g } CO_2}{1 \text{ mol}} = 1.83 \text{ g } CO_2\text{/L}$$

A balloon filled with helium therefore weighs significantly less than the air it displaces, and it floats. A balloon filled with $CO_2$, on the other hand, weighs more than the air it displaces, so it sinks.

**CHECKPOINT**

Use the difference between the average molar mass of air (29.0 g/mol) and the molar mass of water (18.0 g/mol) to explain why 1 L of dry air weighs more than 1 L of air that has been saturated with water vapor.

## 4.13 IDEAL GAS CALCULATIONS: PART II

Gas law problems often ask you to predict what happens when one or more changes are made in the variables that describe the gas. There are two ways of working these problems. One approach was used in Exercises 4.3 through 4.5. A more powerful approach is based on the fact that the ideal gas constant is in fact a constant.

We start by solving the ideal gas equation for the ideal gas constant:

$$R = \frac{PV}{nT}$$

We then note that the ratio of $PV/nT$ at any time must be equal to this ratio at any other time:

$$\frac{P_1V_1}{n_1T_1} = \frac{P_2V_2}{n_2T_2}$$

We then substitute the known values of pressure, temperature, volume, and amount of gas into this equation and solve for the appropriate unknown. This approach has two advantages. First, only one equation has to be remembered. Second, it can be used to handle problems in which more than one variable changes at a time.

**EXERCISE 4.14**

A sample of $NH_3$ gas fills a 27.0-L container at −15°C and 2.58 atm. Calculate the volume of this gas at 21°C and 751 mmHg.

**SOLUTION** Let's start by listing what we know and what we don't know about this system:

| *Initial conditions* | *Final conditions* |
|---|---|
| $P_1$ = 2.58 atm | $P_2$ = 751 mmHg |
| $V_1$ = 27.0 L | $V_2$ = ? |
| $T_1$ = −15°C | $T_2$ = 21°C |
| $n_1$ = ? | $n_2$ = ? |

Now we convert these data to a consistent set of units for use in the ideal gas equation:

| *Initial conditions* | *Final conditions* |
|---|---|
| $P_1 = 2.58$ atm | $P_2 = 0.988$ atm |
| $V_1 = 27.0$ L | $V_2 = ?$ |
| $T_1 = 258$ K | $T_2 = 294$ K |
| $n_1 = ?$ | $n_2 = ?$ |

We can then substitute this information into the following equation:

$$\frac{P_1V_1}{n_1T_1} = \frac{P_2V_2}{n_2T_2}$$

When this is done we get one equation with three unknowns: $n_1$, $n_2$, and $V_2$:

$$\frac{(2.58 \text{ atm})(27.0 \text{ L})}{(n_1)(258 \text{ K})} = \frac{(0.988 \text{ atm})(V_2)}{(n_2)(294 \text{ K})}$$

**Problem-Solving Strategy**

It is impossible to solve one equation with three unknowns. But a careful reading of the problem suggests that the number of moles of $NH_3$ is the same before and after the temperature and pressure change:

$$n_1 = n_2$$

Substituting this equality into the unknown equation gives the following:

$$\frac{(2.58 \text{ atm})(27.0 \text{ L})}{(n_1)(258 \text{ K})} = \frac{(0.988 \text{ atm})(V_2)}{(n_1)(294 \text{ K})}$$

Multiplying both sides of this equation by $n_1$ gives one equation with one unknown, which can be solved for that unknown:

$$\frac{(2.58 \text{ atm})(27.0 \text{ L})}{(258 \text{ K})} = \frac{(0.988 \text{ atm})(V_2)}{(294 \text{ K})}$$

$$V_2 = \frac{(2.58 \cancel{\text{atm}})(27.0 \text{ L})(294 \cancel{\text{K}})}{(258 \cancel{\text{K}})(0.988 \cancel{\text{atm}})} = \textbf{80.3 L}$$

According to this calculation, the volume of the gas increases from 27.0 L at −15°C and 2.58 atm to 80.3 L at 21°C and 751 mmHg.

## 4.14 DALTON'S LAW OF PARTIAL PRESSURES

The *CRC Handbook of Chemistry and Physics* describes the atmosphere as 78.084% $N_2$, 20.946% $O_2$, 0.934% Ar, and 0.033% $CO_2$ by volume when the water vapor has been removed. What image does this description evoke in your mind? Does this mean that only 20.463% of the room in which you are sitting contains $O_2$? Or is the atmosphere in your room a more or less homogeneous mixture of these gases?

Section 4.2 argued that gases expand to fill their containers. The volume of $O_2$ in your room is therefore the same as the volume of $N_2$. (Both gases expand to fill the room.) When we describe the atmosphere as 20.946% $O_2$ by volume, we mean that the volume of the atmosphere in a flexible container, such as a balloon, would shrink by 20.946% if the $O_2$ were removed.

What about the pressure of the different gases in your room? Is the pressure of the

$O_2$ in the atmosphere the same as the pressure of the $N_2$? We can answer this question by rearranging the ideal gas equation as follows:

$$P = n \times \frac{RT}{V}$$

According to this equation, the pressure of a gas is proportional to the number of moles of gas, if the temperature and volume are held constant. Because the temperature and volume of the $O_2$ and $N_2$ in the atmosphere are the same, the pressure of each gas must be proportional to the number of the moles of the gas. Because there is more $N_2$ in the atmosphere than $O_2$, the contribution to the total pressure of the atmosphere from $N_2$ is larger than the contribution from $O_2$.

John Dalton was the first to recognize that the total pressure of a mixture of gases is the sum of the contributions of the individual components of the mixture. By convention, the part of the total pressure of a mixture that results from one component is called the **partial pressure** of that component. **Dalton's law of partial pressures** states that the total pressure of a mixture of gases is the sum of the partial pressures of the various components:

$$\boldsymbol{P_T = P_1 + P_2 + P_3 + \cdots}$$

## EXERCISE 4.15

Calculate the total pressure of a mixture that contains 1.00 g of $H_2$ and 1.00 g of He in a 5.00-L container at 21°C.

**SOLUTION** We can start by listing what we know about the conditions in the problem:

$$V = 5.00 \text{ L} \qquad T = 21°\text{C}$$

We also know the masses of $H_2$ and He in the flask, however, so we can calculate the number of moles of each gas in the container:

$$1.00 \text{ g } H_2 \times \frac{1 \text{ mol } H_2}{2.016 \text{ g } H_2} = 0.496 \text{ mol } H_2$$

$$1.00 \text{ g He} \times \frac{1 \text{ mol He}}{4.003 \text{ g He}} = 0.250 \text{ mol He}$$

Because we know the volume, temperature, and number of moles of each component of the mixture, we can calculate the partial pressures of $H_2$ and He. We start by rearranging the ideal gas equation as follows:

$$P = \frac{nRT}{V}$$

We then use this equation to calculate the partial pressures:

$$P_{\text{hydrogen}} = \frac{(0.496 \text{ mol } H_2)(0.08206 \text{ L-atm/mol-K})(294 \text{ K})}{(5.00 \text{ L})} = 2.39 \text{ atm}$$

$$P_{\text{helium}} = \frac{(0.250 \text{ mol He})(0.08206 \text{ L-atm/mol-K})(294 \text{ K})}{(5.00 \text{ L})} = 1.21 \text{ atm}$$

The total pressure in this mixture is the sum of the partial pressures of the two components:

$$P_T = P_{\text{hydrogen}} + P_{\text{helium}} = \mathbf{3.60\ atm}$$

Dalton derived the law of partial pressures from his work on the amount of water vapor that could be absorbed by air at different temperatures. It is therefore fitting that this law is used most often to correct for the amount of water vapor picked up when a gas is collected by displacing water. Suppose, for example, that we want to collect a sample of $O_2$ prepared by heating potassium chlorate until it decomposes:

$$2\ KClO_3(s) \longrightarrow 2\ KCl(s) + 3\ O_2(g)$$

The gas given off in this reaction can be collected by filling a flask with water, inverting the flask in a trough, and then letting the gas bubble into the flask as shown in Figure 4.12.

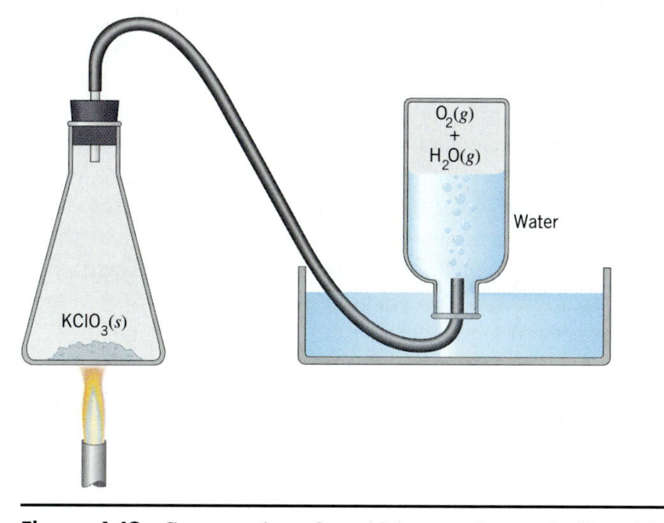


**Figure 4.12** Gases, such as $O_2$, which are only marginally soluble in water can be collected by displacing water from a container.

Because some of the water in the flask will evaporate during the experiment, the gas that collects in this flask is going to be a mixture of $O_2$ and water vapor. The total pressure of this gas is the sum of the partial pressures of these two components:

$$P_T = P_{\text{oxygen}} + P_{\text{water}}$$

The total pressure of this mixture must be equal to atmospheric pressure. (If it was any greater, the gas would push water out of the container. If it was any less, water would be forced into the container.) If we had some way to estimate the partial pressure of the water in this system, we could therefore calculate the partial pressure of the oxygen gas.

By convention, the partial pressure of the gas that collects in a closed container above a liquid is known as the **vapor pressure** of the liquid. Table A-4 in the appendix gives the vapor pressure of water at various temperatures. If we know the temperature at which a gas is collected by displacing water, and we assume that the gas is saturated with water vapor at this temperature, we can calculate the partial pressure of the gas by subtracting the vapor pressure of water from the total pressure of the mixture of gases collected in the experiment.

**RESEARCH IN THE 90s**

## GAS-PHASE ION CHEMISTRY

No single advance played as important a role in the birth of modern chemistry at the end of the eighteenth century as the development of techniques for studying gases. Now, almost 200 years later, students often leave the topic of gases with the belief that they are studying ancient history. To some extent, they are right. Discussions of the behavior of gases are designed to help students appreciate how scientists struggled to understand the behavior of something they could not quite ''see,'' even though they knew gases existed. History, however, is cyclic, and gas-phase chemistry has recently become one of the most powerful methods for obtaining the answers to sophisticated questions.

Professor Hilka Kenttämaa (right) and Brandon Beasley (left), an undergraduate research student, adjusting the FT-ICR spectrometer.

The advantage of gas-phase chemistry is the fact that most of the volume of a gas is empty space. Thus, we can study isolated particles that do not interact with their nearest neighbors. By adjusting the pressure of the gas, we can control the number of particles in a given volume and thereby control the average amount of time before one particle collides with another. We can therefore prepare a molecule in an excited state and study what happens as it decays.

John Farrell, Jr., an undergraduate working with Hilkka Kenttämaa, used these techniques to study radical cations of a family of compounds known as organophosphates [*Journal of the American Chemical Society,* **114,** 1205 (1992)]. Their work was done with an instrument known as a ion cyclotron resonance (ICR) spectrometer. The following describes several key characteristics of their experiment:

- The pressure inside the spectrometer before the sample was injected was less than $1 \times 10^{-9}$ mmHg. After the sample was injected, the pressure was approximately $1 \times 10^{-7}$ mmHg.
- The sample was ionized by bombarding it with high-energy electrons, which remove an electron from a neutral molecule to form a positive ion.
- The positive ions, or cations, formed in this reaction were trapped in a magnetic field, where they moved in a cyclic path—thus, the name ''ion cyclotron.''
- The researchers then transferred the ions into an analyzer cell, where these ions collided with neutral argon atoms over a period of about 0.1 second. These collisions caused the ions to dissociate into fragments. The partial pressures of the sample and argon gas ensured that an average of one to three collisions occurred between the cation and an argon atom before the products of these collisions were studied.
- Farrell and Kenttämaa captured the fragments that formed and analyzed them by measuring their mass with a mass spectrometer.

One set of experiments performed by Farrell and Kenttämaa studied the following organophosphate:

$$
\begin{array}{c}
\mathrm{O} \\
| \\
\mathrm{CH_3O{-}P{-}OCH_3} \\
| \\
\mathrm{OCH_3}
\end{array}
$$

When an electron is removed from this molecule, the result is a positively charged ion, or cation. As we will soon see, most of the electrons in a chemical compound exist as pairs of electrons. In this cation, one electron is unpaired. Compounds that contain an unpaired electron are called ''radicals'' because they are often highly reactive.

Farrell and Kenttämaa used the ICR technique to probe whether the unpaired electron and positive charge exists at the same point on the molecule. They found that they do not. The fragmentation products clearly suggest that the radical cation formed when this molecule is ionized rearranges to give the following product:

$$\begin{array}{c} OH \\ | \\ CH_3O\text{—}P^+\text{—}OCH_2\cdot \\ | \\ OCH_3 \end{array}$$

When the radical cation is formed, a hydrogen atom is transferred from one of the —$OCH_3$ groups to the oxygen atom on the phosphorus. The positive charge resides primarily on the phosphorus atom, and the unpaired electron resides on the $CH_2$ group formed when the hydrogen atom is removed.

## EXERCISE 4.16

Calculate the number of grams of $O_2$ that can be collected by displacing water from a 250-mL flask at 21°C and 746.2 mmHg.

**Problem-Solving Strategy**

**SOLUTION** It is often useful to work a problem backwards. In this case, our goal is the mass of $O_2$ in the flask. To reach this goal we need to calculate the number of moles of $O_2$ in the sample. To do this, we need to know the pressure, volume, and temperature of the $O_2$. We already know the volume of the $O_2$ because a gas expands to fill its container. We also know the temperature of the $O_2$. But we don't know the pressure of the $O_2$.

$$V_{oxygen} = 250 \text{ ml} \qquad T_{oxygen} = 21°C \qquad P_{oxygen} = ?$$

All we know is the total pressure of the oxygen plus the water vapor that collects in the flask:

$$P_T = P_{oxygen} + P_{water} = 746.2 \text{ mmHg}$$

Table A-4 gives the vapor pressure of water at 21°C as 18.7 mmHg. If we assume that the gas collected in this experiment is saturated with water vapor, the partial pressure of $O_2$ must be less than the total pressure in the flask by 18.7 mmHg:

$$P_{oxygen} = 746.2 \text{ mmHg} - 18.7 \text{ mmHg} = 727.5 \text{ mmHg}$$

We now know the pressure (727.5 mmHg), volume (250 mL), and temperature (21°C) of the $O_2$ collected in this experiment. Before we can calculate the number of moles of $O_2$, however, we have to convert these measurements to appropriate units. We can then use the ideal gas equation to calculate the amount of $O_2$ collected in this experiment:

$$n = \frac{PV}{RT} = \frac{(0.9572\ \cancel{\text{atm}})(0.250\ \cancel{\text{L}})}{(0.08206\ \cancel{\text{L}}\text{-}\cancel{\text{atm}}/\text{mol-}\cancel{\text{K}})(294\ \cancel{\text{K}})} = 0.009919 \text{ mol } O_2$$

From this information, we can calculate the number of grams of $O_2$ collected:

$$0.009919\ \cancel{\text{mol}}\ O_2 \times \frac{32.00 \text{ g } O_2}{1\ \cancel{\text{mol}}} = 0.317 \text{ g } O_2$$

## 4.15 THE KINETIC MOLECULAR THEORY

The experimental observations about the behavior of gases discussed so far can be explained with a simple theoretical model known as the **kinetic molecular theory.** This theory is based on the following postulates, or assumptions:

1. Gases are composed of a large number of particles that behave like hard, spherical objects in a state of constant, random motion.
2. These particles move in a straight line until they collide with another particle or the walls of the container.
3. These particles are much smaller than the distance between particles. Most of the volume of a gas is therefore empty space.
4. There is no force of attraction between gas particles or between the particles and the walls of the container.
5. Collisions between gas particles or collisions with the walls of the container are perfectly elastic. None of the energy of a gas particle is lost when it collides with another particle or with the walls of the container.
6. The average kinetic energy of a collection of gas particles depends on the temperature of the gas and nothing else.

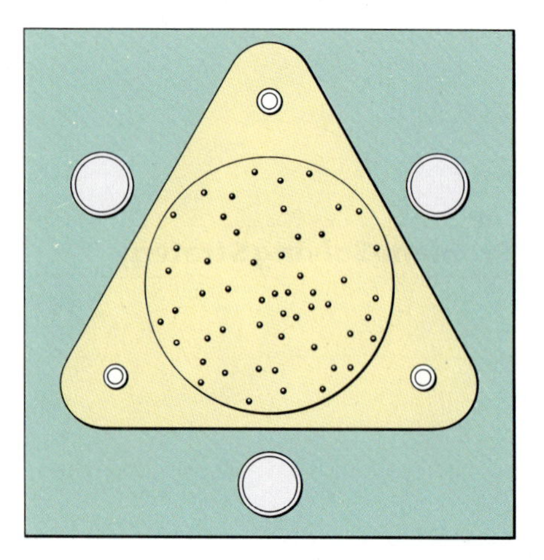

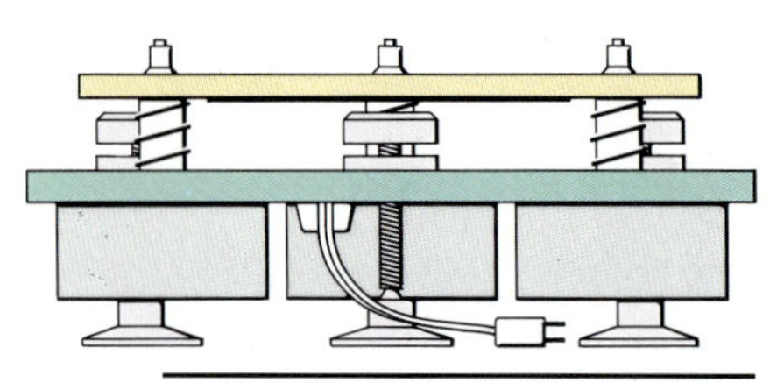

**Figure 4.13** The six postulates of the kinetic molecular theory can be demonstrated with the molecular dynamics simulator shown here.

The assumptions behind the kinetic molecular theory can be illustrated with the apparatus shown in Figure 4.13, which consists of a glass plate surrounded by walls mounted on top of three vibrating motors. A handful of steel ball bearings is placed on top of the glass plate to represent the gas particles.

When the motors are turned on, the glass plate vibrates, which makes the ball bearings move in a constant, random fashion (postulate 1). Each ball moves in a straight line until it collides with another ball or with the walls of the container (postulate 2). Although collisions are frequent, the average distance between the ball bearings is much larger than the diameter of the balls (postulate 3). There is no force of attraction between the individual ball bearings or between the ball bearings and the walls of the container (postulate 4).

The collisions that occur in this apparatus are very different from those that occur when a rubber ball is dropped on the floor. Collisions between the rubber ball and the floor are *inelastic,* as shown in Figure 4.14. A portion of the energy of the ball is lost each time it hits the floor, until it eventually rolls to a stop. In this apparatus, the collisions are perfectly *elastic.* The balls have just as much energy after a collision as before (postulate 5).

Any object in motion has a **kinetic energy** that is defined as one-half of the product of its mass times its velocity squared:

$$KE = \frac{1}{2} mv^2$$

At any time, some of the ball bearings on this apparatus are moving faster than others, but the system as a whole can be described by an *average kinetic energy.* When we increase the "temperature" of the system by increasing the voltage to the motors, we find that the average kinetic energy of the ball bearings increases (postulate 6).

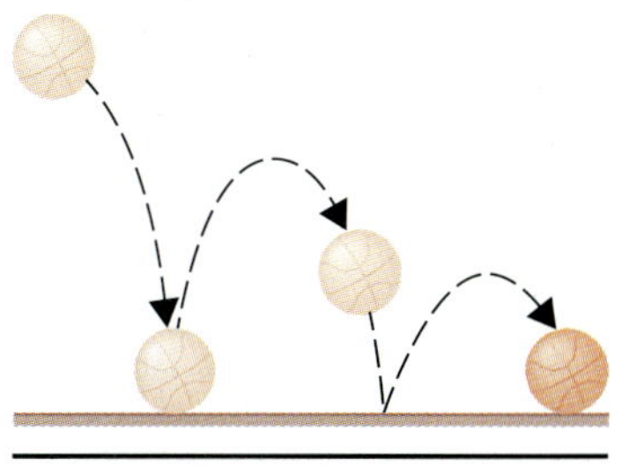

**Figure 4.14** Most collisions are inelastic—some energy is lost each time a ball collides with the floor, for example.

**CHECKPOINT**

The kinetic theory assumes that the temperature of a gas on the macroscopic scale is directly proportional to the average kinetic energy of the particles on the atomic scale. Use this assumption to explain what happens to the particles that form a gas when the temperature of a gas increases.

## 4.16 HOW THE KINETIC MOLECULAR THEORY EXPLAINS THE GAS LAWS

The kinetic molecular theory can be used to explain each of the experimentally determined gas laws.

### THE LINK BETWEEN *P* AND *n*

The pressure of a gas results from collisions between the gas particles and the walls of the container. Each time a gas particle hits the wall, it exerts a force on the wall. An increase in the number of gas particles in the container increases the frequency of collisions with the walls and therefore the pressure of the gas.

### AMONTONS' LAW ($P \propto T$)

The last postulate of the kinetic molecular theory states that the average kinetic energy of a gas particle depends only on the temperature of the gas. Thus, the average kinetic energy of the gas particles increases as the gas becomes warmer. Because the mass of these particles is constant, their kinetic energy can increase only if the average velocity of the particles increases. The faster these particles are moving when they hit the wall, the greater the force they exert on the wall. Since the force per collision becomes larger as the temperature increases, the pressure of the gas must increase as well.

### BOYLE'S LAW ($P = 1/V$)

Gases can be compressed because most of the volume of a gas is empty space. If we compress a gas without changing its temperature, the average kinetic energy of the gas particles stays the same. There is no change in the speed with which the particles move, but the container is smaller. Thus, the particles travel from one end of the container to the other in a shorter period of time. This means that they hit the walls more often. Any increase in the frequency of collisions with the walls must lead to an increase in the pressure of the gas. Thus, the pressure of a gas becomes larger as the volume of the gas becomes smaller.

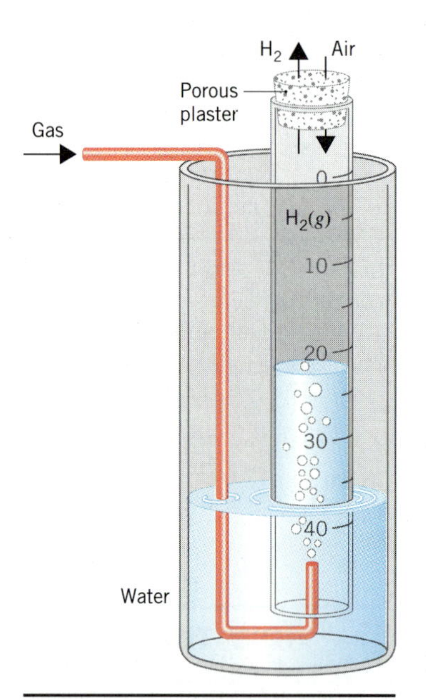


**Figure 4.15** The rate of diffusion, or mixing of a gas with air, can be studied with the apparatus shown here. If the gas escapes from the tube faster than air enters the tube, the amount of water in the tube will increase. If air enters the tube faster than the gas escapes, water will be displaced from the tube.

### CHARLES' LAW ($V \propto T$)

The average kinetic energy of the particles in a gas is proportional to the temperature of the gas. Because the mass of these particles is constant, the particles must move faster as the gas becomes warmer. If they move faster, the particles will exert a greater force on the container each time they hit the walls, which leads to an increase in the pressure of the gas. If the walls of the container are flexible, they will expand until the pressure of the gas once more balances the pressure of the atmosphere. The volume of the gas therefore becomes larger as the temperature of the gas increases.

### AVOGADRO'S HYPOTHESIS ($V \propto n$)

As the number of gas particles increases, the frequency of collisions with the walls of the container must increase. This, in turn, leads to an increase in the pressure of the gas. Flexible containers, such as a balloon, will expand until the pressure of the gas inside the balloon once again balances the pressure of the gas outside. Thus, the volume of the gas is proportional to the number of gas particles.

### DALTON'S LAW OF PARTIAL PRESSURES ($P_T = P_1 + P_2 + P_3 + \cdots$)

Imagine what would happen if six ball bearings of a different size were added to the apparatus in Figure 4.13. The total pressure would increase because there would be more collisions with the walls of the container. But the pressure due to the collisions between the original ball bearings and the walls of the container would remain the same. There is so much empty space in the container that each type of ball bearing hits the walls of the container as often in the mixture as it did when there was only one kind of ball bearing on the glass plate. The total number of collisions with the wall in this mixture is therefore equal to the sum of the collisions that would occur when each size of ball bearing is present by itself. In other words, the total pressure of a mixture of gases is equal to the sum of the partial pressures of the individual gases.

## 4.17 GRAHAM'S LAWS OF DIFFUSION AND EFFUSION

A few of the physical properties of gases depend on the identity of the gas. One of these physical properties can be seen when the movement of gases is studied.

In 1829 Thomas Graham used an apparatus similar to the one shown in Figure 4.15 to study the **diffusion** of gases—the rate at which two gases mix. This apparatus consists of a glass tube sealed at one end with plaster that has holes large enough to allow a gas to enter or leave the tube. When the tube is filled with $H_2$ gas, the level of water in the tube slowly rises because the $H_2$ molecules inside the tube escape through the holes in the plaster more rapidly than the molecules in air can enter the tube. By studying the rate at which the water level in this apparatus changed, Graham was able to obtain data on the rate at which different gases mixed with air.

Graham found that the rates at which gases diffuse is inversely proportional to the square root of their densities:

$$Rate_{\text{diffusion}} \propto \frac{1}{\sqrt{\text{density}}}$$

This relationship eventually became known as **Graham's law of diffusion.**

To understand the importance of this discovery we have to remember that equal volumes of different gases contain the same number of particles. As a result, the number of moles of gas per liter at a given temperature and pressure is constant, which means that the density of a gas is directly proportional to its molecular weight (MW). Graham's law of diffusion can therefore also be written as follows:

$$Rate_{\text{diffusion}} \propto \frac{1}{\sqrt{\text{MW}}}$$

Similar results were obtained when Graham studied the rate of **effusion** of a gas, which is the rate at which the gas escapes through a pinhole into a vacuum. The rate of effusion of a gas is also inversely proportional to the square root of either the density or the molecular weight of the gas:

$$Rate_{\text{effusion}} \propto \frac{1}{\sqrt{\text{density}}} \propto \frac{1}{\sqrt{\text{MW}}}$$

**Graham's law of effusion** can be demonstrated with the apparatus in Figure 4.16. A thick-walled filter flask is evacuated with a vacuum pump. A syringe is filled with 25 mL of gas and the time required for the gas to escape through the syringe needle into the evacuated filter flask is measured with a stop watch. The experimental data in Table 4.6 were obtained by using a special needle with a very small (0.015 cm) hole through which the gas could escape.

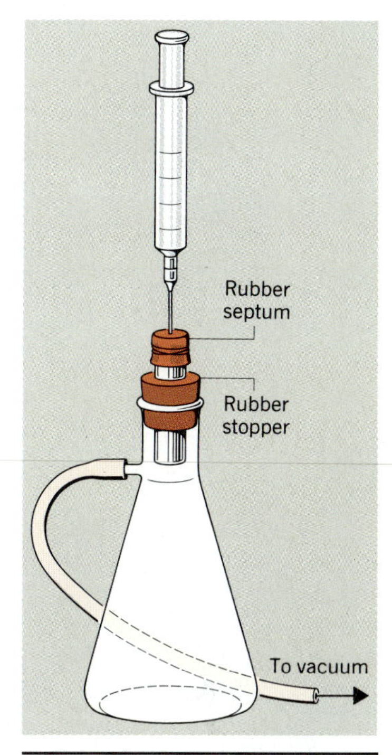


**Figure 4.16** The rate of effusion of a gas can be demonstrated with the apparatus shown here. The time required for the gas in the syringe to escape into an evacuated flask is measured. The faster the gas molecules effuse, the less time it takes for a given volume of the gas to escape into the flask.

**TABLE 4.6 The Time Required for 25-mL Samples of Different Gases to Escape through a 0.015-cm Hole into a Vacuum**

| Compound | Time (s) | Molar Mass (g/mol) |
|---|---|---|
| $H_2$ | 5.1 | 2.02 |
| He | 7.2 | 4.00 |
| $NH_3$ | 14.2 | 17.0 |
| air | 18.2 | 29.0 |
| $O_2$ | 19.2 | 32.0 |
| $CO_2$ | 22.5 | 44.0 |
| $SO_2$ | 27.4 | 64.1 |

As we can see when these data are graphed in Figure 4.17, the *time* required for 25-mL samples of different gases to escape into a vacuum is proportional to the square root of the molecular weight of the gas. The *rate* at which the gases effuse is therefore inversely proportional to the square root of the molecular weight. Graham's observations about the rate at which gases diffuse (mix) or effuse (escape through a pinhole) suggest that relatively light gas particles, such as $H_2$ molecules or He atoms move faster than relatively heavy gas particles, such as $CO_2$ or $SO_2$ molecules.

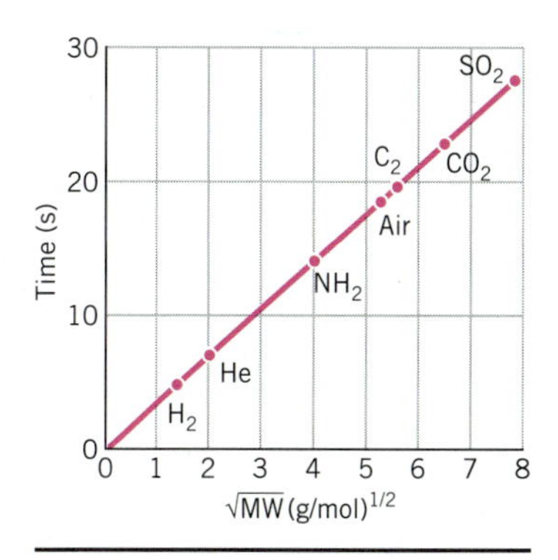


**Figure 4.17** A graph of the time required for 25-mL samples of different gases to escape into an evacuated flask versus the square root of the molecular weight of the gas. Relatively heavy molecules move more slowly, and it takes more time for the gas to escape.

## THE KINETIC MOLECULAR THEORY AND GRAHAM'S LAWS

The kinetic molecular theory can be used to explain the results Graham obtained when he studied the diffusion and effusion of gases. The key to this explanation is the last postulate of this theory, which assumes that the temperature of a system is proportional to the average kinetic energy of its particles and nothing else. In other words, the temperature of a system increases if and only if there is an increase in the average kinetic energy of its particles.

Two gases, such as $H_2$ and $O_2$, at the same temperature, therefore, must have the same average kinetic energy. This can be represented by the following equation:

$$\frac{1}{2}\, m_{\mathrm{H}} v_{\mathrm{H}}^2 = \frac{1}{2}\, m_{\mathrm{O}} v_{\mathrm{O}}^2$$

This equation can be simplified by multiplying both sides by two:

$$m_{\mathrm{H}} v_{\mathrm{H}}^2 = m_{\mathrm{O}} v_{\mathrm{O}}^2$$

It can then be rearranged to give the following:

$$\frac{v_{\mathrm{H}}^2}{v_{\mathrm{O}}^2} = \frac{m_{\mathrm{O}}}{m_{\mathrm{H}}}$$

Taking the square root of both sides of this equation gives a relationship between the ratio of the velocities at which the two gases move and the square root of the ratio of their molecular weights:

$$\boldsymbol{\frac{v_{\mathrm{H}}}{v_{\mathrm{O}}} = \sqrt{\frac{m_{\mathrm{O}}}{m_{\mathrm{H}}}}}$$

This equation is a modified form of Graham's law. It suggests that the velocity (or rate) at which gas molecules move is inversely proportional to the square root of their molecular weights.

## EXERCISE 4.17

Calculate the average velocity of an $H_2$ molecule at 0°C if the average velocity of an $O_2$ molecule at this temperature is 500 m/s.

**SOLUTION** The relative velocities of the $H_2$ and $O_2$ molecules at a given temperature are described by the following equation:

$$\frac{v_{\mathrm{H}}}{v_{\mathrm{O}}} = \sqrt{\frac{m_{\mathrm{O}}}{m_{\mathrm{H}}}}$$

Substituting the molecule weights of $H_2$ and $O_2$ and the average velocity of an $O_2$ molecule into this equation gives the following result:

$$\frac{v_{\mathrm{H}}}{500\ \mathrm{m/s}} = \sqrt{\frac{32.0\ \mathrm{g/mol}}{2.0\ \mathrm{g/mol}}}$$

Solving this equation for the average velocity of an $H_2$ molecule gives a value of 2000 meters per second or about 4500 miles per hour.

Because it is easy to make mistakes when setting up a ratio problem such as this, it is important to check the answer to see whether it makes sense. Graham's law suggests that light molecules move faster on the average than heavy molecules. In this case, the answer makes sense because $H_2$ molecules are much lighter than $O_2$ molecules and they should therefore travel much faster.

**SPECIAL TOPIC**

## DEVIATIONS FROM IDEAL GAS LAW BEHAVIOR: VAN DER WAALS EQUATION

The behavior of real gases usually agrees with the predictions of the ideal gas equation to within ±5% at normal temperatures and pressures. At low temperatures or high pressures, real gases deviate significantly from ideal gas behavior. In 1873, while searching for a way to link the behavior of liquids and gases, Dutch physicist Johannes van der Waals developed an explanation for these deviations and an equation that was able to fit the behavior of real gases over a much wider range of pressures.

Van der Waals realized that two of the assumptions of the kinetic molecular theory were questionable. The kinetic theory assumes that gas particles occupy a negligible fraction of the total volume of the gas. It also assumes that the force of attraction between gas molecules is zero.

The first assumption works at pressures close to 1 atm. But something happens to the validity of this assumption as the gas is compressed. Imagine for the moment that the atoms or molecules in a gas are all clustered in one corner of a cylinder, as shown in Figure 4.18. At normal pressures, the volume occupied by these particles is a negligibly small fraction of the total volume of the gas. But at high pressures, this is no longer true. As a result, real gases are not as compressible at high pressures as an ideal gas. The volume of a real gas is therefore larger than expected from the ideal gas equation at high pressures.

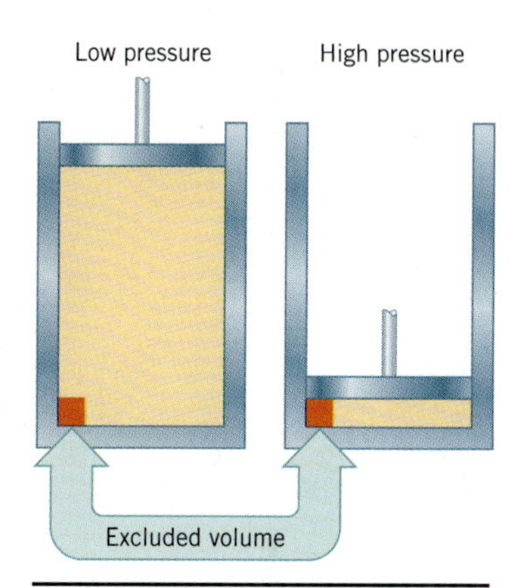


**Figure 4.18** The volume actually occupied by the particles in a gas is relatively small at low pressures, but it can be a significant fraction of the total volume at high pressure. In $O_2$, for example, the gas molecules occupy 0.13% of the total volume at 1.00 atm but 17% of the volume at 100 atm.

Van der Waals proposed that we correct for the fact that the volume of a real gas is too large at high pressures by *subtracting* a term from the volume of the real gas before we substitute it into the ideal gas equation. He therefore introduced a constant ($b$) into the ideal gas equation that was equal to the volume actually occupied by a mole of gas particles. Because the volume of the gas particles depends on the number of moles of gas in the container, the term that is subtracted from the real volume of the gas is equal to the number of moles of gas times $b$:

$$P(V - nb) = nRT$$

When the pressure is relatively small, and the volume is reasonably large, the $nb$ term is too small to make any difference in the calculation. But at high pressures, when the volume of the gas is small, the $nb$ term corrects for the fact that the volume of a real gas is larger than expected from the ideal gas equation.

The assumption that there is no force of attraction between gas particles cannot be true. If it were, gases would never condense to form liquids. In reality, there is a small force of attraction between gas molecules that tends to hold the molecules together. This force of attraction has two consequences: (1) gases condense to form liquids at low temperatures and (2) the pressure of a real gas is sometimes smaller than expected for an ideal gas.

To correct for the fact that the pressure of a real gas is smaller than expected from the ideal gas equation, van der Waals *added* a term to the pressure in this equation. This term contained a second constant ($a$) and has the form: $an^2/V^2$. The complete **van der Waals equation** is therefore written as follows.

$$\left[P + \frac{an^2}{V^2}\right](V - nb) = nRT$$

This equation is something of a mixed blessing. It provides a much better fit with the behavior of a real gas than the ideal gas equation. But it does this at the cost of a loss in

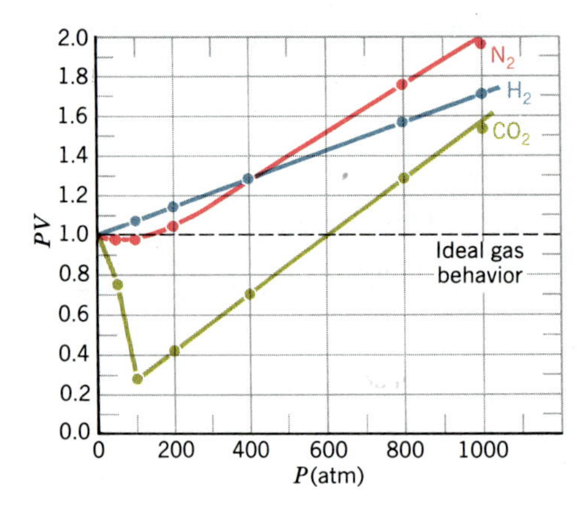


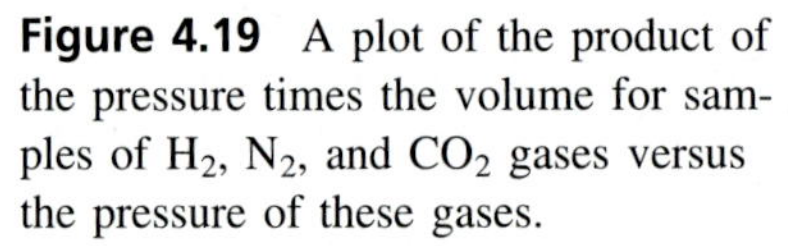
**Figure 4.19** A plot of the product of the pressure times the volume for samples of $H_2$, $N_2$, and $CO_2$ gases versus the pressure of these gases.

generality. The ideal gas equation is equally valid for any gas, whereas the van der Waals equation contains a pair of constants ($a$ and $b$) that change from gas to gas.

The ideal gas equation predicts that a plot of $PV$ versus $P$ for a gas will be a horizontal line because $PV$ should be a constant. Experimental data for $PV$ versus $P$ for $H_2$ and $N_2$ gas at 0°C and $CO_2$ at 40°C are given in Figure 4.19. Values of the van der Waals constants for these and other gases are given in Table 4.7.

**TABLE 4.7 van der Waals Constants for Various Gases**

| Compound | a ($L^2$-atm/$mol^2$) | b (L/mol) |
|---|---|---|
| He | 0.03412 | 0.02370 |
| Ne | 0.2107 | 0.01709 |
| $H_2$ | 0.2444 | 0.02661 |
| Ar | 1.345 | 0.03219 |
| $O_2$ | 1.360 | 0.03803 |
| $N_2$ | 1.390 | 0.03913 |
| CO | 1.485 | 0.03985 |
| $CH_4$ | 2.253 | 0.04278 |
| $CO_2$ | 3.592 | 0.04267 |
| $NH_3$ | 4.170 | 0.03707 |

The magnitude of the deviations from ideal gas behavior can be illustrated by comparing the results of calculations using the ideal gas equation and the van der Waals equation for 1.00 mole of $CO_2$ at 0°C in containers of different volumes. Let's start with a 22.4-L container. According to the ideal gas equation, the pressure of this gas should be 1.00 atm:

$$P = \frac{nRT}{V} = \frac{(1.00\ \cancel{\text{mol}})(0.08206\ \cancel{\text{L}}\text{-atm}/\cancel{\text{mol}}\text{-}\cancel{\text{K}})(273\ \cancel{\text{K}})}{(22.4\ \cancel{\text{L}})} = \mathbf{1.00\ atm}$$

Substituting what we know about $CO_2$ into the van der Waals equation gives a much more complex equation:

$$\left[P + \frac{an^2}{V^2}\right](V - nb) = nRT$$

$$\left[P + \frac{(3.592\ \text{L}^2\text{-atm/mol}^2)(1.00\ \text{mol})^2}{(22.4\ \text{L})^2}\right][22.4\ \text{L} - (1.00\ \text{mol})(0.04267\ \text{L/mol})] = (1.00\ \text{mol})(0.08206\ \text{L-atm/mol-K})(273\ \text{K})$$

This equation can be solved, however, for the pressure of the gas:

$$P = \textbf{0.995 atm}$$

At normal temperatures and pressures, the ideal gas and van der Waals equations give essentially the same results.

Let's now repeat this calculation, assuming that the gas is compressed so that it fills a container that has a volume of only 0.200 liter. According to the ideal gas equation, the pressure would have to be increased to 112 atm to compress 1.00 mole of $CO_2$ at 0°C to a volume of 0.200 L:

$$P = \frac{nRT}{V} = \frac{(1.00\ \text{mol})(0.08206\ \text{L-atm/mol-K})(273\ \text{K})}{(0.200\ \text{L})} = \textbf{112 atm}$$

The van der Waals equation, however, predicts that the pressure will have to increase to only 52.6 atm to achieve the same results:

$$\left[P + \frac{(3.592\ \text{L}^2\text{-atm/mol}^2)(1.00\ \text{mol})^2}{(0.200\ \text{L})^2}\right][0.200\ \text{L} - (1.00\ \text{mol})(0.04267\ \text{L/mol})] = (1.00\ \text{mol})(0.08206\ \text{L-atm/mol-K})(273\ \text{K})$$

$$P = \textbf{52.6 atm}$$

As the pressure of $CO_2$ increases, the van der Waals equation initially gives pressures that are *smaller* than the ideal gas equation, as shown in Figure 4.19, because of the strong force of attraction between $CO_2$ molecules.

Let's now compress the gas even further, raising the pressure until the volume of the gas is only 0.0500 liter. The ideal gas equation predicts that the pressure must increase to 448 atm to condense 1.00 mole of $CO_2$ at 0°C to a volume of 0.0500 L:

$$P = \frac{nRT}{V} = \frac{(1.00\ \text{mol})(0.08206\ \text{L-atm/mol-K})(273\ \text{K})}{(0.050\ \text{L})} = \textbf{448 atm}$$

The van der Waals equation predicts that the pressure will have to reach 1620 atm to achieve the same results:

$$\left[P + \frac{(3.592\ \text{L}^2\text{-atm/mol}^2)(1.00\ \text{mol})^2}{(0.050\ \text{L})^2}\right][0.050\ \text{L} - (1.00\ \text{mol})(0.04267\ \text{L/mol})] = (1.00\ \text{mol})(0.08206\ \text{L-atm/mol-K})(273\ \text{K})$$

$$P = \textbf{1620 atm}$$

The van der Waals equation gives results that are *larger* than the ideal gas equation at very high pressures, as shown in Figure 4.19, because of the volume occupied by the $CO_2$ molecules.

SPECIAL TOPIC

## ANALYSIS OF THE VAN DER WAALS CONSTANTS

The van der Waals equation contains two constants, *a* and *b,* that are characteristic properties of a particular gas. The first of these constants corrects for the force of attraction between gas particles. Compounds for which the force of attraction between particles is strong have large values for *a.* If you think about what happens when a liquid boils, you might expect that compounds with large values of *a* would have higher boiling points. (As the force of attraction between gas particles becomes stronger, we have to go to higher temperatures before we can overcome the force of attraction between the molecules in the liquid to form a gas.) It isn't surprising to find a correlation between the value of the *a* constant in the van der Waals equation and the boiling points of a number of simple compounds, as shown in Figure 4.20.

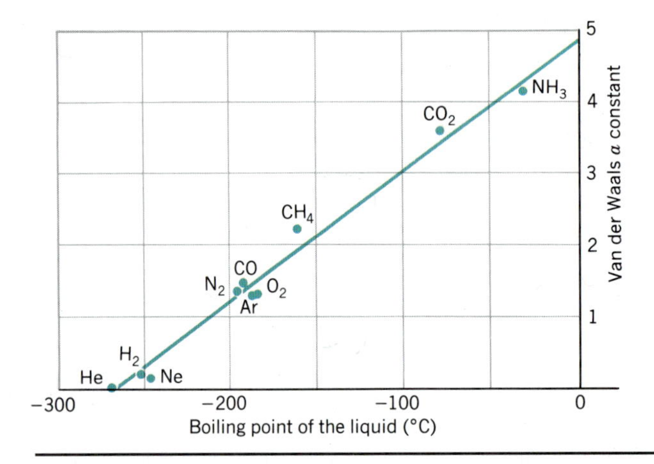


**Figure 4.20** The boiling point of a liquid is an indirect measure of the force of attraction between its molecules. Thus, it isn't surprising to find a correlation between the value of the first van der Waals constant, which measures the force of attraction between gas particles, and the boiling point of a number of simple compounds that are gases at room temperature.

Gases with very small values of *a,* such as $H_2$ and He, must be cooled to almost absolute zero before they condense to form a liquid.

The other van der Waals constant, *b,* is a rough measure of the size of a gas particle. According to the data in Table 4.7, the volume of a mole of argon atoms is 0.03219 liter. This number can be used to estimate the volume of an individual argon atom:

$$\frac{0.03219 \text{ L}}{1 \cancel{\text{mol}}} \times \frac{1 \cancel{\text{mol}}}{6.022 \times 10^{23} \text{ atoms}} = 5.345 \times 10^{-26} \text{ L/atom}$$

The volume of an argon atom then can be converted into cubic centimeters using the appropriate unit factors:

$$5.345 \times 10^{-26} \cancel{\text{L}} \times \frac{1000 \cancel{\text{mL}}}{1 \cancel{\text{L}}} \times \frac{1 \text{ cm}^3}{1 \cancel{\text{mL}}} = 5.345 \times 10^{-23} \text{ cm}^3$$

If we assume that argon atoms are spherical, we can estimate the radius of these atoms. We start by noting that the volume of a sphere is related to its radius by the following formula:

$$V = \tfrac{4}{3}\pi r^3$$

We then assume that the volume of an argon atom is $5.345 \times 10^{-23}$ cm and calculate the radius of the atom:

$$r = \mathbf{2.3 \times 10^{-8}\ cm}$$

According to this calculation, an argon atom has a radius of about $2 \times 10^{-8}$ cm.

## SUMMARY

**1.** The word *gas* comes from the Greek stem *chaos* because a gas is literally the most chaotic of all chemical systems. Gases consist of elementary particles in a state of constant, random motion. The volume of a typical gas at room temperature and atmospheric pressure is about 800 times the volume of the corresponding liquid or solid. As a result, the particles in a gas normally occupy a negligibly small fraction of the total volume of the gas.

**2.** Four characteristic properties are used to describe a gas: pressure, volume, temperature, and the amount of the gas. By holding any two of these properties constant, the relationship between the other two can be determined. The discoveries known as Boyle's law, Charles' law, and so on, are special cases of a more general relationship among these four properties known as the ideal gas equation: $PV = nRT$. This equation summarizes an important characteristic of gases: most of the properties of gases do not depend on the identity of the particles that form the gas.

**3.** One of the properties of a gas that depends on its identity is the average speed with which the gas particles move. Relatively heavy gas particles move slower than lighter gas particles at a given temperature.

**4.** Deviations from ideal gas behavior occur at low temperatures or high pressures. At low temperatures, gases condense to form liquids because the force of attraction between gas particles isn't zero. At high pressures, the volume of a real gas is larger than predicted from ideal gas behavior because the volume of the gas particles is no longer a negligibly small fraction of the total volume of the gas.

## KEY TERMS

**Absolute zero** (p. 121, 123)
**Amontons' law** (p. 120)
**Avogadro's hypothesis** (p. 127)
**Boyle's law** (p. 120)
**Charles' law** (p. 122, 124)
**Dalton's law of partial pressures** (p. 134)
**Diffusion** (p. 140)
**Effusion** (p. 141)
**Force** (p. 115)
**Gay-Lussac's law** (p. 125)
**Graham's laws** (p. 140, 141)
**Ideal gas equation** (p. 128)
**Kinetic energy** (p. 139)
**Kinetic molecular theory** (p. 138)
**Law of combining volumes** (p. 125)
**Partial pressure** (p. 134)
**Pressure** (p. 115)
**State** (p. 112)
**van der Waals equation** (p. 143)
**Vapor pressure** (p. 135)

## KEY EQUATIONS

$$\text{Pressure} = \frac{\text{force}}{\text{area}}$$

$$P_1V_1 = P_2V_2$$

$$\frac{P_1}{P_2} = \frac{T_1}{T_2}$$

$$\frac{V_1}{V_2} = \frac{T_1}{T_2}$$

$$PV = nRT$$

$$P_T = P_1 + P_2 + P_3 + \cdots$$

$$KE = \tfrac{1}{2}mv^2$$

$$\frac{v_H}{v_O} = \sqrt{\frac{m_O}{m_H}}$$

$$\left[P + \frac{an^2}{V^2}\right](V - nb) = nRT$$

## PROBLEMS

### Gases at Room Temperature

**4-1** Which of the following elements and compounds are most likely to be gases at room temperature?

(a) Ar (b) CO (c) $CH_4$ (d) $C_{10}H_{22}$ (e) $Cl_2$
(f) $Fe_2O_3$ (g) Na (h) NaCl (i) Pt (j) $S_8$

### The Properties of Gases

**4-2** Describe in detail how you would measure the following properties of a gas. In each case, describe the equipment you would use and the steps you would have to take to make these measurements.

(a) volume (b) mass (c) pressure (d) density

**4-3** Describe how the behavior of gases can be used to show that fluorine consists of $F_2$ molecules, not F atoms.

**4-4** Imagine two identical flasks at the same temperature. One contains 2 grams of $H_2$ and the other contains 28 grams of $N_2$. Which of the following properties are the same for the two flasks?

(a) pressure (b) average kinetic energy (c) density
(d) the number of molecules per container

**4-5** Explain why the difference between the radius of a helium atom and that of a xenon atom has no effect on the volume of a mole of these gases.

**4-6** Use the fact that the pressure of a gas is the same in all directions to predict what would happen to the weight of an empty metal cylinder when it is filled with helium gas. Will it increase, decrease, or remain the same?

### Pressure versus Force

**4-7** Calculate the force on the bottom of the column of mercury in a barometer when atmospheric pressure is 1.00 atm. Assume that the diameter of the tube is 1.00 cm and the density of mercury is 13.6 g/cm$^3$. Calculate the pressure in pounds per square inch and newtons per square meter.

### Atmospheric Pressure

**4-8** What is the pressure in units of atmospheres when a barometer reads 745.8 mmHg? What is the pressure in units of pascals?

**4-9** One atmosphere pressure will support a column of mercury 760 millimeters tall in a barometer with a tube 1.00 centimeter in diameter. What would be the height of the column of mercury if the diameter of the tube was twice as large?

**4-10** Atmospheric pressure is announced during weather reports in the United States in units of inches of mercury. How many inches of mercury would exert a pressure of 1.00 atm? In Canada, atmospheric pressure is reported in units of pascals. What is 1.00 atm in pascals?

**4-11** If the directions that come with your car tell you to inflate the tires to 200 kPa pressure and you have a tire pressure gauge calibrated in pounds per square inch (psi), what pressure in psi will you use?

**4-12** The vapor pressure of the mercury gas that collects at the top of a barometer is $2 \times 10^{-3}$ mmHg. Calculate the vapor pressure of this gas in atmospheres.

**4-13** Calculate the mass of the atmosphere if atmospheric pressure is 14.7 lb/in$^2$ and the surface area of the planet is $5.1 \times 10^8$ km$^2$.

### Boyle's Law

**4-14** A 425-mL sample of $O_2$ gas was collected at 742.3 mmHg. What would be the pressure in mmHg if this gas was allowed to expand to 975 mL at constant temperature?

**4-15** What would happen to the volume of a balloon filled with 0.357 L of $H_2$ gas collected at 741.3 mmHg if the atmospheric pressure increases to 758.1 mmHg?

**4-16** What is the volume of a scuba tank if it takes 2000 L of air collected at 1 atm to fill the tank to a pressure of 150 atm?

**4-17** Calculate the volume of a balloon that could be filled with the helium in a 2.50-L compressed gas cylinder in which the pressure is 200 atm at 25°C.

### Amontons' Law

**4-18** A butane lighter fluid can is filled with 5.00 atm of $C_4H_{10}$ gas at 21°C. Calculate the pressure in the can when it is stored in a warehouse on a hot summer day when the temperature reaches 38°C.

**4-19** An automobile tire was inflated to a pressure of 32 lb/in$^2$ at 21°C. At what temperature would the pressure reach 60 psi?

**4-20** At 25°C, four-fifths of the pressure of the atmosphere is due to $N_2$ and one-fifth is due to $O_2$. At 100°C, what fraction of the pressure is due to $N_2$?

### Charles' Law

**4-21** Calculate the percent change in the volume of a toy balloon when the gas inside is heated from 22°C to 75°C in a hot-water bath.

**4-22** A sample of $O_2$ gas with a volume of 0.357 liter was collected at 21°C. Calculate the volume of this gas when it is cooled to 0°C.

**4-23** Explain why a balloon filled with $O_2$ at room temperature collapses when cooled in liquid nitrogen, whereas a balloon filled with He does not.

### Gay-Lussac's Law

**4-24** Calculate the ratio of the volumes of sulfur dioxide and oxygen produced when sulfuric acid decomposes:

$$2\,H_2SO_4(aq) \longrightarrow 2\,SO_2(g) + O_2(g) + 2\,H_2O(l)$$

**4-25** Calculate the volume of $H_2$ and $N_2$ gas formed when 1.38 liters of $NH_3$ decompose at a constant temperature and pressure:

$$2\,NH_3(g) \longrightarrow N_2(g) + 3\,H_2(g)$$

**4-26** Ammonia burns in the presence of oxygen to form nitrogen oxide and water:

$$4\,NH_3(g) + 5\,O_2(g) \longrightarrow 4\,NO(g) + 6\,H_2O(g)$$

What volume of NO can be prepared when 15.0 liters of ammonia react with excess oxygen, if all measurements are made at the same temperature and pressure?

**4-27** Acetylene burns in oxygen to form $CO_2$ and $H_2O$:

$$2\,C_2H_2(g) + 5\,O_2(g) \longrightarrow 4\,CO_2(g) + 2\,H_2O(g)$$

Calculate the total volume of the products formed when 15.0 liters of $C_2H_4$ burn in the presence of 15.0 liters of $O_2$, if all measurements are made at the same temperature and pressure.

**4-28** Methane reacts with steam to form hydrogen and carbon monoxide:

$$CH_4(g) + H_2O(g) \longrightarrow CO(g) + 3\,H_2(g)$$

It can also react with steam to form carbon dioxide:

$$CH_4(g) + 2\,H_2O(g) \longrightarrow CO_2(g) + 4\,H_2(g)$$

What is the product of this reaction if 1.50 liters of methane are found by experiment to react with 1.50 liters of water vapor?

### Avogadro's Hypothesis

**4-29** Which weighs more, dry air at 25°C and 1 atm or air at this temperature and pressure that is saturated with water vapor? (Assume that the average molecular weight of air is 29.0 g/mol.)

**4-30** Which of the following samples would have the largest volume at 25°C and 750 mmHg?

(a) 100 g $CO_2$ (b) 100 g $CH_4$ (c) 100 g NO (d) 100 g $SO_2$

**4-31** Nitrous oxide decomposes to form nitrogen and oxygen. Use Avogardro's hypothesis to determine the formula for nitrous oxide if 2.36 liters of this compound decompose to form 2.36 liters of $N_2$ and 1.18 liters of $O_2$.

**4-32** Why does a balloon filled with helium rise? Why does a balloon filled with $CO_2$ sink? (Assume that the average molecular weight of air is 29.0 g/mol.)

### The Ideal Gas Equation

**4-33** Predict the shape of the following graphs for an ideal gas.

(a) pressure versus volume (b) pressure versus temperature (c) volume versus temperature (d) kinetic energy versus temperature (e) pressure versus the number of moles of gas

**4-34** Which of the following graphs does not give a straight line for an ideal gas?

(a) $V$ versus $T$ (b) $T$ versus $P$ (c) $P$ versus $1/V$ (d) $n$ versus $1/T$ (e) $n$ versus $1/P$

**4-35** Which of the following statements is always true for an ideal gas?

(a) If the temperature and volume of a gas both increase at constant pressure, the amount of gas must also increase. (b) If the pressure increases and the temperature decreases for a constant amount of gas, the volume must decrease. (c) If the volume and the amount of gas both decrease at constant temperature, the pressure must decrease.

**4-36** Calculate the value of the ideal gas constant in units of mL-psi/mol-K if 1.00 mole of an ideal gas at 0°C occupies a volume of 22,400 mL at 14.7 lb/in$^2$ pressure.

### Ideal Gas Calculations: Part I

**4-37** Nitrogen gas sells for roughly 50 cents per 100 cubic feet at 0°C and 1 atm. What is the price per gram of nitrogen?

**4-38** Calculate the pressure in atmospheres of 80 grams of $CO_2$ in a 30-liter container at 23°C.

**4-39** Calculate the temperature at which 1.5 grams of $O_2$ have a pressure of 740 mmHg in a 1-liter container.

**4-40** Calculate the number of kilograms of $O_2$ gas that can be stored in a compressed gas cylinder with a volume of 40 L when the cylinder is filled at 150 atm and 21°C.

**4-41** Calculate the pressure in a 250-mL container at 0°C when the $O_2$ in 1 cm$^3$ of liquid oxygen evaporates. Assume that liquid oxygen has a density of 1.118 g/cm$^3$.

**4-42** Assume that a 1.00-L flask was evacuated, 5.0 grams of liquid ammonia were added to the flask, and the flask was sealed with a cork. When the $NH_3$ in the flask warms to 21°C, will the cork be blown out of the mouth of the flask?

### Ideal Gas Calculations: Part II

**4-43** What is the volume of the gas in a balloon at $-195°C$ if the balloon was filled to a volume of 5.0 L at 25°C?

**4-44** $CO_2$ gas with a volume of 25.0 L was collected at 25°C and 0.982 atm. Calculate the pressure of this gas if it is compressed to a volume of 0.15 L and heated to 350°C.

**4-45** Calculate the pressure of 4.80 grams of ozone, $O_3$, in a 2.46-liter flask at 25°C. Assume that the ozone decomposes to molecular oxygen when the temperature is raised to 125°C:

$$2\,O_3(g) \longrightarrow 3\,O_2(g)$$

Calculate the pressure inside the flask once this reaction is complete.

**4-46** One mole of an ideal gas occupies a volume of 22.4 L at 0°C and 1.00 atm. What would be the volume of a mole of an ideal gas at 22°C and 748.8 mmHg?

**4-47** Assume that 10.0 L of $O_2$ gas were collected at 120°C and 749.3 mmHg. Calculate the volume of this gas when it is cooled to 0°C and stored in a container at 1.00 atm.

**4-48** Assume that 5.0 L of $CO_2$ gas were collected at 25°C and 2.5 atm. At what temperature would this gas have to be stored to fill a 10.0-L flask at 0.978 atm?

**4-49** Assume that two 10-L samples of $O_2$ collected at 120°C and 749.3 mmHg were combined and stored in a 1.25-L flask at 27°C. Calculate the pressure of this gas.

### Ideal Gas Calculations Involving the Density of a Gas

**4-50** Calculate the density of $CH_2Cl_2$ in the gas phase at 40°C and 1.00 atm. Compare this with the density of liquid $CH_2Cl_2$ (1.336 g/cm$^3$).

**4-51** Calculate the density of methane gas, $CH_4$, in kilograms per cubic meter at 25°C and 956 mmHg.

**4-52** Calculate the density of helium at 0°C and 1.00 atm and compare this with the density of air (1.29 g/L) at 0°C and 1.00 atm. Explain why 1.00 cubic foot of helium can lift a weight of 0.07 pounds under these conditions.

**4-53** Calculate the ratio of the densities of $H_2$ and $O_2$ at 0°C and 100°C.

**4-54** Which of the rare gases in group VIIIA of the periodic table has a density of 3.7493 g/L at 0°C?

**4-55** Calculate the average molecular weight of air if a sample of air weighs 1.700 times as much as an equivalent volume of ammonia, $NH_3$.

### Dalton's Law of Partial Pressures ($P_T = P_1 + P_2 + \cdots$)

**4-56** Calculate the partial pressure of propane in a mixture that contains equal weights of propane ($C_3H_8$) and butane ($C_4H_{10}$) at 20°C and 746 mmHg.

**4-57** Calculate the partial pressure of helium in a 1.00-liter flask that contains equal numbers of moles of $N_2$, $O_2$, and He at a total pressure of 7.5 atm.

**4-58** Calculate the total pressure in a 10.0-liter flask at 27°C of a sample of gas that contains 6.0 grams of $H_2$, 15.2 grams of $N_2$, and 16.8 grams of He.

**4-59** A 1-L flask is filled with carbon monoxide at 27°C until the pressure is 0.200 atm. Calculate the total pressure after 0.450 gram of carbon dioxide has been added to this flask.

**4-60** A few milliliters of water were added to a 1-L flask at 25°C and the vapor pressure of the water that evaporated was found to be 23.8 mmHg at this temperature. What would be the vapor pressure of the water if this experiment was repeated in a 2-L flask?

**4-61** Calculate the volume of the hydrogen obtained when the water vapor is removed from 289 mL of $H_2$ gas collected by displacing water from a flask at 15°C and 0.988 atm.

### The Kinetic Molecular Theory

**4-62** What would happen to a balloon if the collisions between gas molecules were not perfectly elastic?

**4-63** What would happen to a balloon if the gas molecules were in a state of constant motion, but the motion was not random?

**4-64** Explain why the pressure of a gas is evidence for the assumption that gas particles are in a state of constant random motion.

**4-65** Use the kinetic molecular theory to explain why the pressure of a gas is proportional to the number of gas particles and the temperature of the gas but inversely proportional to the volume of the gas.

**4-66** Which of the following statements explains why a hot-air balloon rises when the air in the balloon is heated? (a) The average kinetic energy of the molecules increases, and the collisions between these molecules and the walls of the balloon make it rise. (b) The pressure of the gas inside the balloon increases, pushing up on the balloon. (c) The gas expands, forcing some of it to escape from the bottom of the balloon, and the decrease in the density of this gas lifts the balloon. (d) The balloon expands, causing it to rise. (e) The hot air rising inside the balloon produces enough force to lift the balloon.

### Graham's Laws of Diffusion and Effusion

**4-67** List the following gases in order of increasing rate of diffusion:

(a) Ar (b) $Cl_2$ (c) $CF_2Cl_2$ (d) $SO_2$ (e) $SF_6$

**4-68** Bromine vapor is roughly five times as dense as oxygen gas. Calculate the relative rates at which $Br_2(g)$ and $O_2(g)$ diffuse.

**4-69** Two flasks with the same volume are connected by a valve. One gram of hydrogen is added to one flask, and 1 gram of oxygen is added to the other. What happens to the weight of the gas in the flask filled with hydrogen when the valve is opened?

**4-70** What happens to the relative amounts of $N_2$, $O_2$, Ar, $CO_2$, and He in air as it diffuses from one flask to another through a pinhole?

**4-71** $N_2O$ and NO are often known by the common names nitrous oxide and nitric oxide. Associate the correct formula with the appropriate common name if nitric oxide diffuses through a pinhole 1.21 times as fast as nitrous oxide.

**4-72** If it takes 6.5 seconds for 25.0 $cm^3$ of helium gas to effuse through a pinhole into a vacuum, how long would it take for 25.0 $cm^3$ of $CH_4$ to escape under the same conditions?

**4-73** Calculate the molecular weight of an unknown gas if it takes 60.0 seconds for 250 $cm^3$ of this gas to escape through a pinhole in a flask into a vacuum and it takes 84.9 seconds for the same volume of oxygen to escape under identical conditions.

**4-74** A lecture hall has 50 rows of seats. If laughing gas ($N_2O$) is released from the front of the room at the same time hydrogen cyanide (HCN) is released from the back of the room, in which row (counting from the front) will students first begin to die laughing? (Assume Graham's law of diffusion is valid.)

**4-75** The atomic weight of radon was first estimated by comparing its rate of diffusion with that of mercury vapor. What is the atomic weight of radon if mercury vapor diffuses 1.082 times as fast?

### Deviations from Ideal Gas Behavior

**4-76** Predict whether the force of attraction between particles makes the volume of a real gas larger or smaller than an ideal gas.

**4-77** Predict whether the fact that the volume of gas particles is not zero makes the volume of a real gas larger or smaller than an ideal gas.

**4-78** Identify the term in the van der Waals equation used to explain why gases become cooler when they are allowed to expand rapidly.

**4-79** Describe the conditions under which significant deviations from ideal gas behavior are observed.

**4-80** Calculate the fraction of empty space in $CO_2$ gas, assuming 1 L of this gas at 0°C and 1.00 atm can be compressed until it changes to a liquid with a volume of 1.26 $cm^3$.

**4-81** The following data were obtained in a study of the pressure and volume of a sample of acetylene:

| $P$(atm): | 1 | 45.8 | 84.2 | 110.5 | 176.0 | 282.2 | 398.7 |
|---|---|---|---|---|---|---|---|
| $V$(L): | 1 | 0.01705 | 0.00474 | 0.00411 | 0.00365 | 0.00333 | 0.00313 |

Calculate the product of the pressure times the volume for each measurement. Plot $PV$ versus $P$ and explain the shape of this curve.

**4-82** Use the van der Waals constants for helium, neon, and argon to calculate the relative size of atoms of these gases.

**4-83** Calculate the pressure of 1.00 mole of $O_2$ at 0°C in 1-L, 0.10-L and 0.05 L containers using both the ideal gas equation and the van der Waals equation.

### Combined Stoichiometry and Gas Law Problems

**4-84** Equal volumes of oxygen and an unknown gas weigh 3.00 and 7.50 grams, respectively. Which of the following is the unknown gas?

(a) CO (b) $CO_2$ (c) NO (d) $NO_2$ (e) $SO_2$ (f) $SO_3$

**4-85** What is the mass of a mole of acetone if 0.520 gram of acetone occupies a volume of 275.5 mL at 100°C and 756 mmHg?

**4-86** Boron forms a number of compounds with hydrogen, including $B_2H_6$, $B_4H_{10}$, $B_5H_9$, $B_5H_{11}$ and $B_6H_{10}$. For which compound would a 1.00-gram sample occupy a volume of 390 $cm^3$ at 25°C and 0.993 atm?

**4-87** Cyclopropane is an anesthetic that is 85.63% carbon and 14.37% hydrogen by mass. What is the molecular formula of this compound if 0.45 L of cyclopropane reacts with excess oxygen at 120°C and 0.72 atm to form 1.35 L of carbon dioxide and 1.35 L of water vapor?

**4-88** Calculate the molecular formula of diazomethane, assuming this compound is 28.6% C, 4.8% H, and 66.6% N by mass and the density of this gas is 1.72 g/L at 25°C and 1 atm.

**4-89** What are the molecular formulas for phosphine, $PH_x$, and diphosphine, $P_2H_y$, if the densities of these gases are 1.517 and 2.944 g/L, respectively, at 0°C and 1 atm?

**4-90** Calculate the mass of magnesium that would be needed to generate 500 mL of hydrogen gas at 0°C and 1.00 atm:

$$Mg(s) + 2\,HCl(aq) \longrightarrow Mg^{2+}(aq) + 2\,Cl^-(aq) + H_2(g)$$

**4-91** Calculate the formula of the oxide formed when 10.0 g of chromium metal react with 6.98 L of $O_2$ at 20°C and 0.994 atm.

**4-92** Calculate the volume of $CO_2$ gas given off at 756 mmHg and 23°C when 150 kg of limestone are heated until they decompose:

$$CaCO_3(s) \longrightarrow CaO(s) + CO_2(g)$$

**4-93** Calculate the volume of $O_2$ that would have to be inhaled at 20°C and 1.00 atm to consume 1.00 kg of fat, $C_{57}H_{110}O_6$:

$$2\,C_{57}H_{110}O_6(s) + 163\,O_2(g) \longrightarrow 114\,CO_2(g) + 110\,H_2O(l)$$

**4-94** Calculate the volume of $CO_2$ gas collected at 23°C and 0.991 atm that can be prepared by reacting 10.0 g of calcium carbonate with excess acid:

$$CaCO_3(aq) + 2\,H^+(aq) \longrightarrow Ca^{2+}(aq) + CO_2(g) + H_2O(l)$$

**4-95** Determine the identity of an unknown metal if 1.00 g of this metal reacts with excess acid according to the following equation to produce 374 mL of $H_2$ gas at 25°C and 1.00 atm:

$$M(s) + 2\,H^+(aq) \longrightarrow M^{2+}(aq) + H_2(g)$$

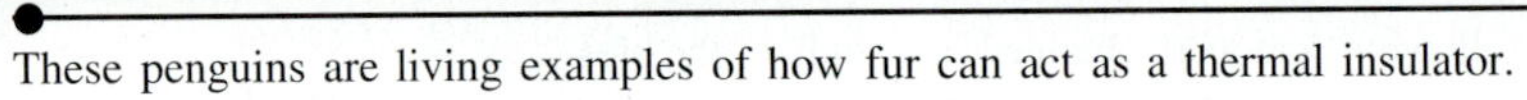

These penguins are living examples of how fur can act as a thermal insulator.

# THERMOCHEMISTRY

This chapter examines the relationships between the terms *temperature, heat, energy,* and *work*. This discussion provides the basis for answering the following questions:

## POINTS OF INTEREST

- Why does the floor feel "cold" but an adjacent rug feels "warm" on a cold winter night, when both objects have the same temperature?
- Why can't you heat the water in an open pot above 100°C, no matter how high you turn the burner on which the pot is placed?
- Why does the tea in "iced tea" become colder when the ice melts?
- Why does it feel good when someone places a wet cloth on your forehead when you have a fever?
- Why does steam cause more serious burns than boiling water?

## 5.1 TEMPERATURE

We don't need to be told that the temperature is 95°F (35°C) on a summer afternoon to know it is hot. Nor do we need to be told that it is −15°F (−26°C) on a clear winter night to realize it is cold. For most purposes, we can rely on our senses to distinguish between hot and cold.

But there are times when our senses can be misled. You probably have had the experience of getting out of bed in the middle of the night and stepping onto a "cold" floor. If you then step onto a small throw rug, your senses tell you that the rug is warmer than the floor. Unfortunately, your senses are wrong. The floor is not colder than the rug. Both objects are just as warm (or just as cold) as the air in the room.

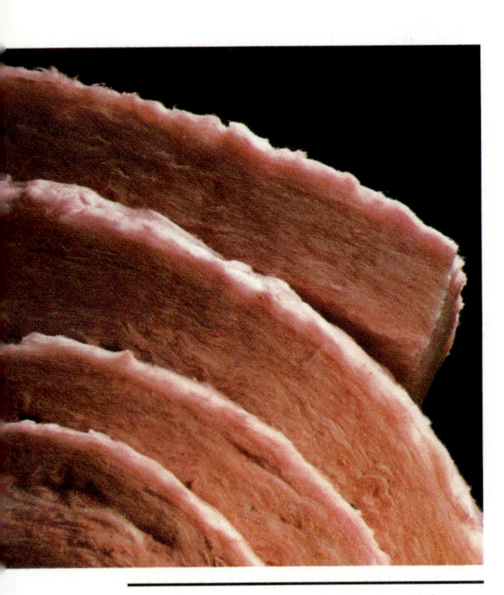

The fiberglass insulation used in houses is a good example of a thermal insulator.

Then why do they feel different? The rug is an example of a **thermal insulator,** which tends to prevent an object from becoming hotter or colder. But the floor is not a thermal insulator. Because the bottoms of your feet are usually warmer than the floor on a cold winter night, your feet become colder for the same reason that a cup of hot coffee left at room temperature gradually cools down.

If you use both metal and plastic ice-cube trays you may encounter a similar phenomenon. When you reach into the freezer, metal ice-cube trays feel colder than plastic trays. But this can't be true; all the trays in the freezer are equally cold. Metals are better **thermal conductors,** however. They are better at conducting heat away from our hands and therefore feel colder.

**CHECKPOINT**

Use the concepts of thermal insulators and thermal conductors to explain why some people cover the seats of their cars with lambskin. Explain why this works equally well in very cold and very hot weather.

Because our senses can be misled, it is useful to have a reliable measure of the degree to which an object is either hot or cold. This quantity is called the **temperature** of the object. Temperature is one of four terms that serve as the basis of this chapter. The others are *heat, energy,* and *work*. All four terms are words that everyone seems to understand, but few can adequately define. As a result, we often use these terms without worrying about exact definitions. Neither the discussion of scales for measuring temperature in Section 1.15, for example, nor the discussion of the effect of temperature on the behavior of gases in Chapter 4 is based on a specific definition of temperature.

This chapter focuses on **thermochemistry,** which is part of the more general subject known as *thermodynamics*. The prefix *thermo-* comes from the Greek word for "heat." The suffix *-dynamics* comes from the Greek word for "force." Thermodynamics literally means an analysis of the force of heat (or other forms of energy). As a result, discussions of thermodynamics require exact definitions of the four terms introduced above. Because the concept of temperature lies at the heart of thermodynamics, we begin our discussion by defining what is meant by the temperature of an object.

## 5.2 TEMPERATURE AS AN INTENSIVE PROPERTY OF MATTER

Temperature is nothing more than a reliable, quantitative measurement of the degree to which an object is either "hot" or "cold." As we saw in Section 1.15, temperature can be measured on either relative or absolute scales (see Figure 5.1). The Celsius (°C) and Fahrenheit (°F) scales measure *relative* temperatures. These scales

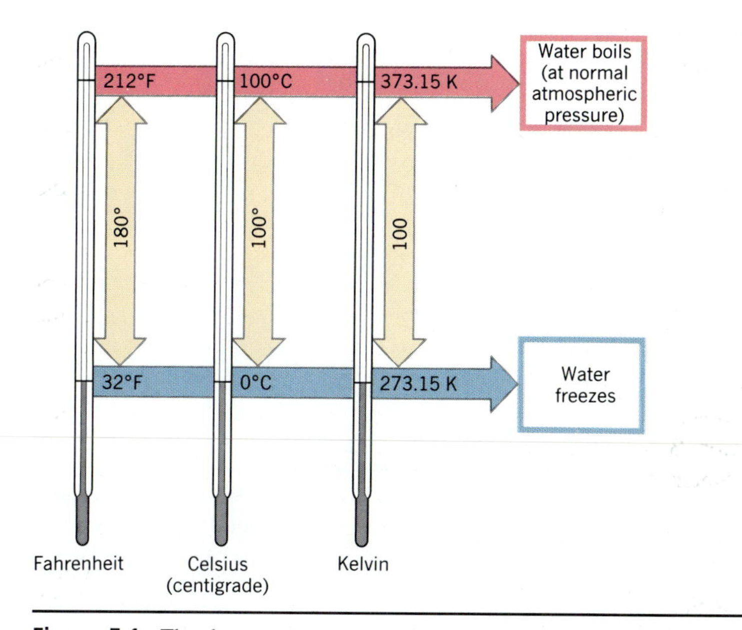


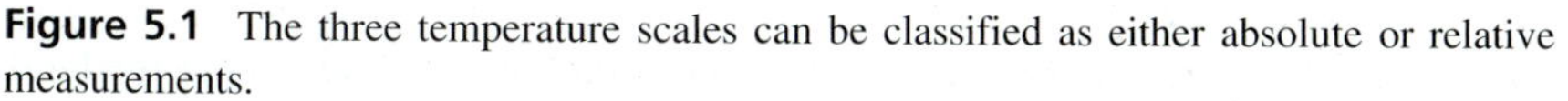

**Figure 5.1** The three temperature scales can be classified as either absolute or relative measurements.

compare the temperature being measured with arbitrary standards, such as boiling water and an ice–water bath. The Kelvin scale measures *absolute* temperatures. On this scale, the temperature of an object is compared with the temperature at which the volume and pressure of an ideal gas extrapolate to zero.

Temperature is one of the seven quantities measured by the base units in the SI system (see Table 1.4). All the other quantities (length, mass, time, electric current, luminous intensity, and amount of substance) are *extensive properties,* which depend on the size of the sample being studied. Temperature is an *intensive property,* which does not depend on the size of the sample. When we mix two 50-mL samples of water at 20°C, the mass doubles and the volume doubles, but the temperature remains the same.

Sec. 1.13

## 5.3 HEAT AND HEAT CAPACITY

As we have seen, temperature is a quantitative measure of whether an object can be labeled hot or cold. **Heat** is a phenomenon that can cause a change in temperature.

To probe the relationship between heat and temperature, it might be useful to consider a series of experiments done by Joseph Black between 1759 and 1762, which studied what happened when liquids at different temperatures were mixed. Black noted that combining equal volumes of water at 100°F and 150°F produced a mixture whose temperature was the average of the two samples (125°F), as shown in Figure 5.2*a*. However, when water at 100°F was mixed with an equal volume of mercury at 150°F, the temperature of the liquids after mixing was only 115°F (see Figure 5.2*b*). The temperature of the mercury fell by 35°F, but the temperature of the water increased by only 15°F.

Black assumed that the heat lost by the mercury as it cooled down was equal to the heat gained by the water as it became warmer. He therefore concluded that the temperature of the water changed by a smaller amount because water had a larger "capacity for heat." In other words, it takes more heat to produce a given change in the temperature of water than it does to produce the same change in the temperature of an equivalent volume of mercury.

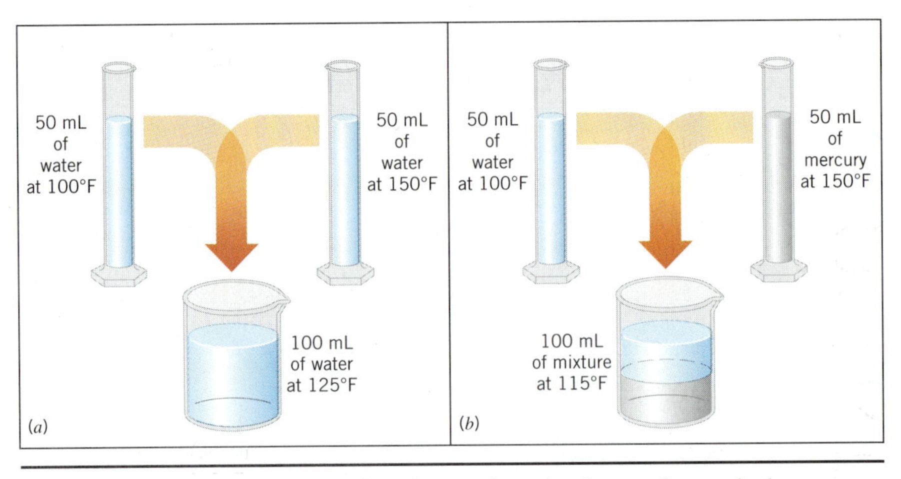


**Figure 5.2** (*a*) The temperature of a mixture of equal volumes of water is the average of the temperatures before the samples were combined. (*b*) When equal volumes of water and mercury are mixed, the temperature of the mixture is much closer to the temperature of the water before mixing.

More than 200 years after Black's experiment, scientists still talk about the **heat capacity** of a substance. Because water has a larger heat capacity than mercury, it takes more heat to raise the temperature of a given mass of water by 1° than it does to raise the temperature of the same mass of mercury by 1°. Any unit used to measure heat must therefore specify not only the size of the sample being heated and the change in the temperature observed but also the identity of the substance being heated.

Sec. 1.13 

The heat capacity of a substance is an extensive quantity, which depends on the size of the sample. There are two ways in which the concept of heat capacity can be turned into an intensive quantity. By convention, the **specific heat** of a substance is defined as the amount of heat needed to raise the temperature of *one gram* by 1°C or 1 K. The ease with which a substance gains or loses heat also can be described in terms of its **molar heat capacity,** which is the amount of heat required to raise the temperature of *one mole* of the substance by either 1°C or 1 K.

When the metric system was first introduced, the calorie was defined so that the specific heat of water was 1. In other words, it takes one calorie of heat to raise the temperature of one gram of water by 1°C. When the SI system was introduced, the approved unit of heat became the *joule* (1 cal = 4.184 J). The units of specific heat in the SI system are therefore J/g-K and the units of molar heat capacities are J/mol-K.

A calorie was once defined as the amount of heat required to raise the temperature of 1 gram of water from 14.5 to 15.5°C.

## EXERCISE 5.1

Use the following equality to calculate the specific heat of water in J/g-K and the molar heat capacity of water in J/mol-K.

$$1 \text{ cal} = 4.184 \text{ J}$$

**SOLUTION** Before the SI system was introduced, the specific heat of water was exactly one calorie per gram per degree Celsius. Because there are 4.184 joules in a calorie, the specific heat of water is 4.184 J/g-°C or 4.184 J/g-K:

$$\frac{1 \cancel{\text{cal}}}{\text{g-K}} \times \frac{4.184 \text{ J}}{1 \cancel{\text{cal}}} = \mathbf{4.184 \text{ J/g-K}}$$

To find the molar heat capacity of water, we multiply the specific heat of water by the mass of a mole. The molar heat capacity of water is therefore 75.376 J/mol-K:

$$\frac{4.184 \text{ J}}{\not{g}\text{-K}} \times \frac{18.015 \not{g} \text{ H}_2\text{O}}{1 \text{ mol}} = \mathbf{75.376 \text{ J/mol-K}}$$

The specific heat and the molar heat capacities of a number of substances are given in Table 5.1. These data suggest that the heat capacity of water is unusually large, regardless of whether we compare specific heats or molar heat capacities. This has very important consequences for our weather. The water that covers so much of the surface of our planet acts as a heat sink, which moderates the changes in temperature that occur as the amount of energy absorbed from the sun varies with the sun's distance from the earth.

**TABLE 5.1 Specific Heats and Molar Heat Capacities of Common Substances**

| Substance | Specific Heat (J/g-K) | Molar Heat Capacity (J/mol-K) |
|---|---|---|
| Al(*s*) | 0.901 | 24.3 |
| C(*s*) | 0.7197 | 8.644 |
| Cu(*s*) | 0.3844 | 24.43 |
| $H_2O(l)$ | 4.184 | 75.376 |
| Fe(*s*) | 0.449 | 25.1 |
| Hg(*l*) | 0.1395 | 27.98 |
| $O_2(g)$ | 0.9172 | 29.35 |
| $N_2(g)$ | 1.040 | 29.12 |
| NaCl(*s*) | 0.8641 | 50.50 |

**CHECKPOINT**

Classify heat, temperature, heat capacity, and specific heat as either extensive or intensive quantities.

## 5.4 LATENT HEAT

The discussion of heat capacity in the previous section was based on the hidden assumption that water always becomes hotter when it is heated. To check whether this is a valid assumption, think about what happens when you heat a pot of water. At first, the water gets hotter, but eventually it starts to boil. From that moment on, the temperature of the water remains the same (100°C), regardless of how fast you heat the water, until the liquid evaporates.

Imagine another experiment in which a thermometer is immersed in a snow bank on a day when the temperature finally gets above 0°C. The snow gradually melts as it gains heat from the air above it. But the temperature remains the same (0°C) until the last snow melts.

Joseph Black was the first to recognize the importance of these observations. He used them to differentiate between heat that you can sense, or detect, and *latent heat.*

The word *latent* is sometimes used to describe things that exist but can't be seen. When a photograph is taken, for example, a latent image is captured. The image exists, but it isn't visible until the film is developed. Mystery novels often talk about latent fingerprints, which can't be seen until the object is sprayed with powder.

There is no change in the temperature of the water as snow melts.

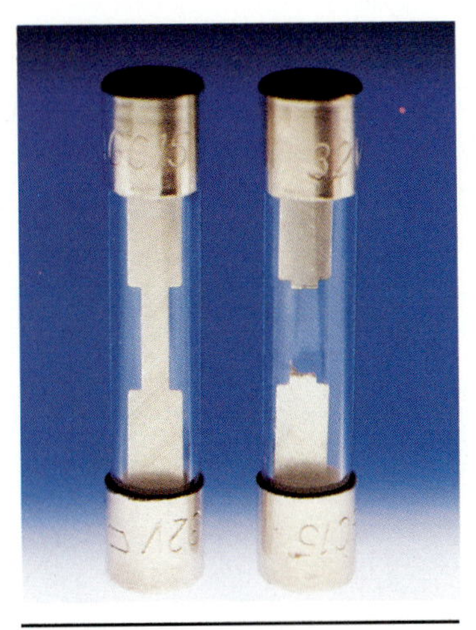

The heat required to melt a substance is known as the latent heat of fusion because the term ''fuse'' originally meant ''to melt.''

Black argued that latent heat is encountered whenever there is a change in the state of matter. Heat can enter or leave a sample without any detectable change in its temperature when a solid melts, when a liquid freezes or boils, or when a gas condenses to form a liquid.

Figure 5.3 shows results obtained when equal amounts of ice and water were heated with a hot plate. Initially, the heat that entered the system was used to melt the ice, and there was no change in the temperature until essentially all of the ice was gone. The amount of heat required to melt the ice is called the **latent heat of fusion.** (The word *fusion* is used because it literally means to melt something. We ''blow a fuse'' in an electric circuit, for example, by melting the thin piece of metal in the center of the fuse.)

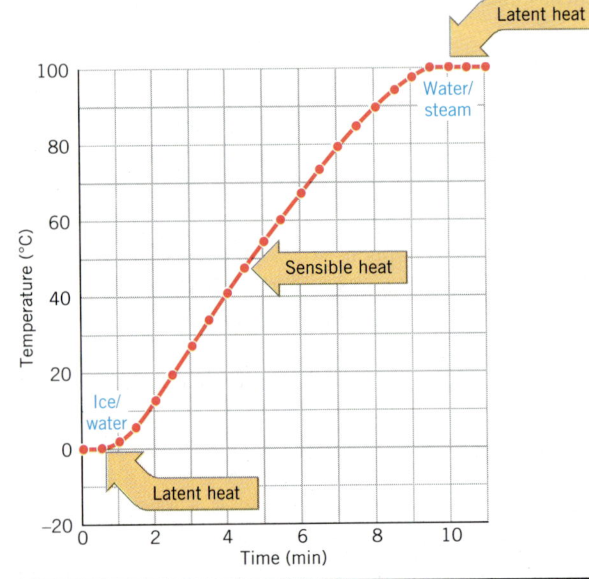


**Figure 5.3** A plot of what happened when equal amounts of ice and water in a 250-mL beaker were heated on a hot plate.

Once the ice melted, the temperature of the water slowly increased from 0°C to 100°C. But once the water started to boil, the heat that entered the sample was used to convert the liquid into a gas and the temperature of the sample remained constant until the liquid evaporated. The amount of heat required to boil, or vaporize, the liquid is called the **latent heat of vaporization.**

## 5.5 THE CALORIC THEORY

Black's experiments served as the basis for the **caloric theory of heat** (from the Latin *calor,* ''heat''). This theory was based on the following assumptions.

1. Heat is a fluid that flows from hot to cold objects.
2. Heat is strongly attracted to matter, which can hold a great deal of heat.
3. Heat is indestructible—it is conserved in all processes.
4. Sensible heat raises the temperature of the object when it flows into the object.
5. Latent heat combines with the particles in matter. (This postulate was used to explain how a solid could be transformed into a liquid, or a liquid into a gas.)
6. Heat has no appreciable weight.

The caloric theory was so powerful that most of the basic structure of thermodynamics was developed during its lifetime. It also served as the basis for much of the language used to discuss heat. Heat is still often described as ''flowing'' from hot to cold objects, as if it were a liquid. We still talk about the ''heat capacity'' of a substance, as if heat were a fluid that could fill its container. Some scientists and engineers even talk about the ''heat content'' of a system, as if a given amount of heat could be stored in the system like a fluid in a container.

Unfortunately there is a problem with the caloric theory. It's wrong. No matter how intuitively satisfying the assumptions listed above may appear, only the last one is valid. Heat is not a fluid. More important, heat is not conserved. The only aspect of the caloric theory that remains intact today is the idea that heat has no weight.

## 5.6 HEAT AND THE KINETIC MOLECULAR THEORY

People who don't understand science often search for a single experiment that proves or disproves a major theory, such as the caloric theory. Some even claim to have found one. There are those, for example, who cite the cannon-boring experiment of Sir Benjamin Thompson (Count Rumford) in 1798 as having sounded the death knell for the caloric theory.

Thompson was an active opponent of the caloric theory. His most famous experiment was done while he was in charge of the military arsenal in Munich. He noticed that the brass barrel of a cannon became hot while it was being bored. Furthermore, the metal chips that were scraped from the barrel as it was hollowed out were often hotter than the temperature of boiling water. Proponents of the caloric theory explained this by assuming that the friction between the boring tool and the brass cannon barrel ''squeezed'' heat out of the brass.

Thompson's opposition to the caloric theory led him to measure the amount of heat given off in an experiment in which the amount of metal removed from the cannon barrel was kept to a minimum. He immersed a cannon in water and used a dull tool to scrape the barrel of the cannon. In a period of 2.5 hours, only 270 grams of metal shavings were scraped from the cannon barrel, but enough heat was generated to boil 26.5 pounds of water. Thompson believed that this experiment showed that friction was an inexhaustible source of heat. If this is true, then heat is not conserved.

According to Thompson, the heat given off in this experiment was not squeezed from the metal. It was generated during the experiment. In hindsight, it is easy to agree with Thompson, but his contemporaries didn't. They were willing to accept that considerable amounts of heat were given off when metal was rubbed or hammered. But they explained Thompson's results by assuming that metals contained so much heat that this experiment gave off only a small fraction of the total. For as much as 50 years after this experiment was done, the caloric theory remained intact.

Sec. 4.15 

Although Thompson's experiments by no means disproved the caloric theory, they marked the starting point in the development of an alternative, the **kinetic theory of heat.** The word "kinetic" is derived from the Greek word, *kinetos,* for motion. The kinetic theory divides the universe into a system and its surroundings. The **system** is that small portion of the universe in which we are interested. It might consist of the water in a beaker or a gas trapped in a piston and cylinder, as shown in Figure 5.4. The **surroundings** are everything else—in other words, the rest of the universe.

One of Rumford's many contributions was the development of a better fireplace. Here we see an etching by Gilray that shows Count Rumford in front of a "Rumford stove."

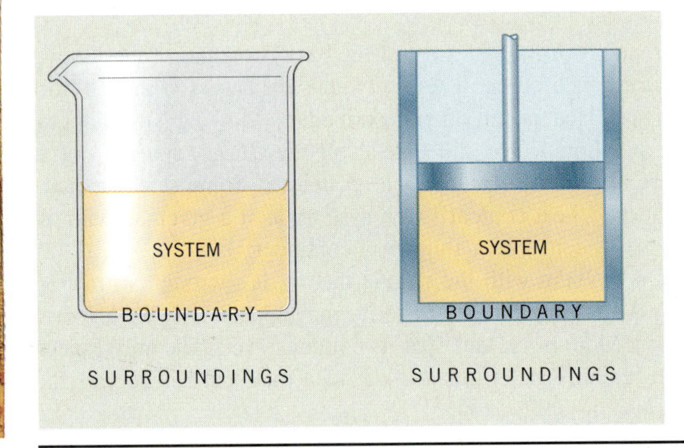


**Figure 5.4** The kinetic theory assumes that heat is transferred across the boundary between a system and its surroundings.

The system and its surroundings are separated by a **boundary.** The boundary can be as real as the glass in a beaker or the walls of a balloon. It also can be imaginary, such as a line 200 nm from the surface of a metal that arbitrarily divides the air close to the metal surface from the rest of the atmosphere. The boundary can be rigid or it can be elastic. It can be assumed to be an ideal thermal conductor or a perfect thermal insulator.

In the kinetic theory, heat is no longer a fluid that can be stored in an object. Heat is something that is transferred across the boundary between a system and its surroundings. In the kinetic theory, heat is no longer conserved; it can be either created or destroyed. The kinetic theory agrees with Thompson's belief that a virtually inexhaustible amount of heat can be generated by the friction of a dull tool rubbing against the barrel of a brass cannon. The last postulate of the kinetic molecular theory provides the basis for understanding the kinetic theory of heat:

Sec. 4.15

**The average kinetic energy of a collection of gas particles depends on the temperature of the gas and nothing else.**

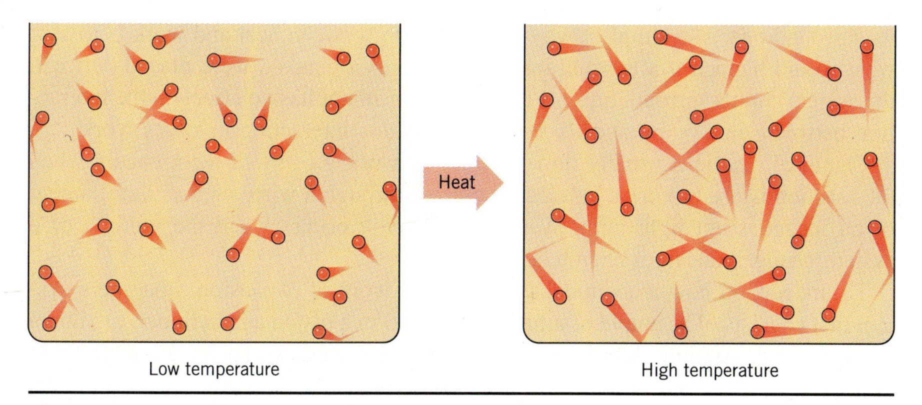


**Figure 5.5** According to the kinetic theory, a system becomes hotter when the average kinetic energy of the particles in the system increases.

According to this postulate, a gas becomes warmer if and only if the average kinetic energy of the gas particles increases. When heat enters a system from its surroundings, the net effect is an increase in the kinetic energy of the atoms and molecules in the system. The kinetic theory of heat therefore can be summarized as follows:

**Heat, when it enters a system, produces an increase in the speed with which the particles of the system move, as shown in Figure 5.5.**

**CHECKPOINT**

Use the kinetic theory to explain what happens to the particles of a gas in a balloon when heat enters the balloon from its surroundings.

## 5.7 WORK

In physics, **work** is done when an object is moved. Two factors determine the amount of work done: (1) the force that must be applied to move the object, and (2) the distance the object is moved:

$$w = F \times d$$

**EXERCISE 5.2**

Calculate the amount of work that has to be done to lift a 10-pound bag of groceries a distance of 2.5 feet from the floor to the top of the kitchen counter.

**SOLUTION** It takes 25 foot-pounds of work to lift a 10-pound bag of groceries a distance of 2.5 feet:

$$w = (10 \text{ lb})(2.5 \text{ ft}) = \mathbf{25 \text{ ft-lb}}$$

In the course of this task, we (the system) do 25 foot-pounds of work on our surroundings.

Heat and work both result from an interaction between a system and its surroundings. Furthermore both are transferred across the boundary between the system and its surroundings. The simplest way to distinguish between heat and work is to ask: what would happen if a thermal insulator, such as a blanket, were placed between the system and its surroundings? If the thermal insulator has no effect on the interaction between the system and its surroundings, this interaction involves work.

Chemical reactions usually do two kinds of work. *Electrical work* occurs when the reaction is used to drive an electric current through a wire, such as the filament of an incandescent light bulb. *Work of expansion* occurs when the volume of a system changes during a chemical reaction.

To understand how a chemical reaction does work of expansion, imagine a system that consists of a sample of ammonia trapped in a piston and cylinder, as shown in Figure 5.6. Assume that the pressure of the gas pushing up on the piston just

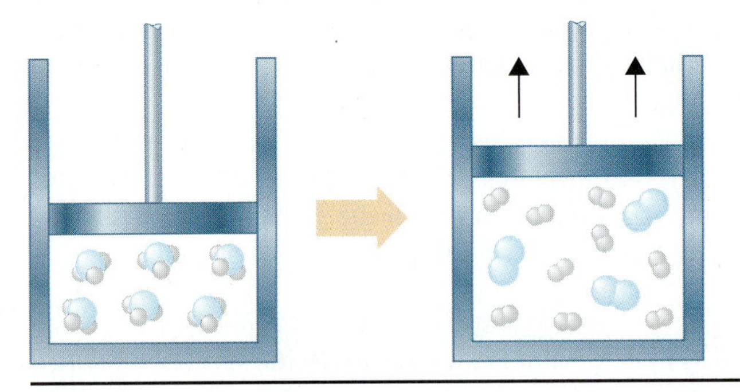

**Figure 5.6** When ammonia decomposes to hydrogen and nitrogen, the number of particles increases. This causes the gas to expand, pushing the piston partway out of the cylinder. The amount of work done by the system is the product of the force applied to the piston times the distance the piston moves.

balances the weight of the piston, so that the volume of the gas is constant. Now assume that the gas decomposes to form nitrogen and hydrogen.

$$2\ NH_3(g) \longrightarrow N_2(g) + 3\ H_2(g)$$

Sec. 4.10

The net effect of this reaction is to increase the number of gas particles in the container. According to Avogadro's hypothesis, if the temperature and pressure of the gas are held constant, the volume of the gas must increase.

The volume of the gas can increase by pushing the piston partway out of the cylinder. This requires moving the piston against the force of gravity, so it involves work. Because the work is generated by the expansion of a gas, it is work of expansion. The amount of work done is equal to the product of the force exerted on the piston times the distance the piston is moved:

$$w = F \times d$$

Sec. 4.4

The pressure ($P$) the gas exerts on the piston is equal to the force ($F$) with which it pushes up on the piston divided by the surface area ($A$) of the piston:

$$P = \frac{F}{A}$$

Thus, the force exerted by the gas is equal to the product of its pressure times the surface area of the piston:

$$F = P \times A$$

Substituting this expression into the equation defining work gives the following result:

$$w = (P \times A) \times d$$

The product of the area of the piston times the distance the piston moves is equal to the change that occurs in the volume of the system when the gas expands. By convention, the change in the volume is represented by the symbol $\Delta V$:

$$\Delta V = A \times d$$

The magnitude of the work (symbolized $|w|$) done when a gas expands is therefore equal to the product of the pressure of the gas times the change in the volume of the gas:

$$|w| = P\Delta V$$

## 5.8 HEAT FROM WORK AND VICE VERSA

Thompson's cannon-boring experiment suggested that heat and work were related and that work could be used to produce heat. In a series of experiments begun 40 years later, James Prescott Joule measured the amount of heat that could be produced from a given amount of work.

Joule's most famous experiment used falling weights to do work by turning a paddle immersed in water, mercury, or oil, as shown in Figure 5.7. By measuring the resulting changes in the temperatures of the liquids, Joule was able to show that 772.5 foot-pounds of work could generate 1 Btu of heat. In recognition of his contri-

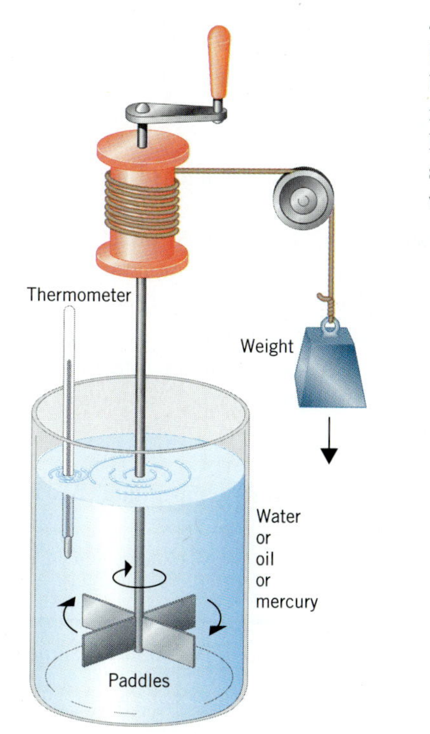


**Figure 5.7** Joule measured the heat that could be generated from a given amount of work by measuring the change in the temperature of a liquid that occurred when the work done by a set of falling weights was used to turn a paddle wheel immersed in the liquid.

butions to this field, the unit of work was eventually named the *joule*. By definition, 1 joule is the work done when a force of 1 newton is used to move an object 1 meter:

$$1 \text{ J} = 1 \text{ N-m}$$

Because work can be converted into heat and vice versa, the SI system uses the joule to measure energy in the form of both heat and work.

## 5.9 THE FIRST LAW OF THERMODYNAMICS: CONSERVATION OF ENERGY

The relationship between heat and work was first established between 1850 and 1851 by the British physicist William Thomson (Lord Kelvin) and his German contemporary Rudolf Clausius. Working independently, Thomson and Clausius concluded that neither heat nor work was conserved; only energy is conserved. Their conclusions are summarized in the **first law of thermodynamics,** which states that energy cannot be created or destroyed.

Energy is neither created or destroyed. The energy absorbed by the system (the ice cubes) in this example is exactly equal to the energy lost by its surroundings (the tea).

A system can gain or lose energy. But any change in the energy of the system must be accompanied by an equivalent change in the energy of its surroundings because the total energy of the universe is constant. By convention, the change in one of the properties of a system or its surroundings is represented by an uppercase Greek delta ($\Delta$) added to the symbol for that property. The first law of thermodynamics therefore can be described by the following equation:

$$\boldsymbol{\Delta E_{univ} = \Delta E_{sys} + \Delta E_{surr} = 0}$$

(The subscripts *univ, sys,* and *surr* stand for the universe, the system, and its surroundings, respectively.)

The energy of a system is often called its **internal energy,** because it is the sum of the kinetic and potential energies of the particles that form the system. For an ideal gas, the internal energy is directly proportional to the temperature of the gas because the potential energy of an ideal gas is zero and the kinetic energy of the gas particles is proportional to the temperature of the gas, as noted in Section 4.15:

$$\boldsymbol{E_{sys} = \tfrac{3}{2} RT}$$

(In this equation, $R$ is the ideal gas constant and $T$ is the temperature of the gas in units of Kelvin.)

The internal energy for more complex systems can't be measured directly. But changes in the internal energy of the system can be detected by monitoring the temperature of the system. If we detect an increase in the temperature of the system, we conclude that the internal energy of the system has increased.

The magnitude of the change in the internal energy of a system is defined as the difference between the initial and final values of this quantity:

$$\Delta E_{sys} = E_{final} - E_{initial}$$

Because the internal energy of a system is proportional to its temperature, $\Delta E$ is positive when the temperature of the system increases.

## 5.10 THE FIRST LAW OF THERMODYNAMICS: INTERCONVERSION OF HEAT AND WORK

The first law of thermodynamics states that the energy of the universe is constant:

$$\Delta E_{univ} = \Delta E_{sys} + \Delta E_{surr} = 0$$

Energy can be transferred between a system and its surroundings, however, as long as the energy gained by one of these components of the universe is equal to the energy lost by the other:

$$\Delta E_{sys} = -\Delta E_{surr}$$

Energy can be transferred between a system and its surroundings in the form of either heat ($q$) or work ($w$). Thus, the change in the internal energy of the system represents a balance between the heat and the work that cross the boundary between the system and its surroundings:

$$\boldsymbol{\Delta E_{sys} = q + w}$$

This equation limits the amount of heat or work that can be extracted from a system. It suggests that an inexhaustible amount of heat can be obtained from the system only if we are willing to do enough work on the system to compensate for the drain on the system's internal energy. Alternatively the system can do an inexhaustible amount of work if we are willing to pump enough heat into it to make up for the drain on its internal energy. The first law of thermodynamics therefore has been described, tongue in cheek, as suggesting that you can't get something for nothing.

---

**CHECKPOINT**

Some textbooks write the following equation for the first law of thermodynamics:

$$\Delta E = 0$$

Others write a different equation:

$$\Delta E = q + w$$

Explain why both equations are correct, if the appropriate subscript is added to the symbol for $\Delta E$.

---

The first law of thermodynamics can be used to explain how the kinetic theory accounts for the phenomenon of latent heat. The first law suggests that heat can do one of two things when it enters a system. It can increase the temperature of the system or it can do work:

$$q = \Delta E_{sys} - w$$

Heat that increases the temperature of the system ($\Delta E_{sys} > 0$) can be detected. It is therefore "sensible" heat. Heat that only does work on the system ($\Delta E_{sys} = 0$) can't be detected. It is therefore "latent" heat.

The sign convention for the relationship between the internal energy of a system and the **heat** that crosses the boundary between the system and its surroundings is given on the right side of Figure 5.8.

1. When the heat that enters a system increases the temperature of the system, the internal energy of the system increases, and $\Delta E$ is positive.
2. When the temperature of the system decreases because heat leaves the system, $\Delta E$ is negative.

The sign convention for the relationship between **work** and the internal energy of a system is shown on the left side of Figure 5.8. This convention can be understood by referring back to the model of a gas trapped in a piston and cylinder shown in Figure 5.6. Work is done by the system when the gas inside the cylinder expands.

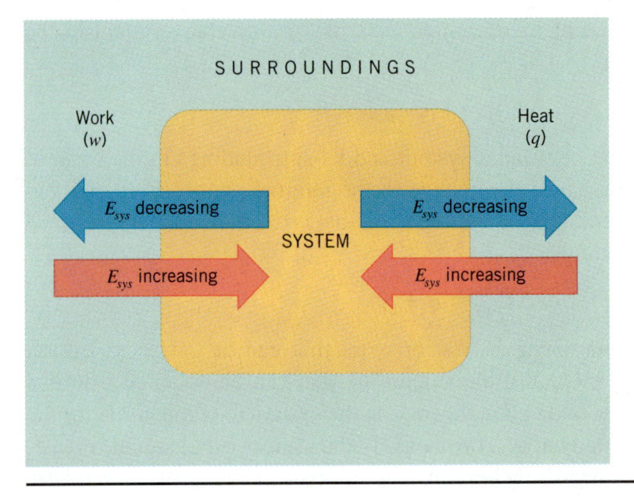


**Figure 5.8** The internal energy of a system decreases ($\Delta E < 0$) when the system loses heat to its surroundings or does work on its surroundings. When heat enters the system from its surroundings or when the surroundings do work on the system, the internal energy of the system increases ($\Delta E > 0$).

When this happens, energy is lost to the surroundings and the system becomes colder.

1. When the system does work on its surroundings, energy is lost, and $\Delta E$ is negative.
2. When the surroundings do work on the system, the internal energy of the system becomes larger, so $\Delta E$ is positive.

In Section 5.7, the relationship between the magnitude of the work done by a system when it expands and the change in the volume of the system was described by the following equation:

$$|w| = P\Delta V$$

Figure 5.8 shows that the sign convention for work of expansion can be included by writing this equation as follows:

$$\mathbf{w = -P\Delta V}$$

## 5.11 STATE FUNCTIONS

Every system can be described by certain measurable properties. A gas, for example, can be described in terms of the number of moles of particles it contains, its temperature, pressure, volume, mass, density, internal energy, and so on. Because these properties describe the **state** of the system, equations that connect two or more of these properties are called **equations of state.** The ideal gas law, for example, is an equation of state:

$$PV = nRT$$

As we have seen, the quantities that describe the state of a system can be classified as either extensive or intensive. They also can be sorted into categories depending on whether or not they are **state functions.**

**By definition, one of the properties of a system is a state function if it depends only on the state of the system, not on the path used to get to that state.**

Assume that $X$ stands for one of the properties of a system. By convention, the change in the value of $X$ when the system undergoes a change in state is equal to the final value of this quantity minus the initial value:

$$\Delta X = X_{\text{final}} - X_{\text{initial}}$$

By definition, $X$ is a state function if and only if the value of $\Delta X$ is the same regardless of the path used to go from the initial to the final state of the system.

The concept of a state function is best illustrated by an analogy. Imagine a trip from New York to Los Angeles, as shown in Figure 5.9. There are an infinite number of routes that can be taken from one of these cities to the other. For the moment, let's compare a more or less direct route between the cities with a route that passes through Bangor, Maine; Miami Beach, Florida; Chicago, Illinois; Dallas, Texas; Missoula, Montana; and Phoenix, Arizona.

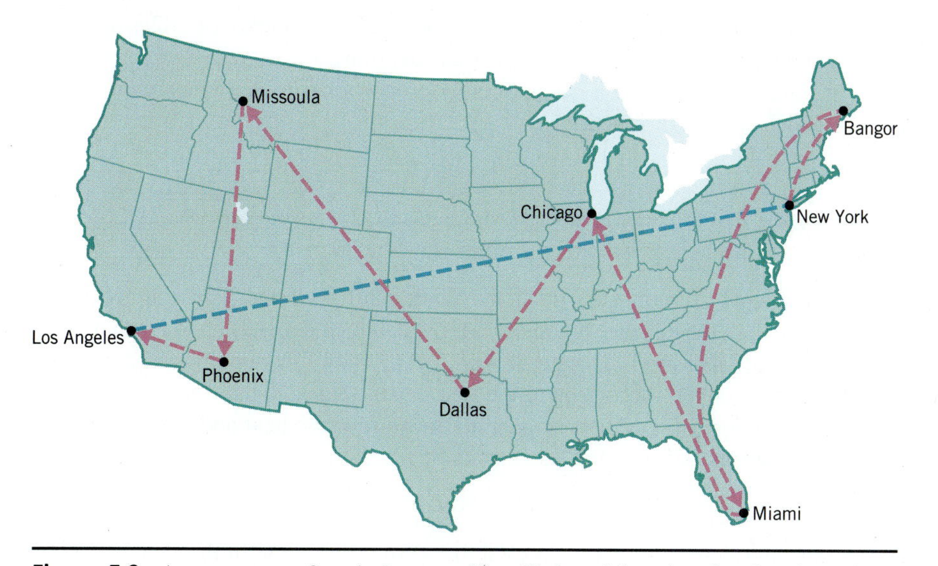


**Figure 5.9** Any property of a trip between New York and Los Angeles that depends on the path used to drive from one city to the other is *not* a state function.

If a property of the trip depends on the route used to go from one city to the other, it is *not* a state function. A number of extensive quantities belong in this category. The cost of the trip obviously depends on the route. So does the amount of gasoline consumed, the amount of tire wear, the distance traveled, and both the length of time and the amount of work it takes to drive this distance.

Any property that has the same value regardless of the route between the cities is a state function. At least three state functions can be used to describe this trip: latitude, longitude, and altitude. The path we take from New York to Los Angeles does not influence the magnitude of the change in any of these quantities.

Another test of a state function involves asking: ''What happens if we return to the initial state?'' If the quantity is a state function, the overall change must be zero because the path takes us back to the same initial state. In other words, when the final state of the system is the same as the initial state, there can be no change in the value of a state function:

$$\Delta X = X_{\text{initial}} - X_{\text{initial}} = 0$$

If tire wear were a state function, for example, rubber would wear off the tires as we drove to Los Angeles and then reappear as we drove back to New York!

**EXERCISE 5.3**

Which of the following properties of a gas are state functions?
(a) Temperature, $T$ (b) Volume, $V$ (c) Pressure, $P$
(d) Number of moles of gas, $n$ (e) Internal energy, $E$

**SOLUTION** Temperature is a state function. No matter how many times we heat, cool, expand, compress, or otherwise change the system, the net change in the temperature only depends on the initial and final states of the system:

$$\Delta T = T_{\text{final}} - T_{\text{initial}}$$

The same can be said for the volume, pressure, and the number of moles of gas in the sample. These quantities are all state functions.

The internal energy of a gas is directly proportional to the temperature of the gas, which is a state function:

$$E = \tfrac{3}{2}RT$$

This means that the internal energy of the gas is also a state function. All of these properties of an ideal gas are therefore state functions.

Heat and work are not state functions. Work can't be a state function because it is proportional to the distance an object is moved, which depends on the path used to go from the initial to the final state. If work isn't a state function, then heat can't be a state function either. According to the first law of thermodynamics, the change in the internal energy of a system is equal to the sum of the heat and the work transferred between the system and its surroundings:

$$\Delta E_{\text{sys}} = q + w$$

If $\Delta E$ does not depend on the path used to go from the initial to the final state, but the amount of work does depend on the path used, the amount of heat given off or absorbed must depend on the path.

The thermodynamic properties of a system that are state functions are usually symbolized by capital letters ($T$, $V$, $P$, $E$, and so on). Thermodynamic properties that are not state functions are often described by lowercase letters ($q$ and $w$).

## 5.12 MEASURING HEAT WITH A CALORIMETER

Thermodynamics can be applied to any system, but the systems of interest in this chapter are chemical reactions. The primary question therefore becomes: "How can we measure the heat given off or absorbed by a chemical reaction?"

The first law of thermodynamics suggests that both heat and work can affect the internal energy of a system:

$$\Delta E_{\text{sys}} = q + w$$

If we can find a way to prevent the system from doing work on its surroundings (or vice versa) the amount of heat given off or absorbed by a chemical reaction will be equal to the change in the system's internal energy:

$$\Delta E = q \qquad \text{(if and only if } w = 0)$$

We can therefore measure the heat given off or absorbed in a reaction by determining the change in temperature that occurs when the reaction is run under conditions in which no work is done by the system on its surroundings, or vice versa.

This raises a related question: ''How can we prevent the system from doing work on its surroundings (or vice versa) while we measure the heat given off or absorbed by a reaction?'' For most chemical reactions, the only kind of work we have to concern ourselves with is work of expansion.

According to Section 5.10, the work done by a system on its surroundings when it expands can be described by the following equation:

$$w = -P\Delta V$$

This equation suggests that we can prevent the system from doing work of expansion by holding the volume of the reaction constant. If there is no change in the volume of the system, there can be no work of expansion.

Heat measured under conditions of constant volume is given the symbol $q_v$. When a chemical reaction is run in a container at constant volume, the change in the internal energy that accompanies the reaction is equal to the heat given off or absorbed by the reaction:

$$\Delta E_{sys} = q_v \qquad \text{(when } V \text{ is constant and therefore } w = 0)$$

These measurements are made in an apparatus known as a **bomb calorimeter,** shown in Figure 5.10. The term *bomb* refers to the fact that the reaction is assumed to take place rapidly, as if the reactants exploded. The apparatus is a *calorimeter* because it is literally a meter for measuring calories (or joules).

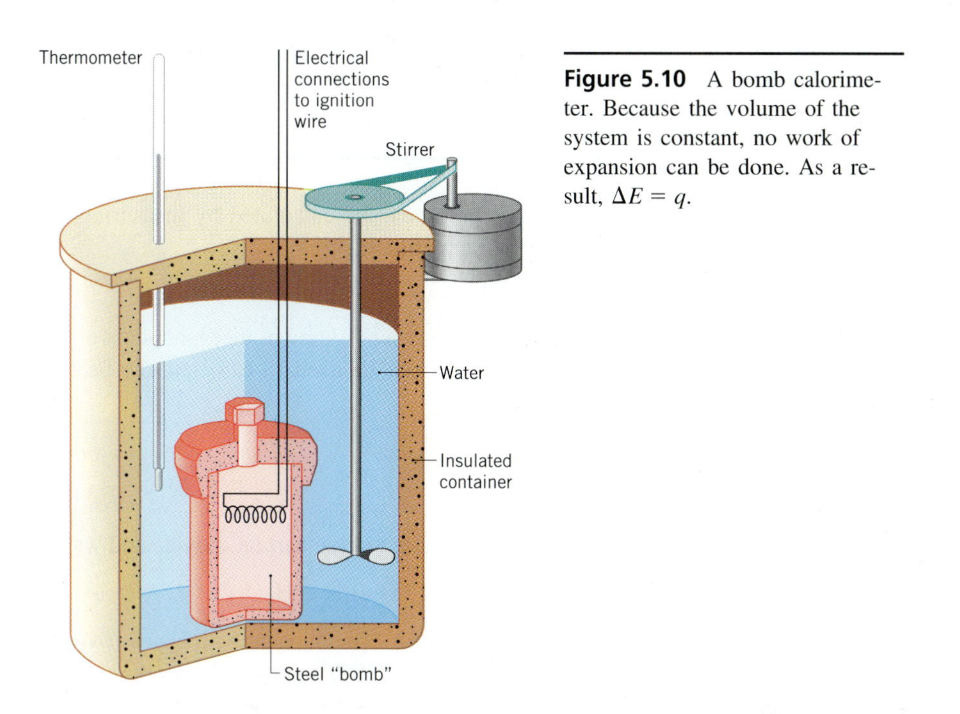


**Figure 5.10** A bomb calorimeter. Because the volume of the system is constant, no work of expansion can be done. As a result, $\Delta E = q$.

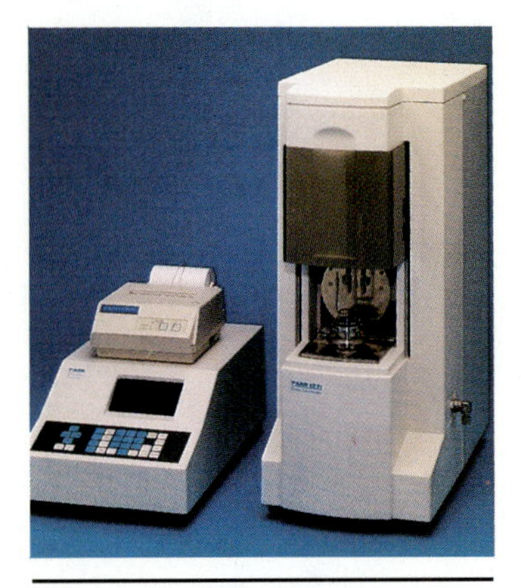

A photograph of a bomb calorimeter.

After the reactants have been added to the bomb, it is sealed and then immersed in a container that holds a known amount of water. The heat given off or absorbed in the reaction is measured by determining what happens to the temperature of the water bath that surrounds the "bomb."

The calorimeter is designed so that the heat given off by the reaction is absorbed by either the bomb in which the reaction is run or the water that surrounds the bomb.

$$q_{\text{reaction}} = -(q_{\text{bomb}} + q_{\text{water}})$$

The magnitude of the heat given off by the reaction is equal to the magnitude of the heat absorbed by the bomb and the water in the calorimeter. The signs of these terms are different, however, because the bomb and the water absorb heat that is given off by the reaction.

The amount of heat given off or absorbed by the water in a calorimeter can be calculated from the molar heat capacity of water. As noted in Section 5.2, it takes 75.376 joules of heat to raise the temperature of 1 mole of water by 1°C or 1 K:

$$C_{\text{water}} = 75.376 \text{ J/mol-K}$$

The units of this constant suggest that the heat given off or absorbed by the water is equal to the product of the number of moles of water in the bath times the molar heat capacity of water times the change in the temperature of the water in kelvin.

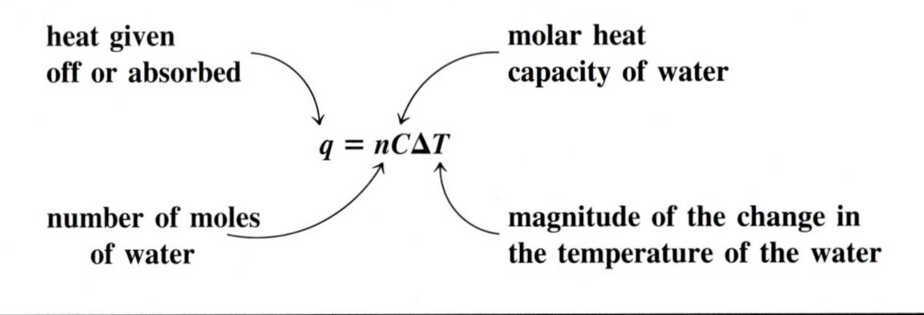


## ▶ CHECKPOINT

Show how the units of molar heat capacity (J/mol-K) can be used to derive the following equation, if you forget the equation.

$$q = nC\Delta T$$

The amount of heat absorbed by the bomb can be calculated by measuring the heat capacity of the bomb in a separate experiment.

$$q_{\text{bomb}} = C_{\text{bomb}}\Delta T$$

Once this characteristic property of the calorimeter is known, measurements of the change in the temperature of the water bath can be used to calculate the heat of reaction.

## EXERCISE 5.4

The natural gas in methane reacts with oxygen to give carbon dioxide and water:

$$CH_4(g) + 2\ O_2(g) \longrightarrow CO_2(g) + 2\ H_2O(g)$$

Calculate the heat given off when 0.160 g of methane reacts with excess oxygen in a bomb calorimeter with a heat capacity of 958 J/°C if the temperature of the 1.000 kg of water in the bath surrounding the bomb increases by 1.56°C.

**SOLUTION** The amount of heat absorbed by the water can be calculated from the number of moles of water that captured this heat ($n$), the molar heat capacity of water ($C$), and the change in the temperature of the water ($\Delta T$). We might therefore start by calculating the number of moles of water in 1.000 kilogram:

$$1000\,\cancel{g} \times \frac{1 \text{ mol } H_2O}{18.02\,\cancel{g}} = 55.49 \text{ mol } H_2O$$

We then substitute the known values for the amount of water in the bath surrounding the calorimeter, the heat capacity of water, and the change in the temperature of water into the following equation:

$$q_{water} = nC\Delta T = (55.49\,\cancel{\text{mol}})(75.376 \text{ J}/\cancel{\text{mol-K}})(1.56\,\cancel{\text{K}}) = 6525 \text{ J}$$

The heat absorbed by the bomb can be calculated from the heat capacity of the bomb and the change in the temperature of the system:

$$q_{bomb} = (958 \text{ J/°C})(1.56\text{°C}) = 1494 \text{ J}$$

The sum of these two calculations gives the total amount of heat absorbed by the calorimeter:

$$q_{bomb} + q_{water} = 8019 \text{ J}$$

We can therefore conclude that the reaction gives off 8.02 kilojoules of energy in the form of heat:

$$q_{reaction} = -8.02 \text{ kJ}$$

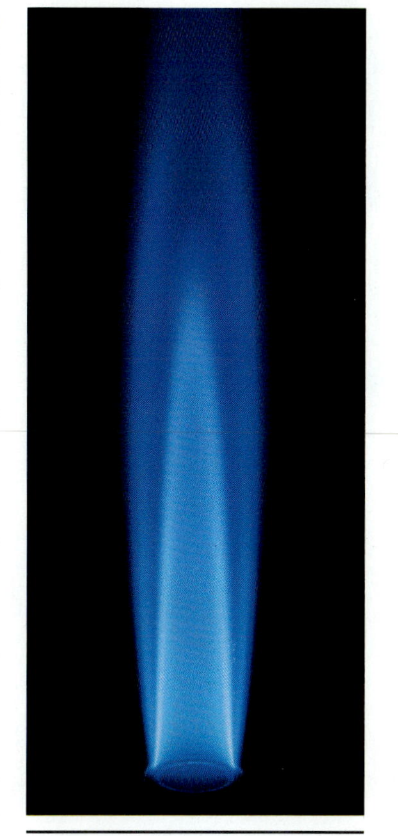

The heat given off when methane burns is an extensive quantity, which obviously depends on the size of the sample consumed. To obtain an intensive quantity, we have to measure the heat given off per gram or per mole of methane consumed.

Exercise 5.4 illustrates an important point: Heat is an *extensive quantity*. The amount of heat given off when methane burns depends on the size of the sample. (In the presence of excess $O_2$, doubling the amount of $CH_4$ in the calorimeter doubles the amount of heat given off.) To compare this reaction with other sources of heat, we have to find a way to convert this measurement into an *intensive quantity*.

We can do this by dividing the heat given off by another extensive quantity that describes the size of the sample. We could calculate the amount of heat given off per gram of $CH_4$ consumed, for example:

$$\frac{8.022 \text{ kJ}}{0.160 \text{ g}} = \mathbf{50.1 \text{ kJ/g}}$$

Among chemists, however, the most common approach for converting measurements of heat into an intensive quantity is to calculate the heat of reaction in units of kilojoules per mole. Because 0.160 grams of methane corresponds to 0.0100 moles, the heat of reaction for the combustion of methane is 802 kJ/mol:

$$\frac{8.022 \text{ kJ}}{0.0100 \text{ mol}} = \mathbf{802 \text{ kJ/mol}}$$

The result of this calculation is a quantity known as the **molar heat of reaction.** By definition, the molar heat of reaction is the heat given off or absorbed by the reaction expressed in units of kilojoules per mole of one of the reagents in the reaction.

## 5.13 ENTHALPY VERSUS INTERNAL ENERGY

The previous section took advantage of the fact that the system can't do work of expansion on its surroundings (or vice versa) when a chemical reaction is run in a closed container at constant volume. The heat given off or absorbed by the reaction is therefore equal to the change in the internal energy of the system.

$$\Delta E_{sys} = q_v \qquad \text{(if and only if } \Delta V = 0\text{)}$$

Most chemical reactions are run in beakers and flasks in which the volume of the system is allowed to change, but the *pressure is constant.*

Chemists, however, seldom run reactions in sealed containers at constant volume. They are more likely to run reactions in an open flask. The system isn't at *constant volume* because gas can enter or leave the flask during the reaction. The system is at *constant pressure,* however, because the total pressure inside the flask is always equal to atmospheric pressure.

If a gas is driven out of the flask during the reaction, the system will do work of expansion on its surroundings. Alternatively, if the reaction pulls a gas into the flask, the surroundings will do work on the system. Either way, the change in the internal energy of the system will be the sum of the heat and the work transferred between the system and its surroundings:

$$\Delta E_{sys} = q + w$$

We can still measure the heat given off or absorbed by the reaction. But it is no longer equal to the change in the system's internal energy.

Fortunately there is a way out of this problem. We start by introducing a new concept known as the **enthalpy (*H*)** of the system. By definition, the enthalpy of the system is the sum of the internal energy of the system plus the product of the pressure of the gas in the system times its volume:

$$\mathbf{H_{sys} = E_{sys} + PV}$$

The change in the enthalpy of a system during a chemical reaction is equal to the change in the sum of the internal energy plus the pressure times the volume of the system:

$$\Delta H_{sys} = \Delta(E_{sys} + PV)$$

This in turn is equal to the change in the internal energy of the system plus the change in the product of the pressure times the volume of the system:

$$\Delta H_{sys} = \Delta E_{sys} + \Delta(PV)$$

If the pressure of the system is constant, the change in the product of the pressure times the volume of the gas in the system is equal to the product of the constant pressure times the change in the volume of the gas. Thus, at constant pressure, the change in the enthalpy of the system is the sum of the change in the internal energy plus the pressure times the change in volume:

$$\Delta H_{sys} = \Delta E_{sys} + P\Delta V \qquad \text{(at constant pressure)}$$

Substituting the first law of thermodynamics into this equation gives the following result:

$$\Delta H_{sys} = (q + w) + P\Delta V \qquad \text{(at constant pressure)}$$

Assuming that the only work done by the reaction is work of expansion gives the following equation:

$$\Delta H_{sys} = (q - \cancel{P\Delta V}) + \cancel{P\Delta V} \qquad \text{(at constant pressure)}$$

Canceling terms in this equation tells us that the heat given off or absorbed by a chemical reaction run under conditions of constant pressure is equal to the change in the enthalpy of the system:

$$\Delta H_{sys} = q_p \qquad \text{(at constant pressure)}$$

For the sake of simplicity, the subscript "sys" will be left off the symbol for both the internal energy of the system and the enthalpy of the system from now on. We will therefore abbreviate the relationship between the change in the enthalpy of the system and the heat given off or absorbed at constant pressure as follows:

$$\Delta H = q_p$$

By definition, $\Delta H$ is the enthalpy of the final state minus the initial state of the system:

$$\Delta H = H_{final} - H_{initial}$$

When this equation is applied to a chemical reaction, the final state corresponds to the products of the reaction and the initial state of the system is the reactants. The change in the enthalpy of the system as the reactants are converted into the products of the reaction is therefore known as the **enthalpy of reaction.**

Our discussion of the relationships between heat, the internal energy of the system, and the enthalpy of the system during a chemical reaction can be summarized as follows:

1. The heat given off or absorbed when a reaction is run at *constant volume* is equal to the change in the internal energy of the system:

$$\Delta E = q_v$$

2. The heat given off or absorbed when a reaction is run at *constant pressure* is equal to the change in the enthalpy of the system:

$$\Delta H = q_p$$

3. The change in the enthalpy of the system during a chemical reaction is equal to the change in the internal energy plus the change in the product of the pressure of the gas in the system times its volume:

$$\Delta H = \Delta E + \Delta(PV)$$

4. The difference between $\Delta H$ and $\Delta E$ for the system is relatively small for reactions that involve only liquids and solids because there is little, if any, change in the volume of the system during the reaction. This difference can be significant, however, for reactions that involve gases if there is a change in the number of moles of gas in the course of the reaction.

## EXERCISE 5.5

For which of the following reactions is $\Delta H$ about the same as $\Delta E$?

(a) $CaCO_3(s) \longrightarrow CaO(s) + CO_2(g)$
(b) $2\ NH_3(g) \longrightarrow N_2(g) + 3\ H_2(g)$
(c) $Fe_2O_3(s) + 2\ Al(s) \longrightarrow Al_2O_3(s) + 2\ Fe(l)$
(d) $CH_4(g) + 2\ O_2(g) \longrightarrow CO_2(g) + 2\ H_2O(g)$

**SOLUTION** $\Delta H$ is roughly the same as $\Delta E$ for reactions (c) and (d). Reaction (c) only involves solids, so there is no significant change in the volume of the system during this reaction. Reaction (d) involves gases, but the number of moles of gas remains constant during the reaction, so the volume of the system doesn't change. The volume of the system increases significantly in reactions (a) and (b), which means that $\Delta H$ for these is significantly larger than $\Delta E$.

## 5.14 ENTHALPIES OF REACTION

Chemical reactions are divided into two classes on the basis of whether they give off or absorb heat from their surroundings. **Exothermic** reactions give off heat to their surroundings. **Endothermic** reactions absorb heat from their surroundings.

The prefix *exo-* is used to mean ''out'' in words such as *exodus* (to go out), *exorcise* (to drive out), and so on. Reactions that give off heat are therefore *exothermic*. The prefix *endo-* comes from the Greek word meaning ''within.'' Thus, we have words such as *endogenous* (to grow from within), *endoskeleton* (the internal bony structure of vertebrates), and *endoplasm* (the inner part of the cytoplasm of a cell). Reactions that absorb, or take in, heat from their surroundings are therefore *endothermic*.

In the previous section, the heat given off or absorbed in a chemical reaction at constant pressure was defined as the enthalpy of reaction. By definition, this is the difference between the sum of the enthalpies of the products of the reaction and the sum of the enthalpies of the starting materials:

$$\Delta H = \Sigma n H_{\text{products}} - \Sigma m H_{\text{reactants}}$$

In this equation, $n$ and $m$ are the stoichiometric coefficients for the reaction.

At constant pressure, when a reaction gives off heat to its surroundings, the enthalpy of the system decreases. Because the sum of the enthalpies of the products is smaller than the sum of the enthalpies of the reactants, exothermic reactions are characterized by negative values of $\Delta H$:

**Exothermic Reactions: $\Delta H$ is negative ($\Delta H < 0$)**

Endothermic reactions, on the other hand, take in heat from their surroundings. As a result, the enthalpy of the system increases. Endothermic reactions are therefore characterized by positive values of $\Delta H$:

**Endothermic Reactions: $\Delta H$ is positive ($\Delta H > 0$)**

A seemingly endless list of chemical reactions could be cited as examples of exothermic reactions. The reaction between sodium and chlorine to form sodium chloride, for example, gives off 411.15 kilojoules of energy per mole of NaCl formed. There are several ways this information can be added to the balanced equation for the reaction. One approach assumes that the balanced equation is written in terms of moles. When two moles of sodium react with a mole of chlorine, for example, two moles of sodium chloride are formed. Thus, a total of 822.30 kilojoules of energy is released:

$$2\,\text{Na}(s) + \text{Cl}_2(g) \longrightarrow 2\,\text{NaCl}(s) \qquad \Delta H = -822.30\text{ kJ}$$

Another approach reports values of the enthalpy of reaction per mole of one of the reactants or products. This approach is indicated as follows:

$$2\,\text{Na}(s) + \text{Cl}_2(g) \longrightarrow 2\,\text{NaCl}(s) \qquad \Delta H = -411.15\text{ kJ/mol NaCl}$$

Most endothermic reactions have to be driven by some external force, much as work has to be done to roll a boulder uphill. An example of this phenomenon is the electrolysis of molten sodium chloride.

$$2\,\text{NaCl}(l) \xrightarrow[\text{current}]{\text{electric}} 2\,\text{Na}(s) + \text{Cl}_2(g)$$

By doing work on the system in the form of passing an electric current through molten sodium chloride, it is possible to drive this reaction "uphill" to form sodium metal and chlorine gas.

A handful of endothermic reactions are spontaneous. One example of a spontaneous endothermic reaction is the basis of a commercial product, an ice pack that doesn't have to be kept in the freezer. These ice packs contain a small quantity of ammonium nitrate ($NH_4NO_3$) or ammonium chloride ($NH_4Cl$), which is separated from a sample of water by a thin membrane. When the pack is struck with the palm of the hand, the membrane is broken, and the salt dissolves in the water.

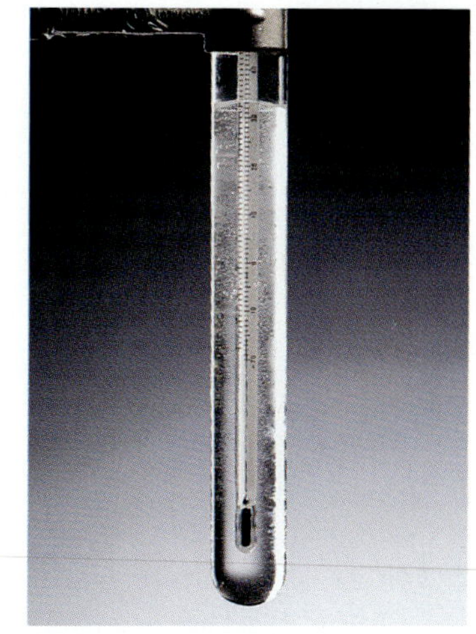

Enough heat is absorbed when $NH_4NO_3$ dissolves in water to lower the temperature of the water from 24°C to 1°C.

$$NH_4NO_3(s) \xrightarrow{H_2O} NH_4^+(aq) + NO_3^-(aq) \qquad \Delta H_{rxn} = 25.7 \text{ kJ/mol}$$

Because the reaction is endothermic, it absorbs heat from its surroundings, and the ice pack can get cold enough to treat minor athletic injuries.

Another spontaneous endothermic reaction can be demonstrated by mixing ammonium thiocyanate, $NH_4SCN$, and barium hydroxide octahydrate, $Ba(OH)_2 \cdot 8\ H_2O$. The starting materials are both solids, but when they react, water is released.

$$2\ NH_4SCN(s) + Ba(OH)_2 \cdot 8\ H_2O(s) \longrightarrow$$
$$Ba^{2+}(aq) + 2\ SCN^-(aq) + 2\ NH_3(aq) + 10\ H_2O(l)$$
$$\Delta H_{rxn} = 130 \text{ kJ/mol } Ba(OH)_2$$

If a little water is poured on the outside of the beaker in which this reaction is run, stirring the contents of the beaker will produce a reaction that absorbs enough heat from its surroundings to freeze the beaker to a plywood board.

The reaction between $NH_4SCN$ and $Ba(OH)_2$ is another example of a spontaneous endothermic reaction. This reaction absorbs so much heat from its surroundings that it can freeze its container to a plywood board if the outside of the container is moistened.

## EXERCISE 5.6

Use your experience with ice, water, and steam to predict which of the following reactions are exothermic and which are endothermic.

(a) $H_2O(s) \longrightarrow H_2O(l)$
(b) $H_2O(l) \longrightarrow H_2O(g)$
(c) $H_2O(g) \longrightarrow H_2O(l)$

**SOLUTION**

(a) This reaction is endothermic because it takes energy in the form of heat to convert any solid into a liquid. As the ice cubes in a glass of iced tea melt, for example, the tea becomes colder. Heat is transferred from the tea (the surroundings) to the ice cubes (the system) as the ice melts.

(b) This reaction is also endothermic because it takes energy to overcome the force of attraction between molecules in a liquid to form a gas. When you place a moist cloth on your forehead on a hot summer day, you feel cooler as the water evaporates. Once again, heat is transferred from the surroundings (your forehead) into the system (the water).

(c) This reaction must be exothermic because it is the opposite of the previous reaction. Steam causes more severe burns than hot water because it also releases heat to the skin as it condenses to form the liquid.

Exercise 5.6 introduces an important point: Reversing the direction in which a reaction is written cannot change the magnitude of the enthalpy of reaction, only the sign of $\Delta H$. It takes heat to convert a liquid into a gas. At 100°C, for example, it takes 40.88 kilojoules of energy in the form of heat to convert a mole of liquid water into a mole of water vapor:

$$H_2O(l) \longrightarrow H_2O(g) \qquad \Delta H_{373} = 40.88 \text{ kJ/mol}$$

Reversing the direction in which the reaction is written changes the sign of $\Delta H$ because the initial and final states of the system have been reversed. Heat is therefore given off when water vapor condenses, and $\Delta H$ for this reaction at 100°C must be $-40.88$ kJ/mol:

$$H_2O(g) \longrightarrow H_2O(l) \qquad \Delta H_{373} = -40.88 \text{ kJ/mol}$$

**CHECKPOINT**

As we have seen, the symbol $\Delta$ is used to describe what happens when the initial value of some quantity is subtracted from its final value. For example,

$$\Delta H = H_f - H_i$$

Use this fact to explain why the sign of $\Delta H$ changes when the direction in which a chemical reaction is written changes but the magnitude of $\Delta H$ stays the same.

## 5.15 STANDARD-STATE ENTHALPIES OF REACTION

The heat given off or absorbed by a chemical reaction depends on the conditions of the reaction. Three factors are particularly important: (1) the concentrations of the reactants and products involved in the reaction, (2) the temperature of the system, and (3) the partial pressure of any gases involved in the reaction. The combustion of methane can be used to illustrate the magnitude of the problem.

Assume that we start with a mixture of $CH_4$ and $O_2$ in which the partial pressure of each gas is 1 atm and the temperature of the system is 25°C. Furthermore assume that we run the following reaction and then let the products cool to 25°C:

$$CH_4(g) + 2\,O_2(g) \longrightarrow CO_2(g) + 2\,H_2O(g)$$

Under these conditions, the reaction gives off a total of 802.4 kilojoules of energy per mole of $CH_4$ consumed. If we start with the reactants at 1000°C and 1 atm pressure, and return the products to the same conditions, the reaction gives off only 792.4 kJ/mol. The difference between these numbers is small (10.0 kJ/mol), but it is still 100 times larger than the experimental error ($\pm$0.1 kJ/mol) with which the measurements were made.

Thermodynamic data are often measured at 25°C (298 K). Measurements taken at other temperatures are identified by adding a subscript specifying the temperature in kelvin. The data collected for the combustion of methane at 1000°C (1273 K), for example, would be reported as: $\Delta H_{1273} = 792.4$ kJ/mol.

The effect of pressure and concentration on thermodynamic data is controlled by defining a set of standard conditions for thermodynamic experiments. By definition, the **standard state** for thermodynamic measurements fulfills the following requirements:

1. The partial pressures of any gas involved in the reaction is 0.1 MPa.
2. The concentrations of all aqueous solutions are 1 *M*.

Measurements done under standard-state conditions are indicated by adding a superscript ° to the symbol of the quantity being reported. The *standard-state* enthalpy of reaction for the combusion of natural gas at 25°C, for example would be reported as: $\Delta H° = -802.4$ kJ/mol $CH_4$.

## 5.16 HESS'S LAW

Section 5.13 defined the enthalpy of a system in terms of the internal energy, pressure, and volume of the gas in the system:

$$H = E + PV$$

Because the internal energy, pressure, and volume of a gas are all state functions, the enthalpy of a system is also a state function. As a result, the difference between the initial and final values of the enthalpy of a system does not depend on the path used to go from one of these states to the other.

In 1840, Germain Henri Hess, professor of chemistry at the University and Artillery School in St. Petersburg, Russia, came to the same conclusion on the basis of experiment and proposed a general rule known as **Hess's law,** which states that the enthalpy of reaction ($\Delta H$) is the same regardless of whether a reaction occurs in one step or in several steps. We can therefore calculate the enthalpy of reaction by adding the enthalpies associated with a series of hypothetical steps into which the reaction can be broken.

**EXERCISE 5.7**

The standard-state molar enthalpies of reaction for the formation of water as both a liquid and a gas have been measured:

$$H_2(g) + \tfrac{1}{2}\,O_2(g) \longrightarrow H_2O(l) \qquad \Delta H^\circ = -285.83\text{ kJ/mol }H_2O$$
$$H_2(g) + \tfrac{1}{2}\,O_2(g) \longrightarrow H_2O(g) \qquad \Delta H^\circ = -241.82\text{ kJ/mol }H_2O$$

Use these data and Hess's law to calculate $\Delta H^\circ$ for the following reaction:

$$H_2O(l) \longrightarrow H_2O(g)$$

**Problem-Solving Strategy**

**SOLUTION** The key to solving this problem is finding a way to combine the two reactions for which experimental data are known to give the reaction for which $\Delta H^\circ$ is unknown. We can do this by reversing the direction in which the first reaction is written and then adding this to the equation for the second reaction. For the sake of argument, we assume that 1 mole of water is decomposed into its elements in the first reaction and a mole of water vapor is formed from its elements in the second reaction, as shown in Figure 5.11.

$$H_2O(l) \longrightarrow \cancel{H_2(g)} + \cancel{\tfrac{1}{2}\,O_2(g)} \qquad \Delta H^\circ = 1\text{ mol }H_2O \times 285.83\text{ kJ/mol}$$
$$\cancel{H_2(g)} + \cancel{\tfrac{1}{2}\,O_2(g)} \longrightarrow H_2O(g) \qquad \Delta H^\circ = 1\text{ mol }H_2O \times -241.82\text{ kJ/mol}$$
$$H_2O(l) \longrightarrow H_2O(g) \qquad \Delta H^\circ = 44.01\text{ kJ}$$

(Note that when we reversed the direction in which the first reaction was written we had to change the sign of $\Delta H^\circ$ for this reaction.)

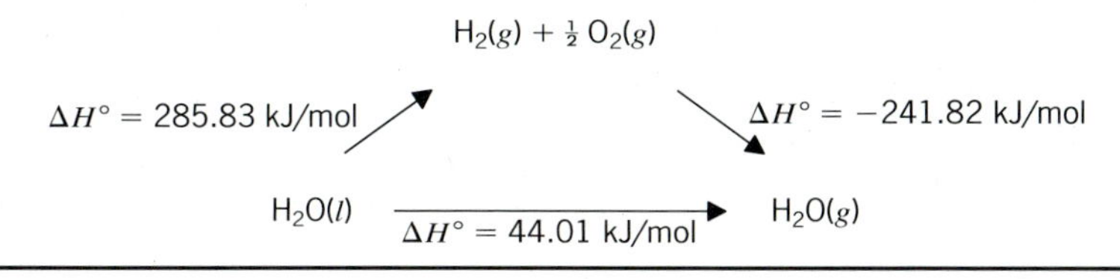

**Figure 5.11** Exercise 5.7

In Section 5.14, the enthalpy of reaction for the following process was given as 40.88 kJ/mol:

$$H_2O(l) \longrightarrow H_2O(g) \qquad \Delta H^\circ_{373} = 40.88\text{ kJ/mol}$$

The calculation in Exercise 5.7 gives a value that is about 8% larger:

$$H_2O(l) \longrightarrow H_2O(g) \qquad \Delta H^\circ = 44.01 \text{ kJ/mol}$$

This discrepancy can be explained by noting that these standard-state enthalpies of reaction were measured at different temperatures. The value of $\Delta H^\circ$ reported in Section 5.14 was measured at 100°C. The value calculated in this exercise was measured at 25°C.

There was just enough information in Exercise 5.7 to solve the problem. The next exercise forces us to choose the reactions that will be combined from a wealth of information.

**EXERCISE 5.8**

Before pipelines were built to deliver natural gas, individual towns and cities contained plants that produced a fuel known as *town gas* by passing steam over red-hot charcoal:

$$C(s) + H_2O(g) \longrightarrow CO(g) + H_2(g)$$

Calculate $\Delta H^\circ$ for this reaction from the following information:

$$C(s) + \tfrac{1}{2} O_2(g) \longrightarrow CO(g) \qquad \Delta H^\circ = -110.53 \text{ kJ/mol CO}$$
$$C(s) + O_2(g) \longrightarrow CO_2(g) \qquad \Delta H^\circ = -393.51 \text{ kJ/mol } CO_2$$
$$CO(g) + \tfrac{1}{2} O_2(g) \longrightarrow CO_2(g) \qquad \Delta H^\circ = -282.98 \text{ kJ/mol } CO_2$$
$$H_2(g) + \tfrac{1}{2} O_2(g) \longrightarrow H_2O(g) \qquad \Delta H^\circ = -241.82 \text{ kJ/mol } H_2O$$

**SOLUTION** The key to success in this calculation is recognizing that we can obtain the desired equation by using the first reaction to generate $CO(g)$ from $C(s)$ and the reverse of the fourth reaction to generate $H_2(g)$ from $H_2O(g)$, as shown in Figure 5.12:

$$C(s) + \cancel{\tfrac{1}{2} O_2(g)} \longrightarrow CO(g) \qquad \Delta H^\circ = 1 \text{ mol CO} \times -110.53 \text{ kJ/mol}$$
$$H_2O(g) \longrightarrow H_2(g) + \cancel{\tfrac{1}{2} O_2(g)} \qquad \Delta H^\circ = 1 \text{ mol } H_2O \times 241.82 \text{ kJ/mol}$$
$$C(s) + H_2O(g) \longrightarrow CO(g) + H_2(g) \qquad \Delta H^\circ = \mathbf{131.29 \text{ kJ}}$$

$$C(s) + H_2O(g)$$
$$\downarrow \Delta H^\circ = 241.82 \text{ kJ/mol}$$
$$C(s) + \tfrac{1}{2} O_2(g) + H_2(g)$$
$$\downarrow \Delta H^\circ = -110.53 \text{ kJ/mol}$$
$$CO(g) + H_2(g)$$

**Figure 5.12** Exercise 5.8

## 5.17 ENTHALPIES OF FORMATION

Hess's law suggests that we can save a great deal of work measuring enthalpies of reaction by using a little imagination in choosing the reactions for which measurements are made. The question is, what is the best set of reactions to study so that we get the greatest benefit from the smallest number of experiments?

Exercise 5.8 predicted $\Delta H^\circ$ for the reaction:

$$C(s) + H_2O(g) \longrightarrow CO(g) + H_2(g)$$

by combining enthalpy of reaction measurements for the following reactions:

$$C(s) + \tfrac{1}{2}\,O_2(g) \longrightarrow CO(g)$$
$$H_2(g) + \tfrac{1}{2}\,O_2(g) \longrightarrow H_2O(g)$$

The reactions used to solve this exercise have one thing in common. Each leads to the formation of a compound from its elements in their most thermodynamically stable form. The enthalpy of reaction for each of these reactions is therefore the **enthalpy of formation** of the compound, $\Delta H_f^\circ$.

**By definition, $\Delta H_f^\circ$ is the enthalpy associated with the reaction that forms a compound from its elements in their most thermodynamically stable states.**

## EXERCISE 5.9

Which of the following equations describes a reaction for which $\Delta H^\circ$ is equal to the enthalpy of formation of a compound, $\Delta H_f^\circ$?

(a) $2\,Mg(s) + O_2(g) \longrightarrow 2\,MgO(s)$

(b) $MgO(s) + CO_2(g) \longrightarrow MgCO_3(s)$

(c) $Mg(s) + C(s) + \tfrac{3}{2}\,O_2(g) \longrightarrow MgCO_3(s)$

**SOLUTION** Equations (a) and (c) describe enthalpy of formation reactions. Each of these reactions results in the formation of a compound from the most thermodynamically stable form of its elements. Equation (b) can't be an enthalpy of formation reaction because the product of this reaction is not formed from its elements.

Hess's law can be used to calculate the enthalpy of reaction for a chemical reaction from the enthalpies of formation of the reactants and products of the reaction.

## EXERCISE 5.10

Use Hess's law to calculate $\Delta H^\circ$ for the reaction:

$$MgO(s) + CO_2(g) \longrightarrow MgCO_3(s)$$

from the following enthalpy of formation data:

$$Mg(s) + \tfrac{1}{2}\,O_2(g) \longrightarrow MgO(s) \qquad \Delta H_f^\circ = -601.70 \text{ kJ/mol MgO}$$
$$C(s) + O_2(g) \longrightarrow CO_2(g) \qquad \Delta H_f^\circ = -393.51 \text{ kJ/mol } CO_2$$
$$Mg(s) + C(s) + \tfrac{3}{2}\,O_2(g) \longrightarrow MgCO_3(s) \qquad \Delta H_f^\circ = -1095.8 \text{ kJ/mol } MgCO_3$$

**SOLUTION** The reaction in which we are interested converts two reactants ($MgO$ and $CO_2$) into a single product ($MgCO_3$). We might therefore start by reversing the direction in which we write the enthalpy of formation reactions for $MgO$ and $CO_2$—thereby decomposing these substances into their elements in their most thermodynamically stable form. We can then add the enthalpy of formation reaction for $MgCO_3$—thereby forming the product from its elements in their most stable form:

$$\begin{array}{ll} MgO(s) \longrightarrow \cancel{Mg(s)} + \cancel{\tfrac{1}{2} O_2(g)} & \Delta H^\circ = 1 \text{ mol MgO} \times 601.70 \text{ kJ/mol} \\ CO_2(g) \longrightarrow \cancel{C(s)} + \cancel{O_2(g)} & \Delta H^\circ = 1 \text{ mol } CO_2 \times 393.51 \text{ kJ/mol} \\ \cancel{Mg(s)} + \cancel{C(s)} + \cancel{\tfrac{3}{2} O_2(g)} \longrightarrow MgCO_3(s) & \Delta H^\circ = 1 \text{ mol } MgCO_3 \times -1095.8 \text{ kJ/mol} \\ \hline MgO(s) + CO_2(g) \longrightarrow MgCO_3(s) & \Delta H^\circ = \mathbf{-100.6\ kJ} \end{array}$$

Adding these three equations gives the desired unknown reaction. $\Delta H^\circ$ for this reaction is therefore the sum of the enthalpies of these three hypothetical steps.

No matter how complex the reaction, the procedure used in Exercise 5.10 works. All we have to do as the reaction becomes more complex is add more intermediate steps. Analysis of the results of Exercise 5.10, however, can be used to construct an algorithm that greatly simplifies problems of this nature.

We obtained the answer to this exercise by adding the enthalpy of formation of each of the products and subtracting the enthalpy of formation of each of the reactants. In general, the enthalpy of reaction for any chemical reaction is equal to the difference between the sum of the enthalpies of formation of the products and the sum of the enthalpies of formation of the reactants:

$$\boldsymbol{\Delta H^\circ = \Sigma n \Delta H^\circ_{f,products} - \Sigma m \Delta H^\circ_{f,reactants}}$$

In this equation, $n$ and $m$ are the stoichiometric coefficients for the reaction. The reaction in Exercise 5.10 obeys a 1:1:1 stoichiometry. The enthalpy of reaction therefore can be calculated from the enthalpy of formation as follows:

$$\begin{aligned} \Delta H^\circ &= [\Delta H^\circ_f\ MgCO_3] - [\Delta H^\circ_f\ MgO + \Delta H^\circ_f\ CO_2] \\ &= [1\ \cancel{mol}\ MgCO_3 \times -1095.8\ kJ/\cancel{mol}] \\ &\qquad - [1\ \cancel{mol}\ MgO \times -601.70\ kJ/\cancel{mol} + 1\ \cancel{mol}\ CO_2 \times -393.51\ kJ/\cancel{mol}] \\ &= \mathbf{-100.6\ kJ} \end{aligned}$$

The formula for calculating $\Delta H^\circ$ for a reaction from enthalpy of formation data works because enthalpy is a state function. Thus, $\Delta H^\circ$ is the same regardless of the path used to get from the starting materials to the products of the reaction. Instead of running the reaction in a single step,

$$MgO(s) + CO_2(g) \longrightarrow MgCO_3(s)$$

we can split it into two steps. In the first step, the starting materials are converted to the elements from which they form in their most thermodynamically stable states:

$$MgO(s) + CO_2(g) \longrightarrow Mg(s) + C(s) + \tfrac{3}{2} O_2(g)$$

In the second step, these elements combine to form the products of the reaction:

$$Mg(s) + C(s) + \tfrac{3}{2} O_2(g) \longrightarrow MgCO_3(s)$$

The first step is the opposite of the enthalpy of formation reactions for the two reactants. The second step corresponds to the enthalpy of formation reaction for the product of the original reaction. Combining these two steps is therefore equivalent to subtracting the sum of the enthalpies of formation of the reactants from the sum of the enthalpies of formation of the products, as shown in Figure 5.13.

Standard-state enthalpy of formation data for a variety of elements and compounds can be found in Table A-15 in the appendix. One more point needs to be understood before this table can be used effectively. By definition, the enthalpy of formation of any element in its most thermodynamically stable form is zero.

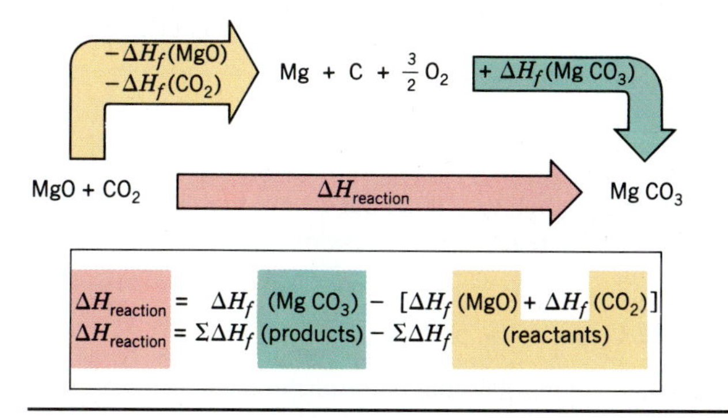


**Figure 5.13** Because enthalpy is a state function, $\Delta H°$ for a reaction is equal to the difference between the sum of the enthalpy of formation of the products and the sum of the enthalpy of formation of the reactants:

$$\Delta H° = \Sigma n\Delta H°_{f,products} - \Sigma m\Delta H°_{f,reactants}$$

$\Delta H°$ is therefore the same regardless of whether the reaction goes directly from the starting materials to the products or through a hypothetical intermediate, which consists of the elements in their most stable states. Converting the reactants into their elements is the same as subtracting $\Delta H_f°$ for these reagents, and converting the elements into their products is the same as adding $\Delta H_f°$ for the products.

Under standard-state conditions, the most thermodynamically stable form of oxygen, for example, is the diatomic molecule in the gas phase: $O_2(g)$. By definition, the enthalpy of formation of this substance is equal to the enthalpy associated with the reaction in which it is formed from its elements in their most thermodynamically stable form. For $O_2$ molecules in the gas phase, $\Delta H_f°$ is therefore equal to the heat given off or absorbed in the following reaction:

$$O_2(g) \longrightarrow O_2(g)$$

Because the initial and final states of this reaction are identical, no heat can be given off or absorbed, so $\Delta H_f°$ for $O_2(g)$ is zero.

## EXERCISE 5.11

Which of the following substances should have a standard-state enthalpy of formation equal to zero?

(a) $Hg(l)$ (b) $Br_2(g)$ (c) $H(g)$

**SOLUTION**

(a) The standard-state enthalpy of formation of liquid mercury is zero, because the most stable form of this element under standard-state conditions is a liquid.

(b) Bromine is a liquid, not a gas, under standard-state conditions. Because it takes heat to boil a liquid, $\Delta H_f°$ for this compound in the gas phase should be positive. (Table A-15 lists a value of 30.91 kJ/mol for the standard-state enthalpy of formation of $Br_2$ as a gas.)

(c) Hydrogen exists as $H_2$ molecules in the gas phase under standard-state conditions. Because it takes energy to break the bonds in a molecule, $\Delta H_f°$ for hydrogen atoms in the gas phase should be positive. (Table A-15 lists a value of 217.65 kJ/mol for the standard-state enthalpy of formation of hydrogen atoms in the gas phase.)

$\Delta H_f°$ for $Br_2$ as a gas is positive because the most stable form of bromine at 25°C and 1 atm is the liquid.

We are now ready to use standard-state enthalpy of formation data to predict enthalpies of reaction.

**EXERCISE 5.12**

Pentaborane(9), $B_5H_9$, was once studied as a potential rocket fuel. Calculate the heat given off when a mole of $B_5H_9$ reacts with excess oxygen according to the following equation:

$$2\ B_5H_9(g) + 12\ O_2(g) \longrightarrow 5\ B_2O_3(s) + 9\ H_2O(g)$$

**SOLUTION** Table A-15 contains the following data for the reactants and products of this reaction:

| *Compound* | $\Delta H_f^\circ$ *(kJ/mol)* |
|---|---|
| $B_5H_9(g)$ | 73.2 |
| $B_2O_3(s)$ | −1272.77 |
| $O_2(g)$ | 0 |
| $H_2O(g)$ | −241.82 |

The standard-state enthalpy of reaction for this reaction is equal to the sum of the standard-state enthalpies of formation of the products minus the sum of the standard-state enthalpies of formation of the reactants:

$$\Delta H^\circ = \Sigma n \Delta H^\circ_{\text{f,products}} - \Sigma m \Delta H^\circ_{\text{f,reactants}}$$

We can therefore calculate $\Delta H^\circ$ by multiplying the enthalpies of formation of the reactants and products by the number of moles of each reactant or product involved in the reaction:

$$\begin{aligned}\Delta H^\circ &= \Sigma n \Delta H^\circ_{\text{f,products}} - \Sigma m \Delta H^\circ_{\text{f,reactants}} \\ &= [5\ \cancel{\text{mol}}\ B_2O_3 \times -1272.77\ \text{kJ}/\cancel{\text{mol}} + 9\ \cancel{\text{mol}}\ H_2O \times -241.82\ \text{kJ}/\cancel{\text{mol}}] \\ &\qquad - [2\ \cancel{\text{mol}}\ B_5H_9 \times 73.2\ \text{kJ}/\cancel{\text{mol}} + 12\ \cancel{\text{mol}}\ O_2 \times 0\ \text{kJ}/\cancel{\text{mol}}] \\ &= \mathbf{-8686.6\ kJ}\end{aligned}$$

According to the balanced equation, this is the energy given off when *two* moles of $B_5H_9$ are consumed. $\Delta H^\circ$ for this reaction is therefore −4343.5 kJ/mol $B_5H_9$. This is five times the molar enthalpy of reaction for the combustion of $CH_4$. On a per gram basis, it is about 40% larger than the energy released when methane burns.

## 5.18 BOND-DISSOCIATION ENTHALPIES

As we have seen, $\Delta H^\circ$ can be calculated with the following formula when the enthalpy of formation is known for all of the reactants and products of a chemical reaction:

$$\Delta H^\circ = \Sigma n \Delta H^\circ_{\text{f,products}} - \Sigma m \Delta H^\circ_{\text{f,reactants}}$$

Only a limited number of enthalpies of formation have been measured, and it is easy to imagine a reaction for which $\Delta H_f^\circ$ data aren't available for one or more reagents. When this happens, we can no longer predict the exact value of $\Delta H^\circ$ for the reaction, but we can estimate the enthalpy of reaction using **bond-dissociation enthalpies.** By definition, the bond-dissociation enthalpy for an *X*—*Y* bond is the enthalpy of the gas-phase reaction in which this bond is broken to give isolated *X* and *Y* atoms:

$$XY(g) \longrightarrow X(g) + Y(g)$$

The bond-dissociation enthalpy for a C—H bond, for example, can be calculated by combining $\Delta H_f^\circ$ data to give a net equation in which the only thing that happens is the breaking of C—H bonds in the gas phase:

$$\begin{array}{ll} CH_4(g) \longrightarrow C(s) + 2\,H_2(g) & \Delta H^\circ = 1 \text{ mol} \times 74.81 \text{ kJ/mol } CH_4 \\ C(s) \longrightarrow C(g) & \Delta H^\circ = 1 \text{ mol} \times 716.68 \text{ kJ/mol C} \\ 2\,H_2(g) \longrightarrow 4\,H(g) & \Delta H^\circ = 4 \text{ mol} \times 217.65 \text{ kJ/mol H} \\ \hline CH_4(g) \longrightarrow C(g) + 4\,H(g) & \Delta H^\circ = 1662.09 \text{ kJ} \end{array}$$

If it takes 1662 kJ/mol to break the 4 moles of C—H bonds in a mole of $CH_4$, the average bond-dissociation enthalpy for a single C—H bond is about 415 kJ/mol.

The results of many calculations such as this are summarized in Table A-14 in the appendix. To use this table correctly, it is important to remember the sign convention for $\Delta H$:

Exothermic reactions: $\Delta H$ is negative

Endothermic reactions: $\Delta H$ is positive

Bond-dissociation enthalpies are always positive numbers because it takes energy to break a bond. When this table is used to estimate the enthalpy associated with the formation of a bond, the sign becomes negative because energy is released when bonds are formed.

## EXERCISE 5.13

Use bond-dissociation enthalpies to estimate $\Delta H^\circ$ for the gas-phase reaction between hydrogen and nitrogen to form ammonia:

$$N_2(g) + 3\,H_2(g) \longrightarrow 2\,NH_3(g)$$

Assume that $N_2$ molecules are held together by N≡N triple bonds.

**Problem-Solving Strategy**

**SOLUTION** The key to this calculation is keeping track of how many bonds have to be broken and how many bonds have to be formed. In this case, we can transform the starting materials into isolated nitrogen and hydrogen atoms by breaking 1 mole of N≡N triple bonds and 3 moles of H—H single bonds (see Figure 5.14).

$$N{\equiv}N + 3\,H{-}H \xrightarrow{\text{Bond Dissociation}} 2\,N + 6\,H \xrightarrow{\text{Bond Formation}} 2\,NH_3$$

**Figure 5.14**

We can estimate the energy it takes to break these bonds by multiplying the number of bonds of each kind by the bond-dissociation enthalpy for that bond in Table A-14:

**Bond Dissociation**

$$\begin{array}{lrl} \text{N}{\equiv}\text{N bond:} & (1\,\text{mol})(946\text{ kJ/mol}) = & 946\text{ kJ} \\ \text{H—H bonds:} & (3\,\text{mol})(435\text{ kJ/mol}) = & 1305\text{ kJ} \\ \hline & \text{total} = & 2251\text{ kJ} \end{array}$$

When these isolated atoms in the gas phase recombine to form the products of the reaction, six moles of N—H single bonds are formed:

**Bond Formation**

$$\text{N—H bonds:} \qquad (6\ \cancel{\text{mol}})(-390\ \text{kJ}/\cancel{\text{mol}}) = -2340\ \text{kJ}$$

Note the sign convention in this calculation. It takes energy to break bonds, which means the bond-dissociation step is endothermic and $\Delta H°$ for this step is positive. Energy is released when bonds form. As a result, the second step is exothermic and $\Delta H°$ is negative.

Adding the energy consumed to break the bonds in the starting materials to the energy given off when the products are formed gives us an estimate for the overall enthalpy of reaction:

$$\begin{array}{r} 2251\ \text{kJ} \\ -2340\ \text{kJ} \\ \hline -89\ \text{kJ} \end{array}$$

Because this is $\Delta H°$ for the formation of 2 moles of ammonia, the molar enthalpy of reaction is half as large: $-45$ kJ/mol. The results of this calculation agree reasonably well with the value of $\Delta H_f°$ for ammonia of $-46.11$ kJ/mol reported in Table A-15.

## EXERCISE 5.14

Oxyacetylene torches, which operate at temperatures as high as 3300°C, are fueled by the combustion of acetylene, $C_2H_2$:

$$2\ C_2H_2(g) + 5\ O_2(g) \longrightarrow 4\ CO_2(g) + 2\ H_2O(g)$$

Assume that acetylene contains C—H single bonds and C≡C triple bonds, that $O_2$ is held together by O═O double bonds, and that $CO_2$ contains C═O double bonds. Estimate $\Delta H°$ for this reaction from the bond-dissociation enthalpies in Table A-14. Compare the results of this calculation with the value of $\Delta H°$ obtained from the enthalpies of formation of the reactants and products.

**SOLUTION** The first step in using bond-dissociation enthalpies to estimate $\Delta H°$ for a reaction involves analyzing the number of bonds that must be broken and then formed to convert the reactants into the products of the reaction. This analysis can be simplified by writing the following skeleton structures for the components of this reaction.

$$2\ \text{H—C≡C—H} + 5\ \text{O═O} \longrightarrow 4\ \text{O═C═O} + 2\ \text{H—O—H}$$

We can transform the starting materials into isolated carbon, hydrogen, and oxygen atoms in the gas phase by breaking 4 moles of C—H single bonds, 2 moles of C≡C triple bonds, and 5 moles of O═O double bonds, as shown in Figure 5.15. We can estimate the energy it takes to break these bonds by multiplying the

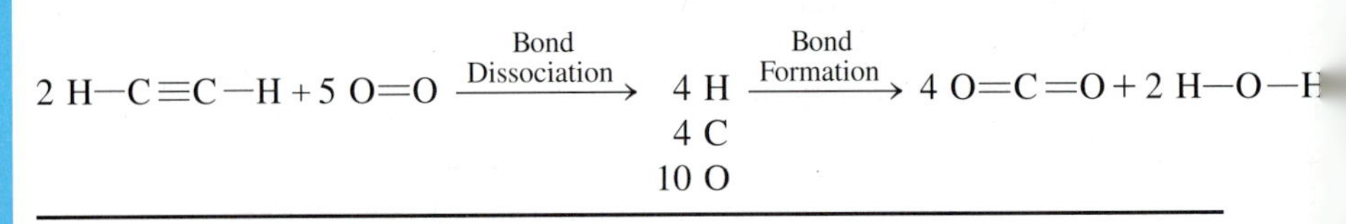

**Figure 5.15** Exercise 5.14

number of bonds of each kind by the bond-dissociation enthalpy for that bond given in Table A-14:

**Bond Dissociation**

| | |
|---|---|
| C—H bonds: | (4 mol)(415 kJ/mol) = 1660 kJ |
| C≡C bond: | (2 mol)(837 kJ/mol) = 1674 kJ |
| O═O bonds: | (5 mol)(498 kJ/mol) = 2490 kJ |
| | total = 5824 kJ |

When these atoms recombine to form the products of the reaction, 8 moles of C═O double bonds and 4 moles of H—O single bonds are formed. We can estimate the energy released when these bonds form by reversing the signs of the bond-dissociation enthalpies for these bonds given in Table A-14:

**Bond Formation**

| | |
|---|---|
| 8 C═O bonds: | (8 mol)(−745 kJ/mol) = −5960 kJ |
| 4 H—O bonds: | (4 mol)(−464 kJ/mol) = −1856 kJ |
| | total = −7816 kJ |

Adding the energy consumed to break the bonds in the starting materials to the energy given off when the products are formed gives an estimate for the overall enthalpy of reaction:

$$\begin{array}{r} 5824 \text{ kJ} \\ -7816 \text{ kJ} \\ \hline -1992 \text{ kJ} \end{array}$$

Because this is the energy given off when 2 moles of acetylene are burned, the estimated value of $\Delta H°$ for this reaction is −996 kJ/mol $C_2H_2$.

Bond-dissociation enthalpies can only give estimated values for $\Delta H°$. In this case, we can calculate the actual value of $\Delta H°$, because the standard-state enthalpies of formation of the reactants and products are all listed in Table A-15.

| *Compound* | $\Delta H_f^\circ$ *(kJ/mol)* |
|---|---|
| $C_2H_2(g)$ | 226.73 |
| $O_2(g)$ | 0 |
| $CO_2(g)$ | −393.51 |
| $H_2O(g)$ | −241.82 |

The standard-state enthalpy of reaction for the combustion of acetylene therefore can be calculated as follows:

$$\begin{aligned} \Delta H° &= \Sigma n\Delta H^\circ_{f,\text{products}} - \Sigma m\Delta H^\circ_{f,\text{reactants}} \\ &= [4 \text{ mol } CO_2 \times -393.51 \text{ kJ/mol} + 2 \text{ mol } H_2O \times -241.82 \text{ kJ/mol}] \\ &\qquad - [2 \text{ mol } C_2H_2 \times 226.73 \text{ kJ/mol} + 5 \text{ mol } O_2 \times 0 \text{ kJ/mol}] \\ &= \mathbf{-2511.14\ kJ} \end{aligned}$$

Because 2 moles of acetylene are consumed in the balanced equation for this reaction, $\Delta H°$ is −1255.6 kJ/mol $C_2H_2$.

In Exercise 5.13, bond-dissociation enthalpies gave an estimate for $\Delta H°$ that was remarkably close to value obtained from $\Delta H_f^\circ$ data. In this case, the bond-dissociation enthalpy calculation gives a result that is about 20% smaller than the value of $\Delta H°$ obtained from enthalpy of formation data. This is an unusually large discrepancy, but it illustrates an important point. Bond-dissociation enthalpies can only

## RESEARCH IN THE 90s

### BIOLOGICAL MICROCALORIMETRY

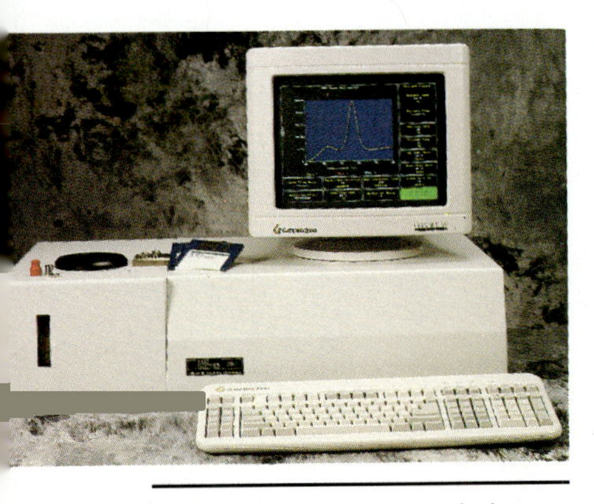

A differential scanning calorimeter for microcalorimetry.

Calorimetry was first applied to biological systems in 1856, when a French scientist studied the production of heat during microbial fermentation in a 21,400-L culture that contained 3.5 tons of sugar [M. Dufrunfaut, *Comptes rendus hebdomadaires des séances de lacademie des sciences, (Paris),* **42,** 945 (1856)]. Since that time, calorimetry has been used to obtain some of the classic data that describe the metabolism and growth of biochemical systems. It was used, for example, to show that different amounts of energy can be extracted from a gram of a simple carbohydrate (glucose: 9.87 kJ/mol), a complex carbohydrate (glycogen: 11.3 kJ/mol), and a fat (trimyristin: 21.4 kJ/mol). More important, it showed that metabolism extracts about 60% of the total energy available in each of these fuels.

Over the years, calorimetry has provided information on such diverse questions as the effect of nutritional factors on the growth of bacteria and the conditions under which proteins undergo denaturation. It also has been used to probe the process by which organisms generate heat (thermogenesis) and thereby to understand the effect of changes in diet or hibernation on the basal metabolism rate. It has been used even to probe membrane–drug interactions, providing information on both the site at which drugs bind and the nature of the changes they produce.

The role of calorimetry in basic research in the 90s is evidenced by a pair of papers that appeared in recent issues of a major biochemistry journal. In the first paper, scientists at the Universität Kaiserslautern in Germany used calorimetry to study the binding of positive ions to the phospholipids that form membranes [A. Blume and J. Tuchtenhagen, *Biochemistry,* **31,** 4636 (1992)].

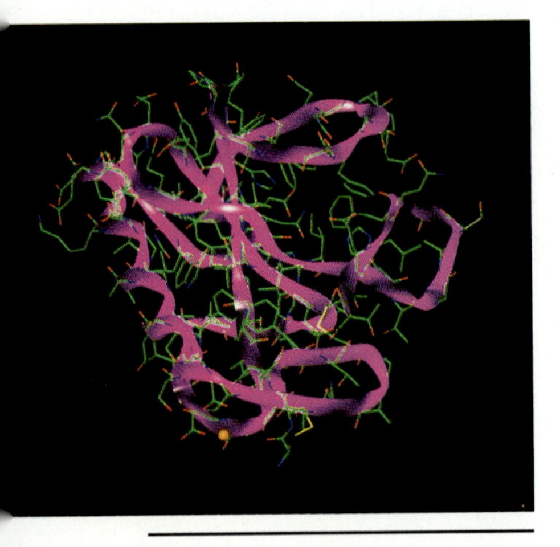

The structure of ribonuclease T1.

In the other, members of the Department of Chemistry at Yale and the Department of Biochemistry at Texas A&M University used calorimetry to study the denaturation of ribonuclease T1 across a range of pH values [C.-Q. Hu, J. M. Sturtevant, J. A. Thomson, R. E. Erickson, and C. N. Pace, *Biochemistry,* **31,** 4876 (1992)]. Ribonuclease T1, like many other globular proteins, can exist in two forms, which are labeled ''folded'' and ''unfolded.'' Research in recent years has shown that the difference in the energies of these two conformations for globular proteins is relatively small—between 20 and 60 kJ/mol. This is important because only the folded protein is active. Pace and co-workers used calorimetery to study the transition between these forms at different values of pH. They found that the folded form of ribonuclease T1 is most stable at pH 5. They also found that the folded protein became even more stable when it bound either $Na^+$ or $Ca^{2+}$ ions. They concluded that the folding of the protein could be affected by the binding of $H^+$ from water or positively charged cations in the solution.

provide an estimate of the value of $\Delta H°$ because they are based on estimates of the strength of an average bond. In this case, the C—H bond in acetylene is significantly weaker than the C—H bonds in $CH_4$ that were used to estimate the strength of these bonds. As a result, the bond-dissociation enthalpy calculation underestimates the energy released when acetylene is burned.

## SUMMARY

**1.** Thermochemistry revolves around four concepts. **Temperature** is a quantitative measure of the degree to which an object is ''cold'' or ''hot.'' **Heat** is a way of transferring energy between a system and its surroundings that often, but not always, changes the temperature of the system. For an ideal gas, the **internal energy** is directly proportional to the temperature of the system: $E = \frac{3}{2} RT$. The internal energy of more complex systems can't be measured, but changes in internal energy can be detected as changes in the temperature of the system. **Work** is defined as the product of the force used to move an object times the distance the object is moved.

**2.** The language used to describe heat was derived from the caloric theory of heat, which assumed that heat was a fluid that could neither be created nor destroyed. The caloric theory has been replaced by a kinetic theory, which recognizes that an inexhaustible amount of heat can be generated by doing work. In the kinetic theory, heat and work are ways of transferring energy across the boundary between a system and its surroundings.

**3.** The amount of heat given off or absorbed in a chemical reaction can be measured with a calorimeter. Because the reaction occurs in a sealed container at constant volume, no work of expansion is done during the reaction. As a result, the heat given off or absorbed by the reaction is equal to the change in the internal energy of the system during the course of the reaction: $\Delta E_{sys} = q_V$. Most chemical reactions are run in open flasks or beakers, however, in which the volume of the system changes but the pressure is constant. By definition, the heat of reaction under conditions of constant pressure is equal to the change in the **enthalpy** of the system during the course of the reaction: $\Delta H_{sys} = q_p$.

**4.** Chemical reactions that give off heat are said to be **exothermic;** those that absorb heat are **endothermic.**

**5.** The value of $\Delta H$ for a reaction depends on the conditions under which the reaction is run. A **standard state** for thermochemical measurements therefore has been defined in which the pressure of any gas is 0.1 MPa and the concentration of any aqueous solution is 1 $M$. Measurements made under standard-state conditions are indicated by adding a superscript ° to the symbol for the quantity being measured. Thus, $\Delta H°$ refers to measurements in which the reactants and products are handled in their standard state.

**6.** Because enthalpy is a state function, the magnitude of $\Delta H°$ for a reaction does not change when a reaction is broken up into a series of small steps. Enthalpy of reaction data can therefore be combined to predict the exact value of $\Delta H°$ for reactions that have not been studied experimentally. By convention, the data most often tabulated for use in predicting $\Delta H°$ for a reaction are standard-state enthalpies of formation, $\Delta H_f°$.

**7.** The enthalpy of formation of a compound is the enthalpy associated with the reaction in which this compound is made from its elements in their most thermodynamically stable states at 1 atm pressure.

**8.** Bond-dissociation enthalpies can be used to estimate the value of $\Delta H°$ for gas-phase reactions in which the value of $\Delta H_f°$ for one or more of the reagents is not known. This involves estimating the enthalpy associated with breaking the bonds in the starting materials and adding the result to the enthalpy associated with the formation of the bonds in the products.

## KEY TERMS

**Bond-dissociation enthalpy** (p. 182)
**Boundary** (p. 160)
**Caloric theory of heat** (p. 159)
**Calorimeter** (p. 169)
**Endothermic** (p. 174)
**Enthalpy (*H*)** (p. 172)
**Enthalpy of formation ($\Delta H_f$)** (p. 179)
**Enthalpy of reaction ($\Delta H$)** (p. 173)
**Equations of state** (p. 166)
**Exothermic** (p. 174)
**First law of thermodynamics** (p. 164)
**Heat** (p. 155)
**Heat capacity** (p. 156)
**Hess's law** (p. 177)
**Internal energy (*E*)** (p. 164)
**Kinetic theory of heat** (p. 160)
**Latent heat of fusion** (p. 158)
**Latent heat of vaporization** (p. 159)
**Molar heat capacity** (p. 156)
**Molar heat of reaction** (p. 171)
**Specific heat** (p. 156)
**Standard state** (p. 176)
**State** (p. 166)
**State functions** (p. 166)
**Surroundings** (p. 160)
**System** (p. 160)
**Temperature** (p. 154)
**Thermal insulators** (p. 154)
**Thermal conductors** (p. 154)
**Thermochemistry** (p. 154)
**Work** (p. 161)

## KEY EQUATIONS

$w = F \times d$

$\Delta E_{univ} = \Delta E_{sys} + \Delta E_{surr} = 0$

$E_{sys} = \frac{3}{2} RT$

$\Delta E_{sys} = q + w$

$w = -P\Delta V$

$q = nC\Delta T$

$H_{sys} = E_{sys} + PV$

$\Delta H = \Sigma nH_{products} - \Sigma mH_{reactants}$

$\Delta H^\circ = \Sigma n\Delta H^\circ_{f,products} - \Sigma m\Delta H^\circ_{f,reactants}$

## PROBLEMS

### Temperature

**5-1** When you touch a parked motorcycle on a hot August day, the metal feels even hotter than the seat. Explain why.

**5-2** It can be a painful experience to sit on the vinyl seats of a car that has been out in the hot summer sun. Explain why it is less painful if someone has left a towel on the seat, or if the seats are covered with lambskin.

**5-3** Plastic ice-cube trays are cheaper than those made from metal. It is also easier to remove ice cubes from plastic trays. In which tray would ice form more quickly: plastic or metal?

### Heat and Heat Capacity

**5-4** Young children often confuse heat and temperature. How would you explain the difference to them?

**5-5** Describe the units for heat capacity when heat is measured in Btu, in calories, and in joules.

**5-6** Calculate the number of joules in 1 Btu of heat.

**5-7** Describe the difference between the specific heat of a substance and its molar heat capacity.

**5-8** Use the heat capacities of mercury and water in Table 5.1 to explain why the temperature of mercury changes more than the temperature of water when equal volumes of the two liquids at different temperatures are mixed.

**5-9** A piece of copper metal weighing 145 grams was heated to 100°C and then dropped in 250 grams of water at 25°C. The copper cooled down and the water became warmer until each had a temperature of 28.8°C. Calculate the amount of heat absorbed by the water. Assuming that the heat lost by the copper was absorbed by the water, what is the molar heat capacity of copper metal?

### Latent Heat

**5-10** Children who are beginning to understand the concepts of heat and temperature often believe that temperature always increases when you heat something. Describe the example you would use to convince them they are wrong.

**5-11** Define the terms *latent heat of fusion* and *latent heat of vaporization,* and classify them as examples of either extensive or intensive properties.

### The Caloric Theory

**5-12** Describe the difference between the model of heat based on the caloric theory and the model based on the kinetic molecular theory.

**5-13** The caloric theory is based on the assumption that heat is neither created nor destroyed. Give examples from daily experience that suggest that heat is conserved. Give examples that violate this hypothesis.

### The Kinetic Theory of Heat

**5-14** What physical properties on both the atomic and macroscopic scales change when a balloon filled with helium is heated? Use the kinetic theory of heat to explain each of these changes.

**5-15** Define the terms *system, surroundings,* and *boundary.* Give three examples of a system separated from its surroundings by a boundary, either real or imaginary.

**5-16** Children often believe that things that are hot contain a lot of heat. Use the thermodynamic concepts of system, surroundings, and boundaries to explain why they are wrong.

### Work

**5-17** Describe the relationship among work, force, and the distance an object is moved.

**5-18** Which of the following reactions could do work of expansion on their surroundings?

(a) $CH_4(g) + 2\ O_2(g) \longrightarrow CO_2(g) + 2\ H_2O(g)$

(b) $CaCO_3(s) \longrightarrow CaO(s) + CO_2(g)$

(c) $2\ CO(g) + O_2(g) \longrightarrow 2\ CO_2(g)$

(d) $2\ N_2O(g) \longrightarrow 2\ N_2(g) + O_2(g)$

### Heat from Work and Vice Versa

**5-19** Describe a simple test that can be used to decide whether the interaction between a system and its surroundings involves the transfer of heat or of work.

**5-20** Describe how each of the following interactions between a system and its surroundings involves the transfer of heat and/or work.

(a) Cooling a glass of lemonade by adding ice cubes.
(b) A balloon expanding as more gas is forced into the system.
(c) A gas in a cylinder being rapidly compressed by a piston.
(d) The combustion of a mixture of gasoline and air forcing the piston out of the cylinder in an internal combustion engine.
(e) An electric current being driven through a thin coil of copper wire.

### The First Law of Thermodynamics

**5-21** What physical property of an ideal gas is directly proportional to the internal energy of the gas? Describe how changes in the internal energy of more complex systems can be detected.

**5-22** The first law of thermodynamics is often stated as **energy is conserved.** Describe why it is incorrect to assume that the first law suggests that the **energy of a *system* is conserved.**

**5-23** Describe what happens to the internal energy of a system when the system does work on its surroundings. What happens to the internal energy of the system when it loses heat to its surroundings?

**5-24** Give examples of a system doing work on its surroundings and a system losing heat to its surroundings. Describe what happens to the temperature of the system in each case.

**5-25** Give examples of a system having work done on it by its surroundings and a system gaining heat from its surroundings. What happens to the temperature of the system?

**5-26** Explain why the first law of thermodynamics is often described as suggesting that there is no such thing as a free lunch.

### State Functions

**5-27** Give examples of at least five physical properties that are state functions.

**5-28** Which of the following descriptions of a trip are state functions?

(a) work done (b) energy expanded (c) cost (d) distance traveled (e) tire wear (f) gasoline consumed (g) location of the car (h) elevation (i) latitude (j) longitude

**5-29** Explain why some of the following physical properties of a system are state functions, but others are not.

(a) temperature (b) internal energy (c) enthalpy (d) pressure (e) volume (f) heat (g) work

**5-30** When $X$ is a state function, $\Delta X$ can be defined as follows.

$$\Delta X = X_f - X_i$$

Explain why there is a unique value for $\Delta X$ for a given set of initial and final states if and only if $X$ is a state function.

### Measuring Heat with a Calorimeter

**5-31** Chemists often determine the heat given off in a chemical reaction by running the reaction in a bomb calorimeter and measuring the change in the temperature of a sample of water surrounding the calorimeter. How much heat is given off by a reaction that raises the temperature of 740.3 grams of water in a calorimeter with a heat capacity of 1.05 kJ/°C by 1.38°C?

**5-32** Calculate the heat given off when 0.01248 mole of $Fe_2O_3$ reacts with an excess of powdered aluminum in a bomb calorimeter with a heat capacity of 980 J/°C if the temperature of the 985 grams of water surrounding the calorimeter increases by 2.58°C.

$$Fe_2O_3(s) + 2\,Al(s) \longrightarrow Al_2O_3(s) + 2\,Fe(l)$$

**5-33** Calculate the molar heat of reaction for the previous problem in units of kilojoules per mole of $Fe_2O_3$ consumed and in units of kilojoules per mole of iron metal produced.

**5-34** Describe the difference between $q_V$ and $q_P$. When chemists measure the heat given off or absorbed in a reaction they often run the reaction in a steel container. Does this experiment measure $q_V$ or $q_P$? When students measure the heat given off or absorbed in a reaction, they often run the reaction in a styrofoam cup. Does their experiment measure $q_V$ or $q_P$?

**5-35** The heat given off in a chemical reaction is an extensive quantity that depends on the amount of reactants used. It is converted to an intensive quantity when it is expressed in units of kilojoules per mole of one of the reactants or products of the reaction. Calculate the molar heat of reaction for the combustion of butane if 45.71 kilojoules of heat are released when 1.000 gram of butane is burned:

$$2\,C_4H_{10}(g) + 13\,O_2(g) \longrightarrow 8\,CO_2(g) + 10\,H_2O(g)$$

**5-36** In theory, $B_5H_9$ should be an excellent rocket fuel because of the enormous amount of energy released when this compound burns:

$$2\,B_5H_9(g) + 12\,O_2(g) \longrightarrow 5\,B_2O_3(s) + 9\,H_2O(g)$$

Calculate the molar heat of reaction for the combustion of $B_5H_9$ if the reaction between 0.100 gram of $B_5H_9$ and excess oxygen in a bomb calorimeter with a heat capacity of 0.920 kJ/°C raises the temperature of the 852 grams of water surrounding the calorimeter by 1.57°C.

### Enthalpy versus Internal Energy

**5-37** When will the change in the enthalpy associated with a chemical reaction ($\Delta H$) be equal to the change in internal energy ($\Delta E$)?

**5-38** For which of the following reactions is $\Delta H$ roughly equal to $\Delta E$?

(a) $2\,H_2(g) + O_2(g) \longrightarrow 2\,H_2O(g)$
(b) $Pb(NO_3)_2(s) + 2\,KI(s) \longrightarrow PbI_2(s) + 2\,KNO_3(s)$
(c) $HCl(aq) + NaOH(aq) \longrightarrow NaCl(aq) + H_2O(l)$
(d) $NaOH(s) + CO_2(g) \longrightarrow NaHCO_3(s)$

**5-39** If the internal energy of a system is directly proportional to the temperature of the system, what is the advantage of introducing the concept of enthalpy?

### Enthalpies of Reaction

**5-40** Which of the following reactions would you expect to be endothermic?

(a) $H_2(g) \longrightarrow 2\ H(g)$
(b) $2\ H_2(g) + O_2(g) \longrightarrow 2\ H_2O(g)$
(c) $H_2O(g) \longrightarrow H_2O(l)$
(d) $HCl(aq) + NaOH(aq) \longrightarrow NaCl(aq) + H_2O(l)$

**5-41** Which of the following reactions would you expect to be endothermic?

(a) $2\ Na(s) + 2\ H_2O(l) \longrightarrow 2\ Na^+(aq) + 2\ OH^-(aq) + H_2(g)$
(b) $2\ Mg(s) + O_2(g) \longrightarrow 2\ MgO(s)$
(c) $2\ NaCl(s) \longrightarrow 2\ Na(s) + Cl_2(g)$
(d) $Na^+(g) + e^- \longrightarrow Na(g)$

**5-42** Humans sweat and dogs pant to keep cool. Explain how these processes help keep them cool.

### Standard-State Enthalpies of Reaction

**5-43** Describe the difference between the enthalpy of reaction ($\Delta H$) and the standard-state enthalpy of reaction ($\Delta H^\circ$) for a chemical reaction.

**5-44** Explain why it is important to have a standard state for thermochemical measurements.

**5-45** How much heat is given off when 1 mole of nitrogen reacts with 2 moles of oxygen to give 2 moles of $NO_2$ gas, if $\Delta H^\circ$ for the following reaction is 33.2 kilojoules per mole of $NO_2$?

$$N_2(g) + 2\ O_2(g) \longrightarrow 2\ NO_2(g)$$

**5-46** Calculate the standard-state molar enthalpy of reaction for the following reaction, if 1.00 gram of magnesium gives off 46.22 kilojoules of heat when it reacts with excess fluorine:

$$Mg(s) + F_2(g) \longrightarrow MgF_2(s)$$

**5-47** Calculate $\Delta H^\circ$ for the following reaction, assuming that 1.00 gram of hydrogen gives off 4.65 kilojoules of heat when it reacts with 1.00 gram of calcium.

$$Ca(s) + H_2(g) \longrightarrow CaH_2(s)$$

### Hess's Law

**5-48** Explain how Hess's law is a direct consequence of the fact that the enthalpy of a system is a state function.

**5-49** Use the following data to calculate $\Delta H^\circ$ for the conversion of graphite into diamond:

$C(graphite) + O_2(g) \longrightarrow CO_2(g)$ $\quad \Delta H^\circ = -393.51$ kJ/mol $CO_2$

$C(diamond) + O_2(g) \longrightarrow CO_2(g)$ $\quad \Delta H^\circ = -395.94$ kJ/mol $CO_2$

**5-50** Use the following data:

$2\ H_2(g) + O_2(g) \longrightarrow 2\ H_2O(l)$ $\quad \Delta H^\circ = -285.83$ kJ/mol $H_2O$

$H_2(g) + O_2(g) \longrightarrow H_2O_2(aq)$ $\quad \Delta H^\circ = -187.8$ kJ/mol $H_2O_2$

to calculate $\Delta H^\circ$ for the decomposition of hydrogen peroxide:

$$2\ H_2O_2(aq) \longrightarrow 2\ H_2O(l) + O_2(g)$$

**5-51** In the presence of a spark, nitrogen and oxygen react to form nitrogen oxide:

$$N_2(g) + O_2(g) \longrightarrow 2\ NO(g)$$

Calculate $\Delta H^\circ$ for this reaction from the following data:

$N_2(g) + 2\ O_2(g) \longrightarrow 2\ NO_2(g)$ $\quad \Delta H^\circ = 33.2$ kJ/mol $NO_2$

$2\ NO(g) + O_2(g) \longrightarrow 2\ NO_2(g)$ $\quad \Delta H^\circ = -57.1$ kJ/mol $NO_2$

**5-52** Enthalpy of reaction data can be combined to determine $\Delta H^\circ$ for reactions that are difficult, if not impossible, to study directly. Nitrogen and oxygen, for example, do not react directly to form dinitrogen pentoxide:

$$2\ N_2(g) + 5\ O_2(g) \longrightarrow 2\ N_2O_5(g)$$

Use the following data to determine $\Delta H^\circ$ for the hypothetical reaction in which nitrogen and oxygen combine to form $N_2O_5$:

$N_2(g) + \frac{3}{2}\ O_2(g) + H_2(g) \longrightarrow 2\ HNO_3(aq)$ $\quad \Delta H^\circ = -207.4$ kJ/mol

$N_2O_5(g) + H_2O(g) \longrightarrow 2\ HNO_3(aq)$ $\quad \Delta H^\circ = 218.4$ kJ/mol

$2\ H_2(g) + O_2(g) \longrightarrow 2\ H_2O(g)$ $\quad \Delta H^\circ = -285.8$ kJ/mol

**5-53** Use the following data:

$3\ C(s) + 4\ H_2(g) \longrightarrow C_3H_8(g)$ $\quad \Delta H^\circ = -103.85$ kJ/mol $C_3H_8$

$C(s) + O_2(g) \longrightarrow CO_2(g)$ $\quad \Delta H^\circ = -393.51$ kJ/mol $CO_2$

$H_2(g) + 2\ O_2(g) \longrightarrow H_2O(g)$ $\quad \Delta H^\circ = -241.83$ kJ/mol $H_2O$

to calculate the heat of combustion of propane, $C_3H_8$:

$$C_3H_8(g) + 5\ O_2(g) \longrightarrow 3\ CO_2(g) + 4\ H_2O(g)$$

**5-54** Use the following data:

$C_4H_9OH(l) + 6\ O_2(g) \longrightarrow 4\ CO_2(g) + 5\ H_2O(g)$ $\quad \Delta H^\circ = -2456.1$ kJ/mol $C_4H_9OH$

$(C_2H_5)_2O(l) + 6\ O_2(g) \longrightarrow 4\ CO_2(g) + 5\ H_2O(g)$ $\quad \Delta H^\circ = -2510.0$ kJ/mol $(C_2H_5)_2O$

to calculate $\Delta H^\circ$ for the following reaction:

$$(C_2H_5)_2O(l) \longrightarrow C_4H_9OH(l)$$

### Enthalpies of Formation

**5-55** For which of the following substances is $\Delta H_f^\circ$ equal to zero?

(a) $P_4(s)$ (b) $H_2O(g)$ (c) $H_2O(l)$ (d) $O_3(g)$
(e) $Cl(g)$ (f) $F_2(g)$ (g) $Na(g)$

**5-56** Use the enthalpy of formation data in the appendix to determine whether heat is given off or absorbed when limestone is converted to lime and carbon dioxide:

$$CaCO_3(s) \longrightarrow CaO(s) + CO_2(g)$$

**5-57** Calculate $\Delta H^\circ$ for the following reaction from the enthalpy of formation data in the appendix:

$$CO(g) + NH_3(g) \longrightarrow HCN(g) + H_2O(g)$$

**5-58** Phosphine ($PH_3$) is a foul-smelling gas, which often burns on contact with air. Use the enthalpy of formation data in the appendix to calculate $\Delta H^\circ$ for this reaction, to obtain an estimate of the amount of energy given off when this compound burns:

$$PH_3(g) + 2\ O_2(g) \longrightarrow H_3PO_4(s)$$

**5-59** Carbon disulfide ($CS_2$) is a useful, but flammable, solvent. Calculate $\Delta H^\circ$ for the following reaction from the enthalpy of formation data in the appendix:

$$CS_2(l) + 3\ O_2(g) \longrightarrow CO_2(g) + 2\ SO_2(g)$$

**5-60** The disposable lighters that so many smokers carry use butane as a fuel:

$$2\ C_4H_{10}(g) + 13\ O_2(g) \longrightarrow 8\ CO_2(g) + 10\ H_2O(g)$$

Calculate $\Delta H^\circ$ for the combusion of butane from the enthalpy of formation data in the appendix.

**5-61** The first step in the synthesis of nitric acid involves burning ammonia:

$$4\ NH_3(g) + 5\ O_2(g) \longrightarrow 4\ NO(g) + 6\ H_2O(g)$$

Calculate $\Delta H^\circ$ for this reaction from the enthalpy of formation data in the appendix.

**5-62** Lavoisier believed that all acids contain oxygen because so many compounds he studied that contained oxygen form acids when they dissolve in water. Calculate $\Delta H^\circ$ for the reaction between tetraphosphorus decaoxide and water to form phosphoric acid from the enthalpy of formation data in the appendix:

$$P_4O_{10}(s) + 6\ H_2O(l) \longrightarrow 4\ H_3PO_4(aq)$$

**5-63** Small quantities of oxygen can be prepared in the laboratory by heating potassium chlorate ($KClO_3$) until it decomposes. Calculate $\Delta H^\circ$ for the following reaction from the enthalpy of formation data in the appendix:

$$2\ KClO_3(s) \longrightarrow 2\ KCl(s) + 3\ O_2(g)$$

**5-64** Use enthalpies of formation to predict which of the following reactions gives off the most heat per mole of aluminum consumed:

$$Fe_2O_3(s) + 2\ Al(s) \longrightarrow 2\ Fe(s) + Al_2O_3(s)$$
$$Cr_2O_3(s) + 2\ Al(s) \longrightarrow 2\ Cr(s) + Al_2O_3(s)$$

### Bond-Dissociation Enthalpies

**5-65** Estimate the enthalpy of formation of Cl atoms in the gas phase from the bond-dissociation enthalpy of the Cl—Cl bond.

**5-66** Use bond-dissociation enthalpies to estimate $\Delta H^\circ$ for the following reaction:

$$H_2C{=}CH_2(g) + H_2(g) \longrightarrow CH_3{-}CH_3{-}CH_3(g)$$

Compare the results of this calculation with the value of $\Delta H^\circ$ calculated from enthalpies of formation.

**5-67** Calculate the bond-dissociation enthalpy for the N≡N triple bond in the $N_2$ molecule from the N—H and H—H bond-dissociation enthalpies and the enthalpy of reaction for the following reaction:

$$N_2(g) + 3\ H_2(g) \longrightarrow 2\ NH_3(g) \qquad \Delta H^\circ = -46.1\ \text{kJ/mol}\ NH_3$$

**5-68** Use bond-dissociation enthalpies to estimate the energy given off when the isooctane used in octane ratings is burned:

$$CH_3{-}CH(CH_3){-}C(CH_3)_2{-}CH_2{-}CH_3 + \tfrac{25}{2}\ O_2 \longrightarrow 8\ CO_2 + 9\ H_2O$$

**5-69** Use bond-dissociation enthalpies to estimate the enthalpy of formation of HCl.

**5-70** Use bond-dissociation enthalpies to estimate the enthalpy of reaction for the combustion of $CS_2$ in the gas phase:

$$CS_2(g) + 3\ O_2(g) \longrightarrow CO_2(g) + 2\ SO_2(g)$$

Assume $CS_2$ contains C=S double bonds, $O_2$ contains O=O double bonds, $CO_2$ contains C=O double bonds, and that $SO_2$ contains S=O double bonds. Compare the results of this calculation with the value of $\Delta H^\circ$ obtained from enthalpy of formation data.

**5-71** Methanol, $CH_3OH$, has been proposed as an alternative to gasoline for use as a fuel. Use bond-dissociation enthalpies to estimate the energy given off when methanol burns in the gas phase. Assume that methanol contains 3 C—H bonds, a C—O bond, and an O—H bond. Assume that $O_2$ contains O=O double bonds and that $CO_2$ contains C=O double bonds.

$$2\ CH_3OH(g) + 3\ O_2(g) \longrightarrow 2\ CO_2(g) + 4\ H_2O(g)$$

### Integrated Problems

**5-72** Calculate the energy needed to convert 18 grams of ice at 0°C to steam at 100°C using some or all of the following information:

$$2\ H_2(g) + O_2(g) \longrightarrow 2\ H_2O(g) \qquad \Delta H^\circ = -241.8\ \text{kJ/mol}\ H_2O$$

$$H_2O(s) \longrightarrow H_2O(l) \qquad \Delta H^\circ = 6.03\ \text{kJ/mol}\ H_2O$$

$$H_2O(l) \longrightarrow H_2O(g) \qquad \Delta H^\circ = 40.67\ \text{kJ/mol}\ H_2O$$

$$H_2O(l,\ 0°C) \longrightarrow H_2O(l,\ 100°C) \qquad \Delta H^\circ = 7.53\ \text{kJ/mol}\ H_2O$$

**5-73** The first ionization energy of an element is defined as the energy it takes to remove the highest energy electron from an atom of the element in the gas phase. The equation for the first ionization energy of hydrogen is written:

$$H(g) \longrightarrow H^+(g) + e^-$$

Use the following data to calculate the first ionization energy of hydrogen in kilojoules per mole:

$H_2(g) + Cl_2(g) \longrightarrow 2\ HCl(g)$
$\Delta H° = -92.31$ kJ/mol HCl

$H_2(g) \longrightarrow 2\ H(g)$
$\Delta H° = 435.94$ kJ/mol $H_2$

$Cl_2(g) \longrightarrow 2\ Cl(g)$
$\Delta H° = 243.36$ kJ/mol $Cl_2$

$Cl(g) + e^- \longrightarrow Cl^-(g)$
$\Delta H° = -348.79$ kJ/mol $Cl^-$

$H^+(g) + Cl^-(g) \longrightarrow HCl(g)$
$\Delta H° = -1395.38$ kJ/mol HCl

**5-74** Calculate the O—H bond-dissociation enthalpy from the H—H and O=O bond-dissociation enthalpies and the enthalpy of formation of $H_2O$ in the gas phase.

$2\ H_2(g) + O_2(g) \longrightarrow 2\ H_2O(g)$
$\Delta H_f° = -241.83$ kJ/mol $H_2O$

**5-75** Use the enthalpy of combustion for methane, given below, to estimate the energy released when 100 cubic feet of natural gas is burned:

$CH_4(g) + 2\ O_2(g) \longrightarrow CO_2(g) + 2\ H_2O(g)$
$\Delta H° = -802.36$ kJ/mol $CH_4$

**5-76** Use the enthalpies of formation of $O_2$, $CO_2$ and $H_2O$ to calculate the enthalpy of formation of ethanol, $CH_3CH_2OH$, if ethanol gives off 1277.41 kJ/mol when burned:

$$CH_3CH_2OH(l) + 3\ O_2(g) \longrightarrow 2\ CO_2(g) + 3\ H_2O(g)$$

**5-77** Calculate the enthalpy of formation of benzoic acid, $C_6H_5CO_2H$, if this compound gives off 3095.0 kJ/mol when burned:

$$2\ C_6H_5CO_2H(s) + 15\ O_2(g) \longrightarrow 14\ CO_2(g) + 6\ H_2O(g)$$

# Intersection

## NATURE ON AN ATOMIC SCALE

When you plug a lamp into an electrical socket and then turn on the lamp, what happens to the electric current that flows from the wall socket after it enters the light bulb? Osborne and co-workers have studied the responses to questions like this from students between the ages of 8 and 18 [R. Osborne and P. Freyberg, *Learning in Science: The Implications of Children's Science,* Heinemann: Auckland, NZ]. Some students argued that the current flowed back to the power source, or continued on around the circuit. Others gave very different responses. They concluded that: ''It is used up to create heat and light.'' Or: ''It gets burnt up.''

The idea that a light bulb consumes electric current, or transforms it into heat and light energy, might explain why half of the 8- to 12-year-old students in another of Osborne's experiments tried to get a small bulb to light by touching the base of the bulb to the top of the battery or by attaching a single wire between the bulb and the top of the battery.

When analogous experiments were done to probe students' concepts of heat and matter, similar results were obtained. In each case, the results were consistent with models of electricity, matter, or heat that assumed that these phenomena were continuous fluids. This isn't surprising, because there is nothing in our daily lives that would lead us to question the notion that electricity, matter, and heat are continuous.

Historians of science have argued that the years between 1803 and 1905 mark a period during which a revolution occurred in our view of the world in which we live. When this period began, it was generally accepted that matter, heat, electricity, and light were all continuous fluids. In 1803, John Dalton proposed the revolutionary notion that matter *was not* continuous. When it was viewed on a small enough scale, he argued, it was composed of individual particles, or atoms.

The next step in this revolution involved a change in our notion of heat from that of a continuous fluid, which was conserved as it was transferred from one body to another, into a model in which heat was carried by particles in a state of constant motion. The third step occurred toward the end of the 19th century, when it became apparent that electricity was carried by particles, which inevitably became known as electrons. In 1905, the revolution became complete when Einstein showed that light was not a continuous wave, as everyone believed, but discrete wave-packets of energy.

Contrary to popular belief, this light bulb doesn't ''consume'' electricity. Every electron that enters the bulb is eventually returned to the power company.

The next three chapters will require a shift in our perspective from the macroscopic scale to the atomic scale. Once this transition has been achieved, the remainder of the text will oscillate between discussions of the results of experiments performed on the macroscopic scale and interpretations of these results based on our model for nature on the atomic scale.

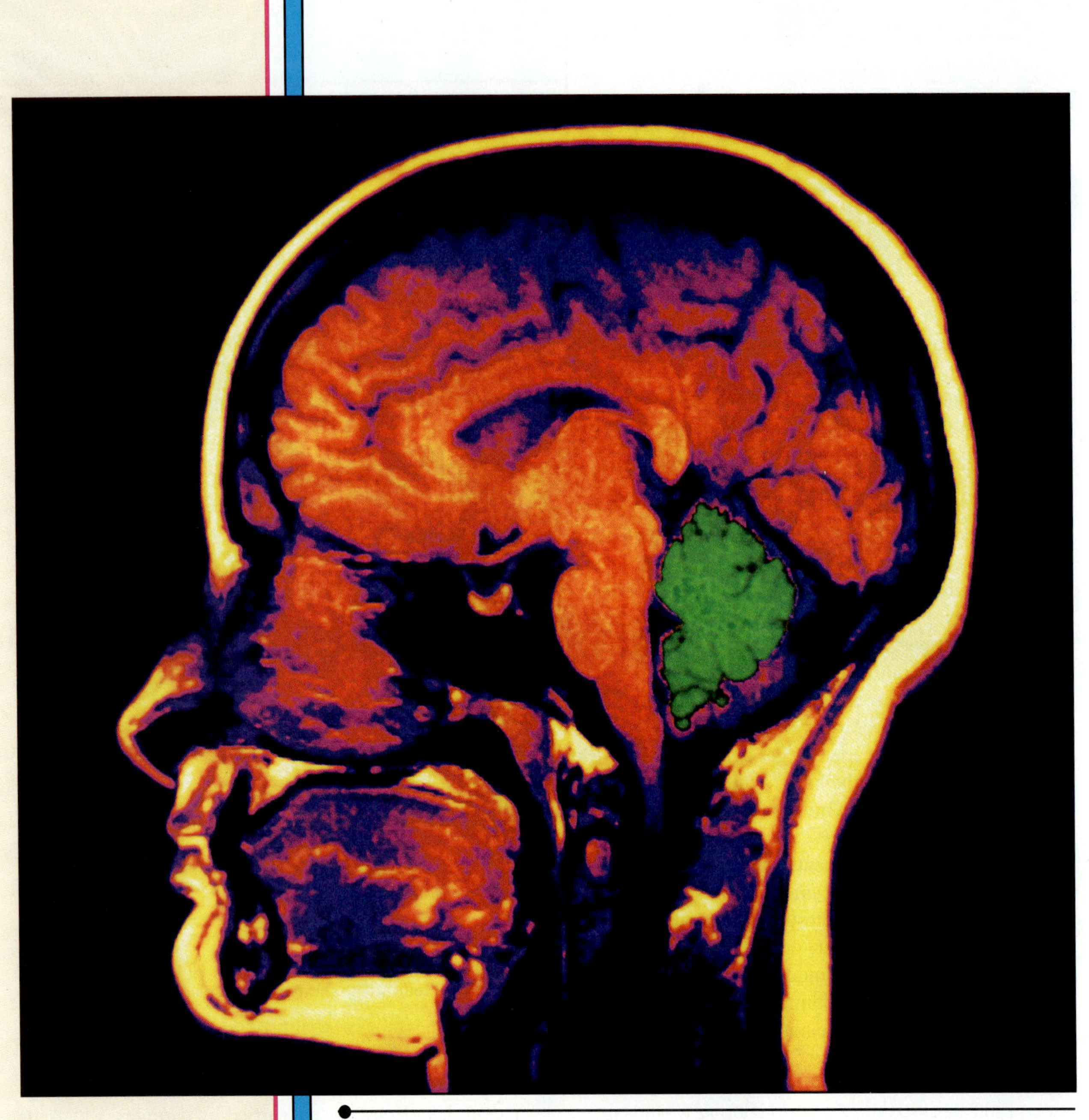

An MRI image of a normal brain. The cerebellum, which is the portion of the brain that coordinates movement of the muscles, is shown here.

CHAPTER 6

# THE STRUCTURE OF THE ATOM

Because it examines the process by which our present understanding of the structure of the atom was achieved, this chapter addresses one of the fundamental questions scientists face: How do we design experiments that can be performed on the macroscopic scale that enable us to probe the structure of matter on a scale so small it cannot be ''seen?'' It also provides the basis for answering the following questions:

## POINTS OF INTEREST

- What does it mean when we say that clothes cling together when removed from the dryer because of a ''static electric charge?''
- Why are computer monitors called ''CRT's?'' Where did the term ''cathode-ray tube'' come from?
- What motivated Hans Geiger to develop the ''Geiger counter.''
- What does it mean when we say that light is simultaneously both a particle and a wave? Why can't we find examples of the same phenomenon with objects large enough to be seen with the naked eye?
- Why do a crucible and the iron triangle on which it rests both give off a red glow when heated with a burner until they are ''red hot?''

## 6.1 THE PROCESS OF DISCOVERY: ELECTRICITY

When amber is rubbed with fur, it attracts small pieces of paper.

As early as the fourth century B.C., Plato noted that a yellow substance then known as *elektron* attracted lightweight objects when rubbed against a piece of fur. (This material is now known as amber, and it has been found to be the fossilized resin from pine trees.) By 1600, William Gilbert was able to show that a similar effect could be obtained by rubbing a number of different "electrics," including diamonds, sapphires, opals, glass, sulfur, and sealing wax.

In 1733 Charles Francois de Cisternay du Fay noticed that objects that had been rubbed sometimes attracted and sometimes repelled each other. He explained this by proposing two different kinds of electricity. *Vitreous electricity* (from the Latin for "glass") is produced when glass or gems were rubbed. *Resinous electricity* (from the Latin for "resin," or "amber") is obtained by rubbing amber, silk, or paper. Du Fay argued that objects with different kinds of electricity attract each other, whereas those with the same kind of electricity repel.

Du Fay's discovery led to a theory of electricity that assumed the existence of two fluids. Objects that were not electrified were assumed to have equal amounts of these fluids, which neutralize each other. Rubbing an object was assumed to remove one of the fluids, leaving an excess of the other.

An opposing school of thought was developed by Benjamin Franklin, who believed there was only one kind of electric fluid, which was present in all bodies. According to this theory, an object picks up an electric charge when some of this fluid is transferred from one body to another. When an object is rubbed, it either gains or loses electricity and thereby picks up either a positive or negative charge. Objects with opposite charges attract each other; those with the same charge repel.

Both du Fay's two-fluid theory of electricity and Franklin's one-fluid model were based on the assumption that it is possible to *charge* an object with these fluids in much the same way that one can load, or charge, a cannon with gunpowder. To this day, we still talk about an object that has been electrified as if it carries an electric "charge." Furthermore, we describe the charge as *static* because it does not move.

## 6.2 THE PROCESS OF DISCOVERY: THE ELECTRON

Many of the early experiments with electricity focused on liquids and solids. By about 1830, Michael Faraday and others had begun to study the effect of an electric current on a gas. Faraday's apparatus consisted of a pair of metal plates sealed in a glass tube as shown in Figure 6.1*a*. The tube was filled with a gas, the metal plates were connected to a series of batteries, and the gas was slowly pumped out of the tube.

Faraday noticed that when the pressure of the gas in the tube became small enough, the gas began to glow. In 1858 Julius Plücker noted that when the residual pressure of the gas inside the tube is very small, the *glass* at one end of the tube emits light. He also found that he could change the position of the patch of glass that glowed by bringing a magnet close to the tube, as shown in Figure 6.1*b*. Plücker interpreted the effect of the magnetic field as evidence that whatever produced this glow is electrically charged.

Much of the language used to describe electricity was proposed by either Benjamin Franklin or William Whewell. Franklin introduced the words *plus, minus, positive, negative, charge,* and *battery*. Whewell coined a number of other terms, including *cation, anion, cathode,* and *anode.*

By convention, the metal plate connected to the negative terminal of the series of batteries that was used to study the effect of an electric current on a gas is called the **cathode.** The other plate is the **anode.**

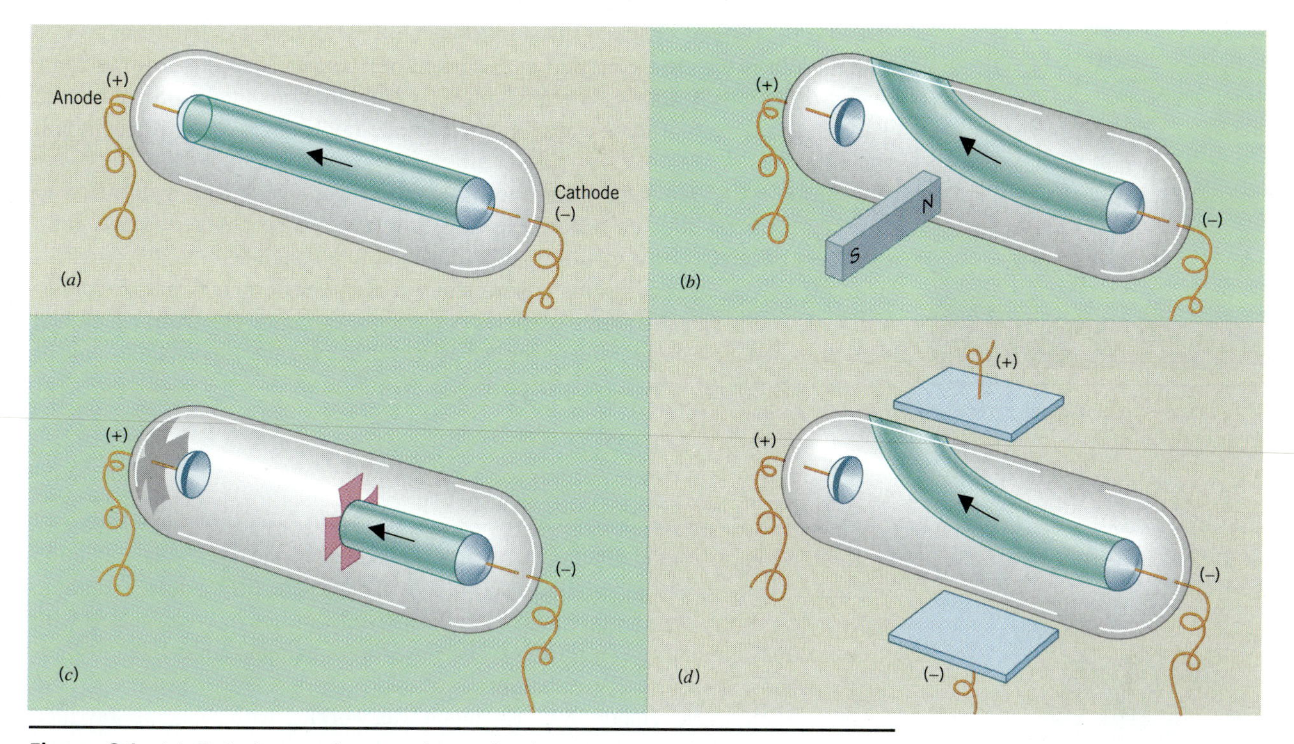


**Figure 6.1** (*a*) Cathode-ray tubes contain a pair of metal plates sealed into a glass tube that has been partially evacuated. If the residual pressure of the gas is small enough, the glass at the end of the tube across from the cathode will glow when the tube is connected to a series of batteries. (*b*) The rays given off by the cathode can be deflected by a magnetic field in a direction which suggests that these cathode rays are negatively charged. (*c*) A solid object placed in the path of the cathode rays casts a shadow on the wall of the tube across from the cathode. (*d*) The cathode rays also can be deflected by an electric field in a direction which suggests they are negatively charged.

In 1869 Johannes Hittorf showed that when a solid object is placed between the cathode and the anode of this apparatus, a shadow is cast on the end of the tube across from the cathode, as shown in Figure 6.1*c*. This suggested that some beam or ray is given off by the cathode, and these tubes soon became known as **cathode-ray tubes.**

The definitive experiments with cathode-ray tubes were done by William Crookes in 1879. Crookes' major contribution was the development of a better vacuum pump that allowed him to produce cathode-ray tubes with a smaller residual gas pressure. Crookes not only confirmed the previous work by Plücker and Hittorf, he was able to show that cathode rays are *negatively* charged by studying the direction in which cathode rays are deflected by a magnetic field.

The key question at this point was whether cathode rays were really ''rays'' (i.e., waves) or whether they are composed of negatively charged particles. The answer to this question was obtained by J. J. Thomson in 1897. Thomson found that the cathode rays could be deflected by an electric field, as shown in Figure 6.1*d*. By balancing the effect of a magnetic field on a cathode-ray beam with an electric field, Thomson was able to show that cathode ''rays'' are actually composed of particles. This experiment also provided an estimate of the ratio of the charge to the mass of these particles.

Neon lights are modern examples of what happens when an electric current is passed through a gas in a cathode-ray tube at low pressure.

J. J. Thompson (left) and Ernst Rutherford (right).

In the SI system, charge is measured in units of coulombs. By definition, 1 coulomb (C) is the charge carried by a current of 1 ampere (amp) that flows for 1 second (s): 1 C = 1 amp-s. When Thomson's data are converted to SI units, the charge-to-mass ratio of the particles in the cathode-ray beam is about $10^8$ coulomb per gram.

Thomson found the same charge-to-mass ratio regardless of the metal used to make the cathode and the anode. He also found the same charge-to-mass ratio regardless of the gas used to fill the tube. He therefore concluded that the particles given off by the cathode in this experiment are a universal component of matter. Although Thomson called these particles *corpuscles,* the name **electron,** which had been proposed by George Stoney several years earlier for the fundamental unit of negative electricity, was soon accepted.

## 6.3 THE DISCOVERY OF X-RAYS

In 1895 William Conrad Roentgen became interested in the ultraviolet radiation emitted by cathode-ray tubes. Because the eye cannot detect ultraviolet radiation, experiments of this type require a UV detector. Roentgen used a screen coated with barium tetracyanoplatinate [$BaPt(CN)_4$] because this compound emits light, or *fluoresces,* when exposed to UV radiation.

One evening in November 1895, Roentgen was working with a cathode-ray tube that had been carefully wrapped with black cardboard. Much to his surprise, the $BaPt(CN)_4$ screen next to the cathode-ray tube gave off light when the tube was switched on. Obviously, something had hit the screen to make it emit light. It was equally obvious that it couldn't have been either UV radiation or cathode rays because neither of these substances could pass through the opaque cardboard.

In an intense series of experiments over a 7-week period, Roentgen found that this new kind of radiation—which he called *x-rays*—passed through solid objects placed between the cathode-ray tube and the detector. He even found that he could see the image of the bones in his hand when he held it between the tube and the screen. Roentgen eventually discovered that he could capture such images on film and one of his first x-ray images is reproduced in Figure 6.2.

**Figure 6.2** One of the earliest x-ray images. There is some reason to believe this is an x-ray of Roentgen's wife's hand.

## 6.4 THE DISCOVERY OF RADIOACTIVITY

The discovery of a form of radiation that could pass through solid matter fired the imagination of a generation of scientists who rushed to study this phenomenon. Roentgen had noted that cathode-ray tubes emit x-rays at the spot that emits light when the cathode rays hit the glass walls of the tube. This observation caught the attention of the French physicist Henri Becquerel.

Becquerel decided to investigate the connection between x-rays and the fluorescence of the glass walls of the cathode-ray tube. He knew that salts of uranium, such as potassium uranyl sulfate [$K_2UO_2(SO_4)_2 \cdot 2\ H_2O$], emitted light when exposed to the UV radiation in sunlight. He therefore wrapped a photographic plate in black paper, placed crystals of this salt on top of the plate, and exposed the crystals to sunlight. When the plates were developed, black spots were found beneath the crystals, suggesting that some form of radiation had been emitted by the uranium salts that passed through the paper and fogged the photographic plate.

Becquerel found the same results, however, when the crystals and photographic plate were prepared and kept in the dark. He also noted that much better images were obtained with pure uranium metal, which did not fluoresce when exposed to sunlight. The radiation given off by uranium metal and its compounds apparently had nothing to do with whether they were exposed to sunlight. It soon became evident that a new form of radiation had been discovered, for which Marie and Pierre Curie suggested the name *radioactivity*.

In 1899 Ernest Rutherford studied the absorption of radioactivity by thin sheets of metal foil and found two components: *alpha* ($\alpha$) radiation, which is absorbed by a few hundredths of a centimeter of metal foil, and *beta* ($\beta$) radiation, which can pass through 100 times as much foil before it is absorbed. Shortly thereafter, a third form of radiation, named *gamma* ($\gamma$) rays, was discovered that can penetrate as much as several centimeters of lead. The three kinds of radiation also differ in the way they are affected by electric and magnetic fields, as shown in Figure 6.3.

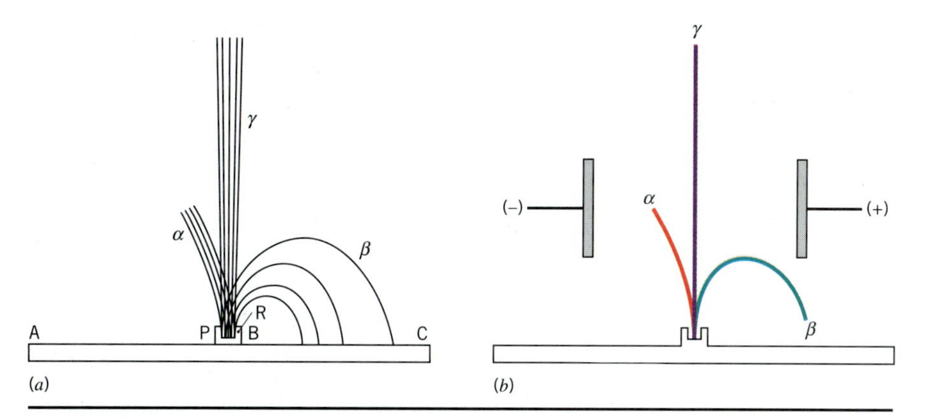


**Figure 6.3** (*a*) In her thesis, Marie Curie reported the drawing on the left which showed the effect of a magnetic field on the three forms of radioactivity. $\alpha$-Particles were deflected more slowly than $\beta$-particles, which suggested that $\alpha$-particles were heavier than $\beta$-particles. $\gamma$-Rays were not affected by a magnetic field. (*b*) The effect of an electric field on the different forms of radioactivity shows that $\alpha$-particles and $\beta$-particles are both electrically charged, but they carry charges with opposite signs. $\gamma$-Rays are not affected by an electric field and therefore have no electrical charge.

**EXERCISE 6.1**

Use Figure 6.3 and the rule that opposite charges attract to determine the charge on $\alpha$-particles, $\beta$-particles, and $\gamma$-rays.

**SOLUTION** $\alpha$-Particles are attracted to the negative pole of an electric field and must therefore carry a positive charge. $\beta$-Particles are attracted to the positive pole of the field and therefore carry a negative charge. Because $\gamma$-rays are not affected by an electric field, they must be electrically neutral.

## 6.5 MILLIKAN'S OIL DROP EXPERIMENT

Thomson's experiment with cathode-ray tubes provided an estimate of the ratio of the charge of the electron divided by its mass. In SI units, this ratio is $10^8$ C/g. This result was disconcerting because it was 1000 times larger than any previous measurement of the charge-to-mass ratio of a particle. Until Thomson's experiment was done, the largest charge-to-mass ratio was found for the $H^+$ ion, for which this ratio is about $10^5$ C/g.

There are two ways of explaining the unusually large charge-to-mass ratio of an electron. We can assume that the charge on the electron is 1000 times larger than the charge on a hydrogen ion, which doesn't make sense. Or we can assume that the magnitude of the charge on the electron is the same as the charge on a hydrogen ion, but the mass of the electron is 1000 times smaller. Thomson preferred the second alternative, believing that the electron was at least 1000 times lighter than the smallest atom.

This hypothesis has important consequences. If electrons are a universal component of matter very much smaller than atoms, Dalton's model of the atom needs to be revised. Atoms are not indivisible. It must be possible to strip one or more negatively charged electrons from an atom when enough energy is applied to the system.

The way to test Thomson's hypothesis was to measure either the charge on an electron or its mass. Between 1908 and 1917, Robert Millikan measured the charge on an electron with the apparatus shown in Figure 6.4. In these experiments, the atomizer from a perfume bottle was used to spray water or oil droplets into a sample chamber. Some of these droplets fell through a pinhole between two plates of an electric field, where they could be observed through a microscope.

A source of x-rays was then used to ionize the air in the chamber by removing electrons from the molecules in the air. Droplets that did not capture one of these electrons fell to the bottom of the chamber due to the force of gravity. Droplets that captured one or more electrons were attracted to the positive plate at the top of the viewing chamber and either fell more slowly or rose toward the top.

By carefully studying individual droplets, Millikan was able to show that the charge on a drop was always an integral multiple of a small, but finite value. When his data are converted to SI units, the charge on a drop is always some multiple of $1.59 \times 10^{-19}$ C. Combining this value for the charge on a single electron with the charge-to-mass ratio for the electron confirms Thomson's hypothesis. The mass of an electron is at least 1000 times smaller than the lightest atom.

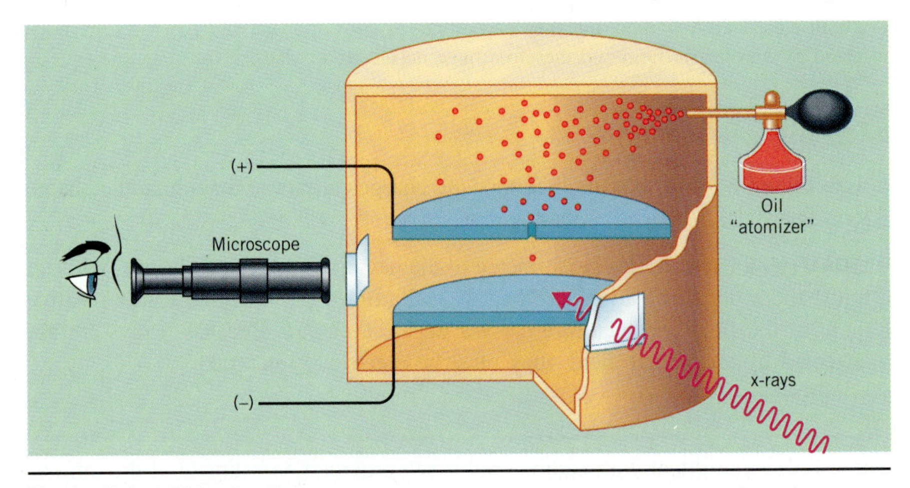


**Figure 6.4** Millikan's oil-drop apparatus.

### ▶ CHECKPOINT

The number of particles in a mole was first estimated by combining the charge on an electron with a constant known as the *Faraday,* which describes the charge on a mole of electrons. Use the following values of the charge on an electron and Faraday's constant to calculate Avogadro's number to the maximum number of significant figures.

| | |
|---|---|
| Charge on an electron: | $e = 1.6021892 \times 10^{-19}$ C |
| Faraday's constant: | $F = 96{,}484.56$ C/mol |

## 6.6 THOMSON'S RAISIN PUDDING MODEL OF THE ATOM

Thomson recognized one of the consequences of the discovery of the electron. Because matter is electrically neutral, there must be a positively charged particle that balances the negative charge on the electrons in an atom. Furthermore, since electrons are very much lighter than atoms, these positively charged particles must carry the mass of the atom. Thomson therefore suggested that atoms were spheres of positive charge in which light, negatively charged electrons were embedded, much as raisins might be embedded in the surface of a pudding (see Figure 6.5).

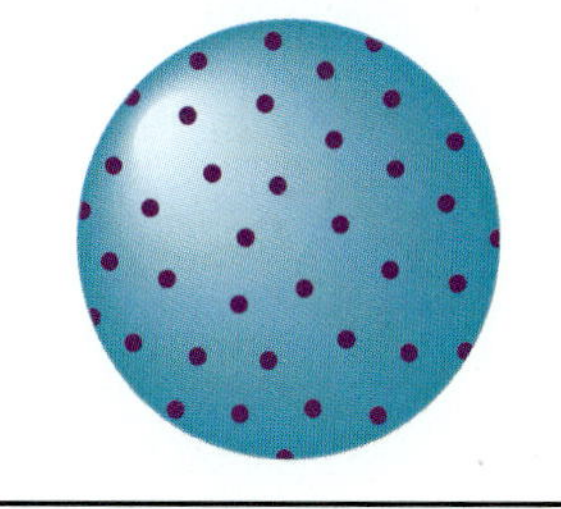

**Figure 6.5** Thomson's raisin pudding model for the structure of an atom.

At the time Thomson proposed this model, evidence for the existence of positively charged particles was available from cathode-ray tube experiments. In 1886 Eugen Goldstein noted that cathode-ray tubes with a perforated cathode (Figure 6.6) emitted a glow from the end of the tube near the cathode. Goldstein concluded that in addition to the electrons, or cathode rays, that travelled from the negatively charged cathode toward the positively charged anode, there was another ray that travelled in the opposite direction, from the anode toward the cathode. Because these rays passed through the holes, or channels, in the cathode, Goldstein called them *canal rays.* By 1898 Wilhelm Wien was able to show that these canal rays could be deflected by both magnetic and electric fields as would be expected for particles that carried an electric charge. They were deflected, however, in the direction expected for positively charged particles.

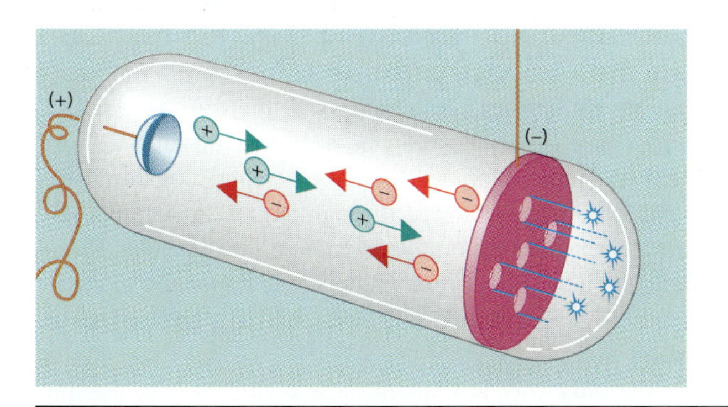


**Figure 6.6** When the cathode of a cathode-ray tube was perforated, Goldstein observed rays he called "canal rays," which passed through the holes, or channels, in the cathode to strike the glass walls of the tube at the end near the cathode. Since these canal rays travel in the opposite direction from cathode rays, they must carry the opposite charge.

These observations can be explained by assuming that the atoms or molecules in the gas in a cathode-ray tube lose one or more electrons when excited by the voltage applied to the electrodes of the tube. The negatively charged electrons flow toward the anode. The positive ions, formed when the atoms or molecules lose electrons, flow in the opposite direction.

The charge-to-mass ratio for the cathode rays is always the same because the rays consist of negatively charged electrons. The charge-to-mass ratio for the positively charged canal rays depends on the gas used to fill the tube because the nature of the ions produced when the atoms or molecules lose electrons depends on the identity of that gas.

## 6.7 THE RUTHERFORD MODEL OF THE ATOM

Thomson's model of the atom was relatively short lived. Within 10 years it was replaced by a model developed by one of his students, Ernest Rutherford. Rutherford began his graduate work by studying the effect of x-rays on various materials. Shortly after the discovery of radioactivity, he turned to the study of the $\alpha$-particles emitted by uranium metal and its compounds.

Before he could study the effect of $\alpha$-particles on matter, Rutherford had to develop a way of counting individual $\alpha$-particles. He found that a screen coated with zinc sulfide emitted a flash of light when it was hit by an $\alpha$-particle. Rutherford and his assistant, Hans Geiger, would sit in the dark until their eyes became sensitive enough. They would then try to count the flashes of light given off by the ZnS screen. (It is not surprising that Geiger was motivated to develop the electronic radioactivity counter that carries his name.)

Rutherford found that a narrow beam of $\alpha$-particles was broadened when it passed through a thin film of mica or metal. He therefore had Geiger measure the angle through which these $\alpha$-particles were scattered by a thin piece of metal foil. Because it is unusually ductile, gold can be made into a foil that is only 0.00004 cm thick. When this foil was bombarded with $\alpha$-particles, Geiger found that the scattering was small, on the order of 1°.

These results were consistent with Rutherford's expectations. He anticipated that virtually all of the $\alpha$-particles would be able to penetrate the metal foil, although they would be scattered slightly by collisions with the atoms through which they passed. In other words, Rutherford expected the $\alpha$-particles to pass through the metal foil the way a rifle bullet would penetrate a bag of sand.

One day, Geiger suggested that a research project should be given to Ernest Marsden, who was working in Rutherford's laboratory. Rutherford responded, "Why not let him see whether any $\alpha$-particles can be scattered through a large angle?" When this experiment was done, Marsden found that a small fraction (perhaps 1 in 20,000) of the $\alpha$-particles were scattered through angles larger than 90° (see Figure 6.7*a*). Many years later, reflecting on his reaction to these results, Rutherford said: "It was quite the most incredible event that has ever happened to me in my life. It was almost as incredible as if you fired a 15-inch shell at a piece of tissue paper and it came back and hit you."

Rutherford concluded that the only way to explain these results was to assume that the positive charge and the mass of an atom were concentrated in a small fraction of the total volume (see Figure 6.7*b*). He then derived mathematical equations for the scattering that would occur. These equations predicted that the number of $\alpha$-particles scattered through a given angle would be proportional to the thickness of the foil and the square of the charge on the nucleus, and inversely proportional to

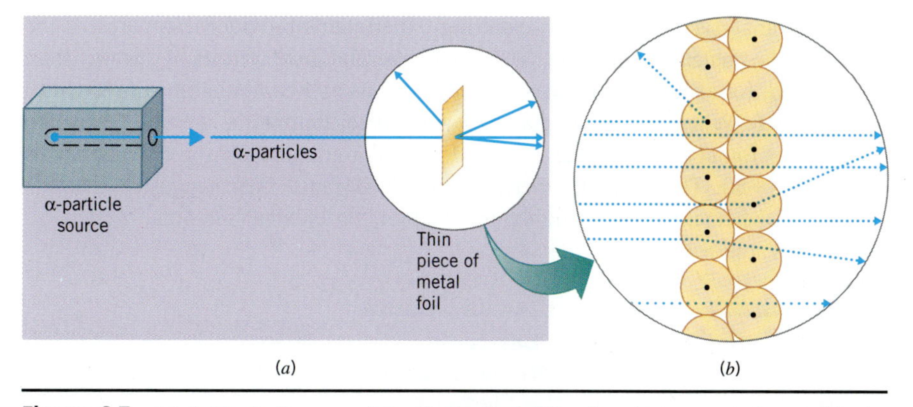


**Figure 6.7** (*a*) A block-diagram of the Rutherford–Marsden–Geiger experiment. Most of the $\alpha$-particles pass through the empty space between nuclei. A few come close enough to be repelled by the nucleus, and these particles are deflected through small angles. Occasionally, an $\alpha$-particle travels along a path that would lead to a direct hit with the nucleus. These particles are deflected through large angles by the force of repulsion between the $\alpha$-particle and the positively charged nucleus of the atom.

the velocity with which the $\alpha$-particles moved raised to the fourth power. In a series of experiments, Geiger and Marsden verified each of these predictions.

When he published the results of these experiments in 1911, Rutherford proposed a model for the structure of the atom that is still accepted today. This model states that all of the positive charge and essentially all of the mass of the atom is concentrated in an infinitesimally small fraction of the total volume of the atom, which Rutherford called the **nucleus** (from the Latin for little nut).

Most of the $\alpha$-particles were able to pass through the gold foil without encountering anything large enough to significantly deflect their paths. A small fraction of the $\alpha$-particles came close to the nucleus of a gold atom as they passed through the foil. When this happened, the force of repulsion between the positively charged $\alpha$-particle and the nucleus deflected the $\alpha$-particle by a small angle. Occasionally, an $\alpha$-particle traveled along a path that would lead to a direct collision with the nucleus of one of the 2000 or so atoms it had to pass through. When this happened, repulsion between the nucleus and the $\alpha$-particle deflected the $\alpha$-particle through an angle of 90° or more.

By carefully measuring the fraction of the $\alpha$-particles deflected through large angles, Rutherford was able to estimate the size of the nucleus. According to his calculations, the radius of the nucleus is at least 10,000 times smaller than the radius of the atom. The vast majority of the volume of an atom is therefore empty space.

### CHECKPOINT

Explain why Rutherford was initially surprised by the results of the $\alpha$-particle scattering experiments.

## 6.8 THE DISCOVERY OF THE PROTON

Rutherford's model assumed that all of the positive charge and most of the mass of the atom was concentrated in the nucleus. The remaining volume of the atom was occupied by lightweight, negatively charged electrons. The problem with this model was that, at the time it was proposed, the number of electrons or positively charged particles in an atom was not known.

Careful analysis of $\alpha$-particle scattering experiments provided estimates of the charge on the nucleus of an atom. When the scattering of $\alpha$-particles by paraffin wax was studied, for example, the charge on the nucleus of a carbon atom was found to be about +6. When this technique was applied to aluminum, the charge on the nucleus was found to be between +13 and +14. For gold, the charge on the nucleus was about +80. These results were interesting because they suggested that the charge on the nucleus was roughly one-half of the element's atomic weight, as shown in Table 6.1.

**TABLE 6.1 The Relationship between Estimates of the Charge on the Nucleus and the Atomic Weight of the Element**

| Element | Charge on the Nucleus | Atomic Weight |
|---|---|---|
| C | +6 | 12 |
| Al | +13-14 | 27 |
| Au | +80 | 197 |

In 1913 A. van den Broek suggested that it was a mistake to compare the charge on the nucleus with the atomic weight of the element. He suggested that the charge should be compared with the **atomic number,** which specifies the position of the element in the periodic table. (Hydrogen has an atomic number of 1, helium is 2, and so on.) Within experimental error, the estimate of the charge on the nucleus obtained from $\alpha$-particle scattering experiments is equal to the atomic number of the element.

Shortly before the first world war, this observation was explained by H. G. J. Moseley, who studied the frequencies of the x-rays given off by cathode-ray tubes when the electrons strike the anode. His results show that the frequencies of these x-rays depend on the metal used to form the anode.

When the elements are arranged in order of increasing atomic weight, aluminum is the first metal. It was therefore the lightest element for which Moseley was able to obtain results. Because it is the thirteenth element in the periodic table, aluminum has an atomic number of 13. (By convention, the symbol for atomic number is a capital Z.) Moseley found a relationship between the frequencies ($v$) of the x-rays given off by a cathode-ray tube and the atomic number of the metal used to form the anode (see Figure 6.8):

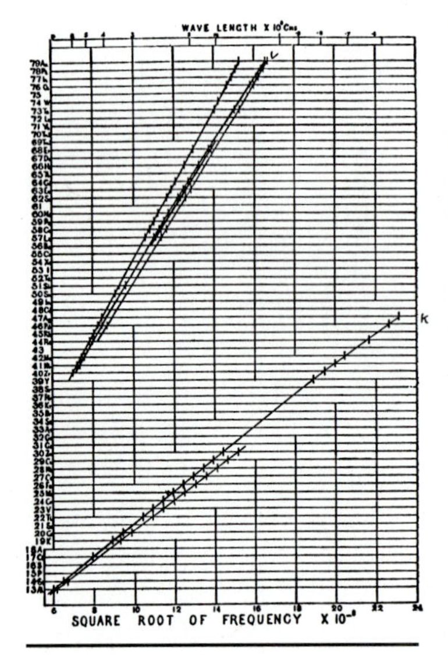


**Figure 6.8** This figure shows the plot of atomic number versus the square root of the frequencies of the x-rays that accompanied the paper Moseley published in the *Philosophical Magazine* in 1914.

$$\nu \propto (Z - 1)^2$$

Moseley argued that the frequencies of the x-rays given off by the anode when it was hit by electrons in the cathode-ray tube should depend on the charge on the nucleus of the atom emitting these x-rays. He therefore concluded that the atomic number of an element was equal to the positive charge on the nucleus of its atoms. Aluminum, for example, must have a net charge of +13 on the nucleus of each atom.

Less than 6 months after completing this work, Moseley volunteered to serve in the British army during World War I and became one of the more than 8.5 million casualties of this war when he was killed during the battle at Gallipoli in August 1915. Shortly after the war, in 1920, Rutherford proposed the name **proton** for the positively charged particles in the nucleus of an atom.

## 6.9 THE DISCOVERY OF THE NEUTRON

According to the model of the atom that resulted from Moseley's work, the nucleus of a hydrogen atom ($Z = 1$) has a charge of +1 and therefore contains one proton. The remaining volume of the atom is occupied by an electron that balances the charge on the nucleus. Because the mass of a hydrogen atom is about 1 amu and the mass of an electron is negligibly small, the mass of a proton must be about 1 amu.

A problem arises when this model is extended to other elements because the mass of most other atoms is at least twice as large as the mass predicted from its atomic number. Aluminum, for example, has an atomic number of 13, but the mass of an aluminum atom is 27 amu. There are two possibilities.

1. We can assume the nucleus of an aluminum atom contains 27 protons and 14 electrons, for a net positive charge of +13. This nucleus then would be surrounded by another 13 electrons that would balance the net positive charge on the nucleus.
2. We can assume the nucleus of an aluminum atom contains 13 protons and 14 other particles that have the same weight as a proton but are electrically neutral.

At the same time that Rutherford proposed the name *proton* for the positively charged particle in the nucleus of an atom, he proposed that the nucleus also contained a neutral particle, eventually named the **neutron.** It was not until 1932, however, that James Chadwick was able to prove that these neutral particles exist. Chemistry textbooks published before 1932 therefore often used the first model of the atom described above, not the second.

Today, we assume that atoms are composed of the three fundamental subatomic particles described in Table 6.2. The protons and neutrons are concentrated in the nucleus of the atom and this positively charged nucleus is surrounded by a sea of negatively charged electrons.

**TABLE 6.2 Fundamental Subatomic Particles**

| Particle | Symbol | Charge | Mass |
|---|---|---|---|
| Electron | $e^-$ | −1 | 0.0005486 amu |
| Proton | $p^+$ | +1 | 1.007276 amu |
| Neutron | $n^\circ$ | 0 | 1.008665 amu |

The number of protons, neutrons, and electrons in an atom can be determined from a set of simple rules:

- The number of protons in the nucleus of the atom is equal to the atomic number ($Z$).
- The number of electrons in a *neutral* atom is equal to the number of protons.
- The mass number ($M$) of the atom is equal to the sum of the number of protons and neutrons in the nucleus.
- The number of neutrons is equal to the difference between the mass number ($M$) of the atom and the atomic number ($Z$).

**EXERCISE 6.2**

Calculate the number of electrons, protons, and neutrons in neutral atoms of the following isotopes.

(a) $^{12}C$ (b) $^{13}C$ (c) $^{14}C$ (d) $^{14}N$

Sec. 3.3

**SOLUTION** The different isotopes of an element are identified by writing the mass number of the atom in the upper left corner of the symbol for the element. $^{12}C$, $^{13}C$, and $^{14}C$ are isotopes of carbon ($Z = 6$) and therefore contain six protons. If the atoms are neutral, they also must contain six electrons. The only difference between these isotopes is the number of neutrons in the nucleus.

$^{12}C$: 6 electrons, 6 protons, and 6 neutrons
$^{13}C$: 6 electrons, 6 protons, and 7 neutrons
$^{14}C$: 6 electrons, 6 protons, and 8 neutrons

Neutral nitrogen atoms ($Z = 7$) contain seven protons and seven electrons. The remaining mass of the atom comes from seven neutrons.

$^{14}N$: 7 electrons, 7 protons, and 7 neutrons

**EXERCISE 6.3**

Calculate the number of electrons in the $Cl^-$ and $Fe^{3+}$ ions.

**SOLUTION** The atomic number of chlorine is 17. The number of protons in both a neutral Cl atom and a negatively charged $Cl^-$ ion is exactly the same: 17. The difference between these particles is the number of electrons that surround the nucleus. The $Cl^-$ ion has one more electron than a neutral Cl atom, for a total of 18 electrons.

Atoms can gain or lose electrons, but the number of protons in the nucleus of the atom cannot change unless the identity of the atom changes. The only way an iron atom ($Z = 26$), with 26 protons and 26 electrons, can form a +3 ion is to lose three electrons. The $Fe^{3+}$ ion therefore has 23 electrons.

This model readily explains the observations made in the early experiments on electricity. Matter can pick up an electrical charge because electrons can be transferred from one neutral atom or molecule to another. Substances that exhibit ''vitreous electricity'' lose electrons when they are rubbed and are therefore left with a net positive charge. Substances that exhibit ''resinous electricity'' gain electrons when rubbed and therefore have a net negative charge.

## 6.10 PARTICLES AND WAVES

Rutherford's model assumes that most of the mass of an atom is concentrated in an infinitesimally small nucleus that is surrounded by a sea of lightweight, negatively charged electrons that occupy the vast majority of the volume of the atom. Our next goal is to develop a picture of how electrons are distributed around the nucleus. Before we do this, however, it is useful to consider the following question: ''What are the characteristic properties of the tools that can be used to probe the distribution of electrons in an atom?'' First, and foremost, they must exist on a scale that is smaller than the size of an atom. Second, they must be capable of being accurately calibrated, so that we know what we are measuring. Finally, they must interact with matter. It therefore isn't surprising that much of what we know about the structure of the electrons in an atom has been obtained by studying the interaction between matter and different forms of **electromagnetic radiation.** Our first goal is therefore to understand what we mean when we say that electromagnetic radiation has some of the properties of a particle and some of the properties of a wave.

Scientists since the time of Galileo have divided matter into two categories: particles and waves. *Particles* are easy to understand, because they have a definite mass and they occupy space. *Waves* are more challenging. They have no mass and yet they carry energy as they travel through space. The best way to demonstrate that waves carry energy is to watch what happens when a pebble is tossed into a lake. As the waves produced by this action travel across the lake, they set in motion any leaves that lie on the surface.

The fact that waves carry energy as they travel through space can be understood by imagining what happens to a leaf on the surface of a pond as the ripples formed when a pebble is dropped into the pond move past the leaf.

In addition to their ability to carry energy, waves have four other characteristic properties: speed, frequency, wavelength, and amplitude. As we watch waves travel across the surface of a lake, we can see that they move at a certain *speed.* By watching waves strike a pier at the edge of the lake, we can see that they are also characterized by a **frequency** ($\nu$), which is the number of waves (or cycles) that hit per unit of time. The frequency of a wave is reported in units of cycles per second ($s^{-1}$) or hertz (Hz).

The idealized drawing of a wave in Figure 6.9 illustrates the definitions of amplitude and wavelength. The **wavelength** ($\lambda$) is the smallest distance between repeating points on the wave. The *amplitude* of the wave is the distance between the highest (or lowest) point on the wave and the center of gravity of the wave.

If we measure the frequency ($\nu$) of a wave in cycles per second and the wavelength ($\lambda$) in meters, the product of these two numbers has the units of meters per second. The product of the frequency ($\nu$) times the wavelength ($\lambda$) of a wave is therefore the speed ($s$) at which the wave travels through space:

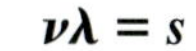

$$\nu\lambda = s$$

To understand the relationship between the speed, frequency, and wavelength of a wave, it might be useful to imagine a concrete example: What is the speed of a wave

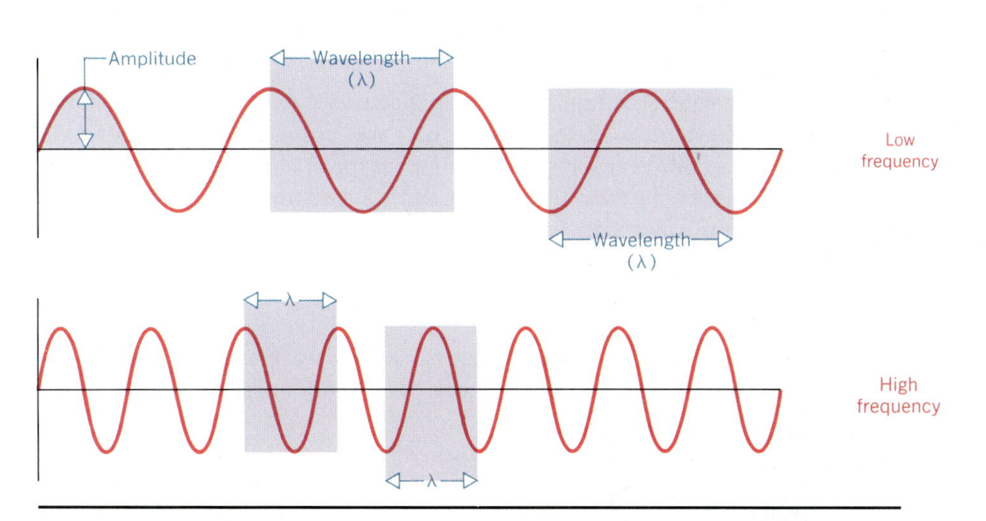


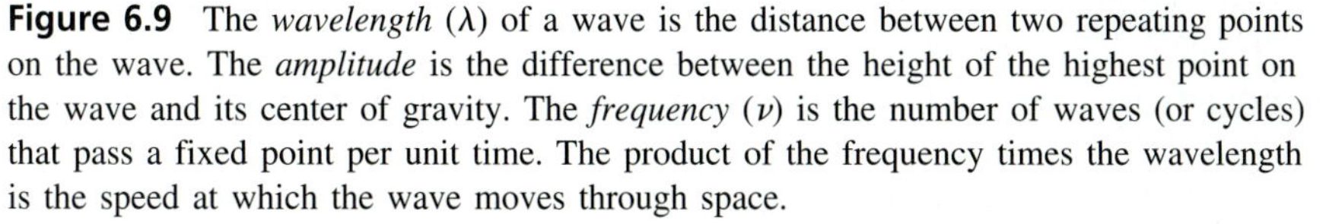

**Figure 6.9** The *wavelength* ($\lambda$) of a wave is the distance between two repeating points on the wave. The *amplitude* is the difference between the height of the highest point on the wave and its center of gravity. The *frequency* ($\nu$) is the number of waves (or cycles) that pass a fixed point per unit time. The product of the frequency times the wavelength is the speed at which the wave moves through space.

that has a wavelength of 1 meter and a frequency of 60 cycles per second? If 60 of these waves pass a fixed point each second, and each wave is 1 meter long, the speed of the wave must be 60 meters per second:

$$60 \frac{1}{s} \times 1\ m = 60\ m/s$$

**EXERCISE 6.4**

Orchestras in the United States tune their instruments to an ''A'' that has a frequency of 440 cycles per second, or 440 Hz. If the speed of sound is 1116 feet per second, what is the wavelength of this note?

**SOLUTION** The product of the frequency times the wavelength of any wave is equal to the speed with which the wave travels through space:

$$(440\ \text{s}^{-1})(\lambda) = 1116\ \text{ft/s}$$

Solving this equation for the wavelength of this note gives a value of 2.54 feet.

## 6.11 LIGHT AND OTHER FORMS OF ELECTROMAGNETIC RADIATION

White light can be separated into a spectrum of colors by passing the light through a glass prism.

In 1865 James Clerk Maxwell proposed that light was a wave with both *electric* and *magnetic* components. It is therefore a form of **electromagnetic radiation.** Because it is a wave, light is bent when it enters a glass prism. When white light is focused on a prism, the light rays of different wavelengths are bent by differing amounts and the light is transformed into a spectrum of colors. Starting from the side of the spectrum where the light is bent by the smallest angle, the colors are red, orange, yellow, green, blue, and violet.

Light contains the narrow band of frequencies and wavelengths in the portion of the electromagnetic spectrum that our eyes can detect. It includes radiation with wavelengths between about 400 nm (violet) and 700 nm (red). Because the wavelength of electromagnetic radiation can be as long as 40 m or as short as $10^{-5}$ nm, the visible spectrum is only a small portion of the total range of electromagnetic radiation.

The electromagnetic spectrum includes radio and TV waves, microwaves, infrared, visible light, ultraviolet, x-rays, $\gamma$-rays, and cosmic rays, as shown in Figure 6.10. These different forms of radiation all travel at the speed of light ($c$). They differ, however, in their frequencies and wavelengths. The product of the frequency times the wavelength of electromagnetic radiation is always equal to the speed of light:

$$\nu\lambda = c$$

As a result, electromagnetic radiation that has a long wavelength has a low frequency, and radiation with a high frequency has a short wavelength.

**EXERCISE 6.5**

Calculate the frequency of red light with a wavelength of 700.0 nm if the speed of light is $2.998 \times 10^8$ m/s.

**SOLUTION** The product of the frequency times the wavelength of any wave is equal to the speed with which the wave travels through space. In this case, the wave travels at the speed of light: $2.998 \times 10^8$ m/s. Before we can calculate the

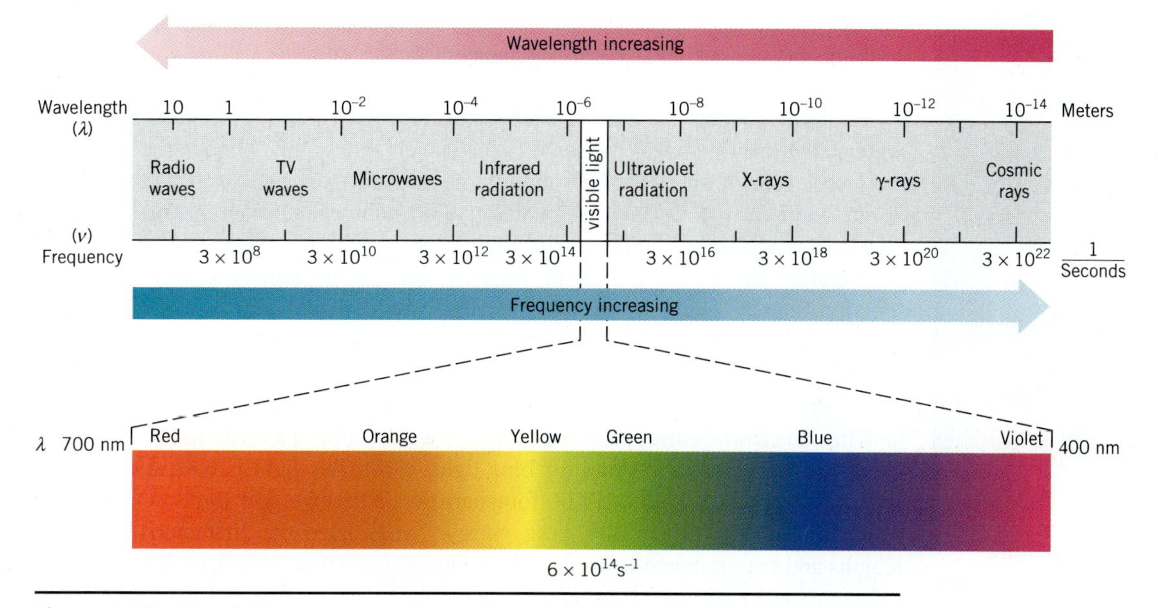


**Figure 6.10** The visible spectrum is the small portion of the total electromagnetic spectrum that our eyes can detect. Other forms of electromagnetic radiation include radio and TV waves, microwaves, infrared and ultraviolet radiation, as well as x-rays, γ-rays and cosmic rays.

frequency of this radiation we have to convert the wavelength into units of meters:

$$700.0 \, \cancel{\text{nm}} \times \frac{1 \text{ m}}{10^9 \, \cancel{\text{nm}}} = 7.000 \times 10^{-7} \text{ m}$$

We can then substitute the known values of the wavelength and the speed of light into the equation $\nu\lambda = c$:

$$\nu(7.000 \times 10^{-7} \text{ m}) = 2.998 \times 10^8 \text{ m/s}$$

We then solve this equation for the frequency of this light in units of cycles per second:

$$\nu = 4.283 \times 10^{14} \text{ s}^{-1}$$

## 6.12 ATOMIC SPECTRA

In 1814 Joseph von Fraunhofer noticed that the light from the sun did not give a continuous spectrum. By using an unusually good prism, Fraunhofer observed a series of dark lines in the sun's spectrum, the more prominent of which he labeled *A* through *H*. Fraunhofer's experiment is an example of an **absorption spectrum.** Some of the light given off by the sun is apparently absorbed before it reaches the earth.

For more than 200 years chemists have known that sodium salts produce a yellow color when added to a flame. Robert Bunsen, however, was the first to systematically study this phenomenon. (Bunsen went so far as to design a new burner that would produce a colorless flame for this work.) Between 1855 and 1860, Bunsen and his colleague Gustav Kirchhoff developed a spectroscope that focused the light from the burner flame onto a prism that separated this light into its spectrum. Using

The characteristic color of the light given off when sodium atoms are excited can be seen by placing crystals of a sodium salt into the flame of a bunsen burner.

this device, Bunsen and Kirchhoff were able to show that the **emission spectrum** of sodium salts consists of two narrow bands of radiation in the yellow portion of the spectrum.

Kirchhoff noticed that the wavelength of the light given off when sodium salts were added to a flame was the same as the wavelength of the *D* line in the Fraunhofer's spectrum of sunlight. He therefore concluded that absorption and emission spectra were related. Certain substances give off light when heated that has the same frequencies and wavelengths as the light they absorb under other conditions.

Chemists and physicists soon began using the spectroscope to catalog the wavelengths of the light either emitted or absorbed by a variety of compounds. These data were then used to detect the presence of certain elements in everything from mineral water to sunlight. No obvious patterns were discovered in these data, however, until 1885, when Johann Jacob Balmer analyzed the spectrum of hydrogen.

When an electric current is passed through a glass tube that contains hydrogen gas at low pressure the tube gives off blue light. When this light is passed through a prism (as shown in Figure 6.11), four narrow bands of bright light are observed against a black background. These narrow bands have the characteristic wavelengths and colors shown in Table 6.3. Balmer noticed that these data fit the following equation to within ±0.02%:

$$\frac{1}{\lambda} = R_H\left[\frac{1}{2^2} - \frac{1}{n^2}\right]$$

**TABLE 6.3 The Characteristic Lines in the Visible Spectrum of Hydrogen**

| Wavelength (nm) | Color |
|---|---|
| 656.2 | red |
| 486.1 | blue |
| 434.0 | blue-violet |
| 410.1 | violet |

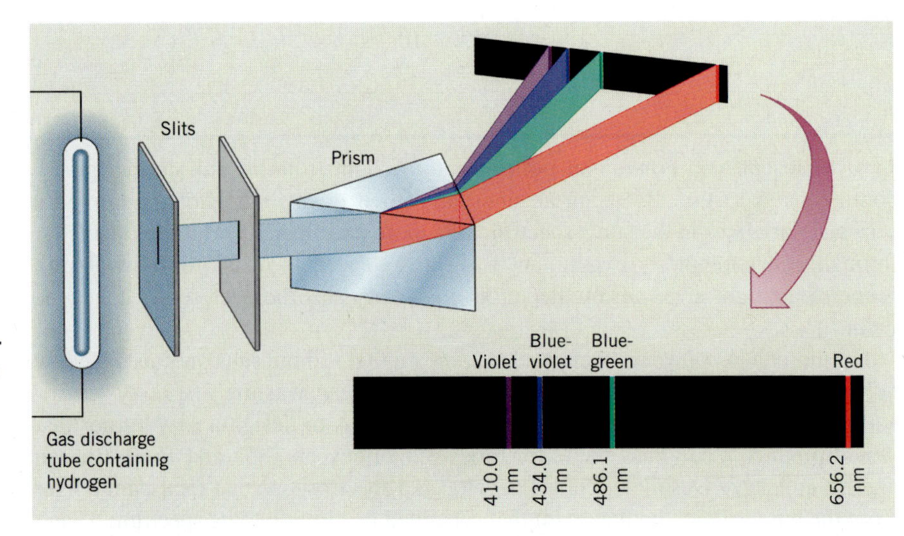


**Figure 6.11** The blue light given off when a tube filled with $H_2$ gas is excited with an electric discharge can be separated into four narrow bands of light when it is passed through a prism.

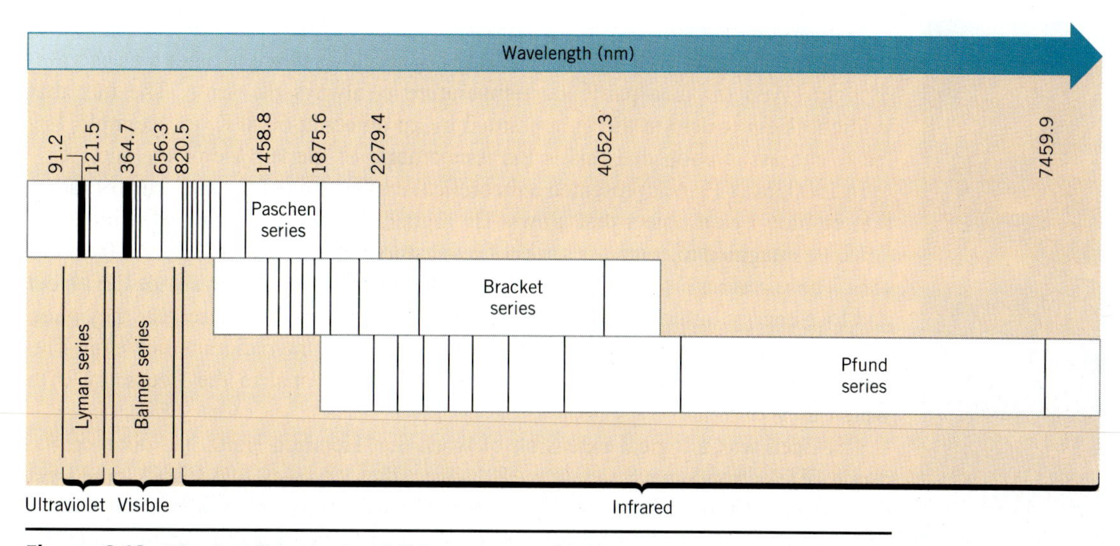


**Figure 6.12** The four lines in the visible spectrum of hydrogen analyzed by Balmer are only a portion of the emission spectrum of this gas. Lyman discovered another series of lines in the UV portion of the spectrum and three series of lines were discovered in the IR spectrum by Paschen, Brackett, and Pfund.

In this equation, $R_H$ is a constant known as the Rydberg constant, which is equal to $1.09678 \times 10^{-2}\ \text{nm}^{-1}$, and $n$ is an integer between 3 and 6.

Between 1906 and 1924, four more series of lines were discovered in the emission spectrum of hydrogen by searching the infrared spectrum at longer wavelengths and the ultraviolet spectrum at shorter wavelengths. Each of these lines fits the same general equation, where $n_1$ and $n_2$ are integers and $R_H$ is $1.09678 \times 10^{-2}\ \text{nm}^{-1}$.

$$\frac{1}{\lambda} = R_H\left[\frac{1}{n_1^2} - \frac{1}{n_2^2}\right]$$

These five series of lines are shown in Figure 6.12 and are referred to by their discoverers' names.

## 6.13 QUANTIZATION OF ENERGY

The spectrum of the hydrogen atom raises several important questions.

1. Why do hydrogen atoms give off only a handful of narrow bands of radiation when they emit light?
2. Why do hydrogen atoms emit light when excited that has the same wavelengths as the light they absorb at room temperature?

Answers to these questions came from work on a related topic.

It is common knowledge that objects give off light when heated. Examples range from the red glow of an electric burner on a stove to the bright white light emitted when the tungsten wire in a light bulb is heated by passing an electric current through the wire. What is less well known is a phenomenon discovered by Thomas Wedgwood in 1792. Wedgwood, whose father started the famous porcelain factory, noticed that many objects give off a red glow when heated to the same temperature. The bottom of a porcelain crucible and the iron triangle on which the crucible rests, for example, both glow red when heated with a bunsen burner to the same temperature.

Wedgwood's observation that objects heated to the same temperature give off light of the same color can be illustrated by comparing the light given off by a crucible and the iron triangle on which it rests when the crucible is heated.

Wedgwood also noticed that the color of the light emitted by an object changes as it is heated to higher temperatures until the object glows white hot, but the spectrum of light given off at a particular temperature is always the same. The fact that sunlight is equivalent to the light emitted by an object at 6000 K, for example, has led to the assumption that this is the temperature of the surface of the sun.

In 1900, Max Planck presented a theoretical explanation of the spectrum of radiation emitted by an object that glows. He argued that the walls of a glowing solid could be imagined to contain a series of resonators that oscillate at different frequencies. These resonators gain energy in the form of heat from the walls of the object and lose energy in the form of electromagnetic radiation. The more heat they gain, the faster they oscillate and the higher the frequency of the radiation they emit. The energy of these resonators at any moment is proportional to the frequency with which they oscillate.

This idea was a logical extension of work that Heinrich Hertz had done on the origin of electromagnetic radiation. However, the only way that Planck could get this model to fit the spectrum of light emitted by a glowing object was to introduce a revolutionary notion. To fit the observed spectrum, Planck had to assume that the energy of these oscillators could take on only a limited number of values. In other words, the spectrum of energies for these oscillators wasn't continuous. Because the number of values of the energy of these oscillators is limited, they are theoretically "countable." The energy of the oscillators in this system is therefore said to be *quantized.*

Planck introduced the notion of quantization to explain how light was emitted. In 1905 Albert Einstein extended Planck's work to the light that had been emitted. At a time when everyone agreed that light was a wave (and therefore continuous), Einstein suggested that it behaved as if it were a stream of small bundles, or packets, of energy. In other words, light was also quantized.

Einstein's model was based on two assumptions. First, he assumed that light was composed of **photons,** which are small, discrete bundles of energy. Second, he assumed that the energy of a photon is proportional to its frequency.

$$E = h\nu$$

In this equation, $h$ is a constant known as Planck's constant, which is equal to $6.626 \times 10^{-34}$ J-s.

## EXERCISE 6.6

Calculate the energy of a single photon of red light with a wavelength of 700.0 nm and the energy of a mole of these photons.

**SOLUTION** In Exercise 6.5 we found that red light with a wavelength of 700.0 nm has a frequency of $4.283 \times 10^{14}$ s$^{-1}$. Substituting this frequency into the Planck–Einstein equation gives the following result:

$$E = (6.626 \times 10^{-34} \text{ J-}\cancel{\text{s}})(4.283 \times 10^{14} \cancel{\text{s}}^{-1}) = 2.838 \times 10^{-19} \text{ J}$$

A single photon of red light carries an insignificant amount of energy. But a mole of these photons carries about 171,000 joules of energy, or 171 kJ/mol:

$$\frac{2.838 \times 10^{-19} \text{ J}}{1 \cancel{\text{ photon}}} \times \frac{6.022 \times 10^{23} \cancel{\text{ photons}}}{\text{mol}} = \mathbf{170.9 \text{ kJ/mol}}$$

Absorption of a mole of photons of red light would therefore provide enough energy to raise the temperature of a liter of water by more than 40°C.

Einstein's hypothesis that light consists of packets of energy had important consequences for any model that tried to explain the absorption or emission spectrum of the hydrogen atom. The fact that hydrogen atoms emit or absorb radiation at a limited number of frequencies implies that these atoms can only absorb radiation with certain energies. This suggests that there are only a limited number of energy levels within the hydrogen atom. If this is true, these energy levels are countable. In other words, the energy levels of the hydrogen atom are quantized.

This hypothesis explains why an object gives off red light when heated until it just starts to glow. Red light has the longest wavelength and the smallest frequency of any form of visible radiation. Because the energy of electromagnetic radiation is proportional to its frequency, red light carries the smallest amount of energy of any form of visible radiation. The light given off when an object just starts to glow therefore has the color characteristic of the lowest energy form of electromagnetic radiation our eyes can detect. As the object becomes hotter it also emits light at higher frequencies, until it eventually appears to give off white light.

## 6.14 THE BOHR MODEL OF THE ATOM

Although he was never able to incorporate electrons into his model of the atom, Rutherford favored a model in which the electrons circled the nucleus in much the same way planets revolve around the sun. Unfortunately, this model violates one of the basic principles of physics. Classical physics suggests that a charged particle, such as an electron, revolving in a circular orbit around the nucleus of an atom would enter into a spiral that would eventually lead to its collapse into the nucleus.

In 1913 one of Rutherford's students, Niels Bohr, proposed a model for the hydrogen atom that was consistent with Rutherford's model and yet also explained the spectrum of the hydrogen atom. The **Bohr model** was based on the following assumptions:

1. The electron in a hydrogen atom travels around the nucleus in a circular orbit.
2. The energy of the electron in an orbit is proportional to its distance from the nucleus. The further the electron is from the nucleus, the more energy it has.
3. Only a limited number of orbits with certain energies are allowed. In other words, the orbits are quantized.
4. The only orbits that are allowed are those for which the angular momentum of the electron is an integral multiple of Planck's constant divided by $2\pi$.
5. Light is absorbed when an electron jumps to a higher energy orbit and emitted when an electron falls into a lower energy orbit.
6. The energy of the light emitted or absorbed is exactly equal to the difference between the energies of the orbits.

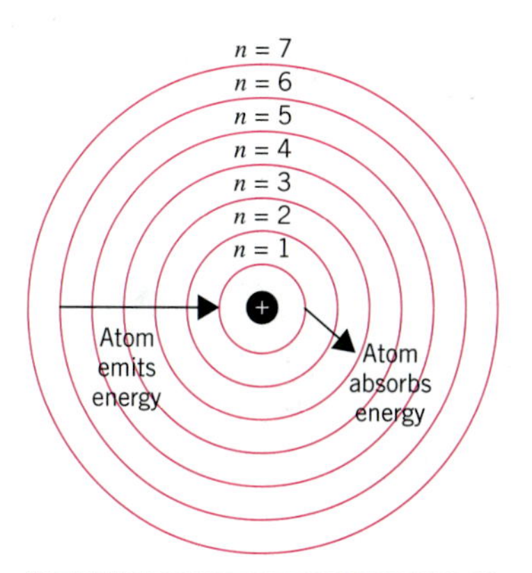


**Figure 6.13** According to the Bohr model, hydrogen atoms absorb light when an electron is excited from a low-energy orbit (such as $n = 1$) into a higher energy orbit (such as $n = 3$). Atoms that have been excited by an electric discharge can give off light when an electron drops from a high-energy orbit (such as $n = 6$) into a lower energy orbit (such as $n = 1$). The energy of the photon absorbed or emitted when the electron moves from one orbit to another is equal to the difference between the energies of the orbits.

Some of the key elements of this hypothesis are illustrated in Figure 6.13. Three points deserve particular attention. First, Bohr recognized that his first assumption violated the principles of classical mechanics. But he knew that it was impossible to explain the spectrum of the hydrogen atom within the limits of classical physics. He was therefore willing to assume that one or more of the principles from classical physics might not be valid on the atomic scale.

Second, he assumed there are only a limited number of orbits in which the electron can reside. He based this assumption on the fact that there are only a limited number of lines in the spectrum of the hydrogen atom and his belief that these lines were the result of light being emitted or absorbed as an electron moved from one orbit to another in the atom.

Finally, Bohr restricted the number of orbits on the hydrogen atom by limiting the allowed values of the angular momentum of the electron. Any object moving along a straight line has a *momentum* equal to the product of its mass ($m$) times the velocity ($v$) with which it moves. An object moving in a circular orbit has an *angular momentum* equal to its mass ($m$) times the velocity ($v$) times the radius of the orbit ($r$). Bohr postulated that the angular momentum of the electron could take on only certain values, equal to an integer times Planck's constant divided by $2\pi$:

$$mvr = n\left[\frac{h}{2\pi}\right] \qquad \text{(where } n = 1, 2, 3, 4, 5, \ldots\text{)}$$

Bohr then used classical physics to show that the energy of an electron in any one of these orbits is inversely proportional to the square of the integer $n$:

$$E \propto -\frac{1}{n^2}$$

The difference between the energies of any two orbits is therefore given by the following equation:

$$\Delta E = R_H\left[\frac{1}{n_1^2} - \frac{1}{n_2^2}\right]$$

In this equation, $n_1$ and $n_2$ are both integers and $R_H$ is the proportionality constant known as the Rydberg constant.

Planck's equation states that the energy of a photon is proportional to its frequency:

$$\boldsymbol{E = h\nu}$$

Substituting the relationship between the frequency, wavelength, and the speed of light into this equation suggests that the energy of a photon is inversely proportional to its wavelength:

$$E = h\left[\frac{c}{\lambda}\right] = \frac{hc}{\lambda}$$

The inverse of the wavelength of electronmagnetic radiation is therefore directly proportional to the energy of this radiation:

$$\frac{1}{\lambda} = \frac{E}{hc}$$

By properly defining the units of the constant, $R_H$, Bohr was able to show that the wavelengths of the light given off or absorbed by a hydrogen atom should be given by the following equation:

$$\frac{1}{\lambda} = R_H\left[\frac{1}{n_1^2} - \frac{1}{n_2^2}\right]$$

Thus, once he introduced his basic assumptions, Bohr was able to derive an equation that matched the relationship obtained from the analysis of the spectrum of the hydrogen atom discussed in Section 6.12. Bohr was able to show that the wavelengths in the UV spectrum of hydrogen discovered by Lyman correspond to transitions from one of the higher energy orbits into the $n = 1$ orbit, as shown in Figure 6.14. The wavelengths in the visible spectrum of hydrogen analyzed by

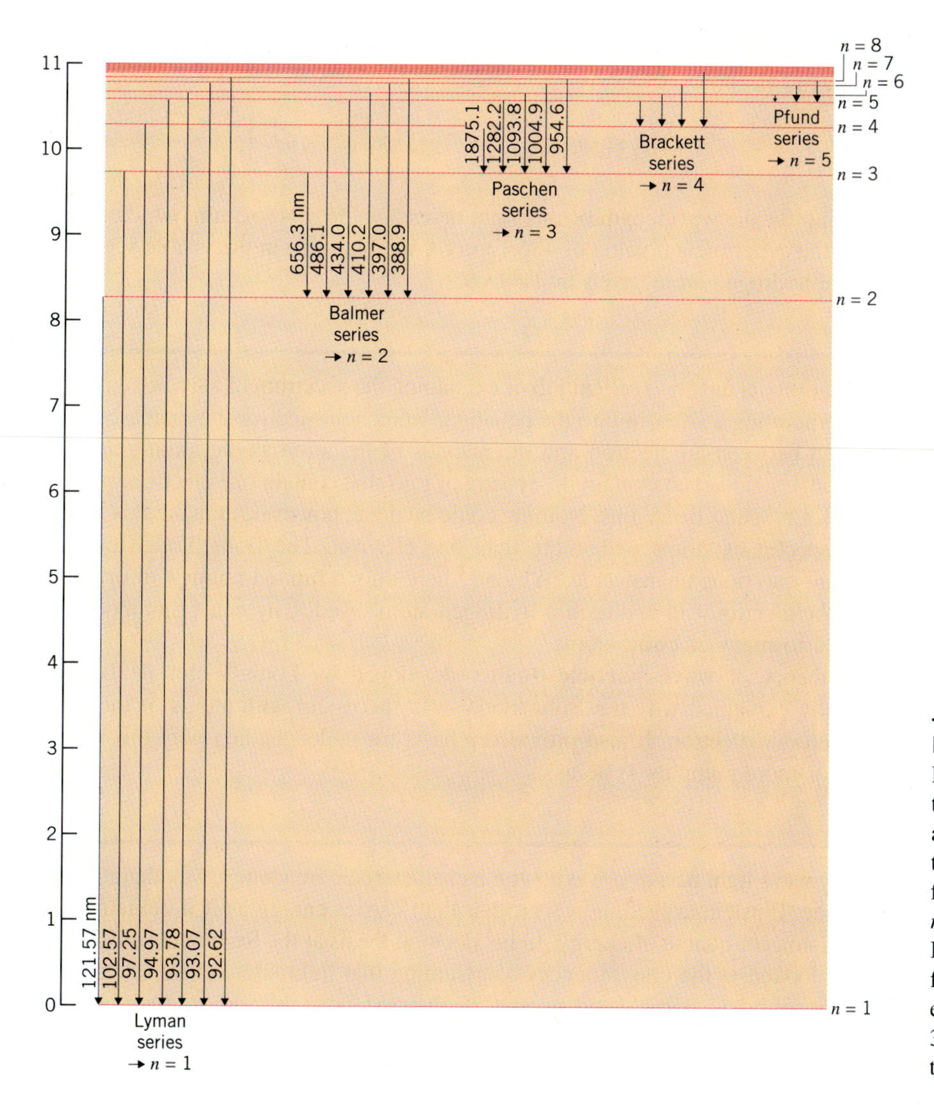


**Figure 6.14** According to the Bohr model, the wavelengths in the UV spectrum of the hydrogen atom discovered by Lyman were the result of electrons dropping from higher energy orbits into the $n = 1$ orbit. The Balmer, Paschen, Brackett, and Pfund series result from electrons falling from high-energy orbits into the $n = 2$, $n = 3$, $n = 4$, and $n = 5$ orbits, respectively.

Balmer are the result of transitions from one of the higher energy orbits into the $n = 2$ orbit. The Paschen, Brackett, and Pfund series of lines in the infrared spectrum of hydrogen result from electrons dropping into the $n = 3$, $n = 4$, and $n = 5$ orbits, respectively.

## EXERCISE 6.7

Calculate the wavelength of the light given off by a hydrogen atom when an electron falls from the $n = 4$ to the $n = 2$ orbit in the Bohr model.

**SOLUTION** According to the Bohr model, the wavelength of the light emitted by a hydrogen atom when the electron falls from a high energy ($n = 4$) orbit into a lower energy ($n = 2$) orbit is given by the following equation:

$$\frac{1}{\lambda} = R_H\left[\frac{1}{n_1^2} - \frac{1}{n_2^2}\right]$$

Substituting the appropriate values of $R_H$, $n_1$, and $n_2$ into this equation gives the following result:

$$\frac{1}{\lambda} = (1.09678 \times 10^{-2}\ \text{nm}^{-1})\left[\frac{1}{2^2} - \frac{1}{4^2}\right]$$

Solving for the wavelength of this light gives a value of 486.3 nm, which agrees with the experimental value of 486.1 nm for the blue line in the visible spectrum of the hydrogen atom given in Table 6.3.

## 6.15 THE SUCCESS AND THE FAILURE OF THE BOHR MODEL

The Bohr model did an excellent job of explaining the spectrum of a hydrogen atom. By incorporating a $Z^2$ term into the equation, which adjusted for the increase in the attraction between an electron and the nucleus of the atom as the atomic number increased, it could even explain the spectra of ions that contain one electron, such as the $He^+$, $Li^{2+}$, and $Be^{3+}$ ions. Nothing could be done, however, to make this model fit the spectra of atoms with more than one electron. The Bohr model left two important questions unanswered. Why are there only a limited number of orbits in which the electron can reside in a hydrogen atom? And, why can't this model be extended to many-electron atoms?

The theory of **wave–particle duality** developed by Louis-Victor de Broglie eventually explained why the Bohr model was successful with atoms or ions that contained one electron. It also provided a basis for understanding why this model failed for more complex systems.

## 6.16 WAVE–PARTICLE DUALITY

In many ways light acts as a wave, with a characteristic frequency, wavelength, and amplitude. Einstein argued, however, that light carries energy as if it contains discrete photons or quanta of energy. In his doctoral thesis at the Sorbonne in 1924, de Broglie looked at the consequences of assuming that light simultaneously has the properties of both a particle and a wave. He then extended this idea to other objects, such as an electron.

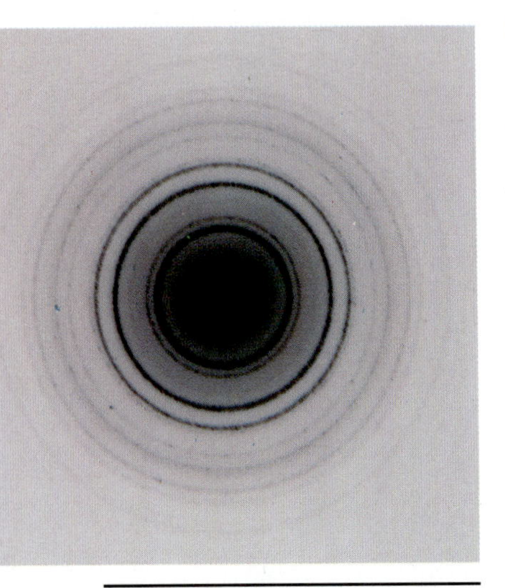

The mass of an electron is small enough that its wave behavior can be detected. This photograph shows the diffraction of electrons by a crystal of cubic zirconia.

When an object behaves as a particle in motion, it has an energy proportional to its mass ($m$) and speed ($s$) with which it moves through space:

$$E = ms^2$$

When it behaves as a wave, however, it has an energy that is proportional to its frequency:

$$E = h\nu = \frac{hs}{\lambda}$$

By simultaneously assuming that an object can be both a particle and a wave, de Broglie set up the following equation:

$$ms^2 = \frac{hs}{\lambda}$$

By rearranging this equation, he derived a relationship between one of the wavelike properties of matter and one of its properties as a particle:

$$\lambda = \frac{h}{ms}$$

As noted in the previous section, the product of the mass of an object times the speed with which it moves is the *momentum* ($p$) of the particle. Thus, the **de Broglie equation** suggests that the wavelength ($\lambda$) of any object in motion is inversely proportional to its momentum:

$$\lambda = \frac{h}{p}$$

De Broglie concluded that most particles are too heavy to observe their wave properties. However when the mass of an object is very small, the wave properties can be detected experimentally. De Broglie predicted that the mass of an electron was small enough to exhibit the properties of both particles and waves. In 1927 this prediction was confirmed when the diffraction of electrons was observed experimentally by C. J. Davisson.

De Broglie applied his theory of wave–particle duality to the Bohr model to explain why only certain orbits are allowed for the electron. He argued that only certain orbits allow the electron to satisfy both its particle and wave properties at the same time because only certain orbits have a circumference that is an integral multiple of the wavelength of the electron, as shown in Figure 6.15.

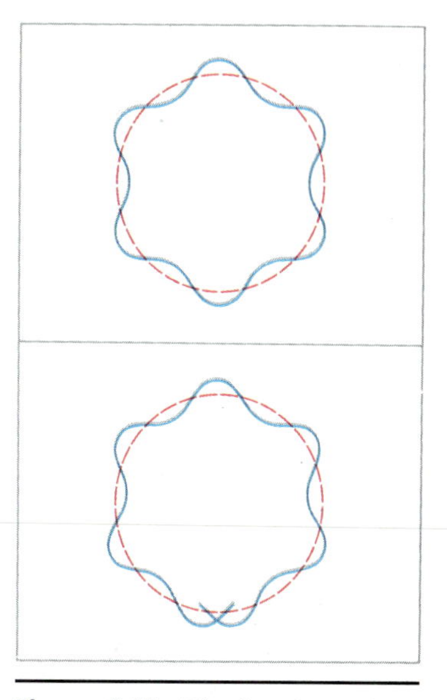

**Figure 6.15** The fact that only certain orbits are allowed in the Bohr model of the atom can be explained by assuming that the circumference of the orbit must be an integral ($n$ = 1, 2, 3, 4 . . . ) multiple of the wavelength of the electron. When it is not, the electron cannot simultaneously satisfy its wave and particle behavior.

**CHECKPOINT**

Use the de Broglie equation to explain why particles that have more mass have less of the character of a wave.

## 6.17 CONSEQUENCES OF THE WAVE PROPERTIES OF ELECTRONS

At first glance, the Bohr model looks like a two-dimensional model of the atom because it restricts the motion of the electron to a circular orbit in a two-dimensional plane. In reality the Bohr model is a one-dimensional model, because a circle can be defined by specifying only one dimension: its radius, $r$. As a result, only one coordinate ($n$) is needed to describe the orbits in the Bohr model.

Unfortunately, electrons aren't particles that can be restricted to a one-dimensional circular orbit. They act to some extent as waves and therefore exist in three-dimensional space. The Bohr model works for one-electron atoms or ions only because certain factors present in more complex atoms are not present in these atoms or ions. To construct a model that describes the distribution of electrons in atoms that contain more than one electron, we have to allow the electrons to occupy three-dimensional space. We therefore need a model that uses three coordinates to describe the distribution of electrons in these atoms.

## 6.18 QUANTUM NUMBERS

Why do we still talk about the Bohr model of the atom 75 years after its discovery, if the only thing this model can do is explain the spectrum of the hydrogen atom? The answer is that it was the last model of the atom for which a simple physical picture can be constructed. It is easy to imagine an atom that consists of solid electrons revolving around the nucleus in circular orbits.

A more powerful model of the atom was developed by Erwin Schrödinger in 1926. Schrödinger combined the equations for the behavior of waves with the de Broglie equation to generate a mathematical model for the distribution of electrons in an atom. The advantage of this model is that it consists of mathematical equations known as *wave functions* that satisfy the requirements placed on the behavior of electrons. The disadvantage is that it is difficult to imagine a physical model of electrons as waves.

The Schrödinger model assumes that the electron is a wave and tries to describe the regions in space, or **orbitals,** where electrons are most likely to be found. Instead of trying to tell us where the electron is at any time, the Schrödinger model describes the probability that an electron can be found in a given region of space at a given time. This model no longer tells us where the electron is; it only tells us where it may be.

The Bohr model was a one-dimensional model that used one quantum number to describe the distribution of electrons in the atom. The only information that was important was the *size* of the orbit, which was described by the $n$ quantum number. Schrödinger's model allowed the electron to occupy three-dimensional space. It therefore required three coordinates, or three **quantum numbers,** to describe the orbitals in which electrons can be found.

The three coordinates that come from Schrödinger's wave equations are the principal ($n$), angular ($l$), and magnetic ($m$) quantum numbers. These quantum numbers describe the size, shape, and orientation in space of the orbitals on an atom.

The **principal quantum number** ($n$) describes the size of the orbital. Orbitals for which $n = 2$ are larger than those for which $n = 1$, for example. Because they have opposite electrical charges, electrons are attracted to the nucleus of the atom. Energy must therefore be absorbed to excite an electron from an orbital in which the electron is close to the nucleus ($n = 1$) into an orbital in which it is further from the nucleus ($n = 2$). The principal quantum number therefore indirectly describes the energy of an orbital.

The **angular quantum number** ($l$) describes the shape of the orbital. Orbitals have shapes that are best described as spherical ($l = 0$), polar ($l = 1$), or cloverleaf ($l = 2$), as shown in Figure 6.16. They can even take on more complex shapes as the value of the angular quantum number becomes larger.

There is only one way in which a sphere ($l = 0$) can be oriented in space. Orbitals that have polar ($l = 1$) or cloverleaf ($l = 2$) shapes, however, can point in different directions. We therefore need a third quantum number, known as the **magnetic quantum number** ($m$), to describe the orientation in space of a particular orbital. (It

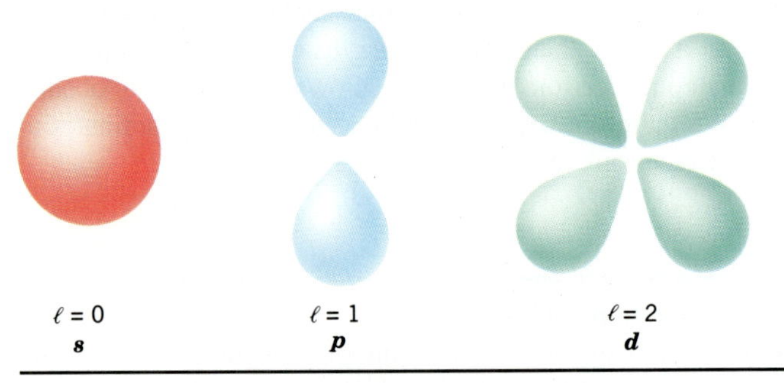


**Figure 6.16** The angular quantum number specifies the shape of the orbital where electrons can be found. When $l = 0$, the orbital is spherical. When $l = 1$, the orbital is polar. When $l = 2$, the orbital typically has the shape of a cloverleaf.

is called the *magnetic* quantum number because the effect of different orientations of orbitals was first observed in the presence of a magnetic field.)

Atomic spectra suggest that the orbitals on an atom are quantized. We therefore need a set of rules that limit the possible combinations of the $n$, $l$, and $m$ quantum numbers:

**RULES GOVERNING THE ALLOWED COMBINATIONS OF QUANTUM NUMBERS**

- The three quantum numbers ($n$, $l$, and $m$) that describe an orbital are integers.
- The principal quantum number ($n$) cannot be zero. The allowed values of $n$ are therefore 1, 2, 3, 4, and so on.
- The angular quantum number ($l$) can be any integer between 0 and $n - 1$. If $n = 3$, for example, $l$ can be either 0, 1, or 2.
- The magnetic quantum number ($m$) can be any integer between $-l$ and $+l$. If $l = 2$, $m$ can be either $-2$, $-1$, 0, $+1$, or $+2$.

## CHECKPOINT

Explain why three quantum numbers are required to describe an orbital. What feature of the three-dimensional structure of an orbital is specified by each of these quantum numbers?

## EXERCISE 6.8

Describe the allowed combinations of the $n$, $l$, and $m$ quantum numbers when $n = 3$.

**SOLUTION** The angular quantum number ($l$) can be any integer between 0 and $n - 1$. Thus, if $n = 3$, $l$ can be either 0, 1, or 2. The magnetic quantum number ($m$) can be any integer between $-l$ and $+l$. Thus, when $l = 0$, $m$ must be 0. When $l = 1$, $m$ can be either $-1$, 0, or $+1$. When $l = 2$, $m$ can be $-2$, $-1$, 0, $+1$, or $+2$. The allowed combinations of $n$, $l$, and $m$ for which $n = 3$ are therefore restricted to the following values:

| $n$ | $l$ | $m$ |
|---|---|---|
| 3 | 0 | 0 |
| 3 | 1 | −1 |
| 3 | 1 | 0 |
| 3 | 1 | +1 |
| 3 | 2 | −2 |
| 3 | 2 | −1 |
| 3 | 2 | 0 |
| 3 | 2 | +1 |
| 3 | 2 | +2 |

## 6.19 SHELLS AND SUBSHELLS OF ORBITALS

Orbitals that have the same value of the principal quantum number form a **shell.** Orbitals within a shell are divided into **subshells** that have the same value of the angular quantum number. Chemists describe the shell and subshell in which an orbital belongs with a two-character code such as $2p$ or $4f$. The first character indicates the shell ($n = 2$ or $n = 4$). The second character identifies the subshell. By convention, the following lowercase letters are used to indicate different subshells.

$$s\colon \; l = 0 \qquad d\colon \; l = 2$$
$$p\colon \; l = 1 \qquad f\colon \; l = 3$$

Although there is no pattern in the first four letters ($s$, $p$, $d$, $f$), the letters progress alphabetically from that point ($g$, $h$, and so on). Some of the allowed combinations of the $n$ and $l$ quantum numbers are shown in Figure 6.17.

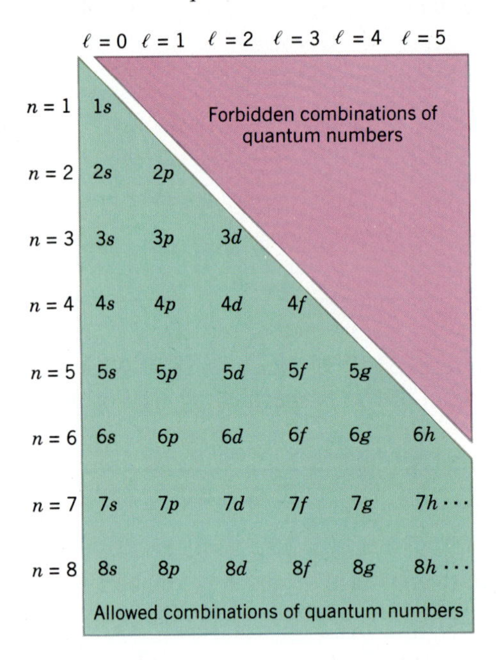


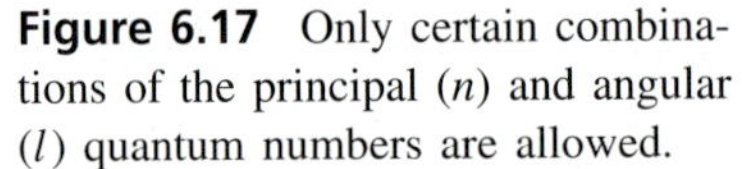

**Figure 6.17** Only certain combinations of the principal ($n$) and angular ($l$) quantum numbers are allowed.

The third rule limiting allowed combinations of the $n$, $l$, and $m$ quantum numbers has an important consequence: The number of subshells in a shell is equal to the principal quantum number for the shell. The $n = 3$ shell, for example, contains three subshells: the $3s$, $3p$, and $3d$ orbitals.

Let's look at some of the possible combinations of the $n$, $l$, and $m$ quantum numbers. There is only one orbital in the $n = 1$ shell because there is only one way in which a sphere can be oriented in space. The only allowed combination of quantum numbers for which $n = 1$ is the following:

| $n$ | $l$ | $m$ | |
|---|---|---|---|
| 1 | 0 | 0 | $1s$ |

There are four orbitals in the $n = 2$ shell:

| $n$ | $l$ | $m$ | |
|---|---|---|---|
| 2 | 0 | 0 | $2s$ |
| 2 | 1 | −1 | $2p$ |
| 2 | 1 | 0 | |
| 2 | 1 | 1 | |

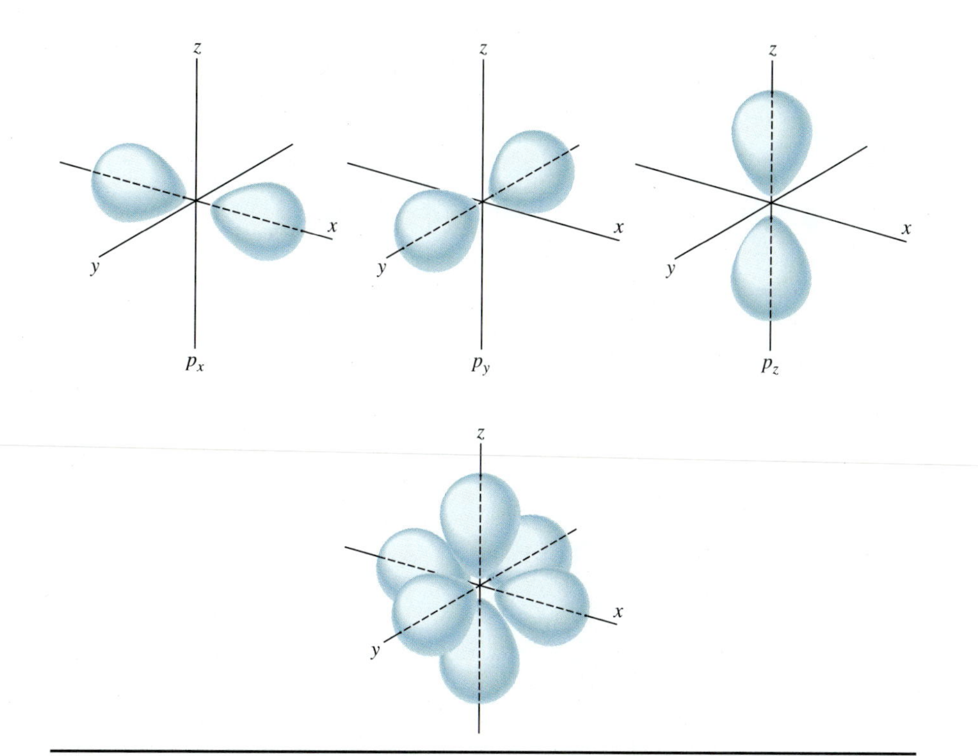


**Figure 6.18** When the angular quantum number ($l$) is 1, there are three possible values of the magnetic quantum number ($m$): $-1$, 0 and $+1$. These three values of $m$ correspond to the three different orientations of these orbitals in space.

There is only one orbital in the 2*s* subshell. But, there are three orbitals in the 2*p* subshell because there are three directions in which a *p* orbital can point. One of these orbitals is oriented along the *x* axis, another along the *y* axis, and the third along the *z* axis of a coordinate system, as shown in Figure 6.18. These orbitals are therefore known as the $2p_x$, $2p_y$, and $2p_z$ orbitals.

There are nine orbitals in the $n = 3$ shell.

| $n$ | $l$ | $m$ | |
|---|---|---|---|
| 3 | 0 | 0 | 3*s* |
| 3 | 1 | $-1$ | |
| 3 | 1 | 0 | 3*p* |
| 3 | 1 | 1 | |
| 3 | 2 | $-2$ | |
| 3 | 2 | $-1$ | |
| 3 | 2 | 0 | 3*d* |
| 3 | 2 | 1 | |
| 3 | 2 | 2 | |

There is one orbital in the 3*s* subshell and three orbitals in the 3*p* subshell. The $n = 3$ shell, however, also includes 3*d* orbitals.

The five different orientations of orbitals in the 3*d* subshell are shown in Figure 6.19. One of these orbitals lies between the axes in the *xy* plane of an *xyz* coordinate system and is called the $3d_{xy}$ orbital. The $3d_{xz}$ and $3d_{yz}$ orbitals have the same shape, but they lie between the axes of the coordinate system in the *xz* and *yz*

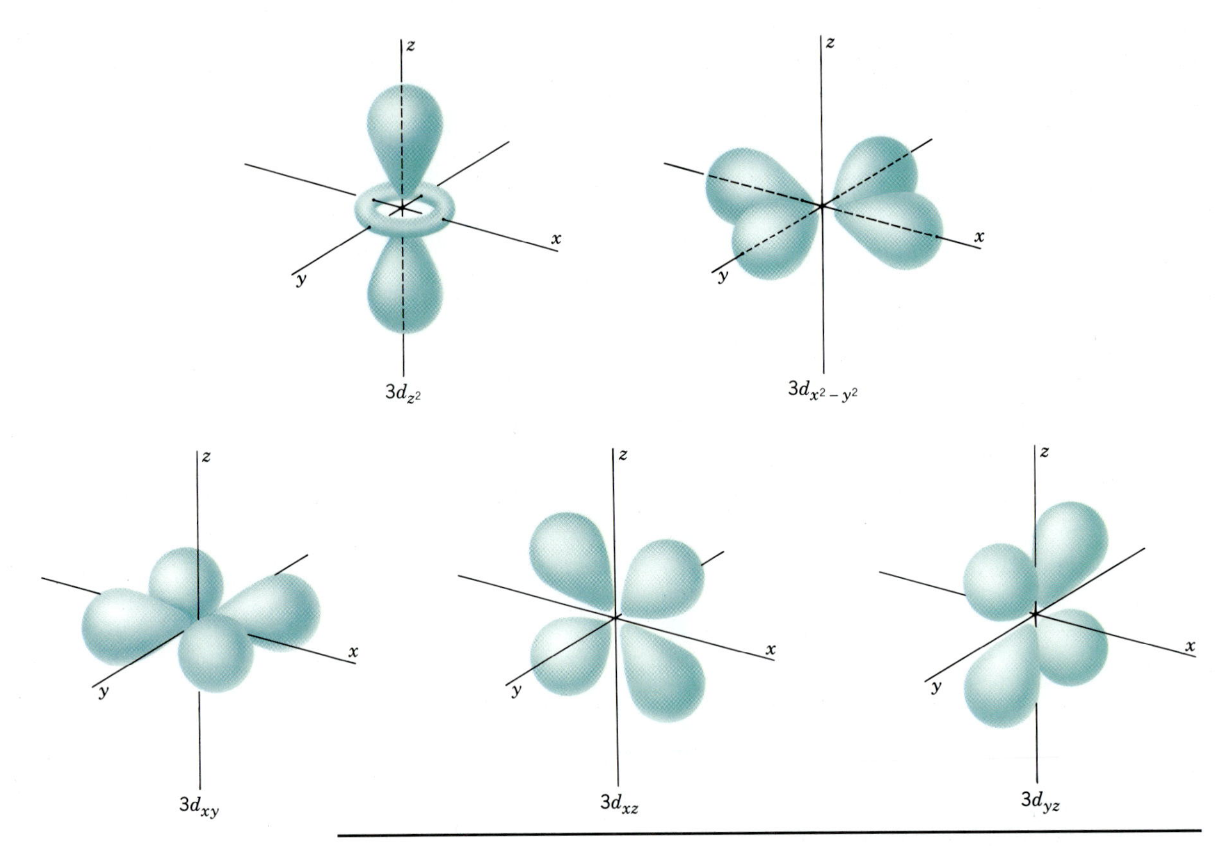


**Figure 6.19** When the angular quantum number ($l$) is 2, there are five allowed values of the magnetic quantum number ($m$) corresponding to the five different orientations of these orbitals in space.

planes. The fourth orbital in this subshell lies along the $x$ and $y$ axes and is called the $3d_{x^2-y^2}$ orbital. Most of the space occupied by the fifth orbital lies along the $z$ axis and this orbital is called the $3d_{z^2}$ orbital.

Exercise 6.9 examines the relationships between orbitals and quantum numbers.

## EXERCISE 6.9

Determine the relationships between: (1) the number of orbitals in a shell and the principal quantum number of the shell, and (2) the number of orbitals in a subshell and the angular quantum number for the subshell.

**SOLUTION** There is one orbital in the $n = 1$ shell, there are four orbitals in the $n = 2$ shell, nine orbitals in the $n = 3$ shell, and so on. The number of orbitals in a shell is therefore the square of the principal quantum number: $1^2 = 1$, $2^2 = 4$, $3^2 = 9$. There is one orbital in an $s$ subshell ($l = 0$), three orbitals in a $p$ subshell ($l = 1$), and five orbitals in a $d$ subshell ($l = 2$). The number of orbitals in a subshell is therefore $2(l) + 1$.

All we have done so far is describe a series of orbitals in which electrons can be placed. Before we can use these orbitals we need to know the number of electrons that can occupy an orbital and how they can be distinguished from one another. Experimental evidence suggests that an orbital can hold no more than two electrons.

To distinguish between the two electrons in an orbital, we need a fourth quantum number. This is called the **spin quantum number** ($s$) because electrons behave as if they are spinning in either a clockwise or a counterclockwise fashion (Figure 6.20). One of the electrons in an orbital is arbitrarily assigned an $s$ quantum number of $+\frac{1}{2}$, the other is assigned an $s$ quantum number of $-\frac{1}{2}$. Thus, it takes three quantum numbers to define an orbital but four quantum numbers to identify one of the electrons that can occupy the orbital.

The allowed combinations of $n$, $l$, and $m$ quantum numbers for the first four shells are given in Table 6.4. For each of these orbitals, there are two allowed values of the spin quantum number, $s$.

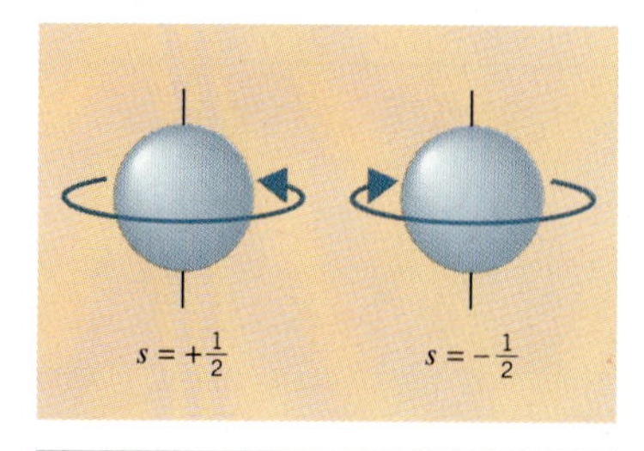


**Figure 6.20** It takes three quantum numbers to describe an orbital. A fourth quantum number is then used to differentiate between the two electrons that can occupy an orbital. Because these electrons behave as if they are spinning in different directions, this fourth quantum number is called the spin ($s$) quantum number. By convention, the allowed values of the spin quantum number are $+\frac{1}{2}$ and $-\frac{1}{2}$.

**TABLE 6.4 Summary of Allowed Combinations of Quantum Numbers**

| $n$ | $l$ | $m$ | Subshell Notation | Number of Orbitals in the Subshell | Number of Electrons Needed to Fill Subshell | |
|---|---|---|---|---|---|---|
| 1 | 0 | 0 | 1$s$ | 1 | 2 | total: 2 |
| 2 | 0 | 0 | 2$s$ | 1 | 2 | |
| 2 | 1 | 1,0,−1 | 2$p$ | 3 | 6 | total: 8 |
| 3 | 0 | 0 | 3$s$ | 1 | 2 | |
| 3 | 1 | 1,0,−1 | 3$p$ | 3 | 6 | |
| 3 | 2 | 2,1,0,−1,−2 | 3$d$ | 5 | 10 | total: 18 |
| 4 | 0 | 0 | 4$s$ | 1 | 2 | |
| 4 | 1 | 1,0,−1 | 4$p$ | 3 | 6 | |
| 4 | 2 | 2,1,0,−1,−2 | 4$d$ | 5 | 10 | |
| 4 | 3 | 3,2,1,0,−1,−2,−3 | 4$f$ | 7 | 14 | total: 32 |

## RESEARCH IN THE 90s

### CONSEQUENCES OF THE SPIN OF SUBATOMIC PARTICLES

Electrons are not the only subatomic particles that behave as if they were spinning in one direction or another—protons and neutrons also have a *spin quantum number* of $\pm\frac{1}{2}$. Because they contain protons and neutrons, many (but not all) nuclei also have a spin quantum number.

Table 6.5 lists the spin quantum number of the common isotopes of the elements in the first and second rows of the periodic table. Several patterns can be seen in these data:

- Nuclei, such as $^{4}$He, $^{12}$C, $^{16}$O, and $^{20}$Ne, that contain an even number of both protons (p) and neutrons (n) have no net spin. (This suggests that both protons and neutrons can pair, in much the same way that electrons pair.)
- Neutrons do not pair with protons. [Deuterium ($^{2}$H), for example, has an unpaired proton and an unpaired neutron, for a total spin quantum number of 1.]
- Protons or neutrons don't always pair when we might expect them to. ($^{7}$Li contains three unpaired protons and $^{10}$B contains three unpaired protons and three unpaired neutrons.)

The spin of a nucleus becomes important in the presence of a magnetic field. Consider a compound that contains hydrogen, for example. One-half of the $^{1}H$ nuclei in the sample will spin in a direction that produces a tiny magnetic field aligned with the external magnetic field. The other half will spin in a direction that generates a tiny magnetic field opposed to the laboratory magnet. The result is a small difference between the energies of the two spin states of these nuclei. In a magnetic field of 2.35 Tesla, the difference between the energies of the $+\frac{1}{2}$ and $-\frac{1}{2}$ spin states of the $^{1}H$ nucleus is equal to the energy carried by radiowaves that have a frequency of 100,000,000 cycles per second, or 100 MHz.

**TABLE 6.5 The Common Isotopes of the First and Second Row Elements**

| Nuclei | Spin Quantum Number | Abundance (%) |
|---|---|---|
| $^{1}H$ (1p) | $\frac{1}{2}$ | 99.985 |
| $^{2}H$ (1p, 1n) | 1 | 0.015 |
| $^{4}He$ (2p, 2n) | 0 | ≈100 |
| $^{7}Li$ (3p, 4n) | $\frac{3}{2}$ | 92.5 |
| $^{9}Be$ (4p, 5n) | $\frac{3}{2}$ | ≈100 |
| $^{10}B$ (5p, 5n) | 3 | 19.9 |
| $^{11}B$ (5p, 6n) | $\frac{3}{2}$ | 80.1 |
| $^{12}C$ (6p, 6n) | 0 | 98.90 |
| $^{13}C$ (6p, 7n) | $\frac{1}{2}$ | 1.10 |
| $^{14}N$ (7p, 7n) | 1 | 99.63 |
| $^{16}O$ (8p, 8n) | 0 | 99.76 |
| $^{19}F$ (9p, 10n) | $\frac{1}{2}$ | ≈100 |
| $^{20}Ne$ (10p, 10n) | 0 | 90.48 |

Let's imagine what would happen if we placed a sample of a compound of hydrogen in a 2.35 Tesla magnet and then irradiated the sample with radiowaves of gradually increasing frequencies. Once the frequency reached 100 MHz, the sample would absorb radiation and the spins of the $^{1}H$ nuclei would flip over, from $+\frac{1}{2}$ to $-\frac{1}{2}$, and vice versa. This phenomenon has three important characteristics. First, the energy is absorbed by the nuclei. Second, the experiment must be done in a magnetic field. Finally, the absorption occurs when the system is in resonance—when the energy of the radiation is exactly equal to the difference between the energies of the two spin states. The experiment is therefore known as *nuclear magnetic resonance* (NMR) spectroscopy.

The NMR experiment can be done with any nucleus that has a spin quantum number that is not zero. At first glance, this experiment would seem to be of interest only to physicists, to probe the properties of a nucleus. Shortly after the NMR phenomenon was observed, however, it was found that the chemical environment of the nucleus influences the frequency at which it absorbs radiation. When the author of this section was a graduate student, the $^{13}C$ NMR spectrum of cholesteryl chloride shown in Figure 6.21 was obtained by his colleagues. This compound contains a total of 27 different environments in which a carbon atom can be found and 25 distinct lines were observed in this $^{13}C$ NMR spectrum. Because of the wealth of information in the $^{13}C$ NMR spectrum, $^{13}C$ NMR soon became one of the most powerful techniques in the chemists' repertoire to identify the product of a chemical reaction or to determine the structure of a substance isolated from natural sources.

In recent years, NMR spectroscopy has led to a revolution in the practice of

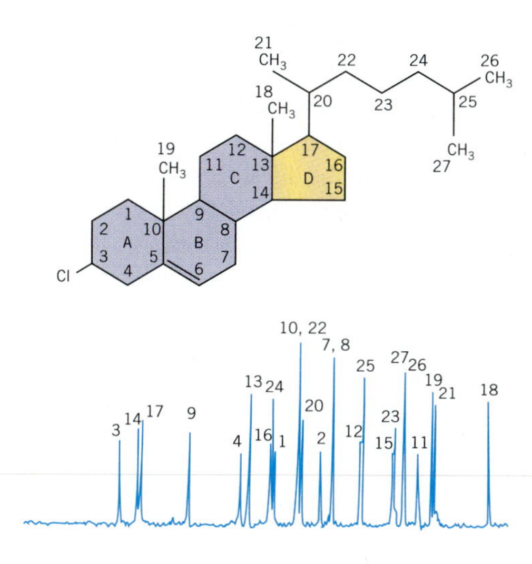


**Figure 6.21** The structure and $^{13}C$ NMR spectrum of cholesteryl chloride.

diagnostic medicine. To understand why, let's consider a sample in which the relative number of $^1H$ nuclei have a spin of $+\frac{1}{2}$ versus $-\frac{1}{2}$. In the absence of a magnetic field, there is no preference for one spin state over another. When the sample is placed in a magnetic field, however, the energies of the two spin states are not quite the same. As a result, the number of $^1H$ nuclei with a spin of $+\frac{1}{2}$ is no longer equal to the number with a spin of $-\frac{1}{2}$. The difference in the population of these states is relatively small—only a few parts per million. At resonance, however, this means that there is a slightly higher probability of a spin flip occurring in one direction than the other. As a result, there is a net absorption of radiation until the population of the two spin states becomes the same. If we turn off the radiation, we can measure how long it takes for the system to *relax*—for the population of the two spin states to return to equilibrium.

In 1971 a remarkable discovery was made: The relaxation time for the $^1H$ nuclei in the water in malignant tissue in rats was significantly longer than normal tissue. As additional experiments were carried out it became apparent that there were differences in the relaxation times of small molecules, such as inorganic salts and water, and large molecules, such as lipids and proteins. These differences were significant enough to differentiate between gray matter and white matter in the brain, which appear virtually identical in x-ray or CAT scans.

Before this technique could be used for medical diagnosis, spectrometers had to be developed that could scan objects as large as the human body. Because it involves using NMR to produce images of the tissue within the body, this technique was initially called *NMR imaging.* It soon became apparent that people were frightened by the term *nuclear* in the phrase *nuclear magnetic resonance imaging.* The technique therefore became known as *MRI.* MRI is now used routinely to obtain the kind of information that previously could be obtained only by exploratory surgery.

The diverse role that MRI will play in biomedical research in the 1990s can be appreciated by noting that MRI recently has been used to measure the effect of age on the rate of flow through the aortic system [*Journal of Applied Physiology,* **74,** 492–497 (1993)], to map the human visual cortex [*Proceedings of the National Academy of Sciences,* **89,** 11069–11073 (1992)], and to study the effect of sensory stimulation on the brain [*Proceedings of the National Academy of Sciences,* **89,** 5951–5955 (1992)].

## 6.20 THE RELATIVE ENERGIES OF ATOMIC ORBITALS

Because of the force of attraction between objects of opposite charge, the most important factor influencing the energy of an orbital is its size and therefore the value of the principal quantum number, $n$. For an atom that contains only one electron, there is no difference between the energies of the different subshells within a shell. The $3s$, $3p$, and $3d$ orbitals, for example, have the same energy in a hydrogen atom. The Bohr model, which specified the energies of orbits in terms of nothing more than the distance between the electron and the nucleus, therefore works for this atom.

The hydrogen atom is unusual, however. As soon as an atom contains more than one electron, the different subshells no longer have the same energy. Within a given shell, the $s$ orbitals always have the lowest energy. The energy of the subshells gradually becomes larger as the value of the angular quantum number becomes larger:

$$\text{Relative energies:} \quad s < p < d < f$$

As a result, two factors control the energy of an orbital for most atoms: the size of the orbital and its shape, as shown in Figure 6.22.

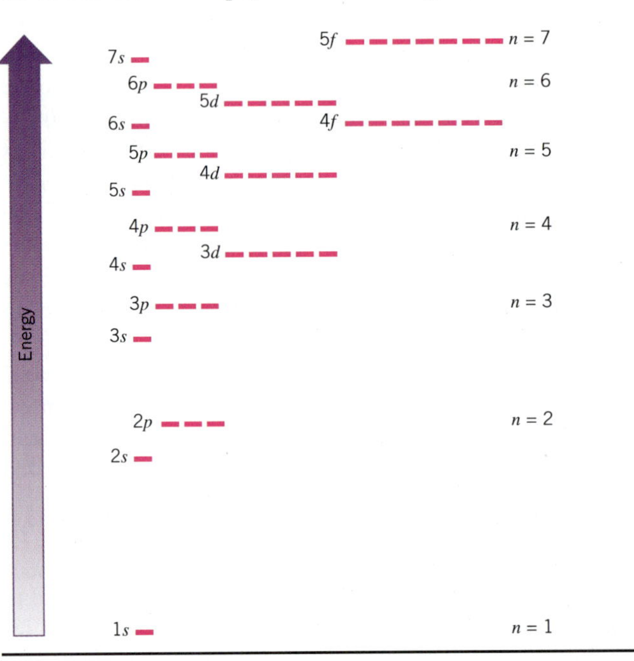


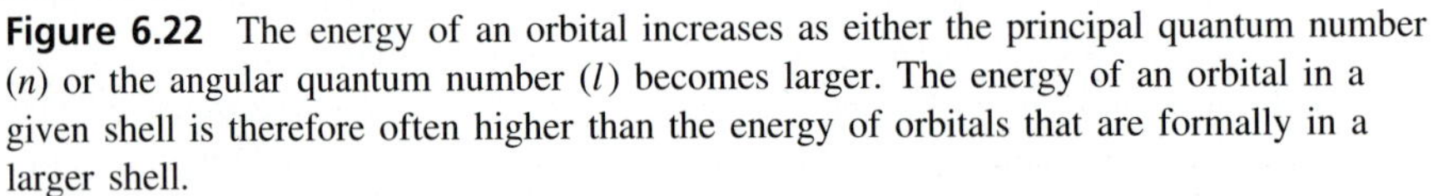

**Figure 6.22** The energy of an orbital increases as either the principal quantum number ($n$) or the angular quantum number ($l$) becomes larger. The energy of an orbital in a given shell is therefore often higher than the energy of orbitals that are formally in a larger shell.

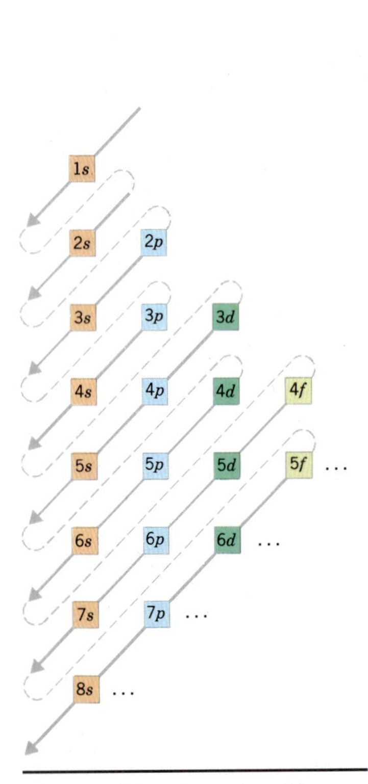


**Figure 6.23** The order of increasing energy for atomic orbitals can be predicted by simply following the arrows in this diagram: $1s < 2s < 2p < 3s < 3p < 4s < 3d < 4p$, and so on.

A very simple device can be constructed to estimate the relative energies of atomic orbitals. The allowed combinations of the $n$ and $l$ quantum numbers are organized in a table, as shown in Figure 6.23, and arrows are drawn at 45° angles pointing toward the bottom left corner of the table. The order of increasing energy of the orbitals is then read off by following these arrows, starting at the top of the first line and then proceeding on to the second, third, fourth lines, and so on. This diagram predicts the following order of increasing energy for atomic orbitals:

$$1s < 2s < 2p < 3s < 3p < 4s < 3d < 4p < 5s < 4d < 5p < 6s < 4f < 5d < 6p < 7s < 5f < 6d < 7p < 8s \ldots$$

## 6.21 ELECTRON CONFIGURATIONS

We are now ready to predict the **electron configuration** of an atom, which describes the orbitals occupied by electrons on the atom. The basis of this prediction is a rule known as the **aufbau principle,** which comes from the German word meaning "building up." The aufbau principle assumes that electrons are added to an atom, one at a time, starting with the lowest energy orbital, until all of the electrons have been placed in an appropriate orbital.

A hydrogen atom ($Z = 1$) has only one electron, which goes into the lowest energy orbital, the $1s$ orbital. This is indicated by writing a superscript "1" after the symbol for the orbital:

$$\text{H } (Z = 1): \qquad 1s^1$$

The next element has two electrons and the second electron fills the $1s$ orbital because there are only two possible values for the spin quantum number used to distinguish between the electrons in an orbital:

$$\text{He } (Z = 2): \qquad 1s^2$$

The third electron goes into the next orbital in the energy diagram, the $2s$ orbital:

$$\text{Li } (Z = 3): \qquad 1s^2\ 2s^1$$

The fourth electron fills this orbital:

$$\text{Be } (Z = 4): \qquad 1s^2\ 2s^2$$

After the $1s$ and $2s$ orbitals have been filled, the next lowest energy orbitals are the three $2p$ orbitals. The fifth electron therefore goes into one of these orbitals:

$$\text{B } (Z = 5): \qquad 1s^2\ 2s^2\ 2p^1$$

When the time comes to add a sixth electron, the electron configuration is obvious:

$$\text{C } (Z = 6): \qquad 1s^2\ 2s^2\ 2p^2$$

But this configuration raises an interesting question. There are three orbitals in the $2p$ subshell. Does the second electron go into the same orbital as the first, or does it go into one of the other orbitals in this subshell?

Before we can answer this question, we need to understand the concept of **degenerate orbitals.**

**By definition, orbitals are *degenerate* when they have the same energy.**

The energy of an orbital depends on both its size and its shape because the electron spends more of its time further from the nucleus of the atom as the orbital becomes larger or the shape becomes more complex. In an isolated atom, however, the energy of an orbital doesn't depend on the direction in which it points in space. Orbitals that differ only in their orientation in space, such as the $2p_x$, $2p_y$, and $2p_z$ orbitals, are therefore degenerate.

Electrons fill degenerate orbitals according to rules first stated by Friedrich Hund. **Hund's rules** can be summarized as follows:

1. One electron is added to each of the degenerate orbitals in a subshell before two electrons are added to any orbital in the subshell.
2. Electrons are added to a subshell with the same value of the spin quantum number until each orbital in the subshell has at least one electron.

When the time comes to place two electrons into the $2p$ subshell we put one electron into each of two of these orbitals. (The choice between the $2p_x$, $2p_y$, and $2p_z$ orbitals is purely arbitrary.)

C (Z = 6): $1s^2\ 2s^2\ 2p_x^1\ 2p_y^1$

The fact that both of the electrons in the $2p$ subshell have the same spin quantum number can be shown by representing an electron for which $s = +\frac{1}{2}$ with an arrow pointing up and an electron for which $s = -\frac{1}{2}$ with an arrow pointing down. The electrons in the $2p$ orbitals on carbon therefore can be represented as follows:

$\underline{\uparrow}\ \underline{\uparrow}\ \underline{\ \ }$

When we get to N ($Z = 7$), we have to put one electron into each of the three degenerate $2p$ orbitals:

N (Z = 7): $1s^2\ 2s^2\ 2p^3$ $\underline{\uparrow}\ \underline{\uparrow}\ \underline{\uparrow}$

Because each orbital in this subshell now contains one electron, the next electron added to the subshell must have the opposite spin quantum number, thereby filling one of the $2p$ orbitals:

O (Z = 8): $1s^2\ 2s^2\ 2p^4$ $\underline{\uparrow\downarrow}\ \underline{\uparrow}\ \underline{\uparrow}$

The ninth electron fills a second orbital in this subshell:

F (Z = 9): $1s^2\ 2s^2\ 2p^5$ $\underline{\uparrow\downarrow}\ \underline{\uparrow\downarrow}\ \underline{\uparrow}$

The tenth electron completes the $2p$ subshell:

Ne (Z = 10): $1s^2\ 2s^2\ 2p^6$ $\underline{\uparrow\downarrow}\ \underline{\uparrow\downarrow}\ \underline{\uparrow\downarrow}$

There is something unusually stable about atoms, such as He and Ne, that have electron configurations with filled shells of orbitals. By convention, we therefore write abbreviated electron configurations in terms of the number of electrons beyond the previous element with a filled-shell electron configuration. Electron configurations of the next two elements in the periodic table, for example, could be written as follows:

Na (Z = 11): [Ne] $3s^1$
Mg (Z = 12): [Ne] $3s^2$

## EXERCISE 6.10

Predict the electron configuration for a neutral tin atom (Sn, $Z = 50$).

**SOLUTION** We start by predicting the order of increasing energy of atomic orbitals from the diagram in Figure 6.23. We then add electrons to these orbitals, starting with the lowest energy orbital, until all 50 electrons have been included.

The key to success with the aufbau process is remembering that each orbital can hold two electrons. Because an $s$ subshell contains only one orbital, only two electrons can be added to this subshell. There are three orbitals in a $p$ subshell, however, so a set of $p$ orbitals can hold up to six electrons. There are five orbitals in a $d$ subshell, so a set of $d$ orbitals can hold up to 10 electrons. The following is the complete electron configuration for tin:

Sn (Z = 50): $1s^2\ 2s^2\ 2p^6\ 3s^2\ 3p^6\ 4s^2\ 3d^{10}\ 4p^6\ 5s^2\ 4d^{10}\ 5p^2$ = [Kr]$5s^2\ 4d^{10}\ 5p^2$

## ▶ CHECKPOINT

Use Hund's rules to explain why iron metal behaves as if its atoms contain more than one unpaired electron.

---

The aufbau process can be used to predict the electron configuration for an element. The actual configuration used by the element has to be determined experimentally. The experimentally determined electron configurations for the elements in the first four rows of the periodic table are given in Table 6.6.

**TABLE 6.6 The Electron Configurations of the First, Second, Third, and Fourth Row Elements**

| Atomic Number | Symbol | Electron Configuration |
|---|---|---|
| 1 | H | $1s^1$ |
| 2 | He | $1s^2$ = [He] |
| 3 | Li | [He] $2s^1$ |
| 4 | Be | [He] $2s^2$ |
| 5 | B | [He] $2s^2\ 2p^1$ |
| 6 | C | [He] $2s^2\ 2p^2$ |
| 7 | N | [He] $2s^2\ 2p^3$ |
| 8 | O | [He] $2s^2\ 2p^4$ |
| 9 | F | [He] $2s^2\ 2p^5$ |
| 10 | Ne | [He] $2s^2\ 2p^6$ = [Ne] |
| 11 | Na | [Ne] $3s^1$ |
| 12 | Mg | [Ne] $3s^2$ |
| 13 | Al | [Ne] $3s^2\ 3p^1$ |
| 14 | Si | [Ne] $3s^2\ 3p^2$ |
| 15 | P | [Ne] $3s^2\ 3p^3$ |
| 16 | S | [Ne] $3s^2\ 3p^4$ |
| 17 | Cl | [Ne] $3s^2\ 3p^5$ |
| 18 | Ar | [Ne] $3s^2\ 3p^6$ = [Ar] |
| 19 | K | [Ar] $4s^1$ |
| 20 | Ca | [Ar] $4s^2$ |
| 21 | Sc | [Ar] $4s^2\ 3d^1$ |
| 22 | Ti | [Ar] $4s^2\ 3d^2$ |
| 23 | V | [Ar] $4s^2\ 3d^3$ |
| 24 | Cr | [Ar] $4s^1\ 3d^5$ |
| 25 | Mn | [Ar] $4s^2\ 3d^5$ |
| 26 | Fe | [Ar] $4s^2\ 3d^6$ |
| 27 | Co | [Ar] $4s^2\ 3d^7$ |
| 28 | Ni | [Ar] $4s^2\ 3d^8$ |
| 29 | Cu | [Ar] $4s^1\ 3d^{10}$ |
| 30 | Zn | [Ar] $4s^2\ 3d^{10}$ |
| 31 | Ga | [Ar] $4s^2\ 3d^{10}\ 4p^1$ |
| 32 | Ge | [Ar] $4s^2\ 3d^{10}\ 4p^2$ |
| 33 | As | [Ar] $4s^2\ 3d^{10}\ 4p^3$ |
| 34 | Se | [Ar] $4s^2\ 3d^{10}\ 4p^4$ |
| 35 | Br | [Ar] $4s^2\ 3d^{10}\ 4p^5$ |
| 36 | Kr | [Ar] $4s^2\ 3d^{10}\ 4p^6$ = [Kr] |

## 6.22 EXCEPTIONS TO PREDICTED ELECTRON CONFIGURATIONS

There are several patterns in the electron configurations listed in Table 6.6. One of the most striking is the remarkable level of agreement between these configurations and the configurations we would predict. There are only two exceptions among the first 40 elements: chromium and copper. These exceptions are interesting, however, because they provide the first glimpse at a factor that will play an important role in the next chapter.

Strict adherence to the rules of the aufbau process would predict the following electron configurations for chromium and copper:

Cr ($Z = 24$): [Ar] $4s^2\ 3d^4$ (predicted electron configurations)
Cu ($Z = 29$): [Ar] $4s^2\ 3d^9$

The experimentally determined electron configurations for these elements are slightly different.

Cr ($Z = 24$): [Ar] $4s^1\ 3d^5$ (actual electron configurations)
Cu ($Z = 29$): [Ar] $4s^1\ 3d^{10}$

In each case, one electron has been transferred from the 4*s* orbital to a 3*d* orbital, even though the 3*d* orbitals are supposed to be at a higher level than the 4*s* orbital.

Once we get beyond atomic number 40, the difference between the energies of adjacent orbitals is small enough that it becomes much easier to transfer an electron from one orbital to another. Most of the exceptions to the electron configuration predicted from the diagram in Figure 6.23 therefore occur among elements with atomic numbers larger than 40. Although it is tempting to focus attention on the handful of elements that have electron configurations that differ from those predicted with Figure 6.23, the amazing thing is that this simple diagram works for so many elements.

## 6.23 ELECTRON CONFIGURATIONS AND THE PERIODIC TABLE

Table 6.6 lists the electron configurations of the elements in order of increasing atomic number. When these data are arranged so that we can compare elements in one of the horizontal rows of the periodic table, we find that these rows typically correspond to the filling of a shell of orbitals. The second row, for example, contains elements in which the orbitals in the $n = 2$ shell are filled:

Li ($Z = 3$): [He] $2s^1$
Be ($Z = 4$): [He] $2s^2$
B ($Z = 5$): [He] $2s^2\ 2p^1$
C ($Z = 6$): [He] $2s^2\ 2p^2$
N ($Z = 7$): [He] $2s^2\ 2p^3$
O ($Z = 8$): [He] $2s^2\ 2p^4$
F ($Z = 9$): [He] $2s^2\ 2p^5$
Ne ($Z = 10$): [He] $2s^2\ 2p^6$

There is an obvious pattern within the vertical columns, or groups, of the periodic table as well. The elements in a group have similar configurations for their outermost electrons. This relationship can be seen by looking at the electron configurations of elements in columns on either side of the periodic table:

| | | | |
|---|---|---|---|
| H | $1s^1$ | | |
| Li | [He] $2s^1$ | F | [He] $2s^2\ 2p^5$ |
| Na | [Ne] $3s^1$ | Cl | [Ne] $3s^2\ 3p^5$ |
| K | [Ar] $4s^1$ | Br | [Ar] $4s^2\ 3d^{10}\ 4p^5$ |
| Rb | [Kr] $5s^1$ | I | [Kr] $5s^2\ 4d^{10}\ 5p^5$ |
| Cs | [Xe] $6s^1$ | At | [Xe] $6s^2\ 4f^{14}\ 5d^{10}\ 6p^5$ |

Figure 6.24 shows the relationship between the periodic table and the orbitals being filled during the aufbau process. The two columns on the left side of the periodic table correspond to the filling of an *s* orbital. The next 10 columns include elements in which the five orbitals in a *d* subshell are filled. The six columns on the right represent the filling of the three orbitals in a *p* subshell. Finally, the 14 columns at the bottom of the table correspond to the filling of the seven orbitals in an *f* subshell.

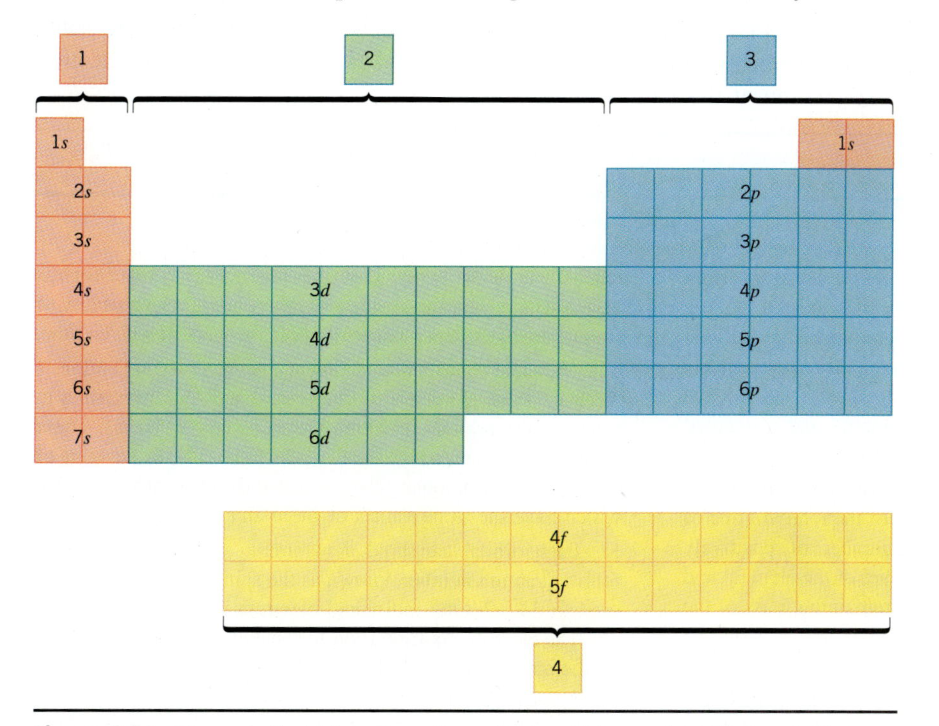


**Figure 6.24** The periodic table reflects the order in which atomic orbitals are filled. The *s* orbitals are filled in the two columns on the far left and the *p* orbitals are filled in the six columns on the right. The *d* orbitals are filled along the transition between the *s* and *p* orbitals. The *f* orbitals are filled in the two long rows of elements at the bottom of the table.

Exercise 6.11 shows how we can combine information from the columns and rows of the periodic table to predict the electron configuration of an atom.

## EXERCISE 6.11

Predict the electron configuration for calcium (Z = 20) and zinc (Z = 30) from their positions in the periodic table.

**SOLUTION** Calcium is in the second column and the fourth row of the table. The second column corresponds to the filling of an *s* orbital. The 1*s* orbital is filled in the first row, the 2*s* orbital in the second row, and so on. By the time we get to the fourth row, we are filling the 4*s* orbital. Calcium therefore has all of the electrons of argon, plus a filled 4*s* orbital:

$$\text{Ca } (Z = 20): \quad [\text{Ar}]\ 4s^2$$

Zinc is the tenth element in the region of the periodic table where *d* orbitals are filled. Zinc therefore has a filled subshell of *d* orbitals. The only question is which set of *d* orbitals are filled. Although zinc is in the fourth row of the periodic table, the first time *d* orbitals occur is in the $n = 3$ shell. The following is therefore the abbreviated electron configuration for zinc:

$$\text{Zn } (Z = 30): \quad [\text{Ar}]\ 4s^2\ 3d^{10}$$

### CHECKPOINT

Theoreticians predict that the element with atomic number 114 will be more stable than the elements with atomic numbers between 103 and 114. Predict the electron configuration of this element.

## SUMMARY

1. At the beginning of the twentieth century, scientists could still argue in good faith about whether atoms existed. As we approach the end of this century, the issue has been resolved. As the result of theoretical and experimental accomplishments of some of the greatest minds of our time, definitive evidence for the existence of atoms has been gathered and a detailed model of their internal structure has been developed.

2. Atoms contain a small, massive positively charged nucleus that occupies a negligibly small fraction of the total volume of the atom. The nucleus contains the number of positively charged protons equal to the atomic number of the element and enough neutral neutrons to make up the remaining mass of the atom.

3. The nucleus is embedded in a sea of lightweight, negatively charged electrons that occupy the remaining space in the atom. These electrons have the properties of both particles and waves, which means that we can't specify their exact location in three-dimensional space. We therefore turn to a quantum mechanical model, which describes the regions in space **(orbitals)** in which electrons reside and talks about the probability of finding an electron at a particular point in space at a given time.

4. The three coordinates chosen to describe the distribution of electrons in an atom specify the size, shape, and orientation in space of the orbitals in which electrons can be found. These orbitals are organized into shells, which have the same value of the *n* quantum number. They are also divided into subshells, which have the same values of the *n* and *l* quantum numbers.

5. Each orbital can hold a maximum of two electrons. A fourth quantum number, known as the spin quantum number, is used to describe the individual electrons in an orbital.

The electron configuration for an element can be predicted by the aufbau process, which involves adding electrons to orbitals starting with the lowest energy orbitals on the atom. The results of these predictions agree to a remarkable extent with the experimentally determined electron configurations of the elements.

## KEY TERMS

**Absorption spectrum** (p. 209)
**Angular quantum number (*l*)** (p. 218)
**Anode** (p. 196)
**Atomic number** (p. 204)
**Aufbau principle** (p. 227)
**Bohr model** (p. 213)
**Cathode** (p. 196)
**Cathode-ray tube** (p. 197)
**de Broglie equation** (p. 217)
**Degenerate orbitals** (p. 227)
**Electromagnetic radiation** (p. 207)
**Electron** (p. 198)

**Electron configuration** (p. 227)
**Emission spectrum** (p. 210)
**Frequency** (p. 207)
**Hund's rules** (p. 227)
**Magnetic quantum number (*m*)** (p. 218)
**Neutron** (p. 205)
**Nucleus** (p. 203)
**Orbitals** (p. 218)
**Photon** (p. 212)
**Principal quantum number (*n*)** (p. 218)
**Proton** (p. 204)
**Quantum number** (p. 218)
**Shell** (p. 220)
**Spin quantum number (*s*)** (p. 223)
**Subshell** (p. 220)
**Wavelength** (p. 207)
**Wave–particle duality** (p. 216)

## KEY EQUATIONS

$$\nu\lambda = s \qquad E = h\nu$$

$$\frac{1}{\lambda} = R_H\left[\frac{1}{n_1^2} - \frac{1}{n_2^2}\right] \qquad \lambda = \frac{h}{p}$$

## PROBLEMS

### Electricity and Matter

**6-1** Use the present model for the structure of the atom to explain static electricity.

**6-2** Describe ways in which our present model for the structure of the atom is consistent with du Fay's two-fluid model of electricity. Describe how it is consistent with Franklin's one-fluid model.

### The Discovery of the Electron

**6-3** Describe the evidence that led to the belief that cathode rays consist of a beam of negatively charged particles that flow from the cathode to the anode of the tube.

**6-4** Use the present model for the structure of the atom to explain the observations that Faraday, Plücker, Hittorf, and Crookes made while working with cathode-ray tubes.

### The Discovery of X-Rays

**6-5** The discovery of x-rays in 1895 set off what has been described as a scientific gold rush to investigate the properties of new forms of radiation. What was it about x-rays that captured the imagination of so many scientists?

**6-6** X-rays and light are both forms of electromagnetic radiation. Describe the difference between x-rays and light that allows one to pass through solid bodies while the other is absorbed?

### Radioactivity

**6-7** Describe the difference between $\alpha$-particles, $\beta$-particles, and $\gamma$-rays.

**6-8** Describe the role that x-rays played in the discovery of radioactivity.

### Millikan's Oil Drop Experiment

**6-9** Use the present model for the structure of the atom to explain why the $H^+$ ion had the largest charge-to-mass ratio of any particle until the electron was discovered.

**6-10** Which of the following particles has the largest charge-to-mass ratio?

(a) a proton (b) a neutron (c) an $\alpha$-particle
(d) an electron (e) a $Li^+$ ion

### The Structure of the Atom

**6-11** What aspects of the chemical behavior of the elements can be attributed to each of the three fundamental subatomic particles?

**6-12** Describe the differences between a hydrogen atom and a proton, between a hydrogen atom and a neutron, and between a proton and a neutron.

**6-13** Describe the present model for the structure of the atom.

**6-14** Choose an example of an element that forms positive ions. Describe the relationship between the atomic number, mass number, number of protons, number of neutrons, and number of electrons in an atom and its positively charged ion. Do the same for an element that forms negative ions.

**6-15** Write the symbol for the atom or ion that contains 24 protons, 21 electrons, and 28 neutrons.

**6-16** Calculate the number of protons and neutrons in the nucleus and the number of electrons surrounding the nucleus of a $^{39}K^+$ ion. What is the atomic number and the mass number of this ion?

**6-17** Calculate the number of protons and neutrons in the nucleus and the number of electrons surrounding the nucleus of a $^{127}I^-$ ion. What is the atomic number and the mass number of this ion?

**6-18** Identify the element that forms atoms with a mass number of 20 that contain 11 neutrons.

**6-19** Give the symbol for the atom or ion that has 34 protons, 45 neutrons, and 36 electrons.

**6-20** Calculate the number of electrons in a $^{134}Ba^{2+}$ ion.

**6-21** A chemistry text published in 1922 proposed a model of the atom based on only two subatomic particles: electrons and protons. All of the protons and some of the electrons were concentrated in the nucleus of the atom. The other electrons revolved around the nucleus. The number of protons in the nucleus was equal to the mass of the atom. The charge on the nucleus was equal to the number of protons minus the number of electrons in the nucleus. Enough electrons were then added to the atom to neutralize the charge on the nucleus. Use this model to calculate the number of protons and electrons in a neutral fluorine atom, F, and a fluoride ion, $F^-$. Compare this calculation with the results obtained by assuming that the nucleus is composed of protons and neutrons.

**6-22** The charge on a single electron is $1.6021892 \times 10^{-19}$ coulomb. Calculate the charge on a mole of electrons and compare the results of this calculation with Faraday's constant, given in Table A-2 in the Appendix.

### Particles and Waves

**6-23** Describe the difference between a particle and a wave.

**6-24** To examine the relationship between the frequency, wavelength, and speed of a wave, imagine that you are sitting at a railroad crossing, watching a train that consists of 45-foot boxcars go by. Furthermore assume it takes 3 seconds for each boxcar to pass in front of your car. Calculate the ''frequency'' of the boxcar and the product of the 45-foot ''wavelength'' times this ''frequency''. Convert the final answer into units of miles per hour.

**6-25** An octave in a musical scale corresponds to a change in the frequency of a note by a factor of 2. If a note with a frequency of 440 hertz is an *A*, then the *A* one octave above this note has a frequency of 880 hertz. What happens to the wavelength of the sound as the frequency increases by a factor of 2? What happens to the speed at which the sound travels to your ear?

**6-26** The human ear is capable of hearing sound waves with frequencies between about 20 and 20,000 hertz. If the speed of sound is 340.3 meters per second at sea level, what is the wavelength in meters of the longest wave the human ear can hear?

### Light and Other Forms of Electromagnetic Radiation

**6-27** Calculate the wavelength in meters of green light that has a frequency of $5.0 \times 10^{14}$ cycles per second.

**6-28** Calculate the frequency of red light that has a wavelength of 700 nanometers.

**6-29** Which has the longer wavelength, red light or blue light?

**6-30** Which has the larger frequency, radio waves or microwaves?

**6-31** In a magnetic field of 2.35 Tesla, $^{13}C$ nuclei absorb electromagnetic radiation that has a frequency of 25.147 megahertz. Calculate the wavelength of this radiation. In which region of the electromagnetic spectrum does this radiation fall?

**6-32** The meter has been defined as 1,650,763.73 wavelengths of the orange-red line of the emission spectrum of $^{86}Kr$. Calculate the frequency of this radiation. In what portion of the electromagnetic spectrum does this radiation fall?

### Atomic Spectra

**6-33** Soap bubbles pick up colors because they reflect light with wavelengths equal to the thickness of the walls of the bubble. What frequency of light is reflected by a soap bubble 6 nanometers thick?

**6-34** Methylene blue, $C_{16}H_{18}ClN_3S$, absorbs light most intensely at wavelengths of 668 and 609 nanometers. What color is the light absorbed by this dye?

**6-35** Sodium salts give off a characteristic yellow-orange light when added to the flame of a bunsen burner. This yellow-orange color is due to two narrow bands of radiation with wavelengths of 588.9953 nanometers and 589.5923 nanometers. Calculate the frequencies of these emission lines.

### Quantization of Energy

**6-36** Which carries more energy, ultraviolet or infrared radiation?

**6-37** Which carries more energy, yellow light with a wavelength of 580 nanometers or green light with a wavelength of 660 nanometers?

**6-38** List the four lines in the emission spectrum of hydrogen in order of increasing energy.

**6-39** Calculate the energy in joules of the radiation in the emission spectrum of the hydrogen atom that has a wavelength of 656.2 nanometers.

**6-40** Calculate the energy in joules of a single particle of radiation broadcast by an amateur radio operator who transmits at a wavelength of 10 meters.

**6-41** $Cl_2$ molecules can dissociate to form chlorine atoms by absorbing electromagnetic radiation. It takes 243.4 kJ of energy to break the bonds in a mole of $Cl_2$ molecules. What is the wavelength of the radiation that has just enough energy to decompose $Cl_2$ molecules to chlorine atoms? In what portion of the spectrum is this wavelength found?

### The Bohr Model of the Atom

**6-42** Calculate the wavelength of the radiation emitted by a hydrogen atom when an electron falls from the $n = 3$ to the $n = 2$ orbit in the Bohr model.

**6-43** Identify the transition between orbits in the Bohr model that gives rise to the line in the emission spectrum of the hydrogen atom that has a wavelength of 410.1 nanometers.

**6-44** Which of the following transitions in the spectrum of the hydrogen atom results in the emission of light with the longest wavelength?

(a) $n = 3$ to $n = 2$ (b) $n = 3$ to $n = 1$
(c) $n = 5$ to $n = 4$ (d) $n = 2$ to $n = 3$
(e) $n = 1$ to $n = 3$ (f) $n = 4$ to $n = 5$

**6-45** Which of the following transitions in a hydrogen atom results in the absorption of a photon with the largest energy?

(a) $n = 2$ to $n = 3$ (b) $n = 2$ to $n = 4$
(c) $n = 1$ to $n = 4$ (d) $n = 3$ to $n = 1$
(e) $n = 3$ to $n = 2$ (f) $n = 4$ to $n = 3$

**6-46** Use Figure 6.14 to determine the region in the electromagnetic spectrum where the Lyman, Balmer, Paschen, Brackett, and Pfund series of lines occur.

**6-47** According to the Bohr model, the energy absorbed when an electron on a hydrogen atom jumps from the $n = 1$ to the $n = 2$ quantum level is $1.63 \times 10^{-18}$ J. Calculate the energy of the photon emitted when an electron falls from the $n = 3$ to the $n = 2$ quantum level.

### Quantum Numbers

**6-48** Describe the function of each of the four quantum numbers: *n, l, m,* and *s*.

**6-49** Describe the selection rules for the *n, l, m,* and *s* quantum numbers.

**6-50** Identify the quantum number that specifies each of the following.

(a) The size of the orbital. (b) The shape of the orbital. (c) The way the orbital is oriented in space. (d) The spin of the electrons that occupy an orbital.

**6-51** Determine the allowed values of the angular quantum number, *l*, when the principal quantum number is 4. Describe the difference between orbitals that have the same principal quantum number and different angular quantum numbers.

**6-52** Determine the allowed values of the magnetic quantum number, *m*, when the angular quantum number is 2. Describe the difference between orbitals that have the same angular quantum number and different magnetic quantum numbers.

**6-53** Determine the allowed values of the spin quantum number, *s*, when $n = 5$, $l = 2$, and $m = -1$.

**6-54** Determine the number of allowed values of the magnetic quantum number when $n = 3$ and $l = 2$.

**6-55** Determine the maximum value for the angular quantum number, *l*, when the principal quantum number is 4.

**6-56** Which of the following is a legitimate set of *n, l, m,* and *s* quantum numbers?

(a) 4, −2, −1, $\frac{1}{2}$ (b) 4, 2, 3, $\frac{1}{2}$ (c) 4, 3, 0, 1
(d) 4, 0, 0, $-\frac{1}{2}$

**6-57** Which of the following is a legitimate set of *n, l, m,* and *s* quantum numbers?

(a) 0, 0, 0, $\frac{1}{2}$ (b) 8, 4, −3, $-\frac{1}{2}$ (c) 3, 3, 2, $+\frac{1}{2}$
(d) 2, 1, −2, $-\frac{1}{2}$ (e) 5, 3, 3, −1

**6-58** Calculate the maximum number of electrons that can have the quantum numbers $n = 4$ and $l = 3$.

**6-59** Calculate the number of electrons in an atom that can simultaneously possess the quantum numbers $n = 4$ and $s = +\frac{1}{2}$.

**6-60** Write the combination of quantum numbers for every electron in the $n = 1$ and $n = 2$ shells using the selection rules outlined in this chapter.

**6-61** Write the combination of quantum numbers for every electron in the $n = 1$ and $n = 2$ shell assuming that the selection rules for assigning quantum numbers are changed to the following:

1. The principal quantum number can be any integer greater than or equal to 1.
2. The angular quantum number can have any value between 0 and *n*.
3. The magnetic quantum number can have any value between 0 and 1.
4. The spin quantum number can have a value of either +1 or −1.

### Shells and Subshells of Orbitals

**6-62** Describe what happens to the difference between the energies of subshells of orbitals as the value of the principal quantum number, *n*, becomes larger.

**6-63** Identify the symbols used to describe orbitals for which $l = 0$, 1, 2, and 3.

**6-64** Determine the number of orbitals in the $n = 3$, $n = 4$, and $n = 5$ shells.

**6-65** Which of the following sets of *n, l, m,* and *s* quantum numbers can be used to describe an electron in a 2*p* orbital?

(a) 2, 1, 0, $-\frac{1}{2}$ (b) 2, 0, 0, $\frac{1}{2}$ (c) 2, 2, 1, $\frac{1}{2}$ (d) 3, 2, 1, $-\frac{1}{2}$ (e) 3, 1, 0, $\frac{1}{2}$

**6-66** Which of the following orbitals cannot exist:

(a) 6*s* (b) 3*p* (c) 2*d* (d) 4*f* (e) 17*f*

**6-67** Calculate the maximum number of electrons in the $n = 1$, $n = 2$, $n = 3$, $n = 4$, and $n = 5$ shells of orbitals.

**6-68** Calculate the maximum number of electrons that can fit into a 4*d* subshell.

**6-69** Calculate the maximum number of unpaired electrons that can be placed in a 5*d* subshell.

**6-70** Explain why the difference between the atomic numbers of pairs of elements in a vertical column, or group, of the periodic table is either 8, 18, or 32.

**6-71** Explain why the number of orbitals in a shell is always equal to $n^2$, where *n* is the principal quantum number.

**6-72** Explain why the maximum number of electrons in a

shell of orbitals is equal to $2n^2$, where $n$ is the principal quantum number.

**6-73** Explain why the number of orbitals in a subshell is equal to $2(l) + 1$, where $l$ is the angular quantum number.

### The Relative Energies of Atomic Orbitals

**6-74** Which pair of quantum numbers determines the energy of an electron in an orbital?

(a) $n$ and $l$ (b) $n$ and $m$ (c) $n$ and $s$ (d) $l$ and $m$ (e) $l$ and $s$ (f) $m$ and $s$

**6-75** Which of the following sets of orbitals is arranged in order of increasing energy?

(a) $3d < 4s < 4p < 5s < 4d$ (b) $3d < 4s < 4p < 4d < 5s$ (c) $4s < 3d < 4p < 5s < 4d$ (d) $4s < 3d < 4p < 4d < 5s$ (e) $3d < 4s < 4p < 4d < 5s$

**6-76** Which of the following orders of increasing energies is incorrect?

(a) $3s < 4s < 5s$ (b) $5s < 5p < 5d$ (c) $5s < 4d < 5p$ (d) $6s < 4f < 5d$ (e) $6s < 5f < 6p$

**6-77** As atomic orbitals are filled according to the aufbau principle, the $6p$ orbitals are filled immediately after which of the following orbitals?

(a) $4f$ (b) $5d$ (c) $6s$ (d) $7s$

### Electron Configurations

**6-78** Which of the following characteristics describes orbitals that are degenerate?

(a) They contain the same number of electrons. (b) They have the same value of the angular quantum number, $l$, and different values of the principal quantum number, $n$. (c) They have the same set of three quantum numbers, $n$, $l$, and $m$, but have different values of the $s$ quantum number. (d) They have the same energy. (e) They contain the same number of unpaired electrons.

**6-79** Describe the regions of the periodic table in which the $s$, $p$, $d$, and $f$ subshells are filled.

**6-80** Write the electron configurations for the elements in the third row of the periodic table.

**6-81** Use straight lines and arrows to draw the orbital diagram for the electron configuration of the nitrogen atom.

**6-82** Use straight lines and arrows to draw the orbital diagram for the electron configuration for nickel.

**6-83** The electron configuration of Si is $1s^2\ 2s^2\ 2p^6\ 3s^2\ 3p^x$, where $x$ is which of the following?

(a) 1 (b) 2 (c) 3 (d) 4 (e) 6

**6-84** Which of the following electron configurations for carbon satisfies Hund's rules?

(a) $1s^2\ 2s^2\ 2p_x^1\ 2p_y^0\ 2p_z^0$ (b) $1s^2\ 2s^2\ 2p_x^1\ 2p_y^1\ 2p_z^0$ (c) $1s^2\ 2s^2\ 2p_x^1\ 2p_y^1\ 2p_z^1$ (d) $1s^2\ 2s^2\ 2p_x^2\ 2p_y^0\ 2p_z^0$

**6-85** Which of the following is the correct electron configuration for the $P^{3-}$ ion?

(a) [Ne] (b) [Ne] $3s^2$ (c) [Ne] $3s^2\ 3p^3$ (d) [Ne] $3s^2\ 3p^6$

**6-86** Which of the following is the correct electron configuration for the bromide ion, $Br^-$?

(a) [Ar] $4s^2\ 4p^5$ (b) [Ar] $4s^2\ 3d^{10}\ 4p^5$ (c) [Ar] $4s^2\ 3d^{10}\ 4p^6$ (d) [Ar] $4s^2\ 3d^{10}\ 4p^6\ 5s^1$ (e) [Ar] $4s^2\ 3d^{10}\ 3p^6$

**6-87** Determine the number of electrons in a vanadium atom for which the principal quantum number, $n$, is 3.

**6-88** Determine the number of electrons in $s$ orbitals in the $V^{2+}$ ion.

**6-89** Which of the following elements has the largest number of electrons for which the principal quantum number is 3?

(a) Na (b) Al (c) Si (d) Cl (e) Ar (f) Zn

**6-90** Which of the following is a possible set of $n$, $l$, $m$, and $s$ quantum numbers for the last electron added to form a gallium atom ($Z = 31$)?

(a) 3, 1, 0, $-\frac{1}{2}$ (b) 3, 2, 1, $\frac{1}{2}$ (c) 4, 0, 0, $\frac{1}{2}$ (d) 4, 1, 1, $\frac{1}{2}$ (e) 4, 2, 2, $\frac{1}{2}$

**6-91** Which of the following is a possible set of $n$, $l$, $m$, and $s$ quantum numbers for the last electron added to form an $As^{3+}$ ion?

(a) 3, 1, $-1$, $\frac{1}{2}$ (b) 4, 0, 0, $-\frac{1}{2}$ (c) 3, 2, 0, $\frac{1}{2}$ (d) 4, 1, $-1$, $\frac{1}{2}$ (e) 5, 0, 0, $\frac{1}{2}$

**6-92** Theoreticians predict that the element with atomic number 114 will be more stable than the elements with atomic numbers between 103 and 114. On the basis of its electron configuration, in which group of the periodic table should Element 114 be placed?

**6-93** What would be the values of the $n$ and $l$ quantum numbers for the last electron added to the element that starts a new inner transition series below the actinide series?

**6-94** Which is the first element to have $4d$ electrons in its electron configuration?

(a) Ca (b) Sc (c) Rb (d) Y (e) La

**6-95** The elements starting with lanthanum (Ce) in the first of the two long rows of elements at the bottom of the periodic table are called "lanthanides." The lanthanides correspond to the filling of orbitals with what value of the angular quantum number?

**6-96** Which of the following neutral atoms has the largest number of unpaired electrons?

(a) Na (b) Al (c) Si (d) P (e) S

**6-97** Which of the following ions has five unpaired electrons?

(a) $Ti^{4+}$ (b) $Co^{2+}$ (c) $V^{3+}$ (d) $Fe^{3+}$ (e) $Zn^{2+}$

### Integrated Problems

**6-98** The most recent estimates give values of about $10^{-10}$ meters for the radius of an atom and $120^{-14}$ meters for the radius of the nucleus of the atom. Calculate the fraction of the total volume of an atom that is essentially empty space.

**6-99** Protons and neutrons are both examples of a category of nuclear particles known as baryons. Both of these baryons contains three quarks. One contains two down quarks and an up quark, the other contains two up quarks and a down quark. An up quark ($u$) has a charge of $+\frac{2}{3}$, whereas a down quark ($d$) has a charge of $-\frac{1}{3}$. Use this information to identify which baryon should be described as *ddu* and which should be described as *duu*.

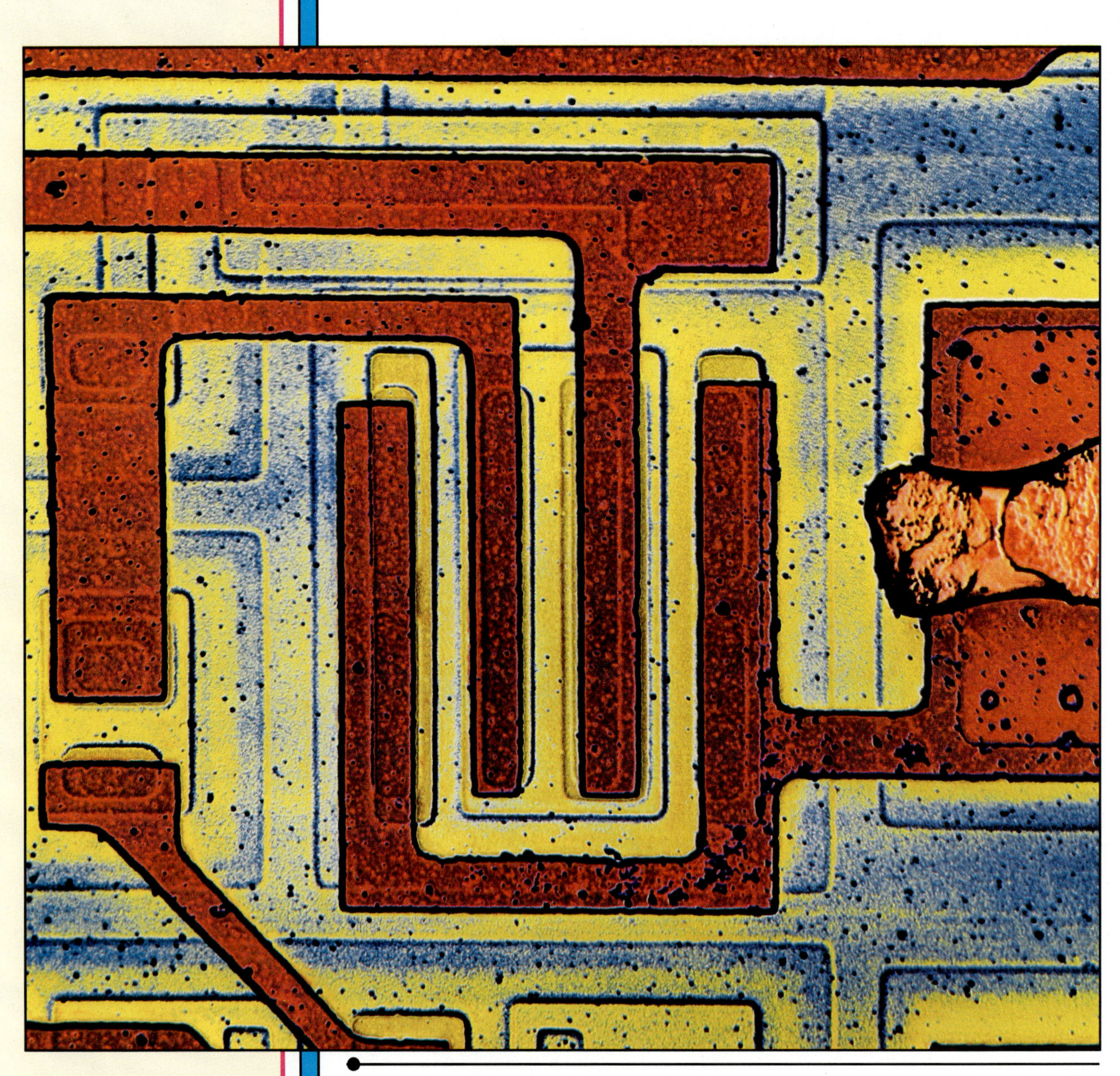

A photomicrograph of an integrated circuit etched onto the surface of a silicon chip. The operation of this circuit depends on the fact that silicon has the properties of neither a metal nor a nonmetal.

# THE PERIODIC TABLE: AN INTRODUCTION TO IONIC COMPOUNDS

This chapter examines the process by which periodic trends in the physical properties of the elements were discovered and uses these trends to explain why ionic compounds exist. It provides the basis for answering the following questions:

## POINTS OF INTEREST

- Why do we need a periodic table, instead of an alphabetical list of elements?
- What differences between the physical properties of sodium and chlorine atoms give rise to the enormous differences between the chemical and physical properties of these elements on the macroscopic scale?
- What's wrong with the popular notion that sodium reacts with chlorine because "chlorine likes electrons more than sodium does?"
- Why does sodium react with chlorine to form NaCl, not $NaCl_2$? Why does magnesium form $MgCl_2$, not MgCl?
- Why is NaCl relatively soluble in water, whereas MgO is not?

## 7.1 THE PROCESS OF DISCOVERY: THE ELEMENTS

In 1661, when Boyle defined an *element* as a substance that couldn't be decomposed into a simpler substance by a chemical reaction, only 13 elements were known: antimony, arsenic, bismuth, carbon, copper, gold, iron, lead, mercury, silver, sulfur, tin, and zinc. By the end of the eighteenth century, when Lavoisier published a list of elements, another 11 had been discovered: chlorine, cobalt, hydrogen, manganese, molybdenum, nickel, nitrogen, oxygen, phosphorus, platinum, and tungsten. Since that time, a new element has been discovered on the average of every two and one-half years (see Figure 7.1).

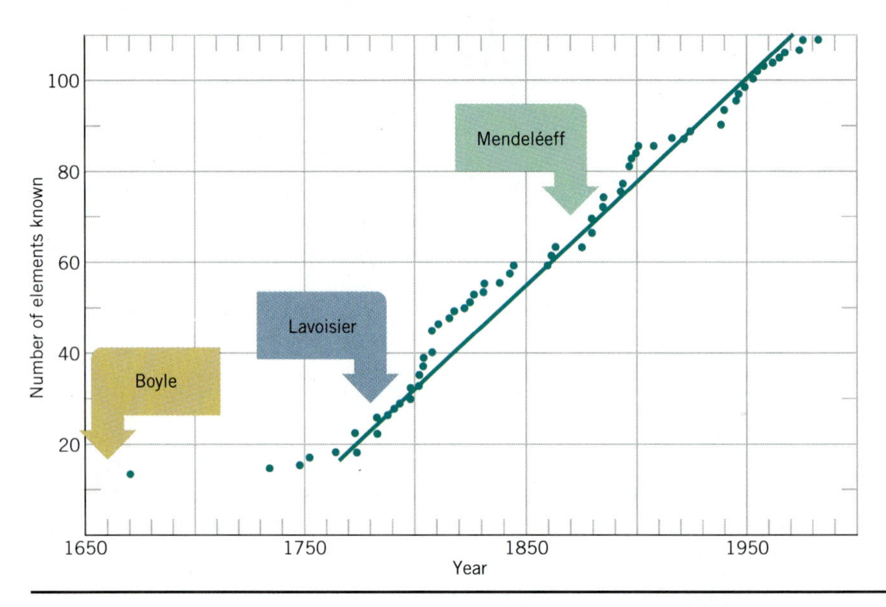


**Figure 7.1** A plot of the number of elements versus time from 1661 to the present.

As the number of elements increased, chemists inevitably began to find patterns in their properties. In 1829 Johann Wolfgang Döbereiner discovered the existence of families of elements with similar chemical properties. Because there always seemed to be three elements in these families, he called them *triads*. Each of the vertical columns in Table 7.1 represents one of these triads.

**TABLE 7.1 Döbereiner's Triads**

| | | | | |
|---|---|---|---|---|
| Li | Ca | S | Cl | Mn |
| Na | Sr | Se | Br | Cr |
| K | Ba | Te | I | Fe |

### CHECKPOINT

Which triads in Table 7.1 contain elements that are still grouped in the same column of the modern periodic table?

Döbereiner's triads grouped elements with similar chemical properties. Consider lithium, sodium, and potassium, for example.

1. These elements all react with water at room temperature.
2. They react with chlorine to form compounds with similar formulas: LiCl, NaCl, and KCl.
3. They combine with hydrogen to form compounds with similar formulas: LiH, NaH, and KH.
4. They form hydroxides with similar formulas: LiOH, NaOH, and KOH.

Döbereiner also found patterns in the physical properties of the elements in a triad. He noted, for example, that the atomic weight of the middle element in each

Based on empty spaces in his periodic table, Mendeléeff predicted the discovery of 10 elements, which he tentatively named by adding the prefix *eka-* to the name of the element immediately above each empty space. He then tried to predict in some detail the properties of four of these elements: **eka-**aluminum, **eka-**boron, **eka-**silicon, and **eka-**tellurium.

The remarkable agreement between Mendeléeff's predictions for **eka-**aluminum and the observed properties of the element gallium, which was discovered in 1875, is shown in Table 7.3. A similar comparison between the predicted properties of

**TABLE 7.3 The Predicted Properties of Eka-Aluminum Compared with the Observed Properties of Gallium**

| Mendeléeff's Predictions | Observed Properties |
|---|---|
| Atomic weight: about 68 | Atomic weight: 69.72 |
| Density: 5.9 | Density: 5.94 |
| Melting point: low | Melting point: 30.15°C |
| Formula of oxide: $Ea_2O_3$ | Formula of oxide: $Ga_2O_3$ |
| Formula of chloride: $EaCl_3$ | Formula of chloride: $GaCl_3$ |
| Chemistry: The hydroxide should dissolve in acids and bases; the metal should form basic salts; the sulfide should precipitate with $H_2S$ or $(NH_4)_2S$; the chloride should be more volatile than $ZnCl_2$. | Chemistry: The hydroxide dissolves in acids and bases; the metal forms basic salts; $Ga_2S_3$ is precipitated by either $H_2S$ or $(NH_4)_2S$; $GaCl_3$ is more volatile than $ZnCl_2$. |
| The element will probably be discovered by spectroscopy. | The element was discovered with the aid of the spectroscope. |

**eka-**silicon and the observed properties of germanium, discovered in 1886, is given in Table 7.4. It was the extraordinary success of Mendeléeff's predictions that led chemists not only to accept the periodic table, but to recognize Mendeléeff more than anyone else as the originator of the concept on which it was based.

**TABLE 7.4 The Predicted Properties of Eka-Silicon Compared with the Observed Properties of Germanium**

| Mendeléeff's Predictions | Observed Properties |
|---|---|
| Atomic weight: 72 | Atomic weight: 72.59 |
| Density: 5.5 | Density: 5.47 |
| Formula of oxide: $EsO_2$ | Formula of oxide: $GeO_2$ |
| Density of oxide: 4.7 | Density of oxide: 4.703 |
| Formula of chloride: $EsCl_4$ | Formula of chloride: $GeCl_4$ |
| Boiling point of $EsCl_4$: $< 100°C$ | Boiling point of $GeCl_4$: 86°C |

## 7.3 MODERN VERSIONS OF THE PERIODIC TABLE

More than 700 versions of the periodic table were proposed in the first 100 years after the publication of the table in Figure 7.2. Some of these tables grouped elements on the basis of similar chemical properties. Copper and silver, for example, can be found in Group I of Mendeléeff's table because they form compounds with oxygen ($Ag_2O$ and $Cu_2O$) and chlorine (AgCl and CuCl) that have the same formula as the compounds formed by other elements in this group, such as lithium ($Li_2O$ and LiCl) or sodium ($Na_2O$ and NaCl). Other versions were based on the fact that

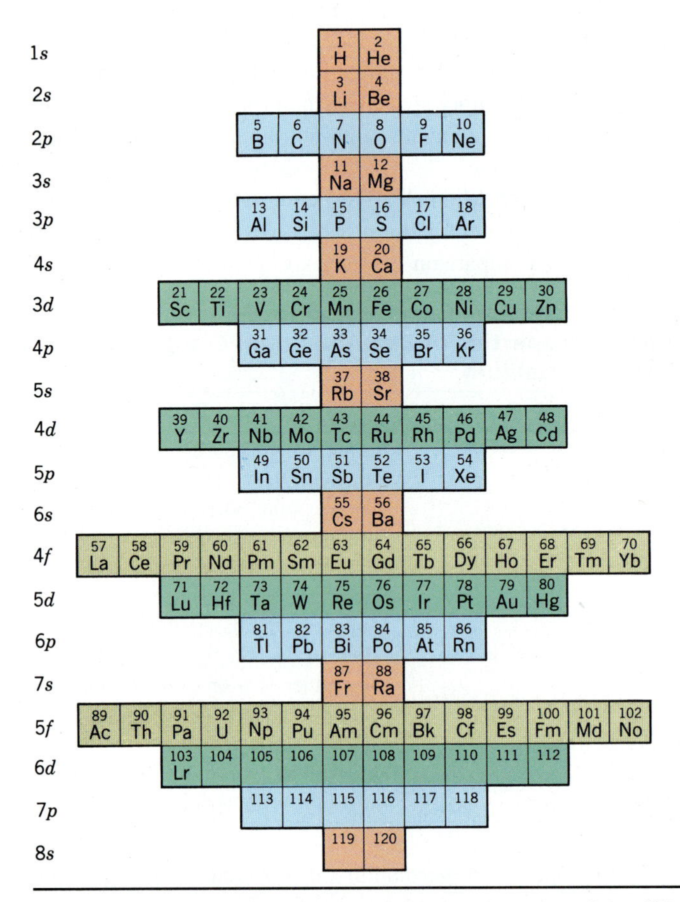


**Figure 7.3** This periodic table emphasizes the pattern of the filling of orbital subshells that goes into determining the electron configuration of an element.

elements with similar chemical properties often have similar electron configurations. The table shown in Figure 7.3, for example, emphasizes the order in which subshells of orbitals are filled.

The most successful tables try to achieve both goals, listing elements in the same group when they have similar chemical properties and similar electron configurations. The most popular version of the periodic table is shown in Figure 7.4. Each group contains compounds with similar chemical properties. In most cases, the elements in a column also have similar electron configurations. The two most important exceptions are hydrogen and helium.

On the basis of electron configuration, hydrogen ($1s^1$) should be in Group 1 (IA), along with lithium ([He] $2s^1$), sodium ([Ne] $3s^1$), potassium ([Ar] $4s^1$), and so on. But the other elements in Group IA are all metals, and hydrogen is not. Thus, it may be appropriate to include hydrogen in Group 17 (VIIA) of the periodic table, with the other nonmetals that have one less electron than a filled-shell configuration. Many periodic tables therefore list hydrogen in both Group 1 and Group 17.

On the basis of electron configurations, helium ($1s^2$) should be placed in Group 2 (IIA) of the table, along with beryllium ([He] $2s^2$), magnesium ([Ne] $3s^2$), calcium ([Ar] $4s^2$), and so on. But the elements in Group 2 are all metals, and helium is not. Helium behaves more like the elements with a filled-shell electron configuration in the last column of the periodic table. Virtually every periodic table therefore includes helium among the elements in Group 18 (VIIIA).

| Periods | 1 IA | 2 IIA | 3 IIIB | 4 IVB | 5 VB | 6 VIB | 7 VIIB | 8 VIIIB | 9 VIIIB | 10 VIIIB | 11 IB | 12 IIB | 13 IIIA | 14 IVA | 15 VA | 16 VIA | 17 VIIA | 18 VIIIA |
|---|---|---|---|---|---|---|---|---|---|---|---|---|---|---|---|---|---|---|
| 1 | 1 H | | | | | | | | | | | | | | | | 1 H | 2 He |
| 2 | 3 Li | 4 Be | | | | | | | | | | | 5 B | 6 C | 7 N | 8 O | 9 F | 10 Ne |
| 3 | 11 Na | 12 Mg | | | | | | | | | | | 13 Al | 14 Si | 15 P | 16 S | 17 Cl | 18 Ar |
| 4 | 19 K | 20 Ca | 21 Sc | 22 Ti | 23 V | 24 Cr | 25 Mn | 26 Fe | 27 Co | 28 Ni | 29 Cu | 30 Zn | 31 Ga | 32 Ge | 33 As | 34 Se | 35 Br | 36 Kr |
| 5 | 37 Rb | 38 Sr | 39 Y | 40 Zr | 41 Nb | 42 Mo | 43 Tc | 44 Ru | 45 Rh | 46 Pd | 47 Ag | 48 Cd | 49 In | 50 Sn | 51 Sb | 52 Te | 53 I | 54 Xe |
| 6 | 55 Cs | 56 Ba | 57 La | 72 Hf | 73 Ta | 74 W | 75 Re | 76 Os | 77 Ir | 78 Pt | 79 Au | 80 Hg | 81 Tl | 82 Pb | 83 Bi | 84 Po | 85 At | 86 Rn |
| 7 | 87 Fr | 88 Ra | 89 Ac | 104 | 105 | 106 | 107 | 108 | 109 | | | | | | | | | |
| | | | | | 58 Ce | 59 Pr | 60 Nd | 61 Pm | 62 Sm | 63 Eu | 64 Gd | 65 Tb | 66 Dy | 67 Ho | 68 Er | 69 Tm | 70 Yb | 71 Lu |
| | | | | | 90 Th | 91 Pa | 92 U | 93 Np | 94 Pu | 95 Am | 96 Cm | 97 Bk | 98 Cf | 99 Es | 100 Fm | 101 Md | 102 No | 103 Lr |

**Figure 7.4** The most popular form of the periodic table separates the elements into three classes: (1) the main-group elements (yellow), (2) the transition metals (red), and (3) the lanthanides and actinides (green).

**CHECKPOINT**

Use the electron configurations of the elements in Group 18 to dispute the common belief that the modern periodic table groups elements in vertical columns that have the same configurations.

## 7.4 THE SIZE OF ATOMS: METALLIC RADII

The relative size of atoms of different elements is an important physical property because it influences their chemical behavior. Unfortunately, the size of an isolated atom cannot be measured directly because we can't determine the location of the electrons that surround the nucleus. We can estimate the size of an atom, however, by assuming that the radius of an atom is equal to one-half the distance between adjacent atoms in a solid. This technique is best suited to elements that are metals, which form solids composed of extended planes of atoms of that element. The results of these measurements are therefore often known as **metallic radii.**

Because more than 75% of the elements are metals, metallic radii are available for most elements in the periodic table. Figure 7.5 shows the relationship between the metallic radii for elements in Groups 1 (IA) and 2 (IIA).

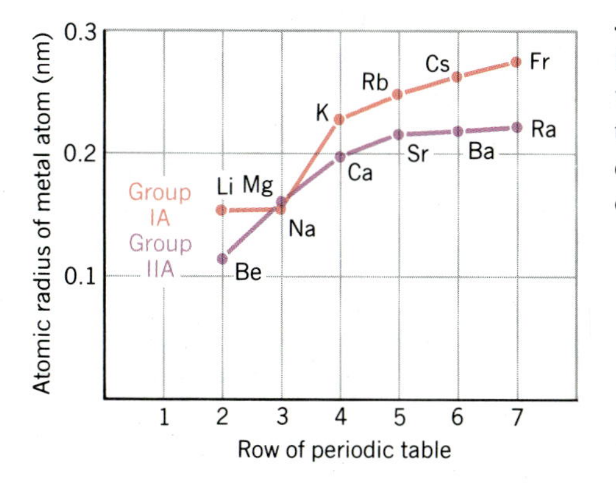


**Figure 7.5** Metallic radii increase as we go down a column of the periodic table. With rare exception, they decrease from left to right across a row of the table.

There are two general trends in these data.

1. The metallic radius increases as we go down a column of the periodic table.
2. The metallic radius decreases as we go from left to right across a row of the periodic table.

The first trend in metallic radii is easy to understand. As we go down a column of the periodic table, electrons are placed in larger orbitals. When this happens, the size of the atom increases.

The second trend is a bit surprising. We might expect that atoms would become larger as we go from left to right across a row of the periodic table because each element has one more electron than the preceding element. But the additional electrons are added to the same shell of orbitals, not to larger orbitals. Because the number of protons in the nucleus also increases as we go across a row of the table, the force of attraction between the nucleus and the electrons that surround it increases. The nucleus therefore tends to hold each electron more tightly and the atoms become smaller.

## 7.5 THE SIZE OF ATOMS: COVALENT RADII

The size of an atom also can be estimated by measuring the distance between adjacent atoms in a covalent compound. The **covalent radius** of a chlorine atom, for example, is half of the distance between the nuclei of the atoms in a $Cl_2$ molecule.

The covalent radii of the main-group elements are given in Figure 7.6. These data confirm the trends observed for metallic radii. Atoms become *larger* as we go down a column of the periodic table, and they become *smaller* as we go across a row of the table.

### CHECKPOINT

Check the validity of the data in Figure 7.6 by comparing the sum of the covalent radii of carbon and chlorine with the length of the C—Cl bond in $CCl_4$, which is 0.1766 nm.

Covalent radius

| | | | | | | | |
|---|---|---|---|---|---|---|---|
| H 0.037 | | | | | | H 0.037 | He |
| Li 0.123 | Be 0.089 | B 0.088 | C 0.077 | N 0.070 | O 0.066 | F 0.064 | Ne |
| Na 0.157 | Mg 0.136 | Al 0.125 | Si 0.117 | P 0.110 | S 0.104 | Cl 0.099 | Ar |
| K 0.203 | Ca 0.174 | Ga 0.125 | Ge 0.122 | As 0.121 | Se 0.117 | Br 0.114 | Kr |
| Rb 0.216 | Sr 0.192 | Im 0.150 | Sn 0.140 | Sb 0.141 | Te 0.137 | I 0.133 | Xe |
| Cs 0.235 | Ba 0.198 | Tl 0155 | Pd 0.154 | Bi 0.152 | Po 0.153 | At — | Rn |
| Fr — | Ra — | | | | | | |

**Figure 7.6** The covalent radii for the main-group elements. Once again, atoms become larger as we go down a column of the periodic table and smaller as we go from left to right across a row of the table.

Table A-5 in the appendix contains covalent and metallic radii for a number of elements. The covalent radius for an element is usually a little smaller than the metallic radius because covalent bonds tend to squeeze the atoms together, as shown in Figure 7.7.

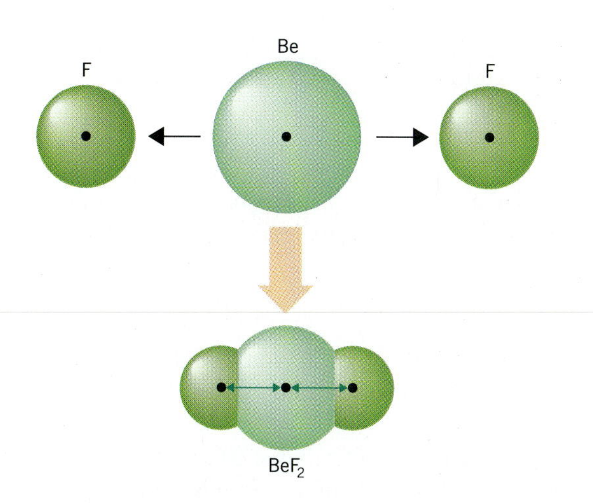


**Figure 7.7** Covalent radii are smaller than metallic radii because the force of attraction between the atoms in a covalent molecule tends to squeeze these atoms together.

## 7.6 THE SIZE OF ATOMS: IONIC RADII

The relative size of atoms also can be studied by measuring the radii of their ions. The first **ionic radii** were obtained by studying the structure of LiI, which contains a relatively small positive ion and a relatively large negative ion. The analysis of the structure of LiI was based on the following assumptions:

1. The relatively small $Li^+$ ions pack in the holes between the much larger $I^-$ ions, as shown in Figure 7.8.
2. The relatively large $I^-$ ions touch one another.
3. The $Li^+$ ions touch the $I^-$ ions.

If these assumptions are valid, the radius of the $I^-$ ion can be estimated by measuring the distance between the nuclei of adjacent iodide ions. The radius of the $Li^+$ ion then can be estimated by subtracting the radius of the $I^-$ ion from the distance between the nuclei of adjacent $Li^+$ and $I^-$ ions.

Unfortunately only two of the three assumptions that were made for LiI are correct. The $Li^+$ ions in this crystal do not quite touch the $I^-$ ions. As a result, this experiment overestimated the size of the $Li^+$ ion. Repeating this analysis with a large number of ionic compounds, however, has made it possible to obtain a set of more accurate ionic radii, which can be found in Table A-5 in the appendix.

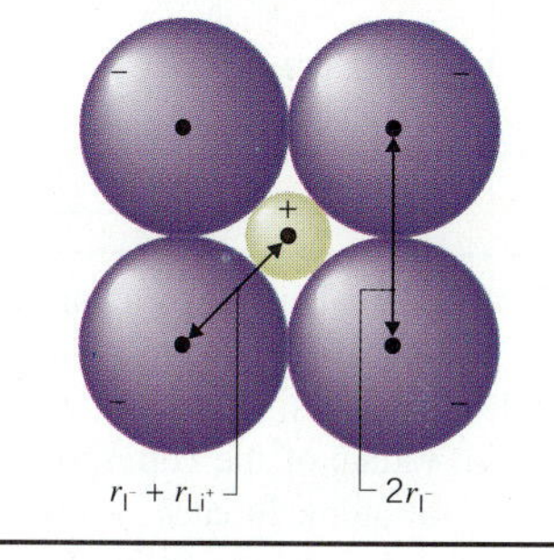


**Figure 7.8** The size of the $Li^+$ and $I^-$ ions in LiI can be calculated from measurements of the distance between the nuclei of adjacent ions if we assume that the $Li^+$ ions are so small that they fit into the holes left when the $I^-$ ions pack together so that adjacent $I^-$ ions touch.

## 7.7 THE RELATIVE SIZE OF ATOMS AND THEIR IONS

Table 7.5 and Figure 7.9 compare the covalent radius of neutral fluorine, chlorine, bromine, and iodine atoms with the radii of their $F^-$, $Cl^-$, $Br^-$, and $I^-$ ions. In each case, the negative ion is much larger than the atom from which it has been formed. In fact, the negative ion can be more than twice as large as the neutral atom.

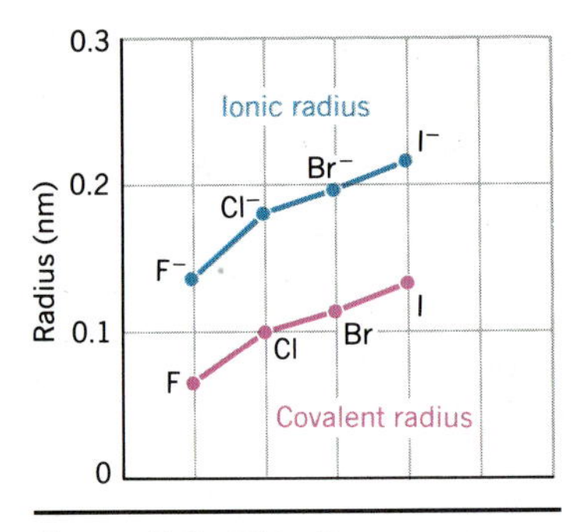


**Figure 7.9** This figure compares the radius of the $F^-$, $Cl^-$, $Br^-$, and $I^-$ ions with the covalent radius of the corresponding neutral atom. In each case, the negatively charged ion is larger than the neutral atom.

**TABLE 7.5 The Covalent Radii of Neutral Group 17 Atoms and the Ionic Radii of Their Negative Ions**

| Element | Covalent Radii (nm) | Ionic Radii (nm) |
|---|---|---|
| F | 0.064 | 0.136 |
| Cl | 0.099 | 0.181 |
| Br | 0.1142 | 0.196 |
| I | 0.1333 | 0.216 |

The only difference between an atom and its ions is the number of electrons that surround the nucleus. A neutral chlorine atom, for example, contains 17 electrons, while a $Cl^-$ ion contains 18 electrons.

$$Cl: \quad [Ne]\ 3s^2\ 3p^5 \qquad Cl^-: \quad [Ne]\ 3s^2\ 3p^6$$

Because the nucleus can't hold the 18 electrons in the $Cl^-$ ion as tightly as the 17 electrons in the neutral atom, the negative ion is significantly larger than the atom from which it forms.

Extending this line of reasoning suggests that positive ions should be smaller than the atoms from which they are formed. The 11 protons in the nucleus of an $Na^+$ ion, for example, should be able to hold the 10 electrons on this ion more tightly than the 11 electrons on a neutral sodium atom. The $Na^+$ ion therefore should be much smaller than a sodium atom. Table 7.6 and Figure 7.10 provide data to test this hypothesis. They compare the covalent radii for neutral atoms of the Group 1 elements with the ionic radii for the corresponding positive ions. In each case, the positive ion is much smaller than the atom from which it forms.

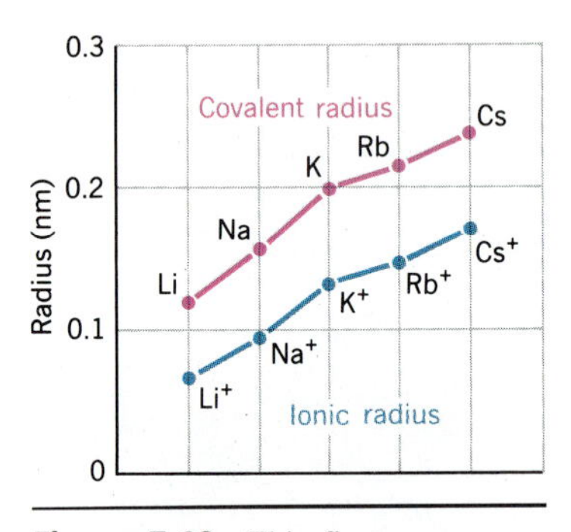


**Figure 7.10** This figure compares the radius of the $Li^+$, $Na^+$, $K^+$, $Rb^+$, and $Cs^+$ ions with the covalent radius of the corresponding neutral atom. In each case, the positively charged ion is smaller than the neutral atom.

**TABLE 7.6 The Covalent Radii of Neutral Group 1 Atoms and the Ionic Radii of Their Positive Ions**

| Element | Covalent Radii (nm) | Ionic Radii (nm) |
|---|---|---|
| Li | 0.123 | 0.068 |
| Na | 0.157 | 0.095 |
| K | 0.2025 | 0.133 |
| Rb | 0.216 | 0.148 |
| Cs | 0.235 | 0.169 |

## EXERCISE 7.1

Compare the sizes of neutral sodium and chlorine atoms and their $Na^+$ and $Cl^-$ ions.

**SOLUTION** A neutral sodium atom is significantly larger than a neutral chlorine atom:

$$Na = 0.157 \text{ nm} \qquad Cl = 0.099 \text{ nm}$$

But an $Na^+$ ion is only one-half the size of a $Cl^-$ ion:

$$Na^+ = 0.095 \text{ nm} \qquad Cl^- = 0.181 \text{ nm}$$

These particles therefore increase in size in the following order:

$$Na^+ \approx Cl < Na < Cl^-$$

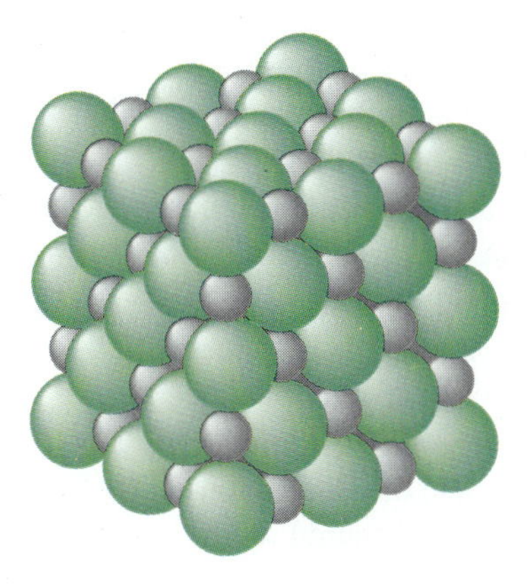

**Figure 7.11** The $Na^+$ ions in NaCl are so much smaller than the $Cl^-$ ions that they fit in the holes between planes of $Cl^-$ ions.

The relative size of positive and negative ions has important implications for the structure of ionic compounds. The positive ions are often so small they pack in the holes between planes of adjacent negative ions. In NaCl, for example, the $Na^+$ ions are so small that the $Cl^-$ ions almost touch, as shown in Figure 7.11.

## 7.8 PATTERNS IN IONIC RADII

The ionic radii in Tables 7.5 and 7.6 confirm one of the patterns observed for both metallic and covalent radii: Atoms become larger as we go down a column of the periodic table. We can examine trends in ionic radii across a row of the periodic table by comparing data for atoms and ions that are **isoelectronic.** By definition, isoelectronic atoms or ions have the same number of electrons. Table 7.7 summarizes data on the radii of a series of isoelectronic ions and atoms of second- and third-row elements.

The data in Table 7.7 are easy to explain if we note that these atoms or ions all have 10 electrons but the number of protons in the nucleus increases from 6 in the $C^{4-}$ ion to 13 in the $Al^{3+}$ ion. As the charge on the nucleus increases, the nucleus can hold a constant number of electrons more tightly. As a result, the atoms or ions become significantly smaller.

**TABLE 7.7 Radii for Isoelectronic Second-Row and Third-Row Atoms or Ions**

| Atom or Ion | Radius (nm) | Electron Configuration |
|---|---|---|
| $C^{4-}$ | 0.260 | $1s^2\ 2s^2\ 2p^6$ |
| $N^{3-}$ | 0.171 | $1s^2\ 2s^2\ 2p^6$ |
| $O^{2-}$ | 0.140 | $1s^2\ 2s^2\ 2p^6$ |
| $F^-$ | 0.136 | $1s^2\ 2s^2\ 2p^6$ |
| Ne | 0.112 | $1s^2\ 2s^2\ 2p^6$ |
| $Na^+$ | 0.095 | $1s^2\ 2s^2\ 2p^6$ |
| $Mg^{2+}$ | 0.065 | $1s^2\ 2s^2\ 2p^6$ |
| $Al^{3+}$ | 0.050 | $1s^2\ 2s^2\ 2p^6$ |

### EXERCISE 7.2

For each of the following pairs of atoms or ions, predict which is larger.

(a) $S^{2-}$ or $O^{2-}$ (b) Na or Al (c) $C^{4-}$ or $F^-$ (d) $P^{3-}$ or P

**SOLUTION**

(a) The $S^{2-}$ ion should be larger than the $O^{2-}$ ion because both atoms and their ions become larger as we go down a column of the periodic table.

(b) A sodium atom should be larger than an aluminum atom because atoms become smaller as we go from left to right across the periodic table.

(c) The $C^{4-}$ ion should be larger than the $F^-$ ion. Both ions have 10 electrons, but the nine protons in the nucleus of the $F^-$ ion should hold these electrons more tightly than the six protons in the nucleus of the $C^{4-}$ ion.

(d) The $P^{3-}$ ion should be significantly larger than the P atom because the 15 protons in the nucleus of both particles can't hold the 18 electrons in the $P^{3-}$ ion as tightly as they can the 15 electrons in the neutral atom.

## 7.9 THE FIRST IONIZATION ENERGY

The magnitude of the first ionization energy of hydrogen ($IE$ = 1312.0 kJ/mol) can be appreciated by comparing it with the energy given off in the thermite reaction, in which $Fe_2O_3$ reacts with powdered aluminum to give aluminum oxide and molten iron ($\Delta H°$ = 850 kJ/mol).

Another physical property of atoms that influences their chemical behavior is the energy needed to remove one or more electrons from a neutral atom to form a positively charged ion. By definition, the **first ionization energy** of an element is the energy needed to remove the outermost, or highest energy, electron from a neutral atom in the gas phase. The process by which the first ionization energy of hydrogen is measured would be represented by the following equation:

$$H(g) \longrightarrow H^+(g) + e^- \qquad \Delta H° = 1312.0 \text{ kJ/mol}$$

**EXERCISE 7.3**

Use the Bohr model to calculate the wavelength and energy of the photon that would have to be absorbed to ionize a neutral hydrogen atom in the gas phase.

**SOLUTION** The Bohr model suggests that the wavelength of radiation required to excite the electron from one orbit to another on a hydrogen atom is given by the following equation:

$$\frac{1}{\lambda} = R_H\left[\frac{1}{n_1^2} - \frac{1}{n_2^2}\right]$$

Ionizing the hydrogen atom is equivalent to exciting the electron from the $n_1 = 1$ into the $n_2 = \infty$ orbit. Substituting the Rydberg constant ($R_H$) and the values of $n_1$ and $n_2$ into this equation gives the following result:

$$\frac{1}{\lambda} = (1.09737 \times 10^{-2} \text{ nm}^{-1})\left[\frac{1}{1^2} - \frac{1}{\infty^2}\right]$$

This equation now can be solved for the wavelength of the photon that would have to be absorbed to ionize the hydrogen atom:

$$\lambda = 91.127 \text{ nm} = 9.1127 \times 10^{-8} \text{ m}$$

We can calculate the frequency of this photon from the relationship between frequency, wavelength, and the speed of light:

$$\nu = \frac{c}{\lambda} = \frac{2.99792 \times 10^8 \cancel{\text{m}}/\text{s}}{9.1127 \times 10^{-8} \cancel{\text{m}}} = 3.2898 \times 10^{15} \text{s}^{-1}$$

We can then use Planck's equation to calculate the energy of this photon:

$$E = h\nu = (6.6262 \times 10^{-34} \text{ J-}\cancel{\text{s}})(3.2898 \times 10^{15} \cancel{\text{s}}^{-1}) = 2.1799 \times 10^{-18} \text{ J}$$

Finally, the number of particles in a mole can be used to calculate the energy of a mole of these photons:

$$\frac{2.1799 \times 10^{-18}\ \text{J}}{\cancel{\text{photon}}} \times \frac{6.0220 \times 10^{23}\ \cancel{\text{photons}}}{\text{mol}} = \mathbf{1312.7\ kJ/mol}$$

A mole of these photons carries 1312.7 kJ of energy. Within experimental error, the results of this calculation agree with the first ionization energy of hydrogen.

The magnitude of the first ionization energy of hydrogen can be brought into perspective by comparing it with the energy given off in a chemical reaction. When we burn natural gas, about 800 kJ of energy is released per mole of methane consumed:

$$CH_4(g) + 2\,O_2(g) \longrightarrow CO_2(g) + 2\,H_2O(g) \qquad \Delta H^\circ = -802.4\ \text{kJ/mol}$$

The thermite reaction, which is used to weld iron rails, gives off about 850 kJ of energy per mole of iron oxide consumed:

$$Fe_2O_3(s) + 2\,Al(s) \longrightarrow Al_2O_3(s) + 2\,Fe(l) \qquad \Delta H^\circ = -851.5\ \text{kJ/mol}$$

The first ionization energy of hydrogen is half again as large as the energy given off in either of these reactions.

## 7.10 PATTERNS IN THE FIRST IONIZATION ENERGIES

The first ionization energy for helium is roughly twice the ionization energy for hydrogen because each electron in helium feels the attractive force of two protons, instead of one:

$$He(g) \longrightarrow He^+(g) + e^- \qquad \Delta H^\circ = 2372.3\ \text{kJ/mol}$$

It takes far less energy, however, to remove an electron from a lithium atom, which has three protons in its nucleus:

$$Li(g) \longrightarrow Li^+(g) + e^- \qquad \Delta H^\circ = 572.3\ \text{kJ/mol}$$

This can be explained by noting that the outermost, or highest energy, electron on a lithium atom is in the 2*s* orbital. Because the electron in a 2*s* orbital is already at a higher energy than the electrons in a 1*s* orbital, it takes less energy to remove this electron from the atom.

The first ionization energies for the main-group elements are given in Figure 7.12 and Figure 7.13.

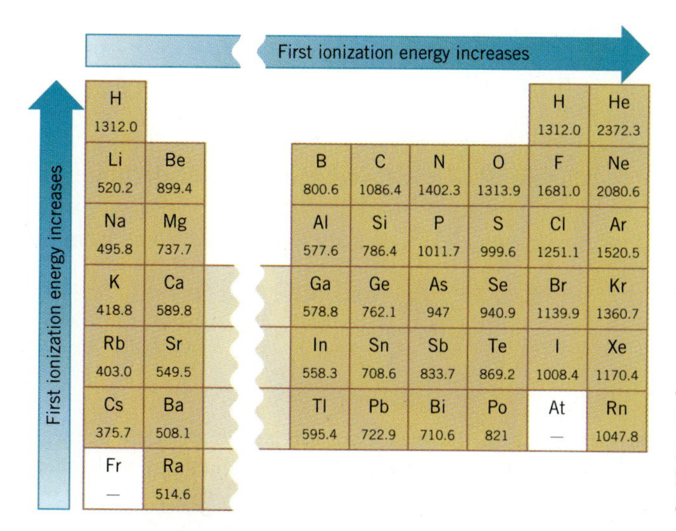


**Figure 7.12** The first ionization energies of the main-group elements. In general, the elements with the largest first ionization energies are found in the upper right-hand corner of the periodic table.

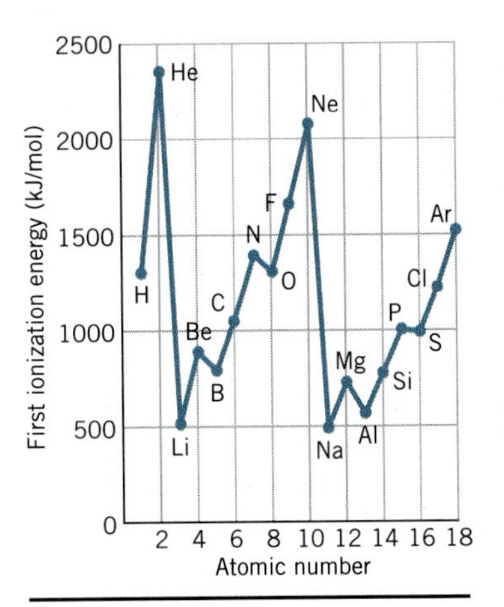


**Figure 7.13** A plot of the first ionization energies of the first 18 elements in the periodic table. There is a gradual increase in the first ionization energy across a row of the periodic table from H to He, from Li to Ne, and from Na to Ar. There is a gradual decrease in the first ionization energy as we go down a column of the periodic table, as can be seen by comparing the values for He, Ne, and Ar.

Two trends are apparent from these data:

1. In general, the first ionization energy increases as we go from left to right across a row of the periodic table.
2. The first ionization energy decreases as we go down a column of the periodic table.

The first trend isn't surprising. We might expect the first ionization energy to increase as we go across a row of the periodic table because the force of attraction between the nucleus and an electron increases as the number of protons in the nucleus of the atom increases.

The second trend results from the fact that the principal quantum number of the orbital holding the outermost electron increases as we go down a column of the periodic table. Although the number of protons in the nucleus also becomes larger, the electrons in smaller shells and subshells tend to screen the outermost electron from some of the force of attraction of the nucleus. Furthermore, the electron being removed when the first ionization energy is measured spends less of its time near the nucleus of the atom, and it therefore takes less energy to remove this electron from the atom.

## 7.11 EXCEPTIONS TO THE GENERAL PATTERN OF FIRST IONIZATION ENERGIES

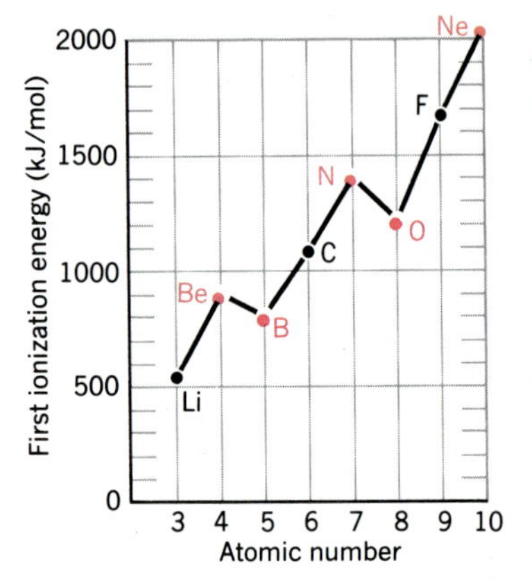


**Figure 7.14** This figure emphasizes the minor exceptions to the general trend of increasing first ionization energy from left to right across a row of the periodic table.

Figure 7.14 shows the first ionization energies for elements in the second row of the periodic table. Although there is a general trend toward an increase in the first ionization energy as we go from left to right across this row, there are two minor inversions in this pattern. The first ionization energy of boron is smaller than that of beryllium, and the first ionization energy of oxygen is smaller than that of nitrogen.

These observations can be explained by looking at the electron configurations of these elements. The electron removed when a beryllium atom is ionized comes from the 2*s* orbital, but a 2*p* electron is removed when boron is ionized:

$$\text{Be:} \quad [\text{He}]\ 2s^2$$
$$\text{B:} \quad [\text{He}]\ 2s^2\ 2p^1$$

The energy of an electron in a 2*p* orbital is larger than that of an electron in a 2*s* orbital (see Figure 6.22). It therefore takes less energy to remove an electron from the 2*p* orbital.

The electrons removed when nitrogen and oxygen are ionized also come from 2*p* orbitals:

$$\text{N:} \quad [\text{He}]\ 2s^2\ 2p^3$$
$$\text{O:} \quad [\text{He}]\ 2s^2\ 2p^4$$

But there is an important difference in the way electrons are distributed in these atoms. Hund's rules predict that the three electrons in the 2*p* orbitals of a nitrogen atom all have the same spin, but electrons are paired in one of the $2p$ orbitals on an oxygen atom:

N: ↑ ↑ ↑
O: ↑↓ ↑ ↑

Hund's rules can be understood by assuming that electrons try to stay as far apart as possible to minimize the force of repulsion between these particles. The three electrons in the $2p$ orbitals on nitrogen therefore enter different orbitals with their spins aligned in the same direction. In oxygen, two electrons must occupy one of the $2p$ orbitals. The force of repulsion between these electrons is minimized to some extent by pairing the electrons. There is still some residual repulsion between these electrons, however, which makes it slightly easier to remove an electron from a neutral oxygen atom than we would expect from the number of protons in the nucleus of the atom.

**EXERCISE 7.4**

Predict which element in each of the following pairs has the larger first ionization energy.

(a) Na or Mg (b) Mg or Al (c) F or Cl

**SOLUTION**

(a) Magnesium, because the first ionization energy tends to increase across a row of the periodic table from left to right.
(b) Magnesium, even though aluminum is to the right of magnesium in the periodic table. When an electron is removed from magnesium, it comes from a $3s$ orbital, but the outermost electron on an aluminum atom is in a $3p$ orbital, so less energy is needed to remove this electron.
(c) Fluorine, because it takes less energy to remove an electron from a $3p$ orbital on chlorine than it does to remove one from a $2p$ orbital on fluorine.

## 7.12 SECOND, THIRD, FOURTH, AND HIGHER IONIZATION ENERGIES

By now we know that sodium forms $Na^+$ ions, magnesium forms $Mg^{2+}$ ions, and aluminum forms $Al^{3+}$ ions. But have you ever wondered why sodium doesn't form $Na^{2+}$ ions, or even $Na^{3+}$ ions? The answer can be obtained from data for the second, third, and higher ionization energies of the element.

The *first ionization energy* of sodium is the energy it takes to accomplish the following reaction:

$$Na(g) + \text{energy} \longrightarrow Na^+(g) + e^-$$

The *second ionization energy* is the energy it takes to remove another electron to form an $Na^{2+}$ ion in the gas phase:

$$Na^+(g) + \text{energy} \longrightarrow Na^{2+}(g) + e^-$$

The *third ionization energy* of sodium can be represented by the following equation:

$$Na^{2+}(g) + \text{energy} \longrightarrow Na^{3+}(g) + e^-$$

The energy required to form a $Na^{3+}$ ion in the gas phase is the sum of the first, second, and third ionization energies of the element.

A complete set of data for the ionization energies (*IE*) of the elements is given in Table A-6 in the appendix. For the moment, let's look at the first, second, third, and fourth ionization energies of sodium, magnesium, and aluminum listed in Table 7.8.

It doesn't take much energy to remove one electron from a sodium atom to form an $Na^+$ ion with a filled-shell electron configuration. Once this is done, however, it takes almost 10 times as much energy to break into this filled-shell configuration to remove a second electron. Because it takes more energy to remove the second electron than is given off in any chemical reaction, sodium can react with other elements to form compounds that contain $Na^+$ ions but not $Na^{2+}$ or $Na^{3+}$ ions.

**TABLE 7.8 First, Second, Third, and Fourth Ionization Energies of Sodium, Magnesium, and Aluminum (kJ/mol)**

| | 1st *IE* | 2nd *IE* | 3rd *IE* | 4th *IE* |
|---|---|---|---|---|
| Na | 495.8 | **4562.4** | **6912** | **9543** |
| Mg | 737.7 | 1450.6 | **7732.6** | **10,540** |
| Al | 577.6 | 1816.6 | 2744.7 | **11,577** |

A similar pattern is observed when the ionization energies of magnesium are analyzed. The first ionization energy of magnesium is larger than sodium because magnesium has one more proton in its nucleus to hold on to the electrons in the 3*s* orbital.

$$\text{Mg:} \quad [\text{Ne}]\ 3s^2$$

The second ionization energy of Mg is larger than the first because it always takes more energy to remove an electron from a positively charged ion than from a neutral atom. The third ionization energy of magnesium is enormous, however, because the $Mg^{2+}$ ion has a filled-shell electron configuration.

The same pattern can be seen in the ionization energies of aluminum. The first ionization energy of aluminum is smaller than magnesium, as predicted in Exercise 7.4. The second ionization energy of aluminum is larger than the first, and the third ionization energy is even larger. Although it takes a considerable amount of energy to remove three electrons from an aluminum atom to form an $Al^{3+}$ ion, the energy needed to break into the filled-shell configuration of the $Al^{3+}$ ion is astronomical. Thus, it would be a mistake to look for an $Al^{4+}$ ion as the product of a chemical reaction.

## EXERCISE 7.5

Predict the group in the periodic table in which an element with the following ionization energies (*IE*) would most likely be found.

1st *IE* = 786 kJ/mol
2nd *IE* = 1577
3rd *IE* = 3232
4th *IE* = 4355
5th *IE* = 16,091
6th *IE* = 19,784

**SOLUTION** The gradual increase in the energy needed to remove the first, second, third, and fourth electrons from this element is followed by an abrupt increase in the energy required to remove one more electron. This is consistent with an element that has four more electrons than a filled-shell configuration, and we can conclude that the element is in Group IVA (Group 14) of the periodic table. These data are in fact the ionization energies of silicon:

$$\text{Si} = [\text{Ne}]\ 3s^2\ 3p^2$$

## EXERCISE 7.6

Use the trends in the ionization energies of the elements to explain the following observations:

(a) Elements on the left side of the periodic table are more likely than those on the right to form positive ions.

(b) The maximum positive charge on an ion is equal to the old-convention group number of the element.

**SOLUTION**

(a) The ionization energies of elements on the left side of the table are much smaller than those of elements on the right. Consider the first ionization energies for sodium and chlorine, for example:

$$\text{Na:} \quad 1\text{st } IE = 495.8 \text{ kJ/mol}$$
$$\text{Cl:} \quad 1\text{st } IE = 1251.1 \text{ kJ/mol}$$

Elements on the left side of the periodic table are therefore more likely to lose electrons to form positive ions.

(b) The electrons that can be removed from an atom in a chemical reaction are the electrons that weren't present in the filled-shell electron configuration of the previous rare gas, such as the $3s^2$ electrons on magnesium. The number of these electrons is equal to the old-convention group number, so the maximum positive charge on an ion is also equal to that group number. Because aluminum is in Group IIIA, for example, it can lose only three electrons before it reaches a filled-shell configuration. Thus, the maximum positive charge on an aluminum ion is +3.

## 7.13 ELECTRON AFFINITY

Ionization energies measure the tendency of a neutral atom to resist the loss of electrons. It takes a considerable amount of energy, for example, to remove an electron from a neutral fluorine atom to form a *positively charged ion:*

$$\text{F}(g) \longrightarrow \text{F}^+(g) + \text{e}^- \qquad \Delta H^\circ = 1681.0 \text{ kJ/mol}$$

The **electron affinity** of an element is the energy given off when a neutral atom in the gas phase gains an extra electron to form a *negatively charged ion.* A fluorine atom in the gas phase, for example, gives off energy when it gains an electron to form a fluoride ion:

$$\mathrm{F}(g) + \mathrm{e}^- \longrightarrow \mathrm{F}^-(g) \qquad \Delta H^\circ = -328.0 \text{ kJ/mol}$$

## CHECKPOINT

Use the chemical equations for the reactions describing the first ionization energy and the electron affinity of fluorine to show that the process by which the electron affinity of an element is measured is not the reverse of the process used to measure the first ionization energy of the element.

---

Electron affinities are more difficult to measure than ionization energies and are usually known to fewer significant figures. The electron affinities of the main-group elements are shown in Figure 7.15. (A more complete set of data can be found in Table A-7 in the appendix.) Several patterns can be found in these data.

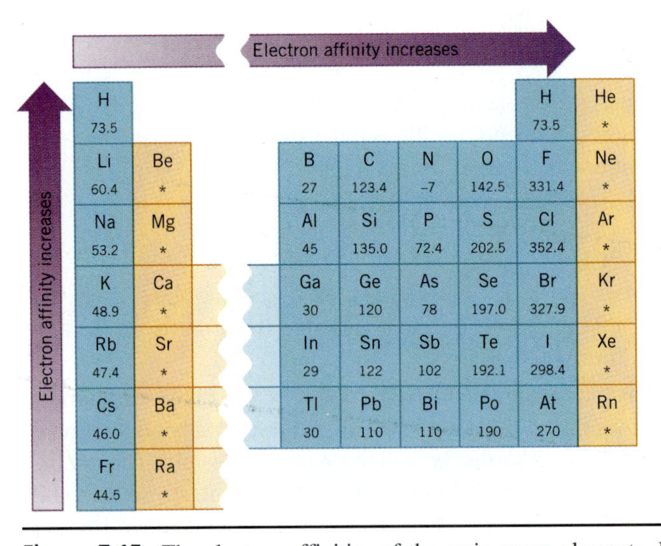


**Figure 7.15** The electron affinities of the main-group elements. Elements with no affinity for extra electrons are marked with an asterisk.

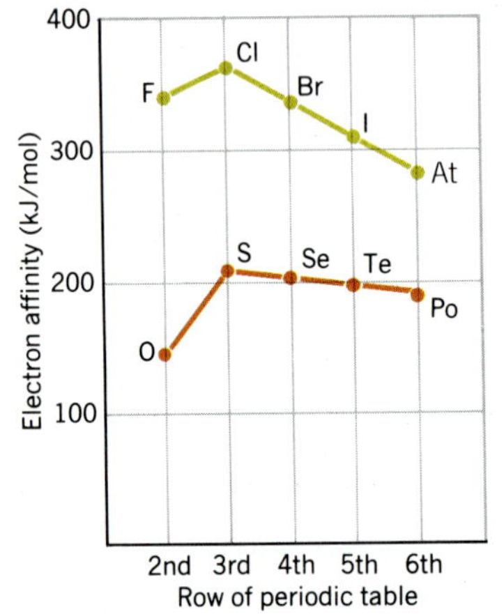


**Figure 7.16** There is a slight increase in the electron affinity among nonmetals as we go from the second to the third row of the periodic table, and then a gradual decrease as we continue down the column.

Electron affinities generally decrease as we go down a column of the periodic table for two reasons. First, the electron being added to the atom is placed in larger orbitals, where it spends less time near the nucleus of the atom. Second, the number of electrons on an atom increases as we go down a column, so the force of repulsion between the electron being added and the electrons already present on a neutral atom becomes larger.

Electron affinity data are complicated by the fact that the repulsion between the electron being added to the atom and the electrons already present on the atom depends on the volume of the atom. Among the nonmetals in Groups VIA and VIIA (Groups 16 and 17), this force of repulsion is largest for the very smallest atoms in these columns: oxygen and fluorine. As a result, these elements have a smaller electron affinity than the elements below them in these groups, as shown in Figure

7.16. From that point on, however, the electron affinities decrease as we continue down these columns.

At first glance, there appears to be no pattern in electron affinity across a row of the periodic table, as shown in Figure 7.17. When these data are listed along with the electron configurations of these elements, as shown in Table 7.9, they begin to make sense. These data can be understood by noting that electron affinities are much smaller than ionization energies. As a result, elements such as helium, beryllium, nitrogen, and neon, which have unusually stable electron configurations, have such small affinities for extra electrons that no energy is given off when a neutral atom of these elements picks up an electron. These configurations are so stable that it actually takes energy to force one of these elements to pick up an extra electron to form a negative ion.

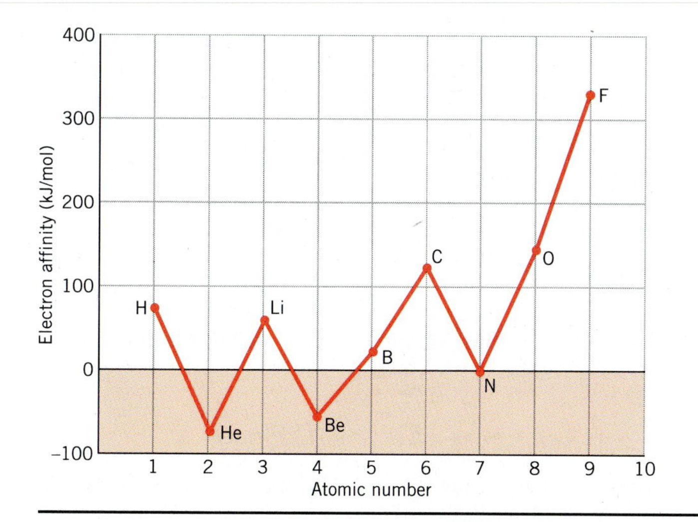


**Figure 7.17** There are a number of exceptions to the general trend toward larger electron affinities as we go across a row of the periodic table. These exceptions occur when the neutral atom has an electron configuration in which all of the subshells are either filled or half-filled.

**TABLE 7.9 Electron Affinities and Electron Configurations for the First 10 Elements in the Periodic Table**

| Element | Electron Affinity (kJ/mol) | Electron Configuration |
|---|---|---|
| H | 72.8 | $1s^1$ |
| He | $<0$ | $1s^2$ |
| Li | 59.8 | [He] $2s^1$ |
| Be | $<0$ | [He] $2s^2$ |
| B | 27 | [He] $2s^2\ 2p^1$ |
| C | 122.3 | [He] $2s^2\ 2p^2$ |
| N | $<0$ | [He] $2s^2\ 2p^3$ |
| O | 141.1 | [He] $2s^2\ 2p^4$ |
| F | 328.0 | [He] $2s^2\ 2p^5$ |
| Ne | $<0$ | [He] $2s^2\ 2p^6$ |

**RESEARCH IN THE 90s**

## MEASURING FUNDAMENTAL PHYSICAL CONSTANTS

Measurements of ionization energies trace back to 1905, when Einstein developed a quantitative relationship between the energy of the photon an atom absorbs and the sum of the ionization energy of the atom and the kinetic energy of the electron ejected when the atom is ionized. Ionization energies are now known for up to seven electrons on an element, and these data are known to as many as seven significant figures. Electron affinities are much more difficult to measure. As recently as 1970, data were available for only eight elements: H, F, Cl, Br, I, C, O, and S.

In recent years, researchers have developed a variety of techniques for measuring the electron affinity of atoms or molecules. One approach is based on advances in the measurement of $\Delta H°$ for gas-phase reactions. During her graduate work at Cal Tech, for example, Amy Stevens Miller measured $\Delta H°$ for the following reaction [*Journal of the American Chemical Society,* **113,** 8765 (1991)]:

$$HMn(CO)_5(g) \longrightarrow H^+(g) + Mn(CO)_5^-(g) \qquad \Delta H° = 1330 \text{ kJ/mol}$$

When Miller obtained these data, the energy required to split the Mn—H bond in this compound to give a *neutral* $Mn(CO)_5$ molecule and an isolated hydrogen atom was already known:

$$HMn(CO)_5(g) \longrightarrow H(g) + Mn(CO)_5(g) \qquad \Delta H° = 248 \text{ kJ/mol}$$

So was the first ionization energy for hydrogen:

$$H(g) \longrightarrow H^+(g) + e^- \qquad \Delta H° = 1312 \text{ kJ/mol}$$

Miller was therefore able to use Hess's law to calculate the electron affinity for $Mn(CO)_5$, as shown in Figure 7.18.

$$Mn(CO)_5(g) + e^- \longrightarrow Mn(CO)_5^-(g) \qquad \Delta H° = -230 \text{ kJ/mol}$$

Because experiments such as this increase our understanding of how and why chemical reactions occur, research that involves the measurement of fundamental physical constants of atoms and molecules will continue well into the next century.

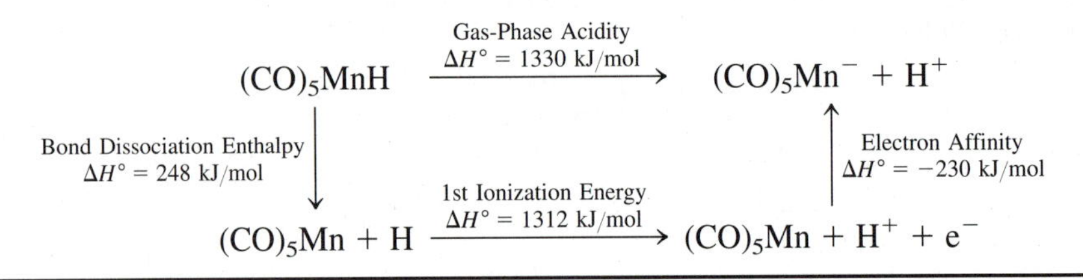


**Figure 7.18** The Hess's law cycle needed to calculate the electron affinity of $(CO)_5Mn$ from measurements of the gas-phase acidity of $(CO)_5MnH$, the energy required to break the Mn—H bond in this compound, and the first ionization energy for hydrogen.

## 7.14 CONSEQUENCES OF THE RELATIVE SIZE OF IONIZATION ENERGIES AND ELECTRON AFFINITIES

Students often believe that sodium reacts with chlorine to form $Na^+$ and $Cl^-$ ions because chlorine atoms ''like'' electrons more than sodium atoms do. There is no doubt that sodium reacts vigorously with chlorine to form NaCl:

$$2\ Na(s) + Cl_2(g) \longrightarrow 2\ NaCl(s)$$

Furthermore, the ease with which solutions of NaCl in water conduct electricity is evidence for the fact that the product of this reaction is a salt, which contains $Na^+$ and $Cl^-$ ions:

$$NaCl(s) \xrightarrow{H_2O} Na^+(aq) + Cl^-(aq)$$

The only question is whether it is legitimate to assume that this reaction occurs because chlorine atoms ''like'' electrons more than sodium atoms do.

According to the data in Tables A-6 and A-7, the first ionization energy for sodium is significantly larger than the electron affinity (*EA*) for chlorine:

$$\text{Na:} \quad \text{1st } IE = 495.8\ \text{kJ/mol}$$
$$\text{Cl:} \quad EA = 328.8\ \text{kJ/mol}$$

Thus, it takes more energy to remove an electron from a neutral sodium atom than is given off when the electron is picked up by a neutral chlorine atom. We obviously will have to find another explanation for why sodium reacts with chlorine to form NaCl. Before we can do this, however, we need to know more about the chemistry of ionic compounds.

## 7.15 LATTICE ENERGIES AND THE STRENGTH OF THE IONIC BOND

The first measurements of the force of attraction between oppositely charged particles were reported by Charles Augustine Coulomb between 1785 and 1791. Coulomb found that the force of attraction was directly proportional to the product of the charges on the two objects ($q_1$ and $q_2$) and inversely proportional to the square of the distance between the objects ($r^2$):

$$F = \frac{q_1 \times q_2}{r^2}$$

The strength of the bond between the ions of opposite charge in an ionic compound therefore depends on the charges on the ions and the distance between the centers of the ions when they pack to form a crystal.

An estimate of the strength of the bonds in an ionic compound can be obtained by measuring the **lattice energy** of the compound, which is the energy given off when oppositely charged ions in the gas phase come together to form a solid. The lattice energy of NaCl, for example, is the energy given off when $Na^+$ and $Cl^-$ ions in the gas phase come together to form the lattice of alternating $Na^+$ and $Cl^-$ ions in the NaCl crystal shown in Figure 7.19:

$$Na^+(g) + Cl^-(g) \longrightarrow NaCl(s) \qquad \Delta H^\circ = -787.3\ \text{kJ/mol}$$

The lattice energies of ionic compounds are relatively large. The lattice energy of NaCl, for example, is $-787.3$ kJ/mol, which is only slightly less than the energy given off when natural gas burns.

The bond between ions of opposite charge is strongest when the ions are small. The lattice energies for the alkali metal halides is therefore largest for LiF and

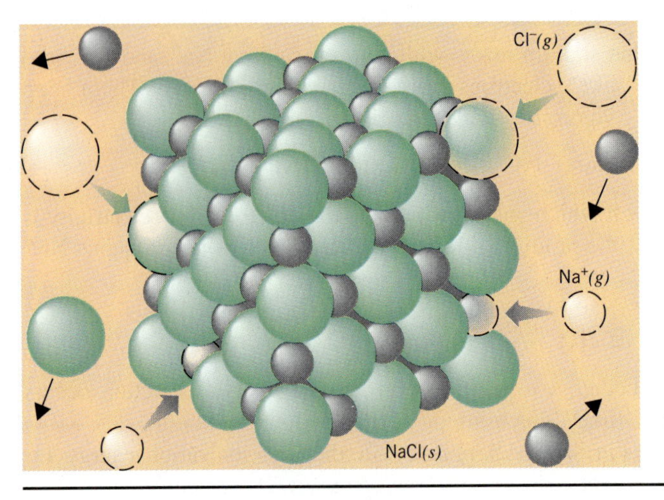


**Figure 7.19** The lattice energy of NaCl is the energy released when $Na^+$ and $Cl^-$ ions in the gas phase come together to form a solid crystalline lattice of alternating $Na^+$ and $Cl^-$ ions.

smallest for CsI, as shown in Table 7.10. The ionic bond should also become stronger as the charge on the ions becomes larger. The data in Table 7.11 show that the lattice energies for salts of the $OH^-$ and $O^{2-}$ ions increase rapidly as the charge on either ion becomes larger.

**TABLE 7.10 The Magnitude of the Lattice Energies of Alkali Metals Halides (kJ/mol)**

| | $F^-$ | $Cl^-$ | $Br^-$ | $I^-$ |
|---|---|---|---|---|
| $Li^+$ | 1036 | 853 | 807 | 757 |
| $Na^+$ | 923 | 787 | 747 | 704 |
| $K^+$ | 821 | 715 | 682 | 649 |
| $Rb^+$ | 785 | 689 | 660 | 630 |
| $Cs^+$ | 740 | 659 | 631 | 604 |

**TABLE 7.11 The Magnitude of the Lattice Energies of Salts of the $OH^-$ and $O^{2-}$ Ions (kJ/mol)**

| | $OH^-$ | $O^{2-}$ |
|---|---|---|
| $Na^+$ | 900 | 2481 |
| $Mg^{2+}$ | 3006 | 3791 |
| $Al^{3+}$ | 5627 | 15,916 |

**CHECKPOINT**

Explain why the magnitude of the lattice energy of sodium salts increases in the following order:

$$\text{NaI} < \text{NaBr} < \text{NaCl} < \text{NaF}$$

## 7.16 LATTICE ENERGIES AND SOLUBILITY

What happens when a salt, such as NaCl, dissolves in water? On the macroscopic scale, the crystal disappears. On the atomic scale, the $Na^+$ and $Cl^-$ ions in the crystal are released into solution:

$$\text{NaCl}(s) \xrightarrow{H_2O} \text{Na}^+(aq) + \text{Cl}^-(aq)$$

The lattice energy of a salt therefore gives a rough indication of the solubility of the salt in water because it reflects the energy needed to separate the positive and negative ions in a salt.

Sodium and potassium salts are soluble in water because they have relatively small lattice energies. Magnesium and aluminum salts are often much less soluble because it takes more energy to separate the positive and negative ions in these salts. NaOH, for example, is very soluble in water (420 g/L), but $Mg(OH)_2$ dissolves in water only to the extent of 0.009 g/L, and $Al(OH)_3$ is essentially insoluble in water.

**CHECKPOINT**

Use general trends in lattice energy to explain why neither CaO nor MgO dissolves readily in water, but CaO is more soluble than MgO.

## 7.17 WHY DOES SODIUM FORM NaCl?

We are now ready to explain why sodium reacts with chlorine to form $Na^+$ ions and $Cl^-$ ions, in spite of the fact that the first ionization energy of sodium is larger than the electron affinity of chlorine. To do this, we need to divide the reaction between sodium and chlorine into a number of hypothetical steps for which we know how much energy is given off or absorbed.

The starting materials for this reaction are solid sodium metal and chlorine molecules in the gas phase, and the product of the reaction is solid sodium chloride:

$$2\,Na(s) + Cl_2(g) \longrightarrow 2\,NaCl(s)$$

Let's imagine that the reaction takes place by the following sequence of steps.

The reaction between sodium metal and chlorine gas to form sodium chloride gives off energy in the form of both heat and light.

1. A mole of sodium is converted from the solid to a gas. As might be expected, this reaction is endothermic:

$$Na(s) \longrightarrow Na(g) \qquad \Delta H^\circ = 107.3 \text{ kJ}$$

2. An electron is then removed from each sodium atom to form a mole of $Na^+$ ions:

$$Na(g) \longrightarrow Na^+(g) \qquad \Delta H^\circ = 495.8 \text{ kJ}$$

The energy consumed in this reaction is equal to the first ionization energy of sodium.

3. A mole of chlorine atoms is formed by breaking the bonds in one-half a mole of chlorine molecules. Like the previous steps, this is an endothermic reaction:

$$\tfrac{1}{2}Cl_2(g) \longrightarrow Cl(g) \qquad \Delta H^\circ = 121.7 \text{ kJ}$$

4. An electron is then added to each chlorine atom to form a $Cl^-$ ion:

$$Cl(g) + e^- \longrightarrow Cl^-(g) \qquad \Delta H^\circ = -348.8 \text{ kJ}$$

This is the first exothermic step in this process, and the energy released is equal to the electron affinity of chlorine.

5. The isolated $Na^+$ and $Cl^-$ ions in the gas phase then come together to form solid NaCl:

$$Na^+(g) + Cl^-(g) \longrightarrow NaCl(s) \qquad \Delta H^\circ = -787.3 \text{ kJ}$$

This is a strongly exothermic reaction, for which $\Delta H^\circ$ is equal to the lattice energy of NaCl.

Sec. 5.16

If we consider just the first four steps in this reaction, Hess's law suggests that it takes 376.0 kJ/mol to form $Na^+$ and $Cl^-$ ions from sodium metal and chlorine gas.

$$Na(s) + \tfrac{1}{2}Cl_2(g) \longrightarrow Na^+(g) + Cl^-(g) \qquad \Delta H^\circ = 376.0 \text{ kJ/mol}$$

When we include the last step in the calculation, the lattice energy of NaCl is large enough to compensate for all of the steps in this reaction that consume energy, as shown in Figure 7.20.

$$Na(s) + \tfrac{1}{2}Cl_2(g) \longrightarrow NaCl(s) \qquad \Delta H^\circ = -411.3 \text{ kJ/mol}$$

The primary driving force behind this reaction is therefore the force of attraction between the $Na^+$ and $Cl^-$ ions formed in the reaction, not the affinity of a chlorine atom for electrons.

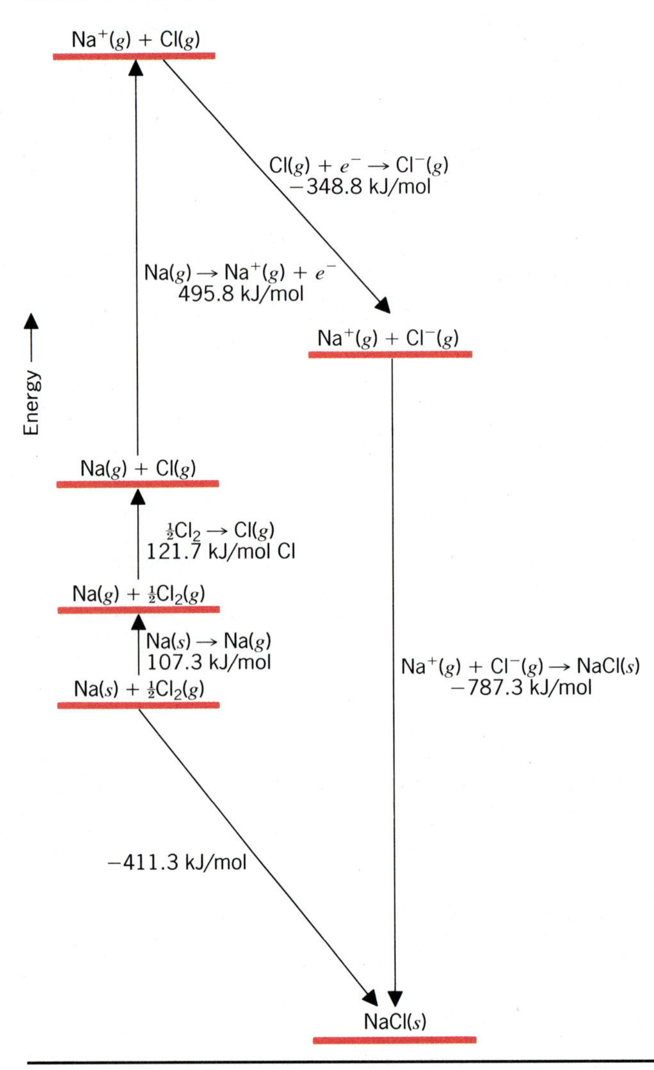


**Figure 7.20** This figure illustrates what happens to the energy of the system as solid sodium metal and chlorine gas are converted step by step into solid sodium chloride.

Why does the reaction stop at NaCl? Why doesn't it keep going to form $NaCl_2$ or $NaCl_3$? The lattice energy would increase as the charge on the sodium atom increased from $Na^+$ to $Na^{2+}$ or $Na^{3+}$. But to form an $Na^{2+}$ ion, we have to remove a second electron from the sodium atom, and the second ionization energy of sodium

(4562.4 kJ/mol) is almost 10 times as large as the first ionization energy. The increase in the lattice energy that would result from forming an $Na^{2+}$ ion can't begin to compensate for the energy needed to break into the filled-shell configuration of the $Na^+$ ion to remove a second electron. The reaction between sodium and chlorine therefore stops at NaCl.

## 7.18 WHY DOES MAGNESIUM FORM $MgCl_2$?

If the reaction between sodium and chlorine stops at NaCl, why does the reaction between magnesium and chlorine go on to form $MgCl_2$? To answer this question, let's break the reaction into the following steps:

**1.** A mole of magnesium is converted from the solid to a gas:

$$Mg(s) \longrightarrow Mg(g) \qquad \Delta H° = 147.7 \text{ kJ}$$

**2.** An electron is removed from each magnesium atom to form a mole of $Mg^+$ ions:

$$Mg(g) \longrightarrow Mg^+(g) \qquad \Delta H° = 737.7 \text{ kJ}$$

**3.** A second electron is then removed to form a mole of $Mg^{2+}$ ions:

$$Mg^+(g) \longrightarrow Mg^{2+}(g) \qquad \Delta H° = 1450.6 \text{ kJ}$$

**4.** Two moles of chlorine atoms are formed by breaking the bonds in a mole of chlorine molecules:

$$Cl_2(g) \longrightarrow 2Cl(g) \qquad \Delta H° = 243.4 \text{ kJ}$$

**5.** An electron is then added to each chlorine atom to form $Cl^-$ ions:

$$2\,Cl(g) + e^- \longrightarrow 2\,Cl^-(g) \qquad \Delta H° = -697.4 \text{ kJ}$$

**6.** The isolated $Mg^{2+}$ and $Cl^-$ ions in the gas phase then come together to form solid $MgCl_2$:

$$Mg^{2+}(g) + 2\,Cl^-(g) \longrightarrow MgCl_2(s) \qquad \Delta H° = -2526 \text{ kJ}$$

Once ignited, magnesium metal is so active that it will burn between two pieces of dry ice to form magnesium oxide and carbon.

$$2\,Mg(s) + CO_2(s) \longrightarrow 2\,MgO(s) + C(s)$$

The lattice energy for $MgCl_2$ is large enough to compensate for the energy it takes to remove the second electron from a magnesium atom because we don't have to break into a filled-shell configuration to form $Mg^{2+}$ ions:

$$Mg(s) + Cl_2(g) \longrightarrow MgCl_2(s) \qquad \Delta H° = -644 \text{ kJ/mol}$$

The reaction stops at $MgCl_2$, however, because it would take an enormous amount of energy to break into the filled-shell configuration of the $Mg^{2+}$ ion to remove another electron.

## 7.19 WHY DO SEMIMETALS EXIST?

The tendency to divide elements into metals, nonmetals, and semimetals on the basis of differences in the chemical and physical properties of these elements was first discussed in Section 2.7. *Metals,* as we have seen, have some or all of the following properties:

1. They have a metallic shine or luster.
2. They are typically solids at room temperature.

3. They are malleable and ductile.
4. They conduct heat and electricity.
5. They exist as extended planes of atoms.
6. They combine with other metals to form alloys, which behave like metals.
7. They form positive ions, such as the $Na^+$, $Mg^{2+}$, $Fe^{3+}$, and $Cu^{2+}$ ions.

*Nonmetals* have the opposite properties:

1. They seldom have a metallic luster.
2. They are often gases at room temperature.
3. They are neither malleable nor ductile.
4. They are poor conductors of both heat and electricity.
5. They often form molecules in their elemental form.
6. They combine with other nonmetals to form covalent compounds.
7. They tend to form negative ions, such as the $Cl^-$, $P^{3-}$, $SO_4^{2-}$, and $PO_4^{3-}$ ions.

The differences in the chemical and physical properties of metals and nonmetals can be traced to differences in their electron configurations, their ionization energies, their electron affinities, and the radii of their atoms and ions. As a rule:

**Metals have relatively few electrons in their outermost shell of orbitals, have lower ionization energies, have smaller electron affinities, and have larger atoms than nonmetals.**

There are no abrupt changes in the physical properties discussed in this chapter as we go across a row of the periodic table or down a column. As a result, the change from metal to nonmetal must be gradual. Instead of arbitrarily dividing elements into metals and nonmetals, it might be better to describe some elements as being more metallic and other elements as more nonmetallic. Metallic character decreases as we go across a row of the periodic table from left to right, while nonmetallic character increases.

**Metallic character decreases** →

Na Mg Al Si P S Cl Ar

**Nonmetallic character increases** →

Diamond is one of the elemental forms of carbon, which is a nonmetal with none of the characteristic properties of a metal.

For the same reasons, elements toward the top of a column of the periodic table are the most nonmetallic, while elements toward the bottom of a column are the most metallic. This is evident in Group 14 (IVA), where a gradual transition is seen from a nonmetallic element (carbon) toward the well-known metals tin and lead:

| | | | | |
|---|---|---|---|---|
| | | C | | |
| | **metallic** | Si | **nonmetallic** | |
| ↓ | **character** | Ge | **character** | ↑ |
| | **increases** | Sn | **increases** | |
| | | Pb | | |

If elements become less metallic and more nonmetallic as we go across a row of the periodic table from left to right, we should encounter elements along each row that have properties that lie between the extremes of metals and nonmetals. The eight elements in this class (B, Si, Ge, As, Sb, Te, Po, and At) often look metallic, but they are brittle, like nonmetals. They are neither good conductors nor good insulators but serve as the basis for the semiconductor industry. These eight elements are often known as *metalloids,* or *semimetals.*

## SUMMARY

**1.** The periodic table provides a basis for predicting general trends in a number of physical properties, including the relative size of atoms and the ions they form and the relative ease with which atoms either gain or lose electrons.

**2.** There is a general trend toward larger atoms or ions as we go down a column of the periodic table. Both atoms and the ions they form become smaller as we go from left to right across a row of the periodic table.

**3.** The number of protons in the nucleus of an atom remains constant as the atom gains or loses electrons to form an ion. As a result, positive ions are typically much smaller than neutral atoms. Negative ions, on the other hand, tend to be much larger than neutral atoms.

**4.** The energy needed to remove electrons from a neutral atom decreases as we go down a column of the periodic table. In general, it increases as we go from left to right across a row of the table. As a result, elements toward the bottom left corner of the periodic table are the most likely to form positive ions.

**5.** As electrons are removed from an atom, the energy needed to remove the next electron always increases. But there is a point at which this energy becomes so large it is impossible to remove another electron in a chemical reaction. This point usually corresponds to an electron configuration for the ion in which all the subshells are filled.

**6.** Because of the repulsion between electrons, the electron affinity of an element is much smaller than its ionization energy. It therefore takes more energy to remove an electron from an atom than we get back when the electron is transferred to another atom.

**7.** Ionic compounds are held together by the force of attraction between ions of opposite charge, which is directly proportional to the charge on the ions and inversely proportional to the distance between the nuclei of the ions.

**8.** The energy given off when isolated ions in the gas phase come together to form an ionic compound is known as the lattice energy of the compound. The lattice energy of an ionic compound more than compensates for the fact that it usually takes more energy to remove one or more electrons to form a positive ion than is given off when these electrons are added to another atom to form a negative ion.

**9.** The physical properties of the atoms play an important role in controlling the macroscopic properties of an element. Elements with relatively small ionization energies, low electron affinities, and large radii tend to be metals.

**10.** There are no abrupt changes in the physical properties of atoms from one end of the periodic table to the other, which means that there is no abrupt change from metal to nonmetal as we go across a row or down a column of the table. As a result, elements that are not either metals or nonmetals can be found along the boundaries between these categories.

## KEY TERMS

**Covalent radius** (p. 246)
**Electron affinity** (p. 255)
**First ionization energy** (p. 250)
**Ionic radii** (p. 247)
**Isoelectronic** (p. 249)
**Lattice energy** (p. 259)
**Metallic radii** (p. 245)

## PROBLEMS

### The Search for Patterns in the Chemistry of the Elements

**7-1** Eleven elements were listed in Section 2.4 as having symbols that were taken from the Latin or German name of the element. Compare the elements on this list with the 13 elements known in 1661 and comment on any pattern you observe.

**7-2** Describe the difference between families, periods, and groups of elements in the periodic table.

**7-3** Use the graph in Figure 7.1 to explain why the first attempts to recognize regular trends or patterns in the properties of the elements did not occur until about 1830.

**7-4** Describe some of the similarities between the chemistries of chlorine, bromine, and iodine that might have led Döbereiner to suggest that these elements form a triad.

### The Development of the Periodic Table

**7-5** Describe the advantages of Mendeléeff's periodic table. What were the keys to the rapid acceptance of this model by his peers?

**7-6** Find the three places in the periodic table where the main-group elements are not listed in order of increasing atomic weights.

**7-7** Mendeléeff placed both silver and copper in the same group as lithium and sodium. Look up the chemistry of these four elements in the *CRC Handbook of Chemistry and Physics.*[2] Describe some of the similarities that allow these elements to be classified in a single group on the basis of their chemical properties.

**7-8** Modern periodic tables no longer place silver and copper in the same group as lithium and sodium. The relationship between the chemical properties of these elements is still retained, however, in periodic tables that use the symbols IA, IIA, and so on to identify groups. Describe this relationship.

**7-9** Use Mendeléeff's periodic table to predict the formulas for the oxides of the following elements.

(a) K (b) Zn (c) In (d) Si (e) V

### Modern Versions of the Periodic Table

**7-10** Describe some of the evidence that could be used to justify the argument that the modern periodic table is based on similarities in the chemical properties of the elements.

**7-11** Describe some of the evidence that could be used to refute the argument that the modern periodic table groups elements with similar electron configurations.

**7-12** Determine the row and column of the periodic table in which you would expect to find the first element to have $4d$ electrons in its electron configuration.

**7-13** Determine the row and column of the periodic table in which you would expect to find the element that has five more electrons than the rare gas krypton.

**7-14** Determine the group of the periodic table in which an element with the following electron configuration belongs: $[X] = 1s^2\, 2s^2\, 2p^6\, 3s^2\, 3p^6\, 4s^2\, 3d^{10}\, 4p^6\, 5s^2\, 4d^{10}\, 5p^3$

**7-15** In which group of the periodic table should Element 119 belong if and when it is discovered?

**7-16** Which of the following ions do not have the electron configuration of argon?

(a) $Ga^{3+}$ (b) $Cl^-$ (c) $P^{3-}$ (d) $Sc^{3+}$ (e) $K^+$

**7-17** Which of the following ions has the electron configuration [Ar] $3d^4$?

(a) $Ca^{2+}$ (b) $Ti^{2+}$ (c) $Cr^{2+}$ (d) $Mn^{2+}$ (e) $Fe^{2+}$

[2] CRC Press, Boca Raton, Florida.

### The Size of Atoms: Metallic Radii

**7-18** Describe what happens to the size of an atom as we go down a column of the periodic table. Explain why.

**7-19** Describe what happens to the size of an atom as we go across a row of the periodic table from left to right. Explain why.

**7-20** At one time, the size of an atom was given in units of angstroms because the radius of a typical atom was about 1 angstrom (Å). Now they are given in a variety of units. If the radius of a gold atom is 1.442 Å, and 1 Å is equal to $10^{-8}$ cm, what is the radius of this atom in units of both nanometers and picometers?

**7-21** Which of the following atoms has the smallest radius?

(a) Na (b) Mg (c) Al (d) K (e) Ca

### The Size of Atoms: Covalent Radii

**7-22** Explain why the covalent radius of an atom is smaller than the metallic radius of the atom.

**7-23** Which of the following atoms has the largest covalent radius?

(a) N (b) O (c) F (d) P (e) S

### The Size of Atoms: Ionic Radii

**7-24** Describe the assumptions that have to be made to determine the size of the $Li^+$ and $I^-$ ions from measurements of the distance between the nuclei of adjacent ions in a lithium iodide crystal.

**7-25** Describe what happens to the radius of an atom when electrons are removed to form a positive ion. Describe what happens to the radius of the atom when electrons are added to form a negative ion.

**7-26** Predict the order of increasing ionic radius for the following ions: $H^-$, $F^-$, $Cl^-$, $Br^-$, and $I^-$. Compare your predictions with the data for these ions in Table A-5 of the appendix. Explain any differences between your predictions and experiment.

### The Relative Size of Atoms and Their Ions

**7-27** Look up the covalent radii for magnesium and sulfur atoms and the ionic radii of $Mg^{2+}$ and $S^{2-}$ ions in the appendix. Explain why $Mg^{2+}$ ions are smaller than $S^{2-}$ ions even though magnesium atoms are larger than sulfur atoms.

**7-28** Explain why the radius of a $Pb^{2+}$ ion (0.120 nm) is very much larger than that of a $Pb^{4+}$ ion (0.084 nm).

**7-29** Predict the relative size of the $Fe^{2+}$ and $Fe^{3+}$ ions found in a variety of proteins, including hemoglobin, myoglobin, and the cytochromes.

### Patterns in Ionic Radii

**7-30** Sort the following atoms or ions into isoelectronic groups.

(a) $N^{3-}$ (b) $O^{2-}$ (c) $F^-$ (d) Ne (e) $Na^+$

(f) $Mg^{2+}$ (g) $Al^{3+}$ (h) $Si^{4+}$ (i) $P^{3-}$ (j) $S^{2-}$
(k) $Cl^-$ (l) Ar (m) $K^+$ (n) $Ca^{2+}$

**7-31** Which of the following contains sets of atoms or ions that are isoelectronic?

(a) $B^{3+}$, $C^{4+}$, $H^+$, He (b) $Na^+$, Ne, $N^{3+}$, $O^{2-}$
(c) $Mg^{2+}$, $F^-$, $Na^+$, $O^{2-}$ (d) Ne, Ar, Xe, Kr
(e) $O^{2-}$, $S^{2-}$, $Se^{2-}$, $Te^{2-}$

**7-32** Predict whether the $Al^{3+}$ or the $Mg^{2+}$ ion is smaller. Explain why.

**7-33** Which of the following ions has the largest radius?

(a) $Na^+$ (b) $Mg^{2+}$ (c) $S^{2-}$ (d) $Cl^-$ (e) $Se^{2-}$

**7-34** Which of the following atoms or ions is the smallest?

(a) Na (b) Mg (c) $Na^+$ (d) $Mg^{2+}$ (e) $O^{2-}$

**7-35** Which of the following ions has the smallest radius?

(a) $K^+$ (b) $Li^+$ (c) $Be^{2+}$ (d) $O^{2-}$ (e) $F^-$

**7-36** Which of the following isoelectronic ions is the largest?

(a) $Mn^{7+}$ (b) $P^{3-}$ (c) $S^{2-}$ (d) $Sc^{3+}$ (e) $Ti^{4+}$

### The First Ionization Energy

**7-37** Write balanced chemical equations for the reactions that occur when the first, second, third, and fourth ionization energies of aluminum are measured.

**7-38** Explain why it takes energy to remove an electron from an isolated atom in the gas phase.

**7-39** Describe the general trend in first ionization energies from left to right across the second row of the periodic table.

**7-40** Describe the general trend in first ionization energies from top to bottom of a column of the periodic table.

**7-41** Explain why the first ionization energy of B is smaller than that of Be and why the first ionization energy of O is smaller than that of N.

**7-42** Explain why the first ionization energy of hydrogen is so much larger than the first ionization energy of sodium.

**7-43** Describe a possible set of *n, l, m,* and *s* quantum numbers for the electron removed when the first ionization energy of sodium is measured.

**7-44** List the following elements in order of increasing first ionization energy:

(a) Li (b) Be (c) F (d) Na (e) Si

**7-45** Which of the following elements should have the largest first ionization energy?

(a) B (b) C (c) N (d) Mg (e) Al

**7-46** Which of the following elements should have the smallest first ionization energy?

(a) Mg (b) Ca (c) Si (d) S (e) Se

### Second, Third, Fourth, and Higher Ionization Energies

**7-47** Which of the following atoms or ions has the largest ionization energy?

(a) P (b) $P^+$ (c) $P^{2+}$ (d) $P^{3+}$ (e) $P^{4+}$

**7-48** Explain why the second ionization energy of sodium is so much larger than the first ionization energy of this element.

**7-49** What is the most probable electron configuration for the element that has the following ionization energies.

1st *IE* = 578 kJ/mol
2nd *IE* = 1817
3rd *IE* = 2745
4th *IE* = 11,577
5th *IE* = 14,831

(a) [Ne] (b) [Ne] $3s^1$ (c) [Ne] $3s^2$ (d) [Ne] $3s^2\,3p^1$
(e) [Ne] $3s^2\,3p^2$ (f) [Ne] $3s^2\,3p^3$

**7-50** Which of the following ionization energies is the largest?

(a) 1st *IE* of Ba (b) 1st *IE* of Mg (c) 2nd *IE* of Ba
(d) 2nd *IE* of Mg (e) 3rd *IE* of Al (f) 3rd *IE* of Mg

**7-51** Which of the following elements should have the largest second ionization energy?

(a) Na (b) Mg (c) Al (d) Si (e) P

**7-52** Which of the following elements should have the largest third ionization energy?

(a) B (b) C (c) N (d) Mg (e) Al

**7-53** List the following elements in order of increasing second ionization energy:

(a) Li (b) Be (c) Na (d) Mg (e) Ne

**7-54** Use the ionization energies in Table A-6 in the appendix to predict the most positive oxidation state of each of the following elements:

(a) C (b) N (c) O (d) Si (e) S (f) Ca

**7-55** Use the ionization energies in Table A-6 in the appendix to predict the most positive oxidation states of the following elements:

(a) Sc (b) Ti (c) V (d) Cr (e) Mn

**7-56** Some elements, such as tin and lead, have more than one common oxidation state. Use the ionization energies in Table A-6 of the appendix to predict the most likely oxidation states of these metals.

**7-57** Explain why the largest oxidation state for vanadium corresponds to the $V^{5+}$ ion, while the largest oxidation state for chromium corresponds to the $Cr^{6+}$ ion.

**7-58** Use the data in Table A-6 to explain why iron forms the $Fe^{2+}$ and $Fe^{3+}$ ions, but not $Fe^{4+}$ ions.

### Electron Affinity

**7-59** Explain why energy is usually released when the electron affinity of an element is measured.

**7-60** Describe in general terms the trends in electron affinity from left to right across a row or down a column of the periodic table.

**7-61** What do He, Be, Ne, Mg, and Ar have in common that explains why the electron affinities of these elements are all less than zero?

**7-62** Compare the magnitudes of the first ionization energies and the electron affinities of a variety of elements in the periodic table. In general, which is larger?

**7-63** The first ionization energy increases more or less regularly across a row of the periodic table, but the electron affinity reaches its peak in the second to the last column. Explain why.

**7-64** Which of the following elements has the largest electron affinity?

(a) Na (b) K (c) N (d) O (e) Ne

**7-65** Which of the following elements has the largest electron affinity?

(a) Li: $1s^2\,2s^1$ (b) Be: $1s^2\,2s^2$ (c) B: $1s^2\,2s^2\,2p^1$ (d) C: $1s^2\,2s^2\,2p^2$ (e) N: $1s^2\,2s^2\,2p^3$

**7-66** Which of the following elements gives off the most energy when its atoms accept an electron in the gas phase?

(a) Na (b) Mg (c) P (d) Cl (e) Ar

**7-67** Use experimental data to either defend or dispute the following statement: "There is a strong correlation between the ionization energies and electron affinities of the main-group elements. Elements that have relatively large ionization energies also have relatively large electron affinities."

### Consequences of the Relative Sizes of Ionization Energies and Electron Affinities

**7-68** What's wrong with the misconception that sodium reacts with chlorine because "chlorine likes electrons more than sodium does"?

**7-69** Compare the energy required to remove three electrons from a neutral aluminum atom to form the $Al^{3+}$ ion with the energy given off when these electrons are added to neutral bromine atoms to form $Br^-$ ions. Is it legitimate to say that aluminum metal reacts with bromine because bromine atoms like electrons more than aluminum atoms do?

### Metals Versus Nonmetals

**7-70** List the elements in the third row of the periodic table in order of decreasing metallic character. List these elements in order of increasing nonmetallic character. Explain the similarities between these trends. Identify each element as either a metal, a nonmetal, or a semimetal.

**7-71** List the elements in Group 15 (VA) of the periodic table in terms of increasing metallic character. Identify each element in this group as a metal, a nonmetal, or a semimetal.

**7-72** Which of the following sets of elements is arranged in order of increasing nonmetallic character?

(a) Sr < Al < Ga < N (b) K < Mg < Rb < Si (c) Ge < P < As < N (d) Al < B < N < F

**7-73** In general, which of the following statements about metals are true?

(a) They have large atomic radii. (b) They have large ionization energies. (c) They have large electron affinities. (d) They are found on the left and toward the bottom of the periodic table.

### Lattice Energies and the Strength of the Ionic Bond

**7-74** Define the term *lattice energy*.

**7-75** The lattice energy of NaCl is the energy given off in which of the following reactions?

(a) $2\,Na(s) + Cl_2(s) \longrightarrow 2\,NaCl(s)$ (b) $Na^+(s) + Cl^-(s) \longrightarrow NaCl(s)$ (c) $Na(g) + Cl(g) \longrightarrow NaCl(g)$ (d) $Na^+(g) + Cl^-(g) \longrightarrow NaCl(g)$ (e) $Na^+(g) + Cl^-(g) \longrightarrow NaCl(s)$

**7-76** Which of the following salts has the largest lattice energy?

(a) LiF (b) LiCl (c) LiBr (d) LiI

**7-77** Which of the following salts has the largest lattice energy?

(a) NaCl (b) NaI (c) KI (d) MgO (e) MgS

**7-78** Explain the following trends in lattice energies:

$$MgF_2 > MgCl_2 > MgBr_2 > MgI_2$$
$$BeF_2 > MgF_2 > CaF_2 > SrF_2 > BaF_2$$

**7-79** Use the *CRC Handbook of Chemistry and Physics*[3] to determine the solubility in water of NaF, NaCl, NaBr, and NaI. Describe the relationship between the solubilities of these salts and their lattice energies.

**7-80** Use lattice energies to explain why MgO is much less soluble in water than CaO.

### Why are NaCl and $MgCl_2$ the Products of Reactions between Na or Mg Metal and $Cl_2$?

**7-81** Explain why sodium reacts with chlorine to form NaCl and not $NaCl_2$ or $NaCl_3$.

**7-82** Explain why magnesium reacts with chlorine to form $MgCl_2$ and not MgCl or $MgCl_3$.

**7-83** Describe why lithium reacts with nitrogen to give $Li_3N$ and not a compound with another formula.

**7-84** Describe the most important factors in determining the formula of an ionic compound such as NaCl or $MgCl_2$.

### Integrated Problems

**7-85** Carl Lineberger and co-workers recently measured the electron affinity of both hydrogen atoms and deuterium atoms [*Physical Review,* **43,** 6104 (1991)]. They reported the results of their experiments as follows: $^1H = 6082.99 \pm 0.15\ cm^{-1}$ and $^2H = 6086.2 \pm 0.6\ cm^{-1}$. Assume that these results are the inverse of the wavelength of the radiation required to remove an electron from the negatively charged ions these atoms form. Calculate the energy of this radiation in units of kJ/mol. You can feel confident in your calculation if you get results similar

[3] CRC Press, Boca Raton, Florida.

to the value of the electron affinity of hydrogen reported in Table 7.9.

**7-86** Lineberger and co-workers also measured the electron affinity of the $NH_2$ molecule [*Journal of Chemical Physics,* **91,** 2762 (1989)]. In their experiment, $NH_3$ molecules were excited with microwave radiation to form $NH_2^-$ ions, which were excited with a laser beam with a wavelength of 363.8 nm. The kinetic energy of the electron ejected when $NH_2^-$ ions absorbed this radiation was then measured. The difference between the energy of the laser radiation (3.408 electron volts) and the kinetic energy of the electron that was ejected was assumed to be equal to the electron affinity of the $NH_2$ molecule: 0.771 eV. Calculate the electron affinity of the $NH_2$ molecule in units of kJ/mol. Assume that you can convert from eV/molecule to J/mol by multiplying the electron affinity (in eV) by the charge on a single electron and the number of electrons in a mole. You can feel confident in your results if you get a value comparable to the electron affinity of the hydrogen atom reported in Table 7.9.

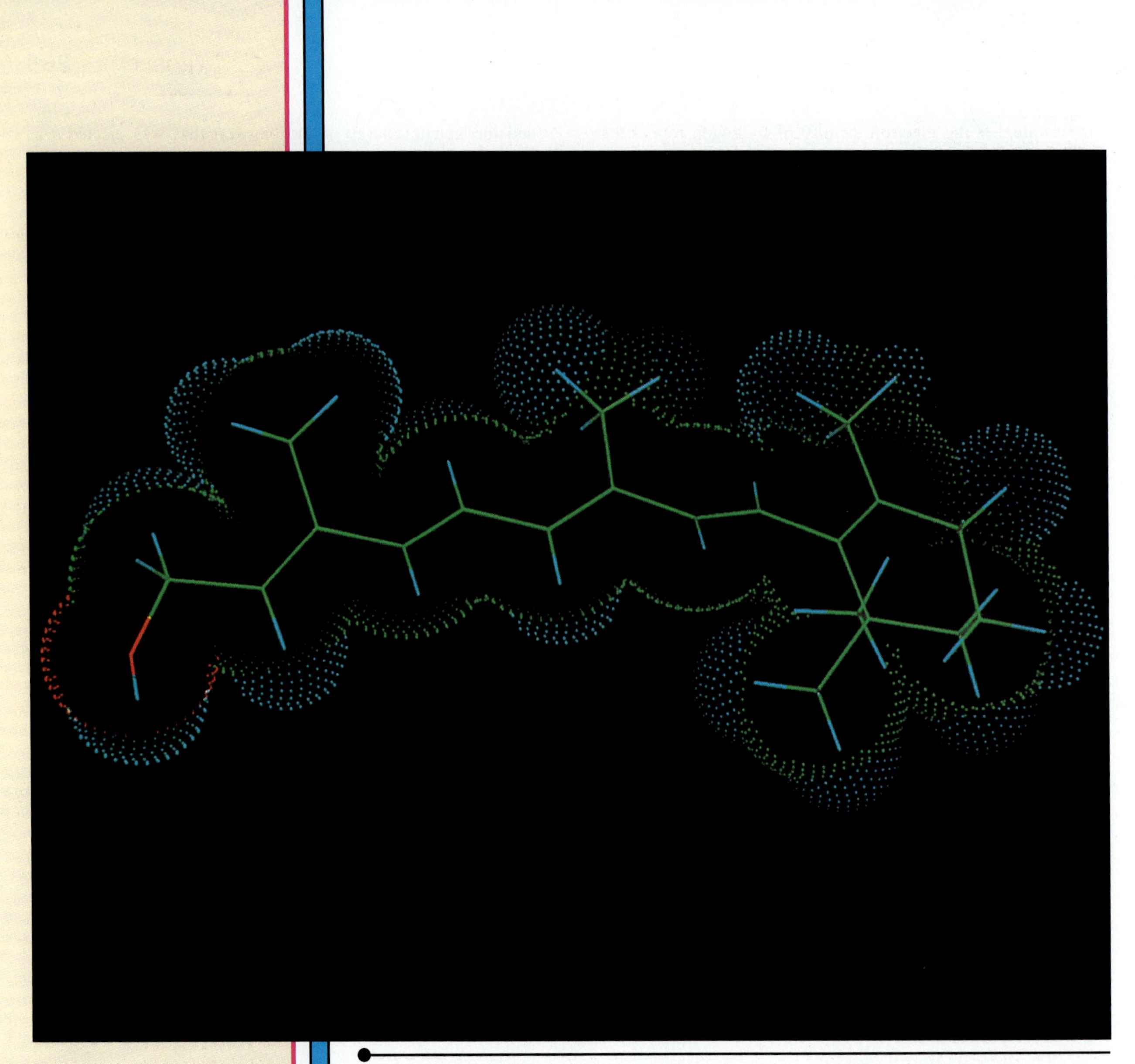

A computer model of the structure of Vitamin A. The net result of all the forces that control the structure around each atom is a molecule that can lie—more or less—in a two-dimensional plane.

# THE COVALENT BOND

Ever since Dalton introduced his atomic theory in 1803, chemists have tried to understand the forces that hold atoms together in chemical compounds. The goal of this chapter is to build a model for bonding that explains the difference between ionic compounds—such as NaCl—which contain positive and negative ions, and covalent compounds—such as $CO_2$—which do not. This chapter provides the basis for answering the following questions:

## POINTS OF INTEREST

- Why is an $H_2$ molecule more stable than a pair of isolated H atoms?
- What holds the atoms together in a symmetric molecule, such as $H_2$ or $O_2$, which can't be held together by the force of attraction between oppositely charged particles?
- Why is NaCl a solid at room temperature, whereas $Cl_2$ is a gas?
- How do we predict the shape of a simple molecule, such as an $H_2O$ molecule, or the geometry around an individual atom in a more complex molecule?
- Why does liquid $O_2$ bridge the gap between the poles of a horseshoe magnet if liquid $N_2$ does not?

## 8.1 THE PROCESS OF DISCOVERY: VALENCE ELECTRONS

| | | | | | | | |
|---|---|---|---|---|---|---|---|
| H | | | | | | H | He |
| Li | Be | B | C | N | O | F | Ne |
| Na | Mg | Al | Si | P | S | Cl | Ar |
| K | Ca | Ga | Ge | As | Se | Br | Kr |
| Rb | Sr | In | Sn | Sb | Te | I | Xe |
| Cs | Ba | Tl | Pb | Bi | Po | At | Rn |
| Fr | Ra | | | | | | |

**Figure 8.1** The model developed by G. N. Lewis was first applied to the main-group elements, which are found on either side of the periodic table.

In 1902, while trying to find a way to explain the periodic table to a beginning chemistry class, G. N. Lewis discovered that the chemistry of the main-group elements shown in Figure 8.1 could be explained with a model in which electrons are arranged in atoms in the form of concentric cubes. This model was based on four assumptions.

1. The number of electrons in the outermost cube on an atom is equal to the number of electrons lost when the atom forms positive ions.
2. Each neutral atom has one more electron in the outermost cube than the atom that precedes it in the periodic table.
3. It takes eight electrons—an **octet**—to complete a cube.
4. Once an atom has an octet of electrons in its outermost cube, this cube becomes part of the core of electrons about which the next cube is built.

Figure 8.2 shows a copy of the notes Lewis used to summarize the key points in this model.

Lewis explained the formulas of simple ionic compounds by assuming that atoms gain electrons if the outermost cube is more than one-half full and lose electrons if the cube is less than one-half full, until the cube is either full or empty. Sodium, for example, loses the only electron in its outermost cube at the same time that chlorine gains the electron it needs to fill its outermost cube.

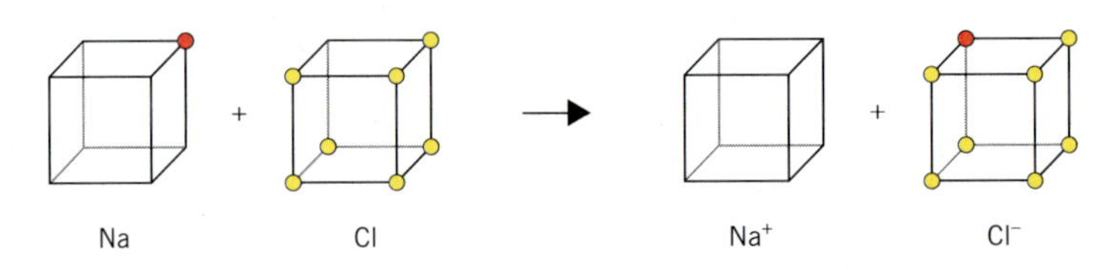


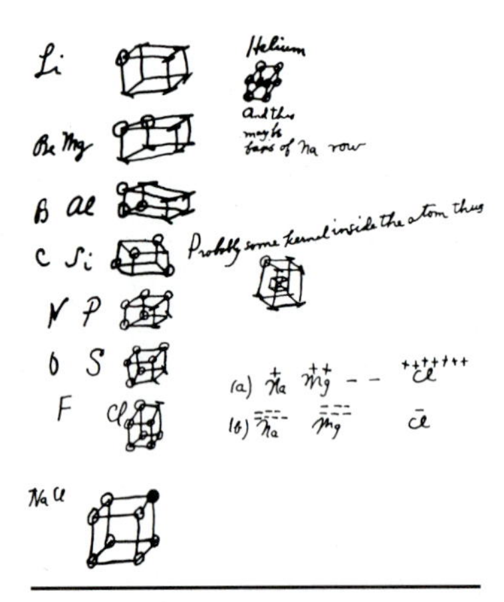


**Figure 8.2** This copy of notes Lewis made while working on his octet theory appeared in his book, *Valence and the Structure of Atoms and Molecules,* published in 1923.

Lewis was the first to describe the compounds formed by the main-group elements in terms of the number of electrons in the outermost shell on an atom. The magnitude of this achievement can be appreciated by noting that this model was generated five years after Thomson's discovery of the electron and nine years *before* Rutherford proposed that the atom consisted of an infinitesimally small nucleus surrounded by a sea of electrons.

As our understanding of the structure of the atom developed, it became apparent why the magic number of electrons for main-group elements was eight. The outermost atomic orbitals for these elements are the $s$ and $p$ orbitals in a given shell, and it takes eight electrons to fill a set of these orbitals. A neutral nitrogen atom, for example, has five electrons in its outermost shell. It therefore must gain three electrons to fill this shell:

$$\text{N:} \quad [\text{He}]\ 2s^2\ 2p^3$$
$$\text{N}^{3-}\text{:} \quad [\text{He}]\ 2s^2\ 2p^6$$

The electrons in the outermost shell eventually became known as the **valence electrons.** This name can be traced to the fact that the number of bonds an element can form is called its *valence* (from the Latin *valens,* "to be strong"). Because the number of electrons in the outermost cube in the Lewis theory controls the number of bonds the atom can form, these outermost electrons are the valence electrons.

The valence electrons are the electrons on an atom that can be gained or lost in a chemical reaction. Since filled $d$ or $f$ subshells are seldom disturbed in a chemical reaction, we can define valence electrons as follows:

**The electrons on an atom that are not present in the previous atom with a filled-shell electron configuration, ignoring filled *d* or *f* subshells.**

Gallium, for example, has the following electron configuration:

$$\text{Ga:} \quad [\text{Ar}]\ 4s^2\ 3d^{10}\ 4p^1$$

The 4*s* and 4*p* electrons can be lost in a chemical reaction, but not the electrons in the filled 3*d* subshell. Gallium therefore has three valence electrons.

**EXERCISE 8.1**

Determine the number of valence electrons in neutral atoms of the following elements:

(a) Si (b) Mn (c) Sb (d) Pb

**SOLUTION** We start by writing the electron configuration for each element:

Si: $[\text{Ne}]\ 3s^2\ 3p^2$
Mn: $[\text{Ar}]\ 4s^2\ 3d^5$
Sb: $[\text{Kr}]\ 5s^2\ 4d^{10}\ 5p^3$
Pb: $[\text{Xe}]\ 6s^2\ 4f^{14}\ 5d^{10}\ 6p^2$

Ignoring filled *d* and *f* subshells, we conclude that neutral atoms of these elements contain the following numbers of valence electrons:

(a) Si = 4 (b) Mn = 7 (c) Sb = 5 (d) Pb = 4

## 8.2 THE PROCESS OF DISCOVERY: THE COVALENT BOND

By 1916 Lewis realized that atoms could also achieve an octet of valence electrons by sharing electrons. Two fluorine atoms, for example, can form a stable $F_2$ molecule in which each atom has an octet of valence electrons by sharing a pair of electrons:

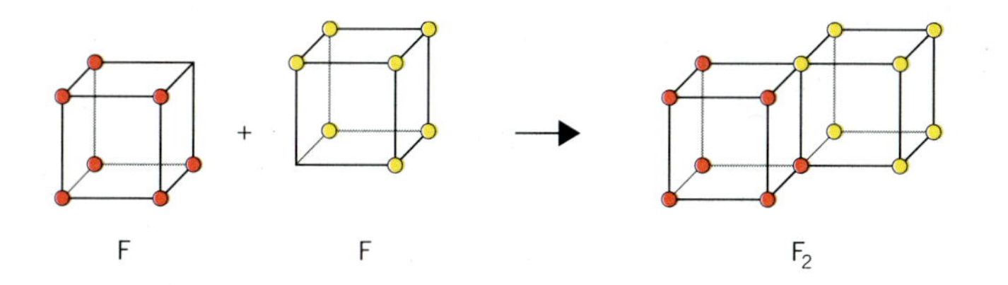


A pair of oxygen atoms can form an $O_2$ molecule in which each atom has a total of eight valence electrons by sharing two pairs of electrons:

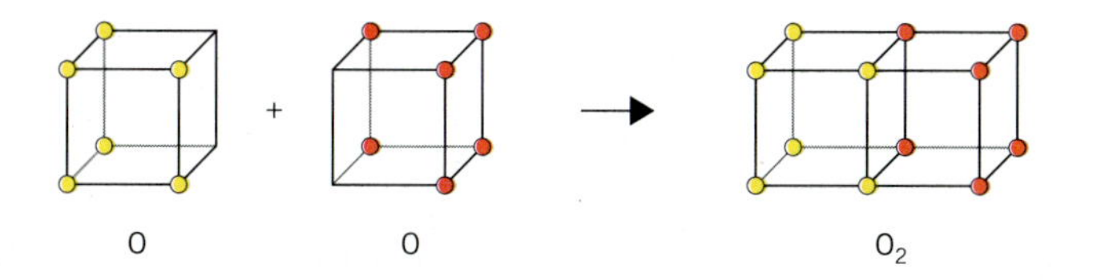


Whenever he applied this model to covalent compounds, Lewis noted that the atoms seem to share pairs of electrons. He also noted that most compounds contain an even number of electrons, which suggests that they exist in pairs. He therefore

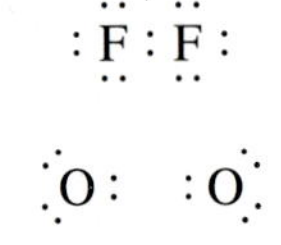

Figure 8.3 The Lewis structures of $F_2$ and $O_2$.

replaced his cubic model of the atom, in which eight electrons are oriented toward the corners of a cube, with a model based on pairs of electrons. In this notation, each atom is surrounded by up to four pairs of dots corresponding to the eight possible valence electrons. The **Lewis structures** of $F_2$ and $O_2$ were therefore written as shown in Figure 8.3. This symbolism is still in use today. The only significant change is the use of lines to indicate covalent bonds formed by the sharing of a pair of electrons.

The prefix *co-* is used to indicate when things are joined or equal (for example, *coalesce, coexist, cooperate,* and *coordinate*). It is therefore appropriate that the term **covalent bond** is used to describe the bonds in compounds that result from the sharing of one or more pairs of electrons.

## 8.3 HOW DOES SHARING OF ELECTRONS BOND ATOMS?

To understand how atoms can be held together by sharing a pair of electrons, let's look at the simplest covalent bond—the bond that forms when two isolated hydrogen atoms come together to form an $H_2$ molecule:

$$\mathrm{H\cdot + \cdot H \longrightarrow H{-}H}$$

An isolated hydrogen atom contains one proton and one electron held together by the force of attraction between oppositely charged particles. The magnitude of this force is equal to the product of the charge on the electron ($q_e$) times the charge on the proton ($q_p$) divided by the square of the distance between these particles ($r^2$):

$$F = \frac{q_e \times q_p}{r^2}$$

When a pair of isolated hydrogen atoms are brought together, a new force of attraction appears (Figure 8.4*a*), which acts twice, because of the attraction between the electron on one atom and the proton on the other. But two forces of repulsion are also created (Figure 8.4*b*) because the negatively charged electrons repel each other, as do the positively charged protons.

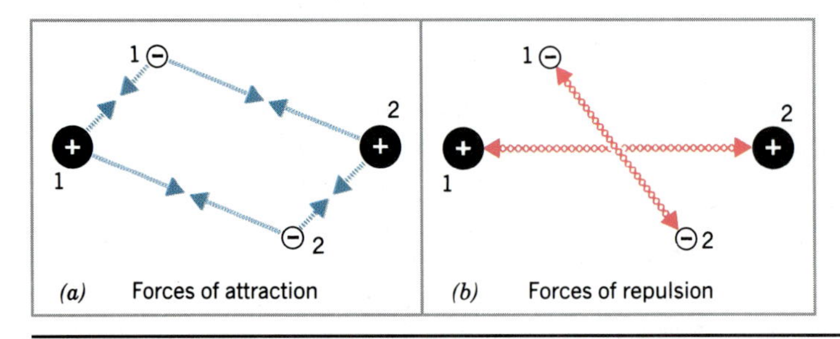


Figure 8.4 (*a*) Two forces of attraction act to bring a pair of hydrogen atoms together—the force of attraction between the electron on each atom and the proton on the other atom. (*b*) Two forces of repulsion drive a pair of hydrogen atoms apart—the repulsion between the two protons and the repulsion between the two electrons.

At first glance, it might seem that the two new repulsive forces would balance the two new attractive forces. If this happened, the $H_2$ molecule would be no more stable than a pair of isolated hydrogen atoms. But there are ways in which the forces of repulsion can be minimized.

As we have seen, electrons behave as if they are tops spinning on an axis. Just as there are two ways in which a top can spin, there are two possible states for the spin of an electron: $s = +\frac{1}{2}$ and $s = -\frac{1}{2}$. When electrons are paired so that they have opposite spins, the force of repulsion between these electrons is minimized.

Sec. 6.19

The force of repulsion between the protons can be minimized if the pair of electrons is placed between the two nuclei. The distance between the electron on one atom and the nucleus of the other is now smaller than the distance between the two nuclei, as shown in Figure 8.5. As a result, the force of attraction between each electron and the nucleus of the other atom is larger than the force of repulsion between the two nuclei, as long as the nuclei are not brought too close together.

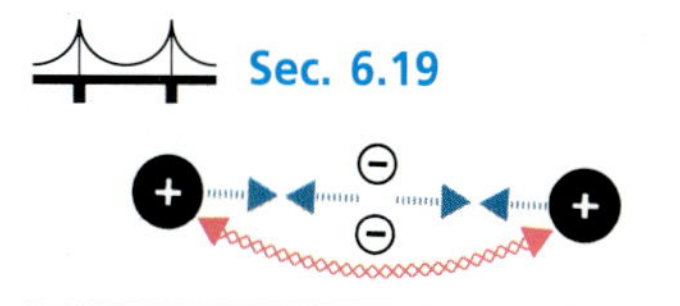

**Figure 8.5** If the electrons are paired and restricted to the region directly between the two nuclei, the attractive forces are larger than the repulsive forces in this molecule. As a result, the molecule is more stable than a pair of isolated atoms.

The net result of pairing the electrons and placing them between the two nuclei is a system that is more stable than a pair of isolated atoms if the nuclei are close enough together to share the pair of electrons, but not so close that repulsion between the nuclei becomes too large. The hydrogen atoms in an $H_2$ molecule are therefore held together by the sharing of a pair of electrons, and this bond is the strongest when the distance between the two nuclei is about 0.074 nm.

## 8.4 SIMILARITIES AND DIFFERENCES BETWEEN IONIC AND COVALENT COMPOUNDS

There is a significant difference between the physical properties of sodium chloride and chlorine, as shown in Table 8.1, which results from the difference between the ionic bonds in NaCl and the covalent bonds in $Cl_2$.

**TABLE 8.1 Some Physical Properties of NaCl and $Cl_2$**

| | NaCl | $Cl_2$ |
|---|---|---|
| Phase at room temperature | Solid | Gas |
| Density | 2.165 g/cm$^3$ | 0.003214 g/cm$^3$ |
| Melting point | 801°C | −100.98°C |
| Boiling point | 1413°C | −34.6°C |
| Ability of aqueous solution to conduct electricity | Conducts | Does not conduct |

Each $Na^+$ ion in NaCl is surrounded by six $Cl^-$ ions, and vice versa, as shown in Figure 8.6*a*. Removing an ion from this compound therefore involves breaking at least six bonds. Some of these bonds would have to be broken to melt NaCl, and they would all have to be broken to boil this compound. As a result, ionic compounds such as NaCl tend to have high melting points and boiling points. Ionic compounds are therefore invariably solids at room temperature.

$Cl_2$ consists of molecules in which one atom is tightly bound to another, as shown in Figure 8.6*b*. The covalent bonds within these molecules are at least as strong as an ionic bond, but we don't have to break these covalent bonds to separate one $Cl_2$ molecule from another. As a result, it is much easier to melt $Cl_2$ to form a liquid or boil it to form a gas, and $Cl_2$ is a gas at room temperature.

The difference between ionic and covalent bonds also explains why aqueous solutions of ionic compounds conduct electricity, while aqueous solutions of cova-

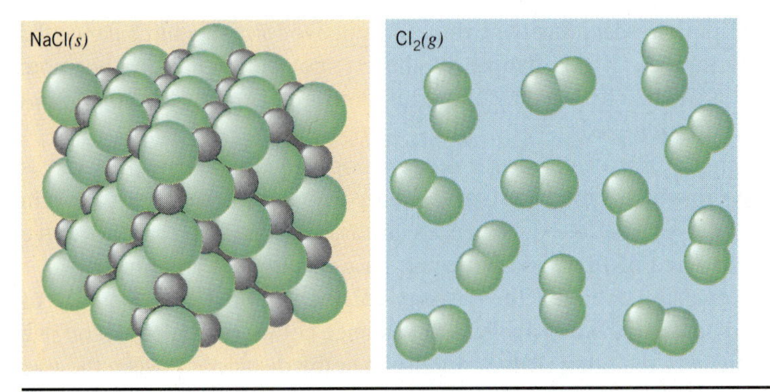


**Figure 8.6** (*a*) Because it takes a great deal of energy to remove the $Na^+$ and $Cl^-$ ions from the crystal, the melting point and boiling point of NaCl are relatively large. (*b*) Although there is a strong force of attraction between the chlorine atoms within a $Cl_2$ molecule, the force of attraction between these molecules is much smaller. As a result, it is relatively easy to separate $Cl_2$ molecules, and this element is a gas at room temperature.

lent compounds often do not. When a salt dissolves in water, the ions are released into solution:

$$NaCl(s) \xrightarrow{H_2O} Na^+(aq) + Cl^-(aq)$$

Aqueous solutions of ionic compounds, such as NaCl, conduct an electric current because these compounds dissociate in water to give positive and negative ions that can carry the current through the solution.

These ions can flow through the solution, producing an electric current that completes the circuit. When a covalent compound dissolves in water, neutral molecules are released into the solution, which cannot carry an electric current:

$$C_{12}H_{22}O_{11}(s) \xrightarrow{H_2O} C_{12}H_{22}O_{11}(aq)$$

The difference between the physical properties of NaCl and $Cl_2$ is so large that it is easy to fall into the trap of believing that the bond between a pair of atoms is either ionic or covalent. Lewis recognized that this was not true. In the paper in which he first described bonds based on the sharing of electrons, Lewis argued that words such as *ionic* and *covalent* referred to the extremes at either end of a continuous spectrum of bonding.

To see how he came to this conclusion, let's compare what happens when covalent and ionic bonds form. When two chlorine atoms come together to form a covalent bond, each atom contributes one electron to form a pair of electrons shared equally by the two atoms, as shown in Figure 8.7. When a sodium atom combines with a chlorine atom to form an ionic bond, each atom still contributes one electron to form a pair of electrons, but this pair of electrons is not shared by the two atoms. The electrons spend most of their time on the chlorine atom.

The Formation of a Covalent Bond

$$:\ddot{\underset{\cdot\cdot}{Cl}}\cdot + \cdot\ddot{\underset{\cdot\cdot}{Cl}}: \longrightarrow :\ddot{\underset{\cdot\cdot}{Cl}}{-}\ddot{\underset{\cdot\cdot}{Cl}}:$$

The Formation of an Ionic Bond

$$Na\cdot + \cdot\ddot{\underset{\cdot\cdot}{Cl}}: \longrightarrow [Na]^+[:\ddot{\underset{\cdot\cdot}{Cl}}:]^-$$

**Figure 8.7** Covalent bonds are formed when electrons are shared more or less equally by the atoms that form the bonds. Ionic bonds occur when there is a net transfer of one or more electrons from one atom to another.

Ionic and covalent bonds differ in the extent to which a pair of electrons is shared by the atoms that form the bond. When one of the atoms is much better at drawing electrons toward itself than the other, the bond is *ionic.* When the atoms are approximately equal in their ability to draw electrons toward themselves, the atoms share the pair of electrons more or less equally, and the bond is *covalent.* The following rule of thumb for predicting whether a compound is ionic or covalent was introduced in Section 2.7:

**Metals often react with nonmetals to form ionic compounds or salts. Nonmetals combine with other nonmetals to form covalent compounds.**

This rule of thumb is useful, but it is also naive, for two reasons.

1. The only way to tell whether a compound is ionic or covalent is to measure the relative ability of the atoms to draw electrons in a bond toward themselves.
2. Any attempt to divide compounds into just two classes (ionic and covalent) is doomed to failure because the bonding in many compounds falls between these two extremes.

The first limitation is the basis of the concept of electronegativity. The second serves as the basis for the concept of polarity.

## 8.5 ELECTRONEGATIVITY AND POLARITY

The relative ability of an atom to draw electrons in a bond toward itself is called the **electronegativity** (*EN*) of the atom. Atoms with large electronegativities attract the electrons in a bond better than those that have small electronegativities. The electronegativities of the main-group elements are given in Figure 8.8. (A complete set of data can be found in Table A-8 in the appendix.) When the magnitude of the electronegativities of the main-group elements is added to the periodic table as a third axis, we get the results shown in Figure 8.9. There are several clear patterns in the data in Figures 8.8 and 8.9.

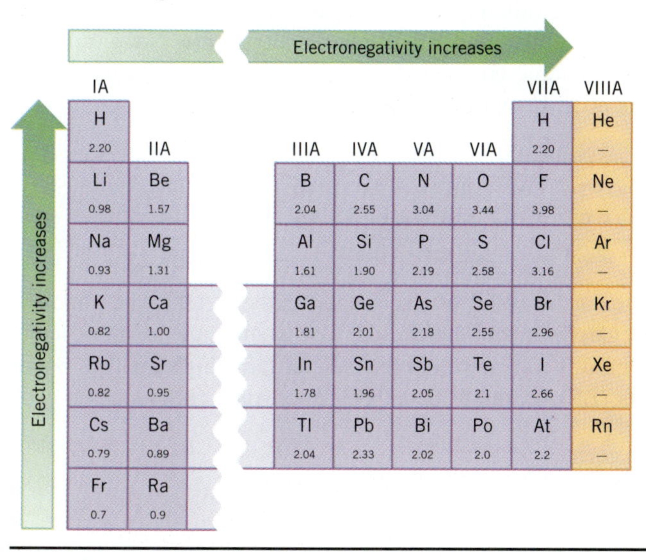


**Figure 8.8** With rare exception, electronegativity increases from left to right across the periodic table and decreases regularly from top to bottom of most columns. Note that elements in the last column of the table have electronegativity of zero.

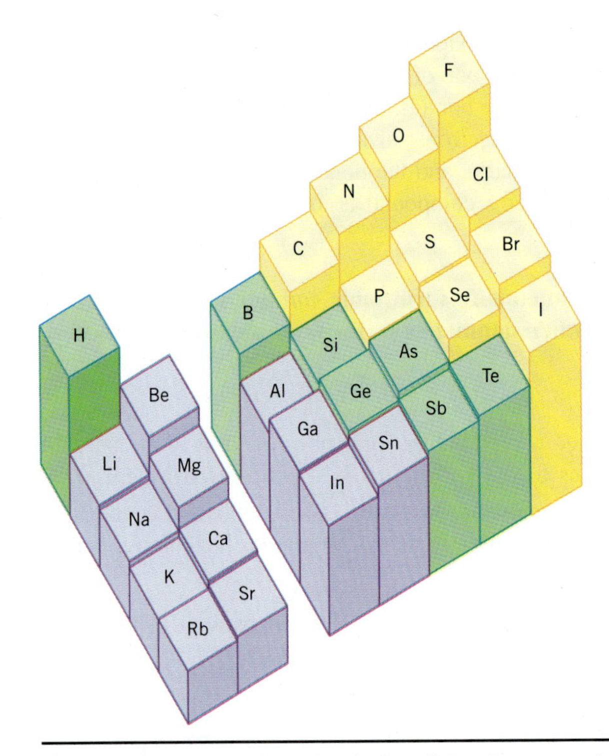


**Figure 8.9** A three-dimensional plot of the electronegativities of the main-group elements versus their position in the periodic table.

1. Electronegativity increases in a regular fashion from left to right across a row of the periodic table.
2. Electronegativity decreases down a column of the periodic table.

When the difference between the electronegativities of the elements in a compound is relatively large, the compound is best classified as *ionic.* NaCl, LiF, and $SrBr_2$ are good examples of ionic compounds. In each case, the electronegativity ($EN$) of the nonmetal is at least two units larger than that of the metal:

| *NaCl* | *LiF* | *$SrBr_2$* |
|---|---|---|
| Cl $EN = 3.16$ | F $EN = 3.98$ | Br $EN = 2.96$ |
| Na $EN = 0.93$ | Li $EN = 0.98$ | Sr $EN = 0.95$ |
| $\Delta EN = 2.23$ | $\Delta EN = 3.00$ | $\Delta EN = 2.01$ |

NaCl: $[Na^+][:\ddot{\underset{..}{Cl}}:^-]$

LiF: $[Li^+][:\ddot{\underset{..}{F}}:^-]$

$SnBr_2$: $[Sn^{2+}][:\ddot{\underset{..}{Br}}:^-]_2$

**Figure 8.10** Lewis structures for NaCl, LiF, and $SrBr_2$.

We can therefore assume a net transfer of electrons from the metal to the nonmetal to form positive and negative ions and write the Lewis structures of these compounds as shown in Figure 8.10. These compounds all have high melting points (MP) and boiling points (BP), as might be expected for ionic compounds:

| | *NaCl* | *LiF* | *$SrBr_2$* |
|---|---|---|---|
| MP | 801°C | 846°C | 657°C |
| BP | 1413°C | 1717°C | 2146°C |

They also dissolve in water to give aqueous solutions of positive and negative ions that conduct electricity, as would be expected.

When the electronegativities of the elements in a compound are about the same, the atoms share electrons, and the substance is *covalent.* Examples of covalent compounds are methane ($CH_4$), nitrogen dioxide ($NO_2$), and sulfur dioxide ($SO_2$):

| $CH_4$ | $NO_2$ | $SO_2$ |
|---|---|---|
| C $EN = 2.55$ | O $EN = 3.44$ | O $EN = 3.44$ |
| H $EN = 2.20$ | N $EN = 3.04$ | S $EN = 2.58$ |
| $\Delta EN = 0.35$ | $\Delta EN = 0.40$ | $\Delta EN = 0.86$ |

These compounds have relatively low melting points and boiling points, as might be expected for covalent compounds, and they are all gases at room temperature:

| | $CH_4$ | $NO_2$ | $SO_2$ |
|---|---|---|---|
| MP | $-182.5$°C | $-163.6$°C | $-75.5$°C |
| BP | $-161.5$° | $-151.8$°C | $-10$°C |

Inevitably, there must be compounds that fall between these extremes. For these compounds, the difference between the electronegativities of the elements is large enough to be significant but not large enough to classify the compound as ionic. Consider water, for example:

| $H_2O$ |
|---|
| O $EN = 3.44$ |
| H $EN = 2.20$ |
| $\Delta EN = 1.24$ |

Water is neither purely ionic nor purely covalent. It does not contain positive and negative ions, as indicated by the Lewis structure on the left in Figure 8.11. But the electrons are not shared equally, as indicated by the Lewis structure on the right in this figure. Water is best described as a **polar compound.** One end, or pole, of the molecule has a partial positive charge ($\delta+$), and the other end has a partial negative charge ($\delta-$).

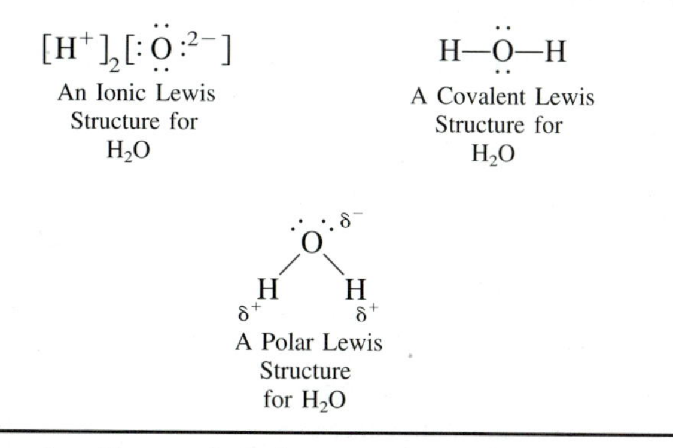


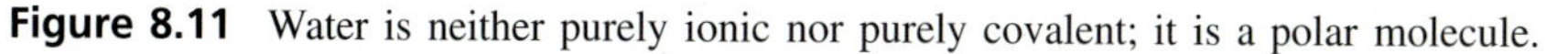

**Figure 8.11** Water is neither purely ionic nor purely covalent; it is a polar molecule.

Ionic and covalent bonds differ in the extent to which electrons are shared by the atoms that form these bonds. When one atom is much more electronegative than the other, it draws the electrons in the bond closer to itself. This produces a separation of charge, which gives rise to positive and negative ions that are held together by an ionic bond. When the atoms in a bond have similar electronegativities, neither atom can draw the electrons away from the other. They therefore share pairs of electrons more or less equally in a covalent bond. When the difference between the electronegativities of the atoms in a bond is neither large enough to form an ionic bond nor small enough to allow the atoms to share electrons equally, the bonds are best described as polar, or polar covalent.

As a rule:

**1.** When the difference between the electronegativities of two elements is less than 1.2, we assume that the bond between atoms of these elements is *covalent.*

**2.** When the difference is larger than 1.8, the bond is assumed to be *ionic.*

**3.** Compounds for which the electronegativity difference is between about 1.2 and 1.8 are best described as *polar,* or *polar covalent:*

| | |
|---|---|
| **Covalent:** | $\Delta EN < 1.2$ |
| **Polar:** | $1.2 \leq \Delta EN \leq 1.8$ |
| **Ionic:** | $\Delta EN > 1.8$ |

### ▶ CHECKPOINT

Use the concept of electronegativity to explain why magnesium burns in the presence of oxygen to form an ionic compound that contains $Mg^{2+}$ and $O^{2-}$ ions, whereas carbon burns to form covalent compounds, such as CO and $CO_2$, in which electrons are shared.

### EXERCISE 8.2

Use electronegativities to decide whether the following compounds are best described as covalent, ionic, or polar:

(a) sodium cyanide (NaCN) (b) tetraphosphorus decasulfide ($P_4S_{10}$)
(c) carbon monoxide (CO) (d) silicon tetrachloride ($SiCl_4$)

**Problem-Solving Strategy**

**SOLUTION**

(a) When there are more than two elements in a compound, it is useful to look at the difference between the least and most electronegative atoms. The difference between sodium and nitrogen is 2.11 electronegativity units. We can therefore assume that this compound is ionic. In this case, it contains the $Na^+$ and $CN^-$ ions.

(b) The difference between the electronegativities of phosphorus and sulfur is only 0.39, and $P_4S_{10}$ is best described as covalent.

(c) The electronegativity difference in CO is slightly larger ($\Delta EN = 0.89$), but this compound is still covalent.

(d) The electronegativity difference in this compound is significant ($\Delta EN = 1.26$) but too small to assume the bonds are ionic. $SiCl_4$ is best described as polar, or polar covalent.

## 8.6 LIMITATIONS OF THE ELECTRONEGATIVITY CONCEPT

Electronegativity is a powerful concept that summarizes the tendency of an element to gain, lose, or share electrons when it combines with another element. But there are limits to the success with which it can be applied. $BF_3$ ($\Delta EN = 1.94$) and $SiF_4$ ($\Delta EN = 2.08$), for example, have electronegativity differences that lead us to expect these compounds to behave as if they were ionic, but both compounds are covalent. They are both gases at room temperature, and their boiling points are −99.9°C and −86°C, respectively.

The source of this difficulty is simple. By convention, each element is assigned an electronegativity value that is used for all of its compounds. But fluorine is less electronegative when it bonds to semimetals (such as B or Si) or nonmetals (such as C) than when it bonds to metals (such as Na or Mg).

This problem surfaces once again when we look at elements that form compounds in more than one oxidation state. $TiCl_2$ and MnO, for example, have many of the properties of ionic compounds. They are both solids at room temperature, and they have very high melting points, as expected for ionic compounds:

| *$TiCl_2$* | *MnO* |
|---|---|
| MP = 1035°C | MP = 1785°C |

$TiCl_4$ and $Mn_2O_7$, on the other hand, are both liquids at room temperature, with melting points below 0°C and relatively low boiling points, as might be expected for covalent compounds:

| *$TiCl_4$* | *$Mn_2O_7$* |
|---|---|
| MP = −24.1°C | MP = −20°C |
| BP = 136.4°C | BP = 25°C |

$TiCl_2$ is an ionic solid, which melts at very high temperatures (*MP* = 1035°C). $TiCl_4$ is a covalent compound, which is a liquid at room temperature (*MP* = −24.1°C). The difference between these compounds can be understood by assuming that the titanium atom becomes more electronegative as the oxidation state of this atom increases.

The principal difference between these compounds is the oxidation state of the metal. As the oxidation state of an atom becomes larger, so does its ability to draw electrons in a bond toward itself. In other words, titanium atoms in a +4 oxidation state and manganese atoms in a +7 oxidation state are more electronegative than titanium and manganese atoms in an oxidation state of +2.

As the oxidation state of the metal becomes larger, the difference between the electronegativities of the metal and the nonmetal with which it combines decreases. The bonds in the compounds these elements form therefore become less ionic (or more covalent).

### CHECKPOINT

It is dangerous to add potassium permanganate to concentrated sulfuric acid. The $MnO_4^-$ ion reacts with strong acid to form $Mn_2O_7$:

$$2\,MnO_4^-(s) + 2\,H^+(aq) \longrightarrow \mathbf{Mn_2O_7(g)} + H_2O(l)$$

which decomposes violently to form $Mn_2O_3$:

$$Mn_2O_7(g) \longrightarrow \mathbf{Mn_2O_3(s)} + 2\,O_2(g)$$

Use the fact that the effective electronegativity of an element increases with oxidation state to explain why $Mn_2O_7$ is a gas at room temperature, whereas $Mn_2O_3$ is a solid.

## 8.7 THE DIFFERENCE BETWEEN POLAR BONDS AND POLAR MOLECULES

The difference between the electronegativities of chlorine ($EN = 3.16$) and hydrogen ($EN = 2.20$) is large enough to assume that the bond in HCl is polar:

$$^{\delta^+}\text{H—Cl}^{\delta^-}$$

Because it contains only this one bond, the HCl molecule is also described as polar.

The polarity of a molecule can be determined by measuring a quantity known as the **dipole moment,** which depends on two factors: (1) the magnitude of the separation of charge and (2) the distance between the negative and positive poles of the molecule. Dipole moments are reported in units of *debye* (*d*). The dipole moment of HCl is small: $\mu = 1.08\ d$. This can be understood by noting that the separation of charge in the HCl bond is relatively small ($\Delta EN = 0.96$) and that the H—Cl bond is relatively short.

C—Cl bonds ($\Delta EN = 0.61$) are not as polar as H—Cl bonds ($\Delta EN = 0.96$), but they are significantly longer. As a result, the dipole moment for $CH_3Cl$ is about the same as HCl: $\mu = 1.10\ d$. At first glance, we might expect a similar dipole moment for carbon tetrachloride ($CCl_4$), which contains four polar C—Cl bonds. The dipole moment of $CCl_4$, however, is 0. This can be understood by considering the structure of $CCl_4$ shown in Figure 8.12. The individual C—Cl bonds in this molecule are polar, but the four C—Cl dipoles cancel each other. Carbon tetrachloride therefore illustrates an important point: Not every molecule that contains polar bonds has a dipole moment.

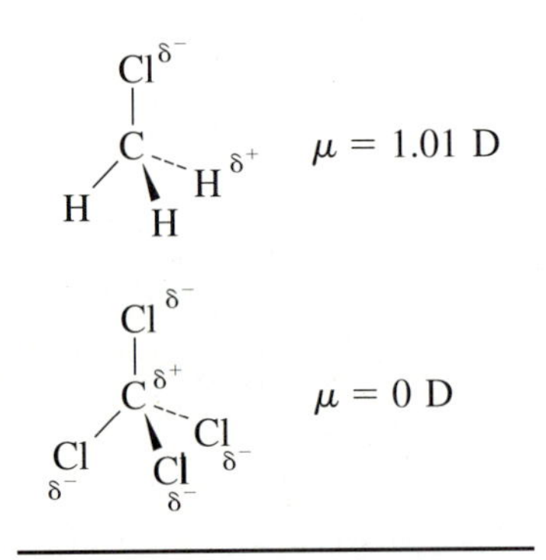


**Figure 8.12** Although the C—Cl bonds in both $CH_3Cl$ and $CCl_4$ are polar, $CCl_4$ has no net dipole moment.

## 8.8 WRITING LEWIS STRUCTURES BY TRIAL AND ERROR

The Lewis structure of a compound can be generated by trial and error. We start by writing symbols that contain the correct number of valence electrons for the atoms in the molecule. We then combine electrons to form covalent bonds until we come up with a Lewis structure in which all of the elements (with the exception of the hydrogen atoms) have an octet of valence electrons.

**▶ CHECKPOINT**

Explain why hydrogen atoms are an exception to the general rule that main-group elements gain, lose, or share electrons until they have eight valence electrons.

Let's apply the trial and error approach to generating the Lewis structure of carbon dioxide, $CO_2$. We start by determining the number of valence electrons on each atom from the electron configurations of the elements. Carbon has four valence electrons, and oxygen has six:

C: [He] $2s^2\ 2p^2$

O: [He] $2s^2\ 2p^4$

We can symbolize this information as shown at the top of Figure 8.13. We now combine one electron from each atom to form covalent bonds between the atoms. When this is done, each oxygen atom has a total of seven valence electrons and the carbon atom has a total of six valence electrons. Because none of these atoms has an octet of valence electrons, we combine another electron on each atom to form two more bonds. The result is a Lewis structure in which each atom has an octet of valence electrons.

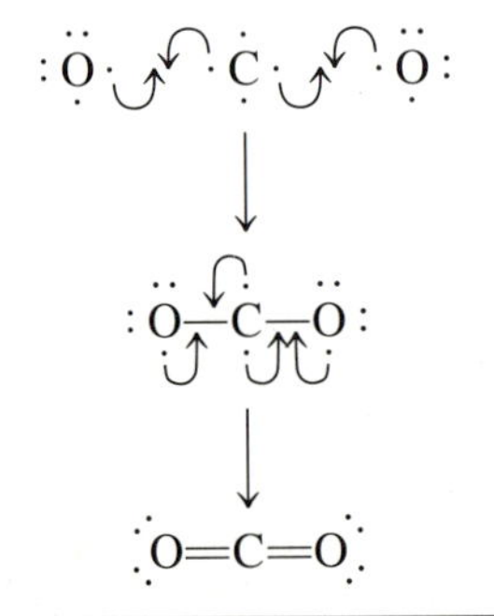

**Figure 8.13** The generation of the Lewis structure of $CO_2$.

## 8.9 DRAWING SKELETON STRUCTURES

The most difficult step in generating the Lewis structure of a molecule is the step in which the skeleton structure of the molecule is written. As a general rule, the less electronegative element is at the center of the molecule. Thus, the formulas of thionyl chloride ($SOCl_2$) and sulfuryl chloride ($SO_2Cl_2$) can be translated into the following skeleton structures:

```
    O            O
    |            |
Cl—S—Cl      Cl—S—Cl
                 |
                 O
```

It is also useful to recognize that the formulas for complex molecules often provide hints at the skeleton structure of the molecule. Dimethyl ether, for example, is often written as $CH_3OCH_3$, which translates into the following skeleton structure:

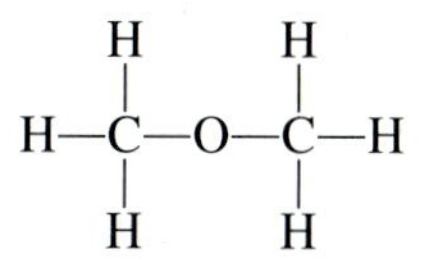

Finally, it is useful to recognize that many compounds that are acids contain O—H bonds. The formula of acetic acid, for example, is often written as $CH_3CO_2H$ because this molecule contains the following skeleton structure:

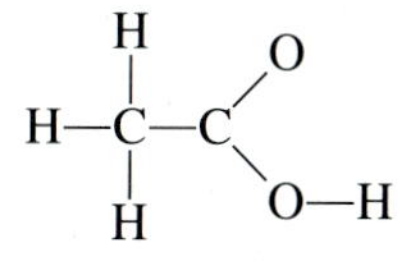

## 8.10 A STEP-BY-STEP APPROACH TO WRITING LEWIS STRUCTURES

The trial-and-error method for writing Lewis structures can be time consuming. For all but the simplest molecules, the following step-by-step process is faster:

**Step 1:** Determine the total number of valence electrons.

**Step 2:** Write the skeleton structure of the molecule.

**Step 3:** Use two valence electrons to form each bond in the skeleton structure.

**Step 4:** Try to satisfy the octets of the atoms by distributing the remaining valence electrons as nonbonding electrons.

The first step in this process involves calculating the number of valence electrons in the molecule or ion. For a neutral molecule this is the sum of the valence electrons on each atom. If the molecule carries an electric charge, we add one electron for each negative charge or subtract an electron for each positive charge.

Consider the chlorate ($ClO_3^-$) ion, for example. A chlorine atom (Group VIIA) has seven valence electrons and each oxygen atom (Group VIA) has six valence electrons. Because the chlorate ion has a charge of $-1$, this ion contains one more electron than a neutral $ClO_3$ molecule. Thus, the $ClO_3^-$ ion has a total of 26 valence electrons:

$$ClO_3^-: \quad 7 + 3(6) + 1 = 26$$

The second step in generating the Lewis structure involves deciding which atoms in the molecule are connected by covalent bonds. As we have seen, the formula of the compound often provides a hint as to the skeleton structure. The formula for the chlorate ion, for example, suggests the following skeleton structure:

```
O—Cl—O
   |
   O
```

The third step assumes that the skeleton structure of the molecule is held together by covalent bonds. The valence electrons are therefore divided into two categories: **bonding electrons** and **nonbonding electrons.** Because it takes two electrons to form a covalent bond, we can calculate the number of nonbonding electrons in the molecule by subtracting two electrons from the total number of valence electrons for each bond in the skeleton structure.

There are three covalent bonds in the most reasonable skeleton structure for the chlorate ion. As a result, six of the 26 valence electrons must be used as bonding electrons. This leaves 20 nonbonding electrons in the valence shell:

$$\begin{array}{rl} & 26 \text{ valence electrons} \\ - & 6 \text{ bonding electrons} \\ \hline & 20 \text{ nonbonding electrons} \end{array}$$

The fourth step in the process by which Lewis structures are generated involves using the nonbonding valence electrons to satisfy the octets of the atoms in the molecule. Each oxygen atom in the $ClO_3^-$ ion already has two electrons—the electrons in the Cl—O covalent bond. Because each oxygen atom needs six nonbonding electrons to satisfy its octet, it takes 18 nonbonding electrons to satisfy the three oxygen atoms. This leaves one pair of nonbonding electrons, which can be used to fill the octet of the central atom.

```
 ..  ..  ..
:O—Cl—O: −
 ¨    |   ¨
     :O:
      ¨
```

## 8.11 MOLECULES THAT DON'T SEEM TO SATISFY THE OCTET RULE

### NOT ENOUGH ELECTRONS

Occasionally we encounter a molecule that doesn't seem to have enough valence electrons. When this happens, we have to remember why atoms share electrons in the first place. If we can't get a satisfactory Lewis structure by sharing a single pair of electrons, it may be possible to achieve this goal by sharing two or even three pairs of electrons.

Consider formaldehyde ($H_2CO$), for example, which contains 12 valence electrons:

$$H_2CO: \quad 2(1) + 4 + 6 = 12$$

The formula of this molecule suggests the following skeleton structure:

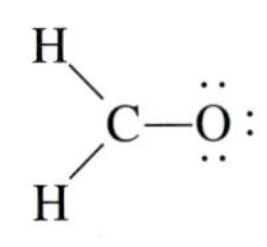

There are three covalent bonds in this skeleton structure, which means that six valence electrons must be used as bonding electrons. This leaves six nonbonding electrons. However, it is impossible to satisfy the octets of the atoms in this molecule with only six nonbonding electrons. When the nonbonding electrons are used to satisfy the octet of the oxygen atom, the carbon atom has a total of only six valence electrons.

We therefore assume that the carbon and oxygen atoms share two pairs of electrons. There are now four bonds in the skeleton structure, which leaves only four nonbonding electrons. This is enough, however, to satisfy the octets of the carbon and oxygen atoms:

H
C=O
H

Every once in a while, we encounter a molecule for which it is impossible to write a satisfactory Lewis structure. Consider boron trifluoride for example, which contains 24 valence electrons:

$$BF_3: \quad 3 + 3(7) = 24$$

There are three covalent bonds in the most reasonable skeleton structure for the molecule. Because it takes six electrons to form the skeleton structure, there are 18 nonbonding valence electrons. But each fluorine atom needs six nonbonding electrons to satisfy its octet. Thus, all of the nonbonding electrons are consumed by the three fluorine atoms. As a result, we run out of electrons while the boron atom has only six valence electrons:

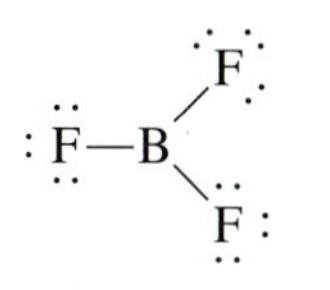

For reasons that we will discuss elsewhere, the elements that form strong double or triple bonds are C, N, O, P, and S. Because neither boron nor fluorine falls in this category, we have to stop with what appears to be an unsatisfactory Lewis structure.

## TOO MANY ELECTRONS

It is also possible to encounter a molecule that seems to have too many valence electrons. When that happens, we expand the valence shell of the central atom. Consider the Lewis structure for sulfur tetrafluoride for example, which contains 34 valence electrons:

$$SF_4: \quad 6 + 4(7) = 34$$

There are four covalent bonds in the skeleton structure for $SF_4$. If eight of the 34 valence electrons are used to form the covalent bonds that hold the molecule together, there are 26 nonbonding valence electrons.

Each fluorine atom needs six nonbonding electrons to satisfy its octet. Because there are four of these atoms, we need 24 nonbonding electrons for this purpose. But there are 26 nonbonding electrons in this molecule. We have already satisfied the octets for all five atoms, and we still have one more pair of valence electrons. We therefore expand the valence shell of the sulfur atom to hold more than eight electrons:

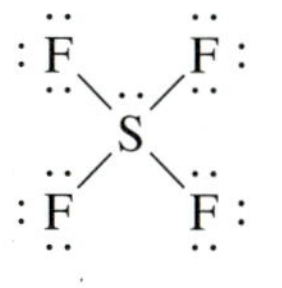

This raises an interesting question: How does the sulfur atom in $SF_4$ hold 10 electrons in its valence shell? The electron configuration for a neutral sulfur atom seems to suggest that only eight electrons will fit in the valence shell of this atom because it takes eight electrons to fill the $3s$ and $3p$ orbitals. But let's look, once again, at the section rules for atomic orbitals. According to these rules, the $n = 3$ shell of orbitals contains $3s$, $3p$, and $3d$ orbitals. Because the $3d$ orbitals on a neutral sulfur atom are all empty, one of these orbitals can be used to hold the extra pair of electrons on the sulfur atom in $SF_4$:

Sec. 6.18

$$\text{S:} \qquad [\text{Ne}]\ 3s^2\, 3p^4\, 3d^0$$

**CHECKPOINT**

Elements in the first and second rows of the periodic table do not have valence shell $d$ orbitals. Use the fact that nitrogen and oxygen do not contain $2d$ orbitals to explain why these elements cannot expand their valence shells.

**EXERCISE 8.3**

Write the Lewis structure for xenon tetrafluoride ($XeF_4$).

**SOLUTION** Xenon (Group VIIIA) has eight valence electrons and fluorine (Group VIIA) has seven. Thus, there are 36 valence electrons in this molecule:

$$XeF_4: \qquad 8 + 4(7) = 36$$

The most reasonable skeleton structure for the molecule contains four covalent bonds. If eight electrons are used to form this skeleton structure, there are 28 nonbonding valence electrons. If each fluorine atom needs six nonbonding electrons, a total of 24 nonbonding electrons are used to complete the octets of these atoms. This leaves four extra nonbonding electrons. Because the octet of each atom appears to be satisfied and we have electrons left over, we expand the valence shell of the central atom until it contains a total of 12 electrons:

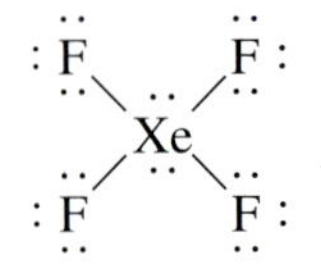


## 8.12 RESONANCE HYBRIDS

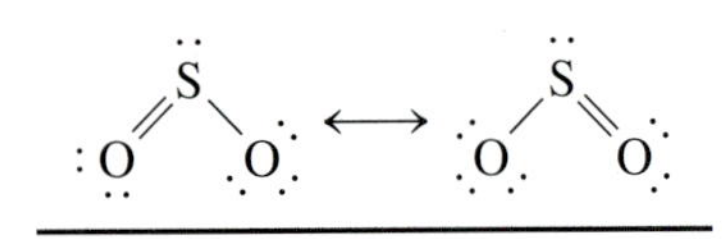


**Figure 8.14** Two equally satisfactory Lewis structures can be written for $SO_2$.

Two Lewis structures can be written for sulfur dioxide, as shown in Figure 8.14. The only difference between these Lewis structures is the identity of the oxygen atom to which the double bond is formed. As a result, they must be equally satisfactory representations of the molecule. This raises an important question: Which of these Lewis structures is correct?

Interesting enough, neither of these structures is correct. The two Lewis structures suggest that one of the sulfur–oxygen bonds is stronger than the other. But there is no difference between the length of the two bonds in $SO_2$, which suggests that the two sulfur–oxygen bonds are equally strong.

When we can write more than one satisfactory Lewis structure, the molecule is an average, or **resonance hybrid,** of these structures. The meaning of the term *reso-*

*nance* can be best understood by an analogy. In music, the notes in a chord are often said to resonate—they mix to give something that is more than the sum of its parts. In a similar sense, the two Lewis structures for the $SO_2$ molecule mix to give a hybrid that is more than the sum of its components. The fact that $SO_2$ is a resonance hybrid of two Lewis structures is indicated by writing a double-headed arrow between these Lewis structures, as shown in Figure 8.14.

The relationship between the $SO_2$ molecule and its Lewis structures can be illustrated by an analogy. Suppose a knight of the Round Table returns to Camelot to describe a wondrous beast encountered during his search for the Holy Grail. He suggests that the animal looks something like a unicorn because it has a large horn in the center of its forehead. But it also looks something like a dragon because it is huge, ugly, and thick-skinned.

The beast was, in fact, a rhinoceros. The animal was real, but it was described as a hybrid of two mythical animals. The $SO_2$ molecule is also real, but we have to use two mythical Lewis structures to describe its properties.

## EXERCISE 8.4

Acetic acid dissociates to some extent in water to give the acetate ion, $CH_3CO_2^-$. Write two alternative Lewis structures for this ion.

**SOLUTION** There are two carbon atoms, three hydrogen atoms, and two oxygen atoms in an acetate ion. Because it has a negative charge, this ion contains a total of 24 valence electrons:

$$CH_3CO_2^-: \qquad 2(4) + 3(1) + 2(6) + 1 = 24$$

The skeleton structure for this molecule contains six covalent bonds, which leaves 12 nonbonding electrons. Unfortunately, it takes all 12 nonbonding electrons to satisfy the octets of the oxygen atoms, which leaves no nonbonding electrons for the carbon atom:

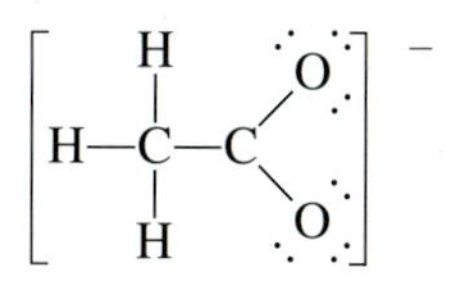

Because there aren't enough electrons to satisfy the octets of the atoms, we assume that there is at least one C=O double bond. There are now seven covalent bonds in the skeleton structure, which leaves only 10 nonbonding electrons. Fortunately, this is enough:

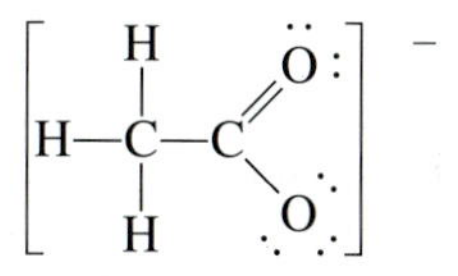

This process gives us one satisfactory Lewis structure for the acetate ion. We can get another by changing the location of the C=O double bond. As a result, the acetate ion is a resonance hybrid of the following Lewis structures:

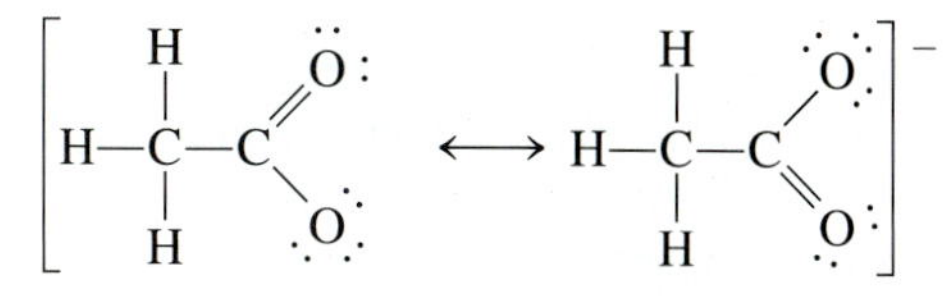

## 8.13 FORMAL CHARGE

It is sometimes useful to calculate the **formal charge** on each atom in a Lewis structure. The first step in this calculation involves dividing the electrons in each covalent bond between the atoms that form the bond. The number of valence electrons formally assigned to each atom is then compared with the number of valence electrons on a neutral atom of the element. If the atom has more valence electrons than a neutral atom, it is assumed to carry a formal negative charge. If it has fewer valence electrons, it is assigned a formal positive charge.

### EXERCISE 8.5

The formula of the amino acid known as glycine is often written $H_3N^+CH_2CO_2^-$. Use the concept of formal charge to explain the meaning of the positive and negative signs in the following Lewis structure:

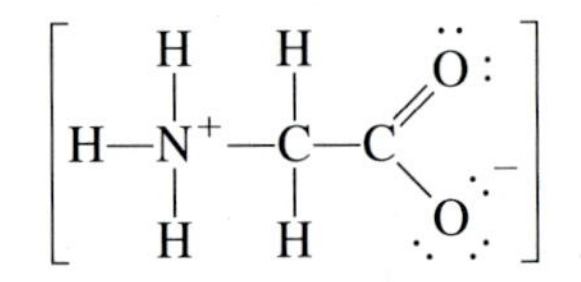

**SOLUTION** We start by arbitrarily dividing pairs of bonding electrons so that each atom in a bond is formally assigned one of these electrons, as shown in Figure 8.15. Once this is done, the nitrogen has four valence electrons, one fewer than a neutral nitrogen atom. The nitrogen therefore carries a formal charge of +1 in this Lewis structure. Both of the carbon atoms formally have four electrons, which is equal to the number on a neutral carbon atom. As a result, neither carbon atom carries a formal charge. The oxygen atom in the C=O double bond formally carries six electrons, which means it has no formal charge. The other oxygen atom, however, carries seven electrons, which means that it formally has a charge of −1.

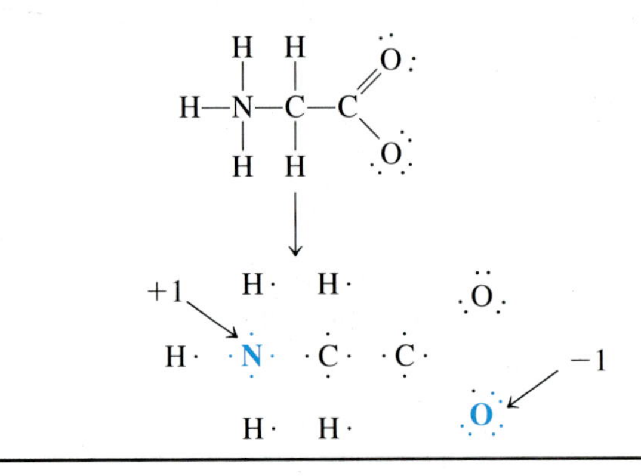


**Figure 8.15** The first step in calculating the formal charge on the nitrogen and oxygen atoms in glycine.

Although the glycine molecule carries no *net* charge, one end of this molecule carries a positive charge, and the other end carries a negative charge.

## 8.14 THE SHAPES OF MOLECULES

The shape of a molecule can play an important role in its chemistry. The changes in the three-dimensional structure of proteins that occur when an egg is heated, for example, are the primary source of the differences between a raw egg and a cooked one. The best illustration of how sensitive proteins are to changes in their three-dimensional structure is provided by the chemistry of hemoglobin, a protein with a molecular weight of 65,000 amu that carries oxygen through the body. This protein contains four chain of amino acids, two $\alpha$ chains and two $\beta$ chains.

The structure of one of the $\beta$ chains in hemoglobin is shown in Figure 8.16. Sickle cell anemia occurs when the identity of a single amino acid among the 146 amino acids on this chain is changed. This substitution of valine for glutamic acid at the sixth position on this chain produces a subtle change in the structure of the hemoglobin, which interferes with its ability to pick up oxygen at low pressures. The result is so severe that children who inherit this disorder from both parents seldom live past the age of two years.

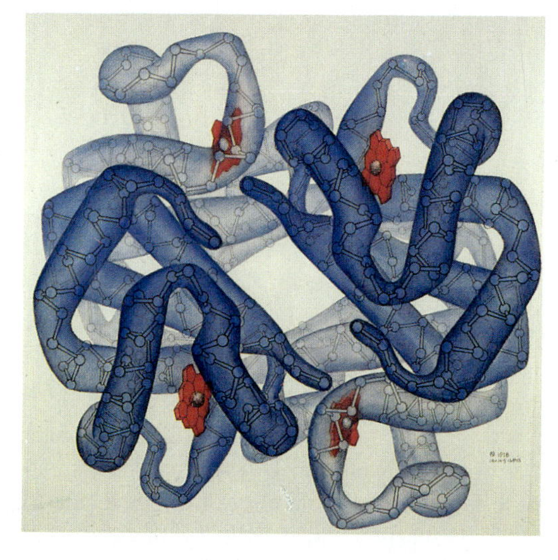

**Figure 8.16** A subtle change in the identity of one of the 146 amino acids on one of the chains in hemoglobin causes a large enough change in the structure of hemoglobin to interfere with its ability to carry oxygen through the blood.

**RESEARCH IN THE 90s**

## THE SHAPES OF MOLECULES

Research in recent years has greatly expanded our understanding of how the shape of a molecule affects the way it binds to certain receptors on the membrane of a cell. It has long been known, for example, that the cholera toxin binds to a receptor, called GM1, on the cells that line the intestine. Once this happens, this toxin can slip through the cell membrane. This initiates a series of reactions that induce these cells to secrete water—as much as 9 gallons a day. Cholera victims die of the debilitating effects of diarrhea.

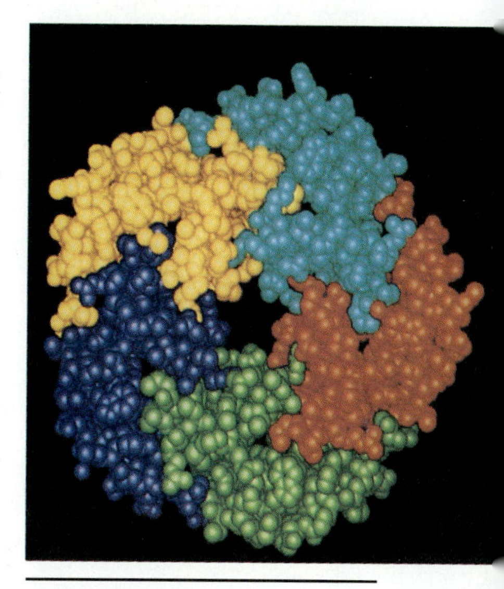

**Figure 8.17** The B subunit of the cholera toxin.

Until recently, the mechanism by which the cholera toxin achieved these results was unknown. In 1991 Rongguang Zhang and Edwin Westbrook at Argonne National Laboratory reported the structure of the cholera toxin [*Science,* **253,** 382–383 (1991)]. This protein contains two subunits. The B subunit consists of five parts—each of which has a molecular weight of about 11,000 amu—that form the doughnut structure shown in Figure 8.17. The A subunit is smaller, with a molecular weight of 29,000 amu. The B subunit apparently binds to the cell and then pushes the smaller A subunit through the membrane of the cell. The A subunit then acts as an enzyme, initiating the reactions that lead to the secretion of water.

In the short term, knowledge of the structure of this protein will drive research toward the discovery of vaccines that can prevent cholera. In the long term, it might provide the basis for the design of a system to deliver drugs across the cell membrane to destroy cancer cells.

Another example of how the structure of a molecule can affect the way it binds to a receptor site is provided by progesterone and the synthetic analog of this steroid shown in Figure 8.18. Progesterone is a hormone that is secreted in the second half of the menstrual cycle. It binds to receptors in the endometrium, preparing the uterus for implantation of a fertilized egg. If pregnancy occurs, continued production of progesterone is essential for maintenance of the embryo and eventually the fetus.

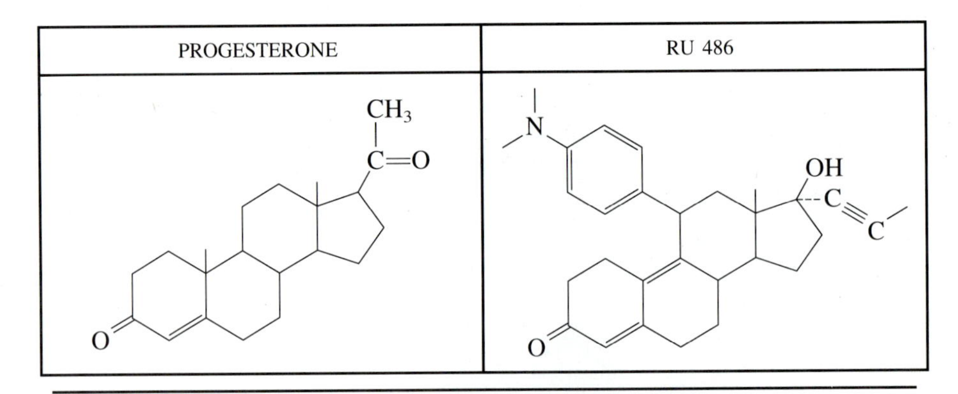


**Figure 8.18** The structures of progesterone and a synthetic analog that is the basis of the drug known as RU-486.

In 1975 Georges Teutsch initiated a project at Roussel-Uclaf to study how small changes in the structure of progesterone might affect the ability of the molecule to bind to the five different kinds of steroid receptors in a cell. Synthetic steroid hormones are divided into two categories. The *agonists* produce the same effect as the natural hormone when they bind. The *antagonists* bind to the receptor but don't switch on the activity that occurs when the natural hormone binds.

Teutsch and co-workers were searching for an antagonist for glucocorticoid receptors, which might increase the rate at which burns and other wounds heal. When they tested a compound originally known as RU-38486, they found that it was a powerful glucocorticoid anatagonist. But it did more than that; it also bound very tightly to the progesterone receptor. As a progesterone antagonist, RU-486, as it soon became known, became a candidate for testing a drug for fertility control.

In September 1988, RU-486 became available in France for the termination of pregnancy. The protocol for its use involves delivering 600 mg of this drug in a single dose followed 36 to 48 hours later with a dose of a prostaglandin to induce the contractions necessary to expel the embryo from the uterus. When this protocol is followed, the rate of its success is similar to that of surgical procedures [*New England Journal of Medicine,* **322**(10), 645–648 (1990)].

Because it binds so tightly to several kinds of hormone receptors, RU-486 has the potential to serve many functions. It could be administered with oxytocin, which is currently used to induce labor at the end of pregnancy, therefore reducing the number of cesarean deliveries that have to be done. It might be used to treat cancers that involve progesterone receptors, such as certain forms of breast cancer. It might also control the growth of certain noncancerous tumors that affect the brain. It might even be used for its original purpose—as a glucocorticoid antagonist. Research is therefore sure to continue on this controversial drug and its potential uses.

## 8.15 PREDICTING THE SHAPES OF MOLECULES

There is no direct relationship between the formula of a compound and the shape of its molecules, as shown by the examples in Figure 8.19. The shapes of these molecules can be predicted from their Lewis structures, however, with a model developed about 30 years ago, known as the **valence-shell electron-pair repulsion (VSEPR) theory.**

| LINEAR | BENT OR ANGULAR |
|---|---|
| F—Be—F | Sn, F, F |

| TRIGONAL PLANAR | TRIGONAL PYRAMIDAL |
|---|---|
| F, Ga—F, F | As, F, F, F |

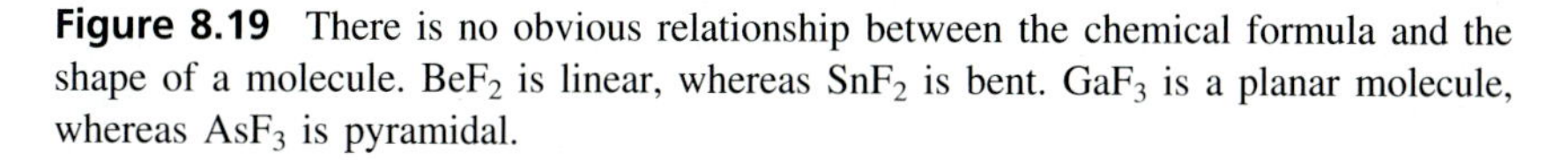

**Figure 8.19** There is no obvious relationship between the chemical formula and the shape of a molecule. $BeF_2$ is linear, whereas $SnF_2$ is bent. $GaF_3$ is a planar molecule, whereas $AsF_3$ is pyramidal.

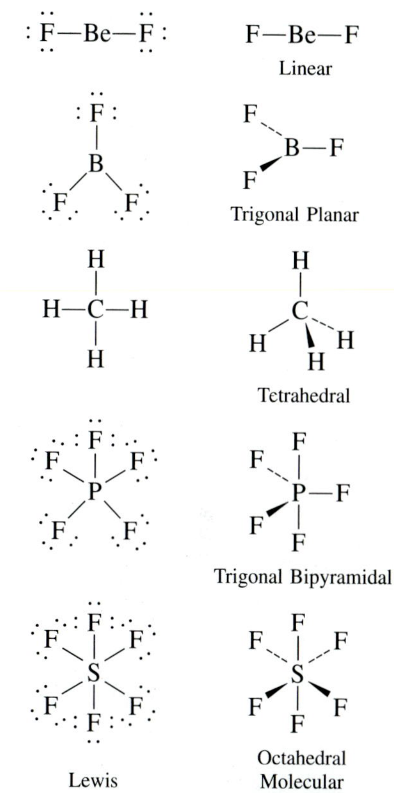


**Figure 8.20** The structures of $BeF_2$, $BF_3$, $CH_4$, $PF_5$, and $SF_6$.

The VSEPR theory predicts the geometry around one atom at a time. This theory assumes that each atom in a molecule will achieve a geometry that minimizes the repulsion between electrons in the valence shell of that atom. The five compounds shown in Figure 8.20 can be used to demonstrate how the VSEPR theory can be applied to simple molecules.

There are only two places in the valence shell of the central atom in $BeF_2$ where electrons can be found. Repulsion between these pairs of electrons can be minimized by arranging them so that they point in opposite directions. Thus, the VSEPR theory predicts that $BeF_2$ should be a **linear** molecule, with a 180° angle between the two Be—F bonds.

There are three places on the central atom in boron trifluoride ($BF_3$) where valence electrons can be found. Repulsion between these electrons can be minimized by arranging them toward the corners of an equilateral triangle. The VSEPR theory therefore predicts a **trigonal planar** geometry for the $BF_3$ molecule, with a F—B—F bond angle of 120°.

$BeF_2$ molecules in the gas phase have a linear structure in which the bond angle is 180°.

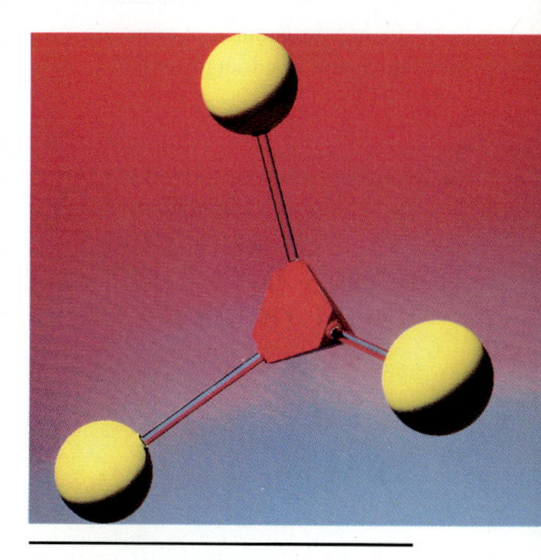

$BF_3$ has a trigonal planar structure in which the bond angles are 120°.

$CH_4$ (*left*) has a tetrahedral structure in which the bond angle is 109°28′, or 109.5°. $PF_5$ (*right*) has a trigonal bipyramidal structure. The five flourine atoms are divided into two groups: those in the equatorial plane of the molecule (*E*) and those along the axis perpendicular to this plane (*A*). The bond angles in this molecule are 90°, 120°, and 180°.

$SF_6$ has an octahedral structure in which the bond angle is 90°.

$BeF_2$ and $BF_3$ are both two-dimensional molecules, in which the atoms lie in the same plane. If we place the same restriction on methane ($CH_4$), we would get a square-planar geometry in which the H—C—H bond angle is 90°. If we let this system expand into three dimensions, however, we end up with a **tetrahedral** molecule in which the H—C—H bond angle is 109°28′.

Repulsion between the five pairs of valence electrons on the phosphorus atom in $PF_5$ can be minimized by distributing these electrons toward the corners of a **trigonal bipyramid.** Three of the positions in a trigonal bipyramid are labeled *equatorial* because they lie along the equator of the molecule. The other two are *axial* because they lie along an axis perpendicular to the equatorial plane. The angle between the three equatorial positions is 120°, while the angle between an axial and an equatorial position is 90°.

There are six places on the central atom in $SF_6$ where valence electrons can be found. The repulsion between these electrons can be minimized by distributing them toward the corners of an **octahedron.** The term *octahedron* literally means "eight sides," but it is the six corners, or vertices, that interest us. To imagine the geometry of an $SF_6$ molecule, locate fluorine atoms on opposite sides of the sulfur atom along the *x*, *y*, and *z* axes of an *xyz* coordinate system.

## 8.16 INCORPORATING DOUBLE AND TRIPLE BONDS INTO THE VSEPR THEORY

Compounds that contain double and triple bonds raise an important point: The geometry around an atom is determined by the number of places in the valence shell of an atom where electrons can be found, not the number of pairs of valence electrons. Consider the Lewis structures of carbon dioxide ($CO_2$) and the carbonate ($CO_3^{2-}$) ion, for example.

$CO_2$: :Ö=C=Ö:

$CO_3^{2-}$: $\left[\begin{array}{c}\ddot{\text{O}}\\ \Vert \\ \text{C} \\ /\ \ \backslash \\ :\ddot{\text{O}}:\ \ :\ddot{\text{O}}:\end{array}\right]^{2-}$

There are four pairs of bonding electrons on the carbon atom in $CO_2$, but only two places where these electrons can be found. (There are electrons in the C=O double bond on the left and electrons in the double bond on the right.) The force of repulsion between these electrons is minimized when the two C=O double bonds are placed on opposite sides of the carbon atom. The VSEPR theory therefore predicts that $CO_2$ will be a linear molecule, just like $BeF_2$, with a bond angle of 180°.

The Lewis structure of the carbonate ion also suggests a total of four pairs of valence electrons on the central atom. But these electrons are concentrated in three places: the two C—O single bonds and the C=O double bond. Repulsions between these electrons are minimized when the three oxygen atoms are arranged toward the corners of an equilateral triangle. The $CO_3^{2-}$ ion therefore should have a trigonal-planar geometry, just like $BF_3$, with a 120° bond angle.

## 8.17 THE ROLE OF NONBONDING ELECTRONS IN THE VSEPR THEORY

The VSEPR theory predicts that the valence electrons on the central atoms in ammonia and water will point toward the corners of a tetrahedron. Because we can't locate the nonbonding electrons with any precision, this prediction can't be tested directly. But the results of the VSEPR theory can be used to predict the positions of the nuclei in these molecules, which can be tested experimentally. If we focus on the positions of the nuclei in ammonia, we predict that the $NH_3$ molecule should have a shape best described as *trigonal pyramidal,* with the nitrogen at the top of the pyramid. Water, on the other hand, should have a shape that can be described as *bent,* or *angular.* Both of these predictions have been shown to be correct, which reinforces our faith in the VSEPR theory.

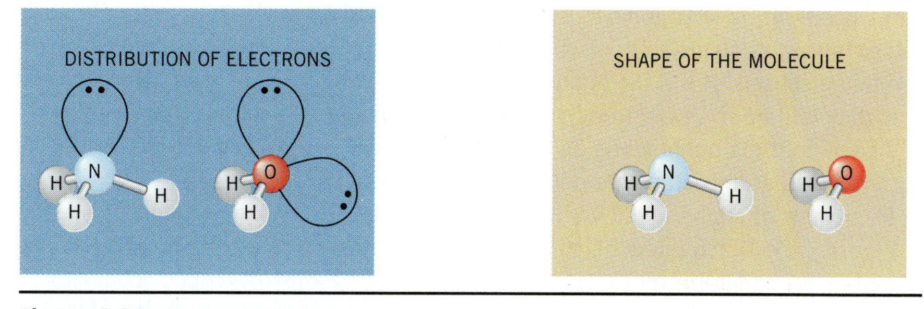


**Figure 8.21** The VSEPR theory predicts that the valence electrons on the central atom in $NH_3$ and $H_2O$ will be oriented toward the corners of a tetrahedron. The shape of these molecules, however, is determined by the positions of their nuclei. Ammonia is therefore described as trigonal pyramidal, and water is described as bent, or angular.

### EXERCISE 8.6

Use the Lewis structure of the $ICl_2^+$ ion shown in Figure 8.22 to predict the shape of this ion.

**SOLUTION** The VSEPR theory predicts that the four pairs of electrons in the valence shell of the iodine atom should be distributed toward the corners of a tetrahedron. When we focus on the positions of the nuclei, however, the ion is best described as *bent,* or *angular.*

$$\left[:\underset{..}{\overset{..}{Cl}}-\underset{..}{\overset{..}{I}}-\underset{..}{\overset{..}{Cl}}:\right]^+$$

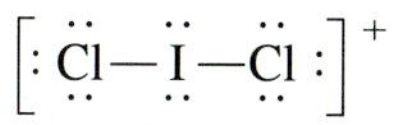

**Figure 8.22** The Lewis structure of the $ICl_2^+$ ion.

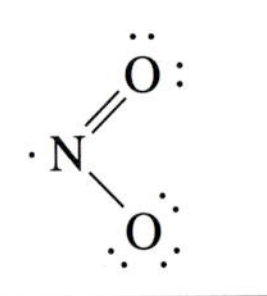

Figure 8.23 The Lewis structure of $NO_2$.

## EXERCISE 8.7

Use the Lewis structure of the $NO_2$ molecule shown in Figure 8.23 to predict the shape of this molecule.

**SOLUTION** The odd number of valence electrons in $NO_2$ makes it impossible to write a Lewis structure in which the electrons are all paired. The Lewis structure in Figure 8.23, however, suggests that there are three places in the valence shell of the nitrogen atom where electrons can be found. There are electrons in the N—O single bond, the N=O double bond, and a single, unpaired, nonbonding electron. The geometry for this molecule is based on arranging these electrons toward the corners of a triangle, and the shape of the molecule is *bent,* or *angular.*

When we extend the VSEPR theory to molecules in which the electrons are distributed toward the corners of a trigonal bipyramid, we run into the question of whether nonbonding electrons should be placed in equatorial or axial positions. Experimentally we find that nonbonding electrons usually occupy equatorial positions in a trigonal bipyramid.

To understand why, we have to recognize that nonbonding electrons take up more space than bonding electrons. Nonbonding electrons need to be close to only one nucleus, and there is a considerable amount of space in which nonbonding electrons can reside and still be near the nucleus of the atom. Bonding electrons, however, must be simultaneously close to two nuclei, and only a small region of space between the nuclei satisfies this restriction.

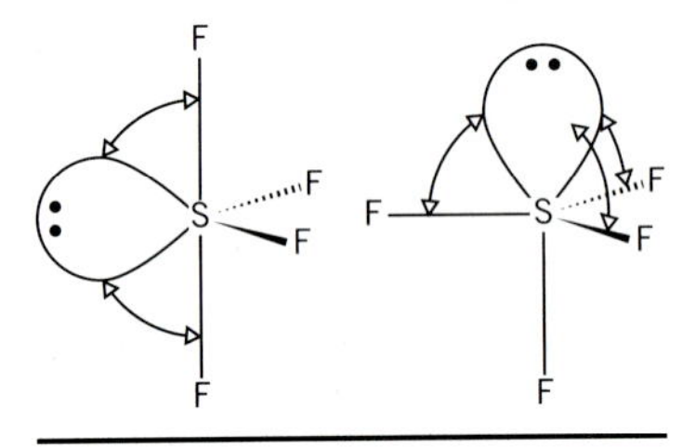

Figure 8.24 The force of repulsion between bonding and nonbonding electrons in $SF_4$ can be minimized by placing the nonbonding electrons in equatorial positions.

Because they occupy more space, the force of repulsion between pairs of nonbonding electrons is relatively large. The force of repulsion between a pair of nonbonding electrons and a pair of bonding electrons is somewhat smaller, and the repulsion between pairs of bonding electrons is even smaller.

Figure 8.24 can help us understand why nonbonding electrons are placed in equatorial positions in a trigonal bipyramid. If the nonbonding electrons in $SF_4$ are placed in an axial position, they will be relatively close (90°) to three pairs of bonding electrons. But if the nonbonding electrons are placed in an equatorial position, they will be 90° away from only two pairs of bonding electrons. As a result, the repulsion between nonbonding and bonding electrons is minimized if the nonbonding electrons are placed in an equatorial position in $SF_4$.

The results of applying the VSEPR theory to $SF_4$, $ClF_3$, and the $I_3^-$ ion are shown in Figure 8.25. When the nonbonding pair of electrons on the sulfur atom in $SF_4$ is placed in an equatorial position, the molecule can be best described as having a *seesaw* or *teeter-totter* shape. Repulsion between valence electrons on the chlorine

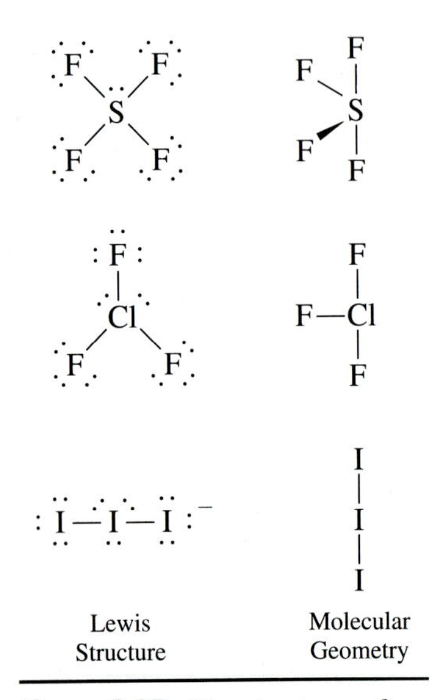


Figure 8.25 The structures of $SF_4$, $ClF_3$ and the $I_3^-$ ion.

The electrons in the valence shell of the sulfur atom in $SF_4$ are oriented toward the corners of a trigonal bipyramid. The geometry of the molecule, however, might be best described as having the shape of a seesaw.

Both pairs of nonbonding electrons (*left*) are placed in equatorial positions in $ClF_3$. The net result is a molecule whose geometry is best described as T-shaped. When all three pairs of nonbonding electrons (*right*) are placed in equatorial positions in the $I_3^-$ ion, the result is a linear molecule.

atom in $ClF_3$ can be minimized by placing both pairs of nonbonding electrons in equatorial positions in a trigonal bipyramid. When this is done, we get a geometry that can be described as T-*shaped.* The Lewis structure of the triiodide ($I_3^-$) ion suggests a trigonal-bipyramidal distribution of valence electrons on the central atom. When the three pairs of nonbonding electrons on this atom are placed in equatorial positions, we get a *linear* molecule.

Molecular geometries based on an octahedral distribution of valence electrons are easier to predict because the corners of an octahedron are all identical.

## EXERCISE 8.8

Use the Lewis structures of $BrF_5$ and $XeF_4$ shown in Figure 8.26 to predict the shape of these molecules.

**SOLUTION** The valence electrons on the bromine atom in $BrF_5$ are distributed toward the corners of an octahedron to form a molecule with a *square-pyramidal* geometry.

When two of the six places in the valence shell of the central atom contain nonbonding electrons, these electrons are kept as far apart as possible to minimize the repulsion between the relatively large nonbonding pairs. The VSEPR theory therefore predicts a *square-planar* structure for $XeF_4$.

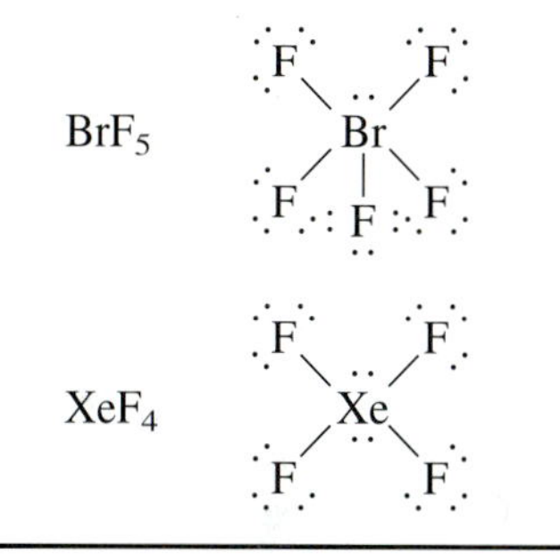


**Figure 8.26** The Lewis structures of $BrF_5$ and $XeF_4$.

The results of this discussion of the VSEPR theory are summarized in Table 8.2.

The structure of $BrF_5$ (*left*) is best described as square pyramidal. The structure of $XeF_4$ (*right*) is square planar.

**TABLE 8.2 The Relationship Between the Number of Places Where Valence Electrons Can Be Found and the Geometry Around an Atom**

| Places Where Electrons Are Found | Places With Bonding Electrons | Places With Nonbonding Electrons | Distribution of Electrons | Molecular Geometry | Examples | |
|---|---|---|---|---|---|---|
| 2 | 2 | 0 | linear | linear | $BeF_2$, $CO_2$ | F—Be—F |
| | 1 | 1 | | linear | CO, $N_2$ | :C≡O: |
| 3 | 3 | 0 | trigonal planar | trigonal planar | $BF_3$, $CO_3^{2-}$ | 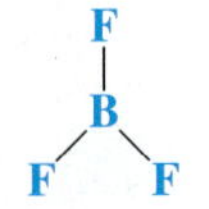 |
| | 2 | 1 | | bent | $O_3$, $SO_2$ | 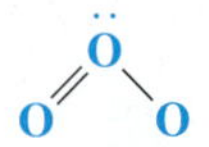 |
| | 1 | 2 | | linear | $O_2$ | 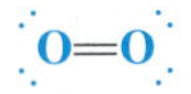 O=O |
| 4 | 4 | 0 | tetrahedral | tetrahedral | $CH_4$, $SO_4^{2-}$ | 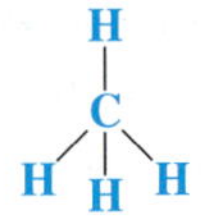 |
| | 3 | 1 | | trigonal pyramidal | $NH_3$, $H_3O^+$ | 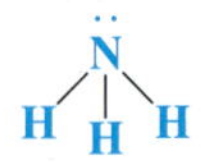 |
| | 2 | 2 | | bent | $H_2O$, $ICl_2^+$ | 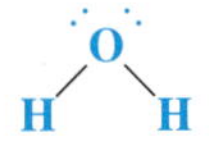 |
| | 1 | 3 | | linear | HF, $OH^-$ | H—F: |
| 5 | 5 | 0 | trigonal bipyramidal | trigonal bipyramidal | $PF_5$ | 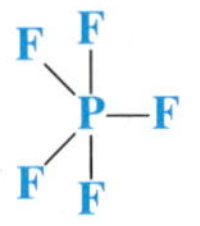 |
| | 4 | 1 | | seesaw | $SF_4$, $TeCl_4$, $IF_4^+$ | 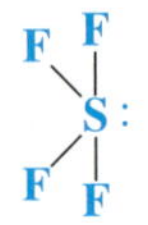 |
| | 3 | 2 | | T-shaped | $ClF_3$ | 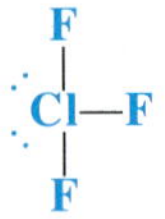 |
| | 2 | 3 | | linear | $I_3^-$, $XeF_2$ |  |
| 6 | 6 | 0 | octahedral | octahedral | $SF_6$, $PF_6^-$, $SiF_6^{2-}$ |  |
| | 5 | 1 | | square pyramidal | $BrF_5$, $SbCl_5^{2-}$ | 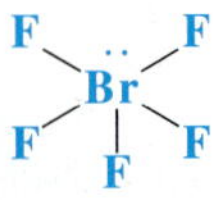 |
| | 4 | 2 | | square planar | $XeF_4$, $ICl_4^-$ | 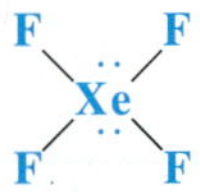 |

SPECIAL TOPIC

## HYBRID ATOMIC ORBITALS

It is difficult to explain the shapes of even the simplest molecules with the atomic orbitals introduced in Chapter 6. A solution to this problem was proposed by Linus Pauling, who argued that the valence orbitals on an atom could be combined to form **hybrid atomic orbitals.**

The geometry of a $BeF_2$ molecule can be explained, for example, by mixing the 2*s* orbital on the beryllium atom with one of the 2*p* orbitals to form a set of *sp* hybrid orbitals that points in opposite directions, as shown in Figure 8.27. One of the valence electrons on the beryllium atom is then placed in each of these orbitals, and these orbitals are allowed to overlap with half-filled 2*p* orbitals on a pair of fluorine atoms to form a linear $BeF_2$ molecule.

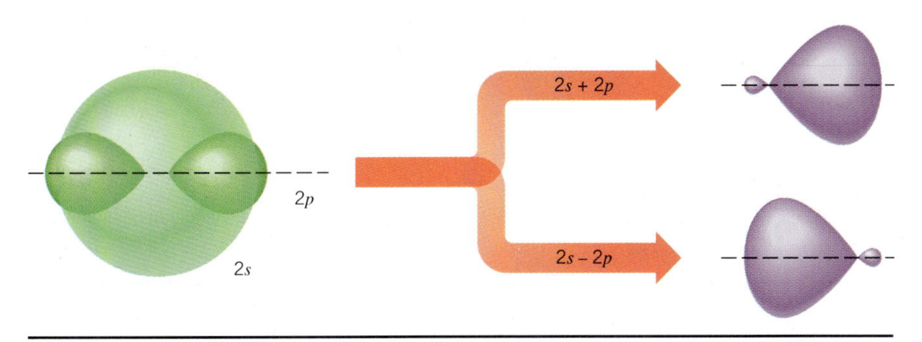


**Figure 8.27** The *sp* hybrid orbitals used by the beryllium atom in the linear $BeF_2$ molecule are formed by combining the wave functions for the 2*s* and 2*p* orbitals on that atom! When these wave functions are added we get an *sp* orbital that points in one direction! When one of these wave functions is subtracted from the other we get an *sp* orbital that points in the opposite direction.

Pauling also showed that the geometry of trigonal planar molecules such as $BF_3$ and the $CO_3^{2-}$ ion could be explained by mixing a 2*s* orbital with both a $2p_x$ and a $2p_y$ orbital on the central atom to form three $sp^2$ hybrid orbitals that point toward the corners of an equilateral triangle. When he mixed a 2*s* orbital with all three 2*p* orbitals ($2p_x$, $2p_y$ and $2p_z$), Pauling obtained a set of four $sp^3$ orbitals oriented toward the corners of a tetrahedron. These $sp^3$ hybrid orbitals are ideal for explaining the geometries of tetrahedral molecules such as $CH_4$ or the $SO_4^{2-}$ ion.

The hybrid atomic orbital model can be extended to molecules whose shapes are based on trigonal-bipyramidal or octahedral distributions of electrons by including valence-shell *d* orbitals. Pauling showed that when the $3d_{z^2}$ orbital is mixed with the 3*s*, $3p_x$, $3p_y$, and $3p_z$ orbitals on an atom, the resulting $sp^3d$ hybrid orbitals point toward the corners of a trigonal bipyramid. When both the $3d_{z^2}$ and $3d_{x^2-y^2}$ orbitals are mixed with the 3*s*, $3p_x$, $3p_y$, and $3p_z$ orbitals, the result is a set of six $sp^3d^2$ hybrid orbitals that point toward the corners of an octahedron.

The geometries of the five different sets of hybrid atomic orbitals ($sp$, $sp^2$, $sp^3$, $sp^3d$, and $sp^3d^2$) are shown in Figure 8.28. The relationship between hybridization and the distribution of electrons in the valence shell of an atom is summarized in Table 8.3.

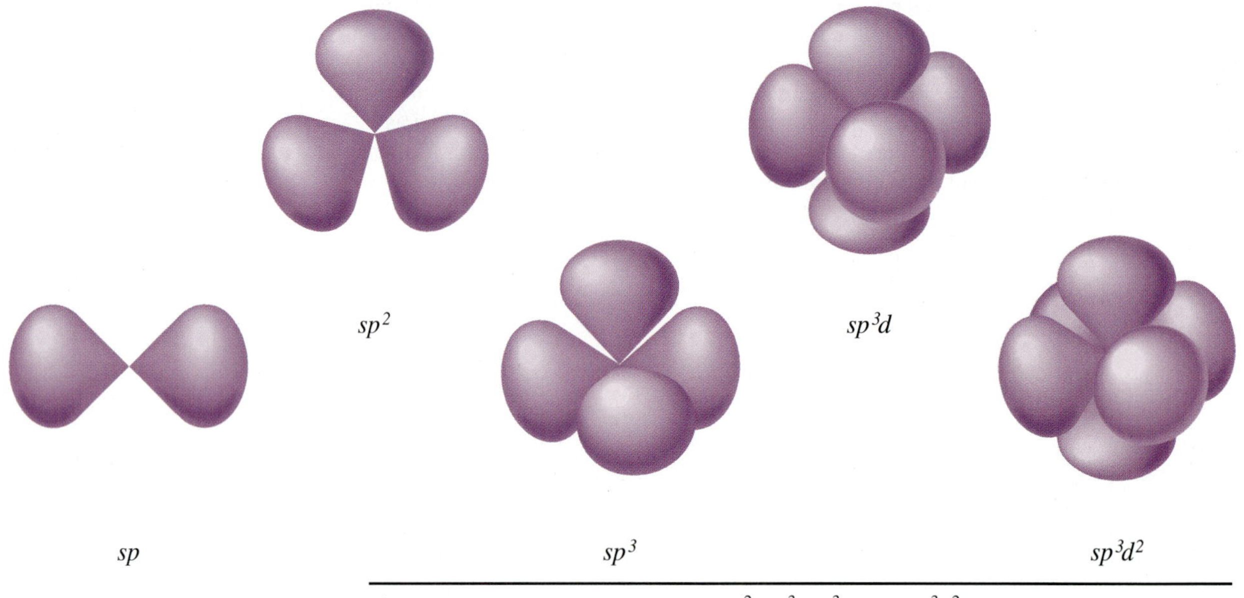


**Figure 8.28** The shapes of the $sp$, $sp^2$, $sp^3$, $sp^3d$, and $sp^3d^2$ hybrid orbitals.

**TABLE 8.3 The Relationship Between the Distribution of Electrons on an Atom and the Hybridization of That Atom**

| Number of Places Where Electrons Are Found | Molecular Geometry | Hybridization | Examples |
|---|---|---|---|
| 2 | linear | $sp$ | $BeF_2$, $CO_2$ |
| 3 | trigonal planar | $sp^2$ | $BF_3$, $CO_3^{2-}$ |
| 4 | tetrahedral | $sp^3$ | $CH_4$, $SO_4^{2-}$ |
| 5 | trigonal bipyramidal | $sp^3d$ | $PF_5$ |
| 6 | octahedral | $sp^3d^2$ | $SF_6$ |

## EXERCISE 8.9

Use Table 8.3 to determine the hybridization of the central atom in the following compounds:

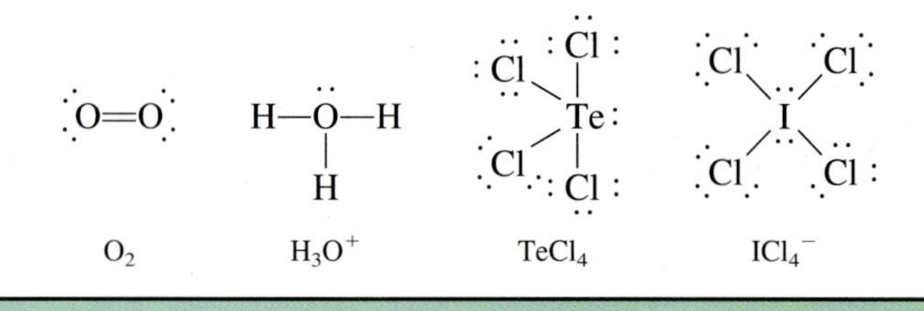

**SOLUTION**

(a) The Lewis structure for the $O_2$ molecule suggests that the distribution of electrons around each oxygen atom should be trigonal planar. Orbitals that point toward the corners of an equilateral triangle can be created when the $2s$, $2p_x$, and $2p_y$ orbitals on each oxygen atom are mixed to form a set of $sp^2$ hybrid orbitals. Thus, the hybridization of the oxygen atoms in the $O_2$ molecule is assumed to be $sp^2$.

(b) The Lewis structure suggests a tetrahedral distribution of electrons around the oxygen atom in the hydronium ion. A set of orbitals that points toward the corners of a tetrahedron can be constructed by mixing the $2s$, $2p_x$, $2p_y$, and $2p_z$ atomic orbitals on the oxygen atom. As a result, the oxygen in this molecule is $sp^3$ hybridized.

(c) The Lewis structure suggests a trigonal-bipyramidal distribution of electrons in the valence shell of the tellurium atom. Orbitals pointing toward the corners of a trigonal bipyramid can be obtained by mixing the $5s$, $5p_x$, $5p_y$, $5p_z$, and $5d_z{}^2$ orbitals on this atom to form an $sp^3d$ hybrid.

(d) The Lewis structure suggests an octahedral distribution of electrons in the valence shell of the iodine atom. Orbitals pointing toward the corners of an octahedron can be obtained when the $5s$, $5p_x$, $5p_y$, $5p_z$, $5d_z{}^2$, and $5d_{x^2-y^2}$ orbitals on iodine are mixed. The iodine atom is therefore $sp^3d^2$ hybridized.

## MOLECULES WITH DOUBLE AND TRIPLE BONDS

The hybrid atomic orbital model can be also used to explain the formation of double and triple bonds. Consider the bonding in formaldehyde ($H_2CO$), for example, which has the following Lewis structure:

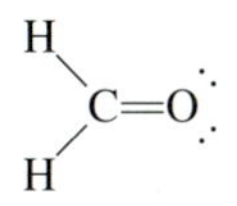

There are three places where electrons can be found in the valence shell of both the carbon and oxygen atoms in this molecule. As a result, the VSEPR theory predicts that the valence electrons on these atoms will be oriented toward the corners of an equilateral triangle. For the sake of argument, let's place the formaldehyde molecule in the $xy$ plane of a coordinate system. We can create a set of $sp^2$ hybrid orbitals on the carbon and oxygen atoms that lie in this plane by mixing the $2s$, $2p_x$, and $2p_y$ orbitals on each atom.

There are four valence electrons on a neutral carbon atom. One of these electrons is placed in each of the three $sp^2$ hybrid orbitals. The fourth electron is placed in the $2p_z$ orbital that was not used during hybridization.

There are six valence electrons on a neutral that point away from the carbon atom oxygen atom. A pair of these electrons is placed in each of the $sp^2$ hybrid orbitals. One electron is then placed in the $sp^2$ hybrid orbital that points toward the carbon atom, and another is placed in the unhybridized $2p_z$ orbital.

The C—H bonds are formed when the electrons in two of the $sp^2$ hybrid orbitals on carbon interact with a $1s$ electron on a hydrogen atom, as shown in Figure 8.29. A C—O bond is formed when the electron in the other $sp^2$ hybrid orbital on carbon interacts with the unpaired electron in the $sp^2$ hybrid orbital on the oxygen atom. These bonds are called **sigma ($\sigma$) bonds** because they look like an *s* orbital when viewed along the bond.

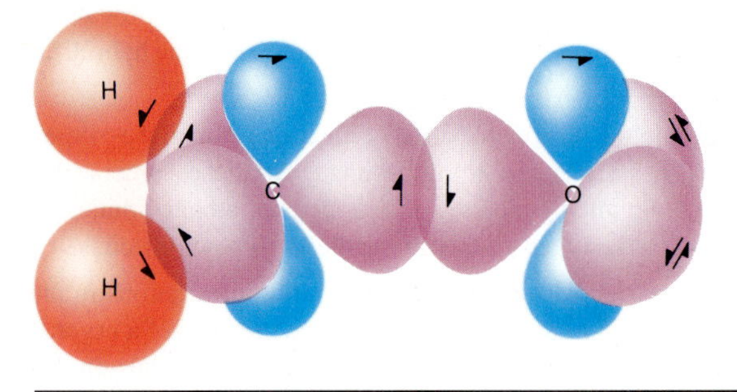


**Figure 8.29** The hybrid atomic orbital picture of the bonding in formaldehyde.

The electron in the $2p_z$ orbital on the carbon atom then interacts with the electron in the $2p_z$ orbital on the oxygen atom to form a second covalent bond between these atoms. This is called a **pi ($\pi$) bond** because it looks like a *p* orbital when viewed along the bond.

Sec. 8.11 

When double bonds were first introduced, we mentioned that they occur most often in compounds that contain C, N, O, P, or S. There are two reasons for this: First, double bonds by their very nature are covalent bonds. They are therefore most likely to be found among the elements that form covalent compounds. Second, the interaction between $2p_z$ orbitals to form a $\pi$ bond requires that the atoms come relatively close together, so these bonds tend to be the strongest for atoms that are relatively small.

**SPECIAL TOPIC**

## MOLECULAR ORBITAL THEORY

Ever since orbitals were introduced in Chapter 6, our discussion has focused on *atomic orbitals.* Because arguments based on these orbitals focus on the bonds formed between valence electrons on an atom, they are often said to involve a *valence-bond* theory.

One limitation of the valence-bond theory was identified in Section 8.12. The valence-bond model of $SO_2$ can't adequately explain the fact that this molecule contains two equivalent bonds with a bond order between that of a S—O single bond and an S=O double bond. The best it can do is suggest that $SO_2$ is a mixture, or hybrid, of the two Lewis structures that can be written for this molecule.

This problem, and many others, can be overcome by using a more sophisticated model of bonding based on **molecular orbitals.** Molecular orbital theory is more powerful than valence-bond theory because the orbitals reflect the geometry of the molecule to which they are applied. But this power carries a significant cost in terms of the ease with which the model can be visualized. We can understand the basics of molecular orbital theory by constructing the molecular orbitals for the simplest possible molecule: a homonuclear diatomic molecule, such as $H_2$, $O_2$, or $N_2$.

Molecular orbitals are obtained by combining the atomic orbitals on the atoms in the molecule. Consider the $H_2$ molecule, for example. One of the molecular orbitals in this molecule is constructed by adding the mathematical functions for the two 1*s* atomic orbitals that come together to form this molecule. Another orbital is formed by subtracting one of these functions from the other, as shown in Figure 8.30.

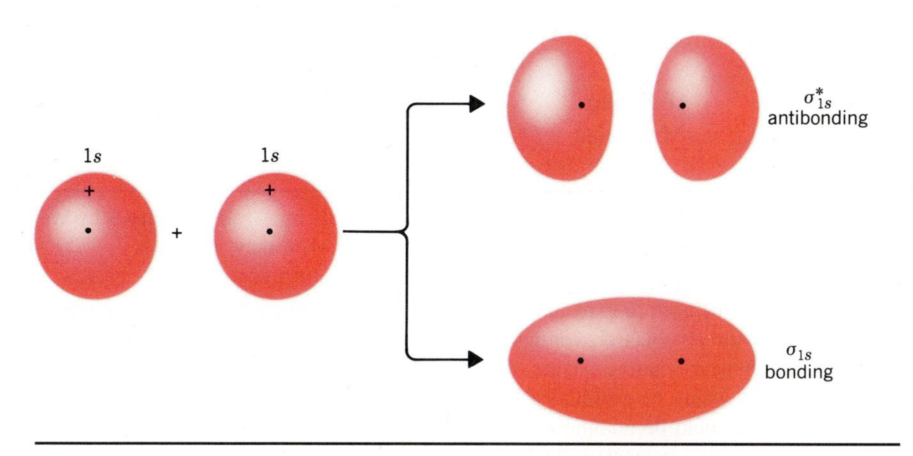


**Figure 8.30** The interaction between a pair of 1*s* atomic orbitals on neighboring atoms leads to the formation of a bonding ($\sigma$) and an antiboding ($\sigma^*$) molecular orbital.

One of these orbitals is called a **bonding molecular orbital** because electrons in this orbital spend most of their time in the region directly between the two nuclei. It is called a **sigma ($\sigma$) molecular orbital** because it looks like an *s* orbital when viewed along the H—H bond. Electrons placed in the other orbital spend most of their time away from the region between the two nuclei. This orbital is therefore an **antibonding,** or **sigma star ($\sigma^*$), molecular orbital.**

The $\sigma$-bonding molecular orbital concentrates electrons in the region directly between the two nuclei. Placing an electron in this orbital therefore stabilizes the $H_2$ molecule. Since the $\sigma^*$-antibonding molecular orbital forces the electron to spend most of its time away from the area between the nuclei, placing an electron in this orbital makes the molecule less stable.

Electrons are added to molecular orbitals, one at a time, starting with the lowest energy molecular orbital. The two electrons associated with a pair of hydrogen atoms are placed in the lowest energy, or $\sigma$-bonding, molecular orbital, as shown in Figure 8.31. This diagram suggests that the energy of an $H_2$ molecule is lower than that of a pair of isolated atoms. As a result, the $H_2$ molecule is more stable than a pair of isolated atoms.

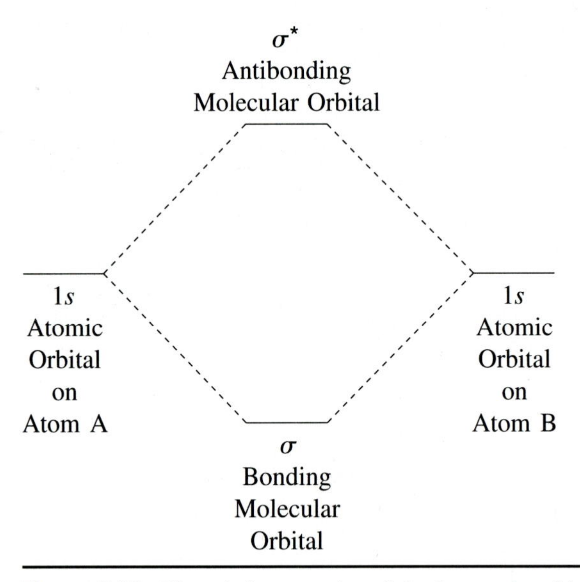


**Figure 8.31** The relative energies of the 1*s* atomic orbitals on a pair of isolated hydrogen atoms and the $\sigma$ and $\sigma^*$ molecular orbitals they form when they interact.

This model can be used to explain why $He_2$ molecules do not exist. Combining a pair of helium atoms with $1s^2$ electron configurations would produce a molecule with a pair of electrons in both the $\sigma$-bonding and the $\sigma^*$-antiboding molecular orbitals. The total energy of an $He_2$ molecule would be essentially the same as the energy of a pair of isolated helium atoms, and there would be nothing to hold the helium atoms together to form a molecule.

The fact that an $He_2$ molecule is neither more nor less stable than a pair of isolated helium atoms illustrates an important principle:

**The core orbitals on an atom make no contribution to the stability of the molecules that contain this atom.**

The only orbitals that are important in our discussion of molecular orbitals are those formed when valence-shell orbitals are combined. The molecular orbital diagram for an $O_2$ molecule would therefore ignore the 1*s* electrons on both oxygen atoms and concentrate on the interactions between the 2*s* and 2*p* valence orbitals.

The $2s$ orbitals on one atom combine with the $2s$ orbitals on another to form a $\sigma_{2s}$ bonding and a $\sigma_{2s}{}^*$ antibonding molecular orbital, just like the $\sigma_{1s}$ and $\sigma_{1s}{}^*$ orbitals formed from the $1s$ atomic orbitals. If we arbitrarily define the $z$ axis of the coordinate system for the $O_2$ molecule as the axis along which the bond forms, the $2p_z$ orbitals on the adjacent atoms will meet head-on to form a $\sigma_{2p}$ bonding and a $\sigma_{2p}{}^*$ antibonding molecular orbital, as shown in Figure 8.32. These are called sigma orbitals because they look like $s$ orbitals when viewed along the oxygen–oxygen bond.

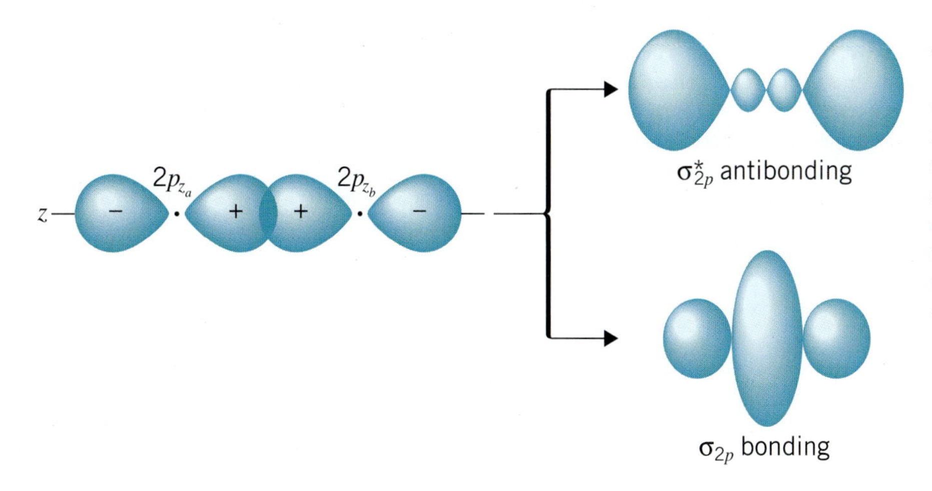


**Figure 8.32** When a pair of $2p$ orbitals are combined so that they meet head-on, bonding ($\sigma_{2p}$) and antibonding ($\sigma_{2p}{}^*$) molecular orbitals are formed. These are called $\sigma$ orbitals because they concentrate the electrons along the axis between the two nuclei, just like the $\sigma_{1s}$ and $\sigma_{2s}$ orbitals.

The $2p_x$ orbital on one atom interacts with the $2p_x$ orbital on the other to form molecular orbitals that have a different shape, as shown in Figure 8.33. These molecular orbitals are called *pi* ($\pi$) orbitals because they look like $p$ orbitals when viewed along the bond. Whereas

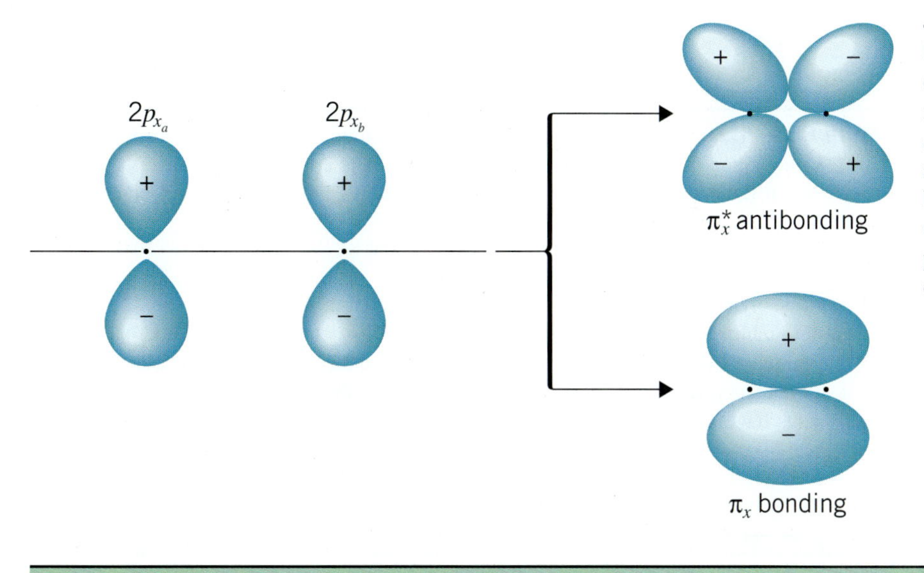


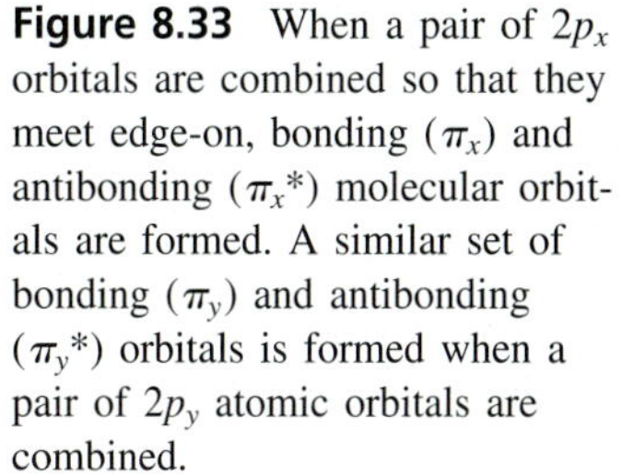

**Figure 8.33** When a pair of $2p_x$ orbitals are combined so that they meet edge-on, bonding ($\pi_x$) and antibonding ($\pi_x{}^*$) molecular orbitals are formed. A similar set of bonding ($\pi_y$) and antibonding ($\pi_y{}^*$) orbitals is formed when a pair of $2p_y$ atomic orbitals are combined.

$\sigma$ and $\sigma^*$ orbitals concentrate the electrons along the axis on which the nuclei of the atoms lie, $\pi$ and $\pi^*$ orbitals concentrate the electrons either above or below this axis.

The $2p_x$ atomic orbitals combine to form a $\pi_x$ bonding molecular orbital and a $\pi_x^*$ antibonding orbital. The same thing happens when the $2p_y$ orbitals interact, only in this case we get a $\pi_y$ bonding molecular orbital and a $\pi_y^*$ antibonding molecular orbital. Because there is no difference between the energies of the $2p_x$ and $2p_y$ atomic orbitals, there is no difference between the energies of the $\pi_x$ and $\pi_y$ or the $\pi_x^*$ and $\pi_y^*$ molecular orbitals.

There is a significant difference between the energies of the 2*s* and 2*p* orbitals on an atom. As a result, the $\sigma_{2s}$ and $\sigma_{2s}^*$ orbitals both lie at lower energies than the $\sigma_{2p}$, $\sigma_{2p}^*$, $\pi_x$, $\pi_y$, $\pi_x^*$, and $\pi_y^*$ orbitals. To sort out the relative energies of the six molecular orbitals formed when the 2*p* atomic orbitals on a pair of atoms are combined, we need to understand the relationship between the strength of the interaction between a pair of orbitals and the relative energies of the molecular orbitals they form.

Because they meet head-on, the interaction between the $2p_z$ orbitals is stronger than the interaction between the $2p_x$ or $2p_y$ orbitals, which meet edge-on. As a result, the $\sigma_{2p}$ orbital lies at a lower energy than the $\pi_x$ and $\pi_y$ orbitals, and the $\sigma_{2p}^*$ orbital lies at higher energy than the $\pi_x^*$ and $\pi_y^*$ orbitals, as shown in Figure 8.34.

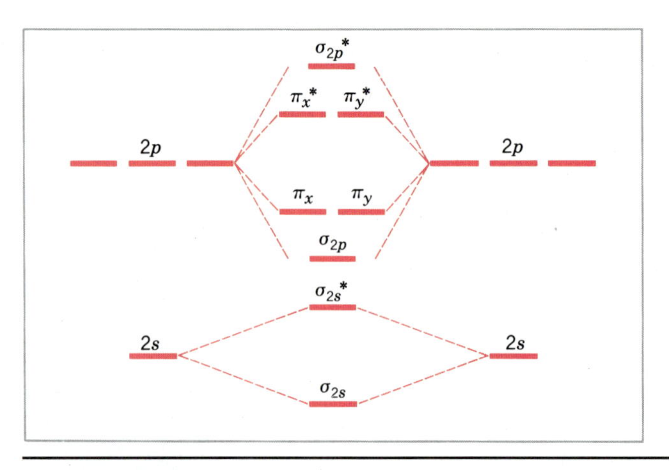


**Figure 8.34** The interaction between $2p_z$ atomic orbitals that meet head-on is stronger than the interaction between $2p_x$ or $2p_y$ orbitals that meet edge-on. As a result, the difference between the energies of the $\sigma_{2p}$ and $\sigma_{2p}^*$ orbitals is larger than the difference between the $\pi_x$ and $\pi_x^*$ or $\pi_y$ and $\pi_y^*$ orbitals.

Unfortunately an interaction is missing from this model. It is possible for the 2*s* orbital on one atom to interact with the $2p_z$ orbital on the other. This interaction introduces an element of *s–p* mixing, or hybridization, into the molecular orbital theory. The result is a slight change in the relative energies of the molecular orbitals, to give the diagram shown in Figure 8.35. Experiments have shown that $O_2$ and $F_2$ are best described by the model in Figure 8.34, but $B_2$, $C_2$, and $N_2$ are best described by a model that includes hybridization, as shown in Figure 8.35.

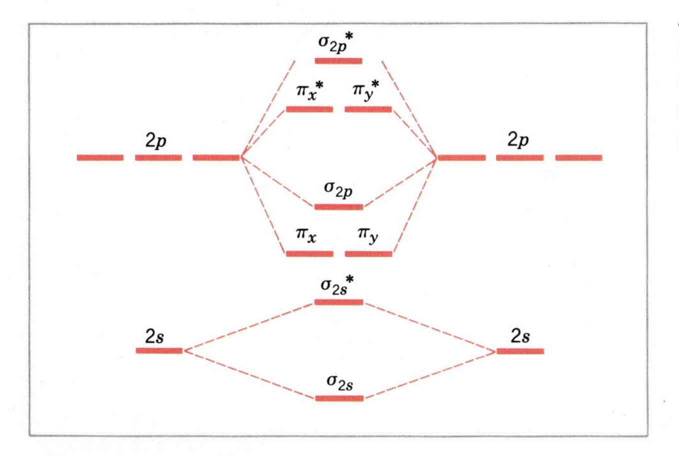


**Figure 8.35** When hybridization is added to the molecular orbital theory, there is a slight change in the energies of the orbitals formed by overlap of the $2p$ atomic orbitals on neighboring atoms.

## EXERCISE 8.10

Construct a molecular orbital diagram for the $O_2$ molecule.

**SOLUTION** There are six valence electrons on a neutral oxygen atom and therefore 12 valence electrons in an $O_2$ molecule. These electrons are added to the diagram in Figure 8.34, one at a time, starting with the lowest energy molecular orbital:

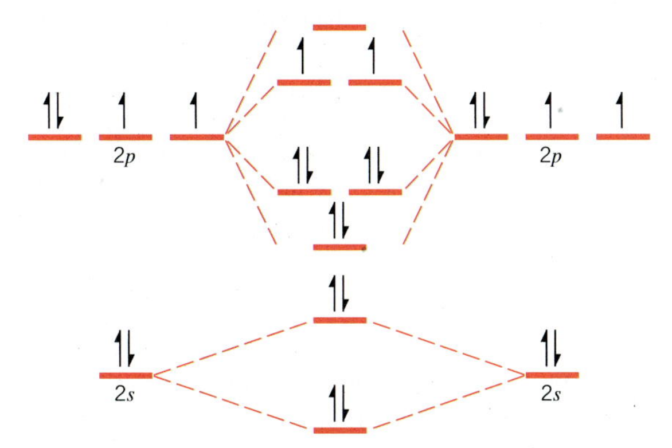


Because Hund's rules apply to the filling of molecular orbitals, molecular orbital theory predicts that there should be two unpaired electrons on this molecule—one electron each in the $\pi_x$* and $\pi_y$* orbitals.

When writing the electron configuration of an atom, we usually list the orbitals in the order in which they fill:

$$\text{Pb:} \quad [\text{Xe}]\ 6s^2\ 4f^{14}\ 5d^{10}\ 6p^2$$

We can write the electron configuration of a molecule by doing the same thing. Concentrating only on the valence orbitals, we write the electron configuration of $O_2$ as follows:

$$O_2: \quad \sigma_{2s}{}^2\sigma_{2s}{}^{*2}\sigma_{2p}{}^2\pi_x{}^2\pi_y{}^2\pi_x{}^{*1}\pi_y{}^{*1}$$

The number of bonds between a pair of atoms is called the **bond order.** Bond orders can be calculated from Lewis structures, which are the heart of the valence-bond model. Oxygen, for example, has a bond order of two:

$$\ddot{\text{O}}{=}\ddot{\text{O}}$$

When there is more than one Lewis structure for a molecule, the bond order is an average of these structures. The bond order in sulfur dioxide, for example, is 1.5, the average of an S—O single bond in one Lewis structure and an S=O double bond in the other:

$$:\ddot{\text{O}}{-}\ddot{\text{S}}{=}\ddot{\text{O}} \longleftrightarrow \ddot{\text{O}}{=}\ddot{\text{S}}{-}\ddot{\text{O}}:$$

The bond order for a homonuclear diatomic molecule can be calculated by assuming that two electrons in a bonding molecular orbital contribute one net bond and that two electrons in an antibonding molecular orbital cancel the effect of one bond. We can calculate the bond order in the $O_2$ molecule by noting that there are eight valence electrons in bonding molecular orbitals and four valence electrons in antibonding molecular orbitals in the electron configuration of this molecule given in Exercise 8.10. Thus, the bond order is two:

$$\text{Bond order} = \frac{\text{bonding electrons} - \text{antibonding electrons}}{2} = \frac{8-4}{2} = 2$$

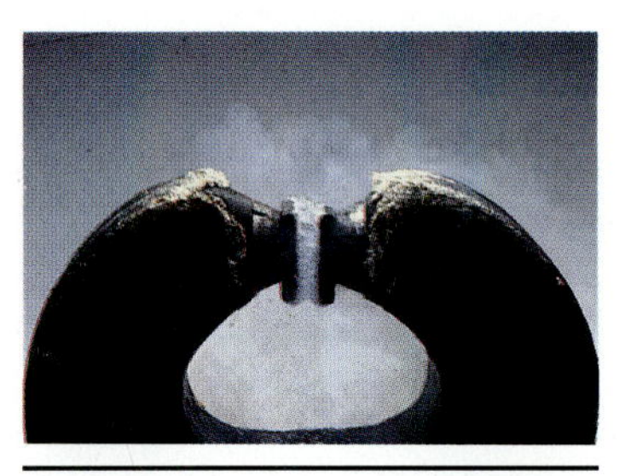

The molecular orbital theory predicts that the $O_2$ molecule will be paramagnetic. This photograph shows that liquid $O_2$ is so strongly attracted to a magnetic field that it will bridge the gap between the poles of a horseshoe magnet.

Although the Lewis structure and molecular orbital models of oxygen yield the same bond order, there is an important difference between these models. The electrons in the Lewis structure are all paired, but there are two unpaired electrons in the molecular orbital description of the molecule. As a result, we can test the predictions of these theories by studying the effect of a magnetic field on oxygen.

Atoms or molecules in which the electrons are paired are **diamagnetic**—repelled by both poles of a magnet. Those that have one or more unpaired electrons are **paramagnetic**—attracted to a magnetic field. Liquid oxygen is attracted to a magnetic field and can actually bridge the gap between the poles of a horseshoe magnet. The molecular orbital model of $O_2$ is therefore superior to the valence-bond model, which cannot explain this property of oxygen.

## EXERCISE 8.11

Use the molecular orbital diagram in Figure 8.35 to calculate the bond order in nitrogen oxide (NO). Compare the results of this calculation with the bond order obtained from the Lewis structure.

**SOLUTION** There are 11 valence electrons in this molecule, and there are two possible Lewis structures, depending on whether we locate the unpaired electron on the nitrogen atom or on the oxygen atom:

$$\dot{\text{N}}{=}\ddot{\text{O}} \longleftrightarrow \text{N}{=}\dddot{\text{O}}$$

Both Lewis structures contain an N=O double bond, however, so the bond order in the valence-bond model for NO is 2.

To calculate the bond order in the molecular orbital model of NO, we have to predict the electron configuration of this molecule by adding electrons, one at a time, to the molecular orbitals in Figure 8.35:

$$\text{NO:} \quad \sigma_{2s}{}^2\sigma_{2s}{}^{*2}\pi_x{}^2\pi_y{}^2\sigma_{2p}{}^2\pi_x{}^{*1}$$

There are eight electrons in bonding molecular orbitals and three electrons in antibonding molecular orbitals in this compound. The bond order in the molecular orbital model of NO is therefore 2.5:

$$\text{Bond order} = \frac{\text{bonding electrons} - \text{antibonding electrons}}{2} = \frac{8 - 3}{2}$$

The strength of the covalent bond in NO has been shown by experiment to be stronger than a typical N=O double bond, in agreement with the predictions of the molecular orbital theory for this molecule.

## SUMMARY

**1.** Ionic and covalent bonds differ in the extent to which electrons are shared by the atoms that form these bonds. When one atom is much more electronegative than the other, it draws the electrons in the bond closer to itself. This produces a separation of charge that gives rise to positive and negative ions that are held together by an ionic bond. When the atoms in a bond have similar electronegativities, neither atom can draw the electrons away from the other. They therefore share pairs of electrons more or less equally in a covalent bond.

**2.** When the difference between the electronegativities of the atoms in a bond is neither large enough to form an ionic bond nor small enough to allow the atoms to share electrons equally, the bonds are best described as polar, or polar covalent.

**3.** At the turn of the century, G. N. Lewis proposed a model for covalent compounds that assumed that the atoms of main-group elements gain, lose, or share electrons until they have an octet of valence-shell electrons. The octet theory is the basis for writing so-called *Lewis structures* of molecules.

**4.** The valence-shell electron-pair repulsion (VSEPR) theory can be used to predict the distribution of electrons in the valence shell of an atom, and therefore the shape of a molecule from the Lewis structure of the molecule.

**5.** The atomic orbitals introduced in Chapter 6 were defined for use with an isolated hydrogen atom in the gas phase. When the time comes to explain the properties of the molecules, it is necessary to modify these orbitals. This can be done by mixing the atomic orbitals on an atom to form a set of *hybrid atomic orbitals.* The hybrid orbitals on one atom then interact with the hybrid orbitals on another to form the covalent bonds in a molecule. Lewis structures and hybrid atomic orbitals are part of a **valence-bond** model of covalent compounds. They focus attention on the bonds formed when electrons in the valence orbitals on different atoms are combined.

**6.** The bonding in covalent compounds also can be explained with **molecular orbital theory,** in which orbitals on the atoms are mixed to form a set of orbitals that is characteristic of a particular molecule. Molecular orbital theory is more powerful than valence-bond theory because it comes closer to predicting the properties of covalent compounds. It isn't as easy to use, however, because a different set of molecular orbitals has to be created for each molecular geometry.

## KEY TERMS

**Antibonding molecular orbital** (p. 301)
**Bond order** (p. 306)
**Bonding electrons** (p. 284)
**Bonding molecular orbital** (p. 301)
**Covalent bond** (p. 274)
**Diamagnetic** (p. 306)
**Dipole moment** (p. 282)
**Electronegativity** (p. 277)
**Formal charge** (p. 288)
**Hybrid atomic orbitals** (p. 297)
**Lewis structure** (p. 274)
**Molecular orbitals** (p. 301)

**Nonbonding electrons** (p. 284)
**Octahedron** (p. 292)
**Octet** (p. 272)
**Paramagnetic** (p. 306)
**Pi ($\pi$) bond** (p. 300)
**Polar compounds** (p. 279)
**Resonance hybrids** (p. 286)
**Sigma ($\sigma$) bonds** (p. 300, 301)
**Tetrahedron** (p. 292)
**Trigonal bipyramid** (p. 292)
**Trigonal planar** (p. 291)
**Valence electrons** (p. 272)
**VSEPR** (p. 291)

## PROBLEMS

### Valence Electrons

**8-1** Define the term *valence electrons.*

**8-2** Describe the relationship between the number of valence electrons on an atom and the most positive oxidation state of the element.

**8-3** Determine the number of valence electrons in neutral atoms of the following elements:

(a) Li (b) C (c) Mg (d) Ar

**8-4** Determine the number of valence electrons in neutral atoms of the following elements:

(a) Fe (b) Cu (c) Bi (d) I

**8-5** Determine the number of valence electrons in the following negative ions and describe any general trends:

(a) $C^{4-}$ (b) $N^{3-}$ (c) $S^{2-}$ (d) $I^-$

**8-6** Determine the number of valence electrons in the following positive ions and describe any general trends:

(a) $Na^+$ (b) $Mg^{2+}$ (c) $Al^{3+}$ (d) $Sc^{3+}$

**8-7** Explain why filled *d* and *f* subshells are ignored when the valence electrons on an atom are counted.

**8-8** Compare the number of valence electrons on a mercury atom in the +1 oxidation state with the number of valence electrons on a neutral hydrogen atom. Use this information to explain why mercury in the +1 oxidation state forms $Hg_2^{2+}$ ions.

**8-9** Predict the group of the periodic table in which an element belongs if its compounds exhibit oxidation states between +6 and −2.

**8-10** Calculate the largest positive oxidation state of a tantalum atom if the electron configuration for a neutral Ta atom is [Xe] $6s^24f^{14}5d^3$

**8-11** Determine the total number of valence electrons in the following molecules or ions:

(a) $BF_3$ (b) $CH_4$ (c) $NH_4^+$ (d) $H_2SO_4$

**8-12** Determine the total number of valence electrons in the following molecules or ions:

(a) $KrF_2$ (b) $SF_4$ (c) $SiF_6^{2-}$ (d) $ZrF_7^{3-}$

**8-13** Which of the following molecules or ions contain the same number of valence electrons?

(a) $CO_2$ (b) $N_2O$ (c) $CNO^-$ (d) $NO_2^+$ (e) $SO_2$ (f) $O_3$ (g) $NO_2^-$

### The Covalent Bond

**8-14** Write electron dot symbols for the elements in the third row of the periodic table.

**8-15** Describe how the sharing of a pair of electrons by two chlorine atoms makes a $Cl_2$ molecule more stable than a pair of isolated Cl atoms.

**8-16** Describe the difference between the ionic bond in NaCl and the covalent bond in $Cl_2$.

**8-17** Students sometimes get the impression that the difference between ionic and covalent compounds is black and white. Explain why chemists believe that ionic and covalent are the two extremes of a continuum of differences in the bonding between two atoms.

### Electronegativity and Polarity

**8-18** Predict whether the following compounds are ionic or covalent by using both the general rule that metals combine with nonmetals to form ionic compounds and the difference in the electronegativities of the elements.

(a) $OF_2$ (b) $CS_2$ (c) MgO (d) ZnS

**8-19** Predict whether the following compounds are ionic or covalent by using both the general rule that metals combine with nonmetals to form ionic compounds and the difference in the electronegativities of the elements.

(a) $IF_3$ (b) $SiCl_4$ (c) $BF_3$ (d) $Na_2S$

**8-20** Describe the trends in the electronegativities of the elements across a row and down a column of the periodic table.

**8-21** Which element in the periodic table has the largest electronegativity? Which has the smallest? Which compound is therefore the most ionic?

**8-22** Which of the following elements is the most electronegative?

(a) S (b) As (c) P (d) Se (e) Cl (f) Br

**8-23** Which of the following series of elements is arranged in order of decreasing electronegativity?

(a) C > Si > P > As > Se (b) O > P > Al > Mg > K
(c) Na > Li > B > N > F (d) K > Mg > Be > O > N
(e) Li > Be > B > C > N

**8-24** Which of the following elements forms bonds with fluorine that are the most covalent?

(a) P (b) Ca (c) Al (d) O (e) Se

**8-25** Which of the following elements forms bonds with oxygen that are the most covalent?

(a) Sr (b) In (c) Sb (d) Te (e) Se

**8-26** Which of the following compounds are best described as polar covalent?

(a) CO (b) $H_2O$ (c) $BeF_2$ (d) $MgBr_2$ (e) $AlI_3$ (f) ZnS

**8-27** Which of the following compounds are best described as polar covalent?

(a) $CaH_2$ (b) $BrF_5$ (c) $NF_3$ (d) $FeCl_4^-$ (e) $UO_2^{2+}$ (f) $SiCl_4$

### Polar Bonds Versus Polar Molecules

**8-28** Explain why $CHCl_3$ is a polar molecule but $CCl_4$ is not.

**8-29** Use the Lewis structure of $CO_2$ to explain why this molecule has no dipole moment.

**8-30** Which of the following molecules should be polar?

(a) $CH_3OH$ (b) $H_2O$ (c) $CH_3OCH_3$ (d) $CH_3CO_2H$

**8-31** Explain why formaldehyde ($H_2CO$) is a polar molecule.

**8-32** Explain why thionyl chloride ($SOCl_2$) is a polar molecule but sulfuryl chloride ($SO_2Cl_2$) is not.

### Limitations of the Electronegativity Concept

**8-33** The electronegativity of an element depends on the oxidation number of the element. Which of the following compounds should be the most covalent?

(a) MnO (b) $Mn_2O_3$ (c) $MnO_2$ (d) $Mn_2O_7$

**8-34** Arrange the following compounds in order of increasing ionic character:

(a) TiO (b) $Ti_2O_3$ (c) $Ti_4O_7$ (d) $TiO_2$

### Writing Lewis Structures

**8-35** Write Lewis structures for the following ionic compounds:

(a) NaCl (b) $CaF_2$ (c) MgO (d) $Na_2S$

**8-36** Write Lewis structures for the following ionic compounds:

(a) $MgSO_4$ (b) $NH_4Cl$ (c) $Na_3PO_4$ (d) $Ca(OCl)_2$

**8-37** Write Lewis structures for the following molecules:

(a) $NH_3$ (b) $CH_3^+$ (c) $H_3O^+$ (d) $BH_4^-$

**8-38** Write Lewis structures for the following ions:

(a) $NO_3^-$ (b) $SO_3^{2-}$ (c) $CO_3^{2-}$ (d) $NO_2^+$

**8-39** Write Lewis structures for the following molecules:

(a) $C_2H_6$ (b) $C_2H_4$ (c) $C_2H_2$ (d) $C_2^{2-}$

**8-40** Write Lewis structures for the following nitrogen compounds:

(a) $N_2O$ (b) NO (c) $NO_2$ (d) $N_2O_3$ (ON—$NO_2$)

**8-41** Write Lewis structures for the following nitrogen compounds.

(a) ClNO (b) $ClNO_2$ (c) $NO^+$ (d) $NO_2^-$ (e) $ONF_3$

**8-42** Explain why a satisfactory Lewis structure for $N_2O_5$ cannot be based on a skeleton structure that contains an N—N bond: $O_2N$—$NO_3$. Show how a satisfactory Lewis structure can be written if we assume that the skeleton structure is $O_2NONO_2$.

**8-43** Write Lewis structures for the following molecules:

(a) $O_2$ (b) $O_3$ (c) $O_2^{2-}$ (d) $O^{2-}$ (e) $O_2^-$

**8-44** Write Lewis structures for the following molecules:

(a) $SO_2$ (b) $SO_3$ (c) $SO_3^{2-}$ (d) $SO_4^{2-}$

**8-45** Write Lewis structures for the following molecules:

(a) $XeF_2$ (b) $XeF_4$ (c) $XeF_3^+$ (d) $OXeF_4$

**8-46** A pair of $NO_2$ molecules can combine to form $N_2O_4$:

$$2\ NO_2(g) \longrightarrow N_2O_4(g)$$

Use the Lewis structures of these compounds to explain why.

**8-47** Which of the following molecules have the same electron configuration as the $N_2$ molecule?

(a) CO (b) NO (c) $CN^-$ (d) $NO^+$ (e) $NO^-$

### Exceptions to the Lewis Octet Rule

**8-48** Which of the following are exceptions to the Lewis octet rule?

(a) $CO_2$ (b) $BeF_2$ (c) $SF_4$ (d) $SO_3$

**8-49** Which of the following are exceptions to the Lewis octet rule?

(a) $BF_3$ (b) $H_2CO$ (c) $XeF_4$ (d) $IF_3$

### Lewis Structures of Molecules with Double or Triple Bonds

**8-50** Which of the following does not contain a double bond?

(a) $N_2$ (b) $CO_2$ (c) $C_2H_4$ (d) $NO_2$ (e) $SO_3^{2-}$

**8-51** Which of the following contains only single bonds?

(a) $CN^-$ (b) $NO^+$ (c) CO (d) $O_2^{2-}$ (e) $Cl_2CO$

**8-52** Ingestion of oxalic acid ($H_2C_2O_4$), which can be found in a variety of plants and vegetables, can produce nausea, vomiting, and diarrhea. When taken in excess, oxalic acid can be toxic. Draw the Lewis structure for this molecule. (Assume that the skeleton structure can be described as $HO_2CCO_2H$.)

### Lewis Structures and Nonbonding Electrons

**8-53** Determine the number of nonbonding pairs of electrons on the iodine atom in the following molecules:

(a) $I_2$ (b) $I_3^-$ (c) $IF_3$ (d) $ICl_4^-$

**8-54** One reason for writing Lewis structures is to determine which ions or molecules can act as bases. Bases invariably have one or more pairs of nonbonding electrons. Which of the following ions or molecules can act as a base?

(a) $OH^-$ (b) $O_2$ (c) $CO_3^{2-}$ (d) $Br^-$ (e) $NH_3$

**8-55** Which of the following cannot act as a base?

(a) $H_2S$ (b) $NH_4^+$ (c) $AlH_3$ (d) $CH_3^-$ (e) $NH_2^-$

### Resonance Hybrids

**8-56** Draw all of the possible Lewis structures for the $CO_3^{2-}$ ion.

**8-57** Draw the three Lewis structures that contribute to the resonance hybrid that describes the $SCN^-$ ion.

**8-58** Which of the following Lewis structures can contribute to the resonance hybrid that describes the electron structure of $N_2O$?

(a) :N≡N—Ȯ: (b) :N=N=Ȯ:

(c) :N̈—N≡O: (d) :N=N—Ö:

(e) :N≡N—Ö:

**8-59** Which of the following molecules exist as resonance hybrids?

(a) $HCO_2^-$ (b) $PH_3$ (c) HCN (d) $SO_4^{2-}$ (e) $NO_2$

**8-60** Which of the following molecules is not a resonance hybrid?

(a) $CO_2$ (b) $NO_2$ (c) $N_2O$ (d) $SO_3$

**8-61** Calculate the average S—O bond order in the $SO_3$ molecule.

**8-62** Calculate the average N—O bond order in $NO_2$, $NO_2^-$, and $NO_3^-$.

### Formal Charge

**8-63** Calculate the formal charge on the sulfur atom in the following molecules:

(a) $SO_2$ (b) $SO_3$ (c) $SO_3^{2-}$ (d) $SO_4^{2-}$

**8-64** Calculate the formal charge on the bromine atom in the following molecules:

(a) HBr (b) $Br_2$ (c) NaBr (d) HOBr (e) $BrF_5$

**8-65** Calculate the formal charge on the nitrogen atom in the following compounds:

(a) $NH_3$ (b) $NH_4^+$ (c) $N_2H_4$ (d) $NH_2^-$ (e) $Mg_3N_2$

**8-66** Calculate the formal charge on the nitrogen atom in the following molecules:

(a) $N_2O$ (b) NO (c) $NO_2$ (d) $N_2O_3$ (e) $N_2O_5$

**8-67** Calculate the formal charge on the boron and nitrogen atoms in $BF_3$, $NH_3$, and the $F_3B—NH_3$ molecule formed when $BF_3$ and $NH_3$ combine.

**8-68** Calculate the formal charge on the two different sulfur atoms in the thiosulfate, $S_2O_3^{2-}$, ion. Assume a skeleton structure that could be described as $S—SO_3$.

### Predicting the Shapes of Molecules

**8-69** Predict the geometry around the central atom in the following molecules:

(a) $PH_3$ (b) $GaH_3$ (c) $ICl_3$ (d) $XeF_3^+$

**8-70** Predict the geometry around the central atom in the following molecules:

(a) $PO_4^{3-}$ (b) $SO_4^{2-}$ (c) $XeO_4$ (d) $MnO_4^-$

**8-71** The same elements often form compounds with very different shapes. Predict the geometry around the central atom in the following molecules:

(a) $SnF_2$ (b) $SnF_3^-$ (c) $SnF_4$ (d) $SnF_6^{2-}$

**8-72** Sulfur reacts with fluorine to form a pair of neutral molecules, $SF_4$ and $SF_6$, that in turn form positive and negative ions. Predict the geometry around the sulfur atom in each of the following:

(a) $SF_3^+$ (b) $SF_4$ (c) $SF_5^-$ (d) $SF_6$

**8-73** Iodine and fluorine combine to form interhalogen compounds that can exist as either neutral molecules, positive ions, or negative ions. Predict the geometry around the iodine atom in each of the following:

(a) $IF_2^-$ (b) $IF_3$ (c) $IF_4^+$ (d) $IF_4^-$ (e) $IF_6^-$

**8-74** Predict the geometry around the central atom in each of the following oxides of nitrogen:

(a) $N_2O$ (b) NO (c) $NO_2$ (d) $NO_2^-$ (e) $NO_3^-$

**8-75** Predict the geometry around the central atom in the following molecules:

(a) $Hg(CH_3)_2$ (b) $Pb(CH_2CH_3)_4$

**8-76** Which of the following compounds is best described as T shaped?

(a) $XeF_3^+$ (b) $NO_3^-$ (c) $NH_3$ (d) $ClO_3^-$ (e) $SF_4$

**8-77** Which of the following molecules have the same shape or geometry?

(a) $NH_2^-$ and $H_2O$ (b) $NH_2^-$ and $BeH_2$ (c) $H_2O$ and $BeH_2$ (d) $NH_2^-$, $H_2O$, and $BeH_2$

**8-78** Which of the following molecules have the same shape or geometry?

(a) $SF_4$ and $CH_4$ (b) $CO_2$ and $H_2O$ (c) $CO_2$ and $BeH_2$ (d) $N_2O$ and $NO_2$ (e) $PCl_4^+$ and $PCl_4^-$

**8-79** Which of the following molecules are best described as bent, or angular?

(a) $H_2S$ (b) $CO_2$ (c) ClNO (d) $NH_2^-$ (e) $O_3$

**8-80** Which of the following molecules are planar?

(a) $SO_3$ (b) $SO_3^{2-}$ (c) $NO_3^-$ (d) $PF_3$ (e) $BH_3$

**8-81** Which of the following molecules are tetrahedral?

(a) $SiF_4$ (b) $CH_4$ (c) $NF_4^+$ (d) $BF_4^-$ (e) $TeF_4$

**8-82** Which of the following molecules are linear?

(a) $C_2H_2$ (b) $CO_2$ (c) $NO_2^-$ (d) $NO_2^+$ (e) $H_2O$

**8-83** Which of the following elements would form a linear compound with the formula $XO_2$?

(a) Li (b) Be (c) C (d) O (e) F

**8-84** Explain why the nonbonding electrons occupy equatorial positions in $ClF_3$, not axial positions.

### Hybrid Atomic Orbitals

**8-85** Determine the hybridization of the central atom in the following molecules:

(a) $NH_3$ (b) $NH_4^+$ (c) NO (d) $NO_2$ (e) $NO_2^+$

**8-86** Determine the hybridization of the central atom in the following molecules:

(a) $CH_4$ (b) $H_2CO$ (c) $HCO_2^-$

**8-87** Determine the hybridization of the central atom in the following molecules:

(a) $SF_4$ (b) $BrO_3^-$ (c) $XeF_3^+$ (d) $Cl_2CO$

### A Hybrid Atomic Orbital Description of Multiple Bonds

**8-88** Write the Lewis structures for molecules of ethylene, $C_2H_4$, and acetylene, $C_2H_2$. Use hybrid atomic orbitals to describe the bonding in these compounds.

**8-89** Write the Lewis structures for carbon monoxide and carbon dioxide. Use hybrid atomic orbitals to describe the bonding in these compounds.

### Molecular Orbital Theory for a Diatomic Molecule

**8-90** Describe the difference between $\sigma$ and $\pi$ molecular orbitals.

**8-91** Describe the difference between bonding and antibonding molecular orbitals.

**8-92** Explain why the difference between the energies of the $\sigma_{2p}$ bonding and $\sigma_{2p}{}^*$ antibonding orbitals is larger than the difference between the $\pi_{2p}$ bonding and $\pi_{2p}{}^*$ antibonding orbitals.

**8-93** Describe the molecular orbitals formed by the overlap of the following atomic orbitals. (Assume that the bond lies along the $z$ axis of the coordinate system.)

(a) $2s + 2s$ (b) $2p_x + 2p_x$ (c) $2p_y + 2p_y$
(d) $2p_z + 2p_z$ (e) $2s + 2p_z$

**8-94** Write the electron configuration for the following diatomic molecules and calculate the bond order in each molecule:

(a) $H_2$ (b) $C_2$ (c) $N_2$ (d) $O_2$ (e) $F_2$

**8-95** Use molecular orbital theory to predict whether the $H_2^+$, $H_2^-$, and $H_2^{2-}$ ions would be more stable or less stable than a neutral $H_2$ molecule.

**8-96** Use molecular orbital theory to explain why the oxygen–oxygen bond is stronger in the $O_2$ molecule than in the $O_2^{2-}$ ion.

**8-97** Use molecular orbital theory to predict whether the bond order in the superoxide ion, $O_2^-$, would be stronger or weaker than the bond order in a neutral $O_2$ molecule.

**8-98** Use molecular orbital theory to predict whether the peroxide ion, $O_2^{2-}$, should be paramagnetic.

**8-99** Write the electron configuration for the following diatomic molecules. Calculate the bond order in each molecule.

(a) HF (b) CO (c) $CN^-$ (d) $ClO^-$ (e) $NO^+$

**8-100** Which of the following molecules is paramagnetic?

(a) HF (b) CO (c) $CN^-$ (d) NO (e) $NO^+$

### Integrated Problems

**8-101** Chemical formulas usually give a good idea of the skeleton structure of a compound, but some formulas are misleading. The formulas of common acids are often written as follows: $HNO_3$, $H_2SO_4$, $H_3PO_4$, $HClO_4$, and so on. Write Lewis structures for these four acids, assuming that their skeleton structures are more appropriately described by the formulas $HONO_2$, $(HO)_2SO_2$, $(HO)_3PO$, and $HOClO_3$.

**8-102** Write Lewis structures for the following compounds and then describe why these compounds are unusual.

(a) NO (b) $NO_2$ (c) $ClO_2$ (d) $ClO_3$

**8-103** Which element forms a compound with the following Lewis structure?

$$\dot{\underset{\cdot}{\text{O}}}\!=\!\ddot{\text{X}}\!-\!\ddot{\underset{\cdot\cdot}{\text{O}}}\!:$$

(a) Al (b) Si (c) P (d) S (e) Cl

**8-104** Which element forms a compound with the following Lewis structure?

$$:\!\ddot{\underset{\cdot\cdot}{\text{Cl}}}\!-\!\ddot{\text{X}}\!=\!\dot{\underset{\cdot}{\text{O}}}$$

(a) Be (b) B (c) C (d) N (e) O (f) F

**8-105** Which element forms an ion with the formula $XF_6^{2-}$ that has no nonbonding electrons in the valence shell of the central atom?

(a) Ca (b) C (c) Si (d) S (e) P

**8-106** In which group of the periodic table does element $X$ belong if there are two pairs of nonbonding electrons in the valence shell of the central atom in the $XF_4^-$ ion?

**8-107** In which group of the periodic table does element $X$ belong if the distribution of electrons in the valence shell of the central atom in the $XCl_4^-$ ion is octahedral?

**8-108** In which group does element $X$ belong if the shape of the $XF_2^-$ ion is linear?

# Intersection

# DESCRIPTIVE CHEMISTRY OF THE ELEMENTS

In 1980, Fred Basolo, of Northwestern University, and Robert Parry, of the University of Utah, proposed the following definition for the term *descriptive chemistry:*

> . . . chemical reactions, syntheses, and commercial processes. It is the chemistry that students can often see, hear, and/or smell in lecture demonstrations and laboratory experiments. . . . It is the explosion which results when a balloon filled with oxygen and hydrogen ignites. It is the pungent odor of burning sulfur, the pink color which appears when Milk of Magnesia is added to an acid solution containing phenolphthalein. [*Journal of Chemical Education,* **57,** 773–777(1980)]

The next three chapters focus on the descriptive chemistry of the main-group elements—the metals and nonmetals at either end of the periodic table. The fourth chapter in this section discusses the chemistry of the transition metals.

The only way in which the human organism can make sense out of the chaos of information impinging on its senses is to search for patterns in this information. There are two ways in which the thousands of chemical reactions of the main-group elements can be categorized. Basolo and Parry proposed a classification system that groups reactions on the basis of similarities in their chemical equations. **Combination reactions,** for example, occur when two elements combine to form a compound:

$$2\,Mg(s) + O_2(g) \longrightarrow 2\,MgO(s)$$
$$P_4(s) + 5\,O_2(g) \longrightarrow P_4O_{10}(s)$$

Or when a pair of compounds combine:

$$MgO(s) + H_2O(l) \longrightarrow Mg(OH)_2(aq)$$
$$P_4O_{10}(s) + 6\,H_2O(l) \longrightarrow 4\,H_3PO_4(aq)$$
$$CaO(s) + SO_2(g) \longrightarrow CaSO_3(s)$$

The reaction between sodium bicarbonate and citric acid that occurs when an Alka-Seltzer™ tablet dissolves in water could be classified as a decomposition reaction, but it is more useful to think of it as a metathesis reaction in which none of the elements undergo a change in oxidation state.

$$NaHCO_3(s) + H^+(aq) \longrightarrow Na^+(aq) + CO_2(g) + H_2O(l)$$

**Decomposition reactions,** as you might expect, are the opposite of combination reactions. They occur when certain compounds are heated:

$$2\,KClO_3(s) \longrightarrow 2\,KCl(s) + 3\,O_2(g)$$
$$CaCO_3(s) \longrightarrow CaO(s) + CO_2(g)$$
$$(NH_4)_2Cr_2O_7(s) \longrightarrow N_2(g) + Cr_2O_3(s) + 4\,H_2O(g)$$

Or when an electric current is passed through a molten sample:

$$2\,H_2O(l) \xrightarrow{\text{electrolysis}} 2\,H_2(g) + O_2(g)$$

**Replacement reactions** involve the displacement of one element by another, such as the displacement of hydrogen by a metal:

$$2\,Na(s) + 2\,H_2O(l) \longrightarrow 2\,NaOH(aq) + H_2(g)$$
$$Mg(s) + 2\,HCl(aq) \longrightarrow MgCl_2(aq) + H_2(g)$$

They also include the displacement of one metal by another. At elevated temperatures, for example, molten sodium metal will react with molten aluminum chloride to give aluminum metal and sodium chloride:

$$3\,Na(l) + AlCl_3(l) \longrightarrow 3\,NaCl(l) + Al(l)$$

The reaction between 1.0 $M$ $Cd(NO_3)_2$ and 1.0 $M$ $Na_2S$ to form a bright yellow precipitate of CdS could be classified as a replacement reaction, but it is more useful to think of it as a metathesis reaction.

$$Cd(NO_3)_2(aq) + Na_2S(aq) \longrightarrow CdS(s) + 2\,NaNO_3(aq)$$

Basolo and Parry's classification system works, but it doesn't probe the processes that occur during the reaction. This text uses a classification system that divides chemical reactions into two categories on the basis of what happens to the oxidation states of the atoms or elements during the reaction. Reactions in which one or more atoms or elements undergo changes in oxidation state are called **oxidation–reduction reactions.** Those in which no change in oxidation state occurs are classified as **metathesis reactions.**

Molten iron being poured into a mold.

CHAPTER 9

# THE MAIN-GROUP METALS

As we have seen, the change from metal to nonmetal must be gradual as we go across a row or down a column of the periodic table. The metallic character of an element decreases as we go from left to right across a row of the periodic table:

Sec. 7.17

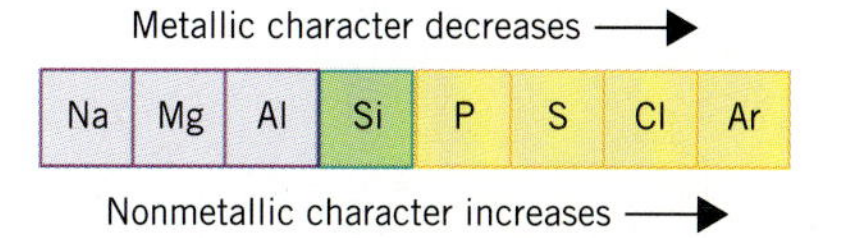


Metallic character increases, however, as we go down a column of the table:

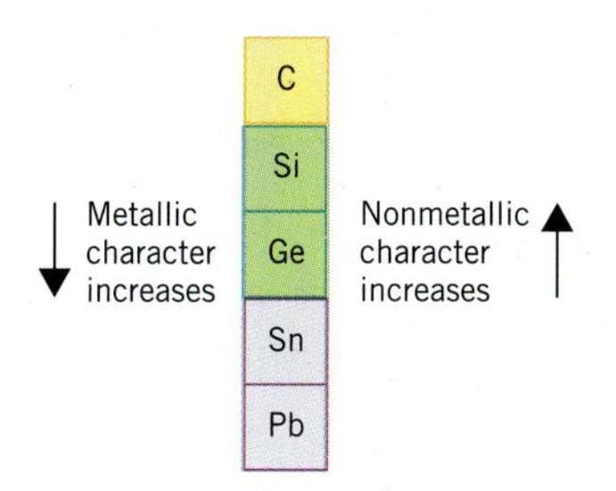


This chapter probes the following questions:

## POINTS OF INTEREST

- Why do some cereals use tiny pieces of iron metal, not $Fe^{3+}$ salts, as the source of the "iron" they provide?
- Why does NaCl taste "salty?"
- When we eat a banana, what kind of "potassium" enters our bodies—potassium metal or $K^+$ ions?
- What is Drano, and how does it work?
- Why is the inside of an aluminum soft-drink can lined with plastic? Why isn't this necessary when "tin" cans are used to package food?

## 9.1 THE ACTIVE METALS

The primary difference between metals is the ease with which they undergo chemical reactions. The elements toward the bottom left corner of the periodic table, which have the most metallic character, are the metals that are the most **active** in the sense of being the most **reactive** (see Figure 9.1). Lithium, sodium, and potassium all react with water, for example. The rate of this reaction increases as we go down this column, however, because these elements become more active as they become more metallic.

Sodium has a shiny, metallic luster when freshly cut.

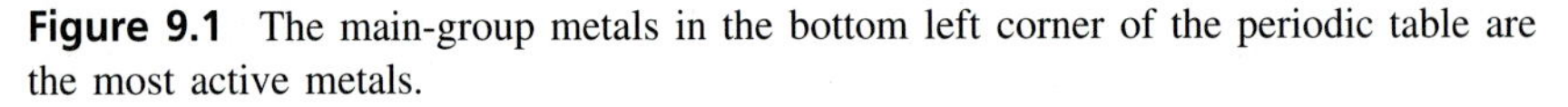


**Figure 9.1** The main-group metals in the bottom left corner of the periodic table are the most active metals.

The metals are often divided into four classes on the basis of their activity, as shown in Table 9.1. The most active metals are so reactive that they readily combine with the $O_2$ and $H_2O$ vapor in the atmosphere and therefore must be stored under an

**TABLE 9.1 Common Metals Divided into Classes on the Basis of Their Activity**

| |
|---|
| *Class I Metals: The Active Metals* |
| Li, Na, K, Rb, Cs (Group IA) |
| Ca, Sr, Ba (Group IIA) |
| *Class II Metals: The Less Active Metals* |
| Mg, Al, Zn, Mn |
| *Class III Metals: The Structural Metals* |
| Cr, Fe, Sn, Pb, Cu |
| *Class IV Metals: The Coinage Metals* |
| Ag, Au, Pt, Hg |

inert liquid, such as mineral oil. These metals are found exclusively in Groups IA (Group 1) and IIA (Group 2) of the periodic table.

Metals in the second class are slightly less active. They don't react with water at room temperature, but they react rapidly with acids. These less active metals can be used for a variety of purposes. Aluminum, for example, is used for alumunim foil and beverage cans. It is important to protect these metals from exposure to acids, however, so aluminum cans are lined with a plastic coating to prevent contact with the acidic soft drinks or fruit juices they contain.

The third class contains metals such as chromium, iron, tin, and lead, which react only with strong acids. It also contains even less active metals such as copper, which only dissolves when treated with acids that can oxidize the metal. Metals in the fourth class are so unreactive they are essentially inert at room temperature. These metals are ideal for making jewelry or coins because they do not react with the vast majority of the substances with which they come into daily contact. As a result, they are often called the "coinage metals."

The reaction between aluminum foil and concentrated hydrochloric acid.

### CHECKPOINT

Use the fact that class III metals dissolve in strong acid to explain why some cereals use tiny pieces of iron metal, not $Fe^{3+}$ salts, as the source of the "iron" they provide. (Hint: Stomach fluid is equivalent to a 1 *M* HCl solution.)

## 9.2 GROUP IA: THE ALKALI METALS

The metals in Group IA include lithium (Li), sodium (Na), potassium (K), rubidium (Rb), cesium (Cs), and francium (Fr). These elements are called the **alkali metals** because they all form hydroxides (such as NaOH) that were once known as *alkalies*. Sodium and potassiuim are relatively common elements. In fact, they are among the eight most abundant elements in the earth's crust, as shown in Table 9.2. The other

**TABLE 9.2 The Percent by Weight of the Most Common Elements in the Earth's Crust**

| Element | Percent by Weight |
|---|---|
| O | 46.60 |
| Si | 27.72 |
| Al | 8.13 |
| Fe | 5.00 |
| Ca | 3.63 |
| Na | 2.83 |
| K | 2.59 |
| Mg | 2.09 |

alkali metals are much less abundant. Discussions of the chemistry of alkali metals therefore focus on sodium and potassium.

Solutions of solvated electrons prepared by dissolving sodium metal in liquid ammonia have a characteristic blue color. More concentrated solutions are so intensely colored they look like metals.

The electron configurations of the alkali metals are characterized by a single valence electron:

| | | | |
|---|---|---|---|
| Li: | [He] $2s^1$ | Rb: | [Kr] $5s^1$ |
| Na: | [Ne] $3s^1$ | Cs: | [Xe] $6s^1$ |
| K: | [Ar] $4s^1$ | Fr: | [Rn] $7s^1$ |

As a result, the chemistry of these elements is dominated by their tendency to lose an electron to form positively charged ions ($Li^+$, $Na^+$, $K^+$). The assumption that these metals form compounds in which they have an oxidation state of +1 was safe enough to be incorporated into the rules for assigning oxidation numbers.

The alkali metals lose electrons so easily that sodium dissolves in liquid ammonia at temperatures below the boiling point of ammonia (−33°C) to give $Na^+$ ions and electrons:

$$Na(s) \xrightarrow{NH_3(l)} Na^+(sol) + e^-(sol)$$

The electrons given off in this reaction are bound to adjacent solvent ($NH_3$) molecules. They are therefore called *solvated* electrons. Dilute solutions of solvated electrons in ammonia have a characteristic blue color.

## 9.3 HALIDES, HYDRIDES, SULFIDES, NITRIDES, AND PHOSPHIDES

The alkali metals react with the nonmetals in Group VIIA ($F_2$, $Cl_2$, $Br_2$, $I_2$, and $At_2$) to form ionic compounds or salts. Chlorine, for example, reacts with sodium metal to produce sodium chloride, table salt:

$$2\,Na(s) + Cl_2(g) \longrightarrow 2\,NaCl(s)$$

The alkali metal halides are colorless or white solids, like the crystal of sodium chloride shown here.

Because they form salts with so many metals, the elements in Group VIIA are known as the **halogens.** This name comes from the Greek word for salt (*hals*) and the Greek word meaning "to produce" (*gennan*). The salts formed by the halogens are called **halides.** These salts include **fluorides** (LiF), **chlorides** (NaCl), **bromides** (KBr), and **iodides** (NaI).

The alkali metals react with hydrogen to form **hydrides,** such as potassium hydride, KH. They react with sulfur to form **sulfides,** such as sodium sulfide, $Na_2S$. Although $N_2$ is virtually inert to chemical reactions at room temperature, the most active metals react with nitrogen to form **nitrides,** such as lithium nitride, $Li_3N$. Elemental phosphorus reacts with the alkali metals to form **phosphides,** such as sodium phosphide, $Na_3P$.

**▶ CHECKPOINT**

Name the following compounds: (a) RbI (b) $Li_2S$ (c) $K_3N$

## RESEARCH IN THE 90s

### THE SEARCH TO UNDERSTAND SALTINESS

In the 1950s Lloyd Beidler of Florida State University showed that the $Na^+$ ions in table salt play a more important role than the $Cl^-$ ions in producing a ''salty'' taste when they encounter the taste cells on the surface of the tongue. The $Na^+$ ions apparently enter these cells through special channels in the cell membrane. This reduces the negative charge within the taste cell, which changes the voltage across the cell membrane, thereby exciting the cell. The excited cells release a neurotransmitter, which stimulates an adjacent nerve cell that begins the process of carrying the signal to the brain.

Beidler found that the chloride ion has an effect on the magnitude of the response of the taste cells to $Na^+$ ions. In 1987, Harry Harper, a physiologist at Stauffer Chemical Company, proposed a possible explanation for the effect of the $Cl^-$ ion on the perception of saltiness. He suggested that the $Cl^-$ ions diffuse through junctions *between* taste cells in the epithelium on the surface of the tongue, as shown in Figure 9.2. By altering the concentration of negative ions that surround the taste cells, these ions modify the cells' response to the $Na^+$ ion.

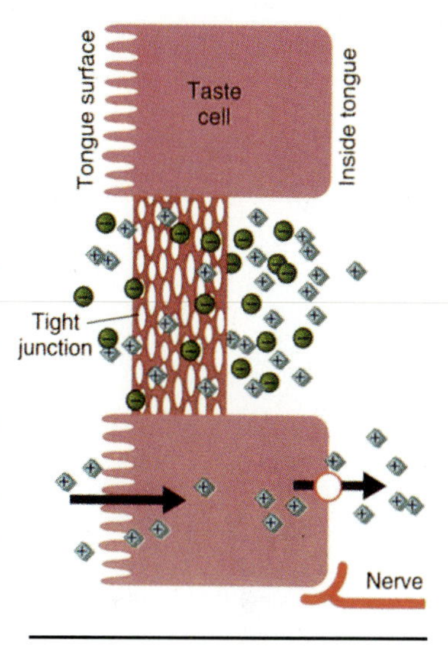


**Figure 9.2** Chloride ions pass easily through the junction between taste cells on the surface of the tongue. Larger ions, such as the acetate ion, cannot pass through this junction.

John DeSimone, Gerald Heck, and Qing Ye of Virginia Commonwealth University recently reported experimental evidence to support Harper's hypothesis [*Science,* **254,** 654 (1991)]. They developed an apparatus that could simultaneously deliver solutions to a rat's tongue and measure the voltage across the surface layer of cells that line the tongue. They found that sodium acetate (NaOAc) increased the voltage across the surface layer of cells but that sodium chloride (NaCl) produced only a small change in this voltage.

These results suggest that both $Na^+$ and $Cl^-$ ions can diffuse across the layer of cells on the tongue into the epithelium. The $OAc^-$ ion, on the other hand, is too large to diffuse into the epithelium and surround the taste cell the way the $Cl^-$ ion can.

When $Na^+$ ions diffuse into the taste cell and $Cl^-$ ions diffuse through the junction between these cells, we get a voltage *across the cell membrane* that excites the cell. When $Na^+$ ions diffuse into the taste cell and $OAc^-$ ions remain behind, we get a voltage *across the epithelium* that interferes with the electrical signal that excites the cell. It therefore takes significantly larger amounts of NaOAc to produce the response to the $Na^+$ ion obtained with a sample of NaCl.

## 9.4 PREDICTING THE PRODUCT OF MAIN-GROUP METAL REACTIONS

Secs. 7.14, 7.17

The product of many reactions between main-group metals and other elements can be predicted from the electron configurations of the elements. Consider the reaction between sodium and chlorine to form sodium chloride, for example. As we have seen, it takes more energy to remove an electron from a sodium atom to form an $Na^+$ ion than we get back when this electron is added to a chlorine atom to form a $Cl^-$ ion. Once these ions are formed, however, the force of attraction between these ions liberates enough energy to make the following reaction exothermic:

$$Na(s) + \tfrac{1}{2}\,Cl_2(g) \longrightarrow NaCl(s) \qquad \Delta H^\circ = -411.3 \text{ kJ/mol}$$

The net effect of this reaction is to transfer one electron from a neutral sodium atom to a neutral chlorine atom to form $Na^+$ and $Cl^-$ ions that have filled-shell configurations:

$$Na\cdot + :\ddot{\underset{..}{Cl}}\cdot \longrightarrow [Na^+][:\ddot{\underset{..}{Cl}}:^-]$$

Potassium and hydrogen have the following electron configurations:

$$K: \quad [Ar]\ 4s^1 \qquad H: \quad 1s^1$$

When these elements react, an electron has to be transferred from one element to the other. We can determine which element will lose an electron by comparing the first ionization energy for potassium (418.8 kJ/mol) with that for hydrogen (1312.0 kJ/mol). Potassium is much more likely to lose an electron in this reaction, which means that hydrogen gains an electron to form $K^+$ and $H^-$ ions.

$$K\cdot + H\cdot \longrightarrow [K^+][H:^-]$$

## EXERCISE 9.1

Write a balanced equation for the following reaction:

$$Li(s) + O_2(s) \longrightarrow$$

**SOLUTION** We can start by comparing the ionization energies of lithium (520.2 kJ/mol) and oxygen (1313.9 kJ/mol) or the electronegativities of lithium (0.98) and oxygen (3.44) to predict which element is likely to lose electrons and which is likely to gain electrons. We then use the electron configurations of the elements to predict the ions that form in this reaction:

$$Li: \quad [He]\ 2s^1 \qquad Li^+: \quad [He]$$
$$O: \quad [He]\ 2s^2\ 2p^4 \qquad O^{2-}: \quad [He]\ 2s^2\ 2p^6 = [Ne]$$

We then combine these ions to predict the product of the reaction, $Li_2O$, and write a balanced equation for this reaction:

$$4\ Li(s) + O_2(g) \longrightarrow 2\ Li_2O(s)$$

## 9.5 OXIDES, PEROXIDES, AND SUPEROXIDES

The method just used to predict the products of reactions of the main-group metals is simple, yet remarkably powerful. Exceptions to its predictions arise, however, when very active metals react with oxygen, which is one of the most reactive nonmetals.

Lithium is well behaved. It reacts with $O_2$ to form an **oxide,** as predicted in Exercise 9.1:

$$4\ Li(s) + O_2(g) \longrightarrow 2\ Li_2O(s)$$

Sodium, however, reacts with $O_2$ under normal conditions to form a compound that contains twice as much oxygen:

$$2\ Na(s) + O_2(g) \longrightarrow Na_2O_2(s)$$

Compounds such as $Na_2O_2$ that are unusually rich in oxygen are called **peroxides.** The prefix *per-* means "above normal" or "excessive." $Na_2O_2$ is a peroxide because it contains more than the usual amount of oxygen.

The difference between oxides and peroxides can be understood by remembering the basic assumption behind the chemistry of the alkali metals: These elements form compounds in which they are present as +1 ions. Oxides, such as $Li_2O$, contain the $O^{2-}$ ion. Peroxides, such as $Na_2O_2$, contain the $O_2^{2-}$ ion.

The formation of sodium peroxide can be explained by assuming that sodium is so reactive that the metal is consumed before each $O_2$ molecule can combine with enough sodium to form $Na_2O$. This explanation is supported by the fact that sodium reacts with $O_2$ in the presence of a large excess of the metal—or a limited amount of $O_2$—to form the oxide expected when this reaction goes to completion:

$$4\ Na(s) + O_2(g) \longrightarrow 4\ Na_2O(s)$$

It is also consistent with the fact that the very active alkali metals—potassium, rubidium, and cesium—react so rapidly with oxygen they form **superoxides,** in which the alkali metal reacts with $O_2$ in a 1:1 mole ratio:

$$K(s) + O_2(g) \longrightarrow KO_2(s)$$

Potassium superoxide forms on the surface of potassium metal, even when the metal is stored under an inert solvent. As a result, old pieces of potassium metal are potentially dangerous. When someone tries to cut the metal, the pressure of the knife pushing down on the area where the superoxide touches the metal can induce the following reaction:

$$KO_2(s) + 3\ K(s) \longrightarrow 2\ K_2O(s)$$

Because potassium oxide is more stable than potassium superoxide, this reaction gives off enough energy to boil potassium metal off the surface, which reacts explosively with the oxygen and water vapor in the atmosphere.

**CHECKPOINT**

Which Group IIA metal, Mg, Ca, or Ba, would be the most likely to form a peroxide?

## 9.6 REACTIONS WITH $H_2O$ AND $NH_3$

### ACTIVE METALS AND WATER

The most common demonstration of the reactivity of the active metals involves dropping pieces of lithium, sodium, and potassium into water. Students who watch this demonstration often notice that lithium reacts slowly with water, that sodium reacts much more rapidly, and that potassium reacts violently. A number of other observations can be gleaned from this demonstration:

1. Each metal floats on the surface of the water, which suggests that these metals have densities less than 1 g/cm$^3$.
2. Bubbles form when lithium reacts with water, which indicates that a gas is given off in this reaction.
3. These bubbles form on the metal surface, which suggests that the reaction occurs between the metal and water and is not the result of the reaction of water with an atom or ion released when the metal dissolves.
4. Sodium reacts so rapidly that no bubbles of gas are observed. Instead, a stream of ''smoke'' can be seen rising off the surface of the liquid.

5. Enough heat is given off when sodium or potassium reacts with water to melt the metal.

Sec. 6.12

6. The gas given off when sodium reacts with water often catches fire. When this happens, the flame has the distinct yellow color of excited sodium atoms.
7. Potassium reacts so violently with water that the metal often explodes off the surface of the liquid. When this happens, the flame has the distinct violet color of excited potassium atoms.

Sodium is so active it reacts with water the moment it comes in contact with it. Enough heat is given off to ignite the hydrogen gas produced in this reaction.

The model used to predict the products of reactions between main-group metals and nonmetals can be extended to predict what will happen when these metals react with covalent compounds. Let's start by applying this model to the reaction between sodium metal and water:

$$Na(s) + H_2O(l) \longrightarrow ?$$

One thing is certain: In the course of this reaction, each sodium atom will lose an electron to form a $Na^+$ ion:

$$Na \longrightarrow Na^+ + e^-$$

Figure 9.3 shows what happens when these electrons are picked up by neighboring water molecules:

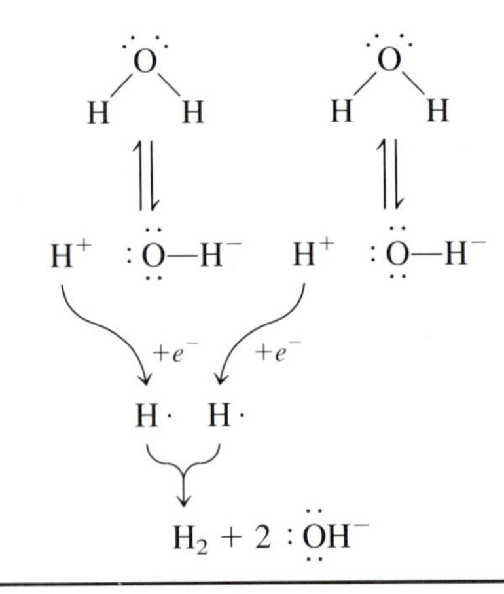


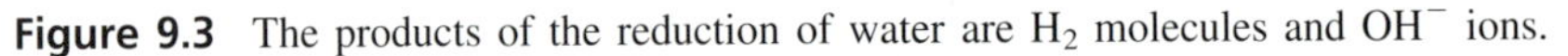

**Figure 9.3** The products of the reduction of water are $H_2$ molecules and $OH^-$ ions.

In a formal sense, hydrogen atoms in a +1 oxidation state pick up electrons to form neutral hydrogen atoms, which combine to form $H_2$ molecules:

$$2\,H^+ + 2\,e^- \longrightarrow H_2$$

Subtracting a positively charged $H^+$ ion from a neutral $H_2O$ molecule leaves an $OH^-$ ion. The reaction that occurs when water molecules gain electrons therefore can be written:

$$2\,H_2O + 2\,e^- \longrightarrow H_2 + 2\,OH^-$$

In retrospect, we can see that this discussion of the reaction between sodium and water divides the reaction into two halves. One half-reaction describes what happens when sodium atoms lose electrons. The other half-reaction describes what happens when water molecules gain these electrons. By combining these half-reactions so that electrons are neither created nor destroyed, we can obtain an overall equation for the reaction:

$$\begin{array}{rcl} 2\,[\mathrm{Na} & \longrightarrow & \mathrm{Na}^+ + \mathrm{e}^-] \\ 2\,\mathrm{H_2O} + 2\,\mathrm{e}^- & \longrightarrow & \mathrm{H_2} + 2\,\mathrm{OH}^- \\ \hline 2\,\mathrm{Na} + 2\,\mathrm{H_2O} & \longrightarrow & 2\,\mathrm{Na}^+ + 2\,\mathrm{OH}^- + \mathrm{H_2} \end{array}$$

The balanced equation for this reaction can be written as follows:

$$2\,\mathrm{Na}(s) + 2\,\mathrm{H_2O}(l) \longrightarrow 2\,\mathrm{Na}^+(aq) + 2\,\mathrm{OH}^-(aq) + \mathrm{H_2}(g)$$

**CHECKPOINT**

Magnesium metal burns rapidly in air to form magnesium oxide. This reaction gives off an enormous amount of energy in the form of light and is used in both flares and fireworks. Use the fact that magnesium is an active metal to predict what happens when someone tries to extinguish a magnesium flare by pouring water on the reaction.

When heated, magnesium metal bursts into flame as it reacts with the $O_2$ and $N_2$ in the atmosphere to form MgO and a trace of $Mg_3N_2$. The reaction gives off both heat and a brilliant white light.

## ACTIVE METALS AND AMMONIA

The same line of reasoning can be used to predict what happens when an alkali metal, such as potassium, reacts with liquid ammonia ($NH_3$). We can start by noting that the potassium atoms will lose one electron each:

$$\mathrm{K} \longrightarrow \mathrm{K}^+ + \mathrm{e}^-$$

Figure 9.4 shows what happens to these electrons.

Hydrogen atoms in a +1 oxidation state gain electrons to form neutral hydrogen atoms, which combine to form $H_2$ molecules:

$$2\,\mathrm{H}^+ + 2\,\mathrm{e}^- \longrightarrow \mathrm{H_2}$$

Removing an $H^+$ ion from an $NH_3$ molecule leaves a negatively charged $NH_2^-$ ion. The following equation therefore describes what happens when ammonia gains electrons:

$$2\,\mathrm{NH_3} + 2\,\mathrm{e}^- \longrightarrow \mathrm{H_2} + 2\,\mathrm{NH_2}^-$$

Combining the two halves of this reaction so that electrons are conserved gives the overall equation for the reaction:

$$\begin{array}{rcl} 2\,[\mathrm{K} & \longrightarrow & \mathrm{K}^+ + \mathrm{e}^-] \\ 2\,\mathrm{NH_3} + 2\,\mathrm{e}^- & \longrightarrow & \mathrm{H_2} + 2\,\mathrm{NH_2}^- \\ \hline 2\,\mathrm{K} + 2\,\mathrm{NH_3} & \longrightarrow & 2\,\mathrm{K}^+ + 2\,\mathrm{NH_2}^- + \mathrm{H_2} \end{array}$$

The balanced equation for this reaction is written:

$$2\,\mathrm{K}(s) + 2\,\mathrm{NH_3}(l) \longrightarrow 2\,\mathrm{K}^+(sol) + 2\,\mathrm{NH_2}^-(sol) + \mathrm{H_2}(g)$$

(The symbol *sol* in this equation indicates that the $K^+$ and $NH_2^-$ ions in this solution are solvated by neighboring $NH_3$ molecules.)

**CHECKPOINT**

Write a balanced equation for the reaction between calcium metal and liquid ammonia.

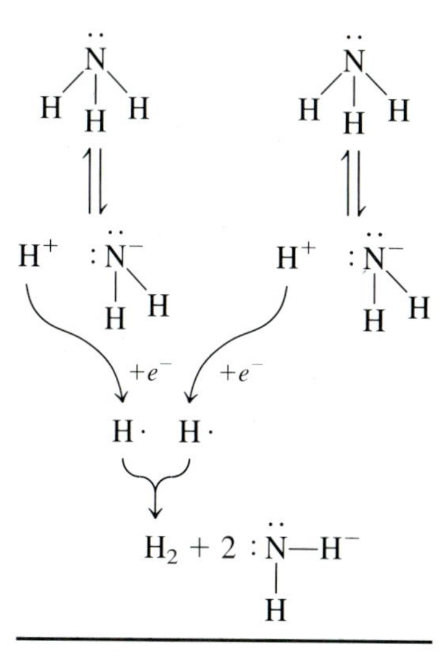


**Figure 9.4** The products of the reduction of ammonia are $H_2$ molecules and $NH_2^-$ ions.

## 9.7 GROUP IIA: THE ALKALINE-EARTH METALS

The elements in Group IIA (Be, Mg, Ca, Sr, Ba, and Ra) are all metals, and all but Be and Mg are active metals. These elements are often called the **alkaline-earth metals.** The term *alkaline* reflects the fact that many compounds of these metals are basic or alkaline. The term *earth* was historically used to describe the fact that many of these compounds are insoluble in water. Most of the chemistry of the alkaline-earth metals can be predicted from the behavior of the alkali metals in Group IA. Three points should be kept in mind, however:

1. The alkaline-earth metals tend to lose two electrons to form $M^{2+}$ ions ($Be^{2+}$, $Mg^{2+}$, $Ca^{2+}$, and so on).
2. These metals are less reactive than the neighboring alkali metals. Magnesium is less active than sodium; calcium is less active than potassium; and so on.
3. These metals become more active as we go down the column. Magnesium is more active than beryllium; calcium is more active than magnesium; and so on.

The alkaline-earth metals react with nonmetals to give the products expected from the electron configurations of the elements:

$$Mg(s) + Cl_2(g) \longrightarrow MgCl_2(s)$$
$$3\,Mg(s) + N_2(g) \longrightarrow Mg_3N_2(s)$$
$$Ca(s) + H_2(g) \longrightarrow CaH_2(s)$$

Because they are not as active as the alkali metals, most of these elements form oxides:

$$2\,Mg(s) + O_2(g) \longrightarrow 2\,MgO(s)$$

Calcium, strontium, and barium can also form peroxides:

$$Ba(s) + O_2(g) \longrightarrow BaO_2(s)$$

The more active members of Group IIA (Ca, Sr, and Ba) react with water at room temperature. The products of these reactions are what we might expect. Calcium, for example, loses two electrons to form $Ca^{2+}$ ions when it reacts with water:

$$Ca \longrightarrow Ca^{2+} + 2\,e^-$$

These electrons are picked up by water molecules to form $H_2$ gas and $OH^-$ ions:

$$2\,H_2O + 2\,e^- \longrightarrow H_2 + 2\,OH^-$$

Combining the two halves of the reaction so that electrons are conserved gives the following result:

$$Ca(s) + 2\,H_2O(l) \longrightarrow Ca^{2+}(aq) + 2\,OH^-(aq) + H_2(g)$$

Although Mg does not react with water at room temperature, it will react with steam. The products of this reaction can't be aqueous $Mg^{2+}$ and $OH^-$ ions because there is no liquid water around to stabilize these ions. The products of this reaction are $H_2$ gas and magnesium oxide, MgO:

$$Mg(s) + H_2O(g) \longrightarrow MgO(s) + H_2(g)$$

### EXERCISE 9.2

Magnesium reacts with hydrogen to form compound *A*, which is a white solid at room temperature. It also reacts with hydrochloric acid to form gas *B* and an aqueous solution of compound *C*. Identify the products of these reactions and write balanced equations for each reaction.

**SOLUTION** Magnesium reacts with hydrogen to form magnesium hydride (*A*), which contains the $Mg^{2+}$ and $H^-$ ions:

$$Mg(s) + H_2(g) \longrightarrow MgH_2(s)$$

Magnesium loses two electrons to form $Mg^{2+}$ ions when it reacts with hydrochloric acid. The electrons given off in this half of the reaction are picked up by the $H^+$ ions from hydrochloric acid to form neutral hydrogen atoms, which combine to give $H_2$ molecules:

$$\begin{array}{rcl} Mg & \longrightarrow & Mg^{2+} + 2\,e^- \\ 2\,H^+ + 2\,e^- & \longrightarrow & H_2 \\ \hline Mg + 2\,H^+ & \longrightarrow & Mg^{2+} + H_2 \end{array}$$

The products of the reaction between magnesium metal and hydrochloric acid are therefore $H_2$ gas (*B*) and an aqueous solution of $Mg^{2+}$ and $Cl^-$ ions (*C*):

$$Mg(s) + 2\,HCl(aq) \longrightarrow \underset{B}{H_2(g)} + \underbrace{Mg^{2+}(aq) + 2\,Cl^-(aq)}_{C}$$

## 9.8 GROUP IIIA: THE CHEMISTRY OF ALUMINUM

The elements in Group IIIA (B, Al, Ga, In, and Tl) can be divided into three classes.

1. Boron is the only element in this group that is not a metal. It behaves like a semimetal or even a nonmetal.
2. Aluminum is the third most abundant element in the earth's crust. It is just slightly less reactive than the active metals.
3. The other three elements in this group are active metals, but they are so scarce they are of limited interest. Gallium, indium, and thallium combined total less than $10^{-10}\%$ of the earth's crust.

Discussions of the Group IIIA metals therefore revolve around the chemistry of aluminum, which can be understood by assuming that the metal reacts to form compounds in which it has an oxidation number of +3.

$$\begin{array}{rcl} 2\,Al(s) + 3\,Br_2(l) & \longrightarrow & Al_2Br_6(s) \\ 4\,Al(s) + 3\,O_2(g) & \longrightarrow & 2\,Al_2O_3(s) \end{array}$$

Aluminum reacts with concentrated acids to give $Al^{3+}$ ions and $H_2$ gas:

$$2\,Al(s) + 6\,H^+(aq) \longrightarrow 2\,Al^{3+}(aq) + 3\,H_2(g)$$

It also reacts with concentrated bases to give $H_2$ gas and the aluminate ion, $Al(OH)_4^-$, in which aluminum is in the +3 oxidation state:

$$2\,Al(s) + 2\,OH^-(aq) + 6\,H_2O(l) \longrightarrow 2\,Al(OH)_4^-(aq) + 3\,H_2(g)$$

You can experience the reaction between aluminum metal and acid by using an aluminum pot with a recipe that calls for large quantities of vinegar. The inside of the pot will look like new when it is washed because a finite amount of the aluminum metal on the inner surface of the pot dissolves when it reacts with the acid in vinegar. The reaction between aluminum metal and concentrated base is even easier to experience. Solid drain cleaners, such as Drano, are mixtures of aluminum metal and sodium hydroxide. When they are mixed with water, reaction between the aluminum metal and the $OH^-$ ion generates $H_2$ gas and gives off a significant amount of energy in the form of heat. The result is a hot, concentrated solution of the $OH^-$ ion, which attacks whatever has plugged the drain.

**CHECKPOINT**

Chemists argue that aluminum is much more reactive than iron. Most people believe the opposite. Describe the common uses of aluminum metal that make most people believe aluminum is less reactive than iron.

## 9.9 MAGNESIUM AND ALUMINUM AS STRUCTURAL METALS

Aluminum is becoming increasingly important as a structural metal. It is used for soft-drink cans, electrical wire, pots and pans, and the siding on houses. It is slowly replacing steel in cars and trucks and is an important metal in commercial aircraft. The property that makes aluminum so attractive as a structural metal is its density ($2.70\ g/cm^3$), which is only about one-third the density of copper ($8.92\ g/cm^3$) or steel ($7.8\ g/cm^3$). Aluminum is an excellent conductor of heat and electricity. It is also unusually malleable and ductile. It can be drawn therefore into electrical wire or rolled into thin sheets of aluminum foil.

Magnesium is even less dense ($1.74\ g/cm^3$). It is therefore used when reducing the weight of an object is particularly important. Magnesium alloys can be found in the frames of luggage, in hand trucks, ladders, camera bodies, airplane components, and the wheels of race cars.

Magnesium is so reactive that a variety of products ranging from flash bulbs to fireworks and military flares are based on the reaction:

$$2\ Mg(s) + O_2(g) \longrightarrow 2\ MgO(s)$$

The reaction between $Fe_2O_3$ and powdered aluminum gives off so much energy it is called the ''thermite'' reaction.

Aluminum has such a high affinity for oxygen that mixtures of $Fe_2O_3$ and powdered aluminum give off enormous amounts of energy when they react to form molten iron metal and $Al_2O_3$ in a reaction known as the *thermite reaction:*

$$Fe_2O_3(s) + 2\ Al(s) \longrightarrow Al_2O_3(s) + 2\ Fe(l) \qquad \Delta H^\circ = -851.5\ kJ$$

This raises an interesting question: If aluminum has a higher affinity for oxygen than iron and magnesium is even more reactive than aluminum, how can these metals be used to make products, such as ladders, pots and pans, aluminum foil, and ''mag'' wheels, that seem to be chemically inert?

The answer is subtle. Magnesium and aluminum are so active that they react with $O_2$ in the air to form a thin coating of MgO or $Al_2O_3$. MgO then slowly reacts with carbon dioxide in the air to form magnesium carbonate:

$$MgO(s) + CO_2(g) \longrightarrow MgCO_3(s)$$

Although the coating may be only 5 nanometers thick, it covers the metal surface so evenly that the metal beneath this layer is protected from further reaction.

If we immerse these metals in an acid, however, the acid washes off the $MgCO_3$ or $Al_2O_3$ coating. This exposes a fresh layer of metal that can react with either oxygen or acid. If the acid is strong enough, these metals will dissolve within a few minutes.

## 9.10 GROUP IVA: TIN AND LEAD

The elements in Group IVA can be divided into three classes: (1) carbon, which is a nonmetal; (2) silicon and germanium, which are semimetals; and (3) tin and lead, which are metals. Tin and lead are among the oldest metals known. Lead was first used to glaze pottery in Egypt more than 7000 years ago, and tin was combined with copper to form bronze as early as 3500 B.C.

Tin and lead are much less reactive than any of the main-group metals discussed so far. According to the argument that elements become more metallic—and therefore more active—as we go down a column of the periodic table, lead should be more reactive than tin.

Tin and lead are less active than aluminum. Lead is a bit more active than tin, however, and finely divided lead metal will glow, or even burst into flame, when it comes in contact with air.

### CHECKPOINT

The inside surfaces of aluminum soft-drink cans are coated with plastic to keep the metal from dissolving on contact with the acid in most soft drinks. The "tin" cans used for packaging food are actually steel cans that have been coated with a thin layer of tin. Use the fact that tin is less active than aluminum to explain why the inner surfaces of tin cans don't have to be coated with plastic.

---

Lead reacts with air to form a thin coating of PbO and/or $PbCO_3$, which protects the metal from further reaction.

$$2\ Pb(s) + O_2(g) \longrightarrow 2\ PbO(s)$$
$$PbO(s) + CO_2(g) \longrightarrow PbCO_3(s)$$

When finely divided, lead is **pyrophoric**—it bursts into flame in the presence of oxygen.

Tin does not react with either air or water at room temperature. When heated until white hot, tin reacts with air to form $SnO_2$:

$$Sn(s) + O_2(g) \longrightarrow SnO_2(s)$$

At high temperatures it also reacts with steam to give $SnO_2$:

$$Sn(s) + 2\ H_2O(g) \longrightarrow SnO_2(s) + 2\ H_2(g)$$

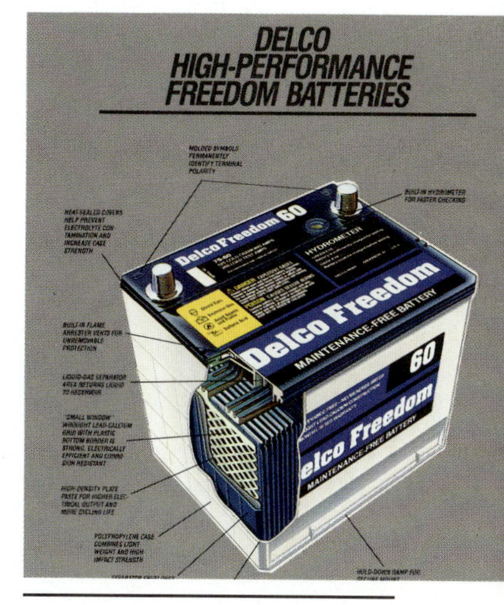


The fact that lead storage batteries contain plates of lead metal immersed in sulfuric acid is evidence for the fact that lead is less reactive than the other main-group metals.

Tin and lead are both less active than aluminum. Neither metal reacts with either dilute hydrochloric acid or dilute sulfuric acid at room temperature. Tin, when heated, reacts with either concentrated hydrochloric acid or concentrated sulfuric acid:

$$Sn(s) + 2\ H^+(aq) \longrightarrow Sn^{2+}(aq) + H_2(g)$$

Lead reacts slowly with hydrochloric acid at room temperature and with concentrated sulfuric acid at temperatures above 200°C:

$$Pb(s) + 2\ H^+(aq) \longrightarrow Pb^{2+}(aq) + H_2(g)$$

Tin and lead are the first main-group metals we have encountered that form compounds in more than one oxidation state. As a rule, the lower oxidation state becomes more stable as we go down a column of the periodic table. Lead, for example, is more likely to be found in the +2 oxidation state than tin. Thus, while tin reacts with oxygen at high temperatures to form $SnO_2$, lead forms PbO.

## 9.11 OXIDATION–REDUCTION REACTIONS

The reactions between a main-group metal and a nonmetallic element or compound described in this chapter are all examples of **oxidation–reduction reactions.**

The term **oxidation** was originally used to describe reactions in which an element combines with oxygen. The reaction between magnesium metal and oxygen to form magnesium oxide, for example, involves the oxidation of magnesium:

$$2\ Mg(s) + O_2(g) \longrightarrow 2\ MgO(s)$$

oxidation

The term **reduction** comes from the Latin stem meaning "to lead back." Anything that leads back to magnesium metal therefore involves reduction. The reaction between magnesium oxide and carbon at 2000°C to form magnesium metal and carbon monoxide is an example of the reduction of magnesium oxide to magnesium metal:

$$MgO(s) + C(s) \longrightarrow Mg(s) + CO(g)$$

reduction

After electrons were discovered, chemists became convinced that oxidation–reduction reactions involved the transfer of electrons from one atom to another. From this perspective, the reaction between magnesium and oxygen is written as follows:

$$2\ Mg + O_2 \longrightarrow 2\ [Mg^{2+}][O^{2-}]$$

In the course of this reaction, each magnesium atom loses two electrons to form a $Mg^{2+}$ ion:

$$Mg \longrightarrow Mg^{2+} + 2\ e^-$$

And, each $O_2$ molecule gains four electrons to form a pair of $O^{2-}$ ions:

$$O_2 + 4\ e^- \longrightarrow 2\ O^{2-}$$

Because electrons are neither created nor destroyed in a chemical reaction, oxidation and reduction are linked. It is impossible to have one without the other, as shown in Figure 9.5.

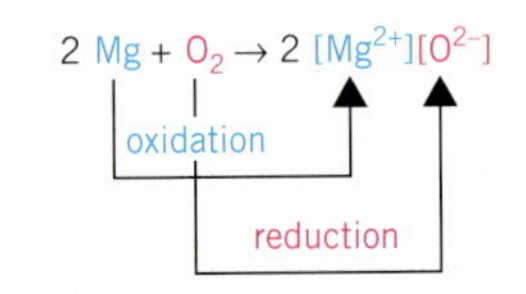


**Figure 9.5** Oxidation cannot occur in the absence of reduction.

## EXERCISE 9.3

Determine which element is oxidized and which is reduced when lithium reacts with nitrogen to form lithium nitride:

$$6\ Li(s) + N_2(g) \longrightarrow 2\ Li_3N(s)$$

**SOLUTION** In the course of this reaction, the lithium atoms lose one electron to form $Li^+$ ions, and the nitrogen atoms gain three electrons to form $N^{3-}$ ions. Lithium is therefore oxidized and nitrogen is reduced:

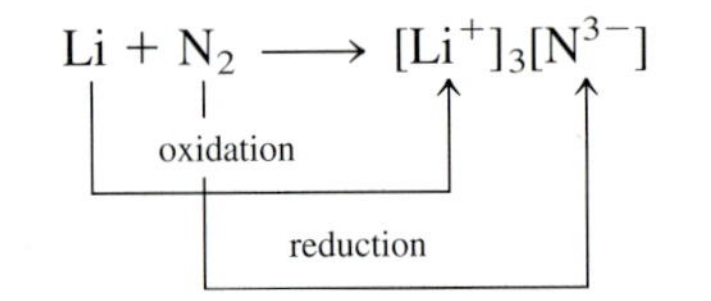


## 9.12 THE ROLE OF OXIDATION NUMBERS IN OXIDATION–REDUCTION REACTIONS

Chemists eventually extended the idea of oxidation and reduction to reactions that do not formally involve the transfer of electrons. Consider the following reaction, for example:

$$CO(g) + H_2O(g) \longrightarrow CO_2(g) + H_2(g)$$

As can be seen in Figure 9.6, the total number of electrons in the valence shell of each atom remains constant in this reaction.

$$:C{\equiv}O: + H{-}\ddot{\underset{..}{O}}{-}H \longrightarrow \ddot{O}{=}C{=}\ddot{O} + H{-}H$$

**Figure 9.6** Oxidation–reduction reactions don't always involve the transfer of electrons. They can also involve the transfer of an atom, such as an oxygen atom.

What changes in this reaction is the oxidation state of these atoms. The oxidation state of carbon increases from +2 to +4, while the oxidation state of the hydrogen decreases from +1 to 0.

$$CO + H_2O \longrightarrow CO_2 + H_2$$

+2 +1 +4 0

Oxidation and reduction are therefore best defined as follows:

***Oxidation* occurs when the oxidation number of an atom becomes larger. *Reduction* occurs when the oxidation number of an atom becomes smaller.**

## EXERCISE 9.4

Determine which atom is oxidized and which is reduced in the following reaction:

$$Sr(s) + 2\ H_2O(l) \longrightarrow Sr^{2+}(aq) + 2\ OH^-(aq) + H_2(g)$$

**SOLUTION** We begin by assigning an oxidation number to each component of the reaction:

$$Sr + 2\ H_2O \longrightarrow Sr^{2+} + 2\ OH^- + H_2$$

0 +1 −2 +2 −2 +1 0

Nothing happens to the oxidation number of the oxygen in this reaction, which means that oxygen is neither oxidized nor reduced. The oxidation number of strontium becomes larger (from 0 to +2), which means Sr is oxidized. The oxidation number of some (but not all) of the hydrogen atoms becomes smaller (from +1 to 0). Hydrogen is therefore reduced in this reaction.

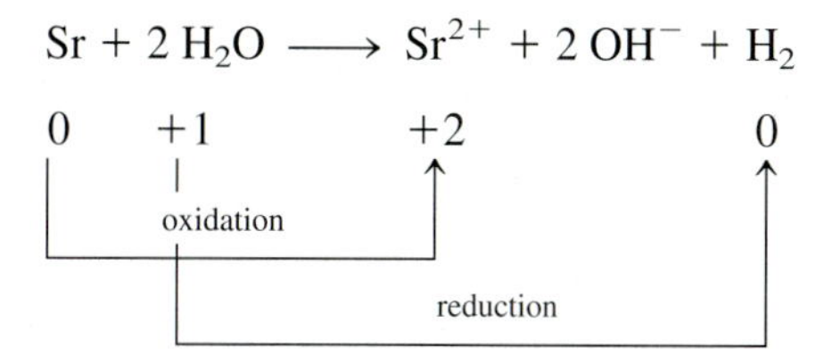


## 9.13 OXIDATION NUMBERS VERSUS THE TRUE CHARGE ON IONS

Discussions of the chemistry of the main-group metals illustrate the importance of understanding the balance between ionic and covalent bonding. In Chapter 8, we noted that the terms *ionic* and *covalent* describe the extremes of a continuum of bonding. There is some covalent character in even the most ionic compounds and vice versa.

Sec. 8.5

As we have seen, it is useful to think about the compounds of the main-group metals as if they contained positive and negative ions. The chemistry of magnesium oxide, for example, is easy to understand if we assume that MgO contains $Mg^{2+}$ and $O^{2-}$ ions. But no compounds are 100% ionic. There is experimental evidence, for example, that the true charge on the magnesium and oxygen atoms in MgO is +1.5 and −1.5.

Oxidation states provide a compromise between a powerful model of oxidation–reduction reactions based on the assumption that these compounds contain ions and our knowledge that the true charge on the ions in these compounds is not as large as this model predicts. By definition, the oxidation state of an atom is the charge that atom would carry if the compound were purely ionic.

For the active metals in Groups IA and IIA, the difference between the oxidation state of the metal atom and the charge on this atom is small enough to be ignored. The main-group metals in Groups IIIA and IVA, however, form compounds that have a significant amount of covalent character. It is misleading, for example, to assume that aluminum bromide contains $Al^{3+}$ and $Br^-$ ions. It actually exists as $Al_2Br_6$ molecules.

This problem becomes even more severe when we turn to the chemistry of the transition metals. MnO, for example, is ionic enough to be considered a salt that contains $Mn^{2+}$ and $O^{2-}$ ions. $Mn_2O_7$, on the other hand, is a covalent compound that boils at room temperature. It is therefore more useful to think about this compound as if it contained manganese in a +7 oxidation state, not $Mn^{7+}$ ions.

## 9.14 OXIDIZING AGENTS AND REDUCING AGENTS

So far we have focused on what happens when a particular element gains or loses electrons. Let's now consider the role that each element plays in the reaction.

When magnesium reacts with oxygen, the magnesium atoms donate electrons to $O_2$ molecules and thereby reduce the oxygen. Magnesium therefore acts as a **reducing agent** in this reaction:

$$\underset{\text{reducing agent}}{2\,Mg} + O_2 \longrightarrow 2\,MgO$$

The $O_2$ molecules, on the other hand, gain electrons from magnesium atoms and thereby oxidize the magnesium. Oxygen is therefore an **oxidizing agent:**

$$2\,Mg + \underset{\text{oxidizing agent}}{O_2} \longrightarrow 2\,MgO$$

Oxidizing and reducing agents can be defined as follows:

***Oxidizing agents* gain electrons. *Reducing agents* lose electrons.**

### EXERCISE 9.5

Identify the oxidizing agent and the reducing agent in the following reaction:

$$Ca(s) + H_2(g) \longrightarrow CaH_2(g)$$

**SOLUTION** We start by assigning oxidation numbers to the reaction:

$$\underset{0}{Ca} + \underset{0}{H_2} \longrightarrow \underset{+2\,-1}{CaH_2}$$

We then decide which element is oxidized and which is reduced. In the course of this reaction, calcium atoms are transformed into $Ca^{2+}$ ions by the loss of a pair of electrons. Calcium metal is therefore oxidized in this reaction. The hydrogen atoms, on the other hand, formally gain an electron to form $H^-$ ions. Hydrogen is therefore reduced.

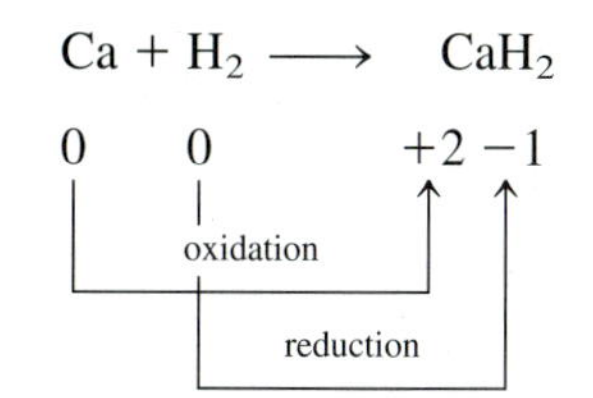


If calcium reduces $H_2$ molecules to $H^-$ ions, calcium metal must be the reducing agent. The $H_2$, on the other hand, oxidizes calcium atoms to form $Ca^{2+}$ ions. $H_2$ is therefore the oxidizing agent:

Reducing agent: Ca
Oxidizing agent: $H_2$

Table 9.3 identifies the reducing agent and the oxidizing agent for some of the reactions discussed in this chapter. One trend is immediately obvious:

**The main-group metals act as reducing agents in all of their chemical reactions.**

**TABLE 9.3 Typical Reactions of Main-Group Metals**

| Reaction | Reducing Agent | Oxidizing Agent |
|---|---|---|
| $2\,Na + Cl_2 \longrightarrow 2\,NaCl$ | Na | $Cl_2$ |
| $2\,K + H_2 \longrightarrow 2\,KH$ | K | $H_2$ |
| $4\,Li + O_2 \longrightarrow 2\,Li_2O$ | Li | $O_2$ |
| $2\,Na + O_2 \longrightarrow Na_2O_2$ | Na | $O_2$ |
| $2\,Na + 2\,H_2O \longrightarrow 2\,Na^+ + 2\,OH^- + H_2$ | Na | $H_2O$ |
| $2\,K + 2\,NH_3 \longrightarrow 2\,K^+ + 2\,NH_2^- + H_2$ | K | $NH_3$ |
| $2\,Mg + O_2 \longrightarrow 2\,MgO$ | Mg | $O_2$ |
| $3\,Mg + N_2 \longrightarrow Mg_3N_2$ | Mg | $N_2$ |
| $Ca + 2\,H_2O \longrightarrow Ca^{2+} + 2\,OH^- + H_2$ | Ca | $H_2O$ |
| $2\,Al + 3\,Br_2 \longrightarrow Al_2Br_6$ | Al | $Br_2$ |
| $Mg + H_2O \longrightarrow MgO + H_2$ | Mg | $H_2O$ |
| $Mg + 2\,H^+ \longrightarrow Mg^{2+} + H_2$ | Mg | $H^+$ |

## 9.15 CONJUGATE OXIDIZING AGENT–REDUCING AGENT PAIRS

Metals act as reducing agents in their chemical reactions. When copper is heated over a flame, for example, the surface slowly turns black as the copper metal reduces oxygen in the atmosphere to form copper(II) oxide:

oxidizing agent

$$2\,Cu + O_2 \longrightarrow 2\,Cu\,O$$

0 0 +2 −2

reducing agent

If we turn off the flame, and blow $H_2$ gas over the hot metal surface, the black CuO that formed on the surface of the metal is slowly converted back to copper metal. In

Copper metal reacts with the oxygen in the atmosphere to form a black film of copper(II) oxide. The thin film of CuO that forms on the surface of the metal can be reduced back to copper metal by turning off the bunsen burner, lowering the funnel shown in this photograph, and blowing $H_2$ gas over the metal surface while it is still hot.

the course of this reaction, CuO is reduced to copper metal. Thus, $H_2$ is the reducing agent in this reaction, and CuO acts as an oxidizing agent:

$$\underset{\substack{+2 \\ \text{oxidizing agent}}}{CuO} + \underset{\substack{0 \\ \text{reducing agent}}}{H_2} \longrightarrow \underset{0}{Cu} + \underset{+1}{H_2O}$$

An important feature of oxidation–reduction reactions can be recognized by examining what happens to the copper in this pair of reactions. The first reaction converts copper metal into CuO, thereby transforming a reducing agent (Cu) into an oxidizing agent (CuO). The second reaction converts an oxidizing agent (CuO) into a reducing agent (Cu). Every reducing agent is therefore linked, or coupled, to a conjugate oxidizing agent, and vice versa.

Every time a reducing agent loses electrons, it forms an oxidizing agent that could gain electrons if the reaction were reversed:

$$\underset{\substack{\text{reducing} \\ \text{agent}}}{Cu} \longrightarrow \underset{\substack{\text{oxidizing} \\ \text{agent}}}{Cu^{2+}} + 2\,e^-$$

Conversely, every time an oxidizing agent gains electrons, it forms a reducing agent that could lose electrons if the reaction went in the opposite direction:

$$\underset{\substack{\text{oxidizing} \\ \text{agent}}}{O_2} + 4\,e^- \longrightarrow \underset{\substack{\text{reducing} \\ \text{agent}}}{2\,O^{2-}}$$

The idea that oxidizing agents and reducing agents are linked, or coupled, is not difficult to understand. But why call them **conjugate** oxidizing agents and reducing agents? Most people associate the term *conjugate* with the process in which the different forms of a verb are systematically linked together. *Conjugate* actually comes from the Latin stem meaning ''to join together.'' It is therefore used to describe things that are linked or coupled, such as oxidizing agents and reducing agents.

The main-group metals are all reducing agents. More importantly, they tend to be ''strong'' reducing agents. The active metals in Group IA, for example, give up electrons better than any other elements in the periodic table. The fact that an active metal such as sodium is a strong reducing agent should tell us something about the relative strength of the $Na^+$ ion as an oxidizing agent. If sodium metal is relatively good at giving up electrons, $Na^+$ ions must be unusually bad at picking up electrons. If Na is a strong reducing agent, the $Na^+$ ion must be a weak oxidizing agent:

$$\underset{\substack{\text{strong} \\ \text{reducing} \\ \text{agent}}}{Na} \longrightarrow \underset{\substack{\text{weak} \\ \text{oxidizing} \\ \text{agent}}}{Na^+} + e^-$$

Conversely, if $O_2$ has such a high affinity for electrons that it is unusually good at accepting them from other elements, it should be able to hang onto these electrons once it picks them up. In other words, if $O_2$ is a strong oxidizing agent, then the $O^{2-}$ ion must be a weak reducing agent:

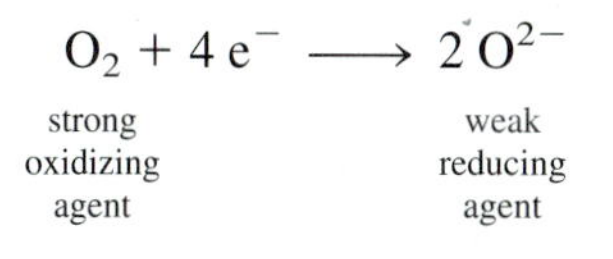


In general, the relationship between conjugate oxidizing and reducing agents can be described as follows:

**Every strong reducing agent (such as Na) has a weak conjugate oxidizing agent (such as the $Na^+$ ion). Every strong oxidizing agent (such as $O_2$) has a weak conjugate reducing agent (such as the $O^{2-}$ ion).**

---

### CHECKPOINT

Use the fact that potassium metal reacts violently with water to predict whether the $K^+$ ion is a strong oxidizing agent or a weak oxidizing agent. Test your prediction by considering what happens when you eat a banana, which is a source of the $K^+$ ion.

---

## 9.16 THE RELATIVE STRENGTHS OF METALS AS REDUCING AGENTS

We can determine the relative strengths of a pair of metals as reducing agents by determining whether a reaction occurs when one of these metals is mixed with a salt of the other. Consider the relative strength of iron and aluminum, for example. Nothing happens when we mix powdered aluminum metal with iron(III) oxide. If we place this mixture in a crucible, however, and get the reaction started by applying a little heat, a vigorous reaction takes place to give aluminum oxide and molten iron metal:

$$2\,Al(s) + Fe_2O_3(s) \longrightarrow Al_2O_3(s) + 2\,Fe(l)$$

By assigning oxidation numbers, we can pick out the oxidation and reduction halves of the reaction:

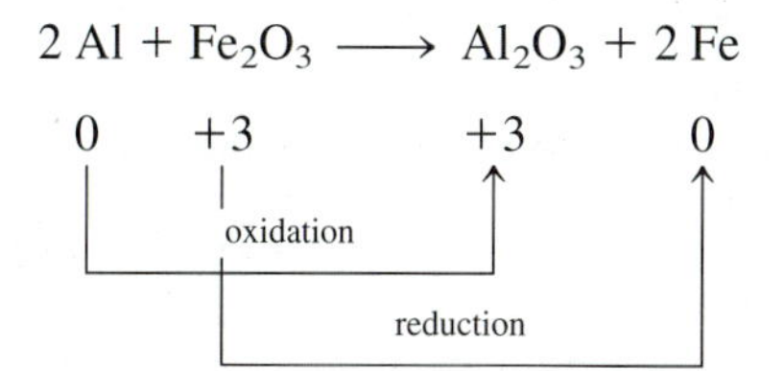


Aluminum is oxidized to $Al_2O_3$ in this reaction, which means that $Fe_2O_3$ must be the oxidizing agent. Conversely, $Fe_2O_3$ is reduced to iron metal, which means that aluminum must be the reducing agent. Because a reducing agent is always transformed into its conjugate oxidizing agent in an oxidation–reduction reaction, the products of this reaction include a new oxidizing agent ($Al_2O_3$) and a new reducing agent (Fe):

$$2\,Al + Fe_2O_3 \longrightarrow Al_2O_3 + 2\,Fe$$

reducing agent — oxidizing agent — oxidizing agent — reducing agent

Since the reaction proceeds in this direction, it seems reasonable to assume that the starting materials contain the stronger reducing agent and the stronger oxidizing agent:

$$\underset{\text{stronger reducing agent}}{2\ Al} + \underset{\text{stronger oxidizing agent}}{Fe_2O_3} \longrightarrow \underset{\text{weaker oxidizing agent}}{Al_2O_3} + \underset{\text{weaker reducing agent}}{2\ Fe}$$

In other words, if aluminum reduces $Fe_2O_3$ to form $Al_2O_3$ and iron metal, aluminum must be a stronger reducing agent than iron.

What can we conclude from the fact that aluminum cannot reduce sodium chloride to form sodium metal?

$$Al(s) + NaCl(s) \not\longrightarrow$$

It seems reasonable to assume that the starting materials in this reaction are the weaker oxidizing agent and the weaker reducing agent:

$$\underset{\text{weaker reducing agent}}{Al} + \underset{\text{weaker oxidizing agent}}{3\ NaCl} \longrightarrow \underset{\text{stronger oxidizing agent}}{AlCl_3} + \underset{\text{stronger reducing agent}}{3\ Na}$$

We can test this hypothesis by asking: What happens when we try to run the reaction in the opposite direction? (Is sodium metal strong enough to reduce a salt of aluminum to aluminum metal?) When this reaction is run, we find that sodium metal can, in fact, reduce aluminum chloride to aluminum metal and sodium chloride when the reaction is run at temperatures hot enough to melt the reactants:

$$3\ Na(l) + AlCl_3(l) \longrightarrow 3\ NaCl(l) + Al(l)$$

If sodium is strong enough to reduce $Al^{3+}$ salts to aluminum metal and aluminum is strong enough to reduce $Fe^{3+}$ salts to iron metal, the relative strengths of these reducing agents can be summarized as follows:

$$Na > Al > Fe$$

## EXERCISE 9.6

Use the following equations to determine the relative strengths of sodium, magnesium, aluminum, and calcium metal as reducing agents.

$$2\ Na + MgCl_2 \longrightarrow 2\ NaCl + Mg$$
$$Al + MgBr_2 \not\longrightarrow$$
$$Ca + MgI_2 \longrightarrow CaI_2 + Mg$$
$$Ca + 2\ NaCl \not\longrightarrow$$

**SOLUTION** The first reaction suggests that sodium is a stronger reducing agent than magnesium. The second reaction suggests that aluminum is not as strong a reducing agent as magnesium or, conversely, that magnesium is a stronger reducing agent than alumunim. The first two reactions therefore give the following sequence:

$$Na > Mg > Al$$

The third and fourth reactions suggest that calcium is a stronger reducing agent than magnesium but a weaker reducing agent than sodium. Thus, calcium has to be inserted into this sequence between sodium and magnesium:

$$Na > Ca > Mg > Al$$

The results of a large number of experiments of this nature are summarized in Table 9.4. By convention, each reaction is written in the direction of reduction. Furthermore, the table is organized so that the strongest reducing agents are found in the upper right corner and the strongest oxidizing agents are in the bottom left corner. The active metals are therefore at the top of the table.

### ▶ CHECKPOINT

Use the fact that potassium metal reacts explosively with water and the fact that gold metal is virtually inert to chemical reactions to explain why there is no need to memorize which end of Table 9.4 contains the strongest reducing agent and which end contains the weakest reducing agent.

**TABLE 9.4 The Relative Strengths of the Common Metals as Reducing Agents**

| Oxidizing agent | Reaction | Reducing agent |
|---|---|---|
| weakest oxidizing agent → | $K^+ + e^- \longrightarrow K$ | ← strongest reducing agent |
| | $Ba^{2+} + 2\,e^- \longrightarrow Ba$ | |
| | $Sr^{2+} + 2\,e^- \longrightarrow Sr$ | |
| | $Ca^{2+} + 2\,e^- \longrightarrow Ca$ | |
| | $Na^+ + e^- \longrightarrow Na$ | |
| ↓ | $Mg^{2+} + 2\,e^- \longrightarrow Mg$ | ↑ |
| | $Al^{3+} + 3\,e^- \longrightarrow Al$ | |
| | $Mn^{2+} + 2\,e^- \longrightarrow Mn$ | |
| | $Ti^{4+} + 4\,e^- \longrightarrow Ti$ | |
| | $Zn^{2+} + 2\,e^- \longrightarrow Zn$ | |
| | $Cr^{3+} + 3\,e^- \longrightarrow Cr$ | |
| | $Fe^{2+} + 2\,e^- \longrightarrow Fe$ | |
| Oxidizing strength of the metal ion increases | $Co^{2+} + 2\,e^- \longrightarrow Co$ | Reducing strength of the metal increases |
| | $Ni^{2+} + 2\,e^- \longrightarrow Ni$ | |
| | $Sn^{2+} + 2\,e^- \longrightarrow Sn$ | |
| | $Pb^{2+} + 2\,e^- \longrightarrow Pb$ | |
| | $Fe^{3+} + 3\,e^- \longrightarrow Fe$ | |
| | $2\,H^+ + 2\,e^- \longrightarrow H_2$ | |
| | $Cu^{2+} + 2\,e^- \longrightarrow Cu$ | |
| | $Ag^+ + e^- \longrightarrow Ag$ | |
| | $Hg^{2+} + 2\,e^- \longrightarrow Hg$ | |
| | $Pt^{2+} + 2\,e^- \longrightarrow Pt$ | |
| strongest oxidizing agent → | $Au^+ + e^- \longrightarrow Au$ | ← weakest reducing agent |

## EXERCISE 9.7

Use Table 9.4 to predict whether copper metal should react with $Ag^+$ ions to form silver metal and $Cu^{2+}$ ions:

$$Cu(s) + 2\,Ag^+(aq) \longrightarrow Cu^{2+}(aq) + 2\,Ag(s)$$

**SOLUTION** According to Table 9.4, copper metal is a better reducing agent than silver metal. The $Ag^+$ ion, on the other hand, is a stronger oxidizing agent than

Copper metal reacts with a solution of the $Ag^+$ ion to form silver metal and a solution of the $Cu^{2+}$ ion, as predicted in Exercise 9.7.

Zinc metal dissolves in concentrated acid, as predicted in Exercise 9.8.

the $Cu^{2+}$ ion. We would therefore expect copper metal to react with $Ag^+$ ions to form silver metal and $Cu^{2+}$ ions:

$$\underset{\substack{\text{stronger}\\\text{reducing}\\\text{agent}}}{Cu} + \underset{\substack{\text{stronger}\\\text{oxidizing}\\\text{agent}}}{2\,Ag^+} \longrightarrow \underset{\substack{\text{weaker}\\\text{oxidizing}\\\text{agent}}}{Cu^{2+}} + \underset{\substack{\text{weaker}\\\text{reducing}\\\text{agent}}}{2\,Ag}$$

**EXERCISE 9.8**

A strong acid can be defined as anything that acts as a good source of the $H^+$ ion. Use Table 9.4 to identify the metals that are not likely to dissolve in strong acids.

**SOLUTION** The $H^+$ ion can be reduced by any metal that is a stronger reducing agent than $H_2$. Zinc, for example, can reduce $H^+$ ions to elemental hydrogen.

$$Zn(s) + 2\,H^+(aq) \longrightarrow Zn^{2+}(aq) + H_2(g)$$

The only metals in Table 9.4 that cannot reduce $H^+$ ions are Cu, Ag, Hg, Pt, and Au. These are the only metals in this table that should not dissolve in a strong acid.

## 9.17 THE PREPARATION OF METALS: CHEMICAL MEANS

In Section 9.1 the metals were divided into four categories on the basis of their activity. These four categories are reflected in the positions of these metals in Table 9.4. The active metals are at the top of the table. The less active metals can be found in the top half. The structural metals are usually found in the lower half of the table and the metals used to make coins and jewelry are at the bottom.

It is also possible to classify metals on the basis of the ease with which they can be prepared from their salts or ores. As we proceed up the column in Table 9.4, it becomes more difficult to prepare the metal from one of its ores.

The first step in preparing most metals involves *concentrating* the ore by separating it from the rock in which it is embedded. The rock is crushed and then ground into fine particles, if necessary. The finely divided rock is then poured into a large tank of water, which contains a wetting agent that adheres to the ore particles. Air blown into the bottom of the tank causes the wetting agent to form bubbles that carry the ore to the surface of the tank.

The next step often involves *sintering,* a procedure in which the ore particles are heated until they come together to form larger particles. Some ores also undergo *calcining* when they are heated, a process in which a gas is given off. Sintering does not change the chemical composition of the ore, but calcining does. An example of an ore that undergoes calcining is bauxite, a mixture of $Al(OH)_3$ and $AlO(OH)$. The $Al(OH)_3$ in bauxite gives off water to form aluminum oxide when the ore is heated:

$$2\ Al(OH)_3(s) \longrightarrow Al_2O_3(s) + 3\ H_2O(g)$$

Many commercially important ores are sulfides, including chalcopyrite ($CuFeS_2$), cinnabar ($HgS$), galena ($PbS$), pyrite ($FeS_2$), and zincblende ($ZnS$). Before these ores can be reduced to the metal, they must be *roasted*. This involves heating the ore in the presence of $O_2$ to convert the sulfide into the corresponding oxide:

$$2\ ZnS(s) + 3\ O_2(g) \longrightarrow 2\ ZnO(s) + 2\ SO_2(g)$$
$$2\ PbS(s) + 3\ O_2(g) \longrightarrow 2\ PbO(s) + 2\ SO_2(g)$$

Cinnabar ($HgS$) is slightly different. When it is roasted, the sulfur is oxidized to $SO_2$ and the mercury is reduced to mercury metal, which boils off:

$$HgS(s) + O_2(g) \longrightarrow Hg(g) + SO_2(g)$$

The steps so far are all designed to prepare the ore for *smelting,* the process in which the ore is reduced to the metal. Copper, iron, lead, tin, and zinc are just a few of the metals produced by smelting metal oxides in a blast furnace. The reducing agent in this process is almost always "coke," an amorphous form of carbon produced when coal is heated in the absence of air. Some metals react directly with the carbon in coke:

$$SnO_2(s) + C(s) \longrightarrow Sn(l) + CO_2(g)$$

Other metals are reduced by the carbon monoxide formed when coke reacts with oxygen:

$$ZnO(s) + CO(g) \longrightarrow Zn(g) + CO_2(g)$$

Note that these reactions are run at such high temperatures that the metal comes off as a liquid or even a gas.

The pouring of a mixture of ore and silica during the smelting of iron.

Figure 9.7 shows a drawing of a typical blast furnace used to make iron metal. The blast furnace is essentially a large chemical reactor in which the temperature varies from about 200°C at the top to 2000°C at the bottom. One of the oxides of iron, such as magnetite ($Fe_3O_4$) or hematite ($Fe_2O_3$), is mixed with coke and loaded at the top of the furnace. Preheated compressed air or $O_2$ is blown into the furnace at the bottom, where it reacts with the coke to produce a mixture of CO and $CO_2$.

At the top of the furnace the iron oxides are reduced to iron metal by carbon monoxide gas:

$$Fe_2O_3(s) + 3\ CO(g) \longrightarrow 2\ Fe(l) + 3\ CO_2(g)$$
$$Fe_3O_4(s) + 4\ CO(g) \longrightarrow 3\ Fe(l) + 4\ CO_2(g)$$

Any iron oxide left as the ore drops through the reactor will eventually reach temperatures above 700°C and react directly with the carbon in coke:

$$Fe_2O_3 + 3\ C(s) \longrightarrow 2\ Fe(l) + 3\ CO(g)$$
$$Fe_3O_4(s) + 4\ C(s) \longrightarrow 3\ Fe(l) + 4\ CO_2(g)$$

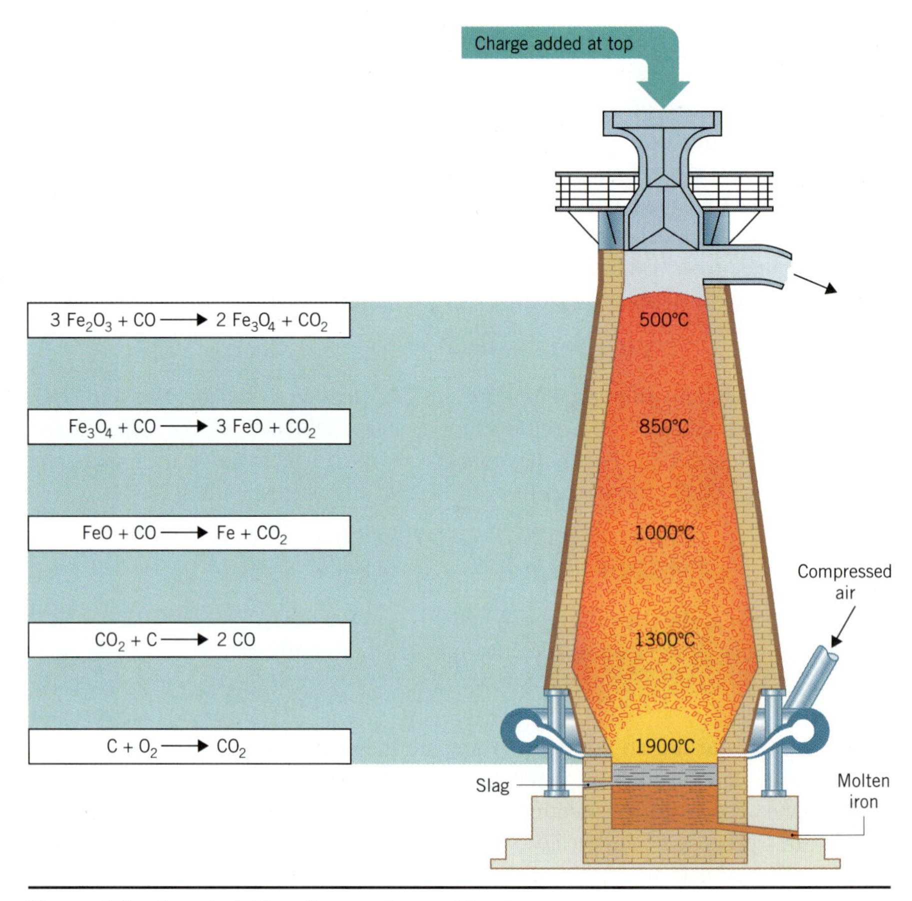


**Figure 9.7** A typical blast furnace for smelting iron.

The mixture of iron oxide and coke is added more or less continuously at the top of the furnace, and molten iron, or pig iron, is drained off at the bottom every few hours.

A wrought-iron fence.

The product of the smelting process contains impurities that must be removed during *refining*. Pig iron, for example, contains carbon (3% to 5%), silicon (0.5% to 4%), manganese (0.15% to 2.5%), phosphorus (0.025% to 2.5%), and sulfur (0.2%). Historically, pig iron was heated until white hot and then worked (or ''wrought'') with a hammer to drive out the impurities. The *wrought iron* produced this way was harder than bronze, and yet it could be hammered into different shapes to make anything from a gate or a railing to a horseshoe. Wrought iron usually has less than 0.1% carbon by weight. Steel, which is harder than wrought iron, can be produced by refining pig iron so that 1% to 2% by weight of carbon remains.

The primary reason for using coke as the reducing agent for the smelting of copper, iron, lead, tin, and zinc is its low cost. It is so cheap that part of the coke is used as a reducing agent and the remainder as a fuel to heat the blast furnace. As we proceed up the column in Table 9.4, we eventually encounter metals whose ores cannot be reduced by coke. Such ores require stronger (and more expensive) reducing agents. Titanium, for example, has many attractive properties as a metal. Its principal drawback is the fact that it is prepared commercially by reducing titanium tetrachloride with magnesium metal, which is relatively expensive:

$$TiCl_4(l) + 2\,Mg(s) \longrightarrow Ti(s) + 2\,MgCl_2(s)$$

Other industrial processes that rely on relatively expensive reducing agents are the reduction of cobalt(III) oxide with $H_2$ gas and the use of powdered aluminum to reduce chromium(III) oxide to chromium:

$$Co_2O_3(s) + 3\ H_2(g) \longrightarrow 2\ Co(s) + 3\ H_2O(g)$$
$$Cr_2O_3(s) + 2\ Al(s) \longrightarrow Al_2O_3(s) + 2\ Cr(l)$$

## 9.18 THE PREPARATION OF ACTIVE METALS: ELECTROLYSIS

Magnesium can be used to reduce $TiCl_4$ to titanium metal, and aluminum can be used to reduce $Cr_2O_3$ to chromium metal. But this leaves us with the task of preparing magnesium and aluminum metal. To do this, we need something that is an even stronger reducing agent.

Magnesium was first prepared in bulk quantities by heating magnesium chloride and potassium metal in a glass tube:

$$MgCl_2(s) + 2\ K(s) \longrightarrow Mg(s) + 2\ KCl(s)$$

Aluminum was first prepared by heating aluminum chloride and potassium metal in a platinum crucible:

$$AlCl_3(s) + 3\ K(s) \longrightarrow Al(s) + 3\ KCl(s)$$

Eventually, it became possible to prepare magnesium and aluminum by using sodium metal, which is cheaper than potassium metal. But this still leaves us with the task of preparing sodium metal.

There are no chemical reducing agents strong enough to prepare sodium metal. We therefore have to resort to physical means, such as passing an electric current through the salt. Humphry Davy first prepared potassium metal in 1807 by connecting a battery to a small piece of ''potash,'' or potassium carbonate ($K_2CO_3$). A few days later he repeated this procedure with crystals of ''caustic soda'' (NaOH) and was able to prepare small quantities of sodium metal.

Only small quantities of metal can be prepared this way because the $Na^+$ or $K^+$ ions are locked in place within the salt crystals. This problem can be overcome by heating the salt until it melts. The ions are then free to move through the molten salt, and significant quantities of the metal eventually collect at the electrode connected to the negative end of the battery.

Figure 9.8 shows a diagram of an electrical cell in which two electrodes connected to a high-voltage battery are immersed in a molten sample of sodium chloride. One of these electrodes gains electrons from the battery and becomes negatively charged. The other electrode loses electrons to the battery and becomes

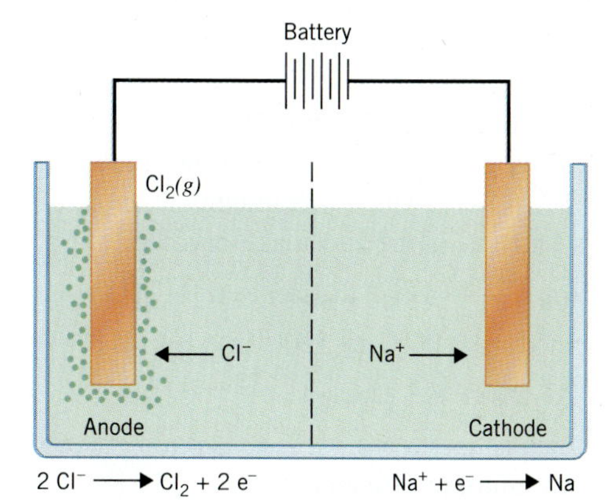


**Figure 9.8** A cell that could be used to prepare sodium metal by the electrolysis of molten NaCl. $Na^+$ ions migrate toward the negative electrode, or cathode, where they are reduced to sodium metal. $Cl^-$ ions migrate toward the positive electrode, or anode, where they are oxidized to form $Cl_2$ gas, which bubbles out of the cell.

positively charged. Because particles of opposite charge attract, the $Na^+$ ions (or *cations*) migrate toward the negatively charged electrode, or **cathode.** Conversely, the $Cl^-$ ions (or *anions*) move toward the positively charged **anode.**

The $Na^+$ ions gain electrons at the cathode to form sodium metal, while the $Cl^-$ ions lose electrons at the anode to form $Cl_2$ gas:

Cathode:

$$Na^+ + e^- \longrightarrow Na \qquad \text{(reduction)}$$

Anode:

$$2\,Cl^- \longrightarrow Cl_2 + 2\,e^- \qquad \text{(oxidation)}$$

Thus, reduction occurs at the cathode and oxidation occurs at the anode. This process is called **electrolysis** (literally, ''splitting with electrons'') because it uses electrons to split a compound into its elements:

$$2\,NaCl(l) \longrightarrow 2\,Na(l) + Cl_2(g)$$

(Note that the sodium produced in this reaction is a liquid because the cell operates at temperatures much higher than the melting point of sodium metal.)

The same process can be used to make magnesium metal. In this case, an electric current is passed through a molten sample of magnesium chloride. The positive $Mg^{2+}$ ions, or cations, are reduced to magnesium metal at the cathode, and the negative $Cl^-$ ions, or anions, are oxidized at the anode:

Cathode:

$$Mg^{2+} + 2\,e^- \longrightarrow Mg \qquad \text{(reduction)}$$

Anode:

$$2\,Cl^- \longrightarrow Cl_2 + 2\,e^- \qquad \text{(oxidation)}$$

The overall electrolysis reaction can therefore be written as follows:

$$MgCl_2(l) \longrightarrow Mg(l) + Cl_2(g)$$

In theory, aluminum metal could be made the same way. In practice, electrolysis of aluminum ores is difficult because they have extremely high melting points. $Al_2O_3$, for example, melts at temperatures above 2000°C. In 1886 Charles Martin Hall, a 21-year-old graduate of Oberlin College, discovered a way to make aluminum by electrolyzing a mixture of $Al_2O_3$ dissolved in cryolite, $Na_3AlF_6$. This mixture melts at a lower temperature than either of the individual ores, which allows the electrolysis to be done at more reasonable temperatures of 960°C to 980°C.

Electrolysis of mixtures of $Al_2O_3$ and $Na_3AlF_6$ is still the principal source of aluminum metal. The $Al_2O_3$ is obtained from the mineral bauxite. The formula for bauxite is often written as $Al_2O_3 \cdot 2\,H_2O$. This is misleading because bauxite does not consist of $Al_2O_3$ with water molecules trapped in holes in the crystal. Bauxite is actually a mixture of two salts: $Al(OH)_3$ and $AlO(OH)$.

The first step in making aluminum is to dissolve bauxite in NaOH to form sodium aluminate, $NaAlO_2$, which is filtered to remove impurities and then allowed to react with water to form aluminum hydroxide, $Al(OH)_3$, which precipitates from solution. Part of the $Al(OH)_3$ formed in this reaction is used to make cryolite:

$$6\,HF(aq) + Al(OH)_3(s) + 3\,NaOH(aq) \longrightarrow Na_3AlF_6(aq) + 6\,H_2O(l)$$

The rest is filtered out, washed and calcined at 1100°C to 1200°C:

$$2\,Al(OH)_3(s) \longrightarrow Al_2O_3(s) + 3\,H_2O(g)$$

A small quantity of $Al_2O_3$ (2% to 8%) is then added to $Na_3AlF_6$, heated to between 960°C and 980°C, and electrolyzed to form aluminum metal.

**SPECIAL TOPIC**

## SALTS OF THE MAIN-GROUP METALS

With the exception of aluminum, tin, and lead, the average citizen never comes in contact with the main-group metals. The salts of these elements, however, have seemingly endless uses in our daily lives.

### GROUP IA

The "lithium" used to treat manic-depressive patients is not lithium metal but a lithium salt, such as lithium carbonate ($Li_2CO_3$).

Sodium bicarbonate ($NaHCO_3$) is sold as baking soda for use in cooking and as bicarbonate of soda for use as an antacid. It is even mixed with kitty litter, in which it serves as a deodorizer. Ordinary table salt (NaCl) is so important that it has entered our language in the form of cliches such as "salt of the earth," and "not worth his salt," and it is the source of the word "salary," which comes from the Latin *salarium*—money given to Roman soldiers to buy salt.

The only potassium salt commonly used in cooking is cream of tartar, or potassium hydrogen tartrate ($KHC_4H_4O_6$), which is used to transform baking soda into baking powder. Potassium salts are essential to plant nutrition, however, and potassium chloride (KCl), potassium nitrate ($KNO_3$), and potassium sulfate ($K_2SO_4$) are used extensively in agriculture.

### GROUP IIA

Beryllium was once known as glucinium because so many of its compounds taste sweet. Exposure to even small quantities of beryllium salts can be fatal, however, so beryllium plays no significant role in our daily lives.

The best known magnesium salts are magnesium hydroxide, $Mg(OH)_2$, and magnesium sulfate, $MgSO_4 \cdot 7\ H_2O$. $Mg(OH)_2$ is used in "milk of magnesia" to treat acid indigestion, and $MgSO_4 \cdot 7\ H_2O$ can be found in "Epsom salts."

There are a number of important calcium salts. Calcium carbonate ($CaCO_3$) is found in chalk, coral, egg shells, limestone, marble, pearl, and clam shells. $CaCO_3$ decomposes on heating to form calcium oxide, or lime (CaO):

$$CaCO_3(s) \xrightarrow{\text{heat}} CaO(s) + CO_2(g)$$

Lime is used to neutralize soil that is too acidic, and it is the principal ingredient of cement.

Calcium sulfate occurs naturally as gypsum, $CaSO_4 \cdot 2\ H_2O$. When heated, gypsum loses water to form plaster of Paris:

$$\underset{\text{gypsum}}{2\ [CaSO_4 \cdot 2\ H_2O(s)]} \longrightarrow \underset{\text{plaster of Paris}}{[(CaSO_4)_2 \cdot H_2O(s)]} + 3\ H_2O(l)$$

When plaster of Paris is mixed with water, the reaction is reversed. This makes plaster of Paris useful for everything from plastering the interior of buildings to making casts to support broken bones.

The principal component of both bone and tooth enamel is a mineral known as hydroxyapatite, $Ca_5(PO_4)_3OH$. The idea behind fluoridating water or adding fluorides to toothpaste is to convert a portion of the hydroxyapatite in tooth enamel into fluoroapatite, $Ca_5(PO_4)_3F$, which is a harder mineral and therefore more resistant to decay.

## GROUP IIIA

Aluminum oxide ($Al_2O_3$) is found in nature as corundum, one of the hardest known minerals. Rubies are crystals of corundum that contain small quantities of chromium impurities and sapphires are crystals of corundum with trace quantities of iron or titanium. An aluminum salt with which many people have daily contact is "aluminum chlorhydrate," or more accurately, aluminum hydroxychloride, $Al_2(OH)_5Cl \cdot 2\,H_2O$. This compound is an astringent, which contracts body tissues. When applied to underarms, its tendency to contract the pores of the skin makes it an antiperspirant.

## GROUP IVA

Tin(II) fluoride ($SnF_2$), or stannous fluoride, was the first source of the fluoride ion used in toothpastes. Tin(IV) oxide, $SnO_2$, is mixed with a variety of transition-metal oxides, such as $V_2O_5$ and $Cr_2O_3$, to form glazes of different colors for use with ceramics. Trialkyltin compounds such as $(C_4H_9)_3SnOH$ are important biocides used to control fungi, bacteria, insects, and weeds. These compounds also have been used to inhibit the formation of barnacles on ships' hulls.

Lead(II) oxide (PbO), or litharge, is often added to glass to increase its density, brilliance, and strength. PbO is also an important component of the lead storage batteries used in cars and trucks. At one time, lead chromate ($PbCrO_4$) and lead dichromate ($PbCr_2O_7$) were important yellow and orange pigments for use in paint. As our awareness of the toxicity of lead compounds has increased, these pigments have been gradually replaced by others. Until recently, tetraethyl lead, $Pb(C_2H_5)_4$, was routinely added to gasoline to increase its octane number. Concern over high levels of lead emissions from automobile exhaust has resulted in a switch to "unleaded" gasolines.

The rubies embedded in this ore consist of aluminum oxide ($Al_2O_3$) with trace amounts of $Cr_2O_3$, which gives rise to their red color.

## SUMMARY

1. Many reactions of the main-group metals can be understood by assuming that these metals lose electrons and the nonmetals with which they react gain electrons until both elements reach a filled-shell electron configuration.

2. Exceptions to the predictions of this model are encountered when the most active main-group metals react with oxygen. These exceptions can be explained by assuming that these metals react so rapidly that the reaction stops prematurely at the stage of either the peroxide (such as $Na_2O_2$) or the superoxide ($KO_2$).

3. The chemistry of the main-group metals revolves around the concept of oxidation–reduction reactions. **Oxidation** occurs when an atom or ion undergoes an increase in its oxidation number. **Reduction** involves a decrease in oxidation number.

4. Metals are oxidized when they react with either nonmetal elements or compounds. They therefore act as reducing agents in these reactions.

5. The stronger of a pair of reducing agents always reacts with the stronger of a pair of oxidizing agents to produce a weaker oxidizing agent and a weaker reducing agent. This phenomenon does not imply that the starting materials are "strong" reducing agents and "strong" oxidizing agents or that the products are "weak" reducing agents and "weak" oxidizing agents. It suggests only that the starting materials are *stronger* oxidizing and reducing agents than the products.

6. Metals become harder to prepare from their salts as we move up the table of relative reducing strengths. Metals at the bottom of this table can sometimes be obtained by simply heating one of their ores. Metals toward the middle of the table are often prepared by reducing one of their ores with carbon. As we move toward the top of the table, it takes stronger (and more expensive) reducing agents to prepare the metal. The active metals at the top of the table can be prepared in useful quantities by passing an electric current through a molten salt.

## KEY TERMS

**Active metal** (p. 316)
**Alkali metal** (p. 317)
**Alkaline-earth metal** (p. 324)
**Anode** (p. 340)
**Cathode** (p. 340)
**Conjugate** (p. 332)
**Electrolysis** (p. 340)
**Halides** (p. 318)
**Halogens** (p. 318)
**Hydrides** (p. 318)
**Nitrides** (p. 318)
**Oxidation** (p. 327)
**Oxides** (p. 320)
**Oxidizing agent** (p. 330)
**Peroxide** (p. 320)
**Phosphides** (p. 318)
**Pyrophoric** (p. 327)
**Reducing agent** (p. 330)
**Reduction** (p. 328)
**Sulfides** (p. 318)
**Superoxide** (p. 321)

## PROBLEMS

### The Active (Reactive) Metals

**9-1** Describe the general trends in the activity, or reactivity, of the main-group metals.

**9-2** What do we mean when we say that sodium is more metallic than lithium?

**9-3** For each of the following pairs of metals, which is more active?

(a) Mg or Ca (b) Na or Mg (c) Na or Ca
(d) K or Mg (e) Mg or Al

**9-4** Which metal in each of the following pairs would you expect to react more rapidly with water?

(a) Na or K (b) Na or Mg (c) Mg or Ca
(d) Ca or Al

### Group IA: The Alkali Metals

**9-5** Define the terms *alkali metal, halide, hydride, sulfide, nitride, phosphide, oxide, peroxide,* and *superoxide*. Give at least one example of each.

**9-6** Write the formulas for the bromide, hydride, sulfide, nitride, phosphide, oxide, and peroxide of rubidium.

**9-7** Describe the difference between the hydrogen atoms in metal hydrides such as LiH and nonmetal hydrides such as $CH_4$ and $H_2O$.

**9-8** White phosphorus bursts spontaneously into flame in the presence of air and is therefore stored under water. Cesium metal can also burst into flame in the presence of air, but it can't be stored under water. Why not?

**9-9** Cesium metal is so active that it reacts explosively with cold water and even with ice at temperatures as low as $-116°C$. Explain why cesium is more active than sodium.

### Group IIA: The Alkaline-Earth Metals

**9-10** Write the formulas for the fluoride, hydride, sulfide, nitride, phosphide, oxide, and peroxide of barium.

**9-11** Write balanced equations for the reactions that lead to the explosions that occur when water is poured onto burning magnesium metal. Write a balanced equation for the reaction that occurs when someone tries to extinguish burning magnesium metal with a $CO_2$ extinguisher.

**9-12** Explain why calcium reacts with water at room temperature, but magnesium only reacts with steam at high temperatures.

**9-13** Explain why $BaO_2$ must be a peroxide, $[Ba^{2+}][O_2{}^{2-}]$, and not an oxide $[Ba^{4+}][O^{2-}]_2$.

### Group IIIA: The Chemistry of Aluminum

**9-14** People often believe that the advantage of replacing steel with aluminum is that aluminum is "lighter." Explain why it is wrong to say that aluminum is lighter than steel.

**9-15** Explain why it often seems that aluminum is less reactive than iron. Explain why chemists think otherwise. Explain how aluminum can be used as a structural metal in spite of its activity.

**9-16** Explain why an even coating of $Al_2O_3$ about 5 nanometers thick protects aluminum from corrosion and makes aluminum seem less active than it really is.

### Group IVA: Tin and Lead

**9-17** The first practical storage battery was built by Gaston Planté in 1860. His battery consisted of two sheets of lead separated by a strip of rubber immersed in a 10% solution of sulfuric acid in water. Use the fact that lead storage batteries can be found in every car and truck on the road to compare the activity of lead with the other main-group metals.

**9-18** The "tin" cans found on the shelves of groceries stores are actually made of steel that has been thinly coated with tin. Explain why these cans are coated with tin. What would happen if Del Monte tried to package fruit cocktail in a steel can?

**9-19** When aluminum is used to make cans, the metal has to be coated with plastic to keep it from dissolving on contact with the acids in food. Why don't tin cans have to be coated with plastic?

**9-20** Children often believe that metals such as iron lose weight when they rust. Explain why they might believe this and why it is a mistake.

### Identifying Main-Group Metals from the Products of Their Reactions

**9-21** Which of the following elements is most likely to form an oxide with the formula $XO$ and a hydride with the formula $XH_2$?

(a) Na (b) Mg (c) Al (d) Si (e) P

**9-22** A main-group metal reacts with hydrogen and oxygen to form compounds with the formulas $XH_4$ and $XO_2$. In which group of the periodic table does this element belong?

**9-23** In which column of the periodic table do we find main-group metals (M) that form sulfides with the formula $M_2S_3$ and react with acid to form $M^{3+}$ ions and $H_2$ gas?

### Predicting the Products of Reactions Between Main-Group Metals and Nonmetal Elements

**9-24** Predict the product of the reaction between aluminum and nitrogen.

**9-25** Predict the product of the reaction between strontium metal and phosphorus.

**9-26** The light meters in automatic cameras are based on the sensitivity of gallium arsenide to light. Predict the formula of gallium arsenide.

**9-27** Write balanced equations for the reactions of sodium metal with the following elements:

(a) $F_2$ (b) $O_2$ (c) $H_2$
(d) $S_8$ (e) $P_4$

**9-28** Write balanced equations for the reactions of calcium metal with the following elements:

(a) $H_2$ (b) $O_2$ (c) $S_8$
(d) $F_2$ (e) $N_2$ (f) $P_4$

**9-29** Predict the product of the reactions of $F_2$ with the following main-group metals:

(a) Zn (b) Al (c) Sn
(d) Mg (e) Bi

### Predicting the Products of Reactions Between Main-Group Metals and Nonmetal Compounds

**9-30** Which of the following describes the products of the reaction of calcium metal and water?

(a) $Ca^{2+}$, $OH^-$, and $H_2$ (b) $Ca^{2+}$, $H^-$, and $O_2$
(c) $Ca^{2+}$, $H^-$, and $OH^-$ (d) $Ca^{2+}$ and $OH^-$
(e) $Ca^{2+}$, $OH^-$, and $H^+$

**9-31** Explain why magnesium reacts with steam to form MgO, not $Mg(OH)_2$.

**9-32** Which of the following is not formed when potassium metal reacts with phosphine?

$$K(s) + PH_3(g) \longrightarrow$$

(a) $K^+$ ions (b) $PH_2^-$ ions (c) $H_2$ gas (d) $H^+$ ions
(e) $OH^-$ ions

**9-33** Predict the products of the following reactions.

(a) $Mg(s) + HCl(aq) \longrightarrow$
(b) $Cu(s) + HCl(aq) \longrightarrow$
(c) $Cs(s) + H_2O(l) \longrightarrow$

**9-34** Which of the following reactions does not make sense?

(a) $Mg(s) + 2\ NH_3(l) \longrightarrow Mg(NH_2)_2(s) + H_2(g)$
(b) $Mg(s) + N_2H_4(g) \longrightarrow Mg(NH_2)_2(s)$
(c) $Mg(s) + 2\ HNO_3(aq) \longrightarrow Mg(NO_3)_2(aq) + H_2(g)$
(d) $Mg(s) + 2\ NO_2(g) + 2\ OH^-(aq) \longrightarrow Mg(NO_3)_2(aq) + 2\ H^+(aq)$

**9-35** Predict the products of the following reactions:

(a) $Mg(s) + I_2(s) \longrightarrow$
(b) $Al(s) + H_2SO_4(aq) \longrightarrow$
(c) $Fe_2O_3(s) + H_2(g) \longrightarrow$
(d) $Mg(s) + TiCl_4(l) \longrightarrow$

**9-36** Predict the products of the following reactions:

(a) $K(s) + H_2O(l) \longrightarrow$
(b) $NaH(s) + H_2O(l) \longrightarrow$
(c) $Ag^+(aq) + Cu(s) \longrightarrow$
(d) $Zn(s) + CuSO_4(aq) \longrightarrow$

### Oxidation–Reduction Reactions

**9-37** Explain why carbon is oxidized when it reacts with sulfur at high temperatures to form carbon disulfide, even though $O_2$ is not involved in the reaction.

$$4\ C(s) + S_8(l) \longrightarrow 4\ CS_2(l)$$

**9-38** The $Ag_2S$ that forms when silver tarnishes can be removed by polishing the silver with a source of the cyanide ion or by wrapping it in aluminum foil and immersing it in salt water:

$$Ag_2S(s) + 4\ CN^-(aq) \longrightarrow 2\ Ag(CN)_2^-(aq) + S^{2-}(aq)$$
$$3\ Ag_2S(s) + 2\ Al(s) \longrightarrow 6\ Ag(s) + Al_2S_3(s)$$

Which of these reactions involves oxidation–reduction?

**9-39** Decide whether each of the following reactions involves oxidation–reduction. If it does, identify what is oxidized and what is reduced.

(a) $CO_2(g) + H_2O(l) \longrightarrow H_2CO_3(aq)$
(b) $Fe_2O_3(s) + 3\ CO(g) \longrightarrow 2\ Fe(s) + 3\ CO_2(g)$
(c) $SiO_2(s) + 3\ C(s) \longrightarrow SiC(s) + 2\ CO(g)$
(d) $CO_2(g) + H_2(g) \longrightarrow CO(g) + H_2O(g)$
(e) $CO(g) + 2\ H_2(g) \longrightarrow CH_3OH(l)$

**9-40** Decide whether each of the following reactions involves oxidation–reduction. If it does, identify what is oxidized and what is reduced.

(a) $Mg(s) + 2\ HCl(aq) \longrightarrow MgCl_2(aq) + H_2(g)$
(b) $I_2(s) + 3\ Cl_2(g) \longrightarrow 2\ ICl_3(l)$
(c) $HCl(aq) + NaOH(aq) \longrightarrow NaCl(aq) + H_2O(l)$
(d) $2\ Na(s) + 2\ H_2O(l) \longrightarrow 2\ NaOH(aq) + H_2(g)$

**9-41** Decide whether each of the following reactions involves oxidation–reduction. If it does, identify what is oxidized and what is reduced.

(a) $Ca_3P_2(s) + 6\ H_2O(l) \longrightarrow 3\ Ca(OH)_2(aq) + 2\ PH_3(g)$
(b) $2\ PH_3(g) + 4\ O_2(g) \longrightarrow 2\ H_3PO_4(s)$
(c) $PH_3(g) + HCl(g) \longrightarrow PH_4Cl(s)$
(d) $P_4(s) + 5\ O_2(g) \longrightarrow P_4O_{10}(s)$

### Reducing Agents and Oxidizing Agents

**9-42** Define the terms *oxidizing agent* and *reducing agent*.

**9-43** Identify the reducing agent and the oxidizing agent in the following reaction:

$$3\ S_8(s) + 16\ KClO_3(s) \longrightarrow 24\ SO_2(g) + 16\ KCl(s)$$

**9-44** Identify the reducing agent and the oxidizing agent in the following reaction:

$$H_2O_2(aq) + 2\ HI(aq) \longrightarrow 2\ H_2O(l) + I_2(s)$$

**9-45** Identify the reducing agent and the oxidizing agent in the following reaction:

$$Cu(s) + 2\ Ag^+(aq) \longrightarrow Cu^{2+}(aq) + 2\ Ag(s)$$

**9-46** Identify the reducing agent and the oxidizing agent in each of the following reactions:

(a) $2\ Al(s) + Cr_2O_3(s) \longrightarrow Al_2O_3(s) + 2\ Cr(s)$
(b) $2\ Al(s) + 6\ H^+(aq) \longrightarrow 2\ Al^{3+}(aq) + 3\ H_2(g)$
(c) $2\ Al(s) + 3\ I_2(s) \longrightarrow 2\ AlI_3(s)$

**9-47** Identify the reducing agent and the oxidizing agent in each of the following reactions.

(a) $2\ Mg(s) + CO_2(g) \longrightarrow 2\ MgO(s) + C(s)$
(b) $Mg(s) + 2\ HCl(aq) \longrightarrow Mg^{2+}(aq) + 2\ Cl^-(aq) + H_2(g)$
(c) $Mg(s) + H_2O(g) \longrightarrow MgO(s) + H_2(g)$
(d) $3\ Mg(s) + N_2(g) \longrightarrow Mg_3N_2(s)$
(e) $3\ Mg(s) + 2\ NH_3(g) \longrightarrow Mg_3N_2(s) + 3\ H_2(g)$

**9-48** Identify the conjugate oxidizing agents for the following reducing agents:

(a) Na (b) Zn (c) $H_2$ (d) $Sn^{2+}$ (e) $H^-$

**9-49** Identify the conjugate reducing agents for the following oxidizing agents:

(a) $Al^{3+}$ (b) $Hg^{2+}$ (c) $H^+$ (d) $H^2$ (e) $Sn^{2+}$

**9-50** Explain why every reducing agent has a conjugate oxidizing agent and why every oxidizing agent has a conjugate reducing agent.

**9-51** Explain the difference between a strong reducing agent such as sodium or potassium metal and a weak reducing agent such as silver or gold.

**9-52** Classify the following substances as either oxidizing agents or reducing agents and predict whether they will be relatively strong or relatively weak.

(a) Mg (b) MgO (c) $AgNO_3$ (d) Cu

**9-53** Explain why the fact that sodium metal is a stronger reducing agent than aluminum metal implies that the $Al^{3+}$ ion must be a stronger oxidizing agent than the $Na^+$ ion.

**9-54** Explain why the fact that iron is a stronger reducing agent than copper does not mean that iron is a ''strong'' reducing agent in an absolute sense.

### The Relative Strengths of Metals as Reducing Agents

**9-55** Archaeologists have recovered samples of copper metal that are at least 5000 years old. What does this say about the activity of copper compared with other metals, such as iron, aluminum, sodium, and magnesium?

**9-56** Use Table 9.4 to predict which of the following elements are strong enough to reduce $Fe_2O_3$ to iron metal:

(a) Na (b) Mg (c) Al (d) Ag (e) $H_2$

**9-57** Use Table 9.4 to predict which of the following oxides can be reduced to the metal with $H_2$:

(a) $Na_2O$ (b) MgO (c) $Al_2O_3$ (d) PbO (e) $Fe_2O_3$
(f) HgO

**9-58** Which of the following elements should be able to reduce $Sn^{2+}$ ions to tin metal?

(a) Na (b) Mg (c) Al (d) Fe (e) Hg

**9-59** Sodium metal reacts with ammonia to form sodium amide and hydrogen gas:

$$2\,Na(s) + 2\,NH_3(l) \longrightarrow 2\,NaNH_2(s) + H_2(g)$$

Use this observation to determine the relative strengths of sodium metal and hydrogen gas as reducing agents.

**9-60** HgO decomposes to mercury metal when heated. CuO does not decompose on heating, but it can be reduced to copper metal with elemental carbon or hydrogen. $Al_2O_3$ cannot be reduced to aluminum metal with either carbon or hydrogen. Arrange the three metals in order of increasing activity.

**9-61** Powdered aluminum reacts with iron oxide to give aluminum oxide and enough heat to melt the iron metal produced in this reaction:

$$Fe_2O_3(s) + 2\,Al(s) \longrightarrow Al_2O_3(s) + 2\,Fe(l)$$

Predict the relative strengths of iron and aluminum as reducing agents.

**9-62** Use Table 9.4 to predict whether powdered aluminum should be able to reduce chromium(III) oxide to chromium:

$$Cr_2O_3(s) + 2\,Al(s) \longrightarrow Al_2O_3(s) + 2\,Cr(l)$$

**9-63** Titanium metal was first produced commercially by the following reaction:

$$TiCl_4(l) + 2\,Mg(s) \longrightarrow Ti(s) + 2\,MgCl_2(s)$$

What does this reaction tell us about the relative strengths of titanium and magnesium metal as reducing agents?

**9-64** Use Table 9.4 to determine which of the following reactions should occur.

(a) $Zn^{2+}(aq) + Cu(s) \longrightarrow Cu^{2+}(aq) + Zn(s)$
(b) $3\,Ag^{+}(aq) + Au(s) \longrightarrow 3\,Ag(s) + Au^{3+}(aq)$
(c) $Cu^{2+}(aq) + Fe(s) \longrightarrow Fe^{2+}(aq) + Cu(s)$
(d) $2\,Cu^{2+}(aq) + Ti(s) \longrightarrow 2\,Cu(s) + Ti^{4+}(aq)$

**9-65** Use the following results to decide where the reduction of $Cd^{2+}$ to Cd metal should be placed in Table 9.4.

$$Cd(s) + Sn^{2+}(aq) \longrightarrow Cd^{2+}(aq) + Sn(s)$$
$$Cd^{2+}(aq) + Fe(s) \longrightarrow Fe^{2+}(aq) + Cd(s)$$
$$Cd(s) + Zn^{2+}(aq) \not\longrightarrow$$
$$Cd^{2+}(aq) + Cr(s) \not\longrightarrow$$

**9-66** Explain why the strongest reducing agents in Table 9.4 are elements toward the bottom left corner of the periodic table.

### The Preparation of Metals

**9-67** On January 21, 1941, Dow Chemical Company produced an ingot of magnesium metal that represented the first commercial ingot of a metal taken from sea water. Speculate on how Dow transformed the $Mg^{2+}$ ions in sea water into magnesium metal.

**9-68** Which of the following metals is not manufactured by electrolysis?

(a) Na (b) Mg (c) Al (d) Fe

**9-69** Metal ores are roasted to convert sulfides into the corresponding oxides:

$$2\,ZnS(s) + 3\,O_2(g) \longrightarrow 2\,ZnO(s) + 2\,SO_2(g)$$

Is this an oxidation–reduction reaction? If so, what is oxidized and reduced?

**9-70** List the metals in Table 9.4 that are prepared by electrolysis. Why are these metals prepared by electrolysis instead of reacting the corresponding salts with metals that are stronger reducing agents?

**9-71** Describe the reactions that take place at the cathode and the anode of an electrolytic cell when $MgCl_2$ is electrolyzed. Which reaction occurs at the cathode of an electrolysis cell, oxidation or reduction?

### Integrated Problems

**9-72** Explain why some, but not all, cans of beer float in water when the cans are made of aluminum.

**9-73** Sodium metal dissolves in lead to form an alloy when these metals are heated. Assume that you have been given a sample of a sodium–lead alloy. Furthermore, assume that the sodium in this alloy undergoes the same reactions as pure sodium metal, although at a slower rate. Finally, assume that the lead in this alloy undergoes the same reactions as pure lead metal. Design an experiment that could be used to determine the percent by weight of sodium in this alloy. (Previous generations of students have discovered at least three different ways in which this experiment can be performed.)

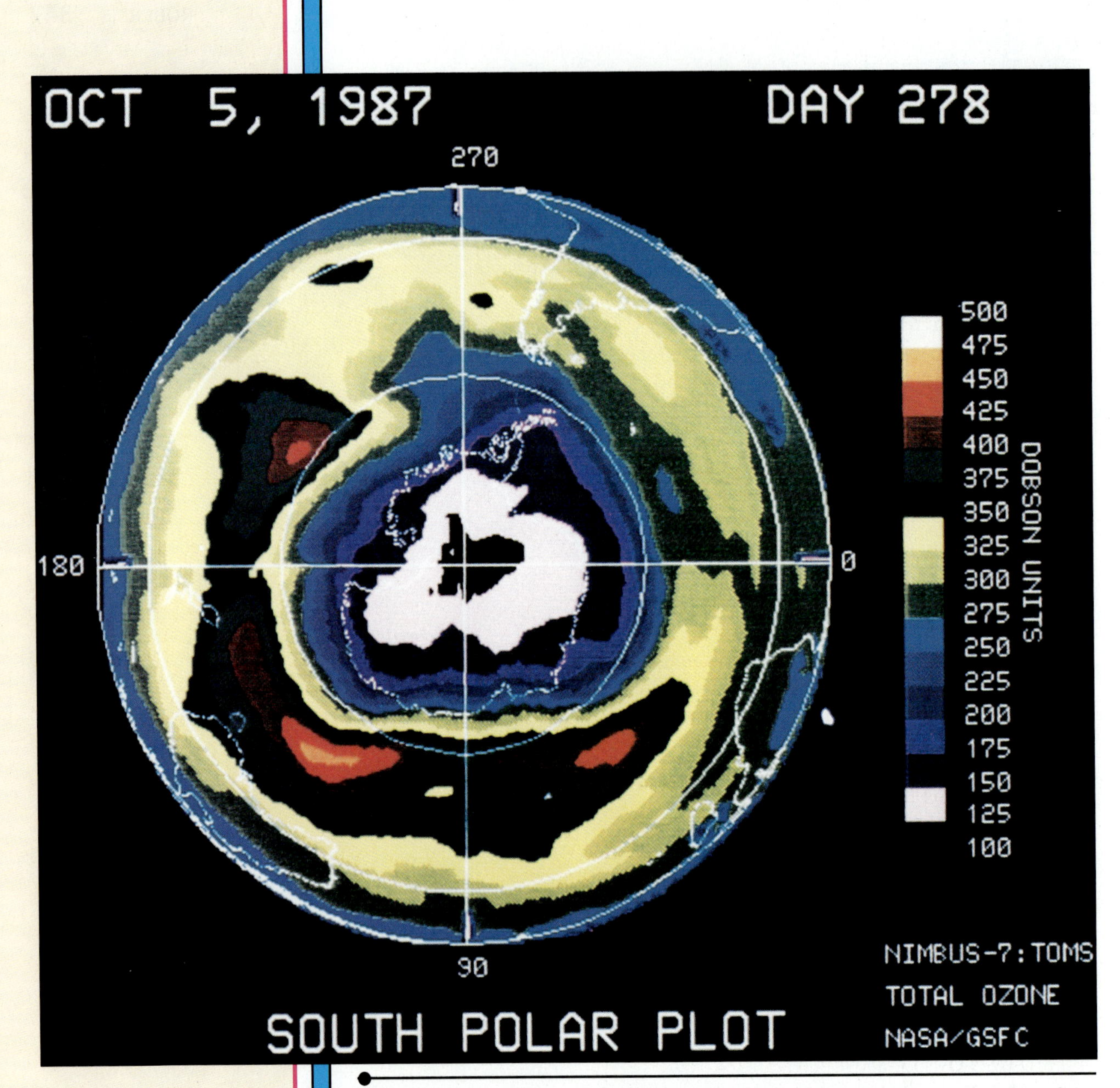


Data taken by the Nimus-7 satellite on September 15, 1987 clearly show depletion of the ozone in the atmosphere above the Antarctic.

CHAPTER 10

# THE CHEMISTRY OF THE NONMETALS

More than 75% of the known elements have the characteristic properties of *metals* (see Figure 10.1). They have a metallic luster; they are malleable and ductile; and they conduct heat and electricity. Eight other elements (B, Si, Ge, As, Sb, Te, Po, and At) are best described as *semimetals* or *metalloids.* They often look like metals, but they tend to be brittle, and they are more likely to be semiconductors than conductors of electricity.

Once the metals and semimetals are removed from the list of known elements, only 17 are left to be classified as *nonmetals.* Six of these elements belong to the family of rare gases in Group VIIIA, most of which are virtually inert to chemical reactions. Discussions of the chemistry of the nonmetals therefore tend to focus on the following elements: H, C, N, O, F, P, S, Cl, Se, Br, I, and Xe. This discussion provides the basis for answering questions such as the following:

## POINTS OF INTEREST

- Why does a balloon filled with a mixture of $H_2$ and $O_2$ explode when you touch it with a match? Why do you need a match to start the reaction? Why doesn't the reaction occur at room temperature?
- Why do periodic tables include hydrogen among both the metals in Group IA and the nonmetals in Group VIIA?
- If nitrogen and oxygen occupy adjacent positions in the periodic table, why is $O_2$ so reactive that it combines with most other elements, whereas $N_2$ is virtually inert to chemical reactions at room temperature?
- How does the ozone in the stratosphere protect us from high-energy solar radiation? What evidence do we have that chlorofluorocarbons present a threat to the ozone layer?
- What is the difference between baking soda and baking powder? Why can we replace baking powder with a mixture of baking soda and sour milk?
- Both graphite and diamond are pure carbon. Why is the former so soft it's used as a lubricant and the latter so hard it is the hardest naturally occurring substance?

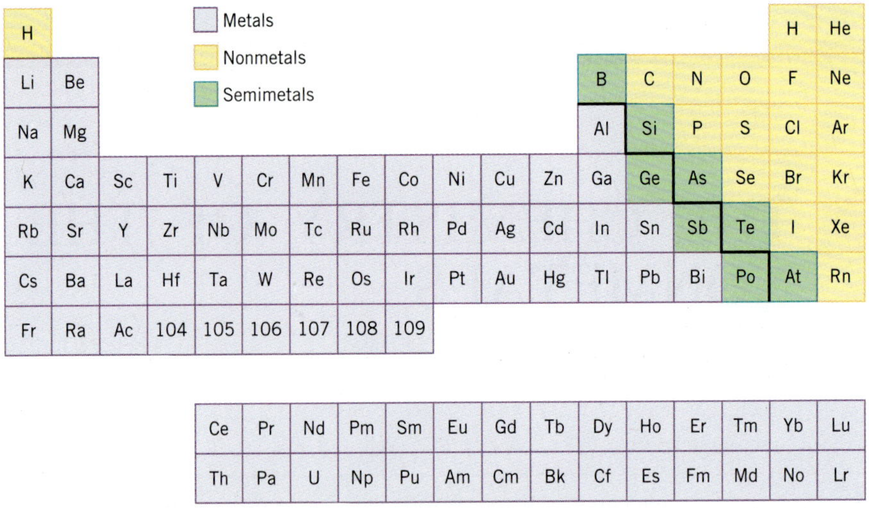

Metals
Nonmetals
Semimetals

| | | | | | | | | | | | | | | | | | |
|---|---|---|---|---|---|---|---|---|---|---|---|---|---|---|---|---|---|
| H | | | | | | | | | | | | | | | | H | He |
| Li | Be | | | | | | | | | | | B | C | N | O | F | Ne |
| Na | Mg | | | | | | | | | | | Al | Si | P | S | Cl | Ar |
| K | Ca | Sc | Ti | V | Cr | Mn | Fe | Co | Ni | Cu | Zn | Ga | Ge | As | Se | Br | Kr |
| Rb | Sr | Y | Zr | Nb | Mo | Tc | Ru | Rh | Pd | Ag | Cd | In | Sn | Sb | Te | I | Xe |
| Cs | Ba | La | Hf | Ta | W | Re | Os | Ir | Pt | Au | Hg | Tl | Pb | Bi | Po | At | Rn |
| Fr | Ra | Ac | 104 | 105 | 106 | 107 | 108 | 109 | | | | | | | | | |

| | | | | | | | | | | | | | |
|---|---|---|---|---|---|---|---|---|---|---|---|---|---|
| Ce | Pr | Nd | Pm | Sm | Eu | Gd | Tb | Dy | Ho | Er | Tm | Yb | Lu |
| Th | Pa | U | Np | Pu | Am | Cm | Bk | Cf | Es | Fm | Md | No | Lr |

**Figure 10.1** The elements can be divided into three classes: metals (shown in violet), semimetals (shown in green), and nonmetals (shown in yellow).

## 10.1 THE NONMETALS

There is a clear pattern in the chemistry of the main-group metals discussed in Chapter 9:

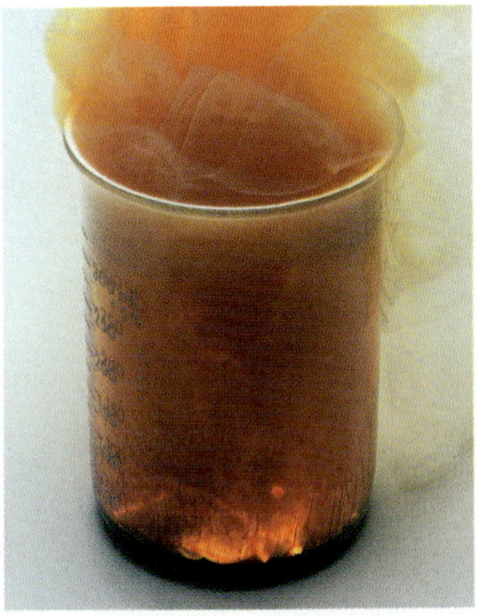

**The main-group metals are oxidized in all of their chemical reactions.**

The reaction between aluminum and bromine gives off enough energy to melt the aluminum metal and boil the liquid bromine.

These metals are oxidized when they react with nonmetal elements. Aluminum, for example, is oxidized by bromine:

$$2\,Al + 3\,Br_2 \longrightarrow Al_2Br_6$$

0 0 +3 −1
oxidation
reduction

The chemistry of the nonmetals is more interesting because these elements can undergo both oxidation and reduction. Phosphorus, for example, is oxidized when it reacts with oxygen to form $P_4O_{10}$.

$$P_4 + 5\,O_2 \longrightarrow P_4O_{10}$$

0 0 +5 −2
oxidation
reduction

But it is reduced when it reacts with calcium to form calcium phosphide:

$$6\,Ca + P_4 \longrightarrow 2\,Ca_3P_2$$

0 0 +2 −3
oxidation
reduction

These reactions can be understood by looking at the relative electronegativities of the elements. Phosphorus ($EN = 2.19$) is less electronegative than oxygen ($EN = 3.44$). When these elements react, the electrons are drawn toward the more electronegative oxygen atoms. Phosphorus is therefore oxidized in this reaction, and oxygen is reduced. Calcium ($EN$ = 1.00), on the other hand, is significantly less electronegative than phosphorus ($EN$ = 2.19). When these elements react, the electrons are drawn toward the more electronegative phosphorus atoms. As a result, calcium is oxidized and phosphorus is reduced.

The behavior of the nonmetals can be summarized as follows:

1. Nonmetals tend to oxidize metals:

$$2\,Mg(s) + O_2(g) \longrightarrow 2\,MgO(s)$$

2. Nonmetals with relatively large electronegativities (such as oxygen and chlorine) oxidize substances with which they react:

$$2\,H_2S(g) + 3\,O_2(g) \longrightarrow 2\,SO_2(g) + 2\,H_2O(g)$$
$$PH_3(g) + 3\,Cl_2(g) \longrightarrow PCl_3(l) + 3\,HCl(g)$$

3. Nonmetals with relatively small electronegativities (such as carbon and hydrogen) can reduce other substances:

$$Fe_2O_3(s) + 3\,C(s) \longrightarrow 2\,Fe(s) + 3\,CO(g)$$
$$CuO(s) + H_2(g) \longrightarrow Cu(s) + H_2O(g)$$

This photograph shows the reaction in which phosphorus is oxidized by oxygen to form $P_4O_{10}$.

The nonmetallic elements can be used as reducing agents. In the manufacture of steel, elemental carbon is used to reduce $Fe_2O_3$ and $Fe_3O_4$ to iron metal.

**EXERCISE 10.1**

Determine which element is oxidized and which is reduced when sulfur vapor reacts with red-hot charcoal to form carbon disulfide:

$$4\,C(s) + S_8(g) \longrightarrow 4\,CS_2(g)$$

**SOLUTION** Sulfur ($EN$ = 2.58) is just slightly more electronegative than carbon ($EN$ = 2.55). We therefore assume that carbon is oxidized in this reaction, and sulfur is reduced:

$$4\,C + S_8 \longrightarrow 4\,CS_2$$

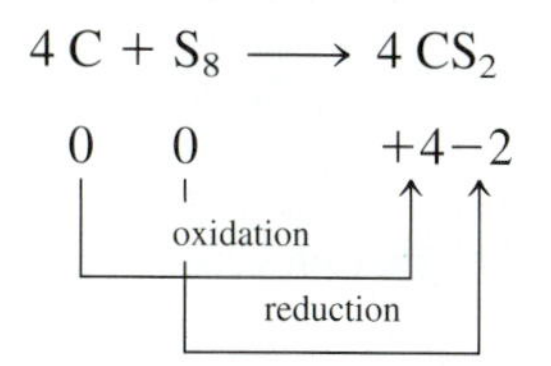

## 10.2 THE CHEMISTRY OF HYDROGEN

Hydrogen is the most abundant element in the universe: 90% of the atoms and 75% of the mass of the universe are hydrogen. Hydrogen is much less abundant on Earth—only 0.15% of the mass of Earth's crust, for example, is hydrogen. Even when the enormous number of hydrogen atoms in the Earth's oceans is included, hydrogen makes up less than 1% of the mass of the planet.

Models for the evolution of the Earth assume that the atmosphere once contained very little oxygen but significant amounts of both elemental hydrogen and compounds of hydrogen such as methane ($CH_4$) and ammonia ($NH_3$). Today, the atmosphere is 21% $O_2$ by volume, with only traces of $CH_4$ (2 ppm) and $H_2$ (0.5 ppm).

What happened to the hydrogen in the atmosphere? As the concentration of $O_2$ in the atmosphere increased, so did the probability of reaction between $H_2$ and $O_2$ to form water:

$$2\,H_2(g) + O_2(g) \longrightarrow 2\,H_2O(g)$$

or the likelihood of reactions between compounds that contain hydrogen and oxygen to produce water:

$$CH_4(g) + 2\,O_2(g) \longrightarrow CO_2(g) + 2\,H_2O(g)$$

The very name *hydrogen* comes from the Greek stems *hydro-*, "water," and *gennan,* "to form or generate." Thus, hydrogen is literally the "water former."

Hydrogen that did not react with oxygen to form the water in the earth's oceans escaped into space. As we have seen, the speed with which gas particles move is inversely proportional to the square root of their atomic or molecular weights. Because they are so light, $H_2$ molecules move faster than any other molecules in the atmosphere. (According to Exercise 4.16, the average velocity of an $H_2$ molecule at 0°C is 2000 m/s.) In spite of this, most of the $H_2$ molecules in the atmosphere are traveling too slowly to escape from the earth's gravity. (The velocity required to escape from the earth's gravitational field is 11.18 km/s.) But a very small fraction of these molecules are moving fast enough to escape. Thus, over geological time periods, a significant amount of the hydrogen that was once in the atmosphere escaped from the planet's surface.

Sec. 4.17

The three main engines of the space shuttle develop 1.1 million pounds of thrust by burning a mixture of liquid hydrogen and liquid oxygen.

Hydrogen combines with every element in the periodic table except the nonmetals in Group VIIIA (He, Ne, Ar, Kr, Xe, and Rn). Although it is often stated that more compounds contain carbon than any other element, this is not necessarily true. Most carbon compounds also contain hydrogen, and hydrogen forms compounds with virtually all the other elements as well. For historical reasons, compounds of hydrogen are frequently called **hydrides,** even though the name "hydride" literally describes compounds that contain an $H^-$ ion. There is a regular trend in the formula of the hydrides across a row of the periodic table, as shown in Figure 10.2. This trend is so regular that the combining power, or *valence,* of an element was once defined as the number of hydrogen atoms bound to the element in its hydride.

| $H_2$ | | | | | | | $H_2$ |
|---|---|---|---|---|---|---|---|
| LiH | $BeH_2$ | | $B_2H_6$ | $CH_4$ | $NH_3$ | $H_2O$ | HF |
| NaH | $MgH_2$ | | $AlH_3$ | $SiH_4$ | $PH_3$ | $H_2S$ | HCl |
| KH | $CaH_2$ | | $GaH_3$ | $GeH_4$ | $AsH_3$ | $H_2Se$ | HBr |
| RbH | $SrH_2$ | | | $SnH_4$ | $SbH_3$ | $H_2Te$ | HI |
| CsH | $BaH_2$ | | | $PbH_4$ | $BiH_3$ | $H_2Po$ | HAt |
| | | | | | | | |

**Figure 10.2** Hydrogen combines with every element in the periodic table except those in Group VIIIA. The formulas of the hydrides of the main-group elements are shown here.

Hydrogen is the only element that forms compounds in which the valence electrons are in the $n = 1$ shell. As a result, hydrogen can have three oxidation states, corresponding to the $H^+$ ion, a neutral H atom, and the $H^-$ ion:

$$H^+ = 1s^0$$
$$H = 1s^1$$
$$H^- = 1s^2$$

Because hydrogen forms compounds with oxidation numbers of both +1 and −1, many periodic tables include this element in both Group IA (with Li, Na, K, Rb, Cs, and Fr) and Group VIIA (with F, Cl, Br, I, and At).

There are many reasons for including hydrogen among the elements in Group IA. It forms compounds (such as HCl and $HNO_3$) that are analogs of alkali metal compounds (such as NaCl and $KNO_3$). Under conditions of very high pressure, it has the properties of a metal. (It has been argued, for example, that any hydrogen present at the center of the planet Jupiter is likely to be a metallic solid.) Finally, hydrogen combines with a handful of metals, such as scandium, titanium, chromium, nickel, or palladium, to form materials that behave as if they were alloys of two metals.

There are equally valid arguments for placing hydrogen in Group VIIA. It forms compounds (such as NaH and $CaH_2$) that are analogs of halogen compounds (such as NaF and $CaCl_2$). It also combines with other nonmetals to form covalent compounds (such as $H_2O$, $CH_4$, and $NH_3$), the way a nonmetal should. Finally, the element is a gas at room temperature and atmospheric pressure, like other nonmetals (such as $O_2$ and $N_2$).

It is difficult to decide where hydrogen belongs in the periodic table because of the physical properties of the element. The first ionization energy of hydrogen (1312 kJ/mol), for example, is roughly halfway between the elements with the largest (2372 kJ/mol) and smallest (376 kJ/mol) ionization energies. Hydrogen also has an electronegativity ($EN = 2.20$) halfway between the extremes of the most electronegative ($EN = 3.98$) and least electronegative ($EN = 0.7$) elements. On the basis of electronegativity, it is tempting to classify hydrogen as a semimetal, as shown in Figure 10.3.

**Figure 10.3** This three-dimensional graph of the electronegativities of the main-group elements helps us understand why it is difficult to classify hydrogen as a metal or nonmetal.

Hydrogen is oxidized by elements that are more electronegative to form compounds in which it has an oxidation number of +1:

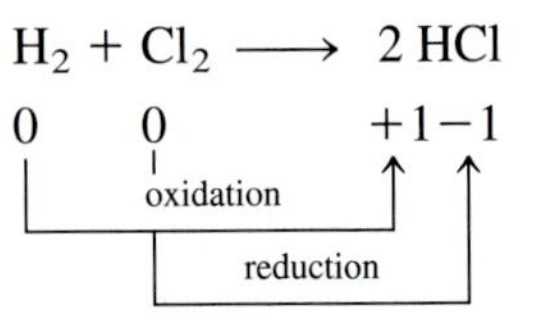


Hydrogen is reduced by elements that are less electronegative to form compounds in which its oxidation number is −1:

$$2\ Na + H_2 \longrightarrow 2\ NaH$$

0 0 +1 −1

oxidation

reduction

At room temperature, hydrogen is a colorless, odorless gas with a density only one-fourteenth the density of air. Small quantities of $H_2$ gas can be prepared in several ways:

1. By reacting an active metal with water:

$$2\,Na(s) + 2\,H_2O(l) \longrightarrow 2\,Na^+(aq) + 2\,OH^-(aq) + H_2(g)$$

2. By reacting a less active metal with a strong acid:

$$Zn(s) + 2\,HCl(aq) \longrightarrow Zn^{2+}(aq) + 2\,Cl^-(aq) + H_2(g)$$

3. By reacting an ionic metal hydride with water:

$$NaH(s) + H_2O(l) \longrightarrow Na^+(aq) + OH^-(aq) + H_2(g)$$

4. By decomposing water into its elements with an electric current:

$$2\,H_2O(l) \xrightarrow{\text{electrolysis}} 2\,H_2(g) + O_2(g)$$

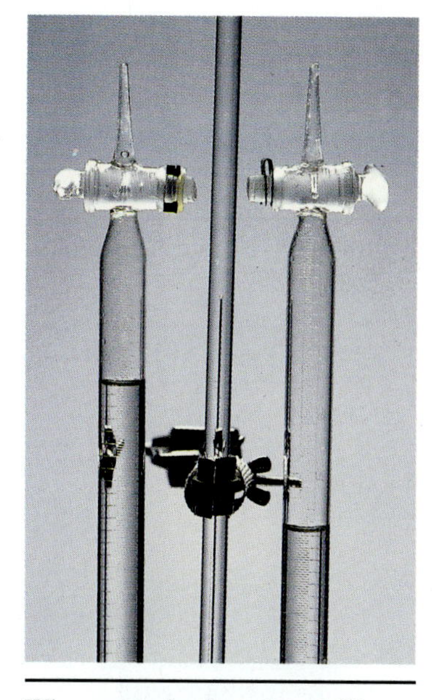

When water is electrolyzed, $H_2$ gas collects above one electrode and $O_2$ gas collects above the other. As expected from the balanced equation for this reaction, twice as much $H_2$ gas collects in one arm of this apparatus as $O_2$ gas in the other arm.

**EXERCISE 10.2**

Use oxidation numbers to determine what is oxidized and what is reduced in the following reactions, which are used to prepare $H_2$ gas.

(a) $Mg(s) + 2\,HCl(aq) \longrightarrow Mg^{2+}(aq) + 2\,Cl^-(aq) + H_2(g)$

(b) $Ca(s) + 2\,H_2O(l) \longrightarrow Ca^{2+}(aq) + 2\,OH^-(aq) + H_2(g)$

**SOLUTION**

(a) Magnesium metal is oxidized in this reaction and the $H^+$ ions from hydrochloric acid are reduced:

$$Mg + 2\,HCl \longrightarrow Mg^{2+} + 2\,Cl^- + H_2$$

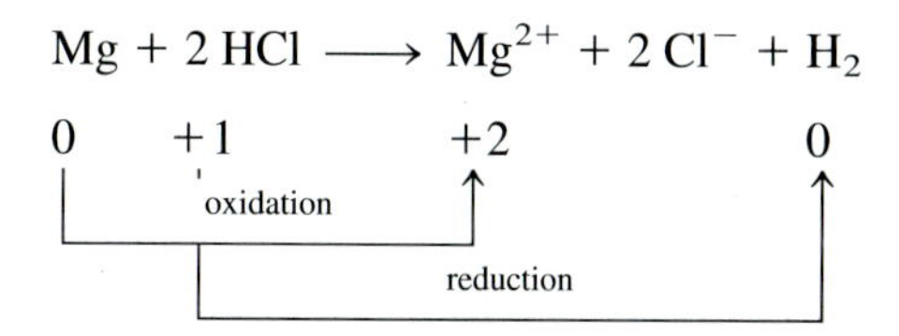


(b) Calcium metal is oxidized in this reaction and the $H^+$ ions from water are reduced:

$$Ca + 2\,H_2O \longrightarrow Ca^{2+} + 2\,OH^- + H_2$$

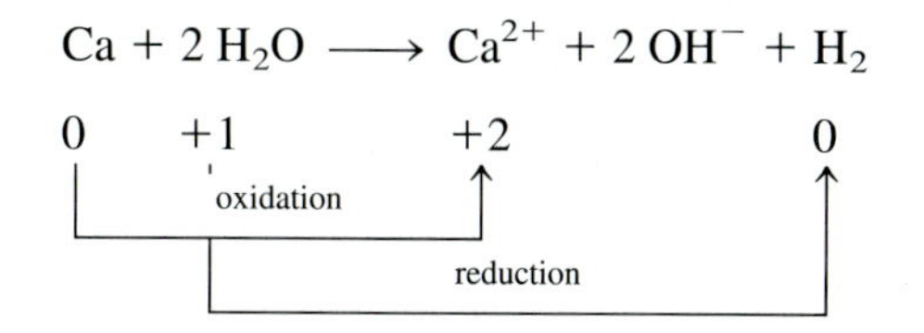


The covalent radius of a neutral hydrogen atom is 0.0371 nm, smaller than that of any other element. Because small atoms can come very close to each other, they tend to form strong covalent bonds. As a result, the bond dissociation enthalpy for the H—H bond is relatively large (435 kJ/mol). $H_2$ therefore tends to be unreactive at room temperature. In the presence of a spark, however, a fraction of the $H_2$ molecules dissociate to form hydrogen atoms that are highly reactive:

$$H_2(g) \xrightarrow{\text{spark}} 2\,H(g)$$

The heat given off when these H atoms react with $O_2$ is enough to catalyze the dissociation of additional $H_2$ molecules. Mixtures of $H_2$ and $O_2$ that are stable at room temperature therefore explode in the presence of a spark or flame.

## 10.3 THE CHEMISTRY OF OXYGEN

**Figure 10.4** The Lewis structure of the $O_2$ molecule.

p. 301

Oxygen is the most abundant element on this planet. The earth's crust is 46.6% oxygen by weight, the oceans are 86% oxygen by weight, and the atmosphere is 21% oxygen by volume. The name *oxygen* comes from the Greek stems *oxys*, "acid," and *gennan*, "to form or generate." Thus, oxygen literally means "acid former." This name was introduced by Lavoisier, who noticed that compounds rich in oxygen, such as $SO_2$ and $P_4O_{10}$, dissolve in water to give acids.

The electron configuration of an oxygen atom, [He] $2s^2\ 2p^4$, suggests that neutral oxygen atoms can achieve an octet of valence electrons by sharing two pairs of electrons to form an O=O double bond, as shown in Figure 10.4. According to this Lewis structure, all of the electrons in the $O_2$ molecule are paired. The compound should therefore be **diamagnetic**—it should be repelled by a magnetic field. Experimentally, $O_2$ is found to be **paramagnetic**—it is attracted to a magnetic field. As noted in Chapter 8, this can be explained by assuming that there are two unpaired electrons in the $\pi^*$ antibonding molecular orbitals of the $O_2$ molecule.

Liquid oxygen has a pale-blue color due to the absorption of photons with a wavelength of 630 nm.

At temperatures below −183°C, $O_2$ condenses to form a liquid with a characteristic light blue color that results from the absorption of light with a wavelength of 630 nm. This absorption is not seen in the gas phase and is relatively weak even in the liquid because it requires that three bodies—two $O_2$ molecules and a photon—collide simultaneously, which is a very rare phenomenon, even in the liquid phase.

### THE CHEMISTRY OF OZONE

The $O_2$ molecule is not the only elemental form of oxygen. In the presence of lightning or another source of a spark, $O_2$ molecules dissociate to form oxygen atoms:

$$O_2(g) \xrightarrow{\text{spark}} 2\ O(g)$$

These O atoms can react with $O_2$ molecules to form ozone, $O_3$, whose Lewis structure is shown in Figure 10.5.

$$O_2(g) + O(g) \longrightarrow O_3(g)$$

Oxygen ($O_2$) and ozone ($O_3$) are examples of **allotropes** (from the Greek meaning "in another manner"). By definition, allotropes are different forms of an element. Because they have different structures, allotropes have different chemical and physical properties (see Table 10.1).

Ozone is an unstable compound with a sharp, pungent odor that slowly decomposes to oxygen:

$$3\ O_3(g) \longrightarrow 3\ O_2(g)$$

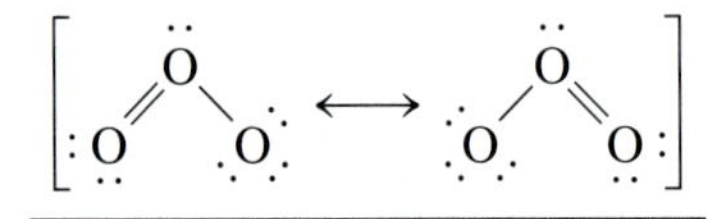

**Figure 10.5** Ozone is a resonance hybrid of two Lewis structures, each of which contains one O=O double bond and one O—O single bond. The valence shell electrons on the central atom are distributed toward the corners of an equilateral triangle. The $O_3$ molecule is therefore bent, with an O—O—O bond angle of 116.5°.

**TABLE 10.1 Properties of Allotropes of Oxygen**

| | Oxygen ($O_2$) | Ozone ($O_3$) |
|---|---|---|
| Melting point | −218.75°C | −192.5°C |
| Boiling point | −182.96°C | −110.5°C |
| Density (at 20°C) | 1.331 g/L | 1.998 g/L |
| O—O bond order | 2 | 1.5 |
| O—O bond length | 0.1207 nm | 0.1278 nm |

At low concentrations, ozone can be relatively pleasant. (The characteristic clean odor associated with summer thunderstorms is due to the formation of small amounts of $O_3$.) Exposure to $O_3$ at higher concentrations leads to coughing, rapid beating of the heart, chest pain, and general body pain. At concentrations above 1 ppm, ozone is toxic.

One of the characteristic properties of ozone is its ability to absorb radiation in the ultraviolet portion of the spectrum ($\lambda \leq 300$ nm), thereby providing a filter that protects us from exposure to high-energy ultraviolet radiation emitted by the sun. We can understand the importance of this filter if we think about what happens when radiation from the sun is absorbed by our skin.

Electromagnetic radiation in the infrared, visible, and low-energy portions of the ultraviolet spectrum ($\lambda \geq 300$ nm) carries enough energy to excite an electron in a molecule into a higher energy orbital. This electron eventually falls back into the orbital from which it was excited and energy is given off to the surrounding tissue in the form of heat. Anyone who has suffered from a sunburn can appreciate the painful consequences of excessive amounts of this radiation.

Radiation in the high-energy portion of the ultraviolet spectrum ($\lambda \leq 300$ nm) has a different effect when it is absorbed. This radiation carries enough energy to ionize atoms or molecules. The ions formed in these reactions have an odd number of electrons and are extremely reactive. They can cause permanent damage to the cell tissue and induce processes that eventually result in skin cancer. Relatively small amounts of this radiation therefore can have drastic effects on living tissue.

In 1974 Molina and Rowland pointed out that chlorofluorocarbons, such as $CFCl_3$ and $CF_2Cl_2$, which had been used as refrigerants and as propellants in aerosol cans, were beginning to accumulate in the atmosphere. In the stratosphere, at altitudes of 10 to 50 km above the earth's surface, chlorofluorocarbons decompose to form Cl atoms and chlorine oxides such as ClO when they absorb sunlight. Cl atoms and ClO molecules have an odd number of electrons, as shown in Figure 10.6. As a result, these substances are unusually reactive. In the atmosphere, they react with ozone or with the oxygen atoms that are needed to form ozone:

$$Cl + O_3 \longrightarrow ClO + O_2$$
$$ClO + O \longrightarrow Cl + O_2$$

Molina and Rowland postulated that these substances would eventually deplete the ozone shield in the stratosphere, with dangerous implications for biological systems that would be exposed to increased levels of high-energy ultraviolet radiation.

$:\ddot{\underset{..}{Cl}}\cdot$

$\cdot\ddot{\underset{..}{Cl}}-\ddot{\underset{..}{O}}:$

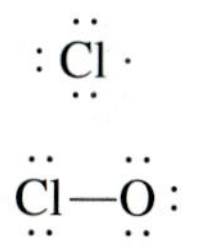

**Figure 10.6** The Lewis structure of the Cl atom and the ClO molecule formed when chlorofluorcarbons in the atmosphere decompose.

**CHECKPOINT**

Use Lewis structures to explain what happens in the following reactions:

$$Cl + O_3 \longrightarrow ClO + O_2$$
$$ClO + O \longrightarrow Cl + O_2$$

## OXYGEN AS AN OXIDIZING AGENT

Fluorine is the only element that is more electronegative than oxygen. As a result, oxygen gains electrons in virtually all its chemical reactions. Each $O_2$ molecule must gain four electrons to satisfy the octets of the two oxygen atoms without

$$:\ddot{O}=\ddot{O}: + 4\,e^- \longrightarrow :\ddot{O}:^{2-} + :\ddot{O}:^{2-}$$

**Figure 10.7** Two $O^{2-}$ ions are formed when an $O_2$ molecule picks up four electrons.

sharing electrons, as shown in Figure 10.7. Oxygen therefore oxidizes metals to form salts in which the oxygen atoms are formally present as $O^{2-}$ ions. Rust forms, for example, when iron reacts with oxygen in the presence of water to give a salt that formally contains the $Fe^{3+}$ and $O^{2-}$ ions, with an average of three water molecules coordinated to each $Fe^{3+}$ ion in this solid:

$$4\,Fe(s) + 3\,O_2(g) \xrightarrow{H_2O} 2\,Fe_2O_3(s) \cdot 3\,H_2O$$

Oxygen also oxidizes nonmetals, such as carbon, to form covalent compounds in which the oxygen formally has an oxidation number of $-2$:

$$C(s) + O_2(g) \longrightarrow CO_2(g)$$

Sec. 9.14

Oxygen is the perfect example of an **oxidizing agent** because it increases the oxidation state of almost any substance with which it reacts. In the course of its reactions, oxygen is reduced. The substances it reacts with are therefore **reducing agents.**

## EXERCISE 10.3

Identify the oxidizing agents and reducing agents in the following reaction:

$$CH_4(g) + 2\,O_2(g) \longrightarrow CO_2(g) + 2\,H_2O(g)$$

**SOLUTION** We start by identifying the oxidation and reduction halves of the reaction:

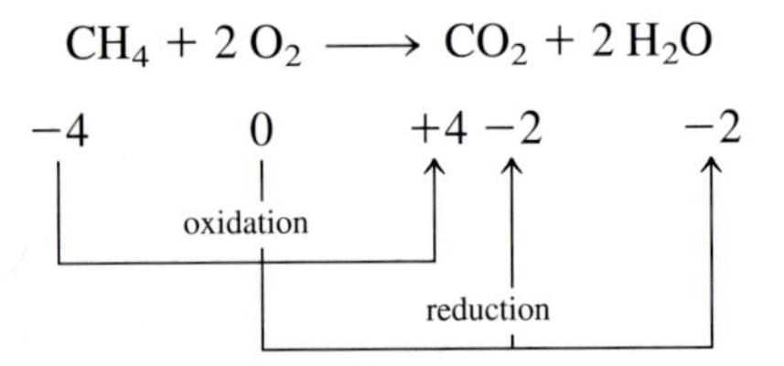


Oxygen oxidizes $CH_4$ in this reaction, so $O_2$ is the oxidizing agent. The oxygen is reduced by $CH_4$, which means that $CH_4$ is the reducing agent. Early models of the atmosphere, which assumed the presence of $CH_4$, $NH_3$, and $H_2$, were called *reducing atmospheres* because they contain large quantities of compounds that are reducing agents.

Each year between 75 and 80 quads, or quadrillion ($10^{15}$) Btu, of energy is consumed in the United States. Less than 10% of this energy is provided by nuclear, solar, geothermal, or hydro power. The rest can be traced to a combustion reaction in which a fuel is oxidized by $O_2$. The cars, trucks, and buses that fill our highways are powered by gasoline engines that burn hydrocarbons such as octane, $C_8H_{18}$, or diesel engines that burn larger hydrocarbons such as cetane, $C_{16}H_{34}$:

$$2\,C_8H_{18}(l) + 25\,O_2(g) \longrightarrow 16\,CO_2(g) + 18\,H_2O(g)$$
$$2\,C_{16}H_{34}(l) + 49\,O_2(g) \longrightarrow 32\,CO_2(g) + 34\,H_2O(g)$$

We heat our homes by burning the methane ($CH_4$) in natural gas, the high molecular weight hydrocarbons in fuel oil, the hydrocarbons in wood, or by using electricity generated in a power plant that burns either oil or coal.

The energy we use to fuel our bodies also comes from combustion reactions. Energy enters our bodies in the form of lipids, proteins, and carbohydrates. These "fuels" are converted into carbohydrates, such as glucose ($C_6H_{12}O_6$), which react with oxygen to produce the energy we need to survive:

$$C_6H_{12}O_6(aq) + 6\,O_2(g) \longrightarrow 6\,CO_2(g) + 6\,H_2O(l) \qquad \Delta H^\circ = -2870 \text{ kJ/mol}$$

About 65% of the energy given off in this reaction is used to synthesize the ATP (adenosine triphosphate) that fuels biological processes. The remaining 35% is released as the heat that keeps our body temperatures higher than the temperature of the surroundings.

There is an ever-growing understanding that Earth contains a finite amount of fossil fuels, such as oil and coal, that will eventually run out. Nuclear, solar, and geothermal power will be increasingly important sources of energy. But they won't replace fossil fuels by themselves because they are used to produce electrical energy, which is difficult to store. One possible solution to this problem has been labeled the *hydrogen economy*.

The first step in the hydrogen economy is to use energy from nuclear, solar, or geothermal power to split water into its elements:

$$2\,H_2O(l) \longrightarrow 2\,H_2(g) + O_2(g)$$

The oxygen is then released to the atmosphere, and the hydrogen is burned as a fuel:

$$2\,H_2(g) + O_2(g) \longrightarrow 2\,H_2O(g) \qquad \Delta H^\circ = -241.83 \text{ kJ/mol}$$

or used to reduce carbon monoxide to methanol or gasoline, which can be stored and later burned as a fuel:

$$CO(g) + 2\,H_2(g) \longrightarrow CH_3OH(l)$$
$$8\,CO(g) + 17\,H_2(g) \longrightarrow C_8H_{18}(l) + 8\,H_2O(l)$$

## PEROXIDES

It takes four electrons to reduce an $O_2$ molecule to a pair of $O^{2-}$ ions. If the reaction stops after the $O_2$ molecule has gained only two electrons, the $O_2^{\ 2-}$ ion shown in Figure 10.8 is produced. This ion has two more electrons than a neutral $O_2$ molecule, which means that the oxygen atoms must share only a single pair of bonding electrons to achieve an octet of valence electrons. The $O_2^{\ 2-}$ ion is called the **peroxide** ion because compounds that contain this ion are unusually rich in oxygen. They are not just oxides—they are (hy-)peroxides.

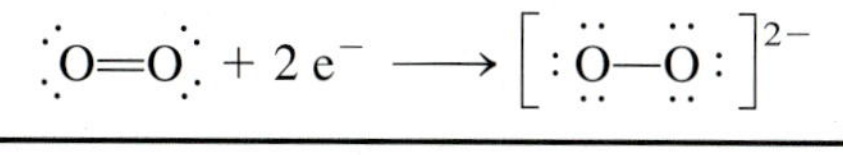

$$\ddot{O}{=}\ddot{O} + 2\,e^- \longrightarrow \left[\,:\ddot{O}{-}\ddot{O}:\,\right]^{2-}$$

**Figure 10.8** An $O_2^{\ 2-}$ (peroxide) ion is formed when an $O_2$ molecule picks up two electrons.

The easiest way to prepare a peroxide is to react sodium or barium metal with oxygen:

$$2\,Na(s) + O_2(g) \longrightarrow Na_2O_2(s)$$
$$Ba(s) + O_2(g) \longrightarrow BaO_2(s)$$

When these peroxides are allowed to react with a strong acid, hydrogen peroxide ($H_2O_2$) is produced:

$$BaO_2(s) + 2\,H^+(aq) \longrightarrow Ba^{2+}(aq) + H_2O_2(aq)$$

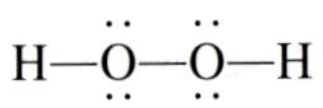

The Lewis structure of hydrogen peroxide contains an O—O single bond, as shown in Figure 10.9. The VSEPR theory predicts that the geometry around each oxygen atom in $H_2O_2$ should be bent. But this theory cannot predict whether the four atoms lie in the same plane or whether the molecule should be visualized as lying in two intersecting planes. The experimentally determined structure of $H_2O_2$ is shown in Figure 10.10. The H—O—O bond angle in this molecule is only slightly larger than the angle between a pair of adjacent $2p$ atomic orbitals on the oxygen atom, and the angle between the planes that form the molecule is slightly larger than the tetrahedral angle.

H—Ö—Ö—H (with lone pairs on each O)

**Figure 10.9** The Lewis structure for hydrogen peroxide.

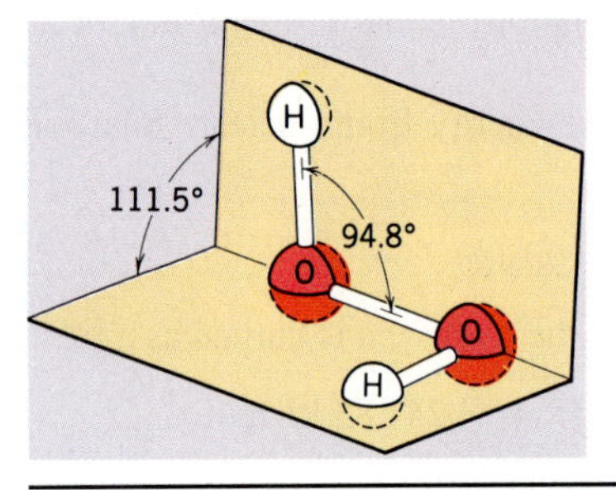


**Figure 10.10** The VSEPR theory predicts that the valence electrons on the oxygen atoms in $H_2O_2$ should be distributed toward the corners of a tetrahedron. The geometry around each oxygen atom is therefore bent, or angular, with an O—O—H bond angle of 94.8°. In order to keep the nonbonding electrons on the oxygen atoms as far apart as possible, the four atoms in this molecule lie in two planes that intersect at an angle of 111.5°.

The oxidation number of the oxygen atoms in hydrogen peroxide is −1. $H_2O_2$ can therefore act as an oxidizing agent and capture two more electrons to form a pair of hydroxide ions, in which the oxygen has an oxidation number of −2.

$$H_2O_2 + 2\,e^- \longrightarrow 2\,OH^-$$

Or, it can act as a reducing agent and lose a pair of electrons to form an $O_2$ molecule.

$$H_2O_2 \longrightarrow O_2 + 2\,H^+ + 2\,e^-$$

### ▶ CHECKPOINT

Use Lewis structures to explain what happens in the following reactions:

$$H_2O_2 + 2\,e^- \longrightarrow 2\,OH^-$$

$$H_2O_2 \longrightarrow O_2 + 2\,H^+ + 2\,e^-$$

Reactions in which a compound simultaneously undergoes both oxidation and reduction are called **disproportionation reactions.** The products of the disproportionation of $H_2O_2$ are oxygen and water:

$$2\,H_2O_2(aq) \longrightarrow O_2(g) + 2\,H_2O(l)$$

**EXERCISE 10.4**

Use the reactions that describe what happens when $H_2O_2$ loses a pair of electrons or picks up a pair of electrons to explain why the disproportionation of hydrogen peroxide gives oxygen and water.

**SOLUTION** Adding the half-reaction for the oxidation of $H_2O_2$ to the half-reaction for the reduction of this compound gives the following results:

$$\begin{array}{rcl} H_2O_2 + 2\,e^- & \longrightarrow & 2\,OH^- \\ H_2O_2 & \longrightarrow & O_2 + 2\,H^+ + 2\,e^- \\ \hline 2\,H_2O_2 & \longrightarrow & O_2 + 2\,H^+ + 2\,OH^- \end{array}$$

The $H^+$ and $OH^-$ ions produced in the two halves of this reaction combine to form water to give the following overall stoichiometry for the reaction:

$$2\,H_2O_2 \longrightarrow O_2 + 2H_2O$$

The disproportionation of $H_2O_2$ is an exothermic reaction:

$$2\,H_2O_2(aq) \longrightarrow O_2(g) + 2\,H_2O(l) \qquad \Delta H^\circ = -94.6\text{ kJ/mol }H_2O$$

This reaction is relatively slow, however, in the absence of a catalyst, such as dust or a metal surface. The principal uses of $H_2O_2$ revolve around its oxidizing ability. It is used in dilute (3%) solutions as a disinfectant. In more concentrated solutions (30%), it is used as a bleaching agent for hair, fur, leather, or the wood pulp used to make paper. In very concentrationed solutions, $H_2O_2$ has been used as rocket fuel because of the ease with which it decomposes to give $O_2$.

## METHODS OF PREPARING $O_2$

Small quantities of $O_2$ gas can be prepared in a number of ways.

1. By decomposing a dilute solution of hydrogen peroxide with dust or a metal surface as the catalyst:

$$2\,H_2O_2(aq) \longrightarrow O_2(g) + 2\,H_2O(l)$$

2. By reacting hydrogen peroxide with a strong oxidizing agent, such as the permanganate ion, $MnO_4^-$:

$$5\,H_2O_2(aq) + 2\,MnO_4^-(aq) + 6\,H^+(aq) \longrightarrow 2\,Mn^{2+}(aq) + 5\,O_2(g) + 8\,H_2O(l)$$

3. By passing an electric current through water:

$$2\,H_2O(l) \xrightarrow{\text{electrolysis}} 2\,H_2(g) + O_2(g)$$

4. By heating potassium chlorate ($KClO_3$) in the presence of a catalyst until it decomposes:

$$2\,KClO_3(s) \xrightarrow{MnO_2} 2\,KCl(s) + 3\,O_2(g)$$

**CHECKPOINT**

Which of the following elements or compounds reacts with water to produce a solution that could be used to produce $O_2$? (a) Na (b) $Na_2O$ (c) $Na_2O_2$ (d) NaOH (e) NaCl

RESEARCH IN THE 90s

## THE CHEMISTRY OF THE ATMOSPHERE

Although major changes occurred in the atmosphere during the early history of our planet, the chemistry of Earth's atmosphere was more or less constant during the time in which the human race evolved. This is no longer true. The amount of methane ($CH_4$) in the atmosphere is increasing at a rate of more than 1% per year. The concentration of carbon dioxide has more than doubled since 1750 and seems to be increasing at an exponential rate (see Figure 10.11). In recent years, color photographs such as the one with which this chapter opened have focused attention on one particular change in the atmosphere, the depletion of ozone above the Antarctic continent.

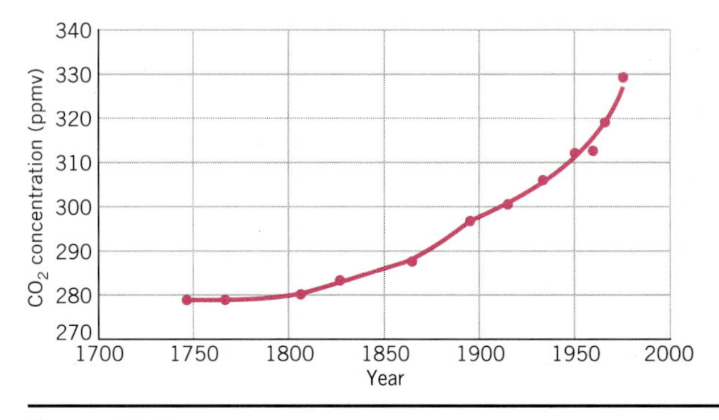


**Figure 10.11** Observed trends in $CO_2$ concentration in the atmosphere from the time of the industrial revolution to present. The units with which these concentrations are reported are parts per million by volume (ppmv).

For over 25 years, a team of scientists collected data on variations in the amount of $O_3$ at different altitudes above the British Antarctic Survey station at Halley Bay. In 1985, they reported that the $O_3$ concentration declined after the return of solar radiation each September. The effect was small when first noticed in the late 1970s but reached such high levels by 1984 that the $O_3$ concentration declined by 30% before the end of October of that year.

The data from Halley Bay sampled only a small portion of the atmosphere over the Antarctic. Once the notion of ozone depletion was reported, however, scientists were able to retrieve data recorded by satellites over a period of years, which confirmed that the same effect occurred over virtually the entire Antarctic continent. Several features of the ozone depletion were particularly interesting:

1. The drop in the ozone concentration occurred very rapidly, within a period of six weeks each spring. (''Spring'' occurs during September and October in the southern hemisphere.)
2. Although the drop in $O_3$ occurs above the Antarctic, it corresponds to a loss of 3% of the total ozone concentration in the planet's atmosphere.
3. The decline in the $O_3$ concentration became more serious each year. By 1989, the $O_3$ concentration during the summer months dropped by 70%.

4. The decline was temporary. During the winter months, the ozone level built back to normal levels.

Several possible explanations were available for the ozone holes. One of the most popular was the suggestion by Molina and Rowland that the ozone in the atmosphere could be destroyed by Cl and ClO radicals created when chlorofluorocarbons (CFCs) in the atmosphere decomposed:

$$Cl + O_3 \longrightarrow ClO + O_2$$
$$ClO + O \longrightarrow Cl + O_2$$

The question is: What evidence can be found to either support or refute this hypothesis? The problem is complex because more than 200 chemical reactions have been included in the models used to explain the chemistry of the atmosphere. It is further complicated by the fact that ClO radicals exist in the atmosphere at concentrations of only about 1 part per trillion by volume (pptv).

In the January 4, 1991, issue of *Science,* James Anderson and co-workers reported data that probed the link between the release of chlorofluorocarbons into the atmosphere and the disappearance of ozone from the stratosphere above the Antarctic each spring. These data were obtained between August 23 and September 22, 1987, using special instruments mounted in high-altitude aircraft. Initial measurements after the plane took off from the tip of Chile (54° south latitude) suggested that the background concentration of ClO radicals was at the threshold of detection: about 1 pptv. As the plane flew toward the south pole, the ClO concentration increased slightly until about 65° latitude, when it rapidly increased.

A plot of the concentration of ClO radicals versus latitude for the September 16 flight is shown in Figure 10.12. By the time the aircraft reached a latitude of 68° the abundance of these radicals had increased by three orders of magnitude, to a level of

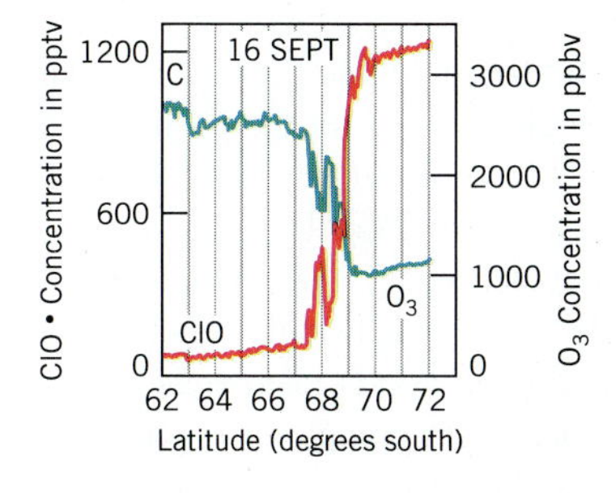


**Figure 10.12** Variations in the ClO· and $O_3$ concentrations in the atmosphere during the September 16, 1987, flight toward the Antarctic continent. ClO· concentrations increased from about 1 part per trillion by volume to 1200 pptv, while $O_3$ concentrations decreased from 2700 parts per billion by volume to 1000 ppbv.

approximately 1200 pptv. These data by themselves are suggestive, but this figure contains more compelling evidence. The plot at the top of Figure 10.12 shows that the concentration of $O_3$ dropped by a factor of about 2.5 at virtually the same time that the ClO concentration increased.

Anderson and co-workers concluded as follows: ''When taken independently, each element in the case contains a segment of the puzzle that in itself is not conclusive. When taken together, however, they provide convincing evidence that the dramatic reduction in . . . $O_3$ over the Antarctic continent would not have occurred had CFCs not been synthesized and then added to the atmosphere.''

## 10.4 THE CHEMISTRY OF SULFUR

Because sulfur is directly below oxygen in the periodic table, these elements have similar electron configurations. As a result, sulfur forms many compounds that are analogs of oxygen compounds, as shown in Table 10.2. Examples in this table show how the prefix *thio-* is used to indicate compounds in which sulfur replaces an oxygen atom.

There are four principal differences between the chemistry of sulfur and the chemistry of oxygen:

1. O=O double bonds are much stronger than S=S double bonds.
2. S—S single bonds are almost twice as strong as O—O single bonds.
3. Sulfur ($EN = 2.58$) is much less electronegative than oxygen ($EN = 3.44$).
4. Sulfur can expand its valence shell to hold more than eight electrons, but oxygen cannot.

These seemingly minor differences have important consequences for the chemistry of these elements.

### THE EFFECT OF DIFFERENCES IN THE STRENGTH OF SINGLE AND DOUBLE BONDS

The radius of a sulfur atom is about 60% larger than that of an oxygen atom:

$$\frac{\text{Covalent radius of sulfur}}{\text{Covalent radius of oxygen}} = \frac{0.104 \text{ nm}}{0.066 \text{ nm}} = 1.58$$

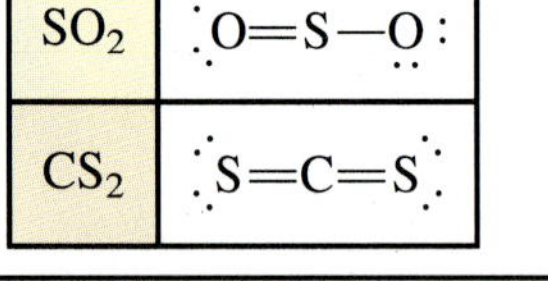


**Figure 10.13** Lewis structures for $SO_2$ and $CS_2$.

As a result, it is harder for sulfur atoms to come close enough together to form $\pi$ bonds. S=S double bonds are therefore much weaker than O=O double bonds.

Double bonds between sulfur and oxygen or carbon atoms can be found in compounds such as $SO_2$ and $CS_2$ (see Figure 10.13). But these double bonds are much weaker than the equivalent double bonds to oxygen atoms in $O_3$ or $CO_2$. The bond dissociation enthalpy for a C=S double bond is 477 kJ/mol, for example, whereas the bond dissociation enthalpy for a C=O double bond is 745 kJ/mol.

Elemental oxygen consists of $O_2$ molecules in which each atom completes its octet of valence electrons by sharing two pairs of electrons with a single neighboring atom. Because sulfur does not form strong S=S double bonds, elemental sulfur usually consists of cyclic $S_8$ molecules in which each atom completes its octet by forming single bonds to two neighboring atoms, as shown in Figure 10.14.

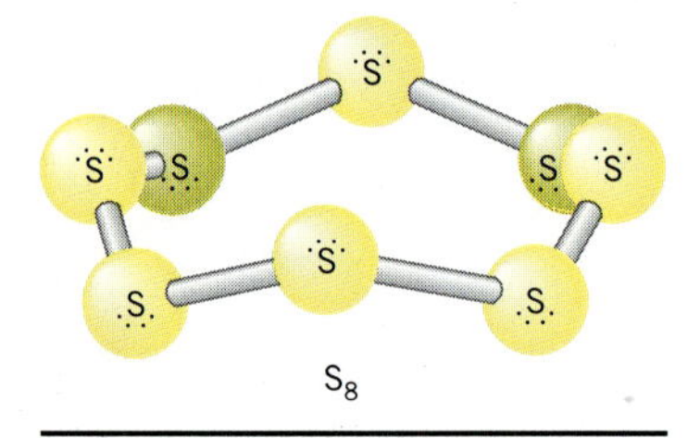


**Figure 10.14** Because they are too large to form strong S=S double bonds, sulfur atoms form single bonds to two different atoms in elemental sulfur. The result is a cyclic $S_8$ molecule.

**TABLE 10.2 Oxygen Compounds and Their Sulfur Analogs**

| Oxygen Compounds | Sulfur Compounds |
|---|---|
| $Na_2O$ (sodium oxide) | $Na_2S$ (sodium sulfide) |
| $H_2O$ (water) | $H_2S$ (hydrogen sulfide) |
| $O_3$ (ozone) | $SO_2$ (sulfur dioxide) |
| $CO_2$ (carbon dioxide) | $CS_2$ (carbon disulfide) |
| $OCN^-$ (cyanate) | $SCN^-$ (thiocyanate) |
| $OC(NH_2)_2$ (urea) | $SC(NH_2)_2$ (thiourea) |

$S_8$ molecules can pack to form more than one crystal. The most stable form of sulfur consists of *orthorhombic* crystals of $S_8$ molecules, which are often found near volcanos. If these crystals are heated until they melt and the molten sulfur is then cooled, an allotrope of sulfur consisting of *monoclinic* crystals of $S_8$ molecules is formed. These monoclinic crystals slowly transform themselves into the more stable orthorhombic structure over a period of time.

The tendency of an element to form bonds to itself is called *catenation* (from the Latin *catena*, "chain"). Because sulfur forms unusually strong S—S single bonds, it is better at catenation than any element except carbon. As a result, the orthorhombic and monoclinic forms of sulfur are not the only allotropes of the element. Allotropes of sulfur also exist that differ in the size of the molecules that form the crystal. Cyclic molecules that contain 6, 7, 8, 10, and 12 sulfur atoms are known.

Sulfur melts at 119.25°C to form a yellow liquid that is less viscous than water. If this liquid is heated to 159°C, it turns into a dark red liquid that cannot be poured from its container. The viscosity of this dark red liquid is 2000 times greater than that of molten sulfur because the cyclic $S_8$ molecules open up and link together to form long chains of as many as 100,000 sulfur atoms.

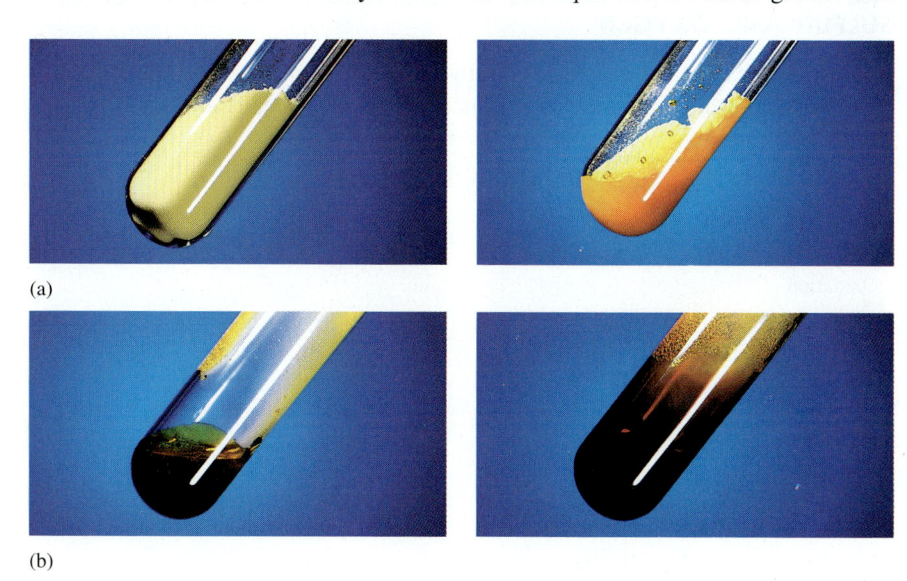


(a) When heated gently, sulfur melts to form a yellow liquid that flows more easily than water.
(b) When heated further, the liquid turns a dark red color, and it becomes much more viscous as the individual $S_8$ molecules link together to form long polymeric chains.

When sulfur reacts with an active metal, it can form the sulfide ion, $S^{2-}$:

$$16\,K(s) + S_8(s) \longrightarrow 8\,K_2S(s)$$

This is not the only product that can be obtained, however. A variety of polysulfide ions with a charge of −2 can be produced that differ in the number of sulfur atoms in the chain:

$$\begin{aligned}
2\,K(s) + S_8(s) \longrightarrow\; & K_2S_2 = [K^+]_2[S{-}S]^{2-} \\
& K_2S_3 = [K^+]_2[S{-}S{-}S]^{2-} \\
& K_2S_4 = [K^+]_2[S{-}S{-}S{-}S]^{2-} \\
& K_2S_5 = [K^+]_2[S{-}S{-}S{-}S{-}S]^{2-} \\
& K_2S_6 = [K^+]_2[S{-}S{-}S{-}S{-}S{-}S]^{2-} \\
& K_2S_8 = [K^+]_2[S{-}S{-}S{-}S{-}S{-}S{-}S{-}S]^{2-}
\end{aligned}$$

Iron pyrite crystals have the empirical formula $FeS_2$. This mineral therefore can be considered to be a salt of the $Fe^{2+}$ and $S_2^{2-}$ ions.

**EXERCISE 10.5**

Use the tendency of sulfur to form polysulfide ions to explain why iron has an oxidation number of +2 in iron pyrite, $FeS_2$, one of the most abundant sulfur ores.

**SOLUTION** If the oxidation number of iron in $FeS_2$ is +2, the sulfur must be present as the $S_2^{2-}$ ion:

$$FeS_2 = [Fe^{2+}][S_2^{2-}]$$

This disulfide ion is the sulfur analog of the peroxide ion, $O_2^{2-}$, and has an analogous Lewis structure.

## THE EFFECT OF DIFFERENCES IN THE ELECTRONEGATIVITIES OF SULFUR AND OXYGEN

Because sulfur is much less electronegative than oxygen, it is more likely to form compounds in which it has a positive oxidation number (see Table 10.3).

**TABLE 10.3 Common Oxidation Numbers for Sulfur**

| Oxidation Number | Examples |
|---|---|
| −2 | $Na_2S$, $H_2S$ |
| −1 | $Na_2S_2$, $H_2S_2$ |
| 0 | $S_8$ |
| +1 | $S_2Cl_2$ |
| +2 | $S_2O_3^{2-}$ |
| $+2\frac{1}{2}$ | $S_4O_6^{2-}$ |
| +3 | $S_2O_4^{2-}$ |
| +4 | $SF_4$, $SO_2$, $H_2SO_3$, $SO_3^{2-}$ |
| +5 | $S_2O_6^{2-}$ |
| +6 | $SF_6$, $SO_3$, $H_2SO_4$, $SO_4^{2-}$ |

In theory, sulfur can react with oxygen to form either $SO_2$ or $SO_3$, whose Lewis structures are given in Figure 10.15. In practice, combustion of sulfur compounds gives $SO_2$, regardless of whether sulfur or a compound of sulfur is burned:

$$S_8(s) + 8\,O_2(g) \longrightarrow 8\,SO_2(g)$$

$$CS_2(l) + 3\,O_2(g) \longrightarrow CO_2(g) + 2\,SO_2(g)$$

$$3\,FeS_2(s) + 8\,O_2(g) \longrightarrow Fe_3O_4(s) + 6\,SO_2(g)$$

Although the $SO_2$ formed in these reactions should react with $O_2$ to form $SO_3$, the rate of this reaction is very slow. The rate of the conversion of $SO_2$ into $SO_3$ can be greatly increased by adding an appropriate catalyst:

$$2\,SO_2(g) \xrightarrow{V_2O_5/K_2O} 2\,SO_3(g)$$

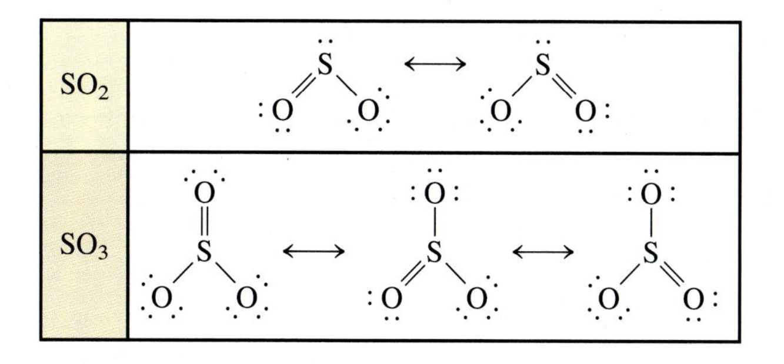


**Figure 10.15** $SO_2$ is a resonance hybrid of two Lewis structures analogous to the structure of ozone in Figure 10.5. $SO_3$ is a resonance hybrid of three Lewis structures. $SO_3$ can be imagined to result from the donation of a pair of nonbonding electrons on the sulfur atom in $SO_2$ to a neutral oxygen atom to form a covalent bond.

Enormous quantities of $SO_2$ are produced by industry each year and then converted to $SO_3$, which can be used to produce sulfuric acid, $H_2SO_4$. In theory, sulfuric acid can be made by dissolving $SO_3$ gas in water:

$$SO_3(g) + H_2O(l) \longrightarrow H_2SO_4(aq)$$

In practice, this is not convenient. Instead, $SO_3$ is absorbed in 98% $H_2SO_4$, where it reacts with the water to form additional $H_2SO_4$ molecules. Water is then added, as needed, to keep the concentration of this solution between 96% and 98% $H_2SO_4$ by weight.

Sulfuric acid is by far the most important industrial chemical. It has even been argued that there is a direct relationship between the amount of sulfuric acid a country consumes and its standard of living. More than 50% of the sulfuric acid produced each year is used to make fertilizers. The rest is used to make paper, synthetic fibers and textiles, insecticides, detergents, feed additives, dyes, drugs, antifreeze, paints and enamels, linoleum, synthetic rubber, printing inks, cellophane, photographic film, explosives, automobile batteries, and metals such as magnesium, aluminum, iron, and steel.

Sulfuric acid dissociates in water to give the $HSO_4^-$ ion, which is known as the hydrogen sulfate, or bisulfate, ion:

$$H_2SO_4(aq) \longrightarrow H^+(aq) + HSO_4^-(aq)$$

Ten percent of these hydrogen sulfate ions dissociate further to give the $SO_4^{2-}$, or sulfate, ion:

$$HSO_4^-(aq) \longrightarrow H^+(aq) + SO_4^{2-}(aq)$$

A variety of salts can be formed by replacing the $H^+$ ions in sulfuric acid with positively charged ions, such as the $Na^+$ or $K^+$ ions:

$$NaHSO_4 = \text{sodium hydrogen sulfate}$$
$$Na_2SO_4 = \text{sodium sulfate}$$

Sulfur dioxide dissolves in water to form sulfurous acid:

$$SO_2(g) + H_2O(l) \longrightarrow H_2SO_3(aq)$$

Sulfurous acid doesn't dissociate in water to as great extent as sulfuric acid, but it is still possible to replace the $H^+$ ions in $H_2SO_3$ with positive ions to form salts:

$$NaHSO_3 = \text{sodium hydrogen sulfite}$$
$$Na_2SO_3 = \text{sodium sulfite}$$

**EXERCISE 10.6**

Which of the following reactions involves a change in the oxidation number of sulfur?

(a) $2\ SO_2(g) + O_2(g) \longrightarrow 2\ SO_3(g)$
(b) $SO_3(g) + H_2O(l) \longrightarrow H_2SO_4(aq)$
(c) $H_2SO_4(aq) \longrightarrow H^+(aq) + HSO_4^-(aq)$

**SOLUTION** Only the first reaction involves a change in the oxidation number of sulfur, from +4 to +6. The others involve compounds in which the oxidation number of sulfur is +6.

Sulfuric acid and sulfurous acid are both examples of a class of compounds known as **oxyacids** because they are literally acids that contain oxygen. Because they are negative ions (or anions) that contain oxygen, the $SO_3^{2-}$ and $SO_4^{2-}$ ions are known as **oxyanions.** The Lewis structures of some of the oxides of sulfur that form oxyacids or oxyanions are given in Figure 10.16. One of these oxyanions deserves special mention. This ion, which is known as the thiosulfate ion, is formed by the reaction between sulfur and the sulfite ($SO_3^{2-}$) ion:

$$8\ SO_3^{2-}(aq) + S_8(s) \longrightarrow 8\ S_2O_3^{2-}(aq)$$

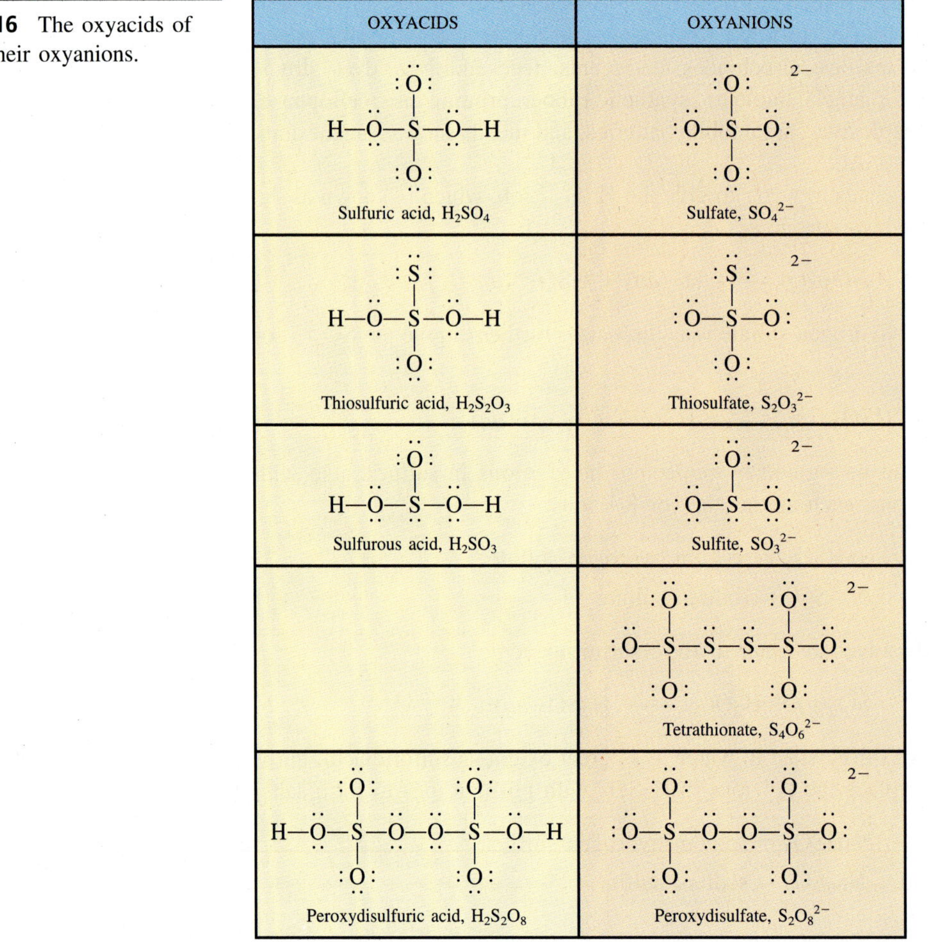


**Figure 10.16** The oxyacids of sulfur and their oxyanions.

## ▶ CHECKPOINT

Use the following Lewis structures to explain why the $S_2O_3^{2-}$ ion is literally a *thio*sulfate:

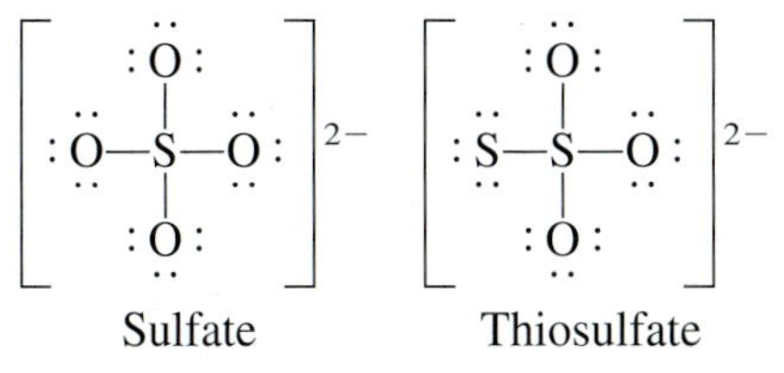


## THE EFFECT OF DIFFERENCES IN THE ABILITIES OF SULFUR AND OXYGEN TO EXPAND THEIR VALENCE SHELLS

The electron configurations of oxygen and sulfur are usually written as follows:

$$\text{O:} \quad [\text{He}]\ 2s^2\ 2p^4$$
$$\text{S:} \quad [\text{Ne}]\ 3s^2\ 3p^4$$

This notation shows the similarity between the configurations of the two elements. However, it hides an important difference that allows sulfur to expand its valence shell to hold more than eight electrons, whereas oxygen cannot.

There are only four orbitals in the valence shell of an oxygen atom: the $2s$, $2p_x$, $2p_y$, and $2p_z$ orbitals. As a result, oxygen can hold no more than eight valence electrons. The valence orbitals of sulfur, however, are in the $n = 3$ shell, which includes $3s$, $3p$, and $3d$ orbitals. Sulfur therefore has empty $3d$ valence orbitals that can be used to expand its valence shell:

$$\text{S:} \quad [\text{Ne}]\ 3s^2\ 3p^4\ 3d^0$$

Oxygen reacts with fluorine to form $OF_2$:

$$O_2(g) + 2\ F_2(g) \longrightarrow 2\ OF_2(g)$$

The reaction stops at this point because oxygen can hold only eight electrons in its valence shell, as shown in Figure 10.17. Sulfur reacts with fluorine to form $SF_4$ and $SF_6$, shown in Figure 10.18, because sulfur can expand its valence shell to hold 10 or even 12 electrons:

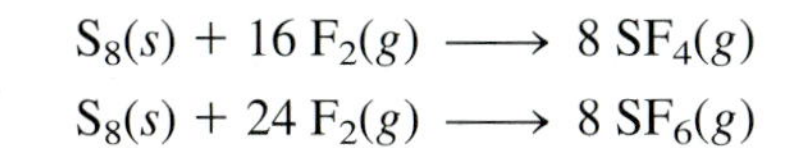

$$S_8(s) + 16\ F_2(g) \longrightarrow 8\ SF_4(g)$$
$$S_8(s) + 24\ F_2(g) \longrightarrow 8\ SF_6(g)$$

:F̤—Ö—F̤:

**Figure 10.17** The Lewis structure for $OF_2$.

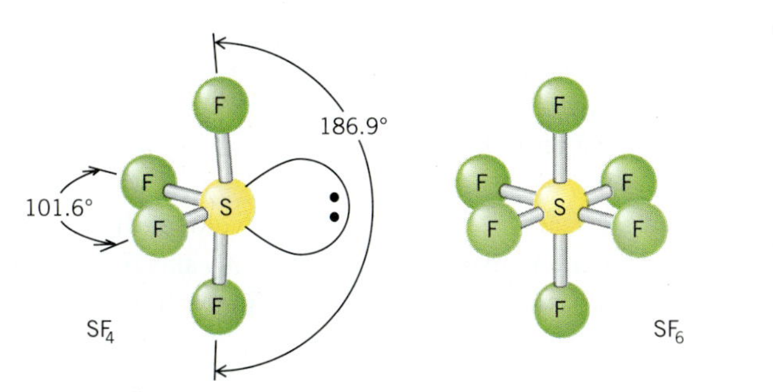


**Figure 10.18** There are 10 valence electrons on the sulfur atom in $SF_4$, so the structure of this molecule is based on a distorted trigonal bipyramid. There are 12 valence electrons on the central atom in $SF_6$, which are distributed toward the corners of an octahedron.

## 10.5 THE CHEMISTRY OF NITROGEN

The chemistry of nitrogen is dominated by the ease with which nitrogen atoms form double and triple bonds. A neutral nitrogen atom contains five valence electrons: $2s^2$ $2p^3$. A nitrogen atom can therefore achieve an octet of valence electrons by sharing three pairs of electrons with another nitrogen atom.

$$:N{\equiv}N:$$

Because the covalent radius of a nitrogen atom is relatively small (only 0.070 nm), nitrogen atoms come close enough together to form very strong $\pi$ bonds. The bond-dissociation enthalpy for the N≡N triple bond is 946 kJ/mol, almost twice as large as that for an O=O double bond.

The strength of the N≡N triple bond makes the $N_2$ molecule very unreactive. $N_2$ is so inert that lithium is one of the few elements with which it reacts at room temperature:

$$6\,Li(s) + N_2(g) \longrightarrow 2\,Li_3N(s)$$

In spite of the fact that the $N_2$ molecule is unreactive, compounds containing nitrogen exist for virtually every element in the periodic table except those in Group VIIIA (He, Ne, Ar, and so on). This can be explained in two ways. First, $N_2$ becomes significantly more reactive as the temperature increases. At high temperatures, nitrogen reacts with hydrogen to form ammonia and with oxygen to form nitrogen oxide:

$$N_2(g) + 3\,H_2(g) \longrightarrow 2\,NH_3(g)$$

$$N_2(g) + O_2(g) \longrightarrow 2\,NO(g)$$

Second, a number of catalysts found in nature overcome the inertness of $N_2$ at low temperature.

### THE SYNTHESIS OF AMMONIA

It is difficult to imagine a living system that does not contain nitrogen, which is an essential component of the proteins, nucleic acids, vitamins, and hormones that make life possible. Animals pick up the nitrogen they need from the plants or other animals in their diet. Plants have to pick up their nitrogen from the soil, or absorb it as $N_2$ from the atmosphere. The concentration of nitrogen in the soil is fairly small, so the process by which plants reduce $N_2$ to $NH_3$—or "fix" $N_2$—is extremely important.

Although 200 million tons of $NH_3$ are produced by nitrogen fixation each year, plants, by themselves, cannot reduce $N_2$ to $NH_3$. This reaction is carried out by blue-green algae and bacteria that are associated with certain plants. The best understood example of nitrogen fixation involves the rhizobium bacteria found in the root nodules of legumes such as clover, peas, and beans. These bacteria contain a nitrogenase enzyme, which is capable of the remarkable feat of reducing $N_2$ to $NH_3$ at room temperature.

Ammonia is made on an industrial scale by a process first developed between 1909 and 1913 by Fritz Haber. In the **Haber process,** a mixture of $N_2$ and $H_2$ gas at 200 to 300 atm and 400°C to 600°C is passed over a catalyst of finely divided iron metal.

$$N_2(g) + 3\,H_2(g) \xrightarrow{Fe} 2\,NH_3(g)$$

Almost 20 million tons of $NH_3$ are produced in the United States each year by this process. About 80% of it, worth more than $2 billion, is used to make fertilizers for

The Haber process uses high temperatures (400–600°C) and high pressures (200–300 atm) to produce $NH_3$. The nitrogenase enzyme in biological systems, such as the clover in this shot, can achieve the same effect at room temperature and atmospheric pressure.

plants that can't fix nitrogen from the atmosphere. On the basis of weight, ammonia is the second most important industrial chemical in the United States. (Only sulfuric acid is produced in larger quantities.)

Two-thirds of the ammonia used for fertilizers is converted into solids such as ammonium nitrate, $NH_4NO_3$; ammonium phosphate, $(NH_4)_3PO_4$; ammonium sulfate, $(NH_4)_2SO_4$; and urea, $H_2NCONH_2$. The other third is applied directly to the soil as **anhydrous** (literally, "without water") ammonia. Ammonia is a gas at room temperature. It can be handled as a liquid when dissolved in water to form an aqueous solution. Alternatively, it can be cooled to temperatures below −33°C, in which case the gas condenses to form the anhydrous liquid, $NH_3(l)$.

## THE SYNTHESIS OF NITRIC ACID

Ammonia produced by the Haber process that is not used as fertilizer is burned in oxygen to generate nitrogen oxide:

$$4\,NH_3(g) + 5\,O_2(g) \longrightarrow 4\,NO(g) + 6\,H_2O(g)$$

Nitrogen oxide—or nitric oxide, as it was once known—is a colorless gas that reacts rapidly with oxygen to produce nitrogen dioxide, a dark brown gas:

$$2\,NO(g) + O_2(g) \longrightarrow 2\,NO_2(g)$$

Nitrogen dioxide dissolves in water to give nitric acid and NO, which can be captured and recycled:

$$3\,NO_2(g) + H_2O(l) \longrightarrow 2\,HNO_3(aq) + NO(g)$$

Thus, by a three-step process developed by Friedrich Ostwald in 1908, ammonia can be converted into nitric acid:

$$\begin{aligned} 4\,NH_3(g) + 5\,O_2(g) &\longrightarrow 4\,NO(g) + 6\,H_2O(g) \\ 2\,NO(g) + O_2(g) &\longrightarrow 2\,NO_2(g) \\ 3\,NO_2(g) + H_2O(l) &\longrightarrow 2\,HNO_3(aq) + NO(g) \end{aligned}$$

The NO produced when ammonia burns is a colorless gas that does not dissolve in water. When it is exposed to air, NO instantly reacts with $O_2$ to form $NO_2$, a dark-brown gas. When the flask containing $NO_2$ is shaken, the brown gas disappears as it dissolves in water to produce a mixture of nitric acid and NO gas.

The Haber process for the synthesis of ammonia combined with the **Ostwald process** for the conversion of ammonia into nitric acid revolutionized the explosives industry. Nitrates have been important explosives in the West ever since Friar Roger Bacon mixed sulfur, saltpeter, and powdered carbon to make gunpowder in 1245:

$$16\ KNO_3(s) + S_8(s) + 24\ C(s) \longrightarrow 8\ K_2S(s) + 24\ CO_2(g) + 8\ N_2(g) \qquad \Delta H^\circ = -571.9\ \text{kJ/mol}\ N_2$$

Before the Ostwald process was developed the only source of nitrates for use in explosives was naturally occurring minerals such as saltpeter, which is a mixture of $NaNO_3$ and $KNO_3$. Once a dependable supply of nitric acid became available from the Ostwald process, a number of nitrates could be made for use as explosives. Combining $NH_3$ from the Haber process with $HNO_3$ from the Ostwald process, for example, gives ammonium nitrate, which is both an excellent fertilizer and a cheap, dependable explosive commonly used in blasting powder:

$$2\ NH_4NO_3(s) \longrightarrow 2\ N_2(g) + O_2(g) + 4\ H_2O(g)$$

The destructive power of ammonium nitrate is apparent from a famous accident at Texas City, Texas. In 1947, a freighter loaded with $NH_4NO_3$ blew up in this harbor, killing nearly 600 people and injuring 4000 others.

**TABLE 10.4 Common Oxidation Numbers for Nitrogen**

| Oxidation Number | Examples |
|---|---|
| −3 | $NH_3$, $NH_4^+$, $NH_2^-$, $Mg_3N_2$ |
| −2 | $N_2H_4$ |
| −1 | $NH_2OH$ |
| $-\frac{1}{3}$ | $NaN_3$, $HN_3$ |
| 0 | $N_2$ |
| +1 | $N_2O$ |
| +2 | NO, $N_2O_2$ |
| +3 | $HNO_2$, $NO_2^-$, $N_2O_3$, $NO^+$ |
| +4 | $NO_2$, $N_2O_4$ |
| +5 | $HNO_3$, $NO_3^-$, $N_2O_5$ |

## INTERMEDIATE OXIDATION NUMBERS

Nitric acid ($HNO_3$) and ammonia ($NH_3$) represent the maximum (+5) and minimum (−3) oxidation numbers for nitrogen. Nitrogen also forms compounds with every oxidation number between these extremes (see Table 10.4).

## NEGATIVE OXIDATION NUMBERS OF NITROGEN BESIDES −3

At about the time that Haber developed the process for making ammonia and Ostwald worked out the process for converting ammonia into nitric acid, Friedrich Raschig developed a process that used the hypochlorite ($OCl^-$) ion to oxidize ammonia to produce hydrazine, $N_2H_4$:

$$2\ NH_3(aq) + OCl^-(aq) \longrightarrow N_2H_4(aq) + Cl^-(aq) + H_2O(l)$$

This reaction can be understood by noting that the $OCl^-$ ion is a two-electron oxidizing agent. The loss of a pair of electrons and a pair of $H^+$ ions by neighboring $NH_3$ molecules would produce a pair of highly reactive $NH_2$ molecules, which would combine to form a hydrazine molecule as shown in Figure 10.19.

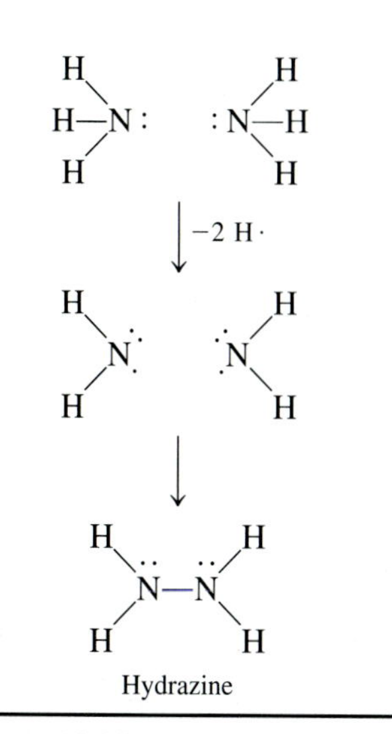


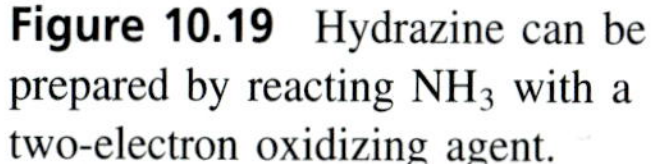
**Figure 10.19** Hydrazine can be prepared by reacting $NH_3$ with a two-electron oxidizing agent.

Hydrazine is a colorless liquid with a faint odor of ammonia that can be collected when this solution is heated until $N_2H_4$ distills out of the reaction flask. Many of the physical properties of hydrazine are similar to those of water:

| | $H_2O$ | $N_2H_4$ |
|---|---|---|
| Density | 1.000 g/cm$^3$ | 1.008 g/cm$^3$ |
| Melting point | 0.00°C | 1.54°C |
| Boiling point | 100°C | 113.8°C |

However, there is a significant difference between the chemical properties of these compounds. Hydrazine burns when ignited in air to give nitrogen gas, water vapor, and large amounts of energy:

$$N_2H_4(l) + O_2(g) \longrightarrow N_2(g) + 2\,H_2O(g) \qquad \Delta H^\circ = -534.3 \text{ kJ/mol } N_2H_4$$

The principal use of hydrazine is as a rocket fuel. It is second only to liquid hydrogen in terms of the number of kilograms of thrust produced per kilogram of fuel burned. Hydrazine has several advantages over liquid $H_2$, however. It can be stored at room temperature, whereas liquid hydrogen must be stored at temperatures below −253°C. Hydrazine is also more dense than liquid $H_2$ and therefore requires less storage space.

Sec. 1.17

Pure hydrazine is seldom used as a rocket fuel, however, because it freezes at the temperatures encountered in the upper atmosphere. Hydrazine is mixed with *N,N*-dimethylhydrazine, $(CH_3)_2NNH_2$, to form a solution that remains a liquid at low temperatures. Mixtures of hydrazine and *N,N*-dimethylhydrazine were used to fuel the Titan II rockets that carried the Project Gemini spacecraft, and the reaction between hydrazine derivatives and $N_2O_4$ is still used to fuel the small rocket engines that enable the space shuttles to maneuver in space.

The products of the combustion of hydrazine are unusual. When carbon compounds burn, the carbon is oxidized to CO or $CO_2$. When sulfur compounds burn, $SO_2$ is produced. When hydrazine is burned, the products of the reaction include $N_2$ because of the unusually strong N≡N triple bond in the $N_2$ molecule:

$$N_2H_4(l) + O_2(g) \longrightarrow N_2(g) + 2\,H_2O(g)$$

Hydrazine reacts with nitrous acid ($HNO_2$) to form hydrogen azide, $HN_3$, in which the nitrogen atom formally has an oxidation state of $-\frac{1}{3}$.

$$N_2H_4(aq) + HNO_2(aq) \longrightarrow HN_3(aq) + 2\,H_2O(l)$$

Pure hydrogen azide is an extremely dangerous substance. Even dilute solutions should be handled with care because of the risk of explosions. Hydrogen azide is best described as a resonance hybrid of the Lewis structures shown in Figure 10.20. The corresponding azide ion, $N_3^-$, is a linear molecule, which is a resonance hybrid of three Lewis structures.

| | |
|---|---|
| $HN_3$ | [H—N̈=N=N̈: ⟷ H—N̈—N≡N:] |
| $N_3^-$ | [:N≡N—N̈: ⟷ :N̈=N=N̈: ⟷ :N̈—N≡N:]$^-$ |

**Figure 10.20** The Lewis structures for hydrogen azide ($HN_3$) and the azide ($N_3^-$) ion.

## POSITIVE OXIDATION NUMBERS FOR NITROGEN: THE NITROGEN HALIDES

Fluorine, oxygen, and chlorine are the only elements more electronegative than nitrogen. As a result, positive oxidation numbers of nitrogen are found in compounds that contain one or more of these elements.

In theory, $N_2$ could react with $F_2$ to form a compound with the formula $NF_3$. In practice, $N_2$ is too inert to undergo this reaction at room temperature. $NF_3$ is made by reacting ammonia with $F_2$ in the presence of a copper metal catalyst:

$$NH_3(g) + 3\ F_2(g) \xrightarrow{Cu} NF_3(g) + 3\ HF(g)$$

The HF produced in this reaction combines with ammonia to form ammonium fluoride. The overall stoichiometry for the reaction is therefore written as follows:

$$4\ NH_3(g) + 3\ F_2(g) \xrightarrow{Cu} NF_3(g) + 3\ NH_4F(s)$$

The Lewis structure of $NF_3$ is analogous to the Lewis structure of $NH_3$, and the two molecules have similar shapes.

Ammonia reacts with chlorine to form $NCl_3$, which seems at first glance to be closely related to $NF_3$. But there is a significant difference between these compounds. $NF_3$ is essentially inert at room temperature, whereas $NCl_3$ is a shock-sensitive, highly explosive liquid that decomposes to form $N_2$ and $Cl_2$:

$$2\ NCl_3(l) \longrightarrow N_2(g) + 3\ Cl_2(g)$$

### EXERCISE 10.7

Use the following enthalpy of formation data to explain why $NF_3$ is essentially inert at room temperature, whereas $NCl_3$ decomposes violently to $N_2$ and $Cl_2$:

| **Compound** | $\Delta H^\circ_f$ |
|---|---|
| $NF_3(g)$ | −124.7 kJ/mol |
| $NCl_3(l)$ | 230 kJ/mol |

**SOLUTION** The enthalpy of formation of $NF_3$ is negative, which means that the compound is more stable than its elements at 25°C and 1 atm. The enthalpy of formation of $NCl_3$ is positive, which means that $NCl_3$ is less stable than its elements. Considering both the sign and magnitude of $\Delta H^\circ_f$ for $NCl_3$, it is not surprising that this compound decomposes to its elements.

Ammonia reacts with iodine to form a solid that is a complex between $NI_3$ and $NH_3$. This material is the subject of a popular, but dangerous, demonstration in which freshly prepared samples of $NI_3$ in ammonia are poured onto filter paper, which is allowed to dry on a ring stand. After the ammonia evaporates, the $NH_3 \cdot NI_3$ crystals are touched with a feather attached to a meter stick, resulting in detonation of this shock-sensitive solid, which decomposes to form a mixture of $N_2$ and $I_2$.

$$2\ NI_3(s) \longrightarrow N_2(g) + 3\ I_2(g)$$

**TABLE 10.5 Enthalpy of Formation Data for the Oxides of Nitrogen**

| Compound | $\Delta H^\circ_f$ (kJ/mol) |
|---|---|
| $N_2O(g)$ | 82.05 |
| $NO(g)$ | 90.25 |
| $NO_2(g)$ | 33.18 |
| $N_2O_3(g)$ | 83.72 |
| $N_2O_4(g)$ | 9.16 |
| $N_2O_5(g)$ | 11.35 |

## POSITIVE OXIDATION NUMBERS FOR NITROGEN: THE NITROGEN OXIDES

Lewis structures for seven oxides of nitrogen with oxidation numbers ranging from +1 to +5 are given in Figure 10.21. These compounds all have two things in common: they contain N=O double bonds and they are less stable than their elements in the gas phase, as shown by the enthalpy of formation data in Table 10.5.

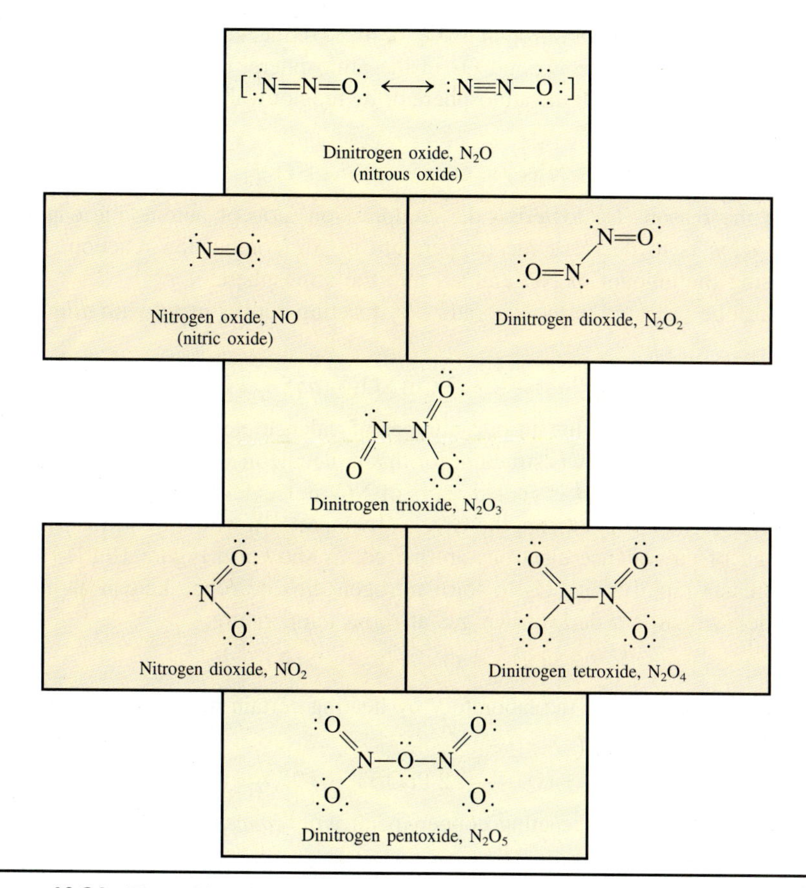


**Figure 10.21** The oxides of nitrogen.

For many years, the endings *-ous* and *-ic* were used to distinguish between the lowest and highest of a pair of oxidation numbers. $N_2O$ is nitrous oxide because the oxidation number of the nitrogen is $+1$. NO is nitric oxide because the oxidation number of the nitrogen is $+2$.

Dinitrogen oxide, $N_2O$, or nitrous oxide, can be prepared by carefully decomposing ammonium nitrate:

$$NH_4NO_3(s) \xrightarrow{170 \text{ to } 200°C} N_2O(g) + 2\,H_2O(g)$$

Nitrous oxide is a sweet-smelling, colorless gas best known to nonchemists as ''laughing gas.'' As early as 1800, Humphry Davy noted that $N_2O$, inhaled in relatively small amounts, produced a state of apparent intoxication often accompanied by either convulsive laughter or crying. When taken in larger doses, nitrous oxide provides fast and efficient relief from pain. $N_2O$ was therefore used as the first anesthetic. Because large doses are needed to produce anesthesia, and continued exposure to the gas can be fatal, $N_2O$ is used today only for relatively short operations.

Nitrous oxide has several other interesting properties. First, it is highly soluble in cream; for that reason, it is used as the propellant in whipped cream dispensers. Second, although it does not burn by itself, it is better than air at supporting the combustion of other objects. This can be explained by noting that $N_2O$ can decompose to form an atmosphere that is one-third $O_2$ by volume, whereas normal air is only 21% oxygen by volume:

$$2\,N_2O(g) \longrightarrow 2\,N_2(g) + O_2(g)$$

The Royal Institute was founded in 1800 to provide a place where lectures on the latest developments in science could be given to ''improve the lot of the working man.'' This etching by James Gillray shows Thomas Garrett and Humphrey Davy administering nitrous oxide or ''laughing gas,'' with unfortunate consequences.

The photograph on the top shows a sealed tube filled with dark-brown $NO_2$ gas. The photograph on the bottom shows how the brown color of this gas disappears as the $NO_2$ dimerizes to form $N_2O_4$ when the tube is cooled in liquid nitrogen.

Enormous quantities of nitrogen oxide, or nitric oxide, are generated each year by the reaction between the $N_2$ and $O_2$ in the atmosphere, catalyzed by a stroke of lightning passing through the atmosphere or by the hot walls of an internal combustion engine:

$$N_2(g) + O_2(g) \longrightarrow 2\,NO(g)$$

One of the reasons for lowering the compression ratio of automobile engines in recent years is to decrease the temperature of the combustion reaction, thereby decreasing the amount of NO emitted into the atmosphere.

NO can be prepared in the laboratory by reacting copper metal with *dilute* nitric acid:

$$3\,Cu(s) + 8\,HNO_3(aq) \longrightarrow 3\,Cu(NO_3)_2(aq) + 2\,NO(g) + 4\,H_2O(l)$$

The NO molecule contains an odd number of valence electrons. As a result, it is impossible to write a Lewis structure for this molecule in which all of the electrons are paired. When NO gas is cooled, pairs of NO molecules combine in a reversible reaction to form a **dimer** (from the Greek, ''two parts''), with the formula $N_2O_2$, in which all of the valence electrons are paired, as shown in Figure 10.21.

NO reacts rapidly with $O_2$ to form nitrogen dioxide (once known as nitrogen peroxide), which is a dark brown gas at room temperature:

$$2\,NO(g) + O_2(g) \longrightarrow 2\,NO_2(g)$$

$NO_2$ can be prepared in the laboratory by heating certain metal nitrates until they decompose:

$$2\,Pb(NO_3)_2(s) \longrightarrow 2\,PbO(s) + 4\,NO_2(g) + O_2(g)$$

It can also be made by reacting copper metal with *concentrated* nitric acid:

$$Cu(s) + 4\,HNO_3(aq) \longrightarrow Cu(NO_3)_2(aq) + 2\,NO_2(g) + 2\,H_2O(l)$$

$NO_2$ also has an odd number of electrons and therefore contains at least one unpaired electron in its Lewis structures. $NO_2$ dimerizes at low temperatures to form $N_2O_4$ molecules, in which all the electrons are paired, as shown in Figure 10.21.

Mixtures of NO and $NO_2$ combine when cooled to form dinitrogen trioxide, $N_2O_3$, which is a blue liquid. The formation of a blue liquid when either NO or $NO_2$ is cooled therefore implies the presence of at least a small portion of the other oxide because $N_2O_2$ and $N_2O_4$ are both colorless.

When a mixture of NO and $NO_2$ is cooled, a deep-blue-colored liquid with the formula $N_2O_3$ is formed.

By carefully removing water from concentrated nitric acid at low temperatures with a dehydrating agent we can form dinitrogen pentoxide:

$$4\ HNO_3(aq) + P_4O_{10}(s) \longrightarrow 2\ N_2O_5(s) + 4\ HPO_3(s)$$

$N_2O_5$ is a colorless solid that decomposes in light or on warming to room temperature. As might be expected, $N_2O_5$ dissolves in water to form nitric acid:

$$N_2O_5(s) + H_2O(l) \longrightarrow 2\ HNO_3(aq)$$

## 10.6 THE CHEMISTRY OF PHOSPHORUS

Phosphorus is the first element whose discovery can be traced to a single individual. In 1669, while searching for a way to convert silver into gold, Hennig Brand obtained a white, waxy solid that glowed in the dark and burst spontaneously into flame when exposed to air. Brand made this substance by evaporating the water from urine and allowing the black residue to putrefy for several months. He then mixed this residue with sand, heated this mixture in the presence of a minimum of air, and collected under water the volatile products that distilled out of the reaction flask.

White phosphorus bursts into flame on contact with air. It does not react with water, however, so it is often stored under water.

Phosphorus forms a number of compounds that are direct analogs of nitrogen-containing compounds. However, the fact that elemental nitrogen is virtually inert at room temperature, whereas elemental phosphorus can burst spontaneously into flame when exposed to air, shows that there are differences between these elements as well. Phosphorus often forms compounds with the same oxidation numbers as the analogous nitrogen compounds, but with different formulas, as shown in Table 10.6.

The same factors that explain the differences between sulfur and oxygen can be used to explain the differences between phosphorus and nitrogen:

1. N≡N triple bonds are much stronger than P≡P triple bonds.
2. P—P single bonds are stronger than N—N single bonds.
3. Phosphorus ($EN = 2.19$) is much less electronegative than nitrogen ($EN = 3.04$).
4. Phosphorus can expand its valence shell to hold more than eight electrons, but nitrogen cannot.

**TABLE 10.6 Nitrogen and Phosphorus Compounds with the Same Oxidation Numbers but Different Formulas**

| Oxidation Number | Nitrogen Compound | Phosphorus Compound |
|---|---|---|
| 0 | $N_2$ | $P_4$ |
| +3 | $HNO_2$ (nitrous acid) | $H_3PO_3$ (phosphorous acid) |
| +3 | $N_2O_3$ | $P_4O_6$ |
| +5 | $HNO_3$ (nitric acid) | $H_3PO_4$ (phosphoric acid) |
| +5 | $NaNO_3$ (sodium nitrate) | $Na_3PO_4$ (sodium phosphate) |
| +5 | $N_2O_5$ | $P_4O_{10}$ |

### THE EFFECT OF DIFFERENCES IN THE SINGLE AND TRIPLE BOND STRENGTHS

The ratio of the radii of phosphorus and nitrogen atoms is the same as the ratio of the radii of sulfur and oxygen atoms, within experimental error:

$$\frac{\text{Covalent radius of phosphorus}}{\text{Covalent radius of nitrogen}} = \frac{0.110\ \text{nm}}{0.070\ \text{nm}} = 1.57$$

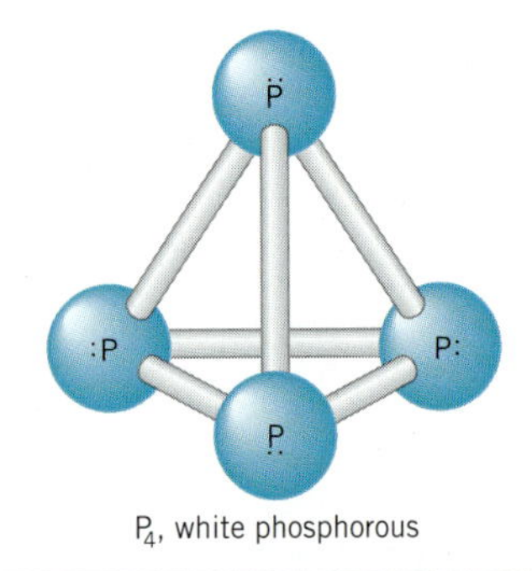


**Figure 10.22** Pure elemental phosphorus is a white, waxy solid that consists of tetrahedral $P_4$ molecules in which the P—P—P bond angle is only 60°.

As a result, P≡P triple bonds are much weaker than N≡N triple bonds, for the same reason that S=S double bonds are weaker than O=O double bonds: phosphorus atoms are too big to come close enough together to form strong $\pi$ bonds.

Each atom in an $N_2$ molecule completes its octet of valence electrons by sharing three pairs of electrons with a single neighboring atom. Because phosphorus does not form strong multiple bonds with itself, elemental phosphorus consists of tetrahedral $P_4$ molecules in which each atom forms single bonds with three neighboring atoms, as shown in Figure 10.22.

Phosphorus is a white solid with a waxy appearance, which melts at 44.1°C and boils at 287°C. It is made by reducing calcium phosphate with carbon in the presence of silica (sand) at very high temperatures:

$$2\,Ca_3(PO_4)_2(s) + 6\,SiO_2(s) + 10\,C(s) \longrightarrow 6\,CaSiO_3(s) + P_4(s) + 10\,CO(g)$$

White phosphorus is stored under water because the element spontaneously bursts into flame in the presence of oxygen at temperatures only slightly above room temperature.

(left) White phosphorus does not react with water, but it can burn under water when $O_2$ gas is bubbled into the solution at temperatures above 70°C. (right) White phosphorus dissolves in $CS_2$ to form solutions that are relatively stable, until the solvent evaporates. When this happens, the phosphorus bursts into flame.

Although phosphorus is insoluble in water, it is very soluble in carbon disulfide. Solutions of $P_4$ in $CS_2$ are reasonably stable. As soon as the $CS_2$ evaporates, however, the phosphorus bursts into flame.

The P—P—P bond angle in a tetrahedral $P_4$ molecule is only 60°. This very small angle produces a considerable amount of strain in the $P_4$ molecule, which can be relieved by breaking one of the P—P bonds. Phosphorus therefore forms other allotropes by opening up the $P_4$ tetrahedron. When white phosphorus is heated to 300°C, one bond inside each $P_4$ tetrahedron is broken, and the $P_4$ molecules link together to form a *polymer* (from the Greek *polys,* "many," and *meros,* "parts") with the structure shown in Figure 10.23. This allotrope of phosphorus is dark red, and its presence in small traces often gives white phosphorus a light yellow color. Red phosphorus is more dense (2.16 g/cm$^3$) than white phosphorus (1.82 g/cm$^3$) and is much less reactive at normal temperatures.

Red phosphorus is much less reactive than white phosphorus.

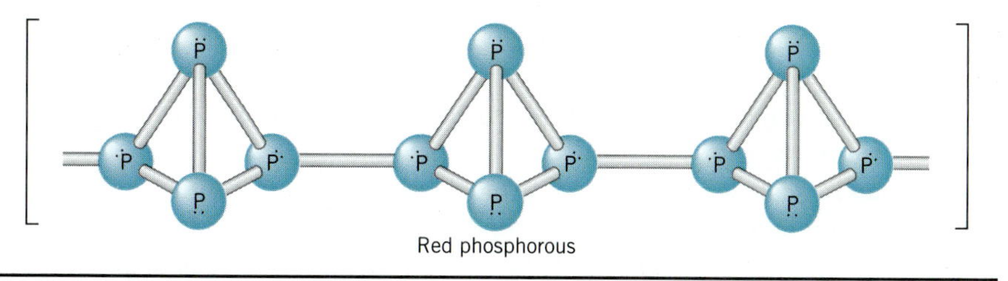


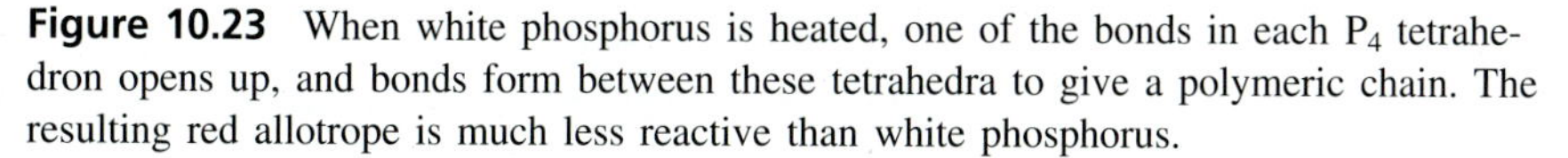

**Figure 10.23** When white phosphorus is heated, one of the bonds in each $P_4$ tetrahedron opens up, and bonds form between these tetrahedra to give a polymeric chain. The resulting red allotrope is much less reactive than white phosphorus.

## THE EFFECT OF DIFFERENCES IN THE STRENGTHS OF P=*X* AND N=*X* DOUBLE BONDS

The size of a phosphorus atom also interferes with its ability to form double bonds to other elements, such as oxygen, nitrogen, and sulfur. As a result, phosphorus tends to form compounds that contain two P—O single bonds where nitrogen would form an N=O double bond. Nitrogen forms the nitrate, $NO_3^-$, ion, for example, in which it has an oxidation number of +5. When phosphorus forms an ion with the same oxidation number, it is the phosphate, $PO_4^{3-}$ ion, as shown in Figure 10.24. Similarly, nitrogen forms nitric acid, $HNO_3$, which contains an N=O double bond, whereas phosphorus forms phosphoric acid, $H_3PO_4$, which contains P—O single bonds, as shown in Figure 10.25.

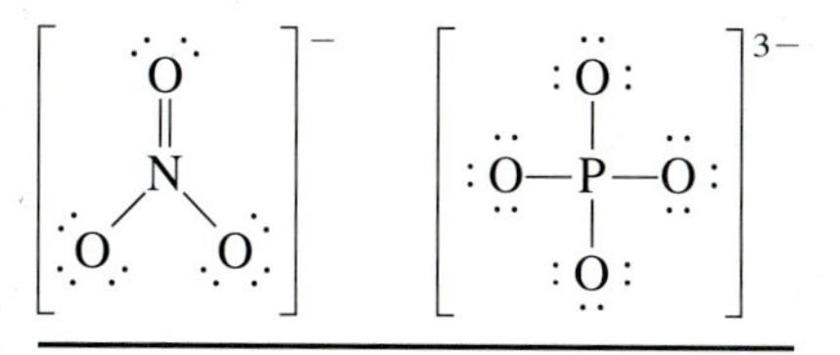

**Figure 10.24** The Lewis structures for the $NO_3^-$ and $PO_4^{3-}$ ions.

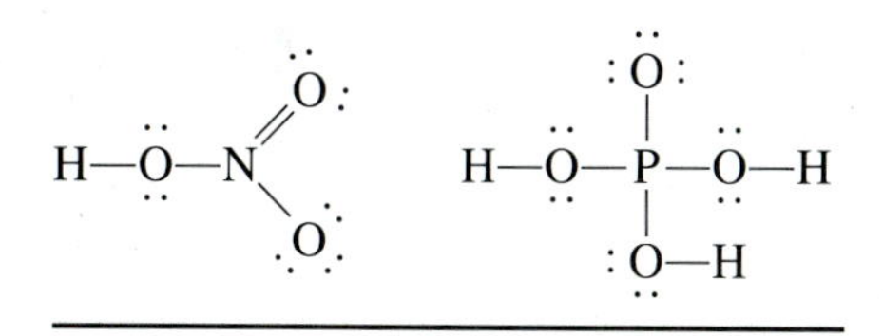

**Figure 10.25** $HNO_3$ and phosphoric acid ($H_3PO_4$).

## THE EFFECT OF DIFFERENCES IN THE ELECTRONEGATIVITIES OF PHOSPHORUS AND NITROGEN

The difference between the electronegativities of phosphorus and nitrogen ($\Delta EN$ = 0.85) is the same as the difference between the electronegativities of sulfur and oxygen ($\Delta EN$ = 0.86), within experimental error. Because it is less electronegative, phosphorus is more likely than nitrogen to exhibit positive oxidation numbers. The most important oxidation numbers for phosphorus are −3, +3, and +5 (see Table 10.7).

Because it is more electronegative than most metals, phosphorus reacts with metals at elevated temperatures to form phosphides, in which it has an oxidation number of −3:

$$6\ Ca(s) + P_4(s) \longrightarrow 2\ Ca_3P_2(s)$$

These metal phosphides react with water to produce a poisonous, highly reactive, colorless gas known as phosphine ($PH_3$), which has the foulest odor the authors have encountered:

$$Ca_3P_2(s) + 6\ H_2O(l) \longrightarrow 2\ PH_3(g) + 3\ Ca^{2+}(aq) + 6\ OH^-(aq)$$

**TABLE 10.7 Common Oxidation Numbers of Phosphorus**

| Oxidation Number | Examples |
|---|---|
| −3 | $Ca_3P_2$, $PH_3$ |
| +3 | $PF_3$, $P_4O_6$, $H_3PO_3$ |
| +5 | $PF_5$, $P_4O_{10}$, $H_3PO_4$ |

Samples of $PH_3$, the phosphorus analog of ammonia, are often contaminated by traces of $P_2H_4$, the phosphorus analog of hydrazine. As if the toxicity and odor of $PH_3$ were not enough, mixtures of $PH_3$ and $P_2H_4$ burst spontaneously into flame in the presence of oxygen.

Compounds (such as $Ca_3P_2$ and $PH_3$) in which phosphorus has a negative oxidation number are far outnumbered by compounds in which the oxidation number of phosphorus is positive. Phosphorus burns in $O_2$ to produce $P_4O_{10}$ in a reaction that gives off extraordinary amounts of energy in the form of heat and light:

$$P_4(s) + 5\,O_2(g) \longrightarrow P_4O_{10}(s) \qquad \Delta H° = -2985 \text{ kJ/mol } P_4$$

When phosphorus burns in the presence of a limited amount of $O_2$, $P_4O_6$ is produced:

$$P_4(s) + 3\,O_2(g) \longrightarrow P_4O_6(s) \qquad \Delta H° = -1640 \text{ kJ/mol } P_4$$

$P_4O_6$ consists of a tetrahedron in which an oxygen atom has been inserted into each P—P bond in the $P_4$ molecule (see Figure 10.26). $P_4O_{10}$ has an analogous structure, with an additional oxygen atom bound to each of the four phosphorus atoms.

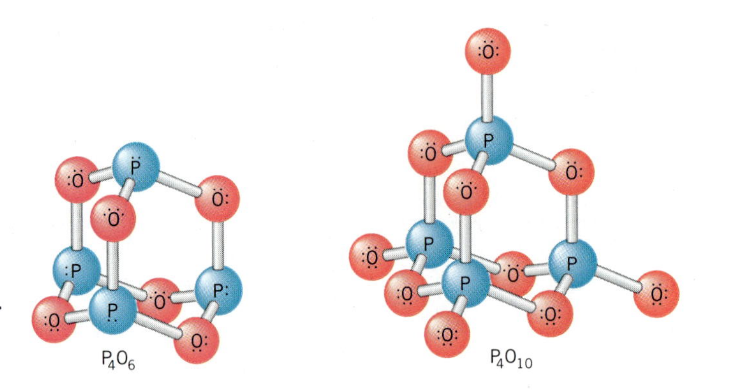


**Figure 10.26** The structures of $P_4O_6$ and $P_4O_{10}$.

$P_4O_6$ and $P_4O_{10}$ react with water to form phosphorous acid, $H_3PO_3$, and phosphoric acid, $H_3PO_4$, respectively:

$$P_4O_6(s) + 6\,H_2O(l) \longrightarrow 4\,H_3PO_3(aq)$$
$$P_4O_{10}(s) + 6\,H_2O(l) \longrightarrow 4\,H_3PO_4(aq)$$

$P_4O_{10}$ has such a high affinity for water that it is commonly used as a dehydrating agent. Phosphorus acid, $H_3PO_3$, and phosphoric acid, $H_3PO_4$, are examples of a large class of oxyacids of phosphorus. Lewis structures for some of these oxyacids and their related oxyanions are given in Figure 10.27.

## THE EFFECT OF DIFFERENCES IN THE ABILITIES OF PHOSPHORUS AND NITROGEN TO EXPAND THEIR VALENCE SHELLS

The reaction between ammonia and fluorine stops at $NF_3$ because nitrogen uses the $2s$, $2p_x$, $2p_y$, and $2p_z$ orbitals to hold valence electrons. Nitrogen atoms can therefore hold a maximum of eight valence electrons. Phosphorus, however, has empty $3d$ atomic orbitals that can be used to expand the valence shell to hold 10 or more electrons. Thus, phosphorus can react with fluorine to form both $PF_3$ and $PF_5$. Phosphorus can even form the $PF_6^-$ ion, in which there are 12 valence electrons on the central atom, as shown in Figure 10.28.

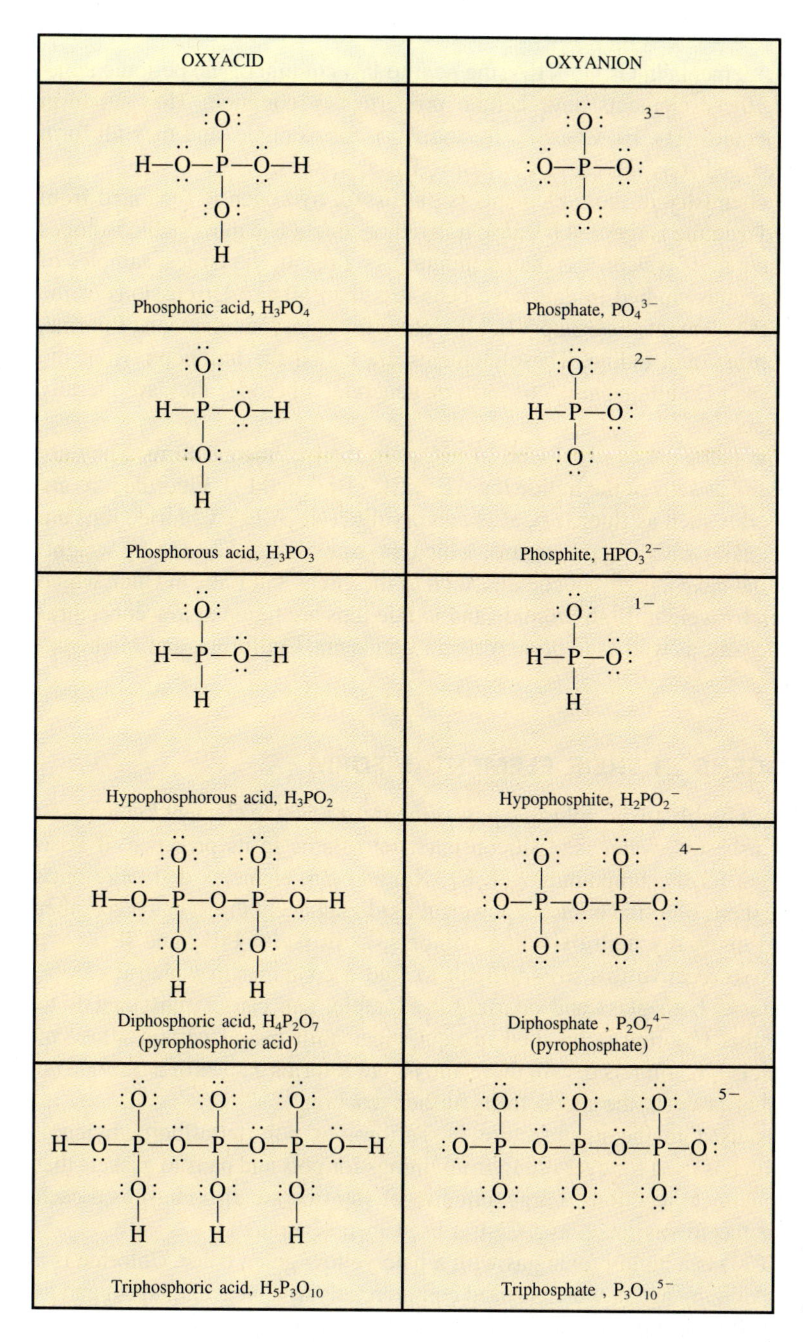


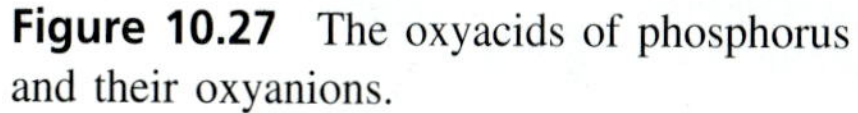

**Figure 10.27** The oxyacids of phosphorus and their oxyanions.

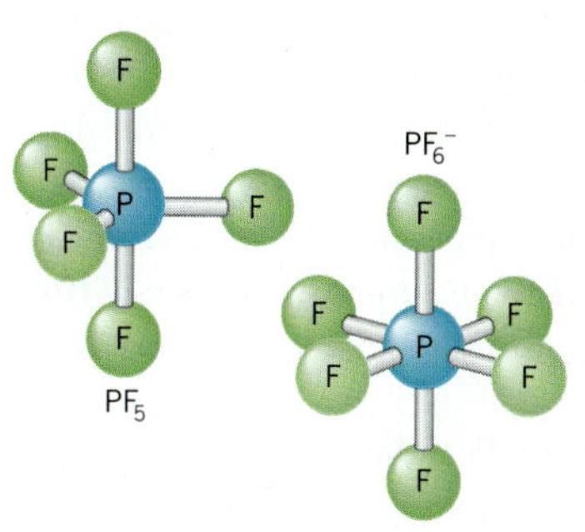


**Figure 10.28** Because it has empty 3*d* orbitals, phosphorus can expand its valence shell to form $PF_5$ molecules or even $PF_6^-$ ions.

## 10.7 THE CHEMISTRY OF THE HALOGENS

There are six elements in Group VIIA, the next to last column of the periodic table. As expected, these elements have certain properties in common. They all form diatomic molecules ($H_2$, $F_2$, $Cl_2$, $Br_2$, $I_2$, and $At_2$), for example, and they all form negatively charged ions ($H^-$, $F^-$, $Cl^-$, $Br^-$, $I^-$, and $At^-$).

When the chemistry of these elements is discussed, hydrogen is separated from the others and astatine is ignored because it is radioactive. (The most stable isotopes of astatine have half-lives of less than a minute. As a result, the largest samples of astatine compounds studied to date have been less than 50 ng.) Discussions of the chemistry of the elements in Group VIIA therefore focus on four elements: fluorine, chlorine, bromine, and iodine. These elements are called the **halogens** (from the Greek *hals*, ''salt,'' and *gennan*, ''to form or generate'') because they are literally the salt formers.

At room temperature, $Cl_2$ is a pale-yellow gas, $Br_2$ is a dark-red liquid, and $I_2$ is a solid that is so intensely colored it looks like a metal.

None of the halogens can be found in nature in their elemental form. They are invariably found as salts of the **halide** ions ($F^-$, $Cl^-$, $Br^-$, and $I^-$). Fluoride ions are found in minerals such as fluorite ($CaF_2$) and cryolite ($Na_3AlF_6$). Chloride ions are found in rock salt (NaCl); in the oceans, which are roughly 2% $Cl^-$ ion by weight; and in lakes that have a high salt content, such as the Great Salt Lake in Utah, which is 9% $Cl^-$ ion by weight. Both bromide and iodide ions are found at low concentrations in the oceans, as well as in brine wells in Louisiana, California, and Michigan.

### THE HALOGENS IN THEIR ELEMENTAL FORM

Fluorine ($F_2$), a highly toxic, colorless gas, is the most reactive element known—so reactive that asbestos, water, and silicon burst into flame in its presence. It is so reactive it even forms compounds with Kr, Xe, and Rn, elements that were once thought to be inert. Fluorine is such a powerful oxidizing agent that it can coax other elements into unusually high oxidation numbers, as in $AgF_2$, $PtF_6$, and $IF_7$.

Fluorine is so reactive that it is difficult to find a container in which it can be stored. $F_2$ attacks both glass and quartz, for example, and causes most metals to burst into flame. Fluorine is handled in equipment built out of certain alloys of copper and nickel. It still reacts with these alloys, but it forms a layer of a fluoride on the surface that protects the metal from further reaction.

Fluorine is used in the manufacture of Teflon—or poly(tetrafluoroethylene), $(C_2F_4)_n$—which is used for everything from linings for pots and pans to gaskets that are inert to chemical reactions. Large amounts of fluorine are also consumed each year to make the freons (such as $CCl_2F_2$) used in refrigerators.

Iodine does not melt when heated; it *sublimes*—it goes directly from the solid to the gas phase.

Chlorine ($Cl_2$) is a highly toxic gas with a pale yellow-green color. Chlorine is a very strong oxidizing agent, which is used commercially as a bleaching agent and as a disinfectant. It is strong enough to oxidize the dyes that give wood pulp its yellow or brown color, for example, thereby bleaching out this color, and strong enough to destroy bacteria and thereby act as a germicide. Large quantities of chlorine are used each year to make solvents such as carbon tetrachloride ($CCl_4$), chloroform ($CHCl_3$), dichloroethylene ($C_2H_2Cl_2$), and trichloroethylene ($C_2HCl_3$).

Bromine ($Br_2$) is a reddish-orange liquid with an unpleasant, choking odor. The name of the element, in fact, comes from the Greek stem *bromos*, ''stench.'' Bromine is used to prepare flame retardants, fire-extinguishing agents, sedatives, antiknock agents for gasoline, and insecticides.

Iodine is an intensely colored solid with an almost metallic luster. This solid is relatively volatile, and it sublimes when heated to form a violet gas. Iodine has been

**TABLE 10.8 Some Properties of $F_2$, $Cl_2$, $Br_2$, and $I_2$**

| | Melting Point (°C) | Boiling Point (°C) | Color | Natural Abundance (ppm) | First Ionization Energy (kJ/mol) | Electron Affinity (kJ/mol) | Ionic Radius (nm) | Density (g/cm$^3$) |
|---|---|---|---|---|---|---|---|---|
| $F_2$ | −218.6 | −188.1 | colorless | 544 | 1680.6 | 322.6 | 0.133 | 1.513 |
| $Cl_2$ | −101.0 | −34.0 | pale green | 126 | 1255.7 | 348.5 | 0.184 | 1.655 |
| $Br_2$ | −7.3 | 59.5 | dark red-brown | 2.5 | 1142.7 | 324.7 | 0.196 | 3.187 |
| $I_2$ | 113.6 | 185.2 | very dark violet, almost black | 0.46 | 1008.7 | 295.5 | 0.220 | 3.960 |

used for many years as a disinfectant in "tincture of iodine." Iodine compounds are used as catalysts, drugs, and dyes. Silver iodide (AgI) plays an important role in the photographic process and in attempts to make rain by seeding clouds. Iodide is also added to salt to protect against goiter, an iodine deficiency disease characterized by a swelling of the thyroid gland.

Some of the chemical and physical properties of the halogens are summarized in Table 10.8. There is a regular increase in many of the properties of the halogens as we proceed down the column from fluorine to iodine, including the melting point, boiling point, intensity of the color of the halogen, the radius of the corresponding halide ion, and the density of the element. On the other hand, there is a regular decrease in the first ionization energy as we go down this column. As a result, there is a regular decrease in the oxidizing strength of the halogens from fluorine to iodine:

$$F_2 > Cl_2 > Br_2 > I_2$$

Oxidizing strength

This trend is mirrored by an increase in the reducing strength of the corresponding halides:

$$I^- > Br^- > Cl^- > F^-$$

Reducing strength

**EXERCISE 10.8**

Use the fact that $Cl_2$ is a stronger oxidizing agent than $Br_2$ to devise a way to prepare elemental bromine from an aqueous solution of the $Br^-$ ion.

**SOLUTION** In Chapter 9 we found that metals can be prepared by reacting one of their salts with a metal that is a stronger reducing agent. Titanium metal, for example, is prepared by reacting $TiCl_4$ with magnesium metal:

$$TiCl_4(l) + 2\,Mg(s) \longrightarrow Ti(s) + 2\,MgCl_2(s)$$

Extending this argument to oxidizing agents suggests that we can produce $Br_2$ by reacting a solution that contains the $Br^-$ ion with something that is an even stronger oxidizing agent than $Br_2$, such as $Cl_2$ dissolved in water.

$$2\,Br^-(aq) + Cl_2(aq) \longrightarrow Br_2(aq) + 2\,Cl^-(aq)$$

## METHODS OF PREPARING THE HALOGENS FROM THEIR HALIDES

The halogens can be made by reacting a solution of the halide ion with any substance that is a stronger oxidizing agent. Iodine, for example, can be made by reacting the iodide ion with either bromine or chlorine:

$$2\,I^-(aq) + Br_2(aq) \longrightarrow I_2(aq) + 2\,Br^-(aq)$$

Bromine was first prepared by A. J. Balard in 1826 by reacting bromide ions with a solution of $Cl_2$ dissolved in water:

$$2\,Br^-(aq) + Cl_2(aq) \longrightarrow Br_2(aq) + 2\,Cl^-(aq)$$

To prepare $Cl_2$, we need a particularly strong oxidizing agent, such as manganese dioxide ($MnO_2$):

$$2\,Cl^-(aq) + MnO_2(aq) + 4\,H^+ \longrightarrow Cl_2(aq) + Mn^{2+}(aq) + 2\,H_2O(l)$$

The synthesis of fluorine escaped the efforts of chemists for almost 100 years. Part of the problem was finding an oxidizing agent strong enough to oxidize the $F^-$ ion to $F_2$. The task of preparing fluorine was made even more difficult by the extraordinary toxicity of both $F_2$ and the hydrogen fluoride (HF) used to make it.

In Chapter 9 we noted that the best way to produce a strong reducing agent is to pass an electric current through a salt of the metal. Sodium, for example, can be prepared by the electrolysis of molten sodium chloride:

$$2\,NaCl(l) \xrightarrow{\text{electrolysis}} 2\,Na(s) + Cl_2(g)$$

In theory, the same process can be used to generate strong oxidizing agents, such as $F_2$.

Attempts to prepare fluorine by electrolysis, however, were initially unsuccessful. Humphry Davy, who prepared potassium, sodium, barium, strontium, calcium, and magnesium by electrolysis, repeatedly tried to prepare $F_2$ by the electrolysis of fluorite ($CaF_2$) and succeeded only in ruining his health. Joseph Louis Gay-Lussac and Louis Jacques Thenard, who prepared elemental boron for the first time, also tried to prepare fluorine and suffered from very painful exposures to hydrogen fluoride. George and Thomas Knox were badly poisoned during their attempts to make fluorine, and both Paulin Louyet and Jerome Nickles died from fluorine poisoning.

Finally, in 1886 Henri Moissan successfully isolated $F_2$ gas from the electrolysis of a mixed salt of KF and HF:

$$2\,KHF_2(s) \xrightarrow{\text{electrolysis}} H_2(g) + F_2(g) + 2\,KF(s)$$

Electrolysis of $KHF_2$ is still used to prepare fluorine today, as shown in Figure 10.29.

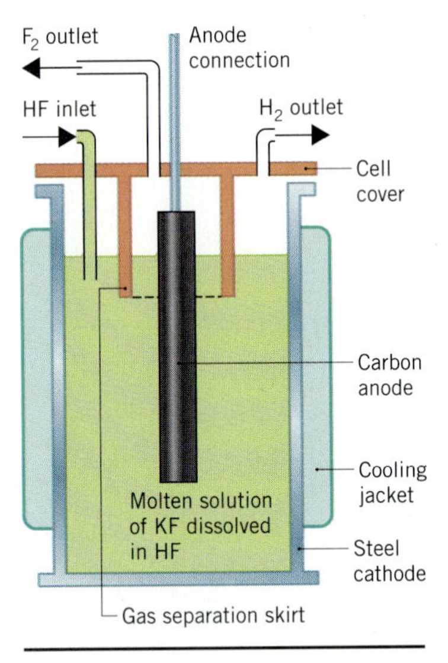


**Figure 10.29** A schematic diagram for the electrolysis of $KHF_2$ to produce $F_2$ gas at the anode and $H_2$ gas at the cathode.

**TABLE 10.9 Common Oxidation Numbers for the Halogens**

| Oxidation Number | Examples |
|---|---|
| −1 | $CaF_2$, HCl, NaBr, AgI |
| 0 | $F_2$, $Cl_2$, $Br_2$, $I_2$ |
| +1 | HClO, ClF |
| +3 | $HClO_2$, $ClF_3$ |
| +5 | $HClO_3$, $BrF_5$, $BrF_6^-$, $IF_5$ |
| +7 | $HClO_4$, $BrF_6^+$, $IF_7$ |

## COMMON OXIDATION NUMBERS FOR THE HALOGENS

Fluorine is the most electronegative element in the periodic table. As a result, it has an oxidation number of −1 in all its compounds. Because chlorine, bromine, and iodine are less electronegative, it is possible to prepare compounds in which these elements have oxidation numbers of +1, +3, +5, and +7, as shown in Table 10.9.

## GENERAL TRENDS IN HALOGEN CHEMISTRY

There are several patterns in the chemistry of the halogens:

1. Neither double nor triple bonds are needed to explain the chemistry of the halogens.

2. The chemistry of fluorine is simplified by the fact it is the most electronegative element in the periodic table and by the fact that it has no *d* orbitals in its valence shell, so it can't expand its valence shell.
3. Chlorine, bromine, and iodine have valence shell *d* orbitals and can expand their valence shells to hold as many as 14 valence electrons.
4. The chemistry of the halogens is dominated by oxidation–reduction reactions.

## THE HYDROGEN HALIDES (HX)

The *hydrogen halides* are compounds that contain hydrogen attached to one of the halogens (HF, HCl, HBr, and HI). These compounds are all colorless gases, which are soluble in water. Up to 512 mL of HCl gas can dissolve in a mL of water at 0°C and 1 atm, for example. Each of the hydrogen halides ionizes to at least some extent when it dissolves in water:

$$HCl(g) \xrightarrow{H_2O} H^+(aq) + Cl^-(aq)$$

**EXERCISE 10.9**

Explain why students often believe that HCl is an ionic compound. Describe the evidence that it is not.

**SOLUTION** HCl is often believed to be an ionic compound because it forms ions when it dissolves in water:

$$HCl(g) \xrightarrow{H_2O} H^+(aq) + Cl^-(aq)$$

This belief is based on the similarity between this reaction and the reaction that occurs when ionic compounds dissolve in water:

$$NaCl(s) \xrightarrow{H_2O} Na^+(aq) + Cl^-(aq)$$

The best evidence that HCl is not an ionic compound is the fact that it is a gas at room temperature, and ionic compounds are invariably solids at room temperature.

Some chemists try to distinguish between the behavior of HCl and NaCl when they dissolve in water as follows. They argue that HCl *ionizes* when it dissolves in water because ions are created by this reaction. NaCl, on the other hand, *dissociates* in water because NaCl already consists of $Na^+$ and $Cl^-$ ions in the solid.

Several of the hydrogen halides can be prepared directly from the elements. Mixtures of $H_2$ and $Cl_2$, for example, react with explosive violence in the presence of light to form HCl:

$$H_2(g) + Cl_2(g) \longrightarrow 2\ HCl(g)$$

Because chemists are usually more interested in aqueous solutions of these compounds than the pure gases, these compounds are usually synthesized in water. Aqueous solutions of the hydrogen halides are often called *mineral acids* because they are literally acids prepared from minerals. Hydrochloric acid is prepared by reacting table salt with sulfuric acid, for example, and hydrofluoric acid is prepared from fluorite and sulfuric acid:

$$2\ NaCl(s) + H_2SO_4(aq) \longrightarrow 2\ HCl(aq) + Na_2SO_4(aq)$$
$$CaF_2(s) + H_2SO_4(aq) \longrightarrow 2\ HF(aq) + CaSO_4(aq)$$

These acids are purified by taking advantage of the ease with which HF and HCl gas boil out of these solutions. The gas given off when one of these solutions is heated is collected and then redissolved in water to give relatively pure samples of the mineral acid.

## THE INTERHALOGEN COMPOUNDS

Interhalogen compounds are formed by reactions between different halogens. All possible interhalogen compounds of the type *XY* are known. Bromine reacts with chlorine, for example, to give BrCl, which is a gas at room temperature:

$$Br_2(l) + Cl_2(g) \longrightarrow 2\ BrCl(g)$$

Interhalogen compounds with the general formulas $XY_3$, $XY_5$, and even $XY_7$ are formed when pairs of halogens react. Chlorine reacts with fluorine, for example, to form chlorine trifluoride:

$$Cl_2(g) + 3\ F_2(g) \longrightarrow 2\ ClF_3(g)$$

These compounds are easiest to form when *Y* is fluorine. Iodine is the only halogen that forms an $XY_7$ interhalogen compound, and it does so only with fluorine.

$ClF_3$ and $BrF_5$ are extremely reactive compounds. $ClF_3$ is so reactive that wood, asbestos, and even water spontaneously burn in its presence. These compounds are excellent fluorinating agents, which tend to react with each other to form positive ions such as $ClF_2^+$ and $BrF_4^+$ and negative ions such as $ClF_4^-$ and $BrF_6^-$:

$$2\ BrF_5(l) \longrightarrow [BrF_4^+][BrF_6^-](s)$$

## NEUTRAL OXIDES OF THE HALOGENS

Under certain conditions, it is possible to isolate neutral oxides of the halogens, such as $Cl_2O$, $Cl_2O_3$, $ClO_2$, $Cl_2O_4$, $Cl_2O_6$, and $Cl_2O_7$. $Cl_2O_7$, for example, can be obtained by dehydrating perchloric acid, $HClO_4$. These oxides are notoriously unstable compounds that explode when subjected to either thermal or physical shock. Some are so unstable they detonate when warmed to temperatures above −40°C.

## OXYACIDS OF THE HALOGENS AND THEIR SALTS

Chlorine reacts with the $OH^-$ ion to form chloride ions and hypochlorite ($OCl^-$) ions:

$$Cl_2(aq) + 2\ OH^-(aq) \longrightarrow Cl^-(aq) + OCl^-(aq) + H_2O(l)$$

This is a disproportionation reaction in which one-half of the chlorine atoms are oxidized to hypochlorite ions and the other half are reduced to chloride ions:

$$Cl_2 + 2\ OH^- \longrightarrow Cl^- + OCl^- + H_2O$$

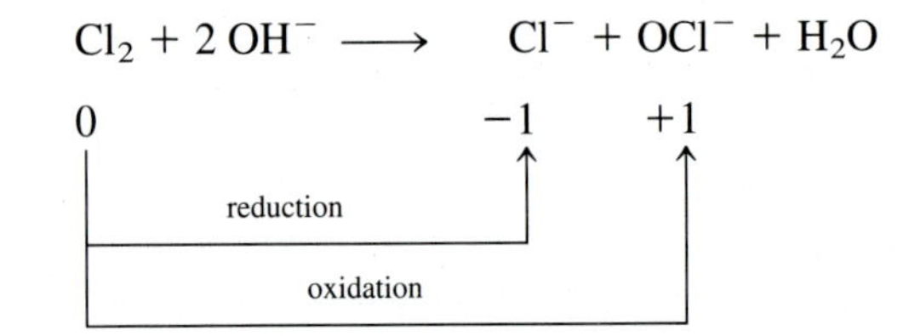


When the solution is hot, this reaction gives a mixture of the chloride and chlorate ($ClO_3^-$) ions:

$$3\ Cl_2(aq) + 6\ OH^-(aq) \longrightarrow 5\ Cl^-(aq) + ClO_3^-(aq) + 3\ H_2O(l)$$

Under carefully controlled conditions, it is possible to convert a mixture of the chlorate and hypochlorite ions into a solution that contains the chlorite ($ClO_2^-$) ion:

$$ClO_3^-(aq) + ClO^-(aq) \longrightarrow \mathbf{2\ ClO_2^-(aq)}$$

The last member of this class of compounds, the perchlorate ion ($ClO_4^-$), is made by electrolyzing solutions of the chlorate ion.

The names of the oxyanions of the halogens use the endings *-ite* and *-ate* to indicate low and high oxidation numbers and the prefixes *hypo-* and *per-* to indicate the very lowest and very highest oxidation numbers, as shown in Table 10.10. Each of these ions can be converted into an oxyacid, which is named by replacing the *-ite* ending with *-ous* and the *-ate* ending with *-ic*.

**TABLE 10.10 Oxyanions and Oxyacids of Chlorine**

| Compound | Name | Compound | Name | Oxidation State of the Chlorine |
|---|---|---|---|---|
| $ClO^-$ | hypochlorite | HOCl | hypochlorous acid | +1 |
| $ClO_2^-$ | chlorite | HOClO | chlorous acid | +3 |
| $ClO_3^-$ | chlorate | $HOClO_2$ | chloric acid | +5 |
| $ClO_4^-$ | perchlorate | $HOClO_3$ | perchloric acid | +7 |

## 10.8 THE CHEMISTRY OF THE RARE GASES

In 1892 Lord Rayleigh found that oxygen was always 15.882 times more dense than hydrogen, no matter how it was prepared. When he tried to extend this work to nitrogen, he found that nitrogen isolated from air was denser than nitrogen prepared from ammonia. William Ramsey attacked this problem by purifying a sample of nitrogen gas to remove any moisture, carbon dioxide, and organic contaminants. He then passed the purified gas over hot magnesium metal, which reacts with nitrogen to form the nitride:

$$3\ Mg(s) + N_2(s) \longrightarrow Mg_3N_2(s)$$

When he was finished, Ramsey was left with a small residue of gas that occupied roughly 1/80 of the original volume. He excited this gas in an electric discharge tube and found that the resulting emission spectrum contained lines that differed from those of all known gases. After repeated discussions of the results of these experiments, Rayleigh and Ramsey jointly announced the discovery of a new element, which they named *argon* from the Greek word meaning the "lazy one" because this gas refused to react with any element or compound they tested.

Argon did not fit into any of the known families of elements in the periodic table, but its atomic weight suggested that it might belong to a new group that could be inserted between chlorine and potassium. Shortly after reporting the discovery of argon in 1894, Ramsey found another unreactive gas when he heated a mineral of uranium. The lines in the spectrum of this gas also occurred in the spectrum of the sun, which led Ramsey to name the element *helium* (from the Greek *helios,* "sun"). Experiments with liquid air led Ramsey to a third gas, which he named *krypton* ("the hidden one"). Experiments with liquid argon led him to a fourth gas, *neon* ("the new one"), and finally a fifth gas, *xenon* ("the stranger").

These elements were discovered between 1894 and 1898. Because Moissan had only recently isolated fluorine for the first time and fluorine was the most active of the known elements, Ramsey sent a sample of argon to Moissan to see whether it

Crystals of the rare gas compound $XeF_4$.

would react with fluorine. It did not. The failure of Moissan's attempts to react argon with fluorine, coupled with repeated failures by other chemists to get the more abundant of these gases to undergo chemical reaction, eventually led to their being labeled *inert gases*. The development of the electronic theory of atoms did little to dispel this notion because it was obvious that these gases had very symmetrical electron configurations. As a result, these elements were labeled "inert gases" in almost every textbook and periodic table until about 30 years ago.

In 1962 Neil Bartlett found that $PtF_6$ was a strong enough oxidizing agent to remove an electron from an $O_2$ molecule:

$$PtF_6(g) + O_2(g) \longrightarrow [O_2^+][PtF_6^-](s)$$

Bartlett realized that the first ionization energy of Xe (1170 kJ/mol) was slightly smaller than the first ionization energy of the $O_2$ molecule (1177 kJ/mol). He therefore predicted that $PtF_6$ might also react with Xe. When he ran the reaction, he isolated the first compound of a Group VIIIA element:

$$Xe(g) + PtF_6(g) \longrightarrow [Xe^+][PtF_6^-](s)$$

A few months later, workers at the Argonne National Laboratory near Chicago found that Xe reacts with $F_2$ to form $XeF_4$. Since that time, more than 200 compounds of Kr, Xe, and Rn have been isolated. No compounds of the more abundant elements in this group (He, Ne, and Ar) have been isolated yet. However, the fact that elements in this family can undergo chemical reactions has led to the use of the term *rare gases* rather than *inert gases* to describe these elements.

Compounds of xenon are by far the most numerous of the **rare-gas** compounds. With the exception of $XePtF_6$, rare-gas compounds have oxidation numbers of +2, +4, +6, and +8, as shown by the examples in Table 10.11.

There is some controvery over whether the rare gases should be viewed as having the outermost shell of electrons filled (in which case they should be labeled Group VIIIA) or empty (in which case they should be labeled Group 0). The authors believe these elements should be labeled Group VIIIA because they behave as if they contribute eight valence electrons when they form compounds. The distribution of electrons in the valence shell of the central atom in a number of xenon compounds is shown in Figure 10.30.

The synthesis of most xenon compounds starts with the reaction between Xe and $F_2$ at high temperatures (250°C to 400°C) to form a mixture of $XeF_2$, $XeF_4$, and $XeF_6$:

$$Xe(g) + F_2(g) \longrightarrow XeF_2(s) + XeF_4(s) + XeF_6(s)$$

**TABLE 10.11 Compounds of Xenon and Their Oxidation Numbers**

| Compound | Oxidation Number | Compound | Oxidation Number |
|---|---|---|---|
| $XeF^+$ | +2 | $XeO_3$ | +6 |
| $XeF_2$ | +2 | $XeOF_4$ | +6 |
| $Xe_2F_3^+$ | +2 | $XeO_2F_2$ | +6 |
| $XeF_3^+$ | +4 | $XeO_3F^-$ | +6 |
| $XeF_4$ | +4 | $XeO_4$ | +8 |
| $XeOF_2$ | +4 | $XeO_6^{4-}$ | +8 |
| $XeF_5^+$ | +6 | $XeO_3F_2$ | +8 |
| $XeF_6$ | +6 | $XeO_2F_4$ | +8 |
| $Xe_2F_{11}^+$ | +6 | $XeOF_5^+$ | +8 |

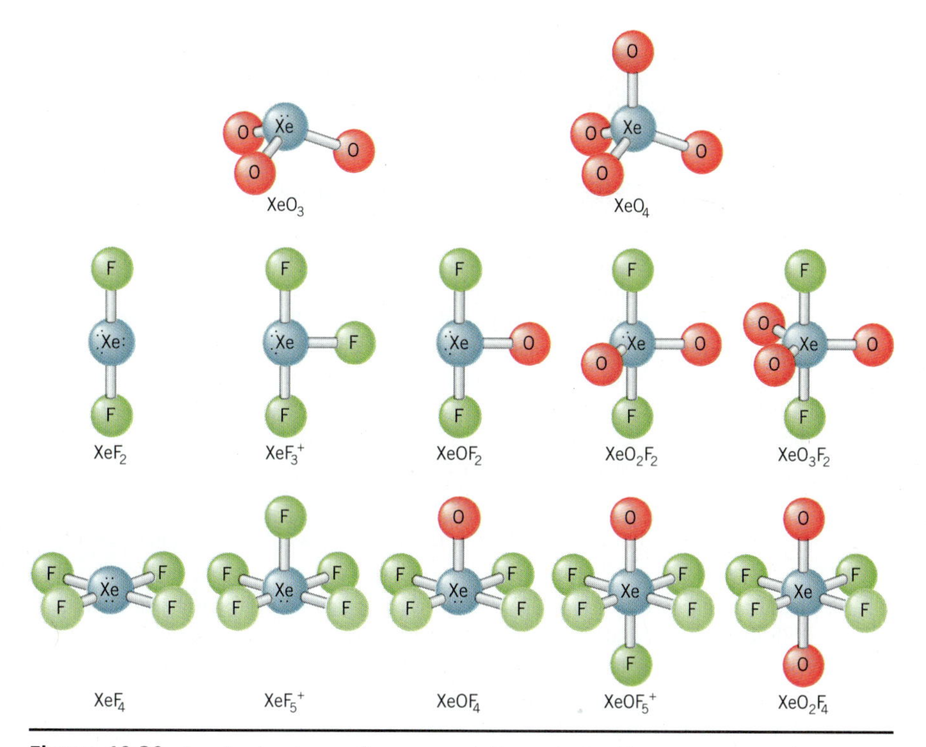


**Figure 10.30** Lewis structures of a representative sample of xenon compounds.

The positively charged $XeF_n^+$ ions are then made by reacting $XeF_2$, $XeF_4$, or $XeF_6$ with either $AsF_5$, $SbF_5$, or $BiF_5$:

$$XeF_2(s) + SbF_5(l) \longrightarrow [XeF^+][SbF_6^-](s)$$
$$2\,XeF_2(s) + AsF_5(g) \longrightarrow [Xe_2F_3^+][AsF_6^-](s)$$
$$XeF_4(s) + BiF_5(s) \longrightarrow [XeF_3^+][BiF_6^-](s)$$
$$2\,XeF_6(s) + AsF_5(g) \longrightarrow [Xe_2F_{11}^+][AsF_6^-](s)$$

Oxides of xenon, such as $XeOF_2$, $XeOF_4$, $XeO_2F_2$, $XeO_3F_2$, $XeO_2F_4$, $XeO_3$, and $XeO_4$, are prepared by reacting $XeF_4$ or $XeF_6$ with water. The $XeO_6^{4-}$ ion, for example, is produced when $XeF_6$ dissolves in strong base:

$$2\,XeF_6(s) + 16\,OH^-(aq) \longrightarrow XeO_6^{4-}(aq) + Xe(g) + O_2(g) + 12\,F^-(aq) + 8\,H_2O(l)$$

Some xenon compounds are relatively stable. $XeF_2$, $XeF_4$, and $XeF_6$, for example, are stable solids that can be purified by sublimation in a vacuum at 25°C. $XeOF_4$ and $Na_4XeO_6$ are also reasonably stable. Others, such as $XeO_3$, $XeO_4$, $XeOF_2$, $XeO_2F_2$, $XeO_3F_2$, and $XeO_2F_4$, are unstable compounds that can decompose violently.

The principal use of rare-gas compounds at present is as the light-emitting component in lasers. Mixtures of 10% Xe, 89% Ar, and 1% $F_2$, for example, can be ''pumped,'' or excited, with high-energy electrons to form excited XeF molecules, which emit a photon with a wavelength of 354 nm.

SPECIAL TOPIC

# THE INORGANIC CHEMISTRY OF CARBON

For more than 200 years, chemists have divided compounds into two categories. Those that were isolated from plants or animals were called *organic,* while those extracted from ores and minerals were *inorganic.* Organic chemistry is often defined as the chemistry of carbon. But this definition would include calcium carbonate ($CaCO_3$) and graphite, which more closely resemble inorganic compounds. We will therefore define organic chemistry as the study of compounds, such as formic acid ($HCO_2H$), methane ($CH_4$), and vitamin C ($C_6H_8O_6$), that contain both carbon and hydrogen. Chapter 24 will be devoted to a discussion of organic chemistry. This section will focus on inorganic carbon compounds.

The chemistry of carbon is dominated by three factors:

1. Carbon forms unusually strong C—C single bonds, C═C double bonds, and C≡C triple bonds.
2. The electronegativity of carbon ($EN = 2.55$) is too small to allow carbon to form $C^{4-}$ ions with most metals and too large for carbon to form $C^{4+}$ ions when it reacts with nonmetals. Carbon therefore forms covalent bonds with many other elements.
3. Carbon forms strong double and triple bonds with a number of other nonmetals, including N, O, P, and S.

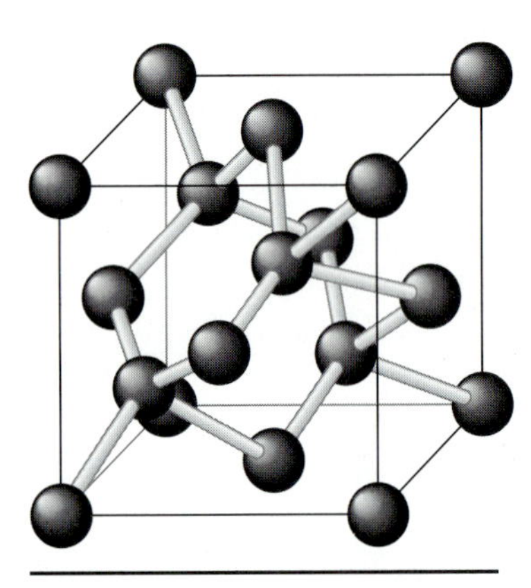

**Figure 10.31** The remarkable properties of diamond can be traced to the fact that each $sp^3$-hybridized carbon atom is tightly bound to four neighboring carbon atoms arranged in a tetrahedral array, so that the crystal can be considered to be a single giant molecule.

## ELEMENTAL FORMS OF CARBON: GRAPHITE, DIAMOND, COKE, AND CARBON BLACK

Carbon occurs as a variety of allotropes. There are two crystalline forms—diamond and graphite—and a number of amorphous (noncrystalline) forms, such as charcoal, coke, and carbon black.

References to the characteristic hardness of diamond (from the Greek *adamas,* ''invincible'') date back at least 2600 years. It was not until 1797, however, that Smithson Tennant was able to show that diamonds consist solely of carbon. The properties of diamond are remarkable. It is among the least volatile substances known (MP = 3550°C, BP = 4827°C), it is also the hardest naturally occurring substance, and it expands less on heating than any other material.

The properties of diamond are a logical consequence of its structure. Carbon, with four valence electrons, forms covalent bonds to four neighboring carbon atoms arranged toward the corners of a tetrahedron, as shown in Figure 10.31. Each of these $sp^3$-hybridized atoms is then bound to four other carbon atoms, which form bonds to four other carbon atoms, and so on. As a result, a perfect diamond can be thought of as a single giant molecule. The strength of the individual C—C bonds and their arrangement in space give rise to the unusual properties of diamond.

The two allotropes of elemental carbon are diamond and graphite.

In some ways, the properties of graphite are like those of diamond. Both compounds boil at 4827°C, for example. But graphite is also very different from diamond. Diamond (3.514 g/cm$^3$) is much denser than graphite (2.26 g/cm$^3$). Whereas diamond is the hardest substance known, graphite is one of the softest. Diamond is an excellent insulator, with little or no tendency to carry an electric current. Graphite is such a good conductor of electricity that graphite electrodes are used in electrical cells.

The physical properties of graphite can be understood from the structure of the solid shown in Figure 10.32. Graphite consists of extended planes of $sp^2$-hybridized carbon atoms in which each carbon is tightly bound to three other carbon atoms. (The strong bonds between

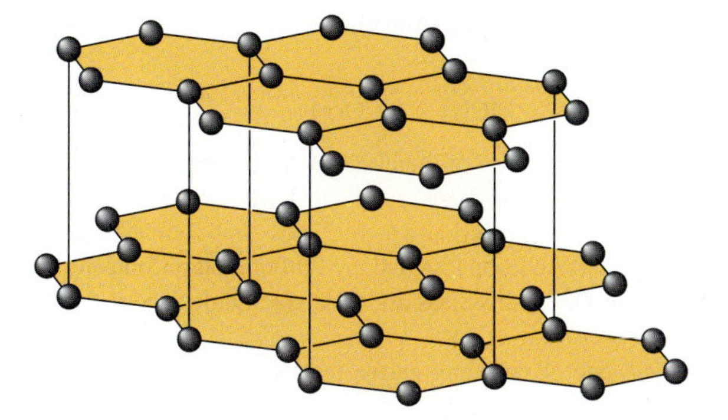

**Figure 10.32** The high melting point and boiling point of graphite can be explained by the very strong bonds within the extended planes of $sp^2$-hybridized carbon atoms. (The bond dissociation enthalpy for the bonds within one of these planes is 477 kJ/mol.) The softness of graphite can be explained by the ease with which these planes slide past one another. (The bond between planes of atoms in graphite has a bond dissociation enthalpy of only 17 kJ/mol.)

carbon atoms *within each plane* explain the exceptionally high melting point and boiling point of graphite.) The distance between these planes of atoms, however, is very much larger than the distance between the atoms within the planes. Because the bonds *between planes* are weak, it is easy to deform the solid by allowing one plane of atoms to move relative to another. As a result, graphite is soft enough to be used in pencils and as a lubricant in motor oil.

''Lead'' pencils do not, incidentally, contain lead. (This is fortunate because many people chew pencils and lead compounds are toxic.) Lead pencils contain graphite, or ''black lead'' as it was once known, which is mixed with clay (20% to 60% by weight) and then baked to form a ceramic rod. Increasing the percentage of clay makes the pencil harder, so that less graphite is deposited on the paper.

The characteristic properties of graphite and diamond might lead you to expect diamond to be more stable than graphite. This is not what is observed experimentally. The standard enthalpy of formation of diamond ($\Delta H°_f$ = 2.425 kJ/mol) is slightly larger than the enthalpy of formation of graphite, which is the most stable form of carbon at 25°C and 1 atm pressure. At very high temperatures and pressures, diamond becomes more stable than graphite. In 1955 General Electric developed a process to make industrial-grade diamonds by treating graphite with a metal catalyst at temperatures of 2000 to 3000 K and pressures above 125,000 atmospheres. Roughly 40% of industrial-quality diamonds are now synthetic. Although gem-quality diamonds can be synthesized, the costs involved are prohibitive.

Both diamond and graphite occur as regularly packed crystals. Other forms of carbon are *amorphous*—they lack a regular structure. Charcoal, carbon black, and coke are all amorphous forms of carbon. *Charcoal* results from heating wood in the absence of oxygen. To make *carbon black,* natural gas or other carbon compounds are burned in a limited amount of air to give a thick, black smoke that contains extremely small particles of carbon, which can be collected when the gas is cooled and passed through an electrostatic precipitator. *Coke* is a more regularly structured material, closer in structure to graphite than either charcoal or carbon black, and is made from coal.

## COVALENT, IONIC, AND INTERSTITIAL CARBIDES

Although carbon is essentially inert at room temperature, it reacts with less electronegative elements at high temperatures to form compounds known as **carbides.** When carbon reacts with an element of similar size and electronegativity, a *covalent carbide* is produced. Silicon

Carborundum is the trade name for finely powdered samples of silicon carbide that are used as an abrasive.

carbide, for example, is made by treating silicon dioxide from quartz with an excess of carbon in an electric furnace at 2300 K:

$$SiO_2(s) + 3\,C(s) \longrightarrow SiC(s) + 2\,CO(g)$$

Covalent carbides have properties similar to those of diamond. Both SiC and diamond are inert to chemical reactions, except at very high temperatures; both have very high melting points; and both are among the hardest substances known. SiC was first synthesized by Edward Acheson in 1891. Shortly thereafter, Acheson founded the Carborundum Company to market this material. Then, as now, materials in this class are most commonly used as abrasives.

Compounds that contain carbon and one of the more active metals are called *ionic carbides:*

$$CaO(s) + 3\,C(s) \longrightarrow CaC_2(s) + CO(g)$$

It is useful to think about these compounds as if they contained negatively charged carbon ions: $[Ca^{2+}][C_2^{2-}]$ or $[Al^{3+}]_4[C^{4-}]_3$. This model is useful because it explains why these carbides burst into flame when added to water. The ionic carbides that formally contain the $C^{4-}$ ion react with water to form methane, which is ignited by the heat given off in this reaction:

$$C^{4-} + 4\,H_2O \longrightarrow CH_4 + 4\,OH^-$$

The ionic carbides that formally contain the $C_2^{2-}$ ion react with water to form acetylene, which is ignited by the heat of reaction:

$$C_2^{2-} + 2\,H_2O \longrightarrow C_2H_2 + 2\,OH^-$$

At one time, miners' lamps were fueled by the combustion of acetylene prepared from the reaction of calcium carbide with water.

Calcium carbide reacts with water to form acetylene, which burns when ignited in the presence of air.

*Interstitial carbides,* such as tungsten carbide (WC), form when carbon combines with a metal that has an intermediate electronegativity and a relatively large atomic radius. In these compounds, the carbon atoms pack in the holes (interstices) between planes of metal atoms. The interstitial carbides, which include TiC, ZrC, and MoC retain the properties of metals. They act as alloys, rather than as either salts or covalent compounds.

## THE OXIDES OF CARBON

Although the different forms of carbon are essentially inert at room temperature, they combine with oxygen at high temperatures to produce a mixture of carbon monoxide and carbon dioxide:

$$2\,C(s) + O_2(g) \longrightarrow 2\,CO(g) \qquad \Delta H^\circ = -110.52\ \text{kJ/mol CO}$$
$$C(s) + O_2(g) \longrightarrow CO_2(g) \qquad \Delta H^\circ = -393.51\ \text{kJ/mol }CO_2$$

CO also can be obtained when red-hot carbon is treated with steam:

$$C(s) + H_2O(g) \longrightarrow CO(g) + H_2(g)$$

Because this mixture of CO and $H_2$ is formed by the reaction of charcoal or coke with water, it is often referred to as *water gas.* It is also known as *town gas* because it was once made by towns and cities for use as a fuel. Water gas, or town gas, was a common fuel for both home and industrial use before natural gas became readily available. The $H_2$ burns to form water, and the CO is oxidized to $CO_2$. Eventually, as our supply of natural gas is depleted, it will become economical to replace natural gas with other fuels, such as water gas, that can be produced from our abundant supply of coal.

CO and $CO_2$ are both colorless gases. CO boils at $-191.5°C$, and $CO_2$ sublimes at $-78.5°C$. Although CO has no odor or taste, $CO_2$ has a faint, pungent odor and a distinctly acidic taste. Both are dangerous substances but at very different levels of exposure. Air contaminated with as little as 2 mg of CO per liter can be fatal because CO binds tightly to the hemoglobin and myoglobin that carry oxygen through the blood. $CO_2$ is not lethal until the concentration in the air approaches 15%. At that point, it has replaced so much oxygen that a person who attempts to breathe this atmosphere suffocates. The danger of $CO_2$ poisoning is magnified by the fact that $CO_2$ is roughly 1.5 times more dense than air. Thus, $CO_2$ can accumulate at the bottom of tanks or wells.

## $CO_2$ IN THE ATMOSPHERE

$CO_2$ influences the temperature of the atmosphere through the greenhouse effect, which works as follows. $CO_2$ absorbs some of the lower energy, longer wavelength, infrared radiation from the sun that would otherwise be reflected back from the surface of the planet. Thus, $CO_2$ in the atmosphere traps heat. Although there are other factors at work, it is worth noting that Venus, whose atmosphere contains a great deal of $CO_2$, has a surface temperature of roughly 400°C, whereas Mars, with little or no atmosphere, has a surface temperature of $-50°C$.

There are many sources of $CO_2$ in the atmosphere. Over geologic time scales, the largest source has been volcanos. Within the last century, the combustion of petroleum, coal, and natural gas has made a significant contribution to atmospheric levels of $CO_2$ (see Figure 10.11). Between 1958 and 1978, the average level of $CO_2$ in the atmosphere increased by 6%, from 315.8 to 334.6 ppm.

At one time, the amount of $CO_2$ released to the atmosphere was not a matter for concern because natural processes that removed $CO_2$ from the atmosphere could compensate for the $CO_2$ that entered the atmosphere. The vast majority of the $CO_2$ liberated by volcanic action, for example, was captured by calcium oxide or magnesium oxide to form calcium carbonate or magnesium carbonate:

$$CaO(s) + CO_2(g) \longrightarrow CaCO_3(s)$$
$$MgO(s) + CO_2(g) \longrightarrow MgCO_3(s)$$

$CaCO_3$ is found as limestone or marble, or mixed with $MgCO_3$ as dolomite. The amount of $CO_2$ in deposits of carbonate minerals is at least several thousand times larger than the amount in the atmosphere.

$CO_2$ also dissolves, to some extent, in water:

$$CO_2(g) \xrightarrow{H_2O} CO_2(aq)$$

It then reacts with water to form carbonic acid, $H_2CO_3$:

$$CO_2(aq) + H_2O(l) \longrightarrow H_2CO_3(aq)$$

As a result of these reactions, the sea contains about 60 times more $CO_2$ than the atmosphere.

Can the sea absorb more $CO_2$ from the atmosphere, or is it near its level of saturation? Is the rate at which the sea absorbs $CO_2$ greater than the rate at which we are adding it to the atmosphere? The observed increase in the concentration of $CO_2$ in recent years suggests pessimistic answers to these two questions. A gradual warming of the earth's atmosphere could result from continued increases in $CO_2$ levels, with adverse effects on the climate and therefore the agriculture of at least the northern hemisphere.

## THE CHEMISTRY OF CARBONATES: $CO_3^{2-}$ AND $HCO_3^-$

Egg shells are almost pure calcium carbonate. $CaCO_3$ can be found also in the shells of many marine organisms and in both limestone and marble. The fact that none of these substances dissolves in water suggests that $CaCO_3$ is normally insoluble in water. Calcium carbonate will dissolve in water saturated with $CO_2$, however, because carbonated water (or carbonic acid) reacts with calcium carbonate to form calcium bicarbonate, which is soluble in water:

$$CaCO_3(s) + H_2CO_3(aq) \longrightarrow Ca^{2+}(aq) + 2\,HCO_3^-(aq)$$

When water rich in carbon dioxide flows through limestone formations, part of the limestone dissolves. If the $CO_2$ escapes from this water, or if some of the water evaporates, solid $CaCO_3$ is redeposited. When this happens as water runs across the roof of a cavern, *stalactites,* which hang from the roof of the cave, are formed. If the water drops before the carbonate reprecipitates, *stalagmites,* which grow from the floor of the cave, are formed.

The chemistry of carbon dioxide dissolved in water is the basis of the soft drink industry. The first artificially carbonated beverages were introduced in Europe at the end of the nineteenth century. Carbonated soft drinks today consist of carbonated water, a sweetening agent (such as sugar, saccharin, or aspartame), an acid to impart a sour or tart taste, flavoring agents, coloring agents, and preservatives. As much as 3.5 liters of gaseous $CO_2$ dissolve in a liter of soft drink. The $CO_2$ contributes the characteristic bite associated with carbonated beverages.

Carbonate chemistry plays an important role in other parts of the food industry as well. *Baking soda,* or bicarbonate of soda, is sodium bicarbonate, $NaHCO_3$, a weak base, which is added to recipes to neutralize the acidity of other ingredients. *Baking powder* is a mixture of baking soda and a weak acid, such as tartaric acid or calcium hydrogen phosphate ($CaHPO_4$). When mixed with water, the acid reacts with the $HCO_3^-$ ion to form $CO_2$ gas, which causes the dough or batter to rise:

$$HCO_3^-(aq) + H^+(aq) \longrightarrow H_2CO_3(aq) \longrightarrow H_2O(l) + CO_2(g)$$

Before commercial baking powders were available, cooks obtained the same effect by mixing roughly a teaspoon of baking soda with a cup of sour milk or buttermilk. The acids that give sour milk and buttermilk their characteristic taste also react with the bicarbonate ion to give $CO_2$.

## RESEARCH IN THE 90s

### FULLERENES

In 1985 Richard Smalley and co-workers at Rice University made a uniquely stable form of carbon by vaporizing graphite with a laser. The apparatus in which this experiment was performed was designed to create small molecules that were clusters of atoms. In this cluster generator, a pulse of helium gas was swept over the surface of the graphite as it was excited with the laser. The mixture of helium and carbon atoms that vaporized from the graphite surface cooled as the gas expanded, and molecules with the formula $C_{60}$ were formed that have a structure that has the symmetry of a soccer ball. Because this structure resembles the geodesic dome invented by R. Buckminster Fuller, $C_{60}$ was named *buckminsterfullerene,* or ''buckyball'' for short.

Although it was formally a new form of pure carbon, $C_{60}$ seemed to be nothing more than a laboratory curiosity until 1990, when Wolfgang Kratschmer and Konsantinos Kostiropoulos, at the Max Planck Institute in Heidelberg, reported that this material could be made by heating a graphite rod in an atmosphere of helium until the graphite evaporated. Once it was known that $C_{60}$ could be synthesized in large quantities, researchers looked for it, and found it, in such common sources as the flame of a sooty candle. It has even been found in the black soot that collects on the glass screen in front of a fireplace.

Some of the excitement chemists experienced when $C_{60}$ was synthesized can be understood by contrasting this form of pure carbon with diamond and graphite. $C_{60}$ is unique because it exists as distinct molecules, not extended arrays of atoms. Equally important, $C_{60}$ can be obtained as a pure substance, whereas diamond and graphite are inevitably contaminated by hydrogen atoms that bind to the carbon atoms on the surface.

$C_{60}$ is now known to be a member of a family of compounds known as the *fullerenes.* Other compounds in this family include $C_{32}$, $C_{44}$, $C_{50}$, $C_{58}$, and $C_{70}$. $C_{60}$ may be the most important of the fullerenes because it is the most perfectly symmetric molecule possible, spinning in the solid at a rate of more than 100 million times per second. Because of their symmetry, $C_{60}$ molecules pack as regularly as Ping-Pong balls. The resulting solid has unusual properties. Initially, it is as soft as graphite, but when compressed by 30%, it becomes harder than diamond. When this pressure is released, the solid springs back to its original volume. $C_{60}$ therefore has the remarkable property that it bounces when shot at a metal surface at high speeds.

$C_{60}$ also has the remarkable ability to form compounds in which it is an insulator, a conductor, a semiconductor, or a superconductor. By itself, $C_{60}$ is a semiconductor. When mixed with just enough potassium to give the empirical formula $K_3C_{60}$, it conducts electricity like a metal. When excess potassium is added, this solid becomes an insulator. When $K_3C_{60}$ is cooled to 18 K, the result is a superconductor. The potential of fullerene chemistry for both practical materials and laboratory curiosities is large enough to explain why this molecule can be described as ''exocharmic''—it exudes charm.

## SUMMARY

**1.** The nonmetals can act as either oxidizing agents or reducing agents. In general, the more electronegative the element, the more likely it is to act as an oxidizing agent to form compounds in which it has a negative oxidation number.

**2.** A neutral hydrogen atom has only one electron. Hydrogen therefore forms compounds in which it has an oxidation number of either +1, 0, or −1. In a formal sense, these oxidation numbers correspond to $H^+$ ions, neutral H atoms, and $H^-$ ions.

**3.** Oxygen is so electronegative it oxidizes virtually every substance with which it reacts. Most of the time, oxygen forms compounds in which it has an oxidation number of −2. Under certain conditions, it forms peroxides, in which it has an oxidation number of −1.

**4.** In many ways, the chemistry of sulfur resembles that of oxygen. There are four major differences: (1) sulfur forms weaker multiple bonds; (2) sulfur forms stronger single bonds; (3) sulfur is more likely to form compounds with positive oxidation numbers; and (4) sulfur can expand its valence shell to hold more than eight electrons.

**5.** The chemistry of nitrogen is dominated by the strength of the double and triple bonds it forms with other nonmetals. The strength of these bonds makes $N_2$ relatively inert at room temperature. It also explains why a number of nitrogen compounds decompose to form $N_2$ or react with other substances to form products that include $N_2$.

**6.** Phosphorus forms compounds that have oxidation numbers analogous to those of nitrogen compounds. The formulas of these compounds are different, however, because phosphorus atoms are too large to form strong double and triple bonds. Where nitrogen compounds form an N=O double bond, for example, the corresponding phosphorus compounds usually contain two P—O single bonds.

**7.** The halogens are relatively strong oxidizing agents. As we go up this column of the periodic table, each succeeding element requires a stronger oxidizing agent to be prepared from one of its compounds. $F_2$ is such a strong oxidizing agent that it must be prepared by electrolysis.

**8.** For many years, chemists assumed that there was no chemistry of the elements in Group VIIIA. For the most part, the chemistry of these rare gases is still restricted to compounds formed by reactions between the heavier members of this group and the most reactive nonmetals, such as fluorine and oxygen.

**9.** The chemistry of carbon is dominated by the strength of the covalent bonds it forms with itself and its ability to form strong double and triple bonds to other nonmetals. As a result, the chemistry of this element is more diverse than any other element.

## KEY TERMS

**Allotropes** (p. 356)
**Anhydrous** (p. 371)
**Carbides** (p. 391)
**Diamagnetic** (p. 356)
**Dimer** (p. 376)
**Disproportionation reactions** (p. 360)
**Haber process** (p. 370)
**Hydrides** (p. 353)
**Ostwald process** (p. 372)
**Oxidizing agent** (p. 358)
**Oxyacids** (p. 368)
**Oxyanions** (p. 368)
**Paramagnetic** (p. 356)
**Peroxide** (p. 359)
**Rare gases** (p. 388)
**Reducing agent** (p. 358)

## PROBLEMS

### Metals, Nonmetals, and Semimetals

**10-1** List the elements that are nonmetals. Describe where these elements are found in the periodic table.

**10-2** Explain why semimetals (such as B, Si, Ge, As, Sb, Te, Po, and At) exist and describe some of their physical properties.

**10-3** Which member of each of the following pairs of elements is more nonmetallic?

(a) As or Bi (b) As or Se (c) As or S
(d) As or Ge (e) As or P

### The Chemistry of the Nonmetals

**10-4** Which of the following elements can exist as a triatomic molecule?

(a) hydrogen (b) helium (c) sulfur (d) oxygen
(e) chlorine

**10-5** Which of the following elements should form compounds with the formulas $Na_2X$, $H_2X$, $XO_2$, and $XF_6$?

(a) B (b) C (c) N (d) O (e) S

**10-6** Which of the following elements should form compounds with the formulas $XH_3$, $XF_3$, and $Na_3XO_4$?

(a) Al (b) Ge (c) As (d) S (e) Cl

**10-7** Which of the following can't be found in nature? Explain why.

(a) $MgCl_2$ (b) $CaCO_3$ (c) $F_2$ (d) $Na_3AlF_6$
(e) NaCl

### The Role of Nonmetal Elements in Chemical Reactions

**10-8** Explain why more electronegative elements tend to oxidize less electronegative elements.

**10-9** Which of the following can be oxidized?

(a) $H_2SO_3$ (b) $P_4$ (c) $Cl^-$ (d) $SiO_2$ (e) $PO_4^{3-}$ (f) $Mg^{2+}$

**10-10** Which of the following can be reduced?

(a) $H_2O$ (b) $H_2SO_3$ (c) HCl (d) $CO_2$ (e) $Mg^{2+}$ (f) Na

### Deciding What Is Oxidized and What Is Reduced

**10-11** For each of the following reactions, identify what is oxidized and what is reduced.

(a) $Fe_2O_3(s) + 3\,CO(g) \longrightarrow 2\,Fe(s) + 3\,CO_2(g)$

(b) $H_2(g) + CO_2(g) \longrightarrow H_2O(g) + CO(g)$

(c) $CH_4(g) + 2\,O_2(g) \longrightarrow CO_2(g) + 2\,H_2O(g)$

(d) $2\,H_2S(g) + 3\,O_2(g) \longrightarrow 2\,SO_2(g) + 2\,H_2O(g)$

**10-12** For each of the following reactions, identify what is oxidized and what is reduced.

(a) $PH_3(g) + 3\,Cl_2(g) \longrightarrow PCl_3(g) + 3\,HCl(g)$

(b) $2\,NO(g) + F_2(g) \longrightarrow 2\,NOF(g)$

(c) $2\,Na(s) + 2\,NH_3(l) \longrightarrow 2\,NaNH_2(s) + H_2(g)$

(d) $3\,NO_2(g) + H_2O(l) \longrightarrow 2\,HNO_3(aq) + NO(g)$

**10-13** Hydrazine is made by a reaction known as the Raschig process:

$$2\,NH_3(aq) + NaOCl(aq) \longrightarrow N_2H_4(aq) + NaCl(aq) + H_2O(l)$$

Decide whether this is an oxidation–reduction reaction. If it is, identify the compound oxidized and the compound reduced.

**10-14** The thiosulfate ion, $S_2O_3^{2-}$, is prepared by boiling solutions of sulfur dissolved in sodium sulfite:

$$8\,SO_3^{2-}(aq) + S_8(s) \longrightarrow 8\,S_2O_3^{2-}(aq)$$

Is this an oxidation–reduction reaction? If it is, identify what is oxidized and what is reduced.

**10-15** Chlorine dioxide, $ClO_2$, is used commercially as a bleach or a disinfectant because of its excellent oxidizing ability. $ClO_2$ is prepared by decomposing chlorous acid.

$$8\,HOClO(aq) \longrightarrow 6\,ClO_2(g) + Cl_2(g) + 4\,H_2O(l)$$

Is this an oxidation–reduction reaction? If it is, identify the oxidizing agent and the reducing agent.

**10-16** Nitric acid sometimes acts as an acid (as a source of the $H^+$ ion) and sometimes as an oxidizing agent. For each of the following reactions, decide whether $HNO_3$ acts as an acid or as an oxidizing agent.

(a) $Na_2CO_3(s) + 2\,HNO_3(aq) \longrightarrow 2\,NaNO_3(aq) + CO_2(g) + H_2O(l)$

(b) $3\,P_4(s) + 20\,HNO_3(aq) + 8\,H_2O(l) \longrightarrow 12\,H_3PO_4(aq) + 20\,NO(g)$

(c) $Al_2O_3(s) + 6\,HNO_3(aq) \longrightarrow 2\,Al(NO_3)_3(aq) + 3\,H_2O(l)$

(d) $3\,Cu(s) + 8\,HNO_3(aq) \longrightarrow 3\,Cu(NO_3)_2(aq) + 2\,NO(g) + 4\,H_2O(l)$

**10-17** Which of the following would you expect to be the best oxidizing agent?

(a) Na (b) $H_2$ (c) $N_2$ (d) $P_4$ (e) $O_2$

**10-18** Which of the following would you expect to be the best reducing agent?

(a) $Na^+$ (b) $F^-$ (c) Na (d) $Br_2$ (e) $Fe^{3+}$

**10-19** For each of the following pairs of elements, determine which is the better reducing agent:

(a) $P_4$ or As (b) As or $S_8$ (c) $P_4$ or $S_8$ (d) $S_8$ or $Cl_2$ (e) C or $O_2$

### Predicting the Products of Chemical Reactions

**10-20** Predict the products of the following reactions:

(a) $Mg(s) + N_2(g) \longrightarrow$

(b) $Li(s) + O_2(g) \longrightarrow$

(c) $Br_2(l) + I^-(aq) \longrightarrow$

**10-21** Predict the products of the following reactions:

(a) $SO_2(g) + H_2O(l) \longrightarrow$

(b) $Cl_2(g) + OH^-(aq) \longrightarrow$

(c) $CO_2(g) + H_2O(l) \longrightarrow$

**10-22** Predict the products of the following reactions:

(a) $HCl(g) + H_2O(l) \longrightarrow$

(b) $P_4O_{10}(s) + H_2O(l) \longrightarrow$

(c) $NO_2(g) + H_2O(l) \longrightarrow$

**10-23** Predict the products of the following reactions:

(a) $S_8(s) + O_2(g) \longrightarrow$

(b) $Al(s) + I_2(s) \longrightarrow$

(c) $P_4(s) + F_2(g) \longrightarrow$

### The Chemistry of Hydrogen

**10-24** Describe three ways of preparing small quantities of $H_2$ in the lab.

**10-25** Explain why it is not a good idea to prepare $H_2$ by reacting sodium metal with a strong acid.

**10-26** Give an example of a compound in which hydrogen has an oxidation number of +1; of 0; of −1.

**10-27** Which of the following reactions produce a compound in which hydrogen has an oxidation number of −1?

(a) $Li + H_2 \longrightarrow$ (b) $O_2 + H_2 \longrightarrow$ (c) $S_8 + H_2 \longrightarrow$ (d) $Cl_2 + H_2 \longrightarrow$ (e) $Ca + H_2 \longrightarrow$

**10-28** Use tables of first ionization energies and electronegativities to explain why it is so difficult to decide whether hydrogen belongs in Group IA or Group VIIA of the periodic table.

**10-29** Which of the following substances can be used as evidence for placing hydrogen in Group IA? Which can be used as evidence for including hydrogen in Group VIIA?

(a) $CaH_2$ (b) $AlH_3$ (c) $H_2S$ (d) $H_3PO_4$ (e) $H_2$

**10-30** The earth's atmosphere once contained significant amounts of $H_2$. Explain why only traces of $H_2$ are left in the earth's atmosphere, whereas the atmospheres of other planets, such as Jupiter, Saturn, and Neptune, contain large quantities of $H_2$.

**10-31** Use Lewis structures to explain what happens in the four reactions described in Section 10.2 that can be used to prepare small quantities of $H_2$ gas.

### The Chemistry of Oxygen and Sulfur

**10-32** Describe three ways of preparing small quantities of $O_2$ in the lab.

**10-33** Describe the relationship among oxygen ($O_2$), the peroxide ion ($O_2^{2-}$), and the oxide ion ($O^{2-}$). Explain why the number of electrons shared by a pair of oxygen atoms decreases as the oxidation number of the oxygen becomes more negative.

**10-34** Which of the following elements or compounds could eventually produce $O_2$ when it reacts with water?

(a) Ba (b) BaO (c) $BaO_2$ (d) $Ba(OH)_2$
(e) $BaNO_3$

**10-35** Explain why the only compounds in which oxygen has a positive oxidation number are compounds, such as $OF_2$, that contain fluorine.

**10-36** Explain why hydrogen peroxide can be either an oxidizing agent or a reducing agent. Describe at least one reaction in which $H_2O_2$ oxidizes another substance and one reaction in which it reduces another substance.

**10-37** Write the Lewis structures for ozone, $O_3$, and sulfur dioxide, $SO_2$. Discuss the relationship between these compounds.

**10-38** Explain why elemental oxygen exists as $O_2$ molecules, whereas elemental sulfur forms $S_8$ molecules.

**10-39** Explain why sulfur forms compounds such as $SF_4$ and $SF_6$, whereas oxygen can only form $OF_2$.

**10-40** Describe the relationship between the thiosulfate and sulfate ions and between the thiocyanate and cyanate ions. Use this relationship to predict the formula of the trithiocarbonate ion.

**10-41** Write the Lewis structures of the following products of the reaction between sodium and sulfur:

(a) $Na_2S$ (b) $Na_2S_2$ (c) $Na_2S_3$ (d) $Na_2S_8$

**10-42** Explain why sulfur readily forms compounds in the +2, +4, and +6 oxidation states, but only a handful of compounds exist in which oxygen is in a positive oxidation state.

**10-43** Explain why sulfur-containing compounds such as $FeS_2$, $CS_2$, and $H_2S$ form $SO_2$ instead of $SO_3$ when they burn.

**10-44** Use Lewis structures to explain what happens when the $SO_3^{2-}$ ion reacts with sulfur to form thiosulfate, $S_2O_3^{2-}$.

**10-45** Use Lewis structures to explain why a two-electron reduction of the $S_2O_6^{2-}$ ion gives the $SO_3^{2-}$ ion.

$$S_2O_6^{2-} + 2\,e^- \longrightarrow 2\,SO_3^{2-}$$

**10-46** Use Lewis structures to explain why a two-electron oxidation of the $S_2O_3^{2-}$ ion gives the $S_4O_6^{2-}$ ion.

**10-47** Which of the following does not have a reasonable oxidation number for sulfur?

(a) $Na_2S$ (b) $H_2S$ (c) $SO_3^{2-}$ (d) $SO_4$ (e) $SF_4$

**10-48** Explain why $SO_2$ plays an important role in the phenomenon known as acid rain.

**10-49** Explain why problems with acid rain would be much more severe if sulfur compounds burned to form $SO_3$ instead of $SO_2$.

### The Chemistry of Nitrogen and Phosphorus

**10-50** Nitrogen has a reasonable oxidation number in all of the following compounds, and yet one of them is still impossible. Which one is impossible?

(a) $NF_3$ (b) $NF_5$ (c) $NO_3^-$ (d) $NO_2^-$ (e) NO

**10-51** Earth's atmosphere contains roughly $4 \times 10^{16}$ tons of nitrogen, and yet the biggest problem facing agriculture in the world today is a lack of "nitrogen." Explain why.

**10-52** Explain why elemental nitrogen is almost inert, but nitrogen compounds such as $NH_4NO_3$, $NaN_3$, nitroglycerine, and trinitrotoluene (TNT) form some of the most dangerous explosives.

**10-53** Which of the following oxides of nitrogen are paramagnetic?

(a) $N_2O$ (b) NO (c) $NO_2$ (d) $N_2O_3$ (e) $N_2O_4$
(f) $N_2O_5$

**10-54** Use Lewis structures to explain what happens in the following reaction:

$$2\,NO + O_2 \longrightarrow 2\,NO_2$$

**10-55** Use Lewis structures to explain why NO reacts with $NO_2$ to form $N_2O_3$ when a mixture of these compounds is cooled.

**10-56** Use the fact that nitrous oxide decomposes to form nitrogen and oxygen to explain why a glowing splint bursts into flame when immersed in a container filled with $N_2O$:

$$2\,N_2O(g) \longrightarrow 2\,N_2(g) + O_2(g)$$

**10-57** Describe ways of preparing small quantities of each of the following compounds in the laboratory:

(a) $N_2O$ (b) NO (c) $NO_2$ (d) $N_2O_4$

**10-58** Describe how to tell the difference between a flask filled with NO gas and a flask filled with $N_2O$.

**10-59** Lightning catalyzes the reaction between nitrogen and oxygen in the atmosphere to form nitrogen oxide, NO:

$$N_2(g) + O_2(g) \longrightarrow 2\,NO(g)$$

Explain why lightning is a source of acid rain.

**10-60** Which of the following elements or compounds is not involved at some stage in the preparation of nitric acid?

(a) $O_2$ (b) $N_2$ (c) NO (d) $NO_2$ (e) $H_2$

**10-61** Explain why phosphorus forms both $PCl_3$ and $PCl_5$ but nitrogen forms only $NCl_3$.

**10-62** Explain why nitrogen is essentially inert at room temperature, but white phosphorus bursts spontaneously into flame when it comes into contact with air.

**10-63** Explain why red phosphorus is much less reactive than white phosphorus.

**10-64** Explain why nitrogen forms extraordinarily stable $N_2$ molecules at room temperature, but phosphorus forms $P_2$ molecules only at very high temperatures.

**10-65** Explain why nitric acid has the formula $HNO_3$ and phosphoric acid has the formula $H_3PO_4$.

**10-66** Write the Lewis structures for phosphoric acid, $H_3PO_4$, and phosphorous acid, $H_3PO_3$. Explain why phosphoric acid can lose three $H^+$ ions to form a phosphate ion, $PO_4^{3-}$, whereas phosphorous acid can lose only two $H^+$ ions to form the $HPO_3^{2-}$ ion.

**10-67** Explain why only two of the four hydrogen atoms in $H_4P_2O_5$ are lost when this oxyacid forms an oxyanion.

**10-68** Describe the role of carbon in the preparation of elemental phosphorus from calcium phosphate.

**10-69** Predict the product of the reaction of phosphorus with excess oxygen and then predict what will happen when the product of this reaction is dissolved in water.

**10-70** Explain why the most common oxidation states of antimony are +3 and +5.

**10-71** Which of the following compounds should not exist?

(a) $Na_3P$ (b) $(NH_4)_3PO_4$ (c) $PO_2$ (d) $PH_3$
(e) $POCl_3$

### The Chemistry of the Halogens

**10-72** Which of the halogens is the most active, or reactive? Explain why.

**10-73** Describe the difference between halogens and halides. Give examples of each.

**10-74** $Fe^{3+}$ ions can oxidize $Br^-$ ions to $Br_2$, but they can't oxidize $Cl^-$ ions to $Cl_2$. Use this information to determine where the $Fe^{3+}$ ion belongs in the following sequence of decreasing oxidizing strength: $F_2 > Cl_2 > Br_2 > I_2$.

**10-75** HBr can be prepared by reacting $PBr_3$ with water:

$$PBr_3(l) + 3\,H_2O(l) \longrightarrow 3\,HBr(aq) + H_3PO_3(aq)$$

Use this information to explain what happens in the following reaction:

$$P_4(s) + 6\,Br_2(s) + 12\,H_2O(l) \longrightarrow 12\,HBr(aq) + 4\,H_3PO_3(aq)$$

**10-76** Explain why chlorine reacts with fluorine to form $ClF_3$ but not $FCl_3$.

**10-77** Chlorine reacts with base to form the hypochlorite ion:

$$Cl_2(aq) + 2\,OH^-(aq) \longrightarrow Cl^-(aq) + OCl^-(aq) + H_2O(l)$$

Use this information to explain why people who make the mistake of mixing Chlorox with toilet bowl cleaners that contain hydrochloric acid often suffer damage to their lungs from breathing chlorine gas.

### The Inorganic Chemistry of Carbon

**10-78** Use the structure of graphite to explain why the bonds between carbon atoms are so strong that it is difficult to boil off individual carbon atoms, and yet the material is so soft it can be used as a lubricant.

**10-79** Explain why silicon forms a covalent carbide, but calcium forms an ionic carbide.

**10-80** Write balanced equations for the combustion of both CO and $H_2$ that explain why a mixture of these gases can be used as a fuel.

**10-81** Use Lewis structures to explain the following reaction:

$$CO_2(g) + H_2O(l) \longrightarrow H_2CO_3(aq)$$

### Integrated Problems

**10-82** A recent catalog listed the following prices: \$21.40 for 450 grams of sodium, \$18.00 for 1 kilogram of zinc, and \$52.80 for 250 grams of sodium hydride. Which reagent would be the least expensive source of one liter of $H_2$ gas?

**10-83** A solution of hydrogen peroxide in water that is 30% hydrogen peroxide by weight sells for \$15.95 per 500 mL, and potassium chlorate sells for \$12.75 per 500 grams. Is it cheaper to generate oxygen by decomposing $H_2O_2$ or by decomposing $KClO_3$?

**10-84** Write a sequence of reactions for the conversion of elemental nitrogen into nitric acid. Calculate the weight of nitric acid that can be produced from a ton of nitrogen gas.

**10-85** At 1700°C, $P_4$ molecules decompose partially to form $P_2$.

$$P_4(g) \longrightarrow 2\,P_2(g)$$

If the average molecular weight of phosphorus at that temperature is 91 g/mol, what fraction of the $P_4$ molecules decomposes?

**10-86** Uranium reacts with fluorine to form $UF_6$, which boils at 51°C. The relative rate of diffusion of $^{235}UF_6$ and $^{238}UF_6$ in the gas phase was used in the Manhattan project to separate the more abundant $^{238}U$ isotope from $^{235}U$. Predict which substance diffuses more rapidly, and calculate the ratio of the rate of diffusion of these compounds.

A spruce forest damaged by acid rain in Norway.

# ACIDS AND BASES

One of the guiding principles behind the development of chemistry was the idea that the behavior of the elements and their compounds could be explained in terms of pairs of opposites, such as metal versus nonmetal, positive versus negative, and so on. This chapter examines one of these pairs of opposites: acid versus base. This discussion answers questions such as the following:

## POINTS OF INTEREST

- Why do lemon juice, vinegar, and sour milk all taste sour?
- Why can we use vinegar, which contains acetic acid, in salad dressing? What would happen if we tried to use hydrochloric acid instead?
- What is the difference between $SO_2$ and CaO that makes one of these compounds an important contributor to acid rain and the other a way of fighting the consequences of acid rain?
- Why does the air in chemistry labs become cloudy when concentrated solutions of HCl and $NH_3$ are used in the same experiment?
- Why does HCl react with NaOH to form NaCl? Why doesn't NaCl dissolve in water to form a mixture of HCl and NaOH when we add table salt to water while cooking?

## 11.1 THE PROCESS OF DISCOVERY: ACIDS AND BASES

For more than 300 years, chemists have classified substances that behave like vinegar as **acids.** Substances that have properties like wood ash, on the other hand, have been classified as **alkalies** (or **bases**). In 1661 Robert Boyle summarized the properties of acids as follows:

1. Acids have a sour taste.
2. Acids are corrosive.
3. Acids change the color of certain vegetable dyes, such as litmus, from blue to red.
4. Acids lose their acidity when they are combined with alkalies.

The name ''acid'' comes from the Latin *acidus,* which means ''sour,'' and refers to the sharp odor and sour taste of many acids. Vinegar, for example, tastes sour because it is a dilute solution of acetic acid in water. Lemon juice tastes sour because it contains citric acid. Milk turns sour when it spoils because lactic acid is formed, and the unpleasant, sour odor of rotten meat or butter can be attributed to compounds such as butyric acid that form when fat spoils.

One of the characteristic properties of acids is their ability to dissolve most metals. Zinc metal, for example, rapidly dissolves in hydrochloric acid to form an aqueous solution of $ZnCl_2$ and hydrogen gas:

$$Zn(s) + 2\,HCl(aq) \longrightarrow Zn^{2+}(aq) + 2\,Cl^-(aq) + H_2(g)$$

Another characteristic property of acids is their ability to change the color of vegetable dyes, such as litmus or ''syrup of violets.'' Litmus is a mixture of blue dyes extracted from several species of lichens native to the Netherlands, which turns red in the presence of acid. Litmus has been used to test for acids for at least 300 years.

In 1661 Boyle summarized the properties of alkalies as follows:

1. Alkalies feel slippery.
2. Alkalies change the color of litmus from red to blue.
3. Alkalies become less alkaline when they are combined with acids.

In essence, Boyle defined alkalies as substances that consume, or neutralize, acids. Acids lose their characteristic sour taste and ability to dissolve metals when they are mixed with alkalies. Alkalies even reverse the change in color that occurs when litmus comes in contact with an acid. Eventually alkalies became known as *bases* because they serve as the ''base'' for making certain salts.

Zinc metal rapidly dissolves in concentrated hydrochloric acid to give an aqueous solution of $Zn^{2+}$ ions and $H_2$ gas.

Vegetable dyes, such as litmus, have been used for more than 300 years to distinguish between acids and bases.

## 11.2 THE ARRHENIUS DEFINITION OF ACIDS AND BASES

Boyle's definitions left an important question unanswered: What factors determine whether a compound is an acid or a base? The first step toward answering this question was taken by Antoine Lavoisier, who found the element oxygen in every acid he analyzed.

Oxygen occurs in a variety of common acids, including acetic acid ($CH_3CO_2H$), boric acid ($H_3BO_3$), carbonic acid ($H_2CO_3$), nitric acid ($HNO_3$), phosphoric acid ($H_3PO_4$), and sulfuric acid ($H_2SO_4$). Lavoisier believed that all acids must contain oxygen and named this element from the Greek stems meaning "acid former."

In 1811 Humphry Davy raised an important objection to Lavoisier's theory of acids when he showed that hydrochloric acid does not contain oxygen. By 1830 at least ten more acids that do not contain oxygen had been discovered, including aqueous solutions of HF, HBr, HI, HCN, HSCN, and $H_2S$.

In 1838 Justig Liebig proposed an alternative to Lavoisier's oxygen theory of acids. Liebig suggested that acids contain one or more hydrogen atoms that can be replaced by metal atoms. According to his theory, HSCN is an acid because it contains a hydrogen atom that can be replaced by a metal atom to form a salt, such as NaSCN. Similarly, $H_2S$ is an acid because it contains hydrogen atoms that can be replaced to form salts such as NaSH or $Na_2S$.

Liebig's theory provided a way to recognize acids, but it didn't explain what gives acids their characteristic properties or why acids have so many properties in common. The first step toward answering these questions occurred in 1884 when Svante Arrhenius suggested that salts such as NaCl dissociate when they dissolve in water to give particles he called **ions:**

$$\mathrm{NaCl}(s) \xrightarrow{\mathrm{H_2O}} \mathrm{Na^+}(aq) + \mathrm{Cl^-}(aq)$$

Three years later Arrhenius extended this theory by suggesting that acids are neutral compounds that *ionize* when they dissolve in water to give $H^+$ ions and a corresponding negative ion. According to his theory, hydrogen chloride is an acid because it ionizes when it dissolves in water to give hydrogen ($H^+$) and chloride ($Cl^-$) ions (see Figure 11.1).

$$\mathrm{HCl}(g) \xrightarrow{\mathrm{H_2O}} \mathrm{H^+}(aq) + \mathrm{Cl^-}(aq)$$

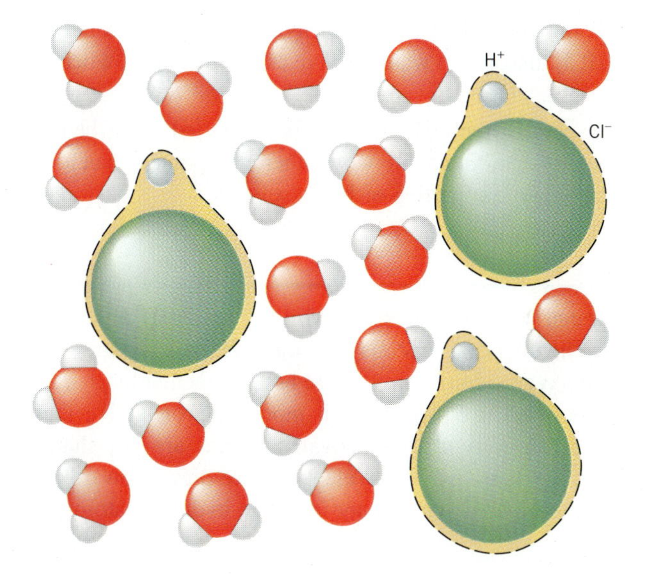


**Figure 11.1** The Arrhenius theory assumes that HCl dissociates into $H^+$ and $Cl^-$ ions when it dissolves in water.

Arrhenius argued that bases are neutral compounds that either dissociate or ionize in water to give $OH^-$ ions and a positive ion. NaOH is an Arrhenius base because it dissociates in water to give the hydroxide ($OH^-$) and sodium ($Na^+$) ions:

$$NaOH(s) \xrightarrow{H_2O} Na^+(aq) + OH^-(aq)$$

An **Arrhenius acid** is therefore any substance that ionizes when it dissolves in water to give the $H^+$, or hydrogen, ion. An **Arrhenius base** is any substance that gives the $OH^-$, or hydroxide, ion when it dissolves in water. Arrhenius acids include compounds such as HCl, HCN, and $H_2SO_4$ that ionize in water to give the $H^+$ ion. Arrhenius bases include ionic compounds that contain the $OH^-$ ion, such as NaOH, KOH, and $Ca(OH)_2$.

**CHECKPOINT**

"Milk of magnesia" is a suspension of MgO in water. It is often used to neutralize excess stomach acid and is therefore a base. In spite of this, MgO does not fit the Arrhenius definition of a base. Explain why.

Phillips Milk of Magnesia.

The Arrhenius theory explains why Liebig's definition of an acid was successful. Compounds that ionize in water to give the $H^+$ ion must contain hydrogen with an oxidation number of +1. These compounds therefore contain a hydrogen atom that can be replaced by a positively charged metal ion, such as the $Na^+$ or $K^+$ ion. This theory also explains why acids have similar properties: The characteristic properties of acids result from the presence of the $H^+$ ion generated when an acid dissolves in water. It also explains why acids neutralize bases and vice versa. Acids provide the $H^+$ ion; bases provide the $OH^-$ ion; and these ions combine to form water:

$$H^+(aq) + OH^-(aq) \longrightarrow H_2O(l)$$

The Arrhenius theory was a major step toward our present understanding of acids and bases, but it has several disadvantages.

1. It can be applied only to reactions that occur in water, because it defines acids and bases in terms of what happens when compounds dissolve in water.
2. It doesn't explain why some compounds in which hydrogen has an oxidation number of +1 (such as HCl) dissolve in water to give acidic solutions, whereas others (such as $CH_4$) do not.
3. Only the compounds that contain the $OH^-$ ion can be classified as Arrhenius bases. The Arrhenius theory can't explain why other compounds (such as $Na_2CO_3$) have the characteristic properties of bases.

## 11.3 AN OPERATIONAL DEFINITION OF ACIDS AND BASES

The Lewis structure of water can help us understand why $H^+$ and $OH^-$ ions play such an important role in the chemistry of aqueous solutions. This Lewis structure suggests that the hydrogen and oxygen atoms in water are bound together by sharing a pair of electrons. But oxygen ($EN = 3.44$) is much more electronegative than hydrogen ($EN = 2.20$), so the electrons in these covalent bonds are not shared equally by the hydrogen and oxygen atoms. These electrons are drawn toward the oxygen atom in the center of the molecule and away from the hydrogen atoms on either end. As a result, the water molecule is **polar.** The oxygen atom carries a partial negative charge ($\delta-$), and the hydrogen atoms carry a partial positive charge ($\delta+$):

Sec. 8.6

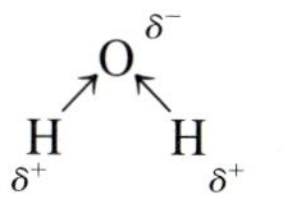


When they dissociate to form ions, water molecules therefore form a positively charged $H^+$ ion and a negatively charged $OH^-$ ion:

$$H{-}\ddot{\underset{\cdot\cdot}{O}}{-}H \rightleftharpoons H^+ + :\ddot{\underset{\cdot\cdot}{O}}{-}H^-$$

The opposite reaction can also occur: $H^+$ ions can combine with $OH^-$ ions to form neutral water molecules:

$$H^+ + OH^- \longrightarrow H_2O$$

The fact that water molecules dissociate to form $H^+$ and $OH^-$ ions, which can then recombine to form water molecules, is indicated by the following equation:

$$H_2O(l) \rightleftharpoons H^+(aq) + OH^-(aq)$$

The pair of arrows that separate the ''reactants'' and the ''products'' of this reaction indicate that the reaction occurs in both directions.

## TO WHAT EXTENT DOES WATER DISSOCIATE TO FORM IONS?

At 25°C, the density of water is 0.9971 g/cm$^3$, or 0.9971 g/mL. The concentration of water is therefore 55.35 *M*:

$$\frac{0.9971\ \cancel{g\ H_2O}}{1\ \cancel{mL}} \times \frac{1000\ \cancel{mL}}{1\ L} \times \frac{1\ mol\ H_2O}{18.015\ \cancel{g\ H_2O}} = \mathbf{55.35\ mol\ H_2O/L}$$

The concentration of the $H^+$ and $OH^-$ ions formed by the dissociation of neutral $H_2O$ molecules at this temperature is only $1.0 \times 10^{-7}$ mol/L. The ratio of the concentration of the $H^+$ (or $OH^-$) ion to the concentration of the neutral $H_2O$ molecules is therefore $1.8 \times 10^{-9}$:

$$\frac{1.0 \times 10^{-7}\ M\ H^+}{55.35\ M\ H_2O} = \mathbf{1.8 \times 10^{-9}}$$

In other words, only about 2 parts per billion (ppb) of the water molecules dissociate into ions at room temperature.

It is difficult to imagine what 2 ppb means. One way to visualize this number is to assume that 2500 letters appear on a typical page of this book. If typographical errors occurred with a frequency of 2 ppb, this book would have to be 400,000 pages long to contain two of these errors. Figure 11.2 shows a model of 20 water molecules, one of which has dissociated to form a pair of $H^+$ and $OH^-$ ions. If this illustration were a very high resolution photograph of the structure of water, we would encounter a pair of $H^+$ and $OH^-$ ions on the average of only once for every 25 million such photographs.

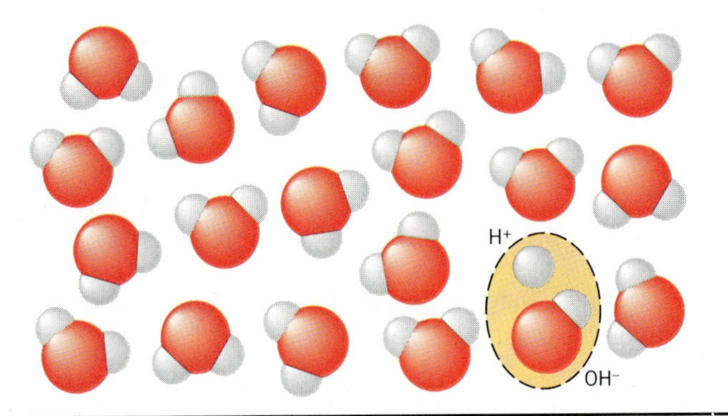


**Figure 11.2** If this figure is thought of as a high-resolution snapshot of 20 $H_2O$ molecules, one of which has dissociated to form $H^+$ and $OH^-$ ions, we would have to look at an average of 25 million such snapshots before we encountered another pair of these ions.

## THE OPERATIONAL DEFINITION OF ACIDS AND BASES

The fact that water dissociates to form $H^+$ and $OH^-$ ions in a reversible reaction is the basis for an operational definition of acids and bases that is more powerful than the definitions proposed by Arrhenius. In an operational sense, an acid is any substance that increases the concentration of the $H^+$ ion when it dissolves in water. A base is any substance that increases the concentration of the $OH^-$ ion when it dissolves in water.

These definitions tie the theory of acids and bases to a simple laboratory test for acids and bases. To decide whether a compound is an acid or a base, we dissolve it in water and test the solution to see whether the $H^+$ or $OH^-$ ion concentration has increased.

$CO_2$ dissolves in water to give a solution that tests acid. Even though CaO and MgO aren't very soluble in water, their solutions still test basic.

There is a subtle difference between the operational definitions of acids and bases and the Arrhenius definitions given in Section 11.2. $CO_2$, for example, is not an Arrhenius acid because it can't dissociate to give the $H^+$ ion in water. $CO_2$ is an acid, however, according to the operational definition, because it reacts with water to form a compound that dissociates to increase the $H^+$ ion concentration:

$$CO_2(g) + H_2O(l) \rightleftharpoons H_2CO_3(aq)$$
$$H_2CO_3(aq) \rightleftharpoons H^+(aq) + HCO_3^-(aq)$$

The operational definition expands the number of compounds that can be called bases to include salts that react with water to form the $OH^-$ ion.

$$CaO(s) + H_2O(l) \rightleftharpoons Ca^{2+}(aq) + 2\,OH^-(aq)$$

## 11.4 TYPICAL ACIDS AND BASES

The chemistry of acids and bases is not the first time we have encountered substances that have opposite properties (see Section 2.7). It is therefore tempting to assume that the properties of acids and bases result from differences between the chemistry of metals and nonmetals. To test this hypothesis, let's look at three categories of compounds: hydrides, oxides, and hydroxides.

Sec. 2.7

Compounds that contain hydrogen bound to a nonmetal are called *nonmetal hydrides.* Because they contain hydrogen in the +1 oxidation state, these compounds can act as a source of the $H^+$ ion in water:

$$HCl(g) \xrightarrow{H_2O} H^+(aq) + Cl^-(aq)$$

$$H_2S(g) \xrightarrow{H_2O} H^+(aq) + HS^-(aq)$$

*Metal hydrides,* on the other hand, contain hydrogen bound to a metal. Because these compounds contain hydrogen in a −1 oxidation state, they dissociate in water to give the $H^-$ (or hydride) ion:

$$NaH(s) \xrightarrow{H_2O} Na^+(aq) + H^-(aq)$$

The $H^-$ ion, with its pair of valence electrons, can abstract an $H^+$ ion from a water molecule.

$$H:^- + H\text{—}\ddot{\underset{..}{O}}\text{—}H \longrightarrow H_2 + :\ddot{\underset{..}{O}}\text{—}H^-$$

Since removing $H^+$ ions from water molecules is one way to increase the $OH^-$ ion concentration in a solution, metal hydrides are bases:

$$NaH(s) + H_2O(l) \longrightarrow Na^+(aq) + OH^-(aq) + H_2(g)$$

$$CaH_2(s) + 2\ H_2O(l) \longrightarrow Ca^{2+}(aq) + 2\ OH^-(aq) + 2\ H_2(g)$$

A similar pattern can be found in the chemistry of the oxides formed by metals and nonmetals. *Nonmetal oxides* dissolve in water to form acids. $CO_2$ dissolves in water to give carbonic acid, $SO_3$ gives sulfuric acid, and $P_4O_{10}$ reacts with water to give phosphoric acid:

$$CO_2(g) + H_2O(l) \longrightarrow H_2CO_3(aq)$$

$$SO_3(g) + H_2O(l) \longrightarrow H_2SO_4(aq)$$

$$P_4O_{10}(s) + 6\ H_2O(l) \longrightarrow 4\ H_3PO_4(aq)$$

*Metal oxides,* on the other hand, are bases. Metal oxides formally contain the $O^{2-}$ ion, which reacts with water to give a pair of $OH^-$ ions:

$$O^{2-}(aq) + H_2O(l) \longrightarrow 2\ OH^-(aq)$$

Metal oxides therefore fit the operational definition of a base:

$$CaO(s) + H_2O(l) \longrightarrow Ca^{2+}(aq) + 2\ OH^-(aq)$$

We see the same pattern in the chemistry of compounds that contain the —OH, or hydroxide, group. *Metal hydroxides,* such as LiOH, NaOH, KOH, and $Ca(OH)_2$, are bases:

$$NaOH(s) \xrightarrow{H_2O} Na^+(aq) + OH^-(aq)$$

*Nonmetal hydroxides,* such as hypochlorous acid (HOCl), are acids:

$$HOCl(aq) \longrightarrow H^+(aq) + OCl^-(aq)$$

Table 11.1 summarizes the trends observed in these three categories of compounds. Metal hydrides, metal oxides, and metal hydroxides are bases. Nonmetal hydrides, nonmetal oxides, and nonmetal hydroxides are acids.

**TABLE 11.1 Typical Acids and Bases**

| Acids | Bases |
|---|---|
| Nonmetal hydrides | Metal hydrides |
| HF, HCl, HBr, HI, | LiH, NaH, KH, |
| HCN, HSCN, $H_2S$ | $MgH_2$, $CaH_2$ |
| Nonmetal oxides | Metal oxides |
| $CO_2$, $SO_2$, $SO_3$, | $Li_2O$, $Na_2O$, $K_2O$, |
| $NO_2$, $P_4O_{10}$ | MgO, CaO |
| Nonmetal hydroxides | Metal hydroxides |
| HOCl, $HONO_2$, | LiOH, NaOH, KOH, |
| $O_2S(OH)_2$, $OP(OH)_3$ | $Ca(OH)_2$, $Ba(OH)_2$ |

The *nonmetal hydroxides* in Table 11.1, which include such common acids as $H_2SO_4$, $H_3PO_4$, and $HNO_3$, are a potential source of confusion. The formulas for these compounds may give the impression that the acidic hydrogen atoms are bound to the nitrogen, sulfur, or phosphorus atoms, but they aren't. In each of these compounds, the acidic hydrogen is attached to an *oxygen* atom. These compounds are therefore all examples of **oxyacids.**

Sec. 10.30

Skeleton structures for eight oxyacids are given in Figure 11.3. In theory, the formulas of these compounds should be written as $O_2S(OH)_2$, $OP(OH)_3$, $HONO_2$, $HOClO_3$, $OC(OH)_2$, and so on, to indicate their skeleton structures. In practice, these skeleton formulas can't compete with the well-established formulas by which

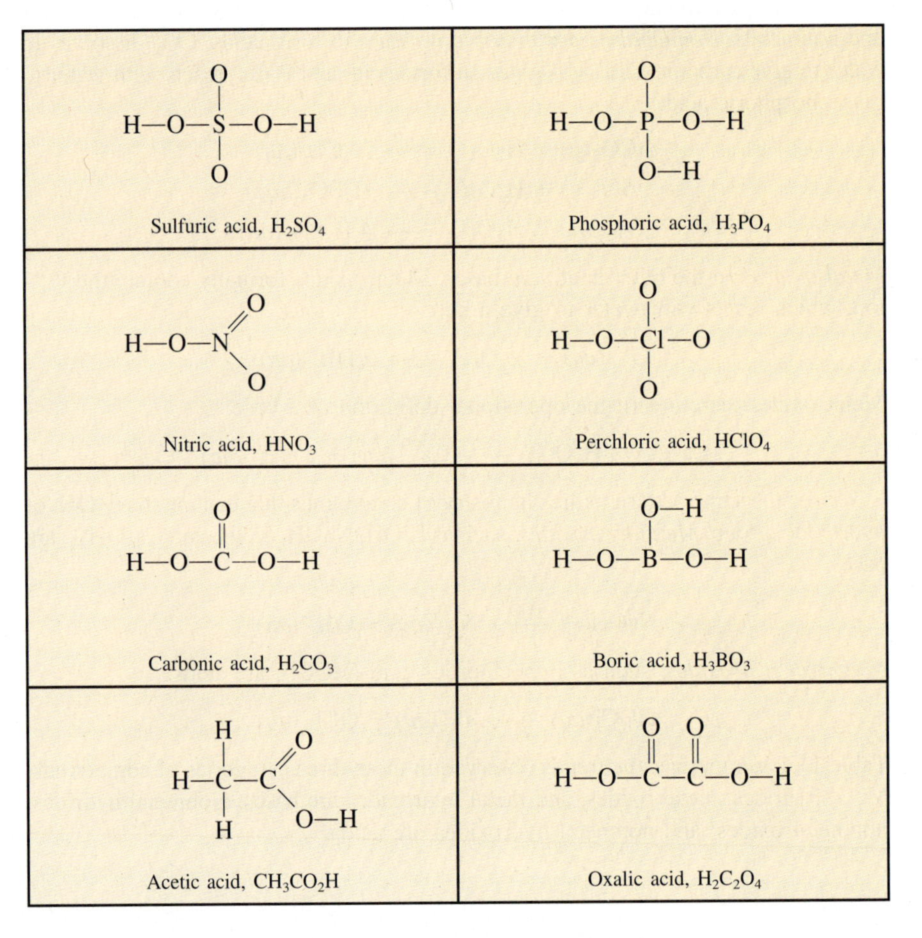


**Figure 11.3** The skeleton structures for some of the common nonmetal hydroxides, or oxyacids.

the acids have long been known: $H_2SO_4$, $H_3PO_4$, $HNO_3$, $HClO_4$, and $H_2CO_3$. As a general rule, acids that contain oxygen have skeleton structures in which the acidic hydrogens are attached to oxygen atoms.

**EXERCISE 11.1**

Use Lewis structures to classify the following acids as either nonmetal hydrides (*X*H) or nonmetal hydroxides (*X*OH):

(a) HCN (b) $HNO_3$ (c) $H_2C_2O_4$ (d) $CH_3CO_2H$

**SOLUTION** HCN is the only nonmetal hydride among these acids:

$$H—C \equiv N:$$

$HNO_3$, $H_2C_2O_4$, and $CH_3CO_2H$ are all nonmetal hydroxides, or oxyacids:

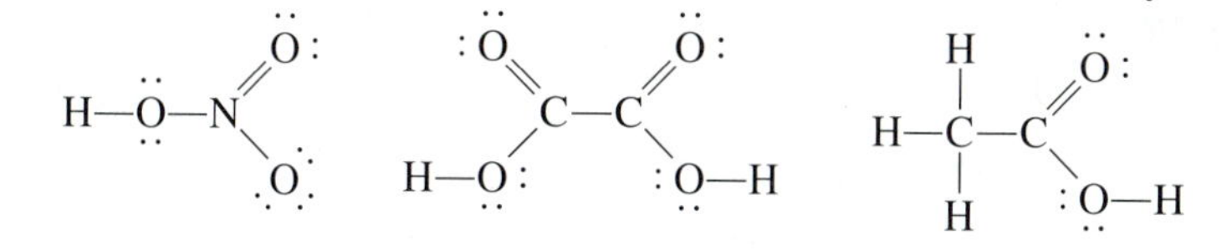

## 11.5 WHY ARE METAL HYDROXIDES BASES AND NONMETAL HYDROXIDES ACIDS?

To understand why nonmetal hydroxides are acids and metal hydroxides are bases, we have to look at the electronegativities of the atoms in these compounds. Let's start with a typical metal hydroxide: sodium hydroxide (see Figure 11.4). The differ-

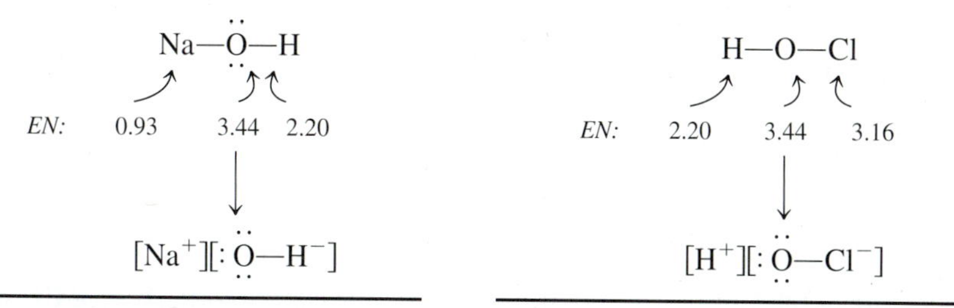


**Figure 11.4** Because the difference between the electronegativities of sodium and oxygen is relatively large, NaOH dissociates to form $Na^+$ and $OH^-$ ions.

**Figure 11.5** The difference between the electronegativities of chlorine and oxygen is very small. As a result, when HOCl dissociates, it forms $H^+$ and $OCl^-$ ions.

ence between the electronegativities of sodium and oxygen is very large ($\Delta EN$ = 2.51). As a result, the electrons in the Na—O bond are not shared equally; these electrons are drawn toward the more electronegative oxygen atom. NaOH therefore dissociates to give $Na^+$ and $OH^-$ ions when it dissolves in water:

$$NaOH(s) \xrightarrow{H_2O} Na^+(aq) + OH^-(aq)$$

We get a very different pattern when we apply the same procedure to hypochlorous acid, HOCl, a typical nonmetal hydroxide (see Figure 11.5). Here, the difference between the electronegativities of the chlorine and oxygen atoms is small ($\Delta EN$ = 0.28). As a result, the electrons in the Cl—O bond are shared more or less equally by the two atoms. The O—H bond, on the other hand, is polar ($\Delta EN$ = 1.24); the electrons in this bond are drawn toward the more electronegative oxygen atom. When this molecule ionizes, the electrons in the O—H bond remain with the oxygen atom, and $OCl^-$ and $H^+$ ions are formed:

$$HOCl(aq) \longrightarrow H^+(aq) + OCl^-(aq)$$

Sec. 7.17

There is no abrupt change from metal to nonmetal across a row or down a column of the periodic table. We should therefore expect to find compounds that lie between the extremes of metal and nonmetal oxides, or metal and nonmetal hydroxides. These compounds, such as $Al_2O_3$ and $Al(OH)_3$, are called **amphoteric** (literally, "either or both") because they can act as either acids or bases. $Al(OH)_3$, for example, acts as an acid when it reacts with a base:

$$\underset{\text{acid}}{Al(OH)_3(s)} + \underset{\text{base}}{OH^-(aq)} \longrightarrow Al(OH)_4^-(aq)$$

Conversely, it acts as a base when it reacts with an acid:

$$\underset{\text{base}}{Al(OH)_3(s)} + \underset{\text{acid}}{3\ H^+} \longrightarrow Al^{3+}(aq) + 3\ H_2O(l)$$

---

**CHECKPOINT**

In the periodic table, where are we most likely to find the elements that form amphoteric hydroxides?

---

## 11.6 THE BRØNSTED DEFINITION OF ACIDS AND BASES

In 1923 Johannes Brønsted and Thomas Lowry independently proposed a more powerful set of definitions of acids and bases. The Brønsted, or Brønsted–Lowry, model is based on a simple assumption:

**Acids donate $H^+$ ions to another ion or molecule, which acts as a base.**

The dissociation of water, for example, involves the transfer of an $H^+$ ion from one water molecule to another to form $H_3O^+$ and $OH^-$ ions:

$$2\ H_2O(l) \rightleftharpoons H_3O^+(aq) + OH^-(aq)$$

According to this model, HCl doesn't dissociate in water to form $H^+$ and $Cl^-$ ions. Instead, an $H^+$ ion is transferred from HCl to a water molecule to form $H_3O^+$ and $Cl^-$ ions, as shown in Figure 11.6:

$$HCl(g) + H_2O(l) \rightleftharpoons H_3O^+(aq) + Cl^-(aq)$$

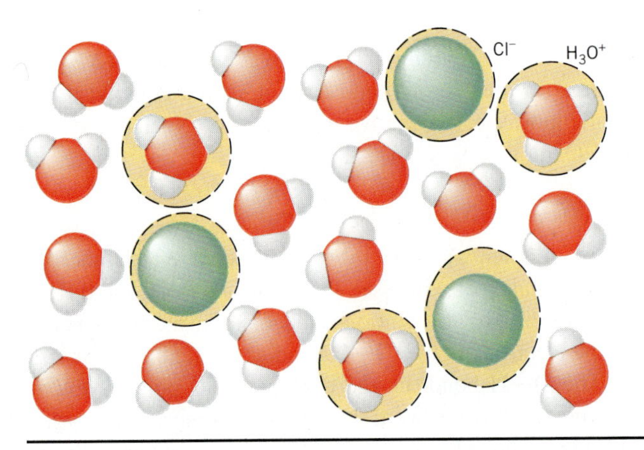


**Figure 11.6** The Brønsted theory assumes that HCl molecules donate an $H^+$ ion to water molecules to form $H_3O^+$ and $Cl^-$ ions when they dissolve in water.

Because it is a proton, an $H^+$ ion is several orders of magnitude smaller than the smallest atom. As a result, the charge on an isolated $H^+$ ion is distributed over such a small amount of space that this $H^+$ ion is attracted toward any source of negative charge that exists in the solution. Thus, the instant that an $H^+$ ion is created in an aqueous solution, it bonds to a water molecule. The Brønsted model, in which $H^+$ ions are transferred from one ion or molecule to another, therefore reflects reality better than the Arrhenius theory, which assumes that $H^+$ ions exist in aqueous solution.

Even the Brønsted model is naive, however. Each $H^+$ ion that an acid donates to water is actually bound to four neighboring water molecules, as shown in Figure 11.7. A more realistic formula for the substance produced when an acid loses an $H^+$ ion is therefore $H(H_2O)_4{}^+$, or $H_9O_4{}^+$. For all practical purposes, however, this substance can be represented as the $H_3O^+$ ion.

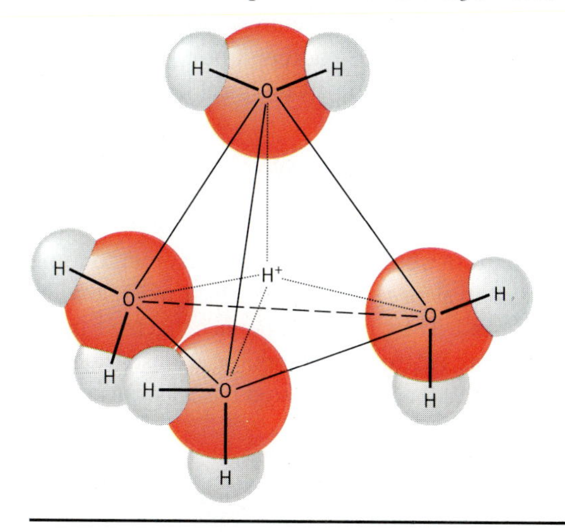


**Figure 11.7** The structure of the $H(H_2O)_4{}^+$ ion formed when an acid reacts with water. For practical purposes, this ion can be thought of as an $H_3O^+$ ion.

The reaction between HCl and water provides the basis for understanding the definitions of a Brønsted acid and a Brønsted base. According to this theory, an $H^+$ ion is transferred from an HCl molecule to a water molecule when HCl dissociates in water:

$$HCl(g) + H_2O(l) \rightleftharpoons H_3O^+(aq) + Cl^-(aq)$$

HCl acts as an $H^+$-ion donor in this reaction, and $H_2O$ acts as an $H^+$-ion acceptor. A **Brønsted acid** is therefore any substance (such as HCl) that can donate an $H^+$ ion to a base. A **Brønsted base** is any substance (such as $H_2O$) that can accept an $H^+$ ion from an acid.

We have two ways of naming the $H^+$ ion. Some chemists call it a hydrogen ion; others call it a proton. As a result, Brønsted acids are known as either **hydrogen-ion donors** or **proton donors.** Brønsted bases are **hydrogen-ion acceptors** or **proton acceptors.**

From the perspective of the Brønsted model, reactions between acids and bases always involve the transfer of an $H^+$ ion from a proton donor to a proton acceptor. Acids can be neutral molecules:

$$\underset{\text{acid}}{HCl(aq)} + \underset{\text{base}}{NH_3(aq)} \rightleftharpoons Cl^-(aq) + NH_4{}^+(aq)$$

They also can be positive ions:

$$\underset{\text{acid}}{NH_4^+(aq)} + \underset{\text{base}}{OH^-(aq)} \rightleftharpoons NH_3(aq) + H_2O(l)$$

or negative ions:

$$\underset{\text{acid}}{H_2PO_4^-(aq)} + \underset{\text{base}}{H_2O(l)} \rightleftharpoons HPO_4^{2-}(aq) + H_3O^+(aq)$$

The Brønsted theory therefore expands the number of potential acids. It also allows us to decide which compounds are acids from their chemical formulas. Any compound that contains hydrogen with an oxidation number of +1 can be an acid. Brønsted acids therefore include HCl, $H_2S$, $H_2CO_3$, $H_2PtF_6$, $NH_4^+$, $HSO_4^-$, and $HMnO_4$.

Brønsted bases can be identified from their Lewis structures. According to the Brønsted model, a base is any ion or molecule that can accept a proton. To understand the implications of this definition, look at how the prototypical base, the $OH^-$ ion, accepts a proton:

$$H^+ + :\ddot{\underset{..}{O}}\text{—}H^- \rightleftharpoons H\text{—}\ddot{\underset{..}{O}}\text{—}H$$

The only way to accept an $H^+$ ion is to form a covalent bond to it. In order to form a covalent bond to an $H^+$ ion, which has no valence electrons, the base must provide both of the electrons needed to form the bond. Thus, only compounds that have pairs of nonbonding valence electrons can act as $H^+$-ion acceptors, or Brønsted bases. The following compounds, for example, can all act as Brønsted bases because they all contain nonbonding pairs of electrons.

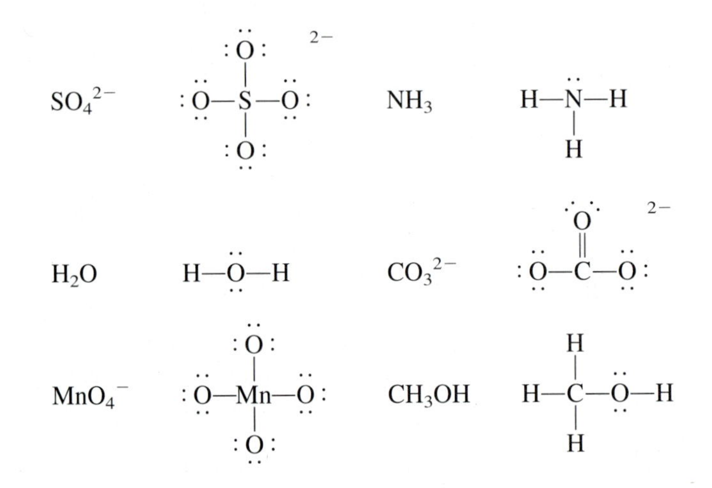


The only bases in the Arrhenius model are compounds such as NaOH or $Ca(OH)_2$ that contain the $OH^-$ ion. The operational definition expands this list to include compounds such as NaH or $Na_2CO_3$ that react with water to give off the $OH^-$ ion. But the Brønsted model expands the list of potential bases to include any ion or molecule that contains one or more pairs of nonbonding valence electrons. The Brønsted definition of a base applies to so many ions and molecules that it is almost easier to count substances, such as the following, that can't be Brønsted bases because they don't have pairs of nonbonding valence electrons:

$BeH_2$ H—Be—H

$H_2$ H—H

$CH_4$ (H—C—H with H above and below C)

$NH_4^+$ [H—N—H with H above and below N]$^+$

Many common cleaning agents, such as Windex™, which is an aqueous $NH_3$ solution, test basic toward litmus.

## EXERCISE 11.2

Which of the following compounds can be Brønsted acids? Which can be Brønsted bases?

(a) $H_2O$ (b) $NH_3$ (c) $HSO_4^-$ (d) $OH^-$

**SOLUTION** All four of these compounds can be Brønsted acids because they all contain hydrogen with an oxidation number of +1. Furthermore, all four compounds can be Brønsted bases because they all contain at least one pair of nonbonding valence electrons:

H—Ö—H (two lone pairs on O)  H—N̈—H with H below N  [H—Ö—S—Ö: with :Ö: above and below S]$^-$  [:Ö—H]$^-$

## 11.7 THE ROLE OF WATER IN THE BRØNSTED THEORY

The Brønsted theory explains water's role in acid–base reactions:

1. Water dissociates to form ions by transferring an $H^+$ ion from one molecule acting as an acid to another molecule acting as a base:

$$\underset{\text{acid}}{H_2O(l)} + \underset{\text{base}}{H_2O(l)} \rightleftharpoons H_3O^+(aq) + OH^-(aq)$$

2. Acids react with water by donating an $H^+$ ion to a neutral water molecule to form the $H_3O^+$ ion:

$$\underset{\text{acid}}{HCl(g)} + \underset{\text{base}}{H_2O(l)} \rightleftharpoons H_3O^+(aq) + Cl^-(aq)$$

3. Bases react with water by accepting an $H^+$ ion from a water molecule to form the $OH^-$ ion:

$$\underset{\text{base}}{NH_3(aq)} + \underset{\text{acid}}{H_2O(l)} \rightleftharpoons NH_4^+(aq) + OH^-(aq)$$

4. Water molecules can act as intermediates in acid–base reactions by gaining $H^+$ ions from the acid:

$$HCl(g) + H_2O(l) \rightleftharpoons H_3O^+(aq) + Cl^-(aq)$$

and then losing these $H^+$ ions to the base:

$$NH_3(aq) + H_3O^+(aq) \rightleftharpoons NH_4^+(aq) + H_2O(l)$$

The Brønsted model can be extended to acid–base reactions in other solvents. For example, there is a small tendency in liquid ammonia for an $H^+$ ion to be transferred from one $NH_3$ molecule to another to form the $NH_4^+$ and $NH_2^-$ ions:

$$2\,NH_3 \rightleftharpoons NH_4^+ + NH_2^-$$

By analogy to the chemistry of aqueous solutions, we conclude that acids in liquid ammonia include any source of the $NH_4^+$ ion and that bases include any source of the $NH_2^-$ ion.

**CHECKPOINT**

Explain why $NH_4Cl$ is an acid and $NaNH_2$ is a base when these compounds are dissolved in liquid ammonia.

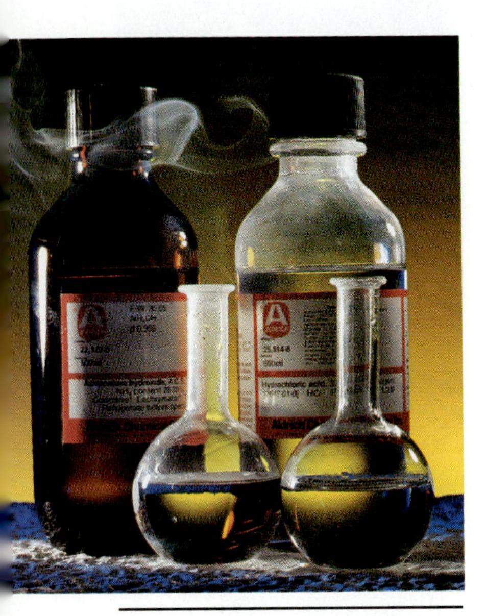

One advantage of the Brønsted theory is the ease with which this model can be extended to include acid-base reactions that do not occur in water, such as the formation of a white cloud of $NH_4Cl(s)$ when a bottle of concentrated $HCl(aq)$ is held close to a bottle of concentrated $NH_3(aq)$.

The Brønsted model can be extended even to reactions that do not occur in solution. A classic example of a gas-phase acid–base reaction is encountered when open containers of concentrated hydrochloric acid and aqueous ammonia are held next to each other. A white cloud of ammonium chloride soon forms as the HCl gas that escapes from one solution reacts with the $NH_3$ gas from the other:

$$HCl(g) + NH_3(g) \rightleftharpoons NH_4Cl(s)$$

This reaction involves the transfer of an $H^+$ ion from HCl to $NH_3$ and is therefore a Brønsted acid–base reaction, even though it occurs in the gas phase.

## 11.8 CONJUGATE ACID–BASE PAIRS

An important consequence of the Brønsted theory is the recognition that acids and bases exist as **conjugate acid–base pairs.** The term *conjugate* comes from the Latin stems meaning ''joined together'' and refers to things that are joined, particularly in pairs. It is an ideal term to describe the relationship between Brønsted acids and bases.

Every time a Brønsted acid acts as an $H^+$-ion donor, it forms a conjugate base. Imagine a generic acid, HA. When this acid donates an $H^+$ ion to water, one product of the reaction is the $A^-$ ion, which is a hydrogen-ion acceptor, or Brønsted base:

$$\underset{\text{acid}}{HA} + H_2O \rightleftharpoons H_3O^+ + \underset{\text{base}}{A^-}$$

Conversely, every time a base gains an $H^+$ ion, the product is a Brønsted acid, HA:

$$\underset{\text{base}}{A^-} + H_2O \rightleftharpoons \underset{\text{acid}}{HA} + OH^-$$

Acids and bases in the Brønsted model therefore exist as conjugate pairs whose formulas are related by the gain or loss of a hydrogen ion.

Our use of the symbols HA and $A^-$ for a conjugate acid–base pair does not mean that all acids are neutral molecules or that all bases are negative ions. It signifies only that the acid contains an $H^+$ ion that isn't present in the conjugate base. As noted earlier, Brønsted acids or bases can be neutral molecules, positive ions, or negative ions. Various Brønsted acids and their conjugate bases are given in Table 11.2.

It is sometimes difficult to appreciate that a compound can be both a Brønsted acid and a Brønsted base. $H_2O$, $OH^-$, $HSO_4^-$, and $NH_3$, for example, can be found in both columns in Table 11.2. Water is the perfect example of this behavior because it simultaneously acts as an acid and as a base when it forms the $H_3O^+$ and $OH^-$ ions:

$$H_2O + H_2O \rightleftharpoons H_3O^+ + OH^-$$

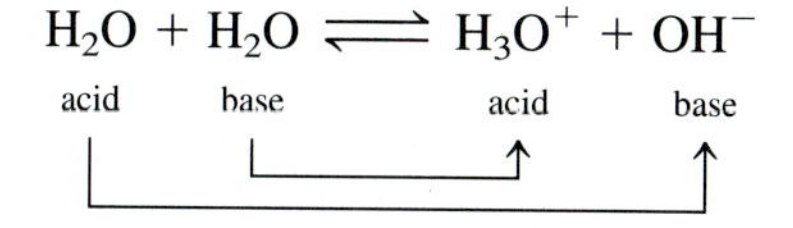


**TABLE 11.2 Typical Brønsted Acids and Their Conjugate Bases**

| Acid | Base |
|---|---|
| $H_3O^+$ | $H_2O$ |
| $H_2O$ | $OH^-$ |
| $OH^-$ | $O^{2-}$ |
| HCl | $Cl^-$ |
| $H_2SO_4$ | $HSO_4^-$ |
| $HSO_4^-$ | $SO_4^{2-}$ |
| $NH_4^+$ | $NH_3$ |
| $NH_3$ | $NH_2^-$ |

## 11.9 THE RELATIVE STRENGTHS OF ACIDS AND BASES

Many hardware stores sell ''muriatic acid,'' a 6 *M* solution of hydrochloric acid HCl(*aq*), to clean bricks and concrete. Grocery stores sell vinegar, which is a 1 *M* solution of acetic acid: $CH_3CO_2H$. Although both substances are acids, you wouldn't use muriatic acid in salad dressing, and vinegar is ineffective in cleaning bricks or concrete.

The difference between these acids is simple: Muriatic acid is a **strong acid** and vinegar is a **weak acid.** Muriatic acid is strong because it is very good at transferring an $H^+$ ion to a water molecule. In a 6 *M* solution of hydrochloric acid, 99.996% of the HCl molecules react with water to form $H_3O^+$ and $Cl^-$ ions:

$$HCl(aq) + H_2O(l) \rightleftharpoons H_3O^+(aq) + Cl^-(aq)$$

Only 0.004%–or 1 out of every 30,000—of the HCl molecules remain in solution. Muriatic acid is a strong acid indeed.

Vinegar is a weak acid because it is not very good at transferring $H^+$ ions to water. In a 1 *M* solution, less than 0.4% of the $CH_3CO_2H$ molecules reacts with water to form $H_3O^+$ and $CH_3CO_2^-$ ions:

$$CH_3CO_2H(aq) + H_2O(l) \rightleftharpoons H_3O^+(aq) + CH_3CO_2^-(aq)$$

More than 99.6% of the acetic acid molecules remain intact.

A bottle of muriatic acid, which is a strong acid.

A bottle of vinegar, which is a weak acid.

It would be useful to have a quantitative measure of the relative strengths of acids to replace the labels *strong* and *weak*. For reasons that are discussed in Chapter 17, the relative strength of acids is described in terms of an **acid-dissociation equilibrium constant,** $K_a$. To understand the nature of this equilibrium constant, let's assume that the reaction between an acid and water can be represented by the following generic equation:

$$HA(aq) + H_2O(l) \rightleftharpoons H_3O^+(aq) + A^-(aq)$$

In other words, we will assume that some of the HA molecules react to form $H_3O^+$ and $A^-$ ions, as shown in Figure 11.8. By convention, the concentrations of these ions in units of moles per liter are represented by their formulas in brackets: $[H_3O^+]$ and $[A^-]$. The concentration of the HA molecules that remain in solution is represented by [HA].

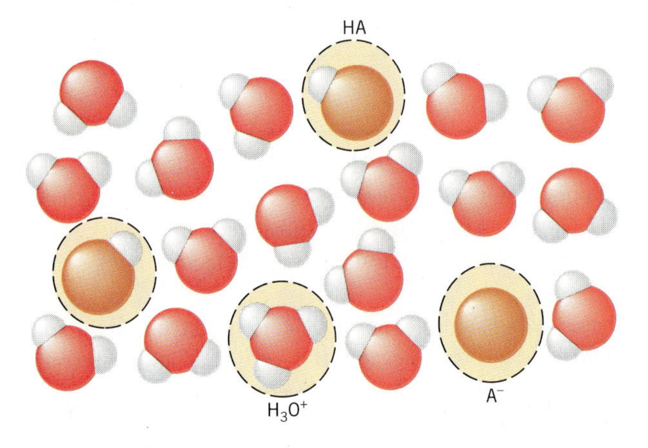


**Figure 11.8** Some, but not all, of the HA molecules in a typical acid react with water to form $H_3O^+$ ions and $A^-$ ions when the acid dissolves in water. In a strong acid, there are very few undissociated HA molecules. In a weak acid, most of the HA molecules remain.

$K_a$ is calculated from the following equation:

$$K_a = \frac{[H_3O^+][A^-]}{[HA]}$$

When a strong acid dissolves in water, the acid reacts extensively with water to form $H_3O^+$ and $A^-$ ions. (Only a small residual concentration of the HA molecules remains in solution.) The product of the concentrations of the $H_3O^+$ and $A^-$ ions is therefore much larger than the concentration of the HA molecules, so $K_a$ for a strong acid is greater than 1. Hydrochloric acid, for example, has a $K_a$ of roughly $1 \times 10^6$:

$$\frac{[H_3O^+][Cl^-]}{[HCl]} = 1 \times 10^6$$

Weak acids, on the other hand, react only slightly with water. The product of the concentrations of the $H_3O^+$ and $A^-$ ions is therefore smaller than the concentration of the residual HA molecules. As a result, $K_a$ for a weak acid is less than 1. Acetic acid, for example, has a $K_a$ of only $1.8 \times 10^{-5}$:

$$\frac{[H_3O^+][CH_3CO_2^-]}{[CH_3CO_2H]} = 1.8 \times 10^{-5}$$

$K_a$ therefore can be used to distinguish between strong acids and weak acids:

Strong acids: $K_a > 1$
Weak acids: $K_a < 1$

A list of common acids and their acid dissociation constants is given in Table 11.3.

**TABLE 11.3 Common Acids and Their Acid-Dissociation Equilibrium Constants**

| Acid Formulas | $K_a$ |
|---|---|
| Strong acids | |
| HI | $3 \times 10^9$ |
| HBr | $1 \times 10^9$ |
| HCl | $1 \times 10^6$ |
| $H_2SO_4$ | $1 \times 10^3$ |
| $HClO_4$ | $1 \times 10^3$ |
| $H_3O^+$ | 55 |
| $HNO_3$ | 28 |
| $H_2CrO_4$ | 10 |
| Intermediate acids | |
| $H_3PO_4$ | $7.1 \times 10^{-3}$ |
| HF | $7.2 \times 10^{-4}$ |
| Weak acids | |
| Citric acid | $7.5 \times 10^{-5}$ |
| $CH_3CO_2H$ | $1.8 \times 10^{-5}$ |
| $H_2S$ | $1.0 \times 10^{-7}$ |
| $H_2CO_3$ | $4.5 \times 10^{-7}$ |
| $H_3BO_3$ | $5.8 \times 10^{-10}$ |
| $H_2O$ | $1.8 \times 10^{-16}$ |

## 11.10 THE RELATIVE STRENGTHS OF CONJUGATE ACID–BASE PAIRS

In order to examine the relationship between the strength of an acid and its conjugate base, consider the implications of the fact that HCl is a strong acid. If HCl is a strong acid, it must be a good proton donor. HCl can only be a good proton donor, however, if the $Cl^-$ ion is a poor proton acceptor. Thus, the $Cl^-$ ion must be a weak base.

$$\underset{\text{strong acid}}{HCl(g)} + H_2O(l) \rightleftharpoons H_3O^+(aq) + \underset{\text{weak base}}{Cl^-(aq)}$$

Let's now consider the relationship between the strength of the ammonium ion ($NH_4^+$) and its conjugate base, ammonia ($NH_3$). The $NH_4^+$ ion is a weak acid because ammonia is a reasonably good base:

$$\underset{\text{weak acid}}{NH_4^+(aq)} + H_2O(l) \rightleftharpoons H_3O^+(aq) + \underset{\text{strong base}}{NH_3(aq)}$$

This discussion is summarized in the following general rules:

> ***Strong acids* have weak conjugate bases. *Strong bases* have weak conjugate acids.**

### EXERCISE 11.3

Use the data in Table 11.3 to predict whether the $CH_3CO_2^-$ ion or the $OH^-$ ion is the stronger base.

**SOLUTION** Table 11.3 contains the following data for the conjugate acids of these bases:

$$CH_3CO_2H: \quad K_a = 1.8 \times 10^{-5}$$
$$H_2O: \quad K_a = 1.8 \times 10^{-16}$$

These data suggest that acetic acid is a stronger acid than water:

$$CH_3CO_2H > H_2O$$

The conjugate base of acetic acid must therefore be weaker than the conjugate base of water:

$$CH_3CO_2^- < OH^-$$

Thus, the $CH_3CO_2^-$, or acetate, ion is a weaker base than the $OH^-$, or hydroxide, ion.

## 11.11 THE RELATIVE STRENGTHS OF PAIRS OF ACIDS AND BASES

Section 11.9 showed how the value of $K_a$ for an acid can be used to decide whether it is a strong acid or a weak acid, in an absolute sense. At times it is also useful to compare the relative strengths of a pair of acids to decide which is stronger. Consider HCl and the $H_3O^+$ ion, for example:

$$HCl: \quad K_a = 1 \times 10^6$$
$$H_3O^+ \quad K_a = 55$$

These $K_a$ values suggest that both are strong acids, but HCl is a stronger acid than the $H_3O^+$ ion.

Sec. 11.9

We can now understand why such a high proportion of the HCl molecules in an aqueous solution reacts with water to form $H_3O^+$ and $Cl^-$ ions. The Brønsted theory suggests that every acid–base reaction converts an acid into its conjugate base and a base into its conjugate acid.

$$\underset{\text{acid}}{HCl(g)} + \underset{\text{base}}{H_2O(l)} \rightleftharpoons \underset{\text{acid}}{H_3O^+(aq)} + \underset{\text{base}}{Cl^-(aq)}$$

Note that there are two acids and two bases in this reaction. The stronger acid, however, is on the left side of the equation:

$$\underset{\substack{\text{stronger}\\\text{acid}}}{HCl(g)} + H_2O(l) \rightleftharpoons \underset{\substack{\text{weaker}\\\text{acid}}}{H_3O^+(aq)} + Cl^-(aq)$$

What about the two bases: $H_2O$ and the $Cl^-$ ion? The general rules given in the previous section suggest that the stronger of a pair of acids must form the weaker of a pair of conjugate bases. The fact that HCl is a stronger acid than the $H_3O^+$ ion implies that the $Cl^-$ ion is a weaker base than water.

$$\text{Acid strength:} \quad HCl > H_3O^+$$
$$\text{Base strength:} \quad Cl^- < H_2O$$

Thus, the equation for the reaction between HCl and water can be written as follows:

$$\underset{\substack{\text{stronger}\\\text{acid}}}{HCl(g)} + \underset{\substack{\text{stronger}\\\text{base}}}{H_2O(l)} \rightleftharpoons \underset{\substack{\text{weaker}\\\text{acid}}}{H_3O^+(aq)} + \underset{\substack{\text{weaker}\\\text{base}}}{Cl^-(aq)}$$

It isn't surprising that 99.996% of the HCl molecules in a 6 $M$ solution react with water to give $H_3O^+$ ions and $Cl^-$ ions. The stronger of a pair of acids should react with the stronger of a pair of bases to form a weaker acid and a weaker base.

---

**CHECKPOINT**

Use the fact that the conjugate base of a strong acid is always a weak base and the conjugate acid of a strong base is always a weak acid to explain why the stronger of a pair of acids and the stronger of a pair of bases always have to be on the same side of the equation.

---

Let's now look at the relative strengths of acetic acid and the $H_3O^+$ ion.

$$CH_3CO_2H: \quad K_a = 1.8 \times 10^{-5}$$
$$H_3O^+ \quad K_a = 55$$

The values of $K_a$ for these acids suggest that acetic acid is a much weaker acid than the $H_3O^+$ ion, which explains why acetic acid is a weak acid in water. Once again, the reaction between the acid and water must convert the acid into its conjugate base and the base into its conjugate acid:

$$\underset{\text{acid}}{CH_3CO_2H(aq)} + \underset{\text{base}}{H_2O(l)} \rightleftharpoons \underset{\text{acid}}{H_3O^+(aq)} + \underset{\text{base}}{CH_3CO_2^-(aq)}$$

But this time, the stronger acid and the stronger base are on the right side of the equation:

$$\underset{\text{weaker acid}}{CH_3CO_2H(aq)} + \underset{\text{weaker base}}{H_2O(l)} \rightleftharpoons \underset{\text{stronger acid}}{H_3O^+(aq)} + \underset{\text{stronger base}}{CH_3CO_2^-(aq)}$$

As a result, only a few of the $CH_3CO_2H$ molecules actually donate an $H^+$ ion to a water molecule to form the $H_3O^+$ and $CH_3CO_2^-$ ions.

## EXERCISE 11.4

$NaNH_2$ reacts with water to give an aqueous solution of NaOH and $NH_3$:

$$NaNH_2(s) + H_2O(l) \rightleftharpoons Na^+(aq) + OH^-(aq) + NH_3(aq)$$

Which is the stronger base, the $NH_2^-$ or the $OH^-$ ion?

**SOLUTION** This reaction involves the transfer of an $H^+$ ion from a water molecule acting as an acid to an $NH_2^-$ ion acting as a base.

$$\underset{\text{base}}{NH_2^-(aq)} + \underset{\text{acid}}{H_2O(l)} \rightleftharpoons \underset{\text{acid}}{NH_3(aq)} + \underset{\text{base}}{OH^-(aq)}$$

If the reaction occurs as written, the reactants must include the stronger acid and the stronger base. Thus, the $NH_2^-$ ion is a stronger base than the $OH^-$ ion.

$$\underset{\text{stronger base}}{NH_2^-(aq)} + \underset{\text{stronger acid}}{H_2O(l)} \rightleftharpoons \underset{\text{weaker acid}}{NH_3(aq)} + \underset{\text{weaker base}}{OH^-(aq)}$$

The magnitude of $K_a$ can be used also to explain why some compounds that qualify as Brønsted acids or bases don't act like acids or bases when they dissolve in water. When the value of $K_a$ for an acid is relatively large, the acid reacts with water until essentially all of the acid molecules have been consumed. Sulfuric acid ($K_a = 1 \times 10^3$), for example, reacts with water until 99.9% of the $H_2SO_4$ molecules in a 1 *M* solution have lost a proton to form $HSO_4^-$ ions:

$$H_2SO_4(aq) + H_2O(l) \rightleftharpoons H_3O^+(aq) + HSO_4^-(aq)$$

As $K_a$ becomes smaller, the extent to which the acid reacts with water decreases.

As long as $K_a$ for the acid is significantly larger than the value of $K_a$ for water, the acid will ionize to some extent. Acetic acid, for example, reacts to some extent with water to form $H_3O^+$ and $CH_3CO_2^-$, or acetate, ions:

$$CH_3CO_2H(aq) + H_2O(l) \rightleftharpoons H_3O^+(aq) + CH_3CO_2^-(aq)$$

As the $K_a$ value for the acid approaches the $K_a$ for water, the compound becomes more like water in its acidity. Although it is still a Brønsted acid, it is so weak that we may be unable to detect this acidity in aqueous solution.

Some potential Brønsted acids are so weak that their $K_a$ values are smaller than water's. Ammonia, for example, has a $K_a$ of only $1 \times 10^{-33}$. Although $NH_3$ can be a Brønsted acid, because it has the potential to act as a hydrogen-ion donor, there is no evidence of this acidity when it dissolves in water.

### THE LEVELING EFFECT OF WATER

All strong acids and bases seem to have the same strength when dissolved in water, regardless of the value of $K_a$. This phenomenon is known as the **leveling effect** of water—the tendency of water to limit the strength of strong acids and bases. We can explain this by noting that strong acids react extensively with water to form the $H_3O^+$ ion. More than 99% of the HCl molecules in hydrochloric acid react with water to form $H_3O^+$ and $Cl^-$ ions, for example,

$$HCl(g) + H_2O(l) \rightleftharpoons H_3O^+(aq) + Cl^-(aq)$$

and more than 99% of the $H_2SO_4$ molecules in a 1 *M* solution react with water to form $H_3O^+$ ions and $HSO_4^-$ ions:

$$H_2SO_4(aq) + H_2O(l) \rightleftharpoons H_3O^+(aq) + HSO_4^-(aq)$$

Thus, the strength of strong acids is limited by the strength of the acid ($H_3O^+$) formed when water molecules pick up an $H^+$ ion.

A similar phenomenon occurs in solutions of strong bases. Strong bases react quantitatively with water to form the $OH^-$ ion. Once this happens, the solution cannot become any more basic. The strength of strong bases is limited by the strength of the base ($OH^-$) formed when water molecules lose an $H^+$ ion.

## 11.12 THE ADVANTAGES OF THE BRØNSTED DEFINITION

The Brønsted definition of acids and bases offers many advantages over the Arrhenius and operational definitions:

1. It expands the list of potential acids to include positive and negative ions, as well as neutral molecules.
2. It expands the list of bases to include any molecule or ion with at least one pair of nonbonding valence electrons.
3. It explains the role of water in acid–base reactions: Water accepts $H^+$ ions from acids to form the $H_3O^+$ ion.
4. It can be expanded to include solvents other than water and reactions that occur in the gas or solid phases.
5. It links acids and bases into conjugate acid–base pairs.
6. It can explain the relationship between the strengths of an acid and its conjugate base.
7. It can explain differences in the relative strengths of a pair of acids or a pair of bases.
8. It can explain the leveling effect of water—the fact that strong acids and bases all have the same strength when dissolved in water.

Because of these advantages, whenever chemists use the words *acid* or *base* without any further description, they are referring to a **Brønsted acid** or a **Brønsted base.**

## RESEARCH IN THE 90s

### SOLID-STATE ACIDS

What are the implications of finding an acid that can overcome the leveling effect of water? In a decade when fuel supply is threatened by international politics and resource scarcity, researchers are investigating ways to create new fuels. "Solid acids," which can be up to $10^7$ times stronger than sulfuric acid, play a key role in this process.

A little background is necessary to explain solid acids. A family of minerals known as the *silicates* make up 95% of the Earth's crust. The structure of these minerals consists of one-, two-, or three-dimensional polymers of tetrahedral $MO_4$ units in which a silicon or aluminum atom is surrounded by four oxygen atoms, as shown in Figure 11.9.

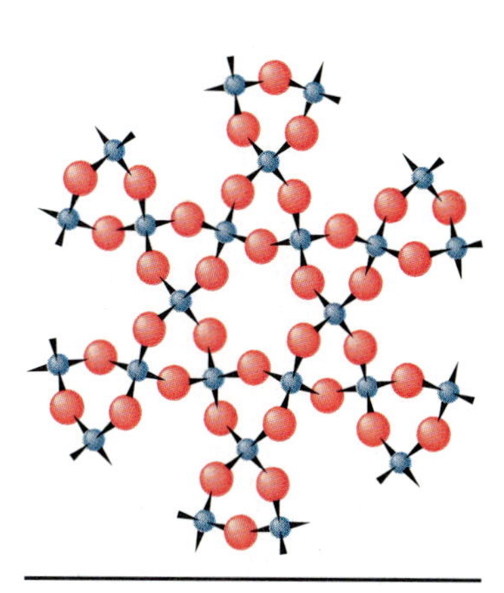

**Figure 11.9** Each of the silicon atoms in quartz is surrounded by four oxygen atoms arranged toward the corners of a tetrahedron. Because each oxygen atom is shared by two silicon atoms, the empirical formula is $SiO_2$. Taken from Wells, *Structural Inorganic Chemistry,* 3rd Ed., Oxford University Press.

A solution of the $SiO_4$ monomer can be obtained by dissolving sodium polysilicate $(Na_3SiO_3)_n$ in water, diluting the solution to $10^{-3}$ *M*, and then carefully lowering the pH to form silicic acid, $Si(OH)_4$, in which the silicon atom is surrounded by a tetrahedron of —OH groups. This solution is unstable, however. If acidified further, it forms a gel, known as *silica gel,* with the formula $SiO_2 \cdot n\,H_2O$. Given enough time and high enough temperatures, silica gel can lose water to form silica $(SiO_2)$. Once again, each silicon atom is surrounded by four oxygen atoms arranged toward the corners of a tetrahedron, but now, each oxygen atom acts as a bridge between two silicon atoms, as shown in Figure 11.9.

The silicates also contain elements other than silicon and oxygen. Feldspar, for example, has the empirical formula $KAlSi_3O_8$. The formula of this mineral can be understood by assuming that one-fourth of the $MO_4$ tetrahedra contain an aluminum atom instead of a silicon atom. Because this formally involves replacing an $Si^{4+}$ ion with an $Al^{3+}$ ion, another positive ion ($K^+$) must be present to keep the charge balanced.

The formula of feldspar raises an interesting question: What would happen if we formed an aluminosilicate in which $H^+$ ions were used to balance the charge that accumulated on the crystal each time an $Al^{3+}$ ion was substituted for $Si^{4+}$? The result would be both a solid and an acid. In theory, *solid acids* could overcome the leveling effect observed when strong acids are dissolved in water. In practice, solid acids have been found that are $10^7$ times stronger than sulfuric acid. Because of the important role strong acids play in catalyzing a variety of chemical reactions, the study of solid acids has become an important area of research in recent years (see J. M. Thomas, *Scientific American,* **1992**(4), 112–117).

Some of the earliest solid acids were based on natural minerals known as *zeolites* (from the Greek *zeo,* "to boil," and *lithos,* "stone"). For more than 200 years, geologists have known that zeolites bubble when they are heated, as if they are literally "boiling stones." We now know that this behavior results from the release of water trapped in the porous channels in the structure of these minerals.

Synthetic zeolites were first used to catalyze the process by which large hydrocarbon molecules are "cracked," or broken into smaller fragments. Other zeolites were then produced that catalyze the process by which the size of hydrocarbon molecules is increased. Zeolite catalysts have therefore significantly increased the amount of gasoline that can be extracted from a barrel of crude oil.

**Figure 11.10** A model of the structure of the ZSM-5 solid acid catalyst.

In recent years chemists at Mobil Oil have found that the zeolite shown in Figure 11.10 (ZSM-5) can be used to transform methanol ($CH_3OH$) into gasoline. In the short run, ZSM-5 will benefit countries such as New Zealand that do not have oil deposits but have an abundant supply of methane ($CH_4$) that can be converted into methanol. In the long run, it may provide the basis for a commercial process in which gasoline is produced from the carbon in coal.

## 11.13 pH AS A MEASURE OF THE CONCENTRATION OF THE $H_3O^+$ ION

Pure water is both a weak acid and a weak base. By itself, water forms only a very small number of the $H_3O^+$ and $OH^-$ ions that characterize aqueous solutions of stronger acids and bases:

$$\underset{\text{base}}{H_2O(l)} + \underset{\text{acid}}{H_2O(l)} \rightleftharpoons \underset{\text{acid}}{H_3O^+(aq)} + \underset{\text{base}}{OH^-(aq)}$$

The concentrations of the $H_3O^+$ and $OH^-$ ions in water can be determined by carefully measuring the ability of water to conduct an electric current. At 25°C, the concentration of these ions in pure water is $1.0 \times 10^{-7}$ moles per liter:

$$[H_3O^+] = [OH^-] = 1.0 \times 10^{-7}\ M \qquad \text{(at 25°C)}$$

Although values of $1.0 \times 10^{-7}$ $M$ for the concentrations of the $H_3O^+$ and $OH^-$ ions in water have been etched into the memories of generations of chemistry students, the concentrations of these ions depend on the temperature of the solution. At 45°C they are about twice as large. In superheated steam, at temperatures of 600°C to 700°C, they are large enough to eventually corrode metal pipes.

When we add a strong acid to water, the concentration of the $H_3O^+$ ion obviously becomes larger:

$$HCl(aq) + H_2O(l) \rightleftharpoons H_3O^+(aq) + Cl^-(aq)$$

At the same time, the $OH^-$ ion concentration becomes smaller because the $H_3O^+$ ions produced in this reaction neutralize some of the $OH^-$ ions in water:

$$H_3O^+(aq) + OH^-(aq) \rightleftharpoons 2\ H_2O(l)$$

Experimentally, we find that the product of the concentrations of the $H_3O^+$ and $OH^-$ ions is constant, no matter how much acid or base is added to water. In pure water at 25°C, the product of the concentration of these ions is $1.0 \times 10^{-14}$:

$$[H_3O^+][OH^-] = 1.0 \times 10^{-14} \qquad \text{(at 25°C)}$$

In 1.0 *M* hydrochloric acid, the $H_3O^+$ ion concentration is 1.0 *M* and the concentration of the $OH^-$ ion is only $1.0 \times 10^{-14}$ *M*. In 1.0 *M* sodium hydroxide, the $OH^-$ ion concentration is 1.0 *M* and the concentration of the $H_3O^+$ ion is only $1.0 \times 10^{-14}$.

The range of concentrations of the $H_3O^+$ and $OH^-$ ions in aqueous solution is so large that it is difficult to work with. Assume, for example, that we want to follow the concentration of the $H_3O^+$ ion as we add an acid or base to water. Furthermore, assume that we want to plot our data on a graph sensitive enough to register the smallest $H_3O^+$ ion concentration. To do this we have to use a graph with intervals of $10^{-14}$ *M* (see Figure 11.11). If we construct this graph so that each change of $10^{-14}$ *M* in the concentration of the $H_3O^+$ ion corresponds to 1 mm on the horizontal axis of the graph, we need a piece of graph paper $10^{14}$ mm long to cover the entire range of $H_3O^+$ ion concentrations. Unfortunately, $10^{14}$ mm is about 60 million miles, or about two-thirds of the distance between the Earth and the Sun.

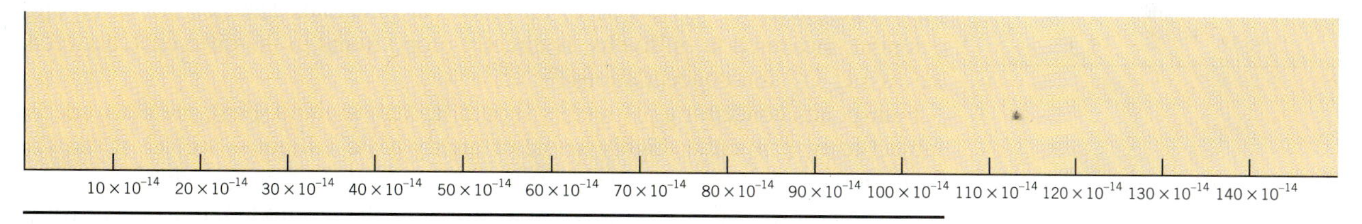


**Figure 11.11** The range of $H_3O^+$ ion concentration in aqueous solution is so large it would take a graph 62 million miles long to reflect changes across the entire range.

In 1909 the Danish biochemist S. P. L. Sørenson suggested reporting the concentration of the $H_3O^+$ ion on a logarithmic scale, which he named the **pH scale.** Because the $H_3O^+$ ion concentration in water is almost always smaller than 1, the log of these concentrations is a negative number. To avoid having to constantly work with negative numbers, Sørenson defined pH as the negative of the log of the $H_3O^+$ ion concentration:

$$\mathbf{pH = -\log [H_3O^+]}$$

**EXERCISE 11.5**

Calculate the pH of Pepsi Cola if the concentration of the $H_3O^+$ ion in this solution is 0.00347 *M*.

**SOLUTION** The pH of a solution is the negative of the log of the $H_3O^+$ ion concentration:

$$\begin{aligned} pH &= -\log [H_3O^+] \\ &= -\log (3.47 \times 10^{-3}) \\ &= -(-2.46) = \mathbf{2.46} \end{aligned}$$

The pH of Pepsi Cola is approximately 2.5.

The concept of pH compresses the range of $H_3O^+$ ion concentrations into a scale that is much easier to handle. As the $H_3O^+$ ion concentration decreases from roughly $10^0$ to $10^{-14}$, the pH of the solution increases from 0 to 14. The relationship

**TABLE 11.4 The Relationship Between the $H_3O^+$ Ion Concentration and the pH of an Aqueous Solution**

| $H_3O^+$ Ion Concentration ($M$) | pH | |
|---|---|---|
| 1.0 | 0 | Acid |
| $1.0 \times 10^{-1}$ | 1 | Acid |
| $1.0 \times 10^{-2}$ | 2 | Acid |
| $1.0 \times 10^{-3}$ | 3 | Acid |
| $1.0 \times 10^{-4}$ | 4 | Acid |
| $1.0 \times 10^{-5}$ | 5 | Acid |
| $1.0 \times 10^{-6}$ | 6 | Acid |
| $1.0 \times 10^{-7}$ | 7 | Neutral |
| $1.0 \times 10^{-8}$ | 8 | Base |
| $1.0 \times 10^{-9}$ | 9 | Base |
| $1.0 \times 10^{-10}$ | 10 | Base |
| $1.0 \times 10^{-11}$ | 11 | Base |
| $1.0 \times 10^{-12}$ | 12 | Base |
| $1.0 \times 10^{-13}$ | 13 | Base |
| $1.0 \times 10^{-14}$ | 14 | Base |

between the $H_3O^+$ ion concentration and the pH of a solution is shown in Table 11.4.

If the concentration of the $H_3O^+$ ion in pure water at 25°C is $1.0 \times 10^{-7}$ $M$, the pH of pure water is 7:

$$\text{pH} = -\log [H_3O^+] = -\log (1.0 \times 10^{-7}) = 7$$

When the pH of a solution is less than 7, the solution is acidic. When the pH is more than 7, the solution is basic.

$$\text{Acidic:} \quad \text{pH} < 7$$
$$\text{Basic:} \quad \text{pH} > 7$$

To decide whether a solution is acidic or basic, we need a way to determine its pH. Historically, this was done with **acid–base indicators.** To a large extent, these indicators have been replaced by pH meters, which are more accurate. The actual measuring device in a pH meter is an electrode that consists of a resin-filled tube with a thin glass bulb at one end. The glass bulb is filled with a 0.1 $M$ HCl solution in contact with a silver wire coated with a thin layer of silver chloride. The electrode is immersed in the solution to be measured, and the difference between the concentration of the $H_3O^+$ ion within the glass bulb and the $H_3O^+$ ion concentration in the surrounding solution gives rise to an electrical potential. The magnitude of this potential, measured in millivolts, is directly proportional to the difference between the two $H_3O^+$ ion concentrations.

Four points concerning pH meters should be kept in mind. First, the electrode is subject to error in either highly acidic or highly basic solutions. Second, it takes a finite amount of time for the electrode to respond to changes in pH. Third, the glass bulb at the tip of the electrode is easily broken if the electrode bumps against the bottom or sides of a beaker. Fourth, and perhaps most important, these instruments can make only relative measurements. All they can do is compare the pH of a solution with a known standard. Thus, pH meters must be calibrated against a solution of known pH before they are used.

Acid–base indicators are still used in the laboratory for rough pH measurements. These indicators are weak acids or weak bases that change color when they gain or lose an $H^+$ ion. The oldest example of an acid–base indicator is litmus, which has been used for more than 300 years. Litmus turns pink in solutions whose pH is below 5 and turns blue when the pH is above 8. Another common indicator, phenolphthalein, is colorless when the pH is less than 8.0 and pink when the pH is above 10.0. More than 200 acid–base indicators, which change color over a broad range of pH values, are known. Table 11.5 lists common indicators, the pH range over which they change color, and their color changes.

Blue litmus paper turns red in the presence of the citric acid in lemons.

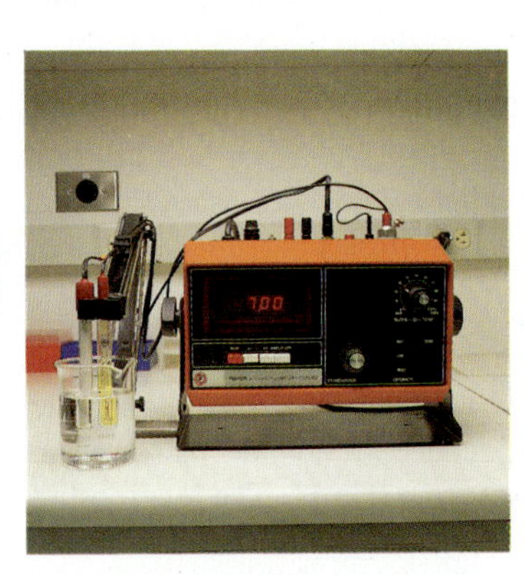

A digital pH meter testing the pH of a neutral solution.

**TABLE 11.5 Acid–Base Indicators**

| Indicator | pH Range | Color Change |
|---|---|---|
| Methyl violet | 0.0–1.6 | yellow to blue-violet |
| Thymol blue | 1.2–2.8 | red to yellow |
| | 8.0–9.6 | yellow to blue |
| Bromophenol blue | 3.0–4.6 | yellow to blue-violet |
| Methyl orange | 3.2–4.4 | red to yellow-orange |
| Methyl red | 4.4–6.2 | red to yellow |
| Litmus | 5–8 | pink to blue |
| Bromocresol purple | 5.2–6.8 | yellow to purple |
| Bromophenol red | 5.2–6.8 | yellow to red |
| Bromothymol blue | 6.2–7.6 | yellow to blue |
| Cresol red | 7.2–8.8 | yellow to red |
| Thymol blue | 8.0–9.6 | yellow to blue |
| Phenolphthalein | 8.0–10.0 | colorless to pink |
| Alizarin yellow | 10.0–12.0 | yellow to red-violet |

The glass electrodes of a pH meter.

By mixing two or more of these compounds, we can develop indicators for use over almost the entire range of pH. A so-called universal indicator can be made from equal volumes of methyl red, methyl orange, phenolphthalein, and bromothymol blue. As the pH of the solution increases, the indicator changes color from red to orange to yellow to green to blue and finally to purple.

The pH of a solution depends on the strength of the acid or base in the solution. Measurements of the pH of dilute solutions are therefore good indicators of the relative strengths of acids and bases. Values of the pH of 0.10 *M* solutions of a number of common acids and bases are given in Table 11.6.

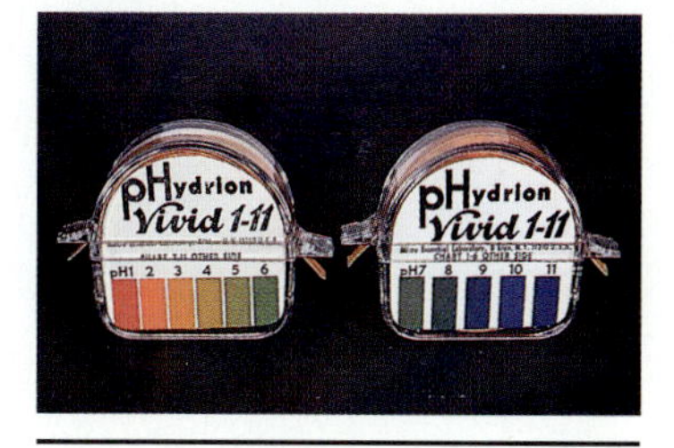


By mixing indicators it is possible to produce a universal indicator pH paper.

**TABLE 11.6 pH of 0.10 *M* Solutions of Common Acids and Bases**

| Compound | pH |
|---|---|
| HCl (hydrochloric acid) | 1.1 |
| $H_2SO_4$ (sulfuric acid) | 1.2 |
| $NaHSO_4$ (sodium hydrogen sulfate) | 1.4 |
| $H_2SO_3$ (sulfurous acid) | 1.5 |
| $H_3PO_4$ (phosphoric acid) | 1.5 |
| HF (hydrofluoric acid) | 2.1 |
| $CH_3CO_2H$ (acetic acid) | 2.9 |
| $H_2CO_3$ (carbonic acid) | 3.8 (saturated solution) |
| $H_2S$ (hydrogen sulfide) | 4.1 |
| $NaH_2PO_4$ (sodium dihydrogen phosphate) | 4.4 |
| $NH_4Cl$ (ammonium chloride) | 4.6 |
| HCN (hydrocyanic acid) | 5.1 |
| $Na_2SO_4$ (sodium sulfate) | 6.1 |
| NaCl (sodium chloride) | 6.4 |
| $NaCH_3CO_2$ (sodium acetate) | 8.4 |
| $NaHCO_3$ (sodium bicarbonate) | 8.4 |
| $Na_2HPO_4$ (sodium hydrogen phosphate) | 9.3 |
| $Na_2SO_3$ (sodium sulfite) | 9.8 |
| NaCN (sodium cyanide) | 11.0 |
| $NH_3$ (ammonia) | 11.1 |
| $Na_2CO_3$ (sodium carbonate) | 11.6 |
| $Na_3PO_4$ (sodium phosphate) | 12.0 |
| NaOH (sodium hydroxide, lye) | 13.0 |

## 11.14 FACTORS THAT CONTROL THE RELATIVE STRENGTHS OF ACIDS AND BASES

Several factors influence the probability of having a heart attack. Heart attacks occur more often among those who smoke than those who don't, among the elderly more often than the young, among those who are overweight more than those who aren't, and among those who never exercise more than those who exercise regularly. It is possible to sort out the relative importance of these factors, however, by trying to keep as many of the other factors constant as possible.

The same approach can be used to sort out the factors that control the relative strengths of acids and bases. Three factors affect the acidity of the *X*—H bond in a nonmetal hydride: (1) the polarity of this bond, (2) the size of the *X* atom, and (3) the charge on the ion or molecule. A fourth factor has to be considered to understand the acidity of the *X*—OH group in an oxyacid: the oxidation state of the *X* atom.

### THE POLARITY OF THE *X*—H BOND

When all other factors are kept constant, acids become stronger as the *X*—H bond becomes more polar. The second-row nonmetal hydrides, for example, become more acidic as the difference between the electronegativity of the *X* and H atoms increases. HF is the strongest of these four acids, and $CH_4$ is one of the weakest Brønsted acids known:

| | | |
|---|---|---|
| HF | $K_a = 7.2 \times 10^{-4}$ | $\Delta EN = 1.8$ |
| $H_2O$ | $K_a = 1.8 \times 10^{-16}$ | $\Delta EN = 1.2$ |
| $NH_3$ | $K_a = 1 \times 10^{-33}$ | $\Delta EN = 0.8$ |
| $CH_4$ | $K_a = 1 \times 10^{-49}$ | $\Delta EN = 0.4$ |

This factor is easy to understand. When these compounds act as an acid, an H—*X* bond is broken to form $H^+$ and $X^-$ ions. The more polar this bond, the easier it is to form these ions. Thus, the more polar the bond, the stronger the acid.

The data in Table 11.6 illustrate the magnitude of this effect. An 0.1 *M* HF solution is moderately acidic. Water is much less acidic, and the acidity of ammonia is so small that the chemistry of aqueous solutions of this compound is dominated by its ability to act as a base.

| | |
|---|---|
| HF | pH = 2.1 |
| $H_2O$ | pH = 7 |
| $NH_3$ | pH = 11.1 |

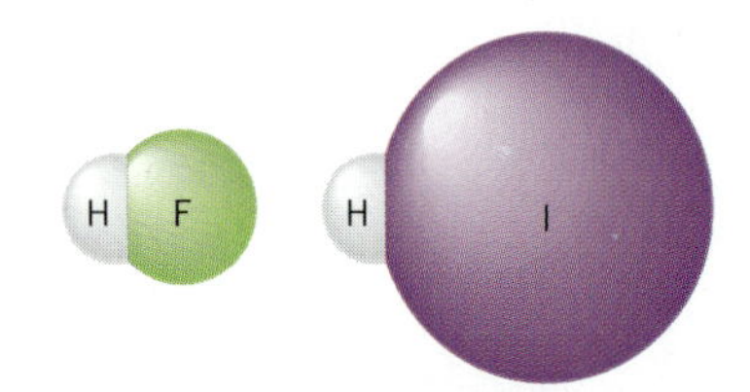


**Figure 11.12** HI is a much stronger acid than HF because of the relative sizes of the fluorine and iodine atoms. The covalent radius of iodine is more than twice as large as fluorine, which means that the charge density on the $F^-$ ion is very much larger than the charge density on the $I^-$ ion. As a result, it is more difficult to separate the atoms in an HF molecule to form $H^+$ and $F^-$ ions than it is to separate the atoms in an HI molecule to form $H^+$ and $I^-$ ions.

### THE SIZE OF THE *X* ATOM

At first glance, we might expect that the hydrogen halides, HF, HCl, HBr, and HI, would become weaker acids as we go down this column of the periodic table because the *X*—H bond becomes less polar. Experimentally, we find the opposite trend. These acids actually become stronger as we go down this column.

This occurs because the size of the *X* atom influences the acidity of the *X*—H bond. Acids become stronger as the *X*—H bond becomes weaker, and bonds generally become weaker as the atoms get larger (see Figure 11.12). The $K_a$ data for HF, HCl, HBr, and HI reflect the fact that the *X*—H bond-dissociation enthalpy (*BDE*) becomes smaller as the *X* atom becomes larger.

| | | |
|---|---|---|
| HF | $K_a = 7.2 \times 10^{-4}$ | $BDE = 569$ kJ/mol |
| HCl | $K_a = 1 \times 10^{6}$ | $BDE = 431$ kJ/mol |
| HBr | $K_a = 1 \times 10^{9}$ | $BDE = 370$ kJ/mol |
| HI | $K_a = 3 \times 10^{9}$ | $BDE = 300$ kJ/mol |

## THE CHARGE ON THE ACID OR BASE

The charge on a molecule or ion can influence its ability to act as an acid or a base. This is clearly shown when the pH of 0.1 *M* solutions of $H_3PO_4$ and the $H_2PO_4^-$, $HPO_4^{2-}$, and $PO_4^{3-}$ ions are compared:

| | |
|---|---|
| $H_3PO_4$ | pH = 1.5 |
| $H_2PO_4^-$ | pH = 4.4 |
| $HPO_4^{2-}$ | pH = 9.3 |
| $PO_4^{3-}$ | pH = 12.0 |

Compounds become less acidic and more basic as the negative charge increases:

Acidity: $H_3PO_4 > H_2PO_4^- > HPO_4^{2-}$

Basicity: $H_2PO_4^- < HPO_4^{2-} < PO_4^{3-}$

The decrease in acidity with increasing negative charge is easy to explain. It is much easier to remove an $H^+$ ion from a neutral $H_3PO_4$ molecule than it is to remove an $H^+$ ion from a negatively charged $H_2PO_4^-$ ion. It is even harder to remove an $H^+$ ion from a negatively charged $HPO_4^{2-}$ ion.

The increase in basicity with increasing negative charge is equally easy to understand. There is a strong force of attraction between the negative charge on a $PO_4^{3-}$ ion and the positive charge on an $H^+$ ion. As a result this ion is a reasonably good base. The force of attraction between the $HPO_4^{2-}$ ion and an $H^+$ ion is smaller because this ion carries a smaller charge. It isn't surprising that the $HPO_4^{2-}$ ion is therefore a weaker base. The charge on the $H_2PO_4^-$ ion is even smaller, so this ion is an even weaker base.

## THE OXIDATION STATE OF THE CENTRAL ATOM

There is no difference in the polarity, size, or charge when we compare oxyacids of the same element, such as $H_2SO_4$ and $H_2SO_3$ or $HNO_3$ and $HNO_2$, yet there is a significant difference in the strengths of these acids. Consider the following $K_a$ data, for example:

| | | | |
|---|---|---|---|
| $H_2SO_4$: | $K_a = 1 \times 10^3$ | $HNO_3$: | $K_a = 28$ |
| $H_2SO_3$: | $K_a = 1.7 \times 10^{-2}$ | $HNO_2$: | $K_a = 5.1 \times 10^{-4}$ |

The acidity of these oxyacids increases significantly as the oxidation state of the central atom becomes larger. $H_2SO_4$ is a much stronger acid than $H_2SO_3$, and $HNO_3$ is a much stronger acid than $HNO_2$. This trend is easiest to see in the four oxyacids of chlorine.

| *Oxyacid* | $K_a$ | *Oxidation number of the chlorine* |
|---|---|---|
| HOCl | $2.9 \times 10^{-8}$ | +1 |
| HOClO | $1.1 \times 10^{-2}$ | +3 |
| $HOClO_2$ | $5.0 \times 10^2$ | +5 |
| $HOClO_3$ | $1 \times 10^3$ | +7 |

Sec. 8.7

Since the atoms in these four compounds are the same size and these compounds all have the same charge, it might seem difficult to explain a factor of $10^{11}$ difference in the value of $K_a$ for hypochlorous acid (HOCl) and perchloric acid ($HOClO_3$). This difference can be traced to a point that was introduced in Section 8.7: There is only one value for the electronegativity of an element, but the tendency

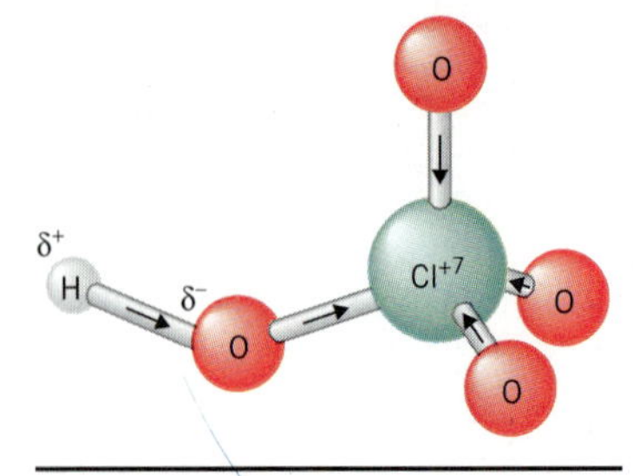


**Figure 11.13** The factor of $10^{11}$ difference between the values of $K_a$ for HOCl and $HOClO_3$ is the result of differences between the electronegativity of the chlorine atoms in these compounds. The increase in the oxidation number of the chlorine from +1 in HOCl to +7 in $HOClO_3$ results in an increase in the effective electronegativity of this atom. As the chlorine atom becomes more electronegative, it pulls electrons in the Cl—O bonds toward itself. The oxygen in the O—H bond can partially compensate for this by pulling the electrons in this bond toward itself. The net result is an increase in the polarity of the O—H bond, which leads to an increase in the acidity of the compound.

of an atom to draw electrons toward itself increases as the oxidation number of the atom increases.

As the oxidation number of the chlorine atom increases, the atom becomes more electronegative. This tends to draw electrons away from the oxygen atoms that surround the chlorine, thereby making the oxygen atoms more electronegative as well, as shown in Figure 11.13. As a result, the O—H bond becomes more polar, and the compound becomes more acidic.

## EXERCISE 11.6

For each of the following pairs, predict which compound is the stronger acid:

(a) $H_2O$ or $NH_3$ (b) $NH_4^+$ or $NH_3$ (c) $NH_3$ or $PH_3$ (d) $H_3PO_4$ or $H_3PO_3$

**SOLUTION**

(a) Oxygen and nitrogen are about the same size, and $H_2O$ and $NH_3$ are both neutral compounds. The only difference between these compounds is the polarity of the *X*—H bond. Because oxygen is more electronegative, the O—H bond is more polar, and $H_2O$ ($K_a = 1.8 \times 10^{-16}$) is a much stronger acid than $NH_3$ ($K_a = 1 \times 10^{-33}$).

(b) The only difference between these compounds is the charge on the acid. The $NH_4^+$ ion is a stronger acid than $NH_3$ because it is easier to remove an $H^+$ ion from an $NH_4^+$ ion than from a neutral $NH_3$ molecule.

(c) $PH_3$ is a stronger acid than $NH_3$ because compounds become more acidic as the size of the atom holding the hydrogen atom increases and the *X*—H bond becomes weaker.

(d) These compounds are both oxyacids, which differ only in the oxidation state of the phosphorus atom. The phosphorus atom in $H_3PO_4$ is more electronegative than the phosphorus atom in $H_3PO_3$, which means that $H_3PO_4$ is the stronger acid.

The relative strengths of Brønsted bases can be predicted from the relative strengths of their conjugate acids combined with the general rule that the stronger of a pair of acids always has the weaker conjugate base.

## EXERCISE 11.7

For each of the following pairs of compounds, predict which compound is the stronger base:

(a) $OH^-$ or $NH_2^-$ (b) $NH_3$ or $NH_2^-$ (c) $NH_2^-$ or $PH_2^-$
(d) $NO_3^-$ or $NO_2^-$

**SOLUTION**

(a) The $OH^-$ ion is the conjugate base of water, and the $NH_2^-$ ion is the conjugate base of ammonia. If $H_2O$ is a stronger acid than $NH_3$, the $OH^-$ ion must be a weaker base than the $NH_2^-$ ion.

(b) Because it carries a negative charge, the $NH_2^-$ ion is a stronger base than $NH_3$.

(c) $PH_3$ is a stronger acid than $NH_3$, which means the $PH_2^-$ ion must be a weaker base than the $NH_2^-$ ion.

(d) $HNO_3$ is a stronger acid than $HNO_2$, which means that the $NO_3^-$ ion is the weaker base.

SPECIAL TOPIC

## TRANSITION-METAL IONS AS BRØNSTED ACIDS

It is easy to understand why aqueous solutions of HCl or $CH_3CO_2H$ are acidic. The following data for the pH of 0.1 *M* solutions of transition-metal ions are a bit harder to explain:

| | |
|---|---|
| $FeCl_3$: | pH = 2.0 |
| $AlCl_3$: | pH = 3.0 |
| $Cu(NO_3)_2$: | pH = 4.0 |

We can't attribute the acidity of these solutions to the $Cl^-$ or $NO_3^-$ ions because these ions are weak bases. The acidity of these solutions must result from the behavior of the $Fe^{3+}$, $Al^{3+}$, and $Cu^{2+}$ ions.

The $Fe^{3+}$, $Al^{3+}$, and $Cu^{2+}$ ions can't be Brønsted acids by themselves. They can only act as proton donors by influencing the ability of the neighboring water molecules to give up $H^+$ ions. They do this by first forming covalent bonds to six water molecules to form a **complex ion,** as shown in Figure 11.14.

$$Al^{3+}(aq) + 6\,H_2O(l) \rightleftharpoons Al(H_2O)_6{}^{3+}(aq)$$
$$Fe^{3+}(aq) + 6\,H_2O(l) \rightleftharpoons Fe(H_2O)_6{}^{3+}(aq)$$
$$Cu^{2+}(aq) + 6\,H_2O(l) \rightleftharpoons Cu(H_2O)_6{}^{2+}(aq)$$

Water molecules covalently bound to one of these metal ions are more acidic than normal. Thus, reactions such as the following occur:

$$Al(H_2O)_6{}^{3+}(aq) + H_2O(l) \rightleftharpoons Al(H_2O)_5(OH)^{2+}(aq) + H_3O^+(aq)$$
$$Fe(H_2O)_6{}^{3+}(aq) + H_2O(l) \rightleftharpoons Fe(H_2O)_5(OH)^{2+}(aq) + H_3O^+(aq)$$

These reactions give rise to a net increase in the $H_3O^+$ ion concentration in these solutions, thereby making the solutions acidic.

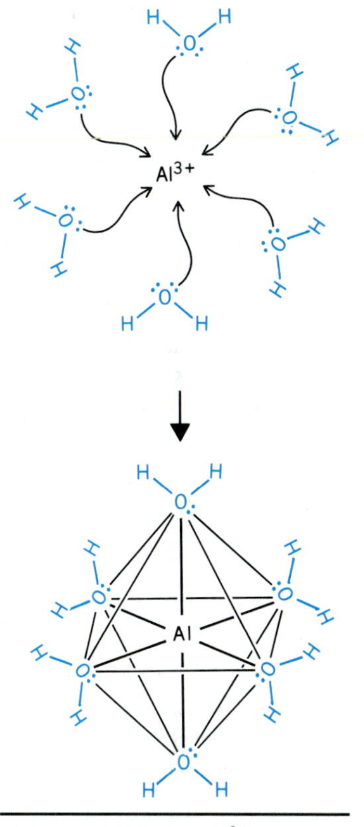


**Figure 11.14** The $Al^{3+}$ ion is a Lewis acid that can accept pairs of nonbonding electrons from six $H_2O$ molecules to form a complex ion with the formula $Al(H_2O)_6{}^{3+}$.

**EXERCISE 11.8**

As many as 25% of the books in the Library of Congress are in brittle condition, in part because of the acidity of the paper on which they were printed. Explain how adding aluminum sulfate to paper when it is manufactured makes paper acidic.

**SOLUTION** At first glance, $Al_2(SO_4)_3$ would appear to be neither an acid nor a base. $Al^{3+}$ ions, however, react with water to form $Al(H_2O)_6{}^{3+}$ complex ions, in which the $H_2O$ molecules bound to the $Al^{3+}$ ion are much more acidic than normal.

**SPECIAL TOPIC**

## THE LEWIS DEFINITIONS OF ACIDS AND BASES

In 1923 G. N. Lewis suggested another way of looking at the reaction between $H^+$ and $OH^-$ ions. In the Brønsted model, the $OH^-$ ion is the active species in this reaction: It accepts an $H^+$ ion to form a covalent bond. In the Lewis model, the $H^+$ ion is the active species: It accepts a pair of electrons from the $OH^-$ ion to form a covalent bond:

$$H^+ \curvearrowleft :\ddot{\underset{..}{O}}—H^- \longrightarrow H—\ddot{\underset{..}{O}}—H$$

In the Lewis theory of acid–base reactions, bases donate pairs of electrons, and acids accept pairs of electrons. A **Lewis acid** is therefore any substance, such as the $H^+$ ion, that can accept a pair of nonbonding electrons. In other words, a Lewis acid is an **electron-pair acceptor.** A **Lewis base** is any substance, such as the $OH^-$ ion, that can donate a pair of nonbonding electrons. A Lewis base is therefore an **electron-pair donor.**

Sec. 9.15 

One advantage of the Lewis theory is the way it complements the model of oxidation–reduction reactions introduced in Chapter 9. Oxidation–reduction reactions involve a transfer of electrons from one atom to another, with a net change in the oxidation number of one or more atoms:

$$Na\cdot + :\ddot{\underset{..}{Cl}}\cdot \longrightarrow [Na^+][:\ddot{\underset{..}{Cl}}:^-]$$

The Lewis theory suggests that acids react with bases to share a pair of electrons, with no change in the oxidation numbers of any atoms. Many chemical reactions can be sorted into one or the other of these classes. Either electrons are transferred from one atom to another, or the atoms come together to share a pair of electrons.

The principal advantage of the Lewis theory is the way it expands the number of acids and therefore the number of acid–base reactions. In the Lewis theory, an acid is any ion or molecule that can accept a pair of nonbonding valence electrons. In the preceding section, we concluded that $Al^{3+}$ ions form bonds to six water molecules to give a complex ion:

$$Al^{3+}(aq) + 6\ H_2O(l) \rightleftharpoons Al(H_2O)_6{}^{3+}(aq)$$

This is an example of a Lewis acid–base reaction. The Lewis structure of water suggests that this molecule has nonbonding pairs of valence electrons and can therefore act as a Lewis base. The electron configuration of the $Al^{3+}$ ion suggests that this ion has empty $3s$, $3p$, and $3d$ orbitals that can be used to hold pairs of nonbonding electrons donated by neighboring water molecules:

$$Al^{3+} = [Ne]\ 3s^0\ 3p^0\ 3d^0$$

Thus, the $Al(H_2O)_6{}^{3+}$ ion is formed when an $Al^{3+}$ ion acting as a Lewis acid picks up six pairs of electrons from neighboring water molecules acting as Lewis bases to give an **acid–base complex,** or **complex ion.**

Lewis acid–base complexes play a vital role in nature. Oxygen is carried from the lungs to the body in the form of a Lewis acid–base complex between $O_2$ molecules and an $Fe^{3+}$ ion in either hemoglobin or myoglobin (see Figure 11.15). The $CO_2$ given off when $O_2$ is used to burn sugars or other carbohydrates is then carried through the body to the lungs in the form of a Lewis acid–base complex with the $Fe^{3+}$ ion in hemoglobin and myoglobin. The nitrogenase enzymes that catalyze the reduction of $N_2$ to $NH_3$ operate by first forming a Lewis acid–base complex between the $N_2$ molecule and a transition-metal ion. Indeed, any protein that carries

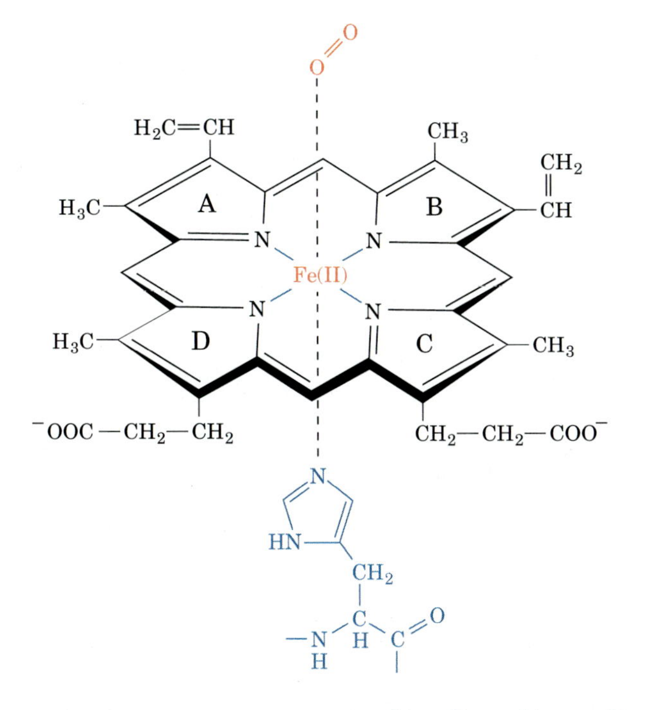


**Figure 11.15** This figure shows a small portion of the active site of the proteins hemoglobin and myoglobin. These proteins contain a *heme* group, which consists of either an $Fe^{2+}$ or $Fe^{3+}$ ion bound to a planar molecule known as a porphyrin. The covalent bonds represented by lines between the iron and nitrogen atoms in this structure result from donating pairs of nonbonding electrons on the nitrogen atoms into empty valence orbitals on the iron atom.

a transition metal ion, such as the $Cu^{2+}$, $Fe^{3+}$, $Mn^{2+}$, or $Zn^{2+}$ ions, operates by forming a Lewis acid–base complex.

Another example of Lewis acid–base chemistry is the use of lime to absorb, or scrub, $SO_2$ and $SO_3$ fumes emitted from smokestacks:

$$CaO(s) + SO_2(g) \rightleftharpoons CaSO_3(s)$$

In the course of this reaction, the $O^{2-}$ ion in CaO acts as an electron-pair donor, or Lewis base, and an $SO_2$ molecule acts as an electron-pair acceptor, or Lewis acid, to form the $SO_3^{2-}$ ion (see Figure 11.16).

$BF_3$ is a trigonal-planar molecule because electrons can be found in only three places in the valence shell of the boron atom. As a result, the boron atom is $sp^2$ hybridized, which leaves an empty $2p_z$ orbital on the boron atom. $BF_3$ can therefore act as an electron-pair acceptor, or Lewis acid. It can use the empty $2p_z$ orbital to pick up a pair of nonbonding electrons from a Lewis base to form a covalent bond. $BF_3$ therefore reacts with Lewis bases such as $NH_3$ to

Sec. 8.14

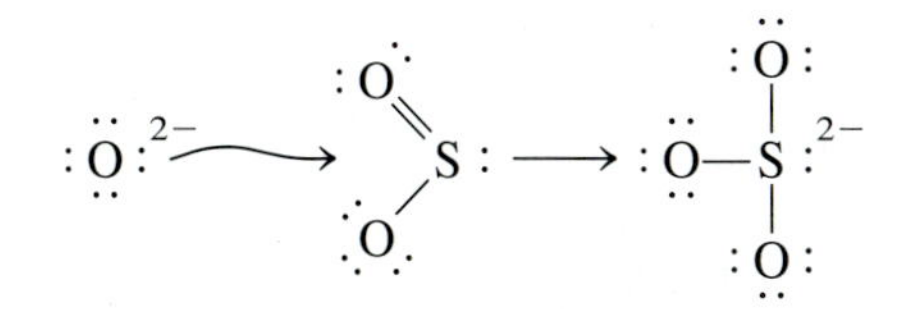


**Figure 11.16** The reaction between CaO and $SO_2$ to form $CaSO_3$ is a Lewis acid–base reaction. The $O^{2-}$ ion in CaO acts as an electron-pair donor, or Lewis base, and $SO_2$ acts as an electron-pair acceptor, or Lewis acid, to form the $SO_3^{2-}$ ion.

form acid–base complexes in which all of the atoms have a filled shell of valence electrons, as shown in Figure 11.17.

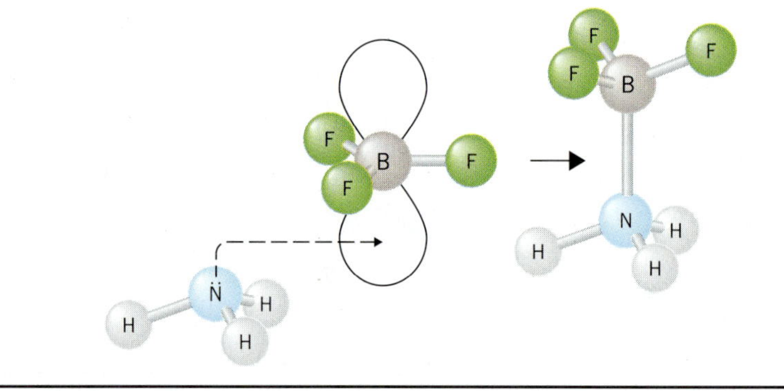


**Figure 11.17** $BF_3$ is a trigonal-planar molecule that contains an $sp^2$-hybridized boron atom with only six valence electrons. The boron atom therefore has an empty $2p_z$ atomic orbital that can accept a pair of nonbonding electrons from an $NH_3$ molecule to form a covalent B—N bond.

The Lewis acid–base theory also can be used to explain why nonmetal oxides such as $CO_2$ dissolve in water to form acids, such as carbonic acid, $H_2CO_3$:

$$CO_2(g) + H_2O(l) \rightleftharpoons H_2CO_3(aq)$$

In the course of this reaction, the water molecule acts as an electron-pair donor, or Lewis base. The electron-pair acceptor is the carbon atom in $CO_2$. When the carbon atom picks up a pair of electrons from the water molecule, it no longer needs to form double bonds with both of the other oxygen atoms (see Figure 11.18).

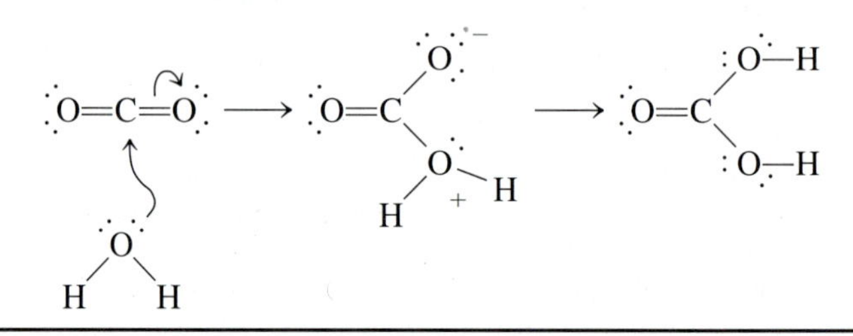

**Figure 11.18** The reaction between $CO_2$ and $H_2O$ to form carbonic acid is another example of a Lewis acid–base reaction. In this case, water is a Lewis base that donates a pair of nonbonding electrons to the carbon atom of carbon dioxide. An $H^+$ ion then migrates from one oxygen atom to another to form a symmetrical $H_2CO_3$ molecule.

Sec. 8.12 

One of the oxygen atoms in the intermediate formed when water is added to $CO_2$ carries a positive charge; another carries a negative charge. After an $H^+$ ion has been transferred from one of these oxygen atoms to the other, all of the oxygen atoms in the compound are electrically neutral. The net result of the reaction between $CO_2$ and water is therefore carbonic acid, $H_2CO_3$.

**EXERCISE 11.9**

Predict whether the following ions or molecules can act as either a Lewis acid or a Lewis base:

(a) $Ag^+$ (b) $NH_3$

**SOLUTION**

(a) The $Ag^+$ ion has empty $5s$ and $5p$ valence orbitals that allow it to act as a Lewis acid, or electron-pair acceptor:

$$Ag^+ = [Kr]\ 4d^{10}\ 5s^0\ 5p^0$$

(b) There are no empty valence orbitals in the Lewis structure of $NH_3$, which suggests that it can't be a Lewis acid. There is a pair of nonbonding electrons, however, which means that ammonia can be an electron-pair donor, or Lewis base.

$$\mathrm{H{-}\ddot{N}{-}H}$$
$$\mathrm{|}$$
$$\mathrm{H}$$

## SUMMARY

**1.** For more than 300 years, chemists have recognized the existence of acids and bases (or alkalies), which combine to form salts. The first step toward explaining this observation was to try to find an element that was present in all acids. Lavoisier believed it was oxygen, but Davy showed this was wrong. Liebig then focused attention on the presence of hydrogen in acids, and Arrhenius proposed a theory that explained why compounds that contain hydrogen are acids.

**2.** According to Arrhenius, acids and bases could be viewed in terms of the ions that form when water dissociates:

$$H_2O(l) \rightleftharpoons H^+(aq) + OH^-(aq).$$

**3.** Brønsted recognized that isolated $H^+$ ions cannot exist in water. He suggested that water molecules react by transferring an $H^+$ ion from one molecule to another to form $H_3O^+$ and $OH^-$ ions: $2\ H_2O(l) \rightleftharpoons H_3O^+(aq) + OH^-(aq)$. According to this model, acid–base reactions involve the transfer of an $H^+$ ion from an acid to a base to form the conjugate base and the conjugate acid.

**4.** The Brønsted definition significantly broadens the category of compounds that can be classified as bases. In the Brønsted model, a base is any compound that can accept an $H^+$ ion. It is therefore any compound with at least one pair of nonbonding valence electrons that can be used to form a bond to an $H^+$ ion.

**5.** Lewis noted that a base must have at least one pair of nonbonding electrons to act as a hydrogen-ion acceptor. Instead of thinking of the base as a *hydrogen-ion acceptor,* he argued that we can think of it as an *electron-pair donor.* This doesn't have much effect on the number of compounds that can be classified as bases, but it greatly expands the number of possible acids. A Lewis acid is any substance that has an empty valence orbital that can be used to accept a pair of electrons.

**6.** By convention, many chemists use the words *acid* and *base* to describe substances that are Brønsted acids and bases. They then use the terms *Lewis acid* and *Lewis base* to describe acid–base reactions that can be understood only in terms of electron-pair donors and acceptors.

## KEY TERMS

**Acids** (p. 402)
**Acid–base complex** (p. 430)
**Acid-dissociation equilibrium constant** ($K_a$) (p. 416)
**Alkalies** (p. 402)
**Amphoteric** (p. 410)
**Arrhenius acid/base** (p. 404)
**Bases** (p. 402)
**Brønsted acid/base** (p. 411, 420)
**Complex ion** (p. 429)
**Conjugate acid–base pairs** (p. 414)
**Electron-pair acceptor/donor** (p. 430)
**Indicators** (p. 424)
**Leveling effect** (p. 420)
**Lewis acid/base** (p. 430)
**Oxyacids** (p. 408)
**pH scale** (p. 423)
**Proton acceptor/donor** (p. 411)
**Strong acid** (p. 415)
**Weak acid** (p. 415)

## PROBLEMS

### Acids and Bases

**11-1** Describe how acids and bases differ. Give examples of substances from daily life that fit each of these categories.

**11-2** Strong acids (such as hydrochloric acid and sulfuric acid) react with strong bases (such as sodium hydroxide and ammonia) to form salts (such as sodium chloride, sodium sulfate, ammonium chloride, and ammonium sulfate). Write balanced equations for these reactions and describe how these salts differ from the acids and bases from which they are made.

### The Arrhenius and Operational Definitions of Acids and Bases

**11-3** Describe the Arrhenius definition of acids and bases, and give examples of acids and bases that fit this definition.

**11-4** Which of the following compounds are Arrhenius acids?

(a) HCl (b) $H_2SO_4$ (c) $H_2O$ (d) $FeCl_3$ (e) $H_2S$

**11-5** Which of the following compounds are Arrhenius bases?

(a) NaOH (b) MgO (c) $Ca(OH)_2$ (d) $H_2O$ (e) $NH_3$

**11-6** The following compounds dissolve in water to give solutions that turn litmus from red to blue. Which of these compounds satisfy the operational definition of a base?

(a) $Na_2O$ (b) $NaHCO_3$ (c) CaO (d) $CaH_2$ (e) $NH_3$

### Typical Acids and Bases

**11-7** Write balanced equations to show what happens when hydrogen bromide dissolves in water to form an acidic solution and when ammonia dissolves in water to form a basic solution.

**11-8** Give an example of an acid whose formula carries a positive charge, an acid that is electrically neutral, and an acid that carries a negative charge.

**11-9** Explain why compounds that contain hydrogen in the $-1$ oxidation state, such as NaH or $CaH_2$, are bases, whereas compounds that contain hydrogen in the $+1$ oxidation state, such as HCl or $HNO_3$, are more likely to be acids.

**11-10** Would you expect metal peroxides, such as $Na_2O_2$ and $BaO_2$, to be acids or bases? What about nonmetal peroxides, such as $H_2O_2$?

**11-11** Which of the following hydrides are acids and which are bases?

(a) $H_2CO_3$ (b) $CH_3CO_2H$ (c) $LiAlH_4$ (d) $H_2S$ (e) $ZnH_2$

**11-12** Which of the following oxides are acids and which are bases?

(a) $CO_2$ (b) $SO_3$ (c) $P_4O_{10}$ (d) MgO

**11-13** Which of the following compounds are acids when dissolved in water?

(a) $SO_2$ (b) $HNO_3$ (c) SrO (d) HI (e) $K_2S$

**11-14** Which of the following compounds are bases when dissolved in water?

(a) $Ca(OH)_2$ (b) $Na_2O$ (c) $NO_2$ (d) $Li_2O$ (e) CsOH

**11-15** Predict the products of the reaction between each of the following compounds and water:

(a) $CO_2$ (b) $P_4O_{10}$ (c) CaO (d) $SO_3$ (e) $Na_2O$

**11-16** Explain why lime (CaO) is used to neutralize soils that are too acidic. Use the fact that microorganisms in soil oxidize sulfur to $SO_2$ and $SO_3$ to explain why sulfur is added to soil that is too basic.

**11-17** Which of the following elements is most likely to form an acidic oxide with the formula $XO_2$ and an acidic hydride with the formula $XH_2$?

(a) Na (b) Mg (c) Al (d) P (e) S (f) Cl

### Why Are Metal Hydroxides Bases and Nonmetal Hydroxides Acids?

**11-18** Which of the following hydroxides are acids and which are bases?

(a) $H_2CO_3$ (b) $HNO_3$ (c) $Ca(OH)_2$ (d) $Zn(OH)_2$ (e) $H_3PO_4$

**11-19** The following compounds are all oxyacids. In each case, the acidic hydrogen atoms are bound to oxygen atoms. Write the skeleton structures for these acids.

(a) $H_3PO_4$ (b) $HNO_3$ (c) $HClO_4$ (d) $H_3BO_3$ (e) $H_2CrO_4$

**11-20** Explain why metal hydroxides, such as LiOH, are bases and nonmetal hydroxides, such as HOBr, are acids.

**11-21** Explain why metal oxides, such as CaO, are bases and nonmetal oxides, such as $CO_2$, are acids.

**11-22** Many insects leave small quantities of formic acid, $HCO_2H$, behind when they bite. Explain why the itching sensation can be relieved by treating the bite with an aqueous solution of ammonia ($NH_3$) or baking soda ($NaHCO_3$).

**11-23** Barium oxide (BaO) and ''phosphorus pentoxide'' ($P_4O_{10}$) are both white solids. What is the easiest way of distinguishing between these compounds?

**11-24** Which of the following compounds is most likely to be amphoteric?

(a) $Na_2O$ (b) CaO (c) $Al_2O_3$ (d) $P_4O_{10}$ (e) $Cl_2O_7$

### The Brønsted Definition of Acids and Bases

**11-25** Describe how the Brønsted definition of an acid can be used to explain why compounds that contain hydrogen with an oxidation number of +1 are often acids.

**11-26** Which of the following compounds can act as both a Brønsted acid and a Brønsted base?

(a) $NaHCO_3$ (b) $Na_2CO_3$ (c) $H_2CO_3$ (d) $CO_2$ (e) $H_2O$

**11-27** Which of the following compounds cannot be Brønsted bases?

(a) $H_3O^+$ (b) $MnO_4^-$ (c) $BH_4^-$ (d) $CN^-$ (e) $S^{2-}$

**11-28** Which of the following compounds cannot be Brønsted bases?

(a) $O_2^{2-}$ (b) $CH_4$ (c) $PH_3$ (d) $SF_4$ (e) $CH_3^+$

**11-29** Label the Brønsted acids and bases in the following reactions:

(a) $HSO_4^-(aq) + H_2O(l) \longrightarrow H_3O^+(aq) + SO_4^{2-}(aq)$

(b) $CH_3CO_2H(aq) + OH^-(aq) \longrightarrow CH_3CO_2^-(aq) + H_2O(l)$

(c) $CaF_2(s) + H_2SO_4(aq) \longrightarrow CaSO_4(aq) + 2\ HF(aq)$

(d) $HNO_3(aq) + NH_3(aq) \longrightarrow NH_4NO_3(aq)$

(e) $LiCH_3(l) + NH_3(l) \longrightarrow CH_4(g) + LiNH_2(s)$

### Conjugate Acid–Base Pairs

**11-30** Identify the conjugate acid–base pairs in the following reactions:

(a) $HCl(aq) + H_2O(l) \longrightarrow H_3O^+(aq) + Cl^-(aq)$

(b) $HCO_3^-(aq) + H_2O(l) \longrightarrow H_2CO_3(aq) + OH^-(aq)$

(c) $NH_3(aq) + H_2O(l) \longrightarrow NH_4^+(aq) + OH^-(aq)$

(d) $CaCO_3(s) + 2\ HCl(aq) \longrightarrow Ca^{2+}(aq) + 2\ Cl^-(aq) + H_2CO_3(aq)$

(e) $CaO(s) + H_2O(l) \longrightarrow Ca^{2+}(aq) + 2\ OH^-(aq)$

**11-31** Which of the following is not a conjugate acid–base pair?

(a) $NH_4^+/NH_3$ (b) $H_2O/OH^-$ (c) $H_3O^+/OH^-$ (d) $CH_4/CH_3^-$

**11-32** Write Lewis structures for the following Brønsted acids and their conjugate bases:

(a) formic acid, $HCO_2H$ (b) methanol, $CH_3OH$

**11-33** Write Lewis structures for the following Brønsted bases and their conjugate acids:

(a) methylamine, $CH_3NH_2$ (b) acetate ion, $CH_3CO_2^-$

**11-34** Identify the conjugate base of each of the following Brønsted acids:

(a) $H_3O^+$ (b) $H_2O$ (c) $OH^-$ (d) $NH_4^+$

**11-35** Identify the conjugate base of each of the following Brønsted acids:

(a) $HPO_4^{2-}$ (b) $Al(H_2O)_6^{3+}$ (c) $HCO_3^-$ (d) $HS^-$

**11-36** Identify the conjugate acid of each of the following Brønsted bases:

(a) $O^{2-}$ (b) $OH^-$ (c) $H_2O$ (d) $NH_2^-$

**11-37** Identify the conjugate acid of each of the following Brønsted bases:

(a) $Na_2HPO_4$ (b) $NaHCO_3$ (c) $Na_2SO_4$ (d) $NaNO_2$

**11-38** Which of the following ions is the conjugate base of a strong acid?

(a) $OH^-$ (b) $HSO_4^-$ (c) $NH_2^-$ (d) $S^{2-}$ (e) $H_3O^+$

**11-39** Use Lewis structures to predict the products of the following acid–base reactions:

$NaOCH_3(aq) + NaHCO_3(aq) \longrightarrow$

$NH_4Cl(aq) + NaSH(aq) \longrightarrow$

**11-40** Write balanced chemical equations for the following acid–base reactions:

(a) $Al_2O_3(s) + HCl(aq) \longrightarrow$

(b) $CaO(s) + H_2SO_4(aq) \longrightarrow$

(c) $CO_2(g) + NaOH(aq) \longrightarrow$

(d) $MgCO_3(s) + HCl(aq) \longrightarrow$
(e) $Na_2O_2(s) + H_3PO_4(aq) \longrightarrow$

### The Relative Strengths of Acids and Bases

**11-41** In the Brønsted model of acid–base reactions, what does the statement that HCl is a stronger acid than $H_2O$ mean in terms of the following reaction?

$$HCl(aq) + H_2O(l) \longrightarrow H_3O^+(aq) + Cl^-(aq)$$

**11-42** In the Brønsted model of acid–base reactions, what does it mean to say that $NH_3$ is a weaker acid than $H_2O$?

**11-43** What can you conclude from experiments that suggest that the following reaction proceeds to the right, as written?

$$HBr(aq) + H_2O(l) \longrightarrow H_3O^+(aq) + Br^-(aq)$$

**11-44** What can you conclude from experiments that suggest that the following reaction proceeds to the right, as written?

$$HCl(aq) + NH_3(aq) \longrightarrow NH_4^+(aq) + Cl^-(aq)$$

**11-45** What can you conclude from experiments that suggest that the following reaction does not proceed to the right, as written?

$$HCO_2^-(aq) + H_2O(l) \not\longrightarrow HCO_2H(aq) + OH^-(aq)$$

**11-46** Describe the relationship between the acid-dissociation equilibrium constant for an acid, $K_a$, and the strength of the acid.

**11-47** Use the acid-dissociation equilibrium constants in Table A-9 in the appendix to classify the following acids as either strong or weak:

(a) acetic acid, $CH_3CO_2H$ (b) boric acid, $H_3BO_3$
(c) chromic acid, $H_2CrO_4$ (d) formic acid, $HCO_2H$
(e) hydrobromic acid, HBr

**11-48** Which of the following is the weakest Brønsted acid?

(a) $H_2S_2O_3$: $K_a = 0.3$ (b) $H_2CrO_4$: $K_a = 9.6$
(c) $H_3BO_3$: $K_a = 7.3 \times 10^{-10}$
(d) $C_6H_5OH$: $K_a = 1.0 \times 10^{-10}$

### Factors That Control the Relative Strengths of Acids and Bases

**11-49** Explain the following observations.

(a) $H_2SO_4$ is a stronger acid than $HSO_4^-$.
(b) $HNO_3$ is a stronger acid than $HNO_2$.
(c) $H_2S$ is a stronger acid than $H_2O$.
(d) $H_2S$ is a stronger acid than $PH_3$.

**11-50** Arrange the following acids in order of increasing strength:

(a) $H_2O$ (b) HCl (c) $CH_4$ (d) $NH_3$

**11-51** Arrange the following bases in order of increasing strength:

(a) $NH_3$ (b) $PH_3$ (c) $H_2O$ (d) $H_2S$

**11-52** Which of the following is the strongest Brønsted acid?

(a) $H_3O^+$ (b) HF (c) $NH_3$ (d) $NaHSO_4$
(e) NaOH

**11-53** Which of the following is the weakest Brønsted acid?

(a) $H_2Se$ (b) $H_2S$ (c) $H_2O$ (d) $SH^-$ (e) $OH^-$

**11-54** Which of the following is the weakest Brønsted acid?

(a) $HClO_4$ (b) $H_3PO_4$ (c) $H_2CO_3$ (d) $H_2SiO_4$
(e) $Al(OH)_3$

**11-55** Which of the following is the strongest Brønsted base?

(a) $NH_2^-$ (b) $PH_2^-$ (c) $CH_3^-$ (d) $SiH_3^-$

**11-56** Which of the following is the strongest Brønsted base?

(a) $H_2O$ (b) $SH^-$ (c) $OH^-$ (d) $S^{2-}$ (e) $O^{2-}$

**11-57** Which of the following has the strongest conjugate base?

(a) $H_2O$ (b) $H_2S$ (c) $NH_3$ (d) $PH_3$ (e) $CH_4$

### pH as a Measure of the Relative Strengths of Acids and Bases

**11-58** What happens to the concentration of the $H_3O^+$ ion when a strong acid is added to water? What happens to the concentration of the $OH^-$ ion? What happens to the pH of the solution?

**11-59** What happens to the $OH^-$ ion concentration when a strong base is added to water? What happens to the concentration of the $H_3O^+$ ion? What happens to the pH of the solution?

**11-60** Which of the following solutions is the most acidic?

(a) 0.10 *M* acetic acid: pH = 2.9 (b) 0.10 *M* hydrogen sulfide: pH = 4.1 (c) 0.10 *M* sodium acetate: pH = 8.4
(d) 0.10 *M* ammonia: pH = 11.1

**11-61** Which of the following solutions is the most acidic?

(a) 0.10 *M* $H_3PO_4$: pH = 1.4 (b) 0.10 *M* $H_2PO_4^-$: pH = 4.4 (c) 0.10 *M* $HPO_4^{2-}$: pH = 9.3 (d) 0.10 *M* $PO_4^{3-}$: pH = 12.0

**11-62** Explain why aqueous solutions of the $Fe^{3+}$ ion are acidic.

**11-63** Which of the following compounds could dissolve in water to give a solution with a pH of about 5?

(a) $NH_3$ (b) NaCl (c) HCl (d) KOH (e) $NH_4Cl$

### The Lewis Definition of Acids and Bases

**11-64** Describe the difference between the Lewis and Brønsted definitions of acids and bases. Describe the advantages of the Lewis definition.

**11-65** Which of the following compounds can be a Lewis acid?

(a) $CH_3^+$ (b) $CH_4$ (c) $CH_3^-$ (d) $BF_3$ (e) $Ag^+$
(f) $Fe^{3+}$

**11-66** Which of the following compounds can be a Lewis base?

(a) $H_2O$ (b) $OH^-$ (c) $O^{2-}$ (d) $O_2^{2-}$ (e) $MnO_4^-$
(f) $CH_4$

### Integrated Problems

**11-67** Write balanced equations for the reaction between lithium and nitrogen to form lithium nitride and the subsequent reaction of lithium nitride with water to form a basic solution.

**11-68** Which of the following substances is the most likely to be a white solid with a high melting point that dissolves in water to form a basic solution?

(a) $O_2$ (b) $CO_2$ (c) $Na_2O$ (d) $P_4O_{10}$ (e) $Cl_2O_7$

**11-69** Explain why rain water has a pH of 5.6. Explain why boiling water to drive off the $CO_2$ raises the pH. Explain why the presence of $SO_2$, $SO_3$, NO, $NO_2$ and similar compounds in the atmosphere gives rise to acid rain.

**11-70** On April 10, 1974, at Pitlochny, Scotland, the rain was found to have a pH of 2.4. Calculate the $H_3O^+$ ion concentration in this rain and compare it with the $H_3O^+$ concentration in 0.10 *M* acetic acid, which has a pH of 2.9.

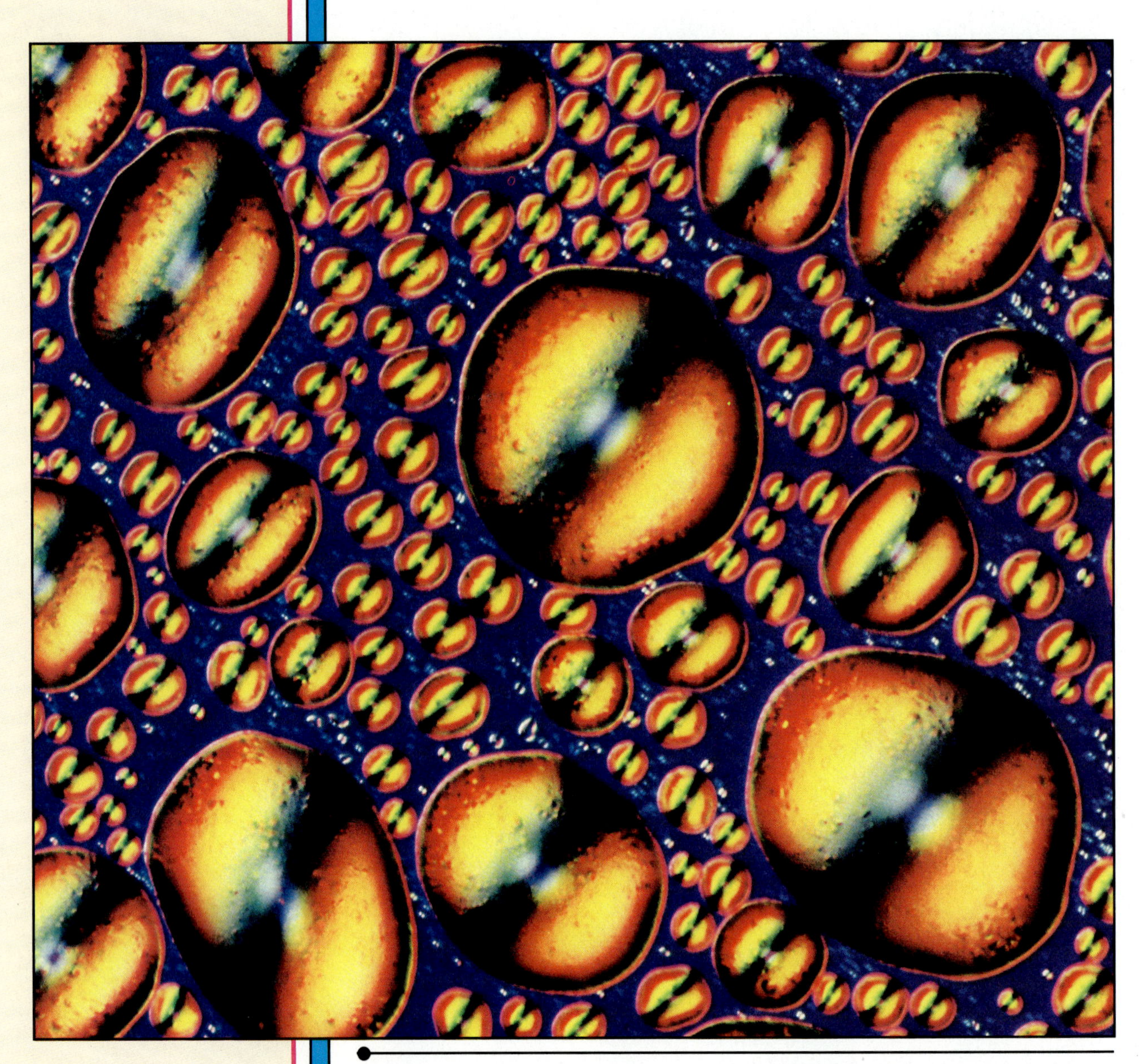

A photomicrograph of crystals of nickel oxide (NiO). These crystals become superconductors when cooled because the bond-oxygen bond is more covalent than it would be in an oxide of one of the main-group metals. Whereas $Ni^{2+}$ ions are relatively small compared with $O^{2+}$ ions, the covalent radius of nickel is significantly larger than oxygen. Thus, the nickel atoms in this compound are large enough to almost touch in this crystal.

CHAPTER
12

# TRANSITION-METAL CHEMISTRY

The elements in the periodic table are often divided into the four categories shown in Figure 12.1: (1) main-group elements, (2) transition metals, (3) lanthanides, and (4) actinides. The *main-group elements* include the active metals in the two columns on the extreme left of the periodic table and the metals, semimetals, and nonmetals in the six columns on the far right. The **transition metals** are the metallic elements that serve as a bridge, or transition, between the two sides of the table. The *lanthanides* and the *actinides* at the bottom of the table are sometimes known as the *inner transition metals* because they have atomic numbers that fall between the first and second elements in the last two rows of the transition metals. This chapter probes the following questions:

## POINTS OF INTEREST

- Why does cobalt form complexes, such as $CoCl_3 \cdot 4\ NH_3$ and $CoCl_3 \cdot 6\ NH_3$?
- Why does the $CoCl_3 \cdot 6\ NH_3$ complex dissociate in water to give four ions, whereas $CoCl_3 \cdot 4\ NH_3$ gives a solution that only contains two ions?
- Why does the $Cu^{2+}$ ion form covalent bonds to $NH_3$ molecules to give $Cu(NH_3)_4{}^{2+}$ complex ions? Why don't the active metals form similar complexes?
- Why do solutions of the $VO_2{}^+$ ion change from yellow to green to blue and then eventually to violet as this ion is reduced, one electron at a time, to the $V^{2+}$ ion?
- How does nature ensure that vitamin $B_{12}$ molecules contain $Co^{3+}$ ions and hemoglobin molecules contain $Fe^{3+}$ ions, and not vice versa?

Main-group Elements | Transition Metals | Main-group Elements

| | | | | | | | | | | | | | | | | | |
|---|---|---|---|---|---|---|---|---|---|---|---|---|---|---|---|---|---|
| H | | | | | | | | | | | | | | | | H | He |
| Li | Be | | | | | | | | | | | B | C | N | O | F | Ne |
| Na | Mg | | | | | | | | | | | Al | Si | P | S | Cl | Ar |
| K | Ca | Sc | Ti | V | Cr | Mn | Fe | Co | Ni | Cu | Zn | Ga | Ge | As | Se | Br | Kr |
| Rb | Sr | Y | Zr | Nb | Mo | Tc | Ru | Rh | Pb | Ag | Cd | In | Sn | Sb | Te | I | Xe |
| Cs | Ba | La | Hf | Ta | W | Re | Os | Ir | Pt | Au | Hg | Tl | Pb | Bi | Po | At | Rn |
| Fr | Ra | Ac | Rf | Ha | 106 | 107 | 108 | 109 | | | | | | | | | |

| | | | | | | | | | | | | | | |
|---|---|---|---|---|---|---|---|---|---|---|---|---|---|---|
| Lanthanides | Ce | Pr | Nd | Pm | Sm | Eu | Gd | Tb | Dy | Ho | Er | Tm | Yb | Lu |
| Actinides | Th | Pa | U | Np | Pu | Am | Cm | Bk | Cf | Es | Fm | Md | No | Lr |

**Figure 12.1** Elements in the periodic table are divided into four categories: (1) main-group elements, (2) transition metals, (3) lanthanides, and (4) actinides.

## 12.1 THE TRANSITION METALS

There is some controversy about the classification of the elements on the boundary between the main-group and transition-metal elements on the right side of the table (see Figure 12.2). The elements in question are zinc (Zn), cadmium (Cd), and mercury (Hg). The disagreement about whether these elements should be classified as main-group elements or transition metals suggests that the differences between these categories are not clear. Transition metals are like main-group metals in many ways: They look like metals, they are malleable and ductile, they conduct heat and electricity, and they form positive ions. The fact that the two best conductors of electricity are a transition metal (copper) and a main-group metal (aluminum) shows the extent to which the physical properties of main-group metals and transition metals overlap.

Transition metals | Main-group elements

| | | | | | | | | | |
|---|---|---|---|---|---|---|---|---|---|
| | | | | | | | | H | He |
| | | | | B | C | N | O | F | Ne |
| | | | | Al | Si | P | S | Cl | Ar |
| Co | Ni | Cu | Zn | Ga | Ge | As | Se | Br | Kr |
| Rh | Pd | Ag | Cd | In | Sn | Sb | Te | I | Xe |
| Ir | Pt | Au | Hg | Tl | Pb | Bi | Po | At | Rn |

**Figure 12.2** There is some debate about whether zinc, cadmium, and mercury should be classified as transition metals or main-group metals. We will classify them as transition metals, because they occur in the transition between the two ends of the table where *d* orbitals are filled.

There are also differences between these metals. The transition metals are more electronegative than the main-group metals, for example, and are therefore more likely to form covalent compounds.

Another difference between the main-group metals and transition metals can be seen in the formulas of the compounds they form. The main-group metals tend to form salts (such as NaCl, $Mg_3N_2$, and CaS) in which there are just enough negative ions to balance the charge on the positive ions. The transition metals form similar compounds [such as $FeCl_3$, $HgI_2$, or $Cd(OH)_2$], but they are more likely than main-group metals to form complexes, such as the $FeCl_4^-$, $HgI_4^{2-}$, and $Cd(OH)_4^{2-}$ ions, that have an excess number of negative ions.

A third difference between main-group and transition-metal ions is the ease with which they form stable compounds with neutral molecules, such as water or ammonia. Salts of main-group metal ions dissolve in water to form aqueous solutions:

$$NaCl(s) \xrightarrow{H_2O} Na^+(aq) + Cl^-(aq)$$

When we let the water evaporate, we get back the original starting material, NaCl(*s*). Salts of the transition-metal ions can display a very different behavior. Chromium(III) chloride, for example, is a violet compound that dissolves in liquid ammonia to form a yellow compound with the formula $CrCl_3 \cdot 6\ NH_3$, which can be isolated when the ammonia is allowed to evaporate:

$$CrCl_3(s) + 6\ NH_3(l) \longrightarrow CrCl_3 \cdot 6\ NH_3(s)$$

One way transition metals differ from main-group metals is their ability to bind ligands to form compounds with different chemical and physical properties. $CrCl_3$ (shown on the left) dissolves in liquid ammonia to form $CrCl_3 \cdot 6\ NH_3$ (shown on the right).

The $FeCl_4^-$ ion and $CrCl_3 \cdot 6\ NH_3$ are called **coordination compounds** because they contain ions or molecules linked, or coordinated, to a transition metal. They are also known as **complex ions** or **coordination complexes,** because they are Lewis acid–base complexes. The ions or molecules that bind to transition-metal ions to form these complexes are called **ligands** (from Latin, ''to tie or bind''). The number of ligands bound to the transition-metal ion is called the **coordination number.** A transition metal complex with six ligands, for example, is referred to as six-coordinate.

Although coordination complexes are particularly important in the chemistry of the transition metals, some main-group elements also form complexes. Aluminum, tin, and lead, for example, form complexes such as the $AlF_6^{3-}$, $SnCl_4^{2-}$, and $PbI_4^{2-}$ ions.

## 12.2 THE PROCESS OF DISCOVERY: WERNER'S THEORY OF COORDINATION COMPLEXES

Alfred Werner first became interested in coordination complexes in 1892, while preparing lectures for a course on atomic theory. Six months later, he proposed a model for coordination complexes that still serves as the basis for work in this field. Werner then spent the remainder of his life collecting evidence to support his theory. Before we introduce Werner's model of coordination complexes, let's look at some of the experimental data available at the turn of the century.

1. At least three different cobalt(III) complexes can be isolated when $CoCl_2$ is dissolved in aqueous ammonia and then oxidized by air to the +3 oxidation state. A fourth complex can be made by slightly different techniques. These complexes have different colors and different empirical formulas:

| | |
|---|---|
| $CoCl_3 \cdot 6\ NH_3$ | orange-yellow |
| $CoCl_3 \cdot 5\ NH_3 \cdot H_2O$ | red |
| $CoCl_3 \cdot 5\ NH_3$ | purple |
| $CoCl_3 \cdot 4\ NH_3$ | green |

2. The reactivity of the ammonia in these complexes has been drastically reduced. By itself, ammonia reacts rapidly with hydrochloric acid to form ammonium chloride:

$$NH_3(aq) + HCl(aq) \longrightarrow NH_4^+(aq) + Cl^-(aq)$$

These complexes don't react with hydrochloric acid, even at 100°C:

$$CoCl_3 \cdot 6\ NH_3(aq) + HCl(aq) \not\longrightarrow$$

3. Solutions of the $Cl^-$ ion react with $Ag^+$ ion to form a white precipitate of AgCl:

$$Ag^+(aq) + Cl^-(aq) \longrightarrow AgCl(s)$$

When excess $Ag^+$ ion is added to solutions of the $CoCl_3 \cdot 6\ NH_3$ and $CoCl_3 \cdot 5\ NH_3 \cdot H_2O$ complexes, three moles of AgCl are formed for each mole of complex in solution, as might be expected. However, only two of the $Cl^-$ ions in the $CoCl_3 \cdot 5\ NH_3$ complex and only one of the $Cl^-$ ions in $CoCl_3 \cdot 4\ NH_3$ can be precipitated with $Ag^+$ ions.

4. Measurements of the conductivity of aqueous solutions of these complexes suggest that the $CoCl_3 \cdot 6\ NH_3$ and $CoCl_3 \cdot 5\ NH_3 \cdot H_2O$ complexes dissociate in water to give a total of four ions. $CoCl_3 \cdot 5\ NH_3$ dissociates to give three ions, and $CoCl_3 \cdot 4\ NH_3$ dissociates to give only two ions.

Werner explained these observations by suggesting that transition-metal ions such as the $Co^{3+}$ ion have a primary valence and a secondary valence. The *primary valence* is the number of negative ions needed to satisfy the charge on the metal ion. In each of the cobalt(III) complexes previously described, three $Cl^-$ ions are needed to satisfy the primary valence of the $Co^{3+}$ ion.

The *secondary valence* is the number of ions or molecules that are coordinated to the metal ion. Werner assumed that the secondary valence of the transition metal in these cobalt(III) complexes is six. The formulas of these compounds therefore can be written as follows.

| | |
|---|---|
| $[Co(NH_3)_6{}^{3+}][Cl^-]_3$ | orange-yellow |
| $[Co(NH_3)_5(H_2O)^{3+}][Cl^-]_3$ | red |
| $[Co(NH_3)_5Cl^{2+}][Cl^-]_2$ | purple |
| $[Co(NH_3)_4Cl_2{}^+][Cl^-]$ | green |

Brackets are used in these formulas to distinguish between ligands that contribute to the primary and secondary valences of the cobalt atom.

**CHECKPOINT**

Determine the primary and secondary valences of the chromium atom in the $Cr(NH_3)_4(H_2O)_2Cl_3$ complex.

## EXERCISE 12.1

Describe how Werner's theory of coordination complexes explains both the number of ions formed when the following compounds dissolve in water and the number of chloride ions that precipitate when these solutions are treated with $Ag^+$ ions.

(a) $[Co(NH_3)_6]Cl_3$
(b) $[Co(NH_3)_5(H_2O)]Cl_3$
(c) $[Co(NH_3)_5Cl]Cl_2$
(d) $[Co(NH_3)_4Cl_2]Cl$

**SOLUTION** The cobalt ion is coordinated to a total of six ligands in each complex, which satisfies the secondary valence of this ion. Each complex also has a total of three chloride ions that satisfy the primary valence. Some of the $Cl^-$ ions are free to dissociate when the complex dissolves in water. Others are bound to the $Co^{3+}$ ion and neither dissociate nor react with $Ag^+$.

(a) The three chloride ions in $[Co(NH_3)_6]Cl_3$ are free to dissociate when this complex dissolves in water. The complex therefore dissociates in water to give a total of four ions, and all three $Cl^-$ ions are free to react with $Ag^+$ ion:

$$[Co(NH_3)_6]Cl_3(s) \xrightarrow{H_2O} Co(NH_3)_6{}^{3+}(aq) + 3\,Cl^-(aq)$$

(b) Once again, the three $Cl^-$ ions are free to dissociate when $[Co(NH_3)_5(H_2O)]Cl_3$ dissolves in water, and they precipitate when $Ag^+$ ions are added to the solution:

$$[Co(NH_3)_5(H_2O)]Cl_3(s) \xrightarrow{H_2O} Co(NH_3)_5(H_2O)^{3+}(aq) + 3\,Cl^-(aq)$$

(c) One of the chloride ions is bound to the cobalt in the $[Co(NH_3)_5Cl]Cl_2$ complex. Only three ions are formed when this compound dissolves in water, and only two $Cl^-$ ions are free to precipitate with $Ag^+$ ions:

$$[Co(NH_3)_5Cl]Cl_2(s) \xrightarrow{H_2O} Co(NH_3)_5Cl^{2+}(aq) + 2\,Cl^-(aq)$$

(d) Two of the chloride ions are bound to the cobalt in $[Co(NH_3)_4Cl_2]Cl$. Only two ions are formed when this compound dissolves in water, and only one $Cl^-$ ion is free to precipitate with $Ag^+$ ions:

$$[Co(NH_3)_4Cl_2]Cl(s) \xrightarrow{H_2O} Co(NH_3)_4Cl_2{}^+(aq) + Cl^-(aq)$$

Werner also assumed that transition-metal complexes had definite shapes. According to his theory, the ligands in six-coordinate cobalt(III) complexes are oriented toward the corners of an octahedron, as shown in Figure 12.3.

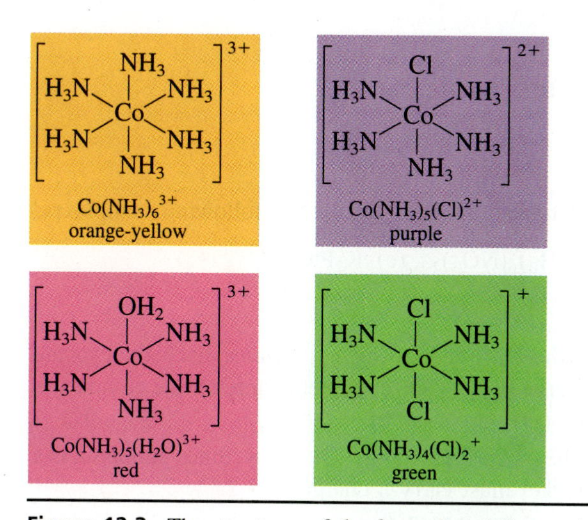


**Figure 12.3** The structures of the four cobalt(III) complexes that played a major role in the development of Werner's theory of coordination complexes.

## 12.3 TYPICAL COORDINATION NUMBERS

Transition-metal complexes have been characterized with coordination numbers that range from 1 to 12, but the most common coordination numbers are 2, 4, and 6. Examples of complexes with these coordination numbers are given in Table 12.1.

**TABLE 12.1 Examples of Common Coordination Numbers**

| Metal Ion | | Ligand | | Complex | Coordination Number |
|---|---|---|---|---|---|
| $Ag^+$ | + | $2\ NH_3$ | $\rightleftharpoons$ | $Ag(NH_3)_2{}^+$ | 2 |
| $Ag^+$ | + | $2\ S_2O_3{}^{2-}$ | $\rightleftharpoons$ | $Ag(S_2O_3)_2{}^{3-}$ | 2 |
| $Ag^+$ | + | $2\ Cl^-$ | $\rightleftharpoons$ | $AgCl_2{}^-$ | 2 |
| $Pb^{2+}$ | + | $2\ OAc^-$ | $\rightleftharpoons$ | $Pb(OAc)_2$ | 2 |
| $Cu^+$ | + | $2\ NH_3$ | $\rightleftharpoons$ | $Cu(NH_3)_2{}^+$ | 2 |
| $Cu^{2+}$ | + | $4\ NH_3$ | $\rightleftharpoons$ | $Cu(NH_3)_4{}^{2+}$ | 4 |
| $Zn^{2+}$ | + | $4\ CN^-$ | $\rightleftharpoons$ | $Zn(CN)_4{}^{2-}$ | 4 |
| $Hg^{2+}$ | + | $4\ I^-$ | $\rightleftharpoons$ | $HgI_4{}^{2-}$ | 4 |
| $Co^{2+}$ | + | $4\ SCN^-$ | $\rightleftharpoons$ | $Co(SCN)_4{}^{2-}$ | 4 |
| $Fe^{2+}$ | + | $6\ H_2O$ | $\rightleftharpoons$ | $Fe(H_2O)_6{}^{2+}$ | 6 |
| $Fe^{3+}$ | + | $6\ H_2O$ | $\rightleftharpoons$ | $Fe(H_2O)_6{}^{3+}$ | 6 |
| $Co^{3+}$ | + | $6\ NH_3$ | $\rightleftharpoons$ | $Co(NH_3)_6{}^{3+}$ | 6 |
| $Ni^{2+}$ | + | $6\ NH_3$ | $\rightleftharpoons$ | $Ni(NH_3)_6{}^{2+}$ | 6 |
| $Fe^{2+}$ | + | $6\ CN^-$ | $\rightleftharpoons$ | $Fe(CN)_6{}^{4-}$ | 6 |

Coordination complexes are often used to measure the concentrations of transition-metal ion solutions. The intensity of the blue color of the $Cu(NH_3)_6{}^{2+}$ ion, for example, is directly proportional to the concentration of this complex.

Note that the charge on the complex is always the sum of the charges on the ions or molecules that form the complex:

$$Cu^{2+} + 4\ NH_3 \rightleftharpoons Cu(NH_3)_4{}^{2+}$$
$$Pb^{2+} + 2\ OAc^- \rightleftharpoons Pb(OAc)_2$$
$$Fe^{2+} + 6\ CN^- \rightleftharpoons Fe(CN)_6{}^{4-}$$

Note also that the coordination number of a complex often increases as the charge on the metal ion becomes larger:

$$Cu^+ + 2\ NH_3 \rightleftharpoons Cu(NH_3)_2{}^+$$
$$Cu^{2+} + 4\ NH_3 \rightleftharpoons Cu(NH_3)_4{}^{2+}$$

### EXERCISE 12.2

Calculate the charge on the transition-metal ion in the following complexes: (a) $Na_2Co(SCN)_4$ (b) $Ni(NH_3)_6(NO_3)_2$ (c) $K_2PtCl_6$

**SOLUTION**

(a) This complex contains the $Na^+$ and $Co(SCN)_4{}^{2-}$ ions. Each thiocyanate ($SCN^-$) ion carries a charge of $-1$. Since the net charge on the complex is $-2$, the cobalt ion must carry a charge of $+2$.

(b) This complex contains the $Ni(NH_3)_6{}^{2+}$ and $NO_3{}^-$ ions. Since ammonia is a neutral molecule, the nickel must carry a charge of $+2$.

(c) This complex contains the $K^+$ and $PtCl_6{}^{2-}$ ions. Each chloride ion carries a charge of $-1$. Since the net charge on the $PtCl_6{}^{2-}$ ion is $-2$, the platinum must carry a charge of $+4$.

## 12.4 THE ELECTRON CONFIGURATION OF TRANSITION-METAL IONS

The electron configurations of ions formed by the main-group metals are easy to understand. Aluminum, for example, loses its three valence electrons when it forms $Al^{3+}$ ions:

$$\text{Al:} \quad [\text{Ne}]\ 3s^2\ 3p^1$$
$$\text{Al}^{3+}\text{:} \quad [\text{Ne}]$$

The relationship between the electron configurations of transition-metal elements and their ions is more complex. Consider the chemistry of cobalt, for example, which forms complexes that contain either $Co^{2+}$ or $Co^{3+}$ ions.

Which valence electrons are removed from a cobalt atom when the $Co^{2+}$ and $Co^{3+}$ ions are formed? The electron configuration of a neutral cobalt atom is written as follows:

$$\text{Co:} \quad [\text{Ar}]\ 4s^2\ 3d^7$$

The discussion of the relative energies of the atomic orbitals suggests that the 4*s* orbitals have lower energy than the 3*d* orbitals. Thus, we might expect cobalt to lose electrons from the higher energy 3*d* orbitals, but this is not what is observed. The $Co^{2+}$ and $Co^{3+}$ ions have the following electron configurations:

Sec. 6.20

$$\text{Co}^{2+}\text{:} \quad [\text{Ar}]\ 3d^7$$
$$\text{Co}^{3+}\text{:} \quad [\text{Ar}]\ 3d^6$$

In general, electrons are removed from the valence-shell *s* orbitals before they are removed from valence *d* orbitals when transition metals are ionized.

Why are electrons removed from 4*s* orbitals before 3*d* orbitals if they are placed in 4*s* orbitals before 3*d* orbitals? There are several ways of answering this question. First, note that the difference between the energies of the 3*d* and 4*s* orbitals is very small. For this reason, the electron configurations for chromium and copper differ slightly from the predictions of the aufbau principle:

Sec. 6.21

$$\text{Cr:} \quad [\text{Ar}]\ 4s^1\ 3d^5$$
$$\text{Cu:} \quad [\text{Ar}]\ 4s^1\ 3d^{10}$$

Note also that the relative energies of the 4*s* and 3*d* orbitals used during the aufbau process hold true for neutral atoms. When transition metals form positive ions, the 3*d* orbitals become more stable than the 4*s* orbitals. As a result, electrons are removed from the 4*s* orbitals before the 3*d* orbitals.

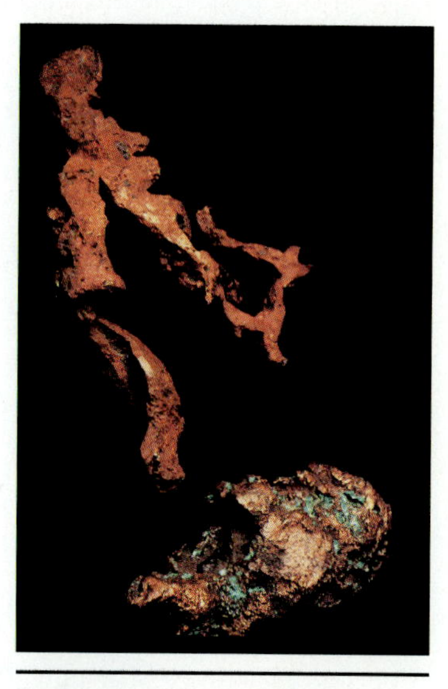

Copper metal can be found in its elemental form in nature. When this happens, it is called "native" copper.

### EXERCISE 12.3

Predict the electron configuration of the $Fe^{3+}$ ion.

**SOLUTION** We start with the configuration of a neutral iron atom:

$$\text{Fe:} \quad [\text{Ar}]\ 4s^2\ 3d^6$$

We then remove three electrons to form the +3 ion. Two of these electrons come from the 4*s* orbital. The other comes from a 3*d* orbital.

$$\text{Fe}^{3+}\text{:} \quad [\text{Ar}]\ 3d^5$$

Because the valence electrons in transition-metal ions are concentrated in *d* orbitals, these ions are often described as having $d^n$ configurations. The $Co^{3+}$ and $Fe^{2+}$ ions, for example, are said to have a $d^6$ configuration.

$$\text{Co}^{3+}\text{:} \quad [\text{Ar}]\ 3d^6$$
$$\text{Fe}^{2+}\text{:} \quad [\text{Ar}]\ 3d^6$$

## 12.5 OXIDATION STATES OF THE TRANSITION METALS

Most transition metals form more than one oxidation state. Manganese, for example, forms compounds in any oxidation state from $-1$ to $+7$. Some oxidation states, however, are more common than others. The most common oxidation states of the first series of transition metals are given in Table 12.2. Efforts to explain the apparent pattern in this table ultimately fail for a combination of reasons. Some of these oxidation states are common because they are relatively stable. Others describe compounds that are not necessarily stable but which react slowly. Still others are common only from an historic perspective.

**TABLE 12.2 Common Oxidation States of the First Series of Transition Metals**

| | Sc | Ti | V | Cr | Mn | Fe | Co | Ni | Cu | Zn |
|---|---|---|---|---|---|---|---|---|---|---|
| +1 | | | | | | | | | $d^{10}$ | |
| +2 | | | $d^3$ | | $d^5$ | $d^6$ | $d^7$ | $d^8$ | $d^9$ | $d^{10}$ |
| +3 | $d^0$ | | | $d^3$ | | $d^5$ | $d^6$ | | | |
| +4 | | $d^0$ | | | $d^3$ | | | | | |
| +5 | | | $d^0$ | | | | | | | |
| +6 | | | | $d^0$ | | | | | | |
| +7 | | | | | $d^0$ | | | | | |

Sec. 8.7

One point about the oxidation states of transition metals deserves particular attention: Transition-metal ions with charges larger than +3 cannot exist in aqueous solution. Consider the following reaction, for example, in which manganese is oxidized from the +2 to the +7 oxidation state:

$$Mn^{2+}(aq) + 4\,H_2O(l) \longrightarrow MnO_4^{-}(aq) + 8\,H^{+}(aq) + 5\,e^{-}$$

When the manganese atom is oxidized, it becomes more electronegative. In the +7 oxidation state, this atom is electronegative enough to react with water to form a covalent oxide, $MnO_4^-$.

It is useful to have a way to distinguish between the charge on a transition-metal ion and the oxidation state of the transition metal. By convention, symbols such as $Mn^{2+}$ refer to ions that carry a +2 charge. Symbols such as Mn(VII) are used to describe compounds in which manganese is in the +7 oxidation state.

Mn(VII) is not the only example of an oxidation state powerful enough to decompose water. As soon as $Mn^{2+}$ is oxidized to Mn(IV), it reacts with water to form $MnO_2$. A similar phenomenon can be seen in the chemistry of both vanadium and chromium. Vanadium exists in aqueous solutions as the $V^{2+}$ ion. But once it is oxidized to the +4 or +5 oxidation state, it reacts with water to form the $VO^{2+}$ or $VO_2^+$ ion. The $Cr^{3+}$ ion can be found in aqueous solution. But once this ion is oxidized to Cr(VI), it reacts with water to form the $CrO_4^{2-}$ and $Cr_2O_7^{2-}$ ions.

## 12.6 LEWIS ACID–LEWIS BASE APPROACH TO BONDING IN COMPLEXES

G. N. Lewis was the first to recognize that the reaction between a transition-metal ion and ligands to form a coordination complex was analogous to the reaction between the $H^+$ and $OH^-$ ions to form water. The reaction between $H^+$ and $OH^-$ ions involves the donation of a pair of electrons from the $OH^-$ ion to the $H^+$ ion to form a covalent bond:

$$H^{+} \; :\ddot{\underset{..}{O}}{-}H^{-} \longrightarrow H{-}\ddot{\underset{..}{O}}{-}H$$

The $H^+$ ion can be described as an **electron-pair acceptor.** The $OH^-$ ion, on the other hand, is an **electron-pair donor.** Lewis argued that any ion or molecule that behaves like the $H^+$ ion should be an acid. Conversely, any ion or molecule that behaves like the $OH^-$ ion should be a base. A **Lewis acid** is therefore any ion or molecule that can accept a pair of electrons. A **Lewis base** is an ion or molecule that can donate a pair of electrons.

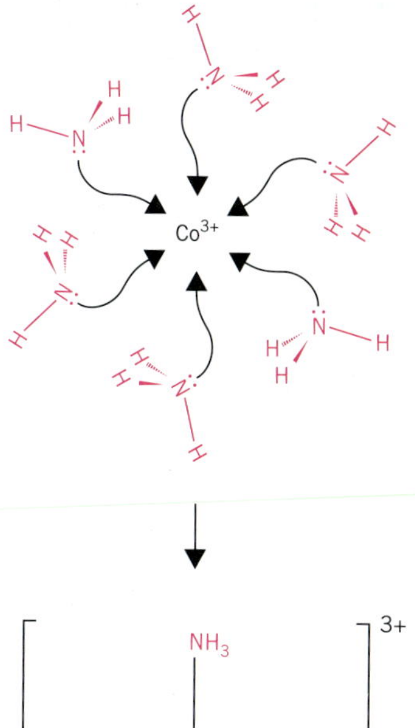


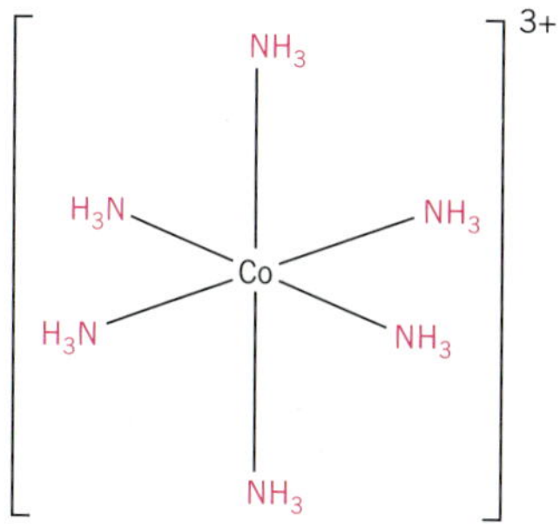


**Figure 12.4** The reaction between $Co^{3+}$ ions and ammonia to form the $Co(NH_3)_6{}^{3+}$ ion is a Lewis acid–base reaction. Each of the six $NH_3$ molecules acts as a Lewis base, donating a pair of nonbonding electrons to the $Co^{3+}$ ion. The $Co^{3+}$ ion acts as a Lewis acid, using empty valence-shell orbitals to pick up pairs of nonbonding electrons to form covalent cobalt–nitrogen bonds.

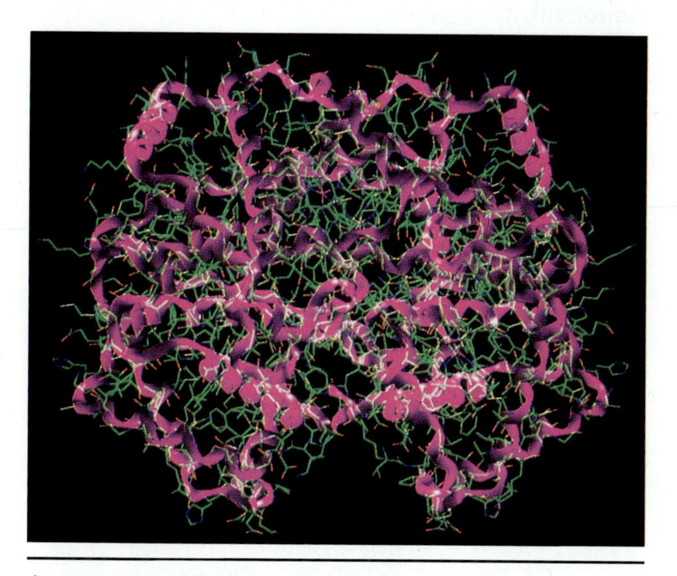

A computer model of the structure of hemoglobin, the oxygen-carrying protein in blood. Oxygen forms a Lewis acid-base complex with the iron atoms represented by the yellow balls in this structure.

When $Co^{3+}$ ions react with ammonia, the $Co^{3+}$ ion accepts pairs of nonbonding electrons from six $NH_3$ ligands to form covalent cobalt–nitrogen bonds (see Figure 12.4). The metal ion is therefore a Lewis acid, and the ligands coordinated to this metal ion are Lewis bases:

$$Co^{3+} + 6\ NH_3 \rightleftharpoons Co(NH_3)_6{}^{3+}$$

| $Co^{3+}$ | $6\ NH_3$ | $Co(NH_3)_6{}^{3+}$ |
|---|---|---|
| electron-pair acceptor (Lewis acid) | electron-pair donor (Lewis base) | acid–base complex |

The $Co^{3+}$ ion is an electron-pair acceptor, or Lewis acid, because it has empty valence-shell orbitals that can be used to hold pairs of electrons. To emphasize these empty valence orbitals, we can write the configuration of the $Co^{3+}$ ion as follows:

$$Co^{3+}: \quad [Ar]\ 3d^6\ 4s^0\ 4p^0$$

There is room in the valence shell of this ion for 12 more electrons. (Four electrons can be added to the $3d$ subshell, two to the $4s$ orbital, and six to the $4p$ subshell.) The $NH_3$ molecule is an electron-pair donor, or Lewis base, because it has a pair of nonbonding electrons on the nitrogen atom.

According to this model, transition-metal ions form coordination complexes because they have empty valence-shell orbitals that can accept pairs of electrons from a Lewis base. Ligands must therefore be Lewis bases: They must contain at least one pair of nonbonding electrons that can be donated to a metal ion.

## 12.7 TYPICAL LIGANDS

Any ion or molecule with a pair of nonbonding electrons can be a ligand. Many ligands are described as **monodentate** (literally, ''one-toothed'') because they ''bite'' the metal in only one place. Typical monodentate ligands are given in Table 12.3.

**TABLE 12.3 Typical Monodentate Ligands**

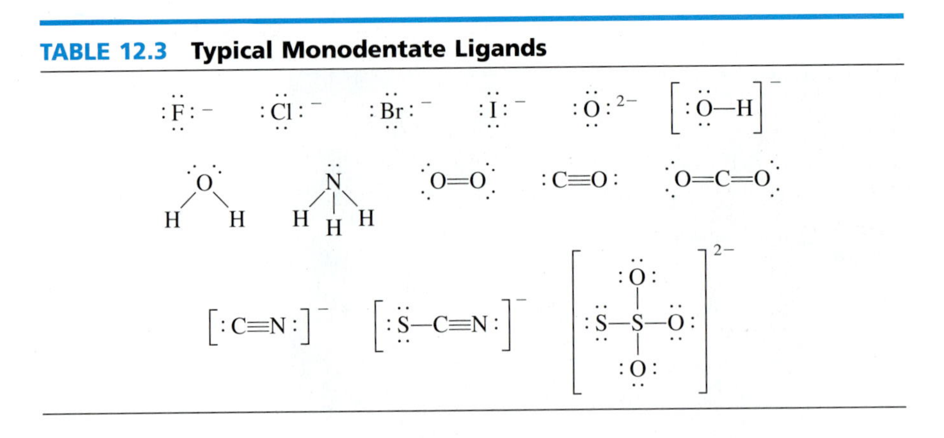

Other ligands can attach to the metal more than once. Ethylenediamine (en) is a typical **bidentate ligand:**

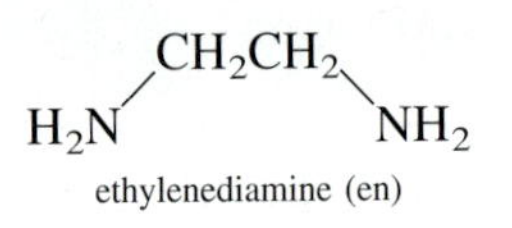


ethylenediamine (en)

Each end of this molecule contains a pair of nonbonding electrons that can form a covalent bond to a metal ion. Ethylenediamine is also an example of a **chelating ligand.** The term *chelate* comes from a Greek stem meaning ''claw.'' It is used to describe ligands that can grab the metal in two or more places, the way a claw would. A typical ethylenediamine complex is shown in Figure 12.5.

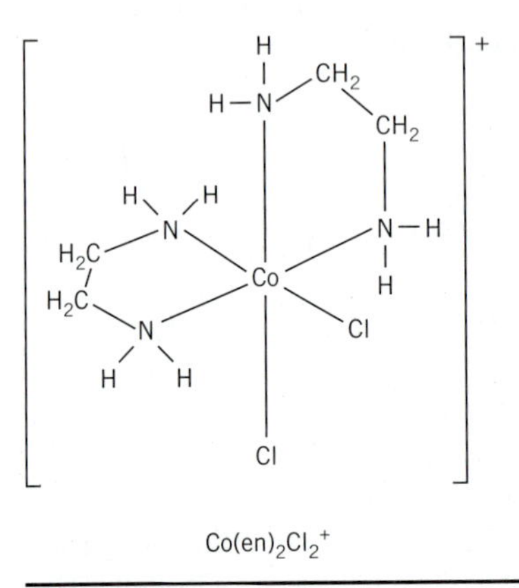

$Co(en)_2Cl_2^+$

**Figure 12.5** Ethylenediamine is a bidentate (literally, ''two-toothed'') ligand, which ''bites'' the metal in two places. Because multidentate ligands hold the metal as if they were claws, these ligands are also known as chelating ligands (from the Greek name for claw).

A number of multidentate ligands are shown in Table 12.4. Linking ethylenediamine fragments gives *tridentate ligands* and *tetradentate ligands,* such as diethylenetriamine (dien) and triethylenetetramine (trien). Adding four $—CH_2CO_2^-$ groups to an ethylenediamine framework gives a *hexadentate ligand,* which can single-handedly satisfy the secondary valence of a transition-metal ion, as shown in Figure 12.6.

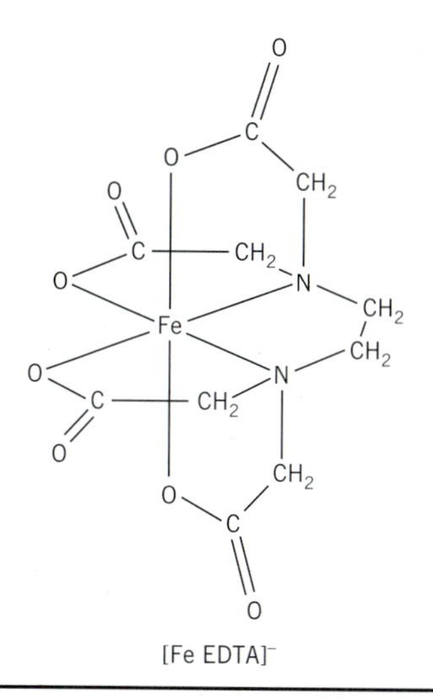


**Figure 12.6** EDTA is a hexadentate ligand that can single-handedly satisfy the secondary valence of a transition-metal ion. EDTA therefore forms very strong complexes with many transition-metal ions.

## ▶ CHECKPOINT

Use Lewis structures to explain why the acetylacetonate ligand (acac) is bidentate and the nitrilotriacetate ligand (NTA) is tetradentate.

**TABLE 12.4 Typical Multidentate Ligands**

| Structure | Name |
|---|---|
| **Bidentate Ligands** | |
| 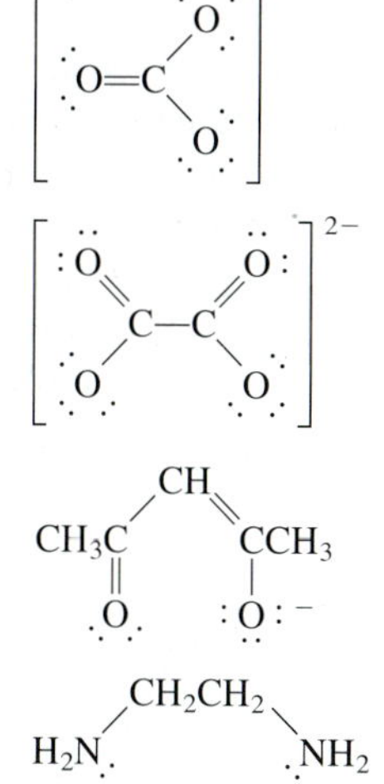 | carbonate |
| | oxalate (ox) |
| | acetylacetonate (acac) |
| $H_2N—CH_2CH_2—NH_2$ | ethylenediamine (en) |
| **Tridentate Ligands** | |
| $H_2NCH_2CH_2NHCH_2CH_2NH_2$ | diethylenetriamine (dien) |
| **Tetradentate Ligands** | |
| 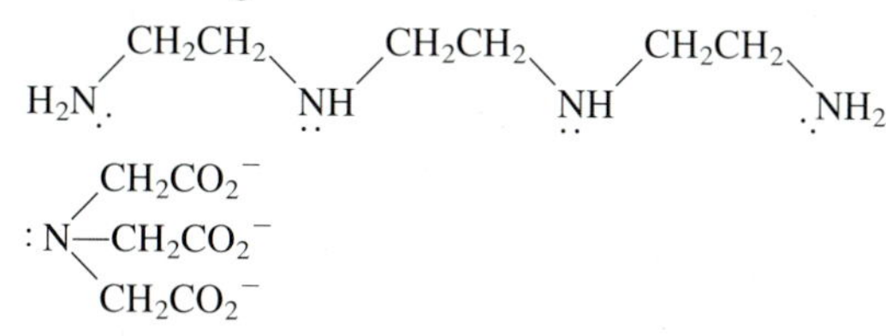 | triethylenetetraamine (trien) |
| | nitrilotriacetate (NTA) |
| **Hexadentate Ligands** | |
| 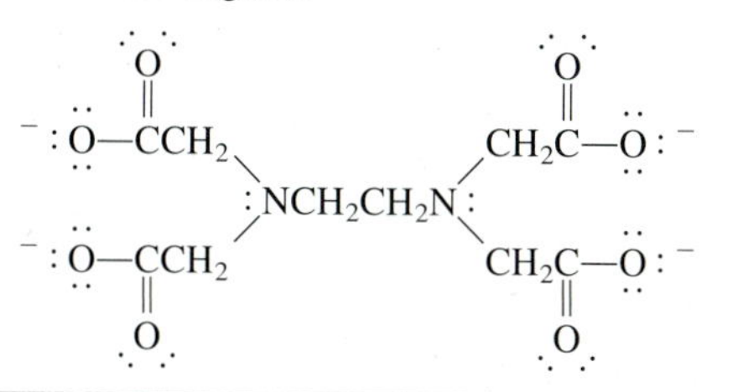 | ethylenediaminetetraacetate (EDTA) |

## RESEARCH IN THE 90s

### NITROGEN FIXATION

Sec. 10.5

Ever since the Haber process was developed, chemists have been fascinated by the contrast between the conditions required to reduce nitrogen to ammonia in the laboratory and the conditions under which this reaction occurs in nature:

$$N_2(g) + 3\,H_2(g) \xrightarrow{Fe} 2\,NH_3(g)$$

In the laboratory, this reaction requires high pressures (200 to 300 atm) and high temperatures (400°C to 600°C). In nature, it occurs at room temperature and atmospheric pressure.

The nitrogenase enzyme that catalyzes this reaction in nature can be separated into two proteins, which contain a total of 34 iron atoms and two molybdenum atoms. One of these proteins contains both Fe and Mo; the other contains only Fe.

The Fe protein contains four iron atoms, in a $Fe_4S_4$ unit, which bind the ATP (adenosine triphosphate) that provides the energy necessary to reduce nitrogen to ammonia. The MoFe protein consists of four subunits ($\alpha_2\beta_2$). Each subunit includes an $Fe_4S_4$ unit that can hold one of the electrons necessary to reduce $N_2$. The remaining metal atoms can be found in a pair of $MoFe_7S_8$ units. One end of these $MoFe_7S_8$ units is bound to the —SH group on the side chain of a residue of the amino acid cysteine in the protein; the other is bound to an analog of the citrate ion known as *homocitrate.*

The Fe protein uses the energy given off when a pair of ATP molecules is hydrolyzed to pump an electron into the MoFe protein, which binds the $N_2$ molecule. Although it takes only six electrons to reduce nitrogen to ammonia, the nitrogenase enzyme consumes a total of eight electrons. (The extra electrons are used to reduce a pair of $H^+$ ions to an $H_2$ molecule.) Thus, the overall stoichiometry of this reaction can be written as follows:

$$N_2 + 8\,H^+ + 8\,e^- \longrightarrow 2\,NH_3 + H_2$$

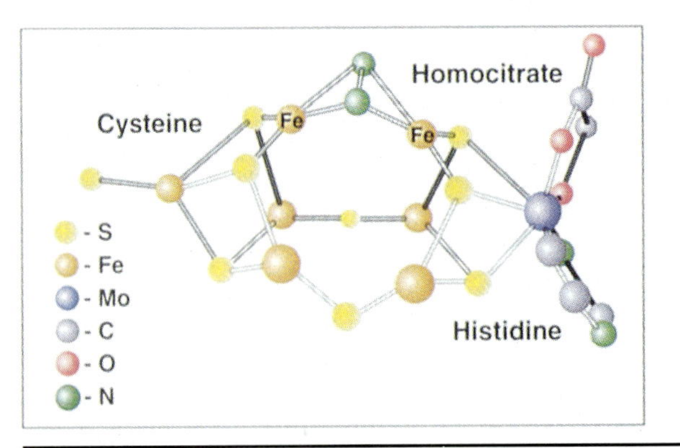


**Figure 12.7** A model for the binding of $N_2$ to the MoFe-protein in the nitrogenase enzyme.

Jongsun Kim and Douglas Rees, of the California Institute of Technology, recently reported x-ray crystal structures of the Fe protein from *Azotobacter vinelandii* and the MoFe protein from *Clostridium pasteurianum* [*Science,* **257,** 1677 (1922)]. Their work suggests that the $N_2$ molecule is bound, side-on, between a pair of iron atoms in the $MoF_7S_8$ (homocitrate) complex, as shown in Figure 12.7.

These results explain why chemists have been frustrated for so many years by their inability to understand the role of the molybdenum atoms in the MoFe protein. Contrary to most researchers' expectations, the $N_2$ molecule is not carried by the molybdenum atoms in this protein. It apparently binds to iron atoms in both the industrial catalyst and the nitrogenase enzyme. The difference between the two catalysts is the ease with which the enzyme system can pump electrons into the metal atom that carries the $N_2$ molecule, thereby reducing it to ammonia.

## 12.8 COORDINATION COMPLEXES IN NATURE

Complex ions play a vital role in the chemistry of living systems, as shown by three biologically important complexes: vitamin $B_{12}$, chlorophyll *a*, and the heme found in the oxygen-carrying proteins hemoglobin and myoglobin.

As early as 1926, it was known that patients who suffered from pernicious anemia became better when they ate liver. It took more than 20 years, however, to determine that the vitamin $B_{12}$ in liver was one of the active factors in relieving the anemia. It took another 10 years to determine the structure of vitamin $B_{12}$ shown in Figure 12.8*a*. Vitamin $B_{12}$ is a six-coordinate $Co^{3+}$ complex. The $Co^{3+}$ ion is coordinated to four nitrogen atoms that lie in a planar molecule known as a *corrin ring*. It is also coordinated to a fifth nitrogen atom and to a sixth group, labeled R in this figure. The R group can be a $CN^-$, $NO_2^-$, $SO_3^{2-}$, or $OH^-$ ion, depending on the source of the vitamin $B_{12}$.

Every carbon atom in our bodies can be traced to a reaction in which plants use the energy in sunlight to make glucose ($C_6H_{12}O_6$) from $CO_2$ and water:

$$6\ CO_2(g) + 6\ H_2O(l) \xrightarrow[\text{chlorophyll}]{\text{light}} C_6H_{12}O_6(aq) + 6\ O_2(g)$$

The fundamental step in the photosynthesis of glucose is the absorption of sunlight by the chlorophyll in higher plants and algae. Chlorophyll *a* (see Figure 12.8*b*) is a complex in which an $Mg^{2+}$ ion is coordinated to four nitrogen atoms in a planar *chlorin ring*. The chlorin ring in chlorophyll is similar to the corrin ring in vitamin $B_{12}$, but not identical. The minor differences between these rings adjust their sizes so that each ring is the right size to hold the correct transition-metal ion.

The coordination complex shown in Figure 12.8*c* is known as a *heme*. Hemes contain an $Fe^{2+}$ or $Fe^{3+}$ ion coordinated to four nitrogen atoms in a *porphyrin ring*. These complexes are found in myoglobin and hemoglobin, which carry oxygen from the lungs to the muscles. The porphyrin ring in a heme is slightly larger than the corrin ring in vitamin $B_{12}$ and slightly smaller than the chlorin ring in chlorophyll *a*, so it is just the right size to hold the $Fe^{2+}$ or $Fe^{3+}$ ion.

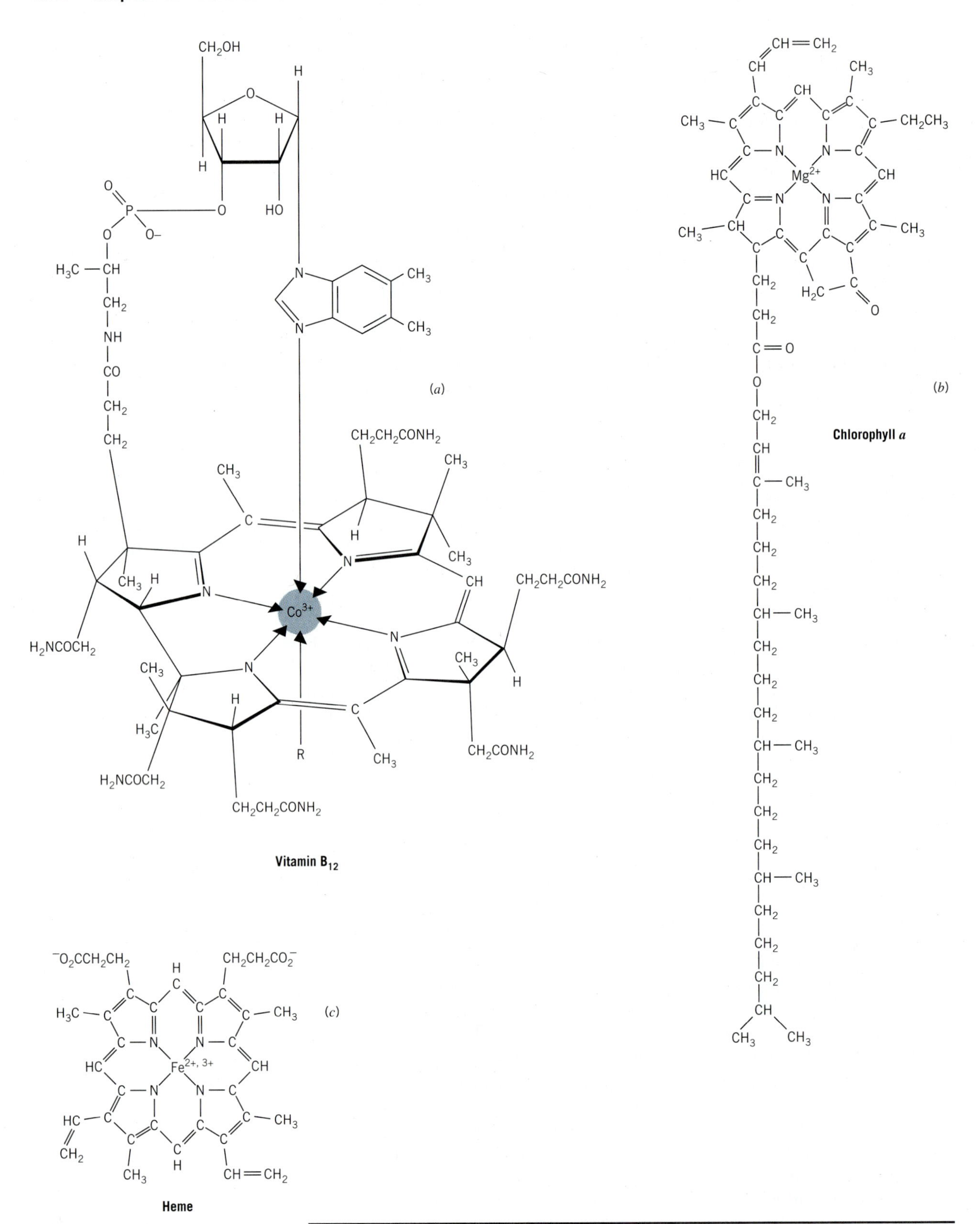


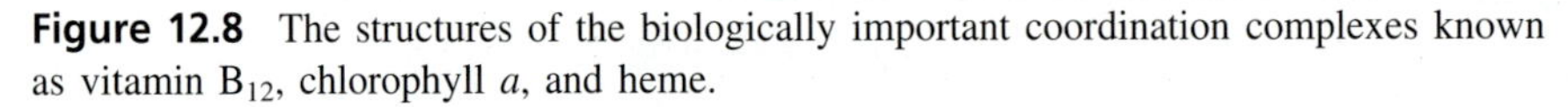

**Figure 12.8** The structures of the biologically important coordination complexes known as vitamin $B_{12}$, chlorophyll *a*, and heme.

## 12.9 NOMENCLATURE OF COMPLEXES

The rules for naming chemical compounds are established by nomenclature committees of the International Union of Pure and Applied Chemistry (IUPAC). The IUPAC nomenclature of coordination complexes is based on the following rules:

**RULES FOR NAMING COORDINATION COMPLEXES**

**1.** The name of the positive ion is written before the name of the negative ion.

**2.** The name of the ligand is written before the name of the metal to which it is coordinated.

**3.** The Greek prefixes *mono-*, *di-*, *tri-*, *tetra-*, *penta-*, *hexa-*, and so on are used to indicate the number of ligands when these ligands are relatively simple. The Greek prefixes *bis-*, *tris-*, and *tetrakis-* are used with more complicated ligands.

**4.** The names of negative ligands always end in *o*, as in *fluoro* ($F^-$), *chloro* ($Cl^-$), *bromo* ($Br^-$), *iodo* ($I-$), *oxo* ($O^{2-}$), *hydroxo* ($OH^-$), and *cyano* ($CN^-$).

**5.** A handful of neutral ligands are given common names, such as *aquo* ($H_2O$), *ammine* ($NH_3$), and *carbonyl* ($CO$).

**6.** Ligands are listed in the following order: negative ions, neutral molecules, and positive ions. Ligands with the same charge are listed in alphabetical order.

**7.** The oxidation number of the metal atom is indicated by a Roman numeral in parentheses after the name of the metal atom.

**8.** The names of complexes with a net negative charge end in *-ate*. $Co(SCN)_4^{2-}$, for example, is the tetrathiocyanatocobaltate(II) ion. When the symbol for the metal is derived from its Latin name, *-ate* is added to the Latin name of the metal. Thus, negatively charged iron complexes are ferrates and negatively charged copper complexes are cuprates.

### EXERCISE 12.4

Name the following coordination complexes:

(a) $K_4Fe(CN)_6$ (d) $[Cr(NH_3)_5(H_2O)](NO_3)_3$
(b) $Fe(acac)_3$ (e) $[Cr(NH_3)_4Cl_2]Cl$
(c) $[Cr(en)_3]Cl_3$

**SOLUTION**

(a) This salt contains the $K^+$ and $Fe(CN)_6^{4-}$ ions. It is therefore potassium hexacyanoferrate(II).
(b) Tris(acetylacetonato)iron(III).
(c) Tris(ethylenediamine)chromium(III) chloride.
(d) Pentaammineaquochromium(III) nitrate.
(e) Dichlorotetraamminechromium(III) chloride.

## 12.10 ISOMERS

### (CIS/TRANS) ISOMERS

An important test of Werner's theory of coordination complexes involved the study of coordination complexes that formed **isomers** (literally, "equal parts"). Isomers are compounds with the same chemical formula but different structures. There are two isomers for the $Co(NH_3)_4Cl_2^+$ complex ion, for example, as shown in Figure 12.9. The structures of these isomers differ in the orientation of the two chloride ions around the $Co^{3+}$ ion. In the **trans** isomer, the chlorides occupy positions across from one another in the octahedron. In the **cis** isomer, they occupy adjacent positions. The difference between cis and trans isomers can be remembered by noting that the prefix *trans* is used to describe things that are on opposite sides, as in *transatlantic* or *transcontinental.*

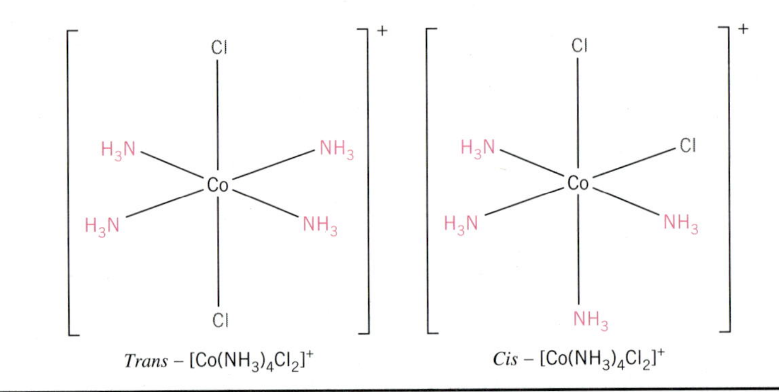


**Figure 12.9** The cis and trans isomers of the $Co(NH_3)_4(Cl)_2^+$ complex ion. In the trans isomer, the two chlorides lie across from each other. In the cis isomer, they occupy adjacent positions in the octahedron.

At the time Werner proposed his theory, only one isomer of the $[Co(NH_3)_4Cl_2]Cl$ complex was known—the green complex described in Section 12.2. Werner predicted that a second isomer should exist and his discovery in 1907 of a purple compound with the same chemical formula was a key step in convincing scientists who were still critical of his model.

Cis/trans isomers are also possible in four-coordinate complexes that have a square-planar geometry. Figure 12.10 shows the structures of the cis and trans isomers of dichlorodiammineplatinum(II). The cis isomer is used as a drug to treat brain tumors, under the trade name Cisplatin. This square-planar complex inserts itself into the grooves in the double-helix structure of the DNA in cells, which inhibits the replication of DNA. This slows down the rate at which the tumor grows, which allows the body's natural defense mechanisms to act on the tumor.

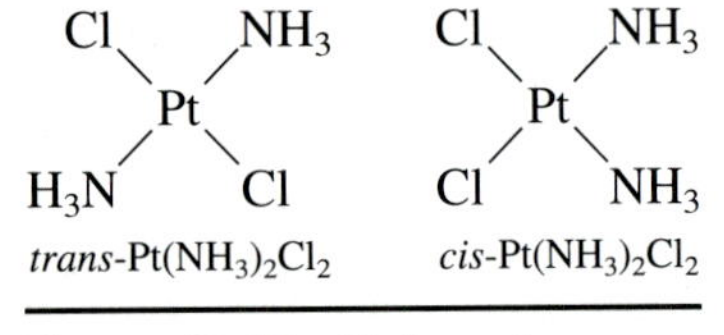


**Figure 12.10** Cis/trans isomers can also exist in square-planar complexes, such as $Pt(NH_3)_2Cl_2$. In the trans isomer, equivalent ligands are across from each other. In the cis isomer, equivalent ligands occupy adjacent positions.

**CHECKPOINT**

Which of the following compounds can form cis/trans isomers?

(a) square-planar $Rh(CO)Cl_3$
(b) trigonal-bipyramidal $Fe(CO)_4(PH_3)$
(c) octahedral $Ni(en)_2(H_2O)_2^{2+}$
(d) tetrahedral $Ni(CO)_3(PH_3)$

## CHIRAL ISOMERS

Another form of isomerism can be understood by considering the difference between gloves and mittens. One glove in each pair fits the left hand, and the other fits the right hand. Mittens usually fit equally well on either hand. To understand why, hold a glove and a mitten in front of a mirror. There is no difference between the mitten shown in Figure 12.11 and its mirror image. There is a difference, however, between the glove and its mirror image. The mirror image of the glove that fits the left hand looks like the glove that fits the right hand, and vice versa.

Each member of a pair of gloves is the mirror image of the other the way the right and left hands are mirror images of each other. Gloves therefore have the same property as the hands on which they are placed. As a result, they are said to be **chiral** (from the Greek *cheir*, ''hand''). By definition, any object that has a mirror image that is different from itself is chiral. The $Co(en)_3^{3+}$ ion is an example of a chiral molecule, which forms a pair of isomers that are mirror images of each other (see Figure 12.12). These isomers have almost identical physical and chemical properties. They have the same melting point, boiling point, density, and color, for example. They differ only in the way they interact with plane-polarized light.

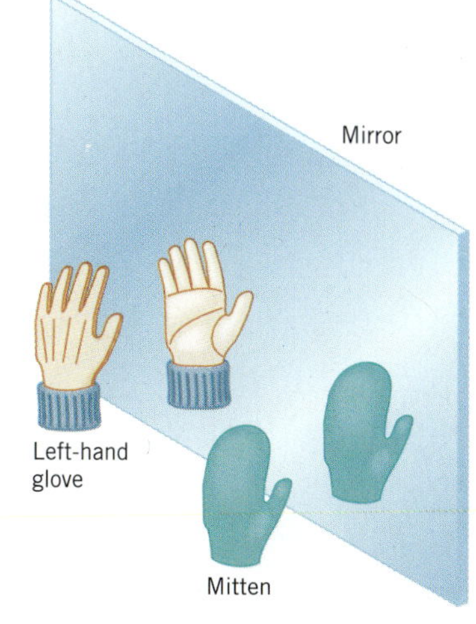


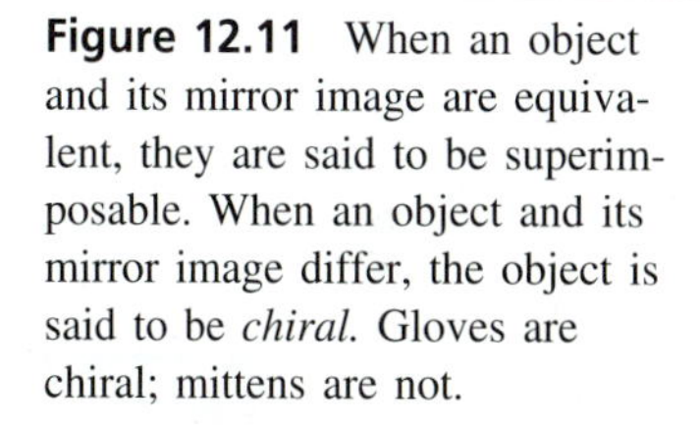

**Figure 12.11** When an object and its mirror image are equivalent, they are said to be superimposable. When an object and its mirror image differ, the object is said to be *chiral*. Gloves are chiral; mittens are not.

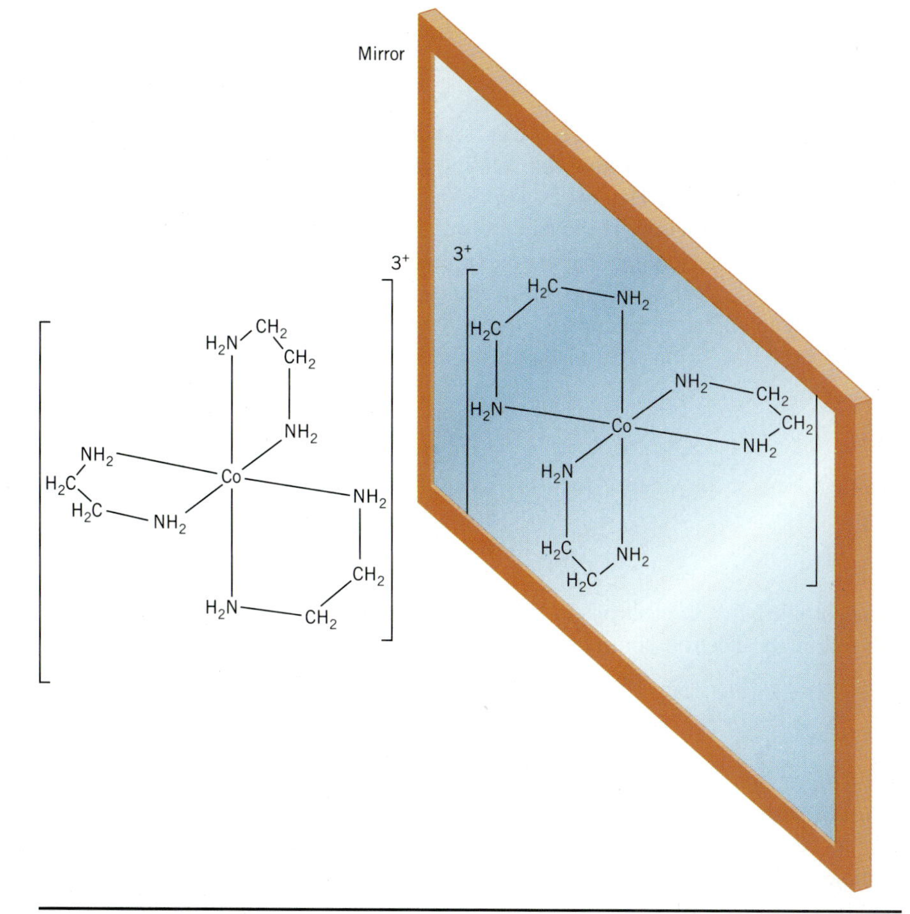


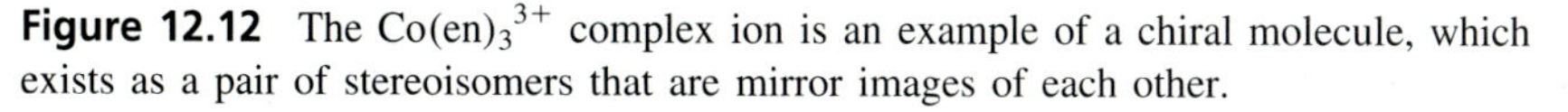

**Figure 12.12** The $Co(en)_3^{3+}$ complex ion is an example of a chiral molecule, which exists as a pair of stereoisomers that are mirror images of each other.

The effect of optically active compounds on light can be demonstrated by filling a large tube with a saturated solution of cane sugar in water and then shining polarized light through the tube to form a spiral of light of different colors along the length of the tube.

Light consists of electric and magnetic fields that oscillate in all directions perpendicular to the path of the light ray. When light is passed through a polarizer, such as a lens in a pair of polarized sunglasses, these oscillations are confined to a single plane. Compounds that can rotate plane-polarized light are said to be **optically active.** Those that rotate the plane of polarization to the right (clockwise) are said to be **dextrorotatory** (from the Latin *dexter*, ''right''). Those that rotate the plane to the left (counterclockwise) are **levorotatory** (from the Latin *laevus*, ''left''). All chiral compounds are optically active; one isomer is dextrorotatory and the other is levorotatory.

**CHECKPOINT**

Which of the following molecules is chiral?

(a) octahedral $Cr(CO)_6$ (b) tetrahedral $Ni(CO)_4$
(c) $SF_4$ (see Figure 8.36) (d) octahedral $Fe(acac)_3$

## 12.11 THE VALENCE-BOND APPROACH TO BONDING IN COMPLEXES

The idea that atoms form covalent bonds by sharing pairs of electrons was first proposed by G. N. Lewis in 1902. It was not until 1927, however, that Walter Heitler and Fritz London showed how the sharing of pairs of electrons holds a covalent molecule together. The Heitler–London model of covalent bonds was the basis of the **valence-bond theory.** The last major step in the evolution of this theory was the suggestion by Linus Pauling that atomic orbitals mix to form hybrid orbitals, such as the *sp*, $sp^2$, $sp^3$, $dsp^3$, and $d^2sp^3$ orbitals discussed in Chapter 8.

Sec. 8.4

It is easy to apply the valence-bond theory to some coordination complexes, such as the $Co(NH_3)_6{}^{3+}$ ion. We start with the electron configuration of the transition-metal ion:

$$Co^{3+}: \quad [Ar]\ 3d^6$$

We then look at the valence-shell orbitals and note that the 4*s* and 4*p* orbitals are empty:

$$Co^{3+}: \quad [Ar]\ 3d^6\ 4s^0\ 4p^0$$

Concentrating the 3*d* electrons in the $d_{xy}$, $d_{xz}$, and $d_{yz}$ orbitals in this subshell gives the following electron configuration:

$$Co^{3+}: \quad \underbrace{\uparrow\downarrow\ \uparrow\downarrow\ \uparrow\downarrow\ \_\ \_}_{3d}\quad \underbrace{\_}_{4s}\quad \underbrace{\_\ \_\ \_}_{4p}$$

The $3d_{x^2-y^2}$, $3d_{z^2}$, 4*s*, $4p_x$, $4p_y$, and $4p_z$ orbitals are then mixed to form a set of empty $d^2sp^3$ orbitals that point toward the corners of an octahedron. Each of these orbitals can accept a pair of nonbonding electrons from a neutral $NH_3$ molecule to form a complex in which the cobalt atom has a filled shell of valence electrons:

$$Co(NH_3)_6{}^{3+}: \quad \underbrace{\uparrow\downarrow\ \uparrow\downarrow\ \uparrow\downarrow\ \uparrow\downarrow\ \uparrow\downarrow}_{3d}\quad \underbrace{\uparrow\downarrow}_{4s}\quad \underbrace{\uparrow\downarrow\ \uparrow\downarrow\ \uparrow\downarrow}_{4p}$$

**EXERCISE 12.5**

Use valence-bond theory to explain why $Fe^{2+}$ ions form the $Fe(CN)_6{}^{4-}$ complex ion.

**SOLUTION** We start with the electron configuration of the transition-metal ion:

$$Fe^{2+}: \quad [Ar]\ 3d^6\ 4s^0\ 4p^0$$

By investing a little energy in the system, we can pair the six $3d$ electrons, thereby creating empty $3d_{x^2-y^2}$ and $3d_{z^2}$ orbitals:

$Fe^{3+}$: ⥮ ⥮ ⥮ __ __ (3d) __ (4s) __ __ __ (4p)

These orbitals are then mixed with the empty $4s$ and $4p$ orbitals to form a set of six $d^2sp^3$ hybrid orbitals, which can accept pairs of nonbonding electrons from the $CN^-$ ligands to form an octahedral complex in which the metal atom has a filled shell of valence electrons:

$Fe(CN)_6{}^{4-}$: ⥮ ⥮ ⥮ ⥮ ⥮ (3d) ⥮ (**4s**) ⥮ ⥮ ⥮ (**4p**)

At first glance, some complexes, such as the $Ni(NH_3)_6{}^{2+}$ ion, seem hard to explain with the valence-bond theory. We start, as always, by writing the configuration of the transition-metal ion:

$$Ni^{2+}: \quad [Ar]\ 3d^8$$

This configuration creates a problem because there are eight electrons in the $3d$ orbitals. Even if we invest the energy necessary to pair the $3d$ electrons, we can't find two empty $3d$ orbitals to use to form a set of $d^2sp^3$ hybrids:

$Ni^{2+}$: ⥮ ⥮ ⥮ ⥮ __ (3d) __ (4s) __ __ __ (4p)

There is a way around this problem. The five $4d$ orbitals on nickel are empty, so we can form a set of empty $sp^3d^2$ hybrid orbitals by mixing the $4d_{x^2-y^2}$, $4d_{z^2}$, $4s$, $4p_x$, $4p_y$, and $4p_z$ orbitals. These hybrid orbitals then accept pairs of nonbonding electrons from six ammonia molecules to form a complex ion:

$Ni(NH_3)_6{}^{2+}$: ⥮ ⥮ ⥮ ⥮ __ (3d) ⥮ (**4s**) ⥮ ⥮ ⥮ (**4p**) ⥮ ⥮ __ __ __ (4d)

The valence-bond theory therefore formally distinguishes between "inner-shell" complexes, which use $3d$, $4s$, and $4p$ orbitals to form a set of $d^2sp^3$ hybrids, and "outer-shell" complexes, which use $4s$, $4p$, and $4d$ orbitals to form $sp^3d^2$ hybrid orbitals.

## 12.12 CRYSTAL-FIELD THEORY

At almost exactly the same time that chemists were developing the valence-bond model for coordination complexes, physicists such as Hans Bethe, John Van Vleck, and Leslie Orgel were developing an alternative known as **crystal-field theory.** This theory tried to describe the effect of the electrical field of neighboring ions on the energies of the valence orbitals of an ion in a crystal. Crystal-field theory was developed by considering two compounds: manganese(II) oxide, MnO, and copper(I) chloride, CuCl.

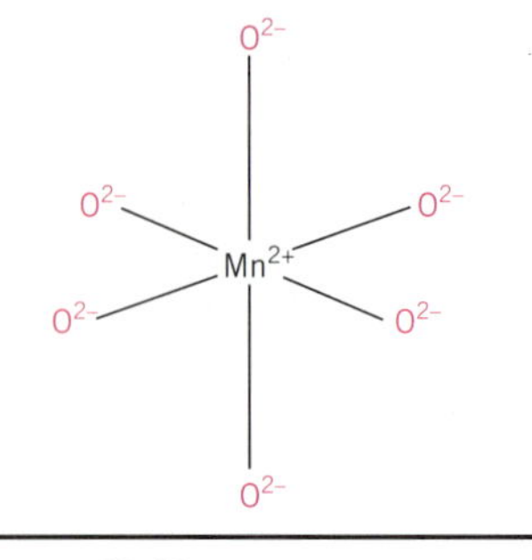


**Figure 12.13** MnO crystallizes in a structure in which each $Mn^{2+}$ ion is surrounded by six $O^{2-}$ ions oriented toward the corners of an octahedron.

## OCTAHEDRAL CRYSTAL FIELDS

Each $Mn^{2+}$ ion in manganese(II) oxide is surrounded by six $O^{2-}$ ions arranged toward the corners of an octahedron, as shown in Figure 12.13. MnO is therefore a model for an *octahedral* complex in which a transition-metal ion is coordinated to six ligands.

What happens to the energies of the 4*s* and 4*p* orbitals on an $Mn^{2+}$ ion when this ion is buried in an MnO crystal? Repulsion between electrons that might be added to these orbitals and the electrons on the six $O^{2-}$ ions that surround the metal ion in MnO increase the energies of these orbitals. The three 4*p* orbitals are still degenerate, however. These orbitals still have the same energy because each 4*p* orbital points toward two $O^{2-}$ ions at the corners of the octahedron.

Repulsion between electrons on the $O^{2-}$ ions and electrons in the 3*d* orbitals on the metal ion in MnO also increases the energy of these orbitals. But the five 3*d* orbitals on the $Mn^{2+}$ ion are no longer degenerate. Let's assume that the six $O^{2-}$ ions that surround each $Mn^{2+}$ ion define an *xyz* coordinate system. Two of the 3*d* orbitals ($3d_{x^2-y^2}$ and $3d_{z^2}$) on the $Mn^{2+}$ ion point directly toward the six $O^{2-}$ ions, as shown in Figure 12.14. The other three orbitals ($3d_{xy}$, $3d_{xz}$, and $3d_{yz}$) lie between the $O^{2-}$ ions.

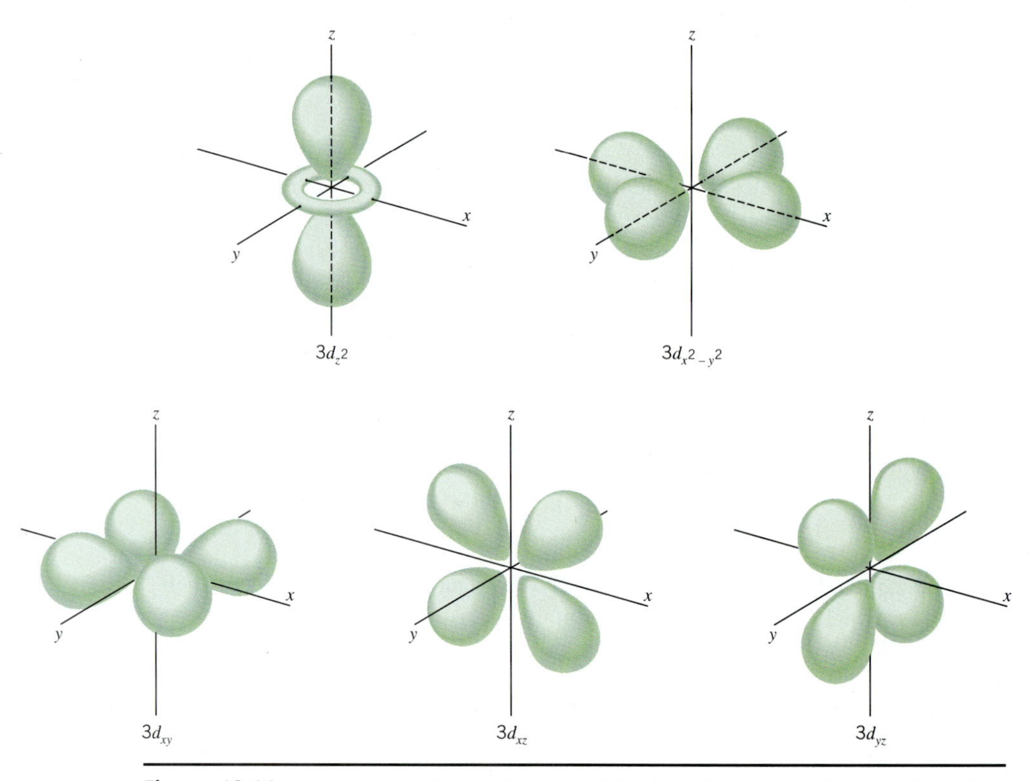


**Figure 12.14** Two of the orbitals ($d_{x^2-y^2}$ and $d_{z^2}$) in the *d* subshell point directly at the ligands in an octahedral complex. The other three ($d_{xy}$, $d_{xz}$, and $d_{yz}$) point between the ligands in these complexes.

The energy of the five 3*d* orbitals increases when the six $O^{2-}$ ions are brought close to the $Mn^{2+}$ ion. However, the energy of two of these orbitals ($3d_{x^2-y^2}$ and $3d_{z^2}$) increases much more than the energy of the other three ($3d_{xy}$, $3d_{xz}$, and $3d_{yz}$), as shown in Figure 12.15. The crystal field of the six $O^{2-}$ ions in MnO therefore

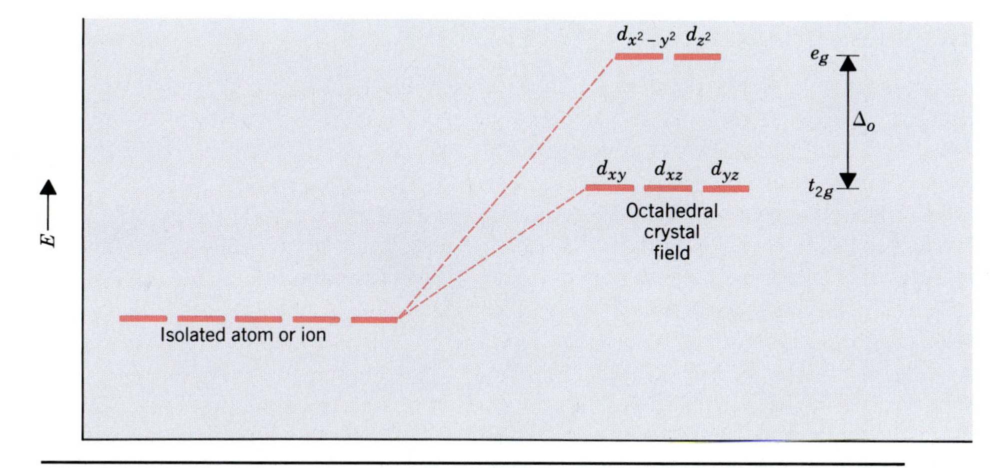


**Figure 12.15** The two $d$ orbitals that point toward the ligands in an octahedral complex are higher in energy than the three $d$ orbitals that lie between these ligands. The five $d$ orbitals are therefore split into $t_{2g}$ and $e_g$ sets of orbitals. The difference between the energies of these orbitals in an octahedral complex is represented by the symbol $\Delta_o$.

splits the degeneracy of the five $3d$ orbitals. Three of these orbitals are now lower in energy than the other two.

By convention, the $d_{xy}$, $d_{xz}$, and $d_{yz}$ orbitals in an octahedral complex are called the $t_{2g}$ orbitals. The $d_{x^2-y^2}$ and $d_{z^2}$ orbitals, on the other hand, are called the $e_g$ orbitals. The easiest way to remember this convention is to note that there are three orbitals in the $t_{2g}$ set:

$$t_{2g}\text{: } d_{xy}, d_{xz}, \text{ and } d_{yz} \qquad e_g\text{: } d_{x^2-y^2} \text{ and } d_{z^2}$$

The difference between the energies of the $t_{2g}$ and $e_g$ orbitals in an octahedral complex is represented by the symbol $\Delta_o$. This splitting of the energy of the $d$ orbitals is not trivial; $\Delta_o$ for the $Ti(H_2O)_6{}^{3+}$ ion, for example, is 242 kJ/mol.

The magnitude of the splitting of the $t_{2g}$ and $e_g$ orbitals changes from one octahedral complex to another. It depends on the identity of the metal ion, the charge on this ion, and the nature of the ligands coordinated to the metal ion.

## TETRAHEDRAL CRYSTAL FIELDS

Each $Cu^+$ ion in copper(I) chloride is surrounded by four $Cl^-$ ions arranged toward the corners of a tetrahedron, as shown in Figure 12.16. CuCl is therefore a model for a *tetrahedral* complex in which a transition-metal ion is coordinated to four ligands.

Once again, the negative ions in the crystal split the energy of the $d$ atomic orbitals on the transition-metal ion. The tetrahedral crystal field splits these orbitals into the same $t_{2g}$ and $e_g$ sets of orbitals as does the octahedral crystal field:

$$t_{2g}\text{: } d_{xy}, d_{xz}, \text{ and } d_{yz} \qquad e_g\text{: } d_{x^2-y^2} \text{ and } d_{z^2}$$

But the two orbitals in the $e_g$ set are now lower in energy than the three orbitals in the $t_{2g}$ set, as shown in Figure 12.17.

To understand the splitting of $d$ orbitals in a tetrahedral crystal field, imagine four ligands lying at alternating corners of a cube to form a tetrahedral geometry, as shown in Figure 12.18. The $d_{x^2-y^2}$ and $d_{z^2}$ orbitals on the metal ion at the center of the cube lie between the ligands, and the $d_{xy}$, $d_{xz}$, and $d_{yz}$ orbitals point toward the

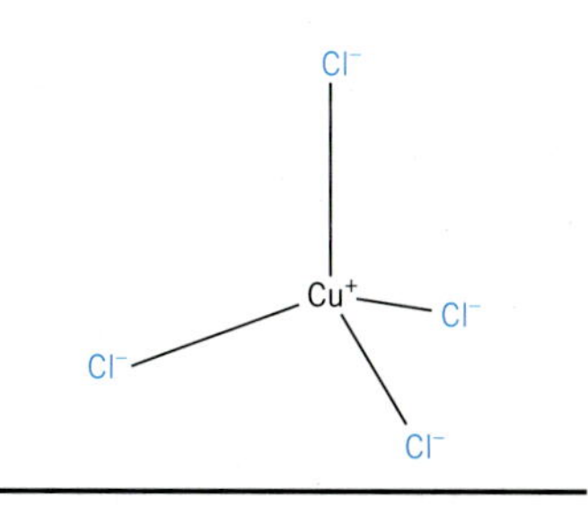


**Figure 12.16** CuCl crystallizes in a structure in which each $Cu^+$ ion is surrounded by four $Cl^-$ ions arranged toward the corners of a tetrahedron.

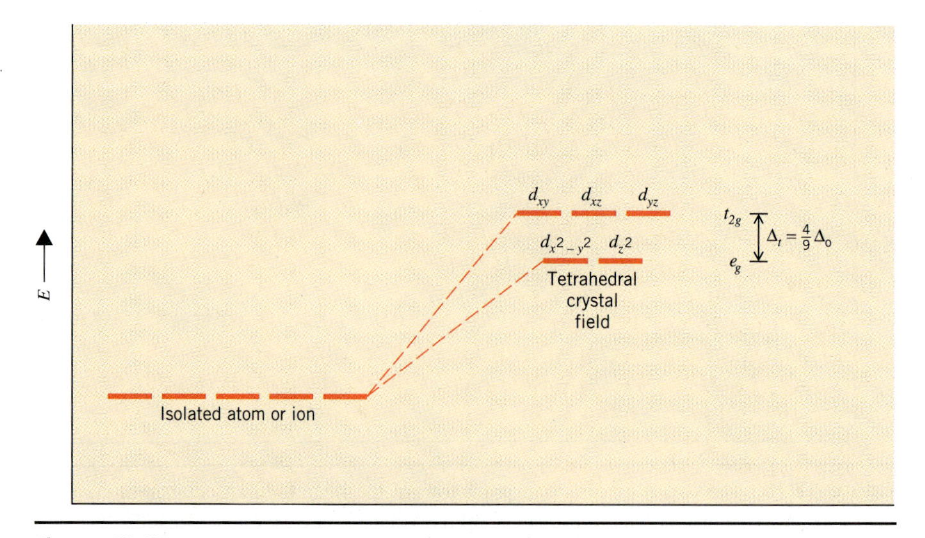


**Figure 12.17** In a tetrahedral complex, the $e_g$ orbitals are lower in energy than the $t_{2g}$ orbitals. The difference between the energies of these orbitals is smaller in tetrahedral complexes than in an equivalent octahedral complex: $\Delta_t = \frac{4}{9}\Delta_o$.

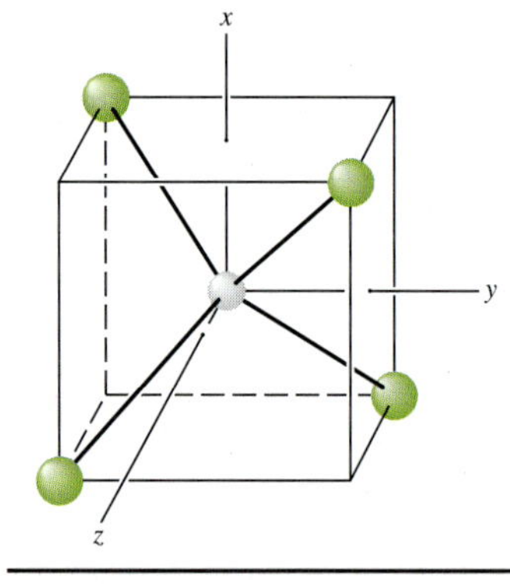


**Figure 12.18** The splitting of the *d* orbitals in a tetrahedral complex is the opposite of the splitting in octahedral complexes. In tetrahedral complexes, the orbitals that lie along the axes of an *xyz* coordination system point between the ligands. The orbitals that lie in the *xy*, *xz*, and *yz* planes point toward these ligands.

ligands. As a result, the splitting observed in a tetrahedral crystal field is the opposite of the splitting in an octahedral complex.

Because a tetrahedral complex has fewer ligands, the magnitude of the splitting is smaller. The difference between the energies of the $t_{2g}$ and $e_g$ orbitals in a tetrahedral complex ($\Delta_t$) is slightly less than half as large as the splitting in analogous octahedral complexes ($\Delta_o$):

$$\Delta_t = \tfrac{4}{9}\Delta_o$$

## SQUARE-PLANAR COMPLEXES

The crystal-field theory can be extended to square-planar complexes, such as $Pt(NH_3)_2Cl_2$. The splitting of the *d* orbitals in these compounds is shown in Figure 12.19.

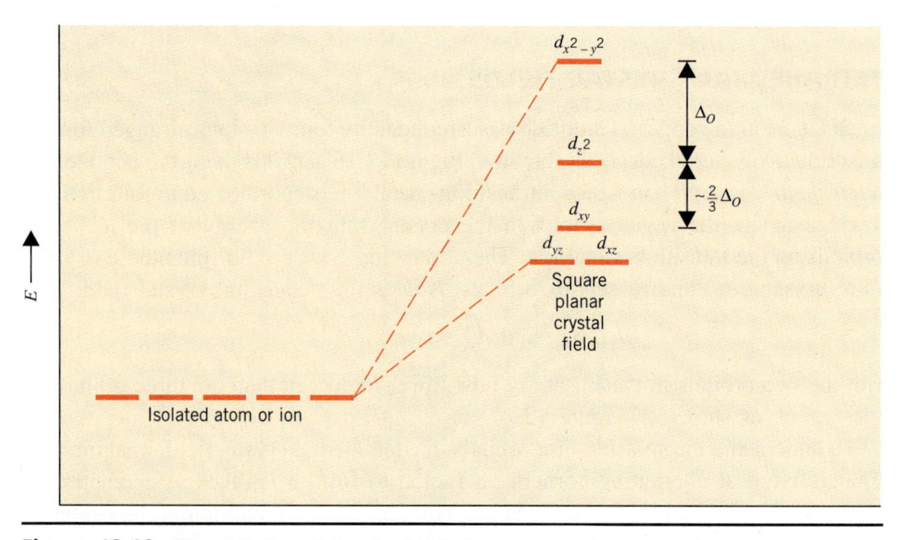


**Figure 12.19** The splitting of the *d* orbitals in a square-planar complex.

## 12.13 THE SPECTROCHEMICAL SERIES

The splitting of $d$ orbitals in the crystal-field model not only depends on the geometry of the complex, it also depends on the nature of the metal ion, the charge on this ion, and the ligands that surround the metal. When the geometry and the ligands are held constant, this splitting decreases in the following order:

$$Pt^{4+} > Ir^{3+} > Rh^{3+} > Co^{3+} > Cr^{3+} > Fe^{3+} > Fe^{2+} > Co^{2+} > Ni^{2+} > Mn^{2+}$$

strong-field ions — weak-field ions

Metal ions at one end of this continuum are called *strong-field ions,* because the splitting due to the crystal field is unusually strong. Ions at the other end are known as *weak-field ions.*

When the geometry and the metal are held constant, the splitting of the $d$ orbitals decreases in the following order:

$$CO \approx CN^- > NO_2^- > NH_3 > H_2O > OH^- > F^- > SCN^- \approx Cl^- > Br^-$$

strong-field ligands — weak-field ligands

Ligands that give rise to large differences between the energies of the $t_{2g}$ and $e_g$ orbitals are called *strong-field ligands.* Those at the opposite extreme are known as *weak-field ligands.*

Because they result from studies of the absorption spectra of transition-metal complexes, these generalizations are known as the **spectrochemical series.** The range of values of $\Delta$ for a given geometry is remarkably large. The value of $\Delta_o$ is 100 kJ/mol in the $Ni(H_2O)_6^{2+}$ ion, for example, and 520 kJ/mol in the $Rh(CN)_6^{3-}$ ion.

## 12.14 HIGH-SPIN VERSUS LOW-SPIN OCTAHEDRAL COMPLEXES

Sec. 6.21

Once we know the relative energies of the $d$ orbitals in a transition-metal complex, we have to worry about how these orbitals are filled. Degenerate orbitals are filled according to Hund's rules:

1. One electron is added to each of the degenerate orbitals in a subshell before a second electron is added to any orbital in the subshell.
2. Electrons are added to a subshell with the same value of the spin quantum number until each orbital in the subshell has at least one electron.

Octahedral transition-metal ions with $d^1$, $d^2$, or $d^3$ configurations therefore can be described by the following diagrams:

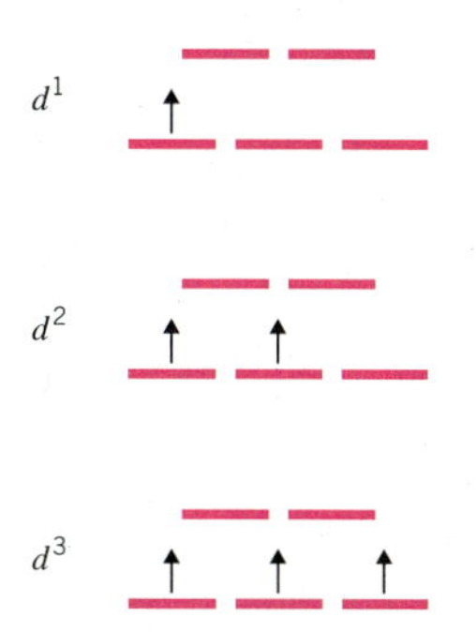

When we try to add a fourth electron, we are faced with a problem. This electron could be used to pair one of the electrons in the lower energy ($t_{2g}$) set of orbitals or it could be placed in one of the higher energy ($e_g$) orbitals. One of these configurations is called **high-spin** because it contains four unpaired electrons with the same spin. The other is called **low-spin** because it contains only two unpaired electrons. The same problem occurs with octahedral $d^5$, $d^6$, and $d^7$ complexes:

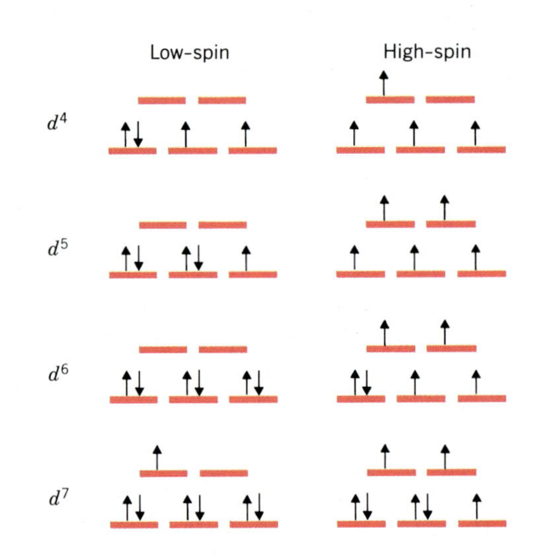


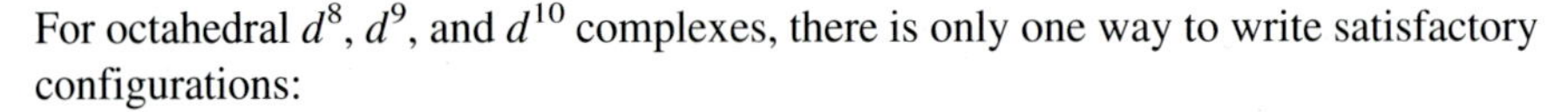

For octahedral $d^8$, $d^9$, and $d^{10}$ complexes, there is only one way to write satisfactory configurations:



As a result, we have to worry about high-spin versus low-spin octahedral complexes only when there are four, five, six, or seven electrons in the $d$ orbitals.

The choice between high-spin and low-spin configurations for octahedral $d^4$, $d^5$, $d^6$, or $d^7$ complexes is easy. All we have to do is compare the energy it takes to pair electrons with the energy it takes to excite an electron to the higher energy orbitals. If it takes less energy to pair the electrons, the complex is low-spin. If it takes less energy to excite the electron, the complex is high-spin.

The amount of energy required to pair electrons in an octahedral complex is more or less constant. The amount of energy needed to excite an electron into the higher energy orbitals, however, depends on the value of $\Delta_o$ for the complex. As a result, we expect to find low-spin complexes among metal ions and ligands that lie toward the high-field end of the spectrochemical series. High-spin complexes are expected among metal ions and ligands that lie toward the low-field end of this series.

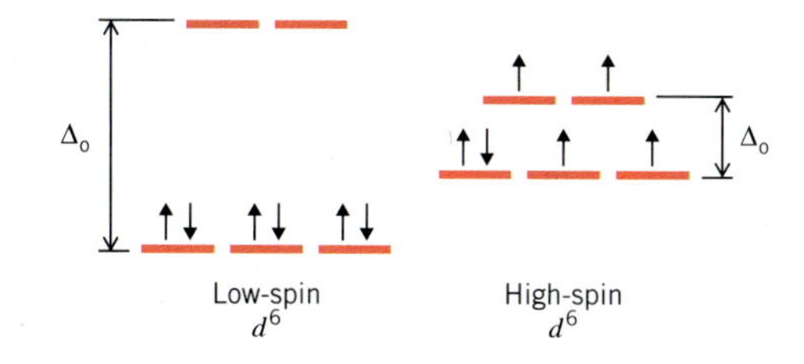


Compounds in which all of the electrons are paired are **diamagnetic**—they are repelled by both poles of a magnet. Compounds that contain one or more unpaired electrons are **paramagnetic**—they are attracted to the poles of a magnet. The force of attraction between paramagnetic complexes and a magnetic field is proportional to the number of unpaired electrons in the complex. We can therefore determine whether a complex is high-spin or low-spin by measuring the strength of the interaction between the complex and a magnetic field.

**EXERCISE 12.6**

Explain why the $Co(NH_3)_6{}^{3+}$ ion is a diamagnetic, low-spin complex, whereas the $CoF_6{}^{3-}$ ion is a paramagnetic, high-spin complex.

**SOLUTION** Both complexes contain the $Co^{3+}$ ion, which lies toward the strong-field end of the spectrochemical series. $NH_3$ also lies toward the strong-field end of this series. As a result, $\Delta_o$ for the $Co(NH_3)_6{}^{3+}$ ion should be relatively large. When the splitting is large, it takes less energy to pair the electrons in the $t_{2g}$ orbitals than it does to excite the electrons to the $e_g$ orbitals. The $Co(NH_3)_6{}^{3+}$ ion is therefore a low-spin $d^6$ complex in which the electrons are all paired, and the ion is diamagnetic.

The $F^-$ ion lies toward the low-field end of the spectrochemical series. As a result, $\Delta_o$ for the $CoF_6{}^{3-}$ ion is much smaller. In this complex, it takes less energy to excite electrons into the $e_g$ orbitals than to pair them to the $t_{2g}$ orbitals. As a result, the $CoF_6{}^{3-}$ ion is a high-spin $d^6$ complex, which contains unpaired electrons and is therefore paramagnetic.

## 12.15 THE COLORS OF TRANSITION-METAL COMPLEXES

One of the joys of working with transition metals is the extraordinary range of colors exhibited by their compounds. By changing the number of ammonia, water, and chloride ligands on a $Co^{3+}$ ion, for example, we can vary the color of the complex over the entire visible spectrum, from red to violet, as shown in Section 12.2. We can also vary the color of the complex by keeping the ligand constant and changing the metal ion. The characteristic colors of aqueous solutions of some common transition-metal ions are given in Table 12.5.

The color of these complexes also changes with the oxidation state of the metal atom. By reducing the $VO_2{}^+$ ion with zinc metal, for example, we can change the color of the solution from yellow to green to blue and then to violet as the vanadium is reduced, one electron at a time, from the +5 to the +2 oxidation state (see Figure 12.20.)

| | | | |
|---|---|---|---|
| $VO_2{}^+$ | yellow | $V^{3+}$ | blue |
| $VO^{2+}$ | green | $V^{2+}$ | violet |

**Figure 12.20** Solutions of the $Cu(NH_3)_4^{2+}$ ion look blue because they absorb yellow light (left). Solutions of the $CrO_4^{2-}$ ion look yellow because they absorb blue light (right).

**TABLE 12.5 Characteristic Colors of Common Transition Metal Ions**

| | | | |
|---|---|---|---|
| $Cr^{2+}$ | blue | $Mn^{2+}$ | very faint pink |
| $Cr^{3+}$ | blue-violet | $Ni^{2+}$ | green |
| $Co^{2+}$ | pinkish red | | |
| $Cu^{2+}$ | light blue | $CrO_4^{2-}$ | yellow |
| $Fe^{3+}$ | yellow | $Cr_2O_7^{2-}$ | orange |
| | | $MnO_4^-$ | deep violet |

To explain why transition-metal complexes are colored, and why the color is so sensitive to changes in the metal, ligand, or oxidation state, we need to understand the physics of color. There are two ways of producing the sensation of color. We can add color where none exists or subtract it from white light. The three primary *additive colors* are red, green, and blue. When all three are present at the same intensity, we get white light. The three primary *subtractive colors* are cyan, magenta, and yellow. When these three colors are absorbed with the same intensity, light is absorbed across the entire visible spectrum. If the object absorbs strongly enough, or if the intensity of the light is dim enough, all of the light can be absorbed.

The additive and subtractive colors are complementary, as shown in Figure 12.21. If we subtract yellow from white light, we get a mixture of cyan and magenta, and the light looks blue. If we subtract blue—a mixture of cyan and magenta—from white light, the light looks yellow. The $Cu(NH_3)_4^{2+}$ complex ion has a blue color because it absorbs light in the yellow portion of the spectrum. The $CrO_4^{2-}$ ion appears yellow because it absorbs blue light. Table 12.6 describes what happens when light in different portions of the visible spectrum is absorbed.

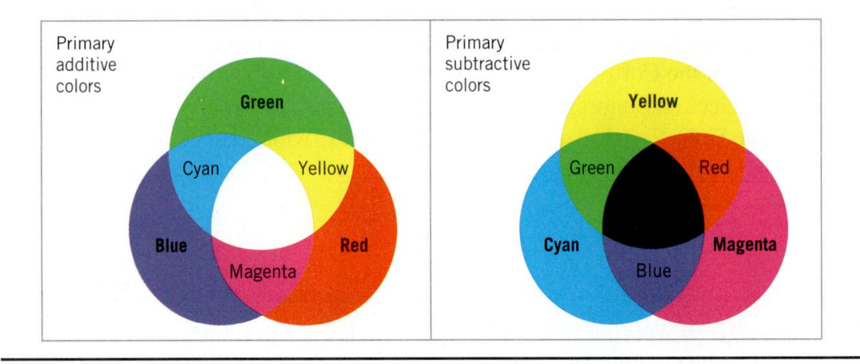


**Figure 12.21** The three primary additive colors are red, blue, and green. The three primary subtractive colors are yellow, cyan, and magenta. The additive colors can be prepared from two of the subtractive colors. Yellow and cyan give green; yellow and magenta give red; cyan and magenta give blue. The subtractive colors can be prepared from two of the additive colors. Green and red give yellow; red and blue give magenta; blue and green give cyan.

Light is absorbed when it carries just enough energy to excite an electron from one orbital to another. Compounds therefore absorb light when the difference between the energy of an orbital that contains an electron and the energy of an empty orbital corresponds to a wavelength in the narrow band of the electromagnetic spectrum that is visible to the naked eye.

**TABLE 12.6 The Relationship Between the Color of Transition-Metal Complexes and the Wavelength of Light Absorbed**

| Wavelength Absorbed (nm) | Color of Light Absorbed | Color of Complex |
|---|---|---|
| 410 | violet | lemon-yellow |
| 430 | indigo | yellow |
| 480 | blue | orange |
| 500 | blue-green | red |
| 530 | green | purple |
| 560 | lemon-yellow | violet |
| 580 | yellow | indigo |
| 610 | orange | blue |
| 680 | red | blue-green |

How much energy is associated with the absorption of a typical photon, such as a photon with a wavelength of 480 nanometers? Since we know the wavelength of this photon, we can calculate its frequency:

$$\nu = \frac{c}{\lambda} = \frac{2.998 \times 10^8 \cancel{\text{m}}/\text{s}}{480 \times 10^{-9} \cancel{\text{m}}} = 6.246 \times 10^{14}\ \text{s}^{-1}$$

We can calculate the energy of this photon from its frequency and Planck's constant:

$$E = h\nu = (6.626 \times 10^{-34}\ \text{J-}\cancel{\text{s}})(6.246 \times 10^{14}\ \cancel{\text{s}}^{-1}) = 4.138 \times 10^{-19}\ \text{J}$$

But this is the energy associated with only a single photon. Multiplying by Avogadro's number gives us the energy in units of kilojoules per mole:

$$\frac{4.138 \times 10^{-19}\ \text{J}}{1\ \cancel{\text{photon}}} \times \frac{6.022 \times 10^{23}\ \cancel{\text{photons}}}{\text{mol}} = \mathbf{249.2\ kJ/mol}$$

Repeating this calculation for all frequencies in the visible portion of the spectrum reveals that the energy associated with a photon of light ranges from 160 to 300 kJ/mol. If we refer back to the discussion of the spectrochemical series in Section 12.13, we find that these energies fall in the range of values of $\Delta$. Anything that changes the difference between the energies of the $t_{2g}$ and $e_g$ orbitals in a transition-metal complex therefore influences the color of the light absorbed by this complex. It is not surprising that the color of the complex is sensitive to factors such as the geometry of the complex, the identity of the metal, the nature of the ligand, and the oxidation state of the metal.

SPECIAL TOPIC

## LIGAND-FIELD THEORY

Secs. 12.11 and 12.12

The valence-bond model described in Section 12.11 and the crystal-field theory described in Section 12.12 explain some aspects of the chemistry of the transition metals, but neither model is good at predicting all of the properties of transition-metal complexes. A third model, based on molecular orbital theory, was therefore developed that is known as **ligand-field theory.** Ligand-field theory is more powerful than either the valence-bond or the crystal-field theories. Unfortunately it is also more abstract.

The ligand-field model for an octahedral transition-metal complex such as the $Co(NH_3)_6^{3+}$ ion assumes that the $3d$, $4s$, and $4p$ orbitals on the metal overlap with one orbital on each of the six ligands to form a total of 15 molecular orbitals, as shown in Figure 12.22. Six of these orbitals are *bonding molecular orbitals,* whose energies are much lower than those of the original atomic orbitals. Another six are *antibonding molecular orbitals,* whose energies are higher than those of the original atomic orbitals. Three are best described as *nonbonding*

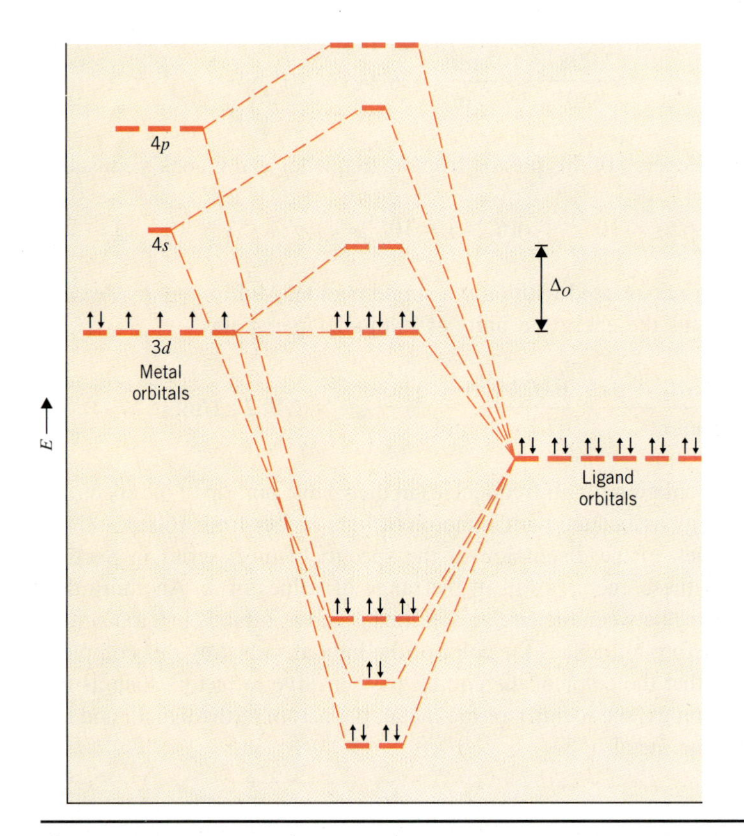


**Figure 12.22** Ligand-field theory is a molecular orbital alternative to the valence-bond and crystal-field models of transition-metal complexes. The molecular orbitals in this diagram were generated by allowing the valence-shell $3d$, $4s$, and $4p$ orbitals on the transition metal to overlap with an orbital on each of the six ligands that contribute nonbonding pairs of electrons to form an octahedral complex.

molecular orbitals, because they have essentially the same energy as the $3d$ atomic orbitals on the metal.

Ligand-field theory enables the $3d$, $4s$, and $4p$ orbitals on the metal to overlap with orbitals on the ligand to form the octahedral covalent bond skeleton that holds this complex together. At the same time, this model generates a set of five orbitals in the center of the diagram that are split into $t_{2g}$ and $e_g$ subshells, as predicted by the crystal-field theory. As a result, we don't have to worry about ''inner-shell'' versus ''outer-shell'' metal complexes. In effect, we can use the $3d$ orbitals in two different ways. We can use them to form the covalent bond skeleton and then use them again to form the orbitals that hold the electrons that were originally in the $3d$ orbitals of the transition metal.

## SUMMARY

**1.** The transition metals serve as a bridge, or transition, between the main-group elements on either side of the periodic table. Although they have many of the chemical and physical properties of the main-group metals, they differ in several ways. First, they tend to be more electronegative than the main-group metals and therefore more likely to form covalent compounds. Second, they are more likely to form coordination complexes, such as the $FeCl_4^-$ and $Co(NH_3)_6^{3+}$ ions.

**2.** Coordination complexes result from reactions between a transition-metal ion acting as a Lewis acid and one or more ligands acting as Lewis bases. In each case, a pair of nonbonding electrons on the ligand (the Lewis base) is donated to an empty orbital on the transition metal (the Lewis acid) to form a covalent bond.

**3.** The coordination number of a transition-metal complex is the number of ligands bound to the transition metal. Common coordination numbers are 2, 4, and 6. Four-coordinate complexes usually have tetrahedral or square-planar structures. Six-coordinate complexes usually have an octahedral structure.

**4.** Transition-metal ions that have a charge of +1, +2, or +3 can exist in water. Ions that have a larger charge are usually so electronegative that they form covalent bonds to oxygen atoms that they strip from neighboring water molecules.

**5.** The valence-bond theory can be applied to coordination complexes. When there are enough empty valence-shell $d$ orbitals, the transition-metal ion is assumed to form a set of empty $dsp^3$ or $d^2sp^3$ hybrid orbitals that can accept pairs of nonbonding electrons from the ligands. When there aren't enough empty valence-shell $d$ orbitals, the transition metal is assumed to use the empty $d$ orbitals in the next shell to form sets of $sp^3d$ or $sp^3d^2$ hybrid orbitals that can accept pairs of electrons from the ligands.

**6.** The crystal-field theory suggests that the five valence-shell $d$ orbitals on the transition metal in a coordination complex are no longer degenerate. In an octahedral complex, they are split into a set of $t_{2g}$ orbitals with relatively low energies and a set of $e_g$ orbitals with higher energies. The $d$ orbitals are also split into a $t_{2g}$ and an $e_g$ set in tetrahedral complexes, but the $e_g$ orbitals now have lower energies than the $t_{2g}$ orbitals.

**7.** When the difference between the energies of the $t_{2g}$ and $e_g$ sets of orbitals is relatively small, electrons are distributed between these orbitals as expected from Hund's rules to give a *high-spin* complex. When the difference between the $t_{2g}$ and $e_g$ sets of orbitals is relatively large, it takes less energy to pair electrons in the lower energy set of orbitals than to satisfy Hund's rules. The result is a *low-spin* complex.

**8.** The splitting of the $t_{2g}$ and $e_g$ sets of orbitals in transition-metal complexes is roughly equal to the energy of the photons in the visible spectrum. Because the magnitude of this splitting depends on the metal ion, the ligands on the metal, and the geometry of the complex, the chemistry of transition-metal complexes is colorful.

## KEY TERMS

**Bidentate ligand** (p. 448)
**Chelating ligand** (p. 448)
**Chiral** (p. 455)
**Cis** (p. 454)

**Complex ions** (p. 441)
**Coordination compounds** (p. 441)
**Coordination number** (p. 441)
**Crystal-field theory** (p. 457)
**Dextrorotatory** (p. 456)
**Diamagnetic** (p. 463)
**Electron-pair acceptor/donor** (p. 447)
**High-spin** (p. 462)
**Isomers** (p. 454)
**Levorotatory** (p. 456)
**Lewis acid/base** (p. 447)
**Ligand-field theory** (p. 466)
**Ligands** (p. 441)
**Low-spin** (p. 462)
**Monodentate** (p. 448)
**Optically active** (p. 456)
**Paramagnetic** (p. 463)
**Spectrochemical series** (p. 461)
**Trans** (p. 454)
**Transition metals** (p. 439)
**Valence-bond theory** (p. 456)

## PROBLEMS

### Transition Metals and Coordination Complexes

**12-1** Identify the subshell of atomic orbitals filled among the transition metals in the fifth row of the periodic table.

**12-2** Describe some of the ways in which transition metals differ from main-group metals such as aluminum, tin, and lead. Describe ways in which they are similar.

**12-3** Use the electron configurations of $Zn^{2+}$, $Cd^{2+}$, and $Hg^{2+}$ to explain why these ions often behave as if they are main-group metals.

**12-4** Use the electronegativities of cobalt and nitrogen to predict whether the Co—N bond in the $Co(NH_3)_6{}^{3+}$ ion is best described as ionic, polar covalent, or covalent.

**12-5** Define the terms *coordination number* and *ligand.*

### Werner's Model of Coordination Complexes

**12-6** Werner wrote the formula of one of his coordination complexes as $CoCl_3 \cdot 6\,NH_3$. Today, we write this compound as $[Co(NH_3)_6]Cl_3$ to indicate the presence of $Co(NH_3)_6{}^{3+}$ and $Cl^-$ ions. Write modern formulas for the compounds Werner described as $CoCl_3 \cdot 5\,NH_3$, $CoCl_3 \cdot 4\,NH_3$, and $CoCl_3 \cdot 5\,NH_3 \cdot H_2O$.

**12-7** Describe the difference between the primary and secondary valences of the $Co^{3+}$ ion in $[Co(NH_3)_6]Cl_3$.

**12-8** Explain why aqueous solutions of $CoCl_3 \cdot 6\,NH_3$ conduct electricity better than aqueous solutions of $CoCl_3 \cdot 4\,NH_3$.

**12-9** Explain why $Ag^+$ ions precipitate three chloride ions from an aqueous solution of $CoCl_3 \cdot 6\,NH_3$, but only one chloride ion from an aqueous solution of $CoCl_3 \cdot 4\,NH_3$.

**12-10** Predict the number of $Cl^-$ ions that could precipitate from an aqueous solution of the $Co(en)_2Cl_2$ complex.

**12-11** Predict the number of ions formed when the $NiSO_4 \cdot 4\,NH_3 \cdot 2\,H_2O$ complex dissociates in water.

### Typical Coordination Numbers

**12-12** Use the examples in Table 12.1 to identify at least one factor that influences the coordination number of a transition-metal ion.

**12-13** Determine both the coordination number and the charge on the transition-metal ion in the following complexes:

(a) $CuF_4{}^{2-}$ (b) $Cr(CO)_6$ (c) $Fe(CN)_6{}^{4-}$
(d) $Pt(NH_3)_2Cl_2$

**12-14** Determine both the coordination number and the charge on the transition-metal ion in the following complexes:

(a) $Co(SCN)_4{}^{2-}$ (b) $Fe(acac)_3$ (c) $Ni(en)_2(H_2O)_2{}^{2+}$
(d) $Co(NH_3)_5(H_2O)^{3+}$

### The Electron Configuration of Transition-Metal Ions

**12-15** Write the electron configuration of the following transition-metal ions:

(a) $V^{2+}$ (b) $Cr^{2+}$ (c) $Mn^{2+}$ (d) $Fe^{2+}$ (e) $Ni^{2+}$

**12-16** Explain why the $Co^{2+}$ ion can be described as a $d^7$ ion.

**12-17** Which of the following ions can be described as $d^5$?

(a) $Cr^{2+}$ (b) $Mn^{2+}$ (c) $Fe^{3+}$ (d) $Co^{3+}$ (e) $Cu^+$

**12-18** Explain the difference between the symbols $Cr^{3+}$ and Cr(VI).

**12-19** Which of the following is not an example of a $d^0$ transition-metal complex?

(a) $TiO_2$ (b) $VO^{2+}$ (c) $Cr_2O_7{}^{2-}$ (d) $MnO_4{}^-$

**12-20** Explain why manganese becomes more electronegative when it is oxidized from $Mn^{2+}$ to Mn(VII).

**12-21** Use the idea that the electronegativity of an element increases with oxidation state to explain why the Mn—O bonds in $MnO_4{}^-$ are covalent in spite of the fact that $\Delta EN = 1.89$.

### Lewis Acid–Lewis Base Approach to Bonding in Complexes

**12-22** Describe what to look for when deciding whether an ion or molecule is a Lewis acid.

**12-23** Explain how Lewis acids, such as the $Co^{3+}$ ion, pick up Lewis bases, or ligands, to form coordination complexes.

**12-24** Which of the following are Lewis acids?

(a) $Fe^{3+}$ (b) $BF_3$ (c) $H_2$ (d) $Ag^+$ (e) $Cu^{2+}$

**12-25** Which of the following are Lewis bases, and therefore potential ligands?

(a) CO (b) $O_2$ (c) $Cl^-$ (d) $N_2$ (e) $NH_3$

**12-26** Which of the following are Lewis bases, and therefore potential ligands?

(a) $CN^-$ (b) $SCN^-$ (c) $CO_3^{2-}$ (d) $NO^+$
(e) $S_2O_3^{2-}$

### Typical Ligands

**12-27** Define the terms *monodentate, bidentate, tridentate,* and *tetradentate.* Give an example of each category of ligand.

**12-28** Use Lewis structures and the VSEPR theory to explain why carbon monoxide could act as a bridge between a pair of transition metals, but it can't be a chelating ligand that coordinates to the same metal twice.

**12-29** Draw the structures of the following coordination complexes:

(a) $Fe(acac)_3$ (b) $Co(en)_3^{3+}$ (c) $Fe(EDTA)^-$
(d) $Fe(CN)_6^{4-}$

### Nomenclature of Complexes

**12-30** Name the following complexes:

(a) $Cu(NH_3)_4^{2+}$ (b) $Mn(H_2O)_6^{2+}$ (c) $Fe(CN)_6^{4-}$
(d) $Ni(en)_3^{2+}$ (e) $Cr(acac)_3$

**12-31** Name the following complexes:

(a) $Pt(NH_3)_2Cl_2$ (b) $Ni(CO)_4$ (c) $Co(en)_3^{3+}$
(d) $Na[Mn(CO)_5]$

**12-32** Name the following complexes:

(a) $Na_3[Co(NO_2)_6]$ (b) $Na_2[Zn(CN)_4]$
(c) $[Co(NH_3)_4Cl_2]Cl$ (d) $[Ag(NH_3)_2]Cl$

**12-33** Write the formulas for the following compounds:

(a) hexamminechromium(III) chloride
(b) chloropentamminechromium(III) chloride
(c) triethylenediamminecobalt(III) chloride
(d) potassium tetranitritodiamminecobaltate(III)

### Isomers

**12-34** Which of the following octahedral complexes can form cis/trans isomers?

(a) $Co(NH_3)_6^{3+}$ (b) $Co(NH_3)_5Cl^{2+}$
(c) $Co(NH_3)_5(H_2O)^{3+}$ (d) $Co(NH_3)_4Cl_2^+$
(e) $Co(NH_3)_4(H_2O)_2^{3+}$

**12-35** Predict the structures of the cis/trans isomers of $Ni(en)_2(H_2O)_2^{2+}$.

**12-36** The octahedral $Mo(PH_3)_3(CO)_3$ complex can exist as a pair of isomers. Predict the structures of these compounds.

**12-37** Which of the following square-planar complexes can form cis/trans isomers?

(a) $Cu(NH_3)_4^{2+}$ (b) $Pt(NH_3)_2Cl_2$ (c) $RhCl_3(CO)$
(d) $IrCl(CO)(PH_3)_2$

**12-38** Explain why square-planar complexes with the generic formula $MX_2Y_2$ can form cis/trans isomers, but tetrahedral complexes with the same generic formula cannot.

**12-39** Explain why square-planar complexes with the generic formula $MX_3Y$ can't form cis/trans isomers.

**12-40** Use the fact that $Rh(CO)(H)(PH_3)_2$ forms cis/trans isomers to predict whether the geometry around the transition metal is square planar or tetrahedral.

**12-41** Compounds are optically active when the mirror image of the compound cannot be superimposed upon itself. Draw the mirror images of the following complex ions and determine which of these ions are chiral.

(a) $Cu(NH_3)_4^{2+}$ (a square-planar complex)
(b) $Co(NH_3)_6^{2+}$ (an octahedral complex)
(c) $Ag(NH_3)_2^+$ (a linear complex)
(d) $Cr(en)_3^{3+}$ (an octahedral complex)

**12-42** Explain why neither of the cis/trans isomers of $Pt(NH_3)_2Cl_2$ is chiral.

**12-43** Explain why the cis isomer of $Ni(en)_2(H_2O)_2^{2+}$ is chiral, but the trans isomer is not.

**12-44** Explain why the cis isomer of $Ni(en)_2(H_2O)_2^{2+}$ is chiral, but the cis isomer of $Ni(NH_3)_4(H_2O)_2^{2+}$ is not.

**12-45** Which of the following octahedral complexes are chiral?

(a) $Cr(acac)_3$ (b) $Cr(C_2O_4)_3^{3-}$ (c) $Cr(CN)_6^{3-}$
(d) $Cr(CO)_4(NH_3)_2$

### The Valence-Bond Approach to Bonding in Complexes

**12-46** Describe the difference between the valence-bond model for the $Co(NH_3)_6^{3+}$ complex ion and the valence-bond model for the $Ni(NH_3)_6^{2+}$ complex ion.

**12-47** Apply the valence-bond model of bonding in transition-metal complexes to $Ni(CO)_4$, $Fe(CO)_5$, and $Cr(CO)_6$. (Hint: Assume that the valence electrons on transition metals in complexes in which the metal is in the zero oxidation states are concentrated in the *d* orbitals.)

**12-48** Use the results of the previous problem to explain why the transition metal is $sp^3$ hybridized in $Ni(CO)_4$, $dsp^3$ hybridized in $Fe(CO)_5$, and $d^2sp^3$ hybridized in $Cr(CO)_6$.

**12-49** Use valence-bond theory to explain the $-2$ charge on the $Fe(CO)_4^{2-}$ ion. Predict the charge on the equivalent $Co(CO)_4^{x-}$ ion. Use acid–base chemistry to predict the charge on the $HFe(CO)_4^{x-}$ ion.

**12-50** Use Lewis structures to explain what happens in the following reaction:

$$2\,CrO_4^{2-}(aq) + 2\,H^+(aq) \rightleftharpoons Cr_2O_7^{2-}(aq) + H_2O(l)$$

Predict the charge on the product of the following reaction.

Explain why this transition-metal complex is a gas at room temperature:

$$2\,MnO_4^{2-}(aq) + 2\,H^+(aq) \rightleftharpoons Mn_2O_7^{x-}(g) + H_2O(l)$$

**12-51** Apply the valence-bond model of bonding in transition-metal complexes to the $Zn(NH_3)_4^{2+}$ and $Fe(H_2O)_6^{3+}$ complex ions.

**12-52** Use the notion that transition metals often pick up enough ligands to fill their valence shells to predict the charge on the $Mn(CO)_5^{x-}$ ion.

**12-53** Use the notion that transition metals often pick up enough ligands to fill their valence shells to predict the charge on the $HgI_4^{x-}$ ion.

**12-54** Use the notion that transition-metals often pick up enough ligands to fill their valence shells to predict the coordination number of the $Cd^{2+}$ ion in the $Cd(OH)_x^{2-}$ ion.

**12-55** Explain why $Zn^{2+}$ ions form both $Zn(CN)_4^{2-}$ and $Zn(NH_3)_4^{2+}$ ions.

### Crystal-Field Theory

**12-56** Describe what happens to the energies of the 3*d* atomic orbitals in an octahedral crystal field.

**12-57** Describe what happens to the energies of the 3*d* atomic orbitals in a tetrahedral crystal field.

**12-58** Which of the 3*d* atomic orbitals in an octahedral crystal field belong to the $t_{2g}$ set of orbitals? Which belong to the $e_g$ set?

**12-59** What do the orbitals in a $t_{2g}$ set have in common? What do the orbitals in the $e_g$ set have in common?

**12-60** The 3*d* orbitals are split into $t_{2g}$ and $e_g$ sets in both octahedral and tetrahedral crystal fields. Is there any difference between the orbitals that go into the $t_{2g}$ set in octahedral and in tetrahedral crystal fields?

**12-61** Explain why $\Delta_t$ for a tetrahedral complex is much smaller than $\Delta_o$ for the analogous octahedral complex.

**12-62** Use the splitting of the 3*d* atomic orbitals in an octahedral crystal field to explain the stability of oxidation states corresponding to $d^3$ and $d^6$ electron configurations in the $Cr(NH_3)_6^{3+}$ and $Fe(H_2O)_6^{2+}$ complex ions.

**12-63** The difference between the energies of the $t_{2g}$ and $e_g$ sets of atomic orbitals in an octahedral or tetrahedral crystal field depends on both the metal atom and the ligands that form the complex. Which of the following metal ions would give the largest difference?

(a) $Rh^{3+}$ (b) $Cr^{3+}$ (c) $Fe^{3+}$ (d) $Co^{2+}$ (e) $Mn^{2+}$

**12-64** Which of the following ligands would give the largest value of $\Delta_o$?

(a) $CN^-$ (b) $NH_3$ (c) $H_2O$ (d) $OH^-$ (e) $F^-$

### High-Spin Versus Low-Spin Complexes

**12-65** Describe the difference between a high-spin and a low-spin $d^6$ complex.

**12-66** What factors determine whether a complex is high-spin or low-spin?

**12-67** Explain why the $Mn(H_2O)_6^{2+}$ ion is a high-spin complex.

**12-68** One of the $Fe(H_2O)_6^{2+}$ and $Fe(CN)_6^{4-}$ complex ions is high-spin and the other is low-spin. Which is which?

**12-69** Use the relative magnitudes of $\Delta_o$ and $\Delta_t$ to explain why there are no low-spin tetrahedral complexes.

**12-70** Compare the positions of the $Co^{2+}$, $Co^{3+}$, $Fe^{2+}$, and $Fe^{3+}$ ions in the spectrochemical series. Explain why the value of $\Delta$ increases with the charge on the transition-metal ion.

**12-71** Compare the positions of the $Co^{3+}$, $Rh^{3+}$, and $Ir^{3+}$ ions in the spectrochemical series. What happens to the value of $\Delta$ as we go down a column among the transition metals?

### The Colors of Transition-Metal Complexes

**12-72** Describe the characteristic colors of aqueous solutions of the following transition-metal ions.

(a) $Cu^{2+}$ (b) $Fe^{3+}$ (c) $Ni^{2+}$ (d) $CrO_4^{2-}$
(e) $MnO_4^-$

**12-73** Explain why so many of the pigments used in oil paints, such as vermilion (HgS), cadmium red (CdS), cobalt yellow [$K_3Co(NO_2)_6$], chrome yellow ($PbCrO_4$), prussian blue ($Fe_4[Fe(CN)_6]_3$), and cobalt blue ($CoO \cdot Al_2O_3$), contain transition-metal ions.

**12-74** Explain why $Cu(NH_3)_4^{2+}$ complexes have a deep-blue color if they don't absorb blue light. What light do they absorb?

**12-75** $CrO_4^{2-}$ ions are bright yellow. In what portion of the visible spectrum do these ions absorb light?

**12-76** When $CrO_4^{2-}$ reacts with acid to form $Cr_2O_7^{2-}$ ions, the color shifts from bright yellow to orange. Does this mean that the light absorbed shifts toward a higher or a lower frequency?

**12-77** $Ni^{2+}$ forms a complex with the dimethylglyoxime (DMG) ligand that absorbs light in the blue-green portion of the spectrum. What is the color of this $Ni(DMG)_2$ complex?

**12-78** Explain why a white piece of paper looks as if it has a faint pink color to a person who has been working for several hours at a computer terminal that has a green screen.

### Ligand-Field Theory

**12-79** Describe how ligand-field theory eliminates the difference between inner-shell complexes, such as the $Co(NH_3)_6^{3+}$ ion, and outer-shell complexes, such as the $Ni(NH_3)_6^{2+}$ ion.

**12-80** Explain how ligand-field theory allows the valence-shell *d* orbitals on the transition metal to be used simultaneously to form the skeleton structure of the complex and to hold the electrons that were originally in the *d* orbitals on the transition metal.

# Intersection

# THE STRUCTURE OF SOLIDS AND LIQUIDS

The kinetic molecular theory explains the characteristic properties of gases by assuming that gas particles are in a state of constant, random motion and that the diameter of these particles is very much smaller than the distance between the particles. Because most of the volume of a gas is empty space, the simplest analogy compares the particles of a gas to fruit flies in a jar.

Many of the properties of solids have been captured in the way the term *solid* is used in English. It describes something that holds its shape, such as a solidly constructed house. It implies continuity, as in a movie that runs for three solid hours. It implies the absence of empty space, as in a solid chocolate Easter bunny. Finally, it describes things that occupy three dimensions, as in solid geometry. A solid may be compared to a brick wall in which the individual bricks form a regular structure and the amount of empty space is kept to a minimum.

Liquids have properties between the extremes of gases and solids. Like gases, they flow to conform to the shape of their containers. Like solids, they cannot expand to fill their containers, and they are very difficult to compress. The structure of a liquid may be compared to a collection of marbles in a bag being shaken back and forth.

Because water is the only substance that we encounter routinely as a solid, a liquid, and a gas, it is useful to consider what happens to water as we change the temperature. At low temperatures, it is a solid in which the individual molecules are locked into a rigid structure. As we raise the temperature, the average kinetic energy of the molecules increases, which increases the rate at which these molecules move.

There are three ways in which a water molecule can move: (1) vibration, (2) rotation, and (3) translation, as shown in art below. Water molecules *vibrate* when H—O bonds are stretched or bent. *Rotation* involves the motion of a molecule around its center of gravity. *Translation* literally means to change from one place to another. It therefore describes the motion of molecules through space.

To understand the effect of this motion, we need to differentiate between intramolecular and intermolecular bonds. The covalent bonds between the hydrogen and oxygen atoms in a water molecule are called **intramolecular bonds.** (The prefix *intra-* comes from the Latin stem meaning "within or inside." Thus, intramural sports match teams from within the same institution.) The bonds between the neighboring water molecules in ice are called **intermolecular bonds,** from the Latin stem *inter* meaning "between." (This far more common prefix is used in words such as *interact, intermediate,* and *international.*)

The *intramolecular* bonds that hold the atoms in $H_2O$ molecules together are almost 25 times as strong as the *intermolecular* bonds between water molecules. (It takes

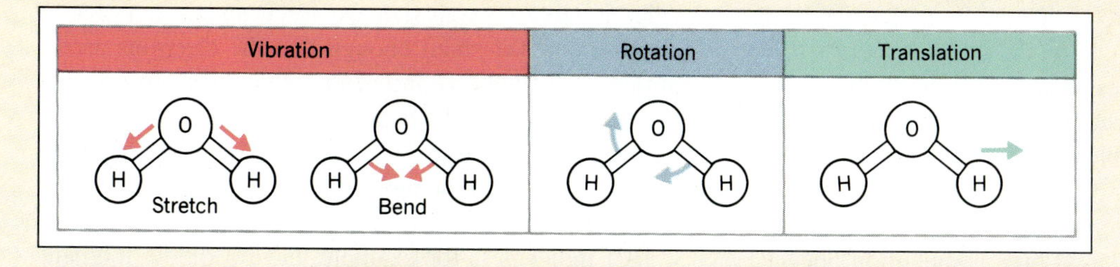


A water molecule can move in three ways: (a) vibration occurs when O—H bonds stretch or bend, (b) rotation involves movement around the center of gravity of the molecule, and (c) translation occurs when a water molecule moves through space.

The next chapters examine the difference between the structures of solids and liquids that explain such simple observations as the difference in the ease with which these two states of matter flow.

464 kJ/mol to break the H—O bonds within a water molecule and only 19 kJ/mol to break the bonds between water molecules.)

All three modes of motion disrupt the bonds between water molecules. As the system becomes warmer, the thermal energy of the water molecules eventually becomes too large to allow these molecules to be locked into the rigid structure of ice. At this point, the solid melts to form a liquid in which intermolecular bonds are constantly broken and reformed as the molecules move through the liquid. Eventually, the thermal energy of the water molecules becomes so large that the lifetime of an intermolecular bond becomes so short that the liquid boils to form a gas in which each particle moves more or less randomly through space.

The difference between solids and liquids, or liquids and gases, is therefore based on a competition between the strength of intermolecular bonds and the thermal energy of the system. At a given temperature, substances that contain strong intermolecular bonds are more likely to be solids. For a given intermolecular bond strength, the higher the temperature, the more likely the substance will be a gas.

The kinetic theory assumes that there is no force of attraction between the particles in a gas. If this assumption were correct, gases would never condense to form liquids and solids at low temperatures. In 1873 Dutch physicist Johannes van der Waals derived an equation that not only included the force of attraction between gas particles but also corrected for the fact that the volume of these particles becomes a significant fraction of the total volume of the gas at high pressures.

The van der Waals equation is used today to give a better fit to the experimental data of real gases than can be obtained with the ideal gas equation. But that was not van der Waals's goal. He was trying to develop a model that would explain the behavior of liquids by including terms that reflected the size of the atoms or molecules in the liquid and the strength of the bonds between these atoms or molecules. The weak intermolecular bonds in liquids and solids are therefore often called **van der Waals forces.** These forces can be divided into three categories: (1) dipole–dipole, (2) dipole-induced dipole, and (3) induced dipole–induced dipole.

## DIPOLE–DIPOLE FORCES

Many molecules contain bonds that fall between the extremes of ionic and covalent bonds. The difference between the electronegativities of the atoms in these molecules is large enough that the electrons are not shared equally, and yet small enough that the electrons aren't drawn exclusively to one of the atoms to form positive and negative ions. The bonds in these molecules are said to be *polar,* because they have positive and negative ends, or poles, and the molecules are often said to have a *dipole moment.*

HCl molecules, for example, have a dipole moment because the hydrogen atom has a slight positive charge and the chlorine atom has a slight negative charge. Because of the force of attraction between oppositely charged particles, there is a small dipole–dipole force of attraction between adjacent HCl molecules:

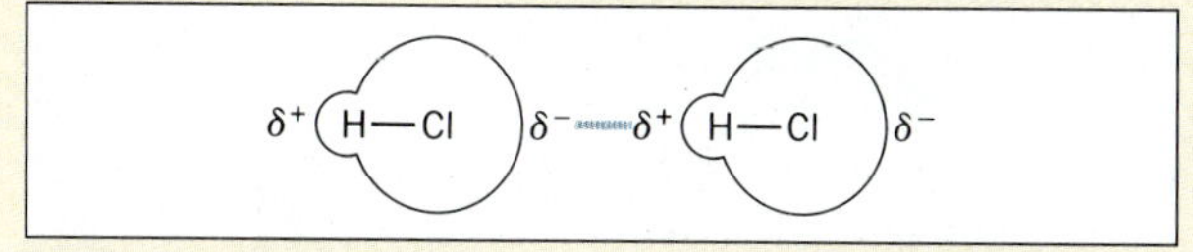


The dipole–dipole interaction in HCl is relatively weak, with a bond dissociation enthalpy of only 3.3 kJ/mol. (The covalent bonds between the hydrogen and chlorine atoms in HCl are 130 times as strong.) The force of attraction between HCl molecules is so small that hydrogen chloride boils at $-85.0$°C.

## DIPOLE-INDUCED DIPOLE FORCES

What would happen if we mixed HCl with argon, which has no dipole moment? The electrons on an argon atom are distributed homogeneously around the nucleus of the atom. But these electrons are in constant motion. When an argon atom comes close to a polar HCl molecule, the electrons can shift to one side of the nucleus to produce a very small dipole moment that lasts for only an instant:

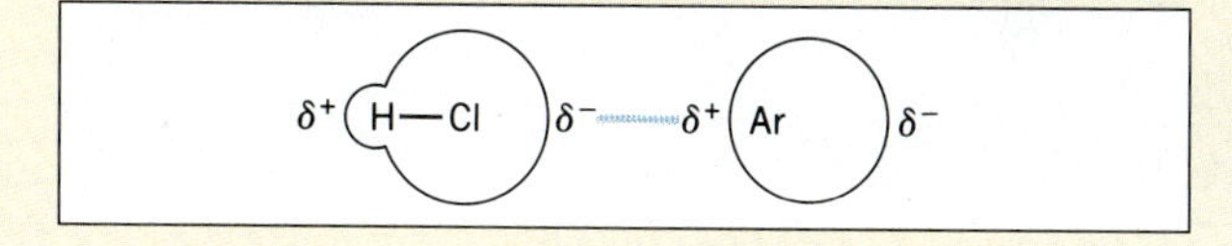


By distorting the distribution of electrons around the argon atom, the polar HCl molecule induces a small dipole moment on this atom, which creates a weak dipole-induced dipole force of attraction between the HCl molecule and the Ar atom. This force is very weak, with a bond energy of about 1 kJ/mol.

## INDUCED DIPOLE–INDUCED DIPOLE FORCES

Neither dipole–dipole nor dipole-induced forces can explain the fact that helium becomes a liquid at temperatures below 4.2 K. By itself, a helium atom is perfectly symmetrical. But movement of the electrons around the nuclei of a pair of neighboring helium atoms can become synchronized so that each atom simultaneously obtains an induced dipole moment:

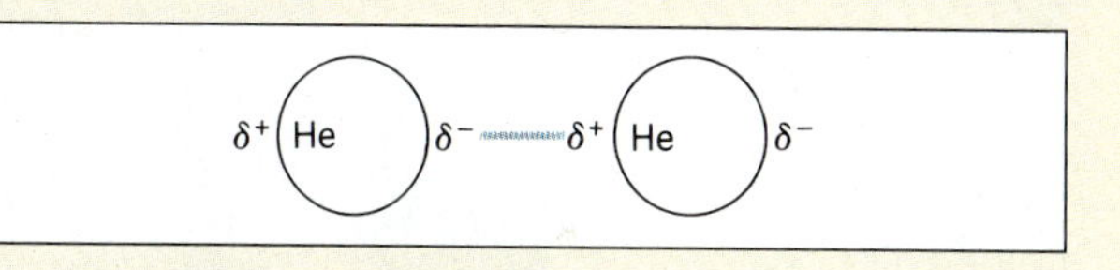


These fluctuations in electron density occur constantly, creating an induced dipole–induced dipole force of attraction between pairs of atoms. As might be expected, this force is relatively weak in helium—only 0.076 kJ/mol. But atoms or molecules become more polarizable as they become larger because there are more electrons to be polarized. It has been argued that the primary force of attraction between molecules in solid $I_2$ and in frozen $CCl_4$ is induced dipole–induced dipole attraction.

It only takes a small amount of an impurity to give glass brilliant color.

CHAPTER 13

# THE STRUCTURE OF SOLIDS

Gases are so disordered that their physical properties usually don't depend on the identity of the gas. Solids represent the other extreme. They are so organized that we have to specify the structure of each substance before we can understand its physical properties. Our discussion of solids will provide the basis for answering questions such as the following:

## POINTS OF INTEREST

- Why does Saran Wrap cling, whereas other plastic wraps do not?
- Why does the metal of a car left in the summer sun feel so much hotter than the grass next to which it was parked?
- Why is steel so much stronger than the iron from which it is made?
- Why do salts with similar formulas—such as NaCl, ZnS, and CsCl—crystallize in different structures?
- How do we know that the radius of a nickel atom is 0.1246 nm, or that the radius of a $Cl^-$ ion is 0.181 nm, if these particles are far too small to be seen?
- Why do metals become better conductors of electricity when cooled? Why do semiconductors become better conductors when heated?
- What is superconductivity? Why do we have to cool substances to very low temperatures before they become superconductors? What progress has been made in recent years toward a "high-temperature" superconductor?

## 13.1 SOLIDS

Solids can be divided into three categories on the basis of how the particles that form the solid pack together. **Crystalline solids** are three-dimensional analogs of a brick wall. They have a regular structure, in which the particles pack in a repeating pattern from one edge of the solid to the other. Diamond is an obvious example of a crystalline solid. **Amorphous solids,** such as glass and many plastics, (literally, "solids without form") have a random structure, with little if any long-range order. Many solids, such as aluminum and steel, have a structure that falls between these extremes. These **polycrystalline** solids are an aggregate of a large number of small crystals or grains in which the structure is regular, but the crystals or grains are arranged in a random fashion.

A polarized-light micrograph of a thin section of brass showing the random distribution of microcrystals that gives rise to the polycrystalline structure of this alloy.

As might be expected, the extent to which a solid is crystalline has important effects on its physical properties. The polyethylene used to make sandwich bags and garbage packs is an amorphous solid that consists of more or less randomly oriented chains of ($—CH_2—CH_2—$) linkages. Milk bottles are made from a more crystalline form of polyethylene, and they have a much more rigid structure.

### ▶ CHECKPOINT

Use the notion that solids become more crystalline when they are cooled to explain why racquetballs immersed in liquid nitrogen at temperatures of −196°C shatter when bounced on the floor.

Solids also can be classified on the basis of the bonds that hold the atoms or molecules together. This approach tends to categorize solids as either molecular, covalent, ionic, or metallic.

Iodine ($I_2$), sugar ($C_{12}H_{22}O_{11}$), and polyethylene are examples of compounds that are **molecular solids** at room temperature. Water and bromine are liquids that form molecular solids when cooled slightly; $H_2O$ freezes at 0°C and $Br_2$ freezes at −7°C. Molecular solids are characterized by relatively strong *intramolecular bonds* between the atoms that form the molecules and much weaker *intermolecular bonds* between these molecules. Because the intermolecular bonds are relatively weak, molecular solids are often soft substances with low melting points.

Dry ice, or solid carbon dioxide, is a perfect example of a molecular solid. The van der Waals forces holding the $CO_2$ molecules together are weak enough that dry ice *sublimes*—it passes directly from the solid to the gas phase—at −78°C.

The $CO_2$ molecules in dry ice are held together by intermolecular van der Waal's forces. This material is known as dry ice because it is cold (like ice) but dry (because it doesn't melt). Dry ice sublimes at −78° C, it goes directly from the solid to the gas phase.

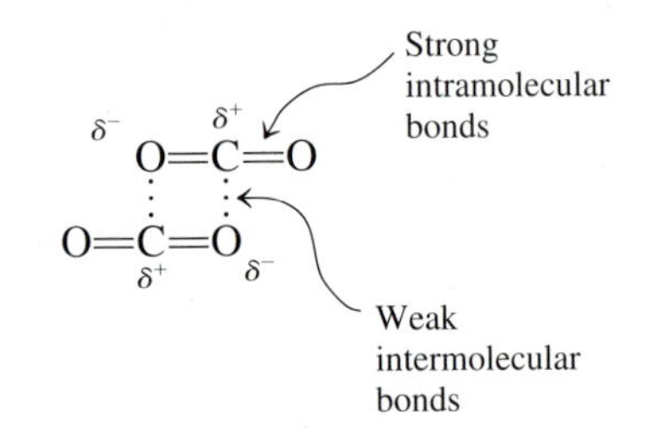


Anything that changes the strength of the van der Waals forces that hold molecular solids together has important consequences for the properties of the solid. Polyethylene $(—CH_2—CH_2—)_n$ is a soft plastic that melts at relatively low temperatures. When one of the hydrogens on every other carbon atom is replaced with a chlorine atom, we get a plastic known as poly(vinyl chloride), or PVC. This plastic is hard enough to be used to make the plastic pipes, which are so strong and long lasting that they are slowly replacing metal pipes for plumbing.

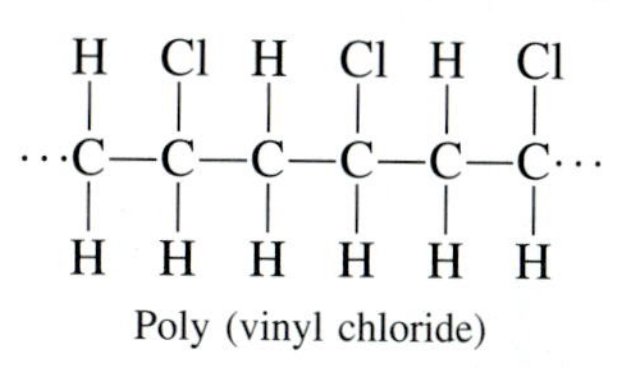


Poly (vinyl chloride)

The strength of PVC can be attributed to an increase in the strength of the van der Waals force of attraction between the chains of $(\text{—}CH_2\text{—}CHCl\text{—})_n$ molecules that form the solid. When poly(vinyl chloride) $(\text{—}CH_2\text{—}CHCl\text{—})_n$ and poly(vinylidene chloride) $(\text{—}CH_2\text{—}CCl_2\text{—})_n$ are copolymerized, a substance is formed that is sold under the trade name Saran. The same increase in the force of attraction between chains that makes PVC harder than polyethylene gives a thin film of Saran a tendency to cling to itself.

**Covalent solids** form crystals that can be viewed as a single giant molecule made up of an almost endless number of covalent bonds. Each carbon atom in diamond, for example, is covalently bound to four other carbon atoms oriented toward the corners of a tetrahedron, as shown in Figure 13.1. Because all of the bonds in this structure are equally strong, covalent solids are often very hard and they are notoriously difficult to melt. Diamond is the hardest natural substance and it melts at 3550°C.

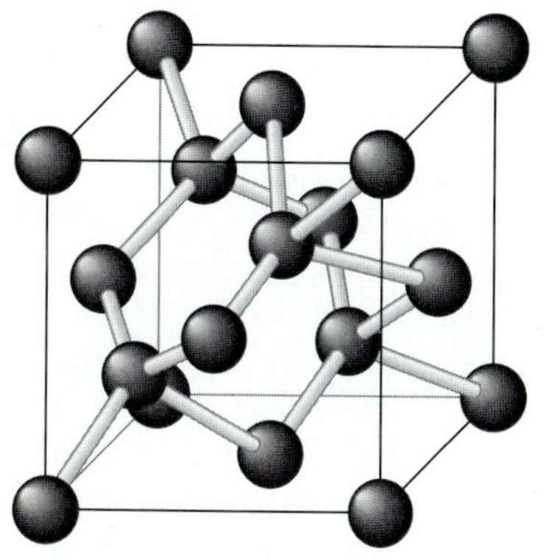

**Figure 13.1** A perfect diamond is a single molecule in which each carbon atom is tightly bound to four neighboring carbon atoms arranged toward the corners of a tetrahedron.

**Ionic solids** are held together by the strong force of attraction between ions of opposite charge:

$$F = \frac{q_1 \times q_2}{r^2}$$

Sec. 7.15

Because this force of attraction depends on the square of the distance between the positive and negative charges, the strength of an ionic bond depends on the radii of the ions that form the solid. As these ions become larger, the bond becomes weaker. But the ionic bond is still strong enough to ensure that salts have relatively high melting points and boiling points.

To understand **metallic solids** we have to amplify our understanding of chemical bonds. Ionic and covalent bonds are often imagined to be opposite ends of a two-dimensional model of bonding, in which compounds that contain polar bonds fall somewhere between these extremes:

ionic ........ polar ........ covalent

In reality, there are three kinds of bonds between adjacent atoms: ionic, covalent, and metallic, as shown in Figure 13.2. The force of attraction between atoms in metals, such as copper and aluminum, or alloys, such as brass and bronze, are metallic bonds.

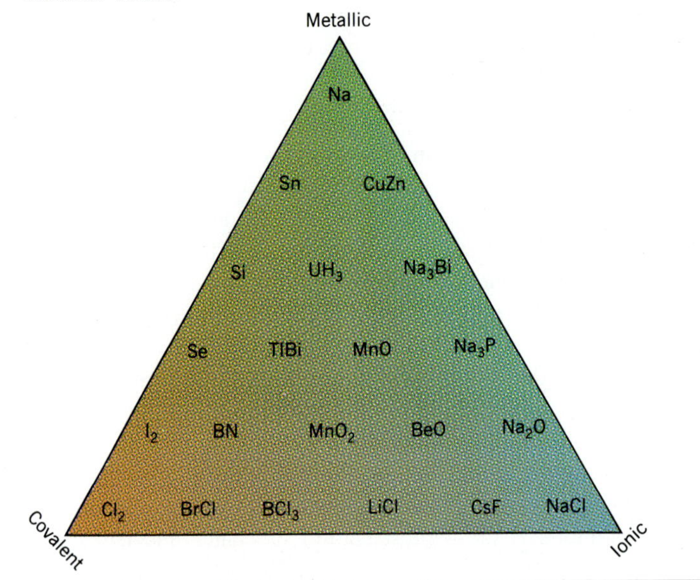


**Figure 13.2** The three different forms of chemical bonds—ionic, covalent, and metallic—form a two-dimensional plane, not a linear continuum.

Molecular, ionic, and covalent solids all have one thing in common. With only rare exceptions, the electrons in these solids are *localized*. They either reside on one of the atoms or ions, or they are shared by a pair of atoms or a small group of atoms.

Metal atoms don't have enough electrons to fill their valence shells by sharing electrons with their immediate neighbors. Electrons in the valence shell are therefore shared by many atoms, instead of just two. In effect, the valence electrons are *delocalized* over many metal atoms. Because these electrons aren't tightly bound to individual atoms, they are free to migrate through the metal. As a result, metals are good conductors of electricity. Electrons that enter the metal at one edge can displace other electrons to give rise to a net flow of electrons through the metal.

## 13.2 THE STRUCTURE OF METALS AND OTHER MONATOMIC SOLIDS

The structures of pure metals are easy to describe because the atoms that form these metals can be pictured as identical perfect spheres. The same can be said about the structure of the rare gases (He, Ne, Ar, and so on) at very low temperatures. These substances all crystallize in one of four basic structures, known as simple cubic (SC), body-centered cubic (BCC), hexagonal closest packed (HCP), and cubic closest packed (CCP).

Solids are very difficult to compress. This means that the amount of space between atoms in a solid must be kept to a minimum. As a rule, the most probable structure for a solid is the structure that makes the most efficient use of space.

To illustrate the principle that the particles that form the solid pack as tightly as possible, picture the best way of packing spheres, such as ping-pong balls, into an empty box.

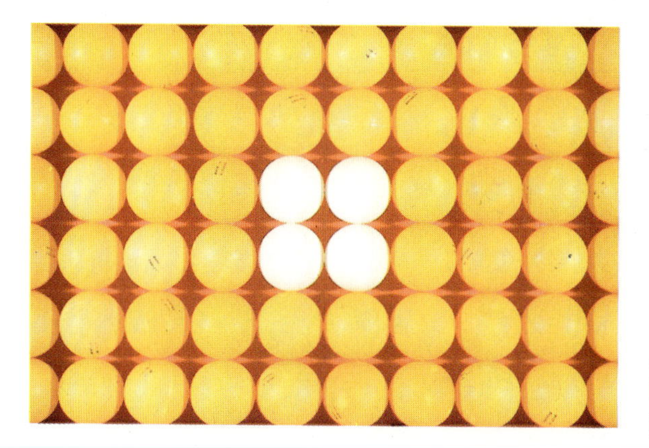

**Figure 13.3** A square-packed plane of spheres.

One approach involves carefully packing the ping-pong balls to form a square-packed plane of spheres, as shown in Figure 13.3. By tilting the box to one side, we can stack a second plane of spheres directly on top of the first. The result is a regular structure in which the simplest repeating unit is a cube of eight spheres, as shown in Figure 13.4. This structure is called **simple cubic packing.** Each sphere in this structure touches four identical spheres in the same plane. It also touches one sphere in the plane above and one in the plane below. Because each atom in this structure can form bonds to its six nearest neighbors, each sphere is said to have a **coordination number** of 6.

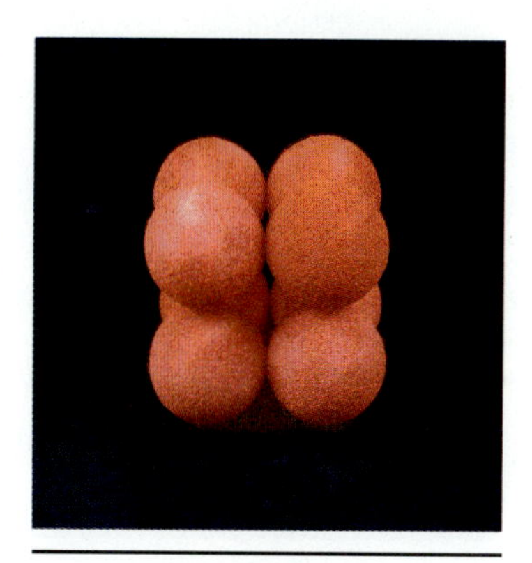

**Figure 13.4** The smallest repeating unit in a simple-cubic structure.

It is fairly easy to show that a simple cubic structure is not an efficient way of using space. Only 52% of the available space is actually occupied by the spheres in a simple cubic structure. The rest is empty space. Because this structure is inefficient, only one element, polonium, actually crystallizes in a simple cubic structure.

This raises the interesting question: How can space be used more efficiently? Suppose we start by separating the spheres to form a square-packed plane in which they do not quite touch each other, as shown in Figure 13.5. The spheres in the second plane pack above the holes in the first plane, as shown in Figure 13.6. Spheres in the third plane pack above holes in the second plane. Spheres in the fourth plane pack above holes in the third plane, and so on. The result is a structure in which the odd-numbered planes of atoms are identical and the even-numbered planes are identical. This *ABABABAB* . . . repeating structure is known as **body-centered cubic packing.**

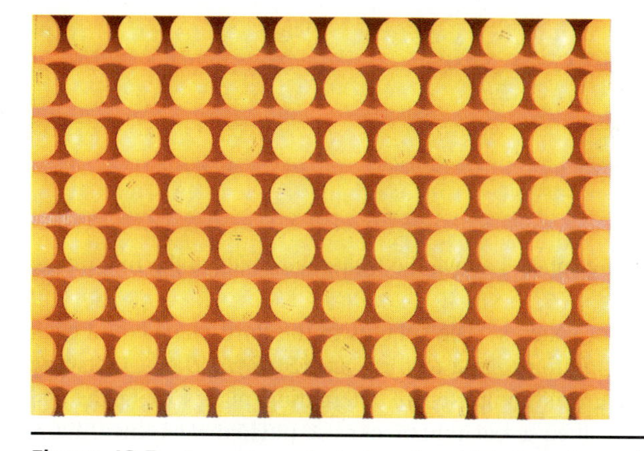

**Figure 13.5** A square-packed plane in which the spheres do not quite touch.

**Figure 13.6** The spheres in the second plane of a body-centered cubic structure pack above the holes in the plane shown in Figure 13.5.

**Figure 13.7** The smallest repeating unit in a body-centered cubic structure.

This structure is called *body-centered cubic* because each sphere touches four spheres in the plane above and four more in the plane below, arranged toward the corners of a cube. Thus, the repeating unit in this structure is a cube of eight spheres with a ninth identical sphere in the center of the body—in other words, a body-centered cube, as shown in Figure 13.7. The coordination number in this structure is 8.

Body-centered cubic packing is a more efficient way of using space than simple cubic packing—68% of the space in this structure is filled. Body-centered cubic packing is an important structure for metals. All of the metals in Group IA (Li, Na, K, and so on), the heavier metals in Group IIA (Ca, Sr, and Ba), and a number of the early transition metals (such as Ti, V, Cr, Mo, W, and Fe) pack in a body-centered cubic structure.

Two structures pack spheres so efficiently they are called **closest packed structures.** Both start by packing the spheres in planes in which each sphere touches six others oriented toward the corners of a hexagon, as shown in Figure 13.8. A second plane is then formed by packing spheres above the triangular holes in the first plane, as shown in Figure 13.9.

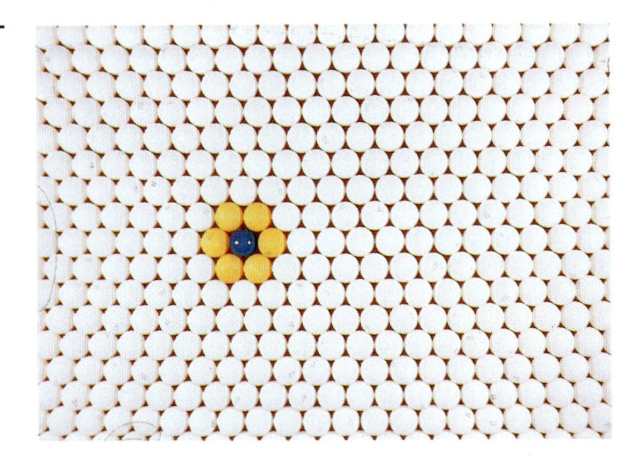

**Figure 13.8** A closest-packed plane in which each sphere touches six others oriented toward the corners of a hexagon.

What about the next plane of spheres? The spheres in the third plane could pack directly *above the spheres in the first plane* to form an *ABABABAB* . . . repeating structure. Because this structure is composed of alternating planes of hexagonal closest packed spheres, it is called a **hexagonal closest packed structure.** Each sphere touches three spheres in the plane above, three spheres in the plane below,

**Figure 13.9** Atoms in the second plane of closest-packed structures pack above the triangular holes in the first plane shown in Figure 13.8.

and six spheres in the same plane, as shown in Figure 13.10. Thus, the coordination number in a hexagonal closest packed structure is 12.

In a hexagonal closest packed structure, 74% of the space is filled. No more efficient way of packing spheres is known, and the hexagonal closest packed structure is important for metals such as Be, Co, Mg, and Zn, as well as the rare gas He at low temperatures.

There is another way of stacking hexagonal closest packed planes of spheres. The atoms in the third plane can be packed *above the holes in the first plane* that were not used to form the second plane. The fourth hexagonal closest packed plane of atoms then packs directly above the first. The net result is an *ABCABCABC* . . . structure, which is called **cubic closest packed.** Each sphere in this structure touches six others in the same plane, three in the plane above, and three in the plane below, as shown in Figure 13.11. Thus, the coordination number is still 12, and, once again, 74% of the space is filled.

The difference between hexagonal and cubic closest packed structures can be understood by comparing Figures 13.10 and 13.11. In the hexagonal closest packed structure, the atoms in the first and third planes lie directly above each other. In the cubic closest packed structure, the atoms in these planes are oriented in different directions.

The cubic closest packed structure is just as efficient as the hexagonal closest packed structure. Many metals, including Ag, Al, Au, Ca, Co, Cu, Ni, Pb, and Pt, crystallize in a cubic closest packed structure. So do most of the rare gases when these gases are cooled to low enough temperatures to solidify.

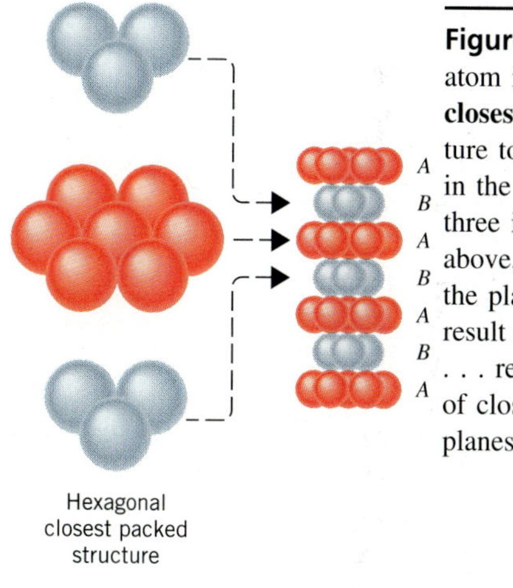


**Figure 13.10** Each atom in a **hexagonal closest packed** structure touches six atoms in the same plane, three in the plane above, and three in the plane below. The result is an *ABABAB* . . . repeating pattern of closest packed planes.

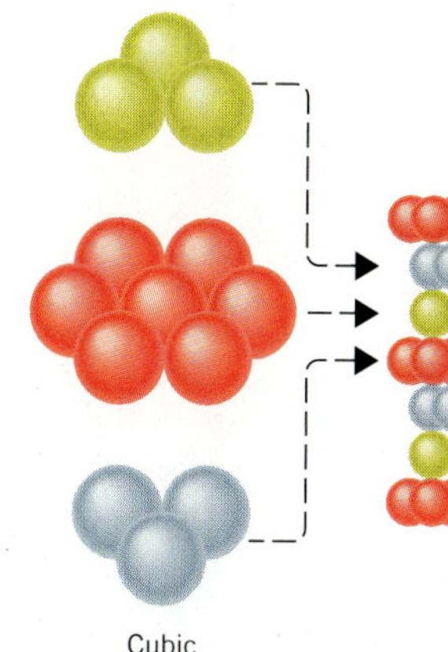


**Figure 13.11** Each atom in a **cubic closest packed** structure also touches six atoms in the same plane, three in the plane above, and three in the plane below. But the atoms in the top plane are rotated by 180° relative to the bottom plane. These planes of atoms therefore form an *ABCABCABC* . . . repeating pattern.

**CHECKPOINT**

Use the fact that the atoms in a metal form strong bonds to up to 12 nearest neighbors to explain why most metals have unusually high melting points and boiling points.

## 13.3 COORDINATION NUMBERS AND THE STRUCTURES OF METALS

The coordination numbers of the four structures described in the preceding section are summarized in Table 13.1. It is easy to understand why metals pack in hexagonal or cubic closest packed structures. Not only do these structures use space as efficiently as possible, they also have the largest possible coordination numbers, which allows each metal atom to form bonds to the largest number of neighboring metal atoms.

**TABLE 13.1 Coordination Numbers for Common Crystal Structures**

| Structure | Coordination Number | Stacking Pattern |
|---|---|---|
| Simple cubic | 6 | *AAAAAAAA* . . . |
| Body-centered cubic | 8 | *ABABABAB* . . . |
| Hexagonal closest packed | 12 | *ABABABAB* . . . |
| Cubic closest packed | 12 | *ABCABCABC* . . . |

It is less obvious why one-third of the metals pack in a body-centered cubic structure in which the coordination number is only 8. The popularity of this structure can be understood by referring to Figure 13.12. The coordination number for body-centered cubic structures given in Table 13.1 counts only the atoms that actually touch a given atom in this structure. Figure 13.12 shows that each atom also *almost touches* four neighbors in the same plane, a fifth neighbor two planes above, and a sixth two planes below. The distance from each atom to the nuclei of these nearby atoms is only 15% larger than the distance to the nuclei of the atoms that it actually touches. Each atom in a body-centered cubic structure can therefore form a total of 14 bonds—eight strong bonds to the atoms that it touches and six weaker bonds to the atoms it almost touches.

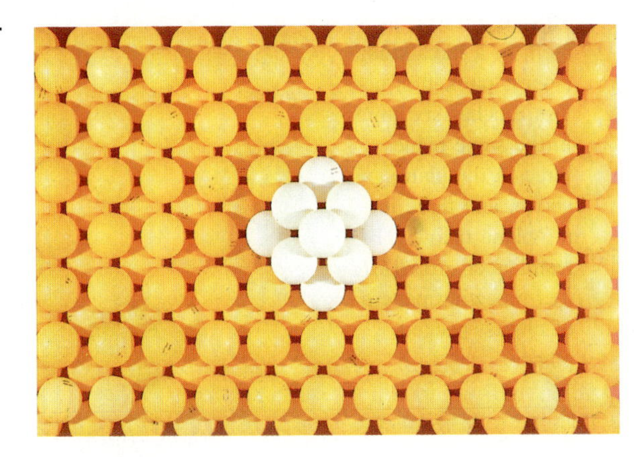

**Figure 13.12** Each atom in a body-centered cubic structure touches four atoms in the plane above and four in the plane below. In addition, each atom almost touches six more atoms.

This makes it easier to understand why atoms in some metals pack in a body-centered cubic structure instead of a hexagonal or cubic closest packed structure. Each metal atom in the closest packed structures can form strong bonds to 12 neighboring atoms. In the body-centered cubic structure, each atom forms a total of 14 bonds to neighboring atoms, although six of these bonds are somewhat weaker than the other eight.

## 13.4 PHYSICAL PROPERTIES THAT RESULT FROM THE STRUCTURE OF METALS

Sec. 2.7

The structures of metals can be used to explain many of the characteristic physical properties of metals.

For example, metals have a characteristic metallic shine, or luster. At first, a reasonable explanation for this may appear to be that metals reflect (literally, throw back) the light that shines on their surface. This implies, in effect, that light bounces off a metal's surface the way a racquetball bounces off the walls of a racquetball court. There is something wrong with this analogy, however, because metals actually absorb a significant fraction of the light that hits their surfaces.

A portion of the energy captured when the metal absorbs light is turned into thermal energy. (You can easily demonstrate this by placing your hand on the surface of a car that has spent several hours in the sun.) The rest of the energy is *reradiated* by a metal as what appears to be reflected light. Silver is better than any other metal at ''reflecting'' light, and yet only 88% of the light that hits the surface of a silver mirror is reradiated.

Solids, such as glass, that do not absorb light can be colored by adding a small quantity of an impurity that absorbs visible light.

This raises the important question: Why do metals absorb and reradiate light when other substances, such as the glass in the car's windows, do not? Light is absorbed when the energy of this radiation is equal to the energy needed to excite an electron to a higher energy excited state or when the energy can be used to move an electron through the solid. Because electrons are delocalized in metals—and therefore free to move through the solid—metals absorb light easily. Other solids, such as glass, don't have electrons that can move through the solid, so they can't absorb light the way metals do. These solids are colorless and can be colored only by adding an impurity in which the energy associated with exciting an electron from one orbital to another falls in the visible portion of the spectrum. Glass is usually colored by adding a small quantity of one of the transition metals. Cobalt produces a blue color, chromium makes the glass appear green, and traces of gold give a deep-red color.

Nonmetals such as hydrogen and oxygen are gases at room temperature because these elements can achieve a filled shell of valence electrons by sharing pairs of electrons to form relatively small molecules, such as $H_2$ and $O_2$, that are moving fast enough at room temperature to escape from the liquid into the gaseous phase. Metals can't do this. There aren't enough electrons on a metal atom to allow it to fill its valence shell by sharing pairs of electrons with one or two nearest neighbors. The only way a metal can obtain the equivalent of a filled shell of valence electrons is by allowing these electrons to be shared by a number of adjacent metal atoms. This is possible only if a large number of metal atoms are kept close together, and metals are therefore solids at room temperature.

Why are metals malleable and ductile? Most metals pack in either body-centered cubic, hexagonal closest packed, or cubic closest packed structures. In theory, changing the shape of the metal is simply a matter of applying a force that makes the

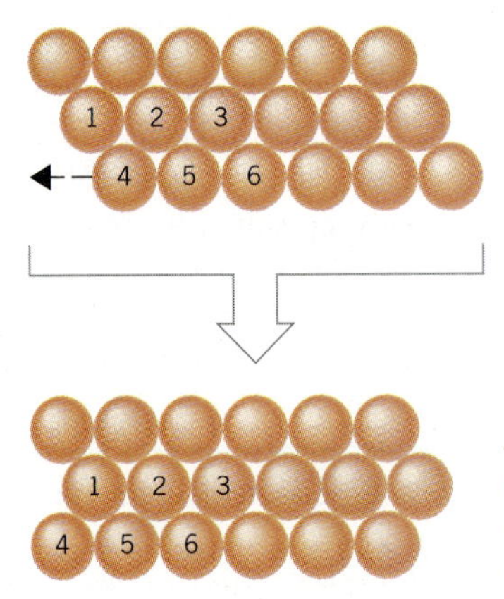


**Figure 13.13** Metals are malleable and ductile because planes of atoms can slip past each other to reach equivalent positions.

atoms in one of the planes slide past the atoms in an adjacent plane, as shown in Figure 13.13. In practice, it is easier to do this when the metal is hot.

Why are metals good conductors of heat and electricity? As we have already seen, the delocalization of valence electrons in a metal allows the solid to conduct an electric current. To understand why metals conduct heat, remember that temperature is a macroscopic property that reflects the kinetic energy of the individual atoms or molecules. The tight packing of atoms in a metal means that kinetic energy can be transferred from one atom to another both rapidly and efficiently.

## 13.5 SOLID SOLUTIONS AND INTERMETALLIC COMPOUNDS

Sec. 3.17

Most of the solutions chemists work with involve a gas (such as HCl) or a solid (such as NaCl) dissolved in a liquid (such as water). It is also possible to prepare solutions in which a gas, a liquid, or a solid dissolves in a solid (see Section 3.17). The most important class of solid solutions are those in which one solid is dissolved in another. Two examples of solid solutions are copper dissolved in aluminum and carbon dissolved in iron.

The solubility of one solid in another usually depends on temperature. At room temperature, for example, copper does not dissolve in aluminum. At 550°C, however, aluminum can form solutions that contain up to 5.6% copper by weight. Aluminum metal that has been saturated with copper at 550°C will try to reject the copper atoms as it cools to room temperature. In theory, the solution could reject copper atoms by forming a polycrystalline structure composed of small crystals of more or less pure aluminum interspersed with small crystals of copper metal. Instead of this, the copper atoms combine with aluminum atoms as the solution cools to form an *intermetallic compound,* with the formula $CuAl_2$, interspersed with the pure aluminum.

$CuAl_2$ is a perfect example of the difference between a mixture (such as a solution of copper dissolved in aluminum) and a compound. The solution can contain varying amounts of copper and aluminum. At 550°C, for example, the solution can contain between 0% and 5.6% copper metal by weight. The intermetallic compound has a fixed composition—$CuAl_2$ is always 49.5% aluminum by weight.

Intermetallic compounds such as $CuAl_2$ are the key to a process known as *precipitation hardening.* Aluminum metal packs in a cubic closest packed structure in which one plane of atoms can slip past another. As a result, pure aluminum metal is too weak to be used as a structural metal in cars or airplanes. Precipitation hardening produces alloys that are five to six times as strong as aluminum and make an excellent structural metal.

The first step in precipitation hardening of aluminum involves heating the metal to 550°C. Copper is then added to form a solution, which is quenched with cold water. The solution cools so fast that the copper atoms can't come together to form microcrystals of copper metal.

Comparing a solid with a brick wall has one major disadvantage. It leads one to believe that atoms can't move through the metal. This is not quite true. Diffusion through the metal can occur, although it occurs slowly. Over a period of time, copper atoms can move through the quenched solution to form microcrystals of the $CuAl_2$ intermetallic compound that are so small they are hard to see with a microscope.

These $CuAl_2$ particles are both hard and strong. So hard they inhibit the flow of the aluminum metal that surrounds them. These microcrystals of $CuAl_2$ strengthen aluminum metal by interfering with the way planes of atoms slip past each other. The result is a metal that is both harder and stronger than pure aluminum.

Alloys of aluminum and copper that have undergone precipitation hardening are much stronger than pure aluminum.

Copper dissolved in aluminum at high temperature is an example of a *substitution solution,* in which copper atoms pack in the positions normally occupied by aluminum atoms. There is another way in which a solid solution can be made. Atoms of one element can pack in the holes, or *interstices,* between atoms of the host element because even the most efficient crystal structures use only 74% of the available space in the crystal. The result is an *interstitial solution.*

Steel at high temperatures is a good example of an interstitial solution. Steel is formed by dissolving carbon in iron. At very high temperatures, iron packs in a cubic closest packed structure that leaves just enough space to allow carbon atoms to fit in the holes between the iron atoms. Below 910°C, iron metal packs in a body-centered cubic structure, in which the holes are too small to hold carbon atoms.

This has important consequences for the properties of steel. At temperatures above 910°C, carbon readily dissolves in iron to form a solid solution that contains as much as 1% carbon by weight. This material is both malleable and ductile, and it can be rolled into thin sheets or hammered into various shapes. When this solution cools below 910°C, the iron changes to a body-centered cubic structure, and the carbon atoms are rejected from the metal. If the solution is allowed to cool gradually, the carbon atoms migrate through the metal to form a compound with the formula $Fe_3C$, which precipitates from the solution. These $Fe_3C$ crystals serve the same role in steel that the $CuAl_2$ crystals play in aluminum—they inhibit the flow of the planes of metal atoms and thereby make the metal stronger.

Steel, which contains up to 1% carbon, is much stronger than iron.

### ▶ CHECKPOINT

It is often difficult to replace the U-trap beneath a sink. Use the fact that atoms can slowly diffuse through a metal to explain what happens when one of these U-traps "freezes."

## 13.6 HOLES IN CLOSEST PACKED AND SIMPLE CUBIC STRUCTURES

Metals are not the only solids that pack in simple cubic, body-centered cubic, hexagonal closest packed, and cubic closest packed structures. A large number of ionic solids use these structures as well.

Sodium chloride (NaCl) and zinc sulfide (ZnS), for example, form crystals that can be thought of as cubic closest packed arrays of negative ions ($Cl^-$ or $S^{2-}$), with positive ions ($Na^+$ or $Zn^{2+}$) packed in holes between the closest packed planes of negative ions. There is a subtle difference between these structures, however, because the $Na^+$ ions in NaCl pack in holes that are different from those used by the $Zn^{2+}$ ions in ZnS.

There are two kinds of holes in a closest packed structure. One kind, called **tetrahedral holes,** is shown in Figure 13.14. The solid lines in this figure represent one plane of closest packed atoms. The dashed lines represent a second plane of atoms, which pack above the holes in the first plane. Each of the holes marked with a *t* touches three atoms in the first plane and one atom in the second plane. They are called tetrahedral holes because positive ions that pack in these holes are surrounded by four negative ions arranged toward the corners of a tetrahedron.

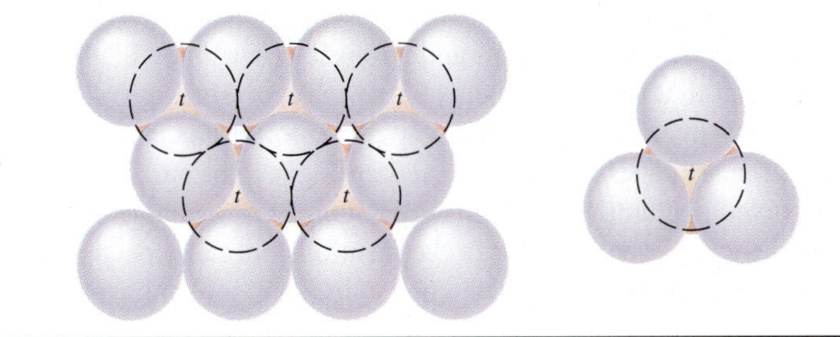

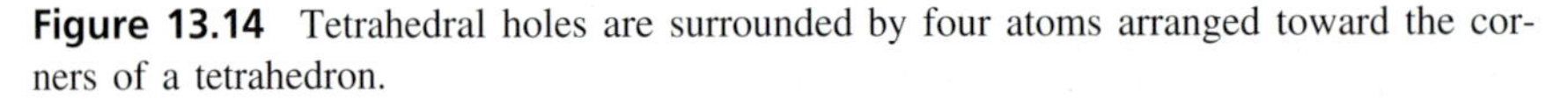

**Figure 13.14** Tetrahedral holes are surrounded by four atoms arranged toward the corners of a tetrahedron.

The second kind, the **octahedral holes** in a closest packed structure, are shown in Figure 13.15. Once again, the solid lines represent a plane of closest packed atoms and the dashed lines correspond to a second plane that packs above the holes in the first plane. Each of the holes marked with an *o* touches three atoms in the first plane and three atoms in the second plane. They are called octahedral holes because positive ions that occupy these holes are surrounded by six negative ions arranged toward the corners of an octahedron.

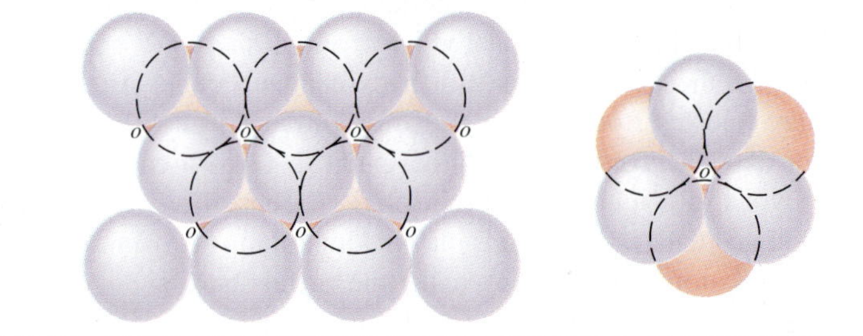

**Figure 13.15** Octahedral holes in a closest packed structure are surrounded by six atoms arranged toward the corners of an octahedron.

Tetrahedral holes are relatively small. The largest atom that can fit into a tetrahedral hole without distorting the tetrahedron has a radius only 0.225 times the radius of the atoms that form the hole. Octahedral holes are almost twice as large as tetrahedral holes. The largest atom that can fit into an octahedral hole has a radius 0.414 times the radius of the atoms that form the hole. The relative size of the atoms or ions that form a crystal therefore dictates whether tetrahedral or octahedral holes are used.

**CHECKPOINT**

Sec. 7.7

Use the discussion of the relative size of atoms and ions in Chapter 7 to explain why simple positive ions can pack in the relatively small holes between planes of closest packed negative ions.

Sometimes positive ions are too big to pack in either tetrahedral or octahedral holes in a closest packed structure of negative ions. When this happens, the negative ions pack in a simple cubic structure, and the positive ions pack in **cubic holes** between the planes of negative ions.

## 13.7 RADIUS RATIO RULES

The discussion of tetrahedral, octahedral, and cubic holes in the previous section suggests that the structure of an ionic solid depends on the relative size of the ions that form the solid. The relative size of these ions is given by the **radius ratio,** which is the radius of the positive ion divided by the radius of the negative ion:

$$\textbf{radius ratio} = \frac{r_+}{r_-}$$

The relationship between the coordination number of the positive ions in ionic solids and the radius ratio of the ions is given in Table 13.2. As the radius ratio increases, the number of negative ions that can pack around each positive ion increases. When the radius ratio is between 0.225 and 0.414, positive ions tend to pack in tetrahedral holes between planes of negative ions in a cubic or hexagonal closest packed structure. When the radius ratio is between 0.414 and 0.732, the positive ions tend to pack in octahedral holes between planes of negative ions in a closest packed structure.

**TABLE 13.2 Radius Ratio Rules**

| Radius Ratio | Coordination Number | Holes in Which Positive Ions Pack |
|---|---|---|
| 0.225–0.414 | 4 | tetrahedral holes |
| 0.414–0.732 | 6 | octahedral holes |
| 0.732–1 | 8 | cubic holes |
| 1 | 12 | closest packed structure |

Table 13.2 suggests that tetrahedral holes aren't used until the positive ion is large enough to touch all four of the negative ions that form this hole. As the radius ratio increases from 0.225 to 0.414, the positive ion distorts the structure of the negative ions toward a structure that might best be described as *closely packed.*

As soon as the positive ion is large enough to touch all six negative ions in an octahedral hole, the positive ions start to pack in octahedral holes. These holes are used until the positive ion is so large that it can't fit into even a distorted octahedral hole.

Eventually a point is reached at which the positive ion can no longer fit into either the tetrahedral or octahedral holes in a closest packed crystal. When the radius ratio is between about 0.732 and 1, ionic solids tend to crystallize in a simple cubic array of negative ions with positive ions occupying some or all of the cubic holes between these planes. When the radius ratio is about 1, the positive ions can be incorporated directly into the positions of the closest packed structure.

**EXERCISE 13.1**

The following compounds have similar empirical formulas. Use the radius ratio rules and the table of ionic radii in the appendix to explain why they have different structures.

(a) NaCl (b) ZnS (c) CsCl

**SOLUTION**

(a) The ionic radii of the $Na^+$ (0.095 nm) and $Cl^-$ (0.181 nm) ions give a radius ratio of 0.52. The $Na^+$ ions should therefore pack in octahedral holes between planes of closest packed $Cl^-$ ions.

(b) The ionic radii of the $Zn^{2+}$ (0.074 nm) and $S^{2-}$ (0.184 nm) ions give a radius ratio of 0.40, which suggests that the $Zn^{2+}$ ions should pack in tetrahedral holes between planes of closest packed $S^{2-}$ ions.

(c) The ionic radii of the $Cs^+$ (0.169 nm) and $Cl^-$ (0.181 nm) ions give a radius ratio of 0.933. The $Cs^+$ ions are too large to fit either tetrahedral or octahedral holes, which means that this substance can't crystallize in a closest packed structure. Instead, it crystallizes as a simple cubic array of $Cl^-$ ions with $Cs^+$ ions in the cubic holes in this structure.

The structure of ionic solids is also affected by differences in the abundance of tetrahedral and octahedral holes in a closest packed structure. There are just as many octahedral holes as there are spheres that form the closest packed structure. Thus, if NaCl is a 1 : 1 salt in which the $Na^+$ ions occupy octahedral holes in a closest packed array of $Cl^-$ ions, all of the octahedral holes in this structure must be filled.

There are twice as many tetrahedral holes as octahedral holes in a closest packed structure. The $Zn^{2+}$ ions in ZnS therefore occupy only half of the tetrahedral holes in a closest packed array of $S^{2-}$ ions.

## 13.8 UNIT CELLS: THE SIMPLEST REPEATING UNIT IN A CRYSTAL

Crystals can be thought of as the three dimensional equivalent of a variety of common two dimensional repeating designs.

So far, our description of solids has focused on the way the particles pack to fill space. The structure of solids also can be described as if they were three-dimensional analogs of a piece of wallpaper. Wallpaper has a regular repeating design that extends from one edge to the other. Crystals have a similar repeating design, but in this case, the design extends in three dimensions from one edge of the solid to the other.

We can unambiguously describe a piece of wallpaper by specifying the size, shape, and contents of the simplest repeating unit in the design. We can describe a three-dimensional crystal by specifying the size, shape, and contents of the simplest repeating unit and the way these repeating units stack to form the crystal.

The simplest repeating unit in a crystal is called a **unit cell.** Each unit cell is defined in terms of **lattice points**—the points in space about which the particles are free to vibrate in a crystal. In 1850, Auguste Bravais showed that crystals could be divided into 14 unit cells that meet the following criteria:

1. The unit cell is the simplest repeating unit in the crystal.
2. Opposite faces of a unit cell are parallel.
3. The edge of the unit cell connects equivalent points.

The 14 Bravais unit cells are shown in Figure 13.16. These unit cells fall into seven categories, which differ in the three unit-cell edge lengths (*a*, *b*, and *c*) and three internal angles ($\alpha$, $\beta$, and $\gamma$), as shown in Table 13.3.

This chapter focuses on the cubic category, which includes the simple cubic, body-centered cubic, and face-centered cubic unit cells shown in Figure 13.17. These unit cells are important for two reasons. First, a number of metals, ionic solids, and intermetallic compounds crystallize in cubic unit cells. Second, it is relatively easy to do calculations with these unit cells because the cell-edge lengths are all the same and the cell angles are all 90°.

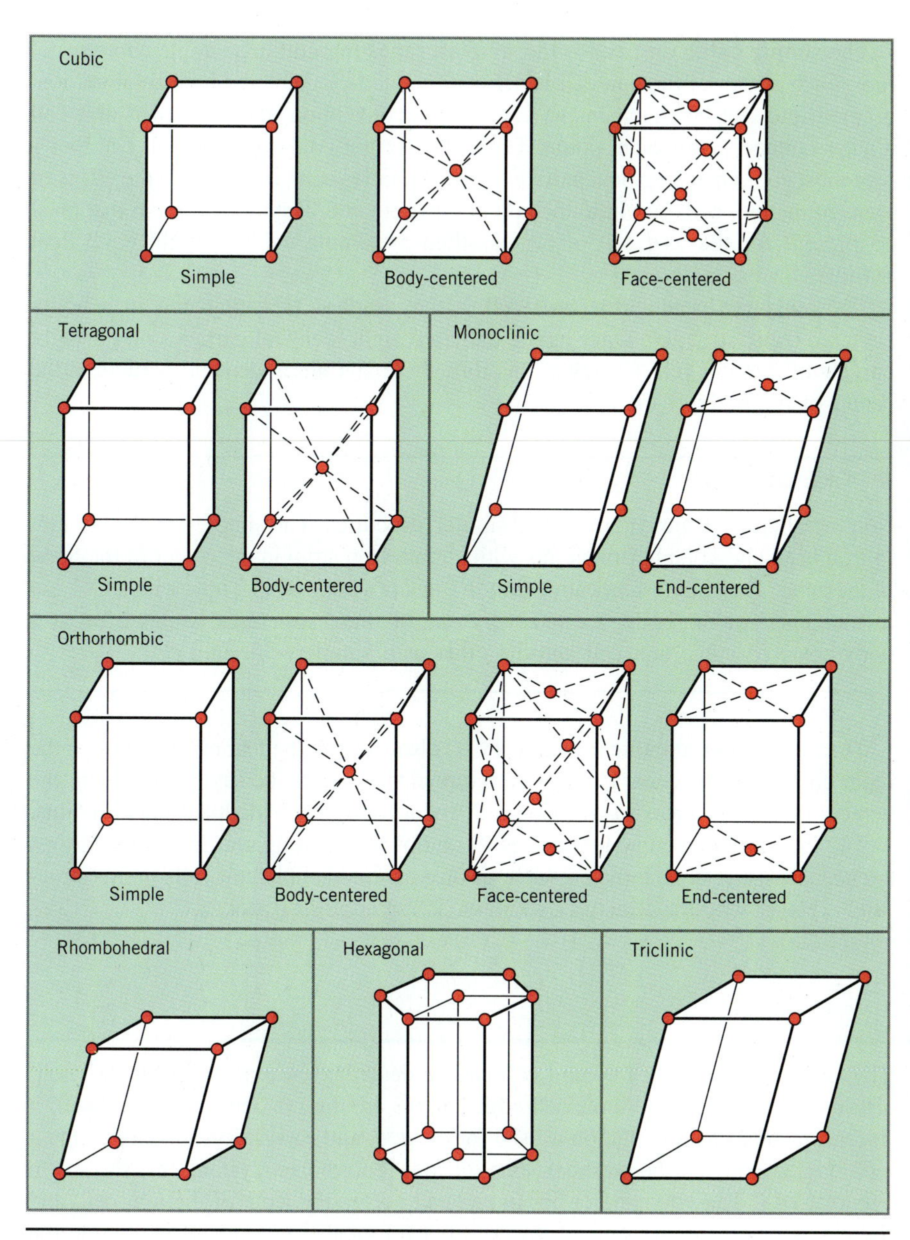


**Figure 13.16** The 14 Bravais unit cells.

**Figure 13.17** Models of simple cubic (top) body-centered cubic (center) and face-centered cubic unit cells (bottom).

**TABLE 13.3 The Seven Categories of Bravais Unit Cells**

| Category | Edge Lengths | Internal Angles |
|---|---|---|
| Cubic | $(a = b = c)$ | $(\alpha = \beta = \gamma = 90°)$ |
| Tetragonal | $(a = b \neq c)$ | $(\alpha = \beta = \gamma = 90°)$ |
| Monoclinic | $(a \neq b \neq c)$ | $(\alpha = \beta = 90° \neq \gamma)$ |
| Orthorhombic | $(a \neq b \neq c)$ | $(\alpha = \beta = \gamma = 90°)$ |
| Rhombohedral | $(a = b = c)$ | $(\alpha = \beta = \gamma \neq 90°)$ |
| Hexagonal | $(a = b \neq c)$ | $(\alpha = \beta = 90°,\ \gamma = 120°)$ |
| Triclinic | $(a \neq b \neq c)$ | $(\alpha \neq \beta \neq \gamma \neq 90°)$ |

The **simple cubic unit cell** is the simplest repeating unit in a simple cubic structure. Each corner of the unit cell is defined by a lattice point at which an atom, ion, or molecule can be found in the crystal. By convention, the edge of a unit cell always connects equivalent points. Each of the eight corners of the unit cell therefore must contain an identical particle. Other particles can be present on the edges or faces of the unit cell, or within the body of the unit cell. But the minimum that must be present for the unit cell to be classified as simple cubic is eight equivalent particles on the eight corners.

The **body-centered cubic unit cell** is the simplest repeating unit in a body-centered cubic structure. Once again, there are eight identical particles on the eight corners of the unit cell. However, this time there is a ninth identical particle in the center of the body of the unit cell.

**CHECKPOINT**

Iron metal and cesium chloride have similar structures. The simplest repeating unit in iron is a cube of eight iron atoms with a ninth iron atom in the center of the body of the cube. The simplest repeating unit in CsCl is a cube of $Cl^-$ ions with a $Cs^+$ ion in the center of the body. Explain why one of these structures is classified as a body-centered cubic unit cell and the other as a simple cubic unit cell.

The **face-centered cubic unit cell** (FCC) also starts with identical particles on the eight corners of the cube. But this structure also contains the same particles in the centers of the six faces of the unit cell, for a total of 14 identical lattice points.

The face-centered cubic unit cell is the simplest repeating unit in a cubic closest packed structure. In fact, the presence of face-centered cubic unit cells in this structure explains why the structure is known as *cubic* closest packed.

## 13.9 UNIT CELLS: A THREE-DIMENSIONAL GRAPH

The lattice points in a cubic unit cell can be described in terms of a three-dimensional graph. Because all three cell-edge lengths are the same in a cubic unit cell, it doesn't matter what orientation is used for the *a, b,* and *c* axes. By convention, the *a* axis is defined as the vertical axis of our coordinate system, as shown in Figure 13.18. The *b* axis describes movement across the front of the unit cell, and the *c* axis represents movement toward the back of the unit cell. If we define the bottom left corner of the unit cell as the origin (0,0,0), the coordinates 1,0,0 indicate a lattice point that is one cell-edge length away from the origin along the *a* axis. Similarly, 0,1,0 and 0,0,1 represent lattice points that are displaced by one cell-edge length from the origin along the *b* and *c* axes, respectively.

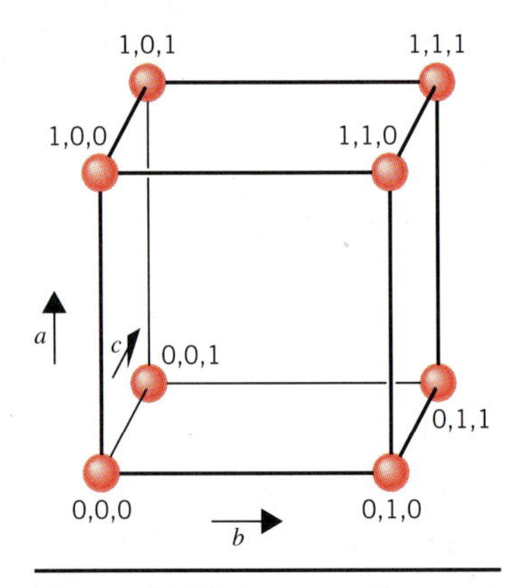


**Figure 13.18** The coordinates of the eight lattice points that define a cubic unit cell.

Thinking about the unit cell as a three-dimensional graph allows us to describe the structure of a crystal with a remarkably small amount of information. We can specify the structure of cesium chloride, for example, with only four pieces of information:

1. CsCl crystallizes in a cubic unit cell.
2. The length of the unit cell edge is 0.4123 nm.
3. There is a $Cl^-$ ion at the coordinates 0,0,0.
4. There is a $Cs^+$ ion at the coordinates $\frac{1}{2},\frac{1}{2},\frac{1}{2}$.

Because the cell edge must connect equivalent lattice points, the presence of a $Cl^-$ ion at one corner of the unit cell (0,0,0) implies the presence of a $Cl^-$ ion at every corner of the cell. The coordinates $\frac{1}{2},\frac{1}{2},\frac{1}{2}$ describe a lattice point at the center of the cell. Because there is no other point in the unit cell that is one cell-edge length away from these coordinates, this is the only $Cs^+$ ion in the cell. CsCl is therefore a simple cubic unit cell of $Cl^-$ ions with a $Cs^+$ in the center of the body of the cell, as shown in Figure 13.19.

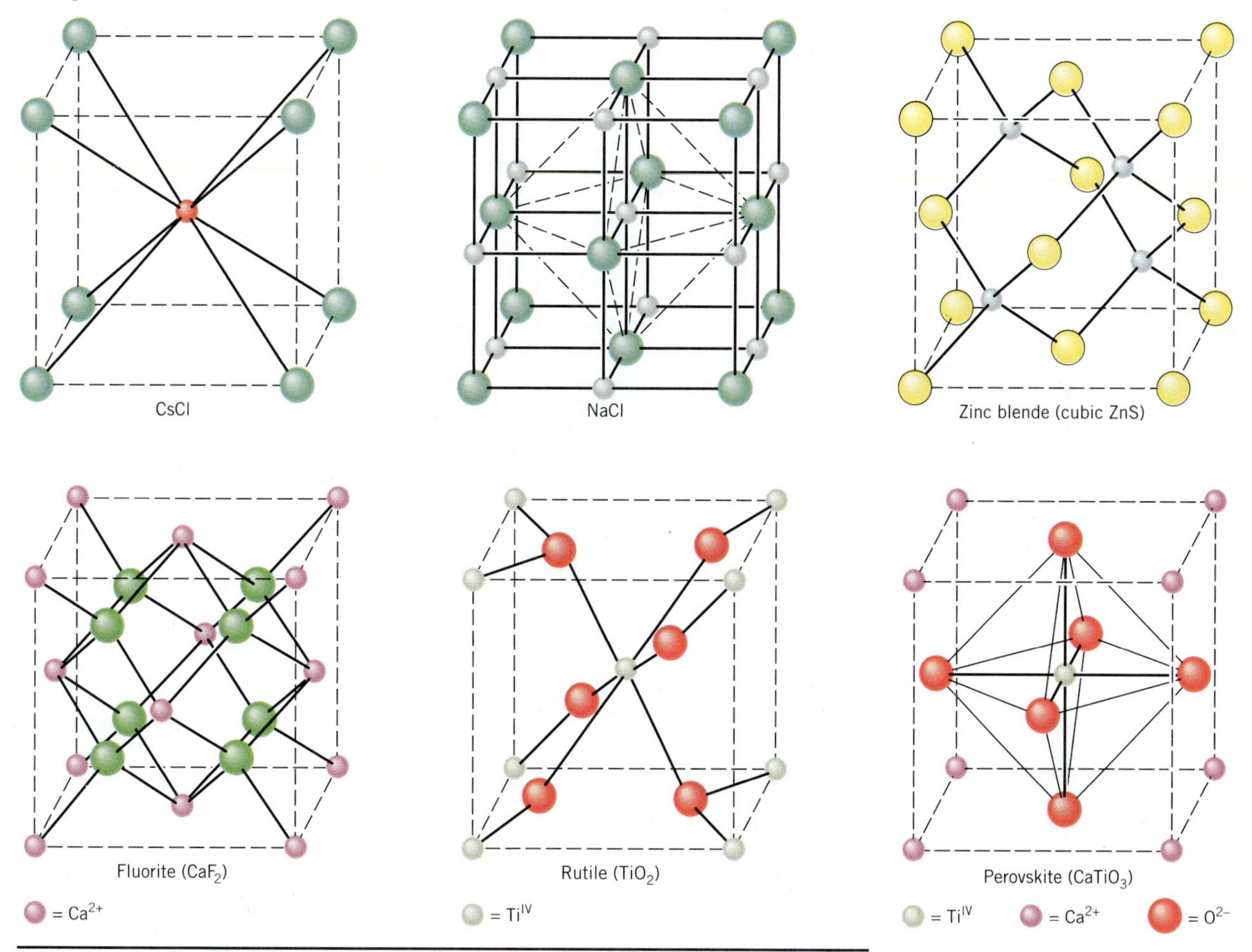


**Figure 13.19** The unit cells of CsCl, NaCl (rock salt), ZnS (zincblende), $CaF_2$ (fluorite), $TiO_2$ (rutile), and $CaTiO_3$ (perovskite).

## 13.10 UNIT CELLS: NaCl AND ZnS

Exercise 13.1 predicted that NaCl should crystallize in a cubic closest packed array of $Cl^-$ ions with $Na^+$ ions in the octahedral holes between planes of $Cl^-$ ions. We can translate this information into a unit-cell model by remembering that the face-centered cubic unit cell is the simplest repeating unit in a cubic closest packed structure.

There are four unique positions in a face-centered cubic unit cell. These positions are defined by the coordinates: 0,0,0; $0,\frac{1}{2},\frac{1}{2}$; $\frac{1}{2},0,\frac{1}{2}$; and $\frac{1}{2},\frac{1}{2},0$. The presence of an particle at one corner of the unit cell (0,0,0) requires the presence of an equivalent particle on each of the eight corners of the unit cell. Because the unit-cell edge connects equivalent points, the presence of a particle in the center of the bottom face

($0,\frac{1}{2},\frac{1}{2}$) implies the presence of an equivalent particle in the center of the top face ($1,\frac{1}{2},\frac{1}{2}$). Similarly, the presence of particles in the center of the $\frac{1}{2},0,\frac{1}{2}$ and $\frac{1}{2},\frac{1}{2},0$ faces of the unit cell implies equivalent particles in the centers of the $\frac{1}{2},1,\frac{1}{2}$ and $\frac{1}{2},\frac{1}{2},1$ faces.

Figure 13.20*a* shows that there is an octahedral hole in the center of a face-centered cubic unit cell, at the coordinates $\frac{1}{2},\frac{1}{2},\frac{1}{2}$. Any particle at this point touches the particles in the centers of the six faces of the unit cell. The other octahedral holes in a face-centered cubic unit cell are on the edges of the cell, as shown in Figure 13.20*b*. If $Cl^-$ ions occupy the lattice points of a face-centered cubic unit cell and all of the octahedral holes are filled with $Na^+$ ions, we get the unit cell shown in Figure 13.19.

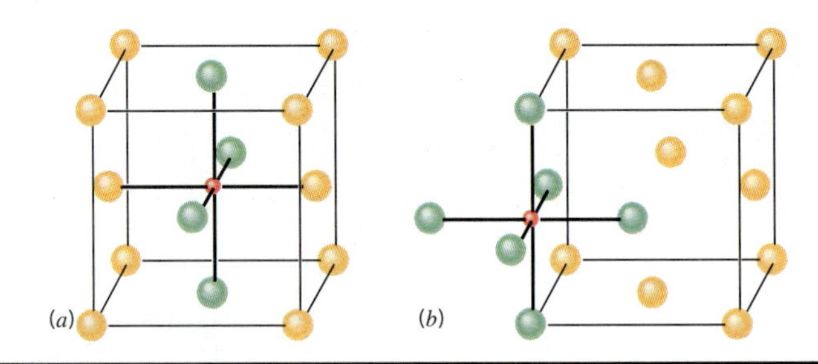


**Figure 13.20** The octahedral holes in a face-centered cubic unit cell are in the center of the body (*a*) and on the edges of the unit cell (*b*).

We can therefore describe the structure of NaCl in terms of the following information:

1. NaCl crystallizes in a cubic unit cell.
2. The cell-edge length is 0.5641 nm.
3. There are $Cl^-$ ions at the positions $0,0,0$; $\frac{1}{2},\frac{1}{2},0$; $\frac{1}{2},0,\frac{1}{2}$; and $0,\frac{1}{2},\frac{1}{2}$.
4. There are $Na^+$ ions at the positions $\frac{1}{2},\frac{1}{2},\frac{1}{2}$; $\frac{1}{2},0,0$; $0,\frac{1}{2},0$; and $0,0,\frac{1}{2}$.

Placing a $Cl^-$ ion at these four positions implies the presence of a $Cl^-$ ion on each of the 14 lattice points that define a face-centered cubic unit. Placing a $Na^+$ ion in the center of the unit cell ($\frac{1}{2},\frac{1}{2},\frac{1}{2}$) and on the three unique edges of the unit cell ($\frac{1}{2},0,0$; $0,\frac{1}{2},0$; and $0,0,\frac{1}{2}$) requires an equivalent $Na^+$ ion in every octahedral hole in the unit cell.

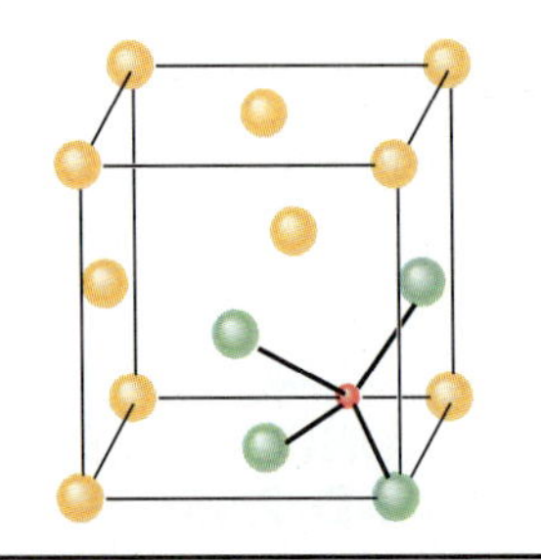

**Figure 13.21** The tetrahedral holes in a face-centered cubic unit cell are found in the eight corners of the unit cell. Each hole is surrounded by the atom on the corner of the unit cell and the three atoms in the centers of the faces that form this corner.

Exercise 13.1 predicted that ZnS would crystallize as a cubic closest packed array of $S^{2-}$ ions with $Zn^{2+}$ ions in tetrahedral holes. The $S^{2-}$ ions in this crystal occupy the same positions as the $Cl^-$ ions in NaCl. The only difference between these crystals is the location of the positive ions. Figure 13.21 shows that the tetrahedral holes in a face-centered cubic unit cell are in the corners of the unit cell, at coordinates such as $\frac{1}{4},\frac{1}{4},\frac{1}{4}$. An atom with these coordinates would touch the atom at this corner as well as the atoms in the centers of the three faces that form this corner. Although it is difficult to see without a three-dimensional model, the four atoms that surround this hole are arranged toward the corners of a tetrahedron.

Because the corners of a cubic unit cell are identical, there must be a tetrahedral hole in each of the eight corners of the face-centered cubic unit cell. If $S^{2-}$ ions occupy the lattice points of a face-centered cubic unit cell and $Zn^{2+}$ ions are packed into every other tetrahedral hole, we get the unit cell of ZnS shown in Figure 13.19. The structure of ZnS therefore can be described as follows:

1. ZnS crystallizes in a cubic unit cell.
2. The cell-edge length is 0.5411 nm.

3. There are $S^{2-}$ ions at the positions 0,0,0; $\frac{1}{2},\frac{1}{2},0$; $\frac{1}{2},0,\frac{1}{2}$; and $0,\frac{1}{2},\frac{1}{2}$.
4. There are $Zn^{2+}$ ions at the positions $\frac{1}{4},\frac{1}{4},\frac{1}{4}$; $\frac{1}{4},\frac{3}{4},\frac{3}{4}$; $\frac{3}{4},\frac{1}{4},\frac{3}{4}$; and $\frac{3}{4},\frac{3}{4},\frac{1}{4}$.

Note that only half of the tetrahedral holes are occupied in this crystal because there are two tetrahedral holes for every $S^{2-}$ ion in a closest packed array of these ions.

## 13.11 UNIT CELLS: MEASURING THE DISTANCE BETWEEN PARTICLES

Sec. 6.11

Nickel was identified in Section 13.1 as one of the metals that crystallizes in a cubic closest packed structure. When you consider that a nickel atom has a mass of only $9.75 \times 10^{-23}$ g and an ionic radius of only $1.24 \times 10^{-10}$ m, it is a remarkable achievement to be able to describe the structure of this metal. The obvious question is: How do we know that nickel packs in a cubic closest packed structure?

The only way to determine the structure of matter on an atomic scale is to use a probe that is even smaller. As we have seen, one of the most useful probes for studying matter on this scale is electromagnetic radiation.

In 1912, Max van Laue found that x-rays that struck the surface of a crystal were diffracted into patterns that resembled the patterns produced when light passes through a very narrow slit. Shortly thereafter, William Lawrence Bragg, who was just completing his undergraduate degree in physics at Cambridge, explained van Laue's results. Bragg argued that x-rays were reflected from planes of atoms near the surface of the crystal, as shown in Figure 13.22.

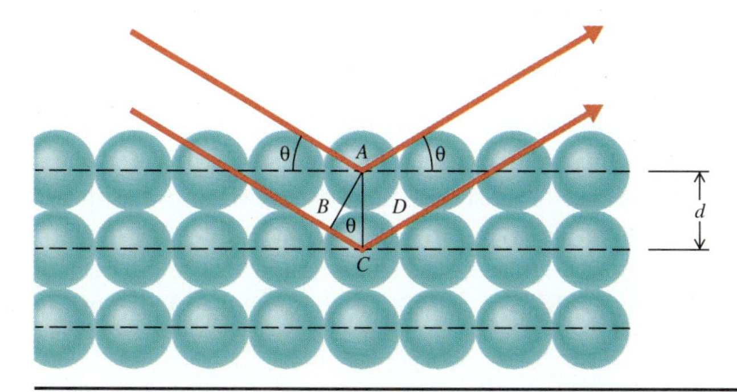


**Figure 13.22** The diffraction of x-rays by the first and second planes in a crystal.

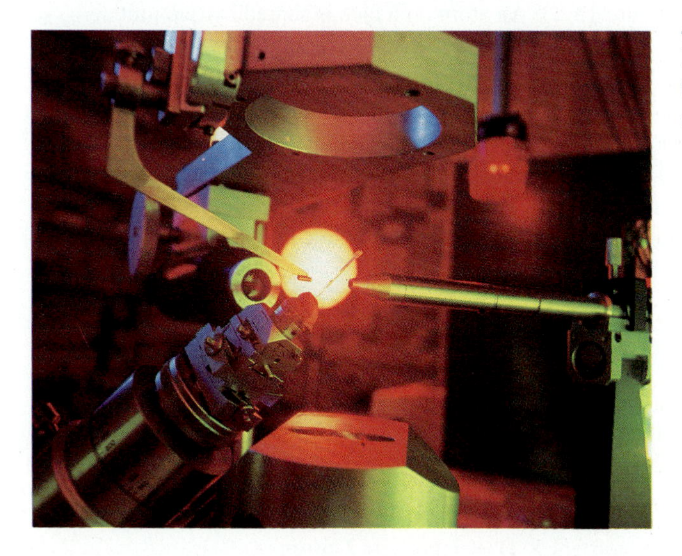

X-ray diffraction crystallography being used to determine the molecular structure of the enzyme methanol dehydrogenase.

X-rays that were reflected from different planes of atoms would be no longer in phase, and they would tend to cancel each other. They could stay in phase only if the extra distance ($BC + CD$) the x-ray moved when it struck the second plane of atoms was equal to an integer ($n$) times the wavelength ($\lambda$) of the radiation. Thus, Bragg argued that the sum of the distances labeled $BC$ and $CD$ in Figure 13.22 was equal to an integer times the wavelength of the radiation that remained in phase:

$$BC + CD = n\lambda$$

Bragg defined the angle between the beam of x-rays and the planes of the crystal as $\Theta$. Because the $BAC$ angle is also equal to $\Theta$, the $BC$ distance is equal to the $AC$ distance times the sine of $\Theta$:

$$BC = AC \sin \Theta$$

Because the $AC$ distance is equal to the distance between planes of atoms, $d$, the relationship between the $BC$ and $AC$ distance can be written as follows:

$$BC = d \sin \Theta$$

Because the $BC$ distance is equal to the $CD$ distance, the following must be true:

$$BC + CD = 2d \sin \Theta$$

Substituting the relationship between $BC + CD$ and $n\lambda$ gives a formula known as the **Bragg equation,** which allows us to calculate the distance between planes of atoms in a crystal from the pattern of diffraction of x-rays of known wavelength:

$$n\lambda = 2d \sin \Theta$$

The pattern by which x-rays are diffracted by nickel metal suggests that this metal packs in a cubic unit cell with a distance between planes of atoms of 0.3524 nm. Thus, the cell-edge length in this crystal must be 0.3524 nm. Knowing that nickel crystallizes in a cubic unit cell is not enough. We still have to decide whether it is a simple cubic, body-centered cubic, or face-centered cubic unit cell. As you will see in the next section, we can do this by measuring the density of the metal.

## 13.12 UNIT CELLS: DETERMINING THE UNIT CELL OF A CRYSTAL

Atoms on the corners, edges, and faces of a unit cell are shared by more than one unit cell, as shown in Figure 13.23. An atom on a face is shared by two unit cells, so only half of the atom belongs to each of these cells. An atom on an edge is shared by four unit cells, and an atom on a corner is shared by eight unit cells. Thus, only one-quarter of an atom on an edge and one-eighth of an atom on a corner can be assigned to each of the unit cells that share these atoms.

If nickel crystallized in a simple cubic unit cell, there would be a nickel atom on each of the eight corners of the cell. Because only one-eighth of these atoms can be assigned to a given unit cell, each unit cell in a simple cubic structure would have one net nickel atom.

*Simple cubic structure:*

8 corners $\times \frac{1}{8}$ = **1 atom**

If nickel formed a body-centered cubic structure, there would be two atoms per unit cell, because the nickel atom in the center of the body wouldn't be shared with any other unit cells.

*Body-centered cubic structure:*

(8 corners $\times \frac{1}{8}$) + 1 body = **2 atoms**

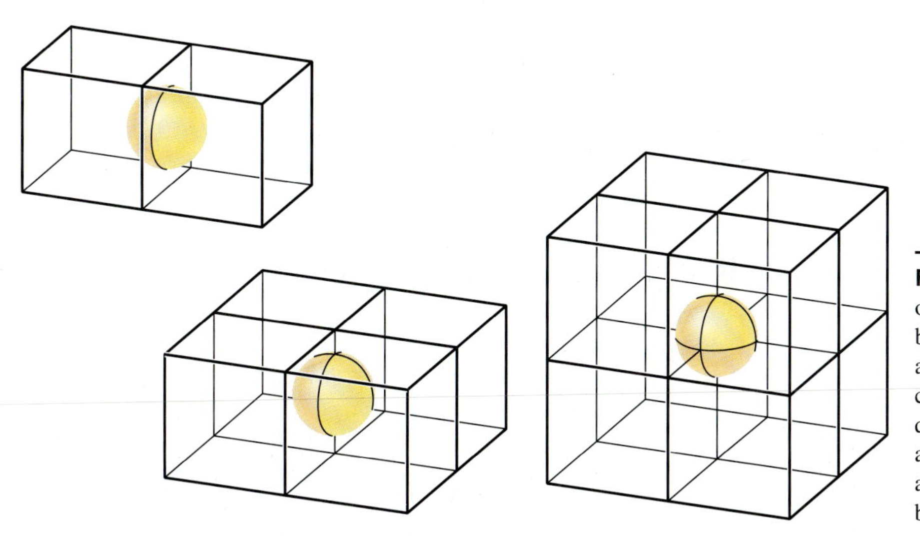

**Figure 13.23** Because an atom on the face of a unit cell is shared by two unit cells, only half of this atom belongs to each of these cells. For similar reasons, one-quarter of an atom on the edge of a unit cell and one-eighth of an atom on the corner of a unit cell belong to this unit cell.

If nickel crystallized in a face-centered cubic structure, the six atoms on the faces of the unit cell would contribute three net nickel atoms, for a total of four atoms per unit cell.

*Face-centered cubic structure:*

$$(8 \text{ corners} \times \tfrac{1}{8}) + (6 \text{ faces} \times \tfrac{1}{2}) = \mathbf{4\ atoms}$$

Because they have different numbers of atoms in a unit cell, each of these structures would have a different density. The first step toward identifying the structure of nickel metal therefore involves calculating the density for nickel for each of these structures. In order to do this, we need to know the volume of the unit cell in cubic centimeters and the mass of a single nickel atom.

The volume ($V$) of the unit cell is equal to the cell-edge length ($a$) cubed:

$$V = a^3 = (0.3524 \text{ nm})^3 = 0.04376 \text{ nm}^3$$

Since there are $10^9$ nm in a meter and 100 cm in a meter, there must be $10^7$ nm in a centimeter:

$$\frac{10^9 \text{ nm}}{1 \cancel{\text{m}}} \times \frac{1 \cancel{\text{m}}}{100 \text{ cm}} = 10^7 \text{ nm/cm}$$

We can therefore convert the volume of the unit cell to $\text{cm}^3$ as follows:

$$4.376 \times 10^{-2} \cancel{\text{nm}^3} \times \left[\frac{1 \text{ cm}}{10^7 \cancel{\text{nm}}}\right]^3 = 4.376 \times 10^{-23} \text{ cm}^3$$

The mass of a nickel atom can be calculated from the atomic weight of this metal and Avogadro's number:

$$\frac{58.69 \text{ g Ni}}{1 \cancel{\text{mol}}} \times \frac{1 \cancel{\text{mol}}}{6.022 \times 10^{23} \text{ atoms}} = 9.746 \times 10^{-23} \text{ g/atom}$$

The density of nickel, if it crystallized in a simple cubic structure, would therefore be 2.23 $\text{g/cm}^3$, to three significant figures.

*Simple cubic structure:*

$$\frac{9.746 \times 10^{-23} \text{ g}/\cancel{\text{unit cell}}}{4.376 \times 10^{-23} \text{ cm}^3/\cancel{\text{unit cell}}} = \mathbf{2.23\ g/cm^3}$$

Because there would be twice as many atoms per unit cell if nickel crystallized in a body-centered cubic structure, the density of nickel in this structure would be twice as large.

*Body-centered cubic structure:*

$$\frac{2(9.746 \times 10^{-23}\ \text{g/}\cancel{\text{unit cell}})}{4.376 \times 10^{-23}\ \text{cm}^3/\cancel{\text{unit cell}}} = \mathbf{4.45\ g/cm^3}$$

There would be four atoms per unit cell in a face-centered cubic structure and the density of nickel in this structure would be four times as large.

*Face-centered cubic structure:*

$$\frac{4(9.746 \times 10^{-23}\ \text{g/}\cancel{\text{unit cell}})}{4.376 \times 10^{-23}\ \text{cm}^3/\cancel{\text{unit cell}}} = \mathbf{8.91\ g/cm^3}$$

The experimental value for the density of nickel is 8.90 g/cm$^3$. The obvious conclusion is that nickel crystallizes in a face-centered cubic unit cell and therefore has a cubic closest packed structure.

## 13.13 UNIT CELLS: CALCULATING METALLIC OR IONIC RADII

Estimates of the radii of most metal atoms can be found in Table A-5 in the appendix. Where do these data come from? How do we know, for example, that the radius of a nickel atom is 0.1246 nm?

We have found, so far, that nickel crystallizes in a face-centered cubic unit cell with a cell-edge length of 0.3524 nm, and this information can be used to calculate the radius of a nickel atom as follows.

One of the faces of a face-centered cubic unit cell is shown in Figure 13.24. According to this figure, the diagonal $d$ across the face of this unit cell is equal to four times the radius $r$ of a nickel atom:

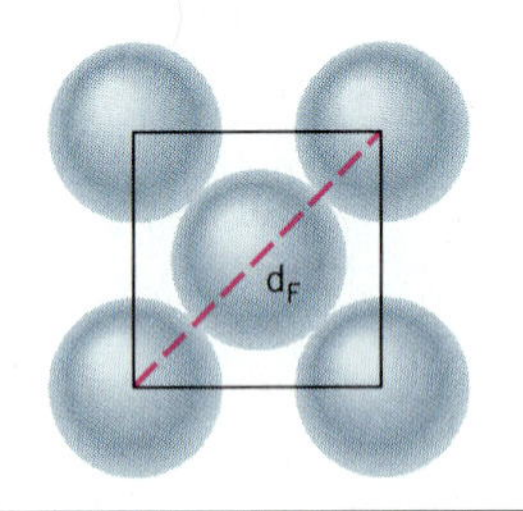


**Figure 13.24** The diagonal across the face of a face-centered cubic unit cell is equal to four times the radius of the atoms that form this cell.

$$d_{\text{face}} = 4\ r_{\text{Ni}}$$

The Pythagorean theorem states that the diagonal across a right triangle is equal to the sum of the squares of the other sides. The diagonal across the face of the unit cell is therefore related to the unit-cell edge length by the following equation:

$$d_{\text{face}}{}^2 = a^2 + a^2$$

Taking the square root of both sides gives the following result:

$$d_{\text{face}} = \sqrt{2}\ a$$

We now substitute into this equation the relationship between the diagonal across the face of this unit cell and the radius of a nickel atom:

$$4\ r_{\text{Ni}} = \sqrt{2}\ a$$

Solving for the radius of a nickel atom gives a value of 0.1246 nm:

$$r_{\text{Ni}} = \frac{\sqrt{2}\ a}{4} = \frac{\sqrt{2}(0.3524\ \text{nm})}{4} = \mathbf{0.1246\ nm}$$

A similar approach can be taken to estimating the size of an ion. Let's start by using the fact that the cell-edge length in cesium chloride is 0.4123 nm to calculate the distance between the centers of the $Cs^+$ and $Cl^-$ ions in CsCl.

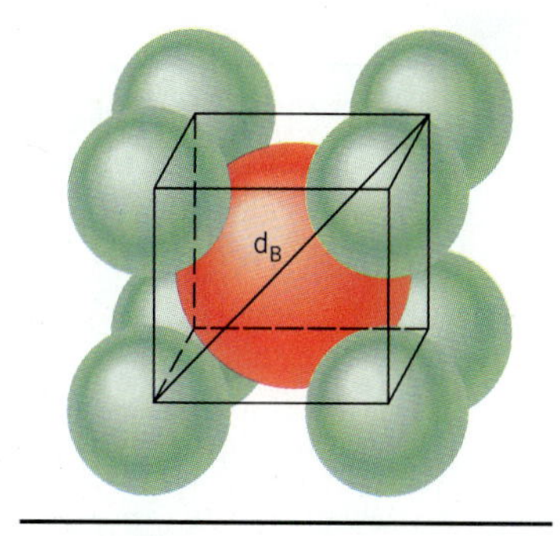


**Figure 13.25** The diagonal across the body of the CsCl unit cell is equal to twice the sum of the radii of the $Cs^+$ and $Cl^-$ ions.

CsCl crystallizes in a simple cubic unit cell of $Cl^-$ ions with a $Cs^+$ ion in the center of the body of the cell, as shown in Figure 13.19. Before we can calculate the distance between the centers of the $Cs^+$ and $Cl^-$ ions in this crystal, however, we have to recognize the validity of one of the simplest assumptions about ionic solids: The positive and negative ions that form these crystals touch.

The diagonal across the body of the CsCl unit cell is therefore the sum of the radii of two $Cl^-$ ions and two $Cs^+$ ions (see Figure 13.25):

$$d_{\text{body}} = 2\,r_{\text{Cs}^+} + 2\,r_{\text{Cl}^-}$$

The three-dimensional equivalent of the Pythagorean theorem suggests that the square of the diagonal across the body of a cube is the sum of the squares of the three sides:

$$d_{\text{body}}^2 = a^2 + a^2 + a^2$$

Taking the square root of both sides of this equation gives the following result:

$$d_{\text{body}} = \sqrt{3}\,a$$

If the cell-edge length in CsCl is 0.4123 nm, the diagonal across the body in this unit cell is 0.7141 nm:

$$d_{\text{body}} = \sqrt{3}\,a = \sqrt{3}(0.4123 \text{ nm}) = 0.7141 \text{ nm}$$

The sum of the ionic radii of $Cs^+$ and $Cl^-$ ions is half this distance, or 0.3571 nm:

$$r_{\text{Cs}^+} + r_{\text{Cl}^-} = \frac{d_{\text{body}}}{2} = \frac{0.7141 \text{ nm}}{2} = \mathbf{0.3571 \text{ nm}}$$

If we had an estimate of the size of either the $Cs^+$ or $Cl^-$ ion, we could use these results to calculate the radius of the other ion. Ever since Chapter 7, we have used a value of 0.181 nm for the ionic radius of the $Cl^-$ ion. Substituting this value into the last equation gives a value of 0.176 nm for the radius of the $Cs^+$ ion:

Sec. 7.6

$$r_{\text{Cs}^+} + r_{\text{Cl}^-} = 0.3571 \text{ nm}$$
$$r_{\text{Cs}^+} + 0.181 \text{ nm} = 0.3571 \text{ nm}$$
$$r_{\text{Cs}^+} = \mathbf{0.176 \text{ nm}}$$

The results of this calculation are in reasonable agreement with the value of 0.169 nm given in Table A-5 for the radius of the $Cs^+$ ion. The discrepancy between these values reflects the fact that ionic radii vary from one crystal to another. The tabulated values are averages of the results of a number of calculations of this type.

SPECIAL TOPIC

## LIQUID CRYSTALS

Molecules that are large, rigid, and linear can form an intermediate phase during the transition between the liquid and solid states. Because it has some of the structure of solids and some of the freedom of motion associated with liquids, this phase is best described as a **liquid crystal.**

Liquid crystals were discovered in 1888, but they were primarily a laboratory curiosity until about 30 years ago. They are now used routinely in the displays of electrical devices such as digital watches, calculators, and computers. These LCD devices take advantage of the fact that the weak bonds that hold molecules together in a liquid crystal are easily affected by changes in pressure, temperature, or electromagnetic fields.

Advances in liquid crystal displays have allowed portable computers to compete with desk-top models.

Liquid crystals are divided into three categories: smectic, nematic, and cholesteric. *Smectic* liquid crystals have a structure that resembles a handful of cigars, as shown in Figure 13.26*a*. Not only do the molecules all point in the same direction, they are so well ordered that they form planes perpendicular to the axes of the molecules. *Nematic* liquid crystals are slightly less well ordered. The molecules still point in the same direction, but they start and stop at different positions within the liquid, as shown in Figure 13.26*b*.

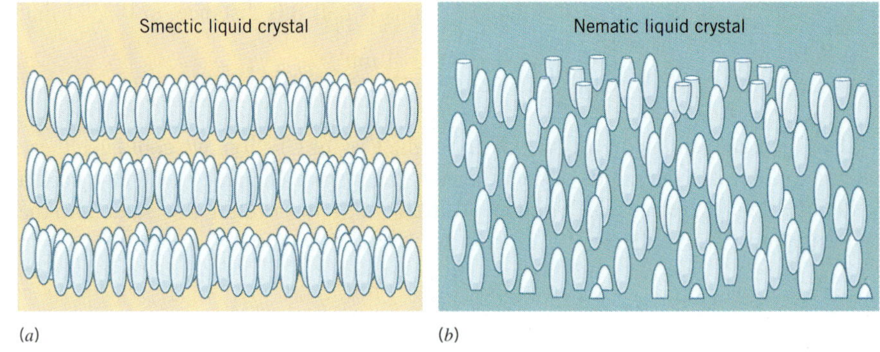


**Figure 13.26** The structures of (*a*) smectic and (*b*) nematic liquid crystals.

*Cholesteric* liquid crystals have a structure similar to nematic liquid crystals, but each plane of molecules is twisted slightly in relation to the plane above or below. These liquid crystals received their name from the fact that many derivatives of cholesterol form this structure. The slight twist in the planes these structures form tends to make these liquid crystals colored, and the fact that changes in the amount of twisting lead to changes in color make these crystals sensitive indicators of changes in temperature or pressure. The most sensitive cholesteric liquid crystals show a detectable color change with temperature changes as small as 0.001°C.

SPECIAL TOPIC

## METALS, SEMICONDUCTORS, AND INSULATORS

A significant fraction of the gross national product (GNP) of the United States, and all of the contribution to the GNP from the high-technology industries, can be traced to efforts to harness differences in the way metals, semiconductors, and insulators conduct electricity. This difference can be expressed in terms of *electrical conductivity* ($\sigma$), which measures the ease with which materials conduct an electric current. It can be expressed also in terms of *electrical resistivity* ($\rho$), the inverse of conductivity, which measures the resistance of a material to carrying an electric charge. The unit of electrical resistance is the ohm, and the unit of resistivity is the ohm-centimeter.

Silver and copper metal are among the best conductors of electricity, with a resistance of only $10^{-6}$ ohm-cm. (This is why copper is the metal most often used in electric wires.) The resistance of semiconductors such as silicon and germanium is $10^8$ to $10^{10}$ times as large. When pure, these semimetals have a resistivity of $10^2$ to $10^4$ ohm-cm. Insulators include glass ($10^{10}$ ohm-cm), diamond ($10^{14}$ ohm-cm), and quartz ($10^{18}$ ohm-cm), which have extremely large resistances to carrying an electric current.

The $10^{24}$-fold range of resistance is not the only difference between metals, semiconductors, and insulators. Metals become better conductors when they are cooled to lower temperatures. Some metals are such good conductors at very low temperatures that they no longer have a measurable resistance and therefore become superconductors. Semiconductors show the opposite behavior—they become much better conductors as the temperature increases. The difference between the temperature dependence of metals and semiconductors is so significant it is often the best criterion for distinguishing between these materials.

**Semiconductors** are very sensitive to impurities. The conductivity of silicon or germanium can be increased by a factor of up to $10^6$ by adding as little as 0.01% of an impurity. Metals, on the other hand, are fairly insensitive to impurities. It takes a lot of impurity to change the conductivity of a metal by as much as a factor of 10; and unlike semiconductors, metals become poorer conductors when impure.

In order to explain the behavior of metals, semiconductors, and insulators, we need to look at what happens during bonding in solids. Because it is the lightest element in the periodic table that is a solid at room temperature, let's start by building a model of what happens when lithium atoms interact. As a first step, we can consider what happens when a pair of lithium atoms with a $1s^2\ 2s^1$ configuration interact to form a hypothetical $Li_2$ molecule.

The molecular orbital diagram for an $Li_2$ molecule is shown in Figure 13.27*a*. The $1s$ orbitals interact to form a pair of $\sigma_{1s}$ and $\sigma_{1s}$* molecular orbitals. The same thing happens to the $2s$ orbitals. In each case, one of the molecular orbitals is lower in energy than the atomic orbitals, and the other molecular orbital is higher in energy than the atomic orbitals.

Now imagine what happens when $10^{21}$ lithium atoms combine to form a tiny (0.01 g) crystal. The $1s$ orbitals on these atoms overlap to form a band of $10^{21}$ orbitals with energies between the extremes of the $\sigma_{1s}$ and $\sigma_{1s}$* molecular orbitals in the $Li_2$ molecule, as shown in Figure 13.27*b*. Similarly, the $2s$ orbitals on these atoms overlap to form a band of $10^{21}$ orbitals with energies between the extremes of the $\sigma_{2s}$ and $\sigma_{2s}$* molecular orbitals.

The $10^{21}$ orbitals in the $1s$ band are filled with electrons. The $2s$ band, however, has only $1 \times 10^{21}$ electrons—it is half-filled. It takes little, if any, energy to excite one of the electrons in the $2s$ band from one orbital to another in this band, so the electrons are free to move from one end of the crystal to the other. This band of orbitals is called a **conduction band** because it enables lithium metal to conduct electricity.

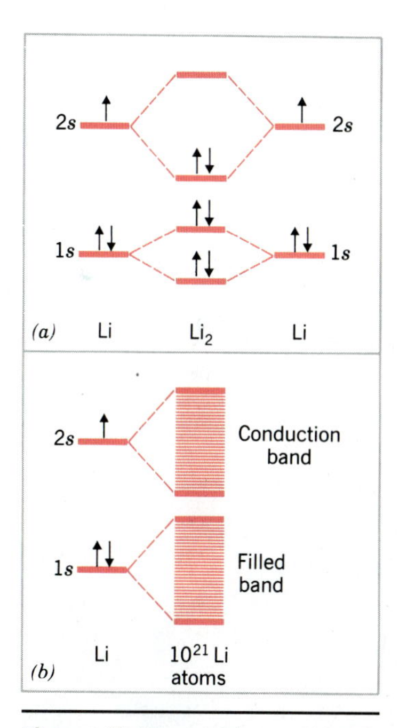


**Figure 13.27** (*a*) The overlap of atomic orbitals on a pair of atoms forms a limited number of molecular orbitals with distinct energies. (*b*) The overlap of atomic orbitals on a large number of atoms forms continuous bands of orbitals.

Let's now turn to magnesium, which has a [Ne] $3s^2$ configuration. The $3s$ orbitals on $10^{21}$ magnesium atoms would overlap to form a band of $10^{21}$ orbitals. But there are two electrons in each $3s$ orbital, so this band of orbitals is totally filled with $2 \times 10^{21}$ electrons. The empty $3p$ orbitals on magnesium, however, also interact to form a band of $3 \times 10^{21}$ orbitals. This empty $3p$ band overlaps the filled $3s$ band in magnesium, so that the combined band is only partially filled, allowing magnesium to conduct electricity.

The differences in the way metals, semiconductors, and insulators conduct electricity can be explained with the diagram in Figure 13.28. Metals have partially filled bands of orbitals that allow electrons to move from one end of the crystal to the other. They therefore conduct an electric current. All of the bands in an insulator are either filled or empty. Furthermore, the gap between the highest energy filled band and the lowest energy empty band in an insulator is so large that it is difficult to excite electrons from one of these bands to the other. As a result, it is difficult to move electrons through an insulator.

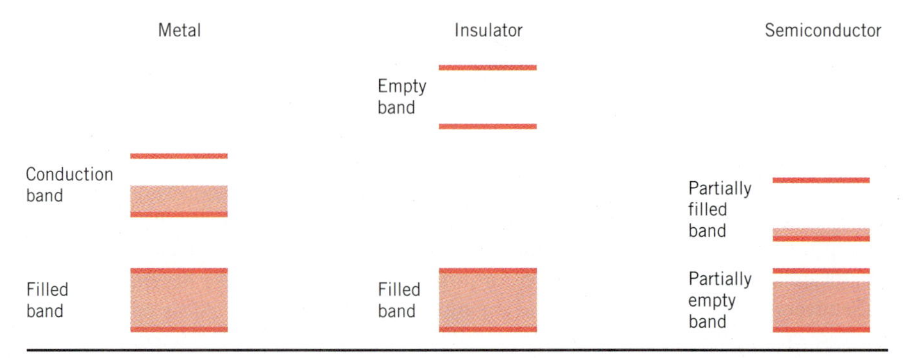


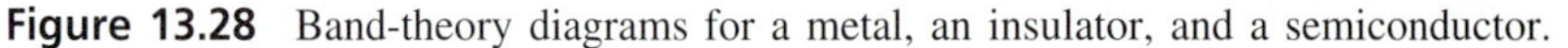

**Figure 13.28** Band-theory diagrams for a metal, an insulator, and a semiconductor.

Semiconductors also have a band structure that consists of filled and empty bands. The gap between the highest energy filled band and the lowest energy empty band is small enough, however, that electrons can be excited into the empty band by the thermal energy these electrons carry at room temperature. Semiconductors therefore fall between the extremes of metals and insulators in their ability to conduct an electric current.

To understand why metals become better conductors at low temperature, it is important to remember that temperature is a macroscopic reflection of the kinetic energy of the individual particles. Much of the resistance of a metal to an electric current at room temperature is the result of scattering of the electrons by the thermal motion of the metal atoms. As the metal is cooled, and the thermal motion slows down, there is less scattering, and the metal becomes a better conductor.

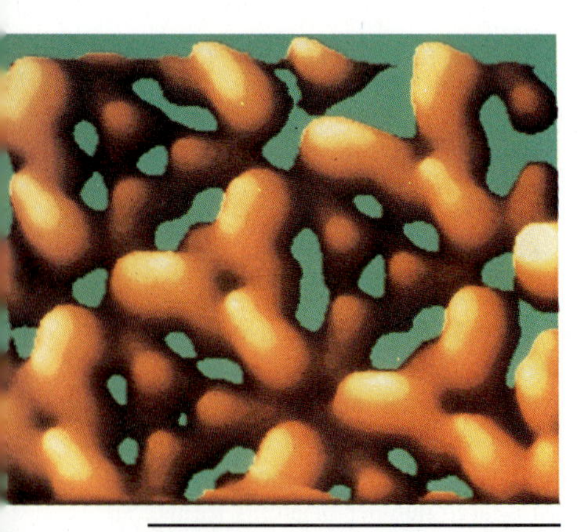

The structure of silicon as seen through a ST Microscope.

Semiconductors become better conductors at high temperatures for exactly the opposite reason. As the temperature increases, the number of electrons with enough thermal energy to be excited from the filled band to the empty band increases and the semiconductor becomes a better conductor of electricity.

To understand why semiconductors are sensitive to impurities, let's look at what happens when we add a small amount of a Group VA element, such as arsenic, to one of the Group IVA semiconductors. Arsenic atoms have one more valence electron than germanium and silicon atoms. Arsenic atoms can therefore lose an electron to form $As^+$ ions, which can occupy some of the lattice points in the crystal where silicon or germanium atoms are normally found.

If the amount of arsenic is kept very small, the distance between these atoms is so large that they do not interact. As a result, the extra electrons from the arsenic atoms occupy orbitals in a very narrow band of energies that lie between the filled and empty bands of the semiconductor, as shown in Figure 13.29. This structure decreases the amount of energy required to excite an electron into the lowest energy empty band in the semiconductor and therefore increases the number of electrons that have enough energy to cross this gap. As a result, this ''doped'' semiconductor becomes a very much better conductor of electricity. Because the electric charge is carried by a flow of *negative* particles, these semiconductors are called *n-type.*

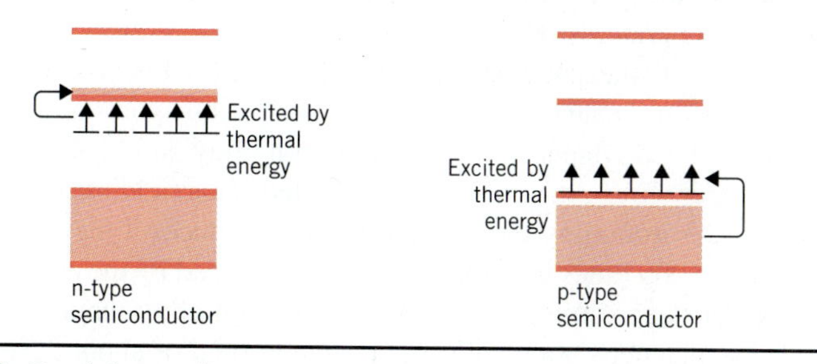

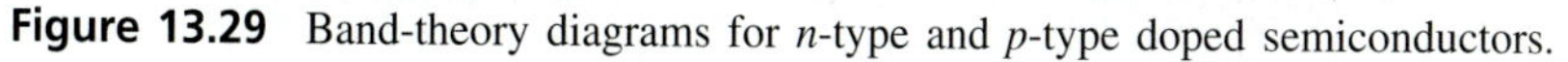

**Figure 13.29** Band-theory diagrams for *n*-type and *p*-type doped semiconductors.

It is also possible to dope a Group IVA semiconductor with one of the elements in Group IIIA, such as indium. Indium atoms have one less valence electron than silicon and germanium atoms, and they can capture electrons from the highest energy filled band of orbitals to form holes in this band. The presence of holes in a filled band has the same effect as the presence of electrons in an empty band: It allows the solid to carry an electric current. The electric charge is now carried by a flow of *positive* particles, or holes, so these semiconductors are called *p-type.*

Because semiconductors tend to carry a much smaller electric current than metals, it is easier to control the current's flow. Furthermore, bringing *n*-type and *p*-type semiconductors together produces a device, the transistor, that has a natural one-directional flow of electrons, which can be stopped by applying a small voltage in the opposite direction. This junction between *n*-type and *p*-type semiconductors was the basis of the revolution in industrial technology that followed the discovery of the transistor by William Shockley, John Bardeen, and Walter Brattain at Bell Laboratories in 1948.

## RESEARCH IN THE 90s

### THE SEARCH FOR HIGH-TEMPERATURE SUPERCONDUCTORS

When an electric current is passed through any material at room temperature, some of the energy of the electrons is dissipated in the form of heat. In metals, as we have seen, this resistance to an electric current decreases as the metal is cooled. In 1911, Heike Onnes found that when mercury was cooled to temperatures below 4.1 K, this resistance fell to zero. Above this temperature, mercury was a conductor of electricity. Below this transition point, it became a **superconductor,** transferring electricity with absolute efficiency. By 1913, Onnes found that tin and lead also became superconductors at temperatures below 4 K.

Onnes recognized the potential of superconductivity for constructing magnets with unusually strong magnetic fields. Standard electromagnets are made by winding a coil of insulated copper wire around an iron-alloy core. As the current flows through the copper wire, a magnetic field is created. This field induces an alignment of electrons in the iron-alloy core, which in turn produces a magnetic field in the core that is up to 1000 times larger than the field produced by the copper wire. There is an upper limit to the strength of the field that iron-alloy magnets can produce, however. These magnets ''saturate'' at a magnetic field of about 2 Tesla, which is 40,000 times larger than the earth's magnetic field.

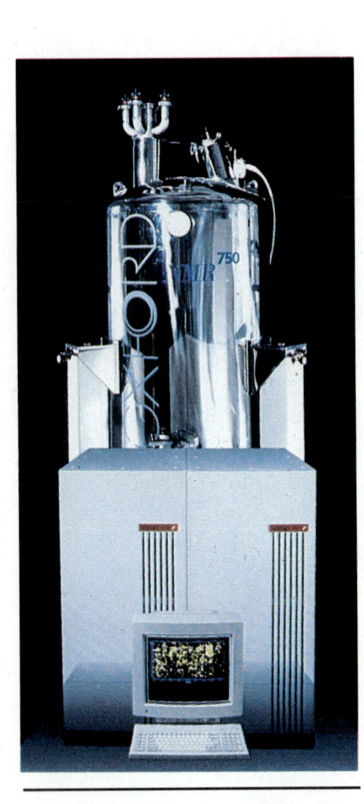

The world's most powerful commercial Nuclear Magnetic Resonance (NMR) Spectrometer.

Onnes believed that superconducting magnets could be produced that achieved much higher fields. Unfortunately, none of the superconducting metals he studied was able to carry enough electric current. It took 50 years before alloys of niobium and tantalum were discovered that could carry the amount of current needed to produce high-field magnets.

Superconducting magnets made from niobium–tantalum alloys became available toward the end of the 1960s. The primary disadvantage of these magnets was the fact that this alloy has to be cooled to the temperature of liquid helium (4.2 K) before it becomes a superconductor. The cost of maintaining one of these magnets could be decreased by as much as a factor of 1000 if it could operate at liquid nitrogen temperatures (77 K).

The search for such ''high-temperature'' superconductors is an important object lesson in the proper role of theory and experiment. At first glance, we might expect $ReO_3$ and $RuO_2$ to be insulators, like other metal oxides. However, these oxides both conduct electricity the way a metal would. In 1964, it was found that other metal oxides, such as NbO and TiO, could conduct electricity so well they become superconductors when cooled to extremely low temperatures (1 K).

NbO and TiO have crystal structures like that of NaCl. It would be a mistake, however, to think of these compounds as $[M^{2+}][O^{2-}]$ salts in which the metal ion is much smaller than the oxide ion. Metal oxides only become conductors of electricity when the metal–oxygen bond has a high degree of covalent character. We therefore have to envision these compounds as containing neutral metal atoms that are much larger than the neutral oxygen atoms. As a result, the interaction between the metal atoms is large enough to enable the solid to behave as a metal.

A major step in the evolution of high-temperature superconductors occurred in 1986, when Alex Müller and Georg Bednorz at the IBM Research Laboratory in

more brittle than plastic milk containers. Use this fact to predict which substance is more crystalline.

### Molecular, Covalent, Ionic, and Metallic Solids

**13-8** Classify the following solids as molecular, covalent, ionic, or metallic:

(a) $BaSO_4$ (b) NaOH (c) Xe (d) $I_2$ (e) aluminum (f) brass (g) $P_4$ (h) $P_4O_{10}$

**13-9** Which of the following solids are held together by an extended network of covalent bonds?

(a) sodium chloride (b) graphite (c) gold (d) calcium carbonate (e) diamond (f) dry ice (solid $CO_2$)

**13-10** Which force makes the most important contribution to the lattice energy of solid $CO_2$?

(a) metallic bonding (b) ionic bonding (c) covalent bonding (d) van der Waals forces

**13-11** Which force makes the most important contribution to the lattice energy of solid argon?

(a) metallic bonding (b) ionic bonding (c) covalent bonding (d) van der Waals forces

**13-12** Which of the following categories is most likely to contain a compound that is a poor conductor of electricity when solid but a very good conductor when heated until it melts?

(a) molecular solids (b) covalent solids (c) ionic solids (d) metallic solids

### The Structure of Metals and Other Monatomic Solids

**13-13** Describe the difference in the way planes of atoms stack to form *hexagonal closest packed, cubic closest packed, body-centered cubic,* and *simple cubic* structures.

**13-14** Explain why the structure of polonium is called *simple cubic;* why the structure of iron is called *body-centered cubic;* and why the structure of copper is called *hexagonal closest packed.*

**13-15** Determine the coordination numbers of the metal atoms in the following structures.

(a) cubic closest packed aluminum (b) hexagonal closest packed magnesium (c) body-centered cubic chromium (d) simple cubic polonium

**13-16** In which of the following structures would a xenon atom form the largest number of induced dipole–induced dipole interactions?

(a) simple cubic (b) body-centered cubic (c) cubic closest packed (d) hexagonal closest packed

**13-17** Explain why a metal such as iron would pack in a body-centered cubic structure, in which the coordination number is eight, instead of a cubic closest packed or hexagonal closest packed structure, in which the coordination number is 12.

**13-18** Sodium crystallizes in a structure in which the coordination number is eight. Which structure best describes this crystal?

(a) simple cubic (b) body-centered cubic (c) cubic closest packed (d) hexagonal closest packed

### Physical Properties That Result from the Structure of Metals

**13-19** Explain why metals are solids at room temperature.

**13-20** Explain why metals are malleable and ductile.

**13-21** Explain why metals conduct heat and electricity.

### Solid Solutions and Intermetallic Compounds

**13-22** Describe the difference between an intermetallic compound, such as $CuAl_2$, and an alloy, such as brass or bronze.

**13-23** Describe how the formation of the intermetallic compounds $CuAl_2$ and $Fe_3C$ hardens aluminum and steel, respectively.

**13-24** Describe the difference between solid solutions and interstitial solutions. Give an example of each. Predict the effect of changes in the relative size of a pair of atoms on their ability to form either of these solutions.

### Holes in Closest Packed and Simple Cubic Structures

**13-25** Tetrahedral holes and octahedral holes can be found in which of the following structures?

(a) simple cubic (b) body-centered cubic (c) cubic closest packed (d) hexagonal closest packed

**13-26** What is the coordination number of a cation packed in each of the following holes?

(a) tetrahedral holes (b) octahedral holes (c) cubic holes

**13-27** Cubic holes can be found in which of the following structures?

(a) simple cubic (b) body-centered cubic (c) cubic closest packed (d) hexagonal closest packed

**13-28** Which is the smallest hole: tetrahedral, octahedral, or cubic? Which is the largest?

**13-29** Prove that the cation that just fits into an octahedral hole formed by six identical anions has a radius 0.414 times the radius of the anions that form the hole.

**13-30** Prove that the cation that just fits into a tetrahedral hole formed by four identical anions has a radius 0.225 times the radius of the anions that form the hole.

### The Structure of Ionic Solids

**13-31** In KF, the $K^+$ and $F^-$ ions are almost exactly the same size; both have ionic radii of approximately 0.134 nm. Which is larger, a neutral potassium atom or a neutral fluorine atom?

**13-32** Rutile is a mineral that contains titanium and oxygen. The crystal structure can be described as a closest packed array of oxygen atoms with titanium atoms in half of the octahedral holes. If the number of octahedral holes in a closest packed array is equal to the number of particles that form this structure, what is the empirical formula for this mineral?

**13-33** Titanium carbide is a covalent carbide that is essentially inert to chemical reactions, has a very high melting point, and is almost as hard as diamond. The crystal structure can be described as a closest packed array of carbon atoms with titanium atoms in all of the octahedral holes. If the number of octahedral holes in a closest packed array is equal to the number of particles that form this structure, what is the empirical formula of this compound?

**13-34** NaCl, AgCl, KH, LiH, MgO, MnS, and CaO all have the same crystal structure—a closest packed array of negative ions with positive ions in all of the octahedral holes. This structure would be predicted for all but one of these compounds on the basis of their chemical formulas and the radius ratios of the ions. For which compound is this structure not expected?

**13-35** Pyrite, $FeS_2$, crystallizes in a structure that can be described as a closest packed array of metal ions with sulfide or polysulfide ions in all of the octahedral holes. If the number of octahedral holes in a closest packed array is equal to the number of particles that forms this array, which formulation describes this mineral better: $[Fe^{4+}][S^{2-}]_2$ or $[Fe^{2+}][S_2{}^{2-}]$?

**13-36** If the number of octahedral holes in a closest packed array is equal to the number of particles that forms this array, what is the formula of an oxide of titanium that crystallizes in a structure that can be described as a closest packed array of oxygen atoms with titanium atoms in two-thirds of the octahedral holes?

**13-37** Use the relative size of the $Ge^{4+}$ and $O^{2-}$ ions to predict the structure of $GeO_2$. What holes are used? What is the coordination number of the $Ge^{4+}$ ions?

**13-38** Zinc telluride crystallizes in a cubic closest packed structure of tellurium atoms with zinc atoms in half of the tetrahedral holes. If the number of tetrahedral holes in a closest packed array is twice as large as the number of particles that form this structure, what is the empirical formula of this compound?

**13-39** Which structure would BeO be expected to most closely resemble?

(a) NaCl (b) CsCl (c) ZnS (d) $CaF_2$

**13-40** An oxide of cobalt crystallizes in a closest packed array of oxygen atoms with cobalt atoms in one-eighth of the tetrahedral holes and one-half of the octahedral holes. What is the oxidation state of the cobalt in this compound? Assume that the number of tetrahedral holes in a closest packed array is twice as large as the number of particles that form this structure.

**13-41** Perovskite is a mineral with the formula $CaTiO_3$. Which of the positive ions in this crystal, $Ti^{4+}$ or $Ca^{2+}$, is more likely to pack in the octahedral holes?

**13-42** Thallium cyanide, $Tl(CN)_x$, crystallizes as a simple cubic array of $CN^-$ ions with thallium ions in all of the cubic holes. If the number of cubic holes in a simple cubic array is equal to the number of particles that form this array, what is the oxidation state of thallium in this compound?

### Unit Cells: The Simplest Repeating Unit in a Crystal

**13-43** Define the term *unit cell*. Describe the common properties of all unit cells.

**13-44** Describe the difference between simple cubic, body-centered cubic, and face-centered cubic unit cells.

**13-45** Simple cubic unit cells are found in simple cubic structures, and body-centered cubic unit cells are found in body-centered cubic structures. Where are face-centered cubic unit cells found: cubic closest packed structures or hexagonal closest packed structures?

**13-46** Sodium hydride crystallizes in a face-centered cubic unit cell of $H^-$ ions with $Na^+$ ions at the center of the unit cell and in the center of each edge of the unit cell. How many $Na^+$ ions does each $H^-$ ion touch? How many $H^-$ ions does each $Na^+$ ion touch?

**13-47** Picture the smallest repeating unit of a simple cubic crystal. Since neighboring atoms touch in this crystal, the length of an edge of this simplest repeating unit is twice the radius of the atoms that form the crystal. If the volume of a sphere is $\frac{4}{3}\pi r^3$, what fraction of this crystal is empty space?

**13-48** Calculate the fraction of empty space in a body-centered cubic unit cell and a face-centered cubic unit cell.

### Unit Cells: A Three-Dimensional Graph

**13-49** The positions of atoms in a unit cell can be described in terms of a three-dimensional graph in which the location 0,0,0 corresponds to one of the corners of the cell. Describe the positions indicated by the following sets of coordinates.

(a) $\frac{1}{2}$,0,0 (b) $\frac{1}{2},\frac{1}{2}$,0 (c) $\frac{1}{2},\frac{1}{2},\frac{1}{2}$ (d) $\frac{1}{4},\frac{1}{4},\frac{1}{4}$

**13-50** Explain why stating that there is a $Cs^+$ ion at the coordinates 0,0,0 in the unit cell of CsCl implies that there must be an equivalent $Cs^+$ ion on each of the other corners of the unit cell.

**13-51** Explain why stating that there are $Cl^-$ ions in the centers of three of the faces of the unit cell of NaCl at the coordinates 0,$\frac{1}{2},\frac{1}{2}$; $\frac{1}{2}$,0,$\frac{1}{2}$; and $\frac{1}{2},\frac{1}{2}$,0 implies that there must be an equivalent $Cl^-$ ion in the center of each of the other faces of the unit cell.

**13-52** At very low temperatures, argon crystallizes in a structure in which Ar atoms are located at the following positions: 0,0,0; 0,$\frac{1}{2},\frac{1}{2}$; $\frac{1}{2}$,0,$\frac{1}{2}$; $\frac{1}{2},\frac{1}{2}$,0. Is this unit cell simple cubic, body-centered cubic, or face-centered cubic?

**13-53** The mineral cuprite crystallizes in a structure in which oxygen atoms are located at the coordinates 0,0,0 and $\frac{1}{2},\frac{1}{2},\frac{1}{2}$ and copper atoms are located at $\frac{1}{4},\frac{1}{4},\frac{1}{4}$; $\frac{1}{4},\frac{3}{4},\frac{3}{4}$; $\frac{3}{4},\frac{1}{4},\frac{3}{4}$; and $\frac{3}{4},\frac{3}{4},\frac{1}{4}$. Is this unit cell simple cubic, body-centered cubic, or face-centered cubic?

### Unit Cells: Determining the Unit Cell of a Crystal

**13-54** Calculate the number of chloride and ammonium ions per unit cell for $NH_4Cl$ if this salt crystallizes in a structure that can be described as a simple cubic unit cell of $NH_4{}^+$ ions with a $Cl^-$ ion in the center of the cell.

**13-55** Gallium arsenide is a photosensitive semiconductor that is used in the light meters in automatic cameras. It crystallizes in a structure in which there are gallium atoms at 0,0,0; $\frac{1}{2},\frac{1}{2},0$; $\frac{1}{2},0,\frac{1}{2}$; and $0,\frac{1}{2},\frac{1}{2}$ and arsenic atoms at $\frac{1}{4},\frac{1}{4},\frac{1}{4}$; $\frac{1}{4},\frac{3}{4},\frac{3}{4}$; $\frac{3}{4},\frac{1}{4},\frac{3}{4}$; and $\frac{3}{4},\frac{3}{4},\frac{1}{4}$. Describe the unit cell of this compound, the kind of holes in which the arsenic atoms are found, and the empirical formula of the compound.

**13-56** The mineral perovskite crystallizes in a cubic unit cell in which there is a titanium atom at 0,0,0, a calcium atom at $\frac{1}{2},\frac{1}{2},\frac{1}{2}$ and oxygen atoms at $\frac{1}{2},0,0$; $0,\frac{1}{2},0$; and $0,0,\frac{1}{2}$. Describe the unit cell and calculate the empirical formula of this compound.

### Unit Cells: Calculating Metallic or Ionic Radii of Atoms and Ions

**13-57** Chromium metal ($d = 7.20$ g/cm$^3$) crystallizes in a body-centered cubic unit cell. Calculate the volume of this unit cell and the radius of a chromium atom.

**13-58** The atomic radius of a titanium atom is 0.1448 nm. What is the density of titanium if this metal crystallizes in a body-centered cubic unit cell?

**13-59** Calculate the atomic radius of an Ar atom, assuming that argon crystallizes at low temperature in a face-centered cubic unit cell with a density of 1.623 g/cm$^3$.

**13-60** Silver crystallizes in a face-centered cubic unit cell with an edge length of 0.40862 nm. Calculate the density of this metal in grams per cubic centimeter.

**13-61** Potassium crystallizes in a cubic unit cell with an edge length of 0.5247 nm. The density of potassium is 0.856 g/cm$^3$. Determine whether this element crystallizes in a simple cubic, body-centered cubic, or face-centered cubic unit cell.

**13-62** Determine whether calcium crystallizes in a simple cubic, body-centered cubic, or face-centered cubic unit cell. Assume that the cell-edge length is 0.5582 nm and the density of this metal is 1.55 g/cm$^3$.

**13-63** Determine whether molybdenum crystallizes in a simple cubic, body-centered cubic, or face-centered cubic unit cell. Assume that the cell-edge length is 0.3147 nm and the density of this metal is 10.2 g/cm$^3$.

**13-64** Which of the following metals would crystallize in a face-centered cubic unit cell with an edge length of 0.3608 nm and a density of 8.95 g/cm$^3$?

(a) Na (b) Ca (c) Tl (d) Cu (e) Au

**13-65** Barium crystallizes in a body-centered cubic structure in which the cell-edge length is 0.5025 nm. Calculate the shortest distance between neighboring barium atoms in this crystal.

**13-66** NaH crystallizes in a structure similar to that of NaCl. If the cell-edge length in this crystal is 0.4880 nm, what is the average length of the Na—H bond?

**13-67** TlI crystallizes in a structure similar to that of CsCl with a cell-edge length of 0.4198 nm. Calculate the average Tl—I bond length in this crystal. If the ionic radius of an $I^-$ ion is 0.216 nm, what is the ionic radius of the $Tl^+$ ion?

**13-68** Calculate the ionic radius of the $Cs^+$ ion if the cell-edge length for CsCl is 0.4123 nm and the ionic radius of a $Cl^-$ ion is 0.181 nm.

### Metals, Semiconductors, and Insulators

**13-69** Describe the differences between metals, semiconductors, and insulators.

**13-70** Explain how metals conduct an electric current.

**13-71** Explain why metals become better conductors of electricity as the temperature decreases but semiconductors become better conductors of electricity as the temperature increases.

**13-72** Explain why adding small quantities of arsenic or gallium increases the conductivity of silicon.

**13-73** Describe the difference between *n*-type and *p*-type semiconductors.

### Integrated Problems

**13-74** Explain why $La_2CuO_4$ is an insulator, but $La_{1.8}Sr_{0.2}CuO_4$ conducts electricity so well it is a superconductor at low temperatures.

**13-75** Describe how the formula of $YBa_2Cu_3O_7$ can be used to explain the fact that this 1-2-3 superconductor behaves as if positive charge was delocalized over the three copper atoms in the empirical formula.

**13-76** Diamond crystallizes in a cubic unit cell in which there are carbon atoms at the positions 0,0,0; $\frac{1}{2},\frac{1}{2},0$; $\frac{1}{2},0,\frac{1}{2}$; $0,\frac{1}{2},\frac{1}{2}$; $\frac{1}{4},\frac{1}{4},\frac{1}{4}$; $\frac{1}{4},\frac{3}{4},\frac{3}{4}$; $\frac{3}{4},\frac{1}{4},\frac{3}{4}$; and $\frac{3}{4},\frac{3}{4},\frac{1}{4}$. What is the cell-edge length of this crystal if the density of diamond is 3.515 g/cm$^3$?

**13-77** CdO crystallizes in a cubic unit cell with a cell-edge length of 0.4695 nm. Calculate the number of $Cd^{2+}$ and $O^{2-}$ ions per unit cell if the density of this crystal is 8.15 g/cm$^3$.

**13-78** LiF crystallizes in a cubic unit cell with a cell-edge length of 0.4017 nm. Calculate the number of $Li^+$ and $F^-$ ions per unit cell if the density of this salt is 2.640 g/cm$^3$.

**13-79** Iron ($d = 7.86$ g/cm$^3$) crystallizes in a body-centered cubic unit cell at room temperature. Calculate the radius of an iron atom in this crystal. At temperatures above 910°C, iron prefers a face-centered cubic structure. If we assume that the change in the size of the iron atom is negligible when the metal is heated to 910°C, what is the density of iron in the face-centered cubic structure? Use this calculation to predict whether iron should expand or contract when it changes from the body-centered cubic to the face-centered cubic structure.

Any model for the structure of liquids has to explain why they flow so easily.

# LIQUIDS

The discussion of liquids in this chapter provides the basis for answering a variety of questions based on observations of the world in which we live:

## POINTS OF INTEREST

- Why do liquids in an open container evaporate at temperatures far below their boiling points?
- Why does a steel sewing needle float on top of a beaker of water, whereas a steel ball bearing falls to the bottom of the beaker?
- Why does rain break up into drops, or beads, when it falls on the surface of a freshly waxed car?
- If you open a butane lighter-fluid can with the nozzle pointed up, you can hear a gas escape. If you do this with the nozzle pointed down, you can see a liquid escape. How can butane exist as both a liquid and a gas at room temperature?
- Why will gasoline flow through a pinhole in a tire, when water molecules—which are significantly smaller—will not?
- Why is water a liquid at room temperature, not a gas or a solid?

## 14.1 THE STRUCTURE OF LIQUIDS

The differences between the structures of gases, liquids, and solids can be best understood by comparing the densities of substances that can exist in all three phases. As shown in Table 14.1, the density of a typical solid is about 20% greater than the corresponding liquid, while the liquid is roughly 800 times as dense as the gas.

**TABLE 14.1 Densities of Solid, Liquid, and Gaseous Forms of Three Elements**

| | Solid ($g/cm^3$) | Liquid ($g/cm^3$) | Gas ($g/cm^3$) |
|---|---|---|---|
| Ar | 1.65 | 1.40 | 0.001784 |
| $O_2$ | 1.426 | 1.149 | 0.001429 |
| $N_2$ | 1.026 | 0.8081 | 0.001251 |

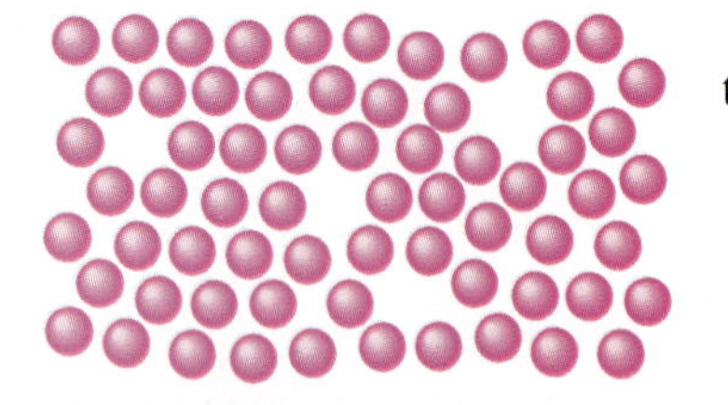

**Figure 14.1** The particles in a liquid do not pack as tightly as they do in a solid because they have more thermal energy and are therefore moving faster in their lattice positions. The structure of liquids also contains small, molecule-sized holes that enable the liquid to flow so that it can conform to the shape of its container.

Figure 14.1 shows a model for the structure of a liquid that is consistent with these data. The key points of this model are summarized as follows:

- The particles that form a liquid are relatively close together, but not as close together as the particles in the corresponding solid.
- The particles in a liquid have more kinetic energy than the particles in the corresponding solid.
- As a result, the particles in a liquid move faster in terms of vibration, rotation, and translation.
- Because they are moving faster, the particles in the liquid occupy more space, and the liquid is less dense than the corresponding solid.
- Differences in kinetic energy alone cannot explain the relative densities of liquids and solids. This model therefore assumes that there are small, particle-sized holes randomly distributed through the liquid.
- Particles that are close to one of these holes behave in much the same way as particles in a gas; those that are far from a hole act more like the particles in a solid.

## 14.2 WHAT KINDS OF MATERIALS FORM LIQUIDS AT ROOM TEMPERATURE?

Three factors determine whether a substance is a gas, a liquid, or a solid at room temperature and atmospheric pressure: (1) the strength of the bonds between the particles that form the substance, (2) the atomic or molecular weight of these particles, and (3) the shape of these particles.

When the forces of attraction between the particles are relatively weak, the substance is likely to be a gas at room temperature. When the forces of attraction are strong, it is more likely to be a solid. As might be expected, a substance is a liquid at room temperature when the intermolecular forces are neither too strong nor too weak.

Sec. 4.15

The role of atomic or molecular weights in determining the state of a substance at room temperature can be understood in terms of the kinetic molecular theory, which includes the following assumption:

Sec. 4.17

**The average kinetic energy of a collection of gas particles depends on the temperature of the gas, and nothing else.**

This means that the average velocity at which different molecules move at the same temperature is inversely proportional to the square root of their molecular weights:

$$\frac{v_A}{v_B} = \sqrt{\frac{MW_B}{MW_A}}$$

Relatively light molecules move so rapidly at room temperature that they can easily break the bonds that hold them together in a liquid or solid. Heavier molecules must be heated to a higher temperature before they can move fast enough to escape from the liquid. They therefore tend to have higher boiling points and are more likely to be liquids at room temperature.

The relationship between the molecular weight of a compound and its boiling point is shown in Table 14.2. The compounds in this table all have the same generic formula: $C_nH_{2n+2}$. The only difference between the compounds is their size and therefore their molecular weights. As shown by Figure 14.2, the relationship between the molecular weights of these compounds and their boiling points is not a straight line, but it is a remarkably smooth curve.

**TABLE 14.2 Melting Points and Boiling Points of Compounds with the Generic Formula $C_nH_{2n+2}$**

| Compound | Melting Point (°C) | Boiling Point (°C) | |
|---|---|---|---|
| $CH_4$ | −182 | −164 | Gases at room temperature |
| $C_2H_6$ | −183.3 | −88.6 | |
| $C_3H_8$ | −189.7 | −42.1 | |
| $C_4H_{10}$ | −138.4 | −0.5 | |
| $C_5H_{12}$ | −130 | 36.1 | Liquids at room temperature |
| $C_6H_{14}$ | −95 | 69 | |
| $C_7H_{16}$ | −90.6 | 98.4 | |
| $C_8H_{18}$ | −56.8 | 125.7 | |
| $C_9H_{20}$ | −51 | 150.8 | |
| $C_{10}H_{22}$ | −29.7 | 174.1 | |

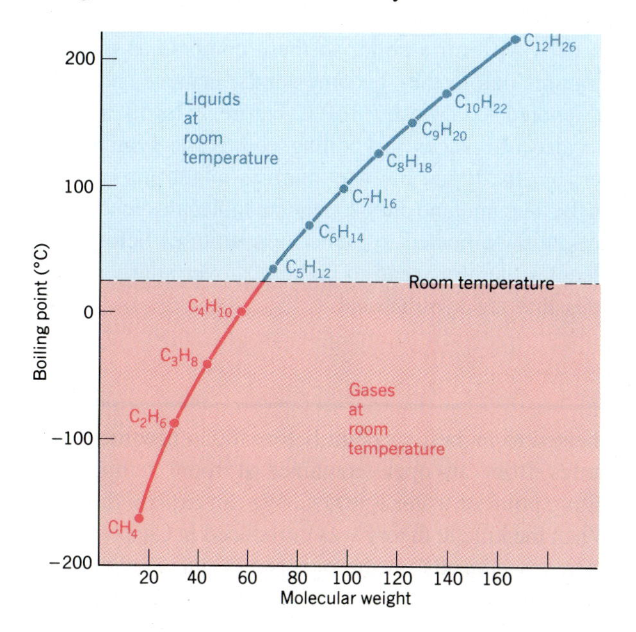


**Figure 14.2** There is a gradual increase in the boiling point of compounds with the generic formula $C_nH_{2n+2}$ as the molecules in this series become heavier.

The data in Figure 14.3 show how the shape of a molecule influences the melting point and boiling point of a compound and therefore the probability that the compound is a liquid at room temperature. The three compounds in this figure are

| COMPOUND | MELTING POINT (°C) | BOILING POINT (°C) |
|---|---|---|
| $CH_3-CH_2-CH_2-CH_2-CH_3$ (zigzag) Pentane | −130 | 36.1 |
| $(CH_3)_2CH-CH_2-CH_3$ Isopentane | −159.9 | 27.8 |
| $C(CH_3)_4$ Neopentane | −16.5 | 9.5 |

**Figure 14.3** Melting points and boiling points for the three isomers with the formula $C_5H_{12}$.

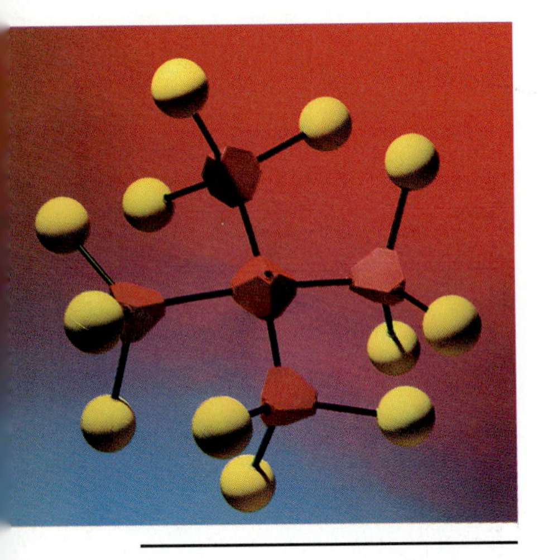

Neopentane.

*isomers* (literally, ''equal parts''). They all have the same chemical formula, but different structures. One of these isomers—neopentane—is a very symmetrical molecule with four identical $CH_3$ groups arranged in a tetrahedral pattern around a central carbon atom. This molecule is so symmetrical that it easily packs to form a solid. Neopentane therefore has to be cooled to only −16.5°C before it crystallizes.

Pentane and isopentane molecules have zigzag structures, which differ only in terms of whether the chain of C—C bonds is linear or branched. These less symmetrical molecules are harder to pack to form a solid, so these compounds must be cooled to much lower temperatures before they become solids. Pentane freezes at −130°C. Isopentane must be cooled to almost −160°C before it forms a solid.

The shape of the molecule also influences the boiling point. The symmetrical neopentane molecules escape from the liquid the way marbles might pop out of a box when it is shaken vigorously. The pentane and isopentane molecules tend to get tangled, like coat hangers, and must be heated to higher temperatures before they can boil. Unsymmetrical molecules therefore tend to be liquids over a larger range of temperatures than molecules that are symmetrical.

## 14.3 VAPOR PRESSURE

Sec. 4.15

A liquid doesn't have to be heated to its boiling point before it can become a gas. Water, for example, evaporates from an open container at room temperature (≈20°C), even though the boiling point of water is 100°C. We can explain this with the diagram in Figure 14.4. When the kinetic theory was introduced in Chapter 4, we noted that the temperature of a gas depends on the *average kinetic energy* of its particles. We had to include the term *average* in this statement because there is an enormous range of kinetic energies for the particles in a gas.

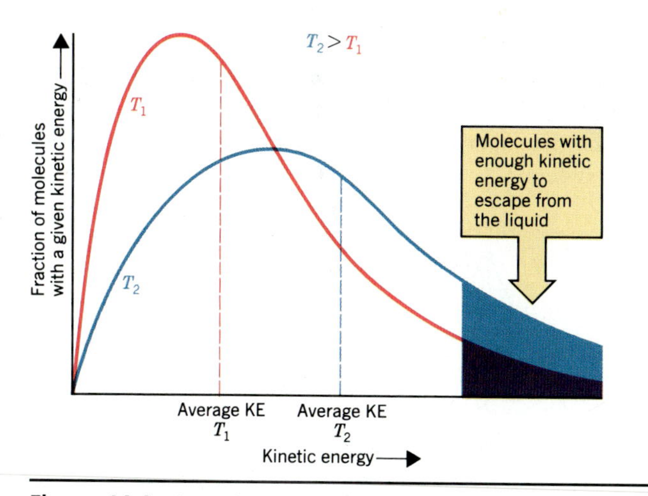


**Figure 14.4** At a given temperature some of the particles in a liquid have enough thermal energy to form a gas. As the temperature increases, the fraction of the molecules moving fast enough to escape from the liquid increases. As a result, the vapor pressure of the liquid also increases.

Liquids and solids are more complex; but the kinetic energy of their particles still depends on the temperature, and some particles still have more kinetic energy than others. Even at low temperatures well below the boiling point of the liquid, some of the particles are moving fast enough to escape from the liquid.

When this happens, the average kinetic energy of the liquid increases. As a result, the liquid becomes cooler. It therefore absorbs energy from its surroundings until it returns to thermal equilibrium. But as soon as this happens, some of the water molecules once again have enough energy to escape from the liquid. In an open container, this process continues until all of the water evaporates.

What happens in a closed container? Once again, some of the molecules escape from the surface of the liquid to form a gas as shown in Figure 14.5. Eventually the rate at which the liquid evaporates to form a gas becomes equal to the rate at which the gas condenses to form the liquid. At this point, the system is said to be in **equilibrium** (from the Latin, "a state of balance"). The space above the liquid is saturated with water vapor, and no more water evaporates.

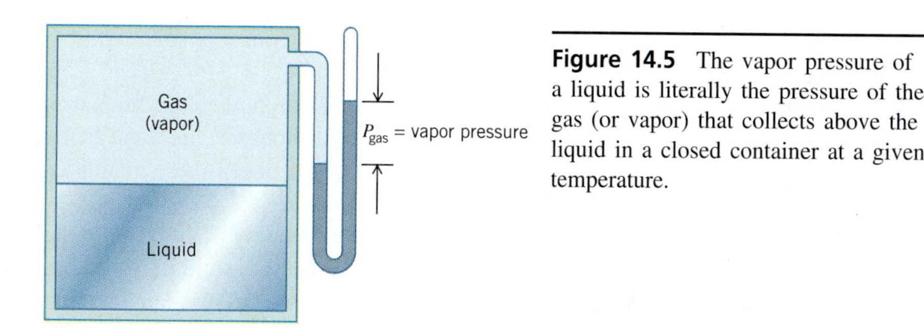


**Figure 14.5** The vapor pressure of a liquid is literally the pressure of the gas (or vapor) that collects above the liquid in a closed container at a given temperature.

The pressure ($P$) of the water vapor in a closed container of water at equilibrium is called the **vapor pressure** of water. The kinetic molecular theory suggests that the vapor pressure of a liquid depends on its temperature. As can be seen in Figure 14.4, the fraction of the molecules that have enough energy to escape from a liquid increases with the temperature of the liquid. As a result, the vapor pressure of a liquid also increases with temperature.

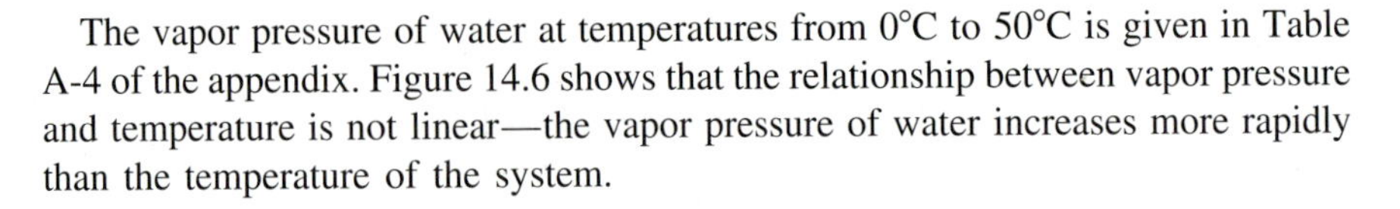

The vapor pressure of water at temperatures from 0°C to 50°C is given in Table A-4 of the appendix. Figure 14.6 shows that the relationship between vapor pressure and temperature is not linear—the vapor pressure of water increases more rapidly than the temperature of the system.

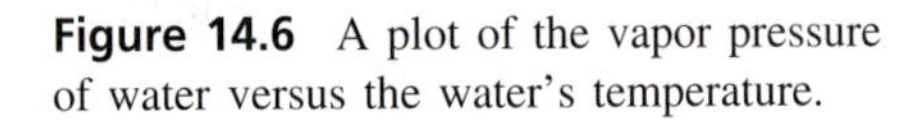

**Figure 14.6** A plot of the vapor pressure of water versus the water's temperature.

Dew drops forming on a blade of grass.

**EXERCISE 14.1**

The dew point is the temperature at which air is saturated with water vapor. If the temperature drops below this point, dew forms. What is the dew point on a day when the relative humidity is 46% and the temperature is 21°C?

**SOLUTION** According to Table A-4, the vapor pressure of water at 21°C is 18.7 mmHg. If the humidity is 46%, the partial pressure of water in the atmosphere on that day is 46% of this value, or 8.6 mmHg. Once again referring to Table A-4, we find that air is saturated with water vapor at a pressure of 8.6 mmHg when the temperature is 9°C. The dew point is therefore 9°C.

## 14.4 MELTING POINT AND FREEZING POINT

Sec. 5.4

It only takes a few seconds for a supersaturated solution of sodium acetate to crystallize after a seed crystal has been added.

Pure, crystalline solids have a characteristic **melting point,** which is the temperature at which the solid melts to become a liquid at atmospheric pressure. The transition between the solid and the liquid is so sharp for small samples of a pure substance that melting points can be measured to ±0.1°C. The melting point of solid oxygen, for example, is −218.4°C.

Liquids have a characteristic temperature at which they turn into solids, known as their **freezing point.** In theory, the melting point of a solid should be the same as the freezing point of the liquid. In practice, small differences between these quantities can be observed.

It is difficult, if not impossible, to heat a solid above its melting point because the heat that enters the solid at its melting point is used to convert the solid into a liquid. It is possible, however, to cool some liquids to temperatures below their freezing points without forming a solid. When this is done, the liquid is said to be *supercooled.*

An example of a supercooled liquid can be made by heating solid sodium acetate trihydrate ($NaCH_3CO_2 \cdot 3\ H_2O$). When this solid melts, the sodium acetate dissolves in the water that was trapped in the crystal to form a solution. When the solution cools to room temperature it should solidify, but it often doesn't. If a small crystal of sodium acetate trihydrate is added to the liquid, however, the contents of the flask solidify within seconds.

Why can a liquid become supercooled? The particles in a solid are packed in a regular structure that is characteristic of that particular substance. Some of these solids form very easily; others do not. Some need a particle of dust, or a seed crystal, to act as a site on which the crystal can grow. In order to form crystals of sodium acetate trihydrate, $Na^+$ ions, $CH_3CO_2^-$ ions, and water molecules must come together in the proper orientation. It is difficult for these particles to organize themselves, but a seed crystal can provide the framework on which the proper arrangement of ions and water molecules can grow.

Because it is difficult to heat solids to temperatures above their melting points, and because pure solids tend to melt over a very small temperature range, melting points are often used to help identify compounds. We can distinguish between the three sugars known as *glucose* (MP = 150°C), *fructose* (MP = 103°C to 105°C), and *sucrose* (MP = 185°C to 186°C), for example, by determining the melting point of a small sample.

Measurements of the melting point of a solid can also provide information about the purity of the substance. Pure, crystalline solids melt over a very narrow range of temperatures, whereas mixtures melt over a broad temperature range. Mixtures also tend to melt at temperatures below the melting points of the pure solids.

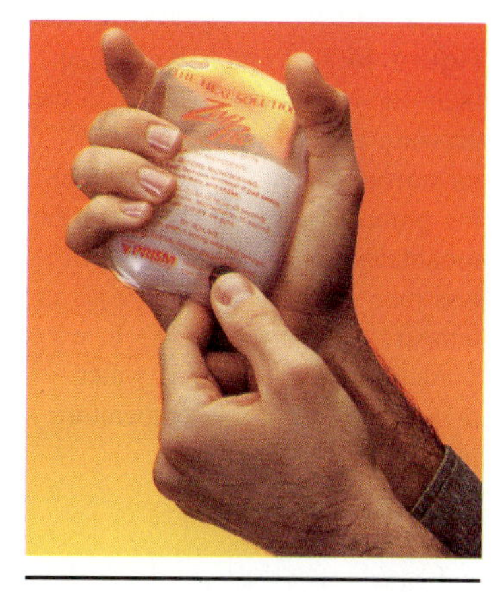

The fact that heat is given off when a supersaturated solution of sodium acetate crystallizes has been used as the basis for a commercial hot pack.

## 14.5 BOILING POINT

When a liquid is heated, it eventually reaches a temperature at which the vapor pressure is large enough that bubbles form inside the body of the liquid. This temperature is called the **boiling point.** Once the liquid starts to boil, this temperature remains constant until all of the liquid has been converted to a gas.

The normal boiling point of water is 100°C. But if you try to cook an egg in boiling water while camping at an elevation of 10,000 feet in the Rocky Mountains, you will find that it takes longer than it does at home because water boils at only 90°C at the higher elevation.

In theory, you shouldn't be able to heat a liquid to temperatures above its normal boiling point. Before microwave ovens became popular, however, pressure cookers were used to decrease the amount of time it took to cook food. In a typical pressure cooker, water can remain a liquid at temperatures as high as 120°C, and food cooks in as little as one-third the normal time.

To understand why water boils at 90°C in the mountains and 120°C in a pressure cooker, even though the normal boiling point of water is 100°C, we have to appreciate why a liquid boils.

**By definition, a liquid boils when the vapor pressure of the gas escaping from the liquid is equal to the pressure exerted on the liquid by its surroundings, as shown in Figure 14.7.**

The normal boiling point of water is 100°C because this is the temperature at which the vapor pressure of water is 760 mmHg, or 1 atm. Under normal conditions, when the pressure of the atmosphere is approximately 760 mmHg, water boils at 100°C. At 10,000 feet above sea level, the pressure of the atmosphere is only 526 mmHg. At these elevations, water boils when its vapor pressure is 526 mmHg, which occurs at a temperature of 90°C.

Pressure cookers are equipped with a valve that lets gas escape when the pressure inside the pot exceeds some fixed value. This valve is often set at 15 psi, which means that the water vapor inside the pot must reach a pressure of 2 atm before it can escape. Because water doesn't reach a vapor pressure of 2 atm until the temperature is 120°C, it boils in this container at 120°C.

Bubbles rising toward the surface of a liquid as it starts to boil.

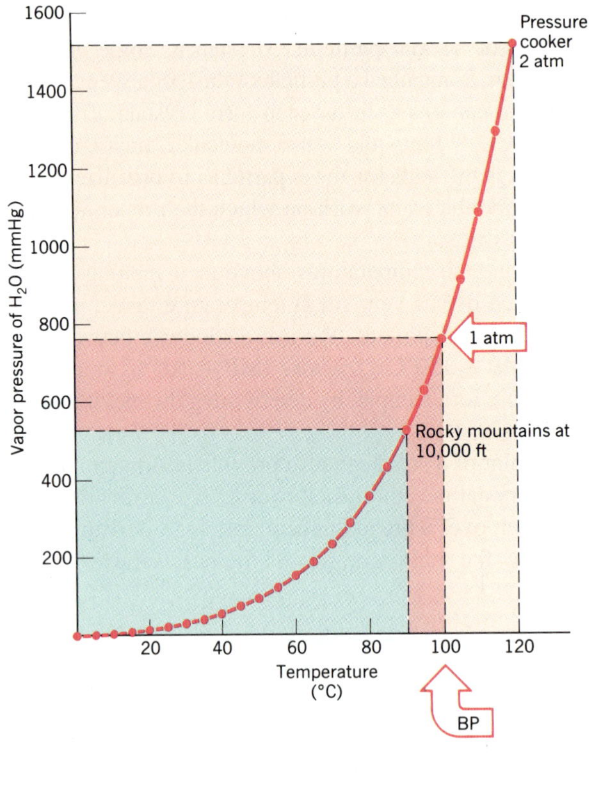


**Figure 14.7** Liquids boil when their vapor pressure is equal to the pressure exerted on the liquid by its surroundings. The normal boiling point of water is 100°C. In the mountains, atmospheric pressure is less than 1 atm, and water boils at temperatures below 100°C. In a pressure cooker at 2 atm, water doesn't boil until the temperature reaches 120°C.

Water boils at room temperature when the pressure in its container is reduced to less than 20 mmHg.

## EXERCISE 14.2

Use the definition of boiling point and the data in Table A-4 to describe how to make water boil at room temperature.

**SOLUTION** Water boils when the vapor pressure of the gas escaping from the liquid is equal to the pressure exerted on the liquid by its surroundings. The vapor pressure of water is roughly 20 mmHg at room temperature. We can therefore make water boil at room temperature by reducing the pressure in its container to less than 20 mmHg.

Liquids often boil in an uneven fashion, or *bump*. They tend to bump when there aren't any scratches on the walls of the container where bubbles can form. Bumping is easily prevented by adding a few boiling chips to the liquid, which provide a rough surface upon which bubbles can form. When boiling chips are used, essentially all of the bubbles that rise through the solution form on the surface of these chips.

## 14.6 CRITICAL TEMPERATURE AND CRITICAL PRESSURE

The obvious way to turn a gas into a liquid is to cool it to a temperature below its boiling point. There is another way of condensing a gas to form a liquid, however, which involves raising the pressure on the gas. Liquids boil at the temperature at which the vapor pressure is equal to the pressure on the liquid from its surroundings. Raising the pressure on a gas therefore effectively increases the boiling point of the liquid.

Suppose that we have water vapor (or steam) in a closed container at 120°C and 1 atm. Since the temperature of the system is above the normal boiling point of water, there is no reason for the steam to condense to form a liquid. Nothing happens as we slowly compress the container—thereby raising the pressure on the gas—until the pressure reaches 2 atm. At this point, the system is at the boiling point of water, and some of the gas will condense to form a liquid. As soon as the pressure on the gas exceeds 2 atm, the vapor pressure of water at 120°C is no longer large enough for the liquid to boil. The gas therefore condenses to form a liquid, as shown in Figure 14.8.

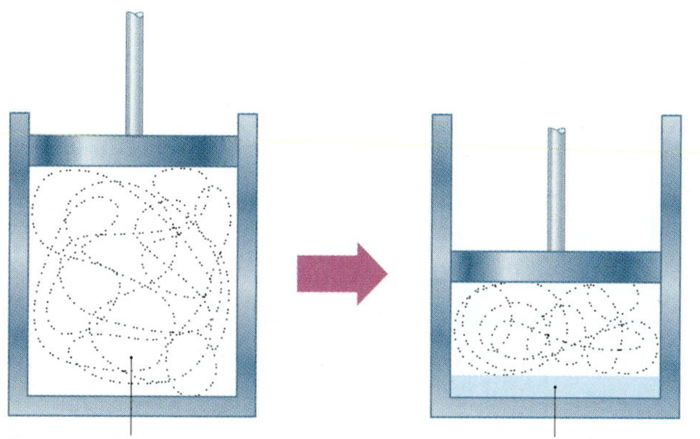

**Figure 14.8** Steam at 120°C and 1 atm pressure will condense to form water when compressed until the pressure on the steam is larger than 2 atm.

In theory, we should be able to predict the pressure at which a gas condenses at a given temperature by consulting a plot of vapor pressure versus temperature, such as the one shown in Figure 14.7. In practice, every compound has a **critical temperature** ($T_c$). If the temperature of the gas is above the critical temperature, the gas can't be condensed, regardless of the pressure applied.

The existence of a critical temperature was discovered by Thomas Andrews in 1869 while studying the effect of temperature and pressure on the behavior of carbon dioxide. Andrews found that he could condense $CO_2$ gas into a liquid by raising the pressure on the gas, as long as he kept the temperature below 31.0°C. At 31.0°C, for example, it takes a pressure of 72.85 atm to liquefy $CO_2$ gas. Andrews found that it was impossible to turn $CO_2$ into a liquid above this temperature, no matter how much pressure was applied.

Gases can't be liquefied at temperatures above the critical temperature because at this point the properties of gases and liquids become the same, and there is no basis on which to distinguish between gases and liquids. The vapor pressure of a liquid at the critical temperature is called the **critical pressure** ($P_c$). The vapor pressure of a liquid never gets larger than this critical pressure.

The critical temperatures, critical pressures, and boiling points of a number of gases are given in Table 14.3. There is an obvious correlation between the critical temperature and boiling points of these gases. These properties are related because they are both indirect measures of the force of attraction between particles in the gas phase.

The experimental values of the critical temperature and pressure of a substance are used to calculate the $a$ and $b$ constants in the van der Waals equation. The values of these constants are calculated from the following equations:

p. 144

$$a = \frac{27R^2T_c^2}{64P_c} \qquad b = \frac{RT_c}{8P_c}$$

**TABLE 14.3 Critical Temperatures, Critical Pressures, and Boiling Points of Common Gases**

| Gas | $T_c$ (°C) | $P_c$ (atm) | BP (°C) |
|---|---|---|---|
| He | −267.96 | 2.261 | −268.94 |
| $H_2$ | −240.17 | 12.77 | −252.76 |
| Ne | −228.71 | 26.86 | −246.1 |
| $N_2$ | −146.89 | 33.54 | −195.81 |
| CO | −140.23 | 34.53 | −191.49 |
| Ar | −122.44 | 48.00 | −185.87 |
| $O_2$ | −118.38 | 50.14 | −182.96 |
| $CH_4$ | −82.60 | 45.44 | −161.49 |
| $CO_2$ | 31.04 | 72.85 | −78.44 |
| $NH_3$ | 132.4 | 111.3 | −33.42 |
| $Cl_2$ | 144.0 | 78.1 | −34.03 |

## 14.7 SURFACE TENSION

Why do wet sheets of paper stick together? Why does a steel sewing needle float on top of a beaker of water, whereas a steel ball bearing falls to the bottom of the beaker? Why does water curve upward in a small-diameter glass tube, whereas mercury curves downward? Why does liquid mercury break up into drops, or beads, when it spills on the floor? Why does rain form similar beads on the surface of a freshly waxed car?

The answers to these questions all can be traced to the force of attraction between the molecules in liquids and the fact that liquids can flow until they take on the shape that maximizes this force of attraction. Below the surface of the liquid, the force of **cohesion** (literally, ''sticking together'') between molecules is the same in all directions, as shown in Figure 14.9. Molecules on the surface of the liquid, however, feel a net force of attraction that pulls them back into the body of the liquid. As a result, the liquid tries to take on the shape that has the smallest possible surface area—the shape of a sphere. The magnitude of the force that controls the shape of the liquid is called the **surface tension.** The stronger the bonds between the molecules in the liquid, the larger the surface tension.

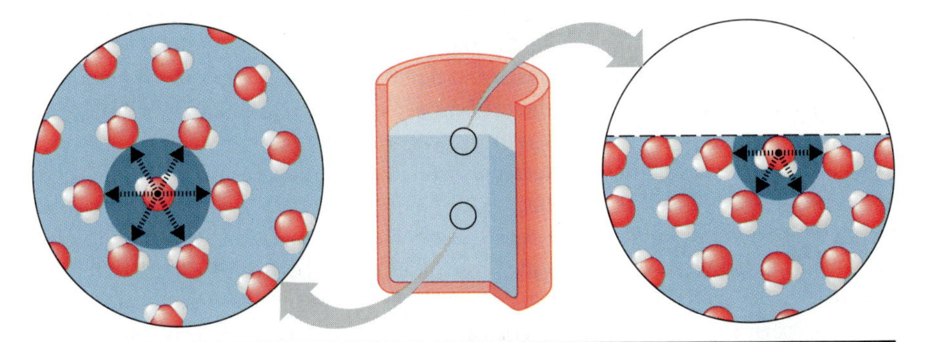

**Figure 14.9** Molecules in a liquid feel a force of *cohesion* that pulls them into the body of the liquid. Molecules that lie along one of the walls of the container also feel a force of *adhesion* that binds them to the walls. When the force of adhesion is much smaller than the force of cohesion, the liquid will pull away from the walls of the container. If the force of adhesion is more than half the force of cohesion, the liquid will ''wet'' the walls of the container.

There is also a force of **adhesion** (literally, ''sticking to'') between a liquid and the walls of the container. When the force of adhesion is more than half as large as the force of cohesion between the liquid molecules, the liquid is said to ''wet'' the solid. A good example of this phenomenon is the wetting of paper by water. The force of adhesion between paper and water combined with the force of cohesion between water molecules explains why sheets of wet paper stick together.

Water wets glass because of the force of adhesion that results from interactions between the positive ends of the polar water molecules and the negatively charged oxygen atoms in glass. As a result, water forms a **meniscus** that curves upward in a small-diameter glass tube, as shown in Figure 14.10. (The term *meniscus* comes from the Greek word for ''moon'' and is used to describe any object that has a crescent shape.) The meniscus that water forms in a buret results from a balance between the force of adhesion pulling up on the column of water to wet the walls of the glass tube and the force of gravity pulling down on the liquid.

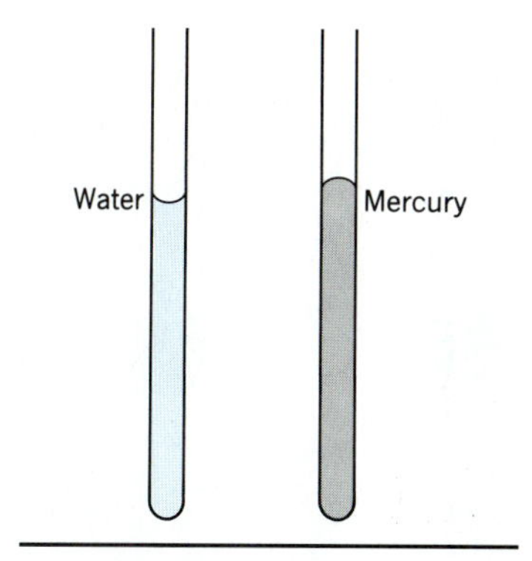


**Figure 14.10** Water climbs the walls of a small-diameter tube to form a meniscus that curves upward, whereas mercury forms a meniscus that curves downward.

The force of *adhesion* between water and wax is very small compared to the force of *cohesion* between water molecules. As a result, rain doesn't adhere to wax. It tends to form beads, or drops, with the smallest possible surface area, thereby maximizing the force of cohesion between the water molecules. The same thing happens when mercury is spilled on glass or poured into a narrow glass tubing. The force of cohesion between mercury atoms is so much larger than the force of adhesion between mercury and glass that the area of contact between mercury and glass is kept to a minimum, with the net result being the meniscus shown in Figure 14.10.

**CHECKPOINT**

Explain why the meniscus curves upward in a water-filled buret and downward in a mercury-filled barometer.

## RESEARCH IN THE 90s

### THE SEARCH FOR SYNTHETIC FIBERS

Synthetic fibers have been produced since 1885, when the French Count Hilaire de Chardonnet received a patent for a synthetic silk. The first step in making Chardonnet silk, as it was known, involved dissolving cellulose from wood pulp in nitric acid to form cellulose nitrate. Anyone who has watched what happens when a flame is held close to a ping-pong ball should understand that cellulose nitrate, by itself, is far too flammable to be used as a fiber. The next step in the production of Chardonnet silk therefore involved decomposing cellulose nitrate back to cellulose under conditions that generated a continuous fiber. This was achieved by extruding a viscous solution of cellulose nitrate in alcohol through small holes into water.

In 1903, a slightly different process was patented in the United Kingdom. It involved dissolving cellulose from wood pulp in a mixture of caustic soda and carbon disulfide ($CS_2$). The resulting viscous material was then extruded through small holes into a solution of sulfuric acid to form a fiber that was sold under the trade name of Rayon. By 1980, the demand for cellulose-based fibers had reached the point where more than 3,250,000 metric tons were produced each year.

Scanning electron microscope images of the structure of the thin fibers of Pirmaloft$^{TM}$. (Photographs provided by Albany International Research Co. Pirmaloft is a registered trademark of Albany International Corp).

More than 100 years after the development of the first synthetic fiber, the most economically important fiber is still cotton, which accounts for almost 50% of the total world production of textile fibers. Recent advances in synthetic fibers, however, have generated materials that not only compete with, but surpass, many natural fibers.

Until recently, the standard against which all fibers were compared as potential insulators was the soft, fluffy down that can be obtained from ducks and geese. Down is simultaneously lightweight and an excellent thermal insulator. Unfortunately, it also readily absorbs water, and, when wet, loses much of its ability to act as an insulator. Thus, even if the supply were plentiful—which it is not—and if down were inexpensive—which it is not—there would be a potential demand for a synthetic fiber that had the insulating properties of down but did not ''wet.''

Several years ago, a synthetic insulator known as *primaloft* was prepared under a contrast issued by the U.S. Army Research, Development and Engineering Center at Natick [*Chemical and Engineering News,* Oct. 16, **1989,** p. 25]. Scanning electron microscopy has shown that natural down is a mixture of relatively large-diameter fibers, which make it stiff, and very thin fibers that ensure the presence of many small pockets of air. The net result is a lightweight, but stiff material that is a good insulator. Primaloft achieves the same effect by combining a small number of large-diameter polyester fibers with many more small fibers. The small fibers have a diameter of 7 $\mu$m, roughly one-fourth the diameter of a human hair. These fibers can be made into a product that is just as ''warm'' as down and even has the same feel when enclosed in a jacket or parka. More importantly, primaloft absorbs much less water when wet, and therefore does not lose its insulating capacity in the rain. Furthermore, it is considerably less expensive than down.

A variety of esoteric measurements are done to compare different textile fibers. Questions that are asked include: Does the fabric swell when the fibers absorb water? Does the fabric have a tendency to build up static electric charge (''static cling'')? Is the fiber even, or does the yarn vary in diameter as it is spun? Does the fiber shrink on exposure to water? How well does the fiber ''breathe?'' (Is it permeable to air?) What is the bursting strength of the fiber? What is its tear strength? How well does it resist snag? How well does it stand up to abrasion and other forms of wear? How well does it conduct heat? (Is it a good insulator, like natural wool?) How well does it pass moisture in the form of water vapor? Is it water repellent, or, at least, can it be made water repellent? How well does it stretch? How well does it accept various dyes? Is it colorfast, once dyed? And, of course, perhaps as important as any other property—is it flammable?

Research in the 90s will continue the search for new fabrics, such as those recently formulated by the Hoechst Celanese Corporation for athletic clothes that have the remarkable ability to pass water vapor but not liquid water through the fabric. These fabrics allow perspiration to evaporate and still provide the protection against rain or snow expected for water-repellant fabrics.

The search for synthetic fibers has produced fabrics that easily pass water vapor, but do not ''wet.'' These fabrics are therefore ideal for making athletic clothes that ''breathe'' but are still water repellant.

## 14.8 VISCOSITY

One way of testing for a slow leak in a tire involves immersing the tire in a barrel of water and looking for the place where bubbles form. In theory, it should be possible to fill the tire with water and see where the water escapes. In practice, this doesn't work because air escapes through holes that are much too small to pass water. It might be tempting to explain this observation by arguing that water molecules are bigger than the $O_2$ or $N_2$ molecules in air. But this can't be correct because gasoline, which contains molecules that are much larger than water molecules, can leak through holes that are too small to pass water.

These observations are the result of differences in the viscosities of air, water, and gasoline. **Viscosity** is a measure of the resistance to flow. Motor oils are more viscous than gasoline, for example, and the maple syrup used on pancakes is more viscous than the vegetable oils used in salad dressings.

Viscosity is measured by determining the rate at which a liquid or gas flows through a small-diameter glass tube. In 1844 Jean Louis Marie Poiseuille showed that the volume of fluid ($V$) that flows down a small-diameter capillary tube per unit of time ($t$) is proportional to the radius of the tube ($r$), the pressure pushing the fluid down the tube ($P$), the length of the tube ($l$), and the viscosity of the fluid ($\eta$):

$$\frac{V}{t} = \frac{\pi r^2 P}{8 \eta l}$$

Viscosity is reported in units called *poise* (pronounced ''pwahz''). The viscosity of water at room temperature is roughly 1 centipoise (1 cP). Gasoline has a viscosity between 0.4 and 0.5 cP; the viscosity of air is 0.018 cP.

Because the molecules closest to the walls of a small-diameter tube adhere to the glass, viscosity measures the rate at which molecules in the middle of the stream of liquid or gas flow past this outer layer of more or less stationary molecules. Viscosity therefore depends on any factor that can influence the ease with which molecules slip past each other. Liquids tend to become more viscous as the molecules become larger, or as the amount of intermolecular bonding increases. They become less viscous as the temperature increses. The viscosity of water, for example, decreases from 1.77 cP at 0°C to 0.28 cP at 100°C.

### ▶ CHECKPOINT

Explain why maple syrup becomes less viscous when heated.

## 14.9 HYDROGEN BONDING AND THE ANOMALOUS PROPERTIES OF WATER

We are so familiar with the properties of water that it is difficult to appreciate the extent to which its behavior is unusual:

- Most solids expand when they *melt*. Water expands when it *freezes*.
- Most solids are more dense than their corresponding liquids. Ice ($d$ = 0.917 g/cm$^3$) is not as dense as water.
- Water has a melting point at least 100°C higher than expected on the basis of the melting points of $H_2S$, $H_2Se$, and $H_2Te$.
- Water has a boiling point almost 200°C higher than expected from the boiling points of $H_2S$, $H_2Se$, and $H_2Te$.
- Water has the largest surface tension of any common liquid except liquid mercury.
- Water has an unusually large viscosity.
- Water is an excellent solvent. It can dissolve compounds, such as NaCl, that are insoluble or only slightly soluble in other liquids.
- Water has an unusually high heat capacity. It takes more heat to raise the temperature of 1 g of water by 1°C than any other liquid.

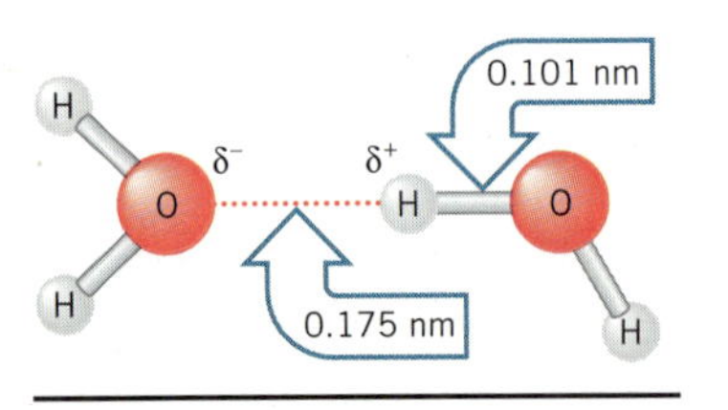


**Figure 14.11** The dipole–dipole attraction between neighboring water molecules is known as *hydrogen bonding*. Because of the polarity of water molecules, this is an unusually strong example of an intermolecular bond.

These anomalous properties all result from the strong intermolecular bonds in water.

In Chapter 8 we concluded that water is best described as a polar molecule in which there is a partial separation of charge to give positive and negative poles. The force of attraction between a positively charged hydrogen atom on one water molecule and the negatively charged oxygen atom on another gives rise to an intermolecular bond, as shown in Figure 14.11. This dipole–dipole interaction between water molecules is known as a **hydrogen bond.**

Hydrogen bonds are separated from other examples of van der Waals forces because they are unusually strong: 20 to 25 kJ/mol. The hydrogen bonds in water are particularly important because of the dominant role that water plays in the chemistry of living systems. Hydrogen bonds are not limited to water, however.

Hydrogen-bond donors include substances that contain relatively polar H—$X$ bonds, such as $NH_3$, $H_2O$, and HF. Hydrogen-bond acceptors include substances that have nonbonding pairs of valence electrons. The H—$X$ bond must be polar to create the partial positive charge on the hydrogen atom that allows dipole–dipole interactions to exist. As the $X$ atom in the H—$X$ bond becomes less electronegative, hydrogen bonding between molecules becomes less important. Hydrogen bonding in HF, for example, is much stronger than in either $H_2O$ or HCl.

### CHECKPOINT

Explain why hydrogen bonds between $NH_3$ molecules are weaker than those between $H_2O$ molecules.

The hydrogen bonds between water molecules in ice produce the open structure shown in Figure 14.12. When ice melts, some of these bonds are broken, and this structure collapses to form a liquid that is about 10% denser. This unusual property of water has several important consequences. The expansion of water when it freezes is responsible for the cracking of concrete, which forms potholes in streets

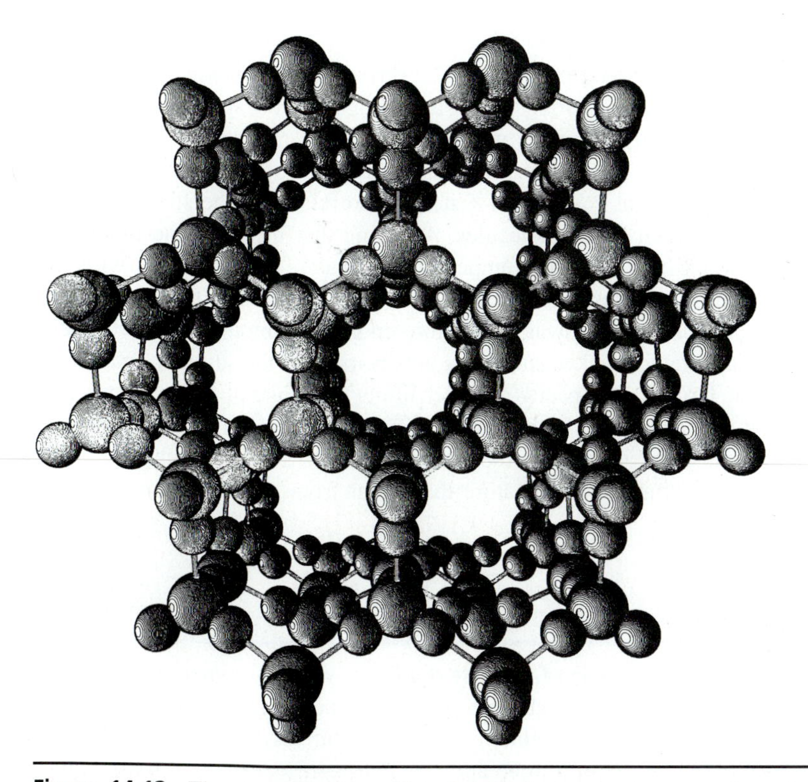

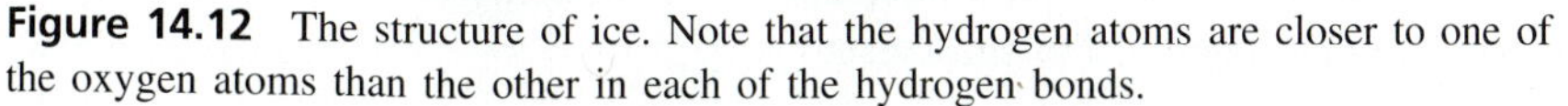

**Figure 14.12** The structure of ice. Note that the hydrogen atoms are closer to one of the oxygen atoms than the other in each of the hydrogen bonds.

and highways. But it also means that ice floats on top of rivers and streams. The ice that forms each winter therefore has a chance to melt during the summer.

Sec. 14.2

Figure 14.13 shows another consequence of the strength of the hydrogen bonds in water. There is a steady increase in boiling point in the series $CH_4$, $GeH_4$, $SiH_4$, and $SnH_4$, as expected from the discussion in Section 14.2. The boiling points of $H_2O$ and HF, however, are anomalously large because of the strong hydrogen bonds between molecules in these liquids. If this doesn't seem important, try to imagine what life would be like if water boiled at $-80°C$.

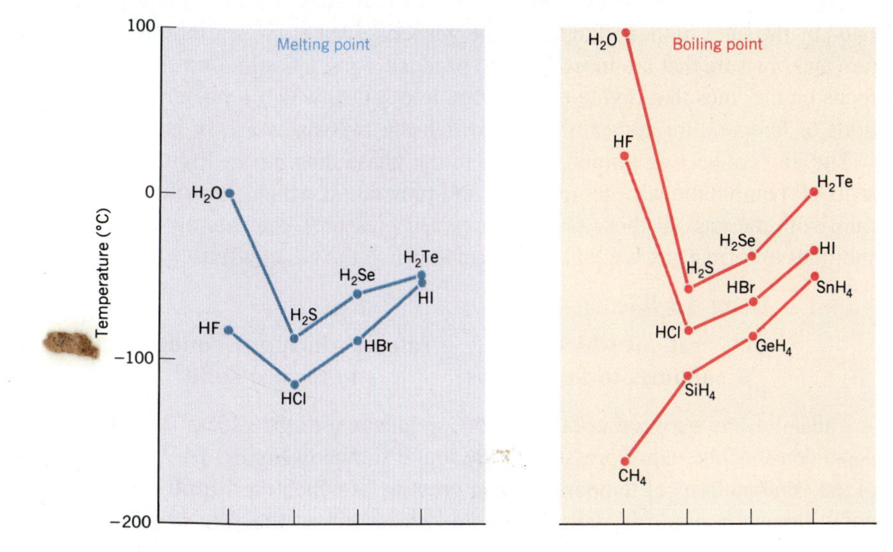


**Figure 14.13** A plot of the melting points and boiling points of hydrides of elements in Groups IVA, VA, and VIA. The melting points and boiling points of HF and $H_2O$ are anomalously large because of the strength of the hydrogen bonds between molecules in these compounds.

The surface tension and viscosity of water are also related to the strength of the hydrogen bonds between water molecules. The surface tension of water is responsible for the capillary action that brings water up through the root systems of plants. It is also responsible for the efficiency with which the wax that coats the surface of leaves can protect plants from excessive loss of water through evaporation.

The unusually large heat capacity of water is also related to the strength of the hydrogen bonds between water molecules. Anything that increases the motion of water molecules, and therefore the temperature of water, must interfere with the hydrogen bonds between these molecules. The fact that it takes so much energy to disrupt these bonds means that water can store enormous amounts of thermal energy. Although the water in lakes and rivers gets warmer in the summer and cooler in the winter, the large heat capacity of water limits the range of temperatures that would otherwise threaten the life that flourishes in this environment. The heat capacity of water is also responsible for the ocean's ability to act as a thermal reservoir that moderates the swings in temperature that occur from winter to summer.

## 14.10 PHASE DIAGRAMS

Figure 14.14 shows an example of a **phase diagram,** which summarizes the effect of temperature and pressure on a substance in a closed container. Every point in this diagram represents a possible combination of temperature and pressure for the system. The diagram is divided into three areas, which represent the solid, the liquid, and the gaseous states of the substance.

The best way to remember which area corresponds to each of these states is to remember the conditions of temperature and pressure that are most likely to be associated with a solid, a liquid, and a gas. Low temperatures and high pressures favor the formation of a solid. Gases, on the other hand, are most likely to be found at high temperatures and low pressures. Liquids lie between these extremes.

You can test whether you have correctly labeled a phase diagram by drawing a line from left to right across the top of the diagram, which corresponds to an increase in the temperature of the system at constant pressure. When a solid is heated at constant pressure, it melts to form a liquid, which eventually boils to form a gas.

Phase diagrams can be used in several ways. We can focus on the regions separated by the lines in these diagrams, and get some idea of the conditions of temperature and pressure that are most likely to produce a gas, a liquid, or a solid. Or we can focus on the lines that divide the diagram into states, which represent the combinations of temperature and pressure at which the two states are in equilibrium.

The line connecting points *A* and *B* in the phase diagram in Figure 14.14 represents all combinations of temperature and pressure at which the solid is in equilibrium with the gas. At these temperatures and pressures, the rate at which the solid sublimes to form a gas is equal to the rate at which the gas condenses to form a solid.

**Along *AB* line:**

$$\textbf{rate at which solid sublimes to form a gas} = \textbf{rate at which gas condenses to form a solid}$$

The solid line between points *B* and *C* is identical to the plot of the temperature dependence of the vapor pressure of the liquid shown in Figure 14.7. It contains all of the combinations of temperature and pressure at which the liquid boils. At every point along this line, the liquid boils to form a gas at the rate at which the gas condenses to form a liquid.

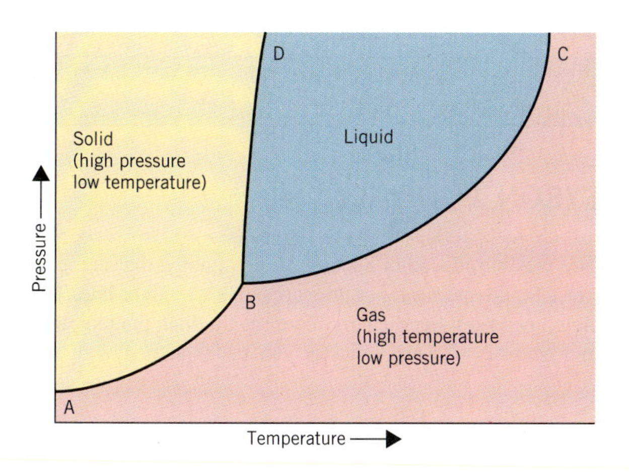


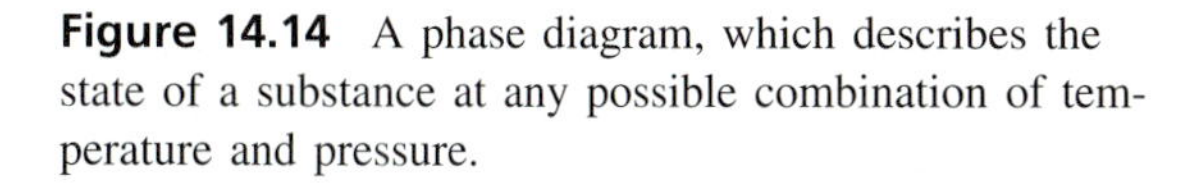
**Figure 14.14** A phase diagram, which describes the state of a substance at any possible combination of temperature and pressure.

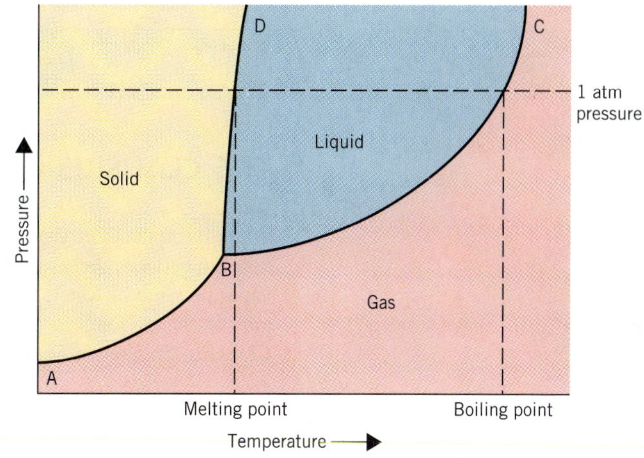


**Figure 14.15** A horizontal line across a phase diagram at a pressure of 1 atm crosses the curve separating solids and liquids at the melting point of the solid, and it crosses the curve separating liquids and gases at the boiling point of the liquid.

**Along *BC* line:**

$$\textbf{rate at which liquid boils to form a gas} = \textbf{rate at which gas condenses to form a liquid}$$

The solid line between points *B* and *D* contains the combinations of temperature and pressure at which the solid and liquid are in equilibrium. At every point along this line, the solid melts at the same rate at which the liquid freezes.

**Along *BD* line:**

$$\textbf{rate at which solid melts to form a liquid} = \textbf{rate at which liquid freezes to form a solid}$$

The *BD* line is almost vertical because the melting point of a solid is not very sensitive to changes in pressure. For most compounds, this line has a small positive slope, as shown in Figure 14.14. The slope of this line is slightly negative for water, however. As a result, ice can melt at temperatures near its freezing point when pressure is applied. The ease with which ice skaters glide across a frozen pond can be explained by the fact that the pressure exerted by their skates melts a small portion of the ice that lies beneath the blades.

Point *B* in this phase diagram represents the only combination of temperature and pressure at which a pure substance can exist simultaneously as a solid, a liquid, and a gas. It is therefore called the **triple point** of the substance, and it represents the only point in the phase diagram in which all three states are in equilibrium. Point *C* is the critical point of the substance, which is the highest temperature and pressure at which a gas and a liquid can coexist at equilibrium.

Because the *BD* line in the phase diagram for water slopes slightly to the left, ice often melts when enough pressure is put upon it.

Figure 14.15 shows what happens when we draw a horizontal line across a phase diagram at a pressure of exactly 1 atm. This line crosses the line between points *B* and *D* at the melting point of the substance because solids normally melt at the temperature at which the solid and liquid are in equilibrium at 1 atm pressure. The line crosses the line between points *B* and *C* at the boiling point of the substance because the normal boiling point of a liquid is the temperature at which the liquid and gas are in equilibrium at 1 atm pressure and the vapor pressure of the liquid is therefore equal to 1 atm.

**SPECIAL TOPIC**

## THE CLAUSIUS-CLAPEYRON EQUATION

When the concept of vapor pressure was introduced in Section 14.3, we noted that the relationship between the temperature of a liquid and its vapor pressure is not a straight line. The vapor pressure of water, for example, increases significantly more rapidly than the temperature of the system, as shown in Figure 14.6. This behavior can be explained with the **Clausius-Clapeyron equation.**

$$\frac{d \ln P}{dT} = \frac{\Delta H_{vap}}{RT^2}$$

According to this equation, the infinitesimally small change in the natural logarithm of the vapor pressure of a liquid ($d \ln P$) that occurs during an infinitesimally small change in the temperature of the system ($dT$) is determined by the molar enthalpy of vaporization of the liquid ($\Delta H_{vap}$), the ideal gas constant ($R$), and the temperature of the system ($T$). If we assume that $\Delta H_{vap}$ does not depend on temperature, the Clausius-Clapeyron equation can be written in the following integrated form, where $C$ is a constant:

$$\ln P = -\frac{\Delta H_{vap}}{RT} + C$$

This form of the Clausius-Clapeyron equation is used to determine the enthalpy of vaporization of a liquid from plots of the natural log of its vapor pressure versus temperature. For our purposes, it would be more useful to take advantage of logarithmic mathematics to write this equation as follows.

$$\ln P = \frac{RT}{\Delta H_{vap}} + C$$

Because the molar enthalpy of vaporization of a liquid is always positive, the natural logarithm of the vapor pressure will increase as the temperature of the system increases. Since the vapor pressure of the liquid increases much more rapidly than its natural logarithm, we get the behavior shown in Figure 14.6.

## SUMMARY

1. Many of the properties of liquids can be explained in terms of a simple model, which assumes that liquids do not pack as tightly as solids because the particles have more thermal energy and are therefore moving faster around their lattice positions. This model also assumes that liquids contain small, particle-sized holes that allow the liquid to flow so that it can conform to the shape of its container.

2. The **vapor pressure** of a liquid is the partial pressure of the gas that collects in a closed container at a particle temperature. The **normal boiling point** of the liquid is the temperature at which the vapor pressure is equal to atmospheric pressure. The **melting point** or **freezing point** of a substance is the temperature at which the solid and the liquid are in equilibrium at atmospheric pressure.

3. Three factors determine whether a substance is a solid, a liquid, or a gas at room temperature: (1) the strength of the interactions between the particles that form the substance, (2) the atomic or molecular weight of these particles, and (3) the shapes of these particles.

4. The relationship between the gaseous, liquid, and solid states of a substance can be summarized in a phase diagram that describes the state of the substance in a closed container under all possible combinations of temperature and pressure.

## KEY TERMS

**Adhesion** (p. 519)
**Boiling point** (p. 515)
**Clausius-Clapeyron equation** (p. 526)
**Cohesion** (p. 518)
**Critical pressure** (p. 517)
**Critical temperature** (p. 517)
**Equilibrium** (p. 513)
**Freezing point** (p. 514)
**Hydrogen bond** (p. 522)
**Melting point** (p. 514)
**Meniscus** (p. 519)
**Phase diagram** (p. 524)
**Surface tension** (p. 518)
**Triple point** (p. 525)
**Vapor pressure** (p. 513)
**Viscosity** (p. 521)

## KEY EQUATIONS

$$\frac{V}{t} = \frac{\pi r^4 P}{8\eta l}$$

Along *AB* line:

$$\text{rate at which solid sublimes to form a gas} = \text{rate at which gas condenses to form a solid}$$

Along *BC* line:

$$\text{rate at which liquid boils to form a gas} = \text{rate at which gas condenses to form a liquid}$$

Along *BD* line:

$$\text{rate at which solid melts to form a liquid} = \text{rate at which liquid freezes to form a solid}$$

## PROBLEMS

### The Structure of Liquids

**14-1** Describe how assuming that there are holes in the structure of a liquid explains the ease with which liquids flow.

**14-2** Explain why liquids are usually less dense than the corresponding solid.

**14-3** Use the data in Table 14.1 to calculate the number of molecules in 1 $cm^3$ of $O_2$ in the solid, liquid, and gaseous phases. Use the results of these calculations to probe the assumption that 99.9% of the volume of a gas is empty space.

### What Materials Form Liquids at Room Temperature?

**14-4** Explain why salts are solids at room temperature.

**14-5** Predict the order in which the boiling points of the following compounds should increase if the intermolecular forces in these compounds are about the same:

(a) $CH_4$ (b) $SiH_4$ (c) $GeH_4$ (d) $SnH_4$

**14-6** Which compound would you expect to have the highest boiling point?

(a) Methane, $CH_4$ (b) Chloromethane, $CH_3Cl$
(c) Dichloromethane, $CH_2Cl_2$ (d) Chloroform, $CHCl_3$
(e) Carbon tetrachloride, $CCl_4$

**14-7** Explain why the boiling points of hydrocarbons that have the generic formula $C_nH_{2n+2}$ increase with molecular weight.

**14-8** Explain why propane ($C_3H_8$) is a gas but pentane ($C_5H_{12}$) is a liquid at room temperature.

**14-9** Explain why methane ($CH_4$) is only a liquid over a very narrow range of temperatures.

**14-10** Explain why pentane is a liquid over a much larger range of temperatures than neopentane.

### Vapor Pressure

**14-11** Explain why it is important to specify the temperature at which the vapor pressure of a liquid is measured.

**14-12** Explain why water eventually evaporates from an open container at room temperature ($\approx$20°C), even though it normally boils at 100°C.

**14-13** Explain why the last postulate of the kinetic theory of gases suggests that the temperature of a gas is directly proportional to the *average kinetic energy* of the particles in the gas.

**14-14** Explain why the vapor pressure of water increases as the temperature of the water increases.

**14-15** Explain why a cloth soaked in water feels cool when placed on your forehead.

**14-16** What would happen to the vapor pressure of liquid bromine, $Br_2$, at 20°C if the liquid was transferred from a narrow 10-mL graduated cylinder into a wide petri dish or crystallizing dish? Would it increase, decrease, or remain the same? What would happen to the vapor pressure in a closed container if more liquid was added to the container? Would it increase, decrease, or remain the same?

**14-17** Explain what it means to say that the liquid and vapor in a closed container are in *equilibrium.*

**14-18** Explain why each of the following increases the rate at which water evaporates from an open container:

(a) Increasing the temperature of the water
(b) Increasing the surface area of the water
(c) Blowing air over the surface of the water
(d) Decreasing atmospheric pressure on the water

**14-19** Calculate the relative humidity when the temperature is

31°C and the partial pressure of water vapor in the atmosphere is 29.8 mmHg.

**14-20** Calculate the dew point on a day when the humidity is 85% and the temperature is 60°F.

### Melting Point, Freezing Point, and Boiling Point

**14-21** Explain why measurements of the melting point of a solid are sometimes more accurate than measurements of the freezing point of a liquid.

**14-22** Explain why a "3-minute egg" cooked while camping in the Rocky Mountains does not taste as good as it does when cooked while camping near the Great Lakes.

**14-23** Explain why it takes less time to cook food in a pressure cooker.

**14-24** Explain why water boils when the pressure on the system is reduced.

**14-25** At what temperature does water boil when the pressure is 50 mmHg?

**14-26** What pressure has to be achieved before water can boil at 20°C?

**14-27** Increasing the temperature of a liquid will do which of the following?

(a) Increase its boiling point (b) Increase its melting point (c) Increase its vapor pressure (d) Increase the amount of heat required to boil a mole of the liquid (e) All of the above

**14-28** Liquid air is composed primarily of liquid oxygen (BP = −183°C) and liquid nitrogen (BP = −196°C). It can be purified by increasing the temperature until one of these gases boils off. Which gas boils off first?

**14-29** Use their vapor pressures at 0°C to predict which of the following liquids has the lowest boiling point.

(a) Acetone: $P = 67$ mmHg (b) Benzene: $P = 24.5$ mmHg (c) Ether: $P = 183$ mmHg (d) Methyl alcohol: $P = 30$ mmHg (e) Water: $P = 4.6$ mmHg

**14-30** According to the data in Table 14.2, butane ($C_4H_{10}$) should be a gas at room temperature (BP = −0.5°C). Use this to explain why you can hear a gas escape when a butane lighter-fluid can is opened with the nozzle pointed up. If you shake one of these cans, however, you can hear a liquid bounce against either end of the can. Furthermore, when you open the can with the nozzle pointed down, you can see a liquid escape. Explain how butane is stored as a liquid in these cans at room temperature.

### Critical Temperature and Critical Pressure

**14-31** Explain why water is a gas at 120°C and 1 atm but some of the water vapor in a closed container condenses to form a liquid at 120°C when the pressure on the vapor becomes larger than 2 atm.

**14-32** What happens to the critical temperature of a series of compounds as the force of attraction between the particles increases?

**14-33** The properties of a liquid become more like those of a gas as the temperature of the liquid increases. The properties of a gas become more like those of a liquid as the pressure on the gas increases. Use these observations to explain why it is impossible to distinguish between a liquid and a gas when the pressure on the system is equal to the critical pressure and the temperature is above the critical temperature.

### Surface Tension

**14-34** The force of cohesion between mercury atoms is much larger than the force of cohesion between water molecules. Conversely, the force of adhesion between water molecules and glass is much larger than the force of adhesion between mercury atoms and glass. Use these observations to explain why mercury forms small drops when it spills rather than a single large puddle.

**14-35** Describe how the surface tension of water can be used to explain the fact that a steel sewing needle floats on the surface of water.

**14-36** If a steel sewing needle floats on the surface of water, why does a steel ball bearing sink to the bottom?

**14-37** Explain why a drop of water seems to spread out on the surface of a car that has not been waxed for many years, but beads up on the surface of a freshly waxed car.

**14-38** Explain the advantages and disadvantages to a plant of having leaves that have a large surface area. Explain why plants coat the surface of leaves with wax. Explain why plants that grow in arid climates seldom have leaves as broad as those found on maple trees.

### Viscosity

**14-39** Gasoline molecules are very much larger than water molecules, and yet gasoline can escape through holes that are much too small to pass water. Explain why.

**14-40** Use the viscosities of water, gasoline, and air to predict the order of increasing force of adhesion between these compounds and glass.

**14-41** What aspect of the structure of water on the molecular scale makes water more viscous than gasoline, which is a mixture of hydrocarbons?

**14-42** Explain why liquids become less viscous when they are heated and more viscous when they are cooled.

### Hydrogen Bonding and the Anomalous Properties of Water

**14-43** Explain why the boiling point and melting point of water are much higher than you would expect from the boiling points and melting points of $H_2S$, $H_2Se$, and $H_2Te$.

**14-44** Explain why hydrogen bonding is very strong in HF and $H_2O$. Explain why hydrogen bonding is much weaker in HCl and $H_2S$.

**14-45** What makes a compound a good hydrogen-bond donor? What makes a compound a good hydrogen-bond acceptor?

**14-46** Explain why the strength of hydrogen bonds decreases in the following order:

$$HF > H_2O > NH_3$$

**14-47** Explain why water has an unusually large heat capacity.

### Phase Diagrams

**14-48** Explain why the *BC* line in Figure 14.14 describes the combinations of temperature and pressure at which a liquid and a gas are in equilibrium.

**14-49** Explain why the *BD* line in Figure 14.14 describes the combinations of temperature and pressure at which a solid and a liquid are in equilibrium.

**14-50** Use a phase diagram to illustrate why a liquid boils when the pressure on the liquid is reduced.

**14-51** Explain what happens as we go from left to right across a phase diagram at a constant pressure of 1 atm.

**14-52** Explain what happens as we go from top to bottom of the phase diagram of water at a constant temperature of 100°C.

**14-53** Use a phase diagram to explain why a gas condenses to form a liquid when the pressure on the gas is increased.

**14-55** Use a phase diagram to explain why most liquids become solids at temperatures near their freezing points when the pressure on the liquid is increased. Explain why ice melts at temperatures near its melting point when the pressure is increased.

**14-56** Use a phase diagram to explain why the boiling point of a liquid is relatively sensitive to changes in pressure, but the melting point is not.

### Integrated Problems

**14-57** Use the radius of an argon atom (0.174 nm), the formula for the volume of a sphere ($\frac{4}{3}\pi r^3$), and the atomic weight of argon (39.948 g/mol) to calculate the density of an argon atom in units of g/cm$^3$. Compare the results of this calculation with the density of argon as a solid, a liquid, and a gas to estimate the percent of the empty space in each phase.

**14-58** Young children sometimes believe that water decomposes into its elements when it boils. How would you explain to a bright middle school student the difference between the bonds that are broken when NaCl or diamond boils and the bonds broken when water boils? How would you explain the difference between the boiling points of NaCl (BP = 1465°C), diamond (BP = 4827°C), and water (BP = 100°C)?

**14-59** Because they are isomers, ethanol ($CH_3CH_2OH$) and dimethyl ether ($CH_3OCH_3$) have the same molecular weight. Explain why ethanol (BP = 78.5°C) has a much higher boiling point than dimethyl ether (BP = −23.6°C).

**14-60** The following compounds have the same molecular weight. Explain why one of these compounds has a higher boiling point than the other.

| $CH_3NCH_3$ (with $CH_3$ on N) | $CH_3CH_2NH$ (with $CH_3$ on N) |
| --- | --- |
| BP = 3°C | BP = 35°C |

The ocean, the rocks against which it breaks, and the sky above it are all examples of a solution.

# SOLUTIONS

Chapter 3 introduced three closely related concepts: **solutes, solvents,** and **solutions.** By definition, one or more *solutes* dissolve in a *solvent* to form a mixture known as the *solution.* This chapter provides the basis for answering questions about solutions such as the following:

## POINTS OF INTEREST

- What do nitrous oxide ($N_2O$) and halothane ($C_2HF_3ClBr$) have in common that allow these compounds to act as anesthetics?
- Why does $KMnO_4$ dissolve in water, but not in $CCl_4$? Why does $I_2$, on the other hand, dissolve in $CCl_4$ much better than it dissolves in water?
- What is the difference between "hard" water and water that has been "softened?"
- What do we mean when we say that something has been "dry-cleaned?"
- How do we determine the molecular weight of a compound? How do we know, for example, that the molecular weight of the *p*-dichlorobenzene in moth balls is 147.01 g/mol?
- How do we separate one chemical compound from another? For example, how do we decaffeinate coffee?

## 15.1 SOLUTIONS: LIKE DISSOLVES LIKE

The photographs that accompany this section illustrate what happens when we add a pair of solutes to a pair of solvents.

Solutes: $I_2$ and $KMnO_4$
Solvents: $H_2O$ and $CCl_4$

The solutes are both solids, and they both have a deep-violet or purple color. The solvents are both colorless liquids, which do not mix with each other.

The difference between the solutes is easy to understand. Iodine consists of individual $I_2$ molecules held together by relatively weak intermolecular bonds. Potassium permanganate consists of $K^+$ and $MnO_4^-$ ions held together by the strong force of attraction between ions of opposite charge. It is therefore much easier to separate the $I_2$ molecules in iodine than it is to separate the ions in $KMnO_4$.

There is also a significant difference between the solvents $CCl_4$ and $H_2O$. The difference between the electronegativities of the carbon and chlorine atoms in $CCl_4$ is so small ($\Delta EN = 0.56$) that there is relatively little ionic character in the C—Cl bonds. Even if there were some separation of charge in these bonds, the $CCl_4$ molecule wouldn't be polar because it has a symmetrical shape in which the four chlorine atoms point toward the corners of a tetrahedron, as shown in Figure 15.1. $CCl_4$ is therefore best described as a **nonpolar solvent.**

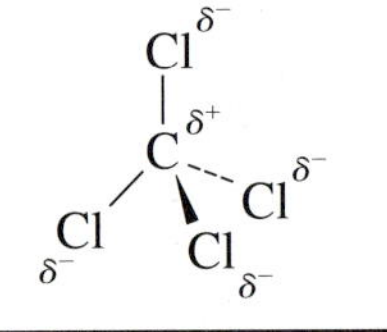


**Figure 15.1** Although there is some separation of charge within the individual bonds in $CCl_4$, the symmetrical shape of the $CCl_4$ molecule ensures that there is no net dipole moment. $CCl_4$ is therefore a *nonpolar solvent.*

The difference between the electronegativities of the hydrogen and oxygen atoms in water is relatively large ($\Delta EN = 1.24$), which means that there is much more charge separation and the H—O bonds in this molecule are therefore polar. If the $H_2O$ molecule were linear, the polarity of the two O—H bonds would cancel, and the molecule would have no net dipole moment. Water molecules are not linear, however; they have a bent, or angular shape. As a result, water molecules have distinct positive and negative poles, and water is a polar molecule, as shown in Figure 15.2. Water is therefore classified as a **polar solvent.**

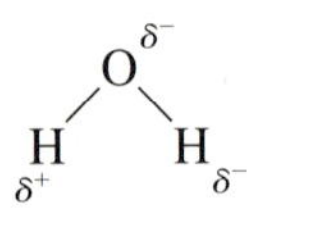


**Figure 15.2** Because water molecules are bent, or angular, they have distinct negative and positive poles. $H_2O$ is therefore an example of a *polar solvent.*

Sec. 8.15

Because the solvents do not mix, when water and carbon tetrachloride are added to a separatory funnel, two separate liquid phases are clearly visible. We can use the relative densities of $CCl_4$ (1.594 g/cm$^3$) and $H_2O$ (1.0 g/cm$^3$) to decide which phase is water and which is carbon tetrachloride. The denser $CCl_4$ settles to the bottom of the funnel.

When a few crystals of iodine are added to the separatory funnel and the contents of the funnel are shaken, the $I_2$ dissolves in the $CCl_4$ layer to form a violet colored solution. The water layer stays essentially colorless, which suggests that little if any $I_2$ dissolves in water.

When this experiment is repeated with potassium permanganate, the water layer picks up the characteristic purple color of the $MnO_4^-$ ion, and the $CCl_4$ layer remains colorless. This suggests that $KMnO_4$ dissolves in water but not in carbon tetrachloride. The results of this experiment are summarized in Table 15.1.

Why does $KMnO_4$ dissolve in water, but not carbon tetrachloride? Why does $I_2$ dissolve in carbon tetrachloride, but not water?

**TABLE 15.1 Solubilities of $I_2$ and $KMnO_4$ in $CCl_4$ and Water**

| | $H_2O$ | $CCl_4$ |
|---|---|---|
| $I_2$ | Insoluble | Very soluble |
| $KMnO_4$ | Very soluble | Insoluble |

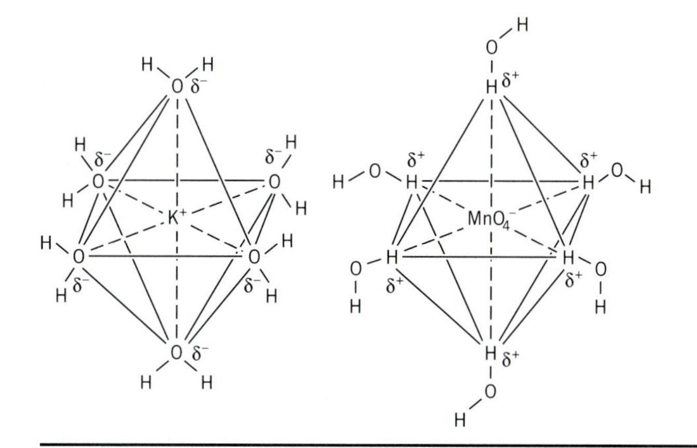


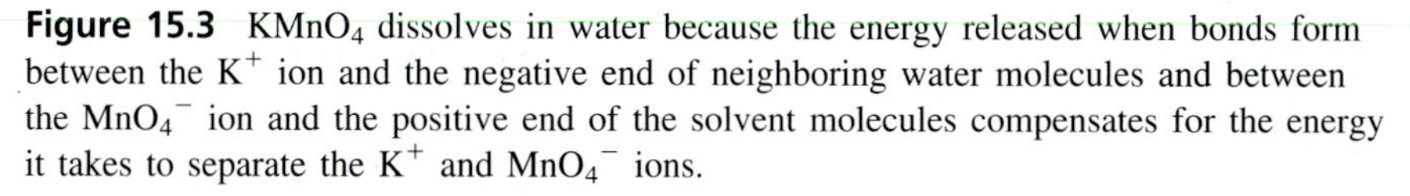

**Figure 15.3** $KMnO_4$ dissolves in water because the energy released when bonds form between the $K^+$ ion and the negative end of neighboring water molecules and between the $MnO_4^-$ ion and the positive end of the solvent molecules compensates for the energy it takes to separate the $K^+$ and $MnO_4^-$ ions.

Water and carbon tetrachloride form two separate liquid phases in a separatory funnel (center); $KMnO_4$ dissolves in the water layer on the top of the separatory funnel to form an intensely colored solution (left); $I_2$ dissolves in the $CCl_4$ layer on the bottom of the separatory funnel to form an intensely colored solution (right).

It takes a lot of energy to separate the $K^+$ and $MnO_4^-$ ions in potassium permanganate. But these ions can form weak bonds with neighboring water molecules, as shown in Figure 15.3. The energy released when these bonds form compensates for the energy that must be invested to rip apart the $KMnO_4$ crystal. No such bonds can form between the $K^+$ or $MnO_4^-$ ions and the nonpolar $CCl_4$ molecules. As a result, $KMnO_4$ can't dissolve in $CCl_4$.

The $I_2$ molecules in iodine and the $CCl_4$ molecules in carbon tetrachloride are both held together by weak intermolecular bonds. Similar intermolecular bonds can form between $I_2$ and $CCl_4$ molecules in a solution. $I_2$ therefore readily dissolves in $CCl_4$. The molecules in water are held together by hydrogen bonds that are stronger than most intermolecular bonds. No interaction between $I_2$ and $H_2O$ molecules is strong enough to compensate for the hydrogen bonds that have to be broken to dissolve iodine in water, so relatively little $I_2$ dissolves in $H_2O$.

We can summarize the results of this experiment by noting that *nonpolar solutes* (such as $I_2$) dissolve in *nonpolar solvents* (such as $CCl_4$), whereas *polar solutes* (such as $KMnO_4$) dissolve in *polar solvents* (such as $H_2O$). As a general rule, we can conclude that:

**Like dissolves like.**

## EXERCISE 15.1

Section 10.6 noted that elemental phosphorus is often stored under water because it doesn't dissolve in water. Elemental phosphorus is very soluble in carbon disulfide, however. Explain why $P_4$ is soluble in $CS_2$ but not in water.

Sec. 10.6

**SOLUTION** The structure of the $P_4$ molecule was shown in Figure 10.22. This molecule is a perfect example of a nonpolar solute. It is therefore more likely to be soluble in nonpolar solvents than in polar solvents such as water.

The Lewis structure of $CS_2$ suggests that this molecule is linear:

$$\ddot{S}=C=\ddot{S}$$

Thus, even if there were some separation of charge in the C=S double bond, the molecule would have no net dipole moment because of its symmetry. The electronegativities of carbon ($EN$ = 2.55) and sulfur ($EN$ = 2.58), however, suggest that the C=S double bonds are almost perfectly covalent. $CS_2$ is therefore a nonpolar solvent, which should readily dissolve $P_4$.

## EXERCISE 15.2

The iodide ion reacts with iodine in aqueous solution to form the $I_3^-$, or triiodide, ion.

$$I^-(aq) + I_2(aq) \longrightarrow I_3^-(aq)$$

What would happen if $CCl_4$ were added to an aqueous solution that contained a mixture of KI, $I_2$, and $KI_3$?

**SOLUTION** KI and $KI_3$ are both salts. One contains the $K^+$ and $I^-$ ions; the other contains the $K^+$ and $I_3^-$ ions. These salts are more soluble in polar solvents, such as water, than in nonpolar solvents, such as $CCl_4$. KI and $KI_3$ would therefore remain in the aqueous solution. $I_2$ is a nonpolar molecule, which is more soluble in a nonpolar solvent, such as carbon tetrachloride. The $I_2$ would therefore leave the aqueous layer and enter the $CCl_4$ layer, where it would exhibit the characteristic violet color of solutions of molecular iodine.

## 15.2 HYDROPHILIC AND HYDROPHOBIC MOLECULES

Section 14.2 described some of the physical properties of the family of compounds known as the *hydrocarbons,* which contain only carbon and hydrogen. Because the difference between the electronegativities of carbon and hydrogen is small ($\Delta EN$ = 0.40), hydrocarbons are nonpolar. As a result, they do not dissolve in polar solvents such as water. Hydrocarbons are therefore described as **immiscible** (literally, "not mixable") in water.

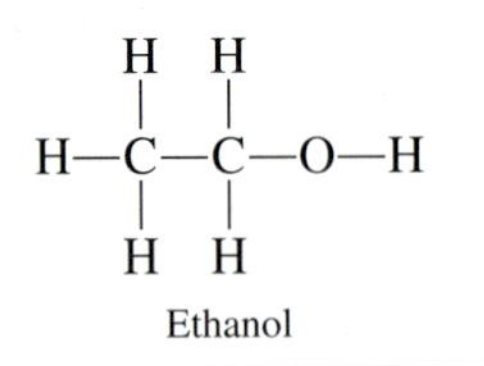


**Figure 15.4** The structure of the alcohol known as ethanol.

When one of the hydrogen atoms in a hydrocarbon is replaced with an —OH group, the compound is known as an **alcohol,** as shown in Figure 15.4. As might be expected, alcohols have properties between the extremes of hydrocarbons and water. When the hydrocarbon chain is short, the alcohol is soluble in water. Methanol ($CH_3OH$) and ethanol ($CH_3CH_2OH$) are infinitely soluble in water, for example. There is no limit on the amount of these alcohols that can dissolve in a given quantity of water. The alcohol in beer, wine, and hard liquors is ethanol, and mixtures of ethanol and water can have any concentration between the extremes of pure alcohol (200 proof) and pure water (0 proof).

As the hydrocarbon chain becomes longer, the alcohol becomes less soluble in water, as shown in Table 15.2. One end of the alcohol molecules has so much

**TABLE 15.2 Solubilities of Alcohols in Water**

| Formula | Name | Solubility in Water (g/100 g) |
|---|---|---|
| $CH_3OH$ | methanol | infinitely soluble |
| $CH_3CH_2OH$ | ethanol | infinitely soluble |
| $CH_3(CH_2)_2OH$ | propanol | infinitely soluble |
| $CH_3(CH_2)_3OH$ | butanol | 9 |
| $CH_3(CH_2)_4OH$ | pentanol | 2.7 |
| $CH_3(CH_2)_5OH$ | hexanol | 0.6 |
| $CH_3(CH_2)_6OH$ | heptanol | 0.18 |
| $CH_3(CH_2)_7OH$ | octanol | 0.054 |
| $CH_3(CH_2)_9OH$ | decanol | insoluble in water |

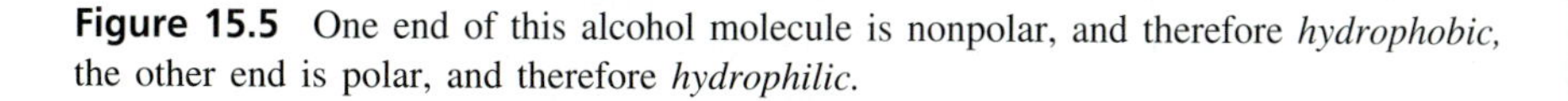


**Figure 15.5** One end of this alcohol molecule is nonpolar, and therefore *hydrophobic,* the other end is polar, and therefore *hydrophilic.*

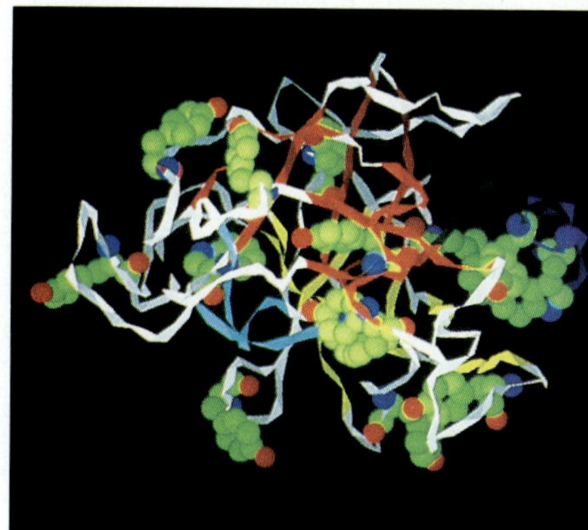

A computer graphics model of the structure of the protein TRYPSIN. The structure of a protein is influenced by the tendency to keep hydrophobic side chains towards the center and hydrophylic side chains pointing toward the water that surrounds the protein.

nonpolar character it is called **hydrophobic** (literally, ''water hating''), as shown in Figure 15.5. The other end contains an —OH group that can form hydrogen bonds to neighboring water molecules and is therefore said to be **hydrophilic** (literally, ''water loving''). As the hydrocarbon chain becomes longer, the hydrophobic character of the molecule increases, and the solubility of the alcohol in water gradually decreases until it becomes essentially insoluble in water.

It is sometimes difficult, encountering the terms *hydrophilic* and *hydrophobic* for the first time to remember which stands for water hating and which stands for water loving. If you remember that Hamlet's girlfriend was named Ophelia (not Ophobia), it will be easier to remember that the prefix *philo-* is commonly used to describe love—for example, in *philanthropist, philharmonic, philosopher,* and so on.

### CHECKPOINT

Amino acids are classified as either hydrophilic or hydrophobic on the basis of their side chains. Use the structure of the side chains of the following amino acids to justify the classifications shown below.

| | *Amino Acid* | *Side Chain* |
|---|---|---|
| **Hydrophobic:** | alanine | $—CH_3$ |
| | cysteine | $—CH_2CH_2SCH_3$ |
| **Hydrophilic:** | lysine | $—CH_2CH_2CH_2CH_2NH_3^+$ |
| | serine | $—CH_2OH$ |

The data in Table 15.2 show one consequence of the general rule that like dissolves like. As molecules become more nonpolar, they become less soluble in water. Table 15.3 shows another example of this rule. NaCl is relatively soluble in water. As the solvent becomes more nonpolar, the solubility of this polar solute decreases.

**TABLE 15.3 Solubility of Sodium Chloride in Water and in Alcohols**

| Formula of Solvent | Solvent Name | Solubility of NaCl (g/100 g solvent) |
|---|---|---|
| $H_2O$ | water | 35.92 |
| $CH_3OH$ | methanol | 1.40 |
| $CH_3CH_2OH$ | ethanol | 0.065 |
| $CH_3(CH_2)_2OH$ | propanol | 0.012 |
| $CH_3(CH_2)_3OH$ | butanol | 0.005 |
| $CH_3(CH_2)_4OH$ | pentanol | 0.0018 |

## 15.3 SOAPS, DETERGENTS, AND DRY-CLEANING AGENTS

The chemistry behind the manufacture of soap hasn't changed since it was made from animal fat and the ash from wood fires almost 5000 years ago. Solid animal fats (such as the tallow obtained from the butchering of sheep or cattle) and liquid plant oils (such as palm oil and coconut oil) are still heated in the presence of a strong base to form a soft, waxy material that enhances the ability of water to wash away the grease and oil that forms on our bodies and our clothes.

Animal fats and plant oils contain compounds known as *fatty acids.* Fatty acids, such as stearic acid (see Figure 15.6), have small, polar, hydrophilic heads attached

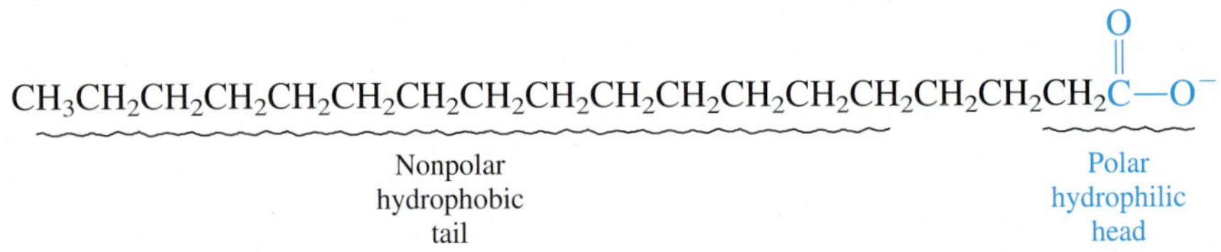

$$\underbrace{CH_3CH_2CH_2CH_2CH_2CH_2CH_2CH_2CH_2CH_2CH_2CH_2CH_2CH_2CH_2CH_2}_{\text{Nonpolar hydrophobic tail}}\underbrace{CH_2\overset{\overset{\displaystyle O}{\|}}{C}{-}O^-}_{\text{Polar hydrophilic head}}$$

**Figure 15.6** The hydrocarbon chain on one end of a fatty acid molecule is nonpolar and hydrophobic, whereas the $—CO_2H$ group on the other end of the molecule is polar and hydrophilic.

to long, nonpolar, hydrophobic tails. Fatty acids are seldom found by themselves in nature. They are usually bound to molecules of glycerol ($HOCH_2CHOHCH_2OH$) to form triglycerides, such as the one shown in Figure 15.7. These triglycerides break

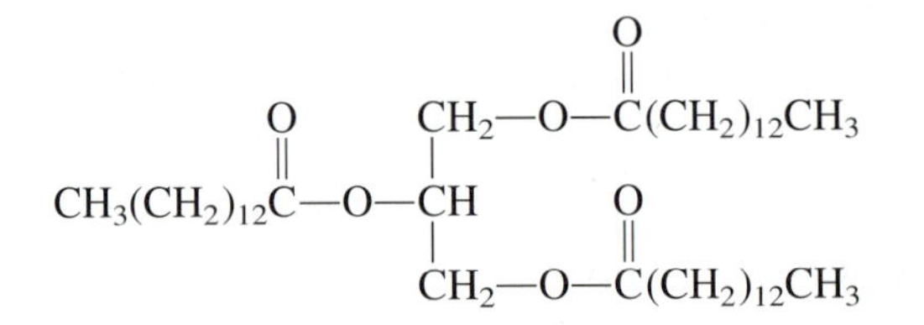

**Figure 15.7** The structure of the triglyceride known as *trimyristin,* which can be isolated in high yield from nutmeg.

down in the presence of a strong base to form the $Na^+$ or $K^+$ salt of the fatty acid, as shown in Figure 15.8. This reaction is called *saponification,* which literally means "the making of soap."

Part of the cleaning action of soap results from the fact that soap molecules are *surfactants*—they tend to concentrate on the surface of water. They cling to the

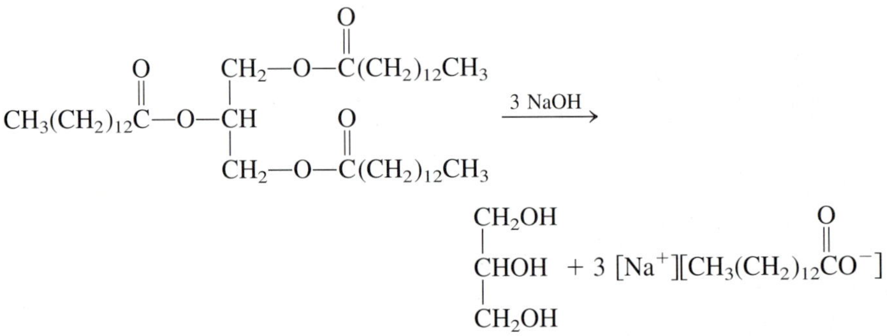

$$\begin{array}{l} \qquad\qquad\qquad\quad CH_2{-}O{-}\overset{\overset{O}{\|}}{C}(CH_2)_{12}CH_3 \\ CH_3(CH_2)_{12}\overset{\overset{O}{\|}}{C}{-}O{-}CH \\ \qquad\qquad\qquad\quad CH_2{-}O{-}\overset{\overset{O}{\|}}{C}(CH_2)_{12}CH_3 \end{array} \xrightarrow{3\ NaOH}$$

$$\begin{array}{l} CH_2OH \\ | \\ CHOH \\ | \\ CH_2OH \end{array} + 3\ [Na^+][CH_3(CH_2)_{12}\overset{\overset{O}{\|}}{C}O^-]$$

**Figure 15.8** The saponification of the trimyristin extracted from nutmeg.

surface because they try to orient their polar $CO_2^-$ heads toward water molecules and their nonpolar $CH_3CH_2CH_2 \ldots$ tails away from neighboring water molecules.

Water can't wash the soil out of clothes by itself because the soil particles that cling to textile fibers are covered by a layer of nonpolar grease or oil molecules that repels water. The nonpolar tails of the soap molecules on the surface of water dissolve in the grease or oil that surrounds a soil particle, as shown in Figure 15.9. The soap molecules therefore disperse, or *emulsify,* the soil particles, which makes it possible to wash these particles out of the clothes.

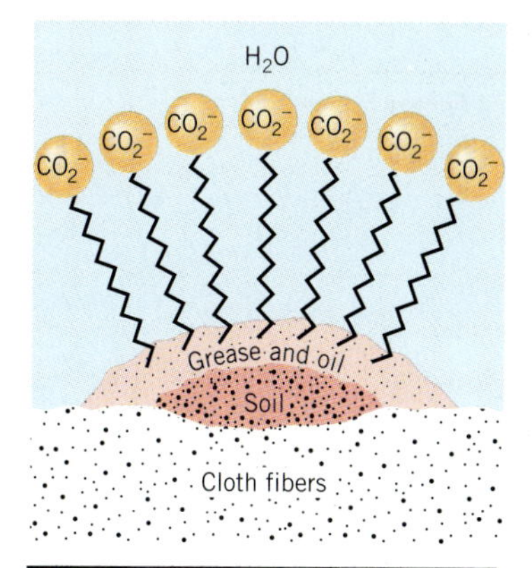


**Figure 15.9** Soap molecules disperse, or emulsify, soil particles coated with a layer of nonpolar grease or oil molecules.

Most soaps are more dense than water. They can be made to float, however, by incorporating air into the soap during its manufacture. Most soaps are also opaque; they absorb rather than transmit light. Translucent soaps can be made by adding alcohol, sugar, and glycerol, which slow down the growth of soap crystals while the soap solidifies. Liquid soaps are made by replacing the sodium salts of the fatty acids with the more soluble $K^+$ or $NH_4^+$ salts.

A translucent soap.

Forty years ago, more than 90% of the cleaning agents sold in the United States were soaps. Today soap represents less than 20% of the market for cleaning agents. The primary reason for the decline in the popularity of soap is the reaction between soap and ''hard'' water. The most abundant positive ions in tap water are $Na^+$, $Ca^{2+}$, and $Mg^{2+}$ ions. Water that is particularly rich in $Ca^{2+}$, $Mg^{2+}$, or $Fe^{3+}$ ions is said to be hard. Hard water interferes with the action of soap because these ions combine with soap molecules to form insoluble precipitates that have no cleaning power. These salts not only decrease the concentration of the soap molecules in solution, they actually bind soil particles to clothing, leaving a dull, gray film.

One way around this problem is to ''soften'' the water by replacing the $Ca^{2+}$ and $Mg^{2+}$ ions with $Na^+$ ions. Many water softeners are filled with a resin that contains $—SO_3^-$ ions attached to a polymer, as shown in Figure 15.10. The resin is treated with NaCl until each $—SO_3^-$ ion picks up an $Na^+$ ion. When hard water flows over this resin, $Ca^{2+}$ and $Mg^{2+}$ ions bind to the $—SO_3^-$ ions on the polymer chain and $Na^+$ ions are released into solution. Periodically, the resin becomes saturated with $Ca^{2+}$ and $Mg^{2+}$ ions. When this happens, it has to be regenerated by being washed with a concentrated solution of NaCl.

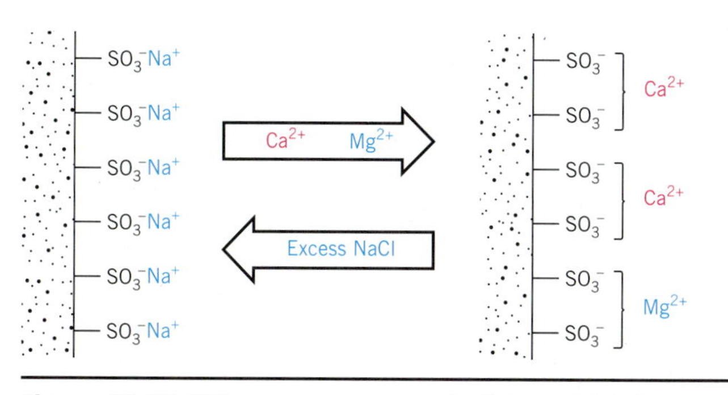


**Figure 15.10** When a water softener is ''charged,'' it is washed with a concentrated NaCl solution until all of the $—SO_3^-$ ions pick up an $Na^+$ ion. The softener then picks up $Ca^{2+}$ and $Mg^{2+}$ ions from hard water, replacing these ions with $Na^+$ ions.

There is another way around the problem of hard water. Instead of removing $Ca^{2+}$ and $Mg^{2+}$ ions from water, we can find a cleaning agent that doesn't form insoluble salts with these ions. Synthetic detergents are examples of such cleaning agents. Detergent molecules consist of long, hydrophobic hydrocarbon tails attached to polar, hydrophilic $—SO_3^-$ or $—OSO_3^-$ heads, as shown in Figure 15.11. By

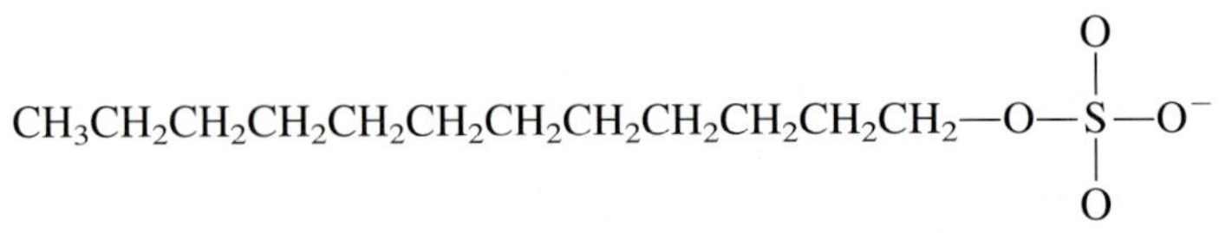

$$CH_3CH_2CH_2CH_2CH_2CH_2CH_2CH_2CH_2CH_2CH_2CH_2{-}O{-}\overset{\overset{\displaystyle O}{|}}{\underset{\underset{\displaystyle O}{|}}{S}}{-}O^-$$

**Figure 15.11** The structure of one of the components of a synthetic detergent.

themselves, detergents don't have the cleaning power of soap. ''Builders'' are therefore added to synthetic detergents to increase their strength. These builders are often salts of highly charged ions, such as the triphosphate ($P_3O_{10}{}^{5-}$) ion.

Cloth fibers swell when they are washed in water. This leads to changes in the dimensions of the cloth that can cause wrinkles—which are local distortions in the structure of the fiber—or even more serious damage, such as shrinking. These problems can be avoided by ''dry cleaning,'' which uses a nonpolar solvent that does not adhere to, or wet, the cloth fibers. The nonpolar solvents used in dry cleaning dissolve the nonpolar grease or oil layer that coats soil particles, freeing the soil particles to be removed by detergents added to the solvent, or by the tumbling action inside the machine. Dry cleaning has the added advantage that it can remove oily soil at lower temperatures than soap or detergent dissolved in water, so it is safer for delicate fabrics.

Sec. 14.7

When dry cleaning was first introduced in the United States between 1910 and 1920, the solvent was a mixture of hydrocarbons isolated from petroleum when gasoline was refined. Over the years, these flammable hydrocarbon solvents have been replaced by halogenated hydrocarbons, such as trichloroethane ($Cl_3C{-}CH_3$), trichloroethylene ($Cl_2C{=}CHCl$), and perchloroethylene ($Cl_2C{=}CCl_2$).

## 15.4 UNITS OF CONCENTRATION: MOLARITY, MOLALITY, AND MOLE FRACTION

The **concentration** of a solution is defined as the amount of solute dissolved in a given amount of solvent or solution:

$$\textbf{Concentration} = \frac{\textbf{amount of solute}}{\textbf{amount of solvent or solution}}$$

Sec. 3.17

There are many ways in which the concentration of a solution can be described. So far we have focused on two: *molarity* and *mass percent.*

The **molarity** ($M$) of a solution is defined as the ratio of the number of moles of solute in the solution divided by the volume of the solution in liters:

$$M = \frac{\textbf{moles of solute}}{\textbf{liters of solution}}$$

**EXERCISE 15.3**

At 25°C, a saturated solution of chlorine in water can be prepared by dissolving 5.77 grams of $Cl_2$ gas in enough water to give a liter of solution. Calculate the molarity of this solution.

**SOLUTION** The key to solving this problem is recognizing that we need to simultaneously know the number of moles of solute in a solution and the volume of the solution before we can calculate the molarity of the solution. The number of moles of solute in this solution is easy to obtain:

$$5.77 \text{ g } Cl_2 \times \frac{1 \text{ mol } Cl_2}{70.91 \text{ g } Cl_2} = \mathbf{0.0814 \text{ mol } Cl_2}$$

We now know the number of moles of solute (0.0814 mol $Cl_2$) and the volume of the solution (1 L). The concentration of $Cl_2$ in this solution is therefore 0.0814 moles per liter, or 0.0814 *molar.*

$$\frac{0.0814 \text{ mol } Cl_2}{1 \text{ L}} = 0.0814 \, M$$

A saturated solution of $Cl_2$ in water can be prepared by bubbling $Cl_2$ gas through a fritted funnel into water.

*Mass percent* is literally the percentage of the total mass of solute in a solution:

$$\textbf{Mass percent} = \frac{\textbf{mass of solute}}{\textbf{mass of solution}} \times 100\%$$

A 3.5% solution of hydrochloric acid, for example, has 3.5 g of HCl in every 100 g of solution. In Exercise 3.14 we saw how the concentration of a solution in units of moles per liter can be calculated from the mass percent and density of the solution.

It is also possible to describe the concentration of a solution in terms of the *volume percent.* This unit is used to describe solutions of one liquid dissolved in another or mixtures of gases:

$$\textbf{Volume percent} = \frac{\textbf{volume of solute}}{\textbf{volume of solution}} \times 100\%$$

Wine labels, for example, describe the alcoholic content as 12% by volume because 12% of the total volume is alcohol.

Molarity is the concentration unit most commonly used by chemists. It has one disadvantage, however. It tells us how much *solute* we need to make a solution, and it gives us the volume of the *solution* produced, but it doesn't tell us how much *solvent* will be required to prepare the solution. We make a 0.100 *M* solution of $CuSO_4$, for example, not by dissolving the $CuSO_4$ in a liter of water, but by dissolving 0.100 mol of $CuSO_4 \cdot 5\,H_2O$ in enough water to give 1 L of solution. But how much water is enough? Because the $CuSO_4 \cdot 5\,H_2O$ crystals occupy some volume, it takes less than a liter of water, but we have no idea how much less.

When it is important to know exactly how much solute and solvent are present in a solution, chemists use two other concentration units: *molality* and *mole fraction.* The **molality** (*m*) of a solution is defined as the number of moles of solute in the solution divided by the mass in kilograms of the solvent used to make the solution:

$$\textbf{Molality } (m) = \frac{\textbf{moles of solute}}{\textbf{kilograms of solvent}}$$

A 0.100 *m* solution of $CuSO_4$, for example, can be prepared by dissolving 0.100 mol of $CuSO_4$ in 1 kg of water. Because the density of water is about 1 g/$cm^3$, or 1 g/mL, the volume of water used to prepare this solution will be approximately 1 L. The total volume of the solution, however, will be larger than 1 L because the $CuSO_4 \cdot 5\,H_2O$ crystals will undoubtedly occupy some volume. As a result, a 0.100 *m* solution is slightly more dilute than a 0.100 *M* solution of the same solute.

## EXERCISE 15.4

A saturated solution of hydrogen sulfide in water can be prepared by bubbling $H_2S$ gas into water until no more dissolves. Calculate the molality of this solution if 0.385 grams of $H_2S$ gas dissolve in 100 grams of water at 20°C and 1 atm.

**Problem-Solving Strategy**

**SOLUTION** The key to solving this problem is recognizing that we can only calculate the molality of a solution if we simultaneously know the number of moles of solute and the number of kilograms of solvent used to prepare the solution.

The number of moles of solute in this solution is easy to obtain:

$$0.385 \text{ g } H_2S \times \frac{1 \text{ mol } H_2S}{34.08 \text{ g } H_2S} = 0.0113 \text{ mol } H_2S$$

Now all we have to do is divide the number of moles of solute by the mass of the solvent in kilograms to find that this solution is 0.113 molal:

$$\frac{0.0113 \text{ mol } H_2S}{0.100 \text{ kg } H_2O} = \mathbf{0.113}\ \boldsymbol{m}$$

Molality has an important advantage over molarity. The molarity of an aqueous solution changes with temperature because the density of water is sensitive to temperature. Because molality is defined in terms of the mass of the solvent, not its volume, the molality of a solution does not change with temperature.

The ratio of solute to solvent in a solution can be described also in terms of the mole fraction of the solute or the solvent in a solution. By definition, the **mole fraction** of any component of a solution is the fraction of the total number of moles of solute and solvent that come from that component. The symbol for mole fraction is a Greek lower case letter chi: $\chi$. The mole fraction of the *solute* is defined as the number of moles of solute divided by the total number of moles of solute and solvent:

$$\chi_{\text{solute}} = \frac{\textbf{moles of solute}}{\textbf{moles of solute + moles of solvent}}$$

Conversely, the mole fraction of the *solvent* is the number of moles of solvent divided by the total number of moles of solute and solvent:

$$\chi_{\text{solvent}} = \frac{\textbf{moles of solvent}}{\textbf{moles of solute + moles of solvent}}$$

**Problem-Solving Strategy**

In a solution that contains a single solute dissolved in a solvent, the sum of the mole fraction of the solute and the solvent must be equal to 1:

$$\chi_{\text{solute}} + \chi_{\text{solvent}} = 1$$

**EXERCISE 15.5**

Calculate the mole fractions of both the solute and the solvent in a saturated solution of hydrogen sulfide in water at 20°C and 1 atm.

**SOLUTION** In the previous exercise, we calculated the number of moles of $H_2S$ in this solution.

$$0.385 \text{ g } H_2S \times \frac{1 \text{ mol } H_2S}{34.08 \text{ g } H_2S} = 0.0113 \text{ mol } H_2S$$

To determine the mole fraction of the solute and solvent, we also need to know the number of moles of water in this solution.

$$100 \text{ g } H_2O \times \frac{1 \text{ mol } H_2O}{18.02 \text{ g } H_2O} = 5.55 \text{ mol } H_2O$$

The mole fraction of the solute is the number of moles of $H_2S$ divided by the total number of moles of both $H_2S$ and $H_2O$.

$$\chi_{\text{solute}} = \frac{(0.0113 \text{ mol } H_2S)}{(0.0113 \text{ mol } H_2S + 5.55 \text{ mol } H_2O)} = 0.002$$

The mole fraction of the solvent is the number of moles of $H_2O$ divided by the moles of both $H_2S$ and $H_2O$.

$$\chi_{\text{solvent}} = \frac{(5.55 \text{ mol } H_2O)}{(0.0113 \text{ mol } H_2S + 5.55 \text{ mol } H_2O)} = 0.998$$

Note that the sum of the mole fractions of the two components of this solution is 1.

$$\chi_{\text{solute}} + \chi_{\text{solvent}} = 0.002 + 0.998 = 1$$

## 15.5 COLLIGATIVE PROPERTIES: VAPOR PRESSURE DEPRESSION

Ever since Chapter 1, we have divided physical properties into two categories. *Extensive properties* (such as mass and volume) depend on the size of the sample. *Intensive properties* (such as density and concentration) are characteristic properties of the substance; they do not depend on the size of the sample being studied. This section introduces a third category that is a subset of the intensive properties of a system. This third category, known as **colligative properties,** only applies to solutions. By definition, one of the properties of a solution is a *colligative property* if it depends only on the ratio of the number of particles of solute and solvent in the solution, not the identity of the solute.

Sec. 1.13

Very few of the physical properties of a solution are colligative properties. As our first example of this limited set of physical properties, let's consider what happens to the **vapor pressure** of the solvent when we add a solute to form a solution. We'll define $P^o$ as the vapor pressure of the pure liquid—the solvent—and $P$ as the vapor pressure of the solvent escaping from the solution after a solute has been added:

$P^o$ = vapor pressure of the pure solvent

$P$ = vapor pressure of the solvent in a solution

### EXERCISE 15.6

When the temperature of a liquid is below its boiling point, we can assume that the only molecules that can escape from the liquid to form a gas are those that lie near the surface of the liquid. Use this assumption to predict whether the vapor pressure of a solvent should increase, decrease, or remain the same when a solute is added to the solvent.

**SOLUTION** When a solute is added to the solvent, some of the solute molecules occupy the space near the surface of the liquid, as shown in Figure 15.12. This has no effect on the rate at which solvent molecules in the gas phase condense to form a liquid. But it decreases the rate at which the solvent molecules in the liquid can escape into the gas phase. As a result, the vapor pressure of the solvent escaping from a solution should be smaller than the vapor pressure of the pure solvent:

$$P < P^o$$

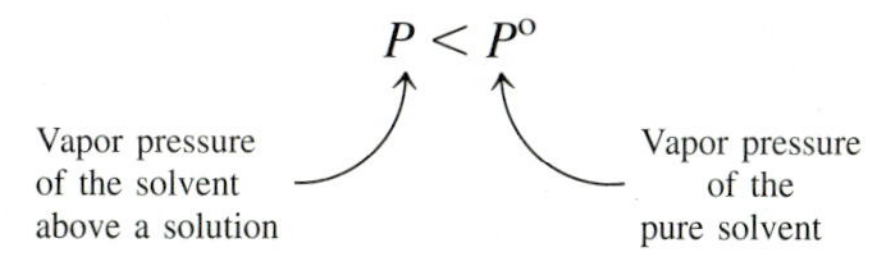


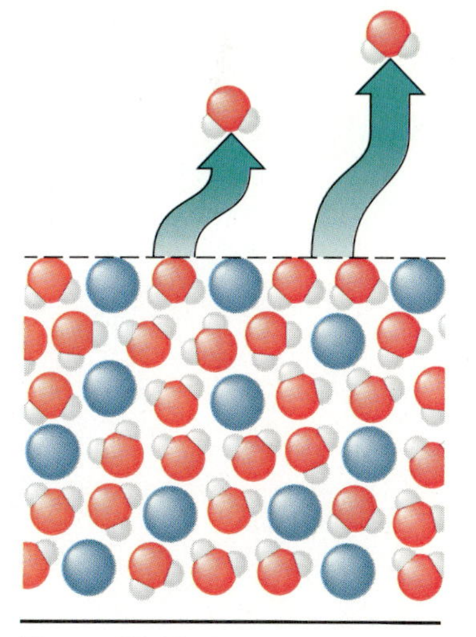

**Figure 15.12** When a solute is dissolved in a solvent, the number of solvent molecules near the surface decreases, and the vapor pressure of the solvent decreases.

Between 1887 and 1888, Francois-Marie Raoult showed that the vapor pressure of the solvent escaping from a solution is equal to the mole fraction of the solvent times the vapor pressure of the pure liquid:

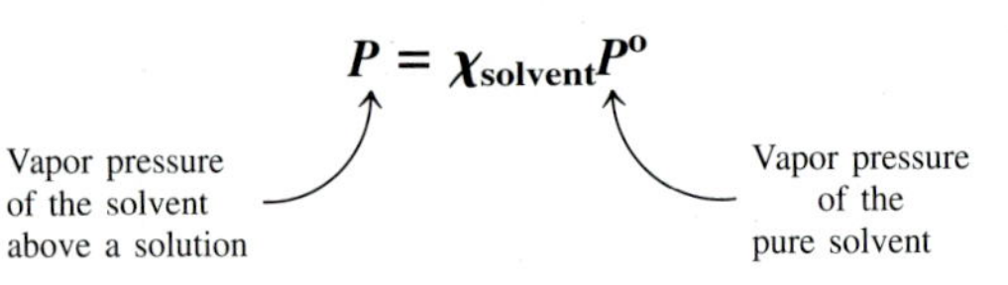


This equation, which is known as **Raoult's law,** is easy to understand. When the solvent is pure, and the mole fraction of the solvent is equal to 1, $P$ is equal to $P^o$. As the mole fraction of the solvent becomes smaller, the vapor pressure of the solvent escaping from the solution also becomes smaller.

Let's assume, for the moment, that the solvent is the only component of the solution that is volatile enough to have a measureable vapor pressure at room temperature. If this is true, the vapor pressure of the solution will be equal to the vapor pressure of the solvent escaping from the solution. Raoult's law suggests that the difference between the vapor pressure of the pure solvent and the solution increases as the mole fraction of the solvent decreases.

The *change in the vapor pressure* that occurs when a solute is added to a solvent is therefore a colligative property. If it depends on the mole fraction of the solute, then it must depend on the ratio of the number of particles of solute to solvent in the solution but not the identity of the solute.

Calculations for solutions that contain a volatile solute are slightly more complex. Consider, for example, a solution prepared by mixing 500 mL of ethanol (EtOH) and 500 mL of water at 25°C. The vapor pressure of pure water at this temperature is 23.76 mmHg and the vapor pressure of ethanol at this temperature is 59.76 mmHg. In order to predict the vapor pressure of the solution we need to know the mole fraction of both components. We start by calculating the number of moles of both ethanol and water in this solution from the densities of the two substances and their molecular weights.

$$500\,\cancel{\text{mL}}\ H_2O \times \frac{0.9971\,\cancel{\text{g}}\ H_2O}{1\,\cancel{\text{mL}}\ H_2O} \times \frac{1\text{ mol } H_2O}{18.02\,\cancel{\text{g}}\ H_2O} = 27.67\text{ mol } H_2O$$

$$500\,\cancel{\text{mL}}\ \text{EtOH} \times \frac{0.786\,\cancel{\text{g}}\ \text{EtOH}}{1\,\cancel{\text{mL}}\ \text{EtOH}} \times \frac{1\text{ mol EtOH}}{46.06\,\cancel{\text{g}}\ \text{EtOH}} = 8.532\text{ mol EtOH}$$

We then calculate the mole fractions of the two components of this solution.

$$\chi_{\text{EtOH}} = \frac{8.532\text{ mol}}{(27.67\text{ mol} + 8.532\text{ mol})} = 0.2356$$

$$\chi_{H_2O} = \frac{27.67\text{ mol}}{(27.67\text{ mol} + 8.532\text{ mol})} = 0.7644$$

According to Raoult's law, the partial pressure of the water escaping from this solution is equal to the product of the mole fraction of water and the vapor pressure of pure water.

$$P_{H_2O} = \chi_{H_2O}P^o{}_{H_2O} = (0.7644)(23.76\text{ mmHg}) = 18.16\text{ mmHg}$$

The partial pressure of the ethanol is equal to the mole fraction of this component times the vapor pressure of the pure alcohol.

$$P_{\text{EtOH}} = \chi_{\text{EtOH}}P^o{}_{\text{EtOH}} = (0.2356)(59.76\text{ mmHg}) = 14.08\text{ mmHg}$$

The total pressure of the gases escaping from the solution is the sum of the partial pressures of these two gases.

$$P_T = P_{H_2O} + P_{EtOH} = 32.24 \text{ mmHg}$$

Although the mixture is 50:50 by volume, slightly more than three-quarters of the particles in this solution are water molecules. As a result, the total pressure of this solution more closely resembles the vapor pressure of pure water than it does pure ethyl alcohol.

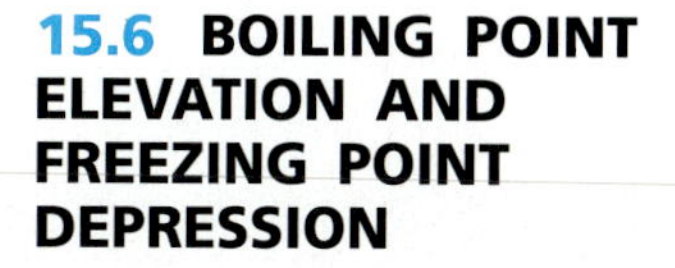

## 15.6 BOILING POINT ELEVATION AND FREEZING POINT DEPRESSION

Figure 15.13 shows the consequences of the fact that solutes lower the vapor pressure of a solvent. The solid line connecting points *B* and *C* in this phase diagram contains the combinations of temperature and pressure at which the pure solvent and its vapor are in equilibrium. Each point on this line therefore describes the vapor pressure of the pure solvent at that temperature. The dotted line in this figure describes the properties of a solution obtained by dissolving a solute in the solvent. At any given temperature, the vapor pressure of the solvent escaping from the solution is smaller than the vapor pressure of the pure solvent. The dotted line therefore lies below the solid line.

According to this figure, the solution can't boil at the same temperature as the pure solvent. If the vapor pressure of the solvent escaping from the solution is smaller than the vapor pressure of the pure solvent at any given temperature, the solution must be heated to a higher temperature before it boils. The lowering of the vapor pressure of the solvent that occurs when it is used to form a solution therefore increases the boiling point of the liquid.

Sec. 14.10

When phase diagrams were introduced in Chapter 14, the triple point was defined as the only combination of temperature and pressure at which the gas, liquid, and solid could exist at the same time. Figure 15.13 shows that the triple point of the

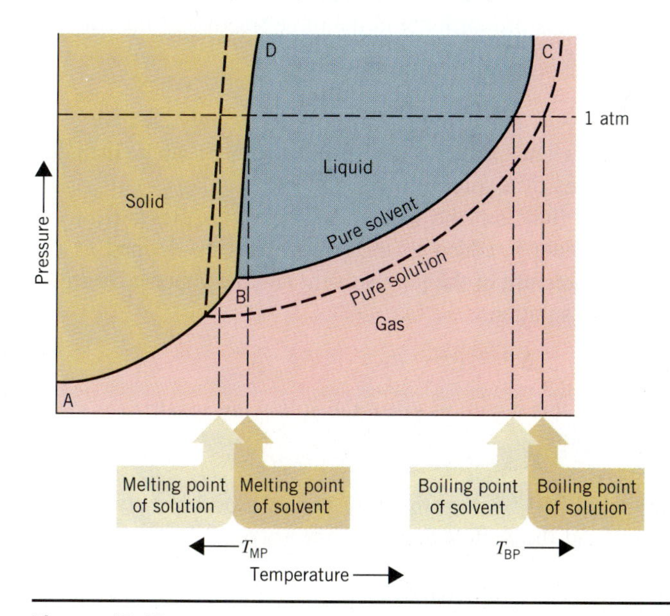


**Figure 15.13** The decrease in the vapor pressure of the solvent that occurs when a solute is added to the solvent causes an increase in the boiling point and a decrease in the melting point of the solution.

solution occurs at a lower temperature than the triple point of the pure solvent. By itself, the change in the triple point is not important. But it results in a change in the temperature at which the solution freezes or melts. To understand why, we have to look carefully at the line that separates the solid and liquid regions in the phase diagram. This line is almost vertical because the melting point of a substance is not very sensitive to pressure.

Adding a solute to a solvent doesn't change the way the melting point depends on pressure. The line that separates the solid and liquid regions of the solution is therefore parallel to the line that serves the same function for the pure solvent. This line must pass through the triple point for the solution, however. The decrease in the triple point that occurs when a solute is dissolved in a solvent therefore decreases the melting point of the solution.

Figure 15.13 shows how the change in vapor pressure that occurs when a solute dissolves in a solvent leads to changes in the melting point and the boiling point of the solvent as well. Because the change in vapor pressure is a colligative property, which depends only on the relative number of solute and solvent particles, the changes in the boiling point and the melting point of the solvent are also colligative properties.

**CHECKPOINT**

Explain why it is wrong to assume that the boiling point and freezing point of a liquid are colligative properties. Explain why only the *change* that occurs in these quantities can be a colligative property.

## 15.7 COLLIGATIVE PROPERTIES CALCULATIONS

The best way to demonstrate the importance of colligative properties is to examine the consequences of Raoult's law. Raoult found that the vapor pressure of the solvent escaping from a solution is proportional to the mole fraction of the solvent.

$$P = \chi_{\text{solvent}} P^{\circ}$$

But the vapor pressure of a solvent is not a colligative property. Only the *change in the vapor pressure* that occurs when a solute is added to the solvent can be included among the colligative properties of a solution.

Sec. 5.11

Because pressure is a state function, the change in the vapor pressure of the solvent that occurs when a solute is added to the solvent can be defined as the difference between the vapor pressure of the pure solvent and the vapor pressure of the solvent escaping from the solution:

$$\Delta P = P^{\circ} - P$$

Substituting Raoult's law into this equation gives the following result:

$$\Delta P = P^{\circ} - \chi_{\text{solvent}} P^{\circ} = (1 - \chi_{\text{solvent}}) P^{\circ}$$

This equation can be simplified by remembering the relationship between the mole fraction of the solute and the mole fraction of the solvent:

$$\chi_{\text{solute}} + \chi_{\text{solvent}} = 1$$

Sec. 15.4

Substituting this relationship into the equation that defines $\Delta P$ gives another form of Raoult's law:

$$\mathbf{\Delta P = \chi_{solute} P^{\circ}}$$

This equation reminds us that *the change in the vapor pressure of the solvent* that occurs when a solute is added to the solvent is proportional to the mole fraction of the solute. As more solute is dissolved in the solvent, the vapor pressure of the solvent decreases, and the change in the vapor pressure of the solvent increases.

Because changes in the boiling point of the solvent ($\Delta T_{BP}$) that occur when a solute is added to a solvent result from changes in the vapor pressure of the solvent, the magnitude of the change in the boiling point is also proportional to the mole fraction of the solute:

$$\Delta T_{BP} = k_b \chi_{solute}$$

In dilute solutions, the mole fraction of the solute is proportional to the molality of the solution, as shown in Figure 15.14. The equation that describes the magnitude of the boiling point elevation that occurs when a solute is added to a solvent is therefore often written as follows:

$$\Delta T_{BP} = k_b m$$

Here, $\Delta T_{BP}$ is the **boiling point elevation**—the change in the boiling point that occurs when a solute dissolves in the solvent—and $k_b$ is a proportionality constant known as the *molal boiling point elevation constant* for the solvent.

A similar equation can be written to describe what happens to the freezing point (FP) (or melting point, MP) of a solvent when a solute is added to the solvent:

$$\Delta T_{FP} = -k_f m$$

In this equation, $\Delta T_{FP}$ is the **freezing point depression**—the change in the freezing point that occurs when the solute dissolves in the solvent—and $k_f$ is the *molal freezing point depression constant* for the solvent. A negative sign is used in this equation to indicate that the freezing point of the solvent *decreases* when a solute is added.

Values of $k_f$ and $k_b$ as well as the freezing points and boiling points for a number of pure solvents are given in Table 15.4. Exercises 15.7 and 15.8 provide an answer to a question that was left unanswered in Chapter 3: How can we determine the molecular weight of a substance? How do we know, for example, that the most common form of sulfur consists of $S_8$ molecules?

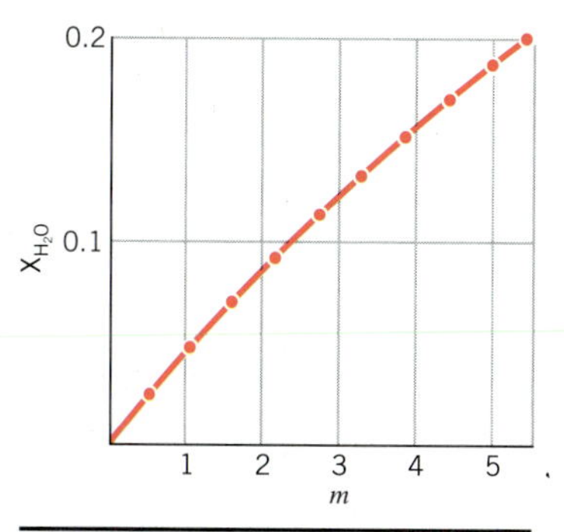


**Figure 15.14** A graph of the mole fraction versus molality for solutions prepared by dissolving increasing amounts of water in ethanol. For dilute solutions, the molality of the solution is directly proportional to the mole fraction of the solute.

## EXERCISE 15.7

Calculate the molecular weight of sulfur if 35.5 g of sulfur dissolve in 100.0 g of $CS_2$ to produce a solution that has a boiling point of 49.48°C.

**TABLE 15.4 Freezing Point Depression and Boiling Point Elevation Constants**

| Compound | Freezing Point (°C) | $k_f$ (°C/$m$) | Compound | Boiling Point (°C) | $k_b$ (°C/$m$) |
|---|---|---|---|---|---|
| Water | 0 | 1.853 | Water | 100 | 0.515 |
| Acetic acid | 16.66 | 3.90 | Ethyl ether | 34.55 | 1.824 |
| Benzene | 5.53 | 5.12 | Carbon disulfide | 46.23 | 2.35 |
| *p*-Xylene | 13.26 | 4.3 | Benzene | 80.10 | 2.53 |
| Naphthalene | 80.29 | 6.94 | Carbon tetrachloride | 76.75 | 4.48 |
| Cyclohexane | 6.54 | 20.0 | Camphor | 207.42 | 5.611 |
| Carbon tetrachloride | −22.95 | 29.8 | | | |
| Camphor | 178.75 | 37.7 | | | |

**Problem-Solving Strategy**

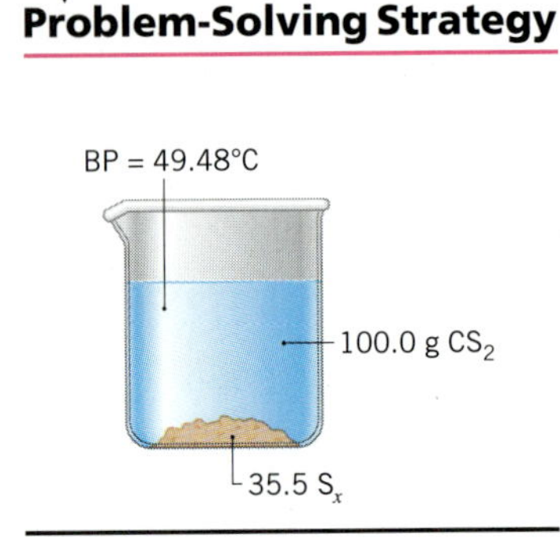


**Figure 15.15** Exercise 15.7

**SOLUTION** The relationship between the boiling point of the solution and the molecular weight of sulfur is not immediately obvious. We therefore start by asking: What do we know about this problem? We might start by drawing a figure, such as Figure 15.15, which helps us organize the information in this problem.

We know the boiling point of the solution, so we might start by looking up the boiling point of the pure solvent in order to calculate the change in the boiling point that occurs when the sulfur is dissolved in $CS_2$:

$$\Delta T_{\text{BP}} = 49.48°\text{C} - 46.23°\text{C} = 3.25°\text{C}$$

We also know that the change in the boiling point is proportional to the molality of the solution:

$$\Delta T_{\text{BP}} = k_b m$$

Since we know the change in the boiling point ($\Delta T_{\text{BP}}$) and the boiling point elevation constant for the solvent ($k_b$) can be looked up in a table, we might decide to calculate the molality of the solution.

$$m = \frac{\Delta T_{\text{BP}}}{k_b} = \frac{3.25°\text{C}}{2.35°\text{C}/m} = 1.38\ m$$

In the search for the solution to a problem, it is useful periodically to consider what we have achieved so far. At this point, we know the molality of the solution and the mass of the solvent used to prepare the solution. We can therefore calculate the number of moles of sulfur present in 100.0 g of carbon disulfide:

$$\frac{1.38 \text{ mol sulfur}}{1000 \text{ g } CS_2} \times 100.0 \text{ g } CS_2 = 0.138 \text{ mol sulfur}$$

We now know the number of moles of sulfur in this solution and the mass of the sulfur. We can therefore calculate the number of grams per mole of sulfur:

$$\frac{35.5 \text{ g}}{0.138 \text{ mol}} = 257 \text{ g/mol}$$

From the periodic table we know the atomic mass of sulfur is about 32 amu. So, the only way to explain this molecular weight is to assume that each sulfur molecule contains eight sulfur atoms:

$$\frac{257 \text{ g/mol}}{32 \text{ g/mol}} = 8$$

Elemental sulfur therefore behaves as if it contains $S_8$ molecules.

## EXERCISE 15.8

Determine the molecular weight of acetic acid if a solution that contains 30.0 g of acetic acid per kilogram of water freezes at −0.93°C. Do these results agree with the assumption that acetic acid has the formula $CH_3CO_2H$?

**SOLUTION** The freezing point depression for this solution is equal to the difference between the freezing point of the solution (−0.93°C) and the freezing point of pure water (0°C):

$$\Delta T_{\text{FP}} = -0.93°\text{C} - 0.0°\text{C} = -0.93°\text{C}$$

We now turn to the equation that defines the relationship between the freezing point depression and the molality of the solution:

$$\Delta T_{FP} = -k_f m$$

Since we know the change in the freezing point, and we can look up the freezing point depression constant in a table, we have enough information to calculate the molality of the solution:

$$m = \frac{-\Delta T_{FP}}{k_f} = \frac{0.93°\text{C}}{1.853°\text{C}/m} = 0.50\ m$$

**Problem-Solving Strategy**

At this point, we might return to the statement of the problem, to see if we are making any progress toward an answer. According to this calculation, there are 0.50 mol of acetic acid per kilogram of water in this solution. The problem stated that there were 30.0 g of acetic acid per kilogram of water in the solution. Since we now know the number of grams and the number of moles of acetic acid in the same mass of water in this sample, we can calculate the molecular weight of acetic acid:

$$\frac{30.0\ \text{g}}{0.50\ \text{mol}} = 60\ \text{g/mol}$$

The results of this experiment are in good agreement with the molecular weight (60.05 g/mol) expected if the formula for acetic acid is $CH_3CO_2H$.

What would happen if the calculation in Exercise 15.8 were repeated with a stronger acid, such as hydrochloric acid?

## EXERCISE 15.9

Explain why an 0.100 $m$ solution of HCl dissolved in water has a freezing point depression ($\Delta T_{FP}$ = 0.352°C) that is twice as large as might be expected from the molality of the solution.

**SOLUTION** We can predict the change in the freezing point that should occur in these solutions from the freezing point depression constant for the solvent and the molality of the solution. The freezing point depression for an 0.100 m solution should be −0.185°C.

$$\Delta T_{FP} = -k_f m = -(1.853°\text{C}/m)(0.100\ m) = -0.185°\text{C}$$

In order to explain these results, it is important to remember that colligative properties depend on the relative number of solute particles in a solution, not their identity. If the acid dissociates to an appreciable extent, the solution will contain more solute particles than we might expect from its molality.

If HCl dissociates completely in water, the total concentration of solute particles ($H_3O^+$ and $Cl^-$ ions) in the solution will be twice as large as the molality of the solution. The freezing point depression for this solution therefore will be twice as large as the change that would be observed if HCl did not dissociate:

$$HCl(g) + H_2O(l) \longrightarrow H_3O^+(aq) + Cl^-(aq)$$

If we assume that 0.100 $m$ HCl dissociates to form $H_3O^+$ and $Cl^-$ ions in water, the freezing point depression for this solution should be −0.371°C, which is slightly larger than what is observed experimentally:

$$\Delta T_{FP} = -k_f m = -(1.853°\text{C}/m)(2 \times 0.100\ m) = -0.371°\text{C}$$

This exercise suggests that HCl does not dissociate into ions when it dissolves in benzene, but dilute solutions of HCl dissociate more or less quantitatively in water.

In 1884 Jacobus Henricus van't Hoff introduced another term into the freezing point depression and boiling point elevation expressions to explain the colligative properties of solutions of compounds that dissociate when they dissolve in water:

$$\Delta T_{FP} = -k_f(i)m$$

Substituting the experimental value for the freezing point depression of an 0.100 *m* HCl solution into this equation gives a value for the *i* term of 1.89. If HCl did not dissociate in water, *i* would be 1. If it dissociates completely, *i* would be 2. The experimental value of 1.89 suggests at least 95% of the HCl molecules dissociate in this solution.

**EXERCISE 15.10**

Explain why 0.60 grams of acetic acid dissolve in 200 grams of benzene to form a solution that lowers the freezing point of benzene to 5.40°C.

**SOLUTION** Because pure benzene freezes at 5.53°C, the freezing point depression in this experiment is −0.13°C:

$$\Delta T_{FP} = 5.40°C - 5.53°C = -0.13°C$$

Once again, we can start with the relationship between the freezing point depression and the molality of the solution:

$$\Delta T_{FP} = -k_f m$$

We then use the molal freezing point depression constant for benzene to calculate the molality of the solution:

$$m = \frac{-\Delta T_{FP}}{k_f} = \frac{0.13\text{°}\cancel{C}}{5.12\text{°}\cancel{C}/m} = 0.025\ m$$

Multiplying the molality of the solution by the amount of solvent used to make the solution gives the number of moles of solute particles in the solution:

$$\frac{0.025 \text{ mol solute}}{1 \cancel{\text{kg benzene}}} \times 0.200 \cancel{\text{kg benzene}} = 0.0050 \text{ mol solute}$$

Since the solution contains 0.60 g of acetic acid, the molecular weight of acetic acid in this experiment is 120 g/mol:

$$\frac{0.60 \text{ g solute}}{0.0050 \text{ mol solute}} = \mathbf{120\ g/mol}$$

The molecular weight of acetic acid in benzene is twice as large as what is expected from the molecular formula, $CH_3CO_2H$. This can be explained by assuming that acetic acid molecules associate in benzene to form dimers that are held together by hydrogen bonds, as shown in Figure 15.16.

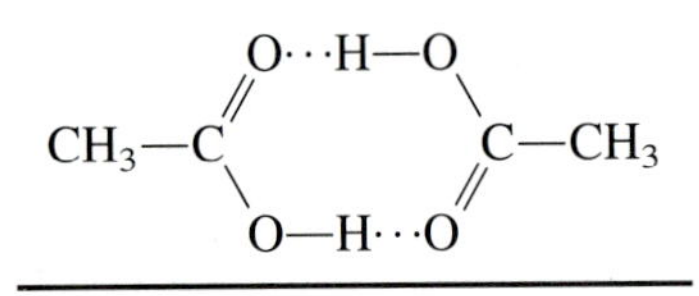


**Figure 15.16** The dimerization of acetic acid that occurs when this compound is dissolved in a nonpolar solvent, such as benzene.

## 15.8 OSMOTIC PRESSURE

In 1784, French physicist and clergyman Jean Antoine Nollet discovered that a pig's bladder filled with a concentrated solution of alcohol in water expanded when it was immersed in water. The bladder acted as a *semipermeable membrane,* which allowed water molecules to enter the solution but kept alcohol molecules from moving in the other direction. Movement of one component of a solution through a membrane to dilute the solution is called **osmosis,** and the pressure this produces is called the **osmotic pressure ($\pi$).**

Osmotic pressure can be demonstrated with the apparatus shown in Figure 15.17. A semipermeable membrane is tied across the open end of a thistle tube. The tube is then partially filled with a solution of sugar or alcohol in water and immersed in a beaker of water. Water will flow into the tube until the pressure on the column of water due to the force of gravity balances the osmotic pressure driving water through the membrane.

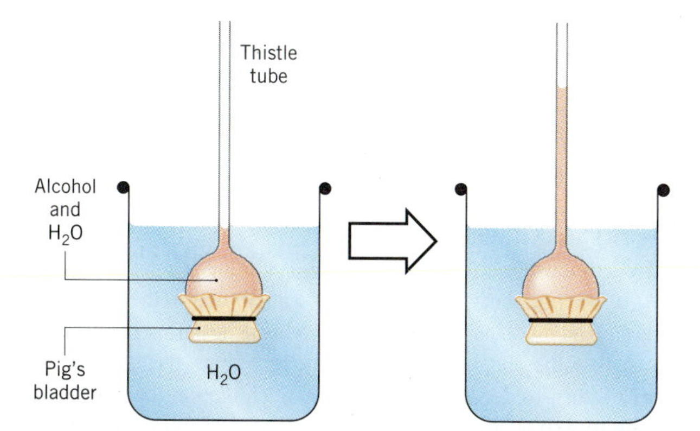


**Figure 15.17** Water flows through the semipermeable membrane to dilute the alcohol solution until the force of gravity pulling down on the column of this solution balances the osmotic pressure pushing the water through the membrane.

A more interesting demonstration of osmotic pressure uses the membrane that surrounds an egg as the semipermeable membrane. The egg is first soaked overnight in acetic acid, which softens the $CaCO_3$ in the egg shell. After the shell is peeled from the egg, the egg is weighed on an analytical balance. It is then immersed in distilled water for 20 to 30 minutes, and then reweighed. The difference between the concentration of the solution within the egg and the distilled water in the beaker generates an osmotic pressure that forces water to pass through the semipermeable membrane that lies beneath the shell of the egg, thereby increasing the mass of the egg by as much as 10%.

This egg initially weighed 72.50 grams. After it was immersed in distilled water for 30 minutes, the weight of this egg increased to 78.14 grams.

The same year that Raoult discovered the relationship between the vapor pressure of a solution and the vapor pressure of a pure solvent, Jacobus Henricus van't Hoff found that the osmotic pressure of a dilute solution ($\pi$) obeyed an equation analogous to the ideal gas equation:

$$\pi = \frac{nRT}{V}$$

This equation suggests that osmotic pressure is another example of a colligative property, because this pressure depends on the ratio of the number of solute particles to the volume of the solution, $n/V$, not the identity of the solute particles. It also reminds us of the magnitude of osmotic pressure. According to this equation, a 1.00 $M$ solution has an osmotic pressure of 22.4 atm at 0°C:

$$\pi = \frac{(1.00 \cancel{\text{mol}})(0.08206 \cancel{\text{L}}\,\text{atm}/\cancel{\text{mol}}\,\cancel{\text{K}})(273 \cancel{\text{K}})}{(1.00 \cancel{\text{L}})} = \mathbf{22.4\ atm}$$

This means that a 1.00 $M$ solution should be able to support a column of mercury 670 in, or almost 56 ft, tall!

Biologists and biochemists often take advantage of osmotic pressure when they isolate the components of a cell. When a cell is added to an aqueous solution that contains a much higher concentration of ions than the liquid within the cell, water leaves the cell by flowing through the cell membrane until the cell shrinks so much that the membrane breaks. Alternatively, when a cell is placed in a solution that has a much smaller ionic strength, water pours into the cell, and the cell expands until the cell membrane bursts.

SPECIAL TOPIC

## CHEMICAL SEPARATIONS

One of the most important—and time-consuming—activities in chemistry involves isolating, separating, and purifying chemical compounds. **Extraction** (literally, "taking out by force") is a useful technique for separating compounds, such as $I_2$ and $KMnO_4$, that have different polarities. The compounds to be separated are treated with a mixture of a polar solvent (such as $H_2O$) and a nonpolar solvent (such as $CCl_4$). The $I_2$ will dissolve in the $CCl_4$, whereas the $KMnO_4$ will dissolve in the $H_2O$. By separating these two phases and allowing the solvents to evaporate, we can cleanly separate $I_2$ and $KMnO_4$.

This technique also can be used to extract solutes from a solid. Coffee, for example, is decaffeinated by extraction. Most of the popularity of coffee can be attributed to the stimulating effect of caffeine, which makes up as much as 2.5% of the dry weight of coffee beans. But many people who have grown to like the flavor of coffee want to reduce the amount of caffeine they consume and therefore turn to decaffeinated coffee. Caffeine can be extracted with hot water, but this also tends to extract the oils that give coffee its flavor. For many years, therefore, coffee beans were decaffeinated by treating them with a nonpolar solvent, such as dichloromethane ($CH_2Cl_2$). This dissolves most of the caffeine without destroying the flavor of the coffee. (The caffeine is then sold to manufacturers of cola drinks, who add it to their products.)

**Distillation** is the technique used most frequently to purify liquids. At its simplest, it involves heating the liquid, as shown in Figure 15.18, until it boils. The vapor that escapes is passed through a water-cooled condenser, where it condenses to form a liquid, which collects in a clean flask.

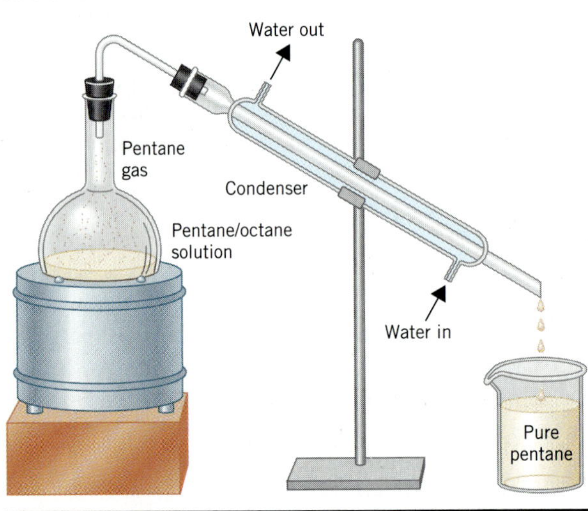


**Figure 15.18** A distillation apparatus.

**EXERCISE 15.11**

Assume that a mixture that contains equal amounts of pentane ($C_5H_{12}$) and octane ($C_8H_{18}$) is distilled. Describe the difference between the composition of the liquid in the distillation flask and the vapor given off when this liquid starts to boil.

**SOLUTION** According to the data in Table 14.2, pentane (BP = 36.1°C) is much more volatile than octane (BP = 125.7°C). Pentane therefore begins to distill from this mixture long before octane, and the partial pressure of the pentane that collects above the solution will be much larger than the partial pressure of the octane. Although the liquid contains equal amounts of the two hydrocarbons, the vapor is much richer in pentane.

Sec. 14.2

We can translate the results of Exercise 15.11 into a general rule for distillations by constructing a phase diagram that describes the mixtures under all possible combinations of temperature and percent composition, as shown in Figure 15.19. The points labeled $T_p$ and $T_o$ in this diagram represent the boiling points of pure pentane and pure octane. These points are connected by two curves. At temperatures below the bottom curve, mixtures of pentane and octane are liquids. At temperatures above the top curve, these mixtures are gases. Points that lie between the curves describe systems that contain both a liquid phase and a gas phase with the same composition. Exercise 15.11 concluded, however, that the liquid and the gas should not have the same composition at a given temperature. Points between the two curves are therefore unstable. Any system at point *B*, for example, would come to equilibrium by forming a vapor with the composition described by point *A* and a liquid with the composition described by point *C*.

The vertical line in Figure 15.19 shows what happens to the system when a 50:50 mixture of pentane and octane is heated. Nothing changes until the temperature reaches point *C*—the temperature at which the liquid starts to boil. The vapor that collects above the boiling liquid will have the composition described by point *A* in this diagram. It will therefore be much richer in pentane than the liquid from which it boils. If this vapor escapes, more pentane

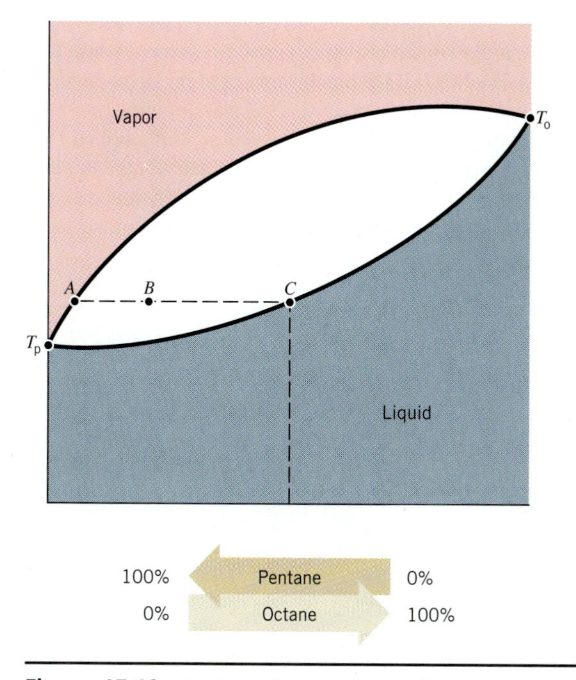


**Figure 15.19** A phase diagram for a mixture of pentane and octane that shows the effect of changes in the composition of the mixture on the boiling point of the solution.

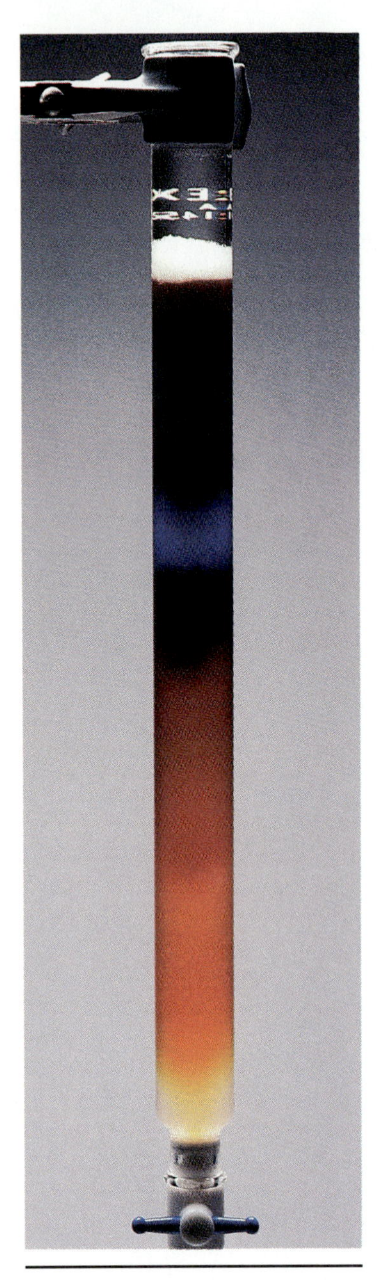

Chromatography literally means "writing with color." The name refers to the fact that this technique was first used to separate the colored pigments in plants.

leaves the system than octane, and the percent by mass of octane in the system increases. This increases the boiling point of the mixture, as well as gradually increasing the amount of octane that collects in the vapor that escapes from the liquid.

Figure 15.19 therefore provides the basis for separating a mixture by distillation. We start by heating the mixture and collecting the first few portions, or fractions, that distill over. These fractions will be rich in the component that has the lower boiling point. We might then discard the next few fractions, which contain a mixture of the low- and high-boiling compounds, and save the last few fractions, which should be rich in the compound that has the higher boiling point.

Distillation works best for purifying compounds that are liquids at room temperature. Solids are more likely to be purified by **recrystallization,** which takes advantage of differences in the solubility of various components of a mixture in different solvents. We start by dissolving as much of the solid as possible in a hot sample of the solvent. We then cool the solution until some of the solid crystallizes back out of solution. Each time the sample is dissolved and recrystallized, it is purer than before. This was the technique used by Marie and Pierre Curie to isolate the first minuscule sample of radium from several tons of the mineral pitchblende.

One of the fastest growing methods of separating compounds was developed by a Russian botanist named Mikhail Tsvet in 1906. This technique is called **chromatography** (literally, "writing with color") because it was first used to separate the colored pigments in plants. The basic principle behind chromatography is simple. A gas or liquid is allowed to flow over a solid support. Compounds that have a high affinity for the solvent are carried along as it moves past the stationary support. Compounds that have a higher affinity for the solid move more slowly.

In *thin-layer chromatography* (TLC), a glass plate or plastic support is coated with a thin layer of a solid such as alumina ($Al_2O_3$) or silica ($SiO_2$). Small samples of the compounds to be separated are then placed on the silica or alumina, and the TLC plate is immersed in a solvent until the solvent line is just below the point where the samples were applied (see Figure 15.20). The solvent slowly moves up the plate, carrying the components of the mixture with it. Some of these components have a high affinity for the $Al_2O_3$ or $SiO_2$ support, and they move very slowly. Others have a high affinity for the solvent, and they move more

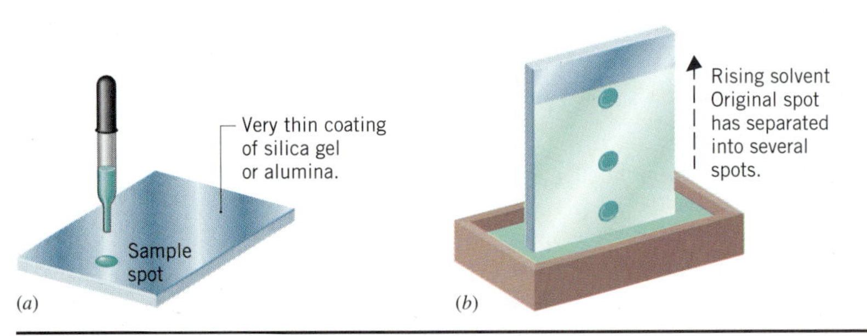


**Figure 15.20** Thin-layer chromatography (TLC). (*a*) The solution is applied near one end of a plate coated with silica or alumina. (*b*) The plate is then immersed in a solvent, which rises up the plate by capillary action.

An ultra pure single-crystal ingot of the semimetal silicon.

rapidly. The net result is a separation of the mixture into its components based on their relative affinities for the stationary solid phase and the mobile liquid phase.

*Column chromatography* extends the principles of TLC to the large-scale separation of mixtures. This technique involves filling a glass column with a solid support, applying up to several grams of the mixture to the top of the column, and then slowly washing the column with solvent.

Mixtures also can be separated by *gas-phase chromatography,* which takes advantage of the relative affinity of the different components in a mixture for the stationary support when the mixture is heated until there is an equilibrium between its gas and liquid phases.

*Zone refining* is used to obtain very pure (99.999%) samples of metals and semimetals. The starting material is drawn into a thin rod which is inserted into a heating coil that slowly moves along the length of the rod. As the coil moves, it constantly brings a new portion of the rod to the point at which the solid just melts. Because solutions melt at lower temperatures than pure solids, it is possible to concentrate the impurities in the small sample of liquid that moves just ahead of the heating coil, and to crystallize almost perfectly pure metal or semimetal just behind the coil.

## RESEARCH IN THE 90s

### SUPERCRITICAL FLUID EXTRACTION

Caffeine is not the only substance removed when coffee beans are extracted with dichloromethane. This nonpolar solvent also removes a significant fraction of the essential oils that give coffee its aroma and taste.

In 1978, a new process for removing caffeine from coffee was developed in Germany (Günther Wilke, *Angewandte Chemie International Edition,* **17,** 702 (1978)]. The best results are achieved when 1 kg of activated charcoal is mixed with 3 kg of coffee beans and this mixture is extracted with $CO_2$ at 90°C and 220 atm pressure for 5 hours.

Sec. 14.6 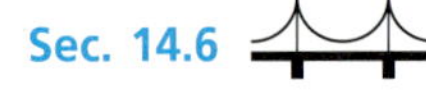

According to the data in Table 14.3, it is possible to condense $CO_2$ gas to form a liquid by raising the pressure on the gas if the temperature is kept *below* 31°C. Because the temperature in this process is kept above the critical temperature, the $CO_2$ never condenses to form a liquid. It is therefore a *supercritical fluid.*

The density of a gas depends on its temperature and pressure. At room temperature and atmospheric pressure, the density of $CO_2$ is 0.002 g/cm$^3$. At its critical point, the density of $CO_2$ is 0.468 g/cm$^3$. At 220 atm, the density of supercritical $CO_2$ can reach values as high as 0.6 g/cm$^3$. The densities of these supercritical fluids therefore begin to compare with those of the nonpolar solvents typically used to do extractions.

There are several advantages to supercritical fluid extraction.

- Because they are gases, these fluids have very low viscosities. As a result, the flow of material through supercritical fluids is relatively fast. Extraction into the fluid is therefore rapid, and so is the rate at which the solid precipitates when the supercritical fluid is allowed to evaporate.
- Because they are gases, supercritical fluids are easy to recover. Thus, very little solvent is lost during the extraction process. Equally important, virtually no solvent residue is left behind.
- The material that has been extracted can be isolated by simply reducing the pressure on the supercritical fluid once the extraction is complete.
- Supercritical fluid extraction is often highly selective. (It is selective enough that decaffeinated coffee today differs from untreated coffee in only one way, the pharmacological effect.)
- There are no potential negative health side effects of supercritical fluid processing, which means it is ideally suited for use in the pharmaceutical and food industries.

Research so far has shown that supercritical fluid extraction can be used to isolate soybean oil from soybeans, or to isolate essential oils from spices. It also can be even used to extract alkaloids, such as nicotine, from plants. Research in the 90s will undoubtedly pursue the promise of supercritical fluid extraction as a means of removing fats and oils from food. It will also examine more efficient large-scale processes for isolating pharmaceutical products from nature.

## SUMMARY

**1.** Solvents can be divided into two categories: **polar** and **nonpolar.** As a rule, **like dissolves like.** Nonpolar compounds dissolve in nonpolar solvents, and polar compounds dissolve in polar solvents.

**2.** There are many ways of describing the concentration of a solution, including **molarity, molality, mole fraction, mass percent,** and **volume percent.** The unit of concentration used most commonly by chemists is *molarity* ($M$)—the number of moles of solute per liter of solution.

**3.** At times it is useful to know the ratio of the amount of solute to the amount of solvent. Under these circumstances, chemists turn to units of *mole fraction* or *molality* ($m$). These units are particularly important in discussions of colligative properties, which depend on the ratio of the number of particles of solute and solvent, but not the identity of the solute.

**4.** Colligative properties include the change in the vapor pressure, the change in the boiling point, and the change in the freezing point of a solvent that occur when a solute is dissolved in the solvent. Adding a solute to a solvent lowers the vapor pressure of the solvent. This results in an elevation in the boiling point of the solution and a depression in its freezing point.

**5.** Another colligative property of a solution is osmotic pressure, which causes the solvent to flow across a semipermeable membrane when the concentrations of solute on either side of the membrane are unequal.

**6.** Because colligative properties depend only on the number of solute particles in a solution, not their identity, these properties can be used to measure the molecular weight of a solute.

## KEY TERMS

**Alcohol** (p. 534)
**Boiling point elevation** (p. 545)
**Chromatography** (p. 552)
**Colligative property** (p. 541)
**Concentration** (p. 538)
**Distillation** (p. 550)
**Extraction** (p. 550)
**Freezing point depression** (p. 545)
**Hydrophilic** (p. 535)
**Hydrophobic** (p. 535)
**Immiscible** (p. 534)
**Molality** (p. 539)
**Molarity** (p. 538)
**Mole fraction** (p. 540)
**Nonpolar solvents** (p. 532)
**Osmosis** (p. 548)
**Osmotic pressure** (p. 548)
**Polar solvents** (p. 532)
**Raoult's law** (p. 542)
**Recrystallization** (p. 552)
**Solute** (p. 531)
**Solution** (p. 531)
**Solvent** (p. 531)
**Vapor pressure** (p. 541)

## KEY EQUATIONS

$$\text{Concentration} = \frac{\text{amount of solute}}{\text{amount of solvent or solution}}$$

$$\text{Molarity: } M = \frac{\text{moles of solute}}{\text{liters of solution}}$$

$$\text{Mass percent} = \frac{\text{mass of solute}}{\text{mass of solution}} \times 100\%$$

$$\text{Volume percent} = \frac{\text{volume of solute}}{\text{volume of solution}} \times 100\%$$

$$\text{Molality: } m = \frac{\text{moles of solute}}{\text{kilograms of solvent}}$$

Mole fraction:

$$\chi_{\text{solute}} = \frac{\text{moles of solute}}{\text{moles of solute + moles of solvent}}$$

$$\chi_{\text{solvent}} = \frac{\text{moles of solvent}}{\text{moles of solute + moles of solvent}}$$

Raoult's Law: $P = \chi_{\text{solvent}} P^{\circ}$

$\Delta P = \chi_{\text{solute}} P^{\circ}$

Freezing point depression: $\Delta T_{\text{FP}} = -k_f m$

Boiling point elevation: $\Delta T_{\text{BP}} = k_b m$

$$\text{Osmotic pressure: } \pi = \frac{nRT}{V}$$

## PROBLEMS

### Solutions: Like Dissolves Like

**15-1** Define the terms *solution, solvent,* and *solute* and give an example of each term.

**15-2** Use a drawing to describe what happens when $I_2$ molecules dissolve in $CCl_4$ and when $KMnO_4$ dissolves in water.

**15-3** One way of screening potential anesthetics involves testing whether the compound dissolves in olive oil, because all common anesthetics, including nitrous oxide ($N_2O$), cyclopropane ($C_3H_6$), and halothane ($C_2HF_3ClBr$), are soluble in olive oil. What property do these compounds have in common?

**15-4** Carboxylic acids with the general formula $CH_3(CH_2)_nCO_2H$ have a nonpolar $CH_3CH_2$ . . . tail and a polar . . . $CO_2H$ head. What effect does increasing the value of $n$ have on the solubility of these acids in polar solvents, such as water? What is the effect on their solubility in nonpolar solvents, such as $CCl_4$?

**15-5** Which of the following compounds would be the most soluble in a nonpolar solvent, such as $CCl_4$?

(a) $H_2O$ (b) $CH_3OH$ (c) $CH_3CH_2CH_2OH$
(d) $CH_3CH_2CH_2CH_2CH_2OH$
(e) $CH_3CH_2CH_2CH_2CH_2CH_2CH_2OH$

**15-6** Potassium iodide reacts with iodine in aqueous solution to form potassium triiodide:

$$KI(aq) + I_2(aq) \longrightarrow KI_3(aq)$$

What would happen if we added $CCl_4$ to this reaction mixture?

(a) The KI and $KI_3$ would dissolve in the $CCl_4$ layer. (b) The $I_2$ would dissolve in the $CCl_4$ layer. (c) Both KI and $I_2$, but not $KI_3$, would dissolve in the $CCl_4$ layer. (d) Neither KI, $KI_3$, nor $I_2$ would dissolve in the $CCl_4$ layer. (e) No distinct $CCl_4$ layer would form because $CCl_4$ is soluble in water.

**15-7** Phosphorus pentachloride can react with itself in a reversible reaction to form a salt that contains the $PCl_4^+$ and $PCl_6^-$ ions:

$$2\ PCl_5 \rightleftharpoons PCl_4^+ + PCl_6^-$$

The extent to which this reaction occurs depends on the solvent in which it is run. Predict whether a nonpolar solvent, such as $CCl_4$, favors the products or the reactants of this reaction. Predict what would happen to this reaction if we used a polar solvent, such as acetonitrile ($CH_3CN$).

### Units of Concentration: Molarity and Molality

**15-8** Define *molarity* and *molality*. How are these units of concentration similar? How are they different?

**15-9** Explain why the molality of a solution is proportional to the mole fraction of the solute for dilute solutions.

**15-10** If liquids expand as the temperature increases, what effect should an increase in the temperature of a solution have on the molarity, molality, and mole fraction of the solute and solvent?

**15-11** When asked to prepare a liter of 1.00 $M$ $K_2CrO_4$, a novice student weighed out exactly 1.00 mol of $K_2CrO_4$ and added this solid to 1.00 L of water in a volumetric flask. What did the student do wrong? Did the student get a solution that was more concentrated than 1.00 $M$ or less concentrated than 1.00 $M$? How would you prepare the solution?

**15-12** Calculate the molality and molarity of an aqueous solution of ammonia that is 28.0% $NH_3$ by mass and has a density of 0.90 g/cm$^3$. Explain why you can calculate both quantities for this solution but not for most solutions.

**15-13** What is the molarity of an aqueous solution that is 18.0% $Pb(NO_3)_2$ by mass if the density of this solution is 1.18 g/cm$^3$?

**15-14** Benzene was once used as a common solvent in chemistry. Its use is now restricted because it can cause cancer. The recommended limit of exposure to benzene is 3.2 milligrams per cubic meter of air (3.2 mg/m$^3$). Calculate the molarity of this solution. Calculate the number of parts per million by mass of benzene in this solution.

**15-15** You can make the chromic acid bath commonly used to clean glassware in the lab by dissolving 92 g of sodium dichromate ($Na_2Cr_2O_7 \cdot 2\ H_2O$) in enough water to give 458 mL of solution and then adding 800 mL of concentrated sulfuric acid. Calculate the molarity of the $Cr_2O_7^{2-}$ ion in this solution.

**15-16** People who have smoked marijuana can be detected by testing their urine for the tetrahydrocannabinols (THC) that are the active ingredient in marijuana. The present limit on detection of THC in urine is 20 nanograms of THC per milliliter of urine (20 ng/mL). Calculate the molarity of this solution at the limit of detection if the molecular weight of THC is 315 g/mol.

### Units of Concentration: Mole Fraction

**15-17** Calculate the mole fraction of $N_2$ and $O_2$ in air, if air is 78% nitrogen and 21% oxygen by volume.

**15-18** Calculate the mole fraction of water in a solution that contains 10.0 g of glucose ($C_6H_{12}O_6$) dissolved in 100 g of water.

**15-19** Calculate the mole fraction of water in a solution prepared by adding 15 g of water to 1000 g of ethanol ($C_2H_6O$). Calculate the molality of this solution. Explain why it is impossible to determine the molarity of the solution.

**15-20** Chlorine gas can be detected in air at a concentration of 3.5 parts per million by mass. Calculate the mole fraction of chlorine in air at this concentration.

**15-21** Nichrome is an alloy that is 60% Ni, 24% Fe, 16% Cr, and 0.1% C by mass. Calculate the mole fraction of each element in this solution.

**15-22** Babbitt metal is an alloy that is 69% Zn, 19% Sn, 4% Cu, 3% Sb, and 5% Pb by mass. Calculate the mole fraction of each metal in this alloy. Calculate the sum of the mole fractions of all five metals. Explain why the sum of these mole fractions is equal to 1, within experimental error.

**15-23** Calculate the mole fraction of both the solute and the solvent in a 0.100 *m* $K_2Cr_2O_7$ solution. Explain why you don't need to know the volume of the solution to do this calculation. Explain why you can't perform the same calculation for a 0.100 *M* $K_2Cr_2O_7$ solution.

**15-24** At high temperatures and pressures, ammonia decomposes to nitrogen and hydrogen according to the following equation:

$$2\ NH_3(g) \longrightarrow N_2(g) + 3\ H_2(g)$$

What is the mole fraction of $N_2$ when one-half of the $NH_3$ has decomposed?

**15-25** What is the density, molarity, molality, and mole fraction of a solution prepared by dissolving 5.0 g of NaCl in 65 g of water if the total volume of the solution is 68 mL?

### Units of Concentration: Mass Percent

**15-26** Concentrated acetic acid is called glacial acetic acid because of the ease with which this solution freezes. (The term *glacial* comes from the Latin stem meaning "ice.") What is the mass percent of the $CH_3CO_2H$ molecules in glacial acetic acid if this solution is 17.4 *M* and its density is 1.05 g/cm$^3$?

**15-27** Calculate the molarity of concentrated hydrochloric acid, if this solution is 38% HCl by mass and its density is 1.1977 g/cm$^3$.

**15-28** The solubility of $NH_3$ gas in water is 33.1% by mass. Calculate the molality of a saturated solution of ammonia in water.

**15-29** Describe the assumption that has to be made in order to estimate the molality of an aqueous solution that is 1.5% sodium chloride by mass and then do this calculation.

**15-30** Calculate the molality of carbon in austenite, one of the crystal forms of steel, if austenite is 0.8% C by mass dissolved in iron.

### Colligative Properties: Vapor Pressure Depression

**15-31** Explain why dissolving a solute in a solvent leads to a decrease in the vapor pressure of the solvent.

**15-32** Explain why the pressure of the solvent escaping from a solution is equal to the mole fraction of the solvent times the vapor pressure of the pure solvent.

**15-33** Predict what will happen to the rate at which water evaporates from an open flask when salt is dissolved in the water, and explain why the rate of evaporation changes.

**15-34** If you place a beaker of pure water (I) and a beaker of a saturated solution of sugar in water (II) in a sealed bell jar, the level of water in beaker I will slowly decrease, and the level of the sugar solution in beaker II will slowly increase. Explain why.

**15-35** Explain why the vapor pressure of a liquid at a particular temperature is an intensive property but the change in the vapor pressure of the liquid when a solute is added is a colligative property.

### Colligative Properties: Boiling Point Elevation and Freezing Point Depression

**15-36** Explain how the decrease in the vapor pressure of a solvent that occurs when a solute is added to the solvent leads to an increase in the solvent's boiling point.

**15-37** Explain how the decrease in the vapor pressure of the solvent that occurs when a solute is added to the solvent leads to a decrease in the solvent's melting point.

**15-38** Predict the shape of a plot of the freezing point of a solution versus the molality of the solution.

**15-39** Explain why salt is added to the ice that surrounds the container in which ice cream is made.

**15-40** Explain why many cities and states spread salt on icy highways. Provide a mechanical explanation for how the salt dissolves in the ice, and then predict what effect this has on the physical properties of the ice.

**15-41** Compare the values of $k_f$ and $k_b$ for water, benzene, carbon tetrachloride, and camphor. Explain why measurements of molecular weight based on freezing point depression might be more accurate than those based on boiling point elevation.

### Colligative Properties: Calculations

**15-42** What is the freezing point of a saturated solution of caffeine ($C_8H_{10}O_2N_4 \cdot H_2O$) in water if it takes 45.6 g of water to dissolve 1.00 g of caffeine?

**15-43** A 0.100 *m* solution of sulfuric acid in water freezes at −0.371°C. Which of the following statements is consistent with this observation?

(a) $H_2SO_4$ does not dissociate in water. (b) $H_2SO_4$ dissociates into $H_3O^+$ and $HSO_4^-$ ions in water. (c) $H_2SO_4$ dissociates in water to form two $H_3O^+$ ions and one $SO_4^{2-}$ ion. (d) $H_2SO_4$ associates in water to form $(H_2SO_4)_2$ molecules.

**15-44** The "Tip of the Week" in a local newspaper suggested using a fertilizer such as ammonium nitrate or ammonium sulfate instead of salt to melt snow and ice on sidewalks because salt can damage lawns. Which of the following compounds would give the largest freezing point depression when 100 g are dissolved in 1 kg of water?

(a) NaCl (b) $NH_4NO_3$ (c) $(NH_4)_2SO_4$

**15-45** A 5% solution of sucrose in water at 20°C has a density of 1.0179 g/cm$^3$. If this solution exhibits an osmotic pressure of 3.57 atm, what is the molecular weight of sucrose?

**15-46** *p*-Dichlorobenzene (PDCB) is replacing naphthalene as the active ingredient in moth balls. Calculate the value of $k_f$ for camphor if a 0.260 *m* solution of PDCB in camphor decreases the freezing point of camphor by 9.8°C.

**15-47** If an aqueous solution boils at 100.50°C, at what temperature does it freeze?

**15-48** What is the boiling point of a solution of 10.0 g of $P_4$ in 25.0 mL of carbon disulfide?

**15-49** Calculate the boiling point of a solution of 4.39 g of naphthalene, $C_{10}H_8$, in 99.5 g of carbon tetrachloride.

**15-50** We usually assume that salts such as KCl dissociate completely when they dissolve in water:

$$KCl(s) + H_2O(l) \longrightarrow K^+(aq) + Cl^-(aq)$$

What percent of the KCl actually dissociates in water if the freezing point of a 0.100 *m* solution of this salt in water is −0.345°C?

**15-51** Calculate the freezing point of a 0.100 *m* solution of acetic acid in water if the $CH_3CO_2H$ molecules are 1.33% ionized in this solution.

**15-52** What fraction of chlorous acid, $HClO_2$, dissociates in water if a solution that contains 3.22 g of $HClO_2$ in 47.0 g of water has a freezing point of 271.10 K?

### Colligative Properties: Osmotic Pressure

**15-53** Use the equation that describes the osmotic pressure of a solution to predict what happens to the osmotic pressure when the solution is cooled.

**15-54** Osmotic pressure is much more sensitive than freezing point depression or boiling point elevation. As a result, it can be used to measure the molecular weights of molecules as large as the hemoglobin that carries oxygen through the blood. The concentration of hemoglobin in blood is roughly 15 g/100 mL. Assume that the osmotic pressure of a solution that contains 15 g of hemoglobin dissolved in 100 g of water is 0.056 atm at 25°C. What is the molecular weight of hemoglobin?

### Chemical Separations

**15-55** Organic compounds are often purified by distillation. Describe the equipment you would use to distill a compound that boils at 156°C.

**15-56** How would you distill a compound that has a boiling point of 156°C if the compound decomposes at temperatures above 100°C?

**15-57** Use Figure 15.19 to explain why a 50:50 mixture of pentane and octane is unstable at temperatures between the boiling point of pentane and octane. Predict what would happen to this mixture.

**15-58** Use Figure 15.19 to explain why the vapor that boils out of a mixture of pentane and octane contains more pentane than octane.

**15-59** Describe the basic principle behind the separation of compounds by thin-layer, column, and gas-phase chromatography.

**15-60** Predict which of the compounds that form the three dots in the TLC plate in Figure 15.20 has the highest affinity for the inert support upon which the chromatography is done. Which of these compounds has the highest affinity for the mobile solvent phase?

### Integrated Problems

**15-61** In large quantities, the nicotine in tobacco is a deadly poison. Calculate the molecular formula of nicotine, $C_xH_yH_z$, if 4.38 mg of this compound burns to form 11.9 mg of $CO_2$ and 3.41 mg of water, and 3.62 g of this poison changes the freezing point of 73.4 g of water by 0.563°C.

**15-62** Dimethylsulfoxide (DMSO) once received attention as a possible treatment for arthritis. Its major advantage is its ability to soak through the skin. DMSO can therefore be applied directly to the joints that are affected by arthritis. Elemental analysis suggests that DMSO is 30.75% C, 7.74% H, 41.03% S, and 20.48% O by mass. What is the molecular formula of this compound if 5.00 g of DMSO dissolved in 100.0 g of dioxane freezes at 8.8°C? (Dioxane: $T_{FP}$ = 11.8°C; $k_f$ = 4.63°C/*m*)

# Intersection

## REACTIONS THAT DO NOT GO TO COMPLETION

Suppose that you were asked to describe the steps involved in calculating the mass of the finely divided white solid produced when a 2.00-g strip of magnesium metal is burned. You might organize your work as follows:

- Start by assuming that the magnesium reacts with oxygen in the atmosphere when it burns.
- Predict that the formula of the product is MgO.
- Use this formula to generate the following balanced equation:

$$2\ Mg(s) + O_2(g) \longrightarrow 2\ MgO(s)$$

- Use the atomic weight of magnesium to convert grams of magnesium into moles of magnesium:

$$2.00\ \cancel{g\ Mg} \times \frac{1\ mol\ Mg}{24.31\ \cancel{g\ Mg}} = 0.08227\ mol\ Mg$$

- Use the balanced equation to convert moles of magnesium into moles of magnesium oxide:

$$0.08227\ \cancel{mol\ Mg} \times \frac{2\ mol\ MgO}{2\ \cancel{mol\ Mg}} = 0.08227\ mol\ MgO$$

- Use the mass of a mole of magnesium oxide to convert moles of MgO into grams of MgO:

$$0.08227\ \cancel{mol\ MgO} \times \frac{40.30\ g\ MgO}{1\ \cancel{mol\ MgO}} = 3.32\ g\ MgO$$

Before reading any further, ask yourself the following questions: ''How confident am I in this answer? Is there anything else that has to be done before I can safely conclude that I have the 'correct' answer to this calculation?''

Before we can trust this answer, we have to consider whether there are any hidden assumptions behind the calculation and then check the validity of these assumptions. Three assumptions were made in this calculation.

- We assumed that the metal contained 2.00 g of magnesium. In other words, we assumed that the strip was pure magnesium.
- We assumed that the magnesium reacted only with the oxygen in the atmosphere to form MgO—ignoring the possibility that some of the magnesium might react with the nitrogen in the atmosphere to form $Mg_3N_2$.
- We assumed that the reaction didn't stop until all of the magnesium metal had been consumed.

It is relatively easy to correct for the fact that the starting material is actually 99% magnesium by mass. We can also correct for the fact that as much as 5% of the product of this reaction is $Mg_3N_2$ instead of MgO. But it is the third assumption that is of particular importance for the discussion in the next three chapters.

At some point in their study of chemistry, most students acquire the impression that all chemical reactions go to completion. This belief is fostered by calculations, such as predicting the amount of MgO produced by burning a known amount of magnesium metal, which assume that the reaction goes to completion. It is reinforced by demonstrations such as the reaction in which a copper penny dissolves in concentrated nitric acid, which seems to continue until the copper penny disappears.

Many students are therefore surprised to learn that chemical reactions don't always go to completion. The following equation provides an example of a chemical reaction that seems to stop prematurely:

$$2\ NO_2(g) \rightleftharpoons N_2O_4(g)$$

At 25°C, when 1 mole of $NO_2$ is added to a 1.00-L flask, this reaction seems to stop when 95% of the $NO_2$ has been converted into $N_2O_4$. Once it has reached this point, the reaction doesn't go any further. As long as the reaction is left at 25°C, about 5% of the $NO_2$ that was present initially will remain in the flask. Reactions that stop before the limiting reagent is consumed are said to reach **equilibrium.**

It is useful to recognize the difference between reactions that come to equilibrium and those that stop when they run out of the limiting reagent. The reaction between a copper penny and nitric acid is an example of a reaction that continues until it has essentially run out of the limiting reagent. We indicate this by writing the equation for the reaction with a single arrow pointing from the reactants to the products:

$$Cu(s) + 4\ HNO_3(aq) \longrightarrow Cu^{2+}(aq) + 2\ NO_3^-(aq) + 2\ NO_2(g) + 2\ H_2O(l)$$

We indicate that a reaction comes to equilibrium by writing a pair of arrows pointing in opposite directions between the two sides of the equation:

$$2\ NO_2(g) \rightleftharpoons N_2O_4(g)$$

To work with reactions that come to equilibrium, we need a way to specify the amount of each reactant or product that is present in the system at any moment in time. By convention, this is done by specifying the concentration of each component of the system in units of moles per liter. This quantity is indicated by a symbol that consists of the formula for the reactant or product written in parentheses. For example,

**$(NO_2)$ = concentration of $NO_2$ in moles per liter at some moment in time**

We then need a way to describe the system when it is at equilibrium. By convention, this is done by writing the symbols for each component of the system in square brackets:

**$[NO_2]$ = concentration of $NO_2$ in moles per liter if and only if the reaction is at equilibrium**

The fact that some reactions come to equilibrium raises a number of interesting questions:

1. Why do these reactions seem to stop prematurely?
2. What is the difference between reactions that seem to go to completion and reactions that reach equilibrium?
3. Is there any way to predict whether a reaction will go to completion or reach equilibrium?
4. How does a change in the conditions of the reaction influence the amount of product formed?

Before we can understand how and why a chemical reac-

Many reactions seem to continue until they run out of the limiting reagent. The reaction between a copper penny and concentrated nitric acid, for example, seems to continue until all the copper metal is gone.

tion comes to equilibrium, we will have to build a model of the factors that influence the rate of a chemical reaction. We will then apply this model to the simplest of all chemical reactions, those that occur in the gas phase. Once we understand gas-phase reactions, we will include interactions between the components of the reaction and the solvent in which it is run. By the end of this section, we will be able to calculate the concentration of any reagent in a system in which many reactions occur simultaneously.

Atmospheric pollution is the result of literally hundreds of gas-phase reactions that occur after the source of this pollution is emitted into the air.

# GAS-PHASE REACTIONS: AN INTRODUCTION TO KINETICS AND EQUILIBRIA

By examining why some reactions stop before they run out of the limiting reagent, this chapter provides the basis for answering questions such as the following:

## POINTS OF INTEREST

- Why does the rate of the following reaction slow down as the reactants are converted into the products of this reaction?

$$ClNO_2(g) + NO(g) \rightleftharpoons NO_2(g) + ClNO(g)$$

- Why does this reaction stop before the limiting reagent has been consumed?
- What does it mean to say that a reaction comes to *equilibrium?*
- How can we determine whether a reaction is at equilibrium at a given moment in time?
- How can we decide in which direction a reaction must shift in order to reach equilibrium?
- How can we use assumptions to transform difficult problems into relatively simple ones? How do we test the validity of these assumptions after they have been made?
- Why does the following reaction shift to the left when we raise the temperature at which the reaction is run?

$$2\ NO_2(g) \rightleftharpoons N_2O_4(g)$$

Why do other reactions shift to the right when we raise the temperature?

## 16.1 GAS-PHASE REACTIONS

The simplest chemical reactions are those that occur in the gas phase in a single step, such as the reaction between $ClNO_2$ to NO to form $NO_2$ and ClNO:

$$ClNO_2(g) + NO(g) \rightleftharpoons NO_2(g) + ClNO(g)$$

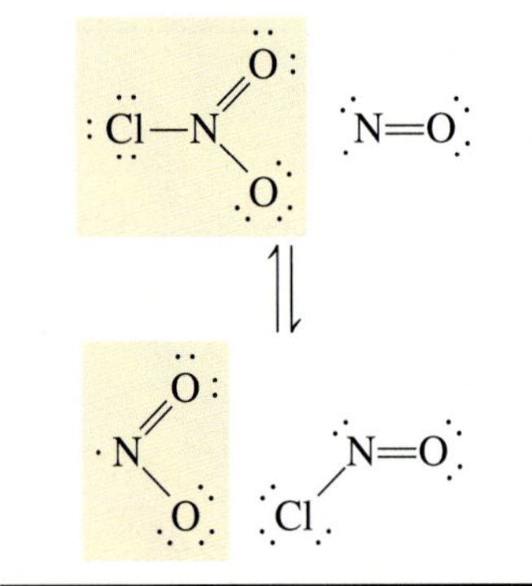


**Figure 16.1** The reaction between $ClNO_2$ and NO to form $NO_2$ and ClNO is a simple, one-step reaction that involves the transfer of a chlorine atom.

This reaction can be understood by writing the Lewis structures for all four components of the reaction. Both NO and $NO_2$ contain an odd number of electrons. Both NO and $NO_2$ can therefore combine with a neutral chlorine atom to form a molecule in which all of the electrons are paired. This reaction therefore involves the transfer of a chlorine atom from one molecule to another, as shown in Figure 16.1.

Figure 16.2 combines a plot of the disappearance of the $ClNO_2$ consumed in this reaction with a plot of the appearance of $NO_2$ formed in the reaction. One of the goals of collecting these data is to describe the **rate of reaction,** which is the rate at which the reactants are transformed into the products of the reaction. The mathematical equation that describes the rate of a chemical reaction is called the **rate law** for the reaction. The data in Figure 16.2 are consistent with the following rate law for this reaction:

$$\text{Rate} = k(ClNO_2)(NO)$$

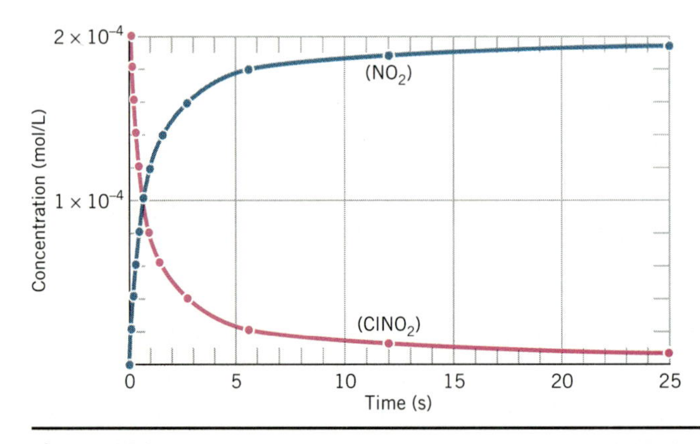


**Figure 16.2** A plot of the change in the concentration of $ClNO_2$ superimposed on a plot of the change in the concentration of $NO_2$ as $ClNO_2$ reacts with NO to produce $NO_2$ and ClNO.

According to this rate law, the rate at which $ClNO_2$ and NO are converted into $NO_2$ and ClNO is proportional to the product of the concentrations of the two reactants. Initially, the rate of reaction is fast. As the reactants are converted into products, however, the $ClNO_2$ and NO concentrations become smaller, and the reaction slows down.

We might expect the reaction to stop when it runs out of either $ClNO_2$ or NO. In practice, the reaction stops before this happens. This is a very fast reaction—the concentration of $ClNO_2$ drops by a factor of two in less than a second. And yet, no matter how long we wait, some residual $ClNO_2$ and NO remains in the reaction flask.

Figure 16.3 divides the plot of the change in the concentrations of $NO_2$ and ClNO into a **kinetic region** and an **equilibrium region.** By definition, the kinetic region is the period during which the concentrations of the components of the reaction are constantly changing. The equilibrium region is the period after which the reaction seems to stop, when there is no further change in the concentrations of the components of the reaction.

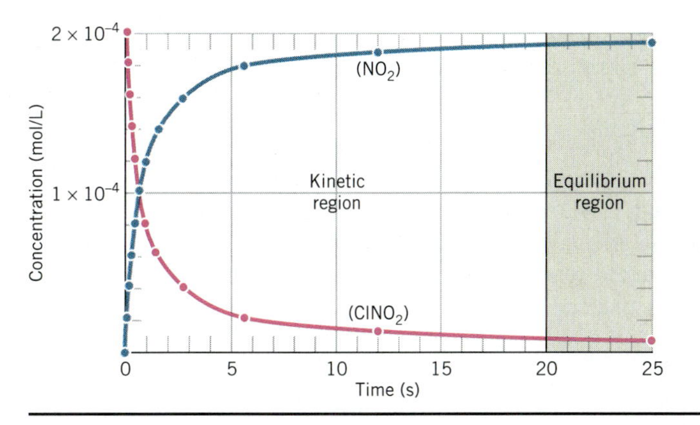


**Figure 16.3** A graph of concentration versus time can be divided into a kinetic and an equilibrium region.

## 16.2 A COLLISION THEORY MODEL FOR GAS-PHASE REACTIONS

The fact that the following reaction

$$ClNO_2(g) + NO(g) \rightleftharpoons NO_2(g) + ClNO(g)$$

seems to stop before all of the reactants are consumed can be explained with a model of chemical reactions known as the **collision theory.** This model assumes that $ClNO_2$ and NO molecules must collide before a chlorine atom can be transferred from one molecule to the other.

This assumption explains why the rate of the reaction is proportional to the concentration of both $ClNO_2$ and NO:

$$\text{Rate} = k(ClNO_2)(NO)$$

The number of collisions per second between $ClNO_2$ and NO molecules depends on their concentrations. As $ClNO_2$ and NO are consumed in the reaction, the number of collisions per second between these molecules decreases, and the reaction slows down.

Suppose that we start with a mixture of $ClNO_2$ and NO, but no $NO_2$ or ClNO. The only reaction that can occur at first is the transfer of a chlorine atom from $ClNO_2$ to NO:

$$ClNO_2(g) + NO(g) \longrightarrow NO_2(g) + ClNO(g)$$

Eventually, $NO_2$ and ClNO build up in the reaction flask, and these molecules begin to collide as well. Collisions between these molecules can result in the transfer of a chlorine atom in the reverse direction:

$$ClNO_2(g) + NO(g) \longleftarrow NO_2(g) + ClNO(g)$$

The collision theory model of chemical reactions assumes that the rate of a simple, one-step reaction is proportional to the product of the concentrations of the ions or molecules consumed in that reaction. The rate of the forward reaction is therefore proportional to the product of the concentrations of the two ''reactants'':

$$\text{Rate}_{\text{forward}} = k_f(ClNO_2)(NO)$$

The rate of the reverse reaction, on the other hand, is proportional to the concentrations of the ''products'' of the reaction:

$$\text{Rate}_{\text{reverse}} = k_r(NO_2)(ClNO)$$

Initially, the rate of the forward reaction is much larger than the rate of the reverse reaction because the system contains $ClNO_2$ and NO, but virtually no $NO_2$ and ClNO:

$$\textbf{Initially:} \qquad \text{rate}_{\text{forward}} \gg \text{rate}_{\text{reverse}}$$

As $ClNO_2$ and NO are consumed, the rate of the forward reaction slows down. At the same time, $NO_2$ and ClNO accumulate, and the reverse reaction speeds up.

If the forward reaction gradually slows down and the reverse reaction speeds up, the system eventually has to reach a point at which the rates of the forward and reverse reactions are the same:

$$\textbf{Eventually:} \qquad \text{rate}_{\text{forward}} = \text{rate}_{\text{reverse}}$$

At this point, the reaction will seem to stop. $ClNO_2$ and NO will be consumed in the forward reaction at the rate at which they are produced in the reverse reaction. The same thing will happen to $NO_2$ and ClNO. When the rates of the forward and reverse reactions are the same, there is no longer any change in the concentrations of the reactants or products of the reaction. In other words, the reaction is at equilibrium.

We can now see that there are two definitions of **equilibrium.**

1. A system in which there is no apparent change in the concentrations of the reactants and products of a reaction.
2. A system in which the rates of the forward and reverse reactions are equal.

The first definition is based on the results of experiments that tell us that some reactions seem to stop prematurely—they reach a point at which no more reactants are converted into products before the limiting reagent is consumed. The other definition is based on a theoretical model of chemical reactions that explains why reactions reach equilibrium.

We can now distinguish between reactions that go to completion and those that reach equilibrium. Reactions that aren't reversible, or that strongly favor the products, are assumed to go to completion and are represented by equations that contain a single arrow:

$$2\ Mg(s) + O_2(g) \longrightarrow 2\ MgO(s)$$

Reversible reactions that reach equilibrium are indicated by a pair of arrows between the two sides of the equation:

$$ClNO_2(g) + NO(g) \rightleftharpoons NO_2(g) + ClNO(g)$$

## 16.3 EQUILIBRIUM CONSTANT EXPRESSIONS

Reactions don't stop when they come to equilibrium. But, the forward and reverse reactions are in balance at equilibrium, so there is no net change in the concentrations of the reactants or products, and the reaction appears to stop on the macroscopic scale. Chemical equilibrium is an example of a *dynamic* balance between opposing actions—the forward and reverse reactions—not a static balance.

Let's look at the logical consequences of the assumption that the reaction between $ClNO_2$ and NO eventually reaches equilibrium:

$$ClNO_2(g) + NO(g) \rightleftharpoons NO_2(g) + ClNO(g)$$

We have argued that the rates of the forward and reverse reactions are the same when this system is at equilibrium:

$$\textbf{At equilibrium:} \qquad \text{rate}_{\text{forward}} = \text{rate}_{\text{reverse}}$$

Substituting the rate laws for the forward and reverse reactions into this equality gives the following result:

$$\text{At equilibrium:} \qquad k_f(ClNO_2)(NO) = k_r(NO_2)(ClNO)$$

But this equation is only valid when the system is at equilibrium, so we should replace the ($ClNO_2$), (NO), ($NO_2$), and (ClNO) terms with symbols that indicate that the reaction is at equilibrium. By convention, we use square brackets for this purpose. The equation describing the balance between the forward and reverse reactions when the system is at equilibrium therefore should be written as follows:

$$\textbf{At equilibrium:} \qquad k_f[ClNO_2][NO] = k_r[NO_2][ClNO]$$

Rearranging this equation gives the following result.

$$\frac{k_f}{k_r} = \frac{[NO_2][ClNO]}{[ClNO_2][NO]}$$

Since $k_f$ and $k_r$ are constants, the ratio of $k_f$ divided by $k_r$ must also be a constant. This ratio is the **equilibrium constant** for the reaction, $K_c$. The ratio of the concentrations of the reactants and products is known as the **equilibrium constant expression:**

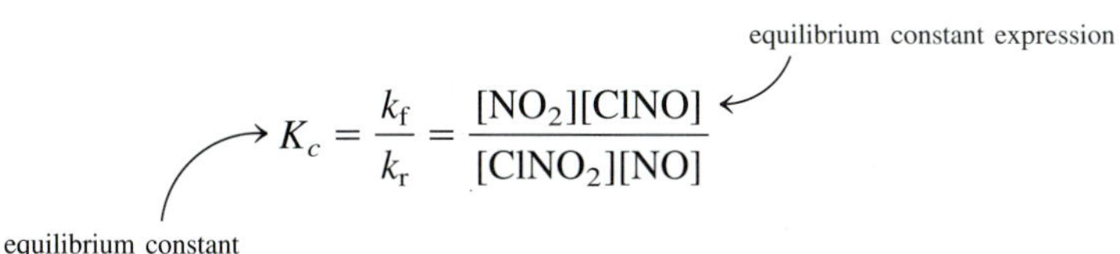


$$K_c = \frac{k_f}{k_r} = \frac{[NO_2][ClNO]}{[ClNO_2][NO]}$$

No matter what combination of concentrations of reactants and products we start with, the reaction will reach equilibrium when the ratio of the concentrations defined by the equilibrium constant expression is equal to the equilibrium constant for the reaction. We can start with a lot of $ClNO_2$ and very little NO, or a lot of NO and very little $ClNO_2$. It doesn't matter.

**At constant temperature, when the reaction reaches equilibrium, the relationship between the concentrations of the reactants and products described by the equilibrium constant expression will always be the same.**

At 25°C, this reaction always reaches equilibrium when the ratio of these concentrations is $1.3 \times 10^4$:

$$K_c = \frac{[NO_2][ClNO]}{[ClNO_2][NO]} = 1.3 \times 10^4$$

## ▶ CHECKPOINT

Explain why the equilibrium constant is larger than 1 for reactions in which the rate constant for the forward reaction is larger than the rate constant for the reverse reaction.

The procedure used in this section to derive the equilibrium constant expression only works with reactions that occur in a single step, such as the transfer of a chlorine atom from $ClNO_2$ to NO. Many reactions take a number of steps to convert

Chemical reactions provide an example of a *dynamic* equilibrium, in which processes are going in opposite directions, which is very different from the *static* equilibrium found in other systems, such as riding a bicycle, where equilibrium involves a balance between opposing forces.

reactants into products. But any reaction that reaches equilibrium, no matter how simple or complex, has an equilibrium constant expression that satisfies the following rules.

**RULES FOR WRITING EQUILIBRIUM CONSTANT EXPRESSIONS**

- Even though chemical reactions that reach equilibrium occur in both directions, the reagents on the right side of the equation are assumed to be the ''products'' of the reaction and the reagents on the left side of the equation are assumed to be the ''reactants.''
- The products of the reaction are always written above the line—in the numerator.
- The reactants are always written below the line—in the denominator.
- For homogeneous systems, the equilibrium constant expression contains a term for every reactant and every product of the reaction.
- The numerator of the equilibrium constant expression is the product of the concentrations of the ''products'' of the reaction raised to a power equal to the coefficient for this component in the balanced equation for the reaction.
- The denominator of the equilibrium constant expression is the product of the concentrations of the ''reactants'' raised to a power equal to the coefficient for this component in the balanced equation for the reaction.

## EXERCISE 16.1

Write equilibrium constant expressions for the following reactions:

(a) $2\ NO_2(g) \rightleftharpoons N_2O_4(g)$

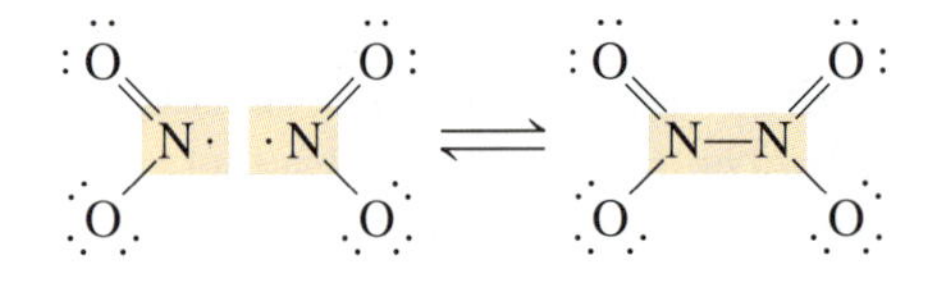

(b) $2\ SO_3(g) \rightleftharpoons 2\ SO_2(g) + O_2(g)$

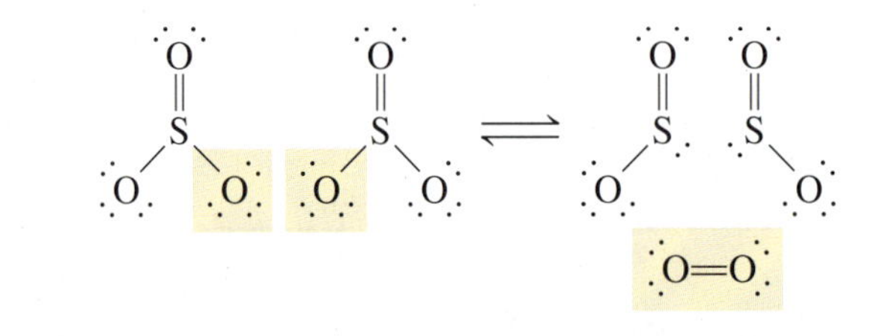

(c) $N_2(g) + 3\ H_2(g) \rightleftharpoons 2\ NH_3(g)$

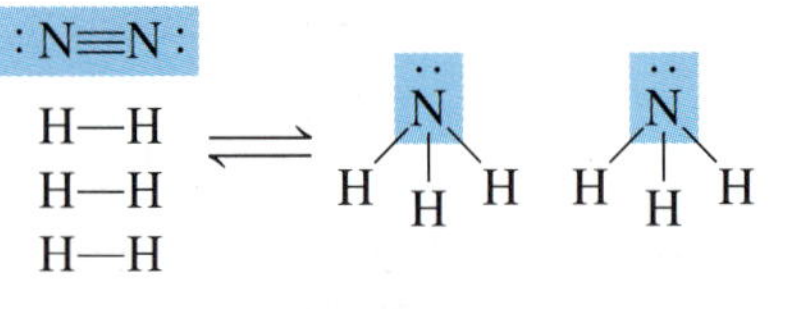

**SOLUTION** In each case, the equilibrium constant expression is the product of the concentrations of the species on the right side of the equation divided by the product of the concentrations of those on the left side of the equation:

$$\text{(a)}\ K_c = \frac{[N_2O_4]}{[NO_2]^2} \quad \text{(b)}\ K_c = \frac{[SO_2]^2[O_2]}{[SO_3]^2} \quad \text{(c)}\ K_c = \frac{[NH_3]^2}{[N_2][H_2]^3}$$

Gas-phase reactions were chosen for this introduction to kinetics and equilibrium because they are among the simplest chemical reactions. Some might therefore question why the equilibrium constant expressions in the preceding exercise are expressed in terms of the concentrations of the gases in units of moles per liter.

Units of concentration were used to emphasize the relationship between chemical equilibria and the rates of chemical reactions, which are reported in terms of the concentrations of the reactants and products. This choice of units is indicated by adding a subscript $c$ to the symbols for the equilibrium constants, to show that they were calculated from the concentrations of the components of the reaction.

## 16.4 ALTERING OR COMBINING EQUILIBRIUM REACTIONS

What happens to the magnitude of the equilibrium constant for a reaction when we turn the equation around? Consider the following reaction, for example:

$$ClNO_2(g) + NO(g) \rightleftharpoons NO_2(g) + ClNO(g)$$

The equilibrium constant expression for this equation is written as follows:

$$K_c = \frac{[NO_2][ClNO]}{[ClNO_2][NO]} = 1.3 \times 10^4 \qquad \text{(at 25°C)}$$

Because this is a reversible reaction, it also can be represented by an equation written in the opposite direction:

$$NO_2(g) + ClNO(g) \rightleftharpoons ClNO_2(g) + NO(g)$$

The equilibrium constant expression is now written as follows:

$$K_c' = \frac{[ClNO_2][NO]}{[NO_2][ClNO]}$$

Each of these equilibrium constant expressions is the inverse of the other. We can therefore calculate $K_c'$ by dividing $K_c$ into 1:

$$K_c' = \frac{1}{K_c} = \frac{1}{1.3 \times 10^4} = 7.7 \times 10^{-5}$$

We can also calculate equilibrium constants by combining two or more reactions for which the value of $K_c$ is known. Assume, for example, that we know the equilibrium constants for the following gas-phase reactions at 200°C:

$$N_2(g) + O_2(g) \rightleftharpoons 2\ NO(g) \qquad K_{c1} = 2.3 \times 10^{-19}$$

$$2\ NO(g) + O_2(g) \rightleftharpoons 2\ NO_2(g) \qquad K_{c2} = 3 \times 10^6$$

We can combine these reactions to obtain an overall equation for the reaction between $N_2$ and $O_2$ to form $NO_2$.

$$\begin{array}{l} N_2(g) + O_2(g) \rightleftharpoons \cancel{2\ NO(g)} \\ +\cancel{2\ NO(g)} + O_2(g) \rightleftharpoons 2\ NO_2(g) \\ \hline N_2(g) + 2\ O_2(g) \rightleftharpoons 2\ NO_2(g) \end{array}$$

It is easy to show that the equilibrium constant expression for the overall reaction is equal to the product of the equilibrium constant expressions for the two steps in this reaction.

$$\frac{[NO_2]^2}{[N_2][O_2]^2} = \frac{\cancel{[NO]^2}}{[N_2][O_2]} \times \frac{[NO_2]^2}{\cancel{[NO]^2}[O_2]}$$

The equilibrium constant for the overall reaction is therefore equal to the product of the equilibrium constants for the individual steps.

$$K_c = K_{c1} \times K_{c2} = (2.3 \times 10^{-19})(3 \times 10^6) = 7 \times 10^{-13}$$

## 16.5 REACTION QUOTIENTS: A WAY TO DECIDE WHETHER A REACTION IS AT EQUILIBRIUM

We now have a model that describes what happens when a reaction reaches equilibrium: At the molecular level, the rate of the forward reaction is equal to the rate of the reverse reaction. Since the reaction proceeds in both directions at the same rate, there is no apparent change in the concentrations of the reactants or the products on the macroscopic scale—the level of objects visible to the naked eye. This model also can be used to predict the direction in which a reaction must shift to reach equilibrium.

If the concentrations of the reactants are too large for the reaction to be at equilibrium, the rate of the forward reaction will be faster than the reverse reaction, and some of the reactants will be converted into products until equilibrium is achieved. Conversely, if the concentrations of the reactants are too small, the rate of the reverse reaction will exceed that of the forward reaction, and the reaction will convert some of the excess products back into reactants until the system reaches equilibrium.

We can determine the direction in which a reaction has to shift to reach equilibrium by calculating the **reaction quotient** ($Q_c$) for the reaction. The reaction quotient is defined as the product of the concentrations of the products of the reaction divided by the product of the concentration of the reactants at any moment in time:

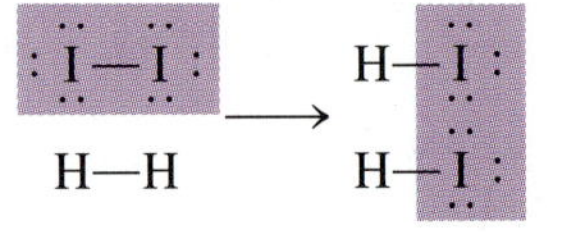


To illustrate how the reaction quotient is used, let's consider the following gas-phase reaction:

$$H_2(g) + I_2(g) \rightleftharpoons 2\ HI(g)$$

The equilibrium constant expression for this reaction is written as follows.

$$K_c = \frac{[HI]^2}{[H_2][I_2]} = 60 \qquad \text{(at 350°C)}$$

By analogy, we can write the expression for the reaction quotient as follows.

$$Q_c = \frac{(HI)^2}{(H_2)(I_2)}$$

$Q_c$ can take on any value between zero and infinity. If the system contains a great deal of HI and very little $H_2$ and $I_2$, the reaction quotient is very large. If the system contains relatively little HI and a great deal of $H_2$ and $I_2$, the reaction quotient is very small.

At any moment in time, there are three possibilities.

1. **$Q_c$ is smaller than $K_c$.** The system contains too much reactant and not enough product to be at equilibrium. The value of $Q_c$ must increase in order for the reaction to reach equilibrium. Thus, the reaction has to convert some of the reactants into products to come to equilibrium.
2. **$Q_c$ is equal to $K_c$.** If this is true, then the reaction is at equilibrium.
3. **$Q_c$ is larger than $K_c$.** The system contains too much product and not enough reactant to be at equilibrium. The value of $Q_c$ must become smaller before the reaction can come to equilibrium. Thus, the reaction must convert some of the products into reactants to reach equilibrium.

**CHECKPOINT**

Explain why there is only one value of the equilibrium constant ($K_c$) for a reaction at a given temperature, but there are an infinite number of possible values for the reaction quotient ($Q_c$).

**EXERCISE 16.2**

Assume that the concentrations of $H_2$, $I_2$, and HI can be measured for the following reaction at any moment in time:

$$H_2(g) + I_2(g) \rightleftharpoons 2\ HI(g) \qquad K_c = 60 \text{ (at 350°C)}$$

For each of the following sets of concentrations, determine whether the reaction is at equilibrium. If it isn't, decide in which direction it must go to reach equilibrium.

(a) $(H_2) = (I_2) = (HI) = 0.010\ M$
(b) $(HI) = 0.30\ M$; $(H_2) = 0.010\ M$; $(I_2) = 0.15\ M$
(c) $(H_2) = (HI) = 0.10\ M$; $(I_2) = 0.0010\ M$

**SOLUTION**

(a) The only way to decide whether the reaction is at equilibrium is to compare the reaction quotient with the equilibrium constant for the reaction:

$$Q_c = \frac{(HI)^2}{(H_2)(I_2)} = \frac{(0.010)^2}{(0.010)(0.010)} = \mathbf{1 < K_c}$$

The reaction quotient in this case is smaller than the equilibrium constant. The only way to get this system to equilibrium is to increase the magnitude of the reaction quotient. This can be done by converting some of the $H_2$ and $I_2$ into HI. The reaction therefore has to shift to the right to reach equilibrium.

(b) The reaction quotient for this set of concentrations is equal to the equilibrium constant for the reaction:

$$Q_c = \frac{(HI)^2}{(H_2)(I_2)} = \frac{(0.30)^2}{(0.010)(0.15)} = \mathbf{60 = K_c}$$

The reaction is therefore at equilibrium.

(c) The reaction quotient for this set of concentrations is larger than the equilibrium constant for the reaction:

$$Q_c = \frac{(HI)^2}{(H_2)(I_2)} = \frac{(0.10)^2}{(0.10)(0.0010)} = 100 > K_c$$

In order to reach equilibrium, the concentrations of the reactants and products must be adjusted until the reaction quotient is equal to the equilibrium constant. This involves converting some of the HI back into $H_2$ and $I_2$. In other words, the reaction has to shift to the left to reach equilibrium.

## 16.6 CHANGES IN CONCENTRATION THAT OCCUR AS A REACTION COMES TO EQUILIBRIUM

The relative size of $Q_c$ and $K_c$ for a reaction tells us whether the reaction is at equilibrium at any particular time. If it isn't, the relative size of $Q_c$ and $K_c$ tell us the direction in which the reaction must shift to reach equilibrium. Now we need a way of predicting how far the reaction has to go to reach equilibrium. Suppose that you are faced with the following problem:

> Phosphorus pentachloride decomposes to phosphorus trichloride and chlorine when heated.
>
> $$PCl_5(g) \rightleftharpoons PCl_3(g) + Cl_2(g)$$
>
> The equilibrium constant for this reaction is 0.030 at 250°C. Assume that the initial concentration of $PCl_5$ is 0.100 moles per liter and there is no $PCl_3$ or $Cl_2$ in the system when we start. Calculate the concentrations of $PCl_5$, $PCl_3$, and $Cl_2$ at equilibrium.

**Problem-Solving Strategy**

The first step toward solving this problem involves organizing the information so that it provides clues as to how to proceed. The problem contains four chunks of information: (1) a balanced equation, (2) an equilibrium constant for the reaction, (3) a description of the initial conditions, and (4) an indication of the goal of the calculation.

The following format offers a useful way to summarize this information.

| | $PCl_5(g)$ | $\rightleftharpoons$ | $PCl_3(g)$ | + | $Cl_2(g)$ | $K_c = 0.030$ |
|---|---|---|---|---|---|---|
| Initial: | 0.100 *M* | | 0 | | 0 | |
| Equilibrium: | ? | | ? | | ? | |

**Problem-Solving Strategy**

We start with the balanced equation and the equilibrium constant for the reaction and then add what we know about the initial and equilibrium concentrations of the various components of the reaction. Initially, the flask contains 0.100 moles per liter of $PCl_5$ and no $PCl_3$ or $Cl_2$. Our goal is to calculate the equilibrium concentrations of these three substances.

Before we do anything else, we have to decide whether the reaction is at equilibrium. We can do this by comparing the reaction quotient for the initial conditions with the equilibrium constant for the reaction:

$$Q_c = \frac{(PCl_3)(Cl_2)}{(PCl_5)} = \frac{(0)(0)}{(0.100)} = 0 < K_c$$

Although the equilibrium constant is small ($K_c = 3.0 \times 10^{-2}$), the reaction quotient is even smaller ($Q_c = 0$). The only way for this reaction to get to equilibrium is for some of the $PCl_5$ to decompose into $PCl_3$ and $Cl_2$.

Since the reaction isn't at equilibrium, one thing is sure—the concentrations of $PCl_5$, $PCl_3$, and $Cl_2$ will all change as the reaction comes to equilibrium. Because the reaction must shift to the right to reach equilibrium, the $PCl_5$ concentration will become smaller, while the $PCl_3$ and $Cl_2$ concentration will become larger.

At first glance, this problem appears to have three unknowns: the equilibrium concentrations of $PCl_5$, $PCl_3$, and $Cl_2$. Because it is difficult to solve a problem in three unknowns, we should look for relationships that can reduce the problem's complexity. One way of achieving this goal is to look at the relationship between the changes that occur in the concentrations of $PCl_5$, $PCl_3$, and $Cl_2$ as the reaction approaches equilibrium.

## EXERCISE 16.3

Calculate the increase in the $PCl_3$ and $Cl_2$ concentrations that occur as the following reaction comes to equilibrium if the concentration of $PCl_5$ decreases by 0.042 moles per liter.

$$PCl_5(g) \rightleftharpoons PCl_3(g) + Cl_2(g)$$

**SOLUTION** The decomposition of $PCl_5$ has a 1 : 1 : 1 stoichiometry, as shown in Figure 16.4:

$$PCl_5(g) \rightleftharpoons PCl_3(g) + Cl_2(g)$$

For every mole of $PCl_5$ that decomposes, we get 1 mole of $PCl_3$ and 1 mole of $Cl_2$. Thus, the change in the concentration of $PCl_5$ that occurs as the reaction comes to equilibrium is equal to the change in the $PCl_3$ and $Cl_2$ concentrations. If 0.042 moles per liter of $PCl_5$ are consumed as this reaction comes to equilibrium, 0.042 moles per liter of both $PCl_3$ and $Cl_2$ must be formed at the same time.

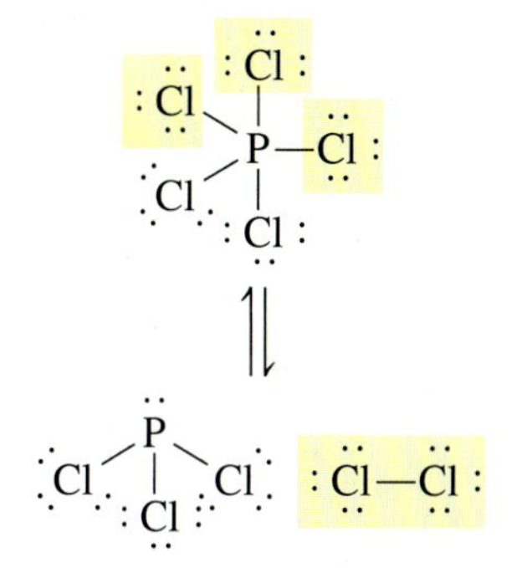

**Figure 16.4** The decomposition of $PCl_5$ to form $PCl_3$ and $Cl_2$ is a reversible reaction with a 1 : 1 : 1 stoichiometry.

This exercise raises an important point. There is a simple relationship between the *change in the concentrations* of the three components of the reaction as it comes to equilibrium because of the stoichiometry of the reaction.

It would be useful to have a symbol to represent the change that occurs in the concentration of one of the components of a reaction as it goes from the initial conditions to equilibrium. In Chapter 5, we introduced the concept of a *state function* as a property of a system whose value only depends on the state of the system. We then argued that the change in the value of a state function was defined by the following equation:

Sec. 5.11

$$\Delta X = X_{\text{final}} - X_{\text{initial}}$$

We can extend this argument to discussions of chemical reactions that come to equilibrium by defining $\Delta(X)$ as the magnitude of the change that occurs in the concentration of $X$ as the reaction comes to equilibrium. We can define $\Delta(PCl_5)$, for example, as the magnitude of the change in the concentration of $PCl_5$ that occurs as this compound decomposes to form $PCl_3$ and $Cl_2$:

$$\underbrace{\Delta(PCl_5)}_{\substack{PCl_5 \text{ consumed} \\ \text{as the reaction} \\ \text{comes to equilibrium}}} = \underbrace{(PCl_5)}_{\substack{\text{initial} \\ \text{concentration}}} - \underbrace{[PCl_5]}_{\substack{\text{concentration at} \\ \text{equilibrium}}}$$

Rearranging this equation, we find that the concentration of $PCl_5$ at equilibrium is equal to the initial concentration of $PCl_5$ minus the amount of $PCl_5$ consumed as the reaction comes to equilibrium:

$$\underset{\text{concentration at equilibrium}}{[PCl_5]} = \underset{\text{initial concentration}}{(PCl_5)} - \underset{\text{PCl}_5\text{ consumed as reaction comes to equilibrium}}{\Delta(PCl_5)}$$

We can then define $\Delta(PCl_3)$ and $\Delta(Cl_2)$ as the changes that occur in the $PCl_3$ and $Cl_2$ concentrations as the reaction comes to equilibrium. The concentrations of both these substances at equilibrium will be larger than their initial concentrations:

$$[PCl_3] = (PCl_3) + \Delta(PCl_3)$$
$$[Cl_2] = (Cl_2) + \Delta(Cl_2)$$

According to Exercise 16.3, the magnitude of the changes in the concentrations of these three substances as the reaction comes to equilibrium are the same. Because of the 1:1:1 stoichiometry of the reaction, the magnitude of the change in the concentration of $PCl_5$ as the reaction comes to equilibrium is equal to the magnitude of the change in the concentrations of $PCl_3$ and $Cl_2$:

$$\Delta(PCl_5) = \Delta(PCl_3) = \Delta(Cl_2)$$

We can therefore rewrite the equations that define the equilibrium concentrations of $PCl_5$, $PCl_3$, and $Cl_2$ in terms of a single unknown: $\Delta$.

$$[PCl_5] = (PCl_5) - \Delta$$
$$[PCl_3] = (PCl_3) + \Delta$$
$$[Cl_2] = (Cl_2) + \Delta$$

Substituting what we know about the initial concentrations of $PCl_5$, $PCl_3$, and $Cl_2$ into these equations gives the following result.

$$[PCl_5] = 0.100 - \Delta$$
$$[PCl_3] = [Cl_2] = 0 + \Delta$$

We can now summarize what we know about this reaction as follows.

| | $PCl_5(g)$ $\rightleftharpoons$ | $PCl_3(g)$ + | $Cl_2(g)$ |
|---|---|---|---|
| Initial: | 0.100 $M$ | 0 | 0 |
| Equilibrium: | $0.100 - \Delta$ | $\Delta$ | $\Delta$ |

We now have only one unknown, $\Delta$, and we need only one equation to solve for one unknown. The obvious equation to turn to is the equilibrium constant expression for this reaction:

$$K_c = \frac{[PCl_3][Cl_2]}{[PCl_5]} = 0.030$$

Substituting what we know about the equilibrium concentrations of $PCl_5$, $PCl_3$, and $Cl_2$ into this equation gives the following result:

$$\frac{[\Delta][\Delta]}{[0.100 - \Delta]} = 0.030$$

This equation can be expanded and then rearranged to give a quadratic equation:

$$\Delta^2 + 0.030\,\Delta - 0.0030 = 0$$

which can be solved with the quadratic formula:

$$\Delta = \frac{-b \pm \sqrt{b^2 - 4ac}}{2a} = \frac{-(0.030) \pm \sqrt{(0.030)^2 - 4(1)(-0.0030)}}{2(1)}$$

$$\Delta = 0.042 \text{ or } -0.072$$

Although two answers come out of this calculation, only the positive root makes any physical sense because we can't have a negative concentration. Thus, the magnitude of the change in the concentrations of $PCl_5$, $PCl_3$, and $Cl_2$ as this reaction comes to equilibrium is 0.042 moles per liter:

$$\Delta = 0.042\ M$$

Plugging this value of $\Delta$ back into the equations that define the equilibrium concentrations of $PCl_5$, $PCl_3$, and $Cl_2$ gives the following results:

$$[PCl_5] = 0.100 - 0.042 = 0.058\ M$$

$$[PCl_3] = [Cl_2] = 0 + 0.042 = 0.042\ M$$

In other words, slightly less than half of the $PCl_5$ present initially decomposes into $PCl_3$ and $Cl_2$ when this reaction comes to equilibrium.

To check whether the results of this calculation represent legitimate values for the equilibrium concentrations of the three components of this reaction, we can substitute these values into the equilibrium constant expression:

$$K_c = \frac{[PCl_3][Cl_2]}{[PCl_5]} = \frac{[0.042][0.042]}{[0.058]} = 0.030$$

These results must be legitimate because the equilibrium constant calculated from these concentrations is equal to the value of $K_c$ given in the problem within experimental error.

## EXERCISE 16.4

Assume the following initial concentrations: $(PCl_5) = 0.100\ M$ and $(Cl_2) = 0.020\ M$. Calculate the equilibrium concentrations of $PCl_5$, $PCl_3$, and $Cl_2$ at 250°C if the equilibrium constant for the decomposition of $PCl_5$ at this temperature is 0.030.

**SOLUTION** We start by representing the problem as follows.

| | $PCl_5(g)$ $\rightleftharpoons$ | $PCl_3(g)$ + | $Cl_2(g)$ | $K_c = 0.030$ |
|---|---|---|---|---|
| Initial: | 0.100 $M$ | 0 | 0.020 $M$ | |
| Equilibrium: | ? | ? | ? | |

Before doing anything else, we need to compare the reaction quotient for the initial conditions with the equilibrium constant for the reaction:

$$Q_c = \frac{(PCl_3)(Cl_2)}{(PCl_5)} = \frac{(0)(0.020)}{(0.100)} = 0 < K_c$$

Once again, the reaction quotient ($Q_c = 0$) is smaller than the equilibrium constant ($K_c = 0.030$). Some of the $PCl_5$ present initially therefore has to decompose to $PCl_3$ and $Cl_2$ before the reaction can come to equilibrium. Because of the stoichiometry of this reaction, the magnitude of the change in the $PCl_5$ concentration as the reaction comes to equilibrium is equal to the magnitude of the changes in the concentrations of $PCl_3$ and $Cl_2$. The problem can therefore be represented as follows:

| | $PCl_5(g)$ $\rightleftharpoons$ | $PCl_3(g)$ + | $Cl_2(g)$ | $K_c = 0.030$ |
|---|---|---|---|---|
| Initial: | 0.100 $M$ | 0 | 0.020 $M$ | |
| Change: | $-\Delta$ | $+\Delta$ | $+\Delta$ | |
| Equilibrium: | $0.100 - \Delta$ | $\Delta$ | $0.020 + \Delta$ | |

Substituting what we know about the concentrations of $PCl_5$, $PCl_3$, and $Cl_2$ into the equilibrium constant expression for the reaction gives the following result:

$$K_c = \frac{[PCl_3][Cl_2]}{[PCl_5]} = \frac{[\Delta][0.020 + \Delta]}{[0.100 - \Delta]} = 0.030$$

Expanding this gives the following quadratic equation:

$$\Delta^2 + 0.050\,\Delta - 0.0030 = 0$$

Solving this equation with the quadratic formula gives the following positive root:

$$\Delta = 0.035\ M$$

This value of $\Delta$ can be used to calculate the equilibrium concentrations of the three components of the reaction:

$$[PCl_5] = 0.100 - \Delta = 0.065\ M$$
$$[PCl_3] = 0 + \Delta = 0.035\ M$$
$$[Cl_2] = 0.020 + \Delta = 0.055\ M$$

In this case, only about one-third of the $PCl_5$ present initially decomposes as the reaction comes to equilibrium.

We can check these results by substituting them into the equilibrium constant expression:

$$K_c = \frac{[PCl_3][Cl_2]}{[PCl_5]} = \frac{[0.035][0.055]}{[0.065]} = 0.030$$

The values for the equilibrium concentrations of $PCl_5$, $PCl_3$, and $Cl_2$ in this calculation must be legitimate because the results of this calculation once again agree with the equilibrium constant for the reaction within experimental error.

## 16.7 HIDDEN ASSUMPTIONS THAT MAKE EQUILIBRIUM CALCULATIONS EASIER

Suppose that we are asked to solve a slightly more difficult problem:

Sulfur trioxide decomposes to give sulfur dioxide and oxygen with an equilibrium constant of $1.6 \times 10^{-10}$ at 300°C:

$$2\ SO_3(g) \rightleftharpoons 2\ SO_2(g) + O_2(g)$$

Calculate the equilibrium concentrations of the three components of this system if the initial concentration of $SO_3$ is 0.100 $M$.

Once again, the first step in this problem involves building a representation of the information in the problem:

| | $2\ SO_3(g)$ $\rightleftharpoons$ | $2\ SO_2(g)$ + | $O_2(g)$ | $K_c = 1.6 \times 10^{-10}$ |
|---|---|---|---|---|
| Initial: | 0.100 *M* | 0 | 0 | |
| Equilibrium: | ? | ? | ? | |

We then compare the reaction quotient for the initial conditions with the equilibrium constant for the reaction:

$$Q_c = \frac{(SO_2)^2(O_2)}{(SO_3)^2} = \frac{(0)^2(0)}{(0.100)^2} = \mathbf{0 < K_c}$$

Because the initial concentrations of $SO_2$ and $O_2$ are zero, the reaction has to shift to the right to reach equilibrium. As might be expected, some of the $SO_3$ must decompose to $SO_2$ and $O_2$.

The stoichiometry of this reaction is more complex than the reaction in the previous section, but the changes in the concentrations of the three components of the reaction are still related. For every 2 moles of $SO_3$ that decompose we get 2 moles of $SO_2$ and 1 mole of $O_2$, as shown in Figure 16.5. We can incorporate this relationship into the format we used earlier by using the balanced equation for the reaction as a guide.

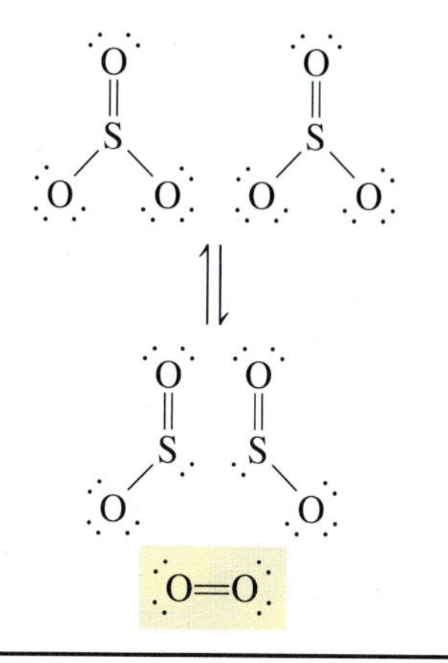

**Figure 16.5** The stoichiometry of this reaction requires that the change in the concentrations of both $SO_3$ and $SO_2$ must be twice as large as the change in the concentration of $O_2$ that occurs as this reaction comes to equilibrium.

The signs of the Δ terms in this problem are determined by the fact that the reaction has to shift from left to right to reach equilibrium. The coefficients in the Δ terms mirror the coefficients in the balanced equation for the reaction. Because twice as many moles of $SO_2$ are produced as moles of $O_2$, the change in the concentration of $SO_2$ as the reaction comes to equilibrium must be twice as large as the change in the concentration of $O_2$. Because 2 moles of $SO_3$ are consumed for every mole of $O_2$ produced, the change in the $SO_3$ concentration must be twice as large as the change in the concentration of $O_2$:

| | $2\ SO_3(g)$ $\rightleftharpoons$ | $2\ SO_2(g)$ + | $O_2(g)$ | $K_c = 1.6 \times 10^{-10}$ |
|---|---|---|---|---|
| Initial: | 0.100 *M* | 0 | 0 | |
| **Change:** | **−2Δ** | **+2Δ** | **+Δ** | |
| Equilibrium: | 0.100 − 2Δ | 2Δ | Δ | |

Substituting what we know about the problem into the equilibrium constant expression for the reaction gives the following equation:

$$K_c = \frac{[SO_2]^2[O_2]}{[SO_3]^2} = \frac{[2\Delta]^2[\Delta]}{[0.100 - 2\Delta]^2} = 1.6 \times 10^{-10}$$

This equation is a bit more of a challenge to expand, but it can be rearranged to give the following cubic equation:

$$4\Delta^3 - 6.4 \times 10^{-10}\Delta^2 + 6.4 \times 10^{-11}\Delta - 1.6 \times 10^{-12} = 0$$

Solving cubic equations is difficult, however. This problem is therefore an example of a family of problems that are difficult, if not impossible, to solve exactly. These problems are solved with a general strategy that consists of making an assumption or approximation that turns them into simpler problems. The following general rules will guide our discussion of **methods of approximation.**

**RULES FOR USING APPROXIMATION METHODS**

1. There is nothing wrong with making an assumption.
2. There are two cardinal sins:
   (a) Forgetting what assumptions have been made.
   (b) Forgetting to check whether the assumptions are valid.

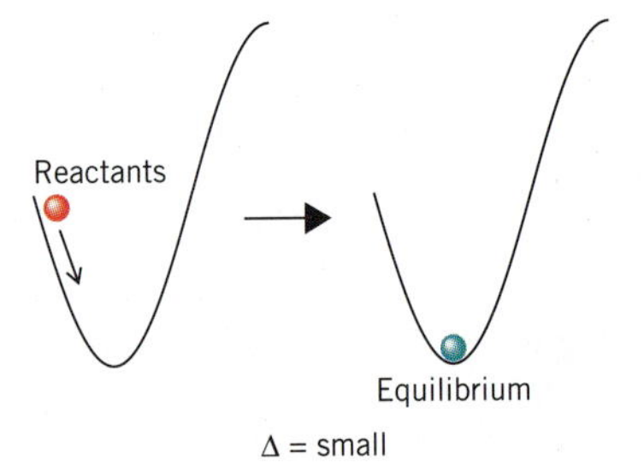


**Figure 16.6** When the initial conditions are close to equilibrium, the changes in the concentrations of the components of the reaction are often small enough compared with the initial concentrations to be ignored.

What assumption can be made to simplify this problem? Let's go back to the first thing we did after building a representation for the problem. We started our calculation by comparing the reaction quotient for the initial concentrations with the equilibrium constant for the reaction.

$$Q_c = \frac{(SO_2)^2(O_2)}{(SO_3)^2} = \frac{(0)^2(0)}{(0.100)^2} = 0 < K_c$$

We then concluded that the reaction quotient ($Q_c = 0$) was smaller than the equilibrium constant ($K_c = 1.6 \times 10^{-10}$) and decided that some of the $SO_3$ would have to decompose in order for this reaction to come to equilibrium.

But what about the relative sizes of the reaction quotient and the equilibrium constant for the reaction? The initial values of $Q_c$ and $K_c$ are both relatively small, which means that the initial conditions are reasonably close to equilibrium, as shown in Figure 16.6. As a result, the reaction does not have far to go to reach equilibrium. It is therefore reasonable to assume that Δ is relatively small in this problem.

**Problem-Solving Strategy**

It is essential to understand the nature of the assumption being made. We are not assuming that Δ is zero. If we did that, all of the unknowns would disappear from the equation! We are only assuming that Δ is small—so small compared with the initial concentration of $SO_3$ that it doesn't make a significant difference when 2Δ is subtracted from this number. We can write this assumption as follows:

$$0.100 - 2\Delta \approx 0.100$$

Let's now go back to the equation we are trying to solve:

$$\frac{[2\Delta]^2[\Delta]}{[0.100 - 2\Delta]^2} = 1.6 \times 10^{-10}$$

By making the assumption that 2Δ is very much smaller than 0.100, we can replace this equation with the following approximate equation:

$$\frac{[2\Delta]^2[\Delta]}{[0.100]^2} \approx 1.6 \times 10^{-10}$$

Expanding this equation gives an equation that is much easier to solve for Δ:

$$4\Delta^3 \approx 1.6 \times 10^{-12}$$

$$\Delta \approx 7.4 \times 10^{-5}\ M$$

Before we can go any further, we have to check our assumption that 2Δ is so small compared with 0.100 that it doesn't make a significant difference when it is subtracted from this number. Is this assumption valid? Is 2Δ small enough compared with 0.100 to be ignored?

$$0.100 - 2(0.000074) \approx 0.100$$

Yes, 2Δ is an order of magnitude smaller than the experimental error involved in the measurement of the initial concentration of $SO_3$.

We can therefore use this approximate value of Δ to calculate the equilibrium concentrations of $SO_3$, $SO_2$, and $O_2$:

$$[SO_3] = 0.100 - 2\Delta \approx \mathbf{0.100}\ M$$
$$[SO_2] = 2\Delta \approx \mathbf{1.5 \times 10^{-4}}\ M$$
$$[O_2] = \Delta \approx \mathbf{7.4 \times 10^{-5}}\ M$$

The equilibrium between $SO_3$ and mixtures of $SO_2$ and $O_2$ therefore strongly favors $SO_3$, not $SO_2$.

We can check the results of our calculation by substituting these results into the equilibrium constant expression for the reaction:

$$K_c = \frac{[SO_2]^2[O_2]}{[SO_3]^2} = \frac{[1.5 \times 10^{-4}]^2[7.4 \times 10^{-5}]}{[0.100]^2} = 1.6 \times 10^{-10}$$

The value of the equilibrium constant that comes out of this calculation agrees with the value given in the problem within experimental error. Our assumption that 2Δ is negligibly small compared with the initial concentration of $SO_3$ is therefore valid, and we can feel confident in the answers it provides.

## 16.8 A RULE OF THUMB FOR TESTING THE VALIDITY OF ASSUMPTIONS

There was no doubt about the validity of the assumption that Δ was small compared with the initial concentration of $SO_3$ in the preceding section. The value of Δ was so small that 2Δ was an order of magnitude smaller than the experimental error involved in measuring the initial concentration of $SO_3$.

In general, we can get some idea of whether Δ might be small enough to be ignored by comparing the initial reaction quotient with the equilibrium constant for the reaction. If $Q_c$ and $K_c$ are both much smaller than 1, or both much larger than 1, the reaction doesn't have very far to go to reach equilibrium, and the assumption that Δ is small enough to be ignored is probably legitimate.

This raises an interesting question: How do we decide whether it is valid to assume that Δ is small enough to be ignored? The answer to this question depends on how much error we are willing to let into our calculation before we no longer trust the results. As a rule of thumb, chemists typically assume that Δ is negligibly small as long as what is added to or subtracted from the initial concentrations of the reactants or products is less than 5% of the initial concentrations. The best way to decide whether the assumption meets this rule of thumb in a particular calculation is to try it and see if it works.

### EXERCISE 16.5

Ammonia is made from nitrogen and hydrogen by the following reversible reaction:

$$N_2(g) + 3\ H_2(g) \rightleftharpoons 2\ NH_3(g)$$

Assume that the initial concentrations of both $N_2$ and $H_2$ are 0.100 moles per liter. Calculate the equilibrium concentrations of the three components of this reaction at 500°C if the equilibrium constant for the reaction at this temperature is 0.040.

**SOLUTION** We start, as always, by building a representation for the problem based on the following general format:

| | $N_2(g)$ | $+ 3\ H_2(g)$ | $\rightleftharpoons 2\ NH_3(g)$ | $K_c = 0.040$ |
|---|---|---|---|---|
| Initial: | 0.100 *M* | 0.100 *M* | 0 | |
| Equilibrium: | ? | ? | ? | |

We then calculate the initial reaction quotient and compare it with the equilibrium constant for the reaction:

$$Q_c = \frac{(NH_3)^2}{(N_2)(H_2)^3} = \frac{(0)^2}{(0.100)(0.100)^3} = 0 < K_c$$

The reaction quotient ($Q_c = 0$) is smaller than the equilibrium constant ($K_c = 0.040$), so the reaction has to shift to the right to reach equilibrium. This will result in a decrease in the concentrations of $N_2$ and $H_2$ and an increase in the $NH_3$ concentration.

The relationship between the changes in the concentrations of $N_2$, $H_2$, and $NH_3$ as this reaction comes to equilibrium is determined by the stoichiometry of the reaction, as shown in Figure 16.7.

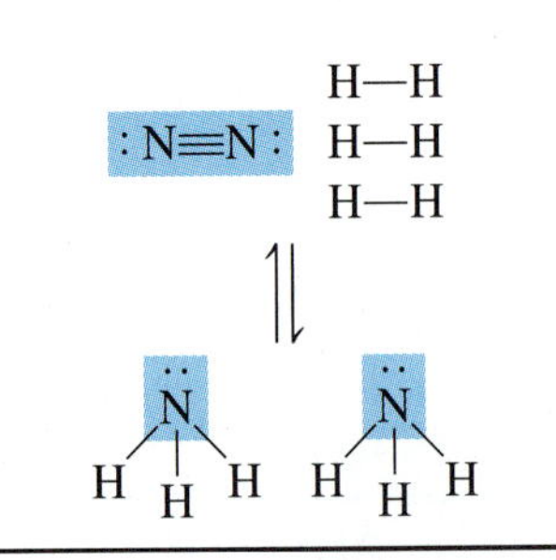


**Figure 16.7** The stoichiometry of this reaction determines the relationship between the magnitude of the changes in the concentrations of $N_2$, $H_2$, and $NH_3$ as this reaction comes to equilibrium.

| | $N_2(g)$ | $+\ 3\ H_2(g)$ | $\rightleftharpoons 2\ NH_3(g)$ | $K_c = 0.040$ |
|---|---|---|---|---|
| Initial: | 0.100 *M* | 0.100 *M* | 0 | |
| Change: | $-\Delta$ | $-3\Delta$ | $+2\Delta$ | |
| Equilibrium: | $0.100 - \Delta$ | $0.100 - 3\Delta$ | $2\Delta$ | |

Substituting this information into the equilibrium constant expression for the reaction gives the following equation:

$$K_c = \frac{[NH_3]^2}{[N_2][H_2]^3} = \frac{[2\Delta^2]}{[0.100 - \Delta][0.100 - 3\Delta]^3} = 0.040$$

Since $Q_c$ for the initial concentrations and $K_c$ for the reaction are both smaller than 1, let's try the assumption that $\Delta$ is small enough that subtracting it from 0.100—or even subtracting $3\Delta$ from 0.100—doesn't make a significant change. This assumption gives the following approximate equation:

$$\frac{[2\Delta]^2}{[0.100][0.100]^3} \approx 0.040$$

Solving this equation for $\Delta$ gives:

$$\boldsymbol{\Delta \approx 1.0 \times 10^{-3}\ M}$$

Now we have to check our assumptions. Is $\Delta$ significantly smaller than 0.100? Yes, $\Delta$ is about 1% of the initial concentration of $N_2$:

$$\frac{0.001}{0.100} \times 100\% = \mathbf{1\%}$$

Is $3\Delta$ significantly smaller than 0.100? Once again, the answer is yes, $3\Delta$ is only about 2% of the initial concentration of $H_2$:

$$\frac{3(0.001)}{0.100} \times 100\% = \mathbf{3\%}$$

We can therefore use this approximate value of $\Delta$ to determine the equilibrium concentrations of $N_2$, $H_2$, and $NH_3$:

$$[NH_3] = 2\Delta \approx \mathbf{0.0020}\ M$$
$$[N_2] = 0.100 - \Delta \approx \mathbf{0.099}\ M$$
$$[H_2] = 0.100 - 3\Delta \approx \mathbf{0.097}\ M$$

Note that only 1% of the nitrogen is converted into ammonia under these conditions.

We can check the validity of our calculation by substituting this information back into the equilibrium constant expression:

$$K_c = \frac{[NH_3]^2}{[N_2][H_2]^3} \approx \frac{[0.0020]^2}{[0.099][0.097]^3} = 0.044$$

Once again, we have reason to accept the assumption that $\Delta$ is small compared with the initial concentrations because the equilibrium constant calculated from these data agrees with the value of $K_c$ given in the problem within experimental error.

## 16.9 WHAT DO WE DO WHEN THE APPROXIMATION FAILS?

It is easy to envision a problem in which the assumption that $\Delta$ is small compared with the initial concentrations can't possibly be valid. All we have to do is construct a problem in which there is a large difference between the values of $Q_c$ for the initial concentrations and $K_c$ for the reaction at equilibrium. Consider the following problem, for example:

Nitrogen oxide reacts with oxygen to form nitrogen dioxide:

$$2\ NO(g) + O_2(g) \rightleftharpoons 2\ NO_2(g)$$

The equilibrium constant for this reaction is $3 \times 10^6$ at 200°C. Assume initial concentrations of 0.100 *M* for NO and 0.050 *M* for $O_2$. Calculate the concentrations of the three components of this reaction at equilibrium.

We start, once again, by representing the information in the problem as follows.

| | $2\ NO(g)$ | $+\ O_2(g)$ | $\rightleftharpoons 2\ NO_2(g)$ | $K_c = 3 \times 10^6$ |
|---|---|---|---|---|
| Initial: | 0.100 *M* | 0.050 *M* | 0 | |
| Equilibrium: | ? | ? | ? | |

The first step is always the same: Compare the initial value of the reaction quotient with the equilibrium constant:

$$Q_c = \frac{(NO_2)^2}{(NO)^2(O_2)} = \frac{(0)^2}{(0.100)^2(0.050)} = 0 < K_c$$

The relationship between the initial reaction quotient ($Q_c = 0$) and the equilibrium constant ($K_c = 3 \times 10^6$) tells us something we may already have suspected, the reaction must shift to the right to reach equilibrium.

Some might ask: ''Why calculate the initial value of the reaction quotient for this reaction? Isn't it obvious that the reaction has to shift to the right to produce at least some $NO_2$?'' Yes, it is. But calculating the value of $Q_c$ for the reaction does more than tell us in which *direction* it has to shift to reach equilibrium. It also gives us an indication of *how far the reaction has to go* to reach equilibrium.

In this case, $Q_c$ is so very much smaller than $K_c$ for the reaction that we have to conclude that the initial conditions are very far from equilibrium. It is therefore unlikely that $\Delta$ will be small enough to be ignored.

**Problem-Solving Strategy**

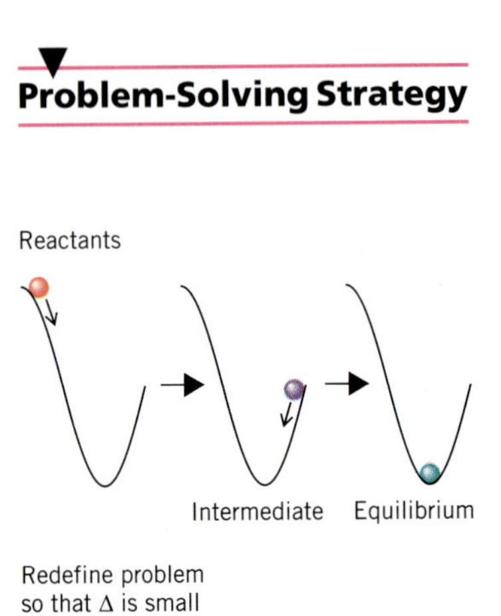


**Figure 16.8** When the initial conditions are very far from equilibrium, it is often useful to redefine the problem. This involves driving the reaction as far as possible in the direction favored by the equilibrium constant. When the reaction returns to equilibrium from these intermediate conditions, changes in the concentrations of the components of the reaction are often small enough compared with the initial concentration to be ignored.

We can't assume that $\Delta$ is negligibly small in this problem, but we can redefine the problem so that this assumption becomes valid. The key to achieving this goal is to remember the conditions under which we can assume that $\Delta$ is small enough to be ignored. This assumption is only valid when $Q_c$ is of the same order of magnitude as $K_c$. (When $Q_c$ and $K_c$ are both much larger than 1 or much smaller than 1.) We can solve problems for which $Q_c$ isn't close to $K_c$ by redefining the initial conditions so that $Q_c$ becomes close to $K_c$ (see Figure 16.8). To show how this is done, let's return to the problem given at the start of this section.

The equilibrium constant for the reaction between NO and $O_2$ to form $NO_2$ is much larger than 1 ($K_c = 3 \times 10^6$). This means that the equilibrium favors the products of the reaction. The best way to handle this problem is to drive the reaction as far as possible to the right, and then let it come back to equilibrium. Let's therefore define an intermediate set of conditions that correspond to what would happen if we push the reaction as far as possible to the right:

| | $2\ NO(g)$ | $+\ O_2(g)$ | $\rightleftharpoons 2\ NO_2(g)$ | $K_c = 3 \times 10^6$ |
|---|---|---|---|---|
| Initial: | $0.100\ M$ | $0.050\ M$ | $0$ | |
| **Intermediate:** | **0** | **0** | **$0.100\ M$** | |

We can see where this gets us by calculating the reaction quotient for the intermediate conditions:

$$Q_c = \frac{(NO_2)^2}{(NO)^2(O_2)} = \frac{(0.100)^2}{(0)^2(0)} = \infty > K_c$$

The reaction quotient is now larger than the equilibrium constant, and the reaction has to shift back to the left to reach equilibrium. Some of the $NO_2$ must now decompose to form NO and $O_2$. The relationship between the changes in the concentrations of the three components of this reaction is determined by the stoichiometry of the reaction, as shown in Figure 16.9.

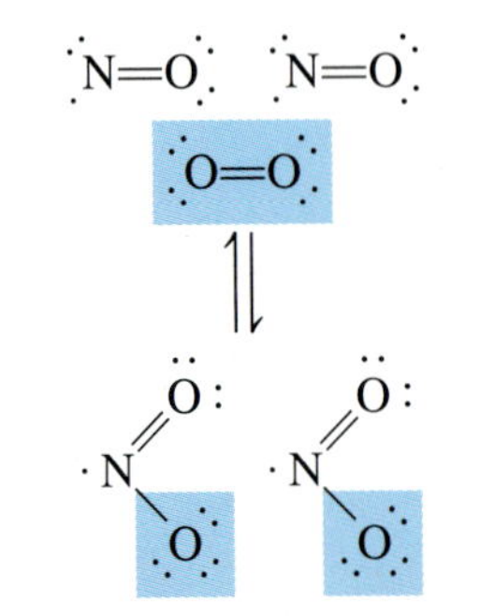

**Figure 16.9** Once again, the stoichiometry of the reaction determines the relationship between the magnitude of the changes in the concentrations of the three components of this reaction as it comes to equilibrium.

| | $2\ NO(g)$ | $+\ O_2(g)$ | $\rightleftharpoons 2\ NO_2(g)$ | $K_c = 3 \times 10^6$ |
|---|---|---|---|---|
| Intermediate: | $0$ | $0$ | $0.100\ M$ | |
| **Change:** | **$+2\Delta$** | **$+\Delta$** | **$-2\Delta$** | |
| Equilibrium: | $2\Delta$ | $\Delta$ | $0.100 - 2\Delta$ | |

We now substitute what we know about the reaction into the equilibrium constant expression:

$$K_c = \frac{[NO_2]^2}{[NO]^2[O_2]} = \frac{[0.100 - 2\Delta]^2}{[2\Delta]^2[\Delta]} = 3 \times 10^6$$

Because the reaction quotient for the intermediate conditions and the equilibrium constant are both relatively large, we can assume that the reaction doesn't have very far to go to reach equilibrium. In other words, we assume that $2\Delta$ is small compared with the intermediate concentration of $NO_2$ and derive the following approximate equation:

$$\frac{[0.100]^2}{[2\Delta]^2[\Delta]} \approx 3 \times 10^6$$

We then solve this equation for an approximate value of $\Delta$:

$$\Delta \approx 9 \times 10^{-4}\ M$$

We now check our assumption that $2\Delta$ is small enough compared with the intermediate concentration of $NO_2$ to be ignored:

$$\frac{2(0.0009)}{0.100} \times 100\% = 1.8\%$$

The value of $2\Delta$ is less than 2% of the intermediate concentration of $NO_2$, which means that it can be legitimately ignored in this calculation.

Since the approximation is valid, we can use the new value of $\Delta$ to calculate the equilibrium concentrations of NO, $NO_2$, and $O_2$:

$$[NO_2] = 0.100 - 2\Delta \approx \mathbf{0.098\ M}$$
$$[NO] = 2\Delta \approx \mathbf{0.0018\ M}$$
$$[O_2] = \Delta \approx \mathbf{0.0009\ M}$$

The results of this calculation provide insight into a chemistry of the pollutants formed by an internal combustion engine. When a mixture of gasoline and air is burned, the $N_2$ and $O_2$ in air react to form NO, which can then react with oxygen to form $NO_2$:

$$N_2(g) + O_2(g) \rightleftharpoons 2\ NO(g)$$
$$2\ NO(g) + O_2(g) \rightleftharpoons 2\ NO_2(g)$$

Although the product of these reactions is often described as $NO_x$—to indicate that it is a mixture of NO and $NO_2$—this calculation suggests that the dominant product of the reaction would be $NO_2$, if this reaction comes to equilibrium.

We can check our calculations by substituting these concentrations back into the equilibrium constant expression:

$$K_c = \frac{[NO_2]^2}{[NO]^2[O_2]} = \frac{[0.098]^2}{[0.0018]^2[0.0009]} = 3 \times 10^6$$

Once again, we can accept the validity of the assumption we had to make to get these equilibrium concentrations because the value of the equilibrium constant that comes out of this calculation agrees with the value of $K_c$ given in the problem within experimental error.

In general, the assumption that $\Delta$ is small compared with the initial concentrations of the reactants or products works best under the following conditions:

1. When $K_c \ll 1$ and we approach equilibrium from left to right. (We start with excess reactants and form some products.)
2. When $K_c \gg 1$ and we approach equilibrium from right to left. (We start with excess products and form some reactants.)

## 16.10 EQUILIBRIA EXPRESSED IN PARTIAL PRESSURES

Chemists usually study gas-phase equilibria by following the partial pressures of the gases in the reaction. We can understand why this is reasonable by rearranging the ideal gas equation to give the following relationship between the pressure of a gas and its concentration in moles per liter:

$$P = \left[\frac{n}{V}\right] \times RT$$

We can therefore characterize the following reaction:

$$N_2(g) + 3\ H_2(g) \rightleftharpoons 2\ NH_3(g)$$

with an equilibrium constant defined in terms of units of concentration:

$$K_c = \frac{[NH_3]^2}{[N_2][H_2]^3}$$

or an equilibrium constant defined in terms of partial pressures:

$$K_p = \frac{P_{NH_3}{}^2}{P_{N_2}P_{H_2}{}^3}$$

What is the relationship between $K_p$ and $K_c$ for a gas-phase reaction? According to the rearranged version of the ideal gas equation, the pressure of a gas is equal to the concentration of the gas times the product of the ideal gas constant and the temperature in units of kelvin:

$$P = \left[\frac{n}{V}\right] \times RT$$

We can therefore calculate the value of $K_p$ for a reaction by multiplying each of the terms in the $K_c$ expression by $RT$:

$$K_p = \frac{P_{NH_3}{}^2}{P_{N_2}P_{H_2}{}^3} = \frac{([NH_3] \times RT)^2}{([N_2] \times RT)([H_2] \times RT)^3}$$

Collecting terms in this example gives the following result:

$$K_p = K_c \times (RT)^{-2}$$

In general, the value of $K_p$ for a reaction can be calculated from $K_c$ with the following equation:

$$\mathbf{K_p = K_c \times (RT)^{\Delta n}}$$

In this equation, $\Delta n$ is the difference between the number of moles of products and the number of moles of reactants in the balanced equation.

## EXERCISE 16.6

Calculate the value of $K_p$ for the following reaction at 500°C if the value of $K_c$ at this temperature is 0.040.

$$N_2(g) + 3\ H_2(g) \rightleftharpoons 2\ NH_3(g)$$

**SOLUTION** The value of $K_p$ for a reaction can be calculated from $K_c$ with the following equation:

$$K_p = K_c \times (RT)^{\Delta n}$$

In order to use this equation, we need to know the value of $\Delta n$ for the reaction. The balanced equation for this reaction generates 2 moles of products for every 4 moles of reactants consumed. The value of $\Delta n$ for this reaction is therefore 2 − 4, or −2:

$$K_p = K_c \times (RT)^{-2}$$

We now use the known value of the ideal gas constant and the temperature in Kelvin to calculate the value of $K_p$ for the reaction at this temperature:

$$K_p = 0.040 \times [(0.08206\ \text{L-atm/mol-}\cancel{K})(773\ \cancel{K})]^{-2} = \mathbf{1.0 \times 10^{-5}}$$

The techniques for working problems using $K_p$ expressions are the same as those described for $K_c$ problems, except that partial pressures are used instead of concentrations to represent the amounts of starting materials and products that are present both initially and at equilibrium.

## 16.11 THE EFFECT OF TEMPERATURE ON A CHEMICAL REACTION

Whenever an equilibrium constant has been given in this chapter, the temperature at which the reaction was run also has been given. If the equilibrium constant is really constant, why do we have to worry about the temperature of the reaction?

The answer is simple. Both $K_c$ and $K_p$ for a reaction are constants at a given temperature, but they can change with temperature. Consider the equilibrium between $NO_2$ and its dimer, $N_2O_4$, for example:

$$2\ NO_2(g) \rightleftharpoons N_2O_4(g)$$

Figure 16.10 shows the effect of temperature on this equilibrium. When we cool a sealed tube containing $NO_2$ in a dry ice–acetone bath at −78°C, the intensity of the brown color of $NO_2$ gas decreases significantly. If we warm the tube in a hot-water bath, the brown color becomes more intense than it was at room temperature.

**Figure 16.10** *Left*: A sealed tube filled with the dark-brown $NO_2$ gas at room temperature. *Right*: When the tube is cooled to −196°C in liquid nitrogen, the following equilibrium shifts so far toward the dimer that the brown color virtually disappears.

$$2\ NO_2(g) \rightleftharpoons N_2O_4(g)$$

The equilibrium constant for this reaction changes with temperature, as shown in Table 16.1. At low temperatures, the equilibrium favors the dimer, $N_2O_4$. At high temperatures, the equilibrium favors $NO_2$. The fact that equilibrium constants are temperature dependent explains why you may find different values for the equilibrium constant for the same chemical reaction.

**TABLE 16.1 The Temperature Dependence of the Equilibrium Constant for the Dimerization of $NO_2$**

| Temperature (°C) | $K_p$ | $K_c$ |
|---|---|---|
| 100 | 0.067 | 2.1 |
| 25 | 7.1 | 170 |
| 0 | 63 | 1400 |
| −78 | 25,000,000 | 400,000,000 |

## 16.12 LE CHÂTELIER'S PRINCIPLE

In 1884 the French chemist and engineer Henry-Louis Le Châtelier proposed one of the central concepts of chemical equilibria. **Le Châtelier's principle** can be stated as follows:

**A change in one of the variables that describe a system at equilibrium produces a shift in the position of the equilibrium that counteracts the effect of this change.**

Our attention so far has been devoted to describing what happens as a system comes to equilibrium. Le Châtelier's principle describes what happens to a system when something momentarily takes it away from equilibrium. This section focuses on three ways in which we can change the conditions of a chemical reaction at equilibrium: (1) changing the concentration of one of the components of the reaction, (2) changing the pressure on the system, and (3) changing the temperature at which the reaction is run.

### CHANGES IN CONCENTRATION

To illustrate what happens when we change the concentration of one of the reactants or products of a reaction at equilibrium, let's consider the following system at 500°C:

| | $N_2(g)$ | $+ 3\ H_2(g)$ | $\rightleftharpoons 2\ NH_3(g)$ | $K_c = 0.040$ |
|---|---|---|---|---|
| Initial: | $0.100\ M$ | $0.100\ M$ | 0 | |
| Equilibrium: | $0.100 - \Delta$ | $0.100 - 3\ \Delta$ | $2\ \Delta$ | |

According to Exercise 16.5, we obtain the following results when we solve this problem:

$$[NH_3] = 2\Delta \approx 0.0020\ M$$
$$[N_2] = 0.100 - \Delta \approx 0.099\ M$$
$$[H_2] = 0.100 - 3\Delta \approx 0.097\ M$$

The fact that $\Delta$ is small compared with the initial concentrations of $N_2$ and $H_2$ makes this calculation relatively easy to do. But it implies that very little ammonia is actually produced in the reaction. According to this calculation, only 1% of the nitrogen initially present is converted into ammonia.

What would happen if we added enough $N_2$ to increase the initial concentration by a factor of 10? The reaction can't be at equilibrium any more because there is far too much $N_2$ in the system. Adding an excess of one of the reactants therefore places a stress on the system. The system responds to minimizing the effect of this stress—by shifting the equilibrium toward the products. The reaction comes back to equilibrium when the concentrations of the three components reach the following values:

$$[NH_3] = 2\Delta \approx 0.0055\ M$$
$$[N_2] = 1.00 - \Delta \approx 1.00\ M$$
$$[H_2] = 0.100 - 3\Delta \approx 0.092\ M$$

By comparing the new equilibrium concentrations with those obtained before excess $N_2$ was added to the system, we can see the magnitude of the effect of adding the excess $N_2$:

| *Before* | *After* |
|---|---|
| $\mathbf{[NH_3] \approx 0.0020\ M}$ | $\mathbf{[NH_3] \approx 0.0055\ M}$ |
| $[N_2] \approx 0.099\ M$ | $[N_2] \approx 1.00\ M$ |
| $[H_2] \approx 0.097\ M$ | $[H_2] \approx 0.092\ M$ |

Increasing the amount of $N_2$ in the system by a factor of 10 leads to an increase in the amount of $NH_3$ at equilibrium by a factor of about 3. Adding an excess of one of the products would have the opposite effect; it would shift the equilibrium toward the reactants.

**CHECKPOINT**

Explain why we can drive a reaction toward completion by removing one of the products of the reaction.

## CHANGES IN PRESSURE

The effect of changing the pressure on a gas-phase reaction depends on the stoichiometry of the reaction. We can demonstrate this by looking at the result of compressing the following reaction at equilibrium.

$$N_2(g) + 3\ H_2(g) \rightleftharpoons 2\ NH_3(g)$$

Let's start with a system that initially contains 2.5 atm of $N_2$ and 7.5 atm of $H_2$ at 500°C, where $K_p$ is $1.4 \times 10^{-5}$, allow the reaction to come to equilibrium, and then compress the system by a factor of 10. When this is done, we get the following results:

| *Before Compression* | *After Compression* |
|---|---|
| $P_{NH_3} = 0.12$ atm | $P_{NH_3} = 8.4$ atm |
| $P_{N_2} = 2.4$ atm | $P_{N_2} = 21$ atm |
| $P_{H_2} = 7.3$ atm | $P_{H_2} = 62$ atm |

Before the system was compressed, the partial pressure of $NH_3$ was only about 1% of the total pressure. After the system is compressed, the partial pressure of $NH_3$ is almost 10% of the total.

These data provide another example of Le Châtelier's principle. A reaction at equilibrium was subjected to a stress—an increase in the total pressure on the system. The reaction then shifted in the direction that minimized the effect of this stress. The reaction shifted toward the products because this reduced the number of particles in the gas, thereby decreasing the total pressure on the system, as shown in Figure 16.11.

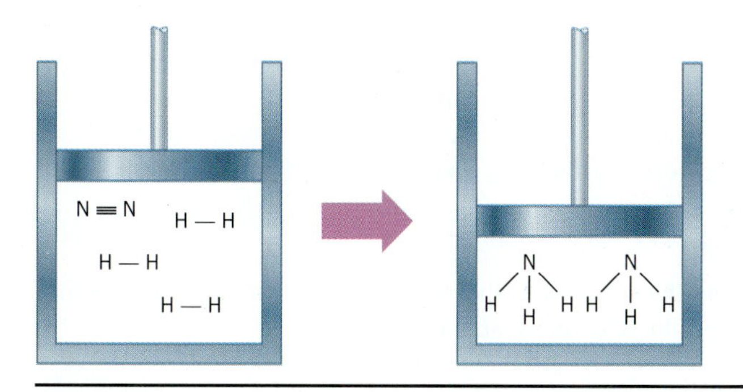


**Figure 16.11** Because the total number of molecules in the system decreases when $N_2$ reacts with $H_2$ to form $NH_3$, shifting this equilibrium toward $NH_3$ decreases the total pressure on the system.

## CHANGES IN TEMPERATURE

Changes in the concentrations of the reactants or products of a reaction shift the position of the equilibrium but do not change the equilibrium constant for the reaction. Similarly, a change in the pressure on a gas-phase reaction shifts the position of the equilibrium without changing the magnitude of the equilibrium constant. Changes in the temperature of the system, however, affect the position of the equilibrium by changing the magnitude of the equilibrium constant for the reaction.

Chemical reactions either give off heat to their surroundings or absorb heat from their surroundings. If we consider heat to be one of the reactants or products of a reaction, we can understand the effect of changes in temperature on the equilibrium. Increasing the temperature of a reaction that gives off heat is the same as adding more of one of the products of the reaction. It places a stress on the reaction, which must be alleviated by converting some of the products back to reactants.

Sec. 5.14

The reaction in which $NO_2$ dimerizes to form $N_2O_4$ provides an example of the effect of changes in temperature on the equilibrium constant for a reaction. This reaction is exothermic:

$$2\ NO_2(g) \rightleftharpoons N_2O_4(g) \qquad \Delta H^\circ = -57.20\ \text{kJ}$$

Thus, raising the temperature of this system is equivalent to adding excess product to the system. The equilibrium constant therefore decreases with increasing temperature, as was shown in Table 15.2.

### EXERCISE 16.7

Predict the effect of the following changes on the reaction in which $SO_3$ decomposes to form $SO_2$ and $O_2$:

$$2\ SO_3(g) \rightleftharpoons 2\ SO_2(g) + O_2(g) \qquad \Delta H^\circ = 197.78\ \text{kJ}$$

(a) Increasing the temperature of the reaction
(b) Increasing the pressure on the reaction
(c) Adding more $O_2$ when the reaction is at equilibrium
(d) Removing $O_2$ from the system when the reaction is at equilibrium

**SOLUTION**

(a) Because this is an endothermic reaction, which absorbs heat from its surroundings, an increase in the temperature of the reaction leads to an increase in the equilibrium constant and therefore a shift in the position of the equilibrium toward the products.

(b) There is a net increase in the number of molecules in the system as the reactants are converted into products, which leads to an increase in the pressure of the system. The system can minimize the effect of an increase in pressure by shifting the position of the equilibrium toward the reactants, thereby converting some of the $SO_2$ and $O_2$ into $SO_3$.

(c) Adding more $O_2$ to the system will shift the position of the equilibrium toward the reactants.

(d) Removing $O_2$ from the system has the opposite effect; it shifts the equilibrium toward the products of the reaction.

**SPECIAL TOPIC**

## SUCCESSIVE APPROXIMATIONS

Our approach to solving equilibrium problems can be summarized as follows:

**Problem-Solving Strategy**

**1.** If the difference between $Q_c$ and $K_c$ is small, the reaction doesn't have far to go to reach equilibrium. We therefore assume that $\Delta$ is small compared with the initial concentrations and solve for an approximate value of $\Delta$.

**2.** If the difference between $Q_c$ and $K_c$ is large, we drive the reaction to completion and then let the reaction come back to equilibrium. In essence, this means redefining the problem so that $Q_c$ is close to $K_c$.

What do we do when the difference between $Q_c$ and $K_c$ is neither relatively small nor relatively large? Consider the following problem, for example:

> Assume that the initial concentration of $N_2$ is 0.250 moles per liter and the initial concentration of $H_2$ is 0.750 moles per liter. Calculate the equilibrium concentrations of $N_2$, $H_2$, and $NH_3$ at 500°C if the equilibrium constant for the following reaction at this temperature is 0.040:
>
> $$N_2(g) + 3\ H_2(g) \rightleftharpoons 2\ NH_3(g)$$

We start, as always, by setting up the problem as follows:

| | $N_2(g)$ | $+\ 3\ H_2(g)$ | $\rightleftharpoons\ 2\ NH_3(g)$ |
|---|---|---|---|
| Initial: | 0.250 *M* | 0.750 *M* | 0 |
| Equilibrium: | ? | ? | ? |

We then calculate the reaction quotient for the initial concentrations:

$$Q_c = \frac{(NH_3)^2}{(N_2)(H_2)^3} = \frac{(0)^2}{(0.250)(0.750)^3} = 0 < K_c$$

The reaction quotient is smaller than the equilibrium constant, so the reaction must shift to the right to produce $NH_3$, as we would expect:

| | $N_2(g)$ | $+\ 3\ H_2(g)$ | $\rightleftharpoons\ 2\ NH_3(g)$ | $K_c = 0.040$ |
|---|---|---|---|---|
| Initial: | 0.250 *M* | 0.750 *M* | 0 | |
| Change | $-\Delta$ | $-3\Delta$ | $+2\Delta$ | |
| Equilibrium: | $0.250 - \Delta$ | $0.750 - 3\Delta$ | $2\Delta$ | |

When we substitute this information into the equilibrium constant expression for the reaction, we obtain the following equation:

$$K_c = \frac{[NH_3]^2}{[NH_2][H_2]^3} = \frac{[2\Delta]^2}{[0.250 - \Delta][0.750 - 3\Delta]^3} = 0.040$$

Because $Q_c$ and $K_c$ are both smaller than 1, we can try the assumption that $\Delta$ is small compared with the initial concentration of $N_2$ and that $3\Delta$ is small compared with the initial concentration of $H_2$:

$$\frac{[2\Delta]^2}{[0.250][0.750]^3} \approx 0.040$$

Solving this equation for Δ gives the following result:

$$\Delta \approx 0.032$$

Is our approximation valid? No, Δ is more than 10% of the initial concentration of $N_2$ and 3Δ is more than 10% of the initial concentration of $H_2$:

$$\frac{0.032}{0.250} \times 100\% = 12.8\% \qquad \frac{3(0.032)}{0.750} \times 100\% = 12.8\%$$

Can we redefine the problem to make Δ small? No, the initial conditions are already on the side of the equilibrium favored by the equilibrium constant. When the equilibrium constant for a reaction is small, we can set up the problem so that $Q_c$ is small and therefore Δ is small. When the equilibrium constant is large, we can drive the reaction toward the products so that $Q_c$ is also large and therefore Δ is small. But there is no way to redefine problems so that Δ is small when $K_c$ is close to 1.

**Problem-Solving Strategy**

Problems like this can be solved by a technique known as **successive approximations.** We set up the problem as usual. We then assume that Δ is small, even if we suspect that this assumption is not valid, and calculate a first approximation for the value of Δ. In this case, as we have already seen, we get the following result:

$$\Delta \approx 0.032$$

We now substitute this approximation value of Δ back into the equilibrium constant expression and solve this equation for a second approximation:

$$\frac{[2\Delta']^2}{[0.250 - 0.032][0.750 - 3(0.032)]^3} = 0.040$$

$$\Delta' \approx 0.025$$

We then substitute this approximate value of Δ′ back into the equation and solve for a third approximation:

$$\frac{[2\Delta'']^2}{[0.250 - 0.025][0.750 - 3(0.025)]^3} = 0.040$$

$$\Delta'' \approx 0.026$$

We then substitute this approximate value of Δ back into the equation and solve for a fourth approximation:

$$\frac{[2\Delta''']^2}{[0.250 - 0.026][0.750 - 3(0.026)]^3} = 0.040$$

$$\Delta''' \approx 0.026$$

The difference between the successive values of Δ obtained by this technique keeps getting smaller until it is eventually smaller than the number of significant figures to which we are entitled.

We therefore assume that this calculation converges to a value of Δ equal to 0.026 and use this value to calculate the concentrations of $N_2$, $H_2$, and $NH_3$ at equilibrium:

$$[N_2] = 0.250 - \Delta = 0.224\ M$$

$$[H_2] = 0.750 - 3\Delta = 0.672\ M$$

$$[NH_3] = 2\Delta = 0.052\ M$$

We can check the validity of these values by substituting them into the equilibrium constant expression for the reaction:

$$K_c = \frac{[NH_3]^2}{[N_2][H_2]^3} = \frac{[0.052]^2}{[0.224][0.672]^3} = 0.040$$

The value of $K_c$ obtained from this calculation agrees with the experimental value of $K_c$ within experimental error. We can therefore feel confident that successive approximations has converged on the correct value of $\Delta$.

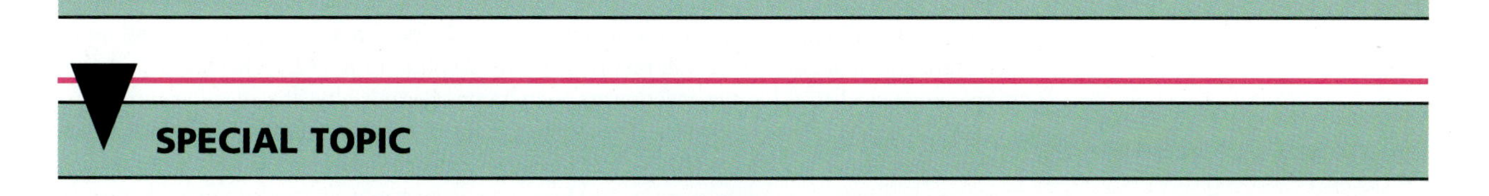

## SPECIAL TOPIC

# LE CHÂTELIER'S PRINCIPLE AND THE HABER PROCESS

Ammonia has been produced commercially from $N_2$ and $H_2$ ever since 1913, when Badische Anilin und Soda Fabrik (BASF) built a plant that used the **Haber process** to make 30 metric tons of synthetic ammonia per day:

$$N_2(g) + 3\ H_2(g) \rightleftharpoons 2\ NH_3(g) \qquad \Delta H^\circ = -92.2\ \text{kJ}$$

Until that time, the principal source of nitrogen for use in farming had been animal and vegetable waste. Today, almost 20 million tons of ammonia worth 2.5 billion dollars is produced in the United States each year, about 80% of which is used for fertilizers. Ammonia is usually applied directly to the fields as a liquid at or near its boiling point of −33.35°C. By using this so-called ''anhydrous ammonia,'' farmers can apply a fertilizer that contains 82% nitrogen by weight.

The Haber process was the first example of the use of Le Châtelier's principle to optimize the yield of an industrial chemical. An increase in the pressure at which this reaction is run favors the products of the reaction because there is a net reduction in the number of molecules in the system as $N_2$ and $H_2$ combine to form $NH_3$. Because the reaction is exothermic, the equilibrium constant decreases as the temperature of the reaction increases.

Table 16.2 shows the mole percent of $NH_3$ at equilibrium when the reaction is run at different combinations of temperature and pressure. The mole percent of $NH_3$ under a partic-

This is a photograph of the first high pressure reactor for the synthesis of ammonia by Haber process.

**TABLE 16.2 Mole Percent of $NH_3$ at Equilibrium**

| | Pressure | | | |
|---|---|---|---|---|
| Temperature (°C) | 200 atm | 300 atm | 400 atm | 500 atm |
| 400 | 38.74 | 47.85 | 58.87 | 60.61 |
| 450 | 27.44 | 35.93 | 42.91 | 48.84 |
| 500 | 18.86 | 26.00 | 32.25 | 37.79 |
| 550 | 12.82 | 18.40 | 23.55 | 28.31 |
| 600 | 8.77 | 12.97 | 16.94 | 20.76 |

Sec. 16.12 

ular set of conditions is equal to the number of moles of $NH_3$ at equilibrium divided by the total number of moles of all three components of the reaction. As the data in Table 16.2 demonstrate, the best yields of ammonia are obtained at low temperatures and high pressures.

Unfortunately, low temperatures slow down the rate of this reaction, and the cost of building plants rapidly escalates as the pressure at which the reaction is run is increased. When commercial plants are designed, a temperature is chosen that allows the reaction to proceed at a reasonable rate without decreasing the equilibrium concentration of the product by too much. The pressure is also adjusted so that it favors the production of ammonia without excessively increasing the cost of building and operating the plant. The optimum conditions for running this reaction at present are a pressure between 140 and 340 atm and a temperature between 400°C and 600°C.

Despite all efforts to optimize reaction conditions, the percentages of hydrogen and nitrogen converted to ammonia are still relatively small. Another form of Le Châtelier's principle is therefore used to drive the reaction to completion. Periodically, the reaction mixture is cycled through a cooling chamber. The boiling point of ammonia (BP = −31°C) is much higher than either hydrogen (BP = −252.8°C) or nitrogen (BP = −195.8°C). Ammonia therefore condenses out of the reaction mixture and can be removed. The remaining hydrogen and nitrogen gases are then recycled through the reaction chamber, where they react to produce more ammonia.

## SUMMARY

**1.** The collision theory model of chemical reactions assumes that the rate of any step in a reaction is proportional to the product of the concentrations of the molecules that must collide in order for the reaction to occur. Initially, the rate of the forward reaction is much faster than the rate of the reverse reaction. As the starting materials are converted into the products of the reaction, the forward reaction slows down and the reverse reaction becomes faster. When the reaction proceeds at the same rate in both directions, it reaches equilibrium. At equilibrium, there is no change in the concentration of the starting materials or the products of the reaction.

**2.** We can decide whether a reaction is at equilibrium at any moment in time by comparing the reaction quotient ($Q_c$) at that moment with the equilibrium constant ($K_c$) for the reaction. If $Q_c$ is smaller than $K_c$, the system contains too much reactant and not enough product to be at equilibrium. The reaction therefore has to shift to the right—converting some of the reactants into products—to reach equilibrium. If $Q_c$ is larger than $K_c$, the system contains too much product and not enough reactant. It therefore has to shift to the left to reach equilibrium.

**3.** The stoichiometry of the reaction dictates the relationship between the changes in the concentrations of the components of the reaction as it comes to equilibrium. This relationship allows us to write an algebraic expression that relates the equilibrium concentration of any component of the reaction to the initial concentration of this substance and a single variable: $\Delta$.

**4.** Calculations of the concentrations of the components of a reaction at equilibrium can be simplified by assuming that $\Delta$ is small compared with the initial concentrations to which it is added or from which it is subtracted. By convention, this assumption is legitimate when $\Delta$ is less than 5% of the initial concentrations with which it is compared.

**5.** The assumption that $\Delta$ is relatively small is valid when $Q_c$ is relatively close to $K_c$. When $Q_c$ is very far from $K_c$, we can redefine the problem to bring $Q_c$ close to $K_c$. This involves creating a set of intermediate conditions in which the reaction is pushed as far as possible in the direction favored by the equilibrium constant.

**6.** At a given temperature, the partial pressure of a gas is directly proportional to its concentration in units of moles per liter. Gas-phase equilibrium constant expressions therefore can be written in terms of either the concentrations of the components of the reaction at equilibrium ($K_c$) or their partial pressures ($K_p$).

**7.** The three kinds of stress that can be applied to a chemical reaction at equilibrium include changes in the temperature, the pressure, or the concentration of the components of the reaction. Le Châtelier's principle states that a chemical reaction at equilibrium responds to these changes by shifting the position of the equilibrium in the direction that minimizes the effect of the changes.

## KEY TERMS

**Approximation methods** (p. 577)
**Collision theory** (p. 565)
**Equilibrium** (p. 566)
**Equilibrium constant, $K_c$** (p. 567)
**Equilibrium constant expression** (p. 567)
**Equilibrium region** (p. 564)
**Haber process** (p. 591)
**Kinetic region** (p. 564)
**Le Châtelier's principle** (p. 586)
**Rate of reaction** (p. 564)
**Rate law** (p. 564)
**Rate law** (p. 564)
**Reaction quotient, $Q_c$** (p. 570)
**Successive approximation** (p. 590)

## KEY EQUATIONS

$$K_c = K_{c1} \times K_{c2}$$
$$K_p = K_c \times (RT)^{\Delta n}$$

## PROBLEMS

### Chemical Reactions Don't Always Go to Completion

**16-1** Describe the difference between reactions that go to completion and reactions that come to equilibrium.

**16-2** Define the terms *equilibrium, equilibrium constant, equilibrium constant expression,* and *reaction quotient.*

**16-3** Describe the meaning of the symbol $[NO_2]$.

### The Rates of Chemical Reactions

**16-4** Describe how the rate of a chemical reaction is analogous to other rate processes, such as the rate at which a car travels or the rate of inflation.

**16-5** Translate the following equation into an English sentence that carries the same meaning: Rate of reaction = $X)/\Delta t$.

**16-6** Sketch a graph of what happens to the concentrations of $N_2$, $H_2$, and $NH_3$ versus time as the following reaction comes to equilibrium. Assume that the initial concentrations of $N_2$ and $H_2$ are both 1.00 moles per liter and that no $NH_3$ is present initially. Label the kinetic and the equilibrium regions of this graph.

$$N_2(g) + 3\ H_2(g) \rightleftharpoons 2\ NH_3(g)$$

### A Collision Theory Model for Gas-Phase Reactions

**16-7** Describe how the collision theory model can be used to explain the fact that the rate at which $ClNO_2$ reacts with NO to form ClNO and $NO_2$ is proportional to the product of the concentrations of $ClNO_2$ and NO.

**16-8** Use the fact that the rate of a chemical reaction is proportional to the product of the concentrations of the reagents consumed in that reaction to explain why reversible reactions inevitably come to equilibrium.

### Writing Equilibrium Constant Expressions

**16-9** Which of the following is the correct equilibrium constant expression for the reaction:

$$Cl_2(g) + 3\ F_2(g) \rightleftharpoons 2\ ClF_3(g)$$

(a) $K_c = \dfrac{2[ClF_3]}{[Cl_2] + 3\ [F_2]}$

(b) $K_c = \dfrac{[Cl_2] + 3\ [F_2]}{2[ClF_3]}$

(c) $K_c = \dfrac{[ClF_3]}{[Cl_2][F_2]}$

(d) $K_c = \dfrac{[ClF_3]^2}{[Cl_2][F_2]^3}$

(e) $K_c = \dfrac{[Cl_2][F_2]^3}{[ClF_3]^2}$

**16-10** Which of the following is the correct equilibrium constant expression for the reaction:

$$2\ NO_2(g) \rightleftharpoons 2\ NO(g) + O_2(g)$$

(a) $K_c = \dfrac{[NO_2]}{[NO][O_2]}$

(b) $K_c = \dfrac{[NO][O_2]}{[NO_2]}$

(c) $K_c = \dfrac{[NO_2]^2}{[NO]^2[O_2]}$

(d) $K_c = \dfrac{[NO]^2[O_2]}{[NO_2]^2}$

(e) $K_c = \dfrac{[2\ NO]^2[O_2]}{[2\ NO_2]^2}$

**16-11** Write equilibrium constant expressions for the following reactions:

(a) $O_2(g) + 2\ F_2(g) \rightleftharpoons 2\ OF_2(g)$

(b) $2\ SO_2(g) + O_2(g) \rightleftharpoons 2\ SO_3(g)$

(c) $2\ SO_3(g) + 2\ Cl_2(g) \rightleftharpoons 2\ SO_2Cl_2(g) + O_2(g)$

**16-12** Write equilibrium constant expressions for the following reactions:

(a) $2\ NO(g) + 2\ H_2(g) \rightleftharpoons N_2(g) + 2\ H_2O(g)$

(b) $2\ NOCl(g) \rightleftharpoons 2\ NO(g) + Cl_2(g)$

(c) $2\ NO(g) + O_2(g) \rightleftharpoons 2\ NO_2(g)$

**16-13** Write equilibrium constant expressions for the following reactions:

(a) $2\ CO(g) + O_2(g) \rightleftharpoons 2\ CO_2(g)$

(b) $CO_2(g) + H_2(g) \rightleftharpoons CO(g) + H_2O(g)$

(c) $CO(g) + 2\ H_2(g) \rightleftharpoons CH_3OH(g)$

### Calculating Equilibrium Constants

**16-14** Calculate $K_c$ for the following reaction at 400 K if 1.00 mole per liter of NOCl decomposes at this temperature to give equilibrium concentrations of 0.0222 *M* NO, 0.0111 *M* $Cl_2$, and 0.989 *M* NOCl:

$$2\ NOCl(g) \rightleftharpoons 2\ NO(g) + Cl_2(g)$$

**16-15** Taylor and Crist [*Journal of the American Chemical Society,* **63,** 1381 (1941)] studied the reaction between hydrogen and iodine to form hydrogen iodide:

$$H_2(g) + I_2(g) \rightleftharpoons 2\ HI(g)$$

They obtained the following data for the concentrations of $H_2$, $I_2$, and HI at equilibrium in units of moles per liter:

| Trial | $[H_2]$ | $[I_2]$ | [HI] |
|---|---|---|---|
| I | 0.0032583 | 0.0012949 | 0.015869 |
| II | 0.0046981 | 0.0007014 | 0.013997 |
| III | 0.0007106 | 0.0007106 | 0.005468 |

Calculate the value of $K_c$ for each of these trials. Is $K_c$ a constant for this reaction, within the limits of experimental error?

### Combining Equilibrium Constant Expressions

**16-16** Write equilibrium constant expressions for the following reactions:

(a) $2\ NO_2(g) \rightleftharpoons 2\ NO(g) + O_2(g)$

(b) $2\ NO(g) + O_2(g) \rightleftharpoons 2\ NO_2(g)$

Calculate the value of $K_c$ at 500 K for reaction (*a*) if the value of $K_c$ for reaction (*b*) is $6.2 \times 10^5$.

**16-17** Write equilibrium constant expressions for the following reactions:

(a) $N_2(g) + 3\ H_2(g) \rightleftharpoons 2\ NH_3(g)$

(b) $2\ NH_3(g) \rightleftharpoons N_2(g) + 3\ H_2(g)$

(c) $NH_3(g) \rightleftharpoons \frac{1}{2}\ N_2(g) + \frac{3}{2}\ H_2(g)$

Calculate the value of $K_p$ at 500°C for reaction (*b*) if $K_p$ at this temperature for reaction (*a*) is $1 \times 10^{-5}$. What is the value of $K_p$ for reaction (*c*) at the same temperature?

**16-18** Use the equilibrium constants for reactions (*a*) and (*b*) at 200°C to calculate the equilibrium constant for reaction (*c*) at this temperature:

(a) $2\ NO(g) \rightleftharpoons N_2(g) + O_2(g)$ $\quad K_c = 4.3 \times 10^{18}$

(b) $2\ NO_2(g) \rightleftharpoons 2\ NO(g) + O_2(g)$ $\quad K_c = 3.4 \times 10^{-7}$

(c) $2\ NO_2(g) \rightleftharpoons N_2(g) + 2\ O_2(g)$ $\quad K_c = ?$

**16-19** Use the equilibrium constants for reactions (*a*) and (*b*) below at 1000 K to calculate the equilibrium constant for reaction (*c*), the water-gas shift reaction, at this temperature:

(a) $CO(g) + \frac{1}{2}\ O_2(g) \rightleftharpoons CO_2(g)$ $\quad K_c = 1.1 \times 10^{11}$

(b) $H_2O(g) \rightleftharpoons H_2(g) + \frac{1}{2}\ O_2(g)$ $\quad K_c = 7.1 \times 10^{-12}$

(c) $CO(g) + H_2O(g) \rightleftharpoons CO_2(g) + H_2(g)$ $\quad K_c = ?$

### Reaction Quotient: A Way to Decide Whether a Reaction Is at Equilibrium

**16-20** Suppose that the reaction quotient ($Q_c$) for the following reaction at some moment in time is $1.0 \times 10^{-8}$ and the equilibrium constant for this reaction ($K_c$) at the same temperature is $3 \times 10^{-7}$.

$$2\ NO_2(g) \rightleftharpoons 2\ NO(g) + O_2(g)$$

Which of the following is a valid conclusion? (a) The reaction is at equilibrium. (b) The reaction must shift toward the products to reach equilibrium. (c) The reaction must shift toward the reactants to reach equilibrium.

**16-21** Which of the following statements correctly describes a system for which $Q_c$ is larger than $K_c$? (a) The reaction is at equilibrium. (b) The reaction must shift to the right to reach equilibrium. (c) The reaction must shift to the left to reach equilibrium. (d) The reaction can never reach equilibrium.

**16-22** Under which set of conditions will the following reaction shift to the right to reach equilibrium?

$$2\ SO_2(g) + O_2(g) \rightleftharpoons 2\ SO_3(g)$$

(a) $K_c < 1$ (b) $K_c > 1$ (c) $Q_c < K_c$ (d) $Q_c = K_c$ (e) $Q_c > K_c$

**16-23** Carbon monoxide reacts with chlorine to form phosgene:

$$CO(g) + Cl_2(g) \rightleftharpoons COCl_2(g)$$

The equilibrium constant, $K_p$, for this reaction is 320 at 300°C. Is the system at equilibrium at the following partial pressures: 0.050 atm $COCl_2$, 0.010 atm CO, and 0.0050 atm $Cl_2$? If not, in which direction does the reaction have to shift to reach equilibrium?

### Changes in Concentration That Occur as a Reaction Comes to Equilibrium

**16-24** Describe the relationship between the initial concentration of a reactant $(X)$, the concentration of this reactant at equilibrium $[X]$, and the change in the concentration of $X$ that occurs as the reaction comes to equilibrium $\Delta(X)$.

**16-25** Explain why the change in the $N_2$ concentration that occurs when the following reaction comes to equilibrium is related to the change in the $H_2$ concentration.

$$N_2(g) + 3\ H_2(g) \rightleftharpoons 2\ NH_3(g)$$

Derive an equation that describes the relationship between the changes in the concentrations of these two reagents.

**16-26** When confronted with the task in the previous problem, one might mistakenly write the following equation:

$$\Delta(N_2) = 3\Delta(H_2)$$

Explain why this equation is wrong. Write the correct form of this relationship.

**16-27** Calculate the changes in the CO and $Cl_2$ concentrations that occur if the concentration of $COCl_2$ decreases by 0.250 moles per liter as the following reaction comes to equilibrium:

$$COCl_2(g) \rightleftharpoons CO(g) + Cl_2(g)$$

**16-28** Calculate the changes in the $N_2$ and $H_2$ concentrations that occur if the concentration of $NH_3$ decreases by 0.234 moles per liter as the following reaction comes to equilibrium:

$$2\ NH_3(g) \rightleftharpoons N_2(g) + 3\ H_2(g)$$

**16-29** Which of the following equations describes the relationship between the magnitude of the changes in the $NO_2$ and $O_2$ concentrations as the following reaction comes to equilibrium?

$$2\ NO(g) + O_2(g) \rightleftharpoons 2\ NO_2(g)$$

(a) $\Delta(NO_2) = \Delta(O_2)$ (b) $\Delta(NO_2) = 2\ \Delta(O_2)$
(c) $\Delta(O_2) = 2\ \Delta(NO_2)$

**16-30** Which of the following equations correctly describes the relationship between the changes in the $Cl_2$ and $F_2$ concentrations as the following reaction comes to equilibrium?

$$Cl_2(g) + 3\ F_2(g) \rightleftharpoons 2\ ClF_3(g)$$

(a) $\Delta(Cl_2) = \Delta(F_2)$ (b) $\Delta(Cl_2) = 2\ \Delta(F_2)$
(c) $\Delta(Cl_2) = 3\ \Delta(F_2)$ (d) $\Delta(F_2) = 2\ \Delta(Cl_2)$
(e) $\Delta(F_2) = 3\ \Delta(Cl_2)$

**16-31** Which of the following describes the change that occurs in the concentration of $H_2O$ when ammonia reacts with oxygen to form nitrogen oxide and water according to the following equation if the change in the $NH_3$ concentration is $\Delta$?

$$4\ NH_3(g) + 5\ O_2(g) \rightleftharpoons 4\ NO(g) + 6\ H_2O(g)$$

(a) $\Delta$ (b) $1.5\Delta$ (c) $2\Delta$ (d) $4\Delta$ (e) $6\Delta$

**16-32** Calculate the concentrations of $H_2$ and $NH_3$ at equilibrium if a reaction that initially contained 1.00 *M* concentrations of both $N_2$ and $H_2$ is found to have an $N_2$ concentration of 0.922 *M* at equilibrium:

| | $N_2(g)$ | $+\ 3\ H_2(g)$ | $\rightleftharpoons 2\ NH_3(g)$ |
|---|---|---|---|
| Initial: | 1.00 *M* | 1.00 *M* | 0 *M* |
| Equilibrium: | 0.922 *M* | ? | ? |

**16-33** Calculate the equilibrium constant for the reaction in the previous problem.

## Hidden Assumptions That Make Equilibrium Calculations Easier

**16-34** Some students have described the technique used in this chapter to simplify equilibrium problems as follows: ''Assume that $\Delta$ is zero.'' Explain why they are wrong. What is the correct way of describing the assumption?

**16-35** Describe the advantage of setting up equilibrium problems so that $\Delta$ is small compared with the initial concentrations.

**16-36** Describe what happens if you make the assumption that $\Delta$ is zero in the following equation:

$$\frac{[0.125 - \Delta][2.40 - 2\ \Delta]^2}{[0.200 + 2\ \Delta]^2} = 1.3 \times 10^{-8}$$

Explain how to get around this problem.

## A Rule of Thumb for Testing the Validity of Assumptions

**16-37** Describe how to test whether $\Delta$ is small enough compared with the initial concentrations to be legitimately ignored.

**16-38** Explain why $\Delta$ is relatively small when the reaction quotient ($Q_c$) is reasonably close to the equilibrium constant for the reaction ($K_c$).

**16-39** Explain why making the assumption that $\Delta$ is small compared with the initial concentrations of the reactants and products is doomed to failure when the reaction quotient ($Q_c$) is very different from the equilibrium constant for the reaction ($K_c$).

## What Do We Do When the Approximation Fails?

**16-40** Describe the technique used to solve problems for which the reaction quotient is very different from the equilibrium constant.

**16-41** Before we can solve the following problem, we have to define a set of intermediate conditions under which the concentration of one of the reactants or products is zero.

| | $2\ NO_2(g)$ | $\rightleftharpoons 2\ NO(g)$ | $+\ O_2(g)$ | $K_c = 5.3 \times 10^{-6}$ |
|---|---|---|---|---|
| Initial: | 0.10 *M* | 0.10 *M* | 0.005 *M* | |

Which of the following goals determines whether we push the reaction as far as possible to the right or as far as possible to the left? (a) To make both $Q_c$ and $K_c$ large. (b) To make both $Q_c$ and $K_c$ small. (c) To bring $Q_c$ as close as possible to $K_c$. (d) To make the difference between $Q_c$ and $K_c$ as large as possible.

## Gas-Phase Equilibrium Problems

**16-42** Calculate the concentrations of $PCl_5$, $PCl_3$, and $Cl_2$ that are present when the following gas-phase reaction comes to equilibrium. Calculate the percent of the $PCl_5$ that decomposes when the reaction comes to equilibrium. Explain the difference between the results of this calculation and the percent decomposition obtained in Section 16.6.

$$PCl_5(g) \rightleftharpoons PCl_3(g) + Cl_2(g) \qquad K_c = 0.0013 \text{ (at 450 K)}$$

| | $PCl_5(g)$ | $PCl_3(g)$ | $Cl_2(g)$ |
|---|---|---|---|
| Initial: | 1.00 *M* | 0 | 0 |

**16-43** Calculate the concentrations of $PCl_5$, $PCl_3$, and $Cl_2$ present when the following gas-phase reaction comes to equilibrium. Calculate the percent decomposition in this reaction and explain any difference between the results of this calculation and the results obtained in Section 16.8.

$$PCl_5(g) \rightleftharpoons PCl_3(g) + Cl_2(g) \qquad K_c = 0.0013 \text{ (at 450 K)}$$

Initial: 1.00 $M$ 0 0.20 $M$

**16-44** Calculate the concentrations of NO, $NO_2$, and $O_2$ present when the following gas-phase reaction reaches equilibrium:

$$2\ NO_2(g) \rightleftharpoons 2\ NO(g) + O_2(g) \qquad K_c = 3.4 \times 10^{-7} \text{ (at 200°C)}$$

Initial: 0.10 $M$ 0 0

**16-45** Calculate the concentrations of NO, $NO_2$, and $O_2$ present when the following gas-phase reaction reaches equilibrium:

$$2\ NO_2(g) \rightleftharpoons 2\ NO(g) + O_2(g) \qquad K_c = 3.4 \times 10^{-7} \text{ (at 200°C)}$$

Initial: 0 0.10 $M$ 0.10 $M$

**16-46** Calculate the equilibrium concentrations of $SO_3$, $SO_2$, and $O_2$ present when 0.100 mol of $SO_3$ in a 250-mL flask at 300°C decomposes to form $SO_2$ and $O_2$:

$$2\ SO_3(g) \rightleftharpoons 2\ SO_2(g) + O_2(g) \qquad K_c = 1.6 \times 10^{-10} \text{ (at 300°C)}$$

**16-47** Calculate the equilibrium concentrations of $SO_3$, $SO_2$, and $O_2$ present when a mixture of 0.100 mol of $SO_2$ and 0.050 mol of $O_2$ in a 250-mL flask at 300°C combine to form $SO_3$:

$$2\ SO_2(g) + O_2(g) \rightleftharpoons 2\ SO_3(g) \qquad K_c = 6.3 \times 10^{9} \text{ (at 300°C)}$$

**16-48** Calculate the equilibrium concentration of $NO_2$ present when 0.100 $M$ $N_2O_4$ decomposes to form $NO_2$ at 25°C:

$$N_2O_4(g) \rightleftharpoons 2\ NO_2(g) \qquad K_c = 5.8 \times 10^{-5}$$

**16-49** Calculate the equilibrium concentration of $NO_2$ present when 1.00 $M$ $NO_2$ dimerizes to form $N_2O_4$ at 25°C:

$$N_2O_4(g) \rightleftharpoons 2\ NO_2(g) \qquad K_c = 5.8 \times 10^{-5}$$

**16-50** Calculate the equilibrium concentrations of $N_2$, $H_2$, and $NH_3$ present when a mixture that was initially 0.10 $M$ $N_2$, 0.10 $M$ $H_2$, and 0.10 $M$ $NH_3$ comes to equilibrium at 500°C:

$$N_2(g) + 3\ H_2(g) \rightleftharpoons 2\ NH_3(g) \qquad K_c = 0.040 \text{ (at 500°C)}$$

**16-51** Calculate the equilibrium concentrations of CO, $H_2O$, $CO_2$, and $H_2$ present in the water-gas shift reaction at 800°C if the initial concentrations of CO and $H_2O$ are 1.00 $M$:

$$CO(g) + H_2O(g) \rightleftharpoons CO_2(g) + H_2(g) \qquad K_c = 0.72 \text{ (at 800°C)}$$

**16-52** What initial concentrations of CO and $H_2O$ would be needed to reach an equilibrium concentration of 1.00 $M$ $CO_2$ in the water-gas shift reaction described Problem 16-51?

**16-53** Calculate the equilibrium concentrations of $N_2$, $O_2$, and NO present when a mixture that was initially 0.100 $M$ in $N_2$ and 0.090 $M$ in NO comes to equilibrium at 600°C:

$$N_2(g) + O_2(g) \rightleftharpoons 2\ NO(g) \qquad K_c = 3.3 \times 10^{-10}$$

**16-54** Sulfuryl chloride decomposes to sulfur dioxide and chlorine. Calculate the concentrations of the three components of this system at equilibrium if 6.75 g of $SO_2Cl_2$ in a 1.00-L flask decomposes at 25°C:

$$SO_2Cl_2(g) \rightleftharpoons SO_2(g) + Cl_2(g) \qquad K_c = 1.4 \times 10^{-5}$$

**16-55** Calculate the concentrations of NO, $Cl_2$, and NOCl at equilibrium if a mixture that was initially 0.50 $M$ in NO and 0.10 $M$ in $Cl_2$ combined to form nitrosyl chloride:

$$2\ NO(g) + Cl_2(g) \rightleftharpoons 2\ NOCl(g) \qquad K_c = 2.1 \times 10^{3} \text{ (at 500 K)}$$

### Equilibria Expressed in Partial Pressures

**16-56** Explain why pressure can be used instead of concentration to describe equilibrium constant expressions for gas-phase reactions.

**16-57** Which equation correctly describes the relationship between $K_p$ and $K_c$ for the following reaction?

$$Cl_2(g) + 3\ F_2(g) \rightleftharpoons 2\ ClF_3(g)$$

(a) $K_p = K_c$ (b) $K_p = K_c \times (RT)^{-1}$
(c) $K_p = K_c \times (RT)^{-2}$ (d) $K_p = K_c \times (RT)$
(e) $K_p = K_c \times (RT)^2$

**16-58** Which equation correctly describes the relationship between $K_p$ and $K_c$ for the following reaction?

$$2\ NO_2(g) \rightleftharpoons 2\ NO(g) + O_2(g)$$

(a) $K_p = K_c$ (b) $K_p = 1/K_c$ (c) $K_p = K_c(RT)$
(d) $K_p = K_c(RT)^{-1}$

**16-59** Calculate $K_p$ for the decomposition of NOCl at 500 K if 27.3% of a 1.00-atm sample of NOCl decomposes to NO and $Cl_2$ at equilibrium:

$$2\ NOCl(g) \rightleftharpoons 2\ NO(g) + Cl_2(g)$$

**16-60** Calculate the partial pressures of phosgene, carbon monoxide, and chlorine at equilibrium when a system that was initially 0.124 atm in $COCl_2$ decomposes at 300°C according to the following equation:

$$COCl_2(g) \rightleftharpoons CO(g) + Cl_2(g) \qquad K_p = 3.2 \times 10^{-3}$$

**16-61** Calculate the partial pressures of $SO_3$, $SO_2$, and $O_2$ that would be present at equilibrium when a mixture that was initially 0.490 atm in $SO_2$ and 0.245 atm in $O_2$ comes to equilibrium at 700 K:

$$2\ SO_2(g) + O_2(g) \rightleftharpoons 2\ SO_3(g) \qquad K_p = 6.7 \times 10^{4}$$

**16-62** Calculate the partial pressures of $SO_3$, $SO_2$, and $O_2$ that would be present at equilibrium when a mixture that was initially 0.490 atm in $SO_2$ and 0.345 atm in $O_2$ reacts at 700 K:

$$2\ SO_3(g) \rightleftharpoons 2\ SO_2(g) + O_2(g) \qquad K_p = 1.5 \times 10^{-5}$$

**16-63** Calculate the partial pressures of $N_2$, $H_2$, and $NH_3$ present at equilibrium when a mixture that was initially 0.50 atm in $N_2$, 0.60 atm in $H_2$, and 0.20 atm in $NH_3$ comes to equilibrium at a temperature at which $K_p$ for the following reaction is $1.0 \times 10^{-5}$:

$$N_2(g) + 3\ H_2(g) \rightleftharpoons 2\ NH_3(g)$$

**16-64** Calculate the partial pressures of $NH_3$, $O_2$, $NO_2$, and $H_2O$ present at equilibrium when a mixture that was initially 0.50 atm in each of the four gases comes to equilibrium at 1500°C:

$$4\ NO_2(g) + 6\ H_2O(g) \rightleftharpoons 4\ NH_3(g) + 7\ O_2(g) \qquad K_p = 1.8 \times 10^{-28}$$

**16-65** Calculate the partial pressures of $N_2$, $H_2$, and $NH_3$ present at equilibrium when 4.00 atm of $NH_3$ decomposes at 1000 K if $K_p$ for this reaction is $1.3 \times 10^6$:

$$2\ NH_3(g) \rightleftharpoons N_2(g) + 3\ H_2(g)$$

**16-66** Calculate the concentrations of $N_2$, $O_2$, and NO present when a mixture that was initially 0.40 *M* $N_2$ and 0.60 *M* $O_2$ reacts to form NO at 700°C. (Hint: Look carefully at the symbol for the equilibrium constant for this reaction.)

$$N_2(g) + O_2(g) \rightleftharpoons 2\ NO(g) \qquad K_p = 4.3 \times 10^{-9} \text{ (at 700°C)}$$

**16-67** Industrial chemicals can be made from coal by a process that starts with the reaction between red-hot coal and steam to form a mixture of CO and $H_2$. This mixture, which is known as synthesis gas, can be converted to methanol ($CH_3OH$) in the presence of a ruthenium–cobalt catalyst. The methanol produced in this reaction then can be converted into a host of other products, ranging from acetic acid to gasoline. Calculate the percent yield of methanol when a mixture of 1.00 atm CO and 1.00 atm $H_2$ comes to equilibrium at 65°C:

$$CO(g) + 2\ H_2(g) \rightleftharpoons CH_3OH(g) \qquad K_p = 2.3 \times 10^4 \text{ (at 65°C)}$$

## Le Châtelier's Principle

**16-68** Le Châtelier's principle has been applied to many fields, ranging from economics to psychology to political science. Give an example of Le Châtelier's principle in a field outside the physical sciences.

**16-69** Predict the effect of increasing the pressure on the following reactions at equilibrium:

(a) $2\ SO_3(g) + 2\ Cl_2(g) \rightleftharpoons 2\ SO_2Cl_2(g) + O_2(g)$

(b) $O_2(g) + 2\ F_2(g) \rightleftharpoons 2\ OF_2(g)$

(c) $2\ NO(g) + O_2(g) \rightleftharpoons 2\ NO_2(g)$

**16-70** Predict the effect of decreasing the pressure on the following reactions at equilibrium:

(a) $N_2O_4(g) \rightleftharpoons 2\ NO_2(g)$

(b) $N_2(g) + O_2(g) \rightleftharpoons 2\ NO(g)$

(c) $NO(g) + NO_2(g) \rightleftharpoons N_2O_3(g)$

**16-71** Predict the effect of increasing the concentration of the indicated reagent on each of the following reactions at equilibrium:

(a) $\mathbf{2\ NO_2(g)} \rightleftharpoons N_2O_4(g)$

(b) $2\ SO_3(g) \rightleftharpoons 2\ SO_2(g) + \mathbf{O_2(g)}$

(c) $\mathbf{PF_5(g)} \rightleftharpoons PF_3(g) + F_2(g)$

**16-72** Predict the effect of decreasing the concentration of the underlined reagent on each of the following reactions at equilibrium:

(a) $N_2(g) + O_2(g) \rightleftharpoons \mathbf{2\ NO(g)}$

(b) $\mathbf{3\ O_2(g)} \rightleftharpoons 2\ O_3(g)$

(c) $Cl_2(g) + \mathbf{3\ F_2(g)} \rightleftharpoons 2\ ClF_3(g)$

**16-73** Use Le Châtelier's principle to predict the effect of an increase in pressure on the solubility of a gas in water.

**16-74** List as many ways as possible of increasing the yield of ammonia in the Haber process:

$$N_2(g) + 3\ H_2(g) \rightleftharpoons 2\ NH_3(g)$$

**16-75** Explain why an increase in pressure favors the formation of ammonia in the Haber process.

**16-76** Predict how an increase in the volume of the container by a factor of 2 would affect the concentrations of ammonia and oxygen in the following reaction:

$$4\ NH_3(g) + 5\ O_2(g) \rightleftharpoons 4\ NO(g) + 6\ H_2O(g)$$

## Successive Approximations

**16-77** Describe the kind of problem whose solution is most likely to require the use of successive approximations.

**16-78** Describe the procedure you would use if you decided to solve an equilibrium problem by successive approximations.

**16-79** Use successive approximations to check the calculations in the *before* and *after* tables in Section 16.12.

The acid-base reaction between the vapors emitted by concentrated solutions of hydrochloric acid and ammonium hydroxide.

# ACID–BASE EQUILIBRIA

Because water is the solvent for so many reactions, this chapter focuses on what happens when the components of a reaction interact with water to provide the basis for answering questions such as the following:

## POINTS OF INTEREST

- Why are $NH_3$ and HCl much more soluble in water than $O_2$?
- Why is the pH of water so sensitive that adding a single drop of 2 *M* HCl to 100 mL lowers the pH by four units?
- Why does $H_2SO_4$ lose only one $H^+$ ion when it dissociates in water? Why does it lose two $H^+$ ions when it reacts with a base such as ammonia?
- Why is it a mistake to call a solution of $NH_3$ dissolved in water "ammonium hydroxide?"
- Why does the strength of a base depend on the strength of its conjugate acid, and vice versa?
- What is a *buffer,* and how does it resist changes in pH?
- Why does the pH remain more or less constant during a titration until we get close to the endpoint of the titration? Why does the pH rise so rapidly at the endpoint?

## 17.1 THE ROLE OF SOLVENTS IN CHEMICAL REACTIONS

Gases, and the gas-phase reactions discussed in the previous chapter, played such a vital role in the history of chemistry that the development of the apparatus shown in Figure 17.1 to collect, measure, and identify gases has been described as the most important factor in turning chemistry into a science. In a practical sense, however, gases are difficult to handle because useful quantities occupy such an enormous amount of space. Gases are therefore stored at high pressures in compressed gas cylinders or they are dissolved in an appropriate solvent, such as water.

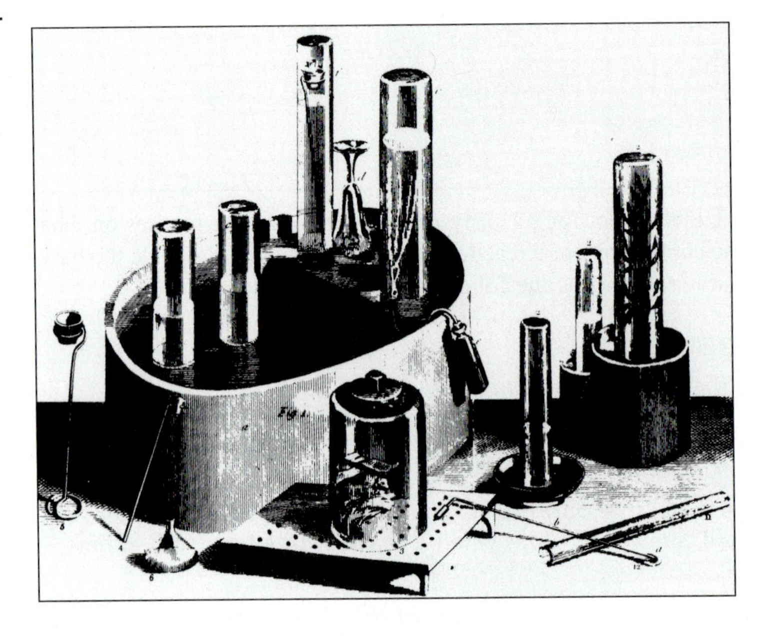

**Figure 17.1** This etching shows some of the apparatus used by Joseph Priestley for studying gases. It is reproduced from his book *Experiments and Observations on Different Kinds of Air,* Vol. 1, published in 1774.

Anyone who has worked in a chemistry laboratory when concentrated solutions of both ammonia and hydrochloric acid are being used has experienced the gas-phase reaction between HCl and $NH_3$ to form finely divided $NH_4Cl$, which collects in the atmosphere:

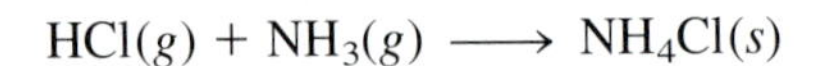

$$HCl(g) + NH_3(g) \longrightarrow NH_4Cl(s)$$

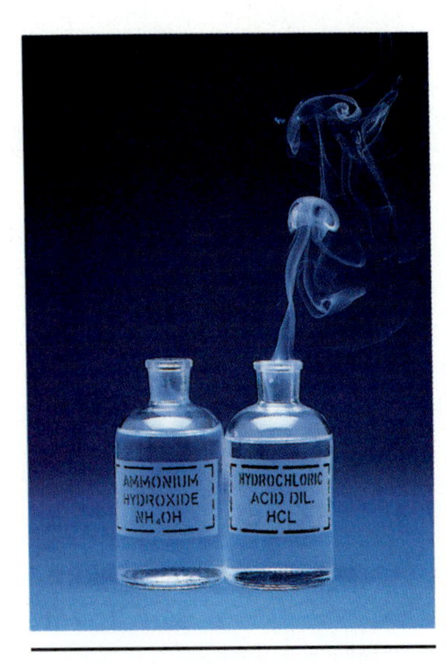


The gas-phase reaction between HCl and $NH_3$ to form $NH_4Cl$.

Outside of the laboratory, however, the only exposure that most people get to ammonia and hydrogen chloride is as aqueous solutions. The best way to clean glass, for example, is to spray a dilute aqueous solution of ammonia, $NH_3(aq)$, on the glass and then rub it with a paper towel or a crumpled piece of newspaper. Few chemists work with hydrogen chloride as a gas, but as an aqueous solution, $HCl(aq)$, it is one of the first chemicals we encounter in our exposure to chemistry laboratories.

Three factors control the solubility of a gas in a solvent: (1) the temperature of the system, (2) the pressure pushing the gas into the solvent, and (3) interactions between particles of the gas and the solvent in which is dissolves. The effect of temperature can be predicted by imagining what would happen when a warm can of a carbonated beverage is opened—the solubility of the $CO_2$ dissolved in the beverage decreases significantly with temperature. The effect of pressure on the solubility of a gas was first studied by William Henry in 1803. His results are summarized in **Henry's law,** which states the mass ($m$) of a gas that dissolves in a given volume of solvent, at constant temperature, is proportional to the pressure ($P$) of the gas with which it is in equilibrium.

$$m = kP$$

**TABLE 17.1 The Solubilities of Common Gases in Water**

| Gas | Solubility (liters of gas per liter of $H_2O$ at 0°C and 1 atm) |
|---|---|
| He | 0.0094 |
| $H_2$ | 0.02148 |
| $N_2$ | 0.02354 |
| CO | 0.03537 |
| $O_2$ | 0.04889 |
| $CH_4$ | 0.05563 |
| $CO_2$ | 1.713 |
| $H_2S$ | 4.670 |
| $SO_2$ | 79.8 |
| HCl | 512 |
| $NH_3$ | 1130 |

By increasing the pressure on a gas, it is possible to increase the amount of gas that dissolves. When the pressure is released (in this case, when the bottle is opened), gas rapidly bubbles out of solution.

To probe the role of the interactions between particles of a gas and the solvent it might be useful to examine the data in Table 17.1, which gives the solubility of common gases in terms of the number of liters of gas that will dissolve in a liter of water at 0°C and 1 atm. Most of these gases are only marginally soluble in water. Others, such as $NH_3$ and HCl, are much more soluble.

The solubility of ammonia in water is the basis of a popular demonstration. A large round-bottomed flask is filled with $NH_3$ gas, sealed with a rubber stopper, and turned upside down. A second flask is then filled with water, a few drops of phenolphthalein are added to the water, and the two flasks are connected with tygon tubing. The demonstration starts when a few drops of water are forced into the upper flask, which contains $NH_3$. All the $NH_3$ in the upper flask rapidly dissolves in this small quantity of water. This creates a vacuum that causes water to rush into this flask with a fountain-like effect.

The ammonia fountain demonstration.

The solubility of $NH_3$ in water can be explained, in part, by the hydrogen bonds that form between ammonia and water shown in Figure 17.2. These bonds effectively pull $NH_3$ into solution, driving the following equilibrium to the right:

$$NH_3(g) \overset{H_2O}{\rightleftharpoons} NH_3(aq)$$

$O_2$ is much less soluble in water because it does not form hydrogen bonds to water.

The color that forms in the upper flask in the ammonia-fountain demonstration reminds us that ammonia not only dissolves in water, it also reacts with water. About 1% of the $NH_3$ molecules in this solution pick up $H^+$ ions from neighboring water molecules to form ammonium ions and hydroxide ions:

$$NH_3(aq) + H_2O(l) \rightleftharpoons NH_4^+(aq) + OH^-(aq)$$

It is the $OH^-$ ions formed in this reaction that turn phenolphthalein pink.

A similar demonstration can be done in which HCl gas dissolves in water that contains an indicator. Once again, all the gas dissolves in a few drops of water that have been forced into the upper flask.

$$HCl(g) \overset{H_2O}{\rightleftharpoons} HCl(aq)$$

This creates a vacuum, which forces water to gush into the upper flask. Once again,

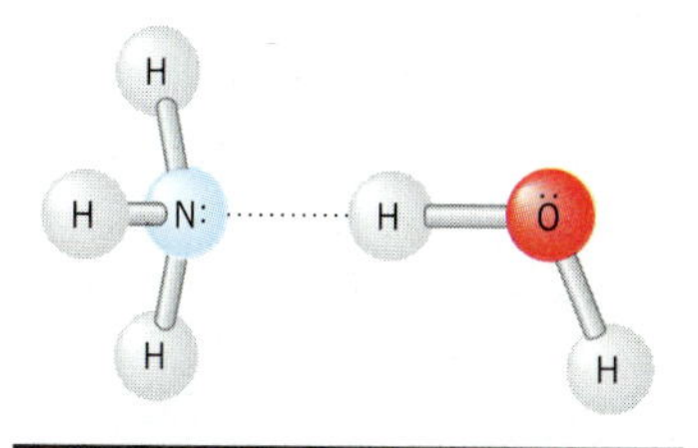


**Figure 17.2** $NH_3$ is soluble in water because it can form hydrogen bonds to $H_2O$ molecules.

the color of the indicator changes, this time because the HCl reacts with water to give hydronium ions and chloride ions:

$$HCl(aq) + H_2O(l) \longrightarrow \mathbf{H_3O^+(aq)} + Cl^-(aq)$$

The remainder of this chapter is devoted to building a model that allows us to calculate the concentrations of the various ions formed when an acid such as HCl or a base such as $NH_3$ reacts with water.

## 17.2 THE ACID–BASE CHEMISTRY OF WATER

The chemistry of aqueous solutions is dominated by the equilibrium between neutral water molecules and the ions they form:

$$2\,H_2O(l) \rightleftharpoons H_3O^+(aq) + OH^-(aq)$$

Strict application of the rules for writing equilibrium constant expressions to this reaction produces the following equation:

$$K_c = \frac{[H_3O^+][OH^-]}{[H_2O]^2}$$

This is a legitimate equilibrium constant expression, but it fails to take into account the enormous difference between the concentrations of neutral $H_2O$ molecules and $H_3O^+$ and $OH^-$ ions at equilibrium.

Measurements of the ability of water to conduct an electric current suggest that pure water at 25°C contains $1.0 \times 10^{-7}$ moles per liter of each of these ions:

$$[H_3O^+] = [OH^-] = 1.0 \times 10^{-7}\,M$$

At the same temperature, the concentration of neutral $H_2O$ molecules is 55.35 molar:

$$\frac{0.9971\ \cancel{g\ H_2O}}{1\ \cancel{mL}} \times \frac{1000\ \cancel{mL}}{1\ L} \times \frac{1\ mol\ H_2O}{18.015\ \cancel{g\ H_2O}} = \mathbf{55.35\ mol\ H_2O/L}$$

The ratio of the concentration of the $H^+$ (or $OH^-$) ion to the concentration of the neutral $H_2O$ molecules is therefore $1.8 \times 10^{-9}$:

$$\frac{1.0 \times 10^{-7}}{55.35} = \mathbf{1.8 \times 10^{-9}}$$

In other words, only about 2 parts per billion (ppb) of the water molecules dissociate into ions at room temperature.

The equilibrium concentration of $H_2O$ molecules is so much larger than the concentrations of the $H_3O^+$ and $OH^-$ ions that it is effectively constant. We therefore build the $[H_2O]$ term into the equilibrium constant for the reaction and thereby greatly simplify equilibrium calculations. We start by rearranging the equilibrium constant expression for the dissociation of water to give the following equation:

$$[H_3O^+][OH^-] = K_c \times [H_2O]^2$$

We then replace the term on the right side of this equation with a constant known as the **water-dissociation equilibrium constant,** $K_w$:

$$\mathbf{[H_3O^+][OH^-] = K_w}$$

In pure water, at 25°C, the $[H_3O^+]$ and $[OH^-]$ ion concentrations are $1.0 \times 10^{-7}$ $M$. The value of $K_w$ at 25°C is therefore $1.0 \times 10^{-14}$:

*Pure Water*

$$[1.0 \times 10^{-7}][1.0 \times 10^{-7}] = \mathbf{1.0 \times 10^{-14}} \qquad \text{(at 25°C)}$$

Although $K_w$ is defined in terms of the dissociation of water, this equilibrium constant expression is equally valid for solutions of acids and bases dissolved in water. Regardless of the source of the $H_3O^+$ and $OH^-$ ions, the product of the concentrations of these ions at equilibrium at 25°C is always $1.0 \times 10^{-14}$.

What happens to the $H_3O^+$ and $OH^-$ concentrations when we add a strong acid to water? Suppose, for example, that we add enough acid to a beaker of water to raise the $H_3O^+$ concentration to 0.010 $M$. According to Le Châtelier's principle, this should drive the equilibrium between water and its ions to the left, reducing the number of $H_3O^+$ and $OH^-$ ions in the solution:

$$2\,H_2O(l) \rightleftharpoons H_3O^+(aq) + OH^-(aq)$$

Because there are so many $H_3O^+$ ions in this solution, the change in the concentration of this ion is too small to notice. When the system returns to equilibrium, the $H_3O^+$ ion concentration is still about 0.010 $M$. Furthermore, when the reaction returns to equilibrium, the product of the $H_3O^+$ and $OH^-$ ion concentrations is once again equal to $K_w$:

$$[H_3O^+][OH^-] = 1 \times 10^{-14}$$

The solution therefore comes back to equilibrium when the dissociation of water is so small that the $OH^-$ concentration is only $1.0 \times 10^{-12}$ $M$:

$$[OH^-] = \frac{K_w}{[H_3O^+]} = \frac{1.0 \times 10^{-14}}{[0.010]} = \mathbf{1.0 \times 10^{-12}\ M}$$

Adding an acid to water therefore has an effect on the concentration of both the $H_3O^+$ and the $OH^-$ ions. Because it is a source of this ion, adding an acid to water increases the concentration of the $H_3O^+$ ion. Adding an acid to water, however, *decreases* the extent to which water dissociates. It therefore leads to a significant decrease in the concentration of the $OH^-$ ion.

As might be expected, the opposite effect is observed when a base is added to water. Because we are adding a base, the $OH^-$ ion concentration increases. Once the system returns to equilibrium, the product of the $H_3O^+$ and $OH^-$ ion concentrations is once again equal to $K_w$. The only way this can be achieved, of course, is by a decrease in the concentration of the $H_3O^+$ ion.

## 17.3 pH AND pOH

Adding an acid to water increases the $H_3O^+$ ion concentration and decreases the $OH^-$ ion concentration, whereas adding a base does the opposite. Regardless of what is added to water, however, the product of the concentrations of these ions at equilibrium is always $1.0 \times 10^{-14}$ at 25°C:

$$[H_3O^+][OH^-] = 1.0 \times 10^{-14}$$

Table 17.2 lists pairs of $H_3O^+$ and $OH^-$ ion concentrations that can coexist at equilibrium in water at 25°C. These concentrations vary over an enormous range (from about 1 $M$ to $10^{-14}$ $M$). The factor of $10^{14}$ that separates one end of this range from the other is difficult, if not impossible, to conceptualize. The difference is comparable to the difference between the value of three pennies and a national debt

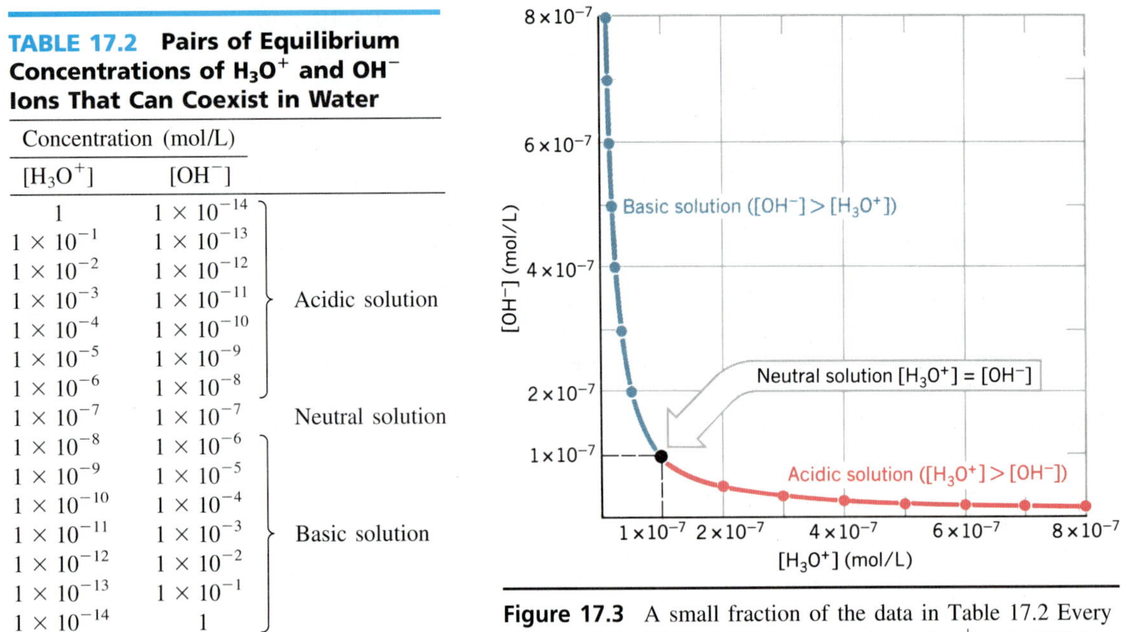


**TABLE 17.2 Pairs of Equilibrium Concentrations of $H_3O^+$ and $OH^-$ Ions That Can Coexist in Water**

| Concentration (mol/L) | | |
|---|---|---|
| $[H_3O^+]$ | $[OH^-]$ | |
| 1 | $1 \times 10^{-14}$ | Acidic solution |
| $1 \times 10^{-1}$ | $1 \times 10^{-13}$ | |
| $1 \times 10^{-2}$ | $1 \times 10^{-12}$ | |
| $1 \times 10^{-3}$ | $1 \times 10^{-11}$ | |
| $1 \times 10^{-4}$ | $1 \times 10^{-10}$ | |
| $1 \times 10^{-5}$ | $1 \times 10^{-9}$ | |
| $1 \times 10^{-6}$ | $1 \times 10^{-8}$ | |
| $1 \times 10^{-7}$ | $1 \times 10^{-7}$ | Neutral solution |
| $1 \times 10^{-8}$ | $1 \times 10^{-6}$ | Basic solution |
| $1 \times 10^{-9}$ | $1 \times 10^{-5}$ | |
| $1 \times 10^{-10}$ | $1 \times 10^{-4}$ | |
| $1 \times 10^{-11}$ | $1 \times 10^{-3}$ | |
| $1 \times 10^{-12}$ | $1 \times 10^{-2}$ | |
| $1 \times 10^{-13}$ | $1 \times 10^{-1}$ | |
| $1 \times 10^{-14}$ | 1 | |

**Figure 17.3** A small fraction of the data in Table 17.2 Every point on the solid line represents a pair of $H_3O^+$ and $OH^-$ ion concentrations for which the solution is at equilibrium.

of $3 trillion, or the difference between the radius of a gold atom and a distance of 8 miles.

Sec. 11.13

Data from Table 17.2 are plotted in Figure 17.3 over a narrow range of concentrations between $1 \times 10^{-7}$ *M* and $1 \times 10^{-6}$ *M*. The point at which the concentrations of the $H_3O^+$ and $OH^-$ ions are equal is called the *neutral* point. Solutions in which the concentration of the $H_3O^+$ ion is larger than $1 \times 10^{-7}$ *M* are described as *acidic.* Those in which the concentration of the $H_3O^+$ ions is smaller than $1 \times 10^{-7}$ *M* are *basic.*

It is impossible to construct a graph that includes all the data from Table 17.2. In 1909, Danish biochemist S. P. L. Sørenson proposed a way around this problem. Sørenson worked at a laboratory set up by the Carlsberg brewery to apply scientific methods to the study of the fermentation reactions in brewing beer. Faced with the task of constructing graphs of the activity of the malt versus the $H_3O^+$ ion concentration, Sørenson proposed using logarithmic mathematics to condense the range of $H_3O^+$ and $OH^-$ concentrations to a more convenient scale. By definition, the logarithm of a number is the power to which a base must be raised to obtain that number. The logarithum to the base 10 of $10^{-7}$ for example, is $-7$.

$$\log(10^{-7}) = -7$$

Since the concentration of the $H_3O^+$ and $OH^-$ ions in aqueous solutions are usually less than 1 *M*, the logarithms of these concentrations are negative numbers. Because he considered positive numbers more convenient, Sørenson suggested that the sign of the logarithm should be changed after it had been calculated. He therefore introduced the symbol p to indicate the negative of the logarithm of a number. Thus, **pH** is the negative of the logarithm of the $H_3O^+$ ion concentration:

$$\mathbf{pH = -\log [H_3O^+]}$$

Similarly, **pOH** is the negative of the logarithm of the $OH^-$ ion concentration:

$$\mathbf{pOH = -\log [OH^-]}$$

The relationship between the pH and pOH of an aqueous solution can be derived by taking the logarithm of both sides of the $K_w$ expression:

$$\log([H_3O^+][OH^-]) = \log(10^{-14})$$

The log of the product of two numbers is equal to the sum of their logs. Thus, the sum of the log of the $H_3O^+$ and $OH^-$ ion concentrations is equal to the log of $10^{-14}$:

$$\log[H_3O^+] + \log[OH^-] = -14$$

Both sides of this equation can now be multiplied by $-1$:

$$-\log[H_3O^+] - \log[OH^-] = 14$$

Substituting the definitions of pH and pOH into this equation gives the following:

$$\mathbf{pH + pOH = 14}$$

This equation can be used to convert from pH to pOH, or vice versa, for any aqueous solution at 25°C, regardless of how much acid or base has been added to the solution. By converting the $H_3O^+$ and $OH^-$ ion concentrations in Table 17.2 into pH and pOH data, we can fit the entire range of concentrations onto a single graph, as shown in Figure 17.4

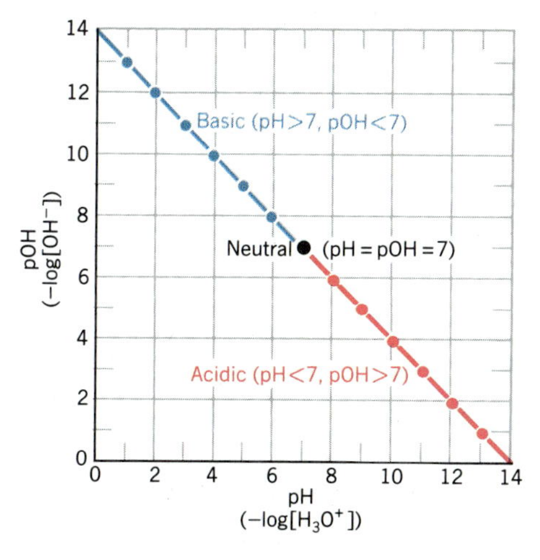


**Figure 17.4** We can fit the entire range of $H_3O^+$ and $OH^-$ ion concentrations in Table 17.2 on a single graph by plotting pH versus pOH. Any point on the solid line corresponds to a solution at equilibrium.

### ▶ CHECKPOINT

Describe what happens to the pH of a solution as the concentration of the $H_3O^+$ ion becomes larger.

## 17.4 ACID-DISSOCIATION EQUILIBRIUM CONSTANTS

Sec. 11.9

Anyone who has had the misfortune of spilling hydrochloric acid on his or her clothes knows that there is a big difference between this strong acid and weak acids such as the acetic acid in vinegar. Both compounds satisfy the Brønsted definition of an acid. (They are both $H^+$ ion, or proton, donors.) But they differ in the extent to which they donate $H^+$ ions to water.

By definition, a strong acid is any substance that is good at donating an $H^+$ ion to water. As we have seen, 99.996% of the HCl molecules in a 6 $M$ solution dissociate when the following reaction comes to equilibrium. This equilibrium lies so far to the right that we write the equation for the reaction with a single arrow, suggesting that hydrochloric acid dissociates completely in aqueous solution:

$$\underset{\text{0.004\% at equilibrium}}{HCl(aq) + H_2O(l)} \longrightarrow \underset{\text{99.996\% at equilibrium}}{\underbrace{H_3O^+(aq) + Cl^-(aq)}}$$

Weak acids are relatively poor $H^+$ ion donors. Acetic acid, for example, is a Brønsted acid because it can donate an $H^+$ ion to water. But it isn't a very good $H^+$ ion donor. Only about 1.3% of the acetic acid molecules in an 0.10 *M* solution lose a proton to water.

$$\underset{\text{98.7\% at equilibrium}}{CH_3CO_2H(aq)} + H_2O(l) \rightleftharpoons \underset{\text{1.3\% at equilibrium}}{\underbrace{H_3O^+(aq) + CH_3CO_2^-(aq)}}$$

It is easy to develop a qualitative feeling for what is meant by the terms *strong acid* and *weak acid.* Strong acids dissociate more or less completely in aqueous solution; weak acids dissociate only slightly. Only a few acids are strong; most acids are weak.

A quantitative feeling for the difference between strong acids and weak acids can be obtained from the equilibrium constants for the reaction between acids and water. Because it is time consuming to write the formula $CH_3CO_2H$ for acetic acid, chemists commonly abbreviate this formula as HOAc and describe the dissociation of the acid as follows.

$$HOAc(aq) + H_2O(l) \rightleftharpoons H_3O^+(aq) + OAc^-(aq)$$

Using this convention, the equilibrium constant expression for the reaction between acetic acid and water might be written as follows:

$$K_c = \frac{[H_3O^+][OAc^-]}{[HOAc][H_2O]}$$

Like the equilibrium constant expression for the dissociation of water, this is a legitimate equation. But most acids are weak, so the equilibrium concentration of $H_2O$ is effectively the same after dissociation as it was before the acid was added. Because the $[H_2O]$ term has no effect on the equilibrium, it is built into the equilibrium constant for the reaction as follows:

$$\frac{[H_3O^+][OAc^-]}{[HOAc]} = K_c \times [H_2O]$$

Sec. 11.9

The result is an equilibrium constant for this equation known as the **acid-dissociation equilibrium constant,** $K_a$. For this reaction:

$$K_a = \frac{[H_3O^+][OAc^-]}{[HOAc]}$$

In general, for any acid HA:

$$\mathbf{K_a = \frac{[H_3O^+][A^-]}{[HA]}}$$

Values of $K_a$ can be used to estimate the relative strengths of acids. The larger the value of $K_a$, the stronger the acid. By definition, a compound is classified as a strong acid when $K_a$ is larger than 1. Weak acids have values of $K_a$ that are smaller than 1.

A list of the acid-dissociation equilibrium constants for some common acids is given in Table 17.3. A more extensive list can be found in Table A-9 in the appendix.

There is a $10^{58}$-fold difference between the values of $K_a$ for the strongest (HI: $K_a = 3 \times 10^9$) and weakest ($CH_4$: $K_a = 10^{-49}$) Brønsted acids. To put this factor of $10^{58}$ into perspective, we should note that the radius of a typical atom ($10^{-8}$ cm) and the radius of that portion of the universe that can be seen with the strongest telescopes (about 10 billion light-years) are separated by a factor of only $10^{36}$.

Table 17.3 provides us with the basis for understanding the difference between strong acids and weak acids. Think about the reaction between a very strong acid and water:

$$\underset{K_a = 10^6}{HCl(aq)} + H_2O(l) \rightleftharpoons \underset{K_a = 55}{H_3O^+(aq)} + Cl^-(aq)$$

HCl is a much stronger acid than the $H_3O^+$ ion, which means that $H_2O$ is a stronger base than the $Cl^-$ ion. It therefore isn't surprising to find that the stronger of a pair of acids reacts with the stronger of a pair of bases to give a weaker acid and a weaker base.

What about the reaction between acetic acid and water?

Sec. 11.10

$$\underset{K_a = 1.8 \times 10^{-5}}{HOAc(aq)} + H_2O(l) \rightleftharpoons \underset{K_a = 55}{H_3O^+(aq)} + OAc^-(aq)$$

In this case, the reaction tries to convert the weaker of a pair of acids and the weaker of a pair of bases into a stronger acid and a stronger base. It isn't surprising to find that this reaction occurs to only a minor extent.

**TABLE 17.3 Values of $K_a$ for Common Acids**

| Strong Acids | $K_a$ |
|---|---|
| Hydrochloric acid (HCl) | $1 \times 10^6$ |
| Sulfuric acid ($H_2SO_4$) | $1 \times 10^3$ |
| **Hydronium ion ($H_3O^+$)** | **55** |
| Nitric acid ($HNO_3$) | 28 |
| **Weak Acids** | **$K_a$** |
| Phosphoric acid ($H_3PO_4$) | $7.1 \times 10^{-3}$ |
| Citric acid ($C_6H_7O_8$) | $7.5 \times 10^{-4}$ |
| Acetic acid ($CH_3CO_2H$) | $1.8 \times 10^{-5}$ |
| Boric acid ($H_3BO_3$) | $7.3 \times 10^{-10}$ |
| Water ($H_2O$) | $1.8 \times 10^{-16}$ |

Of course, as the value of $K_a$ decreases further, the extent to which the acid will react with water must decrease as well. Inevitably, we will encounter acids that are so weak they can't compete with water as a source of the $H_3O^+$ ion.

## CHECKPOINT

By definition, nitric acid is a "strong" acid because $K_a$ is larger than 1. Explain why concentrated nitric acid (15 $M$) does not dissociate completely:

$$\underset{\approx 30\% \text{ at equilibrium}}{HNO_3(aq)} + H_2O(l) \rightleftharpoons \underbrace{H_3O^+(aq) + NO_3^-(aq)}_{\approx 70\% \text{ at equilibrium}}$$

## RESEARCH IN THE 90s

### GAS-PHASE ACID–BASE CHEMISTRY

Until recently, most of the information chemists had about acid–base reactions was obtained from experiments done with solutions, in which the solvent plays an important role. About 20 years ago, papers started to appear in the chemical literature that described techniques that can be used to measure the acidity or basicity of individual molecules in the gas phase.

The importance of gas-phase measurements of this sort can be appreciated by looking at the rate at which publications on the subject have increased. A recent compilation of gas-phase ion data listed all references to such measurements from the 1920s through most of the 1980s [S. G. Lias, J. E. Bartmess, J. F. Liebman, J. L. Holmes, R. D. Levin, and W. G. Mallard, *Journal of Physical Chemistry Reference Data,* **17** (1988)]. Prior to 1960, 11 references could be found. In the 1960s, 89 papers were published on this topic. In the 1970s, more than 600 papers appeared. Between 1980 and 1988, another 1700 papers appeared.

By definition, the *gas-phase acidity* of an HA molecule is the enthalpy ($\Delta H_{HA}$) of the following reaction:

$$HA(g) \longrightarrow H^+(g) + A^-(g)$$

Some of the most accurately determined gas-phase acidities are given in Table 17.4.

**TABLE 17.4 Gas-Phase Acidities**

| Compound | $\Delta H_{HA}$(kJ/mol) |
|---|---|
| $CH_4$ | 1743 |
| $NH_3$ | 1689 |
| $H_2O$ | 1635 |
| HF | 1552 |
| HCl | 1395 |
| HBr | 1354 |
| HI | 1315 |

As might be expected, there is a good correlation ($r = -0.94$) between gas-phase acidity data and the acid-dissociation equilibrium constant. Because of the enormous range of values of $K_a$ for Brønsted acids, it is useful to compare the gas-phase acidity data with values of p$K_a$ for these acids:

$$pK_a = -\log K_a$$

Figure 17.5 shows the correlation between gas-phase and solution acidity across the entire spectrum of Brønsted acids.

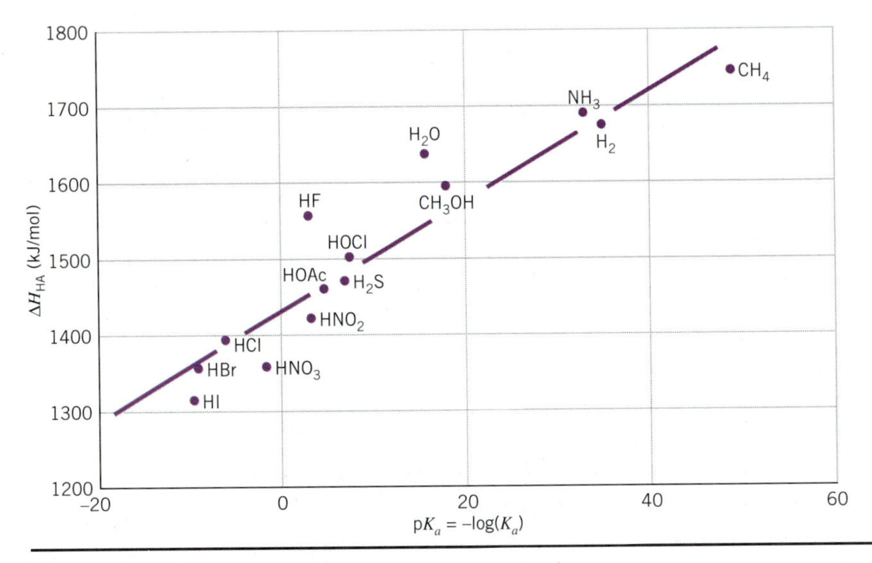


**Figure 17.5** A plot of gas-phase acidity versus p$K_a$ data.

Two gases in Figure 17.5 give results that lie particularly far from the line that describes the best fit to these data: HF and $H_2O$. These compounds seem to be better acids in aqueous solution than would be expected from their gas-phase acidities. It is tempting to relate some of the deviation from linearity to the stabilization of the relatively small $F^-$ and $OH^-$ ions that occurs when these ions interact with the polar solvent.

## 17.5 STRONG ACIDS

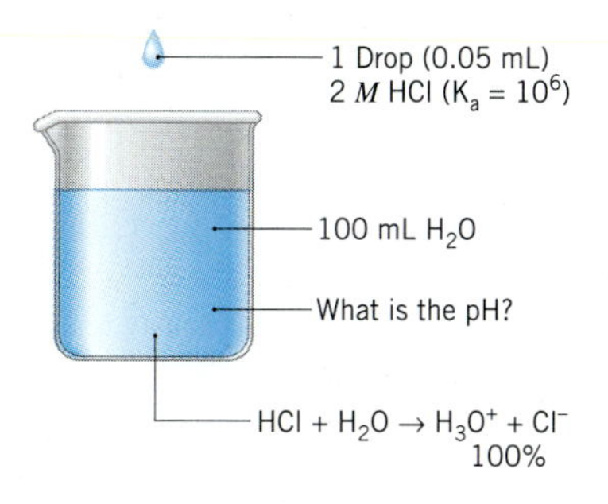


The simplest acid–base equilibria are those in which a strong acid (or base) is dissolved in water. Consider the calculation of the pH of a solution formed by adding a single drop of 2 *M* hydrochloric acid to 100 mL of water, for example. The key to this calculation is remembering that HCl is a strong acid ($K_a = 10^6$) and that acids as strong as this can be assumed to dissociate completely in water:

$$HCl(aq) + H_2O(l) \longrightarrow H_3O^+(aq) + Cl^-(aq)$$

The $H_3O^+$ ion concentration at equilibrium is therefore essentially equal to the initial concentration of the acid.

A useful rule of thumb says that there are about 20 drops in each milliliter. One drop of 2 *M* HCl therefore has a volume of 0.05 mL, or $5 \times 10^{-5}$ L.

The number of moles of HCl added to the water in this calculation can be obtained from the volume and concentration of the hydrochloric acid:

$$\frac{2 \text{ mol HCl}}{\cancel{\text{L}}} \times 5 \times 10^{-5} \cancel{\text{L}} = 1 \times 10^{-4} \text{ mol HCl}$$

The initial concentration of HCl is equal to the number of moles of HCl added to the beaker divided by the volume of water to which the HCl has been added.

$$\frac{1 \times 10^{-4} \text{ mol HCl}}{0.100 \text{ L}} = \mathbf{1 \times 10^{-3}\ \mathit{M}\ HCl}$$

According to this calculation, the initial concentration of HCl is $1 \times 10^{-3}$ *M*. If we assume that this acid dissociates completely, the $[H_3O^+]$ ion concentration at equilibrium is $1 \times 10^{-3}$ *M*. The solution prepared by adding one drop of 2 *M* HCl to 100 mL of water would therefore have a pH of 3:

$$\begin{aligned} pH &= -\log [H_3O^+] \\ &= -\log [1 \times 10^{-3}] = -(-3) = 3 \end{aligned}$$

## 17.6 WEAK ACIDS

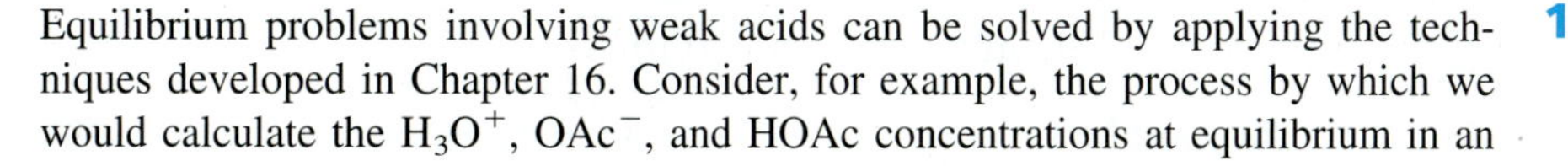
Equilibrium problems involving weak acids can be solved by applying the techniques developed in Chapter 16. Consider, for example, the process by which we would calculate the $H_3O^+$, $OAc^-$, and HOAc concentrations at equilibrium in an

**Problem-Solving Strategy**

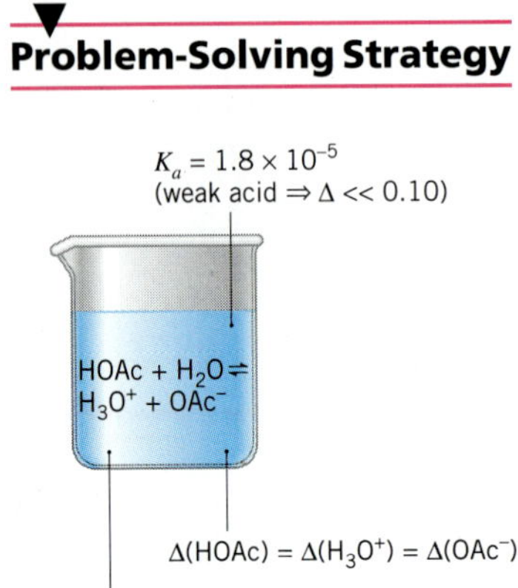


0.10 *M* solution of acetic acid in water. We start this calculation, as always, by building a representation of what we know about the reaction:

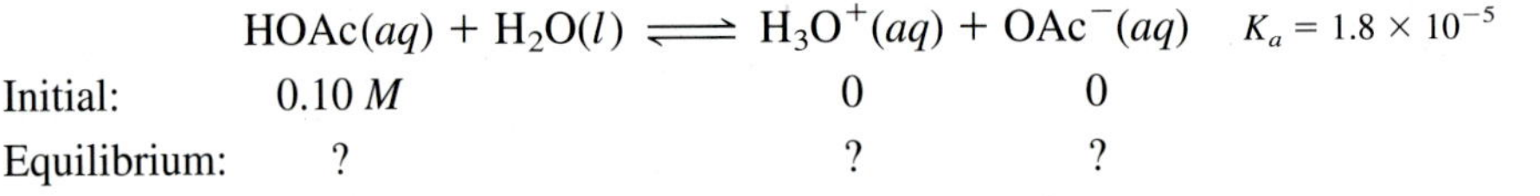

| | $HOAc(aq) + H_2O(l) \rightleftharpoons$ | $H_3O^+(aq)$ + | $OAc^-(aq)$ | $K_a = 1.8 \times 10^{-5}$ |
|---|---|---|---|---|
| Initial: | 0.10 *M* | 0 | 0 | |
| Equilibrium: | ? | ? | ? | |

We then compare the initial reaction quotient ($Q_a$) with the equilibrium constant ($K_a$) for the reaction and reach the obvious conclusion that the reaction must shift to the right to reach equilibrium:

$$Q_a = \frac{(H_3O^+)(OAc^-)}{(HOAc)} = \frac{(0)(0)}{(0.10)} = 0 < K_a$$

Recognizing that we get one $H_3O^+$ ion and one $OAc^-$ ion each time an HOAc molecule dissociates allows us to write equations for the equilibrium concentrations of the three components of the reaction.

| | $HOAc(aq) + H_2O(l) \rightleftharpoons$ | $H_3O^+(aq)$ + | $OAc^-(aq)$ | $K_a = 1.8 \times 10^{-5}$ |
|---|---|---|---|---|
| Initial: | 0.10 *M* | 0 | 0 | |
| Equilibrium: | 0.10 − Δ | Δ | Δ | |

Substituting what we know about the system at equilibrium into the $K_a$ expression gives the following equation:

$$K_a = \frac{[H_3O^+][OAc^-]}{[HOAc]} = \frac{[\Delta][\Delta]}{[0.10 - \Delta]} = 1.8 \times 10^{-5}$$

Although we could rearrange this equation and solve it with the quadratic formula, it is much simpler to apply the reasoning we used in Chapter 16 to test the assumption that Δ is small compared with the initial concentration of acetic acid:

$$\frac{[\Delta][\Delta]}{[0.10]} \approx 1.8 \times 10^{-5}$$

We then solve this approximate equation for the value of Δ:

$$\Delta \approx 0.0013$$

Is Δ small enough to be ignored in this problem? Yes, because it is less than 5% of the initial concentration of acetic acid:

$$\frac{0.0013}{0.10} \times 100\% = 1.3\%$$

We can therefore use this value of Δ to calculate the equilibrium concentrations of $H_3O^+$, $OAc^-$, and HOAc:

$$[HOAc] = 0.10 - \Delta \approx \mathbf{0.10\ M}$$
$$[H_3O^+] = [OAc^-] = \Delta \approx \mathbf{0.0013\ M}$$

We can confirm the validity of these results by substituting these concentrations into the expression for $K_a$.

$$K_a = \frac{[H_3O^+][OAc^-]}{[HOAc]} = \frac{[0.0013][0.0013]}{[0.10]} = 1.7 \times 10^{-5}$$

Our calculation must be valid because the ratio of these concentrations agrees with the value of $K_a$ for acetic acid within experimental error.

## 17.7 HIDDEN ASSUMPTIONS IN WEAK-ACID CALCULATIONS

How many assumptions were made in the calculation in Section 17.6? It may have appeared that we made one assumption—that Δ is small compared with the initial concentration of HOAc. In fact, two assumptions were made. The other assumption was hidden in the way the problem was set up:

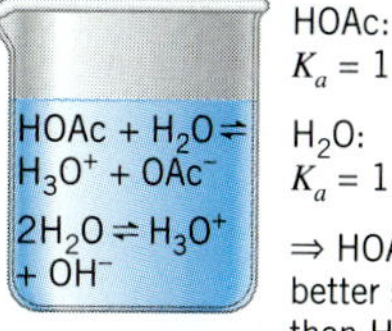


HOAc: $K_a = 1.8 \times 10^{-5}$

$H_2O$: $K_a = 1.8 \times 10^{-16}$

⇒ HOAc is a much better source of $H_3O^+$ than $H_2O$ is.

| | $HOAc(aq) + H_2O(l) \rightleftharpoons$ | $H_3O^+(aq)$ | $+ OAc^-(aq)$ | $K_a = 1.8 \times 10^{-5}$ |
|---|---|---|---|---|
| Initial: | 0.10 *M* | 0 | 0 | |
| Equilibrium: | 0.10 − Δ | Δ | Δ | |

The amount of $H_3O^+$ ion in water is so small that it is common to assume that the initial concentration of this ion is zero, which isn't quite true.

It is important to remember that there are two sources of the $H_3O^+$ ion in this solution. We get $H_3O^+$ ions from the dissociation of acetic acid:

$$HOAc(aq) + H_2O(l) \rightleftharpoons \mathbf{H_3O^+(aq)} + OAc^-(aq)$$

But we also get $H_3O^+$ ions from the dissociation of water:

$$2\,H_2O(l) \rightleftharpoons \mathbf{H_3O^+(aq)} + OH^-(aq)$$

Because the initial concentration of the $H_3O^+$ ion is not quite zero, it might be a better idea to write "≈ 0" beneath the $H_3O^+$ term when we describe the initial conditions of the reaction.

Before we can trust the results of the calculation in Section 17.6, we have to check both of the assumptions made in this calculation:

1. The assumption that the amount of acid that dissociates is small compared with the initial concentration of the acid.
2. The assumption that enough acid dissociates to allow us to ignore the dissociation of water.

We have already confirmed the validity of the first assumption. (Only 1.3% of the acetic acid molecules dissociate in this solution.) Let's now check the second assumption.

According to the calculation in the previous section, the concentration of the $H_3O^+$ ion from the dissociation of acetic acid is 0.0013 *M*. The $OH^-$ ion concentration in this solution is therefore:

$$[OH^-] = \frac{K_w}{[H_3O^+]} = \frac{1.0 \times 10^{-14}}{1.3 \times 10^{-3}} = \mathbf{7.7 \times 10^{-12}}$$

All of the $OH^-$ ion in this solution comes from the dissociation of water. Since we get one $H_3O^+$ ion for each $OH^-$ ion when water dissociates, the contribution to the total $H_3O^+$ ion concentration from the dissociation of water must be $7.7 \times 10^{-12}$ *M*. In other words, only about 6 parts per billion of the $H_3O^+$ ions in this solution come from the dissociation of water:

$$\frac{7.7 \times 10^{-12}}{1.3 \times 10^{-3}} \approx \mathbf{6 \times 10^{-9}}$$

The second assumption is therefore valid in this calculation. For all practical purposes, we can assume that virtually none of the $H_3O^+$ ion in this solution comes from the dissociation of water. As might be expected, this assumption only fails for dilute solutions of very weak acids.

## 17.8 IMPLICATIONS OF THE ASSUMPTIONS IN WEAK-ACID CALCULATIONS

The two assumptions that are made in weak-acid equilibrium problems can be restated as follows.

1. The dissociation of the acid is *small enough* that the change in the concentration of the acid as the reaction comes to equilibrium can be ignored.
2. The dissociation of the acid is *large enough* that the $H_3O^+$ ion concentration from the dissociation of water can be ignored.

In other words, the acid must be *weak enough* that Δ is small compared with the initial concentration of the acid. But it also must be *strong enough* that the $H_3O^+$ ions from the acid overwhelm the dissociation of water. This invokes an image out of the story of Goldilocks and the Three Bears. In order for the approach taken to the calculation for acetic acid to work, the acid has to be "just right." If it's too strong, Δ won't be small enough to be ignored. If it's too weak, the dissociation of water will have to be included in the calculation.

Fortunately, many acids are "just right." To illustrate this point, the next section will use both assumptions in a series of calculations designed to identify the factors that influence the $H_3O^+$ ion concentration in aqueous solutions of weak acids.

## 17.9 FACTORS THAT INFLUENCE THE $H_3O^+$ ION CONCENTRATION IN WEAK-ACID SOLUTIONS

The following exercise examines the relationship between the $H_3O^+$ ion concentration at equilibrium and the acid-dissociation equilibrium constant for the acid.

**EXERCISE 17.1**

Calculate the pH of 0.10 *M* solutions of the following acids.

(a) Hypochlorous acid, HOCl $K_a = 2.9 \times 10^{-8}$

(b) Hypobromous acid, HOBr $K_a = 2.4 \times 10^{-9}$

(c) Hypoiodous acid, HOI $K_a = 2.3 \times 10^{-11}$

**SOLUTION** The first calculation can be set up as follows:

| | $HOCl(aq) + H_2O(l)$ | $\rightleftharpoons$ | $H_3O^+(aq)$ | $+ OCl^-(aq)$ |
|---|---|---|---|---|
| Initial: | 0.10 *M* | | ≈0 | 0 |
| Equilibrium: | 0.10 − Δ | | Δ | Δ |

Substituting this information into the $K_a$ expression gives the following equation:

$$K_a = \frac{[H_3O^+][OCl^-]}{[HOCl]} = \frac{[\Delta][\Delta]}{[0.10 - \Delta]} = 2.9 \times 10^{-8}$$

We then try the assumption that Δ is small compared with the initial concentration of the acid:

$$\frac{[\Delta][\Delta]}{[0.10]} \approx 2.9 \times 10^{-8}$$

Solving this equation for Δ gives the following result:

$$\Delta \approx 5.4 \times 10^{-5}\, M$$

Both assumptions made in this calculation are legitimate. Δ is small compared with the initial concentration of HOCl, but it is several orders of magnitude larger than the $H_3O^+$ ion concentration from the dissociation of water. We can therefore use this value of $\Delta$ to calculate the pH of the solution:

$$\text{pH} = -\log [H_3O^+] = -\log [5.4 \times 10^{-5}] = 4.27$$

Repeating this calculation with hypobromous and hypoodous acid gives the following results:

| | | |
|---|---|---|
| HOCl: | $[H_3O^+] = 5.4 \times 10^{-5}\,M$ | **pH = 4.27** |
| HOBr: | $[H_3O^+] = 1.5 \times 10^{-5}\,M$ | **pH = 4.82** |
| HOI: | $[H_3O^+] = 1.5 \times 10^{-6}\,M$ | **pH = 5.82** |

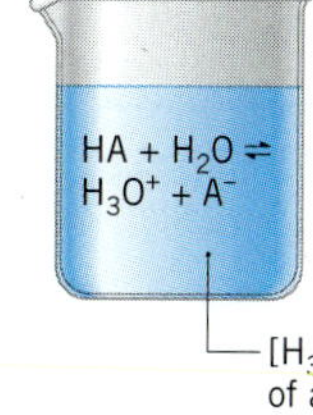


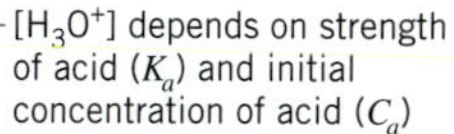


As expected, the $H_3O^+$ ion concentrations at equilibrium—and therefore the pH of the solution—depends on the value of $K_a$ for the acid. The $H_3O^+$ ion concentration decreases and the pH of the solution increases as the value of $K_a$ becomes smaller. The next exercise shows that the $H_3O^+$ ion concentration at equilibrium also depends on the initial concentration of the acid.

## EXERCISE 17.2

Calculate the $H_3O^+$ ion concentration and the pH of acetic acid solutions with the following concentrations: 1.0 *M*, 0.10 *M*, and 0.01 *M*.

**SOLUTION** The first calculation can be set up as follows:

| | $HOAc(aq) + H_2O(l)$ | $\rightleftharpoons$ | $H_3O^+(aq)$ | $+\ OAc^-(aq)$ | $K_a = 1.8 \times 10^{-5}$ |
|---|---|---|---|---|---|
| Initial: | 1.0 *M* | | $\approx 0$ | 0 | |
| Equilibrium: | $1.0 - \Delta$ | | $\Delta$ | $\Delta$ | |

Substituting this information into the $K_a$ expression gives the following equation:

$$K_a = \frac{[H_3O^+][OAc^-]}{[HOAc]} = \frac{[\Delta][\Delta]}{[1.0 - \Delta]} = 1.8 \times 10^{-5}$$

We now assume that $\Delta$ is small compared with the initial concentration of acetic acid:

$$\frac{[\Delta][\Delta]}{[1.0]} \approx 1.8 \times 10^{-5}$$

We then solve this equation for an approximate value of $\Delta$:

$$\Delta \approx 0.0042$$

Both assumptions are legitimate in this calculation. $\Delta$ is small compared with the initial concentration of the acid, but large compared with the concentration of the $H_3O^+$ ion from the dissociation of water. We can therefore use this value of $\Delta$ to calculate the pH of the solution:

$$\text{pH} = -\log [H_3O^+] = -\log [4.2 \times 10^{-3}] = 2.4$$

Repeating this calculation for the different initial concentrations gives the following results:

| | | |
|---|---|---|
| 1.0 *M* HOAc: | $[H_3O^+] = 4.2 \times 10^{-3}\,M$ | **pH = 2.4** |
| 0.10 *M* HOAc: | $[H_3O^+] = 1.3 \times 10^{-3}\,M$ | **pH = 2.9** |
| 0.010 *M* HOAc: | $[H_3O^+] = 4.2 \times 10^{-4}\,M$ | **pH = 3.4** |

The concentration of the $H_3O^+$ ion in an aqueous solution gradually decreases and the pH of the solution increases as the solution becomes more dilute. The results of Exercises 17.1 and 17.2 provide a basis for constructing a model that allows us to predict when we can ignore the dissociation of water in equilibrium problems involving weak acids. Two factors must be built into this model: (1) the strength of the acid as reflected by the value of $K_a$, and (2) the strength of the solution as reflected by the initial concentration of the acid.

## 17.10 NOT-SO-WEAK ACIDS

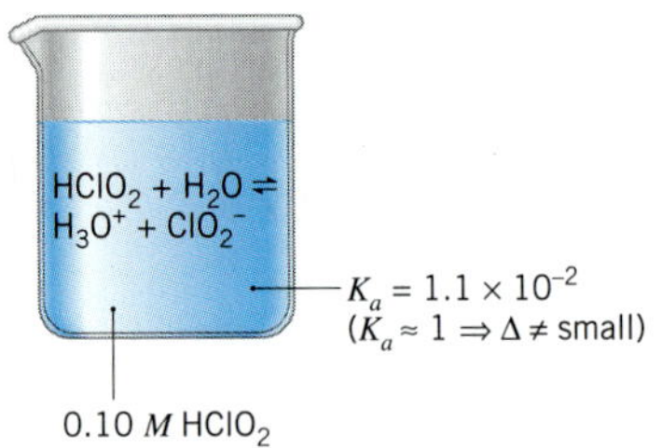


We now have techniques for solving equilibrium problems that involve strong acids or weak acids that are ''just right.'' But we need to develop techniques to handle problems for which one or the other of our assumptions is not valid.

In this section, we will consider acid solutions that aren't weak enough to ignore the value of Δ. Let's start by calculating the $H_3O^+$, $HClO_2$, and $ClO_2^-$ concentrations at equilibrium in an 0.10 *M* solution of chlorous acid ($K_a = 1.1 \times 10^{-2}$):

$$HClO_2(aq) + H_2O(l) \rightleftharpoons H_3O^+(aq) + ClO_2^-(aq)$$

The first step, as always, involves building a representation of the problem:

| | $HClO_2(aq) + H_2O(l) \rightleftharpoons$ | $H_3O^+(aq)$ + | $ClO_2^-(aq)$ | $K_a = 1.1 \times 10^{-2}$ |
|---|---|---|---|---|
| Initial: | 0.10 *M* | ≈0 | 0 | |
| Equilibrium: | 0.10 − Δ | Δ | Δ | |

We then substitute this information into the $K_a$ expression:

$$K_a = \frac{[H_3O^+][ClO_2^-]}{[HClO_2]} = \frac{[\Delta][\Delta]}{[0.10 - \Delta]} = 1.1 \times 10^{-2}$$

The value of $K_a$ for this acid is close enough to 1 to make us suspicious of the assumption that Δ is small compared with the initial concentration of the acid. There is nothing wrong with trying this assumption, however, even if we suspect it isn't valid:

$$\frac{[\Delta][\Delta]}{[0.10]} \approx 1.1 \times 10^{-2}$$

Solving this approximate equation gives a value for Δ that is 33% of the initial concentration of chlorous acid:

$$\Delta \approx 0.033\ M$$

The assumption that Δ is small therefore fails miserably.

**Problem-Solving Strategy**

There are two ways out of this difficulty, as detailed in Chapter 16. We can expand the original equation and solve it by the quadratic formula. Or we can use successive approximations to solve the problem. Both techniques give the following value of Δ for this problem:

$$\Delta = 0.028$$

Using this value of Δ gives the following results.

$$[HClO_2] = 0.10 - \Delta = \mathbf{0.072\ M}$$
$$[H_3O^+] = [ClO_2^-] = \Delta = \mathbf{0.028\ M}$$

Chlorous acid does not belong among the class of strong acids that dissociate more or less completely. Nor does it fit in the category of weak acids, which dissociate only to a negligible extent. Since the amount of dissociation in this solution is about 28%, it might be classified as a "not-so-weak acid."

## 17.11 VERY WEAK ACIDS

It is more difficult to solve equilibrium problems when the acid is too weak to ignore the dissociation of water. Deriving an equation that can be used to solve this class of problems is therefore easier than solving them one at a time. To derive such an equation, we start by assuming that we have a generic acid that dissolves in water. We therefore have two sources of the $H_3O^+$ ion:

$$HA(aq) + H_2O(l) \rightleftharpoons H_3O^+(aq) + A^-(aq)$$
$$2\,H_2O(l) \rightleftharpoons H_3O^+(aq) + OH^-(aq)$$

Because we get one $H_3O^+$ ion for each $OH^-$ ion when water dissociates, the concentration of the $H_3O^+$ ion from the dissociation of water is always equal to the amount of $OH^-$ ion from this reaction.

$$[H_3O^+]_W = [OH^-]_W$$

The total $H_3O^+$ ion concentration in an acid solution is equal to the sum of the $H_3O^+$ ion concentrations from the two sources of this ion, the acid and the water:

$$[H_3O^+]_T = [H_3O^+]_{HA} + [H_3O^+]_W$$

We now write three more equations that describe this system. The first equation is the equilibrium constant expression for this reaction:

$$K_a = \frac{[H_3O^+]_T[A^-]}{[HA]}$$

The second equation summarizes the relationship between the total $H_3O^+$ ion concentration in the solution and the $OH^-$ ion concentration from the dissociation of water:

$$[H_3O^+]_T[OH^-]_W = K_w$$

The third equation summarizes the relationship between the positive and negative ions produced by the two reactions that occur in this solution:

$$[H_3O^+]_T = [A^-] + [OH^-]_W$$

(This equation simply states that the sum of the positive ions formed by the dissociation of the acid and water is equal to the sum of the negative ions produced by these reactions.)

We now substitute the second equation into the third:

$$[H_3O^+]_T = [A^-] + \frac{K_w}{[H_3O^+]_T}$$

We then solve this equation for the $[A^-]$ term:

$$[A^-] = [H_3O^+]_T - \frac{K_w}{[H_3O^+]_T}$$

We then substitute this equation into the equilibrium constant expression:

$$K_a = \frac{[H_3O^+]_T\left([H_3O^+]_T - \dfrac{K_w}{[H_3O^+]_T}\right)}{[HA]}$$

Rearranging this equation by combining terms gives:

$$K_a[HA] = [H_3O^+]_T^2 - K_w$$

We then solve this equation for the $H_3O^+$ ion concentration and take the square root of both sides:

$$[H_3O^+]_T = \sqrt{K_a[HA] + K_w}$$

We can generate a more useful version of this equation by remembering that we are trying to solve equilibrium problems for acids that are so weak we can't ignore the dissociation of water. We can therefore assume that $\Delta$ is small compared with the initial concentration of the acid.

By convention, the symbol used to represent the initial concentration of the acid is $C_a$. If $\Delta$ is small compared with the initial concentration of the acid, then the concentration of HA when this reaction reaches equilibrium will be virtually the same as the initial concentration:

$$[HA] \approx C_a$$

Substituting this approximation into the equation derived in this section gives an equation that can be used to calculate the pH of a solution of a very weak acid:

$$\mathbf{[H_3O^+]_T = \sqrt{K_aC_a + K_w}}$$

**EXERCISE 17.3**

Calculate the $H_3O^+$ concentration in an 0.0001 *M* solution of hydrocyanic acid (HCN):

$$HCN(aq) + H_2O(l) \rightleftharpoons H_3O^+(aq) + CN^-(aq) \qquad K_a = 6 \times 10^{-10}$$

**SOLUTION** This is a dilute solution of a fairly weak acid. It is therefore the kind of solution for which the dissociation of water is likely to make an important contribution to the total $H_3O^+$ concentration. We know $K_a$ for the acid, as well as the initial concentration. We can therefore use the equation derived in this section to calculate the total $H_3O^+$ ion concentration from the dissociation of both hydrocyanic acid and water:

$$[H_3O^+]_T = \sqrt{K_aC_a + K_w}$$
$$= \sqrt{[(6 \times 10^{-10})(1 \times 10^{-4}) + (1 \times 10^{-14})]} = \mathbf{3 \times 10^{-7}\ \mathit{M}}$$

## 17.12 SUMMARIZING THE CHEMISTRY OF WEAK ACIDS

This section compares the way in which the $H_3O^+$ concentration is calculated for pure water, a weak acid, and a very weak acid.

### PURE WATER

The product of the concentrations of the $H_3O^+$ and $OH^-$ ions in pure water is equal to $K_w$:

$$[H_3O^+][OH^-] = K_w$$

But the $H_3O^+$ and $OH^-$ ion concentrations in pure water are the same:

$$[H_3O^+] = [OH^-]$$

Substituting the second equation into the first gives the following result:

$$[H_3O^+]^2 = K_w$$

The $H_3O^+$ ion concentration in pure water is therefore equal to the square root of $K_w$:

$$[H_3O^+] = \sqrt{K_w}$$

## WEAK ACIDS

The generic equilibrium constant expression for a weak acid is written as follows.

$$K_a = \frac{[H_3O^+][A^-]}{[HA]}$$

If the acid is strong enough to ignore the dissociation of water, the $H_3O^+$ ion and $A^-$ ion concentrations in this solution are about equal:

$$[H_3O^+] \approx [A^-]$$

Substituting this information into the acid-dissociation equilibrium constant expression gives the following result:

$$K_a = \frac{[H_3O^+]^2}{[HA]}$$

The concentration of the HA molecules at equilibrium is equal to the initial concentration of the acid minus the amount that dissociates, $\Delta$:

$$K_a = \frac{[H_3O^+]^2}{[C_a - \Delta]}$$

If $\Delta$ is small compared with the initial concentration of the acid, we get the following approximate equation:

$$K_a \approx \frac{[H_3O^+]^2}{C_a}$$

Rearranging this equation and taking the square root of both sides gives the following result:

$$[H_3O^+] \approx \sqrt{K_a C_a}$$

## VERY WEAK ACIDS

When the acid is so weak that we can't ignore the dissociation of water, we use the following equation, derived in Section 17.11, to calculate the concentration of the $H_3O^+$ ion at equilibrium:

$$[H_3O^+] = \sqrt{K_a C_a + K_w}$$

The equations used to calculate the $H_3O^+$ ion concentration in these solutions are summarized below.

| | |
|---|---|
| **Pure water:** | $[H_3O^+] = \sqrt{K_w}$ |
| **Weak acid:** | $[H_3O^+] \approx \sqrt{K_a C_a}$ |
| **Very weak acid:** | $[H_3O^+] = \sqrt{K_a C_a + K_w}$ |

The first and second equations are nothing more than special cases of the third. When we can ignore the dissociation of the acid—because there is no acid in the solution—we get the first equation. When we can ignore the dissociation of water, we get the second equation. When we can't ignore the dissociation of either the acid or water, we have to use the last equation.

This discussion gives us a basis for deciding when we can ignore the dissociation of water. Remember our rule of thumb: We can ignore anything that makes a contribution of less than 5% to the total. Now compare the most inclusive equation for the $H_3O^+$ ion concentration,

$$[H_3O^+] = \sqrt{K_aC_a + K_w}$$

with the equation that assumes that the dissociation of water can be ignored:

$$[H_3O^+] \approx \sqrt{K_aC_a}$$

The only difference is the $K_w$ term, which is under the square root sign.

If the $H_3O^+$ ion concentration were directly proportional to the sum of $K_aC_a$ plus $K_w$ we could assume that the error would be less than 5% when $K_aC_a$ was at least 20 times larger than $K_w$. This is not a linear relationship, however. Because the $H_3O^+$ ion concentration is proportional to the square root of $K_aC_a$ plus $K_w$, $K_aC_a$ only has to be about 10 times larger than $K_w$ to give an error of less than 5% in the concentration of the $H_3O^+$ ion. Stated differently, this means that the dissociation of water can be ignored whenever $K_aC_a$ is at least 10 times larger than $K_w$.

As a rule:

**We can ignore the dissociation of water when $K_aC_a$ for a weak acid is larger than $1.0 \times 10^{-13}$. When $K_aC_a$ is smaller than $1.0 \times 10^{-13}$, the dissociation of water must be included in the calculation.**

## EXERCISE 17.4

Calculate the pH of an 0.023 $M$ solution of saccharin (HSc), if $K_a$ is $2.1 \times 10^{-12}$ for this artificial sweetener.

**SOLUTION** We can decide whether the dissociation of water can be ignored by comparing the product of $K_a$ for the acid times its initial concentration with the value of $K_w$ for water:

$$K_aC_a = (2.1 \times 10^{-12})(0.023) = 4.8 \times 10^{-14} < 1 \times 10^{-13}$$

According to this calculation, the product of $K_a$ times $C_a$ is too small for the dissociation of water to be ignored. Thus, the $H_3O^+$ ion concentration in this solution has to be calculated from the following equation:

$$[H_3O^+] = \sqrt{K_aC_a + K_w} = \sqrt{[(2.1 \times 10^{-12})(0.023) + 1.0 \times 10^{-14}}$$
$$= 2.4 \times 10^{-7}\ M$$

Once we know the concentration of the $H_3O^+$ ion, we can calculate the pH of the solution:

$$\text{pH} = -\log\,[2.4 \times 10^{-7}] = 6.6$$

## 17.13 BASES



With minor modifications, the techniques applied to equilibrium calculations for acids are valid for solutions of bases in water. Consider, for example, the calculation of the pH of an 0.10 $M$ $NH_3$ solution. We can start by writing an equation for the reaction between ammonia and water:

$$NH_3(aq) + H_2O(l) \rightleftharpoons NH_4^+(aq) + OH^-(aq)$$

Strict adherence to the rules for writing equilibrium constant expressions leads to the following equation for this reaction:

$$K_c = \frac{[NH_4^+][OH^-]}{[NH_3][H_2O]}$$

But, taking a lesson from our experience with acid-dissociation equilibria, we can build the $[H_2O]$ term into the value of the equilibrium constant. Reactions between a base and water are therefore described in terms of a **base-ionization equilibrium constant,** $K_b$.

$$K_b = \frac{[NH_4^+][OH^-]}{[NH_3]} = K_c \times [H_2O]$$

The values of $K_b$ for a limited number of bases are given in Table A-10 in the appendix. For $NH_3$, $K_b$ is $1.8 \times 10^{-5}$.

We can organize what we know about this equilibrium with the format we used for equilibria involving acids:

| | $NH_3(aq) + H_2O(l)$ | $\rightleftharpoons$ $NH_4^+(aq)$ | $+ OH^-(aq)$ | $K_b = 1.8 \times 10^{-5}$ |
|---|---|---|---|---|
| Initial: | 0.10 $M$ | 0 | $\approx 0$ | |
| Equilibrium: | $0.10 - \Delta$ | $\Delta$ | $\Delta$ | |

Substituting this information into the equilibrium constant expression gives the following equation:

$$K_b = \frac{[NH_4^+][OH^-]}{[NH_3]} = \frac{[\Delta][\Delta]}{[0.10 - \Delta]} = 1.8 \times 10^{-5}$$

$K_b$ for ammonia is small enough to consider the assumption that $\Delta$ is small compared with the initial concentration of the base:

$$\frac{[\Delta][\Delta]}{[0.10]} \approx 1.8 \times 10^{-5}$$

Solving this approximate equation gives the following result:

$$\Delta \approx 1.3 \times 10^{-3}$$

This value of $\Delta$ is small enough compared with the initial concentration of $NH_3$ to be ignored and yet large enough compared with the $OH^-$ ion concentration in water to ignore the dissociation of water. It can therefore be used to calculate the pOH of the solution:

$$\text{pOH} = -\log(1.3 \times 10^{-3}) = 2.89$$

Which, in turn, can be used to calculate the pH of the solution:

$$\text{pH} = 14 - \text{pOH} = 11.11$$

**Problem-Solving Strategy**

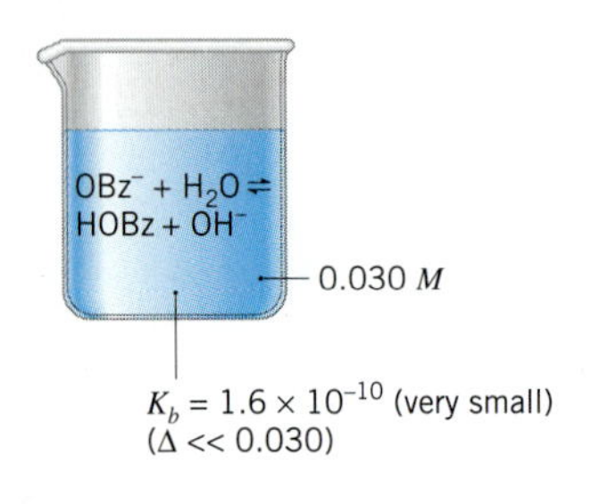


Equilibrium problems involving bases are relatively easy to solve if the value of $K_b$ for the base is known. Values of $K_b$ are listed in Table A-10 for only a limited number of compounds, however. The first step in many base equilibrium calculations therefore involves determining the value of $K_b$ for the reaction from the value of $K_a$ for the conjugate acid. To see how this is done, let's calculate the pH of a 0.030 *M* solution of sodium benzoate ($C_6H_5CO_2Na$) in water from the value of $K_a$ for benzoic acid ($C_6H_5CO_2H$):

$$K_a = 6.3 \times 10^{-5}$$

Benzoic acid and sodium benzoate are members of a family of food additives whose ability to retard the rate at which food spoils has helped produce a 10-fold decrease in the incidence of stomach cancer. To save time and space, we'll abbreviate benzoic acid as HOBz and sodium benzoate as NaOBz. Benzoic acid, as its name implies, is an acid. Sodium benzoate is a salt of the conjugate base, the $OBz^-$ or benzoate ion.

Whenever sodium benzoate dissolves in water, it dissociates into its ions:

$$\text{NaOBz}(aq) \xrightarrow{H_2O} \text{Na}^+(aq) + \text{OBz}^-(aq)$$

The benzoate ion then acts as a base toward water, picking up a proton to form the conjugate acid and a hydroxide ion:

$$\text{OBz}^-(aq) + \text{H}_2\text{O}(l) \rightleftharpoons \text{HOBz}(aq) + \text{OH}^-(aq)$$

The base-ionization equilibrium constant expression for this reaction is therefore written:

$$K_b = \frac{[\text{HOBz}][\text{OH}^-]}{[\text{OBz}^-]}$$

The next step in solving the problem involves calculating the value of $K_b$ for the $OBz^-$ ion from the value of $K_a$ for HOBz.

The $K_a$ and $K_b$ expressions for benzoic acid and its conjugate base both contain the ratio of the equilibrium concentrations of the acid and its conjugate base. $K_a$ is proportional to $[OBz^-]$ divided by [HOBz], and $K_b$ is proportional to [HOBz] divided by $[OBz^-]$.

$$K_a = \frac{[\text{H}_3\text{O}^+][\text{OBz}^-]}{[\text{HOBz}]} \qquad K_b = \frac{[\text{HOBz}][\text{OH}^-]}{[\text{OBz}^-]}$$

Two changes have to made to derive the $K_b$ expression from the $K_a$ expression: We need to remove the $[H_3O^+]$ term and introduce an $[OH^-]$ term. We can do this by multiplying the top and bottom of the $K_a$ expression by the $OH^-$ ion concentration:

**Problem-Solving Strategy**

$$K_a = \frac{[\text{H}_3\text{O}^+][\text{OBz}^-]}{[\text{HOBz}]} \times \frac{[\text{OH}^-]}{[\text{OH}^-]}$$

Rearranging this equation gives:

$$K_a = \frac{[\text{OBz}^-]}{[\text{HOBz}][\text{OH}^-]} \times [\text{H}_3\text{O}^+][\text{OH}^-]$$

The terms on the right side of this equation should look familiar. The first is the inverse of the $K_b$ expression, the second is the expression for $K_w$:

$$K_a = \frac{1}{K_b} \times K_w$$

Rearranging this equation gives the following result:

$$\mathbf{K_a \times K_b = K_w}$$

According to this equation, the value of $K_b$ for the reaction between the benzoate ion and water can be calculated from $K_a$ for benzoic acid:

$$K_b = \frac{K_w}{K_a} = \frac{1.0 \times 10^{-14}}{6.3 \times 10^{-5}} = \mathbf{1.6 \times 10^{-10}}$$

---

**CHECKPOINT**

Use the relationship between the values of $K_a$ for an acid and $K_b$ for its conjugate base to explain the following rules from Chapter 11.

Sec. 11.8

**Strong acids have weak conjugate bases.**
**Strong bases have weak conjugate acids.**

---

Now that we know $K_b$ for the benzoate ion, we can calculate the pH of an 0.030 *M* NaOBz solution with the techniques used to handle weak-acid equilibria. We start, once again, by building a representation for the problem:

| | $OBz^-(aq) + H_2O(l) \rightleftharpoons$ | $HOBz(aq)$ + | $OH^-(aq)$ | $K_b = 1.6 \times 10^{-10}$ |
|---|---|---|---|---|
| Initial: | 0.030 *M* | 0 | ≈0 | |
| Equilibrium: | 0.030 − Δ | Δ | Δ | |

We then substitute this information into the $K_b$ expression:

$$K_b = \frac{[HOBz][OH^-]}{[OBz^-]} = \frac{[\Delta][\Delta]}{[0.030 - \Delta]} = 1.6 \times 10^{-10}$$

Because $K_b$ is relatively small, we assume that Δ is small compared with 0.030:

$$\frac{[\Delta][\Delta]}{[0.030]} \approx 1.6 \times 10^{-10}$$

We then solve the approximate equation for the value of Δ:

$$\Delta \approx 2.2 \times 10^{-6}$$

The assumption that Δ is small is obviously valid. We can therefore use Δ to calculate the pOH of the solution:

$$pOH = -\log[2.2 \times 10^{-6}] = 5.66$$

The problem asked for the pH of the solution, however, so we use the relationship between pH and pOH to calculate the pH:

$$pH = 14 - pOH = \mathbf{8.34}$$

Two assumptions were made in this calculation.

1. We assumed that $\Delta$ was small enough compared with the initial concentration of the base that it could be ignored in the $[0.030 - \Delta]$ term.
2. We assumed that all of the $OH^-$ ion at equilibrium came from the reaction between the benzoate ion and water. (In other words, we ignored the contribution to the $OH^-$ ion concentration from the dissociation of water.)

We have already confirmed the validity of the first assumption. What about the second? The $OH^-$ ion concentration obtained from this calculation is $2.1 \times 10^{-6}$ *M*, which is 21 times the $OH^-$ ion concentration in pure water. According to Le Châtelier's principle, however, the addition of a base suppresses the dissociation of water. This means that the dissociation of water makes a contribution of significantly less than 5% to the total $OH^-$ ion concentration in this solution. It therefore can be legitimately ignored.

By analogy with our discussion of acids in Section 17.9, we can conclude that two factors affect the $OH^-$ ion concentration in aqueous solutions of bases: $K_b$ and $C_b$. By analogy with the discussion of weak acids in Section 17.12, we can develop the following rule:

**We can ignore the dissociation of water when $K_bC_b$ for a weak base is larger than $1.0 \times 10^{-13}$. When $K_bC_b$ is smaller than $1.0 \times 10^{-13}$, we have to include the dissociation of water in our calculations.**

## EXERCISE 17.5

Calculate the HOAc, $OAc^-$, and $OH^-$ concentrations at equilibrium in an 0.10 *M* NaOAc solution. (HOAc: $K_a = 1.8 \times 10^{-5}$)

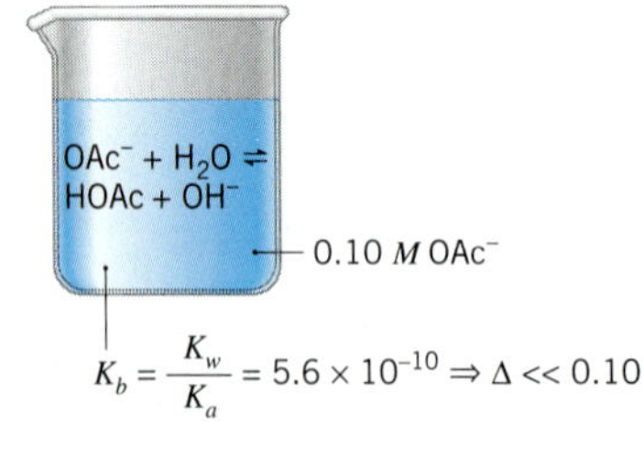


**SOLUTION** Because we start with a solution of a base, NaOAc, dissolved in water, we can summarize the problem we are trying to solve as follows:

| | $OAc^-(aq) + H_2O(l)$ | $\rightleftharpoons$ | $HOAc(aq)$ | $+\ OH^-(aq)$ | $K_b = ?$ |
|---|---|---|---|---|---|
| Initial: | 0.10 *M* | | 0 | $\approx 0$ | |
| Equilibrium: | $0.10 - \Delta$ | | $\Delta$ | $\Delta$ | |

The only other information we need to solve this problem is the value of $K_b$ for the reaction between the acetate ion and water, which can be calculated from the value of $K_a$ for acetic acid:

$$K_b = \frac{K_w}{K_a} = \frac{1.0 \times 10^{-14}}{1.8 \times 10^{-5}} = 5.6 \times 10^{-10}$$

At this point we should test whether the solution is basic enough to allow us to ignore the dissociation of water:

$$K_bC_b = (5.6 \times 10^{-10})(0.10) = 5.6 \times 10^{-11} > 1 \times 10^{-13}$$

Because the product $K_b$ times the initial concentration of the base is large enough to ignore the dissociation of water, we can set up the calculation as follows:

$$K_b = \frac{[HOAc][OH^-]}{[OAc^-]} = \frac{[\Delta][\Delta]}{[0.10 - \Delta]} = 5.6 \times 10^{-10}$$

The value of $K_b$ for this base is relatively small, so we can feel reasonably confident that $\Delta$ is small compared with the initial concentration of base:

$$\frac{[\Delta][\Delta]}{[0.10]} \approx 5.6 \times 10^{-10}$$

We can then solve this approximate equation.

$$\Delta \approx 7.5 \times 10^{-6}$$

We have already confirmed the validity of the assumption that the dissociation of water can be ignored in this problem. We now test the assumption that $\Delta$ is small compared with the initial concentration of the base and find this assumption is also valid. We can therefore use this value of $\Delta$ to calculate the equilibrium concentrations of HOAc, $OAc^-$, and $OH^-$:

$$[OAc^-] = 0.10 - \Delta \approx \mathbf{0.10}\ M$$
$$[HOAc] = [OH^-] = \Delta \approx \mathbf{7.5 \times 10^{-6}}$$

## EXERCISE 17.6

Salts of the fluoride ion are used to help prevent tooth decay. Calculate the pH of an 0.0010 $M$ solution of NaF in water. (HF: $K_a = 7.2 \times 10^{-4}$)

**SOLUTION** The first step in this calculation is to derive the value $K_b$ for the $F^-$ ion from $K_a$ for hydrofluoric acid:

$$K_b = \frac{K_w}{K_a} = \frac{1.0 \times 10^{-14}}{7.2 \times 10^{-4}} = 1.4 \times 10^{-11}$$

The product of $K_b$ times the initial concentration of the $F^-$ ion then can be compared with $1.0 \times 10^{-13}$:

$$K_bC_b = (1.4 \times 10^{-11})(0.0010) = 1.4 \times 10^{-14} < 1 \times 10^{-13}$$

The $F^-$ ion is too weak a base, and the solution is too dilute, for the dissociation of water to be ignored. The $OH^-$ ion concentration in this solution is therefore calculated from the following equation:

$$[OH^-] = \sqrt{K_bC_b + K_w} = \sqrt{(1.4 \times 10^{-11})(0.0010) + 1.0 \times 10^{-14}}$$
$$= 1.5 \times 10^{-7}\ M$$

The pOH of this solution is therefore 6.8 and the pH is 7.2:

$$pOH = -\log [1.5 \times 10^{-7}] = 6.8$$
$$pH = 14 - pOH = \mathbf{7.2}$$

Dilute solutions of NaF in water are such weak bases that NaF is often described as a *neutral salt.*

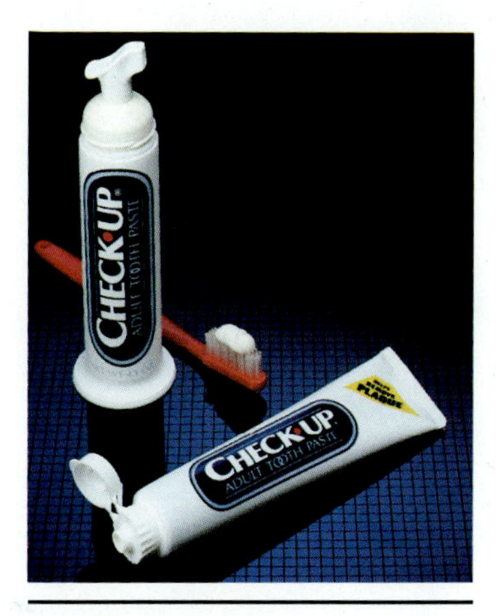

A flouride toothpaste.

## 17.14 MIXTURES OF ACIDS AND BASES

So far, we have examined 0.10 $M$ solutions of both acetic acid and sodium acetate:

| | | |
|---|---|---|
| 0.10 $M$ HOAc: | $[H_3O^+] = 1.3 \times 10^{-3}\ M$ | pH = 2.9 |
| 0.10 $M$ NaOAc: | $[OH^-] = 7.5 \times 10^{-6}\ M$ | pH = 8.9 |

What would happen if we added enough sodium acetate to an acetic acid solution so that the solution is simultaneously 0.10 $M$ in both HOAc and NaOAc?

**Problem-Solving Strategy**

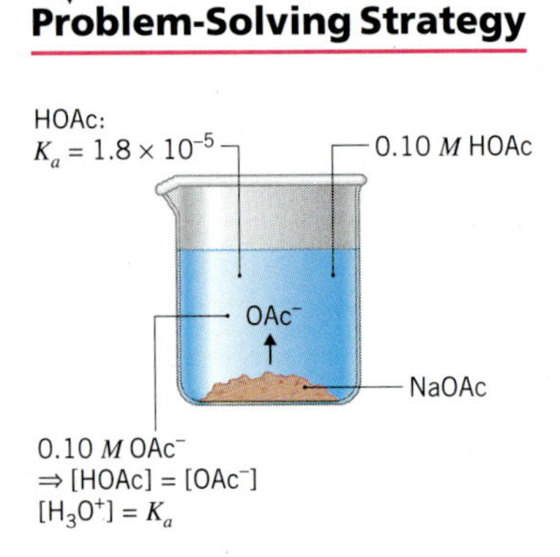


The first step toward answering this question is recognizing that there are two sources of the $OAc^-$ ion in this solution. Acetic acid, of course, dissociates to give the $H_3O^+$ and $OAc^-$ ions:

$$HOAc(aq) + H_2O(l) \rightleftharpoons H_3O^+(aq) + \mathbf{OAc^-(aq)}$$

Sodium acetate, however, also dissociates in water to give the $OAc^-$ ion.

$$NaOAc(s) \xrightarrow{H_2O} Na^+(aq) + \mathbf{OAc^-(aq)}$$

These reactions share a common ion: the $OAc^-$ ion. Le Châtelier's principle predicts that adding NaOAc will shift the equilibrium between HOAc and the $H_3O^+$ and $OAc^-$ ions to the left. Thus, adding NaOAc reduces the extent to which HOAc dissociates. This mixture is therefore less acidic than 0.10 *M* HOAc, by itself.

We can quantitate this discussion by setting up the problem as follows:

| | $HOAc(aq) + H_2O(l) \rightleftharpoons$ | $H_3O^+(aq)$ | $+ OAc^-(aq)$ | $K_a = 1.8 \times 10^{-5}$ |
|---|---|---|---|---|
| Initial: | 0.10 *M* | ≈0 | 0.10 *M* | |
| Equilibrium: | ? | ? | ? | |

We don't have any basis for predicting whether the dissociation of water can be ignored in this calculation, so let's make this assumption and check its validity later.

If most of the $H_3O^+$ ion concentration at equilibrium comes from the dissociation of the acetic acid, the reaction has to shift to the right to reach equilibrium:

| | $HOAc(aq) + H_2O(l) \rightleftharpoons$ | $H_3O^+(aq)$ | $+ OAc^-(aq)$ | $K_a = 1.8 \times 10^{-5}$ |
|---|---|---|---|---|
| Initial: | 0.10 *M* | ≈0 | 0.10 *M* | |
| Equilibrium: | $0.10 - \Delta$ | $\Delta$ | $0.10 + \Delta$ | |

Substituting this information into the $K_a$ expression gives the following result:

$$K_a = \frac{[H_3O^+][OAc^-]}{[HOAc]} = \frac{[\Delta][0.10 + \Delta]}{[0.10 - \Delta]} = 1.8 \times 10^{-5}$$

By now, the next step should be obvious: We assume that $\Delta$ is small compared with the initial concentrations of HOAc and $OAc^-$:

$$\frac{[\Delta][0.10]}{[0.10]} \approx 1.8 \times 10^{-5}$$

We then solve the approximate equation this assumption generates:

$$\Delta \approx 1.8 \times 10^{-5}$$

Are our two assumptions legitimate? Is $\Delta$ small enough compared with 0.10 to be ignored? Is $\Delta$ large enough so that the dissociation of water can be ignored? The answer to both questions is yes. We can therefore use this value of $\Delta$ to calculate the $H_3O^+$ ion concentration at equilibrium and the pH of the solution:

$$pH = -\log [1.8 \times 10^{-5}] = \mathbf{4.74}$$

This solution is therefore acidic, but less acidic than HOAc by itself.

## 17.15 BUFFERS

Sec. 17.5

Mixtures of a weak acid and its conjugate base, such as HOAc and the $OAc^-$ ion, are called **buffers.** The term *buffer* usually means "to lessen or absorb shock." These solutions are buffers because they lessen or absorb the drastic change in pH that occurs when small amounts of acids or bases are added to water. In Section 17.5 we concluded that adding a single drop of 2 *M* hydrochloric acid to 100 mL of pure water changes the pH from 7 to 3. Let's see what happens when we add a drop of 2 *M* HCl to a buffer solution.

### EXERCISE 17.7

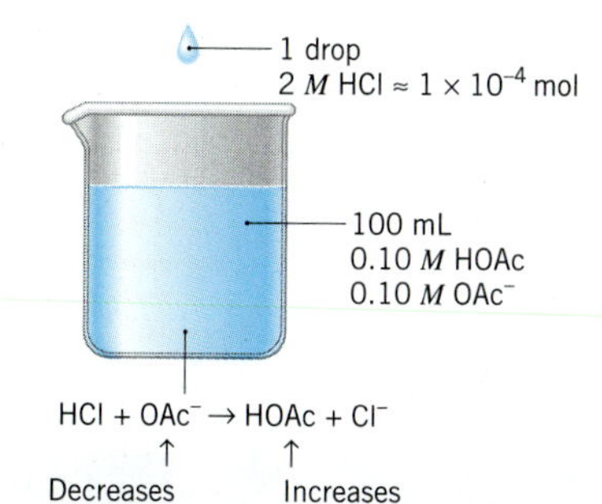

Calculate the effect of adding one drop of 2 *M* HCl to 100 mL of a buffer solution that is 0.100 *M* in both acetic acid and sodium acetate.

**SOLUTION** Before the acid is added, the concentrations of HOAc and the $OAc^-$ ion in this solution are the same, 0.100 *M*:

$$K_a = \frac{[H_3O^+][OAc^-]}{[HOAc]} = \frac{[H_3O^+]\cancel{[0.100]}}{\cancel{[0.100]}} = 1.8 \times 10^{-5}$$

The initial $H_3O^+$ ion concentration in the buffer is therefore equal to $K_a$ for the acid:

$$[H_3O^+] = 1.8 \times 10^{-5}$$

The pH of the buffer solution before the acid is added is therefore 4.74:

$$\text{pH} = -\log\,[1.8 \times 10^{-5}] = 4.74$$

In section 17.5 we concluded that adding a single drop of 2 *M* HCl to 100 mL of pure water was equivalent to adding $1 \times 10^{-3}$ mol of HCl per liter of solution. Because HCl dissociates more or less completely in water, this generates an $H_3O^+$ ion concentration that is $1 \times 10^{-3}$ *M*. The $H_3O^+$ concentration is now too large for the solution to be at equilibrium:

$$Q_a = \frac{(H_3O^+)(OAc^-)}{(HOAc^-)} = \frac{(1.0 \times 10^{-3})(0.100)}{(0.100)} = 1.0 \times 10^{-3} > K_a$$

The reaction will therefore have to shift toward HOAc to reach equilibrium.

For the sake of argument, let's assume that all of the $H_3O^+$ ion from the hydrochloric acid is consumed to convert $OAc^-$ ions into acetic acid:

$$OAc^-(aq) + H_3O^+(aq) \longrightarrow HOAc(aq) + H_2O(l)$$

Because we added 0.001 mol/L of HCl to this solution, the reaction is characterized by the following intermediate conditions:

| | $HOAc(aq) + H_2O(l)$ | $\rightleftharpoons$ | $H_3O^+(aq)$ | + | $OAc^-(aq)$ | $K_a = 1.8 \times 10^{-5}$ |
|---|---|---|---|---|---|---|
| Initial: | 0.100 *M* | | 0.001 *M* | | 0.100 *M* | |
| Change: | +0.001 *M* | | −0.001 *M* | | −0.001 *M* | |
| Intermediate: | 0.101 *M* | | ≈0 | | .099 *M* | |

In order for the reaction to come to equilibrium, some of the acetic acid in these intermediate conditions has to dissociate to form $H_3O^+$ and $OAc^-$ ions:

| | $HOAc(aq) + H_2O(l)$ | $\rightleftharpoons$ | $H_3O^+(aq)$ | + | $OAc^-(aq)$ | $K_a = 1.8 \times 10^{-5}$ |
|---|---|---|---|---|---|---|
| Intermediate: | 0.101 *M* | | ≈0 | | 0.099 *M* | |
| Equilibrium: | $0.101 - \Delta$ | | $\Delta$ | | $0.099 + \Delta$ | |

We then substitute this information into the $K_a$ expression:

$$K_a = \frac{[H_3O^+][OAc^-]}{[HOAc]} = \frac{[\Delta][0.099 + \Delta]}{[0.101 - \Delta]} = 1.8 \times 10^{-5}$$

and assume that Δ is relatively small:

$$\frac{[\Delta][0.099]}{[0.101]} \approx 1.8 \times 10^{-5}$$

When we solve this equation for Δ, we find that there has been no significant change in the $H_3O^+$ ion concentration:

$$\Delta \approx 1.8 \times 10^{-5}$$

As a result, there is no significant change in the pH of this solution:

$$pH = -\log [H_3O^+] = \mathbf{4.74}$$

Although a single drop of 2 *M* HCl drops the pH of 100 mL of water by 4 units (from 7 to 3), there is no change in the pH of the buffer when a drop of 2 *M* HCl is added to 100 mL of this buffer solution. Thus, the pH of the buffer solution is truly "buffered" against the effect of small amounts of acid or base.

Buffers can be made from a weak acid and its conjugate base, such as acetic acid and a salt of the acetate ion:

$$\underset{\text{weak acid}}{\mathbf{HOAc}(aq)} + H_2O(l) \rightleftharpoons H_3O^+(aq) + \underset{\text{conjugate base}}{\mathbf{OAc^-}(aq)}$$

Buffers can also be made from a weak base and its conjugate acid, such as ammonia and a salt of the ammonium ion:

$$\underset{\text{conjugate acid}}{\mathbf{NH_4^+}(aq)} + H_2O(l) \rightleftharpoons H_3O^+(aq) + \underset{\text{weak base}}{\mathbf{NH_3}(aq)}$$

HOAc/$OAc^-$ buffers test acidic to litmus, whereas $NH_3/NH_4^+$ buffers test basic.

There is an important difference between these buffers. Mixtures of HOAc and the $OAc^-$ ion form an **acidic buffer,** with a pH below 7. Mixtures of $NH_3$ and the $NH_4^+$ ion form a **basic buffer,** with a pH above 7.

We can predict whether a buffer will be acidic or basic by comparing the values of $K_a$ and $K_b$ for the conjugate acid–base pair. $K_a$ for acetic acid is significantly larger than $K_b$ for the acetate ion. We therefore expect mixtures of this conjugate acid–base pair to be acidic:

$$\text{HOAc:} \quad K_a = 1.8 \times 10^{-5}$$
$$\text{OAc}^-\text{:} \quad K_b = 5.6 \times 10^{-10}$$

$K_b$ for ammonia, on the other hand, is much larger than $K_a$ for the ammonium ion. Mixtures of this acid–base pair are therefore basic:

$$NH_3\text{:} \quad K_b = 1.8 \times 10^{-5}$$
$$NH_4^+\text{:} \quad K_a = 5.6 \times 10^{-10}$$

In general, if $K_a$ for the acid is larger than $1 \times 10^{-7}$, the buffer will be acidic. If $K_b$ is larger than $1 \times 10^{-7}$, the buffer is basic.

## 17.16 BUFFER CAPACITY AND pH TITRATION CURVES

The most important property of a buffer is its ability to resist changes in pH when small quantities of acid or base are added to the solution. We can develop a model for the capacity of a buffer to absorb acid or base by looking at how the buffer resists changes in pH. Consider an $HOAc/OAc^-$ buffer, for example:

$$K_a = \frac{[H_3O^+][OAc^-]}{[HOAc]} = 1.8 \times 10^{-5}$$

Rearranging the $K_a$ expression for this buffer gives the following equation:

$$[H_3O^+] = K_a \times \frac{[HOAc]}{[OAc^-]}$$

According to this equation, the $H_3O^+$ ion concentration, and therefore the pH of the solution, will remain essentially constant as long as the ratio of the concentrations of HOAc and the $OAc^-$ ion is more or less constant.

Suppose we start with a solution that contains equal amounts of HOAc and the $OAc^-$ ion. When we add small amounts of acid to this solution, some of the acetate ions are converted into acetic acid:

$$OAc^-(aq) + H_3O^+(aq) \longrightarrow HOAc(aq) + H_2O(l)$$

But the ratio of these concentrations doesn't change by much, so the pH stays the same. When we add small amounts of base to the solution, some of the acetic acid is converted into acetate ions:

$$HOAc(aq) + OH^-(aq) \longrightarrow OAc^-(aq) + H_2O(l)$$

But, once again, the ratio of these concentrations stays more or less the same, and so does the pH.

As long as the concentrations of HOAc and the $OAc^-$ ion in the buffer are larger than the amount of acid or base added to the solution, the pH remains constant. Table 17.5 compares the effects of adding different amounts of hydrochloric acid to water and a pair of buffer solutions. The first column gives the number of moles of HCl added per liter of solution. The second column shows the effect of this much acid on the pH of water. The third column shows the effect on a buffer solution that contains 0.10 *M* HOAc and 0.10 *M* NaOAc. The fourth column shows the results of adding the acid to a buffer that contains twice as much acetic acid and twice as much sodium acetate.

**TABLE 17.5 The Effect of Adding Hydrochloric Acid to Water and Two Buffer Solutions**

| Moles HCl per Liter of Solution | pH of Water | pH of Buffer 1 (0.10 *M* HOAc/ 0.10 *M* $OAc^-$) | pH of Buffer 2 (0.20 *M* HOAc/ 0.20 *M* $OAc^-$) |
|---|---|---|---|
| 0 | 7 | 4.74 | 4.74 |
| 0.000001 | 6 | 4.74 | 4.74 |
| 0.00001 | 5 | 4.74 | 4.74 |
| 0.0001 | 4 | 4.74 | 4.74 |
| 0.001 | 3 | 4.74 | 4.74 |
| 0.01 | 2 | 4.66 | 4.70 |
| 0.1 | 1 | 2.72 | 4.27 |

Even when as much as 0.01 moles per liter of acid is added to the first buffer, there is very little change in the pH of the buffer solution, as shown in Figure 17.6. By definition, the **buffer capacity** of a buffer is the amount of acid or base that can be added before the pH of the solution changes significantly. Note that the capacity of the second buffer solution in Table 17.5 is even greater than the first. This buffer protects against quantities of acid or base as large as 0.1 moles per liter.

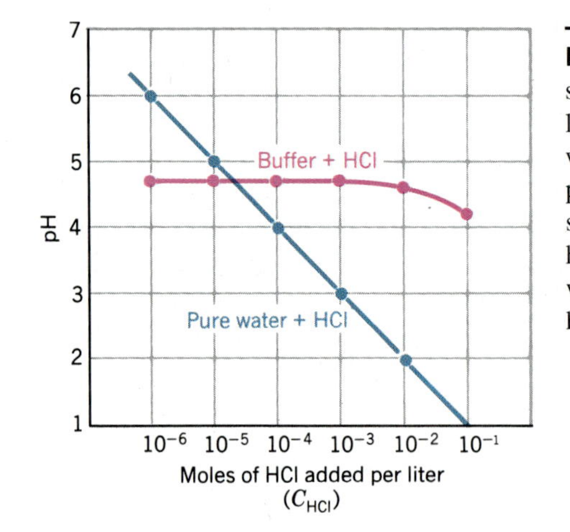


**Figure 17.6** The pH of water is very sensitive to hydrochloric acid. Adding as little as $10^{-3}$ mol of HCl to a liter of water can drop the pH by 4 units. The pH of the buffer, however, remains constant until relatively large amounts of hydrochloric acid have been added, at which point the buffer becomes exhausted.

Buffers and buffering capacity have a lot to do with the shape of the titration curve in Figure 17.7, which shows the pH of a solution of acetic acid as it is titrated with a strong base, sodium hydroxide. Four points (*A, B, C,* and *D*) on this curve will be discussed in some detail.

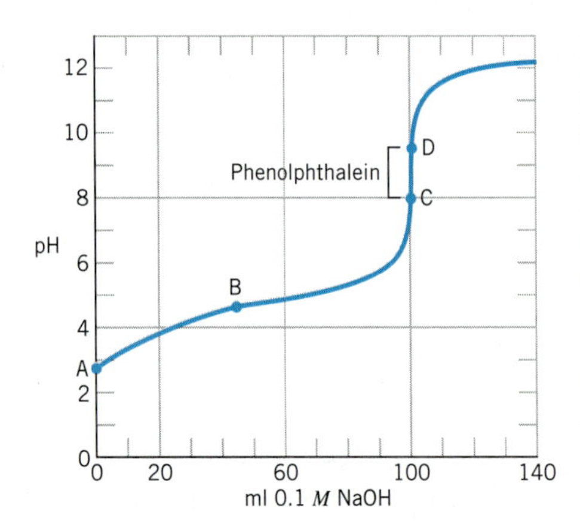


**Figure 17.7** The pH of an 0.10 *M* solution of acetic acid as it is titrated with 0.10 *M* sodium hydroxide. The pH increases at first, as some of the HOAc is converted into $OAc^-$ ions. This creates a buffer, which resists changes in pH until it has become exhausted. At that point, the pH rises rapidly, then slowly levels off as the solution begins to behave more like an 0.10 *M* NaOH solution. Points *A–D* are explained in the text.

Point *A* represents the pH at the start of the titration. The pH at this point is 2.9, the pH of an 0.10 *M* HOAc solution. As NaOH is added to this solution, some of the acid is converted to its conjugate base, the $OAc^-$ ion:

$$HOAc(aq) + OH^-(aq) \longrightarrow OAc^-(aq) + H_2O(l)$$

Point *B* is the point at which exactly half of the HOAc molecules have been converted to $OAc^-$ ions. At this point, the concentration of the HOAc molecules is equal to the concentration of the $OAc^-$ ions. Because the ratio of the HOAc and $OAc^-$ concentrations is 1:1, the concentration of the $H_3O^+$ ion at Point *B* is equal to the value of $K_a$ for the acid:

$$\text{Point } B: \quad [H_3O^+] = K_a$$

This provides us with a way of measuring $K_a$ for an acid. We can titrate a sample of the acid with a strong base, plot the titration curve, and then find the point along this curve at which exactly half of the acid has been consumed. The $H_3O^+$ ion concentration at this point will be equal to the value of $K_a$ for the acid.

Point $C$ is the **equivalence point** for the titration—the point at which enough base has been added to the solution to consume the acid present at the start of the titration. The goal of any titration is to find the equivalence point. What is actually observed is the **endpoint** of the titration—Point $D$—the point at which the indicator changes color.

Every effort is made to bring the endpoint as close as possible to the equivalence point of a titration. Suppose that phenolphthalein is used as the indicator for this titration, for example. Phenolphthalein turns from colorless to pink as the pH of the solution changes from 8 to 10. If we wait until the phenolphthalein permanently turns pink, the endpoint will fall beyond the equivalence point for the titration. So we try to find the point at which adding one drop of base turns the entire solution to a pink color that fades in about 10 seconds while the solution is stirred. In theory, this is one drop before the endpoint of the titration and therefore closer to the equivalence point.

Note the shape of the pH titration curve in Figure 17.7. The pH rises rapidly at first, because we are adding a strong base to a weak acid and the base neutralizes some of the acid. The curve then levels off, and the pH remains more or less constant as we add base because some of the HOAc present initially is converted into $OAc^-$ ions to form a buffer solution. The pH of this buffer solution stays relatively constant until most of the acid has been converted to its conjugate base. At that point the pH rises rapidly because essentially all of the HOAc in the system has been converted into $OAc^-$ ions and the buffer is exhausted. The pH then gradually levels off as the solution begins to look like a 0.10 $M$ NaOH solution.

The titration curve for a weak base, such as ammonia, titrated with a strong acid, such as hydrochloric acid, would be analogous to the curve in Figure 17.7. The principal difference is fact that the pH value is high at first and decreases as acid is added.

## 17.17 DIPROTIC ACIDS

The acid equilibrium problems discussed so far have focused on a family of compounds known as **monoprotic acids.** Each of these acids has a single $H^+$ ion, or proton, it can donate when it acts as a Brønsted acid. Hydrochloric acid (HCl), acetic acid ($CH_3CO_2H$ or HOAc), nitric acid ($HNO_3$), and benzoic acid ($C_6H_5CO_2H$) are all monoprotic acids.

Several important acids can be classified as **polyprotic acids,** which can lose more than one $H^+$ ion when they act as Brønsted acids. **Diprotic acids,** such as sulfuric acid ($H_2SO_4$), carbonic acid ($H_2CO_3$), hydrogen sulfide ($H_2S$), chromic acid ($H_2CrO_4$), and oxalic acid ($H_2C_2O_4$) have two acidic hydrogen atoms. **Triprotic acids,** such as phosphoric acid ($H_3PO_4$) and citric acid ($C_6H_8O_7$), have three.

There is usually a large difference in the ease with which these acids lose the first and second (or second and third) protons. When sulfuric acid is classified as a strong acid, a novice might assume that it loses both of its protons when it reacts with water. But this isn't a correct assumption. Sulfuric acid is a strong acid because $K_a$ for the loss of the first proton is much larger than 1. We can therefore assume that essentially all the $H_2SO_4$ molecules in an aqueous solution lose the *first* proton to form the $HSO_4^-$, or hydrogen sulfate, ion:

$$H_2SO_4(aq) + H_2O(l) \longrightarrow H_3O^+(aq) + HSO_4^-(aq) \qquad K_{a1} = 1 \times 10^3$$

But $K_a$ for the loss of the second proton is only $10^{-2}$ and only 10% of the $H_2SO_4$ molecules in a 1 *M* solution lose a second proton:

$$HSO_4^-(aq) + H_2O(l) \rightleftharpoons H_3O^+(aq) + SO_4^{2-}(aq) \qquad K_{a2} = 1.2 \times 10^{-2}$$

$H_2SO_4$ only loses both $H^+$ ions when it reacts with a base, such as ammonia.

Table 17.6 gives values of $K_a$ for some common polyprotic acids. The large difference between the values of $K_a$ for the sequential loss of protons by a polyprotic acid is important because it means we can assume that these acids dissociate one step at a time.

**TABLE 17.6 Acid-Dissociation Equilibrium Constants for Common Polyprotic Acids**

| Acid | $K_{a1}$ | $K_{a2}$ | $K_{a3}$ |
|---|---|---|---|
| Sulfuric acid ($H_2SO_4$) | $1.0 \times 10^3$ | $1.2 \times 10^{-2}$ | |
| Chromic acid ($H_2CrO_4$) | 9.6 | $3.2 \times 10^{-7}$ | |
| Oxalic acid ($H_2C_2O_4$) | $5.4 \times 10^{-2}$ | $5.4 \times 10^{-5}$ | |
| Sulfurous acid ($H_2SO_3$) | $1.7 \times 10^{-2}$ | $6.4 \times 10^{-8}$ | |
| Phosphoric acid ($H_3PO_4$) | $7.1 \times 10^{-3}$ | $6.3 \times 10^{-8}$ | $4.2 \times 10^{-13}$ |
| Glycine ($C_2H_6NO_2$) | $4.5 \times 10^{-3}$ | $2.5 \times 10^{-10}$ | |
| Citric acid ($C_6H_8O_7$) | $7.5 \times 10^{-4}$ | $1.7 \times 10^{-5}$ | $4.0 \times 10^{-7}$ |
| Carbonic acid ($H_2CO_3$) | $4.5 \times 10^{-7}$ | $4.7 \times 10^{-11}$ | |
| Hydrogen sulfide ($H_2S$) | $1.0 \times 10^{-7}$ | $1.3 \times 10^{-13}$ | |

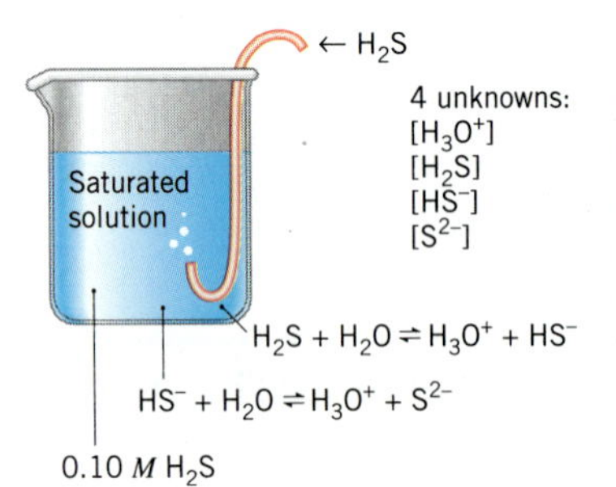


Let's look at the consequence of the **stepwise dissociation** of polyprotic acids by examining the chemistry of a saturated solution of $H_2S$ in water. Hydrogen sulfide is the foul-smelling gas that gives rotten eggs their unpleasant odor. It is an excellent source of the $S^{2-}$ ion, however, and is therefore commonly used in introductory chemistry laboratories. $H_2S$ is a weak acid that dissociates in steps. Some of the $H_2S$ molecules lose a proton in the first step to form the $HS^-$, or hydrogen sulfide, ion:

$$\text{First step:} \qquad H_2S(aq) + H_2O(l) \rightleftharpoons H_3O^+(aq) + HS^-(aq)$$

A small fraction of the $HS^-$ ions formed in this reaction then go on to lose another $H^+$ ion in a second step.

$$\text{Second step:} \qquad HS^-(aq) + H_2O(l) \rightleftharpoons H_3O^+(aq) + S^{2-}(aq)$$

Since there are two steps in this reaction, we can write two equilibrium constant expressions:

$$K_{a1} = \frac{[H_3O^+][HS^-]}{[H_2S]} = 1.0 \times 10^{-7}$$

$$K_{a2} = \frac{[H_3O^+][S^{2-}]}{[HS^-]} = 1.3 \times 10^{-13}$$

Although each of these equations contains three terms, there are only four unknowns—$[H_3O^+]$, $[H_2S]$, $[HS^-]$, and $[S^{2-}]$—because the $[H_3O^+]$ and $[HS^-]$ terms appear in both equations. The $[H_3O^+]$ term represents the total $H_3O^+$ ion concentration from both steps and therefore must have the same value in both equations. Similarly, the $[HS^-]$ term, which represents the balance between the $HS^-$ ions formed in the first step and the $HS^-$ ions consumed in the second step, must have the same value for both equations.

Four equations are needed to solve for four unknowns. We already have two equations: the $K_{a1}$ and $K_{a2}$ expressions. We are going to need either two more equations or a pair of assumptions that can generate two equations. We can base one of these on the fact that the value of $K_{a1}$ for this acid is almost a million times larger than the value of $K_{a2}$:

$$K_{a1} >> K_{a2}$$

This means that only a small fraction of the $HS^-$ ions formed in the first step go on to dissociate in the second step. If this is true, most of the $H_3O^+$ ions in this solution come from the dissociation of $H_2S$, and most of the $HS^-$ ions formed in this reaction remain in solution. As a result, we can assume that the $H_3O^+$ and $HS^-$ ion concentrations are more or less equal.

First assumption: $[H_3O^+] \approx [HS^-]$

**Problem-Solving Strategy**

We need one more equation, and therefore one more assumption. Note that $H_2S$ is a weak acid ($K_{a1} = 1.0 \times 10^{-7}$, $K_{a2} = 1.3 \times 10^{-13}$). Thus, we can assume that most of the $H_2S$ that dissolves in water still will be present when the solution reaches equilibrium. In other words, we can assume that the equilibrium concentration of $H_2S$ is approximately equal to the initial concentration.

Second assumption: $[H_2S] \approx C_{H_2S}$

We now have four equations in four unknowns:

$$K_{a1} = \frac{[H_3O^+][HS^-]}{[H_2S]} = 1.0 \times 10^{-7}$$

$$K_{a2} = \frac{[H_3O^+][S^{2-}]}{[HS^-]} = 1.3 \times 10^{-13}$$

$$[H_3O^+] \approx [HS^-]$$

$$[H_2S] \approx C_{H_2S}$$

Since there is always a unique solution to four equations in four unknowns, we are now ready to calculate the $H_3O^+$, $H_2S$, $HS^-$, and $S^{2-}$ concentrations at equilibrium in a saturated solution of $H_2S$ in water. All we need to know is that a saturated solution of $H_2S$ in water has an initial concentration of about 0.10 *M*.

**Problem-Solving Strategy**

Because $K_{a1}$ is so much larger than $K_{a2}$ for this acid, we can work with the equilibrium expression for the first step without worrying about the second step for the moment. We therefore start with the expression for $K_{a1}$ for this acid:

$$K_{a1} = \frac{[H_3O^+][HS^-]}{[H_2S]} = 1.0 \times 10^{-7}$$

We then invoke one of our assumptions:

$$[H_2S] \approx C_{H_2S} \approx 0.10\ M$$

Substituting this approximation into the $K_{a1}$ expression gives:

$$\frac{[H_3O^+][HS^-]}{[0.10]} \approx 1.0 \times 10^{-7}$$

We then invoke the other assumption:

$$[H_3O^+] \approx [HS^-] \approx \Delta$$

Substituting this approximation into the $K_{a1}$ expression gives the following result:

$$\frac{[\Delta][\Delta]}{[0.10]} \approx 1.0 \times 10^{-7}$$

We can now solve this approximate equation for $\Delta$.

$$\Delta \approx 1.0 \times 10^{-4}$$

If our two assumptions are valid, we are three-fourths of the way to our goal. We know the $H_2S$, $H_3O^+$, and $HS^-$ concentrations:

$$[H_2S] \approx 0.10\ M$$
$$[H_3O^+] \approx [HS^-] \approx 1.0 \times 10^{-4}\ M$$

Having extracted the values of three unknowns from the first equilibrium expression, we turn to the second equilibrium expression:

$$K_{a2} = \frac{[H_3O^+][S^{2-}]}{[HS^-]} = 1.3 \times 10^{-13}$$

Substituting the known values of the $H_3O^+$ and $HS^-$ ion concentrations into this expression gives the following equation:

$$\frac{[1.0 \times 10^{-4}][S^{2-}]}{[1.0 \times 10^{-4}]} = 1.3 \times 10^{-13}$$

Because the equilibrium concentrations of the $H_3O^+$ and $HS^-$ ions are more or less the same, the $S^{2-}$ ion concentration at equilibrium is approximately equal to the value of $K_{a2}$ for this acid:

$$[S^{2-}] \approx \mathbf{1.3 \times 10^{-13}}\ M$$

It is now time to check our assumptions. Is the dissociation of $H_2S$ small compared with the initial concentration? Yes. The $HS^-$ and $H_3O^+$ ion concentrations obtained from this calculation are $1.0 \times 10^{-4}\ M$, which is 0.1% of the initial concentration of $H_2S$. The following assumption is therefore valid:

$$[H_2S] \approx C_{H_2S} \approx 0.10\ M$$

Is the difference between the $S^{2-}$ and $HS^-$ ion concentrations large enough to assume that essentially all of the $H_3O^+$ ions at equilibrium are formed in the first step and that essentially all of the $HS^-$ ions formed in this step remain in solution? Yes. The $S^{2-}$ ion concentration obtained from this calculation is $10^9$ times smaller than the $HS^-$ ion concentration. Thus, our other assumption is also valid:

$$[H_3O^+] \approx [HS^-]$$

We can therefore summarize the concentrations of the various components of this equilibrium as follows:

$$[H_2S] \approx \mathbf{0.10}\ M$$
$$[H_3O^+] \approx [HS^-] \approx \mathbf{1.0 \times 10^{-4}}\ M$$
$$[S^{2-}] \approx \mathbf{1.3 \times 10^{-13}}\ M$$

## 17.18 DIPROTIC BASES

The techniques we have used with diprotic acids can be extended to diprotic bases. The only challenge is calculating the values of $K_b$ for the base. Suppose we are given the following problem:

> Calculate the $H_2CO_3$, $HCO_3^-$, $CO_3^{2-}$, and $OH^-$ concentrations at equilibrium in a solution that is initially 0.10 *M* in $Na_2CO_3$ ($H_2CO_3$: $K_{a1} = 4.5 \times 10^{-7}$; $K_{a2} = 4.7 \times 10^{-11}$).

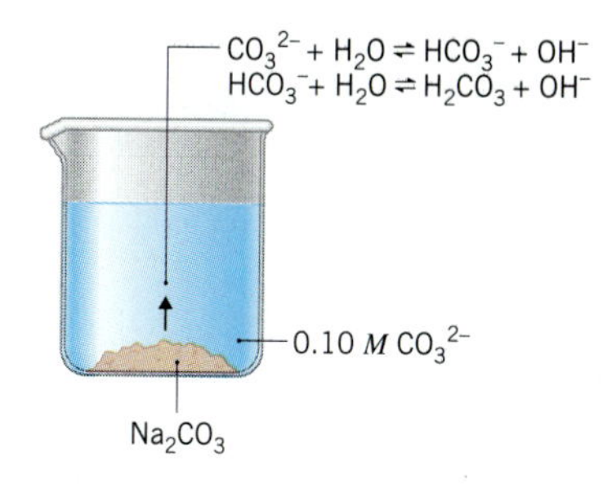


Because it is a salt, sodium carbonate dissociates into its ions when it dissolves in water:

$$Na_2CO_3(aq) \xrightarrow{H_2O} 2\,Na^+(aq) + CO_3^{2-}(aq)$$

The carbonate ion then acts as a base toward water, picking up a pair of protons (one at a time) to form the bicarbonate ion, $HCO_3^-$ ion, and then eventually carbonic acid, $H_2CO_3$.

$$CO_3^{2-}(aq) + H_2O(l) \rightleftharpoons HCO_3^-(aq) + OH^-(aq) \qquad K_{b1} = ?$$
$$HCO_3^-(aq) + H_2O(l) \rightleftharpoons H_2CO_3(aq) + OH^-(aq) \qquad K_{b2} = ?$$

The first step in solving this problem involves determining the values of $K_{b1}$ and $K_{b2}$ for the carbonate ion. We start by comparing the $K_b$ expressions for the carbonate ion with the $K_a$ expressions for carbonic acid:

$$K_{b1} = \frac{[HCO_3^-][OH^-]}{[CO_3^{2-}]} \qquad K_{a2} = \frac{[H_3O^+][CO_3^{2-}]}{[HCO_3^-]}$$
$$K_{b2} = \frac{[H_2CO_3][OH^-]}{[HCO_3^-]} \qquad K_{a1} = \frac{[H_3O^+][HCO_3^-]}{[H_2CO_3]}$$

The expressions for $K_{b1}$ and $K_{a2}$ have something in common—they both depend on the concentrations of the $HCO_3^-$ and $CO_3^{2-}$ ions. The expressions for $K_{b2}$ and $K_{a1}$ also have something in common—they both depend on the $HCO_3^-$ and $H_2CO_3$ concentrations. We can therefore calculate $K_{b1}$ from $K_{a2}$ and $K_{b2}$ from $K_{a1}$.

We start by multiplying the top and bottom of the $K_{a1}$ expression by the $OH^-$ ion concentration to introduce the $[OH^-]$ term:

$$K_{a1} = \frac{[H_3O^+][HCO_3^-]}{[H_2CO_3]} \times \frac{[OH^-]}{[OH^-]}$$

We then group terms in this equation as follows:

$$K_{a1} = \frac{[HCO_3^-]}{[H_2CO_3][OH^-]} \times [H_3O^+][OH^-]$$

The first term in this equation is the inverse of the $K_{b2}$ expression, and the second term is the $K_w$ expression:

$$K_{a1} = \frac{1}{K_{b2}} \times K_w$$

Rearranging this equation gives the following result:

$$K_{a1}K_{b2} = K_w$$

Similarly, we can multiply the top and bottom of the $K_{a2}$ expression by the $OH^-$ ion concentration:

$$K_{a2} = \frac{[H_3O^+][CO_3^{2-}]}{[HCO_3^-]} \times \frac{[OH^-]}{[OH^-]}$$

Collecting terms gives the following equation:

$$K_{a2} = \frac{[CO_3^{2-}]}{[HCO_3^-][OH^-]} \times [H_3O^+][OH^-]$$

The first term in this equation is the inverse of $K_{b1}$, and the second term is $K_w$.

$$K_{a2} = \frac{1}{K_{b1}} \times K_w$$

This equation therefore can be rearranged as follows:

$$\mathbf{K_{a2}K_{b1} = K_w}$$

We can now calculate the values of $K_{b1}$ and $K_{b2}$ for the carbonate ion:

$$K_{b1} = \frac{K_w}{K_{a2}} = \frac{1.0 \times 10^{-14}}{4.7 \times 10^{-11}} = \mathbf{2.1 \times 10^{-4}}$$

$$K_{b2} = \frac{K_w}{K_{a1}} = \frac{1.0 \times 10^{-14}}{4.5 \times 10^{-7}} = \mathbf{2.2 \times 10^{-8}}$$

We are finally ready to do the calculations. We start with the $K_{b1}$ expression because the $CO_3^{2-}$ ion is the strongest base in this solution and therefore the best source of the $OH^-$ ion:

$$K_{b1} = \frac{[HCO_3^-][OH^-]}{[CO_3^{2-}]} = 2.1 \times 10^{-4}$$

The difference between $K_{b1}$ and $K_{b2}$ for the carbonate ion is large enough to suggest that most of the $OH^-$ ions come from this step and most of the $HCO_3^-$ ions formed in this reaction remain in solution:

$$[OH^-] \approx [HCO_3^-] \approx \Delta$$

The value of $K_{b1}$ is small enough to assume that $\Delta$ is small compared with the initial concentration of the carbonate ion. If this is true, the concentration of the $CO_3^{2-}$ ion at equilibrium will be roughly equal to the initial concentration of $Na_2CO_3$:

$$[CO_3^{2-}] \approx C_{Na_2CO_3}$$

Substituting this information into the $K_{b1}$ expression gives the following result:

$$\frac{[\Delta][\Delta]}{[0.10]} \approx 2.1 \times 10^{-4}$$

This approximate equation can now be solved for $\Delta$:

$$\Delta \approx 0.0046\ M$$

We then use this value of $\Delta$ to calculate the equilibrium concentrations of the $OH^-$, $HCO_3^-$, and $CO_3^{2-}$ ions:

$$[CO_3^{2-}] = 0.10 - \Delta \approx 0.095\ M$$
$$[OH^-] \approx [HCO_3^-] \approx \Delta \approx 0.0046\ M$$

We now turn to the $K_{b2}$ expression.

$$K_{b2} = \frac{[H_2CO_3][OH^-]}{[HCO_3^-]}$$

Substituting what we know about the $OH^-$ and $HCO_3^-$ ion concentrations into this equation gives:

$$\frac{[H_2CO_3]\cancel{[0.0046]}}{\cancel{[0.0046]}} \approx 2.2 \times 10^{-8}$$

According to this equation, the $H_2CO_3$ concentration at equilibrium is approximately equal to $K_{b2}$ for the carbonate ion:

$$[H_2CO_3] \approx 2.2 \times 10^{-8}\ M$$

Summarizing the results of our calculations allows us to test the assumptions made generating these results:

$$[CO_3^{2-}] \approx \mathbf{0.095\ M}$$
$$[OH^-] \approx [HCO_3^-] \approx \mathbf{4.6 \times 10^{-3}\ M}$$
$$[H_2CO_3] \approx \mathbf{2.2 \times 10^{-8}\ M}$$

All of our assumptions are valid. The extent of the reaction between the $CO_3^{2-}$ ion and water to give the $HCO_3^-$ ion is less than 5% of the initial concentration of $Na_2CO_3$. Furthermore, most of the $OH^-$ ion comes from the first step, and most of the $HCO_3^-$ ion formed in this step remains in solution.

## SPECIAL TOPIC

### TRIPROTIC ACIDS

Our techniques for working diprotic acid or diprotic base equilibrium problems can be applied to triprotic acids and bases as well. To illustrate this, let's calculate the $H_3O^+$, $H_3PO_4$, $H_2PO_4^-$, $H_2PO_4^{2-}$, and $PO_4^{3-}$ concentrations at equilibrium in a 0.10 $M$ $H_3PO_4$ solution for which $K_{a1} = 7.1 \times 10^{-3}$, $K_{a2} = 6.3 \times 10^{-8}$, and $K_{a3} = 4.2 \times 10^{-13}$.

Let's assume that this acid dissociates by steps and analyze the first step—the most extensive reaction:

$$K_{a1} = \frac{[H_3O^+][H_2PO_4^-]}{[H_3PO_4]} = 7.1 \times 10^{-3}$$

We now assume that the difference between $K_{a1}$ and $K_{a2}$ is large enough that most of the $H_3O^+$ ions come from this first step and most of the $H_2PO_4^-$ ions formed in this step remain in solution:

$$[H_3O^+] \approx [H_2PO_4^-] \approx \Delta$$

Substituting this assumption into the $K_{a1}$ expression gives the following equation:

$$\frac{[\Delta][\Delta]}{[0.10 - \Delta]} = 7.1 \times 10^{-3}$$

The assumption that $\Delta$ is small compared with the initial concentration of the acid fails in this problem. But we don't really need this assumption because we can use the quadratic formula or successive approximations to solve the equation. Either way, we obtain the same answer:

$$\Delta = 0.023\ M$$

We can then use this value of $\Delta$ to obtain the following information:

$$[H_3PO_4] \approx 0.10 - \Delta \approx \mathbf{0.077\ M}$$
$$[H_3O^+] \approx [H_2PO_4^-] \approx \mathbf{0.023\ M}$$

We now turn to the second strongest acid in this solution:

$$K_{a2} = \frac{[H_3O^+][HPO_4^{2-}]}{[H_2PO_4^-]} = 6.3 \times 10^{-8}$$

Substituting what we know about the $H_3O^+$ and $H_2PO_4^-$ ion concentrations into this expression gives the following equation:

$$\frac{\cancel{[0.023]}[HPO_4^{2-}]}{\cancel{[0.023]}} = 6.3 \times 10^{-8}$$

If our assumptions so far are correct, the $HPO_4^{2-}$ ion concentration in this solution is equal to $K_{a2}$:

$$[HPO_4^{2-}] \approx \mathbf{6.3 \times 10^{-8}}$$

We have only one more equation, the equilibrium expression for the weakest acid in the solution:

$$K_{a3} = \frac{[H_3O^+][PO_4^{3-}]}{[HPO_4^{2-}]} = 4.2 \times 10^{-13}$$

Substituting what we know about the concentrations of the $H_3O^+$ and $HPO_4^{2-}$ ions into this expression gives:

$$\frac{[0.023][PO_4^{3-}]}{[6.3 \times 10^{-8}]} \approx 4.2 \times 10^{-13}$$

This equation can be solved for the phosphate ion concentration at equilibrium:

$$[PO_4^{3-}] \approx \mathbf{1.2 \times 10^{-18}}\ M$$

Summarizing the results of the calculations helps us check the assumptions made along the way:

$$[H_3PO_4] \approx \mathbf{0.077}\ M$$
$$[H_3O^+] \approx [H_2PO_4^-] \approx \mathbf{0.023}\ M$$
$$[HPO_4^{2-}] \approx \mathbf{6.3 \times 10^{-8}}\ M$$
$$[PO_4^{3-}] \approx \mathbf{1.2 \times 10^{-18}}\ M$$

The only approximation used in working this problem was the assumption that the acid dissociates one step at a time. Is the difference between the concentrations of the $H_2PO_4^-$ and $HPO_4^{2-}$ ions large enough to justify the assumption that essentially all of the $H_3O^+$ ions come from the first step? Yes. Is it large enough to justify the assumption that essentially all of the $H_2PO_4^-$ formed in the first step remains in solution? Yes.

You may never encounter an example of a polyprotic acid for which the differences between successive values of $K_a$ are too small to allow you to assume stepwise dissociation. This assumption works even when we might expect it to fail.

## EXERCISE 17.8

Calculate the $H_3O^+$, $H_3Cit$, $H_2Cit^-$, $HCit^{2-}$, and $Cit^{3-}$ concentrations (where $Cit^{3-}$ stands for the citrate ion) in a 1.00 *M* solution of citric acid. ($H_3Cit$: $K_{a1} = 7.5 \times 10^{-4}$, $K_{a2} = 1.7 \times 10^{-5}$, $K_{a3} = 4.0 \times 10^{-7}$)

**SOLUTION** At first glance, $K_{a1}$, $K_{a2}$, and $K_{a3}$ for this acid might seem to be too close to each other to assume stepwise dissociation. The only way to tell for sure is to try the assumption and see whether it works. We start with the expression for the first step:

$$K_{a1} = \frac{[H_3O^+][H_2Cit^-]}{[H_3Cit]} = 7.5 \times 10^{-4}$$

Even though we may be suspicious of its validity, let's make the assumption that most of the $H_3O^+$ ions at equilibrium come from the citric acid. Let's also assume that most of the $H_2Cit^-$ ions formed in the first step remain in solution.

If these assumptions are valid, the concentrations of the $H_3O^+$ and $H_2Cit^-$ ions at equilibrium are equal:

$$[H_3O^+] \approx [H_2Cit^-] \approx \Delta$$

Substituting this assumption into the $K_{a1}$ expression gives:

$$\frac{[\Delta][\Delta]}{[H_3Cit]} \approx 7.5 \times 10^{-4}$$

Because citric acid is a weak acid, we can assume that the equilibrium concentration of the undissociated $H_3Cit$ molecules is roughly equal to the initial concentration of this acid:

$$\mathbf{[H_3Cit] \approx C_{H_3Cit}}$$

Substituting this assumption into the $K_{a1}$ expression gives the following result:

$$\frac{[\Delta][\Delta]}{[1.00]} \approx 7.5 \times 10^{-4}$$

We can now solve this approximate equation for $\Delta$:

$$\Delta \approx 0.027\ M$$

This value of $\Delta$ now can be used to calculate the following concentrations:

$$[H_3Cit] \approx 1.00 - \Delta \approx \mathbf{0.97}\ M$$

$$[H_3O^+] \approx [H_2Cit^-] \approx \mathbf{0.027}\ M$$

We can now turn to the $K_{a2}$ expression:

$$K_{a2} = \frac{[H_3O^+][HCit^{2-}]}{[H_2Cit^-]}$$

Substituting what we know about the $H_3O^+$ and $H_2Cit^-$ concentrations into this expression gives the following equation:

$$\frac{\cancel{[0.027]}[HCit^{2-}]}{\cancel{[0.027]}} \approx 1.7 \times 10^{-5}$$

If our assumptions are valid, the $HCit^{2-}$ ion concentration is equal to $K_{a2}$ for this acid:

$$[HCit^{2-}] \approx \mathbf{1.7 \times 10^{-5}}\ M$$

We now turn to the $K_{a3}$ expression:

$$K_{a3} = \frac{[H_3O^+][Cit^{3-}]}{[HCit^{2-}]} = 4.0 \times 10^{-7}$$

Substituting the approximate values for the $H_3O^+$ and $HCit^{2-}$ ion concentrations into this expression gives the following equation:

$$\frac{[0.027][Cit^{3-}]}{[1.7 \times 10^{-5}]} \approx 4.0 \times 10^{-7}$$

We now solve for the $Cit^{3-}$ ion concentration:

$$[Cit^{3-}] \approx \mathbf{2.5 \times 10^{-10}}\ M$$

Summarizing the results of this calculation allows us to test the assumptions made in deriving them:

$$[H_3Cit] \approx \mathbf{0.097}\ M$$

$$[H_3O^+] \approx [H_2Cit^-] \approx \mathbf{0.027}\ M$$

$$[HCit^{2-}] \approx \mathbf{1.7 \times 10^{-5}}\ M$$

$$[Cit^{3-}] \approx \mathbf{2.5 \times 10^{-10}}\ M$$

Is it legitimate to assume that the amount of $H_3Cit$ that dissociates in the first step is small compared with the initial concentration of this acid? Yes. Is it legitimate to assume that most of the $H_3O^+$ ion comes from the first step? Yes. Is it legitimate to assume that most of the $H_2Cit^-$ ion formed in the first step remains in solution? Yes. Since all of the assumptions made in this calculation are valid, the results are also valid.

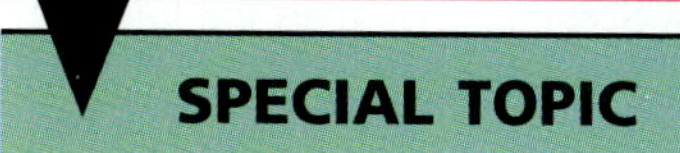


## COMPOUNDS THAT COULD BE EITHER ACIDS OR BASES

Sometimes the hardest part of a calculation is deciding whether the compound is an acid or a base. Consider sodium bicarbonate, for example, which dissolves in water to give the bicarbonate ion:

$$NaHCO_3(s) \xrightarrow{H_2O} Na^+(aq) + HCO_3^-(aq)$$

In theory, the bicarbonate ion can act as both a Brønsted acid and a Brønsted base toward water:

$$HCO_3^-(aq) + H_2O(l) \rightleftharpoons H_3O^+(aq) + CO_3^{2-}(aq)$$
$$HCO_3^-(aq) + H_2O(l) \rightleftharpoons H_2CO_3(aq) + OH^-(aq)$$

Which reaction predominates? Is the $HCO_3^-$ ion more likely to act as an acid or as a base?

We can answer this last question by comparing the equilibrium constants for these reactions. The equilibrium in which $HCO_3^-$ ion acts as a Brønsted acid is described by $K_{a2}$ for carbonic acid:

$$K_{a2} = \frac{[H_3O^+][CO_3^{2-}]}{[HCO_3^-]} = 4.7 \times 10^{-11}$$

The equilibrium in which the $HCO_3^-$ acts as a Brønsted base is described by $K_{b2}$ for the carbonate ion:

$$K_{b2} = \frac{[H_2CO_3][OH^-]}{[HCO_3^-]} = 2.2 \times 10^{-8}$$

Since $K_{b2}$ is significantly larger than $K_{a2}$, the $HCO_3^-$ ion is a stronger base than it is an acid.

### EXERCISE 17.9

Phosphoric acid, $H_3PO_4$, is obviously an acid. The phosphate ion, $PO_4^{3-}$, is a base. Predict whether solutions of the $H_2PO_4^-$ and $HPO_4^{2-}$ ions are more likely to be acidic or basic.

**SOLUTION** Let's start by looking at the stepwise dissociation of phosphoric acid:

$$H_3PO_4(aq) + H_2O(l) \rightleftharpoons H_3O^+(aq) + H_2PO_4^-(aq) \qquad K_{a1} = 7.1 \times 10^{-3}$$
$$H_2PO_4^-(aq) + H_2O(l) \rightleftharpoons H_3O^+(aq) + HPO_4^{2-}(aq) \qquad K_{a2} = 6.3 \times 10^{-8}$$
$$HPO_4^{2-}(aq) + H_2O(l) \rightleftharpoons H_3O^+(aq) + PO_4^{3-}(aq) \qquad K_{a3} = 4.2 \times 10^{-13}$$

We can then look at the steps by which the $PO_4^{3-}$ ion picks up protons from water to form phosphoric acid:

$$PO_4^{3-}(aq) + H_2O(l) \rightleftharpoons HPO_4^{2-}(aq) + OH^-(aq) \qquad K_{b1} = ?$$
$$HPO_4^{2-}(aq) + H_2O(l) \rightleftharpoons H_2PO_4^-(aq) + OH^-(aq) \qquad K_{b2} = ?$$
$$H_2PO_4^-(aq) + H_2O(l) \rightleftharpoons H_3PO_4(aq) + OH^-(aq) \qquad K_{b3} = ?$$

Applying the procedure used for sodium carbonate in Section 17.18 gives the following relationships among the values of these six equilibrium constants:

$$K_{a1}K_{b3} = K_w$$
$$K_{a2}K_{b2} = K_w$$
$$K_{a3}K_{b1} = K_w$$

We can predict whether the $H_2PO_4^-$ ion should be an acid or a base by looking at the values of the $K_a$ and $K_b$ constants that characterize its reactions with water:

$$H_2PO_4^-(aq) + H_2O(l) \rightleftharpoons H_3O^+(aq) + HPO_4^{2-}(aq) \qquad K_{a2} = 6.3 \times 10^{-8}$$
$$H_2PO_4^-(aq) + H_2O(l) \rightleftharpoons H_3PO_4(aq) + OH^-(aq) \qquad K_{b3} = 1.4 \times 10^{-12}$$

Because $K_{a2}$ is significantly larger than $K_{b3}$, solutions of the $H_2PO_4^-$ ion in water should be acidic.

We can use the same method to decide whether the $HPO_4^{2-}$ ion is an acid or a base when dissolved in water. We start by looking at the values of $K_a$ and $K_b$ that characterize its reactions with water:

$$HPO_4^{2-}(aq) + H_2O(l) \rightleftharpoons H_3O^+(aq) + PO_4^{3-}(aq) \qquad K_{a3} = 4.2 \times 10^{-13}$$
$$HPO_4^{2-}(aq) + H_2O(l) \rightleftharpoons H_2PO_4^-(aq) + OH^-(aq) \qquad K_{b2} = 1.6 \times 10^{-7}$$

Because $K_{b2}$ is significantly larger than $K_{a3}$, solutions of the $HPO_4^{2-}$ ion in water should be basic.

Sec. 11.13

Measurements of the pH of 0.10 $M$ solutions of these compounds yield the following results, which are consistent with the predictions of this exercise:

| $H_3PO_4$ | $H_2PO_4^-$ | $HPO_4^{2-}$ | $PO_4^{3-}$ |
|---|---|---|---|
| pH = 1.6 | pH = 4.1 | pH = 10.1 | pH = 12.7 |

## SUMMARY

**1.** An $H_3O^+$ ion is created for each $OH^-$ ion when water dissociates:

$$2\,H_2O(l) \rightleftharpoons H_3O^+(aq) + OH^-(aq)$$

The concentrations of these ions in pure water are therefore the same:

$$[H_3O^+] = [OH^-] = 1.0 \times 10^{-7}\,M.$$

**2.** When an acid is added to water, the concentration of the $H_3O^+$ ion increases. This suppresses the dissociation of water, which decreases the $OH^-$ ion concentration. The opposite occurs when a base dissolves in water.

**3.** The tendency of an acid to dissociate in water is often expressed in terms of the acid-dissociation equilibrium constant, $K_a$:

$$K_a = \frac{[H_3O^+][A^-]}{[HA]}$$

When $K_a > 1$, the acid is a strong acid. When $K_a > 55$, we can assume that the acid dissociates more or less completely.

**4.** For weak acids, we often assume that the amount of dissociation is small compared with the initial concentration of the acid, but large enough to allow us to ignore the dissociation of water:

$$[H_3O^+] \approx \sqrt{K_aC_a}$$

For very weak acids, we have to include the dissociation of water in our calculations:

$$[H_3O^+] = \sqrt{K_aC_a + K_w}$$

**5.** The reaction between a base and water is described in terms of a base-ionization equilibrium constant expression:

$$K_b = \frac{[BH^+][OH^-]}{[B]}$$

When the value of $K_b$ is known, these equilibria are handled with the same techniques as solutions of acids in water. When $K_b$ is not known, it is calculated from the value of $K_a$ for the conjugate acid:

$$K_aK_b = K_w$$

**6.** Mixtures of weak acids and their conjugate bases, or weak bases and their conjugate acids, are known as *buffers.* These solutions have the remarkable ability to resist changes in pH when small amounts of acid or base are added to the solution.

**7.** The shape of a pH titration curve is dictated by the chemistry of buffers. When a weak acid is titrated with a strong base, for example, the pH initially rises as some of the acid is neutralized. This creates a buffer, which resists further changes in pH until essentially all of the acid has been consumed. At that point, the pH rises rapidly. This rapid change in pH therefore indicates the point at which equivalent amounts of acid and base have been added to the solution.

**8.** Calculations involving diprotic and triprotic acids are greatly simplified by the fact that equilibria in which these acids lose protons can be handled one at a time.

## KEY TERMS

**Acid-dissociation equilibrium constant, $K_a$** (p. 606)
**Acidic buffer** (p. 626)
**Base-ionization equilibrium constant, $K_b$** (p. 619)
**Basic buffer** (p. 626)
**Buffer** (p. 625)
**Buffer capacity** (p. 628)
**Diprotic acid** (p. 629)
**Endpoint** (p. 629)
**Equivalence point** (p. 629)
**Henry's law** (p. 600)
**Monoprotic acid** (p. 629)
**pH** (p. 604)
**pOH** (p. 604)
**Polyprotic acid** (p. 629)
**Stepwise dissociation** (p. 630)
**Water-dissociation equilibrium constant, $K_w$** (p. 602)
**Triprotic acid** (p. 629)

## KEY EQUATIONS

$[H_3O^+][OH^-] = K_w = 1.0 \times 10^{-14}$ (at 25°C)

$pH = -\log [H_3O^+]$

$pOH = -\log [OH^-]$

$pH + pOH = 14$

$$K_a = \frac{[H_3O^+][A^-]}{[HA]}$$

$$K_b = \frac{[HB^+][OH^-]}{[B]}$$

Pure water: $[H_3O^+] = \sqrt{K_w}$

Weak acid: $[H_3O^+] \approx \sqrt{K_a C_a}$

Very weak acid: $[H_3O^+] = \sqrt{K_a C_a + K_w}$

$K_a \times K_b = K_w$

## PROBLEMS

### The Acid–Base Chemistry of Water

**17-1** Define the following terms; *acid, base, monoprotic, diprotic, triprotic, weak acid, strong acid, weak base,* and *strong base.*

**17-2** Describe the difference between strong acids (such as hydrochloric acid) and weak acids (such as acetic acid) and the difference between strong bases (such as sodium hydroxide) and weak bases (such as ammonia).

**17-3** Explain why the concentrations of the $H_3O^+$ ion and the $OH^-$ ion in pure water are the same.

**17-4** The dissociation of water is an endothermic reaction:

$$2\,H_2O(l) \rightleftharpoons H_3O^+(aq) + OH^-(aq) \qquad \Delta H^\circ = 55.84 \text{ kJ}$$

Use Le Châtelier's principle to predict what should happen to the fraction of water molecules that dissociate into ions as the temperature of water increases.

**17-5** Use Le Châtelier's principle to explain why adding either an acid or a base to water suppresses the dissociation of water.

**17-6** Explain why it is impossible for water to be at equilibrium when it contains large quantities of both the $H_3O^+$ and $OH^-$ ions.

### pH and pOH

**17-7** Calculate the number of $H_3O^+$ and $OH^-$ ions in 1.00 mL of pure water.

**17-8** Calculate the pH and pOH of an 0.035 *M* HCl solution.

**17-9** Calculate the pH and pOH of a solution that contains 0.568 g of HCl per 250 mL of solution.

**17-10** Calculate the $H_3O^+$ and $OH^-$ ion concentrations in a solution that has a pH of 3.72.

**17-11** Explain why the pH of a solution decreases when the $H_3O^+$ ion concentration increases.

### Acid-Dissociation Equilibrium Constants

**17-12** Which of the following factors influences the value of $K_a$ for the dissociation of formic acid?

$$HCO_2H(aq) + H_2O(l) \rightleftharpoons HCO_2^-(aq) + H_3O^+(aq)$$

(a) temperature (b) pressure (c) pH (d) the initial concentration of $HCO_2H$ (e) the initial concentration of the $HCO_2^-$ ion

**17-13** Which of the following solutions is the most acidic?

(a) 0.10 *M* $CH_3CO_2H$: $K_a = 1.8 \times 10^{-5}$
(b) 0.10 *M* $HCO_2H$: $K_a = 1.8 \times 10^{-4}$
(c) 0.10 *M* $ClCH_2CO_2H$: $K_a = 1.4 \times 10^{-3}$
(d) 0.10 *M* $Cl_2CHCO_2H$: $K_a = 5.1 \times 10^{-2}$

**17-14** Which of the following ions is the strongest base?

(a) $CH_3CO_2^-$ ($CH_3CO_2H$: $K_a = 1.8 \times 10^{-5}$)
(b) $HCO_2^-$ ($HCO_2H$: $K_a = 1.8 \times 10^{-4}$)
(c) $ClCH_2CO_2^-$ ($ClCH_2CO_2H$: $K_a = 1.4 \times 10^{-3}$)
(d) $Cl_2CHCO_2^-$ ($Cl_2CHCO_2H$: $K_a = 5.1 \times 10^{-2}$)

**17-15** List acetic acid, chlorous acid, hydrofluoric acid, and nitrous acid in order of increasing strength if 0.10 *M* solutions of these acids contain the following equilibrium concentrations.

| *Acid* | HA | $A^-$ | $H_3O^+$ |
|---|---|---|---|
| HOAc | 0.099 *M* | 0.0013 *M* | 0.0013 *M* |
| HOClO | 0.072 *M* | 0.028 *M* | 0.028 *M* |
| HF | 0.092 *M* | 0.0081 *M* | 0.0081 *M* |
| $HNO_2$ | 0.093 *M* | 0.0069 *M* | 0.0069 *M* |

### Strong Acids

**17-16** Explain why it is wrong to assume that the pH of a $10^{-8}$ *M* HCl solution is 8?

**17-17** Explain why the $H_3O^+$ ion concentration in a strong acid solution depends on the concentration of the solution but not the value of $K_a$ for the acid.

**17-18** Calculate the pH of an 0.056 *M* solution of hydrochloric acid.

**17-19** Nitric acid is often grouped with sulfuric acid and hydrochloric acid as one of the strong acids. Calculate the pH of 0.10 *M* nitric acid, assuming that it is a strong acid that dissociates completely. Calculate the pH of this solution using the value of $K_a$ for the acid. ($HNO_3$: $K_a = 28$)

### Hidden Assumptions in Weak-Acid Calculations

**17-20** Describe the two assumptions that are commonly made in weak-acid equilibrium problems. Describe how you can test whether these assumptions are valid for a particular calculation.

**17-21** Explain why the techniques used to calculate the equilibrium concentrations of the components of a weak-acid solution can't be used for either strong acids or very weak acids.

### Factors That Influence the $H_3O^+$ Ion Concentration in Weak-Acid Solutions

**17-22** Explain why the $H_3O^+$ ion concentration in a weak-acid solution depends on both the value of $K_a$ for the acid and the concentration of the acid.

**17-23** Which of the following solutions has the largest $H_3O^+$ ion concentration?

(a) 0.10 *M* HOAc (b) 0.010 *M* HOAc
(c) 0.0010 *M* HOAc

### Weak Acids

**17-24** Calculate the percent of HOAc molecules that dissociate in 0.10 *M*, 0.010 *M*, and 0.0010 *M* solutions of acetic acid. What happens to the percent ionization as the solution becomes more dilute? (HOAc: $K_a = 1.8 \times 10^{-5}$)

**17-25** Formic acid ($HCO_2H$) was first isolated from ants. In fact, the name comes from the Latin word for ants, *formi.* Calculate the $HCO_2H$, $HCO_2^-$, and $H_3O^+$ concentrations in an 0.10 *M* solution of formic acid in water. ($HCO_2H$: $K_a = 1.8 \times 10^{-4}$)

**17-26** Hydrogen cyanide (HCN) is a gas that dissolves in water to form hydrocyanic acid. Calculate the $H_3O^+$, HCN, and $CN^-$ concentrations in an 0.174 *M* solution of hydrocyanic acid. (HCN: $K_a = 6 \times 10^{-10}$)

**17-27** The first disinfectant used by Joseph Lister was called *carbolic acid.* This substance is now known as *phenol* (PhOH). Calculate the $H_3O^+$ ion concentration in a 0.0167 *M* solution of phenol. (PhOH: $K_a = 1.0 \times 10^{-10}$)

**17-28** Calculate the concentration of acetic acid that would give an $H_3O^+$ ion concentration of $2.0 \times 10^{-3}$ *M.* (HOAc: $K_a = 1.8 \times 10^{-5}$)

**17-29** Calculate the value of $K_a$ for ascorbic acid (vitamin C) if 2.8% of the ascorbic acid molecules in a 0.10 *M* solution dissociate.

**17-30** Calculate the value of $K_a$ for nitrous acid ($HNO_2$) if an 0.10 *M* solution is 7.1% dissociated at equilibrium.

### Not-So-Weak Acids

**17-31** Calculate the $H_3O^+$ ion concentration in 0.10 *M* solutions of acetic acid ($CH_3CO_2H$: $K_a = 1.8 \times 10^{-5}$), chloroacetic acid ($ClCH_2CO_2H$: $K_a = 1.4 \times 10^{-3}$), and dichloroacetic acid ($Cl_2CHCO_2H$: $K_a = 5.1 \times 10^{-2}$). What happens to the $H_3O^+$ ion concentration as $K_a$ increases?

**17-32** Trichloroacetic acid ($Cl_3CCO_2H$: $K_a = 0.22$) is a much stronger acid than acetic acid. Calculate the concentrations of $Cl_3CCO_2H$, $Cl_3CCO_2^-$, and $H_3O^+$ in an 0.250 *M* solution of trichloroacetic acid.

### Very Weak Acids

**17-33** Under which of the following conditions can we ignore the contribution to the total $H_3O^+$ ion concentration from the dissociation of water?

(a) When $K_aC_a < 1.0 \times 10^{-13}$ (b) When $K_aC_a > 1.0 \times 10^{-13}$ (c) Only when $K_aC_a = 1.0 \times 10^{-13}$

**17-34** Calculate the $H_3O^+$ ion concentration in a solution that is 0.01 $M$ in hydrogen peroxide. ($H_2O_2$: $K_a = 2.2 \times 10^{-12}$)

**17-35** Calculate the $H_3O^+$ ion concentration in an 0.010 $M$ $NH_4NO_3$ solution. ($NH_4^+$: $K_a = 5.8 \times 10^{-10}$)

**17-36** Boric acid ($H_3BO_3$) is a weak acid found in many first-aid products, such as eye drops. Calculate the $H_3O^+$ and $OH^-$ ion concentrations in a 0.0024 $M$ solution of boric acid in water. Clearly state and justify any assumptions you make. ($H_3BO_3$: $K_a = 7.3 \times 10^{-10}$)

### Bases

**17-37** Which of the following equations correctly describes the relationship between $K_b$ for the formate ion ($HCO_2^-$) and $K_a$ for formic acid ($HCO_2H$)?

(a) $K_b = K_w \times K_a$ (b) $K_b = K_a/K_w$ (c) $K_b = K_w/K_a$ (d) $K_b = K_w + K_a$ (e) $K_b = K_w - K_a$

**17-38** Use the relationship between $K_a$ for an acid and $K_b$ for its conjugate base to explain why strong acids have weak conjugate bases and weak acids have strong conjugate bases.

**17-39** Calculate the $HCO_2H$, $OH^-$, and $HCO_2^-$ ion concentrations in a solution that contains 0.020 mol of sodium formate ($NaHCO_2$) in 250 mL of solution. ($HCO_2H$: $K_a = 1.8 \times 10^{-4}$)

**17-40** Calculate the $OH^-$, HOBr, and $OBr^-$ ion concentrations in a solution that contains 0.050 mol of sodium hypobromite (NaOBr) in 500 mL of solution. (HOBr: $K_a = 2.4 \times 10^{-9}$)

**17-41** Calculate the pH of a 0.756 $M$ solution of NaOAc. (HOAc: $K_a = 1.8 \times 10^{-5}$)

**17-42** A solution of $NH_3$ dissolved in water is known as both aqueous ammonia and ammonium hydroxide. Use the value of $K_b$ for the following reaction to explain why aqueous ammonia is the better name.

$$NH_3(aq) + H_2O(l) \rightleftharpoons NH_4^+(aq) + OH^-(aq) \qquad K_b = 1.8 \times 10^{-5}$$

**17-43** At 25°C, a 0.10 $M$ aqueous solution of methylamine ($CH_3NH_2$) is 6.8% ionized.

$$CH_3NH_2(aq) + H_2O(l) \rightleftharpoons CH_3NH_3^+(aq) + OH^-(aq)$$

Calculate $K_b$ for methylamine. Is methylamine a stronger base or a weaker base than ammonia?

**17-44** What is the molarity of an aqueous ammonia solution that has an $OH^-$ ion concentration of $1.0 \times 10^{-3}$ $M$?

**17-45** Proteins contain nitrogen. When they are digested, they are converted to carbohydrates, which do not contain nitrogen. The nitrogen in proteins therefore has to be excreted from the organism. Mammals excrete this nitrogen as urea, $H_2NCONH_2$. Calculate the pH of a 0.10 $M$ solution of urea. ($H_2NCONH_2$: $K_b = 1.5 \times 10^{-4}$)

**17-46** Calculate $K_b$ for hydrazine ($H_2NNH_2$) if the pH of a 0.10 $M$ aqueous solution of this rocket fuel is 10.54.

**17-47** Calculate the pH of a 0.016 $M$ aqueous solution of calcium acetate, $Ca(OAc)_2$. (HOAc: $K_a = 1.8 \times 10^{-5}$)

### Buffers, Buffering Capacity, and pH Titration Curves

**17-48** Explain how buffers resist changes in pH.

**17-49** Explain why a mixture of HOAc and NaOAc is an acidic buffer, but a mixture of $NH_3$ and $NH_4Cl$ is a basic buffer. (HOAc: $K_a = 1.8 \times 10^{-5}$, $NH_4^+$: $K_a = 5.6 \times 10^{-10}$)

**17-50** Which of the following mixtures would make the best buffer?

(a) HCl and NaCl (b) NaOAc and $NH_3$ (c) HOAc and $NH_4Cl$ (d) NaOAc and $NH_4Cl$ (e) $NH_3$ and $NH_4Cl$

**17-51** Which of the following solutions is an acidic buffer?

(a) 0.10 $M$ HCl and 0.10 $M$ NaOH (b) 0.10 $M$ HCl and 0.10 $M$ NaCl (c) 0.10 $M$ $HCO_2H$ and 0.10 $M$ $NaHCO_2$ (d) 0.10 $M$ $NH_3$ and 0.10 $M$ $NH_4Cl$

**17-52** Which of the following solutions is a basic buffer?

(a) 0.10 $M$ HCl and 0.10 $M$ NaOH (b) 0.10 $M$ HCl and 0.10 $M$ NaCl (c) 0.10 $M$ $HCO_2H$ and 0.10 $M$ $NaHCO_2$ (d) 0.10 $M$ $NH_3$ and 0.10 $M$ $NH_4Cl$

**17-53** What is the best way to increase the capacity of a buffer made by dissolving $NaHCO_2$ in an aqueous solution of $HCO_2H$?

(a) Increase the concentration of $HCO_2H$. (b) Increase the concentration of $NaHCO_2$. (c) Increase the concentrations of both $HCO_2H$ and $NaHCO_2$. (d) Increase the ratio of the concentration of $HCO_2H$ to the concentration of $NaHCO_2$. (e) Increase the ratio of the concentration of $NaHCO_2$ to the concentration of $HCO_2H$.

**17-54** Describe in detail the experiment you would use to measure the value of $K_a$ for formic acid, $HCO_2H$.

**17-55** Sketch a titration curve for 0.10 $M$ $NH_3$ ($K_b = 1.8 \times 10^{-5}$) reacting with an 0.10 $M$ HCl solution. Label the equivalence point, the endpoint, and the point at which the $OH^-$ ion concentration is equal to $K_b$ for the base.

### Buffer Calculations

**17-56** Calculate the pH of a solution that contains 0.010 $M$ of proprionic acid (HOPr) and 0.080 $M$ of potassium proprionate (KOPr). (HOPr: $K_a = 1.3 \times 10^{-5}$)

**17-57** Calculate the pH of a solution prepared by dissolving 0.040 mol of sodium nitrite ($NaNO_2$) in 200 mL of 0.10 $M$ nitrous acid. ($HNO_2$: $K_a = 5.1 \times 10^{-4}$)

**17-58** Calculate the $H_3O^+$ ion concentration and the pH of a buffer solution prepared by dissolving 0.218 mol of sodium acetate in 500 mL of 0.100 *M* acetic acid. (HOAc: $K_a = 1.8 \times 10^{-5}$)

**17-59** Calculate the $H_3O^+$ ion concentration in a solution that is 0.050 *M* in acetic acid and 0.10 *M* in sodium acetate. What happens when $1 \times 10^{-3}$ mol of HCl are added to this solution? (HOAc: $K_a = 1.8 \times 10^{-5}$)

**17-60** How many moles of sodium formate ($NaHCO_2$) would have to be added to 500 mL of 0.100 *M* formic acid ($HCO_2H$) to give a solution buffered at pH 4.1? ($HCO_2H$: $K_a = 1.8 \times 10^{-4}$)

**17-61** Calculate the ratio of concentrations of hypochlorous acid (HOCl) and sodium hypochlorite (NaOCl) needed to produce a buffer solution with a pH of 7.60. (HOCl: $K_a = 2.9 \times 10^{-8}$)

**17-62** How many grams of NaOH must be added to 500 mL of 0.100 *M* $NaH_2PO_4$ to give a buffer with a pH of 8.10? ($H_2PO_4^-$: $K_a = 6.3 \times 10^{-8}$)

**17-63** How much $NaHCO_2$ would you have to add to 0.10 *M* $HCO_2H$ to get a buffer solution with a pH of 3.4? ($HCO_2H$: $K_a = 1.8 \times 10^{-4}$)

**17-64** Most bacteria will not grow in solutions that are more acidic than pH = 4.5. What ratio of concentrations of HOAc and $OAc^-$ would you need to prepare a buffer with a pH of 4.5?

**17-65** What ratio of concentrations of $HPO_4^{2-}$ and $H_2PO_4^-$ would you need to prepare a buffer with a pH of 7.00? ($H_2PO_4^-$: $K_a = 6.3 \times 10^{-8}$)

**17-66** How many grams of $NH_4Cl$ must be added to 100 mL of 0.300 *M* $NH_3$ to obtain a solution with a pH of 9.00?

### Polyprotic Acids

**17-67** Calculate the $H_3O^+$, $CO_3^{2-}$, $HCO_3^-$, and $H_2CO_3$ concentrations at equilibrium in a solution that initially contained 0.10 mol of carbonic acid per liter. ($H_2CO_3$: $K_{a1} = 4.5 \times 10^{-7}$, $K_{a2} = 4.7 \times 10^{-11}$)

**17-68** Calculate the equilibrium concentrations of the important components of an 0.25 *M* malonic acid ($HO_2CCH_2CO_2H$) solution. Use the symbol $H_2M$ as an abbreviation for malonic acid and assume stepwise dissociation of this acid. ($H_2M$: $K_{a1} = 1.4 \times 10^{-5}$, $K_{a2} = 2.1 \times 10^{-8}$)

**17-69** Check the validity of the assumption in the previous problem that malonic acid dissociates in a stepwise fashion by comparing the concentrations of the $HM^-$ and $M^{2-}$ ions obtained in this problem. Is this assumption valid?

**17-70** Glycine, the simplest of the amino acids found in proteins, is a diprotic acid with the formula $HO_2CCH_2NH_3^+$. If we symbolize glycine as $H_2G^+$, we can write the following equations for the stepwise dissociation of this amino acid:

$$H_2G^+(aq) + H_2O(l) \rightleftharpoons HG(aq) + H_3O^+(aq) \qquad K_{a1} = 4.5 \times 10^{-3}$$

$$HG(aq) + H_2O(l) \rightleftharpoons G^-(aq) + H_3O^+ \qquad K_{a2} = 2.5 \times 10^{-10}$$

Calculate the concentrations of $H_3O^+$, $G^-$, HG, and $H_2G^+$ in a 2.0 *M* solution of glycine in water.

**17-71** Which of the following equations accurately describes a 2.0 *M* solution of glycine ($H_2G^+$)?

(a) $[H_3O^+] \approx [H_2G^+]$ (b) $[H_3O^+] < [H_2G^+]$
(c) $[H_3O^+] > [HG]$ (d) $[H_3O^+] \approx [HG]$
(e) $[H_3O^+] < [HG]$

**17-72** Which of the following equations results from the fact that glycine ($H_2G^+$) is a weak diprotic acid?

(a) $[G^-] \approx C_{H_2G^+}$ (b) $[G^-] \approx K_{a2}$ (c) $[G^-] \approx [HG]$
(d) $[G^-] > [HG]$ (e) $[G^-] \approx [H_3O^+]$

**17-73** Oxalic acid ($H_2C_2O_4$) has been implicated in diseases such as gout and kidney stones. Calculate the $H_3O^+$, $H_2C_2O_4$, $HC_2O_4^-$, and $C_2O_4^{2-}$ concentrations in a 1.25 *M* solution of oxalic acid:

$$H_2C_2O_4(aq) + H_2O(l) \rightleftharpoons H_3O^+(aq) + HC_2O_4^-(aq) \qquad K_{a1} = 5.4 \times 10^{-2}$$

$$HC_2O_4^-(aq) + H_2O(l) \rightleftharpoons H_3O^+(aq) + C_2O_4^{2-}(aq) \qquad K_{a2} = 5.4 \times 10^{-5}$$

### Polyprotic Bases

**17-74** Which of the following sets of equations can be used to calculate $K_{b1}$ and $K_{b2}$ for sodium oxalate ($Na_2C_2O_4$) from $K_{a1}$ and $K_{a2}$ for oxalic acid ($H_2C_2O_4$)?

(a) $K_{b1} = K_w \times K_{a1}$ and $K_{b2} = K_w \times K_{a2}$ (b) $K_{b1} = K_w \times K_{a2}$ and $K_{b2} = K_w \times K_{a1}$ (b) $K_{b1} = K_w/K_{a1}$ and $K_{b2} = K_w/K_{a2}$ (c) $K_{b1} = K_w/K_{a2}$ and $K_{b2} = K_w/K_{a1}$ (e) $K_{b1} = K_{a1}/K_w$ and $K_{b2} = K_{a2}/K_w$

**17-75** Calculate the pH of a 0.028 *M* solution of sodium oxalate ($Na_2C_2O_4$) in water:

$$H_2C_2O_4(aq) + H_2O(l) \rightleftharpoons H_3O^+(aq) + HC_2O_4^-(aq) \qquad K_{a1} = 5.4 \times 10^{-2}$$

$$HC_2O_4^-(aq) + H_2O(l) \rightleftharpoons H_3O^+(aq) + C_2O_4^{2-}(aq) \qquad K_{a2} = 5.4 \times 10^{-5}$$

**17-76** Calculate the $H_3O^+$, $OH^-$, $H_2CO_3$, $HCO_3^-$, and $CO_3^{2-}$ concentrations in an 0.150 *M* solution of sodium carbonate ($Na_2CO_3$) in water. ($H_2CO_3$: $K_{a1} = 4.5 \times 10^{-7}$; $K_{a2} = 4.7 \times 10^{-11}$)

### Compounds That Could Be Either Acids or Bases

**17-77** According to the data in Table 11.6, the pH of an 0.10 *M* $NaHCO_3$ solution is 8.4. Explain why the $HCO_3^-$ ion forms solutions that are basic, not acidic.

**17-78** Which of the following statements concerning sodium hydrogen sulfide (NaSH) is correct? ($H_2S$: $K_{a1} = 1.0 \times 10^{-7}$, $K_{a2} = 1.3 \times 10^{-13}$)

(a) NaSH is an acid because $K_{a1}$ for $H_2S$ is much larger than $K_{a2}$. (b) NaSH is an acid because $K_{a1}$ for $H_2S$ is smaller than $K_{b1}$ for $Na_2S$. (c) NaSH is a base because $K_{b1}$ for $Na_2S$ is much larger than $K_{b2}$. (d) NaSH is a base because $K_{b2}$ for $Na_2S$ is larger than $K_{a2}$ for $H_2S$.

**17-79** Predict whether an aqueous solution of sodium hydrogen sulfate ($NaHSO_4$) will be acidic, basic, or neutral. ($H_2SO_4$: $K_{a1} = 1 \times 10^3$, $K_{a2} = 1.2 \times 10^{-2}$)

**17-80** Predict whether an aqueous solution of sodium hydrogen sulfite ($NaHSO_3$) will be acidic, basic, or neutral. ($H_2SO_3$: $K_{a1} = 1.7 \times 10^{-2}$, $K_{a2} = 6.4 \times 10^{-8}$)

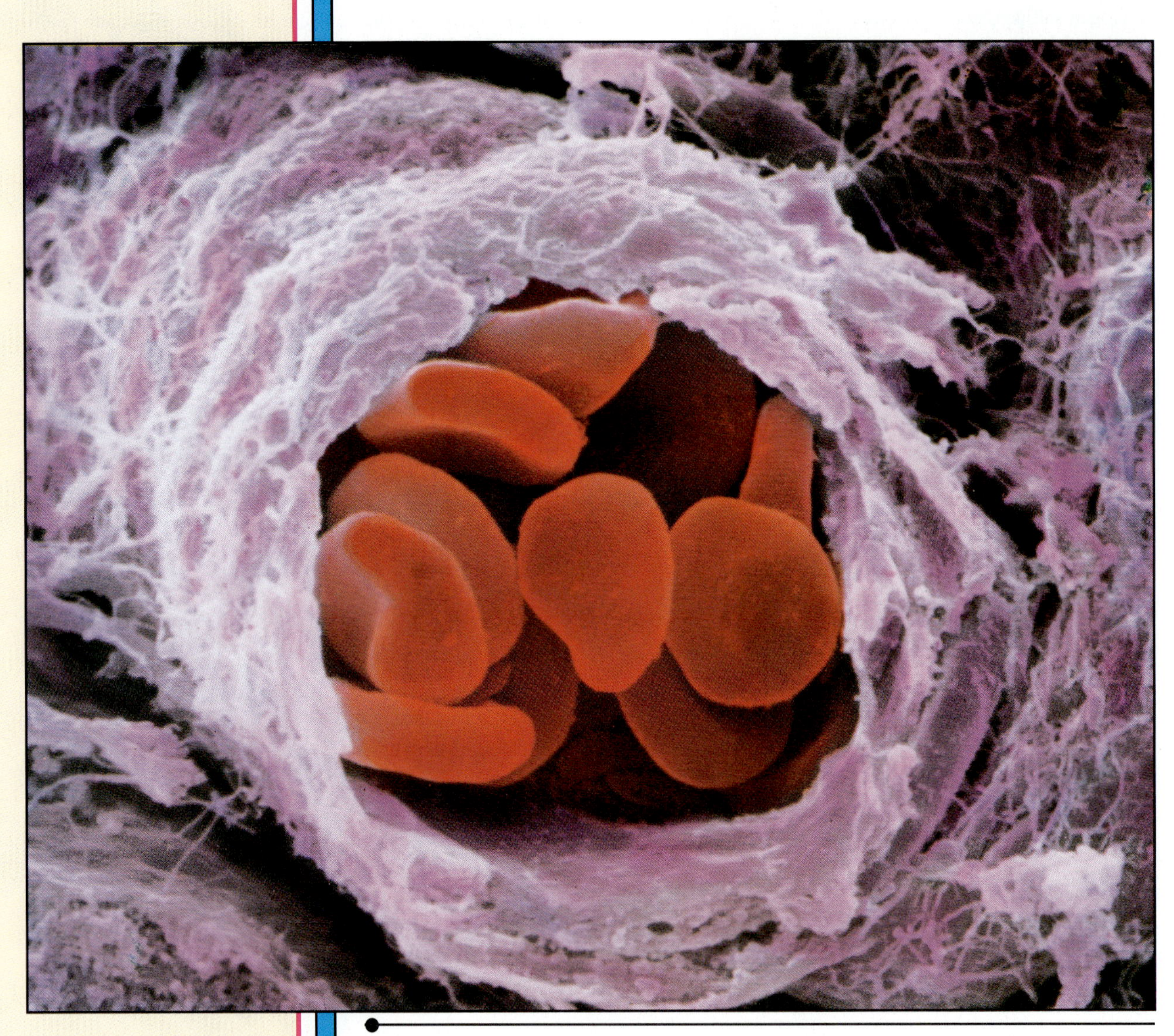

Transition-metal complexes play a vital role in the process by which the blood cells in this blood vessel carry oxygen and carbon dioxide through the blood stream.

CHAPTER 18

# SOLUBILITY AND COMPLEX-ION EQUILIBRIA

This chapter extends the concept of equilibrium to reactions in which there is a change in the state or phase of one or more components. Particular attention will be paid to the question of what happens when a solid dissolves in water. This discussion will provide the basis for answering questions such as the following:

## POINTS OF INTEREST

- Why is NaCl soluble in water, but not AgCl?
- If AgCl is "insoluble" in water, what does it mean to say that a solution of AgCl in water is *saturated?*
- What does it mean to say that two reactions share a *common ion?* What effect does the presence of a common ion have on the solubility of a salt?
- Why is it impossible to prepare an 0.10 *M* $Cr^{3+}$ ion solution by dissolving a chromium(III) salt in water? What do we have to do to prepare this solution?
- Why does a light-blue precipitate form when a small amount of $NH_3$ is added to an aqueous $Cu^{2+}$ ion solution? Why does this precipitate dissolve to form a dark-blue solution when excess $NH_3$ is added?

## 18.1 WHY DO SOME SOLIDS DISSOLVE IN WATER?

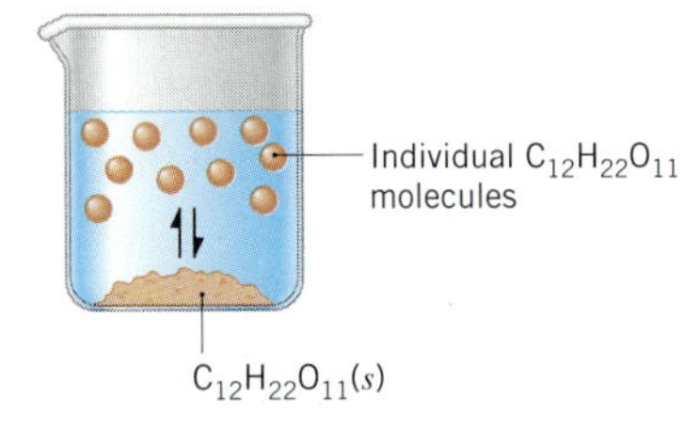


When asked to explain what happens when sugar ($C_{12}H_{22}O_{11}$) dissolves in water, students in introductory chemistry courses often write equations such as the following:

$$C_{12}H_{22}O_{11}(s) \xrightarrow{H_2O} C_{12}H_{22}O_{11}(aq)$$

Although there is nothing wrong with this equation, it doesn't explain what happens at the molecular level when sugar dissolves. Nor does it provide any hints about why sugar dissolves, but other solids do not. To understand what happens at the molecular level when a solid dissolves, we have to refer back to the discussion of molecular and ionic solids in Chapter 13.

The sugar we use to sweeten coffee or tea is a *molecular solid,* in which the individual molecules are held together by relatively weak intermolecular forces. When sugar dissolves in water, the weak bonds between the individual sucrose molecules are broken, and these $C_{12}H_{22}O_{11}$ molecules are released into solution.

It takes energy to break the bonds between the $C_{12}H_{22}O_{11}$ molecules in sucrose. It also takes energy to break the hydrogen bonds in water that must be disrupted to insert one of these sucrose molecules into solution. Sugar dissolves in water because energy is given off when the slightly polar sucrose molecules form intermolecular bonds with the polar water molecules. The weak bonds that form between the solute and the solvent compensate for the energy needed to disrupt the structure of both the pure solute and the solvent. In the case of sugar and water, this process works so well that up to 1800 g of sucrose can dissolve in a liter of water.

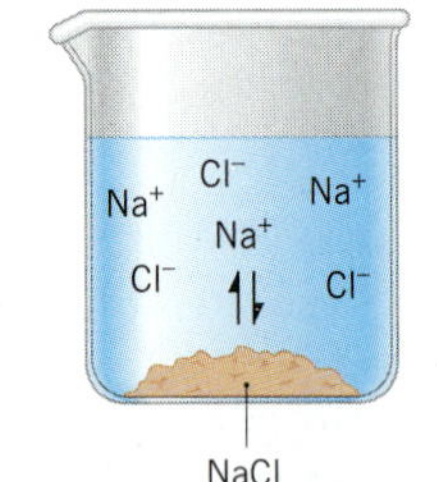


*Ionic solids* (or salts) contain positive and negative ions that are held together by the strong force of attraction between particles with opposite charges. When one of these solids dissolves in water, the ions that form the solid are released into solution, where they become associated with the polar solvent molecules:

$$NaCl(s) \xrightarrow{H_2O} Na^+(aq) + Cl^-(aq)$$

As a result, we can generally assume that salts dissociate into their ions when they dissolve in water. Ionic compounds dissolve in water if the energy given off when the ions interact with water molecules compensates for the energy needed to break the ionic bonds in the solid and the energy required to separate the water molecules so that the ions can be inserted into solution.

A saturated solution of sodium chloride in water.

It takes an enormous amount of energy to rip apart an ionic crystal (see Section 7.15). The lattice energy of sodium chloride, for example, is 787.3 kJ/mol. This means that 787.3 kJ of energy is released when a mole of $Na^+$ and $Cl^-$ ions in the gas phase come together to form solid NaCl. But it also means that it takes 787.3 kJ/mol to transform solid NaCl into $Na^+$ and $Cl^-$ ions in the gas phase:

$$NaCl(s) \longrightarrow Na^+(g) + Cl^-(g) \qquad \Delta H° = 787.3 \text{ kJ/mol}$$

It takes so much energy to separate the $Na^+$ and $Cl^-$ ions in NaCl that we might not expect this compound to dissolve in water. The force of attraction between $Na^+$ ions and water molecules is so large, however, that 783.5 kJ/mol of energy is released when the $Na^+$ and $Cl^-$ ions interact with water molecules:

$$Na^+(g) + Cl^-(g) \xrightarrow{H_2O} Na^+(aq) + Cl^-(aq) \qquad \Delta H° = -783.5 \text{ kJ/mol}$$

The overall enthalpy of reaction for the process in which solid NaCl dissolves in water is therefore close to zero:

$$NaCl(s) \xrightarrow{H_2O} Na^+(aq) + Cl^-(aq) \qquad \Delta H° = 3.8 \text{ kJ/mol}$$

Because $\Delta H°$ for this reaction is very small, thermodynamics favors neither the reactants nor the products. We might therefore expect NaCl to be moderately soluble in water. In fact, up to 360 g of NaCl dissolve in a liter of water at room temperature, to give a solution with a concentration of about 6 *M*.

Now that we have some idea of why NaCl dissolves in water we can understand why silver chloride does not. The lattice energy for AgCl is very large:

$$AgCl(s) \longrightarrow Ag^+(g) + Cl^-(g) \qquad \Delta H° = 915.7 \text{ kJ/mol}$$

So is the energy released when the $Ag^+$ and $Cl^-$ ions interact with water:

$$Ag^+(g) + Cl^-(g) \xrightarrow{H_2O} Ag^+(aq) + Cl^-(aq) \qquad \Delta H° = -850.2 \text{ kJ/mol}$$

But the energy released when the ions interact with water is not quite large enough to compensate for the energy needed to separate the ions in the crystal. As a result, the overall enthalpy of reaction is unfavorable:

$$AgCl(s) \xrightarrow{H_2O} Ag^+(aq) + Cl^-(aq) \qquad \mathbf{\Delta H° = 65.5 \text{ kJ/mol}}$$

We would therefore expect relatively little AgCl to dissolve in water. In fact, less than 0.002 g of AgCl dissolves in a liter of water at room temperature. The solubility of silver chloride in water is so small that it is often said to be ''insoluble'' in water, even though this term is misleading.

## 18.2 SOLUBILITY EQUILIBRIA

Discussions of solubility equilibria are based on the following assumption:

**When solids dissolve in water, they dissociate to give the elementary particles from which they are formed.**

Thus, molecular solids dissociate to give individual molecules:

$$C_{12}H_{22}O_{11}(s) \xrightarrow{H_2O} C_{12}H_{22}O_{11}(aq)$$

and ionic solids dissociate to give solutions of the positive and negative ions they contain:

$$NaCl(s) \xrightarrow{H_2O} Na^+(aq) + Cl^-(aq)$$

We can detect the presence of $Na^+$ and $Cl^-$ ions in an aqueous solution with the conductivity apparatus shown in Figure 18.1. This apparatus consists of a light bulb connected to a pair of metal wires that can be immersed in a beaker of water. The circuit in this conductivity apparatus is not complete. In order for the light bulb to glow when the apparatus is plugged into an electrical outlet, there must be a way for electrical charge to flow through the solution from one of these metal wires to the other.

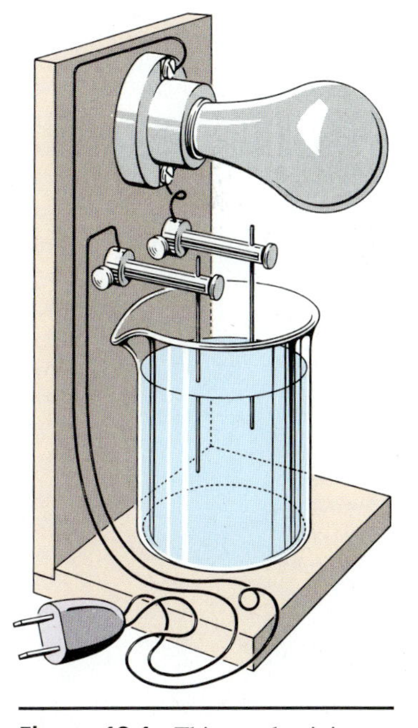

**Figure 18.1** This conductivity apparatus can be used to demonstrate the difference between aqueous solutions of ionic and covalent compounds.

### CHECKPOINT

Use the discussion of the dissociation of water in the previous chapter to explain why pure water is a poor conductor of electricity, at best.

When an ionic solid dissolves in water, the positive and negative ions are free to move through this solution. The net result is a flow of electric charge through the

The light bulb glows brightly because NaCl dissociates in water to form $Na^+$ and $Cl^-$ ions that carry the electric current through the solution.

solution that completes the circuit, and lets the light bulb glow. As might be expected, the brightness of the bulb is proportional to the concentration of the ions in the solution. Slightly soluble salts such as calcium sulfate make the light bulb glow dimly. When the wires are immersed in a solution of a very soluble salt, such as NaCl, the light bulb glows brightly.

The conductivity apparatus in Figure 18.1 only gives us qualitative information about the relative concentrations of the ions in different solutions. It is possible to build a more sophisticated instrument that gives quantitative measurements of the conductivity of a solution, which is directly proportional to the concentration of the ions in the solution. Figure 18.2 shows what happens to the conductivity of water as we gradually add infinitesimally small amounts of AgCl to water and wait for the solid to dissolve before taking measurements.

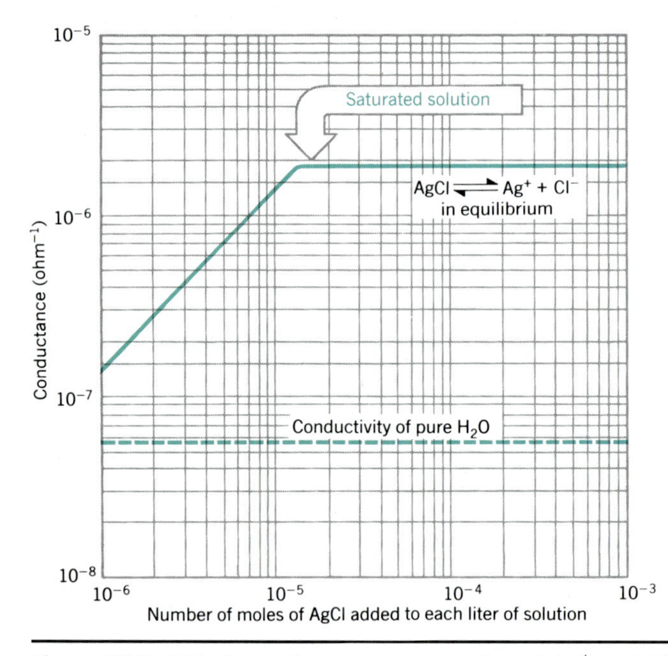


**Figure 18.2** This figure shows the concentration of $Ag^+$ and $Cl^-$ ions versus time as AgCl is added to water.

The system conducts an electric current even before any AgCl is added because of the small quantity of $H_3O^+$ and $OH^-$ ions in water. The solution becomes a slightly better conductor when AgCl is added, because some of this salt dissolves to give $Ag^+$ and $Cl^-$ ions, which can carry an electric current through the solution. The conductivity continues to increase as more AgCl is added, until about 0.002 g of this salt has dissolved per liter of solution.

The fact that the conductivity does not increase after the solution has reached this concentration tells us that there is a limit on the solubility of this salt in water. Once the solution reaches this limit, no more AgCl dissolves, regardless of how much solid we add to the system. This is exactly what we would expect if the solubility of AgCl is controlled by an equilibrium. Once the solution reaches equilibrium, the rate at which AgCl dissolves to form $Ag^+$ and $Cl^-$ ions is equal to the rate at which these ions recombine to form AgCl.

$$AgCl(s) \xrightleftharpoons{H_2O} Ag^+(aq) + Cl^-(aq)$$

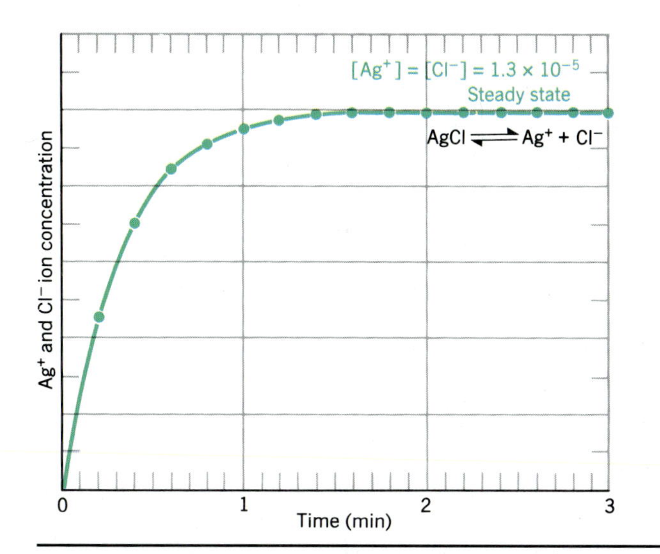


**Figure 18.3** The conductivity of a solution of AgCl in water increases at first as the AgCl dissolves and dissociates into $Ag^+$ and $Cl^-$ ions. Once the solution has become saturated with AgCl, the conductivity remains the same no matter how much solid is added.

Evidence to support this conclusion comes from Figure 18.3, which shows what happens to the conductivity of water when a large excess of solid silver chloride is added to the water. When the salt is first added, it dissolves and dissociates rapidly. The conductivity of the solution therefore increases rapidly at first:

$$\mathrm{AgCl}(s) \xrightarrow[\text{dissociate}]{\text{dissolve}} \mathrm{Ag}^+(aq) + \mathrm{Cl}^-(aq)$$

The concentrations of these ions soon become large enough that the reverse reaction starts to compete with the forward reaction, which leads to a decrease in the rate at which $Ag^+$ and $Cl^-$ ions enter the solution:

$$\mathrm{Ag}^+(aq) + \mathrm{Cl}^-(aq) \xrightarrow[\text{precipitate}]{\text{associate}} \mathrm{AgCl}(s)$$

Eventually, the $Ag^+$ and $Cl^-$ ion concentrations become large enough that the rate at which precipitation occurs exactly balances the rate at which AgCl dissolves. Once that happens, there is no change in the concentration of these ions with time and the reaction is at equilibrium. When this system reaches equilibrium it is called a **saturated solution,** because it contains the maximum concentration of ions that can exist in equilibrium with the solid salt at that temperature. The amount of salt that must be added to a given volume of solvent to form a saturated solution is called the **solubility** of the salt.

## 18.3 SOLUBILITY RULES

The data needed to do the thermodynamic calculations in Section 18.1 are available for only a few compounds. These calculations are therefore more useful for explaining the results of measurements of the solubility of salts than for predicting what happens when a salt is added to water.

**TABLE 18.1 Solubility Rules for Ionic Compounds in Water**

Soluble Salts

1. The $Na^+$, $K^+$, and $NH_4^+$ ions form *soluble salts.* Thus, NaCl, $KNO_3$, $(NH_4)_2SO_4$, $Na_2S$, and $(NH_4)_2CO_3$ are soluble.
2. The nitrate ($NO_3^-$) ion forms *soluble salts.* Thus, $Cu(NO_3)_2$ and $Fe(NO_3)_3$ are soluble.
3. The chloride ($Cl^-$), bromide ($Br^-$), and iodide ($I^-$) ions generally form *soluble salts.* Exceptions to this rule include salts of the $Pb^{2+}$, $Hg_2^{2+}$, $Ag^+$, and $Cu^+$ ions. $ZnCl_2$ is soluble, but CuBr is not.
4. The sulfate ($SO_4^{2-}$) ion generally forms *soluble salts.* Exceptions include $BaSO_4$, $SrSO_4$, and $PbSO_4$, which are insoluble, and $Ag_2SO_4$, $CaSO_4$, and $Hg_2SO_4$, which are slightly soluble.

Insoluble Salts

1. Sulfides ($S^{2-}$) are usually *insoluble.* Exceptions include $Na_2S$, $K_2S$, $(NH_4)_2S$, MgS, CaS, SrS, and BaS.
2. Oxides ($O^{2-}$) are usually *insoluble.* Exceptions include $Na_2O$, $K_2O$, SrO, and BaO, which are soluble, and CaO, which is slightly soluble.
3. Hydroxides ($OH^-$) are usually *insoluble.* Exceptions include NaOH, KOH, $Sr(OH)_2$, and $Ba(OH)_2$, which are soluble, and $Ca(OH)_2$, which is slightly soluble.
4. Chromates ($CrO_4^{2-}$) are usually *insoluble.* Exceptions include $Na_2CrO_4$, $K_2CrO_4$, $(NH_4)_2CrO_4$, and $MgCrO_4$.
5. Phosphates ($PO_4^{3-}$) and carbonates ($CO_3^{2-}$) are usually *insoluble.* Exceptions include salts of the $Na^+$, $K^+$, and $NH_4^+$ ions.

There are a number of patterns in the data obtained from measuring the solubility of different salts. These patterns form the basis for the rules outlined in Table 18.1, which can guide predictions of whether a given salt will dissolve in water. These rules are based on the following definitions of the terms *soluble, insoluble,* and *slightly soluble.*

- A salt is soluble if it dissolves in water to give a solution with a concentration of at least 0.1 *M* at room temperature.
- A salt is insoluble if the concentration of an aqueous solution is less than 0.001 *M* at room temperature.
- Slightly soluble salts give solutions that fall between these extremes.

## 18.4 THE SOLUBILITY PRODUCT EXPRESSION

Silver chloride is so insoluble in water (≈0.002 g/L) that a saturated solution contains only about $1.3 \times 10^{-5}$ mol of AgCl per liter of water:

$$AgCl(s) \xrightleftharpoons{H_2O} Ag^+(aq) + Cl^-(aq)$$

Strict adherence to the rules for writing equilibrium constant expressions for this reaction gives the following:

$$K_c = \frac{[Ag^+][Cl^-]}{[AgCl]}$$

(Water is not included in the equilibrium constant expression because it is neither consumed nor produced in this reaction, even though it is a vital component of the system.)

Two of the terms in this expression are easy to interpret. The $[Ag^+]$ and $[Cl^-]$ terms represent the concentrations of the $Ag^+$ and $Cl^-$ ions in moles per liter when this solution is at equilibrium. The third term, [AgCl], is more ambiguous. It doesn't represent the concentration of AgCl dissolved in water because we assume that AgCl dissociates into $Ag^+$ ions and $Cl^-$ ions when it dissolves in water. It can't represent the amount of solid AgCl in the system because the equilibrium is not affected by the amount of excess solid added to the system. The [AgCl] term has to be translated quite literally as the number of moles of AgCl in a liter of solid AgCl.

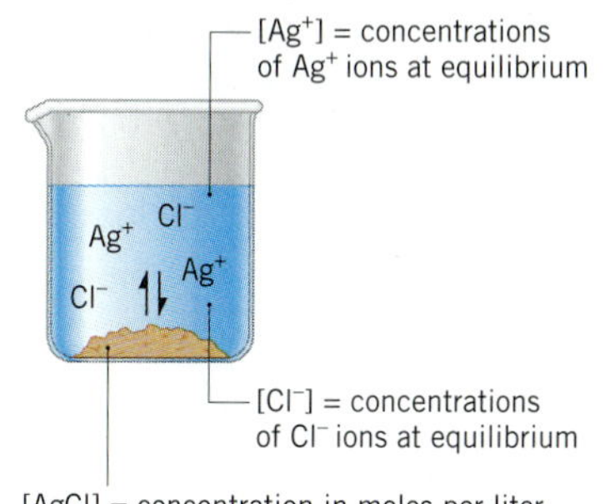


The concentration of solid AgCl can be calculated from its density and the mass of a mole of AgCl:

$$\frac{5.56 \text{ g AgCl}}{1 \text{ cm}^3} \times \frac{1 \text{ cm}^3}{1 \text{ mL}} \times \frac{1000 \text{ mL}}{1 \text{ L}} \times \frac{1 \text{ mol AgCl}}{143.34 \text{ g AgCl}} = 38.8 \text{ mol AgCl/L}$$

This quantity is a constant, however. The number of moles per liter in solid AgCl is the same at the start of the reaction as it is when the reaction reaches equilibrium.

Since the [AgCl] term is a constant, which has no effect on the equilibrium, it is built into the equilibrium constant for the reaction:

$$[Ag^+][Cl^-] = K_c \times [AgCl]$$

This equation suggests that the product of the equilibrium concentrations of the $Ag^+$ and $Cl^-$ ions in this solution is equal to a constant. Since this constant is proportional to the solubility of the salt, it is called the **solubility product equilibrium constant** for the reaction, or $K_{sp}$:

$$K_{sp} = [Ag^+][Cl^-]$$

The $K_{sp}$ expression for a salt is the product of the concentrations of the ions, with each concentration raised to a power equal to the coefficient of that ion in the balanced equation for the solubility equilibrium. Solubility product constants for a number of so-called insoluble salts are given in Table A.11 in the appendix.

## EXERCISE 18.1

Write the $K_{sp}$ expression for a saturated solution of $CaF_2$ in water.

**SOLUTION** We start with a balanced equation for the equilibrium we want to describe:

$$CaF_2(s) \overset{H_2O}{\rightleftharpoons} Ca^{2+}(aq) + 2\,F^-(aq)$$

Because three ions are produced when this salt dissolves in water, the $K_{sp}$ expression contains three terms:

$$K_{sp} = [Ca^{2+}][F^-][F^-] = [Ca^{2+}][F^-]^2$$

## 18.5 THE RELATIONSHIP BETWEEN $K_{SP}$ AND THE SOLUBILITY OF A SALT

$K_{sp}$ is called the solubility product because it is literally the product of the solubilities of the ions in moles per liter. The solubility product of a salt therefore can be calculated from its solubility, or vice versa.

Photographic films are based on the sensitivity of AgBr to light. When light hits a crystal of AgBr, a small fraction of the $Ag^+$ ions are reduced to silver metal. The rest of the $Ag^+$ ions in these crystals are reduced to silver metal when the film is developed. AgBr crystals that do not absorb light are then removed from the film to "fix" the image. Let's calculate the solubility of AgBr in water in grams per liter, to see whether AgBr can be removed by simply washing the film.

We start with the balanced equation for the equilibrium:

$$AgBr(s) \underset{}{\overset{H_2O}{\rightleftharpoons}} Ag^+(aq) + Br^-(aq)$$

We then write the solubility product expression for this reaction:

$$K_{sp} = [Ag^+][Br^-] = 5.0 \times 10^{-13}$$

**Problem-Solving Strategy**

One equation can't be solved for two unknowns—the $Ag^+$ and $Br^-$ ion concentrations. We can generate a second equation, however, by noting that one $Ag^+$ ion is released for every $Br^-$ ion when silver bromide dissolves in water. Because there is no other source of either ion in this solution, the concentrations of these ions at equilibrium must be the same:

$$\mathbf{[Ag^+] = [Br^-]}$$

Substituting this equation into the $K_{sp}$ expression gives the following result:

$$[Ag^+]^2 = 5.0 \times 10^{-13}$$

Taking the square root of both sides of this equation gives the equilibrium concentrations of the $Ag^+$ and $Br^-$ ions:

$$[Ag^+] = [Br^-] = \mathbf{7.1 \times 10^{-7}\ M}$$

Once we know how many moles of AgBr dissolve in a liter of water, we can calculate the solubility in grams per liter:

$$\frac{7.1 \times 10^{-7} \text{ mol AgBr}}{1 \text{ L}} \times \frac{187.8 \text{ g AgBr}}{1 \text{ mol AgBr}} = \mathbf{1.3 \times 10^{-4}\ g\ AgBr/L}$$

The solubility of AgBr in water is only 0.00013 g/L. It therefore isn't practical to try to wash the unexposed AgBr off photographic film with water. In the second half of this chapter we will encounter a more efficient way of removing the unexposed AgBr from film when it is processed.

Solubility product calculations with 1 : 1 salts such as AgBr are relatively easy to perform. In order to extend such calculations to compounds with more complex formulas we need to understand the relationship between the solubility of a salt and the concentrations of its ions at equilibrium. In the previous chapter, we used the symbols $C_a$ and $C_b$ to describe the initial concentration of acids and bases that were added to an aqueous solution. In this chapter, we will use the symbol $C_s$ to describe the amount of a salt that dissolves in water.

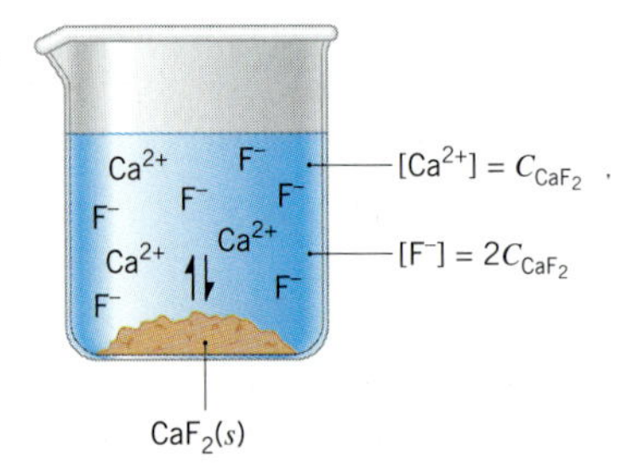


### EXERCISE 18.2

Several compounds were studied as possible sources of the fluoride ion for use in toothpaste. Write equations that describe the relationship between the solubility of $CaF_2$ and the equilibrium concentrations of the $Ca^{2+}$ and $F^-$ ions in a saturated solution as a first step toward evaluating its use as a fluoridating agent.

**SOLUTION** As always, we start with the balanced equation for the reaction:

$$CaF_2(s) \xrightleftharpoons{H_2O} Ca^{2+}(aq) + 2\,F^-(aq)$$

Salts dissociate into their ions when they dissolve in water. For every mole of $CaF_2$ that dissolves, we get a mole of $Ca^{2+}$ ions. The equilibrium concentration of the $Ca^{2+}$ ion is therefore equal to the solubility of this compound in moles per liter:

$$[Ca^{2+}] = C_s$$

For every mole of $CaF_2$ that dissolves, we get twice as many moles of $F^-$ ions. The $F^-$ ion concentration at equilibrium is therefore equal to twice the solubility of the compound in moles per liter:

$$[F^-] = 2\,C_s$$

**EXERCISE 18.3**

Use the $K_{sp}$ for calcium fluoride to calculate its solubility in grams per liter. Comment on the potential of $CaF_2$ to act as a fluoridating agent. ($CaF_2$: $K_{sp} = 4.0 \times 10^{-11}$)

**SOLUTION** According to Exercise 18.1, the solubility product expression for $CaF_2$ is written as follows:

$$K_{sp} = [Ca^{2+}][F^-]^2$$

Exercise 18.2 gave us the following equations for the relationship between the solubility of this salt and the concentrations of the $Ca^{2+}$ and $F^-$ ions.

$$[Ca^{2+}] = C_s \qquad [F^-] = 2\,C_s$$

Substituting this information into the $K_{sp}$ expression gives the following result:

$$[C_s][2C_s]^2 = 4.0 \times 10^{-11}$$
$$4C_s^3 = 4.0 \times 10^{-11}$$

This equation can be solved for the solubility of $CaF_2$ in units of moles per liter:

$$C_s = 2.2 \times 10^{-4}$$

Once we know how many moles of $CaF_2$ dissolve in a liter, we can calculate the solubility in units of grams per liter:

$$\frac{2.2 \times 10^{-4}\ \cancel{\text{mol CaF}_2}}{1\text{ L}} \times \frac{78.1\text{ g CaF}_2}{1\ \cancel{\text{mol CaF}_2}} = 0.017\text{ g CaF}_2\text{/L}$$

The solubility of calcium fluoride is fairly small: 0.017 g/L. Stannous fluoride, or tin(II) fluoride, is over 10,000 times as soluble, so $SnF_2$ was chosen as the first compound used in fluoride toothpastes.

A single crystal of fluorite ($CaF_2$).

The techniques used in the preceding exercise are also valid for salts that contain more positive ions than negative ions.

**EXERCISE 18.4**

Calculate the solubility in grams per liter of silver sulfide in order to decide whether it is accurately labeled when described as an insoluble salt. ($Ag_2S$: $K_{sp} = 6.3 \times 10^{-50}$)

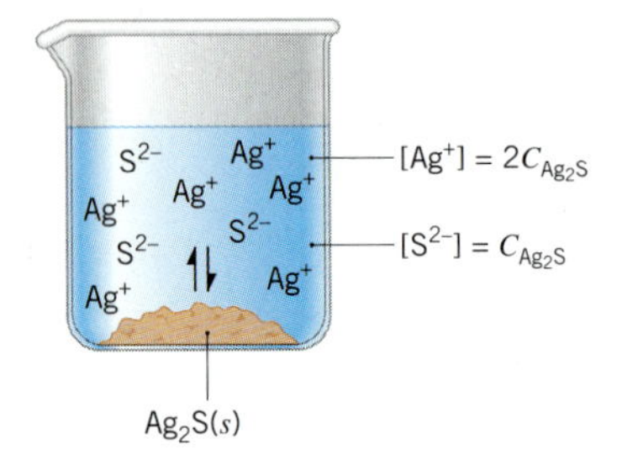


**SOLUTION** Silver sulfide is another example of a 2:1 salt, for which the following solubility product equilibrium expression can be written:

$$K_{sp} = [Ag^+]^2[S^{2-}]$$

The concentration of the $S^{2-}$ ion in a saturated solution is equal to the solubility of this salt. The equilibrium concentration of the $Ag^+$ ion is twice as large:

$$[S^{2-}] = C_s \qquad [Ag^+] = 2\,C_s$$

Substituting this information into the $K_{sp}$ expression for $Ag_2S$ gives the following equation:

$$[2\,C_s]^2[C_s] = 6.3 \times 10^{-50}$$
$$4\,C_s^3 = 6.3 \times 10^{-50}$$

This equation can be solved for the solubility of $Ag_2S$ in units of moles per liter:

$$C_s = 2.5 \times 10^{-17}\,M$$

The number of grams of $Ag_2S$ that dissolve in a liter of water can be then calculated from the solubility in moles per liter and the mass of a mole of $Ag_2S$:

$$\frac{2.5 \times 10^{-17}\ \text{mol Ag}_2\text{S}}{1\ \text{L}} \times \frac{247.8\ \text{g Ag}_2\text{S}}{1\ \text{mol Ag}_2\text{S}} = 6.2 \times 10^{-15}\ \text{g Ag}_2\text{S/L}$$

This calculation suggests that 6.2 femtograms of $Ag_2S$ dissolve in a liter of water. This is $10^{10}$ times smaller than the smallest sample that can be weighed on an analytical balance. Thus, it is not surprising that this salt is described as ''insoluble'' in water.

## 18.6 COMMON MISCONCEPTIONS ABOUT SOLUBILITY PRODUCT CALCULATIONS

Let's focus on one step in the preceding exercise. We started with the solubility product expression for $Ag_2S$:

$$K_{sp} = [Ag^+]^2[S^{2-}]$$

We then substituted the relationship between the concentrations of these ions and the solubility of the salt into this equation:

$$[2\,C_s]^2[C_s] = 6.3 \times 10^{-50}$$

When they see this for the first time, students often ask: ''Why did you double the $Ag^+$ ion concentration and then square it? Aren't you counting this term twice?''

This question results from confusion about the symbols used in the calculation. Remember that the symbol $C_s$ in this equation stands for the solubility of $Ag_2S$ in moles per liter. Since we get two $Ag^+$ ions for each $Ag_2S$ formula unit that dissolves in water, the $Ag^+$ ion concentration at equilibrium is twice the solubility of the salt, or 2 $C_s$. We square the $Ag^+$ ion concentration term because the equilibrium constant expression for this reaction is proportional to the product of the concentrations of the three products of the reaction:

$$K_{sp} = [Ag^+][Ag^+][S^{2-}]$$

It is just more convenient to write this equation in the condensed form:

$$K_{sp} = [Ag^+]^2[S^{2-}]$$

Another common stumbling block in solubility product calculations occurs when writing an equation to describe the relationship between the concentrations of the $Ag^+$ and $S^{2-}$ ions in a saturated $Ag_2S$ solution. It is all too easy to look at the formula for this compound, $Ag_2S$, and then write the following equation:

$$[S^{2-}] = 2\,[Ag^+]$$

This seems reasonable to some, who argue that there are twice as many $Ag^+$ ions as $S^{2-}$ ions in the compound. But the equation is wrong. Because two $Ag^+$ ions are produced for each $S^{2-}$ ion, there are twice as many silver ions as sulfide ions in this solution. This solution is correctly described by the following equation:

$$[Ag^+] = 2\,[S^{2-}]$$

How can you avoid making this mistake? After you write the equation that you think describes the relationship between the concentrations of the ions, try it to see if it works. Suppose just enough $Ag_2S$ dissolved in water to give two $S^{2-}$ ions. How many $Ag^+$ ions would you get? Four. If you get the right answer when you substitute this concrete example into your equation, it must be written correctly.

## 18.7 USING $K_{SP}$ AS A MEASURE OF THE SOLUBILITY OF A SALT

The value of $K_a$ for an acid is proportional to the strength of the acid:

$$K_a = \frac{[H_3O^+][A^-]}{[HA]}$$

If we find the following $K_a$ values in a table, we can immediately conclude that formic acid is a stronger acid than acetic acid:

Formic acid ($HCO_2H$): $K_a = 1.8 \times 10^{-4}$
Acetic acid ($CH_3CO_2H$): $K_a = 1.8 \times 10^{-5}$

The same can be said about values of $K_b$:

$$K_b = \frac{[HB^+][OH^-]}{[B]}$$

The following base-ionization equilibrium constants imply that methylamine is a stronger base than ammonia:

Methylamine ($CH_3NH_2$): $K_b = 4.8 \times 10^{-4}$
Ammonia ($NH_3$): $K_b = 1.8 \times 10^{-5}$

Unfortunately, there is no simple way to predict the relative solubilities of salts from their $K_{sp}$'s if the salts produce different numbers of positive and negative ions when they dissolve in water.

### EXERCISE 18.5

Which salt—$CaCO_3$ or $Ag_2CO_3$—is more soluble in water in units of moles per liter? ($CaCO_3$: $K_{sp} = 2.8 \times 10^{-9}$; $Ag_2CO_3$: $K_{sp} = 8.1 \times 10^{-12}$)

**SOLUTION** We might expect $CaCO_3$ to be more soluble than $Ag_2CO_3$ because it has a larger $K_{sp}$. The only way to test this prediction is to calculate the solubilities of both compounds.

Stalactites are calcium carbonate deposits that hang from the top of limestone caverns. Stalagmites are deposits that build up from the bottom of the cavern.

The solubility product expression for $CaCO_3$ has the following form:

$$K_{sp} = [Ca^{2+}][CO_3^{2-}]$$

The concentrations of the $Ca^{2+}$ and $CO_3^{2-}$ ions in a saturated solution of this salt are both equal to the solubility of the salt: $C_s$:

$$[C_s][C_s] = 2.8 \times 10^{-9}$$

Taking the square root of both sides of this equation gives the solubility of this salt:

$$C_s = \mathbf{5.3 \times 10^{-5}}\ M$$

$Ag_2CO_3$ is a 2:1 salt, for which the following solubility product expression is written:

$$K_{sp} = [Ag^+]^2[CO_3^{2-}]$$

The $CO_3^{2-}$ ion concentration is equal to the solubility of the salt, but the $Ag^+$ ion concentration is twice as large:

$$[CO_3^{2-}] = C_s$$
$$[Ag^+] = 2C_s$$

Substituting this information into the $K_{sp}$ expression gives the following results:

$$[2C_s]^2[C_s] = 8.1 \times 10^{-12}$$
$$4C_s^3 = 8.1 \times 10^{-12}$$

This equation can be solved for the solubility of $Ag_2CO_3$:

$$C_s = \mathbf{1.3 \times 10^{-4}}\ M$$

In spite of the fact that $K_{sp}$ for $CaCO_3$ is larger than $K_{sp}$ for $Ag_2CO_3$, $CaCO_3$ is less soluble than $Ag_2CO_3$.

$$Ag_2CO_3: \quad C_s = \mathbf{0.00013}\ M$$
$$CaCO_3: \quad C_s = \mathbf{0.000053}\ M$$

## 18.8 THE ROLE OF THE ION PRODUCT ($Q_{SP}$) IN SOLUBILITY CALCULATIONS

Consider a saturated solution of AgCl in water:

$$AgCl(s) \overset{H_2O}{\rightleftharpoons} Ag^+(aq) + Cl^-(aq)$$

Because AgCl is a 1:1 salt, the concentrations of the $Ag^+$ and $Cl^-$ ions in this solution are equal.

**Saturated solution of AgCl in water: $[Ag^+] = [Cl^-]$**

Imagine what happens when a few crystals of solid $AgNO_3$ are added to this saturated solution of AgCl in water. According to the rules in Table 18.1, silver nitrate is a soluble salt. It therefore dissolves and dissociates into $Ag^+$ and $NO_3^-$ ions. As a result, there are two sources of the $Ag^+$ ion in this solution:

$$AgNO_3(s) \longrightarrow \mathbf{Ag^+(aq)} + NO_3^-(aq)$$

$$AgCl(s) \overset{H_2O}{\rightleftharpoons} \mathbf{Ag^+(aq)} + Cl^-(aq)$$

Adding $AgNO_3$ to a saturated AgCl solution therefore increases the $Ag^+$ ion concentration. When this happens, the solution is no longer at equilibrium because the product of the concentrations of the $Ag^+$ and $Cl^-$ ions is too large. In more formal terms, we can argue that the **ion product** ($Q_{sp}$) for the solution is larger than the solubility product ($K_{sp}$) for AgCl:

$$Q_{sp} = (Ag^+)(Cl^-) > K_{sp}$$

The ion product is literally the product of the concentrations of the ions at any moment in time. When it is equal to the solubility product for the salt, the system is at equilibrium.

The reaction eventually comes back to equilibrium after the excess ions precipitate from solution as solid AgCl. When equilibrium is reestablished, however, the concentrations of the $Ag^+$ and $Cl^-$ ions won't be the same. Because there are two sources of the $Ag^+$ ion in this solution, there will be more $Ag^+$ ion at equilibrium than $Cl^-$ ion:

**Saturated solution of AgCl to which $AgNO_3$ has been added: $[Ag^+] > [Cl^-]$**

Now imagine what happens when a few crystals of NaCl are added to a saturated solution of AgCl in water. There are two sources of the chloride ion in this solution:

$$NaCl(s) \xrightarrow{H_2O} Na^+(aq) + Cl^-(aq)$$

$$AgCl(s) \overset{H_2O}{\rightleftharpoons} Ag^+(aq) + Cl^-(aq)$$

Once again, the ion product is larger than the solubility product:

$$Q_{sp} = (Ag^+)(Cl^-) > K_{sp}$$

This time, when the reaction comes back to equilibrium, there will be more $Cl^-$ ion in the solution than $Ag^+$ ion:

**Saturated solution of AgCl to which NaCl has been added: $[Ag^+] < [Cl^-]$**

Figure 18.4 shows a small portion of the possible combinations of the $Ag^+$ and $Cl^-$ ion concentrations in an aqueous solution. Any point along the curved line in this graph corresponds to a system at equilibrium because the product of the $Ag^+$ and $Cl^-$ ion concentrations for these solutions is equal to $K_{sp}$ for AgCl.

Point *A* represents a solution at equilibrium that could be produced by dissolving two sources of the $Ag^+$ ion—such as $AgNO_3$ and AgCl—in water. Point *B* represents a saturated solution of AgCl in pure water, in which the $[Ag^+]$ and $[Cl^-]$ terms are equal. Point *C* describes a solution at equilibrium that was prepared by dissolving two sources of the $Cl^-$ ion in water, such as NaCl and AgCl.

Any point that is not along the solid line in Figure 18.4 represents a solution that is not at equilibrium. Any point *below* the solid line (such as point *D*) represents a solution for which the ion product is smaller than the solubility product:

$$\text{Point } D: \quad Q_{sp} < K_{sp}$$

If more AgCl were added to the solution at Point *D*, it would dissolve:

$$\text{If } Q_{sp} < K_{sp}: \quad AgCl(s) \longrightarrow Ag^+(aq) + Cl^-(aq)$$

This photograph shows tiny crystals of $PbI_2$ settling out of a saturated solution that has been shaken.

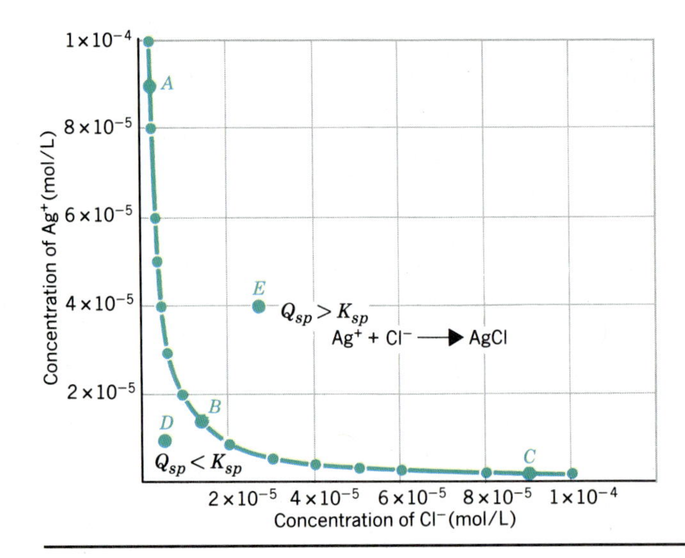


**Figure 18.4** The solid line in this graph is called the saturation curve for AgCl. Every point along this line represents a solution at equilibrium. Any point below the saturation curve represents a solution for which $Q_{sp}$ is smaller than $K_{sp}$. Any point above the saturation curve describes a solution for which $Q_{sp}$ is larger than $K_{sp}$; AgCl will have to precipitate from this solution before it can reach equilibrium.

Points *above* the solid line (such as point *E*) represent solutions for which the ion product is larger than the solubility product:

$$\text{Point } E: \qquad Q_{sp} > K_{sp}$$

The solution described by point *E* will eventually come to equilibrium after enough solid AgCl has precipitated:

$$\text{If } Q_{sp} > K_{sp}: \qquad Ag^+(aq) + Cl^-(aq) \longrightarrow AgCl(s)$$

## 18.9 THE COMMON-ION EFFECT

When $AgNO_3$ is added to a saturated solution of AgCl, it is often described as a source of a common ion, the $Ag^+$ ion. By definition, a *common ion* is an ion that enters the solution from two different sources. Solutions to which both NaCl and AgCl have been added also contain a common ion; in this case, the $Cl^-$ ion. This section focuses on the effect of common ions on solubility product equilibria.

### EXERCISE 18.6

Calculate the solubility of AgCl in pure water. (AgCl: $K_{sp} = 1.8 \times 10^{-10}$)

**SOLUTION** The solubility product expression for AgCl would be written as follows:

$$K_{sp} = [Ag^+][Cl^-] = 1.8 \times 10^{-10}$$

Because there is only one source of the $Ag^+$ and $Cl^-$ ions in this solution, the concentrations of these ions at equilibrium must be the same. Furthermore, they are both equal to the solubility of AgCl in units of moles per liter, $C_s$:

$$[Ag^+] = [Cl^-] = C_s$$

Substituting this information into the solubility product expressions leads to the conclusion that the solubility of AgCl is equal to the square root of $K_{sp}$ for this salt:

$$[C_s][C_s] = 1.8 \times 10^{-10}$$
$$[C_s] = 1.3 \times 10^{-5}\ M$$

The **common-ion effect** can be understood by considering the following question: What happens to the solubility of AgCl when we dissolve this salt in a solution that is already 0.10 $M$ NaCl? As a rule, we can assume that salts dissociate into their ions when they dissolve. A 0.10 $M$ NaCl solution therefore contains 0.10 mol of the $Cl^-$ ion per liter of solution. Because the $Cl^-$ ion is one of the products of the solubility equilibrium, Le Châtelier's principle leads us to expect that AgCl will be less soluble in an 0.10 $M$ $Cl^-$ solution than it is in pure water.

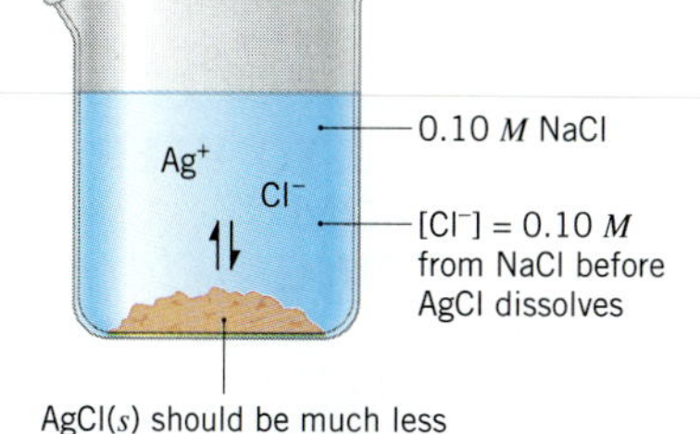

AgCl($s$) should be much less soluble than in pure $H_2O$

## EXERCISE 18.7

Calculate the solubility of AgCl in 0.10 $M$ NaCl. (AgCl: $K_{sp} = 1.8 \times 10^{-10}$)

**SOLUTION** The $Ag^+$ and $Cl^-$ ion concentrations at equilibrium will no longer be the same because there are two sources of the $Cl^-$ ion in this solution: AgCl and NaCl.

$$[Ag^+] \neq [Cl^-]$$

The best way to set up this problem is the format used in Chapters 16 and 17. Initially, there is no $Ag^+$ ion in the solution, but the $Cl^-$ ion concentration is 0.10 $M$. As the reaction comes to equilibrium, some of the AgCl will dissolve and the concentrations of both the $Ag^+$ and $Cl^-$ ions will increase. Both concentrations will increase by an amount equal to the solubility of AgCl in this solution, $C_s$:

| | $AgCl(s) \rightleftharpoons$ | $Ag^+(aq)$ + | $Cl^-(aq)$ | $K_{sp} = 1.8 \times 10^{-10}$ |
|---|---|---|---|---|
| Initial: | | 0 | 0.10 $M$ | |
| Equilibrium: | | $C_s$ | $0.10 + C_s$ | |

We now write the solubility product expression for this reaction:

$$K_{sp} = [Ag^+][Cl^-]$$

We then substitute what we know about the equilibrium concentrations of the $Ag^+$ and $Cl^-$ ions into this equation:

$$[C_s][0.10 + C_s] = 1.8 \times 10^{-10}$$

We could expand the equation and solve it with the quadratic formula, but that would involve a lot of work. Let's see if we can find an assumption that makes the calculation easier.

What do we know about $C_s$? In pure water, the solubility of AgCl is 0.000013 $M$. In this solution, we expect it to be even smaller. It therefore seems reasonable to expect that $C_s$ should be small compared with the initial concentration of the $Cl^-$ ion:

$$[C_s][0.10] \approx 1.8 \times 10^{-10}$$

Solving this approximate equation gives the following result:

$$C_s \approx 1.8 \times 10^{-9}\ M$$

Note that the assumption used to generate the approximate equation is valid. (The $Cl^-$ ion concentration from the dissociation of AgCl is about 50 million times smaller than the initial $Cl^-$ ion concentration.) This assumption works very well with common-ion problems involving insoluble salts because the $K_{sp}$ values for these salts are so small.

Let's compare the results of Exercises 18.6 and 18.7:

In pure water: $C_s = 1.3 \times 10^{-5}\ M$
In 0.10 $M$ NaCl: $C_s = 1.8 \times 10^{-9}\ M$

These calculations show how the common-ion effect can be used to make an "insoluble" salt even less soluble in water.

## 18.10 HOW TO KEEP A SALT FROM PRECIPITATING

The common-ion effect also can be used to prevent a salt from precipitating from solution. Instead of adding a source of a common ion, we add a reagent that removes the common ion from solution.

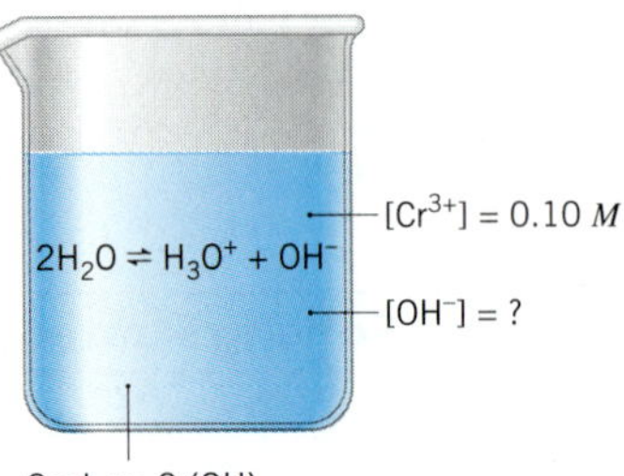


### EXERCISE 18.8

Calculate the pH at which $Cr(OH)_3$ just starts to precipitate from an 0.10 $M$ $Cr^{3+}$ solution. Use the results of this calculation to explain why it is impossible to prepare a 0.10 $M$ $Cr^{3+}$ solution at neutral pH. [$Cr(OH)_3$: $K_{sp} = 6.3 \times 10^{-31}$]

**SOLUTION** Chromium(III) hydroxide starts to precipitate from this solution when the ion product is equal to the solubility product. We can therefore begin by writing the $K_{sp}$ expression for this salt:

$$K_{sp} = [Cr^{3+}][OH^-]^3 = 6.3 \times 10^{-31}$$

We then substitute the desired concentration of the $Cr^{3+}$ ion into this equation and solve for the $OH^-$ ion concentration at which $Cr(OH)_3$ just starts to precipitate:

$$[0.10][OH^-]^3 = 6.3 \times 10^{-31}$$
$$[OH^-] = 1.8 \times 10^{-10}\ M$$

We then calculate the pOH of the solution and use this information to calculate the pH:

$$\text{pOH} = -\log [OH^-] = 9.74$$
$$\text{pH} = 14 - \text{pOH} = 4.26$$

According to this calculation, the $OH^-$ ion concentration is too small for $Cr(OH)_3$ to precipitate from a 0.10 $M$ $Cr^{3+}$ solution if the pH of the solution is kept *below* 4.26.

In this exercise, the common ion is the $OH^-$ ion, which is involved in the following equilibria:

$$Cr(OH)_3(s) \xrightleftharpoons{H_2O} Cr^{3+}(aq) + 3\ OH^-(aq)$$
$$2\ H_2O(l) \rightleftharpoons H_3O^+(aq) + OH^-(aq)$$

We can keep $Cr(OH)_3$ from precipitating by adding enough acid to keep the pH below 4.26. $Cr(OH)_3$ would therefore precipitate long before the $OH^-$ ion concentration could be raised to the point at which the solution is neutral (pH ≈ 7).

Electrolysis was used to chrome plate parts of this Harley.

The results of similar calculations for a number of different $Cr^{3+}$ ion concentrations are given in Table 18.2. Only a small fraction of these data will fit on a normal graph, such as Figure 18.5. We can overcome this problem by plotting the log of the $Cr^{3+}$ ion concentrations versus the log of the $OH^-$ ion concentrations. The log $[Cr^{3+}]$ term for the data in Table 18.2 ranges from 0 to −7, and the log $[OH^-]$ term ranges from −7.7 to −10.1. Another way to overcome the problem involves plotting the $Cr^{3+}$ and $OH^-$ concentrations on log–log graph paper. Figure 18.6 shows a log–log plot of the data in Table 18.2.

The points along the solid line in Figure 18.6 represent combinations of $Cr^{3+}$ and $OH^-$ ion concentrations at which the solution is saturated, and therefore at equilib-

**TABLE 18.2 The Inverse Relationship Between the Equilibrium Concentrations of the $Cr^{3+}$ and $OH^-$ Ions**

| $[Cr^{3+}]$ (mol/L) | $[OH^-]$ (mol/L) |
|---|---|
| 1.0 | $8.6 \times 10^{-11}$ |
| $1 \times 10^{-1}$ | $1.8 \times 10^{-10}$ |
| $1 \times 10^{-2}$ | $4.0 \times 10^{-10}$ |
| $1 \times 10^{-3}$ | $8.6 \times 10^{-10}$ |
| $1 \times 10^{-4}$ | $1.8 \times 10^{-9}$ |
| $1 \times 10^{-5}$ | $4.0 \times 10^{-9}$ |
| $1 \times 10^{-6}$ | $8.6 \times 10^{-9}$ |
| $1 \times 10^{-7}$ | $1.8 \times 10^{-8}$ |

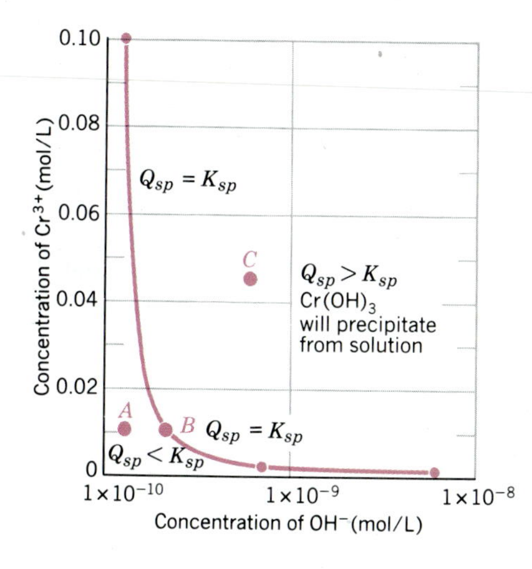


**Figure 18.5** The saturation curve for $Cr(OH)_3$ over a narrow range of $Cr^{3+}$ and $OH^-$ concentrations. The solid line describes the combinations of $Cr^{3+}$ and $OH^-$ concentrations for which the solution is at equilibrium.

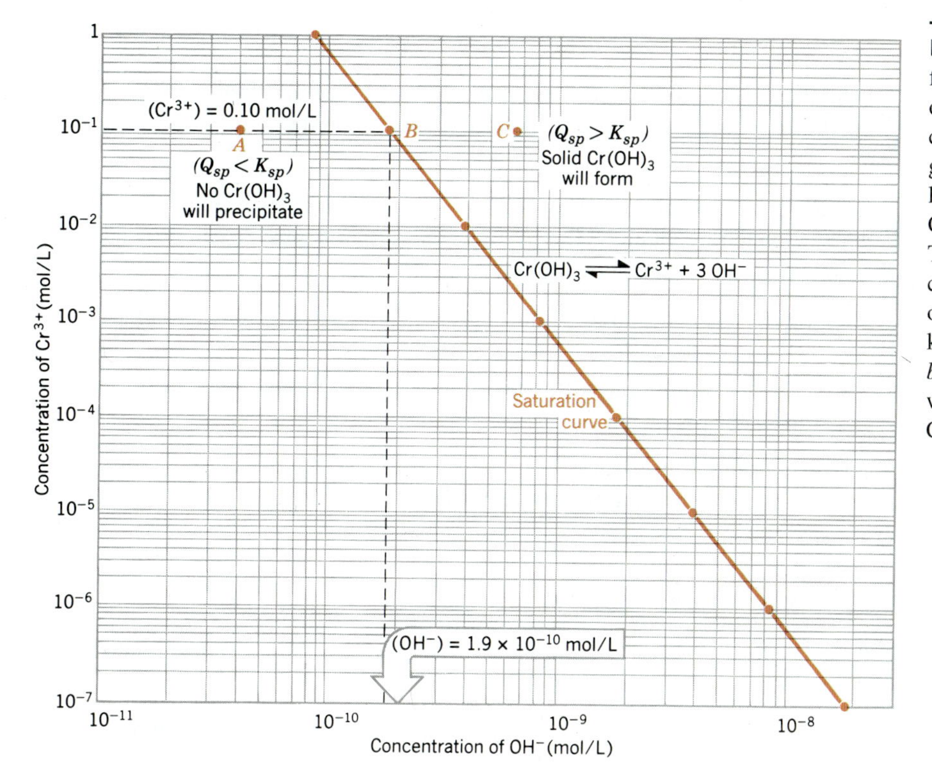


**Figure 18.6** The saturation curve for $Cr(OH)_3$ over the entire range of concentrations in Table 18.2 can be plotted by using log–log graph paper. The horizontal dashed line at the upper left represents a $Cr^{3+}$ ion concentration of 0.10 *M*. This line intersects the saturation curve at an $OH^-$ ion concentration of $1.9 \times 10^{-10}$ *M*. As long as we keep the $OH^-$ ion concentration *below* $1.9 \times 10^{-10}$ *M*, no $Cr(OH)_3$ will precipitate from a 0.10 *M* $Cr^{3+}$ ion solution.

rium. Point *A* represents a solution in which the $OH^-$ ion concentration is too small for $Cr(OH)_3$ to precipitate when the $Cr^{3+}$ ion concentration is 0.10 *M*. Point *B* corresponds to the set of conditions calculated in Exercise 18.9. Point *C* describes a pair of $Cr^{3+}$ and $OH^-$ ion concentrations for which the ion product is too large. $Cr(OH)_3$ would precipitate from any solution that momentarily contained the $Cr^{3+}$ and $OH^-$ ion concentrations that correspond to point *C*.

## 18.11 HOW TO SEPARATE IONS BY SELECTIVE PRECIPITATION

The last few sections have shown how the concentration of an ion can be controlled to either prevent a solid from dissolving or to keep it in solution. The technique known as **selective precipitation** combines these processes to separate a mixture of two or more ions. It involves adding a reagent that selectively brings one of the ions out of solution as a precipitate while leaving the other ions in the solution.

Some mixtures can be separated on the basis of the solubility rules outlined in Table 18.1. We can separate the $Ag^+$ ion from a solution that contains the $Cu^{2+}$ ion, for example, by adding a source of the $Cl^-$ ion. Silver chloride is an insoluble salt ($K_{sp} = 1.8 \times 10^{-10}$), which will precipitate from solution, whereas copper(II) chloride is soluble in water.

Selective precipitation becomes more of a challenge when the ions to be separated form salts with similar solubilities. $Mn^{2+}$ and $Ni^{2+}$ ions, for example, both form insoluble sulfides:

$$\text{MnS:} \quad K_{sp} = 3 \times 10^{-13}$$

$$\text{NiS:} \quad K_{sp} = 3.2 \times 10^{-19}$$

The $10^6$-fold difference in the values of $K_{sp}$ for these salts is large enough, however, to allow us to precipitate the less soluble NiS salt selectively from a mixture of these ions, without precipitating MnS as well.

### EXERCISE 18.9

Describe the conditions under which $Ni^{2+}$ ions can be quantitatively precipitated as NiS from a solution that is 0.10 *M* in the $Ni^{2+}$ and $Mn^{2+}$ ions, while the $Mn^{2+}$ ions are left in solution.

**SOLUTION** We can precipitate $Ni^{2+}$ from solution as NiS by adding a source of the $S^{2-}$ ion to this solution. The $S^{2-}$ ion concentration must be carefully adjusted, however, to meet the following criteria:

1. The $S^{2-}$ ion concentration must be large enough to precipitate as much of the $Ni^{2+}$ as possible.
2. The $S^{2-}$ ion concentration must be small enough that no MnS precipitates from solution.

The second criterion in Exercise 18.9 is easier to test than the first. It is relatively easy to determine when the $S^{2-}$ ion concentration is too large—and MnS starts to precipitate.

### EXERCISE 18.10

Calculate the $S^{2-}$ ion concentration at which MnS will begin to precipitate from a solution that is 0.10 *M* in $Mn^{2+}$ ions. (MnS: $K_{sp} = 3 \times 10^{-13}$)

**SOLUTION** We start with the solubility product expression for MnS:

$$K_{sp} = [Mn^{2+}][S^{2-}]$$

We then substitute what we know about the $Mn^{2+}$ ion concentration into this equation and calculate the $S^{2-}$ ion concentration at which MnS just starts to precipitate:

$$[0.10][S^{2-}] = 3 \times 10^{-13}$$
$$[S^{2-}] = \mathbf{3 \times 10^{-12}} \, M$$

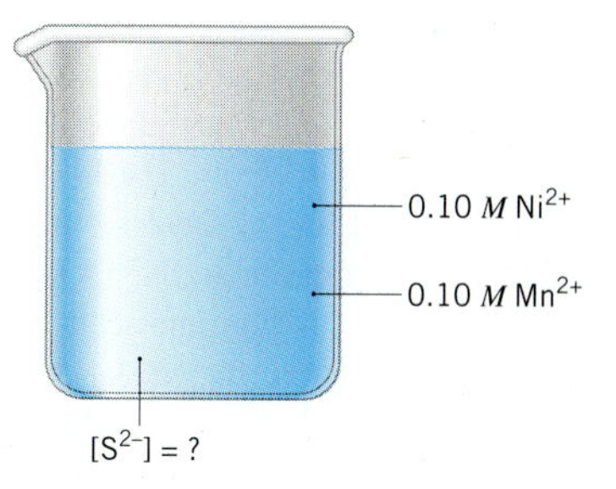


Goal:
Precipitate as much $Ni^{2+}$ as possible without precipitating any $Mn^{2+}$

We can therefore keep $Mn^{2+}$ ions from precipitating from a 0.10 *M* solution if we can keep the $S^{2-}$ ion concentration smaller than $3 \times 10^{-12}$ *M*. Let's see if this $S^{2-}$ ion concentration is large enough to effectively remove $Ni^{2+}$ ions from the mixture.

## EXERCISE 18.11

Calculate the $Ni^{2+}$ ion concentration in a solution to which enough $S^{2-}$ ion has been added to raise the concentration of this ion to $3 \times 10^{-12}$ *M*. (NiS: $K_{sp} = 3.2 \times 10^{-19}$)

**SOLUTION** If we control the rate at which $S^{2-}$ ion is added to the solution, all of this ion initially combines with $Ni^{2+}$ ions to form NiS, which precipitates from the solution. Eventually, the $Ni^{2+}$ ion concentration becomes so small that $S^{2-}$ ion starts to accumulate in the solution. At what point does the $Ni^{2+}$ ion concentration become small enough that the $S^{2-}$ ion concentration at equilibrium is $3 \times 10^{-12}$ *M*?

We start with the solubility product expression for NiS:

$$K_{sp} = [Ni^{2+}][S^{2-}]$$

We then substitute the known $S^{2-}$ ion concentration into this equation and solve for the $Ni^{2+}$ ion concentration at equilibrium:

$$[Ni^{2+}][3 \times 10^{-12}] = 3.2 \times 10^{-19}$$
$$[Ni^{2+}] = \mathbf{1 \times 10^{-7}} \, M$$

In other words, if $S^{2-}$ ion is added very slowly, the $S^{2-}$ concentration at equilibrium can't get as large as $3 \times 10^{-12}$ *M* until the $Ni^{2+}$ concentration has been reduced by a factor of $10^6$, from 0.10 *M* to $1 \times 10^{-7}$ *M*.

The $10^6$-fold difference between the solubility products for MnS and NiS is large enough to allow us to separate the $Mn^{2+}$ and $Ni^{2+}$ ions in a mixture that is initially 0.10 *M* in both ions. Would this technique work equally well if the task involved separating the $Mn^{2+}$ and $Ni^{2+}$ ions when both concentrations were 0.010 *M*? Or when the $Ni^{2+}$ ion was present at a 0.10 *M* concentration, but the $Mn^{2+}$ concentration was only 0.0010 *M*? Instead of repeating the calculations in Exercises 18.10 and 18.11 for each set of initial concentrations, we can construct a graph that allows us to answer this question for almost any combination of $Mn^{2+}$ and $Ni^{2+}$ ion concentrations.

Figure 18.7 shows the saturation curves for NiS and MnS plotted on the same piece of log–log graph paper. The solid line at the left describes pairs of $Ni^{2+}$ and $S^{2-}$ ion concentrations at which NiS is in equilibrium with these ions. The solid line at the right does the same for the equilibrium between MnS and the $Mn^{2+}$ and $S^{2-}$ ions. The horizontal line at the bottom of the graph represents an $Ni^{2+}$ ion concentration of $1 \times 10^{-6}$ *M*. This line intersects the saturation curve for NiS at an $S^{2-}$ ion concentration of $3.2 \times 10^{-13}$ *M*. The vertical line that represents this $S^{2-}$ ion concentration doesn't intersect the MnS saturation curve until the $Mn^{2+}$ ion concentration is 1 *M*. This graph therefore tells us that as long as the initial concentrations of the $Ni^{2+}$ and $Mn^{2+}$ ions are less than 1 *M*, we can reduce the $Ni^{2+}$ ion concentration to $1 \times 10^{-6}$ *M* by adding $S^{2-}$ ions without precipitating MnS.

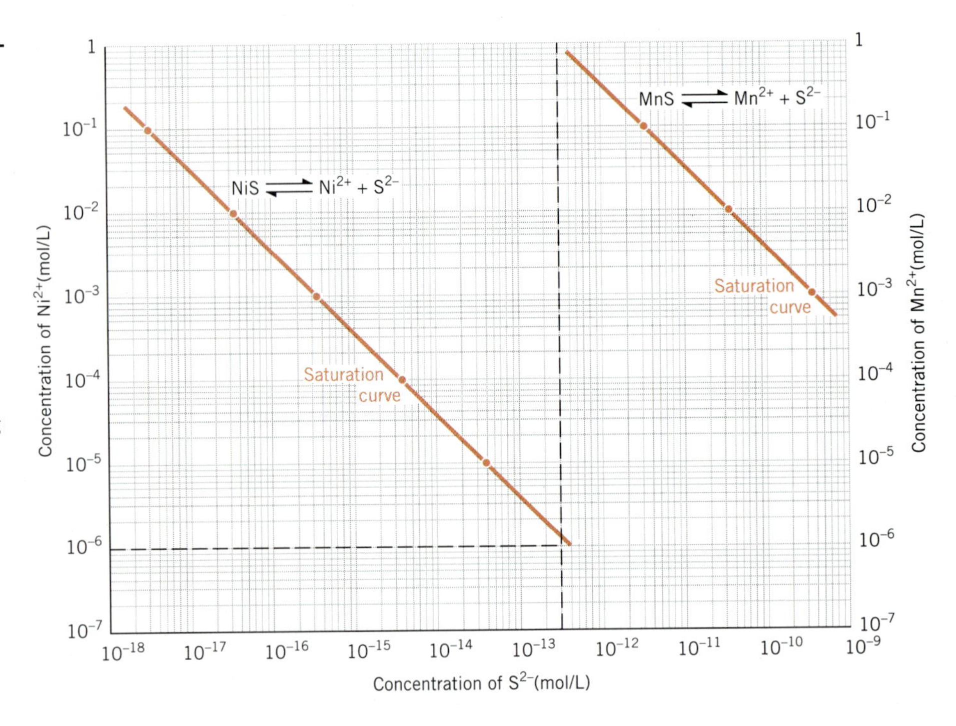


**Figure 18.7** The saturation curves for NiS and MnS plotted on log–log graph paper. The horizontal, dashed line at the bottom of this figure represents $[Ni^{2+}] = 1 \times 10^{-6}\ M$. This line intersects the saturation curve for NiS at $[S^{2-}] = 3.2 \times 10^{-13}\ M$. A vertical line corresponding to this $S^{2-}$ ion concentration intersects the MnS saturation curve at $[Mn^{2+}] = 1\ M$. We can therefore precipitate NiS from solution without precipitating MnS as long as $[Ni^{2+}] > 1 \times 10^{-6}\ M$ and $[Mn^{2+}] < 1\ M$.

## 18.12 HOW TO ADJUST THE CONCENTRATION OF AN ION

The preceding section leaves an important question unanswered: How do we adjust the $S^{2-}$ ion concentration in a solution so that it approaches but does not exceed $3 \times 10^{-12}\ M$?

The foundation for controlling the concentration of ions that are conjugate bases of weak acids was laid in Chapter 17. We start with an aqueous solution of the weak acid—in this case, hydrogen sulfide dissolved in water:

$$H_2S(aq) + H_2O(l) \rightleftharpoons H_3O^+(aq) + HS^-(aq) \qquad K_{a1} = 1.0 \times 10^{-7}$$
$$HS^-(aq) + H_2O(l) \rightleftharpoons H_3O^+(aq) + S^{2-}(aq) \qquad K_{a2} = 1.3 \times 10^{-13}$$

Sec. 17.17

At room temperature, hydrogen sulfide is a gas that is only marginally soluble in water. (A saturated solution of this gas has a concentration of about 0.10 $M$ at room temperature.) Using the techniques introduced in Section 17.17, we can calculate the $S^{2-}$ ion concentration in a 0.10 $M$ $H_2S$ solution.

$$[S^{2-}] = K_{a2} = 1.3 \times 10^{-13}\ M$$

The $S^{2-}$ ion concentration in a saturated solution of $H_2S$ is therefore close to $3 \times 10^{-12}\ M$. How could we bring it even closer?

Le Châtelier's principle suggests that we should be able to either increase or decrease the dissociation of this acid by adjusting the pH of the solution. If we add a strong acid, the amount of $S^{2-}$ ion should decrease as the acid-dissociation equilibria are driven toward the left. If we add a strong base, the $S^{2-}$ ion concentration should increase as the equilibria are pulled to the right.

$H_2S + H_2O \rightleftharpoons HS^- + H_3O^+$ — 0.10 $M$ $H_2S$

$HS^- + H_2O \rightleftharpoons H_3O^+ + S^{2-}$

Question:
What is the effect on $[S^{2-}]$ of adding a strong acid or a strong base to this solution?

In pure water, $H_2S$ dissociates by losing one proton at a time. But when we add either a strong acid or a strong base to this solution, the equilibria shift so much in one direction or the other that we can treat $H_2S$ as if it dissociates by losing two protons in a single step:

$$H_2S(aq) + 2\ H_2O(l) \rightleftharpoons 2\ H_3O^+(aq) + S^{2-}(aq)$$

The equilibrium constant expression for the overall reaction is equal to the product of the equilibrium constant expressions for the individual steps:

$$\frac{[H_3O^+][\cancel{HS^-}]}{[H_2S]} \times \frac{[H_3O^+][S^{2-}]}{[\cancel{HS^-}]} = \frac{[H_3O^+]^2[S^{2-}]}{[H_2S]}$$

The equilibrium constant for the overall reaction is therefore equal to the product of $K_{a1}$ times $K_{a2}$:

$$\frac{[H_3O^+]^2[S^{2-}]}{[H_2S]} = K_{a1} \times K_{a2} = \mathbf{1.3 \times 10^{-20}}$$

A saturated solution of $H_2S$ has an initial concentration of 0.10 $M$. Because $H_2S$ is a weak acid, we can assume that the concentration of this acid at equilibrium is approximately equal to its initial concentration.

$$\frac{[H_3O^+]^2[S^{2-}]}{[0.10]} \approx 1.3 \times 10^{-20}$$

We can then solve this approximate equation for the concentration of the $H_3O^+$ ion:

$$[H_3O^+]^2 \approx \frac{1.3 \times 10^{-20} \times [0.10]}{[S^{2-}]}$$

$$[H_3O^+] \approx \sqrt{\frac{1.3 \times 10^{-21}}{[S^{2-}]}}$$

**EXERCISE 18.12**

Calculate the pH to which a saturated solution of $H_2S$ in water must be adjusted for the $S^{2-}$ ion concentration to be $3 \times 10^{-12}$ $M$.

**SOLUTION** The key to this problem is recognizing the relationship between the $S^{2-}$ and $H_3O^+$ ion concentrations in a saturated $H_2S$ solution to which an acid or a base has been added:

$$[H_3O^+] \approx \sqrt{\frac{1.3 \times 10^{-21}}{[S^{2-}]}}$$

Once we have this equation, we can calculate the $H_3O^+$ ion concentration needed to force the $S^{2-}$ ion concentration to take on a particular value. For example, the $H_3O^+$ ion concentration must be $2 \times 10^{-5}$ $M$ if we want the $S^{2-}$ ion concentration to be equal to $3 \times 10^{-12}$ $M$:

$$[H_3O^+] \approx \sqrt{\frac{1.3 \times 10^{-21}}{3 \times 10^{-12}}} = 2 \times 10^{-5}\ M$$

Once we know the $H_3O^+$ ion concentration needed, we can calculate the pH of this solution:

$$\text{pH} = -\log [H_3O^+] = \mathbf{4.7}$$

Combining the results of this exercise with the calculations in the preceding section leads to the conclusion that we can separate the $Ni^{2+}$ ion from the $Mn^{2+}$ ion by adding a saturated solution of $H_2S$ that has been buffered at a pH of about 4.7. Most of the $Ni^{2+}$ ion will precipitate as NiS, which can be collected by filtration. All of the $Mn^{2+}$ ion will remain in solution.

**RESEARCH IN THE 90s**

## SOLUBILITY EQUILIBRIA

A computer search of the engineering literature produced 140 references to papers published between 1989 and 1992 that included the key word *solubility product.* As might be expected, the research reported used solubility equilibria to either prevent a substance from dissolving or to purify a substance.

Examples of the first category were provided by research that examined the suitability of either clay or concrete as one of the barriers for the long-term storage of radioactive waste. One study, for example, showed that clay can be used to protect against the leaching of $^{14}C$ into groundwater if the clay contains natural buffers that can keep the pH high enough that $CaCO_3$ does not dissolve [R. Dayal and E. J. Reardon, *Waste Management,* **12,** 189 (1992)]. Another found that water that percolates through clay becomes buffered at a sufficiently high pH to ensure that the solubility of waste materials that contain $^{137}Cs$, $^{90}Sr$, $^{60}Co$, and $^{65}Zn$ is negligibly small [U. Bartl and K. A. Czurda, *Applied Clay Science,* **6,** 195 (1991)]. A third paper measured the data necessary to predict whether cement with the empirical formula $[(CaO)_3(Al_2O_3)(H_2O)_6]$ provides a stable environment for $^{125}I$ wastes [M. J. Atkins, D. Macphee, A. Kindness, and F. P. Glasser, *Cement and Concrete Research,* **21,** 991 (1991)].

Examples of the second category can be found in research in the field of biotechnology. As one paper noted, crystallization is one of the most frequently utilized techniques for purification in the pharmaceutical industry [C. J. Orella and D. J. Kirwan, *Biotechnology Progress,* **5,** 89 (1989)]. This paper showed how amino acids can be purified by taking advantage of differences in their solubility in pure water versus water–alcohol mixtures. (The solubility of alanine, for example, decreases by a factor of 1000 as the mole fraction of alcohol in this mixture increases from 0 to about 0.9.) Another paper studied the effect of adding supercritical $CO_2$ to a solution [C. J. Chang, A. D. Randolph, and N. E. Craft, *Biotechnology Progress,* **7,** 275 (1991)]. When $CO_2$ at temperatures slightly above the critical temperature and pressures well above the critical pressure was added to a solution of $\beta$-carotene in an organic solvent, the supercritical fluid acted as an ''antisolvent,'' quantitatively precipitating pure $\beta$-carotene from the solution.

## 18.13 COMPLEX IONS

The basic assumption behind the discussion of solubility equilibria is the idea that salts dissociate into their ions when they dissolve in water. Copper sulfate, for example, dissociates into the $Cu^{2+}$ and $SO_4^{2-}$ ions in water:

$$CuSO_4(s) \xrightleftharpoons{H_2O} Cu^{2+}(aq) + SO_4^{2-}(aq)$$

If we add 2 $M$ $NH_3$ to this solution, the first thing we notice is the formation of a light-blue, almost bluish-white, precipitate. This can be explained by combining what we know about acid–base and solubility equilibria. Ammonia acts as a base toward water to form a mixture of the ammonium and hydroxide ions:

$$NH_3(aq) + H_2O(l) \rightleftharpoons NH_4^{+}(aq) + OH^{-}(aq) \qquad K_b = 1.8 \times 10^{-5}$$

The $OH^-$ ions formed in this reaction combine with $Cu^{2+}$ ions in the solution to form a $Cu(OH)_2$ precipitate:

$$Cu^{2+}(aq) + 2\ OH^-(aq) \rightleftharpoons Cu(OH)_2(s) \qquad K_{sp} = 2.2 \times 10^{-20}$$

In theory, the $OH^-$ ion concentration should increase when more base is added to the solution. As a result, more $Cu(OH)_2$ should precipitate from the solution. At first, this is exactly what happens. In the presence of excess ammonia, however, the $Cu(OH)_2$ precipitate dissolves, and the solution turns deep blue. This raises an important question: ''Why does the $Cu(OH)_2$ precipitate dissolve in excess ammonia?''

The first step toward answering this question involves writing the electron configuration of copper metal and its $Cu^{2+}$ ion:

$$Cu = [Ar]\ 4s^1\ 3d^{10} \qquad Cu^{2+} = [Ar]\ 3d^9$$

It is sometimes useful to think about the electron configuration of the $Cu^{2+}$ ion in terms of the entire set of valence-shell orbitals. In addition to the nine electrons in the 3*d* subshell, this ion has an empty 4*s* orbital and a set of three empty 4*p* orbitals:

$$Cu^{2+} = [Ar]\ 4s^0\ 3d^9\ 4p^0$$

The $Cu^{2+}$ ion can therefore pick up pairs of nonbonding electrons from four $NH_3$ molecules to form covalent Cu—N bonds, as shown in Figure 18.8.

$$Cu^{2+}(aq) + 4\ NH_3(aq) \rightleftharpoons Cu(NH_3)_4{}^{2+}(aq)$$

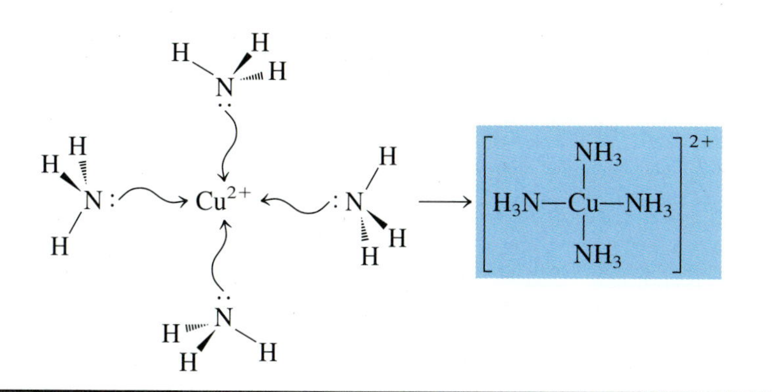


**Figure 18.8** The $Cu^{2+}$ ion has four empty valence-shell orbitals—4*s*, $4p_x$, $4p_y$, and $4p_z$—that can accept pairs of nonbonding electrons from $NH_3$ molecules to form covalent Cu-N bonds.

G. N. Lewis was the first to recognize the similarity between this reaction and the acid–base reaction in which an $H^+$ ion combines with an $OH^-$ ion to form a water molecule. As we discovered in Chapter 11, both reactions involve the transfer of a pair of nonbonding electrons from one atom to an empty orbital on another atom to form a covalent bond. Both reactions can be interpreted therefore in terms of an electron-pair acceptor combining with an electron-pair donor.

p. 430

Lewis suggested that we could expand our definition of acids by assuming that an acid is any substance that acts like the $H^+$ ion to accept a pair of nonbonding electrons. By definition, a Lewis acid is therefore an electron-pair acceptor. A Lewis base, on the other hand, is any substance that acts like the $OH^-$ ion to donate a pair of nonbonding electrons. A Lewis base is therefore an electron-pair donor.

**CHECKPOINT**

Use Lewis structures to predict which of the following can act as an electron-pair donor, or Lewis base.

(a) $H_2O$ (b) CO (c) $O_2$ (d) $H^-$ (e) $BH_3$

The product of the reaction of a Lewis acid with a Lewis base is an **acid–base complex.** When the $Cu^{2+}$ ion reacts with four $NH_3$ molecules, the product of this reaction is called a **complex ion:**

$$\underset{\substack{\text{electron-pair}\\ \text{acceptor}\\ \text{(Lewis acid)}}}{Cu^{2+}(aq)} + \underset{\substack{\text{electron-pair}\\ \text{donor}\\ \text{(Lewis base)}}}{4\,NH_3(aq)} \rightleftharpoons \underset{\substack{\text{acid–base complex}\\ \text{or}\\ \text{complex ion}}}{Cu(NH_3)_4{}^{2+}(aq)}$$

Any atom, ion, or molecule with at least one empty valence-shell orbital can be a Lewis acid. Any atom, ion, or molecule that contains at least one pair of nonbonding electrons is a Lewis base. All of the substances whose Lewis structures are shown in Figure 18.9, for example, can act as Lewis bases to form complex ions.

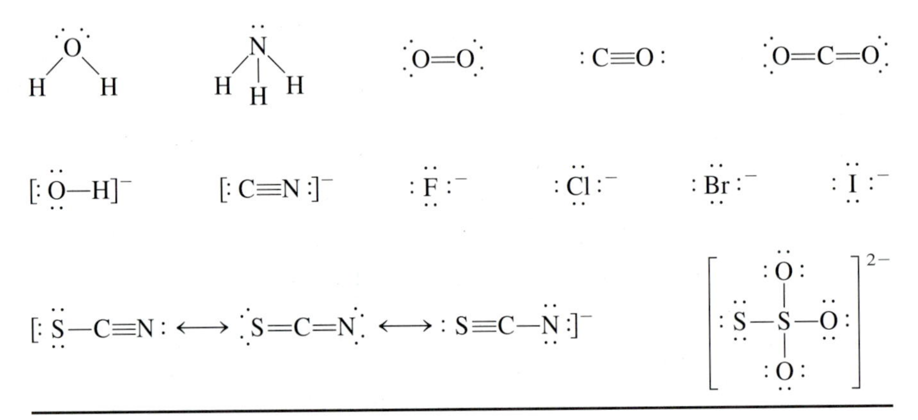

**Figure 18.9** Lewis structures of some potential Lewis bases.

## 18.14 THE STEPWISE FORMATION OF COMPLEX IONS

When a transition-metal ion binds Lewis bases to form a coordination complex, or complex ion, it picks up these ligands one at a time. The $Ag^+$ ion, for example, combines with $NH_3$ in a two-step reaction. It first picks up one $NH_3$ molecule to form a one-coordinate complex:

$$Ag^+(aq) + NH_3(aq) \rightleftharpoons Ag(NH_3)^+(aq)$$

This intermediate then picks up a second $NH_3$ molecule in a separate step:

$$Ag(NH_3)^+(aq) + NH_3(aq) \rightleftharpoons Ag(NH_3)_2{}^+(aq)$$

It is possible to write equilibrium constant expressions for each step in these complex-ion formation reactions. The equilibrium constants for these reactions are known as **complex-formation equilibrium constants, $K_f$.**

The equilibrium constant expressions for the two steps in the formation of the $Ag(NH_3)_2{}^+$ complex ion are written as follows:

$$K_{f1} = \frac{[Ag(NH_3)^+]}{[Ag^+][NH_3]}$$

$$K_{f2} = \frac{[Ag(NH_3)_2{}^+]}{[Ag(NH_3)^+][NH_3]}$$

When polyprotic acids were introduced in Section 17.17, we noted that the difference between the ease with which these acids lose the first and second protons is relatively large:

Sec. 17.17

$$H_2S(aq) + H_2O(l) \rightleftharpoons H_3O^+(aq) + HS^-(aq) \qquad K_{a1} = 1.0 \times 10^{-7}$$

$$HS^-(aq) + H_2O(l) \rightleftharpoons H_3O^+(aq) + S^{2-}(aq) \qquad K_{a2} = 1.3 \times 10^{-13}$$

As a result, calculations for polyprotic acids are based on the assumption of stepwise dissociation.

This assumption is harder to justify for complex formation equilibria. The difference between $K_{f1}$ and $K_{f2}$ for the complexes between $Ag^+$ and ammonia, for example, is only a factor of 4:

$$Ag^+(aq) + NH_3(aq) \rightleftharpoons Ag(NH_3)^+(aq) \qquad K_{f1} = 1.7 \times 10^3$$

$$Ag(NH_3)^+(aq) + NH_3(aq) \rightleftharpoons Ag(NH_3)_2{}^+(aq) \qquad K_{f2} = 6.5 \times 10^3$$

This means that most of the $Ag^+$ ions that pick up one $NH_3$ molecule to form the $Ag(NH_3)^+$ complex ion are likely to pick up another $NH_3$ to form the two-coordinate $Ag(NH_3)_2{}^+$ complex ion. Table 18.3 summarizes the concentrations of the $Ag^+$, $Ag(NH_3)^+$, and $Ag(NH_3)_2{}^+$ ions over a range of $NH_3$ concentrations. These data are plotted in Figure 18.10.

**TABLE 18.3 The Effect of Changes in the $NH_3$ Concentration on the Fraction of Silver Present as the $Ag^+$, $Ag(NH_3)^+$, or $Ag(NH_3)_2{}^+$ Ions**

| $[NH_3]$ ($M$) | $Ag^+$ (%) | $Ag(NH_3)^+$ (%) | $Ag(NH_3)_2{}^+$ (%) |
|---|---|---|---|
| $10^{-6}$ | 99.8 | 0.2 | 0.001 |
| $10^{-5}$ | 98.2 | 1.7 | 0.1 |
| $10^{-4}$ | 78.1 | 13.3 | 8.6 |
| $10^{-3}$ | 7.3 | 12.4 | 80.4 |
| $10^{-2}$ | 0.09 | 1.5 | 98.4 |
| $10^{-1}$ | 0.0009 | 0.2 | 99.8 |

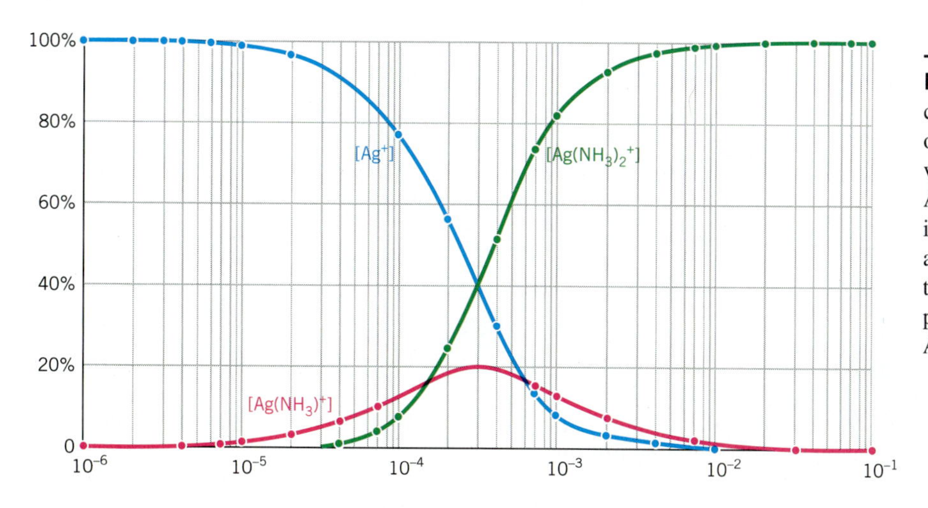


**Figure 18.10** The effect of changes in the $NH_3$ concentration on the proportion of the total silver ion concentration present as $Ag^+$, $Ag(NH_3)^+$, and $Ag(NH_3)_2{}^+$ ions. Even at $NH_3$ concentrations as low as $3.2 \times 10^{-3}$ $M$, more than 95% of the silver ions are present as two-coordinate $Ag(NH_3)_2{}^+$ complex ions.

Essentially all of the silver is present as the $Ag^+$ ion at very low concentrations of $NH_3$. As the $NH_3$ concentration increases, the dominant species soon becomes the two-coordinate $Ag(NH_3)_2{}^+$ ion. Even at $NH_3$ concentrations as small as 0.0010 *M*, most of the silver is present as the $Ag(NH_3)_2{}^+$ ion.

The concentration of the one-coordinate $Ag(NH_3)^+$ intermediate is never very large. Either the $NH_3$ concentration is so small that most of the silver is present as the $Ag^+$ ion or it is large enough that essentially all of the silver is present as the two-coordinate $Ag(NH_3)_2{}^+$ complex ion.

If the only important components of this equilibrium are the free $Ag^+$ ion (at low $NH_3$ concentrations) and the two-coordinate $Ag(NH_3)_2{}^+$ complex ion (at moderate to high $NH_3$ concentrations), we can collapse the individual steps in this reaction into the following overall equation:

$$Ag^+(aq) + 2\,NH_3(aq) \rightleftharpoons Ag(NH_3)_2{}^+(aq)$$

The overall complex formation equilibrium constant expression for this reaction is written as follows:

$$K_f = \frac{[Ag(NH_3)_2{}^+]}{[Ag^+][NH_3]^2}$$

This expression is equal to the product of the equilibrium constant expressions for the individual steps in the reaction:

$$\frac{[Ag(NH_3)_2{}^+]}{[Ag^+][NH_3]^2} = \frac{[\cancel{Ag(NH_3)^+}]}{[Ag^+][NH_3]} \times \frac{[Ag(NH_3)_2{}^+]}{[\cancel{Ag(NH_3)^+}][NH_3]}$$

The overall complex formation equilibrium constant is therefore equal to the product of the $K_f$ values for the individual steps:

$$\boldsymbol{K_f = K_{f1} \times K_{f2} \times K_{f3} \cdots}$$

For the $Ag(NH_3)_2{}^+$ ion, for example:

$$K_f = K_{f1} \times K_{f2} = \mathbf{1.1 \times 10^7}$$

Overall complex formation equilibrium constants for common complex ions can be found in Table A.12 in the appendix.

Dilute solutions of the $Fe^{3+}$ ion (left) are essentially colorless. When the $SCN^-$ ion (right) is added to these solutions, the $Fe(SCN)^{2+}$ and $Fe(SCN)_2{}^+$ complex ions are formed, which gives the solutions a blood-red color.

## EXERCISE 18.13

Calculate the complex formation equilibrium constant for the two-coordinate $Fe(SCN)_2{}^+$ complex ion from the following data:

$$Fe^{3+}(aq) + SCN^-(aq) \rightleftharpoons Fe(SCN)^{2+}(aq) \qquad K_{f1} = 890$$

$$Fe(SCN)^{2+}(aq) + SCN^-(aq) \rightleftharpoons Fe(SCN)_2{}^+(aq) \qquad K_{f2} = 2.6$$

**SOLUTION** The difference between the stepwise formation equilibrium constants for this complex is relatively small:

$$K_{f1} = \frac{[Fe(SCN)^{2+}]}{[Fe^{3+}][SCN^-]} = 890$$

$$K_{f2} = \frac{[Fe(SCN)_2{}^+]}{[Fe(SCN)^{2+}][SCN]} = 2.6$$

Solutions of these complex ions are therefore best described in terms of an overall complex formation equilibrium:

$$Fe^{3+}(aq) + 2\,SCN^-(aq) \rightleftharpoons Fe(SCN)_2{}^+(aq)$$

The equilibrium constant expression for the overall reaction is equal to the product of the expressions for the individual steps in the reaction:

$$\frac{[Fe(SCN)_2^+]}{[Fe^{3+}][SCN^-]^2} = \frac{[\cancel{Fe(SCN)^{2+}}]}{[Fe^{3+}][SCN^-]} \times \frac{[Fe(SCN)_2^+]}{[\cancel{Fe(SCN)^{2+}}][SCN^-]}$$

The overall equilibrium constant is therefore the product of $K_{f1}$ and $K_{f2}$:

$$K_f = K_{f1} \times K_{f2} = \mathbf{2.3 \times 10^3}$$

## 18.15 COMPLEX-DISSOCIATION EQUILIBRIUM CONSTANTS

Complex ions also can be described in terms of complex-dissociation equilibria. We can start by assuming, for example, that most of the silver ions in an aqueous solution are present as the two-coordinate $Ag(NH_3)_2^+$ complex ion. We then assume that some of these ions dissociate to form $Ag(NH_3)^+$ complex ions and then eventually $Ag^+$ ions:

$$Ag(NH_3)_2^+(aq) \rightleftharpoons Ag(NH_3)^+(aq) + NH_3(aq)$$
$$Ag(NH_3)^+(aq) \rightleftharpoons Ag^+(aq) + NH_3(aq)$$

A **complex-dissociation equilibrium constant ($K_d$)** expression, which is the inverse of $K_f$, can be written for each of these reactions. Our problem has two reactions and therefore two $K_d$'s:

$$K_{d1} = \frac{[Ag(NH_3)^+][NH_3]}{[Ag(NH_3)_2^+]}$$

$$K_{d2} = \frac{[Ag^+][NH_3]}{[Ag(NH_3)^+]}$$

Alternatively, the individual steps in this reaction can be collapsed into an overall equation, which can be described by an overall equilibrium constant expression:

$$Ag(NH_3)_2^+(aq) \rightleftharpoons Ag^+(aq) + 2\ NH_3(aq)$$

$$K_d = \frac{[Ag^+][NH_3]^2}{[Ag(NH_3)_2^+]}$$

**EXERCISE 18.14**

Calculate the complex dissociation equilibrium constant for the $Cu(NH_3)_4^{2+}$ ion from the value of $K_f$ for this complex. $[Cu(NH_3)_4^{2+}$: $K_f = 2.1 \times 10^{13})$

**SOLUTION** The equilibrium expression for the complex formation reaction is written as follows:

$$K_f = \frac{[Cu(NH_3)_4^{2+}]}{[Cu^{2+}][NH_3]^4}$$

The equilibrium constant expression for the complex dissociation reaction is nothing more nor less than the inverse of the complex formation equilibrium expression:

$$K_d = \frac{[Cu^{2+}][NH_3]^4}{[Cu(NH_3)_4^{2+}]}$$

The value of $K_d$ is therefore equal to the inverse of $K_f$:

$$K_d = \frac{1}{K_f} = \mathbf{4.8 \times 10^{-14}}$$

## 18.16 APPROXIMATE COMPLEX-ION CALCULATIONS

Sec. 16.9

The techniques introduced during the discussion of gas-phase equilibria in Chapter 16 can be used to solve complex-ion equilibrium problems. For example, consider the seemingly complex question:

> What fraction of the total iron(III) concentration is present as the $Fe^{3+}$ ion in a solution that was initially 0.10 *M* $Fe^{3+}$ and 1.0 *M* $SCN^-$? [$Fe(SCN)_2^+$: $K_f = 2.3 \times 10^3$]

Complex-formation equilibria provide another example of the general rule that it is useful to begin equilibrium calculations by comparing the reaction quotient (here designated $Q_f$) for the initial conditions with the equilibrium constant for the reaction. The initial conditions can be summarized as follows:

$$Fe^{3+}(aq) + 2\,SCN^-(aq) \rightleftharpoons Fe(SCN)_2^+(aq) \qquad K_f = 2.3 \times 10^3$$

| | $Fe^{3+}$ | $SCN^-$ | $Fe(SCN)_2^+$ |
|---|---|---|---|
| Initial: | 0.10 *M* | 1.0 *M* | 0 |

The initial value of the reaction quotient is therefore equal to zero:

$$Q_f = \frac{(Fe(SCN)_2^+)}{(Fe^{3+})(SCN^-)^2} = \frac{(0)}{(0.10)(1.0)^2} = 0 << 2.3 \times 10^3$$

**Problem-Solving Strategy**

Because the reaction quotient is very much smaller than the equilibrium constant for the reaction, it would be absurd to assume that the reaction is close to equilibrium. Section 16.9 introduced a way around this problem: We define a set of intermediate conditions in which we drive the reaction as far as possible toward the right:

$$Fe^{3+}(aq) + 2\,SCN^-(aq) \rightleftharpoons Fe(SCN)_2^+(aq) \quad K_f = 2.3 \times 10^3$$

| | $Fe^{3+}$ | $SCN^-$ | $Fe(SCN)_2^+$ |
|---|---|---|---|
| Initial: | 0.10 *M* | 1.0 *M* | 0 |
| Change: | −0.10 *M* | −2(0.010) *M* | +0.10 *M* |
| Intermediate: | 0 | 0.8 *M* | 0.10 *M* |

We then assume that the reaction comes to equilibrium from these intermediate conditions:

$$Fe^{3+}(aq) + 2\,SCN^-(aq) \rightleftharpoons Fe(SCN)^{2+}(aq) \quad K_f = 2.3 \times 10^3$$

| | $Fe^{3+}$ | $SCN^-$ | $Fe(SCN)^{2+}$ |
|---|---|---|---|
| Intermediate: | 0 | 0.8 *M* | 0.10 *M* |
| Equilibrium: | Δ | 0.8 + 2 Δ | 0.10 − Δ |

$Fe^{3+} + SCN^- \rightleftharpoons Fe(SCN)^{2+}$

$Fe(SCN)^{2+} + SCN^- \rightleftharpoons Fe(SCN)_2^+$

Initial:
$[Fe^{3+}] = 0.10\ M$
$[SCN^-] = 1.0\ M$

$\Delta(SCN^-) = 2\Delta(Fe^{3+})$
As we form the $Fe(SCN)_2^+$ complex

Substituting this information into the equilibrium expression for the reaction gives the following equation:

$$\frac{[0.10 - \Delta]}{[\Delta][0.8 + 2\,\Delta]^2} = 2.3 \times 10^3$$

We are now ready to assume that Δ is relatively small:

$$\frac{[0.10]}{[\Delta][0.8]^2} \approx 2.3 \times 10^3$$

Solving this approximate equation gives the following result:

$$\Delta \approx 6.8 \times 10^{-5}$$

The assumption that Δ is relatively small is valid because the problem was defined so that it would be valid. We can now use this value of Δ to answer the original question:

$$[Fe^{3+}] = 6.8 \times 10^5\ M$$

Even though the complex formation equilibrium constant is not very large, essentially all of the iron in this solution is present as the $Fe(SCN)_2^+$ complex ion. Only a negligible fraction is present as the $Fe^{3+}$ ion:

$$\frac{6.8 \times 10^{-5}}{0.10} \times 100\% = \mathbf{0.068\%}$$

The following exercise illustrates a calculation when $Q_f$ is much greater than $K_f$.

**EXERCISE 18.15**

Calculate the concentration of the $Cu^{2+}$ ion in a solution that is initially 0.10 $M$ $Cu^{2+}$ and 1.0 $M$ $NH_3$. [$Cu(NH_3)_4^{2+}$: $K_f = 2.1 \times 10^{13}$]

**SOLUTION** We can set up this calculation as follows:

$$Cu^{2+}(aq) + 4\,NH_3(aq) \rightleftharpoons Cu(NH_3)_4^{2+}(aq) \qquad K_f = 2.1 \times 10^{13}$$

| | $Cu^{2+}$ | $NH_3$ | $Cu(NH_3)_4^{2+}$ |
|---|---|---|---|
| Initial: | 0.10 $M$ | 1.0 $M$ | 0 |

Because $K_f$ for this complex is very large, essentially all of the $Cu^{2+}$ ions should form $Cu(NH_3)_4^{2+}$ ions at equilibrium. We therefore define a set of intermediate conditions in which we shift the reaction to the right until all of the $Cu^{2+}$ ion is converted into $Cu(NH_3)_4^{2+}$ complex ions:

$$Cu^{2+}(aq) + 4\,NH_3(aq) \rightleftharpoons Cu(NH_3)_4^{2+}(aq) \qquad K_f = 2.1 \times 10^{13}$$

| | $Cu^{2+}$ | $NH_3$ | $Cu(NH_3)_4^{2+}$ |
|---|---|---|---|
| Initial: | 0.10 $M$ | 1.0 $M$ | 0 |
| Change: | −0.10 $M$ | −4(0.10 $M$) | +0.10 $M$ |
| Intermediate: | 0 | 0.6 $M$ | 0.10 $M$ |

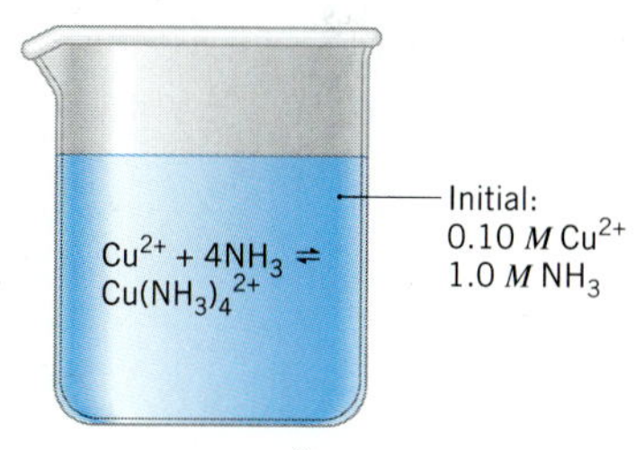


$\Delta(NH_3) = 4\Delta(Cu^{2+})$
As we form the $Cu(NH_3)_4^{2+}$ complex

We then let the reaction come to equilibrium from these intermediate conditions:

$$Cu^{2+}(aq) + 4\,NH_3(aq) \rightleftharpoons Cu(NH_3)_4^{2+}(aq) \qquad K_f = 2.1 \times 10^{13}$$

| | $Cu^{2+}$ | $NH_3$ | $Cu(NH_3)_4^{2+}$ |
|---|---|---|---|
| Intermediate: | 0 | 0.6 $M$ | 0.10 $M$ |
| Equilibrium: | $\Delta$ | $0.6 + 4\,\Delta$ | $0.10 - \Delta$ |

The next step in solving this problem involves writing the equilibrium constant expression for the reaction:

$$K_f = \frac{[Cu(NH_3)_4^{2+}]}{[Cu^{2+}][NH_3]^4}$$

We then substitute what we know about the equilibrium concentrations of the three components of this reaction into this equation:

$$\frac{[0.10 - \Delta]}{[\Delta][0.6 + 4\,\Delta]^4} = 2.1 \times 10^{13}$$

The value of $K_f$ is so large that very little $Cu(NH_3)_4^{2+}$ complex ion dissociates as the reaction comes to equilibrium. It is therefore reasonable to assume that $\Delta$ is relatively small:

$$\frac{[0.10]}{[\Delta][0.6]^4} \approx 2.1 \times 10^{13}$$

Solving this approximate equation gives the following result:

$$\Delta \approx 3.7 \times 10^{-14}$$

The assumption that Δ is small is valid. We can therefore use the results of this calculation to determine the concentration of the free (uncomplexed) $Cu^{2+}$ ion in this solution:

$$[Cu^{2+}] = 3.7 \times 10^{-14}\ M$$

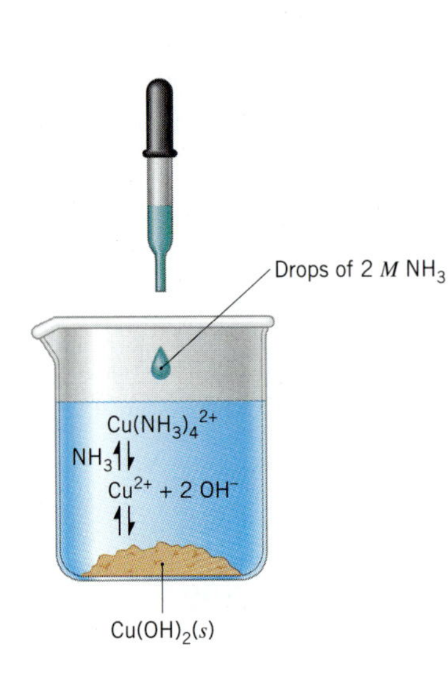


The results of the preceding exercise can be used to explain why $Cu(OH)_2$ dissolves in excess ammonia. Before we can do this, however, we need to understand why $Cu(OH)_2$ precipitates in the first place. Ammonia acts as a base toward water:

$$K_b = \frac{[NH_4^+][OH^-]}{[NH_3]} = 1.8 \times 10^{-5}$$

Even fairly dilute solutions of the $OH^-$ ion have more than enough $OH^-$ ion to precipitate $Cu(OH)_2$ from an 0.10 *M* $Cu^{2+}$ ion solution. As the amount of $NH_3$ added to the solution increases, the concentration of the $OH^-$ ion increases. But it doesn't become very much larger. The $OH^-$ ion concentration in a 0.001 *M* $NH_3$ solution is $1.3 \times 10^{-4}$ *M*. By the time the $NH_3$ concentration reaches 0.10 *M*, the $OH^-$ ion concentration has increased by only a factor of 30, to $4.2 \times 10^{-3}$ *M*.

As the amount of $NH_3$ added to the solution increases, the concentration of $Cu^{2+}$ ions rapidly decreases because these ions are tied up as $Cu(NH_3)_4^{2+}$ complex ions. According to the preceding exercise, the $Cu^{2+}$ ion concentration in 1.0 *M* $NH_3$ is only $3.7 \times 10^{-14}$ *M*. Thus, the ion product for $Cu(OH)_2$ under these conditions is about the same size as the solubility product for this compound.

$$Q_{sp} = (Cu^{2+})(OH^-)^2 = (3.7 \times 10^{-14})(4.2 \times 10^{-3})^2 = 6.5 \times 10^{-19} \approx K_{sp}$$

As soon as the $NH_3$ concentration exceeds 1 *M*, the $Cu^{2+}$ ion concentration becomes so small that the ion product for $Cu(OH)_2$ is smaller than $K_{sp}$, and the $Cu(OH)_2$ precipitate dissolves.

## 18.17 USING COMPLEX-ION EQUILIBRIA TO DISSOLVE AN INSOLUBLE SALT

Sec. 18.5

The key to using complex-ion equilibria to dissolve an insoluble salt is simple: Choose a complex for which $K_f$ is large enough that the concentration of the uncomplexed metal ion is too small for the ion product to exceed the solubility product. To show how this is done, let's return to a topic introduced in Section 18.5.

An essential step in processing photographic film involves removing the AgBr crystals that do not capture light. Because this step permanently fixes the image onto the film, the reagent used to achieve it is called a ''fixer.'' We can decide whether a particular complexing agent is strong enough to be used as a fixer by calculating the solubility of AgBr in an aqueous solution of this reagent.

### EXERCISE 18.16

Calculate the solubility of AgBr in 1 *M* $S_2O_3^{2-}$. (AgBr: $K_{sp} = 5 \times 10^{-13}$; $Ag(S_2O_3)_2^{3-}$: $K_f = 2.9 \times 10^{13}$)

**SOLUTION** Two equilibria must be considered in this problem, the solubility product for AgBr and the complex formation equilibrium for the $Ag(S_2O_3)_2^{3-}$ complex ion:

$$AgBr(s) \rightleftharpoons Ag^+(aq) + Br^-(aq)$$

$$Ag^+(aq) + 2\ S_2O_3^{2-}(aq) \rightleftharpoons Ag(S_2O_3)_2^{3-}(aq)$$

AgBr will dissolve if the $Ag(S_2O_3)_2^{3-}$ complex is strong enough to reduce the $Ag^+$ ion concentration to the point at which the product of the $Ag^+$ and $Br^-$ ion concentrations at equilibrium is smaller than the $K_{sp}$ of AgBr.

**Problem-Solving Strategy**

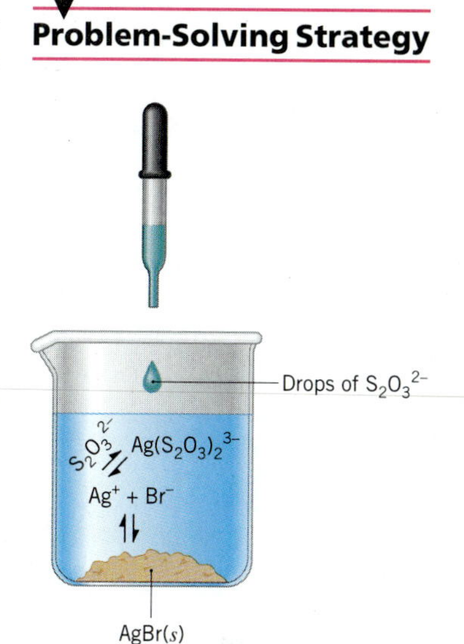


One way to decide whether $K_f$ for this complex ion is large enough to overcome the solubility product equilibrium involves combining the equilibria in this solution to give the following overall equation:

$$AgBr(s) + 2\ S_2O_3^{2-}(aq) \rightleftharpoons Ag(S_2O_3)_2^{3-}(aq) + Br^-(aq)$$

The equilibrium constant for this reaction is the product of $K_f$ times $K_{sp}$:

$$K = K_f \times K_{sp} = (2.9 \times 10^{13})(5 \times 10^{-13}) = \mathbf{15}$$

Because the equilibrium constant is larger than 1, a significant amount of AgBr should dissolve in this solution:

$$\frac{[Ag(S_2O_3)_2^{3-}][Br^-]}{[S_2O_3^{2-}]^2} = 15$$

When AgBr dissolves in pure water, the concentration of the $Ag^+$ and $Br^-$ ions must be the same.

$$\textbf{In pure water:} \quad [Ag^+] = [Br^-] = C_s$$

In 1 $M$ $S_2O_3^{2-}$ essentially all of the silver ion will be present as the two-coordinate $Ag(S_2O_3)_2^{3-}$ ion:

$$\textbf{In 1 } M\ S_2O_3^{2-}\textbf{:} \quad [Ag(S_2O_3)_2^{3-}] = [Br^-] = C_s$$

The concentration of the $S_2O_3^{2-}$ ion at equilibrium will be equal to the initial concentration of this ion minus the amount consumed when the $Ag(S_2O_3)_2^{3-}$ complex is formed:

$$[S_2O_3^{2-}] = 1 - 2C_s$$

Substituting this information into the equilibrium constant expression for the overall reaction gives the following equation:

$$\frac{[C_s][C_s]}{[1 - 2C_s]^2} = 15$$

Solving this equation with the quadratic formula gives the following result:

$$C_s = \mathbf{0.44\ M}$$

The $Ag(S_2O_3)_2^{3-}$ complex ion is strong enough to dissolve up to 0.44 moles of AgBr per liter of solution. It isn't surprising that the thiosulfate ion is used as the fixer in the processing of virtually all commercial photographic films.

## RESEARCH IN THE 90s

### BIOCHEMICAL COMPLEXES

Recent research has greatly expanded our understanding of the interactions that give rise to complexes. For purposes of discussion, it is useful to distinguish between complexes that involve the formation of covalent bonds and those that do not.

The complex formation equilibria discussed so far all involve the formation of a covalent bond between a transition-metal ion and a Lewis base such as $NH_3$ or the $SCN^-$ ion. A similar phenomenon occurs when either hemoglobin or myoglobin picks up oxygen. (A covalent bond is formed between one of the atoms in the $O_2$ molecule and an $Fe^{3+}$ ion in the protein.) Covalent bonds are also formed by the antitumor agent known as cisplatin. This *cis*-$PtCl_2(NH_3)_2$ complex first inserts itself into a groove in the structure of DNA. It then forms covalent bonds to the DNA, which interfere with the reproduction of DNA and thereby slow down the rate at which the tumor grows.

There are four ways in which complexes can form that do not involve the creation of a covalent bond:

1. The complex can be held together by *hydrogen bonds* between donor substituents on one component and acceptors on another.
2. The complex can involve *hydrophobic interactions,* in which hydrophobic substituents on neighboring molecules interact to form a region in space from which water can be excluded.
3. The complex can result from *electrostatic interactions* between groups on neighboring components of the system that carry opposite electrical charges.
4. The complex can be held together by *van der Waals forces* between neighboring groups that possess a dipole moment or groups in which a dipole moment can be induced.

Jacqueline Barton and co-workers have used a variety of techniques to probe the interaction between transition-metal ions and DNA. A recent series of papers used $^1$H NMR spectroscopy to study what happens when the complexes shown in Figure 18.11 are added to a synthetic strand of DNA that contains six base pairs:

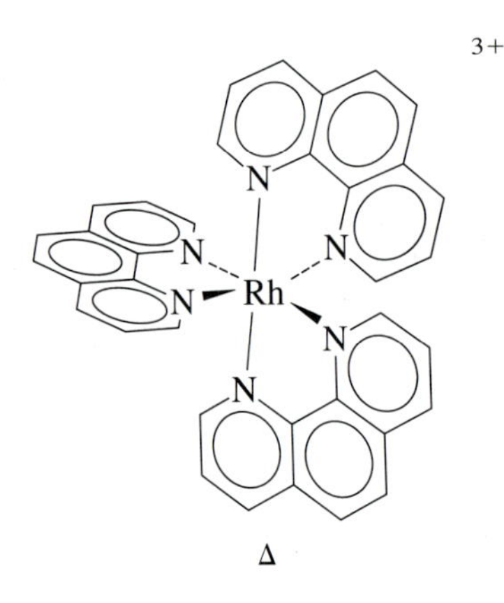


**Figure 18.11** The two stereoisomers of the $Rh(phen)_3{}^{3+}$ complex ion.

$d$(GTGCAC)$_2$ [Jill P. Rehmann and Jacqueline K. Barton, *Biochemistry,* **29,** 1701–1716 (1990)].

The Rh(phen)$_3^{3+}$ complex used in this study exists as a pair of stereoisomers that are mirror images of each other. Barton and co-workers found that in spite of similarities in their structures, these complexes bind at different points on DNA. The Δ isomer binds to the minor groove in the structure of the oligonucleotide used in this experiment. The Λ isomer, on the other hand, binds to the surface of the DNA, as shown in Figure 18.12.

Sec. 12.10

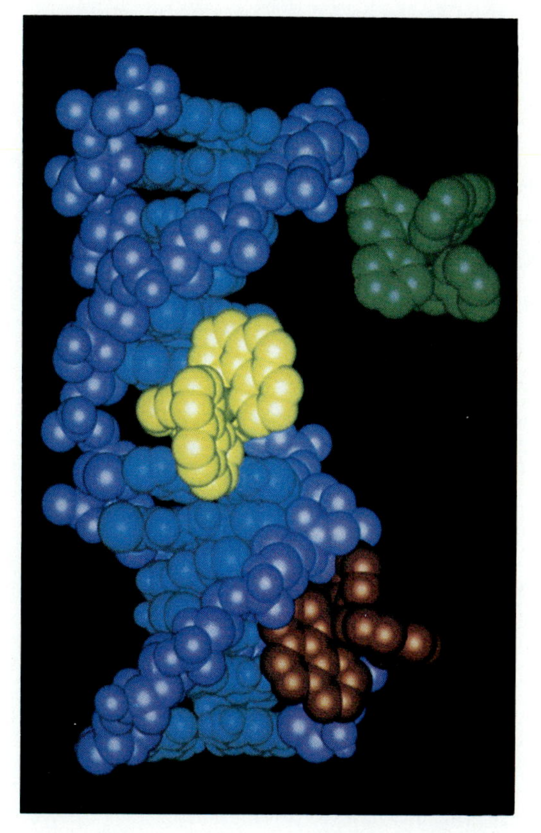

**Figure 18.12** A computer model of the binding of M(bpy)$_3^{2+}$ (green), Λ-M(phen)$_3^{2+}$ (yellow), and Δ-M(phen)$_3^{2+}$ (red) complexes bound to a small segment of double-stranded DNA.

This work suggests that hydrogen bonding does not have to play a role in the binding of transition-metal complexes to DNA, because the Rh(phen)$_3^{3+}$ complex is neither a hydrogen-bond donor nor a hydrogen-bond acceptor. The difference between the sites at which the two isomers bind emphasizes the role that the shape of a substrate has on its binding. The fact that the Δ isomer binds preferentially to the minor groove in this DNA sequence suggests that the site at which binding occurs is not determined exclusively by the ease with which it can approach DNA. (If accessibility were the driving force behind the binding, the Δ isomer would bind in the major groove in the DNA structure.) This work suggests that electrostatic interactions between the positive transition-metal complex and the negative phosphate groups on the backbone of the DNA and van der Waals forces are important components in determining the site at which interactions with DNA occur.

SPECIAL TOPIC

## A QUALITATIVE VIEW OF COMBINED EQUILIBRIA

Most of the discussion so far has focused on individual equilibria. This section will examine how Le Châtelier's principle can be applied to systems in which many chemical equilibria exist simultaneously.

Consider what happens when solid $CuSO_4$ dissolves in an aqueous $NH_3$ solution. If we start with enough copper sulfate to form a saturated solution, the following solubility product equilibrium will exist in this solution:

$$CuSO_4(s) \rightleftharpoons Cu^{2+}(aq) + SO_4^{2-}(aq)$$

The $SO_4^{2-}$ ion formed when $CuSO_4$ dissolves in water is a weak Brønsted base that can pick up an $H^+$ ion to form the hydrogen sulfate and hydroxide ions:

$$SO_4^{2-}(aq) + H_2O(l) \rightleftharpoons HSO_4^-(aq) + OH^-(aq)$$

There are other sources of the $OH^-$ ion in this solution. Water, of course, dissociates to some extent to form the $OH^-$ ion:

$$2\ H_2O(l) \rightleftharpoons H_3O^+(aq) + OH^-(aq)$$

Ammonia also reacts with water, to some extent, to form the $NH_4^+$ and $OH^-$ ions:

$$NH_3(aq) + H_2O(l) \rightleftharpoons NH_4^+(aq) + OH^-(aq)$$

The $Cu^{2+}$ ion released into solution when $CuSO_4$ dissolves reacts with ammonia to form a series of complex ions:

$$Cu^{2+}(aq) + NH_3(aq) \rightleftharpoons Cu(NH_3)^{2+}(aq)$$
$$Cu(NH_3)^{2+}(aq) + NH_3(aq) \rightleftharpoons Cu(NH_3)_2^{2+}(aq)$$
$$Cu(NH_3)_2^{2+}(aq) + NH_3(aq) \rightleftharpoons Cu(NH_3)_3^{2+}(aq)$$
$$Cu(NH_3)_3^{2+}(aq) + NH_3(aq) \rightleftharpoons Cu(NH_3)_4^{2+}(aq)$$

If the concentration of the $OH^-$ ion is large enough, a second solubility product equilibrium will exist in this solution:

$$Cu(OH)_2(s) \rightleftharpoons Cu^{2+}(aq) + 2\ OH^-(aq)$$

The simple process of dissolving copper(II) sulfate in aqueous ammonia therefore can involve nine simultaneous equilibria. In theory, we must consider each of these reactions if we want to predict what will happen under a particular set of experimental conditions. In practice, we can make at least one simplifying assumption. We can assume that the complex ion equilibria in this system can be represented by a single equation in which the $Cu^{2+}$ ion combines with four $NH_3$ molecules to form the four-coordinate $Cu(NH_3)_4^{2+}$ ion. We can therefore construct a fairly complete model of what happens in this solution if we take into account the equilibria summarized in Figure 18.13.

### EXERCISE 18.17

Use Le Châtelier's principle to predict the effect of adding each of the following substances to a saturated solution of $CuSO_4$ in aqueous $NH_3$:

(a) $CuSO_4(s)$ (b) $HNO_3(aq)$ (c) $NaOH(aq)$ (d) $Na_2SO_4(aq)$ (e) $NH_3(aq)$

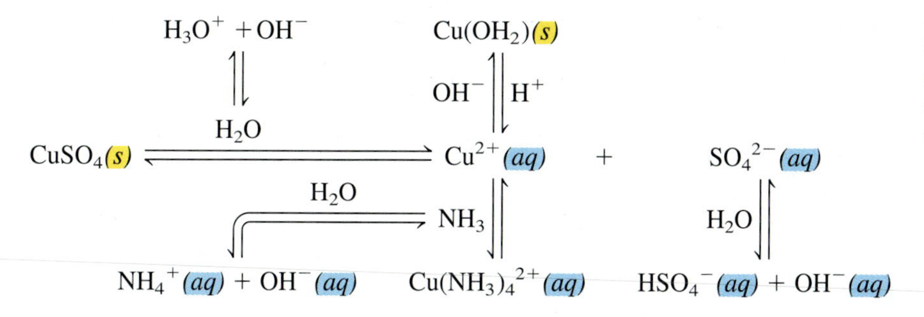


**Figure 18.13** The relationship among the equilibria that exist in a saturated solution of $CuSO_4$ in ammonia.

**SOLUTION** We can predict the effect of adding each of these reagents by applying Le Châtelier's principle to Figure 18.13.

(a) Adding excess $CuSO_4$ to a saturated solution has no effect on any of the equilibria shown in the figure because it has no effect on the concentration of either the $Cu^{2+}$ or $SO_4^{2-}$ ions.

(b) Nitric acid is a strong acid that should convert most of the $NH_3$ into $NH_4^+$ ions. Anything that removes $NH_3$ from the solution tends to destroy the $Cu(NH_3)_4^{2+}$ complex ion. Adding nitric acid therefore tends to increase the $Cu^{2+}$ ion concentration, which should cause $CuSO_4$ to precipitate from solution. We don't expect $Cu(OH)_2$ to precipitate, however, because the concentration of the $OH^-$ ion in a strong acid solution is much too small.

(c) Sodium hydroxide is a strong base. In theory, it should react with the $NH_4^+$ ion in this solution to form more $NH_3$. In practice, there isn't very much $NH_4^+$ ion in the solution to begin with, so adding NaOH has little effect on most of the equilibria in the solution. The presence of excess $OH^-$ ion, however, is likely to precipitate some of the $Cu^{2+}$ ion in solution as $Cu(OH)_2$.

(d) Sodium sulfate is a source of the $SO_4^{2-}$ ion. The common-ion effect therefore predicts that some additional $CuSO_4$ will precipitate if $Na_2SO_4$ is added to this solution.

(e) Any increase in the $NH_3$ concentration should increase the amount of the $Cu^{2+}$ ion tied up as $Cu(NH_3)_4^{2+}$ complex ions. This reduces the concentration of the free $Cu^{2+}$ ion, which makes the ion product for $CuSO_4$ smaller than the solubility product. As a result, adding $NH_3$ can cause more $CuSO_4$ to dissolve. It is also likely to dissolve any residual $Cu(OH)_2$ that might be present.

Another example of combined equilibria revolves around the aqueous chemistry of the $Fe^{3+}$ ion. Dilute solutions of this ion prepared by dissolving an iron(III) salt in perchloric acid are essentially colorless. In the presence of the thiocyanate ion, however, a blood-red solution is formed. This solution contains a pair of complex ions:

$$Fe^{3+}(aq) + SCN^-(aq) \rightleftharpoons Fe(SCN)^{2+}(aq) \qquad K_f = 890$$
$$Fe(SCN)^{2+}(aq) + SCN^-(aq) \rightleftharpoons Fe(SCN)_2^+(aq) \qquad K_f = 2.6$$

The $Fe^{3+}$ ion also forms a complex with the citrate ion ($Cit^{3-}$):

$$Fe^{3+}(aq) + Cit^{3-}(aq) \rightleftharpoons Fe(Cit)(aq) \qquad K_f = 6.3 \times 10^{11}$$

This neutral coordination complex has a pale yellow color.

**Figure 18.14** The aqueous solutions identified in Exercise 18.18.

## EXERCISE 18.18

Figure 18.14 shows a photograph of the following aqueous solutions, reading from left to right.

(a) A dilute aqueous solution of the $Fe^{3+}$ ion

(b) A mixture of the $Fe^{3+}(aq)$ and $SCN^-(aq)$ ions

(c) A mixture of the $Fe^{3+}(aq)$ and $SCN^-(aq)$ ions to which a strong acid has been added

(d) A mixture of the $Fe^{3+}(aq)$ and $Cit^{3-}(aq)$ ions

(e) A mixture of the $Fe^{3+}(aq)$ and $Cit^{3-}(aq)$ ions to which a strong acid has been added

(f) A mixture of the $Fe^{3+}(aq)$, $SCN^-(aq)$, and $Cit^{3-}(aq)$ ions

(g) A mixture of the $Fe^{3+}(aq)$, $SCN^-(aq)$, and $Cit^{3-}(aq)$ ions to which a strong acid has been added.

Explain the color (or lack thereof) for each of these solutions.

### SOLUTION

(a) The color of dilute solutions of the $Fe^{3+}$ ion is so weak they are essentially colorless.

(b) When the $SCN^-$ ion is added to an aqueous solution of the $Fe^{3+}$ ion, the $Fe(SCN)^{2+}$ and $Fe(SCN)_2{}^+$ complex ions are formed, and the solution turns blood red.

(c) Nothing happens when a strong acid is added to a mixture of the $Fe^{3+}$ and $SCN^-$ ions. This tells us something about the strength of the conjugate acid of the $SCN^-$ ion—thiocyanic acid, HSCN. If HSCN were a weak acid, adding a strong acid to the solution would tie up the $SCN^-$ ion as HSCN. If this happened, the $SCN^-$ ion would no longer be free to form a complex with the $Fe^{3+}$ ion. This would decrease the amount of the $Fe(SCN)^{2+}$ and $Fe(SCN)_2{}^+$ complex ions in solution, thereby decreasing the intensity of the color of this solution. Since this is not observed, HSCN must be a relatively strong acid. This conclusion is consistent with the $K_a$ for thiocyanic acid found in Table A.9 in the appendix.

(d) Mixing the $Fe^{3+}(aq)$ and $Cit^{3-}(aq)$ ions forms the Fe(Cit) complex, which is pale yellow.

(e) Because citric acid is a relatively weak acid, the $Cit^{3-}$ ion is a reasonably good base. Adding a strong acid therefore destroys the Fe(Cit) complex by converting the citrate ion into its conjugate acid: citric acid ($H_3Cit$). When this happens, the pale-yellow color disappears.

(f) The $K_f$ for the Fe(Cit) complex is much larger than $K_f$ for the $Fe(SCN)_2{}^+$ complex ion. Given a choice between $SCN^-$ and $Cit^{3-}$ ions, the $Fe^{3+}$ ions form complexes with $Cit^{3-}$ ions. A solution containing a mixture of these three ions therefore has the characteristic color of the Fe(Cit) complex.

(g) When a strong acid is added to a mixture of the $Fe^{3+}$, $Cit^{3-}$, and $SCN^-$ ions, the $Cit^{3-}$ ions are converted into citric acid ($H_3Cit$). When the $Cit^{3-}$ ions are removed from solution, the only complexing agent left in the solution is the $SCN^-$ ion. The solution therefore turns the blood-red color of the $Fe(SCN)^{2+}$ and $Fe(SCN)_2{}^+$ complex ions.

**SPECIAL TOPIC**

## A QUANTITATIVE VIEW OF COMBINED EQUILIBRIA

We can perform quantitative calculations for solutions of combined equilibria if we keep in mind how these equilibria are coupled.

### EXERCISE 18.19

Calculate the solubility of zinc sulfide in a solution buffered at pH 10.00 that is initially 0.10 *M* in $H_2S$.

**SOLUTION** We start by building a representation of the equilibria that must be considered in order to do this calculation, such as the one shown in Figure 18.15. We then calculate the $H_3O^+$ ion concentration in this buffer solution:

$$[H_3O^+] = 10^{-pH} = 10^{-10.00} = 1.0 \times 10^{-10}\ M$$

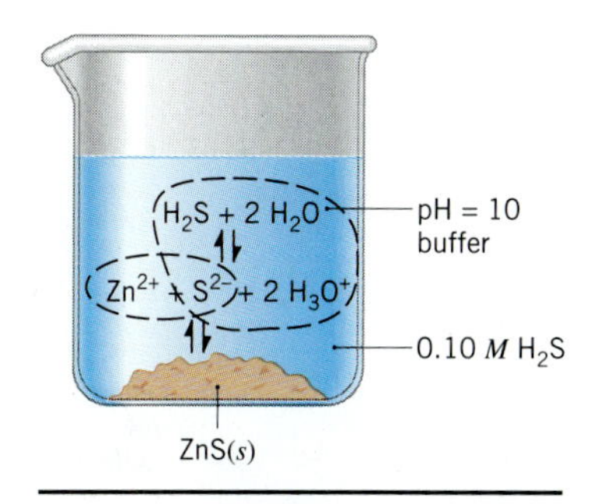


**Figure 18.15** The combined equilibria for the calculation in Exercise 18.19.

We then consider the effect this pH would have on the dissociation of $H_2S$:

$$H_2S(aq) + 2\ H_2O(l) \rightleftharpoons 2\ H_3O^+(aq) + S^{2-}(aq) \qquad K_c = 1.3 \times 10^{-20}$$

Because it is a weak acid, we can assume that the concentration of $H_2S$ at equilibrium is more or less equal to its initial concentration:

$$[H_2S] \approx 0.10\ M$$

We now know the $H_2S$ and $H_3O^+$ concentrations at equilibrium, which means that we can calculate the concentration of the $S^{2-}$ ion at equilibrium before any ZnS dissolves:

$$\frac{[H_3O^+]^2[S^{2-}]}{[H_2S]} \approx \frac{[1.0 \times 10^{-10}]^2[S^{2-}]}{[0.10]} \approx 1.3 \times 10^{-20}$$

$$[S^{2-}] \approx 0.13\ M$$

When ZnS dissolves, we get more $S^{2-}$ ion. The amount of ZnS that dissolves can be calculated from the following equation, where $C_s$ is the solubility of ZnS:

$$[Zn^{2+}][S^{2-}] = K_{sp}$$

$$[C_s][C_s + 0.13] = 1.6 \times 10^{-24}$$

If we have to, we can always solve this equation with the quadratic formula. Before we do this, however, we can look for an assumption that makes the problem easier to solve. In this case, it seems reasonable to assume that the solubility of ZnS is so small that most of the $S^{2-}$ ion at this pH comes from the dissociation of $H_2S$:

$$C_s << 0.13\ M$$

When this assumption is made, we get the following approximate equation:

$$[C_s][0.13] \approx 1.6 \times 10^{-24}$$

When we solve this equation, we get a solubility of ZnS that is so small that we have to conclude that the assumption used to generate this equation is valid:

$$C_s \approx 1.2 \times 10^{-23}$$

**Problem-Solving Strategy**

There is no magic formula that can help us divide problems such as Exercise 18.19 into steps, or decide the order in which steps should be handled. The following general rules, however, might guide us through these problems:

1. Identify the equilibria that must be included in the model of the solution.
2. Draw a figure that shows how these equilibria are coupled.
3. Find the simplest equilibrium—the one for which all of the necessary data are available—and solve this part of the problem.
4. Ask: ''Where did this get me?'' ''What can I do with this information?''
5. Never lose track of what was accomplished in a previous step.
6. Use the results of one step to solve another until all of the equilibria have been utilized.

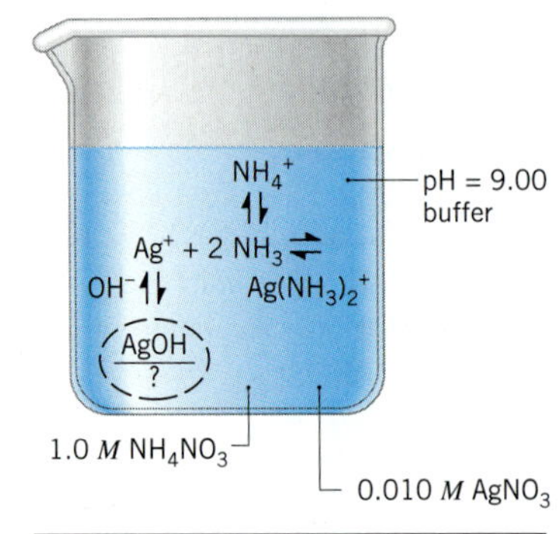


**Figure 18.16** The combined equilibria for the calculation in Exercise 18.20.

## EXERCISE 18.20

Predict whether AgOH will precipitate from a solution buffered at pH 9.00 that is initially 0.010 $M$ $AgNO_3$ and 1.0 $M$ $NH_4NO_3$. (AgOH: $K_{sp} = 2.0 \times 10^{-8}$; $NH_3$: $K_b = 1.8 \times 10^{-5}$; $Ag(NH_3)_2^+$: $K_f = 1.1 \times 10^7$; $H_2O$: $K_w = 1.0 \times 10^{-14}$)

**SOLUTION** We might start by noting that silver nitrate and ammonium nitrate are both soluble salts.

$$AgNO_3(s) \xrightarrow{H_2O} Ag^+(aq) + NO_3^-(aq)$$

$$NH_4NO_3(s) \xrightarrow{H_2O} NH_4^+(aq) + NO_3^-(aq)$$

We then build a representation of the equilibria present in this solution such as the one in Figure 18.16.

We can solve some problems by looking at the initial conditions and working toward the final answer. Others are so complex it is useful to look at the goal and then work backwards. The goal in this problem is to decide whether AgOH precipitates from this solution. In order to make this decision, we need two pieces of information: the $Ag^+$ and $OH^-$ ion concentrations at equilibrium. We can therefore divide the problem into two parts whose goals consist of determining the values of the $[Ag^+]$ and $[OH^-]$ terms when all other reactions in this system are at equilibrium.

It seems reasonable to start with the goal that is easiest to achieve: finding the $OH^-$ ion concentration. Because we know the pH of this buffer solution, we can calculate the $H_3O^+$ ion concentration at equilibrium:

$$[H_3O^+] = 10^{-pH} = 10^{-9.00} = 1.0 \times 10^{-9}\ M$$

We can then use this $H_3O^+$ ion concentration to calculate the $OH^-$ ion concentration in the buffer solution:

$$[OH^-] = \frac{K_w}{[H_3O^+]} = \frac{1.0 \times 10^{-14}}{1.0 \times 10^{-9}} = \mathbf{1.0 \times 10^{-5}\ \mathit{M}}$$

We now turn to the second part of the problem, finding the concentration of the $Ag^+$ ion at equilibrium. We know that this ion is in equilibrium with the $Ag(NH_3)_2^+$ complex ion:

$$K_f = \frac{[Ag(NH_3)_2^+]}{[Ag^+][NH_3]^2}$$

**Problem-Solving Strategy**

If we knew the concentrations of $NH_3$ and the $Ag(NH_3)_2^+$ complex ion, we could calculate the $Ag^+$ ion concentration at equilibrium. This gives us two new subgoals: determining the $Ag(NH_3)_2^+$ and $NH_3$ concentrations.

Determining the $Ag(NH_3)_2^+$ ion concentration is relatively easy. We know that the initial concentration of the $Ag^+$ ion in this solution is 0.010 *M*. We also know that this ion forms a strong complex with $NH_3$. We can therefore assume that essentially all of the silver ions in this solution will be present as $Ag(NH_3)_2^+$ complex ions:

$$[Ag(NH_3)_2^+] \approx \mathbf{0.010\ \mathit{M}}$$

Now all we need is the $NH_3$ concentration. We know that ammonia is in equilibrium with the $NH_4^+$ ion:

$$K_b = \frac{[NH_4^+][OH^-]}{[NH_3]} = 1.8 \times 10^{-5}$$

And we already know the $OH^-$ ion concentration in this solution:

$$\frac{[NH_4^+][1.0 \times 10^{-5}]}{[NH_3]} = 1.8 \times 10^{-5}$$

Rearranging this equation gives:

$$\frac{[NH_4^+]}{[NH_3]} = 1.8$$

The concentration of the $NH_4^+$ ion at equilibrium is therefore 1.8 times the concentration of $NH_3$:

$$\mathbf{[NH_4^+] = 1.8\ [NH_3]}$$

We also know that the solution was initially 1.0 *M* in $NH_4^+$ ions. These ions are now present as $NH_3$ molecules, as $NH_4^+$ ions, or as part of the $Ag(NH_3)_2^+$ complex ions:

$$[NH_3] + [NH_4^+] + 2\ [Ag(NH_3)_2^+] = 1.0\ M$$

We can ignore the last term in this equation because the concentration of the $Ag(NH_3)_2^+$ ion can't be larger than the initial concentration of the $Ag^+$ ion: 0.010 *M*. We can therefore assume that only a negligibly small fraction of the $NH_4^+$ that was added to the solution is present in the $Ag(NH_3)_2^+$ complex ion:

$$[NH_3] + [NH_4^+] \approx 1.0\ M$$

We now have two equations in two unknowns:

$$[NH_4^+] = 1.8\ [NH_3]$$
$$[NH_3] + [NH_4^+] = 1.0\ M$$

Substituting the first equation into the second allows us to solve for the $NH_3$ and $NH_4^+$ concentrations at equilibrium:

$$[NH_3] = 0.36\ M$$
$$[NH_4^+] = 0.64\ M$$

We can now return to the complex formation equilibrium:

$$K_f = \frac{[Ag(NH_3)_2^+]}{[Ag^+][NH_3]^2}$$

and substitute into this equation the approximate values for the concentrations of $NH_3$ and the $Ag(NH_3)_2^+$ complex ion at equilibrium:

$$\frac{[0.010]}{[Ag^+][0.36]^2} \approx 1.1 \times 10^7$$

We can then solve this equation for the $Ag^+$ ion concentration at equilibrium:

$$[Ag^+] \approx 7.0 \times 10^{-9}\ M$$

We can now look at the product of the $Ag^+$ and $OH^-$ ion concentrations when this system is at equilibrium.

$$[Ag^+][OH^-] = [7.0 \times 10^{-9}][1.0 \times 10^{-5}] = 7.0 \times 10^{-14}$$

The ion product for this solution ($7.0 \times 10^{-14}$) is very much smaller than the solubility product for AgOH ($2.0 \times 10^{-8}$), which means that AgOH won't precipitate from solution.

## SUMMARY

**1.** There is a limit to the amount of any solid that will dissolve in a given volume of a solvent. A solution is saturated when this limit is reached. In a **saturated solution,** an equilibrium exists between the solid and the particles released into the solution when the solid dissolves. At equilibrium, the solid dissolves and dissociates at the same rate at which these particles come together to form additional solid that precipitates from solution.

**2.** The basic assumption behind the discussion of the solubility of salts is the notion that salts dissociate into their ions when they dissolve in water. The equilibrium between a salt and its ions in a saturated solution can be described with a **solubility product** equilibrium constant expression, $K_{sp}$. The concentration of water is not included in this expression because water is neither consumed nor produced in these reactions. The concentration of the solid is also left out of the expression because it is a constant.

**3.** The $K_{sp}$ for a salt is equal to the product of the concentrations of the ions formed when the salt dissolves in water raised to a power equal to the coefficient of that ion in a balanced equation for the reaction.

**4.** Adding a source of a common ion to a saturated solution of a salt produces a solution for which the **ion product** ($Q_{sp}$) is larger than the solubility product ($K_{sp}$). This shifts the equilibrium, making the salt less soluble in water.

**5.** The **common-ion effect** also can be used to increase the solubility of a salt. We can achieve this goal by adding a reagent to the solution that decreases the concentration of one of the ions formed when the salt dissolves. Adding an acid, for example, can reduce the $OH^-$ ion concentration in water to a point at which insoluble hydroxides no longer precipitate from solution.

**6.** Anything that can donate a pair of nonbonding electrons is a Lewis base. Anything with at least one empty valence-shell orbital that can accept a pair of nonbonding electrons is a Lewis acid. Lewis acids combine with Lewis bases to form acid–base complexes held together by the sharing of a pair of electrons in a covalent bond.

**7.** Because the values of $K_f$ for the individual steps in the formation of a complex ion are often about the same, complex ion equilibria are typically collapsed into a single overall equation. Complex ion equilibria calculations often involve the construction of an intermediate stage, in which the reaction is driven as far as possible toward the complex. A small amount of the complex ion is then allowed to dissociate as the reaction comes to equilibrium.

**8.** Complex ion equilibria are particularly important in solutions in which more than one equilibrium occurs at the same time. They are therefore an essential part of any discussion of combined equilibria. A qualitative understanding of combined equilibria can be obtained by application of Le Châtelier's principle.

## KEY TERMS

**Acid–Base Complexes** (p. 670)
**Common-ion effect** (p. 661)
**Complex ion** (p. 670)
**Complex-dissociation equilibrium constant ($K_d$)** (p. 673)
**Complex-formation equilibrium constant ($K_f$)** (p. 670)
**Ion product ($Q_{sp}$)** (p. 659)
**Saturated solution** (p. 651)
**Selective precipitation** (p. 664)
**Solubility** (p. 651)
**Solubility product equilibrium constant ($K_{sp}$)** (p. 653)

## KEY EQUATIONS

$$K_{sp} = [Ag^+][Cl^-]$$
$$K_f = K_{f1} \times K_{f2} \times K_{f3} \cdots$$

## PROBLEMS

### What Happens When Solids Dissolve In Water?

**18-1** Both molecular and covalent solids contain covalent bonds. Explain why molecular solids are often soluble in water but covalent solids are not.

**18-2** Explain why NaCl is soluble in water but AgCl is not.

**18-3** Use an example to explain what is meant by the general rule that salts dissociate when they dissolve in water.

### Solubility Equilibria

**18-4** Explain why the bulb in the conductivity apparatus in Figure 18.1 glows more brightly when the wires are immersed in a solution of NaCl than when the wires are immersed in tap water.

**18-5** Explain why the addition of a few small crystals of silver chloride makes water a slightly better conductor of electricity. Explain why the conductivity gradually increases as more AgCl is added, until it eventually reaches a maximum. Describe what is happening in the solution when its conductivity reaches the maximum.

### Solubility Rules

**18-6** Which of the following salts is (are) *insoluble* in water?
(a) $Ba(NO_3)_2$ (b) $BaCl_2$ (c) $BaCO_3$ (d) BaS
(e) $Ba(OAc)_2$

**18-7** Which of the following salts is (are) *insoluble* in water?
(a) $(NH_4)_2SO_4$ (b) $K_2CrO_4$ (c) $Na_2S$ (d) $Pb(NO_3)_2$
(e) $Cr(OH)_3$

**18-8** Which of the following salts is (are) *soluble* in water?
(a) PbS (b) PbO (c) $PbCrO_4$ (d) $PbCO_3$
(e) $Pb(NO_3)_2$

### The Solubility Product Expression

**18-9** Explain why the $[Ag^+]$ and $[Cl^-]$ terms are variables but the [AgCl] term is a constant no matter how much AgCl is added to a saturated solution of silver chloride in water.

**18-10** What is the correct solubility product expression for the following reaction?

$$Ca_3(PO_4)_2(s) \rightleftharpoons 3\ Ca^{2+}(aq) + 2\ PO_4^{3-}(aq)$$

(a) $K_{sp} = \dfrac{[Ca^{2+}][PO_4^{3-}]}{[Ca_3(PO_4)_2]}$ (b) $K_{sp} = \dfrac{[Ca^{2+}]^3[PO_4^{3-}]^2}{[Ca_3(PO_4)_2]}$

(c) $K_{sp} = [Ca^{2+}][PO_4^{3-}]$ (d) $K_{sp} = [Ca^{2+}]^3[PO_4^{3-}]^2$

(e) $K_{sp} = [Ca^{2+}]^2[PO_4^{3-}]^3$

**18-11** Which of the following is the correct solubility product expression for $Al_2(SO_4)_3$?

(a) $K_{sp} = [Al^{3+}][SO_4^{2-}]$ (b) $K_{sp} = [2\ Al^{3+}][3\ SO_4^{2-}]$

(c) $K_{sp} = [Al^{3+}]^2[SO_4^{2-}]^3$

(d) $K_{sp} = [2\ Al^{3+}]^2[3\ SO_4^{2-}]^3$

**18-12** Write the solubility product expression for the following salts:

(a) $BaCrO_4$ (b) $CaCO_3$ (c) $PbF_2$ (d) $Ag_2S$

### The Relationship Between $K_{sp}$ and the Solubility of a Salt

**18-13** Write an equation that describes the relationship between the concentrations of the $Ag^+$ and $CrO_4^{2-}$ ions in a saturated solution of $Ag_2CrO_4$.

**18-14** Write an equation that describes the relationship between the concentrations of the $Bi^{3+}$ and $S^{2-}$ ions in a saturated solution of $Bi_2S_3$.

**18-15** Which of the following equations describes the relationship between the solubility product for $MgF_2$ and the solubility of this compound?

(a) $K_{sp} = 2C_s$ (b) $K_{sp} = C_s^2$ (c) $K_{sp} = 2C_s^2$

(d) $K_{sp} = C_s^3$ (e) $K_{sp} = 4C_s^3$

**18-16** $Hg_2Cl_2$ contains the $Hg_2^{2+}$ and $Cl^-$ ions. Which of the following equations describes the relationship between the solubility product and the solubility of this compound?

(a) $K_{sp} = C_s$ (b) $K_{sp} = C_s^3$ (c) $K_{sp} = 4C_s^3$

(d) $K_{sp} = C_s^4$ (e) $K_{sp} = 16C_s^4$

**18-17** Which is more soluble, $Ag_2S$ or HgS? ($Ag_2S$: $K_{sp} = 6.3 \times 10^{-50}$; HgS: $K_{sp} = 4 \times 10^{-53}$)

**18-18** Which is more soluble, $PbSO_4$ or $PbI_2$? ($PbSO_4$: $K_{sp} = 1.6 \times 10^{-8}$; $PbI_2$: $K_{sp} = 7.1 \times 10^{-9}$)

**18-19** Mercury forms salts that contain either the $Hg^{2+}$ ion or the $Hg_2^{2+}$ ion. Which is more soluble, HgS or $Hg_2S$? (HgS: $K_{sp} = 4 \times 10^{-53}$; $Hg_2S$: $K_{sp} = 1.0 \times 10^{-47}$)

**18-20** What is the concentration of the $CN^-$ ion in a saturated solution of zinc cyanide dissolved in water if the $Zn^{2+}$ ion concentration is $4.0 \times 10^{-5}$ *M*?

**18-21** What is the concentration of the $CrO_4^{2-}$ ion in a saturated solution of silver chromate dissolved in water if the $Ag^+$ ion concentration is $1.3 \times 10^{-4}$ *M*?

**18-22** What is the solubility product for strontium fluoride if the solubility of $SrF_2$ in water is 0.107 g/L?

**18-23** Silver acetate, $Ag(CH_3CO_2)$ is marginally soluble in water. What is the $K_{sp}$ for silver acetate if 1.190 g of $Ag(CH_3CO_2)$ dissolves in 99.40 mL of water?

**18-24** Lithium salts, such as lithium carbonate, are used to treat manic-depressives. What is the solubility product for lithium carbonate if 1.36 g of $Li_2CO_3$ dissolves in 100 mL of water?

**18-25** Magnesia (MgO) is only marginally soluble in water. When it does dissolve, however, it forms magnesium hydroxide—''milk of magnesia.'' What is the $K_{sp}$ for $Mg(OH)_2$ if a saturated solution in water has a pH of 10.22?

**18-26** People who have the misfortune of going through a series of x-rays of the gastrointestinal tract are often given a suspension of solid barium sulfate in water to drink. $BaSO_4$ is used instead of other $Ba^{2+}$ salts, which also reflect x-rays, because it is relatively insoluble in water. (Thus the patient is exposed to the minimum amount of toxic $Ba^{2+}$ ion.) What is the solubility product for barium sulfate if 1 g of $BaSO_4$ dissolves in 400,000 g of water?

**18-27** Lime (CaO) is only slightly soluble in water. When it does dissolve, however, it forms calcium hydroxide. What is the $K_{sp}$ for $Ca(OH)_2$ if it takes 11.0 mL of a 0.0010 *M* HCl solution to neutralize a liter of a saturated $Ca(OH)_2$ solution?

**18-28** What is the solubility of silver sulfide in water in grams per 100 mL if the solubility product for $Ag_2S$ is $6.3 \times 10^{-50}$?

**18-29** What is the solubility in water for each of the following salts in grams per 100 mL?

(a) $Hg_2S$ ($K_{sp} = 1.0 \times 10^{-47}$)

(b) HgS ($K_{sp} = 4 \times 10^{-53}$)

**18-30** What is the solubility in water for each of the following salts in grams per 100 mL?

(a) $Ca_3(PO_4)_2$ ($K_{sp} = 2.0 \times 10^{-29}$) (b) $Pb_3(PO_4)_2$ ($K_{sp} = 8.0 \times 10^{-43}$) (c) $Ag_3PO_4$ ($K_{sp} = 1.4 \times 10^{-16}$)

**18-31** List the following salts in order of increasing solubility in water:

(a) $Ag_2S$ ($K_{sp} = 6.3 \times 10^{-50}$) (b) $Bi_2S_3$ ($K_{sp} = 1 \times 10^{-97}$) (c) CuS ($K_{sp} = 6.3 \times 10^{-36}$) (d) HgS ($K_{sp} = 4 \times 10^{-53}$)

### The Common-Ion Effect

**18-32** Define the term *common-ion effect.* Describe how Le Châtelier's principle can be used to explain the common-ion effect.

**18-33** Describe what happens to the equilibrium concentrations of the $Ag^+$ and $Cl^-$ ions when 10 g of NaCl are added to a liter of a saturated solution of silver chloride in water.

**18-34** Which of the following statements is true?

(a) $MgF_2$ is more soluble in 0.100 *M* NaF than in pure water. (b) $MgF_2$ is less soluble in 0.100 *M* NaF than in pure water. (c) $MgF_2$ is just as soluble in 0.100 *M* NaF as in pure water.

**18-35** In which of the following solutions would $Ag_2S$ be least soluble? ($Ag_2S$: $K_{sp} = 6.3 \times 10^{-50}$; $H_2S$: $K_{a1} = 1.0 \times 10^{-7}$, $K_{a2} = 1.3 \times 10^{-13}$)

(a) Pure water (b) 0.0010 *M* $Na_2S$ (c) 0.10 *M* $H_2S$

**18-36** Calculate the solubility of $Al(OH)_3$ in moles per liter in a solution buffered at pH 9.31. [$Al(OH)_3$: $K_{sp} = 1.3 \times 10^{-33}$]

### How To Keep a Salt from Dissolving or Precipitating

**18-37** Calculate the equilibrium concentration of the $Ag^+$ ion in a solution prepared by dissolving 3.21 g of potassium iodide in 350 mL of water and then adding silver iodide until the solution is saturated with AgI. (AgI: $K_{sp} = 8.3 \times 10^{-17}$)

**18-38** How many grams of silver sulfide will dissolve in 500 mL of an 0.050 *M* $S^{2-}$ solution? ($Ag_2S$: $K_{sp} = 6.3 \times 10^{-50}$)

**18-39** In which of the following solutions is $Pb(OH)_2$ most soluble? [$Pb(OH)_2$: $K_{sp} = 1.2 \times 10^{-15}$]

(a) Pure water (b) 0.010 *M* NaOH (c) 0.010 *M* HCl

**18-40** Calculate the solubility of $Co(OH)_3$ in both pure water and a pH 9 solution. [$Co(OH)_3$: $K_{sp} = 1.6 \times 10^{-44}$]

**18-41** At what pH does $Cr(OH)_3$ just start to precipitate from a solution that is 0.025 *M* in $Cr(NO_3)_3$? [$Cr(OH)_3$: $K_{sp} = 6.3 \times 10^{-31}$]

### How To Separate Ions by Selective Precipitation

**18-42** Start with a solution that contains 7.33 g of $K_2CrO_4$ per 500 mL. Assume that you can add $Ag^+$ ions to this solution without significantly changing its volume. What is the $Ag^+$ ion concentration when 99.9% of the $CrO_4^{2-}$ has been precipitated as $Ag_2CrO_4$? ($Ag_2CrO_4$: $K_{sp} = 1.1 \times 10^{-12}$)

**18-43** Both barium sulfite, $BaSO_3$, and barium sulfate, $BaSO_4$, are insoluble in water. $BaSO_3$ dissolves in acid, however, whereas $BaSO_4$ does not. Explain why. ($BaSO_3$: $K_{sp} = 8 \times 10^{-7}$; $BaSO_4$: $K_{sp} = 1.1 \times 10^{-10}$; $H_2SO_3$: $K_{a1} = 1.7 \times 10^{-2}$, $K_{a2} = 6.4 \times 10^{-8}$; $H_2SO_4$: $K_{a1} = 10^3$, $K_{a2} = 1.2 \times 10^{-2}$)

**18-44** Which ion, $Pb^{2+}$ or $Zn^{2+}$, precipitates first when we slowly add $S^{2-}$ ions to a solution that is $1.0 \times 10^{-4}$ *M* in both $Pb^{2+}$ and $Zn^{2+}$ ions? What is the concentration of this ion when the second ion starts to precipitate? (PbS: $K_{sp} = 8.0 \times 10^{-28}$; ZnS: $K_{sp} = 1.6 \times 10^{-24}$)

**18-45** Which halide, AgCl or AgI, precipitates first when we slowly add $Ag^+$ ions to a solution that is $1.0 \times 10^{-5}$ *M* in both $Cl^-$ and $I^-$ ions? What is the concentration of this ion when the second ion starts to precipitate? (AgCl: $K_{sp} = 1.8 \times 10^{-10}$; AgI: $K_{sp} = 8.3 \times 10^{-17}$)

**18-46** Describe how you would separate $Pb^{2+}$ from $Hg^{2+}$ in a solution that was 0.10 *M* in both ions. (Hint: PbS: $K_{sp} = 8.0 \times 10^{-28}$; HgS: $K_{sp} = 4 \times 10^{-53}$)

**18-47** Which of the following statements is true?

(a) CdS is less soluble than CuS. (b) CdS is about $10^9$ times more soluble than CuS. (c) $Ag_2S$ is less soluble than either CdS or CuS. (d) $Ag_2S$ is about 10 times more soluble than CuS.

(CdS: $K_{sp} = 8 \times 10^{-27}$, CuS: $K_{sp} = 6.3 \times 10^{-36}$, $Ag_2S$: $K_{sp} = 6.3 \times 10^{-50}$)

**18-48** What happens when $H_2S$ is added very slowly to a solution that contains 0.10 *M* concentrations of the $Cu^{2+}$ and $Cd^{2+}$ ions?

(a) CdS will precipitate first. (b) CuS will precipitate first. (c) Most of the $Cu^{2+}$ will still be in solution when most of the $Cd^{2+}$ has been precipitated. (d) Both CdS and CuS will precipitate at the same time.
(e) Neither CdS nor CuS will precipitate.

(CdS: $K_{sp} = 8 \times 10^{-27}$, CuS: $K_{sp} = 6.3 \times 10^{-36}$)

**18-49** A solution is prepared from small amounts of CdS and CuS added to pure water. What happens when a strong acid such as HCl is added to this solution?

(a) Only CdS will dissolve at first. (b) Both CdS and CuS will dissolve at the same time. (c) Some CdS will dissolve, but more CuS will precipitate. (d) More of both CdS and CuS will precipitate.

(CdS: $K_{sp} = 8 \times 10^{-27}$; CuS: $K_{sp} = 6.3 \times 10^{-36}$)

### How To Adjust the Concentration of an Ion

**18-50** Hydrogen sulfide, $H_2S$, is a weak acid in water.

$$H_2S(aq) + H_2O(l) \rightleftharpoons H_3O^+(aq) + HS^-(aq) \qquad K_{a1} = 1.0 \times 10^{-7}$$

$$HS^-(aq) + H_2O(l) \rightleftharpoons H_3O^+(aq) + S^{2-}(aq) \qquad K_{a2} = 1.3 \times 10^{-13}$$

If your goal is to dissolve a ZnS precipitate, would it be better to add a strong acid or a strong base? Use chemical equations to explain your answer. (ZnS: $K_{sp} = 1.6 \times 10^{-24}$)

**18-51** Explain why iron(III) hydroxide precipitates when aqueous solutions of iron(III) chloride and sodium carbonate are mixed. [$Fe(OH)_3$: $K_{sp} = 4 \times 10^{-38}$]

**18-52** What will precipitate from a solution that is simultaneously 0.10 *M* in $Cd^{2+}$, $Fe^{3+}$, HCl, and $H_2S$?

**18-53** Which of the following cations will precipitate as the sulfide from a solution that is 0.10 *M* in $H_2S$, 0.3 *M* in HCl, and $10^{-3}$ *M* in each of the following ions?

(a) $Fe^{3+}$ (b) $Ag^+$ (c) $Cu^{2+}$ (d) $Ni^{2+}$

### Complex Ion Equilibria

**18-54** Define the following terms: *Lewis acid, Lewis base, complex ion,* and *coordination number.*

**18-55** Use examples to explain the difference between Lewis acids and Brønsted acids, and between Lewis bases and Brønsted bases.

**18-56** Which of the following is a Lewis acid?

(a) CO (b) $C_2H_2$ (c) $BeF_2$ (d) $CH_4$ (e) $NF_3$

**18-57** Which of the following is a Lewis acid?

(a) $CH_3^+$ (b) $CH_4$ (c) $NH_3$ (d) $BF_4^-$ (e) $O^{2-}$

**18-58** Which of the following is not a Lewis acid?

(a) $H^+$ (b) $BF_3$ (c) CO (d) $Cu^{2+}$ (e) $Fe^{3+}$

**18-59** Which of the following is not a Lewis base?

(a) $NH_4^+$ (b) $OH^-$ (c) $Cl^-$ (d) $O_2$ (e) $SCN^-$

**18-60** Which of the following Lewis acids is not a Brønsted acid?

(a) HF (b) HOAc (c) $H_3PO_4$ (d) $NH_3$ (e) $BF_3$

**18-61** Explain why the charges on the $Fe(SCN)^{2+}$ and $Fe(SCN)_2^+$ ions are +2 and +1, respectively.

**18-62** Calculate the charge on the transition metal ions in the $Zn(NH_3)_4^{2+}$, $Fe(SCN)_2^+$, $Sn(OH)_3^-$, $Co(SCN)_4^{2-}$, and $Ag(S_2O_3)_2^{3-}$ complex ions.

### The Stepwise Formation of Complex Ions

**18-63** Explain the difference between polyprotic acids and complex ions that allows us to assume that polyprotic acids dissociate in steps, whereas the dissociation of complex ions can be collapsed into a single overall reaction.

**18-64** $Cu^{2+}$ forms a four-coordinate complex with ammonia:

$$Cu^{2+}(aq) + 4\ NH_3(aq) \rightleftharpoons Cu(NH_3)_4^{2+}(aq)$$

What is the relationship between the overall complex formation equilibrium constant for this reaction, $K_f$, and the stepwise formation constants, $K_{f1}$, $K_{f2}$, $K_{f3}$ and $K_{f4}$?

**18-65** Calculate the complex formation equilibrium constant, $K_f$, for the following overall reaction:

$$Ag^+(aq) + 2\ S_2O_3^{2-}(aq) \rightleftharpoons Ag(S_2O_3)_2^{3-}(aq)$$

from the values of the stepwise formation constants.

$$Ag^+(aq) + S_2O_3^{2-}(aq) \rightleftharpoons Ag(S_2O_3)^-(aq) \qquad K_{f1} = 6.6 \times 10^5$$

$$Ag(S_2O_3)^-(aq) + S_2O_3^{2-}(aq) \rightleftharpoons Ag(S_2O_3)_2^{3-}(aq) \qquad K_{f2} = 4.4 \times 10^4$$

**18-66** Calculate the complex formation equilibrium constant, $K_f$, for the following overall reaction:

$$Cd^{2+}(aq) + 4\ CN^-(aq) \rightleftharpoons Cd(CN)_4^{2-}(aq)$$

from the values of the stepwise formation constants:

$$Cd^{2+}(aq) + CN^-(aq) \rightleftharpoons Cd(CN)^+(aq) \qquad K_{f1} = 3.0 \times 10^5$$

$$Cd(CN)^+(aq) + CN^-(aq) \rightleftharpoons Cd(CN)_2(aq) \qquad K_{f2} = 1.3 \times 10^5$$

$$Cd(CN)_2(aq) + CN^-(aq) \rightleftharpoons Cd(CN)_3^-(aq) \qquad K_{f3} = 4.3 \times 10^4$$

$$Cd(CN)_3^-(aq) + CN^-(aq) \rightleftharpoons Cd(CN)_4^{2-}(aq) \qquad K_{f4} = 3.5 \times 10^3$$

**18-67** Which of the following solutions has the smallest concentration of the $Ag^+$ ion?

(a) 0.10 *M* $Ag^+$ and 1.0 *M* $Cl^-$ ($AgCl_2^-$: $K_f = 1.1 \times 10^5$)

(b) 0.10 *M* $Ag^+$ and 1.0 *M* $NH_3$ [$Ag(NH_3)_2^+$: $K_f = 1.1 \times 10^7$]

(c) 0.10 *M* $Ag^+$ and 1.0 *M* $S_2O_3^{2-}$ [$Ag(S_2O_3)_2^{3-}$: $K_f = 2.9 \times 10^{13}$]

(d) 0.10 *M* $Ag^+$ and 1.0 *M* $CN^-$ [$Ag(CN)_2^-$: $K_f = 1.3 \times 10^{21}$]

**18-68** Which of the following solutions has the largest concentration of the $Hg^{2+}$ ion?

(a) 0.10 *M* $Hg^{2+}$ and 1.0 *M* $Cl^-$ ($HgCl_4^{2-}$: $K_f = 1.2 \times 10^{15}$)

(b) 0.10 *M* $Hg^{2+}$ and 1.0 *M* $Br^-$ ($HgBr_4^{2-}$: $K_f = 1 \times 10^{21}$)

(c) 0.10 *M* $Hg^{2+}$ and 1.0 *M* $I^-$ ($HgI_4^{2-}$: $K_f = 6.8 \times 10^{29}$)

(d) 0.10 *M* $Hg^{2+}$ and 1.0 *M* $CN^-$ [$Hg(CN)_4^{2-}$: $K_f = 3 \times 10^{41}$]

### Complex Dissociation Equilibrium Constants

**18-69** Which of the following equations describes the relationship between the complex formation equilibrium constant, $K_f$, and the complex dissociation equilibrium constant, $K_d$, for the $Fe(CN)_6^{3-}$ complex ion?

(a) $K_f = K_d$ (b) $K_f = K_w/K_d$ (c) $K_f = K_d/K_w$
(d) $K_f = K_w \times K_d$ (e) $K_f = 1/K_d$

**18-70** Derive the relationships among $K_{f1}$, $K_{f2}$, $K_{d1}$, and $K_{d2}$ for the $Fe(SCN)^{2+}$ and $Fe(SCN)_2^+$ complex ions.

### Approximate Complex-Ion Calculations

**18-71** Calculate the $Fe^{3+}$ ion concentration at equilibrium in a solution prepared by adding 0.100 mol of $SCN^-$ to 250 mL of 0.0010 *M* $Fe(NO_3)_3$:

$$Fe^{3+}(aq) + 2\ SCN^-(aq) \rightleftharpoons Fe(SCN)_2^+(aq) \qquad K_f = 2.3 \times 10^3$$

**18-72** Calculate the $Cu^{2+}$ ion concentration at equilibrium in a solution that is initially 0.10 *M* in $Cu^{2+}$ and 4 *M* in $NH_3$. [$Cu(NH_3)_4^{2+}$: $K_f = 2.1 \times 10^{13}$]

**18-73** Calculate the $Zn^{2+}$ ion concentration at equilibrium in a solution prepared by dissolving 0.220 mol of $ZnCl_2$ in 500 mL of 2.0 *M* ammonia. [$Zn(NH_3)_4^{2+}$: $K_f = 2.9 \times 10^9$]

**18-74** Calculate the $Sb^{3+}$ ion concentration at equilibrium in an 0.10 *M* $Sb^{3+}$ ion solution that has been buffered at pH 8.00. [$Sb(OH)_4^-$: $K_f = 2 \times 10^{35}$]

**18-75** Calculate the $Sb^{3+}$ ion concentration at equilibrium in a solution that is initially 0.10 *M* in $Sb^{3+}$ and 6 *M* in HCl. ($SbCl_4^-$: $K_f = 5.2 \times 10^4$)

**18-76** Calculate the $Co^{3+}$ ion concentration at equilibrium in a solution that is initially 0.10 *M* in $Co^{3+}$ and 1.0 *M* in $SCN^-$. [$Co(SCN)_4^-$: $K_f = 1 \times 10^3$]

**18-77** Calculate the $Cd^{2+}$ ion concentration at equilibrium in a solution prepared by adding 10.0 mL of 15 *M* aqueous ammonia to 100 mL of a solution of $7.00 \times 10^{-3}$ grams of $CdCl_2$ in water. [$Cd(NH_3)_4^{2+}$: $K_f = 1.3 \times 10^7$]

### Using Complex-Ion Equilibria To Dissolve an Insoluble Salt

**18-78** AgCl is virtually insoluble in water, but it is reasonably soluble in 4 *M* $NH_3$. How many moles of AgCl will dissolve in 1.00 L of 4.00 *M* $NH_3$? [AgCl: $K_{sp} = 1.8 \times 10^{-10}$; $Ag(NH_3)_2^+$: $K_f = 1.1 \times 10^7$]

**18-79** AgBr is less soluble than AgCl, so we need to raise the concentration of ammonia in the solution in order to get a reasonable amount to dissolve. How many grams of AgBr will dissolve in 250 mL of 6.0 *M* $NH_3$? [$Ag(NH_3)_2^+$: $K_f = 1.1 \times 10^7$; AgBr: $K_{sp} = 5.0 \times 10^{-13}$]

**18-80** $K_f$ for the $Ag(NH_3)_2^+$ complex ion is not large enough to allow silver bromide to dissolve in 4 *M* $NH_3$. How many moles of AgBr will dissolve in 1 L of 15 *M* $NH_3$? [AgBr: $K_{sp} = 5.0 \times 10^{-13}$; $Ag(NH_3)_2^+$: $K_f = 1.1 \times 10^7$]

**18-81** $K_f$ for the thiosulfate complex is not large enough to allow silver iodide to dissolve in $S_2O_3^{2-}$ solutions but the value of $K_f$ for the cyanide complex ion, $Ag(CN)_2^-$, is much larger. How many moles of AgI will dissolve in a liter of 4 *M* $CN^-$? [AgI: $K_{sp} = 8.3 \times 10^{-17}$; $Ag(CN)_2^-$: $K_f = 1 \times 10^{21}$]

### A Qualitative View of Combined Equilibria

**18-82** $Fe^{3+}$ forms blood-red complexes with the thiocyanate ion, $SCN^-$. What is the best way to increase the concentration of these complex ions in a solution that contains the following equilibria?

$$2\,H_2O(l) \rightleftharpoons H_3O^+(aq) + OH^-(aq) \qquad K_w = 1.0 \times 10^{-14}$$

$$Fe(OH)_3(s) \rightleftharpoons Fe^{3+}(aq) + 3\,OH^-(aq) \qquad K_{sp} = 4 \times 10^{-38}$$

$$Fe^{3+} + SCN^-(aq) \rightleftharpoons Fe(SCN)^{2+}(aq) \qquad K_{f1} = 890$$

$$Fe(SCN)^{2+}(aq) + SCN^-(aq) \rightleftharpoons Fe(SCN)_2^+(aq) \qquad K_{f2} = 2.6$$

$$SCN^-(aq) + H_2O(l) \rightleftharpoons HSCN(aq) + OH^-(aq) \qquad K_a = 71$$

(a) Add $HNO_3$. (b) Add NaOH. (c) Add NaSCN. (d) Add $Fe(OH)_3$.

**18-83** What is the effect of adding a strong acid to a solution that contains the $Zn(CN)_4^{2-}$ complex ion if HCN is a weak acid ($K_a = 6 \times 10^{-10}$)?

(a) The $Zn^{2+}$ ion concentration increases. (b) The $Zn^{2+}$ ion concentration decreases. (c) The $Zn^{2+}$ ion concentration remains the same. (d) The $Zn^{2+}$ and $CN^-$ ion concentrations both increase. (e) There is no way of predicting what will happen to the $Zn^{2+}$ ion concentration.

**18-84** What is one way to increase the concentration of the $Cu^{2+}$ ion in a saturated solution of $CuSO_4$ in ammonia in which the following equilibria occur?

$$CuSO_4(s) \rightleftharpoons Cu^{2+}(aq) + SO_4^{2-}(aq)$$

$$Cu^{2+}(aq) + 4\,NH_3(aq) \rightleftharpoons Cu(NH_3)_4^{2+}(aq)$$

$$Cu^{2+}(aq) + 2\,OH^-(aq) \rightleftharpoons Cu(OH)_2(s)$$

$$NH_3(aq) + H_2O(l) \rightleftharpoons NH_4^+(aq) + OH^-(aq)$$

$$2\,H_2O(l) \rightleftharpoons H_3O^+(aq) + OH^-(aq)$$

$$SO_4^{2-}(aq) + H_2O(l) \rightleftharpoons HSO_4^-(aq) + OH^-(aq)$$

(a) Add an acid, such as $HNO_3$. (b) Add a base, such as NaOH. (c) Increase the ammonia concentration. (d) Add more $CuSO_4$. (e) None of these increases the $Cu^{2+}$ ion concentration in this solution.

### Integrated Problems

**18-85** What concentration of $NH_3$ must be present in a 0.10 *M* $AgNO_3$ solution to prevent AgCl from precipitating when 4.0 g of sodium chloride are added to 250 mL of this solution? [$Ag(NH_3)_2^+$: $K_f = 1.1 \times 10^7$; AgCl: $K_{sp} = 1.8$ $10^{-10}$]

**18-86** Will $Co(OH)_3$ precipitate from a solution that is initially 0.10 *M* in $Co^{3+}$ and 1.0 *M* in $SCN^-$ if this solution is buffered at pH 7.00? [$Co(OH)_3$: $K_{sp} = 1.6 \times 10^{-44}$; $Co(SCN)_4^-$: $K_f = 1 \times 10^3$]

**18-87** It is possible to keep $Co(OH)_3$ from precipitating from a 0.010 *M* $CoCl_3$ solution by buffering the solution at pH 9.10 with a buffer that contains $NH_3$ and the $NH_4^+$ ion. How much 6 *M* $NH_3$ and 6 *M* HCl must be added per liter of this solution to prevent $Co(OH)_3$ from precipitating? [$NH_3$: $K_b = 1.8 \times 10^{-5}$; $Co(OH)_3$: $K_{sp} = 1.0 \times 10^{-43}$; $Co(NH_3)_6^{3+}$: $K_f = 2 \times 10^{35}$]

**18-88** Calculate the $CO_3^{2-}$ ion concentration in a 0.10 *M* $HCO_3^-$ solution buffered with equal numbers of moles of $NH_3$ and $NH_4^+$. Is this $CO_3^{2-}$ concentration large enough to precipitate $BaCO_3$ when the solution is mixed with an equal volume of a 0.10 *M* $Ba^{2+}$ ion solution? ($BaCO_3$: $K_{sp} = 5.1 \times 10^{-9}$; $H_2CO_3$: $K_{a1} = 4.5 \times 10^{-7}$, $K_{a2} = 4.7 \times 10^{-11}$; $NH_3$: $K_b = 1.8 \times 10^{-5}$)

**18-89** Research has shown that enough $CO_3^{2-}$ ion can be leached out of clay to buffer groundwater at a pH of about 8. Assume that the total concentration of the $HCO_3^-$ and $CO_3^{2-}$ ions in this solution is 0.10 *M*. Calculate the maximum concentration of $^{60}Co$ that could leach into the groundwater if clay were used as a barrier to store this radioactive isotope. ($CoCO_3$: $K_{sp} = 1.4 \times 10^{-13}$; $Co(OH)_3$: $K_{sp} = 1.6 \times 10^{-44}$; $H_2CO_3$: $K_{a1} = 4.5 \times 10^{-7}$, $K_{a2} = 4.7 \times 10^{-11}$)

# Intersection

## OXIDATION–REDUCTION REACTIONS

In 1818, Mary Wollstonecraft Shelley wrote a novel—*Frankenstein; or, the Modern Prometheus*—in which Dr. Frankenstein attempts to create life using lightning, the fire stolen from heaven. But he finds he cannot control the magic he steals. In the end, the monster extracts revenge by killing his creator. Not for creating him, which was difficult enough to forgive, but for abandoning him after he was created.

Although few people still read the book, almost everyone is familiar with one or more of the movie versions. The image of Colin Clive as Dr. Henry Frankenstein bringing life to the monster played by Boris Karloff in the original 1931 film, or Gene Wilder as a descendant of the original Frankenstein bringing life to a monster played by Peter Boyle, is easy for many of us to resurrect from our memories.

What is more difficult to imagine is the climate in which the book was written, 175 years ago. The dark castle, a stormy night, an incompetent assistant, a manic scientist, and an experiment to sustain or reanimate life that went astray are characteristic elements of a Gothic novel. But some of these plot elements go beyond this; they reflect a genuine interest at the turn of the nineteenth century in the implications of early experiments being done with electricity that carried far beyond the group doing the experiments, into the population of scholars in all areas, including the arts and humanities.

Interviews with individuals who have survived being hit by lightning, however, show the error in the metaphor Mary Shelley used to reanimate life. A stroke of lightning carries enough voltage to disrupt the nervous system, causing residual pain that can last for weeks,

months, or even years. There is an element of truth, however, in the assumption that life and electric current are interconnected. When the potential behind the flow of electrons is a fraction of a volt, not tens of thousands of volts, and when the current is relatively small, the flow of electrons through the electron transport system is an essential component of the process by which living systems capture energy from the food they consume.

The next two chapters will focus on the interaction between the flow of electrons and chemical systems. Chapter 19 examines reactions that involve the transfer of electrons and atom-transfer reactions that can be interpreted as if they involved the flow of electrons. Chapter 20 will focus on two questions: How can reactions that involve the transfer of electrons be used to do work? and How can work be done on a chemical system to induce the transfer of electrons?

The three previous chapters have shown how equilibrium constants can be used to determine the concentrations or partial pressures of the components of a chemical reaction at equilibrium. Chapter 20 will introduce a technique for measuring equilibrium constants. When asked: "Where do equilibrium constants come from?" you will no longer have to answer: "From the back of the textbook!"

Cars were painted at first for only one reason—to slow down the rate at which the steel rusted.

# OXIDATION–REDUCTION REACTIONS

The concepts of oxidation and reduction were originally introduced to explain what happens in reactions that involve the transfer of one or more electrons. This chapter provides the basis for answering questions such as the following:

## POINTS OF INTEREST

- What do the following reactions have in common that allow both to be classified as oxidation–reduction reactions?

$$2\,Mg(s) + O_2(g) \longrightarrow 2\,MgO(s)$$
$$CO_2(g) + H_2(g) \longrightarrow CO(g) + H_2O(g)$$

- What does it mean when we say that $O_2$ is an *oxidizing agent* and zinc metal is a *reducing agent*?

- How do we predict the correct stoichiometry for reactions, such as the following, for which it seems that more than one balanced equation can be written?

$$2\,MnO_4^-(aq) + H_2O_2(aq) + 6\,H^+(aq) \longrightarrow 2\,Mn^{2+}(aq) + 3\,O_2(g) + 4\,H_2O(l)$$
$$2\,MnO_4^-(aq) + 3\,H_2O_2(aq) + 6\,H^+(aq) \longrightarrow 2\,Mn^{2+}(aq) + 4\,O_2(g) + 6\,H_2O(l)$$
$$2\,MnO_4^-(aq) + 5\,H_2O_2(aq) + 6\,H^+(aq) \longrightarrow 2\,Mn^{2+}(aq) + 5\,O_2(g) + 8\,H_2O(l)$$
$$2\,MnO_4^-(aq) + 7\,H_2O_2(aq) + 6\,H^+(aq) \longrightarrow 2\,Mn^{2+}(aq) + 6\,O_2(g) + 10\,H_2O(l)$$

- Why does the following reaction occur?

$$Cu^{2+}(aq) + Zn(s) \longrightarrow Cu(s) + Zn^{2+}(aq)$$

  Why won't the reverse reaction occur to an appreciable extent?

$$Zn^{2+}(aq) + Cu(s) \longrightarrow Cu^{2+}(aq) + Zn(s)$$

- How do we predict whether the following oxidation–reduction reaction will occur?

$$2\,Fe^{3+}(aq) + 2\,I^-(aq) \longrightarrow 2\,Fe^{2+}(aq) + I_2(aq)$$

## 19.1 THE PROCESS OF DISCOVERY: OXIDATION AND REDUCTION

The first step toward a theory of chemical reactions was taken by Georg Ernst Stahl in 1697 when he proposed the *phlogiston* theory, which was based on the following observations:

- Metals have many properties in common.
- Metals often produce a ''calx'' when heated. (The *calx* is the crumbly residue left after a mineral or metal is roasted.)
- These calxes are not as dense as the metals from which they are produced.
- Some of these calxes form metals when heated with charcoal.
- With only a few exceptions, the calx is found in nature, not the metal.

These observations led Stahl to the following conclusions.

- Phlogiston (from the Greek *phlogistos,* ''to burn'') is given off whenever something burns.
- Wood and charcoal are particularly rich in phlogiston because they leave very little ash when they burn. (Candles must be almost pure phlogiston because they leave no ash.)
- Because they are found in nature, calxes must be simpler than metals.
- Metals form a calx by giving off phlogiston:

$$\text{metal} \longrightarrow \text{calx} + \text{phlogiston}$$

- Metals can be made by adding phlogiston to the calx:

$$\text{calx} + \text{phlogiston} \longrightarrow \text{metal}$$

- Because charcoal is rich in phlogiston, heating calxes in the presence of charcoal sometimes produces metals.

This model was remarkably successful. It explained why metals have similar properties—they all contained phlogiston. It explained the relationship between metals and their calxes—they were related by the gain or loss of phlogiston. It even explained why a candle goes out when placed in a bell jar—the air eventually becomes saturated with phlogiston.

### CHECKPOINT

Use the phlogiston theory to explain what is observed when a strip of magnesium metal is ignited.

---

Initially, there was only one problem with the phlogiston theory. As early as 1630, Jean Rey noted that tin gains weight when it forms a calx. (The calx is about 25% heavier than the metal.) From our point of view, this seems to be a fatal flaw: If phlogiston is given off when a metal forms a calx, why does the calx weigh more than the metal? This observation didn't bother proponents of the phlogiston theory. Stahl explained it by suggesting that the weight increased because air entered the metal to fill the vacuum left after the phlogiston escaped.

The phlogiston theory was the basis for research in chemistry for most of the eighteenth century. It was not until 1772 that Antoine Lavoisier noted that nonmetals gain enormous amounts of weight when burned in air. (The weight of phosphorus, for example, increases by a factor of about 2.3.) The magnitude of this change led Lavoisier to conclude that phosphorus must combine with something in air when it burns. This conclusion was reinforced by the observation that the volume of air decreases by a factor of one-fifth when phosphorus burns in a limited amount of air.

Lavoisier proposed the name *oxygene* (literally, the ''acid former'') for the substance absorbed from air when a compound burns because the products of the combustion of nonmetals such as phosphorus are acids when they dissolve in water:

$$P_4(s) + 5\,O_2(g) \longrightarrow P_4O_{10}(s)$$

$$P_4O_{10}(s) + 6\,H_2O(l) \longrightarrow 4\,H_3PO_4(aq)$$

Lavoisier's oxygen theory of combustion was eventually accepted and chemists began to describe any reaction between an element or compound and oxygen as **oxidation.** The reaction between magnesium metal and oxygen, for example, involves the oxidation of magnesium:

$$2\,Mg(s) + O_2(g) \longrightarrow 2\,MgO(s)$$

By the turn of the twentieth century, it seemed that all oxidation reactions had one thing in common—oxidation always seemed to involve the loss of electrons. Chemists therefore developed a model for these reactions that focused on the transfer of electrons. Magnesium metal, for example, was thought to lose electrons to form $Mg^{2+}$ ions when it reacted with oxygen. By convention, the element or compound that gained these electrons was said to undergo **reduction.** In this case, $O_2$ molecules were said to be reduced to form $O^{2-}$ ions.

$$2\,Mg + O_2 \longrightarrow 2\,[Mg^{2+}][O^{2-}]$$

oxidation (Mg → $Mg^{2+}$)

reduction ($O_2$ → $O^{2-}$)

A classic demonstration of oxidation–reduction reactions involves placing a piece of copper wire into an aqueous solution of the $Ag^+$ ion. The reaction involves the net transfer of electrons from copper metal to $Ag^+$ ions to produce whiskers of silver metal that grow out from the copper wire and $Cu^{2+}$ ions.

$$Cu(s) + 2\,Ag^+(aq) \longrightarrow Cu^{2+}(aq) + 2\,Ag(s)$$

Whiskers of silver metal grow on the surface of a copper wire when it is immersed in a solution of $Ag^+$ ions. The fact that $Cu^{2+}$ ions are also formed in this reaction can be demonstrated by adding $NH_3(aq)$ to this solution.

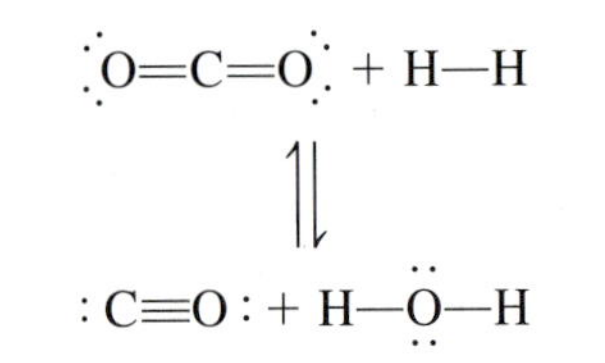


**Figure 19.1** Oxidation–reduction reactions don't always involve the transfer of electrons, they also include atom-transfer reactions such as this.

The $Cu^{2+}$ ions formed in this reaction are responsible for the light-blue color of the solution. Their presence can be confirmed by adding ammonia to this solution to form the deep-blue $Cu(NH_3)_4{}^{2+}$ complex ion.

Chemists eventually recognized that oxidation–reduction reactions don't always involve the transfer of electrons. There is no change in the number of valence electrons on any of the atoms in the following reaction, for example, as shown in Figure 19.1.

$$CO_2(g) + H_2(g) \rightleftharpoons CO(g) + H_2O(g)$$

Chemists therefore developed the concept of **oxidation number** to extend the idea of oxidation and reduction to reactions in which electrons are not really gained or lost.

Sec. 9.12

The most powerful model of oxidation–reduction reactions is based on the following definitions, first introduced in Section 9.12.

**Oxidation involves an increase in the oxidation number of an atom. Reduction occurs when the oxidation number of an atom decreases.**

According to this model, $CO_2$ is reduced when it reacts with hydrogen because the oxidation number of the carbon decreases from +4 to +2. Hydrogen is oxidized in this reaction because its oxidation number increases from 0 to +1.

$$CO_2(g) + H_2(g) \rightleftharpoons CO(g) + H_2O(g)$$

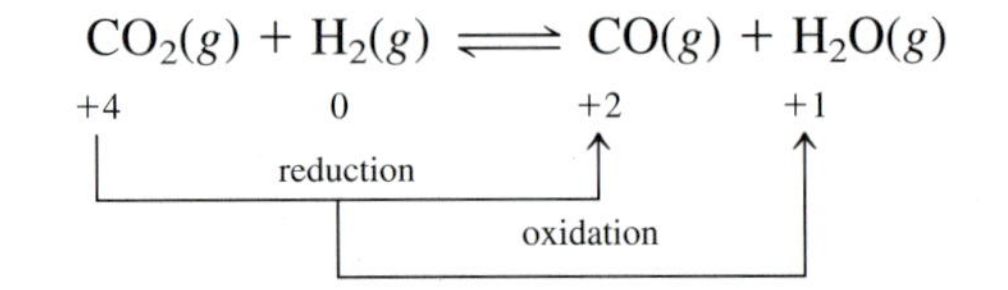


## 19.2 OXIDATION–REDUCTION REACTIONS

We find examples of oxidation–reduction or **redox reactions** almost every time we analyze the reactions used as sources of either heat or work. When natural gas burns, for example, an oxidation–reduction reaction occurs that releases more than 800 kJ/mol of energy:

$$CH_4(g) + 2\,O_2(g) \longrightarrow CO_2(g) + 2\,H_2O(g)$$

Within our bodies, a sequence of oxidation–reduction reactions burns sugars, such as glucose ($C_6H_{12}O_6$), and the fatty acids in the fats we eat:

$$C_6H_{12}O_6(aq) + 6\,O_2(g) \longrightarrow 6\,CO_2(g) + 6\,H_2O(l)$$
$$CH_3(CH_2)_{16}CO_2H(aq) + 26\,O_2(g) \longrightarrow 18\,CO_2(g) + 18\,H_2O(l)$$

The burning of natural gas in a petroleum flare is an example of an oxidation-reduction reaction.

We don't have to restrict ourselves to reactions that can be used as a source of energy, however, to find examples of oxidation–reduction reactions. Silver metal, for example, is oxidized when it comes in contact with trace quantities of $H_2S$ or $SO_2$ in the atmosphere, or foods, such as eggs, that are rich in sulfur compounds.

$$4\,Ag(s) + 2\,H_2S(g) + O_2(g) \longrightarrow 2\,Ag_2S(s) + 2\,H_2O(g)$$

Fortunately, the film of $Ag_2S$ that collects on the metal surface forms a protective coating that slows down further oxidation of the silver metal.

The tarnishing of silver is just one example of a broad class of oxidation–reduction reactions that fall under the general heading of **corrosion.** Another exam-

Silver tarnishes because of an oxidation-reduction reaction.

Rust forms on iron metal that is exposed to both oxygen and water.

ple is the series of reactions that occurs when iron or steel rusts. When heated, iron reacts with oxygen to form a mixture of iron(II) and iron(III) oxides:

$$2\ \mathrm{Fe}(s) + \mathrm{O_2}(g) \longrightarrow 2\ \mathrm{FeO}(s)$$
$$2\ \mathrm{Fe}(s) + 3\ \mathrm{O_2}(g) \longrightarrow 2\ \mathrm{Fe_2O_3}(s)$$

Molten iron even reacts with water to form an aqueous solution of $Fe^{2+}$ ions and $H_2$ gas:

$$\mathrm{Fe}(l) + 2\ \mathrm{H_2O}(l) \longrightarrow \mathrm{Fe^{2+}}(aq) + 2\ \mathrm{OH^-}(aq) + \mathrm{H_2}(g)$$

At room temperature, however, all three of these reactions are so slow they can be ignored.

Iron only corrodes at room temperature in the presence of both oxygen and water. In the course of this reaction, the iron is oxidized to give a hydrated from of iron(II) oxide:

$$2\ \mathrm{Fe}(s) + \mathrm{O_2}(aq) + 2\ \mathrm{H_2O}(l) \longrightarrow 2\ \mathrm{FeO} \cdot \mathrm{H_2O}(s)$$

Because this compound has the same empirical formula as $Fe(OH)_2$, it is often mistakenly called iron(II), or ferrous, hydroxide. The $FeO \cdot H_2O$ formed in this reaction is further oxidized by $O_2$ dissolved in water to give a hydrated form of iron(III), or ferric, oxide:

$$4\ \mathrm{FeO} \cdot \mathrm{H_2O}(s) + \mathrm{O_2}(aq) + 2\ \mathrm{H_2O}(l) \longrightarrow 2\ \mathrm{Fe_2O_3} \cdot 3\ \mathrm{H_2O}(s)$$

To further complicate matters, $FeO \cdot H_2O$ formed at the metal surface combines with $Fe_2O_3 \cdot 3\ H_2O$ to give a hydrated form of magnetic iron oxide ($Fe_3O_4$):

$$\mathrm{FeO} \cdot \mathrm{H_2O}(s) + \mathrm{Fe_2O_3} \cdot 3\ \mathrm{H_2O}(s) \longrightarrow \mathrm{Fe_3O_4} \cdot \mathrm{n}\ \mathrm{H_2O}(s)$$

Because these reactions only occur in the presence of both water and oxygen, cars tend to rust where water collects. Furthermore, because the simplest way of preventing iron from rusting is to coat the metal so that it doesn't come in contact with water, cars were originally painted for only one reason—to slow down the formation of rust.

## 19.3 ASSIGNING OXIDATION NUMBERS

The key to identifying oxidation–reduction reactions is recognizing when a chemical reaction leads to a change in the oxidation number of one or more atoms. It is therefore a good idea to take another look at the rules for assigning oxidation numbers. By definition, the oxidation number of an atom is equal to the charge that would be present on the atom if the compound were composed of ions. If we assume that $CH_4$ contains $C^{4-}$ and $H^+$ ions, for example, the oxidation numbers of the carbon and hydrogen atoms would be −4 and +1.

Sec. 2.12

Note that it doesn't matter whether the compound actually contains ions. The oxidation number is the charge an atom would have if the compound were ionic. The concept of oxidation number is nothing more than a bookkeeping system used to keep track of electrons in chemical reactions. This system is based on a series of rules, summarized in Table 19.1.

**TABLE 19.1 Rules for Assigning Oxidation Numbers**

- The oxidation number of an atom is zero in a neutral substance that contains atoms of only one element. Thus, the atoms in $O_2$, $O_3$, $P_4$, $S_8$, and aluminum metal all have an oxidation number of 0.
- The oxidation number of monatomic ions is equal to the charge on the ion. The oxidation number of sodium in the $Na^+$ ion is +1, for example, and the oxidation number of chlorine in the $Cl^-$ ion is −1.
- The oxidation number of hydrogen is +1 when it is combined with a *nonmetal*. Hydrogen is therefore in the +1 oxidation state in $CH_4$, $NH_3$, $H_2O$, and HCl.
- The oxidation number of hydrogen is −1 when it is combined with a *metal*. Hydrogen is therefore in the −1 oxidation state in LiH, NaH, $CaH_2$, and $LiAlH_4$.
- The metals in Group IA form compounds (such as $Li_3N$ and $Na_2S$) in which the metal atom is in the +1 oxidation state.
- The elements in Group IIA form compounds (such as $Mg_3N_2$ and $CaCO_3$) in which the metal atom is in the +2 oxidation state.
- Oxygen usually has an oxidation number of −2. Exceptions include molecules and polyatomic ions that contain O—O bonds, such as $O_2$, $O_3$, $H_2O_2$, and the $O_2^{2-}$ ion.
- The nonmetals in Group VIIA often form compounds (such as $AlF_3$, HCl, and $ZnBr_2$) in which the nonmetal is in the −1 oxidation state.
- The sum of the oxidation numbers of the atoms in a molecule is equal to the charge on the molecule.
- The most electronegative element in a compound has a negative oxidation number.

Any set of rules, no matter how good, will only get you so far. You then have to rely on a combination of common sense and prior knowledge. Questions to keep in mind while assigning oxidation numbers include the following: Are there any recognizable ions hidden in the molecule? Does the oxidation number make sense in terms of the known electron configuration of the atom?

## EXERCISE 19.1

Determine the oxidation number of each element in the following compounds:

(a) $BaO_2$ (b) $(NH_4)_2MoO_4$ (c) $Na_3Co(NO_2)_6$ (d) $CS_2$

**SOLUTION**

(a) If the oxidation number of the oxygen in $BaO_2$ were −2, the oxidation number of the barium would have to be +4. But elements in Group IIA can't form +4 ions. This compound must be barium peroxide, $[Ba^{2+}][O_2^{2-}]$. Barium is therefore +2 and oxygen is −1.

(b) $(NH_4)_2MoO_4$ contains the $NH_4^+$ ion, in which hydrogen is +1 and nitrogen is −3. Because there are two $NH_4^+$ ions, the other half of the compound must be an $MoO_4^{2-}$ ion, in which molybdenum is +6 and oxygen is −2.

(c) Sodium is in the +1 oxidation state in all of its compounds. This compound therefore contains the $Co(NO_2)_6^{3-}$ complex ion. This complex ion contains six $NO_2^-$ ions in which the oxidation number of nitrogen is +3 and oxygen is −2. The oxidation state of the cobalt atom is therefore +3.

(d) The most electronegative element in a compound always has a negative oxidation number. Since sulfur tends to form −2 ions, the oxidation number of the sulfur in $CS_2$ is −2 and the carbon is +4.

## 19.4 RECOGNIZING OXIDATION–REDUCTION REACTIONS

Chemical reactions are often divided into two categories: oxidation–reduction reactions and metathesis reactions. Metathesis reactions include acid–base reactions that involve the transfer of an $H^+$ ion from a Brønsted acid to a Brønsted base:

$$\underset{\text{Brønsted acid}}{CH_3CO_2H(aq)} + \underset{\text{Brønsted base}}{OH^-(aq)} \rightleftharpoons \underset{\text{Brønsted base}}{CH_3CO_2^-(aq)} + \underset{\text{Brønsted acid}}{H_2O(l)}$$

They can also involve the sharing of a pair of electrons by an electron-pair donor (Lewis base) and an electron-pair acceptor (Lewis acid):

$$\underset{\text{Lewis acid}}{Co^{3+}(aq)} + \underset{\text{Lewis base}}{6\ NO_2^-(aq)} \rightleftharpoons Co(NO_2)_6^{3-}(aq)$$

Oxidation–reduction reactions can involve the transfer of one or more electrons:

$$Cu(s) + 2\ Ag^+(aq) \longrightarrow Cu^{2+}(aq) + 2\ Ag(s)$$

They can also occur by the transfer of oxygen, hydrogen, or halogen atoms:

$$CO_2(g) + H_2(g) \rightleftharpoons CO(g) + H_2O(g)$$
$$SF_4(g) + F_2(g) \longrightarrow SF_6(g)$$

Fortunately, there is an almost foolproof method of distinguishing between metathesis and redox reactions. Reactions in which none of the atoms undergoes a change in oxidation number are **metathesis** reactions. There is no change in the oxidation number of any atom in either of the metathesis reactions in Figure 19.2, for example. The word *methathesis* literally means "interchange" or "transposition," and it is used to describe changes that occur in the order of letters or sounds in a word as a language develops. Metathesis occurred, for example, when the Old English word *brid* became *bird.* In chemistry, methathesis is used to describe reactions that interchange atoms or groups of atoms between molecules.

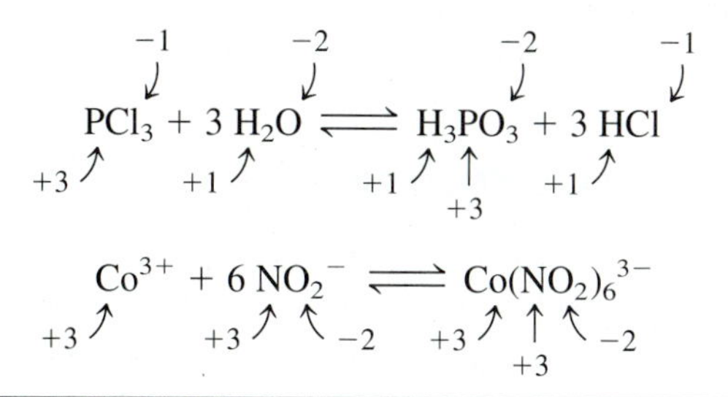


**Figure 19.2** Examples of metathesis reactions.

When at least one atom undergoes a change in oxidation state, the reaction is an oxidation–reduction reaction. Each of the reactions in Figure 19.3 is therefore an example of an oxidation–reduction reaction.

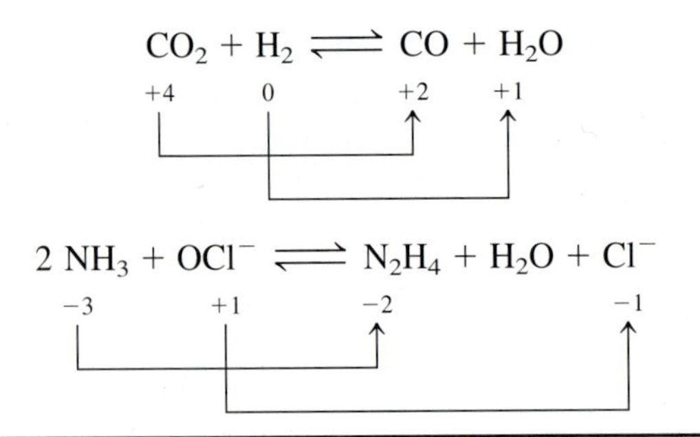


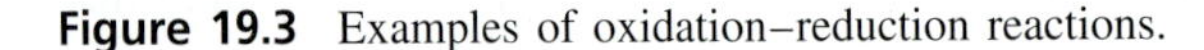

**Figure 19.3** Examples of oxidation–reduction reactions.

**EXERCISE 19.2**

Classify each of the following as either a metathesis or an oxidation–reduction reaction. Note that mercury usually exists in one of three oxidation states: mercury metal, $Hg_2^{2+}$ ions, or $Hg^{2+}$ ions.

(a) $Hg_2^{2+}(aq) + 2\,OH^-(aq) \longrightarrow Hg_2O(s) + H_2O(l)$
(b) $Hg_2^{2+}(aq) + Sn^{2+}(aq) \longrightarrow 2\,Hg(l) + Sn^{4+}(aq)$
(c) $Hg_2^{2+}(aq) + H_2S(aq) \longrightarrow Hg(l) + HgS(s) + 2\,H^+(aq)$
(d) $Hg_2CrO_4(s) + 2\,OH^-(aq) \longrightarrow Hg_2O(s) + CrO_4^{2-}(aq) + H_2O(l)$

**SOLUTION**

(a) Metathesis. Mercury is in the +1 oxidation state in both the $Hg_2^{2+}$ ion and in $[Hg_2^{2+}][O^{2-}]$.

(b) Oxidation–reduction. Mercury is reduced from the +1 to the 0 oxidation state, while tin is oxidized from +2 to +4.

(c) Oxidation–reduction. This is a disproportionation reaction (Section 10.3) in which mercury is simultaneously reduced from +1 to 0 and oxidized from +1 to +2.

(d) Metathesis. Mercury is in the +1 oxidation state in both $[Hg_2^{2+}][CrO_4^{2-}]$ and $[Hg_2^{2+}][O^{2-}]$.

## 19.5 BALANCING OXIDATION–REDUCTION EQUATIONS

A trail-and-error approach to balancing chemical equations was introduced in Section 3.13. This approach involves playing with the equation—adjusting the ratio of the reactants and products—until the following goals have been achieved.

**GOALS FOR BALANCING CHEMICAL EQUATIONS**

**1.** The same number of atoms of each element is found on both sides of the equation and therefore mass is conserved.

**2.** The sum of the positive and negative charges is the same on both sides of the equation and therefore charge is conserved. (Charge is conserved because electrons are neither created nor destroyed in a chemical reaction.)

There are two situations in which relying on trial and error can get you into trouble. Sometimes the equation is too complex to be solved by trial and error within a reasonable amount of time. Consider the following reaction, for example:

$$3\,Cu(s) + 8\,HNO_3(aq) \longrightarrow 3\,Cu^{2+}(aq) + 2\,NO(g) + 6\,NO_3^-(aq) + 4\,H_2O(l)$$

Other times, more than one equation can be written that seems to be balanced. The following are just a few of the balanced equations that can be written for the reaction between the permanganate ion and hydrogen peroxide, for example:

$$2\,MnO_4^-(aq) + H_2O_2(aq) + 6\,H^+(aq) \longrightarrow 2\,Mn^{2+}(aq) + 3\,O_2(g) + 4\,H_2O(l)$$
$$2\,MnO_4^-(aq) + 3\,H_2O_2(aq) + 6\,H^+(aq) \longrightarrow 2\,Mn^{2+}(aq) + 4\,O_2(g) + 6\,H_2O(l)$$
$$2\,MnO_4^-(aq) + 5\,H_2O_2(aq) + 6\,H^+(aq) \longrightarrow 2\,Mn^{2+}(aq) + 5\,O_2(g) + 8\,H_2O(l)$$
$$2\,MnO_4^-(aq) + 7\,H_2O_2(aq) + 6\,H^+(aq) \longrightarrow 2\,Mn^{2+}(aq) + 6\,O_2(g) + 10\,H_2O(l)$$

Equations such as these have to be balanced by a more systematic approach than trial and error.

## 19.6 THE HALF-REACTION METHOD OF BALANCING REDOX EQUATIONS

A powerful technique for balancing oxidation–reduction equations involves dividing these reactions into separate oxidation and reduction half-reactions. We then balance the half-reactions, one at a time, and combine them so that electrons are neither created nor destroyed in the reaction.

The steps involved in the half-reaction method for balancing equations can be illustrated by considering the reaction used to determine the amount of the triiodide ion ($I_3^-$) in a solution by titration.

**Step 1: Write a skeleton equation for the reaction.** The skeleton equation for the reaction on which this titration is based can be written as follows:

$$I_3^- + S_2O_3^{2-} \longrightarrow I^- + S_4O_6^{2-}$$

**Step 2: Assign oxidation numbers to atoms on both sides of the equation.** The negative charge in the $I_3^-$ ion is formally distributed over the three iodine atoms, which means that the average oxidation state of the iodine atoms in this ion is $-\frac{1}{3}$. In the $S_4O_6^{2-}$ ion, the total oxidation state of the sulfur atoms is +10. The average oxidation state of the sulfur atoms is therefore $+2\frac{1}{2}$.

$$\underset{-\frac{1}{3}}{I_3^-} + \underset{+2\,-2}{S_2O_3^{2-}} \longrightarrow \underset{-1}{I^-} + \underset{+2\frac{1}{2}\,-2}{S_4O_6^{2-}}$$

**Step 3: Determine which atoms are oxidized and which are reduced.**

$$\underset{-\frac{1}{3}}{I_3^-} + \underset{+2}{S_2O_3^{2-}} \longrightarrow \underset{-1}{I^-} + \underset{+2\frac{1}{2}}{S_4O_6^{2-}}$$

reduction ($I_3^-$ → $I^-$)

oxidation ($S_2O_3^{2-}$ → $S_4O_6^{2-}$)

**Step 4: Divide the reaction into oxidation and reduction half-reactions and balance these half-reactions one at a time.** This reaction can be arbitrarily divided into two half-reactions. One half-reaction describes what happens during oxidation.

$$\text{Oxidation: } \underset{+2}{S_2O_3^{2-}} \longrightarrow \underset{+2\frac{1}{2}}{S_4O_6^{2-}}$$

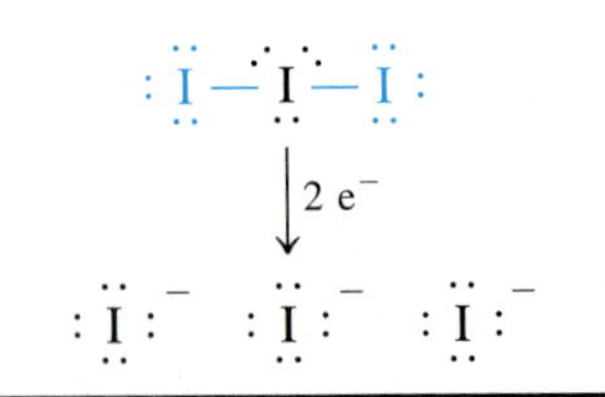

**Figure 19.4** Adding a pair of electrons to the central atom in an $I_3^-$ ion produces a system that consists of three $I^-$ ions.

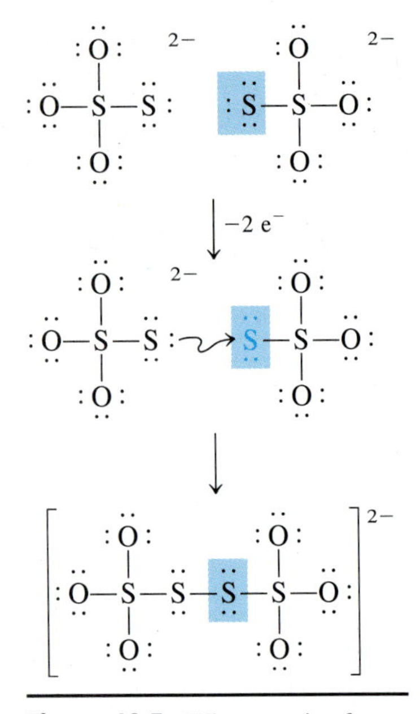

**Figure 19.5** When a pair of electrons is removed from an $S_2O_3^{2-}$ ion we get a neutral $S_2O_3$ molecule with an empty orbital on the terminal sulfur atom. This molecule can accept a pair of nonbonding electrons from the terminal sulfur atom on a neighboring $S_2O_3^{2-}$ ion to form an $S_4O_6^{2-}$ ion.

The other describes the reduction half of the reaction.

$$\text{Reduction: } \underset{-\frac{1}{3}}{I_3^-} \longrightarrow \underset{-1}{I^-}$$

It doesn't matter which half-reaction we balance first, so let's start with the reduction half-reaction. Our goal is to balance this half-reaction in terms of both charge and mass. It seems reasonable to start by balancing the number of iodine atoms on both sides of the equation:

$$\text{Reduction: } I_3^- \longrightarrow \mathbf{3\ I^-}$$

We then balance the charge by noting that two electrons must be added to an $I_3^-$ ion to produce 3 $I^-$ ions, as can be seen from the Lewis structures of these ions shown in Figure 19.4.

$$\text{Reduction: } I_3^- + \mathbf{2\ e^-} \longrightarrow 3\ I^-$$

We now turn to the oxidation half-reaction. The Lewis structures of the starting material and the product of this half-reaction suggest that we can get an $S_4O_6^{2-}$ ion by removing two electrons from a pair of $S_2O_3^{2-}$ ions, as shown in Figure 19.5.

$$\text{Oxidation: } 2\ S_2O_3^{2-} \longrightarrow S_4O_6^{2-} + \mathbf{2\ e^-}$$

**Step 5: Combine these half-reactions so that electrons are neither created nor destroyed.** Two electrons are given off in the oxidation half-reaction and two electrons are picked up in the reduction half-reaction. We can therefore obtain a balanced chemical equation by simply combining these half-reactions:

$$\begin{array}{l} (2\ S_2O_3^{2-} \longrightarrow S_4O_6^{2-} + 2\ e^-) \\ +\ (I_3^- + 2\ e^- \longrightarrow 3\ I^-) \\ \hline I_3^- + 2\ S_2O_3^{2-} \longrightarrow 3\ I^- + S_4O_6^{2-} \end{array}$$

**Step 6: Balance the remainder of the equation by inspection, if necessary.** Since the overall equation is already balanced in terms of both charge and mass, we simply introduce the symbols describing the states of the reactants and products:

$$I_3^-(aq) + 2\ S_2O_3^{2-}(aq) \longrightarrow 3\ I^-(aq) + S_4O_6^{2-}(aq)$$

## 19.7 REDOX REACTIONS IN ACIDIC SOLUTIONS

Some might argue that we didn't need to use half-reactions to balance the equation in the previous section because it can be balanced by trial and error. The half-reaction technique becomes indispensable, however, in balancing reactions such as the oxidation of sulfur dioxide by the dichromate ion in acidic solution:

$$SO_2(aq) + Cr_2O_7^{2-}(aq) \xrightarrow{H^+} SO_4^{2-}(aq) + Cr^{3+}(aq)$$

The reason why this equation is inherently more difficult to balance has nothing to do with the ratio of moles of $SO_2$ to moles of $Cr_2O_7^{2-}$. It is because the solvent takes an active role in both half-reactions.

## EXERCISE 19.3

Use half-reactions to balance the equation for the reaction between sulfur dioxide and the dichromate ion in acidic solution.

**SOLUTION**

**Step 1: Write a skeleton equation for the reaction.**

$$SO_2 + Cr_2O_7^{2-} \longrightarrow SO_4^{2-} + Cr^{3+}$$

**Step 2: Assign oxidation numbers to atoms on both sides of the equation.**

$$\underset{+4-2}{SO_2} + \underset{+6\ -2}{Cr_2O_7^{2-}} \longrightarrow \underset{+6-2}{SO_4^{2-}} + \underset{+3}{Cr^{3+}}$$

**Step 3: Determine which atoms are oxidized and which are reduced.**

$$SO_2 + Cr_2O_7^{2-} \longrightarrow SO_4^{2-} + Cr^{3+}$$

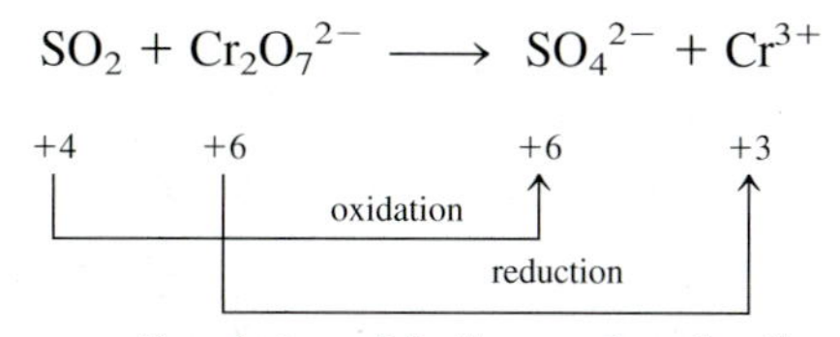


**Step 4: Divide the reaction into oxidation and reduction half-reactions and balance these half-reactions.** This reaction can be divided into the following half-reactions:

$$\text{Oxidation: } \underset{+4}{SO_2} \longrightarrow \underset{+6}{SO_4^{2-}}$$

$$\text{Reduction: } \underset{+6}{Cr_2O_7^{2-}} \longrightarrow \underset{+3}{Cr^{3+}}$$

It doesn't matter which half-reaction we balance first, so let's start with reduction. Because the $Cr_2O_7^{2-}$ ion contains two chromium atoms that must be reduced from the +6 to the +3 oxidation state, six electrons are consumed in this half-reaction:

$$\text{Reduction: } Cr_2O_7^{2-} + 6\,e^- \longrightarrow 2\,Cr^{3+}$$

This raises an interesting question: What happens to the oxygen atoms when the chromium atoms are reduced? The seven oxygen atoms in the $Cr_2O_7^{2-}$ ions are formally in the −2 oxidation state. If these atoms were released into the solution in the −2 oxidation state when the chromium was reduced, they would be present as $O^{2-}$ ions. But it doesn't make sense to write this half-reaction as follows:

$$\text{Reduction: } Cr_2O_7^{2-} + 6\,e^- \longrightarrow 2\,Cr^{3+} + 7\,O^{2-}$$

The reaction is being run in an acidic solution, and the $O^{2-}$ ion is a very strong base that would immediately react with the $H^+$ ions in this solution to form water, as shown in Figure 19.6. The following is therefore a more realistic equation for this half-reaction.

$$\text{Reduction: } Cr_2O_7^{2-} + 14\,H^+ + 6\,e^- \longrightarrow 2\,Cr^{3+} + 7\,H_2O$$

We can now turn to the oxidation half-reaction, and start by noting that two electrons are given off when sulfur is oxidized from the +4 to the +6 oxidation state.

The characteristic yellow-orange color of aqueous solutions of the $Cr_2O_7^{2-}$ ion gradually fades when $SO_2$ is bubbled through this solution.

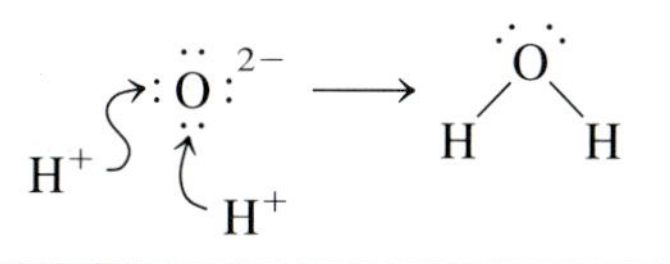


**Figure 19.6** When six electrons are added to a $Cr_2O_7^{2-}$ ion, two $Cr^{3+}$ ions and seven $O^{2-}$ ions are released into the solution. This reaction is run in an acidic solution, however, which means that the $O^{2-}$ ions will instantly react with the acid to form seven $H_2O$ molecules.

$$\text{Oxidation: } SO_2 \longrightarrow SO_4^{2-} + 2\,e^-$$

The key to balancing the charge on both sides of this equation is remembering that the reaction is run in acid, which contains $H^+$ ions and $H_2O$ molecules. We can therefore add $H^+$ ions or $H_2O$ molecules to either side of the equation, as needed. The only way to balance the charge on both sides of this equation is to add four $H^+$ ions to the products of the reaction:

$$\text{Oxidation: } SO_2 \longrightarrow SO_4^{2-} + 2\,e^- + 4\,H^+$$

We can then balance the number of hydrogen and oxygen atoms on both sides of this equation by adding a pair of $H_2O$ molecules to the reactants.

$$\text{Oxidation: } SO_2 + 2\,H_2O \longrightarrow SO_4^{2-} + 2\,e^- + 4\,H^+$$

This equation can be understood in terms of the Lewis structures shown in Figure 19.7.

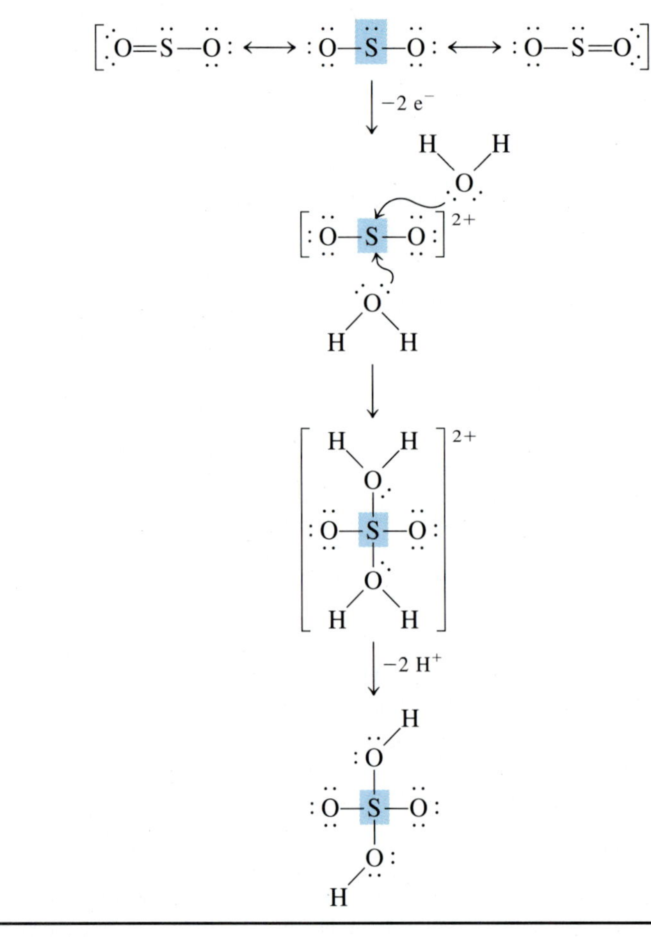


**Figure 19.7** The two-electron oxidation of $SO_2$ in the presence of water leads to the formation of an $SO_4^{2-}$ ion and four $H^+$ ions.

**Step 5: Combine the two half-reactions so that electrons are neither created nor destroyed:** Six electrons are consumed in the reduction half-reaction and two electrons are given off in the oxidation half-reaction. We can combine these half-reactions so that electrons are conserved by multiplying the reduction half-reaction by 3:

$$\begin{array}{l} (Cr_2O_7^{2-} + 14\,H^+ + 6\,e^- \longrightarrow 2\,Cr^{3+} + 7\,H_2O) \\ +\,3(SO_2 + 2\,H_2O \longrightarrow SO_4^{2-} + 2\,e^- + 4\,H^+) \\ \hline Cr_2O_7^{2-} + 3\,SO_2 + 14\,H^+ + 6\,H_2O \longrightarrow \\ \qquad 2\,Cr^{3+} + 3\,SO_4^{2-} + 12\,H^+ + 7\,H_2O \end{array}$$

**Step 6: Balance the remainder of the equation by inspection, if necessary.** Although the equation appears balanced, we are not quite finished with it. We can simplify the equation by subtracting 12 $H^+$ ions and 6 $H_2O$ molecules from each side to generate the following balanced equation:

$$Cr_2O_7^{2-}(aq) + 3\,SO_2(aq) + 2\,H^+(aq) \longrightarrow 2\,Cr^{3+}(aq) + 3\,SO_4^{2-}(aq) + H_2O(l)$$

### CHECKPOINT

On an exam, students were asked to use half-reactions to write a balanced equation for the reaction between $SO_2$ and the $Cr_2O_7^{2-}$ ion in the presence of a strong acid:

$$SO_2(g) + Cr_2O_7^{2-}(aq) \xrightarrow{H^+} SO_4^{2-}(aq) + Cr^{3+}(aq)$$

One student wrote the following half-reactions:

Oxidation: $SO_2 + 4\,OH^- \longrightarrow SO_4^{2-} + 2\,H_2O + 2\,e^-$

Reduction: $Cr_2O_7^{2-} + 7\,H_2O + 6\,e^- \longrightarrow 2\,Cr^{3+} + 14\,OH^-$

The student then wrote the following overall equation:

$$3\,SO_2(g) + Cr_2O_7^{2-}(aq) + H_2O(l) \longrightarrow 3\,SO_4^{2-}(aq) + 2\,Cr^{3+}(aq) + 2\,OH^-(aq)$$

Explain why the instructor gave no credit for this answer, even though the overall equation is balanced.

## EXERCISE 19.4

We can determine the concentration of an acidic permanganate ion solution by titrating this solution with a known amount of oxalic acid until the characteristic purple color of the $MnO_4^-$ ion disappears:

$$H_2C_2O_4(aq) + MnO_4^-(aq) \xrightarrow{H^+} CO_2(g) + Mn^{2+}(aq)$$

Use the half-reaction method to write a balanced equation for this reaction.

**SOLUTION**

**Step 1: Write a skeleton equation for the reaction.**

$$H_2C_2O_4 + MnO_4^- \longrightarrow CO_2 + Mn^{2+}$$

**Step 2: Assign oxidation numbers to atoms on both sides of the equation.**

$$\underset{+1\,+3\,-2}{H_2C_2O_4} + \underset{+7\;-2}{MnO_4^-} \longrightarrow \underset{+4\,-2}{CO_2} + \underset{+2}{Mn^{2+}}$$

**Step 3: Determine which atoms are oxidized and which are reduced.** The

The concentration of a solution of the $MnO_4^-$ ion can be determined by titrating with an oxalic acid solution until the characteristic color of the $MnO_4^-$ ion disappears.

only elements that undergo a change in oxidation state are the carbon atoms in oxalic acid and the manganese atom in the $MnO_4^-$ ion:

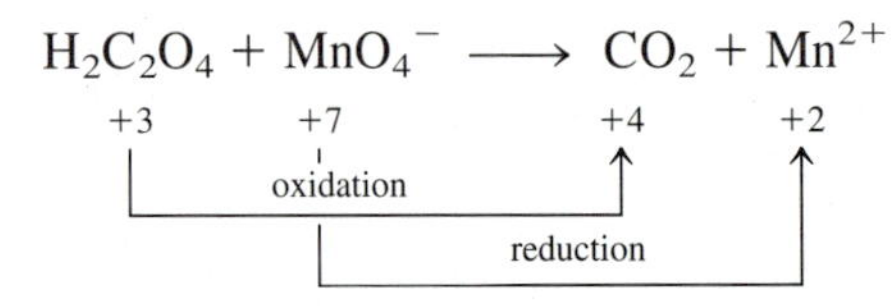


**Step 4: Divide the reaction into oxidation and reduction half-reactions and balance these half-reactions.** This reaction can be divided into the following half-reactions.

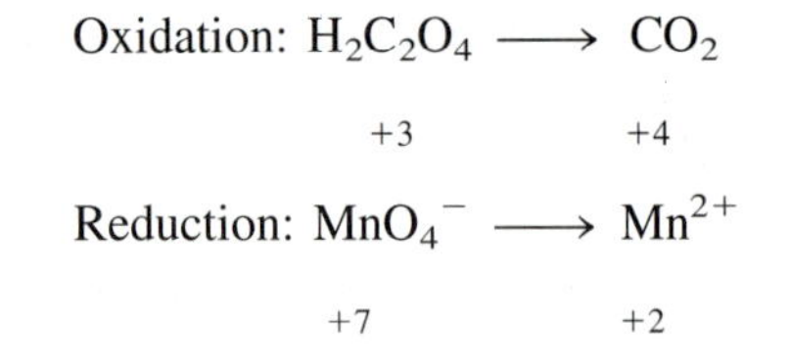


We'll balance the reduction half-reaction first. It takes five electrons to reduce manganese from +7 to +2.

$$\text{Reduction: } MnO_4^- + 5\,e^- \longrightarrow Mn^{2+}$$

Because the reaction is run in acid, we can add $H^+$ ions or $H_2O$ molecules to either side of the equation, as needed. There are two hints that tell us which side of the equation gets $H^+$ ions and which side gets $H_2O$ molecules. The only way to balance the charge on both sides of this equation is to add eight $H^+$ ions to the left side of the equation:

$$\text{Reduction: } MnO_4^- + 8\,H^+ + 5\,e^- \longrightarrow Mn^{2+}$$

The only way to balance the number of oxygen atoms is to add four $H_2O$ molecules to the right side of the equation:

$$\text{Reduction: } MnO_4^- + 8\,H^+ + 5\,e^- \longrightarrow Mn^{2+} + 4\,H_2O$$

We can now turn to the oxidation half-reaction. We might start by assuming that both of the carbon atoms in oxalic acid end up as carbon dioxide. This is therefore a two-electron oxidation half-reaction:

$$\text{Oxidation: } \underset{+3}{H_2C_2O_4} \longrightarrow 2\,\underset{+4}{CO_2} + 2\,e^-$$

We can balance both charge and mass by noting that two $H^+$ ions are given off when oxalic acid is oxidized to carbon dioxide, as shown in Figure 19.8:

$$\text{Oxidation: } H_2C_2O_4 \longrightarrow 2\,CO_2 + 2\,e^- + 2\,H^+$$

**Step 5: Combine the two half-reactions so that electrons are neither created nor destroyed.** Five electrons are consumed in the reduction half-reaction and two electrons are given off in the oxidation half-reaction. We can combine these half-reactions so that electrons are conserved by using the lowest common multiple of 5 and 2:

$$\begin{array}{l} 2(MnO_4^- + 8\,H^+ + 5\,e^- \longrightarrow Mn^{2+} + 4\,H_2O) \\ +\,5(H_2C_2O_4 \longrightarrow 2\,CO_2 + 2\,e^- + 2\,H^+) \\ \hline 2\,MnO_4^- + 16\,H^+ + 5\,H_2C_2O_4 \longrightarrow \\ \qquad 10\,CO_2 + 2\,Mn^{2+} + 8\,H_2O + 10\,H^+ \end{array}$$

**Problem-Solving Strategy**

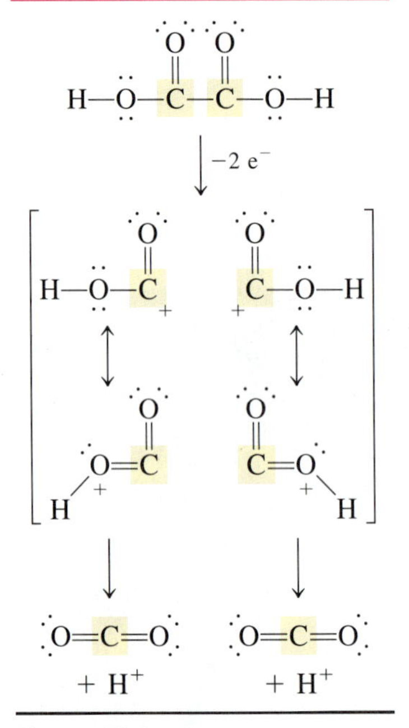


**Figure 19.8** The two electrons in the C—C bond are lost when oxalic acid is oxidized. This leaves us with a pair of ions that have a resonance structure that corresponds to a protonated $CO_2$ molecule. A two-electron oxidation of oxalic acid therefore gives a pair of $CO_2$ molecules and a pair of $H^+$ ions.

**Step 6: Balance the remainder of the equation by inspection, if necessary.** We can generate the simplest balanced equation for this reaction by subtracting 10 $H^+$ ions from both sides of the equation derived in the previous step:

$$2\,MnO_4^-(aq) + 5\,H_2C_2O_4(aq) + 6\,H^+(aq) \longrightarrow 10\,CO_2(g) + 2\,Mn^{2+}(aq) + 8\,H_2O(l)$$

The reaction between oxalic acid and potassium permanganate in acidic solution is a classical technique for standardizing solutions of the $MnO_4^-$ ion. These solutions need to be standardized before they can be used because it is difficult to obtain pure potassium permanganate. There are three sources of error.

- Samples of $KMnO_4$ are usually contaminated by $MnO_2$.
- Some of the $KMnO_4$ reacts with trace contaminants when it dissolves in water, even when distilled water is used as the solvent.
- The presence of traces of $MnO_2$ in this system catalyzes the decomposition of $MnO_4^-$ ion on standing.

Solutions of this ion therefore have to be standardized by titration just before they are used. A sample of reagent-grade sodium oxalate ($Na_2C_2O_4$) is weighed out, dissolved in distilled water, acidified with sulfuric acid, and then stirred until the oxalate dissolves. The resulting oxalic acid solution is then used to titrate $MnO_4^-$ to the endpoint of the titration, which is the point at which the last drop of $MnO_4^-$ ion is decolorized and a faint pink color persists for 30 seconds.

Solutions of the $MnO_4^-$ ion that have been standardized against oxalic acid, using the equation balanced in Exercise 19.4, can be used to determine the concentration of aqueous solutions of hydrogen peroxide, using the equation balanced in the next exercise.

## EXERCISE 19.5

As we have seen, an endless number of balanced equations can be written for the reaction between the permanganate ion and hydrogen peroxide in acidic solution to form the manganese(II) ion and oxygen:

$$MnO_4^-(aq) + H_2O_2(aq) \xrightarrow{H^+} Mn^{2+}(aq) + O_2(g)$$

Use the half-reaction method to determine the correct stoichiometry for this reaction.

**SOLUTION**

**Step 1: Write a skeleton equation for the reaction.**

$$MnO_4^- + H_2O_2 \longrightarrow Mn^{2+} + O_2$$

**Step 2: Assign oxidation numbers to atoms on both sides of the equation.**

$$\underset{+7\ -2}{MnO_4^-} + \underset{+1\ -1}{H_2O_2} \longrightarrow \underset{+2}{Mn^{2+}} + \underset{0}{O_2}$$

**Step 3: Determine which atoms are oxidized and which are reduced.**

$$MnO_4^- + H_2O_2 \longrightarrow Mn^{2+} + O_2$$

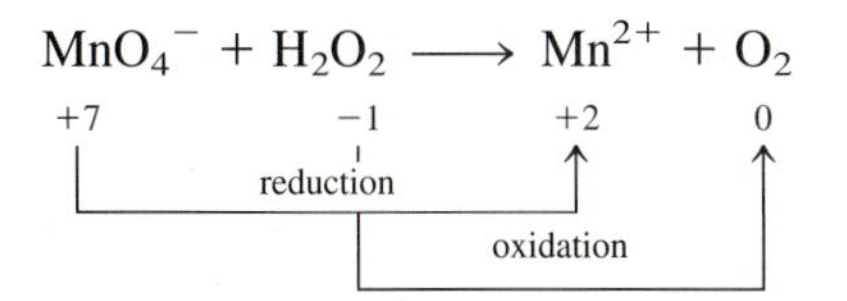

When $H_2O_2$ reacts with the $MnO_4^-$ ion, $O_2$ gas bubbles out of the solution and the characteristic purple color of the $MnO_4^-$ ion disappears.

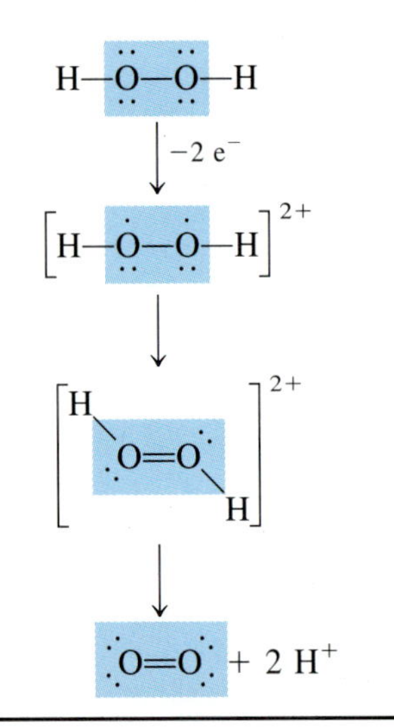


**Figure 19.9** Let's assume that $H_2O_2$ is oxidized by removing one electron from each oxygen atom. The unpaired electrons could then interact to form a doubly protonated $O_2$ molecule. The two-electron oxidation of $H_2O_2$ therefore gives an $O_2$ molecule and a pair of protons.

**Step 4: Divide the reaction into oxidation and reduction half-reactions and balance these half-reactions.** We balanced the reduction half-reaction in the previous exercise:

$$\text{Reduction: } MnO_4^- + 8\,H^+ + 5\,e^- \longrightarrow Mn^{2+} + 4\,H_2O$$

To balance the oxidation half-reaction, we have to remove two electrons from a pair of oxygen atoms in the −1 oxidation state to form a neutral $O_2$ molecule:

$$\text{Oxidation: } H_2O_2 \longrightarrow O_2 + 2\,e^-$$

We can then add a pair of $H^+$ ions to the products to balance both charge and mass in this half-reaction (see Figure 19.9):

$$\text{Oxidation: } H_2O_2 \longrightarrow O_2 + 2\,H^+ + 2\,e^-$$

**Step 5: Combine the two half-reactions so that electrons are neither created nor destroyed.** Two electrons are given off during oxidation and five electrons are consumed during reduction. We can combine these half-reactions so that electrons are conserved by using the lowest common multiple of 5 and 2:

$$\begin{array}{l} 2(MnO_4^- + 8\,H^+ + 5\,e^- \longrightarrow Mn^{2+} + 4\,H_2O) \\ + 5(H_2O_2 \longrightarrow O_2 + 2\,H^+ + 2\,e^-) \\ \hline 2\,MnO_4^- + 5\,H_2O_2 + 16\,H^+ \longrightarrow \\ \qquad 2\,Mn^{2+} + 5\,O_2 + 10\,H^+ + 8\,H_2O \end{array}$$

**Step 6: Balance the remainder of the equation by inspection, if necessary.** The simplest balanced equation for this reaction is obtained when 10 $H^+$ ions are subtracted from each side of the equation derived in the previous step:

$$2\,MnO_4^-(aq) + 5\,H_2O_2(aq) + 6\,H^+(aq) \longrightarrow 2\,Mn^{2+}(aq) + 5\,O_2(g) + 8\,H_2O(l)$$

## 19.8 REDOX REACTIONS IN BASIC SOLUTIONS

Half-reactions are also valuable for balancing equations in basic solutions. The key to success with these reactions is recognizing that basic solutions contain $H_2O$ molecules and $OH^-$ ions. We can therefore add water molecules or hydroxide ions to either side of the equation, as needed.

In the previous section, we obtained the following equation for the reaction between the permanganate ion and hydrogen peroxide in an acidic solution:

$$2\,MnO_4^-(aq) + 5\,H_2O_2(aq) + 6\,H^+(aq) \longrightarrow 2\,Mn^{2+}(aq) + 5\,O_2(g) + 8\,H_2O(l)$$

It might be interesting to see what happens when this reaction occurs in a basic solution.

### EXERCISE 19.6

Write a balanced equation for the reaction between the permanganate ion and hydrogen peroxide in a basic solution to form manganese dioxide and oxygen:

$$MnO_4^-(aq) + H_2O_2(aq) \xrightarrow{OH^-} MnO_2(s) + O_2(g)$$

**SOLUTION**

**Step 1: Write a skeleton equation for the reaction.**

$$MnO_4^- + H_2O_2 \longrightarrow MnO_2 + O_2$$

**Step 2: Assign oxidation numbers to atoms on both sides of the equation.**

$$MnO_4^- + H_2O_2 \longrightarrow MnO_2 + O_2$$

+7 −2    +1 −1    +4 −2    0

**Step 3: Determine which atoms are oxidized and which are reduced.**

$$MnO_4^- + H_2O_2 \longrightarrow MnO_2 + O_2$$

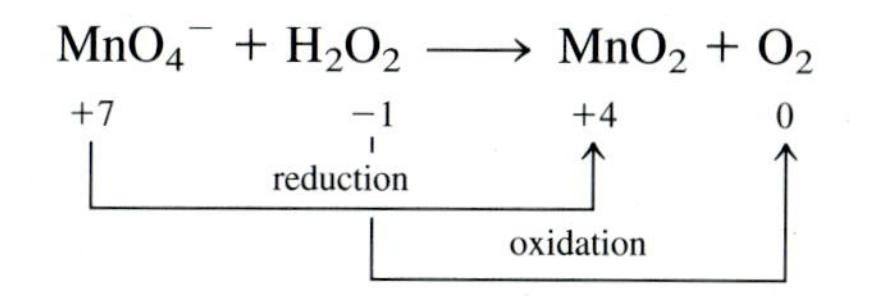


**Step 4: Divide the reaction into oxidation and reduction half-reactions and balance these half-reactions.** This reaction can be divided into the following half-reactions:

$$\text{Reduction: } MnO_4^- \longrightarrow MnO_2$$

+7    +4

$$\text{Oxidation: } H_2O_2 \longrightarrow O_2$$

−1    0

Let's start by balancing the reduction half-reaction. It takes three electrons to reduce manganese from the +7 to the +4 oxidation state:

$$\text{Reduction: } MnO_4^- + 3\,e^- \longrightarrow MnO_2$$

We now try to balance either the number of atoms or the charge on both sides of the equation. Because the reaction is run in basic solution, we can add either $OH^-$ ions or $H_2O$ molecules to either side of the equation, as needed. The key to deciding which side of the equation gets each of these reagents is simple: The only way to balance the net charge of −4 on the left side of the equation is to add four $OH^-$ ions to the products:

$$\text{Reduction: } MnO_4^- + 3\,e^- \longrightarrow MnO_2 + 4\,OH^-$$

We can then balance the number of hydrogen and oxygen atoms by adding two $H_2O$ molecules to the reactants:

$$\text{Reduction: } MnO_4^- + 3\,e^- + 2\,H_2O \longrightarrow MnO_2 + 4\,OH^-$$

We now turn to the oxidation half-reaction. Two electrons are lost when hydrogen peroxide is oxidized to form $O_2$ molecules:

$$\text{Oxidation: } H_2O_2 \longrightarrow O_2 + 2\,e^-$$

We can balance the charge on this half-reaction by adding a pair of $OH^-$ ions to the reactants:

$$\text{Oxidation: } H_2O_2 + 2\,OH^- \longrightarrow O_2 + 2\,e^-$$

The only way to balance the number of hydrogen and oxygen atoms is to add two $H_2O$ molecules to the products, as shown in Figure 19.10:

$$\text{Oxidation: } H_2O_2 + 2\,OH^- \longrightarrow O_2 + 2\,H_2O + 2\,e^-$$

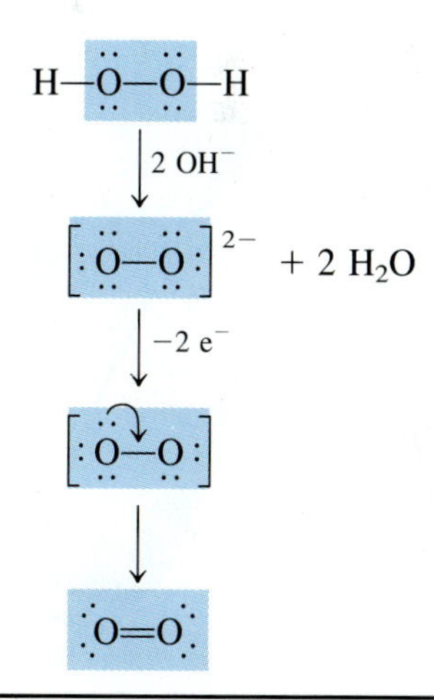


**Figure 19.10** The products of the oxidation of $H_2O_2$ in the presence of base can be understood by assuming that $H_2O_2$ reacts with a pair of $OH^-$ ions to form an $O_2^{2-}$ ion, which loses a pair of electrons to form an $O_2$ molecule.

**Step 5: Combine the two half-reactions so that electrons are neither created nor destroyed.** Two electrons are given off during the oxidation half-reaction and three electrons are consumed in the reduction half-reaction. We can therefore combine these half-reactions by using the lowest common multiple of 2 and 3:

$$\begin{array}{l} 2(MnO_4^- + 3\,e^- + 2\,H_2O \longrightarrow MnO_2 + 4\,OH^-) \\ +\,3(H_2O_2 + 2\,OH^- \longrightarrow O_2 + 2\,H_2O + 2\,e^-) \\ \hline 2\,MnO_4^- + 3\,H_2O_2 + 6\,OH^- + 4\,H_2O \longrightarrow \\ \qquad 2\,MnO_2 + 3\,O_2 + 8\,OH^- + 6\,H_2O \end{array}$$

**Step 6: Balance the remainder of the equation by inspection, if necessary.** The simplest balanced equation is obtained by subtracting four $H_2O$ molecules and six $OH^-$ ions from each side of the equation derived in the previous step:

$$2\,MnO_4^-(aq) + 3\,H_2O_2(aq) \longrightarrow 2\,MnO_2(s) + 3\,O_2(g) + 2\,OH^-(aq) + 2\,H_2O(l)$$

Sec. 10.3

Note that the ratio of moles of $MnO_4^-$ to moles of $H_2O_2$ consumed is different in acidic and basic solutions. This difference results from the fact that $MnO_4^-$ is reduced all the way to $Mn^{2+}$ in acid, but the reaction stops at $MnO_2$ in base.

Reactions in which a single reagent undergoes both oxidation and reduction are called **disproportionation reactions.** Bromine, for example, disproportionates to form bromide and bromate ions when a strong base is added to an aqueous bromine solution:

$$Br_2 \xrightarrow{OH^-} Br^- + BrO_3^-$$

## EXERCISE 19.7

Write a balanced equation for the disproportionation of bromine in the presence of a strong base:

**SOLUTION**

**Step 1: Write a skeleton equation for the reaction.**

$$Br_2 + OH^- \longrightarrow Br^- + BrO_3^-$$

**Step 2: Assign oxidation numbers to atoms on both sides of the equation.**

$$\underset{0}{Br_2} + \underset{-2\,+1}{OH^-} \longrightarrow \underset{-1}{Br^-} + \underset{+5\,-2}{BrO_3^-}$$

**Step 3: Determine which atoms are oxidized and which are reduced.**

$$\underset{0}{Br_2} + OH^- \longrightarrow \underset{-1}{Br^-} + \underset{+5}{BrO_3^-}$$

reduction (0 → −1)

oxidation (0 → +5)

**Step 4: Divide the reaction into oxidation and reduction half-reactions and balance these half-reactions.** Bromine is oxidized from the 0 to the +5 oxidation state in this reaction:

$$\text{Oxidation: } \underset{0}{Br_2} \longrightarrow \underset{+5}{BrO_3^-}$$

In the presence of strong base, the characteristic red-orange color of an aqueous $Br_2$ solution gradually fades as the bromine disproportionates to form the $Br^-$ and $BrO_3^-$ ions.

Bromine is also reduced in this reaction, from the 0 to the −1 oxidation state:

$$\text{Reduction: } \underset{0}{Br_2} \longrightarrow \underset{-1}{Br^-}$$

If we assume that two $Br^-$ ions are produced for each molecule of $Br_2$ reduced, the reduction half-reaction would be written as follows:

$$\text{Reduction: } Br_2 + \mathbf{2\ e^-} \longrightarrow 2\ Br^-$$

If we assume that two $BrO_3^-$ ions are formed each time a $Br_2$ molecule is oxidized, the oxidation half-reaction involves the loss of 10 electrons:

$$\text{Oxidation: } Br_2 \longrightarrow 2\ BrO_3^- + \mathbf{10\ e^-}$$

Because the reaction occurs in basic solution we can add $OH^-$ ions or $H_2O$ molecules to either side of the equation, as needed. The trick, as always, is deciding which side of the equation gets the $OH^-$ ions and which side gets the $H_2O$ molecules. In this case, the only way to balance the charge on both sides of the equation is to add 12 $OH^-$ ions to the reactants:

$$\text{Oxidation: } Br_2 + \mathbf{12\ OH^-} \longrightarrow 2\ BrO_3^- + 10\ e^-$$

Six of the oxygen atoms in the $OH^-$ ions end up in the $BrO_3^-$ ions. What happens to the other six oxygen atoms and the 12 hydrogen atoms? The most reasonable answer is that they combine to form six $H_2O$ molecules:

$$\text{Oxidation: } Br_2 + 12\ OH^- \longrightarrow 2\ BrO_3^- + 10\ e^- + \mathbf{6\ H_2O}$$

**Step 5: Combine the two half-reactions so that electrons are neither created nor destroyed.** Ten electrons are given off in the oxidation half-reaction and two electrons are consumed in the reduction half-reaction. We can therefore combine these half-reactions as follows:

$$\begin{array}{l} 5(Br_2 + 2\ e^- \longrightarrow 2\ Br^-) \\ \underline{+\ (Br_2 + 12\ OH^- \longrightarrow 2\ BrO_3^- + 10\ e^- + 6\ H_2O)} \\ 6\ Br_2 + 12\ OH^- \longrightarrow 2\ BrO_3^- + 10\ Br^- + 6\ H_2O \end{array}$$

**Step 6: Balance the remainder of the equation by inspection, if necessary.** The equation generated in the previous step is balanced, but it is not quite correct. The best answer to this exercise would be the equation with the smallest possible coefficients, so we divide all of the coefficients by 2:

$$3\ Br_2 + 6\ OH^- \longrightarrow BrO_3^- + 5\ Br^- + 3\ H_2O$$

We now complete the exercise by identifying the states of the reactants and products, as follows:

$$3\ Br_2(aq) + 6\ OH^-(aq) \longrightarrow BrO_3^-(aq) + 5\ Br^-(aq) + 3\ H_2O(l)$$

## 19.9 MOLECULAR REDOX REACTIONS

Lewis structures can play a vital role in understanding oxidation–reduction reactions with complex molecules. Consider the following reaction, for example, which is used in the Breathalyzer to determine the amount of ethyl alcohol or ethanol on the breath of individuals who are suspected of driving while under the influence:

p. 95

$$3\ CH_3CH_2OH(g) + 2\ Cr_2O_7^{2-}(aq) + 16\ H^+(aq) \longrightarrow 3\ CH_3CO_2H(aq) + 4\ Cr^{3+}(aq) + 11\ H_2O(l)$$

We could balance the oxidation half-reaction in terms of the molecular formulas of the starting material and the product of this half-reaction:

$$\text{Oxidation: } C_2H_6O \longrightarrow C_2H_4O_2$$

It is easier to understand what happens in this reaction, however, if we assign oxidation numbers to each of the carbon atoms in the Lewis structures of the components of this reaction, as shown in Figure 19.11.

The carbon atom in the $CH_3$— group in ethanol is assigned an oxidation state of −3 so that it can balance the oxidation states of the three substituents it carries. Applying the same technique to the —$CH_2OH$ group in the starting material gives an oxidation state of −1.

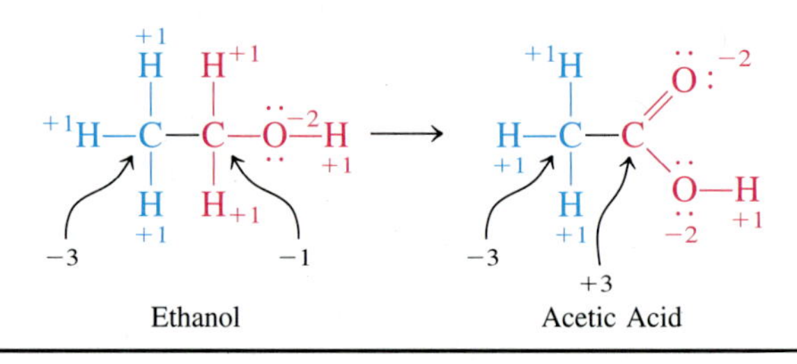


**Figure 19.11** Oxidation of ethanol to acetic acid.

The carbon in the $CH_3$— group in the acetic acid formed in this reaction has the same oxidation state as it did in the starting material: −3. There is a change in the oxidation number of the other carbon atom, however, from −1 to +3. The oxidation half-reaction therefore formally corresponds to the loss of four electrons by one of the carbon atoms:

$$\text{Oxidation: } CH_3CH_2OH \longrightarrow CH_3CO_2H + 4\,e^-$$

Because this reaction is run in acidic solution, we can add $H^+$ and $H_2O$ molecules as needed to balance the equation:

$$\text{Oxidation: } CH_3CH_2OH + H_2O \longrightarrow CH_3CO_2H + 4\,e^- + 4\,H^+$$

The other half of this reaction involves a six-electron reduction of the $Cr_2O_7^{2-}$ ion in acidic solution to form a pair of $Cr^{3+}$ ions:

$$\text{Reduction: } Cr_2O_7^{2-} + 6\,e^- \longrightarrow 2\,Cr^{3+}$$

Adding $H^+$ ions and $H_2O$ molecules as needed gives the following balanced equation for this half-reaction:

$$\text{Reduction: } Cr_2O_7^{2-} + 14\,H^+ + 6\,e^- \longrightarrow 2\,Cr^{3+} + 7\,H_2O$$

We are now ready to combine the two half-reactions by assuming that electrons are neither created nor destroyed in this reaction:

$$3(CH_3CH_2OH + H_2O \longrightarrow CH_3CO_2H + 4\,e^- + 4\,H^+)$$
$$2(Cr_2O_7^{2-} + 14\,H^+ + 6\,e^- \longrightarrow 2\,Cr^{3+} + 7\,H_2O)$$
$$3\,CH_3CH_2OH + 2\,Cr_2O_7^{2-} + 28\,H^+ + 3\,H_2O \longrightarrow 3\,CH_3CO_2H + 4\,Cr^{3+} + 12\,H^+ + 14\,H_2O$$

Simplifying this equation by removing 3 $H_2O$ molecules and 12 $H^+$ ions from both sides of the equation gives the balanced equation for this reaction:

$$3\,CH_3CH_2OH(g) + 2\,Cr_2O_7^{2-}(aq) + 16\,H^+(aq) \longrightarrow 3\,CH_3CO_2H(aq) + 4\,Cr^{3+}(aq) + 11\,H_2O(l)$$

## EXERCISE 19.8

Methyllithium ($CH_3Li$) can be used to form bonds between carbon and either main-group metals or transition metals:

$$HgCl_2(s) + 2\ CH_3Li(l) \longrightarrow Hg(CH_3)_2(l) + 2\ LiCl(s)$$
$$WCl_6(s) + 6\ CH_3Li(l) \longrightarrow W(CH_3)_6(l) + 6\ LiCl(s)$$

It can be used also to form bonds between carbon and other nonmetals:

$$PCl_3(l) + 3\ CH_3Li(l) \longrightarrow P(CH_3)_3(l) + 3\ LiCl(s)$$

or between carbon atoms:

$$CH_3Li(l) + H_2CO(g) \longrightarrow [CH_3CH_2OLi] \xrightarrow{H^+} CH_3CH_2OH(l)$$

Use Lewis structures to explain the stoichiometry of the following oxidation–reaction, which is used to synthesize methyllithium:

$$CH_3Br(l) + 2\ Li(s) \longrightarrow CH_3Li(l) + LiBr(s)$$

**SOLUTION** The active metals are reducing agents in all of their chemical reactions. In this reaction, lithium metal reduces a carbon atom from an oxidation state of −2 to −4, as shown in Figure 19.12:

Sec. 9.14

$$\text{Reduction: } CH_3Br + 2\ e^- \longrightarrow CH_3Li$$

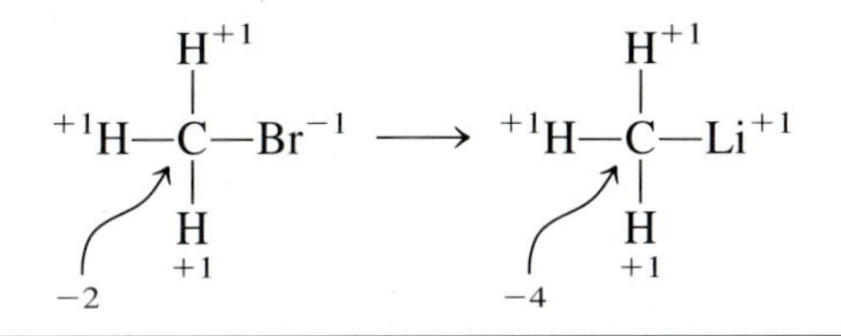


**Figure 19.12** Reduction of $CH_3Br$ with lithium metal.

The other half of the reaction therefore must involve the net loss of a total of two electrons:

$$\text{Oxidation: } 2\ Li \longrightarrow 2\ Li^+ + 2\ e^-$$

Two moles of lithium are therefore consumed for each mole of $CH_3Br$ in this reaction. One of the $Li^+$ ions formed in this reaction combines with the negatively charged carbon atom to form methyllithium and the other precipitates from solution with the $Br^-$ ion as LiBr:

$$CH_3Br(l) + 2\ Li(s) \longrightarrow CH_3Li(l) + LiBr(s)$$

## 19.10 COMMON OXIDIZING AGENTS AND REDUCING AGENTS

Section 9.14 introduced another way of looking at oxidation–reduction reactions, which focuses on the role played by a particular reactant in a chemical reaction. What is the role of the permanganate ion in the following reaction, for example?

$$2\ MnO_4^-(aq) + 5\ H_2C_2O_4(aq) + 6\ H^+(aq) \longrightarrow 10\ CO_2(g) + 2\ Mn^{2+}(aq) + 8\ H_2O(l)$$

Sec. 9.14

According to Exercise 19.4, oxalic acid is oxidized to carbon dioxide in this reaction and the permanganate ion is reduced to the $Mn^{2+}$ ion:

$$\text{Oxidation: } \underset{+3}{H_2C_2O_4} \longrightarrow \underset{+4}{CO_2}$$

$$\text{Reduction: } \underset{+7}{MnO_4^-} \longrightarrow \underset{+2}{Mn^{2+}}$$

The permanganate ion removes electrons from oxalic acid molecules and thereby oxidizes the oxalic acid. Thus, the $MnO_4^-$ ion acts as an **oxidizing agent** in this reaction. Oxalic acid, on the other hand, is a **reducing agent** in this reaction. By giving up electrons, it reduces the $MnO_4^-$ ion to $Mn^{2+}$.

Atoms, ions, and molecules that have an unusually large affinity for electrons tend to be good oxidizing agents. Elemental fluorine, for example, is the strongest common oxidizing agent. $F_2$ is such a good oxidizing agent that metals, quartz, asbestos, and even water burst into flame in its presence. Other good oxidizing agents include $O_2$, $O_3$, and $Cl_2$, which are the elemental forms of the second and third most electronegative elements, respectively.

Sec. 8.6

Another place to look for good oxidizing agents is among compounds with unusually large oxidation states, such as the permanganate ($MnO_4^-$), chromate ($CrO_4^{2-}$), and dichromate ($Cr_2O_7^{2-}$) ions, as well as nitric acid ($HNO_3$), perchloric acid ($HClO_4$), and sulfuric acid ($H_2SO_4$). These compounds are strong oxidizing agents because elements become more electronegative as the oxidation states of their atoms increase.

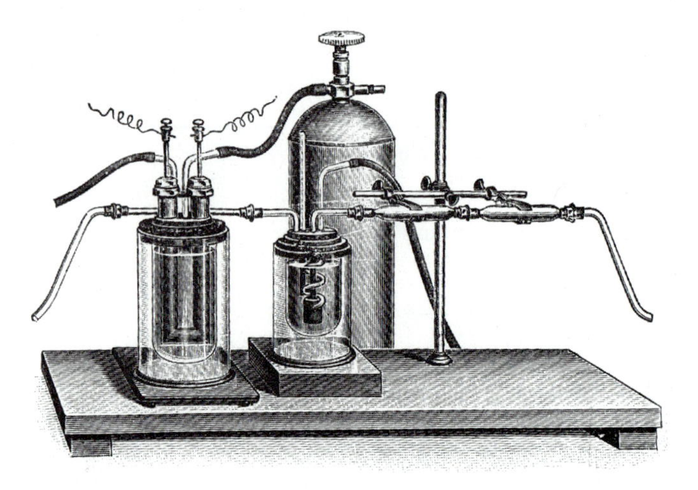

Elemental fluorine is so extremely reactive that it was difficult to prepare. This woodcut shows the apparatus in which Moissan first prepared elemental fluorine by the electrolysis of $KHF_2$.

Good reducing agents include the active metals, such as sodium, magnesium, aluminum, and zinc, which have relatively small ionization energies and low electronegativities. Metal hydrides, such as NaH, $CaH_2$, and $LiAlH_4$, which formally contain the $H^-$ ion, are also good reducing agents.

Some compounds can act as either oxidizing agents or reducing agents. One example is hydrogen gas, which acts as an oxidizing agent when it combines with metals and as a reducing agent when it reacts with nonmetals:

$$2\,Na(s) + H_2(g) \longrightarrow 2\,NaH(s)$$

$$H_2(g) + Cl_2(g) \longrightarrow 2\,HCl(g)$$

Another example is hydrogen peroxide, in which the oxygen atom is in the $-1$ oxidation state. Because this oxidation state lies between the extremes of the more common 0 and $-2$ oxidation states of oxygen, $H_2O_2$ can act as either an oxidizing agent or a reducing agent.

## 19.11 THE RELATIVE STRENGTHS OF OXIDIZING AND REDUCING AGENTS

Sec. 9.16

As we have seen, spontaneous oxidation–reduction reactions convert the stronger of a pair of oxidizing agents and the stronger of a pair of reducing agents into a weaker oxidizing agent and a weaker reducing agent. The fact that the following reaction occurs, for example, suggests that copper metal is a stronger reducing agent than silver metal and that the $Ag^+$ ion is a stronger oxidizing agent than the $Cu^{2+}$ ion:

$$\underset{\text{stronger reducing agent}}{Cu(s)} + \underset{\text{stronger oxidizing agent}}{2\,Ag^+(aq)} \longrightarrow \underset{\text{weaker oxidizing agent}}{Cu^{2+}(aq)} + \underset{\text{weaker reducing agent}}{2\,Ag(s)}$$

On the basis of many such experiments, the common oxidation–reduction half-reactions have been organized into a table in which the strongest reducing agents are at one end and the strongest oxidizing agents are at the other, as shown in Table 19.2. By convention, all of the half-reactions are written in the direction of reduction. Furthermore, by convention, the strongest reducing agents are usually found at the top of the table.

Fortunately, you don't have to memorize these conventions. All you have to do is remember that the active metals, such as sodium and potassium, are excellent reducing agents and look for these entries in the table. The strongest reducing agents will be found at the corner of the table where sodium and potassium metal are listed.

### EXERCISE 19.9

Arrange the following oxidizing and reducing agents in order of increasing strength:

Reducing agents: $Cl^-$, Cu, $H_2$, $H^-$, HF, Pb, and Zn

Oxidizing agents: $Cr^{3+}$, $Cr_2O_7^{2-}$, $Cu^{2+}$, $H^+$, $O_2$, $O_3$, and $Na^+$

**SOLUTION** According to Table 19.2, these reducing agents become stronger in the following order:

$$HF < Cl^- < Cu < H_2 < Pb < Zn < H^-$$

The oxidizing agents become stronger in the following order.

$$Na^+ < Cr^{3+} < H^+ < Cu^{2+} < O_2 < Cr_2O_7^{2-} < O_3$$

### EXERCISE 19.10

Use Table 19.2 to predict whether the following oxidation–reduction reactions should occur as written:

(a) $2\,Ag(s) + S(s) \longrightarrow Ag_2S(s)$

(b) $2\,Ag(s) + Cu^{2+}(aq) \longrightarrow 2\,Ag^+(aq) + Cu(s)$

(c) $MnO_4^-(aq) + 3\,Fe^{2+}(aq) + 2\,H_2O(l) \longrightarrow MnO_2(s) + 3\,Fe^{3+}(aq) + 4\,OH^-(aq)$

(d) $MnO_4^-(aq) + 5\,Fe^{2+}(aq) + 8\,H^+(aq) \longrightarrow Mn^{2+}(aq) + 5\,Fe^{3+}(aq) + 4\,H_2O(l)$

**TABLE 19.2 The Relative Strengths of Common Oxidizing Agents and Reducing Agents**

| | | |
|---|---|---|
| | $K^+ + e^- \rightleftharpoons K$ | Best |
| | $Ba^{2+} + 2\,e^- \rightleftharpoons Ba$ | reducing |
| | $Ca^{2+} + 2\,e^- \rightleftharpoons Ca$ | agents |
| | $Na^+ + e^- \rightleftharpoons Na$ | |
| | $Mg^{2+} + 2\,e^- \rightleftharpoons Mg$ | |
| | $H_2 + 2\,e^- \rightleftharpoons 2\,H^-$ | |
| | $Al^{3+} + 3\,e^- \rightleftharpoons Al$ | |
| | $Mn^{2+} + 2\,e^- \rightleftharpoons Mn$ | |
| | $Zn^{2+} + 2\,e^- \rightleftharpoons Zn$ | |
| | $Cr^{3+} + 3\,e^- \rightleftharpoons Cr$ | |
| | $S + 2\,e^- \rightleftharpoons S^{2-}$ | |
| | $2\,CO_2 + 2\,H^+ + 2\,e^- \rightleftharpoons H_2C_2O_4$ | |
| | $Cr^{3+} + e^- \rightleftharpoons Cr^{2+}$ | |
| | $Fe^{2+} + 2\,e^- \rightleftharpoons Fe$ | |
| | $Co^{2+} + 2\,e^- \rightleftharpoons Co$ | |
| | $Ni^{2+} + 2\,e^- \rightleftharpoons Ni$ | |
| | $Sn^{2+} + 2\,e^- \rightleftharpoons Sn$ | |
| | $Pb^{2+} + 2\,e^- \rightleftharpoons Pb$ | |
| | $Fe^{3+} + 3\,e^- \rightleftharpoons Fe$ | |
| | $2\,H^+ + 2\,e^- \rightleftharpoons H_2$ | |
| | $S_4O_6{}^{2-} + 2\,e^- \rightleftharpoons 2\,S_2O_3{}^{2-}$ | |
| | $Sn^{4+} + 2\,e^- \rightleftharpoons Sn^{2+}$ | |
| | $Cu^{2+} + e^- \rightleftharpoons Cu^+$ | |
| | $O_2 + 2\,H_2O + 4\,e^- \rightleftharpoons 4\,OH^-$ | |
| | $Cu^+ + e^- \rightleftharpoons Cu$ | |
| | $I_2 + 2\,e^- \rightleftharpoons 2\,I^-$ | |
| Oxidizing | $MnO_4{}^- + 2\,H_2O + 3\,e^- \rightleftharpoons MnO_2 + 4\,OH^-$ | ↑ |
| power | $O_2 + 2\,H^+ + 2\,e^- \rightleftharpoons H_2O_2$ | Reducing |
| increases | $Fe^{3+} + e^- \rightleftharpoons Fe^{2+}$ | power |
| ↓ | $Hg_2{}^{2+} + 2\,e^- \rightleftharpoons 2\,Hg$ | increases |
| | $Ag^+ + e^- \rightleftharpoons Ag$ | |
| | $Hg^{2+} + 2\,e^- \rightleftharpoons Hg$ | |
| | $H_2O_2 + 2\,e^- \rightleftharpoons 2\,OH^-$ | |
| | $HNO_3 + 3\,H^+ + 3\,e^- \rightleftharpoons NO + 2\,H_2O$ | |
| | $Br_2 + 2\,e^- \rightleftharpoons 2\,Br^-$ | |
| | $2\,IO_3{}^- + 12\,H^+ + 10\,e^- \rightleftharpoons I_2 + 6\,H_2O$ | |
| | $CrO_4{}^{2-} + 8\,H^+ + 3\,e^- \rightleftharpoons Cr^{3+} + 4\,H_2O$ | |
| | $Pt^{2+} + 2\,e^- \rightleftharpoons Pt$ | |
| | $MnO_2 + 4\,H^+ + 2\,e^- \rightleftharpoons Mn^{2+} + 2\,H_2O$ | |
| | $O_2 + 4\,H^+ + 4\,e^- \rightleftharpoons 2\,H_2O$ | |
| | $Cr_2O_7{}^{2-} + 14\,H^+ + 6\,e^- \rightleftharpoons 2\,Cr^{3+} + 7\,H_2O$ | |
| | $Cl_2 + 2\,e^- \rightleftharpoons 2\,Cl^-$ | |
| | $PbO_2 + 4\,H^+ + 2\,e^- \rightleftharpoons Pb^{2+} + 2\,H_2O$ | |
| | $MnO_4{}^- + 8\,H^+ + 5\,e^- \rightleftharpoons Mn^{2+} + 4\,H_2O$ | |
| | $Au^+ + e^- \rightleftharpoons Au$ | |
| | $H_2O_2 + 2\,H^+ + 2\,e^- \rightleftharpoons 2\,H_2O$ | |
| | $Co^{3+} + e^- \rightleftharpoons Co^{2+}$ | |
| Best | $S_2O_8{}^{2-} + 2\,e^- \rightleftharpoons 2\,SO_4{}^{2-}$ | |
| oxidizing | $O_3 + 2\,H^+ + 2\,e^- \rightleftharpoons O_2 + H_2O$ | |
| agents | $F_2 + 2\,H^+ + 2\,e^- \rightleftharpoons 2\,HF$ | |

**SOLUTION**

(a) No. The $S^{2-}$ ion is a better reducing agent than Ag and the $Ag^+$ ion is a better oxidizing agent than S. Silver doesn't tarnish because it reduces sulfur; it tarnishes because it reacts with sulfur compounds in the presence of oxygen and the oxygen is reduced.

(b) No. Cu is a better reducing agent than Ag and the $Ag^+$ ion is a better oxidizing agent than $Cu^{2+}$ ions.

(c) No. $MnO_4^-$ in base is not a strong enough oxidizing agent to oxidize $Fe^{2+}$ to $Fe^{3+}$.

(d) Yes. $MnO_4^-$ in acid is a strong enough oxidizing agent to oxidize $Fe^{2+}$ to $Fe^{3+}$.

## EXERCISE 19.11

Which of the following pairs of ions cannot exist simultaneously in aqueous solutions?

(a) $Cu^+$ and $Fe^{3+}$ (b) $Fe^{3+}$ and $I^-$ (c) $Al^{3+}$ and $Co^{2+}$

**SOLUTION**

(a) According to Table 19.2, the $Cu^+$ ion is a significantly better reducing agent than the $Fe^{2+}$ ion and the $Fe^{3+}$ ion is a better oxidizing agent than $Cu^{2+}$. Solutions that contain both $Cu^+$ and $Fe^{3+}$ ions therefore should undergo a spontaneous oxidation–reduction reaction. As a result, these ions can't exist simultaneously in the same solution:

$$Cu^+(aq) + Fe^{3+}(aq) \longrightarrow Cu^{2+}(aq) + Fe^{2+}(aq)$$

(b) According to Table 19.2, the $Fe^{3+}$ ion is strong enough to oxidize $I^-$ ions to $I_2$. These ions therefore cannot exist in the same solution:

$$2\,Fe^{3+}(aq) + 2\,I^-(aq) \longrightarrow 2\,Fe^{2+}(aq) + I_2(aq)$$

(c) The $Al^{3+}$ ion is one of the weakest oxidizing agents in Table 19.2 and the $Co^{2+}$ ion is one of the weakest reducing agents. These ions can't undergo an oxidation–reduction reaction and can therefore coexist in aqueous solution:

$$Al^{3+}(aq) + Co^{2+}(aq) \longrightarrow \text{N.R.}$$

## SUMMARY

**1.** Oxidation–reduction reactions can be systematically balanced by a method which balances the two halves of the reaction and then combines them so that electrons are conserved.

**2.** Spontaneous oxidation–reduction reactions convert the stronger of a pair of oxidizing agents and the stronger of a pair of reducing agents into a weaker oxidizing agent and a weaker reducing agent. The fact that copper metal can reduce silver nitrate to silver metal, for example, suggests that copper is a stronger reducing agent than silver.

**3.** The results of a large number of observations of spontaneous chemical reactions are summarized in a table of the relative strengths of common oxidizing agents and reducing agents that can be used to predict whether oxidation–reduction should occur.

## KEY TERMS

**Corrosion** (p. 698)
**Disproportionation reactions** (p. 712)
**Metathesis** (p. 701)
**Oxidation** (p. 697)
**Oxidation number** (p. 698)
**Oxidizing agent** (p. 716)
**Redox reactions** (p. 698)
**Reducing agent** (p. 716)
**Reduction** (p. 697)

## PROBLEMS

### Evolution of the Theory of Oxidation–Reduction Reactions

**19-1** Describe the phlogiston theory of chemical reactions.

**19-2** Summarize the fatal flaw in the phlogiston theory, which inevitably led to its collapse.

**19-3** Organic chemists often discuss oxidation in terms of the loss of a pair of hydrogen atoms. Ethyl alcohol, $CH_3CH_2OH$, is oxidized to acetaldehyde, $CH_3CHO$, for example, by removing a pair of hydrogen atoms. Show that this reaction obeys the general rule that oxidation is any process in which the oxidation number of an atom increases.

**19-4** Organic chemists also often discuss reduction in terms of the gain of a pair of hydrogen atoms. Ethylene, $C_2H_4$, is reduced when it reacts with $H_2$ to form ethane, $C_2H_6$, for example. Show that this reaction obeys the general rule that reduction is any process in which the oxidation number of an atom decreases.

### Oxidation–Reduction Reactions

**19-5** Our bodies convert the carbohydrates, fats, and proteins we consume into simple sugars such as glucose, $C_6H_{12}O_6$, which we burn to form $CO_2$ and $H_2O$. Show that the combustion of glucose is an oxidation–reduction reaction.

**19-6** Organisms that can live in the absence of oxygen are called *anaerobic*. They obtain their energy by converting glucose, $C_6H_{12}O_6$, into lactic acid, $C_3H_6O_3$. Show that these organisms capture energy from their environment with no net oxidation or reduction.

**19-7** Write chemical equations for the corrosion of silver and iron and show that corrosion results from oxidation–reduction reactions.

### Assigning Oxidation States

**19-8** An area of active research in recent years has involved compounds such as $Re_2Cl_8^{2-}$, $Cr_2Cl_9^{3-}$, and $Mo_2Cl_8^{4-}$ that contain metal–metal bonds. Calculate the oxidation state of the metal atom in each of these compounds.

**19-9** Calculate the oxidation state of the aluminum atom in the following compounds:

(a) $LiAlH_4$ (b) $Al(H_2O)_6^{3+}$ (c) $Al_2O_3$
(d) $Al(OH)_4^-$ (e) $NaAl(OH)_2CO_3$ (the active ingredient in Rolaids)

**19-10** Which of the following compounds contain hydrogen in a negative oxidation state?

(a) $H_2S$ (b) $NH_3$ (c) $PH_4^+$ (c) $LiAlH_4$ (e) $CaH_2$
(f) $CH_4$

**19-11** Arrange the following compounds in order of increasing oxidation state of the carbon atom:

(a) C (b) CO (c) $CO_2$ (d) $H_2CO$ (e) $CH_3OH$
(f) $CH_4$

**19-12** Carbon can have any oxidation state between $-4$ and $+4$. Calculate the oxidation state of carbon in the following compounds:

(a) $CCl_4$ (b) $COCl_2$ (c) CO (d) $CO_2$ (e) $CH_3Li$
(f) $CH_4$ (g) $H_2CO$ (h) $Na_2CO_3$ (i) $HCO_2H$

**19-13** Sulfur can have any oxidation state between $+6$ and $-2$. Calculate the oxidation state of sulfur in the following compounds:

(a) $S_8$ (b) $H_2S$ (c) ZnS (d) $SO_2$ (e) $SO_3$
(f) $SO_3^{2-}$ (g) $SO_4^{2-}$ (h) $S_2O_3^{2-}$

**19-14** Calculate the oxidation state of manganese in the following compounds:

(a) MnO (b) $Mn_2O_3$ (c) $MnO_2$ (d) $Mn_2O_7$
(e) $Mn(OH)_2$ (f) $HMnO_4$ (g) $CaMnO_3$ (h) $MnSO_4$
(i) $NaMn(CO)_5$

**19-15** Prussian blue is a pigment with the formula $Fe_4[Fe(CN)_6]_3$. If this compound contains the $Fe(CN)_6^{4-}$ ion, what is the oxidation state of the other four iron atoms? Turnbull's blue is a pigment with the formula $Fe_3[Fe(CN)_6]_2$. This compound contains the $Fe(CN)_6^{3-}$ ion. What is the oxidation state of the other three iron atoms?

### Recognizing Oxidation–Reduction Reactions

**19-16** Describe the difference between metathesis reactions and oxidation–reduction reactions. Give examples of both.

**19-17** Which statement correctly describes the following reaction?

$$3\,Sn^{2+}(aq) + Cr_2O_7^{2-}(aq) + 14\,H^+(aq) \longrightarrow 3\,Sn^{4+}(aq) + 2\,Cr^{3+}(aq) + 7\,H_2O(l)$$

(a) Both the $Sn^{2+}$ and $H^+$ ions are oxidizing agents.

(b) The $Cr_2O_7^{2-}$ ion is the oxidizing agent. (c) The $Sn^{2+}$ ion is reduced. (d) The $Sn^{4+}$ ion must be a weak oxidizing agent. (e) None of the above is true.

**19-18** We can remove the $Ag_2S$ that forms when silver tarnishes by polishing the object with a compound that contains cyanide ions or by wrapping the object in aluminum foil and immersing it in salt water. Which of these reactions involves oxidation–reduction?

$$Ag_2S(s) + 4\,CN^-(aq) \longrightarrow 2\,Ag(CN)_2^-(aq) + S^{2-}(aq)$$

$$3\,Ag_2S(s) + 2\,Al(s) \longrightarrow 6\,Ag(s) + Al_2S_3(s)$$

**19-19** The following reactions involve either carbon monoxide or carbon dioxide. Decide whether each of these reactions involves oxidation–reduction. If it does, identify what is oxidized and what is reduced.

(a) $CO_2(g) + H_2O(l) \rightleftharpoons H_2CO_3(aq)$

(b) $Fe_2O_3(s) + 3\,CO(g) \longrightarrow 2\,Fe(s) + 3\,CO_2(g)$

(c) $SiO_2(s) + 3\,C(s) \longrightarrow SiC(s) + 2\,CO(g)$

(d) $CO(g) + 2\,H_2(g) \longrightarrow CH_3OH(l)$

**19-20** Decide whether each of the following reactions of phosphorus and its compounds involves oxidation–reduction. If it does, identify what is oxidized and what is reduced.

(a) $Ca_3P_2(s) + 6\,H_2O(l) \longrightarrow 3\,Ca(OH)_2(aq) + 2\,PH_3(g)$

(b) $2\,PH_3(g) + 4\,O_2(g) \longrightarrow 2\,H_3PO_4(s)$

(c) $PH_3(g) + HCl(g) \longrightarrow PH_4Cl(s)$

(d) $P_4(s) + 5\,O_2(g) \longrightarrow P_4O_{10}(s)$

**19-21** Use Lewis structures to explain what happens when $CO_2$ reacts with water to form carbonic acid, $H_2CO_3$.

**19-22** Use Lewis structures to explain why reduction of the $Hg^{2+}$ ion to the +1 oxidation state gives a diatomic $Hg_2^{2+}$ ion.

**19-23** Use Lewis structures to explain what happens in the following reaction:

$$NO_3^-(aq) + 4\,H^+(aq) + 3\,e^- \rightleftharpoons NO(g) + 2\,H_2O(l)$$

**19-24** Use Lewis structures to explain what happens in the following reaction:

$$2\,SO_3^{2-}(aq) + O_2(g) \rightleftharpoons 2\,SO_4^{2-}(aq)$$

### The Half-Reaction Method of Balancing Redox Equations

**19-25** The chemistry of nitrogen is unusual because so many of its compounds undergo reactions that give nitrogen in its elemental form. Balance the following oxidation–reduction reactions in which ammonia is converted into molecular nitrogen:

(a) $CuO(s) + NH_3(g) \longrightarrow Cu(s) + N_2(g) + H_2O(l)$

(b) $NH_3(aq) + Cl_2(g) \longrightarrow N_2(g) + NH_4Cl(aq)$

**19-26** Reactions in which a reagent undergoes both oxidation and reduction are known as disproportionation reactions. Write balanced equations for the following disproportionation reactions:

(a) $H_2O_2(aq) \longrightarrow H_2O(l) + O_2(g)$

(b) $NO_2(g) + H_2O(l) \longrightarrow HNO_3(aq) + NO(g)$

**19-27** The term *conproportionation* has been proposed to describe reactions that are the opposite of disproportionation reactions. Write balanced equations for the following conproportionation reactions:

(a) $HIO_3(aq) + HI(aq) \longrightarrow I_2(aq)$

(b) $SO_2(g) + H_2S(g) \longrightarrow S_8(s) + H_2O(l)$

**19-28** Acids that are also oxidizing agents play an important role in the preparation of small quantities of the halogens in the laboratory. Write balanced equations for the following oxidation–reduction equations:

(a) $HCl(aq) + HNO_3(aq) \longrightarrow NO(g) + Cl_2(g)$

(b) $HBr(aq) + H_2SO_4(aq) \longrightarrow SO_2(g) + Br_2(aq)$

(c) $HCl(aq) + MnO_4^-(aq) \longrightarrow Cl_2(g) + Mn^{2+}(aq)$

**19-29** The product of the reaction between copper metal and nitric acid depends on the concentration of the acid. In dilute nitric acid, NO is formed. Concentrated nitric acid produces $NO_2$. Write balanced equations for the following oxidation–reduction reactions between copper metal and nitric acid:

(a) $Cu(s) + HNO_3(aq) \longrightarrow Cu^{2+}(aq) + NO(g)$

(b) $Cu(s) + HNO_3(aq) \longrightarrow Cu^{2+}(aq) + NO_2(g)$

**19-30** Use half-reactions to balance the following oxidation–reduction equation, which seems to give only a single product:

$$H_2S(g) + H_2O_2(aq) \longrightarrow S_8(s)$$

**19-31** Some of the earliest matches consisted of white phosphorus and potassium chlorate glued to wooden sticks. Write a balanced equation for the following oxidation–reduction reaction, which occurs when such a match is rubbed against a hard surface:

$$P_4(s) + KClO_3(s) \longrightarrow P_4O_{10}(s) + KCl(s)$$

### Redox Reactions in Acidic Solutions

**19-32** Write balanced equations for the following reactions in which an acidic solution of the $CrO_4^{2-}$ or $Cr_2O_7^{2-}$ ion is used as an oxidizing agent:

(a) $I^-(aq) + CrO_4^{2-}(aq) \longrightarrow I_2(aq) + Cr^{3+}(aq)$

(b) $Fe^{2+}(aq) + Cr_2O_7^{2-}(aq) \longrightarrow Fe^{3+}(aq) + Cr^{3+}(aq)$

(c) $H_2S(g) + CrO_4^{2-}(aq) \longrightarrow S_8(s) + Cr^{3+}(aq)$

**19-33** Write balanced equations for the following oxidation–reduction reactions that involve an acidic solution of the $MnO_4^-$ ion:

(a) $Fe^{2+}(aq) + MnO_4^-(aq) \longrightarrow Fe^{3+}(aq) + Mn^{2+}(aq)$

(b) $S_2O_3^{2-}(aq) + MnO_4^-(aq) \longrightarrow SO_4^{2-}(aq) + Mn^{2+}(aq)$

(c) $Mn^{2+}(aq) + PbO_2(s) \longrightarrow MnO_4^-(aq) + Pb^{2+}(aq)$

(d) $SO_2(g) + MnO_4^-(aq) \longrightarrow H_2SO_4(aq) + Mn^{2+}(aq)$

**19-34** Write a balanced equation for the following reaction, which occurs in acidic solution:

$$Br_2(aq) + SO_2(g) \longrightarrow H_2SO_4(aq) + HBr(aq)$$

**19-35** The two most common oxidation states of chromium are Cr(III) and Cr(VI). Balance the following oxidation–reduction equations that involve one of these oxidation states of chromium. Assume that these reactions occur in acidic solution.

(a) $Cr(s) + O_2(g) + H^+(aq) \longrightarrow Cr^{3+}(aq)$
(b) $Fe^{2+}(aq) + Cr_2O_7^{2-}(aq) \longrightarrow Fe^{3+}(aq) + Cr^{3+}(aq)$
(c) $BrO_3^-(aq) + Cr^{3+}(aq) \longrightarrow Br_2(aq) + HCrO_4^-(aq)$

### Redox Reactions in Basic Solutions

**19-36** Balance the following oxidation–reduction equations that occur in basic solution.

(a) $NO(g) + MnO_4^-(aq) \longrightarrow NO_3^-(aq) + MnO_2(s)$
(b) $NH_3(aq) + MnO_4^-(aq) \longrightarrow N_2(g) + MnO_2(s)$

**19-37** Most organic compounds react with the $MnO_4^-$ ion in the presence of acid to form $CO_2$ and water. $MnO_4^-$ in the presence of base, however, is a weaker and therefore more gentle oxidizing agent. Balance the following redox reactions in which the $MnO_4^-$ ion is used to oxidize an organic compound:

(a) $CH_3OH(aq) + MnO_4^-(aq) \xrightarrow{H^+} CO_2(g) + Mn^{2+}(aq)$
(b) $CH_3OH(aq) + MnO_4^-(aq) \xrightarrow{OH^-} HCO_2H(aq) + MnO_2(s)$

**19-38** The $Cr^{3+}$ ion is not soluble in basic solutions because it precipitates as $Cr(OH)_3$. Chromium metal will dissolve in excess base, however, because it can form a soluble $Cr(OH)_4^-$ complex ion. $Cr(OH)_3$ also can be dissolved by oxidizing the chromium to the +6 oxidation state. Write balanced equations for the following reactions that occur in basic solution:

(a) $Cr(s) + OH^-(aq) \longrightarrow Cr(OH)_4^-(aq) + H_2(g)$
(b) $Cr(OH)_3(s) + H_2O_2(aq) \longrightarrow CrO_4^{2-}(aq)$

**19-39** Write balanced equations for the following reactions to see whether there is any difference between the stoichiometry of the reaction in which the sulfite ion is oxidized to the sulfate ion when this reaction is run in acidic versus basic solution:

$$SO_3^{2-}(aq) + O_2(g) \xrightarrow{H^+} SO_4^{2-}(aq)$$
$$SO_3^{2-}(aq) + O_2(g) \xrightarrow{OH^-} SO_4^{2-}(aq)$$

**19-40** Use half-reactions to balance the following disproportionation reactions that occur in basic solution:

(a) $Cl_2(g) + OH^-(aq) \longrightarrow Cl^-(aq) + OCl^-(aq)$
(b) $Cl_2(g) + OH^-(aq) \longrightarrow Cl^-(aq) + ClO_3^-(aq)$
(c) $P_4(s) + OH^-(aq) \longrightarrow PH_3(g) + H_2PO_2^-(aq)$
(d) $H_2O_2(aq) \longrightarrow O_2(g) + H_2O(l)$

### Industrial Redox Reactions

**19-41** Write a balanced equation for the oxidation–reduction reaction used to extract gold from its ores:

$Au(s) + CN^-(aq) + O_2(g) \longrightarrow Au(CN)_2^-(aq) + OH^-(aq)$

**19-42** Write a balanced equation for the Raschig process, in which ammonia reacts with the hypochlorite ion in a basic solution to form hydrazine:

$$NH_3(aq) + OCl^-(aq) \longrightarrow N_2H_4(aq) + Cl^-(aq)$$

### Common Oxidizing Agents and Reducing Agents

**19-43** What chemical and physical properties make $Cl_2$, $O_2$, $CrO_4^{2-}$, and $MnO_4^-$ good oxidizing agents?

**19-44** What chemical and physical properties make Na, Al, and Zn good reducing agents?

**19-45** Which of the following can't be an oxidizing agent?

(a) $Cl^-$ (b) $Br_2$ (c) $Fe^{3+}$ (d) Zn (e) $CaH_2$

**19-46** Which of the following can't be a reducing agent?

(a) $H_2$ (b) $Cl_2$ (c) $Fe^{3+}$ (d) Al (e) LiH

**19-47** Which of the following can be both an oxidizing agent and a reducing agent?

(a) $H_2$ (b) $I_2$ (c) $H_2O_2$ (d) $P_4$ (e) $S_8$

**19-48** Identify the oxidizing agents on both sides of the following equation:

$$CrO_4^{2-}(aq) + PH_3(g) \xrightarrow{OH^-} Cr(OH)_4^-(aq) + P_4(aq)$$

### The Relative Strengths of Oxidizing and Reducing Agents

**19-49** Which of the following transition metals is the strongest reducing agent?

(a) Cr (b) Mn (c) Fe (d) Co (e) Ni

**19-50** Which of the following solutions is the strongest oxidizing agent?

(a) $MnO_4^-$ in acid (b) $MnO_4^-$ in base (c) $MnO_2$ in base (d) $CrO_4^{2-}$ in acid (e) $Cr_2O_7^{2-}$ in acid

**19-51** Which of the following oxidation–reduction reactions should occur as written?

(a) $Co(s) + 2\,Cr^{3+}(aq) \longrightarrow Co^{2+}(aq) + 2\,Cr^{2+}(aq)$
(b) $H_2(g) + Zn^{2+}(aq) \longrightarrow Zn(s) + 2\,H^+(aq)$
(c) $Sn(s) + 2\,H^+(aq) \longrightarrow Sn^{2+}(aq) + H_2(g)$
(d) $3\,Na(l) + AlCl_3(l) \longrightarrow 3\,NaCl(l) + Al(l)$

**19-52** Which of the following oxidation–reduction reactions should occur as written?

(a) $PbO_2(s) + Mn^{2+}(aq) \longrightarrow Pb^{2+}(aq) + MnO_2(s)$
(b) $5\,CrO_4^{2-}(aq) + 3\,Mn^{2+}(aq) + 16\,H^+(aq) \longrightarrow 5\,Cr^{3+}(aq) + 3\,MnO_4^-(aq) + 8\,H_2O(l)$
(c) $2\,H_2O_2(aq) \longrightarrow 2\,H_2O(l) + O_2(g)$

**19-53** Use Table 19.2 to predict the products of the following oxidation–reduction reactions:

(a) $HNO_3(aq) + Mn(s) \longrightarrow$
(b) $Mg(s) + CrCl_3(s) \longrightarrow$
(c) $Cr^{3+}(aq) + O_2(aq) + H^+(aq) \longrightarrow$
(d) $F_2(g) + Au(s) \longrightarrow$
(e) $2\,H_2O_2(aq) \longrightarrow$

**19-54** Which of the following pairs of ions can't coexist in aqueous solution?

(a) $Cr^{2+}$ and $MnO_4^-$ (b) $Fe^{3+}$ and $Cr_2O_7^{2-}$
(c) $Cr^{2+}$ and $I^-$ (d) $Mn^{2+}$ and $Cl^-$

**19-55** Which of the following pairs of ions can't coexist in aqueous solution?

(a) $Na^+$ and $S_2O_8^{2-}$ (b) $Hg^{2+}$ and $Cl^-$
(c) $Cr^{2+}$ and $I_3^-$ (d) $Cu^{2+}$ and $I^-$

**19-56** Solutions of the $Sn^{2+}$ ion are unstable because they are slowly oxidized by air to the $Sn^{4+}$ ion. Explain why adding small pieces of tin metal to a solution of the $Sn^{2+}$ can produce a solution with a reasonable shelf life.

**19-57** On an exam, students were asked to use half-reactions to write a balanced equation for the following reaction:

$$NH_3(aq) + MnO_4^-(aq) \xrightarrow{OH^-} N_2(g) + MnO_2(s)$$

One student wrote the following half-reactions:

Oxidation: $2\ NH_3 \longrightarrow N_2 + 6\ e^- + 6\ H^+$

Reduction: $MnO_4^- + 3\ e^- + 4\ H^+ \longrightarrow MnO_2 + 2\ H_2O$

and then wrote the following overall equation:

$$2\ NH_3(aq) + 2\ MnO_4^-(aq) + 2\ H^+(aq) \longrightarrow N_2(g) + 2\ MnO_2(s) + 4\ H_2O(l)$$

Explain why this answer is incorrect.

## Integrated Problems

**19-58** Oxalic acid reacts with the chromate ion in acidic solution to give $CO_2$ and $Cr^{3+}$ ions:

$$H_2C_2O_4(aq) + CrO_4^{2-}(aq) \xrightarrow{H^+} CO_2(g) + Cr^{3+}(aq) + H_2O(l)$$

If 10.0 mL of oxalic acid consume 40.0 mL of 0.0250 *M* $CrO_4^{2-}$ solution, what is the molarity of the oxalic acid solution?

**19-59** A 2.50-g sample of bronze was dissolved in sulfuric acid:

$$Cu(s) + 2\ H_2SO_4(aq) \longrightarrow CuSO_4(aq) + SO_2(g) + 2\ H_2O(l)$$

The $CuSO_4$ formed in this reaction was mixed with KI to form CuI:

$$2\ CuSO_4(aq) + 5\ I^-(aq) \longrightarrow 2\ CuI(s) + I_3^-(aq) + 2\ SO_4^{2-}(aq)$$

The $I_3^-$ formed in this reaction was then titrated with $S_2O_3^{2-}$:

$$I_3^-(aq) + 2\ S_2O_3^{2-}(aq) \longrightarrow 3\ I^-(aq) + S_4O_6^{2-}(aq)$$

If 31.5 mL of 1.00 *M* $S_2O_3^{2-}$ ion was consumed in this titration, what was the percent by weight of copper in the original sample of bronze?

**19-60** The percent zinc in an impure sample of the metal can be determined by titrating the sample with a potassium bromate solution. The unbalanced equation for this reaction could be written as follows:

$$Zn(s) + BrO_3^-(aq) \longrightarrow Zn^{2+}(aq) + Br_2(aq)$$

Assume that a 2.45-g sample of zinc consumed 40.4 mL of an 0.247 *M* $KBrO_3$ solution. Furthermore, assume that zinc is the only component of this sample that is a strong enough reducing agent to reduce the $BrO_3^-$ ion to $Br_2$. Calculate the percent zinc in this sample.

**19-61** A 20.00-mL sample of a $K_2C_2O_4$ solution was acidified and then titrated with an acidic 0.256 *M* $KMnO_4$ solution. What is the molarity of the oxalate solution if it took 14.6 mL of the $MnO_4^-$ solution to reach the endpoint of the titration?

**19-62** Calculate the number of grams of ferrous chloride ($FeCl_2$) that can be oxidized by 3.2 mL of 3.0 *M* $KMnO_4$.

**19-63** The stoichiometry of the reaction between hydrogen peroxide and the $MnO_4^-$ ion depends on the pH of the solution in which the reaction is run.

$$2\ MnO_4^-(aq) + 5\ H_2O_2(aq) + 6\ H^+(aq) \longrightarrow 2\ Mn^{2+}(aq) + 5\ O_2(g) + 8\ H_2O(l)$$

$$2\ MnO_4^-(aq) + 3\ H_2O_2(aq) \longrightarrow 2\ MnO_2(s) + 3\ O_2(g) + 2\ OH^-(aq) + 2\ H_2O(l)$$

Assume that it took 32.45 mL of 0.145 *M* $KMnO_4$ to titrate 28.46 mL of 0.248 *M* $H_2O_2$. Was the reaction run in acidic or basic solution?

**19-64** A 1.893-g sample of oxalic acid ($H_2C_2O_4 \cdot 2\ H_2O$) was dissolved in about 20 mL of water, acidified, and then diluted to a total volume of 50.00 mL. It took 19.35 mL of an acidic $MnO_4^-$ ion solution to titrate a 25.00-mL aliquot of the oxalic acid solution to the $MnO_4^-$ endpoint. The $MnO_4^-$ solution was then used to standardize an $H_2O_2$ solution of unknown concentration. If it took 15.37 mL of the $MnO_4^-$ ion solution to titrate a 25.00-mL aliquot of the $H_2O_2$ solution, what was the concentration of the hydrogen peroxide solution?

The bright shine on many parts of this Harley-Davidson is achieved by using electrolysis to plate a layer of chromium onto a metal surface.

# ELECTROCHEMISTRY

This chapter examines what happens when oxidation–reduction reactions are separated into a pair of half-reactions. It provides the basis for answering questions such as the following:

## POINTS OF INTEREST

- What do we mean when we say that the following reaction is *spontaneous?*

$$Cu^{2+}(aq) + Zn(s) \longrightarrow Cu(s) + Zn^{2+}(aq)$$

- How do we use a spontaneous reaction to build a *battery* and why is this term used to describe a source of electrical energy?
- Why does the voltage of a battery remain more or less constant for most of its life? What happens when the battery "runs down?"
- How can we use the fact that the voltage of a battery eventually reaches zero to measure the equilibrium constant for a reaction?
- What happens when metals corrode? How can we protect metals from corrosion?

## 20.1 ELECTRO-CHEMICAL REACTIONS

The Statue of Liberty, standing more than 300 ft tall from the base of her pedestal to the tip of her torch, was a gift to the United States from the people of France. Designed and built in Paris by the sculptor Frederic Bartholdi, she consists of 300 shaped copper plates weighing 80 tons supported by an iron skeleton designed by Alexandre Eiffel.

The Statue of Liberty consists of copper plates supported on an iron skeleton. Where the two metals came in contact, the iron skeleton has corroded.

The iron skeleton consists of more than 1300 iron bars bent to follow the contours of the copper skin and then connected to the skin with copper bands or saddles. Because the iron bars are flat, they are free to move through the saddles, thereby allowing the skin to expand or contract as the temperature changes. Asbestos strips were originally inserted between the iron bars and the copper saddles to keep the two metals from touching. Over the years, these insulating strips wore away, and the two metals came into contact. When this happened, the iron bars started to corrode.

It isn't surprising that iron corrodes when exposed to salt water. What is surprising is the fact that it corrodes as much as 1000 times faster when it comes in contact with copper metal. Some of the iron bars in the Statue of Liberty lost two-thirds of their cross-sectional area to corrosion in less than 100 years.

To understand why iron corrodes so much faster when it comes in contact with copper we have to build a model for **electrochemical reactions**—chemical reactions that involve the flow of electrons. Once established, this model will not only explain the corrosion of metals, but also the reactions that fuel the batteries that drive modern society. It will also provide a basis for understanding how electrolysis can be used to prepare strong oxidizing agents (such as $F_2$ and $Cl_2$) and strong reducing agents (such as Na, K, and Al).

### EXERCISE 20.1

Predict what will happen when iron nails are used to attach a copper roof to a building.

**SOLUTION** Iron nails corrode so fast when in contact with copper metal that the roof falls off, often within 6 months to 2 years.

## 20.2 ELECTRICAL WORK FROM SPONTANEOUS OXIDATION–REDUCTION REACTIONS

The following rule can be used to predict when an oxidation–reduction reaction should occur.

**Oxidation–reduction reactions convert the stronger of a pair of oxidizing agents and the stronger of a pair of reducing agents into a weaker oxidizing agent and a weaker reducing agent.**

Sec. 19.11

### EXERCISE 20.2

Use Table 19.2 to predict whether zinc metal should dissolve in acid.

**SOLUTION** Zinc will dissolve in acid if the metal is a strong enough reducing agent to reduce $H^+$ ions to $H_2$ gas:

$$Zn(s) + 2\,H^+(aq) \longrightarrow Zn^{2+}(aq) + H_2(g)$$

According to Table 19.2, zinc is a stronger reducing agent than $H_2$, and the $H^+$ ion is a stronger oxidizing agent than the $Zn^{2+}$ ion:

$$\underset{\text{stronger reducing agent}}{Zn(s)} + \underset{\text{stronger oxidizing agent}}{2\,H^+(aq)} \longrightarrow \underset{\text{weaker oxidizing agent}}{Zn^{2+}(aq)} + \underset{\text{weaker reducing agent}}{H_2(g)}$$

Zinc therefore should dissolve in acid.

Zn reacting with HCl.

We can test this prediction by adding a few chunks of mossy zinc to a beaker of concentrated hydrochloric acid. Within a few minutes, the zinc metal dissolves and significant amounts of hydrogen gas are liberated.

This reaction has some of the characteristic features of oxidation–reduction reactions.

- It is exothermic, in this case giving off 153.89 kJ per mole of zinc consumed.
- The equilibrium constant for the reaction is very large ($K_c = 6 \times 10^{25}$), and chemists often write the equation for this reaction as if essentially all of the reactants were converted to products:

$$Zn(s) + 2\,H^+(aq) \longrightarrow Zn^{2+}(aq) + H_2(g)$$

- It can be formally divided into separate oxidation and reduction half-reactions:

$$\text{Oxidation:} \quad Zn \longrightarrow Zn^{2+} + 2\,e^-$$
$$\text{Reduction:} \quad 2\,H^+ + 2\,e^- \longrightarrow H_2$$

- The two half-reactions can be separated so that the energy given off by this reaction can be used to do work.

According to the first law of thermodynamics, the energy given off in a chemical reaction can be converted into heat, work, or a mixture of heat and work. When we drop pieces of zinc metal into an acidic solution, most of the energy is given off as heat, although some work is done to expand the hydrogen gas formed in this reaction. If we run the half-reactions in separate containers, we can force the electrons to flow from the oxidation to the reduction half-reaction through an external wire, which allows us to capture the energy given off in the reaction as electrical work.

We can start by immersing a strip of zinc metal into a 1 *M* $Zn^{2+}$ ion solution, as shown in Figure 20.1. We then immerse a piece of platinum wire in a second beaker filled with 1 *M* HCl and bubble $H_2$ gas over the Pt wire. Finally, we connect the zinc metal and platinum wire to form an electric circuit.

We've now made a system in which electrons can flow from one half-reaction, or **half-cell,** to another. The same driving force that makes zinc metal react with acid when the two are in contact should operate in this system. Zinc atoms on the metal surface lose electrons to form $Zn^{2+}$ ions, which go into solution:

$$\text{Oxidation:} \quad Zn \longrightarrow Zn^{2+} + 2\,e^-$$

The electrons given off in this half-reaction flow through the circuit and eventually accumulate on the platinum wire to give this wire a net negative charge. The $H^+$ ions from the hydrochloric acid are attracted to this negative charge and migrate toward the platinum wire. When the $H^+$ ions touch the platinum wire, they pick up electrons to form hydrogen atoms, which combine to form $H_2$ molecules:

$$\text{Reduction:} \quad 2\,H^+ + 2\,e^- \longrightarrow H_2$$

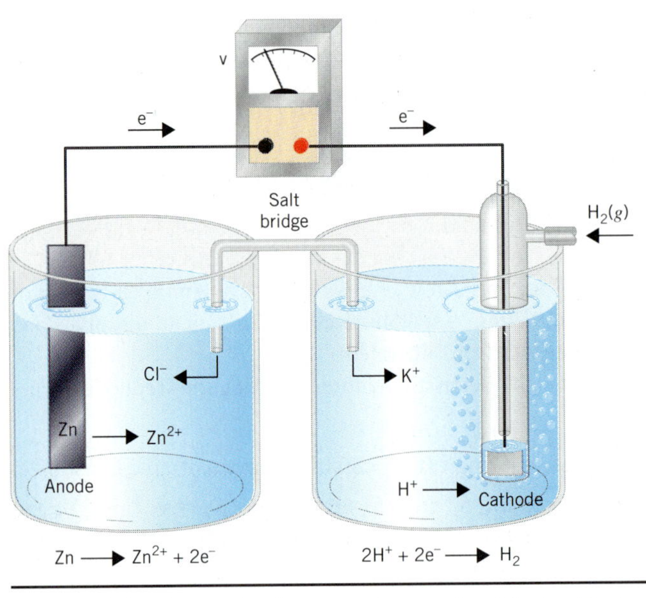


**Figure 20.1** The reaction between zinc metal and hydrochloric acid has the potential to do work if the electrons transferred in this reaction are forced to flow through an external wire. $K^+$ ions flow from the salt bridge into the cathode half-cell to compensate for the $H^+$ ions consumed at this electrode, and $Cl^-$ ions flow into the anode half-cell to balance the charge on the $Zn^{2+}$ ions produced at this electrode.

The oxidation of zinc metal releases $Zn^{2+}$ ions into the $Zn/Zn^{2+}$ half-cell. This half-cell therefore picks up a positive charge that interferes with the transfer of more electrons. The reduction of $H^+$ ions in the $H_2/H^+$ half-cell leads to a net negative charge as these $H^+$ ions are removed from the solution, which also interferes with the transfer of more electrons.

To overcome this problem, we complete the circuit by adding a U-tube filled with a saturated solution of a soluble salt, such as KCl. Negatively charged $Cl^-$ ions flow out of one end of the U-tube to balance the positive charge on the $Zn^{2+}$ ions created in one half-cell. Positively charged $K^+$ ions flow out of the other end of the tube to replace the $H^+$ ions consumed in the other half cell. The U-tube is called a **salt bridge** because it contains a solution of a salt that literally serves as a bridge to complete the electric circuit.

## 20.3 VOLTAIC CELLS

In 1791, an Italian professor of anatomy named Luigi Galvani noticed that the leg muscles of a freshly dissected frog contracted when they were connected to one of the frog's nerves with a circuit composed of two metals. This work soon came to the attention of an Italian professor of physics, Alessandro Volta, who eventually concluded that the frog's leg was a detector of the electric current generated by the bimetallic circuit.

Further experiments led Volta to distinguish between two classes of materials that conduct electric currents. One class contains metals, such as copper, zinc, and silver, that generate an electric current when brought into contact. The other class contains substances, such as aqueous salt solutions, that carry the electric current. By assembling a cell consisting of alternating silver and zinc metal discs separated by moist cardboard, Volta was able to construct the first continuous source of electric current.

In recognition of the contributions of Luigi Galvani and Alessandro Volta, electrochemical cells that use an oxidation–reduction reaction to generate an electric current are known as **galvanic** or **voltaic cells.** Because the potential of these cells to do work by driving an electric current through a wire is measured in units of *volts,* we will refer to these cells from now on as *voltaic cells.*

Let's take another look at the voltaic cell in Figure 20.1. Within each half-cell, reaction occurs on the surface of the metal electrode. At the zinc electrode, zinc atoms are oxidized to form $Zn^{2+}$ ions, which go into solution. The electrons liberated in this reaction flow through the zinc metal until they reach the wire that connects the zinc electrode to the platinum wire. They then flow through the platinum wire, where they eventually reduce a $H^+$ ion in the neighboring solution to a hydrogen atom, which combines with another hydrogen atom to form an $H_2$ molecule.

The electrode at which oxidation takes place in a electrochemical cell is called the **anode.** The electrode at which reduction occurs is called the **cathode.** The identity of the cathode and anode can be remembered by recognizing that positive ions, or **cations,** flow toward the cathode, while negative ions, or **anions,** flow toward the anode. In the voltaic cell shown in Figure 20.1, $H^+$ ions flow toward the cathode, where they are reduced to $H_2$ gas. On the other side of the cell, $Cl^-$ ions are released from the salt bridge and flow toward the anode, where the zinc metal is oxidized.

$H_2$ electrode in which $H_2$ gas is bubbled over a carbonized platinum wire.

## 20.4 STANDARD-STATE CELL POTENTIALS FOR VOLTAIC CELLS

The **cell potential** for a voltaic cell is literally the potential of the cell to do work on its surroundings by driving an electric current through a wire. By definition, one joule (J) of energy is produced when one coulomb (C) of electrical charge is transported across a potential of one volt (V):

$$1 \text{ V} = \frac{1 \text{ J}}{1 \text{ C}}$$

The potential of a voltaic cell depends on the concentrations of any species present in solution, the partial pressures of any gases involved in the reaction, and the temperature at which the reaction is run. To provide a basis for comparing the results of one experiment with another, the following set of **standard-state conditions** for electrochemical measurements has been defined.

1. All solutions are 1 *M*.
2. All gases have a partial pressure of 1 MPa (0.9869 atm).

Although standard-state measurements can be made at any temperature, they are often taken at 25°C.

Cell potentials measured under standard-state conditions are represented by the symbol $E°$. The **standard-state cell potential,** $E°$, measures the strength of the driving force behind the chemical reaction. The larger the difference between the oxidizing and reducing strengths of the reactants and products, the larger the cell potential. To obtain a relatively large cell potential, we have to react a strong reducing agent with a strong oxidizing agent.

The experimental value for the standard-state cell potential for the reaction between zinc metal and acid is 0.76 V:

$$Zn(s) + 2\,H^+(aq) \longrightarrow Zn^{2+}(aq) + H_2(g) \qquad E° = 0.76 \text{ V}$$

The cell potential for this reaction measures the *relative* reducing power of zinc metal compared with hydrogen gas. But it doesn't tell us anything about the *absolute* value of the reducing power for either zinc metal or $H_2$.

Sec. 5.15

We encountered a similar problem with enthalpies of reaction in Chapter 5. There, we solved the problem by arbitrarily defining the enthalpy of formation of an element in its most stable state at 1 atm as 0 kJ/mol. We then defined all other enthalpies of reaction in terms of this arbitrary standard.

We can do the same thing with cell potentials. We will arbitrarily define the standard-state potential for the reduction of $H^+$ ions to $H_2$ gas as exactly zero volts:

$$2\,H^+ + 2\,e^- \longrightarrow H_2 \qquad E° = 0.000\ldots\text{ V}$$

We will then use this reference point to calibrate the potential of any other half-reaction.

The key to using this reference point is recognizing that the overall cell potential for a reaction must be the sum of the potentials for the oxidation and reduction half-reactions:

$$\mathbf{E°_{overall} = E°_{ox} + E°_{red}}$$

If the overall potential for the reaction between zinc and acid is 0.76 V, and the half-cell potential for the reduction of $H^+$ ions is 0 V, then the half-cell potential for the oxidation of zinc metal must be 0.76 V:

$$\begin{array}{lll} \text{Zn} \longrightarrow \text{Zn}^{2+} + \cancel{2e^-} & & E°_{ox} = 0.76\text{ V} \\ + \; 2\,\text{H}^+ + \cancel{2e^-} \longrightarrow \text{H}_2 & & E°_{red} = 0.00\text{ V} \\ \hline \text{Zn} + 2\,\text{H}^+ \longrightarrow \text{Zn}^{2+} + \text{H}_2 & & E° = E°_{ox} + E°_{red} = 0.76\text{ V} \end{array}$$

## 20.5 PREDICTING SPONTANEOUS REDOX REACTIONS FROM THE SIGN OF *E°*

The *magnitude* of the cell potential is a measure of the driving force behind a reaction. The larger the value of the cell potential, the further the reaction is from equilibrium. The *sign* of the cell potential tells us the direction in which the reaction must shift to reach equilibrium.

Consider the reaction between zinc and acid, for example:

$$\text{Zn}(s) + 2\,\text{H}^+(aq) \longrightarrow \text{Zn}^{2+}(aq) + \text{H}_2(g) \qquad E° = 0.76\text{ V}$$

The fact that $E°$ is positive tells us that when this system is present at standard-state conditions, it has to shift to the right to reach equilibrium. Reactions for which $E°$ is positive therefore have equilibrium constants that favor the products of the reaction. It is therefore tempting to describe these reactions as ''spontaneous.''

**EXERCISE 20.3**

Use the overall cell potentials to predict which of the following reactions are spontaneous:

(a) $\text{Cu}(s) + 2\,\text{Ag}^+(aq) \longrightarrow \text{Cu}^{2+}(aq) + 2\,\text{Ag}(s) \qquad E° = 0.46\text{ V}$

(b) $2\,\text{Fe}^{3+}(aq) + 2\,\text{Cl}^-(aq) \longrightarrow 2\,\text{Fe}^{2+}(aq) + \text{Cl}_2(g) \qquad E° = -0.59\text{ V}$

(c) $2\,\text{Fe}^{3+}(aq) + 2\,\text{I}^-(aq) \longrightarrow 2\,\text{Fe}^{2+}(aq) + \text{I}_2(aq) \qquad E° = 0.24\text{ V}$

(d) $2\,\text{H}_2\text{O}_2(aq) \longrightarrow 2\,\text{H}_2\text{O}(l) + \text{O}_2(aq) \qquad E° = 1.09\text{ V}$

(e) $\text{Cu}(s) + 2\,\text{H}^+(aq) \longrightarrow \text{Cu}^{2+}(aq) + \text{H}_2(g) \qquad E° = -0.34\text{ V}$

**SOLUTION** Any reaction for which the overall cell potential is positive is spontaneous. Reactions (a), (c), and (d) are therefore spontaneous.

What happens to the cell potential when we reverse the direction in which a reaction is written?

**EXERCISE 20.4**

Use the standard-state cell potential for the following reaction:

$$Cu(s) + 2\,H^+(aq) \longrightarrow Cu^{2+}(aq) + H_2(g) \qquad E^\circ = -0.34\text{ V}$$

to predict the standard-state cell potential for the opposite reaction:

$$Cu^{2+}(aq) + H_2(g) \longrightarrow Cu(s) + 2\,H^+(aq) \qquad E^\circ = ?$$

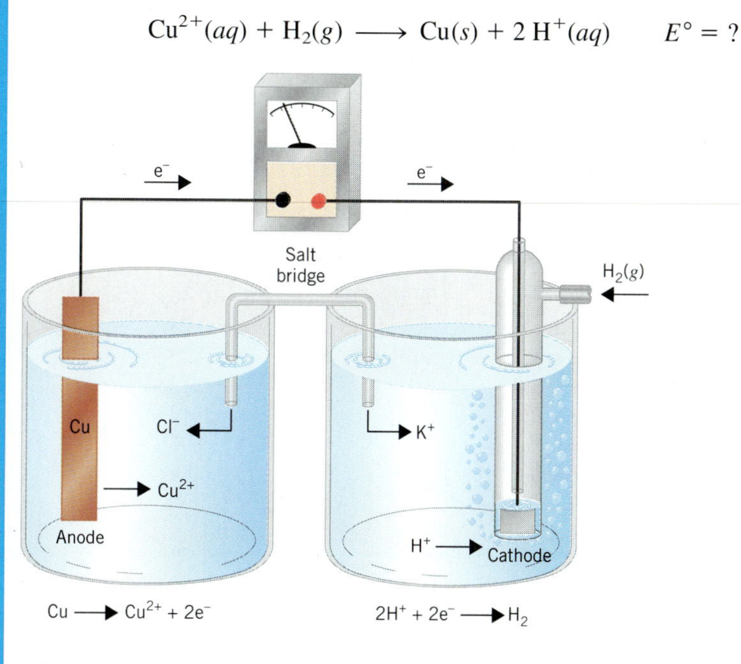


**SOLUTION** Turning the reaction around doesn't change the relative strengths of $Cu^{2+}$ and $H^+$ ions as oxidizing agents, or copper metal and $H_2$ as reducing agents. The *magnitude* of the potential therefore must remain the same. But turning the equation around changes the *sign* of the cell potential, and can therefore turn an unfavorable reaction into one that is spontaneous, or vice versa. The standard-state cell potential for the reduction of $Cu^{2+}$ ions by $H_2$ gas is therefore +0.34 V:

$$Cu^{2+}(aq) + H_2(g) \longrightarrow Cu(s) + 2\,H^+(aq) \qquad E^\circ = -(-0.34\text{ V}) = +0.34\text{ V}$$

## 20.6 STANDARD-STATE REDUCTION HALF-CELL POTENTIALS

Sec. 9.11

The standard-state cell potentials for some common half-reactions are given in Table 20.1; a more complete table can be found in Appendix A-13. Ever since oxidation and reduction were introduced in Chapter 9, we have encountered tables of relative oxidizing and reducing power that list the strongest reducing agents in the upper right corner and the strongest oxidizing agents in the lower left corner. This particular version has the advantage that it gives a quantitative measure of the difference between the strengths of a pair of reducing or oxidizing agents.

There is no need to remember that reducing agents become stronger toward the upper right corner of this table, or that the strength of the oxidizing agents increases toward the bottom left corner. All you have to do is remember some of the chemistry of the elements at the top and bottom of this table.

Take a look at the half-reaction at the top of the table:

$$K^+ + e^- \rightleftharpoons K \qquad E^\circ_{red} = -2.924\text{ V}$$

**TABLE 20.1 Standard-State Reduction Potentials, $E°_{red}$**

| | Half-Reaction | $E°_{red}$ | |
|---|---|---|---|
| | $K^+ + e^- \rightleftharpoons K$ | −2.924 | Best |
| | $Ba^{2+} + 2\,e^- \rightleftharpoons Ba$ | −2.90 | reducing |
| | $Ca^{2+} + 2\,e^- \rightleftharpoons Ca$ | −2.76 | agents |
| | $Na^+ + e^- \rightleftharpoons Na$ | −2.7109 | |
| | $Mg^{2+} + 2\,e^- \rightleftharpoons Mg$ | −2.375 | |
| | $H_2 + 2\,e^- \rightleftharpoons 2\,H^-$ | −2.23 | |
| | $Al^{3+} + 3\,e^- \rightleftharpoons Al$ | −1.706 | |
| | $Mn^{2+} + 2\,e^- \rightleftharpoons Mn$ | −1.04 | |
| | $Zn^{2+} + 2\,e^- \rightleftharpoons Zn$ | −0.7628 | |
| | $Cr^{3+} + 3\,e^- \rightleftharpoons Cr$ | −0.74 | |
| | $S + 2\,e^- \rightleftharpoons S^{2-}$ | −0.508 | |
| | $2\,CO_2 + 2\,H^+ + 2\,e^- \rightleftharpoons H_2C_2O_4$ | −0.49 | |
| | $Cr^{3+} + e^- \rightleftharpoons Cr^{2+}$ | −0.41 | |
| | $Fe^{2+} + 2\,e^- \rightleftharpoons Fe$ | −0.409 | |
| | $Co^{2+} + 2\,e^- \rightleftharpoons Co$ | −0.28 | |
| | $Ni^{2+} + 2\,e^- \rightleftharpoons Ni$ | −0.23 | |
| | $Sn^{2+} + 2\,e^- \rightleftharpoons Sn$ | −0.1364 | |
| | $Pb^{2+} + 2\,e^- \rightleftharpoons Pb$ | −0.1263 | |
| | $Fe^{3+} + 3\,e^- \rightleftharpoons Fe$ | −0.036 | |
| | $2\,H^+ + 2\,e^- \rightleftharpoons H_2$ | 0.0000 . . . | |
| | $S_4O_6^{2-} + 2\,e^- \rightleftharpoons 2\,S_2O_3^{2-}$ | 0.0895 | |
| Oxidizing | $Sn^{4+} + 2\,e^- \rightleftharpoons Sn^{2+}$ | 0.15 | ↑ |
| power | $Cu^{2+} + e^- \rightleftharpoons Cu^+$ | 0.158 | Reducing |
| increases | $Cu^+ + 2\,e^- \rightleftharpoons Cu$ | 0.3402 | power |
| ↓ | $O_2 + 2\,H_2O + 4\,e^- \rightleftharpoons 4\,OH^-$ | 0.401 | increases |
| | $Cu^+ + e^- \rightleftharpoons Cu$ | 0.522 | |
| | $I_3^- + 2\,e^- \rightleftharpoons 3\,I^-$ | 0.5338 | |
| | $MnO_4^- + 2\,H_2O + 3\,e^- \rightleftharpoons MnO_2 + 4\,OH^-$ | 0.588 | |
| | $O_2 + 2\,H^+ + 2\,e^- \rightleftharpoons H_2O_2$ | 0.682 | |
| | $Fe^{3+} + e^- \rightleftharpoons Fe^{2+}$ | 0.770 | |
| | $Hg_2^{2+} + 2\,e^- \rightleftharpoons Hg$ | 0.7961 | |
| | $Ag^+ + e^- \rightleftharpoons Ag$ | 0.7996 | |
| | $Hg^{2+} + 2\,e^- \rightleftharpoons Hg$ | 0.851 | |
| | $H_2O_2 + 2\,e^- \rightleftharpoons 2\,OH^-$ | 0.88 | |
| | $HNO_3 + 3\,H^+ + 3\,e^- \rightleftharpoons NO + 2\,H_2O$ | 0.96 | |
| | $Br_2 + 2\,e^- \rightleftharpoons 2\,Br^-$ | 1.087 | |
| | $2\,IO_3^- + 12\,H^+ + 10\,e^- \rightleftharpoons I_2 + 6\,H_2O$ | 1.19 | |
| | $CrO_4^{2-} + 8\,H^+ + 3\,e^- \rightleftharpoons Cr^{3+} + 4\,H_2O$ | 1.195 | |
| | $Pt^{2+} + 2\,e^- \rightleftharpoons Pt$ | 1.2 | |
| | $MnO_2 + 4\,H^+ + 2\,e^- \rightleftharpoons Mn^{2+} + 2\,H_2O$ | 1.208 | |
| | $O_2 + 4\,H^+ + 4\,e^- \rightleftharpoons 2\,H_2O$ | 1.229 | |
| | $Cr_2O_7^{2-} + 14\,H^+ + 6\,e^- \rightleftharpoons 2\,Cr^{3+} + 7\,H_2O$ | 1.33 | |
| | $Cl_2 + 2\,e^- \rightleftharpoons 2\,Cl^-$ | 1.3583 | |
| | $PbO_2 + 4\,H^+ + 2\,e^- \rightleftharpoons Pb^{2+} + 2\,H_2O$ | 1.467 | |
| | $MnO_4^- + 8\,H^+ + 5\,e^- \rightleftharpoons Mn^{2+} + 4\,H_2O$ | 1.491 | |
| | $Au^+ + e^- \rightleftharpoons Au$ | 1.68 | |
| | $H_2O_2 + 2\,H^+ + 2\,e^- \rightleftharpoons 2\,H_2O$ | 1.776 | |
| | $Co^{3+} + e^- \rightleftharpoons Co^{2+}$ | 1.842 | |
| Best | $S_2O_8^{2-} + 2\,e^- \rightleftharpoons 2\,SO_4^{2-}$ | 2.05 | |
| oxidizing | $O_3 + 2\,H^+ + 2\,e^- \rightleftharpoons O_2 + H_2O$ | 2.07 | |
| agents | $F_2 + 2\,H^+ + 2\,e^- \rightleftharpoons 2\,HF$ | 3.03 | |

What do we know about potassium metal? Potassium is one of the most reactive metals—it bursts into flame when added to water, for example. Furthermore, we know that metals are reducing agents in all of their chemical reactions. When we find potassium in this table, we can therefore conclude that it is listed among the strongest reducing agents.

Conversely, look at the last reaction in the table:

$$F_2 + 2\,e^- \rightleftharpoons 2\,F^- \qquad E^\circ_{red} = 3.03\text{ V}$$

Fluorine is the most electronegative element in the periodic table. It shouldn't be surprising to find that $F_2$ is the strongest oxidizing agent in Table 20.1.

Referring to either end of this table can also help you remember the sign convention for cell potentials. The previous section introduced the following rule:

**Oxidation–reduction reactions that have a positive overall cell potential are spontaneous.**

This is consistent with the data in Table 20.1. We know that fluorine easily gains electrons to form fluoride ions, and the half-cell potential for this reaction is positive:

$$F_2 + 2\,e^- \rightleftharpoons 2\,F^- \qquad E^\circ_{red} = 3.03\text{ V}$$

We also know that potassium is an excellent reducing agent. Thus, the potential for the reduction of $K^+$ ions to potassium metal is negative:

$$K^+ + e^- \rightleftharpoons K \qquad E^\circ_{red} = -2.924\text{ V}$$

but the potential for the oxidation of potassium metal to $K^+$ ions is positive:

$$K \rightleftharpoons K^+ + e^- \qquad E^\circ_{red} = -(-2.924\text{ V}) = 2.924\text{ V}$$

### EXERCISE 20.5

Which of the following is the strongest oxidizing agent?

(a) $H_2O_2$ in acid (b) $H_2O_2$ in base (c) $MnO_4^-$ in acid
(d) $MnO_4^-$ in base (e) $CrO_4^{2-}$ in acid

**SOLUTION** We start by searching Table 20.1 for the reduction half-cell potential for the oxidizing agent in each of these solutions.

(a) $H_2O_2 + 2\,H^+ + 2\,e^- \rightleftharpoons 2\,H_2O \qquad E^\circ_{red} = 1.776\text{ V}$

(b) $H_2O_2 + 2\,e^- \rightleftharpoons 2\,OH^- \qquad E^\circ_{red} = 0.88\text{ V}$

(c) $MnO_4^- + 8\,H^+ + 5\,e^- \rightleftharpoons Mn^{2+} + 4\,H_2O \qquad E^\circ_{red} = 1.491\text{ V}$

(d) $MnO_4^- + 2\,H_2O + 3\,e^- \rightleftharpoons MnO_2 + 4\,OH^- \qquad E^\circ_{red} = 0.588\text{ V}$

(e) $CrO_4^{2-} + 8\,H^+ + 3\,e^- \rightleftharpoons Cr^{3+} + 4\,H_2O \qquad E^\circ_{red} = 1.195\text{ V}$

According to these data, the strongest oxidizing agent is $H_2O_2$ in acid.

## 20.7 PREDICTING STANDARD-STATE CELL POTENTIALS

In 1836, a chemistry professor from London named John Frederic Daniell developed the first voltaic cell stable enough to be used as a battery. This Daniell cell contained a zinc rod coated with mercury immersed in dilute sulfuric acid inside a porous cup, which was in turn immersed in a solution of copper(II) sulfate in contact

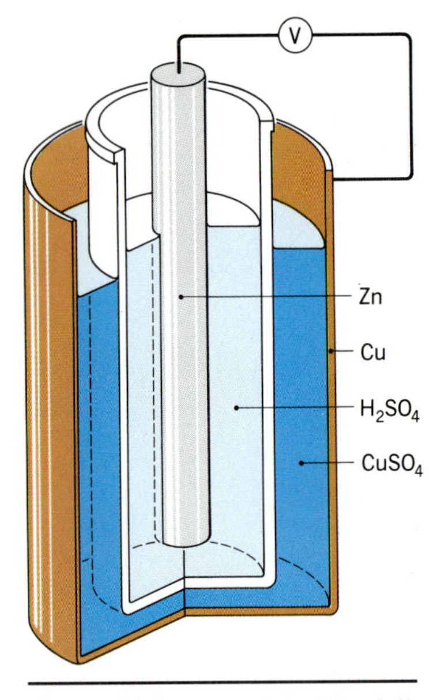


**Figure 20.2** The original Daniell cell.

with a copper container as shown in Figure 20.2. For our purposes, it is easier to work with the idealized Daniell cell in Figure 20.3.

We can use the known values of the standard-state reduction potentials for the $Cu/Cu^{2+}$ and $Zn/Zn^{2+}$ half-cells to predict the overall potential for the Daniell cell and to determine which electrode is the anode and which is the cathode.

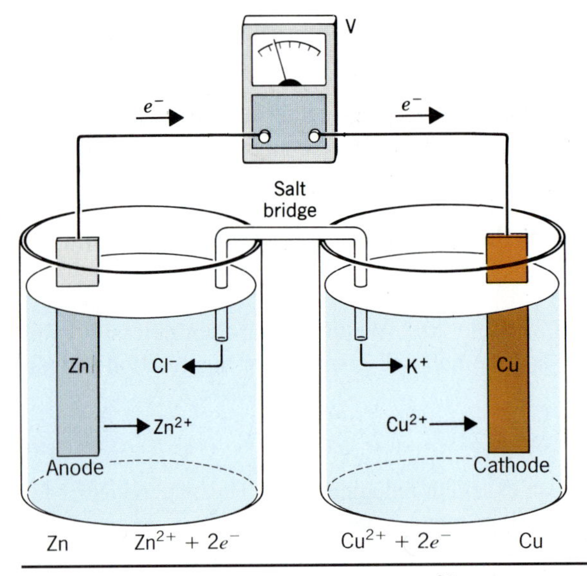


**Figure 20.3** Zinc metal is oxidized to $Zn^{2+}$ ions at the anode, and $Cu^{2+}$ ions are reduced to copper metal at the cathode of a Daniell cell.

We start by writing a balanced chemical equation for the reaction that occurs in this cell. Table 20.1 suggests that zinc is a better reducing agent than copper and that the $Cu^{2+}$ ion is a better oxidizing agent than the $Zn^{2+}$ ion. The overall reaction therefore involves the reduction of $Cu^{2+}$ ions by zinc metal:

$$Zn(s) + Cu^{2+}(aq) \longrightarrow Zn^{2+}(aq) + Cu(s)$$

We then divide the reaction into separate oxidation and reduction half-reactions:

$$\begin{array}{ll} \text{Reduction:} & Cu^{2+} + 2\,e^- \longrightarrow Cu \\ \text{Oxidation:} & Zn \longrightarrow Zn^{2+} + 2\,e^- \end{array}$$

The potential for the reduction of $Cu^{2+}$ ions to copper metal can be found in Table 20.1. To find the potential for the oxidation of zinc metal, we have to reverse the sign on the potential for the $Zn/Zn^{2+}$ couple in this table:

$$\begin{array}{lll} \text{Reduction:} & Cu^{2+} + 2\,e^- \longrightarrow Cu & E^\circ_{red} = 0.34\text{ V} \\ \text{Oxidation:} & Zn \longrightarrow Zn^{2+} + 2\,e^- & E^\circ_{ox} = -(-0.76\text{ V}) = 0.76\text{ V} \end{array}$$

The overall potential for this cell is the sum of the potentials for the two half-cells:

$$\begin{array}{ll} Cu^{2+} + \cancel{2e^-} \longrightarrow Cu & E^\circ_{red} = 0.34\text{ V} \\ +\ Zn \longrightarrow Zn^{2+} + \cancel{2e^-} & E^\circ_{ox} = 0.76\text{ V} \\ \hline Zn + Cu^{2+} \longrightarrow Zn^{2+} + Cu & E^\circ = E^\circ_{red} + E^\circ_{ox} = \mathbf{1.10\text{ V}} \end{array}$$

We now note that oxidation always occurs at the anode and reduction always occurs at the cathode of an electrochemical cell. The $Zn/Zn^{2+}$ half-cell is therefore the anode, and the $Cu^{2+}/Cu$ half-cell is the cathode, as shown in Figure 20.3.

## EXERCISE 20.6

Use cell potential data to explain why copper metal does not dissolve in a typical strong acid, such as hydrochloric acid:

$$Cu(s) + 2\,H^+(aq) \not\longrightarrow$$

but will dissolve in 1 *M* nitric acid:

$$3\,Cu(s) + 2\,HNO_3(aq) + 6\,H^+(aq) \longrightarrow 3\,Cu^{2+}(aq) + 2\,NO(g) + 4\,H_2O(l)$$

**SOLUTION** Copper does not dissolve in a typical strong acid because the overall cell potential for the oxidation of copper metal to $Cu^{2+}$ ions coupled with the reduction of $H^+$ ions to $H_2$ is negative:

$$\begin{array}{lll} & Cu \longrightarrow Cu^{2+} + 2\,e^- & E^\circ_{ox} = -(0.34\text{ V}) \\ + & 2\,H^+ + 2\,e^- \longrightarrow H_2 & E^\circ_{red} = 0.00\ldots\text{ V} \\ \hline & Cu + 2\,H^+ \longrightarrow Cu^{2+} + H_2 & E^\circ = E^\circ_{ox} + E^\circ_{red} = -0.34\text{ V} \end{array}$$

Copper dissolves in nitric acid because the reaction at the cathode now involves the reduction of nitric acid to NO gas, and the potential for this half-reaction is strong enough to overcome the half-cell potential for oxidation of copper metal to $Cu^{2+}$ ions:

$$\begin{array}{lll} & 3\,(Cu \longrightarrow Cu^{2+} + \cancel{2e^-}) & E^\circ_{ox} = -(0.34\text{ V}) \\ + & 2(HNO_3 + 3\,H^+ + \cancel{3e^-} \longrightarrow NO + 2\,H_2O) & E^\circ_{red} = 0.96\text{ V} \\ \hline & 3\,Cu + 2\,HNO_3 + 6\,H^+ \longrightarrow & E^\circ = E^\circ_{ox} + E^\circ_{red} = 0.62\text{ V} \\ & \qquad 3\,Cu^{2+} + 2\,NO + 4\,H_2O & \end{array}$$

The last step in Exercise 20.6 illustrates an important point. Note that the units of half-cell potential are volts, not volts per mole or volts per electron. All we do when combining half-reactions is add the two half-cell potentials.

## EXERCISE 20.7

Use cell potentials to explain why hydrogen peroxide disproportionates to form oxygen and water:

$$2\,H_2O_2(aq) \longrightarrow 2\,H_2O(l) + O_2(g)$$

**SOLUTION** Because the oxygen atoms in hydrogen peroxide are in the −1 oxidation state, $H_2O_2$ can be either oxidized to $O_2$ or reduced to water:

$$\begin{array}{lll} & H_2O_2 + 2\,H^+ + 2\,e^- \longrightarrow 2\,H_2O & E^\circ_{red} = 1.776\text{ V} \\ + & H_2O_2 \longrightarrow O_2 + 2\,H^+ + 2\,e^- & E^\circ_{ox} = -(0.682\text{ V}) \\ \hline & 2\,H_2O_2 \longrightarrow 2\,H_2O + O_2 & E^\circ = E^\circ_{red} + E^\circ_{ox} = 1.094\text{ V} \end{array}$$

The overall cell potential for the disproportionation of $H_2O_2$ is positive, which means this should be a spontaneous oxidation–reduction reaction.

$H_2O_2$ reacts with itself to give $O_2$, which bubbles out of solution, and water.

## 20.8 LINE NOTATION FOR VOLTAIC CELLS

Voltaic cells can be described by a **line notation** based on the following conventions:

- A single vertical line indicates a change in state or phase.
- Within a half-cell, the reactants are listed before the products.

- Concentrations of aqueous solutions are written in parentheses after the symbol for the ion or molecule.
- A double vertical line is used to indicate the junction between the half-cells.
- The line notation for the anode (oxidation) is written before the line notation for the cathode (reduction).

The line notation for a standard-state Daniell cell is written as follows:

$$\underset{\substack{\text{anode}\\\text{(oxidation)}}}{Zn|Zn^{2+}(1.0\ M)}\,||\,\underset{\substack{\text{cathode}\\\text{(reduction)}}}{Cu^{2+}(1.0\ M)|Cu}$$

Electrons flow from the anode to the cathode in a voltaic cell. (They flow from the electrode at which they are given off to the electrode at which they are consumed.) Reading from left to right, this line notation therefore corresponds to the direction in which electrons flow.

**EXERCISE 20.8**

Write the line notation for the cell shown in Figure 20.1.

**SOLUTION** The cell in Figure 20.1 is based on the following half-reactions:

$$\text{Oxidation:} \quad Zn \longrightarrow Zn^{2+} + 2\,e^-$$
$$\text{Reduction:} \quad 2\,H^+ + 2\,e^- \longrightarrow H_2$$

Since we list the reactants before the products in a half-cell, we write the following line notation for the anode:

$$\text{anode:} \quad Zn|Zn^{2+}(1.0\ M)$$

The line notation for the cathode has to indicate that $H^+$ ions are reduced to $H_2$ gas on a platinum metal surface. This can be done as follows:

$$\text{cathode:} \quad H^+(1.0\ M)|H_2(1\ \text{atm})|Pt$$

Since line notation is read in the direction in which electrons flow, we place the notation for the anode before the cathode, as follows:

$$\underset{\text{anode}}{Zn|Zn^{2+}(1.0\ M)}\,||\,\underset{\text{cathode}}{H^+(1.0\ M)|H_2(1\ \text{atm})|Pt}$$

## 20.9 THE NERNST EQUATION

What happens when the Daniell cell in Figure 20.3 is used to do work?

- The zinc electrode becomes lighter as zinc atoms are oxidized to $Zn^{2+}$ ions, which go into solution.
- The copper electrode becomes heavier as $Cu^{2+}$ ions in the solution are reduced to copper metal.
- The concentration of $Zn^{2+}$ ions at the anode increases and the concentration of the $Cu^{2+}$ ions at the cathode decreases.
- Negative ions flow from the salt bridge toward the anode to balance the charge on the $Zn^{2+}$ ions produced at this electrode.
- Positive ions flow from the salt bridge toward the cathode to compensate for the $Cu^{2+}$ ions consumed in the reaction.

An important property of the cell is missing from this list: Over a period of time, the cell runs down, and eventually has to be replaced. Let's assume that our cell is initially a standard-state cell in which the concentrations of the $Zn^{2+}$ and $Cu^{2+}$ ions are both 1 M:

$$Zn|Zn^{2+}(1.0\ M)||Cu^{2+}(1.0\ M)|Cu$$

As the reaction goes forward—as zinc metal is consumed and copper metal is produced—the driving force behind the reaction must become weaker. Therefore, the cell potential must become smaller.

This raises an interesting question: When does the cell potential become zero? In Section 20.5, we argued that the cell potential measures the driving force behind a reaction. The cell potential is therefore zero if and only if the reaction is at equilibrium. When the reaction is at equilibrium, there is no net change in the amount of zinc metal or copper ions in the system, so no electrons flow from the anode to the cathode. If there is no longer a net flow of electrons, the cell can no longer do electrical work. Its potential for doing work therefore must be zero.

Sec. 20.5

In 1889 Hermann Walther Nernst showed that the potential for an electrochemical reaction is described by the following equation:

$$E = E° - \frac{RT}{nF} \ln Q_c$$

In the **Nernst equation,** $E$ is the cell potential at some moment in time, $E°$ is the cell potential when the reaction is at standard-state conditions, $R$ is the ideal gas constant in units of joules per mole, $T$ is the temperature in kelvin, $n$ is the number of moles of electrons transferred in the balanced equation for the reaction, $F$ is the charge on a mole of electrons, and $Q_c$ is the reaction quotient at that moment in time. The symbol ln indicates a natural logarithm, or log to the base $e$, where $e$ is an irrational number equal to 2.71828 . . . .

Three terms in the Nernst equation are constants: $R$, $T$, and $F$. The ideal gas constant is 8.314 J/mol-K. The temperature is usually 25°C. The charge on a mole of electrons can be calculated from Avogadro's number and the charge on a single electron:

Sec. 6.5

$$\frac{6.022045 \times 10^{23}\ \cancel{e^-}}{1\ \text{mol}} \times \frac{1.6021892 \times 10^{-19}\ \text{C}}{1\ \cancel{e^-}} = \mathbf{96{,}484.56\ C/mol}$$

Substituting this information into the Nernst equation gives the following equation:

$$E = E° - \frac{0.02568}{n} \ln Q_c$$

Three of the remaining terms in this equation are characteristics of a particular reaction: $n$, $E°$, and $Q_c$. Consider the Daniell cell, for example.

According to Section 20.7, the standard-state potential for this cell is 1.10 V. Two moles of electrons are transferred from zinc metal to $Cu^{2+}$ ions in the balanced equation for this reaction, so $n$ is 2 for this cell. Because we never include the concentrations of solids in either reaction quotient or equilibrium constant expressions, $Q_c$ for this reaction is equal to the concentration of the $Zn^{2+}$ ion divided by the concentration of the $Cu^{2+}$ ion:

$$Q_c = \frac{(Zn^{2+})}{(Cu^{2+})}$$

Substituting what we know about the Daniell cell into the Nernst equation gives the following result, which represents the cell potential for the Daniell cell at 25°C at any moment in time:

$$E = E° - \frac{0.02568}{2} \ln \left[\frac{(Zn^{2+})}{(Cu^{2+})}\right]$$

Figure 20.4 shows a plot of the potential for the Daniell cell as a function of the natural logarithm of the reaction quotient. When the reaction quotient is very small, the cell potential is positive and relatively large. This isn't surprising because the reaction is far from equilibrium and the driving force behind the reaction should be relatively large. When the reaction quotient is very large, the cell potential is negative. This means that the reaction would have to shift back toward the reactants to reach equilibrium.

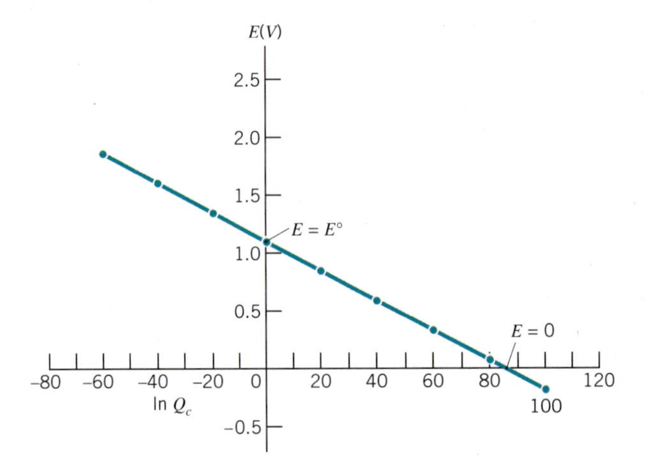


**Figure 20.4** A plot of ln $Q_c$ versus the cell potential for the Daniell cell. The cell potential is equal to $E°$ when the cell is at standard-state conditions. The cell potential is equal to zero if and only if the reaction is at equilibrium.

**EXERCISE 20.9**

Calculate the potential in the following cell when 99.99% of the $Cu^{2+}$ ions have been consumed:

$$Zn|Zn^{2+}(1.00\ M)||Cu^{2+}(1.00\ M)|Cu$$

**SOLUTION** The initial concentrations of the $Zn^{2+}$ and $Cu^{2+}$ ions are both 1.00 *M*. When 99.99% of the $Cu^{2+}$ ions have been consumed, the concentration of the $Cu^{2+}$ ion will be $1.0 \times 10^{-4}$ *M* and the $Zn^{2+}$ ion concentration will be effectively 2.00 *M*. Even though only 0.01% of the limiting reactant is left, the cell potential is still 0.97 V:

$$E = 1.10 \text{ V} - \frac{0.0257}{2} \ln \left[\frac{(2.00)}{(1 \times 10^{-4})}\right] = \mathbf{0.97\ V}$$

Exercise 20.9 raises an important point. The cell potential depends on the logarithm of the ratio of the concentrations of the products and the reactants. As a result, the potential of a cell or battery is more or less constant until virtually all of the reactants have been converted into products.

The Nernst equation can be used to calculate the potential of a cell that operates at non-standard-state conditions.

**EXERCISE 20.10**

Calculate the potential at 25°C for the following cell:

$$Cu|Cu^{2+}(0.024\ M)||Ag^{+}(0.0048\ M)|Ag$$

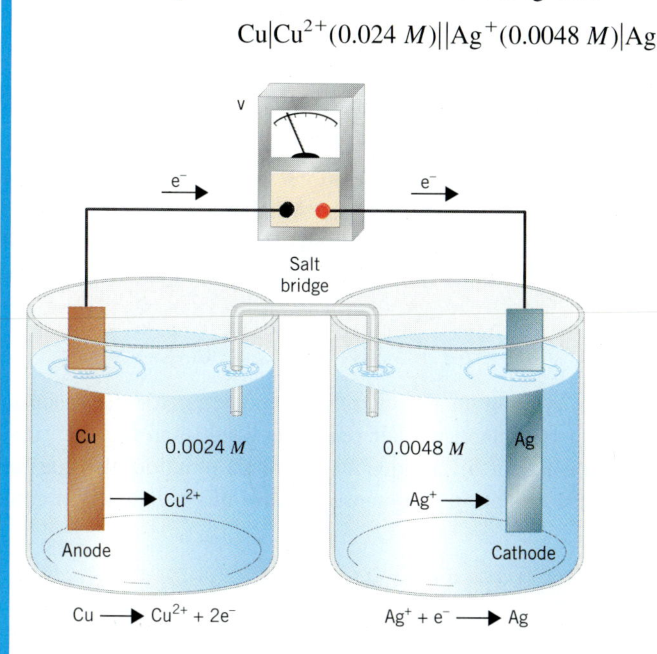


**SOLUTION** We start by translating the line notation into an equation for the reaction:

$$Cu(s) + 2\ Ag^{+}(aq) \longrightarrow Cu^{2+}(aq) + 2\ Ag(s)$$

We then look up the standard-state potentials for the two half-reactions and calculate the standard-state cell potential:

$$\begin{array}{lll} \text{Oxidation:} & Cu \longrightarrow Cu^{2+} + 2\,e^{-} & E^{\circ}_{ox} = -(-0.3402\ V) \\ \text{Reduction:} & Ag^{+} + e^{-} \longrightarrow Ag & E^{\circ}_{red} = 0.7996\ V \\ & & E^{\circ} = E^{\circ}_{red} + E^{\circ}_{ox} = 0.4594\ V \end{array}$$

We now set up the Nernst equation for this cell, noting that $n$ is 2 because two electrons are transferred in the balanced equation for the reaction:

$$E = E^{\circ} - \frac{0.02568}{2} \ln \left[ \frac{(Cu^{2+})}{(Ag^{+})^{2}} \right]$$

Substituting the concentrations of the $Cu^{2+}$ and $Ag^{+}$ ions into this equation gives the value for the cell potential:

$$E = 0.4594\ V - \frac{0.02568}{2} \ln \left[ \frac{(0.024)}{(0.0048)^{2}} \right] = 0.3702\ V$$

The Nernst equation can also help us understand a popular demonstration that uses glucose, $C_6H_{12}O_6$, to reduce $Ag^{+}$ ions to silver metal, thereby forming a silver mirror on the inner surface of the container. The standard-state half-cell potential for the reduction of $Ag^{+}$ ions is approximately 0.800 volt:

$$Ag^{+} + e^{-} \longrightarrow Ag \qquad E^{\circ}_{red} = 0.800\ V$$

The standard-state half-cell potential for the oxidation of glucose is −0.050 V:

$$C_6H_{12}O_6 + H_2O \longrightarrow C_6H_{12}O_7 + 2\,H^+ + 2\,e^- \qquad E^\circ_{ox} = -0.050\text{ V}$$

The overall standard-state cell potential for this reaction is therefore favorable:

$$\begin{array}{lr} 2\,(Ag^+ + \cancel{e^-} \longrightarrow Ag) & E^\circ_{red} = 0.800\text{ V} \\ + \; C_6H_{12}O_6 + H_2O \longrightarrow C_6H_{12}O_7 + 2\,H^+ + \cancel{2\,e^-} & E^\circ_{ox} = -0.050\text{ V} \\ \hline 2\,Ag^+ + C_6H_{12}O_6 + H_2O \longrightarrow 2\,Ag + C_6H_{12}O_7 + 2\,H^+ & E^\circ = 0.750\text{ V} \end{array}$$

The reaction isn't run under standard-state conditions, however. It takes place in a solution to which aqueous ammonia has been added. Most of the silver is therefore present as the $Ag(NH_3)_2^+$ complex ion. This is important because the half-cell potential for the reduction of this complex is considerably smaller than the potential for the reduction of the $Ag^+$ ion:

$$Ag(NH_3)_2^+ + e^- \longrightarrow Ag + 2\,NH_3 \qquad E^\circ_{red} = 0.373\text{ V}$$

This leads to a significant *decrease* in the overall cell potential for the reaction because the $Ag(NH_3)_2^+$ ion is a much weaker oxidizing agent than the $Ag^+$ ion.

The fact that this reaction is run in an aqueous ammonia solution also has an effect on the potential for the oxidation of glucose because this half-reaction contains a pair of $H^+$ ions:

$$C_6H_{12}O_6 + H_2O \longrightarrow C_6H_{12}O_7 + 2\,H^+ + 2\,e^-$$

The half-cell potential for this reaction therefore depends on the pH of the solution. Because two $H^+$ ions are given off when glucose is oxidized, the reaction quotient for this reaction depends on the square of the $H^+$ ion concentration. A change in this solution from standard-state conditions (pH = 0) to the pH of an aqueous ammonia solution (pH ≈ 11) therefore results in an *increase* of 0.650 V in the half-cell potential for this reaction.

The increase in the reducing strength of glucose when the reaction is run at pH 11 more than compensates for the decrease in oxidizing strength that results from the formation of the $Ag(NH_3)_2^+$ complex ion. Thus, the overall cell potential for the reduction of silver ions to silver metal is actually more favorable in aqueous ammonia than under standard-state conditions.

$$\begin{array}{lr} 2\,(Ag(NH_3)_2^+ + \cancel{e^-} \longrightarrow Ag + 2\,NH_3) & E^\circ_{red} = 0.373\text{ V} \\ + \; C_6H_{12}O_6 + H_2O \longrightarrow C_6H_{12}O_7 + 2\,H^+ + \cancel{2\,e^-} & E^\circ_{ox} = 0.6000\text{ V} \\ \hline 2\,Ag(NH_3)_2^+ + C_6H_{12}O_6 + H_2O \longrightarrow & E^\circ = 0.973\text{ V} \\ \qquad 2\,Ag + C_6H_{12}O_7 + 2\,NH_4^+ + 2\,NH_3 & \end{array}$$

## 20.10 USING THE NERNST EQUATION TO MEASURE EQUILIBRIUM CONSTANTS

The Nernst equation can be used to measure the equilibrium constant for a reaction. To understand how this is done, we have to understand what happens to the cell potential as an oxidation–reduction reaction comes to equilibrium. As the reaction approaches equilibrium, the driving force behind the reaction decreases, and the cell potential approaches zero:

$$\text{At equilibrium:} \qquad E = 0$$

What implications does this have for the Nernst equation?

$$E = E^\circ - \frac{RT}{nF}\ln Q_c$$

At equilibrium, the reaction quotient is equal to the equilibrium constant ($Q_c = K_c$) and the overall cell potential for the reaction is zero ($E = 0$).

$$\text{At equilibrium:} \qquad 0 = E^\circ - \frac{RT}{nF}\ln K_c$$

Rearranging this equation gives the following result:

$$\text{At equilibrium:} \qquad nFE^\circ = RT\ln K_c$$

According to this equation, we can calculate the equilibrium constant for any oxidation–reduction reaction from its standard-state cell potential:

$$\boldsymbol{K_c = e^{nFE^\circ/RT}}$$

or, in a more useful arrangement:

$$\boldsymbol{\ln K_c = \frac{nFE^\circ}{RT}}$$

This photograph shows the silver mirror than can be plated out on the inside of a flask by reduction of $Ag^+$ ions with glucose in an aqueous ammonia solution.

## EXERCISE 20.11

Calculate the equilibrium constant at 25°C for the reaction between zinc metal and acid:

$$Zn(s) + 2\,H^+(aq) \longrightarrow Zn^{2+}(aq) + H_2(g)$$

**SOLUTION** We start by calculating the overall standard-state cell potential for this reaction:

$$\begin{array}{llr} Zn \longrightarrow Zn^{2+} + 2\,e^- & & E^\circ_{ox} = -(-0.7628\text{ V}) \\ +\ 2\,H^+ + 2\,e^- \longrightarrow H_2 & & E^\circ_{red} = 0.0000\text{ V} \\ \hline Zn + 2\,H^+ \longrightarrow Zn^{2+} + H_2 & & E^\circ = \mathbf{0.7628\text{ V}} \end{array}$$

We then substitute into the Nernst equation the implications of the fact that the reaction is at equilibrium: $Q_c = K_c$ and $E = 0$.

$$0 = E^\circ - \frac{RT}{nF}\ln K_c$$

We then rearrange this equation:

$$nFE^\circ = RT\ln K_c$$

and solve for the natural logarithm of the equilibrium constant:

$$\ln K_c = \frac{nFE^\circ}{RT}$$

Substituting what we know about the reaction into this equation gives the following result:

$$\ln K_c = \frac{(2)(96{,}485\text{ C})(0.7628\text{ V})}{(8.314\text{ J/mol-K})(298\text{ K})} = \mathbf{59.41}$$

$$K_c = e^{59.41} = \mathbf{6.3 \times 10^{25}}$$

This is a very large equilibrium constant, which means that equilibrium lies heavily on the side of the products. The equilibrium constant is so large that the equation for this reaction is written as if it proceeds to completion:

$$Zn(s) + 2\,H^+(aq) \longrightarrow Zn^{2+}(aq) + H_2(g)$$

This technique can even be used to calculate equilibrium constants for reactions that don't seem to involve oxidation–reduction.

## EXERCISE 20.12

Use the following standard-state cell potentials to calculate the complex formation equilibrium constant for the $Zn(NH_3)_4^{2+}$ complex ion:

$$Zn(NH_3)_4^{2+} + 2\,e^- \rightleftharpoons Zn + 4\,NH_3 \qquad E^\circ_{red} = -1.04\ V$$
$$Zn^{2+} + 2\,e^- \rightleftharpoons Zn \qquad E^\circ_{red} = -0.7628\ V$$

**SOLUTION** By reversing the first half-reaction and adding it to the second, we can obtain an overall equation that corresponds to the complex formation equilibrium:

$$\cancel{Zn} + 4\,NH_3 \rightleftharpoons Zn(NH_3)_4^{2+} + \cancel{2e^-} \qquad E^\circ_{ox} = 1.04\ V$$
$$+\ Zn^{2+} + \cancel{2e^-} \rightleftharpoons \cancel{Zn} \qquad E^\circ_{red} = -0.7628\ V$$
$$Zn^{2+} + 4\,NH_3 \rightleftharpoons Zn(NH_3)_4^{2+} \qquad E^\circ = \mathbf{0.28\ V}$$

When this reaction comes to equilibrium, the cell potential is zero and the reaction quotient is equal to the complex formation equilibrium constant for the $Zn(NH_3)_4^{2+}$ complex ion:

$$0 = E^\circ - \frac{RT}{nF}\ln K_f$$

We now solve this equation for the natural logarithm of the equilibrium constant and calculate the value of $K_f$ for this reaction:

$$\ln K_f = \frac{nFE^\circ}{RT} = \frac{(2)(96{,}485\ C)(0.28\ V)}{(8.314\ J/mol\text{-}K)(298\ K)} = \mathbf{21.8}$$
$$K_f = e^{21.8} = \mathbf{2.9 \times 10^9}$$

The value of $K_f$ for the $Zn(NH_3)_4^{2+}$ complex ion obtained from this calculation is the same as the one given in Table A.12 in the appendix.

## EXERCISE 20.13

Use the following standard-state cell potentials to calculate the solubility product at 25°C for $Mg(OH)_2$.

$$Mg(OH)_2 + 2\,e^- \rightleftharpoons Mg + 2\,OH^- \qquad E^\circ_{red} = -2.69\ V$$
$$Mg^{2+} + 2\,e^- \rightleftharpoons Mg \qquad E^\circ_{red} = -2.375\ V$$

Problem-Solving Strategy

**SOLUTION** This time we have to reverse the second half-reaction to obtain an overall equation that corresponds to the appropriate equilibrium expression:

$$Mg(OH)_2 + \cancel{2e^-} \rightleftharpoons \cancel{Mg} + 2\,OH^- \qquad E^\circ_{red} = -2.69\ V$$
$$+\ \cancel{Mg} \rightleftharpoons Mg^{2+} + \cancel{2e^-} \qquad E^\circ_{ox} = 2.375\ V$$
$$Mg(OH)_2 \rightleftharpoons Mg^{2+} + 2\,OH^- \qquad E^\circ = \mathbf{-0.32\ V}$$

Once again, we start with the Nernst equation and assume that the reaction is at equilibrium:

$$0 = E^\circ - \frac{RT}{nF}\ln K_{sp}$$

We then rearrange this equation, substitute what we know about the reaction into this equation, and calculate the value of $K_{sp}$ for this compound:

$$\ln K_{sp} = \frac{nFE^\circ}{RT} = \frac{(2)(96{,}485\text{ C})(-0.32\text{ V})}{(8.314\text{ J/mol-K})(298\text{ K})} = -24.9$$

$$K_{sp} = 1.5 \times 10^{-11}$$

The value of $K_{sp}$ obtained from this calculation matches the one given in Table A.11 of the appendix, within experimental error.

## 20.11 ELECTROLYTIC CELLS

The cells discussed so far have one thing in common: They use a spontaneous chemical reaction to drive an electric current through an external circuit. These voltaic cells are important because they are the basis for the batteries that fuel modern society. But they are not the only kind of electrochemical cell. It is also possible to construct a cell that does work on a chemical system by driving an electric current through the system, breaking down, or lyzing, a substance to produce charged products. These cells are called **electrolytic cells.** Electrolysis is used to drive an oxidation–reduction reaction in a direction in which it does not occur spontaneously.

## 20.12 THE ELECTROLYSIS OF MOLTEN NaCl

An idealized cell for the electrolysis of sodium chloride is shown in Figure 20.5. A source of direct current is connected to a pair of inert electrodes immersed in molten sodium chloride. Because the salt has been heated until it melts, the $Na^+$ ions flow toward the **negative electrode** (or cathode) and the $Cl^-$ ions flow toward the **positive electrode** (or anode).

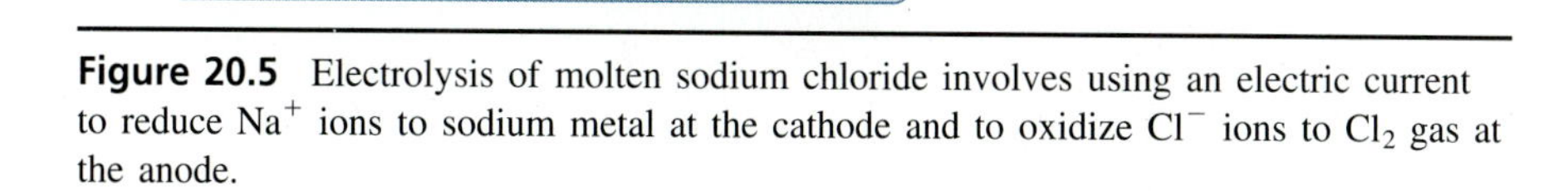


**Figure 20.5** Electrolysis of molten sodium chloride involves using an electric current to reduce $Na^+$ ions to sodium metal at the cathode and to oxidize $Cl^-$ ions to $Cl_2$ gas at the anode.

When $Na^+$ ions collide with the negative electrode, the battery carries a large enough potential to force these ions to pick up electrons to form sodium metal:

$$\text{Negative electrode (cathode):} \qquad Na^+ + e^- \longrightarrow Na$$

Electrolysis is used to prepare either sodium, aluminum or magnesium metal.

Reduction therefore occurs at the cathode of this cell. The $Cl^-$ ions that collide with the positive electrode are oxidized to $Cl_2$ gas, which bubbles off at this electrode:

$$\text{Positive electrode (anode):} \qquad 2\,Cl^- \longrightarrow Cl_2 + 2\,e^-$$

Oxidation therefore occurs at the anode. The net effect of passing an electric current through the molten salt in this cell is to decompose sodium chloride into its elements, sodium metal and chlorine gas:

Electrolysis of NaCl:

$$\text{Cathode } (-): \qquad Na^+ + e^- \longrightarrow Na$$

$$\text{Anode } (+): \qquad 2\,Cl^- \longrightarrow Cl_2 + 2\,e^-$$

The potential required to oxidize $Cl^-$ ions to $Cl_2$ is $-1.36$ V and the potential needed to reduce $Na^+$ ions to sodium metal is $-2.7$ V. The battery used to drive this reaction therefore must have a potential of at least 4.07 V.

This example explains why the process is called **electrolysis.** The suffix *-lysis* comes from the Greek stem meaning ''to loosen or split up.'' Electrolysis literally uses an electric current to split a compound into its elements:

$$2\,NaCl(l) \xrightarrow{\text{electrolysis}} 2\,Na(l) + Cl_2(g)$$

This example also illustrates the difference between voltaic cells and electrolytic cells. Voltaic cells use the energy given off in a spontaneous reaction to do electrical work. Electrolytic cells use electrical work as source of energy to drive the reaction in the opposite direction.

The dotted vertical line in the center of Figure 20.5 represents a diaphragm that keeps the $Cl_2$ gas produced at the anode from coming into contact with the sodium metal generated at the cathode. The function of this diaphragm can be understood by turning to a more realistic drawing of the commercial Downs cell used to electrolyze sodium chloride shown in Figure 20.6.

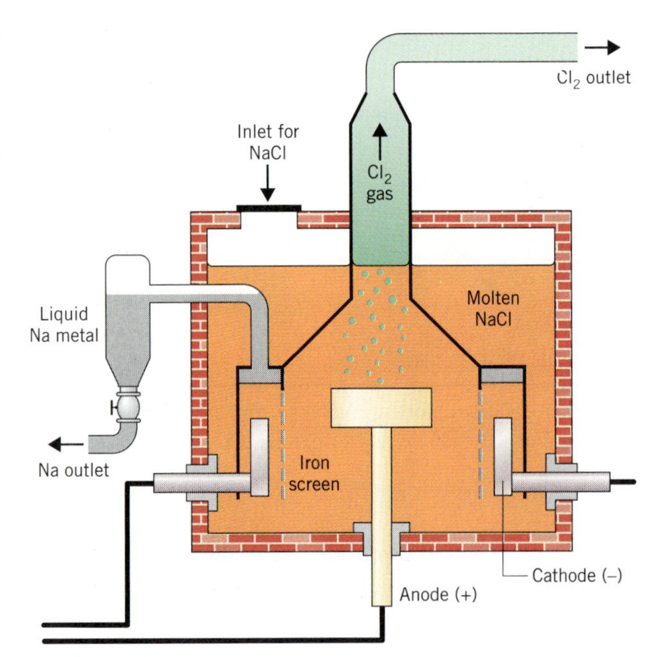


**Figure 20.6** A cross-section of the Downs cell used for the electrolysis of a molten mixture of calcium chloride and sodium chloride.

Chlorine gas that forms on the graphite anode inserted into the bottom of this cell bubbles through the molten sodium chloride into a funnel at the top of the cell. Sodium metal that forms at the cathode floats up through the molten sodium chloride into a sodium-collecting ring, from which it is periodically drained. The diaphragm that separates the two electrodes is a screen of iron gauze, which prevents the explosive reaction that would occur if the products of the electrolysis reaction came in contact with each other.

The feedstock for the Downs cell is a 3:2 mixture by mass of $CaCl_2$ and NaCl. This mixture is used because it has a melting point of 580°C, whereas pure sodium chloride has to be heated to more than 800°C before it melts.

## 20.13 THE ELECTROLYSIS OF AQUEOUS NaCl

Figure 20.7 shows an idealized drawing of a cell in which an aqueous solution of sodium chloride is electrolyzed. Once again, the $Na^+$ ions migrate toward the negative electrode and the $Cl^-$ ions migrate toward the positive electrode. But, now there

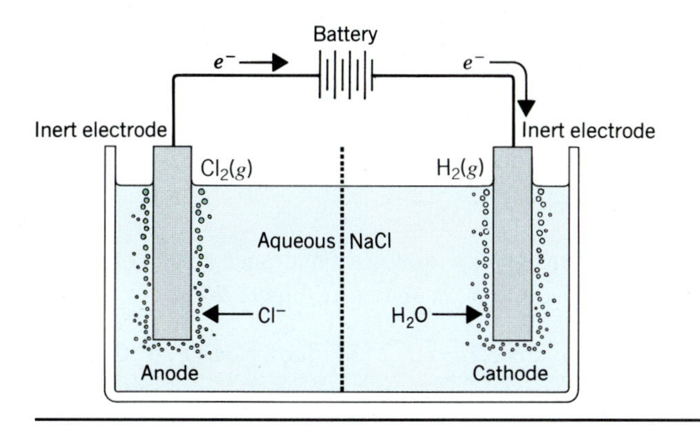


**Figure 20.7** Electrolysis of aqueous sodium chloride results in reduction of water to form $H_2$ gas at the cathode and oxidation of $Cl^-$ ions to form $Cl_2$ gas at the anode.

are two substances that can be reduced at the cathode: $Na^+$ ions and water molecules.

Cathode (−):

$$Na^+ + e^- \longrightarrow Na \qquad E^\circ_{red} = -2.71\ V$$

$$2\,H_2O + 2\,e^- \longrightarrow H_2 + 2\,OH^- \qquad E^\circ_{red} = -0.83\ V$$

Because it is much easier to reduce water than $Na^+$ ions, the only product formed at the cathode is hydrogen gas:

$$\text{Cathode } (-): \qquad 2\,H_2O + 2\,e^- \longrightarrow H_2 + 2\,OH^-$$

There are also two substances that can be oxidized at the anode: $Cl^-$ ions and water molecules.

Anode (+):

$$2\,Cl^- \longrightarrow Cl_2 + 2\,e^- \qquad E^\circ_{ox} = -1.36\ V$$

$$2\,H_2O \longrightarrow O_2 + 4\,H^+ + 4\,e^- \qquad E^\circ_{ox} = -1.23\ V$$

The standard-state potentials for these half-reactions are so close to each other that we might expect to see a mixture of $Cl_2$ and $O_2$ gas collect at the anode. In practice, the only product of this reaction is $Cl_2$:

$$\text{Anode } (+): \qquad 2\,Cl^- \longrightarrow Cl_2 + 2\,e^-$$

At first glance, it would seem easier to oxidize water ($E^\circ_{ox} = -1.23$ V) than $Cl^-$ ions ($E^\circ_{ox} = -1.36$ V). We must remember, however, that these are standard-state half-cell potentials, and in practice, this cell is never allowed to reach standard-state conditions. The solution is typically 25% NaCl by mass, which significantly decreases the potential required to oxidize the $Cl^-$ ion. The pH of the cell is also kept very high, which decreases the oxidation potential for water. The deciding factor is a phenomenon known as **overvoltage,** which is the extra voltage that must be applied to a reaction to get it to occur at the rate at which it would occur in an ideal system.

Under ideal conditions, a potential of 1.23 V is large enough to oxidize water to $O_2$ gas. Under real conditions, however, it can take a much larger voltage to initiate this reaction. (The overvoltage for the oxidation of water can be as large as 1 V.) By carefully choosing the electrode to maximize the overvoltage for the oxidation of water and then carefully controlling the potential at which the cell operates, we can ensure that only chlorine is produced in this reaction.

In summary, electrolysis of aqueous solutions of sodium chloride does not give the same products as electrolysis of molten sodium chloride. Electrolysis of molten NaCl decomposes this compound into its elements:

$$2\,NaCl(l) \xrightarrow{\text{electrolysis}} 2\,Na(l) + Cl_2(g)$$

Electrolysis of aqueous NaCl solutions gives a mixture of hydrogen and chlorine gas and an aqueous sodium hydroxide solution, as shown in Figure 20.8:

$$2\,NaCl(l) + 2\,H_2O(l) \xrightarrow{\text{electrolysis}} 2\,Na^+(aq) + 2\,OH^-(aq) + H_2(g) + Cl_2(g)$$

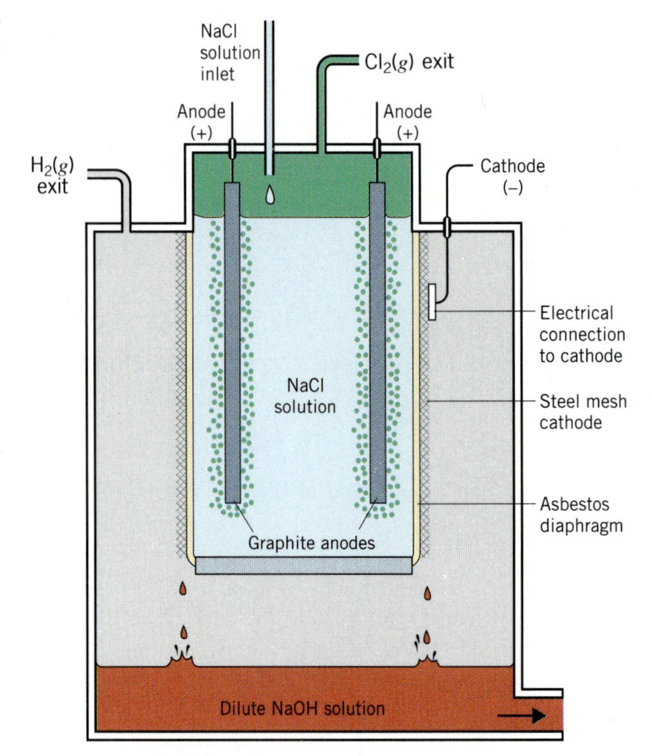


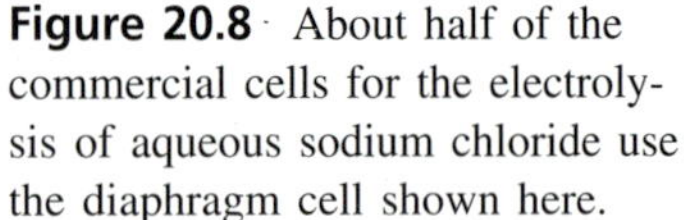
**Figure 20.8** About half of the commercial cells for the electrolysis of aqueous sodium chloride use the diaphragm cell shown here.

Because the demand for chlorine is much larger than the demand for sodium, electrolysis of aqueous sodium chloride is a more important process commercially. Electrolysis of an aqueous NaCl solution has two other advantages. It produces $H_2$ gas at the cathode, which can be collected and sold. It also produces NaOH, which can be drained from the bottom of the electrolytic cell and sold.

The dotted vertical line in Figure 20.7 represents a diaphragm that prevents the $Cl_2$ produced at the anode in this cell from coming into contact with the NaOH that accumulates at the cathode. When this diaphragm is removed from the cell, the products of the electrolysis of aqueous sodium chloride react to form sodium hypochlorite, which is the first step in the preparation of hypochlorite bleaches, such as Chlorox:

$$Cl_2(g) + 2\,OH^-(aq) \longrightarrow Cl^-(aq) + OCl^-(aq) + H_2O(l)$$

## 20.14 ELECTROLYSIS OF WATER

A standard apparatus for the electrolysis of water is shown in Figure 20.9.

$$2\,H_2O(l) \xrightarrow{\text{electrolysis}} 2\,H_2(g) + O_2(g)$$

A pair of inert electrodes is sealed in opposite ends of a container designed to collect the $H_2$ and $O_2$ gas given off in this reaction. The electrodes are then connected to a battery or another source of electric current.

By itself, water is a very poor conductor of electricity. We therefore add an electrolyte to water to provide ions that can flow through the solution, thereby completing the electric circuit. The electrolyte must be soluble in water. It also should be relatively inexpensive. Most importantly, it must contain ions that are harder to oxidize or reduce than water:

$$2\,H_2O + 2\,e^- \longrightarrow H_2 + 2\,OH^- \qquad E^\circ_{red} = -0.83\text{ V}$$

$$2\,H_2O \longrightarrow O_2 + 4\,H^+ + 4\,e^- \qquad E^\circ_{ox} = -1.23\text{ V}$$

According to Table 20.1, the following cations are harder to reduce than water: $Li^+$, $Rb^+$, $K^+$, $Cs^+$, $Ba^{2+}$, $Sr^{2+}$, $Na^+$, and $Mg^{2+}$. Two of these cations are more likely candidates than the others because they form inexpensive, soluble salts: $Na^+$ and $K^+$.

Table 20.1 suggests that the $SO_4{}^{2-}$ ion might be the best anion to use because it is the most difficult anion to oxidize. The potential for oxidation of this ion to the peroxydisulfate ion is $-2.05$ V:

$$2\,SO_4{}^{2-} \longrightarrow S_2O_8{}^{2-} + 2\,e^- \qquad E^\circ_{ox} = -2.05\text{ V}$$

When an aqueous solution of either $Na_2SO_4$ or $K_2SO_4$ is electrolyzed in the apparatus shown in Figure 20.9, $H_2$ gas collects at one electrode and $O_2$ gas collects at the other.

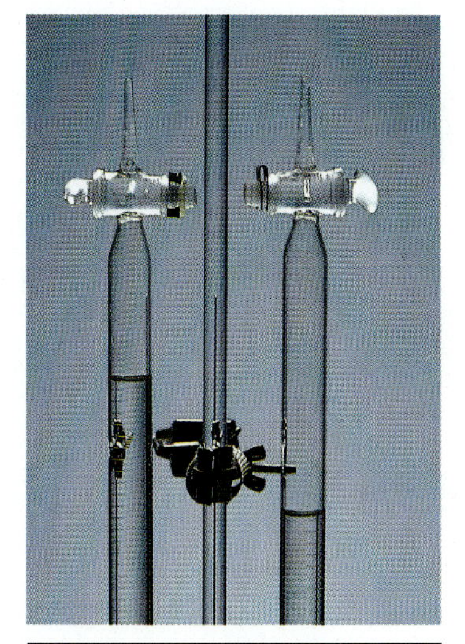

**Figure 20.9** Electrolysis of an aqueous $Na_2SO_4$ solution results in the reduction of water to form $H_2$ gas at the cathode, and oxidation of water to form $O_2$ gas at the anode.

What would happen if we added an indicator such as bromothymol blue to this apparatus? Bromothymol blue turns yellow in acidic solutions (pH $< 6$) and blue in basic solutions (pH $> 7.6$). According to the equations for the two half-reactions, the indicator should turn yellow at the anode and blue at the cathode:

$$\text{Cathode }(-):\quad 2\,H_2O + 2\,e^- \longrightarrow H_2 + 2\,OH^-$$

$$\text{Anode }(+):\quad 2\,H_2O \longrightarrow O_2 + 4\,H^+ + 4\,e^-$$

## 20.15 FARADAY'S LAW

Humphry Davy's experiments with electrochemistry led to his discovery of sodium, potassium, magnesium, calcium, barium, and strontium. It has been said, however, that his greatest discovery was Michael Faraday. Faraday was apprenticed to a bookbinder in London at the age of 13. His exposure to works brought into the shop for binding excited his interest in science, and he started attending lectures on science and doing experiments in chemistry and electricity on his own. He was eventually hired by Davy as a secretary and scientific assistant, and thereby began a lifelong commitment to research in both chemistry and physics.

Faraday's early research on electrolysis led him to propose a relationship between the amount of current passed through a solution and the mass of the substance decomposed or produced by this current. **Faraday's law** of electrolysis can be stated as follows:

> **The amount of a substance consumed or produced at one of the electrodes in an electrolytic cell is directly proportional to the amount of electricity that passes through the cell.**

In order to use Faraday's law we need to recognize the relationship between current, time, and the amount of electric charge that flows through a circuit. By definition, one coulomb of charge is transferred when a one-ampere current flows for one second:

$$1\ \text{C} = 1\ \text{amp-s}$$

To illustrate how Faraday's law can be used, let's calculate the number of grams of sodium metal that will form at the cathode when a 10.0-amp current is passed through molten sodium chloride for a period of 4.00 hours.

We start by calculating the amount of electric charge that flows through the cell:

$$10.0\ \cancel{\text{amp}} \times 4.00\ \cancel{\text{h}} \times \frac{60\ \cancel{\text{min}}}{1\ \cancel{\text{h}}} \times \frac{60\ \cancel{\text{s}}}{1\ \cancel{\text{min}}} \times \frac{1\ \text{C}}{1\ \cancel{\text{amp-s}}} = \mathbf{144{,}000\ C}$$

Before we can use this information, we need a bridge between this macroscopic quantity and the phenomenon that occurs on the atomic scale. This bridge is represented by **Faraday's constant,** which describes the number of coulombs of charge carried by a mole of electrons:

Sec. 20.9

$$\frac{6.022045 \times 10^{23}\ \cancel{e^-}}{1\ \text{mol}} \times \frac{1.6021892 \times 10^{-19}\ \text{C}}{1\ \cancel{e^-}} = \mathbf{96{,}484.56\ C/mol}$$

Thus, the number of moles of electrons transferred when 144,000 coulombs of electric charge flow through the cell can be calculated as follows:

$$144{,}000\ \cancel{\text{C}} \times \frac{1\ \text{mol e}^-}{96{,}485\ \cancel{\text{C}}} = \mathbf{1.49\ mol\ e^-}$$

According to the balanced equation for the reaction that occurs at the cathode of this cell, we get 1 mole of sodium for every mole of electrons:

$$\text{Cathode } (-): \qquad Na^+ + e^- \longrightarrow Na$$

Thus, we get 34.3 g of sodium in 4.00 hours.

$$1.49\ \cancel{\text{mol Na}} \times \frac{22.99\ \text{g Na}}{1\ \cancel{\text{mol Na}}} = \mathbf{34.3\ g\ Na}$$

The implications of this calculation are interesting: We would have to run this electrolysis for more than 2 days to prepare a pound of sodium.

## EXERCISE 20.14

Calculate the volume of $H_2$ gas at 25°C and 1.00 atm that will collect at the cathode when an aqueous solution of $Na_2SO_4$ is electrolyzed for 2.00 hours with a 10.0-amp current.

**SOLUTION** We start by calculating the amount of electrical charge that passes through the solution:

$$10.0 \text{ amp} \times 2.00 \text{ h} \times \frac{60 \text{ min}}{1 \text{ h}} \times \frac{60 \text{ s}}{1 \text{ min}} \times \frac{1 \text{ C}}{1 \text{ amp s}} = 72{,}000 \text{ C}$$

We then calculate the number of moles of electrons that carry this charge:

$$72{,}000 \text{ C} \times \frac{1 \text{ mol e}^-}{96{,}485 \text{ C}} = 0.746 \text{ mol e}^-$$

The balanced equation for the reaction that produces $H_2$ gas at the cathode indicates that we get a mole of $H_2$ gas for every 2 mole of electrons:

$$\text{Cathode } (-): \qquad 2\,H_2O + 2\,e^- \longrightarrow H_2 + 2\,OH^-$$

We therefore get 1 mole of $H_2$ gas at the cathode for every 2 mole of electrons that flow through the cell:

$$0.746 \text{ mol e}^- \times \frac{1 \text{ mol } H_2}{2 \text{ mol e}^-} = 0.373 \text{ mol } H_2$$

We now have the information we need to calculate the volume of the gas produced in this reaction:

$$V = \frac{nRT}{P} = \frac{(0.373 \text{ mol})(0.08206 \text{ L-atm/mol-K})(298 \text{ K})}{(1 \text{ atm})} = 9.12 \text{ L}$$

We can extend the general pattern outlined in this section to answer questions that might seem impossible at first glance.

## EXERCISE 20.15

Determine the oxidation number of the chromium in an unknown salt if electrolysis of a molten sample of this salt for 1.50 hours with a 10.0-amp current deposits 9.71 grams of chromium metal at the cathode.

**SOLUTION** We start, as before, by calculating the number of moles of electrons that passed through the cell during electrolysis:

$$10.0 \text{ amp} \times 1.50 \text{ h} \times \frac{60 \text{ min}}{1 \text{ h}} \times \frac{60 \text{ s}}{1 \text{ min}} \times \frac{1 \text{ C}}{1 \text{ amp-s}} = 54{,}000 \text{ C}$$

$$54{,}000 \text{ C} \times \frac{1 \text{ mol e}^-}{96{,}485 \text{ C}} = 0.560 \text{ mol e}^-$$

But we don't know the balanced equation for the reaction at the cathode in this cell. Thus, it doesn't seem obvious how we are going to use this information.

We therefore write down this result in a conspicuous location, and go back to the original statement of the question, to see what else can be done.

The problem tells us the mass of chromium deposited at the cathode. By now it is obvious that we need to transform this information into the number of moles of chromium metal generated:

$$9.71\ \cancel{\text{g Cr}} \times \frac{1\ \text{mol Cr}}{52.0\ \cancel{\text{g Cr}}} = \mathbf{0.187\ mol\ Cr}$$

**Problem-Solving Strategy**

We now know the number of moles of chromium metal produced and the number of moles of electrons it took to produce this metal. We might therefore look at the relationship between the moles of electrons consumed in this reaction and the moles of chromium produced:

$$\frac{0.560\ \text{mol e}^-}{0.187\ \text{mol Cr}} = \frac{3}{1}$$

Three moles of electrons are consumed for every mole of chromium metal produced. The only way to explain this is to assume that the net reaction at the cathode involves reduction of $Cr^{3+}$ ions to chromium metal:

$$\text{Cathode } (-): \quad Cr^{3+} + 3\ e^- \longrightarrow Cr$$

Thus, the oxidation number of chromium in the unknown salt must be +3.

## SPECIAL TOPIC

### BATTERIES

The term **battery** is used to describe a set of similar or connected items, such as an artillery battery or a battery of telephones. How, then, did it also come to mean a device that converts chemical energy into electrical energy? In Section 20.7, we noted that the first voltaic cell stable enough to be used in a battery was constructed by John Frederic Daniell. This cell was based on the reaction between zinc metal and $Cu^{2+}$ ions and it produced a standard-state cell potential of 1.10 V. When a number of these cells are connected in series we get a ''battery'' of cells with a voltage equal to 1.10 V times the number of cells.

Primary batteries.

#### PRIMARY BATTERIES

In the mid 1800s, batteries were the primary source of electrical energy. Batteries that are meant to be discarded when they lose their charge are still called **primary batteries.** Until recently, the market for disposable batteries was dominated by batteries based on the *Leclanché cell,* which was introduced in 1860. The cathode (positive electrode) in this cell is a mixture of solid $MnO_2$ and carbon. The anode (negative electrode) is a sheet of zinc metal that is amalgamated with a trace of mercury. The electrolytic solution is a mixture of $NH_4Cl$ and $ZnCl_2$. The cell reactions that occur during the discharge of a manganese dioxide–zinc primary battery can be written as follows:

$$\text{Cathode } (+): \quad 3\ MnO_2 + 2\ H_2O + 4\ e^- \longrightarrow Mn_3O_4 + 4\ OH^-$$

$$\text{Anode } (-): \quad Zn + 2\ OH^- \longrightarrow ZnO + H_2O + 2\ e^-$$

$$\text{Overall reaction:} \quad 2\ Zn(s) + 3\ MnO_2(s) \longrightarrow Mn_3O_4(s) + 2\ ZnO(s)$$

In the 1880s, either gelatin, flour, or starch was added to the electrolyte of the Leclanché cell to make the battery spillproof. These batteries therefore soon became known as *dry cells* (see Figure 20.10). The initial cell potential of a single dry cell is 1.54 V. The popularity of this cell is based on the fact that it is relatively inexpensive, available in many voltages and sizes, and suitable for intermittent use.

In 1949, the first ''alkaline'' dry cell battery was produced. These cells also use a mixture of $MnO_2$ and carbon as the cathode and amalgamated zinc as the anode. They use KOH, however, as the electrolyte. Alkaline dry cells (see Figure 20.11) have several advantages. They can be used over a wider range of temperatures because the electrolyte is more stable. They also require very little electrolyte and therefore can be very compact. More importantly, the voltage discharge curve for these batteries is almost flat. This means that they maintain a constant voltage for a longer period of time and, therefore, last longer.

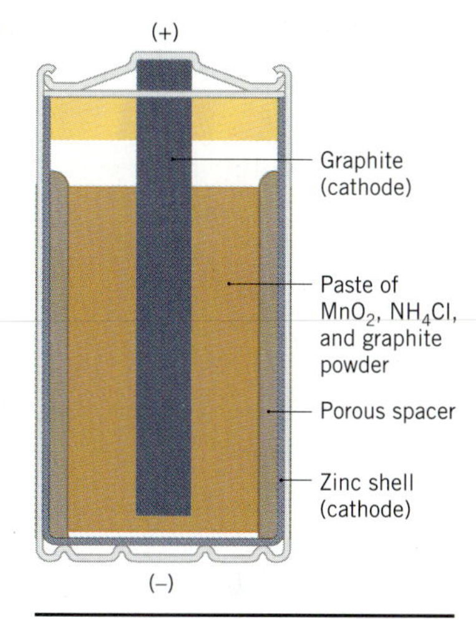


**Figure 20.10** A cross-section of a standard ''D'' cell battery.

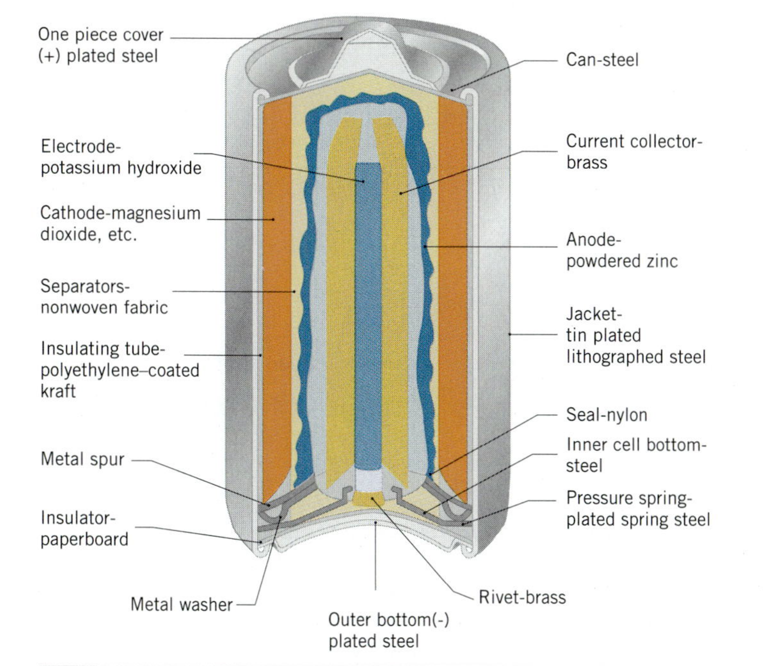


**Figure 20.11** A cutaway drawing of a typical alkaline battery.

## FUEL CELLS

Ever since the start of the space program, enthusiasm has run high for batteries that are **fuel cells.** By definition, a fuel cell is an electrochemical cell in which a combustion reaction is used to generate electrical energy. Fuel cells therefore require the continuous feed of a combustible material and oxygen and produce a low-voltage direct current as long as fuel is provided.

So far, it has not been practical to burn hydrocarbons, such as natural gas, in a fuel cell. The

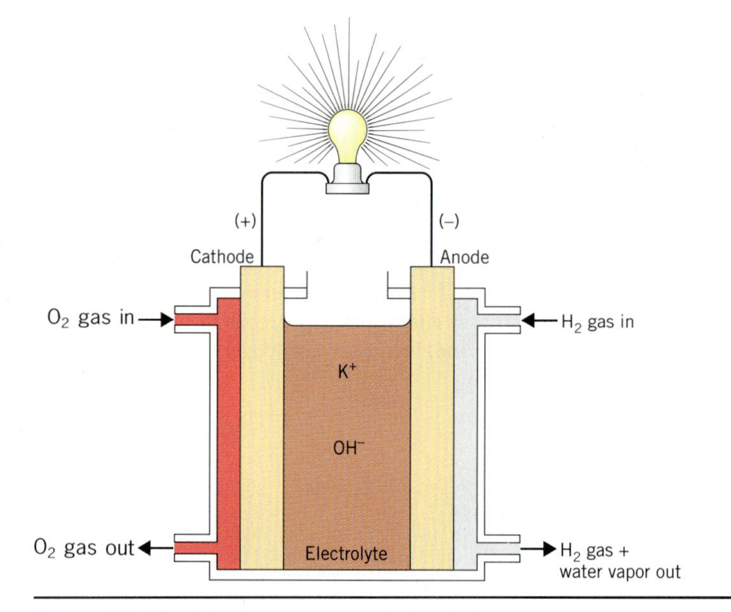


**Figure 20.12** A hydrogen-oxygen fuel cell.

fuel cells used in the space program burn hydrogen gas to give water, with a cell potential of 1.10 V, as shown in Figure 20.12.

$$2\ H_2(g) + O_2(g) \longrightarrow 2\ H_2O(g) \qquad E° = 1.10\ V$$

Fuel cells also can be constructed to burn either ammonia or hydrazine:

$$4\ NH_3(g) + 3\ O_2(g) \longrightarrow 2\ N_2(g) + 6\ H_2O(g) \qquad E° = 1.17\ V$$

## SECONDARY BATTERIES

Batteries that are used to store electrical energy are called **secondary batteries.** The first working model of a secondary battery was demonstrated to the French Academy of Sciences by Gaston Planté in 1860. It contained nine cells in parallel and was able to deliver extraordinarily large currents. These cells were constructed from lead plates separated by layers of flannel that were immersed in a 10% $H_2SO_4$ solution. As might be expected, this system soon became known as the *lead–acid battery.*

The first lead–acid batteries were charged by connecting them to a primary battery. The battery was then discharged, allowed to rest, and the charge–discharge cycle was continued. During the charging process, some of the lead in one plate of the cell was oxidized to $PbO_2$. Over a period that took as long as a year, Planté was able to build up enough $PbO_2$ on these plates to hold a significant charge. Production of the first lead–acid storage batteries therefore was a time-consuming process. As a result, these batteries remained a laboratory curiosity until 1873, when Siemens invented the dynamo, which produced a dependable source of electric current to charge these batteries.

In spite of many studies of the charging and discharging of lead electrodes in acid solution in the last 130 years, there is still some doubt about the exact mechanism of the reactions in

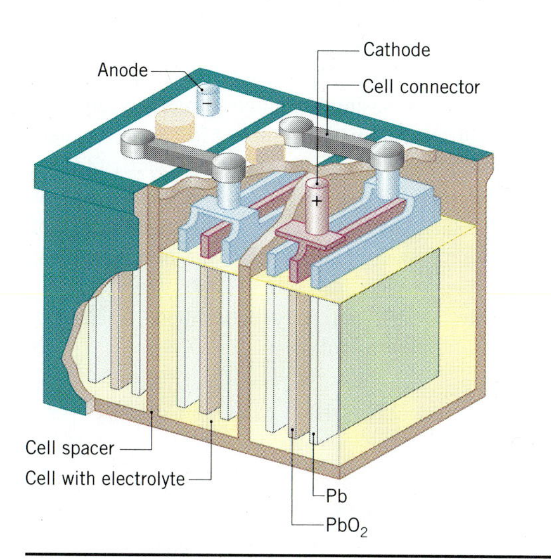


**Figure 20.13** A cutaway drawing of a typical lead–acid battery.

the lead–acid storage battery shown in Figure 20.13. The most popular model suggests that the following reaction occurs at the $PbO_2$ plate:

$$PbO_2 + 4\,H^+ + SO_4^{2-} + 2\,e^- \underset{\text{charge}}{\overset{\text{discharge}}{\rightleftharpoons}} PbSO_4 + 2\,H_2O$$

When the battery discharges, $PbO_2$ is reduced to $PbSO_4$. When it is charged, the reaction is reversed. The standard-state electrode potential for this half-reaction is 1.682 V.

The reaction at the lead metal plate can be described by the following reaction, for which the standard-state potential is 0.3858 V:

$$Pb + SO_4^{2-} \underset{\text{charge}}{\overset{\text{discharge}}{\rightleftharpoons}} PbSO_4 + 2\,e^-$$

The total cell reaction therefore can be written as follows:

$$PbO_2 + Pb + 4\,H^+ + 2\,SO_4^{2-} \underset{\text{charge}}{\overset{\text{discharge}}{\rightleftharpoons}} 2\,PbSO_4 + 2\,H_2O$$

The magnitude of the standard-state potential for a single cell in these batteries is slightly larger than 2 V, although the sign of this potential depends on whether it is being charged or discharged. The typical 12-V lead storage battery therefore contains six 2-V cells.

During the charging of a lead–acid storage battery, water can be decomposed into its elements:

$$2\,H_2O(l) \longrightarrow 2\,H_2(g) + O_2(g)$$

Lead storage batteries therefore represent a potential threat of hydrogen explosions. Because these explosions often spray the 10% sulfuric acid electrolyte onto the individual working on the battery, safety goggles always should be worn when working with these batteries.

## NICAD BATTERIES

A different type of secondary battery has become increasingly popular in recent years. The history of this battery traces back to the turn of the twentieth century. After approximately 10,000 experiments, Thomas Edison found that he could make a rechargeable battery from a nickel–iron cell. The active materials in this cell were $Ni(OH)_2$ and iron metal that was partially oxidized to $Fe^{2+}$. In the mid-1920s, a German manufacturer found that the risk of hydrogen explosions in these batteries was significantly reduced when the iron metal was replaced with cadmium.

The chemistry of nickel–cadmium batteries is still not fully understood. A charge–discharge mechanism for this system might be written as follows:

$$NiO(OH) \cdot x\,H_2O + Cd + 2\,H_2O \underset{\text{charge}}{\overset{\text{discharge}}{\rightleftharpoons}} 2\,Ni(OH)_2 \cdot y\,H_2O + Cd(OH)_2$$

When these batteries discharge, the nickel is reduced from +3 to +2 and the cadmium is oxidized from 0 to +2, with a net potential of 1.29 V. Both the oxidized and reduced forms of nickel contain water trapped in the crystals, but the amount of water of hydration in these two forms is different.

SPECIAL TOPIC

## GALVANIC CORROSION AND CATHODIC PROTECTION

The model for electrochemical reactions developed in this chapter can be used to explain the behavior observed when the corrosion of iron metal is studied:

- **Iron rusts very slowly in contact with dry air.** Iron metal rusts in contact with dry air because iron atoms on the surface of the metal slowly react with oxygen in the atmosphere to form a mixture of oxides.

$$2\,Fe(s) + O_2(g) \longrightarrow 2\,FeO(s)$$
$$4\,Fe(s) + 3\,O_2(g) \longrightarrow 2\,Fe_2O_3(s)$$
$$3\,Fe(s) + 2\,O_2(g) \longrightarrow Fe_3O_4(s)$$

  This reaction is uneven, and there are a number of holes in the oxide layer that forms on the surface of the metal. Oxygen atoms migrate through these holes to the metal surface below the oxide layer. Corrosion therefore continues slowly, but surely. At sites where the metal was deformed when it was worked or shaped, the iron oxide often breaks off, exposing a fresh metal surface to further corrosion and pitting.

- **Iron corrodes on contact with strong acids.** Iron metal dissolves in strong acid, as would be expected from its position in Table 20.1. If the acid is boiled to remove any dissolved $O_2$ gas, it reacts with iron to form solutions of the $Fe^{2+}$ ion. This can be understood by comparing the overall potential for reaction between iron metal and acid to form $Fe^{2+}$ ions with the potential for the reaction to form $Fe^{3+}$ ions:

$$Fe(s) + 2\,H^+(aq) \longrightarrow Fe^{2+}(aq) + H_2(g) \qquad E° = 0.409\text{ V}$$
$$2\,Fe(s) + 6\,H^+(aq) \longrightarrow 2\,Fe^{3+}(aq) + 3\,H_2(g) \qquad E° = 0.036\text{ V}$$

If $O_2$ is present, the $Fe^{2+}$ formed when the iron dissolves is oxidized to $Fe^{3+}$:

$$4\,Fe^{2+}(aq) + O_2(g) + 4\,H^+(aq) \longrightarrow 4\,Fe^{3+}(aq) + 2\,H_2O(l) \qquad E° = 0.46\text{ V}$$

- **Iron does not rust in contact with water that does not contain dissolved $O_2$ gas.** The standard-state potential for the reaction between iron metal and the $H^+$ ion is favorable:

$$Fe + 2\,H^+ \longrightarrow Fe^{2+} + H_2 \qquad E° = 0.409\text{ V}$$

But the concentration of the $H^+$ ion in water is far from the standard state of 1 *M*. Substituting the $1 \times 10^{-7}$ *M* $H^+$ ion concentration in water into the Nernst equation for this reaction lowers the cell potential for the reaction by 0.414 V. This is a large enough change to make the reaction slightly unfavorable.

- **Iron rusts more rapidly in contact with water that contains dissolved $O_2$ than it does in dry air.** When water and $O_2$ are both present, the oxidation and reduction reactions no longer have to occur at the same point on the metal surface. The iron metal now simultaneously acts as the anode, cathode, and the circuit through which electrons flow, as shown in Figure 20.14. At one point on the metal surface, iron metal is oxidized to $Fe^{2+}$ ions:

$$\text{Anode:} \qquad Fe \longrightarrow Fe^{2+} + 2\,e^- \qquad E°_{ox} = 0.409\text{ V}$$

The electrons released in this reaction flow through the metal to another point, at which $O_2$ dissolved in the water is reduced:

$$\text{Cathode:} \qquad O_2 + 2\,H_2O + 4\,e^- \longrightarrow 4\,OH^- \qquad E°_{red} = 0.401\text{ V}$$

The net result is the formation of an aqueous solution of iron(II) hydroxide:

$$2\,Fe(s) + O_2(g) + 2\,H_2O(l) \longrightarrow 2\,Fe^{2+}(aq) + 4\,OH^-(aq) \qquad E° = 0.810\text{ V}$$

The $Fe^{2+}$ and $OH^-$ ions produced in these half-reactions diffuse toward each other and eventually combine to form a hydrated iron(II) oxide that undergoes further oxidation to form rust, $Fe_2O_3 \cdot 3\,H_2O$.

- **Iron rusts even more rapidly when it comes in contact with copper metal.** A voltaic cell is created at the point at which the two metals touch, as shown in Figure 20.15. The iron metal acts as the anode, giving up electrons to form $Fe^{2+}$ ions:

$$\text{Anode:} \qquad Fe \longrightarrow Fe^{2+} + 2\,e^- \qquad E°_{ox} = 0.409\text{ V}$$

These electrons flow to the copper metal, which becomes the cathode at which $O_2$ dissolved in water is reduced:

$$\text{Cathode:} \qquad O_2 + 2\,H_2O + 4\,e^- \longrightarrow 4\,OH^- \qquad E°_{red} = 0.401\text{ V}$$

Once again, the $Fe^{2+}$ ions produced at the anode diffuse toward the $OH^-$ ions given off at the cathode and are eventually oxidized to form rust. This process is called *galvanic corrosion* because the overall reaction is similar to what would happen if the reaction were run in a galvanic cell.

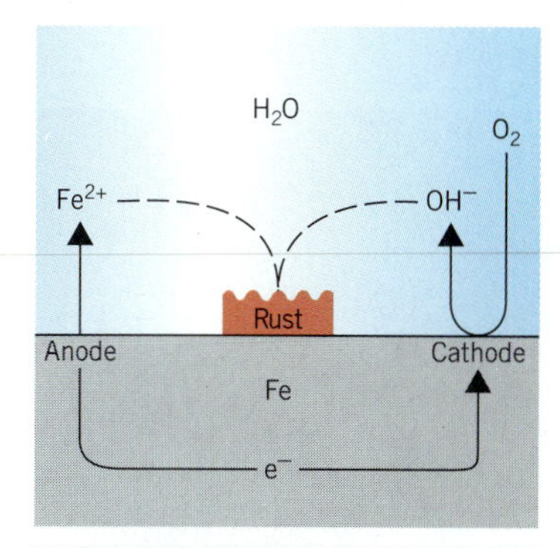


**Figure 20.14** When iron metal is immersed in water that contains dissolved oxygen, oxidation of the iron and reduction of $O_2$ can occur at different points on the metal surface. The net effect of this reaction is the formation of $Fe^{2+}$ ions and $OH^-$ ions, which diffuse toward each other to form a complex that is further oxidized to form rust, $Fe_2O_3 \cdot 3\,H_2O$.

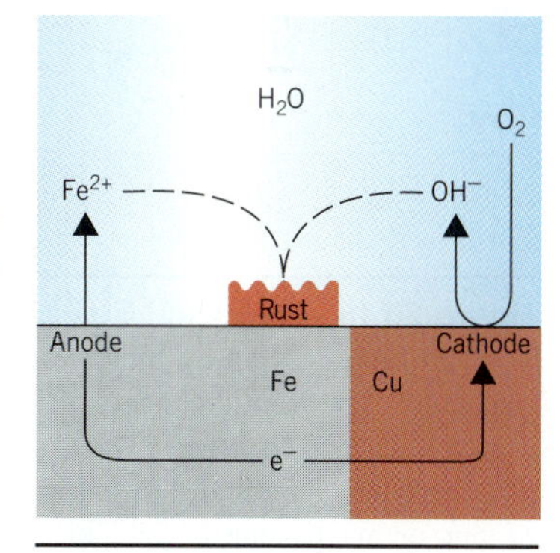


**Figure 20.15** When iron comes in contact with copper metal, a galvanic cell is created in which the iron is oxidized to $Fe^{2+}$ ions. The electrons given off are transferred to the copper metal, where they are used to reduce $O_2$ to $OH^-$ ions.

Connecting an iron pipe to a bar of magnesium protects the iron from corrosion.

Copper metal doesn't corrode when it touches iron because copper metal is a weaker reducing agent than iron. We can see this by imagining the voltaic cell that would be created if copper acted as the anode and the electrons flowed to iron, which would act as the cathode at which $O_2$ was reduced:

$$\text{Anode:} \quad Cu \longrightarrow Cu^{2+} + 2\,e^- \qquad E^\circ_{ox} = -0.340\text{ V}$$

$$\text{Cathode:} \quad O_2 + 2\,H_2O + 4\,e^- \longrightarrow 4\,OH^- \qquad E^\circ_{red} = 0.401\text{ V}$$

The overall potential for this reaction is much smaller than the potential for the reaction that occurs when iron acts as the anode:

$$Cu(s) + O_2(g) + 2\,H_2O(l) \longrightarrow Cu^{2+}(aq) + 4\,OH^-(aq) \qquad E^\circ = 0.061\text{ V}$$

$$Fe(s) + O_2(g) + 2\,H_2O(l) \longrightarrow Fe^{2+}(aq) + 4\,OH^-(aq) \qquad E^\circ = 0.810\text{ V}$$

The iron therefore corrodes before the copper.

An interesting question is raised when Figures 20.14 and 20.15 are compared. The overall reaction for the corrosion of iron is the same, regardless of whether or not copper metal is in contact with the iron metal. Thus, the overall cell potential for these reactions must be the same. Why, then, does iron corrode so much faster when it comes in contact with copper metal?

The answer seems to rest with the concept of overvoltage introduced in Section 20.13. The actual potential for the reduction of $O_2$ on a metal surface is usually larger than the value reported in tables of standard-state cell potentials. The overvoltage for the reduction of $O_2$ on copper metal, however, is significantly smaller than the overvoltage on iron. Since the overvoltage is defined as the extra voltage that must be applied to a reaction to get it to occur at the same rate, oxidation of iron metal in contact with copper should occur much more rapidly than the corresponding reaction on an isolated iron metal surface.

- **Iron does not rust when it is in contact with zinc or magnesium metal.** Iron can be protected from corrosion if it is allowed to stay in contact with a metal that is a better reducing agent, as shown in Figure 20.16. This process is called *cathodic protection* because it involves making iron the cathode of a voltaic cell. The zinc or magnesium metal acts as the anode of a voltaic cell:

$$\text{Anode:} \quad Zn \longrightarrow Zn^{2+} + 2\,e^- \qquad E^\circ_{ox} = 0.763\text{ V}$$

The electrons given off in this reaction flow through the wire to the iron metal, which acts as the cathode at which $O_2$ is reduced:

$$\text{Cathode:} \quad O_2 + 2\,H_2O + 4\,e^- \longrightarrow 4\,OH^- \qquad E^\circ_{red} = 0.401\text{ V}$$

The overall potential for this reaction is much larger than the potential for the reaction that would occur if iron was the anode:

$$2\,Zn(s) + O_2(g) + 2\,H_2O(l) \longrightarrow 2\,Zn^{2+}(aq) + 4\,OH^-(aq) \qquad E^\circ = 1.164\text{ V}$$

Thus, oxidation of the iron won't occur until the anode is either completely consumed or the electrical contact between the iron and zinc metal is broken. Cathodic protection is the theory behind ''galvanized'' iron, which is iron metal coated with a thin layer of zinc. Until virtually all of the zinc has corroded, the iron metal that provides the structure for this material remains intact.

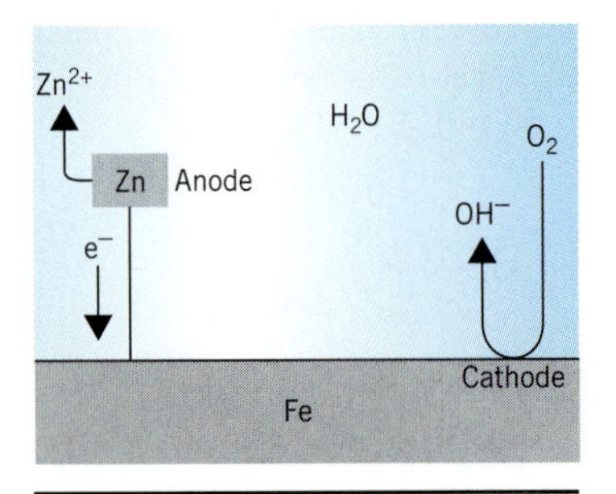


**Figure 20.16** The corrosion of iron can be prevented by cathodic protection. The iron is either coated with zinc to form galvanized iron or connected to a piece of zinc or magnesium metal. Zinc and magnesium are both stronger reducing agents than iron, so they are oxidized instead of the iron. The iron therefore becomes the cathode at which $O_2$ is reduced, which protects the iron from corrosion.

## SUMMARY

**1.** By physically separating the oxidation and reduction halves of a reaction, it is possible to force the electrons that flow from one of these half-reactions to the other to pass through an external wire. The result is an electrochemical cell that can do electrical work. Cells of this sort are commonly known as either **galvanic** or **voltaic cells.**

**2.** The cell potential for a spontaneous oxidation–reduction reaction is a measure of the driving force behind the reaction. A large cell potential is observed when a strong reducing agent reacts with a strong oxidizing agent because the driving force behind this reaction is relatively large. By convention, the sign of the cell potential is positive for a spontaneous oxidation–reduction reaction.

**3.** The magnitude of the potential for an oxidation–reduction reaction depends on the conditions under which the measurement is made. Under **standard-state conditions,** all solutions have a concentration of 1 $M$ and all gases have a partial pressure of 1 MPa. The standard-state cell potential for an oxidation–reduction reaction is represented by the symbol $E°$.

**4.** The standard-state cell potential for the reduction of $H^+$ ions to $H_2$ gas is arbitrarily defined as exactly zero volts. This convention allows us to use measurements of the cell potentials of oxidation–reduction reactions to generate a table of **standard-state half-cell potentials.**

**5.** The standard-state cell potential for an oxidation–reduction reaction is the sum of the standard-state half-cell potentials for the oxidation and reduction halves of the reaction.

**6.** When an oxidation–reduction reaction is at equilibrium, there is no longer any driving force pushing the reaction forward. As a result, $E$ for the cell at equilibrium must be zero. Rewriting the Nernst equation to reflect this condition allows us to calculate the equilibrium constant for an oxidation–reduction reaction from the value of $E°$ for the reaction. The larger the value of $E°$, the larger the equilibrium constant because the standard-state conditions for reactions that have large values of $E°$ are very far from equilibrium.

**7.** **Voltaic cells** use a spontaneous oxidation–reduction reaction to drive an electric current through a wire. **Electrolytic cells** use an electric current to drive an oxidation–reduction reaction in the opposite direction.

## KEY TERMS

**Anion** (p. 729)
**Anode** (p. 729)
**Battery** (p. 750)
**Cathode** (p. 729)
**Cation** (p. 729)
**Cell potential** (p. 729)
**Electrochemical reactions** (p. 726)
**Electrolysis** (p. 744)
**Electrolytic cell** (p. 743)
**Faraday's constant** (p. 748)
**Faraday's law** (p. 748)
**Fuel cell** (p. 751)
**Half-cell** (p. 727)
**Line notation** (p. 735)
**Negative electrode** (p. 743)
**Nernst equation** (p. 737)
**Overvoltage** (p. 746)
**Positive electrode** (p. 743)
**Primary battery** (p. 750)
**Salt bridge** (p. 728)
**Secondary battery** (p. 752)
**Standard-state cell potentials** (p. 729)
**Standard-state conditions** (p. 729)
**Voltaic cell** (p. 729)

## KEY EQUATIONS

$$E°_{overall} = E°_{ox} + E°_{red}$$

$$E = E° - \frac{RT}{nF} \ln Q_c \quad \text{(Nernst equation)}$$

$$K_c = c^{nFE°/RT}$$

$$\ln K_c = \frac{nFE°}{RT}$$

## PROBLEMS

### Electrical Work from Spontaneous Oxidation–Reduction Reactions

**20-1** Define the terms *coulomb, volt, ampere,* and *faraday.*

**20-2** Describe an experiment that would allow you to determine the relative strengths of zinc, copper, silver, and iron metal as reducing agents.

**20-3** Explain why oxidation and reduction half-reactions have to be physically separated for an oxidation–reduction reaction to do work.

**20-4** Describe the function of a salt bridge in an electrochemical cell. Explain what happens when the salt bridge is removed from the system and why.

**20-5** Humphrey Davy described a "living salt bridge," which consists of the first and second fingers of the hand inserted into the two half-cells of a voltaic cell. Explain how this accomplishes the same function as a U-tube filled with a saturated KCl solution.

### Voltaic Cells from Spontaneous Oxidation–Reduction Reactions

**20-6** Describe the relationships between pairs of the following terms: *cathode, anode, cation,* and *anion.*

**20-7** Explain why cations flow towards the cathode and anions towards the anode of an electrochemical cell, regardless of whether it is a voltaic or electrolytic cell.

**20-8** Explain why oxidation occurs at the anode and reduction occurs at the cathode of both voltaic and electrolytic cells.

### Standard-State Cell Potentials for Voltaic Cells

**20-9** Describe the difference between $E$ and $E°$ for a cell.

**20-10** Describe the conditions under which the cell potential must be determined in order for the measurement to be equal to $E°$.

**20-11** Describe what the magnitude of $E°$ for an oxidation–reduction reaction tells us about the reaction.

**20-12** Explain why cell potentials measure the relative strengths of a pair of oxidizing agents and the relative strengths of a pair of reducing agents but not their absolute strengths.

**20-13** Describe the arbitrary convention that is used to turn measurements of the standard-state cell potential for a reaction into a table of half-cell reduction potentials.

### Predicting Spontaneous Oxidation–Reduction Reactions from the Sign of $E°$

**20-14** Describe what the sign of $E°$ for an oxidation–reduction reaction tells us about the reaction.

**20-15** Describe what happens to the sign and magnitude of $E°$ for an oxidation–reduction reaction when the direction in which the reaction is written is reversed.

**20-16** Which of the following oxidation–reduction reactions should occur as written when run under standard-state conditions?

(a) $Al(s) + Cr^{3+}(aq) \longrightarrow Al^{3+}(aq) + Cr(s)$ $E° = 0.966$ V

(b) $3\,Cr^{2+}(aq) \longrightarrow Cr(s) + 2\,Cr^{3+}(aq)$ $E° = 0.33$ V

(c) $Fe(s) + Cr^{3+}(aq) \longrightarrow Cr(s) + Fe^{3+}(aq)$ $E° = -0.70$ V

(d) $3\,H_2(g) + 2\,Cr^{3+}(aq) \longrightarrow 2\,Cr(s) + 6\,H^+(aq)$ $E° = -0.74$

**20-17** Which of the following oxidation–reduction reactions should occur as written when run under standard-state conditions?

(a) $2\,Fe^{2+}(aq) + H_2O_2(aq) \longrightarrow 2\,Fe^{3+}(aq) + 2\,OH^-(aq)$ $E° = 0.11$ V

(b) $2\,Fe^{2+}(aq) + Cl_2(aq) \longrightarrow 2\,Fe^{3+}(aq) + 2\,Cl^-(aq)$ $E° = 0.588$ V

(c) $2\,Fe^{2+}(aq) + Br_2(aq) \longrightarrow 2\,Fe^{3+}(aq) + 2\,Br^-(aq)$ $E° = 0.317$ V

(d) $2\,Fe^{2+}(aq) + I_2(aq) \longrightarrow 2\,Fe^{3+}(aq) + 2\,I^-(aq)$ $E° = -0.235$ V

**20-18** Use the results of the previous problem to determine the relative strengths of $H_2O_2$, $Cl_2$, $Br_2$, $I_2$, and the $Fe^{3+}$ ion as oxidizing agents.

### Standard-State Reduction Half-Cell Potentials

**20-19** What do the following half-cell reduction potentials tell us about the relative strengths of zinc and copper metal as reducing agents? What do they tell us about the relative strengths of $Zn^{2+}$ and $Cu^{2+}$ ions as oxidizing agents?

$$Zn^{2+} + 2\,e^- \rightleftharpoons Zn \qquad E° = -0.7628 \text{ V}$$
$$Cu^{2+} + 2\,e^- \rightleftharpoons Cu \qquad E° = 0.3402 \text{ V}$$

### Predicting Standard-State Cell Potentials

**20-20** Which of the following reactions should occur spontaneously as written when run under standard-state conditions?

(a) $Zn(s) + 2\,H^+(aq) \longrightarrow Zn^{2+}(aq) + H_2(g)$
(b) $Cr(s) + 3\,Fe^{3+}(aq) \longrightarrow Cr^{3+}(aq) + 3\,Fe^{2+}(aq)$
(c) $Mn(s) + Mg^{2+}(aq) \longrightarrow Mn^{2+}(aq) + Mg(s)$

**20-21** Which of the following reactions should occur spontaneously as written when run under standard-state conditions?

(a) $NO_2^-(aq) + ClO^-(aq) \longrightarrow NO_3^-(aq) + Cl^-(aq)$
(b) $2\,ClO_2^-(aq) \longrightarrow ClO^-(aq) + ClO_3^-(aq)$
(c) $3\,Cu(s) + 2\,HNO_3(aq) + 6\,H^+(aq) \longrightarrow 3\,Cu^{2+}(aq) + 2\,NO(g) + 4\,H_2O(l)$

**20-22** Describe what happens inside a Daniell cell when it does work:

$$Zn(s) + Cu^{2+}(aq) \longrightarrow Zn^{2+}(aq) + Cu(s)$$

**20-23** Calculate $E°$ for the following reaction and predict whether the reaction should occur spontaneously as written when run under standard-state conditions:

$$6\,Fe^{2+}(aq) + Cr_2O_7^{2-}(aq) + 14\,H^+(aq) \longrightarrow 6\,Fe^{3+}(aq) + 2\,Cr^{3+}(aq) + 7\,H_2O(l)$$

**20-24** Calculate $E°$ for the following reaction and predict whether the reaction should occur spontaneously as written when run under standard-state conditions. Use the results of this calculation to explain what happens during the thermite reaction:

$$Al^{3+}(aq) + Fe(s) \longrightarrow Fe^{3+}(aq) + Al(s)$$

**20-25** Some metals react with water at room temperature,

others do not. Calculate $E°$ for the following reaction and predict whether the reaction should occur spontaneously as written when run under standard-state conditions.

$$Ca(s) + 2\,H_2O(l) \longrightarrow Ca^{2+}(aq) + 2\,OH^-(aq) + H_2(g)$$

**20-26** Calculate $E°$ for the following reaction to predict whether the $Cu^+$ ion should spontaneously undergo disproportionation under standard-state conditions:

$$2\,Cu^+(aq) \longrightarrow Cu(s) + Cu^{2+}(aq)$$

**20-27** Use standard-state half-cell reduction potentials to predict whether a 1.00 $M$ $Fe^{2+}(aq)$ solution should react with a 1.00 $M$ $H^+(aq)$ solution to produce $Fe^{3+}(aq)$ and $H_2(g)$.

### Line Notation for Voltaic Cells

**20-28** Write the line notation for the following voltaic cell under standard-state conditions:

$$2\,Ag^+(aq) + Cu(s) \longrightarrow Cu^{2+}(aq) + 2\,Ag(s)$$

**20-29** Write the line notation for the voltaic cell in which AgCl is reduced at a silver metal electrode and copper metal is oxidized at the other electrode to form $Cu^{2+}$ ions under standard-state conditions:

$$2\,AgCl(s) + Cu(s) \longrightarrow 2\,Ag(s) + Cu^{2+}(aq) + 2\,Cl^-(aq)$$

**20-30** Determine $E°$ for the following cells and decide in which direction these reactions are spontaneous:

$$Zn|Zn^{2+}(1.0\,M)||Fe^{2+}(1.0\,M)|Fe$$

$$Pt|Mn^{2+}(1.0\,M)|MnO_4^-(1.0\,M)||Fe^{2+}(1.0\,M)|Fe$$

**20-31** Draw a diagram of the following cell, label the anode and cathode, and calculate $E°$ for the cell:

$$Al|Al^{3+}(1.0\,M)||H^+(1.0\,M)|H_2(1\text{ atm})|Pt$$

**20-32** Draw a diagram of the following cell, label the anode and cathode, and calculate $E°$ for the cell:

$$Zn|Zn^{2+}(1.0\,M)||MnO_4^-(1.0\,M)|Mn^{2+}(1.0\,M)|Pt$$

### Using Standard-State Half-Cell Potentials to Understand Chemical Reactions

**20-33** Use half-cell potentials to determine whether the following pairs of ions can coexist in aqueous solution.

(a) $Na^+$, $Sn^{2+}$ (b) $Zn^{2+}$, $I^-$ (c) $Hg_2^{2+}$, $F^-$
(d) $Fe^{2+}$, $Hg^{2+}$ (e) $Ag^+$, $Hg^{2+}$

**20-34** Use half-cell potentials to determine whether the following pairs of ions can coexist in aqueous solution.

(a) $Sn^{2+}$, $Fe^{2+}$ (b) $Au^+$, $Br^-$ (c) $Fe^{3+}$, $CrO_4^{2-}$
(d) $Fe^{3+}$, $SO_4^{2-}$ (e) $MnO_4^-$, $I^-$

**20-35** Explain why copper, gold, mercury, platinum, and silver can be found in the metallic state in nature.

**20-36** Which of the following is the strongest reducing agent?

(a) Zn (b) Fe (c) $H_2$ (d) Cu (e) Ag

**20-37** Which of the following is the strongest oxidizing agent?

(a) $H_2O_2$ in $OH^-$ (b) $H_2O_2$ in $H^+$ (c) Na
(d) $O_2$ in $H^+$ (e) Al

**20-38** Which of the following is the strongest reducing agent?

(a) $H^+$ (b) $H_2$ (c) $H^-$ (d) $H_2O$ (e) $O_2$

**20-39** Which is the better oxidizing agent?

(a) $H^-$ or K (b) Sn or $Fe^{2+}$ (c) $Ag^+$ or $Au^+$

### The Nernst Equation

**20-40** Explain why there is only one value of $E°$ for a cell at a given temperature but many different values of $E$.

**20-41** Explain why the cell potential becomes smaller as the cell comes closer to equilibrium.

**20-42** What does it mean when $E$ for a cell is equal to zero? What does it mean when $E°$ is equal to zero?

**20-43** Which of the following describes an oxidation–reduction reaction at equilibrium?

(a) $E = 0$ (b) $E° = 0$ (c) $E = E°$ (d) $Q_c = 0$
(e) $\ln K_c = 0$

**20-44** Write the Nernst equation for the following half-cell and calculate the half-cell potential, assuming that the $H^+$ ion concentration is $10^{-7}$ $M$ and that the partial pressure of the $H_2$ gas is 1 atm:

$$2\,H^+(aq) + 2\,e^- \rightleftharpoons H_2(g)$$

**20-45** Describe how increasing the pH of the following half-reaction affects the half-cell reduction potential:

$$MnO_4^-(aq) + 8\,H^+(aq) + 5\,e^- \longrightarrow Mn^{2+}(aq)$$

Does the permanganate ion become a stronger oxidizing agent or a weaker oxidizing agent as the solution becomes more basic?

**20-46** Write the Nernst equation for the following reaction and calculate the cell potential if the $Al^{3+}$ ion concentration is 1.2 $M$ and the $Fe^{3+}$ ion concentration is 2.5 $M$:

$$Al^{3+}(aq) + Fe(s) \longrightarrow Al(s) + Fe^{3+}(aq)$$

**20-47** Calculate $E°$ for the following reaction:

$$2\,Fe^{2+}(aq) + H_2O_2(aq) \longrightarrow 2\,Fe^{3+}(aq) + 2\,OH^-(aq)$$

Write the Nernst equation for this reaction and calculate the cell potential for a system in a pH 10 buffer for which all other components are present at standard-state conditions.

**20-48** Assume that we start with a Daniell cell at standard-state conditions:

$$Zn|Zn^{2+}(1\,M)||Cu^{2+}(1\,M)|Cu$$

Calculate the cell potential under the following sets of conditions:

(a) Ninety-nine percent of the zinc metal and $Cu^{2+}$ ions have been consumed. (b) The reaction has reached 99.99% completion. (c) The reaction has reached 99.9999% completion. (d) The $Cu^{2+}$ ion concentration is only $1 \times 10^{-8}$ $M$. (e) The reaction has reached equilibrium.

**20-49** Calculate the standard-state cell potential for the voltaic cell built around the following oxidation–reduction reaction:

$$2\,MnO_4^-(aq) + 5\,H_2O_2(aq) + 6\,H^+(aq) \longrightarrow 2\,Mn^{2+}(aq) + 8\,H_2O(l) + 5\,O_2(g)$$

Use the Nernst equation to predict the effect on the cell potential of an increase in the pH of the solution. Predict the effect of an increase in the $H_2O_2$ concentration.

**20-50** What would be the effect of an increase in temperature on the potential of the Daniell cell? Would it increase, decrease, or remain the same?

**20-51** Predict the cell potential for the following reaction at 25°C and 250°C:

$$Zn|Zn^{2+}(1.0\ M)||Fe^{2+}(1.0\ M)|Fe$$

Use the results of these calculations to comment on the sensitivity of cell potentials to changes in temperature.

**20-52** The standard-state cell potential, $E°$, is zero for any cell in which the reaction at the cathode is the opposite of the reaction at the anode:

$$Cu|Cu^{2+}(1.0\ M)||Cu^{2+}(1.0\ M)|Cu \qquad E° = 0$$

A small voltage can be obtained, however, from a cell in which the $Cu^{2+}$ ion concentration is different in the two half-cells. Determine the anode and the cathode of the following cell and calculate the overall cell potential for this concentration cell:

$$Cu|Cu^{2+}(2.50\ M)||Cu^{2+}(0.18\ M)|Cu$$

### Using the Nernst Equation to Measure Equilibrium Constants

**20-53** Calculate the equilibrium constant at 25°C for the Daniell cell. Use the results of this calculation to explain why a single arrow rather than double arrows is used in the following equation:

$$Zn(s) + Cu^{2+}(aq) \longrightarrow Zn^{2+}(aq) + Cu(s)$$

**20-54** Use the standard-state reduction potentials for the following half-cells to calculate the solubility product for AgI:

$$Ag^+ + e^- \rightleftharpoons Ag \qquad E° = 0.7996\ V$$

$$AgI + e^- \rightleftharpoons Ag^+ + I^- \qquad E° = -0.164\ V$$

**20-55** Use the following data to calculate the complex formation equilibrium constant for the $Ag(NH_3)_2^+$ complex ion:

$$Ag^+ + e^- \rightleftharpoons Ag \qquad E° = 0.7996\ V$$

$$Ag(NH_3)_2^+ + e^- \rightleftharpoons Ag + 2\ NH_3 \qquad E° = 0.373\ V$$

**20-56** Use the data in Table A.13 to calculate the solubility product of AgCl.

**20-57** Use the data in Table A.13 to calculate the solubility product of $Hg_2Cl_2$.

**20-58** Use the data in Table A.13 to calculate the complex formation constant for the $Zn(CN)_4^{2-}$ and $Zn(NH_3)_4^{2+}$ complex ions.

**20-59** Use the data in Table A.13 to calculate the complex formation constants for the $Ni(NH_3)_6^{2+}$ and $Co(NH_3)_6^{2+}$ complex ions.

**20-60** Calculate $E°$ for the following reaction:

$$Cu(s) + 2\ Ag^+(aq) \longrightarrow Cu^{2+}(aq) + 2\ Ag(s)$$

What would happen to the cell potential if enough NaCl were added to increase the $Cl^-$ ion concentration to 1.0 $M$? (AgCl: $K_{sp} = 1.8 \times 10^{-10}$)

### Electrolytic Cells

**20-61** Explain why electrolysis of aqueous NaCl produces $H_2$ gas, not sodium metal at the cathode, and $Cl_2$ gas, not $O_2$ gas, at the anode.

**20-62** Calculate the standard-state cell potential necessary to electrolyze molten NaCl to form sodium metal and chlorine gas.

**20-63** Calculate the standard-state cell potential necessary to electrolyze an aqueous solution of NaCl.

**20-64** Which of the following statements best describes what happens when $MgCl_2$ is electrolyzed?

(a) Mg metal forms at the anode. (b) $Mg^{2+}$ ions are oxidized at the cathode. (c) $Cl_2$ gas is formed at the anode by the oxidation of $Cl^-$. (d) $Cl^-$ ions flow toward the cathode.

**20-65** Explain why an electrolyte, such as $Na_2SO_4$, has to be added to water before the water can be electrolyzed to $H_2$ and $O_2$.

**20-66** Describe what would happen if $NiSO_4$ were used instead of $Na_2SO_4$ when water is electrolyzed.

**20-67** Why does the solution around the cathode become basic when an aqueous solution of $Na_2SO_4$ is electrolyzed?

**20-68** What are the most likely products of the electrolysis of an aqueous solution of KBr?

(a) $K^+(aq)$ and $Br^-(aq)$ (b) $K(s) + Br_2(l)$
(c) $K^+(aq) + OH^-(aq) + H_2(g) + Br_2(l)$
(d) $K^+(aq) + OH^-(aq) + O_2(g) + Br_2(l)$

**20-69** Describe what happens during the electrolysis of the following solutions:

(a) $FeCl_3(aq)$ (b) $NaI(aq)$ (c) $K_2SO_4(aq)$
(d) $H_2SO_4(aq)$

**20-70** Why can't aluminum metal be prepared by the electrolysis of an aqueous solution of the $Al^{3+}$ ion?

**20-71** Which of the following reactions occurs at the anode during electrolysis of an aqueous $Na_2SO_4$ solution?

(a) $SO_4^{2-}(aq) \longrightarrow SO_2(g) + O_2(g) + 2\ e^-$
(b) $2\ H_2O(l) \longrightarrow O_2(g) + 4\ H^+(aq) + 4\ e^-$
(c) $2\ H_2O(l) + O_2(g) + 4\ e^- \longrightarrow 4\ OH^-(aq)$
(d) $SO_4^{2-}(aq) + 4\ H^+(aq) + 2\ e^- \longrightarrow SO_2(g) + H_2O(l)$

### Faraday's Law

**20-72** Describe an experiment that uses Faraday's law to determine the strength of an electric current.

**20-73** Calculate the mass of silver metal that can be prepared with 1 C of electric charge.

**20-74** The rate at which electric power is delivered is measured in units of kilowatts (kw). One watt is the power delivered by a 1-amp current at a potential of 1 V. Calculate the number of kilowatt-hours of electricity it takes to prepare a mole of sodium metal by electrolysis of molten NaCl.

**20-75** If a 5-amp current was used, how long would it take to prepare a mole of sodium metal? Of magnesium metal? Of aluminum metal?

**20-76** Predict the products of the electrolysis of an aqueous solution of LiBr, and calculate the number of grams of each product formed by electrolysis of this solution for 1.0 hour with a 2.5-amp current.

**20-77** Calculate the amount of electric current necessary to produce a metric tonne (1000 kg) of $Cl_2$ gas.

**20-78** Calculate the molarity of 1.00 L of 2.0 *M* $CuSO_4$ after the solution has been electrolyzed for 2.5 hours with a 4.5-amp current.

**20-79** Calculate the ratio of the mass of $O_2$ to the mass of $H_2$ produced when an aqueous $Na_2SO_4$ solution is electrolyzed for 2.56 hours with a 1.34-amp current.

**20-80** Under a particular set of conditions, electrolysis of an aqueous $AgNO_3$ solution generated 1.00 g of silver metal. How many grams of $I_2$ would be produced by the electrolysis of an aqueous solution of NaI under the same condition?

**20-81** Calculate the ratio of the mass of $Br_2$ that collects at the anode to the mass of aluminum metal plated out at the cathode when molten $AlBr_3$ is electrolyzed for 10.5 hours with a 20.0-amp current.

**20-82** Assume that aqueous solutions of the following salts are electrolyzed for 20.0 min with a 10.0-amp current. Which solution will deposit the most grams of metal at the cathode?

(a) $ZnCl_2$ (b) $ZnBr_2$ (c) $WCl_6$ (d) $ScBr_3$
(e) $HfCl_4$

**20-83** Which of the following compounds will give the most grams of metal when a 10.0-amp current is passed through molten samples of these salts for 2.0 hours?

(a) KCl (b) $CaCl_2$ (c) $ScCl_3$

**20-84** An electric current is passed through a series of three cells filled with aqueous solutions of $AgNO_3$, $Cu(NO_3)_2$, and $Fe(NO_3)_3$. How many grams of each metal will be deposited by a 1.58-amp current flowing for 3.54 hours?

**20-85** What are the values of $x$ and $y$ in the formula $Ce_xCl_y$ if electrolysis of an aqueous solution of this salt for 16.5 hours with a 1.00-amp current deposits 21.6 g of cerium metal?

**20-86** What is the oxidation number of the manganese in an unknown salt if electrolysis of an aqueous solution of this salt for 30 min with a 10.0-amp current generates 2.56 g of manganese metal?

**20-87** Gold forms compounds in the +1 and +3 oxidation states. What is the oxidation number of gold in a compound that deposits 1.53 g of gold metal when electrolyzed for 15 min with a 2.50-amp current?

### Galvanic Corrosion and Cathodic Protection

**20-88** Use cell potentials to explain why iron slowly oxidizes to form $Fe_2O_3$.

**20-89** Use cell potentials to explain why iron dissolves in strong acids.

**20-90** Use the Nernst equation to explain why the $H^+$ ion concentration in water is too small for iron to dissolve in water from which all oxygen has been removed.

**20-91** Use cell potentials to explain why iron very slowly dissolves in water that contains oxygen.

**20-92** Explain why it is a mistake to describe rust as iron(III) oxide, or $Fe_2O_3$.

**20-93** When iron rusts, the metal surface acts as the anode of a voltaic cell. Explain why the iron metal becomes the cathode of a voltaic cell when it is connected to a piece of magnesium or zinc.

**20-94** Explain why galvanized iron, which contains a thin coating of zinc metal, corrodes much more slowly than iron by itself.

## Intersection

# THE FORCES THAT CONTROL A CHEMICAL REACTION

The first half of this book divided chemical reactions into two categories: those that occurred and those that did not. We have noted that sodium reacts with water, for example,

$$2\,Na(s) + H_2O(l) \longrightarrow 2\,Na^+(aq) + 2\,OH^-(aq) + H_2(g)$$

whereas magnesium metal does not:

$$Mg(s) + H_2O(l) \not\longrightarrow$$

The second half of the book has introduced a more sophisticated view of chemical reactions. By introducing the idea that chemical reactions can come to equilibrium, we can construct a continuum of chemical reactions. Toward one end are those reactions, such as the dissociation of acetic acid, which lie very heavily on the side of the reactants:

$$HOAc(aq) + H_2O(l) \rightleftharpoons H_3O^+(aq) + OAc^-(aq) \qquad K_a = 1.8 \times 10^{-5}$$

Toward the other end are reactions that proceed almost to completion.

$$Cu^{2+}(aq) + 4\,NH_3(aq) \rightleftharpoons Cu(NH_3)_4{}^{2+}(aq) \qquad K_f = 2.1 \times 10^{13}$$

We are now left with the question: Why do some reactions occur, and not others? In the next chapter, we will see that **thermodynamics** is a powerful tool for predicting what should (or should not) happen in a chemical reaction. It is ideally suited for answering questions that begin, "What if . . . ?" Thermodynamics predicts, for example, that hydrogen should react with oxygen to form water:

$$2\,H_2(g) + O_2(g) \longrightarrow 2\,H_2O(g)$$

It also predicts that iron(III) oxide should react with aluminum metal to form aluminum oxide and iron metal:

$$Fe_2O_3(s) + 2\,Al(s) \longrightarrow Al_2O_3(s) + 2\,Fe(s)$$

The burning of coal in a power plant is an example of kinetic versus thermodynamic control. In the absence of a flame, the coal does not burn to form the more thermodynamically stable products: $CO_2$ and $H_2O$.

Both of these reactions can, in fact, occur. We can experience the first by touching a balloon filled with $H_2$ gas with a candle tied to the end of a meter stick. The other can be demonstrated by igniting a mixture of $Fe_2O_3$ and powdered aluminum.

This doesn't mean that $H_2$ and $O_2$ burst spontaneously into flame the instant these gases are mixed. In the absence of a spark, mixtures of $H_2$ and $O_2$ can be stored for years with no detectable change. Nor does it mean that scrupulous attention must be paid to prevent $Fe_2O_3$ from coming in contact with aluminum metal, lest some violent reaction take place. In the absence of a source of heat, mixtures of $Fe_2O_3$ and powdered aluminum are so stable they are sold in 50-lb lots under the trade name "Thermite."

It is a mistake to lose sight of the fact that thermodynamics can never tell us what will happen. It can only tell us what should or might happen. Reactions that behave as thermodynamics predicts they should are said to be under **thermodynamic control.** It's not surprising from a thermodynamic point of view, for example, that potassium metal bursts into flame when it comes in contact with water:

$$2\,K(s) + 2\,H_2O(l) \longrightarrow 2\,K^+(aq) + 2\,OH^-(aq) + H_2(g)$$

Other reactions are said to be under **kinetic control.** Thermodynamics predicts that these reactions should occur, but the rate of the reactions is negligibly slow. Examples of reactions under kinetic control include the reactions in which elemental sulfur and its compounds burn. Thermodynamics predicts that these reactions should form $SO_3$, not $SO_2$, because $SO_3$ is more stable than $SO_2$. Under normal conditions, however, these reactions invariably stop at $SO_2$ because the reaction between $SO_2$ and $O_2$ to form $SO_3$ is extremely slow:

$$S_8(s) + 8\,O_2(g) \longrightarrow 8\,\mathbf{SO_2}(g)$$

$$CS_2(g) + 3\,O_2(g) \longrightarrow CO_2(g) + 2\,\mathbf{SO_2}(g)$$

$$2\,H_2S(g) + 3\,O_2(g) \longrightarrow 2\,H_2O(g) + 2\,\mathbf{SO_2}(g)$$

In order to understand chemical reactions fully, we need to combine the predictions of thermodynamics with studies of the factors that influence the rates of chemical reactions. These factors fall under the general heading of **chemical kinetics,** which are discussed in the second chapter in this section.

The beauty of the Bryce Canyon National Park in Utah provides an example of the balance between forces that create rocks and those that gradually erode them to form a system at equilibrium.

# CHEMICAL THERMODYNAMICS

This chapter explores one of the most fundamental questions in chemistry: What determines the direction in which a reaction occurs? It will also provide the basis for answering the following related questions:

## POINTS OF INTEREST

- What do we mean when we say that a cup of hot coffee *spontaneously* cools down?
- Why we would be surprised if ice spontaneously formed inside a glass of lemonade on a hot summer's day?
- Why does a deck of cards become more disordered when it is shuffled?
- Why does the equilibrium constant for the following reaction change with temperature?

$$2\ NO_2(g) \rightleftharpoons N_2O_4(g)$$

  How do we predict whether the equilibrium constant for this reaction increases or decreases as we raise the temperature of the system?
- How can we measure the equilibrium constant for this reaction?

## 21.1 CHEMICAL THERMODYNAMICS

In Chapter 5, **thermodynamics** was defined as the branch of science that deals with the relationship between heat and other forms of energy, such as work. It is frequently summarized as three laws that describe restrictions on how different forms of energy can be interconverted. Chapter 5 introduced the first law of thermodynamics.

**First law: Energy is conserved; it can be neither created nor destroyed.**

This chapter introduces the second and third laws of thermodynamics.

**Second law: In an isolated system, natural processes are spontaneous when they lead to an increase in disorder, or entropy.**

**Third law: The entropy of a perfect crystal is zero when the temperature of the crystal is equal to absolute zero (0 K).**

There have been many attempts to build a device that violates the laws of thermodynamics. All have failed. Thermodynamics is one of the few areas of science in which there are no exceptions. At a time when all other areas of physics seemed open to question, Einstein wrote about thermodynamics as follows:

> Thermodynamics is the only science about which I am firmly convinced that, within the framework of the applicability of its basic principles, it will never be overthrown.

The birth of thermodynamics is often traced to the work of a French scientist, Sadi Carnot, who analyzed the factors that control the amount of work that can be done by a steam engine in his book, *Reflections on the Motive Power of Fire,* published in 1824. Much of thermodynamics is still more important to physicists and engineers than chemists. This chapter therefore focuses on **chemical thermodynamics,** the portion of thermodynamics that pertains to chemical reactions.

## 21.2 THE FIRST LAW OF THERMODYNAMICS

One of the basic assumptions of thermodynamics is the idea that we can arbitrarily divide the universe into a **system** and its **surroundings.** The **boundary** between the system and its surroundings can be as real as the walls of a beaker that separates a solution from the rest of the universe (as in Figure 21.1). Or it can be as imaginary as the set of points that divide the air just above the surface of a metal from the rest of the atmosphere (as in Figure 21.2).

One of the thermodynamic properties of a system is its **internal energy,** $E$, which is the sum of the kinetic and potential energies of the particles that form the system. The internal energy of a system can be understood by examining the simplest possible system: an ideal gas. Because the particles in an ideal gas do not interact, this system has no potential energy. The internal energy of an ideal gas is therefore the sum of the kinetic energies of the particles in the gas.

Sec. 4.15 

The kinetic molecular theory assumes that the temperature of a gas is directly proportional to the average kinetic energy of its particles, as shown in Figure 21.3. The internal energy of an ideal gas is therefore directly proportional to the temperature of the gas:

Sec. 5.9

$$E_{sys} = \tfrac{3}{2} RT$$

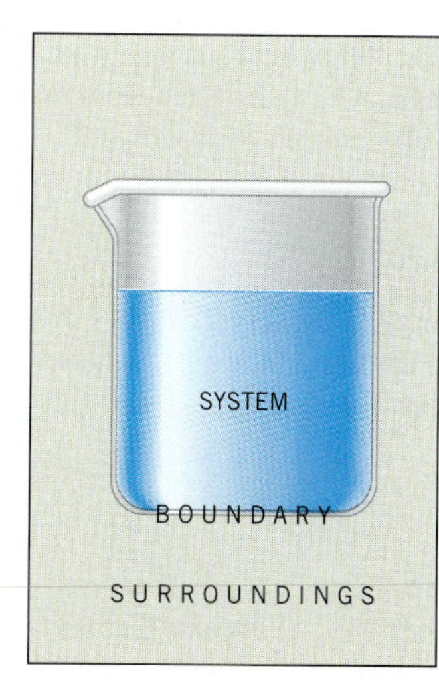


**Figure 21.1** The boundary between the system and its surroundings can be as real as the walls of a beaker that separate a solution (the system) from the rest of the universe (the surroundings).

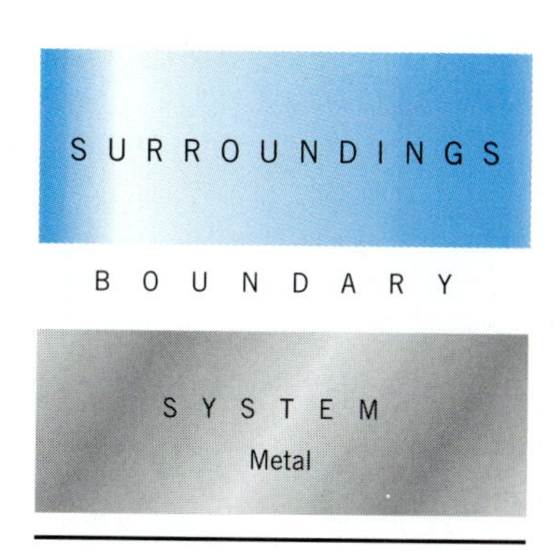


**Figure 21.2** The boundary between the system and its surroundings can be as imaginary as an arbitrary set of points that separates the air just above a metal surface (the system) from the rest of the atmosphere (the surroundings).

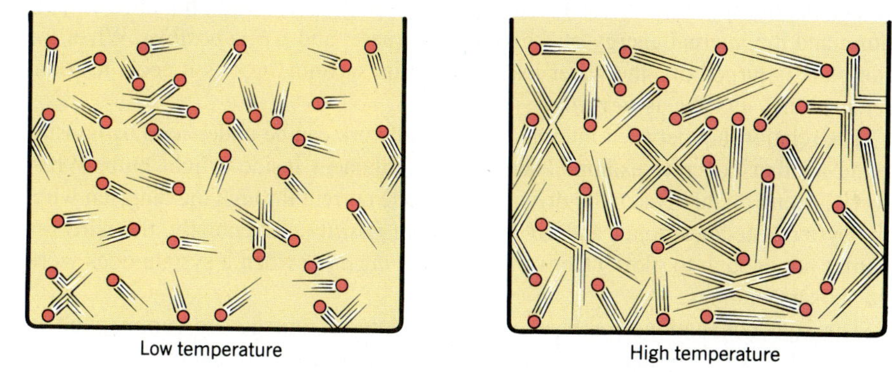


**Figure 21.3** The kinetic theory assumes that the temperature of a system increases if and only if the kinetic energy of the particles in the system increases.

In this equation, $R$ is the ideal gas constant in joules per mole kelvin (J/mol-K) and $T$ is the temperature in kelvin.

The internal energy of systems that are more complex than an ideal gas can't be measured directly. But the internal energy of the system is still proportional to its temperature. We can therefore monitor changes in the internal energy of a system by watching what happens to the temperature of the system. Whenever the temperature of the system increases, we can conclude that the internal energy of the system has also increased.

Assume, for the moment, that a thermometer immersed in a beaker of water on a hot plate reads 73.5°C, as shown in Figure 21.4. This measurement can only describe the state of the system at that moment in time. It can't tell us whether the water was heated directly from room temperature to 73.5°C or heated from room temperature to 100°C and then allowed to cool.

Temperature is therefore a **state function.** It depends only on the state of the system at any moment in time, not the path used to get the system to that state. Because the internal energy of the system is proportional to its temperature, internal energy is also a state function. Any change in the internal energy of the system is equal to the difference between its initial and final values:

$$\Delta E_{sys} = E_f - E_i$$

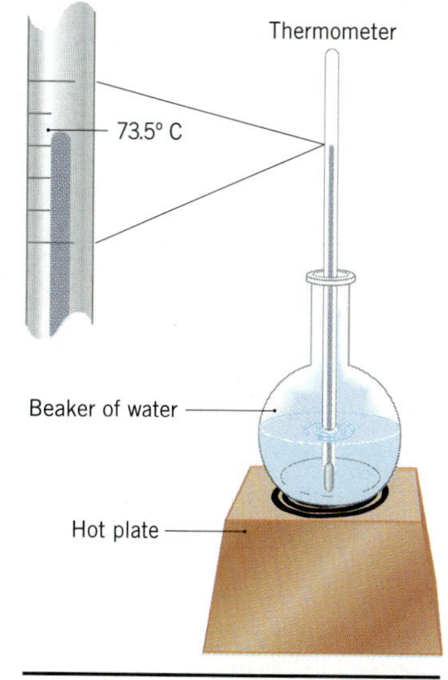


**Figure 21.4** The temperature of a system is a state function, which only reflects the state of the system at that moment in time.

The first law of thermodynamics can be captured in the following equation, which states that the energy of the universe is constant. Energy can be transferred from the system to its surroundings, or vice versa, but it can't be created or destroyed.

**First Law of Thermodynamics:**

$$\Delta E_{univ} = \Delta E_{sys} + \Delta E_{surr} = 0$$

A more useful form of the first law describes how energy is conserved. It says that the change in the internal energy of a system is equal to the sum of the heat gained or lost by the system (q) and the work (w) done by or on the system.

**First Law of Thermodynamics:**

$$\Delta E_{sys} = q + w$$

The sign convention for the relationship between the internal energy of a system and the heat gained or lost by the system can be understood by thinking about a concrete example, such as a beaker of water on a hot plate. When the hot plate is turned on, the system gains heat from its surroundings. As a result, both the temperature and the internal energy of the system increase, and $\Delta E$ is *positive.* When the hot plate is turned off, the water loses heat to its surroundings as it cools to room temperature, and $\Delta E$ is *negative.*

The relationship between internal energy and work can be understood by considering another concrete example: the tungsten filament inside a light bulb. When work is done on this system by driving an electric current through the tungsten wire, the system becomes hotter and $\Delta E$ is therefore *positive.* (Eventually, the wire becomes hot enough to glow.) Conversely, $\Delta E$ is *negative* when a system does work on its surroundings.

The sign conventions for heat, work, and internal energy are summarized in Figure 21.5. The internal energy and temperature of a system decrease ($\Delta E < 0$) when the system either loses heat or does work on its surroundings. Conversely, the internal energy and temperature increase ($\Delta E > 0$) when the system gains heat from its surroundings or when the surroundings do work on the system.

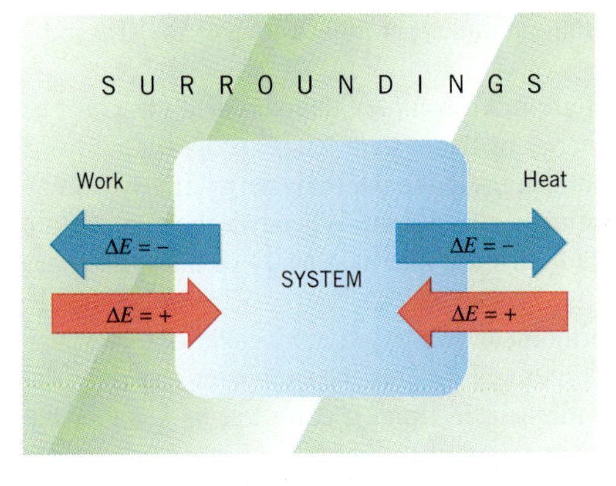


**Figure 21.5** The sign conventions for heat and work are similar when the first law of thermodynamics is written $\Delta E = q + w$. The internal energy of the system decreases when the system loses heat to its surroundings or does work on its surroundings. Conversely, the internal energy increases when the system gains heat from its surroundings or when work is done on the system.

## 21.3 ENTHALPY VERSUS INTERNAL ENERGY

In this chapter, the **system** is usually defined as the chemical reaction and the **boundary** is the container in which the reaction is run. In the course of the reaction, heat is either given off or absorbed by the system. Furthermore, the system either does work on its surroundings or has work done on it by its surroundings. Either of these interactions can affect the internal energy of the system:

$$\Delta E_{sys} = q + w$$

Two kinds of work are normally associated with a chemical reaction: *electrical work* and *work of expansion*. Chemical reactions can do work on their surroundings by driving an electric current through an external wire. Reactions also do work on their surroundings when the volume of the system expands during the course of the reaction. The amount of work of expansion done by the reaction is equal to the product of the pressure against which the system expands times the change in the volume of the system:

$$w = -P\Delta V$$

Sec. 5.7

The sign convention for this equation reflects the fact that the internal energy of the system *decreases* when the system does work on its surroundings.

## CHECKPOINT

Use the sign convention for work of expansion to explain why gases usually become cooler when they expand.

What would happen if we created a set of conditions under which no work is done by the system on its surroundings, or vice versa, during a chemical reaction? Under these conditions, the heat given off or absorbed by the reaction would be equal to the change in the internal energy of the system:

$$\Delta E_{sys} = q \qquad \text{(if and only if } w = 0\text{)}$$

The easiest way to achieve these conditions is to run the reaction at constant volume, where no work of expansion is possible. At constant volume, the heat given off or absorbed by the reaction ($q_V$) is equal to the change in the internal energy that occurs during the reaction:

$$\Delta E_{sys} = q_V \qquad \text{(at constant volume)}$$

Figure 21.6 shows a calorimeter in which reactions can be run at constant volume. Most reactions, however, are run in open flasks and beakers. When this is done, the

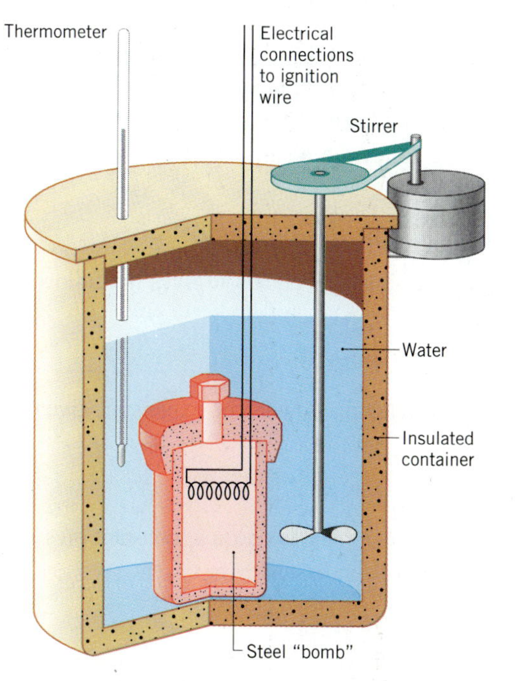


**Figure 21.6** The heat given off when a reaction is run in a sealed container, such as this bomb calorimeter, is equal to the change in the internal energy of the system during the reaction: $\Delta E = q_V$.

volume of the system is not constant because gas can either enter or leave the container during the reaction. The system is at constant pressure, however, because the total pressure inside the container is always equal to atmospheric pressure.

If a gas is driven out of the flask during the reaction, the system does work on its surroundings. If the reaction pulls a gas into the flask, the surroundings do work on the system. We can still measure the amount of heat given off or absorbed during the reaction, but it is no longer equal to the change in the internal energy of the system because some of the heat has been converted into work:

$$\Delta E_{sys} = q + w$$

Sec. 5.13

We can get around this problem by introducing the concept of **enthalpy (*H*)**, which is the sum of the internal energy of the system plus the product of the pressure of the gas in the system times the volume of the system:

$$H_{sys} = E_{sys} + PV$$

For the sake of simplicity, the subscript sys will be left off the symbol for both the internal energy of the system and the enthalpy of the system from now on. We will therefore abbreviate the relationship between the enthalpy of the system and the internal energy of the system as follows:

$$H = E + PV$$

The change in the enthalpy of the system during a chemical reaction is equal to the change in its internal energy plus the change in the product of the pressure times the volume of the system:

$$\Delta H = \Delta E + \Delta(PV)$$

**CHECKPOINT**

Describe the conditions under which $\Delta H$ for a reaction will be approximately equal to $\Delta E$ for the reaction.

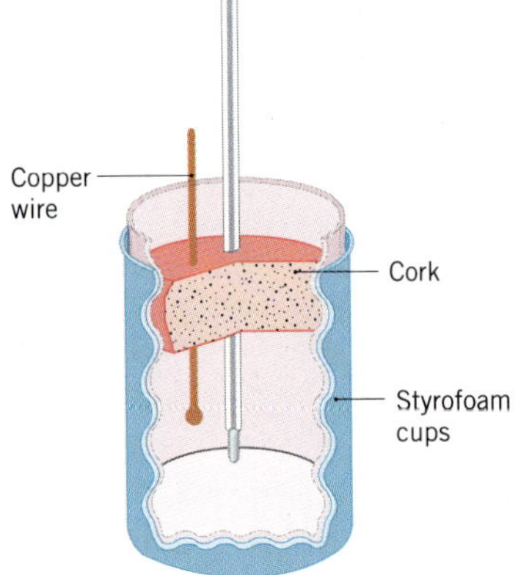

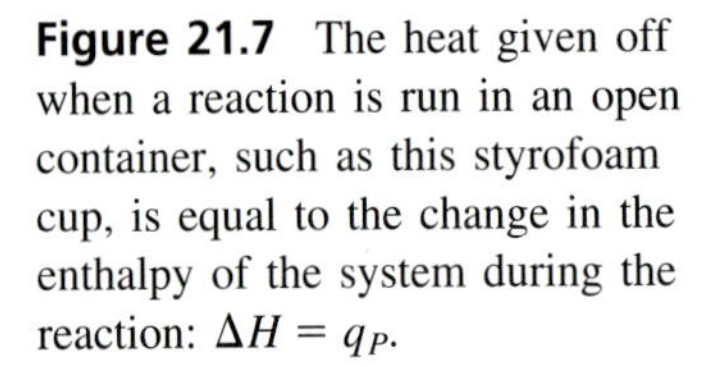

**Figure 21.7** The heat given off when a reaction is run in an open container, such as this styrofoam cup, is equal to the change in the enthalpy of the system during the reaction: $\Delta H = q_P$.

Let's assume that the reaction is run in a styrofoam cup, as shown in Figure 21.7. Because the reaction is run at constant pressure, the change in the enthalpy that occurs during the reaction is equal to the change in the internal energy of the system plus the product of the constant pressure times the change in the volume of the system:

$$\Delta H = \Delta E + P\Delta V \qquad \text{(at constant pressure)}$$

Substituting the first law of thermodynamics into this equation gives the following result:

$$\Delta H = (q_P + w) + P\Delta V$$

Assuming that the only work done by the reaction is work of expansion gives an equation in which the $P\Delta V$ terms cancel:

$$\Delta H = (q_P - \cancel{P\Delta V}) + \cancel{P\Delta V}$$

Thus, the heat given off or absorbed during a chemical reaction at constant pressure ($q_P$) is equal to the change in the enthalpy of the system:

$$\Delta H = q_P \qquad \text{(at constant pressure)}$$

The relationship between the change in the internal energy of the system during a chemical reaction and the enthalpy of reaction can be summarized as follows.

1. The heat given off or absorbed when a reaction is run at *constant volume* is equal to the change in the internal energy of the system:

$$\Delta E_{sys} = q_V$$

2. The heat given off or absorbed when a reaction is run at *constant pressure* is equal to the change in the enthalpy of the system:

$$\Delta H_{sys} = q_P$$

3. The change in the enthalpy of the system during a chemical reaction is equal to the change in the internal energy plus the change in the product of the pressure of the gas in the system and its volume:

$$\Delta H_{sys} = \Delta E_{sys} + \Delta(PV)$$

4. The difference between $\Delta E$ and $\Delta H$ for the system is small for reactions that involve only liquids and solids because there is little if any change in the volume of the system during the reaction. The difference can be relatively large, however, for reactions that involve gases if there is a change in the number of moles of gas in the course of the reaction.

**EXERCISE 21.1**

Which of the following processes are run at constant volume and which are run at constant pressure?

(a) An acid–base titration.

(b) Decomposing $CaCO_3$ by heating limestone in a crucible with a bunsen burner.

(c) The reaction between zinc metal and an aqueous solution of $Cu^{2+}$ ions to form copper metal and $Zn^{2+}$ ions.

(d) Measuring the calories in a 1-oz serving of a breakfast cereal by burning the cereal in a bomb calorimeter.

**SOLUTION** Processes (a), (b), and (c) are all run under conditions of constant pressure. Only (d) is run at constant volume.

## 21.4 SPONTANEOUS CHEMICAL REACTIONS

The first law of thermodynamics suggests that we can't get something for nothing. It allows us to build an apparatus that does work, but it places important restrictions on that apparatus. It says that we have to be willing to pay a price in terms of a loss of either heat or internal energy for any work we ask the system to do. It also puts a limit on the amount of work we can get for a given investment of either heat or internal energy.

The first law allows us to convert heat into work, or work into heat. It also allows us to change the internal energy of a system by transferring either heat or work between the system and its surroundings. But it doesn't tell us whether one of these changes is more easy to achieve than another. Our experiences, however, tell us that there is a preferred direction to many natural processes. We aren't surprised when a cup of coffee gradually loses heat to its surroundings as it cools, for example, or

**Figure 21.8** There is a preferred direction to most chemical reactions. It isn't surprising to find that magnesium metal dissolves in concentrated hydrochloric acid to form $H_2$ gas, which bubbles out of the solution. But it would be surprising to find bubbles of $H_2$ appearing at the surface of the solution and then sinking through the solution until they vanished, while a piece of magnesium metal mysteriously appeared.

when the ice in a glass of lemonade absorbs heat as it melts. But we would be surprised if a cup of coffee suddenly grew hotter until it boiled or the water in a glass of lemonade froze on a hot summer day, even though neither process violates the first law of thermodynamics.

Similarly, we aren't surprised to see a piece of zinc metal dissolve in a strong acid to give bubbles of hydrogen gas, as shown in Figure 21.8.

$$\mathrm{Zn}(s) + 2\,\mathrm{H}^+(aq) \longrightarrow \mathrm{Zn}^{2+}(aq) + \mathrm{H_2}(g)$$

But if we saw a film in which $H_2$ bubbles formed on the surface of a solution and then sank through the solution until they disappeared, while a strip of zinc metal formed in the middle of the solution, we would conclude that the film was being run backward.

Many chemical and physical processes are reversible and yet tend to proceed in a direction in which they are said to be **spontaneous.** This raises the question: What makes a reaction spontaneous? What drives the reaction in one direction and not the other?

So many spontaneous reactions are exothermic that it is tempting to assume that one of the driving forces that determines whether a reaction is spontaneous is a tendency to give off energy. The following are all examples of spontaneous chemical reactions that are exothermic:

$$2\,\mathrm{Al}(s) + 3\,\mathrm{Br_2}(l) \longrightarrow 2\,\mathrm{AlBr_3}(s) \qquad \Delta H^\circ = -511\ \mathrm{kJ/mol\ AlBr_3}$$

$$2\,\mathrm{H_2}(g) + \mathrm{O_2}(g) \longrightarrow 2\,\mathrm{H_2O}(g) \qquad \Delta H^\circ = -241.82\ \mathrm{kJ/mol\ H_2O}$$

$$\mathrm{P_4}(s) + 5\,\mathrm{O_2}(g) \longrightarrow \mathrm{P_4O_{10}}(s) \qquad \Delta H^\circ = -2984\ \mathrm{kJ/mol\ P_4O_{10}}$$

There are also spontaneous reactions, however, that absorb energy from their surroundings. At 100°C, water boils spontaneously even though the reaction is endothermic:

$$\mathrm{H_2O}(l) \longrightarrow \mathrm{H_2O}(g) \qquad \Delta H^\circ = 40.88\ \mathrm{kJ/mol}$$

Ammonium nitrate dissolves spontaneously in water, even though energy is absorbed when this reaction takes place:

$$\mathrm{NH_4NO_3}(s) \xrightarrow{\mathrm{H_2O}} \mathrm{NH_4}^+(aq) + \mathrm{NO_3}^-(aq) \qquad \Delta H^\circ = 28.05\ \mathrm{kJ/mol}$$

Thus, the tendency of a spontaneous reaction to give off energy can't be the only driving force behind a chemical reaction. There must be another factor that helps determine whether a reaction is spontaneous. This factor, known as *entropy,* is a measure of the disorder of the system.

## 21.5 ENTROPY AS A MEASURE OF DISORDER

Perhaps the best way to understand entropy as a driving force in nature is to conduct a simple experiment with a new deck of cards. Open the deck, remove the jokers, and then turn the deck so that you can read the cards. The top card will be the ace of spades, followed by the two, three, and four of spades, and so on, as shown in Figure 21.9. Now divide the cards in half, shuffle the deck, and note that the deck becomes more disordered. The more often the deck is shuffled, the more disordered it becomes. What makes a deck of cards become more disordered when shuffled?

In 1877 Ludwig Boltzmann provided a basis for answering this question when he introduced the concept of the **entropy** of a system as a measure of the amount of disorder in the system. A deck of cards fresh from the manufacturer is perfectly

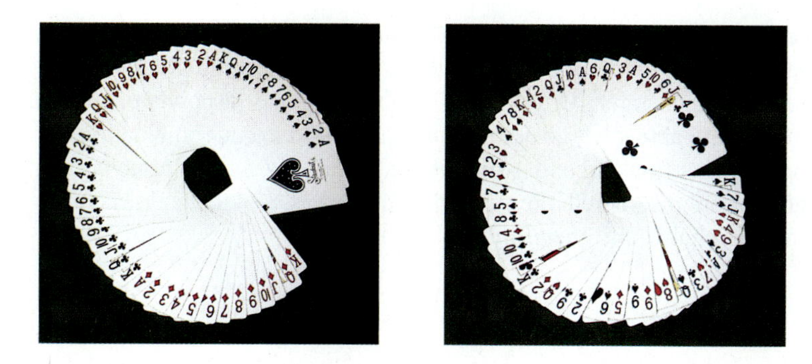

**Figure 21.9** A deck of cards, fresh from the manufacturer, is a perfectly ordered system. When the cards are shuffled, they become more and more disordered. The driving force behind this process is a natural tendency for systems to move toward greater disorder.

ordered and the entropy of this system is zero. When the deck is shuffled, the entropy of the system increases as the deck becomes more disordered.

There are $8.066 \times 10^{67}$ different ways of organizing a deck of cards. The probability of obtaining any particular sequence of cards when the deck is shuffled is therefore 1 part in $8.066 \times 10^{67}$. In theory, it is possible to shuffle a deck of cards until the cards fall into perfect order. But it isn't very likely!

Boltzmann introduced the following equation to describe the relationship between entropy and the amount of disorder in a system:

$$S = k \ln W$$

In this equation, $S$ is the entropy of the system, $k$ is a proportionality constant equal to the ideal gas constant divided by Avogadro's constant, ln represents a logarithm to the base $e$, and $W$ is the number of equivalent ways of describing the state of the system. According to this equation, the entropy of a system increases as the number of equivalent ways of describing the state of the system increases.

The relationship between the number of equivalent ways of describing a system and the amount of disorder in the system can be demonstrated with another analogy

**TABLE 21.1 Number of Equivalent Combinations for Various Types of Poker Hands**

| Hand | $W$ | $\ln W$ |
|---|---|---|
| Royal flush (AKQJ10 in one suit) | 4 | 1.39 |
| Straight flush (five cards in sequence in one suit) | 36 | 3.58 |
| Four of a kind | 624 | 6.44 |
| Full house (three of a kind plus a pair) | 3,744 | 8.23 |
| Flush (five cards in the same suit) | 5,108 | 8.54 |
| Straight (five cards in sequence) | 10,200 | 9.23 |
| Three of a kind | 54,912 | 10.91 |
| Two pairs | 123,552 | 11.72 |
| One pair | 1,098,240 | 13.91 |
| No pairs | 1,302,540 | 14.08 |
| Total | 2,598,960 | |

based on a deck of cards. There are 2,598,960 different hands that could be dealt in a game of five-card poker. More than half of these hands are essentially worthless. Winning hands are much rarer. Only 3,744 combinations correspond to a "full house," for example. Table 21.1 gives the number of equivalent combinations of cards for each category of poker hand, which is the value of *W* for this category. As the hand becomes more disordered, the value of *W* increases, and the hand becomes intrinsically more probable and therefore less valuable.

## 21.6 ENTROPY AND THE SECOND LAW OF THERMODYNAMICS

The vapor that escapes from a liquid is more disordered than the liquid.

The second law of thermodynamics describes the relationship between entropy and the spontaneity of natural processes:

**Second law: In an isolated system, natural processes are spontaneous when they lead to an increase in disorder, or entropy.**

This statement is restricted to **isolated systems** to avoid having to worry about whether the reaction is exothermic or endothermic. By definition, neither heat nor work can be transferred between an isolated system and its surroundings.

We can apply the second law of thermodynamics to chemical reactions by noting that the **entropy** of a system is a state function that is directly proportional to the **disorder** of the system:

$\Delta S_{sys} > 0$ implies that the system becomes *more disordered* during the reaction.

For an isolated system, any process that leads to an increase in the disorder of the system will be spontaneous. The following generalizations can help us decide when a chemical reaction leads to an increase in the disorder of the system.

- Solids have a much more regular structure than liquids. Liquids are therefore more disordered than solids.
- The particles in a gas are in a state of constant, random motion. Gases are therefore more disordered than the corresponding liquids.
- Any process that increases the number of particles in the system increases the amount of disorder.

### EXERCISE 21.2

Which of the following processes will lead to an increase in the entropy of the system?

(a) $N_2(g) + 3\,H_2(g) \rightleftharpoons 2\,NH_3(g)$

(b) $H_2O(l) \rightleftharpoons H_2O(g)$

(c) $CaCO_3(s) \longrightarrow CaO(s) + CO_2(g)$

(d) $NH_4NO_3(s) \xrightarrow{H_2O} NH_4^+(aq) + NO_3^-(aq)$

**SOLUTION**

(a) Because the total number of molecules decreases in this reaction, the system becomes more ordered. The entropy of the system therefore decreases.

(b) Gases are more disordered than the corresponding liquids, so the entropy of the system increases.

(c) Reactions in which a compound decomposes into two products lead to an increase in entropy because the system becomes more disordered. The increase in entropy in this reaction is even larger because the starting material is a solid and one of the products is a gas.

(d) The $NH_4^+$ and $NO_3^-$ ions are free to move in a random fashion through the aqueous solution, whereas these ions are locked into position in the crystal. As a result, the entropy of the system increases in this reaction.

We noted in Section 21.4 that the sign of $\Delta H$ for a chemical reaction affects the direction in which the reaction occurs. As a rule:

**Spontaneous reactions often, but not always, give off energy.**

This section has introduced the notion that the sign of $\Delta S$ for a reaction can also determine the direction of the reaction:

**In an isolated system, chemical reactions occur in the direction that leads to an increase in the disorder of the system.**

In order to decide whether a reaction is spontaneous, it is therefore important to consider the effect of changes in both enthalpy and entropy that occur during the reaction.

## EXERCISE 21.3

Use the Lewis structures of $NO_2$ and $N_2O_4$ and the stoichiometry of the following reaction to decide whether $\Delta H$ and $\Delta S$ favor the reactants or the products of this reaction:

$$2\,NO_2(g) \rightleftharpoons N_2O_4(g)$$

**SOLUTION** $NO_2$ has an unpaired electron in its Lewis structure:

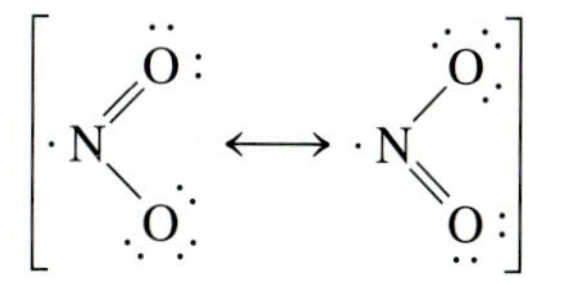

When two $NO_2$ molecules collide in the proper orientation, a covalent bond can form between the nitrogen atoms to produce the dimer, $N_2O_4$:

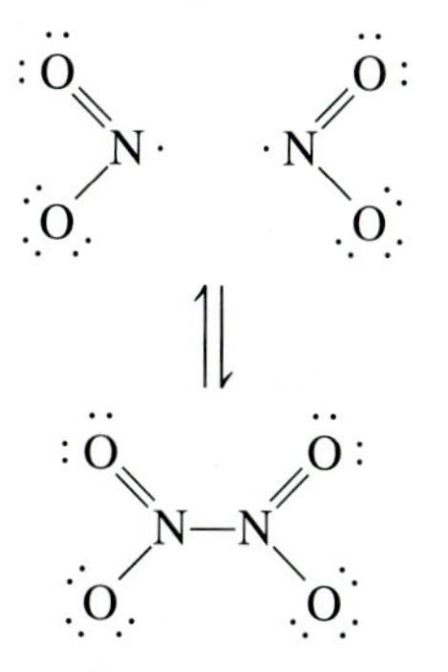

Because it forms a new bond, this reaction should be exothermic. $\Delta H$ for this reaction therefore favors the products.

Sec. 10.5 

The reaction leads to a decrease in the entropy of the system because the system becomes less disordered when two molecules come together to form a larger molecule. $\Delta S$ for this reaction therefore favors the reactants.

## 21.7 THE THIRD LAW OF THERMODYNAMICS

The third law of thermodynamics defines absolute zero on the entropy scale:

**Third law: The entropy of a perfect crystal is zero when the temperature of the crystal is equal to absolute zero (0 K).**

The crystal must be perfect, or else there will be some inherent disorder. It also must be at 0 K; otherwise there will be thermal motion within the crystal, which leads to disorder.

As the crystal warms to temperatures above 0 K, the particles in the crystal start to move, generating some disorder. The entropy of the crystal gradually increases with temperature as the average kinetic energy of the particles increases. At the melting point, the entropy of the system increases abruptly as the compound is transformed into a liquid, which is not as well ordered as the solid. The entropy of the liquid gradually increases as the liquid becomes warmer because of the increase in the vibrational, rotational, and translational motion of the particles. At the boiling point, there is another abrupt increase in the entropy of the substance as it is transformed into a random, chaotic gas.

**TABLE 21.2 The Entropies of Solid, Liquid, and Gaseous Forms of Sulfur Trioxide**

| Compound | $S°$ (J/mol-K) |
|---|---|
| $SO_3(s)$ | 70.7 |
| $SO_3(l)$ | 113.8 |
| $SO_3(g)$ | 256.76 |

Table 21.2 provides an example of the difference between the entropies of a substance in its solid, liquid, and gaseous phases. Note that the units of entropy are joules per mole kelvin (J/mol-K). A plot of the entropy of this system versus temperature is shown in Figure 21.10.

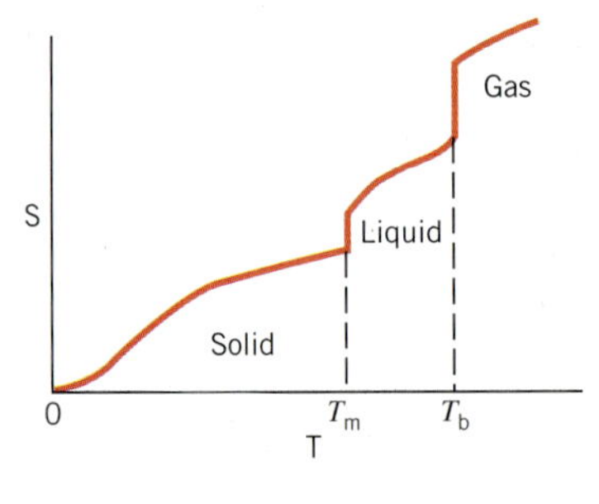


**Figure 21.10** There is a gradual increase in the entropy of a solid when it is heated until the temperature of the system reaches the melting point of the solid. At this point, the entropy of the system increases abruptly as the solid melts to form a liquid. The entropy of the liquid gradually increases until the system reaches the boiling point of the liquid, at which point it increases abruptly once more. The entropy of the gas then increases gradually with temperature.

## 21.8 STANDARD-STATE ENTROPIES OF REACTION

Because entropy is a state function, the change in the entropy of the system that accompanies any process can be calculated by subtracting the initial value of the entropy of the system from the final value:

$$\Delta S = S_f - S_i$$

$\Delta S$ for a chemical reaction is therefore equal to the difference between the sum of the entropies of the reactants and the products of the reaction:

$$\Delta S = \Sigma S_{\text{products}} - \Sigma S_{\text{reactants}}$$

When this difference is measured under standard-state conditions, the result is the **standard-state entropy of reaction, $\Delta S°$**:

$$\Delta S° = \Sigma S°_{\text{products}} - \Sigma S°_{\text{reactants}}$$

By convention, the standard state for thermodynamic measurements is characterized by the following conditions:

> **Standard-state conditions:**
> **All solutions have concentrations of 1 *M*. All gases have partial pressures of 0.1 MPa (0.9869 atm).**

Although standard-state entropies can be measured at any temperature, they are often measured at 25°C. Standard-state entropies at 25°C for a variety of compounds are given in Table A.15 in the appendix.

## EXERCISE 21.4

Calculate the standard-state entropy of reaction for the following reactions and explain the sign of $\Delta S°$ for each reaction:

(a) $Hg(l) \rightleftharpoons Hg(g)$
(b) $NaCl(s) \overset{H_2O}{\rightleftharpoons} Na^+(aq) + Cl^-(aq)$
(c) $2\,NO_2(g) \rightleftharpoons N_2O_4(g)$
(d) $N_2(g) + O_2(g) \rightleftharpoons 2\,NO(g)$

**SOLUTION**

(a) Table A.15 contains the following standard-state entropy data for this reaction:

| Compound | $S°$ (J/mol-K) |
|---|---|
| $Hg(l)$ | 76.02 |
| $Hg(g)$ | 174.96 |

The balanced equation states that 1 mol of mercury vapor is produced for each mole of liquid mercury that boils. The standard-state entropy of reaction is therefore calculated as follows:

$$\begin{aligned}\Delta S° &= \Sigma S°_{\text{products}} - \Sigma S°_{\text{reactants}}\\ &= [1 \text{ mol } Hg(g) \times 174.96 \text{ J/mol-K}] - [1 \text{ mol } Hg(l) \times 76.02 \text{ J/mol-K}]\\ &= \mathbf{98.94 \text{ J/K}}\end{aligned}$$

The sign of $\Delta S°$ is positive because this process transforms a liquid into a gas, which is inherently more disordered.

(b) Table A.15 contains the following standard-state entropy data for this reaction:

| Compound | $S°$ (J/mol-K) |
|---|---|
| $NaCl(s)$ | 72.13 |
| $Na^+(aq)$ | 59.0 |
| $Cl^-(aq)$ | 56.5 |

In this reaction, 1 mole of $Na^+$ ions and 1 mole of $Cl^-$ ions are produced

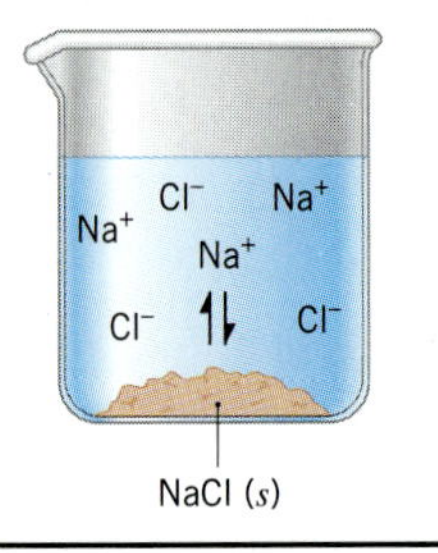


**Figure 21.11** The entropy of the system increases when NaCl dissolves in water.

for each mole of NaCl that dissolves. The standard-state entropy of reaction is therefore calculated as follows:

$$\begin{aligned}\Delta S^\circ &= \Sigma S^\circ_{\text{products}} - \Sigma S^\circ_{\text{reactants}} \\ &= (1 \text{ mol Na}^+ \times 59.0 \text{ J/mol-K} + 1 \text{ mol Cl}^- \times 56.5 \text{ J/mol-K}) \\ &\quad - (1 \text{ mol NaCl} \times 72.13 \text{ J/mol-K}) \\ &= \mathbf{43.4 \text{ J/K}}\end{aligned}$$

The sign of $\Delta S^\circ$ for this reaction is positive because the $Na^+$ and $Cl^-$ ions are free to move through the aqueous solution, as shown in Figure 21.11. The solution is therefore inherently more disordered than solid NaCl, in which these ions are locked into position.

(c) Table A.15 contains the following standard-state entropy data for this reaction:

| Compound | $S^\circ$ (J/mol-K) |
|---|---|
| $NO_2(g)$ | 240.06 |
| $N_2O_4(g)$ | 304.29 |

In this equation, 1 mole of $N_2O_4$ is formed for every 2 mole of $NO_2$ consumed, and the value of $\Delta S^\circ$ is calculated as follows:

$$\begin{aligned}\Delta S^\circ &= \Sigma S^\circ_{\text{products}} - \Sigma S^\circ_{\text{reactants}} \\ &= (1 \text{ mol N}_2\text{O}_4 \times 304.29 \text{ J/mol-K}) - (2 \text{ mol NO}_2 \times 240.06 \text{ J/mol-K}) \\ &= \mathbf{-175.83 \text{ J/K}}\end{aligned}$$

The sign of $\Delta S^\circ$ is negative because two molecules combine in this reaction to form a larger, more ordered product.

(d) Table A.15 contains the following standard-state entropy data for this reaction:

| Compound | $S^\circ$ (J/mol-K) |
|---|---|
| $NO(g)$ | 210.76 |
| $N_2(g)$ | 191.61 |
| $O_2(g)$ | 205.14 |

The balanced equation for this reaction indicates that 2 mole of NO are produced when 1 mole of $N_2$ reacts with 1 mole of $O_2$. Thus, the standard-state entropy of reaction is calculated as follows:

$$\begin{aligned}\Delta S^\circ &= \Sigma S^\circ_{\text{products}} - \Sigma S^\circ_{\text{reactants}} \\ &= (2 \text{ mol NO} \times 210.76 \text{ J/mol-K}) \\ &\quad - (1 \text{ mol N}_2 \times 191.61 \text{ J/mol-K} + 1 \text{ mol O}_2 \times 205.14 \text{ J/mol-K}) \\ &= \mathbf{24.77 \text{ J/K}}\end{aligned}$$

$\Delta S^\circ$ for this reaction is small but positive because the product of the reaction (NO) is slightly more disordered than the reactants ($N_2$ and $O_2$).

## 21.9 THE DIFFERENCE BETWEEN ENTHALPY OF REACTION AND ENTROPY OF REACTION CALCULATIONS

At first glance, tables of thermodynamic data seem inconsistent. Consider the data in Table 21.3, for example. The enthalpy data in this table are given in terms of the standard-state enthalpy of formation of each substance, $\Delta H_f^\circ$. This quantity is the heat given off or absorbed when the substance is made from its elements in their most thermodynamically stable state at 0.1 MPa. The enthalpy of formation of $AlCl_3$, for example, is the heat given off in the following reaction:

$$2\,\text{Al}(s) + 3\,\text{Cl}_2(g) \longrightarrow 2\,\text{AlCl}_3(s) \qquad \Delta H_f^\circ = -704.2 \text{ kJ/mol}$$

**TABLE 21.3 Thermodynamic Data for Aluminum and Its Compounds**

| Substance | $\Delta H_f^\circ$ (kJ/mol) | $S^\circ$ (J/mol-K) |
| --- | --- | --- |
| $Al(s)$ | 0 | 28.33 |
| $Al(g)$ | 326.4 | 164.54 |
| $Al_2O_3(s)$ | −1675.7 | 50.92 |
| $AlCl_3(s)$ | −704.2 | 110.67 |

The enthalpy data in this table are therefore relative numbers, which compare each compound with its elements.

Enthalpy data are listed as relative measurements because there is no absolute zero on the enthalpy scale. All we can measure is the heat given off or absorbed by a reaction. All we can determine, then, is the difference between the enthalpies of the reactants and the products of a reaction. We therefore define the enthalpy of formation of the elements in their most thermodynamically stable states as zero and report all compounds as either more or less stable than their elements.

Entropy data are different. The third law defines absolute zero on the entropy scale. As a result, the absolute entropy of any element or compound can be measured by comparing it with a perfect crystal at absolute zero. The entropy data are therefore given as absolute numbers, $S^\circ$, not entropies of formation, $\Delta S_f^\circ$.

$$AlCl_3(s) \qquad S^\circ = 110.67 \text{ J/mol-K}$$

## 21.10 GIBBS FREE ENERGY

Section 21.6 noted that some reactions are spontaneous because they give off energy in the form of heat ($\Delta H^\circ < 0$). Others are spontaneous because they lead to an increase in the disorder of the system ($\Delta S^\circ > 0$). The following exercise shows how calculations of $\Delta H^\circ$ and $\Delta S^\circ$ can be used to probe the driving force behind a particular reaction.

### EXERCISE 21.5

Calculate $\Delta H^\circ$ and $\Delta S^\circ$ for the following reaction and decide in which direction each of these factors will drive the reaction:

$$N_2(g) + 3\,H_2(g) \rightleftharpoons 2\,NH_3(g)$$

**SOLUTION** Table A.15 contains the following data for this reaction:

| Compound | $\Delta H_f^\circ$ (kJ/mol) | $S^\circ$ (J/mol-K) |
| --- | --- | --- |
| $N_2(g)$ | 0 | 191.61 |
| $H_2(g)$ | 0 | 130.68 |
| $NH_3(g)$ | −46.11 | 192.45 |

The reaction is exothermic ($\Delta H^\circ < 0$), which means that the enthalpy of reaction favors the products of the reaction:

$$\begin{aligned}\Delta H^\circ &= \Sigma H_f^\circ(\text{products}) - \Sigma H_f^\circ(\text{reactants}) \\ &= (2 \text{ mol } NH_3 \times -46.11 \text{ kJ/mol}) \\ &\quad - (1 \text{ mol } N_2 \times 0 \text{ kJ/mol} + 3 \text{ mol } H_2 \times 0 \text{ kJ/mol}) \\ &= -92.22 \text{ kJ}\end{aligned}$$

The entropy of reaction is unfavorable, however, because there is a significant increase in the order of the system ($\Delta S° < 0$) when $N_2$ and $H_2$ combine to form $NH_3$, as shown in Figure 21.12:

$$\begin{aligned}\Delta S° &= \Sigma S°(\text{products}) - \Sigma S°(\text{reactants}) \\ &= (2 \text{ mol } NH_3 \times 192.45 \text{ J/mol-K}) \\ &\quad - (1 \text{ mol } N_2 \times 191.61 \text{ J/mol-K} + 3 \text{ mol } H_2 \times 130.68 \text{ J/mol-K}) \\ &= \mathbf{-198.75 \text{ J/K}}\end{aligned}$$

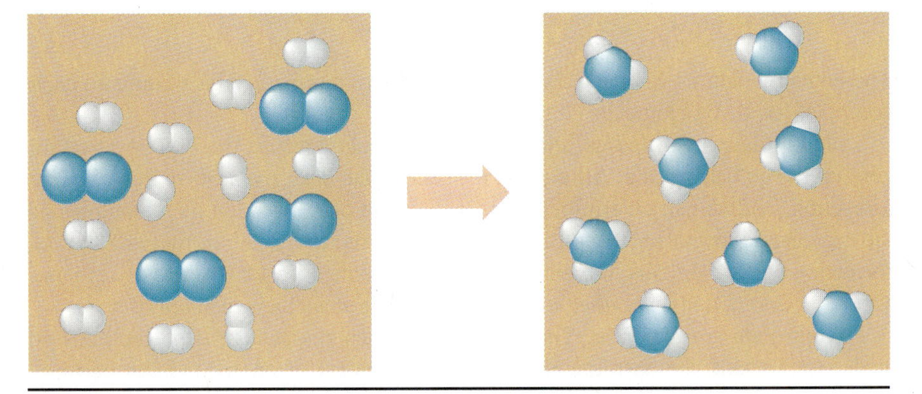

**Figure 21.12** The entropy of the system decreases when $N_2$ and $H_2$ react to form $NH_3$.

This exercise raises an important question: What happens when one of the potential driving forces behind a chemical reaction is favorable and the other is not? We can answer this question by defining a new quantity known as the **Gibbs free energy** ($G$) of the system, which reflects the balance between these forces. It is called the *Gibbs* free energy of the system to recognize the contributions of J. Willard Gibbs, a professor of mathematical physics at Yale from 1871 until the turn of the century, who is considered by some to be the greatest scientist produced by the United States. It is called *free energy* because, as we will soon see, it is the energy that is free to do work.

The Gibbs free energy of a system at any moment in time is defined as the enthalpy of the system minus the product of the temperature times the entropy of the system:

$$G = H - TS$$

The Gibbs free energy of the system is a state function because it is defined in terms of thermodynamic properties that are state functions. The change in the Gibbs free energy of the system that occurs during a reaction is therefore equal to the change in the enthalpy of the system minus the change in the product of the temperature times the entropy of the system:

$$\Delta G = \Delta H - \Delta(TS)$$

If the reaction is run at constant temperature, this equation can be written as follows.

$$\Delta G = \Delta H - T\Delta S$$

The change in the free energy of a system that occurs during a reaction can be measured under any set of conditions. If the data are collected under standard-state conditions, the result is the **standard-state free energy of reaction** ($\Delta G°$):

$$\mathbf{\Delta G° = \Delta H° - T\Delta S°}$$

The beauty of the equation defining the free energy of a system is its ability to determine the relative importance of the enthalpy and entropy terms as driving forces behind a particular reaction. The change in the free energy of the system that occurs during a reaction measures the balance between the two driving forces that determine whether a reaction is spontaneous. As we have seen, the enthalpy and entropy terms have different sign conventions:

| Favorable | Unfavorable |
|---|---|
| $\Delta H^\circ < 0$ | $\Delta H^\circ > 0$ |
| $\Delta S^\circ > 0$ | $\Delta S^\circ < 0$ |

The entropy term is therefore subtracted from the enthalpy term when calculating $\Delta G^\circ$ for a reaction.

Because of the way the free energy of the system is defined, $\Delta G^\circ$ is negative for any reaction for which $\Delta H^\circ$ is negative and $\Delta S^\circ$ is positive. $\Delta G^\circ$ is therefore negative for any reaction that is favored by both the enthalpy and entropy terms. We can therefore conclude that any reaction for which $\Delta G^\circ$ is negative should be favorable, or spontaneous:

$$\text{Favorable, or spontaneous reactions: } \Delta G^\circ < 0$$

Conversely, $\Delta G^\circ$ is positive for any reaction for which $\Delta H^\circ$ is positive and $\Delta S^\circ$ is negative. Any reaction for which $\Delta G^\circ$ is positive is therefore unfavorable:

$$\text{Unfavorable, or non-spontaneous reactions: } \Delta G^\circ > 0$$

Reactions are classified as either **exothermic** ($\Delta H^\circ < 0$) or **endothermic** ($\Delta H^\circ > 0$) on the basis of whether they give off or absorb heat. Reactions also can be classified as **exergonic** ($\Delta G^\circ < 0$) or **endergonic** ($\Delta G^\circ > 0$) on the basis of whether the free energy of the system decreases or increases during the reaction.

When a reaction is favored by both enthalpy ($\Delta H^\circ < 0$) and entropy ($\Delta S^\circ > 0$), there is no need to calculate the value of $\Delta G^\circ$ to decide whether the reaction should proceed. The same can be said for reactions favored by neither enthalpy ($\Delta H^\circ > 0$) nor entropy ($\Delta S^\circ < 0$). Free energy calculations become important for reactions favored by only one of these factors.

## EXERCISE 21.6

Calculate $\Delta H^\circ$ and $\Delta S^\circ$ for the following reaction:

$$NH_4NO_3(s) \xrightarrow{H_2O} NH_4^+(aq) + NO_3^-(aq)$$

Use the results of this calculation to determine the value of $\Delta G^\circ$ for this reaction at 25°C, and explain why $NH_4NO_3$ spontaneously dissolves in water at room temperature.

**SOLUTION** Table A.15 contains the following data for this reaction:

| Compound | $\Delta H_f^\circ$ (kJ/mol) | $S^\circ$ (J/mol-K) |
|---|---|---|
| $NH_4NO_3(s)$ | −365.56 | 151.08 |
| $NH_4^+(aq)$ | −132.51 | 113.4 |
| $NO_3^-(aq)$ | −205.0 | 146.4 |

This reaction is endothermic, and the enthalpy of reaction is therefore unfavorable:

$$\Delta H^\circ = \Sigma\Delta H_f{}^\circ_{\text{products}} - \Sigma\Delta H_f{}^\circ_{\text{reactants}}$$
$$= (1 \text{ mol } NH_4{}^+ \times -132.51 \text{ kJ/mol} + 1 \text{ mol } NO_3{}^- \times -205.0 \text{ kJ/mol})$$
$$- (1 \text{ mol } NH_4NO_3 \times -365.56 \text{ kJ/mol})$$
$$= \mathbf{28.05\ kJ}$$

The reaction leads to a significant increase in the disorder of the system, however, and is therefore favored by the entropy of reaction:

$$\Delta S^\circ = \Sigma S^\circ_{\text{products}} - \Sigma S^\circ_{\text{reactants}}$$
$$= (1 \text{ mol } NH_4{}^+ \times 113.4 \text{ J/mol-K} + 1 \text{ mol } NO_3{}^- \times 146.4 \text{ J/mol-K})$$
$$- (1 \text{ mol } NH_4NO_3 \times 151.08 \text{ J/mol-K})$$
$$= \mathbf{108.7\ J/K}$$

To decide whether $NH_4NO_3$ should dissolve in water at 25°C we have to compare the $\Delta H^\circ$ and $T\Delta S^\circ$ terms to see which is larger. Before we can do this, we have to convert the temperature from °C to kelvin:

$$T_K = 25°C + 273.15 = 298.15 \text{ K}$$

We also have to recognize that the units of $\Delta H^\circ$ for this reaction are kilojoules and the units of $\Delta S^\circ$ are joules per kelvin. At some point in this calculation, we therefore have to convert these quantities to a consistent set of units. Perhaps the easiest way of doing this is to convert $\Delta H^\circ$ to joules. We then multiply the entropy term by the absolute temperature and subtract this quantity from the enthalpy term:

$$\Delta G^\circ = \Delta H^\circ - T\Delta S^\circ$$
$$= 28{,}050 \text{ J} - (298.2 \text{ K} \times 108.7 \text{ J/K})$$
$$= 28{,}050 \text{ J} - 32{,}410 \text{ J} = \mathbf{-4360\ J}$$

At 25°C, the standard-state free energy for this reaction is negative because the entropy term at this temperature is larger than the enthalpy term:

$$\Delta G^\circ = \mathbf{-4.4\ kJ}$$

The reaction is therefore spontaneous at room temperature.

## 21.11 THE EFFECT OF TEMPERATURE ON THE FREE ENERGY OF A REACTION

The balance between the contributions from the enthalpy and entropy terms to the free energy of a reaction depends on the temperature at which the reaction is run.

### EXERCISE 21.7

Use the values of $\Delta H^\circ$ and $\Delta S^\circ$ calculated in Exercise 21.5 to predict whether the following reaction is spontaneous at 25°C:

$$N_2(g) + 3\ H_2(g) \rightleftharpoons 2\ NH_3(g)$$

**SOLUTION** According to Exercise 21.5, this reaction is favored by enthalpy but not by entropy:

$$\Delta H^\circ = -92.22 \text{ kJ} \qquad \text{(favorable)}$$
$$\Delta S^\circ = -198.75 \text{ J/K} \qquad \text{(unfavorable)}$$

Before we can compare these terms to see which is larger, we have to incorporate into our calculation the temperature at which the reaction is run:

$$T_K = 25°C + 273.15 = 298.15 \text{ K}$$

We then multiply the entropy of reaction by the absolute temperature and subtract the $T\Delta S°$ term from the $\Delta H°$ term:

$$\begin{aligned} \Delta G° &= \Delta H° - T\Delta S° \\ &= (-92{,}220 \text{ J}) - (298.15 \text{ K} \times -198.75 \text{ J/K}) \\ &= (-92{,}220 \text{ J}) + 59{,}260 \text{ J} \\ &= \mathbf{-32{,}960 \text{ J}} \end{aligned}$$

According to this calculation, the reaction should be spontaneous at 25°C:

$$\Delta G° = \mathbf{-32.96 \text{ kJ}}$$

What happens as we raise the temperature of the reaction? The equation used to define free energy suggests that the entropy term will become more important as the temperature increases:

$$\Delta G° = \Delta H° - T\Delta S°$$

Since the entropy term is unfavorable, the reaction should become less favorable as the temperature increases.

## EXERCISE 21.8

Predict whether the following reaction is still spontaneous at 500°C:

$$N_2(g) + 3\ H_2(g) \rightleftharpoons 2\ NH_3(g)$$

Assume that the values of $\Delta H°$ and $\Delta S°$ used in Exercise 21.7 are still valid at this temperature.

**SOLUTION** Before we can decide whether the reaction is still spontaneous we need to calculate the temperature on the kelvin scale:

$$T_K = 500°C + 273 = 773 \text{ K}$$

We then multiply the entropy term by this temperature and subtract the result from the value of $\Delta H°$ for the reaction:

$$\begin{aligned} \Delta G°_{773} &= \Delta H°_{298} - T\Delta S°_{298} \\ &= (-92{,}220 \text{ J}) - (773 \text{ K} \times -198.75 \text{ J/K}) \\ &= (-92{,}220 \text{ J}) - (-153{,}600 \text{ J}) = 61{,}380 \text{ J} \\ &= \mathbf{61.4 \text{ kJ}} \end{aligned}$$

Because the entropy term becomes larger as the temperature increases, the reaction changes from one which is favorable at low temperatures to one that is unfavorable at high temperatures.

## 21.12 BEWARE OF OVERSIMPLIFICATIONS

The calculation of $\Delta G°$ for the following reaction at 25°C in Exercise 21.7 was perfectly legitimate:

$$N_2(g) + 3\ H_2(g) \rightleftharpoons 2\ NH_3(g)$$

The enthalpy and entropy data used in this calculation were standard-state measurements made at 25°C, which can be legitimately used to predict the standard-state free energy of reaction at this temperature. The calculation of $\Delta G^\circ_{773}$ for this reaction at 500°C in Exercise 21.8, however, was based on the assumption that the enthalpy and entropy data measured at 25°C are still valid at 500°C.

This is a useful assumption, but it isn't always a valid one. Changes in $\Delta H^\circ$ and $\Delta S^\circ$ are often small over moderate temperature ranges. When the temperature range over which these data are extrapolated is large, as it is in this case, the results of the calculation should be taken with a grain of salt. They represent an estimate, not a prediction, of the magnitude of the effect of the change in temperature. In this case, the assumption that $\Delta H^\circ$ and $\Delta S^\circ$ are constant underestimates the change in the magnitude of $\Delta G$ by about 20%. $\Delta G^\circ_{773}$ for this reaction is 73.2 kJ, not 61.4 kJ.

## 21.13 STANDARD-STATE FREE ENERGIES OF REACTION

$\Delta G^\circ$ for a reaction also can be calculated from tabulated standard-state free energy data. Since there is no absolute zero on the free-energy scale, the easiest way to tabulate such data is in terms of **standard-state free energies of formation,** $\Delta G_f^\circ$. As might be expected, the standard-state free energy of formation of a substance is the difference between the free energy of the substance and the free energies of its elements in their thermodynamically most stable states at 1 atm, all measurements being made under standard-state conditions.

$\Delta G^\circ$ for any reaction can be calculated from $\Delta G_f^\circ$ for the reactants and products in the same way that $\Delta H^\circ$ for a reaction is calculated from $\Delta H_f^\circ$ data. The standard-state free energy of reaction is equal to the sum of the free energies of formation of the products of the reaction minus the sum of the free energies of formation of the reactants:

$$\Delta G^\circ = \Sigma \Delta G^\circ_{f\,\text{products}} - \Sigma \Delta G^\circ_{f\,\text{reactants}}$$

## 21.14 INTERPRETING STANDARD-STATE FREE ENERGY OF REACTION DATA

We are now ready to ask the obvious question: What does the value of $\Delta G^\circ$ tell us about the following reaction?

$$N_2(g) + 3\,H_2(g) \rightleftharpoons 2\,NH_3(g) \qquad \Delta G^\circ = -32.96 \text{ kJ}$$

By definition, the value of $\Delta G^\circ$ for a reaction measures the difference between the free energies of the reactants and products *when all components of the reaction are present at standard-state conditions.*

$\Delta G^\circ$ therefore describes this reaction only when all three components are present at 1 atm pressure:

$$\text{Standard state:} \quad Q_P = \frac{P^2_{NH_3}}{P_{N_2}P^3_{H_2}} = \frac{(1)^2}{(1)(1)^3} = 1$$

The *sign* of $\Delta G^\circ$ tells us the direction in which the reaction has to shift to come to equilibrium. The fact that $\Delta G^\circ$ is negative for this reaction at 25°C means that a system under standard-state conditions at this temperature would have to shift to the right, converting some of the reactants into products, before it can reach equilibrium. The *magnitude* of $\Delta G^\circ$ for a reaction tells how far the standard state is from equilibrium. The larger the value of $\Delta G^\circ$, the further the reaction has to go from the standard-state conditions to get to equilibrium.

Assume, for example, that we start with the following reaction under standard-state conditions, as shown in Figure 21.13:

$$N_2(g) + 3\,H_2(g) \rightleftharpoons 2\,NH_3(g)$$

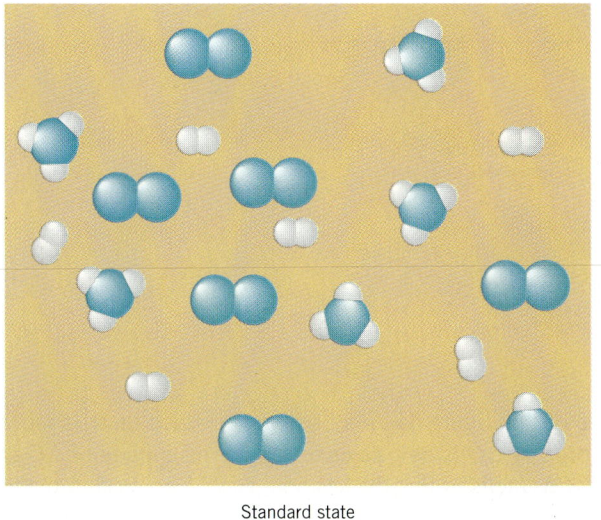


**Figure 21.13** At standard-state conditions, the amount of $N_2$, $H_2$, and $NH_3$ in this system is exactly the same. Thus $Q_P$ for this system is equal to 1.

The value of $\Delta G$ at that moment will be equal to the standard-state free energy for this reaction, $\Delta G°$:

$$\text{When } Q_P = 1: \qquad \Delta G = \Delta G°$$

As the reaction gradually shifts to the right, converting $N_2$ and $H_2$ into $NH_3$, the value of $\Delta G$ for the reaction will decrease. If we could find some way to harness the tendency of this reaction to come to equilibrium, we could get the reaction to do work. The free energy of a reaction at any one moment is therefore said to be a measure of the energy available to do work.

## 21.15 THE RELATIONSHIP BETWEEN FREE ENERGY AND EQUILIBRIUM CONSTANTS

When a reaction leaves the standard state because of a change in the ratio of the concentrations of the products to the reactants, we have to describe the system in terms of non-standard-state free energies of reaction. The difference between $\Delta G°$ and $\Delta G$ for a reaction is important. There is only one value of $\Delta G°$ for a reaction at a given temperature, but there is an infinite number of possible values of $\Delta G$.

Figure 21.14 shows the relationship between $\Delta G$ for the following reaction and the logarithm to the base $e$ of the reaction quotient:

$$N_2(g) + 3\,H_2(g) \rightleftharpoons 2\,NH_3(g)$$

Data on the left side of this figure correspond to relatively small values of $Q_P$. They therefore describe systems in which there is far more reactant than product. The sign of $\Delta G$ for these systems is negative and the magnitude of $\Delta G$ is large. The system is therefore relatively far from equilibrium and the reaction must shift to the right to reach equilibrium.

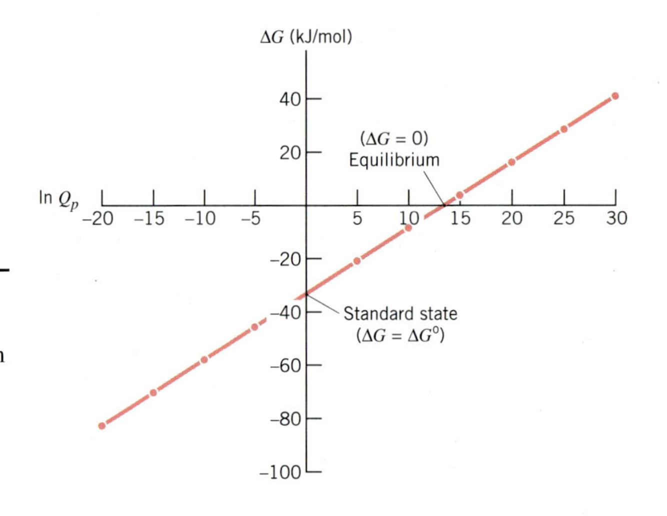


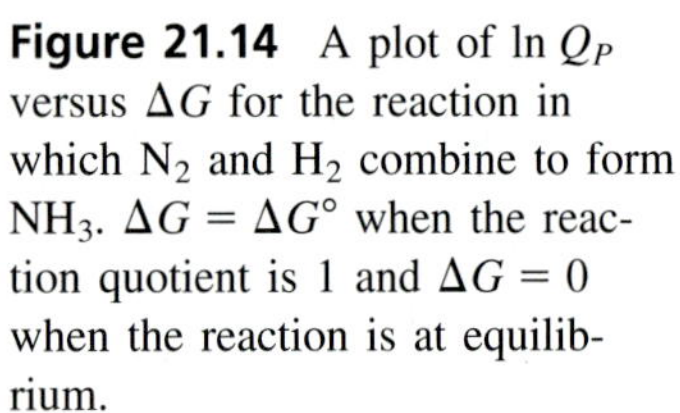
**Figure 21.14** A plot of ln $Q_P$ versus $\Delta G$ for the reaction in which $N_2$ and $H_2$ combine to form $NH_3$. $\Delta G = \Delta G°$ when the reaction quotient is 1 and $\Delta G = 0$ when the reaction is at equilibrium.

Data on the far right side of this figure describe systems in which there is more product than reactant. The sign of $\Delta G$ is now positive and the magnitude of $\Delta G$ is moderately large. The sign of $\Delta G$ tells us that the reaction would have to shift to the left to reach equilibrium. The magnitude of $\Delta G$ tells us that we don't have quite as far to go to reach equilibrium.

The points at which the straight line in Figure 21.14 cross the horizontal and vertical axes of this diagram are particularly important. The straight line crosses the vertical axis when the **reaction quotient,** $Q$ for the system is equal to 1. This point therefore describes the standard-state conditions, and the value of $\Delta G$ at this point is equal to the standard-state free energy of reaction, $\Delta G°$:

$$\text{When } Q = 1: \qquad \Delta G = \Delta G°$$

The point at which the straight line crosses the horizontal axis describes a system for which $\Delta G$ is equal to zero. Because there is no driving force behind the reaction, the system must be at equilibrium:

$$\text{When } Q = K: \qquad \Delta G = 0$$

The relationship between the free energy of reaction at any moment in time ($\Delta G$) and the standard-state free energy of reaction ($\Delta G°$) is described by the following equation:

$$\mathbf{\Delta G = \Delta G° + RT \ln Q}$$

In this equation, $R$ is the ideal gas constant in units of J/mol-K, $T$ is the temperature in kelvin, ln represents a logarithm to the base $e$, and $Q$ is the reaction quotient at that particular moment.

As we have seen, the driving force behind a chemical reaction is zero ($\Delta G = 0$) when the reaction is at equilibrium ($Q = K$):

$$0 = \Delta G° + RT \ln K$$

We can therefore solve this equation for the relationship between $\Delta G°$ and $K$:

$$\mathbf{\Delta G° = -RT \ln K}$$

This equation allows us to calculate the equilibrium constant for any reaction from the standard-state free energy of reaction, or vice versa.

The key to understanding the relationship between $\Delta G°$ and $K$ is recognizing that the magnitude of $\Delta G°$ tells us how far the standard-state is from equilibrium. The smaller the value of $\Delta G°$, the closer the standard-state is to equilibrium. The larger the value of $\Delta G°$, the further the reaction has to go to reach equilibrium. The relationship between $\Delta G°$ and the equilibrium constant $K$ for a chemical reaction is illustrated by the data in Table 21.4.

**TABLE 21.4 Values of $\Delta G°$ and $K$ for Common Reactions at 25°C**

| Reaction | $\Delta G°$ (kJ) | $K$ |
| --- | --- | --- |
| $2\,SO_3(g) \rightleftharpoons 2\,SO_2(g) + O_2(g)$ | 141.7 | $1.4 \times 10^{-25}$ |
| $H_2O(l) \rightleftharpoons H^+(aq) + OH^-(aq)$ | 79.9 | $1.0 \times 10^{-14}$ |
| $AgCl(s) \xrightarrow{H_2O} Ag^+(aq) + Cl^-(aq)$ | 55.6 | $1.8 \times 10^{-10}$ |
| $HOAc(aq) \xrightarrow{H_2O} H^+(aq) + OAc^-(aq)$ | 27.1 | $1.8 \times 10^{-5}$ |
| $N_2(g) + 3\,H_2(g) \rightleftharpoons 2\,NH_3(g)$ | −32.9 | $5.8 \times 10^{5}$ |
| $HCl(aq) \xrightarrow{H_2O} H^+(aq) + Cl^-(aq)$ | −34.2 | $1 \times 10^{6}$ |
| $Cu^{2+}(aq) + 4\,NH_3(aq) \rightleftharpoons Cu(NH_3)_4^{2+}(aq)$ | −76.0 | $2.1 \times 10^{13}$ |
| $Zn(s) + Cu^{2+}(aq) \longrightarrow Zn^{2+}(aq) + Cu(s)$ | −211.8 | $1.4 \times 10^{37}$ |

## EXERCISE 21.9

Use the value of $\Delta G°$ obtained in Exercise 21.7 to calculate the equilibrium constant for the following reaction at 25°C:

$$N_2(g) + 3\,H_2(g) \rightleftharpoons 2\,NH_3(g)$$

**SOLUTION** Exercise 21.7 gave the following value for $\Delta G°$ for this reaction at 25°C:

$$\Delta G° = -32.96 \text{ kJ}$$

We now turn to the relationship between $\Delta G°$ for a reaction and the equilibrium constant for the reaction:

$$\Delta G° = -RT \ln K$$

Solving for the natural log of the equilibrium constant gives the following equation:

$$\ln K = -\frac{\Delta G°}{RT}$$

Substituting the known value of $\Delta G°$, $R$, and $T$ into this equation gives the following result:

$$\ln K = -\frac{-32{,}960 \text{ J/mol}}{(8.314 \text{ J/mol-K})(298 \text{ K})} = 13.3$$

The equilibrium constant for this reaction at 25°C is therefore $6.0 \times 10^5$:

$$K = e^{13.3} = 6.0 \times 10^5$$

**Problem-Solving Strategy**

It is easy to make mistakes when handling the sign of the relationship between $\Delta G°$ and $K$. It is therefore a good idea to check the final answer to see whether it makes sense. $\Delta G°$ for the reaction in this exercise is negative. The reaction is therefore spontaneous, and the equilibrium should lie on the side of the products. The equilibrium constant therefore should be much larger than 1, which it is.

Sec. 16.10

The nature of the equilibrium constant obtained in Exercise 21.9 should be clarified. As we saw in Chapter 16, the equilibrium constant for this reaction can be expressed in two ways: $K_c$ and $K_p$. We can write equilibrium constant expressions in terms of the partial pressures of the reactants and products, or in terms of their concentrations in units of moles per liter:

$$K_p = \frac{P_{NH_3}^2}{P_{N_2}P_{H_2}^3} \qquad K_c = \frac{[NH_3]^2}{[N_2][H_2]^3}$$

For gas-phase reactions, such as the one in Exercise 21.9, the equilibrium constant obtained from $\Delta G°$ is based on the partial pressures of the gases ($K_p$). For reactions in solution, the equilibrium constant that comes from the calculation is based on concentrations ($K_c$).

## EXERCISE 21.10

Use the following standard-state free energy of formation data to calculate the acid–dissociation equilibrium constant ($K_a$) at 25°C for formic acid.

| Compound | $\Delta G_f°$ (kJ/mol) |
|---|---|
| $HCO_2H(aq)$ | −372.3 |
| $H^+(aq)$ | 0.00 |
| $HCO_2^-(aq)$ | −351.0 |

**SOLUTION** We can start by writing the equation that corresponds to the acid–dissociation equilibrium for formic acid:

$$HCO_2H(aq) \rightleftharpoons H^+(aq) + HCO_2^-(aq)$$

We then calculate the value of $\Delta G°$ for this reaction:

$$\begin{aligned}\Delta G° &= \Sigma\Delta G_{f\,\text{products}}° - \Sigma\Delta G_{f\,\text{reactants}}° \\ &= (1 \text{ mol } H^+ \times 0.00 \text{ kJ/mol} + 1 \text{ mol } HCO_2^- \times -351.0 \text{ kJ/mol}) \\ &\quad - (1 \text{ mol } HCO_2H \times -372.3 \text{ kJ/mol}) \\ &= \mathbf{21.3 \text{ kJ}}\end{aligned}$$

We now turn to the relationship between $\Delta G°$ and the equilibrium constant for the reaction:

$$\Delta G° = -RT \ln K$$

and solve for the natural logarithm of the equilibrium constant:

$$\ln K = -\frac{\Delta G°}{RT}$$

Substituting the known values of $\Delta G°$, $R$, and $T$ into this equation gives the following result:

$$\ln K = -\frac{21{,}300 \text{ J/mol}}{(8.314 \text{ J/mol-K})(298 \text{ K})} = \mathbf{-8.60}$$

We can now calculate the value of the equilibrium constant:

$$K = e^{-8.60} = \mathbf{1.8 \times 10^{-4}}$$

The value of $K_a$ obtained from this calculation agrees with the value for formic acid given in Table A.9, within experimental error.

**RESEARCH IN THE 90s**

## THE THERMODYNAMICS OF BIOLOGICAL SYSTEMS

The only way for living systems to cope with a universe that is driven toward maximum disorder is to find a constant source of free energy. These systems survive by extracting free energy from sunlight or from the *exergonic* ($\Delta G < 0$) reactions in which they oxidize carbohydrates, lipids, or proteins. They use this free energy to synthesize adenosine triphosphate (ATP) from adenosine diphosphate (ADP):

$$ADP^{3-}(aq) + PO_3{}^{3-}(aq) + 2\,H^+(aq) \rightleftharpoons ATP^{4-}(aq) + H_2O(l)$$

$$\Delta G° = 30.5 \text{ kJ/mol} \qquad (\text{at pH} = 7)$$

They then use the decomposition of ATP to ADP to drive the *endergonic* ($\Delta G > 0$) processes required to keep the organism alive. ATP is consumed, for example, when cells do mechanical work, when they transport ions and molecules across a membrane, or when they synthesize the complex molecules upon which life depends.

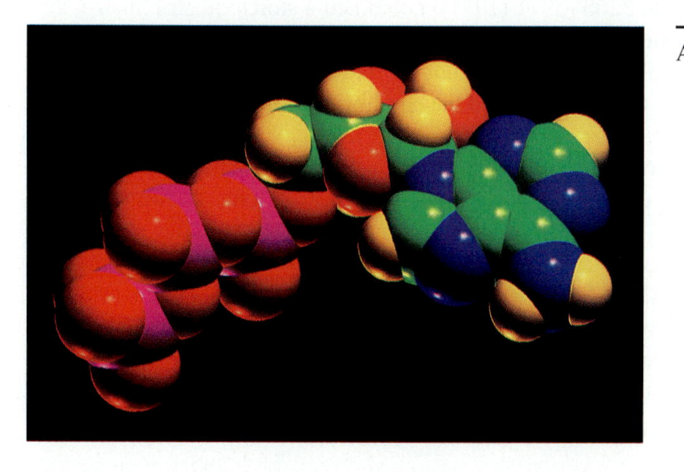

A computer model of the structure of ATP.

In theory, we should be able to calculate the equilibrium constant for the reaction between ADP and the phosphate ion to form ATP from the value of $\Delta G°$ for this reaction:

$$\Delta G° = -\text{RT} \ln K_c$$

We should then be able to calculate the equilibrium concentrations of the various components of this reaction from estimates of the initial concentrations of these ions. In practice, the results of these calculations would be misleading because this system is far more complex than it seems. There are, in fact, at least 12 simultaneous equilibria that have to be considered to build an adequate model of this system.

Both ATP and ADP carry a net negative charge and therefore act as a base toward water:

$$ATP^{4-}(aq) + H_2O(l) \rightleftharpoons HATP^{3-}(aq) + OH^-(aq)$$

$$HATP^{3-}(aq) + H_2O(l) \rightleftharpoons H_2ATP^{2-}(aq) + OH^-(aq)$$

$$ADP^{3-}(aq) + H_2O(l) \rightleftharpoons HADP^{2-}(aq) + OH^-(aq)$$

$$HADP^{2-}(aq) + H_2O(l) \rightleftharpoons H_2ADP^-(aq) + OH^-(aq)$$

The $PO_4^{3-}$ ion is also a base toward water:

$$PO_4^{3-}(aq) + H_2O(l) \rightleftharpoons HPO_4^{2-}(aq) + OH^-(aq)$$
$$HPO_4^{2-}(aq) + H_2O(l) \rightleftharpoons H_2PO_4^-(aq) + OH^-(aq)$$

The system is complicated by the fact that the $ATP^{4-}$, $ADP^{3-}$, $HATP^{3-}$, $HADP^{2-}$, and $HPO_4^{2-}$ ions form complexes with the $Mg^{2+}$ ions present in living systems:

$$ATP^{4-}(aq) + Mg^{2+}(aq) \rightleftharpoons MgATP^{2-}(aq)$$
$$ADP^{3-}(aq) + Mg^{2+}(aq) \rightleftharpoons MgADP^-(aq)$$
$$HATP^{3-}(aq) + Mg^{2+}(aq) \rightleftharpoons MgHATP^-(aq)$$
$$HADP^{2-}(aq) + Mg^{2+}(aq) \rightleftharpoons MgHADP(aq)$$
$$HPO_4^{2-}(aq) + Mg^{2+}(aq) \rightleftharpoons MgHPO_4(aq)$$

These reactions involve a total of 17 species that are important in the pH range between 3 and 10: $H^+$, $OH^-$, $H_2O$, $Mg^{2+}$, $ATP^{4-}$, $ADP^{3-}$, $HATP^{3-}$, $HADP^{2-}$, $H_2ATP^{2-}$, $H_2ADP^-$, $MgHPO_4$, $MgATP^{2-}$, $MgADP^-$, $MgHATP^-$, $MgHADP$, $HPO_4^{2-}$, and $H_2PO_4^-$. A recent paper [R. A. Alberty, *Proceedings of the National Academy of Science,* **88,** 3268–3271 (1991)] combined a stoichiometric matrix that described the contribution of each of these 17 species to the 12 reactions with values of $\Delta G°$ for these reactions to provide enough free energy data to calculate the equilibrium constant of each reaction and the equilibrium concentrations for each of the 17 components of these reactions.

The data necessary to do such calculations are seldom available for biochemical systems because of the complexity of these systems. Research has shown, however, that once values of $\Delta G_f^\circ$ are known for enough compounds, it is possible to calculate the effect of various groups in the structure of each of these compounds on the free energy of formation [M. L. Mavrovouniotis, *Biotechnology and Bioengineering,* **36,** 1070–82 (1990)]. When this approach was used to predict the value of $\Delta G°$ for the following reaction, a value was obtained that agreed with the measured value for $\Delta G°$ within experimental error:

$$ADP^{3-}(aq) + PO_3^{3-}(aq) + 2\,H^+(aq) \rightleftharpoons ATP^{4-}(aq) + H_2O(l)$$

Research throughout the coming decades will see an ever-increasing depth of understanding of the chemical reactions that drive biological systems, some of which will come from advances in measurements and calculations of biothermodynamic data.

## 21.16 THE TEMPERATURE DEPENDENCE OF EQUILIBRIUM CONSTANTS

When equilibrium constants were introduced in Chapter 16, we noted that they are not strictly constant because they change with temperature. We are now ready to understand why.

The standard-state free energy of reaction is a measure of how far the standard-state is from equilibrium:

$$\Delta G° = -RT \ln K$$

Sec. 16.11

But the magnitude of $\Delta G°$ depends on the temperature of the reaction:

$$\Delta G° = \Delta H° - T\Delta S°$$

As a result, the equilibrium constant must depend on the temperature of the reaction.

A good example of this phenomenon is the reaction in which $NO_2$ dimerizes to form $N_2O_4$:

$$2\,NO_2(g) \rightleftharpoons N_2O_4(g)$$

In Exercise 21.3 we noted that this reaction is favored by enthalpy because it forms a new bond, which makes the system more stable. We also noted that the reaction is not favored by entropy because it leads to a decrease in the disorder of the system.

$NO_2$ is a brown gas and $N_2O_4$ is colorless. We can therefore monitor the extent to which $NO_2$ dimerizes to form $N_2O_4$ by examining the intensity of the brown color in a sealed tube of this gas. What should happen to the equilibrium between $NO_2$ and $N_2O_4$ as the temperature is lowered?

For the sake of argument, let's assume that there is no significant change in either $\Delta H°$ or $\Delta S°$ as the system is cooled. The contribution to the free energy of the reaction from the enthalpy term is therefore constant, but the contribution from the entropy term becomes smaller as the temperature is lowered:

$$\Delta G° = \Delta H° - T\Delta S°$$

As the tube is cooled, and the entropy term becomes less important, the net effect is a shift in the equilibrium toward the right. Figure 21.15 shows what happens to the intensity of the brown color when a sealed tube containing $NO_2$ gas is immersed in liquid nitrogen. There is a drastic decrease in the amount of $NO_2$ in the tube as it is cooled to −196°C.

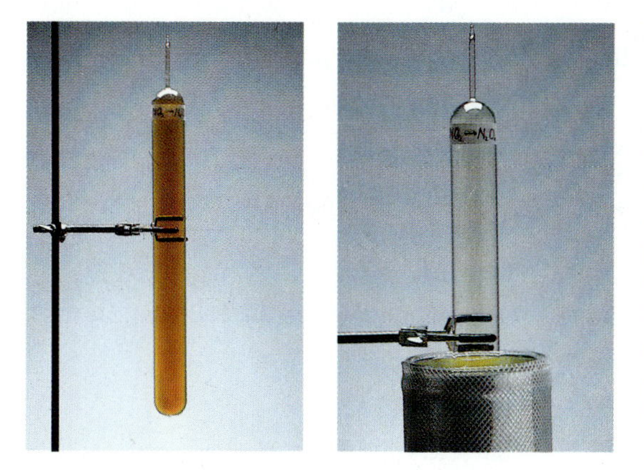

**Figure 21.15** The equilibrium between $NO_2$ and $N_2O_4$ is influenced by two factors. $\Delta H°$ for this reaction favors $N_2O_4$. But $\Delta S°$ favors $NO_2$, which is more disordered. The contribution to $\Delta G°$ from the enthalpy term is more or less constant regardless of the temperature of the system. But the contribution from the entropy term is very sensitive to temperature. At low temperatures, the enthalpy term is larger than the entropy term, and the equilibrium strongly favors $N_2O_4$. As the temperature increases, the equilibrium shifts toward $NO_2$. Thus, the characteristic brown color of $NO_2$ gas disappears as a tube containing $NO_2$ is moved into a liquid nitrogen bath (−196°C).

## EXERCISE 21.11

Use values of $\Delta H°$ and $\Delta S°$ for the following reaction at 25°C to estimate the equilibrium constant for this reaction at the temperature of boiling water (100°C), ice (0°C), a dry ice–acetone bath (−78°C), and liquid nitrogen (−196°C):

$$2\,NO_2(g) \rightleftharpoons N_2O_4(g)$$

**SOLUTION** Table A.15 contains the following information for this reaction:

| Compound | $\Delta H_f°$ (kJ/mol) | $S°$ (J/mol-K) |
|---|---|---|
| $NO_2(g)$ | 33.18 | 240.06 |
| $N_2O_4(g)$ | 9.16 | 304.29 |

According to these data, the reaction is favored by enthalpy:

$$\Delta H° = (1\text{ mol } N_2O_4 \times 9.16\text{ kJ/mol}) - (2\text{ mol } NO_2 \times 33.18\text{ kJ/mol})$$
$$= \mathbf{-57.20\text{ kJ}}$$

But it is not favored by entropy:

$$\Delta S° = (1 \text{ mol } N_2O_4 \times 304.29 \text{ J/mol-K}) - (2 \text{ mol } NO_2 \times 240.06 \text{ J/mol-K})$$
$$= \mathbf{-175.83 \text{ J/K}}$$

If we assume that these values of $\Delta H°$ and $\Delta S°$ are still valid at 100°C, the value of $\Delta G°$ at this temperature is 8400 J:

$$\Delta G°_{373} = \Delta H°_{298} - T\Delta S°_{298}$$
$$= -57{,}200 \text{ J} - (373 \text{ K})(-175.83 \text{ J/K}) = \mathbf{8400 \text{ J}}$$

Repeating this calculation at the other temperatures gives the following results:

| | |
|---|---|
| 100°C: | $\Delta G° = 8.4$ kJ |
| 0°C: | $\Delta G° = -9.2$ kJ |
| −78°C: | $\Delta G° = -22.9$ kJ |
| −196°C: | $\Delta G° = -43.7$ kJ |

We now write the equation for the relationship between $\Delta G°$ and the equilibrium constant for the reaction:

$$\Delta G° = -RT \ln K$$

We then solve this equation for the natural logarithm of the equilibrium constant:

$$\ln K = -\frac{\Delta G°}{RT}$$

Let's start by calculating the value of $\ln K_p$ when the reaction is at 100°C:

$$\ln K = -\frac{8{,}400 \text{ J/mol}}{(8.314 \text{ J/mol-K})(373 \text{ K})} = \mathbf{-2.71}$$

The equilibrium constant at this temperature is therefore 0.067:

$$K = e^{-2.71} = \mathbf{0.067}$$

Repeating this calculation at the other temperatures gives the following results:

| | |
|---|---|
| 100°C: | $K_p = 0.067$ |
| 0°C: | $K_p = 58$ |
| −78°C: | $K_p = 1.4 \times 10^6$ |
| −196°C: | $K_p = 3.8 \times 10^{29}$ |

At 100°C, the unfavorable entropy term is relatively important, and the equilibrium lies on the side of $NO_2$. As the reaction is cooled, the entropy term becomes less important, and the equilibrium shifts toward $N_2O_4$. At the temperature of liquid nitrogen, essentially all of the $NO_2$ condenses to form $N_2O_4$.

## 21.17 THE RELATIONSHIP BETWEEN FREE ENERGY AND CELL POTENTIALS

The value of $\Delta G$ for a reaction at any moment in time tells us two things. The sign of $\Delta G$ tells us in what direction the reaction has to shift to reach equilibrium. The magnitude of $\Delta G$ tells us how far the reaction is from equilibrium at that moment.

In Chapter 20, we saw that the potential of an electrochemical cell is a measure of how far an oxidation–reduction reaction is from equilibrium. In that chapter, we used the Nernst equation to describe the relationship between the cell potential at any one moment and the standard-state cell potential:

Sec. 20.10

$$E = E° - \frac{RT}{nF} \ln Q$$

Let's rearrange this equation as follows:

$$nFE = nFE° - RT \ln Q$$

We can now compare it with the equation used to describe the relationship between the free energy of reaction at any one moment and the standard-state free energy of reaction:

$$\Delta G = \Delta G° + RT \ln Q$$

These equations are similar because the Nernst equation is a special case of the more general free-energy relationship. We can convert one of these equations to the other by taking advantage of the following relationships between the free energy of a reaction and the cell potential of the reaction when it is run as an electrochemical cell:

$$\Delta G = -nFE \qquad \Delta G° = -nFE°$$

**EXERCISE 21.12**

Use the relationship between $\Delta G°$ and $E°$ for an electrochemical reaction to derive the relationship between the standard-state cell potential and the equilibrium constant for the reaction.

**SOLUTION** We can start with the equation that describes the relationship between the standard-state free energy of reaction and the standard-state cell potential for an electrochemical cell:

$$\Delta G° = -nFE°$$

We can then write the relationship between the standard-state free energy of reaction and the equilibrium constant for the reaction:

$$\Delta G° = -RT \ln K$$

Substituting one of these equations into the other gives:

$$-nFE° = -RT \ln K$$

We now multiply both sides of this equation by $-1$:

$$nFE° = RT \ln K$$

Solving for the ln $K$ term gives the equation used in Chapter 20 to calculate the equilibrium constants for electrochemical reactions from their standard-state-state cell potentials:

$$\ln K = \frac{nFE°}{RT}$$

## SUMMARY

**1.** The first law of thermodynamics describes the relationship between heat, work, and the internal energy of a system. According to the first law, the change in the internal energy of the system is equal to the sum of the heat transferred from the system to its surroundings (or vice versa) and the work done by the system on its surroundings (or vice versa):

$$\Delta E = q + w$$

**2.** When a reaction is run in a sealed container at constant volume, the change in the internal energy of the system is equal to the heat given off or absorbed during the reaction:

$$\Delta E = q_V$$

**3.** When the reaction is run in an open container at constant pressure, the heat of reaction is equal to the change in the enthalpy of the system:

$$\Delta H = q_P$$

**4.** Two factors determine whether a reaction will occur: the change in the enthalpy and the change in the entropy of the system that occur during the reaction. Reactions that give off energy tend to occur spontaneously. Reactions that lead to an increase in the disorder of the system—and therefore an increase in its entropy—also tend to occur spontaneously.

**5.** The change in the Gibbs free energy of a system during a reaction is a measure of the balance between the two forces that drive the reaction. The Gibbs free energy is the sum of the enthalpy of the system plus the product of the absolute temperature times the entropy of the system:

$$G = H - TS$$

**6.** The *sign* of $\Delta G$ for a reaction indicates the direction in which the reaction has to shift to reach equilibrium. When $\Delta G$ is negative, the reaction has to shift to the right—toward the products—to reach equilibrium. When $\Delta G$ is positive, the reaction has to shift to the left, toward the reactants. The *magnitude* of $\Delta G$ indicates how far the reaction is from equilibrium at any moment in time.

**7.** There are an infinite number of possible values of $\Delta G$ for a reaction but only one value of $\Delta G^\circ$ at a given temperature, which describes the results of measurements made under standard-state conditions.

**8.** The value of $\Delta G^\circ$ for a reaction indicates how far the standard-state conditions are from equilibrium. As a reaction moves from standard-state conditions toward equilibrium, the value of $\Delta G$ for the reaction gradually decreases until it reaches zero when the reaction is at equilibrium. The relationship between the free energy of reaction and the standard-state free energy of reaction is given by the following equation:

$$\Delta G = \Delta G^\circ + RT \ln Q$$

**9.** When the reaction is at equilibrium, $\Delta G$ is zero and the reaction quotient is equal to the equilibrium constant:

$$0 = \Delta G^\circ + RT \ln K$$

Solving for $\Delta G^\circ$ gives the following equation:

$$\Delta G^\circ = -RT \ln K$$

The equilibrium constant for a reaction therefore can be calculated from the standard-state free energy of reaction (or vice versa).

**10.** The value of $\Delta G^\circ$ for a reaction depends on the temperature of the reaction:

$$\Delta G^\circ = \Delta H^\circ - T\Delta S^\circ$$

Since the equilibrium constant depends on the value of $\Delta G^\circ$, the equilibrium constant also depends on the temperature at which the reaction is run.

**11.** The value of $\Delta G^\circ$ for an oxidation–reduction reaction is proportional to the overall cell potential for the reaction:

$$\Delta G^\circ = -nFE^\circ$$

The standard-state free energy of reaction ($\Delta G^\circ$) for an oxidation–reduction therefore can be calculated from the standard-state cell potential ($E^\circ$), or vice versa.

## KEY TERMS

**Boundary** (p. 766)
**Chemical thermodynamics** (p. 766)
**Disorder** (p. 774)
**Endergonic** (p. 781)
**Endothermic** (p. 781)
**Enthalpy ($H$)** (p. 770)
**Entropy ($S$)** (p. 772)
**Exergonic** (p. 781)
**Exothermic** (p. 781)
**First law of thermodynamics** (p. 766)
**Gibbs free energy ($G$)** (p. 780)
**Internal energy ($E$)** (p. 766)
**Isolated system** (p. 773)
**Reaction quotient ($Q$)** (p. 786)
**Second law of thermodynamics** (p. 766)
**Spontaneous** (p. 772)
**Standard-state entropy of reaction ($S^\circ$)** (p. 777)
**Standard-state free energy of formation ($\Delta G_f^\circ$)** (p. 784)

**Standard-state free energy of reaction ($\Delta G°$)** (p. 780)
**Standard-state conditions** (p. 777)
**State function** (p. 767)
**Surroundings** (p. 766)
**System** (p. 766)
**Thermodynamics** (p. 766)
**Third law of thermodynamics** (p. 766)

## KEY EQUATIONS

$E_{sys} = \frac{3}{2}RT$

$\Delta E_{univ} = \Delta E_{sys} + \Delta E_{surr} = 0$

$\Delta E_{sys} = q + w$

$w = -P\Delta V$

$H = E + PV$

$S = k \ln W$

$\Delta G° = \Delta H° - T\Delta S°$

$\Delta G° = \Sigma \Delta G°_{f\,products} - \Sigma \Delta G°_{f\,reactants}$

$\Delta G = \Delta G° + RT \ln Q$

$\Delta G° = -RT \ln K$

## PROBLEMS

### The First Law of Thermodynamics

**21-1** Define the following terms: *heat, work, internal energy, enthalpy, entropy,* and *free energy.*

**21-2** Describe the first law of thermodynamics in terms of both a verbal definition and a mathematical equation.

**21-3** Describe what happens to the internal energy of the system when an exothermic reaction is run under conditions of constant volume.

**21-4** Calculate the heat absorbed by 3 moles of an ideal gas when the temperature of the gas increases by 3.5°C.

**21-5** Calculate the temperature after 3 moles of an ideal gas at 25°C expand from a volume of 1 L to a volume of 5 L at a constant pressure of 1 atm if no heat is gained or lost during this process.

**21-6** Calculate the change in the internal energy of a system that does 1246 J of work on its surroundings and at the same time absorbs 456 J of heat from its surroundings. Predict whether the temperature of the system will increase or decrease when this happens.

**21-7** Calculate the amount of heat that must be added to an ideal gas that expands from 1.25 L to 8.00 L against a constant pressure of 5.00 atm in order to keep the temperature of the gas constant.

**21-8** Work of expansion can be expressed in units of liter-atmospheres (L-atm). A piston with a diameter of 25.0 cm that moves 15.0 cm under a constant pressure of 9.00 atm does 66.3 L-atm worth of work. Use the two values of the ideal gas constant ($R$ = 0.08206 L-atm/mol-K and $R$ = 8.314 J/mol-K) to convert 66.3 L-atm of work into joules.

### Enthalpy Versus Internal Energy

**21-9** Describe the difference between $\Delta E$ and $\Delta H$ for a chemical reaction.

**21-10** Under what conditions is the heat given off or absorbed by a reaction equal to the change in the internal energy of the system? Under what conditions is it equal to the enthalpy of reaction?

**21-11** For which of the following reactions is $\Delta E$ roughly equal to $\Delta H$?

(a) $2\,H_2(g) + O_2(g) \rightleftharpoons 2\,H_2O(g)$

(b) $Zn(s) + Cu^{2+}(aq) \rightleftharpoons Zn^{2+}(aq) + Cu(s)$

(c) $Zn(s) + 2\,H^+(aq) \rightleftharpoons Zn^{2+}(aq) + H_2(g)$

(d) $NH_3(g) + HCl(g) \rightleftharpoons NH_4Cl(s)$

(e) $HCl(aq) + NaOAc(aq) \rightleftharpoons HOAc(aq) + NaCl(aq)$

**21-12** Explain why there is only one value of $\Delta H°$ for a reaction at 25°C, but many values of $\Delta H$.

### Entropy as a Measure of Disorder

**21-13** Describe the relationship between the entropy of a system and the number of equivalent ways in which the state of the system can be described.

**21-14** What happens to the ability of a poker hand to win a game of poker as the entropy of the hand becomes larger?

**21-15** At first glance, the number of possible poker hands might be expected to be $52 \times 51 \times 50 \times 49 \times 48$, or 311,875,200. Why is the actual number of different hands only 2,598,960, which is 5! times smaller?

**21-16** Give examples of natural processes that spontaneously lead to an increase in the entropy of the system.

### Entropy and the Second Law of Thermodynamics

**21-17** Which of the following reactions leads to an increase in the entropy of the system?

(a) $H_2(g) + Cl_2(g) \rightleftharpoons 2\,HCl(g)$

(b) $2\,NO_2(g) \rightleftharpoons N_2O_4(g)$

(c) $CaO(s) + CO_2(g) \rightleftharpoons CaCO_3(s)$

(d) $2\,H_2(g) + O_2(g) \rightleftharpoons 2\,H_2O(g)$

(e) $2\,NH_3(g) \rightleftharpoons N_2(g) + 3\,H_2(g)$

**21-18** In which of the following processes is $\Delta S^\circ$ negative?

(a) $2\ H_2O_2(aq) \rightleftharpoons 2\ H_2O(l) + O_2(g)$

(b) $CO_2(s) \rightleftharpoons CO_2(g)$

(c) $H_2O(l) \rightleftharpoons H_2O(g)$

(d) $4\ Al(s) + 3\ O_2(g) \longrightarrow 2\ Al_2O_3(s)$

(e) $NaCl(s) \xrightleftharpoons{H_2O} Na^+(aq) + Cl^-(aq)$

**21-19** In which of the following processes is $\Delta S^\circ$ positive?

(a) $2\ NO(g) + Cl_2(g) \rightleftharpoons 2\ NOCl(g)$

(b) $NaCl(s) \rightleftharpoons NaCl(l)$

(c) $3\ O_2(g) \rightleftharpoons 2\ O_3(g)$

(d) $C_2H_4(g) + H_2(g) \longrightarrow C_2H_6(g)$

**21-20** Which of the following processes should have the most positive value of $\Delta S^\circ$?

(a) $N_2(g) + O_2(g) \rightleftharpoons 2\ NO(g)$

(b) $3\ C_2H_2(g) \longrightarrow C_6H_6(l)$

(c) $H_2O(l) \rightleftharpoons H_2O(g)$

(d) $4\ Al(s) + 3\ O_2(g) \longrightarrow 2\ Al_2O_3(s)$

**21-21** Which of the following processes should have the most positive value of $\Delta S^\circ$?

(a) $H_2O(l) \rightleftharpoons H_2O(s)$

(b) $NaNO_3(s) \xrightleftharpoons{H_2O} Na^+(aq) + NO_3^-(aq)$

(c) $2\ HCl(g) \rightleftharpoons H_2(g) + Cl_2(g)$

(d) $2\ H_2(g) + O_2(g) \longrightarrow 2\ H_2O(g)$

### The Third Law of Thermodynamics

**21-22** Explain why the standard-state entropy of a solid increases with temperature.

**21-23** Explain why the standard-state entropy of a compound increases as the compound is converted from a solid to a liquid, or from a liquid to a gas.

### Standard-State Entropies of Reaction

**21-24** One of the key steps in transforming coal into a liquid fuel involves reducing carbon monoxide with $H_2$ gas to form methanol. Calculate $\Delta S^\circ$ for the reaction to determine whether a favorable change in the entropy of the system might be a driving force behind this reaction. Predict the effect of an increase in temperature on the equilibrium constant for this reaction:

$$CO(g) + 2\ H_2(g) \rightleftharpoons CH_3OH(l)$$

**21-25** Tetraphosphorus decaoxide is often used as a dehydrating agent because of its tendency to pick up water. Calculate $\Delta S^\circ$ for the following reaction to determine whether entropy might be the primary driving force behind this reaction:

$$P_4O_{10}(s) + 6\ H_2O(l) \rightleftharpoons 4\ H_3PO_4(aq)$$

**21-26** Calculate $\Delta S^\circ$ for the following reaction. Comment on both the sign and the magnitude of $\Delta S^\circ$. Assume that $S^\circ$ for ammonium nitrite is 138.1 J/mol-K.

$$NH_4NO_2(s) \longrightarrow N_2(g) + 2\ H_2O(g)$$

**21-27** Show how $\Delta S^\circ$ for the following reaction can be used to show that $O_2$ is more disordered than $O_3$, even though $S^\circ$ for $O_3$ is larger than $S^\circ$ for $O_2$:

$$3\ O_2(g) \rightleftharpoons 2\ O_3(g)$$

**21-28** Compare $S^\circ$ for the various forms of elemental phosphorus. Explain why $S^\circ$ increases in the order $P < P_2 < P_4$, even though the system becomes more ordered.

| Compound | $S^\circ$ (J/mol-K) |
|---|---|
| $P(s)$ | 41.09 |
| $P(g)$ | 163.193 |
| $P_2(g)$ | 218.129 |
| $P_4(g)$ | 279.98 |

**21-29** Look up $S^\circ$ data for aqueous solutions of the following ions: $H^+$, $Li^+$, $Na^+$, $K^+$, $Fe^{2+}$, $Fe^{3+}$, $Mg^{2+}$, $Ba^{2+}$, $Ca^{2+}$, and $Al^{3+}$. Use these data to determine two factors that lead to an increase in the standard-state entropy of an aqueous solution of an ion.

**21-30** The $S^\circ$ data for ions in aqueous solutions are confusing. For all other substances, the zero point on the entropy scale is a perfect crystal at 0 K. For ionic solutions, the zero point is defined as the entropy of a 1 $M$ $H^+$ ion solution. Explain why the existence of two different reference points for the entropy scale does not influence the value of $\Delta S^\circ$ for the following reaction, which involves solids and gases as well as aqueous solutions:

$$Zn(s) + 2\ H^+(aq) \rightleftharpoons Zn^{2+}(aq) + H_2(g)$$

**21-31** Explain why enthalpy and free-energy data are given as enthalpies of formation ($\Delta H_f^\circ$) and free energies of formation ($\Delta G_f^\circ$), but entropy data are not given as entropies of formation ($\Delta S_f^\circ$).

### Gibbs Free Energy

**21-32** Explain the difference between $\Delta G$ and $\Delta G^\circ$ for a chemical reaction.

**21-33** What does it mean when $\Delta G$ for a reaction is zero? Can $\Delta G^\circ$ for a reaction be zero?

**21-34** Which of the following combinations of $\Delta H^\circ$ and $\Delta S^\circ$ always indicates a spontaneous reaction?

(a) $\Delta H^\circ > 0$, $\Delta S^\circ < 0$ (b) $\Delta H^\circ < 0$, $\Delta S^\circ > 0$

(c) $\Delta H^\circ > 0$, $\Delta S^\circ > 0$ (d) $\Delta H^\circ < 0$, $\Delta S^\circ < 0$

(e) $\Delta H^\circ = 0$, $\Delta S^\circ = 0$

**21-35** Predict the signs of $\Delta H^\circ$ and $\Delta S^\circ$ for the following reactions without referring to a table of thermodynamic data, and explain your predictions:

(a) $2\ H_2(g) + O_2(g) \rightleftharpoons 2\ H_2O(g)$

(b) $2\ Na(s) + Cl_2(g) \longrightarrow 2\ NaCl(s)$

(c) $N_2(g) + 3\ H_2(g) \rightleftharpoons 2\ NH_3(g)$

(d) $2\ Cu(NO_3)_2(s) \longrightarrow 2\ CuO(s) + 4\ NO_2(g) + O_2(g)$

**21-36** Predict the signs of $\Delta H^\circ$ and $\Delta S^\circ$ for the following reaction without referring to a table of thermodynamic data, and explain your predictions:

$$NH_3(g) \overset{H_2O}{\rightleftharpoons} NH_3(aq)$$

Explain why the odor of $NH_3$ gas that collects above this solution becomes more intense as the temperature increases.

**21-37** Limestone decomposes to form lime and carbon dioxide when it is heated. Calculate $\Delta H°$ and $\Delta S°$ for the decomposition of limestone to identify the driving force behind this reaction:

$$CaCO_3(s) \rightleftharpoons CaO(s) + CO_2(g)$$

**21-38** Calculate $\Delta H°$ and $\Delta S°$ for the thermite reaction to identify the driving force behind this reaction:

$$Fe_2O_3(s) + 2\ Al(s) \longrightarrow Al_2O_3(s) + 2\ Fe(s)$$

**21-39** The first step in extracting iron ore from pyrite, $FeS_2$, involves roasting the ore in the presence of oxygen to form iron(III) oxide and sulfur dioxide:

$$4\ FeS_2(s) + 11\ O_2(g) \longrightarrow 2\ Fe_2O_3(s) + 8\ SO_2(g)$$

Calculate $\Delta H°$ and $\Delta S°$ for this reaction and determine the driving force behind the reaction.

**21-40** Calculate $\Delta H°$ and $\Delta S°$ for the following reaction. What is the major driving force behind this reaction?

$$2\ KMnO_4(s) + 5\ H_2O_2(aq) + 6\ H^+(aq) \longrightarrow 2\ K^+(aq) + 2\ Mn^{2+}(aq) + 5\ O_2(g) + 8\ H_2O(l)$$

**21-41** It is possible to make hydrogen chloride by reacting phosphorus pentachloride with water and boiling the HCl out of this solution. Calculate $\Delta H°$ and $\Delta S°$ for this reaction. What is the driving force behind this reaction: the enthalpy of reaction, the entropy of reaction, or Le Châtelier's principle?

$$PCl_5(g) + 4\ H_2O(l) \longrightarrow H_3PO_4(aq) + 5\ HCl(g)$$

**21-42** Which of the following processes is spontaneous at 25°C?

(a) $CH_3OH(l) \rightleftharpoons CH_3OH(g)$
(b) $CH_3OH(l) \rightleftharpoons HCHO(g) + H_2(g)$
(c) $2\ CH_3OH(l) \longrightarrow 2\ CH_4(g) + O_2(g)$
(d) $CH_3OH(l) \rightleftharpoons CO(g) + 2\ H_2(g)$

**21-43** Calculate $\Delta H°$ and $\Delta S°$ for the following reaction. Explain why it is a mistake to use water to put out a fire that contains white-hot iron metal.

$$3\ Fe(s) + 4\ H_2O(l) \longrightarrow Fe_3O_4(s) + 4\ H_2(g)$$

**21-44** Use thermodynamic data to explain why sodium reacts with water, but silver metal does not:

$$2\ Na(s) + 2\ H_2O(l) \rightleftharpoons 2\ Na^+(aq) + 2\ OH^-(aq) + H_2(g)$$
$$2\ Ag(s) + 2\ H_2O(l) \not\longrightarrow 2\ Ag^+(aq) + 2\ OH^-(aq) + H_2(g)$$

**21-45** Use thermodynamic data to explain why zinc reacts with 1 *M* acid, but not with water:

$$Zn(s) + 2\ H^+(aq) \rightleftharpoons Zn^{2+}(aq) + H_2(g)$$
$$Zn(s) + 2\ H_2O(l) \not\longrightarrow Zn^{2+}(aq) + 2\ OH^-(aq) + H_2(g)$$

**21-46** Calculate $\Delta H°$ and $\Delta S°$ for the following reaction. Which of these factors explains why ammonium nitrate is a potential explosive?

$$2\ NH_4NO_3(s) \longrightarrow 2\ N_2(g) + O_2(g) + 4\ H_2O(g)$$

**21-47** Silane, $SiH_4$, decomposes to form elemental silicon and hydrogen. Calculate $\Delta H°$ and $\Delta S°$ for this reaction and predict the effect on the reaction of an increase in the temperature at which it is run:

$$SiH_4(s) \longrightarrow Si(s) + 2\ H_2(g)$$

**21-48** For which of the following reactions would you expect $\Delta H°$ and $\Delta G°$ to be about the same?

(a) $4\ Fe(s) + 3\ O_2(g) \longrightarrow 2\ Fe_2O_3(s)$
(b) $2\ Na(s) + 2\ H_2O(l) \longrightarrow 2\ Na^+(aq) + 2\ OH^-(aq) + H_2(g)$
(c) $Fe_2O_3(s) + 2\ Al(s) \longrightarrow Al_2O_3(s) + 2\ Fe(s)$
(d) $N_2O_4(g) \rightleftharpoons 2\ NO_2(g)$
(e) $CaC_2(s) + 2\ H_2O(l) \rightleftharpoons Ca^{2+}(aq) + 2\ OH^-(aq) + C_2H_2(g)$

## The Effect of Temperature on the Free Energy of Chemical Reactions

**21-49** For which of the following reactions will conditions change from unfavorable to favorable as the temperature increases?

(a) $2\ CO(g) + O_2(g) \rightleftharpoons 2\ CO_2(g)$
$\Delta H° = -565.97$ kJ/mol, $\Delta S° = -173.00$ J/mol-K
(b) $2\ H_2O(g) \rightleftharpoons 2\ H_2(g) + O_2(g)$
$\Delta H° = 483.64$ kJ/mol, $\Delta S° = 90.01$ J/mol-K
(c) $2\ N_2O(g) \rightleftharpoons 2\ N_2(g) + O_2(g)$
$\Delta H° = -164.1$ kJ/mol, $\Delta S° = 148.66$ J/mol-K
(d) $PbCl_2(s) \overset{H_2O}{\rightleftharpoons} Pb^{2+}(aq) + 2\ Cl^-(aq)$
$\Delta H° = 23.39$ kJ/mol, $\Delta S° = -12.5$ J/mol-K

**21-50** Explain why the equilibrium constant for the following reaction decreases as the temperature increases:

$$N_2(g) + 3\ H_2(g) \rightleftharpoons 2\ NH_3(g)$$

**21-51** Which of the following diagrams best describes the relationship between $\Delta G°$ and temperature for the following reaction?

$$2\ H_2(g) + O_2(g) \rightleftharpoons 2\ H_2O(g)$$

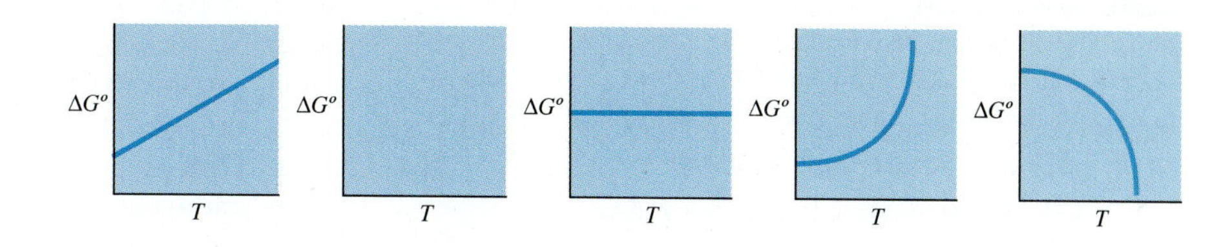

### Standard-State Free Energies of Reaction

**21-52** For which of the following substances is the standard-state free energy of formation equal to zero?

(a) $H(g)$ (b) $Hg(g)$ (c) $H_2O(l)$ (d) $O_2(g)$
(e) $O_3(g)$

**21-53** Calculate $\Delta G°$ for the following reaction from $\Delta H°$ and $\Delta S°$ data for this reaction. Compare the result of this calculation with the value obtained from free energy of formation data.

$$CS_2(l) + 3\,O_2(g) \longrightarrow CO_2(g) + 2\,SO_2(g)$$

**21-54** Adding salt to water does not change $\Delta H°$ for the process in which ice melts:

$$H_2O(s) \rightleftharpoons H_2O(l)$$

But it increases $\Delta S°$ for this process because it increases the entropy of the solution without changing the entropy of the solid. Show how this can be used to explain the fact that adding salt to water lowers its melting point.

**21-55** Adding salt to water does not change $\Delta H°$ for the process in which water boils:

$$H_2O(l) \rightleftharpoons H_2O(g)$$

But it decreases $\Delta S°$ for this process because it increases the entropy of the liquid without changing the entropy of the gas. Show how this can be used to explain the fact that adding salt to water raises its boiling point.

**21-56** Use the concept of standard-state free energies of reaction, $\Delta G°$, to explain why the vapor pressure of water increases as the temperature of the water increases.

### The Relationship Between Free-Energy and Equilibirum Constants

**21-57** Which of the following correctly describes a reaction at equilibrium?

(a) $\Delta G = 0$ (b) $\Delta G° = 0$ (c) $\Delta G = \Delta G°$
(d) $Q = 0$ (e) $\ln K = 0$

**21-58** Calculate the equilibrium constant at 25°C for the following reaction from the standard-state free energies of formation of the reactants and products:

$$CO(g) + 2\,H_2(g) \rightleftharpoons CH_3OH(g)$$

**21-59** Calculate $K_p$ at 1000 K for the following reaction. Describe the assumptions you had to make to do this calculation.

$$2\,CH_4(g) \rightleftharpoons C_2H_6(g) + H_2(g)$$

**21-60** Calculate the equilibrium constant at 25°C for the following reaction:

$$2\,HI(g) + Cl_2(g) \rightleftharpoons 2\,HCl(g) + I_2(s)$$

**21-61** Calculate $\Delta H°$, $\Delta S°$, and $\Delta G°$ for the following reactions. Use these data to calculate the values of $K_a$ for hydrochloric and acetic acid. Compare the results of these calculations with the data in Table A.9.

$$HCl(aq) \rightleftharpoons H^+(aq) + Cl^-(aq)$$
$$CH_3CO_2H(aq) \rightleftharpoons H^+(aq) + CH_3CO_2^-(aq)$$

**21-62** Calculate $\Delta H°$, $\Delta S°$, and $\Delta G°$ for the following reaction. Use these data to calculate the value of $K_b$ for ammonia. Compare the result of this calculation with the value of $K_b$ in Table A.10.

$$NH_3(aq) + H_2O(l) \rightleftharpoons NH_4^+(aq) + OH^-(aq)$$

**21-63** Calculate $\Delta H°$, $\Delta S°$, and $\Delta G°$ for the following reactions. Use these data to calculate values of $K_{a1}$ and $K_{a2}$ for carbonic acid. Compare the results of these calculations with the data in Table A.9.

$$H_2CO_3(aq) \rightleftharpoons H^+(aq) + HCO_3^-(aq)$$
$$HCO_3^-(aq) \rightleftharpoons H^+(aq) + CO_3^{2-}(aq)$$

**21-64** Calculate $\Delta H°$, $\Delta S°$, and $\Delta G°$ for the following reaction. Use these data to calculate the solubility product constant for HgS. Compare the result of this calculation with the data in Table A.11.

$$HgS(s) \xrightleftharpoons{H_2O} Hg^{2+}(aq) + S^{2-}(aq)$$

**21-65** Calculate $\Delta H°$, $\Delta S°$, and $\Delta G°$ for the following reaction. Use these data to calculate the solubility product for silver chloride. Compare the result of this calculation with the data in Table A.11.

$$AgCl(s) \xrightleftharpoons{H_2O} Ag^+(aq) + Cl^-(aq)$$

**21-66** Use the data in Table A.15 to predict which of the following salts should be less soluble in water: PbS or $PbSO_4$.

**21-67** Use thermodynamic data to calculate the solubility of LiF in water. Compare the result of this calculation with the solubility reported in the *CRC Handbook of Chemistry and Physics:* 0.12 g of LiF per 100 g of water.

**21-68** Use $\Delta G_f°$ for the $Ag^+(aq)$ and $S^{2-}(aq)$ ions and the solubility product constant for $Ag_2S$ to estimate $\Delta G_f°$ for $Ag_2S(s)$.

**21-69** Calculate $K_f$ for the $Zn(NH_3)_4^{2+}$ complex ion if $\Delta G°$ for the following reaction is $-54.0$ kJ/mol:

$$Zn^{2+}(aq) + 4\,NH_3(aq) \rightleftharpoons Zn(NH_3)_4^{2+}(aq)$$

**21-70** Use $\Delta G_f°$ for the $Fe^{3+}(aq)$ and $SCN^-(aq)$ ions and the complex formation equilibrium constant for the following reaction to estimate $\Delta G_f°$ for the $Fe(SCN)^{2+}$ complex ion:

$$Fe^{3+}(aq) + SCN^-(aq) \rightleftharpoons Fe(SCN)^{2+} \qquad K_f = 8.9 \times 10^2$$

### The Temperature Dependence of Equilibrium Constants

**21-71** Synthesis gas can be made by reacting red-hot coal with steam. Estimate the temperature at which the equilibrium constant for this reaction is equal to 1.

$$C(s) + H_2O(g) \rightleftharpoons CO(g) + H_2(g)$$

**21-72** Calculate $\Delta H°$, $\Delta S°$, $\Delta G°$, and the equilibrium constant at 25°C for the following reaction. Predict the effect of an increase in the temperature of the system on the equilibrium constant for this reaction.

$$CO_2(g) + H_2(g) \rightleftharpoons CO(g) + H_2O(g)$$

**21-73** What happens to the equilibrium constant for the following reaction as the temperature increases?

$$PCl_5(g) \rightleftharpoons PCl_3(g) + Cl_2(g)$$

**21-74** At approximately what temperature does the equilibrium constant for the following reaction become larger than 1?

$$PCl_5(g) \rightleftharpoons PCl_3(g) + Cl_2(g)$$

**21-75** For which of the following reactions would the equilibrium constant increase with increasing temperature?

(a) $2\ H_2(g) + O_2(g) \rightleftharpoons 2\ H_2O(g)$
(b) $2\ HCl(g) \rightleftharpoons H_2(g) + Cl_2(g)$
(c) $2\ NH_3(g) \rightleftharpoons N_2(g) + 3\ H_2(g)$

**21-76** Calculate the equilibrium constant for the following reaction at 25°C, 200°C, 400°C, and 600°C. Explain any trend you observe.

$$2\ SO_3(g) \rightleftharpoons 2\ SO_2(g) + O_2(g)$$

**21-77** Calculate $\Delta H°$ and $\Delta S°$ for the following reaction. Predict what will happen to the equilibrium constant of this reaction as the temperature increases.

$$Cu(s) + 2\ H^+(aq) \rightleftharpoons Cu^{2+}(aq) + H_2(g)$$

### The Relationship Between Free-Energy and Cell Potentials

**21-78** Which of the following best describes the spontaneous reaction between zinc metal and $Cu^{2+}$ ions in the Daniell cell?

$$Zn(s) + Cu^{2+}(aq) \rightleftharpoons Cu(s) + Zn^{2+}$$

(a) $K_c > 1$, $\Delta G° < 0$, and $E° < 0$ (b) $K_c > 1$, $\Delta G° > 0$, and $E° < 0$ (c) $K_c > 1$, $\Delta G° < 0$, and $E° > 0$ (d) $K_c < 1$, $\Delta G° < 0$, and $E° > 0$ (e) $K_c < 1$, $\Delta G° > 0$, and $E° > 0$

**21-79** Calculate $\Delta G°$ for a standard-state Daniell cell from the standard-state cell potential:

$$Zn(s) + Cu^{2+}(aq) \rightleftharpoons Zn^{2+}(aq) + Cu(s) \qquad E° = 1.103\ V$$

**21-80** Calculate $\Delta G°$ for the following reaction from the standard-state cell potential. Predict whether this reaction is spontaneous at room temperature.

$$Cu(s) + 2\ Ag^+(aq) \rightleftharpoons Cu^{2+}(aq) + 2\ Ag(s) \qquad E° = 0.4594\ V$$

**21-81** Calculate the equilibrium constant for the following reaction from the standard-state cell potential for the reaction:

$$Cu(s) + 2\ H^+(aq) \rightleftharpoons Cu^{2+}(aq) + H_2(g) \qquad E° = 0.3402\ V$$

### Integrated Problems

**21-82** $\Delta S°$ for the following reaction is favorable even though $\Delta H°$ is not:

$$CH_3OH(l) \rightleftharpoons CH_3OH(g)$$

Assume that methanol boils at the temperature at which $\Delta G°$ for this reaction is equal to zero. Use the values of $\Delta H°$ and $\Delta S°$ for this reaction to estimate the boiling point of methanol.

**21-83** Use the equation that describes the relationship between $\Delta G$ and $\Delta G°$ for a reaction to construct a graph of $\Delta G$ versus ln $Q_c$ at 25°C for the following reactions over a range of values of $Q_c$ between $10^{-10}$ and $10^{10}$:

$$HOAc(aq) \xrightleftharpoons{H_2O} H^+(aq) + OAc^-(aq) \qquad \Delta G° = 27.1\ kJ$$

$$HCl(aq) \xrightleftharpoons{H_2O} H^+(aq) + Cl^-(aq) \qquad \Delta G° = -34.2\ kJ$$

Explain the significance of the difference between the points at which the straight lines for these data cross the horizontal and vertical axes.

One of the most powerful techniques for dating archaeological sites involves comparing the amount of $^{14}C$ in samples taken from the site with the $^{14}C$ in living tissue. These petroglyphs were painted by the Anasazi, who first settled in northern Arizona and Utah around 100 A.D.

# KINETICS

Thermodynamics tells us whether a reaction *should occur*. But it cannot tell us whether the reaction *will occur*. To answer that question, we need to understand the factors that control the rate of a chemical reaction. This chapter provides the basis for answering questions such as the following:

## POINTS OF INTEREST

- Why are mixtures of $H_2$ and $O_2$ stable for years when stored at room temperature, whereas they react more or less completely within a few days at 300°C, within a few hours at 500°C, and almost instantly at 700°C?

- What happens to the rate of the following reaction as the reactants are converted into products? Does it increase, decrease, or remain the same?

$$CH_3Br + OH^- \rightleftharpoons CH_3OH + Br^-$$

- How do we decide whether a reaction occurs in a single step:

$$CH_3Br + OH^- \rightleftharpoons CH_3OH + Br^-$$

  or in a sequence of steps?

$$CH_3Br \rightleftharpoons CH_3^+ + Br^-$$
$$CH_3^+ + OH^- \rightleftharpoons CH_3OH$$

- What does it mean when we say that the *half-life* for the decay of $^{14}C$ is 5730 years?

- How can we use the fact that the rate of the forward reaction is much faster than the reverse reaction for the following reaction to predict whether the equilibrium constant for this reaction favors the reactants or the products?

$$ClNO_2(g) + NO(g) \rightleftharpoons NO_2(g) + ClNO(g)$$

## 22.1 CHEMICAL KINETICS

By now you should be familiar with acid–base titrations that use phenolphthalein as the endpoint indicator, as shown in Figure 22.1. You might not have noticed, however, what happens when a solution that contains phenolphthalein in the presence of excess base is allowed to stand for a few minutes. Although the solution initially has a pink color, it gradually turns colorless as the phenolphthalein reacts with the $OH^-$ ion in this solution.

Table 22.1 shows what happens to the concentration of phenolphthalein in a solution that was initially 0.005 *M* in phenolphthalein and 0.61 *M* in $OH^-$ ion. As you can see when these data are plotted in Figure 22.2, the phenolphthalein concentration decreases by a factor of 10 over a period of about four minutes.

**Figure 22.1** An acid-base titration using phenolphthalein as the indicator.

**TABLE 22.1 Experimental Data for the Reaction Between Phenolphthalein and Excess Base**

| Concentration of Phenolphthalein (*M*) | Time (s) |
|---|---|
| 0.0050 | 0.0 |
| 0.0045 | 10.5 |
| 0.0040 | 22.3 |
| 0.0035 | 35.7 |
| 0.0030 | 51.1 |
| 0.0025 | 69.3 |
| 0.0020 | 91.6 |
| 0.0015 | 120.4 |
| 0.0010 | 160.9 |
| 0.00050 | 230.3 |
| 0.00025 | 299.6 |
| 0.00015 | 350.7 |
| 0.00010 | 391.2 |

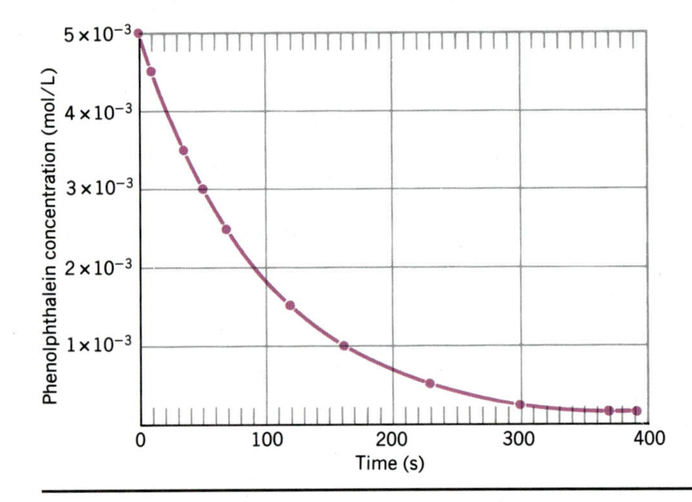


**Figure 22.2** A plot of the concentration of phenolphthalein versus time for the reaction between this indicator and excess $OH^-$ ion.

Experiments such as the one that gave us the data in Table 22.1 are measurements of **chemical kinetics.** One of the goals of these experiments is to describe the rate at which the reactants are transformed into the products of the reaction.

The term *rate* is often used to describe the change in a quantity that occurs per unit of time. The rate of inflation, for example, is the change in the average cost of a collection of standard items per year. The rate at which an object travels through space is the distance traveled per unit of time, such as miles per hour or kilometers per second. In chemical kinetics, the distance traveled is the change in the concentration of one of the components of the reaction. The **rate of a reaction** is therefore the change in the concentration of one of the reactants, $\Delta(X)$, that occurs during a given period of time, $\Delta t$:

$$\text{Rate of reaction} = \frac{\Delta X}{\Delta t}$$

Rate is defined most often as the distance traveled per unit of time.

## EXERCISE 22.1

Use the data in Table 22.1 to calculate the rate at which phenolphthalein reacts with the $OH^-$ ion during each of the following periods.

(a) During the first time interval, when the phenolphthalein concentration falls from 0.0050 *M* to 0.0045 *M*.

(b) During the second interval, when the concentration falls from 0.0045 *M* to 0.0040 *M*.

(c) During the third interval, when the concentration falls from 0.0040 *M* to 0.0035 *M*.

**SOLUTION** The rate of reaction is equal to the change in the phenolphthalein concentration divided by the length of time over which this change occurs.

(a) During the first time period, the rate of the reaction is $4.8 \times 10^{-5}$ moles per liter per second:

$$\text{Rate} = \frac{\Delta(X)}{\Delta t} = \frac{(0.0050\ \text{M}) - (0.0045\ \text{M})}{(10.5\ \text{s} - 0\ \text{s})} = \mathbf{4.8 \times 10^{-5}}\ \boldsymbol{M/\text{s}}$$

(b) During the second time period, the rate of reaction is slightly smaller:

$$\text{Rate} = \frac{\Delta(X)}{\Delta t} = \frac{(0.0045\text{ M}) - (0.0040\text{ M})}{(22.3\text{ s} - 10.5\text{ s})} = \mathbf{4.2 \times 10^{-5}}\ \boldsymbol{M/s}$$

(c) During the third time period, the rate of reaction is even smaller:

$$\text{Rate} = \frac{\Delta(X)}{\Delta t} = \frac{(0.0040\text{ M}) - (0.0035\text{ M})}{(35.7\text{ s} - 22.3\text{ s})} = \mathbf{3.7 \times 10^{-5}}\ \boldsymbol{M/s}$$

## 22.2 INSTANTANEOUS RATES OF REACTION AND THE RATE LAW FOR A REACTION

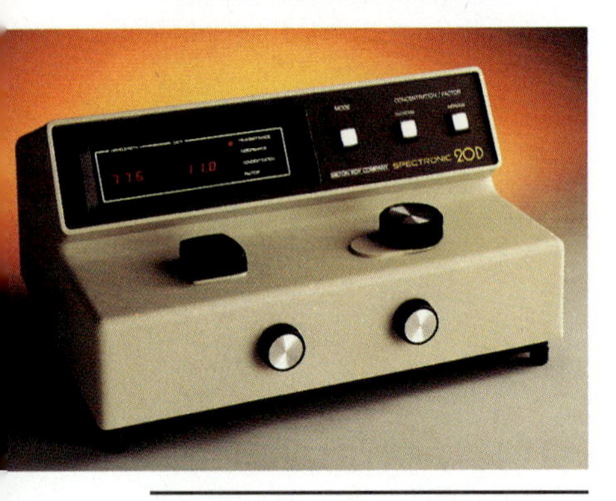


A spectrophotometer can be used to monitor changes in the concentration of a solution that absorbs light. The amount of light absorbed as it passes through the solution is directly proportional to the concentration of the light-absorbing substance.

Exercise 22.1 illustrates an important point: The rate of the reaction between phenolphthalein and the $OH^-$ ion isn't constant; it changes with time. Like most reactions, the rate of this reaction gradually decreases as the reactants are consumed. This means that the rate of reaction changes while it is being measured.

To minimize the error this introduces into our measurements, it seems advisable to measure the rate of reaction over periods of time that are short compared with the time it takes for the reaction to occur. We might try, for example, to measure the infinitesimally small change in concentration, $d(X)$, that occurs over an infinitesimally short period of time, $dt$. The ratio of these quantities is known as the **instantaneous rate of reaction:**

$$\textbf{Rate} = \frac{d(X)}{dt}$$

The instantaneous rate of reaction at any moment in time can be calculated from a graph of the concentration of the reactant (or product) versus time. Figure 22.3 shows how the rate of reaction for the decomposition of phenolphthalein can be calculated from a graph of concentration versus time. The rate of reaction at any moment in time is equal to the slope of a tangent drawn to this curve at that moment.

The instantaneous rate of reaction can be measured at any time between the moment at which the reactants are mixed and the reaction reaches equilibrium. Extrapolating these data back to the instant at which the reagents are mixed gives the *initial instantaneous rate of reaction.*

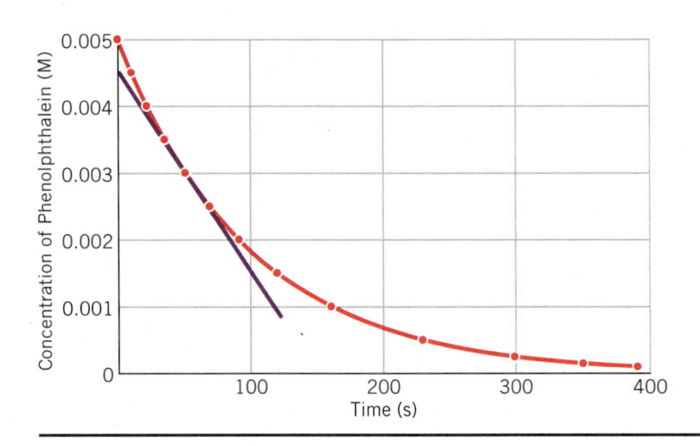


**Figure 22.3** The instantaneous rate of reaction at any moment in time can be calculated from the slope of a tangent to the graph of concentration versus time. At a point 50 s after the beginning of this reaction, for example, the instantaneous rate of reaction is $2.7 \times 10^{-5}$ moles per liter per second.

## 22.3 RATE LAWS AND RATE CONSTANTS

An interesting result is obtained when the instantaneous rate of reaction is calculated at various points along the curve in Figure 22.3. The rate of reaction at every point on this curve is directly proportional to the concentration of phenolphthalein at that moment in time:

$$\text{Rate} = k(\text{phenolphthalein})$$

Because this equation is an experimental law that describes the rate of the reaction, it is called the **rate law** for the reaction. The proportionality constant, $k$, is known as the **rate constant.**

### ▶ CHECKPOINT

Use the rate law for the decomposition of phenolphthalein to explain why the rate of this reaction gradually slows down as phenolphthalein is consumed.

### EXERCISE 22.2

Calculate the rate constant for the reaction between phenolphthalein and the $OH^-$ ion if the instantaneous rate of reaction is $2.5 \times 10^{-5}$ moles per liter per second when the concentration of phenolphthalein is 0.0025 $M$.

**SOLUTION** We start with the rate law for this reaction:

$$\text{Rate} = k(\text{phenolphthalein})$$

We then substitute the known rate of reaction and the known concentration of phenolphthalein into this equation:

$$2.5 \times 10^{-5} \frac{\text{mol/L}}{\text{s}} = k(0.0025 \text{ mol/L})$$

Solving for the rate constant gives the following result:

$$\mathbf{k = 0.010\ s^{-1}}$$

### EXERCISE 22.3

Use the rate constant for the reaction between phenolphthalein and the $OH^-$ ion to calculate the initial instantaneous rate of reaction for the data in Table 22.1.

**SOLUTION** Substituting the rate constant for the reaction from the preceding exercise and the initial concentration of phenolphthalein into the rate law for this reaction gives the following:

$$\text{Rate} = k(\text{phenolphthalein})$$
$$= (0.010 \text{ s}^{-1})(0.0050 \text{ mol/L}) = \mathbf{5.0 \times 10^{-5} \frac{mol/L}{s}}$$

Because the rate of reaction is the change in the concentration of phenolphthalein divided by the time over which this change occurs, it is reported in units of moles per liter per second. Because the number of moles of phenolphthalein per liter is the molarity of this solution, the rate also can be reported in terms of the change in molarity per second: $M$/s.

## 22.4 DIFFERENT WAYS OF EXPRESSING THE RATE OF REACTION

There is usually more than one way to measure the rate of a reaction. We can study the decomposition of hydrogen iodide, for example, by measuring the rate at which $H_2$ or $I_2$ is formed in the following reaction or the rate at which HI is consumed:

$$2\,HI(g) \rightleftharpoons H_2(g) + I_2(g)$$

Experimentally we find that the rate at which $I_2$ is formed is proportional to the square of the HI concentration at any moment in time:

$$\frac{d(I_2)}{dt} = k(HI)^2$$

What would happen if we studied the rate at which $H_2$ is formed? The balanced equation suggests that $H_2$ and $I_2$ must be formed at exactly the same rate:

$$\frac{d(H_2)}{dt} = \frac{d(I_2)}{dt}$$

What would happen, however, if we studied the rate at which HI is consumed in this reaction? Because HI is consumed, the change in its concentration must be a negative number. By convention, the rate of a reaction is always reported as a positive number. We therefore have to change the sign before reporting the rate of reaction for a reactant that is consumed in the reaction:

$$-\frac{d(HI)}{dt} = k'(HI)^2$$

The negative sign does two things. Mathematically, it converts a negative change in the concentration of HI into a positive rate. Physically, it reminds us that the concentration of the reactant decreases with time.

What is the relationship between the rate of reaction ($k$) obtained by monitoring the formation of $H_2$ or $I_2$ and the rate ($k'$) obtained by watching HI disappear? The stoichiometry of the reaction says that two HI molecules are consumed for every molecule of $H_2$ or $I_2$ produced. This means that the rate of decomposition of HI is twice as fast as the rate at which $H_2$ and $I_2$ are formed. We can translate this relationship into a mathematical equation as follows:

$$-\frac{d(HI)}{dt} = 2\left[\frac{d(H_2)}{dt}\right] = 2\left[\frac{d(I_2)}{dt}\right]$$

As a result, the rate constant obtained from studying the rate at which $H_2$ and $I_2$ are formed in this reaction ($k$) is not the same as the rate constant obtained by monitoring the rate at which HI is consumed ($k'$).

### EXERCISE 22.4

Calculate the rate at which HI disappears in the following reaction at the moment when $I_2$ is being formed at a rate of $1.8 \times 10^{-6}$ moles per liter per second:

$$2\,HI(g) \longrightarrow H_2(g) + I_2(g)$$

**SOLUTION** The balanced equation for the reaction shows that 2 moles of HI disappear for every mole of $I_2$ formed. Thus, HI is consumed in this reaction twice as fast as $I_2$ is formed:

$$-\frac{d(HI)}{dt} = 2\left[\frac{d(I_2)}{dt}\right] = 2(1.8 \times 10^{-6})\frac{\text{mol/L}}{\text{s}} = \mathbf{3.6 \times 10^{-6}}\ \boldsymbol{M/s}$$

**Problem-Solving Strategy**

Students sometimes get the wrong answer to this exercise because they become confused about whether the equation for the calculation should be written as:

$$-\frac{d(\mathrm{HI})}{dt} = 2\left[\frac{d(\mathrm{I_2})}{dt}\right]$$

or as:

$$\frac{d(\mathrm{I_2})}{dt} = 2\left[-\frac{d(\mathrm{HI})}{dt}\right]$$

You can avoid mistakes by checking to see whether your answer makes sense. The balanced equation states that 2 moles of HI are consumed for every mole of $I_2$ produced. HI should therefore disappear ($3.6 \times 10^{-6}$ *M*/s) twice as fast as $I_2$ is formed ($1.8 \times 10^{-6}$ *M*/s).

## 22.5 THE RATE LAW VERSUS THE STOICHIOMETRY OF A REACTION

In the 1930s, Sir Christopher Ingold and co-workers at the University of London studied the kinetics of the following substitution reactions.

$$\mathrm{CH_3Br}(aq) + \mathrm{OH^-}(aq) \rightleftharpoons \mathrm{CH_3OH}(aq) + \mathrm{Br^-}(aq)$$

They found that the rate of this reaction is proportional to the concentrations of both reactants:

$$\text{Rate} = k(\mathrm{CH_3Br})(\mathrm{OH^-})$$

When they ran a similar reaction on a slightly different starting material, they obtained similar products:

$$(\mathrm{CH_3})_3\mathrm{CBr}(aq) + \mathrm{OH^-}(aq) \rightleftharpoons (\mathrm{CH_3})_3\mathrm{COH}(aq) + \mathrm{Br^-}(aq)$$

But now the rate of reaction was proportional to the concentration of only one of the reactants:

$$\text{Rate} = k((\mathrm{CH_3})_3\mathrm{CBr})$$

These results illustrate an important point:

**The rate law for a reaction cannot be predicted from the stoichiometry of the reaction; it must be determined experimentally.**

Sometimes, the rate law is consistent with what we expect from the stoichiometry of the reaction:

$$2\,\mathrm{HI}(g) \rightleftharpoons \mathrm{H_2}(g) + \mathrm{I_2}(g) \qquad \text{Rate} = k(\mathrm{HI})^2$$

Often, however, it is not:

$$2\,\mathrm{N_2O_5}(g) \longrightarrow 4\,\mathrm{NO_2}(g) + \mathrm{O_2}(g) \qquad \text{Rate} = k(\mathrm{N_2O_5})$$

As we will see, the rate law for a reaction can differ from what the stoichiometry would lead us to expect when the reaction occurs in more than one step.

## 22.6 ORDER AND MOLECULARITY

Some reactions occur in a single step. The reaction in which a chlorine atom is transferred from $ClNO_2$ to NO to form $NO_2$ and ClNO is a good example of a one-step reaction:

$$ClNO_2(g) + NO(g) \rightleftharpoons NO_2(g) + ClNO(g)$$

Other reactions occur by a series of individual steps. $N_2O_5$, for example, decomposes to $NO_2$ and $O_2$ by a three-step mechanism:

Step 1: $N_2O_5 \rightleftharpoons NO_2 + NO_3$
Step 2: $NO_2 + NO_3 \longrightarrow NO_2 + NO + O_2$
Step 3: $NO + NO_3 \longrightarrow 2\,NO_2$

The steps in a reaction are classified in terms of **molecularity,** which describes the number of molecules consumed. When a single molecule is consumed, the step is called **unimolecular.** When two molecules are consumed, it is **bimolecular.**

### EXERCISE 22.5

Determine the molecularity of each step in the reaction by which $N_2O_5$ decomposes to $NO_2$ and $O_2$.

**SOLUTION** All we have to do is count the number of molecules consumed in each step in this reaction to decide that the first step is unimolecular and the other two steps are bimolecular.

Step 1: $N_2O_5 \rightleftharpoons NO_2 + NO_3$ (unimolecular)
Step 2: $NO_2 + NO_3 \longrightarrow NO_2 + NO + O_2$ (bimolecular)
Step 3: $NO + NO_3 \longrightarrow 2\,NO_2$ (bimolecular)

Reactions also can be classified in terms of their **order.** The decomposition of $N_2O_5$ is a **first-order reaction** because the rate of reaction depends on the concentration of $N_2O_5$ raised to the first power:

$$\text{Rate} = k(N_2O_5)$$

The decomposition of HI is a **second-order reaction** because the rate of reaction depends on the concentration of HI raised to the second power:

$$\text{Rate} = k(\text{HI})^2$$

When the rate of a reaction depends on more than one reagent, we classify the reaction in terms of the order of each reagent.

### EXERCISE 22.6

Classify the order of the reaction between NO and $O_2$ to form $NO_2$:

$$2\,NO(g) + O_2(g) \longrightarrow 2\,NO_2(g)$$

Assume the following rate law for this reaction:

$$\text{Rate} = k(\text{NO})^2(O_2)$$

**SOLUTION** This reaction is first-order in $O_2$, second-order in NO, and third-order overall.

The difference between the molecularity and the order of a reaction is important. The molecularity of a reaction, or a step within a reaction, describes what happens on the molecular level. The order of a reaction describes what happens on the macroscopic scale. We determine the order of a reaction by watching the products of a reaction appear or the reactants disappear. The molecularity of the reaction is something we deduce to explain these experimental results.

## 22.7 A COLLISION THEORY MODEL OF CHEMICAL REACTIONS

Sec. 4.15

The **collision theory model** of chemical reactions can be used to explain the observed rate laws for both one-step and multistep reactions. This model assumes that the rate of any step in a reaction depends on the frequency of collisions between the particles involved in that step.

Figure 22.4 provides a basis for understanding the implications of the collision theory model for simple, one-step reactions, such as the following:

$$ClNO_2(g) + NO(g) \rightleftharpoons NO_2(g) + ClNO(g)$$

The kinetic molecular theory assumes that the number of collisions per second in a gas depends on the number of particles per liter. The rate at which $NO_2$ and ClNO are formed in this reaction therefore should be directly proportional to the concentrations of both $ClNO_2$ and NO:

$$\text{Rate} = k(ClNO_2)(NO)$$

The collision theory model suggests that the rate of any step in a reaction is proportional to the concentrations of the reagents consumed in that step. The rate law for a one-step reaction should therefore agree with the stoichiometry of the reaction. The following reaction, for example, occurs in a single step:

$$CH_3Br(aq) + OH^-(aq) \rightleftharpoons CH_3OH(aq) + Br^-(aq)$$

When these molecules collide in the proper orientation, a pair of nonbonding electrons on the $OH^-$ ion can be donated to the carbon atom at the center of the $CH_3Br$ molecule, as shown in Figure 22.5. When this happens, a carbon–oxygen bond forms at the same time that the carbon–bromine bond is broken. The net result of this reaction is the substitution of an $OH^-$ ion for a $Br^-$ ion. Because the reaction occurs in a single step, which involves collisions between the two reactants, the rate of this reaction is proportional to the concentration of both reactants:

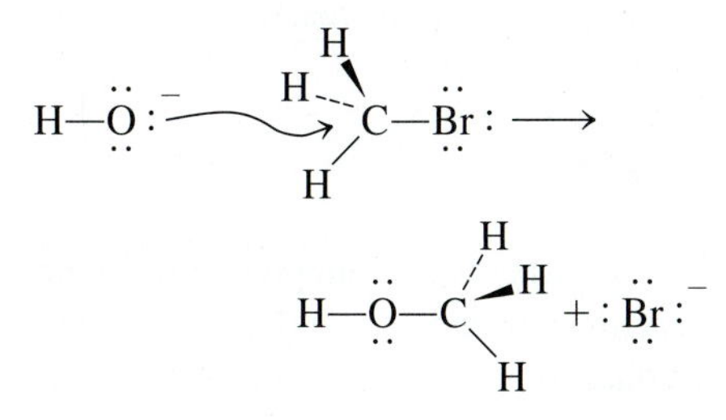

**Figure 22.5** The reaction between $CH_3Br$ and the $OH^-$ ion occurs in a single step, which involves attack by the $OH^-$ ion on the carbon atom. In the course of this reaction, a carbon–oxygen bond forms at the same time that the carbon–bromine bond is broken.

$$\text{Rate} = k(CH_3Br)(OH^-)$$

Not all reactions occur in a single step. The following reaction occurs in three steps, as shown in Figure 22.6.

$$(CH_3)_3CBr(aq) + OH^-(aq) \rightleftharpoons (CH_3)_3COH(aq) + Br^-(aq)$$

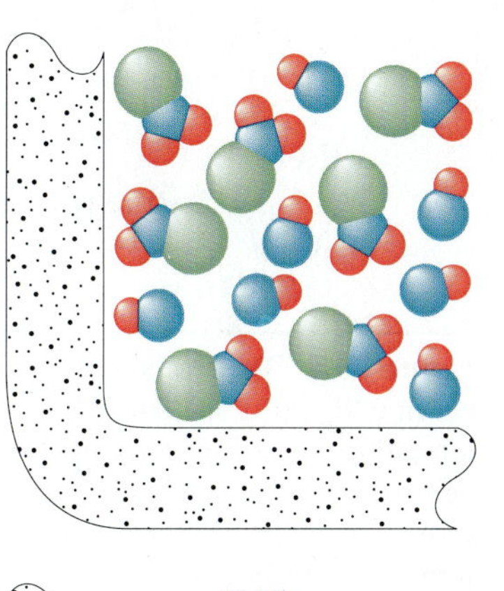

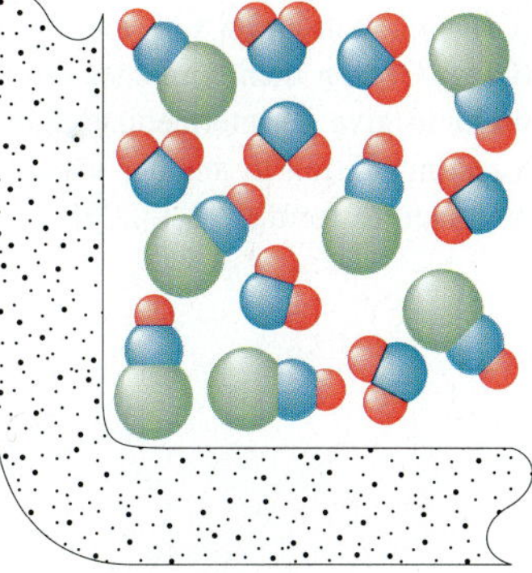

**Figure 22.4** This figure represents a "snapshot" of a small portion of a container in which $ClNO_2$ reacts with NO to form $NO_2$ and ClNO. The collision theory model of reactions assumes that molecules must collide in order to react. Anything that increases the frequency of these collisions increases the rate of reaction, so the rate of reaction must be proportional to the concentration of both of the reactants consumed in this reaction.

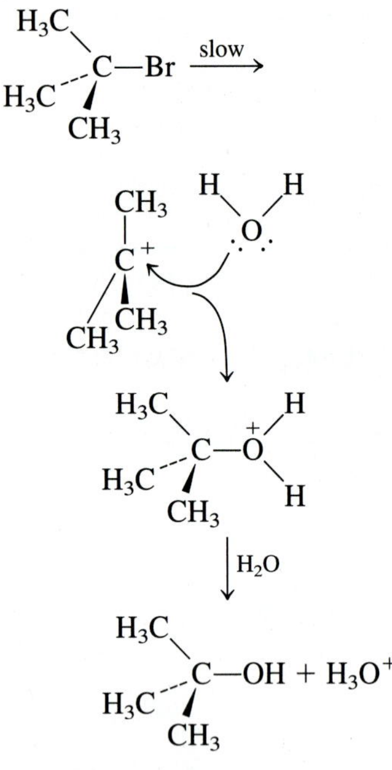


**Figure 22.6** The reaction between $(CH_3)_3CBr$ and the $OH^-$ ion follows a very different mechanism. In the first step, the carbon–bromine bond breaks to form a positively charged $(CH_3)_3C^+$ ion and a $Br^-$ ion. The $(CH_3)_3C^+$ ion then combines with water to give an intermediate that loses an $H^+$ ion to another water molecule to form $(CH_3)_3COH$.

In the first step, the $(CH_3)_3CBr$ molecule dissociates into a pair of ions.

$$\text{First step:} \quad CH_3{-}C(CH_3)_2{-}Br \rightleftharpoons CH_3{-}C^+(CH_3)_2 + Br^-$$

The positively charged $(CH_3)_3C^+$ ion then reacts with water in a second step.

$$\text{Second step:} \quad CH_3{-}C^+(CH_3)_2 + H_2O \rightleftharpoons CH_3{-}C(CH_3)_2{-}OH_2^+$$

The product of this reaction then loses a proton to either the $OH^-$ ion or water in the final step.

$$\text{Third step:} \quad CH_3{-}C(CH_3)_2{-}OH_2^+ + OH^- \rightleftharpoons CH_3{-}C(CH_3)_2{-}OH + H_2O$$

The second and third steps in this reaction are very much faster than first:

$$(CH_3)_3CBr \rightleftharpoons (CH_3)_3C^+ + Br^- \qquad \text{Slow step}$$
$$(CH_3)_3C^+ + H_2O \rightleftharpoons (CH_3)_3COH_2^+ \qquad \text{Fast step}$$
$$(CH_3)_3COH_2^+ + OH^- \rightleftharpoons (CH_3)_3COH + H_2O \qquad \text{Fast step}$$

The overall rate of reaction is therefore more or less equal to the rate of the first step. The first step is called the **rate-limiting step** in this reaction because it literally limits the rate at which the products of the reaction can be formed. Because only one reagent is involved in the rate-limiting step, the overall rate of reaction is proportional to the concentration of only this reagent:

$$\text{Rate} = k((CH_3)_3CBr)$$

The rate law for this reaction therefore differs from what we would predict from the stoichiometry of the reaction. Although the reaction consumes both $(CH_3)_3CBr$ and $OH^-$, the rate of the reaction is proportional only to the concentration of $(CH_3)_3CBr$.

The rate laws for chemical reactions can be explained by the following general rules:

- The rate of any step in a reaction is directly proportional to the concentrations of the reagents consumed in that step.
- The overall rate law for a reaction is determined by the sequence of steps, or the **mechanism,** by which the reactants are converted into the products of the reaction.
- The overall rate law for a reaction is dominated by the rate law for the slowest step in the reaction.

## 22.8 THE MECHANISMS OF CHEMICAL REACTIONS

To understand what happens when the first step in a multistep reaction is not the rate-limiting step, consider the reaction between NO and $O_2$ to form $NO_2$:

$$2\,NO(g) + O_2(g) \rightleftharpoons 2\,NO_2(g)$$

This reaction involves a two-step mechanism. The first step is a relatively fast reaction in which two NO molecules combine to form a dimer, $N_2O_2$. The product of this step then undergoes a much slower reaction in which it combines with $O_2$ to form a pair of $NO_2$ molecules:

$$\begin{array}{lll} \text{Step 1:} & 2\,NO \rightleftharpoons N_2O_2 & \text{(fast step)} \\ \text{Step 2:} & N_2O_2 + O_2 \rightleftharpoons 2\,NO_2 & \text{(slow step)} \end{array}$$

The net effect of these reactions is the transformation of two NO molecules and one $O_2$ molecule into a pair of $NO_2$ molecules:

$$\begin{array}{l} \quad 2\,NO \rightleftharpoons \cancel{N_2O_2} \\ +\,\cancel{N_2O_2} + O_2 \rightleftharpoons 2\,NO_2 \\ \hline \quad 2\,NO_2 + O_2 \rightleftharpoons 2\,NO_2 \end{array}$$

In this reaction, the second step is the rate-limiting step. No matter how fast the first step takes place, the overall reaction cannot proceed any faster than the second step in the reaction. As we have seen, the rate of any step in a reaction is directly proportional to the concentrations of the reactants consumed in that step. The rate law for the second step in this reaction is therefore proportional to the concentrations of both $N_2O_2$ and $O_2$:

$$\text{Step 2:} \qquad \text{Rate}_{2nd} = k(N_2O_2)(O_2)$$

Because the first step in the reaction is much faster, the overall rate of reaction is more or less equal to the rate of this rate-limiting step:

$$\text{Rate} \approx k(N_2O_2)(O_2)$$

This rate law is not very useful because it is difficult to measure the concentrations of intermediates, such as $N_2O_2$, that are simultaneously formed and consumed in the reaction. It would be better to have an equation that related the overall rate of reaction to the concentrations of the original reactants.

Let's take advantage of the fact that the first step in this reaction is reversible.

$$\text{Step 1:} \qquad 2\,NO \rightleftharpoons N_2O_2$$

The rate of the forward reaction in this step depends on the concentration of NO raised to the second power.

$$\text{Step 1:} \qquad \text{Rate}_{forward} = k_f(NO)^2$$

The rate of the reverse reaction depends only on the concentrations of $N_2O_2$.

$$\text{Step 1:} \qquad \text{Rate}_{reverse} = k_r(N_2O_2)$$

Because the first step in this reaction is very much faster than the second, the first step should come to equilibrium. When that happens, the rates of the forward and reverse reactions for the first step are the same:

$$k_f(NO)^2 = k_r(N_2O_2)$$

Let's rearrange this equation to solve for one of the terms that appears in the rate law for the second step in the reaction:

$$(N_2O_2) = \frac{k_f}{k_r}(NO)^2$$

Substituting this equation into the rate law for the second step gives the following result:

$$\text{Rate}_{2nd} = k\left(\frac{k_f}{k_r}\right)(NO)^2(O_2)$$

Since $k$, $k_f$, and $k_r$ are all constants, they can be replaced by a single constant, $k'$, to give the experimental rate law for this reaction described in Exercise 22.6:

$$\text{Rate}_{overall} \approx \text{Rate}_{2nd} = k'(NO)^2(O_2)$$

## 22.9 THE RELATIONSHIP BETWEEN THE RATE CONSTANTS AND THE EQUILIBRIUM CONSTANT FOR A REACTION

There is a simple relationship between the equilibrium constant for a reversible reaction and the rate constants for the forward and reverse reactions *if the mechanism for the reaction involves only a single step*. To understand this relationship, let's turn once more to a reversible reaction that we know occurs by a one-step mechanism:

$$ClNO_2(g) + NO(g) \rightleftharpoons NO_2(g) + ClNO(g)$$

The rate of the forward reaction is equal to the rate constant for this reaction, $k_f$, times the concentrations of the reactants, $ClNO_2$ and NO:

$$\text{Rate}_{forward} = k_f(ClNO_2)(NO)$$

The rate of the reverse reaction is equal to a second rate constant, $k_r$, times the concentrations of the products, $NO_2$ and ClNO:

$$\text{Rate}_{reverse} = k_r(NO_2)(ClNO)$$

This system will reach equilibrium when the rate of the forward reaction is equal to the rate of the reverse reaction:

$$\text{Rate}_{forward} = \text{Rate}_{reverse}$$

Substituting the rate laws for the forward and reverse reactions when the system is at equilibrium into this equation gives the following:

$$[ClNO_2]k_f[NO] = [NO_2]k_r[ClNO]$$

This equation can be rearranged to give the equilibrium constant expression for the reaction:

$$\frac{k_f}{k_r} = \frac{[ClNO][NO_2]}{[NO][ClNO_2]}$$

Thus, the equilibrium constant $K_c$ for a one-step reaction is equal to the forward rate constant divided by the reverse rate constant:

$$K_c = \frac{k_f}{k_r}$$

## EXERCISE 22.7

The rate constants for the forward and reverse reactions in the following equilibrium have been measured. At 25°C, $k_f$ is $7.3 \times 10^3$ liters per mole-second and $k_r$ is 0.55 liters per mole-second. Calculate the equilibrium constant for this reaction:

$$ClNO_2(g) + NO(g) \rightleftharpoons NO_2(g) + ClNO(g)$$

**SOLUTION** We start by assuming that the rates of the forward and reverse reactions at equilibrium are the same.

$$\text{At equilibrium:} \quad \text{rate}_{\text{forward}} = \text{rate}_{\text{reverse}}$$

We then substitute the rate laws for these reactions into this equality:

$$\text{At equilibrium:} \quad k_f[ClNO_2][NO] = k_r[NO_2][ClNO]$$

We then rearrange this equation to get the equilibrium constant expression for the reaction:

$$K_c = \frac{k_f}{k_r} = \frac{[NO_2][ClNO]}{[ClNO_2][NO]}$$

The equilibrium constant for the reaction is therefore equal to the rate constant for the forward reaction divided by the rate constant for the reverse reaction:

$$K_c = \frac{k_f}{k_r} = \frac{7300 \text{ L/mol-s}}{0.55 \text{ L/mol-s}} = 1.3 \times 10^4$$

## 22.10 DETERMINING THE ORDER OF A REACTION FROM RATE OF REACTION DATA

The rate law for a reaction can be determined by studying what happens to the initial instantaneous rate of reaction when we start with different initial concentrations of the reactants. To show how this is done, let's determine the rate law for the decomposition of hydrogen peroxide in the presence of the iodide ion:

$$2\ H_2O_2(aq) \xrightarrow{I^-} 2\ H_2O(l) + O_2(g)$$

Data on initial instantaneous rates of reaction for five experiments run at different initial concentrations of $H_2O_2$ and the $I^-$ ion are given in Table 22.2.

**TABLE 22.2 Rate of Reaction Data for the Decomposition of $H_2O_2$ in the Presence of the $I^-$ Ion**

| | Initial $(H_2O_2)$ $(M)$ | Initial $(I^-)$ $(M)$ | Initial Instantaneous Rate of Reaction $(M/s)$ |
|---|---|---|---|
| Trial 1: | $1.0 \times 10^{-2}$ | $2.0 \times 10^{-3}$ | $2.3 \times 10^7$ |
| Trial 2: | $2.0 \times 10^{-2}$ | $2.0 \times 10^{-3}$ | $4.6 \times 10^7$ |
| Trial 3: | $3.0 \times 10^{-2}$ | $2.0 \times 10^{-3}$ | $6.9 \times 10^7$ |
| Trial 4: | $1.0 \times 10^{-2}$ | $4.0 \times 10^{-3}$ | $4.6 \times 10^7$ |
| Trial 5: | $1.0 \times 10^{-2}$ | $6.0 \times 10^{-3}$ | $6.9 \times 10^7$ |

The only difference between the first three trials is the initial concentration of $H_2O_2$. The difference between trial 1 and trial 2 is a twofold increase in the initial $H_2O_2$ concentration, which leads to a twofold increase in the initial rate of reaction:

$$\frac{\text{Rate for trial 2}}{\text{Rate for trial 1}} = \frac{4.6 \times 10^{-7}\ M/\text{s}}{2.3 \times 10^{-7}\ M/\text{s}} = 2$$

The difference between trial 1 and trial 3 is a threefold increase in the initial $H_2O_2$ concentration, which produces a threefold increase in the initial rate of reaction:

$$\frac{\text{Rate for trial 3}}{\text{Rate for trial 1}} = \frac{6.9 \times 10^{-7}\ M/\text{s}}{2.3 \times 10^{-7}\ M/\text{s}} = 3$$

The only possible conclusion is that the rate of reaction is directly proportional to the $H_2O_2$ concentration.

Experiments 1, 4, and 5 were run at the same initial concentration of $H_2O_2$ but different initial concentrations of the $I^-$ ion. When we compare trials 1 and 4 we see that doubling the initial $I^-$ concentration leads to a twofold increase in the rate of reaction:

$$\frac{\text{Rate for trial 4}}{\text{Rate for trial 1}} = \frac{4.6 \times 10^{-7}\ M/\text{s}}{2.3 \times 10^{-7}\ M/\text{s}} = 2$$

Trials 1 and 5 show that tripling the initial $I^-$ concentration leads to a threefold increase in the initial rate of reaction. We therefore conclude that the rate of the reaction is also directly proportional to the concentration of the $I^-$ ion.

The results of these experiments are consistent with a rate law for this reaction that is first order in both $H_2O_2$ and $I^-$:

$$\text{Rate} = k(H_2O_2)(I^-)$$

**EXERCISE 22.8**

Hydrogen iodide decomposes to give a mixture of hydrogen and iodine:

$$2\ HI(g) \longrightarrow H_2(g) + I_2(g)$$

Use the following data to determine whether the decomposition of HI in the gas phase is first order or second order in hydrogen iodide.

| | **Initial (HI) (*M*)** | **Initial Instantaneous Rate of Reaction (*M*/s)** |
|---|---|---|
| Trial 1: | $1.0 \times 10^{-2}$ | $4.0 \times 10^{-6}$ |
| Trial 2: | $2.0 \times 10^{-2}$ | $1.6 \times 10^{-5}$ |
| Trial 3: | $3.0 \times 10^{-2}$ | $3.6 \times 10^{-5}$ |

**SOLUTION** We can start by comparing trials 1 and 2. When the initial concentration of HI is doubled, the initial rate of reaction increases by a factor of 4:

$$\frac{\text{Rate for trial 2}}{\text{Rate for trial 1}} = \frac{1.6 \times 10^{-5}\ M/\text{s}}{4.0 \times 10^{-6}\ M/\text{s}} = 4$$

Let's now compare Trials 1 and 3. When the initial concentration of HI is tripled, the initial rate increases by a factor of 9:

$$\frac{\text{Rate for trial 3}}{\text{Rate for trial 1}} = \frac{3.6 \times 10^{-5}\ M/\text{s}}{4.0 \times 10^{-6}\ M/\text{s}} = 9$$

The rate of this reaction is proportional to the square of the HI concentration. The reaction is therefore second order in HI, as noted in Section 22.4:

Sec. 22.4

$$\text{Rate} = k(\text{HI})^2$$

## 22.11 THE INTEGRATED FORM OF FIRST-ORDER AND SECOND-ORDER RATE LAWS

The rate law for a reaction is a useful way of probing the mechanism of a chemical reaction but it isn't very useful for predicting how much reactant remains in solution or how much product has been formed in a given amount of time. For these calculations, we use the **integrated form** of the rate law.

Let's start with the rate law for a reaction that is first order in the disappearance of a single reactant, $X$:

$$-\frac{d(X)}{dt} = k(X)$$

When this equation is rearranged and both sides are integrated, we get the following result:

**Integrated form of the first-order rate law:**

$$\ln\left[\frac{(X)}{(X)_o}\right] = -kt$$

In this equation, $(X)$ is the concentration of $X$ at any moment in time, $(X)_o$ is the initial concentration of this reagent, $k$ is the rate constant for the reaction, and $t$ is the time since the reaction started.

To illustrate the power of the integrated form of the rate law for a reaction, let's use this equation to calculate how long it would take for the $^{14}C$ in a piece of charcoal to decay to half of its original concentration. We will start by noting that $^{14}C$ decays by first-order kinetics with a rate constant of $1.21 \times 10^{-4}\ y^{-1}$.

$$-\frac{d(^{14}\text{C})}{dt} = k(^{14}\text{C})$$

The integrated form of this rate law would be written as follows:

$$\ln\left[\frac{(^{14}\text{C})}{(^{14}\text{C})_o}\right] = -kt$$

We are interested in the moment when the concentration of $^{14}C$ in the charcoal is half of its initial value:

$$(^{14}\text{C}) = \tfrac{1}{2}\,(^{14}\text{C})_o$$

Substituting this relationship into the integrated form of the rate law gives the following equation:

$$\ln\left[\frac{\frac{1}{2}\,(^{14}\text{C})_o}{(^{14}\text{C})_o}\right] = -kt$$

We now simplify this equation:

$$\ln\left[\tfrac{1}{2}\right] = -kt$$

Fragments of the Dead Sea Scrolls which were authenticated by $^{14}C$ dating.

and then solve for $t$:

$$t = -\frac{\ln(\frac{1}{2})}{k} = \frac{0.693}{1.21 \times 10^{-4}\ \text{y}^{-1}} = \mathbf{5730\ y}$$

It therefore takes 5730 years for half of the $^{14}C$ in the sample to decay. This is called the **half-life** of $^{14}C$. In general, the half-life for a first-order kinetic process can be calculated from the rate constant as follows:

$$t_{\frac{1}{2}} = -\frac{\ln(\frac{1}{2})}{k} = \frac{0.693}{k}$$

The notion of half-life is used most often to convey the length of time over which radioactive isotopes represent a threat to the environment.

### CHECKPOINT

Which radioactive isotope would have the longer half-life, $^{15}O$ or $^{19}O$?

$$^{15}\text{O:}\quad k = 5.63 \times 10^{-3}\ \text{s}^{-1}$$
$$^{19}\text{O:}\quad k = 2.38 \times 10^{-2}\ \text{s}^{-1}$$

Let's now turn to the rate law for a reaction that is second order in a single reactant, $X$:

$$-\frac{d(X)}{dt} = k(X)^2$$

The integrated form of the rate law for this reaction is written as follows.

**Integrated form of the second-order rate law:**

$$\frac{1}{(X)} - \frac{1}{(X)_o} = kt$$

Once again, $(X)$ is the concentration of $X$ at any moment in time, $(X)_o$ is the initial concentration of $X$, $k$ is the rate constant for the reaction, and $t$ is the time since the reaction started.

## EXERCISE 22.9

Acetaldehyde, $CH_3CHO$, decomposes by second-order kinetics with a rate constant of 0.334 $M^{-1}\ s^{-1}$ at 500°C. Calculate the amount of time it would take for 80% of the acetaldehyde to decompose in a sample that has an initial concentration of 0.00750 $M$.

**SOLUTION** We start with the rate law for the decomposition of acetaldehyde, which follows second-order kinetics:

$$\text{Rate} = k(\text{CH}_3\text{CHO})^2$$

We then write the integrated form of this rate law:

$$\frac{1}{(\text{CH}_3\text{CHO})} - \frac{1}{(\text{CH}_3\text{CHO})_o} = kt$$

The initial concentration of acetaldehyde is 0.00750 $M$. The final concentration is only 20% as large, or 0.00150 $M$:

$$\frac{1}{(0.00150)} - \frac{1}{(0.00750)} = kt$$

Substituting the rate constant for this reaction ($k = 0.334$) into this equation and then solving for $t$ gives the following result:

$$t = 1600 \text{ s}$$

It therefore takes slightly less than one-half of an hour for 80% of the acetaldehyde to decompose at this temperature.

The half-life of a second-order reaction can be calculated from the integrated form of the second-order rate law:

$$\frac{1}{(X)} - \frac{1}{(X)_o} = kt$$

We start by asking: "How long would it take for the concentration of $X$ to decay from its initial value, $(X)_o$, to a value half as large?"

$$\frac{1}{\frac{1}{2}(X)_o} - \frac{1}{(X)_o} = kt_{\frac{1}{2}}$$

The first step in simplifying this equation involves multiplying the top and bottom halves of the first term by 2:

$$\frac{2}{(X)_o} - \frac{1}{(X)_o} = kt_{\frac{1}{2}}$$

Subtracting one term on the left side of this equation from the other gives the following result:

$$\frac{1}{(X)_o} = kt_{\frac{1}{2}}$$

We can now solve this equation for the half-life of the reaction:

$$t_{\frac{1}{2}} = \frac{1}{k(X)_o}$$

There is an important difference between the equations for calculating the half-lives of first-order and second-order reactions. The half-life of a first-order reaction is a constant, which is proportional to the rate constant for the reaction:

**first-order reaction:**

$$t_{\frac{1}{2}} = \frac{0.693}{k}$$

The half-life for a second-order reaction is inversely proportional to both the rate constant for the reaction and the initial concentration of the reactant that is consumed in the reaction:

**second-order reaction:**

$$t_{\frac{1}{2}} = \frac{1}{k(X)_o}$$

Discussions of the half-life of a reaction are therefore usually confined to first-order processes.

## 22.12 DETERMINING THE ORDER OF A REACTION WITH THE INTEGRATED FORM OF RATE LAWS

The integrated form of the rate laws for first- and second-order reactions provides another way of determining the order of a reaction. We can start by assuming, for the sake of argument, that the reaction is first-order in reactant $X$:

$$\text{Rate} = k(X)$$

We then test this assumption by checking concentration versus time data for the reaction to see whether they fit the first-order rate law:

$$\ln\left[\frac{(X)}{(X)_o}\right] = -kt$$

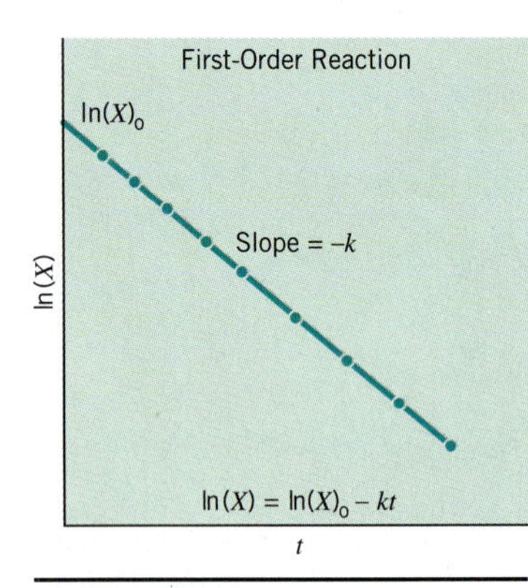


**Figure 22.7** If the rate law for a reaction is first order in $X$, and nothing else, a plot of the natural log of the concentration of $X$ versus the amount of time since the reaction started will be a straight line with a slope equal to the negative of the rate constant.

To see how this is done, let's start by rearranging the integrated form of the first-order rate law as follows:

$$\ln (X) - \ln (X)_o = -kt$$

We then solve this equation for the natural logarithm of the concentration of $X$ at any moment in time:

$$\ln (X) = \ln (X)_o - kt$$

This equation contains two variables, $\ln (X)$ and $t$, and two constants, $\ln (X)_o$ and $k$. It therefore can be set up in terms of the equation for a straight line:

$$y = mx + b$$

$$\ln (X) = -kt + \ln (X)_o$$

If the reaction is first-order in $X$, a plot of the natural logarithm of the concentration of $X$ versus time will be a straight line with a slope equal to $-k$, as shown in Figure 22.7.

If the plot of $\ln (X)$ versus time is not a straight line, the reaction can't be first order in $X$. We therefore assume, for the sake of argument, that it is second order in $X$:

$$\text{Rate} = k(X)^2$$

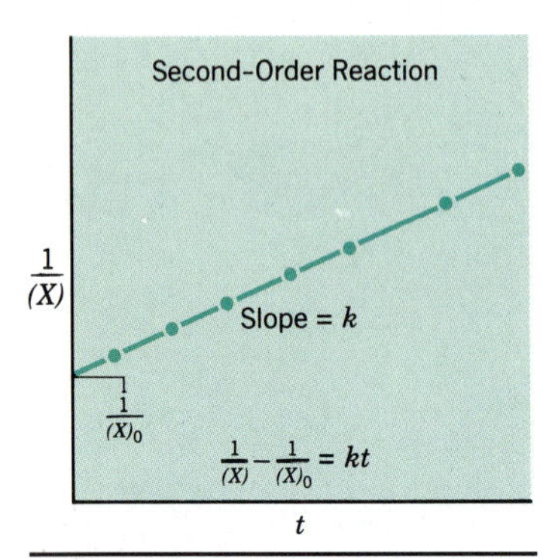


**Figure 22.8** If the rate law for a reaction is second order in $X$, and in nothing else, a plot of the inverse of the concentration of $X$ versus the amount of time since the reaction started will be a straight line with a slope equal to the rate constant, $k$.

We then test this assumption by checking whether the experimental data fit the integrated form of the second-order rate law:

$$\frac{1}{(X)} - \frac{1}{(X)_o} = kt$$

This equation contains two variables, $(X)$ and $t$, and two constants, $(X)_o$ and $k$. Thus, it also can be set up in terms of the equation for a straight line:

$$y = mx + b$$

$$\frac{1}{(X)} = kt + \frac{1}{(X)_o}$$

If the reaction is second order in $X$, a plot of the reciprocal of the concentration of $X$ versus time will be a straight line with a slope equal to $k$, as shown in Figure 22.8.

## EXERCISE 22.10

Use the data in Table 22.1 to determine whether the reaction between phenolphthalein (PHTH) and the $OH^-$ ion is a first-order or a second-order reaction.

**SOLUTION** The first step in solving this problem involves calculating the natural log of the phenolphthalein concentration, ln (PHTH), and the reciprocal of the concentration, 1/(PHTH), for each point at which a measurement was taken:

| (PHTH) (mol/L) | ln (PHTH) | 1/(PHTH) | Time (s) |
|---|---|---|---|
| 0.0050 | −5.30 | 200 | 0.0 |
| 0.0045 | −5.40 | 222 | 10.5 |
| 0.0040 | −5.52 | 250 | 22.3 |
| 0.0035 | −5.65 | 286 | 35.7 |
| 0.0030 | −5.81 | 333 | 51.1 |
| 0.0025 | −5.99 | 400 | 69.3 |
| 0.0020 | −6.21 | 500 | 91.6 |
| 0.0015 | −6.50 | 667 | 120.4 |
| 0.0010 | −6.91 | 1000 | 160.9 |
| 0.00050 | −7.60 | 2000 | 230.3 |
| 0.00025 | −8.29 | 4000 | 299.6 |
| 0.00015 | −8.80 | 6670 | 350.7 |
| 0.00010 | −9.21 | 10,000 | 391.2 |

We then construct graphs of ln (PHTH) versus $t$ (Figure 22.9) and 1/(PHTH) versus $t$ (Figure 22.10). Only one of these graphs, Figure 22.9, gives a straight line. We therefore conclude that these data fit a first-order kinetic equation, as noted in Section 22.1.

Sec. 22.1

$$\text{Rate} = k(\text{phenolphthalein})$$

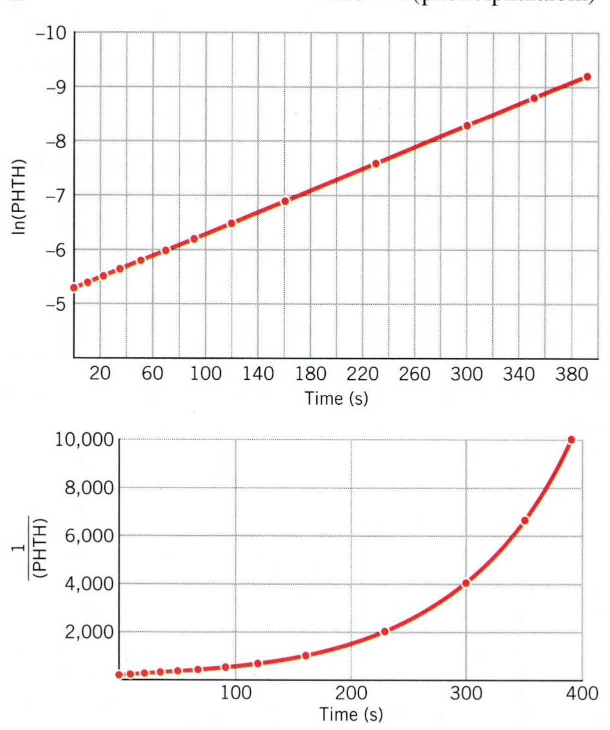



**Figure 22.9** A plot of the natural log of the concentration of phenolphthalein versus time for the reaction between phenolphthalein and excess $OH^-$ ion is a straight line, which shows that this reaction is first order in phenolphthalein.

**Figure 22.10** A plot of the reciprocal of the concentration of phenolphthalein versus time for the reaction between phenolphthalein and the $OH^-$ ion *is not* a straight line, which shows that this reaction is not second order in phenolphthalein.

## 22.13 REACTIONS THAT ARE FIRST ORDER IN TWO REACTANTS

What about reactions that are first order in two reactants, $X$ and $Y$, and therefore second order overall?

$$\text{Rate} = k(X)(Y)$$

A plot of $1/(X)$ versus time won't give a straight line because the reaction is not second order in $X$. Unfortunately, neither will a plot of ln $(X)$ versus time because the reaction is not strictly first order in $X$. It is first order in both $X$ and $Y$.

One way around this problem is to turn the reaction into a **pseudo-first-order reaction** by making the concentration of one of the reactants so large that it is effectively constant. The rate law for the reaction is still first order in both reactants. But the initial concentration of one reactant is so much larger than the other that the rate of reaction seems to be sensitive only to changes in the concentration of the reagent present in limited quantities.

Assume, for the moment, that the reaction is studied under conditions for which there is a large excess of $Y$. If this is true, the concentration of $Y$ will remain essentially constant during the reaction. As a result, the rate of the reaction will not depend on the concentration of the excess reagent. Instead, it will appear to be first order in the other reactant, $X$. A plot of ln $(X)$ versus time will therefore give a straight line:

$$\text{Rate} = k'(X)$$

If there is a large excess of $X$, the reaction will appear to be first order in $Y$. Under these conditions, a plot of log $(Y)$ versus time will be linear:

$$\text{Rate} = k'(Y)$$

The value of the rate constant obtained from either of these equations, $k'$, won't be the actual rate constant for the reaction. It will be the product of the rate constant for the reaction times the concentration of the reagent that is present in excess.

In our discussion of acid–base equilibria, we argued that the concentration of water is so much larger than any other component of these solutions that we can build it into the equilibrium constant expression for the reaction:

$$K_a = \frac{[H_3O^+][A^-]}{[HA]} \qquad K_b = \frac{[BH^+][OH^-]}{[B]}$$

We now understand why this is done. Because the concentration of water is so large, the reaction between an acid or a base and water is a pseudo-first-order reaction that only depends on the concentration of the acid or base.

RESEARCH IN THE 90s

## DETERMINING THE MECHANISM OF CHEMICAL REACTIONS

What do the following processes have in common?

- Determining the amount of copper metal in an alloy or the amount of gold that can be extracted from an ore.
- Stabilizing the potassium iodide added to salt, so that it does not decompose on contact with moist air.
- Measuring the "free available chlorine" (FAC) in the water in a pool or spa.
- Deciding whether the "hypo" used to fix film is still active, or whether it needs to be replaced.
- Determining the concentration of the hydrogen peroxide sold in local drug stores, to see if it is still 3% $H_2O_2$.
- Testing a potential antimicrobial agent to see whether it is a bactericide (that kills bacteria) or just a bacteriostat (that prevents the growth of bacteria).

At some stage, each of these processes involves the reaction between thiosulfate and either iodine:

$$I_2(aq) + 2\,S_2O_3^{2-}(aq) \rightleftharpoons 2\,I^-(aq) + S_4O_6^{2-}(aq)$$

or the triiodide ion:

$$I_3^-(aq) + 2\,S_2O_3^{2-}(aq) \rightleftharpoons 3\,I^-(aq) + S_4O_6^{2-}(aq)$$

Models of the structures of the $S_2O_3^{2}$-ion (left) and the $I_3$-ion (right).

More than 80 years have passed since Friedrich Raschig proposed a mechanism for the reaction between iodine and thiosulfate. Details of this mechanism, however, were not fully understood until 1992, when a technique known as pulsed accelerated flow kinetics was used to study this reaction [W. M. Scheper and D. W. Margerum, *Inorganic Chemistry,* **31,** 5466–5473 (1992)].

The phenomenon that makes these reactions useful for chemical analysis interfered with attempts to elucidate the kinetics of the reactions. These reactions are ideally suited for use in titrations because the overall rate of reaction is rapid. Several steps in these reactions are so fast they are close to the so-called *diffusion limit.* (This is the point at which reaction occurs almost as rapidly as reactant molecules can diffuse through solution.) As a result, the system rapidly comes to equilibrium after each drop of titrant is added.

When the pulsed accelerated flow technique for studying fast reactions was applied to this system, however, enough kinetic data were obtained to confirm the following mechanism. Iodine reacts with the iodide ion in a reversible reaction to form the triiodide ion:

$$I_2 + I^- \underset{k_{-1}}{\overset{k_1}{\rightleftharpoons}} I_3^-$$

The rate constant for the forward reaction ($k_1 = 5.6 \times 10^9\, M^{-1}\, s^{-1}$) is approximately 750 times the rate constant for the reverse reaction ($k_{-1} = 7.5 \times 10^6\, s^{-1}$).

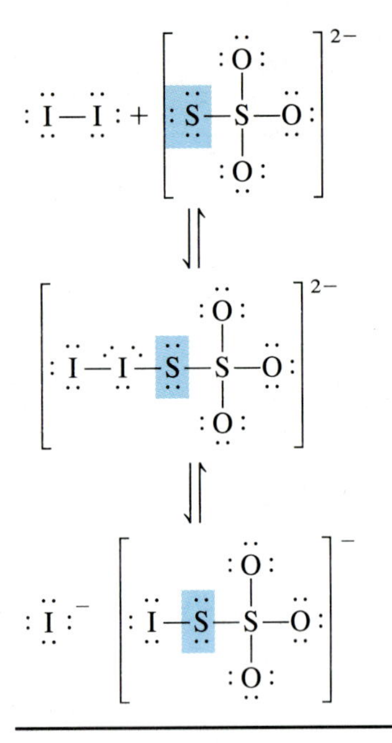

**Figure 22.11** The first step in the reaction between iodine and the thiosulfate ion involves the formation of an $I_2S_2O_3^{2-}$ complex ion that decomposes to form the $I^-$ and $IS_2O_3^-$ ions.

The ratio of these rate constants is therefore consistent with the experimental value of the equilibrium constant for this reaction ($K_1 = 721$).

Iodine in this solution rapidly reacts with thiosulfate to form a complex ion with the structure shown in Figure 22.11. In this case, the rate constant for the forward reaction ($k_2 = 7.8 \times 10^9\ M^{-1}\ s^{-1}$) is so much larger than the rate constant for the reverse reaction ($k_{-2} = 250\ s^{-1}$) that the equilibrium constant for the formation of this complex ion is very large ($K_2 = 3.2 \times 10^7$):

$$I_2 + S_2O_3^{2-} \underset{k_{-2}}{\overset{k_2}{\rightleftharpoons}} I_2S_2O_3^{2-}$$

The triiodide ion undergoes a similar reaction, in which it forms an iodide ion and the $I_2S_2O_3^{2-}$ complex ion. Once again, the rate of the forward reaction ($k_3 = 4.2 \times 10^8\ M^{-1}\ s^{-1}$) is so much larger than the rate of the reverse reaction ($k_{-3} = 9.5 \times 10^3\ M^{-1}\ s^{-1}$) that the equilibrium constant for the reaction is large ($K_3 = 4.4 \times 10^4$):

$$I_3^- + S_2O_3^{2-} \underset{k_{-3}}{\overset{k_3}{\rightleftharpoons}} I_2S_2O_3^{2-} + I^-$$

The $I_2S_2O_3^{2-}$ complex formed in these reactions decomposes to give $I^-$ and an $IS_2O_3^-$ complex ion, as shown in Figure 22.11. The overall rate of reaction is so fast that the rate constants for the reversible reaction in which the $I_2S_2O_3^{2-}$ intermediate is transformed into the $IS_2O_3^-$ intermediate cannot be determined. The equilibrium constant for this reaction, however, has been shown to be close to one ($K_4 = 0.245$):

$$I_2S_2O_3^{2-} \underset{k_{-4}}{\overset{k_4}{\rightleftharpoons}} IS_2O_3^- + I^-$$

The $IS_2O_3^-$ complex ion now reacts with thiosulfate ion to form the iodide and tetrathionate ions ($k_5 = 1.3 \times 10^6\ M^{-1}\ s^{-1}$):

$$IS_2O_3^- + S_2O_3^{2-} \overset{k_5}{\rightleftharpoons} I^- + S_4O_6^{2-}$$

The reaction between $I_2$ or the $I_3^-$ ion and the $S_2O_3^{2-}$ ion illustrates a number of important points about chemical kinetics.

- Reactions for which simple balanced equations can be written often occur by complex mechanisms.
- The overall kinetics of multistep reactions is significantly more complex than the first-order and second-order reactions described in this chapter. The rate at which iodine and triiodide are consumed in this reaction, for example, obeys the following rate law:

$$\text{Rate} = \left(\frac{k_2 + k_3K_1[I^-]}{1 + K_1[I^-]}\right)[S_2O_3^{2-}]([I_2] + [I_3^-]) - \left(\frac{k_{-2}[I^-] + k_{-3}[I^-]^2}{K_4 + [I^-]}\right)([I_2S_2O_3^{2-}] + [IS_2O_3^-])$$

- Oxidation–reduction reactions don't have to involve the direct transfer of electrons, they can also involve the transfer of atoms such as the reactions involved in this mechanism.
- Recent advances in techniques for studying fast reactions provide the basis for significant improvements in our understanding of the mechanisms of reactions through basic research in chemical kinetics.

## 22.14 THE ACTIVATION ENERGY OF CHEMICAL REACTIONS

Only a small fraction of the collisions between reactant molecules convert the reactants into the products of the reaction. This can be understood by turning, once again, to the reaction between $ClNO_2$ and NO:

$$ClNO_2(g) + NO(g) \longrightarrow NO_2(g) + ClNO(g)$$

In the course of this reaction, a chlorine atom is transferred from one nitrogen atom to another. In order for the reaction to occur, the nitrogen atom in NO must collide with the chlorine atom in $ClNO_2$:

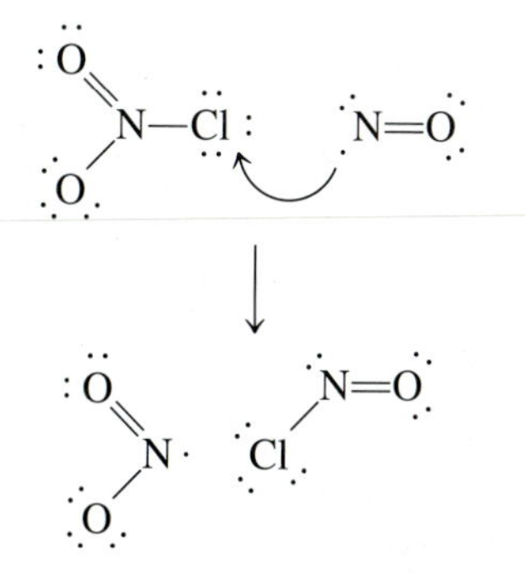

Reaction won't occur if the oxygen end of the NO molecule collides with the chlorine atom on $ClNO_2$:

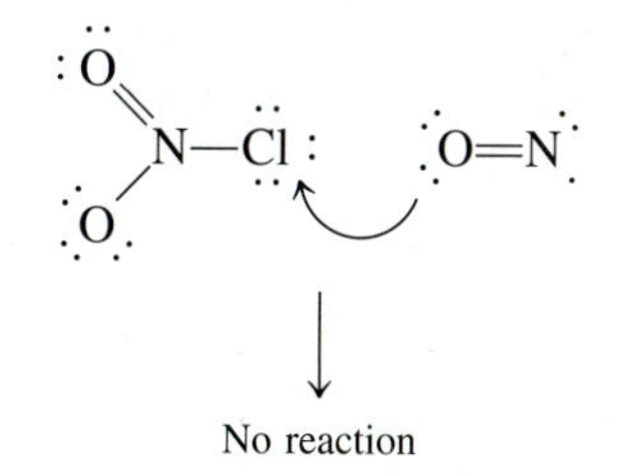


Neither will it occur if one of the oxygen atoms on $ClNO_2$ collides with the nitrogen atom on NO:

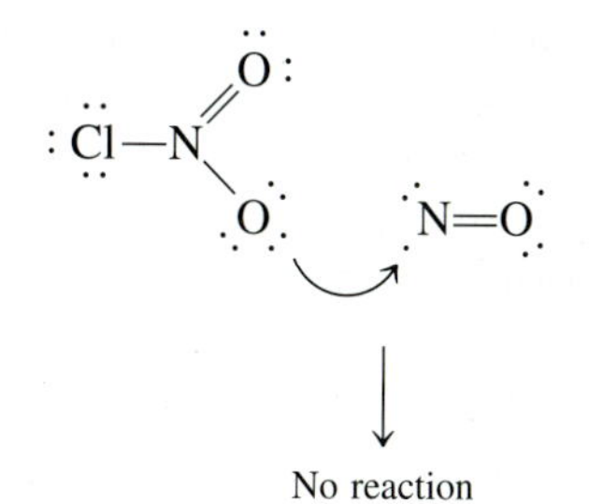


Another factor that influences whether reaction will occur is the energy the molecules carry when they collide. Not all of the molecules have the same kinetic energy, as shown in Figure 22.12. This is important because the kinetic energy molecules carry when they collide is the principal source of the energy that must be invested in a reaction to get it started.

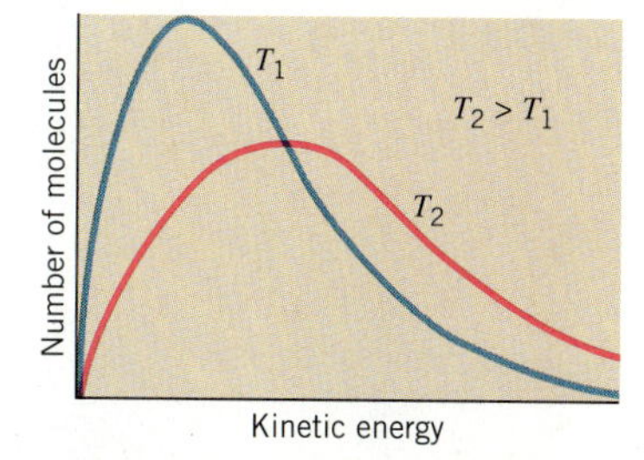


**Figure 22.12** The kinetic molecular theory states that the average kinetic energy of a gas is proportional to the temperature of the gas, and nothing else. At any given temperature, however, some of the gas particles are moving faster than others.

The overall standard free energy for the reaction between $ClNO_2$ and NO is favorable:

$$ClNO_2(g) + NO(g) \rightleftharpoons NO_2(g) + ClNO(g) \qquad \Delta G^\circ = -23.6 \text{ kJ/mol}$$

But, before the reactants can be converted into products, the free energy of the

The effect of activation energy on the rate of a chemical reaction can be understood by remembering that it is easier to traverse a path between mountains than it is to try to climb over these mountains.

system must overcome the **activation energy** for the reaction, as shown in Figure 22.13. The vertical axis in this diagram represents the free energy of a pair of molecules as a chlorine atom is transferred from one to the other. The horizontal axis represents the sequence of infinitesimally small changes that must occur to convert the reactants into the products of this reaction.

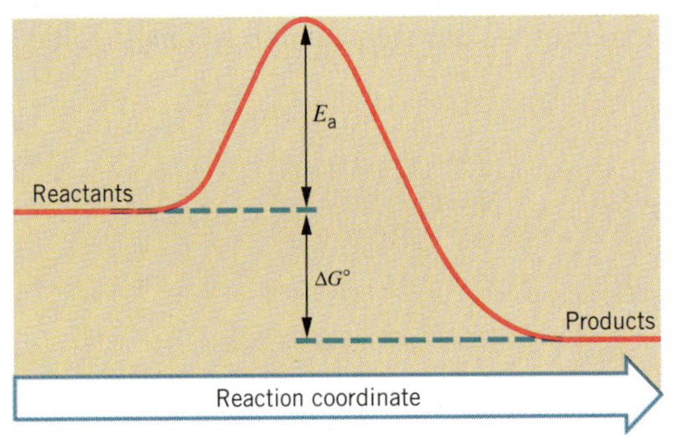


**Figure 22.13** The activation energy for a reaction, $E_a$, is the change in the potential energy of the reactant molecules that must be overcome before the reaction can occur.

To understand why reactions have an activation energy, consider what has to happen in order for $ClNO_2$ to react with NO. First, and foremost, these two molecules have to collide, thereby organizing the system. Not only do they have to be brought together, they have to be held in exactly the right orientation relative to each other to ensure that reaction can occur. Both of these factors raise the free energy of the system by lowering the entropy. Some energy also must be invested to begin breaking the Cl—$NO_2$ bond so that the Cl—NO bond can form.

NO and $ClNO_2$ molecules that collide in the correct orientation, with enough kinetic energy to climb the activation energy barrier, can react to form $NO_2$ and ClNO. As the temperature of the system increases, the number of molecules that carry enough energy to react when they collide also increases. The rate of reaction therefore increases with temperature. As a rule, the rate of a reaction doubles for every 10°C increase in the temperature of the system.

Note that the symbol used to represent the difference between the free energies of the products and the reactants in Figure 22.13 is $\Delta G°$, not $\Delta G°$. A small capital G is used to remind us that this diagram plots the free energy of a pair of molecules as they react, not the free energy of a system that contains many pairs of molecules undergoing collision. If we averaged the results of this calculation over the entire array of molecules in the system, we would get the change in the free energy of the system, $\Delta G°$.

Note also that the symbol used to represent the activation energy is written with a capital *E*. This is unfortunate because it leads students to believe the activation energy is the change in the internal energy of the system, which is not quite true. $E_a$ measures the change in the potential energy of a pair of molecules that is required to begin the process of converting a pair of reactant molecules into a pair of product molecules.

## 22.15 CATALYSTS AND THE RATES OF CHEMICAL REACTIONS

Aqueous solutions of hydrogen peroxide are stable until we add a small quantity of the $I^-$ ion, a piece of platinum metal, a few drops of blood, or a freshly cut slice of turnip, at which point the hydrogen peroxide rapidly decomposes:

$$2\,H_2O_2(aq) \longrightarrow 2\,H_2O(l) + O_2(g)$$

This reaction therefore provides the basis for understanding the effect of a catalyst on the rate of a chemical reaction. Four criteria must be satisfied in order for something to be classified as **catalyst:**

- Catalysts increase the rate of reaction.
- Catalysts are not consumed by the reaction.
- A small quantity of catalyst should be able to affect the rate of reaction for a large amount of reactant.
- Catalysts do not change the equilibrium constant for the reaction.

The first criterion provides the basis for defining a catalyst as something that increases the rate of a reaction. The second reflects the fact that anything consumed in the reaction is a reactant, not a catalyst. The third criterion is a consequence of the second; because catalysts are not consumed in the reaction, they can catalyze the reaction over and over again. The fourth criterion results from the fact that catalysts speed up the rates of the forward and reverse reactions equally, so the equilibrium constant for the reaction remains the same.

Catalysts increase the rates of reactions by providing a new mechanism that has a smaller activation energy, as shown in Figure 22.14. A larger proportion of the collisions that occur between reactants now have enough energy to overcome the activation energy for the reaction. As a result, the rate of reaction increases.

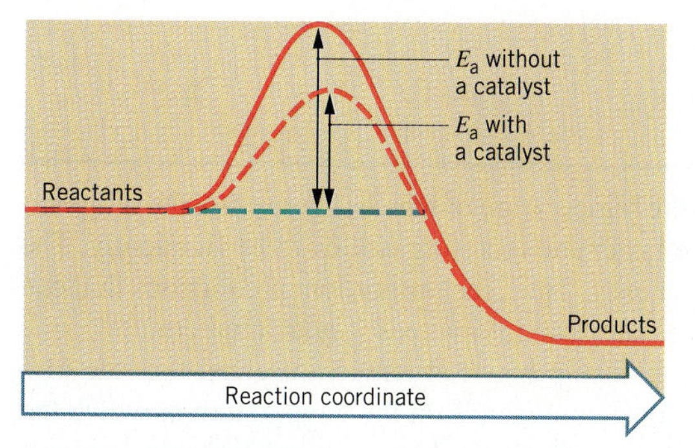


**Figure 22.14** A catalyst increases the rate of a reaction by providing an alternative mechanism that has a smaller activation energy.

The effect of several catalysts on the activation energy for the decomposition of hydrogen peroxide and the relative rate of this reaction is summarized in Table 22.3.

**TABLE 22.3 The Effect of Catalysts on the Activation Energy for the Decomposition of Hydrogen Peroxide**

| Catalyst | Activation Energy (kJ/mol) | Relative Rate of Reaction |
|---|---|---|
| None | 75.3 | 1 |
| $I^-$ | 56.5 | $2.0 \times 10^3$ |
| Pt | 49.0 | $4.1 \times 10^4$ |
| Catalase | 8 | $6.3 \times 10^{11}$ |

Adding a source of the $I^-$ ion to this solution decreases the activation energy by 25%, which increases the rate of the reaction by a factor of 2000. A piece of platinum metal decreases the activation energy even further, and thereby increases the rate of reaction by a factor of 40,000. The *catalase* enzyme in blood, horseradish, or turnips decreases the activation energy by almost a factor of 10, which leads to a 600 billion-fold increase in the rate of reaction.

The bombardier beetle uses the enzyme-catalyzed decomposition of hydrogen peroxide as a defensive mechanism. When attacked, it mixes the contents of a sac that is about 25% $H_2O_2$ with a suspension of a crystalline peroxidase enzyme in a turretlike mixing tube. The rate of reaction is so rapid that the fluid carried from this tube by the $O_2$ liberated in the reaction has a temperature of 100°C.

To illustrate how a catalyst can decrease the activation energy for a reaction by providing another pathway for the reaction, let's look at the mechanism for the decomposition of hydrogen peroxide catalyzed by the $I^-$ ion. In the presence of this ion, the decomposition of $H_2O_2$ doesn't have to occur in a single step. It can occur in two steps, both of which are easier and therefore faster. In the first step, the $I^-$ ion is oxidized by $H_2O_2$ to form the hypoiodite ion, $OI^-$:

$$H_2O_2(aq) + I^-(aq) \longrightarrow H_2O(l) + OI^-(aq)$$

In the second step, the $OI^-$ ion is reduced to $I^-$ by $H_2O_2$:

$$OI^-(aq) + H_2O_2(aq) \longrightarrow H_2O(l) + O_2(g) + I^-(aq)$$

Because there is no net change in the concentration of the $I^-$ ion as a result of these reactions, the $I^-$ ion satisfies the criteria for a catalyst. Because $H_2O_2$ and $I^-$ are both involved in the first step in this reaction, and the first step in this reaction is the rate-limiting step, the overall rate of reaction is first-order in both reagents, as we discovered in Section 22.10.

Sec. 22.10

## 22.16 DETERMINING THE ACTIVATION ENERGY OF A REACTION

The rate of a reaction depends on the temperature at which it is run. As the temperature increases, the molecules move faster and therefore collide more frequently. The molecules also carry more kinetic energy. Thus, the proportion of collisions that can overcome the activation energy for the reaction increases with temperature.

The only way to explain the relationship between temperature and the rate of a reaction is to assume that the rate constant depends on the temperature at which the reaction is run. In 1889, Svante Arrhenius showed that the relationship between temperature and the rate constant for a reaction obeyed the following equation:

$$k = Ze^{-E_a/RT}$$

In this equation, $k$ is the rate constant for the reaction, $Z$ is a proportionality constant that varies from one reaction to another, $E_a$ is the activation energy for the reaction, $R$ is the ideal gas constant in joules per mole kelvin, and $T$ is the temperature in kelvin.

The **Arrhenius equation** can be used to determine the activation energy for a reaction. We start by taking the natural logarithm of both sides of the equation:

$$\ln k = \ln Z - \frac{E_a}{RT}$$

We then rearrange this equation to fit the equation for a straight line:

$$y = mx + b$$

$$\ln k = -\frac{E_a}{R}\left[\frac{1}{T}\right] + \ln Z$$

According to this equation, a plot of ln $k$ versus $1/T$ should give a straight line with a slope of $-E_a/R$, as shown in Figure 22.15.

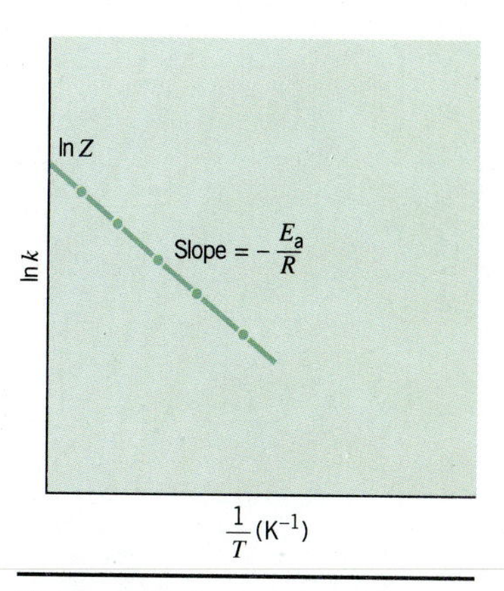


**Figure 22.15** A plot of the natural log of the rate constant for the reaction at different temperatures versus the inverse of the temperature in kelvin is a straight line with a slope equal to $-E_a/R$.

## EXERCISE 22.11

Use the following data to determine the activation energy for the decomposition of HI:

| Temperature (K) | Rate Constant ($M$/s) |
|---|---|
| 573 | $2.91 \times 10^{-6}$ |
| 673 | $8.38 \times 10^{-4}$ |
| 773 | $7.65 \times 10^{-2}$ |

**SOLUTION** We can determine the activation energy for a reaction from a plot of the natural log of the rate constants versus the reciprocal of the absolute temperature. We therefore start by calculating $1/T$ and the natural logarithm of the rate constants:

| ln $k$ | $1/T$ (K$^{-1}$) |
|---|---|
| −12.75 | 0.00175 |
| −7.08 | 0.00149 |
| −2.57 | 0.00129 |

When we construct a graph of these data, we get a straight line with a slope of −22,200 K. According to the Arrhenius equation, the slope of this line is equal to $-E_a/R$:

$$-22{,}200\text{ K} = -\frac{E_a}{8.314\text{ J/mol-K}}$$

When this equation is solved, we get the following value for the activation energy for this reaction.

$$E_a = 183\text{ kJ/mol}$$

By keeping the mathematics of logarithms clearly in mind, it is possible to derive another form of the Arrhenius equation that can be used to predict the effect of a change in temperature on the rate constant for a reaction:

$$\ln\left[\frac{k_1}{k_2}\right] = \frac{E_a}{R}\left[\frac{1}{T_2} - \frac{1}{T_1}\right]$$

## EXERCISE 22.12

Calculate the rate of decomposition of HI at 600°C.

**SOLUTION** We start with the following form of the Arrhenius equation:

$$\ln\left[\frac{k_1}{k_2}\right] = \frac{E_a}{R}\left[\frac{1}{T_2} - \frac{1}{T_1}\right]$$

We then pick any one of the three data points used in the preceding exercise as $T_1$ and allow the value of $T_2$ to be 873 K:

$$T_1 = 573 \text{ K} \qquad k_1 = 2.91 \times 10^{-6}\ M/\text{s}$$
$$T_2 = 873 \text{ K} \qquad k_2 = ?$$

Substituting what we know about the system into the equation given above gives the following result:

$$\ln\left[\frac{(2.91 \times 10^{-6}\ M/\text{s})}{k_2}\right] = \frac{183{,}000 \text{ J/mol}}{8.314 \text{ J/mol-K}}\left[\frac{1}{873 \text{ K}} - \frac{1}{573 \text{ K}}\right]$$

We can simplify the right-hand side of this equation as follows:

$$\ln\left[\frac{(2.91 \times 10^{-6}\ M/\text{s})}{k_2}\right] = -13.20$$

We then take the antilog of both sides of the equation:

$$\frac{(2.91 \times 10^{-6}\ M/\text{s})}{k_2} = 1.85 \times 10^{-6}$$

Solving for $k_2$ gives the rate constant for this reaction at 600°C:

$$k_2 = 1.6\ M/\text{s}$$

Increasing the temperature of the reaction from 573 K to 873 K therefore increases the rate constant for the reaction by a factor of almost a million.

The Arrhenius equation also can be used to calculate what happens to the rate of a reaction when a catalyst lowers the activation energy.

## EXERCISE 22.13

Use the activation energy data in Table 22.3 to predict the effect of adding a source of the $I^-$ ion on the rate of decomposition of $H_2O_2$ at 25°C.

**SOLUTION** We can start by substituting what we know about the reaction in the absence of a catalyst into the Arrhenius equation:

$$k = Ze^{-E_a/RT} = Ze^{(-75{,}300 \text{ J/mol})/(8.314 \text{ J/mol-K})(298 \text{ K})} = Ze^{-30.4}$$

We then repeat this calculation for the reaction in the presence of $I^-$ ions:

$$k_{I^-} = Ze^{-E_a/RT} = Ze^{(-56{,}500 \text{ J/mol})/(8.314 \text{ J/mol-K})(298 \text{ K})} = Ze^{-22.8}$$

We are now ready to calculate the ratio of these rate constants:

$$\frac{k_{I^-}}{k} = \frac{Ze^{-22.8}}{Ze^{-30.4}} = \frac{2000}{1}$$

Adding $I^-$ ion to this system increases the rate of reaction by a factor of 2000, as noted in Table 22.3.

**SPECIAL TOPIC**

## DERIVING THE INTEGRATED RATE LAWS

In order to derive the integrated form of the first-order rate law, we start with the equation that describes the rate law for the reaction:

$$-\frac{d(X)}{dt} = k(X)$$

We then rearrange the equation as follows:

$$\frac{1}{(X)} d(X) = -k\,dt$$

Our goal is to integrate both sides of this equation. Mathematically, this is equivalent to finding the area under the curve that would be produced if this function were graphed. This process is indicated with integral signs, as follows:

$$\int \frac{1}{(X)}\, d(X) = \int -k\,dt$$

We are interested in the area under this curve between the time when the reaction starts ($t = 0$) and some later time ($t$):

$$\int_0^x \frac{1}{(X)}\, d(X) = \int_0^t -k\,dt$$

The integral of $(X)^{-1}\, d(X)$ is equal to the natural logarithm of $(X)$. The integrated form of the first-order rate law therefore can be written as follows:

**Integrated form of the first-order rate law:**

$$\ln\left[\frac{(X)}{(X)_0}\right] = -kt$$

When using this equation, remember that $(X)$ is the concentration of the reactant at any moment in time, $(X)_0$ is the initial concentration of the reactant, $k$ is the rate constant for the reaction, and $t$ is the time since the reaction started.

The derivation of the integrated form of the second-order rate law also starts with the equation that defines the rate law of this reaction:

$$-\frac{d(X)}{dt} = k(X)^2$$

We start by rearranging the equation as follows:

$$-\frac{1}{(X)^2}\, d(X) = k\,dt$$

We then integrate both sides of this equation.

$$\int_0^x -\frac{1}{(X)^2}\, d(X) = \int_0^t k\,dt$$

The integral of $(X)^{-2}d(x)$ is $-(X)^{-1}$. The integrated form of the second-order rate law is therefore written as follows.

**Integrated form of the second-order rate law:**

$$\frac{1}{(X)} - \frac{1}{(X)_o} = kt$$

Once again, the $(X)$ term is the concentration of $X$ at any moment in time, $(X)_o$ is the initial concentration of $X$, $k$ is the rate constant for the reaction, and $t$ is the time since the reaction started.

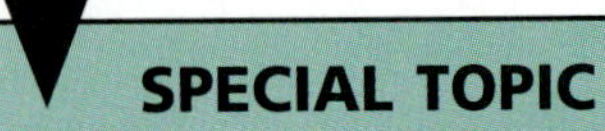


## THE KINETICS OF ENZYME-CATALYZED REACTIONS

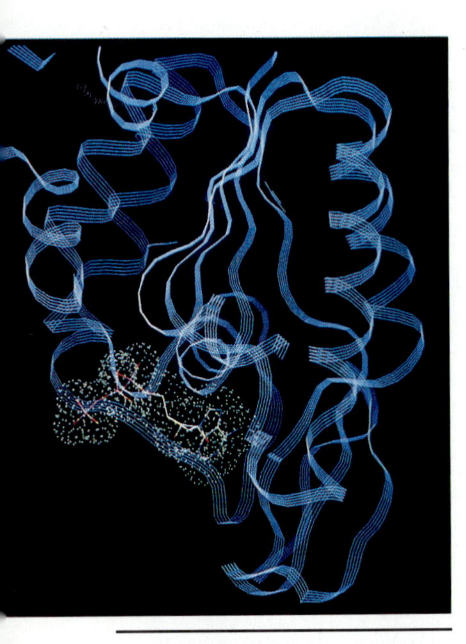

A computer model of the binding of the ATP substrate to an enzyme known as phosphoglycerate kinase.

**Enzymes** are proteins that catalyze reactions in living systems. In 1913, Lenor Michaelis and his student M. L. Menton studied the rate at which an enzyme isolated from yeast catalyzed the hydrolysis of sucrose into fructose and glucose:

$$\text{Sucrose} + H_2O \xrightarrow{\text{enzyme}} \text{fructose} + \text{glucose}$$

At first glance, this reaction seems similar to the reaction between sucrose and acid, which had been studied by Ludwig Wilhemy 60 years earlier:

$$\text{Sucrose} + H_3O^+ \longrightarrow \text{fructose} + \text{glucose}$$

Wilhemy found that the reaction between sucrose and acid was first order in sucrose at constant pH:

$$\text{Rate} = k(\text{sucrose})$$

Michaelis and Menton, however, found that the initial rate of the **enzyme-catalyzed reaction** was first order in sucrose only at low concentrations of sucrose. At high concentrations, the initial rate of reaction did not increase when more sucrose was added to the system.

A reaction that does not depend on the concentration of one of the reactants is said to be **zero order** in that reactant. The enzyme-catalyzed hydrolysis of sucrose apparently changes from a first order reaction to a zero order reaction as the amount of sucrose in the system increases, as shown in Figure 22.16. There is a maximum initial rate of reaction, $\text{rate}_{max}$, for the enzyme-catalyzed reaction that can't be exceeded no matter how much sucrose we add to the solution. Once the reaction reaches $\text{rate}_{max}$, the only way to increase the rate of reaction is to add more enzyme to the solution.

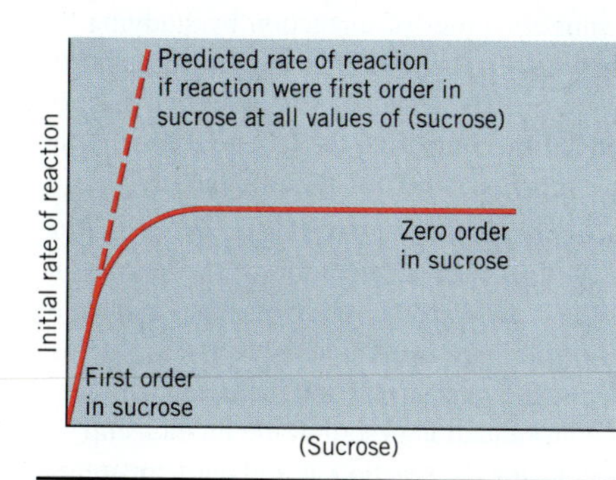


**Figure 22.16** If the reaction in which an enzyme hydrolyzes sucrose to fructose and glucose were first order in the concentration of sucrose, the initial rate of reaction should be directly proportional to the initial concentration of sucrose. Michaelis and Menton, however, showed that this reaction is first order in sucrose at low concentrations of sucrose but zero order in sucrose at high concentrations.

Michaelis and Menton explained this behavior by assuming that the reaction proceeds through a two-step mechanism. In the first step, the enzyme (E) combines with sucrose (S) to form a complex, ES:

$$\mathrm{E} + \mathrm{S} \underset{k_2}{\overset{k_1}{\rightleftharpoons}} \mathrm{ES}$$

The enzyme then converts sucrose into the products (P) of the reaction, which are released in a second step that regenerates the enzyme:

$$\mathrm{ES} \underset{k_4}{\overset{k_3}{\rightleftharpoons}} \mathrm{E} + \mathrm{P}$$

If the first reaction is faster than the second, the overall rate of this reaction is more or less equal to the rate of the second step, which is proportional to the concentration of the enzyme–sucrose complex:

$$\text{Rate}_{\text{overall}} \approx \text{Rate}_{\text{2nd}} = k_3(\text{ES})$$

The maximum rate of reaction will be seen when essentially all of the enzyme is tied up as the ES complex:

$$\text{Rate}_{\text{max}} = k_3(\text{E}_\text{t})$$

In this equation, $E_t$ is the sum of the concentrations of the free enzyme, E, and the enzyme–sucrose complex, ES.

There is a limit to the rate at which the enzyme can consume sucrose. When there is a large excess of sucrose in this solution, every time the enzyme operates on a molecule of sucrose, it will immediately pick up another. No matter how much sucrose is added to the solution, the reaction can't occur any faster. The reaction no longer depends on the sucrose concentration and is therefore zero order in sucrose:

$$\text{Rate} = k(\text{sucrose})^\circ = k$$

## SUMMARY

**1.** The rate of a reaction is the change in the concentration of one of the components of a reaction that occurs in a given period of time: $\Delta(X)/\Delta t$. The instantaneous rate of reaction is the infinitesimally small change in concentration that occurs over an infinitesimally small length of time: $d(X)/dt$.

**2.** The rate law for a one-step reaction is consistent with the stoichiometry of the reaction. For reactions that occur in more than one step, the rate laws may differ from what we would expect from the stoichiometry of the reaction.

**3.** The individual steps in a reaction can be classified in terms of the **molecularity.** Steps may be unimolecular, bimolecular, or, occasionally, trimolecular. The rate law for the reaction is classified in terms of the **order** of the reaction. The molecularity of the steps in a reaction can be deduced from the mechanism proposed for the reaction; the order of the reaction must be determined experimentally.

**4.** We can determine the order of a reaction by studying the effect of changes in the initial concentrations of the reactants on the initial instantaneous rate of reaction. The order of a reaction also can be determined by comparing the kinetic data with the predictions of the integrated forms of the rate laws.

**5.** The rate of a chemical reaction depends on the activation energy for the reaction. As the temperature of the system increases, more of the collisions between reactant molecules carry enough kinetic energy to overcome the activation energy for the reaction. Thus, the rate of reaction increases with temperature. The Arrhenius equation describes the relationship between the rate constant for the reaction at a given temperature and the activation energy for the reaction.

## KEY TERMS

**Activation energy** (p. 824)
**Arrhenius equation** (p. 826)
**Bimolecular** (p. 808)
**Catalyst** (p. 825)
**Chemical kinetics** (p. 803)
**Collision theory model** (p. 809)
**Enzyme-catalyzed reaction** (p. 830)
**First-order reaction** (p. 808)
**Half-life** (p. 816)
**Instantaneous rate of reaction** (p. 804)
**Integrated form of the rate laws** (p. 815)
**Mechanism** (p. 810)
**Molecularity** (p. 808)
**Order** (p. 808)
**Pseudo-first-order reaction** (p. 820)
**Rate of reaction** (p. 803)
**Rate constant** (p. 805)
**Rate law** (p. 805)
**Rate-limiting step** (p. 810)
**Second-order reaction** (p. 808)
**Unimolecular** (p. 808)
**Zero-order reaction** (p. 830)

## KEY EQUATIONS

$$\text{Rate} = \frac{d(X)}{dt}$$

$$K_c = \frac{k_f}{k_r}$$

$$\ln\left[\frac{(X)}{(X)_o}\right] = -kt$$ Integrated form of the first-order rate law

$$\frac{1}{(X)} - \frac{1}{(X)_o} = kt$$ Integrated form of the second-order rate law

$$t_{\frac{1}{2}} = \frac{0.693}{k}$$ (For a first-order reaction)

$$t_{\frac{1}{2}} = \frac{1}{k(X)_o}$$ (For a second-order reaction)

$$k = Ze^{-E_a/RT}$$

$$\ln k = -\frac{E_a}{R}\left[\frac{1}{T}\right] + \ln Z$$

$$\ln\left[\frac{k_1}{k_2}\right] = \frac{E_a}{R}\left[\frac{1}{T_2} - \frac{1}{T_1}\right]$$

## PROBLEMS

### Chemical Kinetics

**22-1** Define the terms *thermodynamic control* and *kinetic control* and give an example of each.

**22-2** Describe the difference between the terms *rate of reaction, rate law,* and *rate constant.* Give an example of each.

**22-3** Describe the difference between the rate of reaction measured over a finite period of time and the instantaneous rate of reaction. Explain the advantage of measuring the instantaneous rate of reaction.

**22-4** Explain why the rate of each step in a reaction is proportional to the concentrations of the reagents consumed in that step.

**22-5** Which of the following graphs best describes the relationship between the rate of a reaction and the temperature of the reaction?

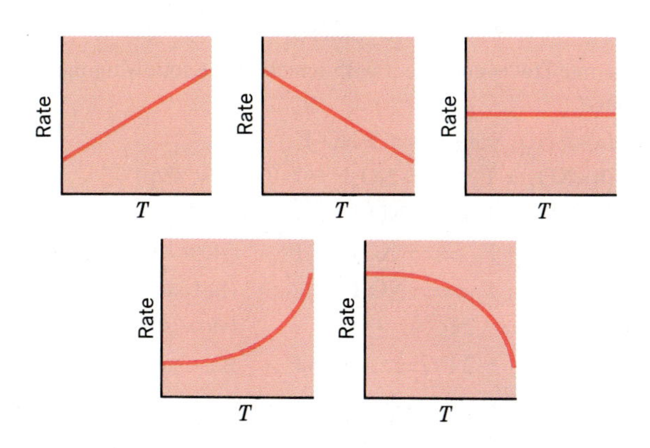


**22-6** Describe what happens to the rate at which a reactant is consumed during the course of a reaction. (Does it increase, decrease, or remain the same?)

**22-7** Explain why mixtures of $H_2$ and $O_2$ gas do not react when stored at room temperature for several years, whereas the reaction is complete within a few days at 300°C, within a few hours at 500°C, and almost instantaneously at 700°C.

### Rate Laws and Rate Constants

**22-8** What are the units of the rate constant for the following reaction?

$$N_2O_4(g) \rightleftharpoons 2\ NO_2(g)$$
$$\text{Rate} = k(N_2O_4)$$

**22-9** What are the units of the rate constant for the following reaction?

$$2\ NO_2(g) \rightleftharpoons N_2O_4(g)$$
$$\text{Rate} = k(NO_2)^2$$

**22-10** What are the units of the rate constant for the following reaction?

$$2\ Br^-(aq) + H_2O_2(aq) + 2\ H^+(aq) \rightleftharpoons Br_2(aq) + 2\ H_2O(aq)$$
$$\text{Rate} = k(Br^-)(H_2O_2)(H^+)$$

**22-11** A reaction is said to be diffusion controlled when it occurs as fast as the reactants diffuse through the solution. The following is a good example of a diffusion-controlled reaction:

$$H_3O^+(aq) + OH^-(aq) \rightleftharpoons 2\ H_2O(l)$$

Given that the rate constant for this reaction is $1.4 \times 10^{11}\ M^{-1}\ s^{-1}$ at 25°C and that the reaction obeys the following rate law:

$$\text{Rate} = k(H_3O^+)(OH^-)$$

calculate the rate of reaction in a neutral solution (pH = 7.00).

### Different Ways of Expressing the Rate of Reaction

**22-12** Which equation describes the relationship between the rates at which $Cl_2$ and $F_2$ are consumed in the following reaction?

$$Cl_2(g) + 3\ F_2(g) \rightleftharpoons 2\ ClF_3(g)$$

(a) $-d(Cl_2)/dt = -d(F_2)/dt$
(b) $-d(Cl_2)/dt = 2[-d(F_2)/dt]$
(c) $2[-d(Cl_2)/dt] = -d(F_2)/dt$
(d) $-d(Cl_2)/dt = 3[-d(F_2)/dt]$
(e) $3[-d(Cl_2)/dt] = -d(F_2)/dt$

**22-13** Which equation describes the relationships between the rates at which $Cl_2$ is consumed and $ClF_3$ is produced in the following reaction?

$$Cl_2(g) + 3\ F_2(g) \rightleftharpoons 2\ ClF_3(g)$$

(a) $-d(Cl_2)/dt = d(ClF_3)/dt$
(b) $-d(Cl_2)/dt = 2[d(ClF_3)/dt]$
(c) $2[-d(Cl_2)/dt] = d(ClF_3)/dt$
(d) $-d(Cl_2)/dt = 3[d(ClF_3)/dt]$
(e) $3[-d(Cl_2)/dt] = d(ClF_3)/dt$

**22-14** Calculate the rate at which $N_2O_4$ is formed in the following reaction at the moment in time when $NO_2$ is being consumed at a rate of 0.0592 *M*/s:

$$2\ NO_2(g) \rightleftharpoons N_2O_4(g)$$

**22-15** Calculate the rate constant for the formation of $I_2$ in the following reaction at 500°C if the rate constant for the disappearance of HI is $0.039\ M^{-1}\ s^{-1}$.

$$2\ HI(g) \rightleftharpoons H_2(g) + I_2(g)$$
$$\text{Rate} = k(HI)^2$$

**22-16** Ammonia burns in the gas phase to form nitrogen oxide and water:

$$4\,NH_3(g) + 5\,O_2(g) \longrightarrow 4\,NO(g) + 6\,H_2O(g)$$

Derive the relationship between the rates at which $NH_3$ and $O_2$ are consumed in this reaction. Derive the relationship between the rates at which $O_2$ is consumed and $H_2O$ is produced.

**22-17** Calculate the rate of formation of NO and the rate of disappearance of $NH_3$ for the following reaction at the moment when the rate of formation of water is 0.040 *M*/s:

$$4\,NH_3(g) + 5\,O_2(g) \longrightarrow 4\,NO(g) + 6\,H_2O(g)$$

**22-18** The instantaneous rate of disappearance of the $MnO_4^-$ ion in the following reaction is $4.56 \times 10^{-3}$ *M*/s at some moment in time:

$$10\,I^-(aq) + 2\,MnO_4^-(aq) + 16\,H^+(aq) \longrightarrow 2\,Mn^{2+}(aq) + 5\,I_2(aq) + 8\,H_2O(l)$$

What is the rate of appearance of $I_2$ at the same moment?

### The Rate Law versus the Stoichiometry of a Reaction

**22-19** Describe the difference between the *stoichiometry* and the *mechanism* of a reaction.

**22-20** Describe the conditions under which the rate law for the reaction is most likely to reflect the stoichiometry of the reaction.

**22-21** Describe one or more factors that can make the rate law for a reaction differ from what the stoichiometry of the reaction leads us to expect.

### Order and Molecularity

**22-22** Describe the difference between unimolecular and bimolecular reactions. Give an example of each.

**22-23** Which graph best describes the rate of the following reaction if this reaction is first order in $N_2O_4$?

$$N_2O_4(g) \rightleftharpoons 2\,NO_2(g)$$

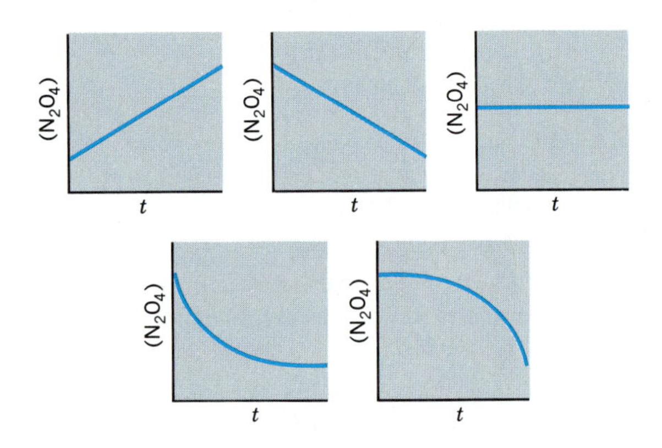


### The Mechanisms of Chemical Reactions

**22-24** Each of the following reactions was found experimentally to be second order overall. Which of them is most likely to be an elementary reaction that occurs in a single step?

(a) $2\,NO_2(g) + Cl_2(g) \longrightarrow 2\,NO_2Cl(g)$

(b) $N_2O_3(g) \longrightarrow NO(g) + NO_2(g)$

(c) $3\,O_2(g) \longrightarrow 2\,O_3(g)$

(d) $2\,NO(g) \longrightarrow N_2(g) + O_2(g)$

**22-25** NO reacts with $H_2$ according to the following equation:

$$2\,NO(g) + 2\,H_2(g) \longrightarrow N_2(g) + 2\,H_2O(g)$$

The mechanism for this reaction involves two steps:

$2\,NO + H_2 \longrightarrow N_2 + H_2O_2$ (slow step)

$H_2O_2 + H_2 \longrightarrow 2\,H_2O$ (fast step)

What is the experimental rate law for this reaction?

**22-26** The following reaction is first order in both $NO_2$ and $F_2$:

$$2\,NO_2(g) + F_2(g) \longrightarrow 2\,NO_2F(g)$$

$$\text{Rate} = k(NO_2)(F_2)$$

This rate law is consistent with which of the following mechanisms?

(a) $2\,NO_2 + F_2 \rightleftharpoons 2\,NO_2F$

(b) $NO_2 + F_2 \rightleftharpoons NO_2F + F$ (fast step)
$NO_2 + F \rightleftharpoons NO_2F$ (slow step)

(c) $NO_2 + F_2 \rightleftharpoons NO_2F + F$ (slow step)
$NO_2 + F \rightleftharpoons NO_2F$ (fast step)

(d) $F_2 \rightleftharpoons 2\,F$ (slow step)
$2\,NO_2 + 2\,F \rightleftharpoons 2\,NO_2F$ (fast step)

**22-27** The disproportionation of NO to $N_2O$ and $NO_2$ is third order in NO:

$$3\,NO(g) \longrightarrow N_2O(g) + NO_2(g)$$

$$\text{Rate} = k(NO)^3$$

This rate law is consistent with which of the following mechanisms?

(a) $NO + NO + NO \rightleftharpoons N_2O + NO_2$ (one-step)

(b) $2\,NO \rightleftharpoons N_2O_2$ (slow step)
$N_2O_2 + NO \rightleftharpoons N_2O + NO_2$ (fast step)

(c) $2\,NO \rightleftharpoons N_2O_2$ (fast step)
$N_2O_2 + NO \rightleftharpoons N_2O + NO_2$ (slow step)

**22-28** Predict the rate law for the oxidation of the iodide ion by hypochlorite:

$$I^-(aq) + OCl^-(aq) \longrightarrow Cl^-(aq) + OI^-(aq)$$

if the reaction proceeds by the following mechanism:

$OCl^- + H_2O \rightleftharpoons HOCl + OH^-$ (fast step)

$I^- + HOCl \longrightarrow HOI + Cl^-$ (slow step)

$HOI + OH^- \longrightarrow OI^- + H_2O$ (fast step)

**22-29** $N_2O_5$ decomposes to form $NO_2$ and $O_2$:

$$2\,N_2O_5(g) \rightleftharpoons 4\,NO_2(g) + O_2(g)$$

The rate law for this reaction is first order in $N_2O_5$:

$$\text{Rate} = k(N_2O_5)$$

Show how this rate law is consistent with the following three-step mechanism for this reaction:

**Step 1:** $N_2O_5 \rightleftharpoons NO_2 + NO_3$ (fast step)

**Step 2:** $NO_2 + NO_3 \longrightarrow NO_2 + NO + O_2$ (slow step)

**Step 3:** $NO + NO_3 \longrightarrow 2\,NO_2$ (fast step)

**22-30** The following is the experimental rate law for the reaction between hydrogen and bromine to form hydrogen bromide, HBr:

$$\text{Rate} = k(H_2)(Br_2)^{\frac{1}{2}}$$

Explain how the following mechanism is consistent with this rate law:

$Br_2 \rightleftharpoons 2\,Br$ (fast step)

$Br + H_2 \longrightarrow HBr + H$ (slow step)

$H + Br_2 \longrightarrow HBr + Br$ (fast step)

$H + HBr \longrightarrow H_2 + Br_2$ (fast step)

### The Relationship Between the Rate Constants and the Equilibrium Constant for a Reaction

**22-31** Describe the relationship between the forward and reverse rate constants and the equilibrium constant for a one-step reaction.

**22-32** The following is a one-step reaction:

$$CH_3Cl(aq) + I^-(aq) \rightleftharpoons CH_3I(aq) + Cl^-(aq)$$

What is the equilibrium constant for this reaction if the rate constant for the forward reaction is $5.2 \times 10^{-7}\,M^{-1}\,s^{-1}$ and the rate constant for the reverse reaction is $1.5 \times 10^{-11}\,M^{-1}\,s^{-1}$?

### Determining the Order of a Reaction from Initial Rates of Reaction

**22-33** Use the following data to determine the rate law for the reaction between nitrogen oxide and chlorine to form nitrosyl chloride:

$$2\,NO(g) + Cl_2(g) \longrightarrow 2\,NOCl(g)$$

| Initial (NO) (*M*) | Initial ($Cl_2$) (*M*) | Initial Rate of Reaction (*M*/s) |
|---|---|---|
| 0.10 | 0.10 | 0.117 |
| 0.20 | 0.10 | 0.468 |
| 0.30 | 0.10 | 1.054 |
| 0.30 | 0.20 | 2.107 |
| 0.30 | 0.30 | 3.161 |

**22-34** Use the results of the preceding problem to determine the rate constant for this reaction. Predict the initial instantaneous rate of reaction when the initial NO and $Cl_2$ concentrations are both 0.50 *M*.

**22-35** Use the following data to determine the rate law for the reaction between nitrogen oxide and oxygen to form nitrogen dioxide:

$$2\,NO(g) + O_2(g) \rightleftharpoons 2\,NO_2(g)$$

| Initial Pressure NO (mmHg) | Initial Pressure $O_2$ (mmHg) | Initial Rate of Reaction (mmHg/s) |
|---|---|---|
| 100 | 100 | 0.355 |
| 150 | 100 | 0.800 |
| 250 | 100 | 2.22 |
| 150 | 130 | 1.04 |
| 150 | 180 | 1.44 |

**22-36** Use the results of the preceding problem to determine the rate constant for this reaction. Predict the initial rate of reaction when the initial pressures for both NO and $O_2$ are 250 mmHg.

**22-37** Use the following data to determine the rate law for the reaction between methyl iodide and the $OH^-$ ion in aqueous solution to form methanol and the iodide ion:

$$CH_3I(aq) + OH^-(aq) \longrightarrow CH_3OH(aq) + I^-(aq)$$

| Initial ($CH_3I$) (*M*) | Initial ($OH^-$) (*M*) | Initial Rate of Reaction (*M*/s) |
|---|---|---|
| 1.35 | 0.10 | $8.78 \times 10^{-6}$ |
| 1.00 | 0.10 | $6.50 \times 10^{-6}$ |
| 0.85 | 0.10 | $5.53 \times 10^{-6}$ |
| 0.85 | 0.15 | $8.29 \times 10^{-6}$ |
| 0.85 | 0.25 | $1.38 \times 10^{-5}$ |

**22-38** Use the results of the preceding question to determine the rate constant for this reaction. Predict the initial instantaneous rate of reaction when the initial $CH_3I$ concentration is 0.10 *M* and the initial $OH^-$ concentration is 0.050 *M*.

### The Integrated Forms of the Rate Laws

**22-39** Describe the kinds of problems that are best solved by using the rate law for a reaction, such as the following:

$$\text{Rate} = k(N_2O_5)$$

Describe the kinds of problems that are best solved with the integrated form of the rate law:

$$\ln\left[\frac{(N_2O_5)}{(N_2O_5)_0}\right] = -kt$$

**22-40** Water was heated in a test tube to 75.0°C and then allowed to cool to room temperature. If this process followed first-order kinetics with a rate constant of $8.0 \times 10^{-4}\,s^{-1}$, what was the temperature of the water after 400 s?

**22-41** In an acceleration test for a BMW 325es, the following speed versus time data were collected after the car shifted into fourth gear:

| Speed (mph): | 80.4 | 83.9 | 87.5 | 91.2 | 95.1 | 99.3 |
|---|---|---|---|---|---|---|
| Time (s): | 16 | 18 | 20 | 22 | 24 | 26 |

Are these data consistent with first-order kinetics? Predict the time at which the speed will reach 100 mph.

**22-42** The reaction in which $NO_2$ forms a dimer is second order in $NO_2$:

$$2\,NO_2(g) \rightleftharpoons N_2O_4(g)$$
$$\text{Rate} = k(NO_2)^2$$

Calculate the rate constant for this reaction if it takes 0.005 seconds for the initial concentration of $NO_2$ to decrease from 0.50 *M* to 0.25 *M*.

**22-43** The decomposition of hydrogen peroxide is first order in $H_2O_2$:

$$2\,H_2O_2(aq) \longrightarrow 2\,H_2O(l) + O_2(g)$$
$$\text{Rate} = k(H_2O_2)$$

How long will it take for half of the $H_2O_2$ in a 10-gal sample to be consumed if the rate constant for this reaction is $5.6 \times 10^{-2}\ s^{-1}$?

### The Integrated Forms of the Rate Laws and Half-Life Calculations

**22-44** Calculate the rate constant for the following acid–base reaction if the half-life for this reaction is 0.0282 s at 25°C and the reaction is first order in the $NH_4^+$ ion:

$$NH_4^+(aq) + H_2O(l) \rightleftharpoons NH_3(aq) + H_3O^+(aq)$$

**22-45** The following reaction is second order in $NO_2$:

$$2\,NO_2(g) \rightleftharpoons N_2O_4(g)$$

What effect would doubling the initial concentration of $NO_2$ have on the half-life for this reaction?

**22-46** The following reaction is first order in both reactants and therefore second order overall:

$$CH_3I(aq) + OH^-(aq) \longrightarrow CH_3OH(aq) + I^-(aq)$$

When the reaction is run in a buffer solution, however, it is pseudo-first order in $CH_3I$:

$$\text{Rate} = k(CH_3I)$$

What is the half-life of this reaction in a pH 10.00 buffer if the rate constant for this pseudo-first-order reaction is $6.5 \times 10^{-9}\ s^{-1}$?

**22-47** The age of a rock can be estimated by measuring the amount of $^{40}Ar$ trapped inside. This calculation is based on the fact that $^{40}K$ decays to $^{40}Ar$ by a first-order process. It also assumes that none of the $^{40}Ar$ produced by this reaction has escaped from the rock since the rock was formed:

$$^{40}K + e^- \longrightarrow {}^{40}Ar \qquad k = 5.81 \times 10^{-11}y^{-1}$$

Calculate the half-life of this radioactive decay.

**22-48** Another way of determining the age of a rock involves measuring the extent to which the $^{87}Rb$ in the rock has decayed to $^{87}Sr$:

$$^{87}Rb \longrightarrow {}^{87}Sr + e^- \qquad k = 1.42 \times 10^{-11}y^{-1}$$

What fraction of the $^{87}Rb$ would still remain in a rock when half of the $^{40}K$ has decayed?

**22-49** $^{14}C$ measurements on the linen wrappings from the Book of Isaiah in the Dead Sea scrolls suggest that these scrolls contain about 79.5% of the $^{14}C$ expected in living tissue. How old are these scrolls, if the half-life for the decay of $^{14}C$ is 5730 years?

**22-50** The Lascaux cave near Montignac in France contains a series of cave paintings. Radiocarbon dating of charcoal taken from this site suggests an age of 15,520 years. What fraction of the $^{14}C$ present in living tissue is still present in this sample? ($^{14}C$: $t_{\frac{1}{2}}$ = 5730 y)

**22-51** A skull fragment found in 1936 at Baldwin Hills, California, was dated by $^{14}C$ analysis. Approximately 100 g of bone was cleaned and treated with 1 *M* HCl(*aq*) to destroy the mineral content of the bone. The bone protein was then collected, dried, and pyrolyzed. The $CO_2$ produced was collected and purified, and the ratio of $^{14}C$ to $^{12}C$ was measured. If this sample contained roughly 5.7% of the $^{14}C$ present in living tissue, how old was the skeleton? ($^{14}C$: $t_{\frac{1}{2}}$ = 5730 y)

**22-52** Charcoal samples from Stonehenge in England emit 62.3% of the disintegrations per gram of carbon per minute expected for living tissue. What is the age of these samples? ($^{14}C$: $t_{\frac{1}{2}}$ = 5730 y)

**22-53** A lump of beeswax was excavated in England near a collection of Bronze Age objects roughly 2500 to 3000 years of age. Radiocarbon analysis of the beeswax suggests an activity equal to roughly 90.3% of the activity observed for living tissue. Determine whether this beeswax was part of the hoard of Bronze Age objects. ($^{14}C$: $t_{\frac{1}{2}}$ = 5730 y)

**22-54** The activity of the $^{14}C$ in living tissue is 15.3 disintegrations per minute per gram of carbon. The limit for reliable determination of $^{14}C$ ages is 0.10 disintegration per minute per gram of carbon. Calculate the maximum age of a sample that can be dated accurately by radiocarbon dating. Assume a half-life of 5730 years.

### Determining the Order of a Reaction with the Integrated Forms of Rate Laws

**22-55** Use the following data to determine the rate law for the decomposition of $N_2O$.

$$2\,N_2O(g) \longrightarrow 2\,N_2(g) + O_2(g)$$

| $(N_2O)$ (*M*): | 0.100 | 0.086 | 0.079 | 0.075 | 0.066 | 0.059 | 0.049 |
|---|---|---|---|---|---|---|---|
| Time (s): | 0 | 80 | 120 | 160 | 240 | 320 | 480 |

**22-56** Use the results of the preceding problem to calculate the rate constant for this reaction. Predict the concentration of $N_2O$ after 900 s.

**22-57** Use the following data to determine the rate law for the hydrolysis of the $BH_4^-$ ion:

$$BH_4^-(aq) + 4\,H_2O(l) \longrightarrow B(OH)_4^-(aq) + 4\,H_2(g)$$

| $(BH_4^-)$ (*M*): | 0.100 | 0.088 | 0.077 | 0.068 | 0.060 | 0.052 | 0.046 |
|---|---|---|---|---|---|---|---|
| Time (h): | 0 | 24 | 48 | 72 | 96 | 120 | 144 |

**22-58** Use the results of the preceding question to calculate the half-life for the reaction.

**22-59** Triphenylphosphine, $PPh_3$, reacts with nickel tetracarbonyl, $Ni(CO)_4$, to displace a molecule of carbon monoxide:

$$Ni(CO)_4 + PPh_3 \longrightarrow Ph_3PNi(CO)_3 + CO$$

The following data were obtained when this reaction was run at 25°C in the presence of a large excess of triphenylphosphine:

| $(Ni(CO)_4)$ ($M$): | 10.0 | 7.6 | 5.8 | 4.4 | 3.3 | 2.5 |
|---|---|---|---|---|---|---|
| Time (s): | 0 | 40 | 80 | 120 | 160 | 200 |

Use these data to determine whether the reaction is first order or second order in $Ni(CO)_4$.

**22-60** The rate of the reaction in the preceding problem does not depend on the concentration of $PPh_3$. Combine this fact with the results of the preceding problem to determine whether the rate law for this reaction is consistent with the following mechanism:

$$Ni(CO)_4 \rightleftharpoons Ni(CO)_3 + CO \qquad \text{(slow step)}$$

$$Ni(CO)_3 + PPh_3 \rightleftharpoons Ph_3PNi(CO)_3 \qquad \text{(fast step)}$$

**22-61** $Cr(NH_3)_5Cl^{2+}$ reacts with the $OH^-$ ion in aqueous solution to displace $Cl^-$ from the complex ion.

$$Cr(NH_3)_5Cl^{2+}(aq) + OH^-(aq) \longrightarrow Cr(NH_3)_5(OH)^{2+}(aq) + Cl^-(aq)$$

The following data were obtained when this reaction was run at 25°C in a buffer solution at constant pH:

| $(Cr(NH_3)_5Cl^{2+})$ ($M$): | 1.00 | 0.81 | 0.66 | 0.53 | 0.43 | 0.35 |
|---|---|---|---|---|---|---|
| Time (min): | 0 | 3 | 6 | 9 | 12 | 15 |

Use these data to determine whether this reaction is first order or second order in $Cr(NH_3)_5Cl^{2+}$.

**22-62** The rate of the reaction in the preceding problem is proportional to the pH of the buffer solution in which the reaction is run. Each time the buffer is changed so that the $OH^-$ ion concentration is doubled, the rate of reaction increases by a factor of 2. Combine this observation with the results of the preceding problem to determine the rate law for this reaction.

**22-63** Show that the rate law derived in the preceding problem is consistent with the following mechanism:

$$Cr(NH_3)_5Cl^{2+} + OH^- \rightleftharpoons Cr(NH_3)_4(NH_2)(Cl)^+ + H_2O \qquad \text{(slow step)}$$

$$Cr(NH_3)_4(NH_2)(Cl)^+ \rightleftharpoons Cr(NH_3)_4(NH_2)^{2+} + Cl^- \qquad \text{(fast step)}$$

$$Cr(NH_3)_4(NH_2)^{2+} + H_2O \rightleftharpoons Cr(NH_3)_5(OH)^{2+} \qquad \text{(fast step)}$$

**22-64** Dimethyl ether, $CH_3OCH_3$, decomposes at high temperatures as shown in the following equation:

$$CH_3OCH_3(g) \longrightarrow CH_4(g) + H_2(g) + CO(g)$$

The following data were obtained when the partial pressure of $CH_3OCH_3$ was studied as this compound decomposed at 500°C. Use these data to determine the order of this reaction.

| $P_{CH_3OCH_3}$ (mmHg): | 312 | 278 | 251 | 227 | 157 |
|---|---|---|---|---|---|
| Time (s): | 0 | 390 | 777 | 1195 | 3155 |

### Reactions That Are First Order in Two Reactants

**22-65** The following reaction is first order in both $CH_3I$ and $OH^-$.

$$CH_3I(aq) + OH^-(aq) \longrightarrow CH_3OH(aq) + I^-(aq)$$

Describe how to turn this reaction into one that is pseudo-first order in $CH_3I$.

### Catalysts and the Rate of Chemical Reactions

**22-66** Describe the four properties of a catalyst. Give an example of a catalyzed reaction and show how the catalyst meets these criteria.

### A Collision Theory Model of Chemical Reactions

**22-67** Describe the factors that determine whether a collision between two molecules will lead to a reaction.

**22-68** Describe the relationship between the rate of a reaction and the activation energy for the reaction.

### Determining the Activation Energy of a Reaction

**22-69** Assume that the activation energy was measured for both the forward ($E_a$ = 120 kJ/mol) and reverse ($E_a$ = 185 kJ/mol) directions of a reversible reaction. What would be the activation energy for the reverse reaction in the presence of a catalyst that decreased the activation energy for the forward reaction to 90 kJ/mol?

**22-70** The rate constant for the decomposition of $N_2O_5$ increases from $1.52 \times 10^{-5}\ s^{-1}$ at 25°C to $3.83 \times 10^{-3}\ s^{-1}$ at 45°C. Calculate the activation energy for this reaction.

**22-71** Calculate the activation energy for the following reaction if the rate constant for this reaction increases from 87.1 $M^{-1}\ s^{-1}$ at 500 K to $1.53 \times 10^3\ M^{-1}\ s^{-1}$ at 650 K:

$$2\ NO_2(g) \longrightarrow 2\ NO(g) + O_2(g)$$

**22-72** Calculate the activation energy for the decomposition of $NO_2$ from the temperature dependence of the rate constant for this reaction:

$$2\ NO_2(g) \longrightarrow N_2(g) + 2\ O_2(g)$$

| Temperature (K): | 319 | 329 | 352 | 381 | 389 |
|---|---|---|---|---|---|
| $k$ ($M^{-1}\ s^{-1}$): | 0.522 | 0.755 | 1.70 | 4.02 | 5.03 |

**22-73** Calculate the rate constant at 780 K for the following reaction if the rate constant for the reaction is $3.5 \times 10^{-7}\ M^{-1}\ s^{-1}$ at 550 K and the activation energy is 188 kJ/mol:

$$2\ HI(g) \rightleftharpoons H_2(g) + I_2(g)$$

**22-74** Calculate the rate constant at 75°C for the following reaction if the rate constant for this reaction is $6.5 \times 10^{-5}\ M^{-1}\ s^{-1}$ at 25°C and the activation energy is 92.9 kJ/mol:

$$CH_3I(aq) + OH^-(aq) \longrightarrow CH_3OH(aq) + I^-(aq)$$

### The Kinetics of Enzyme-Catalyzed Reactions

**22-75** Explain why the rate of the enzyme-catalyzed hydrolysis of sucrose is first order in sucrose at low concentrations of this substance.

**22-76** Explain why the rate of enzyme-catalyzed reactions becomes zero order at very high concentrations of the substrate.

### Integrated Problems

**22-77** A 50-mL Mohr buret was connected to a 1-ft length of capillary tubing. The buret was then filled with water, and the volume of water versus time was monitored as the water gradually flowed through the capillary tubing. Use the following data to determine whether the rate at which water flows through the capillary tubing fits first-order or second-order kinetics. Determine the rate constant for this process.

| $V$ (mL): | 50 | 40 | 30 | 20 | 10 | 0 |
|---|---|---|---|---|---|---|
| $t$ (s): | 0 | 19 | 42 | 72 | 116 | 203 |

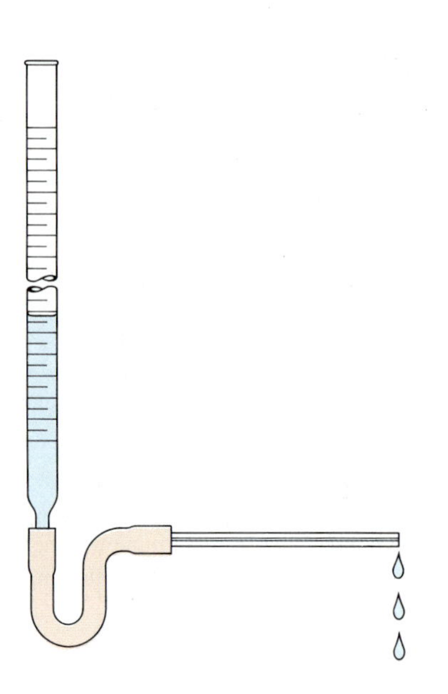

**22-78** Predict the effect of doubling the length of the capillary tubing on the rate constant for the process described in the previous problem. Compare your prediction with the following experimental results:

| $V$ (mL): | 50 | 40 | 30 | 20 | 10 | 0 |
|---|---|---|---|---|---|---|
| $t$ (s): | 0 | 43 | 97 | 171 | 282 | 523 |

**22-79** A piece of glass tubing threaded through a one-hole rubber stopper was inserted in the mouth of the Mohr buret. A source of compressed air was then connected to the glass tubing. When a constant pressure was applied to the top of the buret, the following data were obtained.

| $V$ (mL): | 45 | 40 | 35 | 30 | 25 | 20 | 15 | 10 | 5 | 0 |
|---|---|---|---|---|---|---|---|---|---|---|
| $t$ (s): | 0 | 75 | 150 | 224 | 300 | 376 | 446 | 520 | 595 | 674 |

Determine whether these data fit zero-order, first-order, or second-order kinetics.

# Intersection

## APPLICATION OF CHEMICAL KNOWLEDGE

Chemistry, like any other science, requires a balance between theory and experiment. The search for patterns in experimental data guide the development of theories, which shape the nature of the experiments that are then done. It is not surprising that introductory courses in chemistry therefore oscillate between discussions of experimental facts and the theories developed both to explain these facts and predict the results of experiments that have not yet been done.

Most of the material in this section could be covered at almost any point in the course. But some aspects of the chemistry of nuclear reactions, organic compounds, and both natural and synthetic polymers are best understood after discussions of equilibria, thermodynamics, and kinetics.

This section builds on knowledge constructed throughout the core of this textbook to answer new—and often more sophisticated—questions.

- We know a $^{40}K$ atom contains 19 protons and 11 neutrons. Why is this combination unstable? Why does $^{40}K$ decay to form both $^{40}Ar$ and $^{40}Ca$?
- We know that the energy of electromagnetic radiation depends on its frequency. How can we use this information to explain why high-energy UV radiation is much more dangerous than IR radiation?
- We know that it is possible to transform a gas into a liquid by raising the pressure on the gas. How can we use this information to explain the difference between natural gas and liquified petroleum gases (LPG)?
- How do we explain the fact that theobromine and theophylline have the same molecular formula ($C_7H_8N_4O_2$) and yet the theobromine that can be extracted from chocolate has a very different effect on the central nervous system than the theophylline found in the bronchodilator ''Theo-Dur?''
- The keratins are a family of durable and unreactive proteins that can be found in hair, horn, nails, hoofs, feathers, and even skin. Why does increasing the amount of sulfur in these proteins change the ''soft'' keratins in skin into the ''hard'' keratins in your hair and nails?

The next chapters provide the basis for understanding the difference between the ''soft'' keratins in the fur on these Dall sheep rams and the ''hard'' keratins in their horns.

For some students, the contents of these chapters will be a first step toward answering even more complex questions, such as:

- How can the force of repulsion between objects with the same charge simultaneously be used to explain the $\alpha$-particle scattering observed by Rutherford-Marsden-Geiger and the tendency of a radioactive nuclei to split into two halves of different size?
- How can we use our understanding of solubility, which is often summarized in the general rule that ''like dissolves like,'' to explain diseases such as sickle-cell anemia that are caused by mutations that change the structure of essential proteins such as hemoglobin?
- What is the difference between high-density lipoproteins (HDL) and low-density lipoproteins (LDL)? Why is an excess of LDL a major cause of atherosclerosis? Why do patients suffering from atherosclerosis have a low concentration of HDL? What is the relationship between these lipoproteins and cholesterol? Why are measurements of the cholesterol level no longer as useful as they once were?

Remnants of the Vela supernova.

# NUCLEAR CHEMISTRY

General chemistry courses often begin with discussions of evidence for the existence of atoms and end by describing what happens when changes occur in the structure of atoms. This chapter provides the basis for answering the following questions:

## POINTS OF INTEREST

- Why are some isotopes (such as $^{12}C$ and $^{13}C$) stable, while others (such as $^{14}C$) are not? What information do we need to predict whether a particular isotope is stable? What information do we need to predict how an unstable isotope will decay?
- Why do some nuclei undergo *fusion* to become larger, while others undergo *fission* to become smaller? How can we induce fission in a nucleus that does not do this spontaneously?
- Why do nuclear reactions release far more energy than chemical reactions? Where does this energy come from?
- What are the biological effects of radiation? Why do scientists distinguish between *ionizing* and *nonionizing radiation?*
- At a time when some people fear even the term *nuclear,* can we develop an understanding of the conditions under which such fear is valid, and when it is not?

## 23.1 THE PROCESS OF DISCOVERY: RADIOACTIVITY

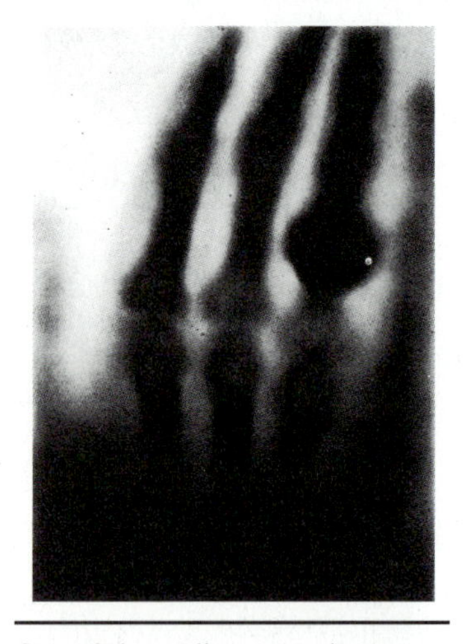

One of the earliest x-ray images. There is reason to believe this is an x-ray of Roentgen's wife's hand.

The discovery of x-rays by William Conrad Roentgen in November of 1895 excited the imagination of a generation of scientists, who rushed to study this phenomenon. Within a few months, Henri Becquerel found that both uranium metal and salts of this element gave off a different form of radiation, which could also pass through solids. By 1898, Marie Curie found that compounds of thorium were also ''radioactive.'' After painstaking effort she eventually isolated two more radioactive elements—polonium and radium—from ores that contained uranium.

In 1899 Ernest Rutherford studied the absorption of radioactivity by thin sheets of metal foil. He found that there were at least two different forms of radioactivity. One form, which he called *alpha (α) particles,* were absorbed by metal foil that was a few hundredths of a centimeter thick. The other, *beta (β) particles,* could pass through 100 times as much metal foil before they became absorbed. Shortly thereafter, a third form of radiation, *gamma (γ) rays,* was discovered that could penetrate as much as several centimeters of lead.

The results of early experiments on the effect of electric fields on these three forms of radiation are shown in Figure 23.1. The direction in which α-particles were deflected suggested that they were positively charged. The magnitude of this deflection suggested that they had the same charge-to-mass ratio as a $He^{2+}$ ion. To test the equivalence between α-particles and $He^{2+}$ ions, Rutherford built an apparatus that allowed α-particles to pass through a very thin glass wall into an evacuated flask that contained a pair of metal electrodes. After a few days, he connected these electrodes to a battery and noted that the gas in the flask did indeed give off the characteristic emission spectrum of helium.

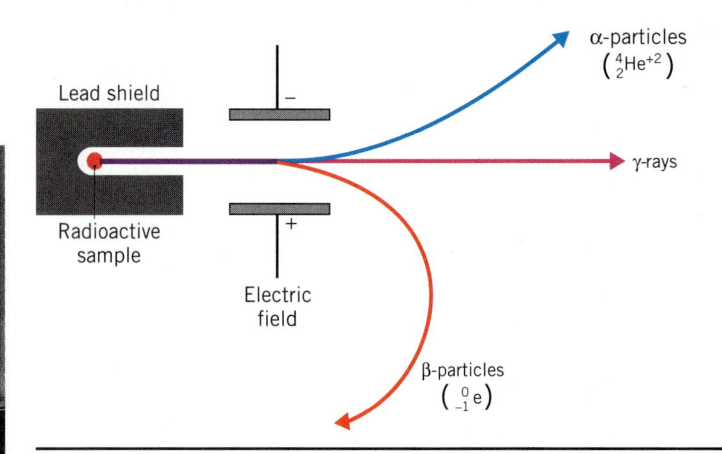


**Figure 23.1** The effect of an electric field on α-, β-, and γ-radiation.

Marie Curie in her laboratory ca. 1905.

Experiments with electric and magnetic fields suggested that β-particles were negatively charged. Furthermore, they had the same charge-to-mass ratio as an electron. To date, no detectable difference has been found between β-particles and electrons. The only reason to retain the name ''β-particle'' is to emphasize the fact that these particles are ejected from the nucleus of an atom when it undergoes radioactive decay.

Sec. 6.11

The fact that γ-rays are not deflected by either electric or magnetic fields suggests that these rays don't carry an electric charge. Since they travel at the speed of light, they are classified as a form of electromagnetic radiation that carries even more energy than x-rays.

At the turn of the century, when radioactivity was discovered, atoms were assumed to be indestructible. Ernest Rutherford and Frederick Soddy, however, found that radioactive substances became less active with time (see Figure 23.2). More importantly, they noticed that radioactivity was always accompanied by the formation of atoms of a different element. By 1903, they concluded that radioactivity was accompanied by a change in the structure of the atom. They therefore assumed that radiation was emitted when an element decayed into a different kind of atom.

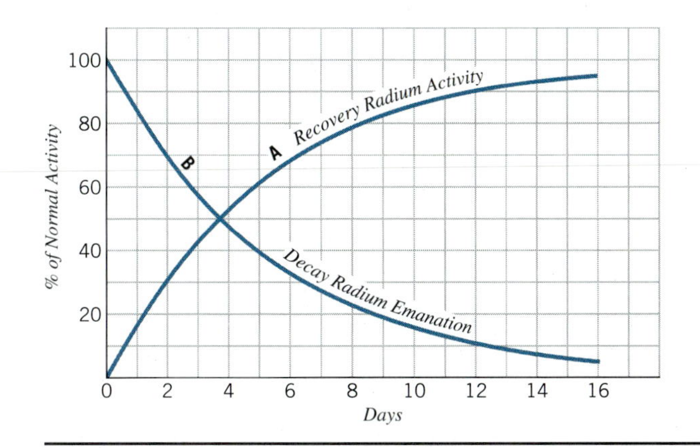


**Figure 23.2** Growth and decay curves reported by Rutherford and Soddy for "uranium X," produced when uranium undergoes radioactive decay. Curve B shows the decay in the activity of this substance after it is extracted from uranium. Curve A shows the growth in the activity of the uranium as "uranium X" is replenished by radioactive decay.

Ernest Rutherford and J. A. Ratcliffe in Rutherford's laboratory.

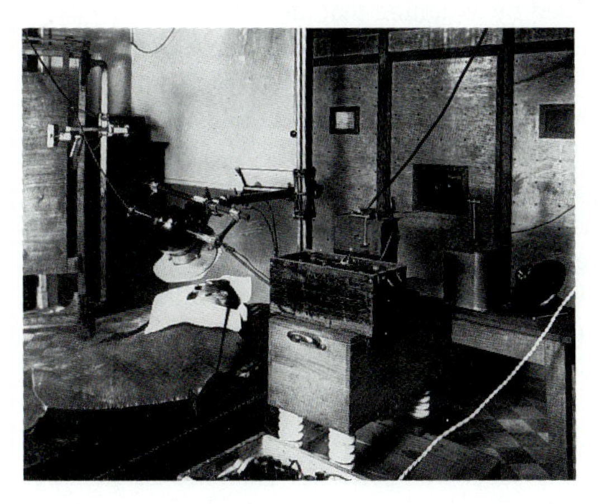

A patient suffering from cancer being treated by the $\alpha$-particle decay of radium in Madame Curie's Institute at the University of Paris. Every part of the patient's body other than the part to be treated is covered with lead which is impervious to the radium rays.

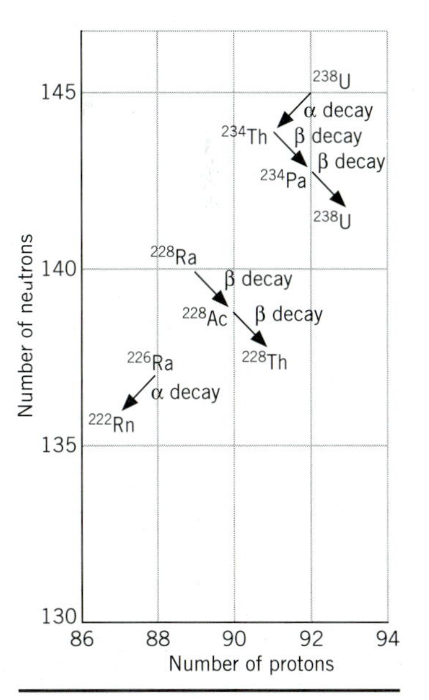


**Figure 23.3** The product of $\alpha$ decay lies two spaces to the left in the periodic table. The product of $\beta$ decay lies one space to the right.

By 1910, at least 40 radioactive elements had been isolated that were associated with the process by which uranium metal decayed to lead. This created a problem, however, because there was space for only 11 elements between lead and uranium. In 1913, Kasimir Fajans and Frederick Soddy proposed an explanation for these results based on the following rules, which are illustrated in Figure 23.3.

**1.** $\alpha$-Particles are emitted when an element is formed that belongs two spaces to the left in the periodic table. Uranium ($Z = 92$), for example, emits an $\alpha$-particle when it decays to form thorium ($Z = 90$).

**2.** $\beta$-Particles are emitted when an element is formed that belongs one space to the right in the periodic table. Actinium ($Z = 89$) emits a $\beta$-particle when it decays to form thorium ($Z = 90$).

**3.** Radioactive elements that fall in the same place in the periodic table are different forms of the same element. The radioactive thorium produced by the $\alpha$-particle decay of uranium is a different form of the element than the radioactive thorium obtained by the $\beta$-particle decay of actinium.

Sec. 3.2

Soddy proposed the name *isotope* to describe different radioactive atoms that occupy the same position in the periodic table. J. J. Thomson and Francis Aston then used a mass spectrometer to show that isotopes are atoms of the same element that have different atomic masses.

## 23.2 THE STRUCTURE OF THE ATOM

The discovery of the electron in 1897 by J. J. Thomson suggested that there was an internal structure to the ''indivisible'' building blocks of matter known as atoms. This, of course, raised the question: How many electrons did an atom contain? By studying the scattering of light, x-rays, and $\alpha$-particles, Thomson concluded that the number of electrons in an atom was between 0.2 and 2 times the weight of the atom.

In 1911, Rutherford concluded that the scattering of $\alpha$-particles by extremely thin pieces of metal foil could be explained by assuming that all of the positive charge and most of the mass of the atom were concentrated in an infinitesimally small fraction of the total volume of the atom, for which he proposed the name *nucleus*. Rutherford's data also suggested that the nucleus of a gold atom carries a positive charge that is about 80 times the charge on an electron.

The discovery of the neutron in 1932 explained the discrepancy between the charge on the nucleus and the mass of an atom. A gold atom that has a mass of 197 amu consists of a nucleus that contains 79 protons and 118 neutrons surrounded by 79 electrons. By convention, this information is specified by the following symbol, which describes the only naturally occurring isotope of gold:

$$^{197}_{79}\text{Au}$$

This convention also can be applied to subatomic particles. The only difference is the use of lowercase letters to identify the particle:

$$\underset{\text{electron}}{^{0}_{-1}\text{e}} \qquad \underset{\text{proton}}{^{1}_{1}\text{p}} \qquad \underset{\text{neutron}}{^{1}_{0}\text{n}}$$

Because anyone with access to a periodic table can find the atomic number of an element, a shorthand notation is often used that reports only the mass number of the atom and the symbol of the element. The shorthand notation for the naturally occurring isotope of gold is $^{197}$Au.

**EXERCISE 23.1**

Determine the number of protons, neutrons, and electrons in a $^{210}Pb^{2+}$ ion.

**SOLUTION** The atomic number of lead is 82, which means this ion contains 82 protons. Since it has a charge of +2, this ion must contain 80 electrons. Because neutrons and protons both have a mass of about 1 amu, the difference between the mass number (210) and the atomic number (82) is equal to the number of neutrons in the nucleus of the atom. This ion therefore contains 128 neutrons.

A particular combination of protons and neutrons is called a **nuclide.** Nuclides with the same number of protons are called **isotopes.** Nuclides with the same mass number are **isobars.** Nuclides with the same number of neutrons are **isotones.**

**EXERCISE 23.2**

Classify the following sets of nuclides as examples of either istopes, isobars or isotones:

(a) $^{12}C$, $^{13}C$, and $^{14}C$ (b) $^{40}Ar$, $^{40}K$, and $^{40}Ca$ (c) $^{14}C$, $^{15}N$, and $^{16}O$

**SOLUTION**

(a) $^{12}C$, $^{13}C$, and $^{14}C$ are *isotopes* because they all contain six protons.
(b) $^{40}Ar$, $^{40}K$, and $^{40}Ca$ are *isobars* because they have the same mass number.
(c) $^{14}C$, $^{15}N$, and $^{16}O$ can't be isotopes because they contain different numbers of protons. They can't be isobars because they have different mass numbers. They all contain eight neutrons, however, so they are examples of *isotones*.

## 23.3 MODES OF RADIOACTIVE DECAY

Early studies of radioactivity indicated that three different kinds of radiation were emitted, symbolized by the first three letters of the Greek alphabet $\alpha$, $\beta$, and $\gamma$. With time, it became apparent that this classification scheme was too simple. The emission of a negatively charged $\beta^-$ particle, for example, is only one example of a family of radioactive transformations known as $\beta$ decay. A fourth category, known as spontaneous fission, also had to be added to describe the process by which certain radioactive nuclides decompose into fragments of different weights.

### ALPHA DECAY

**Alpha decay** is usually restricted to the heavier elements in the periodic table. (Only a handful of nuclei with atomic numbers less than 83 emit an $\alpha$-particle.) The product of $\alpha$ decay is easy to predict if we assume that both mass and charge are conserved in nuclear reactions. Alpha decay of the $^{238}U$ ''parent'' nuclide, for example, produces $^{234}Th$ as the ''daughter'' nuclide:

$$^{238}_{92}U \longrightarrow \, ^{234}_{90}Th + \, ^{4}_{2}He \qquad (\alpha \text{ decay})$$

The sum of the mass numbers of the products (234 + 4) is equal to the mass number of the parent nuclide (238), and the sum of the charges on the products (90 + 2) is equal to the charge on the parent nuclide.

## BETA DECAY

Three different modes of radioactivity fall into the category of **beta decay:** (1) electron ($\beta^-$) emission, (2) electron capture, and (3) positron ($\beta^+$) emission.

In **electron ($\beta^-$) emission,** an electron is ejected from the nucleus. As a result, the charge on the nucleus increases by 1. Electron ($\beta^-$) emitters are found throughout the periodic table, from the lightest elements ($^3$H) to the heaviest ($^{255}$Es). The product of $\beta^-$ emission can be predicted by assuming that mass and charge are conserved in nuclear reactions. If $^{40}$K is a $\beta^-$ emitter, for example, the product of this reaction must be $^{40}$Ca:

$$^{40}_{19}\text{K} \longrightarrow {}^{40}_{20}\text{Ca} + {}^{0}_{-1}\text{e} \qquad \text{[electron } (\beta^-) \text{ emission]}$$

Once again the sum of the mass numbers of the products is equal to the mass number of the parent nuclide, and the sum of the charge on the products is equal to the charge on the parent nuclide.

There is an important difference between the energetics of $\alpha$ and $\beta$ decay. The $\alpha$-particles emitted by $^{238}$U all carry essentially the same energy. The energy of the electrons emitted by $^{40}$K covers a broad spectrum, from slightly more than 1000 kJ/mol to a maximum of 127.8 MJ/mol. In 1930 Wolfgang Pauli explained this by suggesting that the energy given off during this reaction is divided between two particles emitted simultaneously. The first particle, of course, is the $\beta^-$. The second particle is an *antineutrino* ($\overline{\nu}$), which is the antimatter equivalent of the *neutrino* ($\nu$) (from the Italian meaning "a little neutral particle"). The decay of a nuclide by electron emission therefore should be written as follows:

$$^{14}_{6}\text{C} \longrightarrow {}^{14}_{7}\text{N} + {}^{0}_{-1}\text{e} + \overline{\nu}$$

Because it carries neither charge nor mass, we will ignore the antineutrino from now on.

Nuclei can also decay by capturing one of the electrons that surround the nucleus. **Electron capture** leads to a decease of one in the charge on the nucleus. The energy given off in this reaction is carried by an x-ray photon, which is represented by the symbol $h\nu$, where $h$ is Planck's constant and $\nu$ is the frequency of the x-ray. The product of this reaction can be predicted, once again, by assuming that mass and charge are conserved:

$$^{40}_{19}\text{K} + {}^{0}_{-1}\text{e} \longrightarrow {}^{40}_{18}\text{Ar} + h\nu \qquad \text{(electron capture)}$$

The electron captured by the nucleus in this reaction is usually a 1*s* electron because electrons in this orbital are the closest to the nucleus.

A third form of beta decay is called **positron ($\beta^+$) emission.** The positron is the antimatter equivalent of an electron. It has the same mass as an electron, but the opposite charge. Positron ($\beta^+$) decay produces a daughter nuclide with one less positive charge on the nucleus. The energy given off in this reaction is partitioned between the $\beta^+$ and a neutrino (which will be ignored from now on).

$$^{40}_{19}\text{K} \longrightarrow {}^{40}_{18}\text{Ar} + {}^{0}_{+1}\text{e} + \nu \qquad \text{[positron } (\beta^+) \text{ emission]}$$

Positrons have a very short lifetime. They rapidly lose their kinetic energy as they pass through matter. As soon as they come to rest, they combine with an electron to form two $\gamma$-ray photons:

$$^{0}_{+1}\text{e} + {}^{0}_{-1}\text{e} \longrightarrow 2\gamma$$

The three forms of $\beta$ decay for the $^{40}$K nuclide are summarized in Figure 23.4. Note that the mass number of the parent and daughter nuclides are the same for

$$^{40}_{19}\text{K} \longrightarrow {}^{40}_{20}\text{Ca} + {}^{0}_{-1}\text{e}$$

$$^{40}_{19}\text{K} \longrightarrow {}^{40}_{18}\text{Ar} + {}^{0}_{+1}\text{e}$$

$$^{40}_{19}\text{K} + {}^{0}_{-1}\text{e} \longrightarrow {}^{40}_{18}\text{Ar} + h\nu$$

**Figure 23.4** $^{40}$K is an unusual nuclide because it simultaneously decays by all three forms of $\beta$ decay: electron emission, electron capture, and positron emission. The two most common modes of decay for this isotope are electron emission, which increases the atomic number of the nuclide, and electron capture, which decreases the atomic number.

electron emission, electron capture, and position emission. All three forms of $\beta$ decay therefore interconvert isobars.

---

**CHECKPOINT**

Explain why electron capture, electron emission, and positron emission all interconvert nuclides that have the same mass, but different numbers of protons.

---

## GAMMA EMISSION

Daughter nuclides produced by $\alpha$ decay or $\beta$ decay are often emitted in an excited state. The excess energy associated with this excited state is released when the nucleus emits a photon in the $\gamma$-ray portion of the electromagnetic spectrum. Most of the time, the $\gamma$-ray is emitted within $10^{-12}$ seconds after the $\alpha$- or $\beta$-particle. In some cases, gamma decay is delayed, and a short-lived, or **metastable,** nuclide is formed. These metastable nuclides are identified by a small letter m written after the mass number. $^{60m}Co$, for example, is produced by the electron emission of $^{60}Fe$:

$$^{60}_{26}\text{Fe} \longrightarrow {}^{60m}_{27}\text{Co} + {}^{0}_{-1}\text{e}$$

The metastable $^{60m}Co$ nuclide has a half-life of 10.5 min. Since electromagnetic radiation carries neither charge nor mass, the product of $\gamma$-ray emission by $^{60m}Co$ is $^{60}Co$:

$$^{60m}_{27}\text{Co} \longrightarrow {}^{60}_{27}\text{Co} + \gamma \qquad (\gamma\text{-ray emission})$$

## SPONTANEOUS FISSION

Nuclides become less stable as the charge on the nucleus increases. Nuclides with atomic numbers of 90 or more undergo a form of radioactive decay known as **spontaneous fission.** In this reaction, the parent nucleus splits into smaller nuclei. The reaction is usually accompanied by the ejection of one or more neutrons.

$$^{252}_{98}\text{Cf} \longrightarrow {}^{140}_{54}\text{Xe} + {}^{108}_{44}\text{Ru} + 4\ {}^{1}_{0}\text{n}$$

For all but the very heaviest isotopes, spontaneous fission is a very slow reaction. Spontaneous fission of $^{238}U$, for example, is almost 2 million times slower than the rate at which this nuclide undergoes $\alpha$ decay.

### EXERCISE 23.3

Predict the products of the following nuclear reactions:

(a) Electron emission by $^{14}C$
(b) Positron emission by $^{8}B$
(c) Electron capture by $^{125}I$
(d) Alpha emission by $^{210}Rn$
(e) Gamma-ray emission by $^{56m}Ni$

**SOLUTION** We can predict the product of each reaction by writing an equation in which both mass number and charge are conserved:

(a) $^{14}_{6}\text{C} \longrightarrow {}^{14}_{7}\text{N} + {}^{0}_{-1}\text{e}$
(b) $^{8}_{5}\text{B} \longrightarrow {}^{8}_{4}\text{Be} + {}^{0}_{+1}\text{e}$
(c) $^{125}_{53}\text{I} + {}^{0}_{-1}\text{e} \longrightarrow {}^{125}_{52}\text{Te} + h\nu$
(d) $^{210}_{86}\text{Rn} \longrightarrow {}^{206}_{84}\text{Po} + {}^{4}_{2}\text{He}$
(e) $^{56m}_{28}\text{Ni} \longrightarrow {}^{56}_{28}\text{Ni} + \gamma$

## 23.4 NEUTRON-RICH VERSUS NEUTRON-POOR NUCLIDES

In 1934 Enrico Fermi proposed a theory that explained the three forms of beta decay. He argued that a neutron could decay to form a proton by emitting an electron:

$$^{1}_{0}\text{n} \longrightarrow {}^{1}_{1}\text{p} + {}^{0}_{-1}\text{e} \qquad [\text{electron } (\beta^-) \text{ emission}]$$

A proton, on the other hand, could be transformed into a neutron by two pathways. It could capture an electron:

$$^{1}_{1}\text{p} + {}^{0}_{-1}\text{e} \longrightarrow {}^{1}_{0}\text{n} \qquad (\text{electron capture})$$

Or it could emit a positron:

$$^{1}_{1}\text{p} \longrightarrow {}^{1}_{0}\text{n} + {}^{0}_{+1}\text{e} \qquad [\text{positron } (\beta^+) \text{ emission}]$$

According to this theory, electron emission results in an *increase* in the atomic number of the nucleus:

$$^{14}_{6}\text{C} \longrightarrow {}^{14}_{7}\text{N} + {}^{0}_{-1}\text{e}$$

Electron capture, on the other hand, leads to a *decrease* in the atomic number of the nucleus:

$$^{7}_{4}\text{Be} + {}^{0}_{-1}\text{e} \longrightarrow {}^{7}_{3}\text{Li} + h\nu$$

Positron emission also *decreases* the atomic number of the nucleus:

$$^{11}_{6}\text{C} \longrightarrow {}^{11}_{5}\text{B} + {}^{0}_{+1}\text{e}$$

A plot of the number of neutrons versus the number of protons for the stable naturally occurring isotopes is shown in Figure 23.5. Several conclusions can be drawn from this figure.

1. The stable nuclides lie in a very narrow band of neutron-to-proton ratios.
2. The ratio of neutrons to protons in stable nuclides gradually increases as the number of protons in the nucleus increases.
3. Light nuclides, such as $^{12}$C, contain about the same number of neutrons and protons. Heavy nuclides, such as $^{238}$U, contain up to 1.6 times as many neutrons as protons.
4. There are no stable nuclides with atomic numbers larger than 83.
5. This narrow band of stable nuclei is surrounded by a sea of instability.
6. Nuclei that lie above this line have too many neutrons and are therefore **neutron rich.**
7. Nuclei that lie below this line don't have enough neutrons and are therefore **neutron poor.**

### CHECKPOINT

Use the fact that the average mass of a sulfur atom is 32.06 amu to predict whether the $^{35}$S isotope is neutron rich or neutron poor.

The most likely mode of decay for a neutron-rich nucleus is one that converts a neutron into a proton. Every neutron-rich radioactive isotope with an atomic number

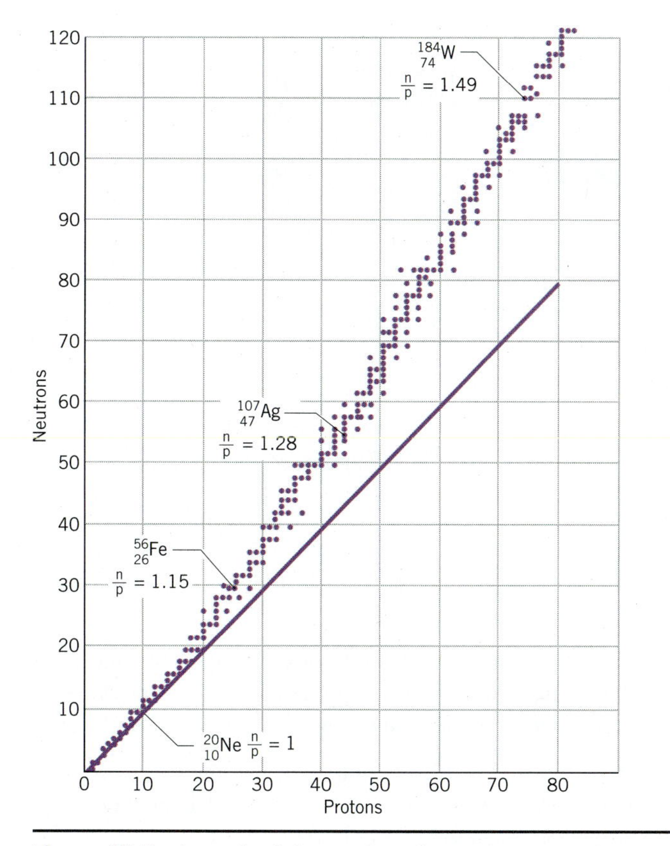


**Figure 23.5** A graph of the number of neutrons versus the number of protons for all stable naturally occurring nuclei. Nuclei that lie to the right of this band of stability are *neutron poor*. Nuclei to the left of the band are *neutron rich*.

smaller than 83 decays by electron ($\beta^-$) emission. $^{14}C$, $^{32}P$, and $^{35}S$ are all examples of neutron-rich nuclei that decay by the emission of an electron:

$$^{35}_{16}S \longrightarrow {}^{35}_{17}Cl + {}^{0}_{-1}e \qquad (\beta^- \text{ emission})$$

Neutron-poor nuclides decay by modes that convert a proton into a neutron. Neutron-poor nuclides with atomic numbers less than 83 tend to decay by either electron capture or positron emission. Many of these nuclides decay by both routes, but positron emission is more often observed in the lighter nuclides, such as $^{22}Na$:

$$^{22}_{11}Na \longrightarrow {}^{22}_{10}Ne + {}^{0}_{+1}e \qquad (\beta^+ \text{ emission})$$

Electron capture is more common among heavier nuclides, such as $^{125}I$, because the 1*s* electrons are held closer to the nucleus as the charge on the nucleus increases:

$$^{125}_{53}I + {}^{0}_{-1}e \longrightarrow {}^{125}_{52}Te + h\nu \qquad (\text{electron capture})$$

A third mode of decay is observed in neutron-poor nuclides that have atomic numbers larger than 83. Although it is not obvious at first, $\alpha$ decay also increases the ratio of neutrons to protons.

$$^{238}_{92}U \longrightarrow {}^{234}_{90}Th + {}^{4}_{2}He \qquad (\alpha \text{ decay})$$

The parent nuclide ($^{238}U$) in this reaction has 92 protons and 146 neutrons, which means that the neutron-to-proton ratio is 1.587. The daughter nuclide ($^{234}Th$) has 90

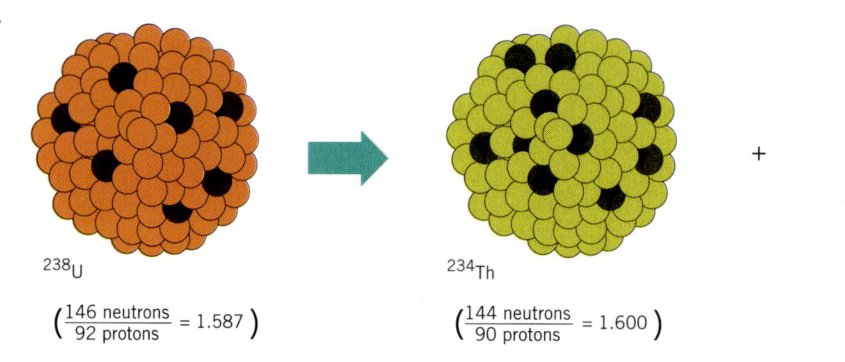


**Figure 23.6** Because the loss of an $\alpha$-particle leads to a small increase in the neutron-to-proton ratio of the nuclide, $\alpha$-particle decay occurs in neutron-poor nuclides such as $^{238}U$.

protons and 144 neutrons, so its neutron-to-proton ratio is 1.600. The daughter nuclide is therefore slightly less likely to be neutron poor, as shown in Figure 23.6.

**EXERCISE 23.4**

Predict the most likely modes of decay and the products of decay of the following nuclides: (a) $^{17}F$ (b) $^{105}Ag$ (c) $^{185}Ta$

**SOLUTION** The first step in predicting the mode of decay of a nuclide is to decide whether the nuclide is neutron rich or neutron poor. This can be done by comparing the mass number of the nuclide with the atomic weight of the element.

(a) The atomic weight of fluorine is 18.998 amu. Because the mass of a $^{17}F$ nuclide is smaller than the average fluorine atom, the $^{17}F$ nuclide must contain fewer neutrons. It is therefore likely to be neutron poor. Because it is a relatively light nuclide, $^{17}F$ might be expected to decay by positron emission:

$$^{17}_{9}F \longrightarrow {}^{17}_{8}O + {}^{0}_{+1}e \qquad (\beta^+ \text{ emission})$$

(b) The atomic weight of silver is 107.868 amu. Because the mass of the $^{105}Ag$ nuclide is smaller than the average silver atom, this nuclide contains fewer neutrons than the stable isotopes of silver. Since $^{105}Ag$ is a relatively heavy neutron-poor nuclide, we expect it to decay by electron capture:

$$^{105}_{47}Ag + {}^{0}_{-1}e \longrightarrow {}^{105}_{46}Pd + h\nu \qquad (\text{electron capture})$$

(c) The atomic weight of tantalum is 180.948 amu. The $^{185}Ta$ isotope is therefore likely to be a neutron-rich isotope, which decays by electron emission.

$$^{185}_{73}Ta \longrightarrow {}^{185}_{74}W + {}^{0}_{-1}e \qquad (\beta^- \text{ emission})$$

## 23.5 BINDING ENERGY CALCULATIONS

We should be able to predict the mass of an atom from the masses of the subatomic particles it contains. A helium atom, for example, contains two protons, two neutrons, and two electrons. The mass of a helium atom therefore should be 4.0329802 amu:

$$\begin{aligned} 2(1.0072765)\text{ amu} &= 2.0145530\text{ amu} \\ 2(1.0086650)\text{ amu} &= 2.0173300\text{ amu} \\ \underline{2(0.0005486)\text{ amu}} &= \underline{0.0010972\text{ amu}} \\ \text{Total mass} &= 4.0329802\text{ amu} \end{aligned}$$

When the mass of a helium atom is measured, we find that the experimental value is smaller than the predicted mass by 0.0303769 amu.

$$\begin{aligned} \text{Predicted mass} &= 4.0329802 \text{ amu} \\ \text{Observed mass} &= \underline{4.0026033 \text{ amu}} \\ \text{Mass defect} &= \mathbf{0.0303769\ amu} \end{aligned}$$

The difference between the mass of an atom and the sum of the masses of its protons, neutrons, and electrons is called the **mass defect.** The mass defect of an atom reflects the stability of the nucleus. It is equal to the energy released when the nucleus is formed from its constituent protons and neutrons. The mass defect is therefore also known as the **binding energy** of the nucleus.

The binding energy serves the function for nuclear reactions that $\Delta H°$ does for a chemical reaction: It measures the difference between the stability of the products of the reaction and the starting materials. The larger the binding energy, the more stable the nucleus. The binding energy also can be viewed as the amount of energy that must be added to take the nucleus apart to form isolated neutrons and protons. It is therefore literally the energy that binds together the neutrons and protons in the nucleus.

The binding energy of a nuclide can be calculated from its mass defect with Einstein's equation that relates mass and energy:

$$E = mc^2$$

To obtain the binding energy in units of joules, we must convert the mass defect from atomic mass units to kilograms:

$$0.0303769 \cancel{\text{amu}} \times \frac{1.6605655 \times 10^{-24} \cancel{\text{g}}}{1 \cancel{\text{amu}}} \times \frac{1 \text{ kg}}{1000 \cancel{\text{g}}} = 5.04428 \times 10^{-29} \text{ kg}$$

Substituting this mass into Einstein's equation gives a binding energy for a single helium atom of $4.53358 \times 10^{-12}$ J.

$$\begin{aligned} E &= (5.04428 \times 10^{-29} \text{ kg})(2.9979246 \times 10^{8} \text{ m/s})^2 \\ &= 4.53358 \times 10^{-12} \text{ J} \end{aligned}$$

Multiplying by the number of atoms in a mole gives a binding energy for helium of $2.730 \times 10^{12}$ J/mol, or 2.730 billion kilojoules per mole:

$$\frac{4.53358 \times 10^{-12} \text{ J}}{1 \cancel{\text{atom}}} \times \frac{6.022 \times 10^{23} \cancel{\text{atoms}}}{1 \text{ mol}} = \mathbf{2.730 \times 10^{12}\ J/mol}$$

This calculation helps us understand the fascination of nuclear reactions. The energy released when natural gas is burned is about 800 kJ/mol. The synthesis of a mole of helium releases 3.4 million times as much energy.

Since most nuclear reactions are carried out on very small samples of material, the mole is not a reasonable basis of measurement. Binding energies are usually expressed in units of electron volts (eV) or million electron volts (MeV) per atom. The binding energy of helium is $28.3 \times 10^6$ eV/atom or 28.3 MeV/atom:

$$\frac{4.53358 \times 10^{-12} \cancel{\text{J}}}{1 \text{ atom}} \times \frac{1 \text{ eV}}{1.6021892 \times 10^{-19} \cancel{\text{J}}} = \mathbf{28.30 \times 10^{6}\ eV/atom}$$

Calculations of the binding energy can be simplified by using the following conversion factor between the mass defect in atomic mass units and the binding energy in million electron volts:

$$\mathbf{1\ amu = 931.5016\ MeV}$$

## EXERCISE 23.5

Calculate the binding energy of $^{235}U$ if the mass of this nuclide is 235.0349 amu.

**SOLUTION** A neutral $^{235}U$ atom contains 92 protons, 92 electrons, and 143 neutrons. The predicted mass of a $^{235}U$ atom is therefore 236.9601 amu, to four significant figures:

$$\begin{aligned} 92(1.00728)\text{ amu} &= 92.6698\text{ amu} \\ 92(0.0005486)\text{ amu} &= 0.0505\text{ amu} \\ 143(1.00867)\text{ amu} &= \underline{144.2398\text{ amu}} \\ \text{Total mass} &= 236.9601\text{ amu} \end{aligned}$$

To calculate the mass defect for this nucleus, we subtract the observed mass from the predicted mass:

$$\begin{aligned} \text{Predicted mass} &= 236.9601\text{ amu} \\ \text{Observed mass} &= \underline{235.0349\text{ amu}} \\ \text{Mass defect} &= \mathbf{1.9252\text{ amu}} \end{aligned}$$

Using the conversion factor that relates the binding energy to the mass defect, we obtain a binding energy for $^{235}U$ of 1793.3 MeV per atom:

$$\frac{1.9252\ \cancel{\text{amu}}}{1\text{ atom}} \times \frac{931.50\text{ MeV}}{1\ \cancel{\text{amu}}} = \mathbf{1793.3\ MeV/atom}$$

Binding energies gradually increase with atomic number, although they tend to level off near the end of the periodic table. A more useful quantity is obtained by dividing the binding energy for a nuclide by the total number of protons and neutrons it contains. This quantity is known as the **binding energy per nucleon.**

The binding energy per nucleon ranges from about 7.5 to 8.8 MeV for most nuclei, as shown in Figure 23.7. It reaches a maximum, however, at an atomic mass of about 60 amu. The largest binding energy per nucleon is observed for $^{56}Fe$, which is the most stable nuclide in the periodic table.

The graph of binding energy per nucleon versus atomic mass explains why energy is released when relatively small nuclei combine to form larger nuclei in **fusion reactions:**

$$^{12}_{6}C + ^{12}_{6}C \longrightarrow ^{24}_{12}Mg \qquad \text{(fusion)}$$

It also explains why energy is released when relatively heavy nuclei split apart in **fission** (literally, "to split or cleave") **reactions:**

$$^{235}_{92}U \longrightarrow ^{139}_{56}Ba + ^{94}_{36}Kr + 2\,^{1}_{0}n \qquad \text{(fission)}$$

There are a number of small irregularities in the binding energy curve at the low end of the mass spectrum, as shown in Figure 23.8. The $^{4}He$ nucleus, for example, is much more stable than its nearest neighbors. The unusual stability of the $^{4}He$ nucleus explains why $\alpha$-particle decay is usually much faster than the spontaneous fission of a nuclide into two large fragments.

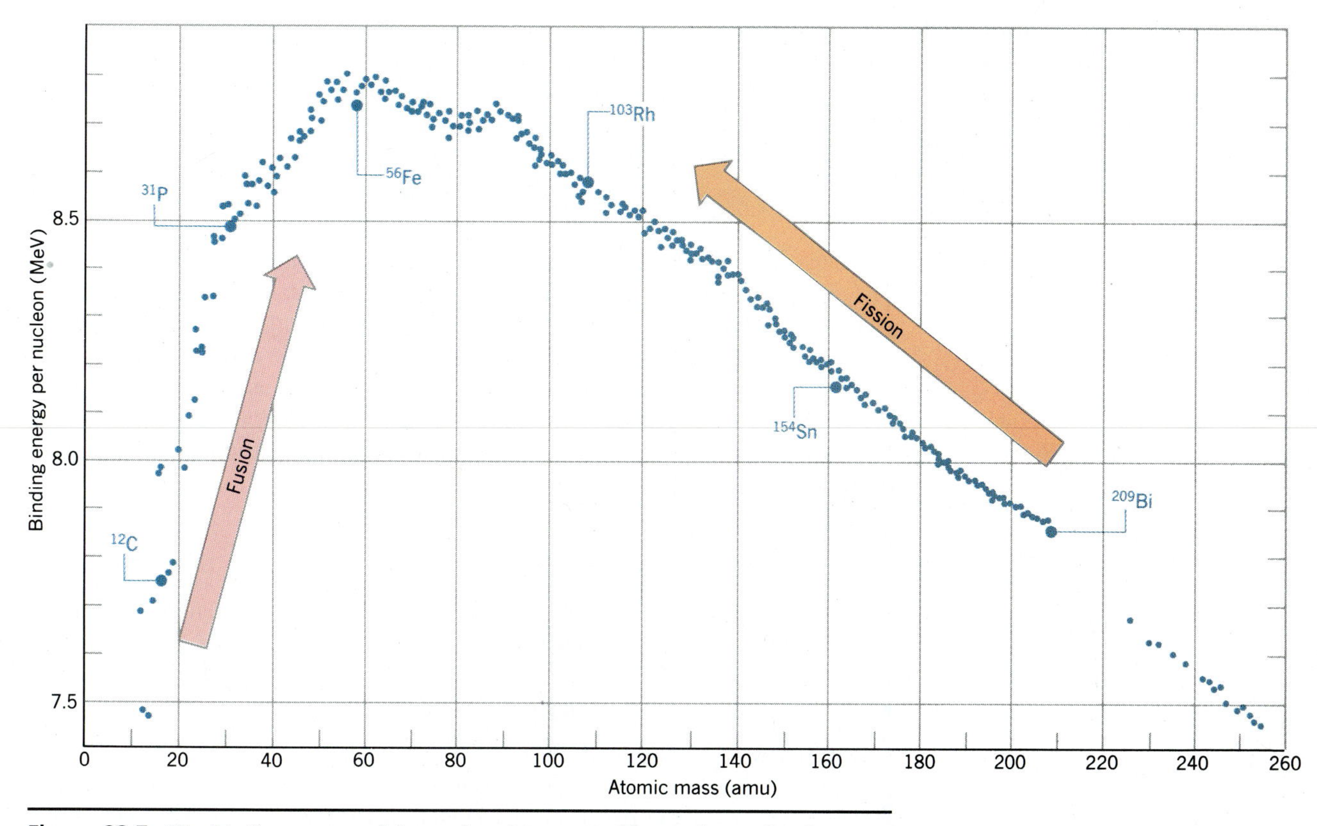


**Figure 23.7** The binding energy of the nucleus increases with atomic number from one end of the periodic table to the other. The binding energy *per nucleon,* however, reaches a maximum at $^{56}$Fe. Nuclei lighter than $^{56}$Fe can become more stable by fusing together; nuclei that are significantly heavier than $^{56}$Fe can become more stable by splitting apart.

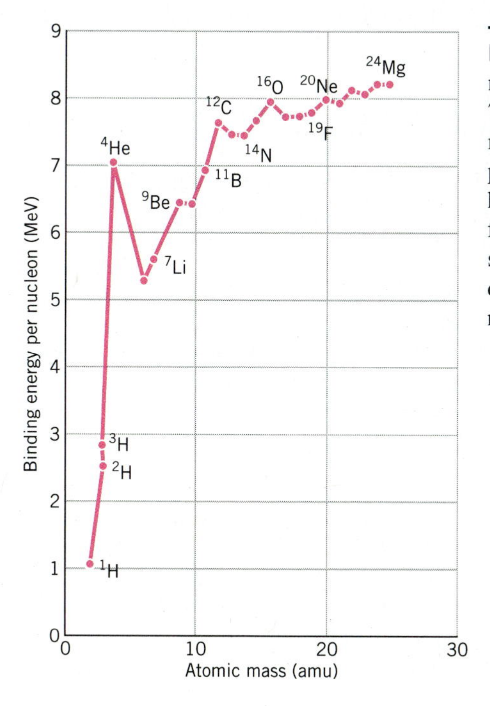


**Figure 23.8** The binding energy per nucleon for most stable nuclei is between 7.8 and 8.8 MeV per nucleon. There is more variability in the binding energy per nucleon for the very light nuclei, however. The binding energy per nucleon for $^{4}$He is particularly large. The unusual stability of this nuclide explains why $\alpha$ decay occurs so often among the heavier nuclides.

## 23.6 THE KINETICS OF RADIOACTIVE DECAY

Radioactive nuclei decay by first-order kinetics. The rate of radioactive decay is therefore the product of a rate constant ($k$) times the number of atoms of the isotope in the sample ($N$):

$$\textbf{Rate} = \frac{-dN}{dt} = kN$$

The rate of radioactive decay does not depend on the chemical state of the isotope. The rate of decay of $^{238}U$, for example, is exactly the same in uranium metal and in uranium hexafluoride, or in any other compound of this element.

The rate at which a radioactive isotope decays is called the **activity** of the isotope. The most common unit of activity is the **curie** (Ci), which was originally defined as the number of disintegrations per second in 1 g of $^{226}Ra$. The curie is now defined as the amount of radioactive isotope necessary to achieve an activity of $3.700 \times 10^{10}$ disintegrations per second.

### EXERCISE 23.6

The most abundant isotope of uranium is $^{238}U$; 99.276% of the atoms in a sample of uranium are $^{238}U$. Calculate the activity of the $^{238}U$ in 1 L of a 1.00 $M$ solution of the uranyl ion, $UO_2^{2+}$. Assume that the rate constant for the decay of this isotope is $4.87 \times 10^{-18}$ disintegrations per second.

**SOLUTION** The rate at which the $^{238}U$ isotope decays depends on the rate constant for this reaction ($k$) and the number of uranium atoms in the sample ($N$). The rate constant for the decay of this nuclide was given in the statement of the problem and we can calculate the number of uranium atoms in the sample by noting that a liter of a 1.00 $M$ $UO_2^{2+}$ solution contains 1 mole of uranium atoms:

$$\text{Rate} = kN = (4.87 \times 10^{-18}\ \text{s}^{-1})(6.02 \times 10^{23}\ \text{atoms}) = 2.93 \times 10^{6}\ \text{atoms/s}$$

To calculate the activity of this sample, we have to convert from disintegrations per second to curies:

$$2.93 \times 10^{6}\ \cancel{\text{atoms/s}} \times \frac{1\ \text{Ci}}{3.700 \times 10^{10}\ \cancel{\text{atoms/s}}} = \mathbf{7.92 \times 10^{-5}\ Ci}$$

The curie is a very large unit of measurement. Activities of samples handled in the laboratory are therefore often reported in millicuries (mCi) or microcuries ($\mu$Ci). This sample has an activity of 79.2 $\mu$Ci.

The relative rates at which radioactive nuclei decay can be expressed in terms of either the rate constants for the decay or the half-lives of the nuclei. We can conclude that $^{14}C$ decays more rapidly than $^{238}U$, for example, by noting that the rate constant for the decay of $^{14}C$ is much larger than that for $^{238}U$:

$$^{14}\text{C}: \quad k = 1.210 \times 10^{-4}\ \text{y}^{-1}$$
$$^{238}\text{U}: \quad k = 1.54 \times 10^{-10}\ \text{y}^{-1}$$

We can reach the same conclusion by noting that the half-life for the decay of $^{14}C$ is much shorter than that for $^{235}U$:

$$^{14}\text{C}: \quad t_{\frac{1}{2}} = 5730\ \text{y}$$
$$^{238}\text{U}: \quad t_{\frac{1}{2}} = 4.51 \times 10^{9}\ \text{y}$$

The **half-life** for the decay of a radioactive nuclide is the length of time it takes for exactly half of the nuclei in the sample to decay. In Chapter 22, we concluded that the half-life of a first-order process is inversely proportional to the rate constant for this process:

Sec. 22.11

$$t_{\frac{1}{2}} = \frac{\ln 2}{k} = \frac{0.693}{k}$$

**EXERCISE 23.7**

Calculate the fraction of the $^{14}C$ that remains in a sample after eight half-lives.

**SOLUTION** Half of the $^{14}C$ present initially decays during the first half-life, half of what is left decays during the second half-life, and so on. The $^{14}C$ left after eight half-lives is equal to one-half raised to the eighth power:

$$(1/2)^8 = 0.00391$$

Less than 0.4% of the original $^{14}C$ is left after eight half-lives.

For more complex calculations, it is easier to convert the half-life of the nuclide into a rate constant and then use the integrated form of the first-order rate law, described in Chapter 22, substituting $N$ for the concentration of $X$:

$$\ln\left(\frac{N}{N_o}\right) = -kt$$

**EXERCISE 23.8**

How long would it take for a sample of $^{222}Rn$ that weighs 0.750 g to decay to 0.100 g? Assume a half-life for $^{222}Rn$ of 3.823 days.

**SOLUTION** We can start by calculating the rate constant for this decay from the half-life:

$$k = \frac{\ln 2}{t_{\frac{1}{2}}} = \frac{0.6931}{3.823 \text{ d}} = 0.1813 \text{ d}^{-1}$$

We then turn to the integrated form of the first-order rate law:

$$\ln\left[\frac{(N)}{(N)_o}\right] = -kt$$

The ratio of the number of atoms that remain in the sample to the number of atoms present initially is the same as the ratio of grams at the end of the time period to the number of grams present initially:

$$\ln\left(\frac{0.100}{0.750}\right) = -(0.1813 \text{ d}^{-1})(t)$$

Solving for $t$, we find that it takes 11.1 days for 0.750 g of $^{222}Rn$ to decay to 0.100 g of this nuclide.

## 23.7 DATING BY RADIOACTIVE DECAY

The earth is constantly bombarded by cosmic rays emitted by the sun. The total energy received in the form of cosmic rays is relatively small—roughly equal to the energy received by the planet from starlight. But the energy of a single cosmic ray is relatively large, on the order of several billion electron volts. These highly energetic rays react with atoms in the atmosphere to produce neutrons that then react with nitrogen atoms in the atmosphere to produce $^{14}C$, as shown in Figure 23.9.

$$^{14}_{7}N + ^{1}_{0}n \longrightarrow ^{14}_{6}C + ^{1}_{1}H$$

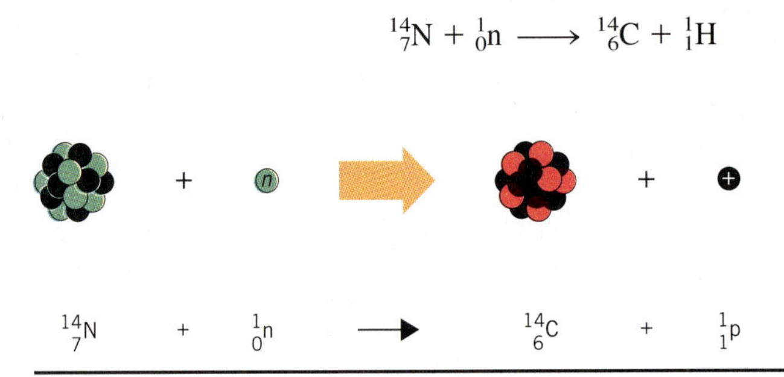


**Figure 23.9** The cycle of reactions on which $^{14}C$ dating is based.

Carbon-14 is a neutron-rich nuclide that decays by electron emission with a half-life of 5730 years:

$$^{14}_{6}C \longrightarrow ^{14}_{7}N + ^{0}_{-1}e$$

Just after World War II, Willard F. Libby proposed a way to use these reactions to estimate the age of carbon-containing substances. The **$^{14}C$ dating** technique for which Libby received the Nobel prize is based on the following assumptions.

1. Carbon-14 is produced in the atmosphere at a more or less constant rate.
2. Carbon atoms circulate between the atmosphere, the oceans, and living organisms at a rate very much faster than they decay. As a result, there is a constant concentration of $^{14}C$ in all living things.
3. After death, organisms no longer pick up $^{14}C$.
4. By comparing the activity of a sample with the activity of living tissue we can estimate how long it has been since the organism died.

The natural abundance of $^{14}C$ is about 1 part in $10^{12}$, and the average activity of living tissue is 15.3 disintegrations per minute per gram of carbon. Typical samples used for radiocarbon dating include charcoal, wood, cloth, paper, sea shells, limestone, flesh, hair, soil, peat, and bone. Since most iron samples also contain carbon, it is possible to estimate the time since iron was last fired by analyzing it for $^{14}C$.

Collection of fossil skulls and fragments of jaw bones from the oldest known group of hominids, the Australopithecines, is too old to be dated by $^{14}C$. Potassium argonne dating, however, suggests dates from 3.7 to 1.6 million years.

**EXERCISE 23.9**

The skin, bone, and clothing of an adult female mummy discovered in Chimney Cave, Lake Winnemucca, Nevada, were dated by radiocarbon analysis. How old is this mummy if the sample retains 73.9% of the activity of living tissue?

**SOLUTION** Because $^{14}C$ decays by first-order kinetics, the log of the ratio of the $^{14}C$ in the sample today ($N$) to the amount that would be present if it was still alive ($N_o$) is proportional to the rate constant for this decay and the time since death.

$$\ln\left[\frac{(N)}{(N)_o}\right] = -kt$$

The rate constant for this reaction can be calculated from the half-life of $^{14}C$, which is 5730 years:

$$k = \frac{\ln 2}{t_{\frac{1}{2}}} = \frac{0.6931}{5730\ \text{y}} = 1.210 \times 10^{-4}\ \text{y}^{-1}$$

If the sample retains 73.9% of its activity, the ratio of the activity today ($N$) to the original activity ($N_o$) is 0.739. Substituting what we know into the integrated form of the first-order rate law gives the following equation:

$$\ln(0.739) = -(1.210 \times 10^{-4}\ \text{y}^{-1})(t)$$

Solving this equation for the unknown gives an estimate of the time since death:

$$t = 2500\ \text{y}$$

One of Libby's assumptions is questionable. The amount of $^{14}C$ in the atmosphere has not been constant with time. Because of changes in solar activity and the earth's magnetic field, it has varied by as much as ±5%. More recently, contamination from the burning of fossil fuels and the testing of nuclear weapons has caused significant changes in the amount of radioactive carbon in the atmosphere. Radiocarbon dates are therefore reported in years before the present era (B.P.). By convention, the present era is assumed to begin in 1950, when $^{14}C$ dating was introduced.

Studies of bristlecone pines allow us to correct for changes in the abundance of $^{14}C$ with time. These remarkable trees, which grow in the White Mountains of California, can live for up to 5000 years. By studying the $^{14}C$ activity of samples taken from the annual growth rings in these trees, researchers have developed a calibration curve for $^{14}C$ dates from the present back to 5145 B.C.

"Earth's oldest living things."

After roughly 45,000 years (eight half-lives), a sample retains only 0.4% of the $^{14}C$ activity of living tissue. At that point it becomes too old to date by radiocarbon techniques. Other radioactive isotopes can be used to date rocks, soils, or archaeological objects that are much older.

Potassium–argon dating, for example, has been used to date samples up to 4.3 billion years old. Naturally occurring potassium contains 0.0118% by weight of the radioactive $^{40}K$ isotope. This isotope decays to $^{40}Ar$ with a half-life of 1.3 billion years. The $^{40}Ar$ produced after a rock crystallizes is trapped in the crystal lattice. It can be released, however, when the rock is melted at temperatures up to 2000°C. By measuring the amount of $^{40}Ar$ released when the rock is melted and comparing it with the amount of potassium in the sample, the time since the rock crystallized can be determined.

## 23.8 IONIZING VERSUS NONIONIZING RADIATION

We live in a sea of radiation. We are exposed to infrared, ultraviolet, visible, and cosmic rays from the sun. We are subjected to radio waves from local radio and television transmitters, microwaves from microwave ovens or cellular telephones, and x-rays produced by the cathode-ray tubes in our television sets. We are also exposed to natural sources of radioactivity, including $\alpha$-particles from trace contam-

The cathode ray tube in a T.V. set gives off a small quantity of x-rays when the electrons given off by the cathode of the CRT strike the metal of the anode. The potential hazard is greatest for young children who often sit too close to the T.V.

Students on spring break exposing themselves to the effects of both nonionizing and ionizing radiation.

inants in brick and clay, $\beta$ emitters in the food chain, and $\gamma$-rays from soils and rocks.

In recent years, people have learned to fear the effects of this radiation. They don't want to live near nuclear reactors. They are frightened by reports of links between excess exposure to sunlight and skin cancer. They are afraid of the leakage from microwave ovens, or the radiation produced by their television sets.

Several factors combine to heighten the public's anxiety about the long-range effects of radiation. Perhaps the most important source of fear is the fact that radiation can't be detected by the average person. Furthermore, the effects of exposure to radiation may not appear for months or even years.

To understand the biological effects of radiation, we must first understand the difference between **ionizing radiation** and **nonionizing radiation.** In general, two things can happen when radiation is absorbed by matter: excitation or ionization. *Excitation* occurs when the radiation excites the motion of the atoms or molecules, or excites an electron from an occupied orbital into an empty, higher energy orbital. *Ionization* occurs when the radiation carries enough energy to remove an electron from an atom or molecule.

Living tissue is 70–90% water by weight. When it is irradiated, most of the energy of the radiation is absorbed by water molecules. The dividing line between radiation that excites electrons and radiation that forms ions therefore falls at an energy equal to the ionization energy of water: 1216 kJ/mol. Radiation that carries less energy can only excite the water molecule. It is therefore called *nonionizing radiation.* Radiation that carries more energy than 1216 kJ/mol can remove an electron from a water molecule, and is therefore called *ionizing radiation.*

Table 23.1 contains estimates of the energies of various kinds of radiation. Radiowaves, microwaves, infrared radiation, and visible light are all forms of nonionizing radiation. X-rays, $\gamma$-rays, and $\alpha$- and $\beta$-particles are forms of ionizing radiation. The

**TABLE 23.1 Energies of Ionizing and Nonionizing Forms of Radiation**

| Radiation | Typical Frequency ($s^{-1}$) | Typical Energy (kJ/mol) | |
|---|---|---|---|
| Particles | | | |
| $\alpha$-Particles | | $4.1 \times 10^8$ | |
| $\beta$-Particles | | $1.5 \times 10^7$ | |
| Electromagnetic radiation | | | |
| Cosmic rays | $6 \times 10^{21}\ s^{-1}$ | $2.4 \times 10^9$ | Ionizing Radiation |
| $\gamma$-Rays | $3 \times 10^{20}\ s^{-1}$ | $1.2 \times 10^8$ | |
| x-Rays | $3 \times 10^{17}\ s^{-1}$ | $1.2 \times 10^5$ | |
| Ultraviolet | $3 \times 10^{15}\ s^{-1}$ | 1200 | |
| Visible | $5 \times 10^{14}\ s^{-1}$ | 200 | Nonionizing Radiation |
| Infrared | $3 \times 10^{13}\ s^{-1}$ | 12 | |
| Microwaves | $3 \times 10^{9}\ s^{-1}$ | $1.2 \times 10^{-3}$ | |
| Radiowaves | $3 \times 10^{7}\ s^{-1}$ | $1.2 \times 10^{-5}$ | |

dividing line between ionizing and nonionizing radiation in the electromagnetic spectrum falls in the ultraviolet portion of the spectrum. It is therefore useful to divide the UV spectrum into: $UV_a$ and $UV_B$. Radiation at the high-energy end of the UV spectrum ($UV_B$) can be as dangerous as x-rays or $\gamma$-rays.

When ionizing radiation passes through living tissue, electrons are removed from neutral water molecules to produce $H_2O^+$ ions.

$$H_2O \longrightarrow H_2O^+ + e^-$$

Between three and four water molecules are ionized for every $1.6 \times 10^{-17}$ J of energy absorbed in the form of ionizing radiation. The $H_2O^+$ ion should not be confused with the $H_3O^+$ produced when acids dissolve in water. The $H_2O^+$ ion is an example of a *free radical,* which contains an unpaired valence-shell electron, as shown in Figure 23.10. Free radicals are extremely reactive. The radicals formed when ionizing radiation passes through water are among the strongest oxidizing agents that can exist in aqueous solution. At the molecular level, these oxidizing agents destroy biologically active molecules by either removing electrons or removing hydrogen atoms. This often leads to damage to the membrane, nucleus, chromosomes, or mitochondria of the cell that either inhibits cell division, results in cell death, or produces a malignant cell.

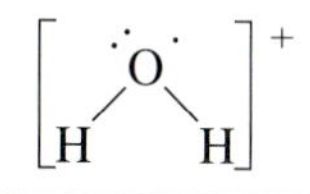

**Figure 23.10** The Lewis structure of the $H_2O^+$ ion. This ion contains an unpaired electron, which makes it extremely reactive.

## 23.9 BIOLOGICAL EFFECTS OF IONIZING RADIATION

From the time that radioactivity was discovered, it was obvious that it caused damage. Glass containers used to store radium compounds, for example, turned a rich purple and eventually cracked because of radiation damage. As early as 1901, Pierre Curie discovered that a sample of radium placed on his skin produced wounds that were very slow to heal. What some find surprising is the magnitude of the difference between the biological effects of nonionizing radiation, such as light and microwaves, and ionizing radiation, such as high-energy ultraviolet radiation, x-rays, $\gamma$-rays, and $\alpha$- or $\beta$-particles.

This dessicator was used to store radium compounds. Radiation emitted by these compounds eventually caused enough damage to the structure of the glass to turn it a deep purple color.

Radiation at the low-energy end of the electromagnetic spectrum excites the movement of atoms and molecules, which is equivalent to heating the sample. Radiation near the visible portion of the spectrum excites electrons into higher energy orbitals. When the electron eventually falls back to a lower energy state, the excess energy is given off to neighboring molecules in the form of heat. The principal effect of nonionizing radiation is therefore an increase in the temperature of the system.

Biological systems are sensitive to heat. We experience this each time we cook with a microwave oven, or spend too long in the sun. But it takes a great deal of nonionizing radiation to reach dangerous levels. We can assume, for example, that absorption of enough radiation to produce an increase of about 6°C in body temperature would be fatal. Since the average 70-kg human is 80% water by weight, we can use the heat capacity of water to calculate that it would take about 1.5 million joules of nonionizing radiation to kill the average human. If this energy were carried by visible light with a frequency of $5 \times 10^{14}\ s^{-1}$, it would correspond to absorption of about 7 moles of photons.

Ionizing radiation is much more dangerous. A dose of only 300 J of x-ray or $\gamma$-ray radiation is fatal for the average human, even though this radiation raises the temperature of the body by only 0.001°C. $\alpha$-Particle radiation is even more dangerous; a dose equivalent to only 15 J is fatal for the average human. Whereas it takes 7 moles of photons of visible light to produce a fatal dose of nonionizing radiation, absorption of only $7 \times 10^{-10}$ mol of the $\alpha$-particles emitted by $^{238}U$ is fatal.

There are three ways of measuring ionizing radiation.

1. Measure the *activity* of the source in units of disintegrations per second or curies, which is the easiest measurement to make.
2. Measure the radiation to which an object is *exposed,* in units of roentgens, by measuring the amount of ionization produced when this radiation passes through a sample of air.

3. Measure the radiation *absorbed* by the object in units of radiation absorbed doses or "rads." This is the most useful quantity, but it is the hardest to obtain.

One **radiation absorbed dose,** or **rad,** corresponds to the absorption of $10^{-5}$ J of energy per gram of body weight. Because this is equivalent to 0.01 J/kg, one rad produces an increase in body temperature of about $2 \times 10^{-6}$°C. At first glance, the rad may seem to be a negligibly small unit of measurement. The destructive power of the radicals produced when water is ionized is so large, however, that cells are inactivated at a dose of 100 rads, and a dose of 400 to 450 rads is fatal for the average human.

Not all forms of radiation have the same efficiency for damaging biological organisms. The faster the energy is lost as the radiation passes through the tissue, the more damage it does. To correct for the differences in **radiation biological effectiveness (RBE)** among various forms of radiation, a second unit of absorbed dose has been defined. The **roentgen equivalent man,** or **rem,** is the product of the absorbed dose in rads times the biological effectiveness of the radiation:

$$\text{rems} = \text{rads} \times \text{RBE}$$

Values for the RBE of different forms of radiation are given in Table 23.2.

**TABLE 23.2 The Radiation Biological Effectiveness of Various Forms of Radiation**

| Radiation | RBE |
|---|---|
| x-Rays and $\gamma$-rays | 1 |
| $\beta$-Particles with energies larger than 0.03 MeV | 1 |
| $\beta$-Particles with energies less than 0.3 MeV | 1.7 |
| Thermal (slow-moving) neutrons | 3 |
| Fast-moving neutrons or protons | 10 |
| $\alpha$-Particles or heavy ions | 20 |

Estimates of the per capita exposure to radiation in the United States are summarized in Table 23.3. These estimates include both external and internal sources of natural background radiation. External sources include cosmic rays from the sun and $\alpha$-particles or $\gamma$-rays emitted from rocks and soil. Internal sources include nuclides that enter the body when we breathe ($^{14}C$, $^{85}Kr$, $^{220}Rn$, and $^{222}Rn$) and through the food chain ($^{3}H$, $^{14}C$, $^{40}K$, $^{90}Sr$, $^{131}I$, and $^{137}Cs$). The actual dose from natural radiation depends on where one lives. People who live in the Rocky Mountains, for example, receive twice as much background radiation as the national average because there is less atmosphere to filter out the cosmic rays from the sun.

**TABLE 23.3 Average Whole-Body Exposure Levels for Sources of Ionizing Radiation**

| Source | Per Capita Dose (rems/y) |
|---|---|
| Natural background | 0.082 |
| Medical x-rays | 0.077 |
| Nuclear test fallout | 0.005 |
| Consumer and industrial products | 0.005 |
| Nuclear power industry | 0.001 |
| | Total: 0.170 |

The average dose from medical x-rays has decreased in recent years because of advances in the sensitivity of the photographic film used for x-rays. Radiation from nuclear test fallout has also decreased as a result of the atmospheric nuclear test ban. The threat of fallout from the testing of nuclear weapons can be appreciated by noting that a Chinese atmospheric test in 1976 led to the contamination of milk in the Harrisburg, PA, vicinity at a level of 300 pCi ($3.00 \times 10^{-10}$ Ci) per liter. This was about eight times the level of contamination (41 pCi per liter) that resulted from the accident at Three Mile Island.

The contribution to the radiation absorbed dose from consumer and industrial products includes radiation from construction materials, x-rays emitted by television sets, and inhaled tobacco smoke. The most recent estimate of the total radiation emitted from the mining and milling of uranium, the fabrication of reactor fuels, the storage of radioactive wastes, and the operation of nuclear reactors is less than 0.001 rem per year.

The total dose from ionizing radiation for the average American is about 0.170 rem per year. The Committee on the Biological Effects of Ionizing Radiation of the National Academy of Sciences recently estimated that an increase in this dose to a level of 1 rem per year would result in 169 additional deaths from cancer per million people exposed. This can be compared with the 170,000 cancer deaths that would normally occur in a population this size that was not exposed to this level of radiation.

Smoking has been estimated to result in the delivery of 8 rads of radiation to the epithelium of the lungs due to $^{210}P_O$ contamination on tobacco leaves.

## EXERCISE 23.10

Calculate the rems of radiation absorbed by the average person from the $^{14}C$ in his or her body. Assume the activity of the $^{14}C$ in the average body is 0.08 $\mu$Ci, the energy of the $\beta$-particles emitted when $^{14}C$ decays is 0.156 MeV, and about one-third of this energy is captured by the body.

**SOLUTION** We can calculate the number of $\beta$-particles emitted per second from the activity of the $^{14}C$ isotope:

$$0.08\ \mu\text{Ci} \times \frac{1\ \text{Ci}}{10^6\ \mu\text{Ci}} \times \frac{3.700 \times 10^{10}\ \text{atoms/s}}{1\ \text{Ci}} = 3000\ \text{atoms/s}$$

To calculate the effect of this radiation over a year, we have to calculate the number of $^{14}C$ nuclei that decay during this time:

$$\frac{3000\ \text{atoms}}{1\ \text{s}} \times \frac{60\ \text{s}}{1\ \text{min}} \times \frac{60\ \text{min}}{1\ \text{h}} \times \frac{24\ \text{h}}{1\ \text{d}} \times \frac{365\ \text{d}}{1\ \text{y}} = 9.5 \times 10^{10}\ \text{atoms/y}$$

We can now calculate the amount of energy absorbed per year from the number of atoms that disintegrate, the energy per disintegration, and the fraction of this energy absorbed by the body:

$$\frac{9.5 \times 10^{10}\ \text{atoms}}{1\ \text{y}} \times \frac{0.156\ \text{MeV}}{1\ \text{atom}} \times \frac{1}{3} = 4.9 \times 10^9\ \text{MeV/y}$$

Before we can go any further, we have to convert this energy from units of million electron volts per year to joules per year:

$$\frac{4.9 \times 10^9\ \text{MeV}}{1\ \text{y}} \times \frac{10^6\ \text{eV}}{\text{MeV}} \times \frac{1.60 \times 10^{-19}\ \text{J}}{1\ \text{eV}} = \mathbf{7.8 \times 10^{-4}\ J/y}$$

Averaged over a 70-kg body weight, this amounts to about $1 \times 10^{-5}$ J per kilogram of body weight per year. If 1 rad is equal to 0.01 J/kg, the radiation absorbed dose from $^{14}C$ decay is 0.001 rad per year. Using an RBE for $\beta$ decay of 1, this corresponds to 0.001 rems, or 1 millirem, per year.

The principal effect of low doses of ionizing radiation is to induce cancers, which may take up to 20 years to develop. What is the effect of high doses of ionizing radiation? Cells that are actively dividing are more sensitive to radiation than cells that are not. Thus, cells in the liver, kidney, muscle, brain, and bone are more resistant to radiation than the cells of bone marrow, the reproductive organs, the epithelium of the intestine, and the skin, which suffer the most damage from radiation. Damage to the bone marrow is the main cause of death at moderately high levels of exposure (200 to 1000 rads). Damage to the gastrointestinal tract is the major cause of death for exposures on the order of 100 to 10,000 rads. Massive damage to the central nervous system is the cause of death from extremely high exposures (over 10,000 rads).

## 23.10 NATURAL VERSUS INDUCED RADIOACTIVITY

### NATURAL RADIOACTIVITY

The vast majority of the nuclides found in nature are stable. If our planet is 4.6 billion years old, the only radioactive isotopes that should remain are members of three classes:

1. Isotopes with half-lives of at least $10^9$ years, such as $^{238}U$.
2. Daughter nuclides produced when long-lived radioactive nuclides decay, such as the $^{234}Th$ ($t_{1/2}$ = 24.1 d) produced by the $\alpha$ decay of $^{238}U$.
3. Nuclides such as $^{14}C$ that are still being synthesized.

Sec. 23.4

In Section 23.4, we encountered one factor that influences the stability of a nuclide: the ratio of neutrons to protons. (Nuclei that contain either too many or too few neutrons are unstable.) Another factor that affects the stability of nuclides can be understood by examining patterns in the numbers of protons and neutrons in stable nuclides.

Half of the elements in the periodic table must have an odd number of protons because atomic numbers that are odd are just as likely to occur as those that are even. In spite of this, about 80% of the stable nuclides have an even number of protons. Very few elements with an odd atomic number have more than one stable isotope. Stable isotopes abound, however, among elements with even atomic numbers. Ten stable isotopes are known for tin ($Z = 50$), for example. It is also interesting to note that 91% of the stable isotopes of elements with an odd number of protons have an even number of neutrons. These observations suggest that certain combinations of protons and neutrons are particularly stable.

In Chapter 6 we found that there are magic numbers of electrons. Electron configurations with 2, 10, 18, 36, 54, and 86 electrons are unusually stable. There also seem to be magic numbers of neutrons and protons. Nuclei with 2, 8, 20, 28, 50, 82, or 126 protons or neutrons are unusually stable. This observation explains the anomalously large binding energies observed in Figure 23.8 for $^{4}He$, $^{16}O$, and $^{20}Ne$. $^{20}Ne$ has a magic number of nucleons when both protons and neutrons are counted. $^{4}He$ and $^{16}O$ have magic numbers of both protons and neutrons.

Sec. 6.21

If nuclei tend to be more stable when they have even numbers of protons and neutrons, it isn't surprising that nuclides with an odd number of both protons and neutrons are unstable. $^{40}K$ is one of only five naturally occurring nuclides that contain both an odd number of protons and an odd number of neutrons. This nuclide simultaneously undergoes the electron capture and positron emission expected for neutron-poor nuclides and the electron emission observed with neutron-rich nuclides, as shown in Figure 23.4.

Only 18 radioactive isotopes with atomic numbers of 80 or less can be found in nature. With the exception of $^{14}C$, which is continuously synthesized in the atmosphere, all these elements have lifetimes longer than $10^9$ years. Although these isotopes all undergo radioactive decay, they decay so slowly that reasonable quantities are still present, 4.6 billion years after the planet was formed.

Another 45 radioactive isotopes found in nature have atomic numbers larger than 80. These nuclides fall into three families, one of which is shown in Figure 23.11.

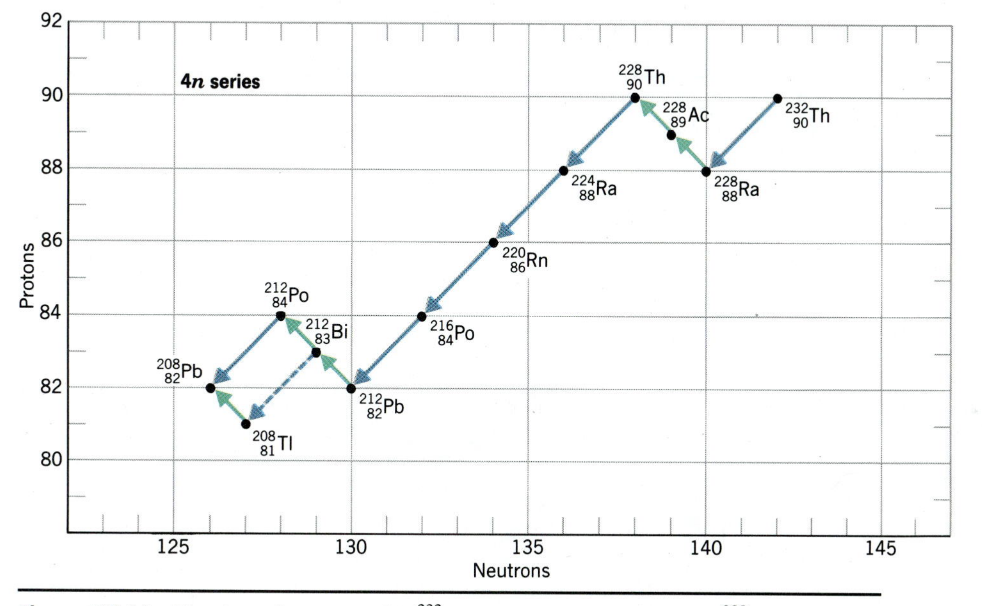


**Figure 23.11** The $4n$ series starts with $^{232}Th$ and eventually decays to $^{208}Pb$.

The parent nuclide is $^{232}Th$, which undergoes $\alpha$ decay to form $^{228}Ra$. The product of this reaction decays by $\beta$ emission to form $^{228}Ac$, which decays to $^{228}Th$, and so on, until the stable $^{208}Pb$ isotope is formed. This family of radionuclides is called the $4n$ series because all its members have a mass number that can be divided by 4.

A second family of radioactive nuclei starts with $^{238}U$ and decays to form the stable $^{206}Pb$ isotope, as shown in Figure 23.12. Every member of this series has a mass number that fits the equation $4n + 2$. The third family, known as the $4n + 3$ series, starts with $^{235}U$ and decays to $^{207}Pb$, as shown in Figure 23.13.

A $4n + 1$ series once existed, which started with $^{237}Np$ and decayed to form the only stable isotope of bismuth, $^{209}Bi$, as shown in Figure 23.14. The half-life of every member of this series is less than $2 \times 10^6$ years, however, so none of the nuclides produced by the decay of neptunium remain in detectable quantities on the earth.

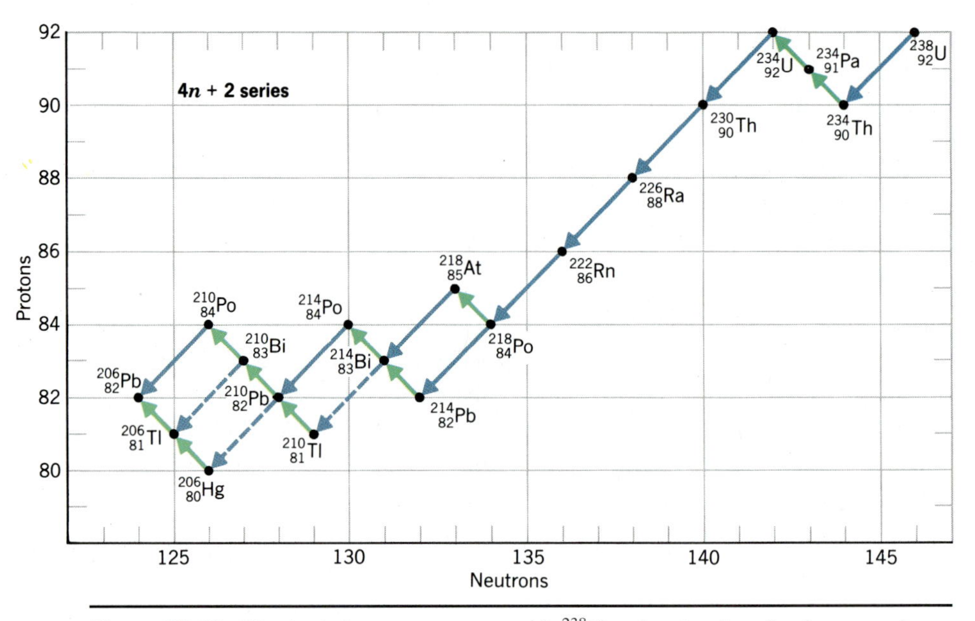


**Figure 23.12** The $4n + 2$ sequence starts with $^{238}U$ and ends when the decay reaches the stable $^{206}Pb$ nuclide.

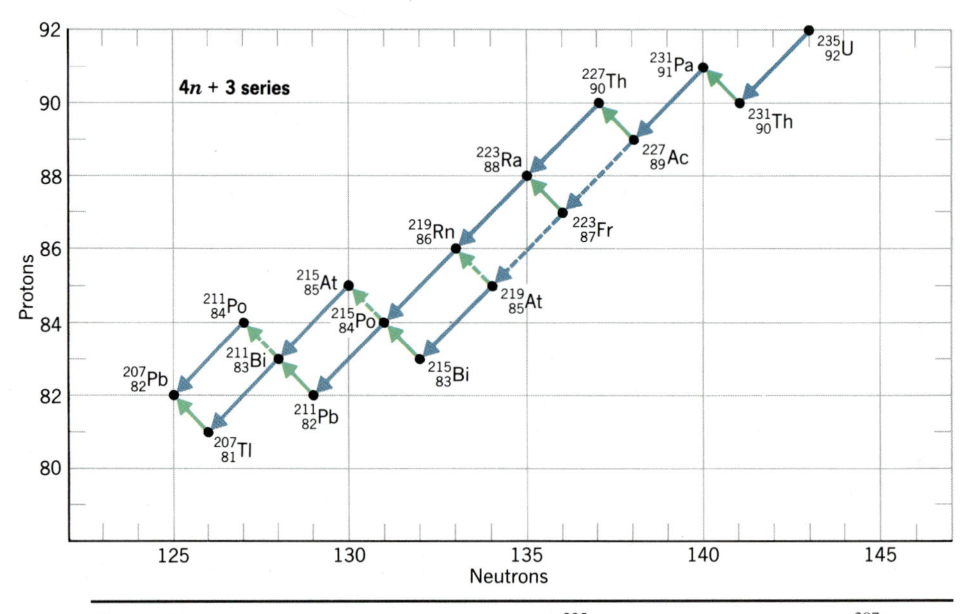


**Figure 23.13** The $4n + 3$ sequence starts with $^{235}U$ and ends with the stable $^{207}Pb$ nuclide.

## INDUCED RADIOACTIVITY

In 1934, Irene Curie, the daughter of Pierre and Marie Curie, and her husband, Frederic Joliot, announced the first synthesis of an artificial radioactive isotope. They bombarded a thin foil of aluminum metal with $\alpha$-particles produced by the decay of polonium and found that the aluminum target became radioactive. Chemical analysis showed that the product of this reaction was an isotope of phosphorus:

$$^{27}_{13}Al + ^{4}_{2}He \longrightarrow ^{30}_{15}P + ^{1}_{0}n$$

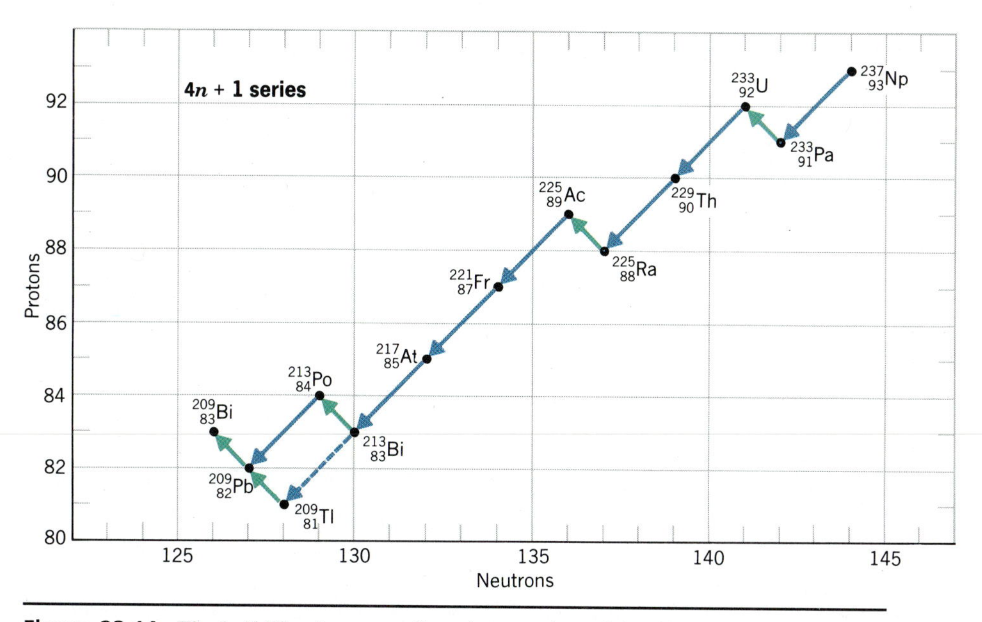


**Figure 23.14** The half-life of every radioactive member of the $4n + 1$ series is so short that none of these nuclides remain in detectable amounts on the earth.

In the next 50 years, more than 2000 other artificial radionuclides were synthesized.

A shorthand notation has been developed for nuclear reactions such as the reaction discovered by Curie and Joliot:

$$^{27}_{13}\mathrm{Al}(\alpha,\mathrm{n})^{30}_{15}\mathrm{P}$$

The parent (or target) nuclide and the daughter nuclide are separated by parentheses that contain the symbols for the particle that hits the target and the particle released in this reaction.

The nuclear reactions used to synthesize artificial radionuclides are characterized by enormous activation energies. Linear accelerators or cyclotrons can be used to overcome these activation energies by exciting charged particles such as protons, electrons, $\alpha$-particles, or even heavier ions, which are then focused on a stationary target. The capture of a positively charged particle usually produces a neutron-poor isotope. The following is an example of a reaction that can be induced by a cyclotron or linear accelerator:

Irene Curie and her husband Frederic Joliot.

$$^{24}_{12}\mathrm{Mg} + ^{2}_{1}\mathrm{H} \longrightarrow ^{22}_{11}\mathrm{Na} + ^{4}_{2}\mathrm{He}$$

Artificial radionuclides are also synthesized in nuclear reactors, which are excellent sources of slow-moving, or **thermal neutrons.** The absorption of a neutron usually results in a neutron-rich isotope. The following neutron absorption reaction occurs in the cooling systems of nuclear reactors cooled with liquid sodium metal:

$$^{23}_{11}\mathrm{Na} + ^{1}_{0}\mathrm{n} \longrightarrow ^{24}_{11}\mathrm{Na} + \gamma$$

In 1940, this technique was used for the first time to synthesize elements with atomic numbers larger than the heaviest naturally occurring element, uranium.

McMillan and Abelson synthesized neptunium and plutonium by irradiating $^{238}U$ with neutrons to form $^{239}U$:

$$^{238}_{92}U + ^{1}_{0}n \longrightarrow ^{239}_{92}U + \gamma \qquad \text{(neutron capture)}$$

which undergoes $\beta^-$ decay to form $^{239}Np$ and then $^{239}Pu$:

$$^{239}_{92}U \longrightarrow ^{239}_{93}Np + ^{0}_{-1}e$$
$$^{239}_{93}Np \longrightarrow ^{239}_{94}Pu + ^{0}_{-1}e$$

Larger bombarding particles were eventually used to produce even heavier transuranium elements:

$$^{253}_{99}Es + ^{4}_{2}He \longrightarrow ^{256}_{101}Md + ^{1}_{0}n$$
$$^{246}_{96}Cm + ^{12}_{6}C \longrightarrow ^{254}_{102}No + 4\,^{1}_{0}n$$

The half-lives for $\alpha$ decay and spontaneous fission decrease as the atomic number of the element increases. Element 104, for example, has a half-life for spontaneous fission of 0.3 seconds. Elements therefore become harder to characterize as the atomic number increases. Recent theoretical work has predicted that a magic number of protons may exist at $Z = 114$. This work suggests that there is an island of stability in the sea of unstable nuclides, as illustrated in Figure 23.15. If this theory is correct, superheavy elements could be formed if we could find a way to cross the gap between elements $Z = 109$ through $Z = 114$.

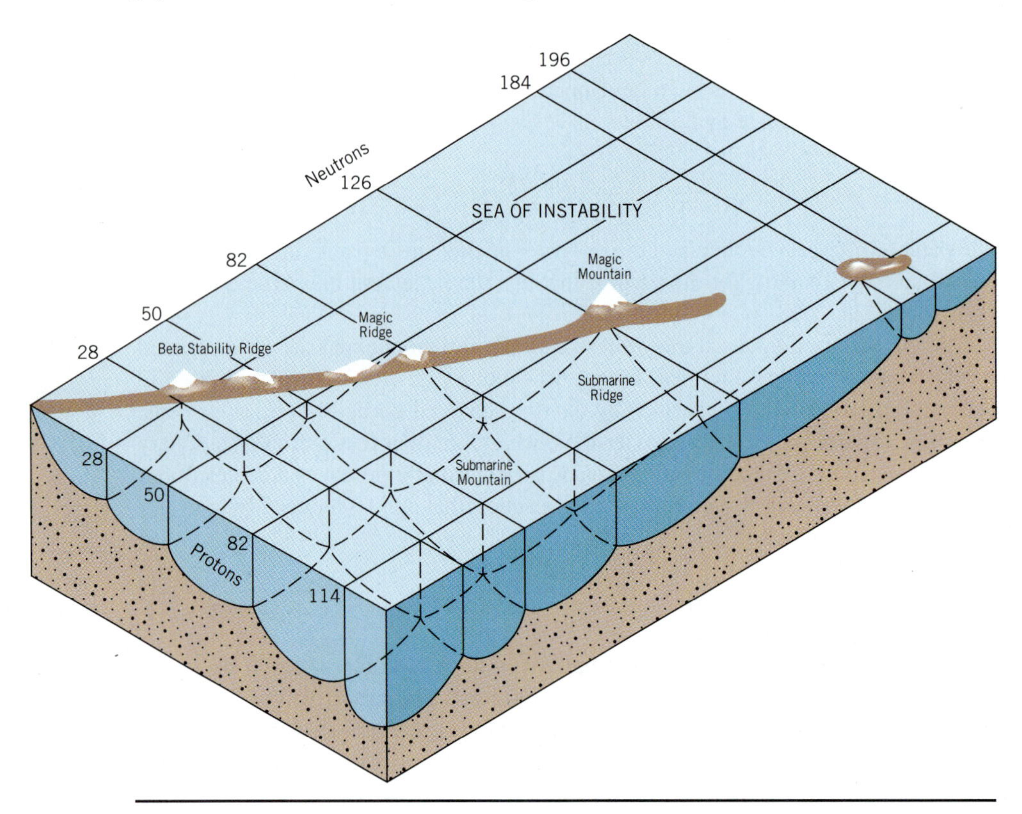


**Figure 23.15** The stable nuclei fall within a narrow band of neutron-to-proton ratios. Glenn Seaborg visualizes this as a ridge of stability in a sea of instability. Theory predicts that an island of stability will exist for superheavy elements with atomic numbers around 114 and atomic masses close to 300 amu. Synthesis of the superheavy elements requires that we jump across the sea of instability in one step.

There is some debate about the number of neutrons needed to overcome the proton–proton repulsion in a nucleus with 114 protons. The best estimates suggest that at least 184 neutrons, and perhaps as many as 196, would be needed. It is not an easy task to bring together two particles that give both the correct total number of protons and the necessary neutrons to produce a nuclide with a half-life long enough to be detected. If we start with a relatively long-lived parent nuclide, such as $^{251}$Cf ($t_{1/2}$ = 800 y) and bombard this nucleus with a heavy ion, such as $^{32}$S, we can envision producing a daughter nuclide with the correct atomic number, but the mass number would be too small by at least 16 amu:

$$^{251}_{98}\text{Cf} + ^{32}_{16}\text{S} \longrightarrow ^{282}_{114}\text{X} + ^{1}_{0}\text{n}$$

## EXERCISE 23.11

Predict the group in the periodic table in which element 114 belongs. What oxidation states are expected for this element?

**SOLUTION** We can use the aufbau principle to predict the following electron configuration for element 114:

$$1s^2 2s^2 2p^6 3s^2 3p^6 4s^2 3d^{10} 4p^6 5s^2 4d^{10} 5p^6 6s^2 4f^{14} 5d^{10} 6p^6 7s^2 5f^{14} 6d^{10} 7p^2$$

This configuration can be abbreviated as follows:

$$X = [\text{Rn}]\ 7s^2 5f^{14} 6d^{10} 7p^2$$

In Section 8.2, we concluded that filled $d$ and $f$ subshells are ignored when valence electrons on an atom are counted. Thus, the valence electrons for this element are the $7s^2$ and $7p^2$ electrons. The element therefore belongs in Group IVA, directly below Pb. Possible oxidation states are +2 and +4.

Sec. 8.2

An expanded periodic table for elements up to $Z$ = 168 is shown in Figure 23.16.

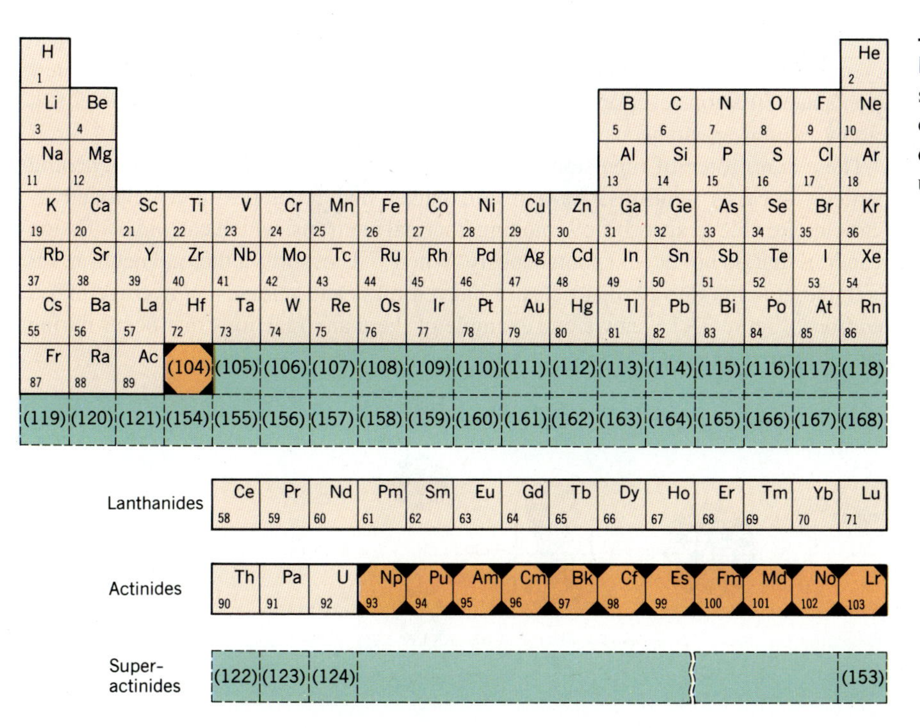


**Figure 23.16** An extended version of the periodic table developed by Glenn Seaborg that predicts the positions for all elements up to atomic number 168.

Elements 104 through 112 are transition metals that fill the 6*d* orbitals. Elements 113 through 120 are main-group elements in which the 7*p* and 8*s* orbitals are filled. The next subshell is the 5*g* atomic orbital, which can hold up to 18 electrons. There is reason to believe that the 5*g* and 6*f* orbitals will be filled at the same time. The next 32 elements are therefore grouped into a so-called superactinide series.

## 23.11 NUCLEAR FISSION

The graph of binding energy per nucleon in Figure 23.7 suggests that nuclides with a mass larger than about 130 amu should spontaneously split apart to form lighter, more stable, nuclides. Experimentally, we find that spontaneous fission reactions occur for only the very heaviest nuclides—those with mass numbers of 230 or more. Even when they do occur, these reactions are often very slow. The half-life for the spontaneous fission of $^{238}U$, for example, is $10^{16}$ years, or about two million times longer than the age of our planet. These observations raise an interesting question: Why are spontaneous fission reactions so slow?

The Rutherford–Geiger–Marsden experiment, which provided the first evidence for the existence of nuclei, suggested that the deflection of the $\alpha$-particles as they passed through the gold foil could be explained by the force of repulsion between the positively charged $\alpha$-particle and the positively charged nuclei of the gold atoms. Because it occurs between the nucleus and a particle that is outside the nucleus, this is an *external* force of repulsion. Theory suggests that there is also an *internal* barrier, which gives rise to an activation energy that opposes the ejection of a positively charged particle from the nucleus of an atom. As long as this activation energy is large, the spontaneous fission of nuclei such as $^{238}U$ will be slow.

Sec. 6.7

The force of repulsion between particles of the same charge depends on the charges ($Z_1$ and $Z_2$) and the distance between the particles ($r$):

$$F = \frac{Z_1 \times Z_2}{r^2}$$

It is therefore easier for a nucleus such as $^{238}U$ to eject an $\alpha$-particle ($Z_1 = 2$ and $Z_2 = 90$) than it is for this nucleus to split in half ($Z_1 = 46$ and $Z_2 = 46$) in a spontaneous fission reaction. This explains why so many of the heavier nuclei that cannot exhibit spontaneous fission can undergo alpha decay.

There is no force of repulsion between a positively charged nucleus and neutral particles such as neutrons. As a result, nuclei that repel positively charged $\alpha$-particles are able to absorb slow-moving, thermal neutrons, which can *induce* fission reactions that do not occur spontaneously. When $^{235}U$ absorbs a thermal neutron, for example, it splits into two particles of uneven mass and releases an average of 2.5 neutrons. Figure 23.17 shows one example of the many reactions that occur during the induced fission of $^{235}U$.

$$^{235}_{92}U + ^{1}_{0}n \longrightarrow ^{139}_{56}Ba + ^{94}_{36}Kr + 3\ ^{1}_{0}n$$

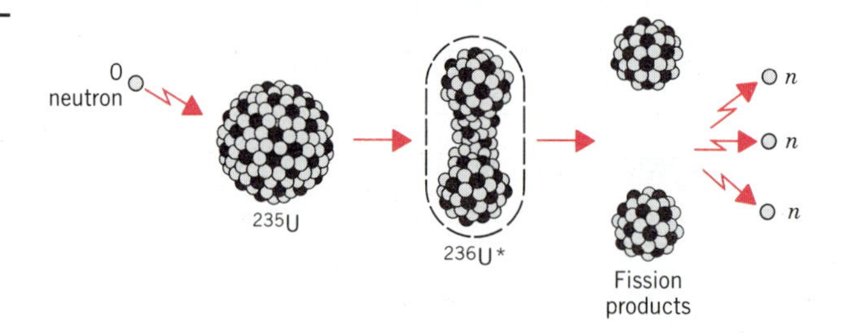


**Figure 23.17** The absorption of a neutron by $^{235}U$ induces oscillations in the nucleus that deform it until it splits into two fragments the way a drop of liquid might break into smaller droplets.

More than 370 daughter nuclides with atomic masses between 72 and 161 amu are formed in the thermal-neutron-induced fission of $^{235}U$. A plot of the relative frequency versus atomic mass of the daughter nuclides produced in this reaction is shown in Figure 23.18.

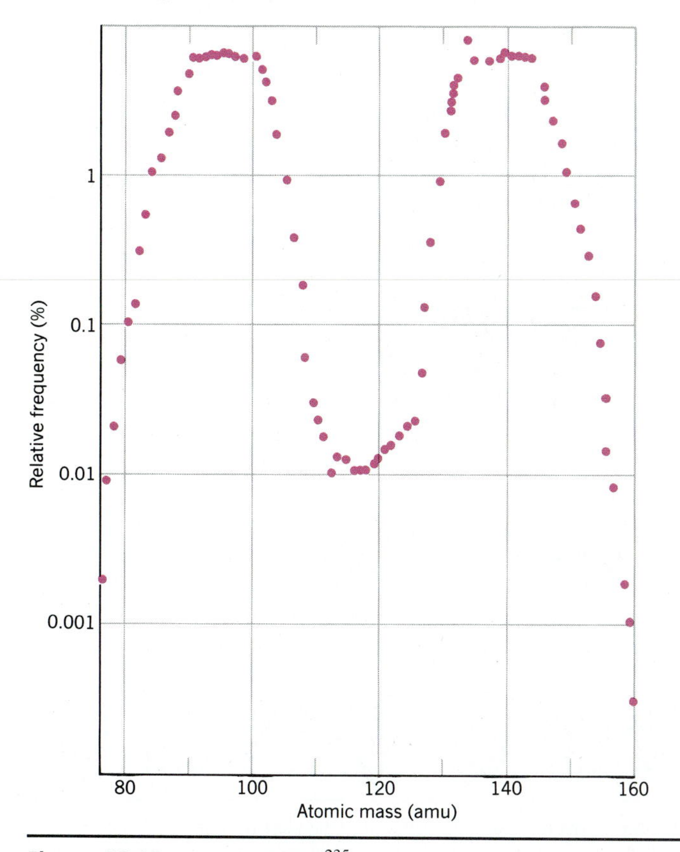


**Figure 23.18** The fission of $^{235}U$ seldom produces two fragments of roughly the same mass. These fission reactions tend to produce one fragment that is roughly 1.4 times heavier than the other.

Several isotopes of uranium undergo induced fission. But the only naturally occurring isotope in which we can induce fission with *thermal* neutrons is $^{235}U$, which is present at an abundance of only 0.72%. The induced fission of this isotope releases an average of 200 MeV per atom, or 80 million kilojoules per gram of $^{235}U$. The attraction of nuclear fission as a source of power can be understood by comparing this figure with the 50 kJ/g released when natural gas is burned.

Pitchblende and yellow cake, which are about 75% $U_3O_8$.

The first artificial nuclear reactor was built by Enrico Fermi and co-workers beneath the University of Chicago's football stadium and brought on line on December 2, 1942. This reactor, which produced several kilowatts of power, consisted of a pile of graphite blocks weighing 385 tons stacked in layers around a cubical array of 40 tons of uranium metal and uranium oxide. Spontaneous fission of $^{238}U$ or $^{235}U$ in this reactor produced a very small number of neutrons. But enough uranium was present so that one of these neutrons induced the fission of a $^{235}U$ nucleus, thereby releasing an average of 2.5 neutrons, which catalyzed the fission of additional $^{235}U$ nuclei in a chain reaction, as shown in Figure 23.19. The amount of fissionable material necessary for the chain reaction to sustain itself is called the **critical mass.**

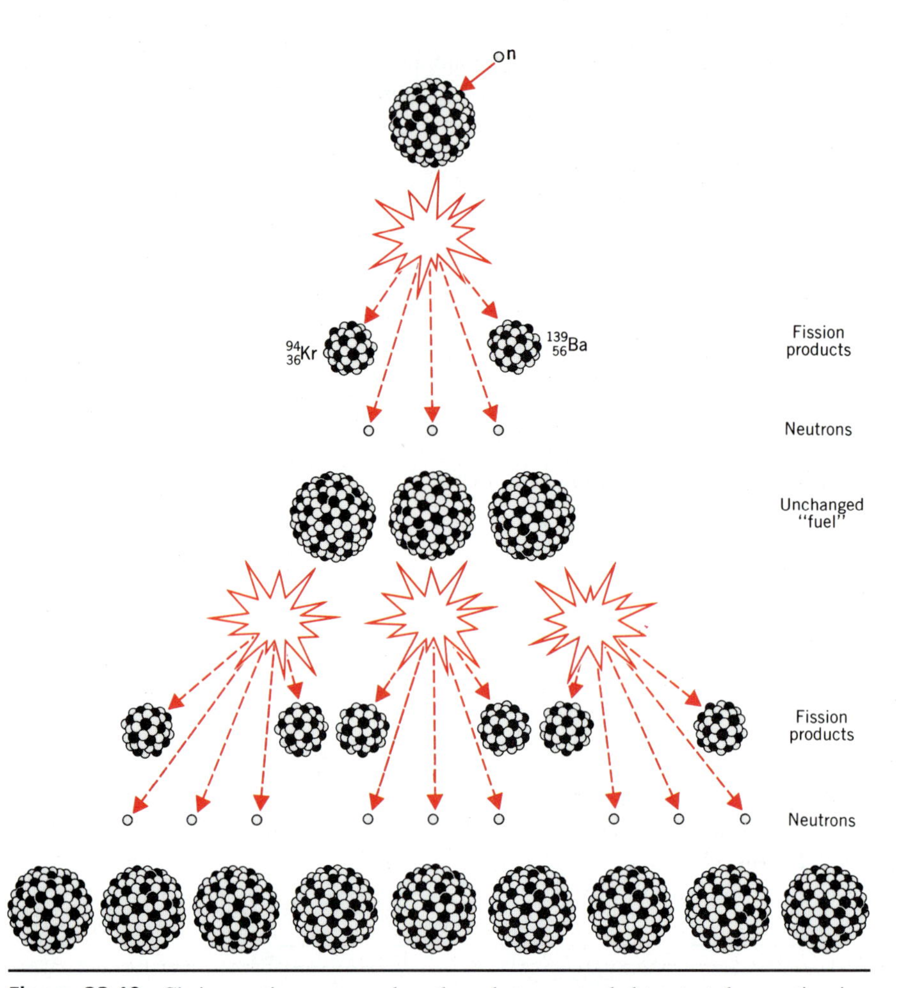


**Figure 23.19** Chain reactions occur when the substance needed to start the reaction is regenerated in the course of the reaction. Induced fission reactions not only regenerate the neutron required to continue the reaction, they produce enough neutrons to initiate new reaction chains. Nuclear reactors therefore contain control rods filled with materials that are good neutron absorbers in order to make sure that the reaction does not get out of control.

The Fermi reactor at Chicago served as a prototype for larger reactors constructed in 1943 at Oak Ridge, Tennessee, and Hanford, Washington, to produce $^{239}$Pu for one of the atomic bombs dropped on Japan at the end of World War II. Some of the neutrons released in the chain reaction are absorbed by $^{238}$U to form $^{239}$U:

$$^{238}_{92}\text{U} + ^{1}_{0}\text{n} \longrightarrow ^{239}_{92}\text{U} + \gamma$$

As we have seen, the product of this reaction undergoes decay by the successive loss of two $\beta^-$-particles to form $^{239}$Pu:

$$^{239}_{92}\text{U} \longrightarrow ^{239}_{93}\text{Np} + ^{0}_{-1}\text{e}$$
$$^{239}_{93}\text{Np} \longrightarrow ^{239}_{94}\text{Pu} + ^{0}_{-1}\text{e}$$

$^{238}$U is therefore an example of a **fertile nuclide.** It does not undergo fission with thermal neutrons, but it can be converted to $^{239}$Pu, which does undergo thermal-neutron-induced fission. $^{232}$Th is another fertile nuclide. It captures a neutron to

A painting of Fermi's first reactor at the University of Chicago.

form $^{233}Th$, which undergoes $\beta$ decay to form $^{233}U$, which is also capable of induced fission:

$$^{232}_{90}Th + ^{1}_{0}n \longrightarrow ^{233}_{90}Th$$
$$^{233}_{90}Th \longrightarrow ^{233}_{91}Pa + ^{0}_{-1}e$$
$$^{233}_{91}Pa \longrightarrow ^{233}_{92}U + ^{0}_{-1}e$$

Fission reactors can be designed to handle naturally abundant $^{235}U$, as well as fuels described as slightly enriched (2 to 5% $^{235}U$), highly enriched (20 to 30% $^{235}U$), or fully enriched (>90% $^{235}U$). Heat generated in the reactor core is transferred to a cooling agent in a closed system. The cooling agent is then passed through a series of heat exchangers in which water is heated to steam. The steam produced in these exchangers then drives a turbine that generates electrical power. There are two ways of specifying the power of such a plant: the thermal energy produced by the reactor or the electrical energy generated by the turbines. The electrical capacity of the plant is usually about one-third of the thermal power.

It takes $10^{11}$ fissions per second to produce one watt of electrical power. As a result, about 1 gram of fuel is consumed per day per megawatt of electrical energy produced. This means that 1 gram of waste products is produced per megawatt per day, which includes 0.5 gram of $^{239}Pu$. These waste products must be either reprocessed to generate more fuel or stored for the tens of thousands of years it takes for the level of radiation to reach a safe limit.

If we design a reactor in which the ratio of the $^{239}Pu$ or $^{233}U$ produced to the $^{235}U$ consumed is greater than 1, the reactor can generate more fuel than it consumes. Such reactors are known as *breeders,* and commercial breeder reactors are now operating in France.

The key to an efficient breeder reactor is a fuel that gives the largest possible number of neutrons released per neutron absorbed. The breeder reactors being built today use a mixture of $PuO_2$ and $UO_2$ as the fuel and fast neutrons to activate fission. Fast neutrons carry an energy of at least several keV and therefore travel 10,000 or more times faster than thermal neutrons. $^{239}Pu$ in the fuel assembly absorbs one of these fast neutrons and undergoes fission with the release of three neutrons. $^{238}U$ in the fuel then captures one of these neutrons to produce additional $^{239}Pu$, thereby producing (or "breeding") fuel faster than it is consumed.

The Superphenix breeder reactor in France.

The advantage of breeder reactors is obvious—they mean a limitless supply of fuel for nuclear reactors. There are significant disadvantages, however. Breeder reactors are more expensive to build. They are also useless without a subsidiary industry to collect the fuel, process it, and ship the $^{239}Pu$ to new reactors.

It is the reprocessing of $^{239}Pu$ that concerns most of the critics of breeder reactors. $^{239}Pu$ is so dangerous as a carcinogen that the nuclear industry places a limit on exposure to this material that assumes workers inhale no more than *0.2 micrograms* of plutonium over their lifetimes. There is also concern that the $^{239}Pu$ produced by these reactors may be stolen and assembled into bombs by terrorist organizations.

The fate of breeder reactors in the United States is linked to economic considerations. Because of the costs of building these reactors and safely reprocessing the $^{239}Pu$ produced, the breeder reactor becomes economical only when the scarcity of uranium drives its price so high that the breeder reactor becomes cost effective by comparison. If nuclear energy is to play a dominant role in the generation of electrical energy in the twenty-first century, breeder reactors eventually may be essential.

The "pile" that Fermi constructed at the University of Chicago in 1942 was the first *artificial* nuclear reactor, but not the first fission reactor. In 1972, a group of French scientists discovered that uranium ore from a deposit in the Oklo mine in Gabon, West Africa, contained 0.4% $^{235}U$ instead of the 0.72% abundance found in all other sources of this ore. Analysis of the trace elements in the ore suggested that natural fission reactors operated in this deposit for a period of 600,000 to 800,000 years about 2 billion years ago.

## 23.12 NUCLEAR FUSION

The graph of binding energy per nucleon in Figure 23.7 suggests another way to obtain useful energy from nuclear reactions: Fusing two light nuclei can liberate as much energy as the fission of $^{235}U$ or $^{239}Pu$. The fusion of four protons to form a helium nucleus, two positrons (and two neutrinos), for example, generates 24.7 MeV of energy:

$$4\ {}^{1}_{1}H \longrightarrow {}^{4}_{2}He + 2\ {}^{0}_{+1}e$$

Most of the energy radiated from the surface of the sun is produced by the fusion of protons to form helium atoms within its core.

Fusion reactions have been duplicated in artificial devices. The enormous destructive power of the $^{235}U$-fueled atomic bomb dropped on Hiroshima on August 6, 1945, which killed 75,000 people, and the $^{239}Pu$-fueled bomb dropped on Nagasaki three days later touched off a violent debate after World War II about the building of the next superweapon—a fusion, or "hydrogen," bomb. Alumni of the Manhattan Project, which had developed the atomic bomb, were divided on the issue. Ernest Lawrence and Edward Teller fought for the construction of the fusion device. J. Robert Oppenheimer and Enrico Fermi argued against it. The decision was made to develop the weapon, and the first artificial fusion reaction occurred when the hydrogen bomb was tested in November 1952.

The history of fusion research is therefore quite different from that of fission research. With fission, the reactor came first, and then the bomb was built. With fusion, the bomb was built long before any progress was made toward the construction of a controlled fusion reactor. More than 40 years after the first hydrogen bomb was exploded, the feasibility of controlled fusion reactions is still open to debate. The reaction that is most likely to fuel the first fusion reactor is the thermonuclear

Hiroshima, shortly after the first atomic bomb fell.

d–t, or deuterium–tritium, reaction. This reaction fuses two isotopes of hydrogen, **deuterium** ($^2$H) and **tritium** ($^3$H), to form helium and a neutron:

$$^2_1\text{H} + ^3_1\text{H} \longrightarrow ^4_2\text{He} + ^1_0\text{n}$$

This ZT-40M Reversed Field Pinch fusion device at the Los Alamos National Laboratory was able to sustain a plasma of 8 million °C for 8 milliseconds in January 1986.

If we consider the implications of this reaction we can begin to understand why it is called a **thermonuclear reaction** and why it is so difficult to produce in a controlled manner. The d–t reaction requires that we fuse two positively charged particles. This means that we must provide enough energy to overcome the force of repulsion between these particles before fusion can occur. To produce a self-sustaining reaction, we have to provide the particles with enough thermal energy so that they can fuse when they collide.

Each fusion reaction is characterized by a specific *ignition temperature,* which must be surpassed before the reaction can occur. The d–t reaction has an ignition temperature above $10^8$ K. In a hydrogen bomb, a fission reaction produced by a small atomic bomb is used to heat the contents to the temperature required to initiate fusion. Obtaining the same result in a controlled reaction is much more challenging.

Any substance at temperatures approaching $10^8$ K will exist as a completely ionized gas, or *plasma*. The goals of fusion research include:

1. To achieve the required temperature to ignite the fusion reaction.
2. To keep the plasma together at this temperature long enough to get useful amounts of energy out of the thermonuclear fusion reactions.
3. To obtain more energy from the thermonuclear reactions than is used to heat the plasma to the ignition temperature.

These are not trivial goals. The only reasonable container for a plasma at $10^8$ K is a magnetic field. Both doughnut-shaped (toroidal) and linear magnetic bottles have been proposed as fusion reactors. But reactors that produce high enough temperatures for ignition are not the same as the reactors that have produced long enough confinement times for the plasma to provide useful amounts of energy.

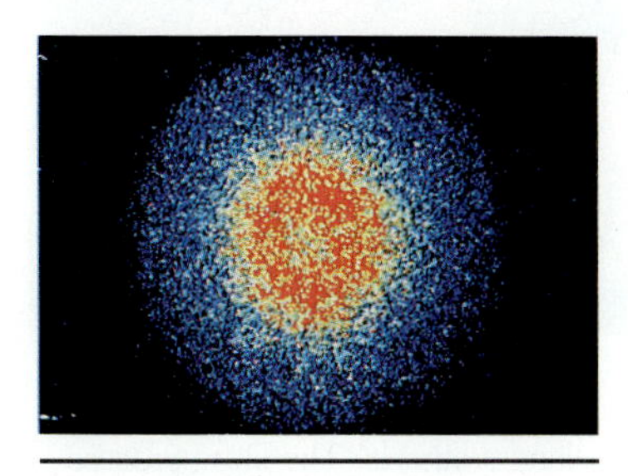

An extremely high-speed x-ray pinhole photo of energy released from a laser fusion target.

A second approach to a controlled fusion reactor involves hitting fuel pellets containing the proper reagents for the thermonuclear reaction with pulsed beams of laser power. If enough power were delivered, the fuel pellets would collapse upon themselves, or implode, to reach densities several orders of magnitude greater than normal. This could produce a plasma both hot enough and dense enough to initiate fusion reactions.

**CHECKPOINT**

Because of the enormous energy released in nuclear reactions, both fission and fusion reactors are possible sources of thermal pollution. Explain why fusion reactors should make smaller contributions to other forms of pollution than the present fission reactors.

## 23.13 NUCLEAR SYNTHESIS

From the most primitive societies to the most complex, people have struggled to explain how the world was created. Within the last 60 years, scientists have generated a new story of creation that offers a model for understanding how nuclei are synthesized.

In 1929, Edwin Hubble first provided evidence to suggest that our universe is expanding. Between 1946 and 1948, George Gamow and co-workers generated a model which assumed that the primordial substance, or *ylem,* from which all other matter was created was an extraordinarily hot, dense singularity, which exploded in a "Big Bang" and has been expanding ever since. This model assumes that neutrons in the ylem were transformed into protons by $\beta$ decay. Neutrons and protons then combined to form $^4$He atoms before the temperature and pressure of the fireball decayed to the point at which no further nuclear reactions were possible.

The first generation stars condensed out of this cloud of hydrogen and helium. As the gas condensed by gravitational attraction, it became warmer. Eventually, the temperature reached $10^7$ K, and first-generation, main-sequence stars were born. The temperatures at the cores of these stars were high enough to ignite the following thermonuclear reactions, which transform hydrogen to helium:

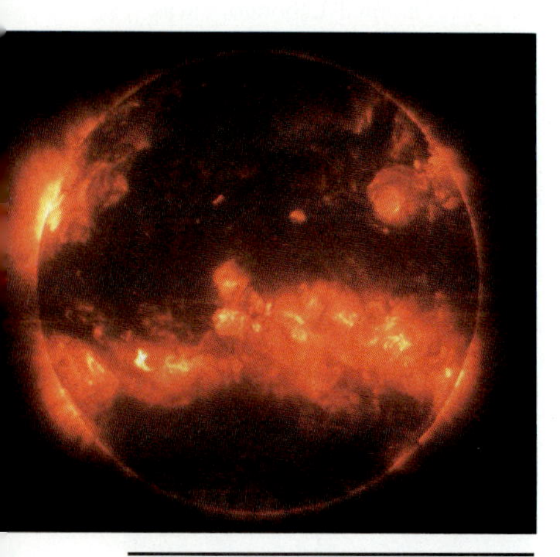

There are two processes for nuclear synthesis. The slow process occurs within a star, such as the sun, during its lifetime. The fast process occurs when the star goes supernova.

$$^{1}_{1}H + ^{1}_{1}H \longrightarrow ^{2}_{1}H + ^{0}_{+1}e$$
$$^{1}_{1}H + ^{2}_{1}H \longrightarrow ^{3}_{2}He$$
$$^{3}_{2}He + ^{3}_{2}He \longrightarrow ^{4}_{2}He + 2\ ^{1}_{1}H$$
$$4\ ^{1}_{1}H \longrightarrow ^{4}_{2}He + 2\ ^{0}_{+1}e$$

The net result of these reactions is the formation of a helium atom from four protons. Eventually the heat generated in this reaction was enough to halt the gravitational collapse of the star, which then entered a stable period during which the energy generated by this reaction balanced the energy radiated at the surface.

The hydrogen-burning reactions in a main-sequence star are concentrated in the core. When enough hydrogen has been consumed, the core begins to collapse, and the temperature of the core rises above $10^8$ K. (The larger the star, the more rapidly it radiates energy from its surface, and the more rapidly it consumes the hydrogen in the core.) As the core collapses, the hydrogen-containing outer shell expands, and the surface of the star cools. Stars that have reached this point in their evolution include the so-called red giants.

Temperatures in the core of these red giants are high enough to ignite further thermonuclear fusion reactions, such as the following:

$$3\ ^{4}_{2}He \longrightarrow ^{12}_{6}C$$

This reaction becomes the principal source of energy in a red giant, although there is undoubtedly some burning of hydrogen to helium in the outer shell of the star. As the amount of $^{12}C$ in the core increases, reactions occur to form $^{16}O$ and $^{20}Ne$:

$$^{12}_{6}C + ^{4}_{2}He \longrightarrow ^{16}_{8}O + \gamma$$
$$^{16}_{8}O + ^{4}_{2}He \longrightarrow ^{20}_{10}Ne + \gamma$$

Eventually, the helium in the core is exhausted, and the core collapses further, reaching temperatures of 6 to 7 $\times$ $10^8$ K. At this point, more complex reactions take place that produce nuclides such as $^{28}Si$ and $^{32}S$:

$$^{12}_{6}C + ^{16}_{8}O \longrightarrow ^{28}_{14}Si + \gamma$$
$$^{16}_{8}O + ^{16}_{8}O \longrightarrow ^{32}_{16}S + \gamma$$

Further gravitational collapse heats the core to temperatures above $10^9$ K, and a complex sequence of reactions takes place to synthesize the nuclei with the highest binding energies, such as Fe and Ni. If the star explodes in a supernova, its contents are ejected across space. Second-generation stars that condense in this region contain not only hydrogen and helium but elements with higher atomic number.

The best estimates of the age of the Milky Way suggest that our galaxy is about 15 billion years old. Our sun and its planets, however, are only 4.6 billion years old. This suggests that the sun is a second-generation star. In such stars, the transformation of hydrogen to helium can be catalyzed by $^{12}C$, as shown in Figure 23.20.

$$^{12}_{6}C + ^{1}_{1}H \longrightarrow ^{13}_{7}N + \gamma$$
$$^{13}_{7}N \longrightarrow ^{13}_{6}C + ^{0}_{+1}e$$
$$^{13}_{6}C + ^{1}_{1}H \longrightarrow ^{14}_{7}N + \gamma$$
$$^{14}_{7}N + ^{1}_{1}H \longrightarrow ^{15}_{8}O + \gamma$$
$$^{15}_{8}O \longrightarrow ^{15}_{7}N + ^{0}_{+1}e$$
$$^{15}_{7}N + ^{1}_{1}H \longrightarrow ^{12}_{6}C + ^{4}_{2}He$$

**Figure 23.20** Second-generation stars do not use the same mechanism to synthesize helium as first-generation stars. They use a sequence of reactions, such as those shown here, that are catalyzed by isotopes heavier than helium.

Two processes can synthesize elements with atomic numbers larger than that of iron. One of them is relatively slow (s process), the other is very rapid (r process). Since the only way to synthesize nuclei with atomic numbers larger than iron is by neutron absorption, both the s process and the r process result from (n,$\gamma$) reactions.

In the **s process,** neutrons are captured one at a time to form a neutron-rich nuclide, which has enough time to undergo $\alpha$ or $\beta^-$ decay before another neutron can be absorbed. An example of an s-process sequence of reactions starts with $^{120}Sn$. The capture of a neutron produces $^{121}Sn$, which undergoes $\beta^-$ decay. If $\beta^-$ decay occurs before this nuclide captures another neutron, a stable isotope of antimony is formed. Eventually, $^{121}Sb$ captures a neutron to produce $^{122}Sb$, which is transformed into $^{122}Te$ by $\beta$ decay. $^{122}Te$ can undergo $\beta^-$ decay to form $^{122}I$, or it can capture a neutron to form $^{123}Te$. With $^{123}Te$, we encounter a series of stable isotopes of tellurium. Thus, neutrons are slowly absorbed, one at a time, until we reach $^{127}Te$, which decays to $^{127}I$, the most abundant isotope of iodine:

$$^{120}_{50}Sn(n,\gamma)^{121}_{50}Sn \xrightarrow{\beta^-} ^{121}_{51}Sb(n,\gamma)^{122}_{51}Sb \xrightarrow{\beta^-} ^{122}_{52}Te(5n,5\gamma)^{127}_{52}Te \xrightarrow{\beta^-} ^{127}_{53}I$$

This slow process can't account for very heavy nuclides, such as $^{232}Th$ and $^{238}U$, because the lifetimes of the intermediate nuclei with atomic numbers between 83 and 90 are too short for this step-by-step absorption of neutrons to proceed. Synthesizing appreciable quantities of uranium and thorium requires a rapid process. In the **r process,** a number of neutrons are captured in rapid succession, before there is time for $\alpha$ or $\beta$ decay to take place. Achieving an r-process reaction, however, requires a very high neutron flux. (These reactions occur during nuclear explosions, for example.) The neutron flux needed to fuel such reactions is not likely to occur in a normal star. During the moment when a star explodes as a supernova, however, the conditions are ripe for r-process reactions. The heavier elements on this planet were produced in a series of supernova explosions that occurred in this portion of the galaxy before our solar system condensed.

## RESEARCH IN THE 90s

### NUCLEAR MEDICINE

Ever since the first x-ray images were obtained by Roentgen in 1895, ionizing radiation and radionuclides have played a vital role in medicine. This work has been so fruitful that a separate field known as nuclear medicine, which focuses on either therapeutic or diagnostic uses of radiation, has developed.

There are three standard approaches to fighting cancer: surgery, chemotherapy, and radiation. Surgery, by its very nature, is invasive. Chemotherapy and classic approaches to radiation therapy are not selective. Research in recent years has therefore examined new approaches to radiation therapy that specifically attack tumor cells, without damaging normal tissue. The technique known as *boron neutron-capture therapy* (BNCT) provides an example of this work [R. F. Barth, A. H. Soloway, and R. G. Fairchild, *Cancer Research,* **50,** 1061 (1990)].

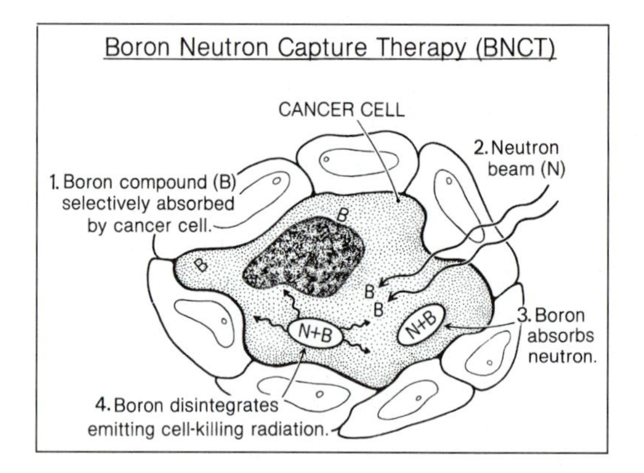


Naturally occurring boron consists of two stable isotopes: $^{10}B$ (19.7%) and $^{11}B$ (80.3%). $^{10}B$ absorbs thermal neutrons to form $^{11}B$ in a nuclear excited state. Although $^{11}B$ in its nuclear ground state is stable, this excited $^{11}B$ nuclide undergoes fission to produce $^{7}Li$ and an alpha particle:

$$^{10}_{5}B + ^{1}_{0}n \longrightarrow ^{7}_{3}Li + ^{4}_{2}He$$

Because the energy of a thermal neutron is only about 0.025 eV, the neutrons that are not absorbed do relatively little damage to the normal tissue. The $\alpha$-particle emitted in this reaction has an energy of 2.79 MeV, however, which makes it an extremely lethal form of radiation. The RBE for $\alpha$-particle radiation is larger than any other particle in Table 23.2 because this relatively massive particle loses energy very efficiently as it collides with matter. Radiation damage from the $\alpha$-particle is therefore restricted to the immediate vicinity of the tissue that absorbed the thermal neutron.

Other common nuclides in living tissue can absorb thermal neutrons. Fortunately, the ability of $^{10}B$ to absorb thermal neutrons is three orders of magnitude larger than these other nuclides. The units with which this measurement is made can be understood by thinking about the area around the nuclide through which the neutron can pass and still be absorbed. The neutron-capture cross-section for a hydrogen atom corresponds to a circle with a radius of about $2 \times 10^{-13}$ cm. For nitrogen atoms, the radius of this circle is about $10^{-12}$ cm. Boron has a neutron-capture cross-section that would be described by a circle with a radius of about $2 \times 10^{-9}$ cm. Virtually all of the neutron capture that occurs is therefore concentrated in the tissue that contains boron.

The potential of boron neutron-capture therapy was recognized as early as 1936. Early clinical trials in the late 1950s and early 1960s failed to prolong the lives of patients suffering from brain tumors because the boron compounds available for testing did not concentrate selectively in the tumor cells. Recent research has discovered boron-labeled compounds that are sufficiently selective for BNCT to become useful as an adjunct to surgery, or in place of surgery with inoperable cases.

## SUMMARY

**1.** We can predict the products of nuclear reactions by assuming that both mass number and charge are conserved. If $^{238}U$ undergoes $\alpha$-particle decay, the product of this reaction must be $^{234}Th$:

$$^{238}_{92}U \longrightarrow ^{234}_{90}Th + ^{4}_{2}He \qquad (\alpha \text{ decay})$$

**2.** There are three forms of beta decay. One corresponds to the emission of an electron:

$$^{40}_{19}K \longrightarrow ^{40}_{19}Ca + ^{0}_{-1}e \qquad (\beta^- \text{ emission})$$

Another involves the emission of a positron:

$$^{40}_{19}K \longrightarrow ^{40}_{18}Ar + ^{0}_{+1}e \qquad (\beta^+ \text{ emission})$$

The third involves the capture of an electron:

$$^{40}_{19}K + ^{0}_{-1}e \longrightarrow ^{40}_{18}Ar + h\nu \qquad (\text{electron capture})$$

**3.** Alpha and beta decay often produce nuclides in an excited state. The daughter nuclide emits its excess energy in the form of a $\gamma$-ray:

$$^{60m}_{27}Co \longrightarrow ^{60}_{27}Co + \gamma \qquad (\gamma\text{-ray emission})$$

**4.** Stable nuclides exist within only a narrow band of neutron-to-proton ratios. When this ratio is too large, the nuclide is said to be **neutron rich.** When the ratio is too small, it is **neutron poor.** Neutron-rich nuclides decay by $\beta^-$ emission. Neutron-poor nuclides decay by $\beta^+$ emission, electron capture, or $\alpha$-particle emission.

**5.** Nuclei can also undergo decay by spontaneous fission. The rate at which a nuclide spontaneously splits into two unequal parts increases with the atomic number of the nuclide. For most nuclides, the rate of spontaneous fission is very slow. The rate of this reaction can be enhanced, however, by irradiating the nuclide with neutrons. The capture of a neutron lowers the activation energy associated with the fission of the nuclide, thereby significantly increasing the rate at which the nuclide undergoes fission.

**6.** The difference between the mass of an atom and the sum of the masses of the electrons, protons and neutrons in the atom is known as the **mass defect.** The mass defect is a direct measure of the **binding energy** that holds the nuclide together.

**7.** The binding energy of nuclides increases with atomic number. When the binding energy per nucleon is calculated, nuclides are found to be most stable at or near the $^{56}Fe$ isotope. Nuclides that are heavier than $^{56}Fe$ can become more stable by undergoing fission reactions. Nuclides that are lighter than $^{56}Fe$ can become more stable by undergoing fusion reactions.

## KEY TERMS

**Activity** (p. 854)
**Alpha decay** (p. 845)
**Beta decay** (p. 846)
**Binding energy** (p. 851)
**Binding energy per nucleon** (p. 852)
**$^{14}C$ dating** (p. 857)
**Critical mass** (p. 869)
**Curie** (p. 854)
**Deuterium** (p. 873)
**Electron capture** (p. 846)
**Electron ($\beta^-$) emission** (p. 846)
**Fertile nuclide** (p. 870)
**Fission reactions** (p. 852)
**Fusion reactions** (p. 852)
**Gamma emission** (p. 847)
**Half-life** (p. 855)

**Induced radioactivity** (p. 864)
**Ionizing radiation** (p. 858)
**Nonionizing radiation** (p. 858)
**Isobar** (p. 845)
**Isotone** (p. 845)
**Isotope** (p. 845)
**Mass defect** (p. 851)
**Metastable** (p. 847)
**Natural radioactivity** (p. 862)
**Neutron-poor nuclides** (p. 848)
**Neutron-rich nuclides** (p. 848)
**Nuclide** (p. 845)
**Positron ($\beta^+$) emission** (p. 846)
**R process** (p. 875)
**Radiation absorbed dose (rad)** (p. 860)
**Radiation biological effectiveness (RBE)** (p. 860)
**Roentgen equivalent man (rem)** (p. 860)
**S process** (p. 875)
**Spontaneous fission** (p. 847)
**Thermal neutron** (p. 865)
**Thermonuclear reaction** (p. 873)
**Tritium** (p. 873)

## KEY EQUATIONS

1 amu = 931.5016 MeV

$$\text{Rate} = -\frac{dN}{dt} = kN$$ Rate of radioactive decay

$$t_{\frac{1}{2}} = \frac{\ln 2}{k} = \frac{0.693}{k}$$ Half-life for radioactive decay

$$\ln\left[\frac{(N)}{(N_o)}\right] = -kt$$

## PROBLEMS

### The Discovery of Radioactivity

**23-1** Identify the particle given off when a nucleus undergoes $\alpha$ decay. Identify the particle given off during $\beta$ decay. Describe the relationship between $\gamma$-rays and other forms of electromagnetic radiation.

**23-2** Describe an experiment that could be used to determine whether an atom emits $\alpha$-particles, $\beta$-particles, or $\gamma$-rays.

### The Structure of the Atom

**23-3** Define the terms *atomic number, mass number, nuclide, nucleon,* and *nucleus.*

**23-4** Define the terms *isotopes, isobars,* and *isotones.* Give examples of each.

**23-5** Calculate the number of electrons, protons, and neutrons in neutral atoms of the following nuclides:

(a) $^{14}C$ (b) $^{75}As$ (c) $^{90}Sr$

**23-6** Calculate the number of electrons, protons, and neutrons in the following ions:

(a) $^{40}K^+$ (b) $^{103}Rh^{3+}$ (c) $^{127}I^-$

### Modes of Radioactive Decay

**23-7** Explain why the three forms of $\beta$ decay interconvert isobars.

**23-8** Describe how electron emission, electron capture, and positron emission differ. Give an example of each.

**23-9** In theory, $\beta$ decay could include reactions in which a positron is captured and the charge on the nucleus increases. Explain why the following positron capture reaction does not occur under normal conditions:

$$^{14}_{6}C + {}^{0}_{+1}e \longrightarrow {}^{14}_{7}N + h\nu$$

**23-10** Which of the following reactions interconvert isotopes? Which interconvert isobars? Which interconvert isotones?

(a) electron emission (b) electron capture
(c) positron emission (d) alpha emission (e) neutron emission (f) neutron absorption (g) alpha emission followed by two $\beta^-$ decays

**23-11** Identify the missing particle in each of the following equations and name the form of radioactive decay:

(a) $^{125}_{53}I + {}^{0}_{-1}e \longrightarrow X$
(b) $^{240}_{94}Pu \longrightarrow X + {}^{144}_{58}Ce + 2\,{}^{1}_{0}n$
(c) $^{90}_{38}Sr \longrightarrow X + {}^{0}_{-1}e$
(d) $^{40}_{19}K \longrightarrow X + {}^{0}_{+1}e$
(e) $^{228}_{90}Th \longrightarrow X + {}^{4}_{2}He$

**23-12** Write balanced equations for the $\beta^-$-particle decay of the following nuclides:

(a) $^{35}S$ (b) $^{52}V$ (c) $^{99}Mo$

**23-13** Write balanced equations for the $\alpha$-particle decay of the following nuclides:

(a) $^{183}Pt$ (b) $^{204}At$ (c) $^{248}Cf$

**23-14** Write balanced equations for the electron capture reactions of the following nuclides:

(a) $^{7}Be$ (b) $^{37}Ar$ (c) $^{62}Cu$

**23-15** Write balanced equations for the positron emission reactions of the following nuclides:

(a) $^{34}Cl$ (b) $^{45}Ti$ (c) $^{90}Mo$

**23-16** Predict the products of the following nuclear reactions:

(a) Electron emission by $^{32}P$ (b) Positron emission by $^{11}C$ (c) Alpha decay by $^{212}Rn$ (d) Electron capture by $^{125}Xe$

**23-17** Predict the products of the following nuclear reactions:

(a) $^{208}_{82}Pb(^{2}_{1}H,n)$__ (b) $^{252}_{98}Cf(^{13}_{6}C,8n)$__
(c) $^{95}_{42}Mo(n,\gamma)$__ (d) $^{202}_{84}Po(^{22}_{10}Ne,4n)$__ (e) $^{14}_{7}N(n,p)$__
(f) $^{63}_{29}Cu(p,2n)$__

**23-18** Identify the missing particle or particles in the following reactions and write a balanced equation for each reaction.

(a) $^{238}_{92}U(\_\_,3n)^{239}_{94}Pu$ (b) __$(\alpha,n)^{242}_{96}Cm$
(c) $^{250}_{98}Cf(^{11}_{5}B,\_\_)^{257}_{103}Lr$ (d) $^{249}_{98}Cf(\_\_,4n)^{257}_{104}Rh$

### Neutron-Rich Versus Neutron-Poor Nuclides

**23-19** Explain why neutron-rich nuclides decay by electron ($\beta^-$) emission.

**23-20** Explain why neutron-poor nuclides decay by either electron capture, positron ($\beta^+$) emission, the emission of an $\alpha$ particle, or spontaneous fission.

**23-21** Which of the following nuclides are most likely to be neutron rich?

(a) $^{14}C$ (b) $^{24}Na$ (c) $^{25}Si$ (d) $^{27}Al$ (e) $^{31}P$

**23-22** Which of the following nuclides are most likely to be neutron poor?

(a) $^{3}H$ (b) $^{11}C$ (c) $^{14}N$ (d) $^{40}K$ (e) $^{61}Cu$

**23-23** Explain why $^{17}Ne$, $^{18}Ne$, $^{19}Ne$ decay by positron emission but $^{23}Ne$ and $^{24}Ne$ decay by electron emission.

**23-24** Which isotope of carbon is most likely to decay by positron emission?

(a) $^{11}C$ (b) $^{12}C$ (c) $^{13}C$ (d) $^{14}C$

**23-25** Which isotope of carbon is most likely to decay by electron emission?

(a) $^{11}C$ (b) $^{12}C$ (c) $^{13}C$ (d) $^{14}C$

### Binding Energy Calculations

**23-26** Define the terms *mass defect* and *binding energy*.

**23-27** Calculate the binding energy of $^{6}Li$ in MeV per atom if the exact mass of this nuclide is 6.01512 amu. Calculate the binding energy per nucleon.

**23-28** Calculate the binding energy of $^{60}Ni$ in MeV per atom if the exact mass is 59.9332 amu. Calculate the binding energy per nucleon.

**23-29** Calculate the exact mass of $^{238}U$ if the binding energy per nucleon is 7.570198 MeV.

**23-30** Which nuclide in Problems 23-27 through 23-29 has the largest binding energy? Which has the largest binding energy per nucleon?

**23-31** Calculate the energy released in the following reaction:

$$^{24}Mg(^{2}H, p)^{25}Mg$$

Use the following data for the masses of the particles involved in the reaction: $^{24}Mg$ = 23.98504 amu; $^{25}Mg$ = 24.98584 amu; $^{2}H$ = 2.0140 amu.

**23-32** Calculate the energy released in the following reaction:

$$^{10}B(n, \alpha)^{7}Li$$

Use the following data for the masses of the particles involved in the reaction: $^{10}B$ = 10.0129 amu; $^{7}Li$ = 7.01600 amu; $^{4}He$ = 4.00260 amu. Explain why the ability of $^{10}B$ to release a high-energy $\alpha$-particle after it absorbs a thermal neutron has generated considerable interest in whether it is possible to get boron compounds to absorb preferentially into the fast-growing tumor in patients who suffer from brain tumors.

### The Kinetics of Radioactive Decay

**23-33** The half-life of $^{32}P$ is 14.3 days. Calculate how long it would take for a 1-g sample of $^{32}P$ to decay to each of the following quantities of $^{32}P$:

(a) 0.500 g (b) 0.250 g (c) 0.125 g

**23-34** Calculate the half-life for the decay of $^{39}Cl$ if a 1-g sample decays to 0.125 g in 165 min.

**23-35** Calculate the rate constants in $s^{-1}$ for the decay of the following nuclides from their half-lives.

(a) $^{18}F$, 110 min (b) $^{54}Mn$, 312 d (c) $^{3}H$, 12.26 y
(d) $^{14}C$, 5730 y (e) $^{129}I$, $1.6 \times 10^{7}$ y

**23-36** A 1-g sample of $^{22}Na$ decays to 0.20 g in 6.04 years. Calculate the half-life for this decay, the rate constant, and the time it would take for this sample to decay to 0.075 g.

**23-37** Calculate the time required for a 2.50-g sample of $^{51}Cr$ to decay to 1.00 g, assuming that the half-life is 27.8 days.

**23-38** A sample of $^{210}Po$ initially weighed 2.000 g. After 25 days, 0.125 g of $^{210}Po$ remained, the rest of the sample having decayed to the stable $^{206}Pb$ isotope. Calculate the half-life of $^{210}Po$ and the weight of $^{206}Pb$ formed.

**23-39** Forgeries that had been accepted by art authorities as paintings by the Dutch artist Vermeer (1632–1675) have been detected by measuring the activity of the $^{210}Pb$ isotope in the lead paints. When lead is extracted from its ores, it is separated from $^{226}Ra$, which is the source of the $^{210}Pb$ isotope. The amount of $^{210}Pb$ in the paint therefore decreases with time. If the half-life for the decay of $^{210}Pb$ is 21 years, what fraction of the $^{210}Pb$ would be present in a 300-year-old painting? What fraction would remain in a 10-year-old forgery?

**23-40** Use the following data for $^{48}Cr$ to calculate the rate constant and the half-life for the decay of this isotope by electron capture.

| Mass (g) | 100 | 80 | 60 | 40 | 20 |
|---|---|---|---|---|---|
| Time (min) | 0 | 444 | 1017 | 1825 | 3205 |

### Units of Activity for Radioactive Decay

**23-41** The threat to people's health from radon in the air trapped in their houses has received attention in recent years. If the average level of radon in a house is approximately 1 picocurie (pCi) per liter of air, how many radon atoms are there per liter? (Assume that $^{222}Rn$ is the principal source of this activity and that the half-life for the decay of this nuclide is 3.823 d.)

**23-42** Calculate the number of disintegrations per minute in a 1.00-mg sample of $^{238}U$, if the half-life is $4.47 \times 10^9$ y.

**23-43** Calculate the activity, in disintegrations per second, for a 1.00-mg sample of the following isotopes of argon:

(a) $^{35}Ar$, $t_{\frac{1}{2}} = 1.85$ min (b) $^{41}Ar$, $t_{\frac{1}{2}} = 1.83$ h
(c) $^{37}Ar$, $t_{\frac{1}{2}} = 35.1$ d (d) $^{39}Ar$, $t_{\frac{1}{2}} = 270$ y

**23-44** Calculate the activity in curies for 1.00-mg samples of the following isotopes of uranium:

(a) $^{238}U$, $t_{\frac{1}{2}} = 9.3$ min (b) $^{230}U$, $t_{\frac{1}{2}} = 20.8$ d
(c) $^{236}U$, $t_{\frac{1}{2}} = 2.39 \times 10^7$ y

**23-45** Calculate the half-life of $^{227}Ac$, if a 0.100-mg sample has an activity of $2.75 \times 10^8$ disintegrations per second.

**23-46** Calculate the weight of 1.00 millicurie (mCi) of $^{14}C$ if the half-life of this nuclide is 5730 y.

### Dating by Radioactive Decay

**23-47** The $^{14}C$ in living matter has an activity of 15.3 disintegrations, or "counts," per minute (cpm). What is the age of an artifact that has an activity of 4 cpm? ($^{14}C$: $t_{\frac{1}{2}} = 5730$ y)

**23-48** A skull fragment found in 1936 at Baldwin Hills, California, was dated by radiocarbon analysis. A 100-g sample of bone was cleaned and treated with 1 *M* hydrochloric acid to destroy the mineral content of the bone. The bone protein was collected, dried, and pyrolyzed. The $CO_2$ produced was collected and purified and the ratio of $^{14}C$ to $^{12}C$ was measured. If this sample contained roughly 5.7% of the $^{14}C$ present in living tissue, how old is the skeleton? ($^{14}C$: $t_{\frac{1}{2}} = 5730$ y)

**23-49** Measurements on the linen wrappings from the Book of Isaiah in the Dead Sea scrolls suggest that the scrolls contain about 79.5% of the $^{14}C$ expected in living tissue. How old are these scrolls? ($^{14}C$: $t_{\frac{1}{2}} = 5730$ y)

**23-50** The Lascaux cave near Montignac in France contains a series of remarkable cave paintings. Radiocarbon dating of charcoal taken from this site suggests an age of 15,520 years. What fraction of the $^{14}C$ present in living tissue is still present in this sample? ($^{14}C$: $t_{\frac{1}{2}} = 5730$ y)

**23-51** Charcoal samples from Stonehenge in England emit 62.3% of the disintegrations per gram of carbon per minute expected for living tissue. What is the age of this charcoal? ($^{14}C$: $t_{\frac{1}{2}} = 5730$ y)

**23-52** A lump of beeswax was excavated in England near a collection of Bronze Age objects that are between 2500 and 3000 years old. Radiocarbon analysis of the beeswax suggests an activity roughly 90.3% of that observed for living tissue. Was this beeswax part of the hoard of Bronze Age objects, or did it date from another period? ($^{14}C$: $t_{\frac{1}{2}} = 5730$ y)

**23-53** The activity of the $^{14}C$ in living tissue is 15.3 disintegrations per minute per gram of carbon. The limit for reliable determination of $^{14}C$ ages is 0.10 disintegration per minute per gram of carbon. Calculate the maximum age of a sample that can be dated accurately by radiocarbon dating if the half-life for the decay of $^{14}C$ is 5730 years.

### Ionizing versus Nonionizing Radiation

**23-54** Use the relationship between the energy and the frequency of a photon (see Section 6.13) to calculate the energy in kilojoules per mole of a photon of blue light that has a frequency of $6.5 \times 10^{14}$ s$^{-1}$. Compare the results of this calculation with the ionization energy of water: 1216 kJ/mol.

**23-55** Calculate the energy in kilojoules per mole for an x-ray that has a frequency of $3 \times 10^{17}$ s$^{-1}$. How do the results of this calculation compare with the ionization energy of water: 1216 kJ/mol?

**23-56** Use Lewis structures to describe the difference between an $H_2O^+$ ion and an $H_3O^+$ ion. If a free radical is an ion or molecule that contains one or more unpaired electrons, which of these ions is a free radical?

### Biological Effects of Ionizing Radiation

**23-57** Radioactivity is measured in units that describe the amount of radiation given off, the amount of radiation to which an object is exposed, the amount of radiation absorbed, or the effectiveness of the radiation as a threat to biological systems. Sort the following units into these categories.

(a) curies (b) rads (c) rems (d) roentgens

**23-58** Explain why sources of $\alpha$-particles are intrinsically more dangerous than sources of $\beta^-$-particles.

### Natural versus Induced Radioactivity

**23-59** Describe the difference between natural and induced radioactivity. Give examples of both processes.

**23-60** The first artificial radioactive elements were synthesized by Irene Curie and Frederic Joliot, who bombarded $^{10}B$ and $^{27}Al$ with $\alpha$-particles to form $^{13}N$ and $^{30}P$. Write balanced equations for these reactions, identify the particle ejected in each reaction, and predict the mode of decay expected for the products of these reactions.

**23-61** Russell, Soddy, and Fajans predicted that the emission of one $\alpha$- and two $\beta^-$-particles by a nuclide would produce an isotope of that nuclide. Which isotope of $^{216}Po$ is produced by such decay? What intermediate nuclides are formed?

**23-62** In the first synthesis of an isotope of mendelevium ($Z = 101$), Ghiorso and co-workers bombarded $^{253}Es$ with $\alpha$-particles. Starting with less than $10^{-12}$ g of einsteinium, one atom of mendelevium was isolated after a period of a few hours. If a neutron was emitted in this reaction, what isotope of Md was produced? Another isotope of mendelevium was produced by bombarding $^{238}U$ with $^{19}F$ atoms. If five neutrons were ejected in this reaction, what isotope of Md was produced?

**23-63** $^{256}Lr$ is produced when $^{243}Am$ is bombarded with $^{18}O$. How many neutrons are emitted in this reaction? $^{256}Lr$ decays by both electron capture and $\alpha$-particle emission. What are the daughter nuclides produced in these reactions?

**23-64** $^{238}_{92}U$ decays by the emission of eight $\alpha$-particles and six $\beta^-$-particles. What stable nuclide is formed at the end of this decay chain?

**23-65** How many alpha and beta particles are emitted when $^{232}_{90}Th$ decays to $^{208}_{82}Pb$?

### Nuclear Fission and Nuclear Fusion

**23-66** Describe the difference between fission and fusion reactions. Give examples of both processes.

**23-67** Explain why relatively light nuclides give off energy when they fuse to form heavier nuclides, whereas relatively heavy nuclides give off energy when they undergo fission.

**23-68** Describe the difference between spontaneous and induced fission reaction. Explain why nuclei undergoing induced fission reactions have much shorter half-lives.

**23-69** Describe the advantages and disadvantages of fusion reactors versus fission reactors.

### Nuclear Synthesis

**23-70** Describe the changes in the nuclear reactions that fuel stars as the stars become older.

**23-71** What evidence do we have that the sun is a second-generation star?

**23-72** Describe the difference between the s process and r process for the synthesis of nuclides. Explain why the s process can't synthesize relatively heavy naturally occurring nuclides, such as $^{238}U$.

A coal fire.

# THE ORGANIC CHEMISTRY OF CARBON

This chapter examines the unique role that carbon plays in the chemistry of the elements. It provides the basis for answering questions such as the following:

## POINTS OF INTEREST

- It is tempting, but misleading, to define organic chemistry as the chemistry of carbon. What do we mean when we say that a compound is *organic* or *inorganic?* What is it that makes a compound *organic?*
- Why were the experiments to test for life on Mars based on the assumption that all living systems must contain carbon?
- What is the difference between ethyl alcohol ($CH_3CH_2OH$) and dimethyl ether ($CH_3OCH_3$)? Why is it useful to introduce functional groups such as *alcohols* and *ethers* to differentiate between these isomers?
- What is the difference between gasoline and kerosene, or kerosene and fuel oil? What does the octane number on the gas pump mean?
- What is the relationship between the active ingredients in coffee and chocolate?
- What is the difference between such simple items as gloves and mittens that make one of these items "handed?" What are the implications of the fact that handedness occurs within organic compounds?

## 24.1 ORGANIC COMPOUNDS

The German chemist Friedrich Wöhler (1800–1882) was a professor of chemistry at the University of Göttingen. He was an outstanding teacher who formed life-long ties with his students and contributed to many different phases of research in chemistry during the nineteenth century. He is remembered today for his synthesis of an organic compound (urea) from inorganic starting materials.

For more than 200 years, chemists divided materials into two categories. Those isolated from plants and animals were classified as *organic,* while those extracted from minerals were *inorganic.* At one time, chemists believed that organic compounds were fundamentally different from those that were inorganic, because organic compounds contained a *vital force* that was only found in living systems.

The synthesis of urea from inorganic starting materials by Friederich Wöhler marked the first step in the decline of the vital force theory. Wöhler was interested in the chemistry of cyanate compounds, which contain the $OCN^-$ ion. In 1828 he tried to synthesize ammonium cyanate ($NH_4OCN$) from silver cyanate ($AgOCN$) and ammonium chloride ($NH_4Cl$). What he expected can be described by the following equation:

$$AgOCN(aq) + NH_4Cl(aq) \longrightarrow AgCl(s) + NH_4OCN(aq)$$

The product he isolated from this reaction had none of the properties of cyanate compounds. It was a white, crystalline material that was identical to urea, $H_2NCONH_2$, which could be isolated from urine.

Neither Wöhler nor his contemporaries claimed that his results disproved the vital force theory. But they set in motion a series of experiments that led to the synthesis of a number of organic compounds from inorganic starting materials. This inevitably led to the removal of vitalism from the list of theories with relevance to chemistry, although it did not lead to the death of the theory, which still had proponents more than 90 years later.

If the difference between organic and inorganic compounds is not the presence of some mysterious vital force required for their synthesis, what is the basis for distinguishing between these classes of compounds? Most compounds extracted from living organisms contain carbon. It is therefore tempting to identify organic chemistry as the chemistry of carbon. But this definition includes compounds such as calcium carbonate ($CaCO_3$), as well as the elemental forms of carbon—diamond and graphite—that are clearly inorganic (see Chapter 10). Perhaps the best definition identifies organic chemistry as the chemistry of compounds that contain both carbon and hydrogen.

p. 390

Even though organic chemistry focuses on compounds that contain two elements, carbon and hydrogen, more than 95% of the compounds that chemists have isolated from natural sources or synthesized in the laboratory are organic. The special role of carbon in the chemistry of the elements is the result of a combination of factors, including the number of valence electrons on a neutral carbon atom, the electronegativity of carbon, and the atomic radius of carbon atoms (see Table 24.1).

Carbon has four valence electrons—$2s^2\,2p^2$—and it must either gain four electrons or lose four electrons to reach a rare-gas configuration. The electronegativity of carbon is too small for carbon to gain electrons from most elements to form $C^{4-}$ ions, and too large for carbon to lose electrons to form $C^{4+}$ ions. Carbon therefore forms covalent bonds with a large number of other elements, including the hydrogen, nitrogen, oxygen, phosphorus, and sulfur found in living systems.

Carbon atoms are relatively small—the covalent radius is only 0.077 nm. As a result, carbon atoms can come close enough together to form strong C=C double bonds or even C≡C triple bonds. Carbon also forms strong double and triple bonds to nitrogen and oxygen. It can even form strong double bonds to elements such as phosphorus or sulfur that don't form strong double bonds to themselves.

View from the Viking II lander on Mars.

Several years ago, an unmanned Viking spacecraft carried out experiments designed to search for evidence of life on Mars. These experiments were based on the assumption that living systems contain carbon, and the absence of any evidence for

**TABLE 24.1 The Physical Properties of Carbon**

| | |
|---|---|
| Electronic configuration | $1s^2\ 2s^2\ 2p^2$ |
| Melting point | |
| Graphite | Sublimes above 3700°C |
| Diamond | Above 3550°C |
| Boiling point | |
| Graphite | 4827°C |
| Diamond | 4827°C |
| Density | |
| Graphite | 2.25 g/cm$^3$ |
| Diamond | 3.514 g/cm$^3$ |
| First ionization energy | 1086.4 kJ/mol |
| Electron affinity | 122.3 kJ/mol |
| Electronegativity | 2.55 |
| C—C single bond length | 0.1544 nm |
| Covalent radius | 0.077 nm |
| C—C bond dissociation enthalpy | 330 kJ/mol |
| Energy to transform C(*s*) into C(*g*) | 716.68 kJ/mol |
| Enthalpy of formation | |
| Graphite | 0.000 kJ/mol |
| Diamond | 2.425 kJ/mol |

carbon-based life on that planet was assumed to mean that no life existed. Several factors make carbon essential to life: (1) the ease with which it forms bonds to itself, (2) its tendency to form multiple bonds to C, N, O, P, and S atoms, and (3) the strength of these covalent bonds. These factors provide an almost infinite variety of potential structures for organic compounds, such as vitamin C shown in Figure 24.1. No other element can provide the variety of combinations and permutations necessary for life to exist.

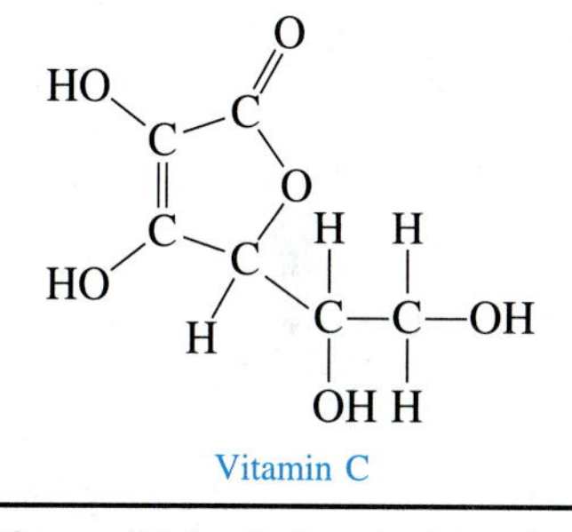


**Figure 24.1** Carbon is the only element that provides the diverse structures, such as the structure of vitamin C shown here, necessary for life to exist.

## 24.2 THE SATURATED HYDROCARBONS: ALKANES AND CYCLOALKANES

Compounds that contain only carbon and hydrogen are known as **hydrocarbons.** Those that contain as many hydrogen atoms as possible are said to be *saturated.* The saturated hydrocarbons are also known as **alkanes.**

The simplest saturated hydrocarbon or alkane is methane: $CH_4$. The Lewis structure of methane can be generated by combining the four electrons in the valence shell of a neutral carbon atom with four hydrogen atoms to form a compound in which the carbon atom shares a total of eight valence electrons:

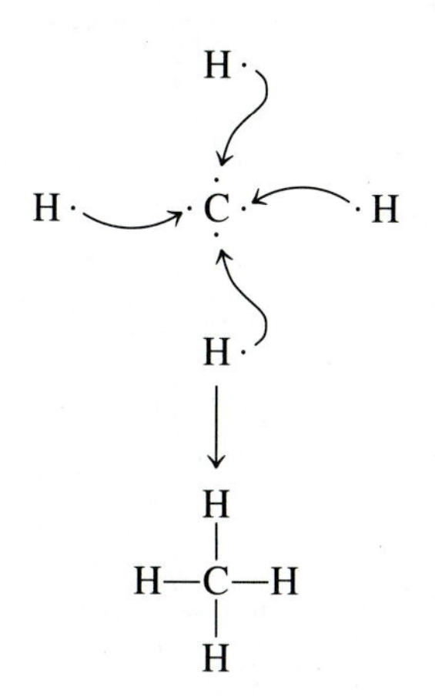

Methane is an example of a general rule that carbon is *tetravalent;* it forms a total of four bonds in virtually all of its compounds.

**EXERCISE 24.1**

Predict the formula of ethane, the alkane that contains two carbon atoms.

**SOLUTION** As a rule, compounds that contain more than one carbon atom are held together by C—C bonds. If we assume that carbon is tetravalent, the formula of this compound must be $C_2H_6$.

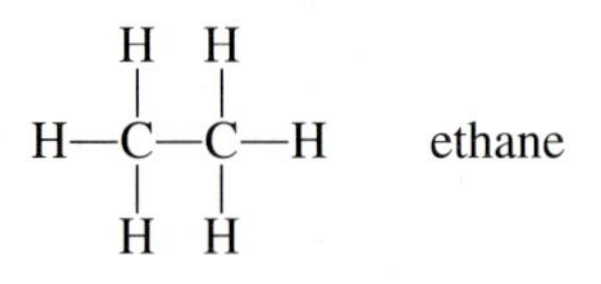


The alkane that contains three carbon atoms is known as propane, which has the formula $C_3H_8$ and the following skeleton structure:

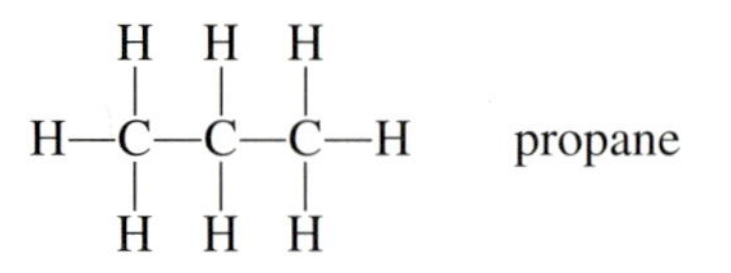


The four-carbon alkane is butane, with the formula $C_4H_{10}$:

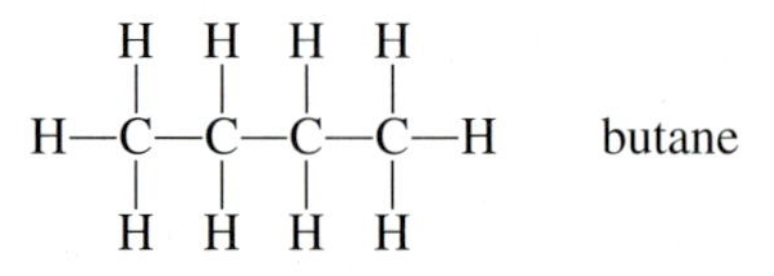


The names, molecular formulas, and physical properties for a number of alkanes are given in Table 24.2. These compounds all have the same generic formula: $C_nH_{2n+2}$. (For every $n$ carbon atoms, there are $2n + 2$ hydrogen atoms.) The boiling points of the alkanes gradually increase with the molecular weight of these compounds. At room temperature, the lighter alkanes are gases; the midweight alkanes are liquids; and the heavier alkanes are solids, or tars.

**TABLE 24.2 The Saturated Hydrocarbons, or Alkanes**

| Name | Molecular Formula | Melting Point (°C) | Boiling Point (°C) | State at 25°C |
|---|---|---|---|---|
| Methane | $CH_4$ | −182.5 | −164 | gas |
| Ethane | $C_2H_6$ | −183.3 | −88.6 | gas |
| Propane | $C_3H_8$ | −189.7 | −42.1 | gas |
| Butane | $C_4H_{10}$ | −138.4 | −0.5 | gas |
| Pentane | $C_5H_{12}$ | −129.7 | 36.1 | liquid |
| Hexane | $C_6H_{14}$ | −95 | 68.9 | liquid |
| Heptane | $C_7H_{16}$ | −90.6 | 98.4 | liquid |
| Octane | $C_8H_{18}$ | −56.8 | 124.7 | liquid |
| Nonane | $C_9H_{20}$ | −51 | 150.8 | liquid |
| Decane | $C_{10}H_{22}$ | −29.7 | 174.1 | liquid |
| Undecane | $C_{11}H_{24}$ | −24.6 | 195.9 | liquid |
| Dodecane | $C_{12}H_{26}$ | −9.6 | 216.3 | liquid |
| Eicosane | $C_{20}H_{42}$ | 36.8 | 343 | solid |
| Triacontane | $C_{30}H_{62}$ | 65.8 | 449.7 | solid |

The alkanes in Table 24.2 are all *straight-chain hydrocarbons,* in which the carbon atoms form a chain that runs from one end of the molecule to the other. Alkanes also form *branched* structures. The smallest hydrocarbon in which a branch can occur has four carbon atoms. This compound has the same formula as butane ($C_4H_{10}$), but a different structure:

$$\begin{array}{c} CH_3 \\ | \\ CH_3{-}CH{-}CH_3 \end{array}$$

isobutane

As noted in earlier chapters, compounds with the same formula and different structures are known as *isomers* (from the Greek *isos,* ''equal,'' and *meros,* ''parts''). The difference between the structures of butane and isobutane is shown in Figure 24.2. These compounds are classified as **constitutional isomers** because they literally differ in their constitution. One contains two $CH_3$— groups and two —$CH_2$— groups; the other contains three $CH_3$— groups and one —CH— group.

Sec. 12.10

There are three constitutional isomers of pentane, $C_5H_{12}$. The first is ''normal'' pentane, or *n*-pentane:

$$CH_3{-}CH_2{-}CH_2{-}CH_2{-}CH_3$$

*n*-pentane

A branched isomer is also possible, which was originally named isopentane.

$$\begin{array}{c} CH_3 \\ | \\ CH_3{-}CH{-}CH_2{-}CH_3 \end{array}$$

isopentane

When a more highly branched isomer was discovered, it was named neopentane (the new isomer of pentane).

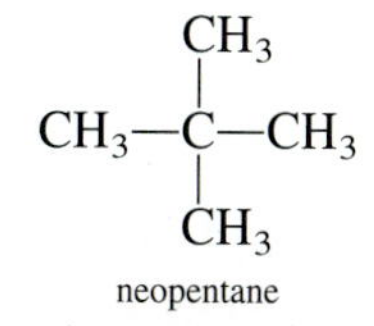

$$\begin{array}{c} CH_3 \\ | \\ CH_3{-}C{-}CH_3 \\ | \\ CH_3 \end{array}$$

neopentane

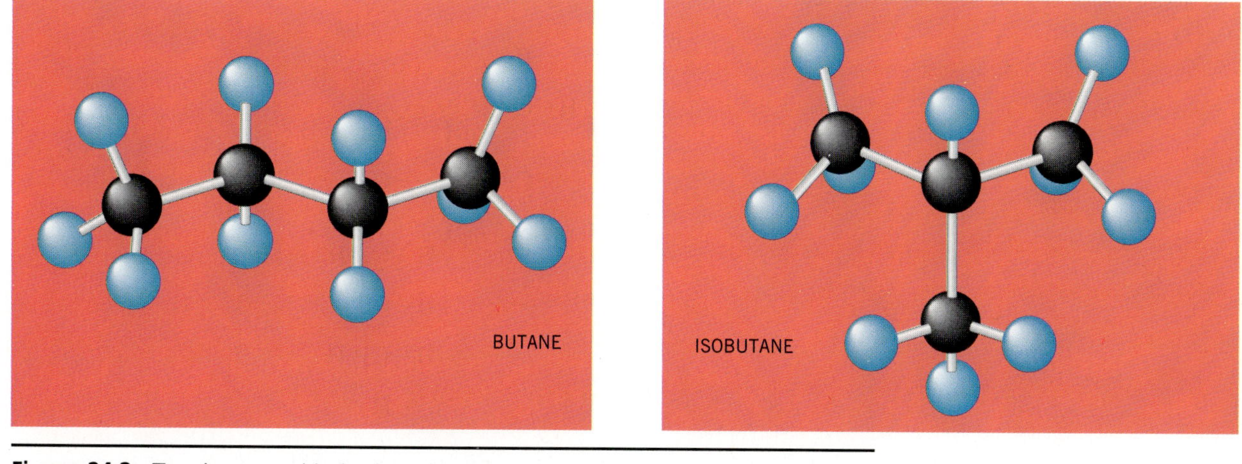


**Figure 24.2** Two isomers with the formula $C_4H_{10}$ are possible. *Butane* is a straight-chain hydrocarbon because the four carbon atoms form a continuous chain. *Isobutane* is a branched hydrocarbon.

## EXERCISE 24.2

Determine the number of constitutional isomers of hexane, $C_6H_{14}$.

**SOLUTION** There are five constitutional isomers of hexane. There is a straight-chain, or normal, isomer:

$$CH_3{-}CH_2{-}CH_2{-}CH_2{-}CH_2{-}CH_3$$

There are two isomers with a single carbon branch:

$$CH_3{-}\underset{}{\overset{CH_3}{\overset{|}{C}H}}{-}CH_2{-}CH_2{-}CH_3 \qquad CH_3{-}CH_2{-}\overset{CH_3}{\overset{|}{C}H}{-}CH_2{-}CH_3$$

And there are two isomers with two branches.

$$CH_3{-}\overset{CH_3}{\overset{|}{C}H}{-}\overset{CH_3}{\overset{|}{C}H}{-}CH_3 \qquad CH_3{-}\underset{CH_3}{\underset{|}{\overset{CH_3}{\overset{|}{C}}}}{-}CH_2{-}CH_3$$

The number of isomers of a compound increases rapidly with the number of carbon atoms. Over four billion isomers are possible for $C_{30}H_{62}$, for example.

If the carbon chain that forms the backbone of a straight-chain hydrocarbon is long enough, we can envision the two ends coming together to form a **cycloalkane.** One hydrogen atom has to be removed from each end of the hydrocarbon chain to form the C—C bond that closes the ring. Cycloalkanes therefore have two less hydrogen atoms than the parent alkane and a generic formula of $C_nH_{2n}$. Cyclohexane, for example, has the formula $C_6H_{12}$.

The International Union of Pure and Applied Chemistry (IUPAC) has developed a systematic approach to naming alkanes and cycloalkanes that is based on the following steps:

- Find the longest continuous chain of carbon atoms in the skeleton structure. Name the compound as a derivative of the alkane with this number of carbon atoms.

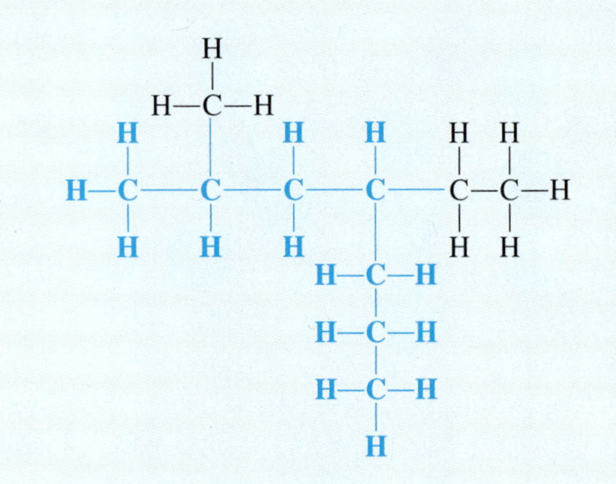

This compound is a derivative of heptane because the longest chain contains seven carbon atoms.

- Name the substituents on the chain. Substituents derived from alkanes are named by replacing the *-ane* ending with *-yl*. This compound contains methyl ($CH_3$—) and ethyl ($CH_3CH_2$—) substituents:

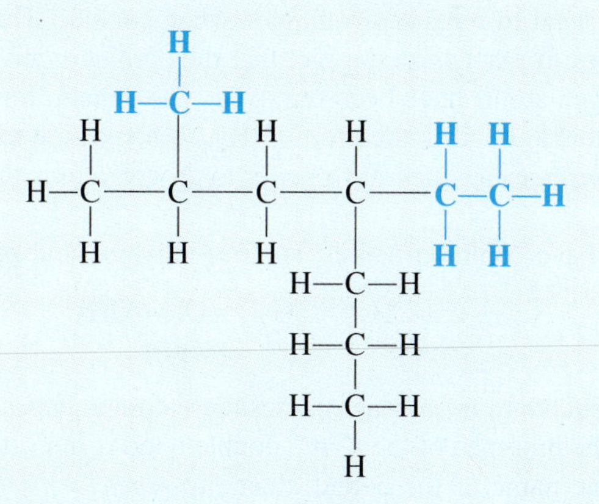

- Number the chain starting at the end nearest the first substituent:

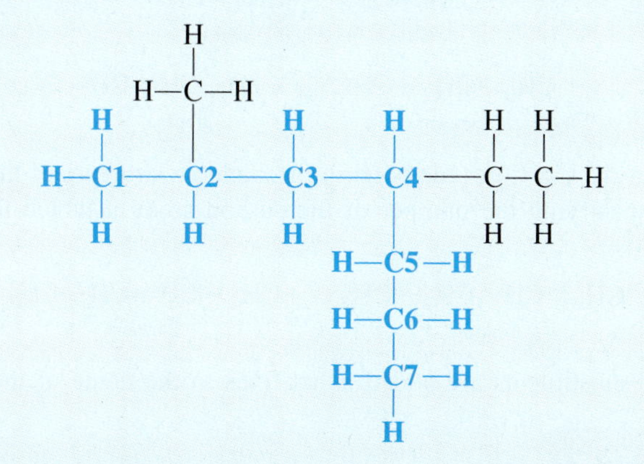

- Use the results of the previous step to specify the carbon atoms on which the substituents are located. This compound, for example, is 4-ethyl-2-methyl-heptane. Use the prefixes *di-*, *tri-*, and *tetra-* to describe substituents that are found two, three, or four times on the same chain of carbon atoms.

## EXERCISE 24.3

Name the following compound.

$$CH_3-\underset{\displaystyle CH_3}{\overset{\displaystyle CH_3}{\underset{|}{\overset{|}{C}}}}-CH_2-\overset{\displaystyle CH_3}{\overset{|}{C}H}-CH_3$$

**SOLUTION** This compound is a derivative of pentane because the longest continuous chain contains five carbon atoms. There are three identical $CH_3$— substituents on the backbone. Two of these methyl groups are on the second carbon, and one is on the fourth carbon. This compound is therefore *2,2,4-trimethylpentane.* Since it contains a total of eight carbon atoms, it is also known by the common name *isooctane.*

## 24.3 THE UNSATURATED HYDROCARBONS: ALKENES AND ALKYNES

Carbon not only forms alkanes with long chains of strong C—C single bonds, it also forms strong C═C double bonds. Compounds that contain C═C double bonds were once known as *olefins* (literally, "to make an oil") because they were hard to crystallize. (They tend to remain oily liquids when cooled.) These compounds are now called **alkenes** to emphasize the fact that they are derivatives of alkanes from which two hydrogen atoms have been removed. The generic formula for an alkene with one C═C double bond is therefore $C_nH_{2n}$. Alkenes are examples of *unsaturated hydrocarbons* because they have fewer hydrogen atoms than the corresponding alkanes.

Many alkenes have common names, such as ethylene and propylene:

| $H_2C{=}CH_2$ | $H_2C{=}CH{-}CH_3$ |
|---|---|
| ethylene | propylene |

The IUPAC nomenclature for alkenes names these compounds as derivatives of the parent alkanes. The presence of the C═C double bond is indicated by changing the *-ane* ending on the name of the parent alkane to *-ene:*

| $CH_3{-}CH_3$ | $CH_2{=}CH_2$ |
|---|---|
| ethane | ethene |
| $CH_3{-}CH_2{-}CH_3$ | $CH_3{-}CH{=}CH_2$ |
| propane | propene |

The location of the C═C double bond in the skeleton structure of the compound is indicated by specifying the number of the carbon atom at which the C═C bond starts:

| $CH_2{=}CH{-}CH_2{-}CH_3$ | $CH_3{-}CH{=}CH{-}CH_3$ |
|---|---|
| 1-butene | 2-butene |

The names of substituents are added as prefixes to the name of the alkene.

### EXERCISE 24.4

Name the following compound.

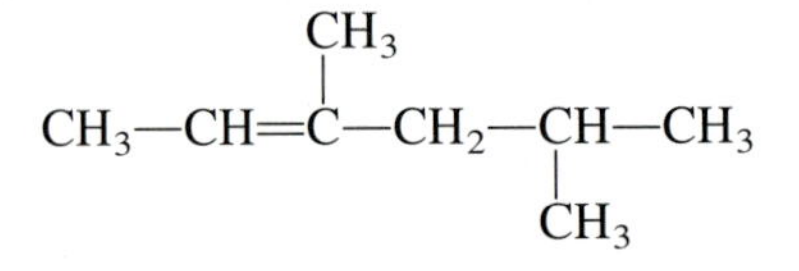


**SOLUTION** This compound is a derivative of hexane because the longest carbon chain contains six carbon atoms. Because it contains a C═C double bond, it is a *hexene.* Because the double bond links the second and third carbon atoms, it is a 2-hexene. The $CH_3$— substituents are on the third and fifth carbon atoms, so the compound is *3,5-dimethyl-2-hexene.*

Sec. 8.16 

The geometry of alkenes differs from that of the parent alkanes. The VSEPR theory predicts that the three regions where electrons are found in the valence shell of the carbon atoms in a C═C double bond should be arranged toward the corners of an equilateral triangle. This prediction is confirmed by experiment, which suggests that the six atoms that form the C═C double bond and its nearest neighbors all lie in the same plane, as shown in Figure 24.3.

The presence of a C═C double bond increases the number of potential isomers of a compound. There are two isomers of butane and three isomers of pentane, for

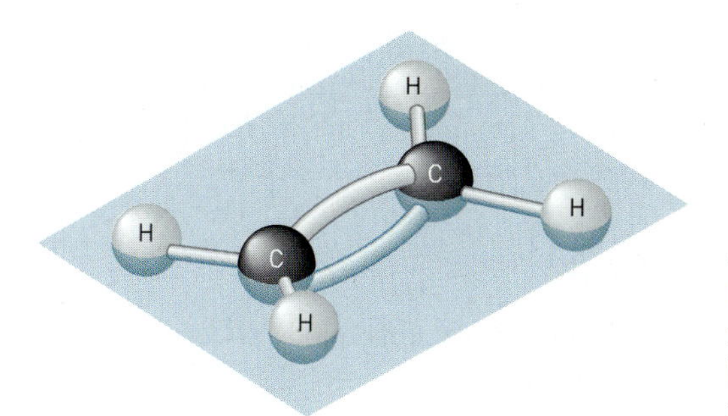


**Figure 24.3** The six atoms that form a C=C double bond and the bond's nearest neighbors must lie in the same plane.

example. But there are four isomers of butene and seven isomers of pentene. Alkenes form constitutional isomers that differ in the location of the C=C bond, such as 1-butene and 2-butene:

$$CH_2{=}CH{-}CH_2{-}CH_3 \qquad CH_3{-}CH{=}CH{-}CH_3$$

1-butene 2-butene

Alkenes also form **stereoisomers** that differ in the way substituents are arranged around the C=C double bond. The isomer with similar substituents on the same side of the double bond is called **cis,** a Latin stem meaning "on this side." The isomer in which similar substituents are across from each other, is called **trans,** a Latin stem meaning "across." The cis isomer of 2-butene, for example, has both $CH_3$— groups on one side of the double bond. In the trans isomer, the $CH_3$— groups are on opposite sides of the double bond:

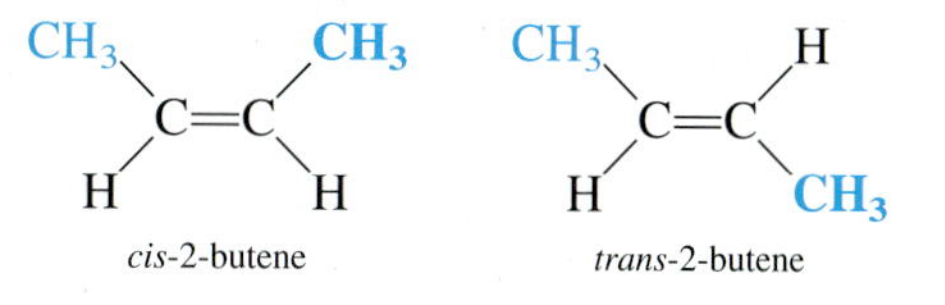


*cis*-2-butene *trans*-2-butene

## EXERCISE 24.5

Name the straight-chain constitutional and stereoisomers of pentene ($C_5H_{10}$).

**SOLUTION** There are two constitutional isomers of pentene in which all the carbon atoms lie in the same chain:

$$CH_2{=}CH{-}CH_2{-}CH_2{-}CH_3 \qquad CH_3{-}CH{=}CH{-}CH_2{-}CH_3$$

1-pentene 2-pentene

There are no cis/trans isomers of 1-pentene because there is only one way of arranging the substituents around the double bond:

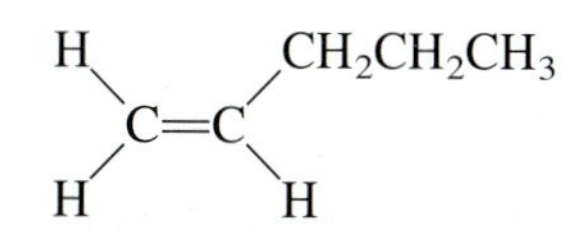


Cis and trans isomers are possible for 2-pentene, however:

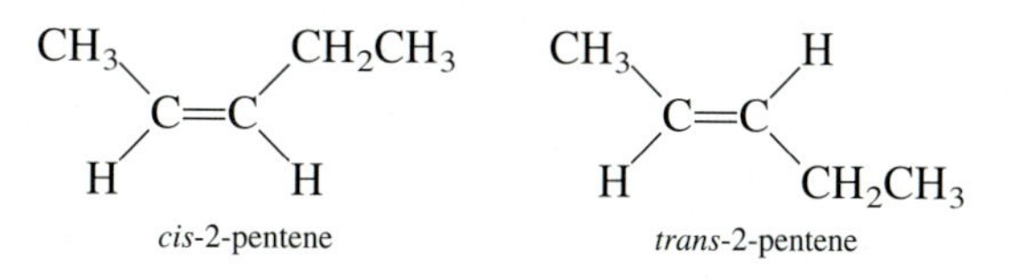


*cis*-2-pentene *trans*-2-pentene

Compounds that contain C≡C triple bonds are called **alkynes.** These compounds have four less hydrogen atoms than the parent alkanes, so the generic formula for an alkyne with one C≡C triple bond is $C_nH_{2n-2}$. The simplest alkyne has the formula $C_2H_2$ and is known by the common name *acetylene:*

$$H{-}C{\equiv}C{-}H$$
acetylene

The systematic nomenclature for alkynes names these compounds as derivatives of the parent alkane, with the ending *-yne* replacing *-ane:*

$$CH_3C{\equiv}CCH_2CH_3 \qquad HC{\equiv}CCH_2CH_3$$
2-pentyne 1-butyne

$$CH_3C{\equiv}CCH_2\underset{}{\overset{CH_3}{\overset{|}{C}}}HCH_2CH_3$$
5-methyl-2-heptyne

In addition to compounds that contain one double bond (*alkenes*) or one triple bond (*alkynes*), we can also envision compounds with two double bonds (*dienes*), three double bonds (*trienes*), or a combination of double and triple bonds:

$$CH_3CH{=}CHCH_2C{\equiv}CH \qquad CH_2{=}CHCH{=}CH_2$$
4-hexen-1-yne 1,3-butadiene

## 24.4 THE REACTIONS OF ALKANES, ALKENES, AND ALKYNES

Alkanes are generally inert to chemical reactions, with two important exceptions. Like all other hydrocarbons, they burn to form $CO_2$ and $H_2O$:

$$C_3H_8(g) + 5\,O_2(g) \longrightarrow 3\,CO_2(g) + 4\,H_2O(g)$$

In the presence of light, or at high temperatures, they also react with halogens to form **alkyl halides:**

$$CH_4(g) + Cl_2(g) \xrightarrow{\text{light}} CH_3Cl(g) + HCl(g)$$

Unsaturated hydrocarbons such as alkenes and alkynes are much more reactive. They react rapidly with bromine, for example, to add a $Br_2$ molecule across the C=C double bond:

$$CH_3CH{=}CHCH_3 + Br_2 \longrightarrow CH_3\overset{Br}{\overset{|}{C}}H\underset{\underset{Br}{|}}{C}HCH_3$$
2,3-dibromobutane

Adding an alkene to a solution of $Br_2$ in $CCl_4$ (left) leads to an almost instantaneous disappearance of the characteristic color of bromine (right).

This reaction provides a way to test for alkenes or alkynes. Solutions of bromine in $CCl_4$ have an intense red-orange color. When $Br_2$ in $CCl_4$ is mixed with a sample of an alkane, no change is initially observed. When it is mixed with an alkene or alkyne, the color of $Br_2$ rapidly disappears.

The reaction between 2-butene and bromine to form 2,3-dibromobutane is just one example of the **addition reactions** of alkenes and alkynes. Hydrogen bromide (HBr) adds across a C=C double bond to form the corresponding alkyl bromide, in which the hydrogen ends up on the carbon atom that had more hydrogen atoms to begin with. Addition of HBr to propene, for example, gives 2-bromopropane:

$$CH_3CH{=}CH_2 + HBr \longrightarrow CH_3\overset{\overset{\displaystyle Br}{|}}{C}HCH_3$$

In the presence of a suitable catalyst, $H_2$ adds across double or triple bonds to convert an alkene or alkyne to the corresponding alkane:

$$CH_3CH{=}CHCH_3 + H_2 \xrightarrow{Pt} CH_3CH_2CH_2CH_3$$

In the presence of an acid catalyst, it is even possible to add a molecule of water across a C=C double bond:

$$CH_3CH{=}CHCH_3 + H_2O \xrightarrow{H_2SO_4} CH_3\overset{\overset{\displaystyle OH}{|}}{C}HCH_2CH_3$$

Addition reactions provide a way to add new substituents to a hydrocarbon chain and thereby produce new derivatives of the parent alkanes.

## 24.5 NATURALLY OCCURRING HYDROCARBONS AND THEIR DERIVATIVES

Complex hydrocarbons and their derivatives are found throughout nature. Natural rubber, for example, is a hydrocarbon that contains long chains of alternating C=C double bonds and C—C single bonds.

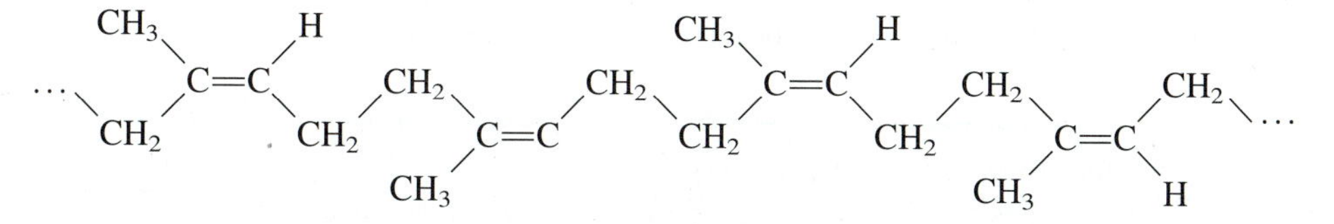

Long hydrocarbon chains are particularly common in *lipids,* the class of biological compounds that includes animal fats and vegetable oils. Another important class of naturally occurring hydrocarbons are the *terpenes,* which can be distilled from plants. Terpenes often have very distinctive odors. β-Pinene (Figure 24.4*a*), for example, is responsible for the characteristic odor of turpentine. Terpenoids are compounds that are derivatives of hydrocarbons, such as citronellal (Figure 24.4*b*), which is one of the compounds that gives rise to the odor of lemons.

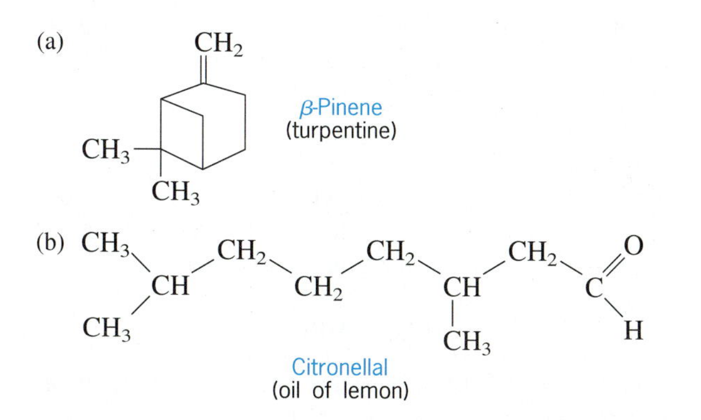


**Figure 24.4** Complex hydrocarbons and their derivatives are found throughout nature. (*a*) β-Pinene is responsible for the characteristic odor of turpentine. (*b*) Citronellal gives rise to the characteristic odor of lemons.

*Vitamins* are important components of nutrition that cannot be made by humans and must therefore be obtained from the diet. The so-called fat-soluble vitamins, such as vitamin $A_1$, $D_2$, and E, have structures based on complex hydrocarbon skeletons (Figure 24.5*a*). The *steroids* (such as cholesterol, shown in Figure 24.5*b*), which are important regulators of biological activity, also have structures based on a complex hydrocarbon skeleton.

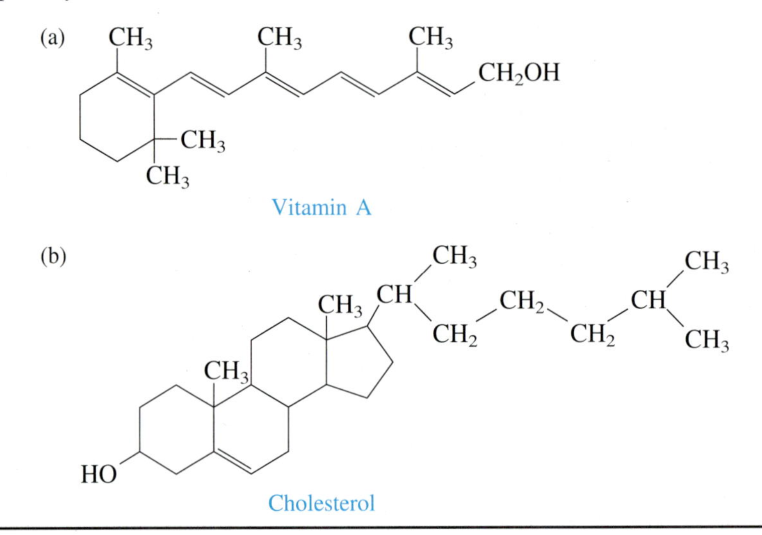


**Figure 24.5** (*a*) Vitamin $A_1$ is a derivative of a complex hydrocarbon that is involved in the process by which our eyes capture images. (*b*) Cholesterol, another complex hydrocarbon derivative, is an important component of cell membranes and is the biosynthetic precursor for the other steroids our bodies produce.

## 24.6 THE AROMATIC HYDROCARBONS AND THEIR DERIVATIVES

At the turn of the nineteenth century, one of the signs of living the good life was having gas lines connected to your house, so that you could use gas lanterns to light the house after dark. The gas burned in these lanterns was called *coal gas* because it was produced by heating coal in the absence of air. The principal component of coal gas was methane, $CH_4$.

In 1825, Michael Faraday analyzed an oily liquid with a distinct odor that collected in tanks used to store coal gas at high pressures. Faraday found that this compound had the empirical formula CH. Ten years later, Eilhardt Mitscherlich produced the same material by heating benzoic acid with lime. Mitscherlich named this substance *benzin,* which became *benzene* when translated into English. He also determined the molecular formula of this compound: $C_6H_6$.

Benzene is obviously an unsaturated hydrocarbon because it has far less hydrogen than the equivalent saturated hydrocarbon: $C_6H_{14}$. But benzene is too stable to be an alkene or alkyne. Alkenes and alkynes react rapidly with potassium permanganate ($KMnO_4$), but benzene does not. They also rapidly add $Br_2$ to the C=C or C≡C bonds, whereas benzene does not react with bromine by itself. It reacts with bromine only in the presence of a $FeBr_3$ catalyst. Furthermore, when it reacts with $Br_2$ in the presence of $FeBr_3$, the product of this reaction is a compound in which a bromine atom has been substituted for a hydrogen atom, not added to the compound:

$$C_6H_6 + Br_2 \xrightarrow{FeBr_3} C_6H_5Br + HBr$$

Other compounds were eventually isolated from coal that had similar properties. Their formulas suggested the presence of multiple C=C bonds, but these compounds were too stable to be alkenes. Since they often have a distinct odor, or aroma, they became known as **aromatic compounds.**

Determining the structure of benzene was a recurring problem throughout most of the nineteenth century. The first step toward solving it was taken by Friedrich August Kekulé in 1865. (Kekulé's interest in the structures of organic compounds may have resulted from the fact that he first enrolled at the University of Giessen as a student of architecture.) One day, while dozing before a fire, Kekulé dreamed of long rows of atoms twisting in a snakelike motion until one of the snakes seized hold of its own tail. This dream led Kekulé to propose that benzene consists of a ring of six carbon atoms with alternating C—C single bonds and C=C double bonds. Since there are two ways in which these bonds can alternate, Kekulé proposed that benzene was a mixture of two compounds in equilibrium.

Kekulé's structure explained the molecular formula of benzene, but it did not explain why benzene did not behave like an alkene. The unusual stability of benzene was not understood until the development of the **theory of resonance.** This theory states that molecules for which two or more satisfactory Lewis structures can be drawn are an average, or hybrid, of these structures. Benzene, for example, is a **resonance hybrid** of the two Kekulé structures, as shown in Figure 24.6.

Sec. 8.11

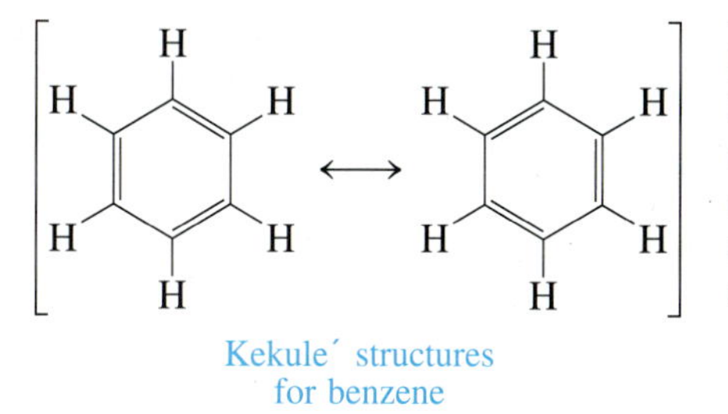


**Figure 24.6** The remarkable stability of benzene is explained by assuming that this compound is a resonance hybrid of a pair of Lewis structures in which the six-membered ring contains alternating C—C single bonds and C=C double bonds.

This model is consistent with the crystal structure of benzene, which shows a planar six-membered ring in which all six carbon–carbon bonds are the same length (0.139 nm). It also explains why the length of these bonds is intermediate between the length of C—C single bonds (0.154 nm) and C=C double bonds (0.133 nm).

Resonance theory suggests that molecules that are hybrids of two or more Lewis structures are more stable than those that are not. It is this extra stability that makes benzene and other aromatic derivatives less reactive than normal alkenes. To emphasize the difference between benzene and a simple alkene, we will replace the Kekulé structure for benzene and its derivatives with the aromatic ring shown in Figure 24.7. The circle in the center of the aromatic ring indicates that the electrons in the ring are delocalized; they are free to move around the ring.

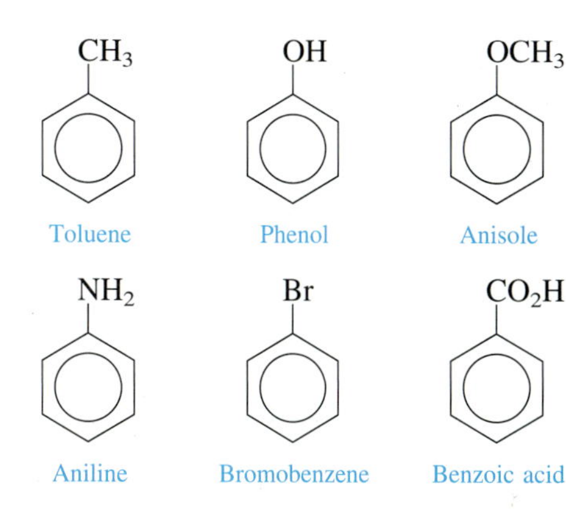


**Figure 24.7** Other aromatic compounds include toluene, phenol, anisole, aniline, bromobenzene, and benzoic acid.

There are three ways in which a pair of substituents can be placed on an aromatic ring. In the *ortho* (*o*) isomer, the substituents are in *adjacent* positions on the ring. In the *meta* (*m*) isomer, they are *separated* by one carbon atom. In the *para* (*p*) isomer, they are on *opposite* ends of the ring. The three isomers of dimethylbenzene, or xylene, are shown in Figure 24.8.

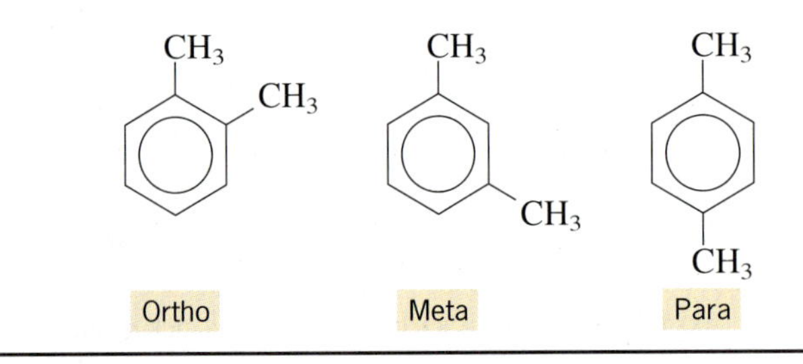


**Figure 24.8** The three isomers of xylene, or dimethylbenzene.

## EXERCISE 24.6

Predict the structure of *para*-dichlorobenzene, one of the active ingredients in moth balls.

**SOLUTION** *Para* isomers of benzene contain two substituents at opposite positions in the six-membered ring. Para-dichlorobenzene therefore has the following structure:

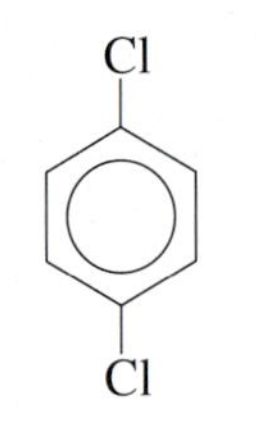


Aromatic compounds can contain more than one six-membered ring. Naphthalene, anthracene, and phenanthrene (see Figure 24.9) are examples of aromatic compounds that contain two or more fused benzene rings.

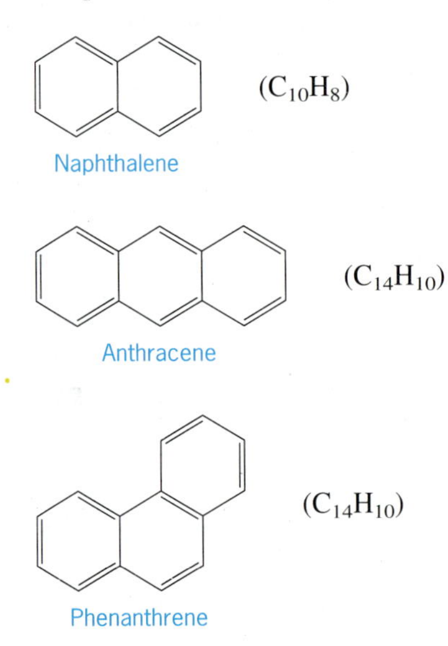


**Figure 24.9** Naphthalene, anthracene, and phenanthrene are complex aromatic compounds that can be obtained by fusing two or more six-membered rings.

## 24.7 THE CHEMISTRY OF PETROLEUM PRODUCTS

The term *petroleum* comes from the Latin stems *petra,* ''rock,'' and *oleum,* ''oil.'' It is used to describe a broad range of fossil hydrocarbons that are found as gases, liquids, or solids beneath the surface of the earth. The two most common forms of petroleum are natural gas and crude oil.

Natural gas is a mixture of lightweight alkanes. The composition of natural gas depends on the source, but a typical sample contains 80% methane ($CH_4$), 7% ethane ($C_2H_6$), 6% propane ($C_3H_8$), 4% butane and isobutane ($C_4H_{10}$), and 3% pentanes ($C_5H_{12}$). The $C_3$, $C_4$, and $C_5$ hydrocarbons are removed before the gas is sold. Commercial natural gas is therefore primarily a mixture of methane and ethane. The propane and butanes removed from natural gas are usually liquefied under pressure and sold as liquefied petroleum gases (LPG).

A plume of natural gas being burned at the Brooklyn Union Gas Company in Brooklyn, New York.

Natural gas was known in England as early as 1659. But it did not replace coal gas as an important source of energy in the United States until after World War II, when a network of gas pipelines was constructed. By 1980, annual consumption of natural gas had grown to more than 55,000 billion cubic feet, which represented almost 30% of total United States' energy consumption.

The first oil well was drilled by Edwin Drake in 1859, in Titusville, PA. It produced up to 800 gallons per day, which far exceeded the demand for this material. By 1980, consumption of oil had reached 2.5 billion gallons per day.

About 225 billion barrels of oil were produced by the petroleum industry between 1859 and 1970. Another 200 billion barrels were produced between 1970 and 1980. The total proven world reserves of crude oil in 1970 were estimated at 546 billion barrels, with perhaps another 800 to 900 billion barrels of oil that remained to be found. It took 500 million years for the petroleum beneath the earth's crust to accumulate. At the present rate of consumption, we may exhaust the world's supply of petroleum by the 200th anniversary of the first oil well.

Crude oil is a complex mixture that is between 50% and 95% hydrocarbon by weight. The first step in refining crude oil involves separating the oil into different hydrocarbon fractions by distillation. A typical set of petroleum fractions is given in Table 24.3. Since there are a number of factors that influence the boiling point of a hydrocarbon, these petroleum fractions are complex mixtures. More than 500 different hydrocarbons have been identified in the gasoline fraction, for example.

p. 550

**TABLE 24.3 Petroleum Fractions**

| Fraction | Boiling Range (°C) | Number of Carbon Atoms |
|---|---|---|
| Natural gas | < 20 | $C_1$ to $C_4$ |
| Petroleum ether | 20–60 | $C_5$ to $C_6$ |
| Gasoline | 40–200 | $C_5$ to $C_{12}$, but mostly $C_6$ to $C_8$ |
| Kerosene | 150–260 | mostly $C_{12}$ to $C_{13}$ |
| Fuel oils | > 260 | $C_{14}$ and higher |
| Lubricants | > 400 | $C_{20}$ and above |
| Asphalt or coke | residue | polycyclic |

Natural gas can be transported by cooling it until it condenses to form a liquid.

About 10% of the product of the distillation of crude oil is a fraction known as *straight-run gasoline,* which served as a satisfactory fuel during the early days of the internal combustion engine. As the automobile engine developed, it was made more powerful by increasing the compression ratio (see Section 4.3). Straight-run gasoline burns unevenly in high-compression engines, which produces a shock wave that causes the engine to ''knock,'' or ''ping.'' As the petroleum industry matured, it

Sec. 4.3

faced two problems: increasing the yield of gasoline from each barrel of crude oil and decreasing the tendency of gasoline to knock when it burned.

The relationship between knocking and the structure of the hydrocarbons in gasoline is summarized in the following general rules:

- Branched alkanes and cycloalkanes burn more evenly than straight-chain alkanes.
- Short alkanes ($C_4H_{10}$) burn more evenly than long alkanes ($C_7H_{16}$).
- Alkenes burn more evenly than alkanes.
- Aromatic hydrocarbons burn more evenly than cycloalkanes.

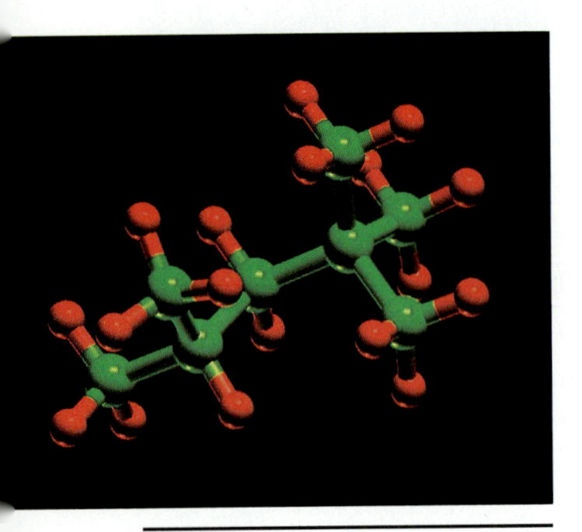

The structure of isooctane

The most commonly used measure of a gasoline's ability to burn without knocking is its *octane number.* Octane numbers compare a gasoline's tendency to knock against the tendency of a blend of two standard hydrocarbons—heptane and 2,2,4-trimethylpentane, or isooctane—to knock. Heptane ($C_7H_{16}$) is a long, straight-chain alkane, which burns unevenly and produces a great deal of knocking. Highly branched alkanes such as 2,2,4-trimethylpentane are more resistant to knocking. Gasolines that match a blend of 87% isooctane and 13% heptane are given an octane number of 87.

There are three ways of reporting octane numbers. Measurements made under conditions of high speed and high temperature are reported as *motor octane numbers.* Measurements taken under relatively mild engine conditions are known as *research octane numbers.* The *road-index octane numbers* reported on gasoline pumps are an average of these two. Road-index octane numbers for a few pure hydrocarbons are given in Table 24.4.

**TABLE 24.4 Hydrocarbon Octane Numbers**

| Hydrocarbon | Road-Index Octane Number |
|---|---|
| Heptane | 0 |
| 2-Methylheptane | 23 |
| Hexane | 25 |
| 2-Methylhexane | 44 |
| 1-Heptene | 60 |
| Pentane | 62 |
| 1-Pentene | 84 |
| Butane | 91 |
| Cyclohexane | 97 |
| 2,2,4-Trimethylpentane (isooctane) | 100 |
| Benzene | 101 |
| Toluene | 112 |

By 1922 a number of compounds had been discovered that could increase the octane number of gasoline. Adding as little as 6 mL of tetraethyllead [$Pb(CH_2CH_3)_4$] to a gallon of gasoline, for example, can increase the octane number by 15 to 20 units. This discovery gave rise to the first "ethyl" gasoline, and enabled the petroleum industry to produce aviation gasolines with octane numbers greater than 100.

Another way to increase the octane number is *thermal reforming.* At high temperatures (500°C to 600°C) and high pressures (25 to 50 atm), straight-chain alkanes isomerize to branched alkanes and cycloalkanes, thereby increasing the octane number of the gasoline. Running this reaction in the presence of hydrogen and a catalyst, such as a mixture of silica ($SiO_2$) and alumina ($Al_2O_3$), results in *catalytic reforming,* which can produce a gasoline with even higher octane numbers. Thermal or catalytic reforming and gasoline additives such as tetraethyllead increase the octane number of the straight-run gasoline obtained from the distillation of crude oil, but neither process increases the yield of gasoline from a barrel of oil.

The data in Table 24.3 suggest that we could increase the yield of gasoline by "cracking" the hydrocarbons that end up in the kerosene or fuel oil fractions into smaller pieces. *Thermal cracking* was discovered as early as the 1860s. At high temperatures (500°C) and high pressures (25 atm), long-chain hydrocarbons break into smaller pieces. A saturated $C_{12}$ hydrocarbon in kerosene, for example, might break into two $C_6$ fragments. The ratio of hydrogen to carbon atoms in the starting material requires that one of the products of this reaction must contain a C=C double bond:

$$CH_3(CH_2)_{10}CH_3 \longrightarrow CH_3CH_2CH_2CH_2CH_2CH_3 + CH_2{=}CHCH_2CH_2CH_2CH_3$$

The presence of alkenes in thermally cracked gasolines increases the octane number (70) relative to that of straight-run gasoline (60), but it also makes thermally cracked gasoline less stable for long-term storage. Thermal cracking therefore has been replaced by *catalytic cracking,* which uses catalysts instead of high temperatures and pressures to crack long-chain hydrocarbons into smaller fragments for use in gasoline.

About 87% of the crude oil refined in 1980 went into the production of fuels such as gasoline, kerosene, and fuel oil. The remainder went for nonfuel uses, such as petroleum solvents, industrial greases and waxes, or as starting materials for the synthesis of *petrochemicals.* Petroleum products are used to produce synthetic fibers such as nylon, Orlon, and Dacron, and other polymers such as polystyrene, polyethylene, and synthetic rubber. They also serve as raw materials in the production of refrigerants, aerosols, antifreeze, detergents, dyes, adhesives, alcohols, explosives, weed killers, insecticides, and insect repellents. The $H_2$ given off when alkanes are converted to alkenes or when cycloalkanes are converted to aromatic hydrocarbons can be used to produce a number of inorganic petrochemicals, such as ammonia, ammonium nitrate, and nitric acid. As a result, most fertilizers as well as other agricultural chemicals are petrochemicals.

Petroleum products.

## 24.8 THE CHEMISTRY OF COAL

*Coal* can be defined as a sedimentary rock that burns. It was formed by the decomposition of plant matter, and it is a complex substance that can be found in many forms. Coal is subdivided into four classes: anthracite, bituminous, subbituminous, and lignite. Elemental analysis gives empirical formulas such as $C_{137}H_{97}O_9NS$ for bituminous coal and $C_{240}H_{90}O_4NS$ for high-grade anthracite. A typical structure for coal is shown in Figure 24.10.

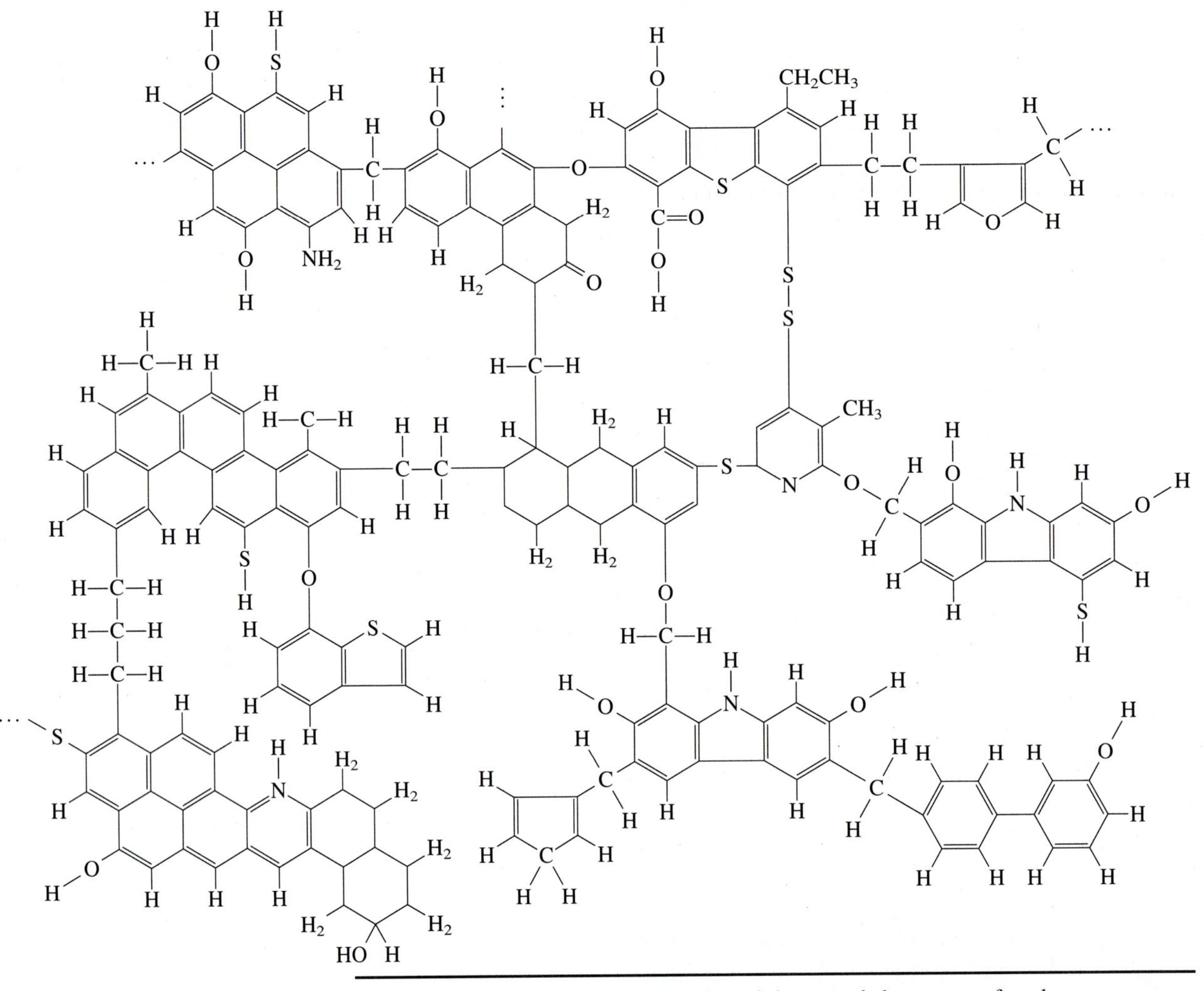

**Figure 24.10** A model for one portion of the extended structure of coal.

*Anthracite coal* is a dense, hard rock with a jet-black color and a metallic luster. It contains between 86% and 98% carbon by weight, and it burns slowly, with a pale blue flame and very little smoke. *Bituminous coal,* or soft coal, contains between 69% and 86% carbon by weight and is the most abundant form of coal. *Subbituminous coal* contains less carbon and more water, and is therefore a less efficient source of heat. *Lignite coal,* or brown coal, is a very soft coal that contains up to 70% water by weight.

The total energy consumption in the United States for 1990 was 86 × $10^{15}$ kJ. Of this total, 41% came from oil, 24% from natural gas, and 23% from coal. Coal is unique as a source of energy in the United States, however, because none of the 2118 billion pounds used in 1990 was imported. Furthermore, the proven reserves are so large we can continue using coal at this level of consumption for at least 2000 years.

At the time this text was written, coal was the most cost-efficient fuel for heating. The cost of coal delivered to the Purdue University physical plant was $1.41 per million kilojoules of heating energy. The equivalent cost for natural gas would have been $5.22 and #2 fuel oil would have cost $7.34. Although coal is cheaper than natural gas and oil, it is more difficult to handle. As a result, there has been a long history of efforts to turn coal into either a gaseous or a liquid fuel.

Coal fire burning.

## COAL GASIFICATION

As early as 1800, *coal gas* was made by heating coal in the absence of air. Coal gas is rich in $CH_4$ and gives off up to 20.5 kJ/L when burned. Coal gas—or *town gas,* as it was also known—became so popular that most major cities and many small towns had a local gashouse, and gas burners were adjusted to burn a fuel that produced 20.5 kJ/L. Gas lanterns, of course, were eventually replaced by electric lights. But coal gas was still used for cooking and heating until the more efficient natural gas (38.3 kJ/L) became readily available.

A slightly less efficient fuel known as *water gas* can be made by reacting the carbon in coal with steam:

$$C(s) + H_2O(g) \longrightarrow CO(g) + H_2(g) \qquad \Delta H^\circ = 118 \text{ kJ/mol}$$

Water gas burns to give $CO_2$ and $H_2O$, releasing roughly 11.2 kJ/L. Note that the enthalpy of reaction for the preparation of water gas is positive, which means that this reaction is endothermic. As a result, the preparation of water gas typically involves alternating blasts of steam and either air or oxygen through a bed of white-hot coal. The exothermic reactions between coal and oxygen to produce CO and $CO_2$ provide enough energy to drive the reaction between steam and coal.

Water gas formed by reacting coal with oxygen and steam is a mixture of CO, $CO_2$, and $H_2$. The ratio of $H_2$ to CO can be increased by adding water to this mixture, to take advantage of a reaction known as the *water-gas shift reaction:*

$$CO(g) + H_2O(g) \longrightarrow CO_2(g) + H_2(g) \qquad \Delta H^\circ = 41 \text{ kJ/mol}$$

The concentration of $CO_2$ can be decreased by reacting $CO_2$ with coal at high temperatures to form CO:

$$C(s) + CO_2(g) \longrightarrow 2\,CO(g) \qquad \Delta H^\circ = 160 \text{ kJ/mol}$$

Water gas from which the $CO_2$ has been removed is called *synthesis gas,* because it can be used as a starting material for a variety of organic and inorganic compounds. It can be used as the source of $H_2$ for the synthesis of ammonia, for example:

$$N_2(g) + 3\,H_2(g) \longrightarrow 2\,NH_3(g)$$

It also can be used to make methyl alcohol, or methanol:

$$CO(g) + 2\,H_2(g) \longrightarrow CH_3OH(l)$$

Methanol then can be used as a starting material for the synthesis of alkenes aromatic compounds, acetic acid, formaldehyde, and ethyl alcohol (ethanol). Synthesis gas also can be used to produce methane, or synthetic natural gas (SNG):

$$CO(g) + 3\,H_2(g) \longrightarrow CH_4(g) + H_2O(g)$$
$$2\,CO(g) + 2\,H_2(g) \longrightarrow CH_4(g) + CO_2(g)$$

## COAL LIQUEFACTION

The first step toward making liquid fuels from coal involves the manufacture of synthesis gas (CO and $H_2$) from coal. In 1925, Franz Fischer and Hans Tropsch developed a catalyst that converted CO and $H_2$ at 1 atm and 250°C to 300°C into liquid hydrocarbons. By 1941, Fischer–Tropsch plants produced 740,000 tons of petroleum products per year in Germany.

Fischer–Tropsch technology is based on a complex series of reactions that use $H_2$ to reduce CO to $-CH_2-$ groups linked to form long-chain hydrocarbons:

$$CO(g) + 2\,H_2(g) \longrightarrow (-CH_2-)_n(l) + H_2O(g) \qquad \Delta H^\circ = -165 \text{ kJ/mol CO}$$

The water produced in this reaction combines with CO in the water-gas shift reaction to form $H_2$ and $CO_2$:

$$CO(g) + H_2O(g) \longrightarrow CO_2(g) + H_2(g) \qquad \Delta H^\circ = 41 \text{ kJ/mol CO}$$

The overall Fischer-Tropsch reaction is therefore described by the following equation.

$$2\,CO(g) + H_2(g) \longrightarrow (-CH_2-)_n(l) + CO_2(g) \qquad \Delta H^\circ = -124 \text{ kJ/mol CO}$$

At the end of World War II, Fischer–Tropsch technology was under study in most industrial nations. The low cost and high availability of crude oil, however, led to a decline in interest in liquid fuels made from coal. The only commercial plants using this technology today are in the Sasol complex in South Africa, which uses 30.3 million tons of coal per year.

Another approach to liquid fuels is based on the reaction between CO and $H_2$ to form methanol, $CH_3OH$:

$$CO(g) + 2\,H_2(g) \longrightarrow CH_3OH(l)$$

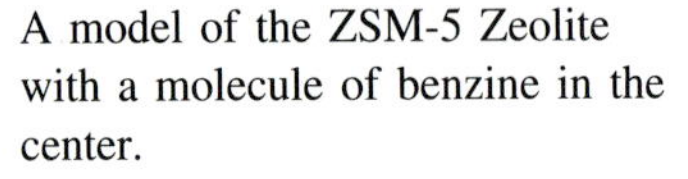
p. 421

Methanol can be used directly as a fuel, or it can be converted into gasoline with catalysts such as the ZSM-5 catalyst developed by Mobil Oil Company.

As the supply of petroleum becomes smaller and its cost continues to rise, a gradual shift may be observed toward liquid fuels made from coal. Whether this takes the form of a return to a modified Fischer–Tropsch technology, the conversion of methanol to gasoline, or other alternatives, only time will tell.

A model of the ZSM-5 Zeolite with a molecule of benzine in the center.

## 24.9 FUNCTIONAL GROUPS

Sodium metal reacts with water to give an aqueous solution of NaOH and $H_2$ gas:

$$2\,Na + 2\,H_2O \longrightarrow 2\,Na^+ + 2\,OH^- + H_2$$

Sodium also reacts with methanol to give $H_2$ gas and a solution of $Na^+$ and $CH_3O^-$ ions dissolved in the alcohol:

$$2\,Na + 2\,CH_3OH \longrightarrow 2\,Na^+ + 2\,CH_3O^- + H_2$$

Instead of trying to memorize both reactions, we can build a general rule that sodium reacts with compounds that contain the —OH functional group to give sodium salts of their conjugate base and $H_2$ gas.

**Functional groups** are atoms or groups of atoms in organic compounds that give these compounds some of their characteristic properties. They provide a way of integrating large quantities of information about the chemistry of organic compounds. We don't have to worry about the difference between 1-butene and 2-methyl-3-hexene, for example. We can focus on the fact that both compounds are alkenes, which add $Br_2$ across the C=C double bond:

$$H_2C{=}CHCH_2CH_3 + Br_2 \longrightarrow BrCH_2CHBrCH_2CH_3$$

$$CH_3\underset{}{\overset{CH_3}{\overset{|}{C}}}HCH{=}CHCH_2CH_3 + Br_2 \longrightarrow CH_3\overset{CH_3}{\overset{|}{C}}HCHBrCHBrCH_2CH_3$$

Some of the important functional groups in organic chemistry are given in Table 24.5.

**TABLE 24.5 Some Important Functional Groups**

| Functional Group | Name | Example |
|---|---|---|
| C—H | alkane | $CH_3CH_2CH_3$ (propane) |
| C=C | alkene | $CH_3CH{=}CH_2$ (propene) |
| C≡C | alkyne | HC≡CH (acetylene) |
| —F, —Cl, —Br, —I | alkyl halide | $CH_3Cl$ (methyl chloride) |
| —OH | alcohol | $CH_3OH$ (methyl alcohol) |
| —O— | ether | $CH_3OCH_3$ (dimethyl ether) |
| $—NH_2$ | amine | $CH_3NH_2$ (methyl amine) |
| $—\overset{O}{\overset{\|}{C}}—H$ | aldehyde | $CH_3\overset{O}{\overset{\|}{C}}H$ (acetaldehyde) |
| $—\overset{O}{\overset{\|}{C}}—$ | ketone | $CH_3\overset{O}{\overset{\|}{C}}CH_3$ (acetone) |
| $—\overset{O}{\overset{\|}{C}}—OH$ | carboxylic acid | $CH_3\overset{O}{\overset{\|}{C}}OH$ (acetic acid) |
| $—\overset{O}{\overset{\|}{C}}—Cl$ | acyl chloride | $CH_3\overset{O}{\overset{\|}{C}}Cl$ (acetyl chloride) |
| $—\overset{O}{\overset{\|}{C}}—O—$ | ester | $CH_3\overset{O}{\overset{\|}{C}}OCH_3$ (methyl acetate) |
| $—\overset{O}{\overset{\|}{C}}—NH_2$ | amide | $CH_3\overset{O}{\overset{\|}{C}}NH_2$ (acetamide) |

**EXERCISE 24.7**

Identify the functional groups in the following compounds:

(a) Vitamin C (Figure 24.1) (b) $\beta$-Pinene (Figure 24.4*a*)
(c) Citronellal (Figure 24.4*b*) (d) Vitamin $A_1$ (Figure 24.5*a*)
(e) Cholesterol (Figure 24.5*b*)

**SOLUTION**

(a) Vitamin C is an alcohol because it contains —OH groups. It is also an alkene because it has a C=C double bond. The most difficult functional group to recognize in this molecule is the ester linkage ($—CO_2—$).
(b) $\beta$-Pinene is an alkene because it contains a C=C double bond.
(c) Citronellal is an aldehyde because it contains a —CHO group.
(d) Vitamin $A_1$ contains both alcohol (—OH) and alkene (C=C) functional groups.
(e) Cholesterol also contains both alcohol (—OH) and alkene (C=C) functional groups.

## 24.10 ALKYL HALIDES

Alkanes react with halogens at high temperatures or in the presence of light to form alkyl halides, as noted in Section 24.4.

$$CH_4(g) + Cl_2(g) \xrightarrow{h\nu} CH_3Cl(g) + HCl(g)$$

The reaction between $CH_4$ and $Cl_2$ has the following characteristic properties:

- It does not take place in the dark, or at low temperatures.
- It occurs in the presence of ultraviolet light, or at temperatures above 250°C.
- Once the reaction gets started, it continues even after the light is turned off.
- The products of the reaction include $CH_2Cl_2$ (dichloromethane), $CHCl_3$ (chloroform), and $CCl_4$ (carbon tetrachloride), as well as $CH_3Cl$ (chloromethane).
- The reaction also produces some $C_2H_6$.

These facts are consistent with a chain-reaction mechanism that involves three processes: chain initiation, chain propagation, and chain termination.

### CHAIN INITIATION

A $Cl_2$ molecule can dissociate into a pair of chlorine atoms by absorbing energy from either ultraviolet light or heat:

$$Cl_2 + \text{energy} \xrightarrow{h\nu} 2\,Cl\cdot \qquad \Delta H° = 243.4 \text{ kJ/mol } Cl_2$$

The chlorine atom produced in this reaction is a *free radical* because it contains an unpaired electron.

### CHAIN PROPAGATION

Free radicals, such as the $Cl\cdot$ atom, are extremely reactive. The $Cl\cdot$ atom can remove a hydrogen atom from $CH_4$ to form HCl and a $CH_3\cdot$ radical. The $CH_3\cdot$

radical then removes a chlorine atom from a $Cl_2$ molecule to form $CH_3Cl$ and a new $Cl\cdot$ radical:

$$CH_4 + Cl\cdot \longrightarrow CH_3\cdot + HCl \qquad \Delta H^\circ = -16 \text{ kJ/mol } CH_4$$
$$CH_3\cdot + Cl_2 \longrightarrow CH_3Cl + Cl\cdot \qquad \Delta H^\circ = -87 \text{ kJ/mol } CH_3Cl$$

Because a $Cl\cdot$ atom is generated in the second reaction for every $Cl\cdot$ atom consumed in the first, this reaction continues in a chainlike fashion until the radicals involved in these chain-propagation steps are destroyed.

### CHAIN TERMINATION

The radicals that keep the reaction going eventually combine in a chain-terminating step. Chain termination can occur in three ways:

$$2\,Cl\cdot \longrightarrow Cl_2 \qquad \Delta H^\circ = -243.4 \text{ kJ/mol } Cl_2$$
$$CH_3\cdot + Cl\cdot \longrightarrow CH_3Cl \qquad \Delta H^\circ = -330 \text{ kJ/mol } CH_3Cl$$
$$2\,CH_3\cdot \longrightarrow CH_3CH_3 \qquad \Delta H^\circ = -350 \text{ kJ/mol } CH_3CH_3$$

### EXERCISE 24.8

Use the free-radical chain-reaction mechanism to explain the five observations listed for the reaction between $CH_4$ and $Cl_2$.

**SOLUTION**

1. The reaction does not occur in the dark or at low temperatures because energy must be absorbed to generate the free radicals that carry the reaction.
2. The reaction occurs in the presence of ultraviolet light because a UV photon has enough energy to dissociate a $Cl_2$ molecule to a pair of $Cl\cdot$ atoms. The reaction occurs at high temperatures because $Cl_2$ molecules can dissociate to form $Cl\cdot$ atoms by absorbing thermal energy.
3. The reaction continues after the light has been turned off because light is needed only to generate the $Cl\cdot$ atoms that start the reaction.
4. The reaction doesn't stop at $CH_3Cl$ because the $Cl\cdot$ atom is so reactive it can abstract additional hydrogen atoms to form $CH_2Cl_2$, $CHCl_3$, and eventually $CCl_4$.
5. The formation of $C_2H_6$ is a clear indication that the reaction proceeds through a free-radical mechanism. When two $CH_3\cdot$ radicals collide, they combine to form a $CH_3CH_3$ molecule.

## 24.11 ALCOHOLS AND ETHERS

**Alcohols** contain an —OH group attached to a saturated carbon. The common names for alcohols are based on the name of the alkyl group.

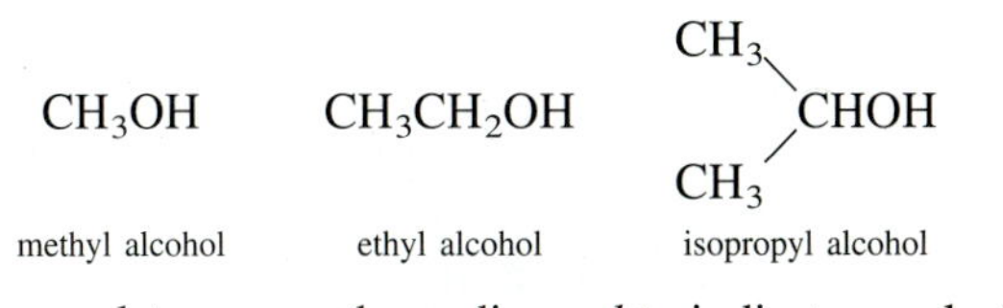


The systematic nomenclature uses the ending *-ol* to indicate an alcohol and a number to identify the carbon that carries the —OH group. (Isopropyl alcohol, for example, is also known as 2-propanol.) The carbon atom that carries the —OH substituent must be a member of the chain upon which the name of the compound is based.

**EXERCISE 24.9**

Use the systematic nomenclature to name the following alcohol:

$$\begin{array}{c} \quad\quad\quad CH_2CH_3 \\ \quad\quad\quad | \\ CH_3CH_2CH_2CHCH_2OH \end{array}$$

**SOLUTION** The longest chain of carbon atoms in this compound contains six atoms. It may be tempting therefore to name this compound as a derivative of hexane:

$$\begin{array}{c} \quad\quad\quad CH_2CH_3 \\ \quad\quad\quad | \\ CH_3CH_2CH_2CHCH_2OH \end{array}$$

But the chain that carries the —OH substituent is only five carbon atoms long. The compound is therefore named as a derivative of pentane:

$$\begin{array}{c} \quad\quad\quad CH_2CH_3 \\ \quad\quad\quad | \\ CH_3CH_2CH_2CHCH_2OH \end{array}$$

2-ethyl-1-pentanol

*Methanol,* or methyl alcohol, is also known as *wood alcohol* because it was originally made by heating wood until a liquid distilled. Methanol is highly toxic, and many people have become blind or died from drinking it. *Ethanol,* or ethyl alcohol, is the alcohol associated with ''alcoholic'' beverages. It has been made for at least 6000 years by adding yeast to solutions that are rich in either sugars or starches. The yeast cells obtain energy from enzyme-catalyzed reactions that convert sugar or starch to ethanol and $CO_2$:

$$C_6H_{12}O_6(aq) \longrightarrow 2\ CH_3CH_2OH(aq) + 2\ CO_2(g)$$

When the alcohol reaches a concentration of 10% to 12% by volume, the yeast cells die. Brandy, rum, gin, and the various whiskeys that have a higher concentration of alcohol are prepared by distilling the alcohol produced by this fermentation reaction.

Ethanol is not as toxic as methanol, but it is still dangerous. Blood alcohol levels of 1.5 to 3 g/L result in intoxication, and levels of 4 to 6 g/L can lead to coma or death. Ethanol is oxidized to $CO_2$ and $H_2O$ by alcohol dehydrogenase enzymes in the body. This reaction gives off 30 kJ/g, which makes ethanol a better source of energy than carbohydrates (17 kJ/g), and almost as good a source of energy as fat (38 kJ/g). An ounce of 80-proof liquor can provide as much as 3% of the average daily caloric intake and drinking alcohol can contribute to obesity. Many alcoholics are malnourished, however, because of the absence of vitamins in the calories they obtain from alcoholic beverages.

Alcohols are classified as *primary* (1°), *secondary* (2°), or *tertiary* (3°) on the basis of their structures. Ethanol is a primary alcohol because there is only one alkyl group attached to the carbon that carries the —OH substituent. The structure of a primary alcohol can be abbreviated as $RCH_2OH$, where R stands for an alkyl group.

$$CH_3CH_2OH$$

1°

Isopropyl alcohol, or rubbing alcohol, is a secondary alcohol, which has two alkyl groups on the carbon atom with the —OH substituent ($R_2CHOH$):

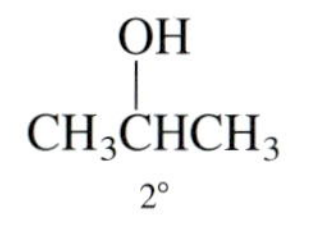


An example of a tertiary alcohol ($R_3COH$) is *t*-butyl alcohol, or 2-methyl-2-propanol:

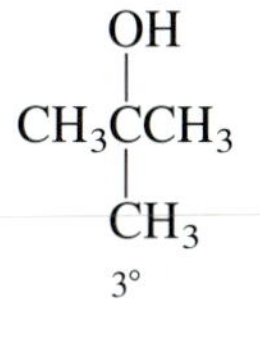


Another class of alcohols are the *phenols,* in which an —OH group is attached to one of the carbon atoms in a benzene ring, as shown in Figure 24.7. Phenols are potent disinfectants. When antiseptic techniques were first introduced to surgery in the 1860s by Joseph Lister, it was phenol (or carbolic acid, as it was then known) that was used. Phenol derivatives are still used in commercial disinfectants such as Lysol.

Water has an unusually high boiling point because of the hydrogen bonds between the $H_2O$ molecules. Alcohols can form similar hydrogen bonds, as shown in Figure 24.11. As a result, alcohols have boiling points that are much higher than alkanes with similar molecular weights. The boiling point of ethanol, for example, is 78.5°C, whereas propane, with about the same molecular weight, boils at −42.1°C.

Sec. 14.9

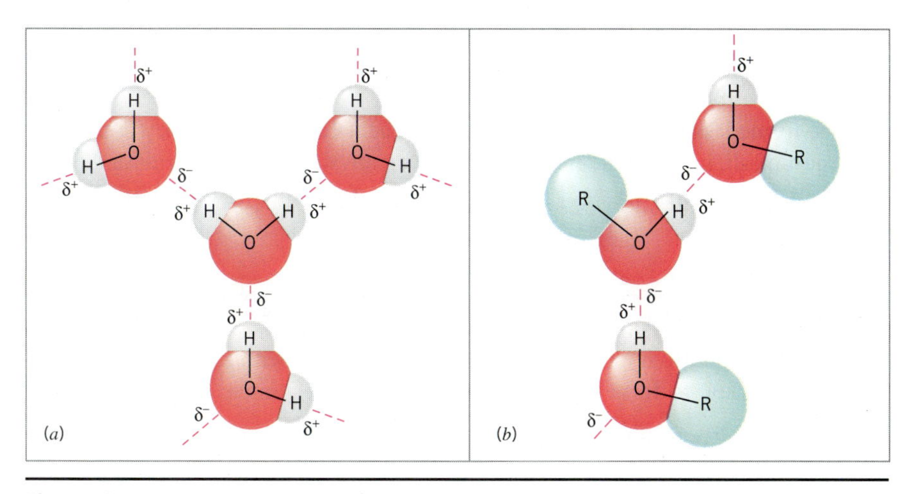


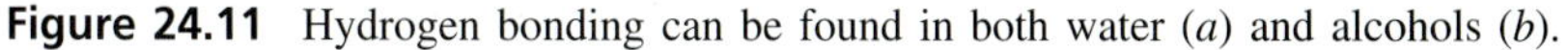

**Figure 24.11** Hydrogen bonding can be found in both water (*a*) and alcohols (*b*).

Alcohols are Brønsted acids in aqueous solution:

$$CH_3CH_2OH(aq) + H_2O(l) \rightleftharpoons H_3O^+(aq) + CH_3CH_2O^-(aq)$$

The acidity of methanol and ethanol is slightly less than that of water. Because of their acidity, alcohols react with sodium metal to produce sodium salts of the corresponding conjugate base:

$$2\,Na + 2\,CH_3OH \longrightarrow 2\,Na^+ + 2\,CH_3O^- + H_2$$

The conjugate base of an alcohol is known as an **alkoxide:**

$$[Na^+][CH_3O^-] \qquad [Na^+][CH_3CH_2O^-]$$

sodium methoxide sodium ethoxide

Alcohols can be prepared by adding water to an alkene in the presence of a strong acid, such as concentrated sulfuric acid:

$$CH_3CH{=}CH_2 + H_2O \xrightarrow{H_2SO_4} CH_3\overset{\overset{\displaystyle OH}{|}}{C}HCH_3$$

Alcohols also can be prepared by substitution reactions between an alkyl halide and the $OH^-$ ion:

$$CH_3CH_2CH_2Br + OH^- \longrightarrow CH_3CH_2CH_2OH + Br^-$$

An **ether** is a compound in which an oxygen atom bridges two alkyl groups. Its name includes the word *ether* attached to the names of the alkyl groups:

$$CH_3CH_2OCH_2CH_3$$

diethyl ether

Diethyl ether, often known by the generic name ''ether,'' was once used extensively as an anesthetic. Because mixtures of diethyl ether and air explode in the presence of a spark, ether has been replaced by safer anesthetics.

Ethers can be synthesized by splitting out a molecule of water between two alcohols in the presence of heat and concentrated sulfuric acid:

$$2\,CH_3CH_2OH \xrightarrow{H_2SO_4} CH_3CH_2OCH_2CH_3$$

They can be formed also by gently reacting a primary alkyl halide with an alkoxide ion:

$$CH_3CH_2CH_2Br + CH_3O^- \longrightarrow CH_3CH_2CH_2OCH_3 + Br^-$$

Once formed, ethers are essentially inert to chemical reactions. They don't react with most oxidizing or reducing agents, and they are stable to most acids and bases, except at high temperatures. They are therefore frequently used as solvents for chemical reactions.

## 24.12 SUBSTITUTION REACTIONS

As noted in the preceding section, alcohols can be made by reacting alkyl halides with the $OH^-$ ion. Both $CH_3Br$ and $(CH_3)_3CBr$ can react with aqueous solutions of the hydroxide ion to form the analogous alcohol, for example:

$$CH_3Br + OH^- \longrightarrow CH_3OH + Br^-$$

$$(CH_3)_3CBr + OH^- \longrightarrow (CH_3)_3COH + Br^-$$

The kinetics of these reactions are different, however. The rate of the reaction between $CH_3Br$ and the $OH^-$ ion is proportional to the concentrations of both reactants:

$$\text{Rate} = k(CH_3Br)(OH^-)$$

Sec. 22.8

But the rate of the reaction between $(CH)_3CBr$ and the $OH^-$ ion is proportional only to the concentration of the alkyl halide:

$$\text{Rate} = k((CH_3)_3CBr)$$

The mechanisms of these reactions therefore also must be different.

The reaction between $CH_3Br$ and the $OH^-$ ion involves attack by the $OH^-$ ion on the carbon atom. Thus, a new carbon–oxygen bond is formed at the same time that the carbon–bromine bond is broken:

$$HO^- + CH_3{-}Br \longrightarrow HO{\cdots}\underset{\substack{|\\ H}}{\overset{H\ \ H}{C}}{\cdots}Br \longrightarrow HO{-}CH_3 + Br^-$$

The $OH^-$ ion acts as a Lewis base in this reaction, which donates a pair of nonbonding electrons to the carbon atom to form a new covalent bond. Organic chemists prefer to call the $OH^-$ ion a **nucleophile** (literally, ‘‘something that loves nuclei’’), but there is no difference between a nucleophile and a Lewis base.

The reaction between $CH_3Br$ and the $OH^-$ ion is an example of a *nucleophilic substitution reaction* because it involves the substitution of one nucleophile (the $OH^-$ ion) for another (the $Br^-$ ion). Since the rate-limiting step in this reaction involves both the $CH_3Br$ and $OH^-$ molecules, it is called a *bimolecular nucleophilic substitution,* or $\mathbf{S_N2}$**,** reaction.

The rate law for the reaction between $(CH_3)_3CBr$ and the $OH^-$ ion is consistent with a multistep mechanism. The first step is a slow, rate-limiting reaction in which the carbon–bromine bond is broken to form a positively charged **carbonium ion** and a negatively charged $Br^-$ ion:

$$CH_3{-}\underset{\substack{|\\ CH_3}}{\overset{\substack{CH_3\\ |}}{C}}{-}Br \rightleftharpoons CH_3{-}\underset{\substack{|\\ CH_3}}{\overset{\substack{CH_3\\ |}}{C^+}} + Br^- \qquad \text{First step}$$

The second step in this reaction is much faster than the first. In this step, the carbonium ion combines with water in the solution:

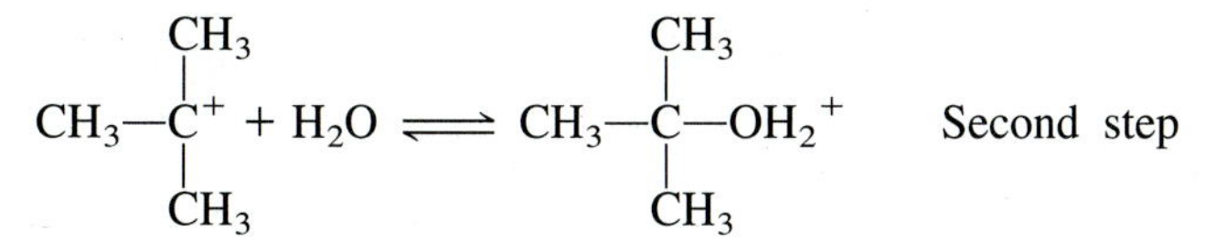

$$CH_3{-}\underset{\substack{|\\ CH_3}}{\overset{\substack{CH_3\\ |}}{C^+}} + H_2O \rightleftharpoons CH_3{-}\underset{\substack{|\\ CH_3}}{\overset{\substack{CH_3\\ |}}{C}}{-}OH_2^+ \qquad \text{Second step}$$

The product of this reaction then loses a proton to either the $OH^-$ ion or water in the final step:

$$CH_3{-}\underset{\substack{|\\ CH_3}}{\overset{\substack{CH_3\\ |}}{C}}{-}OH_2^+ + OH^- \rightleftharpoons CH_3{-}\underset{\substack{|\\ CH_3}}{\overset{\substack{CH_3\\ |}}{C}}{-}OH + H_2O \qquad \text{Third step}$$

Since the rate-limiting step for this reaction involves the dissociation of the $(CH_3)_3CBr$ molecule, it is known as a *unimolecular nucleophilic substitution,* or $\mathbf{S_N1}$**,** reaction.

Why does $CH_3Br$ react with the $OH^-$ ion by the $S_N2$ mechanism if $(CH_3)_3CBr$ does not? The $S_N2$ mechanism requires that the $OH^-$ ion attack the carbon atom. The $OH^-$ ion can get past the small hydrogen atoms in $CH_3Br$ much more easily than it can get past the bulkier $CH_3$— groups in $(CH_3)_3CBr$.

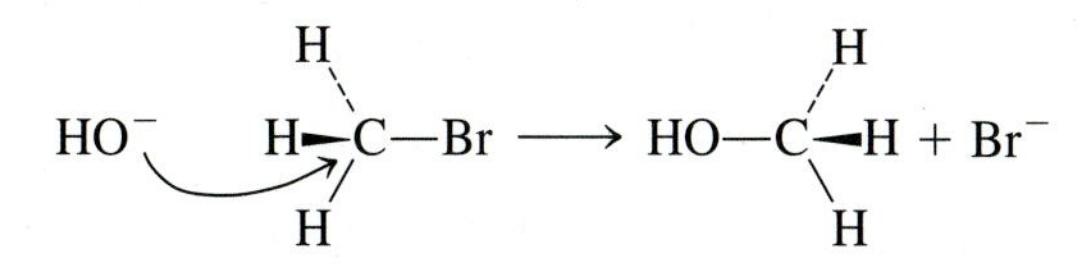

Why does $(CH_3)_3CBr$ react with the $OH^-$ ion by the $S_N1$ mechanism if $CH_3Br$ does not? The $S_N1$ mechanism proceeds through a carbonium ion intermediate, and the stability of carbonium ions decreases in the following order:

$$(CH_3)_3C^+ > (CH_3)_2CH^+ > CH_3CH_2^+ > CH_3^+$$

Tertiary carbonium ions are 343 kJ/mol more stable than primary carbonium ions. As a result, it is much easier for $(CH_3)_3CBr$ to form a carbonium ion intermediate than it is for $CH_3Br$ to undergo a similar reaction.

## 24.13 OXIDATION–REDUCTION REACTIONS

The oxidation numbers of the carbon atoms in a variety of functional groups are given in Table 24.6. These oxidation numbers can be used to classify organic reactions as either oxidation–reduction or metathesis reactions. Three examples of oxidation–reduction reactions of organic compounds are given in Figure 24.12.

**TABLE 24.6 Typical Oxidation Numbers of Carbon**

| Functional Group | Example | Oxidation Number of Carbon |
|---|---|---|
| Alkane | $CH_4$ | −4 |
| Alkene | $H_2C{=}CH_2$ | −2 |
| Alcohol | $CH_3OH$ | −2 |
| Ether | $CH_3OCH_3$ | −2 |
| Alkyl halide | $CH_3Cl$ | −2 |
| Amine | $CH_3NH_2$ | −2 |
| Alkyne | $HC{\equiv}CH$ | −1 |
| Aldehyde | $H_2CO$ | 0 |
| Acid | $HCO_2H$ | 2 |
| Ester | $HCO_2CH_3$ | 2 |
| | $CO_2$ | 4 |

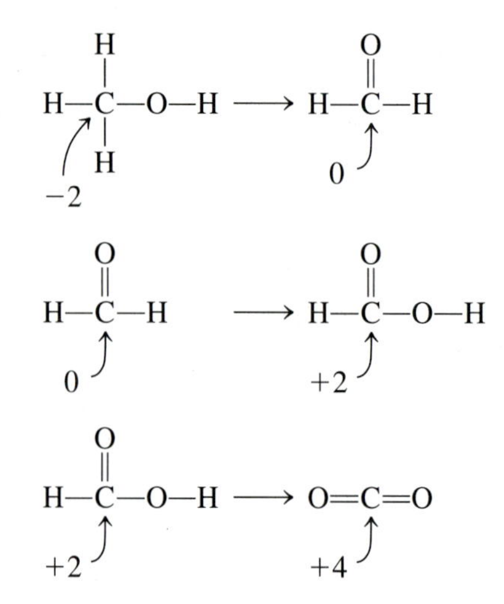


**Figure 24.12** Each of the transformations shown here involves the oxidation of a carbon atom. In the first reaction, the carbon atom is oxidized from the −2 to the 0 oxidation state. In the second reaction, it is oxidized to the +2 state, and the third reaction involves oxidation of the carbon atom to the +4 oxidation state.

## EXERCISE 24.10

Classify the following as either oxidation–reduction or metathesis reactions:

(a) $2\,CH_3OH \xrightarrow{H_2SO_4} CH_3OCH_3 + H_2O$

(b) $HCO_2H + CH_3OH \xrightarrow{H^+} HCO_2CH_3 + H_2O$

(c) $CO + 2\,H_2 \longrightarrow CH_3OH$

**SOLUTION**

(a) There is no change in the oxidation number of the carbon atoms when an alcohol is converted to an ether. This is therefore a metathesis reaction:

$$\underset{-2}{2\,CH_3OH} \longrightarrow \underset{-2\quad -2}{CH_3OCH_3} + H_2O$$

(b) This is also a metathesis reaction because there is no change in the oxidation numbers of the carbon atoms when a carboxylic acid reacts with an alcohol to form an ester:

$$\underset{+2}{HCO_2H} + \overset{-2}{CH_3OH} \longrightarrow \underset{+2}{HCO_2}\overset{-2}{CH_3} + H_2O$$

(c) The carbon atom in carbon monoxide is reduced from +2 to −2 when CO combines with $H_2$ to form methanol, so this is an oxidation–reduction reaction:

$$\underset{+2}{CO} + 2\,H_2 \longrightarrow \underset{-2}{CH_3OH}$$

Assigning oxidation numbers to the individual carbon atoms in complex organic molecules can be difficult. Fortunately, there is another way to recognize **oxidation–reduction** reactions in organic chemistry:

**Oxidation occurs when hydrogen atoms are removed from a carbon atom or when an oxygen atom is added to a carbon atom.**

The following reaction involves oxidation of the carbon atom because a pair of hydrogen atoms are removed from that atom:

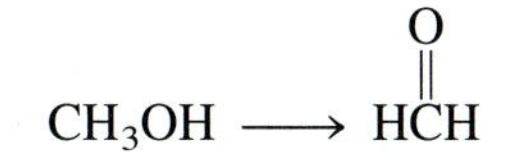


The following reaction involves oxidation of the carbon atom because an oxygen atom is added to the carbon atom:

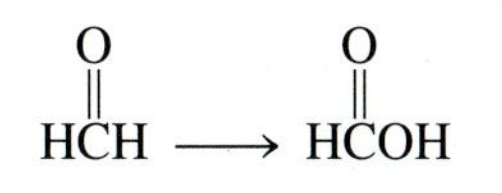


Reduction, on the other hand, occurs when hydrogen atoms are added to a carbon atom or when an oxygen atom is removed from a carbon atom.

## 24.14 ALDEHYDES AND KETONES

The C═O double bond is known as a **carbonyl group.** When at least one of the substituents on the carbonyl is a hydrogen atom, the compound is an **aldehyde.** The common names of aldehydes are derived from the names of the corresponding carboxylic acids:

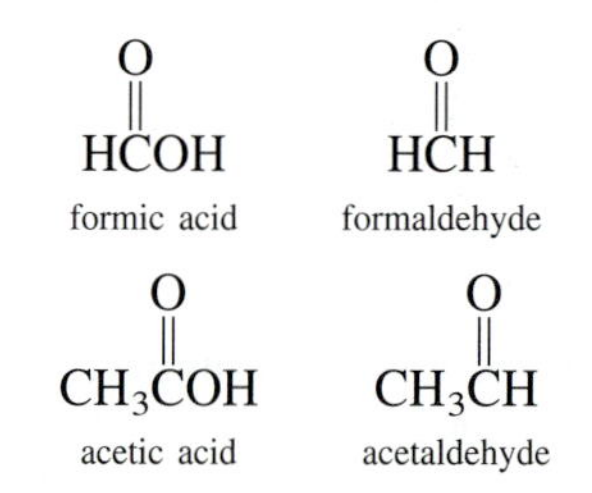


The systematic name for an aldehyde is obtained by adding *-al* to the name of the parent alkane:

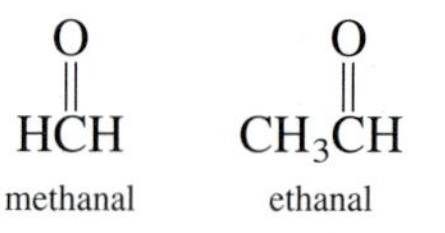


When both substituents on the carbonyl carbon are alkyl groups, the compound is a **ketone.** The common name for these compounds is constructed by adding the word *ketone* to the names of the two alkyl groups. The systematic name is obtained by adding *-one* to the name of the parent alkane and using numbers to indicate the location of the C═O group:

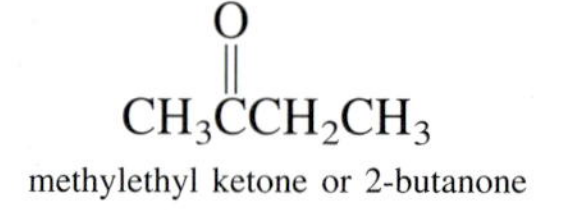


Both aldehydes and ketones can be reduced by $H_2$ in the presence of a catalyst such as platinum metal. When the starting material is an aldehyde, the product is a primary alcohol:

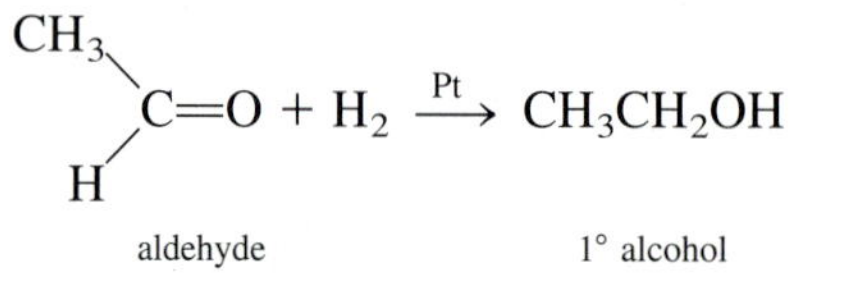


When the starting material is a ketone, the product is a secondary alcohol:

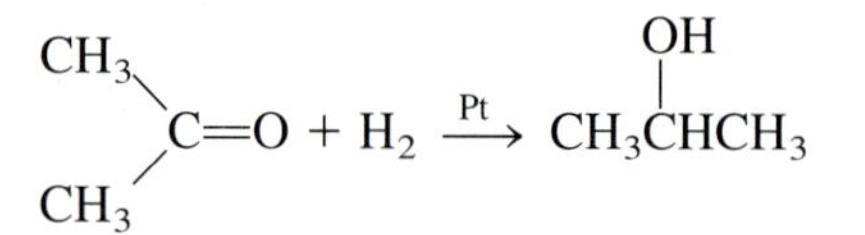

These reactions explain why it is important to distinguish between primary, secondary, and tertiary alcohols. It also provides a hint as to how aldehydes and ketones can be synthesized.

Chromium trioxide ($CrO_3$) in concentrated sulfuric acid is often used to oxidize secondary alcohols to ketones:

$$CH_3CH(OH)CH_3 \xrightarrow{CrO_3/H_2SO_4} CH_3C(=O)CH_3$$

Unfortunately, $CrO_3$ in $H_2SO_4$ is such a good oxidizing agent that it not only oxidizes a primary alcohol to an aldehyde, it oxidizes the aldehyde to the corresponding carboxylic acid:

$$CH_3CH_2OH \xrightarrow{CrO_3/H_2SO_4} [CH_3\overset{\overset{\displaystyle O}{\|}}{C}H] \longrightarrow CH_3\overset{\overset{\displaystyle O}{\|}}{C}OH$$

A weaker oxidizing agent, which is just strong enough to prepare the aldehyde from a primary alcohol, can be obtained by adding $CrO_3$ to a solvent, such as pyridine, $C_6H_5N$, that does not contain water:

$$CH_3CH_2OH \xrightarrow{CrO_3/\text{pyridine}} CH_3\overset{\overset{\displaystyle O}{\|}}{C}H$$

The relationship between $CH_3CH_2OH$ and $CH_3CHO$ explains why the second compound is an *aldehyde*. In the course of this reaction, an alcohol is dehydrogenated. The product of this reaction is therefore quite literally an ''al-dehyd.''

The characteristic orange color of the $CR_2O_7^{2}$ ion rapidly fades and then disappears at the junction between a solution of this ion and an alcohol.

## REACTIONS AT THE CARBONYL GROUP

It is somewhat misleading to write the carbonyl group as a covalent C=O double bond. The difference between the electronegativities of carbon and oxygen is 0.89. As a result, the C=O bond is moderately polar. The C=O bond can be thought of as a hybrid of the following Lewis structures:

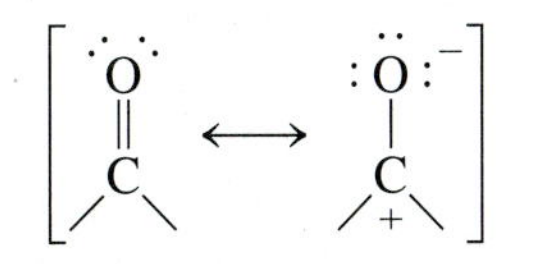

We can represent the polar nature of this hybrid by indicating the presence of a slight negative charge on the oxygen ($\delta^-$) and a slight positive charge ($\delta^+$) on the carbon of the C=O double bond:

$$\overset{\delta^+}{C}=\overset{\delta^-}{O}$$

Reagents that attack the electron-rich $\delta^-$ end of this bond are called **electrophiles** (literally, ''lovers of electrons''). Electrophiles include ions (such as the $H^+$ and $Fe^{3+}$ ions) and neutral molecules (such as $AlCl_3$ and $BF_3$) that are Lewis acids, or electron-pair acceptors. Reagents that attack the electron-poor $\delta^+$ end of the C=O bond are **nucleophiles** (literally, ''lovers of nuclei''). Nucleophiles are Lewis bases (such as $NH_3$ or the $OH^-$ ion).

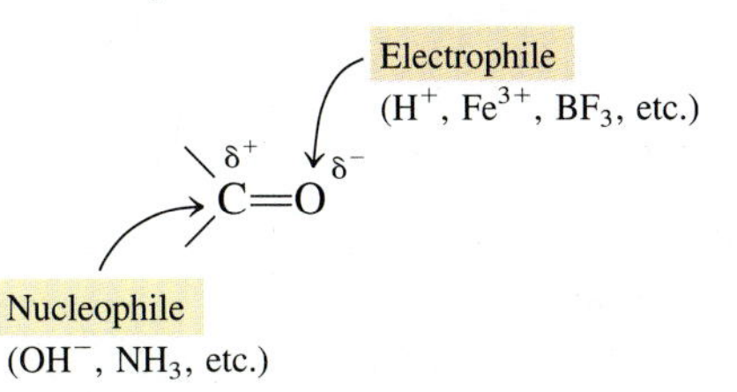


The polarity of the C=O double bond can be used to explain the reactions of carbonyl compounds. Aldehydes and ketones react with a number of hydride ion donors, such as lithium aluminum hydride ($LiAlH_4$) because the $H^-$ ion is a Lewis base, or nucleophile, that attacks the $\delta^+$ end of the C═O bond. When this happens, the two valence electrons on the $H^-$ ion form a covalent bond to the carbon atom. Since carbon is tetravalent, one pair of electrons in the C═O bond is displaced onto the oxygen to form an intermediate with a negative charge on the oxygen atom:

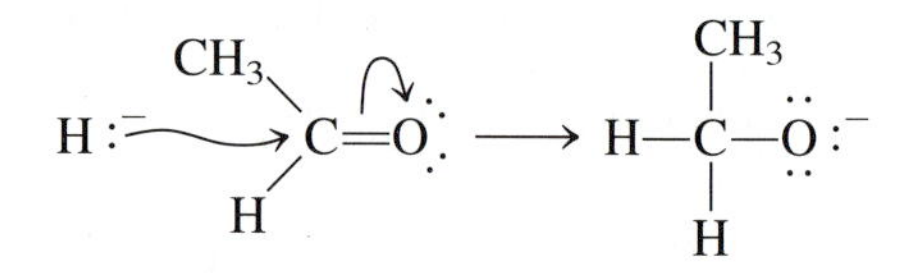

This intermediate ion can then remove an $H^+$ ion from water to form an alcohol:

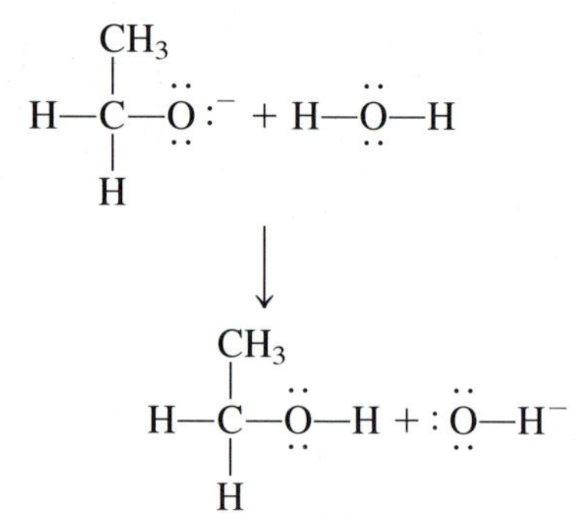

The polar nature of the C═O double bond also explains what happens when carbon dioxide reacts with water to form carbonic acid:

$$CO_2(g) + H_2O(l) \longrightarrow H_2CO_3(aq)$$

$CO_2$ is a linear molecule that contains two C═O double bonds. Water acts as a nucleophile, attacking the $\delta^+$ end of the C═O bond, thereby forming a new carbon–oxygen single bond:

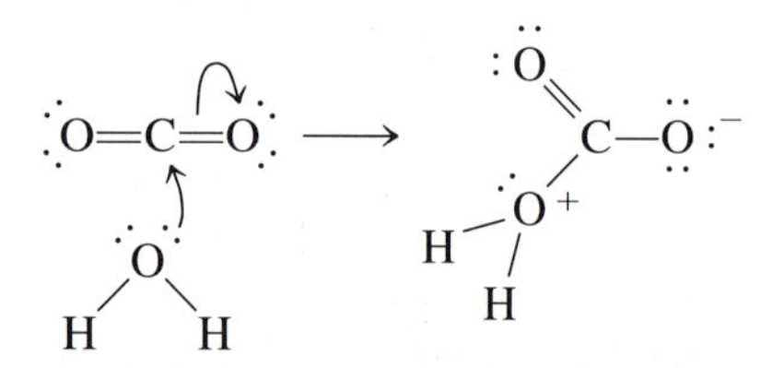

A hydrogen atom then migrates from the positively charged oxygen atom to the negatively charged oxygen atom to form carbonic acid, $H_2CO_3$:

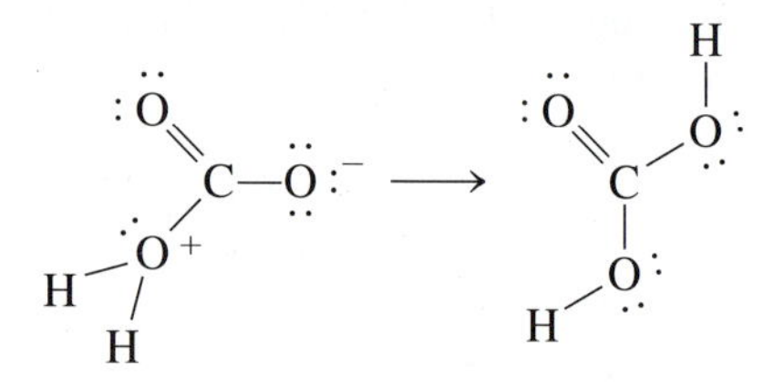

## 24.15 CARBOXYLIC ACIDS AND CARBOXYLATE IONS

When one of the substituents on a carbonyl group is an —OH group, the compound is a **carboxylic acid** with the generic formula $RCO_2H$. These compounds are acids, as the name suggests, which form **carboxylate ions** ($RCO_2^-$) by the loss of an $H^+$ ion. The carboxylate ion is a hybrid of two resonance structures:

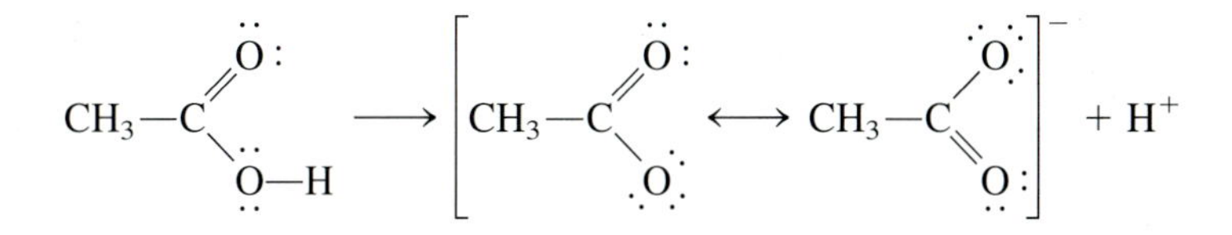

Resonance stabilizes the carboxylate ion and therefore makes carboxylic acids much stronger acids than the analogous alcohols. The value of $K_a$ for a typical carboxylic acid is about $10^{-5}$, whereas alcohols have values of $K_a$ of only $10^{-16}$.

Carboxylic acids were among the first organic compounds to be discovered. They therefore have well-entrenched common names that are often derived from the Latin stems of their sources in nature. Formic acid (Latin *formica,* "an ant") and acetic acid (Latin *acetum,* "vinegar") were first obtained by distilling ants and vinegar, respectively. Butyric acid (Latin *butyrum,* "butter") is found in rancid butter, and caproic, caprylic, and capric acids (Latin *caper,* "goat") are all obtained from goat fat. The systematic names of carboxylic acids are constructed by adding *-oic acid* to the name of the parent alkane. A list of common carboxylic acids is given in Table 24.7.

**TABLE 24.7 Common Carboxylic Acids**

| Compound | Common Name | Solubility in $H_2O$ (g/100 mL) |
|---|---|---|
| Saturated carboxylic acids and fatty acids | | |
| $HCO_2H$ | formic acid | |
| $CH_3CO_2H$ | acetic acid | |
| $CH_3CH_2CO_2H$ | proprionic acid | |
| $CH_3(CH_2)_2CO_2H$ | butyric acid | |
| $CH_3(CH_2)_4CO_2H$ | caproic acid | 0.968 |
| $CH_3(CH_2)_6CO_2H$ | caprylic acid | 0.068 |
| $CH_3(CH_2)_8CO_2H$ | capric acid | 0.015 |
| $CH_3(CH_2)_{10}CO_2H$ | lauric acid | 0.0055 |
| $CH_3(CH_2)_{12}CO_2H$ | myristic acid | 0.0020 |
| $CH_3(CH_2)_{14}CO_2H$ | palmitic acid | 0.00072 |
| $CH_3(CH_2)_{16}CO_2H$ | stearic acid | 0.00029 |
| Unsaturated fatty acids | | |
| $CH_3(CH_2)_7CH{=}CH(CH_2)_7CO_2H$ | oleic acid | |
| $CH_3(CH_2)_4CH{=}CHCH_2CH{=}CH(CH_2)_7CO_2H$ | linoleic acid | |
| $CH_3CH_2CH{=}CHCH_2CH{=}CHCH_2CH{=}CH(CH_2)_7CO_2H$ | linolenic acid | |

Formic acid and acetic acid have sharp, pungent odors. As the length of the alkyl chain increases, the odor of carboxylic acids becomes more unpleasant. Butyric acid, for example, is found in sweat, and the odor of rancid meat is due to carboxylic acids released as the meat spoils.

The data in Table 24.7 show that the solubility of carboxylic acids in water also changes with the length of the alkyl chain. The $—CO_2H$ end of this molecule is polar and therefore soluble in water. As the alkyl chain gets longer, the molecule becomes more nonpolar and less soluble in water.

Sec. 15.1

Compounds that contain two $—CO_2H$ functional groups are known as *dicarboxylic acids.* A number of dicarboxylic acids (see Table 24.8) can be isolated from

**TABLE 24.8 Common Dicarboxylic Acids**

| Compound | Common Name |
|---|---|
| $HO_2CCO_2H$ | oxalic acid |
| $HO_2CCH_2CO_2H$ | malonic acid |
| $HO_2CCH_2CH_2CO_2H$ | succinic acid |
| *cis*-$HO_2CCH{=}CHCO_2H$ | maleic acid |
| *trans*-$HO_2CCH{=}CHCO_2H$ | fumaric acid |
| $HO_2CCH_2\overset{\overset{OH}{\vert}}{C}HCO_2H$ | malic acid |
| $HO_2C\overset{\overset{OH}{\vert}}{C}H\overset{\overset{OH}{\vert}}{C}HCO_2H$ | tartaric acid |
| $HO_2CCH_2\overset{\overset{O}{\Vert}}{C}CO_2H$ | oxaloacetic acid |
| $HO_2C(CH_2)_4CO_2H$ | adipic acid |

natural sources. Tartaric acid, for example, is a byproduct of the fermentation of wine, and succinic, fumaric, malic and oxaloacetic acid are intermediates in the metabolism of sugars to $CO_2$ and $H_2O$.

Several tricarboxylic acids also play an important role in the metabolism of sugar. The most important example of this class of compounds is citric acid:

$$\begin{array}{c} CH_2CO_2H \\ | \\ HO{-}C{-}CO_2H \\ | \\ CH_2CO_2H \end{array}$$

citric acid

## 24.16 ESTERS

Carboxylic acids ($—CO_2H$) can react with alcohols (ROH) in the presence of either acid or base to form **esters** ($—CO_2R$). Acetic acid, for example, reacts with ethanol to form ethyl acetate and water:

$$CH_3\overset{\overset{O}{\Vert}}{C}OH + CH_3CH_2OH \rightleftharpoons CH_3\overset{\overset{O}{\Vert}}{C}OCH_2CH_3 + H_2O$$

In practice, this is not an efficient way of preparing the ester because the equilibrium constant for this reaction is relatively small ($K_c = 3$). Chemists tend to synthesize these esters in a two-step process. They start by reacting the acid with a chlorinating agent such as thionyl chloride to form the corresponding **acyl chloride:**

$$CH_3\overset{\overset{O}{\Vert}}{C}OH + SOCl_2 \longrightarrow CH_3\overset{\overset{O}{\Vert}}{C}Cl + SO_2 + HCl$$

They then react the acyl chloride with the alcohol in the presence of base to form the ester:

$$CH_3\overset{\overset{O}{\Vert}}{C}Cl + CH_3CH_2OH \xrightarrow{\text{base}} CH_3\overset{\overset{O}{\Vert}}{C}OCH_2CH_3 + [\text{base} \cdot H^+]Cl^-$$

The base absorbs the HCl given off in this reaction, thereby driving it to completion.

Esters are named by identifying the alkyl group in the alcohol and then adding *-ate* to the stem of the name of the carboxylic acid:

$$\underset{\text{ethyl acetate}}{CH_3CO_2CH_2CH_3} \qquad \underset{\text{methyl butyrate}}{CH_3CH_2CH_2CO_2CH_3}$$

The term *ester* is commonly used to describe the product of the reaction of any strong acid with an alcohol. Sulfuric acid, for example, reacts with methanol to form a diester known as dimethyl sulfate:

$$HO{-}\overset{\overset{O}{\|}}{\underset{\underset{O}{\|}}{S}}{-}OH + 2\,CH_3OH \longrightarrow CH_3O{-}\overset{\overset{O}{\|}}{\underset{\underset{O}{\|}}{S}}{-}OCH_3$$

Phosphoric acid reacts with alcohols to form triesters such as triethyl phosphate:

$$HO{-}\overset{\overset{O}{|}}{\underset{\underset{OH}{|}}{P}}{-}OH + 3\,CH_3CH_2OH \longrightarrow CH_3CH_2O{-}\overset{\overset{O}{|}}{\underset{\underset{OCH_2CH_3}{|}}{P}}{-}OCH_2CH_3$$

Compounds that contain the $-CO_2R$ functional group therefore should be called **carboxylic acid esters,** to indicate the acid from which they are formed.

Carboxylic acid esters with low molecular weights are colorless, volatile liquids that often have a pleasant odor. They are important components of both natural and synthetic flavors (see Figure 24.13).

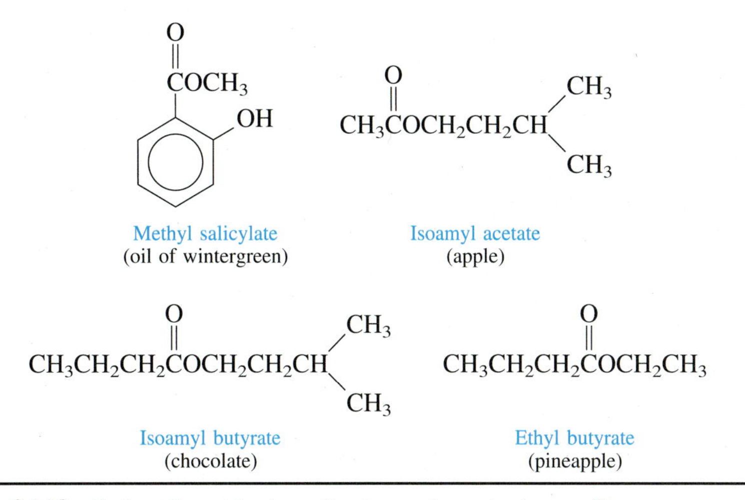

**Figure 24.13** Carboxylic acid esters often have pleasant odors or flavors.

Many of the components of the so called essential oils that give foods their "essence" are esters.

## 24.17 AMINES, ALKALOIDS, AND AMIDES

**Amines** are derivatives of ammonia in which one or more hydrogen atoms are replaced by alkyl groups. We indicate the degree of substitution by labeling the amine as either primary ($RNH_2$), secondary ($R_2NH$), or tertiary ($R_3N$). The common names of these compounds are derived from the names of the alkyl groups:

$$\underset{\text{isopropylamine}}{(CH_3)_2CHNH_2} \qquad \underset{\text{ethylmethylamine}}{CH_3NHCH_2CH_3} \qquad \underset{\text{trimethylamine}}{(CH_3)_3N}$$

The systematic names of primary amines are derived from the name of the parent alkane by adding the prefix *amino* and a number specifying the carbon that carries the $—NH_2$ group:

$$CH_3—CH{=}CH—CH_2—\underset{|}{\overset{NH_2}{CH}}—CH_3$$

5-amino-2-hexene

The chemistry of amines mirrors the chemistry of ammonia. Amines are weak bases that pick up a proton to form ammonium salts. Trimethylamine, for example, reacts with acid to form the trimethylammonium ion:

$$(CH_3)_3N + H^+ \longrightarrow (CH_3)_3NH^+$$

These salts are more soluble in water than the corresponding amines, and this reaction can be used to dissolve otherwise insoluble amines in aqueous solution.

The difference between amines and their ammonium salts plays an important role in both over the counter and illicit drugs. Cocaine, for example, is commonly sold as the hydrochloride salt, which is a white, crystalline solid. By extracting this solid into ether, it is possible to obtain the ''free base.'' A glance at the side panel of almost any over the counter medicine will provide examples of ammonium salts of amines that are used to ensure that the drug dissolves in water (see Figure 24.14).

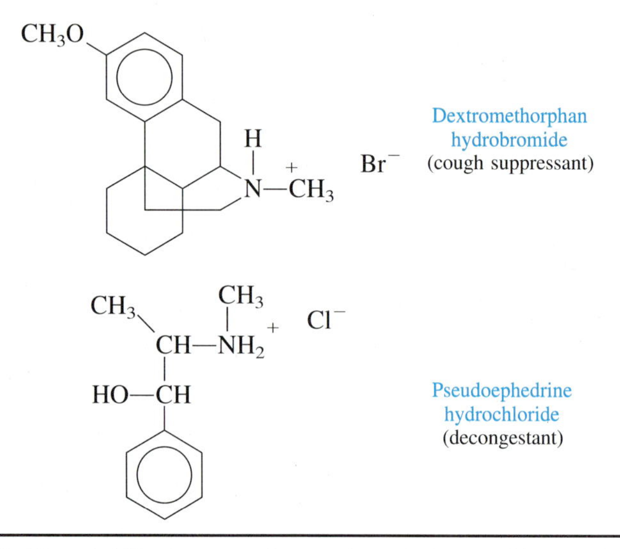


**Figure 24.14** The solubility in water of many drugs is increased by transforming them into amine hydrochloride or hydrobromide salts.

The bark of the chinchona tree, from which quinine was first isolated.

Amines that are isolated from plants are known as *alkaloids*. They include poisons such as nicotine, coniine, and strychnine (shown in Figure 24.15). Nicotine has a pleasant, invigorating effect when taken in minuscule quantities, but is extremely toxic in larger amounts. Coniine is the active ingredient in hemlock, a poison that has been used since the time of Socrates. Strychnine is another toxic alkaloid that has been a popular poison in murder mysteries.

The alkaloids also include a number of drugs, such as morphine, quinine, and cocaine. Morphine is obtained from poppies; quinine can be found in the bark of the chinchona tree; and cocaine is isolated from coca leaves. This family of compounds also includes synthetic analogs of naturally occurring alkaloids, such as heroin and LSD (see Figure 24.16).

encounter a reaction, however, that addresses a basic question: How do we make C—C bonds? One answer resulted from the work that Francois Auguste Victor Grignard started as part of his Ph.D. research at the turn of the century.

Grignard noted that alkyl halides react with magnesium metal in diethyl ether ($Et_2O$) to form compounds that contain a metal–carbon bond. Methyl bromide, for example, forms methylmagnesium bromide:

$$CH_3Br + Mg \xrightarrow{Et_2O} CH_3MgBr$$

Carbon is more electronegative than magnesium ($\Delta EN = 1.24$), and the metal–carbon bond in this compound has a significant amount of ionic character. The alkylmagnesium halides, or **Grignard reagents,** such as $CH_3MgBr$ are best thought of as hybrids of ionic and covalent Lewis structures:

$$\left[CH_3—Mg—\ddot{\underset{..}{Br}}: \longleftrightarrow (CH_3:^-)(Mg^{2+})(:\ddot{\underset{..}{Br}}:^-)\right]$$

Grignard reagents are our first source of **carbanions** (literally, ''anions of carbon''). The Lewis structure of the $CH_3^-$ ion suggests that carbanions can be Lewis bases, or electron-pair donors. Grignard reagents such as methylmagnesium bromide are therefore sources of a nucleophile that can attack the $\delta^+$ end of the C=O double bond in aldehydes and ketones:

$$CH_3:^- + (CH_3)(H)C{=}\ddot{O}: \longrightarrow CH_3—\underset{H}{\overset{CH_3}{C}}—\ddot{\underset{..}{O}}:^-$$

If we treat this intermediate with water, we get an alcohol:

$$CH_3—\underset{H}{\overset{CH_3}{C}}—\ddot{\underset{..}{O}}:^- + H—\ddot{\underset{..}{O}}—H \longrightarrow CH_3—\underset{CH_3}{\overset{CH_3}{C}}—\ddot{\underset{..}{O}}—H + :\ddot{\underset{..}{O}}—H^-$$

By reacting a Grignard reagent with formaldehyde, we can add a single carbon atom to form a primary alcohol:

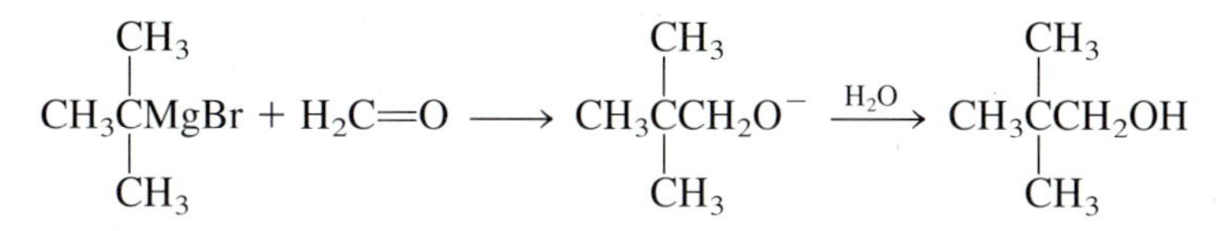

This alcohol then can be oxidized to the corresponding aldehyde:

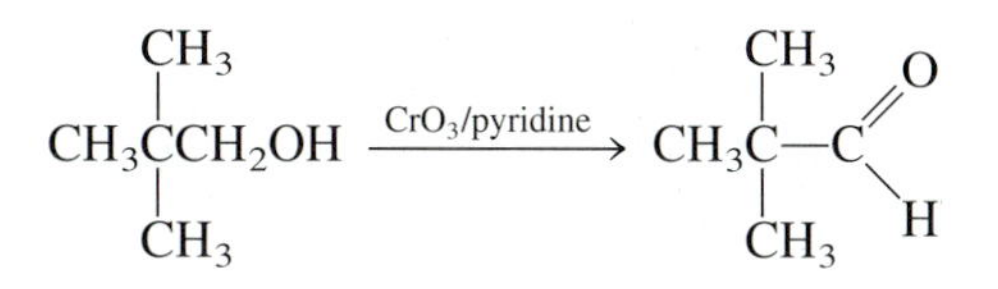

The Grignard reagent therefore provides us with a way of performing the following transformation:

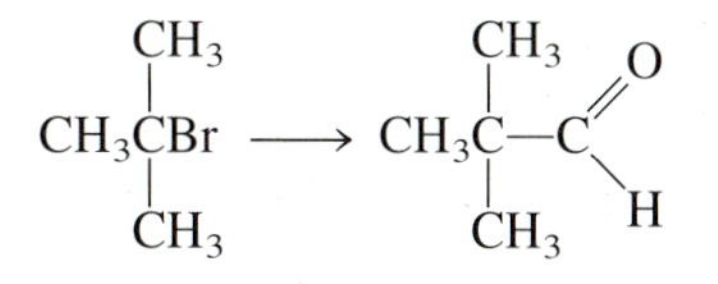

A single carbon atom also can be added if the Grignard reagent is allowed to react with $CO_2$ to form a carboxylic acid:

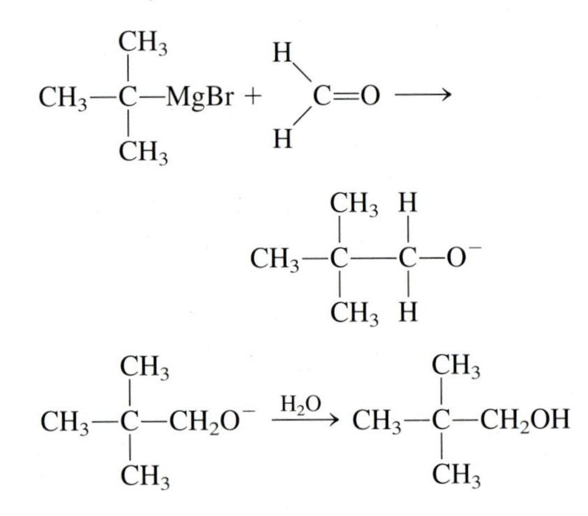

The most important aspect of the chemistry of Grignard reagents is the ease with which this reaction allows us to couple alkyl chains. Isopropylmagnesium bromide, for example, can be used to graft an isopropyl group onto the hydrocarbon chain of an appropriate ketone, as shown in Figure 24.18.

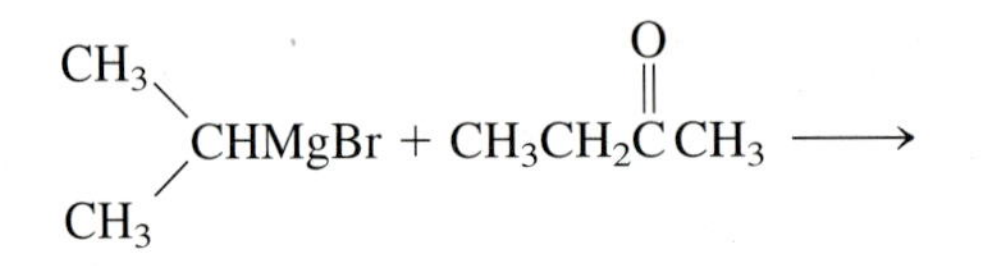

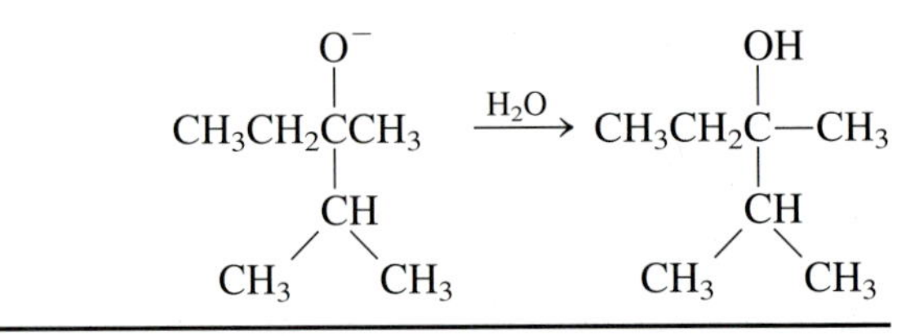

**Figure 24.18** Grignard reagents provide a powerful way of introducing alkyl groups onto a hydrocarbon chain.

## 24.19 OPTICAL ACTIVITY

Section 24.2 described *constitutional isomers* as compounds that have the same formula but differ in the way in which their atoms are joined. Butane and isobutane, for example, are constitutional isomers:

$CH_3CH_2CH_2CH_3$ (butane) $\qquad$ $CH_3\overset{CH_3}{\overset{|}{C}}HCH_3$ (isobutane)

So are ethanol and dimethyl ether:

$CH_3CH_2OH$ (ethanol) $\qquad$ $CH_3OCH_3$ (dimethyl ether)

Constitutional isomers can have very different chemical and physical properties. Ethanol reacts with sodium, but dimethyl ether does not. Ethanol boils at 78.4°C, whereas dimethyl ether boils at −23.7°C.

Sec. 24.3

Section 24.3 defined **stereoisomers** as compounds with similar structures that differ in the way in which the substituents are arranged around a particular functional group. *Cis*- and *trans*-2-butene provide an example of a pair of stereoisomers:

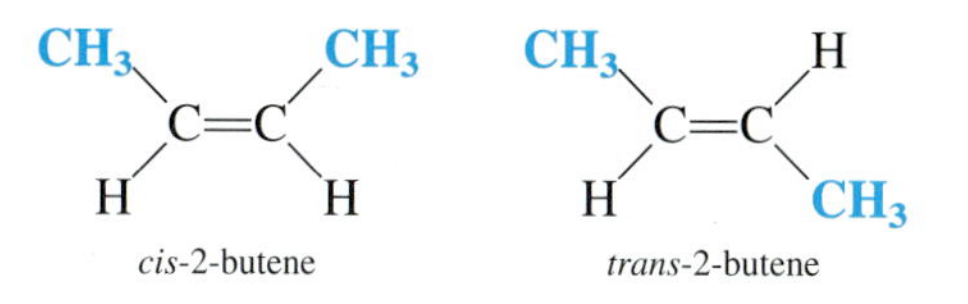


Cis/trans isomers have similar chemical properties, but different physical properties. *cis*-2-Butene, for example, freezes at −138.9°C and *trans*-2-butene freezes at −105.6°C.

This section introduces another form of stereoisomers, that differ only in the way in which they interact with light. Light consists of magnetic and electric fields that oscillate in all directions perpendicular to the path of the light ray, as shown in Figure 24.19. When light is passed through a polarizer, such as the lens of polarized sunglasses, the oscillations are confined to a single plane.

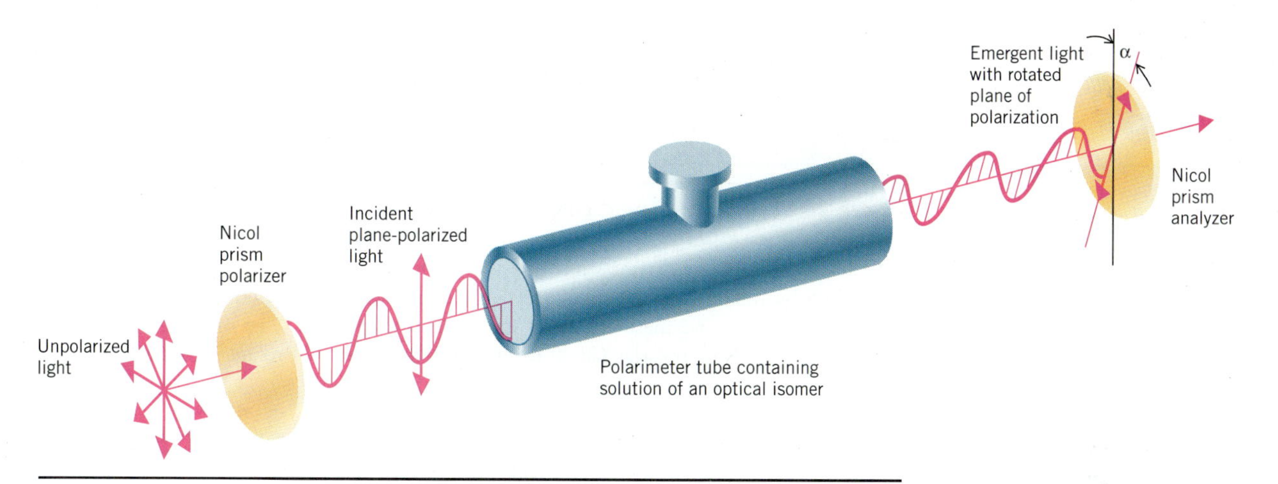


**Figure 24.19** Light consists of magnetic and electric fields that oscillate in all directions perpendicular to the path of the light ray. When light is passed through a polarizer, these oscillations are confined to a single plane. Optically active compounds rotate this plane-polarized light either to the left or to the right.

In 1813 Jean Baptiste Biot noticed that the plane in which polarized light oscillates was rotated either to the right or the left when this light was passed through single crystals of quartz or aqueous solutions of tartaric acid or sugars. Compounds that can rotate plane-polarized light are said to be **optically active.** Those that rotate the plane clockwise (to the right) are said to be **dextrorotatory** (from the Latin *dexter,* ''right''). Those that rotate the plane counterclockwise (to the left) are called **levorotatory** (from the Latin *laevus,* ''left'').

In 1848 Louis Pasteur observed that sodium ammonium tartrate forms two different kinds of crystals that are mirror images of each other, much as the right hand is a mirror image of the left hand (see Figure 24.20). By separating one type of crystal from the other with a pair of forceps he was able to prepare two samples of this compound. One was dextrorotatory when dissolved in aqueous solution, the other was levorotatory. Since the optical activity remained after the compound had been dissolved in water, it could not be the result of macroscopic properties of the crys-

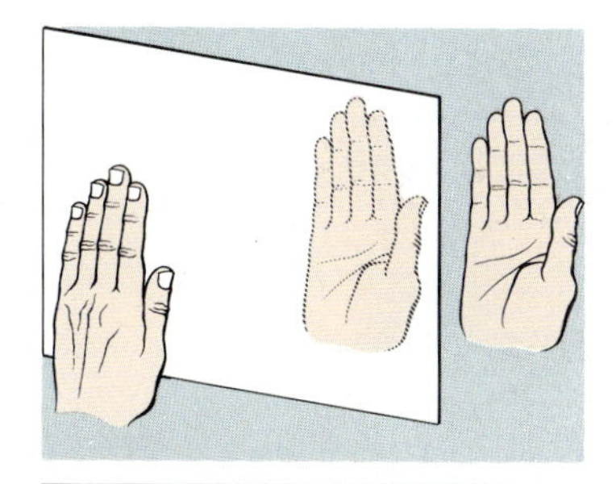

**Figure 24.20** The right hand is the mirror image of the left hand, but neither hand is superimposable on the other. Molecules that have this property are said to be *chiral,* or handed, and they exist as pairs of optical isomers, or stereoisomers.

tals. Pasteur therefore concluded that there must be some asymmetry in the structure of this compound that allowed it to exist in two forms.

In 1874 Jacobus van't Hoff and Joseph Le Bel recognized that a tetrahedral carbon atom with four different substituents could exist in two forms that were mirror images of each other. According to van't Hoff and Le Bel, CHFClBr should be optically active because it contains four different substituents on a tetrahedral carbon atom. Figure 24.21*a* shows one possible arrangement of these substituents and the mirror image of this structure. If we rotate one of these molecules by 180° around the C—H bond we get the structures shown in Figure 24.21*b*. These structures are different, as can be seen in Figure 24.21*c*, because they can't be superimposed on each other.

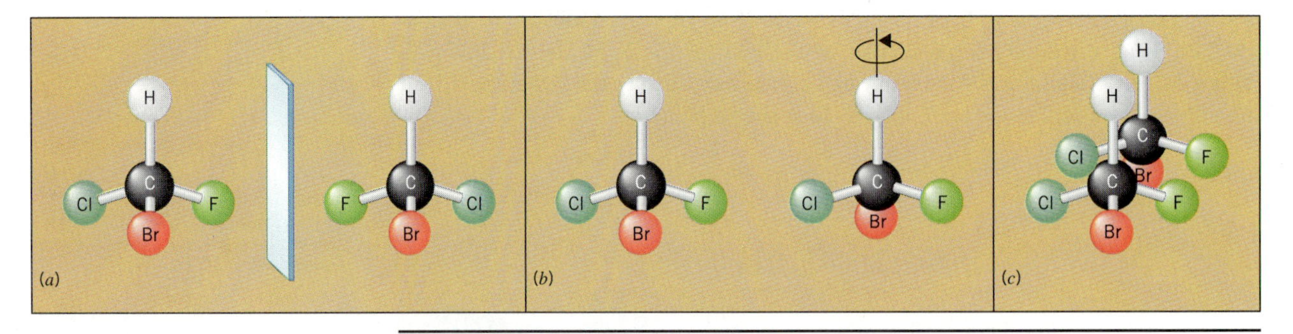


**Figure 24.21** (*a*) This figure shows the structure of one form of the CHFClBr molecule and its mirror image. (*b*) To determine whether a molecule is optically active, we have to decide whether the molecule and its mirror image can be superimposed. We could start by rotating the mirror image by 180° around the C—H bond. (*c*) This compound is chiral because these structures are not superimposable. One of these stereoisomers will rotate plane-polarized light to the right, the other will rotate it to the left.

Sec. 12.10

Any object that is not superimposable on its mirror image is described as **chiral** (from the Greek *cheir,* ''hand''). Hands are chiral. (There is no way to superimpose one hand on the other.) Gloves are also chiral. (Right-hand gloves do not fit on the left hand and vice versa.) Feet and shoes are both chiral, but socks are not.

Compounds, such as CHFClBr, that are chiral are optically active. Chirality is a property of a molecule that results from its structure. Optical activity is a macroscopic property of a collection of these molecules that arises from the way they interact with light. To decide whether a compound should be optically active, we look for centers of chirality—carbon atoms that have four different substituents.

## RESEARCH IN THE 90s

### THE CHEMISTRY OF GARLIC

The volatile materials that could be distilled from plants were named *essential oils* by Paracelsus in the sixteenth century because they were thought to be the quintessence (literally, the ''fifth essence,'' or vital principle) responsible for the odor and flavor of the plants from which they were isolated.

The Egyptians extracted essentials oils from fragrant herbs more than 5000 years ago by pressing the herb or by extracting the fragrant material with olive or palm oil. Some essential oils—such as those isolated from citrus fruit—are still obtained by pressing. Others—including those associated with flowers—are extracted into a nonpolar solvent, such as a petroleum fraction. The most common method for isolating essential oils, however, is steam distillation. The ground botanical is immersed in water which is heated to boiling, or boiling water is allowed to pass through a sample of the ground botanical. The oil and water vapor pass into a condenser, where the oil separates from the water vapor.

The function of essential oils is not fully understood. Some act as attractants for the insects involved in pollination. Most are either bacteriostats or bactericides. In some cases, they can be a source of metabolic energy. In other cases, they appear to be by-products of plant metabolism.

The essential oils are mixtures of up to 200 organic compounds, many of which are either terpenes (with 10 carbon atoms) or sesquiterpenes (with 15 carbon atoms). Although the three compounds shown in Figure 24.22 represent almost 60% of the mass of a sample of rose oil, 50 other components of this essential oil have been identified.

Sec. 24.5

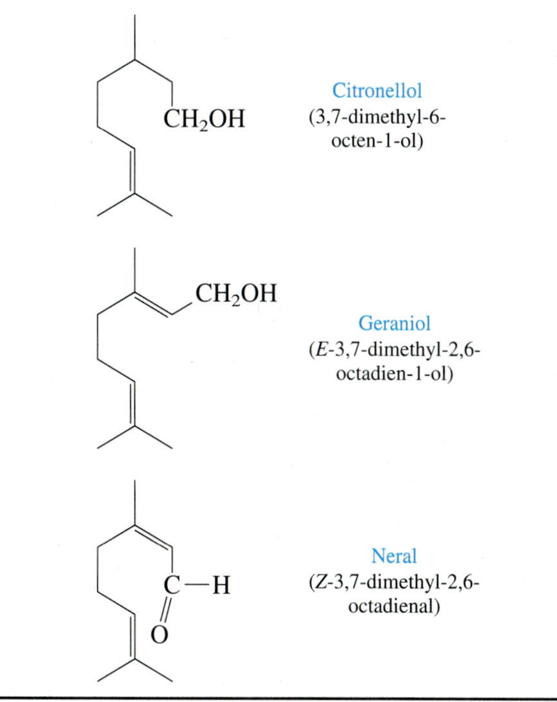


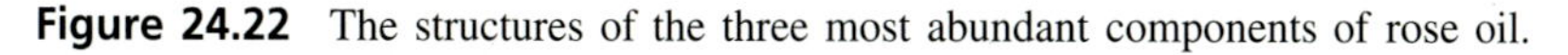

**Figure 24.22** The structures of the three most abundant components of rose oil.

Garlic, onions, and mustard seed differ from most other sources of essential oils. Unlike the scent of a flower, which is continuously released by the bloom in these three cases, the fragrance-producing part of the plant must be crushed before the volatile components are released. For more than 100 years, chemists have known that the principal component of the oil that distills from garlic is diallyl disulfide [F. W. Semmler, *Archiv der Pharmazie,* **230,** 434–448 (1892)]:

$$CH_2{=}CH{-}CH_2{-}S{-}S{-}CH_2{-}CH{=}CH_2$$

diallyl disulfide

Only recently, however, have we learned how this compound is produced when a clove of garlic is crushed [E. Block, *Angewandte Chemie, International Edition in English,* **31,** 1101–1264 (1992)].

Before garlic is crushed, the intact cell contains *S*-2-propenyl-L-cysteine *S*-oxide, or *alliin,* which can be found in the cell cytoplasm:

$$CH_2{=}CH{-}CH_2{-}\overset{\overset{\displaystyle O}{\|}}{S}{-}CH_2{-}\overset{\overset{\displaystyle NH_3^+}{|}}{C}H{-}CO_2^-$$

alliin

Within the cell there are vacuoles that contain an enzyme known as *alliinase.* When the cell is crushed, the enzyme is released. The enzyme transforms the natural product alliin into an intermediate that reacts with itself to form a compound known as *allicin:*

$$CH_2{=}CH{-}CH_2{-}\overset{\overset{\displaystyle O}{\|}}{S}{-}S{-}CH_2{-}CH{=}CH_2$$

allicin

Allicin has been described as an odoriferous, unstable, antibacterial substance that polymerizes easily and must be stored at low temperatures. When heated, it breaks down to give a variety of compounds, including the diallyl disulfide obtained when oil of garlic is distilled from the raw material.

An allinase enzyme also can be found in onions, where it converts an isomer of alliin known as *S*-(*E*)-1-propenyl-L-cysteine *S*-oxide into propanethial *S*-oxide:

$$CH_3{-}CH_2{-}CH{=}S^+{-}O^-$$

propanethial *S*-oxide

The product of this reaction is known as the lachrymator factor of onion because it is the substance primarily responsible for the tears generated when onions are cut.

A great deal of progress has been made in recent years in identifying the various organosulfur compounds formed when garlic or onion are cut and in understanding the process by which these compounds are formed. The structures of some of the principal organosulfur compounds associated with garlic are shown in Figure 24.23. In spite of the progress made so far, much still has to be learned about the compounds that can be isolated from the extracts of the genus *Allium,* which includes both garlic and onion.

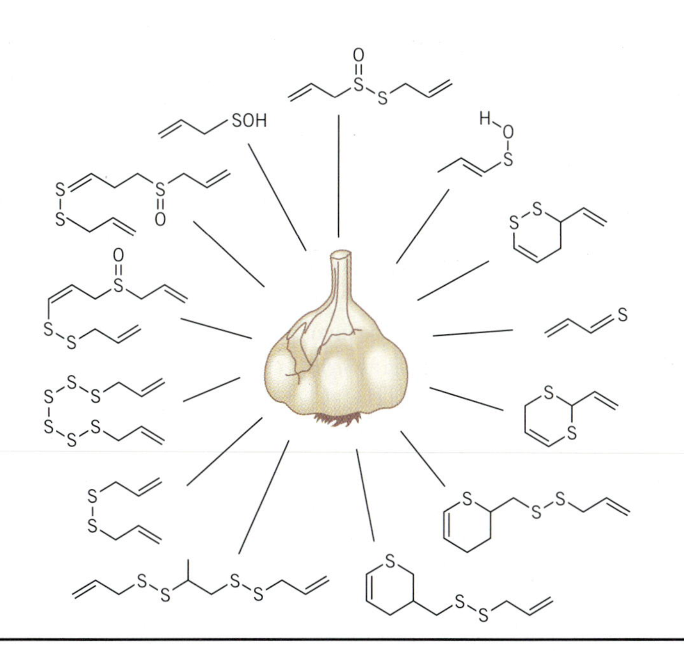

**Figure 24.23** Structures of some of the organosulfur compounds associated with garlic.

## SUMMARY

**1.** Organic chemistry is best defined as the chemistry of compounds that contain both carbon and hydrogen. The chemistry of these compounds is organized around functional groups such as alkanes, alkenes, alkynes, alkyl halides, alcohols, ethers, aldehydes, ketones, carboxylic acids, acyl chlorides, esters, and amides.

**2.** **Alkanes** contain only C—C and C—H single bonds. They are often described as saturated hydrocarbons because they contain as many hydrogen atoms as possible. Unsaturated hydrocarbons include **alkenes** (which contain C═C double bonds), **alkynes** (with C≡C triple bonds), and **aromatic compounds.**

**3.** Constitutional isomers, such as 1-butene and 2-butene or ethanol and dimethyl ether, have the same formula, but they contain different groups of atoms. Stereoisomers, such as *cis*-2-butene and *trans*-2-butene, differ in the geometries around the functional groups. When a compound can't be superimposed on its mirror image, it is said to be chiral, or optically active; it exists as a pair of stereoisomers.

**4.** Organic reactions can be divided into two categories: **metathesis reactions** (such as nucleophilic substitution reactions) and **oxidation–reduction reactions.** Because it is time consuming to assign oxidation numbers to the individual carbon atoms in complex organic compounds, it is useful to remember the following general rules: (1) adding hydrogen atoms to a compound corresponds to the reduction of the carbon atoms, and (2) adding oxygen atoms to a compound involves oxidation of the carbon atom.

## KEY TERMS

**Acyl chloride** (p. 916)
**Addition reaction** (p. 892)
**Alcohol** (p. 905)
**Aldehyde** (p. 912)
**Alkane** (p. 885)
**Alkene** (p. 890)
**Alkoxide** (p. 908)
**Alkyl halide** (p. 892)
**Alkyne** (p. 892)
**Amide** (p. 920)

**Amine** (p. 917)
**Aromatic compound** (p. 895)
**Carbanion** (p. 921)
**Carbonium ion** (p. 909)
**Carbonyl group** (p. 912)
**Carboxylate ion** (p. 914)
**Carboxylic acid** (p. 914)
**Carboxylic acid esters** (p. 917)
**Chiral compound** (p. 924)
**Cis/trans isomers** (p. 891)
**Constitutional isomer** (p. 887)
**Cycloalkane** (p. 888)
**Dextrorotatory** (p. 923)
**Electrophile** (p. 913)
**Ester** (p. 916)
**Ether** (p. 908)
**Functional group** (p. 903)
**Grignard reagent** (p. 921)
**Hydrocarbon** (p. 885)
**Ketone** (p. 912)
**Levorotatory** (p. 923)
**Nucleophile** (pp. 909, 913)
**Optical activity** (p. 923)
**Oxidation reduction** (p. 911)
**Resonance hybrid** (p. 895)
**$S_N1$** (p. 909)
**$S_N2$** (p. 909)
**Stereoisomer** (pp. 891, 923)
**Substitution reaction** (p. 908)
**Theory of resonance** (p. 895)

## PROBLEMS

### The Special Role of Carbon

**24-1** Explain why carbon forms covalent bonds, not ionic bonds, with so many other elements.

**24-2** Explain why carbon forms relatively strong double bonds, not only with itself, but with other nonmetals such as nitrogen, oxygen, phosphorus, and sulfur.

### The Saturated Hydrocarbons: Alkanes and Cycloalkanes

**24-3** Define the terms: *alkane, hydrocarbon, saturated hydrocarbon, straight-chain hydrocarbon, branched hydrocarbon,* and *cyclic hydrocarbon.* Give an example of each.

**24-4** Use the fact that straight-chain alkanes have a $CH_3$— group at either end and a chain of —$CH_2$— groups down the middle to explain why alkanes have the generic formula $C_nH_{2n+2}$. Write the generic formulas for cycloalkanes, alkenes, and alkynes.

**24-5** Describe the difference between *n*-pentane, isopentane, and neopentane. Classify these compounds as either stereoisomers or constitutional isomers.

**24-6** Predict the number of constitutional isomers of heptane, $C_7H_{16}$.

**24-7** Explain why the melting point and boiling point of hydrocarbons increases with molecular weight.

**24-8** Write the molecular formula for the saturated hydrocarbon that has the following carbon skeleton and name this compound:

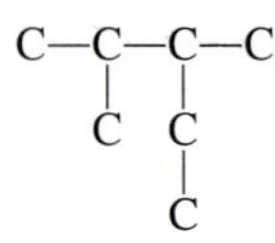

**24-9** Write the molecular formula for the saturated hydrocarbon that has the following carbon skeleton and name this compound:

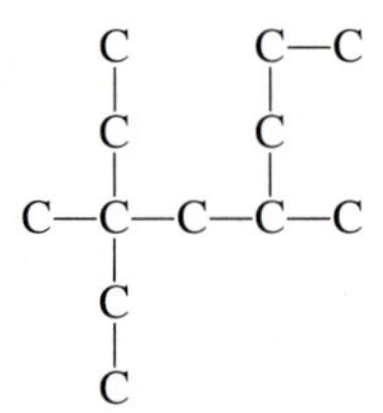

**24-10** Provide the systematic (IUPAC-approved) name for the following compound.

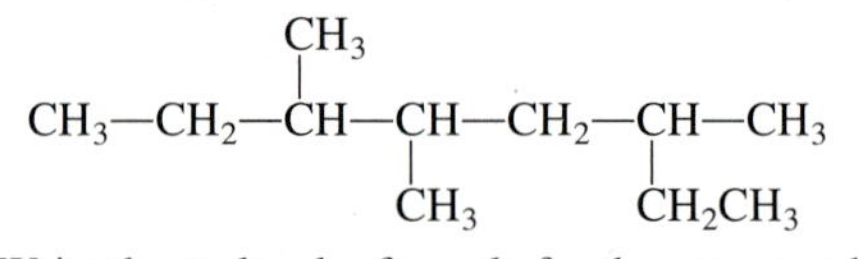

**24-11** Write the molecular formula for the compound that has the following skeleton structure and name this compound:

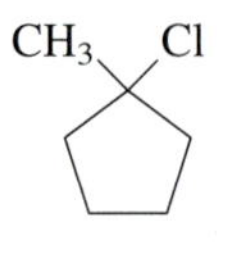


### The Unsaturated Hydrocarbons: Alkenes and Alkynes

**24-12** Define the terms: *alkene, alkyne, olefin,* and *unsaturated hydrocarbon.* Give an example of each.

**24-13** Draw the structures of the alkenes that have the formula $C_6H_{12}$ and name these compounds.

**24-14** Draw the structures of the alkynes that have the formula $C_5H_8$ and name these compounds.

**24-15** Use the VSEPR theory to predict the shape of the ethylene ($C_2H_4$) molecule.

**24-16** Explain why alkenes can form both constitutional and stereoisomers.

**24-17** Explain why alkynes can have constitutional isomers, but not stereoisomers.

**24-18** Describe the hybridization of each of the carbon atoms in the following compound:

$$CH_3{-}CH{=}CH{-}CH_2{-}C{\equiv}CH$$

**24-19** Which of the following compounds does not have the same molecular formula as the others?

(a) cyclopentane (b) methylcyclobutane (c) 1,1-dimethylcyclopropane (d) 1-pentene (e) pentane

**24-20** Which of the following compounds have cis/trans isomers?

(a) $CHCl_3$ (b) $F_2C{=}CF_2$ (c) $Cl_2C{=}CHCH_3$
(d) $FClC{=}CFCl$ (e) $H_2C{=}CHF$

**24-21** Which of the following compounds have cis/trans isomers?

(a) 1-pentene (b) 2-pentene (c) 2-methyl-2-butene
(d) 1-chloro-2-butene (e) 2-pentyne

**24-22** Which of the following pairs of compounds would be constitutional isomers?

(a) $CH_3CH_2OCH_2CH_3$ and $CH_3CH_2CH_2CH_2OH$

(b) $(CH_3)_2CHCH_3$ and $CH_3CH_2CH_2CH_3$

```
                                CH3
                                |
(c) CH3—CH2—CH2—CH3 and CH2—CH2
                                     |
                                     CH3
```

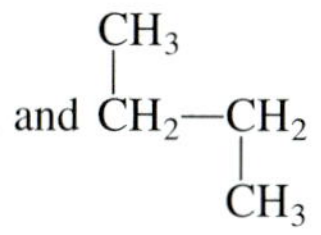

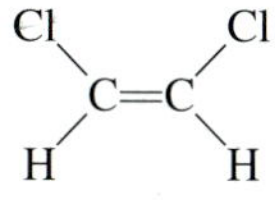

```
      Cl       Cl            Cl       H
        \     /                \     /
(d)      C=C         and        C=C
        /     \                /     \
      H        H             H        Cl
```

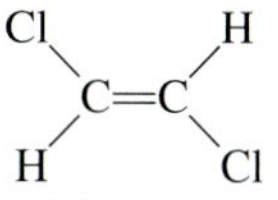

**24-23** Write the molecular formula for the compound that has the following skeleton structure and name this compound:

```
C—C=C—C—C
   |  |
   C  C
```

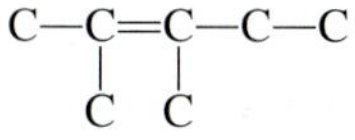

**24-24** Write the molecular formula for the compound that has the following skeleton structure and name this compound:

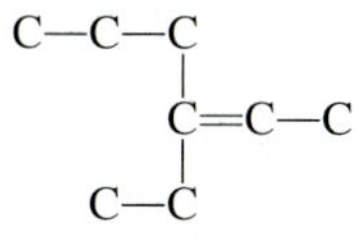

**24-25** Write the molecular formula for the compound that has the following skeleton structure and name this compound:

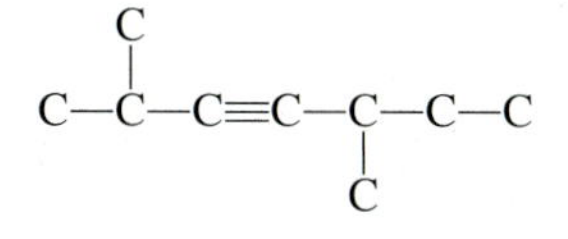

### The Reactions of Alkanes, Alkenes, and Alkynes

**24-26** Write a balanced equation for the reaction in which isooctane, $C_8H_{18}$, burns in oxygen to form $CO_2$ and $H_2O$ vapor.

**24-27** Define the term *addition reaction* and give an example of an addition reaction of an alkene.

**24-28** Predict the product of the addition reaction between $Br_2$ and 3-pentene.

**24-29** Predict the product of the addition reaction between water and 2-butene in the presence of sulfuric acid.

**24-30** Predict the product of the addition reaction between $H_2$ and 3-pentyne in the presence of a platinum metal catalyst.

**24-31** Which of the following is a product of the reaction between chlorine and ethylene?

(a) $CH_3CHCl_2$ (b) $CH_3CH_2Cl$ (c) $ClCH_2CH_2Cl$
(d) $Cl_2CHCHCl_2$

**24-32** Which of the following does not rapidly react with $Br_2$ dissolved in $CCl_4$?

(a) pentane (b) 1-pentene (c) 2-pentene
(d) 1-pentyne (e) 2-pentyne

### The Aromatic Hydrocarbons and Their Derivatives

**24-33** Draw the structures of the following aromatic compounds: aniline, anisole, benzene, and benzoic acid.

**24-34** Draw the structures of the *ortho*-, *meta*-, and *para*-isomers of bromotoluene.

**24-35** TNT is an abbreviation for 2,4,6-trinitrotoluene. Toluene is a derivative of benzene in which a methyl ($CH_3$—) group is substituted for one of the hydrogen atoms. Trinitrotoluene is a derivative of toluene in which —$NO_2$ groups have replaced three more hydrogen atoms on the benzene ring. Draw the structures of all the possible isomers of trinitrotoluene. Label the isomer that is 2,4,6-trinitrotoluene.

**24-36** It might be assumed that benzene is a mixture of two structures that are rapidly being converted from one to the other. What is wrong with this assumption?

### The Chemistry of Petroleum Products and Coal

**24-37** Explain why a mixture of CO and $H_2$ can be used as a fuel.

**24-38** Natural gas, petroleum ether, gasoline, kerosene, and asphalt are all different forms of hydrocarbons that give off energy when burned. Describe how these substances differ. What happens to the boiling points of these mixtures as the average length of the hydrocarbon chain increases?

**24-39** Which of the following compounds has the largest octane number?

(a) *n*-butane (b) *n*-pentane (c) *n*-hexane
(d) *n*-octane

**24-40** Which of the following won't increase the octane number of gasoline?

(a) Increasing the concentration of branched-chain alkanes (b) Increasing the concentration of cycloalkanes (c) Increasing the concentration of aromatic hydrocarbons (d) Increasing the average length of the hydrocarbon chains

**24-41** Describe how thermal cracking and catalytic cracking increase the amount of high-octane gasoline that can be obtained from a barrel of oil. Explain why neither thermal reforming nor catalytic reforming can achieve this.

**24-42** Coal gas can be obtained when coal is heated in the absence of air. Water gas can be obtained when coal reacts with steam. Describe the difference between these gases and explain why much more water gas can be extracted from a ton of coal.

### Functional Groups

**24-43** Give examples of compounds that contain each of the following functional groups.

(a) an alcohol (b) an aldehyde (c) an amine (d) an amide (e) an alkyl halide (f) an alkene (g) an alkyne

**24-44** Describe the difference between the members of each of the following pairs of functional groups.

(a) An alcohol and an alkoxide ion (b) An alcohol and an aldehyde (c) An amine and an amide (d) An ether and an ester (e) An aldehyde and a ketone

**24-45** Classify each of the following compounds as an alcohol, an aldehyde, an ether, or a ketone:

(a) $CH_3CH_2\overset{O}{\overset{\|}{C}}H$

(b) $CH_3CH_2OH$

(c) $CH_3CH_2OCH_2CH_3$

(d) $CH_3CH_2\overset{O}{\overset{\|}{C}}CH_3$

**24-46** Classify each of the following compounds as a primary, secondary, or tertiary alcohol:

(a) $CH_3CH_2OH$ (b) $CH_3CH_2\overset{CH_3}{\overset{|}{C}}HOH$ (c) $CH_3CH_2\underset{CH_3}{\underset{|}{\overset{CH_3}{\overset{|}{C}}}}OH$

(d) $CH_3CH_2\overset{OH}{\overset{|}{C}}HCH_2CH_3$

**24-47** Cortisone is an adrenocortical steroid, which is used as an antiinflammatory agent. Use the fact that carbon is tetravalent to determine the molecular formula of this compound from the following stick structure. Identify the functional groups present in this molecule.

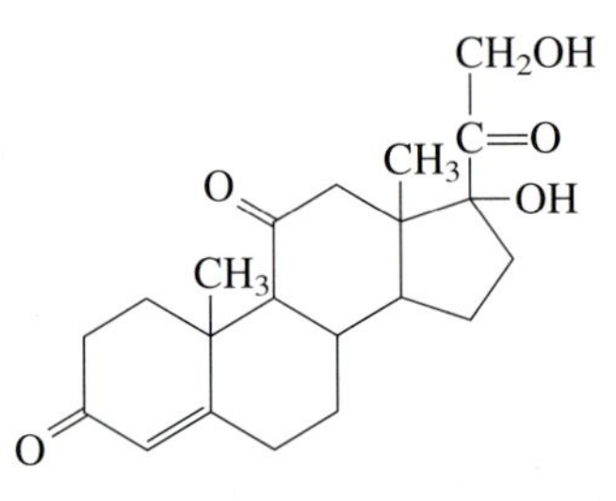


**24-48** Piperine [(*E,E*)-1-[5-(1,3-benzodioxol-5-yl)-1-oxo-2,4-pentadienyl]piperidine] can be extracted from black pepper. Identify the functional groups in this compound.

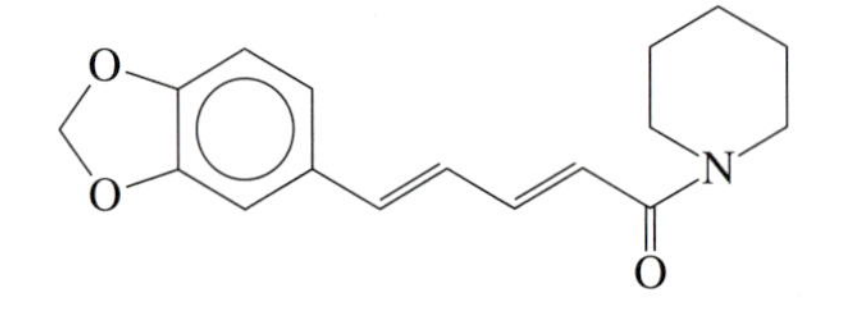


### Alkyl Halides

**24-49** Describe the free-radical chain-reaction mechanism involved in the chlorination of methane to form methyl chloride. Clearly label the chain-initiation, chain-propagation, and chain-termination steps.

**24-50** Use Lewis structures to describe the difference between a $CH_3^+$ ion, a $CH_3\cdot$ free radical, and a $CH_3^-$ ion.

### Alcohols and Ethers

**24-51** Describe the differences between the structures of water, methyl alcohol, and dimethyl ether.

**24-52** Draw the structures of the seven constitutional isomers that have the formula $C_4H_{10}O$. Classify these isomers as either alcohols or ethers.

**24-53** Ethyl alcohol and dimethyl ether have the same chemical formula, $C_2H_6O$. Explain why one of these compounds reacts rapidly with sodium metal but the other does not.

**24-54** Predict the product of the dehydration of ethyl alcohol with sulfuric acid at 130°C.

**24-55** Describe the differences between primary alcohols, secondary alcohols, tertiary alcohols, and phenols. Give an example of each.

**24-56** Draw the structure of *o*-phenylphenol, the active ingredient in Lysol.

**24-57** Explain why alcohols are Brønsted acids in water.

**24-58** Write the structures of the conjugate bases formed when the following alcohols act as a Brønsted acid.

(a) $CH_3CH_2OH$ (ethyl alcohol) (b) $C_6H_5OH$ (phenol)
(c) $(CH_3)_2CHOH$ (isopropyl alcohol)

**24-59** Use the fact that there are no hydrogen bonds between ether molecules to explain why diethyl ether ($C_4H_{10}O$) boils at 34.5°C, whereas its constitutional isomer, 1-butanol ($C_4H_{10}O$), boils at 118°C.

### Substitution Reactions

**24-60** Define the term *substitution reaction* and describe the difference between unimolecular ($S_N1$) and bimolecular ($S_N2$) substitution reactions.

**24-61** Write the mechanism of an $S_N2$ reaction and explain why there is no difference between what organic chemists call a *nucleophile* and what inorganic chemists call a *Lewis base.*

**24-62** Which of the following would not be classified as a *nucleophile?*

(a) $H_2O$ (b) $CH_3O^-$ (c) $NH_4^+$ (d) $CN^-$ (e) $I^-$

**24-63** Assume that the following reaction occurs by an $S_N1$ mechanism:

$$CH_3CH_2\underset{\underset{Br}{|}}{C}HCH_3 + OH^- \longrightarrow CH_3CH_2\underset{\underset{OH}{|}}{C}HCH_3 + Br^-$$

What would be the effect on the rate of reaction of simultaneously increasing the concentration of the $OH^-$ ion by a factor of two and decreasing the concentration of the alkyl halide by a factor of two?

### Oxidation–Reduction Reactions

**24-64** Arrange the following in order of increasing oxidation number of the carbon atom:

(a) C (b) HCHO (c) $HCO_2H$ (d) CO (e) $CO_2$
(f) $CH_4$ (g) $CH_3OH$

**24-65** Classify the following reactions as either metathesis (acid–base) or oxidation–reduction reactions:

(a) $CH_4 + Cl_2 \longrightarrow CH_3Cl + HCl$

(b) $CH_3OH \longrightarrow HCHO$

(c) $HCHO \longrightarrow HCO_2H$

(d) $CH_3OH + HBr \longrightarrow CH_3Br + H_2O$

(e) $CH_3\overset{\overset{O}{\|}}{C}CH_3 \longrightarrow CH_3\overset{\overset{OH}{|}}{C}HCH_3$

**24-66** Classify the following transformations as either oxidation or reduction:

(a) $CH_3CH_2OH \longrightarrow CH_3CHO$

(b) $CH_3CHO \longrightarrow CH_3CO_2H$

(c) $(CH_3)_2C{=}O \longrightarrow (CH_3)_2CHOH$

(d) $CH_3CH{=}CHCH_3 \longrightarrow CH_3CH_2CH_2CH_3$

(e) $(CH_3)_2CHC{\equiv}CH \longrightarrow (CH_3)_2CHCH_2CH_3$

**24-67** Which of the following can be oxidized to form an aldehyde?

(a) $CH_3CH_2OH$ (b) $CH_3CHOHCH_3$ (c) $CH_3OCH_3$
(d) $(CH_3)_2C{=}O$

**24-68** Which of the following should be the most difficult to oxidize?

(a) $CH_3CH_2OH$ (b) $(CH_3)_2CHOH$ (c) $(CH_3)_3COH$
(d) $CH_3CHO$

**24-69** Which of the following can be prepared by reducing $CH_3CHO$?

(a) $CH_3CH_3$ (b) $CH_3CH_2OH$ (c) $CH_3CO_2H$
(d) $CH_3CH_2CO_2H$

**24-70** Which of the following can be prepared by oxidizing of $CH_3CHO$?

(a) $CH_3CH_3$ (b) $CH_3CH_2OH$ (c) $CH_3CO_2H$
(d) $CH_3CH_2CO_2H$

**24-71** Predict the product of the oxidation of 2-methyl-3-pentanol.

### Aldehydes and Ketones

**24-72** Compounds that react with the electron-rich $\delta^-$ end of a carbonyl group are called *electrophiles.* Use Lewis structures to show that there is no difference between an electrophile and a Lewis acid.

**24-73** At which end of a carbonyl group will the following substances react?

(a) $H^+$ (b) $H^-$ (c) $OH^-$ (d) $NH_3$ (e) $BF_3$

**24-74** Explain why mild oxidation of a primary alcohol gives an aldehyde, whereas oxidation of a secondary alcohol gives a ketone.

**24-75** Explain why strong oxidizing agents can't be used to convert a primary alcohol to an aldehyde.

### Carboxylic Acids, Carboxylate Ions, and Carboxylic Acid Esters

**24-76** Explain the difference between a carboxylic acid, a carboxylate ion, and a carboxylic acid ester.

**24-77** What major differences between carboxylic acids (such as butyric acid, $CH_3CH_2CH_2CO_2H$) and esters (such as ethyl butyrate, $CH_3CH_2CH_2CO_2CH_2CH_3$) may help explain why butyric acid gives rise to the odor of rotten meat but ethyl butyrate gives rise to the pleasant odor of pineapple?

**24-78** Explain why fats, which are esters of long-chain carboxylic acids, are insoluble in water. Explain what happens during saponification that makes the resulting compounds marginally soluble in water.

### Amines, Alkaloids, and Amides

**24-79** Classify each of the compounds in Figures 24.15 and 24.16 as either a primary amine, a secondary amine, a tertiary amine, and/or an amide.

### Grignard Reagents

**24-80** Use Lewis structures to decide whether a Grignard reagent is a nucleophile (Lewis base) or an electrophile (Lewis acid). Predict which end of a C=O bond the Grignard reagent will attack.

**24-81** If $CH_3CH_2MgBr$ can be thought of as containing the $CH_3CH_2^-$, $Mg^{2+}$, and $Br^-$ ions, what would be the product of the reaction between this Grignard reagent and dilute aqueous hydrochloric acid?

**24-82** Write a step-by-step mechanism for the reaction between $CH_3MgBr$ and $H_2C{=}O$ to form $CH_3CH_2OH$.

**24-83** Predict the product of the reaction between $(CH_3)_2CHMgBr$ and $CH_3CH_2COCH_3$.

**24-84** If $CH_3CH_2OH$ is used as the only source of carbon, which of the following can be synthesized by a Grignard reaction?

(a) $CH_3CH_2CH_2OH$ (b) $CH_3CH_2CHO$
(c) $CH_3CH_2OCH_2CH_3$ (d) $CH_3CO_2CH_2CH_3$
(e) $CH_3CH_2CH_2CO_2CH_2CH_3$

**24-85** Which of the following reagents could be used to synthesize the following compound?

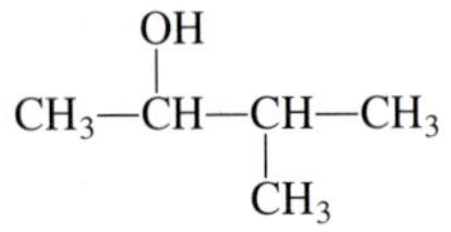

$$CH_3{-}\underset{}{\overset{\displaystyle OH}{\overset{|}{C}}}H{-}\underset{\underset{\displaystyle CH_3}{|}}{C}H{-}CH_3$$

(a) $(CH_3)_2CHMgBr$ (b) $CH_3CHO$ (c) $CH_3MgBr$
(d) $CH_3COCH_2CH_3$ (e) $CH_3COCH_3$
(f) $CH_3CH_2MgBr$ (g) $CH_3COCH_3$

### Optical Activity

**24-86** Describe the difference between constitutional isomers and stereoisomers.

**24-87** Objects that cannot be superimposed on their mirror images are said to be *chiral*, and chiral molecules are optically active. Which of the following molecules are optically active?

(a) $CH_4$ (b) $CH_3Cl$ (c) $CHCl_3$ (d) $CHFCl_2$
(e) CHFClBr

**24-88** Which of the following molecules are optically active?

(a) $C_2H_4$ (b) $C_6H_6$ (c) $C_6H_4Cl_2$ (d) 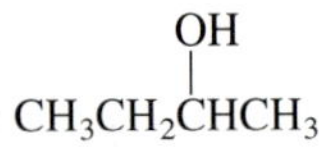 $CH_3CH_2\overset{\displaystyle OH}{\overset{|}{C}}HCH_3$

**24-89** Use the structure of coniine shown in Figure 24.15 to determine whether this compound is optically active.

**24-90** Use the structure of caffeine shown in Figure 24.17 to predict whether this compound is optically active.

**24-91** Determine the number of chiral carbon atoms in the structure of vitamin C shown in Figure 24.1.

### Qualitative Organic Analysis

**24-92** Describe a way of determining whether a compound is an alkane or an alkene.

**24-93** Describe a way of determining whether a compound is a carboxylic acid or an ester.

**24-94** Describe a way of determining whether a compound is an alcohol or an ether.

**24-95** Describe a way of determining whether a compound is a primary or a tertiary alcohol.

**24-96** Describe a way of determining whether a compound is an alcohol or an alkyl halide.

**24-97** Describe a way of determining whether a compound is a carboxylic acid or an amide.

**24-98** Describe a way of determining whether a compound is an alcohol or an amine.

**24-99** Describe a way of determining whether a compound is an aldehyde or ketone.

### Integrated Problems

**24-100** The following Lewis structures were drawn correctly by an organic chemistry student who forgot to indicate whether these molecules carry a positive or a negative charge. Correct his work by specifying whether each molecule is negatively charged, positively charged, or neutral:

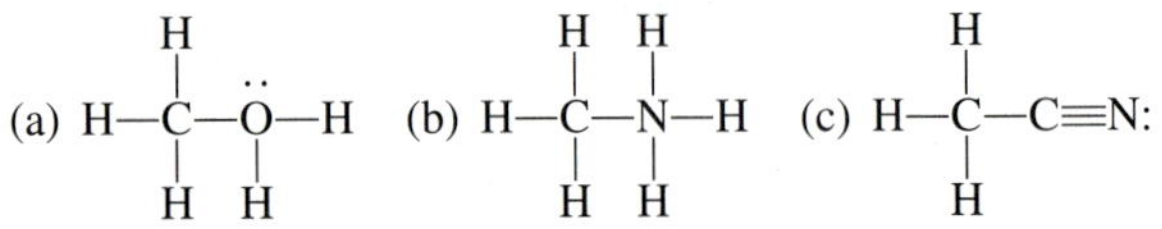

$$\text{(a) } H{-}\underset{\underset{\displaystyle H}{|}}{\overset{\overset{\displaystyle H}{|}}{C}}{-}\underset{\underset{\displaystyle H}{|}}{\dot{\dot{O}}}{-}H \quad \text{(b) } H{-}\underset{\underset{\displaystyle H}{|}}{\overset{\overset{\displaystyle H}{|}}{C}}{-}\underset{\underset{\displaystyle H}{|}}{\overset{\overset{\displaystyle H}{|}}{N}}{-}H \quad \text{(c) } H{-}\underset{\underset{\displaystyle H}{|}}{\overset{\overset{\displaystyle H}{|}}{C}}{-}C{\equiv}N{:}$$

**24-101** Identify the Brønsted acids and the Brønsted bases in the following reaction.

$$CH_3C{\equiv}CH + NH_2^- \longrightarrow CH_3C{\equiv}C^- + NH_3$$

**24-102** Succinic acid plays an important role in the Krebs Cycle, malic acid (*apple acid*) is found in apples, and tartaric acid (*fruit acid*) is found in many fruits. Which of these dicarboxylic acids is chiral? (Hint: See table 24.8 for the structures of these acids.)

**24-103** Predict whether the following compound is chiral:

$$\begin{array}{c} CH_2CH_2CH_2CH_3 \\ | \\ CH_3{-}N^{+}{-}CH_2CH_3 \\ | \\ CH(CH_3)_2 \end{array}$$

**24-104** Which of the following compounds, which play an important role in the chemistry of biological systems, is chiral?

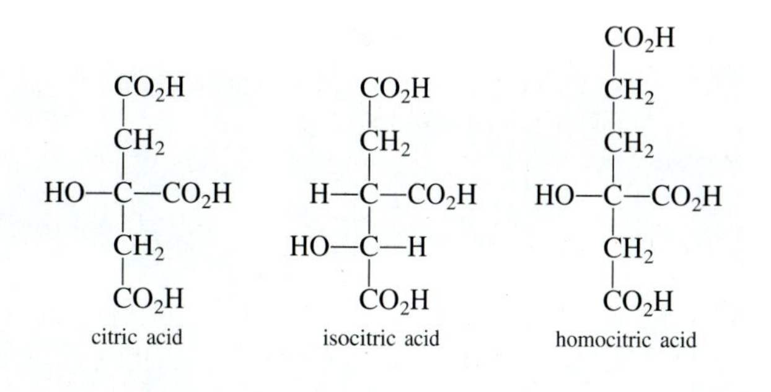

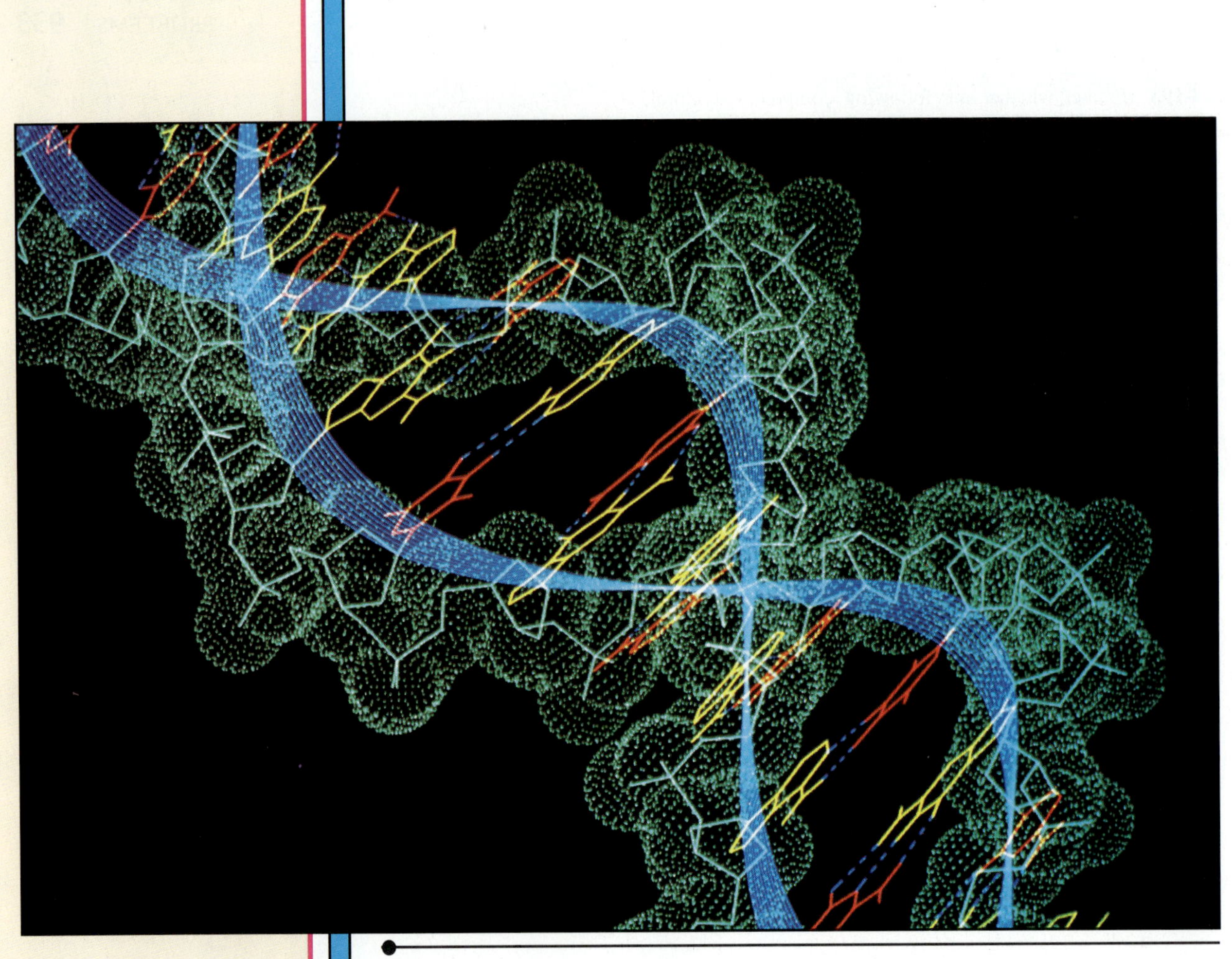

A computer model of the structure of DNA.

CHAPTER 25

# POLYMERS: SYNTHETIC AND NATURAL

This chapter examines how small molecules can be added to a growing chain to form a polymer to provide the basis for answering questions such as the following:

## POINTS OF INTEREST

- What is the difference between the polypropylene used to make milk bottles and the polyethylene in sandwich bags that makes one of these polymers more rigid than the other?
- How does increasing the amount of sulfur in synthetic rubber transform the material used to make automobile tires that absorb energy associated with bumps along the highways, into a material so resilient it bounces to 90% of its initial height when dropped on the floor?
- How can we produce a polymer that absorbs 800 times its weight in distilled water? Why will this polymer absorb only 60 times its weight in urine when it is used in superabsorbent diapers?
- What is the difference between a "fat" and an "oil"? What is the difference between the "oil" we use in an automobile and the "oil" in which we cook food?
- How can thousands of different proteins, with an enormous range of structures and functions, be prepared from a set of only 20 amino acids?
- How can a system of nucleic acids that contains a four-letter alphabet carry enough information to code for the synthesis of the thousands of proteins in a biological system?
- How is the information stored in the structure of nucleic acids used to code for the synthesis of the proteins required for the cell to function?

## 25.1 THE PROCESS OF DISCOVERY: POLYMERS

In 1833 Jöns Jacob Berzelius suggested that compounds with the same formula but different structures should be called *isomers*. He also suggested that compounds with the same empirical formula but different molecular weights should be called **polymers** (literally, ''many parts''). The term *polymer* eventually came to mean compounds with unusually large molecular weights, such as natural rubber. In 1920 Hermann Staudinger proposed the first explanation for why polymer molecules are so heavy. He argued that polymers contain long chains of relatively simple repeating units. Natural rubber, for example, is a polymer that contains large numbers of [$-CH_2C(CH_3){=}CHCH_2-$] units.

The term *rubber* was first used in 1770 by Joseph Priestly to describe the gum from a South American tree that could be used to rub out pencil marks. Natural rubber has only limited uses, however, because it is tacky, strong smelling, perishable, too soft when warm, and too hard when cold. Nathaniel Hayward was the first to note that rubber loses some of its sticky properties when treated with sulfur. It was Charles Goodyear, however, who accidentally dropped a mixture of natural rubber and sulfur onto a hot stove and discovered ''vulcanized'' rubber, which is stable over a wide range of temperatures and far more durable than natural rubber.

The cellulose nitrate that is still used today for ping-pong balls is flammable.

Cellulose is another example of a polymer that contains many copies of a simple repeating unit. Wood is about 50% cellulose by weight; cotton is almost 90% cellulose. Cellulose from wood pulp has been used for centuries to make paper. In the last hundred years, cellulose has also served as the starting material for the synthesis of the first plastics and the first synthetic fibers. Cellulose from wood pulp contains too many impurities to be used to make fibers. It can be purified, however, by dissolving the polymer in a mixture of NaOH and $CS_2$. When this viscous solution is forced through tiny holes in a nozzle into an acid bath, the cellulose fiber known as Rayon is generated. A similar process is used to make a thin film of regenerated cellulose known as *cellophane.*

Each $-C_6H_{10}O_5-$ repeating unit in cellulose contains three $-OH$ groups that can react with nitric acid to form nitrate esters known as *cellulose nitrate.* In 1869 John Wesley Hyatt found that mixtures of cellulose nitrate and camphor dissolve in alcohol to produce a plastic substance he named *celluloid.*

Cellulose nitrate, or celluloid, was used as a substitute for ivory in the manufacture of a variety of items ranging from combs to billiard balls. Cellulose nitrate is extremely flammable, however, and it has been replaced by other plastics for almost all uses except ping-pong balls. No other plastic has been found that has quite the same bounce as celluloid.

## 25.2 DEFINITIONS OF TERMS

### LINEAR, BRANCHED, AND CROSS-LINKED POLYMERS

As we have seen, polymers are compounds with relatively large molecular weights formed by linking together many small monomers. Polyethylene, for example, is formed by *polymerizing* ethylene molecules.

$$\underset{\text{ethylene}}{CH_2{=}CH_2} \longrightarrow \underset{\text{polyethylene}}{[\ldots-CH_2-CH_2-CH_2-CH_2-CH_2-CH_2-\ldots]}$$

Polyethylene is called a *straight-chain polymer* because it consists of a long string of carbon–carbon bonds. This term is somewhat misleading because the geometry around each carbon atom is tetrahedral and the chain is not quite linear, as shown in Figure 25.1. Furthermore, as the chain grows, it folds back on itself in a random fashion to form the structure shown in Figure 25.2.

**Figure 25.1** Each carbon atom in a straight-chain polymer has a tetrahedral geometry.

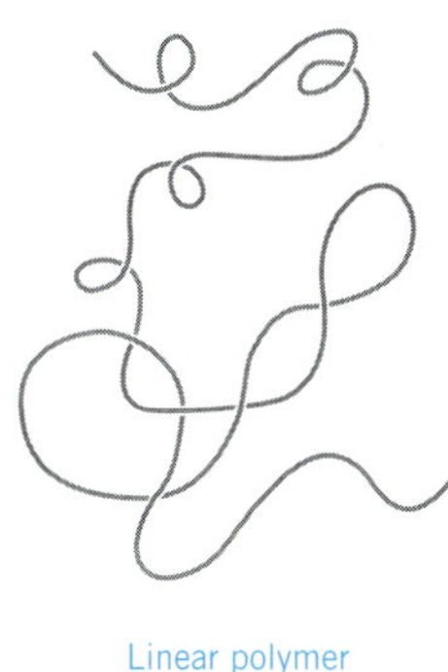


**Figure 25.2** Straight-chain polymers can fold back upon themselves in a random fashion.

Polymers with branches at irregular intervals along the polymer chain are called *branched polymers* (see Figure 25.3). These branches make it difficult for the polymer molecules to pack in a regular array and therefore make the polymer less crystalline.

*Cross-linked polymers* contain branches that connect polymer chains, as shown in Figure 25.4. At first, adding cross-links between polymer chains makes the polymer more elastic. The vulcanization of rubber, for example, results from the introduction of short chains of sulfur atoms that link the polymer chains in the natural rubber. When the number of cross-links is relatively large, however, the polymer becomes more rigid.

Linear and branched polymers form a class of materials known as *thermoplastics.* These materials flow when heated and can be molded into a variety of shapes, which they retain when they cool. Heavy cross-linking produces materials known as *thermoset plastics.* Once the cross-links form, these polymers take on a shape that cannot be changed without destroying the plastic.

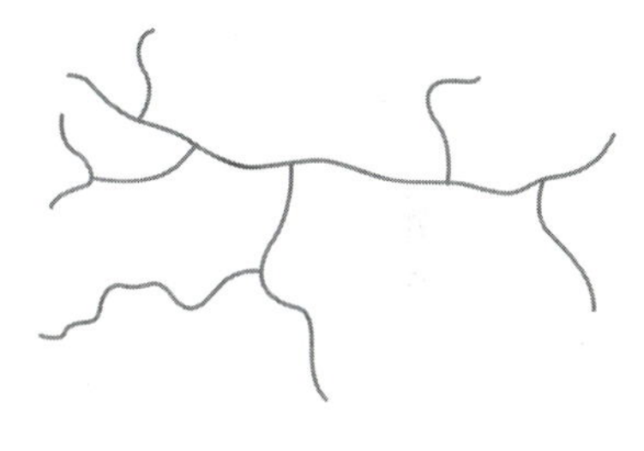


**Figure 25.3** Branched polymers contain short side chains that extend from the main backbone of the polymer.

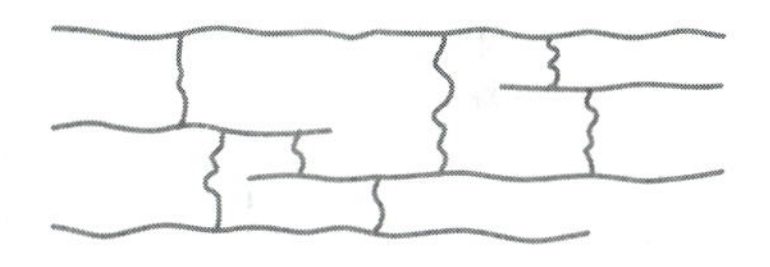


**Figure 25.4** Cross-linked polymers have branches that connect polymer chains.

**EXERCISE 25.1**

Polyethylene can be obtained in two different forms. High-density polyethylene ($0.94\ g/cm^3$) is a linear polymer. Low-density polyethylene ($0.92\ g/cm^3$) is a branched polymer with short side chains on 3% of the atoms along the polymer chain. Explain how the structure of these polymers gives rise to the difference in their densities.

**SOLUTION** Linear polymers are more regular than branched polymers. Linear polymers can therefore pack more tightly, with less wasted space. As a result, linear polymers are slightly more dense than branched polymers.

### HOMOPOLYMERS AND COPOLYMERS

Polyethylene is a *homopolymer,* formed by polymerizing a single monomer. *Copolymers* are formed by polymerizing more than one monomer. Ethylene ($CH_2{=}CH_2$) and propylene ($CH_2{=}CH{-}CH_3$) can be copolymerized, for example, to produce a polymer that has two kinds of repeating units:

$$CH_2{=}CH_2 + CH_2{=}CH{-}CH_3 \longrightarrow$$

$$[\ldots(-CH_2CH_2-)_x(-CH_2-\underset{\displaystyle |}{\overset{\displaystyle CH_3}{}}\!\!\!\!\!\!\!\!\!\!CH-)_y-\ldots]$$

Copolymers are classified on the basis of the way monomers are arranged along the polymer chain, as shown in Figure 25.5. *Random copolymers* contain repeating units arranged in a purely random fashion. *Regular copolymers* contain a sequence of regularly alternating repeating units. The repeating units in *block copolymers* occur in blocks of different lengths. *Graft copolymers* have a chain of one repeating unit grafted onto the backbone of another.

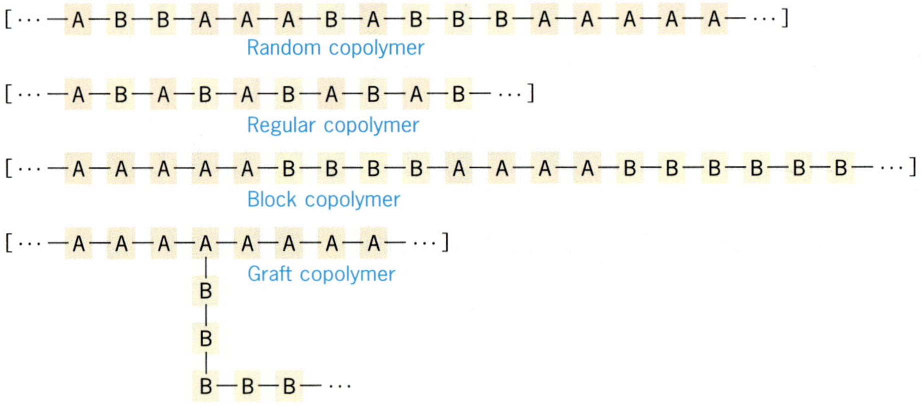

[···—A—B—B—A—A—A—B—A—B—B—B—A—A—A—A—A—···]

Random copolymer

[···—A—B—A—B—A—B—A—B—A—B—···]

Regular copolymer

[···—A—A—A—A—A—B—B—B—B—A—A—A—A—B—B—B—B—B—B—···]

Block copolymer

[···—A—A—A—A—A—A—A—A—···]

(graft branch from the fourth A: B—B—B—B—B—···)

Graft copolymer

**Figure 25.5** Random, regular, block, and graft copolymers.

## TACTICITY

Some monomers form polymers with regular substituents on the polymer chain. These polymers possess a property known as **tacticity** (from the Latin *tacticus,* "fit for arranging"). Tacticity results from the different ways in which these substituents can be arranged on the polymer backbone (see Figure 25.6). When the substituents are arranged in an irregular, random fashion, the polymer is **atactic** (literally, "no

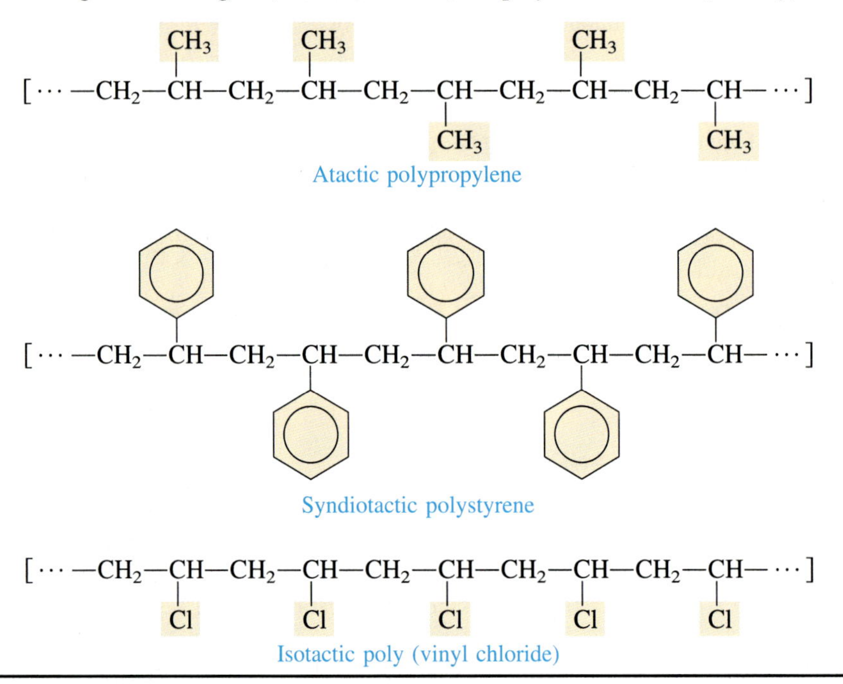


**Figure 25.6** Atactic, syndiotactic, and isotactic polymers.

arrangement''). If the substituents alternate regularly from one side of the chain to the other, the polymer is **syndiotactic.** When the substituents are all on the same side of the chain, the polymer is **isotactic** (literally, ''the same arrangement'').

Plastic chairs made from isotactic polypropylene.

**EXERCISE 25.2**

Atactic polypropylene is a soft, rubbery material with no commercial value. The isotactic polymer is a rigid substance with an excellent resistance to mechanical stress. Explain the difference between these two forms of polypropylene.

**SOLUTION** Isotactic polypropylene is easier to pack in a regular fashion than the atactic form of this polymer. As a result, isotactic polypropylene is more crystalline, which makes the solid more rigid.

## ADDITION VERSUS CONDENSATION POLYMERS

Polyethylene, polypropylene, and poly(vinyl chloride) are examples of **addition polymers,** which are formed by adding monomers to a growing polymer chain. Addition polymers can be recognized by noting that the repeating unit always has the same formula as the monomer from which the polymer is formed:

$$CH_2{=}CH_2 \longrightarrow [\ldots(CH_2CH_2)_n\ldots] \quad \text{polyethylene}$$

$$CH_2{=}CHCH_3 \longrightarrow [\ldots(CH_2\underset{}{\overset{CH_3}{\overset{|}{C}}}H)_n\ldots] \quad \text{polypropylene}$$

$$CH_2{=}CHCl \longrightarrow [\ldots(CH_2\overset{Cl}{\overset{|}{C}}H)_n\ldots] \quad \text{poly(vinyl chloride)}$$

To *condense* means to make something more dense, or compact. Polymers formed when a small molecule condenses out during the polymerization reaction are therefore called **condensation polymers.** $(CH_3)_2Si(OH)_2$ for example, polymerizes to form a condensation polymer known as silicone, as shown in Figure 25.7. Note that the repeating unit in a condensation polymer is inevitably smaller than the monomer from which it is made.

$$H{-}O{-}\underset{CH_3}{\overset{CH_3}{Si}}{-}O{-}H + H{-}O{-}\underset{CH_3}{\overset{CH_3}{Si}}{-}O{-}H + H{-}O{-}\underset{CH_3}{\overset{CH_3}{Si}}{-}O{-}H$$

$$\Big\downarrow -H_2O$$

$$\cdots O{-}\underset{CH_3}{\overset{CH_3}{Si}}{-}O{-}\underset{CH_3}{\overset{CH_3}{Si}}{-}O{-}\underset{CH_3}{\overset{CH_3}{Si}}{-}O\cdots$$

Silicone

**Figure 25.7** Silicone is an example of a condensation polymer. Each time a bond is formed between the monomers that make up the polymer chain, a molecule of water is eliminated.

**EXERCISE 25.3**

Classify the products of the following reactions as either addition or condensation polymers.

(a) poly(methyl methacrylate), sold as Lucite or Plexiglass:

$$CH_2{=}C(CH_3)CO_2CH_3 \longrightarrow [\ldots(-CH_2-C(CH_3)(CO_2CH_3)-)_n\ldots]$$

(b) nylon 6:

$$H_2N(CH_2)_5\overset{O}{\overset{\|}{C}}Cl \longrightarrow [\ldots(-HN(CH_2)_5\overset{O}{\overset{\|}{C}}-)_n\ldots]$$

**SOLUTION**

(a) Poly(methyl methacrylate) is an addition polymer because every atom in the monomer ends up in the repeating unit of the polymer.

(b) Nylon 6 is a condensation polymer because a molecule of HCl condenses from this polymer each time a monomer is added to the chain.

## 25.3 ELASTOMERS

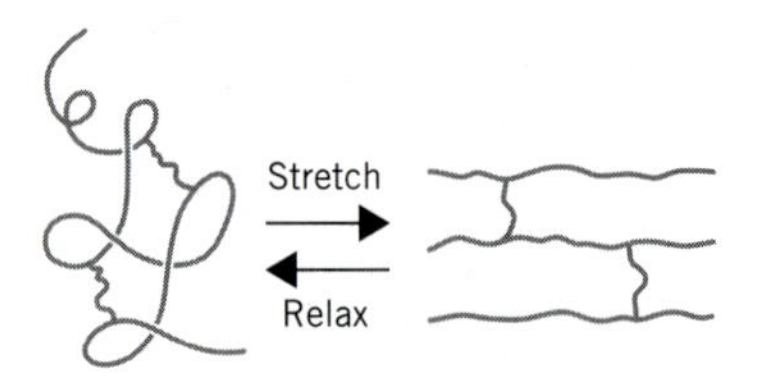


**Figure 25.8** The presence of a few cross-links between polymer chains keeps the chains from slipping past each other when an elastomer is stretched. If the force of attraction between the polymer chains is relatively small, they will coil back into a random shape when the tension is released.

**Elastomers** are polymers that have the characteristic properties of rubber: They are both flexible and elastic. To be elastic, a polymer must meet the following criteria:

- It must contain long, flexible molecules that are coiled in the natural state, which can be stretched without breaking, as shown in Figure 25.8.
- It must contain a few cross-links between polymer chains so that one chain does not slip past another when the substance is stretched.
- It cannot contain too many cross-links, or else it would be too rigid to stretch.
- The intermolecular force of attraction between chains must be relatively small, so that the polymer can curl back into its coiled shape after it has been stretched.

We can understand these requirements by taking a closer look at the chemistry of natural rubber, which is a polymer of a $C_5H_8$ hydrocarbon known as *isoprene:*

$$\underset{\text{isoprene}}{CH_2{=}C(CH_3)-CH{=}CH_2} \longrightarrow \underset{\text{natural rubber}}{[\ldots(-CH_2-C(CH_3){=}CH-CH_2-)_n\ldots]}$$

The double bonds in natural rubber are all in the cis form, which gives rise to long, flexible molecules. The force of attraction between polymer chains is relatively small, so the polymer can curl back into its original shape after the molecules have been oriented by stretching. By adding sulfur to natural rubber, it is possible to introduce a small number of cross-links between these polymer chains that hold these chains together when the polymer is stretched.

At first glance, it might seem easy to make synthetic rubber. All we have to do is find a suitable catalyst that can polymerize isoprene. The task is made more difficult by the fact that the cis isomer of isoprene rearranges into the trans isomer during polymerization and the fact that the trans isomer of polyisoprene, which is known as *gutta percha,* is not elastic. It is therefore important to control the geometry around the C=C double bond during polymerization to make sure that as few of these bonds as possible are converted to the trans geometry. Until recently, this wasn't possible, and other approaches to making synthetic rubber were necessary.

One solution to the problem involved polymerizing derivatives of butadiene such as 2-chloro-1,3-butadiene, which is known as chloroprene, to form the first major synthetic rubber, *neoprene:*

$$\underset{\text{chloroprene}}{CH_2{=}\overset{\overset{\textstyle Cl}{|}}{C}{-}CH{=}CH_2} \longrightarrow \underset{\text{polychloroprene, or neoprene}}{[\ldots(-CH_2-\overset{\overset{\textstyle Cl}{|}}{C}{=}CH-CH_2-]_n\ldots]}$$

This approach is still used to produce a copolymer of 75% butadiene and 25% styrene known as styrene–butadiene rubber (SBR). Roughly 40% of the rubber used in the world today is SBR; another 35% is natural rubber that has been treated with sulfur.

The effect of cross-linking on elastomers can be demonstrated with a pair of rubber balls available from Flinn Scientific Inc. (catalog #AP1971). One of these balls is a polybutadiene rubber that contains an unusually large amount of sulfur. Because the polymer chains are extensively cross-linked, this ball dissipates very little energy in the form of heat when it bounces. It is therefore extremely resilient when bounced on the floor.

The other ball is a styrene–butadiene copolymer with much less cross-linking. When dropped on the floor, the ball seems to ''die.'' This copolymer is used in applications where an energy-absorbing medium is desired, such as automobile tires, which must absorb some of the energy associated with the bumps along the highway.

## 25.4 FREE-RADICAL POLYMERIZATION REACTIONS

It isn't difficult to form addition polymers from monomers containing C=C double bonds. Many of these compounds polymerize spontaneously during storage unless polymerization is actively inhibited. One of the reasons that thermally cracked gasoline was not as good as straight-run gasoline in the early years of the petroleum industry, for example, was the tendency of the alkenes in this gasoline to polymerize.

Addition polymers can be made by a free-radical polymerization reaction, which proceeds by a chain-reaction mechanism that contains three steps: chain initiation, chain propagation, and chain termination.

Most garbage bags are an example of polyethylene made by free radical polymerization.

### CHAIN INITIATION

A source of free radicals is needed to initiate the chain reaction. Benzoyl peroxide is an example of a free-radical initiator that decomposes on heating or in the presence of ultraviolet radiation to form free radicals.

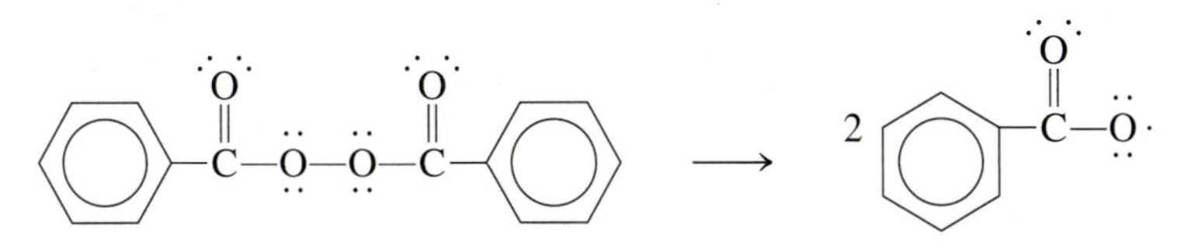

### CHAIN PROPAGATION

The free radical (R · ) produced in the chain-initiation step adds to an alkene to form a new free radical, which then reacts with additional monomers in a chain reaction:

$$R\cdot + H_2C{=}CH_2 \longrightarrow RCH_2CH_2\cdot$$
$$RCH_2CH_2\cdot + CH_2{=}CH_2 \longrightarrow RCH_2CH_2CH_2CH_2\cdot$$

### CHAIN TERMINATION

Whenever pairs of radicals combine to form a covalent bond, the chain reactions carried by these radicals are terminated:

$$R(CH_2CH_2)_nCH_2CH_2\cdot + R\cdot \longrightarrow R(CH_2CH_2)_nCH_2CH_2R$$

## 25.5 IONIC AND COORDINATION POLYMERIZATION REACTIONS

Addition polymers also can be made by chain reactions that proceed through ionic intermediates or by chain reactions in which the alkene is coordinated to a transition-metal catalyst.

### ANIONIC POLYMERIZATION

Anionic polymerization, like free-radical polymerization, takes place by three steps: chain initiation, chain propagation, and chain termination. In this case, the intermediate that carries the chain reaction is a negatively charged ion.

Grignard reagents are hybrids of ionic and covalent Lewis structures:

$$CH_3{-}Mg{-}\ddot{\underset{..}{Br}}:$$
$$\updownarrow$$
$$CH_3:^- \quad Mg^{2+} \quad :\ddot{\underset{..}{Br}}:^-$$

Sec. 24.18 

Methylmagnesium bromide is therefore a good source of the $CH_3^-$ ion. Other sources of negatively charged carbon atoms, or carbanions, include alkyllithium and trialkylaluminum compounds, which can be viewed as hybrids of resonance structures that include carbanions:

$$\left[CH_3Li \longleftrightarrow CH_3:^- \ Li^+\right]$$

$$\left[\begin{array}{c} CH_3{-}\underset{\displaystyle CH_3}{\underset{|}{Al}}{-}CH_3 \end{array} \longleftrightarrow \begin{array}{c} CH_3:^- \ Al^{3+} \ CH_3:^- \\ CH_3:^- \end{array}\right]$$

The $CH_3^-$ ion from one of these metal alkyls can attack an alkene to form a carbon–carbon bond, as shown in Figure 25.9. The product of this reaction is a new carbanion that can attack another alkene in a chain-propagation step. The chain reaction is terminated when the carbanion eventually reacts with water.

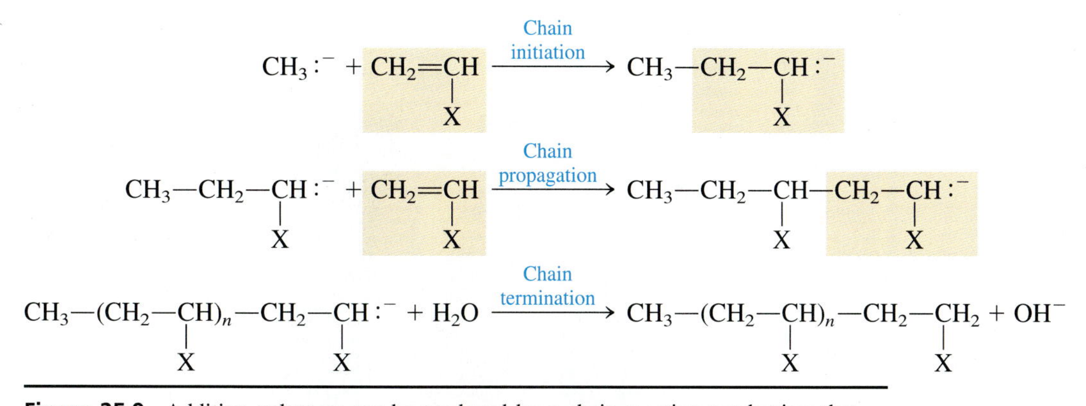


**Figure 25.9** Addition polymers can be produced by a chain-reaction mechanism that results from attack on the C=C double bond in the monomer by a negative ion, such as the $CH_3^-$ ion. Because the chain is initiated and carried by negatively charged ions, this mechanism is known as *anionic polymerization.*

## CATIONIC POLYMERIZATION

The intermediate that carries the chain reaction in polymerization reactions also can be a positive ion, or cation. In this case, the chain reaction is initiated by adding a strong acid to an alkene to form a carbonium ion, as shown in Figure 25.10. The ion produced in this reaction combines with additional monomers to produce a growing polymer chain. The chain reaction is terminated when the carbonium ion reacts with water.

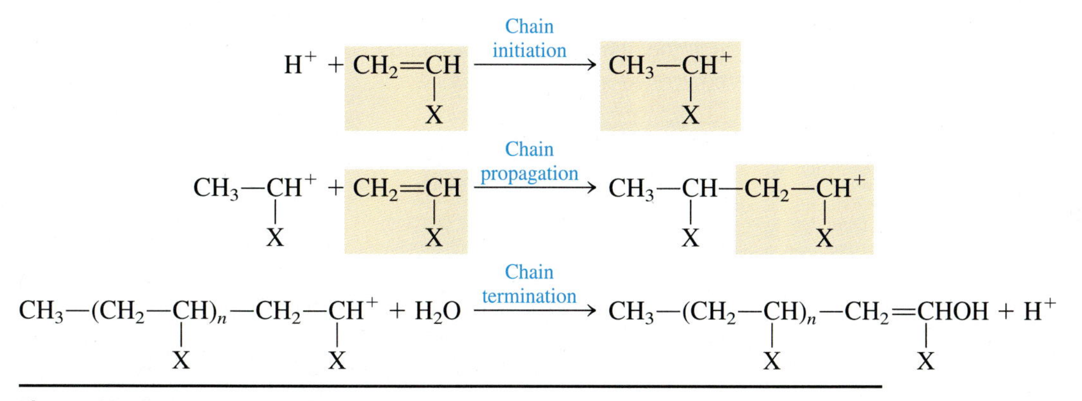


**Figure 25.10** Addition polymers can also result from a *cationic polymerization* mechanism in which the chain is initiated and carried by positively charged ions.

## COORDINATION POLYMERIZATION

In 1963 Karl Ziegler and Giulio Natta received the Nobel prize in chemistry for their work with coordination compound catalysts for addition polymerization reactions. These so-called **Ziegler–Natta catalysts** provide an unprecedented opportunity to control the linearity and tacticity of the polymer.

Free-radical polymerization of ethylene produces a low-density, branched polymer with side chains of one to five carbon atoms on up to 3% of the atoms along the polymer chain. Ziegler–Natta catalysts produce a more linear polymer, which is

High-density polyethylene bottles made with Ziegler-Natta catalysts.

more rigid, with a higher density and a higher tensile strength. Polypropylene produced by free-radical reactions, for example, is a soft, rubbery, atactic polymer with no commercial value. Ziegler–Natta catalysts provide an isotactic polypropylene, which is harder, tougher, and more crystalline.

A typical Ziegler–Natta catalyst would be a mixture of titanium(III) chloride ($TiCl_3$) and triethylaluminum [$Al(CH_2CH_3)_3$]. The first step in this reaction involves the transfer of an ethyl group from aluminum to titanium. An alkene then acts as a Lewis base to form a transition-metal complex, as shown in Figure 25.11. The $Ti—CH_2CH_3$ bond serves as a source of a carbanion, which attacks the alkene to form a carbon–carbon bond. A new alkene then coordinates to the titanium atom and the reaction continues. The titanium atom in this reaction therefore provides a template on which the alkene and the carbanion combine to give a linear polymer with carefully controlled stereochemistry.

$$—Ti—CH_2—CH_3 + CH_2{=}CH_2 \longrightarrow \underset{}{\overset{CH_2=CH_2}{\vdots}}\!\!—Ti—CH_2—CH_3$$

$$\downarrow$$

$$\begin{array}{ccc} CH_2 & — & CH_2 \\ | & & | \\ —Ti & & CH_2—CH_3 \end{array}$$

**Figure 25.11** Ziegler–Natta catalysts provide a mechanism for making addition polymers that proceeds through an intermediate in which the alkene is complexed to a transition metal. This mechanism produces a polymer in which both the linearity and the tacticity can be carefully controlled.

## 25.6 ADDITION POLYMERS

Addition polymers such as polyethylene, polypropylene, poly(vinyl chloride), and polystyrene tend to be linear or branched polymers with little or no cross-linking. As a result, they are thermoplastic materials that flow easily when heated and can be molded into a variety of shapes. The structures, names, and trade names of some common addition polymers are given in Table 25.1.

**TABLE 25.1 Addition Polymers**

| Structure | Chemical Name | Trademark or Common Name |
|---|---|---|
| $(-CH_2-CH_2-)_n$ | polyethylene | |
| $(-CF_2-CF_2-)_n$ | poly(tetrafluoroethylene) | Teflon |
| $(-CH_2-CH(CH_3)-)_n$ | polypropylene | Herculon |
| $(-CH_2-C(CH_3)_2-)_n$ | polyisobutylene | butyl rubber |
| $(-CH_2-CH(C_6H_5)-)_n$ | polystyrene | |
| $(-CH_2-CH(CN)-)_n$ | polyacrylonitrile | Orlon |
| $(-CH_2-CH(Cl)-)_n$ | poly(vinyl chloride) | PVC |
| $(-CH_2-CCl_2-)_n$ | poly(vinylidene chloride) | Saran |
| $(-CH_2-CH(CO_2CH_3)-)_n$ | poly(methyl acrylate) | |
| $(-CH_2-C(CH_3)(CO_2CH_3)-)_n$ | poly(methyl methacrylate) | Plexiglass, Lucite |
| $(-CH_2-CH=CH-CH_2-)_n$ | polybutadiene | |
| $(-CH_2-C(Cl)=CH-CH_2-)_n$ | polychloroprene | neoprene |
| $(-CH_2-C(H)=C(CH_3)-CH_2-)_n$ | poly(*cis*-1,4-isoprene) | natural rubber |
| $(-CH_2-C(H)=C(CH_3)-CH_2-)_n$ | poly(*trans*-1,4-isoprene) | gutta percha |

## POLYETHYLENE

Low-density polyethylene is produced by free-radical polymerization at high temperatures (200°C) and high pressures (above 1000 atm). The high-density polymer results from Ziegler–Natta catalysis at temperatures below 100°C and pressures less than 100 atm. More polyethylene is produced each year than any other plastic.

About 7800 million pounds of low-density and 4400 million pounds of high-density polyethylene were sold in 1980. Polyethylene has no taste or odor and is light-weight, nontoxic, and relatively inexpensive. It is used as a film for packaging food, clothing, and hardware. Most commercial trash bags, sandwich bags, and plastic wrapping are made from polyethylene films. Polyethylene is also used for everything from seat covers to milk bottles, pails, pans, and dishes.

## POLYPROPYLENE

The isotactic polypropylene from Ziegler–Natta-catalyzed polymerization is a rigid, thermally stable polymer with an excellent resistance to stress, cracking, and chemical reaction. Although it costs more per pound than polyethylene, it is much stronger. Thus, bottles made from polypropylene can be thinner, contain less polymer, and often cost less than conventional polyethylene products. The plastic stackable chairs that abound on college campuses are made of polypropylene.

## POLY(TETRAFLUOROETHYLENE)

Tetrafluoroethylene ($CF_2{=}CF_2$) is a gas that boils at −76°C and is therefore stored in cylinders at high pressure. In 1938 Roy Plunkett received a cylinder of tetrafluoroethylene that did not deliver as much gas as it should have. Instead of returning the cylinder, he cut it open with a hacksaw and discovered a white, waxy powder that was the first polytetrafluoroethylene polymer. After considerable effort, a less fortuitous route to this polymer was discovered, and polytetrafluoroethylene, or Teflon, became commercially available.

Teflon is a remarkable substance. It has the best resistance to chemical attack of any polymer, and it can be used at any temperature between −73°C and 260°C with no effect on its properties. It also has a very low coefficient of friction, which makes it waxy or slippery to the touch. Even crude rubber, adhesives, bread dough, and candy won't stick to a Teflon-coated surface.

## POLY(VINYL CHLORIDE) AND POLY(VINYLIDENE CHLORIDE)

Chlorine is one of the top ten industrial chemicals in the United States—more than 20 billion pounds are produced annually. About 20% of this chlorine is used to make the vinyl chloride monomer ($CH_2{=}CHCl$) for the production of poly(vinyl chloride), or PVC. The chlorine substituents on the polymer chain make PVC more fire resistant than polyethylene or polypropylene. They also increase the force of attraction between polymer chains, which increases the hardness of the plastic. The properties of PVC can be varied over a wide range by adding plasticizers, stabilizers, fillers, and dyes, making PVC a remarkably versatile plastic.

A copolymer of vinyl chloride ($CH_2{=}CHCl$) and vinylidene chloride ($CH_2{=}CCl_2$) is sold under the trade name Saran. The same increase in the force of attraction between polymer chains that makes PVC harder than polyethylene gives thin films of Saran a tendency to "cling."

## ACRYLICS

Acrylic acid is the common name for 2-propenoic acid: $CH_2{=}CHCO_2H$. Acrylic fibers such as Orlon are made from a derivative of acrylic acid known as acrylonitrile:

$$\underset{\text{acrylonitrile}}{CH_2{=}CH{-}CN} \longrightarrow \underset{\text{polyacrylonitrile}}{[\ldots(-CH_2-\overset{\overset{\displaystyle CN}{|}}{C}H-)_n\ldots]}$$

Acrylic polymers are usually formed by polymerizing one of the esters of this acid such as methyl acrylate:

$$\underset{\text{methyl acrylate}}{CH_2{=}CH\overset{\overset{\displaystyle O}{\|}}{C}OCH_3} \longrightarrow \underset{\text{poly(methyl acrylate)}}{[\ldots(-CH_2-\overset{\overset{\displaystyle CO_2CH_3}{|}}{C}H-)_n\ldots]}$$

One of the most important acrylic polymers is poly(methyl methacrylate), or PMMA, which is sold under the trade names Lucite and Plexiglass:

$$\underset{\text{methyl methacrylate}}{CH_2{=}\overset{\overset{\displaystyle CH_3}{|}}{C}CO_2CH_3} \longrightarrow \underset{\text{poly(methyl methacrylate), PMMA}}{[\ldots(-CH_2-\overset{\overset{\displaystyle CH_3}{|}}{\underset{\underset{\displaystyle CO_2CH_3}{|}}{C}}H-)_n\ldots]}$$

PMMA is a lightweight, crystal-clear, glasslike polymer used in airplane windows, taillight lenses, and light fixtures. Because it is hard, stable to sunlight, and extremely durable, PMMA is also used to make the reflectors embedded between lanes of interstate highways.

Traffic warnings lights made from PMMA.

The unusual transparency of PMMA makes this polymer ideal for hard contact lenses. Unfortunately, PMMA is impermeable to oxygen and water. Oxygen must therefore be transported to the cornea of the eye in the tears and then passed under the contact lens each time the eye blinks. Soft plastic lenses that pass both oxygen and water are made from poly(2-hydroxyethyl methacrylate), which is cross-linked with ethylene glycol dimethacrylate.

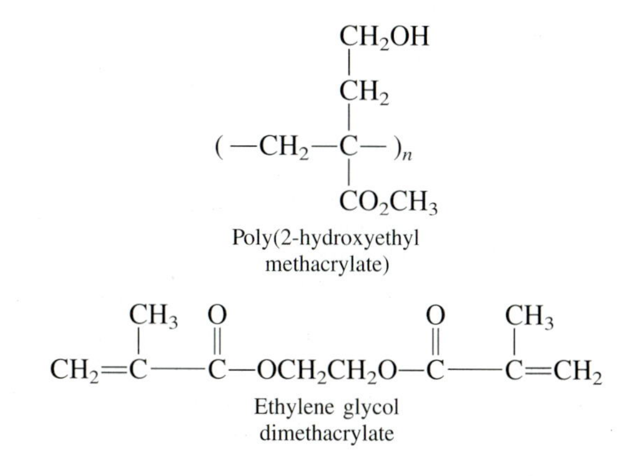

Poly(2-hydroxyethyl methacrylate)

Ethylene glycol dimethacrylate

An interesting polymer can be prepared by copolymerizing a mixture of acrylic acid and the sodium salt of acrylic acid. The product of this reaction has the following structure:

$$[\ldots(\text{—CH}_2\text{—}\underset{\displaystyle\text{CO}_2\text{H}}{\text{CH}}\text{—})_x(\text{—CH}_2\text{—}\underset{\displaystyle\text{CO}_2^-\text{Na}^+}{\text{CH}}\text{—})_y\ldots]$$

sodium polyacrylate

The difference between the $Na^+$ ion concentration inside the polymer network and the solution in which the polymer is immersed generates an osmotic pressure that draws water into the polymer. The amount of liquid that can be absorbed depends on the ionic strength of the solution. This polymer can absorb 800 times its own weight of distilled water, but only 300 times its weight of tap water. Because the ionic strength of urine is equivalent to an 0.1 *M* NaCl solution, this superabsorbent polymer, which can be found in disposable diapers, can absorb 60 times its weight of urine.

## 25.7 CONDENSATION POLYMERS

The Bakelizer steam pressure vessel used by Leo Baekland to commercialize his discovery of bakelite—the world's first synthetic plastic.

The first synthetic plastic was bakelite, developed by Leo Baekland between 1905 and 1914. The synthesis of bakelite starts with the reaction between formaldehyde ($H_2CO$) and phenol ($C_6H_5OH$) to form a mixture of *ortho*- and *para*-substituted phenols. At temperatures above 100°C, these phenols condense to form a polymer in which the aromatic rings are bridged by either $\text{—CH}_2\text{OCH}_2\text{—}$ or $\text{—CH}_2\text{—}$ linkages. The cross-linking in this polymer is so extensive that it is a thermoset plastic. Once it is formed, any attempt to change the shape of this plastic is doomed to failure.

Research started by Wallace Carothers and co-workers at duPont in the 1920s and 1930s eventually led to the discovery of the families of condensation polymers known as polyamides and polyesters. The **polyamides** were obtained by reacting a diacyl chloride with a diamine:

$$\text{H}_2\text{N(CH}_2)_x\text{NH}_2 + \text{Cl}\overset{\displaystyle\text{O}}{\overset{\|}{\text{C}}}(\text{CH}_2)_y\overset{\displaystyle\text{O}}{\overset{\|}{\text{C}}}\text{Cl} \longrightarrow [\ldots(\text{—NH(CH}_2)_x\text{NH}\overset{\displaystyle\text{O}}{\overset{\|}{\text{C}}}(\text{CH}_2)_y\overset{\displaystyle\text{O}}{\overset{\|}{\text{C}}}\text{—})_n\ldots]$$

polyamide

The **polyesters** were made by reacting the diacyl chloride with a dialcohol:

$$\text{HO(CH}_2)_x\text{OH} + \text{Cl}\overset{\displaystyle\text{O}}{\overset{\|}{\text{C}}}(\text{CH}_2)_y\overset{\displaystyle\text{O}}{\overset{\|}{\text{C}}}\text{Cl} \longrightarrow [\ldots(\text{—O(CH}_2)_x\text{O}\overset{\displaystyle\text{O}}{\overset{\|}{\text{C}}}(\text{CH}_2)_y\overset{\displaystyle\text{O}}{\overset{\|}{\text{C}}}\text{—})_n\ldots]$$

polyester

While studying polyesters, Julian Hill found that he could wind a small amount of this polymer on the end of a stirring rod and draw it slowly out of solution as a silky fiber. One day, when Carothers wasn't in the lab, Hill and his colleagues tried to see how long a fiber they could make by stretching a sample of this polymer as they ran down the hall. They soon realized that this playful exercise had oriented the polymer molecules in two dimensions and produced a new material with superior properties.

They immediately tried the same thing with one of the polyamides and produced a sample of the first synthetic fiber: **nylon.** This process can be demonstrated by carefully pouring a solution of hexamethylenediamine in water on top of a solution of adipoyl chloride in $CH_2Cl_2$:

$$\underset{\text{hexamethylene diamine}}{H_2N(CH_2)_6NH_2} + \underset{\text{adipoyl chloride}}{ClC(=O)(CH_2)_4C(=O)Cl} \longrightarrow \underset{\text{nylon 6,6}}{[...(-NH(CH_2)_6NH-C(=O)(CH_2)_4C(=O)-)_n...]}$$

A thin film of polymer forms at the interface between these two phases. By grasping this film with a pair of tweezers, we can draw a continuous string of nylon from the solution. This particular polyamide is known as nylon 6,6 because the polymer is formed from a diamine that has six carbon atoms and a derivative of a dicarboxylic acid that has six carbon atoms.

The effect of pulling on the polymer with the tweezers is much like that of stretching an elastomer—the polymer molecules become oriented in two dimensions. This raises an interesting question: Why don't the polymer molecules return to their original shape when we stop pulling? Secton 25.3 suggested that polymers are elastic when there is no strong force of attraction between the polymer chains. Polyamides and polyesters, however, form strong hydrogen bonds between the polymer chains that keep the polymer molecules oriented as shown in Figure 25.12.

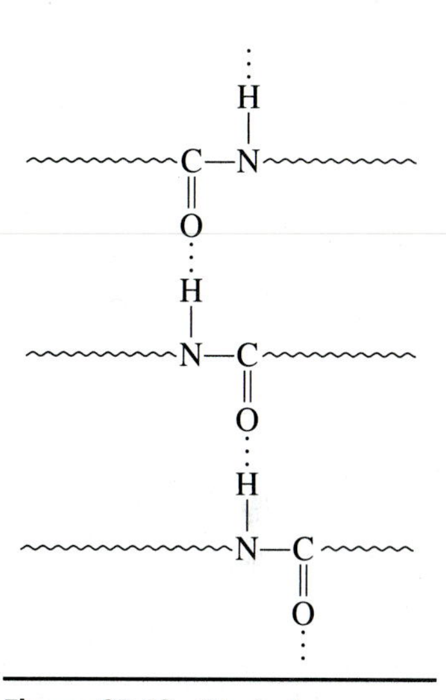

**Figure 25.12** The hydrogen bonds that form between the polymer chains when polyamides and polyesters are stretched help to keep the chains oriented in a two-dimensional fiber.

## EXERCISE 25.4

The synthetic fiber known as nylon 6 has the following structure:

$$[...(-NH(CH_2)_5CO-)_n...]$$

Explain how this polymer is made.

**SOLUTION** This polyamide must be made from a monomer that contains both an $-NH_2$ and a $-COCl$ functional group. This polymer is therefore made by the following condensation reaction:

$$H_2N(CH_2)_5C(=O)Cl + H_2N(CH_2)_5C(=O)Cl \longrightarrow H_2N(CH_2)_5C(=O)-NH(CH_2)_5C(=O)Cl$$

The first synthetic polyester fibers were produced from the reaction between ethylene glycol and either terephthalic acid or one of its esters to give poly(ethylene terephthalate). This polymer is still used to make thin films (Mylar) as well as textile fibers (Dacron and Fortrel).

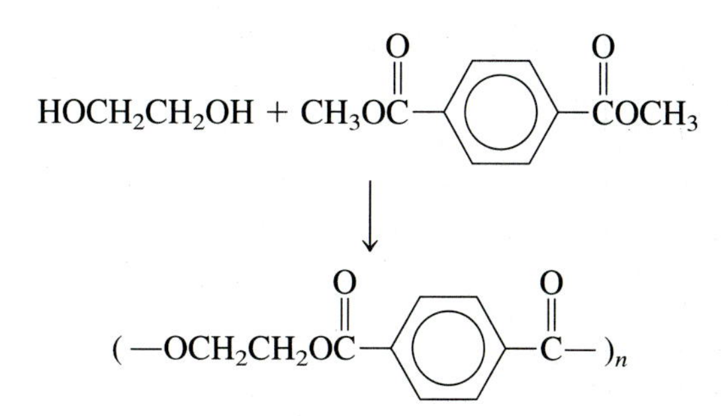

Nylon 6,6 being pulled from the interface between two solutions: adipoyl (or sebacoyl) chloride and hexamethylenediamine.

Phosgene ($COCl_2$) reacts with alcohols to form esters that are analogous to those formed when acyl chlorides react with alcohols:

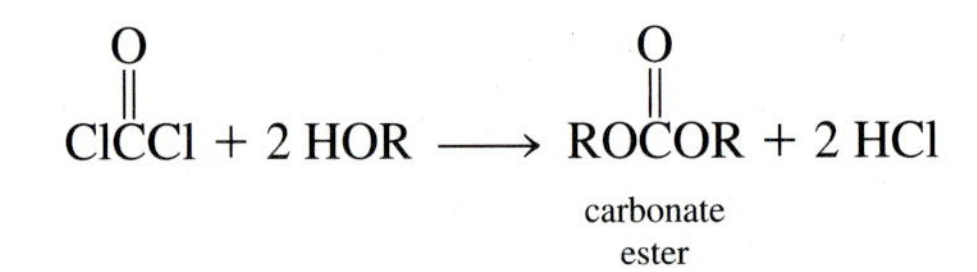


The product of this reaction is called a *carbonate ester* because it is the diester of carbonic acid, $H_2CO_3$. **Polycarbonates** are produced when one of these esters reacts with an appropriate alcohol, as shown in Figure 25.13. The polycarbonate shown in this figure is known as Lexan. It has a very high resistance to impact and is used in safety glass, bulletproof windows, and motorcycle helmets.

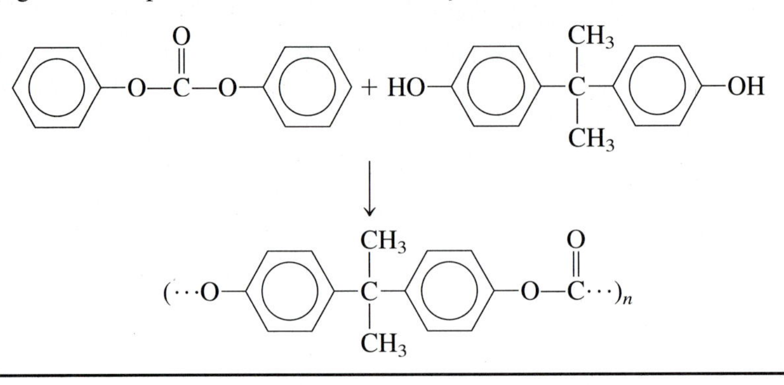


**Figure 25.13** Polycarbonates are condensation polymers formed when a diester of carbonic acid is allowed to react with a dialcohol.

The structures and names of some common condensation polymers are given in Table 25.2.

**TABLE 25.2 Common Condensation Polymers**

| Structure | Trademark or Common Name |
|---|---|
| Polyamides | |
| $(\text{—NH—(CH}_2)_6\text{—NH—C(=O)—(CH}_2)_4\text{—C(=O)—})_n$ | nylon 6,6 |
| $(\text{—NH—(CH}_2)_6\text{—NH—C(=O)—(CH}_2)_3\text{—C(=O)—})_n$ | nylon 6,10 |
| $(\text{—NH—(CH}_2)_5\text{—C(=O)—})_n$ | nylon 6 |
| $(\text{—NH—C}_6\text{H}_{10}\text{—CH}_2\text{—C}_6\text{H}_{10}\text{—NH—C(=O)—(CH}_2)_{10}\text{—C(=O)—})_n$ | Qiana |
| Polyaramides | |
| $(\text{—NH—C}_6\text{H}_4\text{—C(=O)—})_n$ | Kevlar |

**TABLE 25.2 Common Condensation Polymers continued**

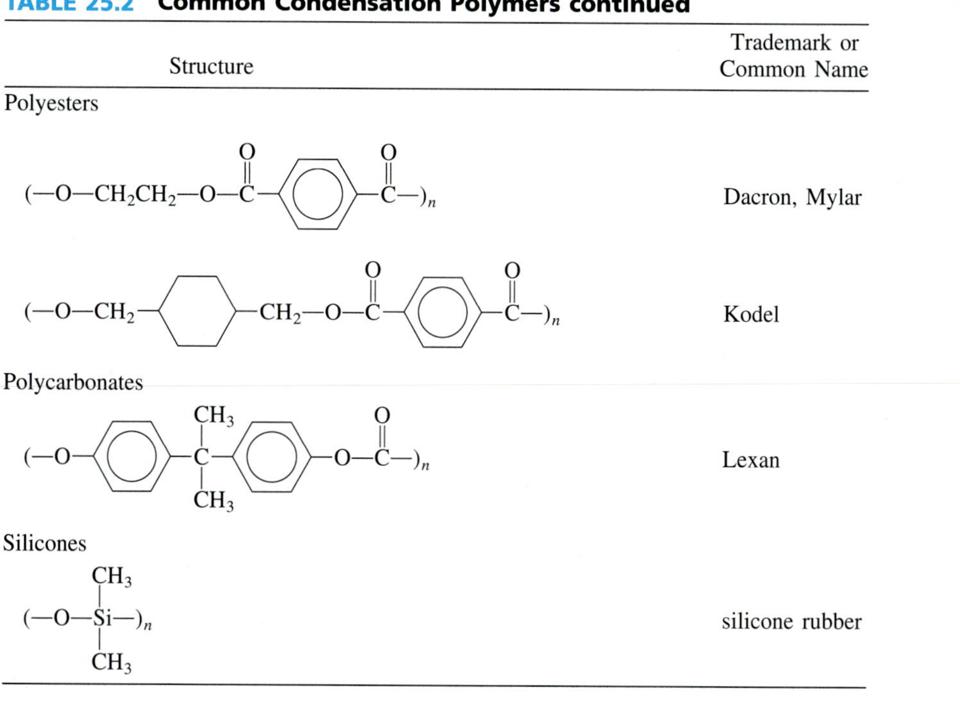

| Structure | Trademark or Common Name |
|---|---|
| Polyesters | |
| (—O—$CH_2CH_2$—O—C(=O)—$C_6H_4$—C(=O)—$)_n$ | Dacron, Mylar |
| (—O—$CH_2$—$C_6H_{10}$—$CH_2$—O—C(=O)—$C_6H_4$—C(=O)—$)_n$ | Kodel |
| Polycarbonates | |
| (—O—$C_6H_4$—C($CH_3$)($CH_3$)—$C_6H_4$—O—C(=O)—$)_n$ | Lexan |
| Silicones | |
| (—O—Si($CH_3$)($CH_3$)—$)_n$ | silicone rubber |

## 25.8 NATURAL POLYMERS

Biochemistry can be defined as the chemistry occurring in living organisms. Biomolecules are often sorted into four categories: proteins, carbohydrates, nucleic acids, and lipids. The name **protein** comes from the Greek stem for "first," which indicates the relative importance of this class of compounds. About half of the dry weight of an organism is made up of proteins, which serve a broader array of functions than any other class of biomolecules:

- *Structure:* The actin and myosin in muscles, the collagen in skin and bone, and the keratins in hair, horn, and hoof are all examples of proteins whose primary function is to provide the structure of the organism.
- *Catalysis:* Most of the chemical reactions in living systems are catalyzed by enzymes, which are proteins.
- *Control:* Many proteins regulate or control biological activity. Insulin, for example, controls the rate at which sugar is metabolized.
- *Energy:* Many proteins, including the casein in milk and the albumin in eggs, are used primarily to store food energy.
- *Transport:* $O_2$ is carried through the bloodstream by the proteins hemoglobin and myoglobin. Other proteins are involved in the transport of sugars, amino acids, and ions across the cell membranes.
- *Protection:* The first line of defense against viruses and bacteria are the antibodies produced by the immune system, which are based on proteins.

The name **carbohydrate** reflects the fact that many of these compounds have the empirical formula $CH_2O$. Carbohydrates are the primary source of food energy for most living systems. They include simple sugars, such as glucose ($C_6H_{12}O_6$) and sucrose ($C_{12}H_{22}O_{11}$), as well as polymers of these sugars, such as starch, glycogen, and cellulose. Carbohydrates are produced from $CO_2$ and $H_2O$ during photosynthesis and are therefore the end products of the process by which plants capture the energy in sunlight.

The name **nucleic acid** was given to a class of relatively strong acids that were found in the nuclei of cells. As monomers, nucleic acids such as adenosine triphosphate (ATP) are involved in the process by which cells capture food energy and make it available to fuel the processes that keep the cell alive. As polymers, they store and process the genetic information that allows an organism to grow and eventually reproduce.

Proteins, carbohydrates, and nucleic acids are all classified on the basis of similarities in their structures and functions. The **lipids** are organized on the basis of one of their physical properties. Any molecule in a biological system that is soluble in nonpolar solvents is classified as a lipid (from the Greek *lipos,* ''fat'').

## 25.9 THE AMINO ACIDS

Proteins are formed by polymerizing monomers that are known as **amino acids** because they contain both an amino ($—NH_2$) and a carboxylic acid ($—CO_2H$) functional group. With only one exception, the amino acids used to synthesize proteins are primary amines with the following generic formula:

$$H_2N—\underset{}{\overset{R}{\overset{|}{C}}}H—CO_2H$$

an amino acid

The chemistry of amino acids is complicated by the fact that the $—NH_2$ group is a base and the $—CO_2H$ group is an acid. In aqueous solution, an $H^+$ ion is therefore transferred from one end of the molecule to the other to form a **zwitterion** (from the German meaning ''mongrel ion'', or hybrid ion):

$$H_2N—\overset{R}{\overset{|}{C}}H—CO_2H \longrightarrow H_3N^+—\overset{R}{\overset{|}{C}}H—CO_2^-$$

Zwitterions are simultaneously electrically charged and electrically neutral. They contain positive and negative charges, but the net charge on the molecule is zero.

### EXERCISE 25.5

Amino acids in aqueous solution at a neutral pH are present as zwitterions with the following generic formula:

$$H_3N^+—\overset{R}{\overset{|}{C}}H—CO_2^-$$

Predict what will happen when a strong acid or a strong base is added to an aqueous solution of an amino acid.

**SOLUTION** In a strongly acidic solution, the $—CO_2^-$ end of this molecule picks up an $H^+$ ion to form a molecule with a net positive charge:

$$H_3N^+—\overset{R}{\overset{|}{C}}H—CO_2H$$

In the presence of a strong base, the $—NH_3^+$ end of the molecule loses an $H^+$ ion, to form a molecule with a net negative charge:

$$H_2N—\underset{\displaystyle |}{\overset{\displaystyle R}{CH}}—CO_2^-$$

The 20 common amino acids used to synthesize proteins are shown in Table 25.3. Most of these amino acids differ only in the nature of the R-group substituent. The common amino acids are therefore classified on the basis of these R groups. Amino acids with nonpolar substituents are said to be *hydrophobic* (water hating). Amino acids with polar R groups that form hydrogen bonds to water are classified as *hydrophilic* (water loving). The remaining amino acids have substituents that carry either negative or positive charges in aqueous solution at neutral pH and are therefore strongly hydrophilic.

Sec. 15.2

## EXERCISE 25.6

Classify the following amino acids.

(a) Valine: $R = —CH(CH_3)_2$
(b) Serine: $R = —CH_2OH$
(c) Aspartic acid: $R = —CH_2CO_2^-$
(d) Lysine: $R = —(CH_2)_4NH_3^+$

**SOLUTION**

(a) Valine is grouped among the amino acids with nonpolar, hydrophobic R groups.
(b) Serine is a member of the class of amino acids with polar, hydrophilic R groups.
(c) Aspartic acid is a hydrophilic amino acid with a negatively charged R group.
(d) Lysine is a hydrophilic amino acid with a positively charged R group.

With the exception of glycine, the common amino acids all contain at least one chiral carbon atom. These amino acids therefore exist as pairs of stereoisomers. The structures of the D and L isomers of alanine are shown in Figure 25.14. Although D-amino acids can be found in nature, they are not used to make proteins. Only the L isomers can be found in proteins.

Sec. 24.19

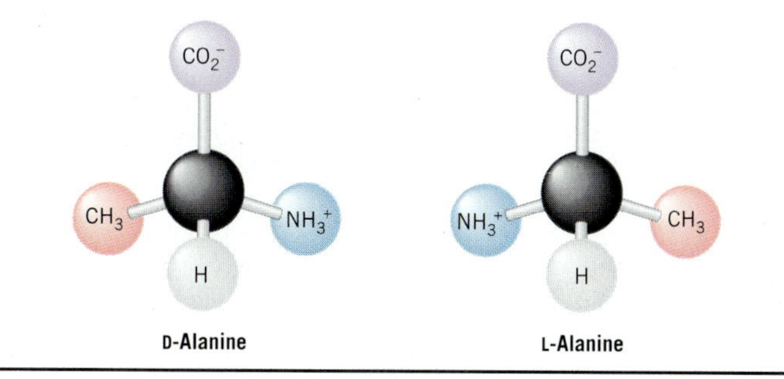


**Figure 25.14** The D and L stereoisomers of the amino acid alanine.

**TABLE 25.3 The 20 Common Amino Acids**

| Name | Structure (at neutral pH) | Name | Structure (at neutral pH) |
|---|---|---|---|
| Nonpolar (Hydrophobic) R Groups | | Polar (Hydrophilic) R Groups | |
| Glycine (GLY) | 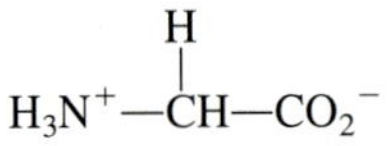 $H$<br>$H_3N^+—CH—CO_2^-$ | Serine (SER) | 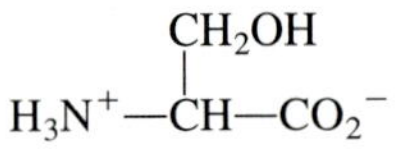 $CH_2OH$<br>$H_3N^+—CH—CO_2^-$ |
| Alanine (ALA) | 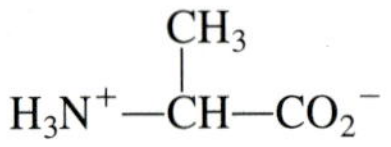 $CH_3$<br>$H_3N^+—CH—CO_2^-$ | Threonine (THR) | 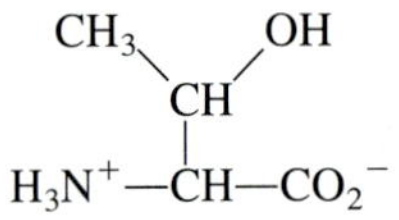 $CH_3 \quad OH$<br>$CH$<br>$H_3N^+—CH—CO_2^-$ |
| Valine (VAL) | 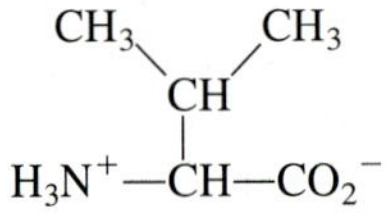 $CH_3 \quad CH_3$<br>$CH$<br>$H_3N^+—CH—CO_2^-$ | Tyrosine (TYR) | 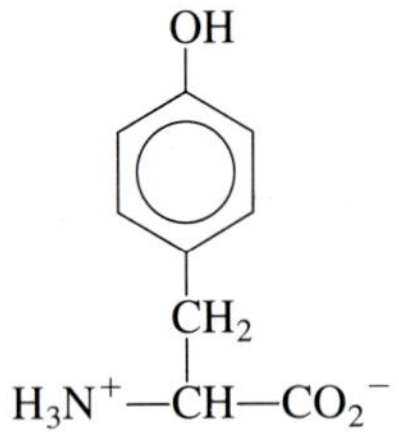 $OH$<br>$CH_2$<br>$H_3N^+—CH—CO_2^-$ |
| Leucine (LEU) | 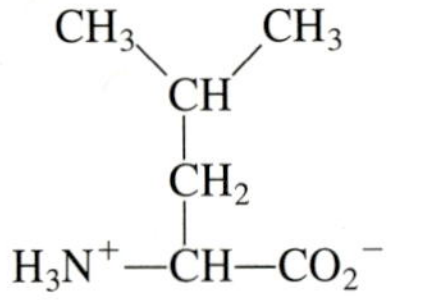 $CH_3 \quad CH_3$<br>$CH$<br>$CH_2$<br>$H_3N^+—CH—CO_2^-$ | Cysteine (CYS) | 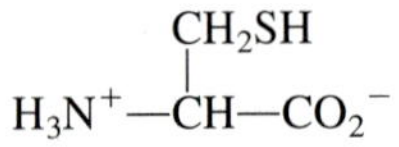 $CH_2SH$<br>$H_3N^+—CH—CO_2^-$ |
| Isoleucine (ILE) | 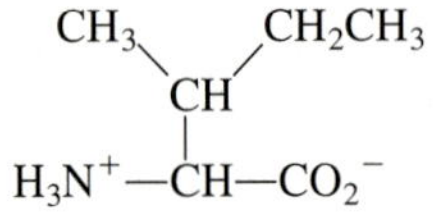 $CH_3 \quad CH_2CH_3$<br>$CH$<br>$H_3N^+—CH—CO_2^-$ | Asparagine (ASN) | 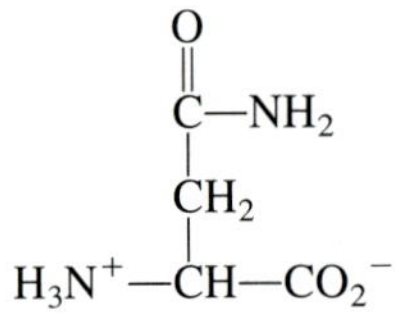 $O$<br>$C—NH_2$<br>$CH_2$<br>$H_3N^+—CH—CO_2^-$ |
| Proline (PRO) | 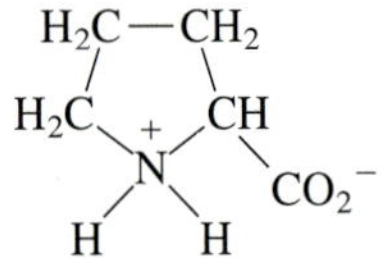 $H_2C—CH_2$<br>$H_2C \quad CH$<br>$N^+ \quad CO_2^-$<br>$H \quad H$ | Glutamine (GLN) | 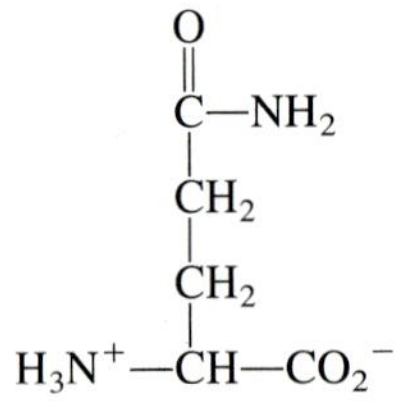 $O$<br>$C—NH_2$<br>$CH_2$<br>$CH_2$<br>$H_3N^+—CH—CO_2^-$ |
| Methionine (MET) | 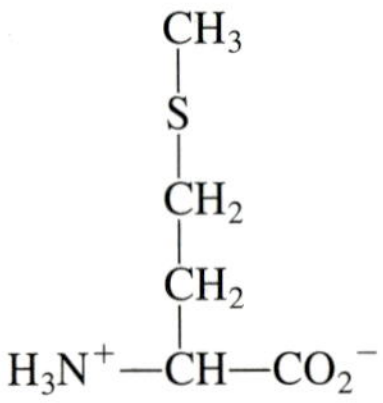 $CH_3$<br>$S$<br>$CH_2$<br>$CH_2$<br>$H_3N^+—CH—CO_2^-$ | Negatively Charged R Groups | |
| Phenylalanine (PHE) | 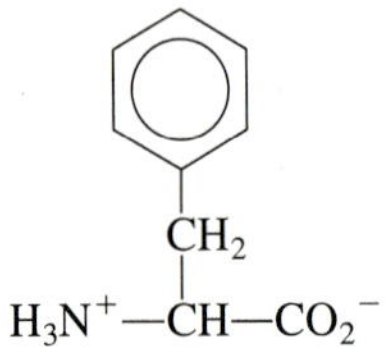 $CH_2$<br>$H_3N^+—CH—CO_2^-$ | Aspartic acid (ASP) | 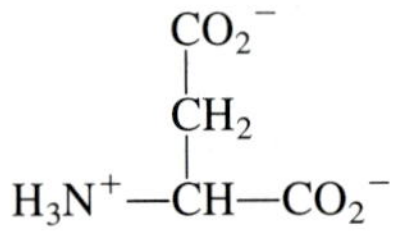 $CO_2^-$<br>$CH_2$<br>$H_3N^+—CH—CO_2^-$ |
| Tryptophan (TRP) | 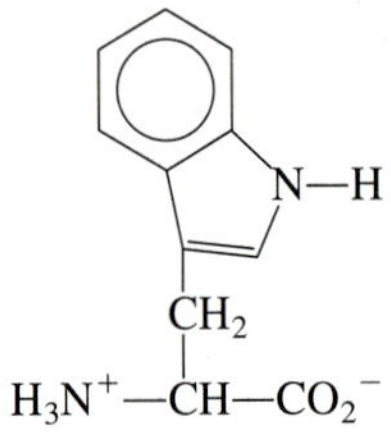 $N—H$<br>$CH_2$<br>$H_3N^+—CH—CO_2^-$ | Glutamic acid (GLU) | 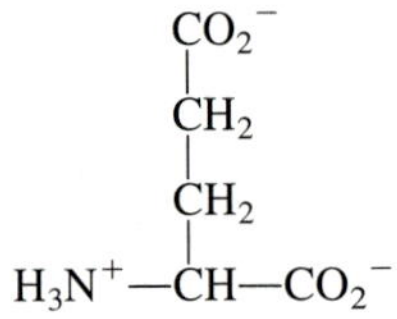 $CO_2^-$<br>$CH_2$<br>$CH_2$<br>$H_3N^+—CH—CO_2^-$ |
| | | Positively Charged R Groups | |
| | | Lysine (LYS) | 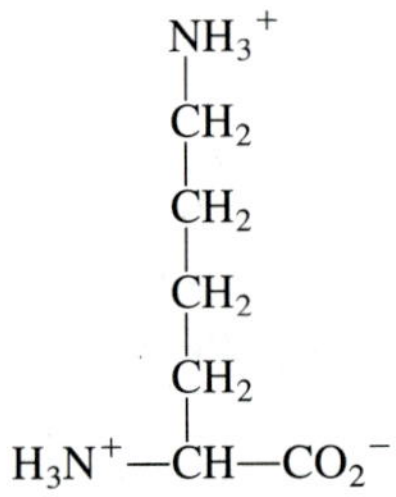 $NH_3^+$<br>$CH_2$<br>$CH_2$<br>$CH_2$<br>$CH_2$<br>$H_3N^+—CH—CO_2^-$ |

**TABLE 25.3 The 20 Common Amino Acids continued**

| Name | Structure (at neutral pH) | Name | Structure (at neutral pH) |
|---|---|---|---|
| Positively Charged R Groups continued | | Positively Charged R Groups continued | |
| Arginine (ARG) | $H_3N^+—CH(—CH_2—CH_2—CH_2—NH—C(=NH_2^+)—NH_2)—CO_2^-$ | Histidine (HIS) | $H_3N^+—CH(—CH_2—C_3H_3N_2H)—CO_2^-$ (imidazole ring: H, N, N—H, H) |

## 25.10 PEPTIDES AND PROTEINS

Proteins are linear polymers of between 40 and 10,000 (or more) amino acids. The average molecular weight of an amino acid is about 110 amu. A modestly sized protein with only 300 amino acids therefore has a molecular weight of 33,000 g/mol, and very large proteins can have molecular weights as high as 1,000,000 g/mol.

Proteins are formed by reacting the $—CO_2H$ end of one amino acid with the $—NH_2$ end of another to form an amide. The —CONH— amide bond between amino acids is also known as a *peptide bond* because relatively short polymers of amino acids are known as **peptides**:

$$H_3N^+—\overset{\displaystyle R}{\overset{|}{C}H}—\overset{\displaystyle O}{\overset{\|}{C}}O^- + H_3N^+—\overset{\displaystyle R}{\overset{|}{C}H}—\overset{\displaystyle O}{\overset{\|}{C}}O^- \longrightarrow H_3N^+—\overset{\displaystyle R}{\overset{|}{C}H}—\overset{\displaystyle O}{\overset{\|}{C}}—NH—\overset{\displaystyle R}{\overset{|}{C}H}—\overset{\displaystyle O}{\overset{\|}{C}}O^-$$

a dipeptide

The same —CONH— bond therefore forms the backbone of both proteins and synthetic fibers such as nylon. This raises an interesting question: How do we explain the enormous range of structures and functions of proteins when nylon has such regular properties?

Nylon has a regular structure that repeats monotonously from one end of the polymer chain to the other because the monomers from which nylon is made are symmetrical. The two ends of an amino acid, on the other hand, are different. Each monomer has both an $—NH_2$ head and a $—CO_2H$ tail. Four different dipeptides can therefore be formed from only two amino acids. Aspartic acid (ASP) could react with phenyl- alanine (PHE), for example, to give two symmetrical dipeptides, PHE-PHE and ASP-ASP, and two unsymmetrical dipeptides, PHE-ASP and ASP-PHE, as shown in Figure 25.15. When the full range of amino acids is considered, we find that it is possible to make 400 different dipeptides, 64 million different hexapeptides, and $10^{52}$ different proteins that contain only 40 amino acids.

### ▶ CHECKPOINT

Show that 27 tripeptides could be formed by combining the amino acids aspartic acid (ASP), phenylalanine (PHE), and cysteine (CYS).

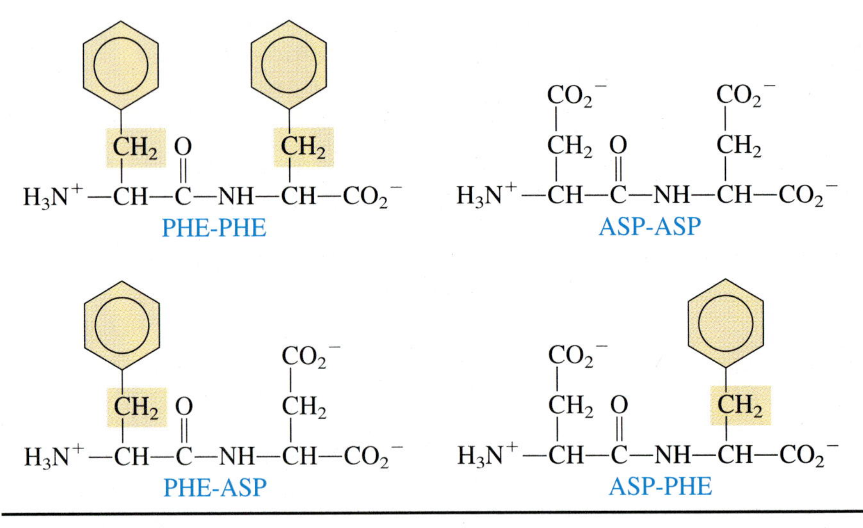


**Figure 25.15** The four dipeptides that can be formed from two amino acids: phenylalanine and aspartic acid.

Differences between the structures of even such closely related dipeptides as ASP-PHE and PHE-ASP give rise to significant differences in their properties as well. The methyl ester of ASP-PHE, for example, has a very sweet taste and is sold as an artificial sweetener under the trade name *aspartame.*

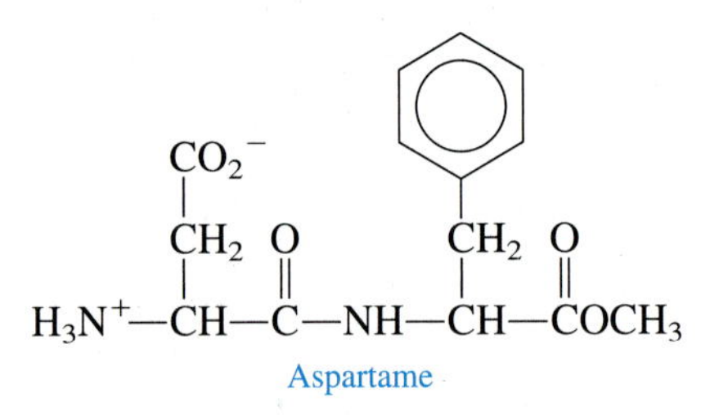


The ester of the dipeptide with the opposite arrangement of amino acids, PHE-ASP, does not taste sweet and has no commercial value. As the length of the polymer chain increases and the number of possible combinations of R groups increases, polymer chains with an almost infinite variety of structures and properties are produced.

In recent years, a group of naturally occurring peptides that mimic painkilling drugs such as morphine has been discovered in human brain cells. These *enkephalins* (from the Greek meaning "in the head") hold the promise of a synthetic painkiller that is both safe and nonaddictive. One of these enkephalins is a pentapeptide that contains four different amino acids—tyrosine (TYR), glycine (GLY), phenylalanine (PHE), and methionine (MET). The first step in describing the structure of this peptide is to list the amino acids in the order in which they are found on the peptide chain: TYR-GLY-GLY-PHE-MET. We then have to identify the amino acid at the —$CO_2H$ end of the chain and the amino acid at the —$NH_2$ end. By convention, proteins are listed from the *N*-terminal amino acid residue toward the *C*-terminal end. The structure of this enkephalin is shown in Figure 25.16.

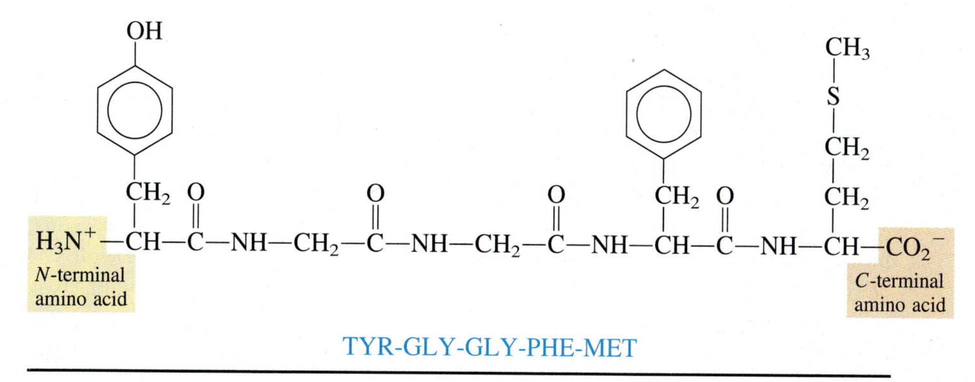


**Figure 25.16** The structure of a naturally occurring pentapeptide that belongs to the family of enkephalins that bind to the same sites in the brain as synthetic painkillers such as morphine.

The pentapetide in Figure 25.16 illustrates the perils that face anyone who tries to synthesize peptides or proteins from amino acids. In order to make this enkephalin in large quantities, we would have to overcome the following problems:

- Forming a peptide bond is an uphill process ($\Delta G° = +17$ kJ/mol), so a way must be found to drive this reaction forward.
- Because the sequence of amino acids is important, the amino acids must be added to the chain one at a time, in a carefully controlled fashion. As a result, a significant entropy factor must be overcome during the synthesis of polypeptides or proteins.
- The R groups of certain amino acids must be protected during polymerization so that no reactions take place on these side chains.

In 1984 R. B. Merrifield received the Nobel Prize in chemistry for developing an automated approach to the synthesis of peptides. The first step involves attaching the amino acid that will become the *C*-terminal residue to an inert, insoluble polystyrene resin. Amino acids are then incorporated, one at a time, by coupling them onto the growing peptide chain. Because the product of each step in this reaction is a solid, it can be easily collected, washed, and purified before the next step in the reaction.

The Merrifield synthesis uses a dehydrating agent known as dicyclohexylcarbodiimide (DCC) to drive the reaction that forms the peptide bond. To prevent reactions at the wrong site, appropriate blocking groups are added to the amino acids before they are polymerized. If we use the symbol *B* to indicate an appropriate blocking group, the synthesis of a dipeptide can be represented by the following equation:

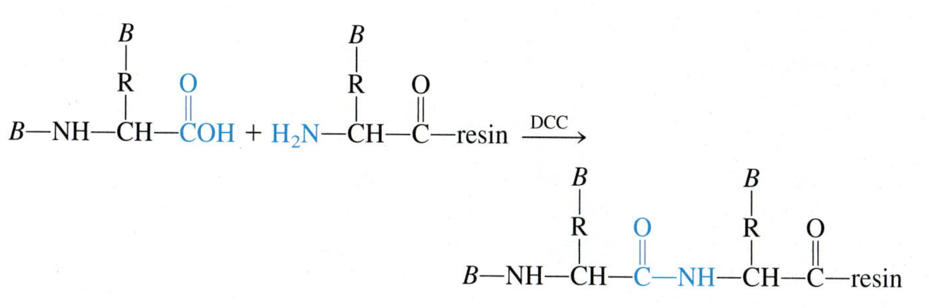

The blocking group on the *N*-terminal end of the dipeptide is then removed, and a third blocked amino acid residue is added to give a tripeptide. This process of adding one amino acid at a time is continued until the polypeptide or protein synthesis is complete. The polypeptide or protein chain is then removed by reacting the resin with HBr in a suitable solvent.

When it was first introduced, this process was automated on an apparatus that required about 4 hours to add an amino acid residue to the peptide chain. Thus, insulin could be synthesized in approximately 8 days, while ribonuclease, with 124 amino acids, required more than a month. The beauty of the Merrifield synthesis is the yield of each step, which is essentially 99%. The synthesis of ribonuclease, for example, took 369 chemical reactions, and 11,931 automated steps, and yet still had an overall yield of 17%.

## 25.11 THE STRUCTURE OF PROTEINS

### THE PRIMARY STRUCTURE OF PROTEINS

The primary structure of a protein is nothing more than the sequence of amino acids, read off one at a time, as if printed on ticker tape. Insulin obtained from cows, for example, consists of two chains (A and B) with the primary structures shown in Figure 25.17. There is more to the structure of a protein, however, than the sequence of amino acids. The polypeptide chain folds back on itself to form a secondary structure. Interactions between amino acid side chains then produce a tertiary structure. For some proteins, interactions between individual polypeptide chains give rise to a quaternary structure.

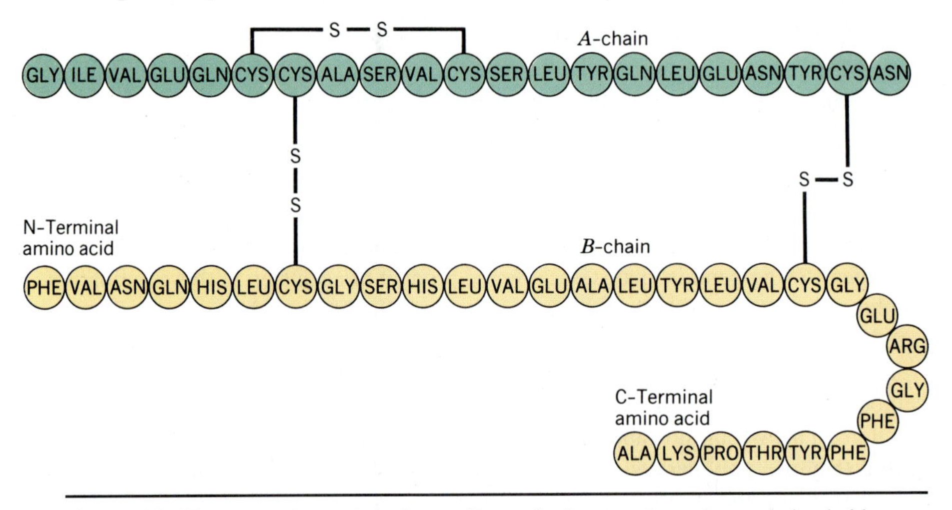


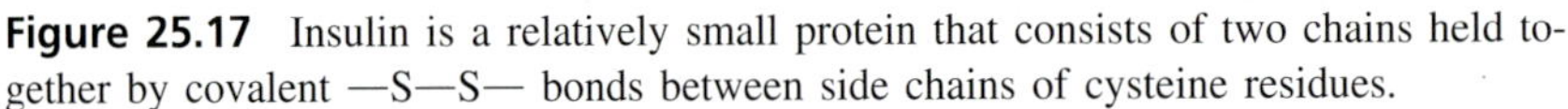
**Figure 25.17** Insulin is a relatively small protein that consists of two chains held together by covalent —S—S— bonds between side chains of cysteine residues.

### THE SECONDARY STRUCTURE OF PROTEINS

The peptide bond in proteins is a resonance hybrid of two Lewis structures:

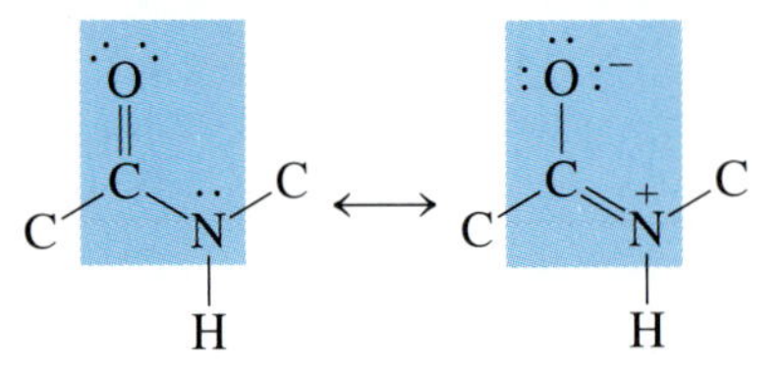

The Lewis structure on the left implies that the geometry around the carbon atom should be trigonal planar and that the carbon atom and its three nearest neighbors should lie in the same plane. The Lewis structure on the right suggests a trigonal-planar geometry for the nitrogen atom as well. Because the peptide bond is a hybrid of these resonance forms, these six atoms must all lie in the same plane.

Since the N—H and C═O bonds are relatively polar, hydrogen bonds form between adjacent peptide chains:

```
        \           /
         N—H···O═C
        /           \
   R—C—H         H—C—R
        \           /
         C═O···H—N
        /           \
```

The fact that the six atoms in the peptide bond must lie in the same plane limits the number of ways in which a polypeptide can be arranged in space. By building models of polypeptides, Linus Pauling and Robert Corey discovered two ways in which an ideal polypeptide chain could fold back on itself to maximize the number of hydrogen bonds between peptides. In one of these structures, the polypeptide chain forms the right-handed $\alpha$-helix shown in Figure 25.18. The other structure is the $\beta$-pleated sheet shown in Figure 25.19.

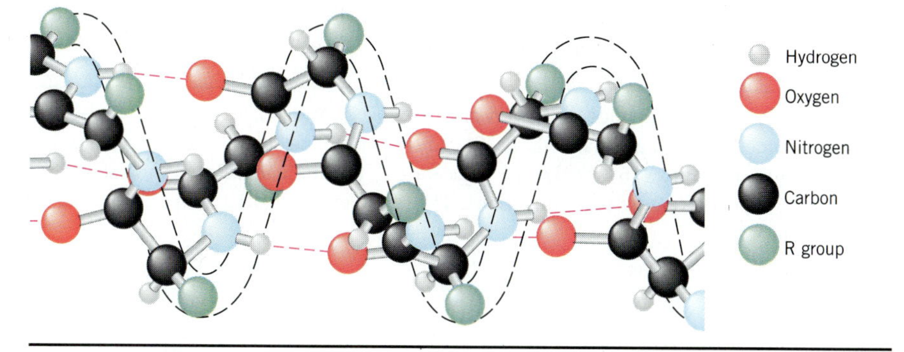


**Figure 25.18** The hydrogen bonds between adjacent peptide bonds allow polypeptides to form a right-handed helix. One 360° turn along this helix contains 3.6 amino acid residues.

## THE TERTIARY STRUCTURE OF PROTEINS

Most proteins have structures that lie between the extremes of ideal $\alpha$-helixes and $\beta$-pleated sheets because other factors influence the way proteins fold to form three-dimensional structures. Particular attention must be paid to interactions between the side chains of the amino acids that form the backbone of the protein. Figure 25.20 shows four ways in which these amino acid side chains can interact to form the tertiary structure of the protein:

*Disulfide (S—S) linkages.* If the folding of a protein brings two cysteine residues together, the two —SH side chains can be oxidized to form a covalent —S—S— bond. These **disulfide bonds** cross-link the polypeptide chain.

*Hydrogen bonding.* In addition to the hydrogen bonds between peptides that gives rise to the secondary structure of the protein, hydrogen bonds can form between amino acid side chains.

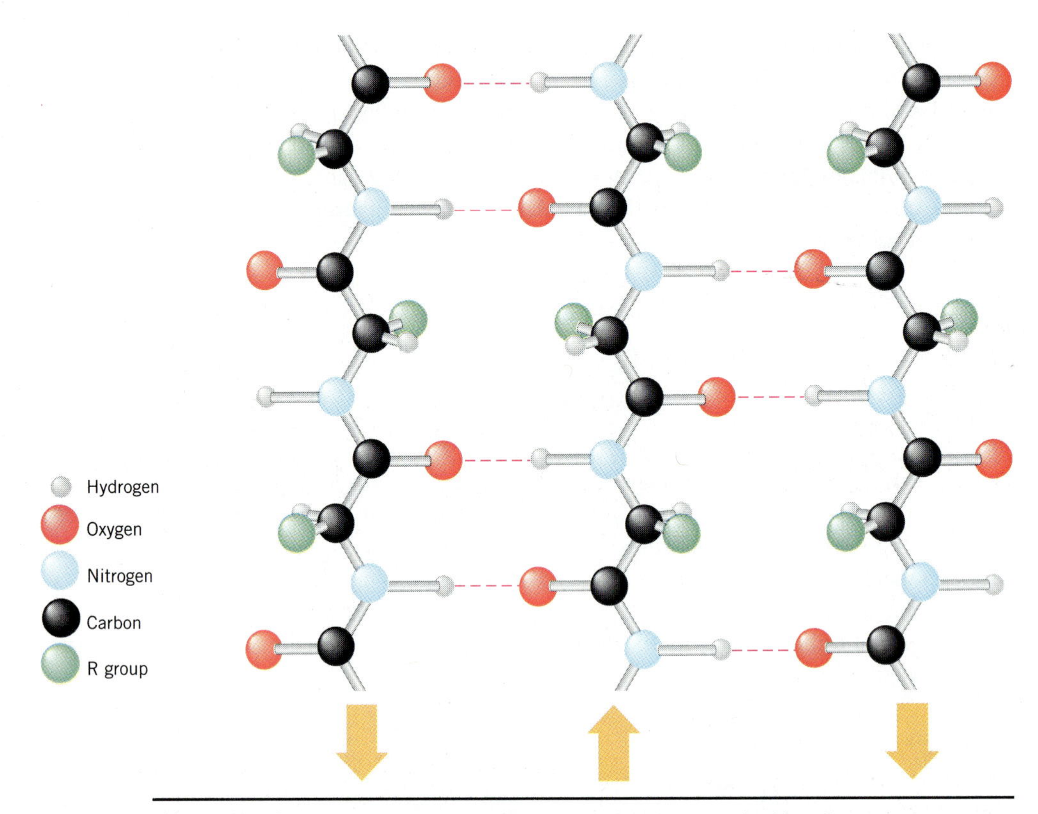


**Figure 25.19** The polypeptide chain can also fold back on itself to form a structure that looks something like a pleated sheet. Two different $\beta$-pleated sheet structures are found in nature that differ in whether the adjacent polypeptide chains run in the same (parallel) or opposite (antiparallel) directions. This drawing shows the antiparallel structure found in silk.

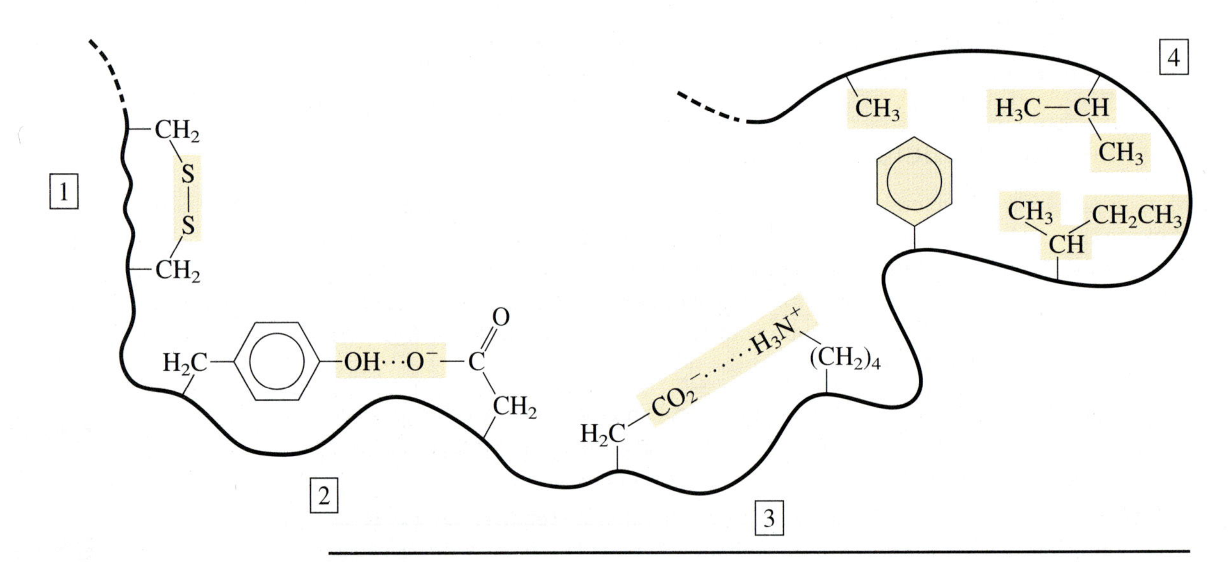


**Figure 25.20** The four factors responsible for the tertiary structure of proteins.

*Ionic bonding.* The structure of a protein can be stabilized by the force of attraction between amino acid side chains of opposite charge, such as the —$NH_3^+$ side chain of LEU and the —$CO_2^-$ side chain of ASP.

*Hydrophobic interactions.* Proteins fold so that the hydrophobic side chains of the amino acids GLY, ALA, VAL, LEU, ILE, PRO, MET, PHE, and TRP are buried within the protein, where they can interact to form hydrophobic pockets. These **hydrophobic interactions** stabilize the structure of the protein.

Human hair contains proteins that are about 14% cysteine. Hair curls as it grows because of the disulfide (—S—S—) links between cysteine residues on adjacent protein molecules. The first step in changing the way hair curls involves shaping the hair to our satisfaction and then locking it into place with curlers. The hair is then treated with a mild reducing agent that reduces the —S—S— bonds to pairs of —SH groups. This relaxes the structure of the proteins in the hair, allowing them to pick up the structure dictated by the curlers. The —SH side chains on cysteine residues that are now adjacent to each other are then oxidized by the $O_2$ in air. New —S—S— linkages form, locking the hair permanently in place—at least until new hair grows.

## THE QUATERNARY STRUCTURE OF PROTEINS

Hemoglobin is the protein that carries $O_2$ through the bloodstream to the muscles. This protein consists of four polypeptide chains—two $\alpha$-chains that contain 141 amino acids and two $\beta$-chains that contain 146 amino acids. Hemoglobin is therefore an example of a protein that has a quaternary structure. It consists of four polymer chains, which must be assembled to form the complete protein.

The polymer chains in a quaternary protein are not linked by covalent bonds such as the S—S bonds that hold together the polypeptide chains in insulin. The primary force of attraction between the $\alpha$- and $\beta$-chains in hemoglobin is the result of interactions between hydrophobic substituents on these polymer chains. In other quaternary proteins, hydrogen bonding or ionic interactions between amino acid side chains on the outer surfaces of adjacent polymer chains also contribute to the process by which the polymer chains are held together.

It would be a mistake to assume that the $\alpha$-chains of hemoglobin are more like an $\alpha$-helix, whereas the $\beta$-chains more closely resemble a $\beta$-pleated sheet. The $\alpha$ and $\beta$ prefixes, which appear with some regularity in biochemistry, refer to the first and second examples discovered of a phenomenon, respectively.

## THE DENATURATION OF PROTEINS

Proteins are fragile molecules that are remarkably sensitive to changes in structure. The replacement of a single glutamic acid by a nonpolar valine at the sixth position on the $\beta$-chains of hemoglobin, for example, produces the disease known as sickle-cell anemia. The introduction of a hydrophobic VAL residue at this position changes the quaternary structure of hemoglobin. The "sticky," nonpolar side chain on a valine residue at this position causes hemoglobin molecules to cluster together in an abnormal fashion, interfering with their function as oxygen-carrying proteins.

Sickle-cell anemia is the result of a change in the way the protein is assembled from amino acids. The structure of a protein also can be changed after it has been

An egg being "denatured" by heating.

made. Anything that causes a protein to leave its normal, or natural, structure is said to **denature** the protein. Factors that can lead to denaturation include the following:

- Heating, which disrupts the secondary and tertiary structure of the protein. The changes we observe when we fry an egg result from denaturation caused by heating.
- Changes in pH that interfere with ionic bonding between amino acid side chains.
- Detergents, which make nonpolar amino acid side chains soluble and thereby destroy the hydrophobic interactions that give rise to the tertiary and quaternary structure of the protein.
- Oxidizing or reducing agents that either create or destroy S—S bonds.
- Reagents such as urea ($H_2NCONH_2$) that disrupt the hydrogen bonds that form the secondary structure of the protein.

## 25.12 CARBOHYDRATES: THE MONOSACCHARIDES

As noted in Chapter 24, the term *carbohydrate* was originally used to describe compounds that were literally "hydrates of carbon" because they had the empirical formula $CH_2O$. In recent years, carbohydrates have been classified on the basis of their structures, not their formulas. They are now defined as *polyhydroxy aldehydes and ketones.* Among the compounds that belong to this family are cellulose, starch, glycogen, and most sugars.

### EXERCISE 25.7

The following compounds are now considered to be carbohydrates because of their structures. Which of these compounds satisfy the literal definition of a carbohydrate?

(a) glucose, $C_6H_{12}O_6$
(b) sucrose, $C_{12}H_{22}O_{11}$
(c) ribose, $C_5H_{10}O_5$

**SOLUTION** Glucose and ribose have the empirical formula $CH_2O$, and therefore meet the literal definition of a carbohydrate. The formula of sucrose suggests that this compound is formed by condensing a pair of $C_6H_{12}O_6$ monomers by eliminating a molecule of water.

There are three important classes of carbohydrates: (1) monosacchrides, (2) disaccharides, and (3) polysaccharides. The **monosaccharides** are white, crystalline solids that contain a single aldehyde or ketone functional group. They are subdivided into two classes—*aldoses* and *ketoses*—on the basis of whether they contain an aldehyde or a ketone. Each is also classified as a triose, tetrose, pentose, hexose, or heptose on the basis of whether it contains three, four, five, six, or seven carbon atoms.

Almost without exception, the monosaccharides are optically active compounds. Although both D and L isomers are possible, most of the monosaccharides found in nature are in the D configuration. Structures for the D and L isomers of the simplest aldose, glyceraldehyde, are shown below:

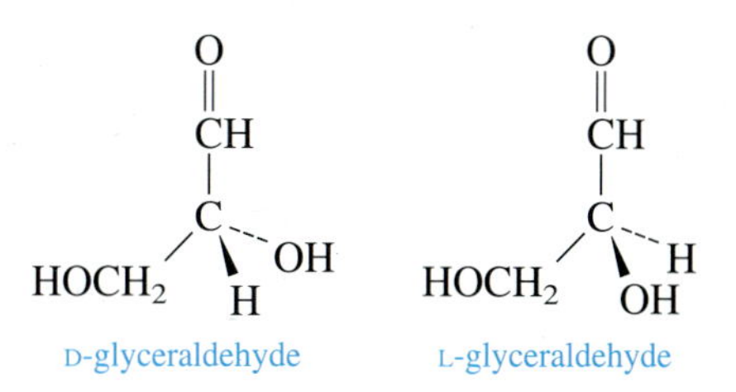


The structures of many monosaccharides were first determined by Emil Fischer in the 1880s and 1890s and are still written according to a convention he developed. The Fischer projection represents what the molecule would look like if its three-dimensional structure were projected onto a piece of paper. By convention, Fischer projections are written vertically, with the aldehyde or ketone at the top. The —OH group on the second-to-last carbon atom is written on the right side of the skeleton structure for the D isomer and on the left for the L isomer. Fischer projections for the two isomers of glyceraldehyde are shown below:

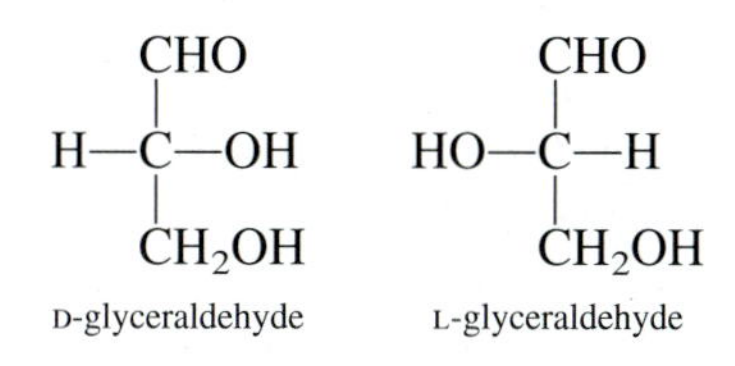


Fischer projections for some of the more common monosaccharides are given in Figure 25.21.

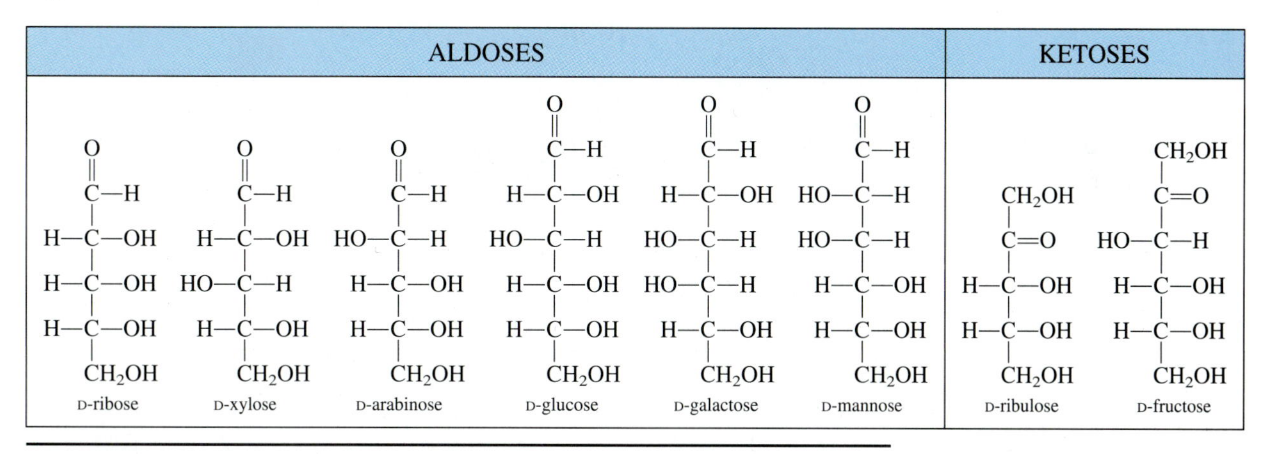


**Figure 25.21** Fischer projections for some of the common monosaccharides.

## EXERCISE 25.8

Glucose and fructose have the same formula: $C_6H_{12}O_6$. Glucose is the sugar with the highest concentration in the bloodstream. Fructose is found in fruit and honey. Use the Fischer projections in Figure 25.21 to explain the difference between the structures of these compounds.

**SOLUTION** Both compounds have the same structure for the third, fourth, fifth, and sixth carbon atoms. The difference between these compounds is the fact that one of them is an aldehyde (glucose) and the other is a ketone (fructose).

If the carbon chain is long enough, the alcohol at one end of a monosaccharide can attack the carbonyl group at the other end to form a cyclic compound. When a six-membered ring is formed, the structure is called a *pyranose* (see Figure 25.22).

**Figure 25.22** The —OH group on the penultimate carbon atom of a glucose molecule can attack the C=O group at the other end to form a six-membered cyclic compound known as a pyranose. Because this reaction introduces another chiral carbon, two isomers are formed, which are known as the $\alpha$- and $\beta$-anomers.

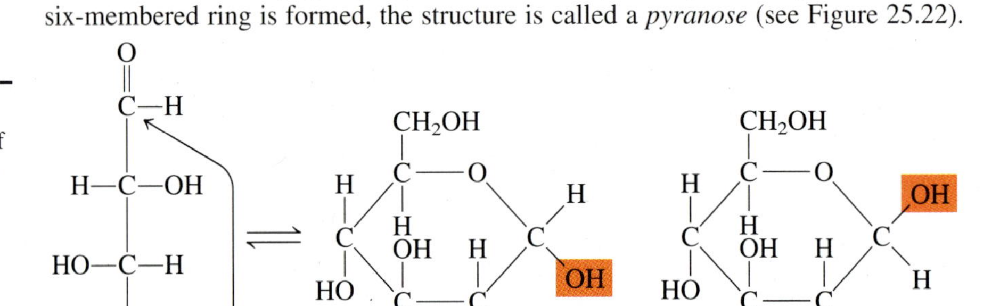


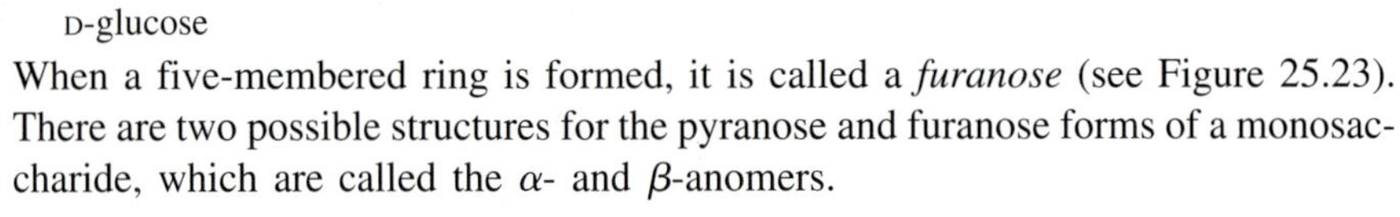


When a five-membered ring is formed, it is called a *furanose* (see Figure 25.23). There are two possible structures for the pyranose and furanose forms of a monosaccharide, which are called the $\alpha$- and $\beta$-anomers.

**Figure 25.23** The $\alpha$- and $\beta$-anomers of the furanose produced when D-ribose and D-fructose form cyclic compounds.

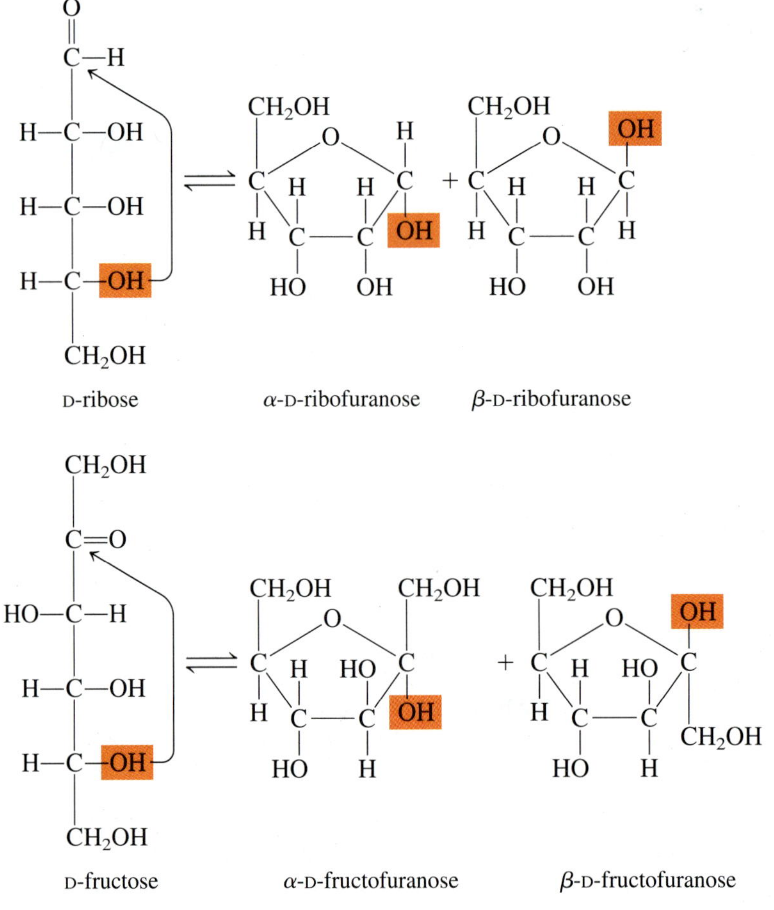

## 25.13 CARBOHYDRATES: THE DISACCHARIDES AND POLYSACCHARIDES

**Disaccharides** are formed by condensing a pair of monosaccharides. The structures of three important disaccharides with the formula $C_{12}H_{22}O_{11}$ are shown in Figure 25.24. *Maltose,* or malt sugar, which forms when starch breaks down, is an important component of the barley malt used to brew beer. *Lactose,* or milk sugar, is a disaccharide found in milk. Very young children have a special enzyme known as lactase that helps digest lactose. As they grow older, many people lose the ability to digest lactose and cannot tolerate milk or milk products. Because human milk has twice as much lactose as milk from cows, young children who develop lactose intolerance while they are being breast-fed are switched to cows' milk or a synthetic formula based on sucrose.

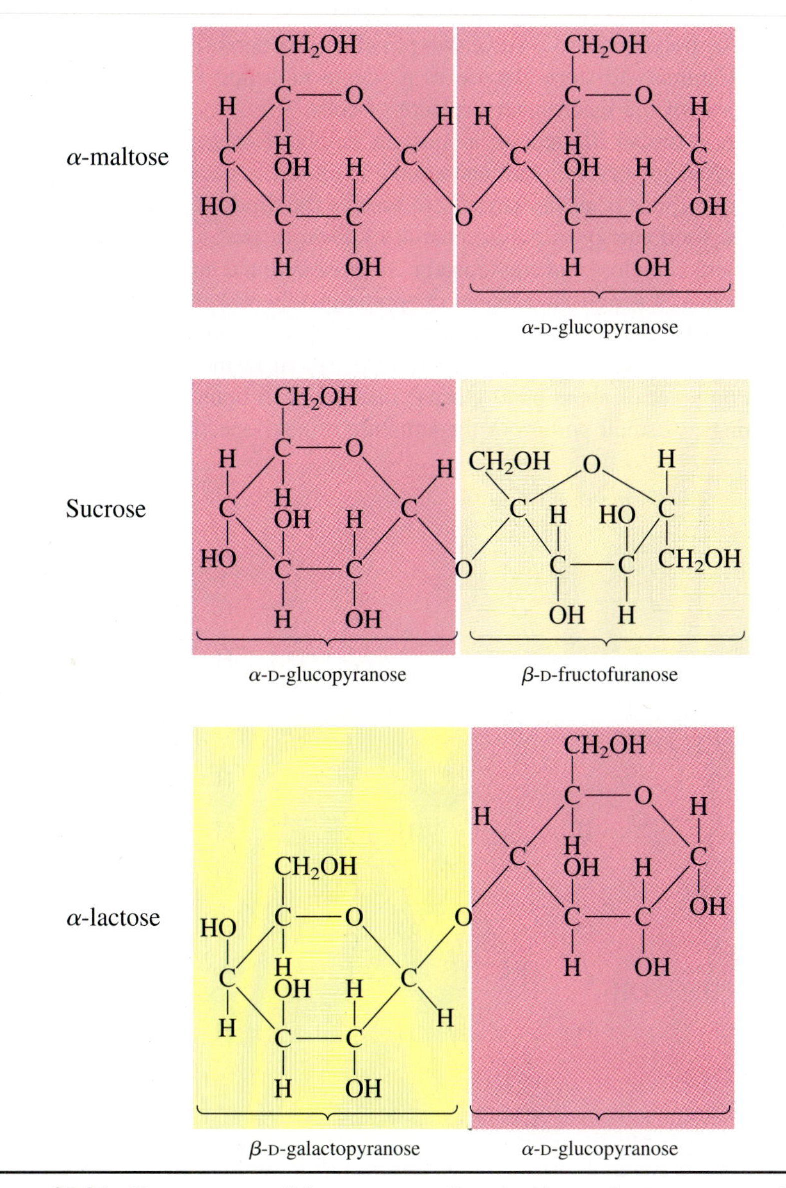


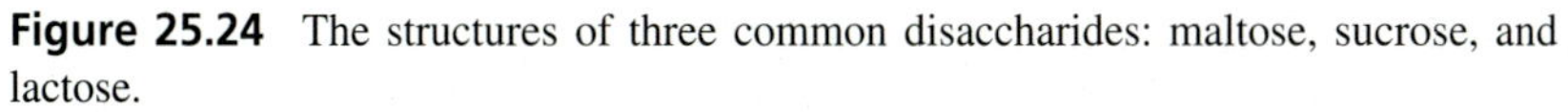

**Figure 25.24** The structures of three common disaccharides: maltose, sucrose, and lactose.

The substance most people refer to as ''sugar'' is the disaccharide *sucrose,* which is extracted from either sugar cane or beets. Sucrose is the sweetest of the disaccharides. It is roughly three times as sweet as maltose and six times as sweet as lactose. In recent years, sucrose has been replaced in many commercial products by corn syrup, which is obtained when the polysaccharides in cornstarch are broken down. Corn syrup is primarily glucose, which is only about 70% as sweet as sucrose. Fructose, however, is about two and a half times as sweet as glucose. A commercial process therefore has been developed that uses an isomerase enzyme to convert about half of the glucose in corn syrup into fructose. This high-fructose corn sweetener is just as sweet as sucrose and has found extensive use in soft drinks.

Monosaccharides and disaccharides represent only a small fraction of the total amount of carbohydrates in the natural world. The great bulk of the carbohydrates in nature are present as **polysaccharides,** which have relatively large molecular weights. The polysaccharides serve two principal functions. They are used by both plants and animals to store glucose as a source of future food energy and they provide some of the mechanical structure of cells.

Very few forms of life receive a constant supply of energy from their environments. In order to survive, plant and animal cells had to develop a way of storing energy during times of plenty in order to survive the times of shortage that follow. Plants store food energy as polysaccharides known as *starch.* There are two basic kinds of starch: amylose and amylopectin. *Amylose* is found in algae and other lower forms of plants. It is a linear polymer of approximately 600 glucose residues whose structure can be predicted by adding α-D-glucopyranose rings to the structure of maltose. *Amylopectin* is the dominant form of starch in the higher plants. It is a branched polymer of about 6000 glucose residues with branches on 1 in every 24 glucose rings. A small portion of the structure of amylopectin is shown in Figure 25.25.

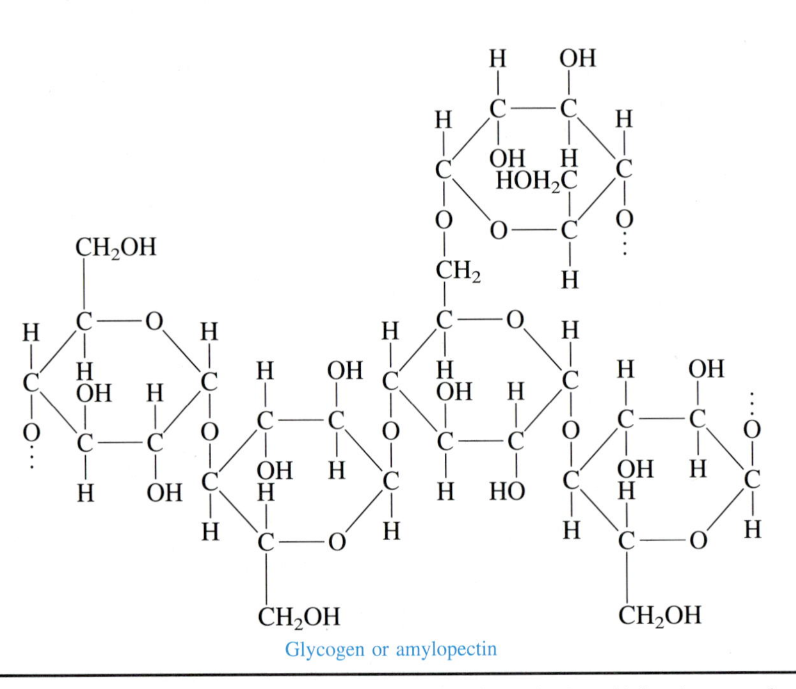


**Figure 25.25** The glycogen in animals and the amylopectin (starch) in plants are both branched polymers of α-D-glucopyranose monomers that are used to store food energy.

The polysaccharide that animals use for the short-term storage of food energy is known as *glycogen.* Glycogen has almost the same structure as amylopectin, with two minor differences. The glycogen molecule is roughly twice as large as amylopectin, and it has roughly twice as many branches.

There is an advantage to branched polysaccharides such as amylopectin and glycogen. During times of shortage, enzymes attack one end of the polymer chain and cut off glucose molecules one at a time. The more branches, the more points at which the enzyme can attack the polysaccharide. Thus, a highly branched polysaccharide is better suited for the rapid release of glucose to the cell than a linear polymer.

Polysaccharides are also used to form the walls of plant and bacterial cells. Cells that don't have a cell wall often break open in solutions whose salt concentrations are either two low (hypotonic) or too high (hypertonic). If the ionic strength of the solution is much smaller than that in the cell, osmotic pressure forces water into the cell to bring the system into balance, which causes the cell to burst. If the ionic strength of the solution is too high, osmotic pressure forces water out of the cell, and the cell breaks open as it shrinks. The cell wall provides the mechanical strength that helps protect plant cells that live in fresh-water ponds (too little salt) or seawater (too much salt) from osmotic shock. The cell wall also provides the mechanical strength that allows plant cells to support the weight of other cells.

Sec. 15.8

The most abundant structural polysaccharide is cellulose. There is so much cellulose in the cell walls of plants that it is the most abundant of all biological molecules. Cellulose is a linear polymer of glucose molecules, with a structure that resembles amylose more closely than amylopectin, as shown in Figure 25.26. The difference between cellulose and amylose can be seen by comparing Figures 25.25 and 25.26. Cellulose is formed by linking $\beta$-glucopyranose rings, instead of the $\alpha$-glucopyranose rings used in starch and glycogen.

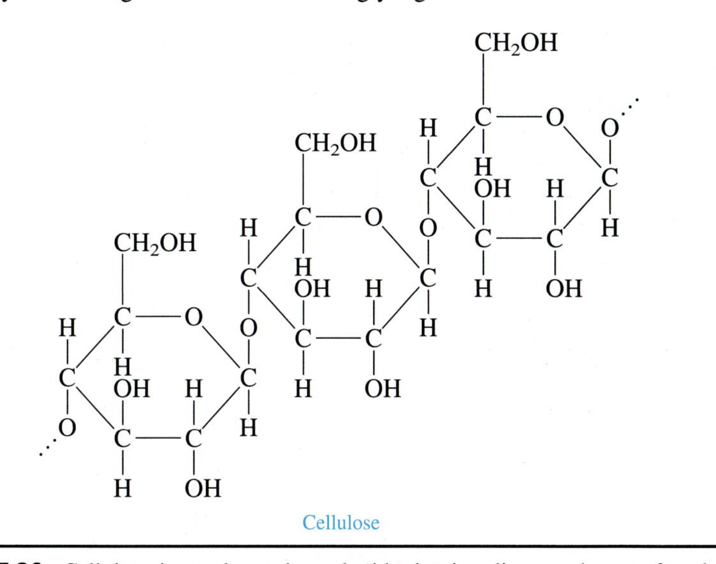


**Figure 25.26** Cellulose is another polysaccharide that is a linear polymer of D-glucopyranose monomers. Cellulose differs from amylose because it is a polymer of the $\beta$ isomer of D-glucopyranose.

The structure of cellulose provides the rigidity plants need.

Cellulose and starch provide an excellent example of the importance of structure in determining the function of biomolecules. At the turn of the century, Emil Fischer suggested that the structure of an enzyme is matched to the substance on which it acts, in much the same way that a lock and key are matched. As a result, the amylase enzymes in saliva that break down the $\alpha$-linkages between glucose molecules in starch cannot act on the $\beta$-linkages in cellulose.

Most animals can't digest cellulose because they don't have an enzyme that can cleave $\beta$-linkages between glucose molecules. Cellulose in their diet therefore serves only as fiber, or roughage. The digestive tracts of some animals, such as cows, horses, sheep, and goats, contain bacteria that have enzymes that cleave these $\beta$-linkages, so these animals can make use of the bacterially digested cellulose.

**EXERCISE 25.9**

Termites provide an example of the symbiotic relationship between bacteria and higher organisms. Termites cannot digest the cellulose in the wood they eat, but their digestive tracts are infested with bacteria that can. Propose a simple way of ridding a house from termites, without killing other insects that might be beneficial.

**SOLUTION** Killing termites is not a difficult task. There are a number of poisons that do an excellent job. Unfortunately, these compounds may also poison other insects or even small animals or children. The simplest way to get rid of termites is to treat the wood with a poison that kills the bacteria that digest the cellulose in wood, so that the termites starve to death.

## 25.14 LIPIDS

Any biomolecule that dissolves in nonpolar solvents—such as chloroform ($CHCl_3$), benzene ($C_6H_6$), or diethyl ether ($CH_3CH_2OCH_2CH_3$)—can be classified as a lipid. Because they are soluble in nonpolar solvents, lipids tend to be insoluble in water and they often feel oily or greasy to the touch.

Water beading on a freshly waxed car.

### NEUTRAL FATS AND OILS

Long-chain carboxylic acids such as stearic acid [$CH_3(CH_2)_{16}CO_2H$] are called **fatty acids** because they can be isolated from animal fats. These fatty acids are divided into two categories on the basis of whether they contain C=C double bonds: *saturated fatty acids* and *unsaturated fatty acids.*

Free fatty acids are seldom found in nature. They are usually tied up with alcohols or amines to form esters ($RCO_2R$) or amides (RCONHR). The most abundant lipids are the triesters formed when a glycerol molecule reacts with three fatty acids, as shown in Figure 25.27. These lipids have been known by a variety of names, including *fat, neutral fat, glyceride, triglyceride,* and *triacylglycerol.*

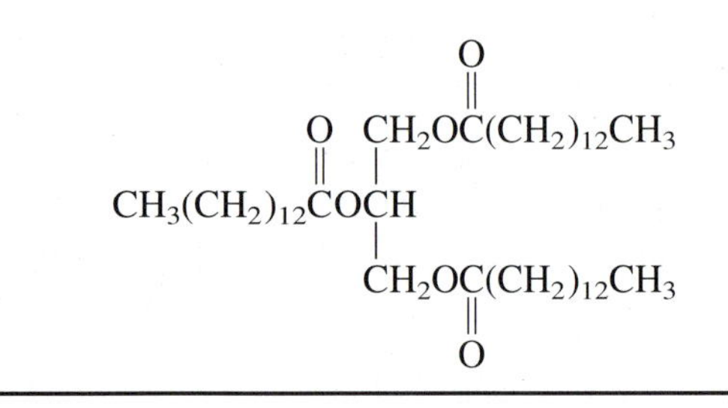


**Figure 25.27** This triester, which is known as trimyristin, can be isolated from nutmeg.

Most animal fats are complex mixtures of different triglycerides. As the percentage of unsaturated fatty acids in these fats becomes larger, the fat melts at lower and lower temperatures until it eventually becomes an oil at room temperature. Beef fat, which is roughly one-third unsaturated fatty acids, is a solid. Olive oil, which is roughly 80% unsaturated, is a liquid.

Fats and oils are used by living cells for only one purpose: to store energy. They are a far more efficient storage system than glycogen or starch because they give off between two and three times as much energy when they are burned. (The metabolism of glycogen releases 15.7 kJ per gram of carbohydrate consumed, whereas the metabolism of lipids gives approximately 40 kJ/g.) This explains why the seeds of many plants are relatively rich in oils, which provide the energy the seed needs to grow until the leaves can begin to produce energy by photosynthesis.

About 15% of the weight of the average human is fat. Although the average human has about 2500 kJ of energy available in the form of glycogen, fat stored in the body represents more than 400,000 kJ. If all of this energy were stored as glycogen, the human body would have to weigh at least one-third more than it does now.

### CHECKPOINT

Several hundred people were poisoned in Spain a few years ago when an unscrupulous merchant diluted olive oil with diesel oil. Explain both the similarities and differences between the structures of the components of these two "oils."

## POLAR LIPIDS

Fats and oils are neutral compounds. When one of the fatty acids in a triacylglycerol is replaced by a phosphate group, a **phospholipid** is obtained that has two nonpolar hydrophobic tails and a charged hydrophilic head, as shown in Figure 25.28. These

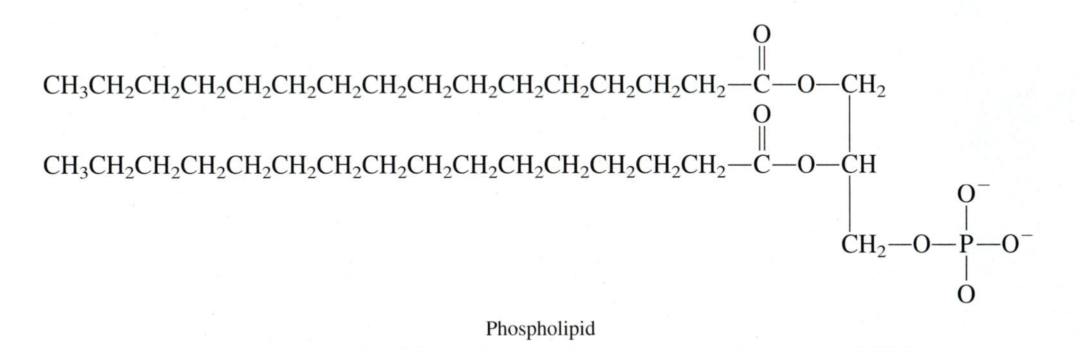


**Figure 25.28** Phospholipids contain two nonpolar hydrophobic tails and a polar hydrophilic head.

phospholipids spontaneously associate to form the membranes (see Figure 25.29) that surround living cells and help subdivide cells into compartments. Although lipids technically don't form polymers like proteins or carbohydrates, they do associate to form structures of macromolecular size and therefore belong in any discussion of polymers.

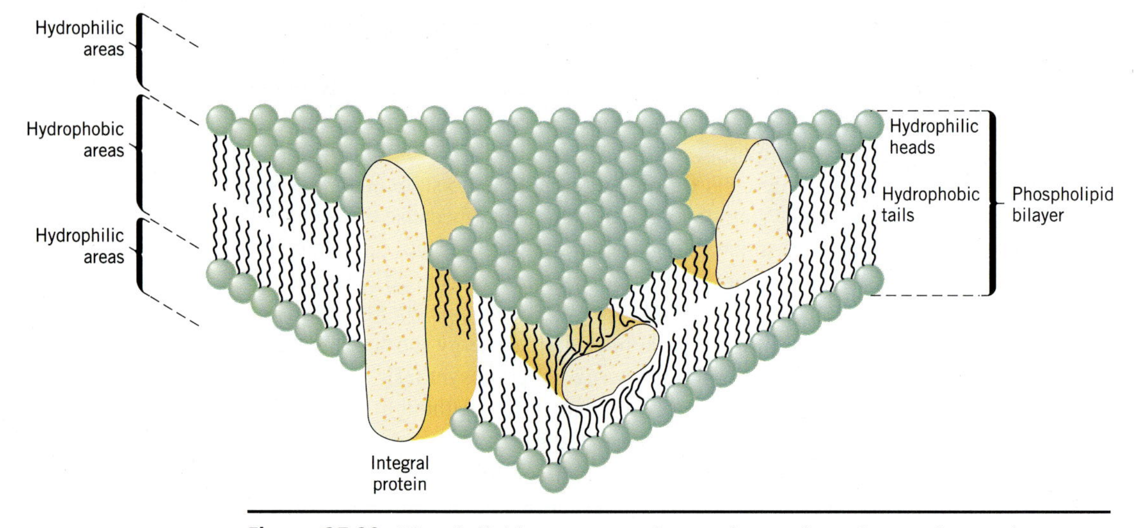


**Figure 25.29** Phospholipids spontaneously associate to form the membranes that separate cells from each other and divide, or compartmentalize, each cell.

## RESEARCH IN THE 90s

### MEDICINAL CHEMISTRY

A photograph of *Salix Alba.*

In 1763, the Reverend Edmund Stone took the first step toward the discovery of one of the most commonly used medicines when he noted that the bark of the English willow was an effective treatment for patients suffering from a fever. Stone explained the effect of willow bark by noting that ". . . many natural maladies carry their cures along with them, or their remedies lie not far from their causes." Thus, the English willow grows in the same moist regions where one was likely to catch the fever treated with its bark.

It took 50 years before the active ingredient in willow bark was isolated and named *salicin,* from the Latin name for the willow (*Salix alba*). Another 50 years elapsed before a large-scale synthesis for this compound was available. By that time, the compound was known as *salicylic acid* because saturated solutions in water are highly acidic (pH = 2.4).

By the end of the nineteenth century, salicylic acid was used to treat rheumatic fever, gout, and arthritis. Many patients treated with this drug complained of chronic stomach irritation because of its acidity and the large doses (6–8 grams per day) required to alleviate the symptoms of these disorders. Because his father was one of these patients, Felix Hoffman searched the chemical literature for a less acidic derivative of salicylic acid. In 1898, Hoffman reported that the acetyl ester of salicylic acid was simultaneously more effective and easier to tolerate than the parent compound. He named this compound *aspirin,* taking the prefix *a-* from the name of the acetyl group and *spirin* from the German name of the parent compound *spirsäure.*

The existence of a drug that reduced both pain and fever initiated a search for other compounds that could achieve the same result. Although it was based on trial and error, this search inevitably produced a variety of substances, such as those in Figure 25.30, that are analgesics, antipyretics, and/or anti-inflammatory agents. Analgesics relieve pain without decreasing sensibility or consciousness. Antipyretics reduce the body temperature when it is elevated. Anti-inflammatory agents counteract swelling or inflammation of the joints, skin, and eyes.

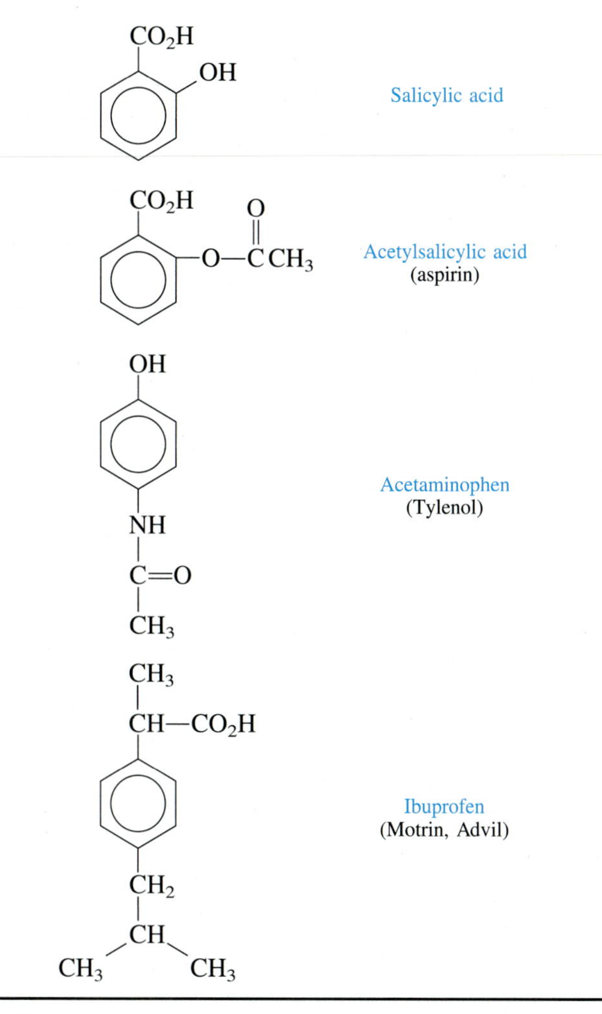


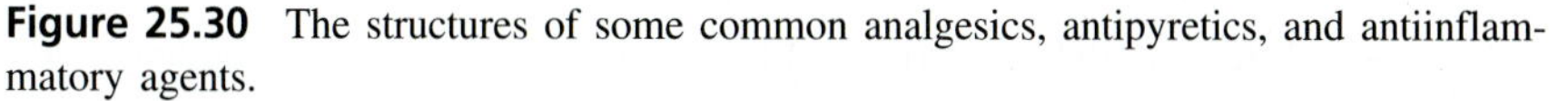

**Figure 25.30** The structures of some common analgesics, antipyretics, and antiinflammatory agents.

The mechanism by which aspirin and similar drugs act was first described in 1971 [J. R. Vane, *Nature,* **231**(25), 232–235 (1971)]. Vane noted that injury to tissue was often followed by the release of a group of hormones known as the *prostaglandins,* which have wide-spread physiological effects at very low concentrations. The prostaglandins regulate blood pressure, mediate the inflammatory response of the joints, induce the process by which blood clots, regulate the sleep/wake cycle, and, when appropriate, induce labor.

Vane suggested that aspirin and other nonsteroidal anti-inflammatory drugs (NSAID's) inhibit the enzyme that starts the process by which prostaglandins such as $PGE_2$ and $PGF_{2\alpha}$ are synthesized from the 20-carbon unsaturated fatty acid known as arachidonic acid shown in Figure 25.31. The steroidal anti-inflammatory drugs (such as hydrocortisone) achieve a similar effect by inhibiting the enzyme that releases arachidonic acid into the cell.

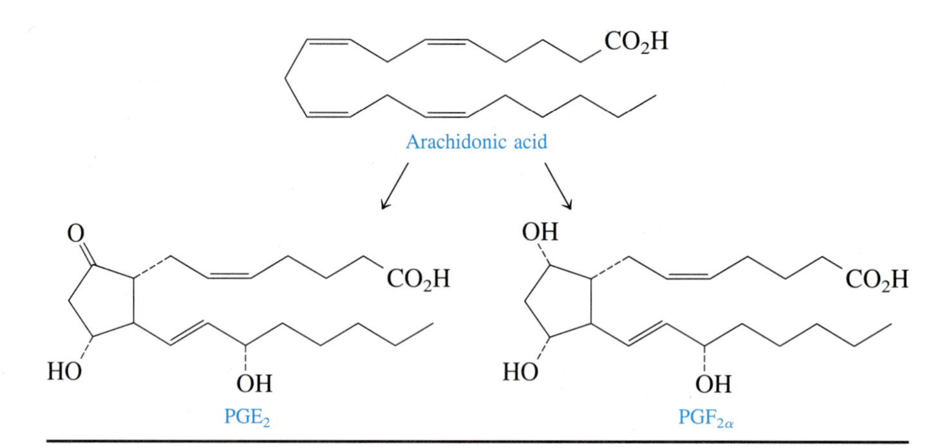


**Figure 25.31** The structures of the $PGE_2$ and $PGF_{2\alpha}$ prostaglandins synthesized from arachidonic acid.

Now that they are beginning to understand the mechanism by which drugs operate, medicinal chemists can approach the design of drugs by a rational process. A recent paper described progress toward the design of a drug to treat several of the debilitating diseases caused by protozoan parasites that afflict millions of people in Latin America, Africa, and Asia [W. N. Hunter, S. Bailey, J. Habash, S. J. Harrop, J. R. Helliwell, T. Aboagye-Kwarteng, K. Smith, and A. H. Fairlamb, *Journal of Molecular Biology,* **227,** 322–333 (1992)]. The potential target for this drug is an enzyme—trypanothione reductase (TR)—that protects the parasite from oxidative damage from the immune system of its mammalian host. Mammalian cells use a similar enzyme, known as glutathione reductase (GR), to protect against damage from oxidation reactions.

Hunter and co-workers found that the human GR enzyme has a smaller, more positively charged active site than the TR enzyme in the parasite. The structural information in this study can now be used to rationally modify a substrate of these enzymes until it possesses the following characteristics.

- The substrate must be too large to bind to the GR enzyme in humans.
- The substrate must have a high affinity for binding to the TR enzyme in the parasite.
- The substrate must inhibit the activity of the TR enzyme, thereby allowing the immune system of the mammalian host to attack and eventually destroy the parasite.

## 25.15 NUCLEIC ACIDS

Nucleic acids were first isolated in 1869. Their name reflects the fact that they are relatively strong acids that were isolated from the nuclei of cells. The nucleic acids are polymers with molecular weights as high as 100,000,000 grams per mole. They can be broken down, or digested, to form monomers known as nucleotides.

Each **nucleotide** is composed of three units: a sugar, an amine, and a phosphate, as shown in Figure 25.32. Nucleic acids are divided into two classes on the basis of the sugar used to form its nucleotides. **Ribonucleic acid (RNA)** is built on a β-D-ribofuranose ring. **Deoxyribonucleic acid (DNA)** contains a modified ribofuranose in which the —OH group on the second carbon atom has been removed, as shown in Figure 25.33.

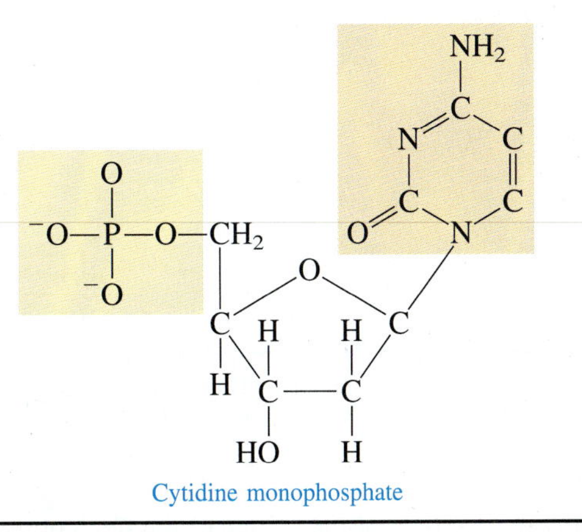


**Figure 25.32** Nucleotides contain a furanose sugar, an amine, and a phosphate ion. In this example, the amine is cytosine and the nucleotide is known as cytidine monophosphate.

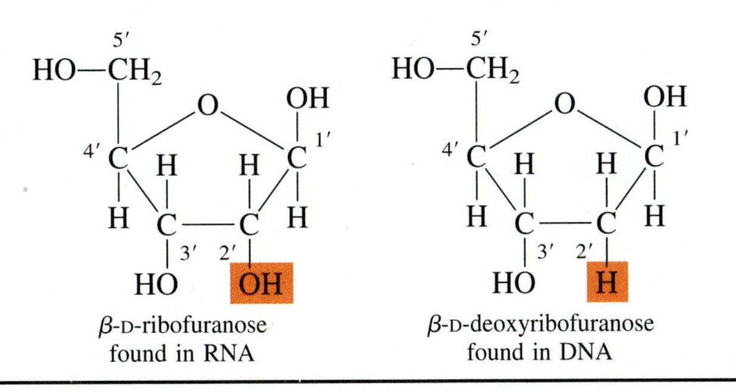


**Figure 25.33** One of the differences between DNA and RNA is the sugar on which the nucleotide is built. RNA uses β-D-ribofuranose. DNA uses a similar sugar in which the —OH group on the 2′ carbon atom has been replaced by a hydrogen.

### CHECKPOINT

Use the structure of a typical nucleotide shown in Figure 25.32 to explain why these compounds are called nucleic *acids*.

The five amines that form nucleic acids fall into two categories: **purines** and **pyrimidines.** There are three pyrimidines—cytosine, thymine and uracil—and two purines—adenine and guanine—as shown in Figure 25.34. DNA and RNA each contain four nucleotides. Both contain the same purines—adenine and guanine—and both also contain the pyrimidine cytosine. But the fourth nucleotide in DNA is thymine, whereas RNA uses uracil to complete its quartet of nucleotides.

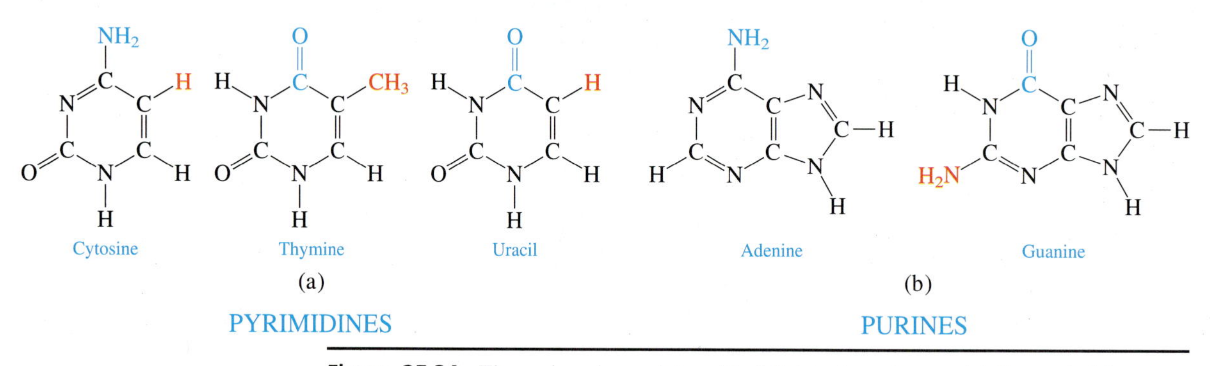


**Figure 25.34** The amines in nucleic acids fall into two classes: (*a*) the *pyrimidines* (uracil, cytosine, and thymine), and (*b*) the *purines* (adenine and guanine).

The carbon atoms in the sugar at the center of the nucleotide are numbered from 1′ to 5′. The —OH group on the 3′ carbon of one nucleotide can react with the phosphate attached to the 5′ carbon of another to form a dinucleotide held together by a phosphate ester bond. As the chain continues to grow, it becomes a **polynucleotide.** A short segment of a DNA chain is shown in Figure 25.35. Reading from the 5′ end of this chain to the 3′ end, this DNA segment contains the following sequence of amine substituents: adenine (A), cytosine (C), guanine (G), and thymine (T).

For many years, the role of nucleic acids in living systems was unknown. In 1944 Oswald Avery presented evidence that nucleic acids were involved in the storage and transfer of the genetic information needed for the synthesis of proteins. This suggestion was actively opposed by many of his contemporaries, who believed that the structure of the nucleic acids was too regular (and therefore too dull) to carry the information that codes for the thousands of different proteins a cell needs to survive.

In retrospect, the first clue about how nucleic acids function was obtained by Erwin Chargaff, who found that DNA always contains the same amounts of certain pairs of bases. There is always just as much adenine as thymine, for example, and just as much guanine as cytosine.

It was not until April 1954, however, that James Watson and Francis Crick proposed a structure for DNA that offered a mechanism for the storage of genetic information. Their structure consisted of two polynucleotide chains running in opposite directions that were linked by hydrogen bonds between a specific purine (A or G) on one strand and a specific pyrimidine (C or T) on the other, as shown in Figure 25.36. These strands form a helix that is not quite as tightly coiled as the $\alpha$-helix Pauling and Corey proposed for proteins.

This structure must be able to explain two processes. There must be some way to make perfect copies of the DNA that can be handed down to future generations **(replication).** There also must be some way to decode the information on the DNA chain **(transcription)** and translate this information into a sequence of amino acids in a protein **(translation).**

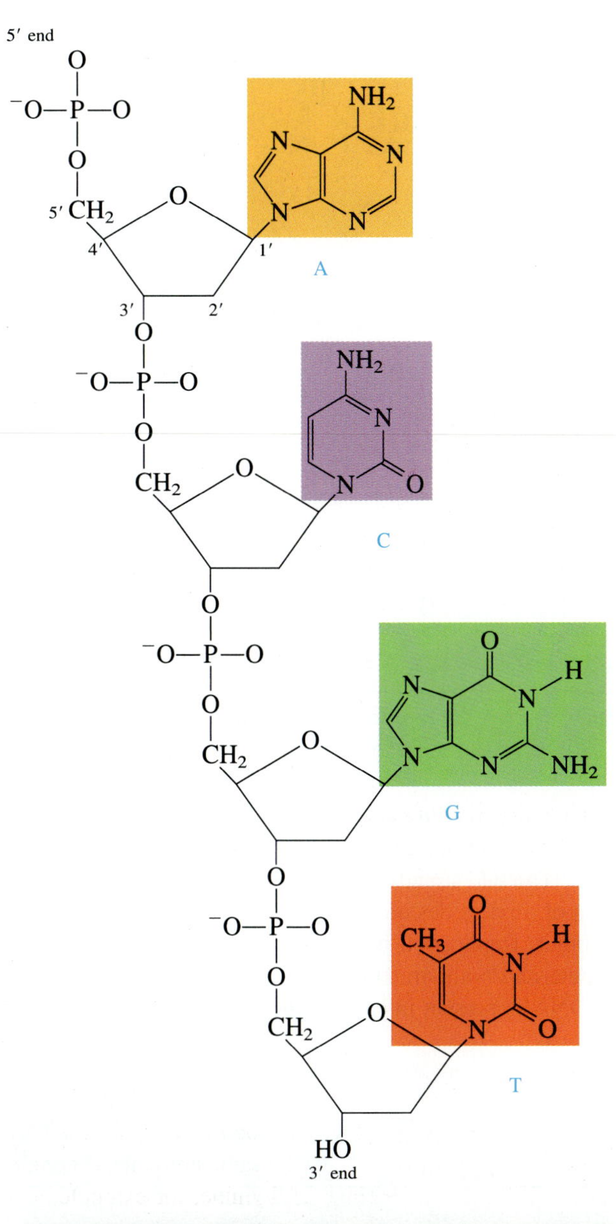


**Figure 25.35** This tetranucleotide provides an example of a short segment of a single strand of DNA.

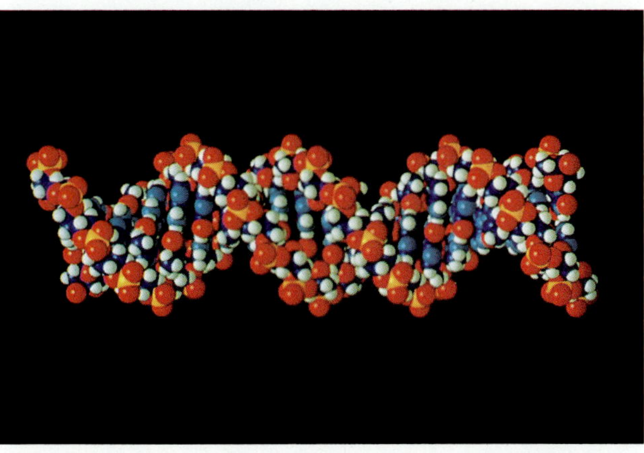

Computer model of the helical structure of the double-stranded DNA.

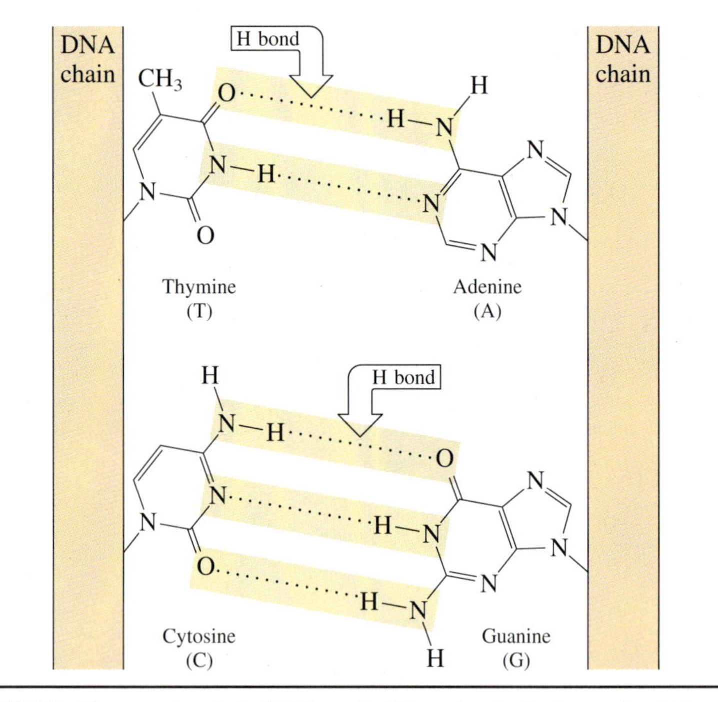


**Figure 25.36** The two strands in DNA are held together by hydrogen bonds between specific pairs of purine and pyrimidine bases. This figure shows the hydrogen bonds between A and T and between G and C.

Replication is easy to understand. According to Watson and Crick, an adenine on one strand of DNA is always paired with a guanine on the other, and a cytosine is always paired with a thymine. The two strands of DNA therefore complement each other perfectly; the sequence of nucleotides on one strand always can be predicted from the sequence on the other. Replication occurs when the two strands of the parent DNA molecule separate and both strands are copied simultaneously. Thus, one strand from the parent DNA is present in each of the daughter molecules produced when a cell divides.

**SPECIAL TOPIC**

## PROTEIN BIOSYNTHESIS

The information that tells a cell how to build the proteins it needs to survive is coded in the structure of the DNA in the nucleus of that cell. This code can't be based on a one-to-one match between nucleotides and amino acids because there are only 4 nucleotides and 20 amino acids must be coded. If the nucleotides are grouped in threes, however, there are 64 possible triplets, or **codons,** which is more than enough combinations to code for the 20 amino acids.

To understand how proteins are made, we have to divide the decoding process into two steps: *transcription* and *translation.* DNA only stores the genetic information; it is not involved in the process by which the information is used. The first step in protein biosynthesis therefore involves transcribing the information in the DNA structure into a useful form. In a separate step, this information is translated into a sequence of amino acids.

## TRANSCRIPTION

Before the information in DNA can be decoded, a small portion of the DNA double helix must be uncoiled, as shown in Figure 25.37. A strand of RNA is then synthesized that is a complementary copy of the DNA.

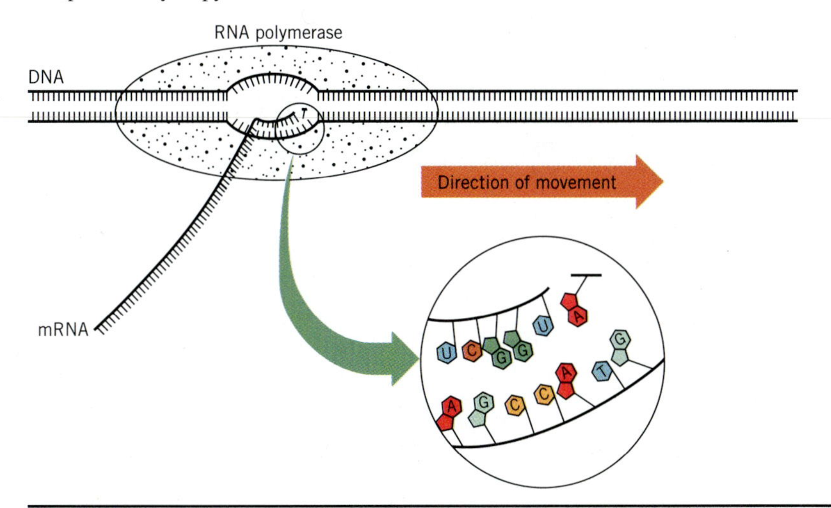


**Figure 25.37** During transcription, an RNA molecule is synthesized that is a complementary copy of one strand of the parent DNA molecule. This RNA now carries the message encoded in the DNA structure and is therefore called messenger RNA.

Assume that the section of the DNA that is copied has the following sequence of nucleotides, starting from the 3′ end:

3′ T-A-C-A-A-G-C-A-G-T-T-G-G-T-C-G-T-G . . . 5′ DNA

When we predict the sequence of nucleotides in the RNA complement, we have to remember that RNA uses U where T would be found in DNA. We also have to remember that base pairing occurs between two chains that run in opposite direction. The RNA complement of this DNA therefore should be written as follows:

3′ T-A-C-A-A-G-C-A-G-T-T-G-G-T-C-G-T-G . . . 5′ DNA
5′ A-U-G-U-U-C-G-U-C-A-A-C-C-A-G-C-A-C . . . 3′ mRNA

Since this RNA strand contains the message that was coded in the DNA, it is called **messenger RNA,** or **mRNA.**

## TRANSLATION

The messenger RNA now binds to a ribosome, where the message is translated into a sequence of amino acids. The amino acids that are incorporated into the protein being synthesized are carried by relatively small RNA molecules known as **transfer RNA,** or **tRNA.** There are at least 60 tRNAs, which differ slightly in their structure in each cell. At one end of each tRNA is a specific sequence of three nucleotides that can bind to the messenger RNA. At the other end there is a specific amino acid. Thus, each three-nucleotide segment of the messenger RNA molecule codes for the incorporation of a particular amino acid. The relationship between these triplets, or codons, and the amino acids is shown in Table 25.4.

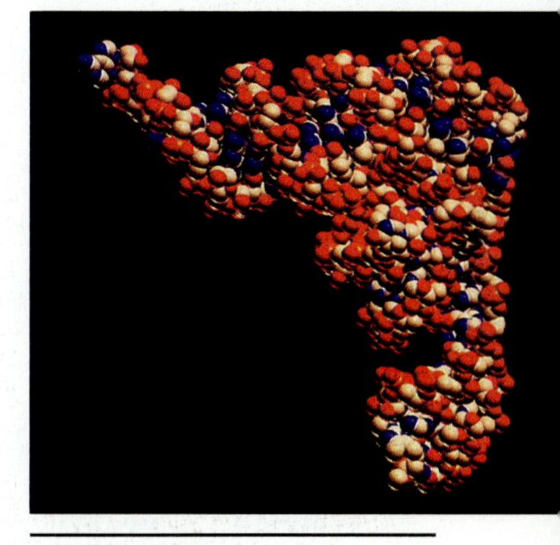

A model of the crystal structure of T-RNA.

**TABLE 25.4 The Genetic Code**

| First Position | Second Position | | | | Third Position |
|---|---|---|---|---|---|
| (5′ End) | *U* | *C* | *A* | *G* | (3′ End) |
| | PHE | SER | TYR | CYS | *U* |
| *U* | PHE | SER | TYR | CYS | *C* |
| | LEU | SER | [a] | [a] | *A* |
| | LEU | SER | [a] | TRP | *G* |
| | LEU | PRO | HIS | ARG | *U* |
| *C* | LEU | PRO | HIS | ARG | *C* |
| | LEU | PRO | GLN | ARG | *A* |
| | LEU | PRO | GLN | ARG | *G* |
| | ILE | THR | ASN | SER | *U* |
| *A* | ILE | THR | ASN | SER | *C* |
| | ILE | THR | LYS | ARG | *A* |
| | MET | THR | LYS | ARG | *G* |
| | VAL | ALA | ASP | GLY | *U* |
| *G* | VAL | ALA | ASP | GLY | *C* |
| | VAL | ALA | GLU | GLY | *A* |
| | VAL | ALA | GLU | GLY | *G* |

[a] Three triplets code for termination of the polypeptide chain: *UAA, UGA,* and *UAG.*

**EXERCISE 25.10**

Assume that the DNA chain that codes for the synthesis of a particular protein contains the triplet A-G-T (reading from the 3′ to the 5′ end). Predict the sequence of nucleotides in the triplet, or codon, that would be built in the messenger RNA constructed on this DNA template. Then predict the amino acid that would be incorporated at this point in the protein.

**SOLUTION** The messenger RNA synthesized when this DNA chain is read as a complement of the DNA. Every time a G appears, a C is placed on the complementary chain, and vice versa. In DNA, A and T are paired, but RNA uses U instead of T:

DNA: 3′ . . . A-G-T . . . 5′
mRNA: 5′ . . . U-C-A . . . 3′

We now look up the U-C-A codon in Table 25.4. According to this table, this triplet on mRNA codes for the amino acid serine.

The signal to start making a polypeptide chain in simple, prokaryotic cells is the triplet AUG, which codes for the amino acid methionine (MET). The synthesis of every protein in these cells therefore starts with a MET residue at the *N*-terminal end of the polypeptide chain. After the tRNA carrying MET binds to the start signal on the messenger RNA, a tRNA carrying the second amino acid binds to the next codon. A dipeptide is synthesized when the MET residue is transferred from the first tRNA to the amino acid on the second tRNA. If the DNA described in this section were translated, the dipeptide would be MET-PHE (reading from the *N*-terminal to the *C*-terminal amino acid).

The messenger RNA now moves through the ribosome, and a tRNA carrying the third amino acid (VAL) binds to the next codon. The dipeptide is then transferred to the amino acid on this third tRNA to form a tripeptide. This sequence of steps continues until one of three codons is encountered: UAA, UGA, or UAG. These codons give the signal for terminating the synthesis of the polypeptide chain, and the chain is cleaved from the last tRNA residue.

The sequence of DNA described in this section would produce the following sequence of amino acids.

MET-PHE-VAL-ASN-GLN-HIS- . . .

This polypeptide is not necessarily an active protein. All proteins in prokaryotic cells start with MET when synthesized, but not all proteins have MET first in their active form. It is often necessary to clip off this MET after the polypeptide has been synthesized to give a protein with a different *N*-terminal amino acid.

Modifications to the polypeptide often have to be made before an active protein is formed. Insulin, for example, consists of two polypeptide chains connected by disulfide linkages. In theory, it would be possible to make these chains one at a time and then try to assemble them to make the final protein. Nature, however, has been more subtle. The polypeptide chain that is synthesized contains a total of 81 amino acids. All of the disulfide bonds that will be present in insulin are present in this chain. The protein is made when a sequence of 30 amino acids is clipped out of the middle of this polypeptide chain.

## SUMMARY

**1.** The term *polymer* has come to mean compounds with large molecular weights that contain long chains of relatively simple repeating units, or monomers. Polymers that are flexible, or elastic, and return to their original shapes when stretched are called **elastomers.** Those that flow when heated are often called **plastics,** or **thermoplastic polymers.** Polymers that take on rigid shapes that can't be changed without destroying the material are called **thermoset plastics.**

**2.** Straight-chain and branched polymers are often plastics. As the amount of cross-linking between polymer chains increases, the polymer initially becomes more elastic. When the cross-linking becomes too extensive, the polymer may become rigid enough to be classified as a thermoset plastic.

**3.** Polymers that contain side chains are classified as either **atactic, syndiotactic,** or **isotactic,** depending on whether the side chains are arranged in random, alternating, or identical directions. Polymers formed by simply adding monomers to a polymer chain are **addition polymers. Condensation polymers** are formed by reactions that condense, or eliminate, a small molecule each time another monomer is added to the polymer chain.

**4.** Amino acids are the building blocks from which proteins are formed. There are 20 common amino acids used to synthesize proteins, all but one of which has the generic formula $H_2N—CHR—CO_2H$. In aqueous solutions at neutral pH, amino acids are present as zwitterions with the generic formula $H_3N^+—CHR—CO_2^-$.

**5.** The amino acids are divided into four categories, depending on whether the R groups are: (1) nonpolar, (2) polar, (3) negatively charged at neutral pH, or (4) positively charged at neutral pH. Nonpolar amino acids have R groups that are hydrophobic; the other three classes have hydrophilic R groups.

**6.** The sequence of amino acids in a protein defines its primary structure. As the polymer chain folds upon itself, hydrogen bonds form between peptide bonds on different portions of the protein, producing the protein's secondary structure. Interactions between the R groups on different portions of the protein give rise to its tertiary structure. Proteins that consist of more than one polymer chain held together by ionic or hydrophobic interactions are said to have a quaternary structure. Anything that disrupts the secondary, tertiary, or quaternary structure of the protein is said to denature the protein.

**7.** Carbohydrates are polyhydroxy aldehydes or ketones. Monosaccharides such as glucose (blood sugar) and fructose (fruit sugar) contain a single aldehyde or ketone group. Disaccharides such as lactose (milk sugar) and sucrose (cane sugar) and polysaccharides, such as starch, glycogen, and cellulose are formed by condensation reactions between monosaccharides.

**8.** Any biologically important compound that is soluble in a nonpolar solvent is classified as a lipid. Lipids are soluble in nonpolar solvents because they have structures that typically include long hydrocarbon chains. They therefore contain carbon in its most highly reduced state, and liberate more energy when burned to form $CO_2$ and $H_2O$ than any other class of biological molecules.

**9.** Nucleic acids are divided into two categories: ribonucleic acids (RNA) and deoxyribonucleic acids (DNA). The individual nucleotides from which nucleic acids are built play an important role in the metabolism of proteins, carbohydrates, and lipids. Polynucleotides are primarily involved in the storage and expression of the genetic information in the cell.

## KEY TERMS

**Addition polymer** (p. 939)
**Amino acid** (p. 952)
**Atactic** (p. 938)
**Carbohydrate** (p. 951)
**Codon** (p. 977)
**Condensation polymer** (p. 939)
**DNA** (p. 973)
**Denaturation** (p. 962)
**Disaccharide** (p. 965)
**Disulfide linkage** (p. 959)
**Elastomer** (p. 940)
**Fatty acid** (p. 968)
**Hydrophobic interaction** (p. 961)
**Isotactic** (p. 939)
**Lipid** (p. 952)
**Messenger RNA (mRNA)** (p. 977)
**Monosaccharide** (p. 962)
**Nucleic acid** (p. 952)
**Nucleotide** (p. 973)
**Nylon** (p. 948)
**Peptide** (p. 955)
**Phospholipid** (p. 969)
**Polyamide** (p. 948)
**Polycarbonate** (p. 950)
**Polyester** (p. 948)
**Polymer** (p. 936)
**Polynucleotide** (p. 974)
**Polysaccharide** (p. 966)
**Protein** (p. 951)
**Purine** (p. 973)
**Pyrimidine** (p. 973)
**Replication** (p. 974)
**RNA** (p. 973)
**Syndiotactic** (p. 939)
**Tacticity** (p. 938)
**Translation** (p. 974)
**Transcription** (p. 974)
**Transfer RNA (tRNA)** (p. 977)
**Ziegler–Natta catalysts** (p. 943)
**Zwitterion** (p. 952)

## PROBLEMS

### Polymers

**25-1** Use Berzelius's definition of a polymer to sort the following compounds into groups of polymers.

(a) formaldehyde, $H_2CO$ (b) ethylene, $C_2H_4$
(c) glyceraldehyde, $C_3H_6O_3$ (d) 2-butene, $C_4H_8$
(e) cyclohexane, $C_6H_{12}$ (f) glucose, $C_6H_{12}O_6$

**25-2** Which of the following substances are polymers?

(a) Teflon (b) proteins (c) starch (d) sugars, such as cane sugar (e) poly(vinyl chloride)
(f) polyethylene (g) cellulose (h) lipids

### Definitions of Terms

**25-3** Describe the differences between linear, branched, and cross-linked polymers.

**25-4** Explain the following observations:

- Linear polymers, such as the polyethylene in garbage bags, tear when stretched.
- Lightly cross-linked polymers, such as rubber, return to their original shape when stretched.
- Highly cross-linked polymers, such as bakelite, break when "stretched."

**25-5** Describe the difference between a homopolymer (such as polystyrene) and a copolymer (such as styrene–butadiene rubber).

**25-6** Describe the difference between random, regular, block, and graft copolymers. Give an example of the sequence of monomer units that would be found in a short length of each of these polymers.

**25-7** Describe the differences in the structures of atactic, isotactic, and syndiotactic forms of poly(vinyl chloride).

**25-8** Calculate the range of molecular weights of polyethylene molecules in a sample in which individual chains contain between 500 and 50,000 ($—CH_2CH_2—$) units.

**25-9** Individual chains in polystyrene polymers typically weigh between 200,000 and 300,000 amu. If the formula for the monomer is $C_6H_5CH{=}CH_2$, how many monomers does the typical chain contain in this addition polymer?

**25-10** Calculate the average molecular weight of the polymer chains in the cellulose nitrate in ping-pong balls if the formula for the monomer is $C_6H_9NO_7$ and the average polymer chain contains 1500 monomers.

**25-11** Explain why it is possible to measure the molecular weight of linear polymers such as polyethylene, but not cross-linked polymers such as those found in rubber.

**25-12** Thermoplastic polymers flow when heated and can be molded into shapes they retain on cooling. Explain how increasing the amount of cross-linking between polymer chains can transform a thermoplastic polymer into a rigid thermoset polymer.

**25-13** Describe the difference between addition and condensation polymers. Give an example of each.

**25-14** Classify the following as either addition or condensation polymers.

(a) Polyethylene, $(—CH_2CH_2—)_n$

(b) Poly(vinyl chloride), $(—CH_2CHCl—)_n$

(c) Nylon, $(—CO(CH_2)_4CONH(CH_2)_6NH—)_n$

(d) Polyester, $(—OCH_2CH_2OCOC_6H_4CO—)_n$

(e) Silicone, $(—OSi(CH_3)_2—)_n$

(f) Saran, $(—CH_2CCl_2—)_n(—CH_2CHCl—)_m$

(g) Plexiglass, $(—CH_2C(CH_3)(CO_2CH_3)—)_n$

### Elastomers

**25-15** What are the characteristic properties of an elastomer?

**25-16** Explain why elastomers must contain long, flexible molecules that are coiled in the natural state. Explain why polymers have to have some cross-links to be elastomers, but can't have too many.

**25-17** Natural rubber becomes too soft when heated and too hard when cooled to be of much use. Explain what happens when natural rubber is treated with sulfur that turns this material into a commercially useful product.

### Free-Radical Polymerization Reactions

**25-18** Describe what happens during the chain-initiation, chain-propagation, and chain-termination steps when propylene ($CH_3CH{=}CH_2$) is polymerized by a free-radical mechanism.

**25-19** Write the Lewis structures of the intermediates in the free-radical polymerization of vinyl chloride, $CH_2{=}CHCl$.

### Ionic and Coordination Polymerization Reactions

**25-20** Describe what happens during the chain-initiation, chain-propagation, and chain-termination steps in the anionic polymerization of vinyl chloride ($CH_2{=}CHCl$) when methyllithium ($CH_3Li$) is used as the chain initiator. Write the Lewis structures of all intermediates.

**25-21** Describe what happens during the chain-initiation, chain-propagation, and chain-termination steps in the cationic polymerization of vinylidene chloride ($CH_2{=}CCl_2$) when hydrobromic acid (HBr) is used to initiate the reaction. Write the Lewis structures of all intermediates.

**25-22** A typical Ziegler–Natta catalyst consists of a mixture of $TiCl_3$ and $Al(C_2H_5)_3$. Explain how the formation of a complex between a molecule of ethylene and the titanium atom in this catalyst can be thought of as an example of a Lewis acid–base reaction.

### Addition Polymers

**25-23** Polypropylene can be made in both atactic and isotactic forms. The atactic form is soft and rubbery, with no commercial value. The isotactic form is much more crystalline—it is hard enough, for example, to be used for furniture. Explain the difference between the physical properties of these two forms of the polymer.

**25-24** Polyethylene can be made in both high-density and low-density forms. One of these polymers has a linear structure; the other is branched, with short side chains of up to five carbon atoms attached to the polymer backbone. Which structure would you expect to give the denser polymer? Which structure would give the more crystalline polymer?

**25-25** Teflon, $(—CF_2CF_2—)_n$, is a waxy polymer to which practically nothing sticks. It is also the most chemically inert of all polymers. Describe at least five ways in which a polymer with these properties can be used.

**25-26** Use the concept of van der Waals forces to explain why plastic wrap made from Saran clings to itself, whereas plastic wrap made from polyethylene does not.

### Condensation Polymers

**25-27** Predict the formula of the repeating unit in the condensation polymers formed by the following reactions:

(a) Dacron: $HOCH_2CH_2OH + ClC(=O)C_6H_4C(=O)Cl \longrightarrow HCl + \ldots$

(b) Nylon 6,6: $H_2N(CH_2)_6NH_2 + ClC(=O)(CH_2)_4C(=O)Cl \longrightarrow HCl + \ldots$

(c) Polycarbonate, $HOCH_2CH_2OH + (CH_3O)_2C{=}O \longrightarrow CH_3OH + \ldots$

(d) Polyaramide, $H_2NC_6H_4C(=O)Cl \longrightarrow HCl + \ldots$

(e) Silicone rubber, $(CH_3)_2Si(OH)_2 \longrightarrow H_2O + \ldots$

**25-28** Explain the difference between the reactions used to prepare nylon 6,6 and nylon 6.

**25-29** Explain why proteins, polysaccharides, and nucleic acids all can be classified as condensation polymers.

### The Amino Acids

**25-30** Explain why amino acids such as glycine ($H_2NCH_2CO_2H$) exist in aqueous solution as zwitterions ($H_3N^+CH_2CO_2^-$).

**25-31** Which of the following forms of the amino acid glycine is present in strongly acidic solutions? In strongly basic solutions?

(a) $H_3N^+CH_2CO_2H$ (b) $H_3N^+CH_2CO_2^-$
(c) $H_2NCH_2CO_2^-$

**25-32** Write the formula for lysine as it would exist in aqueous solution at pH $\approx$ 1.

**25-33** Write the formula for serine as it would exist at pH = 1 and pH = 13.

**25-34** Which of the following amino acids contain side chains that cannot contribute to forming a hydrophobic pocket?

(a) alanine (b) isoleucine (c) methionine (d) serine
(e) lysine

**25-35** The common amino acids in proteins have the generic formula $H_2NCHRCO_2H$. Classify the following amino acids as either hydrophilic or hydrophobic:

(a) Alanine: R = $-CH_3$
(b) Glutamic acid: R = $-CH_2CH_2CO_2H$
(c) Arginine: R = $-CH_2CH_2CH_2NHC(NH_2){=}NH_2^+$
(d) Methionine: R = $-CH_2CH_2SCH_3$
(e) Threonine: R = $-CH(CH_3)OH$

**25-36** Use examples to explain why amino acids with an R group that is either positively charged or negatively charged at neutral pH are hydrophilic.

### Peptides and Proteins

**25-37** How do amino acids, peptides, and proteins differ?

**25-38** How many dipeptides can be formed when alanine, R = $-CH_3$, is condensed with cysteine, R = $-CH_2SH$? What is the difference between these dipeptides?

**25-39** Describe the tripeptides that can form when alanine, R = $-CH_3$, is condensed with cysteine, R = $-CH_2SH$.

**25-40** We can describe peptide chains by listing the amino acids from the *N*-terminal to the *C*-terminal residue. Write the amino acid sequences for all the tetrapeptides containing four different amino acids that can be formed from the amino acids ALA, LYS, SER, and TYR.

### The Structure of Proteins

**25-41** Describe how the primary, secondary, tertiary, and quaternary structures of a protein differ.

**25-42** Describe the four factors that give rise to the tertiary structure of a protein. Give an example of each.

**25-43** List the amino acids that can form hydrogen bonds in the tertiary structure of a protein.

**25-44** List the amino acids that can form ionic bonds in the tertiary structure of a protein.

**25-45** List the amino acids that can form the hydrophobic pockets in the tertiary structure of a protein.

**25-46** Explain why each of the following can denature a protein.

(a) Increasing the temperature (b) Changing the pH
(c) Adding a detergent (d) Adding an oxidizing or reducing agent (e) Adding compounds, such as urea, $(H_2N)_2C{=}O$, that form strong hydrogen bonds

### Carbohydrates

**25-47** Which of the following compounds satisfy the literal definition of the term *carbohydrate?*

(a) glucose, $C_6H_{12}O_6$ (b) sucrose, $C_{12}H_{22}O_{11}$
(c) cellulose, $(C_6H_{10}O_5)_n$ (d) glyceraldehyde, $HOCH_2CHOHCHO$ (e) dihydroxyacetone, $(HOCH_2)_2C{=}O$ (f) ribose, $C_5H_{10}O_5$

**25-48** Explain the difference between an aldose and a ketose and between a triose and a tetrose.

**25-49** Describe the difference between $\alpha$-D-glucopyranose and $\beta$-D-glucopyranose. Describe the difference between $\alpha$-D-glucopyranose and $\alpha$-D-glucofuranose.

**25-50** How do maltose, lactose, and sucrose differ?

**25-51** How do glycogen, amylose, amylopectin, and cellulose differ?

**25-52** Explain why humans are able to digest glycogen, amylose, and amylopectin but not cellulose. What would enable us to digest cellulose?

**25-53** Which of the following compounds is not an aldose?

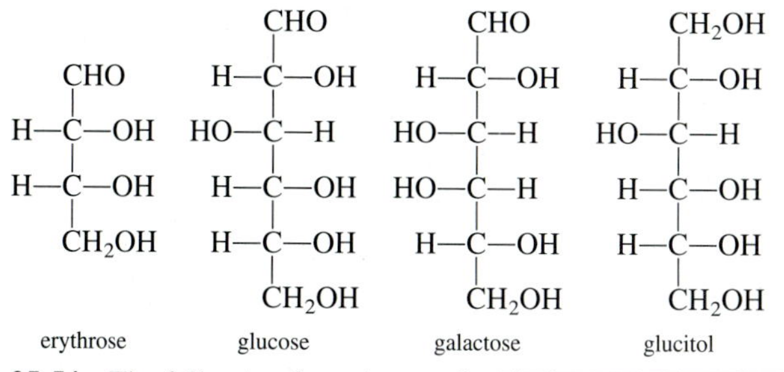


**25-54** The following formulas are the Fischer projections for D-erythrose and D-threose. Draw the Fischer projections for the L isomers of these carbohydrates.

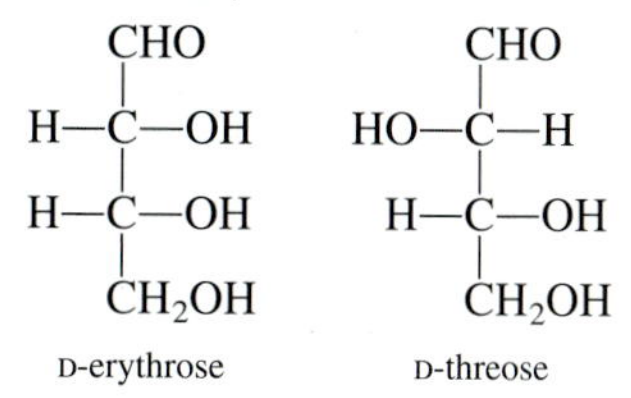


### Lipids

**25-55** Explain why lipids, such as fats and oils, are not soluble in water but are soluble in solvents such as chloroform, $CHCl_3$; benzene, $C_6H_6$, and diethyl ether, $(CH_3CH_2)_2O$.

**25-56** In what ways are corn oil and olive oil similar to petroleum oil? What is the difference between these oils that makes the first two edible but not the third?

**25-57** Draw the structure of the triglyceride formed when stearic acid combines with glycerol.

**25-58** What is the difference between the structures of the fatty acids in fats and oils? How does this difference cause one of these types of compounds to be a solid at room temperature and the other to be a liquid?

**25-59** Calculate the average oxidation state of the carbon atoms in a typical carbohydrate, such as glucose, $C_6H_{12}O_6$, and a typical fatty acid, such as palmitic acid, $C_{16}H_{32}O_2$. Use these oxidation states to explain why fatty acids give off much more energy than carbohydrates are digested.

**25-60** Glucose, $C_6H_{12}O_6$, gives off 2870 kJ/mol when burned, and palmitic acid, $C_{16}H_{32}O_2$, gives off 9790 kJ/mol when burned. Use these data to estimate the energy released per gram when carbohydrates and lipids are burned.

**25-61** Which of the following carboxylic acids is least likely to be found in olive oil?

(a) butyric acid (b) palmitic acid (c) oleic acid

**25-62** The starting material for the synthesis of phospholipids are the phosphatidic acids. Draw the structure of the phosphatidic acid that would be produced by esterifying glycerol with two molecules of stearic acid ($C_{17}H_{35}CO_2H$) and one molecule of phosphoric acid.

### Nucleic Acids

**25-63** Draw the structure of one of the nucleotides found in nucleic acids. Show how the structures of the monomers that form DNA differ from those that form RNA.

**25-64** Classify each of the five common nucleic acids—adenine, cytosine, guanine, thymine, and uracil—as either a purine or a pyrimidine.

**25-65** Explain why studies of nucleic acids invariably find roughly equivalent amounts of purine and pyrimidine nucleotides. Explain why nucleic acids that contain more adenine than guanine also contain more thymine than cytosine.

**25-66** Draw the structures of the side chains on nucleic acids and show how hydrogen bonds form between adenine and thymine and between cytosine and guanine.

**25-67** Describe the difference between replication, transcription, and translation.

### Protein Biosynthesis

**25-68** Describe the difference between the roles played by mRNA and tRNA in protein biosynthesis.

**25-69** Assume that the following sequence of nucleotides on one strand of a DNA molecule is used as the template to code for a protein.

3′ -T-A-C-A-A-G-C-A-G-T-T-G-G-T-C-G-T-G- 5′

Write the sequence of nucleotides, reading from the 5′ to the 3′ end, in the messenger RNA produced when this strand of DNA is transcribed. Describe the polypeptide, starting from the *N*-terminal end, that would be produced during protein biosynthesis.

# TABLES


| | |
|---|---|
| TABLE A.1 | The SI System |
| TABLE A.2 | Values of Selected Fundamental Constants |
| TABLE A.3 | Selected Conversion Factors |
| TABLE A.4 | The Vapor Pressure of Water |
| TABLE A.5 | Radii of Atoms and Ions |
| TABLE A.6 | Ionization Energies |
| TABLE A.7 | Electron Affinities |
| TABLE A.8 | Electronegativities |
| TABLE A.9 | Acid-Dissociation Equilibrium Constants |
| TABLE A.10 | Base-Ionization Equilibrium Constants |
| TABLE A.11 | Solubility Product Equilibrium Constants |
| TABLE A.12 | Complex Formation Equilibrium Constants |
| TABLE A.13 | Standard-State Reduction Potentials |
| TABLE A.14 | Bond-Dissociation Enthalpies |
| TABLE A.15 | Standard-State Enthalpy of Formation, Free Energy of Formation, and Absolute Entropy Data |

**TABLE A.1 The SI System**

Base Units in the SI System

| Physical Quantity | Name of Unit | Symbol |
|---|---|---|
| Length | meter | m |
| Mass | kilogram | kg |
| Time | second | s |
| Temperature | Kelvin | K |
| Electric current | ampere | amp |
| Luminous intensity | candela | cd |
| Amount of substance | mole | mol |

Non-SI Units in Common Use

| Physical Quantity | Name of Unit | Symbol |
|---|---|---|
| Volume | liter | L ($10^{-3}$ $m^3$) |
| Length | angstrom | Å (0.1 nm) |
| Pressure | atmosphere | atm (101.325 kPa) |
| | torr | mmHg (133.32 Pa) |
| Energy | calorie | cal (4.184 J) |
| | electron volt | ev ($1.6022 \times 10^{-19}$ J) |
| Temperature | degree Celsius | °C (K − 273.15) |
| Concentration | molarity | $M$ (mol/L or mol/$dm^3$) |

Common Derived Units in SI

| Physical Quantity | Name of Unit | Symbol |
|---|---|---|
| Energy | joule | J (kg-$m^2$/$s^2$) |
| Frequency | hertz | Hz (cycles/s) |
| Force | newton | N (kg-m/$s^2$) |
| Pressure | pascal | P (N/$m^2$) |
| Power | watt | W (J/s) |
| Electric charge | coulomb | C (amp-s) |
| Electric potential | volt | V (J/C) |
| Electric resistance | ohm | Ω (V/amp) |
| Electric conductance | siemens | S (amp/V) |
| Electric capacitance | farad | F (C/V) |

Fractions and Multipliers for Use in SI

| | | | |
|---|---|---|---|
| exa, E | $10^{18}$ | deci, d | $10^{-1}$ |
| peta, P | $10^{15}$ | centi, c | $10^{-2}$ |
| tera, T | $10^{12}$ | milli, m | $10^{-3}$ |
| giga, G | $10^{9}$ | micro, $\mu$ | $10^{-6}$ |
| mega, M | $10^{6}$ | nano, n | $10^{-9}$ |
| kilo, k | $10^{3}$ | pico, p | $10^{-12}$ |
| hecto, h | $10^{2}$ | femto, f | $10^{-15}$ |
| deka, da | $10^{1}$ | atto, a | $10^{-18}$ |

**TABLE A.2 Values of Selected Fundamental Constants**

| | |
|---|---|
| Speed of light in a vacuum ($c$) | $c = 2.99792458 \times 10^{8}$ m/s |
| Charge on an electron ($q_e$) | $q_e = 1.6021892 \times 10^{-19}$ C |
| Rest mass of an electron ($m_e$) | $m_e = 9.109534 \times 10^{-28}$ g |
| | $m_e = 5.4858026 \times 10^{-4}$ amu |
| Rest mass of a proton ($m_p$) | $m_p = 1.6726485 \times 10^{-24}$ g |
| | $m_p = 1.00727647$ amu |
| Rest mass of a neuron ($m_n$) | $m_n = 1.6749543 \times 10^{-24}$ g |
| | $m_n = 1.008665012$ amu |
| Faraday's constant ($F$) | $F = 96{,}484.56$ C/mol |
| Planck's constant ($h$) | $h = 6.626176 \times 10^{-34}$ J-s |
| Ideal gas constant ($R$) | $R = 0.0820568$ L-atm/mol-K |
| | $R = 8.31441$ J/mol-K |
| Atomic mass unit (amu) | 1 amu $= 1.6605655 \times 10^{-24}$ g |
| Boltzmann's constant ($k$) | $k = 1.380662 \times 10^{-23}$ J/K |
| Avogadro's constant ($N$) | $N = 6.022045 \times 10^{23}$ $mol^{-1}$ |
| Rydberg constant ($R_H$) | $R_H = 1.09737318 \times 10^{7}$ $m^{-1}$ |
| | $= 1.09737318 \times 10^{-2}$ $nm^{-1}$ |
| Heat capacity of water | $C = 75.376$ J/mol-K |

**TABLE A.3 Selected Conversion Factors**

| | |
|---|---|
| Energy | 1 J = 0.2390 cal = $10^7$ erg<br>1 cal = 4.184 J (by definition)<br>1 ev/atom = 1.6021892 × $10^{-19}$ J/atom = 96.484 kJ/mol |
| Temperature | K = °C + 273.15<br>°C = 5/9(°F − 32)<br>°F = 9/5(°C) + 32 |
| Pressure | 1 atm = 760 mmHg = 760 torr = 101.325 kPa |
| Mass | 1 kg = 2.2046 lb<br>1 lb = 453.59 g = 0.45359 kg<br>1 oz = 0.06250 lb = 28.350 g<br>1 ton = 2000 lb = 907.185 kg<br>1 tonne (metric) = 1000 kg = 2204.62 lb |
| Volume | 1 mL = 0.01 L = 1 $cm^3$ (by definition)<br>1 oz (fluid) = 0.031250 qt = 0.029573 L<br>1 qt = 0.946326 L<br>1 L = 1.05672 qt |
| Length | 1 m = 39.370 in.<br>1 mi = 1.60934 km<br>1 in. = 2.540 cm (by definition) |

**TABLE A.4 The Vapor Pressure of Water**

| Temperature (°C) | Pressure (mmHg) | Temperature (°C) | Pressure (mmHg) | Temperature (°C) | Pressure (mmHg) | Temperature (°C) | Pressure (mmHg) |
|---|---|---|---|---|---|---|---|
| 0 | 4.6 | 13 | 11.2 | 26 | 25.2 | 39 | 52.4 |
| 1 | 4.9 | 14 | 12.0 | 27 | 26.7 | 40 | 55.3 |
| 2 | 5.3 | 15 | 12.8 | 28 | 28.3 | 41 | 58.3 |
| 3 | 5.7 | 16 | 13.6 | 29 | 30.0 | 42 | 61.5 |
| 4 | 6.1 | 17 | 14.5 | 30 | 31.8 | 43 | 64.8 |
| 5 | 6.5 | 18 | 15.5 | 31 | 33.7 | 44 | 68.3 |
| 6 | 7.0 | 19 | 16.5 | 32 | 35.7 | 45 | 71.9 |
| 7 | 7.5 | 20 | 17.5 | 33 | 37.7 | 46 | 75.7 |
| 8 | 8.0 | 21 | 18.7 | 34 | 39.9 | 47 | 79.6 |
| 9 | 8.6 | 22 | 19.8 | 35 | 42.2 | 48 | 83.7 |
| 10 | 9.2 | 23 | 21.1 | 36 | 44.6 | 49 | 88.0 |
| 11 | 9.8 | 24 | 22.4 | 37 | 47.1 | 50 | 92.5 |
| 12 | 10.5 | 25 | 23.8 | 38 | 49.7 | | |

**TABLE A.5 Radii of Atoms and Ions**

| Element | Ionic Radius (nm) | Ionic Charge | Covalent Radius (nm) | Metallic Radius (nm) |
|---|---|---|---|---|
| Aluminum | 0.050 | (+3) | 0.125 | 0.1431 |
| Antimony | 0.245 | (−3) | 0.141 | |
| | 0.09 | (+3) | | |
| | 0.062 | (+5) | | |
| Arsenic | 0.222 | (−3) | 0.121 | 0.1248 |
| | 0.058 | (+3) | | |
| | 0.047 | (+5) | | |
| Astatine | 0.227 | (−1) | | |
| | 0.051 | (+7) | | |
| Barium | 0.135 | (+2) | 0.198 | 0.2173 |
| Beryllium | 0.031 | (+2) | 0.089 | 0.1113 |
| Bismuth | 0.213 | (−3) | 0.152 | 0.1547 |
| | 0.096 | (+3) | | |
| | 0.074 | (+5) | | |
| Boron | 0.020 | (+3) | 0.088 | 0.083 |
| Bromine | 0.196 | (−1) | 0.1142 | |
| | 0.039 | (+7) | | |
| Cadmium | 0.097 | (+2) | 0.141 | 0.1489 |
| Calcium | 0.099 | (+2) | 0.174 | 0.1973 |
| Carbon | 0.260 | (−4) | 0.077 | |
| | 0.015 | (+4) | | |
| Cesium | 0.169 | (+1) | 0.235 | 0.2654 |
| Chlorine | 0.181 | (−1) | 0.099 | |
| | 0.026 | (+7) | | |
| Chromium | 0.064 | (+3) | 0.117 | 0.1249 |
| | 0.052 | (+6) | | |
| Cobalt | 0.074 | (+2) | 0.116 | 0.1253 |
| | 0.063 | (+3) | | |
| Copper | 0.096 | (+1) | 0.117 | 0.1278 |
| | 0.072 | (+2) | | |
| Fluorine | 0.136 | (−1) | 0.064 | 0.0717 |
| | 0.007 | (+7) | | |
| Francium | 0.176 | (+1) | | 0.27 |
| Gallium | 0.062 | (+3) | 0.125 | 0.1221 |
| Germanium | 0.272 | (−4) | 0.122 | 0.1225 |
| | 0.053 | (+4) | | |
| Gold | 0.137 | (+1) | 0.134 | 0.1442 |
| | 0.091 | (+3) | | |
| Hydrogen | 0.208 | (−1) | 0.0371 | |
| | $10^{-6}$ | (+1) | | |
| Indium | 0.081 | (+3) | 0.150 | 0.1626 |
| Iodine | 0.216 | (−1) | 0.1333 | |
| | 0.050 | (+7) | | |
| Iron | 0.076 | (+2) | 0.1165 | 0.1241 |
| | 0.064 | (+3) | | |
| Lead | 0.215 | (−4) | 0.154 | 0.1750 |
| | 0.120 | (+2) | | |
| | 0.084 | (+4) | | |
| Lithium | 0.068 | (+1) | 0.123 | 0.152 |
| Magnesium | 0.065 | (+2) | 0.136 | 0.160 |
| Manganese | 0.080 | (+2) | 0.117 | 0.124 |
| | 0.046 | (+7) | | |
| Mercury | 0.127 | (+1) | 0.144 | 0.160 |
| | 0.110 | (+2) | | |
| Molybdenum | 0.062 | (+6) | 0.129 | 0.1362 |
| Nickel | 0.072 | (+2) | 0.115 | 0.1246 |
| Nitrogen | 0.171 | (−3) | 0.070 | |
| | 0.013 | (+3) | | |
| | 0.011 | (+5) | | |
| Oxygen | 0.140 | (−2) | 0.066 | |
| | 0.176 | (−1) | | |
| Phosphorus | 0.212 | (−3) | 0.110 | 0.108 |
| | 0.042 | (+3) | | |
| | 0.034 | (+5) | | |
| Polonium | 0.230 | (−2) | 0.153 | 0.167 |
| | 0.056 | (+6) | | |
| Potassium | 0.133 | (+1) | 0.2025 | 0.2272 |
| Radium | 0.140 | (+2) | | 0.220 |
| Rubidium | 0.148 | (+1) | 0.216 | 0.2475 |
| Scandium | 0.081 | (+3) | 0.144 | 0.1606 |
| Selenium | 0.198 | (−2) | 0.117 | |
| | 0.069 | (+4) | | |
| | 0.042 | (+6) | | |
| Silicon | 0.271 | (−4) | 0.117 | |
| | 0.041 | (+4) | | |
| Silver | 0.126 | (+1) | 0.134 | 0.1444 |
| Sodium | 0.095 | (+1) | 0.157 | 0.1537 |
| Strontium | 0.113 | (+2) | 0.192 | 0.2151 |
| Sulfur | 0.184 | (−2) | 0.104 | |
| | 0.037 | (+4) | | |
| | 0.029 | (+6) | | |
| Tellurium | 0.221 | (−2) | 0.137 | 0.1432 |
| | 0.081 | (+4) | | |
| | 0.056 | (+6) | | |
| Thallium | 0.095 | (+3) | 0.155 | 0.1704 |
| Tin | 0.294 | (−4) | 0.140 | 0.1405 |
| | 0.102 | (+2) | | |
| | 0.071 | (+4) | | |
| Titanium | 0.090 | (+2) | 0.132 | 0.1448 |
| | 0.068 | (+4) | | |
| Tungsten | 0.065 | (+6) | 0.130 | 0.1370 |
| Uranium | 0.083 | (+6) | | 0.1385 |
| Vanadium | 0.059 | (+5) | 0.122 | 0.1321 |
| Xenon | | | 0.209 | |
| Zinc | 0.074 | (+2) | 0.125 | 0.1332 |
| Zirconium | 0.079 | (+4) | 0.145 | 0.167 |

**TABLE A.6 Ionization Energies**

| Element | | I | II | III | IV | V | VI | VII |
|---|---|---|---|---|---|---|---|---|
| 1 | H | 1,312.0 | | | | | | |
| 2 | He | 2,372.3 | 5,250.3 | | | | | |
| 3 | Li | 520.2 | 7,297.9 | 11,814.6 | | | | |
| 4 | Be | 899.4 | 1,757.1 | 14,848.3 | 21,005.9 | | | |
| 5 | B | 800.6 | 2,427.0 | 3,659.6 | 25,025.0 | 32,825.7 | | |
| 6 | C | 1,086.4 | 2,352.6 | 4,620.4 | 6,222.5 | 37,829.4 | 47,275.6 | |
| 7 | N | 1,402.3 | 2,856.0 | 4,578.0 | 7,474.9 | 9,444.7 | 50,370.4 | 64,358.0 |
| 8 | O | 1,313.9 | 3,388.2 | 5,300.3 | 7,469.1 | 10,989.2 | 13,326.1 | 71,332 |
| 9 | F | 1,681.0 | 3,374.1 | 6,050.3 | 8,407.5 | 11,022.4 | 15,163.6 | 17,867.2 |
| 10 | Ne | 2,080.6 | 3,952.2 | 6,122 | 9,370 | 12,177 | 15,238 | 19,998 |
| 11 | Na | 495.8 | 4,562.4 | 6,912 | 9,543 | 13,352 | 16,610 | 20,114 |
| 12 | Mg | 737.7 | 1,450.6 | 7,732.6 | 10,540 | 13,629 | 17,994 | 21,703 |
| 13 | Al | 577.6 | 1,816.6 | 2,744.7 | 11,577 | 14,831 | 18,377 | 23,294 |
| 14 | Si | 786.4 | 1,577.0 | 3,231.5 | 4,355.4 | 16,091 | 19,784 | 23,785 |
| 15 | P | 1,011.7 | 1,903.2 | 2,912 | 4,956 | 6,273.7 | 21,268 | 25,397 |
| 16 | S | 999.58 | 2,251 | 3,361 | 4,564 | 7,012 | 8,495.4 | 27,105 |
| 17 | Cl | 1,251.1 | 2,297 | 3,822 | 5,158 | 6,540 | 9,362 | 11,017.9 |
| 18 | Ar | 1,520.5 | 2,665.8 | 3,931 | 5,771 | 7,238 | 8,780.8 | 11,994.9 |
| 19 | K | 418.8 | 3,051.3 | 4,411 | 5,877 | 7,975 | 9,648.5 | 11,343 |
| 20 | Ca | 589.8 | 1,145.4 | 4,911.8 | 6,474 | 8,144 | 10,496 | 12,320 |
| 21 | Sc | 631 | 1,235 | 2,389 | 7,099 | 8,844 | 10,720 | 13,310 |
| 22 | Ti | 658 | 1,310 | 2,652.5 | 4,174.5 | 9,573 | 11,516 | 13,590 |
| 23 | V | 650 | 1,413 | 2,828.0 | 4,506.5 | 6,294 | 12,362 | 14,489 |
| 24 | Cr | 652.8 | 1,592 | 2,987 | 4,740 | 6,690 | 8,738 | 15,540 |
| 25 | Mn | 717.4 | 1,509.0 | 3,248.3 | 4,940 | 6,990 | 9,220 | 11,508 |
| 26 | Fe | 759.3 | 1,561 | 2,957.3 | 5,290 | 7,240 | 9,600 | 12,100 |
| 27 | Co | 758 | 1,646 | 3,232 | 4,950 | 7,670 | 9,840 | 12,400 |
| 28 | Ni | 736.7 | 1,752.9 | 3,393 | 5,300 | 7,280 | 10,400 | 12,800 |
| 29 | Cu | 745.4 | 1,957.9 | 3,553 | 5,330 | 7,710 | 9,940 | 13,400 |
| 30 | Zn | 906.4 | 1,733.2 | 3,832.6 | 5,730 | 7,970 | 10,400 | 12,900 |
| 31 | Ga | 578.8 | 1,979 | 2,963 | 6,200 | | | |
| 32 | Ge | 762.1 | 1,537.4 | 3,302 | 4,410 | 9,020 | | |
| 33 | As | 947 | 1,797.8 | 2,735.4 | 4,837 | 6,043 | 12,300 | |
| 34 | Se | 940.9 | 2,045 | 2,973.7 | 4,143.4 | 6,590 | 7,883 | 14,990 |
| 35 | Br | 1,139.9 | 2,100 | 3,500 | 4,560 | 5,760 | 8,550 | 9,938 |
| 36 | Kr | 1,350.7 | 2,350.3 | 3,565 | 5,070 | 6,240 | 7,570 | 10,710 |
| 37 | Rb | 403.0 | 2,632 | 3,900 | 5,070 | 6,850 | 8,140 | 9,570 |
| 38 | Sr | 549.5 | 1,064.5 | 4,120 | 5,500 | 6,910 | 8,760 | 10,200 |
| 39 | Y | 616 | 1,181 | 1,980 | 5,960 | 7,430 | 8,970 | 11,200 |
| 40 | Zr | 660 | 1,267 | 2,218 | 3,313 | 7,870 | | |
| 41 | Nb | 664 | 1,382 | 2,416 | 3,960 | 4,877 | 9,899 | 12,100 |
| 42 | Mo | 684.9 | 1,558 | 2,621 | 4,480 | 5,910 | 6,600 | 12,230 |
| 43 | Tc | 702 | 1,472 | 2,850 | | | | |
| 44 | Ru | 711 | 1,617 | 2,747 | | | | |
| 45 | Rh | 720 | 1,744 | 2,997 | | | | |
| 46 | Pd | 805 | 1,874 | 3,177 | | | | |
| 47 | Ag | 731.0 | 2,073 | 3,361 | | | | |
| 48 | Cd | 867.7 | 1,631.4 | 3,616 | | | | |
| 49 | In | 558.3 | 1,820.6 | 2,704 | 5,200 | | | |
| 50 | Sn | 708.6 | 1,411.8 | 2,943.0 | 3,930.2 | 6,974 | | |
| 51 | Sb | 833.7 | 1,595 | 2,440 | 4,260 | 5,400 | 10,400 | |

*(continued)*

**TABLE A.6** **Ionization Energies** (continued)

| Element | | I | II | III | IV | V | VI | VII |
|---|---|---|---|---|---|---|---|---|
| 52 | Te | 869.2 | 1,790 | 2,698 | 3,609 | 5,668 | 6,820 | 13,200 |
| 53 | I | 1,008.4 | 1,845.8 | 3,200 | | | | |
| 54 | Xe | 1,170.4 | 2,046 | 3,100 | | | | |
| 55 | Cs | 375.7 | 2,440 | | | | | |
| 56 | Ba | 502.9 | 965.23 | | | | | |
| 57 | La | 538.1 | 1,067 | 1,850.3 | | | | |
| 58 | Ce | 527.8 | 1,047 | 1,949 | 3,547 | | | |
| 59 | Pr | 523 | 1,018 | 2,086 | 3,761 | 5,543 | | |
| 60 | Nd | 530 | 1,035 | 2,130 | 3,900 | | | |
| 61 | Pm | 535 | 1,052 | 2,150 | 3,970 | | | |
| 62 | Sm | 543 | 1,068 | 2,260 | 3,990 | | | |
| 63 | Eu | 547 | 1,084 | 2,400 | 4,110 | | | |
| 64 | Gd | 592 | 1,167 | 1,990 | 4,250 | | | |
| 65 | Tb | 564 | 1,112 | 2,110 | 3,840 | | | |
| 66 | Dy | 572 | 1,126 | 2,200 | 4,000 | | | |
| 67 | Ho | 581 | 1,139 | 2,203 | 4,100 | | | |
| 68 | Er | 589 | 1,151 | 2,194 | 4,120 | | | |
| 69 | Tm | 596 | 1,163 | 2,285 | 4,120 | | | |
| 70 | Yb | 603 | 1,175 | 2,415 | 4,216 | | | |
| 71 | Lu | 524 | 1,340 | 2,022 | 4,360 | | | |
| 72 | Hf | 642 | 1,440 | 2,250 | 3,210 | | | |
| 73 | Ta | 761 | | | | | | |
| 74 | W | 770 | | | | | | |
| 75 | Re | 760 | 1,260 | 2,510 | 3,640 | | | |
| 76 | Os | 840 | | | | | | |
| 77 | Ir | 880 | | | | | | |
| 78 | Pt | 870 | 1,791.0 | | | | | |
| 79 | Au | 890.1 | 1,980 | | | | | |
| 80 | Hg | 1,007.0 | 1,809.7 | 3,300 | | | | |
| 81 | Tl | 589.3 | 1,971.0 | 2,878 | | | | |
| 82 | Pb | 715.5 | 1,450.4 | 3,081.4 | 4,083 | 6,640 | | |
| 83 | Bi | 703.3 | 1,610 | 2,466 | 4,370 | 5,400 | 8,520 | |
| 84 | Po | 812 | | | | | | |
| 85 | At | | | | | | | |
| 86 | Rn | 1,037.0 | | | | | | |
| 87 | Fr | | | | | | | |
| 88 | Ra | 509.3 | 979.0 | | | | | |
| 89 | Ac | 498.8 | 1,170 | | | | | |
| 90 | Th | 587 | 1,110 | 1,930 | 2,780 | | | |
| 91 | Pa | 568 | | | | | | |
| 92 | U | 584 | | | | | | |
| 93 | Np | 597 | | | | | | |
| 94 | Pu | 585 | | | | | | |
| 95 | Am | 578.2 | | | | | | |
| 96 | Cm | 581 | | | | | | |
| 97 | Bk | 601 | | | | | | |
| 98 | Cf | 608 | | | | | | |
| 99 | Es | 619 | | | | | | |
| 100 | Fm | 627 | | | | | | |
| 101 | Md | 635 | | | | | | |
| 102 | No | 642 | | | | | | |

**TABLE A.7 Electron Affinities**

| Atomic Number | Symbol | Electron Affinity (kJ/mol) | Atomic Number | Symbol | Electron Affinity (kJ/mol) |
|---|---|---|---|---|---|
| 1 | H | 72.8 | 37 | Rb | 46.89 |
| 2 | He | * | 38 | Sr | * |
| 3 | Li | 59.8 | 39 | Y | 0 |
| 4 | Be | * | 40 | Zr | 50 |
| 5 | B | 27 | 41 | Nb | 96 |
| 6 | C | 122.3 | 42 | Mo | 96 |
| 7 | N | −7 | 43 | Tc | 70 |
| 8 | O | 141.1 | 44 | Ru | 110 |
| 9 | F | 328.0 | 45 | Rh | 120 |
| 10 | Ne | * | 46 | Pd | 60 |
| 11 | Na | 52.7 | 47 | Ag | 125.7 |
| 12 | Mg | * | 48 | Cd | * |
| 13 | Al | 45 | 49 | In | 29 |
| 14 | Si | 133.6 | 50 | Sn | 121 |
| 15 | P | 71.7 | 51 | Sb | 101 |
| 16 | S | 200.42 | 52 | Te | 190.15 |
| 17 | Cl | 348.8 | 53 | I | 295.3 |
| 18 | Ar | * | 54 | Xe | * |
| 19 | K | 48.36 | 55 | Cs | 45.49 |
| 20 | Ca | * | 56 | Ba | * |
| 21 | Sc | * | 57–71 | La–Lu | 50 |
| 22 | Ti | 20 | 72 | Hf | * |
| 23 | V | 50 | 73 | Ta | 60 |
| 24 | Cr | 64 | 74 | W | 60 |
| 25 | Mn | * | 75 | Re | 14 |
| 26 | Fe | 24 | 76 | Os | 110 |
| 27 | Co | 70 | 77 | Ir | 150 |
| 28 | Ni | 111 | 78 | Pt | 205.3 |
| 29 | Cu | 118.3 | 79 | Au | 222.74 |
| 30 | Zn | 0 | 80 | Hg | * |
| 31 | Ga | 29 | 81 | Tl | 30 |
| 32 | Ge | 120 | 82 | Pb | 110 |
| 33 | As | 77 | 83 | Bi | 110 |
| 34 | Se | 194.96 | 84 | Po | 180 |
| 35 | Br | 324.6 | 85 | At | 270 |
| 36 | Kr | * | 86 | Rn | * |
| | | | 87 | Fr | 44.0 |

* These elements have negative electron affinities.

**TABLE A.8 Electronegativities**

| Atomic Number | Element | Electro-negativity | Atomic Number | Element | Electro-negativity |
|---|---|---|---|---|---|
| 1 | H | 2.20 | 41 | Nb | 1.6 |
| 2 | He | * | 42 | Mo | 2.16 |
| 3 | Li | 0.98 | 43 | Tc | 1.9 |
| 4 | Be | 1.57 | 44 | Ru | 2.2 |
| 5 | B | 2.04 | 45 | Rh | 2.28 |
| 6 | C | 2.55 | 46 | Pd | 2.20 |
| 7 | N | 3.04 | 47 | Ag | 1.93 |
| 8 | O | 3.44 | 48 | Cd | 1.69 |
| 9 | F | 3.98 | 49 | In | 1.78 |
| 10 | Ne | * | 50 | Sn | 1.96 |
| 11 | Na | 0.93 | 51 | Sb | 2.05 |
| 12 | Mg | 1.31 | 52 | Te | 2.1 |
| 13 | Al | 1.61 | 53 | I | 2.66 |
| 14 | Si | 1.90 | 54 | Xe | * |
| 15 | P | 2.19 | 55 | Cs | 0.79 |
| 16 | S | 2.58 | 56 | Ba | 0.89 |
| 17 | Cl | 3.16 | 57–71 | La–Lu | 1.1–1.3 |
| 18 | Ar | * | 72 | Hf | 1.3 |
| 19 | K | 0.82 | 73 | Ta | 1.3 |
| 20 | Ca | 1.00 | 74 | W | 2.36 |
| 21 | Sc | 1.36 | 75 | Re | 1.9 |
| 22 | Ti | 1.54 | 76 | Os | 2.2 |
| 23 | V | 1.63 | 77 | Ir | 2.20 |
| 24 | Cr | 1.66 | 78 | Pt | 2.28 |
| 25 | Mn | 1.55 | 79 | Au | 2.54 |
| 26 | Fe | 1.83 | 80 | Hg | 2.00 |
| 27 | Co | 1.88 | 81 | Tl | 2.04 |
| 28 | Ni | 1.91 | 82 | Pb | 2.33 |
| 29 | Cu | 1.90 | 83 | Bi | 2.02 |
| 30 | Zn | 1.65 | 84 | Po | 2.0 |
| 31 | Ga | 1.81 | 85 | At | 2.2 |
| 32 | Ge | 2.01 | 86 | Rn | * |
| 33 | As | 2.18 | 87 | Fr | 0.7 |
| 34 | Se | 2.55 | 88 | Ra | 0.9 |
| 35 | Br | 2.96 | 89 | Ac | 1.1 |
| 36 | Kr | * | 90 | Th | 1.3 |
| 37 | Rb | 0.82 | 91 | Pa | 1.5 |
| 38 | Sr | 1.95 | 92 | U | 1.38 |
| 39 | Y | 1.22 | 93–103 | Np–Lw | 1.3 |
| 40 | Zr | 1.33 | | | |

Pauling electronegativities taken from A. L. Allred, *J. Inorg. Nucl. Chem.*, *17,* 215(1961).
* An asterisk indicates a noble gas.

**TABLE A.9 Acid-Dissociation Equilibrium Constants**

| Compound | Dissociation Reaction | $K_a$ | $pK_a$ |
|---|---|---|---|
| Acetic acid | $CH_3CO_2H \rightleftharpoons CH_3CO_2^- + H^+$ | $1.75 \times 10^{-5}$ | 4.757 |
| Ammonium ion | $NH_4^+ \rightleftharpoons NH_3 + H^+$ | $5.8 \times 10^{-10}$ | 9.24 |
| Arsenic acid | $H_3AsO_4 \rightleftharpoons H_2AsO_4^- + H^+$ | $6.0 \times 10^{-3}$ | 2.22 |
| | $H_2AsO_4^- \rightleftharpoons HAsO_4^{2-} + H^+$ | $1.0 \times 10^{-7}$ | 7.00 |
| | $HAsO_4^{2-} \rightleftharpoons AsO_4^{3-} + H^+$ | $3.0 \times 10^{-12}$ | 11.52 |
| Arsenous acid | $H_3AsO_3 \rightleftharpoons H_2AsO_3^- + H^+$ | $6.0 \times 10^{-10}$ | 9.22 |
| | $H_2AsO_3^- \rightleftharpoons HAsO_3^{2-} + H^+$ | $3.0 \times 10^{-14}$ | 13.52 |
| Benzoic acid | $C_6H_5CO_2H \rightleftharpoons C_6H_5CO_2^- + H^+$ | $6.3 \times 10^{-5}$ | 4.20 |
| Boric acid | $H_3BO_3 \rightleftharpoons H_2BO_3^- + H^+$ | $7.3 \times 10^{-10}$ | 9.14 |
| Carbonic acid | $H_2CO_3 \rightleftharpoons HCO_3^- + H^+$ | $4.5 \times 10^{-7}$ | 6.35 |
| | $HCO_3^- \rightleftharpoons CO_3^{2-} + H^+$ | $4.7 \times 10^{-11}$ | 10.33 |
| Chloric acid | $HClO_3 \rightleftharpoons ClO_3^- + H^+$ | $5.0 \times 10^{2}$ | −2.70 |
| Chloroacetic acid | $ClCH_2CO_2H \rightleftharpoons ClCH_2CO_2^- + H^+$ | $1.4 \times 10^{-3}$ | 2.85 |
| Chlorous acid | $HClO_2 \rightleftharpoons ClO_2^- + H^+$ | $1.1 \times 10^{-2}$ | 1.96 |
| Chromic acid | $H_2CrO_4 \rightleftharpoons HCrO_4^- + H^+$ | 9.6 | −0.98 |
| | $HCrO_4^- \rightleftharpoons CrO_4^{2-} + H^+$ | $3.2 \times 10^{-7}$ | 6.50 |
| Citric acid | $H_3Cit \rightleftharpoons H_2Cit^- + H^+$ | $7.5 \times 10^{-4}$ | 3.13 |
| | $H_2Cit^- \rightleftharpoons HCit^{2-} + H^+$ | $1.7 \times 10^{-5}$ | 4.77 |
| | $HCit^{2-} \rightleftharpoons Cit^{3-} + H^+$ | $4.0 \times 10^{-7}$ | 6.40 |
| Dichloroacetic acid | $Cl_2CHCO_2H \rightleftharpoons Cl_2CHCO_2^- + H^+$ | $5.1 \times 10^{-2}$ | 1.29 |
| Formic acid | $HCO_2H \rightleftharpoons HCO_2^- + H^+$ | $1.8 \times 10^{-4}$ | 3.75 |
| Glycine | $H_3N^+CH_2CO_2H \rightleftharpoons H_3N^+CH_2CO_2^- + H^+$ | $4.5 \times 10^{-3}$ | 2.35 |
| | $H_3N^+CH_2CO_2^- \rightleftharpoons H_2NCH_2CO_2^- + H^+$ | $2.5 \times 10^{-10}$ | 9.60 |
| Hydrazoic acid | $HN_3 \rightleftharpoons N_3^- + H^+$ | $1.9 \times 10^{-5}$ | 4.72 |
| Hydrobromic acid | $HBr \rightleftharpoons Br^- + H^+$ | $1 \times 10^{9}$ | −9 |
| Hydrochloric acid | $HCl \rightleftharpoons Cl^- + H^+$ | $1 \times 10^{6}$ | −6 |
| Hydrocyanic acid | $HCN \rightleftharpoons CN^- + H^+$ | $6 \times 10^{-10}$ | 9.22 |
| Hydrofluoric acid | $HF \rightleftharpoons F^- + H^+$ | $7.2 \times 10^{-4}$ | 3.14 |
| Hydroiodic acid | $HI \rightleftharpoons I^- + H^+$ | $3 \times 10^{9}$ | −9.5 |
| Hydrogen peroxide | $H_2O_2 \rightleftharpoons HO_2^- + H^+$ | $2.2 \times 10^{-12}$ | 11.66 |
| Hydrogen selenide | $H_2Se \rightleftharpoons HSe^- + H^+$ | $1.0 \times 10^{-4}$ | 4.00 |
| Hydrogen sulfide | $H_2S \rightleftharpoons HS^- + H^+$ | $1.0 \times 10^{-7}$ | 7.00 |
| | $HS^- \rightleftharpoons S^{2-} + H^+$ | $1.3 \times 10^{-13}$ | 12.89 |
| Hypobromous acid | $HOBr \rightleftharpoons OBr^- + H^+$ | $2.4 \times 10^{-9}$ | 8.62 |
| Hypochlorous acid | $HOCl \rightleftharpoons OCl^- + H^+$ | $2.9 \times 10^{-8}$ | 7.54 |
| Hypoiodous acid | $HOI \rightleftharpoons OI^- + H^+$ | $2.3 \times 10^{-11}$ | 10.64 |
| Iodic acid | $HIO_3 \rightleftharpoons IO_3^- + H^+$ | 0.16 | 0.80 |
| Nitric acid | $HNO_3 \rightleftharpoons NO_3^- + H^+$ | 28 | −1.45 |
| Nitrous acid | $HNO_2 \rightleftharpoons NO_2^- + H^+$ | $5.1 \times 10^{-4}$ | 3.29 |
| Oxalic acid | $H_2C_2O_4 \rightleftharpoons HC_2O_4^- + H^+$ | $5.4 \times 10^{-2}$ | 1.27 |
| | $HC_2O_4^- \rightleftharpoons C_2O_4^{2-} + H^+$ | $5.4 \times 10^{-5}$ | 4.27 |
| Perchloric acid | $HOClO_3 \rightleftharpoons ClO_4^- + H^+$ | $1 \times 10^{8}$ | −8 |
| Periodic acid | $H_5IO_6 \rightleftharpoons H_4IO_6^- + H^+$ | $2.3 \times 10^{-2}$ | 1.64 |
| Phenol | $C_6H_5OH \rightleftharpoons C_6H_5O^- + H^+$ | $1.0 \times 10^{-10}$ | 10.00 |
| Phosphoric acid | $H_3PO_4 \rightleftharpoons H_2PO_4^- + H^+$ | $7.1 \times 10^{-3}$ | 2.15 |
| | $H_2PO_4^- \rightleftharpoons HPO_4^{2-} + H^+$ | $6.3 \times 10^{-8}$ | 7.20 |
| | $HPO_4^{2-} \rightleftharpoons PO_4^{3-} + H^+$ | $4.2 \times 10^{-13}$ | 12.38 |
| Phosphorous acid | $H_3PO_3 \rightleftharpoons H_2PO_3^- + H^+$ | $1.00 \times 10^{-2}$ | 2.00 |
| | $H_2PO_3^- \rightleftharpoons HPO_3^{2-} + H^+$ | $2.6 \times 10^{-7}$ | 6.59 |

*(continued)*

**TABLE A.9 Acid-Dissociation Equilibrium Constants** (continued)

| Compound | Dissociation Reaction | $K_a$ | $pK_a$ |
|---|---|---|---|
| Sulfamic acid | $H_2NSO_3H \rightleftharpoons H_2NSO_3^- + H^+$ | $1.03 \times 10^{-1}$ | 0.987 |
| Sulfuric acid | $H_2SO_4 \rightleftharpoons HSO_4^- + H^+$ | $10^3$ | −3 |
| | $HSO_4^- \rightleftharpoons SO_4^{2-} + H^+$ | $1.2 \times 10^{-2}$ | 1.92 |
| Sulfurous acid | $H_2SO_3 \rightleftharpoons HSO_3^- + H^+$ | $1.7 \times 10^{-2}$ | 1.77 |
| | $HSO_3^- \rightleftharpoons SO_3^{2-} + H^+$ | $6.4 \times 10^{-8}$ | 7.19 |
| Thiocyanic acid | $HSCN \rightleftharpoons SCN^- + H^+$ | 71 | −1.85 |
| Trichloroacetic acid | $Cl_3CCO_2H \rightleftharpoons Cl_3CCO_2^- + H^+$ | 0.22 | 0.66 |
| Water | $H_2O \rightleftharpoons OH^- + H^+$ | $1.8 \times 10^{-16}$ | 15.75 |

**TABLE A.10 Base-Ionization Equilibrium Constants**

| Compound | Ionization Reaction | $K_b$ | $pK_b$ |
|---|---|---|---|
| Ammonia | $NH_3 + H_2O \rightleftharpoons NH_4^+ + OH^-$ | $1.8 \times 10^{-5}$ | 4.76 |
| Aniline | $C_6H_5NH_2 + H_2O \rightleftharpoons C_6H_5NH_3^+ + OH^-$ | $4.0 \times 10^{-10}$ | 9.40 |
| Butylamine | $CH_3(CH_2)_3NH_2 + H_2O \rightleftharpoons CH_3(CH_2)_3NH_3^+ + OH^-$ | $4.0 \times 10^{-4}$ | 3.40 |
| Dimethylamine | $(CH_3)_2NH + H_2O \rightleftharpoons (CH_3)_2NH_2^+ + OH^-$ | $5.9 \times 10^{-4}$ | 3.23 |
| Ethanolamine | $HOCH_2CH_2NH_2 + H_2O \rightleftharpoons HOCH_2CH_2NH_3^+ + OH^-$ | $3.3 \times 10^{-5}$ | 4.50 |
| Ethylamine | $CH_3CH_2NH_2 + H_2O \rightleftharpoons CH_3CH_2NH_3^+ + H_2O$ | $4.4 \times 10^{-4}$ | 3.37 |
| Hydrazine | $H_2NNH_2 + H_2O \rightleftharpoons H_2NNH_3^+ + OH^-$ | $1.2 \times 10^{-6}$ | 5.89 |
| Hydroxylamine | $HONH_2 + H_2O \rightleftharpoons HONH_3^+ + OH^-$ | $1.1 \times 10^{-8}$ | 7.97 |
| Methylamine | $CH_3NH_2 + H_2O \rightleftharpoons CH_3NH_3^+ + OH^-$ | $4.8 \times 10^{-4}$ | 3.32 |
| Pyridine | $C_5H_5NH + H_2O \rightleftharpoons C_5H_5NH_2^+ + OH^-$ | $1.7 \times 10^{-9}$ | 8.77 |
| Trimethylamine | $(CH_3)_3N + H_2O \rightleftharpoons (CH_3)_3NH^+ + OH^-$ | $6.3 \times 10^{-5}$ | 4.20 |
| Urea | $H_2NCONH_2 + H_2O \rightleftharpoons H_2NCONH_3^+ + OH^-$ | $1.5 \times 10^{-14}$ | 13.82 |

**TABLE A.11 Solubility Product Equilibrium Constants**

| Substance | $K_{sp}$ | Substance | $K_{sp}$ | Substance | $K_{sp}$ |
|---|---|---|---|---|---|
| AgBr | $5.0 \times 10^{-15}$ | $\alpha$-CoS | $4.0 \times 10^{-21}$ | $MnCO_3$ | $1.8 \times 10^{-11}$ |
| AgCN | $1.2 \times 10^{-16}$ | $\beta$-CoS | $2.0 \times 10^{-25}$ | $Mn(OH)_2$ | $2 \times 10^{-13}$ |
| $Ag_2CO_3$ | $8.1 \times 10^{-12}$ | $Cr(OH)_3$ | $6.3 \times 10^{-31}$ | MnS | $3 \times 10^{-15}$ |
| AgOH | $2.0 \times 10^{-8}$ | CuBr | $5.3 \times 10^{-9}$ | $NiCO_3$ | $6.6 \times 10^{-9}$ |
| AgOAc | $4.4 \times 10^{-3}$ | CuCl | $1.2 \times 10^{-6}$ | $NiC_2O_4$ | $4 \times 10^{-10}$ |
| $Ag_2C_2O_4$ | $3.4 \times 10^{-11}$ | CuCN | $3.2 \times 10^{-20}$ | $\alpha$-NiS | $3.2 \times 10^{-19}$ |
| AgCl | $1.8 \times 10^{-10}$ | $CuCrO_4$ | $3.6 \times 10^{-6}$ | $\beta$-NiS | $1.0 \times 10^{-24}$ |
| $Ag_2CrO_4$ | $1.1 \times 10^{-12}$ | $CuCO_3$ | $1.4 \times 10^{-10}$ | $\gamma$-NiS | $2.0 \times 10^{-26}$ |
| AgI | $8.3 \times 10^{-17}$ | $Cu(OH)_2$ | $2.2 \times 10^{-20}$ | $PbBr_2$ | $4.0 \times 10^{-5}$ |
| $Ag_2S$ | $6.3 \times 10^{-50}$ | CuI | $1.1 \times 10^{-13}$ | $PbCO_3$ | $7.4 \times 10^{-14}$ |
| AgSCN | $1.0 \times 10^{-12}$ | $Cu_2S$ | $2.5 \times 10^{-48}$ | $PbC_2O_4$ | $4.8 \times 10^{-10}$ |
| $Ag_2SO_4$ | $1.4 \times 10^{-5}$ | CuS | $6.3 \times 10^{-36}$ | $PbCl_2$ | $1.6 \times 10^{-5}$ |
| $Al(OH)_3$ | $1.3 \times 10^{-33}$ | CuSCN | $4.8 \times 10^{-15}$ | $PbCrO_4$ | $2.8 \times 10^{-13}$ |
| AuCl | $2.0 \times 10^{-13}$ | $FeCO_3$ | $3.2 \times 10^{-11}$ | $PbF_2$ | $2.7 \times 10^{-8}$ |
| $AuCl_3$ | $3.2 \times 10^{-23}$ | $Fe_2C_2O_4$ | $3.2 \times 10^{-7}$ | $Pb(OH)_2$ | $1.2 \times 10^{-15}$ |
| AuI | $1.6 \times 10^{-23}$ | $Fe(OH)_2$ | $8.0 \times 10^{-16}$ | $PbI_2$ | $7.1 \times 10^{-9}$ |
| $AuI_3$ | $5.5 \times 10^{-46}$ | $Fe(OH)_3$ | $4 \times 10^{-38}$ | PbS | $8.0 \times 10^{-28}$ |
| $BaCO_3$ | $5.1 \times 10^{-9}$ | FeS | $6.3 \times 10^{-18}$ | $PbSO_4$ | $1.6 \times 10^{-8}$ |
| $BaC_2O_4$ | $2.3 \times 10^{-8}$ | $Hg_2Br_2$ | $5.6 \times 10^{-23}$ | SnS | $1.0 \times 10^{-25}$ |
| $BaCrO_4$ | $1.2 \times 10^{-10}$ | $Hg_2(CN)_2$ | $5 \times 10^{-40}$ | $Sn(OH)_2$ | $1.4 \times 10^{-28}$ |
| $BaF_2$ | $1.0 \times 10^{-6}$ | $Hg_2CO_3$ | $8.9 \times 10^{-17}$ | $Sn(OH)_4$ | $1 \times 10^{-56}$ |
| $Ba(OH)_2$ | $5 \times 10^{-3}$ | $Hg_2(OAc)_2$ | $3 \times 10^{-11}$ | $SrCO_3$ | $1.1 \times 10^{-10}$ |
| $BaSO_4$ | $1.1 \times 10^{-10}$ | $Hg_2C_2O_4$ | $2.0 \times 10^{-13}$ | $SrC_2O_4$ | $1.6 \times 10^{-7}$ |
| $Bi_2S_3$ | $1 \times 10^{-97}$ | $HgC_2O_4$ | $1 \times 10^{-7}$ | $SrCrO_4$ | $2.2 \times 10^{-5}$ |
| $CaCO_3$ | $2.8 \times 10^{-9}$ | $Hg_2Cl_2$ | $1.3 \times 10^{-18}$ | $SrF_2$ | $2.5 \times 10^{-9}$ |
| $CaC_2O_4$ | $4 \times 10^{-9}$ | $Hg_2CrO_4$ | $2.0 \times 10^{-9}$ | $SrSO_4$ | $3.2 \times 10^{-7}$ |
| $CaCrO_4$ | $7.1 \times 10^{-4}$ | $Hg_2I_2$ | $4.5 \times 10^{-29}$ | TlBr | $3.4 \times 10^{-6}$ |
| $CaF_2$ | $4.0 \times 10^{-11}$ | $Hg_2S$ | $1.0 \times 10^{-47}$ | TlCl | $1.7 \times 10^{-4}$ |
| $Ca(OH)_2$ | $5.5 \times 10^{-6}$ | HgS | $4 \times 10^{-53}$ | TlI | $6.5 \times 10^{-8}$ |
| $CdCO_3$ | $5.2 \times 10^{-12}$ | $K_2NaCo(NO_2)_6$ | $2.2 \times 10^{-11}$ | $Zn(CN)_2$ | $2.6 \times 10^{-13}$ |
| $Cd(CN)_2$ | $1.0 \times 10^{-8}$ | $MgCO_3$ | $3.5 \times 10^{-8}$ | $ZnCO_3$ | $1.4 \times 10^{-11}$ |
| $Cd(OH)_2$ | $2.5 \times 10^{-14}$ | $MgC_2O_4$ | $1 \times 10^{-8}$ | $ZnC_2O_4$ | $2.7 \times 10^{-8}$ |
| CdS | $8 \times 10^{-27}$ | $MgF_2$ | $6.5 \times 10^{-9}$ | $Zn(OH)_2$ | $1.2 \times 10^{-17}$ |
| $CoCO_3$ | $1.4 \times 10^{-13}$ | $Mg(OH)_2$ | $1.8 \times 10^{-11}$ | $\alpha$-ZnS | $1.6 \times 10^{-24}$ |
| $Co(OH)_3$ | $1.6 \times 10^{-44}$ | $MgNH_4PO_4$ | $2.5 \times 10^{-13}$ | $\beta$-ZnS | $2.5 \times 10^{-22}$ |

**TABLE A.12 Complex Formation Equilibrium Constants**

| Equilibrium | $K_f$ | Equilibrium | $K_f$ |
|---|---|---|---|
| $AG^+ + 2\,Br^- \rightleftharpoons AgBr_2^-$ | $2.1 \times 10^7$ | $Fe^{2+} + 6\,CN^- \rightleftharpoons Fe(CN)_6^{4-}$ | $1 \times 10^{35}$ |
| $Ag^+ + 2\,Cl^- \rightleftharpoons AgCl_2^-$ | $1.1 \times 10^5$ | $Fe^{3+} + 6\,CN^- \rightleftharpoons Fe(CN)_6^{3-}$ | $1 \times 10^{42}$ |
| $Ag^+ + 2\,CN^- \rightleftharpoons Ag(CN)_2^-$ | $1.3 \times 10^{21}$ | $Fe^{3+} + SCN^- \rightleftharpoons Fe(SCN)^{2+}$ | $8.9 \times 10^2$ |
| $Ag^+ + 2\,I^- \rightleftharpoons AgI_2^-$ | $5.5 \times 10^{11}$ | $Fe^{3+} + 2\,SCN^- \rightleftharpoons Fe(SCN)_2^+$ | $2.3 \times 10^3$ |
| $Ag^+ + 2\,NH_3 \rightleftharpoons Ag(NH_3)_2^+$ | $1.1 \times 10^7$ | $Hg^{2+} + 4\,Br^- \rightleftharpoons HgBr_4^{2-}$ | $1 \times 10^{21}$ |
| $Ag^+ + 2\,SCN^- \rightleftharpoons Ag(SCN)_2^-$ | $3.7 \times 10^7$ | $Hg^{2+} + 4\,Cl^- \rightleftharpoons HgCl_4^{2-}$ | $1.2 \times 10^{15}$ |
| $Ag^+ + 2\,S_2O_3^{2-} \rightleftharpoons Ag(S_2O_3)_2^{3-}$ | $2.9 \times 10^{13}$ | $Hg^{2+} + 4\,CN^- \rightleftharpoons Hg(CN)_4^{2-}$ | $3 \times 10^{41}$ |
| $Al^{3+} + 6\,F^- \rightleftharpoons AlF_6^{3-}$ | $6.9 \times 10^{19}$ | $Hg^{2-} + 4\,I^- \rightleftharpoons HgI_4^{2-}$ | $6.8 \times 10^{29}$ |
| $Al^{3+} + 4\,OH^- \rightleftharpoons Al(OH)_4^-$ | $1.1 \times 10^{33}$ | $I_2 + I^- \rightleftharpoons I_3^-$ | $7.8 \times 10^2$ |
| $Cd^{2+} + 4\,Cl^- \rightleftharpoons CdCl_4^{2-}$ | $6.3 \times 10^2$ | $Ni^{2+} + 4\,CN^- \rightleftharpoons Ni(CN)_4^{2-}$ | $2 \times 10^{31}$ |
| $Cd^{2+} + 4\,CN^- \rightleftharpoons Cd(CN)_4^{2-}$ | $6.0 \times 10^{18}$ | $Ni^{2+} + 6\,NH_3 \rightleftharpoons Ni(NH_3)_6^{2+}$ | $5.5 \times 10^8$ |
| $Cd^{2+} + 4\,I^- \rightleftharpoons CdI_4^{2-}$ | $2.6 \times 10^5$ | $Pb^{2+} + 4\,Cl^- \rightleftharpoons PbCl_4^{2-}$ | $4 \times 10^1$ |
| $Cd^{2+} + 4\,OH^- \rightleftharpoons Cd(OH)_4^{2-}$ | $4.2 \times 10^8$ | $Pb^{2+} + 4\,I^- \rightleftharpoons PbI_4^{2-}$ | $3.0 \times 10^4$ |
| $Cd^{2+} + 4\,NH_3 \rightleftharpoons Cd(NH_3)_6^{2+}$ | $1.3 \times 10^7$ | $Sb^{3+} + 4\,Cl^- \rightleftharpoons SbCl_4^-$ | $5.2 \times 10^4$ |
| $Co^{2+} + 6\,NH_3 \rightleftharpoons Co(NH_3)_6^{2+}$ | $1.3 \times 10^5$ | $Sb^{3+} + 4\,OH^- \rightleftharpoons Sb(OH)_4^-$ | $2 \times 10^{38}$ |
| $Co^{3+} + 6\,NH_3 \rightleftharpoons Co(NH_3)_6^{3+}$ | $2 \times 10^{35}$ | $Sn^{2+} + 4\,Cl^- \rightleftharpoons SnCl_4^{2-}$ | $3.0 \times 10^1$ |
| $Co^{2+} + 4\,SCN^- \rightleftharpoons Co(SCN)_4^{2-}$ | $1 \times 10^3$ | $Zn^{2+} + 4\,CN^- \rightleftharpoons Zn(CN)_4^{2-}$ | $5 \times 10^{16}$ |
| $Cr^{3+} + 4\,OH^- \rightleftharpoons Cr(OH)_4^-$ | $8 \times 10^{29}$ | $Zn^{2+} + 4\,OH^- \rightleftharpoons Zn(OH)_4^{2-}$ | $4.6 \times 10^{17}$ |
| $Cu^{2+} + 4\,OH^- \rightleftharpoons Cu(OH)_4^{2-}$ | $3 \times 10^{18}$ | $Zn^{2+} + 4\,NH_3 \rightleftharpoons Zn(NH_3)_4^{2+}$ | $2.9 \times 10^9$ |
| $Cu^{2+} + 4\,NH_3 \rightleftharpoons Cu(NH_3)_4^{2+}$ | $2.1 \times 10^{13}$ | | |

**TABLE A.13 Standard-State Reduction Potentials**

| Half-Reaction | E° (V) |
|---|---|
| $3\,N_2 + 2\,H^+ + 2\,e^- \rightleftharpoons 2\,HN_3$ | −3.1 |
| $Li^+ + e^- \rightleftharpoons Li$ | −3.045 |
| $Rb^+ + e^- \rightleftharpoons Rb$ | −2.925 |
| $K^+ + e^- \rightleftharpoons K$ | −2.924 |
| $Cs^+ + e^- \rightleftharpoons Cs$ | −2.923 |
| $Ba^{2+} + 2\,e^- \rightleftharpoons Ba$ | −2.90 |
| $Sr^{2+} + 2\,e^- \rightleftharpoons Sr$ | −2.89 |
| $Ca^{2+} + 2\,e^- \rightleftharpoons Ca$ | −2.76 |
| $Na^+ + e^- \rightleftharpoons Na$ | −2.7109 |
| $Mg(OH)_2 + 2\,e^- \rightleftharpoons Mg + 2\,OH^-$ | −2.69 |
| $Mg^{2+} + 2\,e^- \rightleftharpoons Mg$ | −2.375 |
| $H_2 + 2\,e^- \rightleftharpoons 2\,H^-$ | −2.23 |
| $Al^{3+} + 3\,e^- \rightleftharpoons Al$ (0.1 *M* NaOH) | −1.706 |
| $Be^{2+} + 2\,e^- \rightleftharpoons Be$ | −1.70 |
| $Ti^{2+} + 2\,e^- \rightleftharpoons Ti$ | −1.63 |
| $Zn(CN)_4^{2-} + 2\,e^- \rightleftharpoons Zn + 4\,CN^-$ | −1.26 |
| $Zn(NH_3)_4^{2+} + 2\,e^- \rightleftharpoons Zn + 4\,NH_3$ | −1.04 |
| $Mn^{2+} + 2\,e^- \rightleftharpoons Mn$ | −1.029 |
| $SO_4^{2-} + H_2O + 2\,e^- \rightleftharpoons SO_3^{2-} + 2\,OH^-$ | −0.92 |
| $Cr^{2+} + 2\,e^- \rightleftharpoons Cr$ | −0.91 |
| $TiO_2 + 4\,H^+ + 4\,e^- \rightleftharpoons Ti + 2\,H_2O$ | −0.87 |
| $2\,H_2O + 2\,e^- \rightleftharpoons H_2 + 2\,OH^-$ | −0.8277 |
| $Zn^{2+} + 2\,e^- \rightleftharpoons Zn$ | −0.7628 |
| $Cr^{3+} + 3\,e^- \rightleftharpoons Cr$ | −0.74 |
| $2\,SO_3^{2-} + 3\,H_2O + 4\,e^- \rightleftharpoons S_2O_3^{2-} + 6\,OH^-$ | −0.58 |

(*continued*)

**TABLE A.13 Standard-State Reduction Potentials** (continued)

| Half-Reaction | E° (V) |
|---|---|
| $PbO + H_2O + 2\,e^- \rightleftharpoons Pb + 2\,OH^-$ | −0.576 |
| $Ga^{3+} + 3\,e^- \rightleftharpoons Ga$ | −0.560 |
| $S + 2\,e^- \rightleftharpoons S^{2-}$ | −0.508 |
| $2\,CO_2 + 2\,H^+ + 2\,e^- \rightleftharpoons H_2C_2O_4$ | −0.49 |
| $Ni(NH_3)_6{}^{2+} + 2\,e^- \rightleftharpoons Ni + 6\,NH_3$ | −0.48 |
| $Co(NH_3)_6{}^{2+} + 2\,e^- \rightleftharpoons Co + 6\,NH_3$ | −0.422 |
| $Cr^{3+} + e^- \rightleftharpoons Cr^{2+}$ | −0.41 |
| $Fe^{2+} + 2\,e^- \rightleftharpoons Fe$ | −0.409 |
| $Cd^{2+} + 2\,e^- \rightleftharpoons Cd$ | −0.4026 |
| $PbSO_4 + 2\,e^- \rightleftharpoons Pb + SO_4{}^{2-}$ | −0.356 |
| $In^{3+} + 3\,e^- \rightleftharpoons In$ | −0.338 |
| $Tl^+ + e^- \rightleftharpoons Tl$ | −0.3363 |
| $Ag(CN)_2{}^- + e^- \rightleftharpoons Ag + 2\,CN^-$ | −0.31 |
| $Co^{2+} + 2\,e^- \rightleftharpoons Co$ | −0.28 |
| $H_3PO_4 + 2\,H^+ + 2\,e^- \rightleftharpoons H_3PO_3 + H_2O$ | −0.276 |
| $Ni^{2+} + 2\,e^- \rightleftharpoons Ni$ | −0.23 |
| $2\,SO_4{}^{2-} + 4\,H^+ + 2\,e^- \rightleftharpoons S_2O_6{}^{2-} + 2\,H_2O$ | −0.224 |
| $CO_2 + 2\,H^+ + 2\,e^- \rightleftharpoons HCO_2H$ | −0.20 |
| $O_2 + 2\,H_2O + 2\,e^- \rightleftharpoons H_2O_2 + 2\,OH^-$ | −0.146 |
| $Sn^{2+} + 2\,e^- \rightleftharpoons Sn$ | −0.1364 |
| $Pb^{2+} + 2\,e^- \rightleftharpoons Pb$ | −0.1263 |
| $CrO_4{}^{2-} + 4\,H_2O + 3\,e^- \rightleftharpoons Cr(OH)_3 + 5\,OH^-$ | −0.12 |
| $WO_3 + 6\,H^+ + 6\,e^- \rightleftharpoons W + 3\,H_2O$ | −0.09 |
| $Ru^{3+} + e^- \rightleftharpoons Ru^{2+}$ | −0.08 |
| $O_2 + H_2O + 2\,e^- \rightleftharpoons HO_2{}^- + OH^-$ | −0.076 |
| $Fe^{3+} + 3\,e^- \rightleftharpoons Fe$ | −0.036 |
| $2\,H^+ + 2\,e^- \rightleftharpoons H_2$ | 0.0000000... |
| $NO_3{}^- + H_2O + 2\,e^- \rightleftharpoons NO_2{}^- + 2\,OH^-$ | 0.01 |
| $AgBr + e^- \rightleftharpoons Ag + Br^-$ | 0.0713 |
| $S_4O_6{}^{2-} + 2\,e^- \rightleftharpoons 2\,S_2O_3{}^{2-}$ | 0.0895 |
| $Sn^{4+} + 2\,e^- \rightleftharpoons Sn^{2+}$ | 0.15 |
| $Cu^{2+} + e^- \rightleftharpoons Cu^+$ | 0.158 |
| $ClO_4{}^- + H_2O + 2\,e^- \rightleftharpoons ClO_3{}^- + 2\,OH^-$ | 0.17 |
| $SO_4{}^{2-} + 4\,H^+ + 2\,e^- \rightleftharpoons H_2SO_3 + H_2O$ | 0.20 |
| $AgCl + e^- \rightleftharpoons Ag + Cl^-$ | 0.2223 |
| $IO_3{}^- + 3\,H_2O + 6\,e^- \rightleftharpoons I^- + 6\,OH^-$ | 0.26 |
| $Hg_2Cl_2 + 2\,e^- \rightleftharpoons 2\,Hg + 2\,Cl^-$ | 0.2682 |
| $Cu^{2+} + 2\,e^- \rightleftharpoons Cu$ | 0.3402 |
| $ClO_3{}^- + H_2O + 2\,e^- \rightleftharpoons ClO_2{}^- + 2\,OH^-$ | 0.35 |
| $Ag(NH_3)_2{}^+ + e^- \rightleftharpoons Ag + 2\,NH_3$ | 0.373 |
| $O_2 + 2\,H_2O + 4\,e^- \rightleftharpoons 4\,OH^-$ | 0.401 |
| $H_2SO_3 + 4\,H^+ + 4\,e^- \rightleftharpoons S + 3\,H_2O$ | 0.45 |
| $HgCl_4{}^{2-} + 2\,e^- \rightleftharpoons Hg + 4\,Cl^-$ | 0.48 |
| $Cu^+ + e^- \rightleftharpoons Cu$ | 0.522 |
| $I_3{}^- + 2\,e^- \rightleftharpoons 3\,I^-$ | 0.5338 |
| $I_2 + 2\,e^- \rightleftharpoons 2\,I^-$ | 0.535 |
| $MnO_4{}^- + 2\,H_2O + 3\,e^- \rightleftharpoons MnO_2 + 4\,OH^-$ | 0.588 |
| $ClO_2{}^- + H_2O + 2\,e^- \rightleftharpoons ClO^- + 2\,OH^-$ | 0.59 |
| $O_2 + 2\,H^+ + 2\,e^- \rightleftharpoons H_2O_2$ | 0.682 |
| $Fe^{3+} + e^- \rightleftharpoons Fe^{2+}$ | 0.770 |

*(continued)*

**TABLE A.13 Standard-State Reduction Potentials** (continued)

| Half-Reaction | E° (V) |
|---|---|
| $Hg_2^{2+} + 2e^- \rightleftharpoons Hg$ | 0.7961 |
| $Ag^+ + e^- \rightleftharpoons Ag$ | 0.7996 |
| $Hg^{2+} + 2e^- \rightleftharpoons Hg$ | 0.851 |
| $H_2O_2 + 2e^- \rightleftharpoons 2\,OH^-$ | 0.88 |
| $ClO^- + H_2O + 2e^- \rightleftharpoons Cl^- + 2\,OH^-$ | 0.89 |
| $2\,Hg^{2+} + 2e^- \rightleftharpoons Hg_2^{2+}$ | 0.905 |
| $NO_3^- + 3\,H^+ + 2e^- \rightleftharpoons HNO_2 + H_2O$ | 0.94 |
| $ClO_2 + e^- \rightleftharpoons ClO_2^-$ | 0.95 |
| $NO_3^- + 4\,H^+ + 3e^- \rightleftharpoons NO + 2\,H_2O$ | 0.96 |
| $Pd^{2+} + 2e^- \rightleftharpoons Pd$ | 0.987 |
| $HNO_2 + H^+ + e^- \rightleftharpoons NO + H_2O$ | 0.99 |
| $IO_3^- + 6\,H^+ + 6e^- \rightleftharpoons I^- + 3\,H_2O$ | 1.085 |
| $Br_2(aq) + 2e^- \rightleftharpoons 2\,Br^-$ | 1.087 |
| $Cr^{6+} + 3e^- \rightleftharpoons Cr^{3+}$ | 1.10 |
| $ClO_3^- + 2\,H^+ + 2e^- \rightleftharpoons ClO_2^- + H_2O$ | 1.15 |
| $ClO_4^- + 2\,H^+ + 2e^- \rightleftharpoons ClO_3^- + H_2O$ | 1.19 |
| $2\,IO_3^- + 12\,H^+ + 10e^- \rightleftharpoons I_2 + 6\,H_2O$ | 1.19 |
| $HCrO_4^- + 7\,H^+ + 3e^- \rightleftharpoons Cr^{3+} + 4\,H_2O$ | 1.195 |
| $Pt^{2+} + 2e^- \rightleftharpoons Pt$ | 1.2 |
| $MnO_2 + 4\,H^+ + 2e^- \rightleftharpoons Mn^{2+} + 2\,H_2O$ | 1.208 |
| $O_2 + 4\,H^+ + 4e^- \rightleftharpoons 2\,H_2O$ | 1.229 |
| $O_3 + H_2O + 2e^- \rightleftharpoons O_2 + 2\,OH^-$ | 1.24 |
| $Tl^{3+} + 2e^- \rightleftharpoons Tl^+$ | 1.247 |
| $ClO_2 + H^+ + e^- \rightleftharpoons HClO_2$ | 1.27 |
| $2\,HNO_2 + 4\,H^+ + 4e^- \rightleftharpoons N_2O + 3\,H_2O$ | 1.27 |
| $Au^{3+} + 2e^- \rightleftharpoons Au^+$ | 1.29 |
| $Cr_2O_7^{2-} + 14\,H^+ + 6e^- \rightleftharpoons 2\,Cr^{3+} + 7\,H_2O$ | 1.33 |
| $ClO_4^- + 8\,H^+ + 7e^- \rightleftharpoons \frac{1}{2}\,Cl_2 + 4\,H_2O$ | 1.34 |
| $Cl_2(g)2e^- \rightleftharpoons 2\,Cl^-$ | 1.3583 |
| $ClO_4^- + 8\,H^+ + 8e^- \rightleftharpoons Cl^- + 4\,H_2O$ | 1.37 |
| $Au^{3+} + 3e^- \rightleftharpoons Au$ | 1.42 |
| $ClO_3^- + 6\,H^+ + 6e^- \rightleftharpoons Cl^- + 3\,H_2O$ | 1.45 |
| $PbO_2 + 4\,H^+ + 2e^- \rightleftharpoons Pb^{2+} + 2\,H_2O$ | 1.467 |
| $2\,ClO_3^- + 12\,H^+ + 10e^- \rightleftharpoons Cl_2 + 6\,H_2O$ | 1.47 |
| $HClO + H^+ + 2e^- \rightleftharpoons Cl^- + H_2O$ | 1.49 |
| $MnO_4^- + 8\,H^+ + 5e^- \rightleftharpoons Mn^{2+} + 4\,H_2O$ | 1.491 |
| $HClO_2 + 3\,H^+ + 4e^- \rightleftharpoons Cl^- + 2\,H_2O$ | 1.56 |
| $2\,NO + 2\,H^+ + 2e^- \rightleftharpoons N_2O + H_2O$ | 1.59 |
| $2\,HClO_2 + 6\,H^+ + 6e^- \rightleftharpoons Cl_2 + 4\,H_2O$ | 1.63 |
| $2\,HClO + 2\,H^+ + 2e^- \rightleftharpoons Cl_2 + 2\,H_2O$ | 1.63 |
| $HClO_2 + 2\,H^+ + 2e^- \rightleftharpoons HClO + H_2O$ | 1.64 |
| $MnO_4^- + 4\,H^+ + 3e^- \rightleftharpoons MnO_2 + 2\,H_2O$ | 1.679 |
| $Au^+ + e^- \rightleftharpoons Au$ | 1.68 |
| $PbO_2 + SO_4^{2-} + 4\,H^+ + 2e^- \rightleftharpoons PbSO_4 + 2\,H_2O$ | 1.685 |
| $N_2O + 2\,H^+ + 2e^- \rightleftharpoons N_2 + H_2O$ | 1.77 |
| $H_2O_2 + 2\,H^+ + 2e^- \rightleftharpoons 2\,H_2O$ | 1.776 |
| $CO^{3+} + e^- \rightleftharpoons Co^{2+}$ | 1.842 |
| $S_2O_8^{2-} + 2e^- \rightleftharpoons 2\,SO_4^{2-}$ | 2.05 |
| $O_3(g) + 2\,H^+ + 2e^- \rightleftharpoons O_2(g) + H_2O$ | 2.07 |
| $F_2(g) + 2\,H^+ + 2e^- \rightleftharpoons 2\,HF(aq)$ | 3.03 |

## TABLE A.14 Bond-Dissociation Enthalpies

Single-Bond Dissociation Enthalpies (kJ/mol)

| | As | B | Br | C | Cl | F | H | I | N | O | P | S | Si |
|---|---|---|---|---|---|---|---|---|---|---|---|---|---|
| As | 180 | | 255 | 200 | 310 | 485 | 300 | 180 | | 330 | | | |
| B | | 300 | 370 | | 445 | 645 | | 270 | | 525 | | | |
| Br | | | 195 | 270 | 220 | 240 | 370 | 180 | 250 | | 270 | 215 | 330 |
| C | | | | 350 | 330 | 490 | 415 | 210 | 305 | 360 | 265 | 270 | 305 |
| Cl | | | | | 240 | 250 | 431 | 210 | 190 | 205 | 330 | 270 | 400 |
| F | | | | | | 160 | 569 | | 280 | 215 | 500 | 325 | 600 |
| H | | | | | | | 435 | 300 | 390 | 464 | 325 | 370 | 320 |
| I | | | | | | | | 150 | | 200 | 180 | | 230 |
| N | | | | | | | | | 160 | 165 | | | 330 |
| O | | | | | | | | | | 140 | 370 | 423 | 464 |
| P | | | | | | | | | | | 210 | | |
| S | | | | | | | | | | | | 260 | |
| Si | | | | | | | | | | | | | 225 |

Double- and Triple-Bond Dissociation Enthalpies (kJ/mol)

| Bond | kJ/mol | Bond | kJ/mol |
|---|---|---|---|
| C=C | 611 | C=S | 477 |
| C≡C | 837 | N=N | 418 |
| C=O | 745 | N≡N | 946 |
| C≡O | 1075 | N=O | 594 |
| C=N | 615 | O=O | 498 |
| C≡N | 891 | S=O | 523 |

## TABLE A.15 Standard-State Enthalpy of Formation, Free Energy of Formation, and Absolute Entropy Data

| Substance | $\Delta H_f^\circ$ (kJ/mol) | $\Delta G_f^\circ$ (kJ/mol) | $S^\circ$ (J/mol-K) |
|---|---|---|---|
| *Aluminum* | | | |
| $Al(s)$ | 0 | 0 | 28.33 |
| $Al(g)$ | 326.4 | 285.7 | 164.54 |
| $Al^{3+}(aq)$ | −531 | −485 | −321.7 |
| $Al_2O_3(s)$ | −1675.7 | −1582.3 | 50.92 |
| $AlCl_3(s)$ | −704.2 | −628.8 | 110.67 |
| $AlF_3(s)$ | −1504.1 | −1425.0 | 66.44 |
| $Al_2(SO_4)_3(s)$ | −3440.84 | −3099.94 | 239.3 |
| *Antimony* | | | |
| $Sb(s)$ | 0 | 0 | 45.69 |
| $Sb_4(g)$ | 205.0 | 154.8 | 352 |
| $Sb_2O_5(s)$ | −971.9 | −829.2 | 125.1 |
| $Sb_4O_6(s)$ | −1440.6 | −1268.1 | 220.9 |
| $SbH_3(g)$ | 145.105 | 147.75 | 232.78 |
| $SbCl_3(s)$ | −382.17 | −323.67 | 184.1 |
| $SbCl_3(g)$ | −313.8 | −301.2 | 337.80 |
| $SbCl_5(l)$ | −440.2 | −350.1 | 301 |
| $SbCl_5(g)$ | −394.3 | −334.29 | 401.94 |
| $Sb_2S_3(s)$ | −174.9 | −173.6 | 182.0 |

(*continued*)

**TABLE A.15 Standard-State Enthalpy of Formation, Free Energy of Formation, and Absolute Entropy Data** (continued)

| Substance | $\Delta H_f^\circ$ (kJ/mol) | $\Delta G_f^\circ$ (kJ/mol) | $S^\circ$ (J/mol-K) |
|---|---|---|---|
| *Arsenic* | | | |
| $As(s)$ | 0 | 0 | 35.1 |
| $As_4(g)$ | 143.9 | 92.4 | 314 |
| $As_2O_5(s)$ | −924.87 | −782.3 | 105.4 |
| $As_4O_6(s)$ | −1313.94 | −1152.43 | 214.2 |
| $AsH_3(g)$ | 66.44 | 68.93 | 222.78 |
| $AsF_3(g)$ | −785.76 | −770.76 | 289.10 |
| $AsCl_3(g)$ | −261.5 | −248.9 | 327.17 |
| $AsBr_3(g)$ | −130. | −159 | 363.87 |
| $As_2S_3(s)$ | −169 | −168.6 | 163.6 |
| *Barium* | | | |
| $Ba(s)$ | 0 | 0 | 62.8 |
| $Ba(g)$ | 180 | 146 | 170.243 |
| $Ba^{2+}(aq)$ | −537.64 | −560.77 | 9.6 |
| $BaO(s)$ | −553.5 | −525.1 | 70.42 |
| $Ba(OH)_2 \cdot 8\,H_2O(s)$ | −3342.2 | −2792.8 | 427 |
| $BaCl_2(s)$ | −858.6 | −810.4 | 123.68 |
| $BaCl_2(aq)$ | −871.95 | −823.21 | 122.6 |
| $BaSO_4(s)$ | −1473.2 | −1362.2 | 132.2 |
| $Ba(NO_3)_2(s)$ | −992.07 | −769.59 | 213.8 |
| $Ba(NO_3)_2(aq)$ | −952.36 | −783.28 | 302.5 |
| *Beryllium* | | | |
| $Be(s)$ | 0 | 0 | 9.50 |
| $Be(g)$ | 324.3 | 286.6 | 136.269 |
| $Be^{2+}(aq)$ | −382.8 | −379.73 | −129.7 |
| $BeO(s)$ | −609.6 | −508.3 | 14.14 |
| $BeCl_2(s)$ | −490.4 | −445.6 | 82.68 |
| *Bismuth* | | | |
| $Bi(s)$ | 0 | 0 | 56.74 |
| $Bi(g)$ | 207.1 | 168.2 | 187.05 |
| $Bi_2O_3(s)$ | −573.88 | −493.7 | 151.5 |
| $BiCl_3(s)$ | −379.1 | −315.0 | 177.0 |
| $BiCl_3(g)$ | −265.7 | −256.0 | 358.85 |
| $Bi_2S_3(s)$ | −143.1 | −140.6 | 200.4 |
| *Boron* | | | |
| $B(s)$ | 0 | 0 | 5.86 |
| $B(g)$ | 562.7 | 518.8 | 153.45 |
| $B_2O_3(s)$ | −1272.77 | −1193.65 | 53.97 |
| $B_2H_6(g)$ | 35.6 | 86.7 | 232.11 |
| $B_5H_9(l)$ | 42.68 | 171.82 | 184.22 |
| $B_5H_9(g)$ | 73.2 | 175.0 | 275.92 |
| $B_{10}H_{14}(s)$ | −45.2 | 192.3 | 176.56 |
| $H_3BO_3(s)$ | −1094.33 | −968.92 | 88.83 |

*(continued)*

**TABLE A.15 Standard-State Enthalpy of Formation, Free Energy of Formation, and Absolute Entropy Data** (continued)

| Substance | $\Delta H_f^\circ$ (kJ/mol) | $\Delta G_f^\circ$ (kJ/mol) | $S^\circ$ (J/mol-K) |
|---|---|---|---|
| *Boron, continued* | | | |
| $BF_3(g)$ | −1137.00 | −1120.33 | 254.12 |
| $BCl_3(l)$ | −427.2 | −387.4 | 206.3 |
| $B_3N_3H_6(l)$ | −541.00 | −392.65 | 199.6 |
| $B_3N_3H_6(g)$ | −511.75 | −390.00 | 288.68 |
| *Bromine* | | | |
| $Br_2(l)$ | 0 | 0 | 152.231 |
| $Br_2(g)$ | 30.907 | 3.110 | 245.463 |
| $Br(g)$ | 111.884 | 82.396 | 175.022 |
| $HBr(g)$ | −36.40 | −53.45 | 198.695 |
| $HBr(aq)$ | −121.55 | −103.96 | 82.4 |
| $BrF(g)$ | −93.85 | −109.18 | 228.97 |
| $BrF_3(g)$ | −255.60 | −229.43 | 292.53 |
| $BrF_5(g)$ | −428.9 | −350.6 | 320.19 |
| *Calcium* | | | |
| $Ca(s)$ | 0 | 0 | 41.42 |
| $Ca(g)$ | 178.2 | 144.3 | 154.884 |
| $Ca^{2+}(aq)$ | −542.83 | −553.58 | −53.1 |
| $CaO(s)$ | −635.09 | −604.03 | 39.75 |
| $Ca(OH)_2(s)$ | −986.09 | −898.49 | 83.39 |
| $CaCl_2(s)$ | −795.8 | −748.1 | 104.6 |
| $CaSO_4(s)$ | −1434.11 | −1321.79 | 106.7 |
| $CaSO_4 \cdot 2H_2O(s)$ | −2022.63 | −1797.28 | 194.1 |
| $Ca(NO_3)_2(s)$ | −938.39 | −743.07 | 193.3 |
| $CaCO_3(s)$ | −1206.92 | −1128.79 | 92.9 |
| $Ca_3(PO_4)_2$ | −4120.8 | −3884.7 | 236.0 |
| *Carbon* | | | |
| C(graphite) | 0 | 0 | 5.74 |
| C(diamond) | 1.895 | 2.900 | 2.377 |
| $C(g)$ | 716.682 | 671.257 | 158.096 |
| $CO(g)$ | −110.525 | −137.168 | 197.674 |
| $CO_2(g)$ | −393.509 | −394.359 | 213.74 |
| $COCl_2(g)$ | −218.8 | −204.6 | 283.53 |
| $CH_4(g)$ | −74.81 | −50.752 | 186.264 |
| $HCHO(g)$ | −108.57 | −102.53 | 218.77 |
| $H_2CO_3(aq)$ | −699.65 | −623.08 | 187.4 |
| $HCO_3^-(aq)$ | −691.99 | −586.77 | 91.2 |
| $CO_3^{2-}(aq)$ | −677.14 | −527.81 | −56.9 |
| $CH_3OH(l)$ | −238.66 | −166.27 | 126.8 |
| $CH_3OH(g)$ | −200.66 | −161.96 | 239.81 |
| $CCl_4(l)$ | −135.44 | −65.21 | 216.40 |
| $CCl_4(g)$ | −102.9 | −60.59 | 309.85 |
| $CHCl_3(l)$ | −134.47 | −73.66 | 201.1 |
| $CHCl_3(g)$ | −103.14 | −70.34 | 295.71 |
| $CH_2Cl_2(l)$ | −121.46 | −67.26 | 177.8 |

(*continued*)

**TABLE A.15 Standard-State Enthalpy of Formation, Free Energy of Formation, and Absolute Entropy Data** (continued)

| Substance | $\Delta H_f^\circ$ (kJ/mol) | $\Delta G_f^\circ$ (kJ/mol) | $S^\circ$ (J/mol-K) |
|---|---|---|---|
| *Carbon, continued* | | | |
| $CH_2Cl_2(g)$ | −92.47 | −65.87 | 270.23 |
| $CH_3Cl(g)$ | −80.84 | −57.40 | 234.5 |
| $CS_2(l)$ | 89.70 | 65.27 | 151.34 |
| $CS_2(g)$ | 117.36 | 67.12 | 237.84 |
| $HCN(g)$ | 135.1 | 124.7 | 201.78 |
| $CH_3NO_2(l)$ | −113.09 | −14.42 | 171.75 |
| $C_2H_2(g)$ | 226.73 | 209.20 | 200.94 |
| $C_2H_4(g)$ | 52.26 | 68.15 | 219.56 |
| $C_2H_6(g)$ | −84.68 | −32.82 | 229.60 |
| $CH_3CHO(l)$ | −192.30 | −128.12 | 160.2 |
| $CH_3CO_2H(l)$ | −484.5 | −389.9 | 159.8 |
| $CH_3CO_2H(g)$ | −432.25 | −374.0 | 282.5 |
| $CH_3CO_2H(aq)$ | −485.76 | −396.46 | 178.7 |
| $CH_3CO_2^-(aq)$ | −486.01 | −369.31 | 86.6 |
| $CH_3CH_2OH(l)$ | −277.69 | −174.78 | 160.7 |
| $CH_3CH_2OH(g)$ | −235.10 | −168.49 | 282.70 |
| $CH_3CH_2OH(aq)$ | −288.3 | −181.64 | 148.5 |
| $C_6H_6(l)$ | 49.028 | 124.50 | 172.8 |
| $C_6H_6(g)$ | 82.927 | 129.66 | 269.2 |
| *Chlorine* | | | |
| $Cl_2(g)$ | 0 | 0 | 223.066 |
| $Cl(g)$ | 121.679 | 105.680 | 165.198 |
| $Cl^-(aq)$ | −167.159 | −131.228 | 56.5 |
| $ClO_2(g)$ | 102.5 | 120.5 | 256.84 |
| $Cl_2O(g)$ | 80.3 | 97.9 | 266.21 |
| $Cl_2O_7(l)$ | 238 | | |
| $HCl(g)$ | −92.307 | −95.299 | 186.908 |
| $HCl(aq)$ | −167.159 | −131.228 | 56.5 |
| $ClF(g)$ | −54.48 | −55.94 | 217.89 |
| *Chromium* | | | |
| $Cr(s)$ | 0 | 0 | 23.77 |
| $Cr(g)$ | 396.6 | 351.8 | 174.50 |
| $CrO_3(s)$ | −589.5 | | |
| $CrO_4^{2-}(aq)$ | −881.15 | −727.75 | 50.21 |
| $Cr_2O_3(s)$ | −1139.7 | −1058.1 | 81.2 |
| $Cr_2O_7^{2-}(aq)$ | −1490.3 | −1301.1 | 261.9 |
| $(NH_4)_2Cr_2O_7(s)$ | −1806.7 | | |
| $PbCrO_4(s)$ | −930.9 | | |
| *Cobalt* | | | |
| $Co(s)$ | 0 | 0 | 30.04 |
| $Co(g)$ | 424.7 | 380.3 | 179.515 |
| $Co^{2+}(aq)$ | −58.2 | −54.4 | −113 |
| $Co^{3+}(aq)$ | 92 | 134 | −305 |
| $CoO(s)$ | −237.94 | −214.20 | 52.97 |
| $Co_3O_4(s)$ | −891 | −774 | 102.5 |
| $Co(NH_3)_6^{3+}(aq)$ | −584.9 | −157.0 | 146 |

*(continued)*

**TABLE A.15 Standard-State Enthalpy of Formation, Free Energy of Formation, and Absolute Entropy Data** (continued)

| Substance | $\Delta H_f^\circ$ (kJ/mol) | $\Delta G_f^\circ$ (kJ/mol) | $S^\circ$ (J/mol-K) |
|---|---|---|---|
| *Copper* | | | |
| Cu(*s*) | 0 | 0 | 33.150 |
| Cu(*g*) | 338.32 | 298.58 | 166.38 |
| $Cu^+(aq)$ | 71.67 | 49.98 | 40.6 |
| $Cu^{2+}(aq)$ | 64.77 | 65.49 | −99.6 |
| CuO(*s*) | −157.3 | −129.7 | 42.63 |
| $Cu_2O(s)$ | −168.6 | −146.0 | 93.14 |
| $CuCl_2(s)$ | −220.1 | −175.7 | 108.07 |
| CuS(*s*) | −53.1 | −53.6 | 66.5 |
| $Cu_2S(s)$ | −79.5 | −86.2 | 120.9 |
| $CuSO_4(s)$ | −771.36 | −66.69 | 109 |
| $Cu(NH_3)_4{}^{2+}(aq)$ | −348.5 | −111.07 | 273.6 |
| *Fluorine* | | | |
| $F_2(g)$ | 0 | 0 | 202.78 |
| F(*g*) | 78.99 | 61.91 | 158.754 |
| $F^-(aq)$ | −332.63 | −278.79 | −13.8 |
| HF(*g*) | −271.1 | −273.2 | 173.779 |
| HF(*aq*) | −320.08 | −296.82 | 88.7 |
| *Hydrogen* | | | |
| $H_2(g)$ | 0 | 0 | 130.684 |
| H(*g*) | 217.65 | 203.247 | 114.713 |
| $H^+(aq)$ | 0 | 0 | 0 |
| $OH^-(aq)$ | −229.994 | −157.244 | −10.75 |
| $H_2O(l)$ | −285.830 | −237.129 | 69.91 |
| $H_2O(g)$ | −241.818 | −228.572 | 188.25 |
| $H_2O_2(l)$ | −187.78 | −120.35 | 109.6 |
| $H_2O_2(aq)$ | −191.17 | −134.03 | 143.9 |
| *Iodine* | | | |
| $I_2(s)$ | 0 | 0 | 116.135 |
| $I_2(g)$ | 62.438 | 19.327 | 260.69 |
| I(*g*) | 106.838 | 70.50 | 180.791 |
| HI(*g*) | 26.48 | 1.70 | 206.594 |
| IF(*g*) | −95.65 | −118.51 | 236.17 |
| $IF_5(g)$ | −822.49 | −751.73 | 327.7 |
| $IF_7(g)$ | −943.9 | −818.3 | 346.5 |
| ICl(*g*) | 17.78 | −5.46 | 247.551 |
| IBr(*g*) | 40.84 | 3.69 | 258.773 |
| *Iron* | | | |
| Fe(*s*) | 0 | 0 | 27.28 |
| Fe(*g*) | 416.3 | 370.7 | 180.49 |
| $Fe^{2+}(aq)$ | −89.1 | −78.90 | −137.7 |
| $Fe^{3+}(aq)$ | −48.5 | −4.7 | −315.9 |
| $Fe_2O_3(s)$ | −824.2 | −742.2 | 87.40 |
| $Fe_3O_4(s)$ | −1118.4 | −1015.4 | 146.4 |
| $Fe(OH)_2(s)$ | −569.0 | −486.5 | 88 |

(*continued*)

**TABLE A.15 Standard-State Enthalpy of Formation, Free Energy of Formation, and Absolute Entropy Data** (continued)

| Substance | $\Delta H_f^\circ$ (kJ/mol) | $\Delta G_f^\circ$ (kJ/mol) | $S^\circ$ (J/mol-K) |
|---|---|---|---|
| *Iron, continued* | | | |
| $Fe(OH)_3(s)$ | −823.0 | −696.5 | 106.7 |
| $FeCl_3(s)$ | −399.49 | −334.00 | 142.3 |
| $FeS_2(s)$ | −178.2 | −166.9 | 52.93 |
| $Fe(CO)_5(l)$ | −774.0 | −705.3 | 338.1 |
| $Fe(CO)_5(g)$ | −733.9 | −697.21 | 445.3 |
| *Lead* | | | |
| $Pb(s)$ | 0 | 0 | 64.81 |
| $Pb(g)$ | 195.0 | 161.9 | 175.373 |
| $Pb^{2+}(aq)$ | −1.7 | −24.43 | 10.5 |
| $PbO(s)$ | −217.32 | −187.89 | 68.7 |
| $PbO_2(s)$ | −277.4 | −217.33 | 68.6 |
| $PbCl_2(s)$ | −359.41 | −314.10 | 136.0 |
| $PbCl_4(l)$ | −329.3 | | |
| $PbS(s)$ | −100.4 | −98.7 | 91.2 |
| $PbSO_4(s)$ | −919.94 | −813.14 | 148.57 |
| $Pb(NO_3)_2(s)$ | −451.9 | | |
| $PbCO_3(s)$ | −699.1 | −625.5 | 131.0 |
| *Lithium* | | | |
| $Li(s)$ | 0 | 0 | 29.12 |
| $Li(g)$ | 159.37 | 126.66 | 138.77 |
| $Li^{+}(aq)$ | −278.49 | −293.31 | 13.4 |
| $LiH(s)$ | −90.54 | −68.35 | 20.08 |
| $LiOH(s)$ | −484.93 | −438.95 | 42.80 |
| $LiF(s)$ | −615.97 | −587.71 | 35.65 |
| $LiCl(s)$ | −408.61 | −384.37 | 59.33 |
| $LiBr(s)$ | −351.23 | −342.00 | 74.27 |
| $LiI(s)$ | −270.41 | −270.29 | 86.78 |
| $LiAlH_4(s)$ | −116.3 | −44.7 | 78.74 |
| $LiBH_4(s)$ | 190.8 | 125.0 | 75.86 |
| *Magnesium* | | | |
| $Mg(s)$ | 0 | 0 | 32.68 |
| $Mg(g)$ | 147.70 | 113.10 | 148.650 |
| $Mg^{2+}(aq)$ | −466.85 | −454.8 | −138.1 |
| $MgO(s)$ | −601.70 | −569.43 | 26.94 |
| $MgH_2(s)$ | −75.3 | −35.9 | 31.09 |
| $Mg(OH)_2(s)$ | −924.54 | −833.58 | 63.18 |
| $MgCl_2(s)$ | −641.32 | −591.79 | 89.62 |
| $MgCO_3(s)$ | −1095.8 | −1012.1 | 65.7 |
| $MgSO_4(s)$ | −1284.9 | −1170.6 | 91.6 |
| *Manganese* | | | |
| $Mn(s)$ | 0 | 0 | 32.01 |
| $Mn(g)$ | 280.7 | 238.5 | 173.70 |
| $Mn^{2+}(aq)$ | −220.75 | −228.1 | −73.6 |
| $MnO(s)$ | −385.22 | −362.90 | 59.71 |

*(continued)*

**TABLE A.15 Standard-State Enthalpy of Formation, Free Energy of Formation, and Absolute Entropy Data** (continued)

| Substance | $\Delta H_f^\circ$ (kJ/mol) | $\Delta G_f^\circ$ (kJ/mol) | $S^\circ$ (J/mol-K) |
|---|---|---|---|
| *Manganese, continued* | | | |
| $MnO_2(s)$ | −520.03 | −465.14 | 53.05 |
| $Mn_2O_3(s)$ | −959.0 | −881.1 | 110.5 |
| $Mn_3O_4(s)$ | −1387.8 | −1283.2 | 155.6 |
| $KMnO_4(s)$ | −837.2 | −737.6 | 171.76 |
| MnS | −214.2 | −218.4 | 78.2 |
| *Mercury* | | | |
| $Hg(l)$ | 0 | 0 | 76.02 |
| $Hg(g)$ | 61.317 | 31.820 | 174.96 |
| $Hg^{2+}(aq)$ | 171.1 | 164.40 | −32.2 |
| $HgO(s)$ | −90.83 | −58.539 | 70.29 |
| $HgCl_2(s)$ | −224.3 | −178.6 | 146.0 |
| $Hg_2Cl_2(s)$ | −265.22 | −210.745 | 192.5 |
| $HgS(s)$ | −58.2 | −50.6 | 82.4 |
| *Nitrogen* | | | |
| $N_2(g)$ | 0 | 0 | 191.61 |
| $N(g)$ | 472.704 | 455.63 | 153.298 |
| $NO(g)$ | 90.25 | 86.55 | 210.761 |
| $NO_2(g)$ | 33.18 | 51.31 | 240.06 |
| $N_2O(g)$ | 82.05 | 104.20 | 219.85 |
| $N_2O_3(g)$ | 83.72 | 139.46 | 312.28 |
| $N_2O_4(g)$ | 9.16 | 97.89 | 304.29 |
| $N_2O_5(g)$ | 11.35 | 115.1 | 355.7 |
| $NO_3^-(aq)$ | −205.0 | −108.74 | 146.4 |
| $NOCl(g)$ | 51.71 | 66.08 | 261.69 |
| $NO_2Cl(g)$ | 12.6 | 54.4 | 272.15 |
| $HNO_2(aq)$ | −119.2 | −50.6 | 135.6 |
| $HNO_3(g)$ | −135.06 | −74.72 | 266.38 |
| $HNO_3(aq)$ | −207.36 | −111.25 | 146.4 |
| $NH_3(g)$ | −46.11 | −16.45 | 192.45 |
| $NH_3(aq)$ | −80.29 | −26.50 | 111.3 |
| $NH_4^+(aq)$ | −132.51 | −79.31 | 113.4 |
| $NH_4NO_3(s)$ | −365.56 | −183.87 | 151.08 |
| $NH_4NO_3(aq)$ | −339.87 | −190.56 | 259.8 |
| $NH_4Cl(s)$ | −314.43 | −203.87 | 94.6 |
| $N_2H_4(l)$ | 50.63 | 149.34 | 121.21 |
| $N_2H_4(g)$ | 95.40 | 159.35 | 238.47 |
| $HN_3(g)$ | 294.1 | 328.1 | 238.97 |
| *Oxygen* | | | |
| $O_2(g)$ | 0 | 0 | 205.138 |
| $O(g)$ | 249.170 | 231.731 | 161.055 |
| $O_3(g)$ | 142.7 | 163.2 | 238.93 |
| *Phosphorus* | | | |
| P(white) | 0 | 0 | 41.09 |
| $P_4(g)$ | 58.91 | 24.4 | 279.98 |

(continued)

**TABLE A.15 Standard-State Enthalpy of Formation, Free Energy of Formation, and Absolute Entropy Data** (continued)

| Substance | $\Delta H_f^\circ$ (kJ/mol) | $\Delta G_f^\circ$ (kJ/mol) | $S^\circ$ (J/mol-K) |
|---|---|---|---|
| *Phosphorus, continued* | | | |
| $P_2(g)$ | 144.3 | 103.7 | 218.129 |
| $P(g)$ | 314.64 | 278.25 | 163.193 |
| $PH_3(g)$ | 5.4 | 13.4 | 210.23 |
| $P_4O_6(s)$ | −1640.1 | | |
| $P_4O_{10}(s)$ | −2984.0 | −2697.7 | 228.86 |
| $PO_4^{3-}(aq)$ | −1277.4 | −1018.7 | −222 |
| $PF_3(g)$ | −918.8 | −897.5 | 273.24 |
| $PF_5(g)$ | −1595.8 | | |
| $PCl_3(l)$ | −319.7 | −272.3 | 217.1 |
| $PCl_3(g)$ | −287.0 | −267.8 | 311.78 |
| $PCl_5(g)$ | −374.9 | −305.0 | 364.58 |
| $H_3PO_4(s)$ | −1279.0 | −1119.1 | 110.50 |
| $H_3PO_4(aq)$ | −1277.4 | −1018.7 | −222 |
| *Potassium* | | | |
| $K(s)$ | 0 | 0 | 64.18 |
| $K(g)$ | 89.24 | 60.59 | 160.336 |
| $K^+(aq)$ | −252.38 | −283.27 | 102.5 |
| $KOH(s)$ | −424.764 | −379.08 | 78.9 |
| $KCl(s)$ | −436.747 | −409.14 | 82.59 |
| $KNO_3(s)$ | −494.63 | −394.86 | 133.05 |
| $K_2Cr_2O_7(s)$ | −2061.5 | −1881.8 | 291.2 |
| $KMnO_4(s)$ | −837.2 | −737.6 | 171.76 |
| *Silicon* | | | |
| $Si(s)$ | 0 | 0 | 18.83 |
| $Si(g)$ | 455.6 | 411.3 | 167.97 |
| $SiO_2(s)$ | −910.94 | −856.64 | 41.84 |
| $SiH_4(g)$ | 34.3 | 56.9 | 204.62 |
| $SiF_4(g)$ | −1614.94 | −1572.65 | 282.49 |
| $SiCl_4(l)$ | −687.0 | −619.9 | 240. |
| $SiCl_4(g)$ | −657.01 | −616.98 | 330.73 |
| *Silver* | | | |
| $Ag(s)$ | 0 | 0 | 42.55 |
| $Ag(g)$ | 284.55 | 245.65 | 172.97 |
| $Ag^+(aq)$ | 105.579 | 77.107 | 72.68 |
| $Ag(NH_3)_2^+(aq)$ | −111.29 | −17.12 | 245.2 |
| $Ag_2O(s)$ | −31.05 | −11.20 | −121.3 |
| $AgCl(s)$ | −127.068 | −109.789 | 96.2 |
| $AgBr(s)$ | −100.37 | −96.90 | 107.1 |
| $AgI(s)$ | −61.84 | −66.19 | −115.5 |
| *Sodium* | | | |
| $Na(s)$ | 0 | 0 | 51.21 |
| $Na(g)$ | 107.32 | 76.761 | 153.712 |
| $Na^+(aq)$ | −240.13 | −261.905 | 59.0 |
| $NaH(s)$ | −56.275 | −33.46 | −40.016 |

*(continued)*

**TABLE A.15 Standard-State Enthalpy of Formation, Free Energy of Formation, and Absolute Entropy Data** (continued)

| Substance | $\Delta H_f^\circ$ (kJ/mol) | $\Delta G_f^\circ$ (kJ/mol) | $S^\circ$ (J/mol-K) |
|---|---|---|---|
| *Sodium, continued* | | | |
| $NaOH(s)$ | −425.609 | −379.494 | 64.455 |
| $NaOH(aq)$ | −470.114 | −419.150 | 48.1 |
| $NaCl(s)$ | −411.153 | −384.138 | 72.13 |
| $NaCl(g)$ | −176.65 | −196.66 | 229.81 |
| $NaCl(aq)$ | −407.27 | −393.133 | 115.5 |
| $NaNO_3(s)$ | −467.85 | −367.00 | 116.52 |
| $Na_3PO_4(s)$ | −1917.40 | −1788.80 | 173.80 |
| $Na_2SO_3(s)$ | −1123.0 | −1028.0 | 155 |
| $Na_2SO_4(s)$ | −1387.08 | −1270.16 | 149.58 |
| $Na_2CO_3(s)$ | −1130.68 | −1044.44 | 134.98 |
| $NaHCO_3(s)$ | −950.81 | −851.0 | 101.7 |
| $NaOAc(s)$ | −708.81 | −607.18 | 123.0 |
| $Na_2CrO_4(s)$ | −1342.2 | −1234.93 | 176.61 |
| $Na_2Cr_2O_7(s)$ | −1978.6 | | |
| *Sulfur* | | | |
| $S_8(s)$ | 0 | 0 | 31.80 |
| $S_8(g)$ | 102.30 | 49.63 | 430.98 |
| $S(g)$ | 278.805 | 238.250 | 167.821 |
| $S^{2-}(aq)$ | 33.1 | 85.8 | −14.6 |
| $SO_2(g)$ | −296.830 | −300.194 | 248.22 |
| $SO_3(s)$ | −454.51 | −374.21 | 70.7 |
| $SO_3(l)$ | −441.04 | −373.75 | 113.8 |
| $SO_3(g)$ | −395.72 | −371.06 | 256.76 |
| $SO_4^{2-}(aq)$ | −909.27 | −744.53 | 20.1 |
| $SOCl_2(g)$ | −212.5 | −198.3 | 309.77 |
| $SO_2Cl_2(g)$ | −364.0 | −320.0 | 311.94 |
| $H_2S(g)$ | −20.63 | −33.56 | 205.79 |
| $H_2SO_3(aq)$ | −608.81 | −537.81 | 232.2 |
| $H_2SO_4(aq)$ | −909.27 | −744.53 | 20.1 |
| $SF_4(g)$ | −774.9 | −731.3 | 292.03 |
| $SF_6(g)$ | −1209 | −1105.3 | 291.82 |
| $SCN^-(aq)$ | 76.44 | 92.71 | 144.3 |
| *Tin* | | | |
| $Sn(s)$ | 0 | 0 | 44.14 |
| $Sn(g)$ | 302.1 | 267.3 | 168.486 |
| $SnO(s)$ | −285.8 | −256.9 | 56.5 |
| $SnO_2(s)$ | −580.7 | −519.76 | 52.3 |
| $SnCl_2(s)$ | −325.1 | | |
| $SnCl_4(l)$ | −511.3 | −440.21 | 258.6 |
| $SnCl_4(g)$ | −471.5 | −432.2 | 365.8 |
| *Titanium* | | | |
| $Ti(s)$ | 0 | 0 | 30.63 |
| $Ti(g)$ | 469.9 | 425.1 | 180.2 |
| $TiO(s)$ | −519.7 | −495.0 | 34.8 |

(continued)

**TABLE A.15 Standard-State Enthalpy of Formation, Free Energy of Formation, and Absolute Entropy Data** (continued)

| Substancc | $\Delta H_f^\circ$ (kJ/mol) | $\Delta G_f^\circ$ (kJ/mol) | $S^\circ$ (J/mol-K) |
|---|---|---|---|
| *Titanium, continued* | | | |
| $TiO_2(s)$ rutile | −944.8 | −889.5 | 50.33 |
| $TiCl_4(l)$ | −804.2 | −737.2 | 252.3 |
| $TiCl_4(g)$ | −763.2 | −726.8 | 354.8 |
| *Tungsten* | | | |
| $W(s)$ | 0 | 0 | 32.64 |
| $W(g)$ | 849.4 | 807.1 | 173.950 |
| $WO_3(s)$ | −842.87 | −764.083 | 75.90 |
| *Zinc* | | | |
| $Zn(s)$ | 0 | 0 | 41.63 |
| $Zn(g)$ | 130.729 | 95.145 | 160.984 |
| $Zn^{2+}(aq)$ | −153.89 | −147.06 | −112.1 |
| $ZnO(s)$ | −348.28 | −318.30 | 43.64 |
| $ZnCl_2(s)$ | −415.05 | −369.39 | 111.46 |
| $ZnS(s)$ | −205.98 | −201.29 | 57.7 |
| $ZnSO_4(s)$ | −982.8 | −871.5 | 110.5 |

# ANSWERS TO CHECKPOINT QUESTIONS

## CHAPTER 1

**p. 19**

Using a yardstick to obtain the data necessary to calculate the volume of a tin can is a source of *random error* because the smallest subdivision on the yardstick is one-eighth of an inch. The volume calculated from these measurements would sometimes be too small, other times too large. This error affects the precision of the measurement and, indirectly, the accuracy as well.

Using a balance that has not been properly adjusted introduces a *systematic error* into the measurements, which would be systematically either too large or too small. This wouldn't affect the precision of a series of measurements, which would agree with each other. But it would affect the accuracy of the measurements.

**p. 20**

The number *407* is known to $\pm 1$ unit, or one part in 407, which suggests three significant figures. The number *470* is only known to $\pm 10$, or one part in 47, which suggests two significant figures.

**p. 28**

Oil, because it is less dense than vinegar.

**p. 29**

Although the mass of the Titanic was enormous, when this mass is divided by the total volume of the water the ship would displace the average density of this air-filled ''steel balloon'' is less than water. As a result, the ship floated, at least until the ''balloon'' was punctured.

**p. 29**

The densities of charcoal (0.3 to 0.6 g/cm$^3$) and bone (1.7 g/cm$^3$) reported in Table 1.7 are smaller than the density of the solution into which they are dropped. As a result, they float to the top of this solution. Sand, soil, and rocks sink to the bottom of this solution.

**p. 31**

It doesn't matter in which direction you try to do the conversion, from °C to °F or vice versa, $-40$°F is equal to $-40$°C

## CHAPTER 2

**p. 47**

Although an atom is neither strictly indivisible nor indestructible, its identity does not change in the course of a chemical reaction, which means that the intent of the first assumption is valid. Not all atoms of an element are identical because some have more neutrons than others. But, once again, the intent of Dalton's assumption is valid because all atoms of an element have the same number of protons, and it is the number of protons that controls the identity of the element

## CHAPTER 3

**p. 76**

The *molecular weight* of a compound is a weighted average of the mass of the molecules in a sample. Because some of the $C_{12}H_{22}O_{11}$ molecules in a box of sugar contain $^{13}C$, whereas others do not, the molecular weight of this compound lies between the mass of a molecule that contains this isotope and the mass of a molecule that doesn't. Because no single $C_{12}H_{22}O_{11}$ molecule ever has a mass of 342.30 amu, some chemists object to the term *molecular weight.* They prefer to talk about the *molar mass* of $C_{12}H_{22}O_{11}$, which is 342.30 grams per mole.

**p. 82**

The *empirical formula* is the simplest whole-number ratio of the atoms of the different elements in a compound and the *empirical weight* is calculated from this formula. The *molecular formula* describes the actual number of atoms of each element in a molecule and the *molecular weight* is calculated from this formula. The empirical formula of glucose, for example, is $CH_2O$, whereas the molecular formula is $C_6H_{12}O_6$, which is six times the empirical formula. The molecular weight of glucose is therefore 180.2 g/mol, or six times the empirical weight.

**p. 97**

It is easier to measure the volume of a liquid than it is to measure its mass. Thus, it is more useful to know this solution contains 18.0 mol of solute per liter than it is to know that the solution is 96.0% $H_2SO_4$ by mass.

**p. 102**

If the indicator doesn't turn color at the point at which exactly equivalent number of moles of acid and base are present, the endpoint and equivalence point of the titration are different. When phenolphthalein is used, the indicator turns a permanent color one drop beyond the equivalence point of a strong acid–strong base titration.

## CHAPTER 4

**p. 114**

When the densities of the liquid and gas are converted to a common set of units we find that the volume of the gas is approximately 900 times larger than an equivalent amount of the liquid:

$$\frac{NH_3(l)}{NH_3(g)} = \frac{648 \text{ g/L}}{0.719 \text{ g/mL}} = 901$$

**p. 117**

The U-tube in Figure 4.4 is stable when the force of the liquid pushing down in one arm exactly balances the force of the liquid pushing down in the other arm. Because the two liquids have different densities, it takes a taller column of ethanol to balance a given column of water. The same phenomenon occurs in a Toricelli barometer. In this case, the force exerted by the atmosphere on the pool of mercury that surrounds the barometer tube balances the force exerted by a column of mercury (13.6 g/cm$^3$) 760 mm tall inside the tube.

**p. 118**

The vertical height would be the same—760 mm. But, because the barometer was tilted by 30°, the length of the mercury column inside the barometer would have to increase before the vertical height becomes 760 mm.

**p. 119**

When we convert inches to millimeters, we find that the height of the column of water is 10,333 mm, which is 13.6 times the height of a column of mercury 760 mm tall. Thus the product of the height of the water column times its density is equal to the height of the mercury column times its density. In other words, it takes a column of water 33.9 ft tall to exert a force that is equal to the force of a column of mercury 760 mm tall.

**p. 122**

Amonton's law suggests that the pressure of a gas is directly proportional to its temperature. All we have to do to turn this apparatus into a thermometer is calibrate it by checking the pressure of the gas at two reference points, such as 0°C and 100°C.

**p. 128**

The first step in this calculation involves solving the ideal gas equation for the ideal gas constant, $R$:

$$\frac{PV}{nT} = R$$

We then note that 22.414 L is 22,414 mL and that 1 atm pressure is 760 mmHg or 760 torr. The ideal gas constant is therefore:

$$\frac{(760 \text{ torr})(22{,}414 \text{ mL})}{(1.0000 \text{ mol})(273.15 \text{ K})} = 6.236 \times 10^4 \text{ mL-torr/mol-K}$$

**p. 132**

At first glance, it seems that air that is saturated with water vapor would weigh more than dry air. But the molecular weight of water (18 g/mol) is significantly smaller than the molecular weight of the two primary components of air, nitrogen (28 g/mol) and oxygen (32 g/mol). If equal volumes of different gases at the same temperature and pressure contain the same number of particles, the average mass of a sample of air saturated with water vapor must be smaller than the average mass of a sample of dry air.

**p. 139**

The temperature of a gas increases if and only if the average kinetic energy of the particles in the gas increases.

## CHAPTER 5

**p. 154**

Lambskin is a *thermal insulator* that protects you from losing heat to the seat when it is cold or gaining heat from the seat when it is hot.

**p. 157**

Heat is an *extensive quantity* that depends on the size of the sample being studied. Both temperature and heat capacity are *intensive quantities.*

**p. 161**

When heat enters the balloon from its surroundings, the temperature of the gas increases. This can only happen if the average kinetic energy of the gas particles increases.

**p. 165**

The first law of thermodynamics states that energy is neither created nor destroyed. Thus, the total energy of the universe is constant:

$$\Delta E_{\text{univ}} = 0$$

But energy can be transferred from the system to its surroundings, or vice versa. When this happens, the change in the energy of the system depends on the quantity of heat the system gains or loses plus the quantity of the work done by the system or on the system:

$$\Delta E_{\text{sys}} = q + w$$

**p. 170**

If you multiply the heat capacity ($C$) by the number of moles ($n$) in the sample and the change in the temperature of the sample ($\Delta T$), you get the heat in units of joules:

$$\text{J} = \cancel{\text{mol}} \times \frac{(\text{J})}{\cancel{\text{mol}}\text{-}\cancel{\text{K}}} \times \cancel{\text{K}}$$

Thus, $q$ is the product of $n$ times $C$ times $\Delta T$:

$$q = nC\Delta T$$

**p. 176**

The magnitude of $\Delta H$ remains the same because we are still comparing the same states of the system. The sign of $\Delta H$ changes, however, because the identities of the initial and final states of the system are reversed.

## CHAPTER 6

**p. 201**

To seven significant figures, there are $6.022045 \times 10^{23}$ particles in a mole:

$$96{,}484.56 \frac{\cancel{C}}{\text{mol}} \times \frac{1\ \text{e}^-}{1.6021892 \times 10^{-19}\ \cancel{C}} = 6.022045 \times 10^{23} \frac{\text{e}^-}{\text{mol}}$$

**p. 203**

If Thomson's model were correct, the fast-moving $\alpha$-particles would be scattered by small angles as the result of minor collisions as they blasted their way through the diffuse metal atoms in a very thin piece of foil. There was nothing in Thomson's model that was massive enough to scatter an $\alpha$-particle through a large angle, much less scatter some of the $\alpha$-particles back toward the source.

**p. 217**

According to the de Broglie equation, the wavelength of any object in motion is inversely proportional to the momentum of the particle:

$$\lambda = \frac{h}{p}$$

As the mass of a series of particles increases, the wavelength decreases until it is so small that the wave character of the object can no longer be detected.

**p. 219**

It takes three quantum numbers to describe an object that occupies three-dimensional space. The principal ($n$) quantum number describes the size, and therefore indirectly the energy, of an orbital. The angular ($l$) quantum number describes the shape of the orbital. The magnetic ($m$) quantum number describes the orientation of the orbital in space.

**p. 229**

The electron configuration of iron would be written: [Ar] $4s^2\ 3d^6$. According to Hund's rules, the five $3d$ orbitals would be filled as follows:

$$\underline{\uparrow\downarrow}\ \ \underline{\uparrow}\ \ \underline{\uparrow}\ \ \underline{\uparrow}\ \ \underline{\uparrow}$$

Hund's rules therefore predict that an isolated iron atom in the gas phase would behave as if it contained four unpaired electrons in the $3d$ orbitals.

**p. 232**

Element 114 would have the following electron configuration: [Rn] $7s^2\ 5f^{14}\ 6d^{10}\ 7p^2$. This element therefore has four valence electrons ($7s^2\ 7p^2$) and should fall just below lead in Group IVA of the periodic table.

## CHAPTER 7

**p. 240**

The following triads are still grouped in the same columns in the periodic table:

| | | | |
|---|---|---|---|
| Li | Ca | S | Cl |
| Na | Sr | Se | Br |
| K | Ba | Te | I |

**p. 245**

The electron configuration of helium ($1s^2$) is different from that of neon ([He] $2s^2\ 2p^6$) or any other element in Group VIIIA.

**p. 246**

The sum of the covalent radii of carbon (0.077 nm) and chlorine (0.099 nm) is 0.176 nm, which is in excellent agreement with the C—Cl bond length in $CCl_4$.

**p. 256**

The first ionization energy measures the energy required to produce a positively charged ion from a neutral atom in the gas phase:

$$\text{F}(g) \longrightarrow \text{F}^+(g) = \text{e}^-$$

The electron affinity measures the energy given off when a neutral atom picks up an electron to form a negative ion:

$$\text{F}(g) + \text{e}^- \longrightarrow \text{F}^-(g)$$

**p. 260**

As ions become smaller, they can come closer together. This increases the force of attraction between ions of opposite charge and therefore the lattice energy of the salt.

**p. 261**

The lattice energy in a salt that contains $M^{2+}$ and $O^{2-}$ ions would be relatively large, which means that neither MgO nor CaO should readily dissolve in water. However, $Ca^{2+}$ ions (0.099 nm) are considerably larger than $Mg^{2+}$ ions (0.065 nm). As a result, the lattice energy in CaO is significantly smaller than it is in MgO. Because it is easier to separate the $M^{2+}$ and $O^{2-}$ ions in CaO than it is in MgO, CaO is more soluble in water than MgO.

## CHAPTER 8

**p. 280**

The difference between the electronegativities of magnesium and oxygen is relatively large ($\Delta EN = 2.13$), which suggests that MgO should be an ionic compound that contains $Mg^{2+}$ and $O^{2-}$ ions. The electronegativity of carbon is much closer to that of oxygen ($\Delta EN = 0.89$), which suggests that CO and $CO_2$ should be covalent compounds held together by the sharing of pairs of electrons.

**p. 281**

The difference between the electronegativities of manganese and oxygen is large enough to predict that oxides of manganese would be ionic compounds, which are normally solids at room temperature. But, the electronegativity of an element increases with oxidation state. This brings the electronegativity of the Mn(VII) oxidation state close enough to that of oxygen to allow these elements to combine to form covalent compounds, such as $Mn_2O_7$, which is a gas at room temperature.

**p. 282**

It takes only two electrons to fill the valence orbitals on a hydrogen atom.

**p. 286**

Because there are no 2*d* orbitals, nitrogen and oxygen can hold a maximum of eight valence electrons in the 2*s* and 2*p* orbitals.

## CHAPTER 9

**p. 317**

The iron metal in these cereals slowly dissolves in the 1 *M* acid in the stomach to form the $Fe^{3+}$ ion.

**p. 318**

(a) rubidium iodide (b) lithium sulfide (c) potassium nitride

**p. 321**

Barium, which is the most active metal in this series, would be the most likely to form a peroxide.

**p. 323**

Magnesium is so active that it reacts with water to form magnesium oxide and hydrogen gas:

$$Mg(s) + H_2O(g) \longrightarrow MgO(s) + H_2(g)$$

**p. 323**

The stoichiometry of the reaction is described by the following equation:

$$Ca + 2NH_3 \longrightarrow Ca^{2+} + 2NH_2^{-} + H_2$$

**p. 326**

When we use aluminum foil to wrap a sandwich, or fly in an airplane that is made from aluminum metal, it is easy to conclude that aluminum is essentially inert to chemical reactions.

**p. 327**

Aluminum is so reactive toward acid that it must be protected by the thin film of plastic on the inner surface of a softdrink can. Because tin is less active, it doesn't need to be protected. In other words, the rate at which this reaction occurs is so slow that it does not occur to a significant extent within the lifetime of the product stored in the can.

**p. 333**

If potassium metal is a strong enough reducing agent to react violently with water, then the $K^+$ ion produced in this reaction must be a relatively weak oxidizing agent. We know this must be true, because we never feel the $K^+$ ions in a banana oxidizing the cells on our tongues or in our stomach or intestine to form potassium metal, which would then explode on contact with the water in our bodies.

**p. 335**

Locate the portion of the table that contains potassium metal. This is where the strongest reducing agents can be found. The portion that contains gold metal must contain the weakest reducing agents.

## CHAPTER 10

**p. 357**

$$:Cl\cdot + O{=}O{-}O: \longrightarrow \cdot Cl{-}O: + O{=}O$$

$$\cdot Cl{-}O: + \cdot O: \longrightarrow :Cl\cdot + O{=}O$$

**p. 360**

$$H{-}O{-}O{-}H + 2\,e^- \longrightarrow [H{-}O:]^- + [:O{-}H]^-$$

$$H{-}O{-}O{-}H \longrightarrow [H{-}O{-}O{-}H]^{2+} \longrightarrow H^+ + O{=}O + H^+$$

**p. 361**

$Na_2O_2$ reacts with water to produce $H_2O_2$, which can decompose to form water and molecular oxygen.

**p. 369**

If you replace one of the oxygen atoms in the $SO_4^{2-}$ ion with a sulfur atom you can transform the sulfate ion into the *thio*sulfate ion ($S_2O_3^{2-}$).

## CHAPTER 11

**p. 404**

Arrhenius defined a base as a substance that contains the $OH^-$ ion. MgO reacts with water to produce an $OH^-$ ion but doesn't contain this ion itself.

**p. 410**

Metal hydroxides on the left side of the periodic table are basic, whereas the nonmetal oxides on the right side of the table are acidic. Amphoteric oxides should be found toward the middle of this transition. NaOH and $Mg(OH)_2$, for example, are basic, whereas $H_3PO_4$, $H_2SO_4$, and $HClO_4$ are acidic. The amphoteric hydroxides include $Al(OH)_3$ and, perhaps, $Si(OH)_4$.

**p. 414**

When water reacts with itself it produces an acid ($H_3O^+$) and a base ($OH^-$):

$$2\,H_2O \rightleftharpoons H_3O^+ + OH^-$$

Extending the same reasoning to the reaction between a pair of ammonia molecules suggests that the $NH_4^+$ ion is an acid and the $NH_2^-$ ion is a base:

$$2\,NH_3 \rightleftharpoons NH_4^+ + NH_2^-$$

**p. 418**

Consider the following hypothetical acid-base reaction.

$$\underset{\text{acid}}{HA(aq)} + \underset{\text{base}}{B(aq)} \rightleftharpoons \underset{\text{acid}}{HB^+(aq)} + \underset{\text{base}}{A^-(aq)}$$

If HA is a stronger acid than $HB^+$, the $A^-$ ion would have to be a weaker base than B. In other words, the stronger acid and the stronger base must be on the same side of the equation:

$$\underset{\text{stronger acid}}{HA(aq)} + \underset{\text{stronger base}}{B(aq)} \rightleftharpoons \underset{\text{weaker acid}}{HB^+(aq)} + \underset{\text{weaker base}}{A^-(aq)}$$

## CHAPTER 12

**p. 442**

Because this complex contains the $Cr^{3+}$ ion, the primary valence is 3. Because it contains a total of six ligands, the secondary valence is 6.

**p. 449**

The acetylacetonate ligand is bidentate because transition metals can form bonds to both of the oxygen atoms:

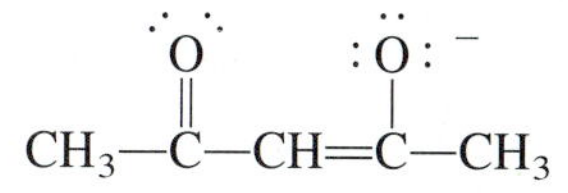

The nitrilotriacetate ion is tetradentate because complexes can form bonds by picking up the pair of nonbonding electrons on the nitrogen or one of the pairs of nonbonding electrons on each of the three $—CO_2^-$ groups.

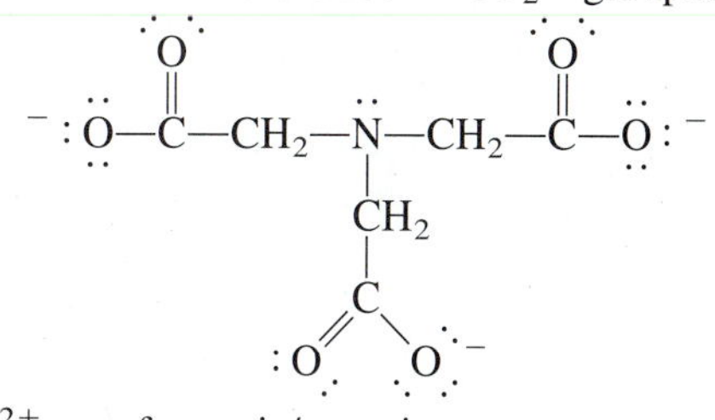

**p. 454**

Only $Ni(en)_2(H_2O)_2^{2+}$ can form cis/trans isomers.

**p. 456**

Only $Fe(acac)_3$ is chiral.

## CHAPTER 13

**p. 476**

When the rubber becomes more crystalline, it becomes more brittle.

**p. 482**

When the bonds between particles are both strong and numerous, it is difficult to disrupt these bonds to form a liquid, much less a gas.

**p. 485**

These traps ''freeze'' when metal atoms migrate from one surface into the other.

**p. 486**

Monatomic positive ions are smaller than the atoms from which they form, whereas monatomic negative ions are larger than the corresponding neutral atom. As a result, simple positive ions are often small enough that they can pack in the holes between planes of closest packed, or at least closely packed, negative ions.

**p. 490**

Iron is body-centered cubic because the lattice point at the center of the unit cell is equivalent to that of the eight corners that define the unit cell. The same cannot be said of cesium chloride, which contains a $Cs^+$ at the center of the cell and $Cl^-$ ions on the eight corners of the cell. CsCl is therefore a simple cubic unit cell of chloride ions with a cesium ion in the center of the cell.

## CHAPTER 14

**p. 519**

The force of adhesion between water and the walls of a buret is a significant fraction of the force of cohesion between water molecules. As a result, water ''wets'' the walls of the buret, and the meniscus curves upward. The force of adhesion between mercury and the walls of a barometer tube is much smaller than the force of cohesion between the mercury atoms. As a result, the mercury tries to minimize contact with the walls of the tube—and maximize contact with itself—and the meniscus curves downward.

**p. 521**

Because the average kinetic energy of a collection of molecules increases with temperature, the velocity at which molecules move down a tube should increase with temperature. As a result, almost any fluid becomes less viscous when heated.

**p. 522**

The water molecule is more polar, which means that the hydrogen bonds between water molecules are stronger.

## CHAPTER 15

**p. 535**

Alanine and cysteine are classified as *hydrophobic* amino acids because they contain hydrocarbon side chains, which have no affinity for water. Lysine and serine are *hydrophilic* amino acids because they have charged side chains or side chains that can form hydrogen bonds to water.

**p. 544**

The boiling point and freezing point of a liquid are characteristic properties of that substance. But the *change* in the freezing point or boiling point that occurs when a solute is dissolved in the liquid depends only on the relative number of solute and solvent, not the identity of the solute particles. As a result, the *change* in the freezing point or boiling point of a liquid are colligative properties.

## CHAPTER 16

**p. 567**

For simple, one-step reactions, the equilibrium constant is equal to the ratio of the rate constants for the forward and reverse reactions:

$$K_c = \frac{k_f}{k_r}$$

When the rate constant for the forward reaction is larger than the rate constant for the reverse reaction, the equilibrium constant for the reaction must be larger than 1.

**p. 571**

At a given temperature, a reaction comes to equilibrium whenever the reaction quotient is equal to the equilibrium constant. Thus, there is only one value of the equilibrium constant for the reaction at a given temperature. The reaction quotient, however, can take on any value between zero and infinity.

**p. 587**

When we remove one of the products of a reaction, we reduce the size of the reaction quotient until it is smaller than the equilibrium constant. The only way the reaction can come back to equilibrium is for the reaction to shift toward the products, converting some of the reactants back into the products of the reaction, thereby increasing the size of the reaction quotient until it is once again equal to the equilibrium constant for the reaction.

## CHAPTER 17

**p. 605**

The pH of a solution is the *negative* of the logarithm of the $H_3O^+$ concentration:

$$pH = -\log [H_3O^+]$$

Thus, as the concentration of the $H_3O^+$ ion increases, the pH of the solution decreases.

**p. 607**

Nitric acid is a strong acid ($K_a = 28$), but it is not a stronger acid than the $H_3O^+$ ion ($K_a = 55$). The stronger acid and stronger base are therefore on the right side of this equation and the reaction cannot go to completion.

**p. 621**

The product of $K_a$ for an acid times $K_b$ for its conjugate base is equal to the dissociation constant for water:

$$K_a \times K_b = K_w = 1 \times 10^{-14}$$

Because $K_w$ is a small number, $K_a$ and $K_b$ can't simultaneously be large numbers. If $K_a$ is large, $K_b$ is small, and vice versa.

## CHAPTER 18

**p. 649**

The concentrations of the $H_3O^+$ and $OH^-$ ions are relatively small, only $1 \times 10^{-7}$ *M* at 25°C. Pure water is therefore a poor conductor of electricity.

**p. 670**

$H_2O$, CO, $O_2$, and the $H^-$ ion can all act as electron-pair donors, or Lewis bases:

$$\mathrm{H{-}\ddot{\underset{..}{O}}{-}H} \qquad \mathrm{:C{\equiv}O:} \qquad \mathrm{\dot{\underset{.}{O}}{=}\dot{\underset{.}{O}}} \qquad \mathrm{H:^-}$$

## CHAPTER 19

**p. 696**

According to the phlogiston theory, light is given off when a strip of magnesium metal is burned because this is the "phlogiston" given off when the metal decomposes to form its calx (MgO).

**p. 707**

Although the equations are balanced in terms of both charge and mass, they are totally unrealistic. The oxidation half-reaction is written as if it consumes the $OH^-$ ion, which will have a minuscule concentration in a strong acid solution, and the reduction half-reaction produces the $OH^-$ ion, which would instantly react with the strong acid to form water. The overall reaction is balanced, but it has no connection to the physical reality of what happens in this reaction.

## CHAPTER 21

**p. 769**

The work done by a system when it expands is defined by the equation:

$$w = -P\Delta V$$

Because the pressure of the gas is positive, and the change in the volume of the system is positive, the system must do work on its surroundings when it expands. If no heat is gained or lost when this happens, the internal energy of the system must decrease:

$$\Delta E_{sys} = q + w$$

Because the temperature of a system is an indirect measure of its internal energy, the system becomes cooler when it expands.

**p. 770**

The relationship between $\Delta H$ and $\Delta E$ for a reaction is $\Delta H = \Delta E + \Delta(PV)$. At constant pressure, this becomes $\Delta H = \Delta E + P\Delta(V)$. Thus, $\Delta H$ is approximately equal to $\Delta E$ when $\Delta V$ is very small. In other words, the change in the enthalpy of the system is approximately equal to the change in the internal energy when there is little, if any, change in the volume of the system. This is true for most reactions that involve only solids and liquids, or reactions that produce exactly the same number of moles of gas particles as they consume. Reactions in which there is a change in the number of moles of gas will have values of $\Delta H$ that are very different from $\Delta E$.

## CHAPTER 22

**p. 805**

The rate at which phenolphthalein reacts with excess base is directly proportional to the concentration of phenolphthalein (PHTH):

$$\text{Rate} = k(\text{PHTH})$$

The rate of this reaction therefore has to slow down as phenolphthalein is consumed in this reaction.

**p. 816**

Because the rate at which $^{15}O$ decays is slower, it has the longer half-life.

## CHAPTER 23

**p. 847**

Electron capture and positron emission transform a proton into a neutron, whereas electron emission converts a neutron into a proton. Because neutrons and protons have roughly the same mass (1 amu), there is no change in the mass number of the nuclide, although the number of protons in the nucleus must change.

**p. 848**

The average mass of a sulfur atom is 32.06 amu because $^{32}S$ is the most common isotope of sulfur. Because the $^{35}S$ isotope contains three more neutrons than normal, it seems reasonable to predict that it is *neutron rich*.

**p. 874**

Fission reactors produce long-lived radioactive byproducts that cannot be released into the environment. If we could harness the fusion reaction used by the sun, which transforms four hydrogen atoms into a helium atom, the product of this reaction would be a stable $^{4}He$ atom, which, in theory, could be released to the environment.

## CHAPTER 25

**p. 955**

The two ends of a peptide are different. Thus, we can use any one of the three amino acids in any one of the three positions. The number of possible tripeptides is therefore $3 \times 3 \times 3$: ASP-ASP-ASP, ASP-ASP-PHE, ASP-PHE-ASP, PHE-ASP-ASP, ASP-ASP-CYS, ASP-CYS-ASP, CYS-ASP-ASP, PHE-PHE-PHE, PHE-PHE-ASP, PHE-ASP-PHE, ASP-PHE-PHE, PHE-PHE-CYS, PHE-CYS-PHE, CYS-PHE-PHE, CYS-CYS-CYS, CYS-CYS-ASP, CYS-ASP-CYS, ASP-CYS-CYS, CYS-CYS-PHE, CYS-PHE-CYS, PHE-CYS-CYS, ASP-CYS-PHE, ASP-PHE-CYS, CYS-PHE-ASP, CYS-ASP-PHE, PHE-ASP-CYS, PHE-CYS-ASP.

**p. 969**

Diesel oil is a mixture of long-chain hydrocarbons, some of which might contain C=C double bonds. Olive oil contains long-chain hydrocarbons *attached to a glycerol backbone by ester linkages.* Similarities in their structures give the two materials similar properties. But the ester linkage at one end of the hydrocarbon chain in the ''oil'' we use to flavor food, rather than the oil we use to fuel our trucks, allows us to digest these compounds.

**p. 973**

Nucleic acids are made by polymerizing nucleotides that are esters of phosphoric acid. Because there are three acidic hydrogen atoms in phosphoric acid, and only two are used to form the ester linkage between nucleotides, the resulting polymer should be about as acidic as phosphoric acid. It isn't surprising that they are called nucleic *acids*.

# APPENDIX C

# ANSWERS TO SELECTED PROBLEMS

**Chapter 1**

**1-3** 40 oz **1-4** 1 y = $3.16 \times 10^7$ s; 1 ton = $3.2 \times 10^4$ oz
**1-5** $1.12 \times 10^5$ grains **1-6** 1.04 oz per fluid oz
**1-7** (a) 0.038126 hogshead (b) 0.5000 kilderkin (c) 0.22222 barrel
**1-8** $2.15 \times 10^3$ in$^3$ **1-9** $3.22 \times 10^{-17}$ light-years
**1-11** (a) $10^{-9}$ (b) $10^{-6}$ (c) $10^{-3}$ (d) $10^{-2}$ (e) $10^3$
**1-12** (a) 43 g (b) 2450 mL (c) 8.14 mL (d) 34.68 cm
**1-13** $1.9 \times 10^{-8}$ cm; 0.19 nm; 190 pm
**1-14** The range of wavelengths are between 0.4 micron and 0.7 micron; 400 millimicron and 700 millimicron; 400 namometer and 700 nanometer; 4000 angstrom and 7000 angstrom
**1-15** (a) m/s (b) m$^2$ (c) m$^3$ (d) kg/m$^3$
**1-18** 2.205 lb **1-19** 3.934 L
**1-20** (a) 8.13 gal (b) 3970 g (c) 0.0839 mi (d) 6.5 L
**1-21** 300 mg **1-22** 20 mL
**1-23** 109.4 yd, therefore 100 m will take longer to run at constant speed.
**1-24** 117 L liquor in a barrel; 159.0 L gasoline or oil in a barrel
**1-25** 757 mL **1-26** 264 gal; 62.4 lb **1-27** 3.99 ft$^3$
**1-28** 1310 cm$^3$
**1-29** 61.02 in$^3$; 0.02832 m$^3$ **1-30** 720 mi/hr **1-31** 29 m/s; 105 km/h; 95 ft/s
**1-32** $2.9979 \times 10^{10}$ cm/s; $6.7085 \times 10^8$ mi/hr; $5.8807 \times 10^{12}$ mi/yr
**1-33** $1.7 \times 10^8$ mi/h
**1-34** human heart: $2.8 \times 10^9$ beats in one lifetime. The Jarvik-7 artificial heart $3.0 \times 10^9$ beats in one lifetime; the Jarvik-7 artificial heart beats 76 beats/min.
**1-35** 0.0471 m$^3$/s **1-36** 1.06 ton/ft$^2$
**1-37** Lightning travels 38892 m/s. Lightning travels approximately 88 times faster than the average velocity of an oxygen molecule at room temperature and approximately 7713 times slower than the speed of light.
**1-38** 744 g/qt **1-39** 0.91 g/cm$^3$ **1-40** 1000 kg/m$^3$
**1-44** (a) three (b) four (c) one (d) five
**1-45** (a) two (b) six (c) four (d) three
**1-46** (a) 475 (b) 0.0680 (c) $9.46 \times 10^{10}$ (d) 30.1
**1-47** 0.4 g/ft$^2$ **1-48** 0.7 kg $H_2O$/kg body wt
**1-49** 4.18783 km
**1-50** $1.8 \times 10^{12}$ tons $CO_2$ in the atmosphere
**1-51** 400 tablets **1-52** 12,000 cans diet soda **1-53** 1 g Ipecac/dose
**1-54** The profit on one 750-mL bottle of Jack Daniels is \$42 − \$8 = \$34.
**1-55** $2 \times 10^{-7}$ cm **1-56** 11 yd$^3$ of topsoil needed
**1-57** 226 lb maximum weight before being considered obese; 6.37 ft, approximately 6 foot 4 inches tall
**1-58** Natural gas is cheaper. **1-59** 3
**1-60** $9.0 \times 10^{13}$ J/g = energy released per gram of solar fuel; $4.8 \times 10^{-10}$ = fraction of sun's energy absorbed by the earth
**1-61** $5 \times 10^{-6}$ cm thick; 178 gold atoms thick
**1-62** (a) $1.198 \times 10^1$ (b) $4.6940 \times 10^{-3}$ (c) $4.679 \times 10^6$
**1-63** (a) $2.126 \times 10^2$ (b) $1.89 \times 10^{-1}$ (c) $1.6221 \times 10^4$
**1-64** (a) 0.00560 (b) 702500 (c) 0.08216
**1-65** (a) 28,093 (b) 33 (c) $3.19 \times 10^4$ (d) $4.31 \times 10^{-2}$
**1-66** (a) 130 (b) 43 (c) $3.90 \times 10^{12}$ (d) $3.13 \times 10^6$ (e) $1.26 \times 10^{10}$
**1-71** Decrease
**1-72** Gasoline = 5.6 lb/gal; water = 8.34 lb/gal. Water is more dense than gasoline.
**1-73** $6.4 \times 10^5$ g peat moss **1-74** $3.907 \times 10^3$ g milk **1-75** $2.5 \times 10^3$ cm$^3$/flask
**1-76** 810 g **1-77** $1.7 \times 10^8$ cm **1-78** $6.0 \times 10^{27}$ g
**1-79** $6 \times 10^{13}$ g/cm$^3$ **1-80** 0.24 g/cm$^3$
**1-81** (a) float (b) sink (c) sink (d) float (e) float
**1-82** 150 g
**1-83** Density cube one and two = 7.19 g/cm$^3$. The cubes were made of the same material.
**1-84** 19.3 g/cm$^3$; The metal is gold. **1-87** 310 K
**1-88** $T_{°C} = T_{°F} = -40$; $T_{°F} = +11.43$, $T_{°C} = -11.43$ **1-89** −6.2°F

**Chapter 2**

**2-3**

| | | |
|---|---|---|
| (a) | diamond | element, nonmetal |
| (b) | brass | mixture, metals |
| (c) | soil | mixture, contains metals and nonmetals |
| (d) | glass | mixture, contains metals and nonmetals |
| (e) | cotton | mixture, nonmetal |
| (f) | milk of magnesium | mixture, contains metals and nonmetals |
| (g) | salt | compound, composed of metal and nonmetal |

(h) iron element, metal
(i) steel mixture, metal

**2-5** (a) Sb (b) Au (c) Fe (d) Hg (e) K (f) Ag (g) Sn (h) W

**2-6** (a) sodium (b) magnesium (c) aluminum (d) silicon (e) phosphorus (f) chlorine (g) argon

**2-7** (a) titanium (b) vanadium (c) chromium (d) manganese (e) iron (f) cobalt (g) nickel (h) copper (i) zinc

**2-8** (a) molybdenum (b) tungsten (c) rhodium (d) iridium (e) palladium (f) platinum (g) silver (h) gold (i) mercury

**2-21** (a) Na, metal (b) Mg, metal (c) Al, metal (d) Si, semimetal (e) P, nonmetal (f) S, nonmetal (g) Cl, nonmetal (h) Ar, nonmetal

**2-22** (a) $S_8$, nonmetal (b) Sb, nonmetal (c) Sc, metal (d) Se, nonmetal (e) Si, semimetal (f) Sm, metal (g) Sn, metal (h) Sr, metal

**2-24** Elements: Group VA: nitrogen, nonmetal; phosphorus, nonmetal; arsenic, semimetal; antimony, semimetal; bismuth, metal. In general, metal-like properties become more pronounced going down the group.

**2-25** The elements of the third period: sodium, metal; magnesium, metal; aluminum, semimetal; silicon, semimetal; phosphorus, nonmetal; sulfur, nonmetal; chlorine, nonmetal; argon, nonmetal. In general, metal-like properties decrease from left to right across the period.

**2-30** (a), (b), (c) **2-31** (a), (b), (e) **2-32** (a), (c), (e)
**2-33** (a), (b) **2-34** (a), (d), (e) **2-38** (a), (b), (d), (e)

**2-40** (a) Mg is in Group IIA, $Mg^{2+}$.
(b) Al is in Group IIIA, $Al^{3+}$.
(c) Si is in Group IVA, $Si^{4+}$.
(d) Cs is in Group IA, $Cs^{+}$.
(e) Ba is in Group IIA, $Ba^{2+}$.

**2-41** (a) C is in Group IVA $4 - 8 = -4$ $C^{4-}$
(b) P is in Group VA $5 - 8 = -3$ $P^{3-}$
(c) S is in Group VIA $6 - 8 = -2$ $S^{2-}$
(d) I is in Group VIIA $7 - 8 = -1$ $I^{-}$

**2-42** −1 **2-43** +2

**2-44** (a) $Mg(NO_3)_2$ (b) $Fe_2(SO_4)_3$ (c) $Na_2CO_3$

**2-45** (a) $Na_2O_2$ (b) $Zn_3(PO_4)_2$ (c) $K_2PtCl_6$

**2-46** $K_3N$; AIN **2-47** $K_2O_2$ **2-48** $x = 6$ **2-50** $Re^{3+}$; $Cr^{3+}$; $Mo^{2+}$

**2-51** (a) 3− (b) 5 + (c) 3 + (d) 2 +

**2-52** (a) 3+ (b) 3+ (c) 3+

**2-53** $Al^{3+}$ **2-54** (e), (g)

**2-55** (a) 0 (b) −1 (c) +1 (d) +3 (e) +5 (f) +7

**2-56** (a) −1 (b) −1 (c) 0 (d) +1 (e) +5 (f) +5 (g) +7 (h) +7

**2-57** (a) +2 (b) +4 (c) +4 (d) +6 (e) +6 (f) +6 (g) +8 (h) +8

**2-59** (a) +4 (b) +4 (c) +2 (d) +4 (e) +4 (f) −4 (g) −4 (h) 0 (i) +4 (j) +2

**2-60** (a) 0 (b) −2 (c) −2 (d) +4 (e) +6 (f) +4 (g) +6 (h) +4 (i) +6 (j) +4 (k) +6

**2-61** (a) +2 (b) +3 (c) +4 (d) +6 (e) +7 (f) +2 (g) +3 (h) +6 (i) +7 (j) +4 (k) +2

**2-62** (a) +2 (b) +4 (c) +3 (d) 4.66 (e) +3 (f) +4 (g) +4 (h) +4 (i) +4

**2-63** Prussian blue, +3; Turnbull's blue, +2

**2-66** (a) $P_4S_3$ (b) $SiO_2$ (c) $CS_2$ (d) $CCl_4$ (e) $PF_5$

**2-67** (a) $SiF_4$ (b) $SF_6$ (c) $OF_2$ (d) $Cl_2O_7$ (e) $ClF_3$

**2-68** (a) $SnCl_2$ (b) $Hg(NO_3)_2$ (c) $SnS_2$ (d) $Cr_2O_3$ (e) $Fe_3P_2$

**2-69** (a) $BeF_2$ (b) $Mg_3N_2$ (c) $Ca_2C$ (d) $BaO_2$ (e) $K_2CO_3$

**2-70** (a) $Co(NO_3)_3$ (b) $Fe_2(SO_4)_3$ (c) $AuCl_3$ (d) $MnO_2$ (e) $WCl_6$

**2-71** (a) potassium nitrate (b) lithium carbonate (c) barium sulfate (d) lead(II) iodide

**2-72** (a) aluminum chloride (b) sodium nitride (c) calcium phosphide (d) lithium sulfide (e) magnesium oxide

**2-73** (a) ammonium hydroxide (b) hydrogen peroxide (c) magnesium hydroxide (d) calcium hypochlorite (e) sodium cyanide

**2-74** (a) antimony(III) sulfide (b) tin(II) chloride (c) sulfur tetrafluoride (d) strontium bromide (e) silicon tetrachloride

**2-75** (a) $CH_3CO_2H$ (b) HCl (c) $H_2SO_4$ (d) $H_3PO_4$ (e) $HNO_3$

**2-76** (a) $H_2CO_3$ (b) HCN (c) $H_3BO_3$ (d) $H_3PO_3$ (e) $HNO_2$

**2-77** sodium bisulfide NaHS
sodium bisulfite $NaHSO_3$

**2-78** $SSO_3^{2-}$ or $S_2O_3^{2-}$

**2-79** (a) fluorite, calcium fluoride (b) galena, lead(II) sulfide
(c) pyrite, iron(II) polysulfide (d) rutile, titanium(IV) oxide
(e) hematite, iron(III) oxide

**2-80** (a) magnetite, iron(II,III) oxide (b) calcite, calcium carbonate
(c) barite, barium sulfate (d) quartz, silicon dioxide

### Chapter 3

**3-1** Chromium **3-2** 1.6816 amu **3-3** Aluminum

**3-4** 79.904 amu

**3-5** 65.3982 amu **3-6** 51.9961 g **3-7** 2:1

**3-8** (a) 12.011 g C (b) 58.70 g Ni (c) 200.59 g Hg

**3-9** \$2.25; 30.027 g **3-10** amu/g = $1.328 \times 10^{21}$

**3-11** $6.022 \times 10^{23}$ atoms O
$6.022 \times 10^{23}$ atoms P
$6.022 \times 10^{23}$ atoms S

**3-12** $5.620 \times 10^{21}$ atoms of Cl

**3-13** (a) $2.41 \times 10^{23}$ atoms O
(b) $7.5 \times 10^{23}$ atoms O
(c) $1.4 \times 10^{24}$ atoms O

**3-14** 106.1 g/mol

**3-15** (d) 4 mole $NH_3$ and 3 mole $N_2H_4$

**3-16** (c) 0.050 moles glucose

**3-17** Formic acid = 46.025 g/mol
Formaldehyde = 30.026 g/mol

**3-18** (a) 16.043 g/mol (b) 180.16 g/mol (c) 74.122 g/mol (d) 75.13 g/mol

**3-19** (a) 444.48 g/mol (b) 46.005 g/mol (c) 97.44 g/mol (d) 158.032 g/mol

**3-20** (a) 220.056 g/mol (b) 241.858 g/mol (c) 294.182 g/mol (d) 310.18 g/mol

**3-21** 162.187 g/mol **3-22** 169.125 g/mol

**3-23** (a) 375.936 g/mol (b) 284.744 g/mol (c) 444.437 g/mol
**3-24** $7.61 \times 10^{-3}$ mol C **2-25** 0.2247 mol P
**3-26** 195.1 g/mol
**3-27** 8.95 g **3-28** 1.587 g/cm$^3$ **3-29** 10 cm$^3$
**3-30** (a) 76.47% (b) 68.42% (c) 52.00%
**3-31** (a) 21.20% N (b) 13.85% N (c) 16.48% N (d) 46.64% N
**3-32** 47.44% C; 2.56% H; 50.00% Cl **3-33** 31.35% Si
**3-34** $CaCO_3$ = 40.04% Ca; $CaSO_4$ = 29.44% Ca; $Ca_3(PO_4)_2$ = 38.76% Ca; Tablets that contain $CaCO_3$ are most efficient in getting $Ca^{+2}$ ions into the body.
**3-35** $SnF_2$
**3-36** The formula of magnetite is $Fe_3O_4$.
**3-37** Pyrolusite has the formula $MnO_2$ (b).
**3-38** Nitrous oxide is $N_2O$.
**3-39** The empirical formula of chalcopyrite is $CuFeS_2$.
**3-40** The empirical formula is $XeF_6$. The oxidation number of Xe is +6.
**3-41** The empirical formula of MDMA is $C_{11}H_{15}NO_2$.
**3-42** The empirical formula is $NO_2$.
**3-43** The empirical formula is CuCl.
**3-44** $C_{40}H_{56}$ **3-45** $C_{10}H_7O_2$ **3-46** $C_8H_{10}N_4O_2$
**3-47** $C_{14}H_{18}N_2O_5$
**3-48** (a) $4\,Cr(s) + 3\,O_2(g) \longrightarrow 2\,Cr_2O_3(s)$
(b) $SiH_4(g) \longrightarrow Si(s) + 2\,H_2(g)$
(c) $2\,SO_3(g) \longrightarrow 2\,SO_2(g) + O_2(g)$
**3-49** (a) $2\,Pb(NO_3)_2(s) \longrightarrow 2\,PbO(s) + 4\,NO_2(g) + O_2(g)$
(b) $NH_4NO_2(s) \longrightarrow N_2(g) + 2\,H_2O(g)$
(c) $(NH_4)_2Cr_2O_7(s) \longrightarrow N_2(g) + Cr_2O_3(s) + 4\,H_2O(g)$
**3-50** (a) $CH_4(g) + 2\,O_2(g) \longrightarrow CO_2(g) + 2H_2O(g)$
(b) $2\,H_2S(g) + 3\,O_2(g) \longrightarrow 2\,H_2O(g) + 2\,SO_2(g)$
(c) $2\,B_5H_9(g) + 12\,O_2(g) \longrightarrow 5\,B_2O_3 + 9\,H_2O(g)$
**3-51** (a) $PF_3(g) + 3\,H_2O(l) \longrightarrow H_3PO_3 + 3\,HF(aq)$
(b) $P_4O_{10}(s) + 6\,H_2O(l) \longrightarrow 4\,H_3PO_4(aq)$
**3-52** (a) $2\,C_3H_8(g) + 10\,O_2(g) \longrightarrow 6\,CO_2(g) + 8\,H_2O(g)$
(b) $C_2H_5OH(l) + 3\,O_2(g) \longrightarrow 2\,CO_2(g) + 3\,H_2O(g)$
(c) $C_6H_{12}O_6(s) + 6\,O_2(g) \longrightarrow 6\,CO_2(g) + 6\,H_2O(aq)$
**3-53** 15.0 mol $O_2$ **3-54** 2.25 mol $O_2$ **3-55** 9 mol CO
**3-56** 2.0 g $O_2$ **3-57** 39.9 g $O_2$; 27.4 g $CO_2$
**3-58** 4.90 lb S **3-59** 9.79 g $O_2$
**3-60** $CrO_3$ **3-61** 0.5 kg $C_2H_5OH$ **3-62** 900 lb Al
**3-63** 3.73 g $PH_3$ **3-64** 12.0 g HCl **3-65** 50 g $N_2$
**3-66** 1000 molecules $H_2O$. If the amount of $O_2$ doubles, the yield remains 500 molecules $H_2O$. If the amount of $H_2$ doubles, then the yield is 1000 molecules $H_2O$.
**3-67** 0.400 mol $P_4S_{10}$ **3-68** 0.18 mol $NO_2$
**3-69** 0.282 mol HCl
**3-70** 67.7 g $Ca_3N_2$ **3-71** 1.137 mol $PF_5$
**3-72** 5.19 g $Al(CH_3)_3$
**3-73** 174 g Fe **3-76** 5.99 M **3-77** $1.3 \times 10^{-5}$ M
**3-78** 14.8 M
**3-79** 8.73 M **3-80** $4.1 \times 10^{-3}$ M **3-81** 12.48 M
**3-82** 0.0814 M
**3-83** 10.7 g $Na_2SO_4$ **3.84** 0.120 M **3-85** 36 mL NaI
**3-86** 172 mL of 17.4 M acetic acid is needed.
**3-87** 2.25 M HCl **3-89** $7.469 \times 10^{-2}$ M acetic acid
**3-90** 0.9631 M NaOH
**3-91** 0.5767 mL $H_2SO_4$ **3-92** 0.376 M $C_6H_{12}O_6$
**3-93** 0.150 M $H_2C_2O_4$
**3-95** $n = 4$ **3-96** $C_{10}H_{14}N_2$ **3-97** 24.286 g/mol; the element is Mg.
**3-98** $C_2HBrClF_3$
**3-99** $2\,KClO_3(s) \longrightarrow 2\,KCl(s) + 3\,O_2(g)$
**3-100** The copper sulfate picks up 5 moles of $H_2O$ or $3.011 \times 10^{24}$ molecules of $H_2O$.
**3-101** $CrO_3$ **3-102** $MnO_2$ **3-104** 66,532 g/mol
**3-105**

| | Carbonate | MW(g/mol) | % $CO_2$ |
|---|---|---|---|
| (a) | $Li_2CO_3$ | 73.89 | 59.56 |
| (b) | $MgCO_3$ | 84.32 | 52.19 |
| (c) | $CaCO_3$ | 100.09 | 43.97 |
| (d) | $ZnCO_3$ | 125.39 | 35.10 |
| (e) | $BaCO_3$ | 197.35 | 22.30 |

**3-106** 39.452 g **3-107** 32.1 g Mg **3-108** 8.00%

## Chapter 4

**4-1** Gasses: Ar, CO, $CH_4$, $Cl_2$
**4-4** (a) pressure
(b) average kinetic energy
(c) number of molecules per container
**4-7** 14.7 lb/in$^2$; $1.01 \times 10^5$ newtons/m$^2$ **4-8** 0.9813 atm; $9.941 \times 10^4$ Pa
**4-10** 1 atm = $1.013 \times 10^5$ Pa **4-11** 29 psi
**4-12** $3 \times 10^{-6}$ atm
**4-13** $1.2 \times 10^{19}$ lb **4-14** 324 mm Hg
**4-15** 0.349 L; The volume of the balloon will decrease.
**4-16** 13 L **4-17** 500 L **4-18** 5.29 atm
**4-19** 550 K = 277°C
**4-21** 18% **4-22** 0.332 L **4-24** 2:1 **4-25** 0.69 L $N_2$; 2.07 L $H_2$
**4-26** 15 L NO **4-27** 18 L total volume
**4-30** (a) 56 L (b) 154 L (c) 82 L (d) 39 L
**4-31** $N_2O$ **4-34** (e) **4-35** (b) **4-36** $1.21 \times 10^3$ mi · psi/mol · K
**4-37** $0.00014 per g $N_2$ **4-38** 1.5 atm **4-39** 253 K
**4-40** 8.0 kg $O_2$ **4-41** 3.13 atm
**4-42** 7.1 atm **4-43** 1.3 L **4-44** 342 atm **4-45** 1.99 atm
**4-46** 24.6 L **4-47** 6.85 L **4-48** 233 K **4-49** 12 atm
**4-50** 0.0389 mol/L; 3.304 g/L **4-51** 0.825 kg/m$^3$
**4-52** 0.179 g/L = density helium at STP **4-53** 0.0631
**4-54** Krypton **4-55** 28.95 g/mol
**4-56** $P_{C_3H_8}$ = 424 mm Hg
**4-57** $P_{He}$ = 2.50 atm **4-58** $P_{Total}$ = 19.0 atm **4-59** $P_{Total}$ = 0.452 atm
**4-67** $SF_6 < CF_2Cl_2 < Cl_2 < SO_2 < Ar$ **4-68** 2.236
**4-72** $x$ = 13.1 s **4-73** 15.98 g/mol
**4-74** The gasses will meet on row 22. **4-75** 234.8 g/mol
**4-82** For He, $r = 2.1 \times 10^{-8}$ cm; for Ne, $r = 1.9 \times 10^{-8}$ cm; for Ar, $r = 2.3 \times 10^{-8}$ cm
**4-83** For V = 1 L, P = 21.92 atm; for V = 0.10 L, P = 225.5 atm; for V = 0.05 L, P = 1327 atm
**4-84** (f) $SO_3$ **4-85** 58.1 g/mol **4-86** The compound is $B_5H_9$.
**4-87** The formula for cyclopropane is $C_3H_6$. **4-88** $CH_2N_2$
**4-89** For $PH_x$, the molecular formula is $PH_3$.
For $P_2H_x$, the molecular formula is $P_2H_4$.
**4-90** 0.542 g Mg **4-91** $CrO_3$ **4-92** $3.66 \times 10^4$ L $CO_2$
**4-93** $2.20 \times 10^3$ L $O_2$
**4-94** 2.45 L $CO_2$ **4-95** Zinc

**Chapter 5**

**5-6** 1054 Joules **5-9** 24.5 J/mol°C **5-18** (b), (d)
**5-28** The state functions of a trip are: (g) location of the car, (h) elevation, (i) latitude, (j) longitude.
**5-29** The state functions are: (a) temperature, (b) internal energy, (c) enthalpy, (d) pressure, (e) volume.
**5-31** −5.72 kJ **5-32** −13.1 kJ **5-33** $-5.25 \times 10^2$ kJ/mol Fe
**5-35** $\frac{2.657 \times 10^3 \text{ kJ}}{\text{mol } C_4H_{10}}$ **5-36** $4.46 \times 10^3$ kJ/mol $B_5H_9$
**5-38** Reactions (b) and (c) would show roughly equivalent changes in enthalpy and internal energy.
**5-40** (a) $H_2(g) \longrightarrow 2H(g)$
**5-41** (c) The separation of NaCl (table salt) into its component elements requires input of significant energy.
**5-45** No heat is given off in this reaction.
**5-46** $\Delta H° = -1123$ kJ/mol Mg **5-47** $\Delta H° = -186$ kJ/mol Ca
**5-49** $\Delta H = 2.43$ kJ/mol $CO_2$ **5-50** $\Delta H°_{rxn} = -196$ kJ
**5-51** $\Delta H°_{rxn} = 180.7$ kJ **5-52** $\Delta H°_f = 11.3$ kJ/mol
**5-53** 2044 kJ/mol $C_3H_8$ **5-54** $\Delta H° = -53.9$ kJ/mol $(C_2H_5)_2O(l)$
**5-55** $F_2(g)$ and $P_4(s)$ **5-56** The reaction is endothermic.
**5-57** The reaction is endothermic. **5-58** −1284 kJ
**5-59** −1075 kJ
**5-60** −5314 kJ **5-61** −905.2 kJ **5-62** −409.2 kJ
**5-63** −91.2 kJ
**5-64** The reaction with iron(III) oxide evolves more heat per mol aluminum consumed.
**5-65** 120 kJ **5-67** 943 kJ **5-68** $\Delta H° = -4127$ kJ
**5-69** $\Delta H° = -93.5$ kJ/mol HCl **5-70** −1104.53 kJ
**5-71** −638.49 kJ/mol $CH_3OH$ **5-72** 54.23 kJ
**5-73** +1312.21 kJ/mol **5-74** 463 kJ **5-75** −101,338 kJ
**5-76** −235.06 kJ **5-77** −385.0 kJ/mol

**Chapter 6**

**6-10** An electron has the largest charge to mass ratio.
**6-15** $^{52}Cr^{3+}$
**6-16** 19 protons, 20 neutrons, 18 electrons, atomic number = Z = 19, mass number = 39
**6-17** 53 protons, 74 neutrons, 54 electrons, Z = 53, mass number = 127
**6-18** $^{20}F$ **6-19** $^{79}Se^{2-}$ **6-20** 54 electrons
**6-22** 96486.1 C/mol
**6-24** 10.2 mi/h **6-26** 17.02 m/cycle **6-27** 600 nm
**6-28** $4.28 \times 10^{14}$ cycles/s **6-31** 11.92 m/cycle; radio wave region
**6-32** $4.949 \times 10^{14}$ cycle/s; visible region **6-33** $5 \times 10^{16}$ cycle/s
**6-35** $5.090 \times 10^{14}$ cycle/s; $5.085 \times 10^{14}$ cycle/s
**6-38** 656.2 nm < 486.1 nm < 434.0 nm < 410.1 nm
**6-39** $3.027 \times 10^{-19}$ J **6-40** $1.986 \times 10^{-26}$ J
**6-41** $3.673 \times 10^{38}$ cycle/s; gamma ray **6-42** 656.467 nm
**6-43** The transition from $n = 6$ to $n = 2$.
**6-44** (a) 656.467 nm
(b) 102.573 nm
(c) 4052.27 nm
(d) not an emission
(e) not an emission
(f) not an emission
The transition from $n = 5$ to $n = 4$ results in the emission of light with the longest wavelength.
**6-45** (a) 656.467 nm
(b) 486.282 nm
(c) 97.2544 nm
(d) not an absorption
(e) not an absorption
(f) not an absorption
The shortest wavelength is the highest energy. The transition from $n = 1$ to $n = 4$ would result in the absorption of a proton with the largest energy.
**6-47** $E = 3.026 \times 10^{-19}$ J **6-50** (a) n (b) l (c) m (d) s
**6-53** +1/2 and −1/2 **6-54** −2, −1, 0, 1, 2 **6-55** 3
**6-56** (d) 4, 0, 0, −1/2 **6-57** (b) 8, 4, 3, −1/2 **6-58** 14 electrons **6-59** 16 electrons
**6-64** for $n = 3$, 9 orbitals; for $n = 4$, 16 orbitals; for $n = 5$, 25 orbitals
**6-65** (a) 2p orbital (b) 2s orbital (c) 2d orbital (d) 3d orbital (e) 3p orbital
**6-66** (c) A 2d orbital cannot exist.
**6-67** $n = 1$, 2 electrons; $n = 2$, 8 electrons; $n = 3$, 18 electrons; $n = 4$, 32 electrons; $n = 5$, 50 electrons
**6-68** 10 electrons **6-69** 5 unpaired electrons **6-74** (a) n and ℓ
**6-75** c is correct. **6-76** e is incorrect. **6-77** 5d orbitals
**6-78** Orbitals are degenerate when (d) they have the same energy.
**6-80** Na [Ne] $3s^1$
Mg [Ne] $3s^2$
Al [Ne] $3s^2\ 3p^1$
Si [Ne] $3s^2\ 3p^2$
P [Ne] $3s^2\ 3p^3$
S [Ne] $3s^2\ 3p^4$
Cl [Ne] $3s^2\ 3p^5$
Ar [Ne] $3s^2\ 3p^6$ = [Ar]
**6-81** N ⇅(1s) ⇅(2s) ↑ ↑ ↑(2p)
**6-82** Ni ⇅(1s) ⇅(2s) ⇅ ⇅ ⇅(2p) ⇅(3s) ⇅ ⇅ ⇅(3p) ⇅(4s) ⇅ ⇅ ⇅ ↑ ↑(3d)
**6-83** $x = 2$ **6-84** (b) **6-85** (b) **6-86** (c)
**6-87** 11 electrons **6-88** 6 electrons **6-89** Zn
**6-90** $n = 4$, $\ell = 1$, $m = 1$, and $s = 1/2$
**6-91** $n = 3$, $\ell = 2$, $m = 0$, and $s = 1/2$
**6-92** Element 114 should be placed in group 14.
**6-93** $n = 5$ $\ell = 4$ **6-94** Y **6-95** $\ell = 3$
**6-96** P has the largest number of unpaired electrons.
**6-97** $Fe^{3+}$ **6-98** $4.19 \times 10^{-27}$ $m^3$ **6-99** ddu = neutron, duu = proton

**Chapter 7**

**7-9** (a) $K_2O$ (b) ZnO (c) $In_2O_3$ (d) $SiO_2$ (e) $V_2O_5$
**7-12** Ytterbium **7-13** Niobium **7-14** Antimony
**7-15** Group 1
**7-16** $Ga^{3+}$ **7-17** $Cr^{+2}$ **7-20** 0.1442 nm; 144.2 pm
**7-21** Aluminum will have the smallest radius (K > Ca > Na > Mg > Al).
**7-23** Phosphorus has the largest covalent radius (P > S > N > O > F).
**7-26** $H^- < F^- < Cl^- < Br^- < I^-$
**7-29** The ion with the fewer electrons will be smaller. $Fe^{3+} < Fe^{2+}$.
**7-30** $N^{3-}$, $O^{2-}$, $F^-$, Ne, $Na^+$, $Mg^{2+}$, $Al^{3+}$, $Si^{4+}$: isoelectronic
$P^{3-}$, $S^{2-}$, $Cl^-$, Ar, $K^+$, $Ca^{2+}$: isoelectronic

**7-31** (a) $B^{3+}$, $C^{4+}$, He are isoelectronic
(b) $N^{3+}$, $Na^+$, Ne, $O^{2-}$ are isoelectronic
(c) $Mg^{2+}$, $F^-$, $Na^+$, $O^{2-}$ are isoelectronic
**7-32** $Al^{3+}$ is smaller than $Mg^{2+}$. **7-33** $Se^{2-}$ **7-34** $Mg^{2+}$
**7-35** $Be^{2+}$ **7-36** $P^{3-}$
**7-37** $Al(g) \longrightarrow Al^{1+}(g) + e^-$
$Al^{1+}(g) \longrightarrow Al^{2+}(g) + e^-$
$Al^{2+}(g) \longrightarrow Al^{3+}(g) + e^-$
$Al^{3+}(g) \longrightarrow Al^{4+}(g) + e^-$
**7-44** F > Be > Si > Li > Na **7-45** Nitrogen
**7-46** Calcium **7-47** $P^{4+}$
**7-49** [Ne] $3s^2$ $3p^1$, aluminum
**7-50** The third ionization energy of Mg
**7-51** Na has the largest second ionization energy.
**7-52** Mg should have the largest third ionization energy.
**7-53** Mg < Be < Ne < Na < Li
**7-54** (a) $C^{4+}$ (b) $N^{5+}$ (c) $O^{6+}$ (d) $Si^{4+}$ (e) $S^{6+}$ (f) $Ca^{2+}$
**7-55** (a) $Sc^{3+}$ (b) $Ti^{4+}$ (c) $V^{5+}$ (d) $Cr^{6+}$ (e) $Mn^{7+}$
**7-56** The most common oxidation states of lead and tin are $Pb^{2+}$, $Pb^{4+}$ and $Sn^{2+}$, $Sn^{4+}$.
**7-64** Oxygen **7-65** Carbon **7-66** Chlorine
**7-70** Decreasing metallic character:
Na > Mg > Al > Si > P > S > Cl > Ar.
Increasing metallic character:
Na < Mg < Al < Si < P < S < Cl < Ar.
**7-71** N < P < As < Sb < Bi
**7-72** In order of increasing non-metallic character:
(a) Sr < Al, Sr < Ga, but Ga < Al
(b) K < Mg but Rb < K
(c) Ge < P but As < P
(d) Al < B < N < F, the correct sequence.
**7-73** Responses (a) and (d) are true of metals.
**7-75** (d) $Na^+(g) + Cl^-(g) \longrightarrow NaCl(s)$
**7-76** LiF **7-77** MgO
**7-85** For $^1H$, E = 72.7685. For $^2H$, E = 72.8071 kJ/mol.
**7-86** The electron affinity of $NH_2$ = 74.4 kJ/mol.

## Chapter 8

**8-3** (a) Li has 1 (b) C has 4
(c) Mg has 2 (d) Ar has 8
**8-4** (a) Fe has 8 (b) Cu has 11
(c) Bi has 5 (d) I has 7
**8-5** All have 8 valence electrons and gain electrons to achieve an octet of electrons.
**8-6** All have 8 valence electrons and lose electrons to achieve an octet of electrons.
**8-7** *d* and *f* subshells seldom take part in the chemical reactivity of the element.
**8-8** A mercury atom in the +1 oxidation state has one valence electron. Hydrogen also has one valence electron.
**8-9** Group 16 **8-10** +5
**8-11** (a) 24 (b) 8
(c) 8 (d) 32
**8-12** (a) 22 (b) 34
(c) 48 (d) 56
**8-13** 16 valence electrons: (a), (b), (c), (d)
18 valence electrons: (e), (f), (*g*)
**8-18** Using the guide when metals combine with nonmetals
(a) covalent (b) covalent
(b) ionic (c) ionic
Using electronegativity differences as a guide
(a) covalent (b) covalent
(b) ionic (c) covalent
**8-19** Using the guide when metals combine with nonmetals
(a) covalent (b) covalent
(b) covalent (c) ionic
Using electronegativity differences as a guide
(a) polar covalent (b) polar covalent
(b) ionic (c) polar covalent
**8-20** Electronegativities increase going from the left to right across a row. They tend to decrease going down a column.
**8-21** Fluorine has the maximum electronegativity, cesium or francium the smallest. The most ionic compound would be FrF.
**8-22** Cl **8-23** (b) **8-24** (d) O **8-25** (e) Se
**8-26** (b) $H_2O$ (d) $MgBr_2$
**8-27** (a) $CaH_2$ (d) $FeCl_4^-$ (f) $SiCl_4$
**8-30** (a) $CH_3OH$, (b) $H_2O$ (d) $CH_3CO_2H$ are polar.
**8-31** It has a dipole moment along the C—O bond.
**8-33** $Mn_2O_7$ **8-34** $TiO_2 < Ti_4O_7 < Ti_2O_3 < TiO$
**8-47** CO, $CN^-$ and $NO^+$ **8-48** (b) $BeF_2$ (c) $SF_4$
**8-49** (a) $BF_3$ (c) $XeF_4$ (d) $IF_3$
**8-50** (a) $N_2$ **8-51** (d) $O_2^{2-}$
**8-53** (a) three pairs (b) three pairs
(c) two pairs (d) two pairs
**8-54** All of the ions and molecules listed can act as Lewis bases.
**8-55** (a) $NH_4^+$ and (c) $AlH_3$ cannot act as Lewis bases.
**8-59** (a), (d) and (e) are resonance hybrids.
**8-60** (a) $CO_2$ **8-61** 1.3
**8-62** 1.5 for $NO_2$; 1.3 for $NO_3^-$
**8-63** (a) + 1 (b) + 2
(c) + 1 (d) + 2
**8-64** (a) 0 (b) 0
(c) −1 (d) 0
(e) 0
**8-65** (a) 0 (b) + 1 (c) 0 (d) −1 (e) −3
**8-66** (a) end (N) = 0; middle (N) = +1 (b) 0 (c) + 1
(d) (N) = 5 valence electrons −4 = +1; (N) = 5 valence $e^-$ −5 = 0 (e) + 1
**8-67** $BF_3$; Formal charge (B) = 0
$NH_3$; Formal charge (N) = 0
$F_3B\text{-}NH_3$; Formal charge (N) = +1; Formal charge (B) = −1
**8-68** end (*S*) = −1; middle (*S*) = +2
**8-69** (a) trigonal pyramidal (b) trigonal planar
(c) T-shaped (d) T-shaped
**8-70** (a) tetrahedral (b) tetrahedral
(c) tetrahedral (d) tetrahedral
**8-71** (a) bent (b) trigonal pyramidal
(c) tetrahedral (d) octahedral
**8-72** (a) trigonal pyramidal (b) trigonal pyramidal
(c) square pyramidal (d) octahedral
**8-73** (a) linear (b) T-shaped
(c) seesaw (d) square planar (e) distorted octahedral
**8-74** (a) linear (b) linear
(c) bent (d) bent (e) trigonal planar
**8-75** (a) linear (b) tetrahedral **8-76** (a) $XeF_3^+$
**8-77** $NH_2^-$ and $H_2O$ **8-78** (c) $CO_2$ and $BeH_2$
**8-79** (a), (c), (d), and (e) **8-80** (a), (c), and (e)

**8-81** (a), (b), (c), and (d) **8-82** (a), (b), and (d)
**8-83** (c) Carbon **8-85** (a) $sp^3$ (b) $sp^3$ (c) $sp^2$ (d) $sp^2$ (e) sp
**8-86** (a) $sp^3$ (b) $sp^2$ (c) $sp^2$
**8-87** (a) $sp^3d$ (b) $sp^3$ (c) $sp^3d$ (d) $sp^2$
**8-98** is not paramagnetic
**8-100** (a) diamagnetic (b) diamagnetic (c) diamagnetic (d)paramagnetic (e) diamagnetic
**8-103** (c) P **8-104** (d) N **8-105** (c) Si
**8-106** Group 17 **8-107** Group 17 **8-108** Group 1

### Chapter 9

**9-3** (a) Ca (b) Na (c) Na (d) K (e) Mg
**9-4** (a) K (b) Na (c) Ca (d) Ca (e) Al
**9-6** RbBr RbH $Rb_2S$ $Rb_3N$ $Rb_3P$ $Rb_2O$ $Rb_2O_2$
**9-10** $BaF_2$, $BaH_2$, BaS, $Ba_3N_2$, $Ba_3P_2$, BaO, $BaO_2$
**9-11** $Mg(s) + H_2O(g) \longrightarrow MgO(s) + H_2(g)$
$MgO(s) + CO_2(g) \longrightarrow MgCO_3(s)$
**9-21** Magnesium **9-22** Group 14 **9-23** Group 13
**9-24** AIN **9-25** $Sr_3P_2$ **9-26** GaAs
**9-27** (a) $2\,Na(s) + F_2(g) \longrightarrow 2\,NaF(s)$ (b) $4\,Na(s) + O_2(g) \longrightarrow 2\,Na_2O(s)$
(c) $2\,Na(s) + H_2(g) \longrightarrow 2\,NaH(s)$ (d) $16\,Na(s) + S_8(s) \longrightarrow 8\,Na_2S(s)$
(e) $12\,Na(s) + P_4(g) \longrightarrow 4\,Na_3P(s)$
**9-28** (a) $Ca(s) + H_2(g) \longrightarrow CaH_2(s)$ (b) $2\,Ca(s) + O_2(g) \longrightarrow 2\,CaO(s)$
(c) $8\,Ca(s) + S_8(s) \longrightarrow 8\,Cas(s)$ (d) $Ca(s) + F_2(g) \longrightarrow CaF_2(s)$
(e) $3\,Ca(s) + N_2(g) \longrightarrow Ca_3N_2(s)$ (f) $6\,Ca(s) + P_4(g) \longrightarrow 2\,Ca_3P_2(s)$
**9-29** (a) $ZnF_2$ (b) $AlF_3$ (c) $SnF_2$ or $SnF_4$ (d) $MgF_2$ (e) $BiF_3$ or $Bif_5$
**9-30** $Ca(s) + 2\,H_2O(l) \longrightarrow Ca^{2+}(aq) + 2\,OH^-(aq) + H_2(g)$
**9-33** (a) $Mg(s) + 2\,HCl(aq) \longrightarrow MgCl_2(aq) + H_2(g)$
(b) $Cu(s) + HCl(aq) \longrightarrow$ no reaction
(c) $2\,Cs(s) + 2\,H_2O(l) \longrightarrow 2\,CsOH(aq) + H_2(g)$
**9-34** Reaction (d)
**9-35** (a) $Mg(s) + I_2(s) \longrightarrow MgI_2(s)$
(b) $2\,Al(s) + 3\,H_2SO_4(aq) \longrightarrow Al_2(SO_4)_3(aq) + 3\,H_2(g)$
(c) $Fe_2O_3(s) + 3\,H_2(g) \longrightarrow 3\,H_2O(l) + 2\,Fe(s)$
(d) $2\,Mg(l) + TiCl_4(g) \longrightarrow 2\,MgCl_2(s) + Ti(s)$
**9-36** (a) $2\,K(s) + 2\,H_2O(l) \longrightarrow H_2(g) + 2\,KOH(aq)$
(b) $NaH(s) + H_2O(l) \longrightarrow NaOH(aq) + H_2(g)$
(c) $Ag^+(aq) + Cu(s) \longrightarrow$ no reaction
(d) $Zn(s) + CuSO_4(aq) \longrightarrow ZnSO_4(aq) + Cu(s)$
**9-38** The second reaction with aluminum involves oxidation.
**9-39** (b) Iron is reduced from +3 in $Fe_2O_3$ to 0 in Fe(*s*). Carbon is oxidized from +2 in CO to +4 in $CO_2$.
(c) Silicon is reduced from +4 in $SiO_2$ to 0 in SiC. Carbon is oxidized from 0 in C(*s*) to +2 in CO.
(d) Carbon is reduced from +4 in $CO_2$ to +2 in CO. Hydrogen is oxidized from 0 in $H_2$ to +1 in $H_2O$.
(e) Carbon is reduced from +2 in CO to −2 in $CH_3OH$. Hydrogen is oxidized from 0 in $H_2$ to +1 in $CH_3OH$.
**9-40** (a) Magnesium is oxidized from 0 in Mg(*s*) to +2 in $MgCl_2$. Hydrogen is reduced from +1 in HCl to 0 in $H_2$.
(b) Iodine is oxidized from 0 in $I_2$ to +3 in $ICl_3$. Chlorine is reduced from 0 in $Cl_2$ to −3 in $ICl_3$.
(d) Sodium is oxidized from 0 in Na(*s*) to +1 in NaOH. Hydrogen is reduced from +1 in $H_2O$ to 0 in $H_2$.
**9-41** (b) The phosphorus is oxidized from a −3 in $PH_3$ to a +5 in $H_3PO_4$. The oxygen is reduced from 0 in $O_2$ to −2 in $H_3PO_4$.
(d) The phosphorus is oxidized from 0 in $P_4$ to +5 in $P_4O_{10}$. The oxygen is reduced from 0 in $O_2$ to −2 in $P_4O_{10}$.
**9-43** $S_8(s)$ reducing agent, $KClO_3$ oxidizing agent
**9-44** $H_2O_2(aq)$ reducing agent, 2HI(*aq*) oxidizing agent
**9-45** Cu(*s*) reducing agent, $Ag^+(aq)$ oxidizing agent
**9-46** (a) Al is the reducing agent, $Cr_2O_3$ the oxidizing agent.
(b) Al is the reducing agent, $H^+$ the oxidizing agent.
(c) Al is the reducing agent, $I_2$ the oxidizing agent.
**9-47** (a) Mg is the reducing agent, $CO_2$ the oxidizing agent.
(b) Mg is the reducing agent, HCl the oxidizing agent.
(c) Mg is the reducing agent, $H_2O$ the oxidizing agent.
(d) Mg is the reducing agent, $N_2$ the oxidizing agent.
(e) Mg is the reducing agent, $NH_3$ the oxidizing agent.
**9-48** (a) $Na \longrightarrow Na^+$ (b) $Zn \longrightarrow Zn^{2+}$
(c) $H_2 \longrightarrow 2\,H^+$ (d) $Sn^{2+} \longrightarrow Sn^{4+}$
(e) $H^- \longrightarrow H$
**9-49** (a) $Al^{3+} \longrightarrow Al$ (b) $Hg^{2+} \longrightarrow Hg$
(c) $H^+ \longrightarrow H$ (d) $H_2 \longrightarrow 2\,H^-$
(e) $Sn^{2+} \longrightarrow Sn$
**9-52** (a) Mg: reducing agent, relatively strong
(b) MgO: oxidation agent, relatively weak
(c) $AgNO_3$: oxidation agent, relatively strong
(d) Cu: reducing agent, relatively weak
**9-56** (a) Na (b) Mg (c) Al
**9-58** (a) Na (b) Mg (c) Al (d) Fe
**9-60** Hg < Cu < Al
**9-64** (a) The reverse reaction is expected to occur.
(b) The reverse reaction is expected to occur.
(c) The reaction is expected to occur as written.
(d) The reaction is expected to occur as written.
**9-68** Iron

### Chapter 10

**10-3** (a) As is more nonmetallic.
(b) Se is more nonmetallic.
(c) S is more nonmetallic.
(d) As is more nonmetallic.
(e) P is more nonmetallic.
**10-4** Ozonc has the molecular formula $O_3$.
**10-5** Sulfur **10-6** Arsenic **10-7** Elemental fluorine, $F_2$
**10-9** (a) $H_2SO_3$ (b) $P_4$ (c) $Cl^-$
**10-10** (b) $H_2SO_3$ (d) $CO_2$ (e) $Mg^{2+}$
**10-11** (a) Iron is reduced. Carbon is oxidized.
(b) Hydrogen is oxidized. Carbon is reduced.
(c) Carbon is oxidized. Oxygen is reduced.
(d) Sulfur is oxidized. Oxygen is reduced.
**10-12** (a) Phosphorus is oxidized. Chlorine is reduced.
(b) Nitrogen is oxidized. Fluorine is reduced.
(c) Sodium is oxidized. Hydrogen is reduced.
(d) Nitrogen is oxidized and reduced.

**10-13** Nitrogen is oxidized. Chlorine is reduced.
**10-14** Sulfur is oxidized from 0 $S_8$ while sulfur is reduced in $SO_3^{2-}$.
**10-16** (a) acid (b) oxidizing agent (c) acid (d) oxidizing agent
**10-17** (e) $O_2$ **10-18** Sodium
**10-19** (a) Arsenic (b) Arsenic (c) Phosphorus (d) Sulfur (e) Carbon
**10-20** (a) $3Mg(s) + N_2(g) \longrightarrow Mg_3N_2(s)$
(b) $4\,Li(s) + O_2(g) \longrightarrow 2\,Li_2O(s)$
(c) $Br_2(l) + 2\,I^-(aq) \longrightarrow I_2(s) + 2Br^-(aq)$
**10-21** (a) $SO_2(g) + H_2O(l) \longrightarrow H_2SO_3(aq)$
(b) $Cl_2(g) + 2\,OH^-(aq) \longrightarrow Cl^-(aq) + OCl^-(aq) + H_2O(l)$
(c) $CO_2(g) + H_2O(l) \longrightarrow H_2CO_3(aq)$
**10-22** (a) $HCl(g) + H_2O(l) \longrightarrow H^+(aq) + Cl^-(aq) + H_2O(l)$
(b) $P_4O_{10}(s) + 6\,H_2O(l) \longrightarrow 4\,H_3PO_4(aq)$
(c) $2NO_2(g) + H_2O(l) \longrightarrow HNO_2(aq) + HNO_3(aq)$
$2HNO_2(aq) \longrightarrow HNO_3(aq) + NO(g)$
**10-23** (a) $S_8(s) + 8\,O_2(g) \longrightarrow 8\,SO_2(g)$
(b) $2\,Al(s) + 3\,I_2(s) \longrightarrow Al_2I_6(s)$
(c) $P_4(s) + 10\,F_2(g) \longrightarrow 5\,PF_5(s)$
**10-27** (a) LiH is formed. (b) $CaH_2$ is formed.
**10-34** (c) $BaO_2$
**10-37** :O=O—O: ⟷ :O—O=O

:O=S—O: ⟷ :O—S=O:
**10-47** (d) $SO_4$
**10-50** (b) $NF_5$
**10-60** All of the elements and compounds listed are involved in the production of nitric acid according to the procedure using the Haber process and the Oswald process.
**10-69** The product of the reaction of elemental phosphorus with excess oxygen is $P_4O_{10}$. This compound reacts with water to form trihydrogen phosphate, $H_3PO_4$, the aqueous solution of which is called phosphoric acid.
**10-71** $PO_2$ should not exist. **10-72** Fluorine
**10-74** $F_2 > Cl_2 > Fe^{3+} > Br_2 > I_2$
**10-80** $2\,CO(g) + O_2(g) \longrightarrow 2\,CO_2(g)$
$2\,H_2(g) + O_2(g) \longrightarrow 2\,H_2O(l)$
**10-82** \$2.19/mol $H_2$ for sodium; \$1.18/mol $H_2$ for zinc; \$5.07/mol $H_2$ for sodium hydride.
**10-83** \$7.23/mol $O_2$ by decomposing $H_2O_2$; \$2.08/mol $O_2$ by decomposing $KCl_3$
**10-84** 4.5 ton $HNO_3$
**10-85** 58% of the tetraphosphorus molecules decompose.
**10-86** $^{235}UF_6$ will diffuse more rapidly.

## Chapter 11

**11-2** $HCl(aq) + NaOH(aq) \longrightarrow NaCl(aq) + H_2O(l)$
$H_2SO_4(aq) + 2\,NaOH(aq) \longrightarrow Na_2SO_4(aq) + 2\,H_2O(l)$
$HCl(aq) + NH_3(aq) \longrightarrow NH_4Cl(aq)$
$H_2SO_4(aq) + 2\,NH_3(aq) \longrightarrow (NH_4)_2SO_4(aq)$
**11-4** (a) HCl (b) $H_2SO_4$ (c) $H_2O$ (e) $H_2S$
**11-5** (a) NaOH (b) $Ca(OH)_2$ (d) $H_2O$
**11-6** All of the listed compounds satisfy the operational definition of a base.
**11-7** $HBr(g) + H_2O(l) \longrightarrow H_3O^+(aq) + Br^-(aq)$
$H_2O(l) + NH_3(g) \longrightarrow OH^-(aq) + NH_4^+(aq)$
**11-11** The acidic hydrides are (a) $H_2CO_3$, (b) $CH_3CO_2H$, and (d) $H_2S$. The basic hydrides are (c) $LiAlH_4$ and (e) $ZnH_2$.
**11-12** The acidic oxides are (a) $CO_2$, (b) $SO_3$, and (c) $P_4O_{10}$. The basic oxide is (d) MgO.
**11-13** (a) $SO_2$ (b) $HNO_3$ (c)HI
**11-14** (a) $Ca(OH)_2$ (b) $Na_2O$ (d) $Li_2O$ (e) CsOH
**11-15** $CO_2(g) + H_2O(l) \longrightarrow H_2CO_3(aq)$ acidic
$6\,H_2O(l) + P_4O_{10}(s) \longrightarrow 4\,H_3PO_4(aq)$ acidic
$CaO(s) + H_2O(l) \longrightarrow Ca(OH)_2(aq)$ basic
$SO_3(g) + H_2O(l) \longrightarrow H_2SO_4(aq)$ acidic
$Na_2O(s) + H_2O(l) \longrightarrow 2\,NaOH(aq)$ basic
**11-17** Sulfur
**11-18** Calcium and zinc hydroxides are the bases. $H_2CO_3$, $HNO_3$ and $H_3PO_4$ are acids.
**11-24** Aluminum
**11-26** (a) $NaHCO_3$ (b) $Na_2CO_3$ (e) $H_2O$
**11-27** $H_3O^+$ **11-28** $CH_4$ and $CH_3^+$
**11-29** (a) $HSO_4^-(aq)$ (Brønsted acid) + $H_2O(l)$ (Brønsted base) $\longrightarrow$ $H_3O^+(aq)$ (Brønsted acid) + $SO_4^{2-}(aq)$ (Brønsted base)
(b) $CH_3CO_2H(aq)$ (Brønsted acid) + $OH^-(aq)$ (Brønsted base) $\longrightarrow$ $CH_3CO_2^-(aq)$ (Brønsted base) + $H_2O(l)$ (Brønsted acid)
(c) $CaF_2$ + $H_2SO_4(aq)$ (Brønsted acid) $\longrightarrow$ $CaSO_4(aq)$ + $2\,HF(aq)$ (Brønsted acid)
The fluoride ion of $CaF_2$ is a Brønsted base.
The $SO_4^{3-}$ ion of $CaSO_4$ is a Brønsted base.
(d) $HNO_3(aq)$ (Brønsted acid) + $NH_3(aq)$ (Brønsted base) $\longrightarrow NH_4NO_3(aq)$
The $NH_4^+$ ion of $NH_4NO_3(aq)$ is a Brønsted acid.
The $NO_3^-$ ion of $NH_4NO_3(aq)$ is a Brønsted base.
(e) $LiCH_3(l) + NH_3(l) \longrightarrow CH_4(g) + LiNH_2(s)$
$CH_3^-$ from $LiCH_3(l)$ is a Brønsted base.
$NH_3(l)$ is a Brønsted acid in this reaction.
**11-30** (a) $HCl(aq)$ (acid) + $H_2O(l)$ (base) $\longrightarrow$ $H_3O^+(aq)$ (acid) + $Cl^-(aq)$ (base)
(b) $HCO_3^-(aq)$ (base) + $H_2O(l)$ (acid) $\longrightarrow$ $OH^-(aq)$ (base) + $CH_3CO_2^-(aq)$ (acid)
(c) $NH_3(aq)$ (base) + $H_2O(l)$ (acid) $\longrightarrow$ $OH^-(aq)$ (base) + $NH_4^+(aq)$ (acid)
(d) $CaCO_3(s)$ (base($co_3^{2-}$)) + $2\,HCl(aq)$ (acid) $\longrightarrow$ $C^{+2}(aq)$ + $2\,Cl^-(aq)$ (base) + $H_2CO_3(aq)$ (acid)
(e) $CaO(s)$ (base($O^{2-}$)) + $H_2O(l)$ (acid) $\longrightarrow$ $Ca^{+2}(aq)$ + $OH^-(aq)$ (base) + $OH^-(aq)$ (acid)
**11-31** Response (c) represents a pair differing by more than one proton.

**11-34** (a) $H_2O$ (b) $^-OH$ (c) $O^{2-}$ (d) $NH_3$
**11-35** (a) $PO_4^{3-}$ (b) $Al(OH)(H_2O)_5^{2+}$ (c) $CO_3^{2-}$ (d) $S^{2-}$
**11-36** (a) $OH^-$ (b) $H_2O$ (c) $H_3O^+$ (d) $NH_3$
**11-37** (a) $NaH_2PO_4$ (b) $H_2CO_3$ (c) $NaHSO_4$ (d) $HNO_2$
**11-38** (b) $HSO_4^-$ (e) $H_3O^+$
**11-40** (a) $Al_2O_3(s) + 6\ HCl(aq) \longrightarrow 2\ AlCl_3(aq) + 3\ H_2O(l)$
(b) $CaO(s) + H_2SO_4(aq) \longrightarrow CaSO_4(s) + H_2O(l)$
(c) $CO_2(g) + 2\ NaOH(aq) \longrightarrow H_2O(l) + Na_2CO_3(aq)$
(d) $MgCO_3 + 2\ HCl(aq) \longrightarrow H_2O(l) + CO_2(g) + MgCl_2(aq)$
(e) $Na_2O_2(s) + H_3PO_4(aq) \longrightarrow H_2O_2(aq) + Na_2HPO_4(aq)$
**11-47** Both chromic acid and hydrobromic acid qualify as "strong acids." The others are classified as weak acids.
**11-48** (d) phenol ($C_6H_5OH$)
**11-50** $CH_4 < NH_3 < H_2O < HCl$
**11-51** $PH_3 < H_2S < H_2O < NH_3$
**11-52** (a) $H_3O^+$ **11-53** $OH^-$ **11-54** $Al(OH)_3$
**11-55** (c) $CH_3^-$ **11-56** (e) the oxidide ion **11-57** methane
**11-60** 0.10 M acetic acid **11-61** $H_3PO_4$ **11-63** (e) $NH_4Cl$
**11-65** (a) $CH_3^+$ (b) $BF_3$ (e) $Ag^+$ (f) $Fe^{3+}$
**11-66** $CH_4$
**11-67** $6\ Li(s) + N_2(g) \longrightarrow 2\ Li_3N(s)$
$Li_3N(s) + 4\ H_2O(l) \longrightarrow 3\ LiOH(aq) + NH_4OH(aq)$
**11-70** The rain in Scotland had $[H_3O^+] = 0.0040$ M. The acetic acid solution has $[H_3O^+] = 0.0013$ M.

## Chapter 12

**12-4** covalent
**12-6** $[Co(NH_3)_5Cl]Cl_2$, $[CO(NH_3)_4Cl_2]Cl$, $[Co(H_2O)(NH_3)_5]Cl_3$
**12-10** This is a $[Co(en)_2Cl_2]^+$ complex from which no $Cl^-$ ions can be precipitated.
**12-11** two ions
**12-13**

| | Coordination Number | Charge on Transition Metal |
|---|---|---|
| (a) $CuF_4^{2-}$ | 4 | +2 |
| (b) $Cr(CO)_6$ | 6 | 0 |
| (c) $Fe(CN_6)^{4-}$ | 6 | +2 |
| (d) $Pt(NH_3)_2Cl_2$ | 4 | +2 |

**12-14**

| | Coordination Number | Charge on Transition Metal |
|---|---|---|
| (a) $Co(SCN)_4^{2-}$ | 4 | +2 |
| (b) $Fe(acac)_3$ | 6 | +3 |
| (c) $[Ni(en)_2(H_2O_2]^{2+}$ | 6 | +2 |
| (d) $[Co(NH_3)_5H_2O)]^{3+}$ | 6 | +3 |

**12-15** (a) $V^{2+}$: $[Ar]3d^3$ (b) $Cr^{2+}$: $[Ar]3d^4$ (c) $Mn^{2+}$: $[Ar]3d^5$
(d) $Fe^{2+}$:$[Ar]3d^6$ (e) $Ni^{2+}$: $[Ar]3d^8$
**12-16** The $Co^{2+}$ ion has an electron configuration of $[Ar]3d^7$.
**12-17** (b) $Mn^{2+}$: $[Ar]3d^5$ and (c) $Fe^{3+}$: $[Ar]3d^5$
**12-19** (b) $VO^{2+}$ is a $d^1$ transition-metal complex.
**12-24** (a) $Fe^{3+}$ (b) $BF_3$ (d) $Ag^+$ (e) $Cu^{2+}$
**12-25** (a) CO (b) $O_2$ (c) $Cl^-$ (d) $N_2$ (e) $NH_3$
**12-26** (a) $CN^-$ (b) $SCN^-$ (c) $CO_3^{2-}$ (d) $NO^+$ (e) $S_2O_3^{2-}$

**12-30**

| | | |
|---|---|---|
| (a) | $Cu(NH_3)_4^{2+}$ | tetramminecopper (II) ion |
| (b) | $Mn(H_2O)_6^{2+}$ | hexaquomanganese (II) ion |
| (c) | $Fe(CN)_6^{4-}$ | hexacyanoiron (II) ion |
| (d) | $Ni(en)_3^{2+}$ | tris(ethylenediamine)nickel (II) ion |
| (e) | $Cr(acac)_3$ | hexacyanoferrate (II) ion |

**12-31**

| | | |
|---|---|---|
| (a) | $Pt(NH_3)_2Cl_2$ | dichlorodiammineplatinum (II) |
| (b) | $Ni(CO)_4$ | tetracarbonylnickel (0) |
| (c) | $Co(en)_3^{3+}$ | tris(ethylenediamine)cobalt (III) ion |
| (d) | $Na[Mn(CO)_5]$ | sodium pentacarbonylmanganate (I) |

**12-32**

| | | |
|---|---|---|
| (a) | $Na_3[Co(NO_2)_6]$ | sodium hexanitritocobaltate (III) |
| (b) | $Na_2[Zn(CN)_4]$ | sodium tetracyanozincate (II) |
| (c) | $[Co(NH_3)_4Cl_2]Cl$ | dichlorotetramminecobalt (III) chloride |
| (d) | $[Ag(NH_3)_2]Cl$ | diamminesilver (I) chloride |

**12-33** (a) $[Cr(NH_3)_6]Cl_3$
(b) $[Cr(NH_3)_5Cl]Cl_2$
(c) $[Co(en)_3]Cl_3$
(d) $K_3[Co(NH_3)_2(NO_2)_4]$
**12-34** (d) $Co(NH_3)_4Cl_2^+$ and (e) $Co(NH_3)_4(H_2O)_2^{3+}$ can form cis/trans isomers.
**12-37** (b) $Pt(NH_3)_2Cl_2$ and (d) $IrCl(CO)(PH_3)_2$
**12-40** square planar **12-41** (a) $Cr(en)_3^{3+}$ chiral
**12-44** The cis isomer of $Ni(NH_3)_4(H_2O)_2^{2+}$ is not chiral.
**12-45** (a) $Cr(acac)_3$ is chiral.
(b) $Cr(C_2O_4)_3^{3-}$ is chiral.
**12-58** See problem 12-56. $t_{2g}$: $d_{xy}$, $d_{yz}$, $d_{xz}$; $e_g$: $d_{x^2-y^2}$, $d_{z^2}$
**12-59** Orbitals in the $t_{2g}$ set are directed between ligands and are at a lower energy, and orbitals in the $e_g$ set are directed at a higher energy.
**12-60** No, the orbitals are the same.
**12-63** (a) $Rh^{3+}$ **12-64** (a) $CN^-$
**12-70** For the ions given, the splitting decreases in the following order: $Co^{3+} > Fe^{3+} > Fe^{2+} > Co^{2+}$
**12-71** In the spectrochemical series, $Ir^{3+} > Rh^{3+} > Co^{3+}$, $\Delta$ increases from top to bottom in a column among the transition metals.
**12-72**

| | | |
|---|---|---|
| (a) | $Cu^{2+}$ | light blue |
| (b) | $Fe^{3+}$ | yellow |
| (c) | $Ni^{2+}$ | green |
| (d) | $CrO_4^{2-}$ | yellow |
| (e) | $MnO_4^-$ | deep violet |

**12-75** The $CrO_4^{2-}$ ion appears yellow because it absorbs blue light.
**12-76** The frequency decreases. **12-77** red

## Chapter 13

**13-8** (a) ionic solid
(b) ionic solid
(c) molecular solid
(d) molecular solid
(e) metallic solid
(f) metallic solid
(g) molecular solid
(h) molecular solid
**13-9** (b) graphite and (e) diamond
**13-10** van der Waals forces
**13-11** van der Waals forces
**13-12** ionic solids
**13-15** Simple cubic packing has a coordination number of 6. Body-centered cubic packing has a coordination number of 8.

Hexagonal closest packing has a coordination number of 12.
Cubic closest packing has a coordination number of 12.

**13-16** Both cubic closest packed and hexagonal closest packed structure types would allow Xenon the largest number of induced dipole - induced dipole interactions.
**13-18** body-centered cubic structure
**13-25** (c) and (d)
**13-26** (a) 4 (b) 6 (c) 8
**13-27** simple cubic structure types
**13-28** The tetrahedral hole is the smallest. The cubic hole is the largest.
**13-31** The neutral potassium atom has the larger size.
**13-32** $TiO_2$ **13-33** TiC **13-34** LiH **13-35** $[Fe^{2+}][S_2^{2-}]$
**13-36** $Ti_2O_3$ **13-38** ZnTe **13-40** +6
**13-41** $Ca^{2+}$ should occupy the larger sized octahedral holes while $Ti^{4+}$ should more likely occupy the tetrahedral holes.
**13-42** +1 **13-45** cubic closest packed **13-47** 0.48
**13-48** For the body-centered cubic unit cell the fraction of empty space is 0.32 For the face-centered cubic unit cell the fraction of empty space is 0.26.
**13-52** face-centered cubic unit cell
**13-53** body-centered cubic
**13-54** There is one chloride and one ammonium ion per unit cell.
**13-55** The empirical formula is GaAs.
**13-56** The empirical formula for the unit cell contents is $CaTiO_3$.
**13-57** The volume of the unit cell = $\frac{2.398 \times 10^{-23}\ \text{cm}^3}{1\ \text{unit cell}}$; $r_{Cr} = 1.933 \times 10^{-8}$ cm
**13-58** d = 4.25 g/cm$^3$ **13-59** $r_{Ar}$ = 193 pm
**13-60** $\frac{10.502\ \text{g}}{1\ \text{cm}^3}$
**13-61** body-centered cubic **13-62** face-centered
**13-63** cubic structure
**13-64** copper **13-65** $r_{Ba} = 4.35 \times 10^{-8}$ cm
**13-66** $r_{Na\text{-}H}$ = 345.1 picometer
**13-67** $r_{Ti}$ = 0.148 nm **13-68** $r_{Cs}$ = 0.176 nm
**13-77** 4 cdo/unit cell
**13-78** 4LiF/unit cell
**13-79** 15.73g Fe/cm$^3$; Ion contracts upon changing from the body-centered cubic structure to the face-centered structure.

**Chapter 14**

**14-3** solid = $2.683 \times 10^{22}$ molecules; liquid = $2.162 \times 10^{22}$ molecules; gas = $2.689 \times 10^{19}$ molecules
**14-5** $CH_4 < SiH_4 < GeH_4 < SnH_4$ **14-6** (e) $CCl_4$
**14-16** The larger surface area will allow a greater total amount of bromine to enter the gas phase, but the actual vapor pressure will remain the same as long as the temperature is not changed. The vapor pressure would not change if more liquid were added or some of the liquid were removed.
**14-19** 88.4% **14-20** 13°C or 55°F **14-25** 38.1°C
**14-26** 17.5 mm Hg
**14-27** (c) increase the vapor pressure of the liquid
**14-28** nitrogen **14-29** ether
**14-56** $d = 3.01\frac{\text{g}}{\text{cm}^3}$; 45.2% empty space in solid; 53.5% empty space in liquid; 99.9% empty space in gas.

**Chapter 15**

**15-5** (e) 1 heptanol
**15-10** Temperature increase will have no effect on the mol fraction or molal concentration of solute. The molar concentration will decrease.
**15-11** The solution was less than 1.00 M.
**15-12** The molarity is 14.8 M. The molality is 22.8 m.
**15-13** $\frac{0.641\ \text{mol}\ Pb(NO_3)_2}{1\ \text{L soln}}$ **15-14** 2.5 ppm $C_6H_6$
**15-15** $\frac{0.34\ \text{mol}\ Cr_2O_7^{-2}}{1\ \text{L soln}}$
**15-16** $6.3 \times 10^{-8}$ M **15-17** $X_{N2} = 0.81$; $X_{O2} = 0.191$
**15-18** $X_{H2O} = 0.989$
**15-19** $X_{H2O} = 0.0369$; molarity = 0.832 m
**15-20** $X_{Cl2} = 1.43 \times 10^{-6}$
**15-21** $X_{Ni} = 0.576$; $X_{Fe} = 0.243$; $X_{Cr} = 0.174$; $X_C = 0.0047$
**15-22** $X_{Pb} = 0.0181$; $X_{Sb} = 0.0185$; $X_{Cu} = 0.0473$; $X_{Sn} = 0.120$
**15-23** $X_{H2O} = 0.998$; $X_{K2Cr2O7} = 0.00180$
**15-24** $X_{N2} = 0.167$
**15-25** $d = \frac{1\ \text{g soln}}{1\ \text{mL soln}}$
The number of mols NaCl in the solution is 0.086 mol NaCl. The molar concentration of the solution is 1.3 m NaCl. The molal concentration of the solution is 1.3 M NaCl. The number of mols $H_2O$ in the solution is 3.6 mol $H_2O$. The mol fraction of NaCl in the solution is 0.023.
**15-26** 99.5% **15-27** $\frac{12\ \text{mol HCl}}{1\ \text{l soln}}$ **15-28** 19.4 m
**15-29** 0.26 m **15-30** 0.671 m
**15-42** Freezing temperature = −0.192°C **15-43** (b)
**15-44** For NaCl = 3.42 mol particles; for $NH_4NO_3$ = 2.50 mol particles; for $(NH_4)_2SO_4$ = 2.27 mol particles. Sodium chloride will give the largest freezing temperature depression per unit mass used.
**15-45** $345\frac{\text{g}}{\text{mol}}$ **15-46** 37.7°C/m **15-47** −1.80°C
**15-49** 78.29°C **15-50** 86% dissociates **15-51** 0.188°C
**15-52** Fraction dissociated .11
**15-54** $\frac{7.4 \times 10^4\ \text{g hemoglobin}}{1\ \text{mol hemoglobin}}$
**15-61** $C_{10}H_{14}N_2$ **15-62** $C_2H_6SO$

**Chapter 16**

**16-9** (d) $K_c = \frac{[ClF_3]^2}{[Cl_2][F_2]}$
**16-10** (d) $K_c = \frac{[NO]^2[O_2]}{[NO_2]}$
**16-11** (a) $K_c = \frac{[OF_2]^2}{[O_2][F_2]^2}$ (b) $K_c = \frac{[SO_3]^2}{[SO_2]^2[O_2]}$

(c) $K_c = \dfrac{[SO_2Cl_2]^2[O_2]}{[SO^3]^2[Cl_2]^2}$

**16-12** (a) $K_c = \dfrac{[N_2][H_2O]^2}{[NO]^2[H_2]^2}$ (b) $K_c = \dfrac{[NO]^2[Cl_2]}{[NOCl]^2}$

(c) $K_c = \dfrac{[NO_2]^2}{[NO]^2[O_2]}$

**16-13** (a) $K_c = \dfrac{[CO_2]^2}{[CO]^2[O_3]}$ (b) $K_c = \dfrac{[CO][H_2O]}{[CO_2][H_2]}$

(c) $K_c = \dfrac{[CH^3OH]}{[CO][H_2]^2}$

**16-14** $K_c = 5.59 \times 10^{-6}$

**16-15** trial I $K_C = 59.686$; trial II $K_c = 59.454$; trial III $K_c = 59.212$; $K_c$ is a constant within the limits of experimental error.

**16-17** $K_c\,(b) = \dfrac{1}{K_c(a)}$ and $K_p\,(b) = \dfrac{1}{K_p(a)} = \dfrac{1}{1 \times 10^{-5}} = 1 \times 10^5$

$K_c\,(c) = (K_c(b))^{1/2}$ and $K_p\,(c) = (K_p)(b))^{1/2} = (1 \times 10^5)^{1/2} = 300$

**16-18** $K_c(c) = 1.5 \times 10^{12}$ **16-19** $K_c(c) = 0.78$

**16-20** (b) **16-21** (c) **16-22** (c)

**16-23** The system is not at equilibrium and must shift to the left to form reactant.

**16-27** $\Delta(CO) = 0.250$ M
$\Delta(Cl_2) = 0.250$ M

**16-28** $\Delta(N_2) = 0.117$
$\Delta(H_2) = 0.351$

**16-29** (b) $\Delta(NO_2) = 2\Delta(O_2)$ **16-30** (e) $\Delta(F_2) = 3\Delta(Cl_2)$

**16-31** (b) $1.5\Delta$

**16-32** $[N_2]_{eq} = 0.922$ M
$[H_2]_{eq} = 0.766$ M
$[NH_3]_{eq} = 0.156$ M

**16-33** $K_c = 0.0587$ **16-41** (c)

**16-42** % decomposition of $PCl_5$ = 3.6%

**16-43** $[PCl_5]_{eq} = 0.994$ M
$[PCl_3]_{eq} = 6.5 \times 10^{-3}$ M
$[ClN_2]_{eq} = 0.2065$ M
% decomposition of $PCl_5$ = 0.65%

**16-44** $[NO_2]_{eq} = 0.098$ M
$[NO]_{eq} = 0.0019$ M
$[O_2]_{eq} = 9.5 \times 10^{-4}$ M

**16-45** $[NO_2]_{eq} = 0.099$ M
$[NO]_{eq} = 2.6 \times 10^{-4}$ M
$[O_2]_{eq} = 0.05013$ M

**16-47** $[SO_3]_{eq} = 0.399$ M
$[SO_2]_{eq} = 3.8 \times 10^{-4}$ M
$[O_2]_{eq} = 1.9 \times 10^{-4}$ M

**16-48** $[N_2O_4]_{eq} = 0.099$ M

**16-49** $[NO_2]_{eq} = 7.6 \times 10^{-3}$ M
$[N_2O_4]_{eq} = 0.99$ M

**16-50** $[N_2] = 0.146$ M
$[H_2] = 0.237$ M
$[NH_3] = 0.0088$ M

**16-51** $[CO]_{eq} = [H_2O]_{eq} = 0.541$ M
$[CO_2]_{eq} = [H_2]_{eq} = 0.459$ M

**16-52** The initial concentration of CO and $H_2O$ is 1.17 M.

**16-53** $[N_2] = 0.1449$ M
$[O_2] = 4.49 \times 10^{-2}$ M
$[NO] = 1.47 \times 10^{-6}$ M

**16-54** $[SO_2Cl_2]_{eq} = 0.0492$ M
$[SO_2]_{eq} = [Cl_2]_{eq} = 8.37 \times 10^{-4}$ M

**16-55** $[NO] = 0.3004$ M
$[Cl_2] = 2.1 \times 10^{-4}$ M
$[NOCl] = 0.199$ M

**16-57** (c) **16-58** (c) **16-59** $K_p = 0.0192$

**16-60** $P_{COCl2} = 0.106$ atm, $P_{CO} = P_{Cl2} = 1.84 \times 10^{-2}$ atm

**16-61** $P_{SO2} = 0.0194$ atm, $P_{O2} = 9.67 \times 10^{-3}$ atm, $P_{SO3} = 0.471$ atm

**16-62** $P_{SO3} = 0.484$ atm, $P_{SO2} = 0.006$ atm, $P_{O2} = 0.103$ atm

**16-63** $P_{N2} = 0.598$ atm, $P_{H2} = 0.897$ atm, $P_{NH3} = 2.08 \times 10^{-3}$ atm

**16-64** $P_{NO2} = 0.790$ atm, $P_{H2O} = 0.929$ atm, $P_{NH3} = 0.210$ atm, $P_{O2} = 2.23 \times 10^{-4}$ atm

**16-65** $P_{NH3} = 1.8 \times 10^{-2}$ atm, $P_{N2} = 1.99$ atm, $P_{H2} = 5.97$ atm

**16-66** $[N_2]_{eq} = 0.399$ M
$[O_2]_{eq} = 0.599$ M
$[NO] = 3.2 \times 10^{-5}$ M

**16-67** % yield $CH_3OH$ = 99.3%

**16-69** (a) The reaction will shift to the right.
(b) The reaction will shift to the right.
(c) The reaction will shift to the right.

**16-70** (a) The reaction will shift to the right.
(b) The reaction will not shift.
(c) The reaction will shift to the left.

**16-71** (a) Increasing the concentration of $NO_2$ will shift the reaction to the right until equilibrium is reached.
(b) Increasing the concentration of $O_2$ will shift the reaction to the left until equilibrium is reached.
(c) Increasing the concentration of $PF_5$ will shift the reaction to the right until equilibrium is reached.

**16-72** (a) Decreasing the concentration of NO will cause the reaction to shift to the right, forming more NO.
(b) Decreasing the concentration of $O_2$ will cause the reaction to shift to the left, forming more $O_2$.
(c) Decreasing the concentration of $F_2(g)$ will cause the reaction to shift to the left, forming more $Cl_2(g)$.

**Chapter 17**

**17-4** The fraction of water molecules that dissociate will increase.

**17-7** $6.022 \times 10^{13}$ ions

**17-8** pH = 1.46 pOH = 12.54

**17-9** pH = 1.21 pOH = 12.79

**17-10** $[H_3O^+] = 1.91 \times 10^{-4}$ M
$[OH^-] = 5.25 \times 10^{11}$ M

**17-12** (a) temperature (c) pH **17-13** (d) $Cl_2CHCO_2H$

**17-14** (a) $CH_3CO_2^-$

**17-15** HOAc < $HNO_2$ < HF < HOClO **17-18** pH = 1.25

**17-19** pH = 1.002 **17-23** (a) 0.10 M HOAc

**17-24** Fraction 0.10 M dissociated = 1.3%
Fraction 0.010 M dissociated = 4.2%
Fraction 0.0010 M dissociated = 13%
As weak acid becomes more dilute, the percent dissociation increases.

**17-25** $[H_3O^+] = [HCO_2^-] = 4.2 \times 10^{-3}$ M
$[HCO_2H] = 9.6 \times 10^{-2}$ M

**17-26** $[H_3O^+] = [CN^-] = 1.0 \times 10^{-5}$ M
$[HCN] = 0.174$ M
**17-27** $[H_3O^+] = 1.3 \times 10^{-6}$ M **17-28** $[HOAc] = 0.22$ M
**17-29** $K_a = 8.1 \times 10^{-5}$
**17-30** $K_a = 5.4 \times 10^{-4}$
**17-31** $[H_3O^+] = 1.3 \times 10^{-3}$ M for acetic acid
$[H_3O^+] = 1.1 \times 10^{-2}$ M for chloroacetic acid
$[H_3O^+] = 5.0 \times 10^{-2}$ M for dichloroacetic acid
As $K_a$ increases, the $[H_3O^+]$ increases.
**17-32** $[H_3O^+] = [Cl_3CCO_2^-] = 0.15$ M
$[Cl_3CCO_2H] = 0.10$ M
**17-33** (b) when $K_aC_a > 1.0 \times 10^{-13}$
**17-34** $[H_3O^+] = 1.1 \times 10^{-7}$ M
**17-35** $[H_3O^+] = 2.4 \times 10^{-6}$ M
**17-36** $[H_3O^+] = 1.3 \times 10^{-6}$ M
$[OH^-] = 7.7 \times 10^{-9}$ M
**17-37** (c) $K_b = K_w/K_a$
**17-39** $[OH^-] = [HCO_2H] = 2.1 \times 10^{-6}$ M
$[HCO_2^-] = 0.08$ M
**17-40** $[OH^-] = 6.5 \times 10^{-4}$ M $= [HOBr]$
$[OBr^-] = 0.10$ M
**17-41** pH = 9.30
**17-43** $K_b = 4.96 \times 10^{-4}$
Methylamine is a stronger base than aqueous ammonia.
**17-44** $C_{NH3} = 0.057$ M **17-45** pH = 11.6
**17-46** $K_b = 1.20 \times 10^{-6}$
**17-47** pH = 8.62 **17-50** (e) $NH_3$ and $NH_4Cl$
**17-51** (c) is an acidic buffer. **17-52** (d) is a basic buffer.
**17-53** (c) Increase $[HCO_2H]$ and $[NaHCO_2]$
**17-56** pH = 5.80 **17-57** pH = 3.59
**17-58** $[H_3O^+] = 4.13 \times 10^{-6}$ M; pH = 5.384
**17-59** $[H_3O^+] = 9.0 \times 10^{-6}$ ; The $[H_3O^+]$ should remain essentially the same.
**17-60** 0.114 mol sodium formate
**17-61** $[HOCl]/[OCl^-] = 0.86$ **17-62** 1.75 g NaOH
**17-63** 3.1 g $NaHCO_2$ **17-64** $[HOAc]/[OAc^-] = 1.8$
**17-65** $[H_2PO_4^-]/[HPO_4^{2-}] = 1.59$ **17-66** 2.88 g $NH_4Cl$
**17-67** $[H_2CO_3] = 0.10$ M
$[H_3O^+] = [HCO_3^-] = 2.1 \times 10^{-4}$ M
$[CO_3^{2-}] = 4.7 \times 10^{-11}$ M
**17-68** $[H_3O^+] = [HM^-] = 1.87 \times 10^{-3}$ M
$[M^{2-}] = 2.1 \times 10^{-8}$ M
**17-70** $[HG] = [H_3O^+] = 9.3 \times 10^{-2}$ M
$[H_2G^+] = 1.9$ M
$[G^-] = 2.5 \times 10^{-10}$ M
**17-71** (b) $9.3 \times 10^{-2}$ M $<$ 1.9 M
(d) $9.3 \times 10^{-2}$ M $\approx 9.3 \times 10^{-2}$ M
**17-72** (b) $[G^-] \approx K_{az}$
**17-73** 0.235 M $= [H_3O^+] = [HC_2O_4^-]$; $[H_2C_2O_4] = 1.02$ M
$[C_2O_4^{2-}] = 5.4 \times 10^{-5}$ M
**17-74** (c) **17-75** pH = 8.36
**17-76** $[CO_3^{2-}] = 5.6 \times 10^{-3}$ M
$[OH^-] = [HCO_3^-] = 5.6 \times 10^{-3}$ M
$[H_3O^+] = 1.8 \times 10^{-12}$ M
$[H_2CO_3] = 2.2 \times 10^{-8}$ M
**17-78** Response (d)
**17-79** pH = 1.6 for $H_3PO_4$
pH = 4.10 for $NaH_2PO_4$
pH = 10.1 for $Na_2HPO_4$
pH = 12.7 for $Na_3PO_4$
**17-80** Acidic **17-81** Acidic

**Chapter 18**

**18-6** (c) $BaCO_3$ **18-7** (e) $Cr(OH)_3$ **18-8** (e) $Pb(NO_3)_2$
**18-10** (d) $K_{sp} = [Ca^{2+}]^3[Po_4^{3-}]^2$
**18-11** (d) $K_{sp} = [Al^{3+}]^2[SO_4^{2-}]^3$
**18-12** (a) $K_{sp} = [Ba^{2+}][CrO_4^{2-}]$
(b) $K_{sp} = [Ca^{2+}][CO_3^{2-}]$
(c) $K_{sp} = [Pb^{2+}][F^-]^2$
(d) $K_{sp} = [Ag^+]^2[S^{2-}]$
**18-13** $K_{sp} = [Ag^+]^2[CrO_4^{2-}]$ **18-14** $K_{sp} = [Bi^{3+}]^2[S^{2-}]^3$
**18-15** $K_{sp} = 4\,C_s^3$
**18-16** $K_{sp} = 4\,C_s^3$ **18-17** $Ag_2S$ **18-18** $PbI_2$ **18-19** $Hg_2S$
**18-20** $[CN^-] = 8.1 \times 10^{-5}\frac{\text{mol}}{\text{L}}$
**18-21** $[CrO_4^{2-}] = 6.5 \times 10^{-5}\frac{\text{mol}}{\text{L}}$
**18-22** $K_{sp} = 2.47 \times 10^{-9}$
**18-23** $K_{sp} = 5.145 \times 10^{-3}$
**18-24** $K_{sp} = 2.49 \times 10^{-2}$
**18-25** $K_{sp} = 2.29 \times 10^{-12}$
**18-26** $K_{sp} = 1.1 \times 10^{-10}$
**18-27** $K_{sp} = 6.7 \times 10^{-16}$
**18-28** $C_s = 6.2 \times 10^{-16}$ g $Ag_2S$ in 100 ml of water
**18-29** (a) $\frac{1.4 \times 10^{-22}\text{ g } Hg_2S}{100\text{ mL soln}}$
(b) $\frac{1.5 \times 10^{-25}\text{ g HgS}}{100\text{ mL soln}}$
**18-30** (a) $\frac{2.2 \times 10^{-5}\text{ g } Ca_3(PO_4)_2}{100\text{ mL soln}}$
(b) $\frac{1.2 \times 10^{-7}\text{ g } Pb_3(PO_4)_2}{100\text{ mL soln}}$ (c) $\frac{2.0 \times 10^{-3}\text{ g } Ag_3PO_4}{100\text{ mL soln}}$
**18-31** $HgS < Bi_2S_3 < CuS < Ag_2S$
**18-34** (b) $MgF_2$ is less soluble in 0.100 M NaF than in pure water.
**18-35** (b) 0.0010 M $Na_2S$ **18-36** $C_s = 1.6 \times 10^{-19}$
**18-37** $[Ag^+] = 1.5 \times 10^{-15}$ M **18-38** $6.8 \times 10^{-23}$g $Ag_2S$
**18-40** In pure water with pH = 7; $C_s = 1.6 \times 10^{-23}$ mol/L
In a solution at pH = 9; $C_s = 1.6 \times 10^{-29}$ mol/L
**18-41** pH = 4.48 **18-42** $[Ag^+] = 1.2 \times 10^{-4}$ mol/L
**18-44** PbS will precipitate first. The lead ion concentration when zinc begins to precipitate is $5.0 \times 10^{-8}$ mol/L.
**18-45** AgI will precipitate first. The corresponding iodide ion concentration is $4.6 \times 10^{-12}$ mol/L.
**18-47** (d) $Ag_2S$ is about 10 times more soluble than CuS.
**18-48** (b) CuS will precipitate first.
**18-49** (a) Only CdS will dissolve at first.
**18-53** (a) $Fe^{3+}$, (b) $Ag^+$, and (c) $Cu^{2+}$ will precipitate first.
**18-56** (c) $BeF_2$ **18-57** (a) $CH_3^+$ **18-58** (c) CO
**18-59** (a) $NH_4^+$ **18-60** (e) $BF_3$
**18-62** $Zn^{2+}$, $Fe^{3+}$, $Sn^{2+}$, $Co^{2+}$, and $Ag^+$
**18-65** $K_f = 2.9 \times 10^{10}$
**18-66** $K_f = 5.9 \times 10^{18}$ **18-67** (d) 0.10 M $Ag^+$ and 1.0 M $CN^-$
**18-68** (a) 0.10 M $Hg^{2+}$ and 1.0 M $Cl^-$
**18-69** (e) $K_f = 1/K_d$
**18-71** $[Fe^{3+}] = 2.8 \times 10^{-6}$ mol/L
**18-72** $[Cu^{2+}] = 2.8 \times 10^{-17}$ mol/L
**18-73** $[Zn^{2+}] = 4.6 \times 10^{-8}$ mol/L

**18-74** $[Sb^{3+}] = 5.0 \times 10^{-13}$ mol/L
**18-75** $[Sb^{3+}] = 2.0 \times 10^{-9}$ mol/L
**18-76** $[Co^{3+}] = 7.7 \times 10^{-4}$ mol/L
**18-77** $[Cd^{2+}] = 7.0 \times 10^{-12}$ mol/L
**18-78** $C_s = 0.83$ mol/L
**18-80** 0.035 mol AgBr **18-81** $C_s = 1.997$ mol/L
**18-82** (c) Add NaSCN **18-83** (a) The $Zn^{2+}$ ion concentration increases.
**18-84** (a) add an acid, such as $HNO_3$
**18-85** $[NH_3] = 13.5$ M
**18-86** Precipitation must occur.
**18-89** The maximum amount of $^{60}Co$ that could leach into the groundwater is $3.0 \times 10^{-10}$ M as $^{60}Co^{2+}$ and $1.6 \times 10^{-26}$ M as $^{60}Co^{3+}$

**Chapter 19**

**19-5** $C_6H_{12}O_6(s) + 6O_2(g) \longrightarrow 6CO_2(g) + H_2O(l)$
**19-7** $4Ag(s) + 2H_2S(g) + O_2(g) \longrightarrow 2Ag_2S(s) + 2H_2O(g)$
$2Fe(s) + 3O_2(g) \longrightarrow 2Fe_2O_3(s)$
**19-8** Re = +3; Cr = +3; Mo = +2
**19-9** +3 in all of the formulas
**19-10** (c) $PH_4^+$ and (e) $CaH_2$
**19-11** $CH_4 < H_3COH < C$ and $H_2CO < CO < CO_2$
**19-12** (a) +4 (b) +4 (c) +2
(d) +4 (e) −4 (f) −4
(g) 0 (h) +4 (i) +2
**19-13** (a) 0 (b) −2 (c) −2
(d) +4 (e) +6 (f) +4
(g) +6 (h) +2
**19-14** (a) +2 (b) +3 (c) +4
(d) +7 (e) +2 (f) +7
(g) +4 (h) +2 (i) +6
**19-15** Prussian blue, +3; Turnbull's blue, +2
**19-17** (b) The $Cr_2O_7^{2-}$ ion is the oxidizing agent.
**19-18** $3Ag_2S(s) + 2Al(s) \longrightarrow 6Ag(s) + Al_2S_3(s)$
**19-19** (b) Fe (+3) is reduced and carbon is oxidized.
(c) Silicon is reduced and carbon is oxidized.
(d) Carbon is reduced and hydrogen is oxidized.
**19-20** (b) Phosphorus is oxidized and oxygen is reduced.
(d) Phosphorus is oxidized and oxygen is reduced.
**19-25** (a) $2NH_3(g) + 3CuO(s) \longrightarrow N_2(g) + 3H_2O(l) + 3Cu(s)$
(b) $3NH_3(aq) + 3Cl_2(g) \longrightarrow N_2(g) + NH_4^+(aq) + 5H^+(aq) + 6Cl^-(aq)$
**19-26** (a) $2H_2O_2(aq) \longrightarrow O_2(g) + 2H_2O(l)$
(b) $3NO_2(g) + H_2O(l) \longrightarrow 2HNO_3(aq) + NO(g)$
**19-27** (a) $HIO_3(aq) + 5HI(aq) \longrightarrow 3I_2(aq) + 3H_2O(l)$
(b) $8SO_2(g) + 16H_2S(g) \longrightarrow 3S_8(s) + 16H_2O(l)$
**19-28** (a) $6HCl(aq) + 2HNO_3(aq) \longrightarrow 2NO(g) + 3Cl_2(g) + 4H_2O(l)$
(b) $2HBr(aq) + H_2SO_4(aq) \longrightarrow SO_2(g) + Br_2(aq) + 2H_2O(l)$
(c) $10HCl(aq) + 6H^+(aq) + 2MnO_4^-(aq) \longrightarrow 5Cl_2(g) + 2Mn^{2+}(aq) + 8H_2O(l)$
**19-30** $8H_2S(g) + 8H_2O_2(aq) \longrightarrow S_8(s) + 16H_2O(l)$
**19-31** $3P_4(s) + 10KClO_3(s) \longrightarrow 3P_4O_{10}(s) + 10KCl(s)$
**19-32** (a) $6I^-(aq) + 2CrO_4^{2-}(aq) + 16H^+(aq) \longrightarrow 3I_2(aq) + 2Cr^{3+}(aq) + 8H_2O(l)$
(b) $6Fe^{2+}(aq) + 14H^+(aq) + Cr_2O_7^{2-}(aq) \longrightarrow 6Fe^{3+}(aq) + Cr^{3+}(aq) + 7H_2O(l)$
(c) $24H_2S(g) + 16CrO_4^{2-}(aq) + 80H^+(aq) \longrightarrow 3S_8(s) + 16Cr^{3+}(aq) + 64H_2O(l)$
**19-34** $Br_2(aq) + SO_2(g) + 2H_2O(l) \longrightarrow H_2SO_4(aq) + 2HBr(aq)$
**19-36** (a) $NO(g) + MnO_4^-(aq) \longrightarrow NO_3^-(aq) + MnO_2(s)$
(b) $2NH_3(g) + 2MnO_4^-(aq) \longrightarrow N_2(g) + 2H_2O(l) + 2MnO_2(s) + 2OH^-(aq)$
**19-38** (a) $2Cr(s) + 2OH^-(aq) + 6H_2O(l) \longrightarrow 2Cr(OH)_4^-(aq) + 3H_2(g)$
(b) $4OH^-(aq) + 2Cr(OH)_3(s) + 3H_2O_2(aq) \longrightarrow 2CrO_4^{2-}(aq) + 8H_2O(l)$
**19-40** (a) $2OH^-(aq) + Cl_2(g) \longrightarrow OCl^-(aq) + H_2O(l) + Cl^-(aq)$
(b) $12OH^-(aq) + 6Cl_2(g) \longrightarrow 2ClO_3^-(aq) + 6H_2O(l) + 10Cl^-(aq)$
(c) $12OH^-(aq) + 4P_4(s) + 12H_2O(l) \longrightarrow 12H_2PO_2^-(aq) + 4PH_3(g)$
(d) $2H_2O_2(aq) \longrightarrow O_2(g) + 2H_2O(l)$
**19-42** $2NH_3(aq) + OCl^-(aq) \longrightarrow N_2H_4(aq) + H_2O(l) + Cl^-(aq)$
**19-45** (a) $Cl^-$, (d) Zn, (e) $CaH_2$ **19-46** (c) $Fe^{3+}$, (b) $Cl_2$
**19-47** All of the substances **19-48** $CrO_4^{2-}(aq)$ and $P_4(s)$
**19-49** (b) Mn **19-50** (a) $MnO_4^-$ in acid
**19-51** (c) $Sn(s) + 2H^+(aq) \longrightarrow Sn^{2+}(aq) + H_2(g)$
(d) $3Na(l) + AlCl_3(l) \longrightarrow 3NaCl(l) + Al(l)$
**19-53** (a) $NO(g)$, $Mn^{2+}(aq)$, $H_2O(l)$
(b) $MgCl_2(s)$ and $Cr(s)$
(c) No reaction
(d) AuF
(e) $O_2$ and $H_2O$
**19-54** (a) $Cr^{2+}$ and $MnO_4^-$ **19-55** (c) $Cr^{2+}$ and $I_3^-$
**19-58** 0.150 M $H_2C_2O_4$ **19-59** 80.1% Cu
**19-60** 66.6% Zn **19-61** 0.47 M $C_2H_2O_4$
**19-62** 6.1 g $FeCl_2$ **19-63** Basic solution
**19-64** 0.2385 M $H_2O_2$

**Chapter 20**

**20-16** (a), (b) **20-17** (a), (b), (c)
**20-18** $Cl_2 > Br_2 > H_2O_2 > I_2$
**20-20** (a), (b) **20-21** (a), (c) **20-23** 0.56 V, yes
**20-24** −1.67 V, no **20-25** 1.93 V, yes
**20-26** 0.364 V, yes **20-27** −0.770 V, no
**20-30** 0.354 V, forward; −1.90 V, reverse
**20-36** (a) **20-37** (b) **20-38** (c)
**20-39** (a) $H^-$ (b) $Fe^{2+}$ **20-43** (a) **20-44** −0.414 V
**20-45** Weaker **20-46** −1.68 V **20-47** 0.35 V
**20-48** (a) 1.03 V (b) 1.00 V (c) 0.914 V (d) 0.855 V (e) 0 V
**20-50** Decrease **20-53** $K = 1.61 \times 10^{37}$
**20-54** $K_{sp} = 5.056 \times 10^{-17}$ **20-55** $K = 1.67 \times 10^7$
**20-56** $K_{sp} = 1.69 \times 10^{-10}$ **20-57** $K_{sp} = 1.4 \times 10^{-18}$
**20-58** $Zn(CN)_4^{2-}$, $K_f = 6.54 \times 10^{16}$; $Zn(NH_3)_4^{2+}$, $K_f = 2.40 \times 10^9$
**20-59** $Ni(NH_3)_6^{2+}$, $K_f = 2.9 \times 10^8$; $Co(NH_3)_6^{2+}$, $K_f = 6.36 \times 10^4$
**20-60** $E° = 0.4594$ V **20-62** $E° = -4.06$ V
**20-63** $E° = -1.36$ V
**20-64** (c) **20-68** (c) **20-71** (b)
**20-73** $1.1180 \times 10^{-3}$ g Ag **20-74** 0.109 kw-h
**20-75** Na = 5.36 h; Mg = 10.72 h; Al = 16.08 h
**20-76** 0.094 g $H_2$; 7.45 g $Br_2$ **20-77** $2.7 \times 10^9$ C
**20-78** 1.79 M $CuSO_4$ **20-79** 8 $O_2$:1 $H_2$
**20-80** 1.18 g $I_2$ **20-82** (e) **20-83** (a)

**20-84** 22.4 g Ag, 6.61 g Cu, 3.87 g Fe
**20-85** $x = 1$, $y = 4$ **20-86** Mn (4+) **20-87** $Au^{3+}$

**Chapter 21**

**21-4** 130.9 J **21-5** 14.2°C
**21-6** −790 J; The temperature will decrease.
**21-7** 3419 J **21-8** 6717 J **21-11** (b) and (e)
**21-17** (e) **21-18** (d) **21-19** (b) **21-20** (c) **21-21** (b)
**21-24** −332.2 J/K **21-25** $-1.54 \times 10^3$ J/K **21-34** (b)
**21-35** (a) $< 0$; (b) $< 0$; (c) $< 0$; (d) $> 0$
**21-36** $\Delta S° < 0$; $\Delta H° < 0$
**21-37** $\Delta H° = 178.4$ kJ; $\Delta S° = 160.6$ J/K; The driving force is entropy.
**21-38** $\Delta H° = -851.5$ kJ; $\Delta S° = -38.6$ J/K The driving force is enthalpy.
**21-39** $\Delta H° = -3310$ kJ; $\Delta S° = -307.7$ J/K The enthalpy is the driving force.
**21-40** $\Delta H°_{rxn} = -602.6$ kJ; $\Delta S°_{rxn} = 579.8$ J/K The major driving force is $\Delta S°$.
**21-41** $\Delta H°_{rxn} = -220.7$ kJ; $\Delta S°_{rxn} = 68.3$ J/K All three are driving forces.
**21-42** None of the reactions is spontaneous.
**21-43** $\Delta H° = 24.9$ kJ; $\Delta S° = 307.7$ J/K
**21-47** $\Delta H° = -34.3$ kJ; $\Delta S° = 75.6$ J/K
**21-48** (c) **21-49** (d) **21-52** (d) **21-53** $-1.060 \times 10^3$ kJ
**21-57** (a) **21-58** $2.20 \times 10^4$ **21-59** $9.29 \times 10^{-5}$
**21-60** $9.75 \times 10^{33}$
**21-61** $\Delta H°_{rxn} = -0.25$ kJ; $\Delta S°_{rxn} = -92.1$ J/K; $\Delta G°_{rxn} = 27.15$ kJ; $K = 1.75 \times 10^{-5}$
**21-62** $\Delta H°_{rxn} = 3.616$ kJ; $\Delta S°_{rxn} = -78.56$ J/K; $\Delta G° = 2.704 \times 10^4$J $K = 1.83 \times 10^{-5}$
**21-63** $K_{a1} = 4.34 \times 10^{-7}$; $K_{a2} = 4.7 \times 10^{-11}$
**21-64** $K_{sp} = 1.99 \times 10^{-53}$ **21-65** $K_{sp} = 1.77 \times 10^{-10}$
**21-66** PbS should be less soluble.
**21-67** 0.113 g LiF/100 g $H_2O$ **21-68** −40.8 kJ/mol
**21-69** $2.89 \times 10^6$ **21-70** 71.18 kJ/mol **21-71** 977 K
**21-74** 516 K **21-75** (c)
**21-77** $\Delta H°_{rxn} = 64.77$ kJ; $\Delta S°_{rxn} = -2.07$ J/K
**21-78** (c) **21-79** −212.8 kJ
**21-80** −88.65 kJ; spontaneous
**21-81** $3.16 \times 10^{-12}$ **21-82** 62.85°C

**Chapter 22**

**22-8** $time^{-1}$ **22-9** $M^{-1}$ $time^{-1}$ **22-10** $M^{-2}$ $time^{-1}$
**22-11** 0.0014 $Ms^{-1}$ **22-12** (e) **22-13** (c)
**22-14** 0.0296 $Ms^{-1}$ **22-15** 0.020 $M^{-1}s^{-1}$
**22-17** 0.0267 $Ms^{-1}$ **22-18** 0.0114 $Ms^{-1}$
**22-25** Rate = $k(NO)^2(H_2)$ **22-26** (c) **22-27** (c)
**22-29** Rate = $k\,K_{eq}\,(N_2O_5)$ **22-32** $K_{eq} = 3.5 \times 10^4$
**22-33** Rate = $k(NO)^2(Cl_2)$ **22-35** Rate = $k(NO)^2(O_2)$
**22-37** Rate = $k(CH_3I)(OH^-)$ **22-38** $3.25 \times 10^{-7}$ $Ms^{-1}$
**22-40** 54.6°C **22-41** 26.38 s **22-43** 12.38 s
**22-44** 24.6 $s^{-1}$ **22-46** $1.1 \times 10^8$ s **22-47** $1.19 \times 10^{10}$ y
**22-48** 0.845 **22-49** 1900 y **22-50** 0.153
**22-51** 23700 y **22-52** 3910 y **22-54** 41811 y
**22-55** $k = 1.49 \times 10^{-3}$ $s^{-1}$ **22-56** $(N_2O) = 0.0274$ M
**22-58** 128 h **22-59** first-order **22-61** first-order
**22-64** first-order **22-70** 218 kJ **22-71** 51.6 kJ
**22-72** 33.3 kJ/mol **22-74** 0.0142 $M^{-1}s^{-1}$
**22-77** 0.0138 mL $s^{-1}$

**Chapter 23**

**23-1** $\alpha = He^{2+}$, $\beta = e^-$, $\gamma$ = electromagnetic radiation
**23-5** (a) 6e 6p 8n
(b) 33e 33p 42n
(c) 38e 38p 52n
**23-6** (a) 18e 19p 21n
(b) 42e 45p 58n
(c) 54e 53p 74n
**23-10** Interconvert isotopes: (e), (f), and (g). Interconvert isobars: (a), (b), and (c). None interconvert isotones.
**23-11** (a) electron capture $^{125}_{52}Te$
(b) nuclear fission $^{94}_{36}Kr$
(c) electron emission $^{90}_{39}Y$
(d) positron emission $^{40}_{18}Ar$
(e) alpha emission $^{224}_{88}Ra$
**23-12** (a) $^{34}_{17}S \longrightarrow {}^{0}_{-1}e + {}^{35}_{17}Cl$
**23-13** (a) $^{183}_{78}Pt \longrightarrow {}^{179}_{76}Os + {}^{4}_{2}He$
**23-14** (a) $^{7}_{4}Be + {}^{0}_{-1}e \longrightarrow {}^{7}_{3}Li$
**23-15** (a) $^{34}_{17}Cl \longrightarrow {}^{34}_{16}S + {}^{0}_{+1}e$
**23-17** (a) $^{209}_{83}Bi$; (b) $^{257}_{104}Unq$; (c) $^{96}_{42}Mo$; (d) $^{220}_{94}Pu$; (e) $^{14}_{6}C$; (f) $^{62}_{30}Zn$
**23-18** (a) $^4He$ $^{238}_{92}U + {}^{4}_{2}He \longrightarrow {}^{239}_{94}Pu + 3{}^{1}_{0}n$
**23-21** (a) $^{14}C$ and (b) $^{24}Na$ **23-22** (b) $^{11}C$ and (e) $^{61}Cu$
**23-24** (a) $^{11}C$ **23-25** (c) $^{13}C$ and (d) $^{14}C$
**23-27** 32.02 MeV/atom; 5.34 MeV/nucleon
**23-28** 524.81 MeV/atom; 8.7 MeV/nucleon
**23-29** 238.0520 amu
**23-30** Largest binding energy per nucleon is $^{60}Ni$. Largest binding energy is $^{238}U$.
**23-31** 5.5 MeV **23-32** 2.79 MeV
**23-33** (a) 14.3 days; (b) 28.6 days; (c) 42.9 days
**23-34** 55.0 min
**23-35** (a) 0.000105 $s^{-1}$
(b) $2.57 \times 10^{-8}$ $s^{-1}$
(c) $1.79 \times 10^{-9}$ $s^{-1}$
(d) $3.84 \times 10^{-12}$ $s^{-1}$
(e) $1.37 \times 10^{-15}$ $s^{-1}$
**23-36** $t_{\frac{1}{2}} = 2.6$ y; $k = 0.27$/y; $t = 9.6$ y **23-37** 36.8 d
**23-38** For $^{210}Po$ $t_{\frac{1}{2}} = 6.25$ d; 1.839 g $^{206}Pb$
**23-39** For 300 years $5.02 \times 10^{-5}$ fraction $^{210}Pb$; for 10 years 0.719 fraction $^{210}Pb$
**23-40** $k = 5.02 \times 10^{-4}$ $min^{-1}$; $t_{\frac{1}{2}} = 1380$ min
**23-41** $1.76 \times 10^4$ atoms/L **23-42** 746 disintegrations/min
**23-43** (a) $1.07 \times 10^{17}$ disintegrations/ s
(b) $1.55 \times 10^{15}$ disintegrations/s
(c) $3.7 \times 10^{12}$ disintegrations/s
(d) $1.26 \times 10^9$ disintegrations/s
**23-44** (a) $8.5 \times 10^4$ Ci (b) 27.3 Ci (c) $6.34 \times 10^{-8}$ Ci
**23-45** $t_{\frac{1}{2}} = 6.68 \times 10^8$ s **23-46** $2.24 \times 10^{-4}$ g $^{14}C$
**23-47** 11100 y **23-48** 23700 y **23-49** 1900 y
**23-50** 0.153 **23-51** 3910 y **23-53** 41600 y
**23-54** 260 kJ/mol **23-55** $1.2 \times 10^5$ kJ/mol
**23-61** $^{216}Po \longrightarrow {}^{212}Pb + {}^{4}He$
$^{214}Pb \longrightarrow {}^{212}Po + 2{}^{0}_{-1}e$
**23-62** $^{253}Es + {}^{4}He \longrightarrow {}^{256}Md + {}^{1}_{0}n$
$^{238}U + {}^{19}F \longrightarrow {}^{252}Md + 5{}^{1}_{0}n$
**23-63** $^{243}Am + {}^{18}O \longrightarrow {}^{256}Lr + 5{}^{1}_{0}n$
$^{256}Lr \longrightarrow {}^{252}Md + {}^{4}He$
$^{256}Lr + {}^{0}_{-1}e \longrightarrow {}^{256}No$
**23-64** $^{238}_{92}U \longrightarrow 8{}^{4}_{2}He + 6{}^{0}_{-1}e + {}^{206}_{82}Pb$
**23-65** $^{232}_{90}Th \longrightarrow {}^{208}_{82}Pb + 6{}^{4}_{2}He + 4{}^{0}_{-1}e$

**Chapter 24**

**24-6** Eight
**24-8** $C_7H_{16}$, 2,3-dimethylpentane
**24-9** $C_{12}H_{26}$, 3-ethyl-3,5-dimethyloctane
**24-10** 3,4,6-trimethyloctane
**24-11** $C_6H_{11}Cl$, 1-chloro-1-methylcyclopentane
**24-13** 1-hexene, 2-hexene, 3-hexene
2-methyl-1-pentene, 2-methyl-2-pentene
4-methyl-2-pentene, 4-methyl-1-pentene
3-methyl-1-pentene, 3-methyl-2-pentene
2-ethyl-1-butene
2,3-dimethyl-1-butene, 2,3-dimethyl-2-butene
3,3-dimethyl-1-butene
**24-14** 1-pentyne, 2-pentyne
3-methyl-1-butyne
**24-19** (e) pentane **24-20** (d) $H_2C = CHF$
**24-21** (b) 2-pentene and (d) 1-chloro-2-butene
**24-22** (a) $CH_3CH_2OCH_2CH_3/CH_3CH_2CH_2CH_2OH$ and
(b) $(CH_3)_2CHCH_3/CH_3CH_2CH_2CH_3$
**24-23** $C_7H_{14}$, 2,3-dimethyl-2-pentene
**24-24** $C_8H_{16}$, 3-ethyl-2-hexene
**24-25** $C_9H_{16}$, 2,5-dimethyl-3-heptyne
**24-26** $2\ C_8H_{18}(g) + 25\ O_2(g) \longrightarrow 16\ CO_2(g) + 18\ H_2O(g)$
**24-28** 2,3-dibromopentane **24-29** 2-butanol
**24-30** pentane **24-31** (c) $ClCH_2CH_2Cl$
**24-32** (a) pentane **24-39** (a) *n*-butane
**24-40** (d) Increasing the average length of the hydrocarbon chains.
**24-45** (a) aldehyde; (b) alcohol; (c) ether; (d) ketone
**24-46** (a) primary; (b) secondary; (c) tertiary; (d) secondary
**24-47** $C_{21}H_{28}O_5$; Functional groups: an alkene, two alcohols, and three ketones.
**24-48** Two ethers, an amide, an alkene, and a benzene ring.
**24-54** diethyl ether **24-62** (c) $NH_4^+$
**24-64** $CH_4 < H_3COH < C$, $H_2CO < HCOOH$, $CO < CO_2$
**24-65** (a) Oxidation–reduction (b) Oxidation–reduction
(c) Oxidation–reduction (d) Metathesis
(e) Oxidation–reduction
**24-66** (a) oxidation; (b) oxidation; (c) reduction;
(d) reduction; (e) reduction
**24-67** (a) $CH_3CH_2OH$ **24-68** (c) $(CH_3)_3COH$
**24-69** (b) $CH_3CH_2OH$
**24-70** (c) $CH_3CO_2H$ **24-71** 2-methyl-3-pentanone
**24-79** Nicotine: tertiary amine
Coniine: secondary amine
Strychnine: tertiary amine and cyclic amide
Lysergic acid dimethyl amide: tertiary amine and amide
The remainder: tertiary amine
**24-83** 3-methyl-3-hexanol
**24-85** Combination of (a) $(CH_3)_2CHMgBr$ and
(b) $CH_3CHO$
**24-87** (e) CHFClBr **24-88** (d) $CH_3CH_2CH(OH)CH_3$
**24-89** yes **24-90** no **24-91** two
**24-100** (a) positive, (b) positive, (c) neutral
**24-102** tartaric acid and malic acid **24-103** chiral
**24-104** both isocitric and homocitric acid

**Chapter 25**

**25-1**

| Group A | Group B |
|---|---|
| (b) ethylene | (a) formaldehyde |
| (d) 2-butene | (c) glyceraldehyde |
| (e) cyclohexane | (f) glucose |

**25-2** All are polymers except (h) lipids.
**25-8** 14,000 amu to 1,400,000 amu
**25-9** If 200,000 amu, 1923 monomers/chain; if 300,000 amu, 2885 monomers/chain
**25-10** 207 amu
**25-14** (a) addition
(b) addition
(c) condensation
(d) condensation
(e) condensation
(f) addition
(g) addition
**25-27** (a) $(—OCH_2CH_2OCOC_6H_4CO—)_n$
(b) $(—NH(CH_2)_6—NHCO—(CH_2)_4CO—)_n$
(c) $(—C_6H_5—(CH_3)_2CC_6H_5—OCOO—)_n$
(d) $(—NH—C_6H_5—NH—CO—C_6H_5—CO—)_n$
(e) $(—O(CH_3)_2—Si)_n$
**25-31** In strongly acidic solution, (a) $+NH_3CH_2CO_2H$
In strongly basic solution, (c) $H_2NCH_2CO_2^-$
**25-34** (d) serine; (e) lysine
**25-35** (a) hydrophobic; (b) hydrophilic; (c) hydrophilic;
(d) hydrophobic; (e) hydrophilic
**25-38** ALA-CYS and CYS-ALA
**25-39** ALA-CYS-ALA, CYS-ALA-CYS
**24-42** (1) Hydrophobic interactions
(2) Hydrogen bonding
(3) Disulfide bonds
(4) Ionic bonding
**25-43** arginine, asparagine, cysteine, glutamine, histidine, lysine, serine, threonine, tyrosine
**25-44** arginine, aspartic acid, histidine, glutamic acid, lysine
**25-45** alanine, valine, isoleucine, leucine, methionine, proline, phenylalanine, tryptophan
**25-47** (a), (d), (e) and (f) and carbohydrates according to the definition.
**25-53** Glucitol is not an aldose.
**25-59** For glucose, the average oxidation state of the carbons is 0 (zero).
For palmitic acid, the average oxidation state is −1.33.
Fatty acids have carbons with a greater capacity to be oxidized.
**25-60** For glucose: 15.9 kJ/g; for palmitic acid: 38.2 kJ/g
**25-61** (a) butyric acid and (b) palmitic acid
**25-64** Purines: adenine and guanine
Pyrimidines: cytosine, thymine and uracil
**25-68** mRNA carries the message (code). tRNA's carry the needed amino acids.
**25-69** 5''-A-U-G-U-U-C-G-U-C-A-A-C-C-A-G-C-A-C-3''
N terminal: MET-PHE-VAL-ASN-GLN-HIS

# GLOSSARY

**Absolute zero** The temperature ($-273.15°C$ or 0 K) at which the volume and pressure of an ideal gas extrapolate to zero.

**Absorption spectrum** The spectrum of dark lines against a light background that results from the absorption of selected frequencies of electromagnetic radiation by an atom or molecule.

**Accuracy** The extent to which a measurement agrees with the true value of the quantity being measured.

**Acid** See **Brønsted acid** or **Lewis acid.**

**Acid–base complex** The product of the reaction between a Lewis acid and a Lewis base. This complex contains a covalent bond formed by donating a pair of nonbonding electrons from the Lewis base to the Lewis acid.

**Acid–base indicator** A weak acid or weak base, such as litmus or phenolphthalein, which changes color when it gains or loses an $H^+$ ion.

**Acid-dissociation equilibrium constant ($K_a$)** A measure of the relative strength of an acid. For a generic acid with the formula HA, $K_a$ can be calculated as:

$$K_a = \frac{[H_3O^+][A^-]}{[HA]}$$

**Acidic buffer** A mixture of a weak acid and its conjugate base that has a pH less than 7 that can resist changes in pH when relatively small amounts of acid or base are added.

**Activation energy** The energy that must be provided to the reactants in a chemical reaction to reach an intermediate or activated state from which the products of the reaction can form.

**Active metal** A metal, such as sodium or potassium, that is unusually reactive.

**Activity** A measure of the number of disintegrations of a radioactive nuclide per unit of time; reported in units of curies or becquerels.

**Acyl chloride** A compound in which a chlorine atom is bound to the carbon atom of a C=O (carbonyl) group, e.g., $CH_3COCl$.

**Addition polymer** A polymer formed without the loss of any atoms in the monomer. Polyethylene $(—CH_2CH_2—)_n$ is an addition polymer of ethylene ($H_2C=CH_2$).

**Addition reaction** A reaction in which a molecule adds across a C=C or C=O double bond.

**Adhesion** The force of attraction between different substances, such as glass and water. See **cohesion.**

**Alcohol** Compounds with an —OH group attached to a carbon atom, such as $CH_3OH$.

**Aldehyde** Literally, an *al*cohol that has been *dehyd*rogenated. Compounds with at least one hydrogen atom attached to a C=O (carbonyl) group, such as formaldehyde ($H_2CO$) or acetaldehyde ($CH_3CHO$)

**Algorithm** A set of rules for calculating something that can be taught to a reasonably intelligent system, such as a computer.

**Alkali** Historically, a compound that neutralizes acids. Now known as a **base.**

**Alkali metal** A metal in Group IA, such as Li, Na, K, and so on.

**Alkaline earth metal** A metal in Group IIA, such as Be, Mg, Ca, and so on.

**Alkaloid** A class of organic compounds that contains nitrogen, isolated from plants.

**Alkane** A hydrocarbon with the generic formula $C_nH_{2n+2}$ that contains only C—C and C—H bonds.

**Alkene** An unsaturated hydrocarbon that contains one or more C=C double bonds.

**Alkoxide** The conjugate base of an alcohol; for example, the $CH_3O^-$ ion.

**Alkyl halide** A derivative of an alkane in which a hydrogen atom has been replaced by a halogen, such as $CH_3Cl$.

**Alkyne** A hydrocarbon that contains one or more C≡C triple bonds.

**Allotropes** Forms of an element with different structures and therefore different chemical and physical properties, such as $O_2$ and $O_3$.

**Alloy** A mixture of two or more elements that acts like a metal. Bronze, for example, is an alloy of copper and tin.

**Alpha particle** A positively charged particle consisting of two protons and two neutrons emitted by one of the radioactive elements. An alpha particle is equivalent to an $He^{2+}$ ion.

**Amide** A compound that can be thought of as the product of the reaction between a carboxylic acid ($RCO_2H$) and an amine, such as $CH_3CONH_2$. This term is also used to describe the $NH_2^-$ ion.

**Amine** Compound with an $—NH_2$, —NHR, or $—NR_2$ substituent attached to a carbon atom.

**Amino acid** One of the essential building blocks from which proteins are made. Amino acids have the generic formula $H_3N^+CHRCO_2^-$.

**Amonton's law** A statement of the relationship between the temperature and pressure of a constant amount of gas at constant volume: $P \propto T$.

**Amorphous** Literally, without form or shape. Used to describe substances that are solids but not crystals.

**Amphoteric** Used to describe a compound, such as $H_2O$ or $Al(OH)_3$, that can act as either an acid or a base.

**Amplitude** The height of a wave. The difference between the center of gravity of the wave and the highest (or lowest) point on the wave.

**Amu** Atomic mass unit. The unit in which the relative masses of atoms are expressed.

**Angular momentum** For a particle in a spherical orbit, the product of the mass of the particle times its velocity times the radius of the orbit.

**Angular quantum number ($l$)** The quantum number used to describe the shape of an atomic orbital.

**Anhydrous** Literally, without water. Used, for example, to differentiate between liquid (anhydrous) ammonia at temperatures below its boiling point (−33°C) and solutions of ammonia dissolved in water.

**Anion** A negatively charged ion, such as the $Cl^-$ ion.

**Anode** The positive end of an electric field. Used to describe the electrode in an electrochemical cell toward which anions flow and the electrode at which oxidation occurs.

**Antibonding molecular orbital** A molecular orbital in which electrons are held in a region of space that does not lie between the atoms.

**Approximation methods** A technique for obtaining approximate answers to calculations that are difficult, if not impossible, to solve exactly.

**Aqueous** Literally, watery. Used to describe solutions of substances dissolved in water.

**Aromatic compound** A compound, such as benzene ($C_6H_6$), that seems to contain C=C double bonds but does not react as an alkene does.

**Arrhenius acid** A substance that dissociates when it dissolves in water to give the $H^+$ ion.

**Arrhenius base** A substance that dissociates when it dissolves in water to give the $OH^-$ ion.

**Arrhenius equation** Describes the relationship between the rate constant for a chemical reaction and the temperature at which the reaction is run.

**Atactic** Used to describe polymers in which the side chains are randomly distributed on either side of the polymer backbone.

**Atom** The smallest particle of an element that retains any of the properties of the element.

**Atomic mass unit** See **amu.**

**Atomic number ($Z$)** The number of protons in the nucleus of an atom.

**Atomic orbital** A region in space where electrons on an atom can be found.

**Atomic weight** The weighted average of the atomic masses of the different isotopes of an element. A single $^{12}C$ atom, for example, has a mass of 12 amu, but naturally occurring carbon also contains a 1.1% $^{13}C$. The atomic weight of carbon is therefore 12.011 amu.

**Aufbau principle** The principle that atomic orbitals are filled one at a time, starting with the orbital that has the lowest energy.

**Avogadro's number** The number of particles in a mole of these particles: $6.0220 \times 10^{23}$.

**Avogadro's hypothesis** The hypothesis that equal volumes of different gases at the same temperature and pressure contain the same number of particles.

**Axial** Describes the two positions in a **trigonal bipyramid** that lie above and below the trigonal plane. See **equatorial.**

**Balanced equation** A symbolic representation of a chemical reaction in which both sides of the equation contain equivalent numbers of atoms of each element. Charge and mass are both conserved in a balanced equation.

**Band of stability** A narrow band of neutron-to-proton ratios that correspond to stable nuclides.

**Base** See **Brønsted base** or **Lewis base.**

**Base-ionization equilibrium constant ($K_b$)** A measure of the relative strength of a base. For the generic base B, $K_b$ can be calculated as:

$$K_b = \frac{[BH^+][OH^-]}{[B]}$$

**Basic buffer** A mixture of a weak acid and its conjugate base that has a pH less than 7 that can resist changes in pH when relatively small amounts of acid or base are added.

**Battery** Literally, a set of similar items, such as an artillery battery. Used to describe a battery of electrochemical cells connected in series.

**Beta decay** A nuclear reaction in which beta particles (electrons, $\beta^-$, or positrons, $\beta^+$) are absorbed by or emitted from the nucleus of an atom.

**Bidentate** A ligand that binds twice. See **chelating ligand.**

**Bimolecular** A step in a chemical reaction in which two molecules are consumed.

**Binding energy** The energy released when the nucleus of an atom is formed by combining neutrons and protons.

**Body-centered cubic** A structure in which the simplest repeating unit consists of nine equivalent lattice points, eight of which are at the corners of a cube and the ninth of which is in the center of the body of the cube.

**Bohr model** A model of the distribution of electrons in an atom based on the assumption that the electron in a hydrogen atom is in one of a limited number of circular orbits.

**Boiling point** The temperature at which the vapor pressure of a liquid is equal to the pressure on the liquid.

**Boiling point elevation** The increase in the boiling point of a solvent that occurs when a solute is added to form a solution.

**Bond-dissociation enthalpy** The energy needed to break an X—Y bond to give X and Y atoms in the gas phase.

**Bond order** The number of bonds between a pair of atoms.

**Bonding electrons** A pair of electrons used to form a covalent bond between adjacent atoms.

**Bonding molecular orbital** A molecular orbital in which the electrons reside in the region in space between the atoms.

**Boundary** In thermodynamics, the boundary separates the system from its surroundings.

**Boyle's law** A statement of the relationship between the pressure and volume of a constant amount of gas at constant temperature: $P \propto 1/V$.

**Bragg equation** A statement of the relationship between the wavelength of the incident x-ray ($\lambda$), the distance between adjacent planes of atoms in a crystal ($d$), and the angle ($\Theta$) between the plane of the incident radiation and the uppermost plane of atoms in a crystal being subjected to x-ray diffraction analysis.

**Breeder reactor** A nuclear reactor that produces more fuel than it consumes.

**British thermal unit** See **Btu.**

**Bromide** Any compound that contains either the $Br^-$ ion or bromine with an oxidation state of $-1$, such as NaBr or HBr.

**Brønsted acid** Any substance that can donate an $H^+$ ion to a base. Brønsted acids are $H^+$-ion or proton donors.

**Brønsted base** Any substance that can accept an $H^+$ ion from an acid. Brønsted bases are $H^+$-ion or proton acceptors.

**Btu (British thermal unit)** The heat needed to raise the temperature of 1 lb of water by 1°F.

**Buffer** A mixture of a weak acid (HA) and its conjugate base ($A^-$) or a weak base (B) and its conjugate acid ($BH^+$). Buffers resist a change in the pH of a solution when small amounts of acid or base are added.

**Buffer capacity** The amount of acid or base a buffer solution can absorb without significant changes in pH.

**Calcining** A process in which an ore loses a gas while being heated.

**Caloric theory** An obsolete theory, which assumed that heat (or caloric) is a fluid that is conserved.

**Calorie** The heat needed to raise the temperature of 1 g of water by 1°C from 14.5°C to 15.5°C.

**Calorimeter** An apparatus used to measure the heat given off or absorbed in a chemical reaction.

**Calx** The powder formed when metals react with air.

**Canal rays** The positively charged particles formed when electrons are removed from the gas particles in a cathode-ray tube. Because they carry a positive charge, canal rays move in the opposite direction from cathode rays.

**$^{14}C$ dating** A technique for estimating the age of a sample that assumes that the $^{14}C$ isotope in the sample decays with a half-life of 5730 years.

**Carbanion** A compound that contains a negatively charged carbon atom, such as the $CH_3^-$ ion.

**Carbide** A compound that contains a negatively charged carbon atom or carbon in a negative oxidation state, such as calcium carbide ($CaC_2$)

**Carbohydrate** Literally, a hydrate of carbon. Originally defined as any compound with an empirical formula of $CH_2O$; now defined as polyhydroxy aldehydes or ketones. Includes starches and sugars. See **monosaccharide** and **polysaccharide.**

**Carbonium ion** A compound that contains a positively charged carbon atom, such as the $CH_3^+$ ion.

**Carbonyl** In organic chemistry, the C=O functional group. In inorganic chemistry, a complex formed when carbon monoxide (CO) binds to a metal.

**Carboxylate ion** The conjugate base of a carboxylic acid, such as the $CH_3CO_2^-$ ion formed when acetic acid loses an $H^+$ ion.

**Carboxylic acid** A compound that contains the $—CO_2H$ functional group.

**Carboxylic acid ester** A compound that can be thought of as the product of the reaction between a carboxylic acid ($RCO_2H$) and an alcohol (R′OH). Any compound that contains the $RCO_2R'$ functional group.

**Catalyst** A substance that increases the rate of a chemical reaction without being consumed in the reaction. A substance that lowers the activation energy for a chemical reaction by providing an alternate pathway for the reaction.

**Catenation** The tendency of an element to form bonds to itself.

**Cathode** The negative end of an electric field. The electrode in an electrochemical cell toward which cations flow and the electrode at which reduction occurs.

**Cathode rays** The negatively charged particles (now recognized as electrons) that travel from the cathode toward the anode in a cathode-ray tube.

**Cathodic protection** The process in which a structural metal, such as iron, is protected from corrosion by connecting it to a metal that has a more negative reduction half-cell potential.

**Cation** A positively charged ion, such as the $Mg^{2+}$ ion.

**Cell potential** A measure of the driving force behind an electrochemical reaction, which is reported in units of volts.

**Chain reaction** A reaction in which one of the starting materials is regenerated in the last step of the reaction.

**Chain-reaction mechanism** A mechanism for a chain reaction that consists of chain-initiating, chain-propagating, and chain-terminating steps.

**Charles' law** A statement of the relationship between the temperature and volume of a constant amount of gas at constant pressure: $V \propto T$.

**Chelating ligand** A ligand that can coordinate to a metal atom more than once.

**Chemical equation** A symbolic representation of the relationship between the reactants and the products of a chemical reaction.

**Chemical kinetics** The study of the rates of chemical reactions.

**Chemiluminescence** A chemical reaction that gives off energy in the form of light instead of heat.

**Chiral** An object whose mirror image is not the same as itself. Compounds with four different substituents on a carbon atom are chiral.

**Chloride** A compound that contains either the $Cl^-$ ion or chlorine with an oxidation state of $-1$, such as HCl.

**Chromatography** A technique for separating the components of a mixture on the basis of differences in their affinity for a stationary and a mobile phase.

**Cis** Literally, on the same side. Describes isomers in which similar substituents are on the same side of a C=C double bond or in adjacent coordination sites on a transition metal. See **trans.**

**Closest packed** The structure in which equivalent particles pack as tightly as possible. Each particle that forms the closest packed structure is surrounded by six nearest neighbors in the same plane arranged toward the corners of a hexagon, three neighbors in the plane above, and three neighbors in the plane below.

**Coal gas** A gas, which is usually rich in methane ($CH_4$), produced when coal is heated in the absence of air.

**Codon** A sequence of three nucleotides on a strand of mRNA that codes for an amino acid.

**Cohesion** The force of attraction between molecules of the same substance. See **adhesion.**

**Colligative property** Any property that depends on the number of solute particles in a solution but not their identity.

**Collision theory model** A model used to explain the rates of chemical reactions, which assumes that molecules must collide in order to react.

**Column chromatography** Chromatography that uses a solid support in a vertical column or tube.

**Combined equilibria** Two or more equilibria that occur simultaneously and involve the same ion or molecule.

**Common-ion effect** The decrease in the solubility of a salt that occurs when the salt is dissolved in a solution that contains another source of one of its ions.

**Complex-dissociation equilibrium constant ($K_d$)** The equilibrium constant for the reaction in which a complex dissociates, such as:

$$Cu(NH_3)_4{}^{2+}(aq) \rightleftharpoons Cu^{2+}(aq) + 4NH_3(aq)$$

**Complex-formation equilibrium constant ($K_f$)** The equilibrium constant for the reaction in which a complex is formed, such as:

$$Cu^{2+}(aq) + 4NH_3(aq) \rightleftharpoons Cu(NH_3)_4{}^{2+}(aq)$$

**Complex ion** An ion in which a ligand is covalently bound to a metal. An ion formed when a Lewis acid such as the $Cu^{2+}$ ion reacts with a Lewis base such as $NH_3$ to form an acid–base complex such as the $Cu(NH_3)_4{}^{2+}$ ion.

**Compound** A substance with a constant composition that contains two or more elements.

**Compressibility** The ability to be compressed. A characteristic of substances, such as gases, that can be compressed to fit into smaller containers.

**Concentration** A measure of the ratio of the amount of solute in a solution to the amount of either solvent or solution. Frequently expressed in units of moles of solute per liter of solution. See **molarity.**

**Condensation polymer** A polymer, such as nylon, that is formed when a small molecule is condensed out or lost during polymerization. See **addition polymer.**

**Conduction band** An energy band in a solid in which electrons are free to move, producing a net transfer of charge.

**Conductivity apparatus** An instrument used to determine whether a substance or solution can conduct an electric current.

**Conjugate acid–base pair** Two substances related by the gain or loss of a proton. Every Brønsted acid has a conjugate Brønsted base. An acid (such as HCl) and its conjugate base (the $Cl^-$ ion), or a base (the $OH^-$ ion) and its conjugate acid ($H_2O$) represent a conjugate acid–base pair.

**Conjugate oxidizing and reducing agents** Two substances related by the gain or loss of electrons. Every oxidizing agent (such as the $Zn^{2+}$ ion) has a conjugate reducing agent (such as zinc metal).

**Constitutional isomers** Two compounds that have the same formula but different constituents. Dimethyl ether ($CH_3OCH_3$) and ethanol ($CH_3CH_2OH$), for example, are constitutional isomers. See **stereoisomers.**

**Coordination compound** A compound in which one or more ligands are coordinated to a metal atom.

**Coordination number** The number of atoms, ions, or molecules to which bonds can be formed.

**Copolymer** A polymer formed from two or more different monomers.

**Corrosion** A process in which a metal is destroyed by a chemical reaction. When the metal is iron, the process is called rusting.

**Covalent bond** A bond between two atoms formed by the sharing of a pair of electrons.

**Covalent compound** A compound, such as water ($H_2O$), composed of neutral molecules in which the atoms are held together by covalent bonds.

**Covalent radius** The radius of an atom in a covalent bond.

**Covalent solid** A solid, such as diamond, in which every atom is covalently bound to its nearest neighbors to form an extended array of atoms rather than individual molecules.

**Critical mass** Spontaneous fission in uranium produces neutrons that induce fission of $^{235}U$, which releases neutrons that can produce a chain reaction. The critical mass is the amount of fissable material necessary for this chain reaction to sustain itself.

**Critical point** The temperature and pressure at which two phases of a substance in equilibrium become identical, forming a single phase.

**Crystal** A three-dimensional solid formed by regular repetition of the packing of atoms, ions, or molecules.

**Crystalline** Used to describe materials that behave as if they are made of regular crystals.

**Crystal-field theory** An extension of the valence-bond theory used to explain transition metal compounds.

**Cubic closest packed** A structure formed by the stacking of closest packed planes of atoms in an *ABCABC* . . . repeating pattern. See **hexagonal closest packed.**

**Cubic hole** A hole in a simple cubic structure, which is surrounded by eight atoms or ions arranged toward the corners of a cube.

**Curie** A unit for measuring the activity of a radioactive nuclide. By definition, 1 $C_i = 3.700 \times 10^{10}$ disintegrations per second.

**Cyclic hydrocarbon** See **cycloalkane.**

**Cycloalkane** An alkane that contains a ring of carbon atoms.

**Dalton's law of partial pressures** A statement of the relationship between the total pressure of a mixture of gases and the partial pressures of the individual components: $P_{total} = P_1 + P_2 + P_3 + \ldots$.

**deBroglie equation** Describes the relationship between the wavelength and momentum of an object. See **wave–particle duality.**

**Degenerate orbitals** Orbitals that have the same energy, such as the three $2p$ atomic orbitals on an isolated atom.

**Dehydrating agent** A reagent, such as $P_4O_{10}$, used to remove water.

**Denaturation** A process that changes the three-dimensional structure of a protein.

**Density** An intensive property of a substance equal to the mass of a sample divided by the volume of the sample.

**Detergent** A synthetic analog of soap that contains a long, hydrophobic tail attached to a hydrophilic $—SO_3^-$ or $—OSO_3^-$ head.

**Dextrorotatory** A compound that rotates plane-polarized light to the right (clockwise) when viewed in the direction of the light source. See **levorotatory.**

**Diamagnetic** A substance repelled by both poles of a magnet. A substance in which the electrons are all paired. See **paramagnetic.**

**Diatomic molecule** A molecule, such as $H_2$ or HCl, which contains two atoms.

**Diffusion** The movement of atoms, ions, or molecules through a gas, liquid, or solid.

**Dilution** The process by which more solvent is added to decrease the concentration of a solution.

**Dimensional analysis** An approach to solving problems that focuses on the way the units of the problem are used to set up the problem.

**Dimer** Literally, two parts. A compound, such as $N_2O_4$, produced by combining two smaller molecules, such as $NO_2$.

**Dipole** Anything with two equal but opposite electrical charges, such as the positive and negative ends of a polar bond or molecule.

**Diprotic acid** An acid, such as $H_2SO_4$, that has the potential to lose two $H^+$ ions.

**Diprotic base** A base, such as the $S^{2-}$ ion, that can pick up two $H^+$ ions.

**Disorder** A measure of the extent to which a system differs from a perfect crystal at 0 K, where there is no disorder.

**Disproportionation** A reaction in which an element or compound simultaneously undergoes both oxidation and reduction, such as the decomposition of $H_2O_2$ to form $H_2O$ and $O_2$.

**Disaccharide** A carbohydrate formed by linking a pair of monosaccharides.

**Dissociation** The process by which salts dissolve in water to give solutions that contain the corresponding ions.

**Dissolve** Literally, to loosen. Used to describe the process in which one substance mixes with another. When a solid dissolves in a liquid, the particles that form the solid are released into solution.

**Distillation** A technique used to separate liquids with different boiling points.

**Disulfide linkage** The —S—S— linkage that can form between the —SH side chains on adjacent cysteine residues in a protein.

**DNA** Deoxyribonucleic acid, the nucleic acid used to store the genetic information that codes for the synthesis of proteins.

**Ductile** Capable of being drawn into thin sheets or wires without breaking.

**Effusion** The process by which a gas escapes through a pinhole into a vacuum.

**Elastic collision** A collision in which no kinetic energy is lost.

**Elastomer** A polymer that snaps back to its original shape after being stretched to at least twice its original length.

**Electrolysis** A process in which an electric current is used to decompose a compound into its elements.

**Electrolytic cell** An electrochemical cell in which electrolysis is done.

**Electromagnetic radiation** Radiation (such as radiowaves, microwaves, infrared rays, light, ultraviolet rays, x-rays, or $\gamma$-rays) that contains both electric and magnetic components and travels at the speed of light.

**Electron** A subatomic particle with a charge of $-1$ and a mass of roughly 0.0005 amu.

**Electron affinity (*EA*)** The energy given off when a neutral atom in the gas phase picks up an electron to form a negatively charged ion.

**Electron capture** A reaction in which the nucleus of an atom captures a $1s$ electron.

**Electron configuration** The arrangement of electrons in atomic orbitals; for example, $1s^2\ 2s^2\ 2p^3$.

**Electron emission** A nuclear reaction in which electrons, or $\beta^-$ particles, are ejected from the nucleus of an atom. See **beta decay.**

**Electron-pair acceptor** See **Lewis acid.**

**Electron-pair donor** See **Lewis base.**

**Electronegativity** The tendency of an atom to draw the electrons in a bond toward itself.

**Electrophile** Literally, something that loves electrons. A Lewis acid that attacks a site rich in electron density.

**Element** A substance that cannot be decomposed into a simpler substance by a chemical reaction. A substance composed of only one kind of atom.

**Elemental analysis** The process by which the percent-by-mass of the elements in a compound is determined.

**Emission spectrum** The spectrum of bright lines against a dark background obtained when an atom or molecule emits radiation when excited by heat or an electric discharge.

**Empirical formula** The simplest formula for a compound. The ratio of the number of atoms of each element in the compound.

**Empirical weight** The weight of the empirical formula, calculated from a table of atomic weights.

**Endergonic** A process that leads to an increase in the free energy of a system and is therefore not spontaneous.

**Endothermic** A process in which the system absorbs heat from the surroundings. See **exothermic.**

**Endpoint** The point at which the indicator of an acid–base titration changes color. See **equivalence point.**

**Enthalpy (*H*)** The sum of the internal energy plus the product of the pressure times the volume of the gas in a system: $H = E + PV$.

**Enthalpy of formation ($\Delta H_f$)** The change in the enthalpy that occurs during a chemical reaction that leads to the formation of a compound from its elements in their most thermodynamically stable states at 1 MPa.

**Enthalpy of reaction ($\Delta H$)** The change in the enthalpy that occurs during a chemical reaction. The difference between the sum of the enthalpies of the reactants and the products of the reaction.

**Entropy (*S*)** A measure of the disorder in a system.

**Enzyme** A protein that catalyzes a biochemical reaction.

**Equality** A symbolic representation of two quantities that are equivalent. For example, 12 inches = 1 foot.

**Equation** A symbolic statement that can be used to do a calculation. For example, $x = 12y$, where $x$ is the number of inches in $y$ feet.

**Equation of state** An equation, such as $PV = nRT$, which relates two or more of the quantities that describe the state of a system.

**Equatorial** The three positions in a **trigonal bipyramid** that lie in the trigonal plane. See **axial.**

**Equilibrium** The point at which there is no longer a change in the concentrations of the reactants and the products of a chemical reaction. The point at which the rates of the forward and the reverse reactions are equal.

**Equilibrium constant ($K_c$ or $K_p$)** The product of the concentrations (or partial pressures) of the products of a reaction divided by the product of the concentrations (or partial pressures) of the reactants.

**Equilibrium constant expression** The expression used to calculate the equilibrium constant for a reaction.

**Equilibrium region** The portion of a plot of the concentration of a substance versus time in which the concentration does not change. The portion of this plot in which the reaction is at equilibrium.

**Equivalence point** The point in an acid–base titration at which equivalent amounts of acid and base have been added to the solution. See **endpoint.**

**Error** The difference between a measurement and the true value of the quantity being measured.

**Ester** See **carboxylic acid ester.**

**Ether** A compound in which an oxygen atom is attached to two carbon atoms, such as diethyl ether, $CH_3CH_2OCH_2CH_3$.

**Excess reagent** In a limiting reagent problem, this is the reactant present in excess. The reaction will stop before all of the excess reagent is consumed.

**Excluded volume** The fraction of the volume of a gas that is not empty space. The volume of the gas actually occupied by gas particles. At room temperature and atmosphere pressure, the excluded volume of a gas is approximately 0.12% of the total volume.

**Exergonic** A process that leads to a decrease in the free energy of the system and is therefore spontaneous.

**Exothermic** A process in which a system gives off heat to the surroundings. See **endothermic.**

**Expandibility** A measure of the ability to expand. A characteristic of substances, such as gases, that can expand to fill their containers.

**Extensive property** A quantity that depends on the size of the sample, such as mass, weight, length, height, and width. See **intensive property.**

**Extraction** A method of separating mixtures based on differences in the solubility of their components in polar versus nonpolar solvents.

**Face-centered cubic** A structure in which the simplest repeating unit consists of fourteen equivalent lattice points, eight of which are at the corners of a cube and another six in the centers of the faces of the cube. Found in cubic closest packed structures.

**Family** A vertical column of elements in the periodic table, such as the elements H, Li, Na, K, and so on.

**Faraday's constant** The charge on a mole of electrons: 96,484.56 C.

**Faraday's law** A statement of the relationship between the amount of product formed during electrolysis and the amount of electric current that passes through the electrolytic cell.

**Fat** A solid triester of glycerol and fatty acids. See **oil.**

**Fatty acid** A carboxylic acid that contains a long, hydrophobic hydrocarbon chain.

**Fertile nuclide** A nuclide that can be converted into one that undergoes spontaneous fission, such as $^{238}U$, which can be converted into $^{239}Pu$.

**Filled-shell configuration** An electron configuration in which a shell of atomic orbitals is filled, such as $1s^2\ 2s^2\ 2p^6$.

**First ionization energy** The energy needed to remove the outermost, or highest energy, electron from a neutral atom in the gas phase.

**First law of thermodynamics** A statement of the relationship between the internal energy of a system and the heat and work transferred from the system to its surroundings, or vice versa: $\Delta E_{sys} = q + w$.

**First-order reaction** A reaction whose rate is proportional to the concentration of a single reactant raised to the first power: rate $= k(X)$

**Fission** A nuclear reaction in which a nuclide splits into two smaller nuclides. See **fusion.**

**Fluoresce** A process in which a compound emits light at one wavelength while being excited by radiation with a shorter wavelength.

**Fluoride** Any compound that contains either the $F^-$ ion or fluorine with an oxidation state of $-1$, such as $CaF_2$.

**Force** The product of the mass of an object times its acceleration.

**Formal charge** The charge on an atom in its Lewis structure. Formal charge is calculated by dividing the electrons in each covalent bond between the atoms in the bond, and then comparing the number of electrons that can be formally assigned to each atom with the number of electrons on a neutral atom of the element.

**Free energy ($G$)** The energy associated with a chemical reaction that can be used to do work. The free energy of a system is the sum of its enthalpy plus the product of the temperature times the entropy of the system: $G = H - TS$.

**Free radical** A neutral atom or molecule that contains an unpaired electron.

**Freezing point** The temperature at which the solid and liquid phases of a substance are in equilibrium at atmospheric pressure. See **melting point.**

**Freezing point depression** The decrease in the freezing point of a solvent that occurs when a solute is added to form a solution.

**Frequency** The number of wave crests or troughs that pass a fixed point per unit time.

**Functional group** An atom or group of atoms in an organic compound that gives the compound some of its characteristic properties, such as the C=O functional group in aldehydes and ketones.

**Fusion** (1) The formation of heavier nuclides by the fusing of two light nuclides. (2) The melting of a solid to form a liquid. See **latent heat of fusion.**

**Galvanic cell** An electrochemical cell that uses a spontaneous chemical reaction to do work. Also known as a **voltaic cell.**

**Galvanic corrosion** Corrosion that occurs when two metals are in contact with each other and with water.

**Gamma ray ($\gamma$)** A high-energy, short-wavelength form of electromagnetic radiation emitted by the nucleus of an atom that carries off some of the energy released in a nuclear reaction.

**Gas** A substance that flows freely, expands to fill its container, and can be compressed to fit into a smaller container.

**Gas-phase chromatography** Chromatography in which the components of a gas are separated as they pass over a solid support.

**Gas-phase reaction** A reaction in which the reactants and products are all gases.

**Gay-Lussac's law** A statement of the fact that the ratio of the volumes of gases consumed or produced in a gas-phase reaction is equal to the ratio of two simple whole numbers. Also known as the **law of combining volumes.**

**Gibbs free energy** The thermodynamic function defined by the equation: $G = H - TS$. See **free energy.**

**Graham's law** The relationship between the rate at which a gas diffuses or effuses and its molecular weight: rate $\propto (MW)^{\frac{1}{2}}$

**Gram** The basic unit of mass in the metric system. A penny weighs roughly 2.5 grams.

**Grignard reagent** An alkylmagnesium halide, such as $CH_3MgBr$. A source of a carbanion (such as the $CH_3^-$ ion) for use in organic synthesis.

**Group** A vertical column, or family, of elements in the periodic table.

**Group number** A number that identifies a group of elements in the periodic table. Until recently, groups were labeled IA, IIA, and so on. A new system has been proposed in which the columns or groups of the periodic table are numbered 1, 2, 3 . . . sequentially from left to right.

**Haber process** The industrial process used to make $NH_3$ from $N_2$ and $H_2$.

**Half-cell** One half of a voltaic cell.

**Half-life** The time required for the amount of a reactant to decrease to half its initial value.

**Half-reaction** The reaction that takes place in a half-cell.

**Halide** A $F^-$, $Cl^-$, $Br^-$, or $I^-$ ion.

**Halogen** $F_2$, $Cl_2$, $Br_2$, or $I_2$.

**Hard water** Water with a high concentration of the $Ca^{2+}$, $Mg^{2+}$, and/or $Fe^{3+}$ ions.

**Heat (*q*)** A form of energy associated with the random motion of the elementary particles in matter.

**Heat of fusion** The heat that must be absorbed to melt a mole of a solid.

**Heat of reaction** The change in the enthalpy of the system that occurs when a reaction is run at constant pressure.

**Heat of vaporization** The heat that must be absorbed to boil a mole of a liquid.

**Heat capacity** The amount of heat required to raise the temperature of a defined amount of a pure substance by one degree. See **Btu, calorie, molar heat capacity,** and **specific heat.**

**Hess's law** A law stating that the heat given off or absorbed in a chemical reaction is the same regardless of whether the reaction occurs in a single step or in many steps.

**Hexadentate** Used to describe a ligand that can bind to a transition metal six times. See **chelating ligand.**

**Hexagonal closest packed** A structure formed by the stacking of closest packed planes of atoms in an ABABAB . . . repeating pattern. See **cubic closest packed.**

**High-spin complex** A transition metal complex in which the difference between the energies of the $t_{2g}$ and $e_g$ sets of orbitals is smaller than the energy it takes to pair two electrons. As a result, the valence-shell *d* electrons on the metal are placed in both the $t_{2g}$ and $e_g$ sets of orbitals. See **low-spin complex.**

**Homonuclear diatomic molecule** A molecule, such as $O_2$ or $F_2$, that contains two atoms of the same element.

**Hund's rules** Rules for adding electrons to degenerate orbitals, which assumes that electrons are added with parallel spins until each of the orbitals has one electron before a second electron is placed in one of these orbitals.

**Hybrid atomic orbitals** Orbitals formed by mixing two or more atomic orbitals.

**Hybridization** A process in which things are mixed. A resonance hybrid is a mixture, or average, of two or more Lewis structures. Hybrid orbitals are formed by mixing two or more atomic orbitals.

**Hydride** Literally, a salt containing the $H^-$ ion, such as NaH. Also used to describe compounds such as HCl that contain hydrogen.

**Hydrocarbon** A compound, such as $CH_4$, that contains only carbon and hydrogen.

**Hydrogen bond** The bond formed when the positive end of one polar molecule, such as water, is attracted to the negative end of another polar molecule.

**Hydrophilic** Literally, water loving. Describes polar groups that attract water molecules.

**Hydrophobic** Literally, water hating. Describes nonpolar groups that repel water molecules.

**Hydroxide** A compound that contains an —OH group.

**Ideal gas** A gas that obeys all the postulates of the kinetic molecular theory. Real gases differ from the expected behavior of an ideal gas for two reasons: (1) the force of attraction between the particles in a gas is not quite zero, and (2) the volume of the particles in a gas is not quite zero.

**Ideal gas equation** The relationship between the pressure, volume, temperature, and amount of an ideal gas: $PV = nRT$. See **equation of state.**

**Immiscible** Liquids, such as gasoline and water, that are not soluble in each other.

**Indicator** A compound, such as phenolphthalein, that changes color at the endpoint of a titration.

**Induced dipole** A short-lived separation of charge, or dipole, created by polarization of a nonpolar atom or molecule.

**Induced fission** Fission of a nuclide that occurs after the nuclide has absorbed another particle. See **spontaneous fission.**

**Inelastic collision** A collision in which at least a portion of the kinetic energy of the colliding particles is lost.

**Inert** Unreactive. Used to describe compounds that do not undergo chemical reactions.

**Initial concentration** The concentration of a reactant before it reacts with either the solvent or another reactant.

**Initial rate of reaction** The rate of a chemical reaction extrapolated back to the instant the reactants were mixed.

**Inner transition metal** Used to describe the elements in the two rows at the bottom of the periodic table.

**Inorganic** Used to describe compounds that do not contain C—H bonds.

**Insoluble** Used to describe a substance that does not dissolve in a solvent to give a reasonable concentration.

**Instantaneous rate of reaction** The rate of a chemical reaction at an instant in time. The infinitesimally small change in the concentration of one of the reactants (or products) that occurs over an infinitesimally small length of time: $d(X)/dt$.

**Integrated form of the rate law** An alternative form of the rate law for a chemical reaction that can be used to predict the concentrations of the reactants (or products) at some moment in time.

**Intensive property** A quantity, such as temperature, density, or pressure, that does not depend on the size of the sample. See **extensive property.**

**Intermetallic compound** A compound, such as $CuAl_2$, with a fixed composition that results from the combination of two or more metals.

**Intermolecular bonds** The weak bonds between molecules in a liquid or solid, such as the hydrogen bonds between water molecules. See **intramolecular bonds.**

**Internal energy ($E$)** The sum of the kinetic and potential energies of the particles in a system. The internal energy of the system is proportional to its temperature.

**Interstitial solution** A solid solution formed when solute atoms are packed in the holes, or interstices, between solvent atoms.

**Intramolecular bonds** The bonds that hold a molecule together, such as the covalent O—H bonds in water. See **intermolecular bonds.**

**Iodide** Any compound that contains either the $I^-$ ion or iodine with an oxidation state of $-1$, such as KI.

**Ion** An atom or molecule that carries an electrical charge, such as the $Na^+$ or $Cl^-$ ions.

**Ion product ($Q_{sp}$)** The product of the concentrations of the ions in a solution at any moment in time.

**Ionic bond** The bond between two ions that results from the force of attraction between particles of opposite charge.

**Ionic compound** A compound that contains ions.

**Ionic radius** The radius of one of the ions in an ionic compound.

**Ionic solid** A solid composed of ions, such as NaCl. Also called a **salt.**

**Ionization** A process in which an ion is created from a neutral atom or molecule by adding or subtracting one or more electrons.

**Ionization energy** See **first** and **second ionization energy.**

**Ionize** A process in which ions are created. HCl, for example, is a covalent compound that ionizes when it dissolves in water.

**Ionizing radiation** Radiation with enough energy to remove an electron from a neutral atom or molecule to produce free radicals. See **nonionizing radiation.**

**Isobars** Nuclides with the same mass number, such as $^{40}K$ and $^{40}Ca$.

**Isoelectronic** Atoms or ions that have the same number of electrons and therefore the same electron configuration, such as $C^{4-}$, $N^{3-}$, $O^{2-}$, $F^-$, and Ne.

**Isomers** Compounds with the same chemical formulas, but different structures, and therefore different chemical or physical properties. See **cis, trans,** etc.

**Isotactic** Used to describe polymers in which all the substituents are oriented on the same side of the polymer backbone.

**Isotones** Nuclides with the same number of neutrons, such as $^{13}C$, $^{14}N$, and $^{15}O$.

**Isotopes** Nuclides with the same number of protons, such as $^{12}C$, $^{13}C$, and $^{14}C$.

**Isotropic** Literally the same in all directions. Used to describe the pressure of a gas.

**Joule** A unit of measurement for both heat and work in the **SI** system. 1 J = 4.184 cal.

**Ketone** A compound in which a C=O (carbonyl) functional group is bound to two carbon atoms, such as acetone: $CH_3COCH_3$.

**Kinetic control** Describes a chemical reaction in which the products are determined by the rate of the reaction. See **thermodynamic control.**

**Kinetic energy** The energy associated with a moving object. The kinetic energy of the object is equal to one-half the product of its mass times its velocity squared ($KE = \frac{1}{2}mv^2$).

**Kinetic molecular theory** The theory that heat is associated with the thermal motion of particles. In the kinetic theory, heat is not conserved—an inexhaustible amount of heat can be created by doing work on a system.

**Kinetic region** The portion of a plot of the concentration of a compound versus time in which the concentration changes. The portion of this plot between the initial conditions and the point at which the system reaches equilibrium.

**Latent heat** Heat that cannot be detected. The heat that enters a system when ice melts to form water or when water boils to form steam is latent heat, because there is no change in the temperature of the system. See **sensible heat.**

**Latent heat of fusion** See **heat of fusion.**

**Latent heat of vaporization** See **heat of vaporization.**

**Lattice energy** The energy given off when oppositely charged ions in the gas phase come together to form a solid. For example, the energy given off in the reaction: $Na^+(g) + Cl^-(g) \rightarrow NaCl(s)$.

**Lattice point** A point about which an atom, ion, or molecule is free to vibrate in a crystal.

**Law of combining volumes** See **Gay-Lussac's law.**

**Law of conservation of matter** The postulate that matter cannot be created or destroyed.

**Law of constant composition** The postulate that a compound always contains the same ratio by mass of its elements, regardless of its source.

**Law of definite proportions** The postulate that the ratio by mass of compounds consumed in a chemical reaction is always the same.

**LeChâtelier's principle** A principle that describes the effect of changes in the temperature, pressure, or concentration of one of the reactants or products of a reaction at equilibrium. It states that when a system at equilibrium is subjected to a stress, it will shift in the direction that minimizes the effect of this stress.

**Leveling effect** The tendency of water to limit the strength of the strongest acids and bases to the strength of the $H_3O^+$ and $OH^-$ ions.

**Levorotatory** Describes compounds that rotate plane-polarized light to the left (counterclockwise) when viewed in the direction of the light source. See **dextrorotatory.**

**Lewis acid** An electron-pair acceptor. A substance that acts in the same way as the $H^+$ ion to accept a pair of electrons.

**Lewis base** An electron-pair donor. A substance that acts in the same way as the $OH^-$ ion to donate a pair of electrons.

**Lewis structure** A symbolic description of the distribution of valence electrons in a molecule. Lewis structures use dots to represent individual electrons and lines to represent covalent bonds.

**Ligand** A Lewis base that can coordinate to a metal atom.

**Ligand-field theory** A molecular orbital description of the bonding in transition metal complexes.

**Limiting reagent** The reactant in a chemical reaction that limits the amount of product that can be formed. The reaction will stop when all the limiting reagent is consumed.

**Line notation** A system for representing electrochemical reactions.

**Lipid** Biologically important molecules that are soluble in nonpolar solvents.

**Liquid** A substance that flows freely, and therefore conforms to the shape of the walls of its container, but cannot expand to fill the container.

**Liquid crystal** A substance that has some of the long-range order of a solid but the freedom of motion of a liquid.

**Liter** The fundamental unit of volume in the metric system. The volume of 1000 grams of water at 4°C.

**Low-spin complex** A transition metal complex in which the difference between the energies of the $t_{2g}$ and $e_g$ orbitals is larger than the energy it takes to pair electrons. As a result, the valence-shell $d$ electrons on the metal are all placed in the lower energy set of orbitals, the $t_{2g}$ set for octahedral complexes and the $e_g$ set for tetrahedral complexes. See **high-spin complex.**

**mRNA** Messenger RNA. The polynucleotide that codes for the synthesis of a protein. mRNA is assembled during transcription of a chain of DNA.

**Macroscopic** Something that can be seen with the naked eye.

**Magnetic quantum number (*m*)** The quantum number used to describe the orientation of an atomic orbital in space.

**Main-group element** An element in one of the groups of the periodic table in which $s$ and $p$ orbitals are filled. Also known as the **representative elements.**

**Malleable** Something that can be hammered, pounded, or pressed into different shapes without breaking.

**Mass** A measure of the amount of matter in an object.

**Mass defect** The difference between the mass of an atom and the sum of the masses of the protons and neutrons that form the nucleus of the atom.

**Mass number (*M*)** An integer equal to the number of protons and neutrons in the nucleus of an atom.

**Measurement** The process by which the amount or quantity of something is measured.

**Mechanism** The steps by which a chemical reaction converts the reactants into the products.

**Melting point** The temperature at which the solid and liquid phases of a substance are in equilibrium at atmospheric pressure.

**Melting point depression** See **freezing point depression.**

**Meniscus** The curved surface of a liquid in a narrow-diameter glass tube.

**Metal** An element that is solid, has a metallic luster, is malleable and ductile, and conducts both heat and electricity.

**Metallic radius** Half the distance between the nuclei of adjacent atoms in a metal.

**Metallic solid** A solid that has the properties of a metal.

**Metalloid** An element with properties that fall between the extremes of metals and nonmetals. See **semimetal.**

**Metastable system** A system that should undergo a spontaneous change, but does not.

**Metathesis reaction** A reaction in which atoms or groups of atoms are interchanged, but none of the atoms undergoes a change in oxidation number. See **oxidation–reduction reaction.**

**Meter** The basic unit of length in both the **metric** and **SI** systems. Currently defined as the distance light travels in 1/299,792,458th of a second.

**Metric system** A system of units introduced in 1790 based on three fundamental quantities: the gram (mass), liter (volume), and meter (length).

**Mixture** A substance that contains two or more elements or compounds that retain their chemical identity and can be separated into the individual components by a physical process. For example, the mixture of $O_2$ and $N_2$ in the atmosphere.

**Molality (*m*)** The number of moles of solute in a solution divided by the number of kilograms of solvent.

**Molar heat capacity (*C*)** The number of joules of heat required to raise the temperature of a mole of a substance by 1 K. See **specific heat.**

**Molar mass** Literally, the mass of a mole. Often used by chemists who are not comfortable using the term **molecular weight** to describe measurements of mass.

**Molarity (*M*)** The number of moles of a solute in a solution divided by the volume of the solution in liters.

**Mole** Literally, a small mass. The amount of any substance that contains the same number of atoms, ions, or molecules as there are $^{12}C$ atoms in exactly 12 g of the $^{12}C$ isotope.

**Mole fraction** Literally, the fraction of the total number of moles in a mixture due to one component of the mixture. The mole fraction of a solute, for example, is the number of moles of solute divided by the total number of moles of solute plus solvent.

**Mole ratio** The ratio of the moles of one reactant or product to the moles of another reactant of product in the balanced equation for a chemical reaction.

**Molecular formula** The formula of a molecule of a compound.

**Molecular geometry** The shape, or geometry, of a molecule as defined by the positions of its nuclei.

**Molecular orbital** A region in space within a molecule where electrons can be found. Molecular orbitals are formed by the overlap, or mixing, of atomic orbitals.

**Molecular solid** A substance that consists of individual molecules held together in the solid by relatively weak intermolecular bonds.

**Molecular weight** The weight of the molecular formula, calculated from a table of atomic weights. The weighted average of the masses of the individual molecules in the substance. Because of the presence of different isotopes, the molecular weight differs from the precise mass of a single molecule.

**Molecularity** Number of molecules consumed in a step of a chemical reaction. See **bimolecular** and **unimolecular.**

**Molecule** The smallest particle that has any of the chemical or physical properties of an element or compound.

**Momentum** The product of the mass times the velocity of an object in motion.

**Monodentate** Literally, ''one toothed.'' Used to describe ligands, such as water or ammonia, that can coordinate to a transition metal only once.

**Monomer** One of the relatively small molecules from which polymers are formed.

**Monoprotic acid (HA)** An acid, such as HCl or HCN, that can lose only one $H^+$ ion or proton.

**Monosaccharide** A carbohydrate that cannot be hydrolyzed to a simpler carbohydrate. See **disaccharide** and **polysaccharide.**

**Natural radioactivity** The radioactive decay that occurs naturally, as opposed to **induced radioactivity.** Also known as **spontaneous fission.**

**Negative electrode** The electrode in an electrochemical cell that carries a negative charge. In an electrolytic cell, it is the **cathode.** In a voltaic cell, it is the **anode.**

**Nernst equation** Describes the relationship between the potential of an electrochemical cell at any moment in time and the standard-state cell potential. Used to understand what happens to the potential of a voltaic cell as it comes to equilibrium.

**Neutrino** A particle with no charge and little or no mass that is ejected from the nucleus at the same time as an electron or positron.

**Neutron** A subatomic particle with a mass of about 1 amu and no charge.

**Neutron-poor nuclide** A nuclide with fewer neutrons than the lightest stable isotope of the element. A nuclide that decays by either electron capture, positron emission, or the emission of an alpha particle.

**Neutron-rich nuclide** A nuclide with more neutrons than the heaviest stable isotope of the element. A nuclide that decays by the emission of an electron.

**Nitride** Any compound that contains either the $N^{3-}$ ion or nitrogen with an oxidation state of $-3$, such as $Li_3N$ or $NH_3$.

**Nomenclature** A systematic way of naming chemical compounds.

**Nonbonding electrons** Electrons in the valence shell of an atom that are not used to form covalent bonds.

**Nonbonding molecular orbital** A molecular orbital whose energy is more or less equal to the energy of the atomic orbitals from which it is formed.

**Nonionizing radiation** Radiation that carries enough energy to excite an atom or molecule, but not enough energy to remove an electron from the atom or molecule. See **ionizing radiation.**

**Nonmetal** An element that lacks the properties generally associated with metals.

**Nonpolar** Used to describe compounds that do not carry a dipole moment.

**Nucleic acid** A molecule of very high molecular weight used to store and process the genetic information in cells.

**Nucleon** A proton or neutron.

**Nucleophile** Literally, something that loves nuclei. All nucleophiles are Lewis bases.

**Nucleophilic substitution reaction** A reaction in which one nucleophile is substituted for another in a molecule.

**Nucleotide** The smallest repeating unit out of which nucleic acids are built.

**Nucleus** Literally, "little nut." It consists of neutrons and protons, occupies an infinitesimally small fraction of the total volume of an atom, and contains almost all of the mass of an atom.

**Nuclide** An atom with a particular combination of protons and neutrons, such as the $^{14}C$ nuclide.

**Octahedral complex** A complex in which six ligands bound to a metal atom are arranged toward the corners of an octahedron.

**Octahedral geometry** A geometry in which six atoms, ions, or molecules are arranged around a central atom in opposite directions along the *x*, *y*, and *z* axes of a coordinate system.

**Octahedral hole** A hole in a closest packed structure surrounded by six atoms or ions arranged toward the corners of a tetrahedron.

**Octahedron** A polyhedron that has eight faces, each of which is an equilateral triangle.

**Octane number** A measure of the antiknock quality of gasoline.

**Octet** Literally, eight objects. Used to describe the tendency of main-group elements to have eight valence-shell electrons in their compounds.

**Oil** One of three kinds of substances: (1) mineral oils, such as crude oil from petroleum, which are mixtures of hydrocarbons; (2) animal and vegetable oils, such as corn oil, which are mixtures of triglycerides; and (3) essential oils or perfumes from plants.

**Olefin** A common name for the class of compounds known as alkenes. Compounds that contain C=C double bonds.

**Optically active** Used to describe the ability of some molecules to rotate plane-polarized light.

**Orbitals** Regions in space where electrons can reside. See **atomic orbitals** and **molecular orbitals.**

**Order** Used to describe the relationship between the rate of a step in a chemical reaction and the concentration of one of the reactants consumed in that step. See **first-order reaction, pseudo-first-order reaction, second-order reaction,** and **zero-order reaction.**

**Organic** Used to describe compounds that contain C—H bonds.

**Osmosis** The process by which one component of a solution passes through a membrane to dilute the solution.

**Osmotic pressure** The pressure exerted by water or other solvents flowing into a solution through a membrane.

**Ostwald process** The industrial process in which ammonia is converted into nitric acid.

**Overlap** The interaction between a pair of atomic orbitals that occurs when the orbitals share space.

**Overvoltage** The voltage over and above the voltage predicted from standard tables of reduction potentials that must be applied before an electrochemical reaction occurs at the expected rate.

**Oxidation** Any process in which the oxidation number of an atom becomes more positive.

**Oxidation number** The charge that would be present on an atom if the element or compound in which the atom is found were ionic.

**Oxidation–reduction reaction** Reaction in which at least one atom undergoes a change in oxidation state. See **metathesis reaction.**

**Oxide** Any compound that contains either the $O^{2-}$ ion or oxygen with an oxidation state of −2, such as $Li_2O$ or $H_2O$.

**Oxidizing agent** An atom, ion, or molecule that gains electrons in a chemical reaction, thereby oxidizing the substance with which it reacts.

**Oxyacid** An acid, such as $H_2SO_4$ or $H_3PO_4$, in which the acidic hydrogen atoms are attached to an oxygen atom.

**Oxyanion** A negatively charged ion, such as the $SO_4^{2-}$ or $PO_4^{3-}$ ion, formed when an oxyacid loses one or more $H^+$ ions.

**pH** A measure of acidity. The negative of the logarithm of the $H_3O^+$ ion concentration: $pH = -\log [H_3O^+]$

**pOH** The negative of the logarithm of the $OH^-$ ion concentration: $pOH = -\log [OH^-]$

**Paramagnetic** A compound that is attracted to a magnetic field. A compound that contains one or more unpaired electrons. See **diamagnetic.**

**Partial pressure** The fraction of the total pressure of a mixture of gases that arises from one component of the mixture.

**Particle** Any object that has mass and therefore occupies space.

**Peptide** A relatively small polymer of amino acids. Peptides are usually too small to have extensive secondary or tertiary structures.

**Peptide bond** The bond between amino acids in a peptide or protein.

**Period** A horizontal row in the periodic table, such as the second period, which contains the elements Li, Be, B, C, N, O, F, and Ne.

**Periodic table** A matrix in which the elements are arranged across rows in order of increasing atomic number so that elements with similar chemical properties fall in the same vertical column.

**Peroxide** Literally, a compound that is rich in oxygen. Used to describe compounds that formally contain the $O_2^{2-}$ ion.

**Phase diagram** A two-dimensional graph that shows the state, or phase, of a substance at any combination of temperature and pressure.

**Phlogiston** An imaginary element once thought to be given off when objects burned.

**Phosphide** Any compound that contains either the $P^{3-}$ ion or phosphorus with an oxidation state of −3, such as $Ca_3P_2$ or $PH_3$.

**Phospholipid** A triester of glycerol with two fatty acids and one phosphate ion. Polar lipids found in biological membranes.

**Photon** The smallest unit of electromagnetic energy.

**Pi bond** A bond formed by the edge-on overlap of *p* atomic orbitals.

**Plastic** Literally, a material that can flow. Used to describe polymers that can be shaped, molded, or milled.

**Poise** The unit of measurement for viscosity.

**Polar** Used to describe compounds that have a dipole moment because they consist of molecules that have negative and positive poles.

**Polar covalent bond** A bond in a molecule, such as $H_2O$, that is neither strictly covalent nor strictly ionic.

**Polarity** The tendency of a molecule to have positive and negative poles because of the unequal sharing of a pair of electrons.

**Polyamide** A polymer, such as nylon or a protein, held together by —CO—NH— bonds.

**Polyatomic ion** An ion that contains more than one atom.

**Polycrystalline solid** A solid composed of many individual crystals or grains arranged in a more or less random order.

**Polyester** A polymer held together by ester linkages between the monomers.

**Polymer** A molecule with a large molecular weight formed by the linking of 30 to 100,000 (or more) repeating units.

**Polynucleotide** A condensation polymer formed by the linking of nucleotides.

**Polyprotic acid** An acid, such as $H_2SO_4$ or $H_3PO_4$, that can lose more than one $H^+$ ion, or proton.

**Polyprotic base** A base, such as the $PO_4^{3-}$ ion, that can accept more than one $H^+$ ion, or proton.

**Polysaccharide** A carbohydrate made by polymerizing many monosaccharide units.

**Positive electrode** The electrode in an electrochemical cell that carries a positive charge. In an electrolytic cell, it is the **anode.** In a voltaic cell, it is the **cathode.**

**Positron** The antimatter analog of an electron ($\beta^+$). Positrons have the same mass as electrons but the opposite charge.

**Positron emission** A nuclear reaction in which a positron is emitted. Positron emission leads to a decrease by one in the atomic number and no change in the mass number of the nuclide.

**Potential** A measure of the driving force behind an electrochemical reaction that is reported in units of volts.

**Precipitation** A process in which positive and negative ions combine to form a salt that separates out of the solution as a solid.

**Precipitation hardening** A process by which an alloy becomes harder when an intermetallic compound, such as $CuAl_2$, is allowed to precipitate from a supersaturated solution of two metals.

**Precision** A measure of the extent to which individual measurements of the same quantity agree.

**Pressure** The force exerted on a surface divided by the area of the surface.

**Primary battery** A battery that cannot be charged. See **secondary battery.**

**Primary structure** The sequence of amino acids in a protein.

**Primary valence** The number of negative ions needed to satisfy the charge on a metal ion. In the $[Co(NH_3)_6]Cl_3$ complex, for example, it is three.

**Principal quantum number (*n*)** The quantum number that describes the size, and therefore the relative energy, of an orbital.

**Product** A substance formed in a chemical reaction.

**Protein** A relatively large polymer of amino acids. A polypeptide that is large enough to have an extensive secondary and tertiary structure.

**Proton** A subatomic particle that has a charge of +1 and a mass of about 1 amu.

**Proton acceptor** An ion or molecule that can gain an $H^+$ ion, or proton. A Brønsted base.

**Proton donor** An ion or molecule that can lose an $H^+$ ion, or proton. A Brønsted acid.

**Pseudo-first-order reaction** A reaction that is formally first order in two reactants, but appears first order in only one of these reactants because it is run in the presence of a large excess of the other reactant.

**Purine** A heterocyclic aromatic compound, containing nitrogen, found in nucleic acids. See **pyrimidine.**

**Pyrimidine** A heterocyclic aromatic compound, containing nitrogen, found in nucleic acids. See **purine.**

**Pyrophoric** Used to describe compounds that burst into flame in the presence of air.

**Quadratic formula** A formula for solving quadratic equations.

**Quantized** Literally, countable. Oranges are quantized; orange juice is not.

**Quantum number** An integer that describes the size, shape, and orientation in space of an atomic orbital.

**Quaternary structure** The ionic or hydrophobic interactions between individual chains of amino acids in some proteins.

**R process** The process by which nuclides are built up by the rapid capture of neutrons. The R process produces heavy nuclides such as $^{238}U$. It requires a large neutron flux, which occurs when a star becomes a supernova. See **S process.**

**Rad** A unit in which the dose of radiation absorbed by an object is reported: 1 rad = $10^{-5}$ J/g.

**Radioactivity** The spontaneous disintegration of an unstable nuclide by a first-order rate law.

**Radius ratio** The radius of the positive ion divided by the radius of the negative ion in a salt.

**Random error** A source of error that limits the precision of a measurement. An error that is equally likely to give results that are too large or too small. See **systematic error.**

**Raoult's law** A law that describes the relationship between the vapor pressure of a solution, the mole fraction of the solute, and the vapor pressure of the solute.

**Rare gases** A group of elements (He, Ne, Ar, Kr, Xe, Rn) that was once erroneously known as the **inert gases.** Also known as the **noble gases.**

**Rate of reaction** The change in the concentration of a compound divided by the amount of time necessary for this change to occur: rate = $d(X)/dt$.

**Rate constant** The proportionality constant in the equation that describes the relationship between the rate of a step in a chemical reaction and the product of the concentrations of the reactants consumed in that step.

**Rate law** An equation that describes how the rate of a chemical reaction depends on the concentrations of the reactants consumed in that reaction.

**Rate-limiting step** The slowest step in a chemical reaction.

**RBE** Radiation biological effectiveness. Used to correct for differences in the effect of equivalent doses of different forms of radiation. See **rad** and **rem.**

**Reactant** One of the starting materials in a chemical reaction.

**Reaction coordinate** The sequence of infinitesimally small steps that must be taken to convert the reactants into the products of a reaction.

**Reaction quotient ($Q_c$ or $Q_p$)** The quotient obtained when the concentrations (or partial pressures) of the products of a reaction are multiplied and the result is divided by the product of the concentrations (or partial pressures) of the reactants. The reaction quotient can have any value between zero and infinity. When the reaction is at equilibrium, the reaction quotient is equal to the equilibrium constant for the reaction.

**Recrystallization** A technique for purifying solids that removes impurities when the solid is dissolved in an appropriate solvent and then allowed to recrystallize.

**Redox** An abbreviation for oxidation–reduction.

**Reducing agent** An atom, ion, or molecule that loses electrons in a chemical reaction, thereby reducing the substance with which it reacts.

**Reduction** Any process that leads to a decrease in the oxidation number of an atom.

**Rem** A unit of radiation absorbed dose equal to the product of the rads of absorbed radiation times the **RBE.**

**Replication** The process by which copies of DNA are made to be passed down to future generations of cells.

**Representative elements** See **main-group elements.**

**Resonance** An averaging or mixing process that occurs when more than one Lewis structure can be written for a molecule.

**Resonance hybrid** A mixture, or average, of two or more Lewis structures.

**RNA** Ribonucleic acid, the nucleic acid involved in transcribing the genetic information for the synthesis of proteins stored in DNA and then translating this information into the sequence of amino acids in the protein.

**Roasting** A process in which a metal ore (such as PbS or ZnS) is transformed into the corresponding oxide (PbO or ZnO) and $SO_2$.

**Rusting** The corrosion of iron or iron-based alloys such as steel. Other metals may corrode, but only iron and steel rust.

**S process** The process by which heavier nuclides are built up by the slow absorption of one neutron at a time. See **R process.**

**Salt** See **ionic compound.**

**Salt bridge** A tube containing a saturated solution of a salt, such as potassium chloride. Used to complete the electrical circuit in a voltaic cell.

**Saponification** The reaction between fats and oils with KOH or NaOH to produce soap.

**Saturated fatty acid** A long-chain carboxylic acid that contains no C=C double bonds. See **unsaturated fatty acid.**

**Saturated hydrocarbon** A hydrocarbon that contains as much hydrogen as possible. A compound with the generic formula $C_nH_{2n+2}$. See **alkane.**

**Saturated solution** A solution that contains as much solute as possible.

**Scientific notation** A system in which a number is expressed as a number between 1 and 10, times 10 raised to an exponent.

**Second ionization energy** The energy needed to remove an electron from an $M^+$ ion in the gas phase.

**Second law of thermodynamics** The notion that natural processes that occur in an isolated system are spontaneous when they lead to an increase in disorder, or entropy.

**Second-order reaction** A reaction whose rate is proportional to the concentration of a single reactant raised to the second power: rate = $k(X)^2$

**Secondary battery** A battery used to store electricity. By definition, these batteries must be capable of being both charged and discharged. See **primary battery.**

**Secondary structure** The hydrogen bonds between adjacent peptide linkages in a protein that stabilize the structure of the protein.

**Secondary valence** The number of ions or molecules that are covalently bound to a transition metal ion. In the $[Co(NH_3)_6]Cl_3$ complex, it is six.

**Selective precipitation** A technique in which one ion is selectively removed from a mixture of ions by precipitation.

**Semiconductor** A material whose ability to carry an electric current falls between those of metals and nonmetals.

**Semimetal** An element, such as silicon or arsenic, with chemical and physical properties that fall between the extremes associated with metals and nonmetals. Also called a **metalloid.**

**Semipermeable membrane** A thin, flexible solid that can pass small molecules such as water but not larger molecules such as sugar or alcohol.

**Sensible heat** Heat that can be sensed, or detected, by a change in the temperature of the system. See **latent heat.**

**Shell** A set of orbitals that have the same principal quantum number.

**SI** The International System of Units, which is based on seven fundamental quantities: the meter (length), kilogram (mass), second (time), kelvin (temperature), ampere (current), mole (amount of substance), and candela (luminous intensity).

**Sigma bond** A cylindrically symmetric bond formed by the overlap of *s* orbitals or the head-on overlap of *p* orbitals.

**Significant figures** The digits in a measurement that are reliable.

**Simple cubic** A structure in which the simplest repeating unit consists of eight equivalent atoms, ions, or molecules at the eight corners of a cube.

**Sintering** A process in which a finely divided ore is heated until it collects to form larger particles.

**Skeleton structure** A representation of the basic structure of a molecule in which lines are used to connect atoms held together by covalent bonds.

**Smelting** A process in which a metal oxide (such as PbO, ZnO, or $Fe_2O_3$) is reduced to the corresponding metal.

**Soap** The sodium or potassium salt of a fatty acid.

**Solid** A substance that does not flow and therefore does not conform to the shape of its container.

**Solubility** The ratio of the maximum amount of solute to the volume of solvent in which this solute can dissolve. Often expressed in units of grams of solute per 100 g of water, or in moles of solid per liter of solution.

**Solubility equilibria** Equilibria that exist in a saturated solution, in which additional solid dissolves at the same rate that particles in solution come together to precipitate more solid.

**Solubility product ($K_{sp}$)** The product of the equilibrium concentrations of the ions in a saturated solution of a salt.

**Solute** The substance that dissolves to form a solution.

**Solution** A mixture of one or more solutes dissolved in a solvent.

**Solvent** The substance in which a solute dissolves.

**Specific heat** The amount of heat required to raise the temperature of 1 g of a substance by either 1°C or 1 K. See **molar heat capacity.**

**Spectrochemical series** A sequence of ligands arranged in order of decreasing magnitude of the splitting of the energies of $e_g$ and $t_{2g}$ orbitals in transition metal complexes.

**Spin quantum number (*s*)** The quantum number used to specify the electrons that occupy an orbital. The spin quantum number can have values of $\pm\frac{1}{2}$.

**Spontaneous fission** Fission of a nuclide that occurs spontaneously. See **induced fission.**

**Spontaneous reaction** A reaction in which the products are favored.

**Square-planar geometry** A geometry in which four atoms, ions, or molecules lying in the same plane are bound to a central atom and arranged toward the corners of a square.

**Square-pyramidal geometry** A geometry in which a fifth atom, ion, or molecule is added to a square-planar geometry along an axis perpendicular to the plane of the other four.

**Standard state** State in which all concentrations are 1 *M* and all partial pressures are 1 MPa.

**Standard-state cell potential** The potential of a cell measured under standard-state conditions.

**Standard-state enthalpy of reaction** The change in the enthalpy that occurs during a chemical reaction that begins and ends under standard-state conditions. See **enthalpy of reaction.**

**Standard-state entropy** The results of measurements of entropy taken under standard-state conditions.

**Standard-state free energy of formation** The results of standard-state measurements of the change in free energy that occurs when a compound is formed from its elements in their most thermodynamically stable states.

**Standard-state free energy of reaction** The energy associated with a chemical reaction that can be used to do work when the reaction begins and ends under standard-state conditions. See **free energy of reaction.**

**State** (1) One of the three states of matter: gas, liquid, or solid. (2) A set of physical properties that describes a system.

**State function** A quantity whose value depends only on the state of the system and not its history; *X* is a state function if and only if the value of $\Delta X$ does not depend on the path used to go from the initial to the final state of the system.

**Stepwise dissociation** The assumption that we can separate the individual steps in a dissociation reaction.

**Stereoisomers** Compounds with the same chemical formula that have different three-dimensional structures. See **cis** and **trans** isomers and **chiral** isomers.

**Stoichiometry** The relationship between the weights of the reactants and the products of a chemical reaction.

**Strong acid** Used to describe acids that dissociate more or less completely in water.

**Sublimation** A process in which a solid goes directly to the gaseous state without passing through an intermediate liquid state.

**Subshell** Atomic orbitals for which the values of the $n$ and $l$ quantum numbers are the same, such as the three $2p$ or five $3d$ atomic orbitals.

**Substitution reaction** A reaction that involves the substitution of one group for another. For example, the reaction between $CH_3Br$ and the $OH^-$ ion to form $CH_3OH$ and the $Br^-$ ion.

**Successive approximations** A technique in which the answer obtained from an approximate calculation is substituted into the original equation to yield a more accurate solution.

**Sulfide** Any compound that contains either the $S^{2-}$ ion or sulfur with an oxidation state of $-2$, such as $CS_2$.

**Superconductor** A substance that has no resistance to conducting an electric current.

**Supercooled liquid** A liquid that has been cooled to a temperature below its freezing point.

**Superoxide** A compound that contains the $O_2^-$ ion, such as $KO_2$.

**Surface tension** The force that controls the shape of a liquid. Surface tension results from the force of cohesion between liquid molecules.

**Surroundings** In thermodynamics, the part of the universe not included in the **system.**

**Syndiotactic** Used to describe polymers in which the substituents alternate regularly between the two sides of the polymer backbone.

**Synthesis gas** A mixture of CO and $H_2$ produced during the gasification of coal that can be used to synthesize a wide range of materials.

**System** In thermodynamics, that small portion of the universe in which we are interested.

**Systematic error** A source of error that limits the accuracy of a measurement, which gives results that are always too small or always too large. See **random error.**

**tRNA** Transfer RNA. The relatively small polynucleotide that carries amino acids and recognizes the codon on mRNA that specifies the amino acid to be incorporated at a particular point on a protein chain.

**T-shaped geometry** A molecular geometry whose shape resembles the letter T.

**Tacticity** Literally, arrangement or system. Used to describe the arrangement of substituents on a polymer chain. See **atactic, isotactic,** and **syndiotactic.**

**Temperature** An intensive property that measures the extent to which an object can be labeled ''hot'' or ''cold.''

**Tertiary structure** The interactions between the side chains on amino acids in a protein that help determine the structure of the protein.

**Tetrahedral complex** A complex in which four ligands are bound to the metal atom and arranged toward the corners of a tetrahedron.

**Tetrahedral hole** A hole in a closest packed structure surrounded by four atoms or ions arranged toward the corners of a tetrahedron.

**Tetrahedron** A geometric shape that resembles a trigonal-based pyramid.

**Tetravalent** Literally, able to form four bonds. Carbon is tetravalent because it forms four bonds in virtually all of its compounds.

**Thermal conductor** An object that conducts heat easily from one side to the other, or from one end to the other.

**Thermal insulator** An object, such as a blanket or a fur coat, that slows down the rate at which heat is transferred from one object to another.

**Thermal neutron** Also known as a *slow neutron.*

**Thermochemistry** The study of the heat given off or absorbed in a chemical reaction.

**Thermodynamic control** Describes a chemical reaction that forms the most stable product, not the product of the fastest reaction. See **kinetic control.**

**Thermodynamics** The study of the relationship between heat, work, and other forms of energy.

**Thermonuclear reaction** A nuclear reaction that only occurs at extremely high temperatures.

**Thin-layer chromatography** Chromatography in which a solvent is passed over a solid support that has been applied as a thin layer to a glass or plastic plate.

**Thio-** A prefix that describes compounds in which a sulfur atom can be found where an oxygen atom is expected. The sulfate ion, for example, has the formula $SO_4^{2-}$, whereas the thiosulfate ion has the formula $S_2O_3^{2-}$.

**Third law of thermodynamics** The postulate that the entropy of a perfect crystal is zero when the temperature of the crystal is equal to absolute zero (0 K).

**Titration** A technique used to determine the concentration of a solute in a solution.

**Torr** A unit of pressure equal to the pressure exerted by a column of mercury 1 mm tall. By definition, 1 torr = 1 mmHg.

**Trans** Literally, across. Describes isomers of compounds in which similar substituents lie on opposite sides of a double bond or on opposite sides of a transition metal. See **cis.**

**Transcription** The process by which the message encoded in DNA is copied to form a sequence of mRNA that can be used as a template to make proteins. See translation.

**Transition metal** Metals in the block of elements that serve as a transition between the two columns on the left side of the table, where *s* orbitals are filled, and the six columns on the right, where *p* orbitals are filled.

**Translation** The process by which the information coded in a sequence of mRNA is transformed into a sequence of amino acids in a protein.

**Translational motion** The net movement of an atom, ion, or molecule through a gas, liquid, or solid.

**Transuranium elements** Elements with atomic numbers larger than that of uranium.

**Triad** Three elements (such as F, Cl, and Br) with similar chemical properties.

**Triglyceride** A triester of glycerol and fatty acids found in fats and oils. Also known as a *triacylglycerol.*

**Trigonal bipyramidal** A geometry in which two atoms, ions, or molecules are added to a **trigonal planar** system along an axis perpendicular to the triangular plane.

**Trigonal planar** A geometry in which three atoms, ions, or molecules are coordinated to a central atom and arranged toward the corners of an equilateral triangle.

**Trigonal-pyramidal** A geometry in which a central atom at the apex of a pyramid is coordinated to three atoms, ions, or molecules to form a pyramid.

**Triple point** The combination of temperature and pressure in a phase diagram at which the substance can simultaneously exist as a gas, a liquid, and a solid.

**Triprotic** Used to describe acids, such as $H_3PO_4$, that can donate three protons or bases, such as the $PO_3^{3-}$ ion, that can accept three protons.

**Uncertainty** The limit on the precision of a measurement. Analytical balances, for example, can weigh an object with an uncertainty of $\pm 0.001$ or $\pm 0.0001$ g.

**Unimolecular** Describes a step in a chemical reaction in which one molecule is consumed.

**Unit** The basis for comparing a measurement with a standard reference, such as the units known as ''inches'' or ''feet.''

**Unit cell** The simplest repeating unit in a crystal.

**Unit factor** A ratio, or factor, equal to one that is constructed from an equality. The fact that 12 inches is equivalent to 1 foot can be used to construct two unit factors: (12 in.)/(1 ft) or (1 ft)/(12 in.). Used in dimensional analysis.

**Unsaturated fatty acid** A long-chain carboxylic acid that contains one or more C=C double bonds. See **saturated fatty acid.**

**Unsaturated hydrocarbon** A hydrocarbon that contains C=C double bonds and/or C≡C triple bonds, and therefore contains less hydrogen than a saturated hydrocarbon.

**Valence band** The highest energy band in a solid that contains electrons.

**Valence-bond theory** A model of the bonding in covalent compounds based on the assumption that atoms in these compounds are held together by the sharing of pairs of valence electrons by adjacent atoms.

**Valence electrons** Electrons in the outermost or highest energy orbitals of an atom. The electrons that are gained or lost in a chemical reaction.

**Valence-shell electron-pair repulsion theory (VSEPR)** A model in which the repulsion between pairs of valence electrons is used to predict the shape or geometry of a molecule.

**van der Waals equation** An equation that corrects for the two erroneous assumptions in the ideal gas equation that the volume occupied by the gas particles is negligibly small and that there is no force of attraction between gas particles.

**van der Waals forces** The weak forces of attraction between atoms or molecules that explain why gases condense to form liquids and solids when cooled.

**Vapor pressure** The pressure of the gas that collects above a liquid in a closed container.

**Vapor pressure depression** The decrease in the vapor pressure of a solvent that occurs when a solute is added to form a solution.

**Vibrational motion** The motion of a molecule that results in the stretching or bending of one or more bonds in the molecule.

**Viscosity** A measure of the resistance of a liquid or gas to flow. Viscous liquids, such as molasses, flow very slowly.

**Vital force** A mythical force that was once assumed to explain the difference between animate (organic) and inanimate (inorganic) objects.

**Vitamin** Literally, vital amines. A wide range of compounds that are either water soluble or fat soluble and are necessary components of the diet of higher organisms, such as mammals.

**Voltaic cell** An electrochemical cell that uses a spontaneous chemical reaction to do work. Also known as a **galvanic cell.**

**Volume percent** The percentage of the total volume of a mixture due to a particular component.

**Water-dissociation equilibrium constant ($K_w$)** The product of the equilibrium concentration of the $H_3O^+$ and $OH^-$ ions in an aqueous solution: $K_w = 1.0 \times 10^{-14}$ at 25°C.

**Water gas** A mixture of CO, $CO_2$, and $H_2$ produced from the reaction of red-hot coal with steam.

**Wave** A phenomenon that does not have mass and therefore does not occupy space. Waves travel through space.

**Wave function** A mathematical equation that describes orbitals in which electrons reside.

**Wavelength** The distance between repeating points on a wave.

**Wave–particle duality** The principle that all objects in motion are simultaneously particles and waves. Most objects have a wavelength that is too small to be detected. Very light objects, such as photons and electrons, can have wavelengths large enough to give these objects some of the properties of waves.

**Weak acid** An acid that dissociates only slightly in water. An acid for which the acid-dissociation equilibrium constant is significantly smaller than one.

**Weight** A measure of the gravitational force of attraction of the Earth acting on an object.

**Weight percent** The percent by weight of an element in a compound, or the percent by weight of one component in a mixture.

**Work (*w*)** Mechanical energy equal to the product of the force applied to an object times the distance the object is moved.

**Work of expansion** The work done when a system that consists of a gas expands against its surroundings.

**X-ray** A high-energy, short-wavelength form of electromagnetic radiation.

**Ylem** The primordial substance out of which the elements were synthesized in the first moments after the creation of the universe.

**Zero-order reaction** A reaction whose rate is not proportional to the concentration of any of the reactants.

**Zeroth law of thermodynamics** A law stating that two objects in thermal equilibrium with a third object are in thermal equilibrium with each other.

**Zwitterion** Literally, a mongrel or hybrid ion. A molecule that contains both negative and positive charges, such as the $H_3N^+CHRCO_2^-$ form of an amino acid.

## PHOTO CREDITS

**Chapter 1** *Opener:* COMSTOCK, Inc. *Page 7:* Smithsonian Institution. *Page 10:* Ken Karp/Omni-Photo Communications, Inc. *Page 12:* Michael Watson. *Page 14:* Andy Washnik. *Page 17:* Andy Sacks/Tony Stone Images. *Page 19:* Courtesy Coin World. *Page 23:* UPI/Bettmann. *Page 25:* Peter Thorne/Science Photo Library/Photo Researchers Inc. *Page 27:* COMSTOCK, Inc. *Page 28:* Andy Washnik. *Page 30:* Barry Robinson/Courtesy The Nature Company.

*Intersection:* NOAA, Coloured by John Wells/Science Photo Library/Photo Researchers.

**Chapter 2** *Opener:* Courtesy Pachyderm Scientific Industries. *Pages 45 and 46 (top):* Andy Washnik. *Page 46 (bottom):* Ken Karp/Omni-Photo Communications, Inc. *Page 49:* Courtesy IBM. *Pages 52 and 53 (top and bottom left):* Andy Washnik. *Page 53 (bottom right):* Reg Watson/Tony Stone Images. *Page 54:* Ken Karp/Omni-Photo Communications, Inc. *Page 55:* Michael Watson. *Page 57:* OPC, Inc. *Page 59:* Ken Karp/Omni-Photo Communications, Inc. *Page 62:* Andy Washnik. *Page 64:* Courtesy Lisa Passmore.

**Chapter 3** *Opener:* Michael Abbey/Photo Researchers. *Page 72:* Courtesy Oesper Collection in the History of Chemistry, University of Cincinnati. *Page 74:* Courtesy Finnigan Corp. *Pages 75 and 76 (top):* Ken Karp/Omni Photo Communications, Inc. *Page 76 (bottom):* Courtesy Domino Sugar Corporation. *Page 80:* David Woodfall/Natural History Photographic Agency. *Page 81:* Andy Washnik. *Pages 82 and 84:* Ken Karp/Omni-Photo Communications, Inc. *Page 85:* Richard Megna/Fundamental Photographs. *Page 87:* Bob Burch/Bruce Coleman, Inc. *Page 88:* Michael Dalton/Fundamental Photographs. *Page 95:* Linda K. Moore. *Page 98:* Andy Washnik. *Page 101:* Michael Watson.

**Chapter 4** *Opener:* Luis Castaneda/The Image Bank. *Pages 101 and 112:* Michael Watson. *Page 113:* Richard Megna/Fundamental Photographs. *Page 115 (top):* Ed Braverman/FPG International. *Page 115 (bottom):* Blair Seitz/Photo Researchers. *Page 118 (top):* Courtesy NASA. *Page 118 (bottom):* Ken Karp/Omni-Photo Communications, Inc. *Page 123:* Bettmann Archive. *Page 124:* Andy Washnik. *Page 126:* Ralph Wetmore/Tony Stone Images. *Page 129:* COMSTOCK, Inc. *Page 132:* Jeff Gnass Photography. *Page 136:* Courtesy Kenttamma Research Group.

**Chapter 5** *Opener:* Terence Harding/Tony Stone Images. *Page 154:* Mark Burnside/AllStock, Inc. *Page 156:* Courtesy Corning. *Page 158 (top):* Sorensen/Bohmer Olse/Tony Stone Images. *Page 158 (left):* Michael Watson. *Page 160:* Courtesy the Old Print Shop, New York City. *Page 164:* Andy Washnik. *Page 170:* Parr Instrument Company. *Page 171:* COMSTOCK, Inc. *Page 172:* Stephen Frisch/Stock, Boston. *Pages 175 and 181:* Andy Washnik. *Page 186 (top):* Hart Scientific Differential Scanning Calorimeter. *Page 186 (bottom):* Tripos Associates.

*Intersection:* David Taylor/Science Photo Library/Photo Researchers.

**Chapter 6** *Opener:* Scott Camazine/Photo Researchers. *Page 196:* Ken Karp/Omni-Photo Communications. *Page 197:* Richard Megna/Fundamental Photographs. *Page 198:* The Bettmann Archive. *Page 207:* Dohrn/Photo Researchers. *Page 208:* Richard Megna/Fundamental Photographs. *Pages 210 and 212:* Ken Karp/Omni-Photo Communications, Inc. *Page 216:* American Institute of Physics. *Page 225:* American Institute of Physics.

**Chapter 7** *Opener:* Dr. Jeremy Burgess/Science Photo Library/Photo Researchers. *Page 242:* Bettmann Archive. *Page 250:* Ken Karp/Omni-Photo Communications, Inc. *Page 258:* Courtesy Amy Stevens Miller. *Pages 261 and 263:* Andy Washnik. *Page 264:* Rosemary Weller/Tony Stone Images.

**Chapter 8** *Opener:* Tripos Associates. *Page 276:* Ken Karp/Omni-Photo Communications, Inc. *Page 281:* Andy Washnik. *Page 289 (bottom):* American Association for the Advancement of Science. *Pages 291–295:* Andy Washnik. *Page 306:* Ken Karp/Omni-Photo Communications, Inc.

*Intersection: Page 312:* Courtesy Miles Laboratories. *Page 313:* Michael Watson.

**Chapter 9** *Opener:* Lester Lefkowitz/Tony Stone Images. *Pages 316 and 317 (top):* Andy Washnik. *Page 318 (top):* Ken Karp/Omni-Photo Communications, Inc. *Page 318 (bottom):* M. Claye/Jacana/Photo Researchers. *Page 319:* American Association for the Advancement of Science. *Page 322:* Andy Washnik. *Page 323:* Richard Megna/Fundamental Photographs. *Pages 326 and 327 (top):* Ken Karp/Omni-Photo Communications, Inc. *Page 327 (bottom):* Courtesy SPO/AC-Delco, General Motors Corp. *Pages 332 and 336 (top):* Michael Watson. *Page 336 (bottom left):* Andy Washnik. *Page 337:* Ray Ellis/Photo Researchers. *Page 338:* Bill Gallery/Stock, Boston. *Page 342:* Karl Hartmann/Sachs/Phototake.

**Chapter 10** *Opener:* Courtesy NASA. *Page 350:* Ken Karp/Omni-Photo Communications, Inc. *Page 351 (top):* Andy Washnik. *Page 351 (bottom):* COMSTOCK, Inc. *Page 352:* Courtesy NASA. *Pages 355 and 356:* Ken Karp/ Omni-Photo Communications, Inc. *Pages 365 (center):* Michael Watson. *Page 365 (bottom right):* COMSTOCK, Inc. *Page 371:* Carr Clifton. *Page 372:* Andy Washnik. *Page 375:* The Royal Institute. *Pages 376–378 (top and bottom left):* Andy Washnik. *Page 378 (top right):* Ken Karp/ Omni-Photo Communications, Inc. *Page 382:* Andy Washnik. *Page 388:* Couresy Argonne National Laboratory. *Page 390:* Paul Silverman/Fundamental Photographs. *Page 392 (top):* Courtesy Norton Company. *Page 392 (bottom):* Andy Washnik.

**Chapter 11** *Opener:* David Woodfall/Tony Stone Images. *Page 402:* Andy Washnik. *Page 404:* Robert Capece. *Page 406:* Andy Washnik. *Page 413:* Courtesy Johnson Wax. *Page 414:* Andy Washnik. *Page 415 (bottom left):* Heinz, U.S.A., a division of the H.J. Heinz Company. *Page 415 (right):* Andy Washnik. *Page 422:* Sir John Meurig Thomas Royal Institution of Great Britain. *Page 423:* Reproduced with permission, PepsiCo, Inc. 1991. *Pages 424 and 425:* Richard Megna/Fundamental Photographs, Inc.

**Chapter 12** *Opener:* Michael W. Davidson/Photo Researchers. *Pages 441 and 444:* Andy Washnik. *Page 445:* Paul Silverman/Fundamental Photographs. *Page 447:* Tripos Associates. *Page 450:* from *Nitrogen Structure: Where to Now?* W. H. Orme-Johnson Science, Vol. 257, Page 1539, 1992. *Page 455:* Ken Karp/Omni-Photo Communications, Inc. *Page 464:* Andy Washnik.

*Intersection:* Willard Clay.

**Chapter 13** *Opener:* Charles Krebs/The Stock Market. *Page 476 (top):* Courtesy Outokumpu American Brass Inc. *Page 476 (bottom):* Andy Washnik. *Pages 479–482:* George Bodner. *Page 483:* Todd Gipstein/Photo Researchers. *Page 485 (top):* David Schultz/Tony Stone Images. *Page 485 (bottom):* Steven Gottlieb/FPG International. *Page 488:* Richard Pasley/Stock, Boston. *Page 489:* George Bodner. *Page 493:* James King-Holmes/Science Photo Library/Photo Researchers. *Page 498:* Courtesy IBM. *Page 500:* Phototake. *Page 502:* Varian Associates. *Page 503 (top):* Chemical Design LTD/Science Photo Library. *Page 503 (bottom):* Tony Stone Images.

**Chapter 14** *Opener:* Jeff Gnass Photography. *Page 512:* Andy Washnik. *Page 514 (top):* Helmut Gritscher/Peter Arnold Inc. *Page 514 (bottom):* Andy Washnik. *Page 515 (top):* Courtesy Prism Technologies, Inc. *Pages 515 (bottom) and 516:* Andy Washnik. *Page 520 (top):* Courtesy Albany International Research Company. *Page 520 (bottom):* Courtesy Hoecht Cellanese. *Page 521:* Pat Lacroix/ The Image Bank. *Page 525:* David Madison/Duomo Photography, Inc.

**Chapter 15** *Opener:* H. Richard Johnston/Tony Stone Images. *Page 533:* Andy Washnik. *Page 535:* Chemical Design/Science Photo Library/Photo Researchers. *Page 537:* Courtesy Neutrogena Corporation. *Page 539:* Andy Washnik. *Pages 549 and 552:* Ken Karp/Omni-Photo Communications, Inc. *Page 553:* John Maher.

*Intersection:* Michael Watson.

**Chapter 16** *Opener:* David Nunuk/Science Photo Library/Photo Researchers. *Page 567:* John Kelly/The Image Bank. *Page 585:* Andy Washnik. *Page 591:* Courtesy BASF Corporation.

**Chapter 17** *Opener:* Andy Washnik. *Page 600 (top):* From *Experiments and Observations on Different Kinds of Air,* Vol. 1, by Joseph Priestley, 1774. *Page 600 (bottom):* Ken Karp/Omni-Photo Communications, Inc. *Page 601 (top):* Paul Silverman/Fundamental Photographs. *Page 601 (bottom):* Andy Washnik. *Page 623:* Courtesy Minnetonka, Inc. *Page 626:* Andy Washnik.

**Chapter 18** *Opener:* Professors P. M. Motta & S. Correr/Science Photo Libary/Photo Researchers. *Page 648:* Andy Washnik. *Page 650:* Ken Karp/Omni-Photo Communications, Inc. *Page 655:* COMSTOCK, Inc. *Page 658:* Richard Megna/Fundamental Photographs. *Page 659:* Andy Washnik. *Page 662:* Gary Russ/The Image Bank. *Page 672:* Andy Washnik. *Page 679:* Reprinted from *Accounts of Chemical Research,* 1991, 24, American Chemical Society and reprinted by permission of Jacqueline K. Barton. *Page 682:* Andy Washnik.

*Intersection: Page 692:* Movie Star News. *Page 693:* Helmut Gritscher/Peter Arnold.

**Chapter 19** *Opener:* David Stoecklein/The Stock Market. *Page 697:* Andy Washnik. *Page 698:* Y. Gladu/Photo Researchers. *Page 699 (left):* James L. Amos/Photo Researchers. *Page 699 (right):* Paul Silverman/Fundamental Photographs. *Page 705:* Ken Karp/Omni-Photo Communications, Inc. *Page 707:* Andy Washnik. *Page 709:* Ken Karp/Omni-Photo Communications, Inc. *Page 712:* Andy Washnik. *Page 716:* Oesper Collection in the History of Chemistry, University of Cincinnati.

**Chapter 20** *Opener:* Gary Ross/The Image Bank. *Page 726:* Henryk Kaiser/Leo de Wys, Inc. *Pages 727, 729, and 735:* Andy Washnik. *Page 741:* John Underwood. *Page 744:* Nikolay Zurek/FPG International. *Page 747:* Ken Karp/Omni-Photo Communications, Inc. *Page 750:* Courtesy Eveready Battery Company, Inc. *Page 756:* Courtesy Magnesium Corporation of America.

*Intersection:* Don Spiro/Tony Stone Images.

**Chapter 21** *Opener:* David Schultz/Tony Stone Images. *Page 766:* Don Spiro/Tony Stone Images. *Page 772:* Ken Karp/Omni-Photo Communications, Inc. *Page 773:* E. R. Degginger. *Page 774:* COMSTOCK, Inc. *Page 789:* Tripos Associates. *Page 791:* Ken Karp/Omni-Photo Communications, Inc.

**Chapter 22** *Opener:* Jeff Gnass/The Stock Market. *Page 802:* Michael Watson. *Page 803:* Rob Gage/FPG International. *Page 804:* Photograph, Courtesy Milton Roy Company, Rochester, NY a subsidiary of Sundstrand Corporation. *Page 815:* Douglas Burrows/Gamma Liaison. *Page 816:* Matt McVay/AllStock, Inc. *Page 821:* Tripos Associates. *Page 824:* Paul Chesley/Tony Stone Images. *Page 830:* Biophysics Laboratory/Science Photo Library/Photo Researchers.

*Intersection:* Tom Ulrich/AllStock, Inc.

**Chapter 23** *Opener:* Royal Observatory, Edinburgh/Science Photo Library/Photo Researchers. *Page 842:* Bettmann Archive. *Page 843 (top):* C. E. Wynn-Williams. *Page 843 (bottom):* UPI/Bettmann Newsphotos. *Page 856:* John Reader/Science Photo Library/Photo Researchers. *Page 857:* Charlie Ott/Photo Researchers. *Page 858 (top):* Steve Starr/SABA. *Page 858 (bottom):* David M. Grossman/Photo Researchers. *Page 859:* Ken Karp/Omni-Photo Communications, Inc. *Page 861:* Noel Quidu/Gamma Liaison. *Page 865:* Bettmann Archive. *Page 869:* Paolo Koch/Photo Researchers. *Page 871 (bottom right):* Maillac/REA/SABA. *Page 871 (top):* Courtesy the Chicago Historical Society. *Page 873 (top):* Official U.S. Air Force Photo. *Page 873 (center right):* U.S. Department of Energy/Science Photo Library/Photo Researchers. *Page 873 (bottom right):* Science Photo Library/Photo Researchers. *Page 874:* Leon Golub/IBM Research and SAO. *Page 876:* Courtesy National Cancer Institute.

**Chapter 24** *Opener:* Colin Bell/Tony Stone Images. *Page 884 (top):* Courtesy The Edgar Fahs Smith collection, taken from *The Development of Modern Chemistry* by Aaron Ihde, Harper & Row, 1964. *Page 884 (bottom):* Astronomical Society of the Pacific. *Page 892:* Andy Washnik. *Page 897 (top):* Rafael Macia/Photo Researchers. *Page 897 (bottom):* Allen Green/Photo Researchers. *Page 898:* Tripos Associates. *Page 899:* Richard Megna/Fundamental Photographs. *Page 901:* Colin Bell/Tony Stone Images. *Page 902:* Courtesy Mobil Oil. *Page 913:* Andy Washnik. *Page 917:* Charles Thatcher/Tony Stone Images. *Page 918:* Walter H. Hodge/Peter Arnold, Inc.

**Chapter 25** *Opener:* Will & Demi McIntyre/Photo Researchers. *Page 936:* E. R. Degginger. *Page 939:* Jean-Marc Barey/Photo Researchers. *Page 941:* Robert E. Daemmrich/Tony Stone Images. *Page 944:* Courtesy Institute of Scrap Recycling Industries, Inc. *Page 947:* Ellis Herwig/Stock, Boston. *Page 948:* National Museum of American History, Smithsonian Institution. *Page 949:* Andy Washnik. *Page 962:* Gary Gay/The Image Bank. *Page 967:* Carr Clifton. *Page 968:* Richard Megna/Fundamental Photographs. *Page 970:* Irene Windridge/A–Z Botanical Collection Ltd. *Page 975:* Nelson Max/LLNL/Peter Arnold, Inc. *Page 977:* Tripos Associates.

# INDEX

Page numbers in italic indicate figures or tables.


Abelson, Philip H., 866
Absolute measurement, 30
Absolute zero, 121, 123–124
Absorption spectrum, 209
Acetic acid, 287, 415
Acetylene, 392, 892
Acheson, Edward, 392
Acid(s), 401–438
 acetic, 287, 415
 amino, 289, 952–955
 Arrhenius, 404, 406
 or bases, compounds that are either, 639
 benzoic, 620
 Brønsted, 410–413, 420, 429, 605
 carboxylic, 914–916
 charge on, 427
 defined, 404–406
 deoxyribonucleic (DNA), 973–979
 dicarboxylic, 915–916
 diprotic, 629–632
 discovery of, 402
 fatty, 536, 968
 hydrochloric, *see* Hydrochloric acid
 Lewis, 429, 430–433, 447, 669–670
 mineral, 385–386
 mixtures of bases and, 623–624
 monoprotic, 629
 muriatic, 415
 nitric, synthesis of, 371–372
 nomenclature of, 65
 nonmetal hydroxides as, 409–410
 not-so-weak, 614–615
 nucleic, 952, 973–979
 operational definition of, 404–406
 oxalic, 102
 oxidation state of central atom of, 427–429
 polyprotic, 629, 630
 properties of, 402
 ribonucleic (RNA), 973
 salicylic, 970
 silicic, 421
 solid, 421
 solid-state, 421–422
 solutions of, oxidation–reduction reactions in, 704–710
 strengths of,
  factors that control, 426–428
  relative to bases, 415–416, 621
  relative to other acids, 417–419
 strong, 415, 606, 609
 succinic, 16
 sulfuric, 97, 101, 367, 368, 421
 sulfurous, 367–368
 triprotic, 629, 636–637
 weak, *see* Weak acid(s)
 weights of, *72*
Acid–base chemistry of water, 602–603
Acid–base complex, 430, 670
Acid–base equilibria, 599–646
Acid–base indicators, 424, *425*
Acid-dissociation equilibrium constant(s) ($K_a$), 416, 605–607; *630,* A-9, A-10
Acidic buffer, 626
Acrylics, 946–948
Actinides, 439
Activation energy, 823–824, 826–828
Active metals, 316–317
 ammonia ($NH_3$) and, 323
 preparation of, 339–340
 water and, 321–323
Activity, of isotope, 854
Acyl chloride, 916
Addition, with significant figures, 21
Addition polymers, 939–940, 944–948
Addition reactions, 892–893
Additive colors, 464
Adhesion, 518, 519
Alar (daminozide), 16–17
Alcohol(s), 905–908
 solubility of, in water, 534–535
Aldehydes, 912–914, 962
Alkali metals, 317–318
Alkaline dry cell, 751
Alkaline–earth metals, 324–325
Alkaloids, 918–920
Alkanes (saturated hydrocarbons), 885–890
 reactions of, 892–893
Alkenes, 890, 892
 reactions of, 892–893
Alkyl halides, 892, 904–905
Alkynes, 892
 reactions of, 892–893
Allotropes, 356
Alloys, 53
Alpha decay, 845
Alpha particles, 842, 844
Alpha radiation, 199
Aluminum, 317, 440
 chemistry of, 325–326
 electrolysis of, 340
 iron oxide ($Fe_2O_3$) combined with, 83, 84
 preparation of, 339
 reaction of, with bromine, 350
 as reducing agent, 333–334
 as structural metal, 326

Aluminum chlorhydrate [$Al_2(OH)_5Cl$], 58
Aluminum oxide ($Al_2O_3$), 83, 84, 342
Aluminum sulfate [$Al_2(SO_4)_3$], 429
Ames, Bruce, 18
Amines, 917–920
Amino acids, 289, 952–955
Ammonia ($NH_3$):
  active metals and, 323
  anhydrous, 371, 591
  burning of, 86
  decomposition of, 162
  Lewis structure of, 374
  liquid versus gaseous state of, 114
  nitrogen reduced to, 450
  oxygen combined with, 90–91
  sodium dissolved in, 318
  solubility of, in water, 600, 601
  synthesis of, 370–372, 450, 591–592
Ammonium chloride ($NH_4Cl$), 175
Ammonium nitrate ($NH_4NO_3$), 175
Amontons, Guillaume, 120
Amontons' law, 120–122, 128, 139
Amorphous carbon, 391
Amorphous solids, 476
Amphoteric compounds, 410
Amplitude of wave, 207
Amu (atomic mass unit), 74, A-2
Amylopectin, 966
Amylose, 966
Analysis:
  elemental, 103–104
  of van der Waals constants, 146–147
Anaximander, 44
Anaximenes, 44
Anderson, James, 363
Andrews, Thomas, 517
Angular molecule, 291, 293
Angular quantum number, 218
Anhydrous ammonia, 371, 591
Anionic polymerization, 942–943
Anions, 340, 729
Anode, 196, 340, 729, 743
Anthracite coal, 900
Antibonding molecular orbital, 301, 466
Antineutrino, 846
Approximation(s):
  failure of, 581–583
  methods of, 577–583
  successive, 589–591
Aqueous solutions, 84
Argon, 387
Aristotle, 44, 47
Aromatic hydrocarbons, 894–896
Arrhenius, Svante, 55, 826
  acid–base theory of, 403–404, 406
Arrhenius acid, 404, 406
Arrhenius base, 404
Arrhenius equation, 826–828
Arrhenius theory, Brønsted theory versus, 411, 412
Aspirin, 81, 970
  analysis of, 103–104
Assumptions:
  failure of, 581–583
  hidden,
    equilibrium calculations and, 576–579
    in weak-acid calculations, 611
  validity of, 579–581
Aston, Francis W., 844
  mass spectrometer invented by, 74–75
Atactic polymer, 938–939
Atmosphere:
  carbon dioxide in, 393–394
  chemistry of, 362–363
Atmospheric pressure, 116–119
Atom(s):
  bonding of, 274–275
  counting of, in solution, 98–100
  defined, 49
  evidence for existence of, 46–48
  formal charge on, 288
  geometry around, valence electrons and, *296*
  hybridization of, *297*
  ions versus, 56–57
  isoelectronic, 249
  mass of, 73–75
  mole of, 75–76
  oxidation number (state) of, 59–61, 329
  physical properties of, 250
  size of, 245–249, *A-4*
  structure of, 56, 195–238, 844–845
  weight, 74–75, 77
Atomic mass, 73–74
  atomic weight versus, 74–75
Atomic mass unit (amu), 74, A-2
Atomic number, 204
Atomic orbitals, 301
  hybrid, 297–300
  relative energies of, 226
Atomic radii, 245–247, *A-4*
Atomic spectra, 209–211
Atomic weight, 77
  atomic mass versus, 74–75
Atomic world, 50
Aufbau principle, 227
Average kinetic energy, 512–523
Avery, Oswald, 974
Avogadro, Amadeo, 126
Avogadro's constant ($N$) (Avogadro's number), 77–78, 201, 737, 773, A-2
Avogadro's hypothesis, 126–127, 128, 140
  work and, 162
Avogadro's number, *see* Avogadro's constant
Azotobacter vinelandii, 451

BAC (blood-alcohol concentration), 95–96
Bacon, Friar Roger, 372
Baekland, Leo, 948
Baking powder, 394
Baking soda, 394
Balancing equations, chemical, 85–87
Balmer, Johann Jacob, 210, 211, 214–215
Bardeen, John, 501
Barium, 324
Barium tetracyanoplatinate [$BaPt(CN_4)$], 198
Barometer, 116–120
Bartholdi, Frederic, 726
Bartlett, Neil, 388
Barton, Jacqueline, 678–679
Base(s), 401–438, 619–623
  or acids, compounds that are either, 639
  Arrhenius, 404
  Brønsted, 410–413, 420
  charge on, 427
  defined, 404–406
  diprotic, 633–635
  discovery of, 402
  Lewis, 430–433, 447, 669–670
  metal hydroxides as, 409–410
  mixtures of acids and, 623–624
  operational definition of, 404–406
  oxidation state of central atom of, 427–429
  solutions of, oxidation–reduction reactions in, 710–713
  strengths of,
    factors that control, 426–428
    relative to acids, 415–416, 621
    relative to other bases, 417–419
  weights of, *72*
Base-ionization equilibrium constant(s) ($K_b$), 619, *A-10*

Basic buffer, 626
Basolo, Fred, 312–313
Battery(ies):
 alkaline, 751
 dry cell, 751
 lead–acid, 752–753
 nickel–cadmium, 754
 primary, 750–751
 secondary, 752–754
 *see also* Cell(s)
Becquerel, Henri, 198–199, 842
Bednorz, Georg, 502–503
Beidler, Lloyd, 319
Bent molecule, 291, 293
Benzoic acid, 620
Beryllium, 324
Berzelius, Jons Jacob, 936
 atomic weights of, 72–73
Beta decay, 846–847
Beta particles, 842, 844
Beta radiation, 199
Bethe, Hans, 457
Bicarbonate of soda, 394
Bidentate ligand, 448
"Big Bang," 874
Bimolecular nucleophilic substitution, 909
Bimolecular reaction, 808
Binding energy, 850–853
Biochemical complexes, 678–679
Biological microcalorimetry, 186
Biological systems, thermodynamics of, 789–790
Biot, Jean Baptiste, 923
Bituminous coal, 900
Black, Joseph, 155, 157–158
Blood-alcohol concentration (BAC), 95–96
BNCT (boron neutron-capture therapy), 876–877
Body-centered cubic packing, 479–480
Bohr, Niels, 213–214
Bohr model, 213–216, 217, 250
 success and failure of, 216
Boiling point elevation, 543–544, 545
Boiling points, *511,* 515–516
Boltzmann, Ludwig, 772, 773
Boltzmann's constant ($k$), 773, A-2
Bomb calorimeter, 169–172, 767
Bond(s):
 covalent, *see* Covalent bond
 disulfide, 959
 double, *see* Double bonds
 hydrogen, 522–524, 678, 679, 959
 intermolecular, 471–472, 476
 ionic, 259–260, 280, 478, 961
 Lewis acid–base approach to, 446–447
 metallic, 478
 peptide, 955
 pi, 300
 polar, 280, 282, 472
 sigma, 300
 single, *see* Single bonds
 triple, *see* Triple bonds
 valence-bond approach to, 456–457
Bond order, 306
Bond-dissociation enthalpies, 182–185, 186, *A-15*
Bonding molecular orbitals, 466
Bonding molecular theory, 301
Borkenstein, R. F., 95
Boron, 325
Boron neutron-capture therapy (BNCT), 876–877
Boundary, 766
 defined, 768
 kinetic molecular theory and, 160
Boyle, Robert, 45, 47, 51, 119
 acid properties summarized by, 402
 alkalies defined by, 402–403
 element defined by, 240
Boyle's law, 2, 119–120, 128, 139
Bragg, William Lawrence, 493–494
Bragg equation, 494
Branched polymers, 937
Branched structures, 887
Brand, Hennig, 377
Brass, 45, 46, 53
Brattain, Walter, 501
Bravais, Auguste, 488
Breathalyzer, 95–96, 713
Breeders, 871
Bromides, 318
Bromine, 382
 reaction of, with aluminum, 350
Brønsted, Johannes, acids and bases defined by, 410–413, 420, 605
Brønsted acid(s), 411, 420
 transition-metal ions as, 429
Brønsted base, 411, 420
Brønsted theory, 410–414
 advantages of, 420
 Arrhenius theory versus, 411, 412
 Lewis model versus, 430–433
 relative strengths of pairs of acids and bases and, 418–419
 role of water in, 413–414
Brønsted–Lowry theory, *see* Brønsted theory
Bronze, 45, 53, 326, 338
Buffer, 625–626
Buffer capacity, 627–629
Bunsen, Robert, 209–210
Butane ($C_4H_{10}$), 88

14C dating, 856
Cadmium, 440
Caffeine, 920
Calcining of ore, 337
Calcium, 324
 oxidation of, 355
 as reducing agent, 331
Calcium carbide, 392
Calcium carbonate ($CaCO_3$), 99
Calcium phosphate [$Ca_3(PO_4)_2$], 58–59
Caloric theory of heat, 159
Calorie, 156
Calorimeter:
 bomb, 169–172, 767
 measuring heat with, 168–171
Calorimetry, 186
Calx, 696
Canal rays, 201
Carbanions, 921
Carbides, 391–392
Carbohydrate(s), 951–952, 962–968
Carbon, 326
 amorphous, 391
 C isotope of, 75
 elemental forms of, 390–391
 inorganic chemistry of, 390–395
 optical activity and, 922–924
 organic chemistry of, 883–934
 oxidation numbers of, *910*
 oxidation state of, 61
 oxides of, 392–393
 physical properties of, *885*
Carbon black, 390, 391
Carbon dioxide ($CO_2$):
 in atmosphere, 393–394
 burning of butane and, 88
 as covalent compound, 54
 Lewis structure of, 282, 292–293
 molecular weight of, 76
 as solid, 476
Carbonate ester, 950
Carbonates, chemistry of, 394
Carbonium ion, 909
Carbonyl group, 912
Carborundum, 392
Carboxylate ions, 914–916
Carboxylic acid, 914–916
Carboxylic acid esters, 917

Carnot, Sadi, 766
Carothers, Wallace, 948
Carr, John Dickson, 40
Catalyst(s):
  criteria for, 825
  rate of chemical reactions and, 825–826
  Ziegler–Natta, 943–944, 945, 946
Catalytic cracking, 899
Catalytic reforming, 899
Catenation, 365
Cathode, 196, 340, 729, 743
Cathode-ray tubes, 197
Cathodic protection, 756
Cationic polymerization, 943
Cations, 340, 729
Cell(s):
  Daniell, 733–734, 736, 737, 738
  Downs, 744, 745
  dry, 751
  dry, alkaline, 751
  electrolytic, 743
  fuel, 751–752
  galvanic, 729
  Leclanche, 750, 751
  voltaic, *see* Voltaic cells
  *see also* Battery(ies)
Cell potential(s), 729–730
  free energy and, 792–793
  magnitude of, 730, 731
  sign of, 730–731
Cellophane, 936
Celluloid, 936
Cellulose nitrate, 936
Celsius, Anders, 30
Celsius scale, 30–31
  relative temperatures measured by, 154–155
Centigrade scale, 30
Ceramics, superconductivity of, 503
Cesium, 317
Chadwick, James, 205
Chain initiation, 941–942
Chain propagation, 942
Chain reactions, 870
Chain termination, 942
Charcoal, 390, 391, 392
Chardonnet, Hilaire, Count de, 519
Chargaff, Erwin, 974
Charge, on electron (*e*), 201, 205, A-2
Charles, Jacques-Alexandre-Cesar, 122
Charles' law, 122–124, 125, 128, 140
Chelating ligand, 448
Chemical equations:
  balancing of, 85–87, 702–703
  chemical reactions represented by, 83–84
  mole ratios and, 87–88
  molecules versus moles in, 84–85
Chemical kinetics, 763, 802–804
Chemical properties, 52–54
Chemical reactions:
  activation energy of, 823–824, 826–828
  chemical equations as representation of, 83–84
  collision theory model of, 809–811
  diffusion limit of, 821
  effect of pressure on, 586–587
  effect of temperature on, 585, 588
  enzyme-catalyzed, 830–831
  forces that control, 762–764
  mechanisms of, 811–812
    determination of, 821–822
  rate-limiting step of, 810
  rates of, catalysts and, 825–826
  spontaneous, 771–772
  *see also* Equilibrium(ia); Reaction(s)
Chemical separations, 550–553
Chemical thermodynamics, *see* Thermodynamics
Chemistry:
  acid–base, of water, 602–603
  of atmosphere, 362–363
  of carbonates, 394
  of coal, 900–902
  descriptive, 312–313
  of garlic, 925–927
  gas-phase, 136–137
  gas-phase acid–base, 608–609
  of halogens, 382–387
  of hydrogen, 352–355
  inorganic, of carbon, 390–395
  knowledge of, application of, 839
  medicinal, 970–972
  of nitrogen, 370–381
  of nonmetals, 349–400
  nuclear, 841–882
  organic, 390
    of carbon, 883–934
  of oxygen, 356–363
  of ozone, 356–357
  of petroleum products, 897–899
  of phosphorus, 377–381
  principles of, 1–3
  of rare gases, 387–389
  of sulfur, 364–369
  transition-metal, 439–470
  of weak acids, 616–618
Chemotherapy, 876
Chiral compounds, 924
Chiral isomers, 455, 456
Chlorides, 318
Chlorin ring, 451
Chlorine, 382
  atomic weight of, 75
  magnesium combined with, 78–79
  oxyanions and oxyacids of, 387
  physical properties of, *275*
Chlorofluorocarbons (CFCs), 363
Chlorophyll *a*, 451, 452
Cholesteric liquid crystals, 498
Chromatography, 552–553
Cinnabar, 46
Cis isomers, 454
Clausius, Rudolf, 164
Clausius–Clapeyron equation, 526
Closest packed structures, 480, 487
  holes in, 485–487
Clostridium pasteurianum, 451
Coal, chemistry of, 900–902
Coal gas, 901
Coal gasification, 901
Coal liquifaction, 902
Cobalt, 441
Cocaine, 918, 919
Cohesion, 518, 519
Coinage metals, 316, 317, 336
Coke, 337, 338, 390, 391, 392
Colligative properties, 541–543, 544–548
Collision theory, 565–566
Collision theory model, 809–811
Color(s):
  additive, 464
  subtractive, 464
  of transition-metal complexes, 463–465
  wavelengths of light and, 465
Column chromatography, 553
Combination reactions, 312
Combined equilibria:
  qualitative view of, 680–682
  quantitative view of, 683–686
Committee on the Biological Effects of Ionizing Radiation, 861
Common ion effect, 660–662
Complex(es):
  acid–base, 430, 670
  biochemical, 678–679
  coordination, 441–443, 451–452
  high-spin, 461–463
  Lewis acid–base approach to bonding in, 446–447
  low-spin, 461–463
  nomenclature of, 453

octahedral, 458–459
high-spin versus low-spin, 461–463
square-planar, 460
tetrahedral, 459–460
transition-metal, colors of, 463–465
valence-bond approach to bonding in, 456–457
Complex ion(s), 430, 441, 668–670
calculations and, 674–676
dissolving of insoluble salt and, 676–677
stepwise formation of, 670–673
Complex-dissociation equilibrium constant ($K_d$), 673
Complex-formation equilibrium constants ($K_f$), 670, *A-12*
Compound(s):
amphoteric, 410
boiling points of, *511*
chiral, 924
coordination, 441
covalent, *see* Covalent compounds
defined, 48
dextrorotatory, 456, 923
diamagnetic, 356, 463
either acid or base, 639
as gas at room temperature, 112
hydrocarbon, *see* Hydrocarbons
inorganic, 884
interhalogen, 386
intermetallic, solid solutions and, 484–485
ionic, *see* Ionic compound(s)
levorotatory, 456, 923
melting points of, *511*
mixture versus, 45–46
nomenclature of, *see* Nomenclature
optically active, 456, 923
organic, 884–885
paramagnetic, 356, 463
polar, 279, 280
polymer, *see* Polymer(s)
rare-gas, 388
tetravalent, 886
Compressibility, of gases, 113–114
Compression ratio, 113
Concentration, 538–541
changes in,
Le Chatelier's principle and, 586–587
upon reaction coming to equilibrium, 572–576
of ion, adjustment of, 666–667
of ore, 336
of solutions, 94–96, 98
Condensation polymers, 939–940, 948–951
Conduction band, 499
Conjugate acid–base pairs, 414–415
relative strengths of, 417, 621
Conjugate oxidizing agent-reducing agent pairs, 331–333
Constant composition, law of, 45
Constant pressure, 172
Constant volume, 172
Constitutional isomers, 887, 922
Conversion factors, selected, *A-3*
Coordination complexes, 441–443
in nature, 451–452
Coordination compounds, 441
Coordination numbers, 441, 444, 479
structures of metals and, 482–483
Coordination polymerization, 943–944
Copolymers, 937–938
Copper, 45, 53, 440
density of, 326
heating of, 331–332
smelting of, 338
Copper sulfate ($CuSO_4$), 98
Corey, Robert, 959, 974
Corrin ring, 451
Corrosion, 358, 698–699, 726
galvanic, 754–756
Coulomb (C), 198, 729, A-2
Covalent bond, 271–311, 478
discovery of, 273–274
polar, 280
*see also* Bond(s); Ionic compounds
Covalent carbides, 391–392
Covalent compounds, 54–56
ionic compounds and, 275–277
nomenclature of, 64
Covalent radii, 246–249
Covalent solids, 477
Crick, Francis, 974, 976
Critical mass, 870
Critical pressure, 516–518
Critical temperature, 516–518
Crookes, William, 197
Cross-linked polymers, 937
Crude oil, 897
Crystal(s):
liquid, 498
monoclinic, 365
orthorhombic, 365
unit cell of, 494–496
Crystal fields:
octahedral, 458–459
tetrahedral, 459–460
Crystal-field theory, 457–460
Crystalline solids, 476
Cubic closest packed structures, 481, 490
Cubic holes, 487
Cubic packing:
body-centered, 479–480
simple, 479, 485–487
Curie, Irene, 864–865
Curie, Marie, 199, 552, 864
Curie, Pierre, 199, 552, 859, 864
Curie (Ci), 854
Cycloalkane, 888

Dalton, John, 47
atomic theory of, 72, 126, 193, 200, 271
law of partial pressures of, 133–138, 140
Daminozide (Alar), 16–17
Daniell, John Frederic, 733–734, 750
Daniell cell, 733–734, 736, 737, 738
Dating:
$^{14}C$, 856
potassium–argon, 857
radioactive decay, 856–857
Davisson, C. J., 217
Davy, Humphry, 403, 748
"laughing gas" and, 375
potassium metal first prepared by, 339, 384
De Broglie, Louis-Victor, 216
De Broglie equation, 217, 218
Debye (*d*), 282
Decay, 846–847
alpha, 845
radioactive,
dating by, 856–857
kinetics of, 854–855
modes of, 845–847
Decomposition reactions, 312
Definite proportions, law of, 47
Degenerate orbitals, 227
Democritus, 47
Density, as intensive property, 27–29
Deoxyribonucleic acid (DNA), 973–979
Department of Agriculture, Alar and, 16–17
Derived SI units, 12
Descriptive chemistry, 312–313
DeSimone, John, 319
Detergents, 536–538
Deuterium, 873
Dextrorotatory compounds, 456, 923
Diamagnetic atoms, 306
Diamagnetic compounds, 356, 463

Diamagnetic molecules, 306
Diamond(s), 390, 391, 392
  as example of crystalline solid, 476
  melting point of, 477
Dicarboxylic acids, 915–916
Dienes, 892
Diffusion, of gases, 140–142
Diffusion limit, 821
Dilution, calculation of, 101–104
Dimensional analysis, 22–25
Dimer, 376
Dioxin (TCDD) (2,3,7,8-tetrachlorodibenzo-*p*-dioxin), 18
Dipole moment, 282, 472
Dipole–dipole forces, 472–473
Dipole-induced dipole forces, 473
Diprotic acids, 629–632
Diprotic bases, 633–635
Disaccharides, 965–966
Disorder, entropy and, 772–774
Disproportionation reactions, 360–361, 712
Dissociation, stepwise, 630
Distillation, 550
Disulfide bonds, 959
Division, with significant figures, 21–22
DNA (deoxyribonucleic acid), 973–979
  biochemical complexes and, 678–679
  transition metal ions and, 678
Dobereiner, Johann Wolfgang, 240–241
Double bonds:
  molecules with, 299–300
  strength of, 364–366, 379
  VSEPR theory and, 292–293
Downs cell, 744, 745
Drake, Edwin, 897
Dry cells, 751
Dry ice, 476
Dry-cleaning agents, 538
Du Fay, Charles Francois de Cisternay, 196
Dynamic equilibrium, 567

Effusion, of gases, 141–142
Eiffel, Alexandre, 726
Einstein, Albert, 193, 212, 213, 216, 258, 766
Elastomers, 940–941
Electrical conductivity, 499–501
Electrical resistivity, 499–501
Electrical work, 162, 726–728, 769
Electricity, 196
Electrochemistry, 725–761
Electrolysis, 339–340
  Faraday's law of, 748–750
  of sodium chloride,
    aqueous, 745–747
    molten, 743–745
  of water, 355, 747
Electrolytic cells, 743
Electromagnetic radiation, 207–209
Electron(s):
  charge on (*e*), 201, 205, A-2
  configuration of, *see* Electron configuration(s)
  discovery of, 196–198, 272
  distribution of, *297*
  localization of, 478
  nonbonding, VSEPR theory and, 293–296
  rest mass of, A-2
  sharing of, 274–275
  solvated, 318
  valence, 272–273, 296, 478
  wave properties of, 217
Electron affinity(ies), 255–257, *A-7*
  relative size of, 259
Electron capture, 846
Electron configuration(s), 227–229
  periodic table and, 230–232
  predicted, exceptions to, 230
  of transition-metal ions, 445
Electron emission, 846
Electron-pair acceptor, 430, 447, 669
Electron-pair donor, 430, 447, 669
Electronegativity(ies), *A-8*
  limitations of, 281
  of main-group metals, 354
  of nitrogen, 379–380
  of oxygen, 366–369
  of phosphorus, 379–380
  polarity and, 277–280
  of sulfur, 366–369
Electrophiles, 913
Electrostatic interactions, 678
Element(s), 45–46
  chemistry of, 51–52
  defined, 45, 48, 240
  as gas at room temperature, 112
  Greek, 44–45
  Newlands' table of, *241*
  *see also* Periodic table
Elemental analysis, 103–104
Emission spectrum, 210
Empedocles, 44
Empirical formulas, 80–82
Empirical weight, 82
Emulsification, 537
Endergonic reactions, 781
Endothermic reactions, 174, 781
Endpoint, in acid–base titration, 102, 629
Energy(ies):
  activation, 823–824, 826–828
  of atomic orbitals, 226
  binding, 850–853
  conservation of, 164–166
  conversion factors of, *A-3*
  free, *see* Free energy
  Gibbs free, 779–782
  internal, *see* Internal energy
  ionization, *see* Ionization energy(ies)
  kinetic, 139, 512–513
  lattice, 259–261
  quantization of, 211–213
  reradiation of, 483
Engine "knock" or "ping," 897
English system of measurement, 8
Enkephalins, 956
Enthalpy(ies) (*H*):
  bond-dissociation, 182–185, 186, *A-15*
  defined, 770
  versus internal energy, 171–173, 768–771
Enthalpy of formation, 178–182, *374*
  standard-state, *see* Standard-state enthalpy(ies) of formation
Enthalpy of reaction, 173, 174–176
  entropy of reaction calculations versus, 778–779
  standard-state, 176
Entropy(ies):
  as measure of disorder, 772–774
  second law of thermodynamics and, 774–776
  standard-state, 777, *A-15–A-24*
Entropy of reaction, 776–779
Environmental Protection Agency (EPA), Alar and, 16–17
Enzyme-catalyzed reactions, 830–831
EPA (Environmental Protection Agency), Alar and, 16–17
Epsom salts, 341
Equality, equation versus, 8–9
Equation(s), 8–9
  Arrhenius, 826–828
  Bragg, 494
  chemical, *see* Chemical equations
  Clausius-Clapeyron, 526
  De Broglie, 217, 218

ideal gas, 128–133, 143–147
Nernst, 736–740, 793
oxidation–reduction, 702–704
state, 166
van der Waals, 143–147, 517
Equilibrium(ia), 513
acid–base, 599–646
acid-dissociation, 416, 605–607, 630
base-ionization, 619
calculations in, 576–583
changes in concentration and, 572–576, 586–587
combined,
qualitative view of, 680–682
quantitative view of, 683–686
complex-dissociation, 673
complex-formation, 670
complex-ion, dissolving of insoluble salt and, 676–677
defined, 566
dynamic, 567
Le Chatelier's principle and, 586–587
partial pressure and, 583–585
solubility, 649–651, 668
solubility product, 653, 654–656, 657–658
water-dissociation, 602
*see also* Chemical reactions
Equilibrium constant(s) ($K$), 567
free energy and, 785–788
Nernst equation and, 740–743
rate constant and, 812–813
temperature dependence of, 790–792
Equilibrium constant expressions, 566–569
Equilibrium reactions, altering or combining of, 569–570
Equilibrium region, 564
Equivalence point, in acid–base titration, 102, 629
Error, in measurement, 6, 14–19
Essential oils, 925
Ester(s), 916–917, 950
Ethanol, 906
Ether, 908
Ethics of science, 2
Ethyl alcohol, 906
Everest, George, 6
Excess reagent, 91
Excitation, 858
Exergonic reactions, 781, 789
Exothermic reactions, 174, 781
Expandability, of gases, 114
Expansion, work of, 162, 769
Experimental law, 2
Extensive properties, 27, 28, 155, 541
Extraction, 550, 554

Face-centered unit cell (FCC), 490
Fahrenheit, Daniel Gabriel, 30
Fahrenheit scale, 30
relative temperatures measured by, 154–155
Fajans, Kasimir, 843
Families (periodic), 51–52
Faraday, Michael, 196, 201, 748, 894
Faraday's constant, 201, 748, A-2
Faraday's law, 748–750
Farrell, John, Jr., 136–137
Fatty acids, 536, 968
FDA, *see* Food and Drug Administration
Feldspar, 421
Fermi, Enrico, 848, 869, 870, 871, 872
Fertile nuclide, 870
First ionization energy, 250–253
First-order rate law, 815–817, 829
First-order reaction, 808
Fischer, Emil, 963
Fischer, Ernst Gottfried, 72
Fischer, Franz, 902
Fission:
nuclear, 868–872
spontaneous, 847
Fission reactions, 852
Fixation, nitrogen, 450–451
Fluorides, 318
Fluorine, 382, 716
Lewis structure of, 274
Food and Drug Administration (FDA):
Alar and, 16–17
pesticide residue analysis by, 18
Force(s):
of adhesion, 518, 519
chemical reactions controlled by, 762–764
of cohesion, 518, 519
dipole–dipole, 472–473
dipole–induced dipole, 473
exerted by gas, versus pressure of gas, 115–116
induced dipole–induced dipole, 473
van der Waals, 472–473, 476, 678, 679
Formal charge, 288
Formaldehyde ($H_2CO$), 299
Formulas:
of common ionic compounds (salts), 57–58
empirical, 80–82
molecular, 81–82
Francium, 317
Franklin, Benjamin, 196
Free energy:
cell potentials and, 792–793
equilibrium constants and, 785–788
Gibbs, 779–782
standard-state,
of reaction, 780–785
Free radical, 859, 904
Free-radical polymerization reactions, 941–942
Freezing point, 514–515
Freezing point depression, 543–544, 545
French Academy of Sciences, 752
Frequency ($\nu$), 207
Fructose, 963, 966
melting point of, 515
Fuel cells, 751–752
Fuller, R. Buckminster, 395
Fullerenes, 395
Functional groups, 903–904
Fundamental constants, selected, values of, *A-2*
Fundamental physical constants, 258
Furanose, 964
Fusion:
latent heat of, 158
nuclear, 872–874
Fusion reactions, 852

Galen, 30
Galileo, 117, 207
first thermoscope constructed by, 29
Gallium, 325
Galvani, Luigi, 728, 729
Galvanic cells, 729
Galvanic corrosion, 754–756
Gamma emission, 847
Gamma particles, 842
Gamma rays, 199
Gamow, George, 874
Garlic, chemistry of, 925–927
Garrett, Thomas, 375
Gas(es), 111–152
Amontons' law and, 120–122, 128, 139
Avogadro's hypothesis and, 126–127, 128, 140, 162

Gas(es) *(Continued)*
  Boyle's law and, 119–120, 128, 139
  Charles' law and, 122–124, 125, 128, 140
  coal, 901
  compressibility of, 113–114
  critical pressure and temperature of, 516–518
  Dalton's law of partial pressures of, 133–138, 140
  diffusion of, 140–142
  effusion of, 141–142
  expandability of, 114
  Gay-Lussac's law and, 125–126
  inert, 388
  kinetic molecular theory of, 138–140, 141–142
  "laughing," 375
  natural, 897
  pressure of,
    versus force exerted by, 115–116
    versus pressure due to weight, 119–120
  properties of, 112–115
  rare, chemistry of, 387–389
  at room temperature, 112, *113*
  solubility of, in solvent, 600–602
  as state of matter, 112
  synthesis, 901
  town, 392–393, 901
  van der Waals constants for, *144*
  van der Waals equation and, 143–147
  volumes of, versus volumes of liquids or solids, 114–115
  water, 392–393, 901
Gas-phase acid–base chemistry, 608–609
Gas-phase acidity, 608
Gas-phase chromatography, 553
Gas-phase ion chemistry, 136–137
Gas-phase reactions, 563–597
  collision theory model for, 565–566
Gay-Lussac, Joseph Louis, 125, 384
Gay-Lussac's law, 125–126
Geiger, Hans, 195, 202–203
Genetic code, *978*
Germanium, 326
Gibbs, Willard, 780
Gibbs free energy, 779–782
Gilbert, William, 196
Glucose, 963
  melting point of, 515
Glycogen, 967
Gold, Lois Swirsky, 18
Goldstein, Eugen, 201
Goodyear, Charles, 936
Graham, John, 17
Graham, Thomas, 140
Graham's laws of diffusion and effusion, 140–142
Grams, conversion of, into moles, 78–79
Graphical treatment of data, 32–35
Graphite, 390–391
  vaporization of, 395
Great Trigonometrical Survey of India, 6–8
Greek elements, 44–45
Green, Laura, 17
Grignard, Francois Auguste Victor, 921
Grignard reagents, 920–922, 942
Group number (periodic), 57
Groups (periodic), 51–52, 57
Gunpowder, 372

Haber, Fritz, 370
Haber process, 370–371, 372, 450
  Le Chatelier's principle and, 591–592
Half-life:
  of $^{14}C$, 816
  defined, 855
Half-reaction method of balancing oxidation–reduction equations, 703–704
Halides, 318
  alkyl, 904–905
  halogen, 382, 384
  hydrogen, 385–386
  nitrogen, 374
Hall, Charles Martin, 340
Halogens, 318
  chemistry of, 382–387
  elemental form of, 382–383
  neutral oxides of, 386
  oxidation numbers for, 384
  preparation of, 384
Hard water, 537
Harper, Harry, 319
Hayward, Nathaniel, 936
Heat, 155
  caloric theory of, 159
  interconversion of work and, 164–166
  kinetic molecular theory and, 159–161
  latent, 157–159
  measurement of, with calorimeter, 168–171
  *see also* Temperature
  specific, 156, *157*
  work from, 163–164
Heat capacity, 156–157
Heck, Gerald, 319
Heitler, Walter, 456
Heitler–London model of covalent bonds, 456
Helium, 387
Heme, 451, 452
Henry, William, 600
Henry's law, 600
Heraclitus, 44
Heroin, 919
Hertz, Heinrich, 212
Hess, Germain Henri, 177
Hess's law, 177–178, 179, 258
Hexadentate ligand, 449
Hexagonal closest packed structures, 480–481
High-spin complexes, 461–463
High-temperature semiconductors, 502–503
Hill, Julian, 948
Hiroshima, 872, 873
Hittorf, Johannes, 197
Hoffman, Felix, 970
Homocitrate, 450, 451
Homopolymers, 937
Hubble, Edwin, 874
Hund, Frederich, 227
Hund's rules, 227–228, 229, 252–253, 461
Hunter, W. N., 972
Hyatt, John Wesley, 936
Hybrid atomic orbitals, 297–300
Hydrazine ($N_2H_4$), 372–373
Hydrides, 318, 353, 407
Hydrocarbons:
  aromatic, 894–896
  immiscibility of, 534
  naturally occurring, 893–894
  saturated (alkanes), 885–890
  straight-chain, 887
  unsaturated, 890–892
Hydrochloric acid (HCl), 99, 415
  metal combined with, 100
  zinc and, 726–728
Hydrogen, 45
  chemistry of, 352–355
  as oxidizing agent, 331
  oxygen combined with, 83, 84–85
  reaction of, with magnesium, 324–325

Hydrogen bonding, 522–524, 678, 679, 959
Hydrogen economy, 359
Hydrogen halides, 385–386
Hydrogen peroxide ($H_2O_2$), 360–361
Hydrogen-ion acceptors, 411
Hydrogen-ion donors, 411
Hydrophilic molecules, 534–535
Hydrophobic interactions, 678, 961
Hydrophobic molecules, 534–535
Hydroxyapatite [$Ca_5(PO_4)_3(OH)$], 58–59
Hypothesis, 2–3

Ice, structure of, 523
ICR (ion cyclotron resonance) spectrometer, 136, 137
Ideal gas calculations, 129–133
Ideal gas constant (*R*), 128, 773, A-2
Ideal gas equation, 128–133
  deviations from, 143–147
Ignition temperature, 873
Immiscible liquids, 534
Indicators, 101–102
Indium, 325
Induced dipole–induced dipole forces, 473
Induced radioactivity, 864–868
Inert gases, 388
Ingold, Sir Christopher, 807
Initial instantaneous rate of reaction, 804
Inner transition metals, 439
Inorganic chemistry, of carbon, 390–395
Inorganic compounds, 884
Insoluble salts, 652
  dissolving of, using complex-ion equilibria, 676–677
Instantaneous rate of reaction, 804–805
Integrated form of rate law, 815–819, 829–830
Intensive properties, 27–31, 154–155, 541
Interhalogen compounds, 386
Intermetallic compounds, solid solutions and, 484–485
Intermolecular bonds, 471–472, 476
Internal energy (*E*), 164, 766
  enthalpy versus, 171–173, 768–771
International System of Units (SI), 12–13, 156, 164, 198, 199–200, *A-2*
International Union of Pure and Applied Chemistry (IUPAC), 888
Interstices, 485
Interstitial carbides, 392
Interstitial solution, 485
Intramolecular bonds, 471–472, 476
Iodides, 318
Iodine, 382–383
Ion(s), 403
  atoms versus, 56–57
  carbonium, 909
  carboxylate, 914–916
  common, 660
  complex, 430, 441, 668–670
    calculations and, 674–676
    dissolving of insoluble salt and, 676–677
    stepwise formation of, 670–673
  concentration of, adjustment of, 666–667
  defined, 56
  formation of, by dissociation of water, 405–406
  isoelectronic, 249
  negative, nomenclature of, 63–64
  oxidation numbers and, 329–330
  polyatomic, 58–59
  positive, nomenclature of, 62–63
  radii of, *A-4*
  separation of, by selective precipitation, 664–666
  strong-field, 461
  transition-metal, 429, 445, 464
  weak-field, 461
Ion cyclotron resonance (ICR) spectrometer, 136, 137
Ion product ($Q_{sp}$), solubility calculations and, 658–660
Ionic bond, 280, 478
  lattice energies and, 259–260
  protein structure and, 961
Ionic carbides, 392
Ionic compound(s), 54–56, 239–269, 278
  covalent compounds and, 275–277
  formulas of, 57–58
  nomenclature of, 62
  in water, solubility rules for, *652*
  *see also* Salt(s); Solubility
Ionic radii, 247–250
Ionic solids, 648
Ionization, 858
Ionization energy(ies), *A-5, A-6*
  first, 250–253
  relative size of, 259
  second, third, fourth, and higher, 253–255
Ionizing radiation, 857–862
Iron, 83, 84
  pig, 338
  as reducing agent, 333–334
  rusting of, 358, 699, 726, 754–756
  smelting of, 338
  wrought, 338
Iron oxide ($Fe_2O_3$), aluminium combined with, 83, 84
Isobars, 845
Isoelectronic atoms or ions, 249
Isolated systems, 774
Isomer(s), 512, 887, 936
  chiral, 455–456
  cis, 454
  constitutional, 887, 922
  meta, 896
  ortho, 896
  para, 896
  trans, 454, 891
Isotactic polymer, 939
Isotones, 845
Isotope(s), 75, 844, 845
  activity of, 854
  common, *225*
IUPAC (International Union of Pure and Applied Chemistry), 888

Joliot, Frederic, 864–865
Joule, James Prescott, 163–164
Joule (J), 156, 164, 729, A-2

Kekule, August, 895
Kelvin, Lord (William Thomson), 30, 164
Kelvin scale, 30–31
  absolute temperatures measured by, 155
Kenttamaa, Hilkka, 136–137
Ketones, 912–914, 962
Kim, Jongsun, 451
Kinetic control, 763
Kinetic energy, 139
Kinetic molecular theory:
  of gases, 138–140, 141–142
  heat and, 159–161
Kinetic region, 564
Kinetics:
  chemical, 763, 802–804
  of enzyme-catalyzed reactions, 830–831
  of radioactive decay, 854–855
  thermodynamics versus, 801

Kirchhoff, Gustav, 209–210
Knox, George and Thomas, 384
Kohn, Alexander, 2
Kostiropoulos, Konstantinos, 395
Kratschmer, Wolfgang, 395
Krypton, 387

Lactose, 965
Lambert, Joseph, 121
Lanthanides, 439
Latent heat, 157–159
Lattice energies, 259–261
Lattice points, 488
"Laughing gas," (nitrous oxide) ($N_2O$), 375
Lavoisier, Antoine, 72, 83, 240, 356
- acid theory of, 403
- oxygen theory of combustion of, 696–697

Law(s):
- Amontons', 120–122, 128, 139
- Boyle's, 2, 119–120, 128, 139
- Charles', 122–124, 125, 128, 140
- of combining volumes, 125–126
- of conservation of energy, 164–166
- of conservation of matter, 83
- of constant composition, 45
- Dalton's, of partial pressures, 133–138, 140
- of definite proportions, 47
- experimental, 2
- Faraday's, 748–750
- Gay-Lussac's, 125–126
- Graham's, 140–142
- Henry's, 600
- Hess's, 177–178, 179, 258
- Raoult's, 542, 544–545
- rate, *see* Rate law(s)
- scientific, 3
- of thermodynamics, *see* Thermodynamics

Lawrence, Ernest, 872
Le Bel, Joseph, 924
Le Chatelier, Henry-Louis, 586
Le Chatelier's principle, 586–588, 603, 622, 680
- haber process and, 591–592

Lead, 326–327
- smelting of, 338

Lead oxide (PbO), 342
Lead-acid battery, 752–753
Leclanche cell, 750, 751
Leucippus, 47
Leveling effect of water, 420
Levorotatory compounds, 456, 923
Lewis, G. N., 446–447, 456
- acids and bases defined by, 430–433, 669
- octet theory of, 272, 273

Lewis acid, 429, 430–433, 447, 669, 670
Lewis acid–base approach to bonding in complexes, 446–447
Lewis base, 430–433, 447, 669, 670
Lewis structures, 274
- bond orders calculated from, 306
- Grignard reagents and, 921, 942
- of ligands, 449
- oxidation–reduction reactions and, 713
- proteins and, 958–959
- theory of resonance and, 895
- writing of, 282–284

Libby, Willard F., 856, 857
Liebig, Justig, 403, 404
Ligand(s), 441, 448–449, 461
Ligand-field theory, 466–467
Light:
- as form of electromagnetic radiation, 208–209
- speed of, 78, 208, A-2
- wavelength of, colors and, 465

Lignite coal, 900
Limiting reagents, 90–94
Line notation, 735–736
Linear molecule, 291, 295
Lipids, 893, 952, 968–970
Liquid(s), 509–530
- boiling point of, 515–516
- critical pressure and temperature of, 516–518
- freezing point of, 514–515
- immiscible, 534
- materials which form at room temperature, 510–512
- melting point and, 514–515
- phase diagrams and, 524–525
- as state of matter, 112
- structure of, 471–473, 510
- supercooling of, 514–515
- surface tension and, 518–519
- vapor pressure and, 512–514
- viscosity of, 521
- volume of, versus volume of gases, 114–115

Liquid crystals, 498
Lister, Joseph, 907
Lithium, 317, 320, 321
- as medication, 341
- nitrogen combined with, 85

Lithium nitride ($Li_3N$), 85, 328
Litmus, 402, 424
Localization, of electrons, 478
London, Fritz, 456
Louyet, Paulin, 384
Low-spin complexes, 461–463
Lowry, Thomas, acids and bases defined by, 410–413
LSD, 919

M (molarity), 98–100
Macroscopic world, 50
Magnesia ($M_gO$), 47
- milk of, 312, 341

Magnesium, 327–328
- burning of, 323
- chlorine combined with, 78–79
- nitrogen combined with, 58
- oxidation of, 355
- oxygen combined with, 87, 93
- preparation of, 339
- reaction of, with hydrogen, 324–325
- as reducing agent, 330, 339
- as structural metal, 326

Magnesium chloride ($MgCl_2$), 78–79
Magnesium hydride ($MgH_2$), 325
Magnesium nitride ($Mg_3N_2$), 58
Magnesium oxide (MgO), 87, 93, 323
Magnetic quantum number, 218–219
Magnetic resonance imaging (MRI), 225
Main-group metals, 315–348
- electronegativities of, 354
- reactions of, *331*, 350–351
  - with other elements, 319–323
- as reducing agents, 331–336
- salts of, 341–342

Maltose, 965
Manhattan Project, 872
Marsden, Ernest, 202–203
Marsh gas (methane) ($CH_4$), 80
Mass:
- atomic, *see* Atomic mass
- conversion factors of, *A-3*
- critical, 870
- as extensive property, 27, 28
- molar, 77
- of products, in chemical reactions, 88–90
- weight versus, 11–12

Mass defect, 851
Mass percent, 539
Mass spectrometer, 73–74
Matter:
- conservation of, 83

properties of, 154–155
states of, 112
Maximum Tolerated Dose (MTD), 17
Maxwell, James Clark, 208
McMillan, Edwin M., 866
Measurement:
absolute, 30
accuracy in, 15–19
electromechanical, 729
elements of, 6
of equilibrium constants, 740–743
error in, 15–19
extensive properties and, 27, 28
of fundamental physical constants, 258
fundamentals of, 5–39
graphical treatment of, 32–35
intensive properties and, 27–31
precision in, 15–19
relative, 30
of risk, 16–18
rounding off in, 22
scientific notation in, 25–26
significant figures in, 19–22
simple unit conversions of, 8–9
uncertainty in, 13–14
in unit cells, 493–494
units of,
conversion of, 8–9, 22–25
English, 8
metric, 10–11, 12, 156
non-SI, 13
SI, 12–13, 156, 164, 198, 199–200, *A-2*
zeros in, 20
*see also* Stoichiometry; Temperature
Mechanisms, of chemical reactions, 811–812, 821–822
Medicinal chemistry, 970–972
Meissner, W., 503
Melting point, *511,* 514–515
Mendeleeff, Dmitri Ivanovitch, 1, 241–243
periodic table of, 2
Meniscus, 519
Menton, M. L., 830, 831
Mercury, 440
in barometer, 116, 117–120
Merrifield, R. B., 957–958
Messenger RNA (mRNA), 977
Meta isomer, 896
Metal hydrides, 407
Metal hydroxides, 407
as bases, 409–410
Metal oxides, 407
Metallic bond, 478
Metallic radii, 245–246
Metallic solids, 477–478
Metalloids (semimetals), 52, 263–265, 349
properties of, 53
Metals, 52
active, 316–317
ammonia ($NH_3$) and, 323
preparation of, 339–340
water and, 321–323
alkali, 317–318
alkaline-earth, 324–325
calx and, 696
coinage, 316, 317, 336
corrosion of, 358, 698–699, 726, 754–756
electrical conductivity of, 499–501
electrical resistivity of, 499–501
hydrochloric acid combined with, 100
main-group, *see* Main-group metals
ore, 336–337
preparation of, by chemical means, 336–339
properties of, 263–264
chemical, 53–54
physical, 53
reaction of, with nonmetals, 277
reactive, 316
reradiation of energy by, 483
roasting of, 696
structural, 326
structure of, 478–482
coordination numbers and, 482–483
physical properties resulting from, 483–484
transition, *see* Transition metals
Metastable nuclide, 847
Metathesis reactions, 313, 701
Methane ($CH_4$), 80, 362
heat given off by, 170–171
Lewis structure of, 885
Methanol ($CH_3OH$), 56, 906
Methods of approximation, 577–583
Methyl alcohol, 56, 906
Metric system, 10–11, 12
calorie defined under, 156
Michaelis, Lenor, 830, 831
Microcalorimetry, biological, 186
Milk of magnesia, 312, 341
Milky Way, 78, 875
Miller, Amy Stevens, 258
Millikan, Robert, 77, 200
Mineral acids, 385–386
Mitscherlich, Eilhardt, 894
Mixture, compound versus, 45–46
Mohr, Hans, 2
Moissan, Henri, 384, 387–388, 716
Molal boiling point elevation constant, 545
Molal freezing point depression constant, 545
Molality (*m*), 539
Molar heat capacity, 156, *157*
Molar heat of reaction, 171
Molar mass, 77
Molar solution, 539
Molarity (*M*), 98–100, 538–539
Mole(s):
chemical equations and, 84–85
as collection of atoms, 75–76
as collection of molecules, 76–77
conversion of grams into, 78–79
defined, 75
number of particles in, 77–78
Mole fraction, 539, 540
Mole ratios, chemical equations and, 87–88
Molecular formulas, 81–82
Molecular orbital theory, 301–307
Molecular oxidation–reduction reactions, 713–715
Molecular solids, 476, 648
Molecular weight, 76, 77, 82
Molecularity, 808–809
Molecule(s):
chemical equations and, 84–85
defined, 49
diamagnetic, 306
with double and triple bonds, 299–300
hydrophilic, 534–535
hydrophobic, 534–535
mole of, 76–77
paramagnetic, 306
polar, 282
shapes of, 288–293
Moment, dipole, 282, 472
Monatomic solids, 478–482
Monoclinic crystals, 365
Monodentate ligands, 448
Monoprotic acids, 629
Monosaccharides, 962–964
Montegolfier, Etienne and Joseph, 122, 123
Morphine, 918, 919
Moseley, H. G. J., 204
Motor octane numbers, 898
MRI (magnetic resonance imaging), 225

mRNA (messenger RNA), 977
MTD (Maximum Tolerated Dose), 17
Muller, Alex, 502–503
Multidentate ligand, 449
Multiplication, with significant figures, 21–22
Muriatic acid, 415

Nagasaki, 872
Naming of compounds, *see* Nomenclature
National Academy of Sciences, 861
National Resource Defense Council (NRDC), Alar and, 16–17
Natta, Giulio, 943
Natural gas, 897
Natural polymers, 951–952
Natural radioactivity, 862–864
Naturally occurring hydrocarbons, 893–894
Nature:
  on atomic scale, 193
  coordination complexes in, 451–452
  on macroscopic scale, 40–42
  of science, 1–3
Negative electrode (cathode), 196, 340, 729, 743
Negative ions, 63–64
Nematic liquid crystals, 498
Neon, 387
Nernst, Hermann Walther, 737
Nernst equation, 736–740, 793
Neuron, rest mass of, A-2
Neutral fats, 968–969
Neutrino, 846
Neutron, 205
  thermal, 865, 869
Newlands, John, 241
Newlands' table of elements, *241*
Nicad batteries, 754
Nickel–cadmium batteries, 754
Nickles, Jerome, 384
Nicotine, 918, 919
Nitrates, 372
Nitric acid ($HNO_3$), 371–372
Nitrides, 318
Nitrogen:
  chemistry of, 370–381
  electronegativities of, 379–380
  lithium combined with, 85
  magnesium combined with, 58
  oxidation numbers for, 372–377
  reduction of, to ammonia, 450
  valence shell of, 380–381
Nitrogen fixation, 450–451
Nitrogen fluoride ($NF_3$), Lewis structure of, 374
Nitrogen halides, 374
Nitrogen oxides, 86, 90–91, 374–377
Nitroglycerin, ($C_3H_5N_3O_9$), 115
Nitrous oxide ($N_2O$) (''laughing gas''), 375
NMR (nuclear magnetic resonance) spectroscopy, 224–225, 502
Nobel, Alfred, 115
Nollet, Jean Antoine, 548
Nomenclature:
  of acids, 65
  of complexes, 453
  of compounds, 62–65
  of covalent compounds, 64
  of ionic compounds (salts), 62
  of negative ions, 63–64
  of positive ions, 62–63
  of salts (ionic compounds), 62
Non-SI units, 13
Nonbonding electrons, VSEPR theory and, 293–296
Nonbonding molecular orbitals, 466–467
Nonionizing radiation, ionizing radiation versus, 857–859
Nonmetal hydrides, 407
Nonmetal hydroxides, 407, 408–410
Nonmetal oxides, 407
Nonmetals, 52
  chemistry of, 349–400
  physical properties of, 53–54, 264
  reactions of, 277
  as reducing agents, 351
Nonpolar solvent, 532
Not-so-weak acids, 614–615
NRDC (National Resource Defense Council), Alar and, 16–17
Nuclear chemistry, 841–882
Nuclear fission, 868–872
Nuclear fusion, 872–874
Nuclear magnetic resonance (NMR) spectroscopy, 224–225, 502
Nuclear medicine, 876–877
Nuclear synthesis, 874–875
Nucleic acids, 952, 973–979
Nucleophiles, 909, 913
Nucleophilic substitution reaction, 909
Nucleotide, 973
Nucleus, 844
Nuclide, 845
  fertile, 870
  metastable, 847
  neutron-rich, 848–850
Nylon, 948–949
Ochsenfeld, R., 503
Octahedral complexes, 458–459
  high-spin versus low-spin, 461–463
Octahedral crystal fields, 458–459
Octahedral holes, 486
Octahedral molecule, 291
Octahedron, 292
Octane number, 898
Octet rule, 272
  molecules that don't satisfy, 284–286
  *see also* Lewis structures
Oil(s), 968–969
  crude, 897
  essential, 925
Olefins, 890
Onnes, Heike, 503
Oppenheimer, J. Robert, 872
Optically active compounds, 456, 923
Orbitals, 218
  antibonding molecular, 466
  atomic, 301
    hybrid, 297–300
    relative energies of, 226
  bonding molecular, 466
  degenerate, 227
  molecular, theory of, 301–307
  nonbonding molecular, 466–467
  shells and subshells of, 220–223
Order, 808–809
  determination of, 813–815, 818–819
  zero, 830
Ore, 336–337
Organic chemistry, 390
  of carbon, 883–934
Organic compounds, 884–885
Orgel, Leslie, 457
Ortho isomer, 896
Orthorhombic crystals, 365
Osmosis, 548
Osmotic pressure, 548–549
Ostwald, Friedrich, 371
Ostwald process, 372
Overvoltage, 746
Oxalic acid ($H_2C_2O_4$), 102
Oxidation, 327, 328–329
  defined, 697
  reduction and, discovery of, 696–698
Oxidation number(s) (oxidation state), 59–61, 328–329, 698
  assignment of, 699–701
  of carbon, *910*
  for halogens, 384
  for nitrogen, 372–377

for nitrogen compounds, 377
for phosphorus, 379
for phosphorus compounds, 377
versus true charge on ions, 329–330
of xenon compounds, 388
Oxidation state(s), of transition metals, 446
Oxidation–reduction reactions, 313, 327–329, 695–724, 910–911
in acidic solutions, 704–710
in basic solutions, 710–713
equations for, balancing of, 702–704
Lewis structures and, 713
metathesis reactions versus, 701
molecular, 713–715
recognition of, 701–702
spontaneous,
electrical work from, 726–728
prediction of, 730–731
Oxides, 320–321
of carbon, 392–393
metal, 407
neutral, of halogens, 386
nitrogen, 374–377
nonmetal, 407
Oxidizing agents, 330–333, 357–361
common, 715–717
relative strengths of, 717–719
Oxyacids, 368, 408
of chlorine, 387
of halogens, 386–387
of phosphorus, 381
Oxyanions, 368, 381
of chlorine, 387
of phosphorus, 381
Oxygen, 45, 46
ammonia combined with, 90–91
burning of sugar and, 88–89
chemistry of, 356–363
electronegativity of, 366–369
hydrogen combined with, 83, 84–85
Lewis structure of, 274, 299
magnesium combined with, 87, 93
methods of preparing, 361
mole versus molecule of, 76–77
oxidation of phosphorus by, 351
as oxidizing agent, 330, 357–361
valence shell of, 369
Oxygene, 697
Ozone ($O_3$), 126
chemistry of, 356–357
depletion of, 362–363
Pace, C. N., 186
Packing, cubic:
body-centered, 479–480
simple, 479, 485–487
Para isomer, 896
Paracelsus, 925
Paramagnetic atoms, 306
Paramagnetic compounds, 356, 463
Paramagnetic molecules, 306
Parmenides, 44
Parry, Robert, 312–313
Partial pressure, equilibria expressed in, 583–585
Particle(s), 206–208
number of, in moles, 77–78
subatomic, spin of, 223–225
in unit cells, 493–494
Pasteur, Louis, 923–924
Pauling, Linus, 297, 456, 959, 974
Pentaborane(9) ($B_5H_9$), 182
Peptide bond, 955
Peptides, 955–958
Periodic table, 239–269
categories of elements in, 52
development of, 241–243
electron configurations and, 230–232
families in, 51–52
groups in, 51–52, 57
Mendeleeff's, 2, 241–243
modern versions of, 243–245
Periods, 52
Peroxides, 320–321, 359–361
Petrochemicals, 899
Petroleum fractions, 897
Petroleum products, chemistry of, 897–899
pH, 603–605
as measure of concentration of $H_3O^+$ ion, 422–425
pH scale, 423
pH titration curves, 627–629
Phase diagrams, 524–525
Phenols, 907
Phlogiston theory, 696
Phosphides, 318
Phospholipids, 969–970
Phosphorus:
chemistry of, 377–381
electronegativities of, 379–380
oxidation of, 351
oxidation numbers for, 379
oxyacids and oxyanions of, 381
valence shell of, 380–381
Photon, 212
Physical properties, 52, 275
of atoms, 250
of carbon, *885*
of metals, 53
of nonmetals, 53
structure of metals and, 483–484
Pi bond, 300
Pig iron, 338
Planck, Max, 212
Planck's constant, 213, 214, 250, 846, A-2
Plante, Gaston, 752
Plastics, thermoset, 937
Platinum, 49
Plato, 44, 196
Plucker, Julius, 196, 197
Plunkett, Roy, 946
PMMA [poly(methyl methacrylate)], 947
pOH, 603–605
Poise, 521
Poiseuille, Jean Louis Marie, 521
Polar bond(s), 472, 280
polar molecules and, 282
Polar compound, 279, 280
Polar covalent bond, 280
Polar covalent compound, 280
Polar lipids, 969–970
Polar molecule, 405
polar bonds and, 282
Polar solvent, 532, 533
Polarity, electronegativity and, 277–280
Poly(methyl methacrylate) (PMMA), 947
Poly(tetrafluoroethylene), 946
Poly(vinyl chloride) (PVC), 946
Poly(vinylidene chloride), 946
Polyamides, 948
Polyatomic ions, 58–59
Polyethylene, 945–946
Polyhydroxy aldehydes, 962
Polymer(s), 378, 935–983
addition, 944–948
condensation versus, 939–940
atactic, 938–939
branched, 937
condensation, 948–951
addition versus, 939–940
cross-linked, 937
discovery of, 936
isotactic, 939
natural, 951–952
straight-chain, 936–937
syndiotactic, 939
tacticity of, 938–939

Polymerization, 942–944, *see also* Polymer(s)
Polynucleotide, 974
Polypropylene, 946
Polyprotic acids, 629, 630
Polysaccharides, 966–968
Porphyrin ring, 451
Positive electrode (anode), 196, 340, 729, 743
Positive ions, nomenclature of, 62–63
Positron emission, 846
Potassium, 317, 320, 321, 322, 324
Potassium dichromate ($K_2Cr_2O_7$), 95
Potassium superoxide ($KO_2$), 321
Potassium-argon dating, 857
Precipitation, 662–666
Precipitation hardening, 484
Precision, in measurement, 15–19
Pressure:
- atmospheric, 116–119
- changes in, Le Chatelier's principle and, 586–587
- constant, 172
- conversion factors of, *A-3*
- critical, 516–518
- of gas,
  - versus force exerted by, 115–116
  - partial, Dalton's law of, 133–138, 140
  - versus pressure due to weight, 119–120
  - vapor, 135
- osmotic, 548–549
- partial, 133–138, 140
  - equilibria expressed in, 583–585
- vapor, 135, 512–514
  - of water, 135, *A-3*

Priestly, Joseph, 936
- apparatus of, for studying gases, *600*

Primary alcohol, 906
Primary batteries, 750–751
Primary valence, 442
Principal quantum number, 218
Products, 83
- mass of, in chemical reactions, 88–90

Propane ($C_3H_8$), 85
Properties:
- of acids, 402
- chemical, 52–54
- colligative, 541–543, 544–548
- extensive, 155, 541
- of gases, 112–115
- intensive, 154–155, 541
- of metals, 53–54, 263–264
- of nonmetals, 53–54, 264
- physical, *see* Physical properties
- of system, 166–168
- of water, 522–524
- of wave, 217

Prostaglandins, 971–972
Protein(s), 951
- denaturation of, 961–962
- peptides and, 955–958
- structure of, 958–962

Protein biosynthesis, 976–979
Proton:
- discovery of, 203–204, 205
- rest mass of, A-2

Proton acceptors, 411
Proton donors, 411
Proust, Joseph Louis, 45, 47
Pseudo-first-order reaction, 820
Pure water, 616–617
Purines, 973, 974
PVC [poly(vinyl chloride)], 946
Pyranose, 964
Pyrimidines, 973–974
Pyrophoric reaction, 327
Pythagorean theorum, 496, 497

Quantization of energy, 211–213
Quantum numbers, 217–219
- angular, 218
- magnetic, 218–219
- principal, 218
- spin, 223–225

R process, 875
Rad (radiation absorbed dose), 860
Radiation:
- alpha, 199
- beta, 199
- diagnostic uses of, 876
- electromagnetic, 207, 208–209
- gamma, 199
- ionizing, 857–862

Radiation absorbed dose (rad), 860
Radiation biological effectiveness (RBE), 860
Radioactive decay:
- dating by, 856–857
- kinetics of, 854–855
- modes of, 845–847

Radioactivity:
- discovery of, 198–199, 842–844
- induced, 864–868
- natural, 862–864

Radius (radii):
- atomic, 245–247, *A-4*
- covalent, 246–249
- ionic, 247–250

Radius ratio rules, 487–488
Ramsey, William, 387–388
Random error, 14–19
Raoult, Francois-Marie, 542, 544, 549
Raoult's law, 542, 544–545
Rare gases, chemistry of, 387–389
Rare-gas compounds, 388
Raschig, Friedrich, 372, 821
Ratcliffe, J. A., 843
Rate:
- defined, 803
- of reaction, 564, 803
  - determining order from, 813–815, 818–819
  - different ways of expressing, 806–807
  - instantaneous, 804–805

Rate constant, 805
- equilibrium constant and, 812–813

Rate law(s), 564, 805
- integrated form of, 815–819
  - derivation of, 829–830
- stoichiometry of a reaction versus, 807

Rate-limiting step, 810
Rayleigh, Lord, 387
Rays:
- alpha, 199
- beta, 199
- canal, 201
- gamma, 199
- x-, 198

RBE (radiation biological effectiveness), 860
Reactants, 83
- mass of, in chemical reactions, 88–90

Reaction(s):
- addition, 892–893
- of alkanes, alkenes, and alkynes, 892–893
- bimolecular, 808
- at carbonyl group, 913–914
- chain, 870
- chemical, *see* Chemical reactions
- combination, 312
- completion not reached in, 559–561
- decomposition, 312
- disproportionation, 360–361, 712
- electrochemical, 726
- electrolysis, 339–340

endergonic, 781
endothermic, 781
enthalpy of, *see* Enthalpy of reaction
entropies of, 776–779
enzyme-catalyzed, 830–831
equilibrium, 569–570
exergonic, 781, 789
exothermic, 781
first-order, 808
pseudo-, 820
fission, 852
free energy of, 780–785
free-radical polymerization, 941–942
fusion, 852
gas-phase, 563–597
of main-group metals, 319–323, *331*, 350–351
metathesis, 313
molar heat of, 171
molecularity of, 808–809
order of, 808–809
determination of, 813–815, 818–819
oxidation–reduction, *see* Oxidation–reduction reactions
pyrophoric, 327
rate of, *see* Rate(s) of reaction
rate constant for, equilibrium constant and, 812–813
rate law for, 805, 807
rate-limiting step of, 810
redox, *see* Oxidation–reduction reactions
replacement, 313
second-order, 808
solvents in, 600–602
spontaneous, 726–728, 730–731, 771–772
substitution, 908–910
thermite, 326
unimolecular, 808
water-gas shift, 901
zero-order, 830
Reaction quotient, 570–572, 786
Reactive metals, 316
Reagent(s):
excess, 91
Grignard, 920–922, 942
limiting, 90–94
Recrystallization, 552
Redox reactions, *see* Oxidation–reduction reactions
Reducing agents, 330–336, 351, 358
common, 715–717
relative strengths of, 717–719
Reducing atmospheres, 358
Reduction, 328–329, 696–698
Rees, Douglas, 451
Refining:
of ore, 338
zone, 553
Relative measurement, 30
Rem (roentgen equivalent man), 860
Replacement reactions, 313
Replication, 974, 976
Reradiation of energy, 483
Research octane numbers, 898
Resinous electricity, 196
Resonance, theory of, 895
Resonance hybrids, 286–287, 895
Rey, Jean, 696
Ribonuclease T1, 186
Ribonucleic acid, *see* RNA
Richter, Jeremias Benjamin, 47, 72
Risk, measurement of, 16–18
RNA (ribonucleic acid), 973
messenger (mRNA), 977
transfer (tRNA), 977
Road-index octane numbers, 898
Roasting, of ore, 337, 696
Roberts, Marc, 17
Rock salt (NaCl), 491–493
Rocket fuel, 182
Roentgen, William Conrad, 198, 842, 876
Roentgen equivalent man (rem), 860
Rotation of water molecules, 471–472
Rounding off, 22
Royal Institute, 375
Royal Society of London, 10
Rubber, 936
Rubidium, 317
Rubies, 342
Rumford, Count (Sir Benjamin Thompson), 159–160
Rust, 358, 699, 726, 754–756
Rutherford, Ernest, 198, 199, 842, 843, 844
atomic model of, 202–203, 206, 213, 272
proton name proposed by, 204, 205
Rutherford–Marsden–Geiger experiment, 202–203, 839, 868
Rydberg constant ($R_H$), 211, 214, 250, A-2

S process, 875
Salicylic acid, 970
Salt(s), 54–56
Epsom, 341
formulas of, 57–58
how to keep from precipitating, 662–664
insoluble, 652
dissolving of, 676–677
of main-group metals, 341–342
nomenclature of, 62
slightly soluble, 652
solubility of, 654–656, 657–658
in water, rules for, *652*
soluble, 652
*see also* Ionic compound(s); Solubility
Saltiness, 319
Saponification, 536
Sapphires, 342
Saturated hydrocarbons (alkanes), 885–890
Scanning tunneling microscopy (STM), 49
Schrodinger, Erwin, 218
Scientific law, 3
Scientific notation, 25–26
Seaborg, Glenn, 866, 867
Second-order rate law, 815–817, 829–830
Second-order reaction, 808
Secondary alcohol, 906
Secondary batteries, 752–754
Secondary valence, 442
Seesaw shaped molecule, 294
Selective precipitation, separation of ions by, 664–666
Semiconductors, 499–503
Semimetals (metalloids), 52, 263–265, 349
properties of, 53
Semipermeable membrane, 548, 549
Shells, 220–223
Shockley, William, 501
SI units of measurement, 12–13, 156, 164, 198, 199–200, *A-2*
Siemens, Werner, 752
Sigma bonds, 300
Sigma molecular orbital, 301
Sigma star molecular orbital, 301
Significant figures, in measurement, 19–22
Silica gel, 421
Silicates, 421
Silicic acid [$Si(OH)_4$], 421
Silicon, 326
Silver, tarnishing of, 698, 699
Simple cubic packing, 479, 485–487

Simple cubic unit cell, 490
Single bonds, strength of, 364–366, 377–379
Sintering, of ore, 337
Slightly soluble salts, 652
Smalley, Richard, 395
Smectic liquid crystals, 498
Smelting, 337–338
Soaps, 536–538
Societe Philomatique, 125
Soddy, Frederick, 75, 843, 844
Sodium, 316, 317, 321–323, 324
  electrolysis of, 339, 340
  sodium chloride formation and, 261–263
Sodium chloride (NaCl), 54, 55, 56, 318
  aqueous, electrolysis of, 745–747
  in endothermic reactions, 174–175
  formation of, 261–263
  lattice energy of, 259
  molten, electrolysis of, 743–745
  physical properties of, *275*
  as rock salt, 491–493
  solubility of, in water and in alcohols, *535*
Sodium hydroxide (NaOH), 102
Solid(s):
  amorphous, 476
  covalent, 477
  crystalline, 476
  dissolving of, in water, 648–649
  ionic, 648
  melting point of, 514–515
  metallic, 477–478
  molecular, 476, 648
  monatomic, structure of, 478–482
  as state of matter, 112
  structure of, 477–508
  volume of, versus volume of gases, 114–115
Solid acids, 421
Solid solutions, intermetallic compounds and, 484–485
Solid-state acids, 421–422
Solubility:
  calculations of, 656–660
  of gas, in solvent, 600–602
  lattice energies and, 260–261
  of salt, 654–656, 657–658
  *see also* Ionic compound(s); Salt(s)
Solubility equilibria, 649–651
Solubility product, 652–653, 654, 668
Solubility product equilibrium constant ($K_{sp}$), 653
  solubility of salt and, 654–656, 657–658, *A-11*
Solubility rules, 651–652
Soluble salts, 652
Solute(s), 96–98, 531, 539, 540
Solution(s), 96–98, 531–558
  acidic, 704–710
  aqueous, 84
  basic, 710–713
  boiling point elevation and, 543–544, 545
  colligative properties of, 541–543, 544–548
  concentration and, 94–96, 538–541
  counting of atoms in, 98–100
  freezing point depression and, 543–544, 545
  immiscibility and, 534
  interstitial, 485
  molality of, 539
  molarity of, 538–539
  osmotic pressure and, 548–549
  solid, 484–485
  substitution, 485
  vapor pressure and, 541–543, 544, 545
Solvated electrons, 318
Solvent(s), 96–98, 531, 539
  nonpolar, 532, 533
  polar, 532, 533
  role of, in chemical reactions, 600–602
Sorensen, S. P. L., pH defined by, 423, 604
Specific heat, 156, *157*
Spectrochemical series, 461
Spectrophotometer, 804
Spectrum (spectra):
  absorption, 209
  atomic, 209–211
  emission, 210
Speed, 207
  of light, 78, 208, A-2
Spin quantum number, 223–225
Spontaneous fission, 847
Spontaneous reactions, 726–728, 730–731, 771–772
Square-planar complexes, 460
Square-planar molecule, 295
Square-pyramidal molecule, 295
Stahl, Georg Ernst, 696
Stalactites, 394
Stalagmites, 394
Standard state, 176
Standard-state cell potential(s) ($E°$), 729–731
  prediction of, 733–735
Standard-state conditions, for electromechanical measurements, 729
Standard-state enthalpies of reaction, 176
Standard-state enthalpy of formation, 180–182, *A-15–A-24*
Standard-state entropies of reaction, 776–778
Standard-state free energies of formation, 784, *A-15–A-24*
Standard-state free energy of reaction, 780–781
  interpreting data of, 784–785
Standard-state reduction potentials, 731–733; *732*, A-12–A-14
Stannous fluoride ($SnF_2$), 342
Starch, 966
State(s):
  equations of, 166
  of matter, 112
  oxidation (oxidation number), 59–61
  standard, 176
State function(s), 166–168, 767
Statue of Liberty, 726
Staudinger, Hermann, 936
Steel, wrought iron versus, 338
Stepwise dissociation, 630
Stereoisomers, 891, 923
Steroids, 894
STM (scanning tunneling microscopy), 49
Stoichiometry, 71–110
  breathalyzer, 95–96
  origins of, 72–73
  of reaction, 807
  *see also* Measurement
Stone, Reverend Edmund, 970
Stoney, George, 198
Straight-chain hydrocarbons, 887
Straight-chain polymer, 936–937
Straight-run gasoline, 897
Strong acid, 415, 606, 609
Strong-field ions, 461
Strong-field ligands, 461
Strontium, 324
Strontium fluoride ($SrF_2$), 56
Structural metals, 326
Strychnine, 918, 919
Subatomic particles, 223–225
Subbituminous coal, 900
Sublimation, 476
Subshells, 220–223
Substitution reactions, 908–910
Substitution solution, 485
Subtraction, with significant figures, 21
Subtractive colors, 464

Succinic acid, 16
Sucrose ($C_{12}H_{22}O_{11}$), 54, 55, 966
  melting point of, 515
  *see also* Sugar
Sugar, 78
  burning of, 88–89
  as covalent compound, 54
  melting point of, 515
  molecular weight of, 76
Sulfides, 318
Sulfur:
  chemistry of, 364–369
  electronegativity of, 366–369
  oxidation numbers of, *366*
  valence shell of, 369
Sulfur dioxide ($SO_2$), Lewis structures of, 286–287
Sulfuric acid ($H_2SO_4$), 97, 101, 367, 368, 421
Sulfurous acid ($H_2SO_3$), 367–368
Superconductors, 502–503
Supercooled liquid, 514–515
Supercritical fluid extraction, 554
Superoxides, 321
Surface tension, 518–519
Surfactants, 536
Surroundings, 766
  kinetic molecular theory and, 160
Symbolic world, 50–51
Syndiotactic polymer, 939
Synthesis gas, 901
Synthetic fibers, 519–520
System(s), 766
  biological, 789–790
  defined, 768
  disorder in, 772–774
  entropy of, 774
  isolated, 774
  kinetic molecular theory and, 160
  properties of, 166–168
Systematic error, 14–19

Tacticity, 938–939
TCDD (dioxin) (2,3,7,8-tetrachlorodibenzo-*p*-dioxin), 18
Teflon, 946
Teeter-totter shaped molecule, 294
Teller, Edward, 872
Temperature:
  absolute, 155
  changes in, Le Chatelier's principle and, 588
  conversion factors of, *A-3*
  critical, 516–518
  effect of,
    on chemical reaction, 585
    on free energy of a reaction, 782–783
  equilibrium constants and, 790–792
  ignition, 873
  as intensive property, 27, 29–31, 154–155
  liquid formation and, 510–512
  relative, 154–155
  scales of, *see* Temperature scale(s)
  as state function, 767
  *see also* Heat
Temperature scale(s):
  Celsius, 30–31, 154–155
  centigrade, 30–31
  conversions among, 30–31
  Fahrenheit, 30–31, 154–155
  Kelvin, 30–31, 155
Tennant, Smithson, 390
Teratogens, 18
Terpenes, 893
Tertiary alcohol, 906
Tetradentate ligand, 449
Tetrahedral crystal fields, 459–460
Tetrahedral holes, 486
Tetrahedral molecule, 292
Tetravalent compounds, 886
Teutsch, Georges, 290
Thales, 44
Thallium, 325
Thenard, Louis Jacques, 384
Theory, 3
Thermal conductors, 154
Thermal cracking, 899
Thermal insulator, 154
Thermal neutrons, 865, 869
Thermal reforming, 898–899
Thermite reaction, 326
Thermochemistry, 153–192
  defined, 154
  *see also* Thermodynamics
Thermodynamic control, 762
Thermodynamics, 765–800
  of biological systems, 789–790
  defined, 154, 766
  first law of, 164–166, 727, 766–768, 771
  kinetics versus, 801
  oversimplications and, 783–784
  second law of, 766
    entropy and, 774–776
  third law of, 766, 776
Thermoplastics, 937
Thermoset plastics, 937
Thin-layer chromatography (TLC), 552
Thompson, Sir Benjamin (Count Rumford), 159–160
Thomson, J. J., 199–200, 844
  electron discovered by, 197–198, 272, 844
  raisin pudding model of atom of, 201–202
Thomson, William (Lord Kelvin), 30, 164
Three Mile Island, 861
Tin, 45, 53, 326–327
  smelting of, 338
Tin fluoride ($SnF_2$), 342
Titanium, 338
Titration, 101–102, 627–629
Torricelli, Evangelista, 117, 119
Toth, Bela, 16–17
Town gas, 392–393, 901
Trans isomers, 454, 891
Transcription, 974, 976–977
Transfer RNA (tRNA), 977
Transition metals:
  chemistry of, 439–470
  colors of complexes of, 463–465
  inner, 439
  ions of, 429, 445
  oxidation states of, 446
Transition-metal complexes, colors of, 463–465
Transition-metal ions:
  as Brønsted acids, 429
  electron configuration of, 445
Translation, 974, 976, 977–979
  of water molecules, 471–472
Triads, 240
Tridentate ligand, 449
Trienes, 892
Trigonal bipyramidal molecule, 292, 294
Trigonal planar molecule, 291
Trigonal pyramidal molecule, 293
Triple bonds:
  molecules with, 299–300
  strength of, 377–379
  VSEPR theory and, 292–293
Triple point, 525
Triprotic acids, 629, 636–637
Tritium, 873
tRNA (transfer RNA), 977
Tropsch, Hans, 902
T-shaped molecule, 295
Tsvet, Mikhail, 552
Tungsten, 49

UDMH (unsymmetric dimethylhydrazine), 16–17
Uncertainty, 6
Unimolecular nucleophilic substitution, 909

Unimolecular reaction, 808
Unit cells, 488–490
  Bravais, 488–489
  calculation of radii of, 496–497
  of crystals, 494–496
  face-centered, 490
  measuring the distance between particles in, 493–494
  of rock salt, 491–493
  simple cubic, 490
  three-dimensional graph of, 490–491
Unit factor, 9
Units, of measurement, 6
Unsaturated hydrocarbons, 890–892
Unsymmetric dimethylhydrazine (UDMH), 16–17
Uranium, 198–199

Valence, 353, 442
Valence electrons, 272–273, 478
  atomic geometry and, *296*
Valence shells, 369, 380–381
Valence-bond approach to bonding in complexes, 456–457
Valence-bond theory, 301, 456
Valence-shell electron-pair repulsion theory, *see* VSEPR theory
Van den Brock, A., 204
van der Waals, Johannes, 143
van der Waals constants, *144,* 146–147
van der Waals equation, 143–147, 517
van der Waals forces, 472–473, 476, 678, 679
Van Laue, Max, 493
Van Vleck, John, 457
Vane, J. R., 971–972
van't Hoff, Jacobus Henricus, 548, 549, 924
Vapor pressure, 512–514
  of water, 135, *A-3*
Vapor pressure depression, 541–543
Vaporization, 159
Very weak acids, 615–616, 617–618
Vibration of water molecules, 471–472
Vinegar, 415
Viscosity, 521
Vitamins, 894
  $B_{12}$, 451, 452
  C ($C_3H_4O_3$), 82
Vitreous electricity, 196
Volt (V), 729, A-2
Volta, Alessandro, 728, 729
Voltaic cell(s), 729, 750
  electrolysis and, 744
  line notation for, 735–736
  standard-state cell potentials for, 729–730
Volume:
  constant, 172
  conversion factors of, *A-3*
  as extensive property, 27, 28
Volume percent, 539
Von Fraunhofer, Joseph, 209, 210
VSEPR theory, 291, 360
  double and triple bonds and, 292–293
  geometry of alkenes and, 890
  nonbonding electrons and, 293–296

Water, 83, 84–85
  acid–base chemistry of, 602–603
  active metals and, 321–323
  Brønsted theory and, 413–414
  as covalent compound, 54
  decomposition of, 45
  dissociation of, 405–406
  dissolving of solids in, 648–649
    rules for, *652*
  electrolysis of, 355, 747
  as Greek element, 44
  hard, 537
  heat capacity of, 155–156, A-2
  ionic compounds in, solubility rules for, *652*
  leveling effect of, 420
  Lewis structure of, 404, 430
  meniscus formation by, 519
  movement of molecules of, 471–472
  as polar solvent, 532, 533
  properties of, 522–524
  pure, 616–617
  shape of molecule of, 293
  solubility(ies),
    of alcohol in, 534–535
    of ammonia in, 600, 601
    of common gases in, *601*
  vapor pressure of, 513–514, *A-3*
Water gas, 392–393, 901
Water-dissociation equilibrium constant ($K_w$), 602
Water-gas shift reactions, 901
Watson, James, 974, 976
Waugh, Sir Andrew, 6
Wave(s), 206–208
  functions of, 218
  properties of, 217
  *see also* Wavelength
Wave–particle duality, 216–217
Wavelength, 207
  of light, colors and, 465
Weak acid(s), 415, 606, 609–610, 617
  calculations regarding, 611–612
  chemistry of, summarized, 616–618
  $H_3O^+$ ion concentration in, 612–614
  not-so-, 614–615
  very, 615–616, 617–618
Weak-field ions, 461
Weak-field ligands, 461
Wedgwood, Thomas, 211–212
Weight:
  atomic, 74–75, 77
  empirical, 82
  mass versus, 11–12
  molecular, 76, 77, 82
Werner, Alfred, 441–443, 454
Westbrook, Edwin, 289
Whewell, William, 196
Wien, William, 201
Wilhemy, Ludwig, 830
Wohler, Friedrich, 884
Wood alcohol, 906
Work, 161–163
  electrical, 162, 726–728, 769
  of expansion, 162, 769
  heat from, 163–164
  interconversion of heat and, 164–166
Wrought iron, 338

Xenon, 387, 388–389
X-rays, 198, 842, 876

Ye, Qing, 319
Ylem, 874

Zeolites, 421–422
Zero(s):
  absolute, 121, 123–124
  as significant figures, 20
Zero-order reaction, 830
Zhang, Rongguang, 289
Ziegler, Karl, 943
Ziegler–Natta catalysts, 943–944, 945, 946
Zinc, 45, 53, 440
  hydrochloric acid and, 726–728
  smelting of, 338
Zincblende (ZnS), 491–493
Zone refining, 553
Zwitterion, 952

## TABLE OF ATOMIC NUMBER AND AVERAGE ATOMIC WEIGHT

| *Element* | *Symbol* | *Atomic Number* | *Mass (amu)* | *Element* | *Symbol* | *Atomic Number* | *Mass (amu)* |
|---|---|---|---|---|---|---|---|
| Actinium | Ac | 89 | 227.0278* | Europium | Eu | 63 | 151.965 |
| Aluminum | Al | 13 | 26.981539 | Fermium | Fm | 100 | 257.0951* |
| Americium | Am | 95 | 243.0614* | Fluorine | F | 9 | 18.9984032 |
| Antimony | Sb | 51 | 121.757 | Francium | Fr | 87 | 223.0197* |
| Argon | Ar | 18 | 39.948 | Gadolinium | Gd | 64 | 157.25 |
| Arsenic | As | 33 | 74.92159 | Gallium | Ga | 31 | 69.723 |
| Astatine | At | 85 | (210)** | Germanium | Ge | 32 | 72.61 |
| Barium | Ba | 56 | 137.327 | Gold | Au | 79 | 196.96654 |
| Berkelium | Bk | 97 | (248)** | Hafnium | Hf | 72 | 178.49 |
| Beryllium | Be | 4 | 9.012182 | Helium | He | 2 | 4.002602 |
| Bismuth | Bi | 83 | 208.98037 | Holmium | Ho | 67 | 164.93032 |
| Boron | B | 5 | 10.811 | Hydrogen | H | 1 | 1.00794 |
| Bromine | Br | 35 | 79.904 | Indium | In | 49 | 114.82 |
| Cadmium | Cd | 48 | 112.411 | Iodine | I | 53 | 126.90447 |
| Calcium | Ca | 20 | 40.078 | Iridium | Ir | 77 | 192.22 |
| Californium | Cf | 98 | (250)** | Iron | Fe | 26 | 55.847 |
| Carbon | C | 6 | 12.011 | Krypton | Kr | 36 | 83.80 |
| Cerium | Ce | 58 | 140.115 | Lanthanum | La | 57 | 138.9055 |
| Cesium | Cs | 55 | 132.9054 | Lawrencium | Lr | 103 | 262.11* |
| Chlorine | Cl | 17 | 35.4527 | Lead | Pb | 82 | 207.2 |
| Chromium | Cr | 24 | 51.9961 | Lithium | Li | 3 | 6.941 |
| Cobalt | Co | 27 | 58.93320 | Lutetium | Lu | 71 | 174.967 |
| Copper | Cu | 29 | 63.546 | Magnesium | Mg | 12 | 24.3050 |
| Curium | Cm | 96 | (247)** | Manganese | Mn | 25 | 54.93805 |
| Dysprosium | Dy | 66 | 162.50 | Mendelevium | Md | 101 | (257)** |
| Einsteinium | Es | 99 | 252.083* | Mercury | Hg | 80 | 200.59 |
| Erbium | Er | 68 | 167.26 | Molybdenum | Mo | 42 | 95.94 |

These data are based on the assumption that the mass of the $^{12}C$ isotope of carbon is exactly 12 amu. The atomic mass of an element is an average of the masses of the isotopes of that element. These data were taken from ''Atomic Weights of the Elements,'' *Pure and Applied Chemistry,* 1991, *63,* 975. The nomenclature of elements with atomic numbers greater than 103 follows the convention described in *Pure and Applied Chemistry,* 1979, *51,* 381.

* The exact mass of the most stable element of this radioactive element.

** The average mass of the stable isotopes of this radioactive element.